Dictionary of
REPORT SERIES CODES
second edition

Dictionary of
REPORT SERIES CODES
second edition

Edited by

Lois E. Godfrey

and

Helen F. Redman

Los Alamos Scientific Laboratory Libraries
(University of California)
Los Alamos, New Mexico

A Project of
SLA Rio Grande Chapter
members and friends

SPECIAL LIBRARIES ASSOCIATION
New York
1973

International Standard Book Number 0-87111-209-4

Library of Congress Catalog Card number 72-87401

© 1973 by Special Libraries Association

235 Park Avenue South, New York 10003

Library of Congress Cataloging in Publication Data
Main entry under title:

Dictionary of report series codes.

    First ed. issued in 1962, by the Report Series Dictionary Committee of Rio Grande Chapter of Special Libraries Association.
    Bibliography: p.
    1. Technical reports—Abbreviations of titles. I. Godfrey, Lois E., ed. II. Redman, Helen F., ed. III. Special Libraries Association. Rio Grande Chapter. Report Series Dictionary Committee. Dictionary of report series codes. IV. Title: Report series codes.

Z6945.A2D5	1973	029.9'6	72-87401
ISBN 0-87111-209-4			

Printed in the United States of America

Preface

In May 1957 the Rio Grande Chapter of Special Libraries Association initiated, as its first Chapter project, the compilation of a *Dictionary of Report Series Codes*. Such a dictionary was felt necessary for bibliographic identification of the hundreds of thousands of technical reports issued since World War II.

These documents are, in general, identified by a prefix consisting of a combination of letters and/or numbers and a serial number. The recurrent letters and numbers used as the prefix are often known as the report series code. It is the purpose of the *Dictionary* to identify as many as possible of these series codes with the agencies originating the reports or assigning the numbers.

In the ten years since the first edition was published, the report literature has grown phenomenally. It now is found increasingly in the social sciences, as well as in the physical sciences and technology. The entries in this edition reflect both trends. The number of entries has doubled; and the corporate names include organizations in the behavioral and social sciences.

The primary sources used in updating the *Dictionary* were the document catalog of the Los Alamos Scientific Laboratory, and published indexes to the report literature. Entries from the first edition, all of which are retained in this, were from the document catalogs of the Air Force Special Weapons Center, Sandia Corporation, and White Sands Missile Range. In addition, the published lists of report series codes used by the United States Atomic Energy Commission, the Armed Services Technical Information Agency, the Office of Technical Services, and the United States Naval Ordnance Laboratory, and lists of codes published by the Special Libraries Association and the Central Air Documents Office were consulted. These are fully identified in the chapter on sources.

The editors sincerely appreciate the support of the Special Libraries Association, the Los Alamos Scientific Laboratory, and all of the Chapter members and friends who assisted, through volunteer participation, with the compilation of the data during the "Dictionary Days" of the Rio Grande Chapter that occurred over the past three years.

Lois E. Godfrey and Helen F. Redman,
Editors

Acknowledgments

Grateful acknowledgment is extended to members of the Computing Science and Service Division of the Los Alamos Scientific Laboratory, especially to Jennie Boring who wrote the computer programs, and to the Central Computing Facility Operations Support Group, whose members prepared the cards and working listings. The Accounting Department prepared the final listings. The compilers also assisted with proofreading.

We are indebted to Margrett Zenich, Director of Libraries, Office of the Chief of Engineers (U.S. Army), who was responsible for prodding us to complete this edition.

We also extend our thanks to the many members of the Science-Technology oriented Divisions of SLA who responded to our 1966 questionnaire, and to our 1972 request for additional codes which was published in *Sci-Tech News*.

Without the hundreds of hours contributed on weekends by the compilers, and the financial support of the Rio Grande Chapter and of SLA, the task would have been impossible.

The Compilers

SLA Rio Grande Chapter Members and Friends

Harold Bellingham	Janet Godfrey	Eleanor Pratt
Ann Beyer	Lois E. Godfrey	Helen F. Redman
Tom Beyer	Dorothy Harrington	Walter Roose
Marti Bogdan	Gertrud Holloway	Gladys Rowe
Irene Brian	Betty S. Jackson	Willie Servis
Betty Burnett	Marjorie Johnson	Charles Snell
Harold Burnett	Helen Keller	Patricia Snell
Madeline Canova	Helen Ketola	Eleanor Standring
Sandra Coleman	Mary King	Adelia Stewart
Calla Ann Crepin	Louise Lewis	Patricia Symes
Lois Dildine	Charles R. Machovec	Frances Szeman
Nina Duncan	Florence Macpherson	Marilyn Treiman
Ruth Farley	M. L. (Sue) Mainord	Lore Watt
William Farrington	Karen Morgan	Marjorie Wilson
Catherine Fifield	Kenneth Olson	Virginia Winsor
Lila Foss	Lee Parman	Kate Young
J. Arthur Freed		

Contents

Introduction to Report Series Codes 1
    Brief History of Report Series Codes 2
    Assignment of Report Designators 6
    Components of Report Series Designators 8
    Evolution and Proliferation of Report Series Codes 10
    Bibliography 12
Explanatory Notes for this Dictionary 14
    Inclusiveness 14
    Form of the Report Series Codes 14
    Arrangement of Entries 14
    Selection of Corporate Entry 15
    Form of Corporate Entry 15
    Parenthetical Notes 16
    Use of Reference Notes 16
If You Don't Find It Here 17
Abbreviations 18
Glossary 20
Sources of Report Series Codes and their Identification 21
Reference Notes 23
    1. Ambiguous Numerical Codes 23
    2. Ambiguous Alphabetical Codes 27
    3. ANP Program 31
    4. Atomic Energy Commission, Operations Offices and Divisions 32
    5. Atomic Energy Commission. Operations Office of Information Services 33
    6. Atomic Energy of Canada Ltd. 33
    7. Atomic Weapon Test Operations 34
    8. Bumblebee Project 36
    9. California. University, Berkeley. Radiation Laboratory 37
    10. Chicago. University. Metallurgical Laboratory 38
    11. Columbia University. Substitute Alloy Materials Laboratories 40
    12. Goodyear Atomic Corporation 41
    13. Joint Armed Services Publications 42
    14. Joint Atomic Weapons Publications 42
    15. Joint Committees of the Atomic Energy Commission and Department
         of Defense 43
    16. Joint Feasibility Study Groups 44
    17. Joint Task Forces 44
    18. Joint Task Groups 45
    19. Joint Working Groups 45
    20. Library of Congress. Technical Information Division and Its Predecessors .. 45
    21. Los Alamos Scientific Laboratory 46
    22. Los Alamos Scientific Laboratory (Incoming Reports) 46
    23. Manhattan District 46
    24. Manhattan District (Declassified Documents) 47
    25. National Advisory Committee for Aeronautics 47
    26. National Aeronautics and Space Administration 48
    27. National Defense Research Committee 49
    28. National Defense Research Committee, Panels and Committees 50
    29. National Research Council of Canada 50
    30. Naval Radiological Defense Laboratory 51
    31. NEPA Project 52
    32. Office of Scientific Research and Development 52
    33. Squid Project 53
    34. Tennessee Eastman Corporation 54

35. Tripartite Nuclear Cross-Sections Committee 55
36. United Kingdom Atomic Energy Authority, Industrial Group and Its
    Predecessors, 1946-55 55
37. United Kingdom Atomic Energy Authority, Industrial Group, 1955-59 59
38. United Kingdom Atomic Energy Authority, Industrial Group, 1959 60
39. United Kingdom Atomic Energy Authority, Industrial Group Successors 61
40. United Kingdom Atomic Energy Authority, Research Group, Atomic Energy
    Research Establishment, and Its Predecessor 61
41. United Kingdom Atomic Energy Authority, Research Group, Radiochemical
    Centre, and Its Predecessor 62
42. TRW Systems 63
43. Vela Uniform Program 63
44. Plowshare Program 64
45. European-American Nuclear Data Committee 65
46. General Electric Company 66
47. The RAND Corporation 66
48. Themis Project 67

Report Series Codes with Related Agencies 69
Corporate Entries with Related Codes 341

Dictionary of
REPORT SERIES CODES
second edition

Introduction to Report Series Codes

Ten years have elapsed since the first edition to this *Dictionary* was published. Reports are still with us, more numerous than ever, and report numbers have proliferated accordingly. This second edition, begun in 1966, has been delayed in part because of the astounding number of codes blossoming each year. All the problems referred to in the introduction to the first edition remain, and new ones have been added. Nevertheless, there are now developments on the horizon that promise improvements.

Compilation of this second edition has brought home to us the essentiality of the report as a means of communication for scholars and engineers. Its beginnings, described later in the history of reports, can be explained by topical events; its continued use can only be ascribed to a need for such a means of communication among the scholarly community. Twenty-five years seems a reasonable proving period. The technical report should now be accorded full recognition as an established form of technical literature, and should be granted the rights therefrom by those handling that literature.

At the time of the publication of the first edition of this *Dictionary*, estimates of the number of reports issued annually varied from 50,000 to 150,000. Since then guesses have ranged as high as 500,000, but no reliable figure has been ascertained. Certainly the number of issuers of reports has more than doubled—as witness this new edition. Furthermore the number of issuers of reports outside the United States has greatly multiplied. Their codes appear frequently in this *Dictionary*. The extension of the report to fields apart from science and technology is also demonstrated here. The social sciences (especially education, housing, urban renewal, and conservation) have begun to use the report as a primary means of communication. Their entry into the report publication field has introduced scores of new report codes.

Along with this proliferation of reports and their codes has come an increased acceptance of them as a respected means of communication. The government abstracting and indexing journals have continued and expanded. *Government Reports Announcements* and its index has evolved from earlier forms, to present to the information and scientific communities a collected announcement and index of the report literature. The earlier journals, *Nuclear Science Abstracts*, *Scientific and Technical Aerospace Reports (STAR)* and the *Technical Abstract Bulletin (TAB)* continue to provide information about reports to those interested in their subject fields. But, representing a major change in the attitude toward reports, many standard abstracting and indexing services are now including them in their coverage. References to reports can be found in almost all the scholarly services. Neglect by them can be ascribed more often to lack of knowledge of the reports than to discrimination against their form of issuance.

With this general acceptance has come a freedom on the part of authors to refer to reports as such in their publications. The "private communication" as a mask for reports is almost dead, and librarians are now presented with both the challenge of tracking down admittedly obscure references to reports and the hope of dealing with a reference that promises something more than total defeat.

Obviously, this greater acceptance of reports has brought solid advantages to the information community. Information about them is far more accessible than formerly. Also, awareness of the need for stricter monitoring of their bibliographic control is growing. Beginning in the middle 1960s the Committee on Scientific and Technical Information (COSATI) of the Federal Council on Science and Technology began to concern itself with the report literature. One of its primary assignments has been to bring the various U.S. government agencies into harmony in their publication practices. In that role it has done much to unify corporate author headings and standardize descriptive cataloging practices, but has so far extended little effort to report numbers. Two COSATI publications (1, 2) have dealt with the general problem, but have not offered specific solutions for it. Nevertheless the very existence of COSATI and its concern with reports has led us to such coordination in their handling as the "Government-wide Index . . . to Federal Research and Development Reports" and its successors.

Another development that holds the greatest hope for organization is the interest of the American National Standards Institute in the technical report literature. The ANSI Z-39 Committee for Library Work, Documentation, and Related Publishing Practices has two subcommittees working on questions relating to technical reports. One has developed a standard for the general format of reports, while the other is concerned specifically with report numbers. The proposals of these subcommittees are discussed later in this Introduction, and hold promise of bringing some logic, restraint and control to the present unmanageable situation.

While recognition of reports is essential to their eventual acceptance and consequent bibliographic control, it has brought with it a proliferation of the problems already existing—more reports, more report series, more untraceable references. It has also created new problems. A serious one is citations to reports, made in all good faith, but containing irrelevant material. An example involving a report from the Los Alamos Scientific Laboratory appears in a French report as this citation:

E. Storm et al. LA-2237—PHYSICS
AND MATHEMATICS
(TID-4500 13th ed. rev.) Nov. 58

The author's name and the report number are both correct, and little harm is done by the remaining bits of information that identify the subject category and distribution list under which initial distribution was made. But other references, made in equally good faith, include the irrelevant material and drop the pertinent items.

Occasionally the confused references are amusing rather than frustrating. One from prestigious *Chemical Abstracts* in 1971 cites a report as "U.S. Clearinghouse Fed. Sci. Tech. Inform., AD 1971, No. 722696." Again no harm is done because any report user can recognize a 6 digit AD number and is unlikely to be misled into suspecting that it was written in 1971 BC.

Purpose of this Dictionary

Report series codes are arbitrarily selected letters and/or numbers used to indicate relationships among individual reports. When combined with other letters and/or numbers specific to those individual reports, they form designators that may identify the reports. They are a device aimed at facilitating report control, both physical and bibliographic.

The development of report codes has followed closely on that of the reports themselves. Through the years codes have become so numerous as to be almost meaningless. Conceived as a shortcut in report reference, they are now often a handicap.

This *Dictionary* is intended as an aid in overcoming that handicap. It attempts to identify or provide an association for most of the codes that have been applied to reports. This introduction endeavors to supply enough background information about the assignment and make-up of report codes to enable the user of the *Dictionary* to interpret these codes and others he may encounter.

Brief History of Report Series Codes

Although reports have been known for many years, their development as a major means of communication dates back only to World War II. The Office of Scientific Research and Development (OSRD) is generally credited with responsibility for their growth. Under the OSRD, government-sponsored research attained the shape it has had ever since. Its outstanding features were decentralization of scientific effort through research contracts, impersonality deriving from the advent of research teams, and problems of security control. Unpublished reports were the most suitable means of recording and disseminating the results of this research.

Another wartime development that produced large numbers of reports was the work of the allied teams that investigated enemy scientific and technical research. They captured many thousands of German and Japanese scientific documents and created thousands of others with reports on their findings and interviews with enemy scientists. The captured documents and the Field Intelligency Agency, Technical (FIAT), British Intelligence Objectives Subcommittee (BIOS), Combined Intelligence Objectives Subcommittee (CIOS), Joint Intelligence Objectives Agency (JIOA) and the Army European Theater of Operations, ALSOS Mission (ALSOS) reports were measured in tons rather than numbers.

Postwar science and technology continued in the pattern of the OSRD. The Atomic Energy Commission (AEC), created by the Atomic Energy Act of 1946, operated its laboratories and promoted nuclear research through contracts. In the military departments there was a tremendous expansion of research. Expenditures for research and development had approximately doubled during the war, and in the decade following they doubled again. During this time the percentage of research money spent by agencies of the federal government increased approximately ten times. By 1950 results of that government-sponsored research were being published in reports at the rate of 75,000 to 100,000 a year.

Obviously there was great need for bibliographic identification and control of these reports. Probably the first attempt was made by the Office of Scientific Research and Development, which adopted a simple numerical series for identifying its reports. Between one-quarter and one-third of its contractors' reports were so numbered, and as Mortimer Taube commented in 1948, "It has been demonstrated over and over again that of all reports issued by OSRD and its contractors, those reports bearing OSRD numbers may be the most conveniently cited and controlled bibliographically." *(3)*

After the disbanding of OSRD, there was no longer a central agency to control or keep track of research reports. In specific areas new agencies were created

INTRODUCTION

whose primary mission was the organization and dissemination of reports. An Executive Order in 1945 established the Publication Board, later known as the Office of Technical Services (OTS), for the purpose of collecting and making available unclassified and declassified reports on war time and postwar research sponsored by the federal government and cooperating foreign governments or captured from enemy countries. The Atomic Energy Commission established a Technical Information Service (TIS), later named the Office of Technical Information (OTI), Division of Technical Information Extension (DTIE), and lastly the Technical Information Center (TIC), which was a continuation and expansion of the central indexing service that the University of Chicago's Metallurgical Laboratory had been providing for the Manhattan District since 1942. Its responsibilities were "the dissemination of scientific and technical information relating to atomic energy," as called for by the Atomic Energy Act of 1946. In 1946/47, the Office of Naval Research (ONR) entered into a contract with the Library of Congress for the establishment of a Science and Technology Project (later known as the Navy Research Section and the Technical Information Division) for servicing technical reports of interest to ONR contractors. In 1948, the Central Air Documents Office (CADO) evolved out of the earlier Air Documents Research Center and Air Documents Division, with the responsibility for organizing all air documents for use by military organizations and their contractors.

These agencies all followed the OSRD example and sought to achieve bibliographic control over the reports in their domains by some sort of report numbering system. Their examples have in turn influenced the agencies they serve, including both originators and users of reports, and report numbers have come to be accepted as the principal device for report control.

Report numbers offer many advantages for bibliographic control. They have simplicity and brevity. They permit open references to security-classified material. When properly used, they provide positive identification of material, and depending on their make-up, may indicate source, date, subject or availability.

Several approaches have been used in the assignment of report numbers. There are simple publication codes with numbers assigned by originating or controlling agencies. These may include a serial number preceded by an indication of the type of publication involved, e.g., Technical Report No. 1, or TR-1, or a serial number preceded by a code for the issuing agency, e.g., LA-2000. Sometimes the serial number is replaced by date of issuance, e.g., ACR, May 1942.

There are accession and distribution codes assigned by recipients of reports. When assigned by individual agencies, they have meaning only within those agencies. When assigned by distribution centers, their meaning resembles publication codes. Many of the best known series have been of this type, e.g.,

PB-	Office of Technical Services
ATI-	Central Air Documents Office
AD-	Defense Documentation Center
U-, R-, C-, S-	Library of Congress, Technical Information Division (*see* Ref. 20)
NP-	Atomic Energy Commission. Technical Information Center
N(YEAR)-	National Aeronautics and Space Administration
X(YEAR)-	National Aeronautics and Space Administration

Although these accession codes are well known and internationally advertised through the report abstracting journals, they still add to the complexities of handling reports. Some feeling for these complexities can be gained from the article by Nancy Boylan *(4)*.

More sophisticated than the simple numerical publication and accession codes are those that reflect the bibliographic description of reports. These codes may include indications of the originating agency, the contract or project under which the work was performed, the type of report, the date of issuance and may or may not contain a serial number, e.g., NACA MR A4L12. It is this type of report code that gave rise to the term "short title" as a synonym for report number.

Finally there are report codes that reflect the subject matter of reports, e.g., RL-25.2.34. These are a natural derivation from traditional library classification schemes and may also be an extension of publication codes, which include symbols for organizational subdivisions of the originating agencies.

Report numbers are put to many different uses, and these uses have influenced the make-up of the codes. When the numbers are used for purposes of property or security control, there is a requirement that they give unique identification of the reports, but only within the property or security control system. There is no advantage to having control numbers indicate anything about the reports or their issuing agencies. When the numbers are used as filing designations, there may be a requirement that the numbers begin with an indication of the originating agency so that the reports will be grouped in the files by source. When the numbers are used in bibliographic references, there is the requirement that they give unique identification, together with the desirability of having the codes indicate the source and availability of the reports and of having them in standard, easily interpreted forms.

One evidence of the increasing acceptance of reports is the role ascribed to them in recent library publications. The 'Standard for Descriptive Cataloging of Government Scientific and Technical Reports" (Rev. No. 1) issued by the Defense Documentation Center in

October 1966 specifies that choice of corporate author should follow the level reflected mnemonically in the report series code *(5)*.

Nancy Boylan *(6)*, in her article, "Technical Reports," that appeared in *RQ* in the fall of 1970 stressed the value of understanding report numbers for librarians engaged in the detective work of reference librarianship. She mentions the first edition of this *Dictionary* as a valuable aid to that detective work.

Through the years since report numbers originated, many problems have arisen in connection with them. These problems were probably inevitable. In addition to the various possible approaches to their assignment and the different uses for which they were intended, there was a lack of precedent in handling them and a lack of understanding about their limitations. Hundreds of agencies were involved in their assignment and use; thousands of different codes in different forms sprang up. There was no way to control them. There was, and is, no way to assess their reliability. Some codes are well-known and indicate the sources of the reports, while others mean nothing. Yet all have been used indiscriminately for reference and control.

As early as 1947, when report numbers were still in their infancy, there was recognition of the problems they could pose. A Conference on the Bibliographic Control of Government Scientific and Technical Reports, held at the Library of Congress in September 1947, determined that the lack of standardization of report numbers detracted from their usefulness and recommended that a committee draw up a proposal for improvement *(7)*. The next year, Mortimer Taube *(8)* proposed that, "What is required is the promulgation by some organization having the requisite authority of a scheme through which every scientific and technical report prepared and issued under a research contract with any Federal agency would bear a symbol which would identify uniquely each report and relate it systematically to all other reports."

In 1949 the Special Committee on Technical Information of the Research and Development Board formed an Ad Hoc Panel on Project Identification Numbering Systems to look into the problem. The Panel was unable to recommend a uniform numbering system but did draw up a list of requirements for satisfactory identification numbers. These included identification of the cognizant government agency, a descriptive classification block and a serial block *(9)*.

By 1950 the situation had become serious enough so that the Central Air Documents Office (CADO) undertook a survey of report numbering systems then in use. The results of that study were issued in June 1950 in a CADO report which listed several hundred different numbering systems *(10)*. This preliminary survey was apparently never followed by a final survey or by an official statement of CADO's stand on report numbers, but it did lead to an article by Major P. K. Sturm in CADO's *Technical Data Digest (11)* in which he pleaded for a standard system of report numbers in which each number could consist of a prefix establishing the source of the report, a series number designating the series or contract number and the serial number of the report, and a suffix giving additional information such as the type of report, its security classification and its special distribution limitations. This article represents one of the few attempts to specify the makeup of report numbers.

Subsequent to the CADO survey no attempts were made to straighten out the confusion of report numbers. The informal g.s.i.s. program (the "group for standardization of information services") made up of representatives of the AEC's Technical Information Service, the Library of Congress' Navy Research Section, the Central Air Documents Office and the National Advisory Committee for Aeronautics' Division of Research Information, attempted to standardize catalog card format, subject headings and corporate author entries but did not concern itself with report numbering systems. At the Institute on the Administration and Use of Technical Research Reports, sponsored by the Science-Technology Division of the Special Libraries Association in 1952, the problem of report numbers was discussed by only one speaker *(12)* and at the Workshop on the Production and Use of Technical Reports, held at the Catholic University of America in 1953, it was not considered at all. Indeed, during the remainder of the 1950s there was only an unhappy resignation to the confused status quo.

In the 1960s that situation changed. Resignation became revulsion and efforts began to be made to find a way of controlling report numbers. The publication of the first edition of this *Dictionary* demonstrated the seriousness of the problem. Its effect was heightened by an article entitled "Report Number Chaos" appearing in *Special Libraries (13)*.

Shortly after publication of the *Dictionary* a report was issued by Herner and Company that gave the results of their findings from research undertaken under contract to the National Science Foundation *(14)*. The purpose of that contract was announced by NSF as the study of categories and code designations of United States government technical reports aimed at analyzing the various categories and code designations assigned to government reports by federal agencies and their contractors, and determining the need for and practicability of developing and adopting a coordinated government-wide system of improved report code designations *(15)*.

Unfortunately the Herner Company found the problem so large that they elected to ignore the portions of report codes acronymic for the issuing bodies, and concentrated instead on those designating the functions of reports. In the introduction to the report they state that useful information was obtained from 160 organizations, but that the basic corps of material

INTRODUCTION

was supplied by 71 organizations and pertained to 245 different report series. Even in so limited a sampling they found so little overlapping of practice from one agency's series to another as to render it insignificant.

In their conclusions about the feasibility of a government-wide controlled system of report numbers they cite several conflicting influences. On the one hand they refer to a steady trend toward the simplification of report numbers and an apparent interest in "order, predictability, mutuality, and continuity" *(16)*, but on the other they quote responses to their questionnaire that show feelings that reports and their designations are the private business of each organization, that local experience is reflected in the series and should be regarded as valuable, that reports are of interest to small circles of users who know and understand the idiosyncracies of the present arrangements, and that administrative controls would make standardization unfeasible.

Despite the negative results of the Herner study, the 1960s showed a series of separate efforts to bring greater appreciation to reports, and through education as well as standards, to provide some measure of bibliographical control over them. Workshops about reports were held by two Chapters of the Special Libraries Association, one by the Rio Grande Chapter in November 1965, and one by the New Jersey Chapter in April 1967. Handbooks on special librarianship began to give more attention to the report literature, and mentioned specifically the problems of dealing with report numbers. Articles about reports appeared in journals dedicated to many different phases of librarianship published in England as well as in the United States. An Italian report issued in 1969 goes so far as to propose international control over report codes and series *(17)*.

In the United States the most definite steps were taken by the Committee on Scientific and Technical Information of the Federal Council for Science and Technology. In 1965 they issued a three volume report entitled "Recommendations for National Document Handling Systems in Science and Technology" that emphasizes the general importance of the report literature *(18)*. This was reinforced the following year by a survey report on "Progress of the United States Government in Scientific and Technical Communication" *(19)*. In October 1966 COSATI issued its "Standard for Descriptive Cataloging of Government Scientific and Technical Reports" that includes precise instructions for ways of citing report numbers, as well as discussing their importance in selecting other descriptive cataloging elements *(20)*. COSATI's "Guidelines to Format Standards for Scientific and Technical Reports Prepared by or for the Federal Government" specifies that "each report shall carry a unique alphanumeric designation" and provides options for a designation provided by the sponsoring agency, one established by the performing organization, or one derived from contract or grant number *(21)*. COSATI also obtained agreement from the major report distribution and indexing agencies to honor one another's report numbers rather than proliferating the numbers applied to each report.

There is evidence that COSATI has given some consideration to the possibilities of national or international control over report series. A report issued in December 1968 by its Task Group on the Role of the Technical Report *(22)* and later revised and published in book form as a portion of "Scientific and Technological Communication" by Sidney Passman contains a reference to a proposal by Dr. C. Sherwin for "a national (and then an international) standard for reporting current efforts [of report identifier control] which would clarify this matter (viz. a Government-wide Form '1498' system." *(23)* His proposal has apparently never been issued by COSATI.

During the 1970s interest in reports and their designators has increased still more. A major tool for untangling report numbers appeared in 1972 with the publication of the *Directory of Engineering Document Sources (24)*. This directory covers approximately 3,600 document identifiers or "initialisms" for "technical/management specifications, standards, reports and related publications" in a broad interpretation of the engineering field. Although it includes only one-sixth the number of entries in this *Dictionary*, it does give additional information about the items covered, such as the indexes in which the document series are covered and the sources from which they are available. The directory also lists government report series arranged by the government branch or department and an organizational index that shows levels and relationships of government bodies. When used in conjunction with this *Dictionary* it should be a helpful reference tool, especially if current plans for making annual revisions are followed.

For the future a promising development is the interest of the American National Standards Institute in technical reports and their identifiers. ANSI Z-39 Committee on Library Work, Documentation and Related Publishing Practices Subcommittee 24 has been working for several years on an American National Standard for Scientific and Technical Reports: Format and Production. The current draft of that standard requires that an identifying report number be assigned to each separately bound document and be shown on all copies.

Another Z-39 subcommittee has been working for the past year on another national standard devoted solely to report numbers. The criteria on which it has proceeded are that every report should have a unique identifying number, that the numbers should be as brief as possible but should designate the issuing organization and date of publication as well as serial number, and that they should be adapted to both machine and manual document systems. Some thought has been given to ways in which the system could later be expanded for international use. The success of achieving unique report identifiers according to the proposed standard is predicated on the designation of a central authority responsible for registering report codes, providing information about assignments, and promoting the use of the standard.

Assignment of Report Designators

The assignment of report designators is indulged in by all who handle reports. Because of the lack of standardization for report designators and the multiple uses to which they may be put, every agency dealing with reports finds at times that it must create a designator or modify the one appearing on the report. Every such creation and modification adds to the likelihood that a subsequent user of the report will be misled by the numbers appearing on it. It is therefore important to know how, why and by whom report designators may be assigned and to recognize the characteristics and limitations of the different types.

The agency in the best position to assign report designators is the one originating or issuing the reports. Designators assigned at the point of origin enjoy the advantage of appearing on every copy of the report, of being accurate (usually), and of being known to the originating and to all distributing agencies. When they contain a designation of the source, they represent the ideal in report designators. Examples of such ideal codes are those applied to formal reports issued by the major contractors of the AEC. The codes are controlled by the AEC, they are brief, they are few in number, and they identify the formal reports completely. Examples of these codes are:

    ANL-   ORNL-
    ISU-   UCRL-

The difficulties with designators assigned by originating agencies arise when the designators 1) attempt to cover additional information about the report origin; 2) contain codes for organizational subdivisions instead of the main agency; or 3) cover only the type of report. Examples of troubles arising from these practices are:

1. Formal reports issued by Goodyear Atomic Corporation carry the code GAT-. Those issued by the divisions have an additional code such as GAT-L- for the Laboratory Division and GAT-DR- for the Engineering Development Division. When those divisions combined and were renamed the Technical Division, a GAT-T- series was created that continues the numbering of the GAT-L- series.

2. Johns Hopkins University reports generally carry the code JHU-. Those of its Solid Propellant Information Agency are designated SPIA-. Although this agency is well known, some recipients of its reports still prefer to prefix JHU- to the SPIA-.

3. The internal technical memoranda of the Naval Torpedo Station, Newport, R.I., carry the series code ITM-. Because this is completely ambiguous, most recipients modify it with a prefix to stand for the Naval Torpedo Station, and the prefixes may vary widely.

Still other difficulties arise when correspondence serials, log numbers, central file or other internal designations are applied to reports by originating agencies. When different correspondence serials are used by each subdivision of an agency, their application to reports can lead to an unexpected multiplicity of report codes. For example, the Los Alamos Scientific Laboratory has officially used only two codes for its reports, LA- and LAMS-, yet over 30 correspondence serials were found listed as report series in the sources used for assembling this *Dictionary*. In contrast to this, there is the complexity of numbers that occurs when the same prefix is used for correspondence as for reports. The General Electric Company's Hanford Atomic Products Operation exemplified this difficulty with its multidigit HW- designators. When file or log numbers are added to reports in addition to the formal report numbers, the result is only further confusion. Reports of Edgerton, Germeshausen & Grier, which carry both report numbers and EG&G Out- numbers, are examples.

After the originating agency, those next best qualified to assign report designators are the contracting or sponsoring agencies. These have cognizance of the work being done and usually set the reporting requirements. Their assignment of report designators is generally done in advance so that, as with the designators of originating agencies, report numbers may appear on the reports when they are issued. In addition their codes enjoy the advantage of being widely known and are frequently more easily recognized than those of small contractors.

Contracting agencies employ several methods of assigning numbers to the reports of their contractors. Some use simple serial systems in which all reports are assigned numbers according to the order of their publication. Reports in the Bumblebee or BB- series of Johns Hopkins University follow this pattern. Others employ subdesignations or suffixes to the series to indicate the various contractors. NEPA- reports used this system. For example, NEPA-893-UAC-5 indicates that NEPA report 893 was the fifth report from United Aircraft Corporation. Still others assign blocks of numbers to different contractors. The Manhattan District and its successor, the Atomic Energy Commission, followed this practice with its A-, M- and P- series.

A practice that has gained popularity in recent years is the inclusion of parts of the contract number in the series designation. In the mid-1960s the Atomic Energy Commission adopted that practice for its short term contractors and those that produce few research reports. Their system is to use the approved alphabetic code of the AEC operations office monitoring the contract, followed by the digits of the sequential identification of the contract and a final serial number.

Problems with designators assigned by contracting or sponsoring agencies arise when a project is sponsored by more than one agency, each with its own numbering system. An extreme example of a project with multiple sponsorship is quoted in the *Directory of Engineering Document Sources* as a single document with seven different numbers assigned as follows (25):

DSAM-4130.1	Defense Supply Agency Manual
AR-708-8	Army Regulation
NAVSUP-PUB-5006	Navy Supply Systems Command Publication
AFM-72-5	Air Force Manual
MCO-P4423.17B	Marine Corps Order
GSA-FSS-4130.1	General Services Administration; Federal Supply Service
DASA1-4100.5A	Defense Atomic Support Agency Instruction

Assignment of report designators by recipients is a very different situation. When the recipient is a national supply center for reports there is no problem. The series like PB-, AD- and N(year)- are well known, and the disadvantage of their assignment after issuance is more than offset by their appearance in standard announcement media and their indication of availability. They are described and explained by Nancy Boylan (26). Nevertheless, they create the necessity for a cross-reference from whatever number the originating agency may have applied first, and thus add a complication to the handling of the report. In the early days of report numbers, the need for numerical cross-references was not realized. Until the publication of *Correlation Index* by Special Libraries Association in 1953 (27), there was no connection between the PB- numbers by which the Office of Technical Services announced and sold reports and the report numbers given them by their originators. Gradually the importance of numerical cross-references has been accepted, even while the nuisance of creating and maintaining them has been increasingly deplored. The complexities are demonstrated most graphically in another article by Nancy Boylan entitled "Identifying Technical Reports through U.S. Government Research Reports and its Published Indexes" (28).

When the recipient is an individual agency the problem is much more acute. Many organizations have adopted control and filing systems based on simple accession numbers. These have the advantages of permitting ready machine control and of requiring added space only at the end of the file. These accession numbers mean nothing to agencies other than the one assigning them, but their appearance in accession lists invites their further use as bibliographic references. When the accession numbers are prefixed with a code indicating the receiving agency, as in the series DUH- and its successor GEH- applied to incoming reports at Hanford Engineer Works, they at least provide an indication of the agency that can identify the reports. When they lack that code, or carry an ambiguous one, they are a source only of confusion.

Almost as bad, and equally prevalent, are codes assigned by recipients in imitation of the codes used by the authoring agencies, or mnemonic for their names. These have the advantage to the assigners of allying reports with others issued by the same agencies, but the hazard for all others is that they treacherously resemble designators which have real meaning to the report's originators.

The most common and least offensive application of this technique involves the addition of a prefix designating source to the series designator used by the originator. This may involve simply the addition of JHU- to an SPIA- report, or it may mean the translation "Quarterly Progress Report Number 5" issued by the Analytical Division of the ABC Corporation into ABC-AD-QPR-5. Although synthetic, such designators are usually translatable by the originating agency.

At the other extreme are the designators that consist of an arbitrary number appended to a source code. So dangerous are these that most agencies assigning them insert a symbol between the source code and the arbitrary number to warn everyone that the numbers are bogus. Such warning symbols are the -Q- used by Sandia Corporation, the -LA- used by the Los Alamos Scientific Laboratory, and the -MP- used by the Air Force Special Weapons Center. In use, such report numbers look like: PADUAU-LA-6 which identifies the sixth report received without its own identifying number from the University of Padua by the Los Alamos Scientific Laboratory in New Mexico.

Agencies committed to systems requiring an identification of source as part of report designators are understandably eager for a standard system that would guarantee this information. Their interests run from a set of rules for assigning report designators to a dictionary of preferred report source codes constructed along the lines of union list library symbols. One example of a set of rules for assigning report numbers was given by W. H. Richardson (29); another comparable set of rules was suggested by C. W. Sargent (30). No list of preferred source codes has been published. If the proposed ANSI standard for report numbers is accepted in its current form one may evolve from the registry maintained by the central monitoring organization.

Because the systems used by agencies originating, sponsoring, distributing and receiving reports may differ and may necessitate the assignment of report numbers by each, multiple report designations occur frequently.

Some reports actually have several numbers printed on their covers; others bear one printed number with several others stamped or handwritten; some have inexplicable combinations of letters and numbers decorating all edges. An example encountered by the reference staff at the Los Alamos Scientific Laboratory recently was identified variously as:

AD 607 380	AFOSR 64 1891
SSL(UC)5-42	N65-13343
IER-64-12	SSL-TN-6

These threaten endless confusion to those handling the reports. The only hope of averting it is the creation of numerous cross-references. The resulting numerical files bristle with strange entries, grow at an alarming rate and irritate all who use them.

Components of Report Series Designators

Report designators are assigned by many people and used for many purposes, and have therefore come to include a large miscellany of information. In their compact form, all particulars of report identification, history and use can be packed into the 80 digit limitation of punched cards for computers. This is an invitation to assigners to load the numbers. A listing of the possible components of report designators in detail would resemble a dictionary of acronyms; a chart of possible arrangements of them would be more like a railroad guide. Neither is possible here, but a generalized listing of the more common components of report series designators follows together with three examples of the ridiculous lengths that can be achieved when any great number of components are combined.

1. Designators for originating or contracting agency

A. Symbols for name of agency: Despite continuing pressure for some sort of guide for selection of source codes, there have never been any generally accepted rules for their choice. Ordinarily initials are selected, or if these are too common or too difficult, acronyms are used. Obviously there is considerable opportunity for duplication. For example, PC- appears in this *Dictionary* as a code for:

Pepperdine College
Philco Corporation
Plasmadyne Corporation
Providence College
Psychological Corporation

Another distressing example is the confusion of the primary portion of the code APL- for Johns Hopkins University's Applied Physics Laboratory and the Air Force Aeropropulsion Laboratory. Their codes interfile to the 6th character, as shown in the sequences appearing in this *Dictionary*:

APL-SOR-MPCS-	APL-TM-
APL-TDR-(YEAR)-	APL-TPR-
APL-TG-	APL-TR-(YEAR)-

This type of confusion accounts for the fact that many codes have been added to this edition of the *Dictionary* with all letters spelled out, whereas in the first edition they were grouped under the general designation (LETTER(S)).

B. Symbols for originating agencies may be supplemented or supplanted by indications of contract, project, assignment, account or work order number. Most usual is the addition of the number or a pertinent part of it following the codes for originating agency. Another approach is the addition or substitution of a name or part of a name for the project. For example, the reports of the Meteor Project at the Massachusetts Institute of Technology are commonly referred to as METEOR- reports, MIT METEOR- reports, or MIT-MR- reports.

C. Occasionally, especially in the early days of reports, report designations were replaced or supplemented by the initials of the individuals responsible for the reports. Examples of this occurred at Argonne National Laboratory, where ANL-WHZ- was used for Dr. Zinn's reports, and ANL-HDY- for Dr. Young's. A recent example appears in this designation for a conference paper, CERN/ECFA-66/WG 2/US-SG4/EE-MP-AJH-1, where the last three groups of letters are the initials of the three authors.

2. Indications of subject classification

Subject classification may be shown by arbitrary symbols either substituting for or augmenting the source codes. There have been numerical codes and alphabetic ones. For example, the Aircraft Laboratory of the Engineering Division at Wright-Patterson Air Force Base employed a numerical code involving three digits for a general research subject and two more for the specific subject. The Radiation Laboratory of the University of California at first based its entire classification system on simple alphabetic abridgements of the subject matter of its research projects and later changed to a duodecimal system representing the same information (*see* Ref. 9). The Metallurgical Laboratory of the University of Chicago employed an alphabetical code system that relied on mnemonic indications of general subject fields. These supplemented, but did not affect, the numerical sequence of the regular series of C- reports (*see* Ref. 10).

3. Designation of form of reports

The most common way of specifying the form of a report is by an alphabetic designation replacing or supplementing the rest of the series code. Many such forms are given in "Ambiguous Alphabetical Codes" (Ref. 2), and are discussed at length in the 1962 Herner report. Other systems frequently employed are suffixes, as with

INTRODUCTION

Sandia Corporation's reports, e.g., SC----(NL) for nomenclature lists; by assignment of blocks of numbers, as with the AEC's Division of Technical Information (*see* Ref. 5); or by a modification of the basic report code, as used to be true of Los Alamos Scientific Laboratory's LA- and LAMS- report series (*see* Ref. 21).

4. Date

Date of writing, date of issuance or date of release is often specified in the report number, either by complete date, by parts of the date or by a date code. The latter practice was employed by the National Advisory Committee for Aeronautics in several of its codes (*see* Ref. 25).

5. Security classification

An indication of classification has often been incorporated into report codes. The most conspicuous use of it was by the Navy Research Section of the Library of Congress, which used the letters U-, R-, C-, and S- as the primary designations for the reports under its cognizance (*see* Ref. 20).

6. Mark showing that the designator was assigned by the recipient

Use of the -Q- by Sandia, the -LA- by Los Alamos and and the -MP- by AFSWC has been discussed in the preceding section.

7. Identifier

The identifier is that portion of a report designation that identifies the report, distinguishing it from all others in the series. Most often it is a serial number or occasionally a serial letter. Sometimes it is the date of issuance of the report. Occasionally it is a very short form of the title, with initial letters for the words of the title and numbers for the date of issuance.

8. Examples of excessively long report numbers, caused by using many of the possible components

AGARD-AC/82-D/4-PT-A SECT 1
        (27 characters and spaces)
CERN/ECFA-66/WG2/US-SG2/TR3
        (27 characters)
WLAD-BL/RP-FVCOG-KPC-70
        (23 characters)

Not only are there all these possible components of report designators, but there are also permutations of them. The ideal situation, a simple prefix, a serial number and possibly a suffix, is subscribed to by most report number assigners, but there is little uniformity as to the substance of the prefix, suffix or even the serial portion. A quick check of CADO's *Preliminary Survey* shows that any one of the various components may serve as the first part of a report designator. For example:

Source code: ISC-69, where ISC- represents Iowa State College.
Subdivisions of source: WD-12064, where WD- stands for Wichita Division of Boeing Airplane Company.
Initials: JRD-18-49R, where JRD- are the initials of J.R. Doe of Phillips Petroleum Company, Research Department, Bartlesville, Oklahoma.
Contract, project or work order: 2070-2, where 2070 is the contract number under the Electronic Subdivision, Engineering Division of the Operations Office at Wright-Patterson Air Force Base; 338-3, where 338 is the laboratory project number under Airborne Instruments Laboratory, Mineola, N.Y.; and 2009-5, where 2009 is the work order number under Aerojet Engineering Corporation, Azusa, California.
Form: E-1000, where E- stands for engineering report under Coles Signal Laboratory, or U-MCREXP-9008, where U- stands for an unsatisfactory report.
Date: 49-1-100, where 49 stands for 1949 under Monsanto Chemical Company, or A-4903-123, where A is the code for January, 49 for 1949, and 03 for the day of the month under Burndy Engineering Company.
Security classification: U-1234, where U- stands for unclassified reports received by the Navy Research Section of the Library of Congress.
Identifier: 123-S, where 123 is the sequence number of a report received by the Menasco Manufacturing Company, Burbank, California.

A similar list of examples could be prepared for components appearing as the last part of designators.

In addition to this bewildering assortment of components arranged in many ways, there is the further complication that some of the components may be ignored for the assignment of serial numbers. For example, in the C- series of reports issued by the Metallurgical Laboratory of the University of Chicago, the second or third letters of the codes were ignored in assignment of the serial number (*see* Ref. 10). In the LA/LAMS- report series of the Los Alamos Scientific Laboratory, the MS- has at times been ignored and at other times treated as indication of a separate series (*see* Ref. 21). Atomic Energy of Canada Ltd. used to maintain one series for its formal reports but inserted a letter for the laboratory of origin between the prefix and the serial number (*see* Ref. 6).

Evolution and Proliferation of Report Series Codes

Naturally, with so many characteristics of report numbers and no control of them, the opportunities for their proliferation are limitless. Many of the changes have been intentional and, however regretted by report users, must be accepted.

Chief among these are the changes in report codes to reflect changes in corporate author. Fox example, when the Manhattan District became the Atomic Energy Commission, its declassified documents were given AECD- numbers instead of MDDC- numbers. After the University of Chicago's Metallurgical Laboratory was named Argonne National Laboratory, its code changed from C- to ANL- but the serial numbers continued. Similarly, when Iowa State College became Iowa State University, its code shifted from ISC- to ISU-.

On the other hand some corporate name changes occur without alteration of the report codes. The University of California's Radiation Laboratory used the code UCRL- through several changes of name including the present Lawrence Berkeley Laboratory and Lawrence Livermore Laboratory.

Occasionally the form of report codes is changed by the originator. An outstanding example is the change of the A.E.R.E.- reports to the simpler AERE-. A.E.R.E. Inf/Bib- reports have become AERE-Bib- reports. More often such changes evolve gradually as references to the report numbers are simplified by recipients. For example, the DEG-INF. SER.- reports are also known as DEGIS- reports (*see* Ref. 39).

Other intentional changes have been perpetrated by the reference tools dealing with reports. For example, the *Directory of Engineering Document Sources* explains in its introduction that "Transcription has been based upon two, often conflicting, criteria: 1) Copying the document identification number directly from the document, and 2) Establishing a coherent format which enables listing by a consistent, left-to-right, character-by-character sequencing system. . . .

"Spaces have been replaced by dashes, particularly between initialisms and ensuing character groupings. Where the original consistently leaves no space throughout the series the original has been followed.

"Periods have been omitted from initialisms and abbreviations; decimal points have been retained within numeric-groupings. . . .

"We have thereby attempted to steer a middle course between the indigestible inconsistencies of the originals and the often unrecognizable interpolations necessitated for certain other special applications, such as logistics cataloging" *(31)*.

The machine produced abstracting and indexing journals follow similar rules in order to attain consistency and to adapt to the limited character sets of their printers. All punctuation is usually converted to hyphens, as are spaces.

The COSATI "Standard for Descriptive Cataloging" actually specifies such practices, calling for insertion of hyphens wherever spaces occur, substitution of slashes for ampersands, changes of Roman to Arabic numerals, and abbreviations of standard functional words to single letter designations *(32)*.

Such alterations of report numbers may offer advantages of consistency, but they also detract from the distinctiveness of a number's appearance. In this *Dictionary* the original form of the number has been preserved wherever possible, with variants included when they are strikingly different in appearance.

A very unsatisfactory, and fortunately relatively uncommon, evolution of codes occurs when report series are changed and instructions are issued for the renumbering of all previously issued reports in the series. Of course these instructions are never followed completely, and all that is accomplished is the necessity for an eternity of cross-references. Outstanding examples of such attempts at change are the Chicago Metallurgical Laboratory reports of the C- series, whose second and third letters were frequently changes (*see* Ref. 10), and the University of California's Radiation Laboratory reports, which were renumbered into the RL- series from an assortment of codes (*see* Ref. 9).

Serious as have been the intentional changes in report numbers, they cannot compare with changes that have occurred through careless references or different interpretations of series designations. Most prevalent, and understandable, has been the varied interpretation of report covers and title pages. For example, there is a confidential report bearing the following identification:

L.O.0510
(Rev. B)
Serial No. 15933
Copy No. -- of 105
ASROC Facility Equipment List
Prepared for U.S. Naval Ordnance Test Station, Pasadena, Calif. by Minneapolis-Honeywell Regulator Co. 1724 S. Mountain Avenue, Duarte, California
8 April 1958

It is conceivable that recipients of this report might in all good faith identify it as:
ASROC-15933
ASROC-L.O. 0510, Rev. B
L.O. 0510 Rev. B
MH-15933
MH-L.O. 0510 Rev. B
MHR-15933
MHR-L.O. 0510 Rev. B
NOTS-15933

INTRODUCTION

NOTS-L.O. 0510 Rev. B
USNOTS-15933
USNOTS-L.O. 0510 Rev. B
MHR-15933-NOTS-58
etc.

For a second example, there is an unclassified report of the (Great Britain) Ministry of Supply, Aeronautical Research Council, Reports and Memoranda, marked:

R. & M. No. 2615
(11,767)
A.R.C. Technical Report

This could be interpreted:

A.R.C. Technical report 11767
ARC-11767
ARC R&M-2615
ARC-TR-11767
R&M 2615
R. & M. No. 2615
etc.

Another source of proliferation in report codes is the abundance of errors appearing in the text and particularly in the indexes of the report abstracting and indexing journals. They seem to arise from faulty keypunching, failure to verify keypunching or to proofread listings, the limitations of optical character recognition devices, and the poor legibility of many computer printouts. These errors have become so numerous in the past few years that the compilation of this *Dictionary* has turned up hundreds of them and has led us to insert a special warning following the Explanatory Notes for the *Dictionary*.

The errors fall into several categories. Here are a few examples, arranged in order of the frequency with which we encountered the categories:

Substitution of the wrong letters

AFFRI-	for	AFRRI-
AFRII-SR-		AFRRI-SR-
AFMIC-TR-		AFMTC-TR-
AP-AN-		HP-AN-
ARCRL-TRANS-		AFCRL-TRANS-
CFA-R-		CEA-R-
CONE-		CONF-
NRE-ISD-TN-		WRE-ISD-TN-

Transposition of letters

AFCRL-PRSP-	for	AFCRL-PSRP-
AIFT-TR-		AFIT-TR-
AMFL-TR-		AFML-TR-
AORD-		AROD-
ARML-TR-		AMRL-TR-
DCLS-		DLCS-

Added letters

CISAV-MEMO-	for	ISAV-MEMO-
DOTS-HS-		DOT-HS-
XX-643-		X-643-

Omitted letters

AFDL-TR-	for	AFFDL-TR-
C/PSEL-TS-		C/PLSEL-TS-

One example that demonstrates the prevalence of these errors was encountered in a single quarterly index to *Government Reports Announcements*. Reports in the AFOSR- series were listed there with these variants:

AFOSR-70-   (intended form)
AFOSR-0-
AF-AFOSR-70-
AFOSR-AFOSR-70-
AFCSR-70-
AFORS-70-
AFSOR-70-

The fact that makes these apparently obvious errors so serious is that by machine sorting, such references are spread out in the report number indexes and may elude the unwary user. Because of that danger, we have elected to include in this *Dictionary* those erroneous codes that we have encountered in the standard abstracting and indexing services. We trust that by so doing we will help reference librarians searching for aberrant report numbers.

Obviously, there is need for some control of report numbers. There must be sets of rules for assigning them, dictionaries of preferred source codes and patterns for arranging their components, all aimed at report designations that will be complete, consistent, concise and unique.

We sincerely hope that this second edition of the *Dictionary of Report Series Codes* will help report users cope with the chaotic situation surrounding report numbers until order can be created by the use of standards such as the one being proposed by the American National Standards Institute.

Literature Cited

1. Committee on Scientific and Technical Information. *Recommendations for National Document Handling Systems in Science and Technology.* (PB 168 267; AD 624 560). Washington, D.C.: November 1965. 3 v. pap.
2. Committee on Scientific and Technical Information. *Progress of the United States Government in Scientific and Technical Communication.* (PB 176 535). Washington, D.C.: 1966. 35p. pap.
3. Taube, Mortimer. Memorandum for a Conference on Bibliographic Control of Government Scientific and Technical

Reports. *Special Libraries* 39 (no. 5): p. 156 (May-Jun 1948).
4. Boylan, Nancy G. Identifying Technical Reports through U.S. Government Research Reports and Its Published Indexes. *College and Research Libraries* 28: p.175-83 (1967).
5. Committee on Scientific and Technical Information. *Standard for Descriptive Cataloging of Government Scientific and Technical Reports.* Revision l. (AD 641 092). Washington, D.C.: October 1966, p.8-9.
6. Boylan, N. G. "Technical Reports". *RQ* 10: p.18-20 (Fall 1970).
7. Library of Congress. Conference on the Bibliographical Control of Government Scientific and Technical Reports. *Minutes of Meetings, September 25-27, 1947.* Washington, D. C.: October 1947, p.15, as quoted in B. M. Fry, *Library Organization and Management of Technical Reports Literature* (Catholic University of America Studies in Library Science, no.1). Washington, D.C.: 1953, p.55.
8. Taube. *loc. cit.*
9. Research and Development Board. *Final Report of the Ad Hoc Panel on Project Identification Numbering System.* Washington, D.C.: January 24, 1949, p.1, as quoted in Fry, *op. cit.*
10. Central Air Documents Office. *Preliminary Survey of Report Numbering Systems* (ATI No. 78633). Wright-Patterson Air Force Base, Dayton, Ohio: June 1950.
11. Sturm, P. K. Some Thoughts Presented on Numbering Systems. *Technical Data Digest* 16 (no. 10): p.5-6 (Sep 1951).
12. Warheit, I. A. Bibliographic Identification and Organization. *American Documentation* 30 (no.2): p. 106-7 (Spring 1952).
13. Redman, H. F. Report Number Chaos. *Special Libraries* 53 (no.10): p.574-8 (Dec 1962).
14. Herner and Company. *Functional Symbols Currently in Use in Government Reports.* prepared by P.R. Mahany et al. Washington, D.C.: December 1962.
15. *Science Information Notes* 3 (no.4): p.1-2 (1961).
16. Herner and Company, *op. cit.*, p.27.
17. Alberani, V. *Rapporto Tecnico e di Ricerca: Considerazioni Generali di Carattere Bibliografico.* (ISS 69/3). Istituto Superiore di Sanita, Roma: Gennaio 1969.
18. Committee on Scientific and Technical Information (PB 168 267) *op. cit.*
19. Committee on Scientific and Technical Information (PB 176 535) *op. cit.*
20. Committee on Scientific and Technical Information. (AD 641 092) *op. cit.*
21. Committee on Scientific and Technical Information. *Guidelines to Format Standards for Scientific and Technical Reports Prepared by or for the Federal Government.* (PB 180 600). Washington, D.C. December 1968. p.4-5.
22. Committee on Scientific and Technical Information. Task Group on the Role of the Technical Report. *The Role of the Technical Report in Scientific and Technological Communication.* (PB 180 944). Washington, D.C.: December 1968.
23. Passman, S. *Scientific and Technological Communication.* Oxford, Pergamon Press: 1969. p.47.
24. Global Engineering and Documentation Services Inc. *Directory of Engineering Document Sources.* D.P. Simonton, comp. Newport Beach, Calif.: 1972.
25. Global Engineering Documentation Services Inc. *op. cit.*, preface.
26. Boylan, *op. cit.*
27. Special Libraries Council of Philadelphia and Vicinity. *Correlation Index: Document Series and PB Reports*, ed. by Gretchen E. Runge. New York: Special Libraries Association, 1953, 271p. pap. litho.
28. Boylan, 1967. *loc. cit.*
29. Richardson, W.H. Report Numbers -- Boon or Bugaboo. *Special Libraries Association, Rio Grande Chapter Bulletin* 4 (no.2): p.3-5 (Sep 1960). Also in *Sci-Tech News* 15 (no. 1): p. 5-6, 9 (Spring 1961).
30. Sargent, C.W. Handling Technical Reports in the Medical Library. *Bulletin of the Medical Library Association* 57 (no.1): p.41-6 (Jan 1969).
31. Global Engineering and Documentation Services Inc. *op. cit.* 1st and 2nd pages of Introduction.
32. Committee on Scientific and Technical Information. *Standard for Descriptive Cataloging. . . op. cit.* p. 21-3.

Bibliography

Alberani, V. Rapporto Tecnico e di Ricerca: Considerazioni Generali di Carattere Bibliografico. (ISS 69/3). Istituto Superiore di Sanita, Roma, Gennaio 1969. 20p. pap.

Barr, K.P. Non-Standard Material at the NLL. *ASLIB Proceedings* 17: p.240-5 (August 1965).

Boylan, Nancy G. Identifying Technical Reports through U.S. Government Research Reports and Its Published Indexes. *College & Research Libraries.* p.175-83 (May 1967).

———. Technical Reports. *RQ* 10: p.18-21 (Fall 1970).

Burkett, J., ed. *Special Library and Information Services in the United Kingdom.* 2nd, rev. ed. London: Library Association, 1965. 366p.

Canter, Louis. The Impact of Documentation on the Research Library. *Special Libraries* 45 (no.10): p.407-13 (Dec 1954).

Carlson, Irving G. Cataloging of Technical Reports in a Military Research Library. *American Documentation* 3 (no.2) p.119-22 (Spring 1952).

Central Air Documents Office. *Preliminary Survey of Report Numbering Systems* (ATI No. 78633). Wright-Patterson Air Force Base, Dayton, Ohio: June 1950. 70p. pap.

Chillag, J.P. Problems with Reports, Particularly Microfiche Reports. *ASLIB Proceedings* 22: p. 201-16 (May 1970).

Committee on Scientific and Technical Information. *Guidelines to Format Standards for Scientific and Technical Reports Prepared by or for the Federal Government* (PB 180 600). Washington, D.C.: December 1968. 16p. pap.

———. *Progress of the United States Government in Scientific and Technical Communication* (PB 176 535). Washington, D.C.: 1966. 35p. pap.

———. *Recommendations for National Document Handling Systems in Science and Technology* (PB 168 267; AD 624 560). Washington, D.C.: November 1965. 3 vols. pap.

———. *Standard for Descriptive Cataloging of Government Scientific and Technical Reports;* Rev. 1 (AD 641 092). Washington, D.C.: October 1966. 50p. pap.

———. Task Group on the Role of the Technical Report. *The Role of the Technical Report in Scientific and Technological Communication* (PB 180 944). Washington, D.C.: December 1968. 112p. pap.

Connor, John M. The Need for Documentation to Government Specifications. *Special Libraries* 47 (no.4): p.152-5 (Apr 1956).

Fry, Bernard M. Cataloging Government Technical Reports. In Weil, B.H. ed. *The Technical Report: Its Preparation, Processing and Use in Industry and Government.* New York: Reinhold, 1954. Chap. 18.

———. *Library Organization and Management of Technical Reports Literature* (Catholic University of America Studies in Library Science, no.1). Washington, D.C.: 1953, 140p. pap. litho.

——— and Kortendick, James J., eds. The Production and Use of Technical Reports: The Proceedings of the Workshop on the Production and Use of Technical Reports, Conducted at the Catholic University of America from April 13 to April 18, 1953. Washington, D.C.: The Catholic University of America Press, 1955. 175p. pap. litho.

Global Engineering Documentation Services Inc. *Directory of Engineering Document Sources.* comp. by D.P. Simonton. Newport Beach, Calif.: 1972. unpaged. pap.

Gray, Dwight E., and Rosenberg, Staffan. Do Technical Reports Become Published Papers? *Physics Today* 10 (no.6): p.18-21 (Jun 1957).

Hall, J. Technical Report Literature. In Ashworth, Wilfred, ed. *Handbook of Special Librarianship and Information Work.* London: ASLIB, 1967. Chap. 7.

Herner and Company. *Functional Symbols Currently in Use in Government Reports*; Final Report to the National Science Foundation. Prep. by Paul R. Mahany et al. Washington, D.C., Herner & Co.: December 1962. 121p.

Houghton, Bernard. *Technical Information Sources:* A Guide to Patents, Standards and Technical Reports Literature. London, Clive Bingley, 1967, Chaps. 9-10.

Jackson, Eugene B. Acquisitions: Sources and Techniques. *American Documentation* 3 (no.2): p.94-100 (Spring 1952).

Kee, Walter A. Availability of Technical Reports. *Sci-Tech News* 16 (no.1): p.7-9 (Spring 1962).

Leondar, Judith C. Workshop Proceedings: Report Literature and Sources of Information. *Special Libraries* 59: p.84-5 (Feb 1968).

Literature of Nuclear Science: Its Management and Use; Proceedings of Conference Held at Division of Technical Information Extension, Oak Ridge, Tennessee, September 11-13, 1962. Oak Ridge, Tenn., USAEC: December 1962. 403p. pap.

Miller, E. Eugene. The Genesis and Characteristics of Report Literature. *American Documentation* 3 (no.2): p.91-4 (Spring 1952).

Passman, Sidney. *Scientific and Technological Communication.* Oxford, Pergamon Press: 1969, Chap. 4.

Pfoutz, Daniel R. Guide to Report Literature. *Library Journal* 84 (no.19): p.3363-6 (Nov 1, 1959).

Proceedings of the Regional Workshop on the Report Literature, Held at the Hilton Hotel, Albuquerque, New Mexico, November 1-2, 1965. Co-sponsored by the Rio Grande Chapter, and Science-Technology Division, Special Libraries Association. North Hollywood, Calif.; Western Periodicals Co. 1966, 150p. pap.

Redman, Helen F. Report Number Chaos. *Special Libraries* 53: p.574-8 (Dec 1962).

Redman, Helen F. Technical Reports: Problems and Predictions. *Arizona Librarian* 23 (no.1): p.11-7 (Winter 1965/66).

Richardson, William H. Report Numbers -- Boon or Bugaboo. *Special Libraries Association, Rio Grande Chapter Bulletin* 4 (no.2): p.3-5 (Sep 1960). Also in *Sci-Tech News* 15 (no.1): p.5-6, 9 (Spring 1961).

Sargent, Charles W. Handling Technical Reports in the Medical Library. *Bulletin of the Medical Library Association* 57 (no.1): p.41-6 (Jan 1969).

Strauss, Lucille J. et al *Scientific and Technical Libraries: Their Organization and Administration.* New York: Interscience, 1964. p.255-7.

Sturm, P.K. Some Thoughts Presented on Numbering Systems. *Technical Data Digest* 16 (no.9): p.5-6 (Sep 1951).

Tallman, Johanna. History and Importance of Technical Reports. *Sci-Tech News*, Part I: vol.15 (no.2): p.44-6 (Summer 1961); Part II: vol.15 (no.4): p.164-5, 168-72 (Winter 1962); Part III: vol.16 (no.1): p.13 (Spring 1962).

Taube, Mortimer. Memorandum for a Conference on Bibliographical Control of Government Scientific and Technical Reports. *Special Libraries* 39 (no.5): p.154-60 (May-Jun 1948).

Warheit, I. A. Bibliographic Identification and Organization. *American Documentation* 3 (no.2): p.105-10 (Spring 1952).

West, Martha W. More on Reports (Ltr.). *RQ* 10: p.286 (Spring 1971).

Explanatory Notes for this Dictionary

Inclusiveness

This *Dictionary* attempts to include all series codes known to have been used in numbering the report literature. It therefore includes codes assigned by agencies other than those issuing the reports. The confusion resulting from code assignment by many agencies is treated fully in the Introduction (*see* p. 6).

The codes included are primarily those used on reports, but codes assigned to other documentary material have been included when they might be confused with report codes. Such codes as those applied to specifications, standards, etc., are therefore partially included.

Correspondence serials are included only when they are known to have been used on reports as well as on correspondence.

Only codes known to have been used on two or more reports are considered to be series codes and, therefore, included.

Because it is not possible for any one group of persons to encounter all codes, there are, undoubtedly, omissions.

Form of the Report Series Codes

The form of the series codes is given as shown on the reports themselves, on published catalog cards for the reports and in published lists of report series codes, except that, because of the computer listing used, upper case letters are used throughout this *Dictionary*, whereas combinations of upper and lower case letters are often encountered elsewhere.

A report series code may be composed of letters and/or numbers in any arrangement. The numbers may be arbitrary or may stand for the year, month, day, organization number, contract number, etc. When the compilers were sure that the latter was the case, such numbers are given in the *Dictionary* as (year), (month), (contract no.), etc. When the use of numbers was obviously arbitrary or was not known to be otherwise, the simple designation (number) or (numbers) is used.

The letters used in codes may also be arbitrary or may stand for the initials of the originating agency, one of its laboratories, a person connected with the project, the designation of a research project, etc. When the letters are obviously arbitrary or were not known to be otherwise, they are given as (letters) in the *Dictionary*. When they were known to be initials, the term (initials) is used. When they were known to stand for a laboratory, research project, etc., they are given as used, with the name in full appearing as a part of or following the corporate entry (*see also* p. 16, Parenthetical Notes).

It was not possible to list every variant form that may be encountered. Possible variants are discussed in the Introduction, p. 10.

In general, a series code is followed by a serial number, which number is the only part of the report designation that is distinctive for a given report. The use of such a distinctive number in the right hand position in a report designation is indicated in this *Dictionary* by hyphen (-) or a slash (/), depending upon which the originating agency uses to set off the distinctive number.

Sometimes the part of the designation that is distinctive is found in some position other than the right end of the designation. In such a case, it is indicated in this *Dictionary* by the fact that there is no hyphen or slash at the end of the series code entry and by the appearance of the designation (number), (numbers), (no.) or (nos.) elsewhere in the series code.

Sometimes, too, the part of the designation that is distinctive is a letter or letters in the series code. This is indicated by the incorporation of the term (letter) or (letters) in the appropriate part of the series code.

Arrangement of Entries

Report series codes are arranged alphabetically letter by letter, disregarding marks of punctuation, until a number is encountered. A number, or a word which stands for a number, as (year), (month), (number), stops the sorting.

The only exception is in codes which start with numbers. Those codes file by the letters in the code, even if the letters are the last of several elements of the code. The difficulty of identifying codes that start with numbers is explained in Ref. 1.

In this edition we have had to ignore punctuation when arranging code entries, because computer manipulation of report numbers by indexing and abstracting services has made punctuation meaningless. Therefore, the alphabetical arrangement of codes in this edition is almost completely different from that of the first edition.

If two or more codes are identical, the arrangement is alphabetical by agency.

EXPLANATORY NOTES FOR THIS DICTIONARY

Selection of Corporate Entry

The corporate entry given for a report series code is either the agency issuing the report or the agency assigning the report number. It is the issuing agency whenever that agency is also the one assigning the number or whenever the code relates solely to that agency, regardless of who assigns it. It is the contracting agency whenever that agency assigns a code that does not relate to the issuing agency. And it is the agency receiving a copy of the report if that agency assigns a code that does not relate to the originator. In this last situation some explanation about the code is given, either in parentheses following the agency's name or in a reference note.

Form of Corporate Entry

The form of corporate entry is the name and location of the agency at the time it assigned the series designation. If the agency had more than one name or location during the period in which it was assigning one series designation, only the latest form of name is used, although more than one location may be shown.

The depth of information about divisions, laboratories, etc. within corporate bodies varied in the sources, and therefore varies in the *Dictionary*.

Approximately half of the entries are from the first edition. We chose to continue the rules used for that rather than use the COSATI rules for corporate entries which have been adopted in the interim.

The form used by the Atomic Energy Commission in its report *Corporate Author Entries Used by the Division of Technical Information in Cataloging Reports* (TID-5059, 6th rev, 1964) has generally been used in this work. An exception is the omission of the name of the city for the corporate entry when the entry includes the name of the city, e.g.:

Springfield Armory, Mass.
Los Alamos Scientific Lab., N. Mex.

Cross-references to later forms of name are given in the Corporate Entries section of this *Dictionary* only when the code(s) used under an early form of the name were continued under a later form. More complete Cross-references to later forms of name are given in COSATI 70-7.

In general, government organizations other than those of the United States are entered under the name of the government, followed by the name and address of the organization. When such an organization starts with the adjectival form of the government's name, as Australian... or Deutsche..., that form is used.

Foreign names are entered under the form first or most frequently encountered by the compilers. Therefore there is a mixture of English and vernacular entries for many foreign countries, e.g.:

Sweden. Flygtekniska Foersoeksanstalten, Stockholm
Sweden. Inst. for Building Research, Stockholm

The following cities in the United States and Canada are given without the state or province designation, following the practice of the Library of Congress and the Division of Technical Information of the Atomic Energy Commission in 1962:

Albany	Nashville
Annapolis	New Haven
Atlanta	New Orleans
Atlantic City	New York
Baltimore	Oklahoma City
Boston	Omaha
Brooklyn	Ottawa
Buffalo	Philadelphia
Chattanooga	Pittsburgh
Chicago	Providence
Cincinnati	Quebec
Cleveland	Richmond
Colorado Springs	St. Augustine
Dallas	St. Louis
Denver	St. Paul
Des Moines	Salt Lake City
Detroit	San Antonio
Duluth	San Francisco
Fort Wayne	Savannah
Grand Rapids	Scranton
Hartford	Seattle
Honolulu	Spokane
Indianapolis	Tacoma
Jersey City	Tallahassee
Los Angeles	Toledo
Memphis	Toronto
Milwaukee	Trenton
Minneapolis	Wheeling
Montreal	

Similarly, the following foreign cities are given without designation of country:

Aachen	Bucharest
Adelaide	Budapest
Amsterdam	Buenos Aires
Antwerp	Cairo
Asunción	Calcutta
Athens	Cape Town
Bangkok	Caracas
Barcelona	Cluj
Basel	Coblenz
Beirut	Cologne
Belfast	Copenhagen
Belgrade	Delhi
Bergen	Dresden
Berlin	Dublin
Bern	Dusseldörf
Bogotá	Edinburgh
Bologna	Erevan
Bombay	Essen
Bonn	Florence
Bordeaux	Frankfurt am Main
Bremen	Freiburg i.B.
Brisbane	Geneva
Brno	Genoa
Brunswick	Ghent
Brussels	Glasgow

Göteborg
Graz
Guatemala
Hague, The
Hamburg
Hanover
Heidelberg
Helsinki
Hiroshima
Innsbruck
Istanbul
Jerusalem
Johannesburg
Kharkov
Kiel
Kiev
Krakow
Kyoto
Lahore
Leiden
Leipzig
Leningrad
Liege
Lima
Lisbon
Liverpool
London
Luxemburg
Lvov
Lyon
Madras
Madrid
Manila
Marseille
Melbourne
Mexico City
Milan
Moscow
Munich
Naples
New Delhi
Nice
Nuremberg
Odessa
Osaka
Oslo
Oxford
Padua
Paris
Prague
Pretoria
Quito
Rio de Janeiro
Rome
Rotterdam
Rouen
Seoul
Sevastopol
Seville
Shanghai
Sofia
Stockholm
Strasbourg
Stuttgart
Sydney
Tashkent
Tokyo
Toulouse
Tunis
Turin
Uppsala
Utrecht
Venice
Vienna
Warsaw
Wiesbaden
Zagreb
Zurich

Parenthetical Notes

All information available to the compilers about the meanings of any sections of the series code is given. If such information is not the name of a division, laboratory, school or other entity within the assigning organization, it is given parenthetically, e.g., (Internal Reports). A listing of frequently-encountered terms like this is given in Ref. 2.

Use of Reference Notes

All long or involved explanations of series designations or of the practice of the assigners in expanding on a series designation are given in the Reference Notes chapter. The notation (Ref. -) refers to these reference notes.

If You Do Not Find It Here. . .

(Hints from the Editors for those seeking to identify report series codes which are not in this Dictionary)

Despite six years of effort, this *Dictionary* is, of course, incomplete and out of date. We are sure that there are many codes that we did not encounter. We know that there were some that we were unable to identify. And, codes are proliferating at a rate exceeding any publishing schedule. Therefore, many are truly missing from this compilation.

But, from our experience in assembling the codes that are here, we suspect that many "missing" codes will turn out to be simple errors. These errors, their causes and forms are discussed in the Introduction, page 11. They fall into several general categories, and may be coped with once they are recognized.

Changes in spacing and punctuation. The arrangement of the codes portion of this *Dictionary* has been changed from that of the first edition to minimize the effects of these alterations. Do not be put off by spaces filled with hyphens, ampersands replaced by slashes, or even spaces closed up. Such changes are made deliberately by many indexing sources, and also occur inadvertently. To compensate for this, ignore all punctuation in looking for the letters in a code.

Substitution of wrong letters. These are the most frequent errors in the indexing journals. Try substituting letter nearby on the keyboard for those you seek: B for V, F for R, S for A, for example. Or, try look-alikes that might have been misread: C for O, P for B, R, or F, I for T; or, a particularly odd one, S for &. If double letters occur in the code you want, try doubling a nearby letter instead of the one that is doubled in the reference you have.

Transposition of letters. Transposed letters occur almost as frequently as wrong letters, and can occur anywhere in the code. Try other arrangements of the letters you are seeking.

Omission of letters. This is a frequent fault and occurs most often in lengthy codes, or codes with double letters. Filling in the missing letter is a real skill that is improved by experience.

Addition of letters. This happens often enough to be troublesome, especially if the added letter is the initial letter. Good guesswork is the only hope.

If detective work from these clues fails to turn up the code you seek, cast about for codes similar in pattern. New codes are being created constantly, but they often follow the pattern of other codes of the same issuing agency.

The usual components and typical patterns of codes are described in the Introduction. By studying that, and with ingenuity and perseverance, you, too, may become a super code sleuth!

Abbreviations

The following abbreviations have been used throughout the *Dictionary*. An asterisk indicates that an abbreviation was used only when it enabled the compilers to avoid running the entry onto another line. Occasionally an abbreviation was written without a period when the single space saved enabled the compilers to condense the entry to one line.

+	and	Conf.	Conference	Gt.	Great
A.B.	Aktiebolag	Conn.	Connecticut	Gt. Brit.	Great Britain
A.F.	Air Force	Constr.	Construction	H.Q.	Headquarters
A.G.	Aktiengesellschaft	Coop.	Cooperative	Hants	Hampshire
Adm.	Administration	Corp.	Corporation	Herts	Hertfordshire
AEC	Atomic Energy Commission	Ctr.	Center, Centre	HQ.	Headquarters
		Cumb.	Cumberland	Ident.	Identification
Aero.	Aeronautical	D.C.	District of Columbia	Ill.	Illinois
AFB	Air Force Base	*DC	District of Columbia	Inc.	Incorporated
Agric.	Agriculture	Del.	Delaware	Incl.	Inclusive
Ala.	Alabama	Dept.	Department	Ind.	Indiana (only at end of entry or when obviously part of address)
Alta.	Alberta	Dev.	Development		
Amer.	American	Dir.	Director		
Anal.	Analytical	Diss.	Dissertation	Ind.	Industrial
Ariz.	Arizona	Div.	Division	Indpls.	Indianapolis
Ark.	Arkansas	Doc(s).	Document(s)	Info.	Information
Assn.	Association	DOD	Department of Defense	Initl(s).	Initial(s)
Assocs.	Associates	E.	East	Inorg.	Inorganic
Asst.	Assistant	E.D.P.	Electronic Data Processing	Inst.(s).	Institute(s), Institut(s)
Atmos.	Atmospheric	E.G.	Exempli gratia	Inst. of Tech.	Institute of Technology
Atomic En.	Atomic Energy	E.V.	Eingetragener Verein		
Auth.	Authority	Edit.	Edition	Instn.	Institution
B.C.	British Columbia	Elec.	Electrical	Instr(s).	Instrument(s)
Bd.	Board	Electrodev.	Electrodevelopment	Instruc. Bk.	Instruction Book
Beds	Bedfordshire	Electrotech.	Electrotechnical	Intell.	Intelligence
Berks	Berkshire	En.	Energy (only used with Atomic...)	Internatl.	International
Bib.	Bibliography			Intl(s).	Initial(s)
Biol.	Biological	Eng.	Engineering	Invest.	Investigation
Bk.	Book	Eng.	England (only at end of entry)	Jnl.	Journal
Bldg.	Building			Jt.	Joint
Br.	Branch	Equip.	Equipment	K.G.	Kommanditgesellschaft
Bros.	Brothers	*Est.	Establishment	Kan.	Kansas
Bucks	Buckinghamshire	Estab.	Establishment	Ky.	Kentucky
Bull(s).	Bulletin(s)	EV	Eingetragener Verein	La.	Louisiana
Bur.	Bureau	Exp.	Experiment	Lab.	Laboratory
*Cal.	California	Exptl.	Experimental	Labs.	Laboratories
Calif.	California	Fdn.	Foundation	Lancs	Lancashire
Can.	Canada	Fla.	Florida	Leics	Leicestershire
CEA	Commissariat a l'Energie Atomique	Ft.	Fort	*Lond.	London
		G.m.b.H.	Gesellschaft mit beschränkter Haftung	*Los Ang.	Los Angeles
Chas.	Charles			Ltd.	Limited
Chem(s).	Chemical(s)	Ga.	Georgia	Ltr(s).	Letter(s)
Ches.	Cheshire	Gd.	Ground	m.b.H.	mit beschränkter Haftung
Cmdr.	Commander	Gen.	General	Mach.	Machinery
Co.	Company	Gloucester	Gloucestershire	Man.	Manitoba
Coll.	College	*GmbH	Gesellschaft mit beschränkter Haftung	Mass.	Massachusetts
Colo.	Colorado			Math.	Mathematics
*Colo. Spgs.	Colorado Springs	Govt.	Government	Md.	Maryland
Comm.	Committee	Gp.	Group	Me.	Maine
Con.	Contract	Grad.	Graduate	Mech.	Mechanical

ABBREVIATIONS

Med.	Medicine	*NY	New York (state)	Spec.	Special
Memo.	Memorandum	*NYC	New York City	Specs.	Specifications
Memos.	Memoranda	Obs.	Observatory	SSR	Soviet Socialist Republic(s)
Met.	Metallurgical	Oct.	October		
Mfg.	Manufacturing	Off.	Office	SSSR	Soyuz Sovetskikh Sotsialisticheskikh Respublik
Mfr(s).	Manufacturer(s)	Okla.	Oklahoma		
Mich.	Michigan	Ont.	Ontario		
Middx.	Middlesex	Ore.	Oregon	St.	Saint
Mil.	Military	Org.	Organization	Sta(s).	Station(s)
Min.	Ministry	*Pa.	Pennsylvania	Std(s).	Standard(s)
Minn.	Minnesota	Penna.	Pennsylvania	Sub-comm.	Sub-committee
Misc.	Miscellaneous	*Phila.	Philadelphia	Subdiv.	Subdivision
Miss.	Mississippi	Phys.	Physics	Subj.	Subject
Mo.	Missouri (only at end of entry or when obviously part of address)	*Pitts.	Pittsburgh	Suff.	Suffolk
		Prelim.	Preliminary	Supp.	Supplement
		Prod(s).	Product(s)	Tech.	Technical
Mo.	Month	Proj(s).	Project(s)	Tech.	Technology (only with Inst. of... or Science and ...)
Mont.	Montana	Protect.	Protection		
Mt.	Mountain	Pt.	Port		
N.	North	Psych.	Psychology	Temp.	Temperature
N.C.	North Carolina	Que.	Quebec	Tenn.	Tennessee
N.D.	North Dakota	R&D	Research and Development	Tex.	Texas
N.H.	New Hampshire	R.I.	Rhode Island	TNO	Toegepastnaturweien-schappelijk Onderzoek
N.J.	New Jersey	R.S.R.	Republici Socialiste Romania		
*N.M.	New Mexico			Trans.	Translation
N. Mex.	New Mexico	Ref.	Reference	Transport.	Transportation
N.S.	Nova Scotia (only at end of entry)	Rept(s).	Report(s)	Trng.	Training
		Res.	Research	U.K.A.E.A.	United Kingdom Atomic Energy Authority
N.S.	Nuclear Ship	Roman No.	Roman Numeral		
N.V.	Naamloze Vernootschap	*Rpt(s).	Report(s)	U.S.	United States
N.W.T.	Northwest Territories	RVO	Rijksverdedigings-organisatie	U.S.S.R.	Union of Soviet Socialist Republics
N.Y.	New York (state)				
N.Y.C.	New York City	S.C.	South Carolina	Univ.	University, Universitaet, Universite, Universidade
NASA	National Aeronautics and Space Administration	S.D.	South Dakota		
		S. Dak.	South Dakota	Univs.	Universities
Natl.	National	S.P.A.	Societa per Azioni	Va.	Virginia
Neb.	Nebraska	*San Fran.	San Francisco	Vict.	Victoria
Nev.	Nevada	Sask.	Saskatchewan	Vt.	Vermont
*NC	North Carolina	Sch.	School	W. Va.	West Virginia
*NJ	New Jersey	Sci.	Scientific	Warwicks	Warwickshire
*NM	New Mexico	Scot.	Scotland	Wash.	Washington
No.	Numeral (only with Roman...)	Sec.	Section	Wilts	Wiltshire
		Secy.	Secretary	Wis.	Wisconsin
No(s).	Number(s)	Ser.	Series	Worcs	Worcestershire
Nr.	Near	Slash	Slash mark (/)	Wyo.	Wyoming
Nuc.	Nuclear	So.	South	Yorks	Yorkshire
NW	Northwest			Yr.	Year

Glossary

Assigning agency: That body responsible for assigning a series code to a given document.

Code: A system of arbitrarily chosen letters and/or numbers used for brevity.

Contracting agency: That body which assigns part or all of the necessary work on a project to another body by contract, while retaining over-all responsibility for the project.

Corporate author: That body responsible for the contents of a given document.

Corporate entry: The form of name used to identify an agency.

Designation: ⎫
Designator: ⎬ That which serves as a means of identification (of a report). This usually includes a report series code and an identifier. It is often called short title or report number.

Document: Any recorded information, regardless of its physical form or characteristics. This includes, but is not limited to, the following: (1) written matter of all kinds, whether handwritten, printed or typed; (2) all painted, drawn or engraved matter; (3) all sound or voice recordings; (4) all printed photographs and exposed or printed film, still or movie; and (5) all reproductions of the foregoing by whatever process reproduced.*

Identifier: ⎫
Indicator: ⎬ That part of a report designator that is distinctive for an individual report. This is usually a serial number.

Issuing agency: ⎫
Originating agency: ⎬ That body responsible for issuing a given document.

Prefix: That code which is put before the serial number of a report.

Reports: Records of research results or of research in progress. They may be written on completed investigations, usually identified as topical or final reports, or they may be periodic progress reports, which summarize the work of a group or the work on a particular project since the preceding report. As records of research investigations, technical reports vary extensively in style, form and method of publication. They range in size from a few pages of brief technical notes or an account of a single experiment to several hundred pages summarizing the research activities of a large-scale project. Typically the technical report is reproduced by a near-print process, such as multilith mat, mimeograph, ditto, etc.†

Report number: See Designator

Report number series: ⎫
Report series abbreviation: ⎬ That system of arbitrarily chosen letters and/or numbers used to identify reports issued successively and in some way related.
Report series code: ⎭

Serial number: An individual number assigned to reports published successively.

Series: Two or more documents in some way related and issued successively.

Short title: See Designator

Sponsoring agency: That body responsible for projects carried out by another body under contract or other assignment.

Atomic Energy Commission. AEC Manual. Washington, D.C.: November 13, 1956, chap. 2105-045.

†Fry, B. M. *Library Organization and Management of Technical Reports Literature* (Catholic University of America Studies in Library Science, no. 1). Washington, D.C.: 1953, p.4

Sources of Report Series Codes and Their Interpretation

Unpublished Sources

The card catalogs of the technical libraries of the following organizations were the primary sources of the material contained in this *Dictionary:*
- Air Force Weapons Laboratory, Kirtland Air Force Base, New Mexico
- Los Alamos Scientific Laboratory of the University of California, Los Alamos, New Mexico
- Sandia Laboratories, Albuquerque, New Mexico
- White Sands Missile Range, New Mexico

Assistance with the entries in the first edition was also received from:
- Army Ballistic Missile Agency, Redstone Arsenal, Alabama
- Defense Atomic Support Agency, Washington, D.C.
- Library of Congress, Science Division, Washington, D.C.

Published Sources

Argonne National Laboratory. *The Classification of Reports by Subject Categories* (CA-1908(Revised)R-C), revision of M-CA-1709, rev. ed., comp. by R. S. Mulliken, rev. by S. R. Schram. Chicago, Ill.: July 29, 1946. 4p. ditto. (Security Classification: Secret)

Argonne National Laboratory. *Guide to Project Information* (CA-3660-C), ed. by M. E. Howe et al., rev. by S. R. Schram. Chicago, Ill.: January 16, 1947. 89p. ditto. (Security Classifcation: Secret)

Armed Services Technical Information Agency. *Corporate Author Headings*, ed. by Eleanor J. Aronson. Arlington Hall Station, Arlington, Va.: 1957. 348p. pap. litho.

Central Air Documents Office. *Preliminary Survey of Report Numbering Systems* (ATI No. 78633). Wright-Patterson Air Force Base, Dayton, Ohio: June 1950. 70p. pap. litho.

Chicago. University. Metallurgical Laboratory. *Bulletin on the Classification of Reports by Subject Categories* (CA-1908 Revised M-C), revision of M-CA-1709, rev. ed., comp. by R. S. Mulliken. Chicago, Ill.: March 27, 1945. 6p. ditto.

Chicago. University. Metallurgical Laboratory. *Guide to Project Information* (M-CA-2096 R), preliminary ed. of an Appendix for the Project Handbook, CL-697, ed. by M. E. Howe et al. Chicago, Ill.: August 30, 1944. 64p. ditto. (Security Classification: Secret)

Committee on Scientific and Technical Information. *Corporate Author Headings* (COSATI 70-7). Washington, D.C.: March 1970. 850p. pap. (Sold by National Technical Information Service)

Defense Documentation Center, Alexandria, Va. *Technical Abstract Bulletin* (and predecessors), 1948-1970.

Division of Technical Information, AEC. *Corporate Author Entries Used by the Division of Technical Information in Cataloging Reports* (TID 5059, 6th Rev.), 6th rev. ed., ed. by Clara L. Fox. Oak Ridge, Tenn.: March 1964. 140p. pap. (Sold by National Technical Information Service)

Division of Technical Information, AEC, *Report Number Codes used by the USAEC Division of Technical Information in Cataloging Reports* (TID-85, 9th Rev.), 9th rev. ed., comp. by Helen W. White and Edna Cockrell, Oak Ridge, Tenn.: September 1970. 181p. pap. (Sold by National Technical Information Service)

Global Engineering Documentation Services Inc. *Directory of Engineering Document Sources*. comp. by D. P. Simonton. Newport Beach, Calif.: 1972. unpaged. pap.

National Technical Information Service, Springfield, Va. *Government Reports Announcements* and *Government Reports Index* (and predecessors), v.1, 1946 - v.71, 1971.

Naval Ordnance Laboratory. *Abbreviation List of Corporate Authors and Report Series* (NAVORD Report 6888), prep. by Eva Liberman. White Oak, Md.: June 1960. 226p. pap. litho. (Security Classification: Confidential)

Office of Information Services, AEC, Washington, D.C. *Nuclear Science Abstracts*, v.1, 1948 - v.25, 1971.

Office of Technical Information, AEC. *Report Number Series Used by the Office of Technical Information in Cataloging Reports; Report Numbering System* (TID-85, 2nd Rev., Suppl.), 2nd rev. ed., suppl., comp. by Frances E. Stratton. Oak Ridge, Tenn.: August 1960. 23p. pap. (Sold by National Technical Information Service)

Office of Technical Services. *Government Document Series Analyzed by OTS*, ed. by Grace Swift. Washington, D.C.: January 1947. 46p. pap. processed.

Scientific and Technical Information Office, NASA,

Washington, D.C. *Scientific and Technical Aerospace Reports*, v.1, 1963 - v.9, 1971.

Special Libraries Council of Philadelphia and Vicinity. *Correlation Index: Document Series and PB Reports*, ed. by Gretchen E. Runge. New York: Special Libraries Association, 1953. 271p. pap. litho.

United Kingdom Atomic Energy Authority. *Guide to U.K.A.E.A. Documents*, 2nd ed., ed. by J. Roland Smith. London: 1960. 32p. pap. (Sold by H. M. Stationery Office)

Reference Notes

Reference 1

AMBIGUOUS NUMERICAL CODES

Many reports are assigned only numbers by their originators; others are assigned combinations of numbers and letters which form series codes. Reports whose series codes start with numbers are the most difficult to identify.

The numbers used in such codes may be arbitrary, or may stand for something. If they stand for something, it may be:
Account number
Assignment number
Command identification
Contract number
Department number
Division number
Group number
Job number
Laboratory project number
Model number
Order number
Part number
Project number
Report number
Report and copy number
Report and project number
Report number and subject
Report number and year
Research project number
Section number
Subject number
Work order number
Year and personnel number
Year, month and day

When the report series code starts with a number, but includes a letter or letters, it is found in this *Dictionary* under the letter(s).

When the code contains no letter, identification is practically hopeless. For this reason, the second edition of the *Dictionary* does not attempt to list any purely numerical codes in the codes section. They are collected below in a list which is not exhaustive but which illustrates the problem.

(SUBJECT NO. +/OR LTR.)-	ADEL PRECISION PRODUCTS CORP., BURBANK, CALIF.
(NUMBERS...)	AEROJET-GENERAL CORP., AZUSA, CALIF.
(WORK ORDER NO.)-	AEROJET-GENERAL CORP., AZUSA, CALIF.
390:64-	AEROJET-GENERAL CORP., AZUSA, CALIF.
(NUMBERS...)	AEROPHYSICS DEVELOPMENT CORP.
(NUMBERS...)	AIR FORCE, WASHINGTON, D.C.
(CONTRACT NUMBER)-	AIR MATERIEL COMMAND. ENGINEERING DIV. ELECTRONIC SUBDIV., WRIGHT-PATTERSON AFB, OHIO
(COMMAND NO.)-(LTRS)...	AIR MATERIEL COMMAND. INTELLIGENCE DEPT. TECH. ANALYSIS DIV., WRIGHT-PATTERSON AFB, OHIO
(BRANCH NO.)-(OTHER NOS.)	AIR PROVING GROUND COMMAND, EGLIN AFB, FLA.
(NUMBERS...)	AIR UNIV., MAXWELL AFB, ALA.
(PROJECT NUMBER-NO.)-	AIRBORNE INSTRUMENTS LAB., INC., MINEOLA, N.Y.
(PROJ. CODE NO.)-	AMERICAN HELICOPTER CO., INC., LOS ANGELES
(NUMBER)	AMERICAN INST. OF AERONAUTICS AND ASTRONAUTICS, N.Y.C.
(NUMBER)	ANEMOSTAT CORP. OF AMERICA, N.Y.C.
(NUMBER)	ARMOUR RESEARCH FOUNDATION, CHICAGO
(NUMBERS...)	ARMY MAP SERVICE, WASHINGTON, D.C.
(NUMBERS...)	ATOMIC ENERGY OF CANADA LTD. CHALK RIVER PROJECT, ONT.
80045-(LETTERS)-	AVCO CORP. ELECTRONICS DIV., CINCINNATI
(NUMBERS...)	AVCO MANUFACTURING CORP. ADVANCED DEVELOPMENT DIV., STRATFORD, CONN.

REFERENCE NOTES

(NUMBER)	BAIRD ASSOCIATES, INC., CAMBRIDGE, MASS.
(NUMBERS...)	BELL AIRCRAFT CORP., BUFFALO
(NUMBERS...)	BOEING AIRPLANE CO., SEATTLE
(NUMBER)	BORG-WARNER CORP. ROY C. INGERSOLL RESEARCH CENTER, DES PLAINES, ILL.
(NUMBER)	BOSTON UNIV. OPTICAL RESEARCH LAB.
(NUMBERS...)	BOYNTON (A.J.) AND CO., CHICAGO
(NUMBERS...)	BUREAU OF AERONAUTICS (NAVY)
(NUMBERS...)	BUREAU OF ORDNANCE (NAVY)
(NUMBER)	CALIFORNIA. UNIV., BERKELEY. DIV. OF ENG. RES.
184-	CALIFORNIA. UNIV., BERKELEY. RADIATION LAB.
(CONTRACT CODE NO.)-	CALIFORNIA INST. OF TECH., PASADENA. JET PROPULSION LAB.
(NUMBERS...)	CANADA. ARMAMENT RESEARCH AND DEVELOPMENT ESTABLISHMENT, VALCARTIER, QUE.
(NUMBER)	CAPEHART-FARNSWORTH CORP., FORT WAYNE
1.(NUMBER).-	CARBIDE AND CARBON CHEMICALS CORP. K-25 PLANT, OAK RIDGE, TENN.
4.(NUMBER).-	CARBIDE AND CARBON CHEMICALS CORP. K-25 PLANT, OAK RIDGE, TENN.
2.(NUMBER).-	CARBIDE AND CARBON CHEMICALS CORP. K-25 PLANT, OAK RIDGE, TENN.
3.(NUMBER).-	CARBIDE AND CARBON CHEMICALS CORP. K-25 PLANT, OAK RIDGE, TENN.
51-(LETTER)-	CHRYSLER CORP., DETROIT
(NUMBERS...)	CHRYSLER CORP. MISSILE OPERATIONS
(NUMBERS...)	COAST AND GEODETIC SURVEY, WASHINGTON, D.C.
(NUMBER)	COLORADO. UNIV., BOULDER. DEPT. OF MECH. ENG.
(NUMBER)	COLORADO. UNIV., BOULDER. DEPT. OF PHYSICS
1-(LETTER)-	COLUMBIA UNIV., N.Y.C. SUBSTITUTE ALLOY MATERIALS LAB.
1(LETTER)-	COLUMBIA UNIV., N.Y.C. SUBSTITUTE ALLOY MATERIALS LAB.
1(LETTER)-(LETTER)-	COLUMBIA UNIV., N.Y.C. SUBSTITUTE ALLOY MATERIALS LAB.
1(LETTER)(NUMBER)-(LTR)-	COLUMBIA UNIV., N.Y.C. SUBSTITUTE ALLOY MATERIALS LAB.
1(LETTER)(NUMBER)(LTR)-	COLUMBIA UNIV., N.Y.C. SUBSTITUTE ALLOY MATERIALS LAB.
1(LETTER)(LETTER)-	COLUMBIA UNIV., N.Y.C. SUBSTITUTE ALLOY MATERIALS LAB.
2-(LETTER)-	COLUMBIA UNIV., N.Y.C. SUBSTITUTE ALLOY MATERIALS LAB.
2(LETTER)-	COLUMBIA UNIV., N.Y.C. SUBSTITUTE ALLOY MATERIALS LAB.
2(LETTER)(NUMBER)-(LTR)-	COLUMBIA UNIV., N.Y.C. SUBSTITUTE ALLOY MATERIALS LAB.
2(LETTER)(NUMBER)(LTR)-	COLUMBIA UNIV., N.Y.C. SUBSTITUTE ALLOY MATERIALS LAB.
2(LETTER)-(LETTER)-	COLUMBIA UNIV., N.Y.C. SUBSTITUTE ALLOY MATERIALS LAB.
2(LETTER)(LETTER)-	COLUMBIA UNIV., N.Y.C. SUBSTITUTE ALLOY MATERIALS LAB.
3-(LETTER)-	COLUMBIA UNIV., N.Y.C. SUBSTITUTE ALLOY MATERIALS LAB.
3(LETTER)-	COLUMBIA UNIV., N.Y.C. SUBSTITUTE ALLOY MATERIALS LAB.
3(LETTER)-(LETTER)-	COLUMBIA UNIV., N.Y.C. SUBSTITUTE ALLOY MATERIALS LAB.
3(LETTER)(LETTER)-	COLUMBIA UNIV., N.Y.C. SUBSTITUTE ALLOY MATERIALS LAB.
4-(LETTER)-	COLUMBIA UNIV., N.Y.C. SUBSTITUTE ALLOY MATERIALS LAB.
4(LETTER)-	COLUMBIA UNIV., N.Y.C. SUBSTITUTE ALLOY MATERIALS LAB.

REFERENCE NOTES

5-(LETTER)-	COLUMBIA UNIV., N.Y.C. SUBSTITUTE ALLOY MATERIALS LAB.
5(LETTER)-	COLUMBIA UNIV., N.Y.C. SUBSTITUTE ALLOY MATERIALS LAB.
5(LETTER)(NUMBER)-(LTR)-	COLUMBIA UNIV., N.Y.C. SUBSTITUTE ALLOY MATERIALS LAB.
5(LETTER)(NUMBER)(LTR)-	COLUMBIA UNIV., N.Y.C. SUBSTITUTE ALLOY MATERIALS LAB.
5(LETTER)-(LETTER)-	COLUMBIA UNIV., N.Y.C. SUBSTITUTE ALLOY MATERIALS LAB.
5(LETTER)(LETTER)-	COLUMBIA UNIV., N.Y.C. SUBSTITUTE ALLOY MATERIALS LAB.
6-(LETTER)-	COLUMBIA UNIV., N.Y.C. SUBSTITUTE ALLOY MATERIALS LAB.
6(LETTER)-	COLUMBIA UNIV., N.Y.C. SUBSTITUTE ALLOY MATERIALS LAB.
6(LETTER)-(LETTER)-	COLUMBIA UNIV., N.Y.C. SUBSTITUTE ALLOY MATERIALS LAB.
6(LETTER)(LETTER)-	COLUMBIA UNIV., N.Y.C. SUBSTITUTE ALLOY MATERIALS LAB.
(ROMAN NUMERAL)/	COMITE CONSULTATIF INTERNATIONAL DE RADIOCOMMUN-ICATIONS, GENEVA
(NUMBER)	CONSOLIDATED VULTEE AIRCRAFT CORP. ORDNANCE AERO-PHYSICS LAB., DAINGERFIELD, TEX.
(JOB NUMBER)	CONSOLIDATED VULTEE AIRCRAFT CORP. SAN DIEGO DIV., CALIF.
(NUMBER-NUMBER-YEAR)	CRUCIBLE STEEL CO. OF AMERICA, PITTSBURGH
(PROJECT NUMBER)-	DESIGNERS FOR INDUSTRY, INC., CLEVELAND
(NUMBER)	DUMONT (ALLEN B.) LABS., INC., CLIFTON, N.J.
(NUMBER)	EASTMAN KODAK CO. NAVY ORDNANCE DIV., ROCHESTER, N.Y.
(YEAR CODE)(MO.NO.)-(DAY)	ECLIPSE-PIONEER DIV., BENDIX AVIATION CORP., TETERBORO, N.J.
(NUMBER)	EDWALD LABORATORIES, INC., RINGWOOD, ILL.
(NUMBERS...)	FANSTEEL METALLURGICAL CORP., NORTH CHICAGO, ILL.
(NUMBER)	FARRAND OPTICAL CO., INC., N.Y.C.
(PROJECT NUMBER)-	FLETCHER AVIATION CORP., PASADENA, CALIF.
(NUMBER)	FLORIDA. UNIV., GAINESVILLE. ENGINEERING AND INDUSTRIAL EXPERIMENT STATION
(CONTRACT ORDER NO.)-	FORMICA CORP., CINCINNATI
(NUMBER)	FORMICA CORP., CINCINNATI
(YEAR)-	GARRETT CORP. AIRESEARCH MFG. DIV., LOS ANG.
27(LETTER)(NUMBERS)	GENERAL DYNAMICS/CONVAIR, SAN DIEGO, CALIF.
55(LETTER)(NUMBERS)	GENERAL DYNAMICS/CONVAIR, SAN DIEGO, CALIF.
69(LETTER)(NUMBERS)	GENERAL DYNAMICS/CONVAIR, SAN DIEGO, CALIF.
(YEAR)(LETTERS)-	GENERAL ELECTRIC CO. (ALL LOCATIONS
(NUMBERS...)	GENERAL ELECTRIC CO. HEAVY MILITARY ELECTRONIC EQUIPMENT DIV., SYRACUSE, N.Y.
(YEAR CODE)-(LETTER)-	GENERAL MOTORS CORP. AC SPARK PLUG DIV., FLINT, MICH.
(NUMBER)	GENERAL TIRE AND RUBBER CO. OF CALIF. RESEARCH AND DEVELOPMENT DIV., PASADENA
(NUMBERS...)	GT. BRIT. ATOMIC WEAPONS RESEARCH ESTABLISHMENT, ALDERMASTON, BERKS, ENGLAND
(NUMBERS...)	GT. BRIT. NATIONAL PHYSICAL LAB., TEDDINGTON, MIDDX., ENGLAND
(PROJECT ACCOUNT NUMBER)-	HALLER, RAYMOND, AND BROWN, INC., STATE COLLEGE, PA.
9479-	HAZELTINE ELECTRONICS CORP., LITTLE NECK, N.Y.
(NUMBERS...)	HOLSTON DEFENSE CORP., KINGSPORT, TENN.
(PROJ. NO.)-(NO.)-(LTR.)	HUGHES TOOL CO. AIRCRAFT DIV., CULVER CITY, CAL.
(PROJECT NUMBER)-	ILLINOIS INST. OF TECH., CHICAGO
(NUMBER)	INSTITUTE OF THE AERONAUTICAL SCIENCES, N.Y.C.
(NUMBERS...)	INTELLIGENCE DIV. (ARMY), WASHINGTON, D.C.
(NUMBERS...)	JOHNS HOPKINS UNIV., BALTIMORE
(NUMBERS...)	KIRTLAND AFB, N. MEX.
(NUMBERS...)	KNOLLS ATOMIC POWER LAB., SCHENECTADY, N.Y.
(NUMBER)	KOLLSMAN INSTRUMENT CO., ELMHURST, N.Y.

REFERENCE NOTES

(NUMBER)	LINDE AIR PRODUCTS CO., TONAWANDA, N.Y.
(NUMBER)	LOCKHEED AIRCRAFT CORP., BURBANK, CALIF.
(NUMBERS...)	LOS ALAMOS SCIENTIFIC LAB., N. MEX.
(NUMBER)	LOUISVILLE, KY. UNIV. INST. OF INDUSTRIAL RES.
(NUMBER)	M.C. MANUFACTURING CO., LAKE ORION, MICH.
(NUMBER)	MC DONNELL AIRCRAFT CORP., ST. LOUIS
(NUMBERS...)	MC GILL UNIV., MONTREAL
(NUMBER)	MARQUARDT AIRCRAFT CO., VAN NUYS, CALIF.
(PROJECT ACCOUNT NO.)-	MARTIN-HUBBARD CORP., BOSTON
(NUMBERS...)	MASSACHUSETTS INST. OF TECH., CAMBRIDGE. DIGITAL COMPUTER LAB.
41-	MASSACHUSETTS INST. OF TECH., CAMBRIDGE. RADIATION LAB.
42-	MASSACHUSETTS INST. OF TECH., CAMBRIDGE. RADIATION LAB.
43-	MASSACHUSETTS INST. OF TECH., CAMBRIDGE. RADIATION LAB.
45-	MASSACHUSETTS INST. OF TECH., CAMBRIDGE. RADIATION LAB.
52-	MASSACHUSETTS INST. OF TECH., CAMBRIDGE. RADIATION LAB.
53-	MASSACHUSETTS INST. OF TECH., CAMBRIDGE. RADIATION LAB.
53/(MO.)/(DAY)/(YR.)	MASSACHUSETTS INST. OF TECH., CAMBRIDGE. RADIATION LAB.
(NUMBERS...)	MASSACHUSETTS INST. OF TECH., LEXINGTON. LINCOLN LAB.
(WORK ORDER NUMBER)-	MAXSON (W.L.) CORP., N.Y.C.
(NUMBERS...)	MICHIGAN. UNIV., ANN ARBOR. ENG. RES. INST.
(NUMBER)	MIDWEST ENGINEERING DEVELOPMENT CO., INC., KANSAS CITY, MO.
(NUMBER)	MINNESOTA. UNIV., MINNEAPOLIS. DEPT. OF AERONAUTICAL ENGINEERING
(NUMBERS...)	MINNESOTA MINING AND MFG. CO., ST. PAUL
(YEAR CODE)-(MONTH NO.)-	MOUND LAB., MIAMISBURG, OHIO
(NUMBERS...)	NATIONAL BUREAU OF STANDARDS, WASH., D.C.
(NUMBER)-(YEAR)	NAVAL AERONAUTICAL ROCKET LAB., LAKE DENMARK, DOVER, N.J.
(NUMBERS...)	NAVAL ORDNANCE TEST STATION, INYOKERN (CHINA LAKE), CALIF.
(NUMBERS...)	NAVAL RADIOLOGICAL DEFENSE LAB., SAN FRANCISCO
(NUMBERS...)	NAVAL RESEARCH LAB., WASHINGTON, D.C.
(NUMBERS...)	NEW YORK UNIV., N.Y.C. COLL. OF ENGINEERING
(NUMBERS...)	NUCLEAR DEV. CORP. OF AMERICA, WHITE PLAINS, N.Y.
(PROJECT NO.)-(SUBJ.NO.)-	PREWITT AIRCRAFT CO., CLIFTON HEIGHTS, PENNA.
(NUMBER)	RAYTHEON MFG. CO., WALTHAM, MASS.
(NUMBER)	REEVES INSTRUMENT CORP., N.Y.C.
(NUMBER)	RHODE ISLAND STATE COLL., KINGSTON. DEPT. OF CHEMISTRY
(NUMBERS...)	ROCHESTER, N.Y. UNIV. INST. OF OPTICS
(NUMBERS...)	ROCKETDYNE DIV., NORTH AMERICAN AVIATION, INC., CANOGA PARK, CALIF.
(NUMBERS...)	SANDIA CORP., ALBUQUERQUE, N. MEX.
(NUMBER)	SHELL OIL CO., N.Y.C.
(NUMBER)	SOLAR AIRCRAFT CO., SAN DIEGO, CALIF.
(NUMBERS...)	SOUTHERN RESEARCH INST., BIRMINGHAM, ALA.
(NUMBER)	STEVENS INST. OF TECH., HOBOKEN, N.J. EXPERIMENTAL TOWING TANK
(NUMBER)	STEWART-WARNER CORP. SOUTH WIND DIV., INDPLS.
(NUMBER)	SYLVANIA ELECTRIC PRODUCTS INC. RADIO TUBE DIV., N.Y.C.
(NUMBERS...)	TEMCO AIRCRAFT CORP., DALLAS
(NUMBERS...)	TRACERLAB, INC., BOSTON
(NUMBERS...)	TRACERLAB, INC., WALTHAM, MASS.
(NUMBERS...)	TRACERLAB, INC. WESTERN DIV., RICHMOND, CAL.
(NUMBER)	TRANSCONTINENTAL + WESTERN AIR, INC., KANSAS CITY, MO.
7000-	TRAVELERS RESEARCH CENTER, INC., HARTFORD (NUMBER MAY BE 7000-7999. IT IS CONTRACT NO.)
(NUMBERS...)	UNDERWATER EXPLOSIONS RESEARCH DIV., PORTSMOUTH, VA.
(NUMBER)	UNITED AIRCRAFT CORP. HAMILTON STANDARD PROPELLERS DIV., EAST HARTFORD, CONN.
(CONTRACT CODE NO.)-(NOS)	UNIVERSITY OF SOUTHERN CALIFORNIA, LOS ANG.
(NUMBER)	WESTINGHOUSE ELECTRIC CORP. SPECIAL PRODUCTS DEVELOPMENT DIV., PITTSBURGH

Reference 2

AMBIGUOUS ALPHABETICAL CODES

These codes are assigned by many agencies to reports they issue or receive. In most cases, they are not sufficiently distinctive to be used by themselves in the bibliographic identification of reports.

Most commonly these codes designate the function of the reports, such as technical reports, internal memoranda, parts lists, etc. These functional codes are discussed in great detail with statistical analysis by Herner and Company in their report listed in the Bibliography (p. 13).

Some of these codes designate subordinate levels within the parent organizations, such as Acoustics Laboratory or Research Laboratory. When these designations appear to us to be limited to a single organization only, they are listed in the main body of this *Dictionary* and are not included in the list below. The organizational designations included here are so common as to be useless in identifying their originators.

A few of the ambiguous alphabetic codes listed below stand for subject matter. Again, they are listed here rather than in the main *Dictionary* only when their commonness destroys their usefulness for identification.

One and two letter codes mnemonic for the issuing organizations are not included in the list below, even though they can now be considered essentially ambiguous. They were once distinctive, but with the great proliferation of codes, have lost much of their meaningfulness. Nevertheless they are still included with their organizations in the main body of the *Dictionary* rather than here.

All the codes listed below become meaningful when they are used in combination with other codes that identify their source, and they are frequently employed in that fashion, indicating the particular series into which a report falls, the subdivision directly responsible for the report, or the subject matter. This list will provide some help for interpreting their meaning when they are employed in that fashion.

A-	Abstracts
AE-	Aeronautical Engineering
AER-	Aeronautical Engineering Report
AL-	Accession List
	Acoustics Laboratory
	Aeronautical Laboratory
AN-	Aerodynamics Note
AP-	Air Publication
APR-	Annual Progress Report
AR-	Aerodynamic Report
	Annual Report
AS-	Administrative and Scientific Progress Report
ASR-	Annual Summary Report
ATR-	Annual Technical Report
B-	Bericht
	Bulletin
BB-	Bibliography
BIB-	Bibliography
BI-MONTHLY PROGRESS RPT.-	
	Bimonthly Progress Report
BM-	Branch Memorandum
BMPR-	Bimonthly Progress Report
BPR-	Bi-weekly Progress Report
	Bimonthly Progress Report
BR-	Branch Report
BTN-	Bulletin
BULL-	Bulletin
BW-	Biological Warfare
BWR-	Bi-weekly Report
C-	Catalog
	Circular
	Communication
	Conference
	Confidential
	Contribution
	Cyclotron
	Technical Compilation
CB-	Confidential Bulletin
CC-	Computation Center
CER-	Civil Engineering Report
CF-	Central Files
	Central Files Memorandum
CL-	Circular Letter
	Cyclotron Lab.
CM-	Confidential Memorandum
CONF-	Conference
CP-	Conference Paper
	Contract Publication
	Current Paper
CPR-	Consolidated Progress Report
CR-	Chemical Report
	Conference Report
	Consolidated Report
	Consulting Report
	Contract Report
	Contractor Report
CSPR-	Consolidated Semiannual Progress Report
CW-	Chemical Warfare

D-	Disclosure	FE-	Field Engineering
	Discussion		Flight Engineering
	Document	FER-	Final Engineering Report
DC-	Declassified Publication Release	FM-	Field Manual
	Development Characteristic	FPL-	Flight Propulsion Lab.
DF-	Data Folder	FPR-	Final Progress Report
DM-	Design Manual	FR-	Field Report
	Design Memorandum		Final Report
DOC-	Document	FS-	Finish Specification
DP-	Design Proposal	FT-	Firing Table
	Discussion Paper		Flight Test
DR-	Data Report		Flight Test Report
	Design Requirement	FTR-	Final Technical Report
	Detailed Report		Flight Test Report
	Development Report		Functional Test Report
	Document Report	FY-	Fiscal Year
DS-	Design Study	G-	Group Report
DT-	Drop Test Report	GM-	Guided Missiles
DWG-	Drawing	GR-	Group Report
E-	Engineering	H-	History
	Engineering Report	HQ-	Headquarters
	Excerpt	HS-	Historical Survey
EDR-	Engineering Dept. Report	I-	Informal Report
	Engineering Division Report		Installation Instructions
EE-	Electrical Engineering		Interim Report
EES-	Engineering Experiment Station		Internal Report
EL-	Electronics Lab.	IB-	Information Bulletin
ELR-	Engineering Lab. Report		Instruction Book
EM-	Education Manual	IC-	Informal Communication
	Electrodeposition Memo		Information Circular
	Engineering Memorandum	IDB-	Internal Distribution Bulletin
	External Memorandum	IDR-	Interim Development Report
EN-	Engineering Note		Internal Development Report
	Engineering Report	IER-	Interim Engineering Report
ENGINEERING REPT.-		II-	Interim Investigation Report
	Engineering Report	IM-	Inspection Memorandum
EO-	Engineering Order		Instructors Manual
EP-	Engineering Paper		Intelligence Memorandum
	External Publication		Interim Memorandum
ER-	Engineering Report		Internal Memorandum
	Evaluation Report	IMR-	Informal Memorandum Report
	Explosives Report	IN-	Instrument Note
	External Report		Internal Note
ERR-	Engineering Research Report	INFO BULL.-	
ES-	Engineering Study		Information Bulletin
	Experimental Station	INSTR-	Instruction
ESR-	Engineering Summary Report	INTERIM ENG. REPT.	
ETM-	External Technical Memorandum		Interim Engineering Report
ETR-	Engineering Test Report	INTERNAL MEMO-	
	External Technical Report		Internal Technical Memo
F-	Final	IOI-	Interim Operating Instructions
	Final Report	IP-	Instrumentation Paper
	Finish Specification	IPR-	Informal Progress Report
	Formal Report		Interim Progress Report
FC-	Flight Charts		Internal Progress Report
FDR-	Final Data Report	IR-	Informal Report

REFERENCE NOTES

	Information Report		Monthly Review
	Instruction Report	MRP-	Monthly Report of Progress
	Instrumentation Report	MRR-	Monthly Review Report
	Intelligence Report	MS-	Material Specifications
	Intelligence Review		Military Specifications
	Interim Report	MT-	Machine Translation
	Internal Report		Missile Test
	Interpretation Report	MTR-	Materials Testing Report
	Interrogation Report	N-	Technical Note
	Invention Report	NB-	Notebook
IS-	Information Series	NM-	Note on Materials
ISR-	Interim Scientific Report	NR-	Report
	Internal Scientific Report	NS-	Nuclear Science
ITM-	Interim Technical Memorandum	NSE-	Nuclear Science and Engineering
	Internal Technical Memorandum	NTIR-	Nontechnical Intelligence Report
ITN-	Interim Technical Note	OA-	Operations Analysis
ITR-	Interim Technical Report	OAR-	Operations Analysis Report
	Interim Test Report	OD-	Ordnance Div.
	Internal Technical Report	OFR-	Open File Report
JPR-	Joint Progress Report	OP-	Operation Plans
L-	Laboratory		Ordnance Pamphlet
	Leaflet	OR-	Open Report
	Lectures		Operational Requirement
	Letter Report		Operations Requirement
	Library		Operations Research
	Library Materials		Ordnance Report
LB-	Laboratory Bulletin		Overall Report
	Library Bulletin	OS-	Ordnance Specification
LETTER-	Informal Technical Report	P-	Pamphlet
LM-	Light Metals		Paper
	Liquid Metals		Periodic Progress Report
LN-	Laboratory Note		Process Report
LOG-	Log Record		Production Report
LR-	Laboratory Report		Progress Report
	Letter Report		Project Report
LS-	Literature Search		Proposal
M-	Manual		Publication
	Memorandum		Reprint/Preprint
	Memorandum Report	PA-	Publication Announcement
	Miscellaneous Project	PAM-	Pamphlet
	Monthly Report	PAPER-	Paper
	Monthly Test Report	PB-	Publication
	Technical Memorandum	PD-	Physics Dept.
MEMO-	Memorandum		Preliminary Design
MIPR-	Monthly Interim Progress Report		Production Dept.
MISCELLANEOUS PAPER-			Purchase Description
	Miscellaneous Paper	PDS-	Preliminary Design Study
MM-	Maintenance Manual	PEM-	Project Engineering Memorandum
MOR-	Monthly Operations Report	PEO-	Production Engineering Order
MP-	Miscellaneous Paper	PEPR-	Periodic Progress Report
	Miscellaneous Publication	PET-	Production Environmental Tests
MPR-	Mine Production Report	PIB-	Product Improvement Bulletin
	Monthly Progress Report	PJM-	Power Jets Memorandum
	Monthly Project Report	PJR-	Power Jets Report
MR-	Memorandum Report	PL-	Parts List
	Miscellaneous Report	PM-	Propulsion Memorandum
	Monthly Report	PMSRP-	Physical and Mathematical Research Paper

PP-	Physical Properties		Research Paper
	Professional Paper		Research Proposal
PR-	Performance Requirement	RPT-	Report
	Plant Report	RR-	Research Report
	Preliminary Report	RS-	Research Study
	Preprint		Research Summary
	Progress Report	RT-	Reduction Tables
	Project Report	RTM-	Research Technical Memorandum
PROC-	Proceedings	RW-	Radiological Warfare
PROJ-	Project	S-	Scientific Progress Report
PS-	Preliminary Study		Special Report
PSR-	Periodic Status Report		Specification
	Progress Summary Report		Staff Report
PSRP-	Physical Science Research Paper		Summary
PTR-	Preliminary Technical Report	SA-	Semi-Annual Report
PUB-	Publication	SAR-	Semi-Annual Report
PUBL-	Publication	SATSR-	Semi-Annual Technical Summary Report
Q-	Quarterly	SB-	Service Bulletin
QC-	Quality Control		Special Bibliography
QCR-	Quality Control Report		Supply Bulletin
QDRI-	Quarterly Development Research Investigation	SCIENTIFIC-	Scientific Report
QI-	Quarterly Index	SCIENTIFIC REPORT-	Scientific Report
QM-	Quality Memorandum	SDP-	Staff Discussion Paper
	Quarterly Memorandum	SM-	Scientific Memorandum
QPR-	Quarterly Progress Report		Special Memorandum
QR-	Quarterly Report		Staff Memorandum
QSR-	Quarterly Status Report		Structure Memorandum
	Quarterly Summary Report		Summary Memorandum
QTR-	Quarterly Technical Report	SN-	Scientific Note
R-	Formal Report	SO-	Shipping Order
	Internal Report	SOP-	Standard Operating Procedure
	Regulation	SP-	Scientific Paper
	Report		Solid Propellants
	Reprint		Special Paper
	Research		Special Project Report
	Research Report		Special Propellants
	Technical Report		Special Publication
RB-	Report Bibliography		Specification
	Research Bulletin		Staff Paper
	Restricted Bulletin	SPEC-	Special
RD-	Research and Development		Specification
RDD-	Research and Development Div.	SPR-	Special Project Report
RDR-	Research Division Report		Special Report
RDS-	Research and Development Survey		Supplementary Progress Report
RDTR-	Research Division Technical Report	SR-	Scientific Report
REPT-	Report		Scientific Research
RI-	Report of Investigation		Section Report
RL-	Radiation Lab.		Simulation Report
	Reading List		Sound Report
	Research Laboratory		Special Regulation
RM-	Memorandum		Special Report
	Reference Material		Staff Report
	Research Memorandum		Standardization Report
RN-	Research Note		Status Report
RP-	Reference Publication		

REFERENCE NOTES

	Summary Report		Training Manual
	Supplemental Report	TMON-	Technical Monograph
SRM-	Staff Research Memorandum	TMR-	Technical Memorandum Report
SRS-	Structural Research Series	TMS-	Technical Manuscript
SS-	Special Series	TN-	Technical Note
	Special Study	TNB-	Technical News Bulletin
SSR-	Special Scientific Report	TO-	Technical Order
ST-	Special Translation	T.O.-	Technical Order
STPR-	Semi-Annual Technical Progress Report	TOR-	Technical Operating Report
STR-	Scientific Technical Report	TP-	Technical Pamphlet
	Summary Technical Report		Technical Paper
SUM-	Summary		Technical Problem
T-	Technical Memorandum		Technical Publication
	Technical Paper		Technical Report
	Technical Report		Technographic Publication
	Test Report		Tentative Pamphlet
	Translation	TPR-	Technical Progress Report
TB-	Technical Bulletin	TR-	Technical Regulation
	Test Bulletin		Test Report
TC-	Technical Communication		Test Run
	Translation Code		Trial Report
TD-	Test Data		Trip Report
TDR-	Technical Data Report	TRANS-	Translation
	Technical Documentary Report	TRN-	Technical Research Note
TECH MEMO-		TRR-	Technical Research Report
	Technical Memorandum		Theoretical Research Report
TECHNICAL PAPER-		TS-	Technical Study
	Technical Paper		Tentative Specification
TECH REPT.-			Test Summary
	Technical Report	TSB-	Technical Service Bulletin
TG-	Technical General Report	TSP-	Technical Specification
THESIS-	Thesis issued as a report	TSR-	Technical Summary Report
TI-	Technical Instruction	TT-	Technical Translation
TIR-	Technical Intelligence Report	VG-	Vu-Graphs
TL-	Technical Letter	WB-	Work Book
	Technical Library	WGR-	Working Group Report
	Title List	WM-	Welding Memorandum
TLM-	Technical Liaison Memo	WP-	Working Paper
TM-	Technical Manual	WT-	Weapon Test Report
	Technical Memorandum		Wind Tunnel
	Technical Minutes	WTD-	Wind Tunnel Data
	Technical Monograph	WTM-	Wind Tunnel Memorandum
	Test Manual	WTN-	Wind Tunnel Note

Reference 3

ANP PROGRAM

ANP-
ANP-DOC.(LETTERS)-
APEX-

CARP-A-
CNLM-
CVC NARF (NOS.-NOS.-LTR.)-

DC-(YEAR)-
DC-(YEAR)-(NUMBER)-
DC-(YEAR)-(MONTH)-

GE-ANP-	GEDR-	NARF-(YEAR)-
GE-ANP-APEX-	GERD-	NARF-PR-OP-
GE-APEX-	GEXDC-(YEAR)-	PREDC-
GE-DC-	GEXDCL-(YEAR)-	PREXDC-
GE-XDC-	NARF-	XDC-(YEAR)-(MONTH)-
GED-		XDCL-

The ANP Program (variously called the Aircraft Nuclear Propulsion Project and the Aircraft Nuclear Propulsion Program) was a continuation, from 1951 through 1961, of the NEPA (Nuclear Energy for the Propulsion of Aircraft) Project.

Reports originating from companies and laboratories participating, by subcontract, in the project were sometimes entered under the name and report series code (ANP-) of the project, but they were also entered under the name and series code of the participants or issuing agencies.

The General Electric Company was the main contractor. Its codes may be found under the corporate entry. The series listed above appear to have been used for ANP Project sponsored reports exclusively. Participants included:

Allison Div., General Motors Corp., Indianapolis
Consolidated Vultee Aircraft Corp., Fort Worth, Tex.
Consolidated Vultee Aircraft Corp., San Diego, Calif.
Convair. Nuclear Aircraft Research Facility, Fort Worth, Tex.
General Dynamics/Fort Worth, Tex.
General Dynamics/Fort Worth, Div. of General Dynamics Corp., Tex.
General Electric Co. Aircraft Gas Turbine Div., Evendale, Ohio
General Electric Co. Aircraft Nuclear Propulsion Dept., Cincinnati
General Electric Co. Aircraft Nuclear Propulsion Project, Cincinnati
General Electric Co. Nuclear Materials and Propulsion Operation, Cincinnati
Lockheed Aircraft Corp., Burbank, Calif.
Oak Ridge National Lab., Tenn.
Pratt & Whitney Aircraft Div., United Aircraft Corp., Middletown, Conn.

Reference 4

ATOMIC ENERGY COMMISSION, OPERATIONS OFFICES AND DIVISIONS

ALO-	HOO-	NVOO-	RLO-	SFO-
BHO-	IDO-	NYDO-	RME-	SO-
COO-	KY-	NYO-	RMO-	SOO-
GJ-	LAO-	NYOO-	RMOO-	SRO-
HASL-	LAR-	ORO-	SAN-	SROO-
HO-	NVO-			

These codes are used to identify technical reports prepared by, or issued through, various operations offices and divisions of the Atomic Energy Commission. Most of the reports bearing these codes have been prepared by small contractors who have no report series codes of their own and have been assigned serial numbers from the series codes of the operations offices or divisions that administer their contracts. The largest number of such reports, prepared by the greatest variety of contractors, carried the code NYO. Occasionally, as with IDO, the operations office code was used on large bodies of reports prepared by a major contractor, which was issued blocks of numbers to be used.

In the mid-1960s the AEC adopted a practice of using the serial portion of the contract number as a second portion of the operations office code to identify the actual issuing installation.

Reference 5

ATOMIC ENERGY COMMISSION. OFFICE OF INFORMATION SERVICES

TID-

This code has been assigned to reports originating with the Office of Information Services, AEC, and its predecessors—Division of Technical Information, Technical Information Service, and Office of Technical Information and their extensions at Oak Ridge—and also to reports received by them from AEC contractors which had not been given distinctive report numbers by the originating agencies.

Since 1952, numbers in this series have been assigned from blocks of numbers that indicate the nature of the reports. The blocks are as follows:

3000-3999	Bibliographies
4000-4999	List of reports received by the division and non-technical reports originating with them
5000-6999	Research and development reports
7000-8999	Special publications
7001-7499	Handbooks, comprehensive surveys, etc.
7500-7999	Compiled papers resulting from technical meetings
8001-8017	AEC monograph series: "The Industrial Atom"
8200-8499	Programmatic or administrative reports
8500-8999	Broad feasibility and study-type reports
9000-9999	Reports connected with the joint AEC-DOD Cooperative Weapon Data Indexing Program
10000-10177	Reports released to the Civilian Application Program with deletions

Reference 6

ATOMIC ENERGY OF CANADA LTD.

B-	ELI-	MET-	PR-REPORTS-
CEI-	ES-	MET-I-	RDI-
CI-	FD-	METI-	REI-
CR-(LTR(S))-	GMI-	NEI-	RLI-
CR (LTR (S))-	GPI-	PBD-	RPI-
CRPR-	HRI-	PBI-	TDS-
DCI-	IOI-	PD-	TDSI-
ED-	IOP-	PR-	TDVI-
EDI-	IPD-	PR-(LTR(S))-	TPI-
EI-	IPI-	PR-(LTR(S))-(NO.)-(LTR(S))-	WD-

Atomic Energy of Canada Ltd., Chalk River Project, generally continued the report numbering system that had been used by the Atomic Energy Project of the National Research Council of Canada. (For NRCC AEP System, see Ref. 29.)

Formal topical reports were assigned numbers in a series beginning with CR and followed by a letter, or letters, standing for the name of the division or branch originating the report. The numbers were assigned serially regardless of the division letters.

For informal reports the CR was dropped and the letters for the originating division, branch, committee or panel were used, usually followed by the letter I, or sometimes by the letters D or P.

For progress reports, the letters PR were used as the main series designation. General laboratory progress reports carried those letters only; division progress reports had the letters for the division separated from the PR by a dash. Summaries of the divisional progress reports were numbered PR-S-, and compilations of periodic lists of reports available from the Scientific Documents Distribution Office were designated PR-REPORTS-.

The letters used at various times to indicate the originating divisions, branches, committees or panels were:

A	Administrative
B	Biology
C	Chemistry
CE	Chemical Engineering
CM	Chemistry and Metallurgy
CRR	Chemistry and Reactor Research
DC	Development Chemistry
E	Engineering Design & Applied Development
ED	Engineering Design; Engineering Development
EL	Electronics
EO	Engineering Operations
ES	Engineering Services
ER	Environmental Research
FD	Fuel Development
GM	General Metallurgy
GP	General Physics
HP	Health Physics
HR	Health and Radiation
IO	Industrial Operations; Operations
IP	(not identified)
MET	Metallurgy
NE	Nuclear Engineering
P	Physics
PB	(not identified)
PD	Nuclear Physics
R	Reactor
RC	Reactor Commissioning
RD	Radiation Dosimetry
RE	Research Engineering
RHC	Radiation Hazards Control
RL	Reactor Loops
RP	Reactor Physics
RRD	Reactor Research and Development
T	Theoretical Physics
TDS	Toronto Design
TDV	Toronto Development
TEC	Technical Physics
TP	Theoretical Physics
W	Whiteshell

Many combinations are possible, as:

CR-MET-	METI-
CRMET-	PR-CRR-(NO.)-MET-
MET-	PR-MET-
MET-I-	PRMET-

Reference 7

ATOMIC WEAPON TEST OPERATIONS

CJTF-	ITR-	SANDSTONE-
CTI-	IVY-	SS-
DASA-POR-	OBJ-	SWPCD-
DASA-POIR-	OC (LETTERS)-	SWPFD-
DWET-	OG-	SWPWT-
GIR-	OS-	UKP-
GREEN-	OUK-	WET-(YEAR)-
GREEN-(PROJECT NO.)-	PILGRIM-	WT-
GREENHOUSE ANNEX-(PROJ. NO)-	POIR-	XR-
HARDTACK-	POR-	XRD-

Formal reports giving the results of atomic weapon test operations are customarily treated as belonging to coherent series for each operation, regardless of the agencies originating the reports.

For the first test operation, Crossroads, no arrangements were made in advance for consolidating reports into a single series, but many recipients of the reports treated them as such and assigned various codes to them. These include:

CJTF-	(Crossroads Joint Task Force)
CTI-	(Crossroads Technical Instrumentation)
OC (LTRS)-	(Operation Crossroads)
SWPCD-	
XR-	

REFERENCE NOTES

Later, in connection with the joint AEC-DOD Cooperative Weapon Data Indexing program, Crossroads reports were assigned sequence numbers in a series designated XRD-.

Operation Sandstone reports were issued as a single series but without a code designation. In the Cooperative Weapon Data Indexing program the prefix SANDSTONE- was assigned to them and their original numbers were retained. Other codes given to the reports by recipients include OS- (Operation Sandstone) and SS-.

With Operation Greenhouse, a formal numbering system was established with the concurrence of the AEC and DOD agencies concerned. A series WT- (standing for "Weapon Test") was created. Numbers in that series are assigned to different operations in blocks of hundreds. One block was assigned retroactively to the reports from Operation Ranger. The other blocks have been assigned in sequence as the operations have occurred.

With Operation Castle, the unified numbering system was extended to include preliminary reports. Before that, preliminary reports had been given numbers in the series of their originating agencies or were designated by series standing for the individual operations. These included:

For Operation Greenhouse
  GREEN-(PROJECT NO.)
  GREENHOUSE ANNEX-(PROJECT NO.)
  OG-
  GIR-   (Greenhouse Interim Report)
For Operation Buster-Jangle
  OBJ-
For Operation Tumbler-Snapper
  OS-   (Operation Snapper)
For Operation Ivy
  IVY-
For Operation Upshot-Knothole
  OUK-
  UKP-   (Upshot-Knothole Preliminary)

The code ITR- was selected for the interim test reports. The serial number assigned is the same as that used in the WT series when the final report is issued. Therefore, WT reports superseded ITR reports bearing the same number.

The schedule of WT numbers is as follows:

0 - 100's	Greenhouse
200's	Ranger
300's - 400's	Buster-Jangle
500's	Tumbler-Snapper
600's	Ivy
700's - 800's	Upshot-Knothole
900's	Castle
1000's	Wigwam
1100's - 1200's	Teapot
1300's	Redwing
1400's - 1500's	Plumbbob
1600's	Hardtack (including Argus)
1700's	Hardtack II
9000's	General summaries covering all operations

Beginning in 1962 the WT series was continued by a series designated POR or POIR. These designators stand for Project Officers' reports, with POIR standing for the interim reports, to be succeeded by final POR reports with the same serial numbers. These reports continued the numbering schedule of the WT reports as follows:

1800's	Nougat and Storax
2000's	Dominic, including Fish Bowl and Christmas series
2100-2149	Prairie Flat
2150-2199	Mine Shaft
2200-2300's	Dominic II
2500's	Roller Coaster
2600's	Back Swing
2700's & 5000-15	Silver Bell
2800-2900's & 6000-10	Whetstone
3000's	Ferris Wheel
4000-49, 5016-5099, 6200-29 6400-6409	Flint Lock
4050-4099	Sailor Hat
6100-6109	Rover
6230-6294	Latchkey
6295-6299	Cyprus
6300's, 6410-29 6440-99	Minute Gun
6500-6569	Mild Wind
6570's	Mighty Mite

In addition to the preliminary and formal reports issued in these series, some reports concerning atomic tests have been assigned numbers in the regular report series of other originating agencies, or in special series of their agencies. Examples of such series are:

HN-(NUMBER)-	Holmes and Harver, Inc., Los Angeles
LA-	Los Alamos Scientific Lab., N. Mex.
LAMS-	Los Alamos Scientific Lab., N. Mex.
UCRL-	California. Univ., Livermore. Lawrence Radiation Lab.
SWPWT-	Armed Forces Special Weapons Project. Washington, D.C.
DWET-	Armed Forces Special Weapons Project. Field Command. Directorate of Weapons Effects Tests, Albuquerque, N. Mex.
WET-(YEAR)-	Armed Forces Special Weapons Project. Field Command. Directorate of Weapons Effects Tests, Albuquerque, N. Mex.

SWPFD- Armed Forces Special Weapons Project. Weapons Effects Div., Washington, D.C. (Operation Wigwam)

HARDTACK- Armed Forces Special Weapons Project. Weapons Effects Program, Washington, D.C. (Operation Hardtack)

PILGRIM- Armed Forces Special Weapons Project. Weapons Effects Program, Washington, D.C. (Operation Plumbbob)

For information on the Joint Task Forces conducting the operations, see Ref. 17.

Reference 8

BUMBLEBEE PROJECT

APL-BB-
APL/JHU-BB-
BB-
BBC-
BUMBLEBEE-
CF-
CM-
JHU-APL-BB-
TG-
TG-(NUMBER)-(NUMBER)-
TG-(NUMBER)(LETTER)-

This was a missile development project conducted by the Applied Physics Laboratory at Johns Hopkins University under contract to the Bureau of Ordnance. Reports on the project were issued by these agencies and also by subcontractors. These include:

Applied Science Corp., Princeton, N.J.
Baird Associates, Inc., Cambridge, Mass.
Bendix Aviation Corp.
Bendix Aviation Corp. Bendix Products Div., Missiles, Mishawaka, Ind.
Bendix Aviation Corp. Pacific Div. Development Labs., North Hollywood, Calif.
Bendix Aviation Corp. Research Labs., Detroit
Convair. Ordnance Aerophysics Lab., Daingerfield, Tex.
Cornell Aeronautical Lab., Inc., Buffalo
Curtiss-Wright Corp. Airplane Div., Columbus, Ohio
Eastman Kodak Co., Rochester, N.Y.
Eclipse-Pioneer Div., Bendix Aviation Corp., Teterboro, N.J.
Esso Research and Engineering Co., Linden, N.J.
Experiment, Inc., Richmond
Farnsworth Electronics Co., Fort Wayne
Goodyear Aircraft Corp.
Hercules Powder Co., Wilmington, Del.
Johns Hopkins Univ., Baltimore
Kellogg (M.W.) Co. Special Products Dept., Jersey City
Little (Arthur D.), Inc., Cambridge, Mass.
McDonnell Aircraft Corp., St. Louis
Michigan. Univ., Ann Arbor. Dept. of Engineering Research
New Mexico State Univ., University Park
North American Aviation, Inc.
Princeton Univ., N.J.
RCA Victor Div., Radio Corp. of America, Camden, N.J.
Sperry Gyroscope Co., Great Neck, N.Y.
Standard Oil Development Co. Esso Labs., Linden, N.J.
Submarine Signal Co.
Texaco Experiment Inc., Richmond
Texas. Univ., Austin. Defense Research Lab.
Texas. Univ., Austin. Electrical Engineering Research Lab.
Virginia. Univ., Charlottesville. Ordnance Research Lab.
Wisconsin. Univ., Madison. Naval Research Lab.

Reports from these agencies were assigned numbers in the BUMBLEBEE (or BB) and the CM series by the Applied Physics Laboratory. In addition, some technical memoranda originating in the Applied Physics Laboratory fall into the regular CF series used for internal reports.

Reference 9

CALIFORNIA. UNIVERSITY, BERKELEY. RADIATION LABORATORY

37 RR-	FUND.-P (MONTH/YEAR)-	SPEC. SEMI.-
37 SM-	FUND.-S-	SUMP.-P (MONTH/YEAR)-
184-	ID-SPEC-	SUMP. SPEC-
ANAL-MISC.-	J-P (NUMBER/YEAR)-	THEO.-P (MONTH/YEAR)-
ASSAY-	J SPEC-	UCRL-BC-
BC-	K-P (MONTH/YEAR)-	UCRL-SPEC-
BETA-	KAPPA-P-	VAC-M-
BETA-S-	MAG. S.-	VAC.-P (MONTH/YEAR)-
BP-	MAG. SPEC-	VAC. SPEC-
C-2-184-M-	MECH. ENG. SPEC.-	VAC-W-
C-2-184-W-	NOMEN.-	XA-1-SM-
CHEM M-	OPER. ANAL.-	XA-2-SM-
CHEM S-	R-1-184-S-	XA-SPEC-
COL. SPEC-	R-1-184-W-	XAC-SPEC-
D1-RR-	R-1-H RR-	XAH-
D1-S-	R1-184-W-	XAH-1-SM-
D1-SM-	R1-M-	XAH-1-SPEC-
D1-SPEC-	R1-RR-	XAH-2.39-
D1M-	R1-S-	XAH-2-RR-
D1W-	R1-SM-	XAH-2-SM-
E-P (MONTH/YEAR)-	R1-SPECIAL-	XAH-SM
E-SPEC-	R1-W-	XAH-SPEC-
EE-	R1H RR-	XAS-
EE-SPEC-	RL-1.7- thru RL-37.9-	XC-
ELEC. ENG. SPEC.-	RL-MISC.-	XL 50.1- thru XL 60.9-
ENG-SPEC.-	RL-NPE-	Z-P (MONTH/YEAR)-
FUND.-184 S2-	SPEC-	Z-SPECIAL-
FUND.-(MONTH/YEAR)-	SPEC SEM-	

For its early reports, the University of California's Radiation Laboratory at Berkeley used a system of report codes based on the subject matter of the reports. The codes were abbreviations or short forms of the names of programs and processes or of the groups working on them. Examples are:

37"	(Magnet)
184"	(Tank)
C-2-184	(Tank)
COL.	(Collection Group)
ID	(Information Division)
KAPPA-	(Process)
MAG	(Unidentified)
R-1-184	(Tank)
R1-	(Tank)
XA-1	(Tank)
XA-2	(Tank)
SUMP-	(Group)

This index heading type of code was frequently followed by another term that designated the type of report, and then by a serial number or by the month and year. On the reports, these codes were frequently shortened to any of several short forms. Thus:
Chemistry Special
Chem. Special
Chem. Spec.
Chem-S-

Later these codes were converted to a duodecimal system consisting of three groups of digits separated by decimal points, the whole prefixed with the letters RL for reports originating with the Radiation Lab. or XL for reports submitted to it. The first digit group indicated the subject of the report, the second group the type of report and the third the sequence number of the specified class and type. The conversion was accomplished by assigning numbers to the various subject classes and types of reports. The decimal-type numbers were assigned to reports that had originally been given numbers based on the index-heading type of codes, but no attempt was made to change the numbers on the reports. Consequently, one report can be referenced by either of two types of numbers. For example:

Old Listings	New Listings
CHEM. SPECIAL NO. 260	RL-4.6.260
BETA SPECIAL NO. 53	RL-3.6.53
RL SM NO. 30	RL-15.4.30
XAH 1 RUN REPORT NO. 34	RL-25.2.34
ELEC. ENG. SPEC. NO. 10	RL-7.6.10

Following are the conversion tables for RL and XL reports:

First part of code of RL generated reports:

Analysis	RL-1
Assay	2
Beta	3
Chemistry	4
D1	5
"E"	6
Electrical Engineering	7
Engineering	8
Fundamental	9
"J" Research	10
"K" Research	11
Mechanical Engineering	12
Nomenclature	13
Operational Analysis	14
RL	15
UCRL Special	16
SUMP	17
Theoretical	18
38"	19
Vacuum	20
XAC	21
XAC-1	22
XAC-2	23
XAH	24
XAH-1	25
XAH-2	26
"Z" Magnetic Group	27
Research Progress Meetings	28
XC	29
"M" Units	30
Focussing Methods	31
C-1	32
C-2	33
Receivers	34
Synchrotron	35
180" Cyclotron	36
Linear Accelerator	37

First part of code for reports generated outside RL:

Alpha	XL-50
Beta	51
Alpha-1	52
Alpha-2	53
General	54
XAX H	55
XAX H-1	56
XAX H-2	57
XBX	58
XBX-1	59
XBX-2	60

Middle number for both RL and XL reports:

Run Reports	1
Daily Reports	2
Weekly Reports	3
Semi-Monthly Reports	4
Monthly & Progress Reports	5
Special Reports	6
Miscellaneous	7
Assay Chemistry	8
Engineering and Plant Development	9

Reports that did not fall in the subject categories of the RL and XL reports were given numbers in the series:

BC-	Berkeley Chemistry
BP-	Berkeley Physics

Reference 10

CHICAGO. UNIVERSITY. METALLURGICAL LABORATORY

The C- reports, published by the Metallurgical Project of the University of Chicago, include those written by members of the Project or its subcontractors or by members of closely associated projects, and a few written elsewhere and republished by the Project. For a short time after the Metallurgical Project became Argonne National Laboratory, the C- series was continued there. Later Argonne reports were assigned ANL- numbers, starting with the number 4000. The C- series reports are divided into subject catego-

REFERENCE NOTES

ries and may or may not include a final letter to indicate the location from which they came.

Each report was assigned a number in consecutive order, without regard for the letters following C-.

At various times reports in these series were renumbered, and the changes made retroactive. Thus, one report may bear the designations CA-166, CC-166, CH-166 and A-239.

The subject categories used at various times (not necessarily simultaneously) may be summarized as:

CATEGORY	SUB-CATEGORY	DESCRIPTION
CA-		Lists of reports, bulletins, or other general material
	M-	Miscellaneous lists and bulletins
	R-	Chemical research, enriched piles and U-233 production
CB-		Chemical research, enriched piles and U-233 production
CC-		Chemical research on miscellaneous materials not included in CB, CK, or CN
	An-	Analytical
	D-	Heavy water
	G-	General
	R-	Radiochemistry
	Rn-	Radiation chemistry
	U-	Basic chemistry of transuranic elements
CE-		Technological research, engineering development
CF-		Fission and fast neutron research; enriched pile research
CH-		Health and protection
CK-		Chemical research. Special chemistry of plutonium
	An-	Analytical
	G-	General
CL-		Lectures and surveys
	(any one of the subject category symbols for sub-category)	
CL-697		Project handbook
CN-		Chemistry of U-235, except for final purification
	An-	Analytical
	G-	General
	H-	Health and protection
	S-	Separation processes
CP-		Physics (except that covered in CF)
	G-	General
	I-	Instruments
	IS-	Instrument specifications
CS-		Policy and administrative material
	I-	Minutes of information meetings
	I-(LETTER)	(the letter depends on the Division reporting at the meeting)
	P-	Minutes of policy meetings
CT-		Metallurgical and technological research
	C-	Coatings and corrosion
	M-	General metallurgy
	W-	Weekly reports
G-(NUMBER)	(one of the subject category symbols after the number)	(translation of German document)
M-	(one of the subject category symbols after the number)	(indicates an informal report or one not considered permanent)
Memo-		(early designation for M-)
W- } X- }		Design blueprints and drawing files, correspondence and reports from Hanford and Oak Ridge

Symbols assigned for the locations from which the reports came and entered as sub-sub entries under the subject categories are:

B-	California. Univ., Berkeley (Latimer)
Ba-	California. Univ., Berkeley (Hamilton)
C-	Argonne, Metallurgical Lab. and Metallurgical Lab. subcontracts
Cr-	Carnegie Inst. of Tech., Pittsburgh
D-	Columbia Univ., N.Y.C.
I-	Iowa State Coll., Ames
O-	Battelle Memorial Inst., Columbus, Ohio
T-	Massachusetts Inst. of Tech., Cambridge
X-	Clinton Labs., Oak Ridge, Tenn.
	Tennessee Eastman Corp., Oak Ridge, Tenn.
	Carbide and Carbon Chemicals Corp., K-25 Plant, Oak Ridge, Tenn.

Reports from these locations were also numbered into the C- series with no indication of the origin given.

Reports from other projects were sometimes assigned numbers by the Metallurgical Laboratory or the Research Control Section of the Manhattan District, and at other times the numbers assigned by the originators were kept but modified to include the Metallurgical Laboratory Project subject classification in the series code. The following symbols are related to sources, as shown:

A- [A]M- M-	Codes assigned by Research Control, Manhattan District to reports from various American sources. The M numbers were sometimes prefixed with an [A] and interfiled with the A- reports.
B- BR-	Various British sources
Beta- Chem S- N	Various American sources
CD-	Clinton Engineer Works, Oak Ridge, Tenn.
LA- LAMS-	Los Alamos Scientific Lab., N. Mex.
MC- ME- MP- MT- MTec- MX-	National Research Council of Canada, Montreal Labs.
Mon- Mon (letter)	Clinton Labs., Oak Ridge, Tenn.

The subject categories were often added after the report number, as MT-9 (CK) for a report designated by the originator as MT-9.

Some of the earlier A and B reports were reissued also as C- reports and were then listed by a three-letter designation, such as CPA- or CPB- in which the first two letters indicate the subject category and the third the origin.

Reference 11

COLUMBIA UNIVERSITY. SUBSTITUTE ALLOY MATERIALS LABORATORIES

O-L- thru O-M-
1-(LETTER)-
1-X(LETTER)-
1(LETTER)-
1(LETTER)-(LETTER)-
1(LETTER)-X(LETTER)-
1(LETTER)(NUMBER)-(LETTER)-
1(LETTER)(NUMBER)-X(LETTER)-
1(LETTER)(NUMBER)(LETTER)-
1(LETTER)(LETTER)-
1X(LETTER)-
2-(LETTER)-
2-X(LETTER)-
2(LETTER)-
2(LETTER)-(LETTER)-
2(LETTER)-X(LETTER)-
2(LETTER)(NUMBER)-(LETTER)-
2(LETTER)(NUMBER)-X(LETTER)-
2(LETTER)(NUMBER)(LETTER)-
2(LETTER)(LETTER)-
2X(LETTER)-
3-(LETTER)-
3-X(LETTER)-
3(LETTER)-
3(LETTER)-(LETTER)-
3(LETTER)-X(LETTER)-

REFERENCE NOTES

3 (LETTER) (LETTER)-
3X (LETTER)-
4- (LETTER)-
4-X (LETTER)-
4 (LETTER)-
4X (LETTER)-
5- (LETTER)-
5-X (LETTER)-
5 (LETTER)-
5 (LETTER)- (LETTER)-
5 (LETTER)-X (LETTER)-
5 (LETTER) (NUMBER)- (LETTER)-
5 (LETTER) (NUMBER)-X (LETTER)-
5 (LETTER) (NUMBER) (LETTER)-
5 (LETTER) (LETTER)-
5X (LETTER)-
6- (LETTER)-
6-X (LETTER)-
6 (LETTER)-
6 (LETTER)- (LETTER)-
6 (LETTER)-X (LETTER)-
6 (LETTER) (LETTER)-
6X (LETTER)-
100-BR- thru 100-ZS

For purposes of accounting for, and controlling access to, classified material issued by or received by Columbia University's Substitute Alloy Materials Laboratories, code designations were assigned to the divisions, groups and sections of the laboratories. The divisions were assigned number codes as follows:

0	Service Division
01	Personnel Division
02	Legal Division
1	Physical Division
2	Chemical Division
3	Engineering Division
4	Theoretical Division
5 & 6	Unidentified divisions

In addition to the numbers listed above, numbers 10-99 were reserved for non-Columbia groups, and the number 100 was assigned to the project headquarters.

The divisions were subdivided into groups, which were designated by letters; the groups were divided into sections, which were designated by numbers. For example, in 2A3-, 2 indicates the Chemical Division, A indicates Group A of the Chemical Division and 3 indicates Section 3 of Group A of the Chemical Division.

Documents issued by the Laboratories received a three-part number consisting of: *1)* the division, group or section symbol; *2)* a letter designating the type of document; and *3)* a serial number, e.g., 2A3-R-100.

Documents received by the Laboratories were assigned a four-part number consisting of: *1)* the division, group or section symbol; *2)* the letter "X" indicating that the number assigned might not necessarily appear on all copies of the document; *3)* a letter designating type of document; and *4)* a serial number, e.g., 2A3-X-R-100.

The letters used to indicate the type of document were:

R	Report	M	Memorandum
L	Letter	S	Special

No consistency was observed in use of dashes to separate the various parts of the numbers.

Reference 12

GOODYEAR ATOMIC CORPORATION

GAT-   GAT-(LETTER(S))-   GAT-DR-   GAT-L-   GAT-T-

Formal reports issued by the Goodyear Atomic Corporation are given serial numbers in the GAT series. Informal reports are assigned numbers in series identified by GAT followed by a letter or letters. The technical divisions issuing reports have been the Laboratory Division (GAT-L-) and the Engineering Development Division (GAT-DR-), which later combined to form the Technical Division (GAT-T-). The numbering of the GAT-T- series begins with GAT-T-501 and is a continuation of the GAT-L- series. The GAT-DR- series died with GAT-DR-300.

Reference 13

JOINT ARMED SERVICES PUBLICATIONS

AN-	JAN-	JANAP-	MS-
AN AB-	JAN-(LETTER)-	JANP-	SPSP-
ANC-	JAN-STD-	JPI-	SPSTP-
AND-	JAN-WM-	MIL-(LETTER)-	WBAN-
JAFNC-	JANAF-	MIL-STD-	

The series codes assigned to joint publications of the armed services sometimes indicate the services involved and may incorporate the letter J, which indicates a joint publication:

AN-	Army-Navy
AN AB-	Army-Navy (Aeronautical Bulletin)
ANC-	Air Force-Navy Civil Committee
AND-	Army-Navy (Drawing)
JAFNC-	Air Force-Navy Committee
JAN-	Army-Navy
JAN-(LETTER)-	Army-Navy (Specification)
JAN-STD-	Army-Navy (Standard)
JAN-WM-	Army-Navy (Work Manual)
JANAF-	Army-Navy-Air Force
JANAP-	Army-Navy-Air Force (Publication)
JANP-	Army-Navy (Publication)
WBAN-	Air Force-Navy Weather Bureau

Other publications applicable to more than one branch of the service are assigned codes that do not indicate the services involved:

JPI-	Army-Navy (joint packaging instruction)
MIL-(LETTER)-	Military Specification
MIL-STD-	Military Standard
MS-	Military Standard
SPSP-	Army-Navy-Air Force Solid Propellant Surveillance Panel
SPSTP-	Army-Navy-Air Force Solid Propellant Rocket Static Test Panel

In assigning designations to military specifications, the Department of Defense follows the JAN- or MIL- with letters that represent the article for which the specification is given, e.g., MIL-T- might be for a tent, tube, etc.

Reference 14

JOINT ATOMIC WEAPONS PUBLICATIONS

AEC-AFSWP-TP-	AIR FORCE T.O. (NUMBERS)-11N	NAVY-SWOP
AEC-DASA-TP-	AIR FORCE T.O.-11N-	SWOP-
AEC-DNA-TP-	ARMY-TM-9-	T.O. (NUMBERS)-11N-
AEC-TP-	ARMY-TM-39-	T.O. 11N-
AF T.O. (NUMBERS)-11N	DASA-TP-	TM-9-
AF T.O.-11N-	DNA-TP-	TM-39-
AF-TO-11N-	JAWPB-	TP-
AFSWP-TP-	JSWPB-	USAF T.O. 11N-
	NAVORD-SWOP-	

This system was administered by the Joint Atomic Weapons Publications Board (previously the Joint Special Weapons Publications Board), which was comprised of representatives from the Defense Atomic Support Agency (previously the Armed Forces Special Weapons Project), Army, Navy, Air Force, Atomic Energy Commission, Sandia Corporation and Los Alamos Scientific Laboratory. The Board has been superseded by an established procedure which is numbered TP-1-1. Two of the agency names have changed. DASA has become Defense Nuclear Agency (DNA) and Sandia Corporation has become Sandia Laboratories. It is designed to provide nonclassified short titles for atomic weapons manuals. Both joint and single service atomic weapons manuals are assigned numbers within this system. Each manual carries as many short titles as services to which it applies. In these short titles the numbers and letters following the prefixes are the same, but the prefixes indicate that they fall within the Tech Manual series of the Army, the Tech Order series of the Air Force and the Special Weapons Ordnance Publications series of the Navy. Thus a single report may bear as many as four short titles, e.g.:

AEC-DASA-TP 0-1   NAVY SWOP 0-1
ARMY TM 39-0-1   AIR FORCE T.O. 0-1-11N

Reference 15
JOINT COMMITTEES OF THE ATOMIC ENERGY COMMISSION AND DEPARTMENT OF DEFENSE

AB-
ACORN-
JC-(LETTERS)-(YEAR)-
JUPITER-
JUPITER-TB-
NZJWCC-
OAK-
PERSHING-
RG-CC-
RG-CC-(YEAR)-
SERGEANT JWCC-
SPRUCE-
SUBROC WJCC-
TABLELEG-
TERRIER CC-

Joint Committees of members from various organizations of the Department of Defense and the Atomic Energy Commission and their contractors, which are formed to consider various aspects of guided missiles and other weapons delivery systems and their warheads, usually issue their minutes and reports under designations beginning with JC- and continue with letters which stand for the missile concerned. For example:

JC-CP-   Corporal Committee
JC-HJ-   Honest John Committee
JC-NB-   Nike B Committee
JC-NH-   Nike Hercules Committee
JC-RS-   Redstone Committee

Occasionally this pattern is broken and the letters JC are omitted. For example:

AB-   Alias Betty Committee
NZJWCC-   Nike Zeus Joint AEC-DOD Warhead Coordinating Committee
RG-CC-   Regulus Coordination Committee

Usually the year of issuance follows the missile designation preceding the serial number.

Sometimes the full name of the missile or an arbitrary name is used. For example:

ACORN-   Acorn Committee
JUPITER-   Jupiter Nose Cone Committee
OAK-   Oak Joint AEC-DOD Warhead Coordinating Committee
PERSHING-   Pershing Joint AEC-DOD Warhead Coordinating Committee
SERGEANT-   Sergeant Joint AEC-DOD Warhead Coordinating Committee
SPRUCE-   Spruce Joint AEC-DOD Warhead Coordinating Committee
SUBROC-   Subroc Joint AEC-DOD Warhead Coordinating Committee
TABLELEG-   Tableleg Committee
TERRIER-   Terrier Coordinating Committee

The reports are usually issued by the organization that hosts the meeting.

Reference 16

JOINT FEASIBILITY STUDY GROUPS

JFSG (LETTERS)-

Joint Feasibility Study Groups were made up of members of various organizations of the Department of Defense and the Atomic Energy Commission and its contractors. Their purpose was the study of some particular phase of a missile or its warhead. The organizations represented usually included:
- Air Force Special Weapons Center, Kirtland AFB, N. Mex.
- California. Univ., Livermore. Lawrence Radiation Lab.

SWC JFSG (LETTERS)-

- Defense Atomic Support Agency. Washington, D.C.
- Defense Atomic Support Agency. Field Command, Albuquerque, N. Mex.
- Los Alamos Scientific Lab., N. Mex.
- Sandia Corp., Albuquerque, N. Mex.

with other organizations sometimes included.

The letters following the JFSG stood for the subject of the study, e.g., JFSG GAR-9.

Reference 17

JOINT TASK FORCES

JF-   JTF-   JTF-(NUMBER)-ATG-   JTF-(NUMBER)-TG-

Joint Task Forces are organizational units made up of members of the Army, Navy, Air Force and other government agencies and their contractors to accomplish specific missions. They are designated by number. Subdivisions of them are called task groups and are designated by a decimal following the number of the task force.

Joint Task Forces that have been responsible for the issuance of many reports are those concerned with the conduct of nuclear tests overseas. Their designations are:

Joint Task Force 1	Operation Crossroads
Joint Task Force 3	Operation Greenhouse
Joint Task Force 7	Operations Sandstone, Castle, Redwing, Hardtack, Wigwam
Joint Task Force 7 Task Group 7.3 (Navy)	Operation Wigwam
Joint Task Force 8	Operation Dominic
Joint Task Force 132	Operation Ivy

Joint Task Force 2 was a special operation at Sandia Base.

Reports issued by Joint Task Forces may be designated by the letters JTF followed by the number of the task force and a serial number. Those issued by task groups bear the JTF designation and number followed by the letters TG (or ATG for Air Task Group), the number of the group and a serial number, for example, JTF-7- and JTF-3-TG-3.4-.

Most reports issued by Joint Task Forces 1, 3, 7 and 132 were given series designations standing for the Operation with which they were concerned (See Ref. 7).

Reference 18

JOINT TASK GROUPS

AFSWP/FC-JTG-(LTRS.-NOS.)   JTG-(LETTERS-NUMBERS)-   MATADOR-JTG-

Joint Task Groups are composed of representatives of the Army, Navy and Air Force for the accomplishment of specific tasks such as the evaluation of atomic warheads or missiles. Their reports are assigned designations made up of the letters JTG, followed by letters and numbers that stand for the subject assigned to the task group.

Occasionally the letters JTG follow the letters standing for the subject, as in MATADOR-JTG-.

Reference 19

JOINT WORKING GROUPS

JWG-   JWG (LETTERS or NUMBERS)-   JWG (LTRS. or NOS.)-(YEAR)-

Joint Working Groups were made up of members from various Department of Defense and Atomic Energy Commission agencies and contractors for the purpose of studying specific missiles, airplanes, etc., and their warheads and payloads. The usual membership was Air Force Special Weapons Center, Defense Atomic Support Agency Field Command and the Sandia Corporation. Their reports and minutes were usually issued by the host organization but were given JWG designations.

The letters and/or numbers following the JWG represented the missile or plane under consideration, e.g., JWG 58-, JWG BOM- and JWG GAR-11.

Sometimes the letters and/or numbers were followed by the year before the serial designation.

Reference 20

LIBRARY OF CONGRESS. TECHNICAL INFORMATION DIVISION—AND ITS PREDECESSORS: NAVY RESEARCH SECTION AND SCIENCE AND TECHNOLOGY PROJECT

C-   R-   S-   U-

Reports received by these divisions of the Library of Congress were assigned accession numbers in four series according to their security classifications. The majority of these reports were issued by naval contractors, but the series were not limited to those. The letters stood for:

   C   Confidential   S   Secret
   R   Restricted   U   Unclassified

Reference 21

LOS ALAMOS SCIENTIFIC LABORATORY

LA-
LA-(NUMBER)-BIB
LA-(NUMBER)-MS
LA-(NUMBER)-PR
LA-(NUMBER)-SOP
LA-(NUMBER)-TR
LA-DC-
LA-DC-(YEAR)-
LA-TR-(YEAR)-
LADC-
LAMS-

Formal reports issued by the Los Alamos Scientific Laboratory are given serial numbers in the LA- series. Less formal reports were once prefixed by LAMS-. Until late 1949 the two series were separately numbered, but after LA-756 and LAMS-953 they were combined. Beginning with -954 only one numerical series was maintained, but the prefix was either LA- or LAMS- as appropriate.

In 1964 this pattern was changed, with the MS relegated to the position of suffix. Subsequently other suffixes were adopted, BIB for bibliographies, PR for progress reports, SOP for standing operating procedures, and TR for translations. Only important translations that have been carefully edited are included in this series.

Two other series are also maintained. LA-TR-(YEAR)- is used for informal translations. The LA-DC- series (formerly LADC- and currently LA-DC-(YEAR)- is used for material released for publication as journal articles, conference papers, books, etc.

Reference 22

LOS ALAMOS SCIENTIFIC LABORATORY (INCOMING REPORTS)

AM-   BM-

These series were assigned by the Los Alamos Scientific Laboratory to miscellaneous reports received from 1946 through 1949, and occasionally thereafter. The choice of designator was determined by the country of origin of the report, e.g., AM- American and BM- British (including Canadian).

Within each series, numbers were assigned in order of accession.

Reference 23

MANHATTAN DISTRICT

A-   M-   P-

These series were assigned by the Manhattan District, and later by the Atomic Energy Commission, to reports issued by their contractors. The A series was used for formal topical reports, the M series for informal reports or memoranda and the P series for progress reports. The numbers were assigned by blocks. Reports issued by the University of Chicago Metallurgical Laboratory were also assigned different numbers in the C-series. Cross references between the two series were compiled and issued in these reports:

CA-555   Chicago. Univ. Metallurgical Lab. Cross index of A and C report numbers based on library information as of April 1, 1943, by Herman H. Fussler, April 6, 1943.

CA-555 (Rev) Technical Information Service, AEC. Cross reference of C and A report numbers by Herman H. Fussler and Charlotte F. Chesnut, December 22, 1948, 29 p.

Reference 24

MANHATTAN DISTRICT (DECLASSIFIED DOCUMENTS)

AECD-

MDDC-

These designate a series of numbers prefixed by MDDC- through number 1779 and by AECD- from then on, assigned by the Manhattan District and later by the Atomic Energy Commission to their declassified reports. Initially numbers in these series were assigned to all reports as they were declassified; later they were used only for reports not previously identified by a distinctive report number. Confusion about these numbers has arisen because later references to the first 1779 declassified reports may give them with an AECD- prefix.

Reference 25

NATIONAL ADVISORY COMMITTEE FOR AERONAUTICS

AC-
ACR-(LETTER)(DATE CODE)-
ARR-(LETTER)(DATE CODE)-
CB-(LETTER)(DATE CODE)-
MR-(LETTER)(DATE CODE)-
NACA-
NACA-(LETTERS)-A(DATE CODE)-
NACA-(LETTERS)-E(DATE CODE)-
NACA-(LETTERS)-H(DATE CODE)-
NACA-(LETTERS)-L(DATE CODE)-
NACA-AC-
NACA-ACR-(DATE)-
NACA-ACR-(LETTER)(DATE CODE)-
NACA-ARR-(DATE)-
NACA-ARR-(LETTER)(DATE CODE)-
NACA-CB-(DATE)-

NACA-CB-(LETTER)(DATE CODE)-
NACA-MR-(DATE)-
NACA-MR-(LETTER)(DATE CODE)-
NACA-RB-(DATE)-
NACA-RB-(LETTER)(DATE CODE)-
NACA-REPORT-
NACA-RM-(LETTER)(DATE CODE)-
NACA-RM-S-
NACA-RM-SE-
NACA-RM-SL-
NACA-TM-
NACA-TN-
NACA-WR-(LETTER)-
RB-(LETTER)(DATE CODE)-
RM-(LETTER)(DATE CODE)-
WR-(LETTER)-

The National Advisory Committee for Aeronautics maintained eleven publication series for its reports, including those issued from Headquarters and from its various laboratories. Reports in four of these series were numbered consecutively without additional identification. These were:

NACA-AC-   Aircraft Circulars
NACA-REPORT-   Reports
NACA-TM-   Technical Memoranda
NACA-TN-   Technical Notes

One series was numbered consecutively but also carried a letter designating the laboratory of origin: NACA-WR-(LETTER)- Wartime Reports.

The other series depended on the date of issuance and later a date symbol combined with a letter for the laboratory of origin for identification. These series were:

NACA-ACR-   Advance Confidential Reports
NACA-ARR-   Advance Restricted Reports
NACA-CB-   Confidential Bulletins
NACA-MR-   Memorandum Reports
NACA-RB-   Restricted Bulletins
NACA-RM-   Research Memoranda

These series were at times referred to without the NACA prefix.

The codes used to indicate the laboratory of origin and the date of issuance were as follows:

Laboratory- A   Ames Aeronautical Lab., Moffett Field, Calif.
            E   Aircraft Engine Research Lab., Cleveland
                Flight Propulsion Research Lab., Cleveland
                Lewis Flight Propulsion Lab., Cleveland
            H   High-Speed Flight Station, Edwards Air Force Base, Calif.
            L   Langley Aeronautical Lab., Langley Field, Va.
                Langley Memorial Aeronautical Lab., Langley Field, Va.

Year-   3   1943
        4   1944
        5   1945 etc.
        50   1950
        51   1951
        52   1952 etc.

Month- A   January   G   July
        B   February   H   August
        C   March   I   September
        D   April   J   October
        E   May   K   November
        F   June   L   December

Day-   01
        02
        03 etc. to 31
        followed by
        a   2nd document issued that date
        b   3rd document issued that date

These codes were assembed in the form:
NACA ACR E4D19
NACA ARR L4K22b
NACA CB E5J11
NACA MR A4L12
NACA RB L5F15

The NACA series were succeeded by the NASA series of the National Aeronautics and Space Administration (See Ref. 26).

Reference 26

NATIONAL AERONAUTICS AND SPACE ADMINISTRATION

NASA-
NASA-(LETTERS)-A-
NASA-(LETTERS)-E-
NASA-(LETTERS)-H-
NASA-(LETTERS)-L-
NASA-(LETTERS)-W-
NASA-M-(DATE)(LETTER)-
NASA-MEMO-(DATE)(LETTER)-

NASA-MEMORANDUM-(DATE)(LTR)-
NASA-RE-(DATE)(LETTER)-
NASA-REPUBLICATION-(DATE)(LTR)-
NASA-TM-X-
NASA-TN-D-
NASA-TR-R-
NASA-TT-F-

The National Aeronautics and Space Administration's publication series supersede those of the National Advisory Committee for Aeronautics (See Ref. 25).

Four series are published, Technical Reports, Technical Notes, Technical Memoranda and Technical Translations (previously called Republications). Until July

1959, reports in these series were identified by date with a suffix letter indicating the source within NASA. These letters were:

A   Ames Research Center, Moffet Field, Calif.
E   Lewis Research Center, Cleveland
H   High-Speed Flight Station, Edwards, Calif.
L   Langley Research Center, Langley Field, Va.
W   NASA Headquarters, Washington, D.C.

Since July 1959 reports in the four series have been numbered consecutively, regardless of place of origin. The form used for the series is:

NASA-TR-R-   Technical Reports
NASA-TN-D-   Technical Notes
NASA-TM-X-   Technical Memoranda
NASA-TT-F-   Technical Translations

Reference 27

NATIONAL DEFENSE RESEARCH COMMITTEE

A-
A-(NUMBER)(LETTER)-
B-
NDRC-1- thru NDRC-19-

NDRC- (1 thru 19) STR-
NDRC-A- thru NDRC-E-
NDRC-DIV.- (1 thru 190)
STR-DIV. (NUMBER)-

The National Defense Research Committee (NDRC) functioned as a subdivision of the Council of National Defense from 1940 to 1941. From 1941 to 1947 it was a subdivision of the Office of Scientific Research and Development. The NDRC supervised contracts with universities and industry for scientific research and development in the interest of the United States government. Reports generated by the contractors were issued with many report series codes in addition to (or instead of) the NDRC- codes. (*See* Ref. 32 for a list of contractors under the parent organization, Office of Scientific Research and Development.)

The NDRC was responsible for Summary Technical Reports (STR-) and for the NDRC- series. Both the NDRC- series and the STR- series were divided by division designation, alphabetical or numerical. The divisions were:

Through Fall 1942
A   Armor and Ordnance (Even in 1943 and later, NDRC-A- was used for reports on the subject of armor and ordnance, although there was no such division at that time.)
B   Bombs, Fuels, Gases and Chemical Problems
C   Communication and Transportation
D   Detection, Controls, and Instruments
E   Patents and Inventions

From December 1942
1   Ballistic Research
2   Effects of Impact and Explosion
3   Rocket Ordnance
4   Ordnance Accessories
5   New Missiles
6   Sub-surface Warfare
7   Fire Control
8   Explosives
9   Chemistry
10   Absorbents and Aerosols
11   Chemical Engineering
12   Transportation
13   Electrical Communication
14   Radar
15   Radio Coordination
16   Optics and Camouflage
17   Physics
18   War Metallurgy
19   Miscellaneous

Reference 28

NATIONAL DEFENSE RESEARCH COMMITTEE, PANELS AND COMMITTEES

AMG-
AMG-B-
AMG-C-
AMG-H-
AMG-IAS-
AMG-N-
AMG-NYU-
AMP-
AMP-MEMO-

AMP-NOTE-
AMP-RPT.-
APP-
NDRC-AMG-
NDRC-AMG-B-
NDRC-AMG-B-MEMO-
NDRC-AMG-C-
NDRC-AMG-C-MEMO-
NDRC-AMG-H-
NDRC-AMG-IAS-

NDRC-AMG-N-
NDRC-AMG-NYU-
NDRC-AMG-NYU-MEMO-
NDRC-AMP-
NDRC-AMP-MEMO-
NDRC-AMP-NOTE-
OSRD-AMP-
OSRD-AMP-NYU-
TDAC-

In addition to its formal divisions, the National Defense Research Committee had the following panels and committees that issued their own report series:
  Applied Mathematics Panel (AMP-)
  Applied Psychology Panel (APP-)
  Committee on Propagation
  Tropical Deterioration Administrative Committee (TDAC-)

The Applied Mathematics Panel and its Applied Mathematics Group worked at and issued reports from many institutions. The report series code often included a letter or letters that indicated the originating institution. These were:
  B-   Brown Univ., Providence
  C-   Columbia Univ., N.Y.C.
  H-   Harvard Univ., Cambridge, Mass.
  IAS- Institute for Advanced Study, Princeton, N.J.
  N-   Northwestern Univ., Evanston, Ill.
  NYU- New York Univ., N.Y.C.

This designation, in combination with other elements of the names of the organizations involved and the types of publications, created many series codes. For example, the following are known to have been used in connection with reports from New York University's Applied Mathematics Group:
  AMG-NYU-
  AMP-MEMO-
  AMP-NOTE-
  AMP-REPORT-
  NDRC-AMG-NYU-
  NDRC-AMG-NYU-MEMO-
  OSRD-AMP-NYU-

Reference 29

NATIONAL RESEARCH COUNCIL OF CANADA

CEI-
CI-
CR-(LETTER(S))-
CR(LETTER(S))-
CRL-
CRP(LETTER(S))-
ED-
EI-
EL-

GPI-
HI-
HRI-
M(LETTER(S))-
MC-
ME-
MI-
MM-
MP-

MP(LETTER(S))-
MT-
MTEC-
MX-
NEI-
NL-
PD-
PR-
PR-(LETTER(S))-

REFERENCE NOTES

PR-(LETTER(S))-(LETTER(S))-
PR-(LTR(S))-(NUMBER)-(LTR(S))-
PR(LETTER(S))-

TEC-PI-
TECPI-

TL-
TPI-
XI-

The Montreal Laboratory and the Chalk River Laboratory of the National Research Council of Canada were predecessors of the Division of Atomic Energy of the Council. After the construction of the Chalk River Laboratories, the entire division was located there, and in 1948 the name was changed to Atomic Energy Project, National Research Council of Canada, which in April 1952 was reorganized as Atomic Energy of Canada Ltd. (*See* Ref. 6).

Reports originating from the Montreal and Chalk River Laboratories were numbered as follows:

The first letter or letters indicated the laboratory of origin: CR for Chalk River and M for Montreal.

The next letters indicated the division, branch, committee or panel issuing the report:

A-	Administration
B-	Biology
C-	Chemistry
CE-	Chemical Engineering
CM-	Chemistry and Metallurgy
DC-	Development Chemistry
E-	Engineering
ED-	Engineering Design
EL-	Electronics
GP-	General Physics
H-	Health
HP-	Health Physics
HR-	Health and Radiation
IO-	Industrial Operations
M-	Medical
MET-	Metallurgy
NE-	Nuclear Engineering
P-	Nuclear Physics
R-	Reactor
RHC-	Radiation Hazards Control
RP-	Reactor Physics
RRD-	Reactor Research & Development
T or TP-	Theoretical Physics
TEC or TECP-	Technical Physics
X-	Extra-Mural

For progress reports, the letter P was inserted between the first letters for the laboratory and the letters for the division, e.g., CRPP, MPT, CRPT, etc., or the letters PR were used as the main series designation, alone for general laboratory progress reports or followed by the letters for the division.

For internal reports, the letters for the laboratory were dropped, and the letter I was added after the letters for divisions.

Lecture reports were numbered as follows:

CRL-	Chalk River Lecture
EL-	Engineering Lecture
NL-	Nuclear Physics Lecture
TL-	Theoretical Physics Lecture

Reference 30

NAVAL RADIOLOGICAL DEFENSE LABORATORY

AD-(NUMBER)-(LETTER)-   AD(LETTER)-   NRDL AD(LETTER)-   USNRDL-

Until 1952 reports issued by the Naval Radiological Defense Laboratory were numbered consecutively and were prefixed with the letters AD. In addition, a letter or letters were used to designate the type of report and/or its subject field. At first these additional letters were placed as suffixes after the serial number; later they followed the AD prefix before the number. The letters used were:

A-	Abstracts, bibliographies, report lists, etc.
B-	Biological studies
C-	Chemical studies
E-	Engineering studies
F-	Fast reactor investigations
H-	Radiological health and safety problems
I-	Instrument research
K-	Special chemistry of the heavy elements
L-	Lectures, seminar notes, manuscripts, translations, mathematical tables, etc.
M-	Medical studies
O-	Classified laboratory memoranda and memoranda for external distribution
P-	Physics
Q-	Scientific Department progress reports
R-	(Unidentified)
S-	Information and/or policy interpretations of

broad phases of the Laboratory's activities
- T- Technological studies
- V- Surveys of the Laboratory's work covering two or more report categories
- X- Atomic bomb detonation studies
- Y- Radiological defense manuals
- Z- Military applications

After report AD-338, the series was continued with the prefix USNRDL.

Reference 31

NEPA PROJECT

IC-
IY-
NCR-
NEPA-
NEPA-(NUMBER)-(LTRS)-(NO.)-
NEPA-IC-(YEAR)-(MONTH)-

NEPA-IY-
NEPA-NCR-
NEPA-SCR-
NEPA-SCRM-
NEPA-SERM-

NEPA-STRM-
SCR-
SCRM-
SERM-
STRM-

The NEPA Project (Nuclear Energy for the Propulsion of Aircraft) under the Fairchild Engine and Airplane Corporation began in May 1946 and ended in April 1951. Associated with Fairchild in the capacity of member companies were:

Allison Div., General Motors Corp., Indianapolis
Continental Aviation and Engineering Corp., Detroit
Flader (Fredric) Inc., North Tonawanda, N.Y.
General Electric Co.
Lycoming Div., Avco Mfg. Corp., Stratford, Conn.
Northrop Aircraft, Inc., Hawthorne, Calif.
United Aircraft Corp., East Hartford, Conn.
Westinghouse Electric Corp., Pittsburgh
Wright Aeronautical Corp.

In addition, many companies and universities participated in the NEPA Project by means of subcontracts.

Reports prepared by all agencies connected with the project were submitted to the NEPA Division of Fairchild for assignment of report number. The number consisted of four parts: 1) NEPA-; 2) consecutive series number-; 3) letters for the department, member company or subcontractor-; and 4) number associated with the department, member company or subcontractor. For example, NEPA-893-UAC-5 is the fifth report submitted under United Aircraft's subcontract to Fairchild and was issued under the consecutive numbering system as NEPA-893. For cataloging purposes, the last two parts of the number may be disregarded.

In addition to the main series, there were several other series with different prefixes. These were:

IC-   NCR-   SERM-
IY-   SCR-   STRM-
        SCRM-

These prefixes were used separately or were in turn prefixed with NEPA. The number following them is not part of the regular consecutive number series.

Reference 32

OFFICE OF SCIENTIFIC RESEARCH AND DEVELOPMENT

CMR-   OSRD-   OSRD (LETTERS)-

The Office of Scientific Research and Development was established in June 1941 and continued until 1947. OSRD subdivisions, including the National Defense Research Committee and the Committee on Medical Research, issued their own reports with distinctive series codes (NDRC- and CMR-).

REFERENCE NOTES

In addition, the OSRD issued its own series and the various contractors (universities and industries) sometimes issued their separate series. Report of OSRD-sponsored research may, therefore, bear one series code and number, or two.

Further, a report written at the California Institute of Technology and recognized by the OSRD as JPC-3 is referred to in other Cal. Tech. reports as CIT/JPC-3.

Some reports of OSRD-sponsored research are known by OSRD-(LETTERS)- in which the letters stand for the contractor or some other identifier readily explained, as:

CF-	Controlled Fragmentation
CIT-	California Inst. of Tech.
DF-	Detonation and Fragmentation
DFA-	Detonation, Fragmentation and Air Blast
EWT-	Effects of Weapons on Targets
FS-	Fundamental Study of Explosives
JP-	Jet Propulsion
JHU-	Johns Hopkins Univ.
LOR-	Liaison Office Research
OD-	National Bureau of Standards, Ordnance Development Div.
OD-SP-	National Bureau of Standards, Ordnance Development Div. Special Projects Group
OD-TB-	National Bureau of Standards, Ordnance Development Div. Special Group on Toss Bombing
ODP-	Organic Development Problems
OTB-	Ordnance and Terminal Ballistics
PT-	Preparation and Testing of Explosives
RL-	Massachusetts Inst. of Tech., Radiation Lab.
RP-	Rocket Propellants
RRC-	RDX and Related Compounds
SC-	Shaped Charges
SP-	Special Propellants
STR-	Summary Technical Reports
TC-	Tracer Compositions
TTM-	Torpedo Technical Memorandum
UE-	Underwater Explosives and Explosions
UERL-	Underwater Explosives Research Lab.

The explanation of the following series is not known to the compilers:

CZC-	JAC-	NDC-	RBC-
DDC-	JBC-	NEC-	RDC-
ERD-	JCC-	NMC-	REI-TMD-
F-	JDC-	NOC-	RP-
LAC-	JEC-	NY-	SM-
IBC-	JFC-	OBC-	SR-
ICC-	JGC-	OD-OAG-	SRG-C-
IDC-	JHC-	OEA-	ST-
IEC-	JIC-	OEC-	TAC-
IFC-	JKC-	OFC-	TPC-
IGC-	JMC-	OGC-	UAC-
IHC-	JNC-	OHC-	UBC-
IIC-	JOC-	OIC-	UEC-
IKC-	JPC-	OMC-	UFC-
ILC-	JQC-	ONC-	UIC-
IMC-	JSC-	OPC-	UM-
INC-	JTC-	ORC-	UM-RPP-
IOC-	K-	OZC-	UMC-
IPC-	LBC-	P-	UNC-
IQC-	LMC-	PMC-	UPC-
ISC-	LP-	PMR-	UZC-
ITC-	MBE-	POM-	W-
IZC-	MTC-	PTM-	ZBC-
J-			

See also Ref. 27 for NDRC-.

Reference 33

SQUID PROJECT

ARC-(NUMBER)-P-
BUM-(NUMBER)-P-
BUM-SQUID-(NUMBER)-
CAL-(NUMBER)(LETTER(S))-
DART-(NUMBER)-(LETTER)-
DART TM-
DEL-(NUMBER)-P-
EXP-(NUMBER)-M-
EXP-(NUMBER)-P-
JHU-(NUMBER)-(LETTER(S))-
JHU-SQUID-
MICH-(NUMBER)-(LETTER)-
MIT-(NUMBER)-M-
MIT-(NUMBER)-P-
MIT-(NUMBER)-R-
MIT-(NUMBER)-T-
NTI-(NUMBER)-(LETTER)-
NYU-(NUMBER)-(LETTER(S))-

PIB-(NUMBER)-(LETTER(S))-
PR-(NUMBER)-M-
PR-(NUMBER)-M-P-
PR-(NUMBER)-P-
PR-(NUMBER)-P-R-
PR-(NUMBER)-R-
PR-(NUMBER)-R-P-
PUR-(NUMBER)-(LETTER(S))-
SQUID-
SQUID-PR-(NUMBER)-C-
SQUID-PR-(NUMBER)-P-
SQUID-TR-(LTRS)-(NO.)-PU-
UCB-(NUMBER)-R-
WIS-(NUMBER)-(LETTER(S))-

Squid Project began as a cooperative program of fundamental research in jet propulsion carried on for the Office of Naval Research of the Navy, the Office of Air Research and the Office of Scientific Research of the Air Force, and the Office of Ordnance Research of the Army by various university, industrial and governmental contractors and agencies. These included:

Aero Chemical Res. Labs., Princeton, N.J.
Atlantic Research Corp., Alexandria, Va.
Brooklyn. Polytechnic Inst.
Bureau of Mines, Pittsburgh
California. Univ., Berkeley
California Inst. of Tech., Pasadena
Catholic Univ. of America, Washington, D.C.
City Coll., NYC
Cornell Aeronautical Lab., Inc., Buffalo
Dartmouth Coll., Hanover, N.H. Thayer School of Engineering
Delaware. Univ., Newark
Dynamic Science, Monrovia, Calif.
Experiment, Inc., Richmond
Illinois. Univ., Urbana
Johns Hopkins Univ., Baltimore
Massachusetts Inst. of Tech., Cambridge
Michigan. Univ., Ann Arbor
Naval Research Lab., Washington, D.C.
New York Univ., N.Y.C.
Northwestern Univ., Evanston, Ill. Technological Inst.
Pennsylvania State Univ., University Park
Polytechnic Research and Development Co., Inc., Brooklyn
Princeton Univ., N.J.
Purdue Univ., Lafayette, Ind.
Stanford Research Inst., Menlo Park, Calif.
Wisconsin. Univ., Madison

Formal reports on the program were given numbers in a series designated SQUID- regardless of the agency issuing them. Informal reports from individual agencies might be placed in the regular series of that agency or given special designations such as BUM-SQUID-(NUMBER)- or JHU-SQUID- by recipients of the reports.

When the regular series code of the agency was used to precede the report number, the number was usually followed by one or more of the letters M, P, R, or T, as in ARC-1-P, BUM-1-P or BUM-2-P, and others shown above.

Recently, since the Headquarters of the Squid Project was established at Purdue University, a different pattern has been developed. Reports are given codes beginning with SQUID-TR- followed by letters indicating the organization at which the work was done, followed by a serial number and the suffix PU.

Reference 34

TENNESSEE EASTMAN CORPORATION

A-0.600- thru A-7.390-
B-0.341- thru B-7.470-
C-0.100- thru C-5.385-
D-1.210- thru D-7.440-
E-0.100- thru E-5.430-
F-3.20- thru F-4.40-
G-1.133- thru G-3.200-
H-0.100- thru H-10.430-

During its operation of the Clinton Laboratories at Oak Ridge, Tennessee, the Tennessee Eastman Corporation used a numbering system for its documents that was designed to indicate the division and group from which the document originated, the subject of the report and its identifying serial number.

A letter was used to designate the division. This was followed by a number that specified the group of the division. The letters and numbers used were:

A-0 thru A-7   Production
B-0 thru B-8   Chemical Production
C-0 thru C-5   Chemical Development

D-0 thru D-9 Process Improvement
E-0 thru E-5 Engineering and Maintenance
F-0 thru F-4 Industrial Relations
G-0 thru G-3 (unidentified)
H-0 thru H-10 Development

These designations were set off from the next two or three digit numbers by a period. The numbers indicated the type or subject of the report as follows:

.01 - .99 Training material
.100 - .199 Periodic, progress, status and summary reports
.200 - .299 Miscellaneous, frequent reports (conferences, seminars, abstracts, indices, etc.)
.300 - .399 Chemical reports
.400 - .499 Development reports
.500 - .599 Engineering and design reports
.600 - .699 Special studies

The final portion of the number was the sequence number, again set off by a period. The form of the complete number was D-1.290.1. This pattern was continued for a short period after the Carbide and Carbon Chemicals Corporation took over the operation of the Oak Ridge installations.

Reference 35

TRIPARTITE NUCLEAR CROSS-SECTIONS COMMITTEE

TNCC-   TNCC(CAN)-   TNCC(UK)-   TNCC(US)-

The Tripartite Nuclear Cross-Sections Committee was comprised of members of the British, Canadian and United States atomic energy agencies. The basic series designation, TNCC, was modified by (UK), (CAN) and (US) to indicate the country that issued the report.

This Committee was discontinued in 1963 and its functions taken over by the European-American Nuclear Data Committee. *(See Reference Note 45.)*

Reference 36

UNITED KINGDOM ATOMIC ENERGY AUTHORITY. INDUSTRIAL GROUP—AND ITS PREDECESSORS, 1946-SEPTEMBER 1955

5001 thru 5256
6001 thru 6286
7001 thru 7008
8001 thru 8151
BP-5001 thru -5256
CENTRAL FILE-5001 thru -5256
CF-5001 thru -5256
CP/M-
CS(SP)-
CSD(CAP.)M-
CSD(CAP.)TM-
CSD(W)-

D(W)MEMO-
DL(S)(LETTERS)-
ENQ-
IDLM-
IG.LIBIB-
IG REPORT-
IG TRANS(R)-
IGC-(LETTER(S))/(LETTER(S))-
MD-
PM(C)-
PM(R)-
PM(S)-

PM(W)-
PRO(TM)-
R.&D.(R)-
R.&D.B.(LTR(S))(LTRS)-
RISLEY-5001 thru -5256
RISLEY-6001 thru -6286
RISLEY-7001 thru -7008
RISLEY-8001 thru -8151
SC-
SCS-(LETTER(S))-
SP-5001 thru -5256

SP (CS)-
SR-5001 thru -5256
SRO-
SRO/ML-
SRR-
SRS-
SRW-
TSM-
VTLM-
WSL-M-
WSL-TM-

Until September 1955, documents in the old Industrial Group series were issued by Great Britain's Division of Atomic Energy (Production), 1946-1953; by the Department of Atomic Energy, Industrial Group, January - July 1954; and by the United Kingdom Atomic Energy Authority, Industrial Group, August 1954 - September 1955.

The various old series are listed below. Those that were continued in later series are so indicated. The later series are explained in Ref. 37.

When received in the United States and distributed by the AEC's Technical Information Service, the straight numerical series were prefixed with the word RISLEY. Both forms of the numerical series are, therefore, included below.

SERIES REFERENCE	DESCRIPTION	SERIES REPLACED BY
5001 thru 5256	Industrial Group; Progress reports	IGM-PR-
	Engineering Branch; Reports	IGE-R-
	Production Branch (later Operations Branch), Chemical Services Dept., Windscale; Reports and other miscellaneous reports	IGO-R/W-
6001 thru 6286	Committees, agenda, minutes and papers	IGC- and initials of Committees' names...
7001 thru 7008	Library Information Section, Risley; Bibliographies	IGRL-IB/R-
8001 thru 8151	Research and Development Branch, (various locations); Reports	IGR-R/...
	Research and Development; Programme and progress reports	IGR-PR/
BP-5001 thru -5256	see 5001 thru 5256 above	
CENTRAL FILE-5001 thru -5256	see 5001 thru 5256 above	
CF-5001 thru -5256	see 5001 thru 5256 above	
CP/M-	see CSD (CAP.) M below	
CS(SP)-	see SCS below	
CSD (CAP.) M-1 thru -143	Production Branch (later Operations Branch), Chemical Service Dept., Capenhurst Works; Methods	IGO-AM/CA-
CSD (CAP.) TM-1 thru -17	Production Branch, (later Operations Branch), Chemical Service Dept., Capenhurst Works; Technical memoranda	IGO-TM/CA-
CSD(W)-	see WSL below	
D(W) MEMO-1 thru -6	Applied Research Dept. (later Research and Dev. Branch), Development Lab., Windscale; Memoranda	R.&D.B.(W) MEMO.-
DL (S) TM-1 thru -3101	Applied Research Dept. (later Research and Dev. Branch), Development Lab., Springfields; Technical memoranda	R.&D.B.(S)TM-

REFERENCE NOTES

SERIES REFERENCE	DESCRIPTION	SERIES REPLACED BY
DL (S) TN-1 thru -2019	Applied Research Dept. (later Research and Dev. Branch), Development Lab., Springfields; Technical notes	R.&D.B.(S)TN-
ENQ-1 thru-23	Production Branch, (later Operations Branch), Courts of Enquiry (accidents)	IGO-AR-
IDLM-1 thru -90	Production Branch, (later Operations Branch), Isotopes Development Lab., Capenhurst; Memoranda	R.&D.B.(CA)TN-
IG. LIBIB-7001-1 thru -2	see 7001 thru 7008 above	
IG REPORT-5001 thru -5256	see 5001 thru 5256 above	
IG TRANS (R)-1 thru -3	Library Information Section, Risley; Translations	IGRL-T/R-
IGC-(LETTER(S))/(LETTER(S))-	IGC apparently stood for reports connected with Committees of the Industrial Group. IGC- series was subdivided by letters that appear to be initials of Committees' names, followed by: /A- Agenda /M- Minutes /Misc.- Miscellaneous /P- Papers /R- Reports There was no obvious indication of the location from which the reports originated.	
MD-1 thru -314	Library Information Section, Risley; Miscellaneous documents	IGRL-IM/R-
PM(C)-1 thru -28	Production Branch (later Operations Branch), Capenhurst Works; Memoranda	IGO-TM/CA-
PM(R)-1 thru -21	Production Branch (later Operations Branch), Risley; Memoranda	
PM(S)-1 thru -30	Production Branch (later Operations Branch), Springfields Works; Memoranda	IGO-TM/S-
PM(W)-1 thru -28	Production Branch (later Operations Branch), Windscale Works; Memoranda	IGO-TM/W-
PRO(TM)-1 thru -5	Production Branch (later Operations Branch), Windscale Works; Technical memoranda	PM(W)-
R.&D.(R)-	see R.&D.B.(R) below	
R.&D.B.(C)REPORT-8001 thru -8151	see 8001 thru 8151 above	
R.&D.B.(C)TM-1 thru -285	Research and Development Branch, Culcheth Labs.; Technical memoranda	IGR-TM/C-
R.&D.B.(C)TN-1 thru -143	Research and Development Branch, Culcheth Labs.; Technical notes	IGR-TN-C-
R.&D.B.(CA)R-8001 thru -8151	see 8001 thru 8151 above	
R.&D.B.(CA)REPORT-8001 thru -8151	see 8001 thru 8151 above	
R.&D.B.(CA)TM-1 thru -73	Research and Development Branch, Capenhurst; Technical memoranda	IGR-TM/CA-
R.&D.B.(CA)TN-1 thru -158	Research and Development Branch, Capenhurst; Technical notes	IGR-TN/CA-
R.&D.B.(CAP.)	see R.&D.B.(CA) above	

REFERENCE NOTES

SERIES REFERENCE	DESCRIPTION	SERIES REPLACED BY
R.&D.B.(R) REPORT -8001 thru -8151	see 8001 thru 8151 above	
R.&D.B.(R)TM-1 thru -92	Research and Development Branch, Risley; Technical memoranda	IGR-TM/R-
R.&D.B.(R)TN-1 thru -61	Research and Development Branch, Risley; Technical notes	IGR-TN/R-
R.&D.B.(S)R-8001 thru -8151	see 8001 thru 8151 above	
R.&D.B.(S)REPORT -8001 thru -8151	see 8001 thru 8151 above	
R.&D.B.(S)TM-1 thru -3490	Research and Development Branch, Springfields; Technical memoranda	IGR-TM/S-
R.&D.B.(S)TN-1 thru -2177	Research and Development Branch, Springfields; Technical notes	IGR-TN/S-
R.&D.B.(W)-8001 thru -8151	see 8001 thru 8151 above	
R.&D.B.(W)MEMO.-1 thru -10	Research and Development Branch, Windscale; Memoranda	R.&D.B.(W)TN-
R.&D.B.(W)REPORT-8001 thru -8151	see 8001 thru 8151 above	
R.&D.B.(W)TM-1 thru -1285	Research and Development Branch, Windscale; Technical memoranda	IGR-TM/W-
R.&D.B.(W)TN-1 thru -225	Research and Development Branch, Windscale; Technical notes	IGR-TN/W-
RD(R)-	(Variant of R.&D.(R), which see above.)	
RDB...-	(Variant of R.&D.B...., which see above.)	
RISLEY 5001-8151	see 5001-, 6001-, etc., above	
SC-	Security Records Office, Capenhurst; Register for miscellaneous unnumbered documents	
SCS-M-1 thru -414	Production Branch (later Operations Branch) Chemical Services Dept., Springfields; Methods	IGO-AM/S-
SCS-MEMO-1 thru -57	Production Branch (later Operations Branch) Chemical Services Dept., Springfields; Memoranda	IGO-TM/S
SCS-R-1 thru -420	Production Branch (later Operations Branch) Chemical Services Dept., Springfields; Reports	IGO-R/S-
SCS-TM-1 thru -14	Production Branch (later Operations Branch) Chemical Services Dept., Springfields; Technical memoranda	IGO-TM/S-
SCS-TN-1 thru -34	Production Branch (later Operations Branch) Chemical Services Dept., Springfields; Technical notes	
SP-5001 thru -5256	see 5001 thru 5256 above	
SP(CS)-	see SCS above	
SR-5001 thru -5256	see 5001 thru 5256 above	
SRO-	Security Records Office, Risley; Register for miscellaneous unnumbered documents	SRR-
SRO/ML-	Security Records Office, Culcheth; Register for miscellaneous unnumbered documents	
SRR-	see SRO above	

SERIES REFERENCE	DESCRIPTION	SERIES REPLACED BY
SRS-	Security Records Office, Springfields; Register for miscellaneous unnumbered documents	
SRW-	Security Records Office, Windscale; Register for miscellaneous unnumbered documents	
TSM-	Engineering Branch, Technical Section, Risley; Memoranda	IGE-TM-
VTLM-1 thru -7	Inspection and Progress Branch, Vacuum Test Lab., Springfields; Memoranda	
WSL-M-1 thru -660A	Production Branch (later Operations Branch) Chemical Services Dept., Windscale; Methods	IGO-AM/W-
WSL-TM-1 thru -351	Production Branch (later Operations Branch) Chemical Services Dept., Windscale; Technical memoranda	IGO-TM/W-

Reference 37

UNITED KINGDOM ATOMIC ENERGY AUTHORITY.

INDUSTRIAL GROUP, SEPTEMBER 1955-JANUARY 1959

IGC-(LETTERS)/(LETTER(S))-
IGD-(LETTER(S))-
IGE-(LETTER(S))-
IGI-(LETTER(S))-
IGM-PR-

IGO-(LETTER(S))/(LETTER(S))-
IGR-(LETTER(S))/(LETTER(S))-
IGRL-(LETTER(S))/(LETTER(S))-
IGS-(LETTER(S))/(LETTER(S))-
IGT-(LETTER(S))-

From about September 1955 to January 1959, most documents issued by the UKAEA's Industrial Group were identified by series codes consisting of several parts. The first part consisted of the letters IG, combined with a letter identifying the main functional branch that was ultimately responsible for the document. Each branch was thus provided with a main series:

- IGE- Industrial Group, Engineering Branch series
- IGI- Industrial Group, Industrial Power Branch series
- IGO- Industrial Group, Operations Branch series
- IGR- Industrial Group, Research and Development Branch series
- IGRL- Industrial Group, Research and Development Branch, Library and Information Dept. series
- IGS- Industrial Group, Health and Safety Branch series (formerly Safety Branch)
- IGT- Industrial Group, Technical Policy Branch series

The branch series then divided into various document sub-series, the code letter(s) of which follow the branch code. These were:

- AM Analytical methods
- AR Accident reports
- CPR Chemical Services progress report
- IB Information bibliography
- IM Information memorandum
- PR Progress report
- R Report
- T Translation
- TM Technical memorandum
- TN Technical note

Following the sub-series code letters (and preceded by a slash) appeared letters indicating at which of the various Industrial Group establishments a document was written. Engineering, Industrial Power and Technical Policy Branches were located only at Risley. Therefore, no location symbol was used for IGE-, IGI- or IGT- reports. The letters used to designate the establishments were:

/C	Culcheth Laboratories
/CA	Capenhurst
/CC	Chapelcross
/CR	Calder Hall
/D	Dounreay
/R	Headquarters, Risley
/S	Springfields
/W	Windscale

Following this is a serial number, which concludes the report number.

At that time (1955-59), other series were also appearing. They included:

1. IGC-, which apparently stood for reports connected with Committees of the Industrial Group. IGC- series was subdivided by letters, which appear to be initials of Committees' names, followed by:

/A-	Agenda
/M-	Minutes
/Misc.-	Miscellaneous
/P-	Papers
/R-	Reports

There was no obvious indication of the location from which the reports originated.

2. IGD-, which appears to have replaced the IGO- and IGR- series for reports written at Dounreay when the name changed from Dounreay Works to Dounreay Experimental Reactor Establishment. IGD- series was subdivided into TM- Technical memoranda and R- Reports.

3. IGM-PR-, which stood for Industrial Group progress reports.

Reference 38

UNITED KINGDOM ATOMIC ENERGY AUTHORITY.

INDUSTRIAL GROUP, JANUARY - JUNE 1959

IG-INF.SER.-(NO.)(LTR/LTR)
IG-MEMO-(NO.)(LTR(S)/LTR(S))
IG-REPORT-(NUMBER)(LTR/LTR)

IGIS-(NUMBER)(LETTER/LTR)
IGM-(NUMBER)(LTR(S)/LTR(S))
IGR-(NUMBER)(LETTER/LETTER)

From January through June 1959, a new publication series with simplified numbering was issued by the UKAEA's Industrial Group. There were three series:

Information Series	IG-INF. SER.-
Memoranda	IG-MEMO-
Reports	IG-REPORT-

The letters of these series are sometimes condensed when written so that, e.g., IG-INF. SER. becomes IGIS.

After the serial number, there is a code in parentheses representing the branch and the establishment that issued the document. This code had no effect on numbering or filing.

(O	Operations Branch		
(RD	Research and Development Branch		
	/C)	Culcheth	
	/CA)	Capenhurst	
	/CR)	Calder Works	
	/D)	Dounreay	
	/R)	Risley	
	/S)	Springfields	
	/W)	Windscale	

Dounreay is sometimes indicated by a (D) with no Branch indication after the number.

Reference 39

UNITED KINGDOM ATOMIC ENERGY AUTHORITY. INDUSTRIAL GROUP SUCCESSORS: DEVELOPMENT AND ENGINEERING GROUP AND PRODUCTION GROUP

DEG-INF. SER.-(NO.)(LTR(S))
DEG-MEMO-(NO.)(LTR(S))
DEG-REPORT-(NUMBER)(LTR(S))
DEGIS-(NUMBER)(LETTER(S))
DEGM-(NUMBER)(LETTER(S))
DEGR-(NUMBER)(LETTER(S))

PG-INF. SER. (NO.)(LTR(S))
PG-MEMO-(NUMBER)(LTR(S))
PG-REPORT-(NUMBER)(LTR(S))
PGIS-(NUMBER)(LETTER(S))
PGM-(NUMBER)(LETTER(S))
PGR-(NUMBER)(LETTER(S))

On July 1, 1959, the UKAEA's Industrial Group ceased to exist and was replaced by two new and independent groups, the Development and Engineering Group and the Production Group, both with headquarters at Risley. Each group issues three publications series:

UKAEA Development and Engineering Group
    Information Series   DEG-INF. SER.-
    Memoranda   DEG-MEMO-
    Reports   DEG-REPORT-

UKAEA Production Group
    Information Series   PG-INF. SER.-
    Memoranda   PG-MEMO-
    Reports   PG-REPORT-

The letters of these series are sometimes condensed when written so that, e.g., DEG-INF. SER. becomes DEGIS.

After the serial number, which follows the publication series code, there is a code in parentheses representing the establishment that issued the document. This code has no effect on numbering or filing. Codes used are:

(C)   Culcheth
(CA)   Capenhurst
(D)   Dounreay
(R)   Risley
(S)   Springfields
(W)   Windscale

Reference 40

UNITED KINGDOM ATOMIC ENERGY AUTHORITY. RESEARCH GROUP. ATOMIC ENERGY RESEARCH ESTABLISHMENT—AND ITS PREDECESSOR: GREAT BRITAIN. ATOMIC ENERGY RESEARCH ESTABLISHMENT

A.E.R.E.-(LETTERS)-
A.E.R.E. (LTRS)/(LTRS)-

AERE-(LETTERS)-
AERE-(LTR(S))/(LTR(S))-

Until 1959 documents issued by the Atomic Energy Research Establishment at Harwell, Berks, were identified by series codes consisting of several parts. The first part was the letters A.E.R.E. The second was a letter or letters designating the division originating the document. The third, separated from the second by a slash, was a letter or letters indicating the type of the document. The final part was the serial number.

The code letters used were as follows:

For Divisions -

C	Chemistry Division
CE	Chemical Engineering Division
D	Director's Office
E	Engineering Division
ED	Engineering Research and Development Division
EL	Electronics Division
ES	Engineering Services Division
G, later GP	General Physics Division
H	Health Division
HP	Health Physics Division
I	Isotope Division
INF	Information Office
LIB	Library
M	Metallurgy Division
MED	Medical Division
MRC	Medical Research Council (Radiobiological Research Unit)
N, later NP	Nuclear Physics Division
R	Reactor Division
RE	Reactor Engineering Division
RP	Reactor Physics Division
RS	Reactor School
T	Theoretical Physics Division
X	Extra-Mural Research

For types of documents -

R	Reports
M	Memoranda
L	Lectures
TN	Technical notes
BIB	Bibliographies
TRANS	Translations

The numbering of the series depended on the type of the document. Reports were in a single numerical series irrespective of the division of origin, e.g., A.E.R.E. C/R 1004; A.E.R.E. N/R 1005; and A.E.R.E. M/R 1006.

Memoranda, lectures and technical notes were divided into sub-series and numbered by division of origin, e.g.:

A.E.R.E. C/M 1, 2, 3, etc.
A.E.R.E. N/M 1, 2, 3, etc.
A.E.R.E. C/L 1, 2, 3, etc.
A.E.R.E. N/L 1, 2, 3, etc.
A.E.R.E. C/TN 1, 2, 3, etc.
A.E.R.E. N/TN 1, 2, 3, etc.

Bibliographies and translations each formed single series, e.g.:

A.E.R.E. INF/BIB 1, 2, 3, etc.
A.E.R.E. LIB/TRANS 1, 2, 3, etc.

In January 1959, a simplified system of numbering was adopted. The letters of the originating divisions were discontinued, and each type of document was put into a single series. The initial letters were run together and followed by a dash. The code letters used for types of documents are:

R	Reports
M	Memoranda
L	Lectures
BIB	Bibliographies
TRANS	Translations
AM	Analytical Methods

The numbering of the new series of reports, bibliographies and translations continues in the same sequence as the former Report, Inf/Bib and Lib/Trans series. The numbering of Memoranda starts at 400 and of Lectures at 100. The series of Analytical Methods starts at 1. Examples of the new series are:

AERE-R-3000   AERE-BIB-130
AERE-M-400   AERE-TRANS-835
AERE-L-100   AERE-AM-1

Reference 41

UNITED KINGDOM ATOMIC ENERGY AUTHORITY. RESEARCH GROUP. RADIOCHEMICAL CENTRE—AND ITS PREDECESSOR: GREAT BRITAIN. RADIOCHEMICAL CENTRE

R.C.C./M-   R.C.C./R-   RCC-M-   RCC-R-

Until 1959, documents issued by the Radiochemical Centre, Amersham, Bucks, were identified by the prefix R.C.C. followed by a slash and the letter R for reports or M for memoranda. The serial numbers were assigned consecutively, regardless of the letter, e.g., R.C.C./R46, R.C.C./M47 and R.C.C./R48.

Beginning in 1959, the two series were separated, the periods were omitted from the prefix and the slash replaced by a dash. The letters R for reports and M for memoranda were retained, e.g., RCC-R-1, 2, 3, etc. and RCC-M-1, 2, 3, etc.

Reference 42

TRW SYSTEMS

(4 DIGITS-NO.-LTRS.)-
(5 DIGITS-NO.-LTRS.)-
STL-(NO.-NO.-LTRS.)-
STL-(NO.-NO.-LTRS)-000

TRW-(NO.-NO.-LTRS.)-
TRW-(NO.-NO.)-R000
TRW-R(5 DIGITS-NO.)-R00-
TRW-TR-(NUMBER.NUMBER)-

TRW Systems and its predecessors, Space Technology Labs., Inc. and TRW Space Technology Labs., Inc., have long used a complicated numbering system for their reports that is based on groupings of digits and letters, with no acronymic relationship to TRW or STL. The format of the number from 1962 to mid-1966 was 0000-0000-XX-000 000 where

the first 4 digits represent the master job order number,

the second four digits represent sequence numbers,

the next two characters include a letter showing type of material and a digit for security classification, and

the two final groups of three digits each indicate supplemental series and copy numbers.

In mid-1966 the system was changed slightly with the replacement of the 4 digit master job order number by a 5 digit sales number.

Because this system, however well constructed internally, does not link TRW Systems' reports to the corporate body by letters recognizable to outside agencies, it is common practice for other libraries to precede them with a STL or TRW prefix in one form or another.

Reference 43

VELA UNIFORM PROGRAM

VESIAC- VUF- VUP-

The Vela Uniform program for the long range detection of nuclear explosions has been conducted under the sponsorship of the Institute for Defense Analyses and the Advanced Research Projects Agency with funding from other government agencies as well.

Work on projects of the program has been carried out by many organizations, including the following:

Air Force Office of Scientific Research, Washington, D.C.
Air Force Technical Applications Center, Washington, D.C.
Air Resources Field Research Office, Las Vegas, Nev.
Allied Research Associates, Inc., Boston, Mass.
Army Geodesy, Intelligence and Mapping Research and Development Agency, Ft. Belvoir, Va.
Association des Amis du Laboratoire de Physique de l'Ecole Normale Superieure, Paris
Barringer Research Ltd., Rexdale, Ont.
Beers (R. F.), Inc., Alexandria, Va.
Blume (John A.) & Associates, San Francisco
Bureau of Mines, Washington, D.C.

Bureau of Mines. Bartlesville Petroleum Research Center, Okla.
California. Univ. Lawrence Livermore Lab.
California Inst. of Tech., Pasadena
California Research Corp., Richmond
Coast and Geodetic Survey, Washington, D.C.
Defense Atomic Support Agency, Washington, D.C.
EG&G, Inc.
Environmental Research Corp., Alexandria, Va.
Engineer Research and Development Lab., Ft. Belvoir, Va.
Environmental Science Services Administration, Boulder, Colo.
Federal Aviation Agency, Salt Lake City
Geological Survey
Geotechnical Corp., Garland, Tex.
Hazelton-Nuclear Science Corp., Palo Alto, Calif.
Holmes & Narver, Inc., Los Angeles
Institute for Defense Analyses, Washington, D.C.
Isotopes, Inc., Palo Alto, Calif.
Isotopes, Inc., Westwood, N.J.
ITEK Corp., Palo Alto, Calif.
Jersey Production Research Co., Tulsa, Okla.
Lamont Geological Observatory, Palisades, N.Y.
Massachusetts Inst. of Tech., Cambridge
Michigan. Univ., Ann Arbor
Naval Radiological Defense Lab., San Francisco
Nevada Bureau of Mines, Reno
Pitkin (Lucius), Inc., N.Y.C.
Public Health Service, Las Vegas, Nev.
Reynolds Electrical and Engineeering Co., Las Vegas, Nev.
Saint Louis Univ.
Sandia Corp., Albuquerque, N. Mex.
Space-General Corp., Glendale, Calif.
Space Technology Labs., Inc., Redondo Beach, Calif.
Stanford Research Inst., Menlo Park, Calif.
Texas Instruments, Inc., Dallas
TRW Systems, San Bernardino, Calif.
United Electrodynamics, Inc., Pasadena, Calif.
United Electrodynamics, Inc. Data Analysis & Technique Development Center, Alexandria, Va.
Waterways Experiment Station, Vicksburg, Miss.
Weather Bureau, Las Vegas, Nev.

These agencies have issued many Vela Uniform reports under their own codes. Some have been assigned VUF and VUP numbers by ARPA. Others have been given VESIAC numbers by the University of Michigan's Vela Seismic Information Analysis Center. Still others are being published by the American Geophysical Union. Many of the reports have been incorporated into two series, one entitled "World Wide Standard Station, Seismic" issued by the Coast and Geodetic Survey in 1962, and the other called "Long Range Seismic Measurements."

Reference 44

PLOWSHARE PROGRAM

AEC PNE-

The Plowshare Program, peaceful uses of nuclear explosives, has been conducted for the Atomic Energy Commission by many government agencies and contractors. Its reports have been gathered into the PNE series, with numbers assigned by the Division of Technical Information Extension in Oak Ridge. Some of the reports carry numbers assigned by their originators as well.

Participating organizations in the Plowshare Program include:
Air Force Weapons Lab., Kirtland AFB, N. Mex.
Alaska. Univ., College
American Science and Engineering Inc., Cambridge, Mass.
Arctic Health Research Center, Anchorage, Alaska

PNE-

Armour Research Foundation, Chicago
Army Engineer Nuclear Cratering Group, Livermore, Calif.
Ballistic Research Labs., Aberdeen Proving Ground, Md.
Battelle-Northwest. Pacific Northwest Lab., Richland, Wash.
Beers (R. F.), Inc., Alexandria, Va.
Blume (John A.) & Associates, San Francisco
Boeing Co. Aero-Space Div., Seattle
Brigham Young Univ., Provo, Utah
Bureau of Mines, Washington, D.C.
California. Univ., Los Angeles
California. Univ. Lawrence Livermore Lab.
CER Geonuclear Corp.

Coast and Geodetic Survey, Washington, D.C.
Corps of Engineers (Army)
Division of Biology and Medicine, AEC
EG&G, Inc.
El Paso Natural Gas Co., Tex.
Environmental Research Corp., Alexandria, Va.
Environmental Science Services Administration, Boulder, Colo.
Federal Aviation Agency, Salt Lake City
Geological Survey, Washington, D.C.
Gt. Brit. Admiralty Research Lab., Teddington, Middx, England
Hanford Atomic Products Operation, Richland, Wash.
Holmes & Narver, Inc., Los Angeles
Isotopes, Inc., Palo Alto, Calif.
Los Alamos Scientific Lab., N. Mex.
Marine Corps Development Center, Quantico, Va.
Maryland. Univ., College Park. Applied Physics Research Lab.
Mathematica, Princeton, N.J.
Naval Radiological Defense Lab., San Francisco
Nevada Operations Office, AEC, Las Vegas
Oak Ridge National Laboratory, Tenn.
Ohio State Univ., Columbus
Panama Canal Co., Canal Zone
Public Health Service, Las Vegas, Nev.
Reynolds Electrical and Engineering Co., Las Vegas, Nev.
Sandia Corp., Albuquerque, N. Mex.
Snow, Ice & Permafrost Research Establishment, Wilmette, Ill.
Southwestern Radiological Health Lab., Las Vegas, Nev.
Space Technology Labs., Inc., Redondo Beach, Calif.
Stanford Research Inst., Menlo Park, Calif.
Washington. Univ., Seattle.
Washington. Univ., Seattle. Applied Fisheries Lab.
Waterways Experiment Station, Vicksburg, Miss.
Weather Bureau, Las Vegas, Nev.

Reports in the PNE series have been assigned numbers according to blocks of hundreds that indicate specific projects within the Plowshare program. The schedule of numbers is:

100's	Gnome
200's	Sedan
300's	Pre-Buggy & Buggy
400's	Chariot
500's	Pre-Schooner & Schooner
600's	Dugout
700's	Sulky
800's	Handcar
900's	Palanquin & Cabriolet
1000's	Gasbuggy
1100's	Pre-Gondola
1200's	Ketch
1300's	Sloop
1400's	Bronco
2000's	Isthmian Canal
3000's	Shoal
5000's	General

Reference 45

EUROPEAN-AMERICAN NUCLEAR DATA COMMITTEE

EANDC-
    EANDC-(NUMBER)(LETTER)

EANDC(LETTER(S))-
    EANDC(LETTER(S))(NUMBER)(LTR.)

The European-American Nuclear Data Committee was formed in 1959 under the auspices of the European Nuclear Energy Agency of the Organization for Economic Cooperation and Development, and consists of 16 individual scientist members representing 21 nations. It is primarily concerned with nuclear cross sections and other fundamental nuclear data that are relevant to nuclear energy development. In 1963 it absorbed the functions of the Tripartite Nuclear Cross Sections Committee.

Its reports are issued in a series prefixed with the letters EANDC. Those bearing no further letters are issued by the secretariat in Paris. Reports issued from member countries carry succeeding letters acronymic for the country or organization names. These include:

AUS	Austria
BZL	Brazil
CAN or CND	Canada
CCP	Union of Soviet Socialist Republics

DEN	Denmark	POL	Poland
E or EUR	Euratom	RUM	Romania
ENEA	European Nuclear Energy Agency	SPN	Spain
GER	German Federal Republic	SWD	Sweden
IAEA	International Atomic Energy Agency	SWT	Switzerland
		UK	United Kingdom
IND	India	US or USA	United States of America
JAP	Japan		
NOR	Norway		
OR	Other participants		

The final letter following the sequential number designates the distribution category.

Reference 46

GENERAL ELECTRIC COMPANY

R(YEAR)(LETTERS)   RM(YEAR)(LETTERS)   TIS(YEAR)(LETTERS)

For years the basic report series of the General Electric Company for all divisions and locations were the R and RM series. The specific divisions and locations were indicated by letters following 2 digits standing for the year of publication of the document. Approximately 250 different groups of letters have been used for this purpose. Many of these are listed individually in this *Dictionary*, either under R(YEAR)..., RM(YEAR)..., or standing alone, together with the identification of the division and location with which they are associated.

The basic pattern of this system is still being continued but the current prefix for the series is TIS, standing for "Technical Information Series." Numbers are assigned by the Technical Information Exchange in Schenectady, N.Y.

Reference 47

THE RAND CORPORATION

D-(NUMBER)-(LETTERS)	R-(NUMBER)-(LETTERS)	RM-(NUMBER)-ARPA
P-(4 DIGITS)-	R-(NUMBER)-AEC	RM-(NUMBER)-ESSA
P-(NO.)-AEC	R-(NUMBER)-PR	RM-(NUMBER)-NRL
P-(NO.)-ARPA	R-(NUMBER)-R	RM-(NUMBER)-RAND
P-(NO.)-(RAND)	RM-(NUMBER)-AEC	

Reports issued in the RAND series D, R and RM, cover work sponsored by other agencies. The identity of these agencies is shown by the letter suffixes following the sequential number. Such suffixes and agencies include:

AEC	Atomic Energy Commission
AID	Agency for International Development
ARPA	Advanced Research Projects Agency
ASDC	Assistant Secretary for Defense/Comptroller
CC	Carnegie Corporation of New York
DASA	Defense Atomic Support Agency
ESSA	Environmental Science Services Administration
FAA	Federal Aviation Agency

FF	Ford Foundation
ISA	Assistant Secretary of Defense/Internal Security Affairs
NASA	National Aeronautics and Space Administration
NIH	National Institutes of Health
NRL	Naval Research Laboratory
NSF	National Science Foundation
PR	U.S. Air Force Project RAND
RAND	RAND Corporation - sponsored research
RC	RAND Corporation - sponsored research
RF	Rockefeller Foundation
TAB	Atomic Energy Commission/Technical Analysis Branch

Papers in the P series supposedly are not reports of research done under or in support of a grant or contract, and therefore should not require suffixes. Nevertheless this *Dictionary* lists three suffixes (AEC, ARPA and RAND) that have appeared in the P series.

Reference 48

THEMIS PROJECT

THEMIS-(YEAR)-
THEMIS-(LETTERS)-
THEMIS-(LETTERS-NO.-YEAR)
THEMIS-(LETTERS-YEAR)-
THEMIS-(LETTERS)-T-
THEMIS-(LETTERS)-T-TR-
THEMIS-(LETTERS)-TR-

THEMIS-A(YEAR)-
THEMIS-B-(YEAR)-
THEMIS-KU-RR-
THEMIS-PROPOSAL-
THEMIS-SR-
THEMIS-TR-
THEMIS-UF-SCIENTIFIC-
THEMIS-UK-RR-

Project THEMIS is a program originated in 1967 by the Department of Defense in an effort to strengthen United States universities, to increase the number of institutions performing high quality research, and to achieve a wider distribution of research funds for science and technology. Funds are made available to universities for projects relevant to the defense mission. Research problems considered relevant to Project THEMIS may fall under the general categories of detection, surveillance, navigation and control; energy and power; information-processing systems; materials sciences; medical sciences; or social and behavioral sciences.

Reports of research work performed under Project THEMIS usually carry report numbers beginning with the word THEMIS and indicating by letters the university at which the work was performed. The pattern for the remainder of the number varies, as is shown in the list above. A few universities do not include an abbreviation for their own names. They have been identified with the codes they use in the main body of the *Dictionary*. Other universities issue their THEMIS-related reports in their own report series.

Universities known to have issued reports bearing THEMIS numbers are the following:

Auburn University, Ala. (AU)
Cincinnati. Univ. Dept. of Aerospace Engineering (AE)
Colorado State Univ., Fort Collins (CER)
Florida. Univ., Gainesville (UF)
Georgia. Univ., Athens (UGA)
Hawaii. Univ., Honolulu
Houston. Univ., Tex. Cullen Coll. of Engineering (RE)
Houston. Univ., Tex. Dept. of Computer Science (RS)
Illinois Inst. of Tech., Chicago
Iowa. Univ., Iowa City (UI)
Kansas State Univ., Manhattan
Kentucky. Univ., Lexington (UK, or as occasionally cited in error, KU)
Lehigh Univ., Bethlehem, Penna. (LU)
Louisiana State Univ., Baton Rouge (LSU, or as occasionally cited in error, LUS)
Massachusetts. Univ., Amherst (UM)
Oklahoma. Univ., Norman. Medical Center (UOMC)
Southern Methodist Univ., Dallas (SMU)
Tennessee. Univ., Knoxville. Dept. of Engineering Mechanics (EM)
Texas A & M Univ., College Station
Texas A & M Univ., College Station. Dept. of Aerospace Engineering (AED)

Texas Tech Univ., Lubbock
University of Notre Dame, South Bend, Ind. (UND)
Utah. Univ., Salt Lake City. Coll. of Engineering (UTEC)

The letters shown in parentheses after the university names are those used in the THEMIS report codes.

Report Series Codes with Related Agencies

Filing starts with the first letter encountered in the code, even though the letter may be preceded by a number.

Codes which contain only numbers are not listed here. They are described and some are listed in Ref. 1.

Code entries are arranged alphabetically, letter by letter, disregarding all marks of punctuation. A number, or a word which stands for a number, as (year), (contract no.), (Roman no.), (no.), (number), stops the alphabetic sorting at that point. The term (letter) or (letters) files before any specific letter.

The agency name is indented under the code to which it is related.

The Explanatory Notes (p. 14) contain the rules for entry and filing. Suggestions for steps to take if you do not find the code you are looking for here are given on p. 17.

A-	(ABSTRACTS) (REF. 2)
A-	ADMIRAL CORP., CHICAGO
A(NO.)-(LTR OR NO.LTR)-	AEROJET-GENERAL CORP. LIQUID ROCKET PLANT, SACRAMENTO, CALIF. (FIRST NO. IS CONTRACT CODE NO.)
A2-ETM-	AEROJET-GENERAL CORP. SOLID ROCKET PLANT, SACRAMENTO, CALIF. (ENVIRONMENTAL TEST MOTOR REPORT)
A (YEAR)-	AEROSPACE CORP., EL SEGUNDO, CALIF. (LIBRARY ACCESSION NUMBER)
A(NUMBER)AW-	AEROSPACE CORP., EL SEGUNDO, CALIF. (ART WORK)
A(YEAR)-CCC(NUMBER)-	AEROSPACE CORP., EL SEGUNDO, CALIF. (INTEROFFICE NO.)
A-	AIR UNIV., MAXWELL AFB, ALA. (ARCTIC REPORTS)
A-	ARMY, WASHINGTON, D.C.
2A-	ARMY ROCKET AND GUIDED MISSILE AGENCY, REDSTONE ARSENAL, ALA.
A-(NO.)(AEI)	ASSOCIATED ELECTRICAL INDUSTRIES, LTD., ALDERMASTON, BERKS., ENGLAND
A-	ATOMIC ENERGY COMMISSION, WASHINGTON, D.C. (REF. 23)
A-	AUTONETICS DIV., NORTH AMERICAN AVIATION, INC., ANAHEIM, CALIF. (NOT USED AFTER 1957)
A-(NUMBER-YEAR)	BELGIUM. INSTITUT D'AERONOMIE SPATIALE DE BELGIQUE, BRUSSELS
A(NUMBER)-	BELL AEROSPACE CO. DIV. OF TEXTRON, BUFFALO
A(NUMBER)-	BELL AEROSPACE CO. DIV. OF TEXTRON.ENVIRONMENTAL DATA COLLECTION AND PROCESSING FACILITY, TUCSON, ARIZ.
A11-	BROWN UNIV., PROVIDENCE. GRAD. DIV. OF APPLIED MATH.
A11-S-	BROWN UNIV., PROVIDENCE. GRAD. DIV. OF APPLIED MATH.
A11-T-	BROWN UNIV., PROVIDENCE. GRAD. DIV. OF APPLIED MATH.
A18-	BROWN UNIV., PROVIDENCE. GRAD. DIV. OF APPLIED MATH.
A-	CARBIDE AND CARBON CHEMICALS CORP. K-25 PLANT, OAK RIDGE, TENN.
A-(NUMBER-NUMBER)	CARNEGIE-MELLON UNIV., PITTSBURGH. DEPT. OF ELEC. ENG
A-	COLUMBIA UNIV., N.Y.C. DIV. OF WAR RESEARCH
A-	DENMARK. TEKNISKE HØJSKOLE, LYNGBY. HYDRO-AND AERODYNAMICS LAB. AERODYNAMICS SECTION
A2-	DOUGLAS AIRCRAFT CO., INC., SANTA MONICA, CALIF.
A(NO.)-	FRANKFORD ARSENAL, PHILADELPHIA
A11-	GENERAL PRECISION LAB., INC., PLEASANTVILLE, N.Y.
A-	GEORGIA INST. OF TECH., ATLANTA
A/(NO.)(S)	GT. BRIT. ADMIRALTY MATERIALS LAB., POOLE, DORSET, ENG
A-	INDIA. CENTRAL MECHANICAL ENG. RES. INST., DURGAPUR
A-	IPSEN INDUSTRIES, INC., ROCKFORD, ILL.
A-1- THRU A-3-	LOS ALAMOS SCIENTIFIC LAB., N. MEX. (INTERNAL CORRESPONDENCE SERIAL USED BY GROUPS IN A DIV.)
A(YEAR)-	LOWELL TECHNOLOGICAL INST. RESEARCH FDN., MASS.
A-59-	MC DONNELL AIRCRAFT CORP., ST. LOUIS
A001 THRU A999	MC DONNELL AIRCRAFT CORP., ST. LOUIS
A-	MANHATTAN DISTRICT. RESEARCH CONTROL (SERIES ASSIGNED TO REPORTS FROM VARIOUS AMERICAN SOURCES)
A(NUMBER)	MASSACHUSETTS INST. OF TECH., CAMBRIDGE
A(NO.)-	MENASCO MANUFACTURING CO., BURBANK, CALIF.
A-	NATIONAL DEFENSE RESEARCH COMMITTEE (REF. 27)
A-(NUMBER)(LETTER)	NATIONAL DEFENSE RESEARCH COMMITTEE (REF. 27)
A-(NUMBER-YEAR)-	NAVAL AIR SYSTEMS COMMAND. ADVANCED SYSTEMS DIV., WASHINGTON, D.C.
A-	NAVAL AIR TURBINE TEST STATION, TRENTON
A-(YEAR)/KL/	NETHERLANDS. CENTRAAL INSTITUUT VOOR VOEDINGSONDERZOEK TNO, ZEIST
A(NO.)-	PHILCO CORP., BLUE BELL, PENNA.
A-59-	PURDUE RESEARCH FOUNDATION, LAFAYETTE, IND.
A-59-	PURDUE UNIV., LAFAYETTE, IND. SCHOOL OF AERO. ENG.
A-(NOS.)(PU)	PURDUE UNIV., LAFAYETTE, IND. SCHOOL OF AERONAUTICS
A-	RADIO CORP. OF AMERICA, CAMDEN, N.J.

A-(NO.LTR.)	REDSTONE ARSENAL, HUNTSVILLE, ALA. (INTERIM AND FINAL REPORT)
2A-	REDSTONE ARSENAL. ORDNANCE MISSILE LABS., HUNTSVILLE, ALA.
A-	RUTGERS UNIV., NEW BRUNSWICK, N.J. SCH. OF CHEMISTRY
A-	SPERRY GYROSCOPE CO., GREAT NECK, N.Y.
A-	SPRAGUE ELECTRIC CO., NORTH ADAMS, MASS.
A-	SWEDEN. FOERSVARETS FORSKNINGSANSTALT, STOCKHOLM
A-(4 DIGITS)-	SWEDEN. FOERSVARETS FORSKNINGSANSTALT, STOCKHOLM
A-	TENNESSEE EASTMAN CORP., OAK RIDGE, TENN.
A-0.600- THRU A-7.390-	TENNESSEE EASTMAN CORP., OAK RIDGE, TENN. (REF. 34)
A-2TD-	TENNESSEE EASTMAN CORP., OAK RIDGE, TENN.
A.-	UNION OF SOUTH AFRICA. MAGNETIC OBS., HERMANUS
A-	UNITED AIRCRAFT CORP., EAST HARTFORD, CONN.
A-	VIRGINIA. UNIV., CHARLOTTESVILLE
A-	WESTINGHOUSE ELECTRIC CORP., BALTIMORE
A-(NO.)(WEC)	WESTINGHOUSE ELECTRIC CORP. AVIATION GAS TURBINE DIV., KANSAS CITY, MO.
A-	WESTINGHOUSE ELECTRIC CORP. AVIATION GAS TURBINE DIV., PHILADELPHIA
AA-	ABT ASSOCIATES, INC., CAMBRIDGE, MASS.
AA-	AIR ASSOCIATES, INC., TETERBORO, N.J.
AA-	AIRCRAFT ARMAMENTS, INC., BALTIMORE
AA-	ANTIAIRCRAFT COMMAND (ARMY), COLORADO SPRINGS
58AA-	ARMED SERVICES TECHNICAL INFO. AGENCY, ARLINGTON, VA.
56AA-	ARMED SERVICES TECHNICAL INFO. AGENCY, ARLINGTON, VA.
57AA-	ARMED SERVICES TECHNICAL INFO. AGENCY, ARLINGTON, VA.
55AA-	ARMED SERVICES TECHNICAL INFO. AGENCY, ARLINGTON, VA.
54AA-	ARMED SERVICES TECHNICAL INFO. AGENCY, ARLINGTON, VA.
AA-	ARMY FIELD FORCES. BOARD NO. 4, FORT BLISS, TEX. (ANTIAIRCRAFT SERVICE TEST SECTION REPORT)
AA-	CORNELL AERONAUTICAL LAB., INC., BUFFALO
AA-	WESTINGHOUSE ELECTRIC CORP., BALTIMORE
AAAS-	AMERICAN ASSN. FOR THE ADVANCEMENT OF SCIENCE, WASH, DC
AAC-	ALASKAN AIR COMMAND, ELMENDORF AFB
AAC-	ALASKAN AIR COMMAND, LADD AFB
AAC-	GARRETT CORP., LOS ANGELES
A/AC.82/INF-	UNITED NATIONS. SCIENTIFIC COMMITTEE ON THE EFFECTS OF ATOMIC RADIATION
A/AC.82/G/1.	UNITED NATIONS. SECRETARIAT
A/AC.82/G/L.-	UNITED NATIONS GENERAL ASSEMBLY, N.Y.C.
A/AC.82/G/R.-	UNITED NATIONS GENERAL ASSEMBLY, N.Y.C.
A/AC.82/R.-	UNITED NATIONS GENERAL ASSEMBLY, N.Y.C.
AACE-	ALFORD (ANDREW) CONSULTING ENGINEER, BOSTON
AAC-OAR-	ALASKAN AIR COMMAND, ELMENDORF AFB (OPERATIONS ANALYSIS REPORT)
AAC OA TM-	ALASKAN AIR COMMAND, ELMENDORF AFB (OPERATIONS ANALYSIS TECHNICAL MEMORANDUM)
AAC OA WP-	ALASKAN AIR COMMAND, ELMENDORF AFB (OPERATIONS ANALYSIS WORKING PAPER)
AAC-WB-	ARMED FORCES SPECIAL WEAPONS PROJECT. FIELD COMMAND. ALBUQUERQUE, N. MEX. (WORKBOOK)
AAD-	WESTINGHOUSE ELECTRIC CORP. AEROSPACE DIV., BALTIMORE
AADB-	ARMY AIR DEFENSE BOARD
AADD-	AVCO MFG. CORP. ADVANCED DEV. DIV., STRATFORD, CONN.
AADD-	BROOKHAVEN NATIONAL LAB. ACCELERATOR DEPT., UPTON, N.Y. (INTERNAL REPORT)
AADS-	ARMY AIR DEFENSE SCHOOL, FORT BLISS, TEX.
AAE-	ILLINOIS. UNIV., URBANA. DEPT. OF AERONAUTICAL AND ASTRONAUTICAL ENGINEERING
AAEC-	AUSTRALIAN ATOMIC ENERGY COMMISSION. RESEARCH ESTABLISHMENT, LUCAS HEIGHTS, NEW SOUTH WALES
AAEC/ARC/P-	UNITED KINGDOM ATOMIC ENERGY AUTHORITY. RESEARCH GP. ATOMIC ENERGY RES. ESTAB., HARWELL, BERKS, ENGLAND
AAEC/E-	AUSTRALIAN ATOMIC ENERGY COMMISSION. RESEARCH ESTABLISHMENT, LUCAS HEIGHTS, NEW SOUTH WALES
AAEC/K-	AUSTRALIAN ATOMIC ENERGY COMMISSION, SYDNEY
AAEC/LIB/BIB-	AUSTRALIAN ATOMIC ENERGY COMMISSION. RESEARCH ESTABLISHMENT, LUCAS HEIGHTS, NEW SOUTH WALES
AAEC-LRP (MONTH/YEAR)	AUSTRALIAN ATOMIC ENERGY COMMISSION. RESEARCH ESTABLISHMENT, LUCAS HEIGHTS, NEW SOUTH WALES (LIST OF REPORT PUBLICATIONS)
AAEC/M-	AUSTRALIAN ATOMIC ENERGY COMMISSION. RESEARCH ESTABLISHMENT, LUCAS HEIGHTS, NEW SOUTH WALES
AAEC/PM-	AUSTRALIAN ATOMIC ENERGY COMMISSION. RESEARCH ESTABLISHMENT, LUCAS HEIGHTS, NEW SOUTH WALES
AAEC/PR(NUMBER)-(LTR.)	AUSTRALIAN ATOMIC ENERGY COMMISSION. RESEARCH ESTABLISHMENT, LUCAS HEIGHTS, NEW SOUTH WALES (PROGRESS REPORT)
AAEC(SP)/(LTR)(NO.)	AUSTRALIAN ATOMIC ENERGY COMMISSION. RESEARCH ESTABLISHMENT, LUCAS HEIGHTS, NEW SOUTH WALES
AAEC/TM-	AUSTRALIAN ATOMIC ENERGY COMMISSION. RESEARCH ESTABLISHMENT, LUCAS HEIGHTS, NEW SOUTH WALES
AAEC/TRANS.-	AUSTRALIAN ATOMIC ENERGY COMMISSION. RESEARCH ESTABLISHMENT, LUCAS HEIGHTS, NEW SOUTH WALES
AAEC/X-	AUSTRALIAN ATOMIC ENERGY COMMISSION. RESEARCH ESTABLISHMENT, LUCAS HEIGHTS, NEW SOUTH WALES
AAEE/	GT. BRIT. AEROPLANE AND ARMAMENT EXPERIMENTAL ESTABLISHMENT, BOSCOMBE DOWN, AMESBURY, WILTS, ENGLAND
AAEE/ATO/	GT. BRIT. AEROPLANE AND ARMAMENT EXPERIMENTAL ESTABLISHMENT, BOSCOMBE DOWN, AMESBURY, WILTS, ENGLAND
AAEE/INST/	GT. BRIT. AEROPLANE AND ARMAMENT EXPERIMENTAL ESTABLISHMENT, BOSCOMBE DOWN, AMESBURY, WILTS, ENGLAND
A/AEE-MEMO-	GT. BRIT. AEROPLANE AND ARMAMENT EXPERIMENTAL ESTABLISHMENT, BOSCOMBE DOWN, AMESBURY, WILTS, ENGLAND
AAEE/NOTE/	GT. BRIT. AEROPLANE AND ARMAMENT EXPERIMENTAL ESTABLISHMENT, BOSCOMBE DOWN, AMESBURY, WILTS, ENGLAND
AAEE/TECH/	GT. BRIT. AEROPLANE AND ARMAMENT EXPERIMENTAL ESTABLISHMENT, BOSCOMBE DOWN, AMESBURY, WILTS, ENGLAND
AAE-N-	ALL AMERICAN ENGINEERING CO., WILMINGTON, DEL.
AAERR-	MISSISSIPPI STATE UNIV., STATE COLLEGE. AGRICULTURAL EXPT. STA.(AEROPHYSICS + AEROSPACE ENG. RES. REPT.)
AAERR-(NO.),JOURNAL-	MISSISSIPPI STATE UNIV., STATE COLLEGE. AGRICULTURAL EXPT. STA.(AEROPHYSICS + AEROSPACE ENG. RES. REPT.)
AA/ES-	PURDUE UNIV., LAFAYETTE, IND. SCHOOL OF AERONAUTICS, ASTRONAUTICS AND ENGINEERING SCIENCES
AA+ES-(YEAR)-	PURDUE UNIV., LAFAYETTE, IND. SCHOOL OF AERONAUTICS, ASTRONAUTICS AND ENGINEERING SCIENCES
AAF-	AIR FORCE, WASHINGTON, D.C.
AAF AAL-	ARCTIC AEROMEDICAL LAB., LADD AFB, ALASKA
AAF AAL PROJ-	ARCTIC AEROMEDICAL LAB., LADD AFB, ALASKA
AAF APP-	AIR MATERIEL COMMAND, WRIGHT-PATTERSON AFB, OHIO (AVIATION PSYCHOLOGY PROGRAM)
AAF APP RR-	ARMY AIR FORCE (AVIATION PSYCHOLOGY PROGRAM)
AAF ARL-	ARMY AIR FORCE. AIRCRAFT RADIATION LAB.
AAF ARL ER-	ARMY AIR FORCE. AIRCRAFT RADIATION LAB. (ENG. RPT.)
AAF ARL MR-	ARMY AIR FORCE. AIRCRAFT RADIATION LAB. (MEMO. RPT.)
AAF ARL TR-	ARMY AIR FORCE. AIRCRAFT RADIATION LAB. (TEST RPT.)
AAF AWAF-	AIR MATERIEL COMMAND. ALL WEATHER FLYING DIV., WRIGHT-PATTERSON AFB, OHIO
AAF AWS-	AIR WEATHER SERVICE, WASHINGTON, D.C.
AAF-AWSM-	AIR WEATHER SERVICE, WASHINGTON, D.C.
AAF-AWSMTR-	AIR WEATHER SERVICE, WASHINGTON, D.C.
AAF AWS R-	AIR WEATHER SERVICE, WASH., D.C. (RUSSIAN REPORT)
AAFB-	ARMY AIR FORCE (BULLETIN)
AAFB-	ARMY AIR FORCE BOARD, ORLANDO, FLA.
AAF BD PROJ-	ARMY AIR FORCE (BOARD PROJECT)

AAF CB-	AIR TECHNICAL SERVICE COMMAND. ENGINEERING DIV. FLYING CLOTHING BR., WRIGHT-PATTERSON AFB, OHIO
AAFCE-	ALLIED AIR FORCES, CENTRAL EUROPE
AAFCE-OA-TM-	ALLIED AIR FORCES, CENTRAL EUROPE (OPERATIONS ANALYSIS TECHNICAL MEMORANDUM)
AAF CRL (NUMBER)-	AIR FORCE CAMBRIDGE RESEARCH LABS., BEDFORD, MASS.
AAF CRL E-	AIR FORCE CAMBRIDGE RESEARCH LABS., BEDFORD, MASS.
AAF DW-	ARMY AIR FORCE. DIRECTORATE OF WEATHER
AAF DW SS-	ARMY AIR FORCE. DIRECTORATE OF WEATHER (SPECIAL SERIES)
AAF ENG-	AIR TECHNICAL SERVICE COMMAND. ENGINEERING DIV., WRIGHT-PATTERSON AFB, OHIO
AAF ENG (NOS.,LETTERS)-	AIR TECHNICAL SERVICE COMMAND. ENGINEERING DIV. WRIGHT-PATTERSON AFB, OHIO
AAF ENG (LETTERS,NOS.)-	AIR TECHNICAL SERVICE COMMAND. ENGINEERING DIV. WRIGHT-PATTERSON AFB, OHIO
AAF ERL-	AIR FORCE CAMBRIDGE RESEARCH LABS., BEDFORD, MASS.
AAF ESE-	AIR TECHNICAL SERVICE COMMAND. ENGINEERING DIV. SERVICE ENG. SUBDIV., WRIGHT-PATTERSON AFB, OHIO
AAF EST-	AIR TECHNICAL SERVICE COMMAND. ENGINEERING DIV. ENG. STDS. SEC., WRIGHT-PATTERSON AFB, OHIO
AAF EU A-2 P-	AIR FORCES IN EUROPE
AAF EXP-	AIR TECHNICAL SERVICE COMMAND. ENGINEERING DIV. EXPTL. ENG. SEC., WRIGHT-PATTERSON AFB, OHIO
AAF EXP-M-(YR-LETTERS)-	AIR TECHNICAL SERVICE COMMAND. ENGINEERING DIV. EXPTL. ENG. SEC., WRIGHT-PATTERSON AFB, OHIO
AAF FCB-	AIR TECHNICAL SERVICE COMMAND. ENGINEERING DIV. FLYING CLOTHING BR., WRIGHT-PATTERSON AFB, OHIO
AAF HRRC-	HUMAN RESOURCES RESEARCH CENTER, LACKLAND AFB, TEXAS
AAF HRRC RB-	HUMAN RESOURCES RESEARCH CENTER, LACKLAND AFB, TEXAS
AAF HRRC RB (NO.)-	HUMAN RESOURCES RESEARCH CENTER, LACKLAND AFB, TEXAS
AAF HRRL-	HUMAN RESOURCES RES. LABS., BOLLING AFB, WASH., D.C.
AAF HRRL (NO.)-	HUMAN RESOURCES RES. LABS., BOLLING AFB, WASH., D.C.
AAF IR-	AIR FORCE (INTELLIGENCE REPORTS)
AAF M-	ARMY AIR FORCE (MANUAL)
AAF MCREE-	AIR MATERIEL COMMAND. ENGINEERING DIV. ELECTRONIC SUBDIV., WRIGHT-PATTERSON AFB, OHIO
AAF MCREXA-	AIR MATERIEL COMMAND. ENGINEERING DIV. AIRCRAFT LAB., WRIGHT-PATTERSON AFB, OHIO (MEMO. REPT.)
AAF MCREXD-	AIR MATERIEL COMMAND. ENGINEERING DIV. AERO-MEDICAL LAB., WRIGHT-PATTERSON AFB, OHIO
AAF NTIR-	AIR FORCE (NON-TECHNICAL INTELLIGENCE REPORTS)
AAF NTIR E-	AIR FORCE (NON-TECHNICAL INTELLIGENCE REPORTS)
AAF OAR-	ARMY AIR FORCE. SECOND. OPERATIONS AND TRAINING DIV. (OPERATIONS ANALYSIS REPORT)
AAF OAR-	OFFICE OF AIR RESEARCH, WRIGHT-PATTERSON AFB, OHIO
AAF-OAR-TR-	OFFICE OF AIR RESEARCH, WRIGHT-PATTERSON AFB, OHIO
AAF P-	ARMY AIR FORCE, WASHINGTON, D.C.
AAF SAM-	SCHOOL OF AVIATION MEDICINE, RANDOLPH AFB, TEX.
AAF SAM PROJ-	SCHOOL OF AVIATION MEDICINE, RANDOLPH AFB, TEX.
AAF SER. X-LM-	ARMY AIR FORCE (LIAISON MEMORANDUM, SERIES X)
AAF SL-	ARMY AIR FORCE (STOCK LIST)
AAF T-2 ATIR-	AIR MATERIEL COMMAND. AIR DOCUMENTS DIV., WRIGHT-PATTERSON AFB, OHIO (AIR TECH. INTELLIGENCE REVIEW)
AAF T-2 INT-	AIR MATERIEL COMMAND. AIR DOCUMENTS DIV., WRIGHT-PATTERSON AFB, OHIO (INTERIM REPORT)
AAF T-2 IR/	AIR MATERIEL COMMAND. AIR DOCUMENTS DIV., WRIGHT-PATTERSON AFB, OHIO (INTELLIGENCE REPORT)
AAF T-2 IRE-	AIR MATERIEL COMMAND. AIR DOCUMENTS DIV., WRIGHT-PATTERSON AFB, OHIO
AAF T-2 MICRO-	AIR MATERIEL COMMAND. AIR DOCUMENTS DIV., WRIGHT-PATTERSON AFB, OHIO (MICROFILM)
AAF T-2 SR/	AIR MATERIEL COMMAND. AIR DOCUMENTS DIV., WRIGHT-PATTERSON AFB, OHIO (SUMMARY REPORT)
AAF T-2 T/	AIR MATERIEL COMMAND. AIR DOCUMENTS DIV., WRIGHT-PATTERSON AFB, OHIO (TRANSLATION)
AAF T-2 TR-	AIR MATERIEL COMMAND. AIR DOCUMENTS DIV., WRIGHT-PATTERSON AFB, OHIO (TECHNICAL REPORT)
AAFT-	ARMY AIR FORCE (TRANSLATION)
AAFTC-	ARMY AIR FORCES TACTICAL CENTER, ORLANDO, FLA.
AAFTC TN (YR.)-	AIR FORCE FLIGHT TEST CENTER, EDWARDS AFB, CALIF.
AAF TIR-	AIR FORCE (TECHNICAL INTELLIGENCE REPORTS)
AAF TIR (LETTER)-	AIR FORCE (TECHNICAL INTELLIGENCE REPORTS)
AAF TN-	AIR TECHNICAL SERVICE COMMAND. ENGINEERING DIV. WRIGHT-PATTERSON AFB, OHIO
AAF TN TSELR-	AIR TECHNICAL SERVICE COMMAND. ENGINEERING DIV. ELECTRONICS SUBDIV., WRIGHT-PATTERSON AFB, OHIO
AAF TO-	ARMY AIR FORCE (TECHNICAL ORDER)
AAFTR-	ARMY AIR FORCE (TECHNICAL REPORT)
AAF-TR-	ARMY AIR FORCE, WASHINGTON, D.C.
AAF TR (NOS.,LETTERS)-	ARMY AIR FORCE, WASHINGTON, D.C.
AAF TSEAA-	AIR TECHNICAL SERVICE COMMAND. ENGINEERING DIV. AERO-MEDICAL LAB., WRIGHT-PATTERSON AFB, OHIO
AAF TSEAA MR-	AIR TECHNICAL SERVICE COMMAND. ENG. DIV. AERO-MEDICAL LAB., WRIGHT-PATTERSON AFB, OHIO (MEMO. RPTS.)
AAF TSEAC-	AIR TECHNICAL SERVICE COMMAND. ENGINEERING DIV. AIRCRAFT LAB., WRIGHT-PATTERSON AFB, OHIO
AAF TSEAC MR-	AIR TECHNICAL SERVICE COMMAND. ENG. DIV. AIRCRAFT LAB., WRIGHT-PATTERSON AFB, OHIO (MEMO. RPTS.)
AAF TSEAL-	AIR TECHNICAL SERVICE COMMAND. ENGINEERING DIV. AERO-MEDICAL LAB., WRIGHT-PATTERSON AFB, OHIO
AAF TSEAL (NO.,LETTER)-	AIR TECHNICAL SERVICE COMMAND. ENGINEERING DIV. AERO-MEDICAL LAB., WRIGHT-PATTERSON AFB, OHIO
AAF TSEAM-	AIR TECHNICAL SERVICE COMMAND. ENGINEERING DIV. MATERIALS LAB., WRIGHT-PATTERSON AFB, OHIO
AAF-TSEAM-M-	AIR TECH. SERVICE COMMAND, WRIGHT-PATTERSON AFB, OHIO
AAF TSEAM MR-	AIR TECHNICAL SERVICE COMMAND. ENG. DIV. MATERIALS LAB., WRIGHT-PATTERSON AFB, OHIO (MEMO. RPTS.)
AAF TSEAP-	AIR TECHNICAL SERVICE COMMAND. ENGINEERING DIV. PERSONAL EQUIPMENT LAB., WRIGHT-PATTERSON AFB, OHIO
AAF TSELA-	AIR TECHNICAL SERVICE COMMAND. ENG. DIV. ELECTRONICS ADMINISTRATIVE SEC., WRIGHT-PATTERSON AFB, OHIO
AAF TSELA(NO.,LTR.,NO.)-	AIR TECHNICAL SERVICE COMMAND. ENG. DIV. ELECTRONICS ADMINISTRATIVE SEC., WRIGHT-PATTERSON AFB, OHIO
AAF TSELC-	AIR TECHNICAL SERVICE COMMAND. ENG. DIV. COMMUNICATION + NAVIGATION LAB., WRIGHT-PATTERSON AFB, OHIO
AAF TSELP-	AIR TECHNICAL SERVICE COMMAND. ENG. DIV. ELECTRONICS PLANS SECTION., WRIGHT-PATTERSON AFB, OHIO
AAF TSELR-	AIR TECHNICAL SERVICE COMMAND. ENGINEERING DIV. ELECTRONICS SUBDIV., WRIGHT-PATTERSON AFB, OHIO
AAF TSELR ESMR (NO.)-	AIR TECHNICAL SERVICE COMMAND. ENGINEERING DIV. ELECTRONICS SUBDIV., WRIGHT-PATTERSON AFB, OHIO
AAF TSELS-	AIR TECHNICAL SERVICE COMMAND. ENGINEERING DIV. SYSTEMS LAB., WRIGHT-PATTERSON AFB, OHIO
AAF TSEPE-	AIR TECHNICAL SERVICE COMMAND. ENGINEERING DIV. EQUIPMENT LAB., WRIGHT-PATTERSON AFB, OHIO
AAF TSEPF-	AIR TECHNICAL SERVICE COMMAND. ENGINEERING DIV. PHOTO LAB., WRIGHT-PATTERSON AFB, OHIO
AAF TSEPL-	AIR TECHNICAL SERVICE COMMAND. ENGINEERING DIV. ENG. PLANS DIV., WRIGHT-PATTERSON AFB, OHIO
AAF TSEPP-	AIR TECHNICAL SERVICE COMMAND. ENGINEERING DIV. POWER PLANT LAB., WRIGHT-PATTERSON AFB, OHIO
AAF TSEPR-	AIR TECHNICAL SERVICE COMMAND. ENGINEERING DIV. PROPELLER LAB., WRIGHT-PATTERSON AFB, OHIO
AAF TSEPS-	AIR TECHNICAL SERVICE COMMAND. ENGINEERING DIV. ARMAMENT LAB., WRIGHT-PATTERSON AFB, OHIO
AAF TSERR-	AIR TECHNICAL SERVICE COMMAND. ENG. DIV. RADIO + RADAR SUBDIV., WRIGHT-PATTERSON AFB, OHIO
AAF TSESA-	AIR TECHNICAL SERVICE COMMAND. ENGINEERING DIV. AIRCRAFT PROJ. SEC., WRIGHT-PATTERSON AFB, OHIO
AAF TSESE-	AIR TECHNICAL SERVICE COMMAND. ENGINEERING DIV. SERVICE ENG. SUBDIV., WRIGHT-PATTERSON AFB, OHIO
AAF TSEST-	AIR TECHNICAL SERVICE COMMAND. ENGINEERING DIV. ENG. STDS. SEC., WRIGHT-PATTERSON AFB, OHIO
AAF U-ENG-	AIR TECHNICAL SERVICE COMMAND. ENGINEERING DIV., WRIGHT-PATTERSON AFB, OHIO

AAF U-EXP-
: AIR TECHNICAL SERVICE COMMAND. ENGINEERING DIV. EXPTL. ENG. SEC., WRIGHT-PATTERSON AFB, OHIO

AAF U-EXP-M-(YR-LTRS)-
: AIR TECHNICAL SERVICE COMMAND. ENGINEERING DIV. EXPTL. ENG. SEC., WRIGHT-PATTERSON AFB, OHIO

AAF U-TSEPL-
: AIR TECHNICAL SERVICE COMMAND. ENGINEERING DIV. ENG. PLANS DIV., WRIGHT-PATTERSON AFB, OHIO

AAF U-TSEPP-
: AIR TECHNICAL SERVICE COMMAND. ENGINEERING DIV. POWER PLANT LAB., WRIGHT-PATTERSON AFB, OHIO

AAF WD-
: ARMY AIR FORCE. WEATHER DIV.

AAF-WD-TR-
: ARMY AIR FORCE. WEATHER DIV.

AAF WIB-
: ARMY AIR FORCE. WEATHER INFORMATION BRANCH

AAF WIS-
: ARMY AIR FORCE. WEATHER INFORMATION SERVICE

AAF WSB-
: ARMY AIR FORCE. WEATHER SERVICE (BULLETIN)

AAF WSM-
: ARMY AIR FORCE. WEATHER SERVICE (MANUAL)

AAF W-U-ENG-
: AIR TECHNICAL SERVICE COMMAND. ENGINEERING DIV., WRIGHT-PATTERSON AFB, OHIO

AAFW-U-EXP-M-(YR)-
: AIR TECHNICAL SERVICE COMMAND. ENGINEERING DIV. EXPTL. ENG. SEC., WRIGHT-PATTERSON AFB, OHIO

AAF W-U-TSEPL-
: AIR TECHNICAL SERVICE COMMAND. ENGINEERING DIV. ENG. PLANS DIV., WRIGHT-PATTERSON AFB, OHIO

AA+GM-
: ANTIAIRCRAFT ARTILLERY AND GUIDED MISSILE CENTER, FORT BLISS, TEXAS

AAI-
: AAI CORP., COCKEYSVILLE, MD.

AAI-ER-
: AAI CORP., COCKEYSVILLE, MD.

AAL-
: NAVAL AIR DEVELOPMENT CENTER. AVIATION ARMAMENT LAB., JOHNSVILLE, PENNA.

AAL-TDR-(YEAR)-
: ARCTIC AEROMEDICAL LAB., FORT WAINWRIGHT, ALASKA

AAL-TDR-
: ARCTIC AEROMEDICAL LAB., LADD AFB, ALASKA

AAL-TN-(YEAR)-
: ARCTIC AEROMEDICAL LAB., LADD AFB, ALASKA

AAL-TR-(YEAR)-
: ARCTIC AEROMEDICAL LAB., FORT WAINWRIGHT, ALASKA

AAL-TR-(YEAR)-
: ARCTIC AEROMEDICAL LAB., LADD AFB, ALASKA

AAM-
: BUREAU OF ORDNANCE (NAVY) (ACOUSTIC ANALYSIS MEMO.)

AAMC-
: ARMY ARTILLERY AND MISSILE CENTER, FORT SILL, OKLA.

AA-N-
: ALL AMERICAN ENGINEERING CO., WILMINGTON, DEL.

AAP-
: GOODYEAR AEROSPACE CORP., AKRON, OHIO

AAP-
: NORTH ATLANTIC TREATY ORGANIZATION. MILITARY AGENCY FOR STANDARDIZATION, PARIS

AAP-PMH-(YEAR)-
: CALIFORNIA. UNIV., LOS ANGELES. CENTER FOR THE HEALTH SCIENCES

AAR-
: BUREAU OF ORDNANCE (NAVY) (ACOUSTIC ANALYSIS REPORT)

AAR-
: RICE UNIV., HOUSTON, TEX.

AAS-
: NAVY (ANTI-AIRCRAFT STUDY)

AASU-
: SOUTHAMPTON, ENGLAND. UNIV., DEPT. OF AERONAUTICS AND ASTRONAUTICS

AASU-TN-
: SOUTHAMPTON, ENGLAND. UNIV., DEPT. OF AERONAUTICS AND ASTRONAUTICS

AATDC-
: UNITED KINGDOM. ARMY AIR TRANSPORT TRAINING AND DEVELOPMENT CENTER, OLD SARUM, WILTS, ENGLAND

A-ATLAS-
: CONSOLIDATED VULTEE AIRCRAFT CORP., SAN DIEGO, CALIF.

AAWSSC-
: ARMY ATOMIC WEAPON SYSTEMS SAFETY COMMITTEE

AAWSSC-
: OFFICE OF THE CHIEF OF ORDNANCE, WASHINGTON, D.C.

AB-
: ALIAS BETTY COMMITTEE (REF. 15)

AB-
: CALIFORNIA. UNIV., BERKELEY. RADIATION LAB.

AB-(NUMBER)-
: GULTON INDUSTRIES, INC. ALKALINE BATTERY DIV., METUCHEN, N.J.

AB-
: REMINGTON ARMS CO., INC., BRIDGEPORT, CONN.

AB-
: SPERRY RAND CORP. SPERRY ELECTRO OPTICS, GT.NECK,N.Y.

AB-
: TENNESSEE EASTMAN CORP., OAK RIDGE, TENN.

ABA-
: AETRON-BLUME-ATKINSON, PALO ALTO, CALIF.

ABBC-
: ATOMIC BOMB CASUALTY COMMISSION, HIROSHIMA

ABCC-
: ATOMIC BOMB CASUALTY COMMISSION, HIROSHIMA

ABCC-(NUMBERS)-(YEAR)
: ATOMIC BOMB CASUALTY COMMISSION, HIROSHIMA

ABCC-AR-
: ATOMIC BOMB CASUALTY COMMISSION, HIROSHIMA (ANNUAL RPT)

ABCC-MANUAL-(NO.)-(YR.)
: ATOMIC BOMB CASUALTY COMMISSION, HIROSHIMA

ABCC-PA-(NUMBER)-(YEAR)
: ATOMIC BOMB CASUALTY COMMISSION, HIROSHIMA

ABCC-TR-(NUMBER)-(NO.)
: ATOMIC BOMB CASUALTY COMMISSION, HIROSHIMA

ABCC-TR-(NUMBER)-(YEAR)
: ATOMIC BOMB CASUALTY COMMISSION, HIROSHIMA

AB-CR-(LETTERS)-
: NATIONAL AERONAUTICS AND SPACE ADM., WASHINGTON, D.C. (TRANSLATOR'S INITIALS) (CATEGORY) (SUBJECT)

ABDL-
: DUMONT (ALLEN B.) LABS., INC., PASSAIC, N.J.

ABEC-
: AMPHENOL-BORG ELECTRONICS CORP., CHICAGO

ABE-NB-
: ARMED FORCES SPECIAL WEAPONS PROJECT. FIELD COMMAND, ALBUQUERQUE, N. MEX. (NOTEBOOK)

ABG-
: NETHERLANDS. ADVIESBUREAU DER GENIE, HAGUE

ABJ-
: BAYLOR UNIV., HOUSTON, TEX. COLL. OF MEDICINE

ABK-
: AMER. BRAKE SHOE CO. AMER. BRAKEBLOK DIV., DETROIT

ABL-
: GOODRICH (B.F.) CO., RIALTO, CALIF.

ABL 1045-
: HARVARD UNIV., CAMBRIDGE, MASS. AMER. BRITISH LAB.

ABL-
: HERCULES POWDER CO. ALLEGANY BALLISTICS LAB., CUMBERLAND, MD.

ABL/AERO-
: AUSTRALIA. AERONAUTICAL RESEARCH LABS., MELBOURNE

ABL/AF/QPR-
: HERCULES POWDER CO. ALLEGANY BALLISTICS LAB., CUMBERLAND, MD.

ABL/ARPA/QTSR-
: HERCULES POWDER CO. ALLEGANY BALLISTICS LAB., CUMBERLAND, MD.

ABL-B-
: HERCULES POWDER CO. ALLEGANY BALLISTICS LAB., CUMBERLAND, MD.

ABL FR-
: HERCULES POWDER CO. ALLEGANY BALLISTICS LAB., CUMBERLAND, MD. (FINAL REPORT)

ABL FR (LETTER)-
: HERCULES POWDER CO. ALLEGANY BALLISTICS LAB., CUMBERLAND, MD. (FINAL REPORT)

ABL-LIBRARY-
: HERCULES POWDER CO. ALLEGANY BALLISTICS LAB., CUMBERLAND, MD.

ABL/MPR-
: HERCULES POWDER CO. ALLEGANY BALLISTICS LAB., CUMBERLAND, MD. (MONTHLY PROGRESS REPORT)

ABL/NASA/QPR-
: HERCULES POWDER CO. ALLEGANY BALLISTICS LAB., CUMBERLAND, MD.

ABL/QPR-
: HERCULES POWDER CO. ALLEGANY BALLISTICS LAB., CUMBERLAND, MD. (QUARTERLY PROGRESS REPORT)

ABL/R-
: HERCULES POWDER CO. ALLEGANY BALLISTICS LAB., CUMBERLAND, MD.

ABL SR-
: HERCULES POWDER CO. ALLEGANY BALLISTICS LAB., CUMBERLAND, MD.

ABL-TR-
: ARMY BIOLOGICAL LABS., FREDERICK, MD.

ABL-TR-(YEAR)-
: HERCULES POWDER CO. ALLEGANY BALLISTICS LAB., CUMBERLAND, MD.

ABL WPR SUPP.-
: HERCULES POWDER CO. ALLEGANY BALLISTICS LAB., CUMBERLAND, MD. (WEEKLY PROGRESS REPT., SUPP.)

ABL/X-
: HERCULES POWDER CO. ALLEGANY BALLISTICS LAB., CUMBERLAND, MD.

ABL/Z-
: HERCULES POWDER CO. ALLEGANY BALLISTICS LAB., CUMBERLAND, MD.

AB-M-
: COLUMBIA UNIV., NYC. SUBSTITUTE ALLOY MATERIALS LABS.

ABMA-
: ARMY BALLISTIC MISSILE AGENCY, REDSTONE ARSENAL, ALA.

ABMA/AFJUPLO CR(NO)(YR)J-
: ARMY BALLISTIC MISSILE AGENCY. CONTROL OFFICE, REDSTONE ARSENAL, ALA. (CONSOLIDATED REPORT)

ABMA/AFJUPLO CR-J-(QTR.)-
: ARMY BALLISTIC MISSILE AGENCY. CONTROL OFFICE, REDSTONE ARSENAL, ALA. (SERIES ENDS WITH CY (FIGURE FOR CALENDAR YEAR)(FIGURE FOR FISCAL YEAR))

ABMA/AFJUPLO CR-J(NO.YR)-
: ARMY BALLISTIC MISSILE AGENCY. CONTROL OFFICE, REDSTONE ARSENAL, ALA. (CONSOLIDATED REPORT)

ABMA ATS-G-
: ARMY BALLISTIC MISSILE AGENCY. GUIDANCE AND CONTROL LAB., REDSTONE ARSENAL, ALA. (ACCEPTANCE TEST SPEC.

ABMA CCTS TR(NO)(NO-LTR)
: ARMY BALLISTIC MISSILE AGENCY. TEST LAB., REDSTONE ARSENAL, ALA. (COLD CALIBRATION TEST STAND TEST REPORT)

ABMA C+F CN-
: ARMY BALLISTIC MISSILE AGENCY. TEST LAB., REDSTONE ARSENAL, ALA. (CALCULATION NOTES)

ABMA C+F EP-
: ARMY BALLISTIC MISSILE AGENCY. TEST LAB., REDSTONE ARSENAL, ALA. (EXPERIMENTAL PROCEDURES)

ABMA C+F F+DC-
　　ARMY BALLISTIC MISSILE AGENCY. TEST LAB., REDSTONE
　　ARSENAL, ALA. (FORMULA AND DATA COMPILATIONS)
ABMA C+F LES-
　　ARMY BALLISTIC MISSILE AGENCY. TEST LAB., REDSTONE
　　ARSENAL, ALA. (LABORATORY EXPERIMENTS SUMMARIES)
ABMA C+F MEMO-
　　ARMY BALLISTIC MISSILE AGENCY. TEST LAB., REDSTONE
　　ARSENAL, ALA.
ABMA C+F NOTE-
　　ARMY BALLISTIC MISSILE AGENCY. TEST LAB., REDSTONE
　　ARSENAL, ALA.
ABMA C+F TEP-
　　ARMY BALLISTIC MISSILE AGENCY. TEST LAB., REDSTONE
　　ARSENAL, ALA. (TEST EVALUATION PROCEDURES)
ABMA-COMM. MIN./
　　ARMY BALLISTIC MISSILE AGENCY, REDSTONE ARSENAL, ALA.
ABMA CR-P(YR)-
　　ARMY BALLISTIC MISSILE AGENCY. CONTROL OFFICE, RED-
　　STONE ARSENAL, ALA. (CONTROL OFFICE PROGRESS RPT.)
ABMA CR-R(NO.)(YR)
　　ARMY BALLISTIC MISSILE AGENCY. CONTROL OFFICE, RED-
　　STONE ARSENAL, ALA. (CONTROL OFFICE REPORT)
ABMA CR-R-(NO)Q(NO)(YR)
　　ARMY BALLISTIC MISSILE AGENCY. CONTROL OFFICE, RED-
　　STONE ARSENAL, ALA. (CONTROL OFFICE QUARTERLY
　　REPORT ON REDSTONE)
ABMA CR-S(NO.)(YR)
　　ARMY BALLISTIC MISSILE AGENCY. CONTROL OFFICE, RED-
　　STONE ARSENAL, ALA. (CONTROL OFFICE SATELLITE RPT.)
ABMA DA-IN-
　　ARMY BALLISTIC MISSILE AGENCY. AEROBALLISTICS LAB.,
　　REDSTONE ARSENAL, ALA. (INTERNAL NOTE)
ABMA DA-M-
　　ARMY BALLISTIC MISSILE AGENCY. AEROBALLISTICS LAB.,
　　REDSTONE ARSENAL, ALA. (MEMORANDUM)
ABMA DA-M(NO.)(YR)
　　ARMY BALLISTIC MISSILE AGENCY. AEROBALLISTICS LAB.,
　　REDSTONE ARSENAL, ALA. (MEMORANDUM)
ABMA DA-M-
　　ARMY BALLISTIC MISSILE AGENCY. COMPUTATION LAB.,
　　REDSTONE ARSENAL, ALA. (MEMORANDUM)
ABMA DA MEMO-
　　ARMY BALLISTIC MISSILE AGENCY. AEROBALLISTICS LAB.,
　　REDSTONE ARSENAL, ALA.
ABMA DA-R-
　　ARMY BALLISTIC MISSILE AGENCY. AEROBALLISTICS LAB.,
　　REDSTONE ARSENAL, ALA. (REPORT)
ABMA DA-TM(NO.)(YR)
　　ARMY BALLISTIC MISSILE AGENCY. AEROBALLISTICS LAB.,
　　REDSTONE ARSENAL, ALA. (TECHNICAL MEMO)
ABMA DA-TN-
　　ARMY BALLISTIC MISSILE AGENCY. AEROBALLISTICS LAB.,
　　REDSTONE ARSENAL, ALA. (TECHNICAL NOTE)
ABMA DA-TN(NO.)(YR)
　　ARMY BALLISTIC MISSILE AGENCY. AEROBALLISTICS LAB.,
　　REDSTONE ARSENAL, ALA.
ABMA DA-TR(NO.)(YR)
　　ARMY BALLISTIC MISSILE AGENCY. AEROBALLISTICS LAB.,
　　REDSTONE ARSENAL, ALA. (TECHNICAL REPORT)
ABMA DA-WTN-
　　ARMY BALLISTIC MISSILE AGENCY. AEROBALLISTICS LAB.,
　　REDSTONE ARSENAL, ALA. (WIND TUNNEL NOTES)
ABMA DCF-TN-
　　ARMY BALLISTIC MISSILE AGENCY. COMPUTATION LAB.,
　　REDSTONE ARSENAL, ALA.
ABMA/DCMR TR-
　　ARMY BALLISTIC MISSILE AGENCY, REDSTONE ARSENAL, ALA.
ABMA DC-TM(NO.)(YR)
　　ARMY BALLISTIC MISSILE AGENCY. COMPUTATION LAB.,
　　REDSTONE ARSENAL, ALA. (TECHNICAL MEMORANDUM)
ABMA DC-TN(NO.)(YR)
　　ARMY BALLISTIC MISSILE AGENCY. COMPUTATION LAB.,
　　REDSTONE ARSENAL, ALA. (TECHNICAL NOTE)
ABMA DC-TR-
　　ARMY BALLISTIC MISSILE AGENCY. COMPUTATION LAB.,
　　REDSTONE ARSENAL, ALA. (TECHNICAL REPORT)
ABMA DFE-TM(NO.)(YR)
　　ARMY BALLISTIC MISSILE AGENCY. FABRICATION + ASSEMBLY
　　ENGINEERING LAB., REDSTONE ARSENAL, ALA.
ABMA DFE-TN(NO.)(YR)
　　ARMY BALLISTIC MISSILE AGENCY. FABRICATION + ASSEMBLY
　　ENGINEERING LAB., REDSTONE ARSENAL, ALA.
ABMA DF-M-
　　ARMY BALLISTIC MISSILE AGENCY. FABRICATION LAB.,
　　REDSTONE ARSENAL, ALA.
ABMA DFR-IN(NO.)(YR)
　　ARMY BALLISTIC MISSILE AGENCY. FABRICATION + ASSEMBLY
　　ENGINEERING LAB., REDSTONE ARSENAL, ALA.
　　(INTERNAL NOTE)
ABMA DFR-TM(NO.)(YR)
　　ARMY BALLISTIC MISSILE AGENCY. FABRICATION + ASSEMBLY
　　ENGINEERING LAB., REDSTONE ARSENAL, ALA.
ABMA DFR-TN(NO.)(YR)
　　ARMY BALLISTIC MISSILE AGENCY. FABRICATION + ASSEMBLY
　　ENGINEERING LAB., REDSTONE ARSENAL, ALA.
ABMA DF-TM(NO.)(YR)
　　ARMY BALLISTIC MISSILE AGENCY. FABRICATION + ASSEMBLY
　　ENGINEERING LAB., REDSTONE ARSENAL, ALA.
ABMA DF-TN(NO.)(YR)
　　ARMY BALLISTIC MISSILE AGENCY. FABRICATION + ASSEMBLY
　　ENGINEERING LAB., REDSTONE ARSENAL, ALA.
ABMA DGIC-TN(NO.)(YR)
　　ARMY BALLISTIC MISSILE AGENCY. GUIDANCE AND CONTROL
　　LAB., REDSTONE ARSENAL, ALA.
ABMA DG-IN(NO.)(YR)
　　ARMY BALLISTIC MISSILE AGENCY. GUIDANCE AND CONTROL
　　LAB., REDSTONE ARSENAL, ALA.
ABMA DGI-TM(NO.)(YR)
　　ARMY BALLISTIC MISSILE AGENCY. GUIDANCE AND CONTROL
　　LAB., REDSTONE ARSENAL, ALA.
ABMA DG-M-
　　ARMY BALLISTIC MISSILE AGENCY. GUIDANCE AND CONTROL
　　LAB., REDSTONE ARSENAL, ALA. (MEMO)
ABMA DG-M(NO.)(YR)
　　ARMY BALLISTIC MISSILE AGENCY. GUIDANCE AND CONTROL
　　LAB., REDSTONE ARSENAL, ALA.
ABMA DGNA(J)(NO.)(YR)
　　ARMY BALLISTIC MISSILE AGENCY. GUIDANCE AND CONTROL
　　LAB., REDSTONE ARSENAL, ALA.
ABMA DGNA-(R)(NO.)/(YR)
　　ARMY BALLISTIC MISSILE AGENCY. GUIDANCE AND CONTROL
　　LAB., REDSTONE ARSENAL, ALA.
ABMA DGNA-(S)(NO.)/(YR)
　　ARMY BALLISTIC MISSILE AGENCY. GUIDANCE AND CONTROL
　　LAB., REDSTONE ARSENAL, ALA.
ABMA DGNC-(J)(NO.)/(YR)
　　ARMY BALLISTIC MISSILE AGENCY. GUIDANCE AND CONTROL
　　LAB., REDSTONE ARSENAL, ALA.
ABMA DGNC-(P)(NO.)/(YR)
　　ARMY BALLISTIC MISSILE AGENCY. GUIDANCE AND CONTROL
　　LAB., REDSTONE ARSENAL, ALA.
ABMA DGND-(J)(NO.)(YR)
　　ARMY BALLISTIC MISSILE AGENCY. GUIDANCE AND CONTROL
　　LAB., REDSTONE ARSENAL, ALA.
ABMA DGNE-(P)(NO.)(YR)
　　ARMY BALLISTIC MISSILE AGENCY. GUIDANCE AND CONTROL
　　LAB., REDSTONE ARSENAL, ALA.
ABMA DGNE-(S)(NO.)(YR)
　　ARMY BALLISTIC MISSILE AGENCY. GUIDANCE AND CONTROL
　　LAB., REDSTONE ARSENAL, ALA.
ABMA DGN-(J)(NO.)(YR)
　　ARMY BALLISTIC MISSILE AGENCY. GUIDANCE AND CONTROL
　　LAB., REDSTONE ARSENAL, ALA.
ABMA DGN-TM(NO.)(YR)
　　ARMY BALLISTIC MISSILE AGENCY. GUIDANCE AND CONTROL
　　LAB., REDSTONE ARSENAL, ALA.
ABMA DG-R-
　　ARMY BALLISTIC MISSILE AGENCY. GUIDANCE AND CONTROL
　　LAB., REDSTONE ARSENAL, ALA.
ABMA DGR-TN(NO.)(YR)
　　ARMY BALLISTIC MISSILE AGENCY. GUIDANCE AND CONTROL
　　LAB., REDSTONE ARSENAL, ALA.
ABMA DG-TM(NO.)(YR)
　　ARMY BALLISTIC MISSILE AGENCY. GUIDANCE AND CONTROL
　　LAB., REDSTONE ARSENAL, ALA.
ABMA DG-TN(NO.)(YR)
　　ARMY BALLISTIC MISSILE AGENCY. GUIDANCE AND CONTROL
　　LAB., REDSTONE ARSENAL, ALA.
ABMA DG-TR(NO.)(YR)
　　ARMY BALLISTIC MISSILE AGENCY. GUIDANCE AND CONTROL
　　LAB., REDSTONE ARSENAL, ALA.
ABMA DI-CC-
　　ARMY BALLISTIC MISSILE AGENCY. WEAPON SYSTEMS
　　INFORMATION OFFICE, REDSTONE ARSENAL, ALA.
ABMA DIR-TN(NO.)(YR)
　　ARMY BALLISTIC MISSILE AGENCY. WEAPON SYSTEMS
　　INFORMATION OFFICE, REDSTONE ARSENAL, ALA.
ABMA DIR-TR(NO.)(YR)
　　ARMY BALLISTIC MISSILE AGENCY. WEAPON SYSTEMS
　　INFORMATION OFFICE, REDSTONE ARSENAL, ALA.
ABMA DI-TM(NO.)(YR)
　　ARMY BALLISTIC MISSILE AGENCY. WEAPON SYSTEMS
　　INFORMATION OFFICE, REDSTONE ARSENAL, ALA.
ABMA DI-TN(NO.)(YR)
　　ARMY BALLISTIC MISSILE AGENCY. WEAPON SYSTEMS
　　INFORMATION OFFICE, REDSTONE ARSENAL, ALA.
ABMA DLE(NO.)(YR)
　　ARMY BALLISTIC MISSILE AGENCY. LAUNCHING + HANDLING
　　EQUIPMENT LAB., REDSTONE ARSENAL, ALA.
ABMA DLEC-TN(NO.)(YR)
　　ARMY BALLISTIC MISSILE AGENCY. SYSTEMS SUPPORT
　　EQUIPMENT LAB., REDSTONE ARSENAL, ALA.
ABMA DLE-TN(NO.)(YR)
　　ARMY BALLISTIC MISSILE AGENCY. SYSTEMS SUPPORT
　　EQUIPMENT LAB., REDSTONE ARSENAL, ALA.
ABMA DL-M-
　　ARMY BALLISTIC MISSILE AGENCY. LAUNCHING + HANDLING
　　EQUIPMENT LAB., REDSTONE ARSENAL, ALA.
ABMA DLMD-TN(NO.)(YR)
　　ARMY BALLISTIC MISSILE AGENCY. SYSTEMS SUPPORT
　　EQUIPMENT LAB., REDSTONE ARSENAL, ALA.
ABMA DLMF-TM(NO.)(YR)
　　ARMY BALLISTIC MISSILE AGENCY. SYSTEMS SUPPORT
　　EQUIPMENT LAB., REDSTONE ARSENAL, ALA.
ABMA DLMF-TN(NO.)(YR)
　　ARMY BALLISTIC MISSILE AGENCY. SYSTEMS SUPPORT
　　EQUIPMENT LAB., REDSTONE ARSENAL, ALA.
ABMA DLMF-TR(NO.)(YR)
　　ARMY BALLISTIC MISSILE AGENCY. SYSTEMS SUPPORT
　　EQUIPMENT LAB., REDSTONE ARSENAL, ALA.
ABMA DLMP-TM(NO.)(YR)
　　ARMY BALLISTIC MISSILE AGENCY. SYSTEMS SUPPORT
　　EQUIPMENT LAB., REDSTONE ARSENAL, ALA.
ABMA DLMP-TN(NO.)(YR)
　　ARMY BALLISTIC MISSILE AGENCY. SYSTEMS SUPPORT
　　EQUIPMENT LAB., REDSTONE ARSENAL, ALA.
ABMA DLM-TN(NO.)(YR)
　　ARMY BALLISTIC MISSILE AGENCY. SYSTEMS SUPPORT
　　EQUIPMENT LAB., REDSTONE ARSENAL, ALA.
ABMA DLMT-TM(NO.)(YR)
　　ARMY BALLISTIC MISSILE AGENCY. WEAPON SYSTEMS
　　INFORMATION OFFICE, REDSTONE ARSENAL, ALA.
ABMA DLMT-TN(NO.)(YR)
　　ARMY BALLISTIC MISSILE AGENCY. SYSTEMS SUPPORT
　　EQUIPMENT LAB., REDSTONE ARSENAL, ALA.

ABMA DLO-TM(NO)(YR)

ABMA DLO-TM(NO.)(YR)
  ARMY BALLISTIC MISSILE AGENCY. SYSTEMS SUPPORT
  EQUIPMENT LAB., REDSTONE ARSENAL, ALA.
ABMA DLS(NO.)(TN)
  ARMY BALLISTIC MISSILE AGENCY. SYSTEMS SUPPORT
  EQUIPMENT LAB., REDSTONE ARSENAL, ALA.
ABMA DLS(NO)(TR)(NO.)(YR)
  ARMY BALLISTIC MISSILE AGENCY. SYSTEMS SUPPORT
  EQUIPMENT LAB., REDSTONE ARSENAL, ALA.
ABMA DLS-TM(NO.)(DATE)
  ARMY BALLISTIC MISSILE AGENCY. SYSTEMS SUPPORT
  EQUIPMENT LAB., REDSTONE ARSENAL, ALA.
ABMA DLS-TN(NO.)(DATE)
  ARMY BALLISTIC MISSILE AGENCY. SYSTEMS SUPPORT
  EQUIPMENT LAB., REDSTONE ARSENAL, ALA.
ABMA DLS-TN(NO.)(YR)
  ARMY BALLISTIC MISSILE AGENCY. SYSTEMS SUPPORT
  EQUIPMENT LAB., REDSTONE ARSENAL, ALA.
ABMA DLS-TR-
  ARMY BALLISTIC MISSILE AGENCY. SYSTEMS SUPPORT
  EQUIPMENT LAB., REDSTONE ARSENAL, ALA.
ABMA DLS-TR(NO.)(YR)
  ARMY BALLISTIC MISSILE AGENCY. SYSTEMS SUPPORT
  EQUIPMENT LAB., REDSTONE ARSENAL, ALA.
ABMA DL-TN-
  ARMY BALLISTIC MISSILE AGENCY. LAUNCHING + HANDLING
  EQUIPMENT LAB., REDSTONE ARSENAL, ALA.
ABMA DL-TN(NO.)(YR)
  ARMY BALLISTIC MISSILE AGENCY. SYSTEMS SUPPORT
  EQUIPMENT LAB., REDSTONE ARSENAL, ALA.
ABMA DL-TR(DATE)
  ARMY BALLISTIC MISSILE AGENCY. LAUNCHING + HANDLING
  EQUIPMENT LAB., REDSTONE ARSENAL, ALA.
ABMA DMCR-TR(NO.)(YR)
  ARMY BALLISTIC MISSILE AGENCY. MISSILE FIRING LAB.,
  CAPE CANAVERAL, FLA.
ABMA DMCR-TR(NO.)(YR)
  ARMY BALLISTIC MISSILE AGENCY. MISSILE FIRING LAB.,
  REDSTONE ARSENAL, ALA.
ABMA DMG-TR(NO.)(YR)
  ARMY BALLISTIC MISSILE AGENCY. MISSILE FIRING LAB.,
  CAPE CANAVERAL, FLA.
ABMA DMG-TR(NO.)(YR)
  ARMY BALLISTIC MISSILE AGENCY. MISSILE FIRING LAB.,
  REDSTONE ARSENAL, ALA.
ABMA DMMA-TR(NO.)(YR)
  ARMY BALLISTIC MISSILE AGENCY. MISSILE FIRING LAB.,
  CAPE CANAVERAL, FLA.
ABMA DMMA-TR(NO.)(YR)
  ARMY BALLISTIC MISSILE AGENCY. MISSILE FIRING LAB.,
  REDSTONE ARSENAL, ALA.
ABMA DMM-TR(NO.)(YR)
  ARMY BALLISTIC MISSILE AGENCY. MISSILE FIRING LAB.,
  CAPE CANAVERAL, FLA.
ABMA DMM-TR(NO.)(YR)
  ARMY BALLISTIC MISSILE AGENCY. MISSILE FIRING LAB.,
  REDSTONE ARSENAL, ALA.
ABMA DMQ-TR(NO.)(YR)
  ARMY BALLISTIC MISSILE AGENCY. MISSILE FIRING LAB.,
  CAPE CANAVERAL, FLA.
ABMA DMQ-TR(NO.)(YR)
  ARMY BALLISTIC MISSILE AGENCY. MISSILE FIRING LAB.,
  REDSTONE ARSENAL, ALA.
ABMA DM-R-
  ARMY BALLISTIC MISSILE AGENCY. MISSILE FIRING LAB.,
  CAPE CANAVERAL, FLA.
ABMA DM-R-
  ARMY BALLISTIC MISSILE AGENCY. MISSILE FIRING LAB.,
  REDSTONE ARSENAL, ALA.
ABMA DM-TM-
  ARMY BALLISTIC MISSILE AGENCY. MISSILE FIRING LAB.,
  CAPE CANAVERAL, FLA.
ABMA DM-TM(NO.)(DATE)
  ARMY BALLISTIC MISSILE AGENCY. MISSILE FIRING LAB.,
  CAPE CANAVERAL, FLA.
ABMA DM-TM-
  ARMY BALLISTIC MISSILE AGENCY. MISSILE FIRING LAB.,
  REDSTONE ARSENAL, ALA.
ABMA DM-TM(NO.)(DATE)
  ARMY BALLISTIC MISSILE AGENCY. MISSILE FIRING LAB.,
  REDSTONE ARSENAL, ALA.
ABMA DM-TR(NO.)(YR)
  ARMY BALLISTIC MISSILE AGENCY. MISSILE FIRING LAB.,
  CAPE CANAVERAL, FLA.
ABMA DM-TR(NO.)(YR)
  ARMY BALLISTIC MISSILE AGENCY. MISSILE FIRING LAB.,
  REDSTONE ARSENAL, ALA.
ABMA DMTR-TM(NO.)(YR)
  ARMY BALLISTIC MISSILE AGENCY. MISSILE FIRING LAB.,
  CAPE CANAVERAL, FLA.
ABMA DMTR-TM(NO.)(YR)
  ARMY BALLISTIC MISSILE AGENCY. MISSILE FIRING LAB.,
  REDSTONE ARSENAL, ALA.
ABMA DMTR-TN(NO.)(YR)
  ARMY BALLISTIC MISSILE AGENCY. MISSILE FIRING LAB.,
  CAPE CANAVERAL, FLA.
ABMA DMTR-TN(NO.)(YR)
  ARMY BALLISTIC MISSILE AGENCY. MISSILE FIRING LAB.,
  REDSTONE ARSENAL, ALA.
ABMA DMTR-TR(NO.)(YR)
  ARMY BALLISTIC MISSILE AGENCY. MISSILE FIRING LAB.,
  CAPE CANAVERAL, FLA.
ABMA DMTR-TR(NO.)(YR)
  ARMY BALLISTIC MISSILE AGENCY. MISSILE FIRING LAB.,
  REDSTONE ARSENAL, ALA.
ABMA D-R-
  ARMY BALLISTIC MISSILE AGENCY. DEVELOPMENT OPERATIONS
  OFFICE, REDSTONE ARSENAL, ALA. (REPORT)
ABMA DRMA-TM(NO.)(YR)
  ARMY BALLISTIC MISSILE AGENCY. SYSTEMS ANALYSIS AND
  RELIABILITY LAB., REDSTONE ARSENAL, ALA.
ABMA DRM-TM(NO.)(YR)
  ARMY BALLISTIC MISSILE AGENCY. SYSTEMS ANALYSIS AND
  RELIABILITY LAB., REDSTONE ARSENAL, ALA.
ABMA DRPG-TM(NO.)(YR)
  ARMY BALLISTIC MISSILE AGENCY. SYSTEMS ANALYSIS AND
  RELIABILITY LAB., REDSTONE ARSENAL, ALA.
ABMA DRP-TM(NO.)(YR)
  ARMY BALLISTIC MISSILE AGENCY. SYSTEMS ANALYSIS AND
  RELIABILITY LAB., REDSTONE ARSENAL, ALA.
ABMA DRR-TM(NO.)(DATE)
  ARMY BALLISTIC MISSILE AGENCY. SYSTEMS ANALYSIS AND
  RELIABILITY LAB., REDSTONE ARSENAL, ALA.
ABMA DRR-TN(NO.)(DATE)
  ARMY BALLISTIC MISSILE AGENCY. SYSTEMS ANALYSIS AND
  RELIABILITY LAB., REDSTONE ARSENAL, ALA.
ABMA DRR-TR(NO.)(DATE)
  ARMY BALLISTIC MISSILE AGENCY. SYSTEMS ANALYSIS AND
  RELIABILITY LAB., REDSTONE ARSENAL, ALA.
ABMA DR-TM(NO.)(YR)
  ARMY BALLISTIC MISSILE AGENCY. SYSTEMS ANALYSIS AND
  RELIABILITY LAB., REDSTONE ARSENAL, ALA.
ABMA DRT-TM(NO.)(DATE)
  ARMY BALLISTIC MISSILE AGENCY. SYSTEMS ANALYSIS AND
  RELIABILITY LAB., REDSTONE ARSENAL, ALA.
ABMA DRT-TN(NO.)(YR)
  ARMY BALLISTIC MISSILE AGENCY. SYSTEMS ANALYSIS AND
  RELIABILITY LAB., REDSTONE ARSENAL, ALA.
ABMA DRW-TN(NO.)(YR)
  ARMY BALLISTIC MISSILE AGENCY. WILLOW PROJECT
  ENGINEERS OFFICE, REDSTONE ARSENAL, ALA.
ABMA DSA-SU MEMO-
  ARMY BALLISTIC MISSILE AGENCY. STRUCTURES AND
  MECHANICS LAB. STRUCTURAL ANALYSIS SECTION, RED-
  STONE ARSENAL, ALA.
ABMA DSA-TB-
  ARMY BALLISTIC MISSILE AGENCY. STRUCTURES AND
  MECHANICS LAB., REDSTONE ARSENAL, ALA.
ABMA DSDA-
  ARMY BALLISTIC MISSILE AGENCY. STRUCTURES AND
  MECHANICS LAB., REDSTONE ARSENAL, ALA.
ABMA DSD-TM(NO.)(YR)
  ARMY BALLISTIC MISSILE AGENCY. STRUCTURES AND
  MECHANICS LAB., REDSTONE ARSENAL, ALA.
ABMA DSD-TN(NO.)(YR)
  ARMY BALLISTIC MISSILE AGENCY. STRUCTURES AND
  MECHANICS LAB., REDSTONE ARSENAL, ALA.
ABMA DSD-TR(NO.)(YR)
  ARMY BALLISTIC MISSILE AGENCY. STRUCTURES AND
  MECHANICS LAB., REDSTONE ARSENAL, ALA.
ABMA DSF-TM(NO.)(YR)
  ARMY BALLISTIC MISSILE AGENCY. STRUCTURES AND
  MECHANICS LAB., REDSTONE ARSENAL, ALA.
ABMA DSF-TN(NO.)(YR)
  ARMY BALLISTIC MISSILE AGENCY. STRUCTURES AND
  MECHANICS LAB., REDSTONE ARSENAL, ALA.
ABMA DSF-TR(NO.)(YR)
  ARMY BALLISTIC MISSILE AGENCY. STRUCTURES AND
  MECHANICS LAB., REDSTONE ARSENAL, ALA.
ABMA DS-IN-
  ARMY BALLISTIC MISSILE AGENCY. STRUCTURES AND
  MECHANICS LAB., REDSTONE ARSENAL, ALA.
ABMA DSL-MEMO-
  ARMY BALLISTIC MISSILE AGENCY. STRUCTURES AND
  MECHANICS LAB., REDSTONE ARSENAL, ALA.
ABMA DSL-TB-
  ARMY BALLISTIC MISSILE AGENCY. STRUCTURES AND
  MECHANICS LAB., REDSTONE ARSENAL, ALA.
ABMA DSL-TB(NO.)(YR)
  ARMY BALLISTIC MISSILE AGENCY. STRUCTURES AND
  MECHANICS LAB., REDSTONE ARSENAL, ALA.
ABMA DSL-TM(NO.)(YR)
  ARMY BALLISTIC MISSILE AGENCY. STRUCTURES AND
  MECHANICS LAB., REDSTONE ARSENAL, ALA.
ABMA DSL-TN(NO.)(YR)
  ARMY BALLISTIC MISSILE AGENCY. STRUCTURES AND
  MECHANICS LAB., REDSTONE ARSENAL, ALA.
ABMA DS-M-
  ARMY BALLISTIC MISSILE AGENCY. STRUCTURES AND
  MECHANICS LAB., REDSTONE ARSENAL, ALA.
ABMA DSN-TM(NO.)(YR)
  ARMY BALLISTIC MISSILE AGENCY. STRUCTURES AND
  MECHANICS LAB., REDSTONE ARSENAL, ALA.
ABMA DSN-TN(NO.)(YR)
  ARMY BALLISTIC MISSILE AGENCY. STRUCTURES AND
  MECHANICS LAB., REDSTONE ARSENAL, ALA.
ABMA DSN-TR(NO.)(YR)
  ARMY BALLISTIC MISSILE AGENCY. STRUCTURES AND
  MECHANICS LAB., REDSTONE ARSENAL, ALA.
ABMA DSP-
  ARMY BALLISTIC MISSILE AGENCY. STRUCTURES AND
  MECHANICS LAB., REDSTONE ARSENAL, ALA.
ABMA DSP-IN-
  ARMY BALLISTIC MISSILE AGENCY. STRUCTURES AND
  MECHANICS LAB., REDSTONE ARSENAL, ALA.
ABMA DSP-TB-
  ARMY BALLISTIC MISSILE AGENCY. STRUCTURES AND
  MECHANICS LAB., REDSTONE ARSENAL, ALA.
ABMA DSP-TM(NO.)(YR)
  ARMY BALLISTIC MISSILE AGENCY. STRUCTURES AND
  MECHANICS LAB., REDSTONE ARSENAL, ALA.
ABMA-DSP-TN-
  ARMY BALLISTIC MISSILE AGENCY. STRUCTURES AND
  MECHANICS LAB., REDSTONE ARSENAL, ALA.
ABMA DSP-TN(NO.)(YR)
  ARMY BALLISTIC MISSILE AGENCY. STRUCTURES AND
  MECHANICS LAB., REDSTONE ARSENAL, ALA.

ABMA DSP-TR(NO.)(YR)
    ARMY BALLISTIC MISSILE AGENCY. STRUCTURES AND
    MECHANICS LAB., REDSTONE ARSENAL, ALA.
ABMA DS-R-
    ARMY BALLISTIC MISSILE AGENCY. STRUCTURES AND
    MECHANICS LAB., REDSTONE ARSENAL, ALA.
ABMA DSRP-TM(NO.)(YR)
    ARMY BALLISTIC MISSILE AGENCY. STRUCTURES AND
    MECHANICS LAB., REDSTONE ARSENAL, ALA.
ABMA DSR-TM(NO.)(YR)
    ARMY BALLISTIC MISSILE AGENCY. STRUCTURES AND
    MECHANICS LAB., REDSTONE ARSENAL, ALA.
ABMA DSS-TB-
    ARMY BALLISTIC MISSILE AGENCY. STRUCTURES AND
    MECHANICS LAB., REDSTONE ARSENAL, ALA.
ABMA DS-TB-
    ARMY BALLISTIC MISSILE AGENCY. STRUCTURES AND
    MECHANICS LAB., REDSTONE ARSENAL, ALA.
ABMA DS-TM-
    ARMY BALLISTIC MISSILE AGENCY. STRUCTURES AND
    MECHANICS LAB., REDSTONE ARSENAL, ALA.
ABMA DS-TM(NO.)(YR)
    ARMY BALLISTIC MISSILE AGENCY. STRUCTURES AND
    MECHANICS LAB., REDSTONE ARSENAL, ALA.
ABMA DS-TN-
    ARMY BALLISTIC MISSILE AGENCY. STRUCTURES AND
    MECHANICS LAB., REDSTONE ARSENAL, ALA.
ABMA DT-EDN-
    ARMY BALLISTIC MISSILE AGENCY. TEST LAB., REDSTONE
    ARSENAL, ALA.
ABMA DT-EM-
    ARMY BALLISTIC MISSILE AGENCY. TEST LAB., REDSTONE
    ARSENAL, ALA.
ABMA DTI-TR(NO.)(YR)
    ARMY BALLISTIC MISSILE AGENCY. TEST LAB., REDSTONE
    ARSENAL, ALA.
ABMA DTM(NO)M1CC
    ARMY BALLISTIC MISSILE AGENCY. TEST LAB., REDSTONE
    ARSENAL, ALA.
ABMA DTM(NO)M1CT
    ARMY BALLISTIC MISSILE AGENCY. TEST LAB., REDSTONE
    ARSENAL, ALA.
ABMA DTM ST-TN(NO.)(YR)
    ARMY BALLISTIC MISSILE AGENCY. TEST LAB., REDSTONE
    ARSENAL, ALA.
ABMA D-TN(NO.)(YR)
    ARMY BALLISTIC MISSILE AGENCY. DEVELOPMENT OPERATIONS
    OFFICE, REDSTONE ARSENAL, ALA. (TECHNICAL NOTE)
ABMA DT-R-
    ARMY BALLISTIC MISSILE AGENCY. TEST LAB., REDSTONE
    ARSENAL, ALA.
ABMA DTR-TM(NO.)(YR)
    ARMY BALLISTIC MISSILE AGENCY. TEST LAB., REDSTONE
    ARSENAL, ALA.
ABMA DTR-TR(NO.)(YR)
    ARMY BALLISTIC MISSILE AGENCY. TEST LAB., REDSTONE
    ARSENAL, ALA.
ABMA DT-TM(NO.)(YR)
    ARMY BALLISTIC MISSILE AGENCY. TEST LAB., REDSTONE
    ARSENAL, ALA.
ABMA DT-TN(NO.)(YR)
    ARMY BALLISTIC MISSILE AGENCY. TEST LAB., REDSTONE
    ARSENAL, ALA.
ABMA DT-TR(NO.)(YR)
    ARMY BALLISTIC MISSILE AGENCY. TEST LAB., REDSTONE
    ARSENAL, ALA.
ABMA DV-TM(NO.)(DATE)
    ARMY BALLISTIC MISSILE AGENCY. RESEARCH PROJECTS
    LAB., REDSTONE ARSENAL, ALA.
ABMA DV-TN(NO.)(DATE)
    ARMY BALLISTIC MISSILE AGENCY. RESEARCH PROJECTS
    LAB., REDSTONE ARSENAL, ALA.
ABMA DV-TN-
    ARMY BALLISTIC MISSILE AGENCY. RESEARCH PROJECTS
    OFFICE, REDSTONE ARSENAL, ALA.
ABMA-DV-TR-
    ARMY BALLISTIC MISSILE AGENCY, REDSTONE ARSENAL, ALA.
ABMA DV-TR(NO.)(YR)
    ARMY BALLISTIC MISSILE AGENCY. RESEARCH PROJECTS
    OFFICE, REDSTONE ARSENAL, ALA.
ABMA DYR-R-
    ARMY BALLISTIC MISSILE AGENCY. TECHNICAL LIAISON
    GROUP OF DOD, REDSTONE ARSENAL, ALA.
ABMA FIELD OFF TLM-
    ARMY BALLISTIC MISSILE AGENCY. FIELD OFFICE, LOS
    ANGELES, CALIF. (TECHNICAL LIAISON MEMO)
ABMA G+C CONTROL NOTE-
    ARMY BALLISTIC MISSILE AGENCY. GUIDANCE AND CONTROL
    LAB., REDSTONE ARSENAL, ALA.
ABMA G+C DEV RPT-
    ARMY BALLISTIC MISSILE AGENCY. GUIDANCE AND CONTROL
    LAB., REDSTONE ARSENAL, ALA.
ABMA G+C-IN-
    ARMY BALLISTIC MISSILE AGENCY. GUIDANCE AND CONTROL
    LAB., REDSTONE ARSENAL, ALA.
ABMA G+C MEMO-
    ARMY BALLISTIC MISSILE AGENCY. GUIDANCE AND CONTROL
    LAB., REDSTONE ARSENAL, ALA.
ABMA G+C MEMO-GS(SP)-T-
    ARMY BALLISTIC MISSILE AGENCY. GUIDANCE AND CONTROL
    LAB., REDSTONE ARSENAL, ALA.
ABMA G+C-TM-
    ARMY BALLISTIC MISSILE AGENCY. GUIDANCE AND CONTROL
    LAB., REDSTONE ARSENAL, ALA.
ABMA G+C-TN-
    ARMY BALLISTIC MISSILE AGENCY. GUIDANCE AND CONTROL
    LAB., REDSTONE ARSENAL, ALA.
ABMA GS(SP)-T-
    ARMY BALLISTIC MISSILE AGENCY. GUIDANCE AND CONTROL
    LAB., REDSTONE ARSENAL, ALA.
ABMA GS-T-
    ARMY BALLISTIC MISSILE AGENCY. GUIDANCE AND CONTROL
    LAB., REDSTONE ARSENAL, ALA.
ABMA HT-
    ARMY BALLISTIC MISSILE AGENCY. LIBRARY, REDSTONE
    ARSENAL, ALA. (ASSIGNED TO REPORTS NOT IDENTIFIED
    BY ANY OTHER SERIES CODE)
ABMA JUPITER TECH BULL-
    ARMY BALLISTIC MISSILE AGENCY, REDSTONE ARSENAL, ALA.
ABMA-M-LOD-
    ARMY BALLISTIC MISSILE AGENCY, REDSTONE ARSENAL, ALA.
ABMA-MTP-LOD-
    ARMY BALLISTIC MISSILE AGENCY, REDSTONE ARSENAL, ALA.
ABMA-OH-(NO.)
    ARMY BALLISTIC MISSILE AGENCY, REDSTONE ARSENAL, ALA.
ABMA ORDAB-CR-
    ARMY BALLISTIC MISSILE AGENCY. CONTROL OFFICE,
    REDSTONE ARSENAL, ALA. (ORDNANCE ARMY BALLISTIC)
    (LETTERS ARE OFFICE SYMBOLS)
ABMA ORDAB-DAA-
    ARMY BALLISTIC MISSILE AGENCY. AEROBALLISTICS LAB.,
    REDSTONE ARSENAL, ALA.
ABMA ORDAB-DAE(NO.)(YR)
    ARMY BALLISTIC MISSILE AGENCY. AEROBALLISTICS LAB.,
    REDSTONE ARSENAL, ALA.
ABMA ORDAB-DAT-
    ARMY BALLISTIC MISSILE AGENCY. AEROBALLISTICS LAB.,
    REDSTONE ARSENAL, ALA.
ABMA ORDAB-DIR-SPU(NO)(YR)
    ARMY BALLISTIC MISSILE AGENCY. TEST LAB., REDSTONE
    ARSENAL, ALA.
ABMA-ORDAB-DM-(LETTERS)-
    ARMY BALLISTIC MISSILE AGENCY. MISSILE FIRING LAB.,
    CAPE CANAVERAL, FLA.
ABMA ORDAB-DR-
    ARMY BALLISTIC MISSILE AGENCY. SYSTEMS ANALYSIS AND
    RELIABILITY LAB., REDSTONE ARSENAL, ALA.
ABMA ORDAB-DSA-DU MEMO-
    ARMY BALLISTIC MISSILE AGENCY. STRUCTURES AND
    MECHANICS LAB. STRUCTURAL ANALYSIS SECTION, RED-
    STONE ARSENAL, ALA.
ABMA ORDAB-DSA MEMO-
    ARMY BALLISTIC MISSILE AGENCY. STRUCTURES AND
    MECHANICS LAB., REDSTONE ARSENAL, ALA.
ABMA ORDAB-DSD-
    ARMY BALLISTIC MISSILE AGENCY. STRUCTURES AND
    MECHANICS LAB., REDSTONE ARSENAL, ALA.
ABMA ORDAB-DSDD-
    ARMY BALLISTIC MISSILE AGENCY. STRUCTURES AND
    MECHANICS LAB., REDSTONE ARSENAL, ALA.
ABMA ORDAB-DSDE-
    ARMY BALLISTIC MISSILE AGENCY. STRUCTURES AND
    MECHANICS LAB., REDSTONE ARSENAL, ALA.
ABMA ORDAB-DSDF-
    ARMY BALLISTIC MISSILE AGENCY. STRUCTURES AND
    MECHANICS LAB., REDSTONE ARSENAL, ALA.
ABMA ORDAB-DSDF(NO.)(YR)
    ARMY BALLISTIC MISSILE AGENCY. STRUCTURES AND
    MECHANICS LAB., REDSTONE ARSENAL, ALA.
ABMA ORDAB-DSDG-
    ARMY BALLISTIC MISSILE AGENCY. STRUCTURES AND
    MECHANICS LAB., REDSTONE ARSENAL, ALA.
ABMA ORDAB-DSE-
    ARMY BALLISTIC MISSILE AGENCY. STRUCTURES AND
    MECHANICS LAB., REDSTONE ARSENAL, ALA.
ABMA ORDAB-DSL-
    ARMY BALLISTIC MISSILE AGENCY. STRUCTURES AND
    MECHANICS LAB., REDSTONE ARSENAL, ALA.
ABMA ORDAB-DSLA-MEMO-
    ARMY BALLISTIC MISSILE AGENCY. STRUCTURES AND
    MECHANICS LAB., REDSTONE ARSENAL, ALA.
ABMA ORDAB-DSLB-MEMO-
    ARMY BALLISTIC MISSILE AGENCY. STRUCTURES AND
    MECHANICS LAB., REDSTONE ARSENAL, ALA.
ABMA ORDAB-DSLC-MEMO-
    ARMY BALLISTIC MISSILE AGENCY. STRUCTURES AND
    MECHANICS LAB., REDSTONE ARSENAL, ALA.
ABMA ORDAB-DSLE-MEMO-
    ARMY BALLISTIC MISSILE AGENCY. STRUCTURES AND
    MECHANICS LAB., REDSTONE ARSENAL, ALA.
ABMA ORDAB-DSL MEMO-
    ARMY BALLISTIC MISSILE AGENCY. STRUCTURES AND
    MECHANICS LAB., REDSTONE ARSENAL, ALA.
ABMA ORDAB-DSL-TB(NO)(YR)
    ARMY BALLISTIC MISSILE AGENCY. STRUCTURES AND
    MECHANICS LAB., REDSTONE ARSENAL, ALA.
ABMA ORDAB-DSM-
    ARMY BALLISTIC MISSILE AGENCY. STRUCTURES AND
    MECHANICS LAB., REDSTONE ARSENAL, ALA.
ABMA ORDAB-DSS-TB-
    ARMY BALLISTIC MISSILE AGENCY. STRUCTURES AND
    MECHANICS LAB., REDSTONE ARSENAL, ALA.
ABMA ORDAB-DTM-SPU-
    ARMY BALLISTIC MISSILE AGENCY. TEST LAB., REDSTONE
    ARSENAL, ALA.
ABMA ORDAB-SE-
    ARMY BALLISTIC MISSILE AGENCY. FIELD SUPPORT
    DIVISION, REDSTONE ARSENAL, ALA.
ABMA ORDAB-SM-MSL-
    ARMY BALLISTIC MISSILE AGENCY. FIELD SUPPORT
    OPERATIONS, REDSTONE ARSENAL, ALA. (MAINTENANCE
    SERVICE LETTER)
ABMA PD-H-
    ARMY BALLISTIC MISSILE AGENCY, REDSTONE ARSENAL, ALA.
    (PURCHASE DESCRIPTION)
ABMA PD-P-
    ARMY BALLISTIC MISSILE AGENCY, REDSTONE ARSENAL, ALA.
ABMA PD-R-
    ARMY BALLISTIC MISSILE AGENCY, REDSTONE ARSENAL, ALA.

ABMA PD-S-
: ARMY BALLISTIC MISSILE AGENCY, REDSTONE ARSENAL, ALA.

ABMA PD-V-
: ARMY BALLISTIC MISSILE AGENCY, REDSTONE ARSENAL, ALA.

ABMA PD-W-
: ARMY BALLISTIC MISSILE AGENCY, REDSTONE ARSENAL, ALA.

ABMA PG-CD-MEMO-
: ARMY BALLISTIC MISSILE AGENCY. STRUCTURES AND MECHANICS LAB., REDSTONE ARSENAL, ALA.

ABMA PG-CE-MEMO-
: ARMY BALLISTIC MISSILE AGENCY. STRUCTURES AND MECHANICS LAB., REDSTONE ARSENAL, ALA.

ABMA PG-CH-MEMO-
: ARMY BALLISTIC MISSILE AGENCY. STRUCTURES AND MECHANICS LAB., REDSTONE ARSENAL, ALA.

ABMA-RA-TR-
: ARMY BALLISTIC MISSILE AGENCY, REDSTONE ARSENAL, ALA.

ABMA R-DS-
: ARMY BALLISTIC MISSILE AGENCY, REDSTONE ARSENAL, ALA.

ABMA-RE-TR-
: ARMY BALLISTIC MISSILE AGENCY, REDSTONE ARSENAL, ALA.

ABMA-RG-TR-
: ARMY BALLISTIC MISSILE AGENCY, REDSTONE ARSENAL, ALA.

ABMA-RJ-TR-
: ARMY BALLISTIC MISSILE AGENCY, REDSTONE ARSENAL, ALA.

ABMA RPO-D-TM-
: ARMY BALLISTIC MISSILE AGENCY. RESEARCH PROJECTS OFFICE, REDSTONE ARSENAL, ALA.

ABMA RPO-P-TN-
: ARMY BALLISTIC MISSILE AGENCY. RESEARCH PROJECTS OFFICE, REDSTONE ARSENAL, ALA.

ABMA RPO-RB-
: ARMY BALLISTIC MISSILE AGENCY. RESEARCH PROJECTS OFFICE, REDSTONE ARSENAL, ALA. (RESEARCH BULLETIN)

ABMA RPO-RS-TN-
: ARMY BALLISTIC MISSILE AGENCY. RESEARCH PROJECTS OFFICE, REDSTONE ARSENAL, ALA.

ABMA RPO-T-TN-
: ARMY BALLISTIC MISSILE AGENCY. RESEARCH PROJECTS OFFICE, REDSTONE ARSENAL, ALA.

ABMA-RP-TR-
: ARMY BALLISTIC MISSILE AGENCY, REDSTONE ARSENAL, ALA.

ABMA-RR-R-DS-
: ARMY BALLISTIC MISSILE AGENCY, REDSTONE ARSENAL, ALA.

ABMA RR-TM(NO.)(YR)
: ARMY BALLISTIC MISSILE AGENCY. RESEARCH LAB., REDSTONE ARSENAL, ALA.

ABMA-RR-TR-
: ARMY BALLISTIC MISSILE AGENCY, REDSTONE ARSENAL, ALA.

ABMA-RS-TN-
: ARMY BALLISTIC MISSILE AGENCY, REDSTONE ARSENAL, ALA.

ABMA-RS-TR-
: ARMY BALLISTIC MISSILE AGENCY, REDSTONE ARSENAL, ALA.

ABMA-RTLA-TM-(YEAR)-...
: ARMY BALLISTIC MISSILE AGENCY, REDSTONE ARSENAL, ALA.

ABMA RT-TM(NO.)(YR)
: ARMY BALLISTIC MISSILE AGENCY. TEST EVALUATION AND FIRING LAB., REDSTONE ARSENAL, ALA.

ABMA-RT-TR-
: ARMY BALLISTIC MISSILE AGENCY, REDSTONE ARSENAL, ALA.

ABMA SA+RL MEMO-
: ARMY BALLISTIC MISSILE AGENCY. SYSTEMS ANALYSIS AND RELIABILITY LAB., REDSTONE ARSENAL, ALA.

ABMA SA+RL-QR-
: ARMY BALLISTIC MISSILE AGENCY. SYSTEMS ANALYSIS AND RELIABILITY LAB., REDSTONE ARSENAL, ALA.

ABMA SM/EA MEMO-
: ARMY BALLISTIC MISSILE AGENCY. STRUCTURES AND MECHANICS LAB., REDSTONE ARSENAL, ALA.

ABMA S+M-TB-G-
: ARMY BALLISTIC MISSILE AGENCY. STRUCTURES AND MECHANICS LAB., REDSTONE ARSENAL, ALA.

ABMA S+M-TDS-
: ARMY BALLISTIC MISSILE AGENCY. STRUCTURES AND MECHANICS LAB., REDSTONE ARSENAL, ALA.

ABMA S+M TEST MEMO-
: ARMY BALLISTIC MISSILE AGENCY. STRUCTURES AND MECHANICS LAB., REDSTONE ARSENAL, ALA.

ABMA S+M TEST REPORT-
: ARMY BALLISTIC MISSILE AGENCY. STRUCTURES AND MECHANICS LAB., REDSTONE ARSENAL, ALA.

ABMA S+M-TM-G-
: ARMY BALLISTIC MISSILE AGENCY. STRUCTURES AND MECHANICS LAB., REDSTONE ARSENAL, ALA.

ABMA S+M-TN-B-
: ARMY BALLISTIC MISSILE AGENCY. STRUCTURES AND MECHANICS LAB., REDSTONE ARSENAL, ALA.

ABMA S+M-TN-C-
: ARMY BALLISTIC MISSILE AGENCY. STRUCTURES AND MECHANICS LAB., REDSTONE ARSENAL, ALA.

ABMA S+M-TN-DDS-
: ARMY BALLISTIC MISSILE AGENCY. STRUCTURES AND MECHANICS LAB., REDSTONE ARSENAL, ALA.

ABMA S+M-TN-DSD-
: ARMY BALLISTIC MISSILE AGENCY. STRUCTURES AND MECHANICS LAB., REDSTONE ARSENAL, ALA.

ABMA S+M-TN-DSP-
: ARMY BALLISTIC MISSILE AGENCY. STRUCTURES AND MECHANICS, REDSTONE ARSENAL, ALA.

ABMA S+M-TN-E-
: ARMY BALLISTIC MISSILE AGENCY. STRUCTURES AND MECHANICS LAB., REDSTONE ARSENAL, ALA.

ABMA S+M-TN-G-
: ARMY BALLISTIC MISSILE AGENCY. STRUCTURES AND MECHANICS LAB., REDSTONE ARSENAL, ALA.

ABMA S+M-TN-H-
: ARMY BALLISTIC MISSILE AGENCY. STRUCTURES AND MECHANICS LAB., REDSTONE ARSENAL, ALA.

ABMA S+M-TN-K-
: ARMY BALLISTIC MISSILE AGENCY. STRUCTURES AND MECHANICS LAB., REDSTONE ARSENAL, ALA.

ABMA S+M-TN-M-
: ARMY BALLISTIC MISSILE AGENCY. STRUCTURES AND MECHANICS LAB., REDSTONE ARSENAL, ALA.

ABMA S+M-TN-N-
: ARMY BALLISTIC MISSILE AGENCY. STRUCTURES AND MECHANICS LAB., REDSTONE ARSENAL, ALA.

ABMA S+M-TN-T-
: ARMY BALLISTIC MISSILE AGENCY. STRUCTURES AND MECHANICS LAB., REDSTONE ARSENAL, ALA.

ABMA SP RPT RP-
: ARMY BALLISTIC MISSILE AGENCY, REDSTONE ARSENAL, ALA.

ABMA TDL BIBLIO-
: ARMY BALLISTIC MISSILE AGENCY. TECHNICAL DOCUMENTS LIBRARY, REDSTONE ARSENAL, ALA.

ABMA TDL-TAL-
: ARMY BALLISTIC MISSILE AGENCY. TECHNICAL DOCUMENTS LIBRARY, REDSTONE ARSENAL, ALA. (TITLE ANNOUNCEMENT LIST)

ABMA TFS-
: ARMY BALLISTIC MISSILE AGENCY. TEST LAB., REDSTONE ARSENAL, ALA.

ABMA TN-DLE(NO.)(YR)
: ARMY BALLISTIC MISSILE AGENCY. SYSTEMS SUPPORT EQUIPMENT LAB., REDSTONE ARSENAL, ALA.

ABMA TN-DLF(NO.)(YR)
: ARMY BALLISTIC MISSILE AGENCY. LAUNCHING + HANDLING EQUIPMENT LAB., REDSTONE ARSENAL, ALA.

ABMA TN-DLG(NO.)(YR)(TN)
: ARMY BALLISTIC MISSILE AGENCY. LAUNCHING + HANDLING EQUIPMENT LAB., REDSTONE ARSENAL, ALA.

ABMA TN-DLM(NO.)(YR)
: ARMY BALLISTIC MISSILE AGENCY. LAUNCHING + HANDLING EQUIPMENT LAB., REDSTONE ARSENAL, ALA.

ABMA TN-DLMF(NO.)(YR)
: ARMY BALLISTIC MISSILE AGENCY. SYSTEMS SUPPORT EQUIPMENT LAB., REDSTONE ARSENAL, ALA.

ABMA TN-DLP(NO.)(YR)
: ARMY BALLISTIC MISSILE AGENCY. LAUNCHING + HANDLING EQUIPMENT LAB., REDSTONE ARSENAL, ALA.

ABMA TN-DLS(NO.)(YR)(TN)
: ARMY BALLISTIC MISSILE AGENCY. SYSTEMS SUPPORT EQUIPMENT LAB., REDSTONE ARSENAL, ALA.

ABMA TN-DSD-
: ARMY BALLISTIC MISSILE AGENCY. STRUCTURES AND MECHANICS, REDSTONE ARSENAL, ALA.

ABMA TN-DSF-
: ARMY BALLISTIC MISSILE AGENCY. STRUCTURES AND MECHANICS LAB., REDSTONE ARSENAL, ALA.

ABMA TP-DLF(NO.)(YR)
: ARMY BALLISTIC MISSILE AGENCY. LAUNCHING + HANDLING EQUIPMENT LAB., REDSTONE ARSENAL, ALA.

ABMA TR(NO.)(YR)
: ARMY BALLISTIC MISSILE AGENCY. SYSTEMS SUPPORT EQUIPMENT LAB., REDSTONE ARSENAL, ALA.

ABMA TR-DLD(NO.)(YR)
: ARMY BALLISTIC MISSILE AGENCY. LAUNCHING + HANDLING EQUIPMENT LAB., REDSTONE ARSENAL, ALA.

ABMA TR-DLE(NO.)(YR)
: ARMY BALLISTIC MISSILE AGENCY. LAUNCHING + HANDLING EQUIPMENT LAB., REDSTONE ARSENAL, ALA.

ABMA TR-DLF(NO.)(YR)
: ARMY BALLISTIC MISSILE AGENCY. LAUNCHING + HANDLING EQUIPMENT LAB., REDSTONE ARSENAL, ALA.

ABMA TR-DLJ(NO.)(YR)
: ARMY BALLISTIC MISSILE AGENCY. LAUNCHING + HANDLING EQUIPMENT LAB., REDSTONE ARSENAL, ALA.

ABMA TR-DLM-
: ARMY BALLISTIC MISSILE AGENCY. LAUNCHING + HANDLING EQUIPMENT LAB., REDSTONE ARSENAL, ALA.

ABMA TR-DLM(NO.)(YR)
: ARMY BALLISTIC MISSILE AGENCY. LAUNCHING + HANDLING EQUIPMENT LAB., REDSTONE ARSENAL, ALA.

ABMA TR-DLMF(NO.)(YR)
: ARMY BALLISTIC MISSILE AGENCY. SYSTEMS SUPPORT EQUIPMENT LAB., REDSTONE ARSENAL, ALA.

ABMA TR-DLMP(NO.)(YR)
: ARMY BALLISTIC MISSILE AGENCY. SYSTEMS SUPPORT EQUIPMENT LAB., REDSTONE ARSENAL, ALA.

ABMA TR-DLMT(NO.)(YR)
: ARMY BALLISTIC MISSILE AGENCY. SYSTEMS SUPPORT EQUIPMENT LAB., REDSTONE ARSENAL, ALA.

ABMA TR-DLP(NO.)(YR)
: ARMY BALLISTIC MISSILE AGENCY. LAUNCHING + HANDLING EQUIPMENT LAB., REDSTONE ARSENAL, ALA.

ABMA TR-DLS(NO.)(YR)
: ARMY BALLISTIC MISSILE AGENCY. LAUNCHING + HANDLING EQUIPMENT LAB., REDSTONE ARSENAL, ALA.

ABMA TR-DLT(NO.)(YR)
: ARMY BALLISTIC MISSILE AGENCY. LAUNCHING + HANDLING EQUIPMENT LAB., REDSTONE ARSENAL, ALA.

ABMA-XMC-
: ARMY BALLISTIC MISSILE AGENCY, REDSTONE ARSENAL, ALA.

ABM-NB-
: ARMED FORCES SPECIAL WEAPONS PROJECT. FIELD COMMAND, ALBUQUERQUE, N. MEX. (NOTEBOOK)

ABN-
: AIRBORNE-QUARTERMASTER RESEARCH + ENG. COMMAND (ARMY)

AB-NP-
: NATIONAL AERONAUTICS AND SPACE ADM., WASHINGTON, D.C. (TRANSLATOR'S INITIALS) (CATEGORY)

AB-R-
: COLUMBIA UNIV., NYC. SUBSTITUTE ALLOY MATERIALS LABS.

ABRL-
: AMERICAN BIOPHYSICS RESEARCH LAB., LANSDALE, PA.

ABS-THH-
: HANOVER. TECHNISCHE HOCHSCHULE

AC-	ACCUMETRICS CORP., CAMBRIDGE, MASS.
AC(NO.)/(CONTRACT NO.)	ADMIRAL CORP., CHICAGO
AC-	AEROCHEM RESEARCH LABS., INC., PRINCETON, N.J.
AC-(NUMBER)-P	AEROCHEM RESEARCH LABS., INC., PRINCETON, N.J. (PROJECT SQUID)
AC-(NUMBER)-PU	AEROCHEM RESEARCH LABS., INC., PRINCETON, N.J.
AC-	AMERICAN BOSCH ARMA CORP. ARMA DIV., GARDEN CITY, N.Y.
AC-	AMHERST COLL., MASS.
AC-	BUREAU OF AERONAUTICS (NAVY). AIRCRAFT DIV.
AC-	CALIFORNIA RESEARCH CORP., RICHMOND
AC-	COLUMBIA UNIV., NYC. SUBSTITUTE ALLOY MATERIALS LABS.
AC(YEAR)-	FEDERAL AVIATION ADM. AERONAUTICAL CTR., OKLAHOMA CITY
AC-(SUBJECT NUMBER)-	FEDERAL AVIATION ADMINISTRATION. FLIGHT STANDARDS SERVICE, WASHINGTON, D.C. (ADVISORY CIRCULAR)
AC-(3 DIGITS-YEAR)	GENERAL DYNAMICS/ELECTRONICS. ACOUSTICS DEPT., ROCHESTER, N.Y.
A.C.-	GT. BRIT. ADVISORY COUNCIL ON SCI. RES. + TECH. DEV.
AC-	GT. BRIT. ADVISORY COUNCIL ON SCI. RES. + TECH. DEV.
AC-	GT. BRIT. ARMAMENT RES. DEPT., FT. HALSTEAD, KENT, ENG.
AC-	GT. BRIT. MINISTRY OF SUPPLY
AC-	NATIONAL ADVISORY COMMITTEE FOR AERONAUTICS (AIRCRAFT CIRCULAR) (REF. 25)
AC-	NAVAL AIR DEV. CTR. AERO. COMPUTER LAB., JOHNSVILLE, PA.
AC-(NUMBER)-PU	PURDUE UNIV., LAFAYETTE, IND.
AC-	TENNESSEE EASTMAN CORP., OAK RIDGE, TENN.
AC-	VIRGINIA. UNIV., CHARLOTTESVILLE
ACA-	ALUMINUM CO. OF AMERICA, NEW KENSINGTON, PA.
ACA-	AUSTRALIAN COUNCIL FOR AERONAUTICS, MELBOURNE
ACA-ARL-QPR-	ALUMINUM CO. OF AMERICA. ALCOA RESEARCH LABS., NEW KENSINGTON, PENNA. (PROGRESS REPORT)
ACA-ARL-TP-	ALUMINUM CO. OF AMERICA. ALCOA RESEARCH LABS., NEW KENSINGTON, PENNA. (TECHNICAL PAPER)
ACAMR-	NATIONAL RESEARCH COUNCIL OF CANADA. ASSOCIATE COMM. ON AVIATION MEDICAL RESEARCH
ACAMR NO. 1 CIU-	NATL. RES. COUNCIL OF CANADA. ASSOCIATE COMM. ON AVIATION MEDICAL RES. NO. 1 CLINICAL INVESTIGATION UNIT (RCAF) (LETTERED REPORTS)
ACB EM-	CAMP COLES SIGNAL LABORATORY. APPLIED COMMUNICATIONS BRANCH, BELMAR, N.J. (ENGINEERING MEMORANDUM)
ACC-	AEROJET-GENERAL CORP., SACRAMENTO, CALIF. (REVISION)
ACC-	AIR COORDINATING COMMITTEE, WASHINGTON, D.C.
ACC-	ALLIED CHEMICAL CORP., MORRISTOWN, N.J.
ACC-	AMERICAN CYANAMID CO., N.Y.C.
ACC-	ARBEITSGEMEINSCHAFT BBC-KRUPP, MANNHEIM, GERMANY
ACC-	NATIONAL RESEARCH COUNCIL OF CANADA. DIV. OF BUILDING RESEARCH, OTTAWA (CBA FRENCH EDITION)
ACC-	NEW YORK UNIV., N.Y.C. INST. FOR MATH. + MECHANICS
ACC-	UNITED KINGDOM. ADMIRALITY CORROSION COMMISSION, LONDON
ACC/H-	UNITED KINGDOM. ADMIRALTY CORROSION COMMISSION, LONDON
ACCL-	AMERICAN COUNCIL OF COMMERCIAL LABS., WASH., D.C.
ACCO-	AMERICAN CYANAMID CO. (ALL LOCATIONS)
ACCO-AR-(YEAR)	AMERICAN CYANAMID CO., STAMFORD, CONN. (ANNUAL RPT.)
ACCO-ISPP-PR-	AMERICAN CYANAMID CO., STAMFORD, CONN. (INTEGRATED SOLID PROPELLANT PROGRAM PROGRESS REPORT)
ACCO-PR-	AMERICAN CYANAMID CO., STAMFORD, CONN.
ACCO-TR-	AMERICAN CYANAMID CO., STAMFORD, CONN. (TECH. REPT.)
ACC-PR-	ALLIED CHEM. CORP. GEN. CHEM. DIV., MORRISTOWN, N.J.
ACC/U.(NUMBER)/(YEAR)	GT. BRIT. COMMITTEE FOR THE PREVENTION OF CORROSION AND FOULING, LONDON
ACD-	DEPARTMENT OF STATE. OFFICE OF EXTERNAL RESEARCH
ACD-	GENERAL ELECTRIC CO. AVIONIC CONTROLS DEPT., BINGHAMTON, N.Y.
ACDA-(NUMBER)-(YEAR)	ARMS CONTROL AND DISARMAMENT AGENCY, WASHINGTON, D.C.
ACDA/E-	ARMS CONTROL AND DISARMAMENT AGENCY, WASHINGTON, D.C.
ACDA-IR-(NUMBERS)	ARMS CONTROL AND DISARMAMENT AGENCY, WASHINGTON, D.C.
ACDA-PUB-	ARMS CONTROL AND DISARMAMENT AGENCY, WASHINGTON, D.C.
ACDA-RR-(YEAR)-	ARMS CONTROL AND DISARMAMENT AGENCY, WASHINGTON, D.C.
ACDA-ST-	ARMS CONTROL AND DISARMAMENT AGENCY, WASHINGTON, D.C.
ACDA/WEC-(YEAR)-	ARMS CONTROL AND DISARMAMENT AGENCY. WEAPONS EVALUATION AND CONTROL BUREAU, WASHINGTON, D.C.
ACDA/WEC/FO-(YEAR)-	ARMS CONTROL AND DISARMAMENT AGENCY. WEAPONS EVALUATION AND CONTROL BUREAU, WASHINGTON, D.C.
ACDCAA-	ARMY COMBAT DEVELOPMENTS COMMAND. AVIATION AGENCY, FORT RUCKER, ALA.
ACDC-CBR-	ARMY COMBAT DEVELOPMENTS COMMAND. CHEMICAL-BIOLOGICAL-RADIOLOGICAL AGENCY, FORT MC CLELLAN, ALA.
ACDC-NG-	ARMY COMBAT DEVS. COMMAND. NUCLEAR GP., FT. BLISS, TEX.
ACDCQA-	ARMY COMBAT DEVELOPMENTS COMMAND. QUARTERMASTER AGENCY, FORT LEE, VA.
ACDC-SWCAG-	ARMY COMBAT DEVELOPMENTS COMMAND. SPECIAL WARFARE AND CIVIL AFFAIRS GROUP, FORT BELVOIR, VA.
AC-DRL-CTN(YEAR)-	AC ELECTRONICS-DEFENSE RES. LABS., SANTA BARBARA, CAL.
AC-DRL-S(YEAR)-	AC ELECTRONICS-DEFENSE RES. LABS., SANTA BARBARA, CAL.
AC-DRL-TR(YEAR)-	AC ELECTRONICS-DEFENSE RES. LABS., SANTA BARBARA, CAL.
ACERC-	ARMY COASTAL ENGINEERING RES. CENTER, WASHINGTON, DC
ACF-	ACF INDUSTRIES, INC. (ALL LOCATIONS)
ACFEL-BL-	ARCTIC CONSTRUCTION AND FROST EFFECTS LAB., BOSTON
ACFEL-MISC.PAPER-	ARCTIC CONSTRUCTION AND FROST EFFECTS LAB., BOSTON
ACFEL-TR-	ARCTIC CONSTRUCTION AND FROST EFFECTS LAB., BOSTON
ACFEL-TRANS-	ARCTIC CONSTRUCTION AND FROST EFFECTS LAB., BOSTON
ACF-EPL-	ACF INDUSTRIES, INC. ELECTRO-PHYSICS LABS., HYATTSVILLE, MD.
ACF-ERR-	NUCLEAR PRODUCTS-ERCO DIV., ACF INDUSTRIES, INC., WASHINGTON, D.C.
ACF-GCPR-	NUCLEAR PRODUCTS-ERCO DIV., ACF INDUSTRIES, INC., WASHINGTON, D.C.
ACFI-	ACF INDUSTRIES, INC., ALBUQUERQUE, N. MEX.
ACFI-C-	ACF INDUSTRIES, INC., ALBUQUERQUE, N. MEX.
ACFI-HATE-	ACF INDUSTRIES, INC., ALBUQUERQUE, N. MEX.
ACF-SA-	ACF INDUSTRIES, INC. ALBUQUERQUE DIV., N. MEX.
ACGSC-	ARMY COMMAND + GEN. STAFF SCH., FT. LEAVENWORTH, KAN.
ACIC-	AERONAUTICAL CHART AND INFORMATION CENTER, ST. LOUIS
ACIC-LD-	AERONAUTICAL CHART AND INFORMATION CENTER, ST. LOUIS
ACIC-REF-PUB-	AERONAUTICAL CHART AND INFORMATION CENTER, ST. LOUIS
ACIC-RM-	AERONAUTICAL CHART AND INFORMATION CENTER, ST. LOUIS (REFERENCE MATERIAL)
ACIC-RP-	AERONAUTICAL CHART AND INFORMATION CENTER, ST. LOUIS (REFERENCE PUBLICATION)
ACIC-SP-	AERONAUTICAL CHART AND INFORMATION CENTER, ST. LOUIS (SPECIAL PUBLICATION)
ACIC-STUDY-	AERONAUTICAL CHART AND INFORMATION CENTER, ST. LOUIS
ACIC-STUDY-TC-	AERONAUTICAL CHART AND INFORMATION CENTER, ST. LOUIS
ACIC-TC-	AERONAUTICAL CHART AND INFORMATION CENTER, ST. LOUIS (TRANSLATION CODE)
ACIC-TECHNICAL PAPER-	AERONAUTICAL CHART AND INFORMATION CENTER, ST. LOUIS
ACIC-TI-	AERONAUTICAL CHART AND INFORMATION CENTER, ST. LOUIS (TECHNICAL INSTRUCTION)
ACIC-TM-	AERONAUTICAL CHART AND INFORMATION CENTER, ST. LOUIS (TECHNICAL MANUAL)
ACIC-TP-	AERONAUTICAL CHART AND INFORMATION CENTER, ST. LOUIS (TECHNICAL PAPER)
ACIC-TR-	AERONAUTICAL CHART AND INFORMATION CENTER, ST. LOUIS (TECHNICAL REPORT)
ACL-	ACTON LABS., INC., MASS.

AC-L-

AC-L-	COLUMBIA UNIV., NYC. SUBSTITUTE ALLOY MATERIALS LABS.
ACL-	NAVY (AVIATION CIRCULAR LETTER)
ACLA-	GENERAL MOTORS CORP. DIGITAL COMPUTER LAB., EL SEGUNDO, CALIF.
ACL MEMO-	ARIZONA. UNIV., TUCSON. ANALOG/HYBRID COMPUTER LAB.
ACM-	ALLIS-CHALMERS MFG. CO., MILWAUKEE
ACM-	BUREAU OF ORDNANCE (NAVY) (ACOUSTIC CALIBRATION MEMO)
AC-M-	COLUMBIA UNIV., NYC. SUBSTITUTE ALLOY MATERIALS LABS.
ACN-	ARMY COMBAT DEVELOPMENTS COMMAND, FORT BELVOIR, VA.
ACNP-	ALLIS CHALMERS CORP. ATOMIC EN. DIV., BETHESDA, MD.
ACNP-ERR-	ALLIS-CHALMERS MFG. CO. ATOMIC ENERGY DIV., WASH., D.C. (ELK RIVER REACTOR)
A/CONF.8/P/	INTERNATIONAL CONFERENCE ON THE PEACEFUL USES OF ATOMIC ENERGY. 1955, GENEVA
A/CONF.15/P/	INTERNATIONAL CONFERENCE ON THE PEACEFUL USES OF ATOMIC ENERGY, 1958, GENEVA
A/CONF.28/P/-	INTERNATIONAL CONFERENCE ON THE PEACEFUL USES OF ATOMIC ENERGY, 1964, GENEVA
A/CONF.49/	INTERNATIONAL CONFERENCE ON THE PEACEFUL USES OF ATOMIC ENERGY, 1971, GENEVA
ACORN-	ACORN COMMITTEE (REF. 15)
ACO/UK/	GT. BRIT. MINISTRY OF DEFENCE, LONDON
ACP-	MASSACHUSETTS INST. OF TECH., CAMBRIDGE. LINCOLN LAB.
ACP-(NUMBER)-	MOTOROLA INC. AEROSPACE CENTER, PHOENIX, ARIZ.
ACP-	OFFICE OF THE JOINT CHIEFS OF STAFF, WASH., D.C.
ACPP-	UNITED KINGDOM ATOMIC ENERGY AUTHORITY. RESEARCH GP. ATOMIC ENERGY RES. ESTAB., HARWELL, BERKS, ENGLAND
ACPP/P-	GT. BRIT. ATOMIC EN. RES. ESTAB., HARWELL, BERKS, ENG
ACR-	ALLIS-CHALMERS MFG. CO. RESEARCH DIV., MILWAUKEE
AC-R-	COLUMBIA UNIV., NYC. SUBSTITUTE ALLOY MATERIALS LABS.
ACR-	DIVISION OF TECHNICAL INFORMATION, AEC (ABSTRACTS OF CLASSIFIED REPORTS)
ACR-	OFFICE OF NAVAL RESEARCH, WASH., D.C. (SERIES FOR PROCEEDINGS OF MEETINGS, CONFERENCES AND SYMPOSIA SPONSORED BY THE OFFICE OF NAVAL RESEARCH)
ACR-(YEAR)-	WEBER AIRCRAFT CORP., BURBANK, CALIF.
ACR-(LETTER)(DATE CODE)	NATIONAL ADVISORY COMMITTEE FOR AERONAUTICS (ADVANCE CONFIDENTIAL REPORT) (REF. 25)
ACRH-	ARGONNE CANCER RESEARCH HOSPITAL, CHICAGO
ACRH-1000-	ARGONNE CANCER RESEARCH HOSPITAL, CHICAGO
ACRL-	AEROCHEM RESEARCH LABS., INC., PRINCETON, N.J.
ACR/NAM-	OFFICE OF NAVAL RESEARCH. NAVAL ANALYSIS GROUP, WASHINGTON, D.C.
ACS-	ALLISON DIV., GENERAL MOTORS CORP., INDIANAPOLIS (APPLIED CHEMISTRY SECTION)
AC-S-	COLUMBIA UNIV., NYC. SUBSTITUTE ALLOY MATERIALS LABS.
ACS-	GT. BRIT. ADMIRALTY COMPUTING SERVICE
ACSD-	ALLIS-CHALMERS MFG. CO. SPACE AND DEFENSE SCIENCES DEPT., MILWAUKEE
ACSDS-	ALLIS-CHALMERS MFG. CO. SPACE AND DEFENSE SCIENCES DEPT., MILWAUKEE
ACSFOR-	ASSISTANT CHIEF OF STAFF FOR FORCE DEVELOPMENT (ARMY) WASHINGTON, D.C.
ACSI-H-	ASSISTANT CHIEF OF STAFF FOR INTELLIGENCE (ARMY), WASHINGTON, D.C.
ACSI-I-	ASSISTANT CHIEF OF STAFF FOR INTELLIGENCE (ARMY), WASHINGTON, D.C.
ACSI-J-	ASSISTANT CHIEF OF STAFF FOR INTELLIGENCE (ARMY), WASHINGTON, D.C.
ACSIL/-	GT. BRIT. ADMIRALTY CENTRE FOR SCIENTIFIC INFORMATION AND LIAISON, LONDON
ACSIL/AD/-	GT. BRIT. ADMIRALTY CENTRE FOR SCIENTIFIC INFORMATION AND LIAISON, LONDON
ACSIL/ADM/-	GT. BRIT. ADMIRALTY CENTRE FOR SCIENTIFIC INFORMATION AND LIAISON, LONDON
ACSIL/LIBY/-	GT. BRIT. ADMIRALTY CENTRE FOR SCIENTIFIC INFORMATION AND LIAISON, LONDON
ACSIL-TRANS-	GT. BRIT. ADMIRALTY CENTRE FOR SCIENTIFIC INFORMATION AND LIAISON, LONDON
AC-SM-	GRUMMAN AEROSPACE CORP. ADVANCED COMPOSITES GROUP, BETHPAGE, N.Y.
AC+SS-	AIR UNIV., MAXWELL AFB, ALA.
ACSU/ADM/-	GT. BRIT. ADMIRALTY MATERIALS LAB., POOLE, DORSET, ENG
ACT-	ADVANCED COMPUTER TECHNIQUES CORP, N.Y.C.
ACTEU-	UNITED KINGDOM. AIR TRAFFIC CONTROL EVALUATION UNIT, BOURNEMOUTH, HANTS, ENGLAND
ACTIAC-	BATTELLE MEMORIAL INST., COLUMBUS, OHIO
ACTIV-	ARMY CONCEPT TEAM IN VIETNAM, SAN FRANCISCO
AC-TR-	AIR MATERIEL COMMAND. ENGINEERING DIV., WRIGHT-PATTERSON AFB, OHIO
ACT-TR-(NO.)-(YR.)	ASTRO CONSULTANTS, PALM SPRINGS, CALIF.
AD-	ABERDEEN PROVING GROUND, MD.
AD-	BUREAU OF AERONAUTICS (NAVY). AIRFRAME DESIGN DIV.
AD-	BUREAU OF AERONAUTICS (NAVY). AVIONICS DIV.
AD-	CHRYSLER CORP., CENTER LINE, MICH.
AD-	CORNELL AERONAUTICAL LAB., INC., BUFFALO
AD-(6 DIGITS)	DEFENSE DOCUMENTATION CENTER, ALEXANDRIA, VA.
AD-	DU PONT DE NEMOURS (E.I.) + CO., WILMINGTON, DEL.
AD-	HUMAN RESOURCES RESEARCH INSTITUTE, MAXWELL AFB, ALA.
AD-	MINNEAPOLIS-HONEYWELL REGULATOR CO., MINNEAPOLIS
AD-(NUMBER)-(LETTER)	NAVAL RADIOLOGICAL DEFENSE LAB., SAN FRAN. (REF. 30)
AD-	OFFICE OF NAVAL RESEARCH (AIR DEFENSE OPERATIONS)
AD-	RESEARCH + DEV. BD., WASH., DC (AIR DEFENCE OPERATIONS)
AD-	SMITH (A.O.) CORP., MILWAUKEE
AD(LETTER)-	NAVAL RADIOLOGICAL DEFENSE LAB., SAN FRAN. (REF. 30)
ADB-	CHRYSLER CORP. MISSILE DIV., DETROIT
ADB-TN-	CHRYSLER CORP., DETROIT
ADC-	ADMIRAL CORP., CHICAGO
ADC-	AEROPHYSICS DEVELOPMENT CORP., SANTA MONICA, CALIF.
ADC-	AIR DEFENSE COMMAND, ENT AFB, COLO.
ADC-	BOSTON UNIV. AREA DEVELOPMENT CENTER
ADC-	NAVAL AIR DEVELOPMENT CENTER, JOHNSVILLE, PENNA.
ADC-	NAVAL ORDNANCE LAB., WHITE OAK, MD. (ACTUATION DATA COMMUNICATION)
ADC-EL-	NAVAL AIR DEVELOPMENT CENTER. AERONAUTICAL ELECTRONIC AND ELECTRIC LAB., JOHNSVILLE, PENNA.
ADC-EL-T-	NAVAL AIR DEVELOPMENT CENTER. AERONAUTICAL ELECTRONIC AND ELECTRIC LAB., JOHNSVILLE, PENNA.
ADC-FC-TP-	FORECAST CENTER, ENT AFB, COLO.
ADC M-	AIR DEFENSE COMMAND, ENT AFB, COLO.
ADCO-	AEROPHYSICS DEVELOPMENT CORP., PACIFIC PALISADES, CAL
ADCO-	AEROPHYSICS DEVELOPMENT CORP., SANTA BARBARA, CALIF.
ADC-OA-TM-	AIR DEFENSE COMMAND, ENT AFB, COLO. (OPERATIONS ANALYSIS TECHNICAL MEMORANDUM)
ADC-OA-WP-	AIR DEFENSE COMMAND, ENT AFB, COLO. (OPERATIONS ANALYSIS WORKING PAPER)
ADCOR-	APPLIED DEVICES CORP., COLLEGE POINT, N.Y.
ADC/ORB-	CANADA. AIR DEFENCE COMMAND. OPERATIONAL RESEARCH BR.
ADC-PR-	AEROPHYSICS DEVELOPMENT CORP., SANTA MONICA, CALIF.
ADC-R-	AEROPHYSICS DEVELOPMENT CORP., SANTA BARBARA, CALIF.
ADCRP-(NUMBER)-	DIRECTORATE OF COMMUNICATIONS AND ELECTRONICS, ENT AFB, COLO.
ADCSB-	AEROPHYSICS DEVELOPMENT CORP., SANTA BARBARA, CALIF.
ADC-SCIENTIFIC RPT.-	ADMIRAL CORP., CHICAGO
ADCSM-	AEROPHYSICS DEVELOPMENT CORP., SANTA MONICA, CALIF.

ADC-SR-	ADMIRAL CORP., CHICAGO
ADC/YAFB/RF-	ROCKET PROFICIENCY CENTER (AIR FORCE), YUMA, ARIZ.
ADD-	AVCO MFG. CORP. ADVANCED DEV. DIV., STRATFORD, CONN.
ADD-	WESTINGHOUSE ELECTRIC CORP. ADVANCED DESIGN AND DEVELOPMENT DEPT., KANSAS CITY, MO.
ADD-(YR)-	WESTINGHOUSE ELECTRIC CORP. ADVANCED DESIGN AND DEVELOPMENT DEPT., KANSAS CITY, MO.
ADD-LJL-	BROOKHAVEN NATIONAL LAB., UPTON, N.Y.
ADD-TR-	GT. BRIT. ARMAMENT DESIGN DEPT., FT.HALSTEAD, KENT, ENG.
ADE-	GT. BRIT. ARMAMENT DESIGN EST., FT.HALSTEAD, KENT, ENG.
ADED-	ARMY NATICK LABS. AIR DELIVERY EQUIPMENT DIV., MASS.
ADF-	ARGONNE NATIONAL LAB., LEMONT, ILL.
ADFSC-01-(NUMBER-NUMBER)	AUTOMATIC DATA FIELD SYSTEMS COMMAND, FT.BELVOIR, VA.
ADFSC-	AUTOMATIC DATA FIELD SYSTEMS COMMAND, FT.BELVOIR, VA.
ADGAS-	AIR DEFENSE COMMAND, ENT AFB, COLO.
ADHAD-	OFFICE OF THE DIRECTOR OF DEFENSE (RES. + ENG.) ADVISORY GROUP ON HIGH ALTITUDE DETECTION, WASH., D.C.
AD HOC-	ARMY ADVANCED MATERIEL CONCEPTS AGENCY, WASHINGTON, DC (AD HOC WORKING GROUP)
ADI-	AMERICAN DOCUMENTATION INSTITUTE, WASHINGTON, D.C.
ADIK-	GT. BRIT. AIR DOCUMENT INTELLIGENCE (CODE LETTER K)
ADI(K)-	GT. BRIT. DIRECTORATE OF INTELLIGENCE
ADL-	LITTLE (ARTHUR D.), INC., CAMBRIDGE, MASS.
ADL-	LITTON PRECISION PRODUCTS, INC. ADVANCED DEVICE LABS. CANOGA PARK, CALIF.
ADL-C-	LITTLE (ARTHUR D.), INC., CAMBRIDGE, MASS.
ADL-LOG-	LITTLE (ARTHUR D.), INC., CAMBRIDGE, MASS.
ADL-NCI-	LITTLE (ARTHUR D.), INC., CAMBRIDGE, MASS.
ADM-	ADMIRAL CORP., CHICAGO
ADM-	AIR MATERIEL COMMAND, WRIGHT-PATTERSON AFB, OHIO
ADM-ENR-SR-	ADMIRAL CORP., CHICAGO (EFFECTS OF NUCLEAR RADIATION. SCIENTIFIC REPORT)
ADM-IM-	ARMED FORCES SPECIAL WEAPONS PROJECT. FIELD COMMAND, ALBUQUERQUE, N. MEX. (INSTRUCTORS MANUAL)
ADN-	CALIFORNIA. UNIV., LIVERMORE. LAWRENCE RADIATION LAB.
ADN-	GRUMMAN AIRCRAFT ENGINEERING CORP., BETHPAGE, N.Y.
ADO/(NOS.)-SER-(LTR.)	NAVAL RADIOLOGICAL DEFENSE LAB., SAN FRANCISCO
ADO-	RESEARCH AND DEVELOPMENT BOARD. AD HOC COMMITTEE ON SCIENTIFIC AND SYNTHETIC ANALYSIS, WASHINGTON, D.C.
ADP-	ADELAIDE. UNIV., AUSTRALIA
ADP-	NAVAL RADIATION LAB., SAN FRANCISCO
ADPUB-	NAVAL ORDNANCE TEST STATION, CHINA LAKE, CALIF. (ADMINISTRATIVE PUBLICATION)
ADR-	(AIR DEFENSE REQUIREMENTS)
ADR-	AVIDYNE RESEARCH, INC., BURLINGTON, MASS.
ADR-	BUREAU OF AERONAUTICS (NAVY). AIRCRAFT DESIGN RESEARCH DIV., WASHINGTON, D.C.
ADR-	DAVID TAYLOR MODEL BASIN, CARDEROCK, MD. (AERO DATA REPORT)
ADR-(NUMBER/YEAR)	GT. BRIT. AEROPLANE AND ARMAMENT EXPTL. EST. ARMAMENT DIV., BOSCOMBE DOWN, AMESBURY, WILTS, ENGLAND
ADR-	GRUMMAN AIRCRAFT ENGINEERING CORP., BETHPAGE, N.Y.
ADR-	LEAR SIEGLER, INC. ASTRONAUTICS DIV., SANTA MONICA, CAL
ADRDE-	GT. BRIT. AIR DEFENCE RESEARCH AND DEVELOPMENT ESTABLISHMENT, MALVERN, WORCS, ENGLAND
ADR-M-	BUREAU OF AERONAUTICS (NAVY). AIRCRAFT DESIGN RESEARCH DIV., WASHINGTON, D.C.
ADR-R-	BUREAU OF AERONAUTICS (NAVY). AIRCRAFT DESIGN RESEARCH DIV., WASHINGTON, D.C.
ADR-REF-	APPLIED DATA RESEARCH, INC., PRINCETON, N.J.
ADS-	FEDERAL AVIATION ADM. AIRCRAFT DEV. SERVICE, WASH., DC
ADS-	NATIONAL AVIATION FACILITIES EXPTL.CTR., ATLANTIC CITY
ADS-	NAVAL AIR DEVELOPMENT CENTER, JOHNSVILLE, PENNA.
ADS-EL-	NAVAL AIR DEVELOPMENT CENTER. AERONAUTICAL ELECTRONIC AND ELECTRIC LAB., JOHNSVILLE, PENNA.
ADSS-	AUSTRALIA. DEFENCE STANDARDS LABS., MARIBYRNONG
ADTC-TR-(YEAR)-	ARMAMENT DEV. AND TEST CTR., EGLIN AFB, FLA.
ADTEC-	ADVANCED TECHNOLOGY CORP., TIMONIUM, MD.
ADTIC-	AIR UNIV., MAXWELL AFB, ALA. ARCTIC, DESERT, TROPIC INFORMATION CENTER
ADTIC-PUB-G-	AIR UNIV., MAXWELL AFB, ALA. ARCTIC, DESERT, TROPIC INFORMATION CENTER
ADVT-	BENDIX CORP. BENDIX SYSTEMS DIV., ANN ARBOR, MICH.
ADWD-	LOS ALAMOS SCIENTIFIC LAB., N. MEX.
AE-	(AERONAUTICAL ENGINEERING) (REF. 2)
AE-	AEROJET-GENERAL CORP. AETRON DIV., COVINA, CALIF.
AE-	AIR VEHICLE ENVIRONMENTAL RESEARCH TEAM, NATICK, MASS
AE-	AKTIEBOLAGET ATOMENERGI, STOCKHOLM
AE-	AKTIEBOLAGET ATOMENERGI, STUDSVIK, SWEDEN
AE-	ARMY NATICK LABS. APPLIED ENTOMOLOGY GROUP, MASS.
AE-	BRITISH AIRCRAFT CORP., PRESTON, LANCS, ENGLAND
AE-	BUREAU OF AERONAUTICS (NAVY). AIRBORNE EQUIP. DIV.
AE(YR)-(NO.)-	CINCINNATI. UNIV.
AE-	CONVAIR, FORT WORTH, TEX.
AE-(NUMBER)-R	GARRETT CORP. AIRESEARCH MFG. DIV., LOS ANGELES
AE-	GARRETT CORP. AIRESEARCH MFG. DIV., LOS ANGELES
AE-(NUMBER)-D	GARRETT CORP. AIRESEARCH MFG. CO. OF ARIZONA, PHOENIX
AE-	GARRETT CORP. AIRESEARCH MFG. DIV., PHOENIX, ARIZ.
AE(YR)-(NO.)	GENERAL DYNAMICS/ASTRONAUTICS, SAN DIEGO, CALIF.
AE-(YEAR)-	GENERAL DYNAMICS/ASTRONAUTICS, SAN DIEGO, CALIF.
AE(YEAR)-	GENERAL DYNAMICS/CONVAIR, SAN DIEGO, CALIF.
AE-(NUMBER)(LETTER)	GEORGE C. MARSHALL SPACE FLIGHT CTR., HUNTSVILLE, ALA.
AE-	INDIAN INST. OF SCIENCE, BANGALORE
AE-	MC GILL UNIV., MONTREAL
AE-	NAVAL AIR DEVELOPMENT CENTER. AERO ELECTRONICS TECHNOLOGY DEPT., JOHNSVILLE, PENNA.
AE-	NORGES TEKNISKE HOEGSKOLE, TRONDHEIM
AE-	STONE + WEBSTER ENGINEERING CORP., N.Y.C.
AE-	TECHNION-ISRAEL INST. OF TECH., HAIFA
AE-	WESTINGHOUSE ELECTRIC CORP., PITTSBURGH
AEB-	MINNESOTA. UNIV., DULUTH
AEC-	AIRSHIP EXPERIMENTAL CENTER, LAKEHURST, N.J.
AEC-	ARGONNE NATIONAL LAB., ILL.
AEC-	ATOMIC ENERGY COMMISSION, WASHINGTON, D.C.
AEC-(NUMBER)-CVAC	CONVAIR, FORT WORTH, TEX.
AEC-	NEW YORK UNIV., N.Y.C. (TECHNICAL PROGRESS REPORT)
A.E.C.-	NEW ZEALAND. DEPT. OF SCIENTIFIC AND INDUSTRIAL RESEARCH. DIV. OF NUCLEAR SCIENCES, LOWER HUTT
AEC-43/-	TECHNICAL COOPERATION PROGRAM, AEC
AEC-	TEXAS. UNIV., AUSTIN. CENTER FOR PARTICLE THEORY
AEC-(LETTERS)-	ATOMIC ENERGY COMMISSION, WASHINGTON, D.C. (FOR EXPLANATION OF THE REMAINDER OF THE CODE, SEE ENTRY UNDER LETTERS FOLLOWING AEC-)
AEC-AFSWP-TP-	JOINT SPECIAL WEAPONS PUBLICATIONS BOARD, ALBUQUERQUE, N. MEX. (REF. 14)
AEC-ALO-DOC-	SANDIA CORP. QUALITY ASSURANCE DEPT., ALBUQUERQUE, NM
AEC-ALO-QAP'S	SANDIA CORP. QUALITY ASSURANCE DEPT., ALBUQUERQUE, NM
AEC-C-	ZENTRALSTELLE FUER ATOMKERNENERGIE-DOKUMENTATION BEIM GMELIN INSTITUT, FRANKFURT AM MAIN
AEC-CTR-	ATOMIC ENERGY COMMISSION, WASHINGTON, D.C. (CONTROLLED THERMONUCLEAR PROGRAM)

AEC-CU-

AEC-CU-	COLUMBIA UNIV., IRVINGTON-ON-HUDSON, N.Y. NEVIS CYCLOTRON LABS.
AECD-	ATOMIC ENERGY COMMISSION, WASHINGTON, D.C. (REF. 24)
AECD-AG-	PAKISTAN. ATOMIC EN. CENTRE. AGRICULTURE DIV., DACCA
AEC-DASA-	SANDIA CORP., ALBUQUERQUE, N. MEX. (TECH. MANUAL)
AEC-DASA-TP-	JOINT ATOMIC WEAPONS PUBLICATIONS BOARD, ALBUQUERQUE, N. MEX. (REF. 14)
AECD/CH/	PAKISTAN. ATOMIC ENERGY CENTRE. CHEMISTRY DIV., DACCA
AECD/EL-	PAKISTAN. ATOMIC ENERGY CENTRE, DACCA
AECD-EP-	PAKISTAN. ATOMIC EN. CENTRE. EXPTL.PHYSICS DIV.,DACCA
AECD/HP/	PAKISTAN. ATOMIC EN. CENTRE.HEALTH PHYSICS DIV.,DACCA
AECD/MISC-	PAKISTAN. ATOMIC ENERGY CENTRE. DACCA
AECD/RB-	PAKISTAN. ATOMIC EN. CENTRE. RADIOBIOLOGY DIV., DACCA
AECD-TN-(YEAR)-	ARNOLD ENGINEERING DEV. CTR., ARNOLD A.F. STA., TENN.
AECD-TP-	PAKISTAN. ATOMIC EN. CENTRE. EXPTL.PHYSICS DIV.,DACCA
AECD-TW-(YEAR)-	ARNOLD ENGINEERING DEV. CTR., ARNOLD A.F. STA., TENN.
AECI-	DIVISION OF INTELLIGENCE, AEC, WASHINGTON, D.C.
AEC/INF/-	UNITED NATIONS ATOMIC ENERGY COMMISSION
AECL-	ATOMIC ENERGY OF CANADA LTD. (ALL LOCATIONS)
AECL-DR-	ATOMIC ENERGY OF CANADA LTD. CHALK RIVER PROJ., ONT.
AECL GPI-	NATIONAL RESEARCH COUNCIL OF CANADA. ATOMIC ENERGY PROJECT, CHALK RIVER, ONT.
AECL-HP-	PAKISTAN. ATOMIC ENERGY CENTRE, LAHORE
AEC-LONDON-	ATOMIC ENERGY COMMISSION, LONDON (INFORMAL SUMMARY)
AECL-PAK-	PAKISTAN. ATOMIC ENERGY COMMISSION, LAHORE
AECL-PAK/LIB-	PAKISTAN. ATOMIC ENERGY CENTRE. LIBRARY DIV., LAHORE
AECL-PAK/RB-	PAKISTAN. ATOMIC EN. CENTRE.RADIOBIOLOGY DIV.,LAHORE
AECL-PUB-	ATOMIC ENERGY OF CANADA LTD. PUBLIC RELATIONS OFFICE, CHALK RIVER, ONT.
AECL-RM-	ATOMIC ENERGY OF CANADA LTD., CHALK RIVER, ONT.
AECOP-	ATOMIC ENERGY COMMISSION COMBINED OPERATIONS PLANNING, OAK RIDGE, TENN.
AEC-ORO-2504-	MARYLAND. UNIV., COLLEGE PARK. DEPT. OF PHYSICS AND ASTRONOMY
AEC-ORO-(CONTRACT NO.)-	OAK RIDGE OPERATIONS OFFICE, AEC, TENN. (REF. 4)
AECPI-	ATOMIC ENERGY COMMISSION, WASHINGTON, D.C. (PROCUREMENT INSTRUCTION)
AEC-PNE-(NUMBER)-	ATOMIC ENERGY COMMISSION, WASHINGTON, D.C. (PEACEFUL USES OF NUCLEAR EXPLOSIVES)
AECPR-	ATOMIC ENERGY COMMISSION, WASHINGTON, D.C. (PROCUREMENT REGULATIONS)
AECR-(LETTERS)-	(AECR- IS ASSIGNED BY VARIOUS RECIPIENTS AS A PREFIX TO CODES FOR REPORTS OF RESEARCH SPONSORED BY THE AEC. FOR EXPLANATION OF THE REMAINDER OF THE CODE, SEE ENTRY UNDER LETTERS FOLLOWING AECR-)
AEC-SP-	DIVISION OF TECHNICAL INFORMATION EXTENSION, AEC, OAK RIDGE, TENN.
AEC-SR-	ATOMIC ENERGY COMMISSION, CHALK RIVER, ONTARIO
AEC-TP-	JOINT ATOMIC WEAPONS PUBLICATIONS BOARD, ALBUQUERQUE, N. MEX. (REF. 14)
AEC-TR-	ATOMIC ENERGY COMMISSION, WASH., D.C. (TRANSLATIONS)
AECTR-	ATOMIC ENERGY COMMISSION, WASH., D.C. (TRANSLATIONS)
AECU-	TECHNICAL INFO. SERVICE, AEC (UNCLASSIFIED RPTS.)
AEC-WO-	ATOMIC ENERGY COMMISSION, WASHINGTON, D.C.
AEC-WT-	ATOMIC ENERGY COMMISSION, WASHINGTON, D.C. (WEAPON TEST REPORT) (REF. 7)
AECWT-	ATOMIC ENERGY COMMISSION, WASHINGTON, D.C. (WEAPON TEST REPORT) (REF. 7)
AED-	GENERAL ELECTRIC CO. AEROSPACE ELECTRONICS DEPT., UTICA, N.Y.
AED-	HUGHES AIRCRAFT CO. AEROSPACE ENGINEERING DIV., CULVER CITY, CALIF.
AED-	ZENTRALSTELLE FUER ATOMKERNENERGIE-DOKUMENTATION BEIM GMELIN INSTITUT, FRANKFURT AM MAIN
AED-A-	ZENTRALSTELLE FUER ATOMKERNENERGIE-DOKUMENTATION BEIM GMELIN INSTITUT, FRANKFURT AM MAIN
AED-AB-(YEAR)-	ZENTRALSTELLE FUER ATOMKERNENERGIE-DOKUMENTATION BEIM GMELIN INSTITUT, FRANKFURT AM MAIN
AED-BRD-	ZENTRALSTELLE FUER ATOMKERNENERGIE-DOKUMENTATION BEIM GMELIN INSTITUT, FRANKFURT AM MAIN
AED-BRD-C-	ZENTRALSTELLE FUER ATOMKERNENERGIE-DOKUMENTATION BEIM GMELIN INSTITUT, FRANKFURT AM MAIN
AEDC-	ARNOLD ENGINEERING DEV. CTR., ARNOLD A.F. STA., TENN.
AEDC-BIB-(YEAR)-	ARNOLD ENGINEERING DEV. CTR., ARNOLD A.F. STA., TENN.
AEDC-GDF-	ARNOLD ENGINEERING DEVELOPMENT CENTER. GAS DYNAMICS FACILITY, ARNOLD AIR FORCE STATION, TENN.
AEDC-GDF-	ARNOLD ENGINEERING DEVELOPMENT CENTER. GROUND DEVELOPMENT FORCES, ARNOLD AIR FORCE STATION, TENN.
AED-CONF.(YEAR)-	ZENTRALSTELLE FUER ATOMKERNENERGIE-DOKUMENTATION BEIM GMELIN INSTITUT, FRANKFURT AM MAIN
AEDC-TDR-(YEAR)-	ARNOLD ENGINEERING DEV. CTR., ARNOLD A.F. STA., TENN. (TECHNICAL DOCUMENTARY REPORT)
AEDC-TM-(YEAR)-	ARNOLD ENGINEERING DEV. CTR., ARNOLD A.F. STA., TENN.
AEDC-TN-(YEAR)-	ARNOLD ENGINEERING DEV. CTR., ARNOLD A.F. STA., TENN.
AEDC-TR-(YEAR)-	ARNOLD ENGINEERING DEV. CTR., ARNOLD A.F. STA., TENN.
AED-D-	ZENTRALSTELLE FUER ATOMKERNENERGIE-DOKUMENTATION BEIM GMELIN INSTITUT, FRANKFURT AM MAIN (THESIS)
AEDG(E)-E-	ARMY RESEARCH AND DEVELOPMENT GROUP (EUROPE), N.Y.C.
AED-M(NUMBER)	ZENTRALSTELLE FUER ATOMKERNENERGIE-DOKUMENTATION BEIM GMELIN INSTITUT, FRANKFURT AM MAIN
AED-R-	RADIO CORP. OF AMERICA. ASTRO-ELECTRONICS DIV., PRINCETON, N.J.
AEDU-	GT. BRIT. ADMIRALTY EXPERIMENTAL DIVING UNIT, PORTSMOUTH, HANTS, ENGLAND
AEDU-	UNITED KINGDOM. ADMIRALTY EXPERIMENTAL DIVING UNIT, PORTSMOUTH, HANTS, ENGLAND
AEEC-	AERONAUTICAL RADIO, INC., WASHINGTON, D.C.
AEEE-	CANADA. ARMY EQUIPMENT ENGINEERING ESTABLISHMENT, OTTAWA
AEEL-	NAVAL AIR DEVELOPMENT CENTER. AERONAUTICAL ELECTRONIC AND ELECTRIC LAB., JOHNSVILLE, PENNA.
AEEP-(NO.)-(NO.)-(YR.)	VIRGINIA. UNIV., CHARLOTTESVILLE. DEPT. OF AEROSPACE ENGINEERING AND ENGINEERING PHYSICS
AEER/HP/SM-	INDIA. ATOMIC ENERGY ESTABLISHMENT, TROMBAY
AEES-	PURDUE UNIV., LAFAYETTE, IND.
AEET-	INDIA. ATOMIC ENERGY ESTABLISHMENT, TROMBAY
AEET/AM/	INDIA. ATOMIC ENERGY ESTABLISHMENT. AIR MONITORING SECTION, TROMBAY
AEET/ANAL/	INDIA. ATOMIC EN. ESTAB. ANALYTICAL DIV., TROMBAY
AEET/CD/	INDIA. ATOMIC EN. ESTAB. CHEMICAL DIV., TROMBAY
AEET/ED/SG/	INDIA. ATOMIC EN. ESTAB. ELECTRONICS DIV., TROMBAY
AEET/HP/R-	INDIA. ATOMIC EN. ESTAB. HEALTH PHYS. DIV., TROMBAY
AEET/HP/TH/	INDIA. ATOMIC EN. ESTAB. HEALTH PHYS. DIV., TROMBAY
AEET/MET/	INDIA. ATOMIC EN. ESTAB. METALLURGY DIV., TROMBAY
AEET/NP-	INDIA. ATOMIC EN. ESTAB. NUCLEAR PHYS. DIV., TROMBAY
AEET/RADIOCHEM/	INDIA. ATOMIC ENERGY ESTABLISHMENT, TROMBAY
AEET-RE-	INDIA. ATOMIC ENERGY ESTABLISHMENT, TROMBAY
AEET-RMS-	INDIA. ATOMIC ENERGY ESTABLISHMENT, TROMBAY
AEET-ROD-	INDIA. ATOMIC ENERGY ESTABLISHMENT, TROMBAY
AEET/SPEC/	INDIA. ATOMIC EN. ESTAB. METALLURGY DIV., TROMBAY
AEET/TP/	INDIA. ATOMIC EN. ESTAB. TECHNICAL PHYS.DIV.,TROMBAY
AEEW-	UNITED KINGDOM ATOMIC ENERGY AUTHORITY. RESEARCH GP. ATOMIC ENERGY ESTAB., WINFRITH, DORSET, ENGLAND
AEEW-AR-	UNITED KINGDOM ATOMIC ENERGY AUTHORITY. RESEARCH GP. ATOMIC ENERGY ESTAB., WINFRITH, DORSET, ENGLAND
AEEW-M-	UNITED KINGDOM ATOMIC ENERGY AUTHORITY. RESEARCH GP. ATOMIC ENERGY ESTAB., WINFRITH, DORSET, ENGLAND
AEEW-R-	UNITED KINGDOM ATOMIC ENERGY AUTHORITY. RESEARCH GP. ATOMIC ENERGY ESTAB., WINFRITH, DORSET, ENGLAND

AEEW-TRANS-	UNITED KINGDOM ATOMIC ENERGY AUTHORITY. RESEARCH GP. ATOMIC ENERGY ESTAB., WINFRITH, DORSET, ENGLAND
AEF-	AKTIEBOLAGET ATOMENERGI, STOCKHOLM
AEFI-	AKTIEBOLAGET ATOMENERGI, STOCKHOLM
AEG-	ALLGEMEINE ELEKTRICITAETS-GESELLSCHAFT, FRANKFURT AM MAIN
AEG/E(NUMBER)/	ALLGEMEINE ELEKTRICITAETS-GESELLSCHAFT, FRANKFURT AM MAIN
AEI-	AMERICAN ELECTRONICS, INC., FULLERTON, CALIF.
AEI-	ASSOCIATED ELECTRICAL INDUSTRIES, LTD., ALDERMASTON, BERKS, ENGLAND
AEI-G-	ASSOCIATED ELECTRICAL INDUSTRIES, LTD. CENTRAL RESEARCH LAB., RUGBY, WARWICK, ENGLAND
AE-INSN-	AKTIEBOLAGET ATOMENERGI, STOCKHOLM (INELASTIC NEUTRON SCATTERING NEWSLETTER)
AEI-T-	ASSOCIATED ELECTRICAL INDUSTRIES, LTD., ALDERMASTON, BERKS, ENGLAND
AEL-	AMERICAN ELECTRONIC LABS., INC., COLMAR, PENNA.
AEL-	AMERICAN ELECTRONIC LABS., INC., PHILADELPHIA
AEL-	ANTENNA ENGINEERING LABS., TORRANCE, CALIF.
AEL-	ARMY ELECTRONICS LABS., FORT MONMOUTH, N.J.
AEL-	GT. BRIT. ADMIRALTY ENGINEERING LAB., WEST DRAYTON, MIDDX, ENGLAND
AEL-	NAVAL AIR EXPERIMENTAL STA. AERO. ENGINE LAB., PHILA.
AEL-	PENNSYLVANIA. UNIV., PHILADELPHIA
AEL-	PRINCETON UNIV., N.J. AERONAUTICAL ENGINEERING LAB.
AEL-M-	GT. BRIT. ADMIRALTY ENGINEERING LAB., WEST DRAYTON, MIDDX, ENGLAND
AELRDL-	ARMY ELECTRONICS RES. + DEV. LAB., FT. MONMOUTH, N.J.
AELRDL-TR-	ARMY ELECTRONICS RES. + DEV. LAB., FT. MONMOUTH, N.J.
AEL-TN-	GT. BRIT. ADMIRALTY ENGINEERING LAB., WEST DRAYTON, MIDDX, ENGLAND
AEL-TR-	AMERICAN ELECTRONIC LABS., INC., PHILADELPHIA
AEL-TR-ECOM-	ARMY ELECTRONICS LABS., FORT MONMOUTH, N.J.
AEM-	AMERICAN ELECTRO METAL CORP., YONKERS, N.Y.
AEMC-	AMERICAN ELECTRO METAL CORP., YONKERS, N.Y.
AEN-	CALIFORNIA. UNIV., SANTA BARBARA. DEPT. OF MECH. ENG.
AEP-	ATOMIC ELECTRIC PROJECT, ST. LOUIS
AE, PARIS-	UNITED STATES EMBASSY, PARIS
AEPC-	NATIONAL RESEARCH COUNCIL OF CANADA. ATOMIC ENERGY PROJECT, CHALK RIVER, ONT.
AEP CRTEC-	NATIONAL RESEARCH COUNCIL OF CANADA. ATOMIC ENERGY PROJECT, CHALK RIVER, ONT.
AEPG-	ARMY ELECTRONIC PROVING GROUND, FORT HUACHUCA, ARIZ.
AEPG-SIG-	ARMY ELECTRONIC PROVING GROUND, FORT HUACHUCA, ARIZ.
AEPI-	AEROPROJECTS, INC., WEST CHESTER, PENNA.
AEPR-	STRATEGIC BOMBING SURVEY (AERO ENGINE PLANT REPORTS)
AEPSC-	AMERICAN ELECTRIC POWER SERVICE CORP., N.Y.C.
AEPSCO-	AMERICAN ELECTRIC POWER SERVICE CORP., N.Y.C.
AER-	(AERONAUTICAL ENGINEERING REPORT) (REF. 2)
AER-	AERONAUTICAL ENGINEERING RESEARCH, INC., PASADENA, CAL
AER-(NO.)-(NO.)-	AMMUNITION PROCUREMENT AND SUPPLY AGENCY. QUALITY EVALUATION DIV., JOLIET, ILL.
AER-	BUREAU OF AERONAUTICS (NAVY)
AER-	NEW MEXICO COLL. OF AGRICULTURE AND MECHANIC ARTS, STATE COLLEGE. PHYSICAL SCIENCE LAB.
AER-(YR)-(LETTER)-	NORTH AMERICAN AVIATION, INC., DOWNEY, CALIF.
AER-	PRINCETON UNIV., N.J. DEPT. OF AERONAUTICAL ENG.
AER-	ROCKETDYNE DIV., NORTH AMERICAN AVIATION, INC., CANOGA PARK, CALIF.
AER-(LTR.)-(NO.)-	CHANCE VOUGHT AIRCRAFT, INC., DALLAS
AERDA-	ARMY ELECTRONICS RESEARCH AND DEVELOPMENT ACTIVITY, WHITE SANDS MISSILE RANGE, N. MEX.
AERDL-	ARMY ENGINEER RESEARCH + DEV. LABS., FT. BELVOIR, VA.
AERDL-T-(NO.)-(YEAR)-	ARMY ELECTRONICS RES. + DEV. LAB., FT. MONMOUTH, N.J.
AERDL-T-	ARMY ENGINEER RESEARCH + DEV. LABS., FT. BELVOIR, VA.
AERDL-TR-	ARMY ELECTRONICS RES. + DEV. LAB., FT. MONMOUTH, N.J.
AERE-9ER/-	GT. BRIT. ATOMIC EN. RES. ESTAB., HARWELL, BERKS, ENG
A.E.R.E.-(LETTERS)-	UNITED KINGDOM ATOMIC ENERGY AUTHORITY. RESEARCH GP. ATOMIC EN. RES. EST., HARWELL, BERKS, ENG. (REF.40)
A.E.R.E.-(LTRS)/(LTRS)-	UNITED KINGDOM ATOMIC ENERGY AUTHORITY. RESEARCH GP. ATOMIC EN. RES. EST., HARWELL, BERKS, ENG. (REF.40)
AERE-AM-	UNITED KINGDOM ATOMIC ENERGY AUTHORITY. RESEARCH GP. ATOMIC ENERGY RES. ESTAB., HARWELL, BERKS, ENGLAND
AERE-A/M-	UNITED KINGDOM ATOMIC EN. AUTH. RES. GROUP. CHEMISTRY DIV., CHATHAM OUTSTATION, KENT, ENGLAND
AERE-A/M-	UNITED KINGDOM ATOMIC ENERGY AUTHORITY. RESEARCH GP. CHEMISTRY DIV., WOOLWICH OUTSTATION, ENGLAND
AERE-A/R-	UNITED KINGDOM ATOMIC ENERGY AUTHORITY. RESEARCH GP. ATOMIC ENERGY RES. ESTAB., HARWELL, BERKS, ENGLAND
AERE-BIB-	UNITED KINGDOM ATOMIC ENERGY AUTHORITY. RESEARCH GP. ATOMIC EN. RES. EST., HARWELL, BERKS, ENG. (REF.40)
AERE-C/(LETTERS)-	UNITED KINGDOM ATOMIC EN. AUTH. RES. GP. CHEMISTRY DIV. ATOMIC ENERGY RES. EST., HARWELL, BERKS, ENG.
AERE-CATALOGUE NO.-	GT. BRIT. ATOMIC EN. RES. ESTAB., HARWELL, BERKS, ENG
AERE-CE/(LETTERS)-	UNITED KINGDOM ATOMIC EN. AUTH. RES. GP. CHEM. ENG. DIV. ATOMIC ENERGY RES. EST., HARWELL, BERKS, ENG.
AERE-CIRCUS/A-	GT. BRIT. ATOMIC EN. RES. ESTAB., HARWELL, BERKS, ENG
AERE-CMND-	UNITED KINGDOM ATOMIC ENERGY AUTHORITY
AERE-D/(LETTERS)-	UNITED KINGDOM ATOMIC EN. AUTH. RES. GP. DIRECTORS OFF. ATOMIC ENERGY RES. EST., HARWELL, BERKS, ENG.
AERE-E.443/N-	GT. BRIT. ATOMIC EN. RES. ESTAB., HARWELL, BERKS, ENG
AERE-E/(LETTERS)-	UNITED KINGDOM ATOMIC EN. AUTH. RES. GP. ENGINEERING DIV. ATOMIC ENERGY RES. EST., HARWELL, BERKS, ENG.
AERE-EL/(LETTERS)-	UNITED KINGDOM ATOMIC EN. AUTH. RES. GP. ELECTRONICS DIV. ATOMIC ENERGY RES. EST., HARWELL, BERKS, ENG.
AERE/EMR/PR-	UNITED KINGDOM ATOMIC ENERGY AUTHORITY. RESEARCH GP. ATOMIC ENERGY RES. ESTAB., HARWELL, BERKS, ENGLAND
AERE-ES/(LETTERS)-	UNITED KINGDOM ATOMIC EN. AUTH. RES. GP. ENG.SERVICES DIV. ATOMIC ENERGY RES. EST., HARWELL, BERKS, ENG.
AERE-FSG-	GT. BRIT. ATOMIC EN. RES. ESTAB., HARWELL, BERKS, ENG
AERE-G/(LETTERS)-	UNITED KINGDOM ATOMIC EN. AUTH. RES. GP. GEN. PHYSICS DIV. ATOMIC ENERGY RES. EST., HARWELL, BERKS, ENG.
AERE-GP/(LETTERS)-	UNITED KINGDOM ATOMIC EN. AUTH. RES. GP. GEN. PHYSICS DIV. ATOMIC ENERGY RES. EST., HARWELL, BERKS, ENG.
AERE-H/(LETTERS)-	UNITED KINGDOM ATOMIC EN. AUTH. RES. GP. HEALTH DIV. ATOMIC ENERGY RES. EST., HARWELL, BERKS, ENG.
AERE-HARD(C)P-	UNITED KINGDOM ATOMIC ENERGY AUTHORITY (HOMOGENE- OUS AQUEOUS REACTOR DESIGN)
AERE-HP/(LETTERS)-	UNITED KINGDOM ATOMIC EN. AUTH. RES. GP. HEALTH PHYS. DIV. ATOMIC ENERGY RES. EST., HARWELL, BERKS, ENG.
AERE-I/(LETTERS)-	UNITED KINGDOM ATOMIC EN. AUTH. RES. GP. ISOTOPES DIV. ATOMIC ENERGY RES. EST., HARWELL, BERKS, ENG.
AER-EIR-	CHANCE VOUGHT CORP., DALLAS
AERE-L-	UNITED KINGDOM ATOMIC ENERGY AUTHORITY. RESEARCH GP. ATOMIC ENERGY RES. ESTAB., HARWELL, BERKS, ENGLAND
AERE-LEO/N-	GT. BRIT. ATOMIC EN. RES. ESTAB., HARWELL, BERKS, ENG
AERE-LEO-RDC-	GT. BRIT. ATOMIC EN. RES. ESTAB., HARWELL, BERKS, ENG
AERE-LMFP/	GT. BRIT. ATOMIC EN. RES. ESTAB., HARWELL, BERKS, ENG
AERE-M-	UNITED KINGDOM ATOMIC ENERGY AUTHORITY. RESEARCH GP. (ALL LOCATIONS) (REF. 40)
AERE-M/(LETTERS)-	UNITED KINDGOM ATOMIC EN. AUTH. RES. GP. METALLURGY DIV. ATOMIC ENERGY RES. EST., HARWELL, BERKS, ENG.
AERE-MED/(LETTERS)-	UNITED KINGDOM ATOMIC EN. AUTH. RES. GP. MEDICAL DIV. ATOMIC ENERGY RES. EST., HARWELL, BERKS, ENG.
AERE-MRC/(LETTERS)-	UNITED KINGDOM ATOMIC ENERGY AUTHORITY. RESEARCH GROUP. MEDICAL RES. COUNCIL. RADIOBIOLOGICAL RES. UNIT. ATOMIC EN. RES. EST., HARWELL, BERKS, ENG.
AERE/MS-	GT. BRIT. ATOMIC EN. RES. ESTAB., HARWELL, BERKS, ENG
AERE-N/(LETTERS)-	UNITED KINGDOM ATOMIC EN. AUTH. RES. GP. NUCLEAR PHYS. DIV. ATOMIC EN. RES. EST., HARWELL,BERKS,ENG.

AERE-NP/(LETTERS)-
: UNITED KINGDOM ATOMIC EN. AUTH. RES. GP. NUCLEAR PHYS. DIV. ATOMIC EN. RES. EST., HARWELL,BERKS,ENG.

AERE-NP-R-
: UNITED KINGDOM ATOMIC ENERGY AUTHORITY. RESEARCH GP. ATOMIC EN. RES. EST., HARWELL, BERKS, ENG. (REF.40)

AER-EOR-
: CHANCE VOUGHT AIRCRAFT, INC., DALLAS

AERE-P-(NO.)WP/P-
: GT. BRIT. ATOMIC EN. RES. ESTAB., HARWELL, BERKS, ENG

AERE-P-(NUMBER)/P-
: UNITED KINGDOM ATOMIC ENERGY AUTHORITY. RESEARCH GP. ATOMIC ENERGY RES. ESTAB., HARWELL, BERKS, ENGLAND

AERE-PGEC/L-
: UNITED KINGDOM ATOMIC ENERGY AUTHORITY. RESEARCH GP. ATOMIC ENERGY RES. ESTAB., HARWELL, BERKS, ENGLAND

AERE-PIPPA-WP/MEMO-
: GT. BRIT. ATOMIC EN. RES. ESTAB., HARWELL, BERKS, ENG

AERE-PR/HPM-
: UNITED KINGDOM ATOMIC ENERGY AUTHORITY. RESEARCH GP. ATOMIC ENERGY RES. ESTAB., HARWELL, BERKS, ENGLAND

AERE-PR/NP-
: UNITED KINGDOM ATOMIC EN. AUTH. RES. GP. NUCLEAR PHYS. DIV. ATOMIC EN. RES. EST., HARWELL,BERKS,ENG.

AERE-PR/SSP-
: UNITED KINGDOM ATOMIC ENERGY AUTHORITY. RESEARCH GP. ATOMIC ENERGY RES. ESTAB., HARWELL, BERKS, ENGLAND

AERE-R-
: UNITED KINGDOM ATOMIC ENERGY AUTHORITY. RESEARCH GP. (ALL LOCATIONS) (REF. 40)

AERE-R/(LETTERS)-
: UNITED KINGDOM ATOMIC EN. AUTH. RES. GP. REACTOR DIV. ATOMIC ENERGY RES. EST., HARWELL, BERKS, ENG.

AERE-R/(LETTERS)-
: UNITED KINGDOM ATOMIC EN. AUTH. RES. GP. REACTOR PHYS. DIV. ATOMIC EN. RES. EST., HARWELL,BERKS,ENG.

AERE-RCDG-
: GT. BRIT. ATOMIC EN. RES. ESTAB., HARWELL, BERKS, ENG

AERE-RE/(LETTERS)-
: UNITED KINGDOM ATOMIC EN. AUTH. RES. GP. REACTOR ENG. DIV. ATOMIC EN. RES. EST., HARWELL,BERKS,ENG.

AERE-REV.-
: GT. BRIT. ATOMIC EN. RES. ESTAB., HARWELL, BERKS, ENG

AERE-RL-
: UNITED KINGDOM ATOMIC EN. AUTH. RES. GP. ATOMIC EN. RES. EST., HARWELL, BERKS, ENG. (READING LIST)

AERE-RP/(LETTERS)-
: UNITED KINGDOM ATOMIC EN. AUTH. RES. GP. REACTOR PHYS. DIV. ATOMIC EN. RES. EST., HARWELL,BERKS,ENG.

AERE-RP/R-
: UNITED KINGDOM ATOMIC ENERGY AUTHORITY. RESEARCH GP. ATOMIC EN. RES. EST., HARWELL, BERKS, ENG. (REF.40)

AERE-RS/(LETTERS)-
: UNITED KINGDOM ATOMIC EN. AUTH. RES. GP. REACTOR SCH. ATOMIC ENERGY RES. EST., HARWELL, BERKS, ENG.

AERE-SC-
: GT. BRIT. NATIONAL INST. FOR RESEARCH IN NUCLEAR SCIENCE, HARWELL, BERKS, ENGLAND

AERE-SC-
: UNITED KINGDOM ATOMIC ENERGY AUTHORITY. SAFETY COMMITTEE, HARWELL,BERKS, ENGLAND

AERE-SPAR-
: GT. BRIT. ATOMIC EN. RES. ESTAB., HARWELL, BERKS, ENG

AERE-SPEC-
: GT. BRIT. ATOMIC EN. RES. ESTAB., HARWELL, BERKS, ENG

AERE-T/(LETTERS)-
: UNITED KINGDOM ATOMIC EN. AUTH. RES. GP. THEORETICAL PHYS. DIV. ATOMIC EN. RES. EST., HARWELL,BERKS,ENG.

AERE-TC-(YEAR)
: UNITED KINGDOM ATOMIC ENERGY AUTHORITY. RESEARCH GROUP. ATOMIC ENERGY RESEARCH ESTABLISHMENT, HARWELL, BERKS, ENG. (TRIPARTITE CONFERENCE)

AERE-TRANS-
: UNITED KINGDOM ATOMIC EN. AUTH. RES. GP. ATOMIC EN. RES. ESTAB., HARWELL, BERKS, ENG. (TRANSLATIONS)

AERE-W/AT-
: UNITED KINGDOM ATOMIC ENERGY AUTHORITY. RESEARCH GP. ATOMIC ENERGY RES. ESTAB., HARWELL, BERKS, ENGLAND

AERE-X/
: UNITED KINGDOM ATOMIC ENERGY AUTHORITY. RESEARCH GP. ATOMIC ENERGY RES. ESTAB., HARWELL, BERKS, ENGLAND

AERE-Z/(LETTERS)-
: UNITED KINGDOM ATOMIC ENERGY AUTHORITY. RESEARCH GP. ATOMIC ENERGY RES. ESTAB., HARWELL, BERKS, ENGLAND

AERF-
: AERONAUTICAL RESEARCH FOUNDATION, CAMBRIDGE, MASS.

AERF-(YR)-(NO)-AROWA
: AEROPHYSICS RESEARCH FOUNDATION, GOLETA, CALIF.

AERI-(LETTERS)-(NO.)-
: KOREA. ATOMIC ENERGY RESEARCH INST., SEOUL

AE-RL-
: AKTIEBOLAGET ATOMENERGI, STOCKHOLM

AERL-
: AVCO CORP. AVCO-EVERETT RESEARCH LAB., EVERETT, MASS.

AERL-(YEAR)-
: AVCO CORP. AVCO-EVERETT RESEARCH LAB., EVERETT, MASS.

AERL-
: PRATT AND WHITNEY AIRCRAFT DIV., UNITED AIRCRAFT CORP., MIDDLETOWN, CONN.

AERL-AMP-
: AVCO CORP. AVCO-EVERETT RESEARCH LAB., EVERETT, MASS.

AERL-BIB-
: AVCO MFG. CORP. AVCO RESEARCH LAB., EVERETT, MASS.

AERL-DSTR-
: AVCO EVERETT RESEARCH LAB., EVERETT, MASS.

AERL-RN-
: AVCO CORP. AVCO-EVERETT RESEARCH LAB., EVERETT, MASS. (RESEARCH NOTE)

AERL-RR-
: AVCO CORP. AVCO-EVERETT RESEARCH LAB., EVERETT, MASS. (RESEARCH REPORT)

AERNU-
: AERONUTRONIC DIV., FORD MOTOR CO., LOS ANGELES

AERNU-S-ITR-
: AERONUTRONIC DIV., FORD MOTOR CO., LOS ANGELES

AERNU-SRD-
: AERONUTRONIC DIV., FORD MOTOR CO., LOS ANGELES

AERO-
: AEROJET-GENERAL CORP., AZUSA, CALIF.

AERO-
: AERONUTRONIC DIV. OF PHILCO CORP., NEWPORT BEACH,CAL.

AERO-
: CRANFIELD INST. OF TECH., BUCKS, ENGLAND

AERO-
: DAVID TAYLOR MODEL BASIN. AERODYNAMICS LAB., WASH.,DC

AERO-
: GT. BRIT. ROYAL AIRCRAFT EST.,FARNBOROUGH, HANTS, ENG

AERO-
: HONEYWELL, INC. AERONAUTICAL DIV., MINNEAPOLIS

AERO-(YEAR)/
: LONDON. CITY UNIV. DEPT. OF AERONAUTICS

AERO-(YEAR)-
: MARYLAND. UNIV., COLLEGE PARK. DEPT. OF AEROSPACE ENGINEERING

AERO-
: NAVY RADIO AND SOUND LAB., SAN DIEGO, CALIF.

AEROBALLISTIC MEMO-
: ARMY BALLISTIC MISSILE AGENCY, HUNTSVILLE, ALA.

AEROBALLISTICS RES. RPT-
: NAVAL ORDNANCE LAB., WHITE OAK, MD.

AERO C-
: AERONUTRONIC DIV. OF PHILCO CORP., NEWPORT BEACH,CAL.

AEROC-
: DE HAVILLAND AIRCRAFT OF CANADA LTD.

AEROCHEM-
: AEROCHEM RESEARCH LABS., INC., PRINCETON, N.J.

AEROCHEM TN-
: AEROCHEM RESEARCH LABS., INC., PRINCETON, N.J.

AEROCHEM TP-
: AEROCHEM RESEARCH LABS., INC., PRINCETON, N.J.

AERO/CONF./
: UNITED KINGDOM ATOMIC ENERGY AUTHORITY. RESEARCH GP. ATOMIC ENERGY RES. ESTAB., HARWELL, BERKS, ENGLAND

AERODYNAMICS RES. REPT.-
: NAVAL ORDNANCE LAB., SILVER SPRING, MD.

AERO. ENG. LAB. NO.-
: PRINCETON UNIV., N.J. DEPT. OF AERONAUTICAL ENG.

AEROJET-
: AEROJET-GENERAL CORP., AZUSA, CALIF.

AEROJET DELFT-
: AEROJET DELFT CORP., AZUSA, CALIF.

AEROJET-L-
: AEROJET-GENERAL CORP., AZUSA, CALIF.

AEROJET LRP-
: AEROJET-GENERAL CORP., AZUSA, CALIF.

AEROJET TR-
: AEROJET-GENERAL CORP., AZUSA, CALIF.

AERO/JHU-
: JOHNS HOPKINS UNIV., BALTIMORE. SCHOOL OF ENGINEERING

AERO/JHU-CM-
: JOHNS HOPKINS UNIV., BALTIMORE. SCHOOL OF ENGINEERING

AEROMEKANO-
: AEROMEKANO, STOCKHOLM

AERONAUTICAL LR-
: NATIONAL RESEARCH COUNCIL OF CANADA, OTTAWA

AERONCA-
: AERONCA MFG. CORP., MIDDLETOWN, OHIO

AERONOMY-
: SMITHSONIAN ASTROPHYSICAL OBS., CAMBRIDGE, MASS.

AERONUTRONIC-C-
: AERONUTRONIC DIV. OF PHILCO CORP., NEWPORT BEACH,CAL.

AERONUTRONIC-S-
: AERONUTRONIC DIV. OF PHILCO CORP., NEWPORT BEACH,CAL.

AERONUTRONIC-U-
: AERONUTRONIC DIV. OF PHILCO CORP., NEWPORT BEACH,CAL.

AEROPHYSICS RES. MEMO-
: GENERAL ELECTRIC CO. MISSILE AND SPACE VEHICLE DEPT., PHILADELPHIA

AEROPHYSICS-RR-
: MISSISSIPPI STATE UNIV., STATE COLLEGE. DEPT. OF AEROPHYSICS

AEROPHYSICS TR-(YR)-
: CURTISS-WRIGHT CORP. AEROPHYSICS DEVELOPMENT DIV., SANTA BARBARA, CALIF.

AERO RPT.-
: DAVID TAYLOR MODEL BASIN. AERODYNAMICS LAB., WASH.,DC

AERO S-
: AERONUTRONIC DIV. OF PHILCO CORP., NEWPORT BEACH,CAL.

AEROSPACE-TR-
: AEROSPACE CORP., EL SEGUNDO, CALIF.

AEROSPACE-TR-
: AEROSPACE CORP., SAN BERNARDINO, CALIF.

AEROTHERM-(YEAR)-
: AEROTHERM CORP., MOUNTAIN VIEW, CALIF.

AEROTHERM-UM-(YEAR)-
: AEROTHERM CORP., MOUNTAIN VIEW, CALIF.

AERO TN-
: SWEDEN. KUNGLIGA TEKNISKA HOEGSKOLAN, STOCKHOLM. INSTITUTIONEN FOER FLYGTEKNIK

AERO U-
: AERONUTRONIC DIV. OF PHILCO CORP., NEWPORT BEACH,CAL.

AER-REP-EL-3-
: NAVAL AIR DEV.CTR. AERO. COMPUTER LAB.,JOHNSVILLE,PA.

AE-RTG-
: AKTIEBOLAGET ATOMENERGI, STOCKHOLM

AE-RTL-
: AKTIEBOLAGET ATOMENERGI, STUDSVIK, SWEDEN

AE-RTR-
    AKTIEBØLAGET ATØMENERGI, STUDSVIK, SWEDEN
AES-
    AKTIEBØLAGET ATØMENERGI, STØCKHØLM
AES-
    NATIONAL DEFENSE RESEARCH COMM. (AIR AND EARTH SHOCK)
A + ES-(NO.)-(NO.)-
    PURDUE UNIV., LAFAYETTE, IND. SCHØØL ØF AERONAUTICAL
        AND ENGINEERING SCIENCES
AES-
    PURDUE UNIV., LAFAYETTE, IND. SCHØØL ØF AERONAUTICAL
        AND ENGINEERING SCIENCES
A + ES-(YEAR)-
    PURDUE UNIV., LAFAYETTE, IND. SCHØØL ØF AERONAUTICAL
        AND ENGINEERING SCIENCES
A/ES-
    PURDUE UNIV., LAFAYETTE, IND. SCHØØL ØF AERONAUTICS,
        ASTRONAUTICS AND ENGINEERING SCIENCES
A/ES-(YEAR)-
    PURDUE UNIV., LAFAYETTE, IND. SCHØØL ØF AERONAUTICS,
        ASTRONAUTICS AND ENGINEERING SCIENCES
AES-
    RCA SERVICE CO., INC., CAMDEN, N.J.
AE-SSI-
    AKTIEBØLAGET ATØMENERGI, STUDSVIK, SWEDEN
AETE-(YEAR)/
    CANADA. AEROSPACE ENGINEERING TEST ESTAB., UPLAND, ONT.
AE-TR-
    AKTIEBØLAGET ATØMENERGI, STØCKHØLM
AEW-(NUMBER)/(YEAR)
    GT. BRIT. ADMIRALTY EXPERIMENT WORKS, HASLAR, ENGLAND
AEW-
    GT. BRIT. ADMIRALTY EXPERIMENT WORKS, HASLAR, ENGLAND
AEWES-(NUMBER)-
    ARMY ENGINEER WATERWAYS EXPERIMENT STATION, VICKSBURG,
        MISS.
AEWES-CR-(NUMBER)-
    ARMY ENGINEER WATERWAYS EXPERIMENT STATION, VICKSBURG,
        MISS. (CONTRACT REPORT)
AEWES-CR-M-
    ARMY ENGINEER WATERWAYS EXPERIMENT STATION, VICKSBURG,
        MISS.
AEWES-CR-N-(YEAR)-
    ARMY ENGINEER WATERWAYS EXPERIMENT STATION, VICKSBURG,
        MISS.
AEWES-CR-S-
    ARMY ENGINEER WATERWAYS EXPERIMENT STATION, VICKSBURG,
        MISS.
AEWES-INSTRUCTION-N-
    ARMY ENGINEER WATERWAYS EXPERIMENT STATION, VICKSBURG,
        MISS.
AEWES-INSTRUCTION-S-(YR)-
    ARMY ENGINEER WATERWAYS EXPERIMENT STATION, VICKSBURG,
        MISS.
AEWES-M-
    ARMY ENGINEER WATERWAYS EXPERIMENT STATION, VICKSBURG,
        MISS.
AEWES-MISC PAPER-(NO.)-
    ARMY ENGINEER WATERWAYS EXPERIMENT STATION, VICKSBURG,
        MISS.
AEWES-MISC. PAPER-M-
    ARMY ENGINEER WATERWAYS EXPERIMENT STATION, VICKSBURG,
        MISS.
AEWES-MISC. PAPER-N-
    ARMY ENGINEER WATERWAYS EXPERIMENT STATION, VICKSBURG,
        MISS.
AEWES-MISC. PAPER-S-
    ARMY ENGINEER WATERWAYS EXPERIMENT STATION, VICKSBURG,
        MISS.
AEWES-N-
    ARMY ENGINEER WATERWAYS EXPERIMENT STATION, VICKSBURG,
        MISS.
AEWES-RR-(LTR.)-(YR.)-
    ARMY ENGINEER WATERWAYS EXPERIMENT STATION, VICKSBURG,
        MISS.
AEWES-S-(YEAR)-
    ARMY ENGINEER WATERWAYS EXPERIMENT STATION, VICKSBURG,
        MISS.
AEWES-TM-(NUMBER)-
    ARMY ENGINEER WATERWAYS EXPERIMENT STATION, VICKSBURG,
        MISS.
AEWES-TR-(NO.)-(NO.)-
    ARMY ENGINEER WATERWAYS EXPERIMENT STATION, VICKSBURG,
        MISS.
AEWES-TR-(NUMBER)-
    ARMY ENGINEER WATERWAYS EXPERIMENT STATION, VICKSBURG,
        MISS.
AEWES-TR-(LTR.)-(YR.)-
    ARMY ENGINEER WATERWAYS EXPERIMENT STATION, VICKSBURG,
        MISS.
AEWES-TRANS-
    ARMY ENGINEER WATERWAYS EXPERIMENT STATION, VICKSBURG,
        MISS.
AEWES-TR-M-
    ARMY ENGINEER WATERWAYS EXPERIMENT STATION, VICKSBURG,
        MISS.
AEWES-TR-N-
    ARMY ENGINEER WATERWAYS EXPERIMENT STATION, VICKSBURG,
        MISS.
AEWES-TR-S-
    ARMY ENGINEER WATERWAYS EXPERIMENT STATION, VICKSBURG,
        MISS.
AEWWS-TR-
    ARMY ENGINEER WATERWAYS EXPERIMENT STATION, VICKSBURG,
        MISS.
AEX-
    NEW MEXICO COLL. OF AGRICULTURE AND MECHANIC ARTS,
        STATE COLLEGE
AF-
    (AIR FACILITIES)
AF-
    AIR FORCE
AF-
    BROWN UNIV., PROVIDENCE. DIV. OF ENGINEERING
AF-
    CANADA. SUFFIELD EXPERIMENTAL STATION, RALSTON, ALBERTA
AF-(NO.)-(LETTER)
    CORNELL AERONAUTICAL LAB., INC., BUFFALO
AF-(NUMBER)Q-(YEAR)
    DOW CHEMICAL CO., MIDLAND, MICH. (ALKYL FUELS)
AF-
    JOHNS HOPKINS UNIV., BALTIMORE. CARLYLE BARTON LAB.
AF-(NUMBER)-
    JOHNS HOPKINS UNIV., BALTIMORE. CARLYLE BARTON LAB.
AF-
    JOHNS HOPKINS UNIV., BALTIMORE. RADIATION LAB.
AF-
    ROHM + HAAS CO. REDSTONE ARSENAL RESEARCH DIV.,
        HUNTSVILLE, ALA.
AF-(LETTERS)-
    AIR FORCE (FOR EXPLANATION OF THE REMAINDER OF THE
        CODE, SEE ENTRY UNDER LETTERS FOLLOWING AF-)
AF-AC-(NUMBER)-
    ANAGRAM CORP., SPRINGFIELD, VA.
AFAC-
    BALDWIN PIANO CO., CINCINNATI
AFACS/SA-
    ASSISTANT CHIEF OF STAFF STUDIES AND ANALYSIS (AIR
        FORCE), WASHINGTON, D.C.
AFAC TM-(YEAR)-
    AIR FORCE ARMAMENT CENTER, EGLIN AFB, FLA.
AFAC TN-(YEAR)-
    AIR FORCE ARMAMENT CENTER, EGLIN AFB, FLA.
AFAC TR-(YEAR)-
    AIR FORCE ARMAMENT CENTER, EGLIN AFB, FLA.
AF-AFOSR-(NUMBER)-(YEAR)
    AIR FORCE OFFICE OF SCIENTIFIC RESEARCH, WASH., D.C.
        (DESIGNATION USED FOR GRANTS)
AFAL-(NUMBERS)-
    AIR FORCE AVIONICS LAB., WRIGHT-PATTERSON AFB, OHIO
AFAL-IR-
    AIR FORCE AVIONICS LAB., WRIGHT-PATTERSON AFB, OHIO
AFAL-TDR-(YEAR)-
    AIR FORCE AVIONICS LAB., WRIGHT-PATTERSON AFB, OHIO
AFAL-TPC (MONTH/YEAR)
    AIR FORCE AVIONICS LAB., WRIGHT-PATTERSON AFB, OHIO
        (TECHNICAL PROGRAMS AND CONTACTS)
AFAL-TR-(YEAR)-
    AIR FORCE AVIONICS LAB., WRIGHT-PATTERSON AFB, OHIO
AFAMA-
    DIRECTORATE OF MANAGEMENT ANALYSIS (AIR FORCE),
        WASHINGTON, D.C.
AF(ANAGRAM)-
    ANAGRAM CORP., SPRINGFIELD, VA.
AFAPL-
    AIR FORCE AERO-PROPULSION LAB., WRIGHT-PATTERSON AFB,
        OHIO
AFAPL-CONF-(YEAR)-
    AIR FORCE AERO-PROPULSION LAB., WRIGHT-PATTERSON AFB,
        OHIO
AFAPL-TDR-(NUMBERS)-
    AIR FORCE AERO-PROPULSION LAB., WRIGHT-PATTERSON AFB,
        OHIO
AFAPL-TR-(YEAR)-
    AIR FORCE AERO-PROPULSION LAB., WRIGHT-PATTERSON AFB,
        OHIO
AF-ARDC-AL-
    BALLISTIC MISSILE DIV. (AIR FORCE) (ACCESSION LIST)
AF-ARDC-BIB-
    BALLISTIC MISSILE DIV. (AIR FORCE)
AF-ARPD-(NUMBER)(LETTER)
    AIR RESEARCH AND DEVELOPMENT COMMAND, BALTIMORE
AFATL-ATRA-TN-(YEAR)-
    AIR FORCE ARMAMENT LAB., EGLIN AFB, FLA.
AFATL-MR-(YEAR)-
    AIR FORCE ARMAMENT LAB., EGLIN AFB, FLA.
AFATL-TR-(YEAR)-
    AIR FORCE ARMAMENT LAB., EGLIN AFB, FLA.
AFBMC-(LETTERS)-(NUMBER)-
    BALLISTIC MISSILES COMM. (AIR FORCE), WASHINGTON, D.C.
AFBMD-
    BALLISTIC MISSILE DIV. (AIR FORCE), INGLEWOOD, CALIF.
AFBMD-DWS-TM-
    BALLISTIC MISSILE DIV. (AIR FORCE), INGLEWOOD, CALIF.
AFBMD-RCS-DD...-
    BALLISTIC MISSILE DIV. (AIR FORCE), INGLEWOOD, CALIF.
AFBMD-TN-(YEAR)-
    BALLISTIC MISSILE DIV. (AIR FORCE), INGLEWOOD, CALIF.
AFBMD-TR-(YEAR)-
    BALLISTIC MISSILE DIV. (AIR FORCE), INGLEWOOD, CALIF.
AFBMD-WDLPM-
    BALLISTIC MISSILE DIV. (AIR FORCE), INGLEWOOD, CALIF.
AFBSD-TDR-(YEAR)-
    BALLISTIC SYSTEMS DIV., NORTON AFB, CALIF.
AFBSD-TN-
    BALLISTIC SYSTEMS DIV., NORTON AFB, CALIF.
AFBSD-TR-
    BALLISTIC SYSTEMS DIV., NORTON AFB, CALIF.
AFC-(YEAR)-SRD-
    AIR FORCE NUCLEAR WEAPON SYSTEM SAFETY GROUP, KIRT-
        LAND AFB, N. MEX.
AFC-(LTRS)-(DAY-MO.-YR)
    WHITE SANDS SIGNAL AGENCY. OFFICE OF THE AREA FRE-
        QUENCY COORDINATOR, WHITE SANDS MISSILE RANGE, N.M.
AFCCDD-
    COMMAND AND CONTROL DEVELOPMENT DIV. (AIR FORCE),
        BEDFORD, MASS.
AFCCDD-TN-(YEAR)-
    COMMAND AND CONTROL DEVELOPMENT DIV. (AIR FORCE),
        BEDFORD, MASS.

AFCCDD-TR-(YEAR)-	COMMAND AND CONTROL DEVELOPMENT DIV. (AIR FORCE), BEDFORD, MASS.
AFC-IM-	ARMED FORCES SPECIAL WEAPONS PROJECT. FIELD COMMAND, ALBUQUERQUE, N. MEX. (INSTRUCTORS MANUAL)
AFCL-	AIR FORCE CAMBRIDGE RESEARCH CENTER, BEDFORD, MASS.
AFC-NB-	ARMED FORCES SPECIAL WEAPONS PROJECT. FIELD COMMAND, ALBUQUERQUE, N. MEX. (NOTEBOOK)
AFCOMB-	DELAWARE. UNIV., NEWARK. DEPT. OF CHEM.
AFCRC-	AIR FORCE CAMBRIDGE RESEARCH CENTER, MASS.
AFCRC-AFP-SR-	AIR FORCE CAMBRIDGE RESEARCH CENTER, BEDFORD, MASS. (AUTOBAROTROPIC FLOW PROJECT, SCIENTIFIC REPORT)
AFCRC AFSG-	AIR FORCE CAMBRIDGE RESEARCH CENTER, MASS. (AIR FORCE SURVEYS IN GEOPHYSICS)
AFCRC-E-	AIR FORCE CAMBRIDGE RESEARCH CENTER. ELECTRONICS RESEARCH DIRECTORATE, MASS.
AFCRC-GRD-	AIR FORCE CAMBRIDGE RESEARCH CENTER. GEOPHYSICS RESEARCH DIRECTORATE, MASS.
AFCRC-GRP-	AIR FORCE CAMBRIDGE RESEARCH CENTER, BEDFORD, MASS. (GEOPHYSICAL RESEARCH PAPER)
AFCRC-TM-(YEAR)-	AIR FORCE CAMBRIDGE RESEARCH CENTER, BEDFORD, MASS.
AFCRC-TN-(YEAR)-	AIR FORCE CAMBRIDGE RESEARCH CENTER, BEDFORD, MASS.
AFCRC-TR-(YEAR)-	AIR FORCE CAMBRIDGE RESEARCH CENTER, BEDFORD, MASS.
AFCRC-TR-(NUMBERS)-	AIR FORCE CAMBRIDGE RESEARCH LABS., BEDFORD, MASS.
AFCRL-	AIR FORCE CAMBRIDGE RESEARCH LABS., BEDFORD, MASS.
AFCRL-(YEAR)-	AIR FORCE CAMBRIDGE RESEARCH LABS., BEDFORD, MASS.
AFCRL-AFSG-	AIR FORCE CAMBRIDGE RESEARCH LABS., BEDFORD, MASS.
AFCRL-AFSIG-	AIR FORCE CAMBRIDGE RESEARCH LABS., BEDFORD, MASS.
AF-CRL-E-	AIR FORCE CAMBRIDGE RESEARCH LABS., BEDFORD, MASS.
AFCRL-ERP-	AIR FORCE CAMBRIDGE RESEARCH LABS., BEDFORD, MASS.
AFCRL-GRD-TR-	AIR FORCE CAMBRIDGE RESEARCH LABS., BEDFORD, MASS.
AFCRL-IP-	AIR FORCE CAMBRIDGE RESEARCH LABS., BEDFORD, MASS. (INSTRUMENTATION PAPER)
AFCRL-PMSRP-	AIR FORCE CAMBRIDGE RESEARCH LABS., BEDFORD, MASS. (PHYSICAL AND MATHEMATICAL RESEARCH PAPER)
AFCRL-PRSP-	AIR FORCE CAMBRIDGE RESEARCH LABS., BEDFORD, MASS. (PHYSICAL SCIENCE RESEARCH PAPER)
AFCRL-PSRP-	AIR FORCE CAMBRIDGE RESEARCH LABS., BEDFORD, MASS. (PHYSICAL SCIENCE RESEARCH PAPER)
AFCRL-SP-	AIR FORCE CAMBRIDGE RESEARCH LABS., BEDFORD, MASS.
AFCRL-SR-	AIR FORCE CAMBRIDGE RESEARCH LABS., BEDFORD, MASS. (SPECIAL REPORT)
AFCRL-TM-	AIR FORCE CAMBRIDGE RESEARCH LABS., BEDFORD, MASS.
AFCRL-TN-(YEAR)-	AIR FORCE CAMBRIDGE RESEARCH LABS., BEDFORD, MASS.
AFCRL-TR-(YEAR)-	AIR FORCE CAMBRIDGE RESEARCH LABS., BEDFORD, MASS.
AFCRL-TRANS-	AIR FORCE CAMBRIDGE RESEARCH LABS., BEDFORD, MASS. (TRANSLATION)
AFCS-	AIR FORCE COMMUNICATIONS SERVICE, SCOTT AFB, ILL.
AFCSA-	ASSISTANT CHIEF OF STAFF STUDIES AND ANALYSIS (AIR FORCE), WASHINGTON, D.C.
AFCS-OPLAN-	AIR FORCE COMMUNICATIONS SERVICE, SCOTT AFB, ILL.
AFCSR-(YEAR)-(NUMBER)TR	AIR FORCE OFFICE OF SCIENTIFIC RESEARCH, WASH., D.C.
AFCS-SS-	AIR FORCE COMMUNICATIONS SERVICE, SCOTT AFB, ILL.
AFCS-UM-	AIR FORCE COMMUNICATIONS SERVICE, SCOTT AFB, ILL.
AFC-VG-	ARMED FORCES SPECIAL WEAPONS PROJECT. FIELD COMMAND, ALBUQUERQUE, N. MEX. (VU-GRAPHS)
AFD-	HYDROGRAPHIC OFFICE, SUITLAND, MD. (AIR FACILITY DIRECTORIES)
AF-DI-	DIRECTORATE OF INTELLIGENCE (AIR FORCE), WASH., D.C.
AF-DI-PVIM-	DIRECTORATE OF INTELLIGENCE (AIR FORCE), WASHINGTON, D.C. (PHYSICAL VULNERABILITY INTERIM MEMORANDUM)
AFDL-TR-(YEAR)-	AIR FORCE FLIGHT DYNAMICS LAB., WRIGHT-PATTERSON AFB, OHIO
AFEEIS-	ALLIED FORCES ENEMY EQUIPMENT INTELLIGENCE SERVICE
AFEO-	(ADVANCED FUZE AND EXPLOSIVES ORDNANCE)
AF-EOAR-	OFFICE OF AEROSPACE RESEARCH (AIR FORCE). EUROPEAN OFFICE
AFEOB-	(ADVANCED FUZE AND EXPLOSIVES ORDNANCE BULLETIN)
AFES-	GT. BRIT. ADMIRALTY FUEL EXPERIMENTAL STATION
AFESD-	ELECTRONIC SYSTEMS DIV. (AIR FORCE), BEDFORD, MASS.
AFESD-TDR-(YEAR)-	ELECTRONIC SYSTEMS DIV. (AIR FORCE), BEDFORD, MASS.
AFESD-TN-(YEAR)-	ELECTRONIC SYSTEMS DIV. (AIR FORCE), BEDFORD, MASS.
AF-ETAC-TN-(YEAR)-	ENVIRONMENTAL TECHNICAL APPLICATIONS CENTER (AIR FORCE), WASHINGTON, D.C.
AFETO-	AMERICAN FORCES EUROPEAN THEATER OF OPERATIONS
AFETO CE-	AMERICAN FORCES EUROPEAN THEATER OF OPERATIONS. CHIEF ENGINEER
AFETO CSO-	AMERICAN FORCES EUROPEAN THEATER OF OPERATIONS. CHIEF SIGNAL OFFICER
AFETO CSO FR-	AMERICAN FORCES EUROPEAN THEATER OF OPERATIONS. CHIEF SIGNAL OFFICER (FIELD REPORT)
AFETO CSO I-	AMERICAN FORCES EUROPEAN THEATER OF OPERATIONS. CHIEF SIGNAL OFFICER (INTERROGATION)
AFETO CSO IR-	AMERICAN FORCES EUROPEAN THEATER OF OPERATIONS. CHIEF SIGNAL OFFICER (INTELLIGENCE REPORT)
AFETR-	AIR FORCE EASTERN TEST RANGE, PATRICK AFB, FLA.
AFETR-NRTOI-	AIR FORCE EASTERN TEST RANGE, PATRICK AFB, FLA.
AFETR-TR-	AIR FORCE EASTERN TEST RANGE, PATRICK AFB, FLA.
AFF-	ARMY FIELD FORCES, FORT BLISS, TEX.
AFFDC-FDCC-TM-	AIR FORCE FLIGHT DYNAMICS LAB. CONTROL CRITERIA BRANCH, WRIGHT-PATTERSON AFB, OHIO
AFFDL-	AIR FORCE FLIGHT DYNAMICS LAB., WRIGHT-PATTERSON AFB, OHIO
AFFDL-FDCL-TM-(YEAR)-	AIR FORCE FLIGHT DYNAMICS LAB., WRIGHT-PATTERSON AFB, OHIO
AFFDL-FDD-(YEAR)-	AIR FORCE FLIGHT DYNAMICS LAB. VEHICLE DYNAMICS DIV., WRIGHT-PATTERSON AFB, OHIO
AFFDL-FDFM-	AIR FORCE FLIGHT DYNAMICS LAB., WRIGHT-PATTERSON AFB, OHIO
AFFDL-FDFR-TM-	AIR FORCE FLIGHT DYNAMICS LAB., WRIGHT-PATTERSON AFB, OHIO
AFFDL-FDTR-TM-	AIR FORCE FLIGHT DYNAMICS LAB., WRIGHT-PATTERSON AFB, OHIO
AFFDL-TDR-(YEAR)-	AIR FORCE FLIGHT DYNAMICS LAB., WRIGHT-PATTERSON AFB, OHIO
AFFDL-TR-	AIR FORCE FLIGHT DYNAMICS LAB., WRIGHT-PATTERSON AFB, OHIO
AFFRI-ARR-	ARMED FORCES RADIOBIOLOGY RES. INST., BETHESDA, MD.
AFFRI-SR-(YEAR)-	ARMED FORCES RADIOBIOLOGY RES. INST., BETHESDA, MD.
AFFTC-SD-	AIR FORCE FLIGHT TEST CENTER, EDWARDS AFB, CALIF.
AFFTC-SP-	AIR FORCE FLIGHT TEST CENTER, EDWARDS AFB, CALIF.
AFFTC-TD-(YEAR)-	AIR FORCE FLIGHT TEST CENTER, EDWARDS AFB, CALIF.
AFFTC-TDR-	AIR FORCE FLIGHT TEST CENTER, EDWARDS AFB, CALIF. (TECHNICAL DOCUMENTARY REPORT)
AFFTC-TIH-	AIR FORCE FLIGHT TEST CENTER, EDWARDS AFB, CALIF.
AFFTC-TM-(YEAR)-	AIR FORCE FLIGHT TEST CENTER, EDWARDS AFB, CALIF.
AFFTC-TN-(YR)-	AIR FORCE FLIGHT TEST CENTER, EDWARDS AFB, CALIF.
AFFTC-TR-	AIR FORCE FLIGHT TEST CENTER, EDWARDS AFB, CALIF.
AFFTC-TR-(YR)-	AIR FORCE FLIGHT TEST CENTER, EDWARDS AFB, CALIF.
AF/GEN/	BRITISH AIRCRAFT CORP., PRESTON, LANCS, ENGLAND
AFGOA-MEMO-(YEAR)-	OPERATIONS ANALYSIS DIV. (AIR FORCE), WASHINGTON,D.C.
AFGOA-TN-(YEAR)-	OPERATIONS ANALYSIS DIV. (AIR FORCE), WASHINGTON,D.C.
AFGWC-TM-	AIR FORCE GLOBAL WEATHER CENTRAL, OFFUTT AFB, NEB.
AFHRL-TR	AIR FORCE HUMAN RESOURCES LAB., WRIGHT-PATTERSON AFB, OHIO
AFIAS-STUDY-	OFFICE OF THE INSPECTOR GENERAL, NORTON AFB, CALIF.
AF-IEPR-	INSTRUMENT PILOT INSTRUCTOR SCHOOL, RANDOLPH AFB, TEX. (INSTRUMENT EVALUATION)
AFIF-	SWITZERLAND. EIDGENOESSISCHE TECHNISCHE HOCHSCHULE, ZURICH

AFINS- NUCLEAR WEAPON SYSTEMS SAFETY GROUP, KIRTLAND AFB, NM
AF-IOI (MONTH/YEAR) AIR FORCE (INTERIM OPERATING INSTRUCTIONS)
AF-IR- DIRECTORATE OF INTELLIGENCE (AIR FORCE), WASH., D.C.
AFISP- INSPECTOR GENERAL OF THE AIR FORCE. DIRECTORATE OF SECURITY POLICE, WASHINGTON, D.C.
AF-IT- AIR FORCE INST. OF TECH., WRIGHT-PATTERSON AFB, OHIO
AFIT- AIR FORCE INST. OF TECH., WRIGHT-PATTERSON AFB, OHIO
AFITB-RPR- AIR FORCE INST. OF TECH., WRIGHT-PATTERSON AFB, OHIO. GRADUATE SCHOOL OF BUSINESS (RESEARCH PROJECT REPT)
AFITB-SS- AIR FORCE INST. OF TECH., WRIGHT-PATTERSON AFB, OHIO. GRADUATE SCHOOL OF BUSINESS (SEMINAR STUDY)
AFIT-CRANFIELD/EE AIR FORCE INST. OF TECH., WRIGHT-PATTERSON AFB, OHIO (WORK DONE AT COLL. OF AERONAUTICS, CRANFIELD, ENG)
AFIT GAE- AIR FORCE INST. OF TECH., WRIGHT-PATTERSON AFB, OHIO
AFIT GAS-(YEAR)- AIR FORCE INST. OF TECH., WRIGHT-PATTERSON AFB, OHIO
AFIT-GNE- AIR FORCE INST. OF TECH., WRIGHT-PATTERSON AFB, OHIO
AFITL-ALR- AIR FORCE INST. OF TECH., WRIGHT-PATTERSON AFB, OHIO. SCHOOL OF LOGISTICS (ADVANCED LOGISTICS REPORT)
AFITL-SRP- AIR FORCE INST. OF TECH., WRIGHT-PATTERSON AFB, OHIO. SCHOOL OF LOGISTICS (ADVANCED LOGISTICS REPORT)
AFIT-SEP- AIR FORCE INST. OF TECH., WRIGHT-PATTERSON AFB, OHIO. SCHOOL OF ENGINEERING
AFIT-SL- AIR FORCE INST. OF TECH., WRIGHT-PATTERSON AFB, OHIO. SCHOOL OF SYSTEMS AND LOGISTICS
AFIT-TR- AIR FORCE INST. OF TECH., WRIGHT-PATTERSON AFB, OHIO. SCHOOL OF ENGINEERING
AFLC-(YEAR)- AIR FORCE LOGISTICS COMMAND, WRIGHT-PATTERSON AFB, OHIO
AFLCP-(NUMBER)- AIR FORCE LOGISTICS COMMAND, WRIGHT-PATTERSON AFB, OHIO
AFLC-TR-(NUMBER-NUMBER) AIR FORCE LOGISTICS COMMAND, WRIGHT-PATTERSON AFB, OHIO
AF-M- AIR FORCE, WASHINGTON, D.C. (MANUAL)
AFM- AIR FORCE, WASHINGTON, D.C. (MANUAL)
AFM- ARMED FORCES MEDICAL POLICY COUNCIL, WASHINGTON, D.C.
AFMDC-(YEAR)- AIR FORCE MACHINABILITY DATA CENTER, CINCINNATI
AFMDC- AIR FORCE MISSILE DEV. CTR., HOLLOMAN AFB, N. MEX.
AFMDC ADJ-(YR)- AIR FORCE MISSILE DEV. CTR., HOLLOMAN AFB, N. MEX.
AFMDC-DAS-(YEAR)- AIR FORCE MISSILE DEV. CTR., HOLLOMAN AFB, N. MEX.
AFMDC HDGR-(YR)- AIR FORCE MISSILE DEV. CTR., HOLLOMAN AFB, N. MEX.
AFMDC MDGR-TM-(YR)- AIR FORCE MISSILE DEV. CTR., HOLLOMAN AFB, N. MEX.
AFMDC-TDR- AIR FORCE MISSILE DEVELOPMENT CENTER, HOLLOMAN AFB, N. MEX. (TECHNICAL DOCUMENTARY REPORT)
AFMDC TEST R- AIR FORCE MISSILE DEVELOPMENT CENTER, HOLLOMAN AFB, N. MEX. (TEST REPORT)
AFMDC-TN-(YR)- AIR FORCE MISSILE DEVELOPMENT CENTER, HOLLOMAN AFB, N. MEX. (TECHNICAL NOTE)
AFMDC-TR-(YR)- AIR FORCE MISSILE DEV. CTR., HOLLOMAN AFB, N. MEX.
AFMIC-TR- AIR FORCE MISSILE TEST CENTER, PATRICK AFB, FLA.
AFML- AIR FORCE MATERIALS LAB., WRIGHT-PATTERSON AFB, OHIO
AFML-AC (MONTH/YEAR) AIR FORCE MATERIALS LAB., WRIGHT-PATTERSON AFB, OHIO (ABSTRACTS OF ACTIVE CONTRACTS)
AFML-IR- AIR FORCE MATERIALS LAB., WRIGHT-PATTERSON AFB, OHIO
AFML-MAN-(YEAR)- AIR FORCE MATERIALS LAB., WRIGHT-PATTERSON AFB, OHIO
AFML-RP-(NUMBER)- AIR FORCE MATERIALS LAB., WRIGHT-PATTERSON AFB, OHIO
AFML-TDR- AIR FORCE MATERIALS LAB., WRIGHT-PATTERSON AFB, OHIO
AFML-TM-MAN-(YEAR)- AIR FORCE MATERIALS LAB., WRIGHT-PATTERSON AFB, OHIO
AFML-TR- AIR FORCE MATERIALS LAB., WRIGHT-PATTERSON AFB, OHIO
AFML-TR-(YEAR)- AIR FORCE MATERIALS LAB., WRIGHT-PATTERSON AFB, OHIO
AFMTC- AIR FORCE MISSILE TEST CENTER, PATRICK AFB, FLA.
AFMTC-FO- AIR FORCE MISSILE TEST CENTER, PATRICK AFB, FLA.
AFMTC-JUPITER-FTDR- AIR FORCE MISSILE TEST CENTER, PATRICK AFB, FLA.
AFMTC-MT-(YEAR)- AIR FORCE MISSILE TEST CENTER, PATRICK AFB, FLA.
AFMTC MTE-TM-(YR)- AIR FORCE MISSILE TEST CENTER, PATRICK AFB, FLA.
AFMTC-MTO- AIR FORCE MISSILE TEST CENTER, PATRICK AFB, FLA.
AFMTC-MTW-TM-(YEAR)- AIR FORCE MISSILE TEST CENTER, PATRICK AFB, FLA.
AFMTC-OD- AIR FORCE MISSILE TEST CENTER, PATRICK AFB, FLA.
AFMTC-OR- AIR FORCE MISSILE TEST CENTER, PATRICK AFB, FLA.
AFMTCP- AIR FORCE MISSILE TEST CENTER, PATRICK AFB, FLA.
AFMTCR- AIR FORCE MISSILE TEST CENTER, PATRICK AFB, FLA.
AFMTC-REDSTONE-FTDR- AIR FORCE MISSILE TEST CENTER, PATRICK AFB, FLA.
AFMTC-SNARK-FT- AIR FORCE MISSILE TEST CENTER, PATRICK AFB, FLA.
AFMTC-SNARK-FTDR- AIR FORCE MISSILE TEST CENTER, PATRICK AFB, FLA.
AFMTC-TDR- AIR FORCE MISSILE TEST CENTER, PATRICK AFB, FLA. (TECHNICAL DOCUMENTARY REPORT)
AFMTC TN-(YR)- AIR FORCE MISSILE TEST CENTER, PATRICK AFB, FLA.
AFMTC-TR-(YR)- AIR FORCE MISSILE TEST CENTER, PATRICK AFB, FLA.
AF-OAD- OPERATIONS ANALYSIS DIV. (AIR FORCE), WASHINGTON, D.C.
AF OA R- OPERATIONS ANALYSIS DIV. (AIR FORCE), WASHINGTON, D.C.
AFOAR-QI-(MO.-MO./YR.) OFFICE OF AEROSPACE RESEARCH (AIR FORCE), WASH., D.C. (QUARTERLY INDEX)
AFOAR-SRP-(YEAR/MONTH) OFFICE OF AEROSPACE RESEARCH (AIR FORCE), WASH., D.C. (STATUS OF RESEARCH PROPOSALS)
AF OA STM- OPERATIONS ANALYSIS DIV. (AIR FORCE), WASHINGTON, D.C.
AFOAT-1- OFFICE FOR ATOMIC ENERGY (AIR FORCE), WASHINGTON, D.C.
AF OA WP- OPERATIONS ANALYSIS DIV. (AIR FORCE), WASHINGTON, D.C.
AF-OMA- OFFICE OF MILITARY ATTACHE (AIR FORCE), LONDON
AFOOA- ARMY AIR FORCE. OFF. OF OPERATIONS ANALYSIS, WASH., DC
AFOOA-S- OPERATIONS ANALYSIS DIV. (AIR FORCE), WASHINGTON, D.C.
AFORS-(YEAR)-(NUMBER)TR AIR FORCE OFFICE OF SCIENTIFIC RESEARCH, WASH., D.C.
AFOSR-(YEAR)- AIR FORCE OFFICE OF SCIENTIFIC RESEARCH, ARLINGTON, VA
AFOSR- AIR FORCE OFFICE OF SCIENTIFIC RESEARCH, WASH., D.C.
AFOSR-(YEAR)-(NUMBER)TR AIR FORCE OFFICE OF SCIENTIFIC RESEARCH, WASH., D.C.
AFOSR-AFOSR-(YR.)-(NO)TR AIR FORCE OFFICE OF SCIENTIFIC RESEARCH, WASH., D.C.
AFOSR/DRA-(YEAR)- DIRECTORATE OF RESEARCH ANALYSES, HOLLOMAN AFB, N.M.
AFOSR-J(NUMBER) AIR FORCE OFFICE OF SCIENTIFIC RESEARCH, WASH., D.C.
AFOSR-O-(NUMBER)TR AIR FORCE OFFICE OF SCIENTIFIC RESEARCH, WASH., D.C.
AFOSR-SR-(YEAR)- AIR FORCE OFFICE OF SCIENTIFIC RESEARCH, WASH., D.C.
AFOSR-TD- AIR FORCE OFFICE OF SCIENTIFIC RESEARCH, WASH., D.C.
AFOSR-TN- AIR FORCE OFFICE OF SCIENTIFIC RESEARCH, WASH., D.C.
AFOSR-TN-(YEAR)- AIR FORCE OFFICE OF SCIENTIFIC RESEARCH, WASH., D.C.
AFOSR-TR- AIR FORCE OFFICE OF SCIENTIFIC RESEARCH, WASH., D.C.
AFOSR-TR-(YEAR)- AIR FORCE OFFICE OF SCIENTIFIC RESEARCH, WASH., D.C.
AFOTC- AIR FORCE OPERATIONAL TEST CENTER, EGLIN AFB, FLA.
AFP- AIR FORCE, WASHINGTON, D.C. (PAMPHLET)
AFPAC- ARMY FORCES IN THE PACIFIC
AFPAC ENG TTS- ARMY FORCES IN THE PACIFIC (ENGINEER, TECHNICAL AND TECHNOLOGICAL SURVEY REPORT)
AFPAC MED- ARMY FORCES IN THE PACIFIC (MEDICAL TECHNICAL INTELLIGENCE FIELD REPORT)
AFPAC MED TIFR- ARMY FORCES IN THE PACIFIC (MEDICAL TECHNICAL INTELLIGENCE FIELD REPORT)
AFPAC ORD- ARMY FORCES IN THE PACIFIC (ORDNANCE TECHNICAL INTELLIGENCE REPORT)
AFPAC ORD TIR- ARMY FORCES IN THE PACIFIC (ORDNANCE TECHNICAL INTELLIGENCE REPORT)
AFPDPL-PR-(YEAR) DIRECTORATE OF PERSONNEL PLANNING (AIR FORCE), WASHINGTON, D.C.
AFPR- STRATEGIC BOMBING SURVEY (AIR FRAMES PLANT REPORTS)
AFPS- ORDNANCE DEPT. ORDNANCE PACKAGING AND CRATING. FIRE CONTROL SUBOFFICE, FRANKFORD ARSENAL, PHILADELPHIA (PACKAGING INSTRUCTIONS)
AFPS-SAAMA- SAN ANTONIO AIR MATERIEL AREA, TEX.
AFPTRC-PL-TM-(NO.)- A.F. PERSONNEL + TRAINING RES. CTR., LACKLAND AFB, TEX.

AFPTRC-TN-(YEAR)-
A.F. PERSONNEL + TRAINING RES. CTR.,LACKLAND AFB,TEX.
AFPTRC-TR-(YEAR)-
A.F. PERSONNEL + TRAINING RES. CTR.,LACKLAND AFB,TEX.
AFR-
AIR FORCE, WASHINGTON, D.C. (REGULATION)
AFRAL-TR-
AIR FORCE ROCKET PROPULSION LAB., EDWARDS AFB, CALIF.
AFRD-(NUMBER)-
AVCO CORP. AVCO-EVERETT RESEARCH LAB., EVERETT, MASS.
AFRDC-
DEPUTY CHIEF OF STAFF. RESEARCH AND DEVELOPMENT
(AIR FORCE), WASHINGTON, D.C.
AFRDC-DPM-(YEAR)-
DEPUTY CHIEF OF STAFF, PLANS AND OPERATIONS (AIR
FORCE), WASHINGTON, D.C.
AFRDC-DPM-(YEAR)-
DEPUTY CHIEF OF STAFF. RESEARCH AND DEVELOPMENT
(AIR FORCE), WASHINGTON, D.C.
AFRDC-DPR-(YEAR)-
DEPUTY CHIEF OF STAFF. PLANS AND OPERATIONS (AIR
FORCE), WASHINGTON, D.C.
AFRDC-DPR-(YEAR)-
DEPUTY CHIEF OF STAFF. RESEARCH AND DEVELOPMENT
(AIR FORCE), WASHINGTON, D.C.
AFRDC-DPR-
OFFICE OF AEROSPACE RESEARCH (AIR FORCE), WASH., D.C.
(DEVELOPMENT PLANNING REPORT)
AFRDC-DPS-(YEAR)-
DEPUTY CHIEF OF STAFF. RESEARCH AND DEVELOPMENT
(AIR FORCE), WASHINGTON, D.C.
AFRDC-DPS-
OFFICE OF AEROSPACE RESEARCH (AIR FORCE), WASH., D.C.
(DEVELOPMENT PLANNING STUDY)
AFRD-QI-(YEAR/MO.-MO.)
RESEARCH DIV. (AIR FORCE), WASHINGTON, D.C.
(QUARTERLY INDEX)
AFRD-SRP-(YEAR/MONTH)
RESEARCH DIV. (AIR FORCE), WASHINGTON, D.C. (STATUS
OF RESEARCH PROPOSALS)
AFRFL-TR-
AIR FORCE ROCKET PROPULSION LAB., EDWARDS AFB, CALIF.
AFRICA-
OHIO UNIV., ATHENS. CENTER FOR INTERNATIONAL STUDIES
AFRII-SR-(YEAR)-
ARMED FORCES RADIOBIOLOGY RES. INST., BETHESDA, MD.
AFRO-
AIR FORCE OFFICE OF SCIENTIFIC RESEARCH, WASH., D.C.
AFRP-
DIRECTORATE OF NUCLEAR SAFETY, KIRTLAND AFB, N. MEX.
AFRPL-PR-(YEAR)-
AIR FORCE ROCKET PROPULSION LAB., EDWARDS AFB, CALIF.
AFRPL-TDR-
AIR FORCE ROCKET PROPULSION LAB., EDWARDS AFB, CALIF.
AFRPL-TM-
AIR FORCE ROCKET PROPULSION LAB., EDWARDS AFB, CALIF.
AFRPL-TR-
AIR FORCE ROCKET PROPULSION LAB., EDWARDS AFB, CALIF.
AFRPL-TR-(YEAR)-
AIR FORCE ROCKET PROPULSION LAB., EDWARDS AFB, CALIF.
AFRRI-
ARMED FORCES RADIOBIOLOGY RES. INST., BETHESDA, MD.
AFRRI-ARR (MO/YR-MO/YR)
ARMED FORCES RADIOBIOLOGY RES. INST., BETHESDA, MD.
(ANNUAL RESEARCH REPORT)
AFRRI-CR-(YEAR)-
ARMED FORCES RADIOBIOLOGY RES. INST., BETHESDA, MD.
AFRRI-SP-(YEAR)-
ARMED FORCES RADIOBIOLOGY RES. INST., BETHESDA, MD.
AFRRI-SR-
ARMED FORCES RADIOBIOLOGY RES. INST., BETHESDA, MD.
AFRRI-SR-(YEAR)-
ARMED FORCES RADIOBIOLOGY RES. INST., BETHESDA, MD.
AFRRI-TN-
ARMED FORCES RADIOBIOLOGY RES. INST., BETHESDA, MD.
AFRRI-TN-(YEAR)-
ARMED FORCES RADIOBIOLOGY RES. INST., BETHESDA, MD.
AFRTW-
ARMY AIR FORCE. WEATHER INFORMATION BRANCH
AFS-
(AIR FACILITIES SERIES)
AFSA-
ARMED FORCES SECURITY AGENCY, WASHINGTON, D.C.
AF-SAC-
STRATEGIC AIR COMMAND, OFFUTT AFB, NEB.
AF-SAM-(YEAR)-
SCHOOL OF AVIATION MEDICINE, BROOKS AFB, TEX.
AF-SAM-
SCHOOL OF AVIATION MEDICINE, RANDOLPH AFB, TEX.
AF-SAM (PROJ. NO.)
SCHOOL OF AVIATION MEDICINE, RANDOLPH AFB, TEX.
AF-SAM-(YR)-
SCHOOL OF AVIATION MEDICINE, RANDOLPH AFB, TEX.
AF-SAM-TDR-
SCHOOL OF AVIATION MEDICINE, BROOKS AFB, TEX. (TECH-
NICAL DOCUMENTARY REPORT)
AFSC-
AIR FORCE SYSTEMS COMMAND, ANDREWS AFB, MD.
AFSC-(NUMBER)-
AIR FORCE SYSTEMS COMMAND, ANDREWS AFB, MD.
AFSCA-E2-
DIRECTORATE OF STUDIES AND ANALYSIS (AIR FORCE),
WASHINGTON, D.C.
AFSC-CP-(NUMBERS)-(LTR)-
AIR FORCE SYSTEMS COMMAND, ANDREWS AFB, MD.
AFSC-CPN-
AIR FORCE SYSTEMS COMMAND, ANDREWS AFB, MD.
AFSC-HISTORICAL-PUB-
AIR FORCE SYSTEMS COMMAND, ANDREWS AFB, MD.

AFSC-HPS-(YEAR)-
AIR FORCE SYSTEMS COMMAND, ANDREWS AFB, MD.
AFSCM-(NUMBER)-
AIR FORCE SYSTEMS COMMAND, ANDREWS AFB, MD.
AFSC/NOR-
ARMED FORCES STAFF COLLEGE, NORFOLK, VA.
AFSC-SPEC-
AIR FORCE SYSTEMS COMMAND, ANDREWS AFB, MD.
AFSC-SPEC-(LTRS.)-
AIR FORCE SYSTEMS COMMAND, ANDREWS AFB, MD.
AFSC-SPEC-DOD-AIMS-(YR)-
AIR FORCE SYSTEMS COMMAND, ANDREWS AFB, MD.
AFSC-TDR-
AIR FORCE SYSTEMS COMMAND, ANDREWS AFB, MD. (TECH.
DOCUMENTARY REPORT)
AFSC-TOD-(NUMBER)(LTR.)
AIR FORCE SYSTEMS COMMAND, ANDREWS AFB, MD.
(TECHNICAL OBJECTIVE)
AFSC-TR-
AIR FORCE SYSTEMS COMMAND, ANDREWS AFB, MD.
AFSG-
AIR FORCE CAMBRIDGE RESEARCH LABS. METEOROLOGY LAB.,
BEDFORD, MASS. (AIR FORCE SURVEYS IN GEOPHYSICS)
AFSIG-
AIR FORCE CAMBRIDGE RESEARCH LABS. METEOROLOGY LAB.,
BEDFORD, MASS. (AIR FORCE SURVEYS IN GEOPHYSICS)
AFSOR-(YEAR)-(NUMBER)TR
AIR FORCE OFFICE OF SCIENTIFIC RESEARCH, WASH., D.C.
AFSWC-
AIR FORCE SPECIAL WEAPONS CENTER, KIRTLAND AFB, N.M.
AFSWC-FTR-
AIR FORCE SPECIAL WEAPONS CENTER, KIRTLAND AFB,
N. MEX. (FUNCTIONAL TEST REPORT)
AFSWC-KAFB-
AIR FORCE SPECIAL WEAPONS CENTER, KIRTLAND AFB, N.M.
AFSWC-NWSSG-(YEAR)-
AIR FORCE NUCLEAR WEAPON SYSTEM SAFETY GROUP, KIRT-
LAND AFB, N. MEX.
AFSWC-OP-(YEAR)-
AIR FORCE SPECIAL WEAPONS CENTER, KIRTLAND AFB,
N. MEX. (OPERATION PLANS)
AFSWC-SWODO-(YEAR)-
AIR FORCE SPECIAL WEAPONS CENTER, KIRTLAND AFB, N.M.
AFSWC-SWOTO-
AIR FORCE SPECIAL WEAPONS CENTER, KIRTLAND AFB, N.M.
AFSWC-SWR-
AIR FORCE SPECIAL WEAPONS CENTER, KIRTLAND AFB, N.M.
AFSWC-SWR-TM-(NUMBER)-
AIR FORCE SPECIAL WEAPONS CENTER, KIRTLAND AFB, N.M.
AFSWC-SWVN-(YEAR)-
AIR FORCE SPECIAL WEAPONS CENTER, KIRTLAND AFB, N.M.
AFSWC-TDR-
AIR FORCE SPECIAL WEAPONS CENTER, KIRTLAND AFB,
N. MEX. (TECHNICAL DOCUMENTARY REPORT)
AFSWC-TM-SWV-(YEAR)-
AIR FORCE SPECIAL WEAPONS CENTER, KIRTLAND AFB,
N. MEX. (TECHNICAL MEMORANDUM)
AFSWC-TN-(YR)-
AIR FORCE SPECIAL WEAPONS CENTER, KIRTLAND AFB, N.M.
AFSWC-TN-SWR-(YEAR)-
AIR FORCE SPECIAL WEAPONS CENTER, KIRTLAND AFB, N.M.
AFSWC-TN-SWW-(YEAR)-
AIR FORCE SPECIAL WEAPONS CENTER, KIRTLAND AFB, N.M.
AFSWC-TR-(YR)-
AIR FORCE SPECIAL WEAPONS CENTER, KIRTLAND AFB, N.M.
AFSWC-TR-SPEC-EC-
AIR FORCE SPECIAL WEAPONS CENTER, KIRTLAND AFB, N.M.
AFSWC-TR-SWR-(YEAR)-
AIR FORCE SPECIAL WEAPONS CENTER, KIRTLAND AFB, N.M.
AFSWC-TS-(YEAR)-
AIR FORCE SPECIAL WEAPONS CENTER, KIRTLAND AFB,
N. MEX. (TEST SUMMARY)
AFSWP-
ARMED FORCES SPECIAL WEAPONS PROJECT, WASHINGTON,D.C.
AFSWP/FC-
ARMED FORCES SPECIAL WEAPONS PROJECT. FIELD COMMAND,
ALBUQUERQUE, N. MEX.
AFSWP/FC (MO.YR.)-
ARMED FORCES SPECIAL WEAPONS PROJECT. FIELD COMMAND,
ALBUQUERQUE, N. MEX.
AFSWP/FC-JTG-(LTRS-NOS.)
ARMED FORCES SPECIAL WEAPONS PROJECT. FIELD COMMAND,
ALBUQUERQUE, N. MEX. (JOINT TASK GP. RPT.) (REF.18)
AFSWP-FS-
ARMED FORCES SPECIAL WEAPONS PROJECT. FIELD COMMAND,
ALBUQUERQUE, N. MEX.
AFSWP-NW-WB-
ARMED FORCES SPECIAL WEAPONS PROJECT. FIELD COMMAND,
ALBUQUERQUE, N. MEX.
AFSWP-SS-(YEAR)-
ARMED FORCES SPECIAL WEAPONS PROJECT. FIELD COMMAND,
ALBUQUERQUE, N. MEX.
AFSWP/SWP-
ARMED FORCES SPECIAL WEAPONS PROJECT, WASHINGTON,D.C.
AFSWP/SWPAN-
ARMED FORCES SPECIAL WEAPONS PROJECT, WASHINGTON,D.C.
AFSWP TL-
ARMED FORCES SPECIAL WEAPONS PROJECT, WASHINGTON,D.C.
(TECHNICAL LETTER SERIES)
AFSWP-TP-
JOINT SPECIAL WEAPONS PUBLICATIONS BOARD, ALBUQUER-
QUE, N. MEX. (REF. 14)
AFSWP-UTG-
ARMED FORCES SPECIAL WEAPONS PROJECT. FIELD COMMAND,
ALBUQUERQUE, N. MEX.
AFSWP-WE-
ARMED FORCES SPECIAL WEAPONS PROJECT. FIELD COMMAND,
ALBUQUERQUE, N. MEX.

AFSWP-WU-	ARMED FORCES SPECIAL WEAPONS PROJECT. FIELD COMMAND, ALBUQUERQUE, N. MEX.
AFTAC-	TACTICAL AIR COMMAND, LANGLEY AFB, VA.
AFTAC/LRSM-	AIR FORCE TECHNICAL APPLICATIONS CENTER, WASH., D.C. (LONG RANGE SEISMIC MEASUREMENTS PROJECT)
AFTAC-TR-(YEAR)-	AIR FORCE TECHNICAL APPLICATIONS CENTER, WASH., D.C.
AF-TDR-	AIR FORCE, WASHINGTON, D.C.
AFTIF-	AIR FORCE (TECHNICAL INFORMATION FILE)
AF-TO-11N-	AIR FORCE (TECHNICAL ORDER) (REF. 14)
AF T.O.(NUMBERS)-11N	JOINT ATOMIC WEAPONS PUBLICATIONS BOARD, ALBUQUERQUE, N. MEX. (REF. 14)
AF T.O.-11N-	JOINT ATOMIC WEAPONS PUBLICATIONS BOARD, ALBUQUERQUE, N. MEX. (REF. 14)
AFTPL-TR-	AIR FORCE ROCKET PROPULSION LAB., EDWARDS AFB, CALIF.
AF-TR-	AIR FORCE, WASHINGTON, D.C.
AFTR-	AIR FORCE, WASHINGTON, D.C. (TECHNICAL REPORT)
AFTRC-	AIR TRAINING COMMAND, SCOTT AFB, ILL.
AFTSW-	AIR WEATHER SERVICE, WASHINGTON, D.C.
AFTSW-	ARMY AIR FORCE. DIRECTORATE OF WEATHER
AFV-	AUSTRALIA. NATIONAL HEALTH AND MEDICAL RES. COUNCIL
AFV-	UNITED KINGDOM. FIGHTING VEHICLES RESEARCH AND DEVELOPMENT ESTAB., CHERTSEY, SURREY, ENGLAND
AF-WDPC-(YEAR)-	BALLISTIC MISSILES COMM. (AIR FORCE), WASHINGTON,D.C.
AFWL-	AIR FORCE WEAPONS LAB., KIRTLAND AFB, N. MEX.
AFWLAA-(YEAR)-	AIR FORCE WEAPONS LAB., KIRTLAND AFB, N. MEX.
AFWLAW-(YEAR)-	AIR FORCE WESTERN TEST RANGE, VANDENBERG AFB, CALIF.
AFWL CR-(MONTH-DAY-YEAR)	AIR FORCE WEAPONS LAB., KIRTLAND AFB, N. MEX.
AFWL-EMP-	AIR FORCE WEAPONS LAB., KIRTLAND AFB, N. MEX. (ELECTROMAGNETIC PULSE NOTE SERIES)
AFWL-TDR-(YEAR)-	AIR FORCE WEAPONS LAB., KIRTLAND AFB, N. MEX.
AFWL-TM-(YEAR)-	AIR FORCE WEAPONS LAB., KIRTLAND AFB, N. MEX.
AFWL-TR-(YEAR)-	AIR FORCE WEAPONS LAB., KIRTLAND AFB, N. MEX.
AFWP-	AIR FORCE, WRIGHT-PATTERSON AFB, OHIO
AFWTR-TR-(YEAR)-	AIR FORCE WESTERN TEST RANGE, VANDENBURG AFB, CALIF.
AFXDC-	DEPUTY CHIEF OF STAFF. PLANS AND OPERATIONS (AIR FORCE), WASHINGTON, D.C.
AFXPD-	DEPUTY CHIEF OF STAFF. PLANS AND OPERATIONS (AIR FORCE), WASHINGTON, D.C.
AG-	AERONAUTICAL STANDARDS GROUP (NAVY)
AG-	AMOCO CHEMICALS CORP., SEYMOUR, IND.
AG-	FEDERAL CIVIL DEFENSE ADMINISTRATION, WASHINGTON,D.C.
AG-	NORTH ATLANTIC TREATY ORGANIZATION. ADVISORY GROUP FOR AEROSPACE RESEARCH AND DEVELOPMENT, PARIS
AGA-	AERO GEO ASTRO CORP., ALEXANDRIA, VA.
AGA-	AMERICAN GAS ASSN., INC., N.Y.C.
AGAAD-	AD HOC GROUP ON HIGH ALTITUDE DETECTION
AGAAD-	OFFICE OF THE DIRECTOR OF DEFENSE (RES+ENG),WASH.,DC
AGAC-	AERO GEO ASTRO CORP., ALEXANDRIA, VA.
AGAC-	AERO GEO ASTRO CORP., CORONA, CALIF.
AGAC-	KELTEC INDUSTRIES INC., CORONA, CALIF.
AGA-RB-	AMERICAN GAS ASSN.TESTING LABS.,CLEVELAND(RES. BULL.)
AGARD-	NORTH ATLANTIC TREATY ORGANIZATION. ADVISORY GROUP FOR AEROSPACE RESEARCH AND DEVELOPMENT, PARIS
AGARD-AC/	NORTH ATLANTIC TREATY ORGANIZATION. ADVISORY GROUP FOR AEROSPACE RESEARCH AND DEVELOPMENT, PARIS
AGARD-ADVISORY-	NORTH ATLANTIC TREATY ORGANIZATION. ADVISORY GROUP FOR AEROSPACE RESEARCH AND DEVELOPMENT, PARIS
AGARD-AG-(NUMBER)-(YEAR)	NORTH ATLANTIC TREATY ORGANIZATION. ADVISORY GROUP FOR AEROSPACE RESEARCH AND DEVELOPMENT, PARIS
AGARD-AR-(NUMBER)-(YR.)	NORTH ATLANTIC TREATY ORGANIZATION. ADVISORY GROUP FOR AEROSPACE RESEARCH AND DEVELOPMENT, PARIS
AGARD-BIB-	NORTH ATLANTIC TREATY ORGANIZATION. ADVISORY GROUP FOR AEROSPACE RESEARCH AND DEVELOPMENT, PARIS
AGARD-BULL-	NORTH ATLANTIC TREATY ORGANIZATION. ADVISORY GROUP FOR AEROSPACE RESEARCH AND DEVELOPMENT, PARIS
AGARD-CP-	NORTH ATLANTIC TREATY ORGANIZATION. ADVISORY GROUP FOR AEROSPACE RESEARCH AND DEVELOPMENT, PARIS
AGARD-ES-	NORTH ATLANTIC TREATY ORGANIZATION. ADVISORY GROUP FOR AEROSPACE RESEARCH AND DEVELOPMENT, PARIS
AGARD-IB-(YEAR)/	NORTH ATLANTIC TREATY ORGANIZATION. ADVISORY GROUP FOR AEROSPACE RESEARCH AND DEVELOPMENT, PARIS
AGARD-IND-(NO./NO.)-(YR)	NORTH ATLANTIC TREATY ORGANIZATION. ADVISORY GROUP FOR AEROSPACE RESEARCH AND DEVELOPMENT, PARIS
AGARD-LS-	NORTH ATLANTIC TREATY ORGANIZATION. ADVISORY GROUP FOR AEROSPACE RESEARCH AND DEVELOPMENT, PARIS
AGARD-MAN-(NUMBER)-(YR.)	NORTH ATLANTIC TREATY ORGANIZATION. ADVISORY GROUP FOR AEROSPACE RESEARCH AND DEVELOPMENT, PARIS
AGARDOGRAPH-	NORTH ATLANTIC TREATY ORGANIZATION. ADVISORY GROUP FOR AEROSPACE RESEARCH AND DEVELOPMENT, PARIS
AGARD-OGRAPH-	NORTH ATLANTIC TREATY ORGANIZATION. ADVISORY GROUP FOR AEROSPACE RESEARCH AND DEVELOPMENT, PARIS
AGARD-R-(NUMBER)-(YEAR)	NORTH ATLANTIC TREATY ORGANIZATION. ADVISORY GROUP FOR AEROSPACE RESEARCH AND DEVELOPMENT, PARIS
AGARD-SPEC-	NORTH ATLANTIC TREATY ORGANIZATION. ADVISORY GROUP FOR AEROSPACE RESEARCH AND DEVELOPMENT, PARIS
AGC-	AEROJET-GENERAL CORP. (ALL LOCATIONS)
AGC-(NUMBERS)-FR	AEROJET-GENERAL CORP., EL MONTE, CALIF.
AGC-(NUMBERS)-SA	AEROJET-GENERAL CORP., EL MONTE, CALIF.
AGC-(NUMBERS)-SUMMARY	AEROJET-GENERAL CORP., EL MONTE, CALIF.
AGC-(NUMBERS)-(LETTERS)	AEROJET-GENERAL CORP., SACRAMENTO, CALIF.
AGC-(NUMBER)-(NUMBER)-	AEROJET-GENERAL CORP. ASTRIONICS DEPT., AZUSA, CALIF.
AGC-(NOS.)-(NOS.)-(NO.)	AEROJET-GENERAL CORP. ORDNANCE DIV., DOWNEY, CALIF.
AGC-(NUMBERS)-	AEROJET-GENERAL CORP. SOLID ROCKET PLANT, SACRAMENTO, CALIF.
AGC-A-	AEROJET-GENERAL CORP. AZUSA, CALIF.
AGC-AE-	AEROJET-GENERAL CORP., AZUSA, CALIF.
AGC-AN-	AEROJET-GENERAL NUCLEONICS, SAN RAMON, CALIF.
AGC-CR-	AEROJET-GENERAL CORP., SACRAMENTO, CALIF.
AGC-FTR-	ANDERSON, GREENWOOD + CO., HOUSTON, TEX.
AGC-L-	AEROJET-GENERAL CORP., AZUSA, CALIF.
AGC-LR-	AEROJET-GENERAL CORP. ADVANCED PROPELLANT DEPT., SACRAMENTO, CALIF.
AGC-LRP-	AEROJET-GENERAL CORP. ADVANCED PROPELLANT DEPT., SACRAMENTO, CALIF.
AGC-LRP-	AEROJET-GENERAL CORP. LIQUID ROCKET PLANT, SACRAMENTO, CALIF.
AGC-M001-	AEROJET NUCLEAR SYSTEMS CO., SACRAMENTO, CALIF.
AGC-NJD-	AEROJET-GENERAL CORP., SACRAMENTO, CALIF.
AGC-Q-	AEROJET-GENERAL CORP. (STATUS REPORT)
AGC-RM-	AEROJET-GENERAL CORP., SACRAMENTO, CALIF.
AGC-RMR-	AEROJET-GENERAL CORP., SACRAMENTO, CALIF.
AGC-RN-	AEROJET-GENERAL CORP., AZUSA, CALIF.
AGC-RN-	AEROJET-GENERAL CORP., SACRAMENTO, CALIF.
AGC-RN-Q-	AEROJET-GENERAL CORP., SACRAMENTO, CALIF.
AGC-RN-S-	AEROJET-GENERAL CORP., AZUSA, CALIF.(DESIGN CRITERIA)
AGC-RN-S-	AEROJET-GENERAL CORP., SACRAMENTO, CALIF.
AGC-RP-DR-	AEROJET-GENERAL CORP., SACRAMENTO, CALIF. (DEV. REPT)
AGC-RP-MR-	AEROJET-GENERAL CORP., SACRAMENTO, CALIF.
AGC-RP-P-	AEROJET-GENERAL CORP., SACRAMENTO, CALIF.
AGC-RP-QR-	AEROJET-GENERAL CORP., SACRAMENTO, CALIF.
AGC-RP-SR-	AEROJET-GENERAL CORP., SACRAMENTO, CALIF.
AGC-RP-TM-	AEROJET-GENERAL CORP., SACRAMENTO, CALIF.
AGC-RP-TR-	AEROJET-GENERAL CORP., SACRAMENTO, CALIF.

AGC-SD-	AEROJET-GENERAL CORP., EL MONTE, CALIF.
AGC-SELT-FR-	AEROJET-GENERAL CORP. PROPULSION DIV., SACRAMENTO, CALIF.
AGC SR-	AEROJET-GENERAL CORP., AZUSA, CALIF.
AGC-SRO-	AEROJET-GENERAL CORP., SACRAMENTO, CALIF. (SECOND QUARTERLY REPORT)
AGC-SRP-	AEROJET-GENERAL CORP. SOLID ROCKET PLANT, SACRAMENTO, CALIF.
AGC-SY-	AEROJET-GENERAL CORP., AZUSA, CALIF.
AGC-TCED-FR-	AEROJET-GENERAL CORP. SOLID ROCKET OPERATIONS, SACRAMENTO, CALIF.
AGC-TCED-QP-	AEROJET-GENERAL CORP., SACRAMENTO, CALIF.
AGC-TCR-	AEROJET-GENERAL CORP., AZUSA, CALIF.
AGC-TM(NO.)-	AEROJET-GENERAL CORP., AZUSA, CALIF. (TECH. MEMO.)
AGC-TM-	AEROJET-GENERAL CORP. LIQUID ROCKET PLANT, SACRAMENTO, CALIF.
AGC-TN-	AEROJET-GENERAL CORP., AZUSA, CALIF.
AGE-	GARRETT CORP. AIRESEARCH MFG. DIV., LOS ANGELES
AGE-	GT. BRIT. ADMIRALTY GUNNERY ESTABLISHMENT
AGET-	DEPT. OF DEFENSE. ADVISORY GROUP ON ELECTRON TUBES
AGF-	CONTINENTAL ARMY COMMAND, FORT MONROE, VA.
AGFSRS-	(AIRCRAFT GROUND FIRE SUPPRESSION AND RESCUE SYSTEM)
AGF-TFF-	ARMY GROUND FORCES (TASK FORCE FRIGID)
AGHAD-	OFFICE OF THE DIRECTOR OF DEFENSE (RES. + ENG.) ADVISORY GROUP ON HIGH ALTITUDE DETECTION, WASH.,D.C.
AGI-	AMERICAN GEOLOGICAL INST. TRANSLATIONS OFF.,WASH.,DC
AGIMRDA-	ARMY ENGINEER GEODESY, INTELLIGENCE AND MAPPING RESEARCH AND DEVELOPMENT AGENCY, FORT BELVOIR, VA.
AGM-	ADELPHI UNIV., GARDEN CITY, N.Y. DEPT. OF GRAD. MATH.
AGM-	AIR FORCE SYSTEMS COMMAND, WRIGHT-PATTERSON AFB, OHIO
AGMB-ST-	ANTIAIRCRAFT + GUIDED MISSILES BRANCH, FT. BLISS, TEX
AG-MR-	DOUGLAS AIRCRAFT CO., INC. (ARMAMENT GROUP MEMO. RPT)
AGN-	AEROJET-GENERAL NUCLEONICS, SAN RAMON, CALIF.
AGN-AN-	AEROJET-GENERAL NUCLEONICS, SAN RAMON, CALIF.
AGN-IDO-	AEROJET-GENERAL NUCLEONICS, SAN RAMON, CALIF.
AGN-RS-	AEROJET-GENERAL NUCLEONICS, SAN RAMON, CALIF.
AGN-TM-	AEROJET-GENERAL NUCLEONICS, SAN RAMON, CALIF.
AGN-TP-	AEROJET-GENERAL NUCLEONICS, SAN RAMON, CALIF.
AGO-	OFFICE OF THE ADJUTANT GENERAL, WASHINGTON, D.C.
AGRC/PWP/P-	UNITED KINGDOM ATOMIC ENERGY AUTHORITY. RESEARCH GP. ATOMIC ENERGY RES. ESTAB., HARWELL, BERKS, ENGLAND
AGREE-	DEPARTMENT OF DEFENSE. ADVISORY GROUP ON RELIABILITY OF ELECTRONIC EQUIPMENT
AGREE-(YR)-(MONTH)-	DEPARTMENT OF DEFENSE. ADVISORY GROUP ON RELIABILITY OF ELECTRONIC EQUIPMENT
AGR-FESG-P(YEAR)-	UNITED KINGDOM ATOMIC ENERGY AUTHORITY. REACTOR GP., RISLEY,LANCS,ENG.(ADVANCED GAS COOLED REACTOR)
AGRIC-	AGRICULTURAL RESEARCH SERVICE, BELTSVILLE, MD.
AGR/OC/P.	UNITED KINGDOM ATOMIC ENERGY AUTHORITY. PRODUCTION GROUP. WINDSCALE WORKS, SELLAFIELD, CUMB., ENGLAND
AGRR-TB-	STUTTGART. TECHNISCHE HOCHSCHULE. INSTITUT FUER AERODYNAMIK UND GASDYNAMIK
AGR/USAEC/	UNITED KINGDOM ATOMIC ENERGY AUTHORITY. INDUSTRIAL GROUP H.Q., RISLEY, LANCS, ENGLAND
AGS-	AMERICAN GEOGRAPHICAL SOCIETY, N.Y.C. (TRANSLATIONS)
AGSCD-	BROOKHAVEN NATIONAL LAB. ACCELERATOR DEPT., UPTON, N.Y. (ALTERNATING-GRADIENT SYNCHROTRON)
AG(SS)-	NAVAL TORPEDO STATION, KEYPORT, WASH.
AGT-	GENERAL ELECTRIC CO. AIRCRAFT GAS TURBINE DIV., CINCINNATI
AGU-	AMERICAN GEOPHYSICAL UNION, WASHINGTON, D.C.
AH-	AMERICAN HELICOPTER CO., INC., MANHATTAN BEACH, CALIF
AH-	AMERICAN HELICOPTER DIV., FAIRCHILD ENGINE AND AIRPLANE CORP., MESA, ARIZ.
AH-	ARMY COMBAT DEVELOPMENTS COMMAND. INST. OF SPECIAL STUDIES, FORT BELVOIR, VA.
AHC-	AMERICAN HELICOPTER CO., INC., MANHATTAN BEACH, CALIF
AHCO-	AMERICAN HELICOPTER CO., INC., MANHATTAN BEACH, CALIF
AHD-B3-	TENNESSEE EASTMAN CORP., OAK RIDGE, TENN.
AHE-IM-	ARMED FORCES SPECIAL WEAPONS PROJECT. FIELD COMMAND, ALBUQUERQUE, N. MEX. (INSTRUCTORS MANUAL)
AHE-NB-	ARMED FORCES SPECIAL WEAPONS PROJECT. FIELD COMMAND, ALBUQUERQUE, N. MEX. (NOTEBOOK)
AHE-VG-	ARMED FORCES SPECIAL WEAPONS PROJECT. FIELD COMMAND, ALBUQUERQUE, N. MEX. (VU-GRAPHS)
AHM-13-	TENNESSEE EASTMAN CORP., OAK RIDGE, TENN.
AHM-IM-	ARMED FORCES SPECIAL WEAPONS PROJECT. FIELD COMMAND, ALBUQUERQUE, N. MEX. (INSTRUCTORS MANUAL)
AHM-NB-	ARMED FORCES SPECIAL WEAPONS PROJECT. FIELD COMMAND, ALBUQUERQUE, N. MEX. (NOTEBOOK)
AHP-	(ALLIED HYDROGRAPHIC PUBLICATION)
59A-HSB-	SHARPLES CORP., PHILADELPHIA
AHSB-	UNITED KINGDOM ATOMIC ENERGY AUTHORITY. AUTHORITY HEALTH AND SAFETY BRANCH, RISLEY, LANCS, ENGLAND
AHSB(A)-R-	UNITED KINGDOM ATOMIC ENERGY AUTHORITY. AUTHORITY HEALTH AND SAFETY BRANCH, HARWELL, BERKS, ENGLAND
AHSB(A)-R-	UNITED KINGDOM ATOMIC ENERGY AUTHORITY. AUTHORITY HEALTH AND SAFETY BRANCH, LONDON
AHSB-MEMO-	UNITED KINGDOM ATOMIC ENERGY AUTHORITY. AUTHORITY HEALTH AND SAFETY BRANCH, RISLEY, LANCS, ENGLAND
AHSB-REPORT-	UNITED KINGDOM ATOMIC ENERGY AUTHORITY. AUTHORITY HEALTH AND SAFETY BRANCH, RISLEY, LANCS, ENGLAND
AHSB(RP)-	UNITED KINGDOM ATOMIC ENERGY AUTHORITY. AUTHORITY HEALTH AND SAFETY BRANCH. RADIOLOGICAL PROTECTION DIV., HARWELL, BERKS, ENGLAND
AHSB (RP) (LETTER)-	UNITED KINGDOM ATOMIC ENERGY AUTHORITY. AUTHORITY HEALTH AND SAFETY BRANCH. RADIOLOGICAL PROTECTION DIV., HARWELL, BERKS, ENGLAND
AHSB/RP/P-	UNITED KINGDOM ATOMIC ENERGY AUTHORITY. RESEARCH GP. ATOMIC ENERGY RES. ESTAB., HARWELL, BERKS, ENGLAND
AHSB/RP/-R-	UNITED KINGDOM ATOMIC ENERGY AUTHORITY. AUTHORITY HEALTH AND SAFETY BRANCH, LONDON
AHSB(RP)R-	UNITED KINGDOM ATOMIC ENERGY AUTHORITY. AUTHORITY HEALTH AND SAFETY BRANCH. RADIOLOGICAL PROTECTION DIV., HARWELL, BERKS, ENGLAND
AHSB (S) (LETTER)-	UNITED KINGDOM ATOMIC ENERGY AUTHORITY. AUTHORITY HEALTH AND SAFETY BRANCH, RISLEY, LANCS, ENGLAND
AHSB (S) R-	UNITED KINGDOM ATOMIC ENERGY AUTHORITY. AUTHORITY HEALTH AND SAFETY BRANCH. SAFEGUARDS DIV., LONDON
AHT-	STANFORD UNIV., CALIF. DEPT. OF MECHANICAL ENG., THERMOSCIENCES DIV.
AI-	(AIRCRAFT INTERCEPTION)
AI-(YEAR)-	ATOMICS INTERNATIONAL, CANOGA PARK, CALIF.
AI-	ATOMICS INTERNATIONAL DIV., NORTH AMERICAN AVIATION, INC., CANOGA PARK, CALIF.
AI-(YEAR)-MEMO-	ATOMICS INTERNATIONAL DIV., NORTH AMERICAN AVIATION, INC., CANOGA PARK, CALIF.
AI-	NAVAL AIR DEV. CTR. AERO. INSTRS. LAB.,JOHNSVILLE,PA.
AI-	SOUTHERN CALIFORNIA EDISON CO.
AIA-	AEROSPACE INDUSTRIES ASSN., LOS ANGELES
AIA-	AIRCRAFT INDUSTRIES ASSOCIATION, WASHINGTON, D.C.
AIAA-	AMERICAN INST. OF AERONAUTICS AND ASTRONAUTICS, N.Y.C. (PAPERS)
AIAA-CP-	AMERICAN INST. OF AERONAUTICS AND ASTRONAUTICS, N.Y.C. (CONFERENCE PUBLICATIONS)
AIAA-PAPER-(YEAR)-	AMERICAN INST. OF AERONAUTICS AND ASTRONAUTICS, N.Y.C. (PAPERS)
AIA-ARTC-	AIRCRAFT INDUSTRIES ASSN., AIRCRAFT RESEARCH AND TESTING COMMITTEE, WASHINGTON, D.C.
AI-AEC-	ATOMICS INTERNATIONAL, CANOGA PARK, CALIF.
AI-AEC-MEMO-	ATOMICS INTERNATIONAL, CANOGA PARK, CALIF.

AIAG-	ALUMINIUM-INDUSTRIE-AKTIEN-GESELLSCHAFT, NEUHAUSEN AM RHEINFALL, SWITZERLAND
AIAR-	AMERICAN INST. FOR AEROLOGICAL RES., PASADENA, CALIF.
AIBS-	AMERICAN INST. OF BIOLOGICAL SCIENCES, WASHINGTON, DC
AIC-	AMERICAN CAR + FOUNDRY CO. AVION DIV., PARAMUS, N.J.
AIC-	ARMY, WASHINGTON, D.C. (AMMUNITION IDENT. CODE)
AIC-	DEPT. OF AGRICULTURE. EASTERN REGIONAL RESEARCH LAB., WYNDMOOR, PENNA. (AGRICULTURAL + IND. CHEMISTRY)
AIC-	GT. BRIT. AIR INTERCEPTION COMMITTEE
AICBM-	ARMED FORCES SPECIAL WEAPONS PROJECT, WASHINGTON, D.C.
AICBM-	NAVAL RESEARCH LAB., WASHINGTON, D.C.
AICE-	AMERICAN INST. OF CROP ECOLOGY, WASHINGTON, D.C.
AI-CE-	ATOMICS INTERNATIONAL, CANOGA PARK, CALIF. (JOINT VENTURE WITH COMBUSTION ENGINEERING, INC. FOR THE HEAVY WATER ORGANIC COOLED REACTOR)
AI-CE-	GEOLOGICAL SURVEY, DENVER
AI-CE-MEMO-	ATOMICS INTERNATIONAL, CANOGA PARK, CALIF. (JOINT VENTURE WITH COMBUSTION ENGINEERING, INC. FOR THE HEAVY WATER ORGANIC COOLED REACTOR)
AI-CE-TDR-	ATOMICS INTERNATIONAL, CANOGA PARK, CALIF. (JOINT VENTURE WITH COMBUSTION ENGINEERING, INC. FOR THE HEAVY WATER ORGANIC COOLED REACTOR)
AIC/P-	UNITED KINGDOM ATOMIC ENERGY AUTHORITY. RESEARCH GP. ATOMIC ENERGY RES. ESTAB., HARWELL, BERKS, ENGLAND
AICS-	INDIA. BHABHA ATOMIC RESEARCH CENTRE, BOMBAY
AID-	GT. BRIT. AERONAUTICAL INSPECTION DIRECTORATE, HAREFIELD, MIDDX, ENGLAND
AID-	LIBRARY OF CONGRESS. AIR INFORMATION DIV., WASH., D.C.
AID-(YR)-(NO.)	LIBRARY OF CONGRESS. AIR INFORMATION DIV., WASH., D.C.
AID/D.(NUMBER/YEAR)	UNITED KINGDOM. AERONAUTICAL INSPECTION DIRECTORATE, HAREFIELD, MIDDX., ENGLAND
AID/DEV/(NUMBER/YEAR)	UNITED KINGDOM. AERONAUTICAL INSPECTION DIRECTORATE, HAREFIELD, MIDDX., ENGLAND
AID-DP-	AGENCY FOR INTERNATIONAL DEVELOPMENT. OFFICE OF PROGRAM AND POLICY COORDINATION, WASHINGTON, D.C.
AID/MET/	GT. BRIT. MINISTRY OF SUPPLY. AID LABS., HAREFIELD, MIDDX, ENGLAND
AID/METRO/	GT. BRIT. AERONAUTICAL INSPECTION DIRECTORATE, HAREFIELD, MIDDX, ENGLAND
AID-METRO-	GT. BRIT. AERONAUTICAL INSPECTION DIRECTORATE, HAREFIELD, MIDDX, ENGLAND
AID-P-	LIBRARY OF CONGRESS. AEROSPACE INFO. DIV., WASH., D.C.
AID-P-(YEAR)-	LIBRARY OF CONGRESS. AIR INFORMATION DIV., WASH., D.C.
AIDR-	AIR FORCE SYSTEMS COMMAND, EGLIN AFB, FLA.
AID-REPORT-	LIBRARY OF CONGRESS. AEROSPACE INFO. DIV., WASH., D.C.
AID REPT. B-	LIBRARY OF CONGRESS. AEROSPACE INFO. DIV., WASH., D.C.
AID REPT. P-	LIBRARY OF CONGRESS. AEROSPACE INFO. DIV., WASH., D.C.
AID REPT. T-	LIBRARY OF CONGRESS. AEROSPACE INFO. DIV., WASH., D.C.
AID-T-(YR)-(NO.)-	LIBRARY OF CONGRESS. AEROSPACE INFO. DIV., WASH., D.C.
AID-T-(YEAR)-	LIBRARY OF CONGRESS. AEROSPACE INFO. DIV., WASH., D.C.
AID-U-(YEAR)-	LIBRARY OF CONGRESS. AEROSPACE INFO. DIV., WASH., D.C.
AID-U-(YR)-(NO.)-	LIBRARY OF CONGRESS. AEROSPACE TECHNOLOGY DIV., WASHINGTON, D.C.
AID-U-	LIBRARY OF CONGRESS. AIR INFORMATION DIV., WASH., D.C.
AIEE-	AMERICAN INST. OF ELECTRICAL ENGINEERS, N.Y.C.
AIEU-	GT. BRIT. ROYAL AIRCRAFT EST. ARMAMENT + INSTR. UNIT
AIGA-	INSPECTORATE GENERAL OF AIR (ARMY) (MANUAL)
AIL-	AIRBORNE INSTRUMENTS LAB., INC., DEER PARK, N.Y.
AIL-	AIRBORNE INSTRUMENTS LAB., INC., MINEOLA, N.Y.
AI-L-(YEAR)AT(NO.)	ATOMICS INTERNATIONAL DIV., NORTH AMERICAN AVIATION, INC., CANOGA PARK, CALIF.
AIL-	COLUMBIA UNIV., N.Y.C. AIRBORNE INSTRUMENTS LAB.
AIL-(NUMBER)-I-	CUTLER-HAMMER, INC., DEER PARK, N.J.
AIL-	NAVAL AIR EXPTL. STA. AERO. INSTRS. LAB., PHILA.
AIL-D/N(NUMBERS)	AIRBORNE INSTRUMENTS LAB., INC., DEER PARK, N.Y.
AIM-	ACADEMY FOR INTERSCIENCE METHODOLOGY, CHICAGO
A-IM-(LETTERS)-	ARMED FORCES SPECIAL WEAPONS PROJECT. FIELD COMMAND, ALBUQUERQUE, N. MEX. (INSTRUCTORS MANUALS)
AI-MEMO-	ATOMICS INTERNATIONAL DIV., NORTH AMERICAN AVIATION, INC., CANOGA PARK, CALIF.
AINA-	ARCTIC INSTITUTE OF NORTH AMERICA, WASHINGTON, D.C.
AI NAA-SR-	ATOMICS INTERNATIONAL DIV., NORTH AMERICAN AVIATION, INC., CANOGA PARK, CALIF.
AINA-RP-	ARCTIC INSTITUTE OF NORTH AMERICA, MONTREAL
AINA-TR-	ARCTIC INSTITUTE OF NORTH AMERICA, WASHINGTON, D.C.
AIP-	AMERICAN INST. OF PHYSICS, N.Y.C. (TRANSLATIONS)
AIP/45 (LETTER)	AMERICAN INST. OF PHYSICS, N.Y.C.
AI-P-	ATOMICS INTERNATIONAL DIV., NORTH AMERICAN AVIATION, INC., CANOGA PARK, CALIF.
AI-PC-	ATOMICS INTERNATIONAL DIV., NORTH AMERICAN AVIATION, INC., CANOGA PARK, CALIF.
AIP-ID-	AMERICAN INST. OF PHYSICS. INFORMATION DIV., N.Y.C.
AIP/SDD-	AMERICAN INST. OF PHYSICS. SYSTEMS DEV. DIV., N.Y.C.
AIP/UDC-	AMERICAN INST. OF PHYSICS, N.Y.C. (UNIVERSAL DECIMAL CLASSIFICATION PROJECT)
AIR-	AMERICAN INST. FOR RESEARCH, PITTSBURGH
AIR-(NUMBER-DATE-LTRS.)-	AMERICAN INSTS. FOR RESEARCH, PITTSBURGH
AIR-	MICHIGAN. UNIV., ANN ARBOR. ENGINEERING RES. INST.
AIR-	NAVAL AIR SYSTEMS COMMAND, WASHINGTON, D.C.
AIR-	SOCIETY OF AUTOMOTIVE ENGINEERS, N.Y.C.
AIRA-	AIRCRAFT ARMAMENTS, INC., COCKEYSVILLE, MD.
AIR-B(NO.-DATE-LTRS.)-	AMERICAN INSTS. FOR RESEARCH, PITTSBURGH
AIR-C(NO.-DATE-LTRS.)-	AMERICAN INSTS. FOR RESEARCH, PITTSBURGH
AIR-C28-(DATE)-TR-	AMERICAN INSTS. FOR RESEARCH IN THE BEHAVIORAL SCIENCES, PALO ALTO, CALIF.
AIRCO-C-	AIR REDUCTION CO., INC. CENTRAL RESEARCH LABS., MURRAY HILL, N.J.
AIRCO-MISC.-	STACK GAS PROBLEM WORKING GROUP, AEC
AIR-D-	AMERICAN INSTS. FOR RESEARCH, PITTSBURGH
AIR-D(NUMBER)B-	AMERICAN INSTS. FOR RESEARCH. INST. FOR PERFORMANCE TECHNOLOGY, PITTSBURGH
AIR-E(NO.)-(DATE)(LTR.)	AMERICAN INST. FOR RESEARCH, PITTSBURGH
AIR-E-	AMERICAN INSTS. FOR RESEARCH, SILVER SPRING, MD.
AIR-E(NUMBER)-	AMERICAN INSTS. FOR RESEARCH, SILVER SPRING, MD.
AI-REPORT-	STOCKHOLM. ARBETSMEDICINSKA INSTITUTET
A.I. REPORT NO.-	NAPLES. UNIVERSITA. ISTITUTO DI AERONAUTICA
AIR-ES-	AMERICAN INSTS. FOR RESEARCH, SILVER SPRING, MD.
AIRESEARCH-F-	GARRETT CORP. AIRESEARCH MFG. CO. OF ARIZONA, PHOENIX
AIRESEARCH-FC-	GARRETT CORP. AIRESEARCH MFG. DIV., LOS ANGELES
AIR-F-(DATE)-TR-	AMERICAN INSTS. FOR RESEARCH, SILVER SPRING, MD.
AIR FORCE T.O.(NOS.)-11N	JOINT ATOMIC WEAPONS PUBLICATIONS BOARD, ALBUQUERQUE, N. MEX. (REF. 14)
AIR FORCE T.O.-11N-	JOINT ATOMIC WEAPONS PUBLICATIONS BOARD, ALBUQUERQUE, N. MEX. (REF. 14)
AIR-G(NO.)-(DATE)(LTR.)	AMERICAN INST. FOR RESEARCH, PITTSBURGH
AIRIC	(AERIAL ICE OBSERVATIONS)
AIR-K-	GARRETT CORP. AIRESEARCH MFG. DIV., LOS ANGELES
AIRL-	NORTHWEST AIRLINES, INC. AERONAUTICAL ICE RESEARCH LABORATORY, MINNEAPOLIS
AIRL-	SMITH, HINCHMAN + GRYLLS, INC. AERONAUTICAL ICING RESEARCH LAB., YPSILANTI, MICH.
AIR-R(NUMBER)-	AMERICAN INSTS. FOR RESEARCH, PITTSBURGH
AIR-RP(NUMBER)-	AMERICAN INSTS. FOR RESEARCH, PITTSBURGH
AIRSOUTH-	ALLIED AIR FORCES SOUTHERN EUROPE, NAPLES
AIRU-	AIR UNIV., MAXWELL AFB, ALA.

AIRU ACSC-	AIR COMMAND AND STAFF COLL., MAXWELL AFB, ALA.
AIRU AM-	SCHOOL OF AVIATION MEDICINE, RANDOLPH AFB, TEX.
AIRU AM SRU-	SCHOOL OF AVIATION MEDICINE, RANDOLPH AFB, TEX.
AIRU-ANPBP-	SCHOOL OF AVIATION MEDICINE, RANDOLPH AFB, TEX. (AIRCRAFT NUCLEAR PROPULSION BIOMEDICAL SUMMARY)
AIRU-AU-	AIR WAR COLL., MAXWELL AFB, ALA. EVALUATION STAFF
AIRU AU-(NUMBER)-ES	AIR WAR COLL., MAXWELL AFB, ALA. EVALUATION STAFF
AIRU-AWC-	AIR WAR COLLEGE, MAXWELL AFB, ALA.
AIS-	ARGUS INFORMATION SERVICE, BALTIMORE (TRANSLATIONS)
AIS-	ARTILLERY INFORMATION SERVICE
AIS-	BUREAU OF ORDNANCE (NAVY) (AD INTERIM SPECIFICATION)
AI-TRANS-	ATOMICS INTERNATIONAL DIV., NORTH AMERICAN AVIATION, INC., CANOGA PARK, CALIF.
AJ-(NUMBER)-	SPERRY GYROSCOPE CO., GREAT NECK, N.Y.
AJA-	AGBABIAN-JACOBSEN ASSOCIATES, LOS ANGELES
AJA-R-(NUMBER)-(NUMBER)	AGBABIAN-JACOBSEN ASSOCIATES, LOS ANGELES
AJBC-	BOYNTON (A.J.) AND CO., CHICAGO
AJEC-	AEROJET-GENERAL CORP., AZUSA, CALIF.
AJEC LRP-	AEROJET-GENERAL CORP. LIQUID ROCKET PLANT, SACRAMENTO, CALIF.
AJM-	TENNESSEE EASTMAN CORP., OAK RIDGE, TENN.
AK-	ADVANCED KINETICS, INC., COSTA MESA, CALIF.
AK-(NUMBER)-	SPERRY GYROSCOPE CO., GREAT NECK, N.Y.
AKBAATC-B-(NO.)-(LTR)-	ARMY AIR DEFENSE CENTER, FORT BLISS, TEX.
AKE-IM-	ARMED FORCES SPECIAL WEAPONS PROJECT. FIELD COMMAND, ALBUQUERQUE, N. MEX. (INSTRUCTORS MANUAL)
AKE-NB-	ARMED FORCES SPECIAL WEAPONS PROJECT. FIELD COMMAND, ALBUQUERQUE, N. MEX. (NOTEBOOK)
AKE-VG-	ARMED FORCES SPECIAL WEAPONS PROJECT. FIELD COMMAND, ALBUQUERQUE, N. MEX. (VU-GRAPHS)
AKM-IM-	ARMED FORCES SPECIAL WEAPONS PROJECT. FIELD COMMAND, ALBUQUERQUE, N. MEX. (INSTRUCTORS MANUAL)
AKM-NB-	ARMED FORCES SPECIAL WEAPONS PROJECT. FIELD COMMAND, ALBUQUERQUE, N. MEX. (NOTEBOOK)
AL-	(ACCESSION LIST) (REF. 2)
AL-	(ACOUSTICS LABORATORY) (REF. 2)
AL-	(AERONAUTICAL LABORATORY) (REF. 2)
AL-	ABBOTT LABS., NORTH CHICAGO, ILL.
AL-(NUMBER)-H	ACF INDUSTRIES, INC., ALBUQUERQUE, N. MEX.
AL-	AERONOMY LAB., BOULDER, COLO.
AL-	AERONUTRONIC, NEWPORT BEACH, CALIF.
AL-	AIR FORCE AVIONICS LAB., WRIGHT-PATTERSON AFB, OHIO
AL-(NO.-LTR-NO.-LTR)	ANDERSEN LABS, INC., BLOOMFIELD, CONN.
AL-	CALIFORNIA. UNIV., BERKELEY
AL-	COLUMBIA UNIV., N.Y.C. ACOUSTICS LAB.
AL-	COLUMBIA UNIV., NYC, SUBSTITUTE ALLOY MATERIALS LABS.
AL-	DAVID TAYLOR MODEL BASIN. AERODYNAMICS LAB., WASH.,DC
AL-	EG+G, INC., ALBUQUERQUE, N. MEX.
AL-	GOODYEAR AEROSPACE CORP., LITCHFIELD PARK, ARIZ.
AL-	NAVAL SHIP RESEARCH AND DEVELOPMENT CENTER. AERODYNAMICS LAB., WASHINGTON, D.C.
AL-1900-	NORTH AMER. AVIATION, INC. AEROPHYSICS DEPT. (NAVAHO)
AL-	NORTH AMERICAN AVIATION, INC. AEROPHYSICS LAB., DOWNEY, CALIF.
AL-	OFFICE OF NAVAL RESEARCH (AIRBORNE LANDING OPERATIONS
AL-	RES. + DEV. BD.,WASH.,DC (AIRBORNE LANDING OPERATIONS
AL-	ROCKETDYNE DIV., NORTH AMERICAN AVIATION, INC., CANOGA PARK, CALIF.
AL(YEAR)L(NO.)	SKF INDUSTRIES, INC. RES. LAB., KING OF PRUSSIA, PA.
AL-(YEAR)Q-	SKF INDUSTRIES, INC. RES. LAB., KING OF PRUSSIA, PA.
AL(YEAR)T(NO.)	SKF INDUSTRIES, INC. RES. LAB., KING OF PRUSSIA, PA.
AL-	SWEDEN. KUNGLIGA TEKNISKA HOEGSKOLAN, STOCKHOLM
ALA-	AMERICAN LIBRARY ASSOCIATION, CHICAGO
ALAKA-	LACQUER + CHEM. CORP. ALAKA RES. LAB., BROOKLYN
AL-ASD-TDR-	GEORGIA INST. OF TECH., ATLANTA
ALASU-	ALASKA. UNIV., COLLEGE
ALAU-	ALABAMA. UNIV., TUSCALOOSA
ALBS-	UNITED STATES LEGATION, BERNE, SWITZERLAND
ALBU-	ALBUQUERQUE URBAN OBSERVATORY, N. MEX.
AL-C-	DAVID TAYLOR MODEL BASIN. AERODYNAMICS LAB., WASH.,DC
ALCC-	ALLIED CHEMICAL CORP., N.Y.C.
ALCO-	AMERICAN LOCOMOTIVE CO., N.Y.C.
ALCOA-	ALUMINUM CO. OF AMERICA. ALCOA RESEARCH LABS., NEW KENSINGTON, PENNA.
ALE-IM-	ARMED FORCES SPECIAL WEAPONS PROJECT. FIELD COMMAND, ALBUQUERQUE, N. MEX. (INSTRUCTORS MANUAL)
ALEUTIAN-	GEOLOGICAL SURVEY, MENLO PARK, CALIF.
ALE-VG-	ARMED FORCES SPECIAL WEAPONS PROJECT. FIELD COMMAND, ALBUQUERQUE, N. MEX. (VU-GRAPHS)
ALFCO-	AMERICAN-LA FRANCE-FOAMITE CORP., ELMIRA, N.Y.
ALFLIN-	FLINT (ALBERT D.) ASSOCIATES, INC.
ALFREDU-	ALFRED UNIV., N.Y.
ALG-	NEW YORK UNIV., N.Y.C.
ALI-	ACTON LABS., INC., MASS.
ALI-	INGALLS (ARTHUR L.), ANN ARBOR, MICH.
ALI-	LITTLE (ARTHUR D.), INC., CAMBRIDGE, MASS.
ALIB-SPEC. BIB.-	ARMY LIBRARY, WASHINGTON, D.C.
ALI-C-	LITTLE (ARTHUR D.), INC., CAMBRIDGE, MASS.
ALL-	ALUMINIUM LABORATORIES LTD., BANBURY, OXFORD.,ENGLAND
ALLISON-EDR-	ALLISON DIV., GENERAL MOTORS CORP., INDIANAPOLIS
ALLOYD-	ALLOYD CORP., CAMBRIDGE, MASS.
ALM-(NUMBER-LETTER)	ARMY LOGISTICS MANAGEMENT CENTER, FORT LEE, VA.
ALM-	NORTH AMERICAN AVIATION, INC. AEROPHYSICS LAB., DOWNEY, CALIF. (MEMO)
ALM-SL-	TENNESSEE EASTMAN CORP., OAK RIDGE, TENN.
ALO-	ALBUQUERQUE OPERATIONS OFFICE, AEC, N. MEX. (REF.4)
ALO-(NUMBERS)-	ALBUQUERQUE OPERATIONS OFFICE, AEC, N. MEX.
ALO-AEC-CMA-	ALBUQUERQUE OPERATIONS OFFICE, AEC, N. MEX.
AL-P-	ATOMICS INTERNATIONAL DIV., NORTH AMERICAN AVIATION, INC., CANOGA PARK, CALIF.
ALPR-	ARGONNE NATIONAL LAB., ILL.
ALR-	ILLINOIS. UNIV., URBANA
ALRAND-(NUMBERS)(LTR.)	NAVAL SUPPLY DEPOT, MECHANICSBURG, PENNA.
ALRC-	AEROJET LIQUID ROCKET CO., SACRAMENTO, CALIF.
ALSKU-	ALASKA. UNIV., COLLEGE
ALSOS-	EUROPEAN THEATER OF OPERATIONS (ARMY). ALSOS MISSION
ALSOS-DDC-	EUROPEAN THEATER OF OPERATIONS (ARMY). ALSOS MISSION
ALSOS-G-	EUROPEAN THEATER OF OPERATIONS (ARMY). ALSOS MISSION (GERLACH FILE)
ALSOS-H-	EUROPEAN THEATER OF OPERATIONS (ARMY). ALSOS MISSION (HOYT FILE)
ALSOS-HA-	EUROPEAN THEATER OF OPERATIONS (ARMY). ALSOS MISSION (HAAGEN FILE)
ALSOS-K-	EUROPEAN THEATER OF OPERATIONS (ARMY). ALSOS MISSION (KLOTTER FILE)
ALSOS-KLI-	EUROPEAN THEATER OF OPERATIONS (ARMY). ALSOS MISSION (KLIEWE FILE)
ALSOS-MF-	EUROPEAN THEATER OF OPERATIONS (ARMY). ALSOS MISSION (MF FILE)
ALSOS MIS-	EUROPEAN THEATER OF OPERATIONS (ARMY). ALSOS MISSION (MIS FILE)

ALSOS O-	EUROPEAN THEATER OF OPERATIONS (ARMY). ALSOS MISSION (OBERTH FILE)
ALSOS RFR-	EUROPEAN THEATER OF OPERATIONS (ARMY). ALSOS MISSION (REICHSFORSCHUNGSRAT FILE)
ALSS-TR-	BENDIX CORP., DETROIT
AL-TDR-(YEAR)	AIR FORCE AVIONICS LAB., WRIGHT-PATTERSON AFB, OHIO
AL-TR-(YEAR)-	AIR FORCE AVIONICS LAB., WRIGHT-PATTERSON AFB, OHIO
AL-TRANS-	ATOMICS INTERNATIONAL DIV., NORTH AMERICAN AVIATION, INC., CANOGA PARK, CALIF.
ALU-	ALFRED UNIV., N.Y.
AM-	ANALYTIC SERVICES, INC., BAILEYS CROSSROADS, VA.
AM-	ARMY ELECTRONIC PROVING GROUND, FORT HUACHUCA, ARIZ.
AM-	ARMY INST. OF PATHOLOGY, WASHINGTON, D.C.
AM-	BALLISTIC RESEARCH LABS., ABERDEEN PROVING GROUND, MD.
AM-(YEAR)-	CALIFORNIA. UNIV., BERKELEY. DIV. OF APPLIED MECHANICS
AM-(YEAR)-	CALIFORNIA. UNIV., BERKELEY. INST. OF ENG. RES.
AM-	COLUMBIA UNIV., NYC. SUBSTITUTE ALLOY MATERIALS LABS.
AM-	CORNELL AERONAUTICAL LAB., INC., BUFFALO
AM-(NUMBER)Q-(YEAR)	DOW CHEMICAL CO., MIDLAND, MICH. (ANTIMALARIAL SYNTHESIS)
AM-(NUMBER)A-(YEAR)	DOW CHEMICAL CO., MIDLAND, MICH.
AM-	ELDORADO MINING + REFINING LTD., SASKATCHEWAN
AM-(YEAR)-	FEDERAL AVIATION ADMINISTRATION. OFFICE OF AVIATION MEDICINE, WASHINGTON, D.C. (CODE USED SINCE 1964)
AM-	GARRETT CORP. AIRESEARCH MFG. DIV., LOS ANGELES
AM-	GERMANY. SCHOOL FOR THEORETICAL PHYSICS, OBERWOLFACH
AM-	GT. BRIT. AIR MINISTRY
AM-	GT. BRIT. ATOMIC WEAPONS RESEARCH ESTABLISHMENT, ALDERMASTON, BERKS, ENGLAND
AM-	LOS ALAMOS SCIENTIFIC LAB., N. MEX. (ASSIGNED BEFORE 1950 TO MISCELLANEOUS AMERICAN REPORTS) (REF. 22)
/A/M-	MANHATTAN DISTRICT. RESEARCH CONTROL (SERIES ASSIGNED TO REPORTS FROM VARIOUS AMERICAN SOURCES)
AM-	NAPLES. UNIVERSITA. ISTITUTO DI FISICA TEORICA
AM-	NATIONAL DEFENSE RESEARCH COMMITTEE. APPLIED MATHEMATICS PANEL (REF. 28)
AM-	NAVAL AIR DEVELOPMENT CENTER. AERO MECHANICS DEPT., JOHNSVILLE, PENNA.
AM-	NEW YORK UNIV., N.Y.C. COLL. OF ENGINEERING
AM-	NORTHROP AIRCRAFT, INC., HAWTHORNE, CALIF. (AERODYNAMICS MEMO)
AM-	OFFICE OF THE ASST. SECY. OF INTERIOR, WATER AND POWER DEV. OFFICE OF SALINE WATER, WASH., D.C.
AM-	PARIS. UNIVERSITE, ORSAY. ECOLE NORMALE SUPERIEURE
AM-(NO.)-	RAMO-WOOLDRIDGE CORP., LOS ANGELES
AM-	RAND CORP., SANTA MONICA, CALIF.
OOAMA-2.5-	AEROJET-GENERAL CORP., SACRAMENTO, CALIF.
AMA-	ASTRO MET ASSOCIATES INC., CINCINNATI
AM AP-	GT. BRIT. AIR MINISTRY (AIR PUBLICATIONS)
AMB-	NAVAL RESEARCH LAB., WASHINGTON, D.C.
AMC-(YEAR)-	AERONUTRONIC DIV. OF PHILCO CORP., NEWPORT BEACH, CAL.
AMC-	AIR MATERIEL COMMAND, WRIGHT-PATTERSON AFB, OHIO
AMC-	ARMY MATERIEL COMMAND, WASHINGTON., D.C.
AMC-	ARMY MISSILE COMMAND, REDSTONE ARSENAL, ALA.
AMC-	NEW MEXICO STATE UNIV., UNIVERSITY PARK. PHYSICAL SCIENCE LAB.
AMCA-	ARMY ADVANCED MATERIEL CONCEPTS AGENCY, WASHINGTON, DC
AMC AFTR-	AIR MATERIEL COMMAND. MATERIEL DIV., WRIGHT-PATTERSON AFB, OHIO
AMC-AF-TSEAA-(NUMBER)-	AIR MATERIEL COMMAND. ENGINEERING DIV., WRIGHT-PATTERSON AFB, OHIO
AMC-AIRL-	AIR MATERIEL COMMAND. AERONAUTICAL ICE RESEARCH LAB.
AMC-ARDC-	AIR MATERIEL COMMAND, WRIGHT-PATTERSON AFB, OHIO
AMC-ARL-	AIR MATERIEL COMMAND. AIRCRAFT RADIATION LAB., WRIGHT-PATTERSON AFB, OHIO
AMC-CD-	AIR MATERIEL COMMAND, WRIGHT-PATTERSON AFB, OHIO
AMC-CFS-	AIR MATERIEL COMMAND. CAMBRIDGE FIELD STATION
AMC E-	AIR MATERIEL COMMAND, WRIGHT-PATTERSON AFB, OHIO (EXCERPT REPORTS)
AMC ED MR-	AIR MATERIEL COMMAND. ENGINEERING DIV., WRIGHT-PATTERSON AFB, OHIO (MEMORANDUM REPORT)
AMCEL-	AMCEL PROPULSION CO., ASHEVILLE, N.C.
AMC-ERC-	AIR MATERIEL COMMAND. CAMBRIDGE FIELD STATION
AMC-ERL-	AIR MATERIEL COMMAND. ELECTRONICS RESEARCH LAB., WRIGHT-PATTERSON AFB, OHIO
AMC-ETO-TM-	ARMY MISSILE COMMAND, REDSTONE ARSENAL, ALA.
AMC F-BB-(NO.)-RE	AIR MATERIEL COMMAND. INTELLIGENCE DIV., WRIGHT-PATTERSON AFB, OHIO (BIBLIOGRAPHY)
AMC F-SU-(NO.)-ND	AIR MATERIEL COMMAND. INTELLIGENCE DIV., WRIGHT-PATTERSON AFB, OHIO (SUMMARY REPORT)
AMC F-TR-(NO.)-ND	AIR MATERIEL COMMAND. INTELLIGENCE DIV., WRIGHT-PATTERSON AFB, OHIO (TECHNICAL REPORT)
AMC F-TS-(NO.)-RE	AIR MATERIEL COMMAND. INTELLIGENCE DIV., WRIGHT-PATTERSON AFB, OHIO (TRANSLATION REPORT)
AMC-GEOD-	AIR MATERIEL COMMAND. GEOPHYSICAL DIVISION, WRIGHT-PATTERSON AFB, OHIO
AMCHITKA-	GEOLOGICAL SURVEY, DENVER
AMC-IRB-	AIR MATERIEL COMMAND. ICE RESEARCH BASE
AMC-IRE-	AIR MATERIEL COMMAND, WRIGHT-PATTERSON AFB, OHIO
AMCM-	AIR MATERIEL COMMAND, WRIGHT-PATTERSON AFB, OHIO (MANUAL)
AMC MCI 102-AE-	AIR MATERIEL COMMAND. INTELLIGENCE DEPT., WRIGHT-PATTERSON AFB, OHIO (STUDY)
AMC MCI 102-EL-	AIR MATERIEL COMMAND. INTELLIGENCE DEPT., WRIGHT-PATTERSON AFB, OHIO (STUDY)
AMC MCREE-(YR.)-	AIR MATERIEL COMMAND. ENGINEERING DIV. ELECTRONIC SUBDIV., WRIGHT-PATTERSON AFB, OHIO
AMC-MCREGA-	AIR MATERIEL COMMAND, WRIGHT-PATTERSON AFB, OHIO
AMC MCREGC-	AIR MATERIEL COMMAND. ENG. DIV. CLIMATIC PROJECTS SEC., WRIGHT-PATTERSON AFB, OHIO (MEMO. REPT.)
AMC MCREXA-	AIR MATERIEL COMMAND. ENGINEERING DIV. AIRCRAFT LAB., WRIGHT-PATTERSON AFB, OHIO (MEMO. REPT.)
AMC MCREXE-	AIR MATERIEL COMMAND. ENGINEERING DIV. EQUIPMENT LAB., WRIGHT-PATTERSON AFB, OHIO (MEMO. REPT.)
AMC MCRFTP-	AIR MATERIEL COMMAND. FLIGHT TEST DIV., WRIGHT-PATTERSON AFB, OHIO (MEMO. REPT.)
AMC/MISS-	ARMY MISSILE COMMAND. REDSTONE ARSENAL, ALA.
AMC/MOB-	ARMY MOBILITY COMMAND. DETROIT ARSENAL
AMC-MR-EXP-M-51/	AIR MATERIEL COMMAND, WRIGHT-PATTERSON AFB, OHIO
AMC-MR-MCREE-	AIR MATERIEL COMMAND, WRIGHT-PATTERSON AFB, OHIO
AMC-MR-MCREXD-	AIR MATERIEL COMMAND. ENGINEERING DIV., WRIGHT-PATTERSON AFB, OHIO
AMC-MR-MCREXE-	AIR MATERIEL COMMAND. ENGINEERING DIV., WRIGHT-PATTERSON AFB, OHIO
AMC-MR-TSEAA-	AIR MATERIEL COMMAND. ENGINEERING DIV., WRIGHT-PATTERSON AFB, OHIO
AMC/MUN-	ARMY MUNITIONS COMMAND. PICATINNY ARSENAL, N.J.
AMCP-	ARMY MATERIEL COMMAND, WASHINGTON, D.C.
AMC-P-	ARMY MISSILE COMMAND, REDSTONE ARSENAL, ALA.
AMC-PAM-	ARMY MATERIEL COMMAND, WASHINGTON, D.C.
AMC-PAM-(NO.)-(NO.)-	ARMY MATERIEL COMMAND, WASHINGTON, D.C.
AMCPM-BR-TR-	PICATINNY ARSENAL. OFFICE OF THE PROJECT MANAGER FOR BOMBS AND RELATED COMPONENTS, DOVER, N.J.
AMC-R-	AIR MATERIEL COMMAND, WRIGHT-PATTERSON AFB, OHIO
AMCR-	ARMY MATERIEL COMMAND, WASHINGTON, D.C.
AMC-RA-RD-TR-(YEAR)-	ARMY MISSILE COMMAND. ADVANCED SYSTEMS LAB., REDSTONE ARSENAL, ALA. (TECHNICAL REPORT)
AMC-RA-RE-TR-(YEAR)-	ARMY MISSILE COMMAND, REDSTONE ARSENAL, ALA.

AMC-RA-RG-TR-(YEAR)-
 ARMY MISSILE COMMAND, REDSTONE ARSENAL, ALA.
AMC-RA-RN-TR-(YEAR)-
 ARMY MISSILE COMMAND, REDSTONE ARSENAL, ALA.
AMC-RA-RR-TR-(YEAR)-
 ARMY MISSILE COMMAND, REDSTONE ARSENAL, ALA.
AMC-RA-RS-TR-(YEAR)-
 ARMY MISSILE COMMAND, REDSTONE ARSENAL, ALA.
AMC-RA-TIR-
 ARMY MISSILE COMMAND, REDSTONE ARSENAL, ALA.
AMC-RC-S-(YEAR)-
 ARMY MISSILE COMMAND. ENGINEERING REQUIREMENTS OFFICE, REDSTONE ARSENAL, ALA.
AMC-RD-TM-(YEAR)-
 ARMY MISSILE COMMAND. ADVANCED SYSTEMS LAB., REDSTONE ARSENAL, ALA. (TECHNICAL MEMORANDUM)
AMC-RD-TR-(YEAR)-
 ARMY MISSILE COMMAND. ADVANCED SYSTEMS LAB., REDSTONE ARSENAL, ALA. (TECHNICAL REPORT)
AMC-RE-TM-(YEAR)-
 ARMY MISSILE COMMAND. ELECTROMAGNETICS LAB., REDSTONE ARSENAL, ALA. (TECHNICAL MEMORANDUM)
AMC-RE-TR-(YEAR)-
 ARMY MISSILE COMMAND. ELECTROMAGNETICS LAB., REDSTONE ARSENAL, ALA. (TECHNICAL REPORT)
AMC-RF-TR-(YEAR)-
 ARMY MISSILE COMMAND. FUTURE MISSILE SYSTEMS DIV., REDSTONE ARSENAL, ALA. (TECHNICAL REPORT)
AMC-RG-TM-(YEAR)-
 ARMY MISSILE COMMAND. ARMY INERTIAL GUIDANCE + CONTROL LAB., REDSTONE ARSENAL, ALA. (TECHNICAL MEMO.)
AMC-RG-TN-(YEAR)-
 ARMY MISSILE COMMAND. ARMY INERTIAL GUIDANCE + CONTROL LAB., REDSTONE ARSENAL, ALA. (TECHNICAL NOTE)
AMC-RG-TR-(YEAR)-
 ARMY MISSILE COMMAND. ARMY INERTIAL GUIDANCE + CONTROL LAB., REDSTONE ARSENAL, ALA. (TECHNICAL REPT.)
AMC-RK-TR-(YEAR)-
 ARMY MISSILE COMMAND, REDSTONE ARSENAL, ALA.
AMC-RL-TR-(YEAR)-
 ARMY MISSILE COMMAND. GROUND SUPPORT EQUIPMENT LAB., REDSTONE ARSENAL, ALA. (TECHNICAL REPORT)
AMC-RN-TR-(YEAR)-
 ARMY MISSILE COMMAND. ARPA DIV., REDSTONE ARSENAL, ALA. (TECHNICAL REPORT)
AMC RPT. CD-
 AIR MATERIEL COMMAND, WRIGHT-PATTERSON AFB, OHIO
AMC-RP-TN-(YEAR)-
 ARMY MISSILE COMMAND, REDSTONE ARSENAL, ALA.
AMC-RR-TM-(YEAR)-
 ARMY MISSILE COMMAND. PHYSICAL SCIENCES LAB., REDSTONE ARSENAL, ALA. (TECHNICAL MEMORANDUM)
AMC-RR-TR-(YEAR)-
 ARMY MISSILE COMMAND. PHYSICAL SCIENCES LAB., REDSTONE ARSENAL, ALA. (TECHNICAL REPORT)
AMC-RSIC-
 REDSTONE SCIENTIFIC INFO. CTR., REDSTONE ARSENAL, ALA
AMC-RS-TM-(YEAR)-
 ARMY MISSILE COMMAND, REDSTONE ARSENAL, ALA.
AMC-RS-TR-(YEAR)-
 ARMY MISSILE COMMAND. STRUCTURES AND MECHANICS LAB., REDSTONE ARSENAL, ALA. (TECHNICAL REPORT)
AMC-RT-TR-(YEAR)-
 ARMY MISSILE COMMAND. TEST AND RELIABILITY EVALUATIONS LAB., REDSTONE ARSENAL, ALA. (TECHNICAL REPT)
AMC-STUDY-(NO.)-
 ARMY MATERIEL COMMAND, WASHINGTON, D.C.
AMCT-
 TEXAS A + M UNIV., COLLEGE STATION
AMC-TIR-
 ARMY MATERIEL COMMAND, WASHINGTON, D.C.
AMC-TIR-(NO.)-(NO.)-
 ARMY MATERIEL COMMAND, WASHINGTON, D.C.
AMC-TIR-
 ARMY MISSILE COMMAND, REDSTONE ARSENAL, ALA.
AMC-TO-
 AIR MATERIEL COMMAND, WRIGHT-PATTERSON AFB, OHIO
AMC-TR-
 AIR MATERIEL COMMAND, WRIGHT-PATTERSON AFB, OHIO
AMC-TR-
 ARMY MISSILE COMMAND, REDSTONE ARSENAL, ALA.
AMC TSEAA (NOS.)(LTRS.)
 AIR MATERIEL COMMAND. ENG. DIV. AERO-MEDICAL LAB., WRIGHT-PATTERSON AFB, OHIO (MEMO. REPT.)
AMC TSELR (NOS.)
 AIR MATERIEL COMMAND. ENG. DIV., AIRCRAFT RADIATION LAB., WRIGHT-PATTERSON AFB, OHIO (MEMO. REPT.)
AMC TSEPE-
 AIR MATERIEL COMMAND. ENGINEERING DIV. EQUIPMENT LAB., WRIGHT-PATTERSON AFB, OHIO (MEMO. REPT.)
AMC-WC-
 WRIGHT AIR DEV. CENTER, WRIGHT-PATTERSON AFB, OHIO
AMC-WCEG-
 WRIGHT AIR DEV. CENTER, WRIGHT-PATTERSON AFB, OHIO
AMC-WCNSW-
 WRIGHT AIR DEV. CENTER, WRIGHT-PATTERSON AFB, OHIO
AMC-WCRR-
 WRIGHT AIR DEV. CENTER, WRIGHT-PATTERSON AFB, OHIO
AMC-WCSG-(YEAR)-
 WRIGHT AIR DEV. CENTER, WRIGHT-PATTERSON AFB, OHIO
AMC/WPN-
 ARMY WEAPONS COMMAND, ROCK ISLAND, ILL.
AMD-
 AEROSPACE MEDICAL DIV., BROOKS AFB, TEX.
AMD-
 APPROVED MARINE DEVICES CO.
AMD-
 DOW CHEMICAL CO. ROCKY FLATS DIV., GOLDEN, COLO.
AMD-AMRL-
 AEROSPACE MEDICAL RESEARCH LABS. (6570TH), WRIGHT-PATTERSON AFB, OHIO
AMD-CR-(NO)-(NO)-(YR.)-
 AEROSPACE MEDICAL DIV., BROOKS AFB, TEX.
AMD-MRL-
 AEROSPACE MEDICAL RESEARCH LABS. (6570TH), WRIGHT-PATTERSON AFB, OHIO
AMDR-
 AEROJET-GENERAL CORP., SACRAMENTO, CALIF.
AMD-TDR-
 AEROSPACE MEDICAL DIV., BROOKS AFB, TEX.
AMD-TR-
 AEROSPACE MEDICAL DIV., BROOKS AFB, TEX.
AMD-TR-(YEAR)-
 AEROSPACE MEDICAL DIV., BROOKS AFB, TEX.
AMD-TT-
 AEROSPACE MEDICAL DIV., BROOKS AFB, TEX.
AME-
 CHRYSLER CORP. MISSILE DIV.
AME-
 GT. BRIT. UNDERWATER COUNTERMEASURES + WEAPONS ESTAB.
AMEE-
 UNITED KINGDOM. ADMIRALTY MARINE ENGINEERING ESTABLISHMENT, GOSPORT, HANTS, ENGLAND
AMEE-TM-(NUMBER/YEAR)
 UNITED KINGDOM. ADMIRALTY MARINE ENGINEERING ESTABLISHMENT, GOSPORT, HANTS, ENGLAND
AMERDC-
 ARMY MOBILITY EQUIPMENT RES. AND DEV. CENTER, FORT BELVOIR, VA.
AMERDL-
 ARMY MEDICAL EQUIP. RES. + DEV. LAB., FT. TOTTEN, NY
AMES-
 NEW YORK OPERATIONS OFFICE, AEC
AMF-
 AMERICAN MACHINE AND FOUNDRY CO.
AMF-ASL-E-(NUMBER)-
 AMERICAN MACHINE AND FOUNDRY CO. ADVANCED SYSTEMS LAB., SANTA BARBARA, CALIF.
AMFC-
 AMERICAN MACHINE AND FOUNDRY CO., CHICAGO
AMFC-
 AMERICAN MACHINE AND FOUNDRY CO. TURBO ENGINE DEPT., PACOIMA, CALIF.
AMFC-
 AMF ATOMICS (CANADA) LTD., PORT HOPE, ONT.
AMFC-(NUMBER)-
 AMF ATOMICS (CANADA) LTD., PORT HOPE, ONT.
AMF-C-
 ATOMIC ENERGY OF CANADA LTD., CHALK RIVER, ONT.
AMF-ER-
 AMERICAN MACHINE AND FOUNDRY CO. GENERAL ENGINEERING LAB., GREENWICH, CONN.
AMF-G-
 AMERICAN MACHINE AND FOUNDRY CO. GENERAL ENGINEERING LAB., GREENWICH, CONN.
AMF-G-
 AMF ATOMICS, INC., GREENWICH, CONN.
AMF-GR-
 AMERICAN MACHINE AND FOUNDRY CO. GENERAL ENGINEERING LAB., GREENWICH, CONN.
AMF-GR-(NUMBER)-(YEAR)
 AMERICAN MACHINE AND FOUNDRY CO. NUCLEAR ENGINEERING LAB., GREENWICH, CONN.
AMF-G-S-
 AMERICAN MACHINE AND FOUNDRY CO. ADVANCED SYSTEMS LAB., SANTA BARBARA, CALIF.
AMF IR-
 AMERICAN MACHINE AND FOUNDRY CO. (INTERIM REPORT)
AMFL-TR-(YEAR)-
 AIR FORCE MATERIALS LAB., WRIGHT-PATTERSON AFB, OHIO
AMF MPR-
 AMERICAN MACHINE + FOUNDRY CO.(MONTHLY PROGRESS RPT.)
AMF MR-
 AMERICAN MACHINE AND FOUNDRY CO., CHICAGO
AMF/MRD TN-
 AMERICAN MACHINE AND FOUNDRY CO., CHICAGO
AMF/TD-
 AMERICAN MACHINE + FOUNDRY CO.TURBO DIV.,PACOIMA,CAL.
AMF-TD-
 SUNDSTRAND TURBO.DIV.OF SUNDSTRAND CORP.,ROCKFORD,ILL
AMF-Y-
 AMF ATOMICS, INC., GREENWICH, CONN.
AMG-
 NATIONAL DEFENSE RESEARCH COMMITTEE. APPLIED MATHEMATICS GROUP (REF. 28)
AMG-B-
 BROWN UNIV., PROVIDENCE. APPLIED MATH. GP. (REF. 28)
AMG-C-
 COLUMBIA UNIV., N.Y.C. APPLIED MATH. GROUP (REF. 28)
AMG-H-
 HARVARD UNIV., CAMBRIDGE, MASS. APPLIED MATHEMATICS GROUP (REF. 28)
AMG-IAS-
 INSTITUTE FOR ADVANCED STUDY. APPLIED MATHEMATICS GROUP, PRINCETON, N.J. (REF. 28)
AMG-M-(NUMBER)-
 SANDIA CORP., ALBUQUERQUE, N. MEX. (MATADOR)
AMG-N-
 NORTHWESTERN UNIV., EVANSTON, ILL. APPLIED MATHEMATICS GROUP (REF. 28)
AMG-NYU-
 NEW YORK UNIV., N.Y.C. APPLIED MATH. GROUP (REF. 28)
AMI-(7 DIGITS)
 AMERICAN MICRO-SYSTEMS, INC., SANTA CLARA, CALIF.
AMINCO-
 AMERICAN INSTRUMENT CO., INC., SILVER SPRING, MD.

AML-	DAVID TAYLOR MODEL BASIN. APPLIED MATHEMATICS LAB., CARDEROCK, MD.
AML-	GT. BRIT. ADMIRALTY MATERIALS LAB., POOLE, DORSET, ENG
AML-	NAVAL AIR ENGINEERING CENTER. AERONAUTICAL MATERIALS LAB., PHILADELPHIA
AML/A/	GT. BRIT. ADMIRALTY MATERIALS LAB., POOLE, DORSET, ENG
AMLC-TR-(YEAR)-	AEROSPACE MEDICAL LAB (CLINICAL), LACKLAND AFB, TEX.
AML-NAM-	NAVAL AIR EXPTL. STA. AERO. MATERIALS LAB., PHILA.
AML-NAM-AE-	NAVAL AIR EXPTL. STA. AERO. MATERIALS LAB., PHILA.
AML/P/	GT. BRIT. ADMIRALTY MATERIALS LAB., POOLE, DORSET, ENG
AM-MEMO-	AUSTRALIA. DEPARTMENT OF CIVIL AVIATION, MELBOURNE
AM MO-	UNITED KINGDOM. AIR MINISTRY. METEOROLOGICAL OFFICE, LONDON
AMMRC-CR-	ARMY MATERIALS + MECHANICS RES. CTR., WATERTOWN, MASS.
AMMRC-CR-(YEAR)-	ARMY MATERIALS + MECHANICS RES. CTR., WATERTOWN, MASS.
AMMRC-MS-	ARMY MATERIALS + MECHANICS RES. CTR., WATERTOWN, MASS.
AMMRC-TR-	ARMY MATERIALS + MECHANICS RES. CTR., WATERTOWN, MASS.
AM.MS-	AMERICAN METEOROLOGICAL SOCIETY, BOSTON
AMNL-	ARMY MEDICAL NUTRITION LAB., DENVER
AMP-	AVCO CORP. AVCO-EVERETT RESEARCH LAB., EVERETT, MASS.
AMP-	NATIONAL DEFENSE RESEARCH COMMITTEE. APPLIED MATHEMATICS PANEL (REF. 28)
AMPEX-	AMPEX CORP., REDWOOD CITY, CALIF.
AMPEX FSB-	AMPEX CORP., REDWOOD CITY, CAL. (FIELD SERVICE BULL.)
AMPEX RB-	AMPEX CORP., REDWOOD CITY, CALIF. (RESEARCH BULL.)
AMPEX RD-	AMPEX CORP., REDWOOD CITY, CALIF. (RESEARCH DOC.)
AMPEX RP-	AMPEX CORP., REDWOOD CITY, CALIF. (RES. PROPOSAL)
AMPEX TM-	AMPEX CORP., REDWOOD CITY, CALIF. (TECHNICAL MEMO.)
AMPEX TR-	AMPEX CORP., REDWOOD CITY, CALIF. (TECHNICAL REPT.)
AMP-MEMO-	NATIONAL DEFENSE RESEARCH COMMITTEE. APPLIED MATHEMATICS PANEL (REF. 28)
AMP-MEMO-	NEW YORK UNIV., N.Y.C. APPLIED MATH. GROUP (REF. 28)
AMP NO. NR-	OFFICE OF NAVAL RESEARCH, WASHINGTON, D.C.
AMP-NOTE-	NATIONAL DEFENSE RESEARCH COMMITTEE. APPLIED MATHEMATICS PANEL (REF. 28)
AMP RPT.-	COLUMBIA UNIV., N.Y.C. APPLIED MATH. GROUP (REF. 28)
AMP-RPT.-	NATIONAL DEFENSE RESEARCH COMMITTEE. APPLIED MATHEMATICS PANEL (REF. 28)
AMQ-(NO.),LOG.NO.-	BENDIX AVIATION CORP. PIONEER-CENTRAL DIV., DAVENPORT, IOWA
AMR-	ADVANCED METALS RESEARCH CORP., BURLINGTON, MASS. (PROGRESS REPORT)
AMR-	ADVANCED METALS RESEARCH CORP., SOMERVILLE, MASS.
AMR-	AMF ATOMICS, INC., YORK, PENNA.
AMR-	APPLIED MECHANICS REVIEWS, SAN ANTONIO, TEX.
AMR-	ARDE-PORTLAND, INC., PARAMUS, N.J.
AMR-	ATLANTIC MISSILE RANGE, PATRICK AFB, FLA.
AM-R(NO.)-	CHRYSLER CORP. MISSILE OPERATIONS (REDSTONE ARTILLERY MISSILE)
AM-R-	COLUMBIA UNIV., NYC. SUBSTITUTE ALLOY MATERIALS LABS.
AMR-	RAYTHEON CO., NORWOOD, MASS.
AMRA-CR-(YEAR)-	ARMY MATERIALS RESEARCH AGENCY, WATERTOWN, MASS. (CONTRACT REPORT)
AMRA-MR-(YEAR)-	ARMY MATERIALS RESEARCH AGENCY, WATERTOWN, MASS. (MEMORANDUM REPORT)
AMRA-MS-(YEAR)-	ARMY MATERIALS RESEARCH AGENCY, WATERTOWN, MASS. (MONOGRAPHS)
AMRA-PS-(YEAR)-	ARMY MATERIALS RESEARCH AGENCY, WATERTOWN, MASS. (SPECIAL PUBLICATION)
AMRA-TR-(YEAR)-	ARMY MATERIALS RESEARCH AGENCY, WATERTOWN, MASS. (TECHNICAL REPORT)
AMRC-	AMERICAN RESEARCH CORP., FULLERTON, CALIF.
A/M-REF-(YEAR)-IT	TEXAS A + M UNIV., COLLEGE STATION. DEPT. OF OCEANOGRAPHY
A/M-REF-	TEXAS A + M UNIV., COLLEGE STATION. RESEARCH FDN.
AMRL-	ARMY MEDICAL RESEARCH LAB., FORT KNOX, KY.
AMRL-MEMO-	AEROSPACE MEDICAL RESEARCH LABS. (6570TH), WRIGHT-PATTERSON AFB, OHIO
AMRL-PUB-	ARMY MEDICAL RES. LAB., FT. KNOX, KY. (PUBLICATIONS)
AMRL R-	ARMY MEDICAL RESEARCH LAB., FORT KNOX, KY.
AMRL-TDR-	AEROSPACE MEDICAL RESEARCH LABS. (6570TH), WRIGHT-PATTERSON AFB, OHIO
AMRL-TR-(YEAR)-	AEROSPACE MEDICAL RESEARCH LABS. (6570TH), WRIGHT-PATTERSON AFB, OHIO
AMRNL-	ARMY MEDICAL RESEARCH AND NUTRITION LAB., DENVER
AMRUE-	ARMY MEDICAL RES. AND DEV. COMMAND, WASHINGTON, D.C.
AMRU-R-(NUMBER-NUMBER)	MC GILL UNIV., MONTREAL. AVIATION MEDICAL RES. UNIT
AMS-	AMERICAN MATHEMATICAL SOCIETY, PROVIDENCE
AMS-	AMERICAN METEOROLOGICAL SOCIETY, BOSTON
AMS-	ARMY MAP SERVICE, WASHINGTON, D.C.
AMS-	MIDWESTERN UNIVERSITIES RESEARCH ASSN., URBANA, ILL.
AMS-	PRINCETON UNIV., N.J. DEPT. OF AEROSPACE AND MECHANICAL SCIENCES
AMS-	PRINCETON UNIV., N.J. GUGGENHEIM LABS. FOR THE AEROSPACE PROPULSION SCIENCES
AMSAA-TM-	ARMY MATERIEL SYSTEMS ANALYSIS AGENCY, ABERDEEN PROVING GROUND, MD.
AMSAA-TR-	ARMY MATERIEL SYSTEMS ANALYSIS AGENCY, ABERDEEN PROVING GROUND, MD.
AMSCD-	ARMY COMBAT DEVELOPMENTS COMMAND. MEDICAL SERVICE AGENCY, FORT SAM HOUSTON, TEX.
AMSCO-	ARMY MEDICAL SERVICE. COMBAT DEV. GROUP, WASH., D.C.
AMSEL-	ARMY ELECTRONICS LABS., FORT MONMOUTH, N.J.
AMS-GEODETIC MEMO-	ARMY MAP SERVICE, WASHINGTON, D.C.
AMSGS-	ARMY MEDICAL SERVICE GRADUATE SCHOOL, WASHINGTON, DC
AMSIS-	GT. BRIT. DIRECTORATE OF INTELLIGENCE
AMS-MURA/-	MIDWESTERN UNIVERSITIES RESEARCH ASSN., URBANA, ILL.
AMST-	AMERICAN MATHEMATICAL SOCIETY, PROVIDENCE (TRANSLATION)
AMS-T-	PRINCETON UNIV., N.J. DEPT. OF AEROSPACE AND MECHANICAL SCIENCES
AMS-TB-	ARMY MAP SERVICE, WASHINGTON, D.C. (TECHNICAL BULL.)
AMS-TI-	ARMY MAP SERVICE, WASHINGTON, D.C. (TECH. INSTRUCTION)
AMS TM-	ARMY MAP SERVICE, WASH., D.C. (TRANSVERSE MERCATOR GRID)
AMS-TM-	ARMY MAP SERVICE, WASHINGTON, D.C. (TECHNICAL MANUAL)
AMS-TR-	AMERICAN MATHEMATICAL SOCIETY, PROVIDENCE
AMS-TR-	ARMY MAP SERVICE, WASHINGTON, D.C. (TECHNICAL REPORT)
AMSWE-CPD-	ARMY WEAPONS COMMAND. COST ANALYSIS OFFICE, ROCK ISLAND, ILL.
AMSWE-OR-	ARMY WEAPONS COMMAND. WEAPONS OPERATIONS RESEARCH OFFICE, ROCK ISLAND, ILL.
AMSWE-RE-	ARMY WEAPONS COMMAND. RESEARCH AND ENGINEERING DIRECTORATE, ROCK ISLAND, ILL.
AMSWE-SM-	ARMY WEAPONS COMMAND. SUPPLY AND MAINTENANCE DIRECTORATE, ROCK ISLAND, ILL.
AMTC-	ARMY MISSILE TEST CTR., WHITE SANDS MISSILE RANGE, NM
AMTC SPEC. RPT.-	ARMY MISSILE TEST CTR., WHITE SANDS MISSILE RANGE, NM
AMTC-TB-	ARMY MISSILE TEST CTR., WHITE SANDS MISSILE RANGE, NM
AMTC-TM-	ARMY MISSILE TEST CTR., WHITE SANDS MISSILE RANGE, NM
AMTC-TR-	ARMY MISSILE TEST CTR., WHITE SANDS MISSILE RANGE, NM
AMTC-WS-TM-	ARMY MISSILE TEST CTR., WHITE SANDS MISSILE RANGE, NM
AMV-R-	VOORHEES (ALAN M.) AND ASSOCIATES, INC., MC LEAN, VA.
AMXFD-AE-	FORT DETRICK, FREDERICK, MD.
AMXFD-AE-T-	FORT DETRICK, FREDERICK, MD.
AMXMR-P-	ARMY MATERIALS RES. AGENCY, WATERTOWN ARSENAL, MASS.

AN

AN-	(AERODYNAMICS NOTE) (REF. 2)
AN-	AEROJET-GENERAL NUCLEONICS, SAN RAMON, CALIF.
AN-	AIR FORCE (AIR NAVIGATION CHARTS)
AN-	ARMY-NAVY (REF. 13)
AN-	ASSOCIATED NUCLEONICS, INC., GARDEN CITY, N.Y.
AN-	LOS ALAMOS SCIENTIFIC LAB., N. MEX.
AN0(NUMBER)	RAYTHEON CO., ANDOVER, MASS.
AN-(LETTER)-	(ARMY-NAVY SPECIFICATION)
ANA-	(AIR FORCE-NAVY AERONAUTICAL BULLETIN)
AN AB-	ARMY-NAVY (AERONAUTICAL BULLETIN) (REF. 13)
ANAF/GMML-	DEPARTMENT OF DEFENSE, WASHINGTON, D.C. (GUIDED MISSILE MAILING LIST)
AN-AGCR-	AEROJET-GENERAL NUCLEONICS, SAN RAMON, CALIF.
ANAL-MISC.	CALIFORNIA. UNIV., BERKELEY. RADIATION LAB. (REF. 9)
ANAL-R-	COLUMBIA UNIV., NYC. SUBSTITUTE ALLOY MATERIALS LABS.
ANAM-	ATOMIC ENERGY OF CANADA LTD. CHALK RIVER PROJECT, ONTARIO. (ALUMINUM-NICKEL ALLOY MEETING)
AN AP-	GT. BRIT. AIR MINISTRY (AIR PUBLICATIONS)
AN-APCSE-(NUMBERS)-VOL-	AEROJET-GENERAL CORP. NUCLEAR ENGINEERING AND MFG. OPERATIONS, SAN RAMON, CALIF.
ANARE-	AUSTRALIA. DEPT. OF SUPPLY. ANTARCTIC DIV., MELBOURNE
ANA-WB-	ARMED FORCES SPECIAL WEAPONS PROJECT. FIELD COMMAND, ALBUQUERQUE, N. MEX. (WORKBOOK)
ANC-	AIR FORCE-NAVY CIVIL COMMITTEE (REF. 13)
ANC-	AMERICAN NUCLEONICS CORP., GLENDALE, CALIF.
ANC-	MUNITIONS BOARD. AIRCRAFT COMMITTEE, WASH., D.C.
ANCP-	CENTRAL UTILITIES ATOMIC POWER ASSOCIATES
ANCR-	AEROJET NUCLEAR CO., IDAHO FALLS
AND-	ARMY-NAVY (DRAWING) (REF. 13)
ANDB-	AIR NAVIGATION DEVELOPMENT BOARD, WASHINGTON, D.C.
AN/EBR-	ARGONNE NATIONAL LAB., IDAHO FALLS (EXPERIMENTAL BREEDER REACTOR)
AN-ENGR-	AEROJET-GENERAL NUCLEONICS, SAN RAMON, CALIF.
ANESB-	ARMY-NAVY EXPLOSIVES SAFETY BOARD, WASHINGTON, D.C.
ANESB TP-	ARMY-NAVY EXPLOSIVES SAFETY BD.,WASH.,DC (TECH.PAPER)
AN/FCC-	LENKURT ELECTRIC CO., INC., SAN CARLOS, CALIF.
AN-FDD-	AEROJET-GENERAL NUCLEONICS, SAN RAMON, CALIF.
AN/GPA-	TACTICAL AIR WARFARE CENTER, EGLIN AFB, FLA.
AN-IDOP-	AEROJET-GENERAL NUCLEONICS, SAN RAMON, CALIF.
ANL-	ARGONNE NATIONAL LAB., ILL. (NOS. START WITH 4000)
ANL-(INITIALS)-	ARGONNE NATIONAL LAB., LEMONT, ILL.
ANL-(INITLS.)-(INITIALS)-	ARGONNE NATIONAL LAB., LEMONT, ILL.
ANL-ACL-	ARGONNE NATIONAL LAB., ILL.
ANLAD-	ARGONNE NATIONAL LAB., ILL.
ANL-C+OM-	ARMY NATICK LABS. CLOTHING AND ORGANIC MATERIALS LAB., MASS.
ANL-C+PLSEL-	ARMY NATICK LABS. CLOTHING AND PERSONAL LIFE SUPPORT EQUIPMENT LAB., MASS.
ANL-EBR-	ARGONNE NATIONAL LAB., ILL.
ANL/ES-CC-	ARGONNE NATIONAL LAB., ILL.
ANL/ES-CEN-	ARGONNE NATIONAL LAB., ILL.
ANL-ES-RPY-	ARGONNE NATIONAL LAB., ILL.
ANL/HEP-	ARGONNE NATIONAL LAB., ILL. (HIGH ENERGY PARTICLES)
ANL/MET-	ARGONNE NATIONAL LAB., ILL.
ANL-MP-(NO.)-(NO.)(FILM)	ARGONNE NATIONAL LAB., ILL. (MOTION PICTURE FILM)
ANL-NDN-	ARGONNE NATIONAL LAB., LEMONT, ILL.
ANL-R.C.-	ARMY NATICK LABS., MASS.
ANL-RE-	ARGONNE NATIONAL LAB., ILL.
ANL/ST-	ARGONNE NATIONAL LAB., ILL. (SODIUM TECHNOLOGY)
ANL-STR-MEMO-	ARGONNE NATIONAL LAB., LEMONT, ILL. (SUBMARINE THERMAL REACTOR)
ANL-TR-(YEAR)-	ARGONNE NATIONAL LAB., ILL.
ANL-TRANS-	ARGONNE NATIONAL LAB., ILL.
AN-MET-	AEROJET-GENERAL NUCLEONICS, SAN RAMON, CALIF.
ANM-WB-	ARMED FORCES SPECIAL WEAPONS PROJECT. FIELD COMMAND, ALBUQUERQUE, N. MEX. (WORKBOOK)
ANNADIV-	NAVAL SHIP RESEARCH AND DEV. CENTER, ANNAPOLIS
ANP-	AIRCRAFT NUCLEAR PROPULSION PROGRAM (REF. 3)
ANP-	CONSOLIDATED VULTEE AIRCRAFT CORP., SAN DIEGO, CALIF. (REF. 3)
ANP-	GENERAL ELECTRIC CO. AIRCRAFT NUCLEAR PROPULSION DEPT., CINCINNATI (REF. 3)
ANP-	OAK RIDGE NATIONAL LAB., TENN. (REF. 3)
ANP DOC. (LETTERS)-	CONSOLIDATED VULTEE AIRCRAFT CORP., FORT WORTH, TEX. (REF. 3)
ANPR-	ARGONNE NATIONAL LAB., ILL.
ANS-	AMERICAN NUCLEAR SOCIETY, HINSDALE, ILL.
ANSC-	AEROJET NUCLEAR SYSTEMS CO., SACRAMENTO, CALIF.
ANSC-	AMERICAN NUCLEAR SCIENCE CORP., N.Y.C.
ANSCO-	ANSCO DIV., GEN. ANILINE + FILM CORP., BINGHAMTON,NY
ANSE-	GENERAL ELECTRIC CO. GEN. ENG. + CONSULTING LAB., SCHENECTADY, N.Y. (ADVANCED NUCLEAR SYSTEMS ENG.)
ANSER-	ANALYTIC SERVICES, INC., BAILEYS CROSSROADS, VA.
ANSER-AR-	ANALYTIC SERVICES, INC., FALLS CHURCH, VA.
ANSER-DBN-(YEAR)-	ANALYTIC SERVICES, INC., FALLS CHURCH, VA.
ANSER-RTBN-(YEAR)-	ANALYTIC SERVICES, INC., FALLS CHURCH, VA.
ANSER-SBN-(YEAR)-	ANALYTIC SERVICES, INC., FALLS CHURCH, VA.
ANSER-SDN-(YEAR)-	ANALYTIC SERVICES, INC., FALLS CHURCH, VA.
ANSER-TBN-(YEAR)-	ANALYTIC SERVICES, INC., FALLS CHURCH, VA.
ANSI-	AMERICAN NATIONAL STANDARDS INST., N.Y.C.
ANSO-	GENERAL ELECTRIC CO. MISSILE AND SPACE DIV., PHILA.
ANSP-	ACADEMY OF NATURAL SCIENCES OF PHILADELPHIA (TRANS-LATIONS)
ANS-RPD-	AMERICAN NUCLEAR SOCIETY, HINSDALE, ILL.
ANS-SD-	AMERICAN NUCLEAR SOCIETY, HINSDALE, ILL.
AN-ST-	SWITZERLAND. EIDGENOESSISCHES INSTITUT FUER REAKTORFORSCHUNG, WUERENLINGEN.
ANTC-	ANTIOCH COLL., YELLOW SPRINGS, OHIO
ANU-	AUSTRALIAN NATIONAL UNIV., CANBERRA
ANU-P-	AUSTRALIAN NATIONAL UNIV., CANBERRA. INST. OF ADVANCED STUDIES.
ANU/P-	AUSTRALIAN NATIONAL UNIV., CANBERRA. RESEARCH SCHOOL OF PHYSICAL SCIENCES
ANWA-PEP-	AEROJET-GENERAL CORP. NUCLEAR SYSTEMS, DOWNEY, CALIF.
ANWB-(LETTER)-	ARMED FORCES SPECIAL WEAPONS PROJECT. FIELD COMMAND, ALBUQUERQUE, N. MEX.
AO-	A.F. PERSONNEL + TRAINING RES. CTR.,LACKLAND AFB,TEX.
AO(NUMBER)-(LETTER)	PHILCO CORP., BLUE BELL, PENNA.
AO-	RES. + DEV. BD., WASH., D.C. (AMPHIBIOUS OPERATIONS)
AOA-	AMERICAN ORDNANCE ASSOCIATION, WASHINGTON, D.C.
AOC-M(YEAR)-	AMERICAN OIL CO. RES. AND DEV. DEPT., WHITING, IND.
AOF-	(AIR OBJECTIVE FOLDER)
AOFP-	(AIR OBJECTIVES FOLDER PROGRAM)
AOG-	SCRIPPS INSTITUTION OF OCEANOGRAPHY. APPLIED OCEAN-OGRAPHY GROUP, LA JOLLA, CALIF.
AOGMS-	ARMY ORDNANCE GUIDED MISSILE SCHOOL
AOG-S-	SCRIPPS INSTITUTION OF OCEANOGRAPHY. APPLIED OCEAN-OGRAPHY GROUP, LA JOLLA, CALIF.
AOG-U-	SCRIPPS INSTITUTION OF OCEANOGRAPHY. APPLIED OCEAN-OGRAPHY GROUP, LA JOLLA, CALIF.

AOK-(YEAR)-	GENERAL DYNAMICS. ADVANCED STUDIES OFFICE, SAN DIEGO, CALIF.
AOL-	GT. BRIT. ADMIRALTY OIL LAB. (ALL LOCATIONS)
AOL-TM-	GT. BRIT. ADMIRALTY OIL LAB. TEDDINGTON, MIDDX, ENG.
AOM-	NATIONAL DEFENSE RES. COMM. (ARMOR + ORDNANCE MEMO.)
AOMC-	ARMY ORDNANCE MISSILE COMMAND, REDSTONE ARSENAL, ALA.
AOMC-RA-TR-(YEAR)-	ARMY ORDNANCE MISSILE COMMAND, REDSTONE ARSENAL, ALA.
AOMC-RE-TR-(YEAR)-	ARMY ORDNANCE MISSILE COMMAND, REDSTONE ARSENAL, ALA.
AOMC-RF-TR-(YEAR)-	ARMY ORDNANCE MISSILE COMMAND, REDSTONE ARSENAL, ALA.
AOMC-RM-TR-(YEAR)-	ARMY ORDNANCE MISSILE COMMAND, REDSTONE ARSENAL, ALA.
AOMC-RR-	ARMY ORDNANCE MISSILE COMMAND, REDSTONE ARSENAL, ALA.
AOMC-RR-TR-	ARMY ORDNANCE MISSILE COMMAND, REDSTONE ARSENAL, ALA.
AOMC-RT-TR-(YEAR)-	ARMY ORDNANCE MISSILE COMMAND, REDSTONE ARSENAL, ALA.
AOML-	ATLANTIC OCEANOGRAPHIC AND METEOROLOGICAL LABS., MIAMI, FLA.
AOO-	ALBUQUERQUE OPERATIONS OFFICE, AEC, N. MEX.
AOO-	RADIO CORP. OF AMERICA, LANCASTER, PENNA.
AOR-	NATIONAL DEFENSE RES. COMM. (ARMOR + ORDNANCE REPORT)
AORC-(NO.)-(NO.)-(N.)-	AMERICAN OPTICAL CO. RESEARCH DIV.,SOUTHBRIDGE, MASS.
AORD-(NO.-NO.-LTR.)	ARMY RESEARCH OFFICE, DURHAM, N.C.
AORE-	UNITED KINGDOM. ARMY OPERATIONAL RESEARCH ESTABLISHMENT, BYFLEET, SURREY, ENGLAND
AORG-	GT. BRIT. ARMY COUNCIL. ARMY OPERATIONAL RESEARCH GROUP, LONDON
AOSWAC-	ARMY ORDNANCE SPECIAL WEAPONS-AMMUNITION COMMAND, DOVER, N.J.
AOW-	ALABAMA ORDNANCE WORKS, CHILDERSBURG
AOWC-	ARMY ORDNANCE WEAPONS COMMAND, ROCK ISLAND, ILL.
AP-	(AIR PUBLICATION) (REF. 2)
AP-	COAST AND GEODETIC SURVEY, WASHINGTON, D.C. (AERONAUTICAL PLANNING CHART)
AP-	ENVIRONMENTAL PROTECTION AGENCY. AIR POLLUTION CONTROL OFFICE, ROCKVILLE, MD.
AP-	GARRETT CORP. AIRESEARCH MFG. DIV., LOS ANGELES
AP-	GARRETT CORP. AIRESEARCH MFG. CO. OF ARIZONA, PHOENIX
AP-(NUMBER)-R	GARRETT CORP. AIRESEARCH MFG. DIV., PHOENIX, ARIZ.
AP-	GT. BRIT. AERONAUTICAL RESEARCH COUNCIL
AP-	GT. BRIT. AIR MINISTRY (AIR PUBLICATIONS)
AP-	GT. BRIT. ATOMIC EN. RES. ESTAB., HARWELL, BERKS, ENG
AP-	MINNESOTA. UNIV., MINNEAPOLIS. SCHOOL OF PHYSICS
AP-	NAVAL AIR DEVELOPMENT CENTER. AERONAUTICAL PHOTOGRAPHIC EXPERIMENT LAB.
AP-	RAMO-WOOLDRIDGE CORP., LOS ANGELES
AP-	WESTINGHOUSE ELECTRIC CORP. ATOMIC POWER DIV., PITTS.
AP-	WESTINGHOUSE ELECTRIC CORP. BETTIS ATOMIC POWER LAB., PITTSBURGH
APA-	JOHNS HOPKINS UNIV., SILVER SPRING, MD. SOLID PROPELLANT INFO. AGENCY. (ABSTRACTS OF MANUSCRIPTS AND INFORMAL REPTS. OF THE JANAF PANEL ON ANAL.CHEMISTRY)
APAE-	ALCO PRODUCTS, INC., FT. BELVOIR, VA.
APAE-	ALCO PRODUCTS, INC., N.Y.C.
APAE-	ALCO PRODUCTS, INC., SCHENECTADY, N.Y. (ATOMIC EN.)
APAE-MEMO-	ALCO PRODUCTS, INC., N.Y.C.
APAE-MEMO-	ALCO PRODUCTS, INC., SCHENECTADY, N.Y.
APC-	AIR FORCE OPERATIONAL TEST CENTER, EGLIN AFB, FLA.
APC-	APPLIED PSYCHOLOGY CORP., WASHINGTON, D.C.
APC-	ATLAS POWDER CO. (ALL LOCATIONS)
APC-	HERCULES POWDER CO. CHEMICAL PROPULSION DIV. BACCHUS WORKS, MAGNA, UTAH
APCC-	AMERICAN POTASH + CHEMICAL CORP., WHITTIER, CALIF.
APCC-QPR-	AMERICAN POTASH + CHEM. CORP. RES.DEPT.,WHITTIER,CAL.
APCI-	AIR PRODUCTS AND CHEMICALS, INC. RESEARCH AND DEVELOPMENT DEPT., ALLENTOWN, PENNA.
APCL-	AIR PRODUCTS AND CHEMICALS, INC., ALLENTOWN, PENNA.
APC-MPR-	ATLAS POWDER CO., TAMAQUA, PENNA.
APC-MPR-	ATLAS POWDER CO., WILMINGTON, DEL.
APCO-	ENVIRONMENTAL PROTECTION AGENCY. AIR POLLUTION CONTROL OFFICE, ROCKVILLE, MD.
APC/R.-	ATOMIC POWER CONSTRUCTIONS LTD.,SUTTON,SURREY,ENG.
APCS-	AIR PHOTOGRAPHIC AND CHARTING SERVICE, WRIGHT-PATTERSON AFB, OHIO
APD-	AUSTRALIA. WEAPONS RESEARCH ESTABLISHMENT, SALISBURY
APD-	GARRETT CORP. AIRESEARCH MFG. CO. OF ARIZONA, PHOENIX
APD-	WESTINGHOUSE ELECTRIC CORP. ATOMIC POWER DIV., PITTS.
APDA-	ATOMIC POWER DEVELOPMENT ASSOCIATES, INC., DETROIT
APDA-CFE-	ATOMIC POWER DEVELOPMENT ASSOCIATES, INC., DETROIT
APDA-FAM-(YEAR)-	ATOMIC POWER DEVELOPMENT ASSOCIATES, INC., DETROIT (ENRICO FERMI ATOMIC POWER PLANT)
APDA-NTS-	ATOMIC POWER DEVELOPMENT ASSOCIATES, INC., DETROIT
APDA-SPEC-	ATOMIC POWER DEVELOPMENT ASSOCIATES, INC., DETROIT
APDA-TM-	ATOMIC POWER DEVELOPMENT ASSOCIATES, INC., DETROIT
APDA-TPI-	ATOMIC POWER DEVELOPMENT ASSOCIATES, INC., DETROIT
APDC-	AEROPHYSICS DEVELOPMENT CORP. (ALL LOCATIONS)
APDR-	ENGLISH ELECTRIC CO., LTD. ATOMIC POWER DIV., WHETSTONE, LEICS, ENGLAND
APE-	GENERAL ELECTRIC CO. ATOMIC PRODUCTS DIV., SCHENECTADY, N.Y.
APEA-	ARMY PRODUCTION EQUIPMENT AGENCY, ROCK ISLAND, ILL.
APED-	GENERAL ELECTRIC CO. NUCLEAR ENERGY DIV.,SAN JOSE,CAL (ATOMIC POWER EQUIPMENT)
APEL-	NAVAL AIR EXPERIMENTAL STATION. AERONAUTICAL PHOTOGRAPHIC EXPERIMENTAL LAB., PHILADELPHIA
APEL-(NO.)-(YR.)-	NAVAL AIR EXPERIMENTAL STATION. AERONAUTICAL PHOTOGRAPHIC EXPERIMENTAL LAB., PHILADELPHIA
APEX-	GENERAL ELECTRIC CO. FLIGHT PROPULSION LAB DEPT., CINCINNATI
APEX-	GENERAL ELECTRIC CO. NUCLEAR MATERIALS AND PROPULSION OPERATION, CINCINNATI (REF. 3)
APF-	AIR POLLUTION FOUNDATION, LOS ANGELES
APF-CR-	SANDERS ASSOCIATES, INC., BEDFORD, MASS.
APG-	ABERDEEN PROVING GROUND, MD.
APG-	AIR PROVING GROUND COMMAND, EGLIN AFB, FLA.
APG/(LTRS.)/	AIR PROVING GROUND COMMAND, EGLIN AFB, FLA. (PROJECT IDENTIFICATION)
APG-AS-	ABERDEEN PROVING GROUND, MD.
APG-BRL-	BALLISTIC RESEARCH LABS.,ABERDEEN PROVING GROUND, MD.
APG-BRLM-	BALLISTIC RESEARCH LABS.,ABERDEEN PROVING GROUND, MD.
APG-BRLR-	BALLISTIC RESEARCH LABS.,ABERDEEN PROVING GROUND, MD.
APGC-	AIR PROVING GROUND CENTER, EGLIN AFB, FLA.
APGC-AIDR(NUMBER-NUMBER)	AIR PROVING GROUND CENTER, EGLIN AFB, FLA.
APG-CCL-	ARMY COATING AND CHEMICAL LAB., ABERDEEN PROVING GROUND, MD.
APGC-PGY-	AIR PROVING GROUND CENTER, EGLIN AFB, FLA.
APGC PROJ-	AIR PROVING GROUND COMMAND, EGLIN AFB, FLA.
APGC-QR-	AIR PROVING GROUND CENTER, EGLIN AFB, FLA.
APG/CSC/-	AIR PROVING GROUND COMMAND, EGLIN AFB, FLA.
APGC-SR-(YEAR)-	AIR PROVING GROUND CENTER, EGLIN AFB, FLA.
APGC-TDR-	AIR PROVING GROUND CENTER, EGLIN AFB, FLA. (TECHNICAL DOCUMENTARY REPORT)
APGC-TM-	AIR PROVING GROUND CENTER, EGLIN AFB, FLA.
APGC TN-	AIR PROVING GROUND CENTER, EGLIN AFB, FLA.
APGC-TN-(YR.)-	AIR PROVING GROUND CENTER, EGLIN AFB, FLA.

APGC-TR-(YEAR)-	AIR PROVING GROUND CENTER, EGLIN AFB, FLA.
APGC-TR-	AIR PROVING GROUND CENTER, EGLIN AFB, FLA.
APG-DPS-	DEVELOPMENT + PROOF SERVICES, ABERDEEN PROVING GD., MD
APG-FR-B-	ABERDEEN PROVING GROUND, MD.
APG-FR-M-	ABERDEEN PROVING GROUND, MD.
APG-FR-S-	ABERDEEN PROVING GROUND, MD.
APG-HEL-...	HUMAN ENGINEERING LABS., ABERDEEN PROVING GROUND, MD.
APG LSD-	LABORATORY SERVICE DIV., ABERDEEN PROVING GROUND, MD.
APG-MISC-	ABERDEEN PROVING GROUND, MD.
APG-MT-	MATERIEL TEST DIRECTORATE, ABERDEEN PROVING GD., MD.
APG OP-	ABERDEEN PROVING GROUND, MD. (ORDNANCE PROGRAM)
APG OP (NUMBER)/	ABERDEEN PROVING GROUND, MD. (ORDNANCE PROGRAM)
APG-OS-	ORDNANCE SCHOOL, ABERDEEN PROVING GROUND, MD.
APG PROJ-REPT-	ORDNANCE DEPT. RESEARCH AND DEVELOPMENT CENTER, ABERDEEN PROVING GROUND, MD.
APG-QPR-	BALLISTIC RESEARCH LABS., ABERDEEN PROVING GROUND, MD.
APG RPT-	ABERDEEN PROVING GROUND, MD.
APG/TAT/-	AIR PROVING GROUND COMMAND, EGLIN AFB, FLA.
APG-TM-	HUMAN ENGINEERING LABS., ABERDEEN PROVING GROUND, MD.
API-	ACADEMIC PRESS, INC., N.Y.C. (TRANSLATIONS)
API-	AEROPROJECTS, INC., WEST CHESTER, PENNA.
API-	AIR PRODUCTS, INC., ALLENTOWN, PENNA.
API-	ALABAMA POLYTECHNIC INSTITUTE, AUBURN
API-	AMERICAN PETROLEUM INSTITUTE, N.Y.C.
APIC-(NO.)SM-	WESTINGHOUSE ELECTRIC CORP. ATOMIC POWER DIV., PITTS.
APID-	ENVIRONMENTAL PROTECTION AGENCY. AIR POLLUTION CONTROL OFFICE, ROCKVILLE, MD.
APIO-	GENERAL ELECTRIC CO. NEUTRON DEVICES DEPT., ST. PETERSBURG, FLA.
APIO-	GENERAL ELECTRIC CO. NUCLEAR ENERGY DIV., SAN JOSE, CAL
API-PR-	AIR PRODUCTS, INC., ALLENTOWN, PENNA.
API-RR-(YEAR)-	AEROPROJECTS, INC., WEST CHESTER, PENNA.
APIT-TM-(YEAR)-	AIR FORCE AERO-PROPULSION LAB., WRIGHT-PATTERSON AFB, OHIO
APJ-	AMERICAN POWER JET CO., MONTCLAIR, N.J.
APJ-	AMERICAN POWER JET CO., RIDGEFIELD, N.J.
APL-	COLORADO. UNIV., BOULDER
APL-	CORNELL AERONAUTICAL LAB., INC., BUFFALO
APL-	JOHNS HOPKINS UNIV., SILVER SPRING, MD. APPLIED PHYSICS LAB.
APL-AQR/(YEAR)-	JOHNS HOPKINS UNIV., SILVER SPRING, MD. APPLIED PHYSICS LAB.
APL-BB-	JOHNS HOPKINS UNIV., SILVER SPRING, MD. APPLIED PHYSICS LAB. (BUMBLEBEE PROJECT) (REF. 8)
APL-BBW-	JOHNS HOPKINS UNIV., SILVER SPRING, MD. APPLIED PHYSICS LAB.
APL-CF-	JOHNS HOPKINS UNIV., SILVER SPRING, MD. APPLIED PHYSICS LAB.
APL-CLB-	JOHNS HOPKINS UNIV., SILVER SPRING, MD. APPLIED PHYSICS LAB.
APL-CM-	JOHNS HOPKINS UNIV., SILVER SPRING, MD. APPLIED PHYSICS LAB.
APL-C-RQR/(YEAR)-	JOHNS HOPKINS UNIV., SILVER SPRING, MD. APPLIED PHYSICS LAB.
APL-DR-	JOHNS HOPKINS UNIV., SILVER SPRING, MD. APPLIED PHYSICS LAB.
APL-E-	JOHNS HOPKINS UNIV., SILVER SPRING, MD. APPLIED PHYSICS LAB.
APL-JHU-	JOHNS HOPKINS UNIV., SILVER SPRING, MD. APPLIED PHYSICS LAB.
APL/JHU-BB-	JOHNS HOPKINS UNIV., SILVER SPRING, MD. APPLIED PHYSICS LAB. (BUMBLEBEE PROJECT) (REF. 8)
APL-JHU-BLP-	JOHNS HOPKINS UNIV., SILVER SPRING, MD. APPLIED PHYSICS LAB.
APL-JHU-BR-	JOHNS HOPKINS UNIV., SILVER SPRING, MD. APPLIED PHYSICS LAB.
APL/JHU-CF-	JOHNS HOPKINS UNIV., SILVER SPRING, MD. APPLIED PHYSICS LAB.
APL-JHU-CLA-RD-	JOHNS HOPKINS UNIV., SILVER SPRING, MD. APPLIED PHYSICS LAB.
APL/JHU-CM-	JOHNS HOPKINS UNIV., SILVER SPRING, MD. APPLIED PHYSICS LAB.
APL/JHU-CPR-	JOHNS HOPKINS UNIV., SILVER SPRING, MD. APPLIED PHYSICS LAB.
APL/JHU-HWL-	JOHNS HOPKINS UNIV., SILVER SPRING, MD. APPLIED PHYSICS LAB.
APL-JHU-PR-	JOHNS HOPKINS UNIV., SILVER SPRING, MD. APPLIED PHYSICS LAB.
APL/JHU-TAT-	JOHNS HOPKINS UNIV., SILVER SPRING, MD. APPLIED PHYSICS LAB.
APL/JHU-TCR-	JOHNS HOPKINS UNIV., SILVER SPRING, MD. APPLIED PHYSICS LAB. (TRANSPORTATION CONTRACTOR REPORT)
APL/JHU-TG-	JOHNS HOPKINS UNIV., SILVER SPRING, MD. APPLIED PHYSICS LAB.
APL/JHU-TPR-	JOHNS HOPKINS UNIV., SILVER SPRING, MD. APPLIED PHYSICS LAB.
APL/JHU-TT-	JOHNS HOPKINS UNIV., SILVER SPRING, MD. APPLIED PHYSICS LAB. (TARTAR MISSILE)
APL-S-	JOHNS HOPKINS UNIV., SILVER SPRING, MD. APPLIED PHYSICS LAB.
APL-SDO-	JOHNS HOPKINS UNIV., SILVER SPRING, MD. APPLIED PHYSICS LAB.
APL-SMS-FS-	JOHNS HOPKINS UNIV., SILVER SPRING, MD. APPLIED PHYSICS LAB.
APL-SOR-(YEAR)-	JOHNS HOPKINS UNIV., SILVER SPRING, MD. APPLIED PHYSICS LAB.
APL-SOR-CAN-(NO.)-(YEAR)	JOHNS HOPKINS UNIV., SILVER SPRING, MD. APPLIED PHYSICS LAB.
APL-SOR-MPCS-	JOHNS HOPKINS UNIV., SILVER SPRING, MD. APPLIED PHYSICS LAB.
APL-TDR-(YEAR)-	AIR FORCE AERO-PROPULSION LAB., WRIGHT-PATTERSON AFB, OHIO
APL-TG-	JOHNS HOPKINS UNIV., SILVER SPRING, MD. APPLIED PHYSICS LAB.
APL-TM-	JOHNS HOPKINS UNIV., SILVER SPRING, MD. APPLIED PHYSICS LAB.
APL-TPR-	JOHNS HOPKINS UNIV., SILVER SPRING, MD. APPLIED PHYSICS LAB. (TRANSPORTATION PROGRESS REPORT)
APL-TR-(YEAR)-	AIR FORCE AERO-PROPULSION LAB., WRIGHT-PATTERSON AFB, OHIO
APL-TRANS-	JOHNS HOPKINS UNIV., SILVER SPRING, MD. APPLIED PHYSICS LAB. (TRANSLATION)
APL-U-RQR/(YEAR)-	JOHNS HOPKINS UNIV., SILVER SPRING, MD. APPLIED PHYSICS LAB.
APL-U-SQR-(YEAR)-	JOHNS HOPKINS UNIV., SILVER SPRING, MD. APPLIED PHYSICS LAB.
APL-UW-	ARMY COLD REGIONS RES. AND ENG. LAB., HANOVER, N.H.
APL-UW-	WASHINGTON. UNIV., SEATTLE. APPLIED PHYSICS LAB.
APL UW/TE-	WASHINGTON. UNIV., SEATTLE. APPLIED PHYSICS LAB.
APL-WQR/(YEAR)-	JOHNS HOPKINS UNIV., SILVER SPRING, MD. APPLIED PHYSICS LAB.
APMD-	AEROJET-GENERAL NUCLEONICS, SAN RAMON, CALIF.
APN-IM-	ARMED FORCES SPECIAL WEAPONS PROJECT. FIELD COMMAND, ALBUQUERQUE, N. MEX. (INSTRUCTORS MANUAL)
AP-NOTE-	ALCO PRODUCTS, INC., SCHENECTADY, N.Y.
APN-VG-	ARMED FORCES SPECIAL WEAPONS PROJECT. FIELD COMMAND, ALBUQUERQUE, N. MEX. (VU-GRAPHS)
APO-	MANNED SPACECRAFT CENTER, HOUSTON, TEX.
APP-	ARMY AIR FORCE (AVIATION PSYCHOLOGY PROGRAM)
APP-	NATIONAL DEFENSE RESEARCH COMMITTEE. APPLIED PSYCHOLOGY PANEL (REF. 28)
APP-	PITTSBURGH. UNIV. ATOMIC AND PLASMA PHYSICS LAB.

APP DP-	SIGNAL CORPS (ARMY). PICTORIAL ENGINEERING AND RESEARCH LAB., LONG ISLAND, N.Y.
APPE-	ATOMIC ENERGY OF CANADA LTD., CHALK RIVER, ONT.
APR-	(ANNUAL PROGRESS REPORT) (REF. 2)
APR-	AMCEL PROPULSION CO., ASHEVILLE, N.C.
APR-	GOODYEAR TIRE AND RUBBER CO. AVIATION PRODUCTION DIV., LITCHFIELD PARK, ARIZ.
APR-	HILLER AIRCRAFT CORP., PALO ALTO, CALIF.
APR-	LOUISVILLE, KY. UNIV.
APR-	NATIONAL RESEARCH COUNCIL OF CANADA. DIV. OF APPLIED PHYSICS, OTTAWA
APR-	NORTHROP CAROLINA, INC., ASHEVILLE, N.C.
APR-	REDEL, INC., ANAHEIM, CALIF.
APR-	TEMPLE UNIV., PHILADELPHIA
APR-	VANDERBILT UNIV., NASHVILLE
APRC-	UNITED KINGDOM. ARMY PERSONNEL RESEARCH COMM., LONDON
APRDC-	ARMY POLAR RESEARCH AND DEV. CTR., FORT BELVOIR, VA.
APRE-	UNITED KINGDOM. ARMY PERSONNEL RESEARCH ESTABLISHMENT, BYFLEET, SURREY, ENGLAND
APRE-RM-Q/	UNITED KINGDOM. ARMY PERSONNEL RESEARCH ESTABLISHMENT, BYFLEET, SURREY, ENGLAND
APRE-TRIAL-MEMO-	UNITED KINGDOM. ARMY PERSONNEL RESEARCH ESTABLISHMENT, BYFLEET, SURREY, ENGLAND
APRF-	AEROPHYSICS RESEARCH FOUNDATION, GOLETA, CALIF.
APRL-	BUREAU OF MINES. APPLIED PHYSICS RESEARCH LAB., COLLEGE PARK, MD.
APRL-(NUMBER)-	BUREAU OF MINES. APPLIED PHYSICS RESEARCH LAB., COLLEGE PARK, MD.
APRL-E-	BUREAU OF MINES. APPLIED PHYSICS RESEARCH LAB., COLLEGE PARK, MD.
APRO-TRN-	ARMY PERSONNEL RESEARCH OFFICE, WASHINGTON, D.C.
APRO-TRW-	ARMY PERSONNEL RESEARCH OFFICE, WASHINGTON, D.C.
APS-	BATTELLE MEMORIAL INST., COLUMBUS, OHIO
APS 71/RSG-	BOEING AIRPLANE CO., SEATTLE
APS-	GARRETT CORP. AIRESEARCH MFG. DIV., LOS ANGELES
APS-	GARRETT CORP. AIRESEARCH MFG. CO. OF ARIZONA, PHOENIX
APS-(NUMBER)-R	GARRETT CORP. AIRESEARCH MFG. DIV., PHOENIX, ARIZ.
APS-CH-	FINNISH ACADEMY OF TECHNICAL SCIENCES, HELSINKI (ACTA POLYTECHNICA SCANDINAVICA. CHEMISTRY)
APS-CI-	FINNISH ACADEMY OF TECHNICAL SCIENCES, HELSINKI (ACTA POLYTECHNICA SCANDINAVICA. CIVIL ENGINEERING)
APS-EI-	FINNISH ACADEMY OF TECHNICAL SCIENCES, HELSINKI (ACTA POLYTECHNICA SCANDINAVICA. ELECTRICAL ENG.)
APS-EL-	FINNISH ACADEMY OF TECHNICAL SCIENCES, HELSINKI (ACTA POLYTECHNICA SCANDINAVICA. ELEC. ENG. SER.)
APS-MA-	FINNISH ACADEMY OF TECHNICAL SCIENCES, HELSINKI (ACTA POLYTECHNICA SCANDINAVICA. MATH. + COMPUTING)
APS-ME-	FINNISH ACADEMY OF TECHNICAL SCIENCES, HELSINKI (ACTA POLYTECHNICA SCANDINAVICA. MECHANICAL ENG.)
APS-PH-	FINNISH ACADEMY OF TECHNICAL SCIENCES, HELSINKI (ACTA POLYTECHNICA SCANDINAVICA. PHYSICS)
APT-	APPLIED PARKING TECHNIQUES, INC., RESTON, VA.
APTD-	ENVIRONMENTAL PROTECTION AGENCY. AIR POLLUTION CONTROL OFFICE, ROCKVILLE, MD. (AIR POLLUTION TECHNICAL DATA)
APTIC-	ENVIRONMENTAL PROTECTION AGENCY. AIR POLLUTION CONTROL OFFICE, ROCKVILLE, MD. (AIR POLLUTION TECHNICAL INFORMATION CENTER)
APU-	GT. BRIT. APPLIED PSYCH. RES. UNIT, CAMBRIDGE, ENG.
APXNR-	NATIONAL RESEARCH COUNCIL OF CANADA. DIV. OF APPLIED PHYSICS, OTTAWA
AQ(NUMBER)D(YEAR)	LITTON SYSTEMS, INC. GUIDANCE AND CONTROL SYSTEMS DIV., WOODLAND HILLS, CALIF.
AQ(NUMBER)C(YEAR)	LITTON SYSTEMS, INC. GUIDANCE AND CONTROL SYSTEMS DIV., WOODLAND HILLS, CALIF.
AQ(NUMBER)M(YEAR)	LITTON SYSTEMS, INC. GUIDANCE SYSTEMS LAB., WOODLAND HILLS, CALIF.
A.Q.D./D(NUMBER/YEAR)	UNITED KINGDOM. AERONAUTICAL QUALITY ASSURANCE DIRECTORATE, HAREFIELD, MIDDX., ENGLAND
AQD/Y/	UNITED KINGDOM. AERONAUTICAL QUALITY ASSURANCE DIRECTORATE, UXBRIDGE, MIDDX., ENGLAND
AQR/(YEAR)-	JOHNS HOPKINS UNIV., SILVER SPRING, MD. APPLIED PHYSICS LAB. (AERONAUTICS DIV. QUARTERLY REPT.)
AR-	(AERODYNAMIC REPORT) (REF. 2)
AR-	(ANNUAL REPORT) (REF. 2)
AR-	ADAMS-RUSSELL CO., INC., WALTHAM, MASS.
AR-	AERONAUTICAL RADIO, INC., WASHINGTON, D.C.
AR-	ANALYTIC SERVICES, INC., FALLS CHURCH, VA.
AR-(YEAR)-	ANALYTIC SERVICES, INC., FALLS CHURCH, VA.
AR-	ARIZONA. UNIV., TUCSON. ENGINEERING EXPERIMENT STA.
AR-	ARMY, WASHINGTON, D.C. (REGULATION)
AR-	AVION ELECTRONICS, INC., PARAMUS, N.J.
AR-	BATTELLE MEMORIAL INST., GENEVA
A-R-	COLLINS RADIO CO., CEDAR RAPIDS, IOWA (APOLLO REPORT)
AR-(1 DIGIT)	COLUMBIA UNIV., NYC. SUBSTITUTE ALLOY MATERIALS LABS.
AR-(NUMBER) Q-(YEAR)	DENVER. UNIV. COLL. OF LAW
	DOW CHEMICAL CO., MIDLAND, MICH. (ADVANCED RES. ON SOLID ROCKET PROPELLANTS. QUARTERLY PROGRESS RPT.)
AR-(NUMBER) S-(YEAR)	DOW CHEMICAL CO., MIDLAND, MICH. (ADVANCED RESEARCH ON SOLID ROCKET PROPELLANTS)
AR(NUMBER)	ERIE TECHNOLOGICAL PRODUCTS, INC., PENNA.
AR-	ETHYL CORP. RESEARCH AND DEVELOPMENT DEPT., FERNDALE, MICH. (AERONAUTICAL RESEARCH)
AR-	FISH AND WILDLIFE SERVICE (INTERIOR), WASHINGTON, D.C. (ADMINISTRATIVE REPORTS)
AR-	FMC CORP., BALTIMORE
AR-	GARRETT CORP. AIRESEARCH MFG. DIV., LOS ANGELES
AR-592-1-	GENERAL DYNAMICS/ASTRONAUTICS, SAN DIEGO, CALIF.
AR-	GT. BRIT. ARMAMENT RES. EST., FT. HALSTEAD, KENT, ENG
AR-	IDAHO. UNIV., MOSCOW
AR-	MAX-PLANCK-INSTITUT FUER KERNPHYSIK, HEIDELBERG
AR-	NAVAL AIR DEVELOPMENT CENTER. AVIATION ARMAMENT LAB., JOHNSVILLE, PENNA.
AR-	NEW HAMPSHIRE. UNIV., DURHAM
AR-	NORTHWESTERN UNIV., EVANSTON, ILL.
AR-	OAK RIDGE NATIONAL LAB., TENN.
AR-	PENNSYLVANIA. UNIV., PHILADELPHIA
AR-	RESEARCH + DEV. BOARD. COMM. ON AERONAUTICS, WASH., DC
AR-	SASKATCHEWAN. UNIV., SASKATOON
AR-	SOUTHWEST RESEARCH INST. DEPT. OF AUTOMOTIVE RES., SAN ANTONIO
AR-(NUMBER)-	SPERRY GYROSCOPE CO., GREAT NECK, N.Y.
AR-	STANFORD RESEARCH INST., MENLO PARK, CALIF.
AR-	UNION OF SOUTH AFRICA. COUNCIL FOR SCIENTIFIC AND INDUSTRIAL RESEARCH, PRETORIA
AR-(YEAR)-	WICHITA STATE UNIV., KANS. DEPT OF AERONAUTICAL ENG.
ARA-	AIRCRAFT RESEARCH ASSN., LTD., BEDFORD, BEDS, ENG.
ARA-	ALLIED RESEARCH ASSOCIATES, INC., CONCORD, MASS.
ARA-	ARA, INC., WEST COVINA, CALIF.
ARA-	AUSTIN RESEARCH ASSOCIATES, INC., TEX.
ARA-C-	AUSTIN RESEARCH ASSOCIATES, INC., TEX.
ARA-F-	ALLIED RESEARCH ASSOCIATES, INC., CONCORD, MASS.
ARAI-	ALLIED RESEARCH ASSOCIATES, INC., BOSTON
ARAI-	AUSTIN RESEARCH ASSOCIATES, INC., TEX.
ARA-LIB-TRANS-	AIRCRAFT RESEARCH ASSN., LTD., BEDFORD, BEDS, ENG.
ARA-M.(NO.)/	AIRCRAFT RESEARCH ASSN., LTD., BEDFORD, BEDS, ENG.
ARA-M-	ALLIED RESEARCH ASSOCIATES, INC., BOSTON

ARAP-

ARAP-	AERONAUTICAL CHART AND INFORMATION CENTER, ST. LOUIS
ARAP-	AERONAUTICAL RESEARCH ASSOCIATES OF PRINCETON,INC.,NJ
ARAP-(YEAR)-	AERONAUTICAL RESEARCH ASSOCIATES OF PRINCETON,INC.,NJ
ARAP-(NUMBER)-P	AERONAUTICAL RESEARCH ASSOCIATES OF PRINCETON,INC.,NJ (PROJECT SQUID)
ARAP-TM-(YEAR)-	AERONAUTICAL RESEARCH ASSOCIATES OF PRINCETON,INC.,NJ
ARA-T-	ALLIED RESEARCH ASSOCIATES, INC., CONCORD, MASS.
ARA-TR-	ALLIED RESEARCH ASSOCIATES, INC., CONCORD, MASS.
ARA-TRANSL-	AIRCRAFT RESEARCH ASSN., LTD., BEDFORD, BEDS, ENG.
ARA-WTN-	AIRCRAFT RESEARCH ASSN., LTD., BEDFORD, BEDS., ENG
ARB-	ARGONNE NATIONAL LAB., IDAHO FALLS
ARB-	DEFENSE DOCUMENTATION CENTER, ALEXANDRIA, VA.
ARB-SC-	DEFENSE DOCUMENTATION CENTER, ALEXANDRIA, VA. (A REPORT BIBLIOGRAPHY, SEARCH CONTROL NUMBER...)
ARB-TN-	UNITED KINGDOM. AIR REGISTRATION BOARD, REIGATE, SURREY, ENGLAND
ARC-	AERONAUTICAL CHART AND INFORMATION CENTER, ST. LOUIS
ARC-(YEAR)-	AMERICAN RESEARCH CORP., FULLERTON, CALIF.
ARC-	AMES RESEARCH CENTER, MOFFETT FIELD, CALIF.
ARC-	ARMED SERVICES TECHNICAL INFORMATION AGENCY. REFERENCE CTR., LIBRARY OF CONGRESS, WASH., D.C.
ARC-	ATLANTIC RESEARCH CORP., ALEXANDRIA, VA.
ARC-(NUMBER)-P	ATLANTIC RESEARCH CORP., ALEXANDRIA, VA. (SQUID PROJECT) (REF. 33)
ARC-	ATLANTIC RES. CORP. SPACE VEHICLES DIV., EL MONTE,CAL
ARC-	CAMBRIDGE UNIV., ENGLAND
ARC-	GT. BRIT. AERONAUTICAL RESEARCH COUNCIL
ARC-	GT. BRIT. NATIONAL GAS TURBINE ESTABLISHMENT, FARNBOROUGH, HANTS, ENGLAND
ARC-	LIBRARY OF CONGRESS. TECH. INFO. DIV., WASH., D.C.
ARC-	LONDON. UNIV.
ARC-	OAK RIDGE NATIONAL LAB., TENN.
ARC-(NUMBER)-PU	PURDUE UNIV., LAFAYETTE, IND.
ARC-	VIRGINIA. UNIV., CHARLOTTESVILLE
ARCAS-	ATLANTIC RESEARCH CORP., ALEXANDRIA, VA.
ARC-C(NUMBER)-(NUMBER)FR	APPLICATION RES. CORP., LOS ANGELES
ARCC-	AUSTRALIA. AERO. RES. LABS.,FISHERMANS BEND, VICTORIA
ARC-COMP-(FM)-	GT. BRIT. AERONAUTICAL RESEARCH COUNCIL, LONDON
ARC-CP-	GT. BRIT. AERONAUTICAL RESEARCH COUNCIL
ARCCP-	GT. BRIT. AERONAUTICAL RES. COUNCIL (CURRENT PAPER)
ARC-FM-	GT. BRIT. AERONAUTICAL RESEARCH COUNCIL
ARC(GB)-	GT. BRIT. AERONAUTICAL RESEARCH COUNCIL
ARCH-	BALLISTIC RESEARCH LABS.,ABERDEEN PROVING GROUND, MD. (TRANSLATIONS OF GERMAN ARCHIVES)
ARC-HMT-	GT. BRIT. AERONAUTICAL RESEARCH COUNCIL, LONDON
ARC-HYP-	AERONAUTICAL CHART AND INFORMATION CENTER, ST. LOUIS
ARCL-	GT. BRIT. AGRICULTURAL RESEARCH COUNCIL. RADIOBIOLOGICAL LAB., GROVE, BERKS, ENGLAND
ARC-LTN-	ASTRO RESEARCH CORP., SANTA BARBARA, CALIF.
ARCO-(4 DIGITS)-	ARCO NUCLEAR CO., LEECHBURG, PENNA.
ARCO-	COMBINED INTELLIGENCE OBJECTIVES SUB-COMMITTEE. AIRCRAFT RESOURCES CONTROL OFFICES
ARCO-	PHILLIPS PETROLEUM CO. ARCO CHEM. PLANT, IDAHO FALLS
ARCO-AM-	APPLIED RADIATION CORP., WALNUT CREEK, CALIF.
ARC-PA-	GT. BRIT. AERONAUTICAL RESEARCH COUNCIL, LONDON
ARC-PERF.-	AERONAUTICAL CHART AND INFORMATION CENTER, ST. LOUIS
ARC-PERF-	GT. BRIT. AERONAUTICAL RESEARCH COUNCIL, LONDON
ARC-PL-	AERONAUTICAL CHART AND INFORMATION CENTER, ST. LOUIS
ARC-QPRS-	ATLANTIC RESEARCH CORP., ALEXANDRIA, VA.
ARC-R-	ASTRO RESEARCH CORP., SANTA BARBARA, CALIF.
ARCRL-	GT. BRIT. AERONAUTICAL RESEARCH COUNCIL
ARCRL-	GT. BRIT. AGRICULTURAL RESEARCH COUNCIL. RADIOBIOLOGICAL LAB., GROVE, BERKS, ENGLAND
ARCRL-TRANS-	AIR FORCE CAMBRIDGE RESEARCH LABS., BEDFORD, MASS.
ARC-R/M-	AERONAUTICAL CHART AND INFORMATION CENTER, ST. LOUIS
ARC-R+M-	GT. BRIT. AERONAUTICAL RESEARCH COUNCIL
ARC RM-	GT. BRIT. AERONAUTICAL RESEARCH COUNCIL
ARC-R/M-	GT. BRIT. AERONAUTICAL RESEARCH COUNCIL
ARC S-	CALIFORNIA. UNIV., BERKELEY. RADIATION LAB.
ARC-S+C-	GT. BRIT. AERONAUTICAL RESEARCH COUNCIL
ARC-T+M-	CAMBRIDGE UNIV., ENGLAND
ARC-T+M-	GT. BRIT. AERONAUTICAL RESEARCH COUNCIL, LONDON
ARC-TN-	GT. BRIT. AERONAUTICAL RESEARCH COUNCIL
ARC-TN-	INDIA. AERONAUTICAL RESEARCH COMMISSION, BANGALORE
ARC-TN-	INDIA. COUNCIL OF SCIENTIFIC AND IND. RES., NEW DELHI
ARC-TP-	GT. BRIT. AERO. RES. COUNCIL. WIND TUNNEL DESIGN COMM
ARC TR-	GT. BRIT. AERONAUTICAL RESEARCH COUNCIL
ARC-TR-	INDIA. COUNCIL OF SCIENTIFIC AND IND. RES., NEW DELHI
ARC-TRDL-	ATLANTIC RESEARCH CORP., ALEXANDRIA, VA.
ARC-TR-PL-	ATLANTIC RESEARCH CORP., ALEXANDRIA, VA.
ARD-	ASSOCIATED RESEARCH DESIGN, ALBUQUERQUE, N. MEX.
ARD-(YEAR)-	EDGERTON, GERMESHAUSEN AND GRIER, INC., BOSTON
ARD-(NUMBER)/	GT. BRIT. ARMAMENT RESEARCH AND DEVELOPMENT ESTABLISHMENT, FORT HALSTEAD, KENT, ENGLAND
ARD-	HILLER AIRCRAFT CORP., PALO ALTO, CALIF.
ARD-(YEAR)-	NORTHROP CORP. NORTRONICS DIV. APPLIED RESEARCH DEPT., NEWBURY PARK, CALIF.
ARD-(YEAR-NUMBER)(LTR.)	NORTRONICS. APPLIED RES. DEPT., NEWBURY PARK, CALIF.
ARD (NUMBER)-	REPUBLIC AVIATION CORP. APPLIED RESEARCH AND DEVELOPMENT MATERIALS LAB., FARMINGDALE, N.Y.
ARD-(LETTERS)-	ASSOCIATED RESEARCH DESIGN, ALBUQUERQUE, N. MEX.
ARDC-	AIR RESEARCH AND DEVELOPMENT COMMAND, ANDREWS AFB, MD
ARDC C(YR)-	AIR RESEARCH AND DEVELOPMENT COMMAND, BALTIMORE
ARDC-CATE-	AIR RESEARCH AND DEVELOPMENT COMMAND, ANDREWS AFB, MD
ARDC-CMCC-GM-	AIR RES.+DEV.COMMAND. WESTERN DEV.DIV.,INGLEWOOD,CAL.
ARDC-DAL-	AIR RES.+DEV.COMMAND. WESTERN DEV.DIV.,INGLEWOOD,CAL.
ARDC-EO-	AIR RESEARCH AND DEV. COMMAND. EUROPEAN OFFICE
ARDC-GM-(YEAR)-	BALLISTIC MISSILE DIV. (AIR FORCE), INGLEWOOD, CALIF.
ARDCLO-	AIR RESEARCH + DEVELOPMENT COMMAND. LIAISON OFFICE, FAR EAST AIR FORCES, TOKYO
ARDC M-	AIR RESEARCH AND DEV. COMMAND, BALTIMORE (MANUAL)
ARDCM-	AIR RESEARCH AND DEVELOPMENT COMMAND, BALTIMORE
ARDC/P/-	UNITED KINGDOM ATOMIC ENERGY AUTHORITY. INDUSTRIAL GROUP. SPRINGFIELDS WORKS, LANCS, ENG.
ARDC-QI-(YEAR/MO.-MO.)	RESEARCH DIV. (AIR FORCE), WASHINGTON, D.C. (QUARTERLY INDEX)
ARDC(RDA-)	AIR RESEARCH AND DEVELOPMENT COMMAND, ANDREWS AFB, MD
ARDC-RM-	AIR RES.+DEV.COMMAND. WESTERN DEV.DIV.,INGLEWOOD,CAL.
ARDC-SR-	AIR RESEARCH AND DEVELOPMENT COMMAND, ANDREWS AFB, MD
ARDC-TM-	ABERDEEN RESEARCH AND DEVELOPMENT CENTER, ABERDEEN PROVING GROUND, MD.
ARDC-TN-(YR)-	AIR RESEARCH AND DEVELOPMENT COMMAND, ANDREWS AFB, MD
ARDC-TR-	ABERDEEN RESEARCH AND DEVELOPMENT CENTER, ABERDEEN PROVING GROUND, MD.
ARDC-TR-	AIR RESEARCH AND DEVELOPMENT COMMAND, ANDREWS AFB, MD
ARDC-TR-(YR)-	AIR RESEARCH AND DEVELOPMENT COMMAND, ANDREWS AFB, MD
ARDC-TRR-	GT. BRIT. ARMAMENT RES. DEPT.,FT.HALSTEAD, KENT, ENG.
ARDC-WDD-	AIR RES.+DEV.COMMAND. WESTERN DEV.DIV.,INGLEWOOD,CAL.

```
ARDC-WDOM-(YEAR)-
            AIR RES.+DEV.COMMAND. WESTERN DEV.DIV.,INGLEWOOD,CAL.
ARDC-WDPCD-(YEAR)-
            AIR RES.+DEV.COMMAND. WESTERN DEV.DIV.,INGLEWOOD,CAL.
ARDC-WDTC-(YEAR)-
            AIR RES.+DEV.COMMAND. WESTERN DEV.DIV.,INGLEWOOD,CAL.
ARDC-WDTP-(YEAR)-
            BALLISTIC MISSILE DIV. (AIR FORCE), INGLEWOOD, CALIF.
ARD-DM-
            GIANNINI CONTROLS CORP. ASTROMECHANICS RESEARCH
               DIV., MALVERN. PENNA.
ARDE-
            GT. BRIT. ARMAMENT RESEARCH AND DEVELOPMENT ESTAB-
               LISHMENT, FORT HALSTEAD, KENT, ENGLAND
ARDE(B)-(NO.)(YR)
            GT. BRIT. ARMAMENT RESEARCH AND DEVELOPMENT ESTAB-
               LISHMENT, FORT HALSTEAD, KENT, ENGLAND
ARDE(M)-(NO.)(YR)
            GT. BRIT. ARMAMENT RESEARCH AND DEVELOPMENT ESTAB-
               LISHMENT, FORT HALSTEAD, KENT, ENGLAND
ARD EM-
            GT. BRIT. ARMAMENT RESEARCH DEPT., FORT HALSTEAD,
               KENT, ENGLAND (ELECTRODEPOSITION MEMO)
ARDE-MEMO-
            GT. BRIT. ARMAMENT RESEARCH AND DEVELOPMENT ESTAB-
               LISHMENT, FORT HALSTEAD, KENT, ENGLAND
ARDE-MEMO(B)-(NO.)(YR)
            GT. BRIT. ARMAMENT RESEARCH AND DEVELOPMENT ESTAB-
               LISHMENT, FORT HALSTEAD, KENT, ENGLAND
ARDE(MX)-(NO.)(YR)
            GT. BRIT. ARMAMENT RESEARCH AND DEVELOPMENT ESTAB-
               LISHMENT, FORT HALSTEAD, KENT, ENGLAND
ARDE-(N)-(NO.)(YR)
            GT. BRIT. ARMAMENT RESEARCH AND DEVELOPMENT ESTAB-
               LISHMENT, FORT HALSTEAD, KENT, ENGLAND
ARDE-S-
            GT. BRIT. ARMAMENT RESEARCH AND DEVELOPMENT ESTAB-
               LISHMENT, HORSHAM, SUSSEX, ENGLAND
ARDE TN-
            ARDE ASSOCIATES, NEWARK, N. J.
ARDE TR-
            ARDE ASSOCIATES, NEWARK, N. J.
ARD EXPLOSIVES REPORT-
            GT. BRIT. ARMAMENT RESEARCH AND DEVELOPMENT ESTAB-
               LISHMENT, FORT HALSTEAD, KENT, ENGLAND
ARD EXP. RPT.(NUMBER)/
            GT. BRIT. ARMAMENT RESEARCH AND DEVELOPMENT ESTAB-
               LISHMENT, FORT HALSTEAD, KENT, ENGLAND
ARD-FR-
            GIANNINI CONTROLS CORP. ASTROMECHANICS RESEARCH
               DIV., MALVERN. PENNA.
ARDG (E)-
            ARMY RESEARCH AND DEVELOPMENT GROUP (EUROPE). N.Y.C.
ARDG(E)-E-
            ARMY RESEARCH AND DEVELOPMENT GROUP (EUROPE), N.Y.C.
ARDG (FE)-
            ARMY RESEARCH AND DEV. GP. (FAR EAST), SAN FRANCISCO
ARDG(FE)-J-(NO.)-
            ARMY RESEARCH AND DEV. GP. (FAR EAST), SAN FRANCISCO
ARDT-
            HILLER AIRCRAFT CORP., PALO ALTO, CALIF.
ARD-TR-
            GIANNINI CONTROLS CORP. ASTROMECHANICS RESEARCH
               DIV., MALVERN, PENNA.
ARD-TRR-
            GT. BRIT. ARMAMENT RESEARCH DEPT., FORT HALSTEAD,
               KENT, ENGLAND (THEORETICAL RESEARCH REPORT)
ARDU-
            AUSTRALIA. AIRCRAFT RES. AND DEV. UNIT, LAVERTON
ARE-
            ATOMIC ENERGY OF CANADA LTD., CHALK RIVER, ONT.
ARE-(NUMBER)/(YEAR)
            UNITED KINGDOM ATOMIC ENERGY AUTHORITY. WEAPONS GROUP
               ATOMIC WEAPONS RES. ESTAB.,ALDERMASTON, BERKS. ENG.
AREA-
            GEOLOGICAL SURVEY, DENVER
AREC-
            AERONAUTICAL BOARD, WASHINGTON, D.C.
ARE-H-(NO.)/(YR)
            GT. BRIT. ARMAMENT RES. EST., FT. HALSTEAD, KENT, ENG
ARE-MEMO-
            GT. BRIT. ARMAMENT RES. EST., FT. HALSTEAD, KENT, ENG
ARE-MR-
            GT. BRIT. ARMAMENT RES. EST., FT. HALSTEAD, KENT, ENG
ARE-NB-
            ARMED FORCES SPECIAL WEAPONS PROJECT. FIELD COMMAND,
               ALBUQUERQUE, N. MEX. (NOTEBOOK)
A/RES/
            UNITED NATIONS GENERAL ASSEMBLY, N.Y.C.
ARE-VG-
            ARMED FORCES SPECIAL WEAPONS PROJECT. FIELD COMMAND,
               ALBUQUERQUE, N. MEX. (VU-GRAPHS)
ARF-
            AEROPHYSICS RESEARCH FOUNDATION, GOLETA, CALIF.
ARF-
            ARMOUR RESEARCH FOUNDATION, CHICAGO
ARF-(LETTER)-
            IIT RESEARCH INST., CHICAGO (PROJECT)
ARF-DCM-
            ARMOUR RESEARCH FOUNDATION, CHICAGO
ARF-DCP-(YEAR)-
            ARMOUR RESEARCH FOUNDATION, CHICAGO
ARF-FR-
            ARMOUR RESEARCH FOUNDATION, CHICAGO
ARF-IR-
            ARMOUR RESEARCH FOUNDATION, CHICAGO
ARF-ORO-
            ARMOUR RESEARCH FOUNDATION, CHICAGO
ARFP-
            A R F PRODUCTS, INC., RATON, N. MEX.
ARF-PR-
            ARMOUR RESEARCH FOUNDATION, CHICAGO
ARF-RP-
            ARMOUR RESEARCH FOUNDATION, CHICAGO
ARF-TM-
            ARMOUR RESEARCH FOUNDATION, CHICAGO
ARF-TM-(LETTER)-
            ARMOUR RESEARCH FOUNDATION, CHICAGO
ARF-TN-(LETTER)-
            ARMOUR RESEARCH FOUNDATION, CHICAGO
ARF-TR-
            ARMOUR RESEARCH FOUNDATION, CHICAGO
ARG(YR)-FR-
            ATMOSPHERIC RESEARCH GROUP, ALTADENA, CALIF.
ARG-
            MASS. INST. OF TECH., CAMBRIDGE. INSTRUMENTATION LAB.
ARG-
            RAYTHEON CO., ANDOVER, MASS.
ARGMA-
            ARMY ROCKET AND GUIDED MISSILE AGENCY, REDSTONE
               ARSENAL, ALA.
ARGMA-MSP-
            ARMY ROCKET AND GUIDED MISSILE AGENCY, REDSTONE
               ARSENAL, ALA.
ARGMA-PUB-
            ARMY ROCKET AND GUIDED MISSILE AGENCY, REDSTONE
               ARSENAL, ALA.
ARGMA-RCS-...
            ARMY ROCKET AND GUIDED MISSILE AGENCY, REDSTONE
               ARSENAL, ALA.
ARGMA-RHA-
            ARMY ROCKET AND GUIDED MISSILE AGENCY, REDSTONE
               ARSENAL, ALA.
ARGMA-RHM-Z-
            ARMY ROCKET AND GUIDED MISSILE AGENCY, REDSTONE
               ARSENAL, ALA.
ARGMA-TM-
            ARMY ROCKET AND GUIDED MISSILE AGENCY, REDSTONE
               ARSENAL, ALA.
ARGMA-TN-
            ARMY ROCKET AND GUIDED MISSILE AGENCY, REDSTONE
               ARSENAL, ALA.
ARGMA-TR-
            ARMY ROCKET AND GUIDED MISSILE AGENCY, REDSTONE
               ARSENAL, ALA.
ARH-
            ATLANTIC RICHFIELD HANFORD CO., RICHLAND, WASH.
ARH-SA-
            ATLANTIC RICHFIELD HANFORD CO., RICHLAND, WASH.
ARI-
            AERONAUTICAL RADIO, INC., WASHINGTON, D.C.
ARI-
            AEROSPACE RESEARCH, INC., BOSTON, MASS.
ARI-
            SYRACUSE UNIV., N.Y. RESEARCH INST.
ARINC-
            ARINC RESEARCH CORP., WASHINGTON, D.C.
AR/INT.SG-
            EUROPEAN ORGANIZATION FOR NUCLEAR RESEARCH, GENEVA
ARIS-
            SWEDEN. FLYGTEKNISKA FOERSOEKSANSTALTEN, STOCKHOLM
ARIZ-U-
            ARIZONA. UNIV., TUCSON
ARJ-
            AEROJET-GENERAL CORP., AZUSA, CALIF.
ARJ-AE-
            AEROJET-GENERAL CORP., AZUSA, CALIF.
ARJ-FR-
            AEROJET-GENERAL CORP., AZUSA, CALIF.
ARJ-IDO-
            AEROJET-GENERAL CORP., AZUSA, CALIF.
ARJ-L-
            AEROJET-GENERAL CORP., AZUSA, CALIF.
ARJ-LRP-
            AEROJET-GENERAL CORP., AZUSA, CALIF.
ARJ-PR-
            AEROJET-GENERAL CORP., AZUSA, CALIF.
ARJ-SR-
            AEROJET-GENERAL CORP., AZUSA, CALIF.
ARJ-TM-
            AEROJET-GENERAL CORP., AZUSA, CALIF.
ARK-
            ARKANSAS. UNIV., LITTLE ROCK. SCHOOL OF MEDICINE
ARK-RE-
            ARKANSAS. UNIV., FAYETTEVILLE
ARK-U-
            ARKANSAS. UNIV., LITTLE ROCK. SCHOOL OF MEDICINE
ARL-
            AACHEN. TECHNISCHE HOCHSCHULE
ARL-
            AEROCHEM RESEARCH LABS., INC., PRINCETON, N.J.
ARL-
            AEROMEDICAL RES. LAB. (6571ST), HOLLOMAN AFB, N. MEX.
ARL-
            AEROSPACE RESEARCH LABS., WRIGHT-PATTERSON AFB, OHIO
ARL-(YR)-(NO.)-
            AEROSPACE RESEARCH LABS., WRIGHT-PATTERSON AFB, OHIO
ARL-
            AIR MATERIEL COMMAND. AIRCRAFT RADIATION LAB.,
               WRIGHT-PATTERSON AFB, OHIO
ARL-
            ARCTIC RESEARCH LABORATORY, POINT BARROW, ALASKA
ARL-
            AUSTRALIA. AERO. RES. LABS.,FISHERMANS BEND, VICTORIA
ARL-(YEAR)-
            AVCO CORP. AVCO MISSILE, SPACE + ELECTRONICS GROUP,
               WILMINGTON, MASS.
ARL-
            AVCO MFG. CORP. AVCO RESEARCH LAB., EVERETT, MASS.
```

ARL-

ARL-	GT. BRIT. ADMIRALTY RES. LAB., TEDDINGTON, MIDDX, ENG
ARL-	OKLAHOMA. UNIV., NORMAN. RESEARCH INST. ATMOSPHERIC RESEARCH LAB.
ARL-	RAMO-WOOLDRIDGE CORP. AERO. RES. LAB., LOS ANGELES
ARL-	THOMPSON PRODUCTS, INC. ANTENNA RESEARCH LAB, COLUMBUS, OHIO
ARL-	VON KARMAN INSTITUTE FOR FLUID DYNAMICS, RHODE-SAINT-GENESE, BELGIUM
ARL-(YEAR)-	YALE UNIV., NEW HAVEN. OBSERVATORY
ARL/A-	AUSTRALIA. AERO. RES. LABS., FISHERMANS BEND, VICTORIA
ARL/AERO-	AUSTRALIA. AERONAUTICAL RESEARCH LABS., MELBOURNE
ARL/AERO-N-	AUSTRALIA. AERONAUTICAL RESEARCH LABS., FISHERMANS BEND, VICTORIA (AERODYNAMICS NOTE)
ARL-B-	UNITED STATES STEEL CORP. APPLIED RESEARCH LAB., MONROEVILLE, PENNA.
ARL/C/R-	GT. BRIT. ADMIRALTY RES. LAB., TEDDINGTON, MIDDX, ENG
ARL/ENG.FAC.-	AUSTRALIA. AERONAUTICAL RESEARCH LABS., MELBOURNE
ARL/F-	AUSTRALIA. AERONAUTICAL RESEARCH LABS., MELBOURNE
ARL/FL-	AUSTRALIA. AERONAUTICAL RESEARCH LABS., MELBOURNE
ARL/G/N(NO.)	GT. BRIT. ADMIRALTY RES. LAB., TEDDINGTON, MIDDX, ENG
ARL-HE-	SMITH (A.O.) CORP. ADMINISTRATION, RESEARCH AND ENGINEERING LABS., MILWAUKEE (HEAT EXCHANGERS)
ARL/H.G.M.-	AUSTRALIA. AERONAUTICAL RESEARCH LABS., MELBOURNE
ARL/H/N(NUMBER)	GT. BRIT. ADMIRALTY RES. LAB., TEDDINGTON, MIDDX, ENG
ARL/I-	AUSTRALIA. AERONAUTICAL RESEARCH LABS., MELBOURNE
ARL/IN-	AUSTRALIA. AERONAUTICAL RESEARCH LABS., FISHERMANS BEND, VICTORIA (INSTRUMENT NOTE)
ARL/L/N(NO.)	GT. BRIT. ADMIRALTY RES. LAB., TEDDINGTON, MIDDX, ENG
ARL/L/R(NO.)	GT. BRIT. ADMIRALTY RES. LAB., TEDDINGTON, MIDDX, ENG
ARL/ME-	AUSTRALIA. AERO. RES. LABS., FISHERMANS BEND, VICTORIA
ARL/ME-	AUSTRALIA. AERONAUTICAL RESEARCH LABS., MELBOURNE
ARL/MET-	AUSTRALIA. AERO. RES. LABS., FISHERMANS BEND, VICTORIA
ARL/M/N(NO.)	GT. BRIT. ADMIRALTY RES. LAB., TEDDINGTON, MIDDX, ENG
ARL/M/P(NUMBER)-	GT. BRIT. ADMIRALTY RES. LAB., TEDDINGTON, MIDDX, ENG
ARL-MR-(YEAR)-	AEROSPACE RESEARCH LABS., WRIGHT-PATTERSON AFB, OHIO
ARL/M/R(NO.)	GT. BRIT. ADMIRALTY RES. LAB., TEDDINGTON, MIDDX, ENG
ARL/N1/R-	GT. BRIT. ADMIRALTY RES. LAB., TEDDINGTON, MIDDX, ENG
ARL/O/N(NUMBER)	GT. BRIT. ADMIRALTY RES. LAB., TEDDINGTON, MIDDX, ENG
ARL/R1/R-	GT. BRIT. ADMIRALTY RES. LAB., TEDDINGTON, MIDDX, ENG
ARL-R2/R-	GT. BRIT. ADMIRALTY RES. LAB., TEDDINGTON, MIDDX, ENG
ARL/R3/E-	GT. BRIT. ADMIRALTY RES. LAB., TEDDINGTON, MIDDX, ENG
ARL/R4/C-	UNITED KINGDOM ATOMIC ENERGY AUTHORITY. WEAPONS GROUP ATOMIC WEAPONS RES. ESTAB., ALDERMASTON, BERKS, ENG.
ARL-S-	UNITED STATES STEEL CORP. APPLIED RESEARCH LAB., MONROEVILLE, PENNA.
ARL/SM-	AUSTRALIA. AERO. RES. LABS., FISHERMANS BEND, VICTORIA
ARL/SM-	AUSTRALIA. AERONAUTICAL RESEARCH LABS., MELBOURNE
ARL-TDR-(YR)-	AEROSPACE RESEARCH LABS., WRIGHT-PATTERSON AFB, OHIO
ARL-TM-	HARVARD UNIV., CAMBRIDGE, MASS. ACOUSTICS RES. LAB.
ARL-TN-(YR)-	AEROSPACE RESEARCH LABS., WRIGHT-PATTERSON AFB, OHIO
ARL-TR-(NUMBERS)-	AEROMEDICAL RES. LAB. (6571ST), HOLLOMAN AFB, N. MEX.
ARL-TR-(YEAR)-	AEROMEDICAL RES. LAB. (6571ST), HOLLOMAN AFB, N. MEX.
ARL-TR-(NO.)(MONTH/YR.)-	AEROSPACE RESEARCH LABS., WRIGHT-PATTERSON AFB, OHIO
ARL-TR-(YR.)-	AEROSPACE RESEARCH LABS., WRIGHT-PATTERSON AFB, OHIO
ARLV-(NUMBER)-(NUMBER)	AIR RESOURCES LAB., LAS VEGAS, NEV.
ARLV-	ATLANTIC-PACIFIC INTEROCEANIC CANAL STUDY COMM., WASHINGTON, D.C.
ARL-WSCI-(YEAR)-	AEROSPACE RESEARCH LABS., WRIGHT-PATTERSON AFB, OHIO
ARM-	ETHYL CORP. RESEARCH AND DEVELOPMENT DEPT., FERNDALE, MICH. (AERONAUTICAL RESEARCH MEMORANDA)
ARM-	GARRETT CORP. AIRESEARCH MFG. CO. OF ARIZONA, PHOENIX
ARM-	NORTHROP AIRCRAFT, INC., HAWTHORNE, CALIF. (ARMAMENT)
ARM-	ROYAL CANADIAN AIR FORCE. CENTRAL EXPERIMENTAL AND PROVING ESTAB., ROCKCLIFFE, ONT.
ARMA-	AMERICAN BOSCH ARMA CORP. ARMA DIV., GARDEN CITY, N.Y.
ARMA-DF-	AMERICAN BOSCH ARMA CORP. ARMA DIV., GARDEN CITY, N.Y.
ARMA-NVOO-NTS-AR-EG+G, INC., SANTA BARBARA, CALIF. (AERIAL RADIOLOGICAL MONITORING SYSTEM, NEVADA TEST SITE)	
ARMC-	GARRETT CORP. AIRESEARCH MFG. DIV., LOS ANGELES
ARM-IM-	ARMED FORCES SPECIAL WEAPONS PROJECT. FIELD COMMAND, ALBUQUERQUE, N. MEX. (INSTRUCTORS MANUAL)
ARML-TR-(YEAR)-	AEROSPACE MEDICAL RESEARCH LABS. (6570TH), WRIGHT-PATTERSON AFB, OHIO
ARM-NB-	ARMED FORCES SPECIAL WEAPONS PROJECT. FIELD COMMAND, ALBUQUERQUE, N. MEX. (NOTEBOOK)
ARMS-(YEAR.NUMBER)-	EG+G, INC., LAS VEGAS, NEV.
ARMS-NVOO-(LTRS)-	EDGERTON, GERMESHAUSEN + GRIER, INC., SANTA BARBARA, CAL
ARMT-	AEROFLEX LABS., INC., LONG ISLAND CITY, N.Y.
ARMTE-IMPROVED HAWK-E-	ARMY MISSILE TEST AND EVALUATION DIRECTORATE, WHITE SANDS MISSILE RANGE, N. MEX.
ARMTE-IMPROVED HAWK-SE-	ARMY MISSILE TEST AND EVALUATION DIRECTORATE, WHITE SANDS MISSILE RANGE, N. MEX.
ARMTE-TOW-	ARMY MISSILE TEST AND EVALUATION DIRECTORATE, WHITE SANDS MISSILE RANGE, N. MEX.
ARMY-AT-LONDON-	ARMY ATTACHE, LONDON
ARMY-TM-	(ARMY TECHNICAL MANUAL)
ARMY-TM-39-	JOINT ATOMIC WEAPONS PUBLICATIONS BOARD, ALBUQUERQUE, N. MEX. (REF. 14)
ARNEC-	ARNOLD ENGINEERING CO., MARENGO, ILL.
ARNEC-	ARNOLD ENGINEERING CO., N.Y.C.
ARO-	ARMY RESEARCH OFFICE, WASHINGTON, D.C.
ARO-	ARO, INC., ARNOLD AIR FORCE STATION, TENN.
ARO-	ARO EQUIPMENT CORP., BRYAN, OHIO
AROD-	ARMY RESEARCH OFFICE, DURHAM, N.C.
AROD-(NUMBERONUMBER)-C	ARMY RESEARCH OFFICE, DURHAM, N.C.
AROD-(NO.-NO.-LTR.)	ARMY RESEARCH OFFICE, DURHAM, N.C.
AROD-I-(NUMBERONO.)-RT	ARMY RESEARCH OFFICE, DURHAM, N.C.
ARODR-	ARMY RESEARCH OFFICE, DURHAM, N.C.
AROD-T-(NUMBERONO.)-RT	ARMY RESEARCH OFFICE, DURHAM, N.C.
ARO-ITR-	ARMY RESEARCH OFFICE, WASHINGTON, D.C.
AROWA-	BUREAU OF AERONAUTICS (NAVY) (AROWA PROJECT)
ARP/(NUMBER)/RJ-	GT. BRIT. ROAD RES. LAB., HARMONDSWORTH, MIDDX., ENG.
ARP-	SOCIETY OF AUTOMOTIVE ENGINEERS, N.Y.C.
ARPA-	ADVANCED RESEARCH PROJECTS AGENCY, WASHINGTON, D.C.
ARPA-AM-	ADVANCED RESEARCH PROJECTS AGENCY, WASHINGTON, D.C.
ARPA-E-	ADVANCED RESEARCH PROJECTS AGENCY, WASHINGTON, D.C.
ARPA/IDA-	INSTITUTE FOR DEFENSE ANALYSES, ARLINGTON, VA.
ARPA/IDA (LETTER)-	INSTITUTE FOR DEFENSE ANALYSES, ARLINGTON, VA.
ARPA-NTDO-(YEAR)-	ADVANCED RESEARCH PROJECTS AGENCY, WASHINGTON, D.C.
ARPA-TIO-(YEAR)-	ADVANCED RESEARCH PROJECTS AGENCY, WASHINGTON, D.C.
ARPA/TR-	AIR RESEARCH AND DEVELOPMENT COMMAND, ANDREWS AFB, MD
ARPA-VUF-	ADVANCED RESEARCH PROJECTS AGENCY, WASHINGTON, D.C. (SERIES USED FOR REPORTS OF VARIOUS CONTRACTORS CONCERNED WITH THE VELA UNIFORM PROGRAM)
ARPC-	AUSTRALIA. RADIO PROPAGATION COMMITTEE
ARPD-	AIR FORCE (APPLIED RESEARCH PLANNING DOCUMENT)
ARR-	BUREAU OF ORDNANCE (NAVY) (ACOUSTIC RANGE REPORT)
ARR-	NAVAL ORDNANCE LAB., WHITE OAK, MD. (AEROBALLISTICS RESEARCH REPORT)
ARR-(LETTER)(DATE CODE)	NATIONAL ADVISORY COMMITTEE FOR AERONAUTICS (ADVANCE RESTRICTED REPORT) (REF. 25)

ARRL-	NAVAL AIR EXPERIMENTAL STATION. AERONAUTICAL RADIO AND RADAR LAB., PHILADELPHIA
ARS-	AMERICAN ROCKET SOCIETY, N.Y.C.
ARS-	DEPARTMENT OF AGRICULTURE. AGRICULTURAL RESEARCH SERVICE, WASHINGTON, D.C.
ARSC-	AIR REDUCTION SALES CO., N.Y.C.
ARSN-	APPLICAZIONI E RICERCHE SCIENTIFICHE, MILAN
AR-STP-	VIENNA. TECHNISCHE HOCHSCHULE
ARTC/R + D-	WRIGHT AIR DEV. DIV., WRIGHT-PATTERSON AFB, OHIO
ARTC-WR-	AIRCRAFT INDUSTRIES ASSN., AIRCRAFT RESEARCH AND TESTING COMMITTEE, WASHINGTON, D.C.
ARTEMIS-	COLUMBIA UNIV., DOBBS FERRY, N.Y. HUDSON LABS.
ARTS-(YEAR),(VOLUME)	ARMY, WASHINGTON, D.C. (ANNUAL RES. TASK SUMMARY)
ARTS-	NAVAL AIR ROCKET TEST STATION, LAKE DENMARK, N.J.
ARU-R(NUMBER)	UNITED KINGDOM. SCIENCE RESEARCH COUNCIL. ASTROPHYSICS RES. UNIT, ABINGDON, BERKS, ENGLAND
ARV(NUMBERS)	ARMY, VIETNAM
ARV (YEAR)-	ITALY. COMITATO NAZIONALE PER L'ENERGIA NUCLEARE, BOLOGNA
ARZ-	ARIZONA STATE COLL., FLAGSTAFF
AS-	(ADMINISTRATIVE AND SCIENTIFIC PROGRESS REPORT)(REF.2
AS-	ABERDEEN PROVING GROUND, MD.
AS-	AEROSPACE CORP., EL SEGUNDO, CALIF.
AS-(YEAR)-(NUMBERS)	AEROSPACE CORP., EL SEGUNDO, CALIF.
AS-	AIR FORCE (SPECIAL AERONAUTICAL CHARTS)
AS-	ALLSTATES DESIGN AND DEVELOPMENT CO., INC., TRENTON
AS-	BROOKHAVEN NATIONAL LAB., UPTON, N.Y. (ANNUAL REPORT)
AS-	CALIFORNIA. UNIV., BERKELEY
AS-(YEAR)-	CALIFORNIA. UNIV., BERKELEY. COLL. OF ENGINEERING
AS-(YEAR)-	CALIFORNIA. UNIV., BERKELEY. DIV. OF AERONAUTICAL SCIENCES
AS-015-N-	EURATOM-BELGIUM FAST REACTOR ASSN.
AS-	ILLINOIS. UNIV., URBANA
AS-	OFFICE OF NAVAL RESEARCH (ANTI-SUBMARINE OPERATIONS)
AS-	QUARTERMASTER RESEARCH AND ENGINEERING COMMAND, NATICK, MASS. (ANALYTICAL SERIES)
AS-	RES. + DEV. BD.,WASH.,DC.(ANTI-SUBMARINE OPERATIONS)
AS(YEAR)-V(NUMBER)-(NO.)	SPACE TECHNOLOGY LABS., INC., LOS ANGELES
AS-	TECHNICAL INDUSTRIAL INTELL. COMM. AERO. SUB-COMM.
ASA-	AERONUTRONIC, NEWPORT BEACH, CALIF.
ASA-	AMERICAN STANDARDS ASSOCIATION, N.Y.C.
ASAE-	AMERICAN-STANDARD. ATOMIC EN. DIV.,MOUNTAIN VIEW,CAL.
ASAE-	AMERICAN-STANDARD. ATOMIC EN. DIV., REDWOOD CITY,CAL.
ASAE-E-	AMERICAN RADIATOR AND STANDARD SANITARY CORP. ATOMIC ENERGY DIV., REDWOOD CITY, CALIF.
ASAE-S-	AMERICAN-STANDARD. ATOMIC EN. DIV.,MOUNTAIN VIEW,CAL.
ASA(0)-	GT. BRIT. AIR MINISTRY
ASARCO-	AMERICAN SMELTING + REFINING CO., N.Y.C.
ASAT-	ARBEITSGEMEINSCHAFT SATELLITENTRAEGER (ASAT), MUNICH
ASATL-	ADVANCED TECHNOLOGY LABS. DIV. OF AMERICAN-STANDARD, MOUNTAIN VIEW, CALIF.
ASB-(YEAR)-	AERONAUTICAL SYSTEMS DIV., WRIGHT-PATTERSON AFB, OHIO
ASBD-	JOINT TACTICAL AIR SUPPORT BOARD, FORT BRAGG, N.C.
ASBES-WP-(YEAR)-	AERONAUTICAL SYSTEMS DIV. MISSION SIMULATION BRANCH, WRIGHT-PATTERSON AFB, OHIO
ASB(S)R-	UNITED KINGDOM ATOMIC ENERGY AUTHORITY. AUTHORITY HEALTH AND SAFETY BRANCH, RISLEY, LANCS, ENGLAND
ASB-TM-(YEAR)-	AERONAUTICAL SYSTEMS DIV., WRIGHT-PATTERSON AFB, OHIO
ASC-	AERO SERVICE CORP., PHILADELPHIA
ASC-	AEROSPACE CORP., EL SEGUNDO, CALIF.
ASC-	AIR TECH. SERVICE COMMAND, WRIGHT-PATTERSON AFB, OHIO
ASC-	APPLIED SCIENCE CORP., PRINCETON, N.J.
ASC-	ARIZONA STATE COLL., TEMPE
ASCC-	AIR STANDARDIZATION COORDINATING COMMITTEE
ASCE-	AMERICAN SOCIETY OF CIVIL ENGINEERS, N.Y.C.
ASC-IR-(NUMBER)-	AERONAUTICAL SYSTEMS DIV., WRIGHT-PATTERSON AFB, OHIO
ASCOP-	APPLIED SCIENCE CORP., PRINCETON, N.J.
ASCP-ASC-	APPLIED SCIENCE CORP., PRINCETON, N.J.
ASC-TR-	AERONAUTICAL SYSTEMS DIV., WRIGHT-PATTERSON AFB, OHIO
ASD-	ADVANCED SYSTEMS DEVELOPMENT DIV., EL SEGUNDO,CALIF.
ASD-2.5-Q-	AEROJET-GENERAL CORP. LIQUID ROCKET PLANT, SACRAMENTO, CALIF. (FINANCIAL MANAGEMENT REPORT)
ASD-	AERONAUTICAL SYSTEMS DIV., WRIGHT-PATTERSON AFB, OHIO
ASD-(NUMBER)-	AERONAUTICAL SYSTEMS DIV., WRIGHT-PATTERSON AFB, OHIO
ASD-	HUGHES AIRCRAFT CO. AERONAUTICAL SYSTEMS DIV., CULVER CITY, CALIF.
ASD-(NUMBER)(LETTER)	HUGHES AIRCRAFT CO. AERONAUTICAL SYSTEMS DIV., CULVER CITY, CALIF.
ASD-	LITTON SYSTEMS, INC. APPLIED SCIENCE DIV.,MINNEAPOLIS
ASD-ASNMH-TM-(YR.)-	AERONAUTICAL SYSTEMS DIV., WRIGHT-PATTERSON AFB, OHIO
ASD-ASTN-	AERONAUTICAL SYSTEMS DIV., WRIGHT-PATTERSON AFB, OHIO
ASD-CR-(YEAR)-	AERONAUTICAL SYSTEMS DIV., WRIGHT-PATTERSON AFB, OHIO (CONTRACT REPORT)
ASDD-RC-	INTERNATIONAL BUSINESS MACHINES CORP. ADVANCED SYSTEMS DEVELOPMENT DIV., YORKTOWN HEIGHTS, N.Y.
ASD-FR-(YEAR)-(NO.)-	AERONAUTICAL SYSTEMS DIV., WRIGHT-PATTERSON AFB, OHIO
ASD-FTR-(YEAR)-	AERONAUTICAL SYSTEMS DIV., WRIGHT-PATTERSON AFB, OHIO
ASD-IR-	AERONAUTICAL SYSTEMS DIV., WRIGHT-PATTERSON AFB, OHIO
ASD-IR-(NUMBER)-	AERONAUTICAL SYSTEMS DIV., WRIGHT-PATTERSON AFB, OHIO
ASD-MC-	AERONAUTICAL SYSTEMS DIV., WRIGHT-PATTERSON AFB, OHIO
ASD-MP-	AERONAUTICAL SYSTEMS DIV., WRIGHT-PATTERSON AFB, OHIO
ASD-QA-	AERONAUTICAL SYSTEMS DIV., WRIGHT-PATTERSON AFB, OHIO
ASDR-	STRATEGIC BOMBING SURVEY. AERO STUDIES DIV.
ASD-TDR-	AERONAUTICAL SYSTEMS DIV., WRIGHT-PATTERSON AFB, OHIO
ASD-TDR-(NUMBER)-	AERONAUTICAL SYSTEMS DIV., WRIGHT-PATTERSON AFB, OHIO
ASD-TDR-(YEAR)-	AERONAUTICAL SYSTEMS DIV., WRIGHT-PATTERSON AFB, OHIO
ASD-TM-(YEAR)-	AERONAUTICAL SYSTEMS DIV., WRIGHT-PATTERSON AFB, OHIO
ASD-TN-(YEAR)-	AERONAUTICAL SYSTEMS DIV., WRIGHT-PATTERSON AFB, OHIO
ASD-TR-	AERONAUTICAL SYSTEMS DIV., WRIGHT-PATTERSON AFB, OHIO
ASD-TR-(NUMBER)-	AERONAUTICAL SYSTEMS DIV., WRIGHT-PATTERSON AFB, OHIO
ASD-TR-(YEAR)-	AERONAUTICAL SYSTEMS DIV., WRIGHT-PATTERSON AFB, OHIO
ASD-TR-(YR.)-(NO)-	AERONAUTICAL SYSTEMS DIV., WRIGHT-PATTERSON AFB, OHIO
ASD-TR-(YEAR)-	PERSONNEL RES. LAB. (6570TH), LACKLAND AFB, TEX.
ASE-	AMERICAN SCIENCE AND ENG. INC., CAMBRIDGE, MASS.
ASE-	GT. BRIT. ADMIRALTY SIGNAL ESTABLISHMENT
ASE-	WESTINGHOUSE ELECTRIC CORP. ADVANCED SYSTEMS ENGINEERING, SUNNYVALE, CALIF.
ASEE-	GT. BRIT. UNDERWATER DETECTION ESTABLISHMENT, PORTLAND, DORSET, ENGLAND
ASEL-	ARMY SIGNAL ENGINEERING LABS., FORT MONMOUTH, N.J.
ASESA-	ARMED SERVICES ELECTRO STANDARDS AGENCY, FORT MONMOUTH, N.J.
ASESB-	ARMED SERVICES EXPLOSIVES SAFETY BOARD, WASH., D.C.
ASETC-	ARMED SERVICES ELECTRO STANDARDS AGENCY, FORT MONMOUTH, N.J.
ASE TN-	AMERICAN SCIENCE AND ENG. INC., CAMBRIDGE, MASS.
ASF-	ARMY SERVICE FORCES
ASF-CAT-ORD-	ORDNANCE DEPT. (ARMY) (CATALOG)
ASF-CAT-TC-	TRANSPORTATION CORPS (ARMY) (CATALOG)

ASF-M-SIG-
    SIGNAL CORPS (ARMY) (MANUAL)
ASG-18-
    HUGHES AIRCRAFT CO., CULVER CITY, CALIF.
ASG-(CONTRACT NO.)
    RCA VICTOR DIV., RADIO CORP. OF AMERICA, CAMDEN, N.J.
ASG-TM-(YEAR)-
    MARQUARDT AIRCRAFT CO. ASTRO SCIENCES GP, VAN NUYS, CAL
ASG-TM-(YEAR)-
    NORTHROP SPACE LABS., HAWTHORNE, CAL. (ASTRO SCIENCES)
ASI-
    AERONUTRONIC SYSTEMS, INC., GLENDALE, CALIF.
ASI AR-
    AERONUTRONIC SYSTEMS, INC., GLENDALE, CALIF.
ASI-C-
    AERONUTRONIC DIV. OF PHILCO CORP., NEWPORT BEACH, CAL.
ASI PUB MPR-(YEAR)-
    AERONUTRONIC SYSTEMS, INC., GLENDALE, CALIF.
ASI PUB U-
    AERONUTRONIC SYSTEMS, INC., GLENDALE, CALIF.
ASI-S-
    AERONUTRONIC DIV. OF PHILCO CORP., NEWPORT BEACH, CAL.
ASI SR-
    AERONUTRONIC SYSTEMS, INC., GLENDALE, CALIF.
ASI SRD-
    AERONUTRONIC SYSTEMS, INC., GLENDALE, CALIF.
ASI-T-
    GT. BRIT. AIR SCIENTIFIC INTELLIGENCE (TECH. TRANS.)
ASI-U-
    AERONUTRONIC DIV. OF PHILCO CORP., NEWPORT BEACH, CAL.
ASL-(NUMBERS)-(YR.)-
    ASTRO-SPACE LABS., INC., HUNTSVILLE, ALA.
AS-L-
    COLUMBIA UNIV., NYC. SUBSTITUTE ALLOY MATERIALS LABS.
ASL-
    NAVAL AIR EXPTL. STA. AERO. STRUCTURES LAB., PHILA.
ASL-
    NORTHROP SPACE LABS., HAWTHORNE, CAL. (ASTRO SCIENCES)
ASL/DR-
    WHITE SANDS MISSILE RANGE. ATMOSPHERIC SCIENCES LAB.,
        N. MEX. (DATA REPORT)
ASL-FE-(YR)-
    ASTRO-SPACE LABS., INC., HUNTSVILLE, ALA.
ASL-FR-(YEAR)-
    ASTRO-SPACE LABS., INC., HUNTSVILLE, ALA.
ASLIB-
    ASSN. OF SPEC. LIBRARIES + INFO. BUREAUX, LONDON
ASL-NAM-
    NAVAL AIR EXPTL. STA. AERO. STRUCTURES LAB., PHILA.
ASL-NAM-AD-
    NAVAL AIR EXPTL. STA. AERO. STRUCTURES LAB., PHILA.
AS-M-
    COLUMBIA UNIV., NYC. SUBSTITUTE ALLOY MATERIALS LABS.
ASM-
    COLUMBIA UNIV., NYC. SUBSTITUTE ALLOY MATERIALS LABS.
ASME-
    AMERICAN SOCIETY OF MECHANICAL ENGINEERS, N.Y.C.
ASMSA-
    ARMY SIGNAL MISSILE SUPPORT AGENCY, WHITE SANDS
        MISSILE RANGE, N. MEX.
ASMSA-MEW-
    ARMY SIGNAL MISSILE SUPPORT AGENCY, WHITE SANDS
        MISSILE RANGE, N. MEX. (MISSILE ELECTRONIC WARFARE)
ASMSA-MGD-
    ARMY SIGNAL MISSILE SUPPORT AGENCY. MISSILE GEO-
        PHYSICS DIV., WHITE SANDS MISSILE RANGE, N. MEX.
ASMSA-MM-
    ARMY SIGNAL MISSILE SUPPORT AGENCY, WHITE SANDS
        MISSILE RANGE, N. MEX. (MISSILE METEOROLOGY)
ASNFS-TM-(YEAR)-
    DIRECTORATE OF AIRFRAME SUBSYSTEMS ENGINEERING,
        WRIGHT-PATTERSON AFB, OHIO
ASNJ-TN-(YEAR)-
    AERONAUTICAL SYSTEMS DIV. DIRECTORATE OF PROPULSION
        AND POWER SUBSYSTEMS, WRIGHT-PATTERSON AFB, OHIO
ASNQ-TM-(YEAR)-
    AERONAUTICAL SYSTEMS DIV. DIRECTORATE OF
        RECONNAISSANCE ENG., WRIGHT-PATTERSON AFB, OHIO
ASO-
    AVIATION SUPPLY OFFICE, PHILADELPHIA
ASP-
    APPLIED SCIENCE CORP., SANTA PAULA, CALIF.
ASPB-
    ARMED SERVICES PETROLEUM BOARD, WASH., D.C.
ASPC-(NUMBER)-
    AEROJET SOLID PROPULSION CO., SACRAMENTO, CALIF.
ASPEN-
    BUREAU OF ORDNANCE (NAVY). ASPEN COMMITTEE
ASPM-
    SYLVANIA ELECTRONIC SYSTEMS-EAST. ADVANCED SYSTEMS
        PLANNING, WALTHAM, MASS.
ASPO-
    MANNED SPACECRAFT CENTER. APOLLO SPECIAL PROJECT
        OFFICE, HOUSTON, TEX.
ASPR
    (ARMED SERVICES PROCUREMENT REGULATIONS)
ASPRL-LN-(NUMBERS)
    ARMAMENT SYSTEMS PERSONNEL RES. LAB., LOWRY AFB, COLO
ASPRL-TM-
    ARMAMENT SYSTEMS PERSONNEL RES. LAB., LOWRY AFB, COLO
ASP-TDR-
    OHIO STATE UNIV. RESEARCH FOUNDATION, COLUMBUS
ASR-
    (ANNUAL SUMMARY REPORT) (REF. 2)
ASR-
    AACHEN. TECHNISCHE HOCHSCHULE
A.S.R.(NUMBER)
    AIX-MARSEILLES UNIV., FRANCE. INSTITUT DE MECANIQUE
        STATISTIQUE DE LA TURBULENCE (ANNUAL SCI. REPT.)
ASR-
    AMERICAN SMELTING + REFINING CO., BARBER, N.J.
ASR-
    ATOMIC ENERGY OF CANADA LTD., CHALK RIVER, ONT.
        (ANNUAL SAFETY REPORT)
AS-R-
    COLUMBIA UNIV., NYC. SUBSTITUTE ALLOY MATERIALS LABS.
ASR-
    GOETTINGEN, GERMANY. UNIV.
ASR-(NUMBER-NUMBER)
    GRUMMAN AIRCRAFT ENGINEERING CORP., BETHPAGE, N.Y.
ASR-
    IIT RESEARCH INST., CHICAGO
ASR-
    ITALY. ISTITUTO UNIVERSITARIO NAVALE, NAPLES
ASR-
    MUNICH. TECHNISCHE HOCHSCHULE
ASR-
    OHIO UNIV., ATHENS
ASR-
    UNION CARBIDE CORP. LINDE DIV., TONAWANDA, N.Y.
ASR-
    ZURICH. UNIVERSITAET
ASRCE-TM-
    AERONAUTICAL SYSTEMS DIV., WRIGHT-PATTERSON AFB, OHIO
ASRDL-
    ARMY SIGNAL RESEARCH + DEV. LAB., FORT MONMOUTH, N.J.
ASRE-
    GT. BRIT. ADMIRALTY SIGNAL AND RADAR ESTABLISHMENT,
        PORTSMOUTH, HANTS, ENGLAND
ASRL-
    MASSACHUSETTS INST. OF TECH., CAMBRIDGE. AERO-ELASTIC
        AND STRUCTURES RESEARCH LAB.
ASRL-LR-(NUMBER)-
    MASSACHUSETTS INST. OF TECH., CAMBRIDGE. AERO-ELASTIC
        AND STRUCTURES RESEARCH LAB.
ASRL-TM-(YEAR)-
    NORTHROP SPACE LABS., HAWTHORNE, CAL. (ASTRO SCIENCES)
ASRL-TR-
    MASSACHUSETTS INST. OF TECH., CAMBRIDGE. AERO-ELASTIC
        AND STRUCTURES RESEARCH LAB.
ASRMDD-
    AERONAUTICAL SYSTEMS DIV. FLIGHT DYNAMICS LAB.,
        WRIGHT-PATTERSON AFB, OHIO
ASRMDD-TM-
    AERONAUTICAL SYSTEMS DIV. FLIGHT DYNAMICS LAB.,
        WRIGHT-PATTERSON AFB, OHIO
ASRMDF-TM-
    AERONAUTICAL SYSTEMS DIV. FLIGHT DYNAMICS LAB.,
        WRIGHT-PATTERSON AFB, OHIO
ASRMDS-TM-
    AIR FORCE FLIGHT DYNAMICS LAB., WRIGHT-PATTERSON AFB,
        OHIO
ASRMPR-TM-
    OHIO PROPULSION LAB., WRIGHT-PATTERSON AFB
ASROC-
    MINNEAPOLIS-HONEYWELL REGULATOR CO., MINNEAPOLIS
ASROC-
    MINNEAPOLIS-HONEYWELL REGULATOR CO., MONROVIA, CALIF
ASROC-D-
    MINNEAPOLIS-HONEYWELL REGULATOR CO., DUARTE, CALIF.
ASRPA-
    ARMY SIGNAL RADIO PROPAGATION AGENCY, FT. MONMOUTH, NJ
ASS-
    ARMY, WASHINGTON, D.C. (ARMY SUBJECT SCHEDULE)
ASSAY-
    CALIFORNIA. UNIV., BERKELEY. RADIATION LAB. (REF. 9)
ASSC-
    ARMY SAFEGUARD SYSTEM COMMAND, HUNTSVILLE, ALA.
ASSDV-
    ARMY ELECTRONIC PROVING GROUND, SPECTRUM SIGNATURE
        LAB., FORT HUACHUCA, ARIZ.
ASS/EPR-R-
    FRANCE. COMMISSARIAT A L'ENERGIE ATOMIQUE. CENTRE
        D'ETUDES NUCLEAIRES, SACLAY
ASSOC.U-
    ASSOCIATED UNIVERSITIES, INC., N.Y.C.
ASSOTW-
    AIR FORCE (AIRFIELD + SEAPLANE STAS. OF THE WORLD)
ASSW-
    (AIRPLANE AND SEAPLANE STATIONS OF THE WORLD)
AST-(NUMBER-NUMBER)-
    VIRGINIA. UNIV., CHARLOTTESVILLE. RESEARCH LABS. FOR
        ENGINEERING SCIENCES
ASTDN-(YEAR)-
    AERONAUTICAL SYSTEMS DIV. FLIGHT TEST ENG. DIV.,
        WRIGHT-PATTERSON AFB, OHIO
AS TDR-
    AEROSPACE CORP., EL SEGUNDO, CALIF.
ASTDV-FTD-(YEAR)-
    DIRECTORATE OF FLIGHT TEST ENGINEERING, WRIGHT-
        PATTERSON AFB, OHIO
ASTDV-FTR-(YEAR)-
    DIRECTORATE OF FLIGHT TEST ENGINEERING, WRIGHT-
        PATTERSON AFB, OHIO
ASTE-
    AMERICAN SOCIETY OF TOOL ENGINEERS, DETROIT
AST/EIR-
    CHANCE VOUGHT AIRCRAFT, INC., DALLAS
ASTF-
    AERONAUTICAL SYSTEMS DIV., WRIGHT-PATTERSON AFB, OHIO
ASTIA-
    ARMED SERVICES TECHNICAL INFO. AGENCY, ARLINGTON, VA.
ASTIA AF TPPD-
    AIR RESEARCH AND DEVELOPMENT COMMAND, ANDREWS AFB, MD
ASTIA ARB-
    ARMED SERVICES TECHNICAL INFO. AGENCY, ARLINGTON, VA.
ASTIA ARC-
    LIBRARY OF CONGRESS. TECH. INFO. DIV., WASH., D.C.

ASTIA BTI-	ARMED SERVICES TECHNICAL INFO. AGENCY, DAYTON, OHIO
ASTIA-CAH-	ARMED SERVICES TECHNICAL INFO. AGENCY, ARLINGTON, VA.
ASTIC-	AEROSPACE TECHNICAL INTELLIGENCE CENTER, WRIGHT-PATTERSON AFB, OHIO
ASTM-	AMERICAN SOCIETY FOR TESTING AND MATERIALS, PHILA.
ASTME-	AMERICAN SOCIETY OF TOOL AND MFG. ENGINEERS, DETROIT
ASTM-STP-	AMERICAN SOCIETY FOR TESTING AND MATERIALS, PHILA.
ASTRA-(NO.)-E-	ADVANCED SCI. TECHNIQUES RES. ASSOCS., MILFORD, CONN.
ASTRA-	ASTRA, INC., MILFORD, CONN.
ASTRA-(3 DIGITS)-	ASTRA, INC., MILFORD, CONN.
ASTRA-	ASTRA, INC., RALEIGH, N.C.
ASTRA-(3 DIGITS)-	ASTRA, INC., RALEIGH, N.C.
ASTRA-G-	ASTRA, INC., RALEIGH, N.C.
ASTRODATA-	ASTRODATA, INC., ANAHEIM, CALIF.
ASTRO-MET-	ASTRO MET ASSOCIATES INC., CINCINNATI
ASTROPOWER-(YEAR)-	AIR FORCE MATERIALS LAB., WRIGHT-PATTERSON AFB, OHIO
ASWDU-	GT. BRIT. AIR/SEA WARFARE DEVELOPMENT UNIT.
ASWE-	UNITED KINGDOM. ADMIRALTY SURFACE WEAPONS ESTABLISHMENT, PORTSMOUTH, HANTS, ENGLAND
ASWE-LAB NOTE-	GT. BRIT. ADMIRALTY SURFACE WEAPONS ESTAB., PORTSMOUTH, HANTS, ENGLAND
ASWE LAB NOTE-XP-(YR.)-	UNITED KINGDOM. ADMIRALTY SURFACE WEAPONS ESTABLISHMENT, PORTSMOUTH, HANTS, ENGLAND
ASWEPS-	NAVAL OCEANOGRAPHIC OFFICE. ANTISUBMARINE WARFARE ENVIRONMENTAL PREDICTION SERVICE, WASHINGTON, D.C.
ASW MEMO.-	ATLANTIC FLEET.ANTISUBMARINE WARFARE FORCE,NORFOLK,VA
ASW/ORT-	CANADA. ANTI-SUBMARINE WARFARE OPERATIONAL RES. TEAM
ASWP ITR-	NAVAL RADIOLOGICAL DEFENSE LAB., SAN FRANCISCO
AT-	AIRTECHNOLOGY CORP., CAMBRIDGE, MASS.
AT-	NAVAL AIR TEST CENTER, PATUXENT RIVER, MD.
ATA-	APPLIED TECHNOLOGY ASSOCIATES, INC., RAMSEY, N.J.
ATAC-	ARMY TANK-AUTOMOTIVE CENTER, WARREN, MICH.
ATACOM-TR-	ARMY TANK-AUTOMOTIVE COMMAND, WARREN, MICH.
ATAC-RRD-	ARMY TANK-AUTOMOTIVE CENTER, WARREN, MICH.
ATAC-TR-	ARMY TANK-AUTOMOTIVE CENTER, WARREN, MICH.
ATB-	ARCTIC TEST BOARD (ARMY)
ATB-	CONTINENTAL ARMY COMMAND. ARCTIC TEST BR., SEATTLE
ATC-	AIR TRAINING COMMAND
ATC-(YR)-	AMERICAN TURBINE CORP., N.Y.C.
ATC-	ASTRO TECHNOLOGY CORP., PALO ALTO, CALIF.
ATCEU-	UNITED KINGDOM. AIR TRAFFIC CONTROL EVALUATION UNIT, BOURNEMOUTH, HANTS, ENGLAND
ATC OJT(LTR.)-(LTRS.)	AIR TRAINING COMMAND, LOWRY AFB, COLO.
ATC OJT(LTR.)-(LTRS.)	AIR TRAINING COMMAND, RANDOLPH AFB, TEXAS
ATD-	LIBRARY OF CONGRESS. AEROSPACE TECHNOLOGY DIV., WASHINGTON, D.C.
ATD-(YEAR)-	LIBRARY OF CONGRESS. AEROSPACE TECHNOLOGY DIV., WASHINGTON, D.C.
ATD-B-(YEAR-NUMBER)	LIBRARY OF CONGRESS. AEROSPACE TECHNOLOGY DIV., WASHINGTON, D.C.
ATDEV-	CONTINENTAL ARMY COMMAND, FORT MONROE, VA.
ATDL-	ATMOSPHERIC TURBULENCE AND DIFFUSION LAB., OAK RIDGE, TENN.
ATDM-	GENERAL ELECTRIC CO., PHILADELPHIA
ATD-P-(YR)-(NO.)	LIBRARY OF CONGRESS. AEROSPACE TECHNOLOGY DIV., WASHINGTON, D.C.
ATD-R-(YEAR)-	LIBRARY OF CONGRESS. AEROSPACE TECHNOLOGY DIV., WASHINGTON, D.C.
ATDU-	GT. BRIT. ROYAL AIRCRAFT EST.,FARNBOROUGH, HANTS, ENG
ATD-U-(YR)-(NO.)	LIBRARY OF CONGRESS. AEROSPACE TECHNOLOGY DIV., WASHINGTON, D.C.
ATE-IM-	ARMED FORCES SPECIAL WEAPONS PROJECT. FIELD COMMAND, ALBUQUERQUE, N. MEX. (INSTRUCTORS MANUAL)
ATE-L-	RADIO CORP. OF AMERICA. AEROSPACE SYSTEMS DIV., BURLINGTON, MASS.
ATE-MS-	RADIO CORP. OF AMERICA. AEROSPACE SYSTEMS DIV., BURLINGTON, MASS.
ATE-MTE-	RADIO CORP. OF AMERICA. DEFENSE ELECTRONIC PRODUCTS, BURLINGTON, MASS. (MULTISYSTEM TEST EQUIPMENT)
ATE-MTE-L-	RADIO CORP. OF AMERICA. AEROSPACE SYSTEMS DIV., BURLINGTON, MASS.
ATE-NB-	ARMED FORCES SPECIAL WEAPONS PROJECT. FIELD COMMAND, ALBUQUERQUE, N. MEX. (NOTEBOOK)
ATE-VG-	ARMED FORCES SPECIAL WEAPONS PROJECT. FIELD COMMAND, ALBUQUERQUE, N. MEX. (VU-GRAPHS)
ATI-	(AIR TECHNICAL INDEX)
ATI-	AIR FORCES IN EUROPE (U.S.), WIESBADEN, GERMANY
ATI-	ARMED SERVICES TECHNICAL INFORMATION AGENCY. DOCUMENT SERVICE CENTER, DAYTON, OHIO
ATI-	CENTRAL AIR DOCUMENTS OFFICE, WRIGHT-PATTERSON AFB, OHIO (ANNOUNCED TECHNICAL INDEX)
ATI-	WAR PRODUCTION BOARD, WASHINGTON, D.C.
ATI-AJA-	AGBABIAN-JACOBSEN ASSOCIATES, LOS ANG. (JT. VENTURE)
ATI-AJA-	APPLIED THEORY, INC., LOS ANGELES (JOINT VENTURE)
ATIC-	AIR TECH. INTELL. CTR., WRIGHT-PATTERSON AFB, OHIO
ATIC 102-AC-	AIR TECHNICAL INTELLIGENCE CENTER, WRIGHT-PATTERSON AFB, OHIO (STUDY)
ATIC 102-AE-	AIR TECHNICAL INTELLIGENCE CENTER, WRIGHT-PATTERSON AFB, OHIO (STUDY)
ATIC 102-EL-	AIR TECHNICAL INTELLIGENCE CENTER, WRIGHT-PATTERSON AFB, OHIO (STUDY)
ATIC FE-(NOS.)-(LTRS.)	AIR TECH. INTELL. CTR., WRIGHT-PATTERSON AFB, OHIO
ATIC F-TS-(NO/ROMAN NO.)	AIR TECHNICAL INTELLIGENCE CENTER, WRIGHT-PATTERSON AFB, OHIO (TRANSLATION REPORT)
ATIC F-TS-(NO.)-RE	AIR TECHNICAL INTELLIGENCE CENTER, WRIGHT-PATTERSON AFB, OHIO (TRANSLATION REPORT)
ATIC T-(YR.)-(NO.)	AIR TECH. INTELL. CTR., WRIGHT-PATTERSON AFB, OHIO
ATIC-T-	ARMY TECHNICAL INTELLIGENCE CENTER, TOKYO
ATIC TIH-(LTRS)(YR)(NO.)	AIR TECH. INTELL. CTR., WRIGHT-PATTERSON AFB, OHIO
ATIC TIR-(LTRS)(YR)(NO.)	AIR TECH. INTELL. CTR., WRIGHT-PATTERSON AFB, OHIO
ATIC-TIR-PR-(YEAR)-	AIR TECH. INTELL. CTR., WRIGHT-PATTERSON AFB, OHIO
ATIC TIS-(LTRS)(YR)(NO.)	AEROSPACE TECHNICAL INTELLIGENCE CENTER, WRIGHT-PATTERSON AFB, OHIO
ATIC-T MAS-	ARMY TECH. INTELL. CTR. MEDICAL ANALYSIS SEC., TOKYO
ATIC-T OS-	ARMY TECHNICAL INTELLIGENCE CENTER. ORDNANCE (ANALYSIS) SECTION, TOKYO
ATIC-T QMAS-	ARMY TECHNICAL INTELLIGENCE CENTER. QUARTERMASTER ANALYSIS SECTION, TOKYO
ATIC-TR-	AIR TECH. INTELL. CTR., WRIGHT-PATTERSON AFB, OHIO
ATIC TR-(LTRS)-(NO.)	AIR TECH. INTELL. CTR., WRIGHT-PATTERSON AFB, OHIO
ATIC-T SAS-	ARMY TECH. INTELL. CTR. SIGNAL ANALYSIS SEC., TOKYO
ATIDG-	(AIR TECHNICAL INDEX DISTRIBUTION GUIDE)
ATIG-	ARMY AIR FORCE. AIR TECHNICAL INTELLIGENCE GROUP
ATIL-	NAVAL EXPLOSIVE ORDNANCE DISPOSAL FACILITY, INDIAN HEAD, MD. (ADVANCED TECHNICAL INFORMATION LETTER)
ATIR-	AIR MATERIEL COMMAND. AIR DOCUMENTS DIV., WRIGHT-PATTERSON AFB, OHIO (AIR TECH. INTELLIGENCE REVIEW)
ATIS-	AIR TECHNICAL INTELLIGENCE SERVICE. FAR EAST COMMAND
ATIS-	SUPREME CMDR. FOR THE ALLIED POWERS. MIL. INTELL. SEC GENL. STAFF ALLIED TRANSLATOR + INTERPRETER SEC.
ATL-(NUMBER)A	ADVANCED TECHNOLOGY LABS. DIV. OF AMERICAN-STANDARD, MOUNTAIN VIEW, CALIF.
ATL-	AIR FORCE ARMAMENT LAB., EGLIN AFB, FLA.
ATL-	GENERAL ELECTRIC CO. ADVANCED TECHNOLOGY LABS., SCHENECTADY, N.Y.
ATL-	WESTINGHOUSE ELECTRIC CORP. AEROSPACE TEST LAB., BALTIMORE

ATL-A-

ATL-A-
    ADVANCED TECHNOLOGY LABS. DIV. OF AMERICAN-STANDARD, MOUNTAIN VIEW, CALIF.
ATLAS-HA-
    GEOLOGICAL SURVEY, DENVER
ATL-D-
    ADVANCED TECHNOLOGY LABS. DIV. OF AMERICAN-STANDARD, MOUNTAIN VIEW, CALIF.
ATLIS-
    OFFICE OF THE CHIEF OF ENGINEERS (ARMY), WASH., D.C.
ATL-TDR-(YEAR)-
    AIR FORCE ARMAMENT LAB., EGLIN AFB, FLA.
ATL-TDR-
    AIR FORCE SYSTEMS COMMAND, EGLIN AFB, FLA.
ATL-TR-
    ADVANCED TECHNOLOGY LABS., INC., JERICHO, N.Y.
ATL-TR-(YEAR)-
    AIR FORCE ARMAMENT LAB., EGLIN AFB, FLA.
ATM-(YEAR)(JOB NO.)-
    AEROSPACE CORP., EL SEGUNDO, CALIF. (AEROSPACE TECHNICAL MEMO)
ATM-(YEAR)A(JOB NO.)-
    AEROSPACE CORP. ATLANTIC MISSILE RANGE OFFICE
ATM-(YEAR)-A(NUMBER)-
    AEROSPACE CORP. EASTERN TEST RANGE OFFICE, PATRICK AFB, FLA.
ATM-(YEAR)S(JOB NO.)-
    AEROSPACE CORP. SAN BERNARDINO OPERATIONS, CALIF.
ATM-
    BENDIX CORP. BENDIX SYSTEMS DIV., ANN ARBOR, MICH.
ATM-
    BUREAU OF AERONAUTICS (NAVY)(ARMAMENT TECHNICAL MEMO)
ATM-IM-
    ARMED FORCES SPECIAL WEAPONS PROJECT. FIELD COMMAND, ALBUQUERQUE, N. MEX. (INSTRUCTORS MANUAL)
ATM-NB-
    ARMED FORCES SPECIAL WEAPONS PROJECT. FIELD COMMAND, ALBUQUERQUE, N. MEX. (NOTEBOOK)
ATMP-
    HYDROGRAPHIC OFFICE, SUITLAND, MD. (AIR TARGET MATERIALS PROGRAM)
ATN-(YEAR)(JOB NO.)-
    AEROSPACE CORP., EL SEGUNDO, CALIF. (AEROSPACE TECHNICAL NOTE)
ATN-(YEAR)A(JOB NO.)-
    AEROSPACE CORP. ATLANTIC MISSILE RANGE OFFICE
ATN-(YEAR)-A(NUMBER)-
    AEROSPACE CORP. EASTERN TEST RANGE OFFICE, PATRICK AFB, FLA.
ATN-(YEAR)S(JOB NO.)-
    AEROSPACE CORP. SAN BERNARDINO OPERATIONS, CALIF.
ATOLL RESEARCH BULL-
    NATIONAL ACADEMY OF SCIENCES-NATIONAL RES. COUNCIL, PACIFIC SCIENCE BOARD, WASHINGTON, D.C.
ATP-
    GARRETT CORP. AIRESEARCH MFG. DIV., LOS ANGELES
ATP-
    GT. BRIT. MINISTRY OF SUPPLY
ATP-
    NAVY, WASHINGTON, D.C. (ALLIED TACTICAL PUBLICATION)
ATPR-(NUMBER-YEAR)-
    GENERAL ELECTRIC CO. MISSILE AND SPACE DIV., PHILA.
ATPR-
    NAVAL ORDNANCE TEST STATION, INYOKERN (CHINA LAKE), CALIF. (ANNUAL TECHNICAL PROGRESS REPORT)
ATR-
    (ANNUAL TECHNICAL REPORT) (REF. 2)
ATR-(YEAR)(JOB NO.)-
    AEROSPACE CORP., EL SEGUNDO, CALIF. (AEROSPACE TECHNICAL REPORT)
ATR-(YEAR)A(JOB NO.)-
    AEROSPACE CORP. ATLANTIC MISSILE RANGE OFFICE
ATR-(YEAR)-A(NUMBER)-
    AEROSPACE CORP. EASTERN TEST RANGE OFFICE, PATRICK AFB, FLA.
ATR-(YEAR)S(JOB NO.)-
    AEROSPACE CORP. SAN BERNARDINO OPERATIONS, CALIF.
ATR-
    AMCEL PROPULSION CO., ASHEVILLE, N.C.
ATR-(YEAR)-
    APPLIED THEORY, INC., LOS ANGELES
ATR-(NUMBERS)
    APPLIED THEORY INC., SANTA MONICA, CALIF.
ATR-
    ARMSTRONG WHITWORTH (SIR W.G.) AIRCRAFT, LTD.
ATR-
    BABCOCK + WILCOX CO. ATOMIC EN. DIV., LYNCHBURG, VA.
ATR-
    BRITISH INTERNAL COMBUSTION ENGINE RESEARCH ASSOCIATION, SLOUGH, BUCKS, ENGLAND
ATR/
    BRITISH NON-FERROUS METALS RESEARCH ASSN., LONDON
ATR-
    NORTHROP CAROLINA, INC., ASHEVILLE, N.C.
ATR-
    UNIV. COLL. OF SOUTH WALES AND MONMOUTHSHIRE, CARDIFF
ATRC-
    AIR TRAINING COMMAND, SCOTT AFB, ILL.
ATRC-
    ATLANTIC RESEARCH CORP., ALEXANDRIA, VA.
ATRCE-
    BABCOCK + WILCOX CO. ATOMIC EN. DIV., LYNCHBURG, VA.
ATRC-SP-QPR-
    ATLANTIC RESEARCH CORP., ALEXANDRIA, VA.
ATR-FE-
    BABCOCK + WILCOX CO. ATOMIC EN. DIV., LYNCHBURG, VA.
ATR-IC-
    BABCOCK + WILCOX CO. ATOMIC EN. DIV., LYNCHBURG, VA. (ADVANCED TEST REACTOR)

ATRS/CONF/
    GT. BRIT. ATOMIC EN. RES. ESTAB., HARWELL, BERKS, ENG
ATR/T-
    GT. BRIT. ADMIRALTY. DEPT. OF SCI. RES. + EXPERIMENT
ATS-
    AEROJET-GENERAL CORP., SACRAMENTO, CALIF. (SPECS.)
ATS-
    ASSOCIATED TECHNICAL SERVICES, INC., EAST ORANGE, N.J. (TRANSLATIONS)
ATS-
    DENMARK. AKADEMIET FOR DE TEKNISKE VIDENSKABER
ATS-(NUMBER-NUMBER)
    FAIRCHILD HILLER CORP., GERMANTOWN, MD. (APPLICATIONS TECHNOLOGY SATELLITES)
ATS-(NUMBER)-
    LOGICON, INC., REDONDO BEACH, CALIF.
ATSC-
    AIR MATERIEL COMMAND, WRIGHT-PATTERSON AFB, OHIO
ATSC-
    AIR TECH. SERVICE COMMAND, WRIGHT-PATTERSON AFB, OHIO
ATSC EEIS
    ALLIED FORCES ENEMY EQUIPMENT INTELLIGENCE SERVICE
ATSC-M-
    AIR TECHNICAL SERVICE COMMAND, WRIGHT-PATTERSON AFB, OHIO (MANUAL)
ATSC-STCO-
    AIR TECHNICAL SERVICE COMMAND, WRIGHT-PATTERSON AFB, OHIO (SUPERVISOR TRAINING CONFERENCE OUTLINE)
ATSR-MEMO-
    ARGONNE NATIONAL LAB., ILL.
ATSTM-LIB-
    ENVIRONMENTAL SCIENCE SERVICES ADM. OFFICE OF ADM. AND TECHNICAL SERVICES, ROCKVILLE, MD. (LIBRARY)
ATS-TR-
    ASSOCIATED TECHNICAL SERVICES, INC., EAST ORANGE, N.J
ATT (NUMBER) G-
    BRITISH INTELLIGENCE OBJECTIVES SUBCOMMITTEE. MILITARY COLL. OF SCIENCE. SCH. OF TANK TECHNOLOGY
ATTI-
    ARIZONA TRANSPORTATION AND TRAFFIC INST., TUCSON
ATTNG-
    CONTINENTAL ARMY COMMAND, FORT MONROE, VA.
ATTS-
    NAVAL AIR TURBINE TEST STATION, TRENTON
ATW-
    TEST WING (DEV.)(6555TH), PATRICK AFB, FLA.
AU-(NUMBER-NUMBER)-ASI
    AIR UNIV., MAXWELL AFB, ALA. AEROSPACE STUDIES INST.
AU-
    AIR UNIV., MAXWELL AFB, ALA. RESEARCH STUDIES INST.
AU-
    AKRON, OHIO. UNIV.
AU-
    ASSOCIATED UNIVERSITIES, INC., N.Y.C.
AUER-(NUMBER)-TR-
    AUERBACH CORP., PHILADELPHIA
AUER-(NUMBER)-FR-
    AUERBACH CORP., PHILADELPHIA
AUERBACH-(NO.)-TR-
    AUERBACH CORP., PHILADELPHIA
AUERBACH-PR-
    AUERBACH CORP., PHILADELPHIA
AU-FR-(YEAR/MONTH)
    ALASKA. UNIV., COLLEGE. GEOPHYSICAL INST.(FINAL RPT.)
AU-GRR-
    ALASKA. UNIV., COLLEGE. GEOPHYSICAL INST. (GEOPHYSICAL RESEARCH REPORT)
AUI-
    ASSOCIATED UNIVERSITIES, INC., N.Y.C.
AUI PER-
    ASSOCIATED UNIVS., INC., N.Y.C. (PROJ. EAST RIVER)
AU-IPS-PR-
    AKRON, OHIO. UNIV. INST. OF POLYMER SCIENCE
AU-IRR-FR-
    AKRON, OHIO. UNIV. INST. OF RUBBER RESEARCH
AU-IRR-PR-
    AKRON, OHIO. UNIV. INST. OF RUBBER RESEARCH
AU-IRR-TR-
    AKRON, OHIO. UNIV. INST. OF RUBBER RESEARCH
AU-ISR-
    ALASKA. UNIV., COLLEGE. GEOPHYSICAL INST. (INTERIM SCIENTIFIC REPORT)
AUL-
    ARENBERG ULTRASONIC LAB., JAMAICA PLAIN, MASS.
AU-QPR-
    ALASKA. UNIV., COLLEGE. GEOPHYSICAL INST. (QUARTERLY PROGRESS REPORT)
AUS ACA R-
    AUSTRALIAN COUNCIL FOR AERONAUTICS, MELBOURNE
AUS ARL/(LETTER)-
    AUSTRALIA. AERO. RES. LABS.,FISHERMANS BEND, VICTORIA
AUS CSIR BR-
    AUSTRALIA. COUNCIL FOR SCIENTIFIC AND INDUSTRIAL RESEARCH. DIVISION OF BUILDING RESEARCH
AUS CSIR-DA-(LETTER)(NO.)
    AUSTRALIA. COUNCIL FOR SCIENTIFIC AND INDUSTRIAL RESEARCH. DIV. OF AERONAUTICS
AUS CSIR DA AN-
    AUSTRALIA. COUNCIL FOR SCIENTIFIC AND INDUSTRIAL RESEARCH. (AERODYNAMICS NOTES)
AUS CSIR DA EN-
    AUSTRALIA. COUNCIL FOR SCIENTIFIC AND INDUSTRIAL RESEARCH. (ENGINES NOTE)
AUS CSIR DA IN-
    AUSTRALIA. COUNCIL FOR SCIENTIFIC AND INDUSTRIAL RESEARCH. (INSTRUMENTS)
AUS CSIR LB A-
    AUSTRALIA. COUNCIL FOR SCIENTIFIC AND INDUSTRIAL RES. SEC. OF LUBRICANTS AND BEARINGS. (A SERIES)

AUS CSIR LB B-	AUSTRALIA. COUNCIL FOR SCIENTIFIC AND INDUSTRIAL RES. SEC. OF LUBRICANTS AND BEARINGS. (BEARINGS)
AUS CSIR LB CF-	AUSTRALIA. COUNCIL FOR SCIENTIFIC + IND. RES. SEC. OF LUBRICANTS AND BEARINGS. (CUTTING FLUIDS)
AUS CSIR LB CF I-	AUSTRALIA. COUNCIL FOR SCI. + IND. RES. SEC. OF LUBRICANTS + BEARINGS. (CUTTING FLUIDS) (INTERIM)
AUS CSIR LB CP-	AUSTRALIA. COUNCIL FOR SCIENTIFIC AND INDUSTRIAL RESEARCH. SECTION OF LUBRICANTS AND BEARINGS. (CYLINDER AND PISTON WEAR IN AERO ENGINES)
AUS CSIR LB CP I-	AUSTRALIA. COUNCIL FOR SCIENTIFIC AND INDUSTRIAL RESEARCH. SEC. OF LUBRICANTS + BEARINGS. (CYLINDER + PISTON WEAR IN AERO ENGINES)(INTERIM)
AUS CSIR LB E-	AUSTRALIA. COUNCIL FOR SCIENTIFIC AND IND. RES. SEC. OF LUBRICANTS AND BEARINGS. (EXPLOSIVES)
AUS CSIR LB LF-	AUSTRALIA. COUNCIL FOR SCI. + IND. RES. SEC. OF LUBRICANTS + BEARINGS. (LUBRICATION + FRICTION)
AUS CSIR LB MV-	AUSTRALIA. COUNCIL FOR SCI. + IND. RES. SEC. OF LUBRICANTS + BEARINGS. (MUZZLE VELOCITY)
AUS CSIR LB PG-	AUSTRALIA. COUNCIL FOR SCIENTIFIC AND IND. RES. SEC. OF LUBRICANTS + BEARINGS. (PRODUCER GAS)
AUS-CSIR-MP-	AUSTRALIA. COUNCIL FOR SCIENTIFIC AND INDUSTRIAL RESEARCH. DIVISION OF BUILDING RESEARCH
AUS CSIR PSS-	AUSTRALIA. COUNCIL FOR SCI. + IND. RES. PHYS. SEC.
AUS CSIR RP-	AUSTRALIA. COUNCIL FOR SCIENTIFIC AND INDUSTRIAL RESEARCH. DIVISION OF RADIOPHYSICS
AUS CSIR T E-	AUSTRALIA. COUNCIL FOR SCIENTIFIC AND INDUSTRIAL RESEARCH. SECTION OF TRIBOPHYSICS. (EXPLOSIVES)
AUS DLNS IWD-	AUSTRALIA. DEPARTMENT OF LABOUR AND NATIONAL SERVICE. INDUSTRIAL WELFARE DIVISION
AUS DLNS IWD TR-	AUSTRALIA. DEPARTMENT OF LABOUR AND NATIONAL SERVICE. INDUSTRIAL WELFARE DIVISION
AUS DM MSL-	AUSTRALIA. MUNITIONS SUPPLY LAB., MARIBYRNONG
AUS DM MSL R-	AUSTRALIA. MUNITIONS SUPPLY LAB., MARIBYRNONG
AUS DM MSL TN-	AUSTRALIA. MUNITIONS SUPPLY LAB., MARIBYRNONG
AUS-MP-TP-	AUSTRALIA. COUNCIL FOR SCIENTIFIC AND INDUSTRIAL RESEARCH. DIVISION OF BUILDING RES. (TECH. PAPER)
AUS-NSL-	AUSTRALIA. NATIONAL STANDARDS LAB., CHIPPENDALE, NEW SOUTH WALES
AU-SPR-	ALASKA. UNIV., COLLEGE. GEOPHYSICAL INST. (SUPPLEMENTARY PROGRESS REPORT)
AU-SR-	ALASKA. UNIV., COLLEGE. GEOPHYSICAL INST. (SCI. RPT.)
AUS TP-	(AUSTRALIAN TECHNICAL PAPERS)
AUS-WTRL-	AUSTRALIA. COUNCIL FOR SCIENTIFIC AND INDUSTRIAL RESEARCH. WOOL TEXTILE RESEARCH LABS.
AU-THESIS-	AIR UNIV., MAXWELL AFB, ALA.
AUTO-	AUTONETICS DIV., NORTH AMER. AVIATION INC., DOWNEY, CAL
AUW-	NAVY, WASHINGTON, D.C. (ATOMIC UNDERWATER WEAPONS)
AUWE-	UNITED KINGDOM. ADMIRALTY UNDERWATER WEAPONS ESTABLISHMENT, PORTLAND, DORSET, ENGLAND
AUWE-TN-	UNITED KINGDOM. ADMIRALTY UNDERWATER WEAPONS ESTABLISHMENT, PORTLAND, DORSET, ENGLAND
AUW/TRI-	GT. BRIT. ADMIRALTY INDEX WORKS
AV-	BUREAU OF AERONAUTICS (NAVY). AVIONICS DIV.
AV-	GRUMMAN AIRCRAFT ENGINEERING CORP., BETHPAGE, N.Y.
A/V-(5 DIGITS)-	MASSACHUSETTS INST. OF TECH., CAMBRIDGE. ACOUSTICS AND VIBRATION LAB.
AVA-	GERMANY. AERODYNAMISCHE VERSUCHSANSTALT, GOETTINGEN
AVAC-	ASOCIACION VENEZOLANA PARA EL AVANCE DE LA CIENCIA, CARACAS
AVA-FB-(YEAR)-	GERMANY. AERODYNAMISCHE VERSUCHSANSTALT, GOETTINGEN
AVAG-	GERMANY. AERODYNAMISCHE VERSUCHSANSTALT, GOETTINGEN
AVA-TB-	GERMANY. AERODYNAMISCHE VERSUCHSANSTALT, GOETTINGEN
AVATD-(NO.)-(YEAR)-RR	AVCO CORP. APPLIED TECHNOLOGY DIV., LOWELL, MASS.
AVATD-(NO.)-(YEAR)-CR	AVCO CORP. APPLIED TECHNOLOGY DIV., LOWELL, MASS.
AVC-	GT. BRIT. ROYAL AIRCRAFT EST., FARNBOROUGH, HANTS, ENG
AVCIR-	ARMY AVIATION MATERIAL LABS., FORT EUSTIS, VA.
AV-CIR-	CORNELL-GUGGENHEIM AVIATION SAFETY CENTER, N.Y.C.
AVCO-	AVCO CORP. (ALL LOCATIONS)
AVCO (NO.)-TM(YR.-NO.)	AVCO MANUFACTURING CORP., LAWRENCE, MASS.
AVCO AADD (NUMBER)-	AVCO MFG. CORP. RESEARCH AND ADVANCED DEVELOPMENT DIV., LAWRENCE, MASS.
AVCO-ADD-	AVCO MFG. CORP. ADVANCED DEV. DIV., STRATFORD, CONN.
AVCO-ADD-	AVCO MFG. CORP. RESEARCH AND ADVANCED DEVELOPMENT DIV., WILMINGTON, MASS. (PROGRESS REPORT)
AVCO-AERL-	AVCO CORP. AVCO-EVERETT RESEARCH LAB., EVERETT, MASS.
AVCO-AMP-	AVCO CORP. AVCO-EVERETT RESEARCH LAB., EVERETT, MASS.
AVCO-ARL-	AVCO MFG. CORP. AVCO RESEARCH LAB., EVERETT, MASS.
AVCO-ARL-RR-	AVCO MFG. CORP. AVCO RESEARCH LAB., EVERETT, MASS.
AVCO-CR-	AVCO CORP. CROSLEY BROADCASTING CORP., CINCINNATI
AVCO-EVERETT RES. REPT.-	AVCO CORP. AVCO-EVERETT RESEARCH LAB., EVERETT, MASS.
AVCO-MPR-(YEAR)-	AVCO MANUFACTURING CORP., EVERETT, MASS.
AVCO-MPR-(YEAR)-	AVCO MANUFACTURING CORP., LAWRENCE, MASS.
AVCOM-TR-	ARMY AVIATION MATERIEL COMMAND, ST. LOUIS
AVCO-QR-	AVCO CORP. RESEARCH AND ADVANCED DEVELOPMENT DIV., WILMINGTON, MASS. (QUARTERLY REPORT)
AVCO-RAD-	AVCO MFG. CORP. RESEARCH AND ADVANCED DEVELOPMENT DIV., LAWRENCE, MASS.
AVCO-RAD-(NO.)-	AVCO MFG. CORP. RESEARCH AND ADVANCED DEVELOPMENT DIV., LAWRENCE, MASS.
AVCO-RAD-(NO.)-TM-	AVCO MFG. CORP. RESEARCH AND ADVANCED DEVELOPMENT DIV., LAWRENCE, MASS.
AVCO-RAD-	AVCO MFG. CORP. RESEARCH AND ADVANCED DEVELOPMENT DIV., WILMINGTON, MASS.
AVCO-RAD-MP-(NO.,YR)-	AVCO MFG. CORP. RESEARCH AND ADVANCED DEVELOPMENT DIV., LAWRENCE, MASS.
AVCO-RAD-MP-(NO.,YR)-	AVCO MFG. CORP. RESEARCH AND ADVANCED DEVELOPMENT DIV., WILMINGTON, MASS.
AVCO-RAD-SA-(NO.)-	AVCO MFG. CORP. RESEARCH AND ADVANCED DEVELOPMENT DIV., LAWRENCE, MASS. (SEMI-ANNUAL RPT.)
AVCO-RAD-SR-	AVCO MFG. CORP. RESEARCH AND ADVANCED DEVELOPMENT DIV., LAWRENCE, MASS.
AVCO-RAD-TM-	AVCO MFG. CORP. RESEARCH AND ADVANCED DEVELOPMENT DIV., LAWRENCE, MASS.
AVCO-RAD-TR-	AVCO MFG. CORP. RESEARCH AND ADVANCED DEVELOPMENT DIV., LAWRENCE, MASS.
AVCO REP SAR (INCL.DATES)	AVCO MFG. CORP. AVCO RESEARCH LAB., EVERETT, MASS. (SEMI-ANNUAL REPORT)
AVCO-RN-	AVCO CORP. AVCO-EVERETT RESEARCH LAB., EVERETT, MASS. (RESEARCH NOTE)
AVCO-RR-	AVCO CORP. AVCO-EVERETT RESEARCH LAB., EVERETT, MASS.
AVCO-SR-(NO.-YR.)-	AVCO MFG. CORP. RESEARCH AND ADVANCED DEVELOPMENT DIV., LAWRENCE, MASS.
AVCO-TR-	AVCO CORP., TULSA, OKLA.
AVCO-TR-(NO.-YR.)-	AVCO MFG. CORP. RESEARCH AND ADVANCED DEVELOPMENT DIV., LAWRENCE, MASS.
AVCSM-(NO.)-(YR.)-CR	AVCO MFG. CORP. SYSTEMS MANAGEMENT (CONTRACT REPORT)
AVCSM-(NO.)-(YR.)-PP	AVCO MFG. CORP. SYSTEMS MANAGEMENT (PROFESSIONAL PAPER)
AVCSM-(NO.)-(YR.)-RM	AVCO MFG. CORP. SYSTEMS MANAGEMENT (RESEARCH REPORT)
AVCSM-(NO.)-(YR.)-RR	AVCO MFG. CORP. SYSTEMS MANAGEMENT (RESEARCH REPORT)
AVD-(NUMBER)-	REPUBLIC AVIATION CORP., FARMINGDALE, N.Y.
AVERT-	AIR VEHICLE ENVIRONMENTAL RESEARCH TEAM, NATICK, MASS
AVIC-	AVION INSTRUMENT CORP., PARAMUS, N.J.
AVIEN-	AVIATION ENGINEERING CORP., WOODSIDE, N.Y.
AVION-	ACF INDUSTRIES, INC. AVION DIV., PARAMUS, N.J.
AVMSD-(NUMBER)-(YR.)-CR	AVCO CORP. AVCO MISSILE, SPACE + ELECTRONICS GROUP, WILMINGTON, MASS.
AVMSD-(NUMBER)-(YR.)-PP	AVCO CORP. AVCO MISSILE SYSTEMS DIV., WILMINGTON, MASS. (PROFESSIONAL PAPER)
AVMSD-(NUMBER)-(YR.)-RM	AVCO CORP. AVCO MISSILE SYSTEMS DIV., WILMINGTON, MASS. (RESEARCH MEMO)
AVMSD-(NUMBER)-(YR.)-RR	AVCO CORP. AVCO MISSILE SYSTEMS DIV., WILMINGTON, MASS. (RESEARCH REPORT)

AVNI-TR-(YEAR)-
: AIR FORCE AVIONICS LAB. RESEARCH AND TECHNOLOGY DIV., HOLLOMAN AFB, N. MEX.

AVR-
: ARBEITSGEMEINSCHAFT VERSUCHS-REAKTOR GMBH, DUESSELDORF

AVR-
: BENDIX AVIATION CORP. RESEARCH LABS. DIV., SOUTHFIELD, MICH.

AVSD-(NUMBER)-(YR.)-RR
: AVCO CORP. AVCO SYSTEMS DIV. AVCO GOVERNMENT PRODUCTS GROUP, WILMINGTON, MASS.

AVSD-(NUMBER)-(YR.)-CR
: AVCO CORP. SYSTEMS DIV., LOWELL, MASS.

AVSER-(YEAR)-
: AVIATION SAFETY ENG. AND RESEARCH, PHOENIX, ARIZ.

AVSSD-(NUMBER)-(YR.)-CR
: AVCO CORP. AVCO SPACE SYSTEMS DIV., LOWELL, MASS.

AVSSD-(NUMBER)-(YR.)-CR
: AVCO CORP. AVCO SPACE SYSTEMS DIV., WILMINGTON, MASS. (CONTRACT REPORT)

AVSSD-(NUMBER)-(YR.)-PP
: AVCO CORP. AVCO SPACE SYSTEMS DIV., WILMINGTON, MASS. (PROFESSIONAL PAPER)

AVSSD-(NUMBER)-(YR.)-RM
: AVCO CORP. AVCO SPACE SYSTEMS DIV., WILMINGTON, MASS. (RESEARCH MEMO)

AVSSD-(NUMBER)-(YR.)-RR
: AVCO CORP. AVCO SPACE SYSTEMS DIV., WILMINGTON, MASS. (RESEARCH REPORT)

AVSSD-(NO.)-(YR.)-
: AVCO CORP. AVCO SPACE SYSTEMS DIV. AVCO GOVERNMENT PRODUCTS GROUP, LOWELL, MASS.

AVSSD-(NUMBER)-(YR.)-PP
: AVCO CORP. SPACE SYSTEMS DIV., LOWELL, MASS.

AVSSD-(LTR)(NO.)-EBB-
: AVCO CORP. AVCO SPACE SYSTEMS DIV. AVCO GOVERNMENT PRODUCTS GROUP, LOWELL, MASS.

AVTM-TR-
: AIR FORCE AVIONICS LAB., WRIGHT-PATTERSON AFB, OHIO

AW-
: AEROSPACE CORP., EL SEGUNDO, CALIF. (ART WORK)

A+W-
: AMMAN + WHITNEY, CONSULTING ENGINEERS, N.Y.C.

AW-
: AMMAN + WHITNEY, CONSULTING ENGINEERS, N.Y.C.

AW-
: BUREAU OF AERONAUTICS (NAVY)

AW-
: GT. BRIT. AERONAUTICAL RESEARCH COUNCIL

AW-
: LOS ALAMOS SCIENTIFIC LAB., N. MEX.

AW-
: NAVAL AIR DEVELOPMENT CENTER. AIR WARFARE RESEARCH DEPT., JOHNSVILLE, PENNA. (USED AFTER JULY 1965)

AW-
: NAVAL AIR DEVELOPMENT CENTER. ANTI-SUBMARINE WARFARE LAB., JOHNSVILLE, PENNA. (USED BEFORE JULY 1965)

AW-
: OFFICE OF NAVAL RESEARCH (ATOMIC WARFARE OPERATIONS)

AW-
: RES. + DEV. BD., WASH., DC. (ATOMIC WARFARE OPERATIONS)

AWAF-
: AIR MATERIEL COMMAND. ALL-WEATHER FLYING DIV., WRIGHT-PATTERSON AFB, OHIO

AWB-G-
: ARMED FORCES SPECIAL WEAPONS PROJECT. FIELD COMMAND, ALBUQUERQUE, N. MEX.

AWB-HJC-
: ARMED FORCES SPECIAL WEAPONS PROJECT. FIELD COMMAND, ALBUQUERQUE, N. MEX. (HONEST JOHN, CORPORAL)

AWB-ME-
: ARMED FORCES SPECIAL WEAPONS PROJECT. FIELD COMMAND, ALBUQUERQUE, N. MEX. (ELECTRICAL)

AWB-MM-
: ARMED FORCES SPECIAL WEAPONS PROJECT. FIELD COMMAND, ALBUQUERQUE, N. MEX. (MECHANICAL)

AWB-NH-
: ARMED FORCES SPECIAL WEAPONS PROJECT. FIELD COMMAND, ALBUQUERQUE, N. MEX. (NIKE HERCULES)

AWB-R-
: ARMED FORCES SPECIAL WEAPONS PROJECT. FIELD COMMAND, ALBUQUERQUE, N. MEX. (REDSTONE)

AWC-
: AIR WAR COLLEGE. MAXWELL AFB, ALA.

AWC-
: ARMY WAR COLL., CARLISLE BARRACKS, PENNA.

AWCC-
: ANACONDA WIRE AND CABLE CO. MAGNET WIRE RESEARCH, MUSKEGON, MICH.

AWCS-CS-
: GENERAL ELECTRIC CO. HEAVY MILITARY ELECTRONICS DEPT., SYRACUSE, N.Y.

AWCS-GCP-
: GENERAL ELECTRIC CO., SYRACUSE, N.Y. (AIR WEAPONS CONTROL SYSTEM, COMPUTER PROGRAM INFORMATION)

AWEC/P-
: GT. BRIT. ARMAMENT RESEARCH AND DEVELOPMENT ESTABLISHMENT, FORT HALSTEAD, KENT, ENGLAND

AWEC/P(YEAR)-
: GT. BRIT. ARMAMENT RESEARCH AND DEVELOPMENT ESTABLISHMENT, FORT HALSTEAD, KENT, ENGLAND

AWEC/P(YEAR)-
: GT. BRIT. ATOMIC WEAPONS RESEARCH ESTABLISHMENT, ALDERMASTON, BERKS, ENGLAND

AWEWS-
: MASSACHUSETTS INST. OF TECH., CAMBRIDGE. SOIL MECHANICS DIV.

AWG-
: SANDIA CORP. AD HOC WORKING GROUP, ALBUQUERQUE, N.M.

AWNW-
: AIR MATERIEL COMMAND. ALL-WEATHER FLYING DIV., WRIGHT-PATTERSON AFB, OHIO

AWNW-
: DIRECTORATE OF FLIGHT AND ALL-WEATHER TESTING, WRIGHT-PATTERSON AFB, OHIO

AWRDM-
: NAVAL AIR DEVELOPMENT CENTER. AIR WARFARE RESEARCH DEPT., JOHNSVILLE, PENNA. (MEMORANDUM)

AWRE-
: UNITED KINGDOM ATOMIC ENERGY AUTHORITY. WEAPONS GROUP ATOMIC WEAPONS RES. ESTAB., ALDERMASTON, BERKS, ENG.

AWRE-DWG-HR/(LTR)(NO.)
: UNITED KINGDOM ATOMIC ENERGY AUTHORITY. WEAPONS GROUP ATOMIC WEAPONS RES. ESTAB., ALDERMASTON, BERKS, ENG. (DRAWING)

AWRE-E(NO.)/(YR.)
: UNITED KINGDOM ATOMIC ENERGY AUTHORITY. WEAPONS GROUP ATOMIC WEAPONS RES. ESTAB., ALDERMASTON, BERKS, ENG.

AWRE-EIVR-
: UNITED KINGDOM ATOMIC ENERGY AUTHORITY. WEAPONS GROUP ATOMIC WEAPONS RES. ESTAB., ALDERMASTON, BERKS, ENG.

AWRE-ERN-
: UNITED KINGDOM ATOMIC ENERGY AUTHORITY. WEAPONS GROUP ATOMIC WEAPONS RES. ESTAB., ALDERMASTON, BERKS, ENG.

AWRE-ERN-(NO.)/(YR.)
: UNITED KINGDOM ATOMIC ENERGY AUTHORITY. WEAPONS GROUP ATOMIC WEAPONS RES. ESTAB., ALDERMASTON, BERKS, ENG.

AWRE-FMP/IMOG/
: UNITED KINGDOM ATOMIC ENERGY AUTHORITY. WEAPONS GROUP ATOMIC WEAPONS RES. ESTAB., ALDERMASTON, BERKS, ENG.

AWRE-GRO/
: UNITED KINGDOM ATOMIC ENERGY AUTHORITY. WEAPONS GROUP ATOMIC WEAPONS RES. ESTAB., ALDERMASTON, BERKS, ENG.

AWRE-HR-
: UNITED KINGDOM ATOMIC ENERGY AUTHORITY. WEAPONS GROUP ATOMIC WEAPONS RES. ESTAB., ALDERMASTON, BERKS, ENG.

AWRE-INSP/
: UNITED KINGDOM ATOMIC ENERGY AUTHORITY. WEAPONS GROUP ATOMIC WEAPONS RES. ESTAB., ALDERMASTON, BERKS, ENG.

AWRE-J(NO.)/(NO.)-
: UNITED KINGDOM ATOMIC ENERGY AUTHORITY. WEAPONS GROUP ATOMIC WEAPONS RES. ESTAB., ALDERMASTON, BERKS, ENG.

AWRE-J(NO.)/(NO.)-UK/
: UNITED KINGDOM ATOMIC ENERGY AUTHORITY. WEAPONS GROUP ATOMIC WEAPONS RES. ESTAB., ALDERMASTON, BERKS, ENG.

AWRE/LIB/BIB/
: UNITED KINGDOM ATOMIC ENERGY AUTHORITY. WEAPONS GROUP ATOMIC WEAPONS RES. ESTAB., ALDERMASTON, BERKS, ENG.

AWRE-NR-(NO.)/(YR.)
: UNITED KINGDOM ATOMIC ENERGY AUTHORITY. WEAPONS GROUP ATOMIC WEAPONS RES. ESTAB., ALDERMASTON, BERKS, ENG.

AWRE-NR/A-(NO.)/(YR.)
: UNITED KINGDOM ATOMIC ENERGY AUTHORITY. WEAPONS GROUP ATOMIC WEAPONS RES. ESTAB., ALDERMASTON, BERKS, ENG.

AWRE-NR/C-(NO.)/(YR.)
: UNITED KINGDOM ATOMIC ENERGY AUTHORITY. WEAPONS GROUP ATOMIC WEAPONS RES. ESTAB., ALDERMASTON, BERKS, ENG.

AWRE-NR/P-(NO.)/(YR.)
: UNITED KINGDOM ATOMIC ENERGY AUTHORITY. WEAPONS GROUP ATOMIC WEAPONS RES. ESTAB., ALDERMASTON, BERKS, ENG.

AWRE-O-
: UNITED KINGDOM ATOMIC ENERGY AUTHORITY. WEAPONS GROUP ATOMIC WEAPONS RES. ESTAB., ALDERMASTON, BERKS, ENG.

AWRE-O-(NO.)/(YR.)
: UNITED KINGDOM ATOMIC ENERGY AUTHORITY. WEAPONS GROUP ATOMIC WEAPONS RES. ESTAB., ALDERMASTON, BERKS, ENG.

AWRE-PAPER-A.-
: UNITED KINGDOM ATOMIC ENERGY AUTHORITY. WEAPONS GROUP ATOMIC WEAPONS RES. ESTAB., ALDERMASTON, BERKS, ENG.

AWRE-R-(NO.)/(YR.)
: UNITED KINGDOM ATOMIC ENERGY AUTHORITY. WEAPONS GROUP ATOMIC WEAPONS RES. ESTAB., ALDERMASTON, BERKS, ENG.

AWRE-SLE-(NO.)/(YR.)
: UNITED KINGDOM ATOMIC ENERGY AUTHORITY. WEAPONS GROUP ATOMIC WEAPONS RES. ESTAB., ALDERMASTON, BERKS, ENG.

AWRE-SSBME-(NO.)/(YR.)
: UNITED KINGDOM ATOMIC ENERGY AUTHORITY. WEAPONS GROUP ATOMIC WEAPONS RES. ESTAB., ALDERMASTON, BERKS, ENG.

AWRE-SSME-TN-(NO.)/(YR.)
: UNITED KINGDOM ATOMIC ENERGY AUTHORITY. WEAPONS GROUP ATOMIC WEAPONS RES. ESTAB., ALDERMASTON, BERKS, ENG.

AWRE-SSPD/USA/
: UNITED KINGDOM ATOMIC ENERGY AUTHORITY. WEAPONS GROUP ATOMIC WEAPONS RES. ESTAB., ALDERMASTON, BERKS, ENG.

AWRE-SWAN-(NO.)/(NO.)
: UNITED KINGDOM ATOMIC ENERGY AUTHORITY. WEAPONS GROUP ATOMIC WEAPONS RES. ESTAB., ALDERMASTON, BERKS, ENG.

AWRE-T-(NO.)/(YR.)
: UNITED KINGDOM ATOMIC ENERGY AUTHORITY. WEAPONS GROUP ATOMIC WEAPONS RES. ESTAB., ALDERMASTON, BERKS, ENG.

AWRE-TERN-(NO.)/(YR.)
: UNITED KINGDOM ATOMIC ENERGY AUTHORITY. WEAPONS GROUP ATOMIC WEAPONS RES. ESTAB., ALDERMASTON, BERKS, ENG.

AWRE-TPN-
: UNITED KINGDOM ATOMIC ENERGY AUTHORITY. WEAPONS GROUP. ATOMIC WEAPONS RESEARCH ESTABLISHMENT, ALDERMASTON, BERKS, ENGLAND (THEORETICAL PHYSICS NOTE)

AWRE-TPN-(NO.)/(YR.)
: UNITED KINGDOM ATOMIC ENERGY AUTHORITY. WEAPONS GROUP ATOMIC WEAPONS RES. ESTAB., ALDERMASTON, BERKS, ENGLAND (THEORETICAL PHYSICS NOTE)

AWRE-TRANS-
: UNITED KINGDOM ATOMIC ENERGY AUTHORITY. WEAPONS GROUP ATOMIC WEAPONS RES. ESTAB., ALDERMASTON, BERKS, ENG.

AWS-
: AIR WEATHER SERVICE, SCOTT AFB, ILL.

AWS-
: AIR WEATHER SERVICE, WASHINGTON, D.C.

AWS-
: AMERICAN WELDING SOCIETY, N.Y.C.
AWS-BIB-
: AIR WEATHER SERVICE, WASHINGTON, D.C. (BIBLIOGRAPHY)
AWSG-(NUMBER)/(YEAR)
: UNITED KINGDOM. ARMY WORK STUDY GROUP, GUILDFORD, SURREY, ENGLAND
AWSM-(NUMBER)-
: AIR WEATHER SERVICE, SCOTT AFB, ILL.
AWSM-
: AIR WEATHER SERVICE, WASHINGTON, D.C. (MEMORANDUM)
AWSM-(NUMBER)-
: AIR WEATHER SERVICE, WASHINGTON, D.C.
AWS MANUAL-
: AIR WEATHER SERVICE, SCOTT AFB, ILL.
AWSM-TR-
: AIR WEATHER SERVICE, WASHINGTON, D.C.
AWSMTR-
: AIR WEATHER SERVICE, WASHINGTON, D.C.
AWSP-(NUMBER)-
: AIR WEATHER SERVICE, SCOTT AFB, ILL.
AWS PAM-
: AIR WEATHER SERVICE, SCOTT AFB, ILL.
AWSR-
: AIR WEATHER SERVICE, WASHINGTON, D.C.
AWS-RP-(NOS. MO./YR.)
: AIR WEATHER SERVICE, SCOTT AFB, ILL. (RECURRING PUBLICATION)
AWS SPECIAL STUDY-
: AIR WEATHER SERVICE, SCOTT AFB, ILL.
AWS-TR-
: AIR WEATHER SERVICE, SCOTT AFB, ILL.
AWS-TR-
: AIR WEATHER SERVICE, WASHINGTON, D.C.
AWTL-
: AIR MATERIEL COMMAND. ALL-WEATHER FLYING DIV. TRAFFIC AND LANDING BRANCH, WRIGHT-PATTERSON AFB, OHIO
AWTR-
: ROBERT A. TAFT WATER RESEARCH CENTER. ADVANCED WASTE TREATMENT RESEARCH LAB., CINCINNATI
AWTSC-
: AUSTRALIA. ATOMIC WEAPONS TESTS COMMITTEE, VICTORIA
AX-
: HUGHES AIRCRAFT CO., CULVER CITY, CALIF.
AXG-(YEAR)-
: GENERAL DYNAMICS/ASTRONAUTICS, SAN DIEGO, CALIF.
AXR-
: COLUMBIA UNIV., NYC. SUBSTITUTE ALLOY MATERIALS LABS.
AY-
: BUREAU OF AERONAUTICS (NAVY). AEROLOGY DIV.
AY(YEAR)-
: GENERAL DYNAMICS/ASTRONAUTICS, SAN DIEGO, CALIF.
AYE-IM-
: ARMED FORCES SPECIAL WEAPONS PROJECT. FIELD COMMAND, ALBUQUERQUE, N. MEX. (INSTRUCTORS MANUAL)
AYE-NB-
: ARMED FORCES SPECIAL WEAPONS PROJECT. FIELD COMMAND, ALBUQUERQUE, N. MEX. (NOTEBOOK)
AZ(LETTER)-
: CONVAIR. ASTRONAUTICS DIV., SAN DIEGO, CALIF.
AZM-
: CONVAIR-ASTRONAUTICS, SAN DIEGO, CALIF.
AZT-(YEAR-NUMBER)-FRLL
: AZTEC SCHOOL OF LANGUAGES, INC., MAYNARD, MASS.
AZT-(YEAR-NUMBER)-RULL
: AZTEC SCHOOL OF LANGUAGES, INC., MAYNARD, MASS.
AZT-(NO.)-(NO.)-GENRL
: AZTEC SCHOOL OF LANGUAGES, INC. RESEARCH TRANSLATION DIV., MC LEAN, VA.
AZT-(YEAR)-
: MASSACHUSETTS INST. OF TECH., LEXINGTON. LINCOLN LAB.
AZU-
: ARIZONA. UNIV., TUCSON
AZU-IAP-
: ARIZONA. UNIV., TUCSON. INST. OF ATMOSPHERIC PHYS.
AZU-PR-
: ARIZONA. UNIV., TUCSON
AZU-SR-
: ARIZONA. UNIV., TUCSON
AZU-TR-
: ARIZONA. UNIV., TUCSON

B-	(BERICHT) (REF. 2)
B-	(BULLETIN) (REF. 2)
B-(3 DIGITS)	AERODYNAMISCHE VERSUCHSANSTALT, GOETTINGEN, GERMANY
B-	AEROJET-GENERAL CORP. SOLID ROCKET PLANT, SACRAMENTO, CALIF. (FLEET BALLISTIC MISSILE PROGRAM)
(NUMBERS)B(NUMBERS)	ATLANTIC FLEET. HUNTER KILLER FORCE, N.Y.
B-	ATOMIC ENERGY OF CANADA LTD. CHALK RIVER PROJECT, ONT. (REF. 6)
B-	BELGIUM. CENTRE D'ETUDE POUR LES APPLICATIONS DE L'ENERGIE NUCLEAIRE, BRUSSELS
B(YEAR)-	BELLCOMM, INC., WASHINGTON, D.C.
B-	BENDIX AVIATION CORP. RESEARCH LABS., SOUTHFIELD, MICH.
B-	BOEING CO., SEATTLE
B11-	BROWN UNIV., PROVIDENCE. GRAD. DIV. OF APPLIED MATH.
B-(NUMBERS)-	BURROUGHS CORP., PAOLI, PENNA.
B-(NUMBER.NUMBER.NUMBER)	CARBIDE AND CARBON CHEMICALS CORP., OAK RIDGE, TENN. (REF. 34)
B-1.1-	CARBIDE AND CARBON CHEMICALS CORP. K-25 PLANT, OAK RIDGE, TENN.
B-7.1-	CARBIDE AND CARBON CHEMICALS CORP. K-25 PLANT, OAK RIDGE, TENN.
B-	CHICAGO. UNIV. METALLURGICAL LAB. (ASSIGNED TO REPORTS FROM VARIOUS BRITISH SOURCES)
B-	CLINTON LABS., OAK RIDGE, TENN.
B-	COLLINS RADIO CO., NEWPORT BEACH, CALIF.
B-	COMPTROLLER GENERAL OF THE U.S., WASHINGTON, D.C.
B-	CZECHOSLOVAKIA. CESKE VYSOKE UCENI TECHNICKE, PRAGUE
B-	DOUGLAS AIRCRAFT CO., INC. PROJECT RAND, SANTA MONICA, CALIF.
B-(4 DIGITS)	EG+G, INC. (ALL LOCATIONS)
B(NO.)-T-	FRANKLIN INST., PHILADELPHIA
B-	FRANKLIN INST. LABS. FOR RES. + DEV., PHILADELPHIA
B-	GENERAL ELECTRIC CO. DASA DATA CENTER, SANTA BARBARA, CALIF.
B1/(NUMBER)/(NUMBER)	GT. BRIT. ARMAMENT RESEARCH AND DEVELOPMENT ESTABLISHMENT, FORT HALSTEAD, KENT, ENGLAND
B-	HYMATIC ENGINEERING CO., LTD., REDDITCH, WORCS, ENG.
B-	INDIA. CENTRAL MECHANICAL ENG. RES. INST., DURGAPUR
B-	LOCKHEED MISSILES AND SPACE CO., SUNNYVALE, CALIF.
B001 THRU B999	MC DONNELL AIRCRAFT CORP., ST. LOUIS
B(NUMBER)-(NUMBER)	MC DONNELL AIRCRAFT CORP., ST. LOUIS
B/68-	MASSACHUSETTS INST. OF TECH., CAMBRIDGE. CENTER FOR INTERNATIONAL STUDIES
1B-	NATIONAL BUREAU OF STANDARDS, WASHINGTON, D.C.
B-	NATIONAL DEFENSE RESEARCH COMMITTEE (REF. 27)
B-	NAVAL TORPEDO STATION, KEYPORT, WASH.
B-	NUTRILITE PRODUCTS, INC., BUENA PARK, CALIF. (BIOL.)
B-	OFFICE OF SCIENTIFIC RESEARCH AND DEVELOPMENT (ASSIGNED TO REPORTS FROM GREAT BRITAIN)
B(NO.)-	PENNSYLVANIA STATE UNIV., UNIVERSITY PARK
B-	PHILCO CORP., BLUE BELL, PENNA.
B(3 DIGITS)-	PHILCO-FORD CORP., BLUE BELL, PENNA.
B-	RAND CORP., SANTA MONICA, CALIF. (BRIEFING)
B(8 DIGITS)	SINGER-GENERAL PRECISION, INC. KEARFOTT DIV., LITTLE FALLS, N.J.
B-	SPERRY GYROSCOPE CO., GREAT NECK, N.Y.
B-	SWEDEN. FOERSVARETS FORSKNINGSANSTALT, STOCKHOLM
B2/(NUMBER)/(YEAR)-	SYLVANIA ELECTRIC PRODUCTS INC., BAYSIDE, N.Y.
B(NO.)/(NO.)/(YR.)-	SYLVANIA ELECTRIC PRODUCTS INC. PRODUCT DEVELOPMENT LABS., FLUSHING, N.Y.
B-	TACTICAL AIR COMMAND, LANGLEY AFB, VA.
B-0.341- THRU B-7.470-	TENNESSEE EASTMAN CORP., OAK RIDGE, TENN. (REF. 34)
B-	UNION OF SOUTH AFRICA. MAGNETIC OBS., HERMANUS
B-	UNITED AIRCRAFT CORP. RES. LABS., EAST HARTFORD, CONN.
(YR.)-9B5-LUBER-R(NO.)	WESTINGHOUSE RESEARCH LABS. INSULATION AND CHEM. TECHNOLOGY DEPT., PITTSBURGH
BA-	BAIRD ASSOCIATES, INC., CAMBRIDGE, MASS.
BA-	BELL AEROSPACE CO. DIV. OF TEXTRON, BUFFALO (LAB.REPTS)
BA-(NUMBER)-	BELL AEROSYSTEMS CO., BUFFALO
B+A	BIOT AND ARNOLD (CONSULTANTS IN APPLIED MATH. + PHYS)
BA-	CARBIDE AND CARBON CHEMICALS CORP., OAK RIDGE, TENN.
BA2-M-	COLUMBIA UNIV., NYC. SUBSTITUTE ALLOY MATERIALS LABS.
BA2-R-	COLUMBIA UNIV., NYC. SUBSTITUTE ALLOY MATERIALS LABS.
BA-	LILIENTHAL-GESELLSCHAFT FUER LUFTFAHRTFORSCHUNG, BERLIN (BERICHT A)
BA-	SYLVANIA ELECTRIC PRODUCTS, INC., BAYSIDE, N.Y.
BAAR-	BOOZ-ALLEN APPLIED RESEARCH, INC., BETHESDA, MD.
BAARINC-	BOOZ-ALLEN APPLIED RESEARCH, INC., BETHESDA, MD.
BAARINC-	BOOZ-ALLEN APPLIED RESEARCH, INC., CHICAGO
BAARINC-PRO-R-	BOOZ-ALLEN APPLIED RESEARCH, INC., BETHESDA, MD.
BAARING-(NUMBER)-	BOOZ-ALLEN APPLIED RESEARCH, INC., ALBUQUERQUE, N.M.
BAARING-(NUMBER)-	BOOZ-ALLEN APPLIED RESEARCH, INC., LOS ANGELES
BAC-	BELL AEROSPACE CO. DIV. OF TEXTRON, BUFFALO
BAC-(NUMBER)-(NUMBER)	BELL AEROSYSTEMS CO., BUFFALO
BAC-	BENDIX AVIATION CORP. (ALL LOCATIONS)
BAC-(NO.)-(NO.)U	BENDIX AVIATION CORP. UTICA DIV., N.Y.
BAC-	BOEING AIRPLANE CO., SEATTLE
BAC-	BRISTOL AEROPLANE CO., LTD., ENGLAND
BAC-AE-	BRITISH AIRCRAFT CORP., PRESTON, LANCS, ENGLAND
BAC-BPD-	BENDIX AVIATION CORP. BENDIX PRODUCTS DIV., MISHAWAKA, IND.
BAC-BPD-MEMO-	BENDIX AVIATION CORP. BENDIX PRODUCTS DIV., MISHAWAKA, IND.
BAC-BSR-	BENDIX AVIATION CORP. BENDIX SYSTEMS DIV., ANN ARBOR, MICH.
BAC-D-	BOEING AIRPLANE CO., SEATTLE
BAC-E-	BENDIX CORP. KANSAS CITY DIV., MO.
BAC-EP-	ECLIPSE-PIONEER DIV., BENDIX AVIATION CORP., TETERBORO, N. J.
BAC-FI-	BENDIX AVIATION CORP. FRIEZ INSTR. DIV., BALTIMORE
BAC/GW/BRISTOL/TR/102	BRISTOL AEROPLANE CO., LTD., ENGLAND
BAC-KC-	BENDIX AVIATION CORP. KANSAS CITY DIV., MO.
BAC-PD-	BENDIX AVIATION CORP. PACIFIC DIV., N. HOLLYWOOD, CAL.
BAC-P-DLM-	BENDIX AVIATION CORP. PACIFIC DIV. DEVELOPMENT LAB., BURBANK, CALIF.
BAC-PR-	BELL AIRCRAFT CORP., BUFFALO
BAC-QR-(NUMBERS)-	BELL AIRCRAFT CORP., BUFFALO (QUARTERLY PROGRESS RPT)
BAC-R-	BENDIX AVIATION CORP. RADIO DIV.
BAC-RDBT-	BENDIX AVIATION CORP. PACIFIC DIV., N. HOLLYWOOD, CAL.
BAC/RL-	BENDIX AVIATION CORP. RESEARCH LABS., DETROIT
BAC/RLD-	BENDIX AVIATION CORP. RESEARCH LABS. DIV., DETROIT
BAC-TM-	BENDIX AVIATION CORP. RESEARCH LABS., DETROIT
BAC-TM-(NO.)-(NO.)-	BENDIX AVIATION CORP. RESEARCH LABS., DETROIT
BA-D-	BELL AEROSPACE CO. DIV. OF TEXTRON, BUFFALO
BAD-	BRITISH ADMIRALTY DELEGATION, WASHINGTON, D. C.
BAEC-	BLASS ANTENNA ELECTRONICS CORP., LONG ISLAND CITY, N.Y.
BA-FTR-(YR./MONTH)	BAIRD ASSOCS., INC., CAMBRIDGE, MASS. (FINAL TECH.RPT)
BAI-	BAIRD ASSOCIATES, INC., CAMBRIDGE, MASS.
BAIR-	BAIRD ASSOCIATES, INC., CAMBRIDGE, MASS.
BAIRD-	BAIRD ASSOCIATES, INC., CAMBRIDGE, MASS.

BALCO-	BALCO RESEARCH LAB., NEWARK, N.J.
BALLISTICS RES. REPT.-	NAVAL ORDNANCE LAB., WHITE OAK, MD.
BAL-TM-	JOHNS HOPKINS UNIV., BALTIMORE. BALLISTIC ANAL. LAB.
BAL-TN-	BALLISTIC RESEARCH LABS., ABERDEEN PROVING GROUND, MD.
BAL-TR-	JOHNS HOPKINS UNIV., BALTIMORE. BALLISTIC ANAL. LAB.
BA-M-	COLUMBIA UNIV., NYC. SUBSTITUTE ALLOY MATERIALS LABS.
BAMIRAC-	MICHIGAN. UNIV., ANN ARBOR. BALLISTIC MISSILE RADIATION ANALYSIS CENTER
B AND W-	BABCOCK + WILCOX CO., N.Y.C.
BAOR-	GT. BRIT. ARMY OF THE RHINE. SPECIAL PROJECTILE OPERATIONS GROUP
BAP-(NUMBER)-	BOEING CO., SEATTLE
BA-QR-	BAIRD ASSOCS., INC., CAMBRIDGE, MASS. (QUARTERLY RPT)
BAR-	BURNS AND ROE, INC., N.Y.C.
BA-R-	CARBIDE AND CARBON CHEMICALS CORP. SUBSTITUTE ALLOY MATERIALS LABS., N.Y.C.
BA-R-	COLUMBIA UNIV., NYC. SUBSTITUTE ALLOY MATERIALS LABS.
BARC-	INDIA. BHABHA ATOMIC RESEARCH CENTRE, BOMBAY
BARC/HP/TM-	INDIA. BHABHA ATOMIC RESEARCH CENTRE, BOMBAY
BARC/I-	INDIA. BHABHA ATOMIC RESEARCH CENTRE, BOMBAY
BARC/INF-	INDIA. BHABHA ATOMIC RESEARCH CENTRE, BOMBAY
BAR-ILANU-	BAR-ILAN UNIVERSITY, RAMAT-GAN, ISRAEL
BARO-	BARODYNAMICS, INC., GEORGETOWN, COLO.
BAROD-(LTRS/LTRS)-	BARODYNAMICS, INC., GEORGETOWN, COLO.
BAR, PASADENA-	BUREAU OF AERONAUTICS REPRESENTATIVE, PASADENA, CAL.
BARRINC-	BOOZ-ALLEN APPLIED RESEARCH, INC., BETHESDA, MD.
BARRY-	BARRY CONTROLS INC., WATERTOWN, MASS.
BASC-	BELL AEROSYSTEMS CO., BUFFALO
BASD-M0-	BENDIX CORP. AEROSPACE SYSTEMS DIV., MISHAWAKA, IND.
BASELU-	BASEL. UNIVERSITAET
BASIC RESEARCH-	GEORGE WASHINGTON UNIV., ARLINGTON, VA. HUMAN RESOURCES RESEARCH OFFICE
BASIC STUDIES-	MONMOUTH COUNTY PLANNING BOARD, N.J.
BASO-	BUREAU OF AERONAUTICS (NAVY) (SHIPPING ORDER)
BAT-	BATTELLE MEMORIAL INST., COLUMBUS, OHIO
BAT-(NOS.)-(NOS.)-	BATTELLE MEMORIAL INST., COLUMBUS, OHIO
BAT-(NUMBER)A-	BATTELLE MEMORIAL INST., COLUMBUS, OHIO
BAT-171-	BATTELLE MEMORIAL INST. REMOTE AREA CONFLICT INFORMATION CENTER, COLUMBUS, OHIO
BAT-G(NUMBER)-	BATTELLE MEMORIAL INST., COLUMBUS, OHIO
BAT-PUB-	BATTELLE MEMORIAL INST., COLUMBUS, OHIO
BAT-RR-	BATTELLE MEMORIAL INST., COLUMBUS, OHIO
BATT-	BATTELLE MEMORIAL INST., COLUMBUS, OHIO
BAT-T(NUMBER)-SR	BATTELLE MEMORIAL INST., WASHINGTON, D.C.
BAW-(NUMBER)-	BABCOCK + WILCOX CO. (ALL LOCATIONS)
BAW-	BABCOCK + WILCOX CO. (ALL LOCATIONS)
BAW-PR-	BABCOCK + WILCOX CO., BARBERTON, OHIO (PROGRESS RPT.)
BAW-RR-	BABCOCK + WILCOX CO. RESEARCH CENTER, ALLIANCE, OHIO
BAW-TM-	BABCOCK + WILCOX CO. ATOMIC EN. DIV., LYNCHBURG, VA.
BAW-TPCC-	BABCOCK + WILCOX CO. ATOMIC EN. DIV., LYNCHBURG, VA.
BAYLOR-	BAYLOR UNIV., HOUSTON, TEX. COLL. OF MEDICINE
BAYLOR-FR/(CONTRACT NO.)	BAYLOR UNIV., HOUSTON, TEX. COLL. OF MEDICINE
BAYLOR-PR/(CONTRACT NO.)	BAYLOR UNIV., HOUSTON, TEX. COLL. OF MEDICINE
BAYLOR UNIV.PR.../MD-(NO)	BAYLOR UNIV., HOUSTON, TEX. COLL. OF MEDICINE
BAYLOR U.PR.../MD-(NC.)	BAYLOR UNIV., HOUSTON, TEX. COLL. OF MEDICINE
BAY U-	BAYLOR UNIV., WACO, TEX.
BB-	(BIBLIOGRAPHY) (REF. 2)
BB-(YEAR)-(NUMBER)-BB	AIR REDUCTION CHEMICAL AND CARBIDE CO., BOUND BROOK, N.J. (USED 1959-64)
BB-	AIR REDUCTION CHEMICAL CO., BOUND BROOK, NJ (1954-60)
BB-	BRUSH BERYLLIUM CO., CLEVELAND
BB-	BUREAU OF ORDNANCE (NAVY) (BUMBLEBEE PROJ.) (REF. 8)
BB-	CARBIDE AND CARBON CHEMICALS CORP., OAK RIDGE, TENN.
BB-	JOHNS HOPKINS UNIV., SILVER SPRING, MD. APPLIED PHYSICS LAB. (BUMBLEBEE PROJECT) (REF. 8)
BBC-	BISSETT-BERMAN CORP., SANTA MONICA, CALIF.
BBC-(YEAR)/	BRITISH BROADCASTING CORP., KINGSWOOD, SURREY, ENG.
BBC-	BRUSH BERYLLIUM CO., CLEVELAND
BBC-(NO.-NO.)X	BRUSH BERYLLIUM CO., CLEVELAND
BBC-	JOHNS HOPKINS UNIV., SILVER SPRING, MD. APPLIED PHYSICS LAB. BUMBLEBEE COMMITTEE (REF. 8)
BBCO-	BRIDGEPORT BRASS CO., CONN.
BBCO-F-	BRIDGEPORT BRASS CO., CONN.
BBC-PR-	BRUSH BERYLLIUM CO., CLEVELAND
BBC-TR-	BRUSH BERYLLIUM CO., CLEVELAND
BBI-	BROWN, BOVERI INC., ZURICH-OERLIKON
BB-M-	COLUMBIA UNIV., NYC. SUBSTITUTE ALLOY MATERIALS LABS.
BB+N-	BOLT, BERANEK, AND NEWMAN, INC., CAMBRIDGE, MASS.
BBN-	BOLT, BERANEK, AND NEWMAN, INC., CAMBRIDGE, MASS.
BBN-	BOLT, BERANEK, AND NEWMAN, INC., LOS ANGELES
BBRL-	GT. BRIT. TELECOMMUNICATIONS RESEARCH ESTABLISHMENT, MALVERN, WORCS, ENGLAND
BBRM-	GT. BRIT. BOMBING RESEARCH MISSION
BC-	BECHTEL CORP., SAN FRANCISCO
BC-	BENDIX CORP., DETROIT
BC-	BENDIX CORP. RESEARCH LABS. DIV., SOUTHFIELD, MICH.
BC-	BERLIN. HAHN-MEITNER-INSTITUT FUER KERNFORSCHUNG (BERICHT, CHEMIE)
BC-	BERYLLIUM CORP., READING, PENNA.
BC-	CALIFORNIA. UNIV., BERKELEY. RADIATION LAB. (REF. 9)
BC-	GENERAL ELECTRIC CO., LTD. RESEARCH LABS., WEMBLEY, MIDDX, ENGLAND
BC-	GT. BRIT. NATL. PHYSICAL LAB., TEDDINGTON, MIDDX., ENG
BC-(NUMBER)N	HYDROGRAPHIC OFFICE, SUITLAND, MD. (BATHYMETRIC CHARTS)
BC-	HYDROGRAPHIC OFFICE, SUITLAND, MD.
BC-(NUMBER)-	MOUND LAB., MIAMISBURG, OHIO
BCC-	ROCKETDYNE DIV., NORTH AMERICAN AVIATION, INC., CANOGA PARK, CALIF.
BCD-	BORDEN CHEMICAL CO., PHILADELPHIA
BCD-MHR-	BENDIX CORP. COMMUNICATIONS DIV., BALTIMORE
BCF-MT-	TENNESSEE EASTMAN CORP., OAK RIDGE, TENN.
BCH/M-	BUREAU OF COMMERCIAL FISHERIES. DIV. OF ECONOMIC RESEARCH, COLLEGE PARK, MD. (MAGNETIC TAPE)
BCI-	BELGIUM. CENTRE D'ETUDE DE L'ENERGIE NUCLEAIRE, BRUSSELS
BCI-	BARRY CONTROLS INC., WATERTOWN, MASS.
BCL-	ROCKETDYNE DIV., NORTH AMERICAN AVIATION, INC., CANOGA PARK, CALIF.
BCL-SED-(NUMBER)-	ILLINOIS. UNIV., URBANA. BIOLOGICAL COMPUTER LAB.
BC-M-	BATTELLE MEMORIAL INST. SYSTEMS AND ELECTRONICS DEPT., COLUMBUS, OHIO
BCM/WP-	COLUMBIA UNIV., NYC. SUBSTITUTE ALLOY MATERIALS LABS.
BC-P-	GT. BRIT. ATOMIC EN. RES. ESTAB., HARWELL, BERKS, ENG
BCP-CR-	BOSE RESEARCH INST., CALCUTTA
BCPG-	SANDERS ASSOCIATES, INC., BEDFORD, MASS.
BCPG-	BECHTEL CORP., SAN FRANCISCO (JOINT VENTURE)
BCPG-	PACIFIC GAS AND ELECTRIC CO., SAN FRANCISCO (JOINT VENTURE)

BCPI-	BECHTEL CORP., SAN FRANCISCO
BCPR-PPS-	BUREAU OF MINES. BARTLESVILLE PETROLEUM RESEARCH CENTER, OKLA.
BCPR-QPR-	BUREAU OF MINES. BARTLESVILLE PETROLEUM RESEARCH CENTER, OKLA.
BC-R-	COLUMBIA UNIV., NYC. SUBSTITUTE ALLOY MATERIALS LABS.
BCR-	FAIRCHILD ENGINE AND AIRPLANE CORP.
BC/RL-	BENDIX CORP. RESEARCH LABS. DIV., SOUTHFIELD, MICH.
BCR-L-	BITUMINOUS COAL RESEARCH, INC., MONROEVILLE, PA.
BC/RLD-	BENDIX CORP. RESEARCH LABS. DIV., SOUTHFIELD, MICH.
BCSO-	BRITISH COMMONWEALTH SCIENTIFIC OFFICE, WASH., D.C.
BCSO (NO.)-(LETTER)-	BRITISH COMMONWEALTH SCIENTIFIC OFFICE, WASH., D.C.
BCSO SCC-	BRITISH COMMONWEALTH SCIENTIFIC OFFICE, WASH., D.C.
BC-TOP-	BECHTEL CORP., SAN FRANCISCO
BCU-	BRITISH COLUMBIA. UNIV., VANCOUVER
BD-	BIO-DYNAMICS, INC., CAMBRIDGE, MASS.
BD-(YEAR)-	BUREAU OF THE CENSUS, WASHINGTON, D.C.
BD-(NUMBER)-	CELESTIAL RESEARCH CORP., SOUTH PASADENA, CALIF.
BD-	CHEMICAL CORPS. METEOROLOGICAL DIV. BIOLOGICAL DEPT., CAMP DETRICK, MD.
BD 1 CN. (NUMBER)-	CONTINENTAL ARMY COMMAND, FORT MONROE, VA.
BD-	SIGNAL CORPS (ARMY). BATTERY DEVELOPMENT SECTION
BD-	UNITED STATES LAKE SURVEY, DETROIT
BD-(NUMBER)-	UNITED STATES LAKE SURVEY, DETROIT
BDC-	BRUSH DEVELOPMENT CO., CLEVELAND
BDC-(YEAR)-TO-	BUREAU OF DOMESTIC COMMERCE, WASHINGTON, D.C.
BDC-QIR-P/P-	BUREAU OF DOMESTIC COMMERCE, WASHINGTON, D.C.
BDDA-	GT. BRIT. DEPT. OF ATOMIC ENERGY (BRITISH DECLASSIFIED DOCUMENTS, ATOMIC)
BDIAC-	BATELLE-DEFENDER INFO. ANALYSIS CTR., COLUMBUS, OHIO
BDM-	BRADDOCK, DUNN AND MC DONALD, INC., EL PASO, TEX.
BDM-(NUMBER-YEAR)-F-	BRADDOCK, DUNN AND MC DONALD, INC., EL PASO, TEX.
BD-M-	COLUMBIA UNIV., NYC. SUBSTITUTE ALLOY MATERIALS LABS.
BDM-W-	BRADDOCK, DUNN AND MC DONALD, INC., MC LEAN, VA.
BD-R-	CARBIDE AND CARBON CHEMICALS CORP. SUBSTITUTE ALLOY MATERIALS LABS., N.Y.C.
BD-R-	COLUMBIA UNIV., NYC. SUBSTITUTE ALLOY MATERIALS LABS.
BDSA-	BUSINESS AND DEFENSE SERVICES ADM., WASHINGTON, D.C.
BDSAF-(NUMBER)(YEAR)	BUREAU OF THE CENSUS, WASHINGTON, D.C.
BDX-	BENDIX CORP. (ALL LOCATIONS)
BE-	AIR FORCE, WASHINGTON, D.C. (BOMBING ENCYCLOPEDIA)
BE-	BARNES ENGINEERING CO., STAMFORD, CONN.
BE-	COMMITTEE ON MEDICAL RESEARCH (BEADLE REPORTS)
BE-	CORNELL AERONAUTICAL LAB., INC., BUFFALO
BE-(NUMBER-LETTER)-	CORNELL AERONAUTICAL LAB., INC., BUFFALO
BE-	OFFICE OF TECH.SERVICES,WASH.,DC.(BOOK EXPLOITATIONS)
BEA-	BRITISH EUROPEAN AIRWAYS
BEAC-ER-	BEECH AIRCRAFT CORP., WICHITA, KAN.
BEAD-ER-	BEECH AIRCRAFT CORP., WICHITA, KAN.
BEAIRA-TR-L/T-	BRITISH ELECTRICAL AND ALLIED INDUSTRIES RESEARCH ASSN., LEATHERHEAD, SURREY, ENGLAND
BEA-SCB-(YEAR)-	BUREAU OF ECONOMIC ANALYSIS, WASHINGTON, D.C. (SURVEY OF CURRENT BUSINESS)
BEB-	OFFICE OF THE CHIEF OF ENGINEERS (ARMY). BEACH EROSION BOARD
BEC-	BABCOCK ELECTRONICS CORP., COSTA MESA, CALIF.
BEC-	BARNES ENGINEERING CO., STAMFORD, CONN.
BEC-(NUMBER)-DSR-	BARNES ENGINEERING CO., STAMFORD, CONN.
BECCO-	BECCO CHEMICAL DIV., FOOD MACHINERY AND CHEMICAL CORP., BUFFALO
BEC-IR-	BELFOUR ENG. CO., SUTTONS BAY, MICH. (INVENTORY REPT)
BEC-R-(NUMBER-NO.-YEAR)	BOWLES ENGINEERING CORP., SILVER SPRING, MD.
BEC-TR-	BARNES ENGINEERING CO., STAMFORD, CONN.
BECV-(YEAR-NUMBER)	CALIFORNIA. UNIV., BERKELEY. LAWRENCE RADIATION LAB.
BEDR-	BURNDY CORP., NORWALK, CONN.
BEECH-	BEECH AIRCRAFT CORP., WICHITA, KAN.
BEECH-	BEECHCRAFT RESEARCH AND DEV., INC., BOULDER, COLO.
BEECH (LETTERS) (LETTERS)	BEECHCRAFT RESEARCH AND DEV., INC., BOULDER, COLO.
BEECHCRAFT-ER-	BEECHCRAFT RESEARCH AND DEV., INC., WICHITA, KAN.
BEECHCRAFT FTRDI-...	BEECHCRAFT RESEARCH AND DEVELOPMENT, INC.
BEECH-ER-	BEECH AIRCRAFT CORP. BOULDER DIV., COLO.
BEECH-TEST REPORT-	BEECH AIRCRAFT CORP. BOULDER DIV., COLO.
BEERS-	BEERS (ROLAND F.), INC., ALEXANDRIA, VA.
BEGE-	BEGE (J.R.M.) CO., ARLINGTON, MASS.
BEGE (NUMBERS)-	BEGE (J.R.M.) CO., ARLINGTON, MASS.
BEI-	BLOCK ENGINEERING INC., CAMBRIDGE, MASS.
BEI-	CALIFORNIA. UNIV., BERKELEY. RADIATION LAB.
BEJG-	SANDIA CORP., ALBUQUERQUE, N. MEX.
BEL-	BREWER ENGINEERING LABS., INC., MARION, MASS.
BELA-	BELL AIRCRAFT CORP., BUFFALO (PROJECT METEOR)
BELFOUR-IR-	BELFOUR ENG. CO. TECH. INFO. SYSTEMS,SUTTONS BAY,MICH
BELFOUR-PR-	BELFOUR ENG. CO. TECH. INFO. SYSTEMS,SUTTONS BAY,MICH
BELL-(NUMBERS)-	BELL AEROSYSTEMS CO., BUFFALO
BELL-	BELL AIRCRAFT CORP., BUFFALO
BE-M-	COLUMBIA UNIV., NYC. SUBSTITUTE ALLOY MATERIALS LABS.
BEND-(NO.)-	BENDIX AVIATION CORP. PACIFIC DIV.,N. HOLLYWOOD, CAL.
BEND-	BENDIX CORP. (ALL LOCATIONS)
BEND (LETTERS, DATES)	BENDIX AVIATION CORP. YORK DIV., PENNA.
BEND BDP-	BENDIX AVIATION CORP. BENDIX PRODUCTS DIV., MISSILES, MISHAWAKA, IND.
BEND BDP-M-	BENDIX AVIATION CORP. BENDIX PRODUCTS DIV., MISSILES, MISHAWAKA, IND.
BEND BSC-	BENDIX CORP. BENDIX SYSTEMS DIV., ANN ARBOR, MICH.
BEND-BSD-	BENDIX CORP. BENDIX SYSTEMS DIV., ANN ARBOR, MICH.
BEND-BSR-	BENDIX CORP. BENDIX SYSTEMS DIV., ANN ARBOR, MICH.
BEND-DLM-	BENDIX CORP. PACIFIC DIV., NORTH HOLLYWOOD, CALIF.
BEND EPD (LTRS)-	ECLIPSE-PIONEER DIV., BENDIX AVIATION CORP., TETERBORO, N. J.
BEND EPD R-	ECLIPSE-PIONEER DIV., BENDIX AVIATION CORP., TETERBORO, N. J.
BEND FID-	BENDIX AVIATION CORP. FRIEZ INSTR. DIV., BALTIMORE
BEND-KCD-(LETTERS)-	BENDIX AVIATION CORP. KANSAS CITY DIV., MO.
BEND-R-	BENDIX AVIATION CORP. RESEARCH LABS., DETROIT
BEND-R-	BENDIX CORP. RED BANK DIV., EATONTOWN, N.J.
BEND/SD-	BENDIX CORP. SCINTILLA DIV., SIDNEY, N.Y.
BEND/SD TR RIG-	BENDIX CORP. SCINTILLA DIV., SIDNEY, N.Y. (RADIO INTERFERENCE GUARD)
BEND TR RIG-	BENDIX CORP. SCINTILLA DIV., SIDNEY, N.Y. (RADIO INTERFERENCE GUARD)
BER-(NUMBER-YEAR)	ALABAMA. UNIV., UNIVERSITY. BUREAU OF ENG. RESEARCH
BE-R-	COLUMBIA UNIV., NYC. SUBSTITUTE ALLOY MATERIALS LABS.
BERCO-	BERYLLIUM CORP., READING, PENNA.
BERGEN U-	BERGEN. UNIV., DEPT. OF APPLIED MATHEMATICS
BERK-	BECKMAN INSTRUMENTS, INC. BERKELEY SCIENCE DIV., RICHMOND, CALIF.
BER R-	BUREAU OF ENGINEERING RESEARCH
BESRL-TRN-	ARMY BEHAVIORAL SCIENCE RES. LAB., WASHINGTON, D.C.

BESRL-TRR-	ARMY BEHAVIORAL SCIENCE RES. LAB., WASHINGTON, D.C.
BETA-	BETA CORP., RICHMOND, VA.
BETA-	CALIFORNIA. UNIV., BERKELEY. RADIATION LAB. (REF. 9)
BETA-	CHICAGO. UNIV. METALLURGICAL LAB. (ASSIGNED TO REPORTS FROM VARIOUS AMERICAN SOURCES)
BETA CD-	TENNESSEE EASTMAN CORP., OAK RIDGE, TENN.
BETA CONF-	TENNESSEE EASTMAN CORP., OAK RIDGE, TENN.
BETA-S-	CALIFORNIA. UNIV., BERKELEY. RADIATION LAB. (REF. 9)
BETA-TR-	BETA CORP., RICHMOND, VA.
BE-TN-RL-	BROWN ENGINEERING CO., INC., HUNTSVILLE, ALA.
BETTIS DESIGNATION-	WESTINGHOUSE ELECTRIC CORP. BETTIS ATOMIC POWER LAB., PITTSBURGH
BEW-	BERLIN. HAHN-MEITNER-INSTITUT FUER KERNFORSCHUNG
BEW RR-	BOARD OF ECONOMIC WARFARE, WASH., D.C. (RES. RPTS.)
BF(NUMBER)-	BENDIX AVIATION CORP. FRIEZ INSTR. DIV., BALTIMORE
BF(NUMBER)-	BENDIX CORP. ENVIRONMENTAL SCIENCE DIV., BALTIMORE
BFC-	BUFFALO FORGE CO.
BFG-CONTROL-(NO.)-A	GOODRICH (B.F.) CO. RESEARCH CENTER, BRECKSVILLE, OHIO
BFG-QR-	GOODRICH (B.F.) CO. RESEARCH CENTER, BRECKSVILLE, OHIO
BFG-SR-	GOODRICH (B.F.) CO., AKRON, OHIO
BF-RATS-	CHICAGO. UNIV. METALLURGICAL LAB.
BG-	MICHIGAN STATE UNIV., EAST LANSING. COLLEGE OF COMMUNICATION ARTS
BGN-	BOARD ON GEOGRAPHIC NAMES
BH-	LITTON SYSTEMS, INC., BEVERLY HILLS, CALIF.
BHC-	BELL HELICOPTER CORP., FORT WORTH, TEX.
BHC-	GENERAL DYNAMICS/ASTRONAUTICS, SAN DIEGO, CALIF.
BHC-TR-	BELL HELICOPTER CORP., FORT WORTH, TEX.
BHC-X/0/	BRITISH HOVERCRAFT CORP. LTD. EXPTL. DEPT., LONDON
BHF-	GERMANY. BEVOLLMAECHTIGTE FUER HOCHFREQUENZFORSCHUNG
BHO-(NUMBER)-	BROOKHAVEN AREA OFFICE, AEC, UPTON, N.Y. (REF. 4)
BH-ONR-	BELL AND HOWELL RESEARCH LABS., PASADENA, CALIF.
BHRA-NEL-JOINT-REPT-	BRITISH HYDROMECHANICS RESEARCH ASSN., CRANFIELD, BEDS, ENGLAND (JOINT VENTURE)
BHRA-NEL-JOINT-REPT-	UNITED KINGDOM. NATIONAL ENGINEERING LAB., GLASGOW (JOINT VENTURE)
BI-	BRANDEIS UNIV., WALTHAM, MASS.
BI(NO.)/(YR)-	GT. BRIT. ARMAMENT RESEARCH AND DEVELOPMENT ESTABLISHMENT, FORT HALSTEAD, KENT, ENGLAND
BI-	GT. BRIT. ATOMIC EN. RES. ESTAB., HARWELL, BERKS, ENG
BI-	IMPERIAL CHEMICAL INDUSTRIES, LTD.
BI (NUMBER) CR-	PENNSYLVANIA. UNIV., PHILA. INST. FOR COOP. RES.
BI(NO.)/(YR)-	SYLVANIA ELECTRIC PRODUCTS INC., BAYSIDE, N.Y.
BIAC INFO. MODULE M(NO.)	AMERICAN INSTITUTE OF BIOLOGICAL SCIENCES. BIO-INSTRUMENTATION ADVISORY COUNCIL, WASHINGTON, D.C.
BIB-	(BIBLIOGRAPHY) (REF. 2)
BIB-	BRITISH HYDROMECHANICS RES. ASSN.,CRANFIELD,BEDS,ENG.
BIB-	NATIONAL RESEARCH COUNCIL OF CANADA. DIV. OF BUILDING RESEARCH, OTTAWA (BIBLIOGRAPHIES)
BIBL./DOK.-	KERNFORSCHUNGSANLAGE JUELICH, GERMANY
BIBLIOGRAPHIC-(YEAR)-	ARMY NATICK LABS. TECHNICAL LIBRARY, MASS.
BIBLIOGRAPHIC LIST-	DEPT. OF TRANSPORTATION. LIBRARY SERVICES DIV., WASHINGTON, D.C.
BIBLIOGRAPHY-	DEPARTMENT OF THE INTERIOR, WASHINGTON, D.C.
BIB-SER-	QUARTERMASTER RESEARCH AND DEV. LABS., PHILADELPHIA
B-ICAS-PAPER-	GERMANY. AERODYNAMISCHE VERSUCHSANSTALT, GOETTINGEN
BIGS-	(BRITISH INTERROGATION OF GERMAN SCIENTISTS)
BIGS-	JOINT INTELLIGENCE OBJECTIVES AGENCY, WASH., D.C.
BIGS-R-	(BRITISH INTERROGATION OF GERMAN SCIENTISTS)
BIHR-	BELTONE INST. FOR HEARING RES.,CHICAGO (TRANSLATIONS)
BII-	BETA INDUSTRIES, INC., DAYTON, OHIO
BII-TR-	BECKMAN INSTRUMENTS, INC., FULLERTON, CALIF.
BILLINGHAM RESEARCH...	IMPERIAL CHEMICAL INDUSTRIES, LTD. BILLINGHAM DIV., DURHAM, ENGLAND
BIMONTHLY PROGRESS RPT.-	(BIMONTHLY PROGRESS REPORT) (REF. 2)
BIN-	HANFORD ATOMIC PRODUCTS OPERATION, RICHLAND, WASH.
BIO/(NUMBER)/(YEAR)	ITALY. COMITATO NAZIONALE PER L'ENERGIA NUCLEARE,ROME
BIOS-	BRITISH INTELLIGENCE OBJECTIVES SUB-COMMITTEE
BIOS B-	BRITISH INTELLIGENCE OBJECTIVES SUB-COMMITTEE
BIOS BAG-	BRITISH INTELL. OBJECTIVES SUB-COMM. (BAG NUMBER)
BIOS CPVA-	BRITISH INTELLIGENCE OBJECTIVES SUB-COMMITTEE
BIOS DOC-	BRITISH INTELLIGENCE OBJECTIVES SUB-COMM. (DOCUMENTS)
BIOS ER-	BRITISH INTELL. OBJECTIVES SUB-COMM. (EVALUATION RPT)
BIOS ERA-	BRITISH INTELLIGENCE OBJECTIVES SUB-COMMITTEE
BIOS FD-	BRITISH INTELL. OBJECTIVES SUB-COMM. (FOREIGN DOCS.)
BIOS FD CRB-	BRITISH INTELLIGENCE OBJECTIVES SUB-COMMITTEE (FOREIGN DOCUMENTS, CENTRAL RADIO BUREAU)
BIOS FD MIRS-	BRITISH INTELLIGENCE OBJECTIVES SUB-COMMITTEE (MILITARY INTELLIGENCE, REPORTS SECTION)
BIOS FR-	BRITISH INTELL. OBJECTIVES SUB-COMM. (FINAL RPTS.)
BIOS GBI-	BRITISH INTELLIGENCE OBJECTIVES SUB-COMMITTEE (GERMAN BAG ISSUE)
BIOS/GP.2/HEC	GT. BRIT. HALSTEAD EXPLOITING CENTRE
BIOS II-	BRITISH INTELLIGENCE OBJECTIVES SUB-COMMITTEE. (INTERIM INVESTIGATION REPORTS)
BIOS IR-	BRITISH INTELLIGENCE OBJECTIVES SUB-COMMITTEE. (INTERROGATION REPORTS)
BIOS IR S-	BRITISH INTELLIGENCE OBJECTIVES SUB-COMMITTEE. (INTERROGATION REPORTS. SUMMARY)
BIOSIS-	BIOSCIENCES INFORMATION SERVICE, PHILADELPHIA
BIOS JAP PR-	BRITISH INTELLIGENCE OBJECTIVES SUB-COMMITTEE. (JAPANESE. PRELIMINARY REPORT)
BIOS M-	BRITISH INTELL. OBJECTIVES SUB-COMM. (MISC. RPTS.)
BIOS-MISC/R-	BRITISH INTELLIGENCE OBJECTIVES SUB-COMMITTEE
BIOS OR-	BRITISH INTELL. OBJECTIVES SUB-COMM. (OVERALL RPT.)
BIOS ORR-	BRITISH INTELLIGENCE OBJECTIVES SUB-COMMITTEE (OVERSEAS RESEARCH REPORTS)
BIOS PR-	BRITISH INTELL. OBJECTIVES SUB-COMM. (PRELIM. RPTS.)
BIOS PR L-	BRITISH INTELLIGENCE OBJECTIVES SUB-COMMITTEE. (PRELIMINARY REPORTS. LISTS)
BIOS R-	BRITISH INTELL. OBJECTIVES SUB-COMM. (FINAL REPORT)
BIOS SO-	BRITISH INTELLIGENCE OBJECTIVES SUB-COMMITTEE
BIOS STT-	BRITISH INTELLIGENCE OBJECTIVES SUBCOMMITTEE. MILITARY COLL. OF SCIENCE. SCH. OF TANK TECHNOLOGY
BIOS TIS-	BRITISH INTELLIGENCE OBJECTIVES SUB-COMMITTEE. TECHNICAL INTELLIGENCE SECTION
BIOS TRIP-	BRITISH INTELL. OBJECTIVES SUB-COMM. (TRIP NO.)
BIR (LTRS)-	BUREAU OF INTELLIGENCE AND RESEARCH, WASHINGTON, D.C.
BIRC PUBLICATION NO.-	CALIFORNIA INST. OF TECH., PASADENA
BIRMINGHAM-	BIRMINGHAM. UNIV., ENGLAND
BIR U-	BIRMINGHAM. UNIV., ENGLAND
BISI-	IRON AND STEEL INSTITUTE, LONDON (TRANSLATIONS)
BISRA-EG/A/(NUMBER/YEAR)	BRITISH IRON AND STEEL RESEARCH ASSN. PLANT ENGINEERING AND ENERGY DIV., LONDON
BISRA-OR/(NO./YR.)	BRITISH IRON AND STEEL RESEARCH ASSN. OPERATIONAL RESEARCH DEPT., LONDON
BISRA-OR/HF/(NO./YR.)	BRITISH IRON AND STEEL RESEARCH ASSN. OPERATIONAL RESEARCH DEPT., LONDON
BISRA-P/(NUMBER/YEAR)	BRITISH IRON AND STEEL RES. ASSN. PHYS. DEPT., LONDON
BISRA-PE/A/(NUMBER/YEAR)	BRITISH IRON AND STEEL RESEARCH ASSN. PLANT ENGINEERING AND ENERGY DIV., LONDON

BISRA-SM/B-

BISRA-SM/B-
  BRITISH IRON AND STEEL RES. ASSN. STEELMAKING DIV., LONDON
BISRA-SM/BE/A/(NO./YR.)
  BRITISH IRON AND STEEL RES. ASSN. STEELMAKING DIV., LONDON
(NUMBER)-BITR
  BELOCK INSTRUMENT CORP., COLLEGE POINT, N.Y.
BIW-
  BOSTON INSULATED WIRE AND CABLE
BJM-
  CALIFORNIA. UNIV., BERKELEY. RADIATION LAB.
BJO-
  BJORKSTEN RESEARCH LABS., INC., MADISON, WIS.
BJ-R-
  COLUMBIA UNIV., NYC. SUBSTITUTE ALLOY MATERIALS LABS.
BJRL-
  BJORKSTEN RESEARCH LABS., INC., MADISON, WIS.
BJSM-
  BRITISH JOINT SERVICES MISSION, WASHINGTON, D.C.
BK-
  CHICAGO. UNIV. METALLURGICAL LAB.
BKC-
  BLAW-KNOX CO., PITTSBURGH
BKC-(NO.)-PE-
  BLAW-KNOX CONSTRUCTION CO., PITTSBURGH
BKI-
  YUGOSLAVIA. INSTITUT ZA NUKLEARNE NAUKE BORIS KIDRIC, BELGRADE
BK-R-
  COLUMBIA UNIV., NYC. SUBSTITUTE ALLOY MATERIALS LABS.
B/L-(NUMBER)-
  BAUSCH AND LOMB, INC., ROCHESTER, N.Y.
B/L-(NUMBER)-IR-
  BAUSCH AND LOMB, INC., ROCHESTER, N.Y.
BL-
  BAUSCH AND LOMB OPTICAL CO., ROCHESTER, N.Y.
BL-
  BOMAC LABS., INC., BEVERLY, MASS.
BL-
  BRUSH LABS. CO., CLEVELAND
BL-
  DU PONT DE NEMOURS (E.I.) + CO. ELASTOMER CHEMICALS DEPT., WILMINGTON, DEL.
BL-(NUMBER) (DATE)
  DU PONT DE NEMOURS (E.I.) + CO. ELASTOMERS DIV., WILMINGTON, DEL.
BL-(NUMBER) (DATE)
  DU PONT DE NEMOURS (E.I.) + CO. RUBBER CHEMICALS DIV., WILMINGTON, DEL.
B/L-
  GT. BRIT. ATOMIC EN. RES. ESTAB., HARWELL, BERKS, ENG
BL-
  GT. BRIT. ROYAL AIRCRAFT EST.,FARNBOROUGH, HANTS, ENG
BL-(3 DIGITS)
  LUNDBERG (BO), BROMMA, SWEDEN
BL-
  STANFORD UNIV., CALIF. BIOPHYSICS LAB.
BL/(NUMBER)/(YEAR)-
  SYLVANIA ELECTRIC PRODUCTS INC., BAYSIDE, N.Y.
BLACK AND VEACH-
  BLACK AND VEACH CONSULTING ENGINEERS, KANSAS CITY, MO
BLC-
  BRUSH LABS. CO., CLEVELAND
BLC-
  NORTHROP AIRCRAFT, INC., HAWTHORNE, CALIF. (BOUNDARY LAYER CONTROL)
BLC-PR-
  BRUSH LABS. CO., CLEVELAND
BLC-TR-
  BRUSH LABS. CO., CLEVELAND
BLEU-
  GT. BRIT. ROYAL AIRCRAFT ESTABLISHMENT. BLIND LANDING EXPERIMENTAL UNIT
BLG-
  BELGIUM. CENTRE D'ETUDE DE L'ENERGIE NUCLEAIRE, BRUSSELS
BLG/B-
  BELGIUM. CENTRE D'ETUDE DE L'ENERGIE NUCLEAIRE, BRUSSELS
BLH-TD-(NUMBERS)-
  BALDWIN-LIMA-HAMILTON CORP., WALTHAM, MASS.
BLI-
  BOMAC LABS., INC., BEVERLY, MASS.
BLIR-
  ARMY CHEMICAL CORPS BIOLOGICAL LABS., CAMP DETRICK, MD. (INTERIM REPORTS)
BL-M-
  COLUMBIA UNIV., NYC. SUBSTITUTE ALLOY MATERIALS LABS.
B-LOC-
  BAUSCH AND LOMB OPTICAL CO., ROCHESTER, N.Y.
BLOC-
  BAUSCH AND LOMB OPTICAL CO., ROCHESTER, N.Y.
BLR-(NUMBER)-
  BELL AEROSPACE CO. DIV. OF TEXTRON,BUFFALO(LAB.REPTS)
BLR-
  BELL AEROSPACE CO. DIV. OF TEXTRON,BUFFALO(LAB.REPTS)
BLR-(YEAR)-(NO.)(LTR.)
  BELL AEROSYSTEMS CO., BUFFALO
BLR-AF-
  TEXACO RESEARCH CENTER, BEACON, N.Y.
B-LRG-
  GT. BRIT. DEPT. OF SCI. + IND. RES. DIRECTORATE OF TUBE ALLOYS (ASSIGNED BY MAJOR GEN. L. R. GROVES)
BLS-
  BUREAU OF LABOR STATISTICS, WASHINGTON, D.C.
BL/TN/
  IMPERIAL COLLEGE OF SCIENCE AND TECHNOLOGY, LONDON. DEPT. OF MECHANICAL ENGINEERING
BL/TN/(LETTER)/(NUMBER)-
  IMPERIAL COLLEGE OF SCIENCE AND TECHNOLOGY, LONDON. DEPT. OF MECHANICAL ENGINEERING
BLW-250-NOTE-
  ATOMIC ENERGY OF CANADA LTD., CHALK RIVER, ONT.
BLWR/N-
  GT. BRIT. ATOMIC EN. RES. ESTAB., HARWELL, BERKS, ENG
BLWT-(NUMBER-YEAR)
  WESTERN ONTARIO UNIV., LONDON. BOUNDARY LAYER WIND TUNNEL LAB.
BM-
  (BRANCH MEMORANDUM) (REF. 2)
B + M-
  BARDWELL AND MC ALISTER, INC., BURBANK, CALIF.
BM-
  BERLIN. HAHN-MEITNER-INSTITUT FUER KERNFORSCHUNG (BERICHT, MATHEMATIK)
BM-(NUMBER-YEAR)-T-
  BRADDOCK, DUNN AND MC DONALD, INC., EL PASO, TEX.
BM-
  BROWN UNIV., PROVIDENCE
BM-
  BUREAU OF MINES
BM-
  GRUMMAN AIRCRAFT ENGINEERING CORP., BETHPAGE, N.Y.
BM-
  LOS ALAMOS SCIENTIFIC LAB., N. MEX. (ASSIGNED BEFORE 1950 TO MISC. BRITISH AND CANADIAN REPORTS)(REF.22)
BM-B-
  BUREAU OF MINES. NW. ELECTRODEV. LAB., ALBANY, ORE.
BM-BULL-(NUMBER)
  BUREAU OF MINES. EXPLOSIVES RES. LAB., PITTSBURGH
BMBW-FBK-(YEAR)-
  GERMANY. BUNDESMINISTERIUM FUER BILDUNG UND WISSENSCHAFT, BONN
BMBW-FB-W-(NO.-NO.)
  GERMANY. BUNDESMINISTERIUM FUER BILDUNG UND WISSENSCHAFT, BONN
BMC-
  BRUBAKER MFG. CO., INC., LOS ANGELES
BMCP-
  HUGHES AIRCRAFT CO., CULVER CITY, CALIF. (BUSINESS MANAGEMENT AND COST PROPOSAL)
BMC/WP-
  GT. BRIT. ATOMIC EN. RES. ESTAB., HARWELL, BERKS, ENG
BMD-
  BALLISTIC MISSILE DIV. (AIR FORCE), INGLEWOOD, CALIF.
BMD (LETTER(S))-
  BALLISTIC MISSILE DIV. (AIR FORCE), INGLEWOOD, CALIF.
BMDD-SAISC-
  TELEDYNE BROWN ENGINEERING. BALLISTIC MISSILE DEFENSE DIV., HUNTSVILLE, ALA.
BMD TN-(YEAR)-
  BALLISTIC MISSILE DIV. (AIR FORCE), INGLEWOOD, CALIF.
BMD TR-(YEAR)-
  BALLISTIC MISSILE DIV. (AIR FORCE), INGLEWOOD, CALIF.
B + MEC-(NO.)-(NO.,LTR)
  BURNS AND MC DONNELL ENGINEERING CO.,KANSAS CITY, MO.
BM/ED-TM-
  BUREAU OF MINES
BM/ED-TN-
  BUREAU OF MINES
BM-EPSD-PR-
  BUREAU OF MINES. EXPLOSIVES AND PHYSICAL SCIENCES DIV., PITTSBURGH
BMET-
  BUREAU OF MINES. DIV. OF EXPLOSIVES TECHNOLOGY, PITTS
BMET-S-
  BUREAU OF MINES. DIV. OF EXPLOSIVES TECHNOLOGY, PITTS
BMI-
  BATTELLE MEMORIAL INST., COLUMBUS, OHIO
BMI-(LETTER)-
  BATTELLE MEMORIAL INST., COLUMBUS, OHIO
BMI-(INITIALS)-
  BATTELLE MEMORIAL INST., COLUMBUS, OHIO
BMI-(INITIALS-LETTER)-
  BATTELLE MEMORIAL INST., COLUMBUS, OHIO
BMI-(INITIALS)-(MEMO)-
  BATTELLE MEMORIAL INST., COLUMBUS, OHIO
BMI-AL-
  BATTELLE MEMORIAL INST., COLUMBUS, OHIO (ACCESSION LIST)
BMI-APDA-(YEAR)-
  ATOMIC POWER DEVELOPMENT ASSOCIATES, INC., DETROIT (JOINT VENTURE)
BMI-APDA-(YEAR)-
  BATTELLE MEMORIAL INST.,COLUMBUS,OHIO (JOINT VENTURE)
BM-IC-
  BUREAU OF MINES (INFORMATION CIRCULARS)
BM-II-
  BUREAU OF MINES. NW. ELECTRODEV. LAB., ALBANY, ORE.
BMI-NLVP-TM-
  BATTELLE MEMORIAL INST., COLUMBUS, OHIO
BMI-NLVP-TR-
  BATTELLE MEMORIAL INST., COLUMBUS, OHIO
BMI-REIC-
  BATTELLE MEMORIAL INST. RADIATION EFFECTS INFORMATION CENTER, COLUMBUS, OHIO
BMI-TML-
  BATTELLE MEMORIAL INST. TITANIUM METALLURGICAL LAB., COLUMBUS, OHIO
BMI-X-
  BATTELLE MEMORIAL INST., COLUMBUS, OHIO
BM-LIV-
  BUREAU OF MINES. EXPLOSIVES RES. LAB., PITTSBURGH
BMN-
  BASEL. UNIVERSITAET
BM-OP-(NUMBER-YEAR)
  BUREAU OF MINES

BM-OPEN FILE-(NO./YR.)	BUREAU OF MINES
BM-OSRD-	BUREAU OF MINES. OIL SHALE RESEARCH BR., LARAMIE, WYO.
BM-PES-	BUREAU OF MINES. PETROLEUM EXPERIMENT STATION, BARTLESVILLE, OKLA.
BMPR-	(BIMONTHLY PROGRESS REPORT) (REF. 2)
BM-PR-	BUREAU OF MINES, SALT LAKE CITY
BM-R-	COLUMBIA UNIV., NYC. SUBSTITUTE ALLOY MATERIALS LABS.
BMR-	HOUGHTON (E.F.) + CO., PHILADELPHIA
BMR-	MICHIGAN. UNIV., ANN ARBOR
BM-RI-	BUREAU OF MINES (REPORTS OF INVESTIGATIONS)
BM-RR-	BUREAU OF MINES, SALT LAKE CITY (RESEARCH REPORT)
BMS-	NATIONAL BUREAU OF STANDARDS, WASHINGTON, D.C.
BMSD-	BENDIX MISSILE SYSTEMS, MISHAWAKA, IND.
BM-SMIB-	BUREAU OF MINES. SPEC. MINERALS INVEST. BR., WASH., D.C
BMS-TII-	MARTIN-MARIETTA CORP. DENVER DIV.
BM-TRANS-	BUREAU OF MINES (TRANSLATIONS)
BMV-	GENERAL DYNAMICS/ASTRONAUTICS, SAN DIEGO, CALIF.
BMVG-FBWT-(YEAR)-	GERMANY. BUNDESMINISTERIUM FUER VERTEIDIGUNG, BONN
BMVTDG-FBWT-(YEAR)-	AVIATEST G.M.B.H., DUESSELDORF
BMW/ET/(NO.)/(YR)	BAYERISCHE MOTOREN WERKE, A. G., SPANDAU, GERMANY
BMW EZV-	BAYERISCHE MOTOREN WERKE. ENTWICKLUNG ZENTRAL-VERWALTUNG
BMWF-	GERMANY. BUNDESMINISTERIUM FUER WISSENSCHAFTLICHE FORSCHUNG, BAD GODESBERG
BMWF-FBK-(YEAR)-	GERMANY. BUNDESMINISTERIUM FUER WISSENSCHAFTLICHE FORSCHUNG, BAD GODESBERG
BMWF-FB-W-(YEAR)-	GERMANY. BUNDESMINISTERIUM FUER WISSENSCHAFTLICHE FORSCHUNG, BAD GODESBERG
BN-	COLUMBIA UNIV., NYC. SUBSTITUTE ALLOY MATERIALS LABS.
BN-	MARYLAND. UNIV., COLLEGE PARK. INST. FOR FLUID DYNAMICS AND APPLIED MATHEMATICS
BN-(4 DIGITS)-	SOCIETE BELGE POUR L'INDUSTRIE NUCLEAIRE, BRUSSELS
BNDC/R-	BRITISH NUCLEAR DESIGN AND CONSTRUCTIONS LTD., WHETSTONE, LEICS, ENGLAND
B-NDV(NUMBER)	BERLIN. HAHN-MEITNER-INSTITUT FUER KERNFORSCHUNG
BNE-(NUMBER)-	CHRYSLER CORP. MISSILE DIV., DETROIT
BNFMRA-	BRITISH NON-FERROUS METALS RESEARCH ASSN., LONDON
BNJV-(LTR.)-(LTR.)-	BOEING CO., SEATTLE (JOINT VENTURE)
BNJV-(LTR.)-(LTR.)-	NORTH AMERICAN AVIATION, INC., LOS ANG.(JT. VENTURE)
BNL-	BROOKHAVEN NATIONAL LAB., UPTON, N.Y.
BNL-(INITIALS)-	BROOKHAVEN NATIONAL LAB., UPTON, N.Y. (AUTHORS INTLS)
BNL-A	BROOKHAVEN NATIONAL LAB., UPTON, N.Y.
BNL-AS-	BROOKHAVEN NATIONAL LAB., UPTON, N.Y.
BNL-C-	BROOKHAVEN NATIONAL LAB., UPTON, N.Y. (CONFERENCES)
BNL-EDC-	BROOKHAVEN NATIONAL LAB., UPTON, N.Y.
BNL-I-	BROOKHAVEN NATIONAL LAB., UPTON, N.Y.
BNL-L-	BROOKHAVEN NATIONAL LAB., UPTON, N.Y.
BNL-LMFR-(DATE)	BROOKHAVEN NATIONAL LAB., UPTON, N.Y. (LIQUID METAL FUEL REACTORS)
BNL-LOG-	BROOKHAVEN NATIONAL LAB., UPTON, N.Y.
BNL-N-	BROOKHAVEN NATIONAL LAB., UPTON, N.Y. (NEUTRON CROSS SECTION EVALUATION GROUP NEWSLETTER)
BNL-P-	BROOKHAVEN NATIONAL LAB., UPTON, N.Y.
BNL-PD-	BROOKHAVEN NATIONAL LAB., UPTON, N.Y.
BNL-PR-	BROOKHAVEN NATIONAL LAB., UPTON, N.Y. (PROGRESS REVIEW)
BNL-RP-	BROOKHAVEN NATIONAL LAB., UPTON, N.Y. (REACTOR PHYS)
BNL-S-	BROOKHAVEN NATIONAL LAB., UPTON, N.Y. (SCIENTIFIC PROGRESS REPORT)
BNL-T-	BROOKHAVEN NATIONAL LAB., UPTON, N.Y.
BNL-TR-	BROOKHAVEN NATIONAL LAB., UPTON, N.Y. (TRANSLATION)
BN-R-	COLUMBIA UNIV., NYC. SUBSTITUTE ALLOY MATERIALS LABS.
BNSA-	BATTELLE-NORTHWEST. PACIFIC NORTHWEST LAB., RICHLAND, WASH.
BNTR-	BATTELLE-NORTHWEST. PACIFIC NORTHWEST LAB., RICHLAND, WASH.
BNW-	BUREAU OF NAVAL WEAPONS
BNW-	MC DONNELL AIRCRAFT CORP., ST. LOUIS
BNW-AL-	BUREAU OF NAVAL WEAPONS
BNWC-	BATTELLE-NORTHWEST. PACIFIC NORTHWEST LAB., RICHLAND, WASH.
BNWL-	BATTELLE-NORTHWEST. PACIFIC NORTHWEST LAB., RICHLAND, WASH. (FORMAL R + D REPORT)
BNWL-B-	BATTELLE-NORTHWEST. PACIFIC NORTHWEST LAB., RICHLAND, WASH.
BNWL-C-	BATTELLE-NORTHWEST. PACIFIC NORTHWEST LAB., RICHLAND, WASH.
BNWL-CC-	BATTELLE-NORTHWEST. PACIFIC NORTHWEST LAB., RICHLAND, WASH. (INFORMAL R + D REPORT)
BNWL-IR-	BATTELLE-NORTHWEST. PACIFIC NORTHWEST LAB., RICHLAND, WASH. (INVENTION)
BNWL-MA-	BATTELLE-NORTHWEST. PACIFIC NORTHWEST LAB., RICHLAND, WASH. (MANUAL)
BNWL-SA-	BATTELLE-NORTHWEST. PACIFIC NORTHWEST LAB., RICHLAND, WASH. (INTENDED FOR JOURNAL PUBLICATION)
BNWL-TR-	BATTELLE-NORTHWEST. PACIFIC NORTHWEST LAB., RICHLAND, WASH. (TRANSLATION)
BOAC-	BOEING CO., SEATTLE
BOAC AR FT-	BOEING AIRPLANE CO., SEATTLE (FLIGHT TEST ANALYSIS REPORT)
BOAC BPR-	BOEING AIRPLANE CO., SEATTLE (BOMARC PROGRESS RPT.)
BOB-(NUMBER-NUMBER)	BUREAU OF THE BUDGET, WASHINGTON, D.C.
BOE-	BOEING AIRPLANE CO., SEATTLE
BOE-D-	BOEING AIRPLANE CO., SEATTLE
BOE-D2-	BOEING AIRPLANE CO., SEATTLE
BOE-D5-	BOEING AIRPLANE CO., SEATTLE
BOE-TD-TOOL DOCUMENT-	BOEING AIRPLANE CO., SEATTLE
BOE-T-TEST-	BOEING AIRPLANE CO., SEATTLE
BOE-WD-	BOEING AIRPLANE CO. WICHITA DIV., KAN.
BOHAC-	BUREAU OF ORDNANCE (NAVY). HYDROBALLISTICS ADVISORY COMMITTEE, WASHINGTON, D. C.
BOHNA-INV-	BOHNA (B.D.) AND CO., INC., SAN FRANCISCO
BOI-	BUREAU OF ORDNANCE (NAVY)
BOLDOW-RF-	BOELKOW ENTWICKLUNGEN K.G., MUNICH
BOLKOW-DSP-	BOELKOW ENTWICKLUNGEN K.G. FUTURE PROJS. DEPT., MUNICH
BOLT B+N-	BOLT, BERANEK, AND NEWMAN, INC., CAMBRIDGE, MASS.
BOM-	BOMAC LABS., INC., BEVERLY, MASS.
BOM-	BUREAU OF MINES
BON-	AIRCRAFT OBSERVER RESEARCH LAB., MATHER AFB, CALIF.
BONNU-PI-	BONN. UNIVERSITAET. PHYSIKALISCHES INSTITUT
BOOK(NUMBER)	AEROJET-GENERAL CORP. SOLID ROCKET PLANT, SACRAMENTO, CALIF (TRAINING MANUAL)
BORDEN-ASR (INCL.DATES)	BORDEN CHEMICAL CO., PHILADELPHIA
BORDEN-PR-	BORDEN CHEMICAL CO., PHILADELPHIA
BORDEN-QPR-	BORDEN CHEMICAL CO., PHILADELPHIA
BOS-	TENNESSEE EASTMAN CORP., OAK RIDGE, TENN.
BOSO-	BUREAU OF ORDNANCE (NAVY) (SHIPMENT ORDER)
BOS-P-	TENNESSEE EASTMAN CORP., OAK RIDGE, TENN.
BOSR-(NUMBER)-	CIVIL AERONAUTICS BOARD, WASHINGTON, D.C.
BOSU-	BOSTON UNIV.
BOSU-ORL-PR-	BOSTON UNIV. OPTICAL RESEARCH LAB.
BOSU-QPR-	BOSTON UNIV.
BOSU-TN-	BOSTON UNIV.

BOTLO-
: BUREAU OF ORDNANCE (NAVY). TECHNICAL LIAISON OFFICE

BOU-
: BOSTON UNIV.

BOU-TN-
: BOSTON UNIV.

BOYNTON-
: BOYNTON ASSOCIATES, LA CANADA, CALIF.

BOYNTON (LTRS,NOS.)
: BOYNTON ASSOCIATES, LA CANADA, CALIF.

BP-
: BECHTEL CORP., SAN FRANCISCO

BP-
: BERLIN. HAHN-MEITNER-INSTITUT FUER KERNFORSCHUNG (BERICHT, PHYSIK)

BP-
: CALIFORNIA. UNIV., BERKELEY. CROCKER LAB.

BP-
: CALIFORNIA. UNIV., BERKELEY. RADIATION LAB. (REF. 9)

BP-
: DU PONT DE NEMOURS (E.I.) + CO., WILMINGTON, DEL.

BP-5001 THRU -5256
: GT. BRIT. DEPT. OF ATOMIC ENERGY. INDUSTRIAL GROUP H.Q., RISLEY, LANCS, ENGLAND (REF. 36)

BP-
: PACIFIC GAS AND ELECTRIC CO., SAN FRANCISCO

BP-
: QUARTERMASTER RESEARCH AND ENGINEERING COMMAND, NATICK, MASS. (BIOPHYSICS)

BPAD-
: BENDIX CORP. BENDIX PRODUCTS AEROSPACE DIV., SOUTH BEND, IND.

BP-AT-
: WESTINGHOUSE ELECTRIC CORP. ATOMIC POWER DEPT.,PITTS.

BPA-TRANS-
: BONNEVILLE POWER ADMINISTRATION, PORTLAND, ORE.

BPC-
: BERMITE POWDER CO., SAUGUS, CALIF.

BPC-
: GT. BRIT. MILITARY PERSONNEL RESEARCH COMMITTEE

BPD-
: BENDIX AVIATION CORP. BENDIX PRODUCTS DIV., MISHAWAKA, IND.

BPE-
: AIR FORCE, WASHINGTON, D.C.

BPI-
: BROOKLYN. POLYTECHNIC INST.

BPI-
: TENNESSEE EASTMAN CORP., OAK RIDGE, TENN.

BPR-
: (BI-WEEKLY PROGRESS REPORT) (REF. 2)

BPR-
: (BIMONTHLY PROGRESS REPORT) (REF. 2)

BPR-
: BUREAU OF PUBLIC ROADS, WASHINGTON, D.C.

BPR-(ROMAN NO.)-
: COLUMBIA UNIV., N.Y.C. ENGINEERING RESEARCH LABS.

BPR-(NUMBER)-
: POLAND. NUCLEAR ENERGY INFORMATION CENTER, WARSAW

BPR-
: VETERANS ADMINISTRATION. PROSTHETIC AND SENSORY AIDS SERVICE, WASHINGTON, D.C.

BPRC-PPS-
: BUREAU OF MINES. BARTLESVILLE PETROLEUM RESEARCH CENTER, OKLA.

BPRC-QPR-
: BUREAU OF MINES. BARTLESVILLE PETROLEUM RESEARCH CENTER, OKLA.

B.P.S.-
: GOODYEAR AIRCRAFT CORP.

BPSN-
: GENERAL ELECTRIC CO. RESEARCH LAB., SCHENECTADY, N.Y.

BPWP/P-
: UNITED KINGDOM ATOMIC ENERGY AUTHORITY. RESEARCH GP. ATOMIC ENERGY RES. ESTAB., HARWELL, BERKS, ENGLAND

BPX-
: AIR FORCE, WASHINGTON, D.C.

BR-
: (BRANCH REPORT) (REF. 2)

BR5-
: AUTONETICS DIV., NORTH AMERICAN AVIATION, INC., ANAHEIM, CALIF.

BR-
: BEECH AIRCRAFT CORP., BOULDER, COLO.

BR-
: BROOKHAVEN NATIONAL LAB., UPTON, N.Y.

BR-
: BRUNSWICK CORP. DEFENSE PRODUCTS DIV., MARION, VA.

BR-
: BUREAU OF ORDNANCE (NAVY) (BUMBLEBEE PROJECT)

BR-
: BUREAU OF RECLAMATION (INTERIOR). COMMISSIONERS OFFICE, DENVER (TRANSLATIONS)

BR-
: CHICAGO. UNIV. METALLURGICAL LAB. (ASSIGNED TO REPORTS FROM VARIOUS BRITISH SOURCES)

BR-
: COMMITTEE ON MEDICAL RESEARCH. (BRITISH REPORTS)

BR-
: GT. BRIT. DEPT. OF ATOMIC ENERGY

BR-
: GT. BRIT. ROAD RES. LAB., HARMONDSWORTH, MIDDX., ENG.

BR-
: NEW YORK UNIV., N.Y.C.

BR-
: NEW YORK UNIV., N.Y.C. INST. OF MATHEMATICAL SCIENCES

BR-
: OFFICE OF NAVAL RESEARCH (BASIC RESEARCH)

BR-
: RAYTHEON CO. (ALL LOCATIONS)

BR-(NUMBER/NUMBER(S))
: SYSTEM DEVELOPMENT CORP., SANTA MONICA, CALIF. (BROCHURE)

BR-(JOB NO.).(TASK NO.)-
: VITRO LABS., SILVER SPRING, MD.

BR-
: WATERLOO. UNIV., ONTARIO

BR-(LTRS.-NO./NO(S).)
: SYSTEM DEVELOPMENT CORP. (LTRS. GIVE LOCATION)

BR-A-
: ASSOCIATED ELECTRICAL INDUSTRIES, LTD., ALDERMASTON, BERKS, ENGLAND

BRA-(NUMBERS)-(YEAR)
: CHRYSLER CORP. MISSILE DIV., DETROIT

BRAB CR-
: NATIONAL RESEARCH COUNCIL. BUILDING RESEARCH ADVISORY BOARD (CONFERENCE REPORT)

BRANCH MEMO B-
: GT. BRIT. ARMAMENT RESEARCH AND DEVELOPMENT ESTABLISHMENT, FORT HALSTEAD, KENT, ENGLAND

BRB-
: BRIDGEPORT BRASS CO., CONN.

BRB-
: GT. BRIT. DEPT. OF SCIENTIFIC AND INDUSTRIAL RESEARCH. (BUILDING RESEARCH BULLETINS)

BRB-
: REACTIVE METALS, INC., SEYMOUR, CONN.

BRC-
: BORAX RESEARCH CORP., ANAHEIM, CALIF.

BRC-
: BROADVIEW RESEARCH CORP., BURLINGAME, CALIF.

BRC-IB-
: BENDIX RADIO CORP. (INSTRUCTION BOOKS)

BRD-
: BARTOL RESEARCH FOUNDATION, SWARTHMORE, PENNA.

BRD-
: BROADVIEW RESEARCH AND DEVELOPMENT, BURLINGAME, CAL.

BRDL-
: BULOVA RESEARCH + DEV. LABS., INC., WOODSIDE, N.Y.

BRDL (LETTERS)-
: BULOVA RESEARCH + DEV. LABS., INC., WOODSIDE, N.Y.

BREA-(NUMBER/NUMBER(S))
: SYSTEM DEVELOPMENT CORP., SANTA MONICA, CALIF. (BROCHURE, EXTERNAL ADMINISTRATION)

BREA-(LTRS.-NO./NO(S).)
: SYSTEM DEVELOPMENT CORP. (LTRS. GIVE LOCATION)

BR-ERA-G/T-
: BRITISH ELECTRICAL AND ALLIED INDUSTRIES RESEARCH ASSN., LONDON

BRF-
: BARTOL RESEARCH FOUNDATION, SWARTHMORE, PENNA.

BRG-(NUMBER/NUMBER(S))
: SYSTEM DEVELOPMENT CORP., SANTA MONICA, CALIF. (BROCHURE, GENERAL)

BRG-(LTRS.-NO./NO(S).)
: SYSTEM DEVELOPMENT CORP. (LTRS. GIVE LOCATION)

BRH/CFS-(YEAR)-
: BUREAU OF RADIOLOGICAL HEALTH, ROCKVILLE, MD.

BRH/DBE-(YEAR)-
: BUREAU OF RADIOLOGICAL HEALTH, ROCKVILLE, MD.

BRH/DEP-(YEAR)-
: BUREAU OF RADIOLOGICAL HEALTH. DIV. OF ELECTRONIC PRODUCTS, ROCKVILLE, MD.

BRH/DER-(YEAR)-
: BUREAU OF RADIOLOGICAL HEALTH. DIV. OF ENVIRONMENTAL RADIATION, ROCKVILLE, MD.

BRH/DMRE-(YEAR)-
: BUREAU OF RADIOLOGICAL HEALTH. DIV. OF MEDICAL RADIATION EXPOSURE, ROCKVILLE, MD.

BRH/NERHL-(YEAR)-
: BUREAU OF RADIOLOGICAL HEALTH. NORTHEASTERN RADIOLOGICAL HEALTH LAB., WINCHESTER, MASS.

BRH/OBD-(YEAR)-
: BUREAU OF RADIOLOGICAL HEALTH, ROCKVILLE, MD.

BRH/OCS-(YEAR)-
: BUREAU OF RADIOLOGICAL HEALTH, ROCKVILLE, MD.

BRH/ORO-(YEAR)-
: BUREAU OF RADIOLOGICAL HEALTH, ROCKVILLE, MD.

BRH/SERHL-(YEAR)-
: BUREAU OF RADIOLOGICAL HEALTH. SOUTHEASTERN RADIOLOGICAL HEALTH LAB., MONTGOMERY, ALA.

BRH/SWRHL-(YEAR)-
: BUREAU OF RADIOLOGICAL HEALTH. SOUTHWESTERN RADIOLOGICAL HEALTH LAB., LAS VEGAS, NEV.

BRI
: BUREAU OF RETIREMENT, INSURANCE AND OCCUPATIONAL HEALTH (CIVIL SERVICE), WASHINGTON, D.C.

BRIA-(NUMBER/NUMBER(S))
: SYSTEM DEVELOPMENT CORP., SANTA MONICA, CALIF. (BROCHURE, INTERNAL ADMINISTRATION)

BRIA-(LTRS.-NO./NO(S).)
: SYSTEM DEVELOPMENT CORP. (LTRS. GIVE LOCATION)

BRJ-
: GENERAL DYNAMICS/ASTRONAUTICS, SAN DIEGO, CALIF.

BRL-
: BALLISTIC RESEARCH LABS.,ABERDEEN PROVING GROUND, MD.

BRL-
: BENDIX CORP. RESEARCH LABS. DIV., SOUTHFIELD, MICH.

BRL-(NO.-LTR.,NO.)
: PARSONS (RALPH M.) CO., PASADENA, CALIF.

BRL-(LETTERS, DATE)
: BJORKSTEN RESEARCH LABS., INC., MADISON, WIS.

BRL-AF-(CONTRACT NO.)
: BALLISTIC RESEARCH LABS.,ABERDEEN PROVING GROUND, MD.

BRL ANNUAL/
: BALLISTIC RESEARCH LABS.,ABERDEEN PROVING GROUND, MD.

BRL BT (NO.,LTRS.)
: BALLISTIC RESEARCH LABS.,ABERDEEN PROVING GROUND, MD.

BRL-CR-
: BALLISTIC RESEARCH LABS.,ABERDEEN PROVING GROUND, MD.

BRLD-	BENDIX CORP. RESEARCH LABS. DIV., SOUTHFIELD, MICH.
BRL DATA RPT.-	BALLISTIC RESEARCH LABS., ABERDEEN PROVING GROUND, MD.
BRL-DR-	BALLISTIC RESEARCH LABS. EXPLOSION KINETICS BRANCH, ABERDEEN PROVING GROUND, MD.
BRL-IBL STATUS-	BALLISTIC RESEARCH LABS., ABERDEEN PROVING GROUND, MD.
BRL-M-	BALLISTIC RESEARCH LABS., ABERDEEN PROVING GROUND, MD.
BRLM-	BALLISTIC RESEARCH LABS., ABERDEEN PROVING GROUND, MD. (MEMORANDUM REPORTS)
BRL-MEMO-	BALLISTIC RESEARCH LABS., ABERDEEN PROVING GROUND, MD.
BRL-MR-	BALLISTIC RESEARCH LABS., ABERDEEN PROVING GROUND, MD. (MEMORANDUM REPORT)
BRL-N-(NUMBERS)	BJORKSTEN RESEARCH LABS., INC., MADISON, WIS.
BRL-OCRR-	BALLISTIC RESEARCH LABS., ABERDEEN PROVING GROUND, MD. (ORDNANCE COMPUTER RES. RPT.)
BRL QPR (NO., LTRS., NOS.)	BALLISTIC RESEARCH LABS., ABERDEEN PROVING GROUND, MD.
BRL-R-	BALLISTIC RESEARCH LABS., ABERDEEN PROVING GROUND, MD. (REPORTS)
BRLR-	BALLISTIC RESEARCH LABS., ABERDEEN PROVING GROUND, MD. (REPORTS)
BRL-TB-	BALLISTIC RESEARCH LABS., ABERDEEN PROVING GROUND, MD.
BRL-TN-	BALLISTIC RESEARCH LABS., ABERDEEN PROVING GROUND, MD. (TECHNICAL NOTES)
BRL-X-	BALLISTIC RESEARCH LABS., ABERDEEN PROVING GROUND, MD.
BR-M-	COLUMBIA UNIV., NYC. SUBSTITUTE ALLOY MATERIALS LABS.
BRN-	BROWN UNIV., PROVIDENCE
BRN-(NUMBER)-P	BROWN UNIV., PROVIDENCE (PROJECT SQUID)
BRN-	CORNELL UNIV., ITHACA, N.Y.
BRN-	NATIONAL RESEARCH COUNCIL OF CANADA. DIV. OF BUILDING RESEARCH, OTTAWA (BLDG. RES. NEWS)
BR-OEO-LN-	BARSS, REITZEL AND ASSOCIATES, INC., CAMBRIDGE, MASS.
BROL-	BROWNING LABS., INC., WINCHESTER, MASS.
BROWN-	BROWN UNIV., PROVIDENCE
BROWN-U-	BROWN UNIV., PROVIDENCE
BROWN U. TR-	BROWN UNIV., PROVIDENCE
BRP-	ATOMIC ENERGY OF CANADA LTD. CHALK RIVER PROJ., ONT.
BRP-	BENDIX CORP. BENDIX RADIO DIV., BALTIMORE, MD.
BRP-	BERLIN. HAHN-MEITNER-INSTITUT FUER KERNFORSCHUNG (BERICHT, REAKTOR)
BRR-	(BRITISH RESEARCH REPORTS)
BR-R-	COLUMBIA UNIV., NYC. SUBSTITUTE ALLOY MATERIALS LABS.
BRR-	NAVAL ORDNANCE LAB., WHITE OAK, MD.
BRS-	GT. BRIT. BLDG. RES. STA., GARSTON, HERTS, ENGLAND
BR-S-	RAYTHEON CO. MISSILE SYSTEMS DIV., BEDFORD, MASS.
BRS-DR-	GT. BRIT. BLDG. RES. STA., GARSTON, HERTS, ENGLAND
BRT-(NUMBER/NUMBER(S))	SYSTEM DEVELOPMENT CORP., SANTA MONICA, CALIF. (BROCHURE, TECHNICAL)
BRT-(LTRS.-NO./NO(S).)	SYSTEM DEVELOPMENT CORP. (LTRS. GIVE LOCATION)
100-BR- THRU 100-ZS-	COLUMBIA UNIV., N.Y.C. SUBSTITUTE ALLOY MATERIALS LAB. (REF. 11)
BR TP-	GT. BRIT. DEPT. OF SCIENTIFIC AND INDUSTRIAL RESEARCH. (BUILDING RESEARCH TECHNICAL PAPERS)
BS-	BELGIUM. SOCIETE D'ETUDES DE RECHERCHES ET D'APPLICATIONS POUR L'INDUSTRIE, BRUSSELS
BS-	BERLIN. HAHN-MEITNER-INSTITUT FUER KERNFORSCHUNG (BERICHT, STRAHLENCHEMIE)
BS-	BRITISH STANDARDS INSTITUTION
BS-	COAST AND GEODETIC SURVEY, WASHINGTON, D.C. (BOTTOM SEDIMENT CHART)
BS-	COMMITTEE ON MEDICAL RES. (NATL. BUR. OF STDS. RPTS.)
BS-	DUBLIN INSTITUTE FOR ADVANCED STUDIES
BS-	EUROPEAN COUNCIL FOR NUCLEAR RESEARCH, GENEVA
BS-	LILIENTHAL-GESELLSCHAFT FUER LUFTFAHRTFORSCHUNG, BERLIN (BERICHT S)
BS/A/(NUMBER-NUMBER)	LIVERPOOL. UNIV. DEPT. OF BUILDING SCIENCE
BSAL-	SANDIA CORP., ALBUQUERQUE, N. MEX. (BASE SPARES ALLOWANCE LIST)
BSC-(NO.)-	BENDIX AVIATION CORP. PIONEER-CENTRAL DIV., DAVENPORT, IOWA
BSC-	BENDIX CORP. BENDIX SYSTEMS DIV., ANN ARBOR, MICH.
BSC-(CONTRACT)-	BETHLEHEM STEEL CO. SHIPBUILDING DIV., QUINCY, MASS.
BSC-(NO.)-	BETHLEHEM STEEL CO. SHIPBUILDING DIV., QUINCY, MASS.
BSCP-(NUMBER-NUMBER)	GEORGE WASHINGTON UNIV., WASHINGTON, D.C. (BIOLOGICAL COMMUNICATIONS PROJECT)
(YEAR)-BSCPS-	BALLISTIC SYSTEMS DIV., NORTON AFB, CALIF.
BSCS-PAMPHLETS-	FLORIDA STATE UNIV., TALLAHASSEE. INSTITUTE OF MOLECULAR BIOPHYSICS
BSD-	BALLISTIC SYSTEMS DIV., NORTON AFB, CALIF.
BSD-	BENDIX AVIATION CORP. BENDIX SYSTEMS DIV., ANN ARBOR, MICH.
BSD(YEAR)-	DUNLAP AND ASSOCIATES, INC., DARIEN, CONN.
BSD-BSSFR-	BALLISTIC SYSTEMS DIV., NORTON AFB, CALIF.
BSD-CR-(YEAR)-	BALLISTIC SYSTEMS DIV. (AIR FORCE), INGLEWOOD, CALIF. (CONTRACT REPORT)
BSD-EXHIBIT-(NO.)-	BALLISTIC SYSTEMS DIV., NORTON AFB, CALIF.
BSD-TDR-(YEAR)-	BALLISTIC SYSTEMS DIV., NORTON AFB, CALIF.
BSD-TR-(YEAR)-	BALLISTIC SYSTEMS DIV., NORTON AFB, CALIF.
BSDU-	GT. BRIT. ROYAL AIR FORCE. BOMBER SUPPORT DEV. UNIT
BSI-BS-	BRITISH STANDARDS INSTITUTION
BSP-	BERLIN. HAHN-MEITNER-INSTITUT FUER KERNFORSCHUNG (BERICHT, STRAHLENPHYSIK)
BSP-IS-(NUMBER-NUMBER)	WISCONSIN. BUREAU OF STATE PLANNING, MADISON (INFORMATION SYSTEM)
BSR-	BENDIX CORP. AEROSPACE SYSTEMS DIV., ANN ARBOR, MICH.
BSR-	BENDIX CORP. BENDIX SYSTEMS DIV., ANN ARBOR, MICH.
BSR-	COMMITTEE ON MEDICAL RES. (BLOOD SUBSTITUTES RPTS.)
BSR-	MASSACHUSETTS INST. OF TECH., CAMBRIDGE
BSRA/	BRITISH SHIPBUILDING RESEARCH ASSN., ATOMIC ENERGY RESEARCH ESTABLISHMENT, WINFRITH, DORSET, ENGLAND
BS-S-	BETHLEHEM STEEL CO., PITTSBURGH
BS-S-	BETHLEHEM STEEL CO. SHIPBUILDING DIV., QUINCY, MASS.
BSS-	NATIONAL BUREAU OF STANDARDS. BUILDING RESEARCH DIV., WASHINGTON, D.C. (BUILDING SCIENCE SERIES)
BSSR-	AIR WEATHER SERVICE, MAC DILL AFB., FLA.
BSSR-	BUREAU OF SOCIAL SCIENCE RESEARCH, INC., WASH., D.C.
BS TC-	AIR WEATHER SERVICE, MAC DILL AFB., FLA.
BSTN-	AIR WEATHER SERVICE, MAC DILL AFB., FLA.
BSWM-SW-	BUREAU OF SOLID WASTE MANAGEMENT, ROCKVILLE, MD.
BT-	BOYCE THOMPSON INST. FOR PLANT RES., INC., YONKERS, N.Y.
BT-	BROWN UNIV., PROVIDENCE
BT-	MARYLAND. UNIV., COLLEGE PARK
BT-	WESTINGHOUSE ELECTRIC CORP. (BETTIS TECHNICAL REVIEW)
BTG-	NORTHROP AIRCRAFT, INC., HAWTHORNE, CALIF.
BTG-	TENNESSEE EASTMAN CORP., OAK RIDGE, TENN.
BTGR-	FRANCE. CENTRE DE RECHERCHES ET D'EXPERIMENTATION DE GENIE RURAL, ANTONY
BTI-	ARMED SERVICES TECHNICAL INFO. AGENCY, ARLINGTON, VA. (BIBLIOGRAPHICAL TECHNICAL INDEX)
BTI-	BIOTECHNOLOGY, INC., ARLINGTON, VA.
BTI-(YEAR)-	BIOTECHNOLOGY, INC., ARLINGTON, VA.
BTI-	BROOKLYN. POLYTECHNIC INST.
BTI-ARC-	ARMED SERVICES TECHNICAL INFORMATION AGENCY. REFERENCE CTR., LIBRARY OF CONGRESS, WASH., D.C.
BTL-	BALLISTIC RESEARCH LABS., ABERDEEN PROVING GROUND, MD.
BTL-	BELL TELEPHONE LABS. (ALL LOCATIONS)

```
BTL-(NUMBER)-(LETTER)
         BELL TELEPHONE LABS., INC., N.Y.C.
BTL-(NO.)(LTR(S))(NO(S))
         BELL TELEPHONE LABS., INC., WHIPPANY, N.J.
BTL MEMO CASE (NO.-NO.)
         BELL TELEPHONE LABS., INC., WHIPPANY, N.J.
BTL/(MISSILE NAME)
         BELL TELEPHONE LABS., INC., WHIPPANY, N.J.
BTL-MM-(YEAR)-
         BELL TELEPHONE LABS., INC., N.Y.C.
BTL-PCP-(YEAR)-
         BELL TELEPHONE LABS., INC., WHIPPANY, N.J.
BTL-QR-
         BELL TELEPHONE LABS., INC., WHIPPANY, N.J.
BTL-TRANS-(INITIAL)-
         BELL TELEPHONE LABS., INC., N.Y.C. (TRANSLATION)
BTN-
         (BULLETIN) (REF. 2)
BTN-
         BUREAU OF ORDNANCE (NAVY) (BALLISTIC TECHNICAL NOTE)
BTN-
         CIVIL SERVICE COMMISSION, WASHINGTON, D.C. (BULLETIN)
BT-R-
         COLUMBIA UNIV., N.Y.C.
BTR-
         THORN-AEI RADIO VALVES AND TUBES, LTD., ROCHESTER, ENG
BTSI-
         BASE TEN SYSTEMS, INC., MONMOUTH JUNCTION, N. J.
BTU-
         BUTLER UNIV., INDIANAPOLIS
BU-
         BOSTON UNIV.
BU-
         BROWN UNIV., PROVIDENCE
BU-
         CORNELL UNIV., ITHACA, N.Y. DEPT. OF PLANT BREEDING
           (BIOMETRICS UNIT)
BU-(LETTER)11-
         BROWN UNIV., PROVIDENCE. GRAD. DIV. OF APPLIED MATH.
BU-A9-T-
         BROWN UNIV., PROVIDENCE. GRADUATE DIVISION OF APPLIED
           MATHEMATICS (TRANSLATION)
BUA-
         BUREAU OF AERONAUTICS (NAVY)
BUAER-
         BUREAU OF AERONAUTICS (NAVY)
BU AER (LETTERS)-
         BUREAU OF AERONAUTICS (NAVY)
BU-ATR-
         BROWN UNIV., PROVIDENCE
BUA-XEL-
         BUREAU OF AERONAUTICS (NAVY) (SPECIFICATIONS)
BU-B-
         BROWN UNIV., PROVIDENCE
BUC-
         BELLARMINE-URSULINE COLL., LOUISVILLE, KY.
BUCI-
         BRUSSELS. UNIVERSITE. INSTITUT DU CANCER
BU DOCKS-
         BUREAU OF YARDS AND DOCKS
BUDOCKS-
         BUREAU OF YARDS AND DOCKS
BU-DR-
         BUREAU OF AERONAUTICS (NAVY)
BU/DTMB/
         BROWN UNIV., PROVIDENCE. DIV. OF ENGINEERING
BU DWG-(NOS.,LTRS,NOS.)
         BOSTON UNIV. (DRAWING NO.)
BU-E-
         BROWN UNIV., PROVIDENCE. DEPT. OF ENGINEERING
BUFFU-
         BUFFALO. UNIV.
BUFFU-PR-
         BUFFALO. UNIV. (PROGRESS REPORT)
BU GDAM-
         BROWN UNIV., PROVIDENCE. GRADUATE DIVISION OF
           APPLIED MATHEMATICS (TRANSLATIONS AND REPORTS)
BULL-
         (BULLETIN) (REF. 2)
BULL-(NUMBER)/
         BELGIUM. ROYAL OBSERVATORY, UCCLE
BULL-
         CLEMSON UNIV., S.C. ENGINEERING EXPERIMENT STATION
BULL-(4 DIGITS)-F
         GEOLOGICAL SURVEY, WASHINGTON, D.C.
BULL-(NUMBER)-PT-
         SHOCK AND VIBRATION INFORMATION CENTER (DEFENSE),
           WASHINGTON, D.C.
BULL-A-
         KODAIKANAL OBSERVATORY, INDIA
BULLETIN HEL-
         LITTLE (ARTHUR D.), INC., CAMBRIDGE, MASS. (HELIUM)
BULLETIN NO.-
         OFFICE OF THE DIRECTOR OF DEFENSE (RES+ENG),WASH.,DC
BULLETIN NOTE NO.-
         BRUSSELS. UNIVERSITE. CENTRE DE PHYSIQUE NUCLEAIRE
BULL-GT-
         LOUISIANA STATE UNIV., BATON ROUGE (JOINT VENTURE)
BULL-GT-
         LOUISIANA WATER RESOURCES RESEARCH INST., BATON ROUGE
           (JOINT VENTURE)
BULL. NO.-
         PUERTO RICO. UNIV., RIO PIEDRAS. COSMIC RAY LAB.
BULL. NOTE-
         BRUSSELS. UNIVERSITE. CENTRE DE PHYSIQUE NUCLEAIRE
BUM-
         BUREAU OF MINES, PITTSBURGH
BUM-(NUMBER)-P
         BUREAU OF MINES, PITTSBURGH (SQUID PROJECT) (REF. 33)

BUMBLEBEE-
         BUREAU OF ORDNANCE (NAVY) (BUMBLEBEE PROJ.) (REF. 8)
BUMBLEBEE-
         CORNELL AERONAUTICAL LAB., INC., BUFFALO
BUMBLEBEE-
         ESSO RESEARCH AND ENGINEERING CO., LINDEN, N.J.
           (BUMBLEBEE PROJECT) (REF. 8)
BUMBLEBEE-
         JOHNS HOPKINS UNIV., BALTIMORE. DEPT. OF AERONAUTICS
BUMBLEBEE-
         JOHNS HOPKINS UNIV., SILVER SPRING, MD. APPLIED
           PHYSICS LAB. (BUMBLEBEE PROJECT) (REF. 8)
BUMBLEBEE-
         LITTLE (ARTHUR D.), INC., CAMBRIDGE, MASS. (BUMBLE-
           BEE PROJECT) (REF. 8)
BUMBLEBEE-
         MICHIGAN. UNIV., ANN ARBOR. OFFICE OF RESEARCH ADM.
BUMBLEBEE-
         STANDARD OIL DEVELOPMENT CO. ESSO LABS., LINDEN, N.J.
           (BUMBLEBEE PROJECT) (REF. 8)
BUMBLEBEE-
         TEXACO EXPERIMENT INC., RICHMOND
BUMBLEBEE-
         TEXAS. UNIV., AUSTIN. DEFENSE RESEARCH LAB.
BUMED-
         BUREAU OF MEDICINE AND SURGERY (NAVY)
BUMED-(NO.)-(NO.)-
         NAVAL RADIOLOGICAL DEFENSE LAB., SAN FRANCISCO
BUMINES-OFR-(NO.-YR.)
         BUREAU OF MINES (OPEN FILE REPT.)
BUM-MPR-
         BUREAU OF MINES. ROLLA METALLURGY RESEARCH CTR., MO.
BUM-PX-
         BUREAU OF MINES, PITTSBURGH
BU-MRL-
         BROWN UNIV., PROVIDENCE. METALS RESEARCH LAB.
BUM-SQUID NO.-
         BUREAU OF MINES, PITTSBURGH (SQUID PROJECT) (REF. 33)
BUM-STR-
         BUREAU OF MINES, PITTSBURGH
BUM-TPR-
         BUREAU OF MINES, PITTSBURGH
BU/NSRDC/(DATE)
         BROWN UNIV., PROVIDENCE
BU/NSRDC/
         BROWN UNIV., PROVIDENCE
BU/NSRDC/(NUMBER-YEAR)
         BROWN UNIV., PROVIDENCE. DIV. OF ENGINEERING
BUO-EAG-
         BUREAU OF ORDNANCE (NAVY). EVALUATION AND ANALYSIS
           STAFF
BUO-EAS-
         BUREAU OF ORDNANCE (NAVY). EVALUATION AND ANALYSIS
           STAFF
BUO OP-
         BUREAU OF ORDNANCE (NAVY)
BU ORD/
         BUREAU OF ORDNANCE (NAVY)
BUORD-
         BUREAU OF ORDNANCE (NAVY)
BUORD BM-
         BUREAU OF ORDNANCE (NAVY)
BUORD BOI-
         BUREAU OF ORDNANCE (NAVY)
BUORD-BOI-(NO.)-(YR.)
         BUREAU OF ORDNANCE (NAVY)(BULLETIN OF ORDNANCE INFO.)
BUORD DRAWINGS-
         BUREAU OF ORDNANCE (NAVY)
BUORD-LWB-(YR.)-
         BUREAU OF ORDNANCE (NAVY) (LIBRARY WEEKLY BULLETIN)
BUORD (NAVY) SER-(NO.)-SP
         BUREAU OF ORDNANCE (NAVY)
BUORD OP-
         BUREAU OF ORDNANCE (NAVY) (ORDNANCE PAMPHLET)
BUORD REM-
         BUREAU OF ORDNANCE (NAVY)
BUORD REXN-
         BUREAU OF ORDNANCE (NAVY)
BUORD-S(NUMBER(S))-(YR.)
         BUREAU OF ORDNANCE (NAVY)
BUORD-SR-
         BUREAU OF ORDNANCE (NAVY)
BUO-REM-
         BUREAU OF ORDNANCE (NAVY)
BU ORL TN-
         BOSTON UNIV.
BU P-
         BROWN UNIV., PROVIDENCE (PROJECTS)
BUPERS-
         BUREAU OF NAVAL PERSONNEL
BUPERS TB-
         NAVAL PERSONNEL RES. FIELD ACTIVITY, SAN DIEGO, CAL.
BU-PRL-TN-
         BOSTON UNIV. PHYSICAL RESEARCH LABS.
BURGESS (LTRS)-QPR-
         BURGESS BATTERY CO., FREEPORT, ILL.
BURK-
         BURKE RESEARCH CO., VAN DYKE, MICH.
BURR-
         BURROUGHS CORP., DETROIT
BURR-
         BURROUGHS CORP., ENCINO, CALIF.
BURR-(CONTRACT NO.)
         BURROUGHS CORP., RANDOR, PENNA.
BURROUGHS(NO.)-MPR-(DATE)
         BURROUGHS CORP. (ALL LOCATIONS)
BURROUGHS-
         BURROUGHS CORP., PAOLI, PENNA.
BURR PROD (LETTERS)-
         BURROUGHS CORP., DETROIT
```

BURR PROD (LETTERS)-	BURROUGHS CORP., ENCINO, CALIF.
BUS-	BUREAU OF SHIPS
BUSANDA-	BUREAU OF SUPPLIES AND ACCOUNTS (NAVY)
BUSHIPS-	BUREAU OF SHIPS
BU SHIPS (NO.-LTR.)-	BUREAU OF SHIPS
BUSHIPS-TM-	BUREAU OF SHIPS
BUSHIPS-TRANS-	BUREAU OF SHIPS (TRANSLATION)
BU-TN-R-	BROWN ENGINEERING CO., INC., HUNTSVILLE, ALA.
BU-TR-	BROWN UNIV., PROVIDENCE
BUWEPS-	BUREAU OF NAVAL WEAPONS
BUWEPS INSTR-	BUREAU OF NAVAL WEAPONS (INSTRUCTION)
BUWEPSREP-	BUREAU OF NAVAL WEAPONS REPRESENTATIVE, SUNNYVALE,CAL
BU-WT-	BROWN UNIV., PROVIDENCE. DIV. OF ENGINEERING
BVCE MP-	BLACK AND VEACH CONSULTING ENGINEERS, KANSAS CITY, MO
BVCP-	BLACK AND VEACH CONSULTING ENGINEERS, KANSAS CITY, MO
BVW-	BRIMAR VALVE WORKS, FOOTSCRAY, KENT, ENGLAND
BW-	(BIOLOGICAL WARFARE) (REF. 2)
BW-	BABCOCK + WILCOX CO. (ALL LOCATIONS, LAB. REPORTS)
B/W-	BABCOCK + WILCOX CO. RESEARCH CENTER, ALLIANCE, OHIO
B+W-	BABCOCK + WILCOX CO. RESEARCH CENTER, ALLIANCE, OHIO
BW-(YEAR)-	BUREAU OF THE CENSUS, WASHINGTON, D.C.
BW-	OFFICE OF NAVAL RESEARCH (BIOL. WARFARE OPERATIONS)
BW-	RES. + DEV. BD.,WASH.,D.C. (BIOL. WARFARE OPERATIONS)
BW-	SYLVANIA ELECTRIC PRODUCTS INC., WOBURN, MASS.
BW-AED-	BABCOCK + WILCOX CO. ATOMIC ENERGY DIV., AKRON, OHIO
BWALR-	DUGWAY PROVING GROUND, TOOELE, UTAH
BWC-	BORG-WARNER CORP., KALAMAZOO, MICH.
BWC-	BORG-WARNER CORP. ROY C. INGERSOLL RESEARCH CENTER, DES PLAINES, ILL.
BW D-	BABCOCK + WILCOX CO. (ALL LOCATIONS, LAB. REPORTS)
BWD-TB-	BENDIX CORP. KANSAS CITY DIV., MO.
BWL-	CHEMICAL CORPS BIOL. WARFARE LABS., CAMP DETRICK, MD.
BWR-	(BI-WEEKLY REPORT) (REF. 2)
BWR-	GT. BRIT. ATOMIC EN. RES. ESTAB., HARWELL, BERKS, ENG
BWRA-	BRITISH WELDING RESEARCH ASSN., CAMBRIDGE, ENGLAND (TRANSLATIONS)
BWRA-	BRITISH WELDING RESEARCH ASSN., LONDON
BW-R/D-	BARRY WRIGHT CORP., WATERTOWN, MASS.
BW-RDE-	BABCOCK + WILCOX CO. (ALL LOCATIONS, LAB. REPORTS)
BW RR-	BABCOCK + WILCOX CO. (ALL LOCATIONS, LAB. REPORTS)
BX-	BENDIX AVIATION CORP. PACIFIC DIV., DEVELOPMENT LABS NORTH HOLLYWOOD, CALIF.
BX-(NUMBER-YEAR)	TRAVENOL LABS., INC., MORTON GROVE, ILL.
BXM-	BENDIX CORP. BENDIX MISHAWAKA DIV., IND.
BXR-	NORTON CO., WORCESTER, MASS.
BY-	CALIFORNIA. UNIV., LIVERMORE. RADIATION LAB.
BYGGFORSKNINGEN-	SWEDEN. INST. FOR BUILDING RESEARCH, STOCKHOLM
BYU-	BRIGHAM YOUNG UNIV., PROVO, UTAH
BYU-(NUMBER)-F	BRIGHAM YOUNG UNIV., PROVO, UTAH
BYU-SR-	BRIGHAM YOUNG UNIV., PROVO, UTAH
BZAH-	BOOZ, ALLEN AND HAMILTON, CHICAGO

C1-

C1-	(ALLIED SERVICES RADIO AND SIGNAL EQUIPMENT SPECS.)
C-	(CATALOG) (REF. 2)
C-	(CIRCULAR) (REF. 2)
C-	(COMMUNICATION) (REF. 2)
C-	(CONFERENCE) (REF. 2)
C-	(CONFIDENTIAL) (REF. 2)
C-	(CONTRACTOR'S REPORT) (REF. 2)
C-	(CONTRIBUTION) (REF. 2)
C-	(CYCLOTRON) (REF. 2)
C-	(TECHNICAL COMPILATION) (REF. 2)
C-	AERONUTRONIC DIV. OF PHILCO CORP., NEWPORT BEACH, CALIF. (CONFIDENTIAL)
C-(NUMBER)-	AIR REDUCTION CO., INC. CENTRAL RESEARCH LABS., MURRAY HILL, N.J.
C2- THRU C9-	AIR RES. + DEV. COMMAND, ANDREWS AFB, MD. (ASSIGNED TO REPORTS ISSUED BY THE COMMAND AND ITS CONTRACTORS C2- IS 1952, C3- 1953, ETC.)
(NUMBER)-C-(YEAR)	ALLISON DIV., GENERAL MOTORS CORP., CLEVELAND
C-	AMPHENOL-BORG ELECTRONICS CORP., CHICAGO
C-	ANSCO DIV., GEN. ANILINE + FILM CORP., BINGHAMTON, NY
C-	ARMOUR RESEARCH FOUNDATION, CHICAGO
C-	ATOMIC ENERGY OF CANADA LTD. CHALK RIVER PROJECT, ONT. (REF. 6)
C(NUMBER)-(NO.)/(NO.)	AUTONETICS, ANAHEIM, CALIF.
C4-	AUTONETICS DIV., NORTH AMERICAN AVIATION, INC., ANAHEIM, CALIF. (CONTRACTUAL REPORT, 1964)
C5-	AUTONETICS DIV., NORTH AMERICAN AVIATION, INC., ANAHEIM, CALIF. (CONTRACTUAL REPORT, 1965)
C6-	AUTONETICS DIV., NORTH AMERICAN AVIATION, INC., ANAHEIM, CALIF. (CONTRACTUAL REPORT, 1966)
C/(NUMBER)/(YEAR)	BELGIUM. CENTRE D'ETUDE DE L'ENERGIE NUCLEAIRE, BRUSSELS
C-	BISSETT-BERMAN CORP., SANTA MONICA, CALIF.
C(NO.)-	BISSETT-BERMAN CORP., SANTA MONICA, CALIF.
C(NUMBER/NUMBER/YEAR)	BRITISH WELDING RESEARCH ASSN., CAMBRIDGE, ENGLAND
C(NO.)-	BROWN UNIV., PROVIDENCE
C11-	BROWN UNIV., PROVIDENCE. GRAD. DIV. OF APPLIED MATH.
C-(YEAR-NUMBER-NUMBER)-	BUNKER-RAMO CORP. DEFENSE SYSTEMS DIV., SILVER SPRING, MD.
C-	BUREAU OF RECLAMATION (INTERIOR). COMMISSIONERS OFFICE, DENVER
C-2-184-M-	CALIFORNIA. UNIV. BERKELEY. RADIATION LAB. (REF. 9)
C-2-184-W-	CALIFORNIA. UNIV. BERKELEY. RADIATION LAB. (REF. 9)
C7-(NUMBER/-)	CALIFORNIA. UNIV., LIVERMORE. LAWRENCE RADIATION LAB.
C-	CAMBRIDGE ACOUSTICAL ASSOCIATES, INC., MASS.
C-	CARBIDE AND CARBON CHEMICALS CORP. K-25 PLANT, OAK RIDGE, TENN.
C1-R-	CARBIDE AND CARBON CHEMICALS CORP. SUBSTITUTE ALLOY MATERIALS LABS., N.Y.C.
C-(NUMBER-YEAR)-	CHARLESTON NAVAL SHIPYARD, S.C.
C-	CHICAGO. UNIV. METALLURGICAL LAB.
C1-L- THRU C1-S-	COLUMBIA UNIV., NYC. SUBSTITUTE ALLOY MATERIALS LABS.
C2-M-	COLUMBIA UNIV., NYC. SUBSTITUTE ALLOY MATERIALS LABS.
C2-R-	COLUMBIA UNIV., NYC. SUBSTITUTE ALLOY MATERIALS LABS.
C2X-M-	COLUMBIA UNIV., NYC. SUBSTITUTE ALLOY MATERIALS LABS.
C-	CONTINENTAL OIL CO., PONCA CITY, OKLA.
C-	COOK ELECTRIC CO. WIRECOM DIV., CHICAGO
C-	CORNING GLASS WORKS, N.Y.
C-	CURTISS-WRIGHT CORP. (ALL LOCATIONS)
C-	DAVID TAYLOR MODEL BASIN, CARDEROCK, MD.
C(NO.)-CR	DEFENSE ATOMIC SUPPORT AGENCY. FIELD COMMAND, ALBUQUERQUE, N. MEX.
C(NO.)-CR-CB(NO.)	DEFENSE ATOMIC SUPPORT AGENCY. FIELD COMMAND, ALBUQUERQUE, N. MEX.
C(NO.)-CR-CN(NO.)	DEFENSE ATOMIC SUPPORT AGENCY. FIELD COMMAND, ALBUQUERQUE, N. MEX.
C(YR.)-(LTR.)-	DUMONT (ALLEN B.) LABS., INC.
C-	FRANKLIN INST. LABS. FOR RES. + DEV., PHILADELPHIA
C-(4 DIGITS)-F	FRANKLIN INST. RESEARCH LABS., PHILADELPHIA
C-(NUMBER-NUMBER)-	FREDERICK RESEARCH CORP., BETHESDA, MD.
C-	GENERAL DYNAMICS CORP. ELECTRIC BOAT DIV., GROTON, CONN
C(NO.)-	GENERAL DYNAMICS CORP. ELECTRIC BOAT DIV., GROTON, CONN
C(NUMBER)-(YEAR)	GENERAL DYNAMICS CORP. ELECTRIC BOAT DIV., GROTON, CONN
C(NUMBER)-(YEAR)-	GENERAL DYNAMICS CORP. ELECTRIC BOAT DIV., SAN DIEGO, CALIF.
C-(NUMBER)-C-	GENERAL ELECTRIC CO. FLIGHT PROPULSION LAB. DEPT., EVENDALE, OHIO
C(YEAR)-	HERCULES POWDER CO., WILMINGTON, DEL.
C-	LAB. FOR ELECTRONICS, INC., ELECTRONICS DIV., BOSTON
C-	LITTLE (ARTHUR D.), INC., CAMBRIDGE, MASS.
C-(NUMBER)-	LITTLE (ARTHUR D.), INC., CAMBRIDGE, MASS.
C-(4 DIGITS)-	LITTON INDUSTRIES. ELECTRON TUBE DIV., SAN CARLOS, CAL
C/(NO.)	LOCKHEED AIRCRAFT CORP., MARIETTA, GA.
C-	LEXINGTON LABS., INC., CAMBRIDGE, MASS.
C-	LIBRARY OF CONGRESS. TECH.INFO.DIV., WASH., DC (REF.20)
C-1- THRU C-8-	LOS ALAMOS SCIENTIFIC LAB., N. MEX. (INTERNAL CORRESPONDENCE SERIAL USED BY GROUPS IN C DIV.)
C/64-	MASSACHUSETTS INST. OF TECH., CAMBRIDGE. CENTER FOR INTERNATIONAL STUDIES
C/65-	MASSACHUSETTS INST. OF TECH., CAMBRIDGE. CENTER FOR INTERNATIONAL STUDIES
C/69-	MASSACHUSETTS INST. OF TECH., CAMBRIDGE. CENTER FOR INTERNATIONAL STUDIES
C-	MINE SAFETY APPLIANCES CO., CALLERY, PENNA.
2C-	NATIONAL BUREAU OF STANDARDS, WASHINGTON, D.C.
C-(NO.-YR)S-	NATL. RES. COUNCIL OF CANADA. NATL. RES. LABS., OTTAWA
C-	NAVAL RESEARCH LAB., WASHINGTON, D.C.
C-(YEAR)-	NETHERLANDS. METAALINSTITUUT TNO, DELFT
C6-(YEAR)/	NORTH AMERICAN AVIATION, INC., ANAHEIM, CALIF.
C-	NORTH AMERICAN AVIATION, INC. OCEAN SYSTEMS OPERATIONS, ANAHEIM, CALIF.
C9-(4 DIGITS)./	NORTH AMERICAN ROCKWELL CORP. AUTONETICS DIV., ANAHEIM, CALIF.
C-	OFFICE OF NAVAL RESEARCH
C-(YEAR)-	OREGON STATE UNIV., CORVALLIS
C-(YEAR)-	OREGON STATE UNIV., CORVALLIS. COMPUTER CENTER
C(NO.)-	PLESSET (E.H.) ASSOCIATES, INC., LOS ANGELES
C-(NUMBER)-(YEAR)-	PLESSET (E.H.) ASSOCIATES, INC., SANTA MONICA, CALIF
C-	RADAR EVALUATION SQUADRON (4754TH), HILL AFB, UTAH
C-	RADIO CORP. OF AMERICA. RCA LABS. DIV., PRINCETON, NJ
C-	REPUBLIC AVIATION CORP., FARMINGDALE, N.Y.
C-	SCHJELDAHL (G.T.) CO., NORTHFIELD, MINN.
C-	STANFORD UNIV., CALIF. MICROWAVE LAB.
C-	STEVENS INST. OF TECH., HOBOKEN, N.J. DEPT. OF METALLURGY
C-	SWEDEN. FOERSVARETS FORSKNINGSANSTALT, STOCKHOLM
C(NUMBER)-	SYLVANIA ELECTRONIC SYSTEMS-CENTRAL, WILLIAMSVILLE, NY
C-(NO.)-	TECHNICAL INDUSTRIAL INTELL. COMM. (COMMUNICATIONS)
C-0.100- THRU C-5.385-	TENNESSEE EASTMAN CORP., OAK RIDGE, TENN. (REF. 34)
C(NO.)-	TEXAS INSTRUMENTS, INC., DALLAS
C(NUMBER)-(NUMBER)-	TEXAS INSTRUMENTS, INC. APPARATUS DIV., DALLAS
C-(YEAR)-	THIOKOL CHEMICAL CORP., HUNTSVILLE, ALA.

C-	THIOKOL CHEMICAL CORP. REDSTONE DIV., HUNTSVILLE, ALA.
C20-0U(NUMBER)-	THOMPSON RAMO WOOLDRIDGE INC., LOS ANGELES
C-(NO.)-B-	TOKYO IMPERIAL UNIV. PATHOLOGICAL INST.
C-	UNDERWATER EXPLOSIONS RESEARCH DIV., PORTSMOUTH, VA.
C-	UNION CARBIDE CORP. PARMA RESEARCH LAB., OHIO
C-	UNION CARBIDE RESEARCH INST., TARRYTOWN, N.Y.
C-	UNITED AIRCRAFT CORP., EAST HARTFORD, CONN.
C(NO.)-	UNITED AIRCRAFT CORP., EAST HARTFORD, CONN.
C-	UNITED AIRCRAFT CORP. RES. LABS., EAST HARTFORD, CONN.
C-	UNITED KINGDOM. ROYAL MILITARY COLL. OF SCIENCE, SHRIVENHAM, WILTS, ENGLAND
C-(NUMBER)(LETTER)-	WESTINGHOUSE ELECTRIC CORP. ELECTRONIC TUBE DIV., BALTIMORE
(YR.)-9C1-ARCS0-R(NO.)	WESTINGHOUSE RESEARCH LABS. QUANTUM ELECTRONICS DEPT., PITTSBURGH
CA-	ARGONNE NATIONAL LAB., LEMONT, ILL. (REF. 10)
CA-	CANADIAN ARSENALS, LTD., TORONTO
CA-	CELESTRON ASSOCIATES, YONKERS, N.Y.
CA-	CHICAGO. UNIV. METALLURGICAL LAB.
CA-	COLLEGE OF AERONAUTICS, CRANFIELD, BUCKS, ENGLAND
CA-(NUMBER)-	COMPUTER ASSOCIATES, INC., ARLINGTON, VA.
CA-	COMPUTER ASSOCIATES, INC., WAKEFIELD, MASS.
CA-	CORNELL AERONAUTICAL LAB., INC., BUFFALO
CA-(NUMBER)-P-	CORNELL AERONAUTICAL LAB., INC., BUFFALO
CA-	GARRETT CORP. AIRESEARCH MFG. CO. OF ARIZONA, PHOENIX
CA-	GT. BRIT. MINISTRY OF CIVIL AVIATION
CA-(NUMBER)	MASSACHUSETTS COMPUTER ASSOCIATES, INC., WAKEFIELD
CA-	OFFICE OF NAVAL RES. (COMBAT AIR SUPPORT OPERATIONS)
CA-	RESEARCH AND DEVELOPMENT BOARD, WASHINGTON, D.C. (COMBAT AIR SUPPORT OPERATIONS)
CA-(NUMBER)-	SPERRY GYROSCOPE CO., GREAT NECK, N.Y.
CA-	SPERRY GYROSCOPE CO. AIR ARMAMENT DIV., GT. NECK, N.Y.
CA-(NUMBER)-	SPERRY GYROSCOPE CO. INERTIAL DIV., GREAT NECK, N.Y.
CA-	SPERRY RAND CORP. FORD INSTRUMENT DIV., LONG ISLAND CITY, N.Y.
CAA-	ARMY BALLISTIC MISSILE AGENCY, REDSTONE ARSENAL, ALA.
CAA-	ARMY ORDNANCE MISSILE COMMAND, REDSTONE ARSENAL, ALA.
CAA-	CIVIL AERONAUTICS ADMINISTRATION, WASHINGTON, D.C.
CAA-D(NUMBER)-(NUMBER)-	CONDUCTRON CORP., ANN ARBOR, MICH.
CAADRP-TR-	GT. BRIT. ROYAL AIRCRAFT EST., FARNBOROUGH, HANTS, ENGLAND (CIVIL AIRCRAFT AIRWORTHINESS DATA RECORDING PROGRAMME)
CAATC-	COLLINS RADIO CO., CEDAR RAPIDS, IOWA (CIVIL AERONAUTICS ADMINISTRATIVE TYPE CERTIFICATE)
CAA-U-	CAMBRIDGE ACOUSTICAL ASSOCIATES, INC., MASS.
CAB-	CIVIL AERONAUTICS BOARD, WASHINGTON, D.C.
CAC-	CHASE AIRCRAFT CO., INC., WEST TRENTON, N.J.
CAC-	COLLEGE OF AERONAUTICS, CRANFIELD, BUCKS, ENGLAND
CAC-	CONTINENTAL AIR COMMAND, MITCHEL AFB, N.Y.
CAC-	CONTINENTAL ARMY COMMAND. COMBAT DEVELOPMENTS SECTION FORT MONROE, VA.
CAC-	OFFICE OF SPECIAL WEAPONS DEVELOPMENTS, FT. BLISS, TEX
CAC (NUMBER)-	OFFICE OF TECHNICAL SERVICES. INDUSTRIAL RESEARCH AND DEVELOPMENT DIV., WASH., D.C. (PROJ. CONTRACT CAC-)
CAC ARTY BD.-	CONTINENTAL ARMY COMMAND. ARTILLERY BOARD, FORT SILL, OKLA.
CAC ATB PROJ.-	CONTINENTAL ARMY COMMAND. ARCTIC TEST BOARD, FORT GREELY, ALASKA
CAC-ATDEV-	CONTINENTAL ARMY COMMAND, FORT MONROE, VA.
CAC ATSWD-(LTR.-YEAR)	CONTINENTAL ARMY COMMAND, FORT MONROE, VA.
CAC BD. 1, PROJ.-	CONTINENTAL ARMY COMMAND. BD. NO. 1, FT. SILL, OKLA.
CAC BD. 4-	CONTINENTAL ARMY COMMAND. BD. NO. 4. FT. BLISS, TEX.
CAC/CORG-	CONTINENTAL ARMY COMMAND, FORT MONROE, VA.
CACI-	CALIFORNIA ANALYSIS CENTER, INC., SANTA MONICA, CALIF
CAC/OSWD-	OFFICE OF SPECIAL WEAPONS DEVELOPMENTS, FT. BLISS, TEX
CAC-OSWDD-	OFFICE OF SPECIAL WEAPONS DEVELOPMENTS, FORT BLISS, TEX. (DOCUMENT)
CAC/SNODGRASS-	CONTINENTAL ARMY COMMAND, FORT MONROE, VA.
CACTQ-	CITIZENS ADVISORY COMMITTEE ON TRANSPORTATION QUALITY, WASHINGTON, D.C.
CAD-LAB-(YEAR-NUMBER)	PENNSYLVANIA STATE UNIV., UNIVERSITY PARK. COMPUTER AIDED DESIGN AND SIMULATION LAB.
CADO-	CENTRAL AIR DOCS. OFF., WRIGHT-PATTERSON AFB, OHIO
CADPL-	ARMY ELECTRONICS COMMAND. COMMUNICATIONS/ADP LAB., FORT MONMOUTH, N.J.
CAE-	CONTINENTAL AVIATION AND ENGINEERING CORP., DETROIT
CAF-	COMMITTEE ON AERONAUTICAL (RES. + DEV.) FACILITIES
CAF-	OFF. OF THE ASST. SECY. OF DEFENSE(RES.+DEV.), WASH, DC
CAF-VJ-	CORNELL AERONAUTICAL LAB., INC., BUFFALO
CAHI-	U.S.S.R. CENTRAL AERO-HYDRODYNAMICAL INST.
CAI-	BOURNS/CAI, INC., BARRINGTON, ILL.
CAI-	CHICAGO AERIAL INDUSTRIES, INC. BARRINGTON, ILL.
CAI-(NUMBER)-(LETTER)	CHICAGO AERIAL INDUSTRIES, INC. BARRINGTON, ILL.
CAI-(NUMBER)-ITRD-	CHICAGO AERIAL INDUSTRIES, INC. BARRINGTON, ILL.
CAI-NY-	COMPUTER APPLICATIONS, INC., N.Y.C.
CAI-R-	PENNSYLVANIA STATE UNIV., UNIVERSITY PARK. COMPUTER ASSISTED INSTRUCTION LAB.
CAI-SED-(NUMBER)-	COMPUTER APPLICATIONS, INC., N.Y.C.
CAI-SYSTEMS MEMO-	FLORIDA STATE UNIV., TALLAHASSEE. COMPUTER-ASSISTED INSTRUCTION CENTER
CAIT-	CASE INST. OF TECH., CLEVELAND
CAI-TM-	FLORIDA STATE UNIV., TALLAHASSEE. COMPUTER-ASSISTED INSTRUCTION CENTER
CAI-TR-	FLORIDA STATE UNIV., TALLAHASSEE. COMPUTER-ASSISTED INSTRUCTION CENTER
CAL-(NUMBER)-(LETTER(S))	CALIFORNIA. UNIV., BERKELEY (SQUID PROJECT) (REF. 33)
CAL-	CANADIAN ARSENALS, LTD., TORONTO
CA-L-	COLUMBIA UNIV., NYC. SUBSTITUTE ALLOY MATERIALS LABS.
CAL-	CORNELL AERONAUTICAL LAB., INC., BUFFALO
CAL-	VIRGINIA. UNIV., CHARLOTTESVILLE
CAL-AA-	CORNELL AERONAUTICAL LAB., INC., BUFFALO
CAL-AC-	CORNELL AERONAUTICAL LAB., INC., BUFFALO
CAL-AD-	CORNELL AERONAUTICAL LAB., INC., BUFFALO
CAL-AF-	CORNELL AERONAUTICAL LAB., INC., BUFFALO
CAL-AG-	CORNELL AERONAUTICAL LAB., INC., BUFFALO
CAL-AI-	CORNELL AERONAUTICAL LAB., INC., BUFFALO
CAL-AM-	CORNELL AERONAUTICAL LAB., INC., BUFFALO
CAL-AN-	CORNELL AERONAUTICAL LAB., INC., BUFFALO
CAL-BB-	CORNELL AERONAUTICAL LAB., INC., BUFFALO
CAL-BE-	CORNELL AERONAUTICAL LAB., INC., BUFFALO
CAL-BM-	CORNELL AERONAUTICAL LAB., INC., BUFFALO
CAL-C-	CORNELL AERONAUTICAL LAB., INC., BUFFALO
CAL-CA-	CORNELL AERONAUTICAL LAB., INC., BUFFALO
CALCH-	CALLERY CHEMICAL CO., PENNA.
CAL-CM-	CORNELL AERONAUTICAL LAB., INC., BUFFALO
CAL-CONTRIB-	COLUMBIA UNIV., N.Y.C. COLUMBIA ASTROPHYSICS LAB.
CAL-DF-	CORNELL AERONAUTICAL LAB., INC., BUFFALO
CAL-DM-	CORNELL AERONAUTICAL LAB., INC., BUFFALO
CAL-FDM-	CORNELL AERONAUTICAL LAB. FLIGHT RES. DEPT., BUFFALO
CAL-FRM-	CORNELL AERONAUTICAL LAB. FLIGHT RES. DEPT., BUFFALO

CAL-GI- CORNELL AERONAUTICAL LAB., INC., BUFFALO
CAL-GM- CORNELL AERONAUTICAL LAB., INC., BUFFALO
CAL-HF- CORNELL AERONAUTICAL LAB., INC., BUFFALO
CAL-HM- CORNELL AERONAUTICAL LAB., INC., BUFFALO
CAL-IG- CORNELL AERONAUTICAL LAB., INC., BUFFALO
CAL-IH- CORNELL AERONAUTICAL LAB., INC., BUFFALO
CAL-IM- CORNELL AERONAUTICAL LAB., INC., BUFFALO
CALIT- CALIFORNIA INST. OF TECH., PASADENA
CAL-JA- CORNELL AERONAUTICAL LAB., INC., BUFFALO
CAL-KA- CORNELL AERONAUTICAL LAB., INC., BUFFALO
CAL-KB- CORNELL AERONAUTICAL LAB., INC., BUFFALO
CAL-KC- CORNELL AERONAUTICAL LAB., INC., BUFFALO
CAL-KD- CORNELL AERONAUTICAL LAB., INC., BUFFALO
CAL-KF- CORNELL AERONAUTICAL LAB., INC., BUFFALO
CAL-KM- CORNELL AERONAUTICAL LAB., INC., BUFFALO
CALLINGS-(YEAR)- CENTER FOR APPLIED LINGUISTICS, WASHINGTON, D.C.
CAL-NM- CORNELL AERONAUTICAL LAB., INC., BUFFALO
CAL-PI- CORNELL AERONAUTICAL LAB., INC., BUFFALO
CAL-QM- CORNELL AERONAUTICAL LAB., INC., BUFFALO
CAL-RA- CORNELL AERONAUTICAL LAB., INC., BUFFALO
CAL-RM- CORNELL AERONAUTICAL LAB., INC., BUFFALO
CAL-SA- CORNELL AERONAUTICAL LAB., INC., BUFFALO
CALT- CALIFORNIA. UNIV., RIVERSIDE
CALT- CALIFORNIA INST. OF TECH., PASADENA
CALT-(NUMBER-NUMBER) CALIFORNIA INST. OF TECH., PASADENA
CALT- CORNELL AERONAUTICAL LAB., INC., BUFFALO
CAL-TB- CORNELL AERONAUTICAL LAB., INC., BUFFALO
CAL-TECH- CALIFORNIA INST. OF TECH., PASADENA
CALTECH- CALIFORNIA INST. OF TECH., PASADENA
CAL-TECH/JPL- CALIF. INST. OF TECH., PASADENA. JET PROPULSION LAB.
CAL-TG- CORNELL AERONAUTICAL LAB., INC., BUFFALO
CAL-TM- CORNELL AERONAUTICAL LAB., INC., BUFFALO
CALU- CALIFORNIA. UNIV. (ALL LOCATIONS)
CAL-U- CALIFORNIA. UNIV. (ALL LOCATIONS)
CAL-UA- CORNELL AERONAUTICAL LAB., INC., BUFFALO
CAL-UB- CORNELL AERONAUTICAL LAB., INC., BUFFALO
CAL-UF- CORNELL AERONAUTICAL LAB., INC., BUFFALO
CAL. U. HE- CALIFORNIA. UNIV., BERKELEY. INST. OF ENG. RES.
CAL. U. IER (NUMBERS) CALIFORNIA. UNIV., BERKELEY. INST. OF ENG. RES.
CALU/LRL(L) CALIFORNIA. UNIV., LIVERMORE. LAWRENCE RADIATION LAB.
CAL-UM- CORNELL AERONAUTICAL LAB., INC., BUFFALO
CAL. U./PROJ. CIVIL- CALIFORNIA. UNIV., RICHMOND. INST. OF ENG. RES.
CAL. U., SERIES- CALIFORNIA. UNIV. (ALL LOCATIONS)
CAL. U.,SERIES-(NO.ISSUE) CALIFORNIA. UNIV. (ALL LOCATIONS)
CAL-VB- CORNELL AERONAUTICAL LAB., INC., BUFFALO
CAL-VC- CORNELL AERONAUTICAL LAB., INC., BUFFALO
CAL-VE- CORNELL AERONAUTICAL LAB., INC., BUFFALO
CAL-VF- CORNELL AERONAUTICAL LAB., INC., BUFFALO
CAL-VG- CORNELL AERONAUTICAL LAB., INC., BUFFALO
CAL-VH- CORNELL AERONAUTICAL LAB., INC., BUFFALO
CAL-VJ- CORNELL AERONAUTICAL LAB., INC., BUFFALO
CAL-VO- CORNELL AERONAUTICAL LAB., INC., BUFFALO
CAL-VS- CORNELL AERONAUTICAL LAB., INC., BUFFALO
CAL-VU- CORNELL AERONAUTICAL LAB., INC., BUFFALO
CAL-VY- CORNELL AERONAUTICAL LAB., INC., BUFFALO
CAL-XA- CORNELL AERONAUTICAL LAB., INC., BUFFALO
CALY- CALIDYNE CO., WINCHESTER, MASS.
CAL-YB- CORNELL AERONAUTICAL LAB., INC., BUFFALO
CAL-YM- CORNELL AERONAUTICAL LAB., INC., BUFFALO
CA-M- ARGONNE NATIONAL LAB., LEMONT, ILL. (REF. 10)
CAM- CAMBRIDGE CORP., BOULDER, COLO.
CAM- CAMBRIDGE CORP., SOMERVILLE, MASS.
CAM- CAMBRIDGE UNIV., ENGLAND
CAM-(NUMBER)- LEHIGH UNIV., BETHLEHEM, PENNA.
CAM- NATIONAL RESEARCH COUNCIL. COMM. ON AVIATION MEDICINE
CAM-(LTRS)- CAMBRIDGE CORP., SOMERVILLE, MASS.
CAMCO- CAMBRIDGE CORP., BOULDER, COLO.
CAMCO- CAMBRIDGE CORP., SOMERVILLE, MASS.
CAMCO-TM- CAMBRIDGE CORP., BOULDER, COLO.
CAM-DAMTP-(YEAR)/(NO.) CAMBRIDGE UNIV., ENGLAND. DEPT. OF APPLIED MATH. AND PHYSICS
CAM EM- CAMBRIDGE CORP., SOMERVILLE, MASS. (ENG. MEMO.)
CAMESA-(NUMBER)A CANADIAN MILITARY ELECTRONICS STANDARDS AGENCY, OTTAWA
CAMESA-L(NUMBER)- CANADIAN MILITARY ELECTRONICS STANDARDS AGENCY, OTTAWA
CAN- ATOMIC ENERGY OF CANADA LTD., CHALK RIVER, ONT.
CAN-(LETTERS)- CANADA. (FOR EXPLANATION OF THE REMAINDER OF THE CODE, SEE ENTRY UNDER LETTERS FOLLOWING CAN-)
CAN-DRB- CANADA. DEFENCE RESEARCH BOARD, OTTAWA
CANEL- PRATT AND WHITNEY AIRCRAFT. CONNECTICUT AIRCRAFT NUCLEAR ENGINE LAB., MIDDLETOWN
CAN-MSER- CANADAIR LTD. MISSILES AND SYSTEMS DIV., MONTREAL
CANO- CANOGA CORP., VAN NUYS, CALIF.
CAN UTIA (NUMBER) TORONTO. UNIV. INST. OF AEROPHYSICS
CAO- NATIONAL RES. COUNCIL. COMM. ON AMPHIBIOUS OPERATIONS
(YEAR)-CAO- VARIAN ASSOCIATES. CALIFORNIA AVENUE OPERATIONS, PALO ALTO, CALIF.
(YEAR)-CAO-(NUMBER)- VARIAN ASSOCIATES. CALIFORNIA AVENUE OPERATIONS, PALO ALTO, CALIF.
CAOPS/ORS- ROYAL CANADIAN A.F. OPERATIONAL RES. SEC., OTTAWA
CAORE- CANADA. ARMY OPERATIONAL RESEARCH ESTABLISHMENT
CAORG- CANADA. ARMY OPERATIONAL RESEARCH ESTABLISHMENT
CAORT- CANADA. ARMY OPERATIONAL RESEARCH ESTABLISHMENT
CAP- UNITED KINGDOM. BOARD OF TRADE, LONDON
CAPE- TECHNICAL INFORMATION SERVICE, AEC (CIVILIAN APPLICATION PROGRAM ENGINEERING DRAWINGS)
CAPENHURST- UNITED KINGDOM ATOMIC ENERGY AUTHORITY. INDUSTRIAL GROUP. CAPENHURST WORKS, CHES., ENGLAND
CAPETOWNU- CAPE TOWN. UNIV.
CA/PMM/(NUMBER/YEAR) NOTTINGHAM, ENGLAND. UNIV. DEPT. OF GEOGRAPHY (CLUSTER ANALYSIS)(AUTHOR'S INITIALS)
CAP-TM- GT. BRIT. CAPENHURST WORKS, CHES., ENGLAND
CA-R- ARGONNE NATIONAL LAB., LEMONT, ILL. (REF. 10)
CAR-(NUMBER)- CARNEGIE INST. OF TECH., PITTSBURGH. DEPT. OF PHYSICS
CAR-17-TR- CARNEGIE INST. OF TECH., PITTSBURGH. METALS RES. LAB.
CAR- CARNEGIE-MELLON UNIV., PITTSBURGH
CA-R- COLUMBIA UNIV., NYC. SUBSTITUTE ALLOY MATERIALS LABS.
CAR- WATERTOWN ARSENAL LAB., MASS. (CAST ARMOR REPORTS)
CARBC- CARBORUNDUM CO., NIAGARA FALLS, N.Y.
CARBC-BI-MONTHLY- CARBORUNDUM CO., NIAGARA FALLS, N.Y.
CARBC-FINAL CARBORUNDUM CO., NIAGARA FALLS, N.Y.
CARBC-PR- CARBORUNDUM CO., NIAGARA FALLS, N.Y. (PROGRESS REPT.)
CARBC-QR- CARBORUNDUM CO., NIAGARA FALLS, N.Y.
CARBC-SR- CARBORUNDUM CO., NIAGARA FALLS, N.Y.
CARBORUNDUM(NO.-NO.)-QPR- CARBORUNDUM CO., NIAGARA FALLS, N.Y.

CARBORUNDUM-QR-	CARBORUNDUM CO., NIAGARA FALLS, N.Y.
CAR-CC-TR-	CARNEGIE INST. OF TECH., PITTSBURGH. COMPUTATION CTR.
CARDE-(NO.)/(YR)-	CANADA. ARMAMENT RES. + DEV. EST., VALCARTIER, QUE.
CARDE-REPRINT-	CANADA. ARMAMENT RES. + DEV. EST., VALCARTIER, QUE.
CARDE-TM-	CANADA. ARMAMENT RES. + DEV. EST., VALCARTIER, QUE.
CARDE-TM-(NUMBER)/(YEAR)-	CANADA. ARMAMENT RES. + DEV. EST., VALCARTIER, QUE.
CARDE-TN-	CANADA. ARMAMENT RES. + DEV. EST., VALCARTIER, QUE.
CARDE-T.N.-	CANADA. ARMAMENT RES. + DEV. EST., VALCARTIER, QUE.
CARDE-TR-	CANADA. ARMAMENT RES. + DEV. EST., VALCARTIER, QUE.
CARI-	CIVIL AEROMEDICAL RESEARCH INSTITUTE, WASHINGTON, D.C. (CODE USED UNTIL 1963)
CARI-	JAPAN. CENTRAL AERONAUTICAL RESEARCH INSTITUTE
CARI (LETTERS)-	JAPAN. CENTRAL AERONAUTICAL RESEARCH INSTITUTE
CARIT-	CARNEGIE INST. OF TECH., PITTSBURGH
CARIW-	CARNEGIE INSTITUTION OF WASHINGTON, D.C.
CARLETONU-	CARLETON UNIV., OTTAWA
CARP-A-	GENERAL ELECTRIC CO. AIRCRAFT NUCLEAR PROPULSION PROJECT, CINCINNATI (REF. 3)
CART-TR-	DARTMOUTH COLL., HANOVER, N.H. DEPT. OF PHYSICS
CAS-	NORTH CAROLINA STATE UNIV., RALEIGH. CENTER FOR ACOUSTICAL STUDIES
CASE-	CASE INST. OF TECH., CLEVELAND
CASE-(YEAR)-	CASE INST. OF TECH., CLEVELAND(LIBRARY ACCESSION NO.)
CASE (NUMBER)-FR-(DATE)	CASE INST. OF TECH., CLEVELAND
CASE-1-(YEAR)-	CASE INST. OF TECH., CLEVELAND. DIGITAL SYSTEMS ENGINEERING GROUP
CASE 400.- THRU 800.-	SANDIA CORP., ALBUQUERQUE, N. MEX.
CASE-CC-	CASE INST. OF TECH., CLEVELAND. COMPUTING CENTER
CASE-FR-	CASE INST. OF TECH., CLEVELAND. DEPT. OF MET. ENG.
CASE-OSR-	CASE INST. OF TECH., CLEVELAND
CASE-SR-	CASE INST. OF TECH., CLEVELAND. METALS RESEARCH LAB.
CASE-TECH. MEMO.-	CASE INST. OF TECH., CLEVELAND. OPERATIONS RES. GP.
CASE-TR-	CASE INST. OF TECH., CLEVELAND
CASIT-	CASE INST. OF TECH., CLEVELAND
CAS-PR-	ARMED FORCES SPECIAL WEAPONS PROJECT. FIELD COMMAND, ALBUQUERQUE, N. MEX.
CATM-	CANADA. ARMY, OTTAWA (TRAINING MANUAL)
CAT. NO. 3-7/	GT. BRIT. ATOMIC EN. RES. ESTAB., HARWELL, BERKS, ENG
CAX-M-	COLUMBIA UNIV., NYC. SUBSTITUTE ALLOY MATERIALS LABS.
CAX-R-	COLUMBIA UNIV., NYC. SUBSTITUTE ALLOY MATERIALS LABS.
CAX-S-	COLUMBIA UNIV., NYC. SUBSTITUTE ALLOY MATERIALS LABS.
CAY MB-	CAYWOOD-SCHILLER, ASSOCIATES, CHICAGO
CB-	(CONFIDENTIAL BULLETIN) (REF. 2)
CB-	AIR TECHNICAL SERVICE COMMAND. ENGINEERING DIV. CLOTHING BRANCH, WRIGHT-PATTERSON AFB, OHIO
CB-	ARGONNE NATIONAL LAB., LEMONT, ILL. (REF. 10)
CB-(YEAR)-	BUREAU OF THE CENSUS, WASHINGTON, D.C. (CURRENT BUSINESS REPORTS)
CB-	CARBIDE AND CARBON CHEMICALS CORP. K-25 PLANT, OAK RIDGE, TENN. (REF. 10)
CB-	CHICAGO. UNIV. METALLURGICAL LAB.
CB-	CONSULTANTS BUREAU, INC., N.Y.C. (TRANSLATIONS)
CB-	FISH AND WILDLIFE SERVICE (INTERIOR), WASHINGTON, D.C. (CONSERVATION BULLETINS)
CB-	GARRETT CORP. AIRESEARCH MFG. DIV., LOS ANGELES
CB-	GT. BRIT. ADMIRALTY
CB-	ITT ELECTRO-PHYSICS LABS., INC., HYATTSVILLE, MD.
CB-	LIBRARY OF CONGRESS. LEGISLATIVE REFERENCE SERVICE, WASHINGTON, D.C.
CB-(YEAR)-	LOCKHEED MISSILES AND SPACE CO., SUNNYVALE, CALIF. (CITATION BIBLIOGRAPHY)
CB-	NATIONAL ADVISORY COMMITTEE FOR AERONAUTICS (CONFIDENTIAL BULLETIN)
CB-	NETHERLANDS. INSTITUUT VOOR TEXTIELREINIGING TNO, DELFT
CB-(LETTER)(DATE CODE)	NATIONAL ADVISORY COMMITTEE FOR AERONAUTICS (CONFIDENTIAL BULLETIN) (REF. 25)
CBA-(NUMBER)-F-	BRADBERRY (CARROLL E.) AND ASSOCIATES,LOS ALTOS, CAL.
CBA-	NATIONAL RESEARCH COUNCIL OF CANADA. DIV. OF BUILDING RESEARCH, OTTAWA (CAN. BLDG. ABSTRACTS)
CBCC-	NATL. RES.COUNCIL. CHEM.-BIOLOGICAL COORDINATION CTR.
CBD-	NATIONAL RESEARCH COUNCIL OF CANADA. DIV. OF BUILDING RESEARCH, OTTAWA (CAN. BLDG. DIGEST)
CBDF-	NATIONAL RESEARCH COUNCIL OF CANADA. DIV. OF BUILDING RESEARCH, OTTAWA (CBD FRENCH EDITION)
CBI-	CHICAGO BRIDGE AND IRON CO., OAK BROOK, ILL.
CBI EEIS SIG PR-	SIGNAL CORPS (ARMY). ENEMY EQUIPMENT INTELLIGENCE DIV. CHINA-BURMA-INDIA THEATER (PRELIMINARY REPORT)
CBI/JHU-	JOHNS HOPKINS UNIV., BALTIMORE. CHESAPEAKE BAY INST.
CBI-TRANS-	CONSULTANTS BUREAU, INC., N.Y.C. (TRANSLATION)
CBL/64/IMA-	JOHNS HOPKINS UNIV., BALTIMORE
CBL-	MARYLAND. DEPT. OF RESEARCH AND EDUCATION. CHESAPEAKE BIOLOGICAL LAB.
CBL-JHU-	JOHNS HOPKINS UNIV., BALTIMORE
CBNM-	EUROPEAN ATOMIC ENERGY COMMUNITY. CENTRAL BUREAU FOR NUCLEAR MEASUREMENTS, GEEL, BELGIUM
CBO-	(COMBINED BOMBING OFFENSIVE SURVEY)
CBRC-	CONTINENTAL BEARING RESEARCH CORP., N.Y.C.
CBRSS-	ARMY COMBAT DEVELOPMENTS COMMAND. CHEMICAL-BIOLOGICAL-RADIOLOGICAL AGENCY, FORT MC CLELLAN, ALA.
CBS-	COLUMBIA BROADCASTING SYSTEM, N.Y.C.
CBS ERD-	COLUMBIA BROADCASTING SYSTEM, N.Y.C. (ENG. RES.+DEV.)
CBS-MLR-	COLUMBIA BROADCASTING SYSTEM, INC. CBS LABORATORIES, STAMFORD, CONN.
CC-	(COMPUTATION CENTER) (REF. 2)
CC-	AMES LAB., IOWA (REF. 10)
CC-	ARGONNE NATIONAL LAB., LEMONT, ILL. (REF. 10)
CC-	CALIFORNIA. UNIV., BERKELEY. RADIATION LAB. (REF. 10)
CC-	CARNEGIE INST. OF TECH., PITTSBURGH (REF. 10)
CC-	CHEMICAL CORPS, ARMY CHEMICAL CENTER, MD.
CC-	CHICAGO. UNIV. METALLURGICAL LAB. (REF. 10)
CC-	CHRYSLER CORP., DETROIT
CC-	CLINTON LABS., OAK RIDGE, TENN. (REF. 10)
CC-(3 DIGITS)-	CONDUCTRON CORP., ANN ARBOR, MICH.
CC-	CONRAC CORP. NEW JERSEY DIV., FAIRFIELD
CC-(NO.)(AEPC)-	NATIONAL RESEARCH COUNCIL OF CANADA. ATOMIC ENERGY PROJECT, CHALK RIVER, ONT.
CC-(NO.)-H-	NATIONAL RESEARCH COUNCIL OF CANADA. ATOMIC ENERGY PROJECT, CHALK RIVER, ONT.
CC-	OFFICE OF RUBBER RESERVE. COMPOUNDERS COMMITTEE
CC-(YEAR)-	OREGON STATE UNIV., CORVALLIS. COMPUTER CENTER
CC-(LETTER(S))-	ARGONNE NATIONAL LAB., LEMONT, ILL. (REF. 10)
CC-AD-	CHRYSLER CORP. ENGINEERING DIV., DETROIT
CC-AME-(LETTERS)-	CHRYSLER CORP. MISSILE DIV.
CCB-	NAVAL RES. LAB. COMBINED COMMUNICATIONS BD.,WASH.,DC
CC-BIB-	CHRYSLER CORP. MISSILE DIV., DETROIT
CCBL-	ARMY CHEMICAL CORPS BIOLOGICAL LABS. PHYSICAL DEFENSE DIV., FORT DETRICK, MD.
CCBL/(LETTERS)-	ARMY CHEMICAL CORPS BIOL. LABS., CAMP DETRICK, MD.
CC BLIR-	ARMY CHEMICAL CORPS BIOLOGICAL LABS., CAMP DETRICK, MD. (INTERIM REPORTS)

CC BL-SPR-

CC BL-SPR-	ARMY CHEMICAL CORPS BIOLOGICAL LABS., CAMP DETRICK, MD. (SPECIAL REPORTS)
CCBP-	WAR DEPT. COMBINED COMMUNICATIONS BOARD (PROJECTS)
CCBP-	WAR DEPT. COMBINED COMMUNICATIONS BOARD (PUBLICATIONS)
CCC(NUMBER)-	AEROSPACE CORP., EL SEGUNDO, CALIF. (INTEROFFICE NO.)
CCC-(NO.)-TR-	BUREAU OF AERONAUTICS (NAVY)
CCC-	CALLERY CHEMICAL CO., PENNA.
CCC-(CONTRACT)	CALLERY CHEMICAL CO., PENNA.
CCC-(CONTRACT)-(LETTERS)-	CALLERY CHEMICAL CO., PENNA.
CCC-	CANADIAN ADMIRAL CORP., LTD., PORT CREDIT, ONT.
CCC7PC(NUMBER)-	CANADIAN COMMERCIAL CORP., OTTAWA
CCC-	CONSOLIDATED CONTROLS CORP., BETHEL, CONN.
CCC-ASR-(DATES)	CALLERY CHEMICAL CO., PENNA. (ANNUAL SUMMARY RPT.)
CC+CC-	COMPUTER COMMAND AND CONTROL CO., PHILADELPHIA
CCCC-(NUMBER-NUMBER)	COMPUTER COMMAND AND CONTROL CO., WASHINGTON, D.C.
CCC-HD-	CONSOLIDATED CONTROLS CORP., BETHEL, CONN.
CCC-CJB-	CLINTON LABS., OAK RIDGE, TENN.
CCCO-	CATALYTIC CONSTRUCTION CO., PHILADELPHIA
CCCO-(CONTRACT)-	CATALYTIC CONSTRUCTION CO., PHILADELPHIA
CCC-P/N-(NO.-NO.-NO.)	CARLETON CONTROLS CORP., EAST AURORA, N.Y.
CCC-QR-(NUMBER)(YEAR)	CALLERY CHEMICAL CO., PENNA. (QUARTERLY REPORT)
CC CRL IR-	CHEMICAL AND RADIOLOGICAL LABS., ARMY CHEMICAL CENTER MD. (INTERIM REPORT)
CC-CRLIR-	CHEMICAL AND RADIOLOGICAL LABS., ARMY CHEM. CTR., MD.
CC CRL R-	CHEMICAL AND RADIOLOGICAL LABS., ARMY CHEM. CTR., MD.
CC-CRLR-	CHEMICAL AND RADIOLOGICAL LABS., ARMY CHEM. CTR., MD.
CCDN-CI/	EUROPEAN NUCLEAR ENERGY AGENCY. NEUTRON DATA COMPILATION CENTRE, GIF-SUR-YVETTE, FRANCE
CCDN-NW-	EUROPEAN NUCLEAR ENERGY AGENCY. NEUTRON DATA COMPILATION CENTRE, GIF-SUR-YVETTE, FRANCE
CCDN/SYS/	EUROPEAN NUCLEAR ENERGY AGENCY. NEUTRON DATA COMPILATION CENTRE, GIF-SUR-YVETTE, FRANCE
CC ED TR-	CHRYSLER CORP. ENGINEERING DIV., DETROIT
CCED TR-	CHRYSLER CORP. ENGINEERING DIV., DETROIT
CC-EL-	CHRYSLER CORP. MISSILE OPERATIONS
CC-FRA-	CHEMICAL CORPS FIELD REQUIREMENTS AGENCY, FORT MCCLELLAN, ALA.
CC-HEC-D(NO.)	CHRYSLER CORP., DETROIT
CCI-	GT. BRIT. MIN. OF SUPPLY. CHEM. INSPECTORATE, LONDON
CCI-	MARTIN CO., BALTIMORE
CCI-	NATIONAL RESEARCH COUNCIL OF CANADA. ATOMIC ENERGY PROJECT, CHALK RIVER, ONT.
CCI-TD-	GT. BRIT. MIN. OF SUPPLY. CHEM. INSPECTORATE, LONDON
C+C-K-	CARBIDE AND CARBON CHEMICALS CORP. K-25 PLANT, OAK RIDGE, TENN.
CCK/TN/	IMPERIAL COLLEGE OF SCIENCE AND TECHNOLOGY, LONDON. DEPT. OF MECHANICAL ENG. (COMBUSTION KINETICS)
CCL-	ARMY COATING AND CHEMICAL LAB., ABERDEEN PROVING GROUND, MD.
CCL-	VIRGINIA. UNIV., CHARLOTTESVILLE. COBB CHEMICAL LAB.
CCMD-	CHRYSLER CORP. MISSILE DIV., DETROIT
CC MD R-	CHEMICAL CORPS. MEDICAL DIV., ARMY CHEMICAL CTR., MD.
CC MD RR-	CHEMICAL CORPS. MEDICAL DIV., ARMY CHEMICAL CENTER, MD. (RESEARCH REPORT)
CC MD SR-	CHEMICAL CORPS. MEDICAL DIV., ARMY CHEMICAL CENTER, MD. (SPECIAL REPORT)
CC-ML-	CHEMICAL CORPS MEDICAL LABS., ARMY CHEM. CTR., MD.
CCML-	CHEMICAL CORPS MEDICAL LABS., ARMY CHEM. CTR., MD.
CCMLR-	CHEMICAL CORPS MEDICAL LABS., ARMY CHEM.CTR.,MD.(RPTS)
CC-MT-	CHRYSLER CORP. MISSILE DIV., DETROIT
CC-MTC-	CHRYSLER CORP. MISSILE DIV., DETROIT
CC-MT-M(NO.)	CHRYSLER CORP. MISSILE DIV., DETROIT
CCN-	MINNEAPOLIS-HONEYWELL REGULATOR CO., ST. PETERSBURG, FLA
CCNDL-	ARMY CHEM.CORPS NUCLEAR DEFENSE LAB., ARMY CHEM.CTR, MD
CCNY-	CITY COLL., NEW YORK
CC-O(NUMBER)-(NUMBER)	CONDUCTRON CORP., ANN ARBOR, MICH.
CC-P-	CONDUCTRON CORP., ANN ARBOR, MICH.
CCP/(NO.)	GT. BRIT. ATOMIC EN. RES. ESTAB., HARWELL, BERKS, ENG
CC/POLARIS-	POLARIS MARK 2 RE-ENTRY BODY COORD. COMM.
CC-PR-	CHRYSLER CORP., DETROIT
CC-PR-(NUMBER,LETTER)-	CHRYSLER CORP. MISSILE DIV., DETROIT
CCR-	COORDINATION CENTER REVIEW
CCR-	WESTINGHOUSE ELECTRIC CORP. RES. LABS., PITTSBURGH
CCRE-	COWLES COMMISSION FOR RESEARCH IN ECONOMICS
CCREL-SR-	ARMY COLD REGIONS RES. AND ENG. LAB., HANOVER, N.H.
CC-RL-	CHRYSLER CORP. MISSILE DIV., DETROIT
CCRPA-M-	CENTRAL CONNECTICUT REGIONAL PLANNING AGENCY, PLAINVILLE
CCRPA-SP-	CENTRAL CONNECTICUT REGIONAL PLANNING AGENCY, PLAINVILLE
CCRTD-	GT. BRIT. COMMITTEE FOR CO-ORDINATION OF CATHODE RAY TUBE DEVELOPMENT
CCS-	CHEMICAL CORPS SCHOOL, ARMY CHEMICAL CENTER, MD.
CCS-	CHEMICAL CORPS SCHOOL, FORT MC CLELLAN, ALA.
CCS-	STANFORD UNIV., CALIF. INST. IN COMPUTER COORDINATED SYSTEMS
CCSI-	CONTINENTAL COPPER AND STEEL INDUSTRIES, INC., N.Y.C.
CCSL-	CAMP COLES SIGNAL LAB., BELMAR, N.J.
CCSL ACB EM-	CAMP COLES SIGNAL LABORATORY. APPLIED COMMUNICATIONS BRANCH, BELMAR, N.J. (ENGINEERING MEMORANDUM)
CCSL EM-	CAMP COLES SIGNAL LAB., BELMAR, N.J. (ENG. MEMO.)
CCSL ER-	CAMP COLES SIGNAL LAB., BELMAR, N.J. (ENG. REPT.)
CCSL MF PROJ-	CAMP COLES SIGNAL LAB., BELMAR, N.J. (MEMO. FILE)
CCSL TM-	CAMP COLES SIGNAL LAB., BELMAR, N.J. (TECH. MEMO.)
CCSL TR-	CAMP COLES SIGNAL LAB., BELMAR, N.J. (TEST REPT.)
CCT-	CONSULTANTS CUSTOM TRANSLATIONS, INC., N.Y.C.
CC-TB-	CHRYSLER CORP., DETROIT
CCTC-	CHEMICAL CORPS. TECH. COMMAND, ARMY CHEMICAL CTR.,MD.
CC TCIR-	CHEMICAL CORPS. TECHNICAL COMMAND, ARMY CHEMICAL CENTER, MD. (INFORMAL REPORT)
CC TCR-	CHEMICAL CORPS. TECH. COMMAND, ARMY CHEMICAL CTR.,MD.
CC/TERRIER-	TERRIER COORD. COMM., VULNERABILITY TEST SUBCOMM.
CCTM-(NUMBER-NUMBER)	ARMY STRATEGIC COMMUNICATIONS COMMAND, FORT HUACHUCA, ARIZ. (TECHNICAL MANUAL)
CC-TM-AME-M(NUMBER)	CHRYSLER CORP., DETROIT
CC-TR-	CHRYSLER CORP., DETROIT
CC-TR-(LTRS)-	CHRYSLER CORP. MISSILE DIV., DETROIT
CCT-TRANS-	CONSULTANTS CUSTOM TRANSLATIONS, INC., N.Y.C.
CC-WEPR-	CHRYSLER CORP. MISSILE OPERATIONS (WEEKLY ENGINEERING PROGRESS REPORT)
C+C-Y-	UNION CARBIDE NUCLEAR CO. Y-12 PLANT, OAK RIDGE, TENN
CD-	ARMY NATICK LABS., MASS.
CD-	CANADA. DEFENCE RESEARCH BOARD, OTTAWA
CD-	CONSOLIDATED VULTEE AIRCRAFT CORP., SAN DIEGO, CALIF.
CD-(YEAR)-	DOW CHEMICAL CO. ROCKY FLATS DIV., GOLDEN, COLO.
CD-(YEAR)-	DOW CHEMICAL CO. ROCKY FLATS PLANT, DENVER
CD/	GT. BRIT. CENTRAL ELECTRICITY GENERATING BOARD, LOND.
CD-	INTERNATL. BUSINESS MACHINES CORP., POUGHKEEPSIE, N.Y.
CD-	NATIONAL RESEARCH COUNCIL OF CANADA. ATOMIC ENERGY PROJECT, CHALK RIVER, ONT.

CD-	NORTH AMERICAN AVIATION, INC., LOS ANGELES
CD-	OAK RIDGE NATIONAL LAB., TENN.
CD-	OFFICE OF RUBBER RESERVE (POLYMER DEVELOPMENT)
CD-	TENNESSEE EASTMAN CORP., OAK RIDGE, TENN.
CD-(NO.)-(INITIALS)-	TENNESSEE EASTMAN CORP., OAK RIDGE, TENN.
CD-	TENNESSEE VALLEY AUTHORITY, OAK RIDGE (CHEMICAL ENGINEERING REPRINT)
CD-(INITIALS)-	TENNESSEE EASTMAN CORP., OAK RIDGE, TENN.
CDA-	CANADA. DEFENCE RESEARCH BOARD, OTTAWA
CD-AC-	TENNESSEE EASTMAN CORP., OAK RIDGE, TENN.
CD-BETA-S-	TENNESSEE EASTMAN CORP., OAK RIDGE, TENN.
CDC-	AEROPHYSICS DEVELOPMENT CORP., SANTA BARBARA, CALIF.
CDC-	BISSETT-BERMAN CORP., SANTA MONICA, CALIF.
CDC-	CHICAGO DEVELOPMENT CORP., RIVERDALE, MD.
CDC-	CONTROL DATA CORP. (ALL LOCATIONS)
CDC-	TENNESSEE EASTMAN CORP., OAK RIDGE, TENN.
CDC-A(NUMBER)-	CONTROL DATA CORP., BETHESDA, MD.
CDCEC-	ARMY COMBAT DEVELOPMENTS COMMAND. EXPERIMENTATION COMMAND, FORT ORD, CALIF.
CD-CHEM-S-	TENNESSEE EASTMAN CORP., OAK RIDGE, TENN.
CDC/HRD-	CONTROL DATA CORP. HOWARD RES. DIV., BETHESDA, MD.
CDC/NG (NUMBERS)	ARMY COMBAT DEVS. COMMAND. NUCLEAR GP., FT. BLISS, TEX
CD-CONF-	TENNESSEE EASTMAN CORP., OAK RIDGE, TENN.
CDC/OSWD (NUMBERS)	ARMY COMBAT DEVELOPMENTS COMMAND. OFFICE OF SPECIAL WEAPONS DEV., FT. BLISS, TEX.
CDD-(NUMBER-NUMBER)	BUREAU OF THE CENSUS, WASHINGTON, D.C.
CDD-	COLLINS RADIO CO., CEDAR RAPIDS, IOWA (DEVELOPMENT DESCRIPTION)
CDE-	FEDERAL PACIFIC ELECTRIC CO. CORNELL DUBILIER ELECTRONICS DIV., NORWOOD, MASS.
CDEC-	ARMY COMBAT DEVELOPMENTS COMMAND. EXPERIMENTATION CENTER, FORT ORD, CALIF.
CDED-	CLEVELAND DIESEL ENGINE DIV., GEN. MOTORS CORP., OHIO
CDEE-	GT. BRIT. CHEMICAL DEFENCE EXPERIMENTAL ESTABLISHMENT, PORTON, WILTS, ENGLAND
CDES-	GT. BRIT. CHEMICAL DEFENCE EXPERIMENTAL ESTABLISHMENT, PORTON, WILTS, ENGLAND
C.D.E.T.N.(NUMBER)	GT. BRIT. CHEMICAL DEFENCE EXPERIMENTAL ESTABLISHMENT, PORTON, WILTS, ENGLAND
CDE-TN-	GT. BRIT. CHEMICAL DEFENCE EXPERIMENTAL ESTABLISHMENT, PORTON, WILTS, ENGLAND
CDE-TP-	GT. BRIT. CHEMICAL DEFENCE EXPERIMENTAL ESTABLISHMENT, PORTON, WILTS, ENGLAND
C.D.E.T.P.(NUMBER)	GT. BRIT. CHEMICAL DEFENCE EXPERIMENTAL ESTABLISHMENT, PORTON, WILTS, ENGLAND
C.D.E.TP(NUMBER)	GT. BRIT. CHEMICAL DEFENCE EXPERIMENTAL ESTABLISHMENT, PORTON, WILTS, ENGLAND
CD-F-	TENNESSEE EASTMAN CORP., OAK RIDGE, TENN.
CDG-	ATOMIC ENERGY OF CANADA LTD., CHALK RIVER, ONT.
CD-GS-	TENNESSEE EASTMAN CORP., OAK RIDGE, TENN.
CDL-	GT. BRIT. CHEMICAL DEFENCE EXPERIMENTAL ESTABLISHMENT, PORTON, WILTS, ENGLAND
C.D.L.ACC/N-	UNITED KINGDOM. CENTRAL DOCKYARD LAB., PORTSMOUTH, HANTS, ENGLAND
CD-MAT-	UNITED KINGDOM. DIRECTORATE OF MATERIALS RESEARCH AND DEVELOPMENT, LONDON
CD-MPR-	TENNESSEE EASTMAN CORP., OAK RIDGE, TENN.
CD-MPR-191-	TENNESSEE EASTMAN CORP., OAK RIDGE, TENN.
CD-MPR-(INITIALS)-	TENNESSEE EASTMAN CORP., OAK RIDGE, TENN.
CDOG-	ARMY COMBAT DEVELOPMENTS COMMAND, FORT BELVOIR, VA. (COMBAT DEVELOPMENTS OBJECTIVES GUIDE)
CDP-	HUGHES AIRCRAFT CO., CULVER CITY, CALIF.
CD-PPO-	TENNESSEE EASTMAN CORP., OAK RIDGE, TENN.
CDP-TR-	RAYTHEON CO., NORWOOD, MASS.
CDR-	COLLINS RADIO CO., CEDAR RAPIDS, IOWA (DEV. REPORT)
CDR-	FAIRCHILD ENGINE AND AIRPLANE CORP. NEPA DIV., OAK RIDGE, TENN.
CDR-(NUMBER)FR(NUMBER)	GEODYNAMICS CORP., SANTA BARBARA, CALIF.
CDR-	GEORGIA. UNIV., ATHENS. (CIVIL DEFENSE RESEARCH)
CDR-	STRATEGIC BOMBING SURVEY (CIVILIAN DEFENSE REPORTS)
CDRA-	TREFILERIES ET LAMINOIRS DU HAVRE. CENTRE DE RECHERCHES, ANTONY, FRANCE
CDRD-(YEAR)-	SUNDSTRAND AVIATION, DENVER
CDRE-	GT. BRIT. CHEM. DEFENCE RES.EST., SUTTON OAK, LANCS, ENG
CDRE-	INDIA. CHEMICAL DEFENCE RESEARCH ESTABLISHMENT
CDR-P-	CALIFORNIA. UNIV., BERKELEY. INST. OF ENG. RES.
CDRP-	CALIFORNIA. UNIV., BERKELEY. INST. OF ENG. RES.
CDS-	COLLINS RADIO CO., CEDAR RAPIDS, IOWA (DESCRIPTIVE SPECIFICATION)
CDS-	TACTICAL AIR COMMAND, LANGLEY AFB, VA.
CD/SA-	GT. BRIT. HOME OFFICE. CIVIL DEFENCE DEPT. SCIENTIFIC ADVISERS BRANCH, LONDON
CDS-LN-(YEAR)-	BROWN UNIV., PROVIDENCE. CENTER FOR DYNAMICAL SYSTEMS
CDSP-	COLLINS RADIO CO., CEDAR RAPIDS, IOWA (DATA SYSTEMS PROPOSAL)
CD-SP-	TENNESSEE EASTMAN CORP., OAK RIDGE, TENN.
CDS-TR-(YEAR)-	BROWN UNIV., PROVIDENCE. CENTER FOR DYNAMICAL SYSTEMS
CDTA-(NUMBER)-	SOCIETE POUR L'ETUDE ET LA REALISATION D'ENGINS BALISTIQUES, COURBEVOIE, FRANCE
CE-(YEAR)-	AIR REDUCTION SALES CO. CRYOGENIC ENG.DEPT., UNION, NJ
CE-	CHICAGO. UNIV. METALLURGICAL LAB. (REF. 10)
CE-	CLINTON LABS., OAK RIDGE, TENN. (REF. 10)
C-E-	COMBUSTION ENGINEERING, INC. MARINE DEPT., WINDSOR, CONN.
CE-(YEAR)-	CONNECTICUT. UNIV., STORRS. DEPT. OF CIVIL ENG.
CE-	CORPS OF ENGINEERS (ARMY), FT. BELVOIR, VA.
CE-(NUMBER)-(LETTER)	COSMODYNE CORP., HAWTHORNE, CALIF.
CE-	CROWN ENGINEERING, ALBUQUERQUE, N. MEX.
CE-	GT. BRIT. ROYAL AIRCRAFT EST., FARNBOROUGH, HANTS, ENG
CE-(NO.)-PR-	MC GILL UNIV., MONTREAL. DEPT. OF CHEMISTRY
CE-(2 DIGITS)(YEAR)-	NEW MEXICO. UNIV., ALBUQUERQUE. BUREAU OF ENG. RES.
CE-	NEW ZEALAND. DOMINION LAB., WELLINGTON
CE-	SYRACUSE UNIV., N.Y. RES. INST. DEPT. OF CIVIL ENG.
CE-	VIRGINIA. UNIV., CHARLOTTESVILLE. RESEARCH LABS. FOR ENGINEERING SCIENCES
CEA-(NO.)(M-H)	CAMBRIDGE ELECTRON ACCELERATOR, MASS.
CEA-	FRANCE. COMMISSARIAT A L'ENERGIE ATOMIQUE (ALL LOCATIONS)
CEA-A-	FRANCE. COMMISSARIAT A L'ENERGIE ATOMIQUE. CENTRE D'ETUDES NUCLEAIRES, SACLAY
CEA-AR-(YEAR)-	FRANCE. COMMISSARIAT A L'ENERGIE ATOMIQUE, PARIS (ANNUAL REPORT)
CEA-ASS/EPR-	FRANCE. COMMISSARIAT A L'ENERGIE ATOMIQUE. SERVICE D'ETUDES DES PILES RAPIDES
CEA-BIB-	FRANCE. COMMISSARIAT A L'ENERGIE ATOMIQUE (ALL LOCATIONS) (BIBLIOGRAPHY)
CEA-CONF-	FRANCE. COMMISSARIAT A L'ENERGIE ATOMIQUE (ALL LOCATIONS) (CONFERENCE PAPERS)
CEAL-	CAMBRIDGE ELECTRON ACCELERATOR, MASS.
CEAL-TM-	CAMBRIDGE ELECTRON ACCELERATOR, MASS.
CEA-(M+H)-	CAMBRIDGE ELECTRON ACCELERATOR, MASS. (MASS. INST. OF TECH. + HARVARD UNIV.)
CEA-N-	FRANCE. COMMISSARIAT A L'ENERGIE ATOMIQUE. CENTRE D'ETUDES NUCLEAIRES (ALL LOCATIONS)
CE AND PE-	ROYAL CANADIAN AIR FORCE. CENTRAL EXPERIMENTAL AND PROVING ESTAB., ROCKCLIFFE, ONT.

CEA-NOTE-
       FRANCE. COMMISSARIAT A L'ENERGIE ATOMIQUE. CENTRE
          D'ETUDES NUCLEAIRES, SACLAY
CEA-PA-
       FRANCE. COMMISSARIAT A L'ENERGIE ATOMIQUE, PARIS
CEA/PA/RT-
       FRANCE. COMMISSARIAT A L'ENERGIE ATOMIQUE, PARIS
CEA-PS-
       FRANCE. COMMISSARIAT A L'ENERGIE ATOMIQUE. CENTRE
          D'ETUDES NUCLEAIRES (PUBLICATIONS SCIENTIFIQUES)
CEA-R-
       FRANCE. COMMISSARIAT A L'ENERGIE ATOMIQUE (ALL
          LOCATIONS)
CEA-S-
       FRANCE. COMMISSARIAT A L'ENERGIE ATOMIQUE. CENTRE
          D'ETUDES NUCLEAIRES, SACLAY
CEA-TP-
       FRANCE. COMMISSARIAT A L'ENERGIE ATOMIQUE, PARIS
CEA-TP-
       FRANCE. COMMISSARIAT A L'ENERGIE ATOMIQUE. CENTRE
          D'ETUDES NUCLEAIRES, SACLAY
CEA-TR-
       FRANCE. COMMISSARIAT A L'ENERGIE ATOMIQUE, PARIS
          (SERIES ASSIGNED BY THE AEC TO TRANSLATIONS
          RECEIVED FROM CEA)
CEA-TR-A-
       FRANCE. COMMISSARIAT A L'ENERGIE ATOMIQUE, PARIS
CEA-TR-R-
       FRANCE. COMMISSARIAT A L'ENERGIE ATOMIQUE, SACLAY
CEA-TR-X-
       FRANCE. COMMISSARIAT A L'ENERGIE ATOMIQUE, SACLAY
CEB-
       TENNESSEE VALLEY AUTHORITY, OAK RIDGE (CHEM.ENG.BULL)
CEC-
       CONSOLIDATED ELECTRODYNAMICS CORP., PASADENA, CALIF.
CEC-
       COOK ELECTRIC CO., MORTON GROVE, ILL.
CEC-
       ITALY. COMITATO NAZIONALE PER L'ENERGIA NUCLEARE,
          BOLOGNA
CEC CRL-(LETTERS)-
       COOK RESEARCH LABS., MORTON GROVE, ILL.
CECD-
       ARMY ENGINEER SCHOOL. COMBAT DEVS. GP., FT.BELVOIR,VA
CECLES-
       EUROPEAN SPACE VEHICLE LAUNCHER DEV. ORG., PARIS
CEC-M(NUMBER)-
       CONSOLIDATED ELECTRODYNAMICS CORP., PASADENA, CALIF.
CECNY-
       CONSOLIDATED EDISON CO. OF NEW YORK, INC.
CEC-PR-
       COOK ELECTRIC CO., MORTON GROVE, ILL.
CEC/SRAP/(INITIALS)-
       FRANCE. CENTRE D'ETUDES CRYOGENIQUES, SASSENAGE
C/ED-
       ARMY NATICK LABS. CLOTHING + EQUIP. DEV. BR., MASS.
C/ED-
       ARMY NATICK LABS. CLOTHING AND ORGANIC MATERIALS
          DIV., MASS.
CEDC-
       ARMY ENGINEER SCHOOL, FORT BELVOIR, VA.
CEE-
       BUREAU OF ORDNANCE (NAVY) (CAPTURED ENEMY EQUIPMENT)
CE EM-
       CORPS OF ENGINEERS (ARMY), FT. BELVOIR, VA.
          (ENGINEERING MANUAL)
C EES B-
       CONNECTICUT. ENG. EXPERIMENT STA. (BULLETIN)
CEF-
       EITEL-MC CULLOUGH, INC., SAN CARLOS, CALIF.
CEGB-RD/B/N-
       GT. BRIT. CENTRAL ELECTRICITY GENERATING BOARD.
          BERKELEY NUCLEAR LABS., BRISTOL, ENGLAND
CEI-
       ATOMIC EN. OF CANADA LTD. CHALK RIVER NUC. LABS., ONT
CEI-
       CENTURY ENGINEERS, INC., BURBANK, CALIF.
CEIS-
       SIGNAL INTELLIGENCE SERVICE, 849TH. CAPTURED ENEMY
          INTELLIGENCE BRANCH
CEL-
       COLLINS RADIO CO., CEDAR RAPIDS, IOWA (ENG. LETTER)
CEL-
       CONTINENTAL ELECTRONICS LTD., BROOKLYN
CELANESE-QR-
       CELANESE CHEMICAL CO. RES. + DEV. DEPT.,CLARKWOOD,TEX
CELESCO BD (NUMBER)-
       CELESTIAL RESEARCH CORP., SOUTH PASADENA, CALIF.
CEM-
       (CAPTURED ENEMY MATERIAL)
CEM-(NUMBER)-
       CENTER FOR THE ENVIRONMENT AND MAN,INC.,HARTFORD,CONN
CEMC-IMP-USP-(YR/YR-NO.)-
       SAO PAULO, BRAZIL. UNIVERSIDADE. CENTER OF STUDIES IN
          CELESTIAL MECHANICS
CEN-
       FRANCE. COMMISSARIAT A L'ENERGIE ATOMIQUE. CENTRE
          D'ETUDES NUCLEAIRES, FONTENAY-AUX-ROSES
CEN-
       NAVAL MISSILE CENTER (CENTER COMMAND)
CENC-
       COMBUSTION ENGINEERING, INC. NUCLEAR COMPONENTS
          ENGINEERING DEPT., CHATTANOOGA
CENCO-
       CENCO EDUCATIONAL FILM CO., CHICAGO
CEND-
       COMBUSTION ENG., INC. NUCLEAR DIV., IDAHO FALLS
CEND-
       COMBUSTION ENG., INC. NUCLEAR DIV., WINDSOR, CONN.

CEND-(NO.)-MD-(NO.)-
       COMBUSTION ENG., INC. NUCLEAR DIV., WINDSOR, CONN.
CEND-(NO.)-RS-(NO.)-
       COMBUSTION ENG., INC. NUCLEAR DIV., WINDSOR, CONN.
CEND/(NUMBER)/TP-
       COMBUSTION ENG., INC. NUCLEAR DIV., WINDSOR, CONN.
CEND-
       UNITED STATES-EURATOM JOINT RESEARCH + DEV. PROGRAM
CEND-PRWA-
       COMBUSTION ENGINEERING, INC., WINDSOR, CONN.
          (JOINT VENTURE)
CEND-PRWA-
       PUERTO RICO WATER RESOURCES AUTHORITY, SAN JUAN
          (JOINT VENTURE)
CEND/PRWRA-
       COMBUSTION ENGINEERING, INC., WINDSOR, CONN.
          (JOINT VENTURE)
CEND/PRWRA-
       PUERTO RICO WATER RESOURCES AUTHORITY, SAN JUAN
          (JOINT VENTURE)
CENDRD-
       COMBUSTION ENGINEERING, INC., WINDSOR, CONN.
CEND-S1C-
       COMBUSTION ENG., INC. NUCLEAR DIV., WINDSOR, CONN.
CENFAM-PV-
       ITALY. CENTRO NAZIONALE DI FISICA DELL ATMOSFERA
          E METEOROLOGIA, ROME
CENFAM-RDP-
       ITALY. CENTRO NAZIONALE DI FISICA DELL ATMOSFERA
          E METEOROLOGIA, ROME
CENFAR-
       FRANCE. COMMISSARIAT A L'ENERGIE ATOMIQUE. CENTRE
          D'ETUDES NUCLEAIRES, FONTENAY-AUX-ROSES
CENG-
       FRANCE. COMMISSARIAT A L'ENERGIE ATOMIQUE. CENTRE
          D'ETUDES NUCLEAIRES, GRENOBLE
CENP-
       COMBUSTION ENGINEERING, INC., N.Y.C.
CENPD-
       COMBUSTION ENGINEERING, INC. NUCLEAR POWER DEPT.,
          WINDSOR, CONN.
CEN-R.-
       BELGIUM. CENTRE D'ETUDE DE L'ENERGIE NUCLEAIRE,
          BRUSSELS
CEN-R-
       FRANCE. COMMISSARIAT A L'ENERGIE ATOMIQUE. CENTRE
          D'ETUDES NUCLEAIRES, SACLAY
CENRD-
       COMBUSTION ENGINEERING, INC. NAVAL REACTORS DIV.,
          WINDSOR, CONN.
CENRD-(NUMBER)-RS-
       COMBUSTION ENGINEERING, INC. NAVAL REACTORS DIV.,
          WINDSOR, CONN.
CENRD/HP-
       COMBUSTION ENGINEERING, INC. NAVAL REACTORS DIV.,
          WINDSOR, CONN.
CENRD-S1C-
       COMBUSTION ENGINEERING, INC. NAVAL REACTORS DIV.,
          WINDSOR, CONN.
CEN-S-PA-
       FRANCE. COMMISSARIAT A L'ENERGIE ATOMIQUE, PARIS
CENS/PA/RT-
       FRANCE. COMMISSARIAT A L'ENERGIE ATOMIQUE, PARIS
CENTRAL FILE NO.-
       (SEE DESIGNATION CF-)
CENTRAL NEVADA-
       GEOLOGICAL SURVEY, DENVER
CEP-
       COLLINS RADIO CO., CEDAR RAPIDS, IOWA (ENG. ARTICLE)
CEPE-
       ROYAL CANADIAN AIR FORCE. CENTRAL EXPERIMENTAL AND
          PROVING ESTAB., ROCKCLIFFE, ONT.
CEPRE-
       GT. BRIT. CLOTHING + EQUIP. PHYSIOLOGICAL RES. ESTAB.
CEPS-
       COMMONWEALTH EDISON CO., CHICAGO (JOINT VENTURE)
CEPS-
       PUBLIC SERVICE OF NORTHERN ILL., CHICAGO (JT.VENTURE)
CER-
       (CIVIL ENGINEERING REPORT) (REF. 2)
CER-
       COLLINS RADIO CO., CEDAR RAPIDS, IOWA (ENG. REPORT)
CER-(NO.)-
       COLORADO STATE UNIV., FORT COLLINS
CER(YR.-YR.)(INITIALS)-
       COLORADO STATE UNIV., FORT COLLINS. DEPT. OF CIVIL
          ENGINEERING
CER(YEAR)(INITIALS)-
       COLORADO STATE UNIV., FORT COLLINS. DEPT. OF CIVIL
          ENGINEERING
CER-
       COLUMBIA UNIV., N.Y.C.
CER-(NUMBER)-
       ROCKETDYNE DIV., NORTH AMERICAN AVIATION, INC.,
          CANOGA PARK, CALIF.
CER-
       TENNESSEE VALLEY AUTHORITY, OAK RIDGE (CHEM.ENG.RPT.)
CERCA-
       UNITED KINGDOM. COMMONWEALTH AND EMPIRE RADIO FOR
          CIVIL AVIATION, LONDON
CERC-MISC PAPER-
       ARMY COASTAL ENGINEERING RES. CENTER, WASHINGTON, DC
CERC-TM-
       ARMY COASTAL ENGINEERING RES. CENTER, WASHINGTON, DC
CER-D(NUMBER)-
       COLLINS RADIO CO., DALLAS
CERD-
       COMBUSTION ENG.,INC. REACTOR DEV. DIV.,WINDSOR,CONN.
CERD-S1C-
       COMBUSTION ENG.,INC. REACTOR DEV. DIV.,WINDSOR,CONN.

CERL-TR-M-	ARMY CONSTRUCTION ENGINEERING RESEARCH LAB., CHAMPAIGN, ILL.
CERN-(NUMBER)/TH.(NUMBER)	EUROPEAN COUNCIL FOR NUCLEAR RESEARCH, GENEVA
CERN-	EUROPEAN ORGANIZATION FOR NUCLEAR RESEARCH, GENEVA
CERN-(YEAR)-	EUROPEAN ORGANIZATION FOR NUCLEAR RESEARCH, GENEVA
CERN-AR-(YEAR)	EUROPEAN ORGANIZATION FOR NUCLEAR RESEARCH, GENEVA
CERN-BIB-	EUROPEAN ORGANIZATION FOR NUCLEAR RESEARCH, GENEVA
CERN-BS-	EUROPEAN ORGANIZATION FOR NUCLEAR RESEARCH, GENEVA
CERN/DD/DH-(YEAR)/(NO.)	EUROPEAN ORGANIZATION FOR NUCLEAR RESEARCH, GENEVA
CERN/DI/HP-	EUROPEAN ORGANIZATION FOR NUCLEAR RESEARCH, GENEVA
CERN/EAG/(NUMBER)	EUROPEAN ORGANIZATION FOR NUCLEAR RESEARCH. EUROPEAN ACCELERATOR STUDY GROUP, GENEVA
CERN/ECFA-(YEAR)/(NUMBER)	EUROPEAN ORGANIZATION FOR NUCLEAR RESEARCH. EUROPEAN COMMITTEE FOR FUTURE ACCELERATORS, GENEVA
CERN/ECFA-(YR)/WG(NO)-	EUROPEAN ORGANIZATION FOR NUCLEAR RESEARCH. EUROPEAN COMMITTEE FOR FUTURE ACCELERATORS, GENEVA (FINAL SECTIONS OF CODE ARE AUTHORS' INITIALS)
CERN/FSG/(NUMBER)	EUROPEAN ORGANIZATION FOR NUCLEAR RESEARCH. EUROPEAN STUDY GROUP ON FUSION, GENEVA
CERN/GEN/-	EUROPEAN COUNCIL FOR NUCLEAR RESEARCH, GENEVA
CERN/HER/(INITLS)(NO)(YR)	EUROPEAN ORGANIZATION FOR NUCLEAR RESEARCH, GENEVA
CERN/HERA/(YEAR)-	EUROPEAN ORGANIZATION FOR NUCLEAR RESEARCH, GENEVA
CERN-INT-MPS/EP-	EUROPEAN ORGANIZATION FOR NUCLEAR RESEARCH, GENEVA
CERN-ISR-MA/(YEAR)-	EUROPEAN ORGANIZATION FOR NUCLEAR RESEARCH, GENEVA
CERN-MPS-(INITLS)-(YEAR)-	EUROPEAN ORGANIZATION FOR NUCLEAR RESEARCH, GENEVA
CERN-NP/(INITLS)-(YEAR)-	EUROPEAN ORGANIZATION FOR NUCLEAR RESEARCH, GENEVA
CERN-NPA/(INITLS)-(YEAR)-	EUROPEAN COUNCIL FOR NUCLEAR RESEARCH, GENEVA
CERN-P-	EUROPEAN COUNCIL FOR NUCLEAR RESEARCH, GENEVA
CERN-PS-(INITIALS)-	EUROPEAN ORGANIZATION FOR NUCLEAR RESEARCH. PROTON-SYNCHROTRON GROUP, GENEVA
CERN/T/(INITIALS)/-	EUROPEAN COUNCIL FOR NUCLEAR RESEARCH, GENEVA
CERN/TC/(INITLS)-(YEAR)-	EUROPEAN ORGANIZATION FOR NUCLEAR RESEARCH, GENEVA
CERN/TC/BEBC-(YR.)-	EUROPEAN ORGANIZATION FOR NUCLEAR RESEARCH, GENEVA
CERN-TH-	EUROPEAN ORGANIZATION FOR NUCLEAR RESEARCH, GENEVA
CERN TRANS (YEAR)-	EUROPEAN ORGANIZATION FOR NUCLEAR RESEARCH, GENEVA
CERPA-ETUDE-	FRANCE. CENTRE D'ETUDES ET DE RECHERCHES. SERVICE DE PSYCHOLOGIE APPLIQUEE, TOULON
CERS-	EUROPEAN SPACE RESEARCH ORGANIZATION, PARIS
CERTS-(NUMBER)-	FRANCE. CENTRE D'ETUDES ET DE RECHERCHES EN TECHNOLOGIE SPATIALE, TOULOUSE
CERTS-NT-(NUMBER)-	FRANCE. CENTRE D'ETUDES ET DE RECHERCHES EN TECHNOLOGIE SPATIALE, TOULOUSE
CERTS-TN-(NUMBER-NUMBER)	FRANCE. CENTRE D'ETUDES ET DE RECHERCHES EN TECHNOLOGIE SPATIALE, TOULOUSE
CES-	BELGIUM. CENTRE D'ETUDES SOCIALES, BRUSSELS
CES-	GT. BRIT. ROYAL NAVAL PERSONNEL RESEARCH COMMITTEE
CES-	OFF. OF THE ASST. SECY. OF DEFENSE(RES.+DEV.),WASH,DC
CES-(LETTER-NUMBER)	INTERNATIONAL BUSINESS MACHINES CORP. CENTER FOR EXPLORATORY STUDIES, CAMBRIDGE, MASS.
CESE-	SIGNAL CORPS (ARMY) (CAPTURED ENEMY SIGNAL EQUIPMENT)
CESE TLI-	SIGNAL CORPS (ARMY) (CAPTURED ENEMY SIGNAL EQUIPMENT. TECHNICAL LIAISON INTELLIGENCE)
CESL-	CAMP EVANS SIGNAL LAB., BELMAR, N.J.
CESL (NO.)(LETTER)/	CAMP EVANS SIGNAL LAB., BELMAR, N.J.
CESL DPB EM-	CAMP EVANS SIGNAL LABORATORY, DEVELOPMENT PLANNING BRANCH, BELMAR, N.J. (ENGINEERING MEMO)
CESL RPSR-	CAMP EVANS SIGNAL LABORATORY. RADIO PROPAGATION BRANCH, BELMAR, N.J.
CESL-SPEC-	CAMP EVANS SIGNAL LAB., BELMAR, N.J. (SPECIFICATION)
CESL TL-UK-	CAMP EVANS SIGNAL LAB., BELMAR, N.J.
CESL TM-	CAMP EVANS SIGNAL LAB., BELMAR, N.J. (TECH. MEMO.)
CESL TR-	CAMP EVANS SIGNAL LAB., BELMAR, N.J. (TECH. REPT.)
CESR-(NUMBER)-(NUMBER)	FRANCE. CENTRE D'ETUDE SPATIALE DES RAYONNEMENTS, TOULOUSE
CESRO-SN-	EUROPEAN SPACE DATA CENTRE, DARMSTADT, GERMANY
CETEC-FR-	CONSOLIDATED ENG. TECHNOLOGY CORP.,MOUNTAIN VIEW,CAL.
CETG-	NEVADA TEST ORGANIZATION. CIVIL EFFECTS TEST GP., AEC
CETIS-	EUROPEAN ATOMIC ENERGY COMMUNITY. EUROPEAN SCIENTIFIC DATA PROCESSING CENTER, ISPRA, ITALY
CE/TN-	AUSTRALIAN ATOMIC ENERGY COMMISSION. RESEARCH ESTABLISHMENT, LUCAS HEIGHTS, NEW SOUTH WALES
CEW-TEC-	CARBIDE AND CARBON CHEMICALS CORP. Y-12 PLANT, OAK RIDGE, TENN.
CEX-	DIVISION OF BIOLOGY AND MEDICINE. CIVIL EFFECTS TEST OPERATIONS, AEC (SERIES ASSIGNED TO REPORTS PREPARED BY VARIOUS ORGANIZATIONS)
CF-	(CENTRAL FILES) (REF. 2)
CF-	(CENTRAL FILES MEMORANDUM) (REF. 2)
CF-	AMERICAN CHAIN AND CABLE CO., INC., DETROIT
CF-	ARGONNE NATIONAL LAB., LEMONT, ILL. (REF. 10)
CF-	BELGIUM. INSTITUT D'AERONOMIE SPATIALE DE BELGIQUE, BRUSSELS
CF-	CHICAGO. UNIV. METALLURGICAL LAB.
CF-(YR-MO)-	CLINTON LABS., OAK RIDGE, TENN.
CF-	GT. BRIT. AERONAUTICAL RESEARCH COUNCIL
CF-5001 THRU -5256	GT. BRIT. DEPT. OF ATOMIC ENERGY. INDUSTRIAL GROUP H.Q., RISLEY, LANCS, ENGLAND (REF. 36)
CF-	GT. BRIT. WINDSCALE WORKS, SELLAFIELD, CUMB., ENGLAND
CF-	JOHNS HOPKINS UNIV., SILVER SPRING, MD. APPLIED PHYSICS LAB. (BUMBLEBEE PROJECT) (REF. 8)
CF-(YEAR)-(MONTH)-	MOUND LAB., MIAMISBURG, OHIO
CF-	NATIONAL DEFENSE RESEARCH COMMITTEE
CF-(YR-MO)-	OAK RIDGE NATIONAL LAB., TENN.
CF-(YEAR)-(MONTH)-	OAK RIDGE SCHOOL OF REACTOR TECHNOLOGY, TENN.
CF-	OFFICE OF SCI. RES. + DEV. (CONTROLLED FRAGMENTATION)
CF-	UNITED KINGDOM ATOMIC ENERGY AUTHORITY. INDUSTRIAL GROUP H.Q., RISLEY, LANCS, ENGLAND (REF. 36)
CF-1498-A THRU -1509	WISCONSIN. UNIV., MADISON. DEPT. OF CHEMISTRY
CFA-R-	FRANCE. COMMISSARIAT A L'ENERGIE ATOMIQUE. CENTRE D'ETUDES NUCLEAIRES, SACLAY
CFB-	BRAUN (C. F.) + CO., ALHAMBRA, CALIF.
CFB-	CHICAGO. UNIV. METALLURGICAL LAB. (ASSIGNED TO REPORTS FROM GREAT BRITAIN) (REF. 10)
CFC-	CAPEHART-FARNSWORTH CORP., FORT WAYNE
CFC-TR-	CERAMIC FINISHING CO., STATE COLLEGE, PENNA.
CFE-	GT. BRIT. CENTRAL FIGHTER ESTABLISHMENT
CFE/GFE POLICY STUDY-	PEAT MARWICK AND LIVINGSTON CO., WASHINGTON, D.C.
CFHQ-TRIAL-	CANADIAN FORCES HEADQUARTERS, OTTAWA
CF+I-	COLORADO FUEL + IRON CORP., PUEBLO
CF/JAD/(NUMBER/YEAR)	NOTTINGHAM, ENGLAND. UNIV. DEPT. OF GEOGRAPHY (CURVE FITTINGS)(AUTHOR'S INITIALS)
CFP-	CANADIAN FORCES HEADQUARTERS, OTTAWA
CF-P-	GT. BRIT. DEPT. OF ATOMIC ENERGY. INDUSTRIAL GROUP H.Q., RISLEY, LANCS, ENGLAND
CFP-CED-GA-(YEAR)	COOSA VALLEY AREA PLANNING + DEVELOPMENT COMMISSION, ROME, GA.
CFR-	BUREAU OF COMMERCIAL FISHERIES, WASHINGTON, D.C. (COMMERCIAL FISHERIES REVIEW)
CFR-	NATIONAL ARCHIVES AND RECORDS SERVICE, WASHINGTON, DC (CODE OF FEDERAL REGULATIONS)
CFRSWP/P(YEAR)(NUMBER)	UNITED KINGDOM ATOMIC ENERGY AUTHORITY. WEAPONS GROUP ATOMIC WEAPONS RES. ESTAB.,ALDERMASTON, BERKS, ENG.
CFS-	AIR FORCE CAMBRIDGE RESEARCH CENTER, MASS.
CFS-	BUREAU OF COMMERCIAL FISHERIES, WASHINGTON, D.C. (CURRENT FISHERIES STATISTICS)
CFSTI-(YEAR)-	CLEARINGHOUSE FOR FEDERAL SCIENTIFIC AND TECHNICAL INFORMATION, SPRINGFIELD, VA.

CFSTI-BIB-(YEAR)-
    CLEARINGHOUSE FOR FEDERAL SCIENTIFIC AND TECHNICAL INFORMATION, SPRINGFIELD, VA.

CG-
    COAST GUARD, WASHINGTON, D.C.

CG-(SUBJ.LTRS)-(EDIT.NO.)
    DIVISION OF CLASSIFICATION, AEC, WASHINGTON, D.C. (CLASSIFICATION GUIDE)

CG-
    NAVAL TORPEDO STATION, KEYPORT, WASH.

CG-
    PHILCO-FORD CORP. SPACE AND RE-ENTRY SYSTEMS DIV., NEWPORT BEACH, CALIF.

CG-
    TENNESSEE EASTMAN CORP., OAK RIDGE, TENN.

CGD-
    BUREAU OF AERONAUTICS (NAVY)

CGE-
    C-G ELECTRONICS, INC., ALBUQUERQUE, N. MEX.

CGE-
    CANADIAN GENERAL ELECTRIC CO., LTD., TORONTO

CGM-
    SPACE TECHNOLOGY LABS., INC., REDONDO BEACH, CALIF.

CGRD-
    COAST GUARD, WASHINGTON, D.C.

CGR-EMC-
    CENTRAL GEEIA REGION, TINKER AFB, OKLAHOMA CITY

CGS-
    CGS LABS., INC., STAMFORD, CONN.

C/GS-
    COAST AND GEODETIC SURVEY, ROCKVILLE, MD.

C+GS-
    NATIONAL OCEAN SURVEY, ROCKVILLE, MD.

CGS-(NUMBER)-
    NATIONAL OCEAN SURVEY, ROCKVILLE, MD.

CGS-
    OFFICE OF THE ASST. SECY. OF DEFENSE (RES. + DEV.) RES. + DEV. COORDINATING COMM. ON GEN. SCIENCES

CGS-
    RESEARCH + DEV. BD. COMM. ON GEN. SCIENCES, WASH.,DC

CGS-
    RES.+ DEV.BD. COMM. ON GEOPHYSICAL SCIENCES, WASH.,DC

CGSC-
    ARMY COMMAND + GEN. STAFF COLL., FT. LEAVENWORTH,KAN.

CGS-C-
    COAST AND GEODETIC SURVEY, LAS VEGAS, NEV.

CGS-C-
    COAST AND GEODETIC SURVEY, ROCKVILLE, MD.

C/GSDR-
    COAST AND GEODETIC SURVEY, FREDERICKSBURG, VA.

CGS-E-
    COAST AND GEODETIC SURVEY, LAS VEGAS, NEV.

CGS-O-
    COAST AND GEODETIC SURVEY, LAS VEGAS, NEV.

CGS-P-
    COAST AND GEODETIC SURVEY, LAS VEGAS, NEV.

CGS-P-
    COAST AND GEODETIC SURVEY, ROCKVILLE, MD.

C/GSTM-
    COAST AND GEODETIC SURVEY, ROCKVILLE, MD. (TECH MEMO)

C+GSTM-
    NATIONAL OCEAN SURVEY, ROCKVILLE, MD.

CGS-TR-
    CGS SCIENTIFIC CORP., WATERTOWN, MASS.

CGTD-
    COAST GUARD, WASHINGTON, D.C.

CG-UF-
    DIVISION OF CLASSIFICATION, AEC, WASHINGTON, D.C.

CGW-
    CORNING GLASS WORKS, N.Y.

CGYD-
    COAST GUARD (YARD)

CH-
    ARGONNE NATIONAL LAB., LEMONT, ILL. (REF. 10)

CH-
    CALIFORNIA. UNIV., BERKELEY. RADIATION LAB. (REF. 10)

CH-23-R-
    CALIFORNIA INST. OF TECH., PASADENA

CH-
    CHICAGO. UNIV.

CH-(NUMBER)(INS)
    CHICAGO. UNIV. INST. FOR NUCLEAR STUDIES

CH-
    CLINTON LABS., OAK RIDGE, TENN. (REF. 10)

CH-(NUMBER)-(ENGLAND)
    HILGER + WATTS LTD., LONDON

CH-
    STAUFFER CHEMICAL CO. CHAUNCEY RESEARCH CENTER, N.Y.

CHABA-
    NATIONAL RESEARCH COUNCIL. COMMITTEE ON HEARING AND BIO-ACOUSTICS, WASHINGTON, D.C.

CHAM-
    CHAMBERLAIN CORP., WATERLOO, IOWA

CHART X-
    CALIFORNIA. UNIV., BERKELEY. RADIATION LAB.

CHASE-
    CHASE AIRCRAFT CO., INC., WEST TRENTON, N.J.

CHAT-
    CHATHAM ELECTRONICS CORP., LIVINGSTON, N.J.

CHATSAV-(YEAR-NUMBER)
    CHATHAM COUNTY-SAVANNAH METROPOLITAN PLANNING COMMISSION, GA.

CH CC-
    CHICAGO. UNIV. METALLURGICAL LAB.

CH-CPR-
    BATTELLE MEMORIAL INST. RADIATION EFFECTS INFORMATION CENTER, COLUMBUS, OHIO

CH-CPR-
    GENERAL ELECTRIC CO. ADVANCED ELECTRONICS CENTER, ITHACA, N.Y.

CH-DM-
    CHICAGO. UNIV. DEPT. OF METEOROLOGY

CHE-
    BUREAU OF RECLAMATION (INTERIOR). CHEMICAL ENG. BR., DENVER

CH-E-
    IIT RESEARCH INST., CHICAGO

CHE (NO.-NO.-LTR.-NO.)
    SYRACUSE UNIV., N.Y. RESEARCH INST.

CHECCHI-
    CHECCHI AND CO., WASHINGTON, D.C.

CHECIT-PL-(YEAR)-
    CALIFORNIA INST. OF TECH., PASADENA. CHEMICAL ENGINEERING-POLYMER LAB.

CHEFINS-(YR)-
    CHICAGO. UNIV. ENRICO FERMI INST. FOR NUCLEAR STUDIES

CH-E/INT.R.-
    KEURING VAN ELECTROTECHNISCHE MATERIALEN, N.V., ARNHEIM, NETHERLANDS

CHEM (NO.-NO.)
    INDIANA UNIV., BLOOMINGTON. CHEMISTRY DEPT.

CHEM-
    TENNESSEE EASTMAN CORP., OAK RIDGE, TENN.

CHEM-(YEAR)
    UNION OF SOUTH AFRICA. COUNCIL FOR SCIENTIFIC AND INDUSTRIAL RESEARCH, PRETORIA

CHEM CD-
    TENNESSEE EASTMAN CORP., OAK RIDGE, TENN.

CHEM/EX-
    GT. BRIT. ADVISORY COUNCIL ON SCI. RES. + TECH. DEV.

CHEMICO-
    CHEMICAL CONSTRUCTION CORP., N.Y.C.

CHEM M-
    CALIFORNIA. UNIV., BERKELEY. RADIATION LAB. (REF. 9)

CHEM S-
    CALIFORNIA. UNIV., BERKELEY. RADIATION LAB. (REF. 9)

CHEM-S-
    CHICAGO. UNIV. METALLURGICAL LAB. (ASSIGNED TO REPORTS FROM VARIOUS AMERICAN SOURCES)

CHEM-S-
    TENNESSEE EASTMAN CORP., OAK RIDGE, TENN.

CHEMSTRAND-QPR-
    CHEMSTRAND RESEARCH CENTER, INC., DURHAM, N.C.

CHIAA-RR-
    CROP HAIL INSURANCE ACTUARIAL ASSN., CHICAGO

CHIC U LAS-TR-
    CHICAGO. UNIV. LABS. FOR APPLIED SCIENCES

CH-IM-
    CHICAGO. UNIV. DEPT. OF METEOROLOGY

CHINA EEIS SIG IC PR-
    SIGNAL CORPS (ARMY). ENEMY EQUIPMENT INTELLIGENCE DIV. CHINA THEATER (PRELIMINARY REPORT)

CH-(INS)-
    CHICAGO. UNIV. INST. FOR NUCLEAR STUDIES

CHIRIKOF-
    GEOLOGICAL SURVEY, DENVER (CHIRIKOF ISLAND, ALASKA)

CH-ISR-
    CHICAGO. UNIV. INST. FOR SYSTEM RESEARCH

CHI U-
    CHICAGO. UNIV.

CHIVE/R-
    CENTRAL INTELLIGENCE AGENCY. OFFICE OF COMPUTER SERVICE, WASHINGTON, D.C.

CH LAS TR-
    CHICAGO. UNIV. LABS. FOR APPLIED SCIENCES

CHN-
    CHICAGO. UNIV. METALLURGICAL LAB. (REF. 10)

CHP-
    CLEAVER-HUME PRESS, LTD., LONDON (TRANSLATIONS)

CHR-
    CHRYSLER CORP. MISSILE OPERATIONS, DETROIT

CHR-
    CONNECTICUT HARD RUBBER CO., NEW HAVEN

CHRISTMAS TREE-
    GEOLOGICAL SURVEY, DENVER

CH-RL-
    CHICAGO. UNIV. AIR FORCE RADIATION LAB.

CHRYSLER-(LETTERS...)
    CHRYSLER CORP., DETROIT

CH-U-
    CHICAGO. UNIV.

CHU-
    CHICAGO. UNIV.

CHU-IM-
    CHICAGO. UNIV. DEPT. OF METEOROLOGY

CHU-INS-
    CHICAGO. UNIV. INST. FOR NUCLEAR STUDIES

CH U/LAS TR-
    CHICAGO. UNIV. LABS. FOR APPLIED SCIENCES

CHU-LMSS-
    CHICAGO. UNIV. LAB. OF MOLECULAR STRUCTURE + SPECTRA

CH USAF R-
    CHICAGO. UNIV. AIR FORCE RADIATION LAB.

CI-
    ATOMIC ENERGY OF CANADA LTD. CHALK RIVER PROJECT, ONT. (REF. 6)

CI-
    CHALMERS TEKNISKA HOEGSKOLA, GOETEBORG

CI-
    CURTISS-WRIGHT CORP.

CI-
    DENMARK. TECHNICAL UNIV.,COPENHAGEN. COASTAL ENG.LAB.

CI-
    GT. BRIT. MIN. OF SUPPLY. CHEM. INSPECTORATE, LONDON

CI-
    IDAHO OPERATIONS OFFICE, AEC, IDAHO FALLS

CI-
    JOHNSON (ARNE) INGENIEURBUERO, STOCKHOLM

CI-
    TORONTO. UNIV.

```
CIA-
      CENTRAL INTELLIGENCE AGENCY, WASHINGTON, D.C.
CIA/BGI-CT-(YEAR-NUMBER)
      CENTRAL INTELLIGENCE AGENCY, WASHINGTON, D.C.
CIACGI-CD-(YEAR-NUMBER)
      CENTRAL INTELLIGENCE AGENCY, WASHINGTON, D.C.
CIA FDD T-
      CENTRAL INTELLIGENCE AGENCY. FOREIGN DOCUMENTS DIV.
      WASHINGTON, D.C. (CONSOLIDATED TRANSLATION SURVEY)
CIA NIE-(DATE)
      CENTRAL INTELLIGENCE AGENCY, WASHINGTON, D.C.
      (NATIONAL INTELLIGENCE ESTIMATE)
CIA NIS-
      CENTRAL INTELLIGENCE AGENCY, WASHINGTON, D.C.
      (NATIONAL INTELLIGENCE SURVEY)
CIA/OSI-(LTR)-(LTRS)(YR)
      CENTRAL INTELL. AGENCY. OFF. OF SCI. INTELL.,WASH.,DC
CIA/SI-(NUMBER-YEAR)
      CENTRAL INTELL. AGENCY. OFF. OF SCI. INTELL.,WASH.,DC
CIC-
      CHEMICAL INST. OF CANADA, OTTAWA
CIC-
      CHESAPEAKE INSTRUMENT CORP., SHADYSIDE, MD.
CIC-(NUMBER.NUMBER)F
      CHESAPEAKE INSTRUMENT CORP., SHADYSIDE, MD.
CIC-E(NUMBER.NUMBER)-
      CALIFORNIA. UNIV., LIVERMORE. LAWRENCE RADIATION LAB.
      COMPUTER INFORMATION CENTER
CIC-MANUAL-
      CALIFORNIA. UNIV., LIVERMORE. LAWRENCE RADIATION LAB.
      COMPUTER INFORMATION CENTER
CIC-MP-
      CALIFORNIA. UNIV., LIVERMORE. LAWRENCE RADIATION LAB.
      COMPUTER INFORMATION CENTER
CIC-N(NUMBER.NUMBER)-
      CALIFORNIA. UNIV., LIVERMORE. LAWRENCE RADIATION LAB.
      COMPUTER INFORMATION CENTER
CIC-P(NUMBER)-
      CALIFORNIA. UNIV., LIVERMORE. LAWRENCE RADIATION LAB.
      COMPUTER INFORMATION CENTER
CIC-TR-
      CLINICAL INVESTIGATION CENTER, OAKLAND, CALIF.
CID-
      HUGHES AIRCRAFT CO., CULVER CITY, CALIF. (COCKPIT
      INSTRUMENT DISPLAYS)
CIDNT-
      POLAND. CENTRALNY INSTYTUT DOKUMENTACJI NAUKOWO-
      TECHNICZEJ, WARSAW (TRANSLATIONS)
CIDS-
      PENNSYLVANIA. UNIV., PHILA. INST. FOR COOP. RES.
CIDS-STATUS-
      ARMY CHEMICAL INFORMATION AND DATA SYSTEM
CIEA-
      MEXICO. INSTITUTO POLITECNICO NACIONAL. CENTRO DE
      INVESTIGACION Y DE ESTUDIOS AVANZADOS, MEXICO CITY
CII-R(NUMBER)
      CONSULTANTS INTERNATIONAL, INC., DALLAS
CIIR-
      NORWAY. SENTRALINSTITUTT FOR INDUSTRIELL FORSKNING,
      OSLO
CIIR-PUB-
      NORWAY. SENTRALINSTITUTT FOR INDUSTRIELL FORSKNING,
      OSLO
CIIR-SI PUBL.-
      NORWAY. SENTRALINSTITJTT FOR INDUSTRIELL FORSKNING,
      OSLO
CIIR-ST.-
      NORWAY. SENTRALINSTITUTT FOR INDUSTRIELL FORSKNING,
      OSLO
CIMA-
      NATIONAL RESEARCH COUNCIL. PACIFIC SCIENCE BOARD
CI/MEMO-
      GT. BRIT. MIN. OF SUPPLY. CHEM. INSPECTORATE, LONDON
CI/METHOD-EA-
      GT. BRIT. MINISTRY OF SUPPLY
CI/METHOD-EB-
      GT. BRIT. MINISTRY OF SUPPLY
CI/METHOD-EC-
      GT. BRIT. MINISTRY OF SUPPLY
CI/METHOD-ED-
      GT. BRIT. MINISTRY OF SUPPLY
CI/METHOD-EE-
      GT. BRIT. MINISTRY OF SUPPLY
CIM-R-
      UNITED KINGDOM ATOMIC ENERGY AUTHORITY. RESEARCH GP.
      CULHAM LAB., ABINGDON, BERKS, ENGLAND
CIN-
      CINCINNATI. UNIV.
CINDA-
      CHRYSLER CORP., NEW ORLEANS (CHRYSLER IMPROVED
      NUMERICAL DIFFERENCING ANALYZER)
CINDA-
      COLUMBIA UNIV., N.Y.C. CINDA CENTRE
CINU-
      CINCINNATI. UNIV.
CIOS-
      COMBINED INTELLIGENCE OBJECTIVES SUB-COMMITTEE
CIOS (ROMAN NUMERAL)-
      COMBINED INTELLIGENCE OBJECTIVES SUB-COMMITTEE
CIOS DOC-
      COMBINED INTELLIGENCE OBJECTIVES SUB-COMM. (DOCUMENT)
CIOS ER-
      COMBINED INTELL. OBJECTIVES SUB-COMM.(EVALUATION RPT)
CIOS-ITEM-
      COMBINED INTELLIGENCE OBJECTIVES SUB-COMMITTEE
CIOS MIRS (LETTER)-
      COMBINED INTELLIGENCE OBJECTIVES SUB-COMMITTEE.
      (MILITARY INTELLIGENCE. REPORTS SECTION)
CIOS OTR-
      COMBINED INTELLIGENCE OBJECTIVES SUB-COMMITTEE.
      (ORDNANCE TARGET REPORTS)
CIOS-STATUS-
      COMBINED INTELLIGENCE OBJECTIVES SUB-COMMITTEE
CIOS T(NO./NO.)
      COMBINED INTELLIGENCE OBJECTIVES SUB-COMM. (TARGET)
CIOS TRIP-
      JOINT INTELLIGENCE OBJECTIVES AGENCY AND BRITISH
      INTELLIGENCE OBJECTIVES SUB-COMM. (COMBINED TRIPS)
CIP/
      GT. BRIT. MIN. OF SUPPLY. CHEM. INSPECTORATE, LONDON
CIR-
      CANADA. DEPT. OF TRANSPORT
CIR-
      CANADA. DEPT. OF TRANSPORT.METEOROLOGICAL BR.,TORONTO
CI-R-
      COLUMBIA UNIV., NYC. SUBSTITUTE ALLOY MATERIALS LABS.
CIR-
      COMMITTEE ON MEDICAL RES. (CLINICAL INVEST. RPTS.)
CIR-
      CORNELL-GUGGENHEIM AVIATION SAFETY CENTER, N.Y.C.
CI/R-
      GT. BRIT. MIN. OF SUPPLY. CHEM. INSPECTORATE, LONDON
CI-R-
      GT. BRIT. SPRINGFIELDS WORKS, LANCS, ENGLAND
CIRA-
      ADVANCE RESEARCH, INC., NEEDHAM HEIGHTS, MASS.
CIRC-
      GEOLOGICAL SURVEY, WASHINGTON, D.C.
CIRC-
      JAPAN. INDUSTRIAL SCIENCE + TECHNOLOGY AGENCY, TOKYO
CIRCULAR LETTERS(FIAT)-
      FIELD INTELLIGENCE AGENCY, TECHNICAL
CIS-
      NUCLEAR DEV. CORP. OF AMERICA, WHITE PLAINS, N.Y.
CISE-
      ITALY. CENTRO INFORMAZIONI STUDI ESPERIENZE, MILAN
CISE-E-
      ITALY. CENTRO INFORMAZIONI STUDI ESPERIENZE, MILAN
CISE-N-
      ITALY. CENTRO INFORMAZIONI STUDI ESPERIENZE, MILAN
CISE-R-
      ITALY. CENTRO INFORMAZIONI STUDI ESPERIENZE, MILAN
CISRC-RT-(YEAR)-
      OHIO STATE UNIV., COLUMBUS. COMPUTER AND INFORMATION
      SCIENCE RESEARCH CENTER
CISRC-TR-(YEAR)-
      OHIO STATE UNIV., COLUMBUS. COMPUTER AND INFORMATION
      SCIENCE RESEARCH CENTER
CI-SST-(NUMBER-NUMBER)
      BROWN ENGINEERING CO., INC., HUNTSVILLE, ALA.
CIT-
      CALIFORNIA INST. OF TECH., PASADENA
CIT-(LETTER(S))-
      CALIFORNIA INST. OF TECH., PASADENA
CIT-(LETTER(S))-
      CARNEGIE INST. OF TECH., PITTSBURGH
CIT-AFBA-
      CARNEGIE INST. OF TECH., PITTSBURGH
CIT AL TR-
      CALIFORNIA INST. OF TECH., PASADENA. ANTENNA LAB.
CITC-
      CASE INST. OF TECH., CLEVELAND
CIT-E-
      CALIFORNIA INST. OF TECH., PASADENA
CIT GAL-
      CALIF. INST. OF TECH., PASADENA. GUGGENHEIM AERO. LAB
CIT-GCLC CON-
      CALIFORNIA INST. OF TECH., PASADENA. GATES AND
      CRELLIN LABS. OF CHEMISTRY (CONTRIBUTION)
CIT GJPC TR-
      CALIFORNIA INST. OF TECH., PASADENA. GUGGENHEIM
      JET PROPULSION CENTER
CIT-HL-
      CALIFORNIA INST. OF TECH.,PASADENA. HYDRODYNAMICS LAB
CIT-IGC-
      CALIFORNIA INST. OF TECH., PASADENA
CIT IPC-
      CALIFORNIA INST. OF TECH., PASADENA
CIT-JPL-
      CALIF. INST. OF TECH., PASADENA. JET PROPULSION LAB.
CIT JPL ABSTRACTS-
      CALIF. INST. OF TECH., PASADENA. JET PROPULSION LAB.
CIT JPL M(NO.)-(NO.)
      CALIFORNIA INST. OF TECH., PASADENA. JET PROPULSION
      LAB. (MEMORANDA)
CIT JPL PR(NO.)-(NO.)
      CALIFORNIA INST. OF TECH., PASADENA. JET PROPULSION
      LAB. (PROGRESS REPORTS)
CIT JPL PUB-
      CALIFORNIA INST. OF TECH., PASADENA. JET PROPULSION
      LAB. (PUBLICATIONS)
CIT JPL R(NO.)-(NO.)
      CALIFORNIA INST. OF TECH., PASADENA. JET PROPULSION
      LAB. (REPORTS)
CIT JPL SR(NO.)-
      CALIFORNIA INST. OF TECH., PASADENA. JET PROPULSION
      LAB. (SECTION REPORTS)
CIT JPL SURVEY-
      CALIF. INST. OF TECH., PASADENA. JET PROPULSION LAB.
CIT-KRL-TR-(YR/MO)-
      CALIFORNIA INST. OF TECH., PASADENA. W.K.KELLOGG
      LAB. OF RADIATION
CIT-M-
      CALIFORNIA INST. OF TECH., PASADENA
CIT-ORD-
      CARNEGIE INST. OF TECH., PITTSBURGH (ORDNANCE)
CIT-ORD-6D-TR-
      CARNEGIE INST. OF TECH., PITTSBURGH
```

CIT-ORD-(LETTER(S))-
  CARNEGIE INST. OF TECH., PITTSBURGH
CIT PL TR-
  CALIFORNIA INST. OF TECH., PASADENA. PLASMA LAB.
CIT QEL TR-
  CALIFORNIA INST. OF TECH., PASADENA. QUANTUM ELECTRONICS LAB.
CIT QESS TR-
  CALIFORNIA INST. OF TECH., PASADENA. QUANTUM ELECTRONICS SOLID STATE
CIT SSEL TR-
  CALIFORNIA INST. OF TECH., PASADENA. SOLID STATE ELECTRONICS LAB.
CIT-S+SN-
  CALIFORNIA INST. OF TECH., PASADENA (SENSOR AND SIMULATION NOTES)
CIT-TR-
  CALIFORNIA INST. OF TECH., PASADENA
CIT-XRR-
  CALIFORNIA INST. OF TECH., PASADENA
1 CIU-
  NATL. RES. COUNCIL OF CANADA. ASSOCIATE COMM. ON AVIATION MEDICAL RES. NO. 1 CLINICAL INVESTIGATION UNIT (RCAF)
CIV-
  GT. BRIT. AERONAUTICAL RESEARCH COUNCIL
CIVO-
  NETHERLANDS. CENTRAAL INSTITUUT VOOR VOEDING-SONDERZOEK TNO, ZEIST
CIVO-R-
  NETHERLANDS. CENTRAAL INSTITUUT VOOR VOEDING-SONDERZOEK TNO, ZEIST
CJ-(NUMBER)-
  SPERRY GYROSCOPE CO., GREAT NECK, N.Y.
CJR-RM-
  NATIONAL BUREAU OF STANDARDS, WASHINGTON, D.C.
CJS-
  KNOLLS ATOMIC POWER LAB., SCHENECTADY, N.Y. (CODE IS AUTHOR'S INITIALS)
CJTF-
  (OPERATION CROSSROADS) (REF. 7)
CK-
  CARBIDE AND CARBON CHEMICALS CORP. K-25 PLANT, OAK RIDGE, TENN. (REF. 10)
CK-
  CHICAGO. UNIV. METALLURGICAL LAB. (REF. 10)
CK-(LETTER(S))-
  CHICAGO. UNIV. METALLURGICAL LAB. (REF. 10)
CKAN-
  CHICAGO. UNIV. METALLURGICAL LAB. (REF. 10)
CKC-
  CARBIDE AND CARBON CHEMICALS CORP. K-25 PLANT, OAK RIDGE, TENN.
CK-KPDO-(YEAR)-
  CONNER (MEL) AND ASSOCIATES, INC., TALLAHASSEE, FLA. (JOINT VENTURE)
CK-KPDO-(YEAR)-
  KENTUCKY PROGRAM DEVELOPMENT OFFICE, FRANKFORT (JOINT VENTURE)
CKN-
  CHICAGO. UNIV. METALLURGICAL LAB. (REF. 10)
CL-
  (CIRCULAR LETTER) (REF. 2)
CL-
  (CYCLOTRON LAB.) (REF. 2)
CL-
  CHICAGO. UNIV. METALLURGICAL LAB. (REF. 10)
CL-
  CLINTON LABS., OAK RIDGE, TENN. (REF. 10)
CL-
  COMMITTEE ON MEDICAL RESEARCH (CLARK REPORTS)
CL-(NUMBER-YEAR)
  COMMUNICATION SATELLITE CORP., WASHINGTON, D.C.
CL-
  CROSBY LABS., INC., MINEOLA, N.Y.
CL-
  FIELD INTELLIGENCE AGENCY, TECH. (CIRCULAR LETTERS)
CL-
  HARVARD UNIV., CAMBRIDGE, MASS. CRUFT LAB.
CL-(YEAR)-
  NETHERLANDS. RIJKSVERDEDIGINGSORGANISATIE, TNO. CHEMICAL LAB., RIJSWIJK
CL-
  NETHERLANDS. TNO. CENTRAAL LABORATORIUM, DELFT.
CL-(YEAR)-
  NETHERLANDS. TNO. CENTRAAL LABORATORIUM, DELFT
CL-(LETTER(S))-
  CHICAGO. UNIV. METALLURGICAL LAB. (REF. 10)
CL-(INITIALS)-
  CLINTON LABS., OAK RIDGE, TENN.
CLA-
  JOHNS HOPKINS UNIV., SILVER SPRING, MD. APPLIED PHYSICS LAB.
CLAB-(NUMBER)-(NUMBER)-
  MATHIESON CHEMICAL CORP., BALTIMORE
CLARKU-
  CLARK UNIV., WORCESTER, MASS. JEPPSON LAB.
CLASS. CONTROL NO. LV-
  EDGERTON, GERMESHAUSEN AND GRIER, INC., BOSTON
CLB-
  JOHNS HOPKINS UNIV., SILVER SPRING, MD. APPLIED PHYSICS LAB.
CLC-
  CLEVITE CORP., CLEVELAND
CLD-
  CBS LABS., STAMFORD, CONN.
(NUMBER)-CLD-
  CBS LABS., STAMFORD, CONN.
CLE-
  NATIONAL RESEARCH COUNCIL OF CANADA. ATOMIC ENERGY PROJECT, CHALK RIVER, ONT.
CLEV-
  CLEVITE CORP., CLEVELAND
CLEVITE-PR-
  CLEVITE CORP., CLEVELAND
CLEVITE-QPR-
  CLEVITE CORP., CLEVELAND
CLEVITE-TR-
  CLEVITE CORP., CLEVELAND
CLM-
  UNITED KINGDOM ATOMIC ENERGY AUTHORITY. CULHAM LAB., ABINGDON, BERKS, ENGLAND
CLM-(INITIALS)-
  CLINTON LABS., OAK RIDGE, TENN.
CLM-AR-
  UNITED KINGDOM ATOMIC ENERGY AUTHORITY. RESEARCH GP. CULHAM LAB., ABINGDON, BERKS, ENGLAND
CLM-BIB-
  UNITED KINGDOM ATOMIC ENERGY AUTHORITY. CULHAM LAB., ABINGDON, BERKS, ENGLAND
CLM-BIB-
  UNITED KINGDOM ATOMIC ENERGY AUTHORITY. RESEARCH GP. CULHAM LAB., ABINGDON, BERKS, ENGLAND
CLM-IIC-
  CLINTON LABS., OAK RIDGE, TENN.
CLM-L-
  UNITED KINGDOM ATOMIC ENERGY AUTHORITY. CULHAM LAB., ABINGDON, BERKS, ENGLAND
CLM-L-
  UNITED KINGDOM ATOMIC ENERGY AUTHORITY. RESEARCH GP. CULHAM LAB., ABINGDON, BERKS, ENGLAND
CLM-LM-(NUMBER/YEAR)
  UNITED KINGDOM ATOMIC ENERGY AUTHORITY. CULHAM LAB., ABINGDON, BERKS, ENGLAND
CLM-M-
  UNITED KINGDOM ATOMIC ENERGY AUTHORITY. CULHAM LAB., ABINGDON, BERKS, ENGLAND
CLM-M-
  UNITED KINGDOM ATOMIC ENERGY AUTHORITY. RESEARCH GP. CULHAM LAB., ABINGDON, BERKS, ENGLAND
CLM-P-
  UNITED KINGDOM ATOMIC ENERGY AUTHORITY. RESEARCH GP. CULHAM LAB., ABINGDON, BERKS, ENGLAND (PREPRINT)
CLM-PR-
  UNITED KINGDOM ATOMIC ENERGY AUTHORITY. CULHAM LAB., ABINGDON, BERKS, ENGLAND
CLM-PR-
  UNITED KINGDOM ATOMIC ENERGY AUTHORITY. RESEARCH GP. CULHAM LAB., ABINGDON, BERKS, ENGLAND
CLM-R-
  UNITED KINGDOM ATOMIC ENERGY AUTHORITY. CULHAM LAB., ABINGDON, BERKS, ENGLAND
CLM-R-
  UNITED KINGDOM ATOMIC ENERGY AUTHORITY. RESEARCH GP. CULHAM LAB., ABINGDON, BERKS, ENGLAND
CLM-TRANS-
  UNITED KINGDOM ATOMIC ENERGY AUTHORITY. CULHAM LAB., ABINGDON, BERKS, ENGLAND
CLN-
  CALIFORNIA. UNIV., LOS ANGELES. DEPT. OF ENGINEERING
CLN AECU-
  CALIFORNIA. UNIV., LOS ANGELES. DEPT. OF ENGINEERING
CLNS-
  CORNELL UNIV., ITHACA, N.Y. LAB. OF NUCLEAR STUDIES
CLOR-(NUMBER)/D
  POLAND. CENTRAL LAB. FOR RADIOLOGICAL PROTECT., WARSAW
CLOR-
  POLAND. CENTRAL LAB. FOR RADIOLOGICAL PROTECT., WARSAW
CLOR-FONTON-
  POLAND. CENTRAL LAB. FOR RADIOLOGICAL PROTECT., WARSAW
CLOR-FOTON-
  POLAND. CENTRAL LAB. FOR RADIOLOGICAL PROTECT., WARSAW
CLOR-GUM-
  POLAND. CENTRAL LAB. FOR RADIOLOGICAL PROTECT., WARSAW
CLOR-I-
  POLAND. CENTRAL LAB. FOR RADIOLOGICAL PROTECT., WARSAW
CLOR-IBJ-
  POLAND. CENTRAL LAB. FOR RADIOLOGICAL PROTECT., WARSAW
CLOR-IO-
  POLAND. CENTRAL LAB. FOR RADIOLOGICAL PROTECT., WARSAW
CL-P-
  OAK RIDGE NATIONAL LAB., TENN.
CL-R-
  COLUMBIA UNIV., NYC. SUBSTITUTE ALLOY MATERIALS LABS.
CLT-
  ATOMIC ENERGY COMMISSION, WASHINGTON, D.C.
CLU-
  CLARK UNIV., WORCESTER, MASS.
CL-V-
  NETHERLANDS. STAATSMIJNEN IN LIMBURG, GELEEN
CM-
  (CONFIDENTIAL MEMORANDUM) (REF. 2)
(YEAR)-(NUMBER)-CM-
  ARMY NATICK LABS., MASS.
CM-
  BUREAU OF AERONAUTICS (NAVY). COMPUTER DIV.
CM-(NUMBER)-D-
  CARBORUNDUM METALS CO., INC., AKRON, N.Y.
CM-
  CHICAGO. UNIV. METALLURGICAL LAB.
CM-
  CONSOLIDATED VULTEE AIRCRAFT CORP. ORDNANCE AEROPHYSICS LAB., DAINGERFIELD, TEX.
CM-(NUMBER)-(YEAR)-
  DOW CHEMICAL CO., MIDLAND, MICH. (CASTABLE MAGNESIUM)
CM-
  GENERAL DYNAMICS/DAINGERFIELD. ORDNANCE AEROPHYSICS LAB., TEX.

CM-	HUGHES AIRCRAFT CO., CULVER CITY, CALIF.
CM-	JOHNS HOPKINS UNIV., BALTIMORE. DEPT. OF AERONAUTICS
CM-	JOHNS HOPKINS UNIV., SILVER SPRING, MD. APPLIED PHYSICS LAB. (BUMBLEBEE PROJECT) (REF. 8)
CM-	MALAKER LABS., INC., HIGH BRIDGE, N.J.
CM-	MICHIGAN. UNIV., ANN ARBOR. DEPT. OF AERONAUTICAL AND ASTRONAUTICAL ENGINEERING
CM-	NATIONAL BUREAU OF STANDARDS. BOULDER LABS., COLO.
CM(NUMBER)-S-	NATIONAL RESEARCH COUNCIL OF CANADA, OTTAWA
CM-	NAVAL RESEARCH LAB., WASHINGTON, D.C.
CM-	RADIO CORP. OF AMERICA. RCA LABS. DIV., PRINCETON, NJ
CM-	TEXAS. UNIV., AUSTIN. ELECTRICAL ENGINEERING RES.LAB.
CM-	WISCONSIN. UNIV., MADISON. NAVAL RESEARCH LAB.
CMB-	LOS ALAMOS SCIENTIFIC LAB., N. MEX.
CMB-1- THRU CMB-14-	LOS ALAMOS SCIENTIFIC LAB., N. MEX. (INTERNAL CORRESPONDENCE SERIAL USED BY GROUPS IN CMB DIV.)
CMC-	CLIMAX MOLYBDENUM CO. OF MICH., DETROIT
CMCC-	RAMO-WOOLDRIDGE CORP., LOS ANGELES
CMCC-GM-	RAMO-WOOLDRIDGE CORP., LOS ANGELES
CMC-MCDA-	CONTINENTAL MOTORS CORP., MUSKEGON, MICH.
CMC-P-	CANADIAN MARCONI CO. AVIONICS DEPT., MONTREAL
CMC-TPL-	CANADIAN MARCONI CO., MONTREAL
CMC-U-(YEAR)-	CARBORUNDUM METALS CO., INC., AKRON, N.Y.
CMEP-M(NUMBER.NUMBER)	FRANCE. CENTRE DE MATHEMATIQUES DE L'ECOLE POLYTECHNIQUE, PARIS
CMERI-A(NUMBER)	INDIA. CENTRAL MECHANICAL ENG. RES. INST., DURGAPUR
CMERI-B(NUMBER)	INDIA. CENTRAL MECHANICAL ENG. RES. INST., DURGAPUR
CMF-	LOS ALAMOS SCIENTIFIC LAB., N. MEX.
CMF-1- THRU CMF-13-	LOS ALAMOS SCIENTIFIC LAB., N. MEX. (INTERNAL CORRESPONDENCE SERIAL USED BY GROUPS IN CMF DIV., WHICH LATER WAS INCORPORATED INTO CMB DIV.)
CML-	CHICAGO. UNIV. CHICAGO MIDWAY LABS.
CML-(YR)-TN-(PROJ.NO.)-	CHICAGO. UNIV. CHICAGO MIDWAY LABS.
CML-(NUMBER)-	CROSS-MALAKER LABS., INC., MOUNTAINSIDE, N.J.
CML-	GT. BRIT. ADMIRALTY CENTRAL METALLURGICAL LAB.
CML-	MULLARD, LTD. CENTRAL MATERIALS LAB., MITCHAM, SURREY, ENGLAND
CMLCD-	CHEMICAL CORPS (ARMY), WASHINGTON, D.C.
CML. C. SCHOOL-	CHEMICAL CORPS SCHOOL, FORT MC CLELLAN, ALA.
CMLEM-52-	CHEMICAL CORPS. MEDICAL DIV., ARMY CHEMICAL CENTER, MD. (RESEARCH REPORT)
CML-L-	CHICAGO. UNIV. CHICAGO MIDWAY LABS.
CML-M-	CHICAGO. UNIV. CHICAGO MIDWAY LABS.
CML-M-(NUMBER)-	CHICAGO. UNIV. CHICAGO MIDWAY LABS.
CMLRE-	CHEMICAL CORPS (ARMY)
CMLRE-ML-	CHEMICAL CORPS MEDICAL LABS., ARMY CHEMICAL CTR., MD.
CML-SR-M-	CHICAGO. UNIV. CHICAGO MIDWAY LABS.
CML-TN-	CHICAGO. UNIV. CHICAGO MIDWAY LABS.
CML-TN-P-	CHICAGO. UNIV. CHICAGO MIDWAY LABS.
CML-TR-P(NUMBER-NUMBER)	CHICAGO. UNIV. CHICAGO MIDWAY LABS.
CMLWD-	CHEMICAL CORPS. SUPPLY AND PROCUREMENT DIV., ARMY CHEMICAL CENTER, MD.
CMM-	CANADA. DEPT. OF TRANSPORT.METEOROLOGICAL BR.,TORONTO
CMMR-	CHEMICAL WARFARE SERVICE, EDGEWOOD ARSENAL, MD. (CAPTURED MATERIEL MEMORANDUM REPORT)
CMN-	SAO PAULO, BRAZIL. UNIVERSIDADE
CMN-PUB-(NUMBER/YEAR)	SAO PAULO, BRAZIL. UNIV. CENTRE DE MEDICINA NUCLEAR
CMP-CF-	ATOMIC ENERGY OF CANADA LTD., CHALK RIVER, ONT.
CMP-FE-	ATOMIC ENERGY OF CANADA LTD., CHALK RIVER, ONT.
CMP-FF-	ATOMIC ENERGY OF CANADA LTD. CHALK RIVER PROJ., ONT.
CMR-	COMMITTEE ON MEDICAL RESEARCH, OFFICE OF SCIENTIFIC RESEARCH AND DEVELOPMENT (REF. 32)
CM-R-	
CMR-	LOS ALAMOS SCIENTIFIC LAB., N. MEX.
CMR-1- THRU CMR-13-	LOS ALAMOS SCIENTIFIC LAB., N. MEX. (INTERNAL CORRESPONDENCE SERIAL USED BY GROUPS IN CMR DIV., WHICH LATER SPLIT INTO CMB AND CMF DIVS.)
CMR-	MASON AND HANGER-SILAS MASON CO. INC. PANTEX ORDNANCE PLANT, AMARILLO, TEX.
CMR A-	COMMITTEE ON MEDICAL RESEARCH (ABBOTT LABS. REPORTS)
CMR-AE-	LOS ALAMOS SCIENTIFIC LAB., N. MEX.
CMR B-	COMMITTEE ON MEDICAL RESEARCH (BACKMANN REPORTS)
CMR BE-	COMMITTEE ON MEDICAL RESEARCH (BEADLE REPORTS)
CMR BR-	COMMITTEE ON MEDICAL RESEARCH. (BRITISH REPORTS)
CMR BS-	COMMITTEE ON MEDICAL RES. (NATL. BUR. OF STDS. RPTS.)
CMR BSR-	COMMITTEE ON MEDICAL RES. (BLOOD SUBSTITUTES RPTS.)
CMR BULL-	COMMITTEE ON MEDICAL RESEARCH (BULLETIN)
CMR C-	COMMITTEE ON MEDICAL RESEARCH (COGHILL REPORTS)
CMR CAM-	COMMITTEE ON MEDICAL RES. COMM. ON AVIATION MED.
CMR CIR-	COMMITTEE ON MEDICAL RES. (CLINICAL INVEST. RPTS.)
CMR CL-	COMMITTEE ON MEDICAL RESEARCH (CLARK REPORTS)
CMR CR-	COMMITTEE ON MEDICAL RESEARCH (CONVALESCENCE AND REHABILITATION REPORTS)
CMR CT-	COMMITTEE ON MEDICAL RESEARCH (CHEMOTHERAPY REPORTS)
CMR CTGC-	COMMITTEE ON MEDICAL RESEARCH. COMMITTEE ON THE TREATMENT OF GAS CASUALTIES
CMR CTGC (LETTER)-	COMMITTEE ON MEDICAL RESEARCH. COMMITTEE ON THE TREATMENT OF GAS CASUALTIES
CMR CU-	COMMITTEE ON MEDICAL RESEARCH (CUTTER REPORTS)
CMR D-	COMMITTEE ON MEDICAL RESEARCH (DU VIGNEAUD REPORTS)
CMR-DO-GS-	LOS ALAMOS SCIENTIFIC LAB., N. MEX.
CMR-DO-TECH-	LOS ALAMOS SCIENTIFIC LAB., N. MEX.
CMREF-PAMPHLET-	COMMITTEE ON MARINE RESEARCH EDUCATION AND FACILITIES WASHINGTON, D.C.
CMR F-	COMMITTEE ON MEDICAL RES. (FOOD + DRUG ADM. RPTS.)
CMR H-	COMMITTEE ON MEDICAL RES. (HEYDEN CHEM. CORP. RPTS.)
CMR ID-	COMMITTEE ON MEDICAL RES. (INFECTIOUS DISEASES RPTS.)
CMR IWB-	COMMITTEE ON MEDICAL RES. (INFECTED WOUNDS AND BURNS)
CMR J-	COMMITTEE ON MEDICAL RESEARCH (JOHNSON REPORTS)
CMR L-	COMMITTEE ON MEDICAL RES. (LILLY RES. LABS. RPTS.)
CMR M-	COMMITTEE ON MEDICAL RESEARCH (MERCK + CO. RPTS.)
CMR-M-	LOS ALAMOS SCIENTIFIC LAB., N. MEX.
CMR MC-	COMMITTEE ON MEDICAL RES. (MISSILE CASUALTIES RPTS.)
CMR ME-	COMMITTEE ON MEDICAL RESEARCH (MEDICINE REPORTS)
CMR MN-	COMMITTEE ON MEDICAL RES. (MEDICAL NUTRITION RPTS.)
CMR MR-	COMMITTEE ON MEDICAL RESEARCH (MALARIA REPORTS)
CMR MR A-	COMMITTEE ON MEDICAL RES. (MALARIA RPTS. ABSTRACT)
CMR N-	COMMITTEE ON MEDICAL RESEARCH (ROCKEFELLER INSTITUTE FOR MEDICAL RESEARCH)
CMR NP-	COMMITTEE ON MEDICAL RESEARCH (NEUROPSYCHIATRY RPTS.)
CMR NS-	COMMITTEE ON MEDICAL RESEARCH (NEUROSURGERY REPORTS)
CMR OL-	COMMITTEE ON MEDICAL RESEARCH (OTOLARYNGOLOGY RPTS.)
CMR P-	COMMITTEE ON MEDICAL RES. (CHAS. PFIZER + CO. RPTS.)
CMR PD-	COMMITTEE ON MEDICAL RES. (PARKE-DAVIS + CO. RPTS.)
CMRR-	CANADA. DEPT. OF TRANSPORT.METEOROLOGICAL BR.,TORONTO
CMR R-	COMMITTEE ON MEDICAL RESEARCH (RANDALL REPORTS)
CMR RS-	COMMITTEE ON MEDICAL RES. (RUSSELL SAGE INST. RPTS.)
CMR S-	COMMITTEE ON MEDICAL RESEARCH (SQUIBB INSTITUTE FOR MEDICAL RESEARCH REPORTS)
CMR SER-	COMMITTEE ON MEDICAL RESEARCH (SANITARY ENG. RPTS.)

CMR SH-

CMR SH-	COMMITTEE ON MEDICAL RES. (SHELL DEV. CO. RPTS.)
CMR SPR-	COMMITTEE ON MEDICAL RESEARCH (SPECIAL REPORTS)
CMR SR-	COMMITTEE ON MEDICAL RESEARCH (SHOCK REPORTS)
CMR-TA-	LOS ALAMOS SCIENTIFIC LAB., N. MEX.
CMR TDR-	COMMITTEE ON MEDICAL RES. (TROPICAL DISEASES RPTS.)
CMR TR-	COMMITTEE ON MEDICAL RESEARCH (TUBERCULOSIS REPORTS)
CMR U-	COMMITTEE ON MEDICAL RESEARCH (UPJOHN CO. REPORTS)
CMR VD-	COMMITTEE ON MEDICAL RES. (VENEREAL DISEASES RPTS.)
CMR W-	COMMITTEE ON MEDICAL RES. (WINTHROP CHEM. CO. RPTS.)
CMR WO-	COMMITTEE ON MEDICAL RESEARCH (WOODWARD REPORTS)
CMTP-CDOT-CAM-	CONNECTICUT. DEPT. OF TRANSPORTATION, HARTFORD (CONN. MASTER TRANSPORTATION PLAN)
CMTR-	CHEMICAL WARFARE SERVICE, EDGEWOOD ARSENAL, MD. (CAPTURED MATERIEL TECHNICAL REPORT)
CM-U-	CARBORUNDUM METALS CO., INC., AKRON, N.Y.
CMU-(NUMBER)-(NUMBER)	CARNEGIE-MELLON UNIV., PITTSBURGH
CMU/PPL-(MONTH/YEAR)	CARNEGIE-MELLON UNIV., PITTSBURGH
CMU-TR-	CARNEGIE-MELLON UNIV., PITTSBURGH
CMX PROGRESS REPORT-	DU PONT DE NEMOURS (E.I.) + CO. ATOMIC ENERGY DIV., WILMINGTON, DEL.
CN-(NUMBER)/D-	AKADEMIYA NAUK SSSR., MOSCOW
CN-	CHICAGO. UNIV. METALLURGICAL LAB. (REF. 10)
CN-	CLINTON LABS., OAK RIDGE, TENN. (REF. 10)
C/N-	COMBINED COMMUNICATIONS BOARD, JOINT CHIEFS OF STAFF, WASHINGTON, D.C.
CN-(NUMBER/LETTER)-	INTERNATIONAL ATOMIC ENERGY AGENCY, VIENNA (CONFERENCE PAPER)
C/N-	UNITED KINGDOM ATOMIC ENERGY AUTHORITY. PRODUCTION GROUP. WINDSCALE WORKS, SELLAFIELD, CUMB., ENGLAND
C/N-	WISCONSIN. UNIV., MADISON
CN-(LETTER(S))-	CHICAGO. UNIV. METALLURGICAL LAB. (REF. 10)
67-CNA-	CANADIAN NUCLEAR ASSN., TORONTO
(YEAR)-CNA-	DIVISION OF RAW MATERIALS, AEC, N.Y.C.
CNAADTRA-	NAVAL AIR ADVANCED TRNG. COMMAND, JACKSONVILLE, FLA.
CNAEM-	TURKEY. AEC. CEKMECE NUCLEAR RESEARCH CENTER, ISTANBUL
CNAM-	TURKEY. AEC. CEKMECE NUCLEAR RESEARCH CENTER, ISTANBUL
CNAOPROFESSIONAL PAPER-	CENTER FOR NAVAL ANALYSES, ARLINGTON, VA.
CNA-PROFESSIONAL PAPER-	CENTER FOR NAVAL ANALYSES, ARLINGTON, VA.
CNA RESEARCH CONTRIB-	CENTER FOR NAVAL ANALYSES, WASHINGTON, D.C.
CNATRA-	NAVAL AIR TRAINING COMMAND, PENSACOLA, FLA.
CNATT-	NAVAL AIR TECHNICAL TRAINING COMMAND, MEMPHIS
CNB-	ITALY. COMITATO NAZIONALE PER L'ENERGIA NUCLEARE. DIVISIONE DI BIOLOGIA E DE PROTEZIONE SANITARIA, ROME
CNBIB-	ITALY. COMITATO NAZIONALE PER LE RICERCHE NUCLEARI, ROME
CNC-	ITALY. COMITATO NAZIONALE PER LE RICERCHE NUCLEARI, ROME
CNC-1- THRU CNC-11-	LOS ALAMOS SCIENTIFIC LAB., N. MEX. (INTERNAL CORRESPONDENCE SERIAL USED BY GROUPS IN CNC DIV.)
CNCS-	GT. BRIT. ROYAL A.F. CENTRAL NAVIGATION + CONTROL SCH
CNEA-	ARGENTINA. COMISION NACIONAL DE ENERGIA ATOMICA, BUENOS AIRES
CNEA-CEC(YEAR)(NUMBER)	ITALY. COMITATO NAZIONALE PER L'ENERGIA NUCLEARE, BOLOGNA
CNEN-	ITALY. COMITATO NAZIONALE PER L'ENERGIA NUCLEARE, ROME
CNEN-IEA-	ITALY. COMITATO NAZIONALE PER L'ENERGIA NUCLEARE. LABORATORI NAZIONALI DI FRASCATI
CNEN-IEA-	SAO PAULO, BRAZIL. UNIVERSIDADE. INSTITUTO DE ENERGIA ATOMICA
CNEN-IN-	ITALY. COMITATO NAZIONALE PER L'ENERGIA NUCLEARE. LABORATORI NAZIONALI DI FRASCATI
CNEN-PRV-R-(YEAR)-	ITALY. COMITATO NAZIONALE PER L'ENERGIA NUCLEARE, BOLOGNA
CNEN-RT/B-	ITALY. COMITATO NAZIONALE PER L'ENERGIA NUCLEARE, ROME
CNEN-RT/BIO-	ITALY. COMITATO NAZIONALE PER L'ENERGIA NUCLEARE, ROME
CNEN-RT/EL-(YEAR)-	ITALY. COMITATO NAZIONALE PER L'ENERGIA NUCLEARE, ROME
CNEN-RT/FI-	ITALY. COMITATO NAZIONALE PER L'ENERGIA NUCLEARE, ROME
CNEN-RT/FIMA-(YEAR)-	ITALY. COMITATO NAZIONALE PER L'ENERGIA NUCLEARE, ROME
CNEN-RT/ING-(YEAR)-	ITALY. COMITATO NAZIONALE PER L'ENERGIA NUCLEARE, ROME
CNEN-RT/MET-(YEAR)-	ITALY. COMITATO NAZIONALE PER L'ENERGIA NUCLEARE, ROME
CNEN-RT/PROT-(YEAR)-	ITALY. COMITATO NAZIONALE PER L'ENERGIA NUCLEARE, ROME
CNEN-RVA-(YEAR)-	ITALY. COMITATO NAZIONALE PER L'ENERGIA NUCLEARE, BOLOGNA
CNEN-RVT-PEC-	ITALY. COMITATO NAZIONALE PER L'ENERGIA NUCLEARE, BOLOGNA
CNEN-RVT-SEC-(YEAR)-	ITALY. COMITATO NAZIONALE PER L'ENERGIA NUCLEARE, BOLOGNA
CNEN-RVT-SIN-(YEAR)-	ITALY. COMITATO NAZIONALE PER L'ENERGIA NUCLEARE, BOLOGNA
CNEN-RVT-SIR-(YEAR)-	ITALY. COMITATO NAZIONALE PER L'ENERGIA NUCLEARE, BOLOGNA
CNES-	FRANCE. CENTRE NATIONAL D'ETUDES SPATIALES, PARIS
CNES-NT-	FRANCE. CENTRE NATIONAL D'ETUDES SPATIALES, PARIS
CNET-	FRANCE. CENTRE NATIONAL D'ETUDES DES TELECOMMUNICATIONS, ISSY LES MOULINEAUX
CNET-NT-EST/APH/	FRANCE. CENTRE NATIONAL D'ETUDES DES TELECOMMUNICATIONS, ISSY LES MOULINEAUX
CNF-	ITALY. COMITATO NAZIONALE PER L'ENERGIA NUCLEARE. LABORATORI NAZIONALI DI FRASCATI
CNG-	ITALY. COMITATO NAZIONALE PER LE RICERCHE NUCLEARI, ROME
CNHW-	CANADA. DEPT. OF NATIONAL HEALTH AND WELFARE. RADIATION PROTECTION DIV., OTTAWA
CNHW(RP-(NUMBER))	CANADA. DEPT. OF NATIONAL HEALTH AND WELFARE. RADIATION PROTECTION DIV., OTTAWA
CNI-	ITALY. COMITATO NAZIONALE PER L'ENERGIA NUCLEARE, ISPRA
CNIE-	ARGENTINA. COMISION NACIONAL DE INVESTIGACIONES ESPACIALES, BUENOS AIRES
CNIE-IC-	ARGENTINA. COMISION NACIONAL DE INVESTIGACIONES ESPACIALES, BUENOS AIRES
CNIE-PE-	ARGENTINA. COMISION NACIONAL DE INVESTIGACIONES ESPACIALES, BUENOS AIRES
CNIE-PT-	ARGENTINA. COMISION NACIONAL DE INVESTIGACIONES ESPACIALES, BUENOS AIRES
CNL-	CLINTON LABS., OAK RIDGE, TENN.
CNL-	NATIONAL DEFENSE RES. COMM. (CONFIDENTIAL NEWS LETTER)
CNLM-	PRATT AND WHITNEY AIRCRAFT DIV., UNITED AIRCRAFT CORP. CONN. AIRCRAFT NUCLEAR ENGINE LAB., MIDDLETOWN (REF. 3)
CNM-R-	ILLINOIS. UNIV., URBANA
CNM-R-	MEXICO. COMISION NACIONAL DE ENERGIA NUCLEAR. CENTRO NUCLEAR DE MEXICO, MEXICO CITY (CONF. PROCEEDINGS)
CNN-	CHICAGO. UNIV. METALLURGICAL LAB. (REF. 10)
CNO-	OFFICE OF THE CHIEF OF NAVAL OPERATIONS, WASH., D.C.
CNP-	CHICAGO. UNIV. METALLURGICAL LAB. (REF. 10)
CNR-	ITALY. CONSIGLIO NAZIONALE DELLE RICERCHE, ROME
CNRC-	NATIONAL RESEARCH COUNCIL OF CANADA, OTTAWA
CNR/PR/	UNITED KINGDOM ATOMIC ENERGY AUTHORITY. WEAPONS GROUP ATOMIC WEAPONS RES. ESTAB., ALDERMASTON, BERKS, ENG.
CNRS-	FRANCE. CENTRE NATIONAL DE LA RECHERCHE SCIENTIFIQUE, PARIS
CNRS-AR-(YEAR)	FRANCE. CENTRE NATIONAL DE LA RECHERCHE SCIENTIFIQUE
CNRS-CPT-(YEAR)-P-	FRANCE. CENTRE NATIONAL DE LA RECHERCHE SCIENTIFIQUE, MARSEILLES + CENTRE DE PHYSIQUE THEORIQUE
CNS-	CHICAGO. UNIV. METALLURGICAL LAB. (REF. 10)
CNSX-	CHICAGO. UNIV. METALLURGICAL LAB. (REF. 10)

CNT-	ITALY. COMITATO NAZIONALE PER L'ENERGIA NUCLEARE. LABORATORI NAZIONALI DI FRASCATI
CNTR-	ITALY. COMITATO NAZIONALE PER LE RICERCHE NUCLEARI, ROME
CO-	AIR RESEARCH AND DEVELOPMENT COMMAND, ANDREWS AFB, MD
CO-	CARTER OBSERVATORY, WELLINGTON, NEW ZEALAND
CO-	CHICAGO OPERATIONS OFFICE, AEC
CO-	COLORADO. UNIV., BOULDER
CO-29(601)-(NOS...)-(NUMBER)-CO-	COLORADO. UNIV., BOULDER. ENGINEERING EXPERIMENT STA.
	GENERAL ELECTRIC CO. ELECTRONICS LAB., SYRACUSE, N.Y.
CO(NUMBER)-(NUMBER)-	GENERAL ELECTRIC CO. HEAVY MILITARY ELECTRONICS DEPT., SYRACUSE, N.Y. (CONFORMAL/PLANAR ARRAY SONAR)
CO-	TENNESSEE EASTMAN CORP., OAK RIDGE, TENN.
COA-	COLLEGE OF AERONAUTICS, CRANFIELD, BUCKS, ENGLAND
COA-AERO-	COLLEGE OF AERONAUTICS, CRANFIELD, BUCKS, ENGLAND
COA-E+C-	COLLEGE OF AERONAUTICS, CRANFIELD, BUCKS, ENGLAND
COA-MAT-	COLLEGE OF AERONAUTICS, CRANFIELD, BUCKS, ENGLAND
COA-MATT-	COLLEGE OF AERONAUTICS, CRANFIELD, BUCKS, ENGLAND
COA-M+P-	COLLEGE OF AERONAUTICS, CRANFIELD, BUCKS, ENGLAND
COA-N-	COLLEGE OF AERONAUTICS, CRANFIELD, BUCKS, ENGLAND
COA-N-MAT.-	COLLEGE OF AERONAUTICS, CRANFIELD, BUCKS, ENGLAND
COA-NOTE-AERO-	COLLEGE OF AERONAUTICS, CRANFIELD, BUCKS, ENGLAND. DEPT. OF AERODYNAMICS
COA-NOTE-E/C-	COLLEGE OF AERONAUTICS, CRANFIELD, BUCKS, ENGLAND. DEPT. OF ELECTRICAL AND CONTROL ENGINEERING
COA-NOTE-MAT-	COLLEGE OF AERONAUTICS, CRANFIELD, BUCKS, ENGLAND
COA-NOTE-M+P-	COLLEGE OF AERONAUTICS, CRANFIELD, BUCKS, ENGLAND
COA-NOTE-M/P-	COLLEGE OF AERONAUTICS, CRANFIELD, BUCKS, ENGLAND
COBSI-WD-	FEDERATION OF AMERICAN SOCIETIES FOR EXPERIMENTAL BIOLOGY. COUNCIL ON BIOLOGICAL SCIENCES INFORMATION, BETHESDA, MD. (WORKING DOCUMENT)
COC-	CALIFORNIA. UNIV., BERKELEY. RADIATION LAB.
COC-	CALIFORNIA. UNIV., LIVERMORE. LAWRENCE RADIATION LAB.
CODE-	CANADIAN OCEANOGRAPHIC DATA CENTRE, OTTAWA
CODE IDENT-	AEROJET-GENERAL CORP., SACRAMENTO, CAL. (TEST PROCEDURE)
COE-	CORPS OF ENGINEERS (ARMY)
COE-EB-	CORPS OF ENGINEERS (ARMY). ENGINEER BOARD
COE-MRDL-	CORPS OF ENGINEERS (ARMY). MISSOURI RIVER DIV. LABS., OMAHA
COE-ORDL	CORPS OF ENGINEERS (ARMY). OHIO RIVER DIV. LABS., CINCINNATI
COEP-	PRECISION TECHNOLOGY, INC., LIVERMORE, CALIF.
COE-UET-	CORPS OF ENGINEERS (ARMY) (UNDERGROUND EXPLOSION TEST)
COHQ-	GT. BRIT. COMBINED OPERATIONS HEADQUARTERS
COI-	AIR FORCE (COMMUNICATIONS OPERATING INSTRUCTIONS)
COI-	CALIFORNIA. UNIV., BERKELEY. RADIATION LAB.
COI-	CALIFORNIA. UNIV., LIVERMORE. RADIATION LAB.
COL-	CALIFORNIA. UNIV., LIVERMORE. LAWRENCE RADIATION LAB.
COLE-	COLEMAN ENGINEERING CO., INC., LOS ANGELES
COLL-	COLLINS RADIO CO., CEDAR RAPIDS, IOWA
COLOGNEU-	COLOGNE. UNIVERSITAT.
COLOU-	COLORADO STATE UNIV., FORT COLLINS
COLR-	COLLINS RADIO CO., CEDAR RAPIDS, IOWA
COLRC-	COLORADO RESEARCH CORP., BROOMFIELD
COL. SERIAL-	COLUMBIA UNIV., N.Y.C.
COL. SPEC-	CALIFORNIA. UNIV., BERKELEY. RADIATION LAB. (REF. 9)
COLU-	COLORADO. UNIV., BOULDER
COLOU-HAO-	HIGH ALTITUDE OBSERVATORY, CLIMAX, COLO.
COLUMBIA SERIAL--	COLUMBIA UNIV., N.Y.C.
C+OM-	ARMY NATICK LABS. CLOTHING AND ORGANIC MATERIALS DIV., MASS.
C/OM-(NUMBER)-	ARMY NATICK LABS. CLOTHING AND ORGANIC MATERIALS DIV., MASS.
C/OM-	ARMY NATICK LABS. CLOTHING AND ORGANIC MATERIALS LAB., MASS.
COM-	BALLISTIC MISSILE DIV. (AIR FORCE), INGLEWOOD, CALIF.
COM-(YEAR)-(5 DIGITS)	NATIONAL TECHNICAL INFORMATION SERVICE, SPRINGFIELD, VA. (ACCESSION NUMBER FOR REPORTS FROM OTHER COMMERCE DEPT. AGENCIES. USED 1971-)
COM-	TENNESSEE EASTMAN CORP., OAK RIDGE, TENN.
COMASWFORLANT-	ATLANTIC FLEET. ANTISUBMARINE WARFARE FORCE, NORFOLK, VA
C/OM-C/ED-	ARMY NATICK LABS. CLOTHING AND ORGANIC MATERIALS DIV., MASS.
C/OM-ER-	ARMY NATICK LABS. CLOTHING AND ORGANIC MATERIALS LAB., MASS.
COMFAIRWINGSLANT-	ATLANTIC FLEET. FLEET AIR WINGS, NORFOLK, VA.
COMFIVE-	FIFTH NAVAL DISTRICT, NORFOLK, VA.
COMHUKFORLAN-	ATLANTIC FLEET. HUNTER KILLER FORCE, NORFOLK, VA.
C/OM-MR/E-(YEAR)-	ARMY NATICK LABS. CLOTHING AND ORGANIC MATERIALS DIV., MASS.
COMMUNICATIONS-	ARIZONA. UNIV., TUCSON. LUNAR AND PLANETARY LAB.
COMNAVEU-	NAVAL FORCES IN EUROPE, U.S.
COMNAVFE-	NAVAL FORCES IN THE FAR EAST, U.S.
COMNAVFORGER-	NAVAL FORCES IN GERMANY, U.S.
COMNAVMISCEN-	NAVAL MISSILE CENTER, POINT MUGU, CALIF.
COMNAVSOUTH-OPS-	ALLIED NAVAL FORCES SOUTHERN EUROPE, MALTA
COMOPDEVFOR-SER-	OPERATIONAL DEV. FORCE, ATLANTIC FLEET, NORFOLK, VA.
COMPLETION-(YEAR)-	MASSACHUSETTS. UNIV., AMHERST. WATER RESOURCES RESEARCH CENTER
COMPT-M(MEFA)-	OFFICE OF THE COMPTROLLER OF THE ARMY. DIRECTOR OF MANAGEMENT, WASHINGTON, D.C.
COMPT-M(MEI)-MPL-	OFFICE OF THE COMPTROLLER OF THE ARMY. DIRECTOR OF MANAGEMENT, WASHINGTON, D.C.
COMSAT-	COMMUNICATION SATELLITE CORP., WASHINGTON, D.C.
COM-SCI-	UNITED KINGDOM. NATIONAL PHYSICAL LAB. DIV. OF COMPUTER SCIENCE, TEDDINGTON, MIDDX., ENGLAND
C/OM-TR-	ARMY NATICK LABS. CLOTHING AND ORGANIC MATERIALS LAB., MASS.
C/OM-TS-	ARMY NATICK LABS. CLOTHING AND ORGANIC MATERIALS LAB., MASS.
COMW-	CALIFORNIA. UNIV., LIVERMORE. LAWRENCE RADIATION LAB.
CON-	CONNECTICUT. UNIV., STORRS. CHEMISTRY DEPT.
CONARC-	CONTINENTAL ARMY COMMAND, FORT MONROE, VA.
CO-NAVAER-	BUREAU OF AERONAUTICS (NAVY)
CON-CIR.-	CONTINENTAL ARMY COMMAND, FORT MONROE, VA.
COND-	CONDUCTRON CORP., ANN ARBOR, MICH.
CONE-(NUMBER)-	DIVISION OF TECHNICAL INFORMATION EXTENSION, AEC, OAK RIDGE, TENN.
CONESCO-	CONESCO INC., CAMBRIDGE, MASS.
CONF-	(CONFERENCE) (REF. 2)
CONF-(6 DIGITS)-	TECHNICAL INFORMATION CENTER, AEC, OAK RIDGE, TENN. (ASSIGNED TO CONF. PAPERS. FIRST 2 DIGITS ARE YEAR, NEXT 2 ARE MONTH, LAST 2 ARE SERIAL NUMBER OF CONF. SERIAL NO. OF INDIVIDUAL PAPER MAY FOLLOW DASH)
CON/HAR/EMR/	IMPERIAL CHEMICAL INDUSTRIES, LTD. METALS DIV., BIRMINGHAM, WARWICK, ENGLAND
CONN-	CONNECTICUT. UNIV., STORRS
CONOCO-	CONTINENTAL OIL CO., PONCA CITY, OKLA.
CON/R-	UNITED KINGDOM ATOMIC ENERGY AUTHORITY. REACTOR GP., RISLEY, LANCS, ENGLAND
CONSECUTIVE-	FOOD + DRUG ADM., WASH., D.C. (INTERBUREAU BY-LINES)
CONSULTEC-	NUCLEAR UTILITY SERVICES, INC., CONSULTEC DIV., WASH, DC
CONTRIB-	COLUMBIA UNIV., DOBBS FERRY, N.Y.

CONTRIB-
CONTRIB-
    FLORIDA STATE UNIV., TALLAHASSEE. GEOPHYSICAL FLUID DYNAMICS INST.
CONTRIB-(YEAR)-
    IOWA STATE UNIV. OF SCIENCE AND TECHNOLOGY, AMES. SOIL RESEARCH LAB.
CONTRIB-
    WASHINGTON. UNIV., SEATTLE. DEPT. OF ATMOSPHERIC SCIENCES
CONVAIR-
    CONVAIR, FORT WORTH, TEX.
CONVAIR-
    GENERAL DYNAMICS/CONVAIR, SAN DIEGO, CALIF.
CONVAIR-
    GENERAL DYNAMICS/POMONA, CALIF.
COO-
    CHICAGO OPERATIONS OFFICE, AEC (REF. 4)
COO-(CONTRACT NUMBER)-
    CHICAGO OPERATIONS OFFICE, AEC (REF. 4)
COOK FPR-
    COOK ELECTRIC CO. COOK RESEARCH LABS., SKOKIE, ILL.
COOK FPR-
    COOK RESEARCH LABS., MORTON GROVE, ILL.
COOS-(CONTRACT NUMBER)-
    CHICAGO OPERATIONS OFFICE, AEC (REF. 4)
COPAC-
    NATIONAL RESEARCH COUNCIL. COMMITTEE ON POLLUTION ABATEMENT AND CONTROL, WASHINGTON, D.C.
COPB-
    CALIFORNIA. UNIV., LIVERMORE. RADIATION LAB.
COPBA-
    CALIFORNIA. UNIV., LIVERMORE. RADIATION LAB.
COPC-
    CALIFORNIA. UNIV., LIVERMORE. RADIATION LAB.
COPCB-
    CALIFORNIA. UNIV., LIVERMORE. RADIATION LAB.
COPENHAGEN-
    COPENHAGEN. UNIVERSITET
COPJ-
    CALIFORNIA. UNIV., BERKELEY. RADIATION LAB.
COPJ-
    CALIFORNIA. UNIV., LIVERMORE. RADIATION LAB.
COPO-
    CALIFORNIA. UNIV., LIVERMORE. RADIATION LAB.
COPP-
    CALIFORNIA. UNIV., LIVERMORE. LAWRENCE RADIATION LAB.
COPV-
    CALIFORNIA. UNIV., LIVERMORE. RADIATION LAB.
CORG-
    CONTINENTAL ARMY COMMAND. COMBAT OPERATIONS RESEARCH GROUP, FORT MONROE, VA.
CORG-(LTRS.)-
    TECHNICAL OPERATIONS, INC. COMBAT OPERATIONS RESEARCH GROUP, FORT MONROE, VA.
CORG-M-
    TECHNICAL OPERATIONS, INC. COMBAT OPERATIONS RESEARCH GROUP, FORT BELVOIR, VA.
CORG-R-
    TECHNICAL OPERATIONS, INC. COMBAT OPERATIONS RESEARCH GROUP, FORT BELVOIR, VA.
CORG-SP-
    TECHNICAL OPERATIONS, INC. COMBAT OPERATIONS RESEARCH GROUP, FORT BELVOIR, VA.
CORN-
    CORNELL UNIV., ITHACA, N.Y.
CORNU-
    CORNELL UNIV., ITHACA, N.Y.
CORNU-EE-
    CORNELL UNIV., ITHACA, N.Y. SCHOOL OF ELECTRICAL ENG.
CORNU-TR-
    CORNELL UNIV., ITHACA, N.Y. (ALL DEPARTMENTS)
CORP-SP-
    TECHNICAL OPERATIONS, INC. COMBAT OPERATIONS RESEARCH GROUP, FORT BELVOIR, VA.
CORU-
    CORNELL UNIV., ITHACA, N.Y.
COS-
    UNITED KINGDOM ATOMIC ENERGY AUTHORITY. WEAPONS GROUP ATOMIC WEAPONS RES. ESTAB., ALDERMASTON, BERKS, ENG.
COSA-SR-(YR.)
    NATIONAL SECURITY AGENCY, WASH., D.C. (SPEC. REPT.)
COSATI-(YEAR)-
    FEDERAL COUNCIL FOR SCIENCE AND TECHNOLOGY. COMMITTEE ON SCIENTIFIC AND TECHNICAL INFORMATION, WASH., DC
COSMIC-
    COSMIC, INC., WASHINGTON, D.C.
COSSACT-
    NAVAL COMMAND SYSTEMS SUPPORT ACTIVITY, WASHINGTON, DC
COSTECH-
    ARMY WEAPONS COMMAND. COST ANALYSIS OFFICE, ROCK ISLAND, ILL.
COT-
    CALIFORNIA. UNIV., BERKELEY. RADIATION LAB.
COT-
    CALIFORNIA. UNIV., LIVERMORE. RADIATION LAB.
COTA-
    CALIFORNIA. UNIV., LIVERMORE. RADIATION LAB.
COUNTING-
    TENNESSEE EASTMAN CORP., OAK RIDGE, TENN.
COURT-
    COURTNEY AND CO., PHILADELPHIA
COVA-
    CALIFORNIA. UNIV., LIVERMORE. LAWRENCE RADIATION LAB.
COWBOY-OO-
    ALBUQUERQUE OPERATIONS OFFICE, AEC, N. MEX.
COWLES FOUNDATION PAPER-
    YALE UNIV., NEW HAVEN. COWLES FOUNDATION FOR RESEARCH IN ECONOMICS
COWL-TM-
    GENERAL DYNAMICS/CONVAIR, SAN DIEGO, CALIF.

COWRR-PUB-
    FEDERAL COUNCIL FOR SCIENCE AND TECHNOLOGY. COMMITTEE ON WATER RESOURCES RESEARCH, WASHINGTON, D.C.
CP-
    (CONFERENCE PAPER) (REF. 2)
CP-
    (CONTRACT PUBLICATION) (REF. 2)
CP-
    (CURRENT PAPER) (REF. 2)
CP-
    AMERICAN INST. OF AERONAUTICS AND ASTRONAUTICS, N.Y.C. (CONFERENCE PUBLICATIONS)
CP-
    ARGONNE NATIONAL LAB., LEMONT, ILL. (REF. 10)
CP-
    CANADA. DEFENCE RESEARCH BOARD, CHIEF OF PERSONNEL BRANCH, OTTAWA
CP-
    GARRETT CORP. AIRESEARCH MFG. DIV., LOS ANGELES
CP-
    GT. BRIT. AERONAUTICAL RESEARCH COUNCIL
CP-(NUMBER/YEAR)
    GT. BRIT. BLDG. RES. STA., GARSTON, HERTS, ENGLAND
CP-
    GT. BRIT. ROYAL NAVAL PERSONNEL RESEARCH COMMITTEE
CP-
    KYOTO UNIV. DEPT. OF AERONAUTICAL ENGINEERING
CP-
    QUARTERMASTER RESEARCH AND ENGINEERING COMMAND, NATICK, MASS. (CHEMICALS AND PLASTICS)
CP-
    TECHNICAL OPERATIONS, INC., ARLINGTON, MASS.
CP-(LETTER(S))-
    CHICAGO. UNIV. METALLURGICAL LAB. (REF. 10)
CPA-
    COLUMBIA UNIV., NYC. SUBSTITUTE ALLOY MATERIALS LABS.
CPA-
    JOHNS HOPKINS UNIV., SILVER SPRING, MD. CHEMICAL PROPULSION INFORMATION AGENCY
CPB-
    CHICAGO. UNIV. METALLURGICAL LAB. (ASSIGNED TO REPORTS FROM GREAT BRITAIN) (REF. 10)
CPC-
    COORS PORCELAIN CO., GOLDEN, COLO.
CPC-(NUMBER)-FR-
    HONEYWELL, INC. CORPORATE PROGRAM CENTER, MINNEAPOLIS
CPC-PR-
    COORS PORCELAIN CO., GOLDEN, COLO.
CPD-
    AUSTRALIA. WEAPONS RESEARCH ESTABLISHMENT
CPD-(YR)-
    AUSTRALIA. WEAPONS RESEARCH ESTABLISHMENT
CPD-
    COLLINS RADIO CO., CEDAR RAPIDS, IOWA (PRODUCT DESCRIPTION)
CPD-(NUMBER)-
    GENERAL ELECTRIC CO. COMMUNICATION PRODUCTS DEPT., LYNCHBURG, VA.
CPDC/H-
    GT. BRIT. ATOMIC EN. RES. ESTAB., HARWELL, BERKS, ENG
CPDC-P-
    UNITED KINGDOM ATOMIC ENERGY AUTHORITY. RESEARCH GP. ATOMIC ENERGY RES. ESTAB., HARWELL, BERKS, ENGLAND
CPDD-(YEAR)-
    CLARKSON COLL. OF TECHNOLOGY, POTSDAM, N.Y.
CPI-
    CLIFTON PRODUCTS, INC., PAINESVILLE, OHIO
CPIA-
    JOHNS HOPKINS UNIV., SILVER SPRING, MD. CHEMICAL PROPULSION INFORMATION AGENCY
CPIA-LTM-
    JOHNS HOPKINS UNIV., SILVER SPRING, MD. CHEMICAL PROPULSION INFORMATION AGENCY
CPIA/M2,REV.-
    JOHNS HOPKINS UNIV., SILVER SPRING, MD. APPLIED PHYSICS LAB.
CPIA-PUB.-
    JOHNS HOPKINS UNIV., SILVER SPRING, MD. CHEMICAL PROPULSION INFORMATION AGENCY
CPK-
    TENNESSEE EASTMAN CORP., OAK RIDGE, TENN.
CPK-E-
    TENNESSEE EASTMAN CORP., OAK RIDGE, TENN.
CPKE-
    TENNESSEE EASTMAN CORP., OAK RIDGE, TENN.
CPL-
    CHICAGO. UNIV. CLOUD PHYSICS LAB.
C/PLSEL-
    ARMY NATICK LABS. CLOTHING AND PERSONAL LIFE SUPPORT EQUIPMENT LAB., MASS.
C/PLSEL-TS-
    ARMY NATICK LABS. CLOTHING AND PERSONAL LIFE SUPPORT EQUIPMENT LAB., MASS.
CPL T-
    GERMANY. OBERKOMMANDO DER KRIEGSMARINE. CHEMISCHES-PHYSIKALISCHES EXPERIMENTELLES LABORATORIUM
CP/M-
    GT. BRIT. CAPENHURST WORKS, CHES., ENGLAND (REF. 36)
CPM-GD/SD-
    NATIONAL MILITARY COMMAND SYSTEM SUPPORT CENTER, WASHINGTON, D.C.
CPN-
    CHICAGO. UNIV. METALLURGICAL LAB. (REF. 10)
CPN/
    UNITED KINGDOM ATOMIC ENERGY AUTHORITY. RESEARCH GP. ATOMIC ENERGY RES. ESTAB., HARWELL, BERKS, ENGLAND
CPNL-WI-
    LOUVAIN, BELGIUM. UNIVERSITE. CENTRE DE PHYSIQUE

CPP-	NATIONAL RESEARCH COUNCIL OF CANADA. ATOMIC ENERGY PROJECT, CHALK RIVER, ONT.
CPP-(YR)-(NO.)-(PPC)	PHILLIPS PETROLEUM CO. ATOMIC EN. DIV., IDAHO FALLS
CPP-(NO.)-(PPC)	PHILLIPS PETROLEUM CO. ATOMIC EN. DIV., IDAHO FALLS
CPP-(YEAR)-	PHILLIPS PETROLEUM CO. ATOMIC EN. DIV., IDAHO FALLS
CPP-	PHILLIPS PETROLEUM CO. ATOMIC EN. DIV., IDAHO FALLS
CPPD-MRR-	CONSUMERS PUBLIC POWER DISTRICT, HALLAM, NEB. (MONTHLY RETIREMENT REPORT)
CPPT-	TEXAS. UNIV., AUSTIN. CENTER FOR PLASMA PHYSICS AND THERMONUCLEAR RESEARCH
CPR-	(CONSOLIDATED PROGRESS REPORT) (REF. 2)
CPR-	AEROJET-GENERAL CORP. LIQUID ROCKET PLANT, SACRAMENTO, CALIF.
CPR-	BUREAU OF ORDNANCE (NAVY) (COMPUTATION PROJECT RPTS.)
CPR-	BUREAU OF SHIPS (COMPUTATION PROJECT REPORTS)
C-PR-	ILLINOIS. UNIV., URBANA
CPR-	JOHNS HOPKINS UNIV., SILVER SPRING, MD. APPLIED PHYSICS LAB.
CPR-	STRATEGIC BOMBING SURVEY (COKING PLANT REPORTS)
CPRC-(YEAR)-	GEORGIA. BUREAU OF STATE PLANNING AND COMMUNITY AFFAIRS, ATLANTA
CPRL-	ILLINOIS. UNIV., URBANA. DEPT. OF ELECTRICAL ENGINEERING, CHARGED PARTICLE RESEARCH LAB.
CPS-	COMMITTEE ON PENICILLIN SYNTHESIS
CPS-(CONTRACT NO.)	ENVIRONMENTAL HEALTH SERVICE, ROCKVILLE, MD.
CPS-	SANDIA CORP., ALBUQUERQUE, N. MEX. (COMMERCIAL PACKAGING SPECIFICATION)
C/PSEL-TS-	ARMY NATICK LABS. CLOTHING AND PERSONAL LIFE SUPPORT EQUIPMENT LAB., MASS.
CPSR-	ATOMIC ENERGY OF CANADA LTD. COMMERCIAL PRODUCTS DIV., OTTAWA (COMMERCIAL PRODUCTS SCIENCE REPORTS)
CPT-	MASSACHUSETTS INST. OF TECH., CAMBRIDGE. CENTER FOR THEORETICAL PHYSICS
CPT-	STANFORD UNIV., CALIF. STANFORD LINEAR ACCELERATOR CENTER
CPT-	TEXAS. UNIV., AUSTIN. CENTER FOR PARTICLE THEORY
CPTEP-	FRANCE. CENTRE DE PHYSIQUE THEORIQUE DE L'ECOLE POLYTECHNIQUE
CPTEP-NO-A(NO.)-	FRANCE. CENTRE DE PHYSIQUE THEORIQUE DE L'ECOLE POLYTECHNIQUE
CPVA-	BRITISH INTELLIGENCE OBJECTIVES SUB-COMMITTEE
CPWA ER-	CANADIAN PRATT AND WHITNEY AIRCRAFT CO., LONGUEUIL, QUE
CPWA TN-	CANADIAN PRATT AND WHITNEY AIRCRAFT CO., LONGUEUIL, QUE
CPX-	CHICAGO. UNIV. METALLURGICAL LAB. (REF. 10)
CQ-	WESTINGHOUSE ELECTRIC CORP. AEROSPACE DIV., BALTIMORE
CR-	(CHEMICAL REPORT) (REF. 2)
CR-	(CONFERENCE REPT.) (REF. 2)
CR-	(CONSOLIDATED REPORT) (REF. 2)
CR-	(CONSULTING REPORT) (REF. 2)
CR-	(CONTRACT REPORT) (REF. 2)
CR-	(CONTRACTOR REPORT) (REF. 2)
CR-(NUMBER)-	AEROJET-GENERAL CORP., SACRAMENTO, CALIF.
CR-	AGRICULTURAL RESEARCH SERVICE, BELTSVILLE, MD.
CR-(YEAR)-J-	ARMY BALLISTIC MISSILE AGENCY, REDSTONE ARSENAL, ALA. (JUPITER PROGRESS REPORTS)
CR-	ARMY ENGINEER WATERWAYS EXPERIMENT STATION, VICKSBURG, MISS.
CR-(YEAR)-(NUMBER)/	ARMY MATERIALS RESEARCH AGENCY, WATERTOWN, MASS.
CR-(NUMBER)-(NO.,LTRS.)	AUTOMETRIC CORP., N.Y.C.
CR-	CALIFORNIA RESEARCH CORP. (ALL LOCATIONS)
CR-(NUMBER)-	CHRYSLER CORP. MISSILE DIV., DETROIT
CR-	COLLINS RADIO CO., CEDAR RAPIDS, IOWA
CR-	COLUMBIA UNIV., N.Y.C.
CR-	COMMITTEE ON MEDICAL RESEARCH (CONVALESCENCE AND REHABILITATION REPORTS)
CR-(NO.)-(NO.)-(NO.)	CONVAIR, POMONA, CALIF.
CR-(NUMBER)-	DEFENSE RESEARCH CORP., SANTA BARBARA, CALIF.
CR-(MONTH)-(YEAR)	DEPT. OF AGRICULTURE. CROPS RES. DIV., WASHINGTON, DC
CR-	FAIRCHILD ENGINE AND AIRPLANE CORP. NEPA DIV., OAK RIDGE, TENN.
CR-	FIELD INTELLIGENCE AGENCY, TECHNICAL (CHEM. RPTS.)
CR-6-(NO.-NO.-NO.)	GENERAL DYNAMICS/POMONA, CALIF.
CR-	MARTIN CO., BALTIMORE
CR-(NO.)-	MARTIN CO., BALTIMORE
CR-(YEAR)-	MARTIN CO., DENVER
CR-(YEAR)-(NO.)-F-	MARTIN CO., DENVER (FINAL REPORT)
CR-(YEAR)-(NO.)-PT-	MARTIN CO., DENVER
CR-	MARTIN-MARIETTA CORP. DENVER DIV.
CR-	MINNESOTA. UNIV., MINNEAPOLIS. COSMIC RAY GROUP
CR-	NEW MEXICO STATE UNIV., UNIVERSITY PARK
CR-	NORTH AMERICAN AVIATION, INC., DOWNEY, CALIF.
CR-	OFFICE OF RUBBER RESERVE (POLYMER RESEARCH, DISCRETION GROUP REPORTS)
CR-(YEAR)-	RADIO CORP. OF AMERICA (ALL LOCATIONS)
CR-	RADIO CORP. OF AMERICA (ALL LOCATIONS)
CR-31-555-	SYLVANIA ELECTRIC PRODUCTS INC., BAYSIDE, N.Y.
CR31-(NUMBER)-	SYLVANIA ELECTRIC PRODUCTS INC. ELECTRONICS DIV., BOSTON
CR-(LETTER(S))-	ATOMIC ENERGY OF CANADA LTD., CHALK RIVER, ONT. (REF.6)
CR(LETTER(S))-	ATOMIC ENERGY OF CANADA LTD., CHALK RIVER, ONT. (REF.6)
CRA-(NUMBER-YEAR)-	CHARLES RIVER ASSOCIATES, INC., CAMBRIDGE, MASS.
CRA-	NATIONAL RESEARCH COUNCIL OF CANADA. ATOMIC ENERGY PROJECT, CHALK RIVER, ONT.
CRAAM-	UNIVERSIDADE MACKENZIE, SAO PAULO, BRAZIL. CENTRO DE RADIO-ASTRONOMIA E ASTROFISICA
CRARE-	ATOMIC ENERGY OF CANADA LTD., CHALK RIVER, ONT.
CR-ARS-(NUMBER-NUMBER)	AGRICULTURAL RESEARCH SERVICE. CROP RESEARCH DIV., WASHINGTON, D.C.
CRB-	GT. BRIT. CENTRAL RADIO BUREAU
CRB-	NATIONAL RESEARCH COUNCIL OF CANADA. ATOMIC ENERGY PROJECT, CHALK RIVER, ONT.
CR-B-RHC-	ATOMIC ENERGY OF CANADA LTD. CHALK RIVER PROJECT, ONT
CRC-	CALIFORNIA RESEARCH CORP., RICHMOND
CRC-	CANADA. COMMUNICATIONS RESEARCH CENTRE, OTTAWA
CRC-	CATALYST RESEARCH CORP., BALTIMORE
CRC-	COLLINS RADIO CO., CEDAR RAPIDS, IOWA
CRC-	COMPUTER RESEARCH CORP., HAWTHORNE, CALIF.
CRC-(YEAR)-	CONCORD RESEARCH CORP., BURLINGTON, MASS.
CRC-	COORDINATING RESEARCH COUNCIL, INC., N.Y.C.
CRC-	NATIONAL RESEARCH COUNCIL OF CANADA. ATOMIC ENERGY PROJECT, CHALK RIVER, ONT.
CRCA-	DIVISION OF TECHNICAL INFORMATION, AEC (CLASSIFIED REPORTS FOR CIVILIAN APPLICATIONS)
CRC-AEC-	CALIFORNIA RESEARCH CORP., RICHMOND
CRCC-	CLEVITE RESEARCH CENTER, CLEVELAND
CRC-CAPA-(NUMBER)-	GENERAL RESEARCH CORP., SANTA BARBARA, CALIF.
CRC-DOC.-	CATALYST RESEARCH CORP., BALTIMORE
CRCE-	NATIONAL RESEARCH COUNCIL OF CANADA. ATOMIC ENERGY PROJECT, CHALK RIVER, ONT.
CR-CH-	FOOD MACHINERY + CHEMICAL CORP., SAN JOSE, CALIF.
CRCI-	COORDINATING RESEARCH COUNCIL, INC., N.Y.C.
CRC-LD-	COORDINATING RESEARCH COUNCIL, INC., N.Y.C.
CRC-NEPA-	CALIFORNIA RESEARCH CORP., RICHMOND

```
CRCO-              CATALYST RESEARCH CORP., BALTIMORE
CRC-PR-            CATALYST RESEARCH CORP., BALTIMORE
CRC-TN-            CANADA. COMMUNICATIONS RESEARCH CENTRE, OTTAWA
CRD-               ACADEMIA R.S.R.INSTITUTUL DE FIZICA ATOMICA,BUCHAREST
CRD-(YEAR)-        AIR REDUCTION CO., INC. CENTRAL RESEARCH DEPT.,
                     MURRAY HILL, N.J. (USED 1961-63)
CRD-               ARMY SCIENTIFIC ADVISORY PANEL
CRD-               ARMY TRANSPORTATION RESEARCH COMMAND, FORT EUSTIS, VA
CRD-               CALIFORNIA RESEARCH AND DEVELOPMENT CO., LIVERMORE
CR-D-              CHRYSLER CORP. MISSILE DIV., DETROIT
CRD-               NATIONAL RESEARCH COUNCIL OF CANADA. ATOMIC ENERGY
                     PROJECT, CHALK RIVER, ONT.
CRD-A(NUMBER)-     CALIFORNIA RESEARCH AND DEVELOPMENT CO., LIVERMORE
CRDARE-            ARMY RESEARCH AND DEVELOPMENT GROUP (EUROPE), N.Y.C.
CRDC-              ATOMIC ENERGY OF CANADA LTD., CHALK RIVER, ONT.
CRDC-              COLUMBIA RESEARCH AND DEV. CORP., COLUMBUS, OHIO
CRDD-              GT. BRIT. MINISTRY OF SUPPLY. CHEMICAL RESEARCH AND
                     DEVELOPMENT DEPT., WALTHAM ABBEY, ESSEX, ENGLAND
CRDL-              ARMY CHEMICAL RES. + DEV. LABS., ARMY CHEM. CTR., MD.
CRDL-              DOW CHEMICAL CO. ROCKY FLATS DIV., GOLDEN, COLO.
CRDL-              DOW CHEMICAL CO. ROCKY FLATS PLANT, GOLDEN, COLO.
CRDLR-             ARMY CHEMICAL RES. + DEV. LABS., ARMY CHEM. CTR., MD.
CRDL-SP-(NUMBER)-  ARMY CHEMICAL RESEARCH AND DEVELOPMENT LABS., ARMY
                     CHEMICAL CENTER, MD. (SPECIAL PUBLICATIONS)
CRDL-TM-(NUMBER)-  ARMY CHEMICAL RES. + DEV. LABS., ARMY CHEM. CTR., MD.
CRD-R-             CALIFORNIA RESEARCH AND DEVELOPMENT CO., LIVERMORE
CRD-T(NUMBER)-     CALIFORNIA RESEARCH AND DEVELOPMENT CO., BERKELEY
CRD-T(NUMBER)-     CALIFORNIA RESEARCH AND DEVELOPMENT CO., LIVERMORE
CRE-               ATOMIC ENERGY OF CANADA LTD., CHALK RIVER, ONT.
CREARE-TN-N-       CREARE INC., HANOVER, N.H.
CREL-              ATOMIC ENERGY OF CANADA LTD., CHALK RIVER, ONT.
CRER-              ATOMIC ENERGY OF CANADA LTD., CHALK RIVER, ONT.
CRES-(NUMBER)-     KANSAS. UNIV., LAWRENCE. CENTER FOR RESEARCH IN
                     ENGINEERING SCIENCE
CRES-REPRINT-      KANSAS. UNIV., LAWRENCE. CENTER FOR RESEARCH IN
                     ENGINEERING SCIENCE
CRESS/CINFAC-R-    AMERICAN UNIV., WASHINGTON, D.C. CENTER FOR RESEARCH
                     IN SOCIAL SYSTEMS
CRES-TR-(NUMBER-NUMBER)
                   KANSAS. UNIV., LAWRENCE. CENTER FOR RESEARCH IN
                     ENGINEERING SCIENCE
CRFD-              ATOMIC ENERGY OF CANADA LTD., CHALK RIVER, ONT.
CRG-               NAVAL RESEARCH LAB. COMBINED RESEARCH GP., WASH.,D.C.
CRGM-              ATOMIC ENERGY OF CANADA LTD., CHALK RIVER, ONT.
CRGP-              ATOMIC ENERGY OF CANADA LTD., CHALK RIVER, ONT.
CRGP-              ATOMIC ENERGY OF CANADA LTD., DEEP RIVER LAB., ONT.
CRHL-              COLORADO STATE UNIV., FORT COLLINS. COLLABORATIVE
                     RADIOLOGICAL HEALTH LAB.
CRHP-              ATOMIC ENERGY OF CANADA LTD., CHALK RIVER, ONT.
CRHR-              NATIONAL RESEARCH COUNCIL OF CANADA. ATOMIC ENERGY
                     PROJECT, CHALK RIVER, ONT.
CRI-               COMMUNICATION RESEARCH INST., MIAMI, FLA.
CRIB-              NATIONAL RESEARCH COUNCIL OF CANADA. ATOMIC ENERGY
                     PROJECT, CHALK RIVER, ONT.
CRIF-EL-           BELGIUM. CENTRE DE RECHERCHES SCIENTIFIQUES ET TECH-
                     NIQUES DE L'INDUSTRIE DES FABRICATION, BRUSSELS
CRIF-MC-           BELGIUM. CENTRE DE RECHERCHES SCIENTIFIQUES ET TECH-
                     NIQUES DE L'INDUSTRIE DES FABRICATION, BRUSSELS
CRIF-MT-           BELGIUM. CENTRE DE RECHERCHES SCIENTIFIQUES ET TECH-
                     NIQUES DE L'INDUSTRIE DES FABRICATION, BRUSSELS
CRIF-PL-           BELGIUM. CENTRE DE RECHERCHES SCIENTIFIQUES ET TECH-
                     NIQUES DE L'INDUSTRIE DES FABRICATION, BRUSSELS
CRIO-              ATOMIC ENERGY OF CANADA LTD., CHALK RIVER, ONT.
CRISP-(YEAR)-      BROOKHAVEN NATIONAL LAB., UPTON, N.Y.
CRIT-              CARNEGIE INST. OF TECH., PITTSBURGH
CRI-TR-(NUMBER)-   CHALLENGER RESEARCH INC., ROCKVILLE, MD.
CR-J-(YEAR)-       ARMY BALLISTIC MISSILE AGENCY, REDSTONE ARSENAL,
                     ALA. (JUPITER PROGRESS REPORTS)
CRK-               KELLEX CORP., N.Y.C.
CRL-               AIR FORCE CAMBRIDGE RESEARCH LABS., BEDFORD, MASS.
CRL-               ATOMIC ENERGY OF CANADA LTD. CHALK RIVER PROJ., ONT.
                     (LECTURE)
CRL-               CAMBRIDGE UNIV., ENGLAND. CAVENDISH LAB.
CRL-               CHEMICAL AND RADIOLOGICAL LABS., ARMY CHEM. CTR., MD.
CRL-               COLUMBIA UNIV., N.Y.C. COLUMBIA RADIATION LAB.
CRL-               COMMUNICATIONS RESEARCH LABS., SANTA ANA, CALIF.
CRL-               COOK RESEARCH LABS., MORTON GROVE, ILL.
CRL-               CREW RESEARCH LAB., RANDOLPH AFB, TEX.
CRL-               NATIONAL BUREAU OF STANDARDS. CENTRAL RADIO PROPA-
                     GATION LAB., WASHINGTON, D.C.
CRL-               NATIONAL RESEARCH COUNCIL OF CANADA. ATOMIC ENERGY
                     PROJECT, CHALK RIVER, ONT. (REF. 29)
CRL-               QUARTERMASTER RESEARCH AND DEVELOPMENT CENTER. CLI-
                     MATIC RESEARCH LAB., LAWRENCE, MASS.
CRL-(LETTERS)-     GT. BRIT. CHEM. RES. LAB., TEDDINGTON, MIDDX., ENG.
CRL/AE-            GT. BRIT. CHEM. RES. LAB., TEDDINGTON, MIDDX., ENG.
CRL-HTC-           SMITH (A.O.) CORP. CERAMIC RES. LAB., MILWAUKEE
CRL-HTC-TOPICAL-   SMITH (A.O.) CORP. CERAMIC RES. LAB., MILWAUKEE
CRLIR-             CHEMICAL AND RADIOLOGICAL LABS.,ARMY CHEMICAL CENTER,
                     MD. (INTERIM REPORTS)
CRL-LN-(YEAR)-     CREW RESEARCH LAB., RANDOLPH AFB, TEX. (LAB. NOTE)
CRL-LN-(YEAR)-     CREW RESEARCH LAB. SURVIVAL FIELD RESEARCH UNIT,
                     STEAD AFB, NEV. (LABORATORY NOTE)
CRL M-             QUARTERMASTER RESEARCH AND DEVELOPMENT CENTER. CLI-
                     MATIC RESEARCH LAB., LAWRENCE, MASS. (MEMO RPTS.)
CRLR-              CHEMICAL AND RADIOLOGICAL LABS.,ARMY CHEMICAL CENTER,
                     MD. (REPORTS)
CRL R (NUMBER(S))- QUARTERMASTER RESEARCH AND DEVELOPMENT CENTER. CLI-
                     MATIC RESEARCH LAB., LAWRENCE, MASS.
CRLR-              QUARTERMASTER RESEARCH AND DEVELOPMENT CENTER. CLI-
                     MATIC RESEARCH LAB., LAWRENCE, MASS.
CRL R (LETTER(S))- QUARTERMASTER RESEARCH AND DEVELOPMENT CENTER. CLI-
                     MATIC RESEARCH LAB., LAWRENCE, MASS.
CRL-S-             JOHNS HOPKINS UNIV., BALTIMORE. INST. FOR COOP. RES.
CRL-TM-(YEAR)-     CREW RESEARCH LAB., RANDOLPH AFB, TEX. (TECH. MEMO)
CRM-               NATIONAL RESEARCH COUNCIL OF CANADA. ATOMIC ENERGY
                     PROJECT, CHALK RIVER, ONT.
CRMET-             ATOMIC ENERGY OF CANADA LTD., CHALK RIVER, ONT.
CRNE-              ATOMIC ENERGY OF CANADA LTD., CHALK RIVER, ONT.
CRNL-              ATOMIC EN. OF CANADA LTD. CHALK RIVER NUC. LABS., ONT
CRNP-              ATOMIC ENERGY OF CANADA LTD., CHALK RIVER, ONT.
CRO-SR-            AIR FORCE CAMBRIDGE RESEARCH LABS., BEDFORD, MASS.
CRP-               ATOMIC ENERGY OF CANADA LTD., CHALK RIVER, ONT.
CRP(LETTER(S))-    NATIONAL RESEARCH COUNCIL OF CANADA. ATOMIC ENERGY
                     PROJECT, CHALK RIVER, ONT. (REF. 29)
CRPA-HOUSING-      CAPITOL REGIONAL PLANNING AGENCY, HARTFORD, CONN.
CRPC-              NATIONAL RESEARCH COUNCIL OF CANADA. ATOMIC ENERGY
                     PROJECT, CHALK RIVER, ONT.
                     PROJECT, CHALK RIVER, ONT.
CRPE-              NATIONAL RESEARCH COUNCIL OF CANADA. ATOMIC ENERGY
                     PROJECT, CHALK RIVER, ONT.
CRPG-              NATIONAL RESEARCH COUNCIL OF CANADA. ATOMIC ENERGY
                     PROJECT, CHALK RIVER, ONT.
CRPL-              ATOMIC ENERGY OF CANADA LTD., CHALK RIVER, ONT.
CRPL-              INSTITUTE FOR TELECOMMUNICATION SCIENCES AND
                     AERONOMY, BOULDER, COLO.
CRPL-              NATIONAL BUREAU OF STANDARDS. CENTRAL RADIO PROPAGA-
                     TION LAB., BOULDER, COLO.
```

CRPL-F-	NATIONAL BUREAU OF STANDARDS. BOULDER LABS., COLO.
CRPP-	NATIONAL RESEARCH COUNCIL OF CANADA. ATOMIC ENERGY PROJECT, CHALK RIVER, ONT.
CRPR-	ATOMIC ENERGY OF CANADA LTD. CHALK RIVER PROJECT, ONT. (REF. 6)
CR-PRG-	NATIONAL RESEARCH COUNCIL OF CANADA. ATOMIC ENERGY PROJECT, CHALK RIVER, ONT.
CR-PRG-(NO.)-TEC.-	NATIONAL RESEARCH COUNCIL OF CANADA. ATOMIC ENERGY PROJECT, CHALK RIVER, ONT.
CRP/RT/(NUMBER/YEAR)	NOTTINGHAM, ENGLAND. UNIV. DEPT. OF GEOGRAPHY (CORRELATION + REGRESSION)(AUTHOR'S INITIALS)
CRPX-	NATIONAL RESEARCH COUNCIL OF CANADA. ATOMIC ENERGY PROJECT, CHALK RIVER, ONT.
C-RQR/(YEAR)-	JOHNS HOPKINS UNIV., SILVER SPRING, MD. APPLIED PHYSICS LAB. (RES. + DEV. QUARTERLY REPT.)
CRR-	ATOMIC ENERGY OF CANADA LTD., CHALK RIVER, ONT.
CRR-	COLLINS RADIO CO., CEDAR RAPIDS, IOWA (RESEARCH RPT.)
CR-R-	COLUMBIA UNIV., NYC. SUBSTITUTE ALLOY MATERIALS LABS.
CRRCE-	ATOMIC ENERGY OF CANADA LTD., CHALK RIVER, ONT.
CRRD-	ATOMIC ENERGY OF CANADA LTD., CHALK RIVER, ONT.
CR-R+DL-	CANADIAN WESTINGHOUSE CO., LTD., HAMILTON, ONT.
CRREL-	ARMY COLD REGIONS RES. AND ENG. LAB., HANOVER, N.H.
CRREL-	ARMY COLD REGIONS RES. + ENG. LAB., WILMETTE, ILL.
CRREL-CRSE-	ARMY COLD REGIONS RES. AND ENG. LAB., HANOVER, N.H.
CRREL-RR-	ARMY COLD REGIONS RES. + ENG. LAB., WILMETTE, ILL. (RESEARCH REPORT)
CRREL-SR-	ARMY COLD REGIONS RES. AND ENG. LAB., HANOVER, N.H. (SPECIAL REPORT)
CRREL-SR-	ARMY COLD REGIONS RES. + ENG. LAB., WILMETTE, ILL. (SPECIAL REPORT)
CRREL-TR-	ARMY COLD REGIONS RES. AND ENG. LAB., HANOVER, N.H. (TECHNICAL REPORT)
CRREL-TR-	ARMY COLD REGIONS RES. + ENG. LAB., WILMETTE, ILL. (TECHNICAL REPORT)
CRREL-TRANS-	ARMY COLD REGIONS RES. + ENG. LAB., WILMETTE, ILL. (TRANSLATION)
CR-RHC-	NATIONAL RESEARCH COUNCIL OF CANADA. ATOMIC ENERGY PROJECT, CHALK RIVER, ONT.
CRRIS-	ATOMIC ENERGY OF CANADA LTD., CHALK RIVER, ONT.
CRRL-	ATOMIC ENERGY OF CANADA LTD., CHALK RIVER, ONT.
CRRM-	ATOMIC ENERGY OF CANADA LTD., CHALK RIVER, ONT.
CRRP-	ATOMIC ENERGY OF CANADA LTD., CHALK RIVER, ONT.
CRS-	ILLINOIS. UNIV., URBANA. DEPT. OF CIVIL ENGINEERING. CONSTRUCTION RESEARCH LAB.
CRS-	NEW JERSEY. CERAMIC RESEARCH STATION, NEW BRUNSWICK
CRSIM-	FRANCE. CENTRE DE RECHERCHES SCIENTIFIQUES, INDUSTRIELLES ET MARITIMES
CRSR-	CORNELL UNIV., ITHACA, N.Y. CENTER FOR RADIOPHYSICS AND SPACE RESEARCH
CRT-	ATOMIC ENERGY OF CANADA LTD., CHALK RIVER, ONT.
CRTC/R-	UNITED KINGDOM ATOMIC ENERGY AUTHORITY. RESEARCH GP. ATOMIC ENERGY RES. ESTAB., HARWELL, BERKS, ENGLAND
CRTEC-	NATIONAL RESEARCH COUNCIL OF CANADA. ATOMIC ENERGY PROJECT, CHALK RIVER, ONT.
CRUFT-TR-	HARVARD UNIV., CAMBRIDGE, MASS. CRUFT LAB.
CRUSTAL STUDIES-	GEOLOGICAL SURVEY, DENVER (CRUSTAL STUDIES)
CRWP-	BRITISH COAL UTILISATION RESEARCH ASSOCIATION
CRWPC-	CANADIAN RADIO WAVE PROPAGATION COMMITTEE
CRWR-	TEXAS. UNIV., AUSTIN. CENTER FOR RESEARCH IN WATER RESOURCES
CRX-	NATIONAL RESEARCH COUNCIL OF CANADA. ATOMIC ENERGY PROJECT, CHALK RIVER, ONT.
CR-/XM-	MIDLAND-WRIGHT, KANSAS CITY, KANS.
CS-	ARGONNE NATIONAL LAB., LEMONT, ILL. (REF. 10)
CS-	CHICAGO. UNIV. METALLURGICAL LAB.
CS-	CLINTON LABS., OAK RIDGE, TENN. (REF. 10)
CS-	COLORADO SCHOOL OF MINES RESEARCH FDN., INC., GOLDEN
C/S-(YEAR)-	COMMUNICATION SYSTEMS INC., FALLS CHURCH, VA.
C/S-(4 DIGITS)-	COMMUNICATIONS AND SYSTEMS, INC., FALLS CHURCH, VA.
C/S-(YEAR)-TR-	COMMUNICATIONS AND SYSTEMS, INC., PARAMUS, N.J.
C/S-(YEAR-3 DIGITS)-	COMMUNICATIONS AND SYSTEMS, INC., PARAMUS, N.J.
CS-	CORNELL UNIV., ITHACA, N.Y. LAB. OF NUCLEAR STUDIES
CS-	DEPARTMENT OF COMMERCE, WASH., D.C. (COMMERCIAL STD.)
CS-	FOREST SERVICE. CENTRAL STATES STATION, BEREA, KY.
CS-	FRANKLIN INST. LABS. FOR RES. + DEV., PHILADELPHIA
CS-(NUMBER)-R(NUMBER)	LOGICON, INC., REDONDO BEACH, CALIF.
CS-(YEAR)-	NCS COMPUTING CORP., ROME, N.Y.
CS-	NETHERLANDS. INSTITUUT VOOR TEXTIELREINIGING TNO, DELFT
CS-	STANFORD UNIV., CALIF. DEPT. OF COMPUTER SCIENCE
CS-	STANLEY AVIATION CORP., DENVER
CS-	TECHNICAL IND. INTELL. COMM. CHEMICALS SUB-COMM.
CS-	TENNESSEE EASTMAN CORP., OAK RIDGE, TENN. (REF. 10)
CS-(2 DIGITS)	TEXAS. UNIV., AUSTIN. CENTER FOR CYBERNETIC STUDIES
CSA-	AIR FORCE, WASHINGTON, D.C.
CSA-	CANADIAN STANDARDS ASSOCIATION, OTTAWA
CSA-	CAYWOOD SCHILLER, ASSOCIATES, CHICAGO
CSA-(YEAR)-P-	CAYWOOD-SCHILLER, ASSOCIATES, CHICAGO
CSA-(YEAR)-F-	CAYWOOD-SCHILLER, ASSOCIATES, CHICAGO
CSA-(YEAR)	SCHOOL OF AVIATION MEDICINE, RANDOLPH AFB, TEX.
CSB-	CLINTON LABS., OAK RIDGE, TENN.
CSC-	CIVIL SERVICE COMMISSION, WASHINGTON, D.C.
CSCA-IR-	CRUCIBLE STEEL CO. OF AMERICA. MIDLAND RES. LAB., PITTSBURGH (INTERIM REPORT)
CSC-FINAL-	COMMERCIAL SOLVENTS CORP., TERRE HAUTE, IND.
CSCHBX-	CIVIL SERVICE COMMISSION, WASHINGTON, D.C. (HANDBOOK)
CS/CJ-	GT. BRIT. CHEM. DEFENCE EXPTL. STA.,PORTON,WILTS,ENG.
CSCJEMPS-	CIVIL SERVICE COMMISSION, WASHINGTON, D.C. (JOB EMPLOYMENT METHOD)
CSCL-	FEDERAL COUNCIL FOR SCIENCE AND TECHNOLOGY. COMMITTEE ON SCIENTIFIC AND TECHNICAL INFORMATION, WASH., DC (SUBJECT CATEGORY LIST)
CSC-M-	COMMERCIAL SOLVENTS CORP., TERRE HAUTE, IND.
CSCO-	CRUCIBLE STEEL CO. OF AMERICA, PITTSBURGH
CSCP-	CIVIL SERVICE COMMISSION, WASHINGTON, D.C. (POSTER)
CSC/P-	UNITED KINGDOM ATOMIC ENERGY AUTHORITY. RESEARCH GP. ATOMIC ENERGY RES. ESTAB., HARWELL, BERKS, ENGLAND
CSC-Q-	COMMERCIAL SOLVENTS CORP., TERRE HAUTE, IND.
CSCRD-	OFFICE OF THE CHIEF OF RESEARCH AND DEVELOPMENT, ARMY RESEARCH OFFICE, WASHINGTON, D.C.
CSC-TR-(YEAR)-	MARYLAND. UNIV., COLLEGE PARK. COMPUTER SCIENCE CTR.
CSD-	NATIONAL ACADEMY OF SCIENCES. COMMITTEE ON SENSORY DEVICES
CSD (CAP.) M-	GT. BRIT. CAPENHURST WORKS, CHES., ENGLAND (REF. 36)
CSD (CAP.) TM-	GT. BRIT. CAPENHURST WORKS, CHES., ENGLAND (REF. 36)
CSD-TR-	PURDUE UNIV.,LAFAYETTE,IND. DEPT. OF COMPUTER SCIENCE
CSD(W)-	GT. BRIT. WINDSCALE WORKS, SELLAFIELD, CUMB., ENGLAND (REF. 36)
CSE-	GT. BRIT. CENTRAL SIGNALS ESTABLISHMENT
CSEE-	GT. BRIT. CLOTHING + EQUIP. PHYSIOLOGICAL RES. ESTAB.
CSE R-	GT. BRIT. CENTRAL SIGNALS ESTABLISHMENT
-CSF-(YEAR)-	COMPAGNIE FRANCAISE THOMSON HOUSTON-HOTCHKISS BRANDT, PARIS
CSGLD-	OFFICE OF ORDNANCE RESEARCH (ARMY), DURHAM, N.C.
CSI-	CASE INST. OF TECH., CLEVELAND

CS-I-	CHICAGO. UNIV. METALLURGICAL LAB. (REF. 10)
CSI-	COMMUNICATION SYSTEMS INC., PARAMUS, N.J.
CSI-(YEAR)-TR	COMMUNICATION SYSTEMS INC., PARAMUS, N.J.
CSI-(YEAR)-	COMPUTER SYMBOLIC, INC., ROME, N.Y.
CSI-(YEAR)-TR-	DEFENSE COMMUNICATIONS AGENCY, WASHINGTON, D.C.
CSIR-	(FOR AUSTRALIAN REPORTS SEE ENTRIES UNDER AUS-CSIR-)
CSIR-	UNION OF SOUTH AFRICA. COUNCIL FOR SCIENTIFIC AND INDUSTRIAL RESEARCH, PRETORIA
CSIR-AR-	UNION OF SOUTH AFRICA. COUNCIL FOR SCIENTIFIC AND INDUSTRIAL RESEARCH, PRETORIA
CSIRC-TR-(YEAR)-	OHIO STATE UNIV., COLUMBUS. COMPUTER AND INFORMATION SCIENCE RESEARCH CENTER.
CSIR-FIS-	UNION OF SOUTH AFRICA. COUNCIL FOR SCIENTIFIC AND INDUSTRIAL RESEARCH, PRETORIA
CSIRO-	AUSTRALIA. COMMONWEALTH SCI. + IND. RES. ORG. INFO. SERVICE, E. MELBOURNE (TRANSLATIONS)
CSIRO-TRANS-	AUSTRALIA. COMMONWEALTH SCIENTIFIC + IND. RES. ORG.
CSIRO/UI-	AUSTRALIA. COMMONWEALTH SCIENTIFIC + IND. RES. ORG.
CSIR-TEL-	UNION OF SOUTH AFRICA. COUNCIL FOR SCIENTIFIC AND INDUSTRIAL RESEARCH, PRETORIA
CSL-	COLES SIGNAL LAB., BELMAR, N.J.
CSL-	MICHIGAN. UNIV., ANN ARBOR.COMMUNICATION SCIENCES LAB
CSL-	SYRACUSE UNIV., N.Y. RESEARCH CORP.
CSL-(YEAR)-	SYRACUSE UNIV., N.Y. RESEARCH CORP.
CSL/ONR-	FLORIDA. UNIV., GAINESVILLE
CSL/ONR-	FLORIDA. UNIV., GAINESVILLE. COMMUNICATIONS SCIENCES LAB.
CSL-R-	ILLINOIS. UNIV., URBANA
CSL-TR-	WESTERN RESERVE UNIV., CLEVELAND
CSM-	COLORADO SCHOOL OF MINES, GOLDEN
CSM-	COLORADO SCHOOL OF MINES RESEARCH FDN., INC., GOLDEN
CSM-(NO.)(LTR.)-	DEFENSE COMMUNICATIONS AGENCY, WASHINGTON, D.C.
CSM-	RAND CORP., SANTA MONICA, CALIF.
CSM-MRL-ONR-	COLORADO SCHOOL OF MINES, GOLDEN. MINING RESEARCH LAB
CSM-PR-	COLORADO SCHOOL OF MINES RESEARCH FDN., INC., GOLDEN
CSM R(YR.)-	COLORADO SCHOOL OF MINES RESEARCH FOUNDATION, INC., GOLDEN. (CONTRIBUTIONS)
CSM-UG-(NO.)-(YR.)-	NATIONAL MILITARY COMMAND SYSTEM SUPPORT CENTER, WASHINGTON, D.C. (COMPUTER SYSTEMS MANUAL)
CSN-	CLINTON LABS., OAK RIDGE, TENN.
CS-P-	CHICAGO. UNIV. METALLURGICAL LAB. (REF. 10)
CSPC-(NUMBER)(LETTER)	GT. BRIT. ATOMIC EN. RES. ESTAB., HARWELL, BERKS, ENG
CSPR-	(CONSOLIDATED SEMIANNUAL PROGRESS REPORT) (REF. 2)
CSPRD-	OFFICE OF ORDNANCE RESEARCH (ARMY), DURHAM, N.C.
C-SQR/(YEAR)-	JOHNS HOPKINS UNIV., SILVER SPRING, MD. APPLIED PHYSICS LAB. (SPACE QUARTERLY REPORT)
CSR-	AEROJET-GENERAL CORP., SACRAMENTO, CALIF. (COMPUTING SCIENCES)
CSR-(CON.NO.)(JOB NO.)-	AEROSPACE CORP., EL SEGUNDO, CALIF.(CONTRACT STATUS)
CSR-(CON.NO.)S(JOB NO.)-	AEROSPACE CORP. SAN BERNARDINO OPERATIONS, CALIF.
C/S-R(NUMBER)-	COMMUNICATION SYSTEMS INC., FALLS CHURCH, VA.
CSR-	DEFENSE RESEARCH CORP., SANTA BARBARA, CALIF.
CSRD-	NAVAL ORDNANCE TEST STA., INYOKERN (CHINA LAKE), CAL.
CSRD-I(RI)(MO./YR.)-	NAVAL ORDNANCE TEST STA., INYOKERN (CHINA LAKE), CAL.
CSRL-MEMO-	ARIZONA. UNIV., TUCSON. COMPUTER SCIENCE RESEARCH LAB
CSRL-RR-	CONVAIR SCI. RES. LAB., SAN DIEGO, CALIF. (RES. RPT.)
CSRP-	CORNELL UNIV., ITHACA, N.Y. (COGNITIVE SYSTEMS RESEARCH PROGRAM)
CSR-P-	MASSACHUSETTS INST. OF TECH., CAMBRIDGE. CENTER FOR SPACE RESEARCH
CSR-P-(YEAR)-	MASSACHUSETTS INST. OF TECH., CAMBRIDGE. CENTER FOR SPACE RESEARCH
CSR-PR-(NUMBERS)-	MASSACHUSETTS INST. OF TECH., CAMBRIDGE. CENTER FOR SPACE RESEARCH (PROGRESS REPORT)
CSR-T-(YR.)-	MASSACHUSETTS INST. OF TECH., CAMBRIDGE. CENTER FOR SPACE RESEARCH (THESIS)
CSR-TN-(YR.)-	MASSACHUSETTS INST. OF TECH., CAMBRIDGE. CENTER FOR SPACE RESEARCH (TECHNICAL NOTE)
CSR-TR-(YR.)-	MASSACHUSETTS INST. OF TECH., CAMBRIDGE. CENTER FOR SPACE RESEARCH (TECHNICAL REPORT)
CSS-(YEAR)-	ROCHESTER, N.Y. UNIV. CENTER FOR SYSTEM SCIENCE
CSSE-	WASHINGTON UNIV., ST. LOUIS. CONTROL SYSTEMS SCIENCE AND ENGINEERING
CS(SP)-	GT. BRIT. SPRINGFIELDS WORKS, LANCS, ENGLAND (REF.36)
CSSR-	NAVY (CONSOLIDATED STOCK STATUS REPORT)
CS/TM-	GT. BRIT. CHEM. DEFENCE EXPTL. STA.,PORTON,WILTS,ENG.
CST-PA-(YEAR)-	MIAMI. UNIV., CORAL GABLES, FLA. CENTER FOR THEORETICAL STUDIES
C/S-TR-(NUMBER)-	COMMUNICATIONS AND SYSTEMS, INC., FALLS CHURCH, VA.
CS-TR-	STANFORD UNIV., CALIF. DEPT. OF COMPUTER SCIENCE
CSU-	COLORADO STATE UNIV., FORT COLLINS
CSUAC-	CORNELL-SYDNEY UNIV., ITHACA, N.Y. ASTRONOMY CENTER
CT-	AMES LAB., IOWA (REF. 10)
CT-	ARGONNE NATIONAL LAB., LEMONT, ILL. (REF. 10)
CT-	BATTELLE MEMORIAL INST., COLUMBUS, OHIO (REF. 10)
CT-	CARBIDE AND CARBON CHEMICALS CORP. K-25 PLANT, OAK RIDGE, TENN. (REF. 10)
CT-	COMMITTEE ON MEDICAL RESEARCH (CHEMOTHERAPY REPORTS)
CT-	DU PONT DE NEMOURS (E.I.) + CO. GRASSELLI CHEMICALS DEPT., WILMINGTON, DEL.
CT-	MASSACHUSETTS INST. OF TECH., CAMBRIDGE (REF. 10)
CT-	MICHIGAN. UNIV., ANN ARBOR. ENGINEERING RES. INST.
CT-	NATIONAL BUREAU OF STANDARDS, WASHINGTON, D.C.
CT-	NAVAL AIR TEST CENTER, PATUXENT RIVER, MD.
CT-	NAVAL MISSILE CENTER (COMPONENT TEST)
CT-	YALE UNIV., NEW HAVEN
CT-	YALE UNIV., NEW HAVEN. DUNHAM LAB.
CTB-	CHICAGO. UNIV. METALLURGICAL LAB. (ASSIGNED TO REPORTS FROM GREAT BRITAIN) (REF. 10)
CTC-	UNION CARBIDE NUCLEAR CO. COMPUTING TECHNOLOGY CENTER, OAK RIDGE, TENN.
CTC-INF-	UNION CARBIDE NUCLEAR CO. COMPUTING TECHNOLOGY CENTER, OAK RIDGE, TENN.
CTD-	BUDD CO. TESTING LABS., PHILADELPHIA
CTD-	COLLINS RADIO CO., CEDAR RAPIDS, IOWA (TECHNICAL DESCRIPTION)
CTD-	NAVAL AIR MISSILE TEST CENTER. COMPONENT TEST DEPT., POINT MUGU, CALIF.
CTDC-TP-	CONTROL DATA CORP., LOS ANGELES
CTDC-V-	COMBAT DEVELOPMENT AND TEST CENTER, VIETNAM
CTE-	CONNECTICUT TELEPHONE + ELECTRIC CORP., MERIDEN
CTEE PTP-	GT. BRIT. CHEMICAL DEFENCE EXPERIMENTAL ESTABLISH-MENT, PORTON, WILTS, ENGLAND
CTG-	TASK GROUP (NO.), FLEET POST OFFICE, SAN FRANCISCO (COMMANDER TASK GROUP)
CTGC-	COMMITTEE ON MEDICAL RESEARCH. COMMITTEE ON THE TREATMENT OF GAS CASUALTIES
CTH-	CHALMERS TEKNISKA HOEGSKOLA, GOETEBORG
CTH-RF-	CHALMERS TEKNISKA HOEGSKOLA, GOETEBORG. INST. FOR REAKTORFYSIK
CTI-	(OPERATION CROSSROADS) (REF. 7)
CTL-	CHROMATIC TELEVISION LABS., INC., N.Y.C.
CTL-(NO.-LETTER-NO.)	CINCINNATI TEST LABORATORY
CTM-	CHICAGO. UNIV. METALLURGICAL LAB. (REF. 10)
CTM-K-	HANFORD WORKS, RICHLAND, WASH.

CTN-	CHICAGO. UNIV. METALLURGICAL LAB.
CTN(YR)-	GENERAL MOTORS CORP., SANTA BARBARA, CALIF.
CTN-	GENERAL MOTORS CORP. DEFENSE RESEARCH LABS., SANTA BARBARA, CALIF.
CTNX-	HANFORD WORKS, RICHLAND, WASH.
CTO-	AKADEMIYA NAUK SSSR. INSTITUT FIZIKI, MOSCOW
CTO-	NAVAL CENTRAL TORPEDO OFFICE
CTO/	UNITED KINGDOM ATOMIC ENERGY AUTHORITY. AUTHORITY HEALTH AND SAFETY BRANCH. RADIOLOGICAL PROTECTION DIV., HARWELL, BERKS, ENGLAND
CTO-	UNITED KINGDOM ATOMIC ENERGY AUTHORITY. CULHAM LAB., ABINGDON, BERKS, ENGLAND (TRANSLATIONS)
CTP-	MASSACHUSETTS INST. OF TECH., CAMBRIDGE. CENTER FOR THEORETICAL PHYSICS
CTR-	AEROJET-GENERAL CORP., SACRAMENTO, CALIF.
CTR-(NO.)-(NO.)-	AUTONETICS, ANAHEIM, CALIF.
CTR-	COLLINS RADIO CO., CEDAR RAPIDS, IOWA (TECH. RPT.)
CT-R-	COLUMBIA UNIV., NYC. SUBSTITUTE ALLOY MATERIALS LABS.
CTR-	COLUMBIA UNIV., NYC. SUBSTITUTE ALLOY MATERIALS LABS.
CTR-	OFFICE OF TECHNICAL SERVICES, WASHINGTON, D.C. (CATALOG OF TECHNICAL REPORTS)
CTR-(NUMBER)-	SYLVANIA ELECTRONIC SYSTEMS-WEST, MOUNTAIN VIEW, CAL.
CTRA-	COAL TAR RESEARCH ASSN., LEEDS, YORKS, ENG. (TRANS.)
CTS-	AERONUTRONIC DIV. OF PHILCO CORP., NEWPORT BEACH, CAL.
CTS-	CENTRAL INTELLIGENCE AGENCY, WASHINGTON, D.C. (CONSOLIDATED TRANSLATION SURVEY)
CTS-	GT. BRIT. DEPT. OF SCIENTIFIC AND INDUSTRIAL RES.
CTS-B-	MIAMI. UNIV., CORAL GABLES, FLA. CENTER FOR THEORETICAL STUDIES
CTS-B-(YEAR)-	MIAMI. UNIV., CORAL GABLES, FLA. CENTER FOR THEORETICAL STUDIES
CTSC/-	GT. BRIT. CAPENHURST WORKS, CHES., ENGLAND
CTSC/P-	UNITED KINGDOM ATOMIC ENERGY AUTHORITY. INDUSTRIAL GROUP. CAPENHURST WORKS, CHES., ENGLAND
CTSG-	NATIONAL RESEARCH COUNCIL OF CANADA. ATOMIC ENERGY PROJECT, CHALK RIVER, ONT.
CTS-H-(YEAR)-	MIAMI. UNIV., CORAL GABLES, FLA. CENTER FOR THEORETICAL STUDIES
CTS-HE-(YEAR)-	MIAMI. UNIV., CORAL GABLES, FLA. CENTER FOR THEORETICAL STUDIES
CTSL-	CALIFORNIA INST. OF TECH., PASADENA. SYNCHROTRON LAB.
CTSL-IR-	CALIFORNIA INST. OF TECH., PASADENA. SYNCHROTRON LAB.
CTS-LN-(YEAR)-	MIAMI. UNIV., CORAL GABLES, FLA. CENTER FOR THEORETICAL STUDIES
CTS-PA-(YR.)-	MIAMI. UNIV., CORAL GABLES, FLA. CENTER FOR THEORETICAL STUDIES
CTS-PHIL. S-(YEAR)-	MIAMI. UNIV., CORAL GABLES, FLA. CENTER FOR THEORETICAL STUDIES
CTS-QED-(YEAR)-	MIAMI. UNIV., CORAL GABLES, FLA. CENTER FOR THEORETICAL STUDIES
CTSRL-(NUMBER-NUMBER)	TEXAS. UNIV., AUSTIN. COMMUNICATIONS THEORY AND SYSTEMS RESEARCH LAB.
CTS-SSP-(YEAR)-	MIAMI. UNIV., CORAL GABLES, FLA. CENTER FOR THEORETICAL STUDIES
CTS-TC-(YEAR)-	MIAMI. UNIV., CORAL GABLES, FLA. CENTER FOR THEORETICAL STUDIES
CTS-T-PHYS-(YEAR)-	MIAMI. UNIV., CORAL GABLES, FLA. CENTER FOR THEORETICAL STUDIES
CTT-	COLUMBIA TECHNICAL TRANSLATIONS, WHITE PLAINS, N.Y.
CU-	CATHOLIC UNIV. OF AMERICA, WASHINGTON, D.C.
CU-	CHICAGO. UNIV.
CU-(4 DIGITS)-	COLUMBIA UNIV., N.Y.C. (4 DIGITS ARE CONTRACT NO.)
CU-(NO)-(YR)-(CON)-(DEPT)	COLUMBIA UNIV., N.Y.C.
CU-2-53-AEC-314-CHEM.	COLUMBIA UNIV., N.Y.C.
CU-(3 DIGITS)-	COLUMBIA UNIV., N.Y.C.
CU-	COLUMBIA UNIV., N.Y.C. (ALL DEPTS.)
CU-(NO.)-(YR)-SC-	COLUMBIA UNIV., N.Y.C. COLUMBIA RADIATION LAB.
CU-(NUMBER-YEAR)-AMC-	COLUMBIA UNIV., N.Y.C. COLUMBIA RADIATION LAB.
CU-(YR.)-(CONTRACT NO.)-	COLUMBIA UNIV., N.Y.C. DEPT. OF ELECTRICAL ENG.
CU-(NO.)-AF-(NO.)-EE	COLUMBIA UNIV., N.Y.C. DEPT. OF ELECTRICAL ENG.
CU-(NUMBER)-ONR-	COLUMBIA UNIV., N.Y.C. DEPT. OF MATHEMATICS
CU-(NO.)-NONR-K-M	COLUMBIA UNIV., N.Y.C. DEPT. OF MATHEMATICS
CU-	COMMITTEE ON MEDICAL RESEARCH (CUTTER REPORTS)
CU-	CORNELL UNIV., ITHACA, N.Y. DEPT. OF CHEMISTRY
CUA-	CATHOLIC UNIV. OF AMERICA, WASHINGTON, D.C.
CUADERNO-	UNIVERSIDAD MAYOR DE SAN ANDRES, LA PAZ, BOLIVIA
CUA-NE-	CATHOLIC UNIV. OF AMERICA, WASHINGTON, D.C.
CUA-PR-	CATHOLIC UNIV. OF AMERICA, WASHINGTON, D.C.
CUA-QPR-	CATHOLIC UNIV. OF AMERICA, WASHINGTON, D.C.
CUA-TR-	CATHOLIC UNIV. OF AMERICA, WASHINGTON, D.C.
CUBIC-	CUBIC CORP., SAN DIEGO, CALIF.
CUBIC-FER-	CUBIC CORP., SAN DIEGO, CALIF.
CUBIC-ITR-	CUBIC CORP., SAN DIEGO, CALIF.
CU-CRL-QPR-	COLUMBIA UNIV., N.Y.C. COLUMBIA RADIATION LAB.
CUD-	COLUMBIA UNIV., N.Y.C.
CUDWR-	COLUMBIA UNIV., N.Y.C. DIV. OF WAR RESEARCH
CUED/A-	CAMBRIDGE UNIV., ENGLAND. DEPT. OF ENGINEERING
CUED/A-TURBO/TR-	CAMBRIDGE UNIV., ENGLAND. DEPT. OF ENGINEERING
CUED/C-	CAMBRIDGE UNIV., ENGLAND. DEPT. OF ENGINEERING
CUED/C-MAT/TR-	CAMBRIDGE UNIV., ENGLAND. DEPT. OF ENGINEERING
CUED/C-STRUCT/TR-	CAMBRIDGE UNIV., ENGLAND. DEPT. OF ENGINEERING
CU-EE-	COLUMBIA UNIV., N.Y.C. DEPT. OF ELECTRICAL ENG.
CUERL-F/	COLUMBIA UNIV., N.Y.C. ELECTRONICS RESEARCH LABS.
CU/F/	CATHOLIC UNIV. OF AMERICA, WASHINGTON, D.C.
CUF-	COLUMBIA UNIV., N.Y.C.
CUG-	COLUMBIA UNIV., N.Y.C.
CUGSAE-	CORNELL UNIV., ITHACA, N.Y. GRAD. SCH. OF AERO. ENG.
CULCHETH-	GT. BRIT. CULCHETH LABS., LANCS, ENGLAND
CU-LGO-	COLUMBIA UNIV., PALISADES, N.Y. LAMONT GEOLOGICAL OBS
CU-LNS-(YEAR)-	OSAKA UNIV. LAB. OF NUCLEAR STUDIES
CU MLR-	CHICAGO. UNIV. METALLURGICAL LAB.
CUN-	COLUMBIA UNIV., N.Y.C.
CUN AIL-	COLUMBIA UNIV., N.Y.C. AIRBORNE INSTRUMENTS LAB.
CUN DWR M-	COLUMBIA UNIV., N.Y.C. DIV. OF WAR RES. (MEMORANDA)
CUN DWR M (LTR.,NO./LTR)-	COLUMBIA UNIV., N.Y.C. DIV. OF WAR RES. (MEMORANDA)
CUN DWR R-	COLUMBIA UNIV., N.Y.C. DIV. OF WAR RESEARCH
CUN DWR R (LTR.,NO./LTR)-	COLUMBIA UNIV., N.Y.C. DIV. OF WAR RESEARCH
CU NEVIS-	COLUMBIA UNIV., IRVINGTON-ON-HUDSON, N.Y. NEVIS LABS.
CUNY-	CITY UNIV., NEW YORK
CUP-	ATOMIC ENERGY OF CANADA LTD. NUCLEAR POWER PLANT DIV. TORONTO (CANADA-U.S. COOP. ATOMIC POWER PROGRAM)
CU/P/	CATHOLIC UNIV. OF AMERICA, WASHINGTON, D.C.
CU PL-	CHICAGO. UNIV. PSYCHOMETRIC LAB.
CU(PNPL)-	COLUMBIA UNIV., N.Y.C. PEGRAM NUCLEAR PHYSICS LABS.
CU-PR-	COLUMBIA UNIV., N.Y.C. COLUMBIA RADIATION LAB.
CURF-	BRITISH COLUMBIA RESEARCH COUNCIL, VANCOUVER
CURF-R-	CANADA. DEPT. OF MINES AND TECH. SURVEYS. MINES BR.
CURL-	CORNELL UNIV., ITHACA, N.Y. NUCLEAR REACTOR LAB.
CURTISS-	CURTISS-WRIGHT CORP. CURTISS DIV., CALDWELL, N.J.
CURTISS-C-	CURTISS-WRIGHT CORP., CALDWELL, N.J.

CURTISS-WRIGHT-	CURTISS-WRIGHT CORP.
CUSE-	COLUMBIA UNIV., N.Y.C. SCHOOL OF ENGINEERING
CUSEE-	CORNELL UNIV., ITHACA, N.Y. SCHOOL OF ELECTRICAL ENG.
CUT-AAV-	CHALMERS TEKNISKA HOEGSKOLA, GOETEBORG. DIV. OF SOLID MECHANICS
CU TLR-	CHICAGO. UNIV. TOXICITY LAB.
CU TR-	COLUMBIA UNIV., N.Y.C.
CU-TR-(NO.)/(CON. NO.)	COLUMBIA UNIV., N.Y.C.
CU-TR-(NUMBER-YEAR)	COLUMBIA UNIV., N.Y.C. (ALL DEPTS.)
CUT-TRANSACTION-	CHALMERS TEKNISKA HOEGSKOLA, GOETEBORG
CUW-	NATIONAL RESEARCH COUNCIL. COMM. ON UNDERSEA WARFARE
CU-WPG-	COLUMBIA UNIV., N.Y.C. WAVE PROPAGATION GROUP
CV-(YR)-(MO)-	CALIFORNIA. UNIV., LIVERMORE. RADIATION LAB.
CV-(3 DIGITS)	CONCHO VALLEY COUNCIL OF GOVERNMENTS, SAN ANGELO, TEX
CVA-	CHANCE VOUGHT AIRCRAFT, INC., DALLAS
CVA-	CONVAIR, SAN DIEGO, CALIF.
CVA AER-(LTR-YR-LTR)-	CHANCE VOUGHT AIRCRAFT, INC. AERONAUTICS DIV., DALLAS
CVA-AST-	VOUGHT ASTRONAUTICS. DIV. OF CHANCE VOUGHT CORP., DALLAS
CVA AST-(LTR-YR-LTR)-	CHANCE VOUGHT AIRCRAFT, INC. ASTRONAUTICS DIV., DALLAS
CVAC-	CONSOLIDATED VULTEE AIRCRAFT CORP., FORT WORTH, TEX.
CVAC-(LETTERS)-	CONVAIR (ALL LOCATIONS)
CVA-E(YEAR)R-	CHANCE VOUGHT AIRCRAFT, INC., DALLAS
CVA-EOR-	CHANCE VOUGHT AIRCRAFT, INC., DALLAS
CVAI-	CHANCE VOUGHT AIRCRAFT, INC., DALLAS
CVAL-	CONSOLIDATED VULTEE AIRCRAFT CORP., SAN DIEGO, CALIF.
CVA-PR-	CHANCE VOUGHT AIRCRAFT, INC., DALLAS
CVA-VE-	VOUGHT ELECTRONICS. DIV. OF CHANCE VOUGHT CORP., DALLAS
CVA-VRS-	VOUGHT RANGE SYSTEMS. DIV. OF CHANCE VOUGHT CORP., DALLAS
CVC-	CHANCE VOUGHT CORP., DALLAS
CVC-	GENERAL DYNAMICS/CONVAIR, SAN DIEGO, CALIF.
CVC-(LTRS,NOS,LTRS)-	CONVAIR, FORT WORTH, TEX.
CVC-(LTRS,NOS,LTRS)-	CONVAIR, SAN DIEGO, CALIF.
CVC/A-	CONVAIR-ASTRONAUTICS, SAN DIEGO, CALIF.
CVC AI-	CONVAIR, SAN DIEGO, CALIF.
CVC AZ (LTR.)-	CONVAIR, SAN DIEGO, CALIF.
CVC CD-	CONVAIR, FORT WORTH, TEX.
CVC CR-(NOS.-NO.)	CONVAIR, POMONA, CALIF.
CVC DC-W-	CONVAIR, SAN DIEGO, CALIF.
CVC ERR-SD-	CONVAIR, SAN DIEGO, CALIF.
CVC FSA (NOS.-NOS.)	CONVAIR, FORT WORTH, TEX.
CVC FSE (NOS.-NOS.)	CONVAIR, FORT WORTH, TEX.
CVC FZA (NOS.-NOS.)	CONVAIR, FORT WORTH, TEX.
CVC FZC (NOS.-NOS.)	CONVAIR, FORT WORTH, TEX.
CVC FZG (NOS.-NOS.)	CONVAIR, FORT WORTH, TEX.
CVC FZK (NOS.-NOS.)	CONVAIR, FORT WORTH, TEX.
CVC FZM (NOS.-NOS.)	CONVAIR, FORT WORTH, TEX.
CVC MRF-	CONVAIR, FORT WORTH, TEX.
CVC MR-N-	CONVAIR, FORT WORTH, TEX.
CVC NARF (NOS.-NOS.-LTR.)	CONVAIR. NUCLEAR AIRCRAFT RESEARCH FACILITY, FORT WORTH, TEX. (REF. 3)
CVC OAL/CM-	CONVAIR. ORDNANCE AEROPHYSICS LAB., DAINGERFIELD,TEX.
CVC OAL-R-	CONVAIR. ORDNANCE AEROPHYSICS LAB., DAINGERFIELD,TEX.
CVC ZA (NO.-NOS.)	CONVAIR, SAN DIEGO, CALIF.
CVC ZC (NO.-NOS.)	CONVAIR, SAN DIEGO, CALIF.
CVC ZD (NO.-NOS.)	CONVAIR, SAN DIEGO, CALIF.
CVC ZJ-	CONVAIR, SAN DIEGO, CALIF.
CVC ZM (NO.-NOS.)	CONVAIR, SAN DIEGO, CALIF.
CVC ZO-P-	CONVAIR, SAN DIEGO, CALIF.
CVC ZP-	CONVAIR, SAN DIEGO, CALIF.
CVC ZPH-	CONVAIR, SAN DIEGO, CALIF.
CVC ZP-M-	CONVAIR, SAN DIEGO, CALIF.
CVC ZR (NOS.-NOS.)	CONVAIR, SAN DIEGO, CALIF.
CVC ZRAI-	CONVAIR, SAN DIEGO, CALIF.
CVC ZRAP-	CONVAIR, SAN DIEGO, CALIF.
CVC ZS-	CONVAIR, SAN DIEGO, CALIF.
CVC ZU-	CONVAIR, SAN DIEGO, CALIF.
CVD-	ASSOCIATED ELECTRICAL INDUSTRIES, LTD. RESEARCH LABS. ALDERMASTON, BERKS, ENGLAND
CVD/(NUMBER)/(YEAR)	THERMAL SYNDICATE LTD., WALLSEND, NORTHUMBERLAND, ENG
CVD/QU-	GT. BRIT. ROYAL NAVAL SCIENTIFIC SERVICE, LONDON
CVD/QU-	QUEENS UNIV., BELFAST
CVD/QUI-	GT. BRIT. ROYAL NAVAL SCIENTIFIC SERVICE, LONDON
CVD/RU-	READING, ENGLAND. UNIV.
CVL-	CALIFORNIA. UNIV., BERKELEY. RADIATION LAB.
CVL-	CALIFORNIA. UNIV., LIVERMORE. LAWRENCE RADIATION LAB.
CVM-	KNOLLS ATOMIC POWER LAB., SCHENECTADY, N.Y.
CVNA-	CAROLINAS-VIRGINIA NUCLEAR POWER ASSOCIATES, INC., CHARLOTTE, N.C.
CVNA-	CAROLINAS-VIRGINIA NUCLEAR POWER ASSOCIATES, INC., COLUMBIA, S.C.
CVNA-	CAROLINAS-VIRGINIA NUCLEAR POWER ASSOCIATES, INC., PARR, S.C.
CW-	(CHEMICAL WARFARE) (REF. 2)
CW-	CURTISS-WRIGHT CORP. (ALL LOCATIONS)
CW-	DUGWAY PROVING GROUND, TOOELE, UTAH
CW-	RES. + DEV. BD.,WASH.,DC. (CHEM. WARFARE OPERATIONS)
CWAED-	CANADIAN WESTINGHOUSE CO., LTD. ATOMIC ENERGY DIV., HAMILTON, ONT.
CWC-ED-	DU PONT DE NEMOURS (E.I.) + CO. ENGINEERING DEPT., WILMINGTON, DEL.
CWC-QPR-	CURTISS-WRIGHT CORP. WRIGHT AERO. DIV., WOOD-RIDGE,NJ
CWD/	UNITED KINGDOM ATOMIC ENERGY AUTHORITY. WEAPONS GROUP ATOMIC WEAPONS RES. ESTAB.,ALDERMASTON, BERKS, ENG.
CWDI-	COOPERATIVE WEAPON DATA INDEXING COMM. (AEC-DOD)
CWDIC-	COOPERATIVE WEAPON DATA INDEXING COMM. (AEC-DOD)
CWH-	CONSOER, WHITE AND HERSHEY, WASH., D.C.
CWL-	CHEMICAL WARFARE LABS., ARMY CHEMICAL CENTER, MD.
CWLR-	CHEMICAL WARFARE LABS., ARMY CHEMICAL CENTER, MD.
CWLSP-	CHEMICAL WARFARE LABS., ARMY CHEMICAL CENTER, MD. (SPECIAL PUBLICATIONS)
CWL-SP-	CHEMICAL WARFARE LABS., ARMY CHEMICAL CENTER, MD. (SPECIAL PUBLICATION)
CWL-TM-	CHEMICAL WARFARE LABS., ARMY CHEMICAL CENTER, MD.
CW-MRJ.OO-	CURTISS-WRIGHT CORP. WRIGHT AERO. DIV., WOOD-RIDGE,NJ
CWO-	CHRYSLER CORP., DETROIT
CW-PR-	CURTISS-WRIGHT CORP. WRIGHT AERO. DIV., WOOD-RIDGE,NJ
CWR-	CURTISS-WRIGHT CORP. (ALL LOCATIONS)
CW-R-	CURTISS-WRIGHT CORP. RESEARCH DIV., CLIFTON, N.J.
CW-R+DL-	CANADIAN WESTINGHOUSE CO., LTD., RES. + DEV. LABS.
CWRL-	CURTISS-WRIGHT CORP. RESEARCH LAB.
CWRR-(NUMBER-NUMBER)	UTAH CENTER FOR WATER RESOURCES RESEARCH, LOGAN
CWRU-	CASE WESTERN RESERVE UNIV., CLEVELAND
CWRU-TR-	CASE WESTERN RESERVE UNIV., CLEVELAND

CWS-
: ALLIED FORCES ENEMY EQUIPMENT INTELLIGENCE SERVICE. INDIA-BURMA THEATER. CHEMICAL WARFARE SECTION

CWS-
: CHEMICAL CORPS, ARMY CHEMICAL CENTER, MD.

CWS-
: CHEMICAL WARFARE SERVICE, EDGEWOOD ARSENAL, MD.

CWS (NO.) CMTR-
: CHEMICAL WARFARE SERVICE. (NO.) CHEMICAL LABORATORY CO. (CAPTURED MATERIEL, TECHNICAL REPORTS)

CWS (NO.) IR-
: CHEMICAL WARFARE SERVICE. (NO.) CHEMICAL LABORATORY CO. (INTELLIGENCE REPORTS)

CWS (NO.) TR-
: CHEMICAL WARFARE SERVICE. (NO.) CHEMICAL LAB. CO.

CWS CMMR-
: CHEMICAL WARFARE SERVICE, EDGEWOOD ARSENAL, MD. (CAPTURED MATERIEL MEMORANDUM REPORT)

CWS CMTR-
: CHEMICAL WARFARE SERVICE, EDGEWOOD ARSENAL, MD. (CAPTURED MATERIEL TECHNICAL REPORT)

CWS CMTR MIT-
: CHEMICAL WARFARE SERVICE, EDGEWOOD ARSENAL, MD. (CAPTURED MATERIEL TECH. RPT. MASS. INST. OF TECH.)

CWS DWG-
: CHEMICAL WARFARE SERVICE, EDGEWOOD ARSENAL, MD. (DRAWING)

CWS FLM-
: CHEMICAL WARFARE SERVICE, EDGEWOOD ARSENAL, MD. (FIELD LABORATORY MEMORANDUM)

CWS FLM (NO.)-(NO.)-
: CHEMICAL WARFARE SERVICE, EDGEWOOD ARSENAL, MD. (FIELD LABORATORY MEMORANDUM)

CWS FMTR-
: CHEMICAL WARFARE SERVICE, EDGEWOOD ARSENAL, MD. (FOREIGN MATERIEL TECHNICAL REPORT)

CWS FMTR M-
: CHEMICAL WARFARE SERVICE, EDGEWOOD ARSENAL, MD. (FOREIGN MATERIEL TECHNICAL REPORT)

CWS FMTR MIT-
: CHEMICAL WARFARE SERVICE, EDGEWOOD ARSENAL, MD. (FOREIGN MATERIEL TECH. RPT. MASS. INST. OF TECH.)

CWS FP-
: CHEMICAL WARFARE SERVICE, EDGEWOOD ARSENAL, MD. (FRED PROJECT)

CWS IDR-
: CHEMICAL WARFARE SERVICE. INTELLIGENCE DIV., EDGEWOOD ARSENAL, MD.

CWS IRC-
: CHEMICAL WARFARE SERVICE, EDGEWOOD ARSENAL, MD. (INSECT AND RODENT CONTROL)

CWS MD (EA)-
: CHEMICAL WARFARE SERVICE. MEDICAL DIV., EDGEWOOD ARSENAL, MD.

CWS MD (EA) MR-
: CHEMICAL WARFARE SERVICE. MEDICAL DIV., EDGEWOOD ARSENAL, MD. (MEMORANDUM REPORTS)

CWS MDR-
: CHEMICAL WARFARE SERVICE. MEDICAL DIV., EDGEWOOD ARSENAL, MD.

CWS MRL (EA)-
: CHEMICAL WARFARE SERVICE. MEDICAL RESEARCH LAB., EDGEWOOD ARSENAL, MD.

CWS-PCS-
: CHEMICAL WARFARE SERVICE, EDGEWOOD ARSENAL, MD.

CWS SPEC-
: CHEMICAL WARFARE SERVICE, EDGEWOOD ARSENAL, MD. (SPECIFICATION)

CWS TCIR-
: CHEMICAL WARFARE SERVICE. TECHNICAL COMMAND, EDGEWOOD ARSENAL, MD. (INFORMAL REPORT)

CWS TCR-
: CHEMICAL CORPS. TECH. COMMAND, ARMY CHEMICAL CTR.,MD.

CWS TDMR-
: CHEMICAL WARFARE SERVICE. TECHNICAL DIV., EDGEWOOD ARSENAL, MD. (MEMORANDUM REPORT)

CWS TIDR-
: CHEMICAL WARFARE SERVICE. TECHNICAL INTELLIGENCE DIV., EDGEWOOD ARSENAL, MD.

CWS TRLR-
: CHEMICAL WARFARE SERVICE. TOXICOLOGICAL RESEARCH LAB., EDGEWOOD ARSENAL, MD.

CWT-
: SOUTHERN COOPERATIVE WIND TUNNEL, PASADENA, CALIF.

CW-WR-(YEAR)-
: CURTISS-WRIGHT CORP., WOOD-RIDGE, N.J.

CX-
: NEW YORK UNIV., N.Y.C. INST. OF MATHEMATICAL SCIENCES

CXRL-
: AUSTRALIA. COMMONWEALTH X-RAY AND RADIUM LAB., MELBOURNE

CY-
: CARBIDE AND CARBON CHEMICALS CORP. Y-12 PLANT, OAK RIDGE, TENN.

CYAP-
: CONNECTICUT YANKEE ATOMIC POWER CO., BOSTON

CYCLOTRON-
: NAVAL RESEARCH LAB. CYCLOTRON BRANCH, WASHINGTON, DC

CZC-
: OFFICE OF SCIENTIFIC RESEARCH AND DEVELOPMENT

D-	(DISCLOSURE) (REF. 2)
D-	(DISCUSSION) (REF. 2)
D-	(DOCUMENT) (REF. 2)
D(NUMBER)	AEROSPACE CORP., EL SEGUNDO, CALIF. (ART WORK)
D-(2 DIGITS)	AKADEMIYA NAUK SSSR., MOSCOW
D-	BELGIUM. CENTRE D'ETUDE DE L'ENERGIE NUCLEAIRE, BRUSSELS
D/(NUMBER)/	BELGIUM. CENTRE D'ETUDE DE L'ENERGIE NUCLEAIRE, BRUSSELS
D(NO.)-	BELL AEROSYSTEMS CO., BUFFALO
D(NO.)-	BELL HELICOPTER CORP., FORT WORTH, TEX.
D-	BERLIN. TECHNISCHE UNIVERSITAET
D-1-(NO.)S	BLAW-KNOX CONSTRUCTION CO., PITTSBURGH
D-2-(NO.)S	BLAW-KNOX CONSTRUCTION CO., PITTSBURGH
D5-	BOEING CO., HUNTSVILLE, ALA. (BOMARC + SATURN)
D6-	BOEING CO., RENTON, WASH.
D203-	BOEING CO., SEATTLE (PROJECT NUMBER, DOCUMENT)
D180-	BOEING CO., SEATTLE (PROJECT NUMBER, DOCUMENT)
D162-	BOEING CO., SEATTLE (PROJECT NUMBER, DOCUMENT)
D2-	BOEING CO. AERO-SPACE GROUP, SEATTLE
D6-	BOEING CO. COMMERCIAL AIRPLANE DIV., SEATTLE
D7-	BOEING CO. SYSTEMS MANAGEMENT DIV., KENT, WASH.
D4-	BOEING CO. TURBINE DIV., SEATTLE
D8-	BOEING CO. VERTOL DIV., MORTON, PENNA.
D3-	BOEING CO. WICHITA DIV., KAN.
D1-	BOEING SCIENTIFIC RESEARCH LABS., SEATTLE (DOCUMENT)
D-1-(NUMBER-NUMBER)	BOEING SCIENTIFIC RESEARCH LABS., SEATTLE
D276-5S-	BUNKER-RAMO CORP. DEFENSE SYSTEMS DIV., CANOGA PARK, CALIF.
D(NO.-NO.)S(NUMBER)-	BUNKER-RAMO CORP. DEFENSE SYSTEMS DIV., CANOGA PARK, CALIF.
D1-RR-	CALIFORNIA. UNIV., BERKELEY. RADIATION LAB. (REF. 9)
D1-S-	CALIFORNIA. UNIV., BERKELEY. RADIATION LAB. (REF. 9)
D1-SM-	CALIFORNIA. UNIV., BERKELEY. RADIATION LAB. (REF. 9)
D1-SPEC-	CALIFORNIA. UNIV., BERKELEY. RADIATION LAB. (REF. 9)
D1M-	CALIFORNIA. UNIV., BERKELEY. RADIATION LAB. (REF. 9)
D1W-	CALIFORNIA. UNIV., BERKELEY. RADIATION LAB. (REF. 9)
D-	CALLERY CHEMICAL CO., PENNA.
D143-(NUMBER)-	CALLERY CHEMICAL CO., PENNA.
D-	CARBIDE AND CARBON CHEMICALS CORP., OAK RIDGE, TENN.
D3-	CARBIDE AND CARBON CHEMICALS CORP. K-25 PLANT, OAK RIDGE, TENN.
D1-M-	COLUMBIA UNIV., NYC. SUBSTITUTE ALLOY MATERIALS LABS.
D1B-M-	COLUMBIA UNIV., NYC. SUBSTITUTE ALLOY MATERIALS LABS.
D1FM-	COLUMBIA UNIV., NYC. SUBSTITUTE ALLOY MATERIALS LABS.
D2L-M-	COLUMBIA UNIV., NYC. SUBSTITUTE ALLOY MATERIALS LABS.
D3-M-	COLUMBIA UNIV., NYC. SUBSTITUTE ALLOY MATERIALS LABS.
D3-X-M-	COLUMBIA UNIV., NYC. SUBSTITUTE ALLOY MATERIALS LABS.
D-	CONDUCTRON CORP., ANN ARBOR, MICH.
D(NUMBER)-	CONDUCTRON CORP., ANN ARBOR, MICH.
D-	CURTISS-WRIGHT CORP. WRIGHT AERO. DIV., WOOD-RIDGE,NJ
D-(3 DIGITS)	DENMARK. TEKNISKE HOJSKOLE, LYNGBY
D-(YR)-(NO.)-	ELDORADO MINING + REFINING LTD. RES.+ DEV.DIV.,OTTAWA
D(NUMBER)-	GENERAL APPLIED SCIENCE LABS., INC., WESTBURY, N.Y.
D(NO.)-	GENERAL DYNAMICS CORP. ELECTRIC BOAT DIV.,GROTON,CONN
D-	GENERAL ELECTRIC CO. AIRCRAFT NUCLEAR PROPULSION DEPT., CINCINNATI
D-	KELLEX CORP., N.Y.C.
D3-	MASSACHUSETTS INST. OF TECH., CAMBRIDGE
D/(YEAR)-	MASSACHUSETTS INST. OF TECH., CAMBRIDGE. CENTER FOR INTERNATIONAL STUDIES
D-	MINNEAPOLIS-HONEYWELL REGULATOR CO. ORDNANCE DIV., DUARTE, CALIF.
D-	MOTOROLA INC. SEMICONDUCTOR PRODS. DIV.,PHOENIX,ARIZ.
D143-(NUMBER)-	MSA RESEARCH CORP., CALLERY, PENNA.
D-	NATIONAL BUREAU OF STANDARDS, WASHINGTON, D.C.
D-	NATIONAL RESEARCH COUNCIL OF CANADA. DIV. OF ATOMIC ENERGY, CHALK RIVER, ONT.
D-3-	NATIONAL RESEARCH COUNCIL OF CANADA. DIV. OF ATOMIC ENERGY, CHALK RIVER, ONT.
D-	NORTH AMERICAN AVIATION, INC., LOS ANGELES
D-	OFFICE OF NAVAL RESEARCH, LONDON
D-	PENNA. STATE UNIV., UNIV. PARK. ORDNANCE RES. LAB.
D-	PRATT AND WHITNEY AIRCRAFT DIV., UNITED AIRCRAFT CORP., HARTFORD, CONN.
D-(YEAR)-	RAMO-WOOLDRIDGE CORP. GUIDED MISSILE RESEARCH DIV., LOS ANGELES
D-	RAND CORP., SANTA MONICA, CALIF.
D-(NUMBER)-(LETTERS)	RAND CORP., SANTA MONICA, CALIF. (REF. 47)
D1-(YEAR)-	SWEDISH DETONIC RESEARCH FOUNDATION, STOCKHOLM
D-(NUMBER/NUMBER(S))	SYSTEM DEVELOPMENT CORP., SANTA MONICA, CALIF. (DOCUMENT)
D-(3 DIGITS)	TELEDYNE CONTINENTAL MOTORS, WARREN, MICH. (DESIGN AND DEVELOPMENT)
D-1.210- THRU D-7.440-	TENNESSEE EASTMAN CORP., OAK RIDGE, TENN. (REF. 34)
D(YEAR)-(NUMBER)	UNIDYNAMICS/PHOENIX. DIV. OF UNIVERSAL MATCH CORP., ARIZ.
D(NO.)-	UNITED AIRCRAFT CORP., EAST HARTFORD, CONN.
D(NUMBERS)-	UNITED AIRCRAFT CORP. RES. LABS., EAST HARTFORD,CONN.
D-	WESTINGHOUSE ELECTRIC CORP. AEROSPACE ELECTRICAL DIV., LIMA, OHIO
DA-	ARMY, WASHINGTON, D.C.
DA-	ARMY BALLISTIC MISSILE AGENCY. DYNAMIC ANALYSIS BRANCH, REDSTONE ARSENAL, ALA.
DA-	BROWN UNIV., PROVIDENCE. GRAD. DIV. OF APPLIED MATH.
DA-	DETROIT ARSENAL, CENTER LINE, MICH.
DA-	DUNLAP AND ASSOCIATES, INC., STAMFORD, CONN.
DA-(YEAR)-	MECHANICS RESEARCH, INC., EL SEGUNDO, CALIF.
DA-	NAVAL AIR STATION, PENSACOLA, FLA.
DA-(NUMBER)-	SPERRY GYROSCOPE CO., GREAT NECK, N.Y.
DA-(LETTERS)-	ARMY, WASHINGTON, D.C. (FOR EXPLANATION OF THE REMAINDER OF THE CODE, SEE ENTRY UNDER LETTERS FOLLOWING DA-)
DA AN-	AUSTRALIA. COUNCIL FOR SCIENTIFIC AND INDUSTRIAL RESEARCH (AERODYNAMICS NOTES)
DABCO-	HOUDRY PROCESS + CHEMICAL CO., MARCUS HOOK, PENNA.
DAC-	DOUGLAS AIRCRAFT CO., INC. (ALL LOCATIONS)
DAC-(NUMBER)-S-	DOUGLAS AIRCRAFT CO., INC., NEWPORT BEACH, CALIF. (SEMIANNUAL REPORT)
DAC-(NUMBER)-(LETTER(S))	DOUGLAS AIRCRAFT CO., INC. ASTROPOWER LAB., NEWPORT BEACH, CALIF.
DAC-(NUMBER)-Q(NUMBER)	DOUGLAS AIRCRAFT CO., INC. ASTROPOWER LAB., NEWPORT BEACH, CALIF.
DAC-	MC DONNELL DOUGLAS ASTRONAUTICS CO. (ALL LOCATIONS)
DAC-	MC DONNELL-DOUGLAS CO., SANTA MONICA, CALIF.
DAC-A(NO.)-(LTRS.)(NO.)	DOUGLAS AIRCRAFT CO., INC. MISSILES AND SPACE SYSTEMS ENGINEERING, SANTA MONICA, CALIF.
DAC-CH-	DOUGLAS AIRCRAFT CO., INC., CHARLOTTE, N.C.
DAC-D(NUMBER)-(NUMBER)	DOUGLAS AIRCRAFT CO., INC. SANTA MONICA DIV., LONG BEACH, CALIF.
DAC-ENG.PAPER-	DOUGLAS AIRCRAFT CO., INC., LONG BEACH, CALIF.
DAC-ENG.PAPER-	DOUGLAS AIRCRAFT CO., INC., SANTA MONICA, CALIF.

DAC-ES-	DOUGLAS AIRCRAFT CO., INC., EL SEGUNDO, CALIF.
DACL-	MASSACHUSETTS INST. OF TECH., CAMBRIDGE. DYNAMIC ANALYSIS AND CONTROL LAB.
DAC-LB-	DOUGLAS AIRCRAFT CO., INC., LONG BEACH, CALIF.
DACL-RM-	MASSACHUSETTS INST. OF TECH., CAMBRIDGE. DYNAMIC ANALYSIS AND CONTROL LAB.
DAC-PAPER-	MC DONNELL DOUGLAS ASTRONAUTICS CO. (ALL LOCATIONS)
DAC R-	DOUGLAS AIRCRAFT CO., INC. MISSILE AND SPACE SYSTEMS DIV., SANTA MONICA, CALIF.
DA-CR-(YR.)-	MARTIN CO., DENVER (CONTRACT REPORT)
DAC-SM-	DOUGLAS AIRCRAFT CO., INC., SANTA MONICA, CALIF.
DAC-SM-	DOUGLAS AIRCRAFT CO., INC. MISSILE AND SPACE SYSTEMS DIV., HUNTINGTON BEACH, CALIF.
DAD-	MC DONNELL DOUGLAS CORP. DOUGLAS AIRCRAFT DIV., LONG BEACH, CALIF.
DAD-	WHITE SANDS MISSILE RANGE. DATA ANALYSIS DIRECTORATE, N. MEX.
DAE-	CINCINNATI. UNIV.
DAE-	LOUISIANA STATE UNIV., BATON ROUGE. DEPT. OF AGRICULTURAL ECONOMICS AND AGRIBUSINESS
DAE (LETTER)-	GT. BRIT. DEPT. OF ATOMIC ENERGY
DAE-AR-	INDIA. DEPT. OF ATOMIC ENERGY, BOMBAY
DA-ED-(YEAR)-	DUNLAP AND ASSOCIATES, INC., DARIEN, CONN.
DAG-	AMERICAN BOSCH ARMA CORP. ARMA DIV., GARDEN CITY, N.Y.
DAI-	DUNLAP AND ASSOCIATES, INC., STAMFORD, CONN.
DAISE-(YEAR)-	ITALY. COMITATO NAZIONALE PER L'ENERGIA NUCLEARE, ROME
DA-L-	COLUMBIA UNIV., NYC. SUBSTITUTE ALLOY MATERIALS LABS.
DAL-	DEUTSCHE AKADEMIE DER LUFTFAHRTFORSCHUNG, BERLIN
DALY-	DALY (LEO A.) CO., OMAHA
DA-M-	COLUMBIA UNIV., NYC. SUBSTITUTE ALLOY MATERIALS LABS.
DAM-	NAVAL ORDNANCE LAB., WHITE OAK, MD. (DESIGN ANALYSIS MEMORANDUM)
DAMP-TM-	RADIO CORP. OF AMERICA, MOORESTOWN, N.J.
DAMTP-(YEAR)/	CAMBRIDGE UNIV., ENGLAND. DEPT. OF APPLIED MATH. AND PHYSICS
DANATOM-	DANISH ASSN. FOR IND. DEV. OF ATOMIC ENERGY, HELLERUP
DANNY-BOY-	GEOLOGICAL SURVEY, DENVER
DAP-	ARMY, WASHINGTON, D.C. (PAMPHLET)
DA-PAM-	ARMY, WASHINGTON, D.C. (PAMPHLET)
DAR-	AIR FORCE OFFICE OF SCIENTIFIC RESEARCH, WASH., D.C.
DA-R-	CARBIDE AND CARBON CHEMICALS CORP. SUBSTITUTE ALLOY MATERIALS LABS., N.Y.C.
DA-R-	COLUMBIA UNIV., NYC. SUBSTITUTE ALLOY MATERIALS LABS.
DARL-	DOUGLAS AIRCRAFT CO., INC. ADVANCED RESEARCH LABS., HUNTINGTON BEACH, CALIF.
DARL-G-	DOUGLAS AIRCRAFT CO., INC. ADVANCED RESEARCH LABS., HUNTINGTON BEACH, CALIF.
DARPD-(NUMBER)-	ARMY, WASHINGTON D.C. (REORGANIZATION PLAN)
DARS-	DETROIT ARSENAL, CENTER LINE, MICH.
DART-	DARTMOUTH COLL., HANOVER, N.H.
DART-(NUMBER)-(LETTER)	DARTMOUTH COLL., HANOVER, N.H. THAYER SCHOOL OF ENGINEERING (SQUID PROJECT) (REF. 33)
DART TM-	DARTMOUTH COLL., HANOVER, N.H. THAYER SCHOOL OF ENGINEERING (SQUID PROJECT) (REF. 33)
DART-TR-	DARTMOUTH COLL., HANOVER, N.H. DEPT. OF PHYSICS
DASA-	DEFENSE ATOMIC SUPPORT AGENCY, WASHINGTON, D.C.
DASA-	DEFENSE ATOMIC SUPPORT AGENCY. FIELD COMMAND, ALBUQUERQUE, N. MEX.
DASA-BIB-	DEFENSE ATOMIC SUPPORT AGENCY, WASHINGTON, D.C.
DASA-DC-BIB-	DEFENSE ATOMIC SUPPORT AGENCY. DATA CENTER, SANTA BARBARA, CALIF. (BIBLIOGRAPHY)
DASA-EM-	DEFENSE ATOMIC SUPPORT AGENCY, WASHINGTON, D.C.
DASA/FC-	DEFENSE ATOMIC SUPPORT AGENCY. FIELD COMMAND, ALBUQUERQUE, N. MEX.
DASA/FC (MO.YR.)-	DEFENSE ATOMIC SUPPORT AGENCY. FIELD COMMAND, ALBUQUERQUE, N. MEX.
DASA-ITR-	DEFENSE ATOMIC SUPPORT AGENCY, WASHINGTON, D.C.
DASA-POIR-	(PROJECT OFFICERS INTERIM REPORT) (REF. 7)
DASA-POR-	(PROJECT OFFICERS REPORT) (REF. 7)
DASAR-	DEFENSE ATOMIC SUPPORT AGENCY. TEST COMMAND, ALBUQUERQUE, N. MEX.
DASA-SC-	DEFENSE ATOMIC SUPPORT AGENCY, WASHINGTON, D.C.
DASA TL-	DEFENSE ATOMIC SUPPORT AGENCY, WASH., DC (TECH. LTR)
DASA-TP-	JOINT ATOMIC WEAPONS PUBLICATIONS BOARD, ALBUQUERQUE, N. MEX. (REF. 14)
DAS-CSI-(LETTER/YR.NO.)	ORGANIZATION FOR ECONOMIC COOPERATION AND DEVELOPMENT. DIRECTORATE OF SCIENTIFIC AFFAIRS, PARIS
DASD-	ARMY, WASHINGTON, D.C. (SHIPPING DOCUMENT)
DASIAC-B(YEAR)-	GENERAL ELECTRIC CO. DASA INFORMATION AND ANALYSIS CENTER, SANTA BARBARA, CALIF.
DASIAC-SB-	GENERAL ELECTRIC CO. DASA INFORMATION AND ANALYSIS CENTER, SANTA BARBARA, CALIF.
DASIAC-SR-	GENERAL ELECTRIC CO. DASA INFORMATION AND ANALYSIS CENTER, SANTA BARBARA, CALIF.
D.ASM-(NUMBER)-CL/CM/PT	COMPAGNIE FRANCAISE THOMSON-HOUSTON. DIV. ACTIVITIES SOUS-MARINES
DAS-MTM-	DOUGLAS AIRCRAFT CO., INC., SANTA MONICA, CALIF.
DASS-(YEAR)-	SOUTHWEST CENTER FOR ADVANCED STUDIES, DALLAS
DAST-	FEDERAL WATER POLLUTION CONTROL ADMINISTRATION, WASH., D.C. (WATER POLLUTION CONTROL RESEARCH SER.)
DATA-	OREGON STATE UNIV., CORVALLIS. DEPT. OF OCEANOGRAPHY
DATDC-	AIR FORCE TECHNICAL APPLICATIONS CENTER, WASH., D.C.
DATDC-	UNITED ELECTRODYNAMICS, INC. DATA ANALYSIS AND TECHNIQUE DEVELOPMENT CENTER, ALEXANDRIA, VA.
DA-TM-(NUMBER)-(NUMBER)	ARMY BALLISTIC MISSILE AGENCY, REDSTONE ARSENAL, ALA.
DA-TN-	ARMY BALLISTIC MISSILE AGENCY, REDSTONE ARSENAL, ALA. (TECHNICAL NOTE)
DATP-	GARRETT CORP. AIRESEARCH MFG. DIV., LOS ANGELES
DA-TR-	ARMY BALLISTIC MISSILE AGENCY, REDSTONE ARSENAL, ALA.
DAV-(NUMBER)-	NAVAL SHIP RESEARCH AND DEVELOPMENT CENTER. DEPT. OF ACOUSTICS AND VIBRATION, WASHINGTON, D.C.
DAYSTROM-	DAYSTROM, INC. MIL. ELECTRONICS DIV., ARCHBALD, PA.
DB(YEAR)D(NUMBER)-	GENERAL ELECTRIC CO. APOLLO SUPPORT DEPT., DAYTONA BEACH, FLA.
DB-(NUMBER)-(YEAR)	MARINE CORPS LANDING FORCE DEV. CENTER, QUANTICO, VA.
DBA-	BROWN (D.) ASSOCIATES, INC., EAU GALLIE, FLA.
DBE-(YEAR)-	BUREAU OF RADIOLOGICAL HEALTH. DIV. OF BIOLOGICAL EFFECTS, ROCKVILLE, MD.
D-BIB-	LOS ALAMOS SCIENTIFIC LAB., N. MEX. (BIBLIOGRAPHIES)
DBIR-	DON BOSCO INST. FOR RESEARCH, RAMSEY, N.J.
DB-M-	COLUMBIA UNIV., NYC. SUBSTITUTE ALLOY MATERIALS LABS.
DBM-	DIVISION OF BIOLOGY AND MEDICINE, AEC, WASH., D.C.
DBM-RIB-	DIVISION OF BIOLOGY AND MEDICINE, AEC. RADIATION INSTRUMENTS BRANCH, WASHINGTON, D.C.
DBO-TR-	DIRECTORATE OF BIOLOGICAL OPERATIONS, PINE BLUFF ARSENAL, ARK.
DB-R-	COLUMBIA UNIV., NYC. SUBSTITUTE ALLOY MATERIALS LABS.
DBR-	NATIONAL RESEARCH COUNCIL OF CANADA. DIV. OF BUILDING RESEARCH, OTTAWA
D-B RPT-	DAIMLER-BENZ A.G., STUTTGART
DB-TR-	PICATINNY ARSENAL, DOVER, N.J.
DB-TR-	PICATINNY ARSENAL. INDUSTRIAL ENG. DIV., DOVER, N.J.
D-B VER-	DAIMLER-BENZ A.G., STUTTGART (VERSUCHSBERICHT)
DC-	(DECLASSIFIED PUBLICATION RELEASE) (REF. 2)
DC-	(DEVELOPMENT CHARACTERISTIC) (REF. 2)
DC-(YEAR)-(NUMBER)-	AIRCRAFT NUCLEAR PROPULSION PROGRAM (REF. 3)

DC-

DC-	ARMY ENGINEER RESEARCH + DEV. LABS., FT. BELVOIR, VA.
DC-(YEAR-NUMBER)-	BABCOCK + WILCOX CO. (ALL LOCATIONS)
DC-	DARTMOUTH COLL., HANOVER, N.H.
DC-(YEAR)-(NUMBER)-	FAIRCHILD ENGINE AND AIRPLANE CORP. NEPA DIV., OAK RIDGE, TENN.
DC-(YEAR)-(NUMBER)-	GENERAL ELECTRIC CO. AIRCRAFT NUCLEAR PROPULSION DEPT., CINCINNATI (REF. 3)
DC-(YEAR)-(NUMBER)-	GENERAL ELECTRIC CO. AIRCRAFT NUCLEAR PROPULSION DEPT., IDAHO FALLS, IDAHO (REF. 3)
DC-	OFFICE OF THE CHIEF OF NAVAL OPERATIONS, WASHINGTON, D.C. (DEVELOPMENT CHARACTERISTIC)
DC-	RAULAND CORP., CHICAGO
DC-	SANDIA CORP., ALBUQUERQUE, N. MEX.
DC-	SANDIA CORP. ALO PRIMARY STANDARDS LAB., ALBUQUERQUE, N. MEX. (DIRECT CURRENT PARAMETERS, RESISTANCE)
DC-(YEAR)-(NUMBER)-	WASHINGTON UNIV., ST. LOUIS
DCA-	BERLIN. TECHNISCHE UNIV. INST. FUER RAUMFAHRTTECHNIK
DCA-(NO.)-(LTR.)-	DEFENSE COMMUNICATIONS AGENCY, WASHINGTON, D.C.
DC-AC-(NUMBER)-	ANAGRAM CORP., SPRINGFIELD, VA.
DCA-CIR-	DEFENSE COMMUNICATIONS AGENCY, WASHINGTON, D.C.
DCAS-	AEROSPACE CORP., EL SEGUNDO, CALIF.
DCASPP-(YEAR)-	AIR FORCE SYSTEMS COMMAND. SPACE SYSTEMS DIV., LOS ANG
DCAS-TDR-	AEROSPACE CORP., EL SEGUNDO, CALIF.
DCAS-TR-	AEROSPACE CORP., EL SEGUNDO, CALIF.
DCA-TR-	DEFENSE COMMUNICATIONS AGENCY, WASHINGTON, D.C.
DCBRE-	CANADA. DEFENCE CHEM. BIOL. + RADIATION LABS., OTTAWA
DCBRL-	CANADA. DEFENCE CHEM. BIOL. + RADIATION LABS., OTTAWA
DCBRL-	ELECTRONIC SYSTEMS DIV. (AIR FORCE). DECISION SCIENCES LAB., BEDFORD, MASS.
DCBRL-TN-	CANADA. DEFENCE CHEM. BIOL. + RADIATION LABS., OTTAWA
DCC-	DAVISON CHEMICAL CORP., BALTIMORE
DCC-	DETROIT CONTROLS CORP., REDWOOD CITY, CALIF.
DCC-	DOW CHEMICAL CO., MIDLAND, MICH.
DCC-RC-	DETROIT CONTROLS CORP. RES. DIV., REDWOOD CITY, CALIF.
DCDE-	DOW CHEMICAL-DETROIT EDISON NUCLEAR POWER DEV. PROJ.
DCD-ER-	WHITE SANDS MISSILE RANGE. DATA COLLECTION DIRECTORATE, N. MEX. (ENGINEERING REPORT)
DCDRD-	GT. BRIT. DIRECTORATE OF CHEMICAL DEFENCE RES. + DEV.
DCE-	ILLINOIS. UNIV., URBANA. DEPARTMENT OF CERAMIC ENG.
DCF-	SYLVANIA-CORNING NUCLEAR CORP., BAYSIDE, N.Y.
DCF-(NUMBER)-CH	SYLVANIA-CORNING NUCLEAR CORP., HICKSVILLE, N.Y.
DCF-(NUMBER)-H	SYLVANIA ELECTRIC PRODUCTS INC., BAYSIDE, N.Y.
DC-FR-	DIKEWOOD CORP., ALBUQUERQUE, N. MEX.
DCH-(LETTERS)-(NUMBER)	SANDIA CORP., ALBUQUERQUE, N. MEX. (DESIGN COORDINATION HANDBOOK)
DCI-	ATOMIC ENERGY OF CANADA LTD., CHALK RIVER, ONT.
DCI-	DEWEY (G.C.) AND CO., INC., N.Y.C.
DCIC-(YEAR)-	BATTELLE MEMORIAL INST., COLUMBUS, OHIO
DCIC-	BATTELLE MEMORIAL INST. DEFENSE CERAMICS INFORMATION CENTER, COLUMBUS, OHIO
DCIC-CAB-	BATTELLE MEMORIAL INST. DEFENSE CERAMICS INFORMATION CENTER, COLUMBUS, OHIO (CURRENT AWARENESS BULLETIN)
DCI DATA/ORD-	DEWEY (G.C.) AND CO., INC., N.Y.C.
DCI-R-	DEWEY (G.C.) AND CO., INC., N.Y.C.
DCL-	GENERAL ELECTRIC CO. AIRCRAFT NUCLEAR PROPULSION DEPT., CINCINNATI
DCL-(YR.)-	GENERAL ELECTRIC CO. AIRCRAFT NUCLEAR PROPULSION DEPT., CINCINNATI
DCL-	LOS ALAMOS SCIENTIFIC LAB., N. MEX.
DCL-	MASS. INST. OF TECH., CAMBRIDGE. DIGITAL COMPUTER LAB
DCLS-	DUQUESNE LIGHT CO., SHIPPINGPORT, PENNA.
DCM-	ARMOUR RESEARCH FOUNDATION, CHICAGO
DCM-(YEAR)-	ARMOUR RESEARCH FOUNDATION, CHICAGO
DC-M-	COLUMBIA UNIV., NYC. SUBSTITUTE ALLOY MATERIALS LABS.
DCM(YEAR)	GENERAL DYNAMICS/CONVAIR, SAN DIEGO, CALIF.
DCM-(YR)-(NO.)-	GENERAL ELECTRIC CO. AIRCRAFT NUCLEAR PROPULSION DEPT., CINCINNATI
DCMR-	ARMOUR RESEARCH FOUNDATION, CHICAGO
DCN-	GRUMMAN AIRCRAFT ENGINEERING CORP., BETHPAGE, N.Y.
DCNO-	OFFICE OF THE CHIEF OF NAVAL OPERATIONS, WASH., D.C.
DCP-(YEAR)	ARMOUR RESEARCH FOUNDATION, CHICAGO
DCP-	ILLINOIS. UNIV., URBANA. ENGINEERING EXPERIMENT STA.
DC/P-	UNITED KINGDOM ATOMIC ENERGY AUTHORITY. PRODUCTION GROUP. WINDSCALE WORKS, SELLAFIELD, CUMB., ENGLAND
DCPWP-P-	UNITED KINGDOM ATOMIC ENERGY AUTHORITY. RESEARCH GP. ATOMIC ENERGY RES. ESTAB., HARWELL, BERKS, ENGLAND
DC-QR-	PICATINNY ARSENAL. AMMUNITION ENG. LAB., DOVER, N.J.
DC-R-	COLUMBIA UNIV., NYC. SUBSTITUTE ALLOY MATERIALS LABS.
DCR-	DYNASCIENCES CORP., BLUE BELL, PENNA.
DCR-	DYNASCIENCES CORP., FORT WASHINGTON, PENNA.
DCR-	GENERAL ELECTRIC CO. AIRCRAFT NUCLEAR PROPULSION DEPT., CINCINNATI
DCRD-	AIR TECHNICAL SERVICE COMMAND. ENG. DIV. AERO-MEDICAL LAB., WRIGHT-PATTERSON AFB, OHIO (MEMO. RPTS.)
DCRDE-	AIR TECHNICAL SERVICE COMMAND. ENG. DIV. AERO-MEDICAL LAB., WRIGHT-PATTERSON AFB, OHIO (MEMO. RPTS.)
DCRDM-	WRIGHT AIR DEVELOPMENT CENTER. AERO MEDICAL LAB., WRIGHT-PATTERSON AFB, OHIO
DCROF-(NUMBER)-	ROYAL ORDNANCE FACTORY, CHORLEY, LANCS, ENGLAND
DCRTT-	WRIGHT AIR DEVELOPMENT CENTER. MATERIALS LAB., WRIGHT-PATTERSON AFB, OHIO
DCS-	ILLINOIS. UNIV., URBANA. DEPT. OF COMPUTER SCIENCE
DC-SR-	DIKEWOOD CORP., ALBUQUERQUE, N. MEX.
DC-TN-	DIKEWOOD CORP., ALBUQUERQUE, N. MEX.
DC-TR-(NUMBER)-(NUMBER)	ARMY BALLISTIC MISSILE AGENCY, REDSTONE ARSENAL, ALA.
DC-TR-(NUMBER)-	DIKEWOOD CORP., ALBUQUERQUE, N. MEX.
DC-TR-(NUMBER)-(NUMBER)	PICATINNY ARSENAL. INDUSTRIAL ENG. DIV., DOVER, N.J.
DC-TR-CSAG-	DIKEWOOD CORP., ALBUQUERQUE, N. MEX.
DCW-	DARTMOUTH COLL., HANOVER, N.H. THAYER SCHOOL OF ENG.
DD(NUMBER)-(YEAR)	COLUMBIA UNIV., ST. DAVID'S, BERMUDA. GEOPHYSICAL FIELD STATION
DD-(NO.)-(LETTER)-	CORNELL AERONAUTICAL LAB., INC., BUFFALO
DD-	NAVAL TORPEDO STATION, KEYPORT, WASH.
DD-	PACKARD MOTOR CAR CO., DETROIT (DESIGN STUDY ON TURBOJET ENGINES)
DDC-	DEFENSE DOCUMENTATION CENTER, ALEXANDRIA, VA.
DDC-	EUROPEAN THEATER OF OPERATIONS (ARMY). ALSOS MISSION
DDC-	OFFICE OF SCIENTIFIC RESEARCH AND DEVELOPMENT
DDC/P-	GT. BRIT. CULCHETH LABS., LANCS, ENGLAND
DDC/P-	GT. BRIT. DIV. OF ATOMIC ENERGY (PRODUCTION), RISLEY, LANCS, ENGLAND
DDC/P-	UNITED KINGDOM ATOMIC ENERGY AUTHORITY. INDUSTRIAL GROUP. CAPENHURST WORKS, CHES., ENGLAND
DDC-TAS-(YEAR)-	DEFENSE DOCUMENTATION CENTER, ALEXANDRIA, VA.
DD-DR+E(M)-	ARMY BALLISTIC MISSILE AGENCY, REDSTONE ARSENAL, ALA.
DDE-	NAVAL TORPEDO STATION, KEYPORT, WASH.
DDI-(NO.-NO.-NO.)	DATA DYNAMICS, INC., LOS ANGELES
DDI (TECH.)-	GT. BRIT. DEPUTY DIRECTORATE OF TECH. INTELLIGENCE
D-DIV-	LOS ALAMOS SCIENTIFIC LAB., N. MEX.
DDL-(NUMBER)-	DATA-DESIGN LABS., ONTARIO, CALIF.
DD-M-	COLUMBIA UNIV., NYC. SUBSTITUTE ALLOY MATERIALS LABS.
D-DOC-	LOS ALAMOS SCIENTIFIC LAB., N. MEX.

DD PROJ KG-	ORDNANCE DEPT. (ARMY). DEVELOPMENT DIV. (PROJECT KG)
DDR-	CLARK BROS. CO. DEVELOPMENT AND PRODUCTION TEST DEPT., OLEAN, N.Y.
DDR + E-	OFFICE OF THE DIRECTOR OF DEFENSE (RES+ENG),WASH,DC
DDRF-	LITTON INDUSTRIES, SAN CARLOS, CALIF.
DE-	AMERICAN OPTICAL CO. J.W. FECKER DIV., PITTSBURGH
DE-	ATOMIC ENERGY OF CANADA LTD. CHALK RIVER PROJ., ONT.
DE-	DYNAMIC ELECTRONICS-NEW YORK, INC., GLENDALE, N.Y.
DE-	NATIONAL RESEARCH COUNCIL OF CANADA. ATOMIC ENERGY PROJECT, CHALK RIVER, ONT.
DEBELL-(NUMBER)-QPR-	DE BELL + RICHARDSON, INC., HAZARDVILLE, CONN. (DECIMAL NUMBERING CODE) DUMONT (ALLEN B.) LABS., INC., CLIFTON, N.J. (REF. 1)
DECO-(NUMBER)-F	WESTINGHOUSE ELECTRIC CORP. DECO COMMUNICATIONS DEPT., LEESBURG, VA.
DED-	CANADA. DIRECTORATE OF ENGINEER DEVELOPMENT
DED-(NUMBER)-MCB-	RUGBY COLL. OF ENG. TECHNOLOGY, WARWICKS, ENGLAND
DE-DC-	DOW CHEMICAL-DETROIT EDISON NUCLEAR POWER DEV. PROJ.
DEE-	GT. BRIT. ADMIRALTY. ELECTRICAL ENGINEERING DEPT.
DEG-	UNITED KINGDOM ATOMIC ENERGY AUTHORITY. DEVELOPMENT AND ENGINEERING GROUP, RISLEY, LANCS, ENGLAND
DEG-(LTRS)-(NO.)(LTR(S))	UNITED KINGDOM ATOMIC ENERGY AUTHORITY. DEVELOPMENT AND ENGINEERING GROUP (REF. 39)
DEG-(LETTERS)-(NO.)(CA)	UNITED KINGDOM ATOMIC ENERGY AUTHORITY. DEVELOPMENT AND ENG. GROUP, CAPENHURST, CHES., ENG. (REF. 39)
DEG-(LETTERS)-(NO.)(R)	UNITED KINGDOM ATOMIC ENERGY AUTHORITY. DEVELOPMENT AND ENG. GROUP, RISLEY, LANCS, ENGLAND (REF. 39)
DEG-INF-SER-	UNITED KINGDOM ATOMIC ENERGY AUTHORITY. DEVELOPMENT AND ENGINEERING GROUP, RISLEY, LANCS, ENGLAND
DEGIS-(NUMBER)(LTR(S))	UNITED KINGDOM ATOMIC ENERGY AUTHORITY. DEVELOPMENT AND ENGINEERING GROUP (REF. 39)
DEGM-(NUMBER)(LETTER(S))	UNITED KINGDOM ATOMIC ENERGY AUTHORITY. DEVELOPMENT AND ENGINEERING GROUP (REF. 39)
DEG-MEMO-	UNITED KINGDOM ATOMIC ENERGY AUTHORITY. DEVELOPMENT AND ENGINEERING GROUP, RISLEY, LANCS, ENGLAND
DEGR-(NUMBER)(LETTER(S))	UNITED KINGDOM ATOMIC ENERGY AUTHORITY. DEVELOPMENT AND ENGINEERING GROUP (REF. 39)
DEG-REPORT-	UNITED KINGDOM ATOMIC ENERGY AUTHORITY. DEVELOPMENT AND ENGINEERING GROUP, RISLEY, LANCS, ENGLAND
DEG/SINR-	FRANCE. COMMISSARIAT A L'ENERGIE ATOMIQUE. CENTRE D'ETUDES NUCLEAIRES, GRENOBLE
DEIS-	DIRECTOR OF ENG. + IND. SERVICES,EDGEWOOD ARSENAL,MD.
DEL-	DEL MAR ENGINEERING LAB., LOS ANGELES
DEL-	DELAWARE. UNIV., NEWARK
DEL-(NUMBER)-P	DELAWARE. UNIV., NEWARK (SQUID PROJECT) (REF. 33)
DEL-(NUMBER)-DD-	GENERAL ELECTRIC CO. ELECTRONICS LAB.,SYRACUSE,N.Y.
DEL-	VIRGINIA. UNIV., CHARLOTTESVILLE
DEL-SPO-(YEAR)-	DELAWARE. STATE PLANNING OFFICE, DOVER
DELTA-	TASK GROUP DELTA, NORFOLK, VA.
DE-M-	COLUMBIA UNIV., NYC. SUBSTITUTE ALLOY MATERIALS LABS.
DEMO-(YEAR)/	GREEK ATOMIC ENERGY COMMISSION. DEMOKRITOS NUCLEAR RESEARCH CENTRE, ATHENS
DEMVPI-(NUMBER-NUMBER)	VIRGINIA POLYTECHNIC INST., BLACKSBURG. DEPT. OF ENGINEERING MECHANICS
DEN-	MARTIN-MARIETTA CORP. DENVER DIV.
DENVERU-	DENVER. UNIV.
DENY-	DYNAMIC ELECTRONICS-NEW YORK, INC., GLENDALE, N.Y.
DEP-(YEAR)-	BUREAU OF RADIOLOGICAL HEALTH. DIV. OF ELECTRONIC PRODUCTS, ROCKVILLE, MD.
DEP-	SYLVANIA ELECTRONIC SYSTEMS, NEEDHAM, MASS.
DEPCO-(LETTER)-	DAYTON ELECTRONIC PRODUCTS CO., INC., OHIO
DEP-SEPP-	FRANCE. COMMISSARIAT A L'ENERGIE ATOMIQUE. CENTRE D'ETUDES NUCLEAIRES, FONTENAY-AUX-ROSES (TRANS.)
(DEPT. NUMBER)-	SPERRY GYROSCOPE CO., GREAT NECK, N.Y. (REF. 1)
DE-R-	COLUMBIA UNIV., NYC. SUBSTITUTE ALLOY MATERIALS LABS.
DER-	MINAS GERAIS UNIV., BELO HORIZONTE, BRAZIL. DIVISAO DE ENGENHARIA DE REATORES
DERE-FRN-(MONTH/YEAR)	UNITED KINGDOM ATOMIC ENERGY AUTHORITY. IND. GP. DOUNREAY EXPTL. REACTOR ESTAB., CAITHNESS, SCOTLAND (FAST REACTOR NEWSLETTER)
DERTS-NT-(NUMBER)-	FRANCE. DEPARTEMENT D'ETUDES DE RECHERCHES EN TECHNOLOGIE SPATIALE, TOULOUSE
DE/SEA-	FRANCE. COMMISSARIAT A L'ENERGIE ATOMIQUE. CENTRE D'ETUDES NUCLEAIRES, SACLAY
DE/SEP/(NUMBER)/	FRANCE. COMMISSARIAT A L'ENERGIE ATOMIQUE. CENTRE D'ETUDES NUCLEAIRES, SACLAY
DESIGN REPORT NO.-	ARGONNE NATIONAL LAB., LEMONT, ILL.
DESPU-	DU PONT DE NEMOURS (E.I.) + CO. SAVANNAH RIVER LAB., AIKEN, S.C.
DESY-	DEUTSCHES ELEKTRONEN-SYNCHROTRON, HAMBURG
DESY-(YEAR)/	DEUTSCHES ELEKTRONEN-SYNCHROTRON, HAMBURG
DESY-A-(NUMBER).(NUMBER)	DEUTSCHES ELEKTRONEN-SYNCHROTRON, HAMBURG
DESY-F-	DEUTSCHES ELEKTRONEN-SYNCHROTRON, HAMBURG
DESY-H-	DEUTSCHES ELEKTRONEN-SYNCHROTRON, HAMBURG
DESY-HERA-	DEUTSCHES ELEKTRONEN-SYNCHROTRON, HAMBURG
DESY-R-	DEUTSCHES ELEKTRONEN-SYNCHROTRON, HAMBURG
DESY-SR-	DEUTSCHES ELEKTRONEN-SYNCHROTRON, HAMBURG
DESY-ST-	DEUTSCHES ELEKTRONEN-SYNCHROTRON, HAMBURG
DET-36-(YEAR)	AIR FORCE. TUSLOG DETACHMENT 36, USAFE, N.Y.C.
DET-36-TR-(YEAR)-	AIR FORCE. TUSLOG DETACHMENT 36, USAFE, N.Y.C.
DET-36-PR-(YEAR)-	AIR FORCE. TUSLOG DETACHMENT 36, USAFE, N.Y.C.
DET 1 APGC-	AIR PROVING GROUND COMMAND, KIRTLAND AFB, N. MEX.
DETRCC-	DETROIT CONTROLS CORP., REDWOOD CITY, CALIF.
DEV-	CONTROL INSTRUMENT CO., INC., BROOKLYN, N.Y.
DEV-	DOUGLAS AIRCRAFT CO., INC., TULSA, OKLA.
DEV-	DOUGLAS AIRCRAFT CO., INC. DEVELOPMENT LABS., SANTA MONICA, CALIF.
DEV-(LETTER)-(NO.)-	ATOMIC ENERGY OF CANADA LTD. CHALK RIVER PROJ., ONT.
DEV-B-	ATOMIC ENERGY OF CANADA LTD. CHALK RIVER PROJ., ONT.
DEV-C-	ATOMIC ENERGY OF CANADA LTD. CHALK RIVER PROJ., ONT.
DEV-F-	ATOMIC ENERGY OF CANADA LTD. CHALK RIVER PROJ., ONT.
DEVF-	CONSOLIDATED VULTEE AIRCRAFT CORP., SAN DIEGO, CALIF.
DEWEY DATA-	DEWEY (G.C.) AND CO., INC., N.Y.C.
DF-	(DATA FOLDER) (REF. 2)
DF-(NUMBER)-O	BOELKOW ENTWICKLUNGEN K.G., MUNICH
D-F-	CALIFORNIA RESEARCH AND DEVELOPMENT CO., LIVERMORE
DF-(YEAR)-CAP-	CANADIAN GENERAL ELECTRIC CO., LTD. CIVILIAN ATOMIC POWER DIV., PETERBOROUGH, ONT.
DF-	COAST AND GEODETIC SURVEY, WASHINGTON, D.C. (RADIO DIRECTION FINDING CHARTS)
DF(NO.)AGT(NO.)	GENERAL ELECTRIC CO., EVENDALE, OHIO
DF-	GENERAL ELECTRIC CO., SCHENECTADY, N.Y. (DATA FOLDER)
DF-(YR LAB.)-	GENERAL ELECTRIC CO., SCHENECTADY, N.Y.
DF-(YR)-(LETTER(S))-	GENERAL ELECTRIC CO., SCHENECTADY, N.Y.
DF (YR.)(LTRS.)(NO.)-	GENERAL ELECTRIC CO. AIRCRAFT GAS TURBINE DIV., EVENDALE, OHIO
DF-(YEAR)-AGT-	GENERAL ELECTRIC CO. AIRCRAFT GAS TURBINE DIV., EVENDALE, OHIO
DF(NO.)(LTRS.)-	GENERAL ELECTRIC CO. ELECTRONICS LAB.,SYRACUSE,N.Y.
DF-(YR)AO-	GENERAL ELECTRIC CO. GENERAL ENGINEERING AND CONSULTING LAB., SCHENECTADY, N.Y.
DF-(YEAR)GL-	GENERAL ELECTRIC CO. GEN. ENG. LAB., SCHENECTADY,N.Y.
DF-(YEAR)-AGT-	GENERAL ELECTRIC CO. JET ENGINE DEPT., CINCINNATI
DF(YEAR)MTG-	GENERAL ELECTRIC CO. MARINE TURBINE AND GEAR DEPT., WEST LYNN, MASS.
DF-	GENERAL ELECTRIC CO. MATERIALS AND PROCESSES LAB., SCHENECTADY, N.Y.

DF(YR)SL-
: GENERAL ELECTRIC CO. MATERIALS AND PROCESSES LAB., SCHENECTADY, N.Y.

DF-(YR)-MTG-
: GENERAL ELECTRIC CO. MEDIUM STEAM TURBINE GENERATOR AND GEAR DEPT., WEST LYNN, MASS.

DF-
: ILLINOIS. UNIV., URBANA. ELECTRICAL ENG. RES. LAB.

DF-
: OFFICE OF SCI. RES. + DEV.(DETONATION+FRAGMENTATION)

DFA-
: OFFICE OF SCIENTIFIC RESEARCH AND DEVELOPMENT (DETONATION, FRAGMENTATION AND AIR BLAST)

DFG-
: DEUTSCHE FORSCHUNGS GESELLSCHAFT

DFI-
: DESIGNERS FOR INDUSTRY, INC., CLEVELAND

DFL-
: DEUTSCHE FORSCHUNGSANSTALT FUER LUFTFAHRT E.V., INSTITUT FUER AERODYNAMIK, BRUNSWICK

DF P-
: GT. BRIT. DEPT. OF FORESTS (PALESTINE REPORTS)

DFP-TP-
: AUSTRALIA. COMMONWEALTH SCIENTIFIC AND INDUSTRIAL RES. ORG. DIV. OF FOREST PRODUCTS, MELBOURNE

DFR-(YEAR)-
: DATA CORP., DAYTON, OHIO

DFS-
: DEUTSCHE FORSCHUNGSANSTALT FUER SEGELFLUG, MUNICH

DFVLR-SONDERDRUCK-
: DEUTSCHE VERSUCHSANSTALT FUER LUFT- UND RAUMFAHRT (ALL LOCATIONS) (REPRINT)

DGAI-
: DANIEL GUGGENHEIM AIRSHIP INST., AKRON, OHIO

DGGW-
: UNITED KINGDOM. DIRECTORATE OF GUIDED WEAPONS RESEARCH AND DEVELOPMENT, LONDON

DGGW-EMR-(YEAR)-
: IMPERIAL COLLEGE OF SCIENCE AND TECHNOLOGY, LONDON

DGLR-(YEAR)-
: DEUTSCHE GESELLSCHAFT FUER LUFT- UND RAUMFAHRT E.V., COLOGNE

DGM-
: BUREAU OF ORDNANCE (NAVY) (DEGAUSSING MEMORANDUM)

DGMS-
: GT. BRIT. ROYAL AIR FORCE. DIRECTOR-GENERAL OF MEDICAL SERVICES

DGN-
: BUREAU OF ORDNANCE (NAVY) (DEGAUSSING NOTE)

DGO-(NUMBER-NUMBER)
: KODAIKANAL OBSERVATORY, INDIA

DG-R-
: ARMY BALLISTIC MISSILE AGENCY, HUNTSVILLE, ALA.

DG-RD-
: NAVAL ORDNANCE LAB. DEGAUSSING RANGE (DATA SHEET)

DGRR/WGLR PAPER-(YR.)-
: DEUTSCHE GESELLSCHAFT FUER RAKETENTECHNIK UND RAUMFAHRT, STUTTGART

DGS-
: OHIO STATE UNIV., COLUMBUS. DEPT. OF GEODETIC SCIENCE

DGSI-
: BUREAU OF ORDNANCE(NAVY)(DEGAUSSING STA. INSTRUCTION)

DG-TR-(NUMBER)-(YEAR)
: ARMY BALLISTIC MISSILE AGENCY, REDSTONE ARSENAL, ALA.

DGWP-
: UNITED KINGDOM. DIRECTORATE OF GUIDED WEAPONS RESEARCH AND DEVELOPMENT, LONDON

DGWRD-
: UNITED KINGDOM. DIRECTORATE OF GUIDED WEAPONS RESEARCH AND DEVELOPMENT, LONDON

DH. (NUMBER)-PH-
: GIRAVIONS DORAND CO., PARIS

DHA-
: DE HAVILLAND AIRCRAFT OF CANADA LTD.

DHC-SP-R.(NO.)-(LTR.)
: DE HAVILLAND AIRCRAFT CO., LTD. SPECIAL PRODUCTS AND APPLIED RESEARCH DIV., MALTON, ONT.

DHC-SP-R.-
: DE HAVILLAND AIRCRAFT OF CANADA LTD. SPECIAL PRODUCTS AND RESEARCH DIV., MALTON, ONT.

DHO-RRL(YEAR)
: CANADA. DEPT. OF HIGHWAYS, DOWNSVIEW

DI-(NUMBER-NUMBER)
: BOEING SCIENTIFIC RESEARCH LABS. PLASMA PHYSICS LAB., SEATTLE

DI-
: DEPARTMENT OF THE INTERIOR. CENTRAL LIBRARY, WASHINGTON, D.C. (TRANSLATIONS)

DI-
: DOCUMENTATION INC., WASHINGTON, D.C.

DI/
: DOCUMENTATION INC., WASHINGTON, D.C.

DIA-
: DEFENSE INTELLIGENCE AGENCY, WASHINGTON, D.C.

DIAM-(YEAR)-
: DEFENSE INTELLIGENCE AGENCY, WASHINGTON, D.C.

DIAS-
: DUBLIN INSTITUTE FOR ADVANCED STUDIES. SCHOOL OF COSMIC PHYSICS

D.I.C.-
: MASSACHUSETTS INST. OF TECH., CAMBRIDGE

DIC-
: MASSACHUSETTS INST. OF TECH., CAMBRIDGE. DIV. OF INDUSTRIAL COOPERATION

DIC-(LETTER(S))-
: MASSACHUSETTS INST. OF TECH., CAMBRIDGE. DIV. OF INDUSTRIAL COOPERATION

DID-
: BUREAU OF ORDNANCE (NAVY) (DESIGN INSPECTION DATA)

DIDAKOMETRY-
: SWEDEN. SCHOOL OF EDUCATION, MALMO. DEPT. OF EDUCATIONAL AND PSYCHOLOGICAL RESEARCH

DIG-
: OFFICE OF THE DEPUTY INSPECTOR GENERAL, NORTON AFB, CALIF.

(YR.)-(3 DIGIT DEPT. NO.)
: INTERNATIONAL BUSINESS MACHINES CORP., OWEGO, N.Y.

(4 DIGITS)
: GT. BRIT. DEPT. OF ATOMIC ENERGY. INDUSTRIAL GROUP H.Q., RISLEY, LANCS, ENGLAND (REFS. 1, 36)

(4 DIGITS)
: GT. BRIT. DIV. OF ATOMIC ENERGY (PRODUCTION), RISLEY, LANCS, ENGLAND (ASSIGNED TO INCOMING MISCELLANEOUS BRITISH REPORTS) (REF. 1)

(4 DIGITS-NO.-LTRS.)-
: TRW SYSTEMS, REDONDO BEACH, CALIF. (REFS. 1, 42)

(5 DIGITS-NO.-LTRS.)-
: TRW SYSTEMS, REDONDO BEACH, CALIF. (REFS. 1, 42)

DI/HP-
: EUROPEAN COUNCIL FOR NUCLEAR RESEARCH, GENEVA

DIL-
: DEFENSE INTELLIGENCE AGENCY, WASH., D.C. (LIBRARY)

DIM-
: NAVAL POWDER FACTORY. ORDNANCE INVESTIGATION LAB., INDIAN HEAD, MD. (DEMOLITION INVEST. MEMORANDA)

DIN-
: GENERAL ELECTRIC CO. RESEARCH LAB., SCHENECTADY, N.Y.

DIN-(LTRS., NUMBERS)-
: GENERAL ELECTRIC CO. MISSILE AND SPACE DIV., PHILA.

DIN-R(YEAR)SD(NO.)
: GENERAL ELECTRIC CO. MISSILE AND SPACE DIV., PHILA.

DINR-E-
: U.S.S.R. JOINT INST. FOR NUCLEAR RESEARCH, DUBNA

DINR-P-
: U.S.S.R. JOINT INST. FOR NUCLEAR RESEARCH, DUBNA

DINR-R-
: U.S.S.R. JOINT INST. FOR NUCLEAR RESEARCH, DUBNA

66-DINS-
: AEROSPACE CORP. ARMS CONTROL PROGRAM OFFICE, SAN BERNARDINO, CALIF.

DIO,(LTRS)(NO)-(LTR)-(YR)
: OFFICE OF THE CHIEF OF NAVAL OPERATIONS, WASH., D.C.

DIR-
: LOS ALAMOS SCIENTIFIC LAB., N. MEX.

DIR.O.P.-(YEAR)/
: NETHERLANDS. DIRECTIE OVERHEIDS-PERSONEELSBELEID, THE HAGUE

DIR/PS/INT.S-
: EUROPEAN ORGANIZATION FOR NUCLEAR RESEARCH, GENEVA

DIR-TR-(NUMBER)-(YEAR)
: ARMY BALLISTIC MISSILE AGENCY, REDSTONE ARSENAL, ALA.

DIRX-
: LOS ALAMOS SCIENTIFIC LAB., N. MEX.

DIS/(NUMBER/YEAR)
: BRITISH STEEL CORP., LONDON

DISCUSSION PAPER-
: ILLINOIS. UNIV., CHICAGO. CENTER FOR URBAN STUDIES

DISS-
: SWITZERLAND. EIDGENOESSISCHE TECHNISCHE HOCHSCHULE, ZURICH

DISS-NR-
: SWITZERLAND. EIDGENOESSISCHE TECHNISCHE HOCHSCHULE, ZURICH

DIT-
: DREXEL INST. OF TECH., PHILADELPHIA

DIT-PC-
: DREXEL INST. OF TECH.,CENTERTON,NJ.LAB.OF CLIMATOLOGY

DI/USAF-
: DIRECTORATE OF INTELLIGENCE (AIR FORCE), WASH., D.C.

DK-
: CORNELL AERONAUTICAL LAB., INC., BUFFALO

DK-(NUMBER.NUMBER.NO.)
: DEUTSCHER WETTERDIENST, OFFENBACH AM MAIN

DKU-
: DUKE UNIV., DURHAM, N.C.

DL-(YEAR)-
: AIR FORCE SYSTEMS COMMAND. DIRECTOR OF LABS., WASHINGTON, D.C.

DL-
: ATOMIC ENERGY OF CANADA LTD., CHALK RIVER, ONT.

DL-(NUMBER-NUMBER)
: BOEING SCIENTIFIC RESEARCH LABS., SEATTLE

D-L-
: COLUMBIA UNIV., NYC. SUBSTITUTE ALLOY MATERIALS LABS.

DL-
: HUGHES AIRCRAFT CO. DEV. LABS., NEWPORT BEACH, CALIF

D(L)-(NUMBER)-AEC
: RAND CORP., SANTA MONICA, CALIF.

DL-(YEAR)-
: SWEDISH DETONIC RESEARCH FOUNDATION, STOCKHOLM

D/L-
: UNITED KINGDOM ATOMIC ENERGY AUTHORITY. RESEARCH GP. ATOMIC ENERGY RES. ESTAB., HARWELL, BERKS, ENGLAND

DLCS-
: DUQUESNE LIGHT CO., SHIPPINGPORT, PENNA.

DLG-
: NAVAL TORPEDO STATION, KEYPORT, WASH.

DLH-
: DEUTSCHE LUFTHANSA A.G. (BERICHT)

DLI-
: BUREAU OF NAVAL WEAPONS

DLI-
: DEVELOPMENT LABS., INC., LOS ANGELES

DLI(YR)-
: DEVELOPMENT LABS., INC., LOS ANGELES

D-LIB-
: LOS ALAMOS SCIENTIFIC LAB., N. MEX.

DLM-
: BENDIX AVIATION CORP. PACIFIC DIV.,N. HOLLYWOOD, CAL.

```
DLMA-(CONTRACT NO.)-
         DEPARTMENT OF LABOR. MANPOWER ADM., WASHINGTON, D.C.
DLNS IWD-
         AUSTRALIA. DEPARTMENT OF LABOUR AND NATIONAL
         SERVICE. INDUSTRIAL WELFARE DIVISION
DL-R-
         CARBIDE AND CARBON CHEMICALS CORP. SUBSTITUTE ALLOY
         MATERIALS LABS., N.Y.C.
DLR-FB-(YEAR)-
         DEUTSCHE GESELLSCHAFT FUER FLUGWISSENSCHAFTEN,
         BRUNSWICK
DLR-FB-(YEAR)-
         DEUTSCHE VERSUCHSANSTALT FUER LUFT- UND RAUMFAHRT
         (ALL LOCATIONS)
DLR-FB-(YEAR)-
         GERMANY. INSTITUT FUER THEORETISCHE GASDYNAMIK,
         AACHEN
DLR-MITT.-(YEAR)-
         DEUTSCHE VERSUCHSANSTALT FUER LUFT- UND RAUMFAHRT
         (ALL LOCATIONS)
DLR-MITT-(YEAR)-
         WISSENSCHAFTLICHE GESELLSCHAFT FUER LUFT- UND RAUM-
         FAHRT, COLOGNE
DL-S-
         DUQUESNE LIGHT CO., SHIPPINGPORT, PENNA.
DL(S)(LETTERS)-
         GT. BRIT. SPRINGFIELDS WORKS, LANCS, ENGLAND (REF.36)
DLSC-MCRL-PTIC-
         DEFENSE LOGISTICS SERVICES CENTER, BATTLE CREEK, MICH.
         (MASTER CROSS REFERENCE LIST)
DL(S)TN-
         GT. BRIT. SPRINGFIELDS WORKS, LANCS, ENGLAND
DL-TR-
         PICATINNY ARSENAL. LIQUID ROCKET PROPULSION LAB.,
         DOVER, N.J.
DLWK-
         DUQUESNE LIGHT CO., SHIPPINGPORT, PENNA.
DLWK-
         KIDDE (WALTER) NUCLEAR LABS., INC., GARDEN CITY, N.Y.
DM-
         (DESIGN MANUAL) (REF. 2)
DM-
         (DESIGN MEMORANDUM) (REF. 2)
DM-(NUMBERS)-
         AEROJET-GENERAL CORP. LIQUID ROCKET OPERATIONS, SAC-
         RAMENTO, CALIF.
DM-
         ATOMIC ENERGY OF CANADA LTD., CHALK RIVER, ONT.
DM-
         BUREAU OF ORDNANCE (NAVY) (DEGAUSSING MEMORANDUM)
D-M-
         COLUMBIA UNIV., NYC. SUBSTITUTE ALLOY MATERIALS LABS.
DM-
         COLUMBIA UNIV., NYC. SUBSTITUTE ALLOY MATERIALS LABS.
DM-
         DORNE AND MARGOLIN, WESTBURY, N.Y.
DM-
         DUMONT (ALLEN B.) LABS., INC., CLIFTON, N.J.
DM-
         FRANCE. COMMISSARIAT A L'ENERGIE ATOMIQUE. CENTRE
         D'ETUDES NUCLEAIRES, SACLAY
DM(YEAR)-
         GENERAL ELECTRIC CO. ADVANCED ENGINE AND TECHNOLOGY
         DEPT., CINCINNATI
DM-
         GENERAL ELECTRIC CO. FLIGHT PROPULSION LAB. DEPT.,
         CINCINNATI
DM-(YEAR)-
         GENERAL ELECTRIC CO. FLIGHT PROPULSION LAB. DEPT.,
         CINCINNATI
DM-
         MARTIN-MARIETTA CORP. DENVER DIV.
DM-
         SANDIA CORP., ALBUQUERQUE, N. MEX. (DESIGN MANUAL)
DM-(YEAR)/(NO.-NO.)
         TEXAS INSTRUMENTS, INC. ELECTRONIC AND OPTICAL
         SYSTEMS DEPT., DALLAS
D-MAT-
         UNITED KINGDOM. DIRECTORATE OF MATERIALS RESEARCH
         AND DEVELOPMENT, LONDON
D-MAT-AVIATION-
         UNITED KINGDOM. DIRECTORATE OF MATERIALS RESEARCH
         AND DEVELOPMENT, LONDON
DMC-
         DUMONT (ALLEN B.) LABS., INC., CLIFTON, N.J.
DMCR-TR-(NO.)-(YEAR)
         ARMY BALLISTIC MISSILE AGENCY, REDSTONE ARSENAL, ALA.
DMCR-TR-(NO.)-(YEAR)
         ARMY BALLISTIC MISSILE AGENCY. MISSILE FIRING LAB.,
         TITUSVILLE, FLA.
DM-CS-
         FRANCE. COMMISSARIAT A L'ENERGIE ATOMIQUE. CENTRE
         D'ETUDES NUCLEAIRES, SACLAY
DMD-
         AEROJET-GENERAL CORP., SACRAMENTO, CALIF. (METALLURG-
         ICAL INVESTIGATIONS)
DMDC-
         NATIONAL BUREAU OF STANDARDS. DIFFUSION IN METALS
         DATA CENTER, WASHINGTON, D.C.
DME-ME-
         NATIONAL RESEARCH COUNCIL OF CANADA, DIVISION OF
         MECHANICAL ENGINEERING, OTTAWA
DME-MER-
         NATIONAL RESEARCH COUNCIL OF CANADA, DIVISION OF
         MECHANICAL ENGINEERING, OTTAWA
DME-MET-
         NATIONAL RESEARCH COUNCIL OF CANADA, DIVISION OF
         MECHANICAL ENGINEERING, OTTAWA
DME-MH-
         NATIONAL RESEARCH COUNCIL OF CANADA, DIVISION OF
         MECHANICAL ENGINEERING, OTTAWA
DME-MISC.-
         NATIONAL RESEARCH COUNCIL OF CANADA. LOW TEMPERATURE
         SECTION, OTTAWA
DME-MK-
         NATIONAL RESEARCH COUNCIL OF CANADA, DIVISION OF
         MECHANICAL ENGINEERING, OTTAWA
DME-ML-
         NATIONAL RESEARCH COUNCIL OF CANADA, DIVISION OF
         MECHANICAL ENGINEERING, OTTAWA
DME-MP-
         NATIONAL RESEARCH COUNCIL OF CANADA, DIVISION OF
         MECHANICAL ENGINEERING, OTTAWA
DME/NAE-(YEAR)-/(NO.)/
         NATIONAL RESEARCH COUNCIL OF CANADA, OTTAWA
DME/NAE-
         NATIONAL RESEARCH COUNCIL OF CANADA. DIVISION OF
         MECHANICAL ENGINEERING, OTTAWA
DMIC-
         BATTELLE MEMORIAL INST. DEFENSE METALS INFORMATION
         CENTER, COLUMBUS, OHIO
DMIC-MEMO-
         BATTELLE MEMORIAL INST. DEFENSE METALS INFORMATION
         CENTER, COLUMBUS, OHIO
DMIC-S-
         BATTELLE MEMORIAL INST. DEFENSE METALS INFORMATION
         CENTER, COLUMBUS, OHIO
DM-M-
         COLUMBIA UNIV., NYC. SUBSTITUTE ALLOY MATERIALS LABS.
DMO-
         OFFICE OF EMERGENCY PREPAREDNESS, WASHINGTON, D.C.
         (DEFENSE MOBILIZATION ORDER)
DMP-
         E.M.I. ELECTRONICS, LTD., HAYES, KENT, ENGLAND
DMP-
         OFFICE OF EMERGENCY PREPAREDNESS, WASHINGTON, D.C.
         (DEFENSE MANPOWER POLICY)
DMP/G-
         BRAZIL. COMISSAO NACIONAL DE ENERGIA NUCLEAR
DM-R-
         ARMY BALLISTIC MISSILE AGENCY, HUNTSVILLE, ALA.
DMR-
         BEECH AIRCRAFT CORP., WICHITA, KAN. (DATA
         MANAGEMENT REPORT)
DMRE-(YEAR)-
         BUREAU OF RADIOLOGICAL HEALTH. DIV. OF MEDICAL RADI-
         ATION EXPOSURE, ROCKVILLE, MD.
DMS-
         GENERAL ELECTRIC CO. SPACE POWER AND PROPULSION
         SECTION, CINCINNATI
DMS-(YEAR)-
         STANFORD UNIV., CALIF. DEPT. OF MATERIALS SCIENCE
DMS-
         STANFORD UNIV., CALIF. DEPT. OF MATERIALS SCIENCE
DMSP-(LETTER)-
         DELFT, NETHERLANDS. TECHNISCHE HOGESCHOOL (DELFT
         MOLTEN SALT PROJECT)
DMSP-G-
         DELFT, NETHERLANDS. TECHNISCHE HOGESCHOOL
DMSRD-
         GT. BRIT. DIRECTORATE OF MATERIALS AND STRUCTURES
         RESEARCH AND DEVELOPMENT
DMXRD-
         GT. BRIT. DIRECTORATE OF MATERIALS AND EXPLOSIVES
         RESEARCH AND DEVELOPMENT
DN-
         CALIFORNIA. UNIV., MERCURY, NEV. LAWRENCE RADIATION
         LAB.
DNA-
         DEFENSE NUCLEAR AGENCY, WASHINGTON, D.C.
DNA-(NUMBER)-F
         DEFENSE NUCLEAR AGENCY, WASHINGTON, D.C.
DNA-(NUMBER)-T
         DEFENSE NUCLEAR AGENCY, WASHINGTON, D.C.
DNC-
         GT. BRIT. ADMIRALTY. NAVAL CONSTRUCTION DEPT.
DND64-
         ONTARIO. HYDRO-ELECTRIC POWER COMMISSION, TORONTO
DNI-
         OFFICE OF NAVAL INTELLIGENCE, WASHINGTON, D.C.
DNPL-
         UNITED KINGDOM. DARESBURY NUCLEAR PHYSICS LAB.,
         CHESHIRE, ENGLAND
DNPL-AR-(YEAR)
         UNITED KINGDOM. DARESBURY NUCLEAR PHYSICS LAB.,
         CHESHIRE, ENGLAND
DNPL/P-
         UNITED KINGDOM. DARESBURY NUCLEAR PHYSICS LAB.,
         CHESHIRE, ENGLAND
DNPL/R-
         UNITED KINGDOM. DARESBURY NUCLEAR PHYSICS LAB.,
         CHESHIRE, ENGLAND
DNS-
         DIRECTORATE OF NUCLEAR SAFETY, KIRTLAND AFB, N. MEX.
DO-
         ACOUSTICA ASSOCIATES, INC., LOS ANGELES
DO-
         BENDIX AVIATION CORP. RESEARCH LABS., DETROIT
DO-(3 DIGITS)
         RADIO TECH. COMMISSION FOR AERONAUTICS, WASH., D.C.
DOC-
         (DOCUMENT) (REF. 2)
DOC.-(YEAR)(LTRS.)(NO.)
         GENERAL ELECTRIC CO., BETHESDA, MD.
DOC-(YEAR)(LTRS.)(NOS.)
         GENERAL ELECTRIC CO., PHILADELPHIA
DOC-(4 DIGITS)
         INTERNATIONAL CIVIL AVIATION ORGANIZATION, MONTREAL
```

DOC-

DOC.-	PORTUGAL. JUNTA DE ENERGIA NUCLEAR. LABORATORIO DE FISICA E ENGENHARIA NUCLEARES, SACAVEM
DOC-B-	SPINDLETOP RESEARCH, INC., LEXINGTON, KY.
DOC-C-	SPINDLETOP RESEARCH, INC., LEXINGTON, KY.
DOC. CONTROL NO.-	AERONUTRONIC SYSTEMS, INC., GLENDALE, CALIF.
DOC-G-	SPINDLETOP RESEARCH, INC., LEXINGTON, KY.
DOCKET-	DIVISION OF REACTOR LICENSING, AEC, WASHINGTON, D.C.
DOC-LFEN-	PORTUGAL. JUNTA DE ENERGIA NUCLEAR. LABORATORIO DE FISICA E ENGENHARIA NUCLEARES, SACAVEM
DOC-S-	SPINDLETOP RESEARCH, INC., LEXINGTON, KY.
DOD-	DEPARTMENT OF DEFENSE, WASHINGTON, D.C.
DOD-(LETTERS)-	DEPARTMENT OF DEFENSE, WASH., D.C. (FOR EXPLANATION OF THE REMAINDER OF THE CODE, SEE LTRS AFTER DOD-)
DOD-AIMS-(YEAR)-	AERONAUTICAL SYSTEMS DIV. AIMS SYSTEM PROGRAM OFFICE, WRIGHT-PATTERSON AFB, OHIO
DODCO TR-	DODCO INC., BLAWENBERG, N.J.
DOD D/GM-	OFFICE OF THE SECRETARY OF DEFENSE. DIRECTOR OF GUIDED MISSILES, WASHINGTON, D.C.
DOD IRIG-(YR)	INTER-RANGE INSTRUMENTATION GROUP, WASHINGTON, D.C.
DOD MLC-	MILITARY LIAISON COMMITTEE, WASHINGTON, D.C.
DOD/R AND E-	OFF. OF THE ASST. SECY. OF DEFENSE(RES.+DEV.),WASH,DC
DOD/R AND E-	OFFICE OF THE DIRECTOR OF DEFENSE (RES+ENG),WASH.,DC
DOD RSB-	RESEARCH AND DEVELOPMENT BOARD, WASHINGTON, D.C.
DOES-	AIR FORCE. 6TH AIR DIV., MAC DILL AFB, FLA.
DOFL-	DIAMOND ORDNANCE FUZE LABS., WASHINGTON, D.C.
DOFL-(NUMBER)-(YEAR)	DIAMOND ORDNANCE FUZE LABS., WASHINGTON, D.C.
DOFL-(YEAR)-T-	DIAMOND ORDNANCE FUZE LABS., WASHINGTON, D.C.
DOFL-FA-	HARRY DIAMOND LABS., WASH.,D.C. (FLUID AMPLIFICATION)
DOFL-IM-	HARRY DIAMOND LABS., WASHINGTON, D.C.
DOFL-LIBRARY-(YEAR)-	DIAMOND ORDNANCE FUZE LABS., WASH., D.C. (LIBRARY)
DOFL-PR-	DIAMOND ORDNANCE FUZE LABS., WASHINGTON, D.C.
DOFL-PR-(YEAR)-	DIAMOND ORDNANCE FUZE LABS., WASHINGTON, D.C.
DOFL-R(NO.)-	DIAMOND ORDNANCE FUZE LABS., WASHINGTON, D.C.
DOFL-TL-RM-	DIAMOND ORDNANCE FUZE LABS., WASHINGTON, D.C.
DOFL-TM-(YEAR)-	DIAMOND ORDNANCE FUZE LABS., WASHINGTON, D.C.
DOFL-TM-	HARRY DIAMOND LABS., WASHINGTON, D.C.
DOFL-TR-	HARRY DIAMOND LABS., WASHINGTON, D.C.
DOF(X)-	GT. BRIT. DIRECTORATE OF ORDNANCE FACTORIES (EXPLOSIVES)
DOL-BLS-B-	BUREAU OF LABOR STATISTICS, WASHINGTON, D.C.
DOMIIT-	IIT RESEARCH INST., CHICAGO
DOMIIT-	ILLINOIS INST. OF TECH., CHICAGO. DEPT. OF MATH.
DOMIIT-	ILLINOIS INST. OF TECH., CHICAGO. DEPT. OF MECHANICS
DOMITT-	IIT RESEARCH INST., CHICAGO
DOR-	GT. BRIT. ADMIRALTY. DEPT. OF OPERATIONAL RESEARCH
DOR-	MASSACHUSETTS INST. OF TECH., LEXINGTON. LINCOLN LAB.
DOR(N)-	CANADA. DIRECTORATE OF OPERATIONAL RESEARCH
DOSIMETRY-TR-(NO.-NO.)	ARMY MISSILE TEST AND EVALUATION DIRECTORATE. NUCLEAR WEAPON EFFECTS DIV. WHITE SANDS MISSILE RANGE, N. MEX.
DOT-CG-	DEPT. OF TRANSPORTATION, WASHINGTON, D.C.
DOT-FRA-OHSGT-	DEPT. OF TRANSPORTATION, WASHINGTON, D.C.
DOT-HS-	DEPT. OF TRANSPORTATION, WASHINGTON, D.C. (HIGHWAY SAFETY)
DOT/HUD-IANAP-(YEAR)-	DEPT. OF TRANSPORTATION, WASH., D.C. (AIRCRAFT NOISE)
DOT/OS-(CONTRACT NO.)-	DEPT. OF TRANSPORTATION, WASHINGTON, D.C.
DOTS-HS-	DEPT. OF TRANSPORTATION, WASHINGTON, D.C. (HIGHWAY SAFETY)
DOT-TD-	DEPT. OF TRANSPORTATION, WASHINGTON, D.C.
DOUGLAS PAPER-	DOUGLAS AIRCRAFT CO., INC. (ALL LOCATIONS)
DOUGLAS PAPER-	DOUGLAS AIRCRAFT CO., INC., SANTA MONICA, CALIF.
DOUGLAS PAPER-	MC DONNELL-DOUGLAS CO., SANTA MONICA, CALIF.
DO VB-	DORNIER-WERKE (VERSUCHSBERICHT)
DOVE	EASTMAN KODAK CO., ROCHESTER, N.Y.
DOW-	DOW CHEMICAL CO., MIDLAND, MICH.
DOW-CD(YEAR)-	DOW CHEMICAL CO. ROCKY FLATS DIV., GOLDEN, COLO.
DOW-CD(YEAR)-	DOW CHEMICAL CO. ROCKY FLATS PLANT, DENVER
DOW-NR-	DOW CHEMICAL CO. NUCLEAR RES. LAB., MIDLAND, MICH.
DOW-QPR-	DOW CHEMICAL CO. POLYMER RES. LAB., MIDLAND, MICH.
DOW-RFP-	DOW CHEMICAL CO. ROCKY FLATS DIV., GOLDEN, COLO.
DOW-RFP-	DOW CHEMICAL CO. ROCKY FLATS PLANT, DENVER
DP-	(DESIGN PROPOSAL) (REF. 2)
DP-	(DISCUSSION PAPER) (REF. 2)
DP-	AEROPROJECTS, INC., WEST CHESTER, PENNA.
DP-	BRIDGEPORT BRASS CO., CONN.
DP-	CALIFORNIA. UNIV., BERKELEY
DP-	DU PONT DE NEMOURS (E.I.) + CO. (ALL LOCATIONS)
DP-(NUMBER)-	EAST COAST AERONAUTICS, INC., MOUNT VERNON, N.Y. (DESIGN PROPOSAL)
DP-	JOINT AIRBORNE TROOP BOARD, FORT BRAGG, N.C.
DP-	MALLINCKRODT CHEMICAL WORKS, ST. LOUIS
DP-	NUCLEAR METALS, INC., CONCORD, MASS.
DP-	UNITED KINGDOM ATOMIC ENERGY AUTHORITY. RESEARCH GP. ATOMIC ENERGY ESTAB., WINFRITH, DORSET, ENGLAND
DP-AD-	DU PONT DE NEMOURS (E.I.) + CO., WILMINGTON, DEL.
DPA.IS/	FRANCE. COMMISSARIAT A L'ENERGIE ATOMIQUE. CENTRE D'ETUDES NUCLEAIRES, SACLAY
DPB EM-	CAMP EVANS SIGNAL LABORATORY. DEVELOPMENT PLANNING BRANCH, BELMAR, N.J. (ENGINEERING MEMO)
DPC-(YEAR)-	DAIRYLAND POWER COOPERATIVE, LA CROSSE, WIS.
DPC-(NUMBER-NUMBER)	DAIRYLAND POWER COOPERATIVE, LA CROSSE, WIS.
DPC-	DU PONT DE NEMOURS (E.I.) + CO., WILMINGTON, DEL.
DPC-CPH-	FRANCE. COMMISSARIAT A L'ENERGIE ATOMIQUE, PARIS
DPC.CPH/	FRANCE. COMMISSARIAT A L'ENERGIE ATOMIQUE. CENTRE D'ETUDES NUCLEAIRES, SACLAY
DPC-CPH-	UNITED KINGDOM ATOMIC ENERGY AUTHORITY. DEVELOPMENT AND ENGINEERING GROUP, RISLEY, LANCS, ENGLAND
DPC-IS-	FRANCE. COMMISSARIAT A L'ENERGIE ATOMIQUE. CENTRE D'ETUDES NUCLEAIRES, SACLAY
DPC.IS/	UNITED KINGDOM ATOMIC ENERGY AUTHORITY. ATOMIC ENERGY RESEARCH ESTABLISHMENT, HARWELL, BERKS, ENGLAND
DPC.P-	UNITED KINGDOM ATOMIC ENERGY AUTHORITY. RESEARCH GP. ATOMIC ENERGY RES. ESTAB., HARWELL, BERKS, ENGLAND
DPC-PCA-	FRANCE. COMMISSARIAT A L'ENERGIE ATOMIQUE. CENTRE D'ETUDES NUCLEAIRES, SACLAY
DPC/PCA/(YEAR)-	FRANCE. COMMISSARIAT A L'ENERGIE ATOMIQUE. CENTRE D'ETUDES NUCLEAIRES, SACLAY
DPCPWP/P.-	UNITED KINGDOM ATOMIC ENERGY AUTHORITY. REACTOR GP., WINFRITH, DORSET, ENGLAND
DPD-(YEAR)-(NUMBER)-(NO.)	DANA AREA OFFICE, AEC
DPD-	DU PONT DE NEMOURS (E.I.) + CO., WILMINGTON, DEL.
DPD-(YR)-	DU PONT DE NEMOURS (E.I.) + CO., WILMINGTON, DEL.
DPD-	DU PONT DE NEMOURS (E.I.) + CO. ATOMIC ENERGY DIV., TERRE HAUTE, IND.
DPD-	MINNESOTA MINING AND MFG. CO. DEFENSE PRODUCTS DEPT., ST. PAUL
DPD-	ORGANIZATION FOR EUROPEAN ECONOMIC COOPERATION, PARIS
DPDTM-	DU PONT DE NEMOURS (E.I.) + CO. ATOMIC ENERGY DIV., WILMINGTON, DEL.
DPE-	DU PONT DE NEMOURS (E.I.) + CO., WILMINGTON, DEL.
DPE-	DU PONT DE NEMOURS (E.I.) + CO. ENGINEERING DEPT., WILMINGTON, DEL.
DPEAER-	AMERICAN MACHINE AND FOUNDRY CO., GREENWICH, CONN.

DR-

DPED-	UNITED KINGDOM ATOMIC ENERGY AUTHORITY. RESEARCH GP. ATOMIC ENERGY RES. ESTAB., HARWELL, BERKS, ENGLAND
DPE/SPE/(YEAR/NUMBER)	FRANCE. COMMISSARIAT A L'ENERGIE ATOMIQUE. CENTRE D'ETUDES NUCLEAIRES. SECTION DE PHYSIQUE ET D'EXPERIMENTATION, SACLAY
DP-ESP-FR-	DU PONT DE NEMOURS (E.I.) + CO. EXPERIMENTAL STATION, WILMINGTON, DEL. (FINAL REPORT)
DPG-(LETTER(S))-	DUGWAY PROVING GROUND, TOOELE, UTAH
DPG-DD-	DUGWAY PROVING GROUND, TOOELE, UTAH
DPGR-	DUGWAY PROVING GROUND, TOOELE, UTAH
DPG SR-	DUGWAY PROVING GROUND, TOOELE, UTAH (SPECIAL REPORTS)
DPGSR-	DUGWAY PROVING GROUND, TOOELE, UTAH
DPG-T(YEAR)-	DUGWAY PROVING GROUND, TOOELE, UTAH
DPG-TEST PLAN-	DUGWAY PROVING GROUND, TOOELE, UTAH
DPGTR-	DUGWAY PROVING GROUND, TOOELE, UTAH (TRIAL RECORD)
DPG-TR-	DUGWAY PROVING GROUND, TOOELE, UTAH (TECH. REPORT)
DPG-TR-	DUGWAY PROVING GROUND, TOOELE, UTAH (TRIAL RECORD)
DPG-TRIAL RECORD-	DUGWAY PROVING GROUND, TOOELE, UTAH
DPG-TR-T(YEAR)-	DUGWAY PROVING GROUND, TOOELE, UTAH
DPH/DOC/(YEAR)-	FRANCE. COMMISSARIAT A L'ENERGIE ATOMIQUE. CENTRE D'ETUDES NUCLEAIRES, SACLAY
DPH-PFC/	FRANCE. COMMISSARIAT A L'ENERGIE ATOMIQUE. CENTRE D'ETUDES NUCLEAIRES, SACLAY
DPH-T/(NO.)-(NO.)	FRANCE. COMMISSARIAT A L'ENERGIE ATOMIQUE. CENTRE D'ETUDES NUCLEAIRES, SACLAY
DPH-T/DOC/	FRANCE. COMMISSARIAT A L'ENERGIE ATOMIQUE. CENTRE D'ETUDES NUCLEAIRES, SACLAY
D/PHYS/R-	CANADA. DIRECTORATE OF PHYSICAL RESEARCH, OTTAWA
D-PHYS-R(G)-HAZEN-	CANADA. DIRECTORATE OF PHYSICAL RESEARCH, OTTAWA (REPORTS FROM HAZEN CAMP, ELLESMERE ISLAND, N.W.T.)
D-PHYS-R(G)-MISC-	CANADA. DIRECTORATE OF PHYSICAL RESEARCH, OTTAWA
DPK-	DU PONT DE NEMOURS (E.I.) + CO., WILMINGTON, DEL.
DPKN-	DU PONT DE NEMOURS (E.I.) + CO. ATOMIC ENERGY DIV., WILMINGTON, DEL.
DPM-	ANALYTIC SERVICES, INC., BAILEYS CROSSROADS, VA.
DPM-	ANALYTIC SERVICES, INC., FALLS CHURCH, VA.
DP-MEMO-	UNITED KINGDOM ATOMIC ENERGY AUTHORITY. RESEARCH GP. ATOMIC ENERGY ESTAB., WINFRITH, DORSET, ENGLAND
DP-MS-(YEAR)-	DU PONT DE NEMOURS (E.I.) + CO. SAVANNAH RIVER LAB., AIKEN, S.C.
DP-MS-(YEAR)	DU PONT DE NEMOURS (E.I.) + CO. SAVANNAH RIVER PLANT, AIKEN, S.C.
DP(NASA)-	DU PONT DE NEMOURS (E.I.) + CO. SAVANNAH RIVER LAB., AIKEN, S.C.
DPNC-	DU PONT DE NEMOURS (E.I.) + CO., WILMINGTON, DEL.
DPNPD-(YEAR)-	ATOMIC ENERGY OF CANADA LTD., CHALK RIVER, ONT.
DPP/MJM/(NUMBER/YEAR)	NOTTINGHAM, ENGLAND. UNIV. DEPT. OF GEOGRAPHY (DATA PROCESSING PACKAGE)(AUTHOR'S INITIALS)
DPR-(YEAR)-	ANALYTIC SERVICES, INC., FALLS CHURCH, VA.
DPR/	LOCKHEED AIRCRAFT CORP. (ALL LOCATIONS) (DEVELOPMENT PLANNING REPORT)
DPR-(NUMBER)-(NUMBER)	MARTIN CO., BALTIMORE
DPR/BWS/-	GT. BRIT. ADMIRALTY. DEPT. OF PHYSICAL RESEARCH
DPR/BWS-(YR.)	GT. BRIT. ADMIRALTY. DEPT. OF PHYSICAL RESEARCH
DPRDD-	UNITED KINGDOM ATOMIC ENERGY AUTHORITY. RESEARCH GP. ATOMIC ENERGY ESTAB., WINFRITH, DORSET, ENGLAND
DP-REPORT-	ARGONNE NATIONAL LAB., ILL.
DP REPORT-	UNITED KINGDOM ATOMIC ENERGY AUTHORITY. RESEARCH GP. (SERIES ASSIGNED TO REPORTS ON DRAGON PROJECT OF ORGANIZATION FOR ECONOMIC COOPERATION AND DEV.)
D.P. REPORT-	UNITED KINGDOM ATOMIC ENERGY AUTHORITY. RESEARCH GP. ATOMIC ENERGY ESTAB., WINFRITH, DORSET, ENGLAND (SERIES ASSIGNED TO REPORTS ON DRAGON PROJECT OF ORGANIZATION FOR ECONOMIC COOPERATION AND DEV.)
DPR/INF/	UNITED KINGDOM ATOMIC ENERGY AUTHORITY, LONDON
DPR/INF/	UNITED KINGDOM ATOMIC ENERGY AUTHORITY. AUTHORITY HEALTH AND SAFETY BRANCH, RISLEY, LANCS, ENGLAND
DPR/JIRC-	GT. BRIT. JOINT INFRA-RED COMMITTEE
D + PS-	DEVELOPMENT + PROOF SERVICES, ABERDEEN PROVING GD., MD
DPS-	DEVELOPMENT + PROOF SERVICES, ABERDEEN PROVING GD., MD
DPS-	MATERIEL TEST DIRECTORATE, ABERDEEN PROVING GD., MD.
DPS-	STANFORD UNIV., CALIF. INST. IN ENG. ECONOMIC SYSTEMS
DPS-AD-	DEVELOPMENT + PROOF SERVICES, ABERDEEN PROVING GD., MD
DPS-FR-B-	DEVELOPMENT + PROOF SERVICES, ABERDEEN PROVING GD., MD
DPS-FR-P-	DEVELOPMENT + PROOF SERVICES, ABERDEEN PROVING GD., MD
DPSO-	PICATINNY ARSENAL. DATA PROCESSING SYSTEMS OFFICE, DOVER, N.J.
DPSO-IR-	PICATINNY ARSENAL. DATA PROCESSING SYSTEMS OFFICE, DOVER, N.J.
DPSO-TM-	PICATINNY ARSENAL. DATA PROCESSING SYSTEMS OFFICE, DOVER, N.J.
DPSP-	DU PONT DE NEMOURS (E.I.) + CO. SAVANNAH RIVER LAB., AIKEN, S.C.
DPSP-	DU PONT DE NEMOURS (E.I.) + CO. SAVANNAH RIVER LAB., AUGUSTA, GA.
DPSP-(YEAR)-(NO.)-(NO.)HW	DU PONT DE NEMOURS (E.I.) + CO. SAVANNAH RIVER LAB., AUGUSTA, GA. (HEAVY WATER)
DPSP-	DU PONT DE NEMOURS (E.I.) + CO. SAVANNAH RIVER PLANT, AIKEN, S.C.
DPSPU-	DU PONT DE NEMOURS (E.I.) + CO. SAVANNAH RIVER LAB., AIKEN, S.C.
DPSPU-	DU PONT DE NEMOURS (E.I.) + CO. SAVANNAH RIVER LAB., AUGUSTA, GA.
DPSPU-(YR.)-(NO.)-	DU PONT DE NEMOURS (E.I.) + CO. SAVANNAH RIVER PLANT, AIKEN, S.C.
DPSPWD-(YEAR)-	DU PONT DE NEMOURS (E.I.) + CO. SAVANNAH RIVER LAB., AUGUSTA, GA.
DPS-SA-(NUMBERS)-	AIR FORCE
DPST-	DU PONT DE NEMOURS (E.I.) + CO. SAVANNAH RIVER LAB., AIKEN, S.C.
DPST-(YEAR)-	DU PONT DE NEMOURS (E.I.) + CO. SAVANNAH RIVER LAB., AIKEN, S.C.
DPST-	DU PONT DE NEMOURS (E.I.) + CO. SAVANNAH RIVER LAB., AUGUSTA, GA.
DPSTM-	DU PONT DE NEMOURS (E.I.) + CO. SAVANNAH RIVER LAB., AIKEN, S.C.
DPSTM-	DU PONT DE NEMOURS (E.I.) + CO. SAVANNAH RIVER LAB., AUGUSTA, GA.
DPSTMWD-	DU PONT DE NEMOURS (E.I.) + CO. SAVANNAH RIVER LAB., AUGUSTA, GA.
DPSTOM-	DU PONT DE NEMOURS (E.I.) + CO. SAVANNAH RIVER LAB., AIKEN, S.C.
DPT-(YEAR)-(MONTH)	CANADA. HYDRO-ELECTRIC POWER COMMISSION OF ONTARIO, ROLPHTON (DOUGLAS POINT GENERATING STA.MONTHLY RPT)
DPTR-	DU PONT DE NEMOURS (E.I.) + CO., WILMINGTON, DEL.
DPW-	DU PONT DE NEMOURS (E.I.) + CO., WILMINGTON, DEL.
DPWD-	DU PONT DE NEMOURS (E.I.) + CO. EXPLOSIVES DEPT., WILMINGTON, DEL.
DPWD-	DU PONT DE NEMOURS (E.I.) + CO. SAVANNAH RIVER LAB., AIKEN, S.C.
DP(WD)-	DU PONT DE NEMOURS (E.I.) + CO. SAVANNAH RIVER LAB., AUGUSTA, GA.
DPW-DUPONT-	DU PONT DE NEMOURS (E.I.) + CO. ATOMIC ENERGY DIV., WILMINGTON, DEL.
DPWR-	HARVEY ALUMINUM, INC., TORRANCE, CALIF.
DPWR-	HORIZONS, INC., CLEVELAND
DPWR-	MICHIGAN. UNIV., ANN ARBOR. SCHOOL OF PUBLIC HEALTH
DPWZ-	DU PONT DE NEMOURS (E.I.) + CO., WILMINGTON, DEL.
DPXN-	DU PONT DE NEMOURS (E.I.) + CO. ATOMIC ENERGY DIV., WILMINGTON, DEL.
DQR/(YEAR)-	JOHNS HOPKINS UNIV., SILVER SPRING, MD. APPLIED PHYSICS LAB.
DR-	(DATA REPORT) (REF. 2)

DR-	(DESIGN REQUIREMENT) (REF. 2)
DR-	(DETAILED REPORT) (REF. 2)
DR-	(DEVELOPMENT REPORT) (REF. 2)
DR-	(DOCUMENT REPORT) (REF. 2)
DR-	AMERICAN BOSCH ARMA CORP. ARMA DIV., GARDEN CITY, N.Y.
DR-	ATMOSPHERIC SCIENCES LAB., WHITE SANDS MISSILE RANGE, NEW MEXICO
DR-	ATOMIC ENERGY OF CANADA LTD. CHALK RIVER PROJ., ONT.
DR-	BUREAU OF AERONAUTICS (NAVY) (DESIGN RESEARCH)
DR-	BUREAU OF AERONAUTICS (NAVY). RESEARCH DIV.
DR-	BUREAU OF NAVAL WEAPONS
DR-	CALIFORNIA. UNIV., BERKELEY
DR-	CANADA. DEFENCE RESEARCH BOARD, OTTAWA
DR-	CANADA. DEPT. OF NATIONAL DEFENCE, OTTAWA
D-R-	CARBIDE AND CARBON CHEMICALS CORP. SUBSTITUTE ALLOY MATERIALS LABS., N.Y.C.
DR-	COLUMBIA UNIV., N.Y.C.
D-R-	COLUMBIA UNIV., N.Y.C. PUPIN PHYSICS LABS.
DR-	COLUMBIA UNIV., NYC. SUBSTITUTE ALLOY MATERIALS LABS.
D-R-	DELCO-REMY DIV., GENERAL MOTORS CORP., MUNCIE, IND.
DR-	DENVER. UNIV. DENVER RESEARCH INST.
DR-	FAIRCHILD ENGINE AND AIRPLANE CORP. NEPA DIV., OAK RIDGE, TENN.
DR-	MC MASTER UNIV., HAMILTON, ONT.
DR-	NATIONAL RESEARCH COUNCIL OF CANADA. ATOMIC ENERGY PROJECT, CHALK RIVER, ONT.
DR-	OREGON STATE UNIV., CORVALLIS
DR-	PHILCO-FORD CORP. SPACE AND RE-ENTRY SYSTEMS DIV., NEWPORT BEACH, CALIF.
DR-	SASKATCHEWAN. UNIV., SASKATOON
DR-	WHITE SANDS MISSILE RANGE, N. MEX.
DRA-	DIRECTORATE OF RESEARCH ANALYSES, HOLLOMAN AFB, N.M.
DRA-(YEAR)-	DIRECTORATE OF RESEARCH ANALYSES, HOLLOMAN AFB, N.M.
DRAE-	CANADA. DEFENCE RESEARCH ANALYSIS ESTAB., OTTAWA
DR AND CWG-	INTER-RANGE INSTRUMENTATION GROUP
DRB-(PROJECT NUMBER)-	CANADA. DEFENCE RESEARCH BOARD, OTTAWA
DRB DR-	CANADA. DEFENCE RESEARCH BOARD, OTTAWA
DRB-REPRINT-	CANADA. DEFENCE RES. TELECOMMUNICATIONS ESTABLISHMENT. RADIO PHYSICS LAB., OTTAWA
DRBS-	CANADA. DEFENCE RESEARCH BOARD, OTTAWA
DRB-S-TP-	CANADA. SUFFIELD EXPERIMENTAL STATION, RALSTON, ALBERTA
DRB-T-(NUMBER)-R	CANADA. DEFENCE RESEARCH BOARD, OTTAWA
DRB-T-	CANADA. DEFENCE RESEARCH BOARD. SCIENTIFIC INFORMATION SERVICE (TRANSLATION)
DRBT (NUMBER)(LETTER)	CANADA. DEFENCE RESEARCH BOARD. SCIENTIFIC INFORMATION SERVICE (TRANSLATION)
DRB-TELS-	CANADA. DEFENCE RESEARCH BOARD, OTTAWA
DRB-TELS-TN-	CANADA. DEFENCE RESEARCH BOARD, OTTAWA
DRC-	DEFENSE RESEARCH CORP., SANTA BARBARA, CALIF.
DRC-(YEAR)-	DEFENSE RESEARCH CORP., SANTA BARBARA, CALIF.
DRC-	OHIO STATE UNIV., COLUMBUS. DISASTER RESEARCH CENTER
DRC-CSR-	DEFENSE RESEARCH CORP., SANTA BARBARA, CALIF.
DRC-CSR-SD-	DEFENSE RESEARCH CORP., SANTA BARBARA, CALIF.
DRC-E-	DYNAMICS RESEARCH CORP., STONEHAM, MASS.
DRC-IMR-	DEFENSE RESEARCH CORP., SANTA BARBARA, CALIF.
DRCL-	CANADA. DEFENCE RESEARCH CHEMICAL LABS., OTTAWA
DRC-MONOGRAPH-	OHIO STATE UNIV., COLUMBUS. DISASTER RESEARCH CENTER
DRCOG-(YEAR)-	DENVER REGIONAL COUNCIL OF GOVERNMENTS
DRC-R-	DYNAMICS RESEARCH CORP., STONEHAM, MASS.
DRC-R-(NUMBER)C	DYNAMICS RESEARCH CORP., STONEHAM, MASS.
DRC-SER-	OHIO STATE UNIV., COLUMBUS. DISASTER RESEARCH CENTER
DRC-TM-	DEFENSE RESEARCH CORP., SANTA BARBARA, CALIF.
DRC-TR-	OHIO STATE UNIV., COLUMBUS. DISASTER RESEARCH CENTER
DR/CWG-	WHITE SANDS MISSILE RANGE, N. MEX.
DRD-	DUNLAP AND ASSOCIATES, INC., DARIEN, CONN.
DRD-	FIRESTONE TIRE AND RUBBER CO. DEFENSE RESEARCH DIV., AKRON, OHIO
DRD-N-	WHITE SANDS MISSILE RANGE, N. MEX.
DRDN-	WHITE SANDS MISSILE RANGE, N. MEX.
DRE-	ISRAEL. DEFENSE RESEARCH ESTABLISHMENT
DREA-	CANADA. DEFENCE RES. ESTAB. ATLANTIC, DARTMOUTH, NS
DREA-(YEAR)-	CANADA. DEFENCE RES. ESTAB. ATLANTIC, DARTMOUTH, NS
DREA-TN-MATH/(YR)/	CANADA. DEFENCE RES. ESTAB. ATLANTIC, DARTMOUTH, NS
DREO-	CANADA. DEFENCE RESEARCH ESTABLISHMENT, OTTAWA
DREO-TN-(YEAR)-	CANADA. DEFENCE RESEARCH ESTABLISHMENT, OTTAWA
DREP-	CANADA. DEFENCE RES. ESTAB. PACIFIC, VICTORIA, B.C.
DREP-TM-	CANADA. DEFENCE RES. ESTAB. PACIFIC, VICTORIA, B.C.
DRES-	CANADA. DEFENCE RES. ESTAB. SUFFIELD, RALSTON, ALTA.
DRES-MEMO-	CANADA. DEFENCE RES. ESTAB. SUFFIELD, RALSTON, ALTA.
DRES-TN-	CANADA. DEFENCE RES. ESTAB. SUFFIELD, RALSTON, ALTA.
DRES-TP-	CANADA. DEFENCE RES. ESTAB. SUFFIELD, RALSTON, ALTA.
DRET-	CANADA. DEFENSE RESEARCH BOARD, OTTAWA
DRET-	CANADA. DEFENCE RES. ESTAB. TORONTO, DOWNSVIEW, ONT.
DRET-RP-	CANADA. DEFENCE RES. ESTAB. TORONTO, DOWNSVIEW, ONT.
DRET-TN-	CANADA. DEFENCE RESEARCH MEDICAL LABS., TORONTO
DREV-	CANADA. DEFENCE RES. ESTAB. VALCARTIER, QUEBEC
DREV-R-	CANADA. DEFENCE RES. ESTAB. VALCARTIER, QUEBEC
DREV-REPRINT-	CANADA. DEFENCE RES. ESTAB. VALCARTIER, QUEBEC
DREV-TL-	CANADA. DEFENCE RES. ESTAB. VALCARTIER, QUEBEC
DREV-TN-	CANADA. DEFENCE RES. ESTAB. VALCARTIER, QUEBEC
DREXELU-TR-	DREXEL UNIV., PHILADELPHIA
DRI-	DENVER. UNIV. DENVER RESEARCH INST.
DRI-(NUMBER)-	DENVER. UNIV. DENVER RESEARCH INST.
DRIBBLE-	GEOLOGICAL SURVEY, DENVER
DRI-LAR-	DENVER. UNIV. DENVER RESEARCH INST.
DRI-MP-	DENVER. UNIV. DENVER RESEARCH INST.
DRI-QR-	DENVER. UNIV. DENVER RESEARCH INST.
DRKL-	CANADA. DEFENCE RESEARCH KINGSTON LAB.
DR-L-	COLUMBIA UNIV., NYC. SUBSTITUTE ALLOY MATERIALS LABS.
D-RL-	LOS ALAMOS SCIENTIFIC LAB., N. MEX.
DRL-	TEXAS. UNIV., AUSTIN. DEFENSE RESEARCH LAB.
DRL-A-	TEXAS. UNIV., AUSTIN. DEFENSE RESEARCH LAB. (ACOUSTICS REPORT)
DRL-TM-	TEXAS. UNIV., AUSTIN
DRL-TM-	TEXAS. UNIV., AUSTIN. DEFENSE RESEARCH LAB.
DRL-TR-	TEXAS. UNIV., AUSTIN. DEFENSE RESEARCH LAB.
DRL/UT-	TEXAS. UNIV., AUSTIN. DEFENSE RESEARCH LAB.
DRL/UT T-	TEXAS. UNIV., AUSTIN. DEFENSE RESEARCH LAB. (TRANS.)
DR-M-	COLUMBIA UNIV., NYC. SUBSTITUTE ALLOY MATERIALS LABS.
DRME-	BRUSSELS. UNIVERSITE
DRML-	CANADA. DEFENCE RESEARCH MEDICAL LABS., TORONTO
DRMLRP-	CANADA. DEFENCE RESEARCH MEDICAL LABS., TORONTO
DRML-TM-	CANADA. DEFENCE RESEARCH BOARD, OTTAWA
DRML-TN-	CANADA. DEFENCE RESEARCH MEDICAL LABS., TORONTO

DRNL-	CANADA. DEFENCE RES. NORTHERN LAB.,FT. CHURCHILL,MAN.
DRNL-TP-	CANADA. DEFENCE RES. NORTHERN LAB.,FT. CHURCHILL,MAN.
DRP/EMTR/CAD.R.(NO.)	FRANCE. COMMISSARIAT A L'ENERGIE ATOMIQUE. CENTRE D'ETUDES NUCLEAIRES, CADARACHE
DRP/EMTR/FAR/R.(NO.)	FRANCE. COMMISSARIAT A L'ENERGIE ATOMIQUE. CENTRE D'ETUDES NUCLEAIRES, SACLAY
DRP/ML/CA ND-	FRANCE. COMMISSARIAT A L'ENERGIE ATOMIQUE. CENTRE D'ETUDES NUCLEAIRES, FONTENAY-AUX-ROSES
DRP/ML/FAR R-	FRANCE. COMMISSARIAT A L'ENERGIE ATOMIQUE. CENTRE D'ETUDES NUCLEAIRES, CADARACHE
DRP/ML/FAR R/(NO.)-	FRANCE. COMMISSARIAT A L'ENERGIE ATOMIQUE. CENTRE D'ETUDES NUCLEAIRES, FONTENAY-AUX-ROSES
DRP/ML/FAR R-	FRANCE. COMMISSARIAT A L'ENERGIE ATOMIQUE. CENTRE D'ETUDES NUCLEAIRES, SACLAY
DRP/MNF/R/	FRANCE. COMMISSARIAT A L'ENERGIE ATOMIQUE, PARIS
DRP/PNR/(LETTERS)/	FRANCE. COMMISSARIAT A L'ENERGIE ATOMIQUE. CENTRE D'ETUDES NUCLEAIRES, CADARACHE
DRPP SRE-	DIRECTOR RESEARCH PROGRAM PLANNING. SCI. RES. ESTAB.
DRPP SRE (LETTER)-	DIRECTOR RESEARCH PROGRAM PLANNING. SCI. RES. ESTAB.
DRP/SCR-	FRANCE. COMMISSARIAT A L'ENERGIE ATOMIQUE. CENTRE D'ETUDES NUCLEAIRES, CADARACHE
DRP/SEMTR/	FRANCE. COMMISSARIAT A L'ENERGIE ATOMIQUE. CENTRE D'ETUDES NUCLEAIRES, CADARACHE
DRP/SEMTR/CAD-	FRANCE. COMMISSARIAT A L'ENERGIE ATOMIQUE. CENTRE D'ETUDES NUCLEAIRES, CADARACHE
DRP/SETR-	FRANCE. COMMISSARIAT A L'ENERGIE ATOMIQUE. CENTRE D'ETUDES NUCLEAIRES, CADARACHE
DRP/SMNF-	FRANCE. COMMISSARIAT A L'ENERGIE ATOMIQUE. CENTRE D'ETUDES NUCLEAIRES, SACLAY
DRS/(NUMBER-NUMBER)TS	COMPAGNIE FRANCAISE THOMSON-HOUSTON, BAGNEUX
DR-SAR-G-	FRANCE. COMMISSARIAT A L'ENERGIE ATOMIQUE. CENTRE D'ETUDES NUCLEAIRES, GRENOBLE
DRSWP/R-	UNITED KINGDOM ATOMIC ENERGY AUTHORITY. RESEARCH GP. ATOMIC ENERGY RES. ESTAB., HARWELL, BERKS, ENGLAND
DRTE-	CANADA. DEFENCE RES. TELECOMMUNICATIONS ESTAB.,OTTAWA
DRTE-(LETTER)-	CANADA. DEFENCE RES. TELECOMMUNICATIONS ESTAB.,OTTAWA
DRTE/EL-	CANADA. DEFENCE RES. TELECOMMUNICATIONS ESTAB.,OTTAWA
DRTE-GEOPHYSICS-	CANADA. DEFENCE RES. TELECOMMUNICATIONS ESTAB.,OTTAWA
DRTE-HAZEN-	CANADA. DEFENCE RES. TELECOMMUNICATIONS ESTAB.,OTTAWA
DRTE-PUBL.-	CANADA. DEFENCE RES. TELECOMMUNICATIONS ESTAB.,OTTAWA
DRUIM-	DEMOLITION RESEARCH UNIT (NAVY) (INTERNAL MEMORANDUM)
DS-	(DESIGN STUDY) (REF. 2)
DS-	AMERICAN BOSCH ARMA CORP. ARMA DIV., GARDEN CITY,N.Y.
DS-	ARMY BALLISTIC MISSILE AGENCY. STRUCTURES AND MECHANICS LAB., REDSTONE ARSENAL, ALA.
DS-(YEAR)-	BOLT, BERANEK, AND NEWMAN, INC., VAN NUYS, CALIF.
DS-	ELECTRONIC PROPERTIES INFO. CTR., CULVER CITY, CALIF
DS-	GARRETT CORP. AIRESEARCH MFG. DIV., LOS ANGELES
DS-	GENERAL ELECTRIC CO. AIRCRAFT NUCLEAR PROPULSION DEPT., CINCINNATI
DS-	HUGHES AIRCRAFT CO., CULVER CITY, CALIF. (DATA SHEET)
DS-	HUGHES AIRCRAFT CO., CULVER CITY, CALIF.(DETAIL SPEC)
DS-(YEAR)-	LIGHTNING AND TRANSIENTS RESEARCH INST., MINNEAPOLIS
DS-	LITTON SYSTEMS, INC., WALTHAM, MASS.
DS-	MARTIN-MARIETTA CORP. DENVER DIV.
DS-(YEAR)-	NAVAL AIR PROPULSION TEST CENTER, PHILADELPHIA
DS-(YEAR)-	UNITED CONTROL CORP., SOUTH EL MONTE, CALIF.
DSA-FSC-	DEFENSE LOGISTICS SERVICES CENTER, BATTLE CREEK,MICH. (FEDERAL SUPPLY CODE)
DSAH-	DEFENSE LOGISTICS SERVICES CENTER, BATTLE CREEK,MICH. (DOD ACTIVITY ADDRESS DIRECTORY)
DSBTF-	DEFENSE SCIENCE BOARD TASK FORCE, WASH., D.C.
DSC-	ARMED SERVICES TECHNICAL INFORMATION AGENCY. DOCUMENT SERVICE CENTER, DAYTON, OHIO
DSC-SN-	DYNAMIC SCIENCE CORP., SOUTH PASADENA, CALIF.
DSD-	GENERAL ELECTRIC CO. TECHNICAL MILITARY PLANNING OPERATION, SANTA BARBARA, CALIF.
DSD-R-	SYRACUSE UNIV., N.Y. RESEARCH CORP.
DSD-R-	SYRACUSE UNIV., N.Y. RES. CORP. DEFENSE SYSTEMS DIV.
DSD-TM-	SYRACUSE UNIV., N.Y. RESEARCH CORP.
DSD-TM-	SYRACUSE UNIV., N.Y. RES. CORP. DEFENSE SYSTEMS DIV.
DSD-TR-(NUMBER)-(YEAR)	ARMY BALLISTIC MISSILE AGENCY, REDSTONE ARSENAL, ALA.
DS/EE/(YEAR)-	AIR FORCE INST. OF TECH., WRIGHT-PATTERSON AFB, OHIO. SCHOOL OF ENGINEERING
DSF-TN-	ARMY BALLISTIC MISSILE AGENCY, REDSTONE ARSENAL, ALA.
DSF-TR-(NUMBER)-(YEAR)	ARMY BALLISTIC MISSILE AGENCY, REDSTONE ARSENAL, ALA.
DSI-	CANADA. DIRECTORATE OF SCIENTIFIC INTELLIGENCE
DSI-(NUMBER)-F	DECISION SCIENCE, INC., SAN DIEGO, CALIF.
DSIF-(NO.)/OPS/(NO.)/	UNION OF SOUTH AFRICA. NATIONAL INST. FOR TELECOMMUNICATIONS RES. DEEP SPACE INSTRUMENTATION LAB., HARTESBEESTHOEK
DSI/JTIC-	GT. BRIT. DIRECTORATE OF SCIENTIFIC INTELLIGENCE. JOINT TECHNICAL INTELLIGENCE COMMITTEE
DSIR-	GT. BRIT. DEPT. OF SCIENTIFIC AND INDUSTRIAL RES.
DSIR-	NEW ZEALAND. DEPT. OF SCI.+ IND. RES. RADIO RES. OFF.
DSIR BRB-	GT. BRIT. DEPT. OF SCIENTIFIC AND INDUSTRIAL RESEARCH. (BUILDING RESEARCH BULLETINS)
DSIR BRD-	GT. BRIT. DEPT. OF SCIENTIFIC AND INDUSTRIAL RESEARCH. (BUILDING RESEARCH DIGEST)
DSIR BR TP-	GT. BRIT. DEPT. OF SCIENTIFIC AND INDUSTRIAL RESEARCH. (BUILDING RESEARCH TECHNICAL PAPERS)
DSIR CR SR-	GT. BRIT. DEPT. OF SCIENTIFIC AND INDUSTRIAL RESEARCH (CHEMISTRY RESEARCH. SPECIAL REPORT)
DSIR FI L-	GT. BRIT. DEPT. OF SCIENTIFIC AND INDUSTRIAL RESEARCH (FOOD INVESTIGATION. LEAFLET)
DSIR FIRE TP-	GT. BRIT. DEPT. OF SCIENTIFIC AND INDUSTRIAL RESEARCH (FIRE RESEARCH. TECHNICAL PAPER)
DSIR FI SR-	GT. BRIT. DEPT. OF SCIENTIFIC AND INDUSTRIAL RESEARCH (FOOD INVESTIGATION. SPECIAL REPORT)
DSIR FI TP-	GT. BRIT. DEPT. OF SCIENTIFIC AND INDUSTRIAL RESEARCH (FOOD INVESTIGATION. TECHNICAL PAPER)
DSIR FPR-	GT. BRIT. DEPT. OF SCIENTIFIC AND INDUSTRIAL RESEARCH. FOREST PRODUCTS RESEARCH LAB. (BULLS.)
DSIR FPR L-	GT. BRIT. DEPT. OF SCIENTIFIC AND INDUSTRIAL RESEARCH. FOREST PRODUCTS RES. LAB. (LEAFLETS)
DSIR FPRR-	GT. BRIT. DEPT. OF SCIENTIFIC AND INDUSTRIAL RESEARCH (FOREST PRODUCTS RESEARCH RECORDS)
DSIR FPR SP-	GT. BRIT. DEPT. OF SCIENTIFIC AND INDUSTRIAL RESEARCH. FOREST PRODUCTS RES. LAB. (SPEC. RPTS.)
DSIR FR SP-	GT. BRIT. DEPT. OF SCIENTIFIC AND INDUSTRIAL RESEARCH (FUEL RESEARCH. SURVEY PAPERS)
DSIR FR TP-	GT. BRIT. DEPT. OF SCIENTIFIC AND INDUSTRIAL RESEARCH (FUEL RESEARCH. TECHNICAL PAPERS)
DSIR L-	GT. BRIT. DEPT. OF SCI. + IND. RES. (LEAFLET)
DSIRLRG-	GT. BRIT. DEPT. OF SCIENTIFIC AND INDUSTRIAL RESEARCH. DIRECTORATE OF TUBE ALLOYS
DSIR NB B-	GT. BRIT. DEPT. OF SCIENTIFIC AND INDUSTRIAL RESEARCH (NATIONAL BUILDING STUDIES, BULLETIN)
DSIR NB SR-	GT. BRIT. DEPT. OF SCIENTIFIC AND INDUSTRIAL RESEARCH (NATIONAL BLDG. STUDIES, SPEC. RPT.)
DSIR NB TP-	GT. BRIT. DEPT. OF SCIENTIFIC AND INDUSTRIAL RESEARCH (NATIONAL BLDG. STUDIES, TECH. PAPERS)
DSIR PWBS-	GT. BRIT. DEPT. OF SCIENTIFIC AND INDUSTRIAL RESEARCH (POST WAR BUILDING STUDIES)
DSIR RAD R SR-	GT. BRIT. DEPT. OF SCIENTIFIC AND INDUSTRIAL RESEARCH (RADIO RESEARCH, SPECIAL REPORT)
DSIR RR-	GT. BRIT. DEPT. OF SCIENTIFIC AND INDUSTRIAL RESEARCH (ROAD RESEARCH BULLETINS)
DSIR RRL RN-	GT. BRIT. ROAD RESEARCH LAB., HARMONDSWORTH, MIDDX., ENGLAND (ROAD NOTE)
DSIR RRL TP-	GT. BRIT. ROAD RESEARCH LAB., HARMONDSWORTH, MIDDX., ENGLAND (TECHNICAL PAPER)

DSIR RR SR-
- GT. BRIT. DEPT. OF SCIENTIFIC AND INDUSTRIAL RESEARCH (ROAD RESEARCH SPECIAL REPORTS)

DSIR RR TP-
- GT. BRIT. DEPT. OF SCIENTIFIC AND INDUSTRIAL RESEARCH (ROAD RESEARCH TECHNICAL PAPER)

DSIR SR-
- GT. BRIT. DEPT. OF SCIENTIFIC AND INDUSTRIAL RESEARCH (SPONSORED RESEARCH REPORTS)

DSIR-TN-
- NEW ZEALAND. DEPT. OF SCIENTIFIC AND INDUSTRIAL RESEARCH. PHYSICS AND ENGINEERING LAB., WELLINGTON

DSIR WPR TP-
- GT. BRIT. DEPT. OF SCIENTIFIC AND INDUSTRIAL RESEARCH (WATER POLLUTION RESEARCH TECH. RPTS.)

DSIS-
- CANADA. DIRECTORATE OF SCI. INFO. SERVICES, OTTAWA

DSIS-B-
- CANADA. DEFENCE SCIENTIFIC INFORMATION SERVICE, OTTAWA

DSIS-T-(NUMBER)-R
- CANADA. DIRECTORATE OF SCI. INFO. SERVICES, OTTAWA

DSI-TRANS-
- CANADA. DIRECTORATE OF SCIENTIFIC INTELLIGENCE

DSL-
- AUSTRALIA. DEFENCE STANDARDS LABS., MARIBYRNONG

DSL-
- DETROIT SIGNAL LABORATORY

DSL AM-
- DETROIT SIGNAL LABORATORY (ASSIGNMENT MEMORANDA)

DSL AM (LETTER(S))-
- DETROIT SIGNAL LABORATORY (ASSIGNMENT MEMORANDA)

DSL EM-
- DETROIT SIGNAL LABORATORY (ENGINEERING MEMORANDA)

DSL EM (LTR.-NO.-LTRS)-
- DETROIT SIGNAL LABORATORY (ENGINEERING MEMORANDA)

DSL-R-
- AUSTRALIA. DEFENCE STANDARDS LABS., MARIBYRNONG

DSL-R-
- SYRACUSE UNIV., N.Y. RES. CORP. DEFENSE SYSTEMS LAB.

DSL-R-
- SYRACUSE UNIV., N.Y. RESEARCH INST. DEFENSE SYSTEMS LAB.

DSL-TM-(NUMBER)-(YEAR)
- ARMY BALLISTIC MISSILE AGENCY, REDSTONE ARSENAL, ALA.

DSL-TM-
- AUSTRALIA. DEFENCE STANDARDS LABS., MARIBYRNONG

DSL-TN-
- AUSTRALIA. DEFENCE STANDARDS LABS., MARIBYRNONG

DSL-TR-
- AUSTRALIA. DEFENCE STANDARDS LABS., MARIBYRNONG

DSM-
- ARMY BALLISTIC MISSILE AGENCY. STRUCTURES AND MECHANICS LAB., REDSTONE ARSENAL, ALA.

DSM-(NUMBER)-(NUMBER)
- DILWORTH, SECORD, MEAGHER AND ASSOCIATES LTD., TORONTO

DS/MC/(YEAR)-
- AIR FORCE INST. OF TECH., WRIGHT-PATTERSON AFB, OHIO. SCHOOL OF ENGINEERING

DSN-TN-(NUMBER)-(YEAR)
- ARMY BALLISTIC MISSILE AGENCY, REDSTONE ARSENAL, ALA.

DSN-TR-(NUMBER)-(YEAR)
- ARMY BALLISTIC MISSILE AGENCY, REDSTONE ARSENAL, ALA.

DSPEA-
- AIR FORCE PACKAGING EVALUATION AGENCY, WRIGHT-PATTERSON AFB, OHIO

DSPEB-
- AIR FORCE PACKAGING EVALUATION AGENCY, WRIGHT-PATTERSON AFB, OHIO

D SPECIAL-
- TENNESSEE EASTMAN CORP., OAK RIDGE, TENN.

DSP-TM-(NUMBER)-(YEAR)
- ARMY BALLISTIC MISSILE AGENCY, REDSTONE ARSENAL, ALA.

DSP-TN-(NUMBER)-(YEAR)
- ARMY BALLISTIC MISSILE AGENCY, REDSTONE ARSENAL, ALA.

DSP-TR-(NUMBER)-(YEAR)
- ARMY BALLISTIC MISSILE AGENCY, REDSTONE ARSENAL, ALA.

DSR-
- MASSACHUSETTS INST. OF TECH., CAMBRIDGE. DIVISION OF SPONSORED RESEARCH

DSR-(NUMBER-NUMBER)
- MASSACHUSETTS INST. OF TECH., CAMBRIDGE. DIVISION OF SPONSORED RESEARCH

DSRC-
- DAVID SARNOFF RESEARCH CENTER, PRINCETON, N.J.

DSRC-QPR-
- DAVID SARNOFF RESEARCH CENTER, PRINCETON, N.J.

DSRC-SAR-
- DAVID SARNOFF RESEARCH CENTER, PRINCETON, N.J.

DSRC-TR-
- DAVID SARNOFF RESEARCH CENTER, PRINCETON, N.J.

DSR-S-
- MARTIN CO., BALTIMORE

DSR SRE (LETTER)-
- DIRECTOR OF SCIENTIFIC RESEARCH. SCI. RES. ESTAB.

DSSC-S-
- WEST VIRGINIA. UNIV., MORGANTOWN. MEDICAL CENTER

DSSM-
- OKLAHOMA. STATE DEPT. OF INSTITUTIONS, SOCIAL AND REHABILITATION SERVICES, OKLAHOMA CITY

DSTI-TRANS-
- GT. BRIT. MINISTRY OF DEFENCE, LONDON

DSU/P-
- UNITED KINGDOM ATOMIC ENERGY AUTHORITY. RESEARCH GP. ATOMIC ENERGY RES. ESTAB., HARWELL, BERKS, ENGLAND

DT-
- (DROP TEST REPORT) (REF. 2)

D-T-
- CALIFORNIA RESEARCH AND DEVELOPMENT CO., LIVERMORE

DT-
- NAVAL MISSILE CENTER. DIRECTOR OF TESTS.

DTA/(NUMBER)/
- SOCIETE POUR L'ETUDE ET LA REALISATION D'ENGINS BALISTIQUES, COURBEVOIE, FRANCE

DTB-
- NAVY (DESTROYER TACTICAL BULLETIN)

DTB REPT-
- CORPS OF ENGINEERS (ARMY). ENGINEER BOARD. DESERT TEST BRANCH

DTC-
- BRAZIL. COMISSAO NACIONAL DE ENERGIA NUCLEAR

DTC-(YEAR)-
- DESERET TEST CENTER, FORT DOUGLAS, UTAH

DTC-
- DESERET TEST CENTER, FORT DOUGLAS, UTAH

DTC-(6 DIGITS)R
- DESERET TEST CENTER, FORT DOUGLAS, UTAH

DTC-(NUMBER-NUMBER)
- DESERET TEST CENTER, FORT DOUGLAS, UTAH

DTC-B-
- DESERET TEST CENTER, FORT DOUGLAS, UTAH

DTC-C-
- DESERET TEST CENTER, FORT DOUGLAS, UTAH

DTC-DR-E(NO.)-VFC
- DESERET TEST CENTER, FORT DOUGLAS, UTAH

DTC-EC-
- DESERET TEST CENTER, FORT DOUGLAS, UTAH

DTC-FIRING RECORD-(YR.)-
- DESERET TEST CENTER, FORT DOUGLAS, UTAH

DTC-FR-E-
- DESERET TEST CENTER, FORT DOUGLAS, UTAH

DTC-IR-E-
- DESERET TEST CENTER, FORT DOUGLAS, UTAH

DTC-LR-E-
- DESERET TEST CENTER, FORT DOUGLAS, UTAH

DTC-R-E-
- DESERET TEST CENTER, FORT DOUGLAS, UTAH

DTC-SPECIAL STUDY-R-
- DESERET TEST CENTER, FORT DOUGLAS, UTAH

DTC-SR-
- DESERET TEST CENTER, FORT DOUGLAS, UTAH

DTC-TB-
- DESERET TEST CENTER, FORT DOUGLAS, UTAH

DTC-TC-
- DESERET TEST CENTER, FORT DOUGLAS, UTAH

DTC-TEST-(YEAR)-
- DESERET TEST CENTER, FORT DOUGLAS, UTAH

DTC-TJ-
- DESERET TEST CENTER, FORT DOUGLAS, UTAH

DTD-
- GARRETT CORP. AIRESEARCH MFG. DIV., LOS ANGELES

DTD-
- GT. BRIT. MINISTRY OF SUPPLY. DEPT. OF TANK DESIGN

DTH-
- DENMARK. TEKNISKE HOJSKOLE, LYNGBY

DTI-
- DIVISION OF TECHNICAL INFORMATION, AEC

DTIE-
- DIVISION OF TECHNICAL INFORMATION EXTENSION, AEC

DTIE-AL-
- DIVISION OF TECH. INFO. EXTENSION, AEC (ACCESSION LIST)

DTI-TR-(NUMBER)-(YEAR)
- ARMY BALLISTIC MISSILE AGENCY, REDSTONE ARSENAL, ALA.

DTL-
- DETROIT TESTING LAB.

DTL-
- FMC CORP. DEFENSE TECHNOLOGY LABS., SANTA CLARA, CALIF.

DTL-(NUMBER)-
- FMC CORP. DEFENSE TECHNOLOGY LABS., SANTA CLARA, CALIF.

DTM-
- CARNEGIE INSTITUTION OF WASHINGTON, D.C. DEPT. OF TERRESTRIAL MAGNETISM

DTMB-
- DAVID TAYLOR MODEL BASIN, CARDEROCK, MD.

DTMB-AERO-
- DAVID TAYLOR MODEL BASIN. AERODYNAMICS LAB., WASH., DC

DTMB-AL-
- DAVID TAYLOR MODEL BASIN, CARDEROCK, MD.

DTMB-AL-C-
- DAVID TAYLOR MODEL BASIN, CARDEROCK, MD.

DTMB-AML-
- DAVID TAYLOR MODEL BASIN. APPLIED MATHEMATICS LAB., CARDEROCK, MD.

DTMB-C-
- DAVID TAYLOR MODEL BASIN, CARDEROCK, MD.

DTMB-R-
- DAVID TAYLOR MODEL BASIN, CARDEROCK, MD.

DTMB-S-
- DAVID TAYLOR MODEL BASIN, CARDEROCK, MD.

DTMB-T-
- DAVID TAYLOR MODEL BASIN, CARDEROCK, MD.

DTMB-TRANS.-
- DAVID TAYLOR MODEL BASIN, CARDEROCK, MD. (TRANSLATION)

DT-R-
- ARMY BALLISTIC MISSILE AGENCY, REDSTONE ARSENAL, ALA.

DTR-(YEAR)-
- DATA CORP., DAYTON, OHIO

DTR-
- UNIDYNAMICS/PHOENIX, ARIZ.

DTS/ES-
- SUD-AVIATION, PARIS

DTW.6161.DK-
- BUREAU OF SHIPS

DTW-
- COMPAGNIE GENERALE DE TELEGRAPHIE SANS FIL, PARIS

DTW.GK.MCD-
- BUREAU OF SHIPS

DU-
- DENVER. UNIV.

DU-
- DENVER. UNIV. DENVER RESEARCH INST.

DU-	DUKE UNIV., DURHAM, N.C.
DU-	WRIGHT AIR DEVELOPMENT CENTER. AERIAL RECONNAISSANCE LAB., WRIGHT-PATTERSON AFB, OHIO
DUCK-	DREXEL UNIV., PHILADELPHIA. COMBUSTION KINETICS LAB.
DU-DRI-	DENVER. UNIV. DENVER RESEARCH INST.
DU DRI (LETTERS)	DENVER. UNIV. DENVER RESEARCH INST.
DU-ER-	DENVER. UNIV. DENVER RESEARCH INST.
DUH-	HANFORD ENGINEER WORKS, RICHLAND, WASH. (ASSIGNED TO INCOMING REPORTS DURING THE PERIOD WHEN HANFORD WAS OPERATED BY E.I. DU PONT DE NEMOURS + CO.)
DU IC-	DUKE UNIV., DURHAM, N.C. (INFORMAL COMMUNICATION)
DUML-	DUMONT (ALLEN B.) LABS., INC., CLIFTON, N.J.
DUMONT-	DUMONT (ALLEN B.) LABS., INC., CLIFTON, N.J.
DU MSB-	DUKE UNIV., BEAUFORT, N.C. MARINE STATION (BULLETIN)
DUN-	DOUGLAS UNITED NUCLEAR, INC., RICHLAND, WASH.
DUN-	DUNLAP AND ASSOCIATES, INC., CLIFTON, N.J.
DUN-AOP-	DOUGLAS UNITED NUCLEAR, INC., RICHLAND, WASH.
DUN-M-	DOUGLAS UNITED NUCLEAR, INC., RICHLAND, WASH.
DUN-SA-	DOUGLAS UNITED NUCLEAR, INC., RICHLAND, WASH.
DUN-TH-	DOUGLAS UNITED NUCLEAR, INC., RICHLAND, WASH.
DU PONT-	DU PONT DE NEMOURS (E.I.) + CO. (ALL LOCATIONS)
DUPONT-	DU PONT DE NEMOURS (E.I.) + CO., WILMINGTON, DEL.
DUPONT-ELAB-(LETTER)-	DU PONT DE NEMOURS (E.I.) + CO. EASTERN LAB., GIBBSTOWN, N.J.
DUPONT-ELAB-(LETTER)-	DU PONT DE NEMOURS (E.I.) + CO. EXPLOSIVES DEPT. EASTERN LAB., GIBBSTOWN, N.J.
DU PONT FOR-	DU PONT DE NEMOURS (E.I.) + CO. (ALL LOCATIONS) (FORMAL REPORTS)
DUPONT-PR-	DU PONT DE NEMOURS (E.I.) + CO. EXPLOSIVES DEPT. EASTERN LAB., GIBBSTOWN, N.J.
DU PONT PRL-	DU PONT DE NEMOURS (E.I.) + CO. PARLIN LAB., N.J.
DUPONT-QPR-(INCL.MOS/YR)	DU PONT DE NEMOURS (E.I.) + CO. PIONEERING RESEARCH DIV., WILMINGTON, DEL.
DUPONT-RR-(YEAR/MONTH)	DU PONT DE NEMOURS (E.I.) + CO. RADIATION PHYSICS LAB., WILMINGTON, DEL. (RESEARCH REPORT)
DU-PR-	DENVER. UNIV. DENVER RESEARCH INST.
DU-TR-	DUKE UNIV., DURHAM, N.C.
DVC-	DALMO VICTOR CO., SAN CARLOS, CALIF.
DVC-	TEXTRON INC. DALMO VICTOR CO., SAN CARLOS, CALIF.
DVC-	TEXTRON INC. DALMO VICTOR CO., SAN CARLOS, CALIF.
DVL-	DEUTSCHE VERSUCHSANSTALT FUER LUFT- UND RAUMFAHRT (ALL LOCATIONS)
DVL-	OCEANICS, INC., PLAINVIEW, N.Y.
DVR-(NUMBERS)-	AEROJET-GENERAL CORP. LIQUID ROCKET OPERATIONS, SACRAMENTO, CALIF.
DVR-	AEROJET-GENERAL CORP. LIQUID ROCKET PLANT, SACRAMENTO, CALIF.
DVR-(YEAR)-	AEROJET-GENERAL CORP. LIQUID ROCKET PLANT, SACRAMENTO, CALIF.
DVRPC-	DELAWARE VALLEY REGIONAL PLANNING COMMISSION, PHILA.
DVT-	MULLARD RADIO VALVE CO., LTD.
DV-TN-(NUMBER)-(YEAR)	ARMY BALLISTIC MISSILE AGENCY, REDSTONE ARSENAL, ALA.
DV-TR-(NUMBER)-(YEAR)	ARMY BALLISTIC MISSILE AGENCY, REDSTONE ARSENAL, ALA.
DW-	ARMY AIR FORCE. DIRECTORATE OF WEATHER
DW-	COAST AND GEODETIC SURVEY, ROCKVILLE, MD. (DENSITY OF SEA WATER)
DW-	PICATINNY ARSENAL. AMMUNITION GROUP, DOVER, N.J.
DW-	PICATINNY ARSENAL. WARHEADS AND SPECIAL PROJECTS LAB., DOVER, N.J.
DW(YR)(NO.)	UNITED KINGDOM ATOMIC ENERGY AUTHORITY. RESEARCH GP. ATOMIC ENERGY RES. ESTAB., HARWELL, BERKS, ENGLAND
DWC-	DIKEWOOD CORP., ALBUQUERQUE, N. MEX.
DWC FR-(NO.)-(NO.)	DIKEWOOD CORP., ALBUQUERQUE, N. MEX. (FINAL REPORT)
DWDL-(NUMBER-NUMBER)	MC DONNELL DOUGLAS CO., RICHLAND, WASH.
DWET-	ARMED FORCES SPECIAL WEAPONS PROJECT. FIELD COMMAND. DIRECTORATE OF WEAPONS EFFECTS TESTS, ALBUQUERQUE, N. MEX. (REF. 7)
DWG-	(DRAWING) (REF. 2)
D (W) MEMO-	GT. BRIT. WINDSCALE WORKS, SELLAFIELD, CUMB., ENGLAND (REF. 36)
DWR-	CALIFORNIA. UNIV., SAN DIEGO. DIV. OF WAR RESEARCH
DWR-	COLUMBIA UNIV., N.Y.C. DIV. OF WAR RESEARCH
DWR(D)-	GT. BRIT. DIRECTORATE OF WEAPON RESEARCH (DEFENCE)
DWS-PM-	TENNESSEE EASTMAN CORP., OAK RIDGE, TENN.
DWS-TM-(YEAR)-	TEST WING (DEV.)(6555TH), PATRICK AFB, FLA.
DWT-	DAVID TAYLOR MODEL BASIN, CARDEROCK, MD.
DWTMB-	DAVID TAYLOR MODEL BASIN, CARDEROCK, MD.
DWTMB (LETTER)-	DAVID TAYLOR MODEL BASIN, CARDEROCK, MD.
DWTMB T/RMB-	DAVID TAYLOR MODEL BASIN, CARDEROCK, MD. (TRANSLATION, RUSSIAN MODEL BASIN)
DXRL-	NEW ZEALAND. DOMINION X-RAY + RADIUM LAB., CHRISTCHURCH
DXRL-(LTRS)-	NEW ZEALAND. DOMINION X-RAY + RADIUM LAB., CHRISTCHURCH
DY-(NUMBER)-(LETTER)	DYNELL ELECTRONICS CORP., PLAINVIEW, N.Y.
DYN-	ALLIED RESEARCH ASSOCIATES, INC. ARADYN DIV., CONCORD, MASS. (VIBRATION ENG. AND TESTING EQUIP.)
DYN-	NORTHROP AIRCRAFT, INC., HAWTHORNE, CALIF. (DYNAMICS)
DYN-	PURDUE UNIV., LAFAYETTE, IND. JET PROPULSION CENTER, SCHOOL OF MECHANICAL ENGINEERING (PROJECT SQUID)
DYNATECH-	DYNATECH CORP., CAMBRIDGE, MASS.
DYR-R-	ARMY BALLISTIC MISSILE AGENCY, HUNTSVILLE, ALA.

E-	(ENGINEERING) (REF. 2)
E-	(ENGINEERING REPORT) (REF. 2)
E-	(EXCERPT) (REF. 2)
E-	AIR FORCE CAMBRIDGE RESEARCH LABS., BEDFORD, MASS.
E-	AMSTERDAM. UNIV.
E-	ARMY SIGNAL ENGINEERING LABS., FORT MONMOUTH, N.J.
E-	ATOMIC ENERGY OF CANADA LTD. CHALK RIVER PROJ., ONT.
E-	BACON (FREDERICK S.) LABS., WATERTOWN, MASS.
E-	BENDIX CORP. KANSAS CITY DIV., MO.
E-	BROWN UNIV., PROVIDENCE. DIV. OF ENGINEERING
E(NUMBER)-	BROWN UNIV., PROVIDENCE. DIV. OF ENGINEERING
E324-	BUNKER-RAMO CORP. DEFENSE SYSTEMS DIV., CANOGA PARK, CALIF.
E-(NUMBER)-	CALIFORNIA. DEPARTMENT OF EMPLOYMENT, SACRAMENTO
E-	CALIFORNIA INST. OF TECH., PASADENA
E-(NUMBER).(NUMBER)	CALIFORNIA INST. OF TECH., PASADENA
E-(NUMBER.NUMBER)	CALIFORNIA INST. OF TECH., PASADENA. HYDRODYNAMICS LAB
E-(YEAR)-	CARLETON UNIV., OTTAWA
E-	CHRYSLER CORP. AMPLEX DIV., DETROIT
E-	COLES SIGNAL LAB., BELMAR, N.J. (ENGINEERING REPT.)
E-	COOK ELECTRIC CO. TECH-CTR. DIV., MORTON GROVE, ILL.
E(NO.)-	DOUGLAS AIRCRAFT CO., INC., CHARLOTTE, N.C.
E-59-	DU PONT DE NEMOURS (E.I.) + CO. JACKSON LAB., WILMINGTON, DEL.
E-	EVANS SIGNAL LAB., BELMAR, N.J. (ENGINEERING REPT.)
E-(YEAR)-R-	EXPLOSIVES CORP. OF AMERICA, ISSAQUAH, WASH.
E-(NUMBER)-	GENERAL ELECTRONIC LABS. INC., BOSTON
E-	GT. BRIT. ATOMIC EN. RES. ESTAB., HARWELL, BERKS, ENG
E.443/N-	GT. BRIT. ATOMIC EN. RES. ESTAB., HARWELL, BERKS, ENG
E-(NUMBER)/(YEAR)	GT. BRIT. ATOMIC WEAPONS RESEARCH ESTABLISHMENT, ALDERMASTON, BERKS, ENGLAND
E-	GT. BRIT. SAFETY IN MINES RESEARCH ESTABLISHMENT, BUXTON, DERBY, ENGLAND
E-	HEBREW UNIV., JERUSALEM
E-(NUMBER)-(YEAR)	HOLMES AND NARVER, INC., LOS ANGELES
E-	HYDRO-ELECTRIC POWER COMMISSION OF ONTARIO, ROLPHTON
E-	IIT RESEARCH INST., CHICAGO
E-(NO.)A	KELLEX CORP., N.Y.C.
E-	LEWIS RESEARCH CENTER, CLEVELAND
E-1- THRU E-5-	LOS ALAMOS SCIENTIFIC LAB., N. MEX. (INTERNAL CORRESPONDENCE SERIAL USED BY GROUPS IN E DIV.)
E001 THRU E999	MC DONNELL AIRCRAFT CORP., ST. LOUIS
E(NUMBER)	MC DONNELL AIRCRAFT CORP., ST. LOUIS
E-	MASSACHUSETTS INST. OF TECH., CAMBRIDGE
E-(4 DIGITS)	MASSACHUSETTS INST. OF TECH., CAMBRIDGE. CHARLES STARK DRAPER LAB.
E-	MASS. INST. OF TECH., CAMBRIDGE. INSTRUMENTATION LAB.
E-	MASS. INST. OF TECH., CAMBRIDGE. SERVOMECHANISMS LAB.
E-	MISSOURI. UNIV., COLUMBIA
E-(YEAR)-	NAVAL ACADEMY, ANNAPOLIS
E7-	NAVAL FLEET MISSILE SYSTEMS, CORONA, CALIF.
E(NUMBER)-	NORTHROP CORP. NORTHROP SPACE LABS., RESEARCH AND ANALYSIS SECTION, HUNTSVILLE, ALA.
E-	NORWAY. FORSVARETS FORSKNINGSINSTITUTT, KJELLER
E-	OXFORD UNIV.
E-	OXFORD UNIV., NUFFIELD, ENGLAND, DEPT. OF ORTHOPAEDIC SURGERY
E-	PUGET SOUND NAVAL SHIPYARD. QUALITY ASSURANCE DIV., BREMERTON, WASH.
E-	SIGNAL CORPS ENG. LABS.,FT. MONMOUTH, NJ (ENG. REPT.)
E-	SQUIER SIGNAL LAB., FORT MONMOUTH, N.J. (ENG. REPT.)
E-	TE CO., SANTA BARBARA, CALIF.
E-0.100- THRU E-5.430-	TENNESSEE EASTMAN CORP., OAK RIDGE, TENN. (REF. 34)
E(NUMBER)-(YEAR)	THIOKOL CHEMICAL CORP. ELKTON DIV., MD.
E-	U.S.S.R. JOINT INST. FOR NUCLEAR RESEARCH, DUBNA
E(NUMBERS)-	UNITED AIRCRAFT CORP. RES. LABS., EAST HARTFORD,CONN.
E(NUMBER)	UNITED KINGDOM. NATIONAL ENGINEERING LAB., GLASGOW
E(NO.)(LETTER)-	VOUGHT AERONAUTICS. DIV. OF CHANCE VOUGHT CORP., DALLAS
(YR.)-(NO.)E2-GASES-P	WESTINGHOUSE RESEARCH LABS. ATOMIC AND MOLECULAR SCIENCES, PITTSBURGH
E-(INITIALS)-	NATIONAL RESEARCH COUNCIL OF CANADA. ATOMIC ENERGY PROJECT, CHALK RIVER, ONT.
EA-(NO.)-	EDGEWOOD ARSENAL, MD.
EA-	ELECTRONIC ASSOCIATES, INC., LONG BRANCH, N.J.
EA-	ELECTRONIC ASSOCIATES INC., PRINCETON, N.J.
EA-	FOREIGN ECONOMIC ADMINISTRATION. ENEMY BRANCH
EA-	GT. BRIT. AERONAUTICAL RESEARCH COUNCIL. ENGINE AERODYNAMICS SUBCOMMITTEE
EA-	OLIN MATHIESON CHEMICAL CORP., EAST ALTON, ILL.
EA-	QUARTERMASTER RESEARCH AND ENGINEERING COMMAND, NATICK, MASS. (ENVIRONMENTAL ANALYSIS)
EA-A(YEAR)-	FRANKFORD ARSENAL. PITMAN-DUNN LAB., PHILADELPHIA
EAC-	CARNEGIE INST. OF TECH., PITTSBURGH
EACD-	CHEMICAL WARFARE SERVICE. CHEMICAL DIV., EDGEWOOD ARSENAL, MD.
EACD PD-	CHEMICAL WARFARE SERVICE. CHEMICAL DIV. PHYSICAL DEPT., EDGEWOOD ARSENAL, MD.
EACRP-A-	EUROPEAN AMERICAN COMMITTEE ON REACTOR PHYSICS, EUROPEAN NUCLEAR ENERGY AGENCY, PARIS
EACRP-L-	EUROPEAN AMERICAN COMMITTEE ON REACTOR PHYSICS, EUROPEAN NUCLEAR ENERGY AGENCY, PARIS
EACRP-U-	EUROPEAN AMERICAN COMMITTEE ON REACTOR PHYSICS, EUROPEAN NUCLEAR ENERGY AGENCY, PARIS
EAD-	AEROJET-GENERAL NUCLEONICS, SAN RAMON, CALIF.
EAES-U.K.-	UNITED KINGDOM ATOMIC ENERGY AUTHORITY. DEVELOPMENT AND ENGINEERING GROUP, RISLEY, LANCS, ENGLAND
EAFB-	AEROJET-GENERAL CORP., SACRAMENTO, CALIF.
EAG-	BUREAU OF ORDNANCE (NAVY). EVALUATION + ANALYSIS GP.
EAH-	GENERAL ELECTRIC CO. GENERAL ENGINEERING AND CONSULTING LAB., SCHENECTADY, N.Y.
EA-L-	COLUMBIA UNIV., NYC. SUBSTITUTE ALLOY MATERIALS LABS.
EAL-	EASTERN AIR LINES, ATLANTA
EA-M-	COLUMBIA UNIV., NYC. SUBSTITUTE ALLOY MATERIALS LABS.
EA-M1F-A(NUMBER)-	EDGEWOOD ARSENAL, MD.(MALFUNCTION INVESTIGATION)
EAMX-	COLUMBIA UNIV., NYC. SUBSTITUTE ALLOY MATERIALS LABS.
EANDC-(NUMBER)(LETTER)	EUROPEAN-AMERICAN NUCLEAR DATA COMMITTEE (REF. 45)
EANDC-	EUROPEAN-AMERICAN NUCLEAR DATA COMMITTEE. SECRETARIAT, OECD, PARIS (REF. 45)
EANDC(LETTER(S))-	EUROPEAN-AMERICAN NUCLEAR DATA COMMITTEE. SECRETARIAT, OECD, PARIS (REF. 45)
EANDC (CAN)(NUMBER)(LTR)	EUROPEAN-AMERICAN NUCLEAR DATA COMMITTEE (REPORTS FROM CANADIAN PARTICIPANTS)
EANDC (E) (NUMBER)(LTR)	EUROPEAN-AMERICAN NUCLEAR DATA COMMITTEE (REPORTS FROM EURATOM MEMBER PARTICIPANTS)
EANDC (EUR)-	EUROPEAN-AMERICAN NUCLEAR DATA COMMITTEE (REPORTS FROM EURATOM MEMBER PARTICIPANTS)
EANDC (OR) (NUMBER)(LTR)	EUROPEAN-AMERICAN NUCLEAR DATA COMMITTEE (REPORTS FROM PARTICIPANTS NOT OTHERWISE DESIGNATED)
EANDC (UK) (NUMBER)(LTR)	EUROPEAN-AMERICAN NUCLEAR DATA COMMITTEE (REPORTS FROM UNITED KINGDOM PARTICIPANTS)
EANDC (US) (NUMBER)(LTR)	EUROPEAN-AMERICAN NUCLEAR DATA COMMITTEE (REPORTS FROM UNITED STATES PARTICIPANTS)
EANS-	GT. BRIT. ROYAL A.F. CENTRAL NAVIGATION + CONTROL SCH

EAP-WQO-(NUMBER)-	ENVIRONMENTAL PROTECTION AGENCY. WATER QUALITY OFFICE, WASHINGTON, D.C.
EA-R-	COLUMBIA UNIV., NYC. SUBSTITUTE ALLOY MATERIALS LABS.
EAR-	FAIRCHILD HILLER CORP. REPUBLIC AVIATION DIV., FARMINGDALE, N.Y.
EAS-	BUREAU OF ORDNANCE (NAVY). EVALUATION AND ANALYSIS STAFF
EAS-	PICATINNY ARSENAL. ENGINEERING ANALYSIS AND SPECIAL AMMUNITION SECTION, DOVER, N.J.
EASP-(NUMBER)-	EDGERTON, GERMESHAUSEN + GRIER,INC.,SANTA BARBARA,CAL
EA-SP-(NO.)-	EDGEWOOD ARSENAL, MD.
EASP-	EDGEWOOD ARSENAL, MD.
EASP-(NUMBER)-	EDGEWOOD ARSENAL, MD.
EAS-TN-	BUREAU OF ORDNANCE (NAVY). EVALUATION AND ANALYSIS STAFF
EATD MR-	CHEMICAL WARFARE SERVICE. TECHNICAL DIV., EDGEWOOD ARSENAL, MD. (MEMORANDA REPORTS)
EA-TM-(NUMBER)-	EDGERTON, GERMESHAUSEN + GRIER,INC.,SANTA BARBARA,CAL
EATM-(NUMBER)-	EDGERTON, GERMESHAUSEN + GRIER,INC.,SANTA BARBARA,CAL
EATM-(NUMBER)-	EDGEWOOD ARSENAL, MD.
EA-TN-	EDGEWOOD ARSENAL, MD.
EATR-	ARMY CHEMICAL CENTER, MD.
EATR-	CHEMICAL WARFARE SERVICE. EDGEWOOD ARSENAL, MD. (TECHNICAL REPORTS)
EATR-	DIRECTOR OF ENG. + IND. SERVICES,EDGEWOOD ARSENAL,MD.
EA-TR-	EDGEWOOD ARSENAL, MD.
EATR-	EDGEWOOD ARSENAL, MD.
EB-	CORPS OF ENGINEERS (ARMY). ENGINEER BOARD
EB-	MARINE CORPS SCHOOLS. EQUIPMENT BOARD, QUANTICO, VA.
EB-(NUMBER)-	SPERRY GYROSCOPE CO., GREAT NECK, N.Y.
EBC-MR-	ENSIGN-BICKFORD CO. RESEARCH DEPT., SIMSBURY, CONN.
EBG/S/(NUMBER/NUMBER)	GENERAL DYNAMICS CORP. ELECTRIC BOAT DIV.,GROTON,CONN
EBL-	ELLIOTT BROS. LTD., LONDON
EBR-	CANADAIR, LTD., MONTREAL
EBR-	GT. BRIT. WATER POLLUTION RES.LAB.,WATFORD,HERTS,ENG.
EBR-	UNITED AIRCRAFT CORP., WINDSOR LOCKS, CONN.
EBR/PA-	BRAZIL. COMISSAO NACIONAL DE ENERGIA NUCLEAR
EBWR-TR-	ARGONNE NATIONAL LAB., ILL.
EC-	BUREAU OF NAVAL WEAPONS REPRESENTATIVE. METROLOGY ENGINEERING CENTER, POMONA, CALIF.
EC-	CARNEGIE INST. OF TECH., PITTSBURGH
EC-	EAU CLAIRE ORDNANCE PLANT
EC-	EMERSON ELECTRIC MFG. CO., ST. LOUIS
EC-	ETHYL CORP., BATON ROUGE, LA.
EC-	GENERAL MOTORS CORP. RESEARCH LABS. DIV., DETROIT (ELECTROCHEMISTRY)
EC-	UTAH. UNIV., SALT LAKE CITY
EC-	WESTINGHOUSE ELECTRIC CORP. DEVELOPMENT ENG. DEPT., LESTER, PENNA.
EC-	WESTINGHOUSE ELECTRIC CORP. STEAM DIV.,LESTER, PENNA.
ECA-	EAST COAST AERONAUTICS, INC., PELHAM MANOR, N.Y.
ECAC-	ELECTROMAGNETIC COMPATIBILITY ANALYSIS CENTER, ANNAPOLIS, MD.
ECAC-D-	ELECTROMAGNETIC COMPATIBILITY ANALYSIS CENTER, ANNAPOLIS, MD. (DIRECTORY, RADAR + COMM. EQUIP.)
ECAC-FAI-	ELECTROMAGNETIC COMPATIBILITY ANALYSIS CENTER, ANNAPOLIS, MD. (FREQUENCY APPLICATION INDEX)
ECAC-FAL-	ELECTROMAGNETIC COMPATIBILITY ANALYSIS CENTER, ANNAPOLIS, MD. (FREQUENCY ALLOCATION LIST)
ECAC-I-	ELECTROMAGNETIC COMPATIBILITY ANALYSIS CENTER, ANNAPOLIS, MD. (SPECTRUM SIGNATURE INDEX)
ECAC-IR-	ELECTROMAGNETIC COMPATIBILITY ANALYSIS CENTER, ANNAPOLIS, MD. (INTERMEDIATE REPORT)
ECAC-IR-J-	ELECTROMAGNETIC COMPATIBILITY ANALYSIS CENTER, ANNAPOLIS, MD.
ECAC-J-	ELECTROMAGNETIC COMPATIBILITY ANALYSIS CENTER, ANNAPOLIS, MD.
ECAC-J-IH-(SS)-	ELECTROMAGNETIC COMPATIBILITY ANALYSIS CENTER, ANNAPOLIS, MD.
ECAC-L-	ELECTROMAGNETIC COMPATIBILITY ANALYSIS CENTER, ANNAPOLIS, MD. (ELECTRONIC ENVIRONMENT LISTING)
ECA-CN-	ELECTRONICS CORP. OF AMERICA, CAMBRIDGE, MASS.
ECAC-PTR-	ELECTROMAGNETIC COMPATIBILITY ANALYSIS CENTER, ANNAPOLIS, MD.
ECAC-R-	ELECTROMAGNETIC COMPATIBILITY ANALYSIS CENTER, ANNAPOLIS, MD.
ECAC-STP-	ELECTROMAGNETIC COMPATIBILITY ANALYSIS CENTER, ANNAPOLIS, MD. (SHORT TERM PROJECT)
ECAC-STP-	IIT RESEARCH INST., CHICAGO
ECAC-TDR-	ELECTROMAGNETIC COMPATIBILITY ANALYSIS CENTER, ANNAPOLIS, MD. (TECHNICAL DOC. REPT.)
ECAC-TDR-	IIT RESEARCH INST., ANNAPOLIS, MD.
ECAC-TM-	ELECTROMAGNETIC COMPATIBILITY ANALYSIS CENTER, ANNAPOLIS, MD. (TECHNICAL MEMORANDUM)
ECAC-TM-	GEORGIA INST. OF TECH., ATLANTA
ECAC-TM-	IIT RESEARCH INST., ANNAPOLIS, MD.
ECAC-TN-	ELECTROMAGNETIC COMPATIBILITY ANALYSIS CENTER, ANNAPOLIS, MD. (TECHNICAL NOTE)
ECAC-TR-	ELECTROMAGNETIC COMPATIBILITY ANALYSIS CENTER, ANNAPOLIS, MD. (TECHNICAL REPORT)
ECAC-TR-	IIT RESEARCH INST., ANNAPOLIS, MD.
ECCO-	ELECTRO-CHEMICAL PRODUCTS CORP., EAST ORANGE, N.J.
ECD-	ENERGY CONVERSION DEVICES, INC., TROY, MICH.
ECI-(NUMBER)-	ELECTRONIC COMMUNICATIONS, INC., ST. PETERSBURG, FLA.
ECI-	ELECTRONIC COMMUNICATIONS INC. RES. DIV.,TIMONIUM,MD.
ECI-I-	ELECTRONIC COMMUNICATIONS, INC., ST. PETERSBURG, FLA.
ECL-(NO.)-(NO.)-(NO.)	AIR FORCE (EQUIPMENT COMPONENT LIST)
ECNG-	EAST CENTRAL NUCLEAR GROUP
ECOLOGY AND EPIZOOLOGY-	UTAH. UNIV., SALT LAKE CITY. ECOLOGY AND EPIZOOLOGY RESEARCH GROUP
ECOM-	ARMY ELECTRONICS COMMAND, FORT MONMOUTH, N.J.
ECOM-(NUMBER)-	ARMY ELECTRONICS COMMAND, FORT MONMOUTH, N.J.
ECOM-(NUMBER)-F	ARMY ELECTRONICS COMMAND, FORT MONMOUTH, N.J.
ECOM-C-	ARMY ELECTRONICS COMMAND, FORT MONMOUTH, N.J.
ECOM-CATE-	ARMY ELECTRONICS COMMAND, FORT MONMOUTH, N.J.
ECOM-TR-(NUMBER)-	ARMY ELECTRONICS COMMAND, FORT MONMOUTH, N.J.
ECONOMIC PLANNING SER-	WYOMING. UNIV., LARAMIE. DIV. OF BUSINESS AND ECONOMIC RESEARCH
ECP-	OFFICE OF THE DIRECTOR OF DEFENSE (RES. + ENG.) ADVISORY GROUP ON ELECTRONIC PARTS AND ELECTRON TUBES (ELECTRONIC COMPONENT PARTS)
EC-PR-	ETHYL CORP. RESEARCH AND DEV. DEPT., FERNDALE, MICH.
ECP-TDR-	EASTMAN CHEMICAL PRODUCTS, INC., KINGSPORT, TENN. (TECHNICAL DATA REPORT)
EC-QR-	ELECTRONA CORP., IRVINGTON, N.J. (QUARTERLY REPORT)
ECR-	DANISH RESEARCH CENTRE FOR APPLIED ELECTRONICS, COPENHAGEN
ECRC/	UNITED KINGDOM. ELECTRICITY COUNCIL RESEARCH CENTRE, CAPENHURST, BERKS, ENGLAND
ECRC/M-	UNITED KINGDOM. ELECTRICITY COUNCIL RESEARCH CENTRE, CAPENHURST, BERKS, ENGLAND
ECRC/N-	UNITED KINGDOM. ELECTRICITY COUNCIL RESEARCH CENTRE, CAPENHURST, BERKS, ENGLAND
ECRC-R-	UNITED KINGDOM. ELECTRICITY COUNCIL RESEARCH CENTRE, CAPENHURST, BERKS, ENGLAND
EC-RD-	BUNKER-RAMO CORP., SILVER SPRING, MD.
EC-RD-	HUGHES AIRCRAFT CO., CULVER CITY, CALIF.
EC-RD-	LOCKHEED ELECTRONICS CO. MILITARY SYSTEMS, PLAINFIELD, N.J.

EC-RD-
: NORTRONICS. APPLIED RES. DEPT., NEWBURY PARK, CALIF.

ECRDC PROJECT NO.-
: CANADA. DEFENCE RES. TELECOMMUNICATIONS ESTAB., OTTAWA

ECS-
: ELECTRONIC CONTROL SYSTEMS, INC., LOS ANGELES

EC.S-
: FRANCE. COMMISSARIAT A L'ENERGIE ATOMIQUE. CENTRE D'ETUDES NUCLEAIRES, SACLAY

EC-SR-
: PENNSYLVANIA RESEARCH ASSOCIATES INC., PHILADELPHIA

ECTF-
: NAVAL ORDNANCE TEST STA., INYOKERN (CHINA LAKE), CAL.

ED-
: ARMY ENGINEER REACTORS GP. ENG. DIV., FT. BELVOIR, VA.

ED-
: ATOMIC ENERGY OF CANADA LTD. CHALK RIVER PROJECT, ONT. (REF. 6)

ED-
: EMPIRE DEVICES, INC., BAYSIDE, N.Y.

ED-
: ESSO RESEARCH AND ENGINEERING CO., LINDEN, N.J.

ED/448/N-
: GT. BRIT. ATOMIC EN. RES. ESTAB., HARWELL, BERKS, ENG

ED-
: MARTIN-MARIETTA CORP. DENVER DIV.

ED-(6 DIGITS)
: OFFICE OF EDUCATION. EDUCATIONAL RESOURCES INFORMATION CENTER, WASHINGTON, D.C. (ERIC DOC.)

ED-(LETTERS)-(NUMBER)
: EUROPEAN COMPANY FOR THE CHEMICAL PROCESSING OF IRRADIATED FUELS, MOL, BELGIUM

EDA-(YEAR)-
: ECONOMIC DEVELOPMENT ADMINISTRATION, WASHINGTON, D.C.

ED-AA-
: FAIRCHILD SPACE AND DEFENSE SYSTEMS, SYOSSET, N.Y.

EDACOMM-(YEAR)-
: ECONOMIC DEVELOPMENT ADMINISTRATION, WASHINGTON, D.C.

ED-AJ-
: FAIRCHILD SPACE AND DEFENSE SYSTEMS, SYOSSET, N.Y.

EDA/OER-(YEAR)-
: ECONOMIC DEVELOPMENT ADMINISTRATION, WASHINGTON, D.C.

ED-AY-
: FAIRCHILD SPACE AND DEFENSE SYSTEMS, SYOSSET, N.Y.

EDC-1.1.(NUMBER)-C
: GILFILLAN BROS., INC., LOS ANGELES

EDC-(GROUP NO.)-(YEAR)-
: CASE INST. OF TECH., CLEVELAND. ENG. DESIGN CENTER

EDCPF-
: BELL AEROSPACE CO. DIV. OF TEXTRON. ENVIRONMENTAL DATA COLLECTION AND PROCESSING FACILITY, TUCSON, ARIZ.

EDF-
: GENERAL MOTORS CORP., MILWAUKEE

EDFB-IBP-(YEAR)-
: OAK RIDGE NATIONAL LAB., TENN. (EASTERN DECIDUOUS FOREST BIOME)

EDG-
: MICHIGAN. UNIV., ANN ARBOR. ENGINEERING RES. INST.

EDH-
: RAYTHEON CO. EQUIPMENT DIV. HDQTRS., WALTHAM, MASS.

EDI-
: ATOMIC ENERGY OF CANADA LTD. CHALK RIVER PROJECT, ONT. (REF. 6)

EDI-
: EDISON (THOMAS A.) INC., ORANGE, N.J.

EDIN-
: EDIN COMPANY, INC., WORCESTER, MASS.

EDL-
: SYLVANIA ELECTRONIC SYSTEMS. ELECTRONIC DEFENSE LABS., MOUNTAIN VIEW, CALIF.

EDL-(LETTER)(NUMBER)
: SYLVANIA ELECTRONIC SYSTEMS-WEST. ELECTRONIC DEFENSE LABS., MOUNTAIN VIEW, CALIF.

EDL-CD-(NUMBER)-M
: SYLVANIA ELECTRONIC SYSTEMS-WEST. ELECTRONIC DEFENSE LABS., MOUNTAIN VIEW, CALIF.

EDL-E(NUMBER)
: SYLVANIA ELECTRONIC SYSTEMS. ELECTRONIC DEFENSE LABS., MOUNTAIN VIEW, CALIF.

EDL-G(NUMBER)
: SYLVANIA ELECTRONIC SYSTEMS. ELECTRONIC DEFENSE LABS., MOUNTAIN VIEW, CALIF.

EDL-M(NUMBER)
: SYLVANIA ELECTRONIC SYSTEMS. ELECTRONIC DEFENSE LABS., MOUNTAIN VIEW, CALIF.

EDO-
: EDO CORP., COLLEGE POINT, N.Y.

EDO-XE-
: EDO (CANADA) LTD., CORNWALL, ONTARIO

EDR-
: (ENGINEERING DEPT. REPORT) (REF. 2)

EDR-
: (ENGINEERING DIV. REPORT) (REF. 2)

EDR-
: ALLISON DIV., GENERAL MOTORS CORP., INDIANAPOLIS

ED-R-
: COLUMBIA UNIV., NYC. SUBSTITUTE ALLOY MATERIALS LABS.

EDR-
: ECLIPSE-PIONEER DIV., BENDIX AVIATION CORP., TETERBORO, N. J.

EDR-
: FAIRCHILD HILLER CORP. REPUBLIC AVIATION DIV., FARMINGDALE, N.Y.

EDR-
: GENERAL MOTORS CORP., INDIANAPOLIS

EDR-CT-(NO.-NO.-NO.)
: MARINE CORPS, WASHINGTON, D.C.

EDR-IT-(NO.-NO.-NO.)
: MARINE CORPS, WASHINGTON, D.C.

EDRL-T-
: ARMY ENGINEER RESEARCH + DEV. LABS., FT. BELVOIR, VA.

EDS-
: ENVIRONMENTAL DATA SERVICE, SILVER SPRING, MD.

EDS-BC-
: ENVIRONMENTAL DATA SERVICE, SILVER SPRING, MD.

EDS-BM-
: ENVIRONMENTAL DATA SERVICE, SILVER SPRING, MD. (BIB)

ED-SR-
: FAIRCHILD HILLER CORP. REPUBLIC AVIATION DIV., FARMINGDALE, N.Y.

EDSTM-
: ENVIRONMENTAL DATA SERVICE, SILVER SPRING, MD.

EDSTM-BC-
: ENVIRONMENTAL DATA SERVICE, SILVER SPRING, MD. (BIB)

EE-
: (ELECTRICAL ENGINEERING) (REF. 2)

EE-
: APPLIED DEVICES CORP., COLLEGE POINT, N.Y.

EE-
: BELOCK INSTRUMENT CORP., COLLEGE POINT, N.Y.

EE-
: CALIFORNIA. UNIV., BERKELEY. RADIATION LAB. (REF. 9)

EE-
: CORNELL UNIV., ITHACA, N.Y. SCHOOL OF ELECTRICAL ENG.

EE-
: DAYTON, OHIO. UNIV.

EE-
: NEW MEXICO. UNIV., ALBUQUERQUE. BUREAU OF ENG. RES.

EE-(NUMBER)/(YEAR)/
: NEW MEXICO. UNIV., ALBUQUERQUE. BUREAU OF ENG. RES.

EE-(3 DIGITS)(YEAR)DC-
: NEW MEXICO. UNIV., ALBUQUERQUE. BUREAU OF ENG. RES.

EE-(3 DIGITS)(YEAR)ONR-
: NEW MEXICO. UNIV., ALBUQUERQUE. BUREAU OF ENG. RES.

EE-(3 DIGITS)(YEAR)HAFB-
: NEW MEXICO. UNIV., ALBUQUERQUE. BUREAU OF ENG. RES.

EE-
: NEW MEXICO. UNIV., ALBUQUERQUE. ENG. EXPERIMENT STA.

EE/(YEAR)/
: SOCIETE NATIONALE D'ETUDE ET DE CONSTRUCTIONS DE MOTEURS D'AVIATION. DIV. ATOMIQUE, SURESNES, FRANCE

EE-(NUMBER)-
: SYRACUSE UNIV., N.Y. DEPT. OF ELECTRICAL ENGINEERING

EE(NUMBER)-(NUMBER)F(NO.)
: SYRACUSE UNIV., N.Y. DEPT. OF ELECTRICAL ENGINEERING

EE(NUMBER-(NUMBER)-T-
: SYRACUSE UNIV., N.Y. DEPT. OF ELECTRICAL ENGINEERING

EE(4 DIGITS)-(YEAR)-
: SYRACUSE UNIV., N.Y. DEPT. OF ELECTRICAL ENGINEERING

EE-
: UNIVAC, ST. PAUL, MINN.

EE-
: UNIVERSITY OF SOUTHERN CALIFORNIA, LOS ANGELES

EE-
: VIRGINIA. UNIV., CHARLOTTESVILLE

EE-(NO.)-(NO.)-(YR.LTR.)
: VIRGINIA. UNIV., CHARLOTTESVILLE. RESEARCH LABS. FOR ENGINEERING SCIENCES

EE-
: WASHINGTON. UNIV., SEATTLE. DEPT. OF ELECTRICAL ENG.

EE-
: WESTINGHOUSE ELECTRIC CORP., BALTIMORE

EE-
: WESTINGHOUSE ELECTRIC CORP. ASTRONUCLEAR LAB., PITTS.

EEC-
: ELECTRONIC ENGINEERING CO. OF CALIF., LOS ANGELES

EEDC-
: NAVAL ORDNANCE LAB. ELECTRICAL EVALUATION DIV. (COMMUNICATION)

EEDTR-T-
: AVCO MFG. CORP. RESEARCH AND ADVANCED DEVELOPMENT DIV., LAWRENCE, MASS.

EEI-
: ENGINEERING ENTERPRISES, INC.

EEIS-
: ALLIED FORCES ENEMY EQUIPMENT INTELLIGENCE SERVICE

EEIS-
: SIGNAL CORPS (ARMY). ENEMY EQUIPMENT INTELLIGENCE DIV. CHINA-BURMA-INDIA THEATER

EEIS CWS PTR-
: ALLIED FORCES ENEMY EQUIPMENT INTELLIGENCE SERVICE. CHEMICAL WARFARE SECTION (PRELIMINARY TECH. REPT.)

EEIS SIG PR-
: ALLIED FORCES ENEMY EQUIPMENT INTELLIGENCE SERVICE. INDIA-BURMA THEATER. SIGNAL SECTION. (PRELIM. RPT.)

EEL-
: ESSO DEVELOPMENT CO., LTD.

EEMC-
: EMERSON ELECTRIC MFG. CO., ST. LOUIS

EEMTIC-(NO.)-
: ELECTRICAL AND ELECTRONIC MEASUREMENT TEST INSTRUMENT CONFERENCE, OTTAWA

EEP-
: ELECTRO-ENGINEERING PRODUCTS CO., CHICAGO

EEP-
: STANFORD UNIV., CALIF. PROJECT IN ENGINEERING ECONOMIC PLANNING

EER-
: OHIO UNIV., ATHENS

EER-(NUMBER-NUMBER)
: OHIO UNIV., ATHENS. DEPT. OF ELECTRICAL ENGINEERING

EER-
: PARKER AIRCRAFT CO., LOS ANGELES

EERC-(YEAR)-
: CALIFORNIA. UNIV., BERKELEY, EARTHQUAKE ENG. RES. CTR

EERD-CRRC-TM-
: AIR FORCE CAMBRIDGE RESEARCH LABS. ELECTRONICS RESEARCH DIRECTORATE, BEDFORD, MASS.

EERI-	EARTHQUAKE ENGINEERING RESEARCH INST., SAN FRANCISCO
EERL-	CORNELL UNIV., ITHACA, N.Y. ELECTRICAL ENG. RES. LAB.
EERL-	MC GILL UNIV., MONTREAL. EATON ELECTRONICS RES. LAB.
EERL-	TEXAS. UNIV., AUSTIN. ELECTRICAL ENGINEERING RES.LAB.
EERL-TR-(YEAR)-	TEXAS. UNIV., AUSTIN. ELECTRICAL ENGINEERING RES.LAB.
EERO-TR-	ARMY ENGINEER EXPLOSIVE EXCAVATION RESEARCH OFFICE, LIVERMORE, CALIF.
EES-	(ENGINEERING EXPERIMENT STATION) (REF. 2)
EES-	NAVAL ENGINEERING EXPERIMENT STATION, ANNAPOLIS
EES-(NO. LETTER)-	NAVAL ENGINEERING EXPERIMENT STATION, ANNAPOLIS
EES-(NUMBER)B-	OHIO UNIV., ATHENS. ENGINEERING EXPERIMENT STATION
EES-(NUMBER-NUMBER)	OHIO UNIV., ATHENS. ENGINEERING EXPERIMENT STATION
EES-B-	NAVAL ENGINEERING EXPERIMENT STATION, ANNAPOLIS
EES-C-	NAVAL ENGINEERING EXPERIMENT STATION, ANNAPOLIS
EE-SPEC-	CALIFORNIA. UNIV., BERKELEY. RADIATION LAB. (REF. 9)
EES-SERIES-	ARIZONA. UNIV., TUCSON. ENGINEERING EXPERIMENT STA.
EES-SR-	ARIZONA. UNIV., TUCSON. ENGINEERING EXPERIMENT STA.
EET-	DAYTON, OHIO. UNIV. DEPT. OF ELECTRICAL ENG.
EET-(NUMBER)/PR-	ELLIOTT ELECTRONIC TUBES LTD.,BOREHAMWOOD, HERTS, ENG
EETC-	AEROSPACE INDUSTRIES ASSN. OF AMERICA, INC., WASHINGTON, D.C.
EE-TR-	KANSAS STATE UNIV., MANHATTAN. DEPT. OF ELECT. ENG.
EEV-	ENGLISH ELECTRIC VALVE CO., LTD.,CHELMSFORD,ESSEX,ENG
EF-	POWER REACTOR DEVELOPMENT CO. ENRICO FERMI ATOMIC POWER PLANT, DETROIT
EF-(NO.)-(LTRS.)-	SOCIETE BELGE POUR L'INDUSTRIE NUCLEAIRE, BRUSSELS
EF-	SPERRY RAND CORP. UNIVAC DIV., PHILADELPHIA
EF-	SPERRY RAND CORP. UNIVAC FEDERAL SYSTEMS DIV., ST. PAUL, MINN.
EF/(LTR.)/(LTR.)/	IMPERIAL COLLEGE OF SCIENCE AND TECHNOLOGY, LONDON. DEPT. OF MECHANICAL ENGINEERING
EF-CRC-	ENGINEERING FOUNDATION, N.Y.C. (COLUMN RES. COUNCIL)
EFI-	CHICAGO. UNIV. ENRICO FERMI INST. FOR NUCLEAR STUDIES
EFI-(YEAR)-	CHICAGO. UNIV. ENRICO FERMI INST. FOR NUCLEAR STUDIES
EFI-ME-(MONTH/YEAR)	U.S.S.R. EREVAN INST. OF PHYSICS
EFINS-	CHICAGO. UNIV. ENRICO FERMI INST. FOR NUCLEAR STUDIES
EFINS-(NUMBER)-	CHICAGO. UNIV. ENRICO FERMI INST. FOR NUCLEAR STUDIES
EFI-TF-(NUMBER-YEAR)	AKADEMIYA NAUK ARMYANSKOI SSR. INSTITUT FIZIKI,EREVAN
EFI-TF-(MONTH/YEAR)	U.S.S.R. EREVAN INST. OF PHYSICS
EFM-	QUARTERMASTER RESEARCH AND ENGINEERING COMMAND, NATICK, MASS. (EXPLOITATION OF FOREIGN MATERIEL)
EFTMR-	FAIRCHILD HILLER CORP. REPUBLIC AVIATION DIV., FARMINGDALE, N.Y.
EF/TN/A/	IMPERIAL COLLEGE OF SCIENCE AND TECHNOLOGY, LONDON. DEPT. OF MECHANICAL ENGINEERING
EF/TN/B/	IMPERIAL COLLEGE OF SCIENCE AND TECHNOLOGY, LONDON
EGCR-	TENNESSEE VALLEY AUTHORITY, OAK RIDGE
EGCR-STUDY-	TENNESSEE VALLEY AUTHORITY, OAK RIDGE
EGEAC TR-	EDISON GENERAL ELECTRIC APPLIANCE CO. (TECH. RPTS.)
E.G.+G.-	EDGERTON, GERMESHAUSEN AND GRIER, INC., BOSTON
EG+G-	EDGERTON, GERMESHAUSEN AND GRIER, INC., BOSTON
EG+G-	EDGERTON, GERMESHAUSEN + GRIER, INC., LAS VEGAS, NEV.
EGG-	EDGERTON, GERMESHAUSEN + GRIER,INC., SANTA BARBARA,CAL
EGG-	EG+G, INC. (ALL LOCATIONS)
EGG-(NUMBER)-	EG+G, INC. (ALL LOCATIONS)
EGG-1183-	EG+G, INC., BOSTON
EG+G-(LETTER(S))-	EDGERTON, GERMESHAUSEN AND GRIER, INC., BOSTON
EG/G-AL-	EG+G, INC., ALBUQUERQUE, N. MEX.
EG + G-B-	EDGERTON, GERMESHAUSEN AND GRIER, INC., BEDFORD, MASS
EG+G-B-	EDGERTON, GERMESHAUSEN AND GRIER, INC., BOSTON
EG/G-B-	EG+G, INC. (ALL LOCATIONS)
EGG-B-	EG+G, INC. SYSTEMS DIV.
EG+G-L-	EDGERTON, GERMESHAUSEN + GRIER, INC., LAS VEGAS, NEV.
EGG-L-	EDGERTON, GERMESHAUSEN + GRIER, INC., LAS VEGAS, NEV.
EG+G-LV-	EDGERTON, GERMESHAUSEN + GRIER, INC., LAS VEGAS, NEV.
EG+G-OUT-	EDGERTON, GERMESHAUSEN AND GRIER, INC., BOSTON
EG+G-PPG-	EDGERTON, GERMESHAUSEN AND GRIER, INC., BOSTON (PACIFIC PROVING GROUND)
EG+G-S-(NO.)-MN	EDGERTON, GERMESHAUSEN + GRIER,INC.,SANTA BARBARA,CAL
EG/G-S-(NUMBER)-R	EDGERTON, GERMESHAUSEN + GRIER,INC.,SANTA BARBARA,CAL
EG/G-S-(NUMBER)-R	EG+G, INC. (ALL LOCATIONS)
EGG-TM-(LETTER)-	EG+G, INC., LAS VEGAS, NEV.
EGG-TR-(LETTER)-	EG+G, INC., LAS VEGAS, NEV.
EG-M-	COLUMBIA UNIV., NYC. SUBSTITUTE ALLOY MATERIALS LABS.
EG-R-	COLUMBIA UNIV., NYC. SUBSTITUTE ALLOY MATERIALS LABS.
EG-REPORT-	UNITED KINGDOM ATOMIC ENERGY AUTHORITY. ENGINEERING GROUP, RISLEY, LANCS., ENGLAND
EH-	GENERAL ELECTRIC CO., SCHENECTADY, N.Y.
EH-	GENERAL ELECTRIC CO., SYRACUSE, N.Y.
EHE-	TEXAS. UNIV., AUSTIN. ENVIRONMENTAL HEALTH ENGINEERING RESEARCH LAB.
EHE-(NUMBER)-	TEXAS. UNIV., AUSTIN. ENVIRONMENTAL HEALTH ENGINEERING RESEARCH LAB.
EHE AT CIT-	CALIFORNIA INST. OF TECH., PASADENA. KECK LAB. OF ENVIRONMENTAL HEALTH ENGINEERING
EHM-	GENERAL ELECTRIC CO. HEAVY MILITARY ELECTRONIC EQUIPMENT DIV., SYRACUSE, N.Y.
EHP-(LETTER NUMBERS...)	PLESSET (E.H.) ASSOCIATES, INC., SANTA MONICA, CALIF
EHPA-(LETTER NUMBERS...)	PLESSET (E.H.) ASSOCIATES, INC., SANTA MONICA, CALIF
EHT/TN/	IMPERIAL COLLEGE OF SCIENCE AND TECHNOLOGY, LONDON
EI-	ATOMIC ENERGY OF CANADA LTD. CHALK RIVER PROJ., ONT.
EI-	EXPERIMENT, INC., RICHMOND
EI-	FINNISH ACADEMY OF TECHNICAL SCIENCES, HELSINKI
EI-	NATIONAL RESEARCH COUNCIL OF CANADA. ATOMIC ENERGY PROJECT, CHALK RIVER, ONT. (REF.29)
EI-	ROYAL CANADIAN AIR FORCE. CENTRAL EXPERIMENTAL AND PROVING ESTAB., ROCKCLIFFE, ONT.
EIA-	ELECTRONIC INDUSTRIES ASSN., WASHINGTON, D.C.
EIB-	NAVY (ELECTRONICS INFORMATION BULLETIN)
EIC-(YEAR)-MECH-	ENGINEERING INST. OF CANADA, MONTREAL
EIC-	MC GILL UNIV., MONTREAL
EID-BP-	DU PONT DE NEMOURS (E.I.) + CO., WILMINGTON, DEL.
EIM-	NAVAL POWDER FACTORY, INDIAN HEAD, MD. (EXPLOSIVES INVESTIGATION MEMORANDUM)
EIMAC-	VARIAN ASSOCIATES. EIMAC DIV., SAN CARLOS, CALIF.
EIMAC-QR-	VARIAN ASSOCIATES. EIMAC DIV., SAN CARLOS, CALIF.
EIMAC-TR-(YEAR)-	EIMAC, SAN CARLOS, CALIF.
EIMAC-TR-	EIMAC, SAN CARLOS, CALIF.
EIN-	OFFICE OF THE CHIEF OF ENGINEERS (ARMY), WASH., D.C.
E.IN.C.-	GT. BRIT. ENGINEER-IN-CHIEF OF THE FLEET
EIR-	BENDIX CORP. FRIEZ INSTRUMENT DIV., BALTIMORE
EIR-	BENDIX CORP. SCIENCE DIV., BALTIMORE
EIR-	EUROPEAN COMPANY FOR THE CHEMICAL PROCESSING OF IRRADIATED FUELS, MOL, BELGIUM
EIR-	SWITZERLAND. EIDGENOESSISCHES INSTITUT FUER REAKTORFORSCHUNG, WUERENLINGEN
EIR-TM-	SWITZERLAND. EIDGENOESSISCHES INSTITUT FUER REAKTORFORSCHUNG, WUERENLINGEN
EIS-	ARMY MAP SERVICE, WASHINGTON, D.C.

EIS-	ENGINEERING INFORMATION SERVICES, PRESTON, LANCS, ENGLAND (TRANSLATIONS)
EIS-	FOREIGN ECONOMIC ADMINISTRATION. LIBERATED AREAS BRANCH (ECONOMIC INSTITUTIONS STAFF REPORTS)
EIST/INST/	FERRANTI, LTD. FLIGHT TRIAL DEPT. (INTERNATIONAL AEROSPACE SYMPOSIUM)
EI-TM-	EXPERIMENT, INC., RICHMOND
EI-TP-	EXPERIMENT, INC., RICHMOND
EIVR-	UNITED KINGDOM ATOMIC ENERGY AUTHORITY. WEAPONS GROUP ATOMIC WEAPONS RES. ESTAB.,ALDERMASTON, BERKS, ENG.
EJ-(6 DIGITS)	OFFICE OF EDUCATION. EDUCATIONAL RESOURCES INFORMATION CENTER, WASHINGTON, D.C. (ERIC JNL.)
EJ-(NUMBER)-	SPERRY UTAH CO., SALT LAKE CITY
EJCC-	EASTERN JOINT COMPUTER CONFERENCE
EK-	EASTMAN KODAK CO., ROCHESTER, N.Y.
EK/ARD-ED-	EASTMAN KODAK CO., ROCHESTER, N.Y.
EKC-	EASTMAN KODAK CO., ROCHESTER, N.Y.
EKCM-	COLE (E.K.), LTD.
EKC-MLR-	EASTMAN KODAK CO., ROCHESTER, N.Y. (MONTHLY LTR. RPT)
EKC-NOD-(LETTERS)-	EASTMAN KODAK CO., ROCHESTER, N.Y.
EK-D-	EASTMAN KODAK CO. NAVY ORDNANCE DIV., ROCHESTER, N.Y.
EK-M-	COLUMBIA UNIV., NYC. SUBSTITUTE ALLOY MATERIALS LABS.
EK/NOD-	EASTMAN KODAK CO. NAVY ORDNANCE DIV., ROCHESTER, N.Y.
EK-R-	COLUMBIA UNIV., NYC. SUBSTITUTE ALLOY MATERIALS LABS.
EKTR-	EASTMAN KODAK CO., ROCHESTER, N.Y.
EL-	(ELECTRONICS LAB.) (REF. 2)
EL-	AEROJET-GENERAL CORP., SACRAMENTO, CALIF.
EL-	BUREAU OF SHIPS. ESSO LAB.
EL-	CANADA. DEFENCE RES. TELECOMMUNICATIONS ESTAB.,OTTAWA
EL-	DOUGLAS AIRCRAFT CO., INC., EL SEGUNDO, CALIF.
EL-(NUMBER) Q-(YEAR)	DOW CHEMICAL CO. SCIENTIFIC PROJS. LAB., MIDLAND,MICH
EL-(2 DIGITS)	FINNISH ACADEMY OF TECHNICAL SCIENCES, HELSINKI (ACTA POLYTECHNICA SCANDINAVICA. ELEC. ENG. SER.)
EL-	GT. BRIT. ROYAL AIRCRAFT EST.,FARNBOROUGH, HANTS, ENG
EL-	NATIONAL RES.COUNCIL OF CANADA. MONTREAL LAB.(REF.29)
EL-	NAVAL AIR DEVELOPMENT CENTER. AERONAUTICAL ELECTRONIC AND ELECTRIC LAB., JOHNSVILLE, PENNA.
EL-	NEW YORK UNIV., N.Y.C. COURANT INSTITUTE OF MATHEMATICAL SCIENCES
EL-	RES. + DEV. BD. COMM. ON ELECTRONICS, WASH., D.C.
EL (NUMBER)/	RES. + DEV. BD. COMM. ON ELECTRONICS, WASH., D.C.
ELAB-AE-	NORGES TEKNISKE HOEGSKOLE, TRONDHEIM
ELAB-B-	DU PONT DE NEMOURS (E.I.) + CO. EASTERN LAB., GIBBSTOWN, N.J.
ELAB-D-	DU PONT DE NEMOURS (E.I.) + CO., GIBBSTOWN, N.J.
ELAB-D-	DU PONT DE NEMOURS (E.I.) + CO. EASTERN LAB., GIBBSTOWN, N.J.
ELAB-IR-	NORGES TEKNISKE HOEGSKOLE, TRONDHEIM
ELAB-IT-	NORGES TEKNISKE HOEGSKOLE, TRONDHEIM
ELAB-TE-	NORGES TEKNISKE HOEGSKOLE, TRONDHEIM
ELAB-TL-	NORGES TEKNISKE HOEGSKOLE, TRONDHEIM
ELAB-TR-	NORGES TEKNISKE HOEGSKOLE, TRONDHEIM
ELAB-VT-	NORGES TEKNISKE HOEGSKOLE, TRONDHEIM
ELAI F/(CONTRACT NO.)	ELECTRONICS ASSOCIATES, INC., PRINCETON, N.J.
ELC-	ELECTRONA CORP., IRVINGTON, N.J.
ELC-	NAVAL AIR STATION, EL CENTRO, CALIF.
ELCA-	ELECTROCHIMICA CORP., MENLO PARK, CALIF.
ELD-	ENJAY LABS., LINDEN, N.J.
ELD-	GENERAL ELECTRIC CO. GEN. ENG. LAB., SCHENECTADY,N.Y.
ELDO-BIB-	EUROPEAN SPACE VEHICLE LAUNCHER DEV. ORG., PARIS
ELDO-LRBA-TN-E-	FRANCE. LABORATOIRE DE RECHERCHES BALISTIQUES ET AERODYNAMIQUES, VERNON
ELDO-M-/(YEAR)/	NETHERLANDS. NATIONAAL LUCHT EN RUIMTEVAART-LABORATORIUM, AMSTERDAM
ELDO-PUBL.-	EUROPEAN SPACE VEHICLE LAUNCHER DEV. ORG., PARIS
ELDO-RFI-	HAWKER SIDDELEY DYNAMICS, LTD.,STEVENAGE, HERTS, ENG.
ELDO-RFI-	SELENIA S.P.A., ROME
ELDO-RFI-GA-	SELENIA S.P.A., ROME
ELDO/SP/(YEAR)/	BOMBRINI PARODI-DELFINO S.P.A., ROME
ELDO/SP/(YEAR)/	FRANCE. COMMISSARIAT A L'ENERGIE ATOMIQUE, SACLAY
ELDO/SP/(YR)/-	FRANCE. OFFICE NATIONAL D'ETUDES ET DE RECHERCHES AEROSPATIALES, PARIS
ELDO/SP/(YEAR)/(NUMBER)	HAWKER SIDDELEY DYNAMICS, LTD.,STEVENAGE, HERTS, ENG.
ELDO/SP/(YEAR)/	SOCIETE D'ETUDE DE LA PROPULSION PAR REACTION, VILLEJUIF, FRANCE
ELDO/SP/(YEAR)/	SOCIETE L'AIR LIQUIDE, SASSENAGE, FRANCE
ELDO/SP/(YEAR)/	SOCIETE POUR L'ETUDE ET LA REALISATION D'ENGINS BALISTIQUES, COURBEVOIE, FRANCE
ELDO-TM-	EUROPEAN SPACE VEHICLE LAUNCHER DEV. ORG., PARIS
ELDO-TM-(LETTER)-	EUROPEAN SPACE VEHICLE LAUNCHER DEV. ORG., PARIS
ELDO-TM/F/	EUROPEAN SPACE VEHICLE LAUNCHER DEV. ORG., PARIS
ELDO-TR-	SELENIA S.P.A., ROME
ELEC-(YEAR)-	MILITARY ACADEMY, WEST POINT, NY. DEPT.OF ELECTRICITY
ELEC. ENG. SPEC.-	CALIFORNIA. UNIV., BERKELEY. RADIATION LAB. (REF. 9)
ELECLAB-	NAVAL SHIP RESEARCH AND DEV. CENTER, ANNAPOLIS
ELECLAB-(NO.)/(YEAR)	NAVAL SHIP RESEARCH AND DEV. CENTER, ANNAPOLIS
ELEX-	NAVAL ELECTRONIC SYSTEMS COMMAND, WASHINGTON, D.C.
ELI-	ATOMIC ENERGY OF CANADA LTD., CHALK RIVER, ONT.
ELR-	(ENGINEERING LAB. REPORT) (REF. 2)
EL-R-	CHRYSLER CORP. MISSILE OPERATIONS
EL-R-	COLUMBIA UNIV., NYC. SUBSTITUTE ALLOY MATERIALS LABS.
ELVPC-(YEAR)-	ELKHART LAKE VILLAGE PLANNING COMMISSION, WIS.
EM-	(EDUCATION MANUAL) (REF. 2)
EM-	(ELECTRODEPOSITION MEMO) (REF. 2)
EM-	(ENGINEERING MEMORANDUM) (REF. 2)
EM-	(EXTERNAL MEMORANDUM) (REF. 2)
EM-(NUMBER)-(NUMBER)	AUTONETICS, DOWNEY, CALIF.
EM-	AUTONETICS DIV., NORTH AMERICAN AVIATION, INC., ANAHEIM, CALIF. (NOT USED AFTER 1963)
EM-	BUREAU OF RECLAMATION (INTERIOR). COMMISSIONERS OFFICE, DENVER
EM-	CORPS OF ENGINEERS (ARMY)
EM-(NO.)-(NO.)-	CORPS OF ENGINEERS (ARMY)
EM-	GENERAL DYNAMICS/ASTRONAUTICS, SAN DIEGO, CALIF.
EM-	GT. BRIT. ARMAMENT RESEARCH DEPT., FORT HALSTEAD, KENT, ENGLAND (ELECTRODEPOSITION MEMO)
EM-	GULTON INDUSTRIES, INC. ENGINEERING MAGNETICS DIV., HAWTHORNE, CALIF.
EM-(YEAR)-	HOLLEY CARBURETOR CO. ELECTRO MECHANICAL DIV.,DETROIT
EM-	NATIONAL RESEARCH COUNCIL OF CANADA. ATOMIC ENERGY PROJECT, CHALK RIVER, ONT.
EM-	NEW YORK UNIV., N.Y.C.
EM-	NEW YORK UNIV., N.Y.C. COURANT INSTITUTE OF MATHEMATICAL SCIENCES
EM-	NEW YORK UNIV., N.Y.C. WASHINGTON SQUARE COLL. OF ARTS AND SCIENCE
EM-(NO.)-(NO.)	SPACE TECHNOLOGY LABS., INC., LOS ANGELES
EM-	TRW SYSTEMS. ENG. MECHANICS LAB., REDONDO BEACH, CAL
EM-	WESTINGHOUSE ELECTRIC CORP. ATOMIC EQUIP. DIV., CHESWICK, PENNA.
EMB-	DAVID TAYLOR MODEL BASIN, CARDEROCK, MD.

EMB-	NAVY (ELECTRONIC MAINTENANCE BOOK)
EMBW-FBK-	HAMBURG. UNIVERSITAET. INST. FUER EXPERIMENTAL PHYSIK
EMC-	ELECTRONICS MAINTENANCE ENGINEERING CTR., NORFOLK,VA.
EMC-(YEAR)-TR-	PHILCO CORP. COMMUNICATIONS+ELECTRONICS DIV., PHILA.
EMCO-	ELECTRO-MECHANICS CO., AUSTIN, TEX.
EMCO-	ELECTRO-MECHANICS CO., DALLAS
EMC/P-	UNITED KINGDOM ATOMIC ENERGY AUTHORITY. INDUSTRIAL GROUP H.Q., RISLEY, LANCS, ENGLAND
EMD-	ELECTRONIQUE MARCEL D'ASSAULT, SAINT-CLOUD, FRANCE
EMD-	MICHIGAN. UNIV., ANN ARBOR
EMD-	MICHIGAN. UNIV., YPSILANTI. WILLOW RUN RESEARCH CTR.
EME-(NO-NO)-(YR.CLASS)-	VIRGINIA. UNIV., CHARLOTTESVILLE. RESEARCH LABS. FOR ENGINEERING SCIENCES
EMEC-	ELECTRONICS MAINTENANCE ENGINEERING CTR., NORFOLK,VA.
EMERSON-	EMERSON ELECTRIC CO. STAR DEPT., SANTA BARBARA, CAL.
EMF-	MICHIGAN. UNIV., YPSILANTI. WILLOW RUN RESEARCH CTR.
EMG-	CONVAIR-ASTRONAUTICS, SAN DIEGO, CALIF.
EMG-	MICHIGAN. UNIV., YPSILANTI. WILLOW RUN RESEARCH CTR.
EMI-	VIRGINIA. UNIV., CHARLOTTESVILLE. RESEARCH LABS. FOR ENGINEERING SCIENCES
EMIC-TR-	E.M.I. COSSOR ELECTRONICS, LTD., DARTMOUTH, N.S.
EMI-DMP-	E.M.I. ELECTRONICS, LTD., HAYES, KENT, ENGLAND
EMI/WB-(MO./YR.)	GERMANY. ERNST-MACH-INSTITUT FREIBURG/BR. DER FRAUNHOFER-GESELLSCHAFT ZUR FORDERUNG DER ANGEWANDTEN FORSCHUNG
EML-	EARTHQUAKE MECHANISM LAB., SAN FRANCISCO
EML-	WHITE SANDS MISSILE RANGE. ELECTRO-MECH. LABS.,N.MEX.
EMM-(YEAR)-	EMMANUEL COLL., BOSTON
EMMT-LAB-TR-	ARIZONA. UNIV., TUCSON. ENERGY MASS AND MOMENTUM TRANSFER LAB.
EMORYU-QR-	EMORY UNIV., ATLANTA
EMP-2-	AIR FORCE WEAPONS LAB., KIRTLAND AFB, N. MEX. (ELECTROMAGNETIC PULSE THEORETICAL NOTES)
EMP-	NEW YORK UNIV., N.Y.C.
EMR-	ARMY MISSILE TEST AND EVALUATION DIRECTORATE, WHITE SANDS MISSILE RANGE, N. MEX.
EM-R-	COLUMBIA UNIV., NYC. SUBSTITUTE ALLOY MATERIALS LABS.
EMR-	GT. BRIT. EXPLOSIVES RESEARCH AND DEVELOPMENT ESTABLISHMENT, WALTHAM ABBEY, ESSEX, ENGLAND
EMRL-	BROWN UNIV., PROVIDENCE. ENG. MATERIALS RES. LAB.
EMRL-	TEXAS. UNIV., AUSTIN. ENGINEERING MECHANICS RES. LAB.
EMRL-RM-	BROWN UNIV., PROVIDENCE. ENG. MATERIALS RES. LAB.
EMRL-RM-	TEXAS. UNIV., AUSTIN. ENGINEERING MECHANICS RES. LAB.
EMRL-TR-	BROWN UNIV., PROVIDENCE. ENG. MATERIALS RES. LAB.
EMRL-TR-	TEXAS. UNIV., AUSTIN. ENGINEERING MECHANICS RES. LAB.
EMR-PD-	ELECTRO-MECHANICAL RESEARCH, INC. PHOTOELECTRIC DIV., PRINCETON, N.J.
EMR-PE-	EMR-PHOTOELECTRIC, PRINCETON, N.J.
EMR-POA-	EMR-PHOTOELECTRIC, PRINCETON, N.J.
EMS-	SHELL DEVELOPMENT CO., EMERYVILLE, CALIF.
EMS-J-	ELECTROMAGNETIC SCIENCES, INC., ATLANTA
EMV-	MICHIGAN. UNIV., YPSILANTI. WILLOW RUN RESEARCH CTR.
EN-	(ENGINEERING NOTE) (REF. 2)
EN-	(ENGINEERING REPORT) (REF. 2)
(NUMBERS)-EN-	ATLANTIC RESEARCH CORP. MISSILE SYSTEMS DIV., COSTA MESA, CALIF.
EN-	CHEMICAL CORPS. ENG. COMMAND, ARMY CHEM. CTR., MD.
EN-	FRANCE. COMMISSARIAT A L'ENERGIE ATOMIQUE. CENTRE E'ETUDES NUCLEAIRES, SACLAY
ENASR-	CHEMICAL CORPS ENG. AGENCY, ARMY CHEM. CTR., MD.
ENATR-	CHEMICAL CORPS ENG. AGENCY, ARMY CHEM. CTR., MD.
ENC-	ENGINEERING COMMAND (ARMY)
ENCR-	CHEMICAL CORPS ENG. AGENCY, ARMY CHEM. CTR., MD.
ENCTR-	CHEM. CORPS. RES. + ENG. COMMAND, ARMY CHEM. CTR.,MD.
ENDF-	NATIONAL NEUTRON CROSS SECTIONS CENTER, BROOKHAVEN NATL. LAB., N.Y. (EVALUATED NUCLEAR DATA FILE)
ENEA-	EUROPEAN NUCLEAR ENERGY AGENCY, PARIS
ENG-(NUMBER)-E-	INTERNATIONAL SILVER CO. TIMES WIRE AND CABLE DIV., WALLINGFORD, CONN.
ENG.-	MALLORY (P.R.) AND CO., INC., INDIANAPOLIS
ENG-(NO. INITIAL(S))	TENNESSEE EASTMAN CORP., OAK RIDGE, TENN.
ENG BEB TM-	OFFICE OF THE CHIEF OF ENGINEERS (ARMY). BEACH EROSION BOARD (TECHNICAL MEMORANDUM)
ENG EB-	AIR MATERIEL COMMAND. ENGINEERING DIV., WRIGHT-PATTERSON AFB, OHIO
ENG EB-	CORPS OF ENGINEERS (ARMY). ENGINEER BOARD
ENG EB CR-	CORPS OF ENGINEERS (ARMY). ENGINEER BOARD
ENG EB DTB RPT-	CORPS OF ENGINEERS (ARMY). ENGINEER BOARD. DESERT TEST BRANCH
ENG EB IR-	CORPS OF ENGINEERS (ARMY). ENGINEER BD.(INTERIM RPT.)
ENG EB MPS-	CORPS OF ENGINEERS (ARMY). ENGINEER BOARD. MAP PROJECTION SECTION
ENG EB PROJ (LETTER(S))-	CORPS OF ENGINEERS (ARMY). ENGINEER BD. (PROJECT)
ENG EB R-	CORPS OF ENGINEERS (ARMY). ENGINEER BOARD
ENG EB REG-	CORPS OF ENGINEERS (ARMY). ENGINEER BD.(REGISTER NO.)
ENG EB REPT-	CORPS OF ENGINEERS (ARMY). ENGINEER BOARD
ENG EN-	CORPS OF ENGINEERS (ARMY). ENGINEER BD. (ENG. NOTE)
ENGG/HT-	PAKISTAN. ATOMIC ENERGY CENTRE. ENG. DIV., LAHORE
ENGINEERING BULL-	RHODE ISLAND. UNIV.,KINGSTON. DIV. OF ENG. RES. + DEV
ENGINEERING CIRCULAR-	BUREAU OF NAVAL WEAPONS REPRESENTATIVE. METROLOGY ENGINEERING CENTER, POMONA, CALIF.
ENGINEERING MONO-	BUREAU OF RECLAMATION (INTERIOR), DENVER
ENGINEERING-NOTES-	CORPS OF ENGINEERS (ARMY). BEACH EROSION BD., WASH,DC
ENGINEERING PAPER-	DOUGLAS AIRCRAFT CO., INC., LONG BEACH, CALIF.
ENGINEERING PAPER-	DOUGLAS AIRCRAFT CO., INC., SANTA MONICA, CALIF.
ENGINEERING REPORT AD-	NAVY, WASHINGTON, D.C.
ENGINEERING RPT.-	(ENGINEERING REPORT) (REF. 2)
ENGINEERING RESEARCH-	PENNSYLVANIA STATE UNIV., UNIVERSITY PARK. COLL. OF ENGINEERING
EN/GI/PS-PW/LP-	ATELIERS DE CONSTRUCTIONS ELECTRIQUES DE CHARLEROI, BELGIUM
ENGORD-	CALIFORNIA INST. OF TECH., PASADENA
ENG. PAPER-	DOUGLAS AIRCRAFT CO., INC., SANTA MONICA, CALIF.
ENG. PROGRESS RPT.-	GENERAL ELECTRIC CO. AIRCRAFT GAS TURBINE DIV.
ENG. PROGRESS RPT.-	PRATT AND WHITNEY AIRCRAFT DIV.,UNITED AIRCRAFT CORP. (ALL LOCATIONS)
ENG. SPEC.-	CALIFORNIA. UNIV., BERKELEY. RADIATION LAB. (REF. 9)
ENG TTS-	ARMY FORCES IN THE PACIFIC (ENGINEER, TECHNICAL AND TECHNOLOGICAL SURVEY REPORT)
ENQ-	GT. BRIT. DEPT. OF ATOMIC ENERGY (REF. 36)
EN-R-	CARBIDE AND CARBON CHEMICALS CORP. SUBSTITUTE ALLOY MATERIALS LABS., N.Y.C.
EN-R-	COLUMBIA UNIV., NYC. SUBSTITUTE ALLOY MATERIALS LABS.
ENRNT-	AIR MATERIEL COMMAND. WATSON LABS., RED BANK, N.J.
ENS-	CALIFORNIA. UNIV., LIVERMORE. LAWRENCE RADIATION LAB.
ENVIRONMENTAL REVIEW-	NATIONAL INSTITUTE OF ENVIRONMENTAL HEALTH SCIENCES, RESEARCH TRIANGLE PARK, N.C.
ENW-	CALIFORNIA. UNIV., LIVERMORE. LAWRENCE RADIATION LAB.
EO-	(ENGINEERING ORDER) (REF. 2)
EO-	ARMY, WASHINGTON, D.C. (ENGINEERING ORDER)
EO-	COOK ELECTRIC CO. TECH-CTR. DIV., MORTON GROVE, ILL.
(NUMBER-NO.)-EO-(NO.)-	GENERAL ELECTRIC CO. MISSILE AND SPACE DIV., PHILA.

EO-(NUMBER)-

E.O.-(NUMBER)-	GOODYEAR AIRCRAFT CORP., AKRON, OHIO
E.O.-(NUMBER)-	MARQUARDT AIRCRAFT CO., VAN NUYS, CALIF.
EO-	MINNESOTA MINING AND MFG. CO., ST. PAUL
EOARDC-	AIR RESEARCH AND DEV. COMMAND. EUROPEAN OFFICE
EODB-	BUREAU OF NAVAL WEAPONS (EXPLOSIVE ORDNANCE DISPOSAL BULLETIN)
EODL-	NAVAL EXPLOSIVE ORDNANCE DISPOSAL FACILITY, INDIAN HEAD, MD. (EXPLOSIVE ORDNANCE DISPOSAL LETTER)
EODR-	NAVAL EXPLOSIVE ORDNANCE DISPOSAL FACILITY, INDIAN HEAD, MD. (EXPLOSIVE ORDNANCE DISPOSAL REPORT)
EODS-	EXPLOSIVE ORDNANCE DISPOSAL SCHOOL (NAVY)
EO-M-	COLUMBIA UNIV., NYC. SUBSTITUTE ALLOY MATERIALS LABS.
EOO-	GTE SYLVANIA, INC. ELECTRO-OPTICS ORGANIZATION, MOUNTAIN VIEW, CALIF.
(YR.)-(NO.)EO-RADIO-	WESTINGHOUSE ELECTRIC CORP., PITTSBURGH
EOS-	ELECTRO-OPTICAL SYSTEMS, INC., PASADENA, CALIF.
EOS-(NUMBER)-(LETTER)	ELECTRO-OPTICAL SYSTEMS, INC., PASADENA, CALIF.
EOS-(NUMBER)-TDR-	ELECTRO-OPTICAL SYSTEMS, INC., PASADENA, CALIF.
EOSI-	ELECTRO-OPTICAL SYSTEMS, INC., PASADENA, CALIF.
EO-TR-	BENDIX CORP. ELECTRO-OPTICS DIV., ANN ARBOR, MICH.
EP-	(ENGINEERING PAPER) (REF. 2)
EP-	(EXTERNAL PUBLICATION) (REF. 2)
EP-	AEROJET-GENERAL CORP., SACRAMENTO, CALIF.
EP-	AMERICAN GEOGRAPHICAL SOCIETY. DEPT. OF EXPLORATION AND FIELD RESEARCH, N.Y.C.
EP-(NUMBER)-0	BOELKOW ENTWICKLUNGEN K.G., MUNICH
E-P(MONTH/YEAR)	CALIFORNIA. UNIV., BERKELEY. RADIATION LAB. (REF. 9)
EP-	CALIF. INST. OF TECH., PASADENA. JET PROPULSION LAB.
EP-	CANADA. DEFENCE RESEARCH MEDICAL LABS., TORONTO
EP-	EAGLE-PICHER CO., JOPLIN, MO.
EP-	FAIRCHILD HILLER CORP. REPUBLIC AVIATION DIV., FARMINGDALE, N.Y.
EP(YEAR)-	GENERAL MOTORS CORP. AC ELECTRONICS DIV., MILWAUKEE
EP-(4 DIGITS)	GENERAL MOTORS CORP. AC ELECTRONICS DIV., MILWAUKEE
EP-	IOWA STATE COLL., AMES. STATISTICAL LAB.
EP-	MISSOURI. UNIV., COLUMBIA
EP-	QUARTERMASTER RESEARCH + DEV. CTR., NATICK, MASS.
EP-	QUARTERMASTER RESEARCH AND ENGINEERING COMMAND, NATICK, MASS. (ENVIRONMENT PROTECTION)
EP-	ROYAL CANADIAN A.F. OPERATIONAL RES. SEC., OTTAWA
EP-	UNITED AIRCRAFT CORP. HAMILTON STANDARD DIV., WINDSOR LOCKS, CONN.
EP-	VIRGINIA. UNIV., CHARLOTTESVILLE. RESEARCH LABS. FOR ENGINEERING SCIENCES.
EPA-CPA-(NUMBER-YEAR)-	ENVIRONMENTAL PROTECTION AGENCY, WASHINGTON, D.C. (CATEGORICAL PROGRAMS)
EPA-SW-(NUMBER-LTRS.)-	ENVIRONMENTAL PROTECTION AGENCY, WASHINGTON, D.C. (SOLID WASTE)
EPA-WQO-(NUMBER)-	ENVIRONMENTAL PROTECTION AGENCY. WATER QUALITY OFFICE, WASHINGTON, D.C.
EPA-WQO-(NO.)-DQH-DATE)	ENVIRONMENTAL PROTECTION AGENCY. WATER QUALITY OFFICE, WASHINGTON, D.C.
EPA-WQO-(NO.)-EBZ-DATE)	ENVIRONMENTAL PROTECTION AGENCY. WATER QUALITY OFFICE, WASHINGTON, D.C.
EPA-WQO-(NO.-LTRS.-DATE)	ENVIRONMENTAL PROTECTION AGENCY. WATER QUALITY OFFICE, WASHINGTON, D.C.
EPC-	EAGLE-PICHER CO., JOPLIN, MO.
EPC-	ENGINEERING PHYSICS CO., ROCKVILLE, MD.
EPCO-(YEAR)-	ENGINEERING PHYSICS CO., ROCKVILLE, MD.
EPD-	CALIF. INST. OF TECH., PASADENA. JET PROPULSION LAB.
EPD-	HUGHES AIRCRAFT CO., NEWPORT BEACH, CALIF. (ELECTRONIC PRODUCTS)
EPE-	ROYAL CANADIAN AIR FORCE. CENTRAL EXPERIMENTAL AND PROVING ESTAB., ROCKCLIFFE, ONT.
EP-F(NUMBER)(LTR.)-(YR)-	VARO, INC., GARLAND, TEX.
EP-F(NUMBER)/I/	VARO, INC., GARLAND, TEX.
EPG-	ARMY ENGINEER RESEARCH + DEV. LABS., FT. BELVOIR, VA.
EPIC-	ENGINE PROGRAM INFORMATION CENTER, WASHINGTON, D.C.
EPIC-	HUGHES AIRCRAFT CO. ELECTRONIC PROPERTIES INFORMATION CENTER, CULVER CITY, CALIF.
EPIC-DS-	HUGHES AIRCRAFT CO. ELECTRONIC PROPERTIES INFORMATION CENTER, CULVER CITY, CALIF.
EPIC-IR-	HUGHES AIRCRAFT CO. ELECTRONIC PROPERTIES INFORMATION CENTER, CULVER CITY, CALIF.
EPIC-S-	HUGHES AIRCRAFT CO. ELECTRONIC PROPERTIES INFORMATION CENTER, CULVER CITY, CALIF.
EPL-	MASSACHUSETTS INST. OF TECH., CAMBRIDGE. ENGINEERING PROJECTS LAB.
EPL-TR-	MICHIGAN. UNIV., ANN ARBOR. ELECTRON PHYSICS LAB.
EPMR-	FAIRCHILD STRATOS CORP., HAGERSTOWN, MD.
EPR-	ARMY MISSILE COMMAND, REDSTONE ARSENAL, ALA.
EPR-	ARMY NATICK LABS. PIONEERING RESEARCH DIV., MASS.
EPRA-	ELECTRONICS PRODUCTION RESOURCES AGENCY, WASH., D.C.
EPRL-	EAGLE-PICHER CO. RESEARCH LABS., JOPLIN, MO.
EPRL-	EAGLE-PICHER CO. RESEARCH LABS., MIAMI, OKLA.
EPRL-(LETTERS)-	EAGLE-PICHER CO. RESEARCH LABS., MIAMI, OKLA.
EP-RR-	AUSTRALIAN NATIONAL UNIV., CANBERRA. DEPT. OF ENGINEERING PHYSICS
EPS-	OFFICE OF THE QUARTERMASTER GENL. MIL. PLANNING DIV. RES. + DEV. BR. (ENVIRONMENTAL PROTECTION SERIES)
EPS-K-	MASS. INST.OF TECH.,OAK RIDGE,TENN. ENG. PRACTICE SCH
EPS-X-	MASS. INST.OF TECH.,OAK RIDGE,TENN. ENG. PRACTICE SCH
EPS-Y-	MASS. INST.OF TECH.,OAK RIDGE,TENN. ENG. PRACTICE SCH
EPT-	ARMY NATICK LABS., MASS.
EP-TR-	CANADA. DIRECTORATE OF BIOSCIENCES RESEARCH. ENVIRONMENTAL PROTECTION SECTION, OTTAWA
EQ-(YEAR)-	STANFORD RESEARCH INST., MENLO PARK, CALIF.
EQR/(YR)-	JOHNS HOPKINS UNIV., SILVER SPRING, MD. APPLIED PHYSICS LAB.
ER-	(ENGINEERING REPORT) (REF. 2)
ER-	(EVALUATION REPORT) (REF. 2)
ER-	(EXPLOSIVES REPORT) (REF. 2)
ER-	(EXTERNAL REPORT) (REF. 2)
ER-	AEROJET-GENERAL CORP., SACRAMENTO, CALIF.
ER-	AERONCA MFG. CORP., MIDDLETOWN, OHIO
ER-	AIRCRAFT ARMAMENTS, INC., COCKEYSVILLE, MD.
ER-	AMERICAN MACHINE AND FOUNDRY CO. GENERAL ENGINEERING LAB., GREENWICH, CONN.
ER-	AMF ATOMICS, INC., N.Y.C.
ER-	BUNKER-RAMO CORP. DEFENSE SYSTEMS DIV., SILVER SPRING, MD.
ER-	BUREAU OF ORDNANCE (NAVY) (EXPLOSIVES RES. RPT.)
ER-	CONN (C.G.) LTD., ELKHART, IND.
ER(NUMBERS)C-	DYNAMICS CORP. OF AMERICA. MASSA DIV., HINGHAM, MASS.
ER-	EMERSON AND CUMING, INC., CANTON, MASS.
ER-	ETHYL CORP. RESEARCH AND DEVELOPMENT DEPT., FERNDALE, MICH. (ENGINEERING RESEARCH)
ER-	FALCON RESEARCH AND DEVELOPMENT CO. TARGET AND VULNERABILITY ANALYSIS LAB., DENVER
ER-	GENERAL DYNAMICS/ELECTRONICS, SAN DIEGO, CALIF.
ER-	GENERAL ELECTRIC CO. AIRCRAFT NUCLEAR PROPULSION PROJECT, CINCINNATI
ER-	GENERAL MOTORS CORP. AEROPRODUCTS DIV., DAYTON, OHIO
ER-	GIANNINI CONTROLS CORP., DUARTE, CALIF.

ER-	INTELECTRON CORP., N.Y.C.
ER-	LINK AVIATION, INC., BINGHAMTON, N.Y.
ER-	LOCKHEED AIRCRAFT CORP. STRUCTURAL RESEARCH LAB., MARIETTA, GA.
ER-	LOCKHEED-GEORGIA CO., MARIETTA
ER-	LOCKHEED NUCLEAR PRODUCTS. GEORGIA DIV., MARIETTA
ER-	LURIA-COURNAND, INC., HAVRE DE GRACE, MD.
ER-(NUMBER-NUMBER)-	MARTIN CO. ELECTRONIC SYSTEMS AND PRODUCT DIV., BALTIMORE
ER-	MARTIN-MARIETTA CORP., BALTIMORE
ER-	MARTIN-MARIETTA CORP. DENVER DIV.
ER-	NAVAL RADIOLOGICAL DEFENSE LAB., SAN FRANCISCO
ER-(NO.)-(NO.)-	OFFICE OF THE CHIEF OF ENGINEERS (ARMY), WASH., D.C.
ER-	SOLAR, SAN DIEGO, CALIF.
ER-(NUMBER)-	SOLAR, SAN DIEGO, CALIF.
ER.(NUMBER)	SYDNEY. UNIV. SCHOOL OF PHYSICS
ER-	THOMPSON RAMO WOOLDRIDGE INC., CLEVELAND
ER-	THOMPSON RAMO WOOLDRIDGE INC. TAPCO DIV., CLEVELAND
ER-	TYCO LABS., INC., CLAREMONT, CALIF.
ERA-	ENGINEERING RESEARCH ASSOCIATES, TORONTO
ERA-	ENGINEERING RESEARCH ASSOCIATES, INC., WASHINGTON, D.C.
ERA-	NATIONAL RESEARCH COUNCIL OF CANADA. RADIO AND ELECTRICAL ENGINEERING DIV., OTTAWA
ER-A-	TENNESSEE EASTMAN CORP., OAK RIDGE, TENN.
ERAC-	SYMPOSIUM ON ELECTRON RING ACCELERATOR, 1968, BERKELEY, CALIF.
ERA-ER-	ENGINEERING RESEARCH ASSOCIATES, INC., WASHINGTON, D.C.
ER-APO-	AMERICAN MACHINE AND FOUNDRY CO., STAMFORD, CONN.
ERA-PR-	ENGINEERING RESEARCH ASSOCIATES, INC., ST. PAUL
ERA-TR-	ENGINEERING RESEARCH ASSOCIATES, INC.
ERB-	NATIONAL RESEARCH COUNCIL OF CANADA. RADIO AND ELECTRICAL ENGINEERING DIV., OTTAWA
ERC-	BUREAU OF MINES. EXPLOSIVES RESEARCH CENTER, PITTSBURGH
ERC-	ENVIRONMENTAL RESEARCH CORP., LAS VEGAS, NEV.
ERC-	ERIE RESISTOR CORP., PENNA.
ERC-	GEORGIA INST. OF TECH., ATLANTA. ENVIRONMENTAL RESOURCES CENTER
ERC-FH-(CONTRACT NO.)-	ARIZONA STATE UNIV., TEMPE. ENGINEERING RESEARCH CTR.
ERCO-	AMERICAN CAR + FOUNDRY CO., RIVERDALE, MD.
ERC-TMR-	ENVIRONMENTAL RESEARCH CORP., ALEXANDRIA, VA.
ERD-	AIR FORCE CAMBRIDGE RESEARCH CENTER. ELECTRONICS RESEARCH DIRECTORATE, MASS.
ERD-	COLUMBIA BROADCASTING SYSTEM, N.Y.C. (ENG. RES.+DEV.)
ERD-	EVANS RESEARCH AND DEVELOPMENT CORP., N.Y.C.
ERD-	OFFICE OF SCIENTIFIC RESEARCH AND DEVELOPMENT
ERDA-	ARMY ELECTRONICS RESEARCH AND DEVELOPMENT ACTIVITY, FORT HUACHUCA, ARIZ.
ERDA-	ARMY ELECTRONICS RESEARCH AND DEVELOPMENT ACTIVITY, WHITE SANDS MISSILE RANGE, N. MEX.
ERDA-	WHITE SANDS MISSILE RANGE. ELECTRONICS RESEARCH AND DEVELOPMENT ACTIVITY, N. MEX.
ERDAA-ELCT-	ARIZONA. UNIV., TUCSON
ERDAA-ELCT-	ARMY ELECTRONICS RESEARCH AND DEVELOPMENT ACTIVITY, FORT HUACHUCA, ARIZ.
ERDAA-MET-	ARMY ELECTRONICS RESEARCH AND DEVELOPMENT ACTIVITY, FORT HUACHUCA, ARIZ.
ERDAW-	ARMY ELECTRONICS RESEARCH AND DEVELOPMENT ACTIVITY, WHITE SANDS MISSILE RANGE, N. MEX.
ERDC-	EVANS RESEARCH AND DEVELOPMENT CORP., N.Y.C.
ERD-CRRD-TM-	AIR FORCE CAMBRIDGE RESEARCH LABS., BEDFORD, MASS.
ERDE-(NO.)/EMR/(YR)	GT. BRIT. EXPLOSIVES RESEARCH AND DEVELOPMENT ESTABLISHMENT, WALTHAM ABBEY, ESSEX, ENGLAND
ERDE-(NO.)/M/(YR)	GT. BRIT. EXPLOSIVES RESEARCH AND DEVELOPMENT ESTABLISHMENT, WALTHAM ABBEY, ESSEX, ENGLAND
ERDE-(NO.)/R/(YR)	GT. BRIT. EXPLOSIVES RESEARCH AND DEVELOPMENT ESTABLISHMENT, WALTHAM ABBEY, ESSEX, ENGLAND
ERDE-TM(NO.)/M/(YR.)	GT. BRIT. EXPLOSIVES RESEARCH AND DEVELOPMENT ESTABLISHMENT, WALTHAM ABBEY, ESSEX, ENGLAND
ERDE-TN-	GT. BRIT. EXPLOSIVES RESEARCH AND DEVELOPMENT ESTABLISHMENT, WALTHAM ABBEY, ESSEX, ENGLAND
ERDE-TR-	GT. BRIT. EXPLOSIVES RESEARCH AND DEVELOPMENT ESTABLISHMENT, WALTHAM ABBEY, ESSEX, ENGLAND
ERDL-	ARMY ENGINEER RESEARCH + DEV. LABS., FT. BELVOIR, VA.
ERDL-(NUMBER)-TR	ARMY ENGINEER RESEARCH + DEV. LABS., FT. BELVOIR, VA.
ERDL R-	ARMY ENGINEER RESEARCH + DEV. LABS., FT. BELVOIR, VA.
ERDL-T-	ARMY ENGINEER RESEARCH + DEV. LABS., FT. BELVOIR, VA. (TRANSLATION)
ERDL-TR-	ARMY ENGINEER RESEARCH + DEV. LABS., FT. BELVOIR, VA.
ERD-ORRD-TM-	AIR FORCE CAMBRIDGE RESEARCH LABS., BEDFORD, MASS.
ERD-PR-	EVANS RESEARCH AND DEVELOPMENT CORP., N.Y.C.
ERD-TN-(YEAR)-	AIR FORCE CAMBRIDGE RESEARCH LABS. ELECTRONICS RESEARCH DIRECTORATE, BEDFORD, MASS.
ERD-TN-	AIR RESEARCH AND DEVELOPMENT COMMAND. ELECTRONICS RESEARCH DIV., BEDFORD, MASS.
EREC-	ESSO RESEARCH AND ENGINEERING CO., LINDEN, N.J.
ERF-	FAIRCHILD HILLER CORP. REPUBLIC AVIATION DIV., FARMINGDALE, N.Y.
ERF-RR-	EYE RESEARCH FOUNDATION OF BETHESDA, MD.
ERG-	EASTERN RESEARCH GROUP, N.Y.C.
ERG/	GT. BRIT. ATOMIC EN. RES. ESTAB., HARWELL, BERKS, ENG
ERI-	ATOMIC ENERGY OF CANADA LTD., CHALK RIVER, ONT.
ERI-	ELECTRONICS RESEARCH, INC., EVANSVILLE, IND.
ERI-	IOWA STATE UNIV. OF SCIENCE AND TECHNOLOGY, AMES. ENGINEERING RESEARCH INST.
ERI-	MICHIGAN. UNIV., ANN ARBOR. ENGINEERING RES. INST.
ERI-(NUMBER)-(NO.)-T	MICHIGAN. UNIV., ANN ARBOR. ENGINEERING RES. INST.
ERIC-	OFFICE OF EDUCATION. EDUCATIONAL RESOURCES INFORMATION CENTER, WASHINGTON, D.C.
ERI/MICH-	MICHIGAN. UNIV., ANN ARBOR. ENGINEERING RES. INST.
ERL-	AIR FORCE CAMBRIDGE RESEARCH LABS., BEDFORD, MASS.
ERL-(NUMBER)-APCL-	ATMOSPHERIC PHYSICS AND CHEMISTRY LAB., BOULDER, COLO.
ERL-	CALIFORNIA. UNIV., BERKELEY. ELECTRONICS RES. LAB.
ERL-(YEAR)-	CALIFORNIA. UNIV., BERKELEY. ELECTRONICS RES. LAB.
ERL-	CALIFORNIA. UNIV., LIVERMORE. COLL. OF ENGINEERING
ERL-(NUMBER)-	MONTANA STATE UNIV., BOZEMAN. ELECTRONICS RES. LAB.
ERL-	RAMO-WOOLDRIDGE CORP. ELECTRONIC RES. LAB., LOS ANG.
ERL-	SMITH (A.O.) CORP., MILWAUKEE
ERL-	SYRACUSE UNIV., N.Y. RES. INST. ELECTRONICS RES. LAB.
ERL-(NUMBER)-WPL-	WAVE PROPAGATION LAB., BOULDER, COLO.
ERL-AOML-	ATLANTIC OCEANOGRAPHIC AND METEOROLOGICAL LABS., MIAMI, FLA.
ERL-APCL-	ATMOSPHERIC PHYSICS AND CHEMISTRY LAB., BOULDER, COLO.
ERL-ARL-	AIR RESOURCES LAB., RESEARCH TRIANGLE PARK, N.C.
ERL-BOMAP-	ENVIRONMENTAL RESEARCH LABS., ROCKVILLE, MD. (BARBADOS OCEANOGRAPHIC + METEOROLOGICAL ANALYSIS PROJ.)
ERL DP-	SIGNAL CORPS (ARMY). PICTORIAL ENGINEERING AND RESEARCH LAB., LONG ISLAND, N.Y. (DIV. REPT.)
ERL-E-	AIR MATERIEL COMMAND. ELECTRONICS RESEARCH LAB., CAMBRIDGE, MASS.
ERL-ESL-	EARTH SCIENCES LABS., BOULDER, COLO.
ERL-LM-	RAMO-WOOLDRIDGE CORP. ELECTRONIC RES. LAB., LOS ANG.
ERL-M-	CALIFORNIA. UNIV., BERKELEY. ELECTRONICS RES. LAB.
ERL-MMTC-	MARINE MINERALS TECHNOLOGY CENTER, TIBURON, CALIF.
ERL-NSSL-	NATIONAL SEVERE STORMS LAB., NORMAN, OKLA.

ERL-OD-

ERL-OD-	ENVIRONMENTAL RESEARCH LABS. OFFICE OF THE DIRECTOR, BOULDER, COLO.
ERL-SEL-	SPACE ENVIRONMENT LAB., BOULDER, COLO.
ERLTM-AOML-	ATLANTIC OCEANOGRAPHIC AND METEOROLOGICAL LABS., MIAMI, FLA.
ERLTM-AL-	AERONOMY LAB., BOULDER, COLO.
ERLTM-AOML-	ATLANTIC OCEANOGRAPHIC AND METEOROLOGICAL LABS., MIAMI, FLA.
ERLTM-ARL-	AIR RESOURCES LAB., RESEARCH TRIANGLE PARK, N.C.
ERLTM-BOMAP-	ENVIRONMENTAL RESEARCH LABS., ROCKVILLE, MD. (BARBADOS OCEANOGRAPHIC + METEOROLOGICAL ANALYSIS PROJ.)
ERLTM-NSSL-	NATIONAL SEVERE STORMS LAB., NORMAN, OKLA.
ERLTM-OD-	ENVIRONMENTAL SCIENCE SERVICES ADM., BOULDER, COLO.
ERLTM-SDL-	SPACE DISTURBANCES LAB., BOULDER, COLO.
ERL-TR-(NO.)-	CALIFORNIA. UNIV., BERKELEY. ELECTRONICS RES. LAB.
ERL-WPL-	WAVE PROPAGATION LAB., BOULDER, COLO.
ERM-	BUREAU OF ORDNANCE (NAVY) (EXPLOSIVES RES. MEMO.)
ERM-	ETHYL CORP. RESEARCH AND DEVELOPMENT DEPT., FERNDALE, MICH. (ENGINEERING RESEARCH MEMORANDA)
ERM-	LOCKHEED AIRCRAFT CORP., BURBANK, CALIF.
ERN-(NUMBER)/(YEAR)	UNITED KINGDOM ATOMIC ENERGY AUTHORITY. WEAPONS GROUP ATOMIC WEAPONS RES. ESTAB.,ALDERMASTON, BERKS, ENG. (EXPLOSIVES RESEARCH NOTE)
ERNA-AB-RTT-	ENTWICKLUNGSRING NORD GMBH, BREMEN
ERP-	AIR FORCE CAMBRIDGE RESEARCH LABS., BEDFORD, MASS.
ERPC-	EMERSON RADIO AND PHONOGRAPH CORP., WASHINGTON, D.C.
ERR-	(ENGINEERING RESEARCH REPORT) (REF. 2)
ERR-	BUREAU OF ORDNANCE (NAVY) (EXPLOSIVES RESEARCH RPT.)
ER-R-	COLUMBIA UNIV., NYC. SUBSTITUTE ALLOY MATERIALS LABS.
ERR-	LITTON INDUSTRIES. AMECOM DIV., SILVER SPRING, MD.
ERR-AN-	GENERAL DYNAMICS/ASTRONAUTICS, SAN DIEGO, CALIF.
ERR-AN-	GENERAL DYNAMICS/CONVAIR, SAN DIEGO, CALIF.
ERR-FW-	GENERAL DYNAMICS/FORT WORTH, TEX.
ERRL-	DEPARTMENT OF AGRICULTURE. EASTERN UTILIZATION RESEARCH BRANCH, WYNDMOOR, PENNA.
ERR-PO-	GENERAL DYNAMICS/POMONA, CALIF.
ERR-SD-	GENERAL DYNAMICS/CONVAIR, SAN DIEGO, CALIF.
ERS-	DEPARTMENT OF AGRICULTURE. ECONOMIC RESEARCH SERVICE, WASHINGTON, D.C.
ERS-(NUMBER)-	FAIRCHILD HILLER CORP. REPUBLIC AVIATION DIV., FARMINGDALE, N.Y.
ERS-	QUARTERMASTER RESEARCH AND ENGINEERING COMMAND, NATICK, MASS. (ENGINEERING RESEARCH STUDY)
ER-SB-	AMERICAN MACHINE AND FOUNDRY CO. ADVANCED PRODUCTS GROUP, SANTA BARBARA, CALIF.
ERTM-	NATIONAL WEATHER SERVICE. EASTERN REGION. GARDEN CITY, N.Y.
ER/TN-	AUSTRALIAN ATOMIC ENERGY COMMISSION. RESEARCH ESTABLISHMENT, LUCAS HEIGHTS, NEW SOUTH WALES
ER-UTC-(YEAR)-	UNITED TECHNOLOGY CENTER, SUNNYVALE, CALIF.
ER-UTC-	UNITED TECHNOLOGY CORP., SUNNYVALE, CALIF.
ER-X	COLUMBIA UNIV., NYC. SUBSTITUTE ALLOY MATERIALS LABS.
ES-	(ENGINEERING STUDY) (REF. 2)
ES-	(EXPERIMENTAL STATION) (REF. 2)
ES-	AMERICAN METAL PRODUCTS CO. ENGINEERING SCIENCE DIV., ANN ARBOR, MICH.
ES-	ARMY NATICK LABS. EARTH SCIENCES DIV., MASS.
ES-	ATOMIC ENERGY OF CANADA LTD. CHALK RIVER PROJECT, ONT. (REF. 6)
ES-	BABCOCK + WILCOX CO., N.Y.C.
ES-	CANADA. DEPT. OF MINES AND TECHNICAL SURVEYS
ES-	DOUGLAS AIRCRAFT CO., INC., SANTA MONICA, CALIF.
ES-	DOUGLAS AIRCRAFT CO., INC. EL SEGUNDO DIV., CALIF.
ES-	ENGINEERING-SCIENCE INC., ARCADIA, CALIF.
ES-	FOREIGN ECONOMIC ADMINISTRATION. ENEMY BRANCH (EXTERNAL ECONOMIC SECURITY STAFF REPORTS)
ES-	ROCK ISLAND ARSENAL LAB., ILL.
ESA-	AIR MATERIEL COMMAND, WRIGHT-PATTERSON AFB, OHIO (EXPERIMENTAL STRESS ANALYSIS)
ESA-MURA-	MIDWESTERN UNIVERSITIES RESEARCH ASSN., URBANA, ILL.
ESAP-(NUMBER)-	EDGEWOOD ARSENAL, MD.
ESB-E-	ELECTRIC STORAGE BATTERY CO. CARL F. NORBERG RESEARCH CENTER, YARDLEY, PENNA.
ESC-	ELECTRONIC SPECIALTY CO., LOS ANGELES
ESC-TR-(YEAR)-	ELECTRONIC SYSTEMS DIV. (AIR FORCE), BEDFORD, MASS.
ESD-	ELECTRONIC SYSTEMS DIV. (AIR FORCE), BEDFORD, MASS.
ESD-(NUMBER)-IR-	GENERAL INSTRUMENT CORP. ELECTRONIC SYSTEMS DIV., HICKSVILLE, N.Y.
ESDAC-	EUROPEAN SPACE DATA CENTRE, DARMSTADT, GERMANY
ESD-CR-	ELECTRONIC SYSTEMS DIV. (AIR FORCE), BEDFORD, MASS.
ESD-TD-	ELECTRONIC SYSTEMS DIV. (AIR FORCE), BEDFORD, MASS.
ESD-TDR-	ELECTRONIC SYSTEMS DIV. (AIR FORCE), BEDFORD, MASS. (TECHNICAL DOCUMENTARY REPORT)
ESD-TDR-(YEAR)-	ELECTRONIC SYSTEMS DIV. (AIR FORCE), BEDFORD, MASS. (TECHNICAL DOCUMENTARY REPORT)
ESD-TM-	NATIONAL WEATHER RECORDS CENTER, ASHEVILLE, N.C.
ESD-TN-	STANFORD RESEARCH INST. EARTH SCIENCES DEPT., MENLO PARK, CALIF.
ESD-TR-	ELECTRONIC SYSTEMS DIV. (AIR FORCE), BEDFORD, MASS.
ESD-TR-(YEAR)-	ELECTRONIC SYSTEMS DIV. (AIR FORCE), BEDFORD, MASS.
ESD-TR-(YEAR)-	INTERNATIONAL BUSINESS MACHINES CORP.,GAITHERSBURG,MD
ESE-	AIR MATERIEL COMMAND. SERVICES ENGINEERING SUBDIV., WRIGHT-PATTERSON AFB, OHIO
ESI-	EBASCO SERVICES, INC., N.Y.C.
E-SIR-	FAIRCHILD HILLER CORP. REPUBLIC AVIATION DIV., FARMINGDALE, N.Y.
ESL-	EARTH SCIENCES LABS., BOULDER, COLO.
ESL-	EATONTOWN SIGNAL LAB., N.J.
ESL-	EVANS SIGNAL LAB., BELMAR, N.J.
ESL-	MASSACHUSETTS INST. OF TECH., CAMBRIDGE. ELECTRONIC SYSTEMS LAB.
ESL-	NAVAL AIR STATION. ELECTRONICS STDS. LAB.,NORFOLK, VA
ESL-	YALE UNIV., NEW HAVEN. EDWARDS STREET LAB.
ESL EM-	EATONTOWN SIGNAL LAB., N.J. (ENGINEERING MEMORANDA)
ESL ER-	EATONTOWN SIGNAL LAB., N.J. (ENGINEERING REPORTS)
ESL-FR-	MASSACHUSETTS INST. OF TECH., CAMBRIDGE. ELECTRONIC SYSTEMS LAB.
ESL-IR-	MASSACHUSETTS INST. OF TECH., CAMBRIDGE. ELECTRONIC SYSTEMS LAB.
ESLI-TM-	ESL, INC., PALO ALTO, CALIF.
ESL-P-	MASSACHUSETTS INST. OF TECH., CAMBRIDGE. ELECTRONIC SYSTEMS LAB.
ESL-QR-	MASSACHUSETTS INST. OF TECH., CAMBRIDGE. ELECTRONIC SYSTEMS LAB.
ESL-R-	MASSACHUSETTS INST. OF TECH., CAMBRIDGE. ELECTRONIC SYSTEMS LAB.
ESL-SR-	MASSACHUSETTS INST. OF TECH., CAMBRIDGE. ELECTRONIC SYSTEMS LAB.
ESL TM-	EATONTOWN SIGNAL LAB., N.J.
ESL-TM-	MASSACHUSETTS INST. OF TECH., CAMBRIDGE. ELECTRONIC SYSTEMS LAB.
ESL TR-	EATONTOWN SIGNAL LAB., N.J.
ESN-	OFFICE OF NAVAL RESEARCH, LONDON
ESO-	ELECTRONIC SUPPLY OFFICE, GREAT LAKES, ILL.
ESP-(YR)-	DU PONT DE NEMOURS (E.I.) + CO. EXPERIMENTAL STATION WILMINGTON, DEL.

ESP-	MOTOROLA INC. MILITARY ELECTRONICS DIV., SCOTTSDALE, ARIZ.
ESP-	NAVAL EXPLOSIVE ORDNANCE DISPOSAL FACILITY, INDIAN HEAD, MD. (EMERGENCY SAFING PROCEDURE)
E-SPEC-	CALIFORNIA. UNIV., BERKELEY. RADIATION LAB. (REF. 9)
ESR-	(ENGINEERING SUMMARY REPORT) (REF. 2)
E.S.R.-	ACADEMIA R.S.R.INSTITUTUL DE FIZICA ATOMICA,BUCHAREST
ESR-	AEROJET-GENERAL CORP., SACRAMENTO, CALIF.
ES-R-	COLUMBIA UNIV., NYC. SUBSTITUTE ALLOY MATERIALS LABS.
ESRD-(NO.)-	REPUBLIC AVIATION CORP., FARMINGDALE, N.Y.
ESRIN-IN-	EUROPEAN SPACE RESEARCH INST., FRASCATI, ITALY
ESRO-	EUROPEAN SPACE RESEARCH ORGANIZATION, PARIS
ESRO-CR-	EUROPEAN SPACE RESEARCH ORGANIZATION, PARIS
ESRO-SM(NUMBER/ESDAC/	EUROPEAN SPACE DATA CENTRE, DARMSTADT, GERMANY
ESRO-SM-	EUROPEAN SPACE RESEARCH ORGANIZATION, PARIS
ESRO-SN-(NO.)/ESDAC/	EUROPEAN SPACE DATA CENTRE, DARMSTADT, GERMANY
ESRO-SN-(NUMBER)/ESLAB/	EUROPEAN SPACE RESEARCH LAB., NOORDWIJK, NETHERLANDS
ESRO-SN-	EUROPEAN SPACE RESEARCH ORGANIZATION, PARIS
ESRO-SN-(NUMBER)/ESLAB	INSTITUTE OF METEOROLOGY, STOCKHOLM
ESRO-SP-	EUROPEAN SPACE RESEARCH ORGANIZATION, PARIS
ESRO-SR-(NO.)/ESDAC/	EUROPEAN SPACE DATA CENTRE, DARMSTADT, GERMANY
ESRO-SR-	EUROPEAN SPACE RESEARCH ORGANIZATION, PARIS
ESRO-TM-	EUROPEAN SPACE RESEARCH ORGANIZATION, PARIS
ESRO-TM/(LETTER)/-	EUROPEAN SPACE TECHNOLOGY CENTER, NOORDWIJK, NETHERLANDS
ESRO-TN-	EUROPEAN SPACE TECHNOLOGY CENTER, DELFT, NETHERLANDS
ESRO-TN-(NUMBER)/ESTEC/	EUROPEAN SPACE TECHNOLOGY CENTER, NOORDWIJK, NETHERLANDS
ESRO-TR-	EUROPEAN SPACE RESEARCH ORGANIZATION, PARIS
ESRP-TN-	STANFORD UNIV., CALIF.
ESSA-EDSTM-	ENVIRONMENTAL DATA SERVICE, SILVER SPRING, MD.
ESSA-ERL-(NUMBER)-APCL-	ATMOSPHERIC PHYSICS AND CHEMISTRY LAB., BOULDER,COLO.
ESSA-ERL-(NO.)-(LTRS.)-	ENVIRONMENTAL SCIENCE SERVICES ADM., BOULDER, COLO.
ESSA-ERL-TM-	ENVIRONMENTAL SCIENCE SERVICES ADM., BOULDER, COLO.
ESSA-ERLTM-ARL-	AIR RESOURCES LAB., RESEARCH TRIANGLE PARK, N.C.
ESSA-IER-(NO.)-ITSA-	INSTITUTE FOR TELECOMMUNICATION SCIENCES AND AERONOMY, BOULDER, COLO.
ESSA-IER-FB-	INSTITUTES FOR ENVIRONMENTAL RESEARCH, BOULDER, COLO.
ESSA-IERTM-EML-	EARTHQUAKE MECHANISM LAB., SAN FRANCISCO
ESSA-SDL-	SPACE DISTURBANCES LAB., BOULDER, COLO.
ESSA-TM-CRTM-	NATIONAL WEATHER SERVICE. CENTRAL REGION, KANSAS CITY, MO.
ESSA-TM-EDSTM-	ENVIRONMENTAL DATA SERVICE, SILVER SPRING, MD.
ESSA-TM-ERLTM-AL-	AERONOMY LAB., BOULDER, COLO.
ESSA-TM-ERLTM-AOML-	ATLANTIC OCEANOGRAPHIC AND METEOROLOGICAL LABS., MIAMI, FLA.
ESSA-TM-ERLTM-NHRL-	NATIONAL HURRICANE RESEARCH LAB., MIAMI, FLA.
ESSA-TM-ERLTM-NSSL-	NATIONAL SEVERE STORMS LAB., NORMAN, OKLA.
ESSA-TM-NESCTM-	NATIONAL ENVIRONMENTAL SATELLITE SERVICE, WASH., D.C.
ESSA-TM-SRTM-	NATIONAL WEATHER SERVICE. SOUTHERN REGION, FT. WORTH, TEX.
ESSA-TM-WBTM-CR-	NATIONAL WEATHER SERVICE. CENTRAL REGION, KANSAS CITY, MO.
ESSA-TM-WBTM-ENG-	NATIONAL WEATHER SERVICE. ENG. DIV., SILVER SPRING,MD
ESSA-TM-WBTM-FCST-	NATIONAL WEATHER SERVICE. WEATHER ANALYSIS + PREDICTION DIV.,SILVER SPRING,MD. (NOTES TO FORECASTERS)
ESSA-TM-WBTM-NMC-	NATIONAL METEOROLOGICAL CENTER, SUITLAND, MD.
ESSA-TM-WBTM-PR-	NATIONAL WEATHER SERVICE. PACIFIC REGION, HONOLULU
ESSA-TM-WBTM-SOS-	NATIONAL WEATHER SERVICE. SPACE OPERATIONS SUPPORT DIV., SILVER SPRING, MD.
ESSA-TM-WBTM-TDL-	NATIONAL WEATHER SERVICE. TECHNIQUES DEVELOPMENT LAB., SILVER SPRING, MD.
ESSA-TM-WBTM-WR-	NATIONAL WEATHER SERVICE. WESTERN REGION, SALT LAKE CITY
ESSA-TM-WRTM-	NATIONAL WEATHER SERVICE. WESTERN REGION, SALT LAKE CITY
ESSA-TR-C+GS	NATIONAL OCEAN SURVEY, ROCKVILLE, MD.
ESSA-TR-EDS-	ENVIRONMENTAL DATA SERVICE, SILVER SPRING, MD.
ESSA-TR-ERL-(NO.)-AL-	AERONOMY LAB., BOULDER, COLO.
ESSA-TR-ERL-(NO.)-APCL-	ATMOSPHERIC PHYSICS AND CHEMISTRY LAB., BOULDER,COLO.
ESSA-TR-ERL-(NO.)-ESL-	EARTH SCIENCES LABS., BOULDER, COLO.
ESSA-TR-ERL-(NO.)-OD-	ENVIRONMENTAL RESEARCH LABS. OFFICE OF THE DIRECTOR, BOULDER, COLO.
ESSA-TR-ERL-	ENVIRONMENTAL SCIENCE SERVICES ADM., BOULDER, COLO.
ESSA-TR-ERL-(NO.)-ITS-	INSTITUTE FOR TELECOMMUNICATION SCIENCES,BOULDER,COLO
ESSA-TR-ERL-(NO.)-POL-	PACIFIC OCEANOGRAPHIC LAB., SEATTLE
ESSA-TR-ERL-(NO.)-SDL-	SPACE DISTURBANCES LAB., BOULDER, COLO.
ESSA-TR-ERL-(NO.)-WPL-	WAVE PROPAGATION LAB., BOULDER, COLO.
ESSA-TR-ERL-APCL-	ATMOSPHERIC PHYSICS AND CHEMISTRY LAB., BOULDER,COLO.
ESSA-TR-ESL-(NO.)-ESL-	EARTH SCIENCES LABS., BOULDER, COLO.
ESSA-TR-IER-	INSTITUTES FOR ENVIRONMENTAL RESEARCH, BOULDER, COLO.
ESSA-TR-NESC-	NATIONAL ENVIRONMENTAL SATELLITE SERVICE, WASH., D.C.
ESSA-TR-WB-	NATIONAL WEATHER SERVICE. OFFICE OF HYDROLOGY, SILVER SPRING, MD.
ESSA-TR-WB-	WEATHER BUREAU, SILVER SPRING, MD.
ESSA-WB-T-	ENVIRONMENTAL DATA SERVICE, SILVER SPRING, MD.
ESSA-WBTM-ER-	NATIONAL WEATHER SERVICE. EASTERN REGION, GARDEN CITY, N.Y.
ESSA-WBTM-SPDD-	NATIONAL WEATHER SERVICE. SYSTEMS PLANS AND DESIGN DIV., SILVER SPRING, MD.
ESSA-WBTM-SR-	NATIONAL WEATHER SERVICE. SOUTHERN REGION, FT. WORTH, TEX.
ESSA-WBTM-TDL-	NATIONAL WEATHER SERVICE. TECHNIQUES DEVELOPMENT LAB., SILVER SPRING, MD.
ESSA-WBTM-T+EL-	NATIONAL WEATHER SERVICE. TEST AND EVALUATION LAB., STERLING, VA.
ESSA-WBTM-WR-	NATIONAL WEATHER SERVICE. WESTERN REGION, SALT LAKE CITY
ESS B-	SUPREME COMMANDER FOR THE ALLIED POWERS. ECONOMIC AND SCI. SEC. RES. + STATISTICS DIV. (BULL.)
ESS/ES-	BRITISH AIRCRAFT CORP., FILTON, GLOUCESTER, ENG.
ESSO-	ESSO RESEARCH AND ENGINEERING CO., LINDEN, N.J.
ESSO-(YEAR)-	ESSO RESEARCH AND ENGINEERING CO., LINDEN, N.J.
ESSO-GR-(NO.)-FBP-(YR.)	ESSO RESEARCH AND ENGINEERING CO., LINDEN, N.J.
ESSO-GR-(LETTERS)-(YR.)	ESSO RESEARCH AND ENGINEERING CO., LINDEN, N.J.
ESSO-MA-	ESSO RESEARCH AND ENGINEERING CO., LINDEN, N.J.
ESS SR-	SUPREME COMMANDER FOR THE ALLIED POWERS. ECONOMIC AND SCIENTIFIC SEC. RES. DIV. (SPEC. RPTS.)
ESS/SS-	BRITISH AIRCRAFT CORP., BRISTOL, GLOUCESTER, ENG.
ES STUDY-	ARMY ENGINEER SCHOOL, FORT BELVOIR, VA.
EST-	AIR MATERIEL COMMAND. ENGINEERING STANDARDS SEC.
ESTEC-	EUROPEAN SPACE TECHNOLOGY CENTRE, DELFT, NETHERLANDS
ESTEC-MT-	EUROPEAN SPACE TECHNOLOGY CENTRE, DELFT, NETHERLANDS
ESTR-	AEROJET-GENERAL CORP., SACRAMENTO, CALIF.
ESYS-	ELECTRONIC SYSTEMS DIV. (AIR FORCE), BEDFORD, MASS.
ET-	AIR FORCE CAMBRIDGE RESEARCH LABS., BEDFORD, MASS.
ET-	AIR FORCE EASTERN TEST RANGE, PATRICK AFB, FLA.
ET-(YEAR)-	AIR FORCE EASTERN TEST RANGE, PATRICK AFB, FLA.
ET(YR.)-	BOEING ATLANTIC TEST CENTER, PATRICK AFB, FLA.
ET-(YEAR)-	BUREAU OF INTERNATIONAL COMMERCE, WASHINGTON, D.C. (ECONOMIC TRENDS)

ET-	NAVAL AIR TEST CENTER. ELECTRONICS TEST DIV., PATUXENT RIVER, MD.
ETAC-IN-	ENVIRONMENTAL TECHNICAL APPLICATIONS CENTER (AIR FORCE), WASHINGTON, D.C.
ETAC-TN-(YEAR)-	ENVIRONMENTAL TECHNICAL APPLICATIONS CENTER (AIR FORCE), WASHINGTON, D.C.
ETC-(NUMBER)-	DUMONT ELECTRON TUBES, CLIFTON, N.J.
E-T-CH-	EMMANUEL COLL., BOSTON
ETDIR	DUGWAY PROVING GROUND, TOOELE, UTAH
ETDR	AEROJET-GENERAL CORP., SACRAMENTO, CALIF.
ETF-	CHEMICAL CORPS, ARMY CHEMICAL CENTER, MD.
ETF-(NUMBER)-	CHEMICAL CORPS, ARMY CHEMICAL CENTER, MD.
E-T-F-	EMMANUEL COLL., BOSTON
E-T-FC-	EMMANUEL COLL., BOSTON
E-T-G-	EMMANUEL COLL., BOSTON
ETH-	SWITZERLAND. EIDGENOESSISCHE TECHNISCHE HOCHSCHULE, ZURICH
ETHYL-RM-	ETHYL CORP. RESEARCH AND DEV. DEPT., FERNDALE, MICH.
ETI-CR-	EFFECTS TECHNOLOGY, INC., SANTA BARBARA, CALIF.
E-T-J-	EMMANUEL COLL., BOSTON
ETL-	AIRTRON, INC., LINDEN, N.J.
ETL-	BUREAU OF MINES. ELECTROTECHNICAL LAB., NORRIS, TENN.
ETL-	FAIRCHILD HILLER CORP. REPUBLIC AVIATION DIV., FARMINGDALE, N.Y.
ETL-	PORTSMOUTH NAVAL SHIPYARD. ELEC. TESTING LAB., N.H.
ETL-ETR-	ARMY ENGINEER TOPOGRAPHIC LABS., FORT BELVOIR, VA.
ETL-N-	PORTSMOUTH NAVAL SHIPYARD. ELEC. TESTING LAB., N.H.
ETLR-	DUGWAY PROVING GROUND, TOOELE, UTAH
ETM-	(EXTERNAL TECHNICAL MEMORANDUM) (REF. 2)
ETM-	ACF INDUSTRIES, INC., HYATTSVILLE, MD.
ETM-	ITT ELECTRO-PHYSICS LABS., INC., HYATTSVILLE, MD.
ETN-	ELLIOTT-AUTOMATION SPACE AND ADVANCED MILITARY SYSTEMS LTD., LONDON
ETO-	AMERICAN FORCES EUROPEAN THEATER OF OPERATIONS
ETO-	GENERAL PRECISION, INC., LITTLE FALLS, N.J.
ETO TM-	ARMY MISSILE COMMAND, REDSTONE ARSENAL, ALA.
ETOUSA ...	AMERICAN FORCES EUROPEAN THEATER OF OPERATIONS
ETP-	NAVAL ORDNANCE TEST STATION, INYOKERN (CHINA LAKE), CALIF. (TERRIER EVALUATION PROGRAM)
ETR-	(ENGINEERING TEST REPORT) (REF. 2)
ETR-	(EXTERNAL TECHNICAL REPORT) (REF. 2)
ETR-	ARMSTRONG WHITWORTH (SIR W.G.) AIRCRAFT, LTD.
(YEAR)-ETR-	BENDIX CORP. ELECTRO-OPTICS DIV., ANN ARBOR, MICH.
ETR-	BENDIX CORP. ELECTRO-OPTICS DIV., ANN ARBOR, MICH.
E-T-R-	EMMANUEL COLL., BOSTON
ETR-	EUROPEAN COMPANY FOR THE CHEMICAL PROCESSING OF IRRADIATED FUELS, MOL, BELGIUM
ETR-	FAIRCHILD HILLER CORP. REPUBLIC AVIATION DIV., FARMINGDALE, N.Y.
ETR-	JOHNS HOPKINS UNIV., SILVER SPRING, MD. APPLIED PHYSICS LAB.
ETR-	NORTHROP AIRCRAFT, INC., HAWTHORNE, CALIF. (ENGINEERING TEST REPORT)
ETR-	UNITED STATES UNDERSEAS CABLE CORP., WASHINGTON, D.C.
E-T-RC-	EMMANUEL COLL., BOSTON
ETR-TR-	RCA SERVICE CO., INC., PATRICK AFB, FLA.
ETS-	EDUCATIONAL TESTING SERVICE, PRINCETON, N.J.
ETS-RB-(YEAR)-	EDUCATIONAL TESTING SERVICE, PRINCETON, N.J.
ETT-	STEVENS INST. OF TECH., HOBOKEN, N.J. EXPERIMENTAL TOWING TANK
ETV-TM-(YR)-(NO)-TM-	PAN AMERICAN WORLD AIRWAYS, INC., PATRICK AFB, FLA.
ETV-TM-(YEAR)-	PAN AMERICAN WORLD AIRWAYS, INC. GUIDED MISSILES RANGE DIV., PATRICK AFB, FLA.
ETW/GT-	GT. BRIT. ROYAL ARMOURED CORPS EQUIPMENT TRIALS WING, WAREHAM, DORSET, ENGLAND
EU-	BERLIN. HAHN-MEITNER-INSTITUT FUER KERNFORSCHUNG
EU-	RESEARCH TRIANGLE INST., DURHAM, N.C.
EU A-2P-	AIR FORCES IN EUROPE
EU-B-	BERLIN. HAHN-MEITNER-INSTITUT FUER KERNFORSCHUNG
EUCOM-	AERONAUTICAL RADIO, INC., WASHINGTON, D.C.
EUR-(4 DIGITS).F	EUROPEAN-AMERICAN NUCLEAR DATA COMMITTEE (REPORTS FROM FRENCH PARTICIPANTS)
EUR-(4 DIGITS).D	EUROPEAN-AMERICAN NUCLEAR DATA COMMITTEE (REPORTS FROM GERMAN PARTICIPANTS)
EUR-(4 DIGITS).E	EUROPEAN-AMERICAN NUCLEAR DATA COMMITTEE (REPORTS FROM EURATOM PARTICIPANTS)
EUR-(4 DIGITS).I	EUROPEAN-AMERICAN NUCLEAR DATA COMMITTEE (REPORTS FROM ITALIAN PARTICIPANTS)
EUR-	EUROPEAN ATOMIC ENERGY COMMUNITY (ALL LOCATIONS)
EURAEC-	UNITED STATES-EURATOM JOINT RESEARCH + DEV. PROGRAM
EURATOM-	EUROPEAN ATOMIC ENERGY COMMUNITY (ALL LOCATIONS)
EUR/C-	EUROPEAN ATOMIC ENERGY COMMUNITY, BRUSSELS
EUR-CEA-	EUROPEAN ATOMIC ENERGY COMMUNITY
EUR-CEA-FC-	EUROPEAN ATOMIC ENERGY COMMUNITY, BRUSSELS
EUR-CEA-FC-(NO.)-TR	EUROPEAN ATOMIC ENERGY COMMUNITY, BRUSSELS (TRANS.)
EURFNR-	UNITED STATES-EURATOM FAST REACTOR EXCHANGE PROGRAM
EURMAR-	GESELLSCHAFT FUER KERNENERGIEVERWERTUNG IN SCHIFFBAU UND SCHIFFAHRT M.B.H., HAMBURG
EURO-	WORLD HEALTH ORGANIZATION. REGIONAL OFFICE FOR EUROPE, COPENHAGEN
EUROCHEMIC-	EUROPEAN COMPANY FOR THE CHEMICAL PROCESSING OF IRRADIATED FUELS, MOL, BELGIUM
EUROSPACE-	EUROPEAN SPACE RESEARCH ORGANIZATION, PARIS
EV-(NUMBER)-0	BOELKOW ENTWICKLUNGEN K.G., MUNICH
EV-	BUREAU OF AERONAUTICS (NAVY). EVALUATION DIV.
EV-	ELECTRO-VOICE, INC., BUCHANAN, MICH.
EV-	WHIRLPOOL CORP. EVANSVILLE ORDNANCE DEPT., IND.
EVANS-	EVANS RESEARCH AND DEVELOPMENT CORP., N.Y.C.
EVE-(NO.)-(YEAR)-	MASSACHUSETTS. UNIV., AMHERST. SCHOOL OF ENGINEERING
EVM-	TENNESSEE EASTMAN CORP., OAK RIDGE, TENN.
EW-	AVCO CORP. CROSLEY DIV. ELECTRONIC RES. LAB., BOSTON
EW-(5 DIGITS)	COMPAGNIE GENERALE DE TELEGRAPHIE SANS FIL, PARIS
EW-	FAIRCHILD HILLER CORP. REPUBLIC AVIATION DIV., FARMINGDALE, N.Y.
EW-	STRATEGIC BOMBING SURVEY (EUROPE)
EW-	WHITE SANDS PROVING GD. ELECTRONIC WARFARE DIV., N.MEX
EWC-	EUREKA WILLIAMS CORP., BLOOMINGTON, ILL.
EWES-LIV-	ARMY ENGINEER WATERWAYS EXPERIMENT STATION, VICKSBURG, MISS.
EWG-JH-(NO.)-(NO.)-(NO.)	EAST-WEST GATEWAY COORDINATING COUNCIL, ST. LOUIS
EWG-KW-(NO.)-(NO.)-(NO.)	EAST-WEST GATEWAY COORDINATING COUNCIL, ST. LOUIS
EW-M-	COLUMBIA UNIV., NYC. SUBSTITUTE ALLOY MATERIALS LABS.
EW-R-	COLUMBIA UNIV., NYC. SUBSTITUTE ALLOY MATERIALS LABS.
EWR-(NUMBER)-(YEAR)-	ENTWICKLUNGSRING SUED GMBH, MUNICH
EWR-E-	ENTWICKLUNGSRING SUED GMBH, MUNICH
EWR-SB-(NUMBER)-	ENTWICKLUNGSRING SUED GMBH, MUNICH
EWT-	OFFICE OF SCIENTIFIC RESEARCH AND DEVELOPMENT (EFFECTS OF WEAPONS ON TARGETS)
EWT-(NUMBER)(LETTER)	OFFICE OF SCIENTIFIC RESEARCH AND DEVELOPMENT (EFFECTS OF WEAPONS ON TARGETS)
EX-(YEAR)-(NUMBERS)	AEROSPACE CORP., EL SEGUNDO, CALIF.
EX-	TENNESSEE EASTMAN CORP., OAK RIDGE, TENN.

EXEP-	DIVISION OF TECHNICAL INFORMATION, AEC, WASH., D.C.
EXEP-	NEW YORK. STATE DEPT. OF HEALTH, ALBANY
EXOTECH-	EXOTECH, INC., ALEXANDRIA, VA.
EXOTECH-TRES-	EXOTECH, INC., WASHINGTON, D.C.
EXP-	AIR MATERIEL COMMAND. EXPERIMENTAL ENG. SEC.
EXP-(NUMBER)-M	EXPERIMENT, INC., RICHMOND (SQUID PROJECT) (REF. 33)
EXP-(NUMBER)-P	EXPERIMENT, INC., RICHMOND (SQUID PROJECT) (REF. 33)
EXP-	TEXACO EXPERIMENT INC., RICHMOND
EXPERIMENT-	RCA VICTOR DIV., RADIO CORP. OF AMERICA, CAMDEN, N.J.
EXP-G-CR-	ATOMIC ENERGY OF CANADA LTD., CHALK RIVER, ONT.
EXPLORATORY STUDY-	GEORGE WASHINGTON UNIV., ARLINGTON, VA. HUMAN RESOURCES RESEARCH OFFICE
EXP-NPD-	ATOMIC ENERGY OF CANADA LTD., CHALK RIVER, ONT.
EXP-NRU-	ATOMIC ENERGY OF CANADA LTD., CHALK RIVER, ONT.
EXP-NRX-	ATOMIC ENERGY OF CANADA LTD., CHALK RIVER, ONT.
EXPR-	ATOMIC ENERGY OF CANADA LTD., CHALK RIVER, ONT.
EXP-TP-	EXPERIMENT, INC., RICHMOND
EXP-WR-	ATOMIC ENERGY OF CANADA LTD. WHITESHELL NUCLEAR RESEARCH ESTABLISHMENT, PINAWA, MANITOBA
EXTRA-MURAL RES. NO.-	GT. BRIT. ATOMIC EN. RES. ESTAB., HARWELL, BERKS, ENG
EZV-	BAYERISCHE MOTOREN WERKE. ENTWICKLUNG ZENTRAL-VERWALTUNG

F- (FINAL) (REF. 2)	
F- (FINAL REPORT) (REF. 2)	
F- (FINISH SPECIFICATION) (REF. 2)	
F- (FORMAL REPORT) (REF. 2)	
F-	AEROJET-GENERAL CORP., SACRAMENTO, CALIF.
F-	AERONAUTICAL SYSTEMS DIV., WRIGHT-PATTERSON AFB, OHIO
F-(NUMBER)/EC-(NO.)H	AIR DEFENSE WEAPONS CENTER, TYNDALL AFB, FLA.
F(NO.)-	ASSOCIATED TESTING LABS, INC., WAYNE, N.J.
F-	BOEING CO., SEATTLE
F(NO.-NO.)U(NUMBER)	BUNKER-RAMO CORP. DEFENSE SYSTEMS DIV., CANOGA PARK, CALIF.
F-(YEAR)-	BURROUGHS CORP. DEFENSE SPACE AND SPECIAL SYSTEMS GROUP, PAOLI, PA.
F-(YEAR)-9-	CANADA. SUFFIELD EXPERIMENTAL STATION, RALSTON, ALBERTA
F-	COLUMBIA UNIV., N.Y.C. ELECTRONICS RESEARCH LABS.
F.991.36-	DANISH DEFENCE RESEARCH BOARD, COPENHAGEN
F-(3 DIGITS)	FEDERAL HIGHWAY ADMINISTRATION. TENN. DIV., NASHVILLE
F-	FRANKLIN INST., PHILADELPHIA
F-(PROJECT)-	FRANKLIN INST. LABS. FOR RES. + DEV., PHILADELPHIA
F-	GARRETT CORP. AIRESEARCH MFG. DIV., LOS ANGELES
F-(NUMBER)-(LETTER)	GARRETT CORP. AIRESEARCH MFG. DIV., LOS ANGELES
F-	GERMANY. LABORATORIUM FUER BETRIEBSFESTIGKEIT, DARMSTADT
F2-	GT. BRIT. ADMIRALTY CENTRE FOR SCIENTIFIC INFORMATION AND LIAISON, LONDON
F.72-	GT. BRIT. ATOMIC EN. RES. ESTAB., HARWELL, BERKS, ENG
F-	LIBRARY OF CONGRESS. LEGISLATIVE REFERENCE SERVICE, WASHINGTON, D.C.
F001 THRU F999	MC DONNELL AIRCRAFT CORP., ST. LOUIS
F(NUMBER)	MC DONNELL ASTRONAUTICS CO., ST. LOUIS
F-	MATHEMATICA, PRINCETON, N.J.
F23-	NAVAL AIR EXPERIMENTAL STA. AERO. ENGINE LAB., PHILA.
F-	NAVAL PROPELLANT PLANT, INDIAN HEAD, MD.
F-	NETHERLANDS. NATL. AEROSPACE LAB., AMSTERDAM
F-	NORWAY. FORSVARETS FORSKNINGSINSTITUTT, KJELLER
F-	OFFICE OF SCIENTIFIC RESEARCH AND DEVELOPMENT
F-20-M-	PRINCETON UNIV., N.J.
F-20-P-	PRINCETON UNIV., N.J.
F-(YEAR)-	PURDUE UNIV., LAFAYETTE, IND.
F-(YR.)-	STANDARD OIL CO. OF INDIANA, CHICAGO
F-	SYLVANIA ELECTRIC PRODUCTS INC., WALTHAM, MASS.
F(NO.)-	SYLVANIA ELECTRIC PRODUCTS INC., WALTHAM, MASS.
F-	SYLVANIA ELECTRONIC SYSTEMS. WALTHAM LABS., MASS.
F(NUMBER)	SYLVANIA ELECTRONIC SYSTEMS-EAST, NEEDHAM, MASS.
F-(NUMBER)-	SYLVANIA ELECTRONIC SYSTEMS-EAST, WALTHAM, MASS.
F-(NUMBER)-	SYLVANIA ELECTRONIC SYSTEMS-EAST. APPLIED RESEARCH LAB., WALTHAM, MASS.
F-3.20- THRU F-4.40-	TENNESSEE EASTMAN CORP., OAK RIDGE, TENN. (REF. 34)
F-(NUMBER)-	UNITED AIRCRAFT CORP. RES. LABS., EAST HARTFORD, CONN.
(YR.)-9F(NO.)-MIOPT-R	WESTINGHOUSE ELECTRIC CORP. RES. + DEV. CTR., PITTS.
(YR.)-9F(NO.)-(NO.)-	WESTINGHOUSE RESEARCH LABS., PITTSBURGH
(YR.)-9F(NO.)-WAVES-P	WESTINGHOUSE RESEARCH LABS., PITTSBURGH
F(LTRS)-	CONSOLIDATED VULTEE AIRCRAFT CORP. FORT WORTH DIV., TEX.
F(LTRS)-(NO.)-	CONSOLIDATED VULTEE AIRCRAFT CORP. FORT WORTH DIV., TEX.
F-(LETTER)(NUMBER)	FRANKLIN INST., PHILADELPHIA
FA-	FAIRCHILD ENGINE AND AIRPLANE CORP. FAIRCHILD AIRCRAFT DIV., HAGERSTOWN, MD.
FA-	FRANKFORD ARSENAL, PHILADELPHIA
F-A-	FRANKLIN INST. LABS. FOR RES. + DEV., PHILADELPHIA
FA-A(YEAR)-	FRANKFORD ARSENAL, PHILADELPHIA
FA-A(YR)-	GT. BRIT. FLYING PERSONNEL RESEARCH COMMITTEE
FAA-(LETTERS)(YEAR)-	GENERAL ELECTRIC CO., CINCINNATI
FAA-AC-(NUMBER/NUMBER)-	FEDERAL AVIATION ADMINISTRATION, WASHINGTON, D.C.
FAA AC(YEAR)-	FEDERAL AVIATION ADM. AERONAUTICAL CTR., OKLAHOMA CITY
FAA-ADS-	FEDERAL AVIATION ADM. AIRCRAFT DEV. SERVICE, WASH., DC
FAA-ADS-	NATIONAL AVIATION FACILITIES EXPTL. CTR., ATLANTIC CITY
FAA-AM-	FEDERAL AVIATION ADMINISTRATION, OKLAHOMA CITY
FAA-AM-(YEAR)-	FEDERAL AVIATION ADMINISTRATION, WASHINGTON, D.C. (AVIATION MEDICINE)
FAA A-RD-	FEDERAL AVIATION ADMINISTRATION, WASHINGTON, D.C.
FAA BIBLIOGRAPHIC LIST-	FEDERAL AVIATION ADMINISTRATION, WASHINGTON, D.C.
FAA-D-	NATIONAL AVIATION FACILITIES EXPTL. CTR., ATLANTIC CITY
FAA-DS-	FEDERAL AVIATION ADM. AIRCRAFT DEV. SERVICE, WASH., DC
FAA-FS-(YEAR)-	FLIGHT SAFETY FOUNDATION, INC., N.Y.C.
FAA-NA-(YEAR)-	NATIONAL AVIATION FACILITIES EXPTL. CTR., ATLANTIC CITY
FAA-RD-(YEAR)-	FEDERAL AVIATION ADMINISTRATION, WASHINGTON, D.C.
FAA-RD-(YEAR)-	FEDERAL AVIATION ADMINISTRATION. SYSTEMS RESEARCH AND DEV. SERVICE, WASHINGTON, D.C. (RESEARCH + DEV.)
FAA-TDR-	FEDERAL AVIATION ADMINISTRATION, WASHINGTON, D.C.
FAC-	NAVAL FACILITIES ENGINEERING COMMAND, WASHINGTON, DC
FA-FCDD-	FRANKFORD ARSENAL, PHILADELPHIA
FA-IEP-(NUMBERS)-	FRANKFORD ARSENAL, PHILADELPHIA
FA-LDN-	FRANKFORD ARSENAL, PHILADELPHIA
FALR-	FRANKFORD ARSENAL, PHILADELPHIA
FALR FC-	FRANKFORD ARSENAL. FIRE CONTROL LAB., PHILADELPHIA
FALR MEMO-	FRANKFORD ARSENAL, PHILADELPHIA
FALR R-	FRANKFORD ARSENAL, PHILADELPHIA
FALR S MEMO-	FRANKFORD ARSENAL, PHILADELPHIA (STATISTICAL MEMO)
FALR T-	FRANKFORD ARSENAL, PHILADELPHIA
FAM-(YEAR)-	ATOMIC POWER DEVELOPMENT ASSOCIATES, INC., DETROIT
FA-M(YEAR)-	FRANKFORD ARSENAL, PHILADELPHIA
FA-M(YEAR)-(NUMBER)-	FRANKFORD ARSENAL, PHILADELPHIA
FAMOS-RR-(NUMBER-YEAR)	BOHAN (WALTER A.) CO., PARK RIDGE, ILL. (PROJ. FAMOS)
FAMOS-RR-(NUMBER-YEAR)	METEOROLOGY INTERNAT'L, INC., MONTEREY, CAL. (PROJ. FAMOS)
FA-MR-M-(NO.)(NO.)(NO.)	FRANKFORD ARSENAL. PHILADELPHIA (MEMO REPORT)
FAO-	FOOD + AGRICULTURE ORG. OF THE UNITED NATIONS, ROME
FAO/DC/AGRIS-	FOOD + AGRICULTURE ORG. OF THE UNITED NATIONS, ROME (PANEL ON INTERNATL. INFO. SYSTEM FOR AGRIC...)
FAO-FID/C-	FOOD + AGRICULTURE ORG. OF THE UNITED NATIONS, ROME
FAO-FID/R-	FOOD + AGRICULTURE ORG. OF THE UNITED NATIONS, ROME
FAO-FRD/T-	FOOD + AGRICULTURE ORG. OF THE UNITED NATIONS, ROME
FAO-FRM/	FOOD + AGRICULTURE ORG. OF THE UNITED NATIONS, ROME
FAO-FRS/C-	FOOD + AGRICULTURE ORG. OF THE UNITED NATIONS, ROME
FAO-FRS/R-	FOOD + AGRICULTURE ORG. OF THE UNITED NATIONS, ROME
FAO-FRS/T-	FOOD + AGRICULTURE ORG. OF THE UNITED NATIONS, ROME
FAO-FRV/T-	FOOD + AGRICULTURE ORG. OF THE UNITED NATIONS, ROME
FAO-P-D-	SWEDEN. FOERSVARETS FORSKNINGSANSTALT, STOCKHOLM
FA-P(YEAR)-	FRANKFORD ARSENAL, PHILADELPHIA
FAP-	RES. + DEV. BD. COMM. ON FUELS + LUBRICANTS, WASH., DC
FA-PAD-	GT. BRIT. FLYING PERSONNEL RESEARCH COMMITTEE
FA-PP-T-	FRANKFORD ARSENAL, PHILADELPHIA
FAR-(5 DIGITS)	DEPT. OF STATE. OFFICE OF RESEARCH AND ANALYSIS FOR THE NEAR EAST AND SOUTH ASIA, WASHINGTON, D.C.
FA-R-	FRANKFORD ARSENAL, PHILADELPHIA

FAR-(NUMBER)-	GODDARD SPACE FLIGHT CENTER, GREENBELT, MD.
FA-R-	GT. BRIT. FLYING PERSONNEL RESEARCH COMMITTEE
FA-R-	NEW YORK EYE AND EAR INFIRMARY, N.Y.C.
FARELF-(NUMBER/YEAR)	FAR EAST LAND FORCES HQ. OPERATIONAL REQUIREMENTS BRANCH, SINGAPORE
FARELF-MEMO(NO./YR.)	FAR EAST LAND FORCES HQ. OPERATIONAL REQUIREMENTS BRANCH, SINGAPORE
FA-RF-(YEAR)-	FRANKFORD ARSENAL, PHILADELPHIA
FAS-	CLEARINGHOUSE FOR FEDERAL SCIENTIFIC AND TECHNICAL INFO.,SPRINGFIELD,VA.(FAST ANNOUNCEMENT SERVICE)
FA-S-	FRANKFORD ARSENAL. PITMAN-DUNN LAB., PHILADELPHIA
FAST-	FIRST ATOMIC SHIP TRANSPORT, INC., N.Y.C.
FA-T(NUMBER)-	FRANKFORD ARSENAL, PHILADELPHIA
FA-TDL-	FRANKFORD ARSENAL, PHILADELPHIA
FA-TEST RPT. T-(NUMBERS)	FRANKFORD ARSENAL, PHILADELPHIA
FA-TN-	FRANKFORD ARSENAL, PHILADELPHIA
FA-TN(YEAR)-	FRANKFORD ARSENAL, PHILADELPHIA
FA/V-	IMPERIAL CHEMICAL INDUSTRIES, LTD., LONDON
FA/V.38-	IMPERIAL CHEMICAL INDUSTRIES, LTD., LONDON
FA/WP/(NO.)/-	EUROPEAN ORGANIZATION FOR NUCLEAR RESEARCH, GENEVA
FB-	BUREAU OF COMMERCIAL FISHERIES, WASHINGTON, D.C. (FISHERY BULLETINS)
F-B(NO.)-	FRANKLIN INST., PHILADELPHIA
F-B(NUMBER)	FRANKLIN INST. RESEARCH LABS., PHILADELPHIA
FB-(NUMBER)-(YEAR)	FRAUNHOFER-GESELLSCHAFT ZUR FORDERUNG DER ANGEWANDTEN FORSCHUNG E.V., DARMSTADT, GERMANY
FB-	GERMANY. LABORATORIUM FUER BETRIEBSFESTIGKEIT, DARMSTADT
FB-	GERMANY. ZENTRALE FUER WISSENSCHAFTLICHES BERICHTS-WESEN DER LUFTFAHRTFORSCHUNG, BERLIN (FORSCHUNGS-BERICHT)
FB-	NAVAL AMMUNITION DEPOT, FALLBROOK, CALIF.
FBFRC-TR-	NAVAL RADIOLOGICAL DEFENSE LAB., SAN FRANCISCO
FBIS-	FOREIGN BROADCAST INFO. SERVICE, WASHINGTON, D.C.
FBM-	GT. BRIT. HALSTEAD EXPLOITING CENTRE
FBRC-TR-	FREDERIC BURK FOUNDATION RESEARCH CTR., SAN FRANCISCO
FC-	(FLIGHT CHARTS) (REF. 2)
FC-	AIR FORCE (FLIGHT CHARTS)
FC/	ARMED FORCES SPECIAL WEAPONS PROJECT. FIELD COMMAND, ALBUQUERQUE, N. MEX.
FC-	BUREAU OF COMMERCIAL FISHERIES, WASHINGTON, D.C. (FISHERY CIRCULARS)
FC-	COAST AND GEODETIC SURVEY, WASH.,D.C. (FLIGHT CHARTS)
FC/	DEFENSE ATOMIC SUPPORT AGENCY. FIELD COMMAND, ALBUQUERQUE, N. MEX.
FC-(8 DIGITS)	DEFENSE ATOMIC SUPPORT AGENCY. FIELD COMMAND, ALBUQUERQUE, N. MEX.
FC-	FIELD COMMAND (ARMY)
FC-	FRANKFORD ARSENAL. FIRE CONTROL LAB., PHILADELPHIA
F-C(NUMBER)	FRANKLIN INST. RESEARCH LABS., PHILADELPHIA
FC-	GARRETT CORP. AIRESEARCH MFG. DIV., LOS ANGELES
FC-	GT. BRIT. FORESTRY COMMISSION
F-C-	WESTINGHOUSE ELECTRIC CORP. BETTIS ATOMIC POWER LAB., PITTSBURGH
FCAG-	JOINT SPEC. WEAPONS PUBLICATIONS BD.,ALBUQUERQUE, N.M
FC B-	GT. BRIT. FORESTRY COMMISSION (BULLETIN)
FCC-	FEDERAL COMMUNICATIONS COMMISSION, WASHINGTON, D.C.
FCCA-	DEFENSE ATOMIC SUPPORT AGENCY. FIELD COMMAND, ALBUQUERQUE, N. MEX.
FCCG-	DEFENSE ATOMIC SUPPORT AGENCY. FIELD COMMAND, ALBUQUERQUE, N. MEX.
FCDA-	FEDERAL CIVIL DEFENSE ADMINISTRATION, WASHINGTON,D.C.
FCDA BULL-	FEDERAL CIVIL DEFENSE ADMINISTRATION, WASHINGTON,D.C.
FCDA H-	FEDERAL CIVIL DEFENSE ADMINISTRATION, WASHINGTON,D.C.
FCDA TB-	FEDERAL CIVIL DEFENSE ADMINISTRATION, WASHINGTON,D.C.
FCDA TM-	FEDERAL CIVIL DEFENSE ADMINISTRATION, WASHINGTON,D.C.
FCDA TR-	FEDERAL CIVIL DEFENSE ADMINISTRATION, WASHINGTON,D.C.
FCDR(NUMBER)-(NUMBER)	ARMED FORCES SPECIAL WEAPONS PROJECT. FIELD COMMAND, ALBUQUERQUE, N. MEX.
FCD/TN-	AUSTRALIAN ATOMIC ENERGY COMMISSION. RESEARCH ESTABLISHMENT, LUCAS HEIGHTS, NEW SOUTH WALES
FC FOS-	GT. BRIT. FORESTRY COMMISSION (FOREST OPERATIONS SER)
FCI-	QUARTERMASTER FOOD AND CONTAINER INSTITUTE FOR THE ARMED FORCES, CHICAGO
FCIC-	FAIRCHILD CAMERA AND INSTRUMENT CORP., JAMAICA, N.Y.
FCL-	MASS. INST. OF TECH., CAMBRIDGE. FLIGHT CONTROL LAB.
FCOD (NUMBER(S))-(DATE)	ARMED FORCES SPECIAL WEAPONS PROJECT. FIELD COMMAND, ALBUQUERQUE, N. MEX.
FCPC-P-PIB-	FLEET COMPUTER PROGRAMMING CENTER, PACIFIC, SAN DIEGO, CALIF. (PROGRAM INFO. BULLETIN)
FCPD-	ARMED FORCES SPECIAL WEAPONS PROJECT. FIELD COMMAND, ALBUQUERQUE, N. MEX.
FCPD-TBR-(NUMBER)-	FULTON COUNTY PLANNING DEPT., JOHNSTOWN, N.Y.
FCPL-A-PIB-	FLEET COMPUTER PROGRAMMING CENTER, ATLANTIC, VIRGINIA BEACH (PROGRAM INFO. BULLETIN)
FC-SWTG-	ARMED FORCES SPECIAL WEAPONS PROJECT. FIELD COMMAND, ALBUQUERQUE, N. MEX.
FCTD-AL-	ARMED FORCES SPECIAL WEAPONS PROJECT. FIELD COMMAND. TRAINING DIV.,ALBUQUERQUE, N. MEX. (ACCESSION LIST)
FCTG-(NUMBER).(NUMBER)	DEFENSE ATOMIC SUPPORT AGENCY. FIELD COMMAND, ALBUQUERQUE, N. MEX.
FCTNG(NO./LTR.-NO.,DATE)	ARMED FORCES SPECIAL WEAPONS PROJECT. FIELD COMMAND, ALBUQUERQUE, N. MEX.
FCTS/(YEAR)-	ARMED FORCES SPECIAL WEAPONS PROJECT. FIELD COMMAND, ALBUQUERQUE, N. MEX.
FCTT/(YEAR)-	ARMED FORCES SPECIAL WEAPONS PROJECT. FIELD COMMAND, ALBUQUERQUE, N. MEX.
FCTT-(LETTERS-LETTERS)-	ARMED FORCES SPECIAL WEAPONS PROJECT. FIELD COMMAND, ALBUQUERQUE, N. MEX.
FCTUG-AFSWP-	ARMED FORCES SPECIAL WEAPONS PROJECT. FIELD COMMAND. SPECIAL WEAPONS UNIT TRAINING GP., ALBUQUERQUE,N.M.
FCWT-	DEFENSE ATOMIC SUPPORT AGENCY. FIELD COMMAND, ALBUQUERQUE, N. MEX. (WEAPONS TESTS)
FD(NUMBER)-	ARMY CHEMICAL CORPS BIOLOGICAL LABS.,FT. DETRICK, MD.
FD-	ARMY NATICK LABS. EARTH SCIENCES DIV., MASS.
FD-	ARMY NATICK LABS. FOOD DIV., MASS.
FD-	ATOMIC ENERGY OF CANADA LTD. CHALK RIVER PROJECT, ONT. (REF. 6)
FD-(NUMBER)Q-(YEAR)	DOW CHEMICAL CO., MIDLAND, MICH. (FLUOROANION DISPLACEMENT)
FD(NUMBER)T(NUMBER)	FAIRCHILD HILLER CORP. AIRCRAFT SERVICE DIV., ST. AUGUSTINE, FLA.
FD-	NATL. RES. COUNCIL. COMM. ON FORTIFICATION DESIGN
FD-	SPERRY RAND CORP. FORD INSTRUMENT DIV., LONG ISLAND CITY, N.Y.
FD-	WISCONSIN. UNIV., MADISON
FDA-NDC-(YEAR)-	FOOD + DRUG ADMINISTRATION. CENTER FOR DRUG INFORMATION, ROCKVILLE, MD.
FDC/ADPO-TM-(YEAR)-	AIR FORCE FLIGHT DYNAMICS LAB., WRIGHT-PATTERSON AFB, OHIO
FDEB-	ABERDEEN PROVING GROUND, MD.
FDL-	AIR FORCE FLIGHT DYNAMICS LAB., WRIGHT-PATTERSON AFB, OHIO
FDL-	NAVY (FAST DEPLOYMENT LOGISTIC PROJECT)
FDLH-	HOLLOMAN AIR DEVELOPMENT CTR., HOLLOMAN AFB, N. MEX.
FDLHO-	HOLLOMAN AIR DEVELOPMENT CTR., HOLLOMAN AFB, N. MEX.
FDLHR-	HOLLOMAN AIR DEVELOPMENT CTR., HOLLOMAN AFB, N. MEX.
FDL-T-	FOOD + DRUG ADMINISTRATION, WASHINGTON, D.C.

FDL-TDR-(YEAR)-	AIR FORCE FLIGHT DYNAMICS LAB., WRIGHT-PATTERSON AFB, OHIO
FDL-TR-(YEAR)-	AIR FORCE FLIGHT DYNAMICS LAB., WRIGHT-PATTERSON AFB, OHIO
FDM-	CORNELL AERONAUTICAL LAB., INC., BUFFALO
FDR-	(FINAL DATA REPORT) (REF. 2)
FDSO-	FRANCE. FACULTE DES SCIENCES D'ORSAY
FDSM-	FRANCE. FACULTE DES SCIENCES DE MARSEILLE
FDTR-TM-(YEAR)-	AIR FORCE SYSTEMS COMMAND, WRIGHT-PATTERSON AFB, OHIO
FDU-(YEAR)-	GENERAL ELECTRIC CO. ORDNANCE DEPT., PITTSFIELD, MASS
FE-	(FIELD ENGINEERING) (REF. 2)
FE-	(FLIGHT ENGINEERING) (REF. 2)
FE-	AEROJET-GENERAL CORP., SACRAMENTO, CALIF.
FE-	CLINTON ENGINEER WORKS, MANHATTAN DISTRICT, OAK RIDGE, TENN.
FE-	DEPARTMENT OF STATE. OFFICE OF FAR EASTERN AFFAIRS, WASHINGTON, D.C.
FE-	GENERAL DYNAMICS/FORT WORTH, TEX.
FE-	ITT FEDERAL ELECTRIC CORP., PARAMUS, N.J.
FE-(NUMBER)-	MARQUARDT CORP. FACILITIES ENG., VAN NUYS, CALIF.
FEA-	FOREIGN ECONOMIC ADMINISTRATION
FEA-	QUARTERMASTER RESEARCH AND ENGINEERING FIELD EVALUATION AGENCY, FORT LEE, VA.
FEA-ABN-	QUARTERMASTER RESEARCH AND ENGINEERING FIELD EVALUATION AGENCY, FORT LEE, VA.
FEAC-	FAIRCHILD ENGINE AND AIRPLANE CORP., HAGERSTOWN, MD.
FEA EA-	FOREIGN ECONOMIC ADMINISTRATION. ENEMY BRANCH
FEA EIS-	FOREIGN ECONOMIC ADMINISTRATION. LIBERATED AREAS BRANCH (ECONOMIC INSTITUTIONS STAFF REPORTS)
FEA ES-	FOREIGN ECONOMIC ADMINISTRATION. ENEMY BRANCH (EXTERNAL ECONOMIC SECURITY STAFF REPORTS)
FEAF-	FAR EAST AIR FORCES, TOKYO
FEAF OA-	FAR EAST AIR FORCES. OPERATIONS ANALYSIS OFF., TOKYO
FEA IND-	FOREIGN ECONOMIC ADM. ENEMY BR. INDUSTRY DIV.
FEA LM-	FOREIGN ECONOMIC ADMINISTRATION. ENEMY BRANCH. LABOR AND MANPOWER DIVISION
FEC-	FAR EAST COMMAND (ARMY)
FEC-	FARNSWORTH ELECTRONICS CO., FORT WAYNE, IND.
FEC-	FEDERAL ELECTRIC CORP., PARAMUS, N.J.
FEC-OPL-	FIELD EMISSION CORP., MC MINNVILLE, ORE.
FED-	AUSTRALIAN ATOMIC ENERGY COMMISSION. RESEARCH ESTABLISHMENT, LUCAS HEIGHTS, NEW SOUTH WALES
FED-	FAIRCHILD ENGINE AND AIRPLANE CORP. FAIRCHILD ENGINE DIV., FARMINGDALE, N.Y.
FEDC-	FROST ENGINEERING DEVELOPMENT CORP., DENVER
FEDC-TR-	FROST ENGINEERING DEVELOPMENT CORP., DENVER
F EES TP-	FLORIDA. ENGINEERING EXPERIMENT STATION, GAINESVILLE (TECHNICAL PAPER)
FEF-	CIVIL SERVICE COMMISSION, WASHINGTON, D.C. (FEDERAL EMPLOYEE FACTS)
FEI-	AKADEMIYA NAUK SSSR. INSTITUT FIZICHESKIKH PROBLEMI, MOSCOW
FEI-	BRITISH WELDING RESEARCH ASSN., LONDON
FEI-	U.S.S.R. GOSUDARSTVENNYI KOMITET PO ISPOL'ZOVANIYU ATOMNOI ENERGII.FIZIKO-ENERGETICHESKII INST.,OBNINSK
F/EIS-	FRANKLIN INST. LABS. FOR RES. AND DEV., PHILADELPHIA (ELECTRIC INITIATOR SYMPOSIUM)
FEL-(NUMBER)-	LEHIGH UNIV., BETHLEHEM, PENNA.
FEL-	LEHIGH UNIV., BETHLEHEM, PENNA. FRITZ LAB.
FER-	(FINAL ENGINEERING REPORT) (REF. 2)
FER/(NUMBER)-	CUBIC CORP., SAN DIEGO, CALIF.
FEWP/P-	GT. BRIT. CENTRAL ELECTRICITY GENERATING BOARD. BERKELEY NUCLEAR LABS., BRISTOL, ENGLAND
FF-115-	FLADER (FREDRIC) INC., NORTH TONAWANDA, N.Y.
FF-180-R-	FLADER (FREDRIC) INC., NORTH TONAWANDA, N.Y.
FF-	NETHERLANDS. NATIONAAL LUCHT EN RUIMTEVAART-LABORATORIUM, AMSTERDAM
FFA-	BOLT, BERANEK, AND NEWMAN, INC., LOS ANGELES
FFA-	SWEDEN. FLYGTEKNISKA FOERSOEKSANSTALTEN, STOCKHOLM
FFA-AU-	SWEDEN. FLYGTEKNISKA FOERSOEKSANSTALTEN, STOCKHOLM
FFAP-	SWEDEN. FLYGTEKNISKA FOERSOEKSANSTALTEN, STOCKHOLM
FFA-R-	SWEDEN. FLYGTEKNISKA FOERSOEKSANSTALTEN, STOCKHOLM
FFGR-	FOUNDATION FOR GLACIER RESEARCH INC., SEATTLE
FFI-	FLADER (FREDRIC) INC., NORTH TONAWANDA, N.Y.
FFI/F-	NORWAY. FORSVARETS FORSKNINGSINSTITUTT, OSLO
FFIF-F-	NORWAY. FORSVARETS FORSKNINGSINSTITUTT, KJELLER
FFIF-INTERN RAPPORT F-	NORWAY. FORSVARETS FORSKNINGSINSTITUTT, KJELLER
FFIF-IR-E-	NORWAY. FORSVARETS FORSKNINGSINSTITUTT, KJELLER
FFIF-IR-F-	NORWAY. FORSVARETS FORSKNINGSINSTITUTT, KJELLER
FFIF-IR-F-	NORWAY. FORSVARETS FORSKNINGSINSTITUTT, OSLO
FFIF-IR-S-	NORWAY. FORSVARETS FORSKNINGSINSTITUTT, KJELLER
FFIF TEKNISK NOTAT F-	NORWAY. FORSVARETS FORSKNINGSINSTITUTT, KJELLER
FFIK-	NORWAY. FORSVARETS FORSKNINGSINSTITUTT, KJELLER
FFIK-IR-K-	NORWAY. FORSVARETS FORSKNINGSINSTITUTT, KJELLER
FFIK TEKNISK-NOTAT-K-	NORWAY. FORSVARETS FORSKNINGSINSTITUTT, KJELLER
FFIR-IR-R-	NORWAY. FORSVARETS FORSKNINGSINSTITUTT, BERGEN
FF-IR-K-	NORWAY. FORSVARETS FORSKNINGSINSTITUTT, KJELLER
FFIS-	NORWAY. FORSVARETS FORSKNINGSINSTITUTT, KJELLER
FFIS-INTERN RAPPORT F-	NORWAY. FORSVARETS FORSKNINGSINSTITUTT, KJELLER
FFIS-IR-S-	NORWAY. FORSVARETS FORSKNINGSINSTITUTT, KJELLER
FFIS-RAPPORT S-	NORWAY. FORSVARETS FORSKNINGSINSTITUTT, KJELLER
FFIU-TEKNISK-NOTATU-	NORWAY. FORSVARETS FORSKNINGSINSTITUTT, OSLO
FFIU-TEKNISK U-	NORWAY. FORSVARETS FORSKNINGSINSTITUTT, KJELLER
FFN-	AKTIEBOLAGET ATOMENERGI, STUDSVIK, SWEDEN
FFP-	UNITED KINGDOM ATOMIC ENERGY AUTHORITY. RESEARCH GP. ATOMIC ENERGY RES. ESTAB., HARWELL, BERKS, ENGLAND
FFR-	DENMARK. FORSVARETS FORSKNINGSRAD, COPENHAGEN
FGE-	LOGICON, INC., SAN PEDRO, CALIF.
FG-E-	OFFICE OF CIVIL DEFENSE, WASHINGTON, D.C.
FGF-	ARGONNE NATIONAL LAB., ILL.
FGMD-	FAIRCHILD GUIDED MISSILES DIV., FARMINGDALE, N.Y.
FGNOC-	FRANKLIN GNO CORP., WEST PALM BEACH, FLA.
FGT-	GENERAL DYNAMICS/FORT WORTH, TEX.
FGZ-	FORSCHUNGSANSTALT GRAF ZEPPELIN
FH(5 DIGITS)	FAIRCHILD HILLER CORP. AIRCRAFT SERVICE DIV., ST. AUGUSTINE, FLA.
FH-EISD-	FAIRCHILD-HILLER CORP. ELECTRONIC AND INFORMATION SYSTEMS DIV., BLADENSBURG, MD.
FHR-	FAIRCHILD HILLER CORP. REPUBLIC AVIATION DIV., FARMINGDALE, N.Y.
FHWA-HP-HSS-(YEAR)-	FEDERAL HIGHWAY ADMINISTRATION, WASHINGTON, D.C.
FI-	FRANKLIN INST., PHILADELPHIA
FI-	PREFORMED LINE PRODUCTS COMPANY. RESEARCH AND ENGINEERING, CLEVELAND (FIELD INSTRUMENTATION)
FI-	SWEDEN. KUNGLIGA TEKNISKA HOEGSKOLAN, STOCKHOLM
FIAT-	FIELD INTELLIGENCE AGENCY, TECHNICAL
FIAT CL-	FIELD INTELLIGENCE AGENCY, TECH. (CIRCULAR LETTERS)
FIAT CR-	FIELD INTELLIGENCE AGENCY, TECHNICAL (CHEM. RPTS.)
FIAT EP-	FIELD INTELLIGENCE AGENCY, TECH. (ENEMY PERSONNEL)
FIAT ER-	FIELD INTELLIGENCE AGENCY, TECH. (EVALUATION REPORTS)
FIAT FR-	FIELD INTELLIGENCE AGENCY, TECHNICAL (FINAL RPTS.)

FIAT IR-	FIELD INTELLIGENCE AGENCY, TECHNICAL (INTELL. RPTS.)
FIAT IR (LTRS/LTRS)/	FIELD INTELLIGENCE AGENCY, TECHNICAL (INTELL. RPTS.)
FIAT MC-	FIELD INTELLIGENCE AGENCY, TECH. (MISC. CHEM. FILE)
FIAT MCF-	FIELD INTELLIGENCE AGENCY, TECH. (MISC. CHEM. FILE)
FIAT MICROFILM-	FIELD INTELLIGENCE AGENCY, TECH. (DOC. ON MICROFILM)
FIAT MM-	FIELD INTELLIGENCE AGENCY, TECHNICAL
FIAT SD-	FIELD INTELL. AGENCY, TECH. (STAATSDRUCKEREI DOCS.)
FIAT-SEN-R-	FIAT. SEZIONE ENERGIA NUCLEARE, TURIN
FIAT TB-	FIELD INTELLIGENCE AGENCY, TECHNICAL (TECH. BULLS.)
FIAT TB T-	FIELD INTELLIGENCE AGENCY, TECHNICAL (TECH. BULLS.)
FI-B(NUMBER)-T-	FRANKLIN INST. LABS. FOR RES. + DEV., PHILADELPHIA
FI-BRF-	BARTOL RESEARCH FOUNDATION, SWARTHMORE, PENNA.
FIC-	FORD INSTRUMENT CO., LONG ISLAND CITY, N.Y.
FICO-	FORD INSTRUMENT CO., LONG ISLAND CITY, N.Y.
FID-	INTERNATIONAL FEDERATION FOR DOCUMENTATION. U.S. NATIONAL COMMITTEE, WASHINGTON, D.C.
FI-F-	FRANKLIN INST., PHILADELPHIA
FI-F-A-	FRANKLIN INST. LABS. FOR RES. + DEV., PHILADELPHIA
FI-FM-	FRANKLIN INST. LABS. FOR RES. + DEV., PHILADELPHIA
FI-FP-	FRANKLIN INST. LABS. FOR RES. + DEV., PHILADELPHIA
FI-I-	FRANKLIN INST. LABS. FOR RES. + DEV., PHILADELPHIA (INTERIM REPORT)
FI-I-A-(NUMBER)-(NUMBER)	FRANKLIN INST. LABS. FOR RES. + DEV., PHILADELPHIA
FIL-	FRANKLIN INST. LABS. FOR RES. + DEV., PHILADELPHIA
FIL-(LTRS)-	FRANKLIN INST. LABS. FOR RES. + DEV., PHILADELPHIA
FILE-(NUMBER)-	NATIONAL TRANSPORTATION SAFETY BOARD, WASHINGTON, DC
FIL-F-	FRANKLIN INST. LABS. FOR RES. + DEV., PHILADELPHIA
FIL-I-	FRANKLIN INST. LABS. FOR RES. + DEV., PHILADELPHIA
FILM-	(ASSIGNED BY VARIOUS RECIPIENTS AS PREFIX FOR MOTION PICTURES. FOR EXPLANATION OF THE REMAINDER OF THE CODE, SEE LETTERS FOLLOWING FILM-)
FIL-MR-	FRANKLIN INST. LABS. FOR RES. + DEV., PHILADELPHIA
FI-MR-	FRANKLIN INST. LABS. FOR RES. + DEV., PHILADELPHIA (MONTHLY REVIEW REPORT)
FI-MU-A-(NUMBER)-(NUMBER)	FRANKLIN INST. LABS. FOR RES. + DEV., PHILADELPHIA
FIN-	JOHNS HOPKINS UNIV., SILVER SPRING, MD. APPLIED PHYSICS LAB.
FINAL REPORT-	MONSANTO CHEMICAL CO., DAYTON, OHIO
FIO-	ARMY TANK-AUTOMOTIVE CENTER. FOREIGN TECHNOLOGY OFFICE, WARREN, MICH.
FI-P-	FRANKLIN INST., PHILADELPHIA
FIPNL-	AKADEMIYA NAUK SSSR. ORDENA LENINA FIZICHESKIYA INST. IMENI P.N. LEBEDEVA
FIPS-PUB-	NATIONAL BUREAU OF STANDARDS. OFFICE OF INFORMATION PROCESSING STANDARDS, WASHINGTON, D.C.
FI-Q-A-(NUMBER)-(NUMBER)	FRANKLIN INST. LABS. FOR RES. + DEV., PHILADELPHIA
FI-Q-B(NUMBER)-	FRANKLIN INST. LABS. FOR RES. + DEV., PHILADELPHIA
FI-QR-	FRANKLIN INST., PHILADELPHIA
FI-QT-	FRANKLIN INST., PHILADELPHIA
FIRENZEU-	FIRENZE UNIV., ITALY. ISTITUTO DI FISICA
FIRL-C(NUMBER)	FRANKLIN INST. RESEARCH LABS., PHILADELPHIA
FIRL-F-B(NUMBER)	FRANKLIN INST. RESEARCH LABS., PHILADELPHIA
FIRL-F-C(NUMBER)	FRANKLIN INST. RESEARCH LABS., PHILADELPHIA
FIRL-I-C(NUMBER)	FRANKLIN INST. RESEARCH LABS., PHILADELPHIA
FI RR-	FRANKLIN INST., PHILADELPHIA (RESEARCH REPORT)
FIS-	UNION OF SOUTH AFRICA. COUNCIL FOR SCIENTIFIC AND INDUSTRIAL RESEARCH, PRETORIA
FI-SA-	FRANKLIN INST. RESEARCH LABS., PHILADELPHIA
FJ-	SPERRY UTAH CO., SALT LAKE CITY
FJSRL-(YEAR)-	AIR FORCE ACADEMY, COLORADO SPRINGS. FRANK J. SEILER RESEARCH LAB.
FK-	GENERAL DYNAMICS/FORT WORTH. NUCLEAR AEROSPACE RESEARCH FACILITY, TEX.
FK-(NUMBER)-	GENERAL DYNAMICS/FORT WORTH. NUCLEAR AEROSPACE RESEARCH FACILITY, TEX.
FKFS RPT-	STUTTGART. TECHNISCHE HOCHSCHULE. FORSCHUNGSINSTITUT FUER KRAFTFAHRWESEN UND FAHRZEUGMOTOREN
FL-	ARMY NATICK LABS. FOOD LAB., MASS.
FL-	BUREAU OF COMMERCIAL FISHERIES, WASHINGTON, D.C. (FISHERY LEAFLETS)
FL-	HARVARD UNIV., CAMBRIDGE, MASS.
FL-	LEHIGH UNIV., BETHLEHEM, PENNA. FRITZ LAB.
FL (NUMBER)/	OFFICE OF THE DIRECTOR OF DEFENSE (RES+ENG), WASH., DC
FL-	RES. + DEV. BD. COMM. ON FUELS + LUBRICANTS, WASH., DC
FL-	SWEDEN. KUNGLIGA TEKNISKA HOEGSKOLAN, STOCKHOLM. FLYGTEKNISKA LABORATINET
FLAU-	FLORIDA. UNIV., GAINESVILLE
FLD-	PRINCETON UNIV., N.J.
FLIGHT SCI. LAB. REPT.-	BOEING SCIENTIFIC RESEARCH LABS. FLIGHT SCIENCES LAB., SEATTLE
FLINDERSU-APT-	FLINDERS UNIV., BEDFORD PARK, SOUTH AUSTRALIA (ASTROPHYSICS AND PLASMA THEORY)
FLM-	CHEMICAL WARFARE SERVICE, EDGEWOOD ARSENAL, MD. (FIELD LABORATORY MEMORANDUM)
FLORIDA-TR-	FLORIDA. UNIV., GAINESVILLE (QUANTUM THEORY PROJECT)
FLORIDA U-(NO.)-QTR-	FLORIDA. UNIV., GAINESVILLE
FLR-	FLUOR CORP., LTD., WHITTIER, CALIF.
FLRL-	SOUTHWEST RESEARCH INST. ARMY FUELS AND LUBRICANTS RESEARCH LAB., SAN ANTONIO
FLUOR-	FLUOR CORP., LTD., LOS ANGELES
FM-	(FIELD MANUAL) (REF. 2)
FM-	AERONAUTICAL CHART AND INFORMATION CENTER, ST. LOUIS
FM-	ARMY, WASHINGTON, D.C. (FIELD MANUAL)
FM-	ARMY COMMAND + GENERAL STAFF COLL., WASHINGTON, D.C.
FM-(NUMBER)-0	BOELKOW ENTWICKLUNGEN K.G., MUNICH
FM-	FACTORY MUTUAL ENGINEERING DIV., BOSTON
FM-	FUNDAMENTAL METHODS ASSOCIATES, N.Y.C.
FM-	GT. BRIT. AERONAUTICAL RESEARCH COUNCIL. ENGINE AERODYNAMICS SUBCOMMITTEE
F.M.-	GT. BRIT. AERO. RES. COUNCIL. FLUID MOTION SUB-COMM.
FM-	GT. BRIT. AERO. RES. COUNCIL. FLUID MOTION SUB-COMM.
FM-	STANFORD UNIV., CALIF. DEPT. OF MECHANICAL ENG.
FMB-(YEAR)-	FORSCHUNGSINSTITUT FUER MILITAERISCHE BAUTECHNIK, ZURICH
FMBR-(NUMBER)/(YEAR)	GERMANY. PHYSIKALISCH-TECHNISCHE BUNDESANSTALT, BRUNSWICK
FMBR-	GERMANY. PHYSIKALISCH-TECHNISCHE BUNDESANSTALT, BRUNSWICK
FMC-	FOOD MACHINERY + CHEMICAL CORP., SAN JOSE, CALIF.
FMC-	FOOTE MINERAL CO., PHILADELPHIA
FMC-BA-	FMC CORP., BALTIMORE
FMC-BA-(NUMBER)-X	FMC CORP. ORGANIC CHEMICALS DIV., BALTIMORE
FMC-DTL-	FMC CORP. DEFENSE TECHNOLOGY LABS., SANTA CLARA, CALIF.
FMC-ORD-	FMC CORP. ORDNANCE ENG. DIV., SAN JOSE., CALIF.
FMC-PCR-	FMC CORP., PRINCETON, N.J.
FMD-	FORT MONMOUTH SIGNAL LAB., N.J.
FMD-	WHITE SANDS MISSILE RANGE. FREQUENCY MANAGEMENT DIV., N. MEX.
FMD-CD-(YR.)-S-(NO.)-R	WHITE SANDS MISSILE RANGE. FREQUENCY MANAGEMENT DIV., N. MEX.
FMD-TM-	WHITE SANDS MISSILE RANGE. FREQUENCY MANAGEMENT DIV., N. MEX.

FMER-
    ARMY CHEM. CTR., MD. (FOREIGN MATERIEL EVALUATION RPT.)
FMFM-
    MARINE CORPS, WASHINGTON, D.C.
FML-(YEAR)-
    MASS. INST. OF TECH., CAMBRIDGE. FLUID MECHANICS LAB.
FML-PUBL-(YEAR-NUMBER)
    MASS. INST. OF TECH., CAMBRIDGE. FLUID MECHANICS LAB.
FMP-(YEAR)/(NUMBER)-MRL
    CANADA. DEPT. OF MINES AND TECHNICAL SURVEYS. FUELS + MINING PRACTICE DIV., OTTAWA
FMPC-
    NATIONAL LEAD CO. OF OHIO, CINCINNATI (FEED MATERIALS PRODUCTION CENTER REPORTS)
FMP/IMOG/
    UNITED KINGDOM ATOMIC ENERGY AUTHORITY. WEAPONS GROUP ATOMIC WEAPONS RES. ESTAB., ALDERMASTON, BERKS, ENG. (FINE MACHINING PANEL)
FMR-(YEAR)-
    GENERAL DYNAMICS/FORT WORTH, TEX.
FMRB-(NUMBER)/(YEAR)
    GERMANY. PHYSIKALISCH-TECHNISCHE BUNDESANSTALT, BRUNSWICK
FMRC-
    FACTORY MUTUAL RESEARCH CORP., NORWOOD, MASS.
FMRD-
    FEDERAL-MOGUL CORP. RES. AND DEV. DIV., ANN ARBOR, MICH.
FMS-
    GENERAL DYNAMICS/FORT WORTH, TEX.
FMSL-
    FORT MONMOUTH SIGNAL LAB., N.J.
FMSL ER-
    FORT MONMOUTH SIGNAL LAB., N.J. (ENGINEERING REPT.)
FMSL SDS EM-
    FORT MONMOUTH SIGNAL LAB. SUPPRESSION DEVELOPMENT SECTION, N.J. (ENGINEERING MEMORANDUM)
FMTR-
    CHEMICAL AND RADIOLOGICAL LABS., ARMY CHEM. CTR., MD.
FMTR-
    CHEMICAL WARFARE SERVICE, EDGEWOOD ARSENAL, MD. (FOREIGN MATERIEL TECHNICAL REPORT)
FMTR-
    PURDUE RESEARCH FOUNDATION, LAFAYETTE, IND.
FMTR-(YEAR)-
    PURDUE UNIV., LAFAYETTE, IND.
FMU-
    STANFORD RESEARCH INST., MENLO PARK, CALIF.
FN-
    ACADEMIA R.S.R. INSTITUTUL DE FIZICA ATOMICA, BUCHAREST
FN-
    NATIONAL ACCELERATOR LAB., OAK BROOK, ILL.
FN-(NUMBER/NUMBER(S))
    SYSTEM DEVELOPMENT CORP., SANTA MONICA, CALIF. (FIELD NOTE)
FN-E-
    FIAT. SEZIONE ENERGIA NUCLEARE, TURIN
FNP-HH-(YEAR)-
    NATIONAL PARK SERVICE. DIV. OF HISTORY, WASH., D.C.
FNRS-
    BELGIUM. FONDS NATIONAL DE LA RECHERCHE SCIENTIFIQUE, BRUSSELS
FNWC-TN-
    FLEET NUMERICAL WEATHER CENTRAL, MONTEREY, CALIF.
FO-(3 DIGITS)
    AMERICAN OPTICAL CO. TECHNICAL PRODUCTS GROUP, SOUTHBRIDGE, MASS.
FOA-
    SWEDEN. FOERSVARETS FORSKNINGSANSTALT, STOCKHOLM
FOA-A-
    SWEDEN. FOERSVARETS FORSKNINGSANSTALT, STOCKHOLM
FOA-P-
    SWEDEN. FOERSVARETS FORSKNINGSANSTALT, STOCKHOLM
FOA-P-C-
    SWEDEN. FOERSVARETS FORSKNINGSANSTALT, STOCKHOLM
FOC-
    FARRAND OPTICAL CO., INC., N.Y.C.
FOCA-OF-
    ARMY FIELD OPERATING COST AGENCY, ALEXANDRIA, VA.
FOCA-WS-
    ARMY FIELD OPERATING COST AGENCY, ALEXANDRIA, VA.
FOCCPAC-
    FLEET OPERATIONS CONTROL CENTER, PACIFIC FLEET, SAN FRANCISCO
FOCCPAC-TN-
    FLEET OPERATIONS CONTROL CENTER, PACIFIC FLEET, SAN FRANCISCO
FOCCPAC-TR-
    FLEET OPERATIONS CONTROL CENTER, PACIFIC FLEET, SAN FRANCISCO
FOI-
    NAVAL EXPLOSIVE ORDNANCE DISPOSAL FACILITY, INDIAN HEAD, MD. (FOREIGN ORDNANCE INFORMATION)
FOIM-
    BUREAU OF ORDNANCE (NAVY)
FOLDER-(NUMBER)-
    AMP INC. CAPITRON DIV., ELIZABETHTOWN, PENNA.
FOM-
    NETHERLANDS. FDN. FOR FUNDAMENTAL RES. OF MATTER
FOOTE-
    FOOTE MINERAL CO., EXTON, PENNA.
FOR-
    BUREAU OF ORDNANCE (NAVY) (FOREIGN ORDNANCE REPORT)
FOR-
    FORDHAM UNIV., N.Y.C.
FOREIGN APPLICATIONS-
    GEOLOGICAL SURVEY, DENVER
FORTIFIKATORISK-NOTAT-
    NORWEGIAN DEFENCE CONSTRUCTION SERVICE. OFFICE OF TEST AND DEVELOPMENT, OSLO

FORU-
    FORDHAM UNIV., N.Y.C.
FORUM REPORT-
    ATOMIC INDUSTRIAL FORUM, INC., N.Y.C.
FOS-
    GT. BRIT. FORESTRY COMMISSION (FOREST OPERATIONS SER)
FOTB-
    AEROJET-GENERAL CORP., SACRAMENTO, CALIF.
FOUNDATION PAPER-
    YALE UNIV., NEW HAVEN. COWLES FOUNDATION FOR RESEARCH IN ECONOMICS
FP-
    CHEMICAL WARFARE SERVICE, EDGEWOOD ARSENAL, MD. (FRED PROJECT)
FP-(NUMBER)Q-(YEAR)
    DOW CHEMICAL CO., MIDLAND, MICH.
FP2/ENEA-
    EUROPEAN SPACE VEHICLE LAUNCHER DEV. ORG., PARIS
FP-
    FURANE PLASTICS, INC., LOS ANGELES
FP-
    GRUMMAN AIRCRAFT ENGINEERING CORP., BETHPAGE, N.Y.
FP-(YEAR)-
    HUGHES AIRCRAFT CO. GROUND SYSTEMS GP., FULLERTON, CAL
FP-
    MASSACHUSETTS INST. OF TECH., CAMBRIDGE. FIBERS AND POLYMERS DIV.
(NUMBER)-FP-
    MATHEMATICA, INC., PRINCETON, N.J.
FP-
    STANFORD UNIV., CALIF. DEPT. OF MECHANICAL ENG.
FPL-
    (FLIGHT PROPULSION LAB.) (REF. 2)
FPL-
    FOREST PRODUCTS LAB., MADISON, WIS.
FPL-
    RES. + DEV. BD. COMM. ON FUELS + LUBRICANTS, WASH., DC
FPL R-
    FOREST PRODUCTS LAB., MADISON, WIS.
FPL-TDR-
    MELPAR, INC., FALLS CHURCH, VA.
FPL-TRANS-
    FOREST PRODUCTS LAB., MADISON, WIS.
FPM-
    CIVIL SERVICE COMMISSION, WASHINGTON, D.C. (FIELD PERSONNEL MANUAL)
FPP-
    GT. BRIT. ATOMIC EN. RES. ESTAB., HARWELL, BERKS, ENG
FPP-
    UNITED KINGDOM ATOMIC ENERGY AUTHORITY. INDUSTRIAL GROUP H.Q., RISLEY, LANCS, ENGLAND
FPP/P-
    GT. BRIT. DIV. OF ATOMIC ENERGY (PRODUCTION), RISLEY, LANCS, ENGLAND
FPR-
    (FINAL PROGRESS REPORT) (REF. 2)
FPR-
    GENERAL DYNAMICS/FORT WORTH, TEX.
FPR-
    GT. BRIT. DEPT. OF SCIENTIFIC AND INDUSTRIAL RESEARCH. FOREST PRODUCTS RESEARCH LABORATORY
FPRC/(NUMBERS)/A
    UNITED KINGDOM. FLYING PERSONNEL RES. COMM., LONDON
FPRC/
    UNITED KINGDOM. FLYING PERSONNEL RES. COMM., LONDON
FPRC/(NUMBERS)/B
    UNITED KINGDOM. FLYING PERSONNEL RES. COMM., LONDON
FPRC/AM-
    UNITED KINGDOM. FLYING PERSONNEL RES. COMM., LONDON
FPRC/MEMO-
    UNITED KINGDOM. FLYING PERSONNEL RES. COMM., LONDON
FPRL-
    FLIGHT PROPULSION RESEARCH LAB., CLEVELAND
FPRR-
    GT. BRIT. DEPT. OF SCIENTIFIC AND INDUSTRIAL RESEARCH (FOREST PRODUCTS RESEARCH RECORDS)
FPS-
    CONVAIR, FORT WORTH, TEX.
FPS-
    GT. BRIT. ATOMIC EN. RES. ESTAB., HARWELL, BERKS, ENG
FPSD/P-
    UNITED KINGDOM ATOMIC ENERGY AUTHORITY. RESEARCH GP. ATOMIC ENERGY RES. ESTAB., HARWELL, BERKS, ENGLAND
FQR/(YEAR)
    JOHNS HOPKINS UNIV., SILVER SPRING, MD. APPLIED PHYSICS LAB.
FR-
    (FIELD REPORT) (REF. 2)
FR-
    (FINAL REPORT) (REF. 2)
FR-
    ABERDEEN PROVING GROUND, MD. (FIRING RECORD)
FR-
    AMERICAN OPTICAL CO. J.W. FECKER DIV., PITTSBURGH
(NUMBER)-FR-
    BAIRD ATOMIC, INC., BEDFORD, MASS.
FR-
    BECKMAN INSTRUMENTS, INC. ADVANCED TECHNOLOGY OPERATIONS, FULLERTON, CALIF.
FR-
    BENDIX AVIATION CORP. FRIEZ INSTR. DIV., BALTIMORE
FR-
    BONN. UNIVERSITAET
FR-
    DATA CORP., DAYTON, OHIO
FR-(YEAR)-
    DATA CORP., DAYTON, OHIO
FR-(NUMBER)-(NUMBER)
    DIKEWOOD CORP., ALBUQUERQUE, N. MEX.

FR-	FRANKLIN INST. LABS. FOR RES. + DEV., PHILADELPHIA
FR-(PROJECT)-	FRANKLIN INST. LABS. FOR RES. + DEV., PHILADELPHIA
FR1-	GENERAL ELECTRIC CO., BETHESDA, MD.
FR-(YEAR)-(NUMBER)/	GHENT RIJKSUNIVERSITEIT. NATUURKUNDIG LABORATORIUM
FR-(3 DIGITS)-(4 DIGITS)	GIANNINI-IRVINE, COSTA MESA, CALIF.
FR-	GIANNINI SCIENTIFIC CORP., SANTA ANA, CALIF.
FR-(NO.)-(CON.NO.)	GIANNINI SCIENTIFIC CORP., SANTA ANA, CALIF.
FR-	GT. BRIT. DEPT. OF SCI. + IND. RES. (FUEL RES.)
FR(YEAR-NUMBER)N	GTE SYLVANIA, INC., NEEDHAM HEIGHTS, MASS.
FR-	HUGHES AIRCRAFT CO., FULLERTON, CALIF.
FR(YEAR)-	HUGHES AIRCRAFT CO., FULLERTON, CALIF.
FR-(YEAR)-(NUMBER)-	HUGHES AIRCRAFT CO. GROUND SYSTEMS GP., FULLERTON, CALIF., (FINAL REPORT)
FR-	JOHNS HOPKINS UNIV., SILVER SPRING, MD. APPLIED PHYSICS LAB.
FR-	NEW YORK UNIV., N.Y.C.
FR-	PHOTOCIRCUITS CORP., GLEN COVE, N.Y.
FR-	PICATINNY ARSENAL, DOVER, N.J.
FR(NO.)	PLASMADYNE CORP., SANTA ANA, CALIF.
FR-(YEAR)-	RAYTHEON CO. AUTOMETRIC OPERATION, ALEXANDRIA, VA.
FR-(YEAR)-	RAYTHEON CO. SPACE + INFO. SYSTEMS DIV., SUDBURY, MASS
FR(YEAR)	SYLVANIA ELECTRONIC SYSTEMS-EAST, NEEDHAM, MASS.
FR-(NUMBER)-(LETTER)	WESTINGHOUSE ELECTRIC CORP. ATOMIC POWER DIV., PITTS.
FR-(NUMBER)-(LETTER)	WESTINGHOUSE ELECTRIC CORP. BETTIS PLANT, PITTSBURGH
FR-(NO.)-B	WESTINGHOUSE ELECTRIC CORP. BETTIS PLANT, PITTSBURGH
FRAN-	FRANKLIN INST., PHILADELPHIA
FRAN (LTR-NO.)	FRANKLIN INST. LABS. FOR RES. + DEV., PHILADELPHIA
FRANCE-(LETTERS)-	(FRANCE- IS ASSIGNED BY VARIOUS RECIPIENTS AS A PREFIX TO CODES FOR REPORTS ORIGINATING IN FRANCE. FOR EXPLANATION OF THE REMAINDER OF THE CODE, SEE ENTRY UNDER LETTERS FOLLOWING FRANCE-)
FRANKFORD-(LETTERS)-	FRANKFORD ARSENAL. PITMAN-DUNN LAB., PHILADELPHIA
FRANKFURTU-	FRANKFURT AM MAIN. UNIVERSITAET
FRAS-(NUMBERS)-(YEAR)	CHRYSLER CORP. MISSILE DIV., DETROIT
FRA-TM-	ARGONNE NATIONAL LAB., ILL.
FRC-	FEDERAL RADIATION COUNCIL, WASHINGTON, D.C.
FRC-(YEAR)-	FREDERICK RESEARCH CORP., BETHESDA, MD.
FRC-	FREED ELECTRONICS + CONTROLS CORP., N.Y.C.
FRDC/P-	GT. BRIT. CAPENHURST WORKS, CHES., ENGLAND
FRDC/P-	GT. BRIT. CULCHETH LABS., LANCS, ENGLAND
FRDC/P-	GT. BRIT. WINDSCALE WORKS, SELLAFIELD, CUMB., ENGLAND
FRDC/P-	UNITED KINGDOM ATOMIC ENERGY AUTHORITY. INDUSTRIAL GROUP H.Q., RISLEY, LANCS, ENGLAND
FRDC/P-	UNITED KINGDOM ATOMIC ENERGY AUTHORITY. RESEARCH GP. ATOMIC ENERGY RES. ESTAB., HARWELL, BERKS, ENGLAND
FRDWP/P-	UNITED KINGDOM ATOMIC ENERGY AUTHORITY. RESEARCH GP. ATOMIC ENERGY RES. ESTAB., HARWELL, BERKS, ENGLAND
FRE-	FAIRCHILD RECORDING EQUIP. CORP., WHITESTONE, N.Y.
FRED-	FREDERICK (CARL L.) AND ASSOCIATES, BETHESDA, MD.
FREL-	PICATINNY ARSENAL. FELTMAN RES. + ENG. LABS., DOVER, NJ
FREL-LS-	PICATINNY ARSENAL. FELTMAN RES. + ENG. LABS., DOVER, NJ
FREWP/P-	GT. BRIT. ATOMIC EN. RES. ESTAB., HARWELL, BERKS, ENG
FRFEWP-	GT. BRIT. ATOMIC EN. RES. ESTAB., HARWELL, BERKS, ENG
FRFEWP/P-	GT. BRIT. CULCHETH LABS., LANCS, ENGLAND
FRFEWP/P-	UNITED KINGDOM ATOMIC ENERGY AUTHORITY. RESEARCH GP. ATOMIC ENERGY RES. ESTAB., HARWELL, BERKS, ENGLAND
FRIGG-	AKTIEBOLAGET ATOMENERGI, STOCKHOLM
FRIGG-NOTE-R-	DANISH ATOMIC ENERGY COMMISSION. RES. ESTAB., RISOE
FRIGG-PM-	DANISH ATOMIC ENERGY COMMISSION. RES. ESTAB., RISOE
FRIT-	LEHIGH UNIV., BETHLEHEM, PENNA. FRITZ LAB.
FRITZ LAB.-	LEHIGH UNIV., BETHLEHEM, PENNA. FRITZ LAB.
FRK-	BOSTON COLL., CHESTNUT HILL, MASS. DEPT. OF CHEMISTRY
FRL-	FABRIC RESEARCH LABS., INC., DEDHAM, MASS.
FRL-	PICATINNY ARSENAL. FELTMAN RESEARCH LABS., DOVER, N.J.
FRL-	UNITED KINGDOM. MINISTRY OF AGRICULTURE, FISHERIES, AND FOOD. FISHERIES RADIOBIOLOGICAL LAB., LOWESTOFT E. SUFFOLK, ENGLAND
FRL/MIT-TR-	MASS. INST. OF TECH., CAMBRIDGE. FUELS RESEARCH LAB.
FRL-TM-	PICATINNY ARSENAL. FELTMAN RESEARCH LABS., DOVER, N.J. (TECHNICAL MEMORANDA)
FRL-TN-	PICATINNY ARSENAL. FELTMAN RESEARCH LABS., DOVER, N.J. (TECHNICAL NOTES)
FRL-TR-	PICATINNY ARSENAL. FELTMAN RESEARCH LABS., DOVER, N.J. (TECHNICAL REPORTS)
FRM-	CORNELL AERONAUTICAL LAB. FLIGHT RES. DEPT., BUFFALO
FRM-	MUNICH. TECHNISCHE HOCHSCHULE. PHYSIKALISCHES INSTITUT
F-RMW-	F-R MACHINE WORKS, INC., LONG ISLAND CITY, N.Y.
FR/N-(YEAR)(NUMBER)	UNITED KINGDOM ATOMIC ENERGY AUTHORITY. REACTOR GP., RISLEY, LANCS, ENGLAND (FAST REACTOR NEWSLETTER)
FRNC-TH-	FRANCE. COMMISSARIAT A L'ENERGIE ATOMIQUE. CENTRE D'ETUDES NUCLEAIRES, SACLAY (THESIS)
FRNOTE-	NATIONAL RESEARCH COUNCIL OF CANADA. DIV. OF BUILDING RESEARCH, OTTAWA (FIRE RESEARCH NOTES)
FR-NPCV-	GENERAL MOTORS CORP. ELECTRO-MOTIVE DIV., DETROIT
FRO-	GENERAL ELECTRIC CO. NUCLEAR ENERGY DIV., SAN JOSE, CAL (FUEL RECOVERY)
FRO-	RESEARCH AND DEVELOPMENT BOARD. PANEL ON FUELS REQUIRING OXIDIZERS, WASHINGTON, D.C.
FROPC/P-	UNITED KINGDOM ATOMIC ENERGY AUTHORITY. RESEARCH GP. ATOMIC ENERGY RES. ESTAB., HARWELL, BERKS, ENGLAND
FROP-C/R-	UNITED KINGDOM ATOMIC ENERGY AUTHORITY. RESEARCH GP. ATOMIC ENERGY RES. ESTAB., HARWELL, BERKS, ENGLAND
FROST-(NUMBER)-	FROST ENGINEERING DEVELOPMENT CORP., ENGLEWOOD, COLO.
FRP-	AIR FORCE CAMBRIDGE RESEARCH LABS., BEDFORD, MASS.
FRPA-FEWP/	UNITED KINGDOM ATOMIC ENERGY AUTHORITY. REACTOR GP., SPRINGFIELDS, LANCS, ENGLAND
FRPC-FEWP/	GT. BRIT. ATOMIC EN. RES. ESTAB., HARWELL, BERKS, ENG
FRPC/RSWP/	UNITED KINGDOM ATOMIC ENERGY AUTHORITY. REACTOR GP., RISLEY, LANCS, ENGLAND
FRPMC/RSWP/	UNITED KINGDOM ATOMIC ENERGY AUTHORITY. REACTOR GP., RISLEY, LANCS, ENGLAND
FRU-	FREIBURG I. B. UNIVERSITAET
FRWP-	UNITED KINGDOM ATOMIC ENERGY AUTHORITY. RESEARCH GP. ATOMIC ENERGY RES. ESTAB., HARWELL, BERKS, ENGLAND
FS-	(FINISH SPECIFICATION) (REF. 2)
FS-(NUMBER)A-(YEAR)	DOW CHEMICAL CO., MIDLAND, MICH.
FS-(NUMBER)Q-(YEAR)	DOW CHEMICAL CO., MIDLAND, MICH.
FS-(NUMBER)-(NUMBER)-	FEDERAL AVIATION ADMINISTRATION, WASHINGTON, D.C.
FS-	FIRTH STERLING, INC., PITTSBURGH
FS-	FIRTH STERLING, INC. AMERICAN ELECTRO METAL DIV., YONKERS, N.Y.
FS-	FOREST SERVICE, WASHINGTON, D.C.
FS-	GENERAL DYNAMICS/FORT WORTH, TEX.
FS-	OFFICE OF SCIENTIFIC RESEARCH AND DEVELOPMENT (FUNDAMENTAL STUDY OF EXPLOSIVES)
FS(YEAR)-	PENNSYLVANIA STATE UNIV., UNIVERSITY PARK. DEPT. OF FUEL SCIENCE
FS-AMD-	FAIRCHILD STRATOS CORP. AIRCRAFT-MISSILES DIV., HAGERSTOWN, MD.
FSC-(NUMBERS)	AEROJET-GENERAL CORP. LIQUID ROCKET OPERATIONS, SACRAMENTO, CALIF.
FSC-(YEAR)	FEDERAL SCIENTIFIC CORP., N.Y.C.
FSC-(YEAR)-	INTERNATIONAL BUSINESS MACHINES CORP., GAITHERSBURG, MD
FSC-	TRG, INC., MELVILLE, N.Y.

FSC-F/	FEDERAL SCIENTIFIC CORP., N.Y.C.
FSC-T-	FEDERAL SCIENTIFIC CORP., N.Y.C.
FSD-	NATIONAL RESEARCH COUNCIL OF CANADA. DIV. OF ATOMIC ENERGY, CHALK RIVER, ONT.
FSD/ING-	ATOMIC ENERGY OF CANADA LTD., CHALK RIVER, ONT.
FSE-	CONVAIR, FORT WORTH, TEX.
FSG-	UNITED KINGDOM ATOMIC ENERGY AUTHORITY. RESEARCH GP. ATOMIC ENERGY RES. ESTAB., HARWELL, BERKS, ENGLAND
FSIB-	MANNED SPACECRAFT CENTER, HOUSTON, TEX. (FLIGHT SAFETY INFORMATION BULLETIN)
FSK-	PRATT AND WHITNEY AIRCRAFT DIV., UNITED AIRCRAFT CORP., HARTFORD, CONN.
FSL-	WHITE SANDS MISSILE RANGE.FLIGHT SIMULATION LAB.,N.M.
FSM-	ARMY AIR FORCE. OFFICE OF FLYING SAFETY (FLYING SAFETY MEMORANDUM)
FSM-	METALS AND CONTROLS CORP., ATTLEBORO, MASS.
FSN-AD(NUMBER)-(NUMBER)	GRUMMAN AIRCRAFT ENGINEERING CORP., BETHPAGE, N.Y.
FSO-AS-	ARTILLERY SCHOOL, FORT SILL, OKLA.
FSP/H-	ATOMIC ENERGY OF CANADA LTD. CHALK RIVER PROJ., ONT.
FSPR-FPL-	FOREST PRODUCTS LAB., MADISON, WIS.
FSPR-PSW-	PACIFIC SOUTHWEST FOREST AND RANGE EXPERIMENT STATION, BERKELEY, CALIF. (FOREST SERVICE RESEARCH PAPER)
FS/Q-TR-	DIRECTORATE OF FLIGHT STANDARDS AND QUALIFICATION RESEARCH (ARMY), ST. LOUIS
FS-R-	COLUMBIA UNIV., NYC. SUBSTITUTE ALLOY MATERIALS LABS.
FSR-	GRUMMAN AIRCRAFT ENGINEERING CORP., BETHPAGE, N.Y.
FSR-	GRUMMAN OPERATIONS ANALYSIS/PLANNING RESEARCH, FUTURE SYSTEMS, BETHPAGE, N.Y.
FSR-(LETTER)(NO.)-	GRUMMAN AIRCRAFT ENGINEERING CORP., BETHPAGE, N.Y.
FSRB-FPL-	FOREST PRODUCTS LAB., MADISON, WIS.
FSRB-SE-	FOREST SERVICE. SOUTHEASTERN STA., ASHEVILLE, N.C.
FSRN-FPL-	FOREST PRODUCTS LAB., MADISON, WIS.
FSRN-NE-	NORTHEASTERN FOREST EXPERIMENT STATION, UPPER DARBY, PENNA. (FOREST SERVICE RESEARCH NOTE)
FSRN-SE-	FOREST SERVICE. SOUTHEASTERN STA., ASHEVILLE, N.C.
FSRP-INT-	FOREST SERVICE. INTERMOUNTAIN STATION, OGDEN, UTAH
FSRP-NE-	FOREST SERVICE. NORTHEASTERN STA., UPPER DARBY, PENNA
FSRP-NE-	NORTHEASTERN FOREST EXP. STA., UPPER DARBY, PENNA.
FSRP-PNW-	FOREST SERVICE. PACIFIC NORTHWEST STA., PORTLAND ORE.
FSRP-RM-	FOREST SERVICE. ROCKY MOUNTAIN STA., FT. COLLINS,COLO
FSR-ST-	GRUMMAN AIRCRAFT ENGINEERING CORP., BETHPAGE, N.Y.
FSSD-SB-	HUGHES AIRCRAFT CO.,CULVER CITY,CAL.(FIELD SERVICE)
FSTC-	ARMY FOREIGN SCIENCE AND TECH. CENTER, WASHINGTON,D.C
FSTC-CS-	ARMY FOREIGN SCIENCE AND TECH. CENTER, WASHINGTON,D.C
FSTC-HF-	ARMY FOREIGN SCIENCE AND TECH. CENTER, WASHINGTON,D.C
FSTC-HT-	ARMY FOREIGN SCIENCE AND TECH. CENTER, WASHINGTON,D.C
FSTC-MP-	ARMY FOREIGN SCIENCE AND TECH. CENTER, WASHINGTON,D.C
FS-TP-	FOREST SERVICE, WASHINGTON, D.C.
F-SU-	AIR MATERIEL COMMAND. AIR DOCUMENTS OFFICE, WRIGHT-PATTERSON AFB, OHIO (TRANSLATION)
FSU-	FLORIDA STATE UNIV., TALLAHASSEE
FSU-HEP-(YEAR)-	FLORIDA STATE UNIV., TALLAHASSEE. HIGH ENERGY PHYSICS LAB.
FSU-M-	FLORIDA STATE UNIV., TALLAHASSEE. DEPT. OF STATISTICS
FSU-SR-	FLORIDA STATE UNIV., TALLAHASSEE
FSU-SR-	FLORIDA STATE UNIV., TALLAHASSEE. DEPT. OF METEOROLOGY (SCIENTIFIC REPORT)
FSU-STATISTICS-	FLORIDA STATE UNIV., TALLAHASSEE. DEPT. OF STATISTICS
FSU-TR-	FLORIDA STATE UNIV., TALLAHASSEE
FT-	(FIRING TABLE) (REF. 2)
FT-	(FLIGHT TEST) (REF. 2)
FT-	(FLIGHT TEST REPORT) (REF. 2)
FT (NUMBER) (YEAR)	ACADEMIA R.S.R.INSTITUTUL DE FIZICA ATOMICA,BUCHAREST
FT-	ARMY, WASHINGTON, D.C. (FIRING TABLES)
FT-	FAIRCHILD ENGINE AND AIRPLANE CORP. FAIRCHILD AIRCRAFT DIV., HAGERSTOWN, MD. (FLIGHT TEST REPORT)
FT-	FARADAY TRANSLATIONS, N.Y.C.
FT-(PROJECT)-	FRANKLIN INST. LABS. FOR RES. + DEV., PHILADELPHIA
FT-	NAVAL AIR TEST CENTER, PATUXENT RIVER, MD.
FT31-	NAVAL AIR TEST CENTER, PATUXENT RIVER, MD.
FT-	NORTHROP AIRCRAFT, INC., HAWTHORNE, CAL.(FLIGHT TEST)
FT-	UNITED KINGDOM. FIGHTING VEHICLES RESEARCH AND DEVELOPMENT ESTAB., CHERTSEY, SURREY, ENGLAND
F.T.-	UNIVERSITATEA DIN TIMISOARA, ROUMANIA
FTAS/TR-(YEAR)-	CASE INST. OF TECH., CLEVELAND. FLUID, THERMAL, AND AEROSPACE SCIENCES GROUP
FTB-	AKTIEBOLAGET ATOMENERGI, STOCKHOLM
FTB-	AKTIEBOLAGET ATOMENERGI, STUDSVIK, SWEDEN
FTC-	AIR FORCE FLIGHT TEST CENTER, EDWARDS AFB, CALIF.
FTC-	FEDERAL TRADE COMMISSION, WASHINGTON, D.C.
FTC-CR-(YR.)-	MARTIN CO., DENVER (CONTRACT REPORT)
FTC-CR-	MARTIN CO., ORLANDO, FLA.
FTC-TDR-	AIR FORCE FLIGHT TEST CENTER, EDWARDS AFB, CALIF.
FTC-TR-(YEAR)-	AIR FORCE FLIGHT TEST CENTER, EDWARDS AFB, CALIF.
FTD-(YEAR)-	FOREIGN TECHNOLOGY DIV., WRIGHT-PATTERSON AFB, OHIO
FTD-(7 DIGITS)	FOREIGN TECHNOLOGY DIV., WRIGHT-PATTERSON AFB, OHIO
FTD-(YEAR)-PHE-	FOREIGN TECHNOLOGY DIV., WRIGHT-PATTERSON AFB, OHIO
FTD-CW-(NUMBERS-YEAR)	FOREIGN TECHNOLOGY DIV., WRIGHT-PATTERSON AFB, OHIO
FTDE-	AIR FORCE EXPERIMENTAL ROCKET ENGINE TEST STATION, EDWARDS AFB, CALIF.
FTD-H-(YEAR)-	FOREIGN TECHNOLOGY DIV., WRIGHT-PATTERSON AFB, OHIO
FTD-HC-(NO.-NO.-YEAR)	FOREIGN TECHNOLOGY DIV., WRIGHT-PATTERSON AFB, OHIO
FTD-HT-(YEAR)-	FOREIGN TECHNOLOGY DIV., WRIGHT-PATTERSON AFB, OHIO
FTDM-	GENERAL DYNAMICS/FORT WORTH, TEX.
FTD-MT-(YEAR)-	FOREIGN TECHNOLOGY DIV., WRIGHT-PATTERSON AFB, OHIO (MACHINE TRANSLATION)
FTD-MT-(NO.-NO.-YEAR)	FOREIGN TECHNOLOGY DIV., WRIGHT-PATTERSON AFB, OHIO
FTD-PHS-(NUMBER-YEAR)-	FOREIGN TECHNOLOGY DIV., WRIGHT-PATTERSON AFB, OHIO
FTD-PWS-(NUMBER-YEAR)-	FOREIGN TECHNOLOGY DIV., WRIGHT-PATTERSON AFB, OHIO
FTD-PWW-IM-(NUMBER)-	FOREIGN TECHNOLOGY DIV., WRIGHT-PATTERSON AFB, OHIO
FTDS-	HUGHES AIRCRAFT CO.,CULVER CITY,CAL.(FLIGHT TESTDATA)
FTD-ST-HB-(NUMBERS)-	FOREIGN TECHNOLOGY DIV., WRIGHT-PATTERSON AFB, OHIO
FTD-TT-(YEAR)-	FOREIGN TECHNOLOGY DIV., WRIGHT-PATTERSON AFB, OHIO
FTD-TT-	FOREIGN TECHNOLOGY DIV., WRIGHT-PATTERSON AFB, OHIO
FTED-	HUGHES AIRCRAFT CO.,CULVER CITY,CAL.(FLIGHT TEST ENG)
FT/F-	CHANCE VOUGHT AIRCRAFT, INC., DALLAS
FTH-	TENNESSEE EASTMAN CORP., OAK RIDGE, TENN.
FTI-(YEAR)-	AKADEMIYA NAUK GRUZINSKOI SSR. FIZIKO-TEKHNICHESKII INSTITUT, SUKHUMI
FTI-	AKADEMIYA NAUK SSSR. ORDENA FIZIKO-TECHNICHESKII INSTITUT IMENI A.F. IOFFE, LENINGRAD
FTI-(NUMBER)/R-	AKADEMIYA NAUK UKRAINSKOI SSR. FIZIKO-TEKHNICHESKII INSTITUT, KHARKOV
FTI-	AKTIEBOLAGET ATOMENERGI, STOCKHOLM
FTI-	U.S.S.R. GOSUDARSTVENNYI KOMITET PO ISPOL'ZOVANIYU ATOMNOI ENERGII. FIZIKO-TEKHNICHESKII INST.,SUKHUMI
FTI-AN-	AKADEMIYA NAUK UKRAINSKOI SSR. FIZIKO-TEKHNICHESKII INSTITUT, KHARKOV
FTID-	HUGHES AIRCRAFT CO., CULVER CITY, CALIF.(FLIGHT TEST)

FTI-LA-	AKADEMIYA NAUK UKRAINSKOI SSR. FIZIKO-TEKHNICHESKII INSTITUT, KHARKOV
FTIO-(NUMBER)-	FOREIGN TECHNICAL INTELLIGENCE OFFICE, ABERDEEN PROVING GROUND, MD.
FTL-5483-QDR-	FEDERAL TELECOMMUNICATION LABS., INC., NUTLEY, N.J.
FTL-	FEDERAL TELEPHONE AND RADIO CORP., CLIFTON, N.J.
FTL-	JOINT PARACHUTE TEST FACILITY, EL CENTRO, CALIF.
FTL-A-A(NUMBER-NUMBER)	SWEDEN. FOERSVARETS TELETEKNISKA LAB., STOCKHOLM
FTL-R-	MASSACHUSETTS INST. OF TECH., CAMBRIDGE. FLIGHT TRANSPORTATION LAB.
FTL-TM-	FEDERAL TELECOMMUNICATION LABS., INC., NUTLEY, N.J.
FTM-	PLANNING RESEARCH CORP., LOS ANGELES
FTMR-	FAIRCHILD HILLER CORP. REPUBLIC AVIATION DIV., FARMINGDALE, N.Y.
FTP-	BENDIX AVIATION CORP. (FLIGHT TEST PROPOSAL)
FTP-	NAVY (FLEET TRAINING PAMPHLET)
FTP-	NAVY (FLEET TRAINING PUBLICATION)
FTR-	(FINAL TECHNICAL REPORT) (REF. 2)
FTR-	(FLIGHT TEST REPORT) (REF. 2)
FTR-	(FUNCTIONAL TEST REPORT) (REF. 2)
FTR-	ACF INDUSTRIES, INC. ALBUQUERQUE DIV., N. MEX.
FTR-	AERONUTRONIC DIV. OF PHILCO CORP., NEWPORT BEACH, CALIF. (FINAL TECHNICAL REPORT)
FTR-	AMERICAN MACHINE AND FOUNDRY CO. GENERAL ENGINEERING LAB., GREENWICH, CONN.
FTR-	ARMY, WASHINGTON, D.C. (FIRING TABLES ROCKETS)
FTR-	AUBURN UNIV., ALA.
F-TR-(NO.)-(LETTERS)	CENTRAL AIR DOCS. OFF., WRIGHT-PATTERSON AFB, OHIO
FTR-	CONVAIR (ALL LOCATIONS) (FLIGHT TEST REPORT)
FTR/	CUBIC CORP., SAN DIEGO, CALIF.
FTR-	FIRESTONE TIRE AND RUBBER CO. ENGINEERING LAB., MONTEREY, CALIF.
FTR-	FLORENCE. UNIVERSITA
FTR-(NUMBER)-	HYDRONAUTICS, INC., LAUREL, MD.
FTR-	ITALY. ISTITUTO NAZIONALE DI ULTRACUSTICA, ROME
FTR-	LEICESTER UNIV., ENGLAND
FTR-	LONDON. UNIV.
FTR-	PICATINNY ARSENAL, DOVER, N.J. (FRAGMENTATION TEST RECORD)
FTRC-	FIRESTONE TIRE AND RUBBER CO., AKRON, OHIO
FTRC RDR-	FARNSWORTH TELEVISION AND RADIO CORP. RESEARCH DEPT., FORT WAYNE
FTRDI-MR(YEAR)-	BEECHCRAFT RESEARCH AND DEVELOPMENT, INC.
FTRM-	MC DONNELL AIRCRAFT CORP., ST. LOUIS (FLIGHT TEST RESEARCH MEMORANDUM)
F-TS-(NUMBER)-RE	AIR TECH. INTELL. CTR., WRIGHT-PATTERSON AFB, OHIO
F-TS-	ARMED SERVICES TECHNICAL INFORMATION AGENCY. DOCUMENT SERVICE CENTER, DAYTON, OHIO
F-TS-	FOREIGN TECHNOLOGY DIV., WRIGHT-PATTERSON AFB, OHIO
FTS-	NORTHROP AIRCRAFT, INC., HAWTHORNE, CALIF. (FLIGHT TEST SUMMARY)
FTSR-	BIOT AND ARNOLD (CONSULTANTS IN APPLIED MATH. + PHYS)
FTSR-	BIRMINGHAM. UNIV., ENGLAND
FT-TM-(NO.)-(YEAR)	NAVAL AIR TEST CENTER, PATUXENT RIVER, MD.
FTZ-	GERMANY. BUNDESPOST. FERNMELDETECHNISCHES ZENTRALAMT DARMSTADT
FTZ-A-(NOS.)-(LTRS.)-	GERMANY. BUNDESPOST. FERNMELDETECHNISCHES ZENTRALAMT DARMSTADT
FU-	FLORIDA. UNIV., GAINESVILLE
FU-	UNDERWRITERS LABS., INC., N.Y.C.
FUND.-184 S2-	CALIFORNIA. UNIV., BERKELEY. RADIATION LAB. (REF. 9)
FUND.-(MONTH/YEAR)	CALIFORNIA. UNIV., BERKELEY. RADIATION LAB. (REF. 9)
FUND.-P(MONTH/YEAR)	CALIFORNIA. UNIV., BERKELEY. RADIATION LAB. (REF. 9)
FUND.-S-	CALIFORNIA. UNIV., BERKELEY. RADIATION LAB. (REF. 9)
FUPH-R-	FLINDERS UNIV., BEDFORD PARK, SOUTH AUSTRALIA. SCHOOL OF PHYSICAL SCIENCES
FU-QPR-(NUMBER)-(YEAR)-	FLORIDA. UNIV., GAINESVILLE. ENGINEERING AND INDUSTRIAL EXPERIMENT STATION
FU-QTR-	FLORIDA. UNIV., GAINESVILLE. ENGINEERING AND INDUSTRIAL EXPERIMENT STATION
FVDD-	UNITED KINGDOM. FIGHTING VEHICLES RESEARCH AND DEVELOPMENT ESTAB., CHERTSEY, SURREY, ENGLAND
FVPE-	UNITED KINGDOM. FIGHTING VEHICLES RESEARCH AND DEVELOPMENT ESTAB., CHERTSEY, SURREY, ENGLAND
FVR-	INTERNATIONAL ATOMIC ENERGY AGENCY. NUCLEAR DATA UNIT, VIENNA
FVRDE-	UNITED KINGDOM. FIGHTING VEHICLES RESEARCH AND DEVELOPMENT ESTAB., CHERTSEY, SURREY, ENGLAND
FVRDE-(LETTERS)-	UNITED KINGDOM. FIGHTING VEHICLES RESEARCH AND DEVELOPMENT ESTAB., CHERTSEY, SURREY, ENGLAND
FW-	FOSTER WHEELER CORP., N.Y.C.
FW-(YEAR)-(NUMBER)	FOSTER WHEELER CORP., N.Y.C.
FWE-	TECHNICAL INFORMATION SERVICE EXTENSION, AEC, OAK RIDGE, TENN. (FOREIGN WEAPONS EFFECTS - ASSIGNED TO UNITED KINGDOM AND CANADIAN REPORTS RELEASED TO THE U. S.)
FWL-(NUMBER-NUMBER)-	GENERAL ELECTRIC CO. FORT WAYNE LAB., IND.
FWM-	BUREAU OF ORDNANCE (NAVY) (FLASHING + WIPING MEMO.)
FWPCA-(NUMBER-DATE)-	FEDERAL WATER POLLUTION CONTROL ADM., WASHINGTON, DC
FWPCA-(NO.)-DWF-(DATE)	FEDERAL WATER POLLUTION CONTROL ADM., WASHINGTON, DC
FWPCA-(NO.-LTRS.-DATE)	FEDERAL WATER POLLUTION CONTROL ADM., WASHINGTON, DC
FWPCA-(NO.)-EAG-(DATE)	FEDERAL WATER POLLUTION CONTROL ADM., WASHINGTON, DC
FWPCA-DAST-	FEDERAL WATER POLLUTION CONTROL ADM., WASHINGTON, DC (DISPOSAL AND SEWAGE TREATMENT)
FWQA-(NUMBER-DATE)	FEDERAL WATER QUALITY ADMINISTRATION, WASHINGTON, DC
FWQA-(NO.)-EAG-(DATE)	FEDERAL WATER QUALITY ADMINISTRATION, WASHINGTON, DC
FWQA-(NO.-LTRS.-DATE)	FEDERAL WATER QUALITY ADMINISTRATION, WASHINGTON, DC
FWQA-(NO.)-DXU-(DATE)	FEDERAL WATER QUALITY ADMINISTRATION, WASHINGTON, DC
FWQO-	FEDERAL WATER QUALITY ADMINISTRATION, WASHINGTON, DC
FWR-	FAIRCHILD STRATOS CORP. AIRCRAFT-MISSILES DIV., HAGERSTOWN, MD.
FW-RDR-	FOSTER WHEELER CORP., LIVINGSTON, N.J.
FWS-DR-	BUREAU OF COMMERCIAL FISHERIES, WASHINGTON, D.C. (FISH AND WILDLIFE DATA RPT.)
FWSGANT-	FLEET WORK STUDY GROUP, ATLANTIC, NORFOLK, VA.
FWSGLANT-	FLEET WORK STUDY GROUP, ATLANTIC, NORFOLK, VA.
FWSGPAC-STUDY-	FLEET WORK STUDY GROUP, PACIFIC, SAN DIEGO, CALIF.
FWWMR-	QUARTERMASTER RESEARCH AND ENG. COMMAND, NATICK, MASS., (FIRE, WATER, WEATHER, MILDEW RESISTANT)
FX/EQ-	GT. BRIT. MARINE AIRCRAFT EXPERIMENTAL ESTABLISHMENT, FELIXSTOWE, SUFF., ENGLAND
FXM-	PRATT AND WHITNEY AIRCRAFT DIV., UNITED AIRCRAFT CORP., HARTFORD, CONN.
FXR-	PRATT AND WHITNEY AIRCRAFT DIV., UNITED AIRCRAFT CORP., HARTFORD, CONN.
FY-	(FISCAL YEAR) (REF. 2)
FY-	SIGNAL CORPS ENG. LABS.,FT.MONMOUTH,N.J.(FISCAL YEAR)
FZA-	GENERAL DYNAMICS/FORT WORTH, TEX.
FZC-	GENERAL DYNAMICS/FORT WORTH, TEX.
FZD-	GENERAL DYNAMICS/FORT WORTH, TEX.
FZE-	GENERAL DYNAMICS/FORT WORTH, TEX.
FZI-	GENERAL DYNAMICS/FORT WORTH, TEX.
FZK-	GENERAL DYNAMICS/FORT WORTH, TEX.
FZM-	GENERAL DYNAMICS/FORT WORTH, TEX.
FZP-	GENERAL DYNAMICS/FORT WORTH, TEX.
FZS-	GENERAL DYNAMICS/FORT WORTH, TEX.
FZT-	GENERAL DYNAMICS/FORT WORTH, TEX.

G-

G-	(GROUP REPORT) (REF. 2)
G-(NUMBER)-(LETTER)-	ADCOM, INC., CAMBRIDGE, MASS.
G(NUMBER)	ADMIRAL CORP., CHICAGO
G-(4 DIGITS)	AMERICAN MACHINE AND FOUNDRY CO. GENERAL ENGINEERING LAB., GREENWICH, CONN.
G-(3 DIGITS)-PR-	ARMOUR RESEARCH FOUNDATION, CHICAGO
G.-	ASSOCIATED ELECTRICAL INDUSTRIES, LTD. RESEARCH LABS. RUGBY, WARWICK, ENGLAND
G-(4 DIGITS)-1-	BATTELLE MEMORIAL INST., COLUMBUS, OHIO
G(4 DIGITS)-R-	BEUKERS (JOHN M.) LABS., INC., STONY BROOK, N.Y.
G101-	BUNKER-RAMO CORP. DEFENSE SYSTEMS DIV., CANOGA PARK, CALIF.
G(NO.-NO.)U(NUMBER)	BUNKER-RAMO CORP. DEFENSE SYSTEMS DIV., CANOGA PARK, CALIF.
G-	BUREAU OF RECLAMATION (INTERIOR). COMMISSIONERS OFFICE, DENVER
G-	CANADA. DEFENCE RESEARCH BOARD, OTTAWA
G-(NUMBER)-C(LETTER)	CHICAGO. UNIV. METALLURGICAL LAB. (REF. 10)
1G-	CHRYSLER CORP. MISSILE DIV., DETROIT
G-	COAST AND GEODETIC SURVEY, ROCKVILLE, MD. (PRELIMINARY GEODETIC PUBL.)
G-	DELAWARE. UNIV., NEWARK
G(4 DIGITS)-A	DORTECH, INC., STAMFORD, CONN.
G-	DOUGLAS AIRCRAFT CO., INC., SANTA MONICA, CALIF.
G-	ENGLEMAN AND CO. INC., WASHINGTON, D.C.
G(YEAR)-	GARRETT CORP. AIRESEARCH MFG. DIV., LOS ANGELES (LIBRARY ACCESSION NUMBER)
G(NO.)(LTRS.)(NO.)	GENERAL ELECTRIC CO. MISSILE + SPACE DIV., CINCINNATI
G(NUMBER)	IIT RESEARCH INST., CHICAGO
G-(NUMBER)/(YEAR)	ISRAEL. NEGEV INST. FOR ARID ZONE RESEARCH. DEPT. OF GEOBOTANY, BEERSHEBA
G-(NUMBER)-	LITTLETON RESEARCH AND ENGINEERING CORP., MASS.
G(NO.).(NO.).(NO.)(LTR.)	LTV ELECTROSYSTEMS, INC., DALLAS
G8406.(NUMBER).(NUMBER)	LTV ELECTROSYSTEMS INC., GREENVILLE, TEX.
G(NO.-NO.-NO.)B	LTV ELECTROSYSTEMS INC., GREENVILLE, TEX.
G(NUMBER)	MC DONNELL AIRCRAFT CORP., ST. LOUIS
G(NUMBER)	MC DONNELL DOUGLAS CORP., ST. LOUIS
G-	MC GILL UNIV., MONTREAL. MAC DONALD PHYSICS LAB.
G-	MARCHETTI (J.W.), INC., NATICK, MASS.
G-	MOTOROLA INC. SEMICONDUCTOR PRODS. DIV., PHOENIX, ARIZ.
G-	NATIONAL OCEANOGRAPHIC DATA CENTER, WASHINGTON, D.C.
G-	NETHERLANDS. NATIONAAL LUCHT EN RUIMTEVAART-LABORATORIUM, AMSTERDAM
G-(NO.)-(YR)	RYAN AERO. CO., LINDBERGH FIELD, SAN DIEGO, CALIF.
G-1.133- THRU G-3.200-	TENNESSEE EASTMAN CORP., OAK RIDGE, TENN. (REF. 34)
G-	UNION OF SOUTH AFRICA. ATOMIC ENERGY BOARD, PRETORIA
G-	UNITED AIRCRAFT CORP., EAST HARTFORD, CONN.
G-91(4 DIGITS)-	UNITED AIRCRAFT CORP. RES. LABS., EAST HARTFORD, CONN.
G-(LETTERS)	(G- IS ASSIGNED BY VARIOUS RECIPIENTS AS A PREFIX TO CODES FOR REPORTS ORIGINATING IN GERMANY. FOR EXPLANATION OF THE REMAINDER OF THE CODE, SEE ENTRY UNDER LETTERS FOLLOWING G-)
GA-	AVCO CORP. RES.+ ADVANCED DEV. DIV., WILMINGTON, MASS.
GA-	CALIFORNIA. UNIV., BERKELEY. RADIATION LAB.
GA-	ETHYL CORP. RESEARCH AND DEVELOPMENT DEPT., FERNDALE, MICH. (GENERAL ADMINISTRATION)
GA-(YEAR)-PR-	GIANNOTTI ASSOCS., BELLPORT, N.Y.
GA-	GULF GENERAL ATOMIC, INC., SAN DIEGO, CALIF.
GA-(NUMBER)	SPERRY GYROSCOPE CO., SYOSSET, N.Y.
GA/(LTRS.)/(YEAR)-	AIR FORCE INST. OF TECH., WRIGHT-PATTERSON AFB, OHIO. SCHOOL OF ENGINEERING (GRADUATE ASTRONAUTICS)
GA-(LETTER(S))-	GENERAL ATOMIC DIV., GEN. DYNAMICS CORP., SAN DIEGO, CAL
GAB-	GABRIEL LABS., NEEDHAM HEIGHTS, MASS.
GAC-	GENERAL ASTROMETALS CORP., YONKERS, N.Y.
GA-C-	GENERAL ATOMIC DIV., GEN. DYNAMICS CORP., SAN DIEGO, CAL
GAC-	GENERAL ATRONICS CORP., PHILADELPHIA
GAC-	GOODYEAR AEROSPACE CORP., AKRON, OHIO
GACA-	AIR FORCE INST. OF TECH., WRIGHT-PATTERSON AFB, OHIO
GACD-	GENERAL ELECTRIC CO., SYRACUSE, N.Y.
GACD-	GULF GENERAL ATOMIC, INC., SAN DIEGO, CALIF.
GACD-6135(MONTH-YEAR)	GULF GENERAL ATOMIC, INC., SAN DIEGO, CALIF. (HTGR BASE PROGRAM PROGRESS REPORT)
GACD-6850(MONTH-YEAR)	GULF GENERAL ATOMIC, INC., SAN DIEGO, CALIF. (PUBLIC SERVICE CO. OF COLO. PLANT DESIGN+CONSTR)
GACD-6900(MONTH-YEAR)	GULF GENERAL ATOMIC, INC., SAN DIEGO, CALIF. (PUBLIC SERVICE CO. OF COLO. RES. + DEV. PROGRAM)
GACD-PR (CONTRACT NO.)	GENERAL ATOMIC DIV., GEN. DYNAMICS CORP., SAN DIEGO, CAL
GAC-GEK-	GOODYEAR AIRCRAFT CORP., AKRON, OHIO
GAC-GER-	GOODYEAR AIRCRAFT CORP., AKRON, OHIO
GACP-	GENERAL ATOMIC DIV., GEN. DYNAMICS CORP., SAN DIEGO, CAL
GACP-	GULF GENERAL ATOMIC, INC., SAN DIEGO, CALIF.
GADC-	GENERAL ATOMIC DIV., GEN. DYNAMICS CORP., SAN DIEGO, CAL
GAE-	AIR FORCE INST. OF TECH., WRIGHT-PATTERSON AFB, OHIO
GAEC-	GRUMMAN AIRCRAFT ENGINEERING CORP., BETHPAGE, N.Y.
GAEC-ENG-POD-(YEAR)-	GRUMMAN AIRCRAFT ENGINEERING CORP., BETHPAGE, N.Y.
GAEC-FSR-S(NUMBER)-	GRUMMAN AIRCRAFT ENGINEERING CORP., BETHPAGE, N.Y.
GAEC-PDR-	GRUMMAN AIRCRAFT ENGINEERING CORP., BETHPAGE, N.Y.
GAEC-PDR-AC-(YEAR)-	GRUMMAN AIRCRAFT ENGINEERING CORP., BETHPAGE, N.Y.
GAEC-RE-	GRUMMAN AIRCRAFT ENGINEERING CORP., BETHPAGE, N.Y.
GA/EE/(YEAR)-	AIR FORCE INST. OF TECH., WRIGHT-PATTERSON AFB, OHIO
GAE/ME/(YEAR)-	AIR FORCE INST. OF TECH., WRIGHT-PATTERSON AFB, OHIO
GAE/PHYS/(YEAR)-	AIR FORCE INST. OF TECH., WRIGHT-PATTERSON AFB, OHIO
GAF-	ELECTRONICS RESEARCH, INC., EVANSVILLE, IND.
GAFB-	ROME AIR DEVELOPMENT CENTER, GRIFFISS AFB, N.Y.
GA+FC-	GENERAL ANILINE AND FILM CORP., EASTON, PENNA.
GAFC (LETTERS)-	GENERAL ANILINE AND FILM CORP.
GAF-US-	GENERAL ANILINE AND FILM CORP., BINGHAMTON, N.Y.
GAG-	MASSACHUSETTS INST. OF TECH., CAMBRIDGE
GAI-	GILBERT ASSOCIATES, INC., WASHINGTON, D.C.
GAI-	GILBERT ASSOCIATES, INC. NUCLEAR ENERGY DEPT., READING, PENNA.
GAI/(LTRS.)/(YEAR)-	AIR FORCE INST. OF TECH., WRIGHT-PATTERSON AFB, OHIO. SCHOOL OF ENGINEERING
GAL-	CALIF. INST. OF TECH., PASADENA. GUGGENHEIM AERO. LAB
GALCIT-	CALIFORNIA INST. OF TECH., PASADENA. GRADUATE AERO-NAUTICAL LABS.
GALCIT-HYPER M-	CALIFORNIA INST. OF TECH., PASADENA. GRADUATE AERO-NAUTICAL LABS. (HYPERSONIC MEMO)
GALCIT JPC TR-	CALIFORNIA INST. OF TECH., PASADENA. GRADUATE AERO-NAUTICAL LABS. (JET PROPULSION CENTER TECH. REPT.)
GALCIT SM TR-	CALIFORNIA INST. OF TECH., PASADENA. GRADUATE AERO-NAUTICAL LABS. (SOLID MECHANICS)
GAM-(YEAR)/(LTRS.)/	AIR FORCE INST. OF TECH., WRIGHT-PATTERSON AFB, OHIO
GAM/(LTRS.)/(YEAR)-	AIR FORCE INST. OF TECH., WRIGHT-PATTERSON AFB, OHIO. SCHOOL OF ENGINEERING (GRADUATE AEROSPACE)
GAMD-	GULF GENERAL ATOMIC, INC., SAN DIEGO, CALIF.
GAMD-TR-	GULF GENERAL ATOMIC, INC., SAN DIEGO, CALIF. (TRANS.)
GA/ME/(YEAR)-	AIR FORCE INST. OF TECH., WRIGHT-PATTERSON AFB, OHIO
GA/MECH/(YEAR)-	AIR FORCE INST. OF TECH., WRIGHT-PATTERSON AFB, OHIO
GAMM-	GESELLSCHAFT FUER ANGEWANDTE MATHEMATIK UND MECHANIK
GAO-	AIR FORCE INST. OF TECH., WRIGHT-PATTERSON AFB, OHIO
GAO-	GENERAL ACCOUNTING OFFICE, WASHINGTON, D.C.

GAO-	OFFICE OF EMERGENCY PREPAREDNESS, WASHINGTON, D.C. (GENERAL ADMINISTRATIVE ORDER)
GA-P-	GENERAL ATOMIC DIV., GEN. DYNAMICS CORP., SAN DIEGO, CAL
GA-P-	GULF GENERAL ATOMIC, INC., SAN DIEGO, CALIF.
GA/PH/(YEAR)-	AIR FORCE INST. OF TECH., WRIGHT-PATTERSON AFB, OHIO
GA/PHYS/(YEAR)-	AIR FORCE INST. OF TECH., WRIGHT-PATTERSON AFB, OHIO
GAR-	(GERMAN AVIATION RESEARCH)
GAR-	GARRETT CORP. AIRESEARCH MFG. DIV., LOS ANGELES
GARD-	GENERAL AMERICAN TRANSPORTATION CORP. GENERAL AMERICAN RESEARCH DIV., NILES, ILL.
GARD-(NUMBER-LETTER)	GENERAL AMERICAN TRANSPORTATION CORP. GENERAL AMERICAN RESEARCH DIV., NILES, ILL.
GARD-(NUMBER-NUMBER-LTR)	GENERAL AMERICAN TRANSPORTATION CORP. GENERAL AMERICAN RESEARCH DIV., NILES, ILL.
GARD-MR-	GENERAL AMERICAN TRANSPORTATION CORP. GENERAL AMERICAN RESEARCH DIV., NILES, ILL.
GAR/LA-(YEAR)-	GARRETT CORP. AIRESEARCH MFG. DIV., LOS ANGELES
GAR/LA-(LETTERS-YEAR)-	GARRETT CORP. AIRESEARCH MFG. DIV., LOS ANGELES
GAR/LA HT-(YR.-NO./NO.)	GARRETT CORP. AIRESEARCH MFG. DIV., LOS ANGELES
GAS-	AIR FORCE INST. OF TECH., WRIGHT-PATTERSON AFB, OHIO
GAS-	SCHWABLAND (GEORGE A.), ALEXANDRIA, VA.
GASBUGGY-	GEOLOGICAL SURVEY, DENVER
GASL-	GENERAL APPLIED SCIENCE LABS., INC., WESTBURY, N.Y.
GASL-	GRUEN APPLIED SCIENCE LABS., INC., WEST HEMPSTEAD, NY
GASL-ITR-	GENERAL APPLIED SCIENCE LABS., INC., WESTBURY, N.Y.
GASL-S-	GENERAL APPLIED SCIENCE LABS., INC., WESTBURY, N.Y.
GASL-TR-	GENERAL APPLIED SCIENCE LABS., INC., WESTBURY, N.Y.
GAT-	GOODYEAR ATOMIC CORP., PIKETON, OHIO
GAT-	GOODYEAR ATOMIC CORP., PORTSMOUTH, OHIO (REF. 12)
GAT-(LETTER(S))-	GOODYEAR ATOMIC CORP., PORTSMOUTH, OHIO (REF. 12)
GATC-	GENERAL AMERICAN TRANSPORT. CORP. GAR DIV., NILES, ILL
GATC-GARD-MR-	GENERAL AMERICAN TRANSPORT. CORP. GAR DIV., NILES, ILL
GATC-MRD-	GENERAL AMERICAN TRANSPORT. CORP. MRD DIV., NILES, ILL
GATC-MRD-TN-	GENERAL AMERICAN TRANSPORT. CORP. MRD DIV., NILES, ILL
GAT-DM-	GOODYEAR ATOMIC CORP., PORTSMOUTH, OHIO
GAT-DR-	GOODYEAR ATOMIC CORP., PORTSMOUTH, OHIO (REF. 12)
GAT-E-	GOODYEAR ATOMIC CORP., PORTSMOUTH, OHIO
GAT-L-	GOODYEAR ATOMIC CORP., PORTSMOUTH, OHIO (REF. 12)
GAT-OR-	GOODYEAR ATOMIC CORP., PORTSMOUTH, OHIO
GAT-P-	GOODYEAR ATOMIC CORP., PIKETON, OHIO
GAT-R-	GOODYEAR ATOMIC CORP., PIKETON, OHIO
GA-TR-	GULF GENERAL ATOMIC, INC., SAN DIEGO, CALIF. (TRANS.)
GAT-SI-	GOODYEAR ATOMIC CORP., PORTSMOUTH, OHIO
GAT-T-	GOODYEAR ATOMIC CORP., PIKETON, OHIO
GATX-	GENERAL AMERICAN TRANSPORT. CORP. GAR DIV., NILES, ILL
GATX-MR-	GENERAL AMERICAN TRANSPORT. CORP. MRD DIV., NILES, ILL
GAT-Z-	GOODYEAR ATOMIC CORP., PIKETON, OHIO
GAW/(LTRS.)/(YEAR)-	AIR FORCE INST. OF TECH., WRIGHT-PATTERSON AFB, OHIO. SCHOOL OF ENGINEERING (GRADUATE AIR WEAPONS)
GB-(4 DIGITS)-	SPERRY RAND CORP. SPERRY SYSTEMS MANAGEMENT DIV., GREAT NECK, N.Y.
GB-ASI-T-	GT. BRIT. AIR SCIENTIFIC INTELLIGENCE (TECH. TRANS.)
GBF-	GOODRICH (B.F.) CO., AKRON, OHIO
GBI-	GILFILLAN BROS., INC., LOS ANGELES
GBR-	UNITED ELECTRODYNAMICS, INC. EARTH SCIENCES DIV., PASADENA, CALIF.
GC-	BEUKERS (JOHN M.) LABS., INC., MEDFORD, N.Y.
GC-	DU PONT DE NEMOURS (E.I.) + CO. GRASSELLI CHEMICALS DEPT. EXPERIMENTAL LAB., CLEVELAND
(YEAR)-GC-	GENERAL ELECTRIC CO. RESEARCH LAB., SCHENECTADY, N.Y.
G + C-	GIBBS AND COX, INC., N.Y.C.
GC-	GLENCO CORP., METUCHEN, N.J.
GC(ROMAN NO.)/INF-	INTERNATIONAL ATOMIC ENERGY AGENCY, VIENNA
GCA-(YEAR NUMBER)	GCA CORP. BEDFORD, MASS. (IN-HOUSE RESEARCH)
GCA-	GEOPHYSICS CORP. OF AMERICA, BEDFORD, MASS.
GCA-	GRAHAM, CROWLEY AND ASSOCIATES, INC., CHICAGO
GCA-TR-(YEAR)-(NO.)-A	GCA CORP. TECHNOLOGY DIV., BEDFORD, MASS. (WORK DONE FOR AIR FORCE)
GCA-TR-(YEAR)-(NO.)-G	GCA CORP. TECHNOLOGY DIV., BEDFORD, MASS. (WORK DONE FOR OTHER AGENCIES)
GCA-TR-(YEAR)-(NO.)-N	GCA CORP. TECHNOLOGY DIV., BEDFORD, MASS. (WORK DONE FOR NASA)
GCC-	GIANNINI CONTROLS CORP., DUARTE, CALIF.
GCC-(YEAR)-(NO.)-	GIANNINI CONTROLS CORP., FAIRFIELD, N.J.
GCC-	GIANNINI CONTROLS CORP., SANTA ANA, CALIF.
GCC-ER-	GIANNINI CONTROLS CORP. (ALL LOCATIONS)
GCD-	ALLIED CHEMICAL CORP. GENERAL CHEMICAL DIV., N.Y.C.
GCD-	DEWEY (G.C.) AND CO., INC., N.Y.C.
GCD-FR-	ALLIED CHEMICAL CORP., N.Y.C.
GCD-MPR-	TENNESSEE EASTMAN CORP., OAK RIDGE, TENN.
GCD-QR-	ALLIED CHEMICAL CORP., N.Y.C.
GCE-(LETTERS)-	ARMED FORCES SPECIAL WEAPONS PROJECT. FIELD COMMAND, ALBUQUERQUE, N. MEX.
GCG-	RES.+ DEV.BD. COMM.ON GEOPHYSICS + GEOGRAPHY, WASH,DC
GCHW-(LETTERS)-	UNITED KINGDOM ATOMIC ENERGY AUTHORITY. RESEARCH GP. ATOMIC ENERGY RES. ESTAB., HARWELL, BERKS, ENGLAND
GCHW/MEMO-	UNITED KINGDOM ATOMIC ENERGY AUTHORITY. RESEARCH GP. ATOMIC ENERGY RES. ESTAB., HARWELL, BERKS, ENGLAND
GCHW-NOTE-	UNITED KINGDOM ATOMIC ENERGY AUTHORITY. RESEARCH GP. ATOMIC ENERGY RES. ESTAB., HARWELL, BERKS, ENGLAND
GC(II)/	INTERNATIONAL ATOMIC ENERGY AGENCY, VIENNA
GCLC CON-	CALIFORNIA INST. OF TECH., PASADENA. GATES AND CRELLIN LABS. OF CHEMISTRY (CONTRIBUTION)
GCM-	ARMY BALLISTIC MISSILE AGENCY, REDSTONE ARSENAL, ALA. (GUIDANCE AND CONTROL MEMORANDUM)
GCMSFC-	GEORGE C. MARSHALL SPACE FLIGHT CTR., HUNTSVILLE, ALA.
GCM/UK/	UNITED KINGDOM ATOMIC ENERGY AUTHORITY. INDUSTRIAL GROUP. CULCHETH LABS., LANCS, ENGLAND
GCM/UK-	UNITED KINGDOM ATOMIC ENERGY AUTHORITY. RESEARCH GP. ATOMIC ENERGY RES. ESTAB., HARWELL, BERKS, ENGLAND
GCM/UK/A-	UNITED KINGDOM ATOMIC ENERGY AUTHORITY. REACTOR GROUP, CULCHETH, LANCS, ENGLAND
GCM/UK/B-	UNITED KINGDOM ATOMIC ENERGY AUTHORITY. RESEARCH GP. ATOMIC ENERGY RES. ESTAB., HARWELL, BERKS, ENGLAND
GCO-(YEAR)-	GCO, INC., ANN ARBOR, MICH.
GC-R(NUMBER)	CHRYSLER CORP. MISSILE OPERATIONS. GUIDANCE AND CONTROL DESIGN DEPT.
GCR-	GRAND CENTRAL ROCKET CO., REDLANDS, CALIF.
GCRD-	TENN. VALLEY AUTH. DIV. OF POWER SUPPLY, CHATTANOOGA
GCR-TR-(YEAR-NUMBER)-N	GEOPHYSICS CORP. OF AMERICA, BEDFORD, MASS.
GCSD-	AEROJET-GENERAL NUCLEONICS, SAN RAMON, CALIF.
G/CSD-AK(NUMBER)	LITTON SYSTEMS, INC. GUIDANCE AND CONTROL SYSTEMS DIV., WOODLAND HILLS, CALIF.
G/CSD-AQ(NUMBER)	LITTON SYSTEMS, INC. GUIDANCE AND CONTROL SYSTEMS DIV., WOODLAND HILLS, CALIF.
G/CSD-AY(NUMBER)	LITTON SYSTEMS, INC. GUIDANCE AND CONTROL SYSTEMS DIV., WOODLAND HILLS, CALIF.
G/CSD-R-	LITTON SYSTEMS, INC. GUIDANCE AND CONTROL SYSTEMS DIV., WOODLAND HILLS, CALIF.
G/CSD-TR-	LITTON SYSTEMS, INC. GUIDANCE AND CONTROL SYSTEMS DIV., WOODLAND HILLS, CALIF.
GCTN-	CALIFORNIA. UNIV., BERKELEY. LAWRENCE RADIATION LAB.
GCTN-	CALIFORNIA. UNIV., LIVERMORE. LAWRENCE RADIATION LAB.

GD-

GD-	COLUMBIA UNIV., NYC. SUBSTITUTE ALLOY MATERIALS LABS.
GD-	GENERAL DYNAMICS CORP. ELECTRIC BOAT DIV., GROTON, CONN
GD-	GENERAL DYNAMICS/ELECTRONICS, ROCHESTER, N.Y.
GD/A-	GENERAL DYNAMICS/ASTRONAUTICS, SAN DIEGO, CALIF.
GDA-	GENERAL DYNAMICS/ASTRONAUTICS, SAN DIEGO, CALIF.
GDA-	GENERAL DYNAMICS/CONVAIR, SAN DIEGO, CALIF.
GDA/A-(YEAR)-	GENERAL DYNAMICS/ASTRONAUTICS, SAN DIEGO, CALIF.
GDA-AA(YEAR)-	GENERAL DYNAMICS/ASTRONAUTICS, SAN DIEGO, CALIF.
GDA-AE-(YEAR)-(NUMBER)-	GENERAL DYNAMICS/ASTRONAUTICS, SAN DIEGO, CALIF.
GDA-AOC-(YEAR)-	GENERAL DYNAMICS/ASTRONAUTICS, SAN DIEGO, CALIF.
GDA-AOH(YEAR)-	GENERAL DYNAMICS/ASTRONAUTICS, SAN DIEGO, CALIF.
GDA-AOJ(YEAR)-	GENERAL DYNAMICS/ASTRONAUTICS, SAN DIEGO, CALIF.
GDA-AY(YEAR)-	GENERAL DYNAMICS/ASTRONAUTICS, SAN DIEGO, CALIF.
GDA-DBE-	GENERAL DYNAMICS/ASTRONAUTICS, SAN DIEGO, CALIF.
GD/A-DBG-(NUMBER)-	GENERAL DYNAMICS/CONVAIR, SAN DIEGO, CALIF.
GD/A-DDB-	GENERAL DYNAMICS/ASTRONAUTICS, SAN DIEGO, CALIF.
GD/A-DDG-	GENERAL DYNAMICS/CONVAIR, SAN DIEGO, CALIF.
GD/A-ERR-	GENERAL DYNAMICS/ASTRONAUTICS, SAN DIEGO, CALIF.
GDA-ERR-	GENERAL DYNAMICS/ASTRONAUTICS, SAN DIEGO, CALIF.
GDA-ERR-AN-	GENERAL DYNAMICS/ASTRONAUTICS, SAN DIEGO, CALIF.
GDA-ESG-	GENERAL DYNAMICS/ASTRONAUTICS, SAN DIEGO, CALIF.
GDAM-	BROWN UNIV., PROVIDENCE. GRADUATE DIVISION OF APPLIED MATHEMATICS (TRANSLATIONS AND REPORTS)
GDAM A-9-M-	BROWN UNIV., PROVIDENCE. GRAD. DIV. OF APPLIED MATH.
GDAM A-9-T-	BROWN UNIV., PROVIDENCE. GRAD. DIV. OF APPLIED MATH.
GDAM A-11-	BROWN UNIV., PROVIDENCE. GRAD. DIV. OF APPLIED MATH.
GDA-MP-	GENERAL DYNAMICS/ASTRONAUTICS, SAN DIEGO, CALIF.
GDAM TR-	BROWN UNIV., PROVIDENCE. GRADUATE DIVISION OF APPLIED MATHEMATICS (TECHNICAL REPORT)
GDA-REL-R-	GENERAL DYNAMICS/ASTRONAUTICS, SAN DIEGO, CALIF.
GDA-VCP-N-	GENERAL DYNAMICS/ASTRONAUTICS, SAN DIEGO, CALIF.
GDA-VTP-E-	GENERAL DYNAMICS/ASTRONAUTICS, SAN DIEGO, CALIF.
GDA-VTP-F-	GENERAL DYNAMICS/ASTRONAUTICS, SAN DIEGO, CALIF.
GDA-VTP-M-	GENERAL DYNAMICS/ASTRONAUTICS, SAN DIEGO, CALIF.
GDA-VTP-P-	GENERAL DYNAMICS/ASTRONAUTICS, SAN DIEGO, CALIF.
GDA-XM-(NO.)-(NO.)(TN)	GENERAL DYNAMICS/ASTRONAUTICS, SAN DIEGO, CALIF.
GDA-ZDU-(NO.)-(NO.)-TN	GENERAL DYNAMICS/ASTRONAUTICS, SAN DIEGO, CALIF.
GDA-ZP-(NO.)-(NO.)-TN	GENERAL DYNAMICS/ASTRONAUTICS, SAN DIEGO, CALIF.
GDA-ZS-(NUMBER)-	GENERAL DYNAMICS/ASTRONAUTICS, SAN DIEGO, CALIF.
GDA-ZT-(NO.)-(NO.)-TN	GENERAL DYNAMICS/ASTRONAUTICS, SAN DIEGO, CALIF.
GD/C-	GENERAL DYNAMICS/CONVAIR, SAN DIEGO, CALIF.
GDC-	GENERAL DYNAMICS/CONVAIR, SAN DIEGO, CALIF.
GDC-(NO.)-(LETTER)-	GENERAL DYNAMICS/CONVAIR, SAN DIEGO, CALIF.
GD-C-(YEAR)-	GENERAL DYNAMICS/CONVAIR, SAN DIEGO, CALIF.
GDC-	GT. BRIT. GERMAN DOCUMENT CENTRE
GDC-	GT. BRIT. MINISTRY OF SUPPLY. TECHNICAL INFO. BUR.
GDC-AAX(YEAR)-	GENERAL DYNAMICS/FORT WORTH. CONVAIR AEROSPACE DIV., TEX.
GDC-ACX(YEAR)-	GENERAL DYNAMICS/CONVAIR, SAN DIEGO, CALIF.
GDC-AWP-(YEAR)-	GENERAL DYNAMICS/CONVAIR, SAN DIEGO, CALIF.
GDC-AWV(YEAR)-	GENERAL DYNAMICS/CONVAIR, SAN DIEGO, CALIF.
GDC-AZD-(YEAR)-	GENERAL DYNAMICS/CONVAIR, SAN DIEGO, CALIF.
GDC-BFK(YEAR)-	GENERAL DYNAMICS/CONVAIR, SAN DIEGO, CALIF.
GDC-BKM(YEAR)-	GENERAL DYNAMICS/CONVAIR, SAN DIEGO, CALIF.
GDC-BNZ(YEAR)-	GENERAL DYNAMICS/CONVAIR, SAN DIEGO, CALIF.
GD/C-BTD(YEAR)-	GENERAL DYNAMICS/CONVAIR, SAN DIEGO, CALIF.
GDC-BTD(YEAR)-	GENERAL DYNAMICS/CONVAIR, SAN DIEGO, CALIF.
GDC/C-ERR-AN-	GENERAL DYNAMICS/CONVAIR, SAN DIEGO, CALIF.
GD/C-CHB-(YEAR)-	GENERAL DYNAMICS/CONVAIR, SAN DIEGO, CALIF.
GD/C-CT-(NO.)B-(NO.)-	GENERAL DYNAMICS/CONVAIR, SAN DIEGO, CALIF.
GDC-CT-(NUMBER-NUMBER)	GENERAL DYNAMICS/CONVAIR, SAN DIEGO, CALIF.
GD/C-DBB-	GENERAL DYNAMICS/CONVAIR, SAN DIEGO, CALIF.
GD/C-DBE-	GENERAL DYNAMICS/CONVAIR, SAN DIEGO, CALIF.
GD/C-DBE(YEAR)-	GENERAL DYNAMICS/CONVAIR, SAN DIEGO, CALIF.
GDC-DBG(YEAR)-	GENERAL DYNAMICS/CONVAIR, SAN DIEGO, CALIF.
GD/C-DCB-	GENERAL DYNAMICS/CONVAIR, SAN DIEGO, CALIF.
GDC-DCB(YEAR)-	GENERAL DYNAMICS/CONVAIR, SAN DIEGO, CALIF.
GDC-DCD(YEAR)-	GENERAL DYNAMICS/CONVAIR, SAN DIEGO, CALIF.
GDC-DDB(YEAR)-	GENERAL DYNAMICS/CONVAIR, SAN DIEGO, CALIF.
GDC-DDC(YEAR)-	GENERAL DYNAMICS/CONVAIR, SAN DIEGO, CALIF.
GDC-DDE(YEAR)-	GENERAL DYNAMICS/CONVAIR, SAN DIEGO, CALIF.
GDC-DDG(YEAR)-	GENERAL DYNAMICS/CONVAIR, SAN DIEGO, CALIF.
GDC-DG-G-	GENERAL DYNAMICS/CONVAIR, SAN DIEGO, CALIF.
GDC-ERR-AN-	GENERAL DYNAMICS/CONVAIR, SAN DIEGO, CALIF.
GDC-ERR-PO-(NO.)-	GENERAL DYNAMICS/POMONA, CALIF.
GDC-ERR-SD-	GENERAL DYNAMICS/CONVAIR, SAN DIEGO, CALIF.
GDC-HST-TR-(NO.)-	GENERAL DYNAMICS/CONVAIR, SAN DIEGO, CALIF.
GDC-OR-P-	GENERAL DYNAMICS/CONVAIR, SAN DIEGO, CALIF.
GDCR-	GENERAL DYNAMICS/POMONA, CALIF.
GD/C-SLV-(NO.)-(NO.)-	GENERAL DYNAMICS/CONVAIR, SAN DIEGO, CALIF.
GDC-ZA-(NUMBER)-	GENERAL DYNAMICS/CONVAIR, SAN DIEGO, CALIF.
GDC-ZC-(NUMBER)-	GENERAL DYNAMICS/CONVAIR, SAN DIEGO, CALIF.
GDC-ZJ-(NO.)-(NO.)-TN	GENERAL DYNAMICS/CONVAIR, SAN DIEGO, CALIF.
GDC-ZM-(NO.)-(NO.)TN	GENERAL DYNAMICS/CONVAIR, SAN DIEGO, CALIF.
GDC-ZP-(NO.)-(NO.)-	GENERAL DYNAMICS/CONVAIR, SAN DIEGO, CALIF.
GDC-ZY-(NUMBER)-	GENERAL DYNAMICS/CONVAIR, SAN DIEGO, CALIF.
GDC-ZZC-(YEAR)-	GENERAL DYNAMICS/CONVAIR, SAN DIEGO, CALIF.
GDF-	ARNOLD ENGINEERING DEVELOPMENT CENTER. GAS DYNAMICS FACILITY, ARNOLD AIR FORCE STATION, TENN.
GD/FW-	GENERAL DYNAMICS/FORT WORTH, TEX.
GD/FW FZK-	GENERAL DYNAMICS/FORT WORTH, TEX.
GD/FW FZM-	GENERAL DYNAMICS/FORT WORTH, TEX.
GD/FW MR-N-	GENERAL DYNAMICS/FORT WORTH, TEX.
GD FZK-	GENERAL DYNAMICS/FORT WORTH, TEX.
GD/LCD-	GENERAL DYNAMICS CORP. LIQUID CARBONIC DIV., CHICAGO
GD MR-N-	GENERAL DYNAMICS/FORT WORTH, TEX.
GD/O AOK-(YEAR)-	GENERAL DYNAMICS/CONVAIR, SAN DIEGO, CALIF.
GD/P-	GENERAL DYNAMICS/POMONA, CALIF.
GD/PAPER-	GENERAL DYNAMICS/CONVAIR, SAN DIEGO, CALIF.
GD/QPR-	GENERAL DYNAMICS/CONVAIR, SAN DIEGO, CALIF.
GD-R-	GENERAL DYNAMICS/ELECTRONICS, SAN DIEGO, CALIF.
GDRL-TR-(YEAR)-	GENERAL MOTORS CORP. DEFENSE RESEARCH LABS., SANTA BARBARA, CALIF.
GD/TDNO-	GENERAL DYNAMICS CORP. ELECTRIC BOAT DIV., GROTON, CONN
GDV-	FOCKE-WULF FLUGZEUGBAU G.M.B.H.
GD ZPH-	GENERAL DYNAMICS/CONVAIR, SAN DIEGO, CALIF.
GD ZR-AP-	GENERAL DYNAMICS/CONVAIR, SAN DIEGO, CALIF.
GE(YEAR)(LETTERS)-	GENERAL ELECTRIC CO. (ALL LOCATIONS)
GE-	GENERAL ELECTRIC CO. (ALL LOCATIONS)
GE(YEAR)ASD(NUMBER)-	GENERAL ELECTRIC CO. APOLLO SUPPORT DEPT., DAYTONA BEACH, FLA.
GE-(NUMBER)-(NUMBER)	GENERAL ELECTRIC CO. HEAVY MILITARY ELECTRONICS DEPT., SYRACUSE, N.Y.

G.E. (DATE)-CAR-
    GENERAL ELECTRIC CO. METALLURGICAL PRODUCTS DEPT.,
      DETROIT (USED 1953-1963)
G.E. (DATE)-MPD-
    GENERAL ELECTRIC CO. METALLURGICAL PRODUCTS DEPT.,
      DETROIT (USED 1964-)
GE-(NUMBER-NUMBER)
    GENERAL ELECTRIC CO. MISSILE AND SPACE VEHICLE DEPT.,
      PHILADELPHIA
GE(YEAR)SD(NUMBER)
    GENERAL ELECTRIC CO. MISSILE AND SPACE VEHICLE DEPT.,
      PHILADELPHIA
GE-(YEAR)-RL-
    GENERAL ELECTRIC CO. RESEARCH LAB., SCHENECTADY, N.Y.
GE-(NUMBERS)SAR-(DATE)
    GENERAL ELECTRIC CO. TECHNICAL MILITARY PLANNING
      OPERATION, SANTA BARBARA, CALIF.
GE (NUMBER/CONTRACT NO.)
    GENERAL ELECTRIC CO. TUBE DEPT., OWENSBORO, KY.
GE-
    GREY (JERRY), LIVINGSTON, N.J.
GE-
    HANFORD ENGINEER WORKS. GE NUCLEONICS PROJECT,
      RICHLAND, WASH.
GE-
    KINETICS CORP., CINCINNATI
GE-
    RESEARCH AND DEVELOPMENT BOARD. COMMITTEE ON GEO-
      GRAPHICAL EXPLORATION, WASHINGTON, D.C.
GE-
    ROYAL CANADIAN AIR FORCE. CENTRAL EXPERIMENTAL AND
      PROVING ESTAB., ROCKCLIFFE, ONT.
GE/(LTRS.)/(YEAR)-
    AIR FORCE INST. OF TECH., WRIGHT-PATTERSON AFB, OHIO.
      SCHOOL OF ENGINEERING (GRADUATE ELECTRONICS)
GE-(INITIALS)-
    GENERAL ELECTRIC CO., SCHENECTADY, N.Y.
GE-(INITIALS)-
    KNOLLS ATOMIC POWER LAB. GE NUCLEONICS PROJECT,
      SCHENECTADY, N.Y.
GEA-
    GENERAL ELECTRIC CO., SCHENECTADY, N.Y.
GE-AEP-(YR)-(LETTERS)-
    KNOLLS ATOMIC POWER LAB. GE NUCLEONICS PROJECT,
      SCHENECTADY, N.Y.
GE-AFM-
    GENERAL ELECTRIC CO. MISSILE AND ORDNANCE SYSTEMS
      DEPT., PHILA. (AERODYNAMICS FUNDAMENTALS MEMO.)
GE-ANP-
    GENERAL ELECTRIC CO. AIRCRAFT NUCLEAR PROPULSION
      PROJECT, CINCINNATI (REF. 3)
GE-ANP-APEX-
    GENERAL ELECTRIC CO. AIRCRAFT NUCLEAR PROPULSION
      PROJECT, CINCINNATI (ENG. PROGRESS RPT.) (REF. 3)
GE-ANPD-DC-
    GENERAL ELECTRIC CO., N.Y.C.
GE-ANSO-(NUMBER)-
    GENERAL ELECTRIC CO. MISSILE AND SPACE DIV., PHILA.
GEAO-
    GENERAL ELECTRIC CO. ADVANCE PRODUCT OPERATION, SAN
      JOSE, CALIF.
GEAP-
    GENERAL ELECTRIC CO. (ALL LOCATIONS)(AEC FUNDED WORK)
GE-APEX-
    GENERAL ELECTRIC CO. AIRCRAFT NUCLEAR PROPULSION
      DEPT., CINCINNATI (REF. 3)
GEATC-
    GENERAL ATRONICS CORP., CONSHOHOCKEN, PENNA.
GE-ATLAS-
    GENERAL ELECTRIC CO. MISSILE AND SPACE VEHICLE DEPT.,
      PHILADELPHIA
GE-C-(NUMBER)-C-(NUMBER)
    GENERAL ELECTRIC CO. FLIGHT PROPULSION LAB. DEPT.,
      EVENDALE, OHIO
GEC-
    GENERAL ELECTRIC CO., LTD. ATOMIC ENERGY DIV., ERITH,
      KENT, ENGLAND
GECO-
    GENERAL ELECTRIC CO. ADVANCE PRODUCT OPERATION, SAN
      JOSE, CALIF. (CUSTOMER REPORT)
GECR-
    GENERAL ELECTRIC CO. NUCLEAR ENERGY DIV.,SAN JOSE,CAL
      (CUSTOMER REPORT)
GED-
    GENERAL ELECTRIC CO. AIRCRAFT NUCLEAR PROPULSION
      DEPT., CINCINNATI (REF. 3)
GE-D-(YEAR)-
    GENERAL ELECTRIC CO. MISSILE AND SPACE DIV., PHILA.
GE-DASIAC-SB-
    GENERAL ELECTRIC CO. DASA INFORMATION AND ANALYSIS
      CENTER, SANTA BARBARA, CALIF. (SPECIAL BIBS.)
GE-DASIAC-SR-
    GENERAL ELECTRIC CO. DASA INFORMATION AND ANALYSIS
      CENTER, SANTA BARBARA, CALIF.
GE-DC-
    GENERAL ELECTRIC CO. AIRCRAFT NUCLEAR PROPULSION
      DEPT., CINCINNATI (REF. 3)
GE-DDC-
    GENERAL ELECTRIC CO. DASA DATA CENTER, SANTA BARBARA,
      CALIF.
GE/DDC-SPEC BIB-
    GENERAL ELECTRIC CO. DASA DATA CENTER, SANTA BARBARA,
      CALIF.
GE/DDC-SPEC RPT-
    GENERAL ELECTRIC CO. DASA DATA CENTER, SANTA BARBARA,
      CALIF.
GE-DIN-(NO.)-(NO.)QPR
    GENERAL ELECTRIC CO. MISSILE AND SPACE VEHICLE DEPT.,
      PHILADELPHIA

GED-O-
    OFFICE OF THE DIRECTOR OF DEFENSE (RES. + ENG.)
      ADVISORY GROUP ON ELECTRON DEVICES, WASH., D.C.
GEDR-
    GENERAL ELECTRIC CO. AIRCRAFT NUCLEAR PROPULSION
      PROJECT, CINCINNATI (REF. 3)
GE-E-
    MASSACHUSETTS INST. OF TECH., CAMBRIDGE
GE/EE/(YEAR)S-
    AIR FORCE INST. OF TECH., WRIGHT-PATTERSON AFB, OHIO
GEEIA-EMC-(YEAR)-
    GROUND ELECTRONICS ENGINEERING INSTALLATION AGENCY,
      GRIFFISS AFB, N.Y.
GEEIA-TR-(YEAR)-
    GROUND ELECTRONICS ENGINEERING INSTALLATION AGENCY,
      GRIFFISS AFB, N.Y.
GE-FWL-
    GENERAL ELECTRIC CO. FORT WAYNE LAB., IND.
GE-GEL-
    GENERAL ELECTRIC CO. GEN. ENG. LAB., SCHENECTADY,N.Y.
GE-GET-
    GENERAL ELECTRIC CO. AERONAUTIC AND ORDNANCE SYSTEMS
      DIV., SCHENECTADY, N.Y.
GE-GS-
    GENERAL ELECTRIC CO. FLIGHT PROPULSION LAB. DEPT.,
      CINCINNATI
GEH-
    HANFORD ENGINEER WORKS, RICHLAND, WASH. (ASSIGNED TO
      INCOMING REPORTS DURING THE PERIOD WHEN HANFORD WAS
      OPERATED BY THE GENERAL ELECTRIC CO.)
GEH-
    HANFORD WORKS, RICHLAND, WASH. (ASSIGNED TO INCOMING
      REPORTS)
GEI-
    GENERAL ELECTRIC CO., SCHENECTADY, N.Y. (INSTRUCTION)
GEIC-
    GENERAL ELECTRIC CO. NUCLEAR ENERGY DIV.,SAN JOSE,CAL
      (IRRADIATION PROCESSING, CUSTOMER REPORT)
GEIR-
    GENERAL ELECTRIC CO. NUCLEAR ENERGY DIV.,SAN JOSE,CAL
      (IRRADIATION PROCESSING, GOVT. CONTRACT WORK)
GE-IY-
    GENERAL ELECTRIC CO., SCHENECTADY, N.Y.
GEJ-
    GENERAL ELECTRIC CO., SCHENECTADY, N.Y.
GE-KAK-
    GENERAL ELECTRIC CO., SCHENECTADY, N.Y.
GE-KHK-
    KNOLLS ATOMIC POWER LAB. GE NUCLEONICS PROJECT,
      SCHENECTADY, N.Y.
GEL-
    GENERAL ELECTRIC CO. GENERAL ENGINEERING AND CON-
      SULTING LAB., SCHENECTADY, N.Y.
GEL-
    GENERAL ELECTRIC CO. GEN. ENG. LAB., SCHENECTADY,N.Y.
GEL-
    GENERAL ELECTRIC CO. RESEARCH LAB., SCHENECTADY, N.Y.
GELAC-ER-
    LOCKHEED-GEORGIA CO., MARIETTA (ENGINEERING REPORTS)
GE-LM-
    GENERAL ELECTRIC CO. LIGHT MILITARY ELECTRONICS
      DEPT., SCHENECTADY, N.Y.
GE-LMED-
    GENERAL ELECTRIC CO. LIGHT MILITARY ELECTRONICS
      DEPT., ITHACA, N.Y.
GE-LMED-
    GENERAL ELECTRIC CO. LIGHT MILITARY ELECTRONICS
      EQUIPMENT DEPT., SCHENECTADY, N.Y.
GEL-PR-
    GENERAL ELECTRIC CO. GEN. ENG. LAB., SCHENECTADY,N.Y.
GEL RL-
    GENERAL ELECTRIC CO., LTD. RESEARCH LABS., WEMBLEY,
      MIDDX, ENGLAND
GE-LT-
    GENERAL ELECTRIC CO., SCHENECTADY, N.Y.
GE-LT-
    KNOLLS ATOMIC POWER LAB. GE NUCLEONICS PROJECT,
      SCHENECTADY, N.Y.
GE-M-
    GENERAL ELECTRIC CO. MALTA TEST STA.,BALLSTON SPA,NY
GEM-
    NATIONAL LEAD CO. OF OHIO, CINCINNATI
GE-MEMO-EN-(YEAR)-
    GENERAL ELECTRIC CO. RESEARCH LAB., SCHENECTADY, N.Y.
GEMP-
    GENERAL ELECTRIC CO. (ALL LOCATIONS)
GE-MPL-
    GENERAL ELECTRIC CO., SUNNYVALE, CALIF.
GEMS-
    GENERAL ELECTRIC CO. MISSILE AND SPACE DIV., PHILA.
GE-MSD-
    GENERAL ELECTRIC CO. MISSILE AND SPACE VEHICLE DEPT.,
      PHILADELPHIA
GEMSIP-
    AEROJET-GENERAL CORP. LIQUID ROCKET OPERATIONS, SAC-
      RAMENTO, CALIF. (GEMINI PROGRAM)
GEMS-NS-
    GENERAL ELECTRIC CO., PHILADELPHIA
GEN-
    ALLIED CHEMICAL CORP. GENERAL CHEMICAL DIV., N.Y.C.
GE-N-
    GENERAL ELECTRIC CO. MISSILE AND SPACE DIV., PHILA.
GEN-
    GENERAL TIRE AND RUBBER CO. OF CALIF., VAN NUYS
GENAPPSL-
    GENERAL APPLIED SCIENCE LABS., INC., WESTBURY, N.Y.
GENAPPSL-TR-
    GENERAL APPLIED SCIENCE LABS., INC., WESTBURY, N.Y.

GEN/DNT-	CONFERENCE ON THE DISCONTINUANCE OF NUCLEAR WEAPON TESTS
GEN/DNT/(LETTERS)	CONFERENCE ON THE DISCONTINUANCE OF NUCLEAR WEAPON TESTS
GENERAL-	BUREAU OF RECLAMATION (INTERIOR). OFFICE OF CHIEF ENGINEER, DENVER
GENIE-P-	CALIFORNIA. UNIV., BERKELEY. ELECTRONICS RES. LAB.
GENIS-	GENISTRON INC. APPLIED RES. DIV., COLLEGE PARK, MD.
GENISTRON-	GENISTRON INC. APPLIED RES. DIV., COLLEGE PARK, MD.
GEN/MISC-	CONFERENCE ON THE DISCONTINUANCE OF NUCLEAR WEAPON TESTS
GE-NMP-	GENERAL ELECTRIC CO. FLIGHT PROPULSION LAB. DEPT., CINCINNATI
GEN/PUB/	INTERNATIONAL ATOMIC ENERGY AGENCY, VIENNA
GE-NUC-LIB-	GENERAL ELECTRIC CO., SCHENECTADY, N.Y.
GE/OD-R-	GENERAL ELECTRIC CO., PITTSFIELD, MASS.
GEOLOGICAL ENGINEERING-	PRINCETON UNIV.,N.J. DEPT. OF CIVIL + GEOLOGICAL ENG.
GEOMET-(NUMBER)-	GEOMET, INC., ROCKVILLE, MD.
GEONUCLEAR-	GEONUCLEAR CORP., LAS VEGAS, NEV.
GEOPHYBDBW-FM-	LUFTWAFFENAMT, PORZ, GERMANY (GEOPHYSIKALISCHER BERATUNGSDIENST DER BUNDESWEHR)
GEOPHYSBDBW-FM/	LUFTWAFFENAMT, PORZ, GERMANY (GEOPHYSIKALISCHER BERATUNGSDIENST DER BUNDESWEHR)
GEOPHYSICS RES. PAPERS-	AIR FORCE CAMBRIDGE RESEARCH CENTER. GEOPHYSICS RESEARCH DIRECTORATE, MASS.
GEO-SUR-TL-	GEOLOGICAL SURVEY, DENVER
GEOTECH-	GEOTECHNICAL CORP., GARLAND, TEX.
GEOTECH-TR-	GEOTECHNICAL CORP., GARLAND, TEX.
GEOTHERMAL ENERGY-	GEOLOGICAL SURVEY, DENVER
GEOU-	GEORGIA. UNIV., ATHENS
GEP-	OFFICE OF THE ASSISTANT SECRETARY OF DEFENSE (RES. + ENG.),WASH.,D.C. ADVISORY GP. ON ELECTRONIC PARTS
GEPM-	GENERAL ELECTRIC CO. RESEARCH LAB., SCHENECTADY, N.Y.
GE-P.O.(NUMBER)-	GENERAL ELECTRIC CO. NUCLEAR MATERIALS AND PROPULSION OPERATION, CINCINNATI
GEPP-	GENERAL ELECTRIC CO. NEUTRON DEVICES DEPT., ST. PETERSBURG, FLA.
GEPP-	GENERAL ELECTRIC CO. PINELLAS PENINSULA PLANT, ST. PETERSBURG, FLA.
GE/PTD(NUMBER)/CON. NO.)	GENERAL ELECTRIC CO. POWER TUBE DEPT., SCHENECTADY,NY
GE-QPR-	GENERAL ELECTRIC CO. SPEC. DEFENSE PROJS.DEPT.,PHILA.
GE-QR-(NUMBER),AEC-(NO.)	GENERAL ELECTRIC CO. MISSILE AND SPACE VEHICLE DEPT., PHILADELPHIA
GE-R(YEAR)(LETTERS)-	GENERAL ELECTRIC CO. (ALL LOCATIONS) (REF. 46)
GER(YEAR)(LTRS.)(NOS.)	GENERAL ELECTRIC CO. (ALL LOCATIONS)
GER-	GENERAL ELECTRIC CO., SCHENECTADY, N.Y.
GE R(YEAR)ELC(NO.-NO.)	GENERAL ELECTRIC CO. ADVANCED ELECTRONICS CENTER, ITHACA, N.Y.
GE-R(YR.)A-	GENERAL ELECTRIC CO. AERONAUTIC AND ORDNANCE SYSTEMS DIV., SCHENECTADY, N.Y. (PROJECT HERMES)
GE-R(YEAR)FPD(NO.)	GENERAL ELECTRIC CO.FLIGHT PROPULSION DIV.,CINCINNATI
GE-R(YR.)A-	GENERAL ELECTRIC CO. GUIDED MISSILES DEPT., SCHENECTADY, N.Y. (PROJECT HERMES)
GE R(YEAR)APS(NO.)(FR)	GENERAL ELECTRIC CO. LIGHT MILITARY ELECTRONICS EQUIPMENT DEPT., SCHENECTADY, N.Y.
GE R(YEAR)SD(NUMBER)	GENERAL ELECTRIC CO. MISSILE AND SPACE VEHICLE DEPT., PHILADELPHIA
GE R(YEAR)TMP-	GENERAL ELECTRIC CO. TECHNICAL MILITARY PLANNING OPERATION, SANTA BARBARA, CALIF.
GER-	GOODRICH (B.F.) CO., AKRON, OHIO
GER-	GOODYEAR AEROSPACE CORP., AKRON, OHIO
GER-(YEAR)-	GOODYEAR AEROSPACE CORP., AKRON, OHIO
GERA-	GOODYEAR AEROSPACE CORP., LITCHFIELD PARK, ARIZ.
GERD-	GENERAL ELECTRIC CO. AIRCRAFT NUCLEAR PROPULSION DEPT., CINCINNATI (REF. 3)
GE-RE-	GENERAL ELECTRIC CO. FLIGHT PROPULSION LAB. DEPT., CINCINNATI
GERL-(YEAR)-GC-	GENERAL ELECTRIC CO. RESEARCH LAB., SCHENECTADY, N.Y.
GERL-(LTRS, NOS.)	GENERAL ELECTRIC CO. RESEARCH LAB., SCHENECTADY, N.Y.
GERL-COMM.-	GENERAL ELECTRIC CO., LTD. RESEARCH LABS., WEMBLEY, MIDDX, ENGLAND
GE-RM(YEAR)(LETTERS)-	GENERAL ELECTRIC CO. (ALL LOCATIONS) (REF. 46)
GE-RS-(YEAR)	GENERAL ELECTRIC CO., PHILADELPHIA
GE-RS-TM-	GENERAL ELECTRIC CO. SPEC. DEFENSE PROJS.DEPT.,PHILA.
GES-	GENERAL ELECTRIC CO., SCHENECTADY, N.Y.
G-ES-	GENERAL ELECTRIC CO. POWER TUBE DEPT., SCHENECTADY,NY
G-E-SD-	GENERAL ELECTRIC CO. TUBE DEPT., SCHENECTADY, N.Y.
GE-SN-	GENERAL ELECTRIC CO., SCHENECTADY, N.Y.
GESP-	GENERAL ELECTRIC CO. NUCLEAR SYSTEMS PROGRAMS, CINCINNATI
GESR-	GENERAL ELECTRIC CO. NUCLEAR ENERGY DIV.,SAN JOSE,CAL
GEST-	GENERAL ELECTRIC CO. NUCLEAR ENERGY DIV.,SAN JOSE,CAL
GET-	GENERAL ELECTRIC CO., SCHENECTADY, N.Y.
GET-	OFFICE OF THE ASSISTANT SECRETARY OF DEFENSE (RES. AND DEV.) ADVISORY GROUP ON ELECTRON TUBES
GE/TD QR-	GENERAL ELECTRIC CO. TUBE DEPT., PALO ALTO, CALIF.
GE/TEMP-	GENERAL ELECTRIC CO. TECHNICAL MILITARY PLANNING OPERATION, SANTA BARBARA, CALIF.
GE-TEMP-(YEAR)TMP-	GENERAL ELECTRIC CO. TECHNICAL MILITARY PLANNING OPERATION, SANTA BARBARA, CALIF.
GE/TEMPO-	GENERAL ELECTRIC CO. TECHNICAL MILITARY PLANNING OPERATION, SANTA BARBARA, CALIF.
GE/TEMPO-DDC-	GENERAL ELECTRIC CO. DASA DATA CENTER, SANTA BARBARA, CALIF.
GE/TEMPO-RM-(YEAR)TMP-	GENERAL ELECTRIC CO. TECHNICAL MILITARY PLANNING OPERATION, SANTA BARBARA, CALIF.
GE-THOR-	GENERAL ELECTRIC CO. MISSILE AND SPACE VEHICLE DEPT., PHILADELPHIA
GE-TM(YEAR)SE-	GENERAL ELECTRIC CO., LYNN, MASS.
GE-TM-	GENERAL ELECTRIC CO. NUCLEAR MATERIALS AND PROPULSION OPERATION, CINCINNATI
GE-TMS-	GENERAL ELECTRIC CO. FLIGHT PROPULSION LAB. DEPT., CINCINNATI
GET/P-	OFFICE OF THE ASSISTANT SECRETARY OF DEFENSE (RES. AND DEV.) ADVISORY GROUP ON ELECTRON TUBES
GE-TR-	GENERAL ELECTRIC CO. GENERAL ENGINEERING AND CONSULTING LAB., SCHENECTADY, N.Y.
GET-REL-	OFFICE OF THE ASSISTANT SECRETARY OF DEFENSE (RESEARCH AND DEVELOPMENT) ADVISORY GROUP ON ELECTRON TUBES. (MILITARY RELIABLE TUBE PROGRAM)
GE-X-	GENERAL ELECTRIC CO. X-RAY DEPT., MILWAUKEE
GEX-	GENERAL ELECTRIC CO. X-RAY DEPT., MILWAUKEE
GE-XDC-	GENERAL ELECTRIC CO. AIRCRAFT NUCLEAR PROPULSION DEPT., CINCINNATI (REF. 3)
GEXDC (YEAR)	GENERAL ELECTRIC CO. AIRCRAFT NUCLEAR PROPULSION PROJECT, CINCINNATI (REF. 3)
GEXDCL (YEAR)-	GENERAL ELECTRIC CO. AIRCRAFT NUCLEAR PROPULSION PROJECT, CINCINNATI (REF. 3)
GEX-R-	FRANCE. COMMISSARIAT A L'ENERGIE ATOMIQUE. CENTRE D'ETUDES NUCLEAIRES, SACLAY
GEZR-	GENERAL ELECTRIC CO. RESEARCH LAB., SCHENECTADY, N.Y.
GF-(NUMBER)Q-(YEAR)	DOW CHEMICAL CO., MIDLAND, MICH. (GYRO FLOTATION FLUIDS)
GF-	ETHYL CORP., FERNDALE, MICH.
GF(NUMBER)-(NUMBER)-	INTERNATIONAL BUSINESS MACHINES CORP., ALBUQUERQUE,NM
GFI-	GLASS FIBERS, INC., TOLEDO
G FILE (ALSOS)-	EUROPEAN THEATER OF OPERATIONS (ARMY). ALSOS MISSION (GERLACH FILE)
GFSC-	GODDARD SPACE FLIGHT CENTER, GREENBELT, MD.
GFW-AB-(YEAR)/	NATIONAL AERONAUTICS AND SPACE ADM., WASHINGTON, D.C. (TRANSLATION)

GFW-TN-(YEAR/NUMBER)	GESELLSCHAFT FUER WELTRAUMFORSCHUNG M.B.H., BAD GODESBERG
GG-	RES.+ DEV.BD. COMM.ON GEOPHYSICS + GEOGRAPHY, WASH,DC
GGA-	GULF GENERAL ATOMIC, INC., SAN DIEGO, CALIF.
GGC/(LTRS.)/(YEAR)-	AIR FORCE INST. OF TECH., WRIGHT-PATTERSON AFB, OHIO. SCHOOL OF ENGINEERING (GRADUATE GUIDANCE + CONTROL)
GHY-	RES.+ DEV.BD. COMM.ON GEOPHYSICS + GEOGRAPHY, WASH,DC
GI-(NO.)-(LETTER)	CORNELL AERONAUTICAL LAB., INC., BUFFALO
GI-	GENERAL ELECTRIC CO. GENERAL ENGINEERING AND CONSULTING LAB., SCHENECTADY, N.Y.
GI-	PENNSYLVANIA STATE UNIV.,UNIVERSITY PARK. GROTH INST.
GI-(NUMBER)-(NUMBER)-	PHILCO CORP. GOVERNMENT AND INDUSTRIAL DIV., PHILA.
GIA-	GIANNINI (G.M.) AND CO., PASADENA, CALIF.
GIC-	GENERAL INSTRUMENT CORP., HICKSVILLE, N.Y.
GIC-	GENERAL INSTRUMENT CORP., NEWARK, N.J. (FINAL REPT.)
GIER-	ILC INDUSTRIES, INC., DOVER, DEL.
GIL-	GILFILLAN BROS., INC., LOS ANGELES
GIMRADA-(NUMBER)-TR	ARMY ENGINEER GEODESY, INTELLIGENCE AND MAPPING RESEARCH AND DEVELOPMENT AGENCY, FORT BELVOIR, VA.
GIMRADA-RN-	ARMY ENGINEER GEODESY, INTELLIGENCE AND MAPPING RESEARCH AND DEVELOPMENT AGENCY, FORT BELVOIR, VA.
GIMRADA-TN-(YEAR)-	ARMY ENGINEER GEODESY, INTELLIGENCE AND MAPPING RESEARCH AND DEVELOPMENT AGENCY, FORT BELVOIR, VA.
GIMRADA-TR-	ARMY ENGINEER GEODESY, INTELLIGENCE AND MAPPING RESEARCH AND DEVELOPMENT AGENCY, FORT BELVOIR, VA.
GIO-(YEAR)/	ITT GILFILLAN, INC., LOS ANGELES
GIPR-	CHALMERS TEKNISKA HOEGSKOLA, GOETEBORG
GI-QRM-	GENERAL INSTRUMENT CORP. CAPACITOR DIV.,DARLINGTON,SC
GIR-	(OPERATION GREENHOUSE) (REF. 7)
GIT-	GEORGIA INST. OF TECH., ATLANTA
GIT-(NUMBER)-PU	PURDUE UNIV., LAFAYETTE, IND.
GIT-(LETTERS)-	GEORGIA INST. OF TECH., ATLANTA. ENG. EXPERIMENT STA.
GIT-A-(NUMBER)-	GEORGIA INST. OF TECH., ATLANTA
GIT-A-(NUMBER)-SR	GEORGIA INST. OF TECH., ATLANTA. ELECTRONICS DIV.
GIT-AER-(YEAR)-	GEORGIA INST. OF TECH., ATLANTA. SCHOOL OF AEROSPACE ENGINEERING
GIT-B-(NUMBER)-	GEORGIA INST. OF TECH., ATLANTA. ENG. EXPERIMENT STA.
GIT-CE-PW-	GEORGIA INST. OF TECH., ATLANTA. SCHOOL OF CIVIL ENG.
GIT/EES-	GEORGIA INST. OF TECH., ATLANTA. ENG. EXPERIMENT STA.
GIT/EES-A(NO.)/T(NO.)	GEORGIA INST. OF TECH., ATLANTA. SCHOOL OF ELECTRICAL ENGINEERING
GIT-GTF-	GEORGIA INST. OF TECH., ATLANTA. ELECTRONICS DIV.
GITIS-(YEAR)-	GEORGIA INST. OF TECH., ATLANTA. SCHOOL OF INFORMATION AND COMPUTER SCIENCE
GIT-QR-	GEORGIA INST. OF TECH., ATLANTA. ENG. EXPERIMENT STA.
GIT-SR-	GEORGIA INST. OF TECH., ATLANTA. ELECTRONICS DIV.
GIT-STR-A-	GEORGIA INST. OF TECH., ATLANTA. ELECTRONICS DIV.
GIT-TR-	GEORGIA INST. OF TECH., ATLANTA. ENG. EXPERIMENT STA.
GIT-TSR-	GEORGIA INST. OF TECH., ATLANTA. ENG. EXPERIMENT STA.
G IVB-	NAVY (WORK IMPROVEMENT PROGRAM, GROUP IV-B. SUPERVISORY CONFERENCE REPORTS)
GJ-	GRAND JUNCTION OPERATIONS OFFICE, AEC, COLO. (REF. 4)
GJ-	SPERRY GYROSCOPE CO., GREAT NECK, N.Y.
GJ-(NUMBER)-	SPERRY GYROSCOPE CO., SYOSSET, N.Y.
GJ-(NUMBER)-	SPERRY RAND CORP., GREAT NECK, N.Y.
GJB/RJL-	ARGONNE NATIONAL LAB., ILL.(CODE IS AUTHOR'S INITLS.)
GJE-NB-	ARMED FORCES SPECIAL WEAPONS PROJECT. FIELD COMMAND, ALBUQUERQUE, N. MEX. (NOTEBOOK)
GJE-VG-	ARMED FORCES SPECIAL WEAPONS PROJECT. FIELD COMMAND, ALBUQUERQUE, N. MEX. (VU-GRAPHS)
GJO-(NUMBER)-	GRAND JUNCTION OFFICE, AEC, COLO. (REF. 4)
GJPC-	CALIFORNIA INST. OF TECH., PASADENA. GUGGENHEIM JET PROPULSION CENTER
G/J-TR-(NUMBER-NUMBER)	GAUTNEY + JONES COMMUNICATIONS, INC., FALLS CHURCH,VA
GK-(NO.)-(NO.)-(NO.)-BK-	SPERRY GYROSCOPE CO., GREAT NECK, N.Y.
GK-	SPERRY RAND CORP. SPERRY SYSTEMS MANAGEMENT DIV., GREAT NECK, N.Y.
GKSS-(YEAR)/E/	GESELLSCHAFT FUER KERNENERGIEVERWERTUNG IN SCHIFFBAU UND SCHIFFAHRT M.B.H., HAMBURG
GL-	GENERAL ELECTRIC CO. GENERAL ENGINEERING AND CONSULTING LAB., SCHENECTADY, N.Y.
(YEAR)-GL-	GENERAL ELECTRIC CO. GEN. ENG. LAB., SCHENECTADY,N.Y.
GL-(YEAR)-	GERMANTOWN LABS., INC., PHILADELPHIA
GLASGOW REPORT-	GT. BRIT. MINISTRY OF SUPPLY
GL L-	GERMANY. LUFTWAFFE (LEHRBILDREIHE)
GLM-	MARTIN CO., BALTIMORE
GLM-(LETTERS)-	MARTIN CO., BALTIMORE
GLM-B-	MARTIN CO., BALTIMORE
GLM-ER-	MARTIN CO., BALTIMORE
GLR-	GEOSCIENCE LTD., LA JOLLA, CALIF.
GLR-	GEOSCIENCE LTD., SAN DIEGO, CALIF.
GLR-	GEOSCIENCE LTD., SOLANA BEACH, CALIF.
GLRC-	GREAT LAKES RESEARCH CORP., ELIZABETHTON, TENN.
GM-	(GUIDED MISSILES) (REF. 2)
GM-	ARMY AIR DEFENSE BOARD, FORT BLISS, TEX.
GM-(NUMBER, YEAR)	ARMY FIELD FORCES, FORT MONROE, VA. (GUIDED MISSILE REPORT)
GM-	ARMY FIELD FORCES. BOARD NO. 4, FORT BLISS, TEX. (GUIDED MISSILE SERVICE TEST SECTION REPORT)
GM-(NUMBER)-(NUMBER)-	BALLISTIC MISSILE DIV. (AIR FORCE), INGLEWOOD, CALIF.
GM-	BUREAU OF AERONAUTICS (NAVY). GUIDED MISSILE DIV.
GM-	CORNELL AERONAUTICAL LAB., INC., BUFFALO
(NUMBER)-GM-	GENERAL ELECTRIC CO. ELECTRONICS LAB.,SYRACUSE,N.Y.
GM-	GENERAL MILLS, INC., MINNEAPOLIS
GM-	NORTHROP AIRCRAFT, INC., HAWTHORNE, CALIF. (GUIDED MISSILE)
GM-	RAMO-WOOLDRIDGE CORP. GUIDED MISSILE RESEARCH DIV., LOS ANGELES
GM-(NUMBER.NUMBER)-	RAMO-WOOLDRIDGE CORP. GUIDED MISSILE RESEARCH DIV., LOS ANGELES
GM-	RES. + DEV. BD. COMM. ON GUIDED MISSILES, WASH., DC
GM-	SPACE TECHNOLOGY LABS., INC., LOS ANGELES
GM-(NUMBER)-(NUMBER)-	SPACE TECHNOLOGY LABS., INC., LOS ANGELES
GM-	WESTINGHOUSE ELECTRIC CORP., EAST PITTSBURGH, PENNA.
GMAC-	GENERAL MOTORS CORP. AC SPARK PLUG DIV., FLINT, MICH.
GMAD-	ALLISON DIV., GENERAL MOTORS CORP., INDIANAPOLIS
GMC-	GENERAL MOTORS CORP. (ALL LOCATIONS)
GMC-	GULTON MFG. CORP., METUCHEN, N.J.
GMC-	JOINT COMMITTEE ON NEW WEAPONS + EQUIPMENT, WASH., DC
GMC-	SANDIA CORP. SPECIAL WEAPONS DEVELOPMENT BOARD. GUIDED MISSILE COMMITTEE, ALBUQUERQUE, N. MEX.
GMC/AD-	ALLISON DIV., GENERAL MOTORS CORP., INDIANAPOLIS
GMCB-	GLADDING, MC BEAN AND CO., LOS ANGELES
GMC EC-	GENERAL MOTORS CORP. ELECTRO CHEMISTRY DEPT.
GMC MR-	GENERAL MOTORS CORP.(ALL LOCATIONS) (MEMORANDUM RPTS)
GMC-MSL-	GENERAL MOTORS CORP. MATERIALS AND STRUCTURES LAB., WARREN, MICH.
GMC PI-	GENERAL MOTORS CORP. RESEARCH LABS. DIV., DETROIT (PHYSICS INSTRUMENTATION)
GMC-S(YEAR)	AC ELECTRONICS-DEFENSE RES. LABS., SANTA BARBARA,CAL.
GMC-TM-	GENERAL MOTORS CORP., SANTA BARBARA, CALIF.

GMC-TR-(YEAR)-
    GENERAL MOTORS CORP. DEFENSE RESEARCH LABS., SANTA
    BARBARA, CALIF.
GMD-
    REPUBLIC AVIATION CORP. GUIDED MISSILE DIV., HICKS-
    VILLE, N.Y.
GMD-(NUMBER)-
    REPUBLIC AVIATION CORP. GUIDED MISSILE DIV.,
    MINEOLA, N.Y.
GM-DRL-TR(YEAR)-
    GENERAL MOTORS CORP. DEFENSE RESEARCH LABS., SANTA
    BARBARA, CALIF.
GME/(LTRS.)/(YEAR)-
    AIR FORCE INST. OF TECH., WRIGHT-PATTERSON AFB, OHIO.
    SCHOOL OF ENGINEERING (GRADUATE MECHANICAL ENG.)
GM EDR-
    ALLISON DIV., GENERAL MOTORS CORP., INDIANAPOLIS
GME-IM-
    ARMED FORCES SPECIAL WEAPONS PROJECT. FIELD COMMAND,
    ALBUQUERQUE, N. MEX. (INSTRUCTORS MANUAL)
GME-VG-
    ARMED FORCES SPECIAL WEAPONS PROJECT. FIELD COMMAND,
    ALBUQUERQUE, N. MEX. (VU-GRAPHS)
GMFT-
    NORTHROP AIRCRAFT, INC., HAWTHORNE, CALIF. (GUIDED
    MISSILE FLIGHT TEST)
GMGC (LTRS.)-
    GIANNINI CONTROLS CORP. SYSTEMS DIV., SANTA ANA, CAL.
GMI-
    ATOMIC ENERGY OF CANADA LTD., CHALK RIVER, ONT.
GMI-
    GENERAL MILLS, INC., MINNEAPOLIS
GMI ERDD-
    GENERAL MILLS, INC. ENG. RES. + DEV.DEPT.,MINNEAPOLIS
GML-
    UNION OF SOUTH AFRICA. GOVERNMENT METALLURGICAL
    LAB., JOHANNESBURG
GMML-
    DEPARTMENT OF DEFENSE, WASHINGTON, D.C. (GUIDED
    MISSILE MAILING LIST)
GM-NED-
    GENERAL MILLS, INC. NUCLEAR EQUIP. DEPT., MINNEAPOLIS
GMOA-
    OPERATIONS ANALYSIS OFF. (AIR FORCE), WASHINGTON,D.C.
GM-P132-(YEAR-NO.-NO.)
    GEORGIA MOUNTAINS PLANNING + DEVELOPMENT COMMISSION,
    GAINESVILLE
GMPG-
    GENERAL MOTORS CORP. PROVING GROUND
GMR-
    GENERAL MOTORS CORP. RESEARCH LABS., WARREN, MICH.
GMRD-R/W-
    RAMO-WOOLDRIDGE CORP. GUIDED MISSILE RESEARCH DIV.,
    LOS ANGELES
GMRI-
    LOCKHEED-GEORGIA CO., MARIETTA
GMRL-
    GENERAL MOTORS CORP. RESEARCH LABS., WARREN, MICH.
GMRWG-
    INTER-BUREAU TECHNICAL COMMITTEE (NAVY). GUIDED
    MISSILE RELAY WORKING GROUP
GMS-
    MC DONNELL AIRCRAFT CORP., ST. LOUIS
GMSFC-(LETTERS)(NUMBERS)
    GEORGE C. MARSHALL SPACE FLIGHT CTR.,HUNTSVILLE, ALA.
GM-TM-
    SPACE TECHNOLOGY LABS., INC. ELECTRONIC LAB,LOS ANG.
GM-TM-
    THOMPSON RAMO WOOLDRIDGE INC., LOS ANGELES
GM-TR-
    BALLISTIC MISSILE DIV. (AIR FORCE), INGLEWOOD, CALIF.
GM TR (YEAR)-
    GENERAL MOTORS CORP. DEFENSE RESEARCH LABS., SANTA
    BARBARA, CALIF.
GM-TR-(NUMBER)-
    RAMO-WOOLDRIDGE CORP. ELECTRONIC RES. LAB., LOS ANG.
GM-TR-
    RAMO-WOOLDRIDGE CORP. GUIDED MISSILE RESEARCH DIV.,
    LOS ANGELES
GM-TR-
    SPACE TECHNOLOGY LABS., INC., EL SEGUNDO, CALIF.
GM-TR-
    SPACE TECHNOLOGY LABS., INC. PHYSICAL RESEARCH LAB.,
    LOS ANGELES
GMX-
    LOS ALAMOS SCIENTIFIC LAB., N. MEX.
GMX-1- THRU GMX-11-
    LOS ALAMOS SCIENTIFIC LAB., N. MEX. (INTERNAL
    CORRESPONDENCE SERIAL USED BY GROUPS IN GMX DIV.)
GN-
    ANSALDO S.P.A., GENOA
GN-
    ELECTRONICS RESEARCH, INC. ANTENNA DIV., EVANSVILLE,
    IND.
GN-
    EUROPEAN ATOMIC ENERGY COMMUNITY
GN-
    ITALY. COMITATO NAZIONALE PER L'ENERGIA NUCLEARE,
    ISPRA
GNB-
    GOULD-NATIONAL BATTERIES, INC., DEPEW, N.Y.
GNBI-
    GOULD-NATIONAL BATTERIES, INC. RES. DIV., MINNEAPOLIS
GNC-
    COAST AND GEODETIC SURVEY, WASHINGTON, D.C. (GLOBAL
    NAVIGATION CHART)
GNC-(NUMBER)(LETTER)-
    GENERAL NUCLEONICS CORP., CLAREMONT, CALIF.
GNE/(LTRS.)/(YEAR)-
    AIR FORCE INST. OF TECH., WRIGHT-PATTERSON AFB, OHIO.
    SCHOOL OF ENGINEERING (GRADUATE NUCLEAR ENG.)
GNEC-
    GENERAL NUCLEAR ENGINEERING CORP., DUNEDIN, FLA.
GNE-PH-
    AIR FORCE INST. OF TECH., WRIGHT-PATTERSON AFB, OHIO.
    SCHOOL OF ENGINEERING
GNE-PHYS/(YEAR)-
    AIR FORCE INST. OF TECH., WRIGHT-PATTERSON AFB, OHIO.
    SCHOOL OF ENGINEERING (GRADUATE NUCLEAR ENG.)
GO-(2 DIGITS)
    GENERAL OCEANOLOGY, INC., CAMBRIDGE, MASS
G.O.-
    ROCKETDYNE DIV., NORTH AMERICAN AVIATION, INC.,
    CANOGA PARK, CALIF.
GO-
    STANFORD RESEARCH INST., MENLO PARK, CALIF.
G.O.-
    WESTINGHOUSE ELECTRIC CORP., BALTIMORE
GOMAC-(YEAR)-
    CONF. ON GOVERNMENT MICROCIRCUIT APPLICATIONS, 1969
GOP-
    MAURER (J.A.), INC., LONG ISLAND, N.Y.
GOR-(LETTERS)-
    MARINE CORPS, WASHINGTON, D.C.
GOR-CC-
    MARINE CORPS, WASHINGTON, D.C.
GOR-CC-
    MARINE CORPS LANDING FORCE DEV. CENTER, QUANTICO, VA.
GO VB-
    GOTHAER WAGGONFABRIK (VERSUCHBERICHT)
GP-
    GENERAL ELECTRIC CO. RESEARCH LAB., SCHENECTADY, N.Y.
GP-
    OFFICE OF NAVAL RESEARCH. GROUP PSYCHOLOGY BRANCH
GPI-
    ATOMIC EN. OF CANADA LTD. CHALK RIVER NUC. LABS., ONT
GPI-
    GENERAL PRECISION, INC., LITTLE FALLS, N.J.
GPI-(NO.)-D-
    GENERAL PRECISION INC., TARRYTOWN, N. J.
GPI-(NO.)-J-
    GENERAL PRECISION INC., TARRYTOWN, N. J.
GPI-
    GENERAL PRECISION, INC., WAYNE, N.J.
GPI-(NO. LTR NO. LTR.)
    GENERAL PRECISION SYSTEMS INC., KEARFOTT SYSTEMS
    INC., WAYNE, N. J.
GPI-
    NATIONAL RESEARCH COUNCIL OF CANADA. ATOMIC ENERGY
    PROJECT, CHALK RIVER, ONT. (REF. 29)
GPL-
    GENERAL PRECISION, INC. GPL DIV., PLEASANTVILLE, N.Y.
GPL-A-
    GENERAL PRECISION, INC. AEROSPACE GP.,LITTLE FALLS,NJ
GPL-A-
    GENERAL PRECISION, INC. GPL DIV., PLEASANTVILLE, N.Y.
GPO-(2 DIGITS)
    GENERAL POST OFFICE, ENG. DEPT., LONDON
GPO-
    GOVERNMENT PRINTING OFFICE, WASHINGTON, D.C.
GPO-RR-
    GENERAL POST OFFICE. ENG. DEPT., LONDON
GPR-
    CANADA. DEPT. OF MINES AND TECH. SURVEYS. MINES BR.
GPSI-(NUMBER)-
    GENERAL PRECISION, INC., LITTLE FALLS, N.J.
GQR/(YR)-
    JOHNS HOPKINS UNIV., SILVER SPRING, MD. APPLIED
    PHYSICS LAB.
GR-
    (GROUP REPORT) (REF. 2)
GR-
    DANISH METEOROLOGICAL INST., CHARLOTTENLUND
GR-(NO.)-APD-(YR.)
    ESSO RESEARCH AND ENGINEERING CO. GOVERNMENT RESEARCH
    LAB., LINDEN, N.J.
GR-(NUMBER)-ASP-
    ESSO RESEARCH AND ENGINEERING CO. GOVERNMENT RESEARCH
    LAB., LINDEN, N.J. (ADVANCED SOLID PROPELLANTS)
GR-(NUMBER)-DCH-
    ESSO RESEARCH AND ENGINEERING CO. GOVERNMENT RESEARCH
    LAB., LINDEN, N.J. (DEV. CATALYST FOR HYDRAZINE)
GR-(NUMBER)-FBP-
    ESSO RESEARCH AND ENG. CO. GOVERNMENT RES. LAB.,
    LINDEN, N.J. (FUNCTIONALITY OF BINDER PREPOLYMERS)
GR-(NUMBER)-MGM-
    ESSO RESEARCH AND ENGINEERING CO. GOVERNMENT RESEARCH
    LAB., LINDEN, N.J. (MAGNETIC GRADIENT MATERIALS)
GR-(NUMBER)-SAM-
    ESSO RESEARCH AND ENGINEERING CO. GOVERNMENT RESEARCH
    LAB., LINDEN, N.J. (SYNTHESIS OF ANTIMALARIALS)
GR-(NUMBER)-SAR-
    ESSO RESEARCH AND ENGINEERING CO. GOVERNMENT RESEARCH
    LAB., LINDEN, N.J. (SYNTHESIS ANTIRADIATION AGENTS)
GR-(NUMBER)-SECA-
    ESSO RES. AND ENG. CO. GOVERNMENT RES. LAB.,
    LINDEN, N.J. (SYNTHESIS + EVAL. OF CURING AGENTS)
GR-(NUMBER)-TSF-
    ESSO RESEARCH AND ENGINEERING CO. GOVERNMENT RESEARCH
    LAB., LINDEN, N.J. (THERMALLY STABLE LIQUID FUELS)
GR-(NUMBER)-VPD-
    ESSO RESEARCH AND ENGINEERING CO. GOVERNMENT RESEARCH
    LAB., LINDEN, N.J. (VAPOR PHASE DEPOSITS)
GR-(NO.)-FRA-(YEAR)
    ESSO RESEARCH AND ENGINEERING CO. GOVERNMENT RESEARCH
    LAB., LINDEN, N.J.

GR-(NO.)-VDP-(YEAR)
    ESSO RESEARCH AND ENGINEERING CO. GOVERNMENT RESEARCH
        LAB., LINDEN, N.J. (VAPOR DEPOSITS)
GR-(NO.)-PAP-(YEAR)
    ESSO RESEARCH AND ENGINEERING CO. GOVERNMENT RESEARCH
        LAB., LINDEN, N.J.
GR-(NO.)-RIR-(YEAR)
    ESSO RESEARCH AND ENGINEERING CO. GOVERNMENT RESEARCH
        LAB., LINDEN, N.J.
GR-(NO.)-OPS-(YEAR)
    ESSO RESEARCH AND ENGINEERING CO. GOVERNMENT RESEARCH
        LAB., LINDEN, N.J. (ORGANOPHOSPHONATES)
GR-(NO.)-MPD-(YEAR)
    ESSO RESEARCH AND ENGINEERING CO. GOVERNMENT RESEARCH
        LAB., LINDEN, N.J.
GR-(YEAR)-
    ETHYL CORP., FERNDALE, MICH.
GR-
    GARRETT CORP. AIRESEARCH MFG. DIV., PHOENIX, ARIZ.
GR-
    GENERAL RADIO CO., CAMBRIDGE, MASS.
GR-
    HONEYWELL RESEARCH CENTER, HOPKINS, MINN.
GR-
    LEAR SIEGLER, INC., GRAND RAPIDS
GR-
    PENNSYLVANIA. UNIV., PHILA. INST. FOR COOP. RES.
GR-
    WESTINGHOUSE ELECTRIC CORP. RES. LABS., E. PITTSBURGH
GRAM-
    UNIVERSIDADE MACKENZIE, SAO PAULO, BRAZIL. GRUPO
        DE RADIO-ASTRONOMIA
GRAZU-
    GRAZ UNIV.
GRC-
    GATES RUBBER CO., DENVER
GRC-
    GENERAL RESEARCH CORP., SANTA BARBARA, CALIF.
GRC-
    GIANNINI RESEARCH LAB., SANTA ANA, CALIF.
GRC-
    GRISCOM-RUSSELL CO., MASSILLON, OHIO
GRC-
    PLASMADYNE CORP., SANTA ANA, CALIF.
GRC-CR-(NUMBER)-
    GENERAL RESEARCH CORP., ARLINGTON, VA.
GRC-CR-(NUMBER)-
    GENERAL RESEARCH CORP., SANTA BARBARA, CALIF.
GRC-CR-(NO.)-(NO.)-F-
    GENERAL RESEARCH CORP., SANTA BARBARA, CALIF.
GRC-IMR-
    GENERAL RESEARCH CORP., SANTA BARBARA, CALIF.
GRCO-
    GATES RUBBER CO., DENVER
GRC-TM-(NUMBER)/(NUMBER)
    GENERAL RESEARCH CORP., SANTA BARBARA, CALIF.
GRD-
    AIR FORCE CAMBRIDGE RESEARCH CENTER. GEOPHYSICS
        RESEARCH DIRECTORATE, MASS.
GR/D-(YEAR)-
    GARRETT RESEARCH + DEVELOPMENT CO., LA VERNE, CALIF.
GRD-
    GRUMMAN AIRCRAFT ENGINEERING CORP., BETHPAGE, N.Y.
GRDC-
    GULF RESEARCH AND DEVELOPMENT CO., PITTSBURGH
GRDC/P-
    UNITED KINGDOM ATOMIC ENERGY AUTHORITY. RESEARCH GP.
        ATOMIC ENERGY RES. ESTAB., HARWELL, BERKS, ENGLAND
GRD-TN-(YEAR)-
    AIR FORCE CAMBRIDGE RESEARCH CENTER. GEOPHYSICS
        RESEARCH DIRECTORATE, BEDFORD, MASS.
GRD-TR-
    AIR FORCE CAMBRIDGE RESEARCH CENTER. GEOPHYSICS
        RESEARCH DIRECTORATE, BEDFORD, MASS.
GRD-TR(YR)-(NO.)-
    AVCO MFG. CORP. RESEARCH AND ADVANCED DEVELOPMENT
        DIV., WILMINGTON, MASS.
GRE/(LTRS.)/(YEAR)-
    AIR FORCE INST. OF TECH., WRIGHT-PATTERSON AFB, OHIO.
        SCHOOL OF ENGINEERING (GRADUATE RELIABILITY ENG.)
GREEN-(PROJECT NUMBER)-
    (OPERATION GREENHOUSE) (REF. 7)
GREEN-
    (OPERATION GREENHOUSE) (REF. 7)
GREENHOUSE ANNEX-(NO.)-
    (OPERATION GREENHOUSE) (REF. 7)
GRE-NB-
    ARMED FORCES SPECIAL WEAPONS PROJECT. FIELD COMMAND,
        ALBUQUERQUE, N. MEX. (NOTEBOOK)
GRE-VG-
    ARMED FORCES SPECIAL WEAPONS PROJECT. FIELD COMMAND,
        ALBUQUERQUE, N. MEX. (VU-GRAPHS)
GRI-(4 DIGITS)
    GAERTNER RESEARCH, INC., STAMFORD, CONN.
GRI-NT-
    FRANCE. GROUPE DE RECHERCHES IONOSPHERIQUES, ISSY-
        LES-MOULINEAUX
GRI-NT-
    FRANCE. GROUPE DE RECHERCHES IONOSPHERIQUES, PARIS
GRI-NTP-
    FRANCE. GROUPE DE RECHERCHES IONOSPHERIQUES, ISSY-
        LES-MOULINEAUX
GRI-NTP-
    FRANCE. GROUPE DE RECHERCHES IONOSPHERIQUES, PARIS
GRI-TP-
    FRANCE. GROUPE DE RECHERCHES IONOSPHERIQUES, PARIS
GRL-TR-
    GIANNINI RESEARCH LAB., SANTA ANA, CALIF.
GRO/(NUMBER)/(NUMBER)/
    UNITED KINGDOM ATOMIC ENERGY AUTHORITY. WEAPONS GROUP
        ATOMIC WEAPONS RES. ESTAB., ALDERMASTON, BERKS, ENG.
GRP-
    AIR FORCE CAMBRIDGE RESEARCH CENTER. GEOPHYSICS
        RES. DIRECTORATE, MASS. (GEOPHYSICAL RES. PAPER)
GRP-(NUMBER)/(YEAR)
    CANADA. DEPT. OF MINES AND TECHNICAL SURVEYS
GRP-
    MICHIGAN. UNIV., ANN ARBOR
GRR-(NO.-LTR.NO.LTR.)
    LEAR SIEGLER, INC. (ALL DIVISIONS)
GRR-
    LEAR SIEGLER, INC., GRAND RAPIDS
GRR-(YEAR)-
    LEAR SIEGLER, INC. INSTRUMENT DIV., GRAND RAPIDS
GRT-
    RES.+ DEV.BD. COMM.ON GEOPHYSICS + GEOGRAPHY, WASH,DC
GR-TR-
    AIR FORCE CAMBRIDGE RESEARCH CENTER, MASS.
G-S-
    AMERICAN MACHINE AND FOUNDRY CO. ADVANCED SYSTEMS
        LAB., SANTA BARBARA, CALIF.
GS-
    BELL AND HOWELL CO., CHICAGO
GS-
    CIVIL SERVICE COMMISSION, WASHINGTON, D.C. (POSITION
        CLASSIFICATION STD.)
GS-
    GEOLOGICAL SURVEY, WASHINGTON, D.C.
GS-
    MELPAR, INC., FALLS CHURCH, VA.
GS-
    RES.+ DEV.BD. COMM. ON GEOPHYSICAL SCIENCES, WASH.,DC
GS/(LTRS.)/(YEAR)-
    AIR FORCE INST. OF TECH., WRIGHT-PATTERSON AFB, OHIO.
        SCHOOL OF ENGINEERING
GSA-
    GENERAL SERVICES ADMINISTRATION, WASHINGTON, D.C.
GSA-
    GRAHAM, SAVAGE, AND ASSOCIATES, INC., JENKINTOWN, PA.
GSA/SM/(YEAR)-
    AIR FORCE INST. OF TECH., WRIGHT-PATTERSON AFB, OHIO
        SCHOOL OF ENGINEERING
GS-B-
    GEOLOGICAL SURVEY, WASHINGTON, D.C. (BIBLIOGRAPHY)
GSB-
    UNITED STATES CONSULATE GENERAL. SCIENCE OFFICE,
        FRANKFURT AM MAIN (GERMAN SCIENCE BULLETIN)
GSC-
    AIR FORCE INST. OF TECH., WRIGHT-PATTERSON AFB, OHIO
GS-C-
    GEOLOGICAL SURVEY, WASHINGTON, D.C. (CIRCULAR)
GSC-
    GEOLOGICAL SURVEY, WASHINGTON, D.C. (CIRCULAR)
GSC-
    GIANNINI SCIENTIFIC CORP., SANTA ANA, CALIF.
GS-CIRC-
    GEOLOGICAL SURVEY, WASHINGTON, D.C. (CIRCULAR)
GSE-
    AEROJET-GENERAL CORP., SACRAMENTO, CALIF.
GSE-
    WESTINGHOUSE ELECTRIC CORP., PITTSBURGH
GSE/MC/(YEAR)-
    AIR FORCE LOGISTICS COMMAND, WRIGHT-PATTERSON AFB, OHIO
GSF-
    GESELLSCHAFT FUER STRAHLENFORSCHUNG M.B.H. INST.
        FUER STRAHLENSCHUTZ, NEUHERBERG BEI MUNICH.
GSF/(LTRS.)/(YEAR)-
    AIR FORCE INST. OF TECH., WRIGHT-PATTERSON AFB, OHIO.
        SCHOOL OF ENGINEERING (GRADUATE SPACE FACILITIES)
GSFC-TECHNICAL ABSTRACTS-
    GODDARD SPACE FLIGHT CENTER, GREENBELT, MD.
GSF-T-
    GESELLSCHAFT FUER STRAHLENFORSCHUNG M.B.H. INST.
        FUER STRAHLENSCHUTZ, NEUHERBERG BEI MUNICH
GSG-
    HUGHES AIRCRAFT CO., CULVER CITY, CALIF.
GSL-TR-(YEAR)-
    NEW YORK UNIV., N.Y.C. GEOPHYSICAL SCIENCES LAB.
GSM/(LTRS.)/(YEAR)-
    AIR FORCE INST. OF TECH., WRIGHT-PATTERSON AFB, OHIO.
        SCHOOL OF ENGINEERING (GRADUATE SYSTEMS MANAGEMENT)
GSM-AED-
    GT. BRIT. GEOLOGICAL SURVEY AND MUSEUM, LONDON
GS-P-(NO.)-A
    GEOLOGICAL SURVEY, WASHINGTON, D.C.
GSP/(LTRS.)/(YEAR)-
    AIR FORCE INST. OF TECH., WRIGHT-PATTERSON AFB, OHIO.
        SCHOOL OF ENGINEERING (GRADUATE SPACE PHYSICS)
GSPP-
    GEOLOGICAL SURVEY, WASH., D.C. (PROFESSIONAL PAPER)
GSR-
    NAVAL AIR MISSILE TEST CENTER, POINT MUGU, CALIF.
GSR-
    SHEFFIELD, ENGLAND. UNIV.
GSR/CSE/
    SNIA VISCOSA, COLLEFERRO, ITALY
GS WSP-
    GEOLOGICAL SURVEY, WASH., D.C. (WATER SUPPLY PAPER)
GT-(NUMBER)-R
    AMERICAN INSTS. FOR RESEARCH, PITTSBURGH
GT-
    GARRETT CORP. AIRESEARCH MFG. CO. OF ARIZONA, PHOENIX
GTC-
    GENERAL TECHNOLOGIES CORP., ALEXANDRIA, VA.
GTC-
    GENERAL TECHNOLOGIES CORP., RESTON, VA.
GTC-
    GENERAL TIME CORP., LA SALLE, ILL.

```
GTD/JWJ-        BROOKHAVEN NATIONAL LAB., UPTON, N.Y. (CODE IS
                AUTHOR'S INITIALS)
GTEL-TR-(YEAR)-...
                GENERAL TELEPHONE AND ELECTRONICS LABS. INC. BAY-
                SIDE LABS., N.Y.
GT/E-TR-(YEAR-NO.-NO.)
                GENERAL TELEPHONE AND ELECTRONICS LABS. INC. BAY-
                SIDE LABS., N.Y.
G-T-G-(YEAR)-
                GEORGIA INST. OF TECH., ATLANTA. ENG. EXPERIMENT STA.
GTI-            (GERMAN TECHNICAL INVESTIGATIONS)
GT I-           (GERMAN TECHNICAL INVESTIGATIONS)
GTI-            GENERAL TELEPHONE AND ELECTRONICS LABS. INC. BAY-
                SIDE LABS., N.Y.
GTP-            AEROJET-GENERAL CORP., SACRAMENTO, CALIF.
GTR-            GENERAL TELEPHONE AND ELECTRONICS LABS. INC. BAY-
                SIDE LABS., N.Y.
GTR-            GENERAL TIRE AND RUBBER CO., AKRON, OHIO
G-T-R-          GEORGIA INST. OF TECH., ATLANTA
GT+R-           GOODYEAR TIRE AND RUBBER CO., AKRON, OHIO
GTR-            HEWLETT-PACKARD LABS., PALO ALTO, CALIF.
GTRC-           GENERAL TIRE AND RUBBER CO., AKRON, OHIO
GTX-(3 DIGITS)
                GENTEX CORP., CARBONDALE, PENNA.
GU-             GEORGIA. UNIV., ATHENS
GU-             GLASGOW. UNIV.
GU-             GULF UNITED NUCLEAR FUELS CORP., ELMSFORD, N.Y.
GUL-            GULTON MFG. CORP., METUCHEN, N.J.
GULF-GA-        GULF GENERAL ATOMIC CO., SAN DIEGO, CALIF.
GULF-RT-        GULF RADIATION TECHNOLOGY, SAN DIEGO, CALIF.
GULF-RT-A-      GULF RADIATION TECHNOLOGY, SAN DIEGO, CALIF.
GUO-            JOINT PUBLICATIONS RESEARCH SERVICE, N.Y.C.
GUPS-MON-       GEORGE WASHINGTON UNIV., WASHINGTON, D.C.
GURC-           GULF UNIVERSITIES RESEARCH SERVICE, WASHINGTON, D.C.
GV-             AEROJET-GENERAL CORP., SACRAMENTO, CALIF.
GV-             ARMOUR RESEARCH FOUNDATION, CHICAGO
GV-             CARBORUNDUM CO. RES. + DEV. DIV., NIAGARA FALLS, N.Y.
GV-             DENVER. UNIV. DENVER RESEARCH INST.
GV-             GENERAL ELECTRIC CO. AIRCRAFT NUCLEAR PROPULSION
                DEPT., CINCINNATI
GV-R-           GIFFELS + ROSSETTI, INC., DETROIT
GW-             DE HAVILLAND PROPELLERS, LTD.
GW-             GEORGE WASHINGTON UNIV., WASHINGTON, D.C.
GW-(4 DIGITS)
                GLEASON WORKS, ROCHESTER, N.Y.
GW/             GT. BRIT. ROYAL AIRCRAFT EST.,FARNBOROUGH, HANTS, ENG
GW-             GT. BRIT. ROYAL AIRCRAFT EST.,FARNBOROUGH, HANTS, ENG
GW-             SHORT BROS. AND HARLAND, LTD.
GWAC-           GT. BRIT. GUIDED WEAPONS ADVISORY COMMITTEE
GWD-            ADVISORY COMMITTEE ON GERMAN WAR DOCUMENTS
GWPS-MON-       GEORGE WASHINGTON UNIV., WASHINGTON, D.C. (POLICY
                STUDIES IN SCIENCE + TECHNOLOGY)
GWPS-OP-        GEORGE WASHINGTON UNIV., WASHINGTON, D.C. (POLICY
                STUDIES IN SCIENCE + TECHNOLOGY)
GWPS-SDP-       GEORGE WASHINGTON UNIV., WASHINGTON, D.C. (POLICY
                STUDIES IN SCIENCE + TECHNOLOGY)
GW/SS-          SMITH (S.) AND SONS, LTD., LONDON
GWT-            GRUMMAN AIRCRAFT ENGINEERING CORP., BETHPAGE, N.Y.
GWTT-           GRUMMAN AIRCRAFT ENGINEERING CORP., BETHPAGE, N.Y.
GWU-            GEORGE WASHINGTON UNIV., WASHINGTON, D.C.
GX-             GT. BRIT. ADMIRALTY SIGNAL AND RADAR ESTABLISHMENT,
                PORTSMOUTH, HANTS, ENGLAND
```

H-	(HISTORY) (REF. 2)
H-	AIR FORCE, WASHINGTON, D.C.
H-	ARMY, WASHINGTON, D.C.
H-	ASSISTANT CHIEF OF STAFF FOR INTELLIGENCE (ARMY), WASHINGTON, D.C.
H(NUMBER)-(NO.)(LTR.)-	BUNKER-RAMO CORP. DEFENSE SYSTEMS DIV., CANOGA PARK, CALIF.
H-(NUMBER).(NO.).(NO.)	CARBIDE AND CARBON CHEMICALS CORP. Y-12 PLANT, OAK RIDGE, TENN.
H.(NUMBER)/(YEAR)	GT. BRIT. ARMAMENT RES. EST., FT. HALSTEAD, KENT, ENG
H-	GT. BRIT. ATOMIC EN. RES. ESTAB., HARWELL, BERKS, ENG
H-	GT. BRIT. DIRECTORATE OF ORDNANCE FACTORIES (EXPLOSIVES)
H.(NO.)/(YR)	GT. BRIT. MINISTRY OF SUPPLY
H-	HANFORD WORKS, RICHLAND, WASH.
H-	HONEYWELL, INC. (ALL LOCATIONS)
(YEAR)H-	HUGHES AIRCRAFT CO., CULVER CITY, CALIF.
H(NO.)MR(YR.)-	LOS ALAMOS SCIENTIFIC LAB., N. MEX.
H-1- THRU H-9-	LOS ALAMOS SCIENTIFIC LAB., N. MEX. (INTERNAL CORRESPONDENCE SERIAL USED BY GROUPS IN H DIV.)
H-	NAVAL OCEANOGRAPHIC OFFICE. HYDROGRAPHIC SURVEYS DEPT., WASHINGTON, D.C.
H-(NUMBER)-(YEAR)	NAVAL OCEANOGRAPHIC OFFICE. HYDROGRAPHIC SURVEYS DEPT., WASHINGTON, D.C.
H-(YEAR-NUMBER)	NETHERLANDS. HOUTINSTITUUT TNO, DELFT
H-	OAK RIDGE NATIONAL LAB., TENN.
H-	OFFICE OF THE ASSISTANT SECRETARY OF DEFENSE (INSTALLATIONS AND LOGISTICS), WASHINGTON, D.C.
H-	PHILCO CORP., LANSDALE, PENNA.
H-	PHILCO CORP., PHILADELPHIA
H-	SPERRY GYROSCOPE CO., GREAT NECK, N.Y.
H-0.100- THRU H-10.430-	TENNESSEE EASTMAN CORP., OAK RIDGE, TENN. (REF. 34)
H-(NUMBER)-	UNITED AIRCRAFT CORP. RES. LABS., EAST HARTFORD, CONN.
H-	WATERTOWN ARSENAL LAB., MASS.
HA-	EUROPEAN THEATER OF OPERATIONS (ARMY). ALSOS MISSION (HAAGEN FILE)
HA-	HARVEY ALUMINUM, INC., TORRANCE, CALIF.
HA-	HARVEY ENGINEERING LABS. FOR RES. + DEV., TORRANCE, CAL
HA-	HITTMAN ASSOCIATES, INC., BALTIMORE
HA-	HONEYWELL, INC. AERONAUTICAL DIV., MINNEAPOLIS
HA-	HUGHES AIRCRAFT CO., CULVER CITY, CALIF.
HAC-	HUGHES AIRCRAFT CO. (ALL LOCATIONS)
HAC-(NO.).(NO.)/(NO.)	HUGHES AIRCRAFT CO. NUCLEAR ELECTRONICS LAB., CULVER CITY, CALIF.
HAC-A(NUMBER)	HUGHES AIRCRAFT CO., CULVER CITY, CALIF.
HAC-ASD-	HUGHES AIRCRAFT CO., CULVER CITY, CALIF.
HAC-ASD/H-	HUGHES AIRCRAFT CO., CULVER CITY, CALIF.
HAC-ATR-	HUGHES RESEARCH LABS. DIV. OF HUGHES AIRCRAFT CO., MALIBU, CALIF.
HAC-B(NUMBER)	HUGHES AIRCRAFT CO., CULVER CITY, CALIF.
HAC-FR-	HUGHES AIRCRAFT CO., FULLERTON, CALIF.
HAC-LS-BIB-	HUGHES AIRCRAFT CO. LIBRARY SERVICES, CULVER CITY, CAL
HAC M-	HUGHES AIRCRAFT CO., CULVER CITY, CALIF.
HAC-MSD-P(YR)-(NUMBER)	HUGHES AIRCRAFT CO., CANOGA PARK, CALIF.
HAC-OP-	HUGHES AIRCRAFT CO., CULVER CITY, CALIF.
HAC-P(YEAR)-	HUGHES AIRCRAFT CO. AERONAUTICAL SYSTEMS DIV., CULVER CITY, CALIF.
HAC-PRI-(NUMBER)Y	HUGHES AIRCRAFT CO. SYSTEMS DEVELOPMENT LABS., CULVER CITY, CALIF.
HAC-PRII-(NUMBER)Y	HUGHES AIRCRAFT CO. SYSTEMS DEVELOPMENT LABS., CULVER CITY, CALIF.
HAC-PUB-	HUGHES AIRCRAFT CO., CULVER CITY, CALIF.
HAC-RR-	HUGHES RESEARCH LABS. DIV. OF HUGHES AIRCRAFT CO., MALIBU, CALIF.
HAC-SDL-	HUGHES AIRCRAFT CO., CULVER CITY, CALIF.
HAC SDN (NOS.)/(NOS.)	HUGHES AIRCRAFT CO., CULVER CITY, CALIF.
HAC SRSM-	HUGHES AIRCRAFT CO., CULVER CITY, CALIF. (SPECIAL RESEARCH STUDY MEMORANDUM)
HAC-SSD-	HUGHES AIRCRAFT CO., CULVER CITY, CALIF.
HAC-TM-	HUGHES AIRCRAFT CO., CULVER CITY, CALIF.
HAC-TOW/H-	HUGHES AIRCRAFT CO., CULVER CITY, CALIF.
HADC-	HOLLOMAN AIR DEVELOPMENT CTR., HOLLOMAN AFB, N. MEX.
HADC (LETTERS)-	HOLLOMAN AIR DEVELOPMENT CTR., HOLLOMAN AFB, N. MEX.
HADC ADJ-(YR.)-	HOLLOMAN AIR DEVELOPMENT CTR., HOLLOMAN AFB, N. MEX.
HADC HDGR-(YR.)-	HOLLOMAN AIR DEVELOPMENT CTR., HOLLOMAN AFB, N. MEX.
HADC-TN-	HOLLOMAN AFB, N. MEX.
HADC-TR-(YEAR)-	HOLLOMAN AIR DEVELOPMENT CTR., HOLLOMAN AFB, N. MEX.
HADES-R-	LITTON SYSTEMS, INC., WOODLAND HILLS, CALIF.
HAD F/(CONTRACT NUMBER)	HUGHES AIRCRAFT CO., NEWPORT BEACH, CALIF.
HAFB-	HOLLOMAN AFB, N. MEX.
HAI-	HARVEY ALUMINUM, INC., TORRANCE, CALIF.
HAIC-	HAYES INTERNATIONAL CORP., BIRMINGHAM, ALA.
HAI-PR-	HARVEY ALUMINUM, INC., TORRANCE, CALIF.
HALC-TR-	BROWN UNIV., PROVIDENCE
HALLIC-	HALLICRAFTERS CO., CHICAGO
HAN-	HANFORD ATOMIC PRODUCTS OPERATION, RICHLAND, WASH.
HAN-	HANFORD OPERATIONS OFFICE, AEC, WASH.
HAN-	HANFORD WORKS, RICHLAND, WASH.
HANFORD-	HANFORD WORKS, RICHLAND, WASH.
HAO-	HIGH ALTITUDE OBSERVATORY, BOULDER, COLO.
HAO-	HIGH ALTITUDE OBSERVATORY, CLIMAX, COLO.
HAO-	MICHIGAN. UNIV., ANN ARBOR. ENGINEERING RES. INST.
HAO-TR-	HIGH ALTITUDE OBSERVATORY, BOULDER, COLO.
HAP-	UNITED KINGDOM ATOMIC ENERGY AUTHORITY. RESEARCH GP. ATOMIC ENERGY RES. ESTAB., HARWELL, BERKS, ENGLAND
HAR-	BROWN-FIRTH RESEARCH LABS., SHEFFIELD, YORKS, ENGLAND
HAR-	GT. BRIT. NATL. CHEMICAL LAB., TEDDINGTON, MIDDX., ENG.
HAR-	HARVEY MACHINE CO., INC., TORRANCE, CALIF.
HAR-	UNITED KINGDOM ATOMIC ENERGY AUTHORITY. RESEARCH GP. ATOMIC ENERGY RES. ESTAB., HARWELL, BERKS, ENGLAND
HARAS-	HUGHES AIRCRAFT CO. HUGHES INTERNATIONAL, CULVER CITY, CALIF.
HARD(LETTER)/(LETTER)-	UNITED KINGDOM ATOMIC ENERGY AUTHORITY. RESEARCH GP. ATOMIC ENERGY RES. ESTAB., HARWELL, BERKS, ENGLAND
HARD(A)/P-	GT. BRIT. CAPENHURST WORKS, CHES., ENGLAND
HARD-(A)/P-	UNITED KINGDOM ATOMIC ENERGY AUTHORITY. RESEARCH GP. ATOMIC ENERGY RES. ESTAB., HARWELL, BERKS, ENGLAND
HARD(B)/P-	UNITED KINGDOM ATOMIC ENERGY AUTHORITY. RESEARCH GP. ATOMIC ENERGY RES. ESTAB., HARWELL, BERKS, ENGLAND
HARD(C)/P-	UNITED KINGDOM ATOMIC ENERGY AUTHORITY. RESEARCH GP. ATOMIC ENERGY RES. ESTAB., HARWELL, BERKS, ENGLAND
HARD/DATA-	UNITED KINGDOM ATOMIC ENERGY AUTHORITY. RESEARCH GP. ATOMIC ENERGY RES. ESTAB., HARWELL, BERKS, ENGLAND
HARD/P-	UNITED KINGDOM ATOMIC ENERGY AUTHORITY. RESEARCH GP. ATOMIC ENERGY RES. ESTAB., HARWELL, BERKS, ENGLAND
HARDTACK-	ARMED FORCES SPECIAL WEAPONS PROJECT. WEAPONS EFFECTS PROGRAM, WASH., D.C. (OPERATION HARDTACK) (REF. 7)
HARE/	GT. BRIT. ATOMIC EN. RES. ESTAB., HARWELL, BERKS, ENG
HARL-	HARVARD UNIV., CAMBRIDGE, MASS. ACOUSTICS RES. LAB.
HA-RL-	HUGHES AIRCRAFT CO. RES. LAB., CULVER CITY, CALIF.
HARVARD U.-(LTRS,NOS.)-	HARVARD UNIV., CAMBRIDGE, MASS.
HASD-	NEW YORK OPERATIONS OFFICE, AEC. HEALTH + SAFETY LAB.

HASL-	NEW YORK OPERATIONS OFFICE, AEC. HEALTH + SAFETY LAB. (REF. 4)
HASL-S-	NEW YORK OPERATIONS OFFICE, AEC. HEALTH + SAFETY LAB.
HASL-SC-	NEW YORK OPERATIONS OFFICE, AEC. HEALTH + SAFETY LAB.
HASL-TM-	NEW YORK OPERATIONS OFFICE, AEC. HEALTH + SAFETY LAB.
HA-SR-	HUGHES AIRCRAFT CO., CULVER CITY, CALIF. (SCI. RPT.)
HAST-	HASTINGS INSTRUMENT CO., INC., HAMPTON, VA.
HAVENS-	BRIDGEPORT BRASS CO., CONN.
HAWAIIU-	HAWAII. UNIV., HONOLULU
HA-WSDL-	HUGHES AIRCRAFT CO. WEAPON SYSTEMS DEVELOPMENT LABS., CULVER CITY, CALIF.
HAWU-	HAWAII. UNIV., HONOLULU
HAYES-ER-	HAYES AIRCRAFT CORP., BIRMINGHAM, ALA.
HB-	AEROJET-GENERAL CORP., SACRAMENTO, CALIF.
HB-	BRUTCHER (HENRY) TECHNICAL TRANSLATIONS, ALTADENA,CAL
HB-(NUMBER)-	SPERRY GYROSCOPE CO., GREAT NECK, N.Y.
HBC-	STANFORD UNIV., CALIF. STANFORD LINEAR ACCELERATOR CENTER (HYDROGEN BUBBLE CHAMBER GROUP)
HBM-	RES. + DEV. BD. COMM. ON HUMAN RESOURCES, WASH., D.C.
HBR-	HYCON MFG. CO., MONROVIA, CALIF.
HBRA-	HYCON MFG. CO., PASADENA, CALIF.
HB-TRANS-	BRUTCHER (HENRY) TECHNICAL TRANSLATIONS, ALTADENA,CAL
HC-	HERLEC CORP., MILWAUKEE
H+C-	HERNER AND CO., WASHINGTON, D.C.
HC-	HUMPHREYS CORP. TAFA DIV., BOW, N.H.
HCC-	HARSHAW CHEMICAL CO., CLEVELAND
HCCC/P-	UNITED KINGDOM ATOMIC ENERGY AUTHORITY. RESEARCH GP. ATOMIC ENERGY RES. ESTAB., HARWELL, BERKS, ENGLAND
HCC-QPR-	HOOKER CHEMICAL CORP., NIAGARA FALLS, N.Y.
HCL-PR-	HARVARD UNIV., CAMBRIDGE, MASS. COMPUTATION LAB.
HCO-	HARVARD UNIV., CAMBRIDGE, MASS. HARVARD COLL. OBSERVATORY (SOLAR REPORT)
HCO-SR-	HARVARD UNIV., CAMBRIDGE, MASS. HARVARD COLL. OBS.
HCO-TR-	HARVARD UNIV., CAMBRIDGE, MASS. HARVARD COLL. OBS.
HCP/P-	UNITED KINGDOM ATOMIC ENERGY AUTHORITY. RESEARCH GP. ATOMIC ENERGY RES. ESTAB., HARWELL, BERKS, ENGLAND
HCRL-	CALIFORNIA. UNIV., LIVERMORE. RADIATION LAB.
HD-	GT. BRIT. ATOMIC EN. RES. ESTAB., HARWELL, BERKS, ENG
HDC-	HANFORD WORKS. DESIGN + CONSTRUCTION DIV., RICHLAND, WASH.
HDL-	HARRY DIAMOND LABS., WASHINGTON, D.C.
HDL-PR-	HARRY DIAMOND LABS., WASHINGTON, D.C.
HDL-RE-	HARRY DIAMOND LABS., WASHINGTON, D.C.
HDL-TM-	HARRY DIAMOND LABS., WASHINGTON, D.C.
HDL-TR-	HARRY DIAMOND LABS., WASHINGTON, D.C.
H-DOC-(NUMBER)-	CONGRESS. HOUSE OF REPRESENTATIVES, WASHINGTON, D.C.
HDRR-	HOLLOMAN AFB, N. MEX. (HOLLOMAN DEV. RESEARCH RPT.)
HDRRM/TM-	HOLLOMAN AIR DEVELOPMENT CTR., HOLLOMAN AFB, N. MEX.
HDS-	GT. BRIT. ATOMIC EN. RES. ESTAB., HARWELL, BERKS, ENG
HDT-	HOLLOMAN AIR DEVELOPMENT CTR., HOLLOMAN AFB, N. MEX.
HDTT-	HOLLOMAN AIR DEVELOPMENT CTR., HOLLOMAN AFB, N. MEX.
HDW/TM-	HOLLOMAN AIR DEVELOPMENT CTR., HOLLOMAN AFB, N. MEX.
HE-116-	CALIFORNIA. UNIV., BERKELEY. FLUID MECHANICS LAB.
HE-150-	CALIFORNIA. UNIV., BERKELEY. INST. OF ENG. RES.
HE-	SWEDEN. FLYGTEKNISKA FOERSOEKSANSTALTEN, STOCKHOLM
HEBREWU-	HEBREW UNIV., JERUSALEM
HEBRU-	HEBREW UNIV., JERUSALEM
HEC-	HAZELTINE ELECTRONICS CORP., LITTLE NECK, N.Y.
HEC-	HERNER AND CO., WASHINGTON, D.C.
HEC-	HOOKER ELECTROCHEMICAL CO., NIAGARA FALLS, N.Y.
HEC-QPR-	HOOKER ELECTROCHEMICAL CO., NIAGARA FALLS, N.Y.
HED-	HAZELTINE SERVICE CORP.
HEDL-SA-	HANFORD ENGINEERING DEVELOPMENT LAB., RICHLAND,WASH.
HEDL-TME-(YEAR)-	HANFORD ENGINEERING DEVELOPMENT LAB., RICHLAND,WASH.
HEE-M-(NUMBER-YEAR)	CHRYSLER CORP. ELECTRICAL AND ELECTRONIC ENG. DEPT., HUNTSVILLE, ALA.
HEHF-	HANFORD OCCUPATIONAL HEALTH FOUNDATION. OCCUPATIONAL HYGIENE DEPT., RICHLAND, WASH.
HEI-	GT. BRIT. ADVISORY COUNCIL ON SCI. RES. + TECH. DEV.
HEIAS-	TUFTS UNIV., MEDFORD, MASS.
HEINZ-	HEINZ (H.J.) CO., PITTSBURGH
HEL-	CALIFORNIA. UNIV., BERKELEY. HYDRAULIC ENG. LAB.
HEL-	HUMAN ENGINEERING LABS., ABERDEEN PROVING GROUND, MD.
HEL-BIB-	HUMAN ENGINEERING LABS., ABERDEEN PROVING GROUND, MD.
HELIODYNE-RR-	HELIODYNE CORP., VAN NUYS, CALIF.
HELPR-	(HANDBOOK OF ELECTRONIC PARTS RELIABILITY)
HEL-S-	HUMAN ENGINEERING LABS., ABERDEEN PROVING GROUND, MD.
HELSINKIU-(NO./YR.)	HELSINKI UNIV.
HEL STANDARD S-	HUMAN ENGINEERING LABS., ABERDEEN PROVING GROUND, MD.
HEL-TM-	HUMAN ENGINEERING LABS., ABERDEEN PROVING GROUND, MD.
HEL-TN-	HUMAN ENGINEERING LABS., ABERDEEN PROVING GROUND, MD.
HEO-(NUMBER)-	WESTINGHOUSE ELECTRIC CORP. HUNTSVILLE ENGINEERING OPERATIONS, ALA.
HEP-(YEAR)-	CAMBRIDGE UNIV., ENGLAND. CAVENDISH LAB.
HEPC-	CANADA. HYDRO-ELECTRIC POWER COMMISSION OF ONTARIO
HEPG-	HAWAII. UNIV., HONOLULU
HEPL-	STANFORD UNIV., CALIF. HIGH ENERGY PHYSICS LAB.
HEP/MISC/	UNITED KINGDOM. RUTHERFORD HIGH ENERGY LAB., CHILTON, DIDCOT, BERKS, ENGLAND
HER-	GT. BRIT. ATOMIC EN. RES. ESTAB., HARWELL, BERKS, ENG
HER-A(NO.)/(YR)	GT. BRIT. ARMAMENT RES. EST., FT. HALSTEAD, KENT, ENG
HER-A-	GT. BRIT. ATOMIC WEAPONS RESEARCH ESTABLISHMENT, ALDERMASTON, BERKS, ENGLAND
HER/AP-(NO.)-(YEAR)	MILAN. UNIV. INSTITUTE OF PHARMACOLOGY
HERCULES MTI-	HERCULES POWDER CO., WILMINGTON, DEL.
HER-H-	GT. BRIT. ATOMIC WEAPONS RESEARCH ESTABLISHMENT, ALDERMASTON, BERKS, ENGLAND
HER-H-	GT. BRIT. MINISTRY OF SUPPLY. MATERIALS RESEARCH DIV.(HIGH EXPLOSIVES RES.), FT. HALSTEAD, KENT, ENG
HERMES-	GENERAL ELECTRIC CO., SCHENECTADY, N.Y.(PROJ. HERMES)
HERO-	(HAZARDS OF ELECTROMAGNETIC RADIATION TO ORDNANCE)
HES-	ILLINOIS. UNIV., URBANA. DEPT. OF CIVIL ENGINEERING
HEW-	HANFORD ENGINEER WORKS, RICHLAND, WASH. (REPORTS ISSUED BY THE GENERAL ELECTRIC CO. JULY - OCT. 1947)
HEX-	HEXCEL PRODUCTS, INC., OAKLAND, CALIF.
HF-	ARMY BOARD FOR AVIATION ACCIDENT RES.,FORT RUCKER,ALA
HF-(YEAR)-	ARMY BOARD FOR AVIATION ACCIDENT RES.,FORT RUCKER,ALA
HF-	CORNELL AERONAUTICAL LAB., INC., BUFFALO
HF-	DENMARK. FORSVARETS FORSKNINGSRAD, COPENHAGEN
HF-	DOW CHEMICAL CO., MIDLAND, MICH.
HF-(NUMBER) M-(YEAR)	DOW CHEMICAL CO. SCIENTIFIC PROJECTS LAB., MIDLAND, MICH. (M IS FOR BIMONTHLY)
HF-(YEAR)-	GENERAL ELECTRIC CO. CAPACITOR DEPT., HUDSON FALLS, N.Y.
(YEAR)HF-	HUGHES AIRCRAFT CO. GROUND SYSTEMS GP., FULLERTON,CAL
HFD-PS-(YEAR)-	CORNELL AERONAUTICAL LAB., INC., BUFFALO
H FILE-	EUROPEAN THEATER OF OPERATIONS (ARMY). ALSOS MISSION (HOYT FILE)

HFO-	AIR RESEARCH AND DEVELOPMENT COMMAND. HUMAN FACTORS OPERATIONS RESEARCH LABS., WASHINGTON, D.C.
HFORL-	AIR RESEARCH AND DEVELOPMENT COMMAND. HUMAN FACTORS OPERATIONS RESEARCH LABS., WASHINGTON, D.C.
HFPS-	OFFICE OF CIVIL DEFENSE, WASHINGTON, D.C. (HOME FALL-OUT PROTECTION SURVEY)
HFR-	KAISER AEROSPACE AND ELECTRONICS CORP., ELECTRONICS PLANT, PALO ALTO, CALIF.
HFS-	HUGHES AIRCRAFT CO., CULVER CITY, CALIF. (FINISH SPECIFICATION)
HFT-	CALIFORNIA. UNIV., BERKELEY. HUMAN FACTORS IN TECHNOLOGY RESEARCH GROUP
HG-	WISCONSIN. UNIV., MADISON
HGG-	GT. BRIT. ATOMIC EN. RES. ESTAB., HARWELL, BERKS, ENG
HGTC-(LETTERS)/	UNITED KINGDOM ATOMIC ENERGY AUTHORITY. RESEARCH GP. ATOMIC ENERGY RES. ESTAB., HARWELL, BERKS, ENGLAND
HH-(NO.)-RR-	HUDSON INST., INC., CROTON-ON-HUDSON, N.Y.
HHFA-	HOUSING AND HOME FINANCE AGENCY, WASHINGTON, D.C.
HHFA RP-	HOUSING AND HOME FINANCE AGENCY, WASHINGTON, D.C. (RESEARCH PAPER)
HHFA TP-	HOUSING AND HOME FINANCE AGENCY, WASHINGTON, D.C. (TECHNICAL PAPER)
HI-	HORIZONS, INC., CLEVELAND
HI-	HUDSON INST., INC., CROTON-ON-HUDSON, N.Y.
HI-(NUMBER)-RR	HUDSON INST., INC., CROTON-ON-HUDSON, N.Y.
HI-(NUMBER)-D	HUDSON INST., INC., CROTON-ON-HUDSON, N.Y. (DRAFT)
HI-	HUDSON INST., INC., HARMON-ON-HUDSON, N.Y.
HI-(NUMBER)-RR	HUDSON INST., INC., HARMON-ON-HUDSON, N.Y.
HI-(YEAR)-	HYDROSYSTEMS, INC., FARMINGDALE, N.Y.
HI-	NATIONAL RESEARCH COUNCIL OF CANADA. ATOMIC ENERGY PROJECT, CHALK RIVER, ONT. (REF. 29)
HI-BN-	HUDSON INST., INC., CROTON-ON-HUDSON, N.Y. (BRIEFING NOTES)
HIC-	HAYES INTERNATIONAL CORP., BIRMINGHAM, ALA.
HIC-	HAYES INTERNATIONAL CORP., HUNTSVILLE, ALA.
HIC-	SHEFFIELD, ENGLAND. UNIV. DEPT. OF FUEL TECHNOLOGY AND CHEMICAL ENGINEERING
HIC-(NUMBER)-APP	SHEFFIELD, ENGLAND. UNIV. DEPT. OF FUEL TECHNOLOGY AND CHEMICAL ENGINEERING
HICKORY TF-	SOUTHEASTERN FOREST EXPERIMENT STATION, ASHEVILLE, NC
HI-C-MEMO-	ARGONNE NATIONAL LAB., ILL.
HICO-	HASTINGS INSTRUMENT CO., INC., HAMPTON, VA.
HI-DP-	HUDSON INST., INC., CROTON-ON-HUDSON, N.Y. (DISCUSSION PAPER)
HIG-	HAWAII INST. OF GEOPHYSICS, HONOLULU
HIG-(YEAR)-	HAWAII INST. OF GEOPHYSICS. HONOLULU
HIGHWAY ENGINEERING SER-	ILLINOIS. UNIV., URBANA. HIGHWAY RESEARCH LAB.
HIM/CONF-	UNITED KINGDOM ATOMIC ENERGY AUTHORITY. RESEARCH GP. ATOMIC ENERGY RES. ESTAB., HARWELL, BERKS, ENGLAND
HI-P-	HUDSON INST., INC., CROTON-ON-HUDSON, N.Y. (PAPER)
HI-PR-	HUDSON INST., INC., CROTON-ON-HUDSON, N.Y.
HIPR-	STRATEGIC BOMBING SURVEY (HEAVY INDUSTRY PLANT RPTS.)
HISTORICAL MONOGRAPH-	EDGEWOOD ARSENAL, MD.
HIT-	HITTMAN ASSOCIATES, INC., BALTIMORE
HI-T-	HUDSON INST., INC., CROTON-ON-HUDSON, N.Y. (TRANSLATION)
HKF-	FERGUSON (H.K.) CO., N.Y.C.
HKF-1942D-	FERGUSON (H.K.) CO., N.Y.C.
HKF-AED-	FERGUSON (H.K.) CO. ATOMIC ENERGY DIV., N.Y.C.
HL(NO.)-	CHICAGO. UNIV.
HL-(NUMBER)-(YEAR)	GENERAL DYNAMICS/ELECTRONICS. HYDROACOUSTICS LAB., ROCHESTER, N.Y.
HL(YEAR/NUMBER)	UNITED KINGDOM ATOMIC ENERGY AUTHORITY, LONDON
HL-(YEAR/NUMBER)	UNITED KINGDOM ATOMIC ENERGY AUTHORITY. RESEARCH GP. ATOMIC ENERGY RES. ESTAB., HARWELL, BERKS, ENGLAND
HL-	VIENNA. TECHNISCHE HOCHSCHULE
HLC-	HALLICRAFTERS CO., ROLLING MEADOWS, ILL.
HLI-	HOFFMAN LABS., INC., LOS ANGELES
HLJ-	JOHNSTON (HERRICK L.) INC., COLUMBUS, OHIO
HM-	CHRYSLER CORP. MISSILE DIV., DETROIT
HM-	CORNELL AERONAUTICAL LAB., INC., BUFFALO
HM-	HANFORD WORKS. DESIGN + CONSTRUCTION DIV., RICHLAND, WASH. (HIGHLY RESTRICTED REPORTS)
HM-	HYDROFOIL CORP., ANNAPOLIS, MD.
HMA/SEE-	GT. BRIT. UNDERWATER DETECTION ESTABLISHMENT
HMC-	HARVEY MACHINE CO., INC., TORRANCE, CALIF.
HMD-	CHRYSLER CORP. HUNTSVILLE SPACE OPERATION, ALA.
HMD-R(NUMBER-YEAR)	CHRYSLER CORP. HUNTSVILLE SPACE OPERATION, ALA.
HMI-B-	BERLIN. HAHN-MEITNER-INSTITUT FUER KERNFORSCHUNG
HML-	CALIF. INST. OF TECH., PASADENA. HYDRAULIC MACH. LAB.
HML-	HAWAII. UNIV., HONOLULU. HAWAII MARINE LAB.
HML ND-	CALIF. INST. OF TECH., PASADENA. HYDRAULIC MACH. LAB.
HMP-	AACHEN. TECHNISCHE HOCHSCHULE
HMP-	RES. + DEV. BD. PANEL ON MANPOWER, WASHINGTON, D.C.
HMS-	HUGHES AIRCRAFT CO.,CULVER CITY,CAL.(MATERIAL SPEC.)
HMSO-	GT. BRIT. HER MAJESTY'S STATIONERY OFFICE, LONDON (TRANSLATIONS)
HMT-	GT. BRIT. AERONAUTICAL RESEARCH COUNCIL
HMUDE-	GT. BRIT. UNDERWATER DETECTION ESTABLISHMENT
HN-	HOLMES AND NARVER, INC., LOS ANGELES
HN-(NUMBER) -	HOLMES AND NARVER, INC., LOS ANGELES
H+N-	HOLMES AND NARVER, INC., LOS ANGELES
HN-	HOLMES AND NARVER, INC. ON-CONTINENT TEST DIV., LAS VEGAS, NEV.
HN-	NATIONAL RESEARCH COUNCIL OF CANADA. DIV. OF BUILDING RESEARCH, OTTAWA (HOUSING NOTES)
HN-CEN-	HOLMES AND NARVER, INC., LOS ANGELES
HN-CHN-	HOLMES AND NARVER, INC., LOS ANGELES
HN-E-(NUMBER)-(YEAR)	HOLMES + NARVER, INC.,LOS ANG. (ENIWETOK PROVING GD.)
HN-EPG-(YEAR)/	HOLMES + NARVER, INC.,LOS ANG. (ENIWETOK PROVING GD.)
H+NID-	HOLMES AND NARVER, INC., LOS ANGELES
HN-JOB-	HOLMES AND NARVER, INC., LOS ANGELES
HN-N-(NUMBER)-(YEAR)	HOLMES + NARVER, INC., LOS ANG. (NEVADA TEST SITE)
HN-NTS-(YEAR)/	HOLMES + NARVER, INC., LOS ANG. (NEVADA TEST SITE)
HN-P/(LETTER)-(NO.)-(YR)	HOLMES AND NARVER, INC., LOS ANG. (PLOWSHARE PROGRAM)
HNS-	BRADBERRY (CARROLL E.) AND ASSOCIATES,LOS ALTOS, CAL.
HNS-(CONTRACT NO.)-	HAZLETON-NUCLEAR SCIENCE CORP., PALO ALTO, CALIF.
HNS-	HAZLETON-NUCLEAR SCIENCE CORP. RADIOCHEMISTRY DEPT., PALO ALTO, CALIF.
HNS-	ISOTOPES, INC., PALO ALTO, CALIF.
H-NSC-	HAZLETON-NUCLEAR SCIENCE CORP., PALO ALTO, CALIF.
HN-SEN-	HOLMES AND NARVER, INC., LOS ANGELES
HNSTR-	HAZLETON-NUCLEAR SCIENCE CORP. EARTH SCIENCES DIV., PALO ALTO, CALIF.
HN-TDP-	HOLMES AND NARVER, INC., LOS ANGELES
HO-	HANFORD OPERATIONS OFFICE, AEC, WASH. (REF. 4)
HO-	HONEYWELL, INC. ORDNANCE DIV., HOPKINS, MINN.
H.O.-	HYDROGRAPHIC OFFICE, SUITLAND, MD.
HO-	HYDROGRAPHIC OFFICE, SUITLAND, MD.
HOBO-	CALIFORNIA. UNIV., LIVERMORE. LAWRENCE RADIATION LAB. (PROJECT HOBO)
HOD-(INITIALS)-	HANFORD OPERATIONS OFFICE, AEC, WASH.

HOD-(INITIALS)-
    HANFORD WORKS, RICHLAND, WASH.
HODS-
    HYDROGRAPHIC OFFICE, SUITLAND, MD. (DATA SHEET)
HOHF-
    HANFORD OCCUPATIONAL HEALTH FDN., RICHLAND, WASH.
HOHF-
    HANFORD OCCUPATIONAL HEALTH FOUNDATION. OCCUPATIONAL HYGIENE DEPT., RICHLAND, WASH.
HOL-
    GT. BRIT. ADMIRALTY MATERIALS LAB., POOLE, DORSET, ENG
H.O. MISC.-
    NAVAL OCEANOGRAPHIC OFFICE, WASHINGTON, D.C. (HYDROGRAPHIC OFFICE MISCELLANEOUS)
HONEY-
    HONEYWELL, INC. (ALL LOCATIONS)
HONEYWELL-(NUMBER)-FRI
    HONEYWELL, INC. SYSTEMS AND RESEARCH DIV., ST. PAUL
HONEYWELL-AERO-
    HONEYWELL, INC., ST. PETERSBURG, FLA.
HOO-
    HANFORD OPERATIONS OFFICE, AEC, WASH. (REF. 4)
H.O. PUB.-
    NAVAL OCEANOGRAPHIC OFFICE, WASHINGTON, D.C. (HYDROGRAPHIC OFFICE PUBLICATION)
HORTICULTURE-PR-
    NEBRASKA. UNIV., LINCOLN. DEPT. OF HORTICULTURE AND FORESTRY
HO/TOTEM-
    GT. BRIT. ATOMIC WEAPONS RESEARCH ESTABLISHMENT, ALDERMASTON, BERKS, ENGLAND
HOT SOURCES-
    CALIFORNIA. UNIV., BERKELEY. RADIATION LAB.
HOUDRY-
    HOUDRY PROCESS AND CHEMICAL CO., LINWOOD, PENNA.
HOUGHTON-QR-
    HOUGHTON (E.F.) + CO., PHILADELPHIA (QUARTERLY REPT)
HP-
    CLINTON LABS., OAK RIDGE, TENN.
HP-
    HARVARD UNIV., CAMBRIDGE, MASS. DIVISION OF ENG. AND APPLIED PHYSICS
HP-
    HEWLETT-PACKARD CO., PALO ALTO, CALIF.
HP-
    HUGHES AIRCRAFT CO., CULVER CITY, CAL. (HUGHES PROCESS)
HP-
    QUARTERMASTER RESEARCH AND ENGINEERING COMMAND, NATICK, MASS. (HIGH POLYMER)
HP-AN-
    HEWLETT-PACKARD CO., COLORADO SPRINGS
HP-AN-
    HEWLETT-PACKARD CO., PALO ALTO, CALIF.
H/PAN-P-
    UNITED KINGDOM ATOMIC ENERGY AUTHORITY. RESEARCH GP. ATOMIC ENERGY RES. ESTAB., HARWELL, BERKS, ENGLAND
HPC-
    GT. BRIT. ATOMIC ENERGY RESEARCH ESTABLISHMENT. HARWELL POWER COMMITTEE, HARWELL, BERKS, ENGLAND
HPC-
    HERCULES POWDER CO., MAGNA, UTAH
HPC-
    HERCULES POWDER CO., WILMINGTON, DEL.
HPC-(YEAR)-
    MONSANTO RESEARCH CORP., ST. LOUIS, MO.
HPC-CPD-
    HERCULES POWDER CO., WILMINGTON, DEL.
HPCO-
    HEWLETT-PACKARD CO., PALO ALTO, CALIF.
HPC-PRM-
    HERCULES POWDER CO., WILMINGTON, DEL.
HPD-
    HELIODYNE CORP., VAN NUYS, CALIF.
HPDI-
    HUGHES AIRCRAFT CO., CULVER CITY, CALIF. (HARD POINT DEFENSE INTERCEPTOR)
HPDS-(LETTERS)-
    HUGHES AIRCRAFT CO., CULVER CITY, CALIF. (HARD POINT DEFENSE SYSTEM)
HP/GEN-
    UNITED KINGDOM ATOMIC ENERGY AUTHORITY. RESEARCH GP. ATOMIC ENERGY RES. ESTAB., HARWELL, BERKS, ENGLAND
HP/M-
    GT. BRIT. MINISTRY OF SUPPLY
HPNS-TR-(NUMBER-YEAR)
    SAN FRANCISCO BAY NAVAL SHIPYARD. HUNTER'S POINT DIV.
HPP-
    HARBOR PROTECTION PROJECT
HPP-
    YALE UNIV., NEW HAVEN. EDWARDS STREET LAB.
HPR-
    ALABAMA STATE HIGHWAY DEPT., MONTGOMERY
HPR-
    NORWAY. INSTITUTT FOR ATOMENERGI. HALDEN REACTOR PROJECT, LILLESTROM
HPRWT-
    HANDLEY PAGE (READING), LTD., ST. ALBANS, HERTS, ENG.
HPS-
    CALIFORNIA. UNIV., BERKELEY
HPS-
    RES. + DEV. BD. COMM. ON HUMAN RESOURCES, WASH., D.C.
HPS-
    RESEARCH AND DEVELOPMENT BOARD. PANEL ON HUMAN ENGINEERING AND PSYCHOPHYSIOLOGY, WASHINGTON, D.C.
HQ-
    (HEADQUARTERS) (REF. 2)
HQ-(NUMBER)/(NUMBER)
    GT. BRIT. ARMAMENT RES. EST., FT. HALSTEAD, KENT, ENG
HQ-
    HARSHAW CHEMICAL CO. CENTRAL RESEARCH LAB., CLEVELAND
HQARDC-
    AIR RESEARCH AND DEVELOPMENT COMMAND, ANDREWS AFB, MD
HQ/IT-
    GT. BRIT. DIRECTORATE OF CHEMICAL DEFENCE RES. + DEV.
HR-
    CANADA. DEFENCE RESEARCH BOARD, OTTAWA
HR-
    CONGRESS. HOUSE OF REPRESENTATIVES, WASHINGTON, D.C.
HR-
    HONEYWELL, INC. CORPORATE RES. CENTER, HOPKINS, MINN.
HR-(YEAR)-
    HONEYWELL, INC. CORPORATE RES. CENTER, HOPKINS, MINN.
HR-
    HUMAN RESOURCES RESEARCH INSTITUTE, MAXWELL AFB, ALA.
HR-
    HYDRAULIC RESEARCH AND MFG. CO., BURBANK, CALIF. HYDRAULIC RESEARCH LAB.
HR-
    RES. + DEV. BD. COMM. ON HUMAN RESOURCES, WASH., D.C.
HRAC-
    NATIONAL RESEARCH COUNCIL OF CANADA. ATOMIC ENERGY PROJECT, CHALK RIVER, ONT.
HRAF-
    (HUMAN RELATIONS AREA FILE)
HRB-
    HALLER, RAYMOND, AND BROWN, INC., STATE COLLEGE, PA.
HRB-
    HRB-SINGER, INC., STATE COLLEGE, PENNA.
HRB-
    NATIONAL RESEARCH COUNCIL. HIGHWAY RESEARCH BOARD
HRB-B-
    NATIONAL RES. COUNCIL. HIGHWAY RES. BD. (BIBLIOGRAPHY)
HRB-BUL-
    NATIONAL RESEARCH COUNCIL. HIGHWAY RES. BD. (BULLETIN)
HRBI-
    HRB-SINGER, INC., STATE COLLEGE, PENNA.
HRB-RR-
    NATIONAL RES. COUNCIL. HIGHWAY RES. BD. (RES. RPT.)
HRB/S-
    HRB-SINGER, INC., STATE COLLEGE, PENNA.
HRB-SINGER-
    HRB-SINGER, INC., STATE COLLEGE, PENNA.
HRB-SR-
    NATIONAL RES. COUNCIL. HIGHWAY RES. BD. (SPEC. RPT.)
HRB-TP-
    HRB-SINGER, INC., STATE COLLEGE, PENNA.
HRC-(YEAR)-
    HERCULES RESEARCH CENTER, WILMINGTON, DEL.
HRC-
    HERCULES RESEARCH CENTER, WILMINGTON, DEL.
HRC-(YR.)-
    HONEYWELL, INC. RADIATION CENTER, BOSTON
HRC-
    HUYCK RESEARCH CENTER, STAMFORD, CONN.
HRC-
    MINNEAPOLIS-HONEYWELL REGULATOR CO., MINNEAPOLIS
HRC-TN-
    HYDROSPACE RES. CORP., ROCKVILLE, MD.
HRC-TR-
    HONEYWELL RESEARCH CENTER, HOPKINS, MINN.
HRC-TR-
    HYDROSPACE RES. CORP., ROCKVILLE, MD.
HRD-
    HAZELTINE CORP. HAZELTINE RES. DIV., LITTLE NECK, NY
HRD-
    HOFF RESEARCH AND DEVELOPMENT LABS., INC., CLEVELAND
HRD-
    ROCKETDYNE DIV., NORTH AMERICAN AVIATION, INC., CANOGA PARK, CALIF.
HRD-AC-(YEAR)-
    GEN. MOTORS CORP. HARRISON RADIATOR DIV., LOCKPORT, NY
HRD-TR-
    ABEX CORP. HYDRODYNAMICS RES. CTR., COLUMBUS, OHIO
HREC-
    LOCKHEED MISSILES AND SPACE CO. HUNTSVILLE RESEARCH AND ENGINEERING CENTER, ALA.
HRGP/P-
    UNITED KINGDOM ATOMIC ENERGY AUTHORITY. RESEARCH GP. ATOMIC ENERGY RES. ESTAB., HARWELL, BERKS, ENGLAND
HR-HML-
    RES. + DEV. BD. COMM. ON HUMAN RESOURCES, WASH., D.C.
HR-HTD-
    RES. + DEV. BD. COMM. ON HUMAN RESOURCES, WASH., D.C.
HRI-
    ATOMIC ENERGY OF CANADA LTD., CHALK RIVER, ONT.
HRI-
    FORDHAM UNIV., N.Y.C.
HRI-(5 DIGITS)-
    HOUSTON RESEARCH INSTITUTE, TEX.
HRI-
    HYDROCARBON RESEARCH, INC., N.Y.C.
HRI-
    NATIONAL RESEARCH COUNCIL OF CANADA. ATOMIC ENERGY PROJECT, CHALK RIVER, ONT. (REF. 29)
HR-ITR-
    HONEYWELL RESEARCH CENTER, HOPKINS, MINN. (TECH.REPT)
HRL-
    HUGHES RESEARCH LABS. DIV. OF HUGHES AIRCRAFT CO., MALIBU, CALIF.
HRM-
    HELIODYNE CORP. BSD/BMRS DATA SERVICES AND ANALYSES, NORTON AFB, CALIF.
HRN-
    HELIODYNE CORP. (ALL LOCATIONS)
HRP-
    NORWAY. INSTITUTT FOR ATOMENERGI, HALDEN
HRRC-
    HUMAN RESOURCES RESEARCH CENTER, LACKLAND AFB, TEXAS
HRRI-
    HUMAN RESOURCES RESEARCH INSTITUTE, MAXWELL AFB, ALA.

HRRI-HR-	HUMAN RESOURCES RESEARCH INSTITUTE, MAXWELL AFB, ALA.
HRRL-	HUMAN RESOURCES RES. LABS., BOLLING AFB, WASH., D.C.
HRRL-MR-	HUMAN RESOURCES RES. LABS., BOLLING AFB, WASH., D.C.
HRS-	HYDRO RESEARCH SCIENCE, SUNNYVALE, CALIF.
HRSCC/P-	UNITED KINGDOM ATOMIC ENERGY AUTHORITY. RESEARCH GP. ATOMIC ENERGY RES. ESTAB., HARWELL, BERKS, ENGLAND
HRSCC/S-	UNITED KINGDOM ATOMIC ENERGY AUTHORITY. RESEARCH GP. ATOMIC ENERGY RES. ESTAB., HARWELL, BERKS, ENGLAND
HRSP/S-	UNITED KINGDOM ATOMIC ENERGY AUTHORITY. RESEARCH GP. ATOMIC ENERGY RES. ESTAB., HARWELL, BERKS, ENGLAND
HR-TR-	HONEYWELL RESEARCH CENTER, HOPKINS, MINN.(TECH.REPT)
HRWP/(NO.)B-	UNITED KINGDOM ATOMIC ENERGY AUTHORITY. INDUSTRIAL GROUP. CAPENHURST WORKS, CHES., ENGLAND
HS-	(HISTORICAL SURVEY) (REF. 2)
HS-	GT. BRIT. ATOMIC EN. RES. ESTAB., HARWELL, BERKS, ENG
HS-	QUARTERMASTER CORPS (HISTORICAL SURVEYS)
HS-	STANFORD UNIV., CALIF. DEPT. OF MECHANICAL ENG.
HS-	UNITED AIRCRAFT CORP. HAMILTON STANDARD DIV., WINDSOR LOCKS, CONN.
HSA-	AUSTRALIA. AERO. RES. LABS.,FISHERMANS BEND, VICTORIA
HSA-	AUSTRALIA. WEAPONS RESEARCH ESTABLISHMENT
HSC-	HAZELTINE SERVICE CORP.
HS/CR-	UNITED KINGDOM ATOMIC ENERGY AUTHORITY. AUTHORITY HEALTH AND SAFETY BRANCH, RISLEY, LANCS, ENGLAND
HSD-TN-	HAWKER SIDDELEY DYNAMICS, LTD., HATFIELD, HERTS, ENG.
HSD-TP-	HAWKER SIDDELEY DYNAMICS, LTD., HATFIELD, HERTS, ENG.
HSE-	SPRINGFIELD ARMORY, MASS.
HSER-	HAMILTON STANDARD. ELECTRONICS DEPT., BROAD BROOK, CONN.
HSER-	HAMILTON STANDARD. ELECTRONICS DEPT., WINDSOR LOCKS, CONN.
HSIC-LAB-(NUMBER)-(YEAR)	HOEGANAES SPONGE IRON CORP., RIVERTON, N.J.
HSIC-PROD HIGH ALLOY-	HOEGANAES SPONGE IRON CORP., RIVERTON, N.J.
HSI-(R)-	ARMY BALLISTIC MISSILE AGENCY, REDSTONE ARSENAL, ALA.
HSIR-	HAMILTON STANDARD, WINDSOR LOCKS, CONN.
HSME-	HAMILTON STANDARD, WINDSOR LOCKS, CONN.
HSM-N(NUMBER)-(YEAR)	CHRYSLER CORP. MISSILE DIV., HUNTSVILLE, ALA.
HSM-R-	CHRYSLER CORP. SPACE DIV., HUNTSVILLE, ALA.
HSP-	UNITED AIRCRAFT CORP. HAMILTON STANDARD DIV., WINDSOR LOCKS, CONN.
HSPH-	HARVARD UNIV., BOSTON. SCHOOL OF PUBLIC HEALTH
HSRD-(YEAR)-	GEORGIA INST. OF TECH., ATLANTA. HEALTH SYSTEMS RESEARCH CENTER
HSR-FRM-	HUMAN SCIENCES RESEARCH, INC., MC LEAN, VA.
HSRI-B(NUMBER)-	MICHIGAN. UNIV., ANN ARBOR. HIGHWAY SAFETY RES. INST.
HSRP-	HARVARD UNIV., CAMBRIDGE, MASS.
HSR RM (YR.)(NO.)(LTR(S))	HUMAN SCIENCES RESEARCH, INC., ARLINGTON, VA.
HSR-RR-(YR.)(NO.)(LTRS)	HUMAN SCIENCES RESEARCH, INC., ARLINGTON, VA.
HSR-RR-	HUMAN SCIENCES RESEARCH, INC., MC LEAN, VA.
HSR TN (YR.)(NO.)(LTR(S))	HUMAN SCIENCES RESEARCH, INC., ARLINGTON, VA.
HSR-TN-(YEAR)-	HUMAN SCIENCES RESEARCH, INC., MC LEAN, VA.
HSSP-	HAMILTON STANDARD, WINDSOR LOCKS, CONN.
HSSTP-TR-	OAK RIDGE NATIONAL LAB., TENN. (HEAVY SECTION STEEL TECHNOLOGY PROGRAM)
HSTP-(YEAR)-	HAMILTON STANDARD, WINDSOR LOCKS, CONN.
HSVA-	HAMBURGISCHE SCHIFFBAU-VERSUCHSANSTALT
HT-	BENDIX AVIATION CORP.
HT-(YEAR)-	GARRETT CORP. AIRESEARCH MFG. DIV., LOS ANGELES
HT-(YEAR)-(NUMBER/NUMBER)	GARRETT CORP. AIRESEARCH MFG. DIV., LOS ANGELES
HT-	GT. BRIT. NATL. PHYSICAL LAB., TEDDINGTON, MIDDX.,ENG
HTC-	HARRIS TRANSDUCER CORP., SOUTHBURY, CONN.
HTC-	HUGHES TOOL CO., HOUSTON, TEXAS
HTC-(YEAR)-	HUGHES TOOL CO. AIRCRAFT DIV., CULVER CITY, CALIF.
HTC-(NO.)-(LETTER(S))-...	HUGHES TOOL CO. AIRCRAFT DIV., CULVER CITY, CALIF.
HTC-1109-(LETTER(S))-	HUGHES TOOL CO. AIRCRAFT DIV., CULVER CITY, CALIF.
HTC-AD-(YEAR)-	HUGHES TOOL CO. AIRCRAFT DIV., CULVER CITY, CALIF.
HTCBN-	HUNGARIAN TRADING CO.FOR BOOKS AND NEWSPAPERS-KULTURA BUDAPEST (TRANSLATIONS)
HTD-GM-	RES. + DEV. BD. COMM. ON GUIDED MISSILES, WASH., DC
HTGC/CSC/P.(NUMBER)	UNITED KINGDOM ATOMIC ENERGY AUTHORITY. RESEARCH GP. ATOMIC ENERGY RES. ESTAB., HARWELL, BERKS, ENGLAND
HTGC-CWP/MEMO-	UNITED KINGDOM ATOMIC ENERGY AUTHORITY. RESEARCH GP. ATOMIC ENERGY RES. ESTAB., HARWELL, BERKS, ENGLAND
HTGC-FEWP/MEMO-	UNITED KINGDOM ATOMIC ENERGY AUTHORITY. RESEARCH GP. ATOMIC ENERGY RES. ESTAB., HARWELL, BERKS, ENGLAND
HTGC/FPWP/P-	UNITED KINGDOM ATOMIC ENERGY AUTHORITY. RESEARCH GP. ATOMIC ENERGY RES. ESTAB., HARWELL, BERKS, ENGLAND
HTGC/FRWP/M	UNITED KINGDOM ATOMIC ENERGY AUTHORITY. RESEARCH GP. ATOMIC ENERGY RES. ESTAB., HARWELL, BERKS, ENGLAND
HTGC/FRWP/MEMO-	UNITED KINGDOM ATOMIC ENERGY AUTHORITY. RESEARCH GP. ATOMIC ENERGY RES. ESTAB., HARWELL, BERKS, ENGLAND
HTGC-MEMO-	UNITED KINGDOM ATOMIC ENERGY AUTHORITY. RESEARCH GP. ATOMIC ENERGY RES. ESTAB., HARWELL, BERKS, ENGLAND
HTGC-MT/MEMO-	UNITED KINGDOM ATOMIC ENERGY AUTHORITY. RESEARCH GP. ATOMIC ENERGY RES. ESTAB., HARWELL, BERKS, ENGLAND
HTGC/P-	UNITED KINGDOM ATOMIC ENERGY AUTHORITY. RESEARCH GP. ATOMIC ENERGY RES. ESTAB., HARWELL, BERKS, ENGLAND
HTGDL-	PURDUE UNIV., LAFAYETTE, IND.
HTL-TR-	MINNESOTA. UNIV., MINNEAPOLIS. HEAT TRANSFER LAB.
HTL-TR-(YEAR)-	MINNESOTA. UNIV., MINNEAPOLIS. HEAT TRANSFER LAB.
HTM-QPR-	HIGH TEMPERATURE MATERIALS, INC., BRIGHTON, MASS.
HTP-	AEROJET-GENERAL CORP., SACRAMENTO, CALIF.
HTR-	AEROJET-GENERAL CORP., SACRAMENTO, CALIF.
HTR/(LETTERS)/P-	UNITED KINGDOM ATOMIC ENERGY AUTHORITY. RESEARCH GP. ATOMIC ENERGY RES. ESTAB., HARWELL, BERKS, ENGLAND
HTR/CWP/P-	UNITED KINGDOM ATOMIC ENERGY AUTHORITY. RESEARCH GP. ATOMIC ENERGY RES. ESTAB., HARWELL, BERKS, ENGLAND
HTRDC/-	UNITED KINGDOM ATOMIC ENERGY AUTHORITY. RESEARCH GP. ATOMIC ENERGY RES. ESTAB., HARWELL, BERKS, ENGLAND
HTRDC/P-	UNITED KINGDOM ATOMIC ENERGY AUTHORITY. RESEARCH GP. ATOMIC ENERGY RES. ESTAB., HARWELL, BERKS, ENGLAND
HTR/FEWP-MEMO-	UNITED KINGDOM ATOMIC ENERGY AUTHORITY. RESEARCH GP. ATOMIC ENERGY RES. ESTAB., HARWELL, BERKS, ENGLAND
HTR-FEWP/P-	UNITED KINGDOM ATOMIC ENERGY AUTHORITY. RESEARCH GP. ATOMIC ENERGY RES. ESTAB., HARWELL, BERKS, ENGLAND
HTR/PWP/MEMO-	UNITED KINGDOM ATOMIC ENERGY AUTHORITY. RESEARCH GP. ATOMIC ENERGY RES. ESTAB., HARWELL, BERKS, ENGLAND
HTR/PWP/P-	UNITED KINGDOM ATOMIC ENERGY AUTHORITY. RESEARCH GP. ATOMIC ENERGY RES. ESTAB., HARWELL, BERKS, ENGLAND
HTVL/LMSC-	LOCKHEED MISSILES AND SPACE CO., HUNTSVILLE, ALA.
HU-	HARVARD UNIV., CAMBRIDGE, MASS.
HU ABL 1045-	HARVARD UNIV., CAMBRIDGE, MASS. AMER. BRITISH LAB.
HU ABL 1045-MR-	HARVARD UNIV., CAMBRIDGE, MASS. AMER. BRITISH LAB.
HU ABL 1045-TM-	HARVARD UNIV., CAMBRIDGE, MASS. AMERICAN BRITISH LAB. (TECHNICAL MEMORANDA)
HU ARL TM-	HARVARD UNIV., CAMBRIDGE, MASS. ACOUSTICS RESEARCH LAB. (TECHNICAL MEMORANDUM)
HU-BHMO-	HARVARD UNIV., CAMBRIDGE, MASS. BLUE HILL METEOR- OLOGICAL OBSERVATORY
HU-CL-PR-	HARVARD UNIV., CAMBRIDGE, MASS. CRUFT LAB.
HU CL TR-	HARVARD UNIV., CAMBRIDGE, MASS. CRUFT LAB.
HU-DEAP-PR-	HARVARD UNIV., CAMBRIDGE, MASS. DIVISION OF ENG. AND APPLIED PHYSICS (PROGRESS REPORT)
HU-DEAP-SPR-	HARVARD UNIV., CAMBRIDGE, MASS. DIVISION OF ENG. AND APPLIED PHYSICS
HU-DEAP-TM-	HARVARD UNIV., CAMBRIDGE, MASS. DIVISION OF ENG. AND APPLIED PHYSICS

HU-DEAP-TR-
    HARVARD UNIV., CAMBRIDGE, MASS. DIVISION OF ENG.
    AND APPLIED PHYSICS (TECHNICAL REPORT)
HU-DL-
    HARVARD UNIV., CAMBRIDGE, MASS. DUNBAR LAB.
HU-DM-
    HARVARD UNIV., CAMBRIDGE, MASS. DEPT. OF MINERALOGY
HUD-MP-
    DEPARTMENT OF HOUSING AND URBAN DEV., WASHINGTON, D.C.
HU-EA-
    HARVARD UNIV., CAMBRIDGE, MASS. DIVISION OF ENG.
    AND APPLIED PHYSICS
HUF-
    PHILLIPS PETROLEUM CO. ATOMIC ENERGY DIV., IDAHO
    FALLS (CODE IS AUTHOR'S INITIALS)
HU FL-
    HARVARD UNIV., CAMBRIDGE, MASS.
HUGHES TM-
    HUGHES AIRCRAFT CO., CULVER CITY, CALIF.
HU-LLP-TR-
    HARVARD UNIV., CAMBRIDGE, MASS. LYMAN LAB. OF PHYSICS
HUM-
    ESSO RESEARCH AND ENGINEERING CO., BAYTOWN, TEX.
HUM-(NUMBER)-P
    ESSO RESEARCH AND ENGINEERING CO., BAYTOWN, TEX.
HUM-
    HUMBLE OIL AND REFINING CO., BAYTOWN, TEX.
HUM-(NUMBER)-P
    HUMBLE OIL AND REFINING CO., BAYTOWN, TEX.
HUM-
    VIRGINIA. UNIV., CHARLOTTESVILLE
HUMBRO-TR-
    GEORGE WASHINGTON UNIV., ARLINGTON, VA. HUMAN
    RESOURCES RESEARCH OFFICE
HUMRRO-
    GEORGE WASHINGTON UNIV., ARLINGTON, VA. HUMAN
    RESOURCES RESEARCH OFFICE
HUMRRO-RBP-D(NO.-YR.)-
    GEORGE WASHINGTON UNIV., ARLINGTON, VA. HUMAN
    RESOURCES RESEARCH OFFICE
HUMRRO-TR-(YEAR)-
    GEORGE WASHINGTON UNIV., ARLINGTON, VA. HUMAN
    RESOURCES RESEARCH OFFICE
HUNTERS PT-TR-(NUMBER-YR)
    SAN FRANCISCO BAY NAVAL SHIPYARD. HUNTER'S POINT DIV.
HU PAL-
    HARVARD UNIV., CAMBRIDGE, MASS. PSYCHO-ACOUSTIC LAB.
HU PAL IC-
    HARVARD UNIV., CAMBRIDGE, MASS. PSYCHO-ACOUSTIC LAB.
    (INFORMAL COMMUNICATION)
HU PAL IR-
    HARVARD UNIV., CAMBRIDGE, MASS. PSYCHO-ACOUSTIC LAB.
    (INVENTION REPORTS)
HU RRL-
    HARVARD UNIV., CAMBRIDGE, MASS. RADIO RESEARCH LAB.
HU RRL 411-IB-
    HARVARD UNIV., CAMBRIDGE, MASS. RADIO RESEARCH LAB.
HU RRL 411-MTR-
    HARVARD UNIV., CAMBRIDGE, MASS. RADIO RESEARCH LAB.
    (MATERIALS TESTING REPORTS)
HU RRL 411-TM-
    HARVARD UNIV., CAMBRIDGE, MASS. RADIO RESEARCH LAB.
    (TECHNICAL MEMORANDA)
HU RRL 411-TR-
    HARVARD UNIV., CAMBRIDGE, MASS. RADIO RESEARCH LAB.
    (TEST REPORTS)
HU RRL D-
    HARVARD UNIV., CAMBRIDGE, MASS. RADIO RESEARCH LAB.
    (DISCLOSURE)
HU RRL TS-
    HARVARD UNIV., CAMBRIDGE, MASS. RADIO RESEARCH LAB.
    (TENTATIVE SPECIFICATION)
HUTR-
    HARVARD UNIV., CAMBRIDGE, MASS. CRUFT LAB.
HU U-
    HULL, ENGLAND. UNIV.
HU USL RPT-
    HARVARD UNIV., CAMBRIDGE, MASS. UNDERWATER SOUND LAB.
HU USL RPT (LETTER)-
    HARVARD UNIV., CAMBRIDGE, MASS. UNDERWATER SOUND LAB.
HUX-(CONTRACT NO.)-
    HARVARD UNIV., CAMBRIDGE, MASS. (MANY DEPTS/SCHOOLS)
HUYCK-
    HUYCK CORP. (PROGRESS REPORT)
HV-
    COAST AND GEODETIC SURVEY, ROCKVILLE, MD. (HOURLY
    VALUES)
HVEC-TN-
    HIGH VOLTAGE ENGINEERING CORP., BURLINGTON, MASS.
HVI-
    HIGH VOLTAGE ENGINEERING CORP., BURLINGTON, MASS.
HVL-(YEAR)-
    GENERAL ELECTRIC CO., PITTSFIELD, MASS.
HW-
    BATTELLE-NORTHWEST. PACIFIC NORTHWEST LAB., RICHLAND,
    WASH.
HW-
    HANFORD ATOMIC PRODUCTS OPERATION, RICHLAND, WASH.
HW-DWL-
    DOUGLAS UNITED NUCLEAR, INC., RICHLAND, WASH.
HWN-
    HANFORD WORKS, RICHLAND, WASH.
HW OPERATING STANDARDS-
    HANFORD WORKS, RICHLAND, WASH.
HW-OREMET-
    OREGON METALLURGICAL CORP., ALBANY
HWR-
    HANFORD WORKS, RICHLAND, WASH.
HW-R-
    HANFORD WORKS, RICHLAND, WASH.

HWR-FEMWG/P-(YEAR)-
    UNITED KINGDOM ATOMIC ENERGY AUTHORITY. REACTOR
    GROUP, CULCHETH, LANCS, ENGLAND
H-WRG-
    TENNESSEE EASTMAN CORP., OAK RIDGE, TENN.
HWS-
    DOUGLAS UNITED NUCLEAR, INC., RICHLAND, WASH.
HWS-
    HANFORD ATOMIC PRODUCTS OPERATION, RICHLAND, WASH.
HW-SA-
    HANFORD ATOMIC PRODUCTS OPERATION, RICHLAND, WASH.
HWT-
    SANDIA CORP., ALBUQUERQUE, N. MEX.
HW TECHNICAL MANUALS-
    HANFORD WORKS, RICHLAND, WASH.
HW-TR-
    HANFORD ATOMIC PRODUCTS OPERATION, RICHLAND, WASH.
HWWP-
    GT. BRIT. ATOMIC EN. RES. ESTAB., HARWELL, BERKS, ENG
HX(NUMBER)-
    PHILCO-FORD CORP. COMMUNICATIONS AND ELECTRONICS
    DIV., WILLOW GROVE, PA.
HY-
    ABEX CORP. HYDRODYNAMICS RES. CTR., COLUMBUS, OHIO
HY-
    DENMARK. TEKNISKE HØJSKOLE, LYNGBY. HYDRO-AND AERO-
    DYNAMICS LAB. HYDRODYNAMICS SECTION
HYC-
    HYCON MFG. CO., PASADENA, CALIF.
HYCON-
    HYCON MFG. CO., PASADENA, CALIF.
HYD-
    BUREAU OF RECLAMATION (INTERIOR). COMMISSIONERS
    OFFICE, DENVER
HYD-
    TEXAS. UNIV., AUSTIN
HYDRA-SUMMARY-
    NAVAL RESEARCH LAB., WASHINGTON, D.C.
HYDROLOGY PAPERS-
    COLORADO STATE UNIV., FORT COLLINS. DEPT. OF CIVIL
    ENGINEERING
HYDRO-TR-
    HYDRONAUTICS, INC., LAUREL, MD.
HYDRO-TR-
    ILLINOIS. STATE WATER SURVEY, URBANA
HYRC-
    HYDROSPACE RES. CORP., ROCKVILLE, MD.
HYTRON-
    CBS-HYTRON, DANVERS, MASS.
HZ-(NUMBER)Q-(YEAR)-
    DOW CHEMICAL CO., MIDLAND, MICH. (LMH-1/HYDRAZINE
    HETEROGENEOUS PROPELLANT DEVELOPMENT)
HZ-
    HORIZONS, INC., CLEVELAND
HZ-
    KERNFORSCHUNGSANLAGE JUELICH. INSTITUT FUER REAKTOR-
    WERKSTOFFE, GERMANY

I-	(INFORMAL REPORT) (REF. 2)
I-	(INSTALLATION INSTRUCTIONS) (REF. 2)
I-	(INTERIM REPORT) (REF. 2)
I-	(INTERNAL REPORT) (REF. 2)
I-(NO.)-(NO.)-/(NO.)	ARMOUR RESEARCH FOUNDATION, CHICAGO
I-	ARMY, WASHINGTON, D.C.
I-	FRANKLIN INST. LABS. FOR RES. + DEV., PHILADELPHIA
I-	FRANKLIN INST. RESEARCH LABS., PHILADELPHIA
I-(NO.)-871A-(NO.)/(NO.)	GT. BRIT. ATOMIC EN. RES. ESTAB., HARWELL, BERKS, ENG
I-	IKOR, INC., WALTHAM, MASS.
I-	ILLINOIS. UNIV., URBANA. COORDINATED SCIENCE LAB.
I-(YEAR)-	PURDUE UNIV., LAFAYETTE, IND.
I-(NUMBER)-	SYLVANIA ELECTRONIC SYSTEMS-EAST. APPLIED RESEARCH LAB., WALTHAM, MASS.
I-(NO.)/(NO.)	TRACERLAB, INC.
I-(NO.)-(NO.)/(NO.)	TRACERLAB, INC.
I(LETTER)C-	OFFICE OF SCIENTIFIC RESEARCH AND DEVELOPMENT
I-A(NUMBER)-	FRANKLIN INST. LABS. FOR RES. + DEV., PHILADELPHIA
IA-	ISRAEL. ATOMIC ENERGY COMMISSION (ALL LOCATIONS)
IA-	NAPLES. UNIVERSITA. ISTITUTO DI AERODINAMICA
IAD-	DEPT. OF STATE. DIV. OF ACQUISITION AND DISTRIBUTION
IAD-	LOCKHEED AIRCRAFT CORP. (ALL LOCATIONS) (INTRA-DEPARTMENTAL COMMUNICATION)
IAE-	U.S.S.R. GOSUDARSTVENNYI KOMITET PO ISPOL'ZOVANIYU ATOMNOI ENERGII. INSTITUT ATOMNOI ENERGII, MOSCOW
IAE-	UNITED KINGDOM ATOMIC ENERGY AUTHORITY. DEVELOPMENT AND ENGINEERING GROUP, CAPENHURST, CHES., ENGLAND
IAEA-	INTERNATIONAL ATOMIC ENERGY AGENCY, VIENNA
IAEA-CN-(NUMBER/NUMBER)	INTERNATIONAL ATOMIC ENERGY AGENCY, VIENNA (CONFERENCE PAPER)
IAEA-INIS-	INTERNATIONAL ATOMIC ENERGY AGENCY. INTERNATIONAL NUCLEAR INFORMATION SYSTEM, VIENNA
IAEA-NPR-	INTERNATIONAL ATOMIC ENERGY AGENCY, VIENNA
IAEA-R-	INTERNATIONAL ATOMIC ENERGY AGENCY, VIENNA
IAEA-SM-	INTERNATIONAL ATOMIC ENERGY AGENCY, VIENNA
IAERU-	RIKKYO UNIV., YOKOSUKA, JAPAN. INST. FOR ATOMIC ENERGY
IAK-	KERNFORSCHUNGSZENTRUM KARLSRUHE. INSTITUT FUER ANGEWANDTE KERNPHYSIK, GERMANY
IAM-	ROYAL CANADIAN AIR FORCE. INST. OF AVIATION MEDICINE
IAN-	COLOMBIA. INSTITUTO DE ASUNTOS NUCLEARES, BOGOTA
IAN-A-	COLOMBIA. INSTITUTO DE ASUNTOS NUCLEARES. DIVISION OF REACTORS, BOGOTA
IAN-ARI-	COLOMBIA. INSTITUTO DE ASUNTOS NUCLEARES, BOGOTA
IAN-B-	COLOMBIA. INSTITUTO DE ASUNTOS NUCLEARES. DIVISION OF MEDICINE AND BIOLOGY, BOGOTA
IAN-F-	COLOMBIA. INSTITUTO DE ASUNTOS NUCLEARES, BOGOTA
IAN-Q-	COLOMBIA. INSTITUTO DE ASUNTOS NUCLEARES, BOGOTA
IAN-RS-ERA-	COLOMBIA. INSTITUTO DE ASUNTOS NUCLEARES. SECCION DE RADIOFISICA SANITARIA, BOGOTA
IAN-SR-	COLOMBIA. INSTITUTO DE ASUNTOS NUCLEARES, DIVISION DE RADIOFISICA SANITARIA, BOGOTA
IAP-	GT. BRIT. ROYAL AIRCRAFT EST., FARNBOROUGH, HANTS, ENG
IARD-	AMERICAN INST. OF PHYSICS. INFORMATION ANALYSIS AND RETRIEVAL DIV., N.Y.C.
IARD-(YEAR)-	AMERICAN INST. OF PHYSICS. INFORMATION ANALYSIS AND RETRIEVAL DIV., N.Y.C.
IAS-	AMERICAN INST. OF AERONAUTICS + ASTRONAUTICS, N.Y.C.
IAS-	INSTITUTE FOR ADVANCED STUDY, PRINCETON, N.J.
IAS-FR-(MONTH)-(YEAR)	INSTITUTE FOR ADVANCED STUDY, PRINCETON, N.J.
IATA-	INTERNATIONAL AIR TRANSPORT ASSOCIATION, CANADA
IAWR-	CHICAGO. UNIV. INST. FOR AIR WEAPONS RESEARCH
IAWR(YR)-	CHICAGO. UNIV. INST. FOR AIR WEAPONS RESEARCH
IB-	(INFORMATION BULLETIN) (REF. 2)
IB-	(INSTRUCTION BOOK) (REF. 2)
I-B(NUMBER)-	FRANKLIN INST. LABS. FOR RES. + DEV., PHILADELPHIA
IB-	GT. BRIT. ADVISORY COUNCIL ON SCI. RES. + TECH. DEV.
IB-	OFFICE OF TECHNICAL SERVICES, WASHINGTON, D.C.
IB-	VETERANS ADMINISTRATION, WASHINGTON, D.C.
IB EEIS-	ALLIED FORCES ENEMY EQUIPMENT INTELLIGENCE SERVICE. INDIA-BURMA THEATER
IB EEIS CWS PTR-	ALLIED FORCES ENEMY EQUIPMENT INTELL. SERVICE. INDIA-BURMA THEATER. CHEM. WARFARE SEC. (PRELIM. TECH. RPT.)
IB EEIS CWS TR-	ALLIED FORCES ENEMY EQUIPMENT INTELL. SERVICE. INDIA-BURMA THEATER. CHEM. WARFARE SEC. (TECH. REPT.)
IB EEIS SIG PR-	ALLIED FORCES ENEMY EQUIPMENT INTELLIGENCE SERVICE. INDIA-BURMA THEATER. SIGNAL SECTION. (PRELIM. RPT.)
IBEG-	ALASKA. UNIV., COLLEGE. INST. OF BUSINESS, GOVERNMENT AND RESEARCH
IBJ-(NO./ROMAN NO./LTR.)	POLAND. INSTITUTE OF NUCLEAR RESEARCH, WARSAW
IBK-	YUGOSLAVIA. INSTITUT ZA NUKLFARNE NAUKE BORIS KIDRIC, BELGRADE
IBM-	GENERAL ELECTRIC CO. AIRCRAFT NUCLEAR PROPULSION DEPT., CINCINNATI
IBM-(YEAR)-	INTERNATIONAL BUSINESS MACHINES CORP. (ALL LOCATIONS)
IBM-	INTERNATIONAL BUSINESS MACHINES CORP. (ALL LOCATIONS)
IBM-(NO. NO. NO.)	INTERNATIONAL BUSINESS MACHINES CORP. (ALL LOCATIONS)
IBMC-	INTERNATIONAL BUSINESS MACHINES CORP., OWEGO, N.Y.
IBMC-	INTERNATIONAL BUSINESS MACHINES CORP. FEDERAL SYSTEMS DIV., OWEGO, N.Y.
IBM-CD-(NO.)-(NO.)-	INTERNATIONAL BUSINESS MACHINES CORP. ELECTRONICS SYSTEMS CENTER, OWEGO, N.Y.
IBM-CD-	INTERNATIONAL BUSINESS MACHINES CORP. FEDERAL SYSTEMS DIV., GAITHERSBURG, MD.
IBM-CD-(NUMBER)-(NO.)-	INTERNATIONAL BUSINESS MACHINES CORP. FEDERAL SYSTEMS DIV., GAITHERSBURG, MD.
IBM-ECBX-	INTERNATIONAL BUSINESS MACHINES CORP. MILITARY PRODUCTS DIV., KINGSTON, N.Y.
IBM ECPX-	INTERNATIONAL BUSINESS MACHINES CORP., KINGSTON, N.Y.
IBM-FSC-(YEAR)-	INTERNATIONAL BUSINESS MACHINES CORP., GAITHERSBURG, MD
IBM-GF(NUMBER)-	INTERNATIONAL BUSINESS MACHINES CORP., ALBUQUERQUE, NM
IBM IER-(NOS.)	INTERNATIONAL BUSINESS MACHINES CORP., ENDICOTT, N.Y. (INTERIM ENGINEERING REPORT)
IBM-NJ-	INTERNATIONAL BUSINESS MACHINES CORP. RESEARCH LAB., SAN JOSE, CALIF.
IBM-NZ-	INTERNATIONAL BUSINESS MACHINES CORP., ZURICH
IBM-RC-	IBM WATSON RESEARCH CENTER, YORKTOWN HEIGHTS, N.Y.
IBM-RJ-	INTERNATIONAL BUSINESS MACHINES CORP. RESEARCH DIV. SAN JOSE LAB., CALIF.
IBM-RW-	IBM WATSON RESEARCH CENTER, YORKTOWN HEIGHTS, N.Y.
IBM TN (NO. NO. NO.)	INTERNATL. BUSINESS MACHINES CORP., POUGHKEEPSIE, N.Y.
IBM-TR-	INTERNATIONAL BUSINESS MACHINES CORP. (ALL LOCATIONS)
IBT-	BUREAU OF SHIPS (INSTRUCTION BOOK, GAROD PART)
IBTC-	INTER-BUREAU TECHNICAL COMMITTEE (NAVY), WASH., D.C.
IC-	(INFORMAL COMMUNICATION) (REF. 2)
IC-	(INFORMATION CIRCULAR) (REF. 2)
I.C.-	BUREAU OF MINES (INFORMATION CIRCULAR)
IC-	BUREAU OF MINES
IC-	FAIRCHILD ENGINE AND AIRPLANE CORP. NEPA DIV., OAK RIDGE, TENN. (REF. 31)
IC-	IMPERIAL COLLEGE OF SCIENCE AND TECHNOLOGY, LONDON. DEPT. OF MECHANICAL ENGINEERING
IC-	IMPERIAL COLLEGE OF SCIENCE AND TECHNOLOGY, LONDON. DEPT. OF PHYSICS
IC/	INTERNATIONAL CENTRE FOR THEORETICAL PHYSICS, TRIESTE, ITALY

IC-(YEAR)-
INTERNATIONAL CENTRE FOR THEORETICAL PHYSICS, TRIESTE, ITALY

IC-(INITIALS)-(YR)-
INTERNUCLEAR CO., INC., CLAYTON, MO.

IC-AERO-(YEAR)-
IMPERIAL COLLEGE OF SCIENCE AND TECHNOLOGY, LONDON. DEPT. OF AERONAUTICS

ICAF-
INDUSTRIAL COLL. OF THE ARMED FORCES, WASHINGTON,D.C.

ICAF-L-(YEAR)-
INDUSTRIAL COLL. OF THE ARMED FORCES, WASHINGTON,D.C.

ICAF-M (YR.)(NO.)
INDUSTRIAL COLL. OF THE ARMED FORCES, WASHINGTON,D.C.

ICAF-TEMPER-
INDUSTRIAL COLL. OF THE ARMED FORCES, WASHINGTON,D.C.

ICAO-
INTERNATIONAL CIVIL AVIATION ORGANIZATION, MONTREAL

IC-AR-(YEAR-YEAR)
INTERNATIONAL CENTRE FOR THEORETICAL PHYSICS, TRIESTE, ITALY

ICAS-
INTERDEPARTMENTAL COMMITTEE FOR ATMOSPHERIC SCIENCES, WASHINGTON, D.C.

ICAS-PAPER-(YEAR)-
INTERDEPARTMENTAL COMMITTEE FOR ATMOSPHERIC SCIENCES, WASHINGTON, D.C.

ICC-
INTERSTATE COMMERCE COMMISSION, WASHINGTON, D.C.

ICC-
NATIONAL RESEARCH COUNCIL. INSECT CONTROL COMMITTEE

ICC AB-
NATIONAL RESEARCH COUNCIL. INSECT CONTROL COMMITTEE (ABSTRACT BULLETIN)

ICD-(NUMBER)-(YEAR)-
AVCO CORP. AVCO MISSILE, SPACE + ELECTRONICS GROUP, WILMINGTON, MASS.

ICE-
CANADA. DEPT. OF TRANSPORT.METEOROLOGICAL BR.,TORONTO

ICF-
IDAHO OPERATIONS OFFICE. HEALTH AND SAFETY DIV., AEC, IDAHO FALLS

IC/HEPR/(YEAR)/
INTERNATIONAL CENTRE FOR THEORETICAL PHYSICS, TRIESTE, ITALY

ICI-
IMPERIAL CHEMICAL INDUSTRIES, LTD., LONDON

ICO PAMPHLET-
INTERAGENCY COMMITTEE ON OCEANOGRAPHY, WASH., D.C.

ICP-
ARMY, WASH., D.C. (ATOMIC INCIDENT CONTROL PLAN)

IC/P-
UNITED KINGDOM ATOMIC ENERGY AUTHORITY. INDUSTRIAL GROUP. CULCHETH LABS., LANCS, ENGLAND

ICR-
JOHNS HOPKINS UNIV., BALTIMORE. INST. FOR COOP. RES.

ICR-
PENNSYLVANIA. UNIV., PHILA. INST. FOR COOP. RES.

ICRPG-
INTERAGENCY CHEMICAL ROCKET PROPULSION GROUP

ICRPG-ROUND ROBIN-
INTERAGENCY CHEMICAL ROCKET PROPULSION GROUP

ICRU-
INTERNATIONAL COMMISSION ON RADIATION UNITS AND MEASUREMENTS, WASHINGTON, D.C.

ICS-
ITT COMMUNICATION SYSTEMS, INC., PARAMUS, N.J.

ICS-(NUMBER)-TR-
ITT COMMUNICATION SYSTEMS, INC., PARAMUS, N.J.

ICS-(YR)-TR-(NO.)-
ITT COMMUNICATION SYSTEMS, INC., PARAMUS, N.J.

ICSC-(NUMBERS-LTRS.-YR)
INTERIM COMMUNICATIONS SATELLITE COMMITTEE

ICSC-
INTERNATIONAL TELECOMMUNICATIONS SATELLITE CONSORTIUM (INTELSAT). INTERIM COMMUNICATIONS SATELLITE COMM.

ICSM (NUMBER)-P-
PANAMA CANAL CO., WASHINGTON, D.C.

ICS MEMO-
PANAMA CANAL CO., WASHINGTON, D.C.

ICS MEMO (NUMBER)-P-
PANAMA CANAL CO., WASHINGTON, D.C.

ICSRD-
INTERDEPARTMENTAL COMM. ON SCI. RES. + DEV., WASH.,DC

ICT-(NUMBER-YEAR)
GERMANY. INSTITUT FUER CHEMIE DER TREIBSTOFFE, BERGHAUSEN

ICTP-
INTERNATIONAL CENTRE FOR THEORETICAL PHYSICS, TRIESTE, ITALY

ID-(YEAR)-
AMERICAN INST. OF PHYSICS. INFORMATION DIV., N.Y.C.

ID-
COMMITTEE ON MEDICAL RES. (INFECTIOUS DISEASES RPTS.)

ID-(NUMBERS)-
NORTHROP AIRCRAFT, INC., HAWTHORNE, CALIF.

ID-
OFFICE OF THE CHIEF OF NAVAL OPERATIONS. INTELL. DIV

IDA-
INSTITUTE FOR DEFENSE ANALYSES, ARLINGTON, VA.

IDA-
OAK RIDGE OPERATIONS OFFICE,AEC. ISOTOPES DIV., TENN.

IDA-ARPA-R-(YEAR)-
INSTITUTE FOR DEFENSE ANALYSES, ARLINGTON, VA.

IDA-ARPA-TR-(YEAR)-
INSTITUTE FOR DEFENSE ANALYSES, ARLINGTON, VA.

IDAC-RT-
ITALY. CONSIGLIO NAZIONALE DELLE RICERCHE, ROME

IDAC-RT-
ITALY. CONSIGLIO NAZIONALE DELLE RICERCHE. ISTITUTO DI ACUSTICA, ROME

IDAC-TR-
ITALY. CONSIGLIO NAZIONALE DELLE RICERCHE, ROME

IDA/HQ-(YEAR)-
INSTITUTE FOR DEFENSE ANALYSES, ARLINGTON, VA.

IDA/HQ-
INSTITUTE FOR DEFENSE ANALYSES, ARLINGTON, VA.

IDA/HQ-JDR-(YEAR)-
INSTITUTE FOR DEFENSE ANALYSES, ARLINGTON, VA. (JOURNAL OF DEFENSE RESEARCH)

IDA-LOG-
INSTITUTE FOR DEFENSE ANALYSES. WEAPONS SYSTEMS EVALUATION DIV., WASHINGTON, D.C.

IDA-R-
INSTITUTE FOR DEFENSE ANALYSES. RESEARCH AND ENGINEERING SUPPORT DIV., ARLINGTON, VA.

IDA-RESG-TR-
INSTITUTE FOR DEFENSE ANALYSES, ARLINGTON, VA.

IDA-RP-P-
INSTITUTE FOR DEFENSE ANALYSES. RESEARCH AND ENGINEERING SUPPORT DIV., ARLINGTON, VA.

IDA-SRD-
INSTITUTE FOR DEFENSE ANALYSES. SCIENCE AND TECHNOLOGY DIV., ARLINGTON, VA.

IDA-TN-
INSTITUTE FOR DEFENSE ANALYSES, ARLINGTON, VA. (TECHNICAL NOTE)

IDB-
(INTERNAL DISTRIBUTION BULLETIN) (REF. 2)

IDB-
OAK RIDGE OPERATIONS OFFICE,AEC. ISOTOPES DIV., TENN.

IDC-
INFORMATION DYNAMICS CORP., READING, MASS.

IDC-
INTERDEPARTMENTAL COMMITTEE FOR THE ACQUISITION OF FOREIGN PUBLICATIONS

IDC-
NAVAL ORDNANCE LAB. INFLUENCE DIV. (COMMUNICATION)

IDC-
OAK RIDGE OPERATIONS OFFICE,AEC. ISOTOPES DIV., TENN.

IDC F-
INTERDEPARTMENTAL COMMITTEE FOR THE ACQUISITION OF FOREIGN PUBLICATIONS (ABSTRACTS ON ENEMY OIL, FUELS AND LUBRICANTS)

IDCK-
NETHERLANDS. TECHNISCH DOCUMENTATIE EN INFORMATIE CENTRUM VOOR DE KRIJGSMACHT, THE HAGUE

IDC R-
INTERDEPARTMENTAL COMMITTEE FOR THE ACQUISITION OF FOREIGN PUBLICATIONS (SUBJECT INDEX)

IDD-
OAK RIDGE OPERATIONS OFFICE,AEC. ISOTOPES DIV., TENN.

IDE-
OAK RIDGE OPERATIONS OFFICE,AEC. ISOTOPES DIV., TENN.

IDEP-
INTERAGENCY DATA EXCHANGE PROGRAM. (SPONSORED BY ARMY, NAVY, AIR FORCE, NASA, AND CANADIAN MILITARY ELECTRONICS STANDARDS AGENCY TO MINIMIZE DUPLICATE TESTS OF PARTS, COMPONENTS AND MATERIALS)

ID GS-
INTELLIGENCE DIV., GENERAL STAFF, WASHINGTON, D.C.

IDIC-
GT. BRIT. FUNGICIDE + INSECTICIDE RESEARCH CO-ORDINATION SERVICE

IDL-
INSTRUMENT DEV. LABS., INC., NEEDHAM HEIGHTS, MASS.

IDLM-
GT. BRIT. CAPENHURST WORKS, CHES., ENGLAND (REF. 36)

ID-M-
COLUMBIA UNIV., N.Y.C.

IDO-
IDAHO OPERATIONS OFFICE, AEC, IDAHO FALLS (REF.4)

IDP-
ATOMIC ENERGY COMMISSION. ISOTOPE DEV.PROGRAM,WASH,DC

IDP-
NAVAL ORDNANCE TEST STATION, CHINA LAKE, CALIF. (INTERNAL DISTRIBUTION PUBLICATION)

IDR-
(INTERIM DEVELOPMENT REPORT) (REF. 2)

IDR-
(INTERNAL DEVELOPMENT REPORT) (REF. 2)

IDR-D(NUMBER)-
COLLINS RADIO CO., DALLAS

IDRES-
FRANKLIN INST. RESEARCH LABS. IDRES INFORMATION CENTER, PHILADELPHIA (INDUSTRIAL RESEARCH)

IDS-
NAVAL AIR DEV. CTR. AERO. INSTRS. LAB.,JOHNSVILLE,PA.

IDS-
NAVAL AIRCRAFT FACTORY. INSTRS. DEV. SEC., PHILA.

ID-SPEC-
CALIFORNIA. UNIV., BERKELEY. RADIATION LAB. (REF. 9)

IE-
GT. BRIT. DIRECTORATE OF CHEMICAL DEFENCE RES. + DEV.

IEA-
JAPAN. INSTITUTE OF PHYSICAL AND CHEMICAL RES., TOKYO

IEA-
SAO PAULO, BRAZIL. UNIVERSIDADE. INSTITUTO DE ENERGIA ATOMICA

IEA-INF-
SAO PAULO, BRAZIL. UNIVERSIDADE. INSTITUTO DE ENERGIA ATOMICA

IEA/RQ-
SAO PAULO, BRAZIL. UNIVERSIDADE. INSTITUTO DE ENERGIA ATOMICA

I+EC-
INDUSTRIAL AND ENGINEERING CHEMISTRY. RESEARCH RESULTS SERVICE, WASHINGTON, D.C.

IEC-	INTERSTATE ENGINEERING CORP., EL SEGUNDO, CALIF.
IED-	PICATINNY ARSENAL, DOVER, N.J.
IED-BP-	DU PONT DE NEMOURS (E.I.) + CO., WILMINGTON, DEL.
IED-G(NUMBER)	NAVAL ORDNANCE LAB., CORONA, CALIF.
IEEE-	INSTITUTE OF ELECTRICAL AND ELECTRONICS ENGINEERS, NYC
IEEE-A(NUMBER)-	INSTITUTE OF ELECTRICAL AND ELECTRONICS ENGINEERS, NYC
IEEE PAPER (NO)-PP-(YR)	INSTITUTE OF ELECTRICAL AND ELECTRONICS ENGINEERS, NYC
IEEI-	INTERNATIONAL ELECTRONICS ENGINEERING, INC., WASH., DC
IEG-	FLORIDA STATE UNIV., TALLAHASSEE
IEG-(NO.) SCI. MEMO-	NATIONAL INSTITUTES OF HEALTH. INFORMATION EXCHANGE GROUP NUMBER- , BETHESDA, MD.
IEI-	COLUMBIA UNIV., N.Y.C.
IEL-	ROHM AND HAAS CO., PHILADELPHIA
IEMC-	INTERNATIONAL ELECTRONICS MANUFACTURING CO., WASH., DC
IEO-	INTERNATIONAL ELECTRIC CORP., PARAMUS, N.J.
IEPR-	EDGEWOOD ARSENAL, MD. (INCENDIARY EVALUATION PROJ.)
IER-	(INTERIM ENGINEERING REPORT) (REF. 2)
IER-	CALIFORNIA. UNIV., BERKELEY. INST. OF ENG. RES.
IER-(NO.)-ITSA-(SAME NO.)	INSTITUTE FOR TELECOMMUNICATION SCIENCES AND AERONOMY, BOULDER, COLO.
IER-	INSTITUTES FOR ENVIRONMENTAL RESEARCH, BOULDER, COLO.
IER-	INSTRUMENT SYSTEMS CORP. TELEPHONICS DIV., HUNTINGTON, N.Y.
IERC-	INTERNATIONAL ELECTRONIC RESEARCH CORP., BURBANK, CAL
IERTM-ATDL-	ATMOSPHERIC TURBULENCE AND DIFFUSION LAB., OAK RIDGE, TENN.
IERTM-EML-	EARTHQUAKE MECHANISM LAB., SAN FRANCISCO
IERTM-ITSA-	INSTITUTE FOR TELECOMMUNICATION SCIENCES AND AERONOMY, BOULDER, COLO.
IERTM-NHRL-	NATIONAL HURRICANE RESEARCH LAB., CORAL GABLES, FLA.
IERTM-NSSL-	NATIONAL SEVERE STORMS LAB., NORMAN, OKLA.
IER-TR-	CALIFORNIA. UNIV., BERKELEY. INST. OF ENG. RES.
IES-	ILLUMINATING ENGINEERING SOCIETY, N.Y.C.
IES-(NUMBER)-	ROCKETDYNE, CANOGA PARK, CALIF.
IESD-AMC-	BROWN ENGINEERING CO., INC. INFORMATION AND ELECTRONIC SYSTEMS DIV., HUNTSVILLE, ALA.
IE-TR-(YEAR)-	ARMY MISSILE COMMAND. ENG. AND QUALITY ASSURANCE DIV. REDSTONE ARSENAL, ALA.
IFA-	ACADEMIA R.S.R. INSTITUTUL DE FIZICA ATOMICA, BUCHAREST
IFA-(YEAR)	AKADEMIYA NAUK UKRAINSKOI SSR. INSTITUT TEORETICHESKOI FIZIKI, KIEV
IFA/A/	ACADEMIA R.S.R. INSTITUTUL DE FIZICA ATOMICA, BUCHAREST
IFA-AL-	ACADEMIA R.S.R. INSTITUTUL DE FIZICA ATOMICA, BUCHAREST
IFA/CO/	ACADEMIA R.S.R. INSTITUTUL DE FIZICA ATOMICA, BUCHAREST
IFA-CP-	ITALY. ISTITUTO DE FISICA DELL'ATMOSFERA, ROME
IFA-CR-	ACADEMIA R.S.R. INSTITUTUL DE FIZICA ATOMICA, BUCHAREST
IFA-CRD-	ACADEMIA R.S.R. INSTITUTUL DE FIZICA ATOMICA, BUCHAREST
IFA/DF/	ACADEMIA R.S.R. INSTITUTUL DE FIZICA ATOMICA, BUCHAREST
IFA-EN-	ACADEMIA R.S.R. INSTITUTUL DE FIZICA ATOMICA, BUCHAREST
IFA-ESR-	ACADEMIA R.S.R. INSTITUTUL DE FIZICA ATOMICA, BUCHAREST
IFA-FN-	ACADEMIA R.S.R. INSTITUTUL DE FIZICA ATOMICA, BUCHAREST
IFA-FR-	ACADEMIA R.S.R. INSTITUTUL DE FIZICA ATOMICA, BUCHAREST
IFA-FR-	AKADEMIYA NAUK SSSR. INSTITUT ATOMNOI ENERGII, MOSCOW
IFA/FT/	ACADEMIA R.S.R. INSTITUTUL DE FIZICA ATOMICA, BUCHAREST
IFA-HE-	ACADEMIA R.S.R. INSTITUTUL DE FIZICA ATOMICA, BUCHAREST
IFA-IS-	ACADEMIA R.S.R. INSTITUTUL DE FIZICA ATOMICA, BUCHAREST
IFA-M-	ACADEMIA R.S.R. INSTITUTUL DE FIZICA ATOMICA, BUCHAREST
IFA-MN-	ACADEMIA R.S.R. INSTITUTUL DE FIZICA ATOMICA, BUCHAREST
IFA-MR-	ACADEMIA R.S.R. INSTITUTUL DE FIZICA ATOMICA, BUCHAREST
IFA-NR-	ACADEMIA R.S.R. INSTITUTUL DE FIZICA ATOMICA, BUCHAREST
IFA-NS-	ACADEMIA R.S.R. INSTITUTUL DE FIZICA ATOMICA, BUCHAREST
IFA-PN-	ACADEMIA R.S.R. INSTITUTUL DE FIZICA ATOMICA, BUCHAREST
IFA-RDP-	ITALY. CONSIGLIO NAZIONALE DELLE RICERCHE. ISTITUTO DI FISICA DELL'ATMOSFERA, TURIN
IFA-SR-	ITALY. ISTITUTO DE FISICA DELL'ATMOSFERA, ROME
IFA-STR-	ITALY. ISTITUTO DE FISICA DELL'ATMOSFERA, ROME
IFA-TR-	ACADEMIA R.S.R. INSTITUTUL DE FIZICA ATOMICA, BUCHAREST
IFA-TR-	ITALY. CONSIGLIO NAZIONALE DELLE RICERCHE. ISTITUTO DI FISICA DELL'ATMOSFERA, TURIN
IFCG-	INTER-RANGE INSTRUMENTATION GROUP
IFH-(NUMBER)-	STUTTGART. TECHNISCHE HOCHSCHULE. INSTITUT FUER HOCHTEMPERATURFORSCHUNG
IFJ-	POLAND. INSTITUTE OF NUCLEAR PHYSICS, KRAKOW
IFL-ITU-OTP-	INTERNATIONAL TELECOMMUNICATIONS UNION, GENEVA (INTERNATIONAL FREQUENCY LIST)
IFP-(NUMBER)-	STUTTGART. TECHNISCHE HOCHSCHULE. INSTITUT FUER PLASMAFORSCHUNG
IFPTH-(MONTH/YEAR)	ITALY. ISTITUTO NAZIONALE DI FISICA NUCLEARE, PADUA
IFPTK-(MONTH/YEAR)	ITALY. ISTITUTO NAZIONALE DI FISICA NUCLEARE, PADUA
IFR-	SIGNAL CORPS (ARMY) (INFRARED REPORT)
IFR/P-	GT. BRIT. ATOMIC EN. RES. ESTAB., HARWELL, BERKS, ENG
IFS-	MARTIN CO., DENVER
IFS-(YEAR)(NUMBER)	NAVY OCEANOGRAPHIC INSTRUMENTATION CTR., WASH., D.C.
IFS-MOL-EFT-	MARTIN CO., DENVER (INTERFACE SPECIFICATION)
IFT-	WARSAW. UNIV. INST. OF THEORETICAL PHYSICS
IFUM-(NUMBER)/FT(R)	MILAN. UNIV. ISTITUTO DI FISICA
IFVE-(LETTERS)-	U.S.S.R. INSTITUT FIZIKI VYSOKIKH ENERGII, SERPUKHOV
IFVE-INZH-(YEAR)-	U.S.S.R. INSTITUT FIZIKI VYSOKIKH ENERGII, SERPUKHOV
IFVE-OP-(YEAR)-	U.S.S.R. INSTITUT FIZIKI VYSOKIKH ENERGII, SERPUKHOV
IFVE-OP-(YEAR)-(NO.)-K	U.S.S.R. INSTITUT FIZIKI VYSOKIKH ENERGII, SERPUKHOV
IFVE-SEF-(YEAR)-	U.S.S.R. INSTITUT FIZIKI VYSOKIKH ENERGII, SERPUKHOV
IFVE-SEF-(YEAR)-(NO.)-K	U.S.S.R. INSTITUT FIZIKI VYSOKIKH ENERGII, SERPUKHOV
IFVE-SEF/(LTRS.)-(YR)-	U.S.S.R. INSTITUT FIZIKI VYSOKIKH ENERGII, SERPUKHOV
IFVE-SKR-(YEAR)-(NO.)-	U.S.S.R. INSTITUT FIZIKI VYSOKIKH ENERGII, SERPUKHOV
IFVE-SKU-(YEAR)-	U.S.S.R. INSTITUT FIZIKI VYSOKIKH ENERGII, SERPUKHOV
IFVE-SPK/(LTRS.)-(YR)-	U.S.S.R. INSTITUT FIZIKI VYSOKIKH ENERGII, SERPUKHOV
IFVE-STF-(YEAR)-(NO.)-	U.S.S.R. INSTITUT FIZIKI VYSOKIKH ENERGII, SERPUKHOV
IFVE-SVM-(YEAR)-	U.S.S.R. INSTITUT FIZIKI VYSOKIKH ENERGII, SERPUKHOV
IG-	CORNELL AERONAUTICAL LAB., INC., BUFFALO
IG-800-S-	CORNELL AERONAUTICAL LAB., INC., BUFFALO
IG-	UNITED KINGDOM ATOMIC ENERGY AUTHORITY. IND. GP. DOUNREAY EXPTL. REACTOR EST., CAITHNESS, SCOT. (REF.37)
IGC-	CALIFORNIA INST. OF TECH., PASADENA
IGC-(LETTER(S))/(LTR(S))-	UNITED KINGDOM ATOMIC ENERGY AUTHORITY. INDUSTRIAL GROUP (COMMITTEES) (REF. 37)
IGC-(LETTER(S))/(LTR(S))-	UNITED KINGDOM ATOMIC ENERGY AUTHORITY. INDUSTRIAL GROUP H.Q., RISLEY, LANCS, ENGLAND (REF. 37)
IGC-ARDC/P-	UNITED KINGDOM ATOMIC EN. AUTH., PRODUCTION GP. CHEMICAL SERVICES DEPT., WINDSCALE, SELLAFIELD, ENG
IGC-PA-(YEAR)-	INFORMATION GENERAL CORP., PALO ALTO, CALIF.
IGD-(LETTER(S))-	UNITED KINGDOM ATOMIC ENERGY AUTHORITY. IND. GP. DOUNREAY EXPTL. REACTOR EST., CAITHNESS, SCOT. (REF.37)
IG DRAWING	I.G. FARBENINDUSTRIE A.G.
IG DRAWING (LETTER(S))-	I.G. FARBENINDUSTRIE A.G.
IGE-	AIRCO INDUSTRIAL GASES DIV., JERSEY CITY, NJ (1965-)
IGE-(LETTER(S))-	UNITED KINGDOM ATOMIC ENERGY AUTHORITY. INDUSTRIAL GROUP H.Q., RISLEY, LANCS, ENGLAND (REF. 37)
IGG-PA-(YEAR)-	INFORMATION GENERAL CORP., PALO ALTO, CALIF.

IGI-(LETTER(S))-
        UNITED KINGDOM ATOMIC ENERGY AUTHORITY. INDUSTRIAL
            GROUP H.Q., RISLEY, LANCS, ENGLAND (REF. 37)
IG-INF.SER-(NO.)(LTR/LTR)
        UNITED KINGDOM ATOMIC ENERGY AUTHORITY. INDUSTRIAL
            GROUP (REF. 38)
IGIS-(NUMBER)(LTR(S)/LTR)
        UNITED KINGDOM ATOMIC ENERGY AUTHORITY. INDUSTRIAL
            GROUP (REF. 38)
IG.LIBIB-
        GT. BRIT. DEPT. OF ATOMIC ENERGY. INDUSTRIAL GROUP
            H.Q., RISLEY, LANCS, ENGLAND (REF. 36)
IGM-(NUMBER)(LTR(S)/LTR)
        UNITED KINGDOM ATOMIC ENERGY AUTHORITY. INDUSTRIAL
            GROUP (REF. 38)
IGM-(NUMBER)(RD/R)
        UNITED KINGDOM ATOMIC ENERGY AUTHORITY. INDUSTRIAL
            GROUP H.Q., RISLEY, LANCS, ENGLAND (REF. 38)
IGM-(NUMBER)(D)
        UNITED KINGDOM ATOMIC ENERGY AUTHORITY. IND. GP.
            DOUNREAY EXPTL.REACTOR EST.,CAITHNESS,SCOT.(REF.38)
IGM-(LETTER(S))/CR-
        UNITED KINGDOM ATOMIC ENERGY AUTHORITY. INDUSTRIAL
            GROUP. CALDER WORKS, CALDERBRIDGE, CUMB., ENGLAND
IG-MEMO-(NO.)(LTR(S)/LTR)
        UNITED KINGDOM ATOMIC ENERGY AUTHORITY. INDUSTRIAL
            GROUP (REF. 38)
I.G. MEMO. (NO.)-(RD/R)
        UNITED KINGDOM ATOMIC ENERGY AUTHORITY. INDUSTRIAL
            GROUP H.Q., RISLEY, LANCS, ENGLAND
IGM-PR-
        UNITED KINGDOM ATOMIC ENERGY AUTHORITY. INDUSTRIAL
            GROUP (PROGRESS REPORT) (REF. 37)
IGN/RT-
        CONFERENCE ON PLASMA DIAGNOSIS, CULHAM, ENGLAND
IGO-(LETTER(S))/(LTR(S))-
        UNITED KINGDOM ATOMIC ENERGY AUTHORITY. INDUSTRIAL
            GROUP (REF. 37)
IGO-(LETTER(S))/R-
        UNITED KINGDOM ATOMIC ENERGY AUTHORITY. INDUSTRIAL
            GROUP H.Q., RISLEY, LANCS, ENGLAND (REF. 37)
IGO-(LETTER(S))/CR-
        UNITED KINGDOM ATOMIC ENERGY AUTHORITY. INDUSTRIAL
            GROUP. CALDER WORKS, CALDERBRIDGE, CUMB., ENGLAND
IGO-(LETTER(S))/CA-
        UNITED KINGDOM ATOMIC ENERGY AUTHORITY. INDUSTRIAL
            GROUP. CAPENHURST WORKS, CHES., ENGLAND (REF. 37)
IGO-(LETTER(S))/CC-
        UNITED KINGDOM ATOMIC ENERGY AUTHORITY. INDUSTRIAL
            GROUP. CHAPELCROSS WORKS, DUMFRIESSHIRE, SCOTLAND
IGO-(LETTER(S))/S-
        UNITED KINGDOM ATOMIC ENERGY AUTHORITY. PRODUCTION
            GROUP, SPRINGFIELDS, LANCS, ENGLAND
IGO-(LETTER(S))/W-
        UNITED KINGDOM ATOMIC EN. AUTH., PRODUCTION GP.
            CHEMICAL SERVICES DEPT., WINDSCALE, SELLAFIELD, ENG
IGPC-
        OHIO STATE UNIV. RESEARCH FOUNDATION, COLUMBUS
IGR-(NUMBER)(LTR(S)/LTR)
        UNITED KINGDOM ATOMIC ENERGY AUTHORITY. INDUSTRIAL
            GROUP (REF. 38)
IGR-(NUMBER)(LTR(S)/R)
        UNITED KINGDOM ATOMIC ENERGY AUTHORITY. INDUSTRIAL
            GROUP H.Q., RISLEY, LANCS, ENGLAND (REF. 38)
IGR-(NUMBER)(LTR(S)/CA)
        UNITED KINGDOM ATOMIC ENERGY AUTHORITY. INDUSTRIAL
            GROUP. CAPENHURST WORKS, CHES., ENGLAND (REF. 38)
IGR-(NUMBER)(LTR(S)/S)
        UNITED KINGDOM ATOMIC ENERGY AUTHORITY. INDUSTRIAL
            GROUP. SPRINGFIELDS WORKS, LANCS, ENG. (REF. 38)
IGR-(NUMBER)(LTR(S)/W)
        UNITED KINGDOM ATOMIC ENERGY AUTHORITY. INDUSTRIAL
            GROUP.WINDSCALE WORKS,SELLAFIELD,CUMB.,ENG.(REF.38)
IGR-(LETTER(S)/CA-
        UNITED KINGDOM ATOMIC ENERGY AUTHORITY. DEVELOPMENT
            AND ENGINEERING GROUP, CAPENHURST, CHES., ENGLAND
IGR-(LETTER(S))/(LTR(S))-
        UNITED KINGDOM ATOMIC ENERGY AUTHORITY. INDUSTRIAL
            GROUP (REF. 37)
IGR-(LETTER(S))/R-
        UNITED KINGDOM ATOMIC ENERGY AUTHORITY. INDUSTRIAL
            GROUP H.Q., RISLEY, LANCS, ENGLAND (REF. 37)
IGR-(LETTER(S))/C-
        UNITED KINGDOM ATOMIC ENERGY AUTHORITY. INDUSTRIAL
            GROUP. CULCHETH LABS., LANCS, ENGLAND (REF. 37)
IGR-(LETTER(S))/D-
        UNITED KINGDOM ATOMIC ENERGY AUTHORITY. INDUSTRIAL
            GROUP. DOUNREAY WORKS, CAITHNESS, SCOT. (REF. 37)
IGR-(LETTER(S))/S-
        UNITED KINGDOM ATOMIC ENERGY AUTHORITY. INDUSTRIAL
            GROUP. SPRINGFIELDS WORKS, LANCS, ENG. (REF. 37)
IGR-(LETTER(S))/W-
        UNITED KINGDOM ATOMIC ENERGY AUTHORITY. INDUSTRIAL
            GROUP.WINDSCALE WORKS,SELLAFIELD,CUMB.,ENG.(REF.37)
IG REPORT-
        (SEE ALSO ENTRIES STARTING IGR-)
IG-REPORT-(NO.)(LTR/LTR)
        UNITED KINGDOM ATOMIC ENERGY AUTHORITY. INDUSTRIAL
            GROUP (REF. 38)
I.G.REPORT (NO.)-(RD/CA)
        UNITED KINGDOM ATOMIC ENERGY AUTHORITY. INDUSTRIAL
            GROUP. CAPENHURST WORKS, CHES., ENGLAND
IGRL-(LTR(S))/(LTR(S))-
        UNITED KINGDOM ATOMIC ENERGY AUTHORITY. INDUSTRIAL
            GROUP (REF. 37)
IGRL-(LETTER(S))/R-
        UNITED KINGDOM ATOMIC ENERGY AUTHORITY. INDUSTRIAL
            GROUP H.Q., RISLEY, LANCS, ENGLAND (REF. 37)
IGRL-(LETTER(S))/CA-
        UNITED KINGDOM ATOMIC ENERGY AUTHORITY. INDUSTRIAL
            GROUP. CAPENHURST WORKS, CHES., ENGLAND (REF. 37)
IGRL-IB/CA-
        GT. BRIT. CAPENHURST WORKS, CHES., ENG. (INFO. BIB.)
IGRL-IB/R-
        GT. BRIT. DEPT. OF ATOMIC ENERGY. INDUSTRIAL GROUP
            H.Q., RISLEY, LANCS, ENGLAND (INFORMATION BIB.)
IGRL-T/W-
        UNITED KINGDOM ATOMIC ENERGY AUTHORITY. INDUSTRIAL
            GROUP H.Q., RISLEY, LANCS, ENGLAND
IGS-(LETTER(S))/(LTR(S))-
        UNITED KINGDOM ATOMIC ENERGY AUTHORITY. INDUSTRIAL
            GROUP (REF. 37)
IGS-(LETTER(S))/R-
        UNITED KINGDOM ATOMIC ENERGY AUTHORITY. INDUSTRIAL
            GROUP H.Q., RISLEY, LANCS, ENGLAND (REF. 37)
IGT-(LETTER(S))-
        UNITED KINGDOM ATOMIC ENERGY AUTHORITY. INDUSTRIAL
            GROUP H.Q., RISLEY, LANCS, ENGLAND (REF. 37)
IG-TRANS-(R)-
        UNITED KINGDOM ATOMIC ENERGY AUTHORITY. INDUSTRIAL
            GROUP H.Q. LIBRARY, RISLEY, LANCS, ENG. (REF. 36)
IGTS-
        (INDUSTRY, GOVERNMENT AND TECHNICAL SERVICE REPORTS)
IGY BULLETIN-
        NATIONAL ACADEMY OF SCIENCES (INTERNATIONAL
            GEOPHYSICAL YEAR BULLETIN)
IGY GRS-
        NATIONAL ACADEMY OF SCIENCES (INTERNATIONAL
            GEOPHYSICAL YEAR. GENERAL REPORT SERIES)
IGY SRS-
        NATIONAL ACADEMY OF SCIENCES (INTERNATIONAL
            GEOPHYSICAL YEAR. SATELLITE REPORT SERIES)
IH-
        CORNELL AERONAUTICAL LAB., INC., BUFFALO
IH-(NUMBER)-F-
        CORNELL AERONAUTICAL LAB., INC., BUFFALO
IHES-A(NUMBER)-
        FRANCE.INSTITUT DES HAUTES ETUDES SCIENTIFIQUES,PARIS
IHES-N-(LETTER)-
        FRANCE.INSTITUT DES HAUTES ETUDES SCIENTIFIQUES,PARIS
IHP-
        PARMA, ITALY. UNIVERSITA. INST. OF HUMAN PHYSIOLOGY
IH/QAS-(YEAR)-
        NAVAL ORDNANCE STATION. QUALITY ASSURANCE DEPT.,
            INDIAN HEAD, MD.
IHR-
        BROOKHAVEN NATIONAL LAB., UPTON, N.Y.
IH/RE-(YEAR)-
        NAVAL ORDNANCE STATION. RESEARCH AND DEVELOPMENT
            DEPT., INDIAN HEAD, MD.
IHT-
        STUTTGART. TECHNISCHE HOCHSCHULE. INSTITUT FUER
            HOCHTEMPERATURFORSCHUNG
IHT-BERICHT-
        STUTTGART. TECHNISCHE HOCHSCHULE. INSTITUT FUER
            HOCHTEMPERATURFORSCHUNG
IH/TFS-(YEAR)-
        JONES + LAUGHLIN STEEL CORP. WIRE ROPE DIV.,MUNCY,PA.
IH/TFS-
        NAVAL ORDNANCE STATION. TEST AND EVALUATION DEPT.,
            INDIAN HEAD, MD.
II-
        (INTERIM INVESTIGATION REPT.) (REF. 2)
II-5-
        GT. BRIT. DEPT. OF SCIENTIFIC AND INDUSTRIAL RE-
            SEARCH. DIRECTORATE OF TUBE ALLOYS
II-188-
        ROCHESTER, N.Y. UNIV.
II-189-
        ROCHESTER, N.Y. UNIV.
IIAE-INF-(NO.)-(YEAR)
        ARGENTINA. INSTITUTO DE INVESTIGACION AERONAUTICA
            Y ESPACIAL, BUENOS AIRES
IIAE-LR-
        ARGENTINA. INSTITUTO DE INVESTIGACION AERONAUTICA
            Y ESPACIAL, BUENOS AIRES
IIAE-SC-LR-
        ARGENTINA. INSTITUTO DE INVESTIGACION AERONAUTICA
            Y ESPACIAL, BUENOS AIRES
IIB-
        GERMANY. BUNDESPOST. IONOSPHAREN-INSTITUT, BREISACH
IICE-
        INSTITUTE FOR INTERNAL COMBUSTION ENGINES
IIF-(NUMBER-YEAR)
        AKADEMIYA NAUK SSSR. INSTITUT YADERNOI FIZIKI,
            NOVOSIBIRSK
IIF-
        AKADEMIYA NAUK SSSR. INSTITUT YADERNOI FIZIKI,
            NOVOSIBIRSK
IIF-LA-
        AKADEMIYA NAUK SSSR. INSTITUT YADERNOI FIZIKI,
            NOVOSIBIRSK
IIHR-
        IOWA. STATE UNIV., IOWA CITY. IOWA INST. OF HYDRAULIC
            RESEARCH
IISL-BIBL.-
        INTERNATIONAL ASTRONAUTICAL FEDERATION, PARIS
IISN-
        BELGIUM. INSTITUT INTERUNIVERSITAIRE DES SCIENCES
            NUCLEAIRES, BRUSSELS
IIT-
        ILLINOIS INST. OF TECH., CHICAGO
IIT-DOMIII-
        ILLINOIS INST. OF TECH., CHICAGO. DEPT. OF MEDICINE
IITRI-
        IIT RESEARCH INST., CHICAGO
IITRI-(NUMBER)-P
        IIT RESEARCH INST., CHICAGO

IITRI-(NO.-LTR.-NO.-NO.)	IIT RESEARCH INST., CHICAGO
IITRI-A-	IIT RESEARCH INST., CHICAGO
IITRI-B-	IIT RESEARCH INST., CHICAGO
IITRI-C-	IIT RESEARCH INST., CHICAGO
IITRI-G-	IIT RESEARCH INST., CHICAGO
IITRI-J-	IIT RESEARCH INST., CHICAGO
IITRI-L-	IIT RESEARCH INST., CHICAGO
IITRI-M-	IIT RESEARCH INST., CHICAGO
IITRI-N-	IIT RESEARCH INST., CHICAGO
IITRI-PROJ-(LTR.)-	IIT RESEARCH INST., CHICAGO
IITRI-T-	IIT RESEARCH INST., CHICAGO
IITRI-U-	IIT RESEARCH INST., CHICAGO
IITRI-V-	IIT RESEARCH INST., CHICAGO
IIT-TR-	ILLINOIS INST. OF TECH., CHICAGO (TECHNICAL REPORT)
IJCPD-	INTERNATIONAL JOINT COMMITTEE ON PSYCHOMETRIC DATA
IJS-	YUGOSLAVIA. INSTITUT JOZEF STEFAN, LJUBLJANA
I-K-	NATIONAL DEFENSE RESEARCH COMMITTEE
IKB-C-B-	BERLIN. HAHN-MEITNER-INSTITUT FUER KERNFORSCHUNG
IKD-	INGERSOLL KALAMAZOO DIV., MICH.
IKF-	FRANKFURT AM MAIN. UNIVERSITAET. INSTITUT FUER KERNPHYSIK.
IKF-D-	FRANKFURT AM MAIN. UNIVERSITAET. INSTITUT FUER KERNPHYSIK
IKO-	AMSTERDAM. INSTITUUT VOOR KERNPHYSISCH ONDERZOEK
IKO-PR-	AMSTERDAM. INSTITUUT VOOR KERNPHYSISCH ONDERZOEK
IL-(NUMBER)-	OFFICE OF SPECIAL WEAPONS DEVELOPMENTS, FT. BLISS,TEX
ILCEP-	INTERLABORATORY COMMITTEE ON EDITING AND PUBLISHING (NAVY), CORONA, CALIF.
ILCEP-EAST MONOGRAPH-	EAST COAST INTERLABORATORY COMMITTEE ON EDITING AND PUBLISHING (NAVY), JOHNSVILLE, PENNA.
ILCF-	INTER-LABORATORY COMMITTEE ON FACILITIES (NAVY)
ILEA-	NORGES TEKNISKE HOEGSKOLE, TRONDHEIM
ILL-	ILLINOIS. UNIV., URBANA
ILL-(NUMBER)-P	ILLINOIS. UNIV., URBANA (PROJECT SQUID)
ILL-(NUMBER)-PU	ILLINOIS. UNIV., URBANA (PROJECT SQUID)
ILL CES SRS-	ILLINOIS. UNIV., URBANA. DEPT. OF CIVIL ENGINEERING (CIVIL ENGINEERING STUDIES. STRUCTURAL RES. SERIES)
ILLCO-	ILLINOIS WATER TREATMENT CO., ROCKFORD
ILL-DCE-	ILLINOIS. UNIV., URBANA. DEPT. OF CIVIL ENGINEERING
ILL EERL TN-	ILLINOIS. UNIV., URBANA. ELECTRICAL ENG. RES. LAB.
ILL EERL TR-	ILLINOIS. UNIV., URBANA. ELECTRICAL ENG. RES. LAB.
ILL-EES	ILLINOIS. UNIV., URBANA. ENGINEERING EXPERIMENT STA.
ILL EES SRS-	ILLINOIS. UNIV., URBANA. ENGINEERING EXPERIMENT STATION. (STRUCTURAL RESEARCH SERIES)
ILLIAC-IV-(NUMBER)-	ILLINOIS. UNIV., URBANA. DEPT. OF COMPUTER SCIENCE
ILL-TECH-	ILLINOIS INST. OF TECH., CHICAGO
ILL.TECH./DOMIIT-	ILLINOIS INST. OF TECH., CHICAGO. DEPT. OF MECHANICS
ILL-(TH)-(YEAR)-	ILLINOIS. UNIV., URBANA. DEPT. OF PHYSICS
ILL U-	ILLINOIS. UNIV., URBANA
ILLU-(NUMBER)-	ILLINOIS. UNIV., URBANA
ILL U.(LETTERS)-	ILLINOIS. UNIV., URBANA
ILL-U-DCE-	ILLINOIS. UNIV., URBANA. DEPT. OF CIVIL ENGINEERING
ILL U.SRS-	ILLINOIS. UNIV., URBANA. COLL. OF ENGINEERING (STRUCTURAL RESEARCH SERIES)
ILL U., STRUC. RES. SER-	ILLINOIS. UNIV., URBANA. COLL. OF ENGINEERING (STRUCTURAL RESEARCH SERIES)
ILL. U. T + AM-	ILLINOIS. UNIV., URBANA. COLL. OF ENGINEERING
ILRS-T-	IDAMI LANGUAGE RESEARCH SECTION, MILAN
ILS-	COAST AND GEODETIC SURVEY, WASHINGTON, D.C. (INSTRUMENT LANDING SYSTEM CHART)
ILT-	ILLINOIS INST. OF TECH., CHICAGO
ILU-	ILLINOIS. UNIV., URBANA
ILU DCE-	ILLINOIS. UNIV., URBANA. DEPARTMENT OF CERAMIC ENG.
ILU DCE MEMO-	ILLINOIS. UNIV., URBANA. DEPARTMENT OF CERAMIC ENG.
ILU DCE MR-	ILLINOIS. UNIV., URBANA. DEPARTMENT OF CERAMIC ENG.
ILU-EES	ILLINOIS. UNIV., URBANA. ENGINEERING EXPERIMENT STA.
ILU EES TR-	ILLINOIS. UNIV., URBANA. ENGINEERING EXPERIMENT STA.
ILU-ES-	ILLINOIS. AGRICULTURAL EXPERIMENT STATION, URBANA
IM-	(INSPECTION MEMORANDUM) (REF. 2)
IM-	(INSTRUCTORS MANUAL) (REF. 2)
IM-	(INTELLIGENCE MEMORANDUM) (REF. 2)
IM-	(INTERIM MEMORANDUM) (REF. 2)
IM-	(INTERNAL MEMORANDUM) (REF. 2)
IM-	BOEING AIRPLANE CO., SEATTLE
IM-(NUMBER)-(LETTER)-	CORNELL AERONAUTICAL LAB., INC., BUFFALO
IM-	NATIONAL RESEARCH COUNCIL OF CANADA. ATOMIC ENERGY PROJECT, CHALK RIVER, ONT.
IMA-	JOHNS HOPKINS UNIV., BALTIMORE. RADIATION LAB.
IMCC-	INTERNATIONAL MINERALS AND CHEMICAL CORP., CHICAGO
IME-	DEPARTMENT OF HOUSING AND URBAN DEVELOPMENT. DIVISION OF INTERNATIONAL AFFAIRS, WASHINGTON, D.C.
IMER-CONTROL-(NO.)-	UTAH. UNIV., SALT LAKE CITY. INST. OF METALS AND EXPLOSIVES RESEARCH
IMF-(NUMBER)/(YEAR)	KERNFORSCHUNGSZENTRUM KARLSRUHE, GERMANY (MATERIAL AND SOLID BODY RESEARCH)
IMI-(NUMBER)-	SANDIA CORP., ALBUQUERQUE, N. MEX. (INSPECTION METHODS INSTRUCTION)
IMIS-(YEAR)-	BUREAU OF INTERNATIONAL COMMERCE, WASHINGTON, D.C. (INTERNATIONAL MARKETING INFORMATION SERVICE)
IMM-	NEW YORK UNIV., N.Y.C. INST. OF MATHEMATICAL SCIENCES
IMM-NYU-	NEW YORK UNIV., N.Y.C. INST. FOR MATH. + MECHANICS
IMM-NYU-	NEW YORK UNIV., N.Y.C. INST. OF MATHEMATICAL SCIENCES
IM-(MODEL NUMBER)-	SANBORN CO., WALTHAM, MASS. (INSTRUCTION MANUAL)
IMOG-	INTERAGENCY MECHANICAL OPERATIONS GP., STEERING COMM.
IMP-	MARYLAND. UNIV.,COLLEGE PARK.INST. OF MOLECULAR PHYS.
IMP-AEC-	MARYLAND. UNIV.,COLLEGE PARK.INST. OF MOLECULAR PHYS.
IMP-NASA-	MARYLAND. UNIV.,COLLEGE PARK.INST. OF MOLECULAR PHYS.
IMP-ONR-	MARYLAND. UNIV.,COLLEGE PARK.INST. OF MOLECULAR PHYS.
IMP-OSR-	MARYLAND. UNIV.,COLLEGE PARK.INST. OF MOLECULAR PHYS.
IMP-TR-	SCRIPPS INSTITUTION OF OCEANOGRAPHY, LA JOLLA, CALIF.
IMR-	(INFORMAL MEMORANDUM REPORT) (REF. 2)
IMR-	DEFENSE RESEARCH CORP., SANTA BARBARA, CALIF.
IMR-	GENERAL MOTORS CORP. DEFENSE RESEARCH LABS., SANTA BARBARA, CALIF.
I.M.R.-(NUMBER/YEAR)	NAVAL OCEANOGRAPHIC OFFICE, WASHINGTON, D.C. (INFORMAL MANUSCRIPT REPORT)
IMR-	NAVAL OCEANOGRAPHIC OFFICE, WASHINGTON, D.C. (INFORMAL MANUSCRIPT REPORTS)
IMR-TR-	CALIFORNIA. UNIV., LA JOLLA. INST.OF MARINE RESOURCES
IMS-(YEAR)-	BELL AEROSYSTEMS CO., BUFFALO
IMS-	UNITED KINGDOM. NATIONAL PHYSICAL LAB. DIV. OF INORG. AND METALLIC STRUCTURE, TEDDINGTON, MIDDX., ENG.
IN-	(INSTRUMENT NOTE) (REF. 2)
IN-	(INTERNAL NOTE) (REF. 2)
IN-	BUREAU OF SHIPS (INDUSTRIAL NOTES)
IN-	IDAHO NUCLEAR CORP., IDAHO FALLS
IN-	NATIONAL REACTOR TESTING STATION, IDAHO FALLS
IN-	VON KARMAN INSTITUTE FOR FLUID DYNAMICS, RHODE-SAINT-GENESE, BELGIUM

INA-

INA-	NATIONAL BUREAU OF STANDARDS, WASHINGTON, D.C.
INAC-PUB-	ITALY. ISTITUTO NAZIONALE PER LE APPLICAZIONI DEL CALCOLO, ROME
INC-	IDAHO NUCLEAR CORP., IDAHO FALLS
INC-	INDUSTRIAL NUCLEONICS CORP., COLUMBUS, OHIO
INC-	ROLLS ROYCE, LTD., DERBY, ENGLAND
INC/(LTR.)(NO.)/(LTR.)-	FEDERAL AVIATION ADMINISTRATION, WASHINGTON, D.C.
INCAP-L-	WORLD HEALTH ORGANIZATION. INSTITUTO DE NUTRICION DE CENTRO AMERICA Y PANAMA, GUATEMALA CITY
INC/C(NUMBER)/P(NUMBER)	AIR FORCE, WASHINGTON, D.C.
INC/C(NO.)/P(NO.)	NATIONAL BUREAU OF STANDARDS, WASHINGTON, D.C.
INC/C(NO.)/P(NO.)	SOCIETY OF AUTOMOTIVE ENGINEERS, N.Y.C.
IND-	FOREIGN ECONOMIC ADM. ENEMY BR. INDUSTRY DIV.
INDC-	HANFORD WORKS. DESIGN + CONSTRUCTION DIV., RICHLAND, WASH. (ASSIGNED TO INCOMING REPORTS)
INDC-	INTERNATIONAL NUCLEAR DATA COMMITTEE, VIENNA
INDC(LETTERS)-(NO./LTR.)	INTERNATIONAL NUCLEAR DATA COMMITTEE, VIENNA
INDC(CCP)-(NUMBER/LTR.)	INTERNATIONAL NUCLEAR DATA COMMITTEE, VIENNA (PAPER FROM U.S.S.R.)
INDC(NDS)-	INTERNATIONAL NUCLEAR DATA COMMITTEE, VIENNA
INDC(US)-(NUMBER/LETTER)	INTERNATIONAL NUCLEAR DATA COMMITTEE, VIENNA (PAPER FROM U.S.A.)
INDEC-	PENNSYLVANIA. UNIV., PHILADELPHIA. INSTITUTE FOR DIRECT ENERGY CONVERSION
INDEC-SR-	PENNSYLVANIA. UNIV., PHILADELPHIA. INSTITUTE FOR DIRECT ENERGY CONVERSION
INDS-	INDUSTRIAL SCIENTIFIC CO., N.Y.C.
INDSWG-	INTERNATIONAL ATOMIC ENERGY AGENCY. INFORMATIONAL NUCLEAR DATA UNIT, VIENNA
IND-TR-	INDIANA UNIV., BLOOMINGTON
INDU-	INDIANA UNIV., BLOOMINGTON
INE/M-	NEW SOUTH WALES, AUSTRALIA. UNIV., KENSINGTON. INST. OF NUCLEAR ENGINEERING
INET-	INDUSTRIAL ELECTRONICS + TRANSFORMER CO., LOS ANG.
INF-(YEAR)-	AKADEMIYA NAUK UKRAINSKOI SSR. INSTITUT TEORETICHESKOI FIZIKI, KIEV
INFCIRC-	INTERNATIONAL ATOMIC ENERGY AGENCY, VIENNA
INFN/AE-(YEAR)/	ITALY. ISTITUTO NAZIONALE DI FISICA NUCLEARE, TRIESTE
INFN/BE-(YEAR)/	ITALY. ISTITUTO NAZIONALE DI FISICA NUCLEARE (ALL LOCATIONS)
INFN-PD-(YEAR)/	ITALY. ISTITUTO NAZIONALE DI FISICA NUCLEARE, PADUA
INFN/RE-	ITALY. ISTITUTO NAZIONALE DI FISICA NUCLEARE, PADUA
INFN/TC-	ITALY. ISTITUTO NAZIONALE DI FISICA NUCLEARE, BOLOGNA
INFN-TC-(YEAR)/	ITALY. ISTITUTO NAZIONALE DI FISICA NUCLEARE, FLORENCE
INFO BULL-	(INFORMATION BULLETIN) (REF. 2)
INFO. BULL. T-	METALLURGICAL ADVISORY COMMITTEE ON TITANIUM
INFO. CIRCULAR-	BUREAU OF MINES
ING ER-	THOMPSON PRODUCTS, INC., CLEVELAND
INIS-MF-	INTERNATIONAL ATOMIC ENERGY AGENCY. INTERNATIONAL NUCLEAR INFORMATION SYSTEM, VIENNA
IN-ITR-	IDAHO NUCLEAR CORP., IDAHO FALLS
INP-(NUMBER)/PH	AKADEMIYA NAUK SSSR. INSTITUT FIZIKI, MOSCOW
INP-(NUMBER)/PH	JAGIELLONIAN UNIV., KRAKOW
INP-(NUMBER)/PS	POLAND. INSTITUTE OF NUCLEAR PHYSICS, KRAKOW
INP-	POLAND. INSTITUTE OF NUCLEAR PHYSICS, KRAKOW
INP-(NUMBER)/PL/PH	POLAND. INSTITUTE OF NUCLEAR PHYSICS, KRAKOW
INP-(NUMBER)-PL	POLAND. INSTITUTE OF NUCLEAR PHYSICS, KRAKOW
INP-(NUMBER)-C	POLAND. INSTITUTE OF NUCLEAR PHYSICS, KRAKOW
INP-(NUMBER)/PS	POLAND. INST. OF NUCLEAR TECHNIQUES, KRAKOW
INPO/TH-	FRANCE. INSTITUT DE PHYSIQUE NUCLEAIRE, ORSAY
IN-P+VE-M-(YEAR)-	GEORGE C. MARSHALL SPACE FLIGHT CTR., HUNTSVILLE, ALA.
INR-	CONVAIR, FORT WORTH, TEX.
INR-	ION PHYSICS CORP., BURLINGTON, MASS.
INR-(NUMBER/YEAR)-	KERNFORSCHUNGSZENTRUM KARLSRUHE. INSTITUT FUER NEUTRONENPHYSIK UND REAKTORTECHNIK, GERMANY
INR-	POLAND. INSTITUTE OF NUCLEAR RESEARCH, WARSAW
INR-(ROMAN NO.)	POLAND. INSTITUTE OF NUCLEAR RESEARCH, WARSAW
INR-(NO./ROMAN NO./LTR.)	POLAND. INSTITUTE OF NUCLEAR RESEARCH, WARSAW
INRCA-QPR-	FRANCE. INSTITUT NATIONAL DE RECHERCHE CHIMIQUE APPLIQUEE, PARIS
INRH-	HUNGARIAN ACADEMY OF SCIENCES. INST. OF NUCLEAR RESEARCH, DEBRECEN
INR-LP-(YEAR)-	POLAND. INSTITUTE OF NUCLEAR RESEARCH, WARSAW
INR-NR-	GERMANY. INSTITUT FUER NEUTRONENPHYSIK UND REAKTORTECHNIK, KARLSRUHE
INR-P-	POLAND. INSTITUTE OF NUCLEAR RESEARCH, WARSAW
INR-P-(NO./ROMAN NO./LTR)	POLAND. INSTITUTE OF NUCLEAR RESEARCH, WARSAW
INR-P-(NUMBER/NO.)/PL	POLAND. INSTITUTE OF NUCLEAR RESEARCH, WARSAW
INR-PT-	POLAND. INSTITUTE OF NUCLEAR RESEARCH, WARSAW
IN-R-QUAL-(YEAR)-	MICHIGAN. UNIV., ANN ARBOR
INS-	CENTER FOR NAVAL ANALYSES. INST. OF NAVAL STUDIES, CAMBRIDGE, MASS.
INS-	NEW ZEALAND. INST. OF NUCLEAR SCIENCES, LOWER HUTT
INS-	NIHON UNIV., TOKYO
INS-	TOKYO UNIV. INST. FOR NUCLEAR STUDIES
INSD-	ISOTOPES, INC. NUCLEAR SYSTEMS DIV., MIDDLE RIVER, MD
INSEAN-QUADERNO-	ITALY. ISTITUTO NAZIONALE PER STUDI ED ESPERIENZE DI ARCHITETTURA NAVALE, ROME
INSJ-	TOKYO UNIV. INST. FOR NUCLEAR STUDIES
INS-PT-	KYOTO UNIV. RESEARCH INST. FOR FUNDAMENTAL PHYSICS
INS-PT-	TOKYO UNIV. INST. FOR NUCLEAR STUDIES
INS-R-	NEW ZEALAND. INST. OF NUCLEAR SCIENCES, LOWER HUTT
INS-RC-	CENTER FOR NAVAL ANALYSES. INST. OF NAVAL STUDIES, CAMBRIDGE, MASS.
INS-RC-	CENTER FOR NAVAL ANALYSES. INST. OF NAVAL STUDIES, WASHINGTON, D.C.
INS-RESEARCH CONTRIB-	CENTER FOR NAVAL ANALYSES. INST. OF NAVAL STUDIES, CAMBRIDGE, MASS.
INS-RESEARCH CONTRIB-	CENTER FOR NAVAL ANALYSES. INST. OF NAVAL STUDIES, WASHINGTON, D.C.
INS-STUDY-	CENTER FOR NAVAL ANALYSES. INST. OF NAVAL STUDIES, CAMBRIDGE, MASS.
INST-	TOKYO UNIV.
INS-TCA-	OSAKA UNIV.
INS-TCA-	TOKYO UNIV. INST. FOR NUCLEAR STUDIES
INS-TH-	TOKYO UNIV. INST. FOR NUCLEAR STUDIES
INS-TL-	TOKYO UNIV. INST. FOR NUCLEAR STUDIES
INSTN-	GT. BRIT. ROYAL AIRCRAFT EST., FARNBOROUGH, HANTS, ENG
INST. OF STATISTICS-MS-	NORTH CAROLINA. UNIV., CHAPEL HILL. DEPT. OF STATISTICS
INSTR-	(INSTRUCTION) (REF. 2)
INSTR-	NAVAL AIR DEV. CTR. AERO. INSTRS. LAB., JOHNSVILLE, PA.
INT-	FOREST SERVICE. INTERMOUNTAIN STATION, OGDEN, UTAH
INT-	NORTHERN FOREST FIRE LAB., MISSOULA, MONT.
INT-(NUMBER)/PS	POLAND. INST. OF NUCLEAR TECHNIQUES, KRAKOW
INT-	UNITED KINGDOM. MINISTRY OF TECHNOLOGY. HYDRAULICS RESEARCH STATION, WALLINGFORD, BERKS, ENGLAND
INTA-	SPAIN. INSTITUTO NACIONAL DE TECNICA AERONAUTICA, MADRID
INTAT-	INTERATOM. INTERNATIONALE ATOMREACTORBAU GMBH, BENSBERG/COLOGNE
INT-DGO-	INTERNUCLEAR CO., INC., CLAYTON, MO.
INT/EL-	FRANCE. COMMISSARIAT A L'ENERGIE ATOMIQUE. CENTRE D'ETUDES NUCLEAIRES, GRENOBLE

INTERAGENCY-
        GEOLOGICAL SURVEY, WASHINGTON, D.C.
INTERIKO-(YEAR/NUMBER)
        NETHERLANDS. INSTITUTE FOR NUCLEAR PHYSICS
        RESEARCH, AMSTERDAM
INTERIM ENG. REPT.-
        (INTERIM ENGINEERING REPORT) (REF. 2)
INTERIM SCIENTIFIC-
        MICHIGAN STATE UNIV., EAST LANSING. DIV. OF
        ENGINEERING RESEARCH
INTERNAL MEMO-
        (INTERNAL TECHNICAL MEMO) (REF. 2)
INTERN RAPPORT-(LTR.)-
        NORWAY. FORSVARETS FORSKNINGSINSTITUTT, KJELLER
INTERNUC-
        INTERNUCLEAR CO., INC., CLAYTON, MO.
INTERNUCLEAR-
        INTERNUCLEAR CO., INC., CLAYTON, MO.
INT.-KRS-MURA-
        WAYNE UNIV., DETROIT
INT/PI (NT)(NO.-NO./YR.)
        FRANCE. COMMISSARIAT A L'ENERGIE ATOMIQUE. CENTRE
        D'ETUDES NUCLEAIRES, GRENOBLE
INT/SPR-
        FRANCE. COMMISSARIAT A L'ENERGIE ATOMIQUE. CENTRE
        D'ETUDES NUCLEAIRES, GRENOBLE
INU-
        INDIANA UNIV., BLOOMINGTON
INUA-RT-N-
        ITALY. ISTITUTO NAZIONALE DI ULTRACUSTICA, ROME
INVENTORY-
        CENTRAL NEW YORK REGIONAL PLANNING AND DEVELOPMENT
        BOARD, SYRACUSE
INVESTIGATION-
        MINNESOTA. DEPT. OF HIGHWAYS. OFFICE OF MATERIALS
IOC-
        ARMED FORCES SPEC. WEAPONS PROJ. FIELD COMMAND.DIREC-
        TORATE OF WEAPONS EFFECTS TESTS, ALBUQUERQUE,N.MEX.
IOC-
        INTERGOVERNMENTAL OCEANOGRAPHIC COMMISSION, PARIS
IOCS-M-ESSA-
        AIR RESOURCES LAB., RESEARCH TRIANGLE PARK, N.C.
IOI
        (INTERIM OPERATING INSTRUCTIONS) (REF. 2)
IOI-
        ATOMIC ENERGY OF CANADA LTD.,CHALK RIVER,ONT. (REF.6)
I.O.M.-(NUMBER/YEAR)
        NAVAL OCEANOGRAPHIC OFFICE, WASHINGTON, D.C.
        (INFORMAL OCEANOGRAPHIC REPORT)
IOM-
        NAVAL OCEANOGRAPHIC OFFICE, WASHINGTON, D.C.
        (INFORMAL OCEANOGRAPHIC REPORT)
ION-
        ION PHYSICS CORP., BURLINGTON, MASS.
IONAC-
        IONAC CHEMICAL CO., BIRMINGHAM, N.J.
IONLAB-P-
        DENMARK. TEKNISKE HOJSKOLE, LYNGBY
IOP-
        ATOMIC ENERGY OF CANADA LTD.,CHALK RIVER,ONT. (REF.6)
IOU-IIHR-
        IOWA. STATE UNIV., IOWA CITY. IOWA INST. OF HYDRAULIC
        RESEARCH
IOWA PERU ES-
        IOWA UNIVERSITIES MISSION TO PERU
IOWA PERU SR-
        IOWA UNIVERSITIES MISSION TO PERU
IP-
        (INSTRUMENTATION PAPER) (REF. 2)
IP-
        AIR FORCE CAMBRIDGE RESEARCH LABS., BEDFORD, MASS.
        (INSTRUMENTATION PAPER)
IP-
        ARMY FIELD FORCES. BOARD NO. 2, FORT KNOX, KY.
        (INFORMAL PROJECT REPORT)
IP-
        BUREAU OF AERONAUTICS (NAVY). IND. PLANNING DIV.
IP-
        FRANCE. COMMISSARIAT A L'ENERGIE ATOMIQUE, PARIS
IP-
        INTERSCIENCE PUBLISHERS, INC., N.Y.C. (TRANSLATIONS)
IP-
        MICHIGAN. UNIV., ANN ARBOR. COLL. OF ENGINEERING
IPAC-(YEAR)-
        NORTHWESTERN UNIV., EVANSTON, ILL. INFORMATION
        PROCESSING AND CONTROL SYSTEMS LAB.
IPAPS-(YEAR/YEAR)-
        CALIFORNIA. UNIV., LA JOLLA. INST. FOR PURE AND
        APPLIED PHYSICAL SCIENCES
IPB-
        NAVY (ILLUSTRATED PARTS BREAKDOWN)
IPC-
        CALIFORNIA INST. OF TECH., PASADENA
IPC-
        INSTITUTE OF PAPER CHEMISTRY, APPLETON, WIS.
IPC-
        INTER-POLYMER CORP., PASSAIC, N.J.
IPD-
        ATOMIC ENERGY OF CANADA LTD., CHALK RIVER, ONT.
IPDC/P-
        GT. BRIT. CAPENHURST WORKS, CHES., ENGLAND
IPF-(YEAR)-
        STUTTGART. UNIVERSITAET. INST. FUER PLASMAFORSCHUNG
IPI-
        ATOMIC ENERGY OF CANADA LTD. CHALK RIVER PROJECT,
        ONT. (REF. 6)
IPIS-RAFB-TN-(YR.)-
        AIR FORCE INSTRUMENT PILOT INSTRUCTOR SCHOOL,
        RANDOLPH AFB, TEXAS

IPK-
        AMERICAN MACHINE AND FOUNDRY CO., ALEXANDRIA, VA.
IPM-
        AKADEMIYA NAUK SSSR. INSTITUT PRIKLADNOI MATHEMATIKI,
        MOSCOW
IPN-
        FRANCE. INSTITUT DE PHYSIQUE NUCLEAIRE, ORSAY
IPNO/LA-
        PARIS. UNIV., ORSAY. INSTITUT DE PHYSIQUE NUCLEAIRE
IPNO/TH-
        FRANCE. INSTITUT DE PHYSIQUE NUCLEAIRE, ORSAY
IPNO/TH-
        PARIS. UNIV., ORSAY. INSTITUT DE PHYSIQUE NUCLEAIRE
IPOR-
        INTERNATIONAL PUBLIC OPINION RESEARCH, INC., N.Y.C.
IPP-
        CESKOSLOVENSKA AKADEMIE VED, USTAV FYZIKY PLAZMATU,
        PRAGUE.
IPP (NUMBER/NUMBER)
        INSTITUT FUER PLASMAPHYSIK, GMBH, GARCHING, GERMANY
IPP-
        INSTITUT FUER PLASMAPHYSIK, GMBH, GARCHING, GERMANY
IPP-AR-
        INSTITUT FUER PLASMAPHYSIK, GMBH, GARCHING, GERMANY
IPPCZ-
        CESKOSLOVENSKA AKADEMIE VED, PRAGUE
IPPJ-
        NAGOYA, JAPAN. UNIV. INST. OF PLASMA PHYSICS
IPPJ-DT-
        NAGOYA, JAPAN. UNIV. INST. OF PLASMA PHYSICS
IPPJ-T-
        NAGOYA, JAPAN. UNIV. INST. OF PLASMA PHYSICS
IPR-
        (INFORMAL PROGRESS REPORT) (REF. 2)
IPR-
        (INTERIM PROGRESS REPORT) (REF. 2)
IPR-
        (INTERNAL PROGRESS REPORT) (REF. 2)
IPR-
        STANFORD RESEARCH INST., MENLO PARK, CALIF.
IPRL-
        INTERCEPTOR PILOT RES. LAB., TYNDALL AFB, FLA.
IPR-SP-
        BUREAU OF MINES. ROLLA METALLURGY RESEARCH CTR., MO.
IPST-
        ISRAEL PROGRAM FOR SCI. TRANS., LTD., JERUSALEM
IPST-CAT.-
        ISRAEL PROGRAM FOR SCI. TRANS., LTD., JERUSALEM
IPT-TM-
        C-E-I-R, INC., ARLINGTON, VA.
IPV-
        SWEDEN. ATOMKOMMITTEN, STOCKHOLM
IR-
        (INFORMAL REPORT) (REF. 2)
IR-
        (INFORMATION REPORT) (REF. 2)
IR-
        (INSTRUCTION REPORT) (REF. 2)
IR-
        (INSTRUMENTATION REPORT) (REF. 2)
IR-
        (INTELLIGENCE REPORT) (REF. 2)
IR-
        (INTELLIGENCE REVIEW) (REF. 2)
IR-
        (INTERIM REPORT) (REF. 2)
IR-
        (INTERNAL REPORT) (REF. 2)
IR-
        (INTERPRETATION REPORT) (REF. 2)
IR-
        (INTERROGATION REPT.) (REF. 2)
IR-
        (INVENTION REPORT) (REF. 2)
IR-(NUMBER)-(NUMBER)
        AIR FORCE MATERIALS LAB., WRIGHT-PATTERSON AFB, OHIO
IR-
        ALABAMA. UNIV., HUNTSVILLE. RESEARCH INSTITUTE
(NUMBER)-IR-
        BAIRD ATOMIC, INC., CAMBRIDGE, MASS.
IR-
        CORNING GLASS WORKS, N.Y.
IR-(YR.)-
        MARTIN CO., DENVER
IR-
        MARTIN-MARIETTA CORP. DENVER DIV.
IR-
        NAVAL OCEANOGRAPHIC OFFICE, WASHINGTON, D.C.
        (SUPERSEDED IMR AND IOM SERIES)
IR-(YEAR)-
        NAVAL OCEANOGRAPHIC OFFICE, WASHINGTON, D.C.
(NUMBER)-IR-
        THIOKOL CHEM. CORP. REACTION MOTORS DIV.,DENVILLE,NJ
IRA-
        ACADEMIA R.S.R.INSTITUTUL DE FIZICA ATOMICA,BUCHAREST
IRA-
        AEROJET-GENERAL CORP., SACRAMENTO, CALIF. (INDE-
        PENDENT RESEARCH AND DEVELOPMENT REPORT)
IRA-
        INFORMATION RESEARCH ASSOCIATES, INC., SAN DIEGO, CAL
IRA-TN-
        INFORMATION RESEARCH ASSOCIATES, INC.,CAMBRIDGE, MASS
IRA-TR-(NUMBER)-(YEAR)
        INFORMATION RESEARCH ASSOCIATES, INC., BERKELEY, CAL.
IRA-TR-
        INFORMATION RESEARCH ASSOCIATES, INC.,LEXINGTON, MASS
IRB-
        KERNFORSCHUNGSZENTRUM KARLSRUHE. INSTITUT FUER
        REAKTORBAUELEMENTE, GERMANY

IRBB-(NO.)(LTS.)(NO.)
    ARMY BALLISTIC MISSILE AGENCY, REDSTONE ARSENAL, ALA.
IRB-TP-(NUMBER)-(YEAR)
    YUGOSLAVIA. INSTITUT RUDJER BOSKOVIC, ZAGREB
IRC-
    CHEMICAL WARFARE SERVICE, EDGEWOOD ARSENAL, MD.
    (INSECT AND RODENT CONTROL)
IRC-
    INTERNATIONAL RESISTANCE CO., PHILADELPHIA
IRCH-(NUMBER/YEAR)-
    KERNFORSCHUNGSZENTRUM KARLSRUHE. INSTITUT FUER
    RADIOCHEMIE, GERMANY
IRCHA-
    FRANCE. INSTITUT NATIONAL DE RECHERCHE CHIMIQUE
    APPLIQUEE, PARIS
IRCO R-(NO.)-
    IRCO CORP., N.Y.C.
IR+D-
    AEROJET-GENERAL CORP., SACRAMENTO, CALIF. (INDE-
    PENDENT RESEARCH AND DEVELOPMENT REPORT)
IRD-(YEAR)-
    INTERNATIONAL RESEARCH + DEVELOPMENT CO. LTD.,
    NEWCASTLE-UPON-TYNE, ENGLAND
IRDC-
    VIRGINIA. UNIV., CHARLOTTESVILLE. INDUSTRIAL
    RESEARCH AND DEVELOPMENT CENTER
IRDD-
    OFFICE OF TECH. SERVICES. IND.RES.+DEV.DIV.,WASH.,DC
IRE-
    AIR MATERIEL COMMAND. AIR DOCUMENTS DIV., WRIGHT-
    PATTERSON AFB, OHIO
IRE-
    INSTITUTE OF RADIO ENGINEERS, N.Y.C.
IRECO-
    INTERMOUNTAIN RES.+ENG.CO.,INC.,SALT LAKE CITY,UTAH
IRI-(NUMBER)-(YEAR)-
    INTERUNIVERSITAIR REACTOR INSTITUUT, DELFT
IRIA-
    MICHIGAN. UNIV., ANN ARBOR. COLL. OF ENGINEERING
IRIG-(NUMBER)-(YEAR)
    WHITE SANDS MISSILE RANGE. INTER-RANGE INSTRUMEN-
    TATION GROUP, N. MEX.
IRIG-
    WHITE SANDS MISSILE RANGE. INTER-RANGE INSTRUMEN-
    TATION GROUP, N. MEX.
IRIG-MWG-
    INTER-RANGE INSTRUMENTATION GROUP. METEOROLOGICAL
    WORKING GROUP, WHITE SANDS MISSILE RANGE, N. MEX.
IRIS SYM-
    INFRARED INFORMATION SYMPOSIA. (PROCEEDINGS)
IRL-
    INDUSTRIAL RESEARCH LABS., BALTIMORE
IRL-
    STANFORD UNIV., CALIF. DEPT. OF GENETICS
IRL-
    STANFORD UNIV., CALIF. INSTRUMENTATION RES. LAB.
IRM-
    OFFICE OF THE CHIEF OF NAVAL OPERATIONS. OPERATIONS
    EVALUATION GROUP, WASHINGTON, D.C.
IRM-
    WHITE SANDS MISSILE RANGE, N. MEX. (INTEGRATED RANGE
    MISSION)
IRMFSG-
    INTER-RANGE MISSILE FLIGHT SAFETY GROUP, WHITE
    SANDS MISSILE RANGE, N. MEX.
IRMGS-
    INTER-RANGE MISSILE GROUND SAFETY GROUP, WHITE
    SANDS MISSILE RANGE, N. MEX.
IRM-NO.-
    UNION CARBIDE RESEARCH INST., TARRYTOWN, N.Y. (INST.
    RESEARCH MEMORANDUM)
IRN-
    FRANCE. INSTITUT DE RECHERCHES NUCLEAIRES, STRASBOURG
IRP-
    AEROJET-GENERAL CORP., SACRAMENTO, CALIF. (INDE-
    PENDENT RESEARCH AND DEVELOPMENT REPORT)
IRP-
    NAVAL APPLIED SCIENCE LAB., BROOKLYN
IRP-(NUMBER/YEAR)
    SWITZERLAND. LABORATOIRE DE RECHERCHES SUR LA
    PHYSIQUE DES PLASMAS, LAUSANNE
IRPL-
    NATIONAL BUREAU OF STANDARDS. CENTRAL RADIO PROPAGA-
    TION LAB., BOULDER, COLO.
IRPL-
    NATIONAL BUREAU OF STANDARDS. INTERSERVICE RADIO
    PROPAGATION LAB., WASH., D.C. (USED 1941-48)
IRPO-(YEAR)(LETTER)-
    PARIS. UNIV., ORSAY. INSTITUT DE PHYSIQUE NUCLEAIRE
IRPO-(YR.)E-
    PARIS. UNIVERSITE, ORSAY. INSTITUT DU RADIUM
IRPO-
    PARIS. UNIVERSITE, ORSAY. INSTITUT DU RADIUM
IRS-(LETTER)-
    TECHNISCHER UEBERWACHUNGS-VEREIN E.V., COLOGNE.
    INSTITUT FUER REAKTORSICHERHEIT
IRSIA-
    BELGIUM. INSTITUT POUR L'ENCOURAGEMENT DE LA
    RECHERCHE SCIENTIFIQUE DANS L'INDUSTRIE ET
    L'AGRICULTURE, BRUSSELS
IRSID-
    FRANCE. INSTITUT DE RECHERCHES DE LA SIDERURGIE,
    SAINT GERMAIN-EN-LAYE
IRS-NO.-
    UNION CARBIDE RESEARCH INST., TARRYTOWN, N.Y. (INST.
    RESEARCH SUMMARY)
IRS-T-
    TECHNISCHER UEBERWACHUNGS-VEREIN E.V., COLOGNE.
    INSTITUT FUER REAKTORSICHERHEIT

IRTWG-
    INTER-RANGE INSTRUMENTATION GROUP
IR-USGS-
    GEOLOGICAL SURVEY, WASHINGTON, D.C. (INTERAGENCY RPT)
IRWA-
    AEROJET-GENERAL CORP., SACRAMENTO, CALIF. (INDE-
    PENDENT RESEARCH AND DEVELOPMENT REPORT)
IS-
    (INFORMATION SERIES) (REF. 2)
IS-
    ACADEMIA R.S.R.INSTITUTUL DE FIZICA ATOMICA,BUCHAREST
IS-
    AMES LAB., IOWA
IS(NO.)M(NO.)P(NO.)
    BUNKER-RAMO CORP., FORT HUACHUCA, ARIZ.
IS-
    IOWA. STATE UNIV., IOWA CITY
IS-
    IOWA STATE UNIV. OF SCIENCE AND TECHNOLOGY, AMES
IS-
    NATIONAL RESEARCH COUNCIL OF CANADA, OTTAWA
IS-(NUMBER)/I-
    SPAIN. JUNTA DE ENERGIA NUCLEAR, MADRID
ISA-
    INSTRUMENT SOCIETY OF AMERICA, PITTSBURGH
ISAS-
    TOKYO UNIV.
ISAV-MEMO-
    SOUTHAMPTON, ENGLAND. UNIV. INSTITUTE OF SOUND AND
    VIBRATION RESEARCH
ISBN-(NO.-NO.-NO.)-
    DANISH METEOROLOGICAL INST., CHARLOTTENLUND
ISC-
    AMES LAB., IOWA
ISC-
    CHRYSLER CORP. MISSILE DIV., DETROIT
ISC-
    IOWA STATE COLL., AMES
ISCE-
    INTERNATIONAL SEISMOLOGICAL CENTRE, EDINBURGH
ISC-FR-
    INDUSTRIAL SCIENTIFIC CO., N.Y.C.
ISC-PTR-(NUMBER).(NUMBER)
    IOWA STATE COLL., AMES. STATISTICAL LAB.
ISCS-
    SANDERS ASSOCIATES, INC., NASHUA, N.H.
ISC-TR-
    IOWA STATE COLL., AMES. STATISTICAL LAB. (TECH. REPT)
ISC-TR-(NUMBER).(NUMBER)
    IOWA STATE COLL., AMES. STATISTICAL LAB.
ISD-
    AUSTRALIA. WEAPONS RESEARCH ESTABLISHMENT
ISD-BIB-
    LOS ALAMOS SCIENTIFIC LAB., N. MEX. (BIBLIOGRAPHY)
ISDO-
    KANSAS STATE UNIV., MANHATTAN. INST. FOR SYSTEMS
    DESIGN AND OPTIMIZATION
ISEG-
    ALASKA. UNIV., COLLEGE. INST. OF SOCIAL, ECONOMIC
    AND GOVERNMENT RESEARCH
ISEGR-
    ALASKA. UNIV., COLLEGE. INST. OF SOCIAL, ECONOMIC
    AND GOVERNMENT RESEARCH
ISL-
    DEUTSCHE VERSUCHSANSTALT FUER LUFTFAHRT
ISL-
    ITEK CORP. INFO. SCIENCES LAB., LEXINGTON, MASS.
ISL-
    MICHIGAN. UNIV., ANN ARBOR. INFORMATION SYSTEMS LAB.
ISL-D-
    INSTITUT FRANCO-ALLEMAND DE RECHERCHES, ST. LOUIS,
    FRANCE
ISL-N-(NUMBER/YEAR)
    INSTITUT FRANCO-ALLEMAND DE RECHERCHES, ST. LOUIS,
    FRANCE
ISL-ST-
    INSTITUT FRANCO-ALLEMAND DE RECHERCHES, ST. LOUIS,
    FRANCE
ISL-T-
    INSTITUT FRANCO-ALLEMAND DE RECHERCHES, ST. LOUIS,
    FRANCE
ISMET-
    GT. BRIT. INTER-SERVICES METALLURGICAL RESEARCH
    COUNCIL, LONDON
ISO-
    INTERNATIONAL ORGANIZATION FOR STANDARDIZATION,GENEVA
ISO-
    ISOCHEM, INC., RICHLAND, WASH.
ISO-
    ISOTOPES, INC., WESTWOOD, N.J.
ISO-SA-
    ISOCHEM, INC., RICHLAND, WASH.
ISOTOPE CATALOG-
    OAK RIDGE NATIONAL LAB., TENN.
ISOTOPES-QR-
    ISOTOPES, INC., WESTWOOD, N.J. (PROGRESS REPORT)
IS/P-
    BROOKHAVEN NATIONAL LAB., UPTON, N.Y.
ISP-
    EUROPEAN ATOMIC ENERGY COMMUNITY. JOINT NUCLEAR
    RESEARCH CENTER, ISPRA, ITALY
IS-PR(MO./YR.-MO./YR.)
    IOWA STATE UNIV. OF SCIENCE AND TECHNOLOGY, AMES
ISPRA-
    EUROPEAN ATOMIC ENERGY COMMUNITY. JOINT NUCLEAR
    RESEARCH CENTER, ISPRA, ITALY
IS-QRMO-
    BROWN ENGINEERING CO., INC. INFO. SYSTEMS DIV.,
    HUNTSVILLE, ALA.

ISR-	(INTERIM SCIENTIFIC REPORT) (REF. 2)
ISR-	(INTERNAL SCIENTIFIC REPORT) (REF. 2)
ISR-	EUROPEAN ORGANIZATION FOR NUCLEAR RESEARCH, GENEVA
ISR-	HUGHES RESEARCH LABS. DIV. OF HUGHES AIRCRAFT CO., MALIBU, CALIF.
ISR-	MICHIGAN. UNIV., ANN ARBOR
IS-RD-	AMES LAB., IOWA (REACTOR MATERIALS DEVELOPMENT)
IS-RIC-	IOWA STATE UNIV. OF SCIENCE AND TECHNOLOGY, AMES. RARE-EARTH INFORMATION CENTER
ISS-	ITALY. ISTITUTO SUPERIORE DI SANITA, ROME
ISS (YEAR/NUMBER)	ITALY. ISTITUTO SUPERIORE DI SANITA, ROME
ISS-LRP-	AEROJET-GENERAL CORP. LIQUID ROCKET PLANT, SACRAMENTO, CALIF.
ISSP-A-	TOKYO UNIV. INST. FOR SOLID STATE PHYSICS
IS-T-	AMES LAB., IOWA
IS-T-	IOWA STATE UNIV. OF SCIENCE AND TECHNOLOGY, AMES
IST-	NATIONAL RESOURCE ANALYSIS CENTER, WASH., D.C.
ISTD-	INTER-SERVICE TOPOGRAPHICAL DEPT.
ISTRACON-	INTERSTATION SUPERSONIC TRACK CONFERENCE. STRUCTURES WORKING GROUP, HOLLOMAN AFB, N.MEX.
IS-TRANS-	IOWA STATE UNIV. OF SCIENCE AND TECHNOLOGY, AMES. INST. FOR ATOMIC RESEARCH
ISTR-TR-	COLORADO. UNIV., CLIMAX. INST. FOR SOLAR TERRESTRIAL RESEARCH
ISU-	CAMBRIDGE LANGUAGE RESEARCH UNIT, ENGLAND
ISU-	IOWA. STATE UNIV., IOWA CITY
ISU-	IOWA STATE UNIV. OF SCIENCE AND TECHNOLOGY, AMES
ISU-EKI-AMES-	IOWA STATE UNIV. OF SCIENCE AND TECHNOLOGY, AMES ENGINEERING RESEARCH INST.
ISVR-	SOUTHAMPTON, ENGLAND. UNIV. INSTITUTE OF SOUND AND VIBRATION RESEARCH
ISVR-TR-	SOUTHAMPTON, ENGLAND. UNIV. INSTITUTE OF SOUND AND VIBRATION RESEARCH
ISWS-	ILLINOIS. STATE WATER SURVEY, URBANA
IT-	ALBRIGHT + WILSON, LTD., OLDBURY, BIRMINGHAM, ENG.
IT-	GENERAL ELECTRIC CO., ITHACA, N.Y.
IT-	STANFORD UNIV., CALIF. DEPT. OF MECHANICAL ENG., THERMOSCIENCES DIV.
ITAC-	PUBLIC HEALTH SERVICE. DIV. OF RADIOLOGICAL HEALTH, WASH., D.C.(INTERLAB TECHNICAL ADVISORY COMM.)
ITAL-	NETHERLANDS. INSTITUUT VOOR TOEPASSING VAN ATOOMENERGIE IN DE LANDBOUW, WAGENINGEN
ITE-	I.T.E. CIRCUIT BREAKER CO., PHILADELPHIA
ITEF-	U.S.S.R. GOSUDARSTVENNYI KOMITET PO ISPOL'ZOVANIYU ATOMNOI ENERGII, MOSCOW
ITEK-	ITEK CORP., LEXINGTON, MASS.
ITEK-	ITEK CORP. VIDYA DIV., PALO ALTO, CALIF.
ITEP-	AKADEMIYA NAUK SSSR. INSTITUT TEORETICHESKOI I EKSPERIMENTAL'NOI FIZIKI, MOSCOW
ITF-(YEAR)-	AKADEMIYA NAUK SSSR. INSTITUT YADERNOI FIZIKI, NOVOSIBIRSK
ITF-(YEAR)-	AKADEMIYA NAUK UKRAINSKOI SSR. FIZIKO-TEKHNICHESKII INSTITUT, KHARKOV
ITF-(YEAR)-	AKADEMIYA NAUK UKRAINSKOI SSR. INSTITUT TEORETICHESKOI FIZIKI, KIEV
ITF-	FOREST SERVICE. INSTITUTE OF TROPICAL FORESTRY, RIO PIEDRAS, PUERTO RICO
ITI-	(ITALIAN TECHNICAL INVESTIGATION)
ITI-	INDUSTRIAL TECTONICS, INC., COMPTON, CALIF.
ITI-BIB-	INFORMATICS TISCO, INC., RIVERDALE, MD.
ITIS-(YEAR)-	GEORGIA INST. OF TECH., ATLANTA. SCHOOL OF INFORMATION AND COMPUTER SCIENCE
ITJ-	POLAND. INST. OF NUCLEAR TECHNIQUES, KRAKOW
ITJ-(NUMBER)/(LTR(S))	POLAND. INST. OF NUCLEAR TECHNIQUES, KRAKOW
ITK-(YEAR)-	AKADEMIYA NAUK UKRAINSKOI SSR. INSTITUT TEORETICHESKOI FIZIKI, KIEV
ITL-	INTERNATIONAL TELECOMMUNICATION LABS., INC., N.Y.C.
ITL-	PHILADELPHIA NAVAL SHIPYARD. INDUSTRIAL TEST LAB.
ITL-SR-	INLAND TESTING LABS., MORTON GROVE, ILL.
ITM-	(INTERIM TECHNICAL MEMORANDUM) (REF. 2)
ITM-	(INTERNAL TECHNICAL MEMORANDUM) (REF. 2)
ITN-	(INTERIM TECHNICAL NOTE) (REF. 2)
ITN/RF-	ANSALDO S.P.A., GENOA
ITP-	STANFORD UNIV., CALIF. INST. OF THEORETICAL PHYSICS
ITR-	(INTERIM TECHNICAL REPORT) (REF. 2)
ITR-	(INTERIM TEST REPORT) (REF. 2 AND REF. 7)
ITR-	(INTERNAL TECHNICAL REPORT) (REF. 2)
ITR-(YEAR)-	LOUISVILLE, KY. UNIV. PERFORMANCE RESEARCH LAB.
ITS-	INSTITUTE FOR TELECOMMUNICATION SCIENCES, BOULDER, COLO
ITSA-	INSTITUTE FOR TELECOMMUNICATION SCIENCES AND AERONOMY, BOULDER, COLO.
ITSA-TM-	NATIONAL BUREAU OF STANDARDS. BOULDER LABS., COLO.
ITT-	ILLINOIS INST. OF TECH., CHICAGO
ITT-	INTERNATIONAL TELEPHONE AND TELEGRAPH CORP., N.Y.C.
ITTCS-	ITT COMMUNICATION SYSTEMS, INC., PARAMUS, N.J.
ITTDC-(YEAR)-	ITT DEFENSE COMMUNICATIONS DIV., NUTLEY, N.J.
ITTE-	CALIFORNIA. UNIV., BERKELEY. INSTITUTE OF TRANSPORTATION AND TRAFFIC ENGINEERING
ITT-EPL-	ITT ELECTRO-PHYSICS LABS., INC., HYATTSVILLE, MD.
ITTFL-	ITT FEDERAL LABS., NUTLEY, N.J.
ITTFL-(YEAR)-	ITT INDUSTRIAL LABS., FORT WAYNE
ITTFL-TD-	IFF FEDERAL LABS., NUTLEY, N.J.
ITTIL-(YEAR)-	ITT INDUSTRIAL LABS., FORT WAYNE
ITT-R-	INSTITUTE OF TEXTILE TECH., CHARLOTTESVILLE, VA.
ITT-TM-	ITT LABS., NUTLEY, N.J.
ITT-TR-	INTERNATIONAL TELEPHONE + TELEGRAPH CORP., NUTLEY, N.J.
IU-	ILLINOIS. UNIV., URBANA
IU-	IOWA. STATE UNIV., IOWA CITY. DEPT. OF PHYSICS AND ASTRONOMY
IU-AL-	ILLINOIS. UNIV., URBANA. ANTENNA LAB.
IU-CL-TR-	ILLINOIS. UNIV., URBANA. CYCLOTRON LAB.
IU DCE-	ILLINOIS. UNIV., URBANA. DEPARTMENT OF CERAMIC ENG.
IU DCE MEMO-	ILLINOIS. UNIV., URBANA. DEPARTMENT OF CERAMIC ENG.
IU-DCL-	ILLINOIS. UNIV., URBANA. DIGITAL COMPUTER LAB.
IU-DCS-	ILLINOIS. UNIV., URBANA. DEPT. OF COMPUTER SCIENCE
IU-EES-	ILLINOIS. UNIV., URBANA. ENGINEERING EXPERIMENT STA.
IU-I-	ILLINOIS. UNIV., URBANA. COORDINATED SCIENCE LAB.
IU-R-	ILLINOIS. UNIV., URBANA. COORDINATED SCIENCE LAB.
IU-RR-(YEAR)-	ILLINOIS. UNIV., URBANA. SMALL HOMES COUNCIL-BUILDING RESEARCH COUNCIL
IUSAFS-	INTER-UNIV SEMINAR ON ARMED FORCES AND SOCIETY, WILMETTE, ILL.
IU-SRS-	ILLINOIS. UNIV., URBANA (STRUCTURAL RESEARCH SERIES)
IU-T+AM-	ILLINOIS. UNIV., URBANA. DEPT. OF THEORETICAL AND APPLIED MECHANICS
IVA-	SWEDEN. INGENIORVETENSKAPSAKADEMIEN, STOCKHOLM
IV-AG-	GENERAL MOTORS CORP., MILWAUKEE
IVA-MEDD-	NETHERLANDS. INSTITUUT TNO VOOR VERPAKKINGEN, DELFT
IVV-	NETHERLANDS. INSTITUUT TNO VOOR VERPAKKINGEN, DELFT
IVY-	(OPERATION IVY) (REF. 7)
IWA-	AEROJET-GENERAL CORP., SACRAMENTO, CALIF. (INDEPENDENT RESEARCH AND DEVELOPMENT REPORT)
IWB-	COMMITTEE ON MEDICAL RES. (INFECTED WOUNDS AND BURNS)

IWL-
- ISOTOPES, INC., WESTWOOD, N.J.

IWP-
- CALIFORNIA. UNIV., BERKELEY. SPACE SCIENCES LAB.

IWR-
- ALASKA. UNIV., COLLEGE. INST. OF WATER RESOURCES

IWTR/
- INDIAN INST. OF SCIENCE, BANGALORE

IWW-SPECIAL STUDY-
- WEATHER WING (1ST), SAN FRANCISCO

IX-QPR-
- COLUMBIA UNIV., N.Y.C. ENGINEERING RESEARCH LABS.

IXR-
- COLUMBIA UNIV., N.Y.C. DIV. OF WAR RESEARCH

IX-TN-
- COLUMBIA UNIV., N.Y.C. ENGINEERING RESEARCH LABS.

IY-
- FAIRCHILD ENGINE AND AIRPLANE CORP. NEPA DIV., OAK RIDGE, TENN. (REF. 31)

IYAF-(NUMBER)-(YEAR)
- AKADEMIYA NAUK SSSR. INSTITUT YADERNOI FIZIKI, NOVOSIBIRSK

IZF-
- NETHERLANDS. INSTITUUT VOOR ZINTUIGFYSIOLOGIE RVO-TNO, SOESTERBERG

J-	ARMY, WASHINGTON, D.C.
J-	ATOMIC ENERGY OF CANADA LTD. CHALK RIVER PROJ., ONT.
J-	COMMITTEE ON MEDICAL RESEARCH (JOHNSON REPORTS)
J-	CZECHOSLOVAKIA. VYZKUMNY USTAV MATEMATICKYCH STROJU, PRAGUE
J-	DYNETICS, INC., MOUNTAIN LAKES, N.J.
J-	JAPAN
J-	JIKEI UNIV., TOKYO. SCHOOL OF MEDICINE
J-	KEIO UNIV., TOKYO. DEPT. OF ANATOMY
J-	KEIO UNIV., TOKYO. SCHOOL OF MEDICINE
J-	LOS ALAMOS SCIENTIFIC LAB., N. MEX.
J-1- THRU J-18-	LOS ALAMOS SCIENTIFIC LAB., N. MEX. (INTERNAL CORRESPONDENCE SERIAL USED BY GROUPS IN J DIV.)
J-(NUMBER)-	MATSUSHITA RES. INST. TOKYO, INC., KAWASAKI, JAPAN
J-	MOTOROLA INC., PHOENIX, ARIZ.
J-	NARA MEDICAL COLL., KASHIHARA CITY, JAPAN. DEPT. OF PHYSIOLOGY
J-	OFFICE OF SCIENTIFIC RESEARCH AND DEVELOPMENT
J-(NO.)(LTR.)-	PYROGENICS, INC., WOODSIDE, N.Y.
J(NUMBER)-FR-	SCIENTIFIC-ATLANTA, INC.
J(NUMBER)-QR-	SCIENTIFIC-ATLANTA, INC.
J-	TECHNE, LTD., CAMBRIDGE, ENGLAND
J-	TECHNICAL INFORMATION SERVICE, AEC (SERIES ASSIGNED TO JAPANESE REPORTS ABOUT HIROSHIMA + NAGASAKI)
J-	TOKYO UNIV.
J-	UNITED AIRCRAFT CORP., EAST HARTFORD, CONN.
J(LETTER)C-	OFFICE OF SCIENTIFIC RESEARCH AND DEVELOPMENT
JA-	CALIFORNIA. UNIV., BERKELEY. DEPT. OF ELEC. ENG.
JA-	DEL MAR ENGINEERING LABS., SANTA MONICA, CALIF.
JA-	MASSACHUSETTS INST. OF TECH., CAMBRIDGE. RESEARCH LAB. OF ELECTRONICS
JA-	MASSACHUSETTS INST. OF TECH., LEXINGTON. LINCOLN LAB.
JA-(NUMBER)-	SPERRY PIEDMONT CO., CHARLOTTESVILLE, VA.
JAB-	BLUME (JOHN A.) + ASSOCS. RES. DIV., SAN FRANCISCO
JAB-(NUMBER)-	BLUME (JOHN A.) + ASSOCS. RES. DIV., SAN FRANCISCO
JAC-	JACOBS INSTRUMENT CO., BETHESDA, MD.
JAC-PAPER-	JOINT AIRWORTHINESS COMMITTEE, LONDON
JADB-	JOINT AIR DEFENSE BOARD, ENT AFB, COLO.
JAEIC-	JOINT ATOMIC ENERGY INTELLIGENCE COMM., WASH. D.C.
JAERI-	JAPAN. ATOMIC ENERGY RESEARCH INSTITUTE, TOKYO
JAERI-MEMO-	JAPAN. ATOMIC ENERGY RESEARCH INSTITUTE, TOKYO
JAFNC-	AIR FORCE-NAVY COMMITTEE (REF. 13)
JAG-	OFFICE OF THE JUDGE ADVOCATE GEN. (NAVY) WASH., D.C.
JAIEG-	JOINT ATOMIC INFORMATION EXCHANGE GROUP, WASH., D.C.
JAJ-CR-	SANDERS ASSOCIATES, INC., BEDFORD, MASS.
JAL-	CALIFORNIA. UNIV., LIVERMORE. RADIATION LAB. ELECTRONIC ENGINEERING DEPT.
JAN-	ARMY-NAVY (REF. 13)
JAN-(LETTER)-	ARMY-NAVY (SPECIFICATION) (REF. 13)
JANAF-	ARMY-NAVY-AIR FORCE (COVERS VARIOUS COMMITTEES AND THEIR PUBLICATIONS) (REF. 13)
JANAF-BUL-	JOINT ARMY-NAVY-AIR FORCE SOLID PROPELLANT GROUP (REF. 13)
JANAF-IP-	JOINT ARMY-NAVY-AIR FORCE IGNITABILITY PANEL (REF.13)
JANAF-ITT-	ARMY-NAVY-AIR FORCE (INTERIM THERMOCHEMICAL TABLES) (REF. 13)
JANAF-SERIAL-	JOINT ARMY-NAVY-AIR FORCE FUZE COMM., WASH.,D.C.
JANAIR-(NUMBER)-(NUMBER)-	JOINT ARMY NAVY AIRCRAFT INSTRUMENTATION RESEARCH PROJECT, WASHINGTON, D.C.
JANAIR-	JOINT ARMY NAVY AIRCRAFT INSTRUMENTATION RESEARCH PROJECT, WASHINGTON, D.C.
JANAIR-D(NUMBER)-	JOINT ARMY NAVY AIRCRAFT INSTRUMENTATION RESEARCH PROJECT, WASHINGTON, D.C.
JANAIR-EL-	JOINT ARMY NAVY AIRCRAFT INSTRUMENTATION RESEARCH PROJECT, WASHINGTON, D.C.
JANAIR-TR-	JOINT ARMY NAVY AIRCRAFT INSTRUMENTATION RESEARCH PROJECT, WASHINGTON, D.C.
JANAIR-TR-D(NUMBERS)-	JOINT ARMY NAVY AIRCRAFT INSTRUMENTATION RESEARCH PROJECT, WASHINGTON, D.C.
JANAP-	OFFICE OF THE JOINT CHIEFS OF STAFF, WASH., D.C. (JOINT ARMY-NAVY-AIR FORCE PUBLICATION) (REF. 13)
JANP-	OFFICE OF THE JOINT CHIEFS OF STAFF, WASH., D.C. (JOINT ARMY-NAVY PUBLICATION) (REF. 13)
JAN-STD-	ARMY-NAVY (STANDARD) (REF. 13)
JAN-STD-	MUNITIONS BOARD STANDARDS AGENCY, WASHINGTON, D.C.
JAN WM-	ARMY-NAVY (WORK MANUAL) (REF. 13)
JAPFNR-	TECHNICAL INFORMATION CENTER, AEC, OAK RIDGE, TENN. (ASSIGNED TO REPORTS FROM THE U.S.-JAPANESE FAST REACTOR EXCHANGE PROGRAM)
JAT-(YEAR)-	RURAL COOPERATIVE POWER ASSN., ELK RIVER, MINN.
JATB-	JOINT AIRBORNE TROOP BOARD, FORT BRAGG, N.C.
JAWPB-	JOINT ATOMIC WEAPONS PUBLICATIONS BOARD, ALBUQUERQUE, N. MEX. (REF. 14)
J/B-(NUMBERS)-	ATLANTIC RESEARCH CORP., ALEXANDRIA, VA.
JB-	LOS ALAMOS SCIENTIFIC LAB., N. MEX.
JBI-	JANSKY + BAILEY, WASHINGTON, D.C.
JBSIP-	JOINT BOARD ON SCIENTIFIC INFORMATION POLICY
JC-	CARBIDE AND CARBON CHEMICALS CORP., OAK RIDGE, TENN.
JC-(LETTERS)-(YEAR)-	JOINT COMMITTEE (AEC-DOD) (REF. 15)
JCB-	OFFICE OF THE JOINT CHIEFS OF STAFF, WASH., D.C.
JCC/ASTOR-	ASTOR JOINT AEC-DOD COORDINATING COMM., SAFETY AND RELIABILITY SUBCOMM.
JCC/HAWK-	HAWK JOINT AEC-DOD COORDINATING COMMITTEE
JCC/SRGNT-	SERGEANT JOINT AEC-DOD WARHEAD COORDINATING COMM.
JCC/TYPHON-	TYPHON JOINT AEC-DOD COORDINATING COMM.
JCH-	TENNESSEE EASTMAN CORP., OAK RIDGE, TENN.
JCL-	JOHN CRERAR LIBRARY, CHICAGO
JC/NHWI-	NIKE-HERCULES SAFETY SUBCOMM. OF THE JOINT COMM. ON NIKE-HERCULES WARHEAD INSTALLATIONS
JCP-	CONGRESS. JOINT COMMITTEE ON PRINTING, WASHINGTON, DC
JCP-(NO.)-(NO.)	DESERET TEST CENTER. JOINT CONTACT POINT DIV., FORT DOUGLAS, UTAH
JCRL-	CALIFORNIA. UNIV., BERKELEY. DEPT. OF PHYSICS
JCS-	OFFICE OF THE JOINT CHIEFS OF STAFF, WASH., D.C.
JCTR-	GENERAL ELECTRIC CO. AVIONIC CONTROLS DEPT., BINGHAMTON, N. Y.
JCU-	JOHN CARROLL UNIV., CLEVELAND
JDK-	BROOKHAVEN NATIONAL LAB., UPTON, N.Y. (CODE IS AUTHOR'S INITIALS)
JDL-	DEUTSCHE LUFTFAHRTFORSCHUNG (JAHRBUCH)
JDT-	JOINT DESIGN TEAM (ARMY), WARREN, MICH.
JDT-S	TENNESSEE EASTMAN CORP., OAK RIDGE, TENN.
JE-(YR)-	ALUMINUM CO. OF AMERICA. ALCOA PROCESS DEVELOPMENT LABS., NEW KENSINGTON, PENNA.
JEIA-	JOINT ELECTRONICS INFORMATION AGENCY, WASH., D.C.
JEN-	SPAIN. JUNTA DE ENERGIA NUCLEAR, MADRID
JEN-(NO.)-DQ/I-	SPAIN. JUNTA DE ENERGIA NUCLEAR. DIRECCION DE QUIMICA E ISOTOPES, MADRID.
JEN-(NO.)-IFIC/I-	SPAIN. JUNTA DE ENERGIA NUCLEAR. DIVISION DE FISICA, MADRID
JEN-(NO.)-DF/I-	SPAIN. JUNTA DE ENERGIA NUCLEAR. DIVISION DE FISICA, MADRID
JEN-(NO.)-DMA/I-	SPAIN. JUNTA DE ENERGIA NUCLEAR. DIVISION DE MATERIALES, MADRID

JEN-(NO.)-DME/I-
: SPAIN. JUNTA DE ENERGIA NUCLEAR. DIVISION DE METALLURGIA, MADRID

JENER-
: JOINT EST. FOR NUCLEAR ENERGY RES., KJELLER, NORWAY

JENER-PUB.-
: JOINT EST. FOR NUCLEAR ENERGY RES., KJELLER, NORWAY

JER-
: THOMPSON PRODUCTS, INC., CLEVELAND

JER-
: THOMPSON PRODUCTS, INC. JET DIV. (ENGINEERING REPORT)

JF-
: JOINT TASK FORCE ... (REF. 17)

JFSG (LETTERS)
: (JOINT FEASIBILITY STUDY GROUP) (REF. 16)

JG-
: MICHIGAN STATE UNIV., EAST LANSING. DEPT. OF STATISTICS

JGC/EAK/AVS-
: BROOKHAVEN NATIONAL LAB., UPTON, N.Y.

JGERA-
: J. G. ENGINEERING RESEARCH ASSOCIATES, BALTIMORE

JH-
: GT. BRIT. MINISTRY OF AVIATION, LONDON

JHJ-CR-
: SANDERS ASSOCIATES, INC., BEDFORD, MASS.

JHRP-
: PURDUE UNIV., LAFAYETTE,IND. JOINT HIGHWAY RES. PROJ.

JHU-
: JOHNS HOPKINS UNIV., BALTIMORE

JHU-(NUMBER)-(LETTER(S))
: JOHNS HOPKINS UNIV., BALTIMORE (SQUID PROJ.)(REF. 33)

JHU-
: VIRGINIA. UNIV., CHARLOTTESVILLE

JHU-AF-
: JOHNS HOPKINS UNIV., BALTIMORE. RADIATION LAB.

JHU-APL-
: JOHNS HOPKINS UNIV., SILVER SPRING, MD. APPLIED PHYSICS LAB.

JHU APL BB-
: JOHNS HOPKINS UNIV., SILVER SPRING, MD. APPLIED PHYSICS LAB. (BUMBLEBEE PROJECT) (REF. 8)

JHU APL CF-
: JOHNS HOPKINS UNIV., SILVER SPRING, MD. APPLIED PHYSICS LAB.

JHU APL CM-
: JOHNS HOPKINS UNIV., SILVER SPRING, MD. APPLIED PHYSICS LAB.

JHU APL SPIA/(L + R)
: JOHNS HOPKINS UNIV., SILVER SPRING, MD. SOLID PROPELLANT INFORMATION AGENCY

JHU APL TG-
: JOHNS HOPKINS UNIV., SILVER SPRING, MD. APPLIED PHYSICS LAB.

JHU-CRSC-
: JOHNS HOPKINS UNIV., BALTIMORE. CENTER FOR RESEARCH IN SCIENTIFIC COMMUNICATION

JHU-DME-
: JOHNS HOPKINS UNIV., BALTIMORE. DEPT. OF MECH. ENG.

JHU-ICR-
: JOHNS HOPKINS UNIV., BALTIMORE. INST. FOR COOP. RES.

JHU(INITIAL)-
: JOHNS HOPKINS UNIV., BALTIMORE

JHU ORO R-
: JOHNS HOPKINS UNIV., CHEVY CHASE, MD. OPERATIONS RESEARCH OFFICE

JHU ORO S-
: JOHNS HOPKINS UNIV., CHEVY CHASE, MD. OPERATIONS RESEARCH OFFICE. (STAFF MEMORANDA)

JHU ORO T-
: JOHNS HOPKINS UNIV., CHEVY CHASE, MD. OPERATIONS RESEARCH OFFICE. (TECHNICAL MEMORANDA)

JHU-PC-
: JOHNS HOPKINS UNIV., SEABROOK, N.J. (PUBLICATIONS IN CLIMATOLOGY)

JHU-PL-
: JOHNS HOPKINS UNIV., BALTIMORE. PSYCHOLOGICAL LAB.

JHU-RL-
: JOHNS HOPKINS UNIV., BALTIMORE. RADIATION LAB.

JHU RL AF-
: JOHNS HOPKINS UNIV., BALTIMORE. RADIATION LAB.

JHU/RL-TR-AF-
: JOHNS HOPKINS UNIV., BALTIMORE. RADIATION LAB.

JHU-SQUID-
: JOHNS HOPKINS UNIV., BALTIMORE (SQUID PROJ.)(REF. 33)

JHU-SR-
: JOHNS HOPKINS UNIV., BALTIMORE

JHU-TR-
: JOHNS HOPKINS UNIV., BALTIMORE

JHUX-
: JOHNS HOPKINS UNIV., BALTIMORE

JHW-
: CALIFORNIA. UNIV., BERKELEY. RADIATION LAB.

JIB-
: GT. BRIT. MINISTRY OF DEFENCE, JOINT INTELL.BUR.,LOND

JILA-
: JOINT INST. FOR LAB. ASTROPHYSICS, BOULDER, COLO.

JINR-(LETTER)(NUMBER)-
: U.S.S.R. JOINT INST. FOR NUCLEAR RESEARCH, DUBNA

JINR-D-
: U.S.S.R. JOINT INST. FOR NUCLEAR RESEARCH, DUBNA

JINR-DC-
: U.S.S.R. JOINT INST. FOR NUCLEAR RESEARCH, DUBNA

JINR-E-
: U.S.S.R. JOINT INST. FOR NUCLEAR RESEARCH, DUBNA

JINR-E(NUMBER)-
: U.S.S.R. JOINT INST. FOR NUCLEAR RESEARCH, DUBNA

JINR-P-
: U.S.S.R. JOINT INST. FOR NUCLEAR RESEARCH, DUBNA

JINR-P(NUMBER)-
: U.S.S.R. JOINT INST. FOR NUCLEAR RESEARCH, DUBNA

JIOA-
: JOINT INTELLIGENCE OBJECTIVES AGENCY, WASH., D.C.

JIOA C-
: TECHNICAL INDUSTRIAL INTELL. COMM. (COMMUNICATIONS)

J.I.R.P.-(NO.)
: (JUNEAU ICE RESEARCH PAPER REPORT)

JIU-
: PORTUGAL. JUNTA DE INVESTIGACOES DO ULTRAMAR, LISBON

JLI-(NUMBER)-
: JOHNSTON LABS., INC., BALTIMORE

JLI-
: JOHNSTON LABS., INC., BALTIMORE

JLI-(NUMBER-NUMBER)-
: JOHNSTON LABS., INC., BALTIMORE

JLI-(NUMBER)-
: JOHNSTON LABS., INC., COCKEYSVILLE, MD.

JLI-
: JOHNSTON (WILLIAM H.) LABS., INC., LAFAYETTE, IND.

JLI-(NO.)-(NO.)-(NO.)-
: JOHNSTON (WILLIAM H.) LABS., INC., LAFAYETTE, IND.

J+M-
: JACKSON AND MORELAND, BOSTON

JM-
: TENNESSEE EASTMAN CORP., OAK RIDGE, TENN.

JMC-
: JOINT METEOROLOGICAL COMMITTEE, WASHINGTON, D.C.

JMC-
: TECHNICAL INFORMATION SERVICE, AEC (JOURNAL OF METALLURGY AND CERAMICS)

JMC/E-
: JOINT METEOROLOGICAL COMMITTEE, WASHINGTON, D.C.

JMC/WC-
: JOINT METEOROLOGICAL COMMITTEE, WASHINGTON, D.C.

JMDR-
: INSTITUTE FOR DEFENSE ANALYSES. RESEARCH AND ENGINEERING SUPPORT DIV., ARLINGTON, VA.

JMG-CD-
: TENNESSEE EASTMAN CORP., OAK RIDGE, TENN.

JMRP-
: UNITED KINGDOM. AIR MINISTRY. METEOROLOGICAL OFFICE, LONDON

JMSAC-
: JOINT METEOROLOGICAL SATELLITE ADVISORY COMMITTEE

JN-
: COAST AND GEODETIC SURVEY, WASHINGTON, D.C. (JET NAVIGATION CHART)

J/N-
: OFFICE OF THE JOINT CHIEFS OF STAFF, WASH., D.C.

JNPC-FEWP(YEAR)/P-
: UNITED KINGDOM ATOMIC ENERGY AUTHORITY. REACTOR GP., WINDSCALE, CUMB., ENGLAND

JNPC-MWP(G)/P-
: UNITED KINGDOM ATOMIC ENERGY AUTHORITY. REACTOR GROUP, CULCHETH, LANCS, ENGLAND

JNPC-MWP-SSG/P(YR)-
: UNITED KINGDOM ATOMIC ENERGY AUTHORITY. REACTOR GROUP, CULCHETH, LANCS, ENGLAND

JNPC/NFISG/-
: UNITED KINGDOM ATOMIC ENERGY AUTHORITY. RESEARCH GP. ATOMIC ENERGY ESTAB., WINFRITH, DORSET, ENGLAND

JNPC-RKWP-
: UNITED KINGDOM ATOMIC ENERGY AUTHORITY. REACTOR GP., RISLEY, LANCS, ENGLAND

JNPC/RPWP/NFISG/P35-
: UNITED KINGDOM ATOMIC ENERGY AUTHORITY. REACTOR GP. CONTROL AND INSTRUMENTATION DIV., ATOMIC ENERGY ESTAB., WINFRITH, DORSET, ENGLAND

JNPC-SSG/P(YEAR)-
: UNITED KINGDOM ATOMIC ENERGY AUTHORITY. REACTOR GROUP, CULCHETH, LANCS, ENGLAND

JNPC/SWP/N-
: GT. BRIT. CENTRAL ELECTRICITY GENERATING BOARD. BERKELEY NUCLEAR LABS., BRISTOL, ENGLAND

JNPC-SWP/P-
: UNITED KINGDOM ATOMIC ENERGY AUTHORITY. REACTOR GP., RISLEY, LANCS, ENGLAND

JNPO-MWP(G)/P-
: GT. BRIT. ATOMIC EN. RES. ESTAB., HARWELL, BERKS, ENG

JNRC-
: EUROPEAN ATOMIC ENERGY COMMUNITY. JOINT NUCLEAR RESEARCH CENTER, ISPRA, ITALY

JNS-
: JAPAN NUCLEAR SHIP DEVELOPMENT AGENCY, TOKYO

J.O.-
: BAIRD ASSOCIATES, INC., CAMBRIDGE, MASS.

JO-
: LOS ALAMOS SCIENTIFIC LAB., N. MEX.

JOB-
: SAN DIEGO COUNTY COMPREHENSIVE PLANNING ORG., CALIF.

JOINT ROAD FRICTION-
: PENNSYLVANIA STATE UNIV., UNIVERSITY PARK. DEPT. OF MECHANICAL ENGINEERING

JOURNAL-
: WEST VIRGINIA. UNIV.,MORGANTOWN. WATER RESEARCH INST.

JOWOG-(NUMBER)/PR-
: UNITED STATES-UNITED KINGDOM JOINT WORKING GROUP

JOWOG-(NUMBER)/(NUMBER)
: UNITED STATES-UNITED KINGDOM JOINT WORKING GROUP

J-P(NUMBER/YEAR)
: CALIFORNIA. UNIV., BERKELEY. RADIATION LAB. (REF. 9)

JP-
: GENERAL ELECTRIC CO. GEN. ENG. LAB., SCHENECTADY,N.Y.

JP-
: MASSACHUSETTS INST.OF TECH.,CAMBRIDGE.NATL.MAGNET LAB

JP-
: NATIONAL DEFENSE RESEARCH COMMITTEE (JET PROPULSION)

JP-
: OFFICE OF SCIENTIFIC RES. + DEV. (JET PROPULSION)

JPBA-	TENNESSEE EASTMAN CORP., OAK RIDGE, TENN.
JPC-	PURDUE UNIV., LAFAYETTE, IND. JET PROPULSION CENTER
JPG-	JEFFERSON PROVING GROUND, MADISON, IND.
JPL-	CALIF. INST. OF TECH., PASADENA. JET PROPULSION LAB.
JPL-(PROJECT NO.)-	CALIF. INST. OF TECH., PASADENA. JET PROPULSION LAB.
JPL-(MONTH)(YEAR)-	CALIF. INST. OF TECH., PASADENA. JET PROPULSION LAB.
JPLAI-	CALIFORNIA INST. OF TECH., PASADENA. JET PROPULSION LAB. (ASTRONAUTICS INFORMATION)
JPLAI/ABSTR. VOL.-	CALIFORNIA INST. OF TECH., PASADENA. JET PROPULSION LAB. (ASTRONAUTICS INFORMATION. ABSTRACTS)
JPLAI/INDEX/	CALIFORNIA INST. OF TECH., PASADENA. JET PROPULSION LAB. (ASTRONAUTICS INFORMATION. INDEX)
JPLAI/LS-	CALIFORNIA INST. OF TECH., PASADENA. JET PROPULSION LAB. (ASTRONAUTICS INFORMATION. LITERATURE SEARCH)
JPLAI-TR-	CALIF. INST. OF TECH., PASADENA. JET PROPULSION LAB.
JPL-BIB-39-	CALIFORNIA INST. OF TECH., PASADENA. JET PROPULSION LAB. (BIBLIOGRAPHY)
JPL-BIBL-	CALIF. INST. OF TECH., PASADENA. JET PROPULSION LAB. (BIBLIOGRAPHY)
JPL-CBS-	CALIFORNIA INST. OF TECH., PASADENA. JET PROPULSION LAB. (COMBINED BIMONTHLY SUMMARY)
JPL-EP-	CALIFORNIA INST. OF TECH., PASADENA. JET PROPULSION LAB. (EXTERNAL PUBLICATION)
JPL-GMS-	CALIFORNIA INST. OF TECH., PASADENA. JET PROPULSION LAB. (GUIDED MISSILE SUMMARY)
JPL/JPS-37-	CALIF. INST. OF TECH., PASADENA. JET PROPULSION LAB.
JPL-MEMO-	CALIF. INST. OF TECH., PASADENA. JET PROPULSION LAB.
JPL-NPS-	CALIFORNIA INST. OF TECH., PASADENA. JET PROPULSION LAB. (NUCLEAR PROPULSION SUMMARY)
JPL-P-	CALIFORNIA INST. OF TECH., PASADENA. JET PROPULSION LAB. (PUBLICATION)
JPL-PR-	CALIFORNIA INST. OF TECH., PASADENA. JET PROPULSION LAB. (PROGRESS REPORT)
JPL-PR-(PROJECT NO.)-	CALIFORNIA INST. OF TECH., PASADENA. JET PROPULSION LAB. (PROGRESS REPORT)
JPL PROPOSAL (NO.)-(NO.)	CALIF. INST. OF TECH., PASADENA. JET PROPULSION LAB.
JPL-PUB-	CALIFORNIA INST. OF TECH., PASADENA. JET PROPULSION LAB. (PUBLICATION)
JPL-QSR-38-	CALIFORNIA INST. OF TECH., PASADENA. JET PROPULSION LAB. (QUARTERLY SUMMARY REPORT)
JPL-RES.SUMM.-	CALIFORNIA INST. OF TECH., PASADENA. JET PROPULSION LAB. (RESEARCH SUMMARY)
JPL-RS-(PROJECT NO.)-	CALIFORNIA INST. OF TECH., PASADENA. JET PROPULSION LAB. (RESEARCH SUMMARY)
JPL-SPEC-(LTR)(NO.)(NO.)	CALIFORNIA INST. OF TECH., PASADENA. JET PROPULSION LAB. (MATERIAL SPECIFICATION)
JPL-SPS-(PROJECT NO.)-	CALIFORNIA INST. OF TECH., PASADENA. JET PROPULSION LAB. (SPACE PROGRAMS SUMMARY)
JPL-SRS-	CALIFORNIA INST. OF TECH., PASADENA. JET PROPULSION LAB. (SPACE RESEARCH SUMMARY)
JPL-STUDY MEMO-	CALIF. INST. OF TECH., PASADENA. JET PROPULSION LAB.
JPL-SUM-	CALIFORNIA INST. OF TECH., PASADENA. JET PROPULSION LAB. (SUMMARY)
JPL-SWT-	CALIFORNIA INST. OF TECH., PASADENA. JET PROPULSION LAB. (SUPERSONIC WIND TUNNEL)
JPL-TECH. RELEASE-(PROJ)-	CALIF. INST. OF TECH., PASADENA. JET PROPULSION LAB.
JPL-TM-(PROJ. NO.)-	CALIFORNIA INST. OF TECH., PASADENA. JET PROPULSION LAB. (TECHNICAL MEMO)
JPL-TR-(NUMBER)-	CALIFORNIA INST. OF TECH., PASADENA. JET PROPULSION LAB.
JPL-TR-(PROJECT NO.)-	CALIFORNIA INST. OF TECH., PASADENA. JET PROPULSION LAB. (TECHNICAL REPORT)
JPR-	(JOINT PROGRESS REPORT) (REF. 2)
JPR-	JERSEY PRODUCTION RESEARCH CO., TULSA
JPRS-	JOINT PUBLICATIONS RESEARCH SERVICE, N.Y.C.
JPRS-GUO-	JOINT PUBLICATIONS RESEARCH SERVICE, N.Y.C.
JPRS-L-	JOINT PUBLICATIONS RESEARCH SERVICE, N.Y.C.
JPRS(NY)-	JOINT PUBLICATIONS RESEARCH SERVICE, N.Y.C.
JPRS-TR-L-	JOINT PUBLICATIONS RESEARCH SERVICE, N.Y.C.
JQMD-	QUARTERMASTER DEPOT, JEFFERSONVILLE, IND.
JRC-	TENNESSEE EASTMAN CORP., OAK RIDGE, TENN.
JRDB-	OFF. OF THE ASST. SECY. OF DEFENSE(RES.+DEV.),WASH,DC
JRL-	IMPERIAL COLL. OF SCI. + TECH., LONDON. JET RES. LAB.
JSEP-TR-	TEXAS. UNIV., AUSTIN. ELECTRONICS RESEARCH CENTER
JSEP-TR-	TEXAS. UNIV., AUSTIN. LABS. FOR ELECTRONICS AND RELATED SCIENCE RESEARCH
JSHP-	ATOMIC ENERGY OF CANADA LTD. CHALK RIVER PROJ., ONT.
JSORT-	CANADA. JOINT SERVICES OPERATIONAL RESEARCH TEAM
J SPEC-	CALIFORNIA. UNIV., BERKELEY. RADIATION LAB. (REF. 9)
JSRP-	JOINT SERVICES READING PANEL, LONDON
JSSG-	JOINT SATELLITE STUDIES GROUP
JSWPB-	JOINT SPECIAL WEAPONS PUBLICATIONS BOARD, ALBUQUERQUE, N. MEX. (REF. 14)
JTETF-	JOINT TEST + EVALUATION TASK FORCE, MAC DILL AFB, FLA
JTF-	CHRYSLER CORP. DEFENSE OPERATIONS DIV., DETROIT
JTF-	JOINT TASK FORCE ... (REF. 17)
JTF-(NUMBER)-ATG-	JOINT TASK FORCE ... (REF. 17)
JTF-(NUMBER)-TG-	JOINT TASK FORCE ... (REF. 17)
JTF2-	JOINT TASK FORCE TWO, SANDIA BASE, N.M.
JTF2-(NO.)-(LTRS.)-	JOINT TASK FORCE TWO, SANDIA BASE, N.M.
JTF2-OAR-	JOINT TASK FORCE TWO, SANDIA BASE, N.M.
JTFI-	JOINT TASK FORCE ONE
JTFMC-TP-	JOINT TASK FORCE SEVEN, PEARL HARBOR
JTF-OPLAN-(NUMBER)-	JOINT TASK FORCE, EGLIN AFB, FLA.
JTG-	JOINT TASK GROUP ... (REF. 18)
JTRU-	AUSTRALIA. JOINT TROPICAL RESEARCH UNIT, INNISFAIL
JUB-BIBL-	KERNFORSCHUNGSANLAGE JUELICH, GERMANY
JUEL-(NUMBER)-(LETTERS)	KERNFORSCHUNGSANLAGE JUELICH, GERMANY
JUL-(NUMBER)-(LETTERS)	KERNFORSCHUNGSANLAGE JUELICH, GERMANY (LETTERS STAND FOR INSTITUTES, WORKING GROUPS, ETC. BUT DO NOT AFFECT THE NUMBERS, WHICH ARE IN ONE SERIES)
JUL-(NUMBER)-	KERNFORSCHUNGSANLAGE JUELICH, GERMANY
JUL-(NUMBER)-NP	KERNFORSCHUNGSANLAGE JUELICH, GERMANY
JUL-(NUMBER)-BO	KERNFORSCHUNGSANLAGE JUELICH. INSTITUT FUER BOTANIK UND MIKROBIOLOGIE, GERMANY
JUL-(NUMBER)-CT	KERNFORSCHUNGSANLAGE JUELICH. INSTITUT FUER CHEMISCHE TECHNOLOGIE, GERMANY
JUL-(NUMBER)-FN	KERNFORSCHUNGSANLAGE JUELICH. INSTITUT FUER FESTKOERPER- UND NEUTRONENPHYSIK, GERMANY
JUL-(NUMBER)-FF	KERNFORSCHUNGSANLAGE JUELICH. INSTITUT FUER FESTKOERPERFORSCHUNG, GERMANY
JUL-(NUMBER)-KP	KERNFORSCHUNGSANLAGE JUELICH. INSTITUT FUER KERNPHYSIK, GERMANY
JUL-(NUMBER)-ME	KERNFORSCHUNGSANLAGE JUELICH. INSTITUT FUER MEDIZIN, GERMANY
JUL-(NUMBER)-PC	KERNFORSCHUNGSANLAGE JUELICH. INSTITUT FUER PHYSIKALISCHE CHEMIE
JUL-(NUMBER)-PP	KERNFORSCHUNGSANLAGE JUELICH. INSTITUT FUER PLASMAPHYSIK, GERMANY
JUL-(NUMBER)-RC	KERNFORSCHUNGSANLAGE JUELICH. INSTITUT FUER RADIOCHEMIE, GERMANY
JUL-(NUMBER)-RB	KERNFORSCHUNGSANLAGE JUELICH. INSTITUT FUER REAKTORBAUELEMENTE, GERMANY
JUL-(NUMBER)-RG	KERNFORSCHUNGSANLAGE JUELICH. INSTITUT FUER REAKTORENTWICKLUNG, GERMANY
JUL-(NUMBER)-RW	KERNFORSCHUNGSANLAGE JUELICH. INSTITUT FUER REAKTORWERKSTOFFE, GERMANY
JUL-(NUMBER)-TP	KERNFORSCHUNGSANLAGE JUELICH. INSTITUT FUER TECHNISCHE PHYSIK, GERMANY
JUL-(NUMBER)-RE	KERNFORSCHUNGSANLAGE JUELICH. ZENTRALABTEILUNG FORSCHUNGSREAKTOREN, GERMANY

```
JUL-(NUMBER)-ST
          KERNFORSCHUNGSANLAGE JUELICH. ZENTRALABTEILUNG
             STRAHLENSCHUTZ, GERMANY
JUL-(NUMBER)-MA
          KERNFORSCHUNGSANLAGE JUELICH. ZENTRALINSTITUT FUER
             ANGEWANDTE MATHEMATIK, GERMANY
JUL-(NUMBER)-RX
          KERNFORSCHUNGSANLAGE JUELICH. ZENTRALINSTITUT FUER
             REAKTOREXPERIMENTE, GERMANY
JUL-(NUMBER)-CA
          KERNFORSCHUNGSANLAGE JUELICH. ZENTRALLABOR FUER
             CHEMISCHE ANALYSE, GERMANY
JUL-(NUMBER)-ZE
          KERNFORSCHUNGSANLAG JUELICH. ZENTRALLABOR FUER
             ELEKTRONIK, GERMANY
JUL-BIBL-
          KERNFORSCHUNGSANLAGE JUELICH, GERMANY
JUL-CONF-
          KERNFORSCHUNGSANLAGE JUELICH, GERMANY
JULKAISU-
          FINLAND. STATE INST. FOR TECHNICAL RESEARCH, HELSINKI
JUNIPER-
          JUNIPER COMMITTEE
JUPITER-
          JUPITER NOSE CONE COMMITTEE (REF. 15)
JUPITER-TB-
          JUPITER NOSE CONE COMMITTEE (TECH. BULL.) (REF. 15)
JWD-
          DU PONT DE NEMOURS (E.I.) + CO. JACKSON LAB.,
             WILMINGTON, DEL.
JWG-
          JOINT WORKING GROUP ... (REF. 19)
JWG (LETTERS OR NUMBERS)-
          JOINT WORKING GROUP ... (REF. 19)
JWG (LTRS OR NOS.)-(YR)-
          JOINT WORKING GROUP ... (REF. 19)
JWGA-
          JOINT WAR GAMES AGENCY, WASHINGTON, D.C.
JWM-
          TENNESSEE EASTMAN CORP., OAK RIDGE, TENN.
JWS/TD-
          JUNGLE WARFARE SCHOOL. TRAIL AND DEVELOPMENTS WING,
             JOHORE BAHRU, MALAYSIA
```

K-(YEAR)-(3 DIGITS)-	CALIFORNIA. UNIV., LIVERMORE. LAWRENCE LIVERMORE LAB.
K-1.0 THRU K-3.--	CALIFORNIA INST. OF TECH., PASADENA. PROJECT CAMEL
K-	CARBIDE AND CARBON CHEMICALS CORP. K-25 PLANT, OAK RIDGE, TENN.
K-1.2.-THRU K-4.76-	CARBIDE AND CARBON CHEMICALS CORP. K-25 PLANT, OAK RIDGE, TENN.
K-	GARRETT CORP. AIRESEARCH MFG. DIV., LOS ANGELES
K-(NUMBER)/MR-(NUMBER)-	HERCULES POWDER CO., KENVIL, N.J.
K-	IIT RESEARCH INST., CHICAGO
K-	KANSAS STATE COLL., MANHATTAN
K-	LOS ALAMOS SCIENTIFIC LAB., N. MEX.
K-1- THRU K-4-	LOS ALAMOS SCIENTIFIC LAB., N. MEX. (INTERNAL CORRESPONDENCE SERIAL USED BY GROUPS IN K DIV.)
K-	NAVAL WEAPONS EVALUATION FACILITY, KIRTLAND AFB, N.M.
K-	NAVAL WEAPONS LAB., DAHLGREN, VA.
K-	OAK RIDGE GASEOUS DIFFUSION PLANT, TENN.
K-	OAK RIDGE NATIONAL LAB., TENN.
K-	OFFICE OF SCIENTIFIC RESEARCH AND DEVELOPMENT
K-	SERCK RADIATORS, LTD., BIRMINGHAM, WAR., ENG.
K60-TR-(YEAR)-	SWEDEN. KIRUNA GEOFYSISKA OBSERVATORIUM
K-	UNION CARBIDE CORP., OAK RIDGE, TENN.
K-	UNION CARBIDE NUCLEAR CO., OAK RIDGE, TENN.
K-(NUMBER)-	UNITED AIRCRAFT CORP. RES. LABS., EAST HARTFORD, CONN.
KA-	CARBIDE AND CARBON CHEMICALS CORP. K-25 PLANT, OAK RIDGE, TENN.
KA-	CORNELL AERONAUTICAL LAB., INC., BUFFALO
KA-(YEAR)-	KAMAN AVIDYNE, BURLINGTON, MASS.
KA-	KINETICS CORP., CINCINNATI
K/AA-	KEYDATA AND ADAMS ASSOCIATES, INC., WATERTOWN, MASS.
KA-AMC-(NUMBER-YEAR)	KAMAN AVIDYNE, BURLINGTON, MASS.
KAC-	KAMAN AIRCRAFT CORP., BLOOMFIELD, CONN.
KAC-(NO.)-(NO.)-PR-	KAMAN AIRCRAFT CORP., BLOOMFIELD, CONN.
KAC KN-(YEAR)-	KAMAN NUCLEAR, COLORADO SPRINGS
KAC-R-	KAMAN AIRCRAFT CORP., BLOOMFIELD, CONN.
KAE-(YEAR)-	ALLMAENNA SVENSKA ELEKTRISKA A.B., VAESTERAES, SWEDEN
KAE-	CARBIDE AND CARBON CHEMICALS CORP. K-25 PLANT, OAK RIDGE, TENN.
KAE-(YEAR)-	NORWAY. INSTITUTT FOR ATOMENERGI, HALDEN
KAFB-	KIRTLAND AFB, N. MEX.
KAFB-OP-	KIRTLAND AFB, N. MEX. (OPERATION PLAN)
KAFB-WCW-	KIRTLAND AFB, N. MEX.
KAI-	CORNELL AERONAUTICAL LAB., INC., BUFFALO
KAMAN-	KAMAN AIRCRAFT CORP., BLOOMFIELD, CONN.
KAMAN-AC-RN-R-(LTR.)(NO.)	KAMAN AIRCRAFT CORP., BLOOMFIELD, CONN.
KAN AF2-TN-	KANSAS. UNIV., LAWRENCE. DEPT. OF MATHEMATICS
KAPL-	KNOLLS ATOMIC POWER LAB., SCHENECTADY, N.Y.
KAPL-A-(LETTERS)-	KNOLLS ATOMIC POWER LAB., SCHENECTADY, N.Y.
KAPL-ADA-	KNOLLS ATOMIC POWER LAB., SCHENECTADY, N.Y. (ADVANCED DEVELOPMENT ACTIVITY)
KAPL-ADM-	KNOLLS ATOMIC POWER LAB., SCHENECTADY, N.Y.
KAPL-A-EOS-	KNOLLS ATOMIC POWER LAB., SCHENECTADY, N.Y. (ELEMENTS OF SPECIFICATIONS)
KAPL-A-S(NO.)-(NO.)-	KNOLLS ATOMIC POWER LAB., SCHENECTADY, N.Y.
KAPL-A-SIR-	KNOLLS ATOMIC POWER LAB., SCHENECTADY, N.Y. (SUBMARINE INTERMEDIATE REACTOR)
KAPL-A-SM-	KNOLLS ATOMIC POWER LAB., SCHENECTADY, N.Y.
KAPL-M-	KNOLLS ATOMIC POWER LAB., SCHENECTADY, N.Y.
KAPL-M-(INITIALS)-	KNOLLS ATOMIC POWER LAB., SCHENECTADY, N.Y.
KAPL-M-AME-	KNOLLS ATOMIC POWER LAB., SCHENECTADY, N.Y.
KAPL-M-AMS-	KNOLLS ATOMIC POWER LAB., SCHENECTADY, N.Y.
KAPL-M-CRDM-	KNOLLS ATOMIC POWER LAB., SCHENECTADY, N.Y. (CONTROL ROD DRIVE MECHANISM STUDY)
KAPL-M-CT-	KNOLLS ATOMIC POWER LAB., SCHENECTADY, N.Y.
KAPL-M-CTU-	KNOLLS ATOMIC POWER LAB., SCHENECTADY, N.Y. (RADIOACTIVE ACCESSIBILITY REPORTS)
KAPL-M-D1G-TD-	KNOLLS ATOMIC POWER LAB., SCHENECTADY, N.Y. (NAVAL REACTOR D1G)
KAPL-M-DNA-	KNOLLS ATOMIC POWER LAB., SCHENECTADY, N.Y. (NAVAL REACTOR D1G)
KAPL-M-EDL-	KNOLLS ATOMIC POWER LAB., SCHENECTADY, N.Y.
KAPL-MEMO-(INITIALS)-	KNOLLS ATOMIC POWER LAB., SCHENECTADY, N.Y.
KAPL-M-HP-	KNOLLS ATOMIC POWER LAB., SCHENECTADY, N.Y. (HEALTH PHYSICS)
KAPL-M-REDOX-	KNOLLS ATOMIC POWER LAB. REDOX GP., SCHENECTADY, N.Y.
KAPL-M-S3G-	KNOLLS ATOMIC POWER LAB., SCHENECTADY, N.Y. (S3G REACTOR)
KAPL-M-SAR-RES-	KNOLLS ATOMIC POWER LAB., SCHENECTADY, N.Y. (SUBMARINE ADVANCED REACTOR)
KAPL-M-SL-	KNOLLS ATOMIC POWER LAB., SCHENECTADY, N.Y. (SHIELDING LAB. EXPERIMENT)
KAPL-M-SMS-	KNOLLS ATOMIC POWER LAB., SCHENECTADY, N.Y.
KAPL-M-SR-	KNOLLS ATOMIC POWER LAB., SCHENECTADY, N.Y.
KAPL-M-SSA-	KNOLLS ATOMIC POWER LAB., SCHENECTADY, N.Y.
KAPL-M-SSD-	KNOLLS ATOMIC POWER LAB., SCHENECTADY, N.Y. (SYSTEM DESCRIPTION)
KAPL-M-TR-	KNOLLS ATOMIC POWER LAB., SCHENECTADY, N.Y. (TEST RUN)
KAPL-P-	KNOLLS ATOMIC POWER LAB., SCHENECTADY, N.Y.
KAPL-PC-	KNOLLS ATOMIC POWER LAB., SCHENECTADY, N.Y.
KAPL-RDTR-	KNOLLS ATOMIC POWER LAB., SCHENECTADY, N.Y.
KAPL-S-	KNOLLS ATOMIC POWER LAB., SCHENECTADY, N.Y.
KAPL-SPEC-(NO.)(LETTERS)-	KNOLLS ATOMIC POWER LAB., SCHENECTADY, N.Y.
KAPL-SPEC-(LTRS)-	KNOLLS ATOMIC POWER LAB., SCHENECTADY, N.Y.
KAPL-SSD-	KNOLLS ATOMIC POWER LAB., SCHENECTADY, N.Y.
KAPL-TRANS-	KNOLLS ATOMIC POWER LAB., SCHENECTADY, N.Y.
KAPPA-P-	CALIFORNIA. UNIV., BERKELEY. RADIATION LAB. (REF. 9)
KA-TR-	KAMAN AVIDYNE, BURLINGTON, MASS.
KB-	CARBIDE AND CARBON CHEMICALS CORP. K-25 PLANT, OAK RIDGE, TENN.
KB-	CORNELL AERONAUTICAL LAB., INC., BUFFALO
KB-826-M-	CORNELL AERONAUTICAL LAB., INC., BUFFALO
KB-935-M-	CORNELL AERONAUTICAL LAB., INC., BUFFALO
KB-	GERMANY. ZENTRALE FUER WISSENSCHAFTLICHES BERICHTSWESEN DER LUFTFAHRTFORSCHUNG, BERLIN (INDUSTRIEKURZBERICHT)
KB-	UNION CARBIDE NUCLEAR CO. K-25 PLANT, OAK RIDGE, TENN
KC-	CARBIDE AND CARBON CHEMICALS CORP. K-25 PLANT, OAK RIDGE, TENN.
KC-	KENYON COLL., GAMBIER, OHIO
KC-	KOPPERS CO., INC., PITTSBURGH
KC-	KORAD CORP., SANTA MONICA, CALIF.
K-C-	OAK RIDGE GASEOUS DIFFUSION PLANT, TENN.
KC-	OAK RIDGE GASEOUS DIFFUSION PLANT, TENN.
KCD-	BENDIX AVIATION CORP., KANSAS CITY, MO.
KCD-E-	BENDIX AVIATION CORP., KANSAS CITY, MO.
KCD-TB-	BENDIX CORP. KANSAS CITY DIV., MO.
KCI-	CARBIDE AND CARBON CHEMICALS CORP., OAK RIDGE, TENN.
KD-	CARBIDE AND CARBON CHEMICALS CORP. K-25 PLANT, OAK RIDGE, TENN.
KD-	CORNELL AERONAUTICAL LAB., INC., BUFFALO
KD-	FLIGHT SAFETY FOUNDATION, INC., N.Y.C.
K-D-	OAK RIDGE GASEOUS DIFFUSION PLANT, TENN.

KD-	UNION CARBIDE NUCLEAR CO. K-25 PLANT, OAK RIDGE, TENN
KDD-	CARBIDE AND CARBON CHEMICALS CORP. K-25 PLANT, OAK RIDGE, TENN.
KDM-AC-	ALLIS-CHALMERS MFG. CO. ATOMIC ENERGY DIV., MILWAUKEE
K-DP-	OAK RIDGE GASEOUS DIFFUSION PLANT, TENN.
KE-	KAISER ENGINEERS DIV., HENRY J. KAISER CO., OAKLAND, CAL
KE-(YR.)-(NO.)-RE	KAISER ENGINEERS DIV., HENRY J. KAISER CO., OAKLAND, CAL
KE-(YEAR)-	KAISER ENGINEERS DIV., HENRY J. KAISER CO., OAKLAND, CAL
KE-	KENTUCKY. UNIV., LEXINGTON
KE-	NETHERLANDS. INSTITUTE FOR NUCLEAR PHYSICS RESEARCH, AMSTERDAM
KEA-	ALLGEMEINE ELEKTRICITAETS-GESELLSCHAFT, FRANKFURT AM MAIN (KERNENERGIEANLAGEN)
KE-ACF-GCPR-	NUCLEAR PRODUCTS-ERCO DIV., ACF INDUSTRIES, INC., WASHINGTON, D.C.
KEAE-	KAISER ENGINEERS DIV., HENRY J. KAISER CO., OAKLAND, CAL
KEARFOTT-	GENERAL PRECISION, INC. KEARFOTT DIV., LITTLE FALLS, NJ
K EE-	KANSAS STATE COLL., MANHATTAN. ENG. EXPERIMENT STA.
KE EES-	KENTUCKY. UNIV., LEXINGTON. ENG. EXPERIMENT STA.
KEES-SR-	KANSAS STATE UNIV., MANHATTAN. ENG. EXPERIMENT STA.
KE-GCPR-	KAISER ENGINEERS DIV., HENRY J. KAISER CO., OAKLAND, CAL
KEK-	KERNFORSCHUNGSZENTRUM KARLSRUHE. INSTITUT FUER KERNVERFAHRENSTECHNIK
KEL-	KELLOGG (M.W.) CO., JERSEY CITY
KELL-	KELLOGG (M.W.) CO., JERSEY CITY
KELLETT-(NO. LTR. NO.)	KELLETT AIRCRAFT CORP., WILLOW GROVE, PENNA.
KELL RDR SPD-	KELLOGG (M.W.) CO. SPECIAL PROJECTS DEPT., JERSEY CITY (RESEARCH AND DEVELOPMENT REPORTS)
KEMA-	KEURING VAN ELECTROTECHNISCHE MATERIALEN, N.V., ARNHEM, NETHERLANDS
KENNI-	KENNAMETAL, INC., LATROBE, PENNA.
KENPS-(YR)-(NO.)-R	KAISER ENGINEERS DIV., HENRY J. KAISER CO., OAKLAND, CAL
KF-	CORNELL AERONAUTICAL LAB., INC., BUFFALO
KF-	KNOLLS ATOMIC POWER LAB., SCHENECTADY, N.Y.
KFA-	KERNFORSCHUNGSANLAGE JUELICH, GERMANY
KFK-(NUMBER/YEAR)-	KERNFORSCHUNGSZENTRUM KARLSRUHE, GERMANY
KFK-	KERNFORSCHUNGSZENTRUM KARLSRUHE, GERMANY
KFK-EXT-(NO./YEAR)-	KERNFORSCHUNGSZENTRUM KARLSRUHE, GERMANY
KFKI-	GERMANY. ZENTRALINSTITUT FUER KERNPHYSIK, DRESDEN
KFKI-	MAGYAR TUDOMANYOS AKADEMIA KOZPONTI FIZIKAI KUTATO INTEZETE, BUDAPEST
KFKI-(YEAR)-(NO.)-(LTRS)	MAGYAR TUDOMANYOS AKADEMIA KOZPONTI FIZIKAI KUTATO INTEZETE, BUDAPEST
KFKI-(NUMBER)/(LETTER)	MAGYAR TUDOMANYOS AKADEMIA KOZPONTI FIZIKAI KUTATO INTEZETE, BUDAPEST
KFKI-(YEAR)-(NUMBER)	MAGYAR TUDOMANYOS AKADEMIA KOZPONTI FIZIKAI KUTATO INTEZETE, BUDAPEST
KFK-IAK-(NUMBER/YEAR)-	KERNFORSCHUNGSZENTRUM KARLSRUHE. INSTITUT FUER ANGEWANDTE KERNPHYSIK, GERMANY
KFK-IAR-(NUMBER/YEAR)-	KERNFORSCHUNGSZENTRUM KARLSRUHE. INSTITUT FUER ANGEWANDTE REAKTORPHYSIK, GERMANY
KFK-IEK-(NUMBER/YEAR)-	KERNFORSCHUNGSZENTRUM KARLSRUHE. INSTITUT FUER EXPERIMENTELLE KERNPHYSIK, GERMANY
KFK-IHC-(NUMBER/YEAR)-	KERNFORSCHUNGSZENTRUM KARLSRUHE. INSTITUT FUER HEISSE CHEMIE, GERMANY
KFK-IMF-(NUMBER/YEAR)-	KERNFORSCHUNGSZENTRUM KARLSRUHE. INSTITUT FUER MATERIAL- UND FESTKOERPERFORSCHUNG, GERMANY
KFK-IMFB-(NUMBER/YEAR)-	KERNFORSCHUNGSZENTRUM KARLSRUHE. INSTITUT FUER MATERIAL- UND FESTKOERPERFORSCHUNG, GERMANY
KFK-INR-(NUMBER/YEAR)-	KERNFORSCHUNGSZENTRUM KARLSRUHE. INSTITUT FUER NEUTRONENPHYSIK UND REAKTORTECHNIK, GERMANY
KFK-IR-(NUMBER/YEAR)-	KERNFORSCHUNGSZENTRUM KARLSRUHE. INSTITUT FUER RADIOCHEMIE, GERMANY
KFK-IR-(NUMBER/YEAR)-	KERNFORSCHUNGSZENTRUM KARLSRUHE. INSTITUT FUER REAKTORENTWICKLUNG, GERMANY
KFK-IRB-(NUMBER/YEAR)-	KERNFORSCHUNGSZENTRUM KARLSRUHE. INSTITUT FUER REAKTORBAUELEMENTE, GERMANY
KFKL-(NO.)/(YR)-	MAGYAR TUDOMANYOS AKADEMIA KOZPONTI FIZIKAI KUTATO INTEZETE, BUDAPEST
KFK-TR-	KERNFORSCHUNGSZENTRUM KARLSRUHE, GERMANY (TRANSLATION)
KFM-	OAK RIDGE GASEOUS DIFFUSION PLANT, TENN.
KFTI-(YEAR/NUMBER)	AKADEMIYA NAUK UKRAINSKOI SSR. FIZIKO-TEKHNICHESKII INSTITUT, KHARKOV
KG-	KOGAKUIN UNIV., TOKYO
KGO-	SWEDEN. KIRUNA GEOFYSISKA OBSERVATORIUM
KG-PROJECT-	ORDNANCE DEPT. (ARMY). DEVELOPMENT DIV. (PROJECT KG)
KHFTI-(YEAR/NUMBER)	AKADEMIYA NAUK UKRAINSKOI SSR. FIZIKO-TEKHNICHESKII INSTITUT, KHARKOV
KH-R-	CALIFORNIA INST. OF TECH., PASADENA. KECK LAB. OF HYDRAULICS AND WATER RESOURCES
KHVL-	SAIGON UNIV., SOUTH VIETNAM. DEPT. OF PHYSICS
KI-	KENNAMETAL, INC., LATROBE, PENNA.
KI-	KOLLSMAN INSTRUMENT CORP., ELMHURST, N.Y.
K-IB-(NO.)(NO.)/	JOHN F. KENNEDY SPACE CENTER, COCOA BEACH, FLA.
KIC-	KOLLSMAN INSTRUMENT CO., ELMHURST, N.Y.
KIC-RD-	KOLLSMAN INSTRUMENT CORP., ELMHURST, N.Y.
KIC-SYS-TDR-	KOLLSMAN INSTRUMENT CORP. SYSTEMS MANAGEMENT DIV., ELMHURST, N.Y.
KIDM-	COLORADO. UNIV., BOULDER
KINGFISHER-	NATIONAL BUREAU OF STANDARDS. CORONA LABS., CALIF.
KIP-	KIP ELECTRONICS CORP., STAMFORD, CONN.
KIP-C-	KIP ELECTRONICS CORP., STAMFORD, CONN.
KIR-	NORWAY. FORSVARETS FORSKNINGSINSTITUTT, OSLO
KIR-(NO.)/(YR.)	NORWAY. FORSVARETS FORSKNINGSINSTITUTT, OSLO
KIR-	NORWAY. INSTITUTT FOR ATOMENERGI, KJELLER
KIR-N-	NORWAY. INSTITUTT FOR ATOMENERGI, KJELLER
KIU-	KYOTO IMPERIAL UNIV. COLLEGE OF AGRICULTURE
KIU C-	KYOTO IMPERIAL UNIV. COLL. OF AGRIC. (CHEM. SERIES)
KIU E-	KYOTO IMPERIAL UNIV. COLLEGE OF AGRICULTURE (ENTOMOLOGICAL SERIES)
KIU G-	KYOTO IMPERIAL UNIV. COLLEGE OF AGRICULTURE (GENETICAL SERIES)
KIU H-	KYOTO IMPERIAL UNIV. COLLEGE OF AGRICULTURE (HORTICULTURAL SERIES)
KIU P-	KYOTO IMPERIAL UNIV. COLLEGE OF AGRICULTURE (PHYTOPATHOLOGICAL SERIES)
KIWI-TNT-	LOS ALAMOS SCIENTIFIC LAB., N. MEX.
KKWP-(NUMBER/YEAR)	KERNKRAFTWERK PLANUNGSGESELLSCHAFT M.B.H., VIENNA
KL-	KELLEX CORP., N.Y.C.
K-L-	OAK RIDGE GASEOUS DIFFUSION PLANT, TENN.
KL-	OAK RIDGE GASEOUS DIFFUSION PLANT, TENN.
K-L-	OAK RIDGE NATIONAL LAB., TENN.
KLD-	OAK RIDGE GASEOUS DIFFUSION PLANT, TENN.
KLG-	KELLOGG (M.W.) CO., JERSEY CITY
KLI-	EUROPEAN THEATER OF OPERATIONS (ARMY). ALSOS MISSION (KLIEWE FILE)
KLI-	OAK RIDGE GASEOUS DIFFUSION PLANT, TENN.
KLO-	CARBIDE AND CARBON CHEMICALS CORP. K-25 PLANT, OAK RIDGE, TENN.
KLX-	VITRO CORP. OF AMERICA, N.Y.C.
KLX-	VITRO LABS., WEST ORANGE, N.J.
KLXS-	VITRO CORP. OF AMERICA. ATOMIC POWER STUDY GROUP, NYC
KM-	CALIFORNIA INST. OF TECH., PASADENA
KM-	KAMAN AIRCRAFT CORP., COLORADO SPRINGS
K-M-	OAK RIDGE GASEOUS DIFFUSION PLANT, TENN.
KMC-	KAISER-FRAZER CORP., WILLOW RUN, MICH.

KMI-	JOHN F. KENNEDY SPACE CENTER, COCOA BEACH, FLA. (KENNEDY MANAGEMENT INSTRUCTIONS)
KMLC-	NAVAL ORDNANCE LAB. KENSINGTON MAGNETIC LAB. (COMMUNICATION)
KMS-(NUMBERS)-(NOS.)(LTR)	KMS INDUSTRIES INC., ANN ARBOR, MICH.
KMSTC-SD-	KMS TECHNOLOGY CENTER, SAN DIEGO, CALIF.
KMT-MAC-	MIDWESTERN UNIVERSITIES RESEARCH ASSN., URBANA, ILL.
KN-	BELL TELEPHONE LABS., INC., MURRAY HILL, N.J.
KN-(YEAR)-(NUMBER)-(LTR.)	KAMAN NUCLEAR, COLORADO SPRINGS
KN-(NO.)-(YEAR)-	KAMAN NUCLEAR, COLORADO SPRINGS
KN-AMC-(YEAR)-	KAMAN NUCLEAR, COLORADO SPRINGS
K-OA-	OAK RIDGE GASEOUS DIFFUSION PLANT, TENN.
KOA-	OAK RIDGE GASEOUS DIFFUSION PLANT, TENN.
KOLL-	KOLLSMAN INSTRUMENT CO., ELMHURST, N.Y.
K-P(MONTH/YEAR)	CALIFORNIA. UNIV., BERKELEY. RADIATION LAB. (REF. 9)
KP-	CARBIDE AND CARBON CHEMICALS CORP. K-25 PLANT, OAK RIDGE, TENN.
KP-	KIP ELECTRONICS CORP., STAMFORD, CONN.
K-P-	OAK RIDGE GASEOUS DIFFUSION PLANT, TENN.
K-PA-	JOHN F. KENNEDY SPACE CENTER, COCOA BEACH, FLA.
KPE-	KNOLLS ATOMIC POWER LAB., SCHENECTADY, N.Y.
KPM-	KNOLLS ATOMIC POWER LAB., SCHENECTADY, N.Y.
KPNO-	KITT PEAK NATIONAL OBSERVATORY, TUCSON, ARIZ.
KPP-	KNOLLS ATOMIC POWER LAB., SCHENECTADY, N.Y.
KPR-	REDEL, INC., ANAHEIM, CALIF.
KPS-	KELLEX CORP., N.Y.C.
KPTR-	KNOLLS ATOMIC POWER LAB., SCHENECTADY, N.Y.
KR-	AMSTERDAM. UNIV.
KR-	NETHERLANDS. REACTOR CENTRUM, PETTEN
KR-	NORWAY. INSTITUTT FOR ATOMENERGI, KJELLER.
KRB-(YEAR/NUMBER)	KERNKRAFTWERK RWE-BAYERNWERK G.M.B.H., GUNDREMMINGEN, GERMANY
KRC-(LTR.)-	INDIA. COUNCIL OF SCIENTIFIC AND IND. RES., NEW DELHI
KRC-(LETTER)(NUMBER)	INDIA. NATIONAL PHYSICAL LAB., NEW DELHI
KRS-MURA-	MIDWESTERN UNIVERSITIES RESEARCH ASSN., URBANA, ILL.
KS-	CARBIDE AND CARBON CHEMICALS CORP. K-25 PLANT, OAK RIDGE, TENN.
KSA-	OAK RIDGE GASEOUS DIFFUSION PLANT, TENN.
KSC-	JOHN F. KENNEDY SPACE CENTER, COCOA BEACH, FLA.
KSC-(LETTERS)-	JOHN F. KENNEDY SPACE CENTER, COCOA BEACH, FLA.
KSC-GP-	JOHN F. KENNEDY SPACE CENTER, COCOA BEACH, FLA. (GENERAL PUBLICATION)
KSC-K-AM-	JOHN F. KENNEDY SPACE CENTER, COCOA BEACH, FLA. (ADMINISTRATIVE MANAGEMENT)
KSSC-	KELLOGG SWITCHBOARD + SUPPLY CO., CHICAGO.
KSU-	KANSAS STATE UNIV., MANHATTAN
KSU-SR-	KANSAS STATE UNIV., MANHATTAN. ENG. EXPERIMENT STA.
KT-	KETT CORP., CINCINNATI
K-T-	MASS. INST.OF TECH.,OAK RIDGE,TENN. ENG. PRACTICE SCH
KT-	MASS. INST.OF TECH.,OAK RIDGE,TENN. ENG. PRACTICE SCH
KT-	NATIONAL BUREAU OF STANDARDS, WASHINGTON, D.C. (KINGFISHER TECH. RPT.)
KT-	UNION CARBIDE NUCLEAR CO., OAK RIDGE, TENN.
KTH-AERO TN-	SWEDEN. KUNGLIGA TEKNISKA HOEGSKOLAN, STOCKHOLM. INSTITUTIONEN FOER FLYGTEKNIK
K-TL-	OAK RIDGE GASEOUS DIFFUSION PLANT, TENN.
KTM-	KETTELLE ASSOCIATES, INC., PAOLI, PENNA.
KTR-	KILGORE, INC. (TECHNICAL REPORT)
K-TRANS-	KYOTO UNIV.
K-TRANS-	OAK RIDGE GASEOUS DIFFUSION PLANT, TENN.
K-TRANS-	SCIENTIFIC TRANSLATION SERVICE, ANN ARBOR, MICH.
KU-	CARBIDE AND CARBON CHEMICALS CORP. K-25 PLANT, OAK RIDGE, TENN.
KU-	KANSAS. UNIV., LAWRENCE.
KUKO-	I.G. FARBENINDUSTRIE A.G. KUNSTSTOFFE KOMMISSION
KUNS-	KYOTO UNIV. DEPT. OF PHYSICS
KURRI-TR-	KYOTO UNIV. RESEARCH REACTOR INST.
K-V-	JOHN F. KENNEDY SPACE CENTER, COCOA BEACH, FLA.
KVT-	KARLSRUHE, GERMANY. TECHNISCHE HOCHSCHULE
KWIS-	MAX-PLANCK-INSTITUT FUER STROEMUNGSFORSCHUNG, GOETTINGEN
KWL-	KERNKRAFTWERK LINGEN G.M.B.H., GERMANY
KWP-	CORNELL AERONAUTICAL LAB., INC., BUFFALO
KX-	KELLEX CORP., N.Y.C.
KX-	OAK RIDGE GASEOUS DIFFUSION PLANT, TENN.
KY-	PADUCAH AREA OFFICE, AEC, KY. (REF. 4)
KY-	PADUCAH GASEOUS DIFFUSION PLANT, KY.
KY-	UNION CARBIDE NUCLEAR CO. PADUCAH PLANT, KY.
KY-(LETTER)-	PADUCAH GASEOUS DIFFUSION PLANT, KY.
KYB-	UNION CARBIDE NUCLEAR CO. PADUCAH PLANT, KY.
KY-D-	PADUCAH GASEOUS DIFFUSION PLANT, KY.
KY-F-	UNION CARBIDE NUCLEAR CO., PADUCAH, KY.
KYL-	CARBIDE AND CARBON CHEMICALS CO., PADUCAH, KY.
KY-L-	PADUCAH GASEOUS DIFFUSION PLANT, KY.
KYM-	UNION CARBIDE NUCLEAR CO. PADUCAH PLANT, KY.
KYUSHU-	KYUSHU UNIV., FUKUOKA. INST. FOR THEORETICAL PHYSICS
KZ-	CARBIDE AND CARBON CHEMICALS CORP. K-25 PLANT, OAK RIDGE, TENN.
KZ-	COLUMBIA UNIV., NYC. SUBSTITUTE ALLOY MATERIALS LABS.

L-	(LABORATORY) (REF. 2)
L-	(LEAFLET) (REF. 2)
L-	(LECTURES) (REF. 2)
L-	(LETTER REPORT) (REF. 2)
L-	(LIBRARY) (REF. 2)
L-	(LIBRARY MATERIALS) (REF. 2)
L-	AEROJET-GENERAL CORP., AZUSA, CALIF.
L-	AEROJET-GENERAL CORP., SACRAMENTO, CALIF.
L(NO.)-	AVCO CORP., WILMINGTON, MASS.
L(NO.)-HKA-(YEAR)-	AVCO CORP. RES.+ ADVANCED DEV. DIV., WILMINGTON, MASS.
L-(2 DIGITS)	BOELKOW ENTWICKLUNGEN K.G., MUNICH
L-4002-C-	BRITISH THOMSON-HOUSTON CO., LTD. RUGBY, WARWICKS, ENG.
L-	EDGERTON, GERMESHAUSEN + GRIER, INC., LAS VEGAS, NEV.
L-	GARRETT CORP. AIRESEARCH MFG. DIV., LOS ANGELES
L(YEAR)-	INDUSTRIAL COLL. OF THE ARMED FORCES, WASHINGTON, D.C.
L-(NUMBER)-(YEAR)-	LOCKHEED MISSILES AND SPACE CO. RESEARCH DIV., PALO ALTO, CALIF.
L-1- THRU L-5-	LOS ALAMOS SCIENTIFIC LAB., N. MEX. (INTERNAL CORRESPONDENCE SERIAL USED BY GROUPS IN L DIV.)
L-	UNION CARBIDE CORP. LINDE DIV., INDIANAPOLIS
L-	UNION CARBIDE CORP. LINDE DIV., NEWARK, N.J.
L-	UNIVERSITY OF NOTRE DAME, IND. DEPT. OF AEROSPACE ENG
L-	UNIV. OF NOTRE DAME, SOUTH BEND, IND. LOBUND INST.
L(LETTER)C-	OFFICE OF SCIENTIFIC RESEARCH AND DEVELOPMENT
LA-	CALIF. INST. OF TECH., PASADENA. JET PROPULSION LAB.
L/A (NO.)-(NO.)-	GENERAL ELECTRIC CO. MISSILE AND ORDNANCE SYSTEMS DEPT., PHILADELPHIA
L/A (NO.)-(NO.)-	GENERAL ELECTRIC CO. MISSILE AND SPACE VEHICLE DEPT., PHILADELPHIA
LA-	GOODYEAR AEROSPACE CORP., AKRON, OHIO
LA-	LOS ALAMOS SCIENTIFIC LAB., N. MEX. (REF. 21)
LA-(NUMBER)-MS	LOS ALAMOS SCIENTIFIC LAB., N. MEX. (REF. 21)
LA-(NUMBER)-SOP	LOS ALAMOS SCIENTIFIC LAB., N.M. (PROCEDURES MANUAL)
LA-(NO.)-UNM	LOS ALAMOS SCIENTIFIC LAB., N. MEX. (WORK DONE AT NEW MEXICO. UNIV., ALBUQUERQUE)
LA-(NUMBER)-PR	LOS ALAMOS SCIENTIFIC LAB., N. MEX. (PROGRESS REPT.)
LA-(NUMBER)-TR	LOS ALAMOS SCIENTIFIC LAB., N. MEX. (TRANSLATION)
LA-(NUMBER)-BIB	LOS ALAMOS SCIENTIFIC LAB., N. MEX. (BIBLIOGRAPHY)
LA(NUMBERS)-(NOS.)(LTR)	NORTH AMERICAN AVIATION, INC. LOS ANGELES DIV.
LA-	NORTHWESTERN UNIV., EVANSTON, ILL.
LAA-	LOS ALAMOS SCIENTIFIC LAB., N. MEX. (RPT. ABSTRACTS)
LAAS-	FRANCE. LABORATOIRE D'AUTOMATIQUE ET DE SES APPLICATIONS SPATIALES, TOULOUSE
LAAS-PUBL-	FRANCE. LABORATOIRE D'AUTOMATIQUE ET DE SES APPLICATIONS SPATIALES, TOULOUSE
LAAS-STI-	FRANCE. LABORATOIRE D'AUTOMATIQUE ET DE SES APPLICATIONS SPATIALES, TOULOUSE
LAB-(LETTERS)-	LOS ALAMOS SCIENTIFIC LAB., N. MEX.
LAB-STU-	ANSALDO S.P.A., GENOA
LAC-	LOCKHEED AIRCRAFT CORP. (ALL LOCATIONS)
LAC-(NUMBER)-LFL-T-	LOCKHEED-CALIFORNIA CO., BURBANK
LAC ANP RE SYM SA-	LOCKHEED AIRCRAFT CORP., MARIETTA, GA. (SEMI-ANNUAL RADIATION EFFECTS SYMPOSIA)
LAC-CMRI-	LOCKHEED-CALIFORNIA CO., BURBANK
LAC-DA-	GENERAL ELECTRIC CO. NUCLEAR ENERGY DIV., SAN JOSE, CAL
LAC DPR-	LOCKHEED AIRCRAFT CORP. (ALL LOCATIONS) (DEVELOPMENT PLANNING REPORT)
LAC-DPTM-	LOCKHEED AIRCRAFT CORP., BURBANK, CALIF. (DEVELOPMENT PLANNING TECHNICAL MEMORANDA)
LAC-ER-	LOCKHEED-GEORGIA CO., MARIETTA (ENGINEERING REPORTS)
LAC-LMSC-	LOCKHEED MISSILES AND SPACE CO., PALO ALTO, CALIF.
LAC-LMSD-	LOCKHEED MISSILES AND SPACE CO., SUNNYVALE, CALIF.
LAC-LR-	LOCKHEED-CALIFORNIA CO., BURBANK
LAC MSD-	LOCKHEED AIRCRAFT CORP. MISSILE SYSTEMS DIV., PALO ALTO, CALIF.
LAC MSD-	LOCKHEED AIRCRAFT CORP. MISSILE SYSTEMS DIV., VAN NUYS, CALIF.
LAC-NR-	LOCKHEED-GEORGIA CO., MARIETTA (NUCLEAR REPORT)
LAC-NR-	LOCKHEED NUCLEAR PRODUCTS, MARIETTA, GA.
LAC-OEA/SST-	LOCKHEED-CALIFORNIA CO., BURBANK
LAC-ORD-	LOCKHEED-GEORGIA CO., MARIETTA
LAC-SB-(YEAR)-	LOCKHEED AIRCRAFT CORP. MISSILES AND SPACE DIV., SUNNYVALE, CALIF. (SPECIAL BIBLIOGRAPHY)
LAC-SRB-(YEAR)-	LOCKHEED AIRCRAFT CORP. MISSILES AND SPACE DIV., SUNNYVALE, CALIF.
LAC-SST/	LOCKHEED-CALIFORNIA CO., BURBANK
LA-DC-	LOS ALAMOS SCIENTIFIC LAB., N. MEX.
LADC-	LOS ALAMOS SCIENTIFIC LAB., N. MEX. (DECLASSIFICATION CODE)
LA-DC-(YEAR)-	LOS ALAMOS SCIENTIFIC LAB., N. MEX. (CODE USED 1972-)
LADC-	MINNESOTA. UNIV., MINNEAPOLIS
LADCO-	LADISH CO., CUDAHY, WIS.
LA-DIR-	LOS ALAMOS SCIENTIFIC LAB., N. MEX.
LAES-	LANDING AIDS EXPERIMENT STATION, ARCATA, CALIF.
LAFE-	BRAZIL. COMISSAO NACIONAL DE ATIVIDADES ESPACIAS. LABORATORIO DE FISICA ESPACIAL, SAO JOSE DOS CAMPOS
LAFE-SCIENTIFIC-	BRAZIL. COMISSAO NACIONAL DE ATIVIDADES ESPACIAS. LABORATORIO DE FISICA ESPACIAL, SAO JOSE DOS CAMPOS
LAFE-TR-	BRAZIL. COMISSAO NACIONAL DE ATIVIDADES ESPACIAS. LABORATORIO DE FISICA ESPACIAL, SAO JOSE DOS CAMPOS
LA-GMX-	LOS ALAMOS SCIENTIFIC LAB., N. MEX.
LA-HU-	HARVARD UNIV., CAMBRIDGE, MASS. DEPT. OF MATHEMATICS
LAI-	LAND-AIR, INC., HOLLOMAN AFB, N. MEX.
L + AI-PR-	LESSELLS AND ASSOCIATES, INC., BOSTON (PROGRESS REPT)
LAIR-	LAND-AIR, INC., HOLLOMAN AFB, N. MEX.
LAIR-	LETTERMAN ARMY INST. OF RESEARCH, SAN FRANCISCO
LA-J-	LOS ALAMOS SCIENTIFIC LAB., N. MEX.
LAKE SUPERIOR STUDIES-	NORTHERN MICHIGAN UNIV., MARQUETTE. DEPT. OF PHYSICS
LAL-	CAGLIARI, ITALY. UNIVERSITA. ISTITUTO DI FISICA
LAL-	PARIS. UNIVERSITE, ORSAY. LABORATOIRE DE L'ACCELERATEUR LINEAIRE
LAL-	STRASBOURG. UNIVERSITE
LAL-RI-	PARIS. UNIVERSITE, ORSAY
LAL-RT-(NUMBER)-(YEAR)	PARIS. UNIVERSITE, ORSAY. LABORATOIRE DE L'ACCELERATEUR LINEAIRE
LAM-	ENGLISH ELECTRIC CO., LTD.
LAMBDA-	LAMBDA CORP., ARLINGTON, VA.
LAMD-	LOS ALAMOS SCIENTIFIC LAB., N. MEX.
LAMS-	LOS ALAMOS SCIENTIFIC LAB., N. MEX. (REF. 21)
LANC-R-	LOS ALAMOS NUCLEAR CORP., N. MEX.
LAND-AIR-	LAND-AIR, INC. WHITE SANDS-HOLLOMAN DIV., HOLLOMAN AFB, N. MEX.
L AND T -	LIGHTNING AND TRANSIENTS RESEARCH INST., MINNEAPOLIS
LAO-	LOCKLAND AREA OFFICE, AEC (REF. 4)
LAO-A-	LOCKLAND AREA OFFICE, AEC
LAP-	LINDE AIR PRODUCTS CO., TONAWANDA, N.Y.
LAP-	PENNSYLVANIA. UNIV., PHILADELPHIA
LAP-	RES. + DEV. BD. COMM. ON ELECTRONICS, WASH., D.C.
LAP-	RESEARCH AND DEVELOPMENT BOARD. PANEL ON ANTENNAS AND PROPAGATION, WASHINGTON, D.C.
LAPC-	LINDE AIR PRODUCTS CO., TONAWANDA, N.Y.

LAPETH-	SWITZERLAND. EIDGENOESSISCHE TECHNISCHE HOCHSCHULE, ZURICH. LABORATORIUM FUER ATOMOSPHAERNPHYSIK
LAP-TR-	CHICAGO. UNIV.
LAR-	DENVER. UNIV. DENVER RESEARCH INST.
LAR-	LOCKLAND AIRCRAFT REACTORS OPERATIONS OFFICE, AEC (REF. 4)
LARS INFORMATION NOTE-	AGRICULTURAL RESEARCH SERVICE, BELTSVILLE, MD.
LARTIUC-	LOS ANGELES REGIONAL TECH. INFO. USERS COUNCIL
LAS-	CHICAGO. UNIV. LABS. FOR APPLIED SCIENCES
LAS/	EUROPEAN SPACE RESEARCH ORGANIZATION, PARIS
LAS-(YEAR)-(LETTERS)-	FRANCE. CENTRE NATIONAL DE LA RECHERCHE SCIENTIFIQUE. LABORATOIRE D'ASTRONOMIE SPATIALE, MARSEILLES
LASL-	LOS ALAMOS SCIENTIFIC LAB., N. MEX.
LAS-L-P-	CHICAGO. UNIV. LABS. FOR APPLIED SCIENCES
LASL-P-	LOS ALAMOS SCIENTIFIC LAB., N. MEX. (PROPOSAL)
LAS-QR-(LTR.)(NO.)-	CHICAGO. UNIV. LABS. FOR APPLIED SCIENCES
LAS-SR-(NUMBER)-(NUMBER)	CHICAGO. UNIV. LABS. FOR APPLIED SCIENCES
LAS-SR-P-	CHICAGO. UNIV. LABS. FOR APPLIED SCIENCES
LAS-TR-(NUMBER)-(NUMBER)	CHICAGO. UNIV. LABS. FOR APPLIED SCIENCES
LAS-TR-(LTR.)(NO.)-	CHICAGO. UNIV. LABS. FOR APPLIED SCIENCES
LAT-	ENGLISH ELECTRIC CO., LTD.
LA-TR-(YEAR)-	LOS ALAMOS SCIENTIFIC LAB., N. MEX. (TRANSLATION) (CODE USED 1964-)
LA-TR-	LOS ALAMOS SCIENTIFIC LAB., N. MEX. (TRANSLATION) (CODE USED BEFORE 1964)
LB-	(LABORATORY BULLETIN) (REF. 2)
LB-	(LIBRARY BULLETIN) (REF. 2)
LB-	DOUGLAS AIRCRAFT CO., INC., LONG BEACH, CALIF.
LB-	DOUGLAS AIRCRAFT CO., INC., NEWPORT BEACH, CALIF.
LB-	MC DONNELL AIRCRAFT CORP., ST. LOUIS (LIBRARY BIB.)
LB-	RADIO CORP. OF AMERICA. RCA LABS. DIV., PRINCETON, NJ
LBF-	GERMANY. LABORATORIUM FUER BETRIEBSFESTIGKEIT, DARMSTADT
LBF-S-	GERMANY. LABORATORIUM FUER BETRIEBSFESTIGKEIT, DARMSTADT
LB/G-	UNITED KINGDOM ATOMIC ENERGY AUTHORITY. RESEARCH GP. ATOMIC ENERGY RES. ESTAB., HARWELL, BERKS, ENGLAND
LBL-	CALIFORNIA. UNIV., BERKELEY. LAWRENCE BERKELEY LAB.
LBSC-	LONG BEACH STATE COLLEGE, CALIF.
LC-	GT. BRIT. BLDG. RES. STA., GARSTON, HERTS, ENGLAND
LC-	LEESONA CORP., PROVIDENCE
LC-	LIBRARY OF CONGRESS, WASHINGTON, D.C.
LC-	NATIONAL BUREAU OF STANDARDS, WASHINGTON, D.C. (LETTER CIRCULAR)
LC-	NATIONAL RESEARCH COUNCIL OF CANADA. ATOMIC ENERGY PROJECT, CHALK RIVER, ONT.
LC-	PATTERSON, MOOS RES. DIV.,LEESONA CORP.,JAMAICA, N.Y.
LC-	RESEARCH AND DEVELOPMENT BOARD, WASHINGTON, D.C. (LANDING COMBAT OPERATIONS)
LCA-	LITHIUM CORP. OF AMERICA, INC., BESSEMER CITY, N.C.
LCA-	LITHIUM CORP. OF AMERICA, INC., MINNEAPOLIS
LCA-	LITHIUM CORP. OF AMERICA, INC., N.Y.C.
LCA/MS-RAE-	FRANCE. LABORATOIRE CENTRAL DE L'ARMEMENT, ARCUEIL
LCC-	LACQUER + CHEM. CORP. ALAKA RES. LAB., BROOKLYN
LCE-TR-(YEAR)-	ARMY MISSILE COMMAND. SYSTEMS ENGINEERING DIV., REDSTONE ARSENAL, ALA.
LCI-	KENT STATE UNIV., OHIO. LIQUID CRYSTAL INST.
LCIE-	LABORATOIRE CENTRAL DES INDUSTRIES ELECTRIQUES, FONTENAY-AUX-ROSES, FRANCE
LCM-	RES. + DEV. BD. COMM. ON ELECTRONICS, WASH., D.C.
LCM-	RES. + DEV. BD. PANEL ON COMMUNICATIONS, WASH., D.C.
LCP-	RESEARCH + DEV. BD. PANEL ON COMPONENTS, WASH., D.C.
LC-PR-	NAVAL ORDNANCE LAB., WHITE OAK, MD.
LC-RB-	LIBRARY OF CONGRESS. NAVY RESEARCH SECTION, WASHINGTON, D.C. (REPORT BIBLIOGRAPHY)
LCS-	RES. + DEV. BD. COMM. ON ELECTRONICS, WASH., D.C.
LC-SL-	LINDE CO. SPEEDWAY LABS., INDIANAPOLIS
LC SP-	GT. BRIT. LEGISLATIVE COUNCIL (SESSIONAL PAPER)
LC-TID-	LIBRARY OF CONGRESS. TECH. INFO. DIV., WASH., D.C.
LC-TRL-	LINDE CO. RESEARCH AND DEV. LAB., TONAWANDA, N.Y.
LD-	THIOKOL CHEMICAL CORP., HUNTSVILLE, ALA.
LDB-	FRANCE. LABORATOIRE D'ETUDES BALISTIQUES DE ST. LOUIS
LDGO-	COLUMBIA UNIV., PALISADES, N.Y. LAMONT-DOHERTY GEOLOGICAL OBSERVATORY
LDGO-	LAMONT-DOHERTY GEOLOGICAL OBSERVATORY, PALISADES, NY
LDL-(4 DIGITS)-	LASER DIODE LABS., INC., METUCHEN, N.J.
LDMR-	NAVAL AIR MISSILE TEST CENTER. LAUNCHER DIV., POINT MUGU, CALIF. (MEMORANDUM REPORT)
LE-	ATOMIC ENERGY OF CANADA LTD. CHALK RIVER PROJ., ONT.
LE-	CALIFORNIA. UNIV., LIVERMORE. LAWRENCE RADIATION LAB. ELECTRONICS ENGINEERING DEPT.
LE-	NATIONAL RESEARCH COUNCIL OF CANADA. ATOMIC ENERGY PROJECT, CHALK RIVER, ONT.
LE-	NAVAL MISSILE CENTER. LABORATORY EVALUATION DEPT., POINT MUGU, CALIF.
LEA-	RES. + DEV. BD. COMM. ON ELECTRONICS, WASH., D.C.
LEA-	RESEARCH + DEV. BD. PANEL ON ACOUSTICS, WASH., D.C.
LEAA-NI-(YEAR)-	DEPARTMENT OF JUSTICE. OFFICE OF LAW ENFORCEMENT ASSISTANCE, WASHINGTON, D.C.
LEAFLET-	DEPARTMENT OF HOUSING AND URBAN DEVELOPMENT. DIVISION OF INTERNATIONAL AFFAIRS, WASHINGTON, D.C.
LEAR-	LEAR, INC., SANTA MONICA, CALIF.
LEAR ADR-	LEAR, INC., SANTA MONICA, CALIF.
LEAR ER-	LEAR, INC., SANTA MONICA, CALIF.
LEAR-ER-GR-	LEAR, INC. ELECTRO-MECHANICAL DIV., GRAND RAPIDS
LEAR SIEGLER-	LEAR SIEGLER, INC., GRAND RAPIDS
LEA-TR-	NORGES TEKNISKE HOEGSKOLE, TRONDHEIM
LEC-	LOCKHEED ELECTRONICS CO., PLAINFIELD, N.J.
LEC-(NUMBER)-(NUMBER)-	LOCKHEED ELECTRONICS CO., PLAINFIELD, N.J.
LEC/HASD-	LOCKHEED ELECTRONICS CO. AEROSPACE SYSTEMS DIV., HOUSTON, TEX.
LEC-TR-(YEAR)-	LOCKHEED ELECTRONICS CO., WHITE SANDS MISSILE RANGE, N. MEX.
LEDEX-	LEDEX INC., DAYTON, OHIO
LEH-	LEHIGH UNIV., BETHLEHEM, PENNA.
LEHIGH-	LEHIGH UNIV., BETHLEHEM, PENNA.
LEHIGH PR-	LEHIGH UNIV., BETHLEHEM, PENNA. FRITZ LAB.
LEHIGH TR-	LEHIGH UNIV., BETHLEHEM, PENNA. FRITZ LAB.
LEHIGH U. PR-	LEHIGH UNIV., BETHLEHEM, PENNA. FRITZ LAB.
LEH TR-	LEHIGH UNIV., BETHLEHEM, PENNA. INST. OF RESEARCH
LEL-	ENGLISH ELECTRIC CO., LTD.
LEL.T-	ENGLISH ELECTRIC CO., LTD.
LEO/A-	GT. BRIT. ATOMIC EN. RES. ESTAB., HARWELL, BERKS, ENG
LEO/N-	GT. BRIT. ATOMIC EN. RES. ESTAB., HARWELL, BERKS, ENG
LEO/RDC-	GT. BRIT. ATOMIC EN. RES. ESTAB., HARWELL, BERKS, ENG
LEO-RDC/P-	UNITED KINGDOM ATOMIC ENERGY AUTHORITY. INDUSTRIAL GROUP H.Q., RISLEY, LANCS, ENGLAND
LER(NO.)	CALIFORNIA. UNIV., LIVERMORE. RADIATION LAB.
LET-	LIGHTNING AND TRANSIENTS RESEARCH INST., MINNEAPOLIS
LET-	OFF. OF THE ASST. SECY. OF DEFENSE(RES.+DEV.),WASH,DC
LET-	RES. + DEV. BD. PANEL ON ELECTRON TUBES, WASH., D.C.

LETOR-	EUROPEAN THEATER OF OPERATIONS (LEATHER REPORT)
LETOR-	QUARTERMASTER CORPS (LEATHER SERIES REPORT)
LETTER-	(INFORMAL TECHNICAL REPORT) (REF. 2)
LEVINTHAL-	LEVINTHAL ELECTRONIC PRODUCTS, INC., PALO ALTO, CAL.
LEXP-	MASS. INST. OF TECH., CAMBRIDGE. LEXINGTON PROJECT
LEY-	PHILLIPS PETROLEUM CO. ATOMIC EN. DIV., IDAHO FALLS
LF-	AEROJET-GENERAL CORP., SACRAMENTO, CALIF.
LF-(NUMBERS)-	AEROJET-GENERAL CORP. EXPERIMENTAL ENGINE DIV., SACRAMENTO, CALIF.
LF-	AEROJET-GENERAL CORP. LIQUID ROCKET OPERATIONS, SACRAMENTO, CALIF.
LF-	GT. BRIT. MINISTRY OF SUPPLY
LF-	LOVELACE FOUNDATION FOR MEDICAL EDUCATION AND RESEARCH, ALBUQUERQUE, N. MEX.
LF-	SWEDEN. FOERSVARETS FORSKNINGSANSTALT, STOCKHOLM
LFA-	LUFTFAHRTFORSCHUNGSANSTALT HERMANN GOERING
LFC-	GT. BRIT. ADVISORY COUNCIL ON SCI. RES. + TECH. DEV.
LFE-	LABORATORY FOR ELECTRONICS, INC., BOSTON
LFEN.(NUMBER)/BETA	OXFORD UNIV.
LFEN-	PORTUGAL. JUNTA DE ENERGIA NUCLEAR. LABORATORIO DE FISICA E ENGENHARIA NUCLEARES, SACAVEM.
LFEN.-	SPAIN. JUNTA DE ENERGIA NUCLEAR, MADRID
LFEN-NI.-	PORTUGAL. JUNTA DE ENERGIA NUCLEAR. LABORATORIO DE FISICA E ENGENHARIA NUCLEARES, SACAVEM
LFEWP-	GT. BRIT. ATOMIC EN. RES. ESTAB., HARWELL, BERKS, ENG
LFF-	CHALMERS TEKNISKA HOEGSKOLA, GOETEBORG
LFF-	SWEDEN. FOERSKNINGRADENS LABORATORIUM, STUDSVIK
LFLWP-	GT. BRIT. ATOMIC EN. RES. ESTAB., HARWELL, BERKS, ENG
LFM-	LUFTFAHRTFORSCHUNG, MUNICH
LFMER-	LOVELACE FOUNDATION FOR MEDICAL EDUCATION AND RESEARCH, ALBUQUERQUE, N. MEX.
LFM-R-(YEAR)-	DENMARK. TECHNICAL UNIV., COPENHAGEN
LFMR-	LOVELACE FOUNDATION FOR MEDICAL EDUCATION AND RESEARCH, ALBUQUERQUE, N. MEX. (INTERNAL REPORTS)
LFR-	BROOKHAVEN NATIONAL LAB., UPTON, N.Y.
LF-TR-	LOVELACE FOUNDATION FOR MEDICAL EDUCATION AND RESEARCH, ALBUQUERQUE, N. MEX. (TRANSLATIONS)
LG-	LILIENTHAL-GESELLSCHAFT FUER LUFTFAHRTFORSCHUNG, BERLIN
LG(NO. LTRS. NO.)	LOCKHEED-GEORGIA CO., MARIETTA (CONTRACT REPORT)
LG-	STANFORD UNIV., CALIF. DEPT. OF MECHANICAL ENG.
LGB-	LILIENTHAL-GESELLSCHAFT FUER LUFTFAHRTFORSCHUNG, BERLIN
LG B (NUMBER)/N-	LILIENTHAL-GESELLSCHAFT FUER LUFTFAHRTFORSCHUNG, BERLIN (BERICHT/NACHTRAG)
LG B-	LILIENTHAL-GESELLSCHAFT FUER LUFTFAHRTFORSCHUNG, BERLIN (BERICHT)
LG B (NUMBER) VB-	LILIENTHAL-GESELLSCHAFT FUER LUFTFAHRTFORSCHUNG, BERLIN (VORBERICHT)
LG BA (NUMBER)/	LILIENTHAL-GESELLSCHAFT FUER LUFTFAHRTFORSCHUNG, BERLIN (BERICHT A)
LG BA-	LILIENTHAL-GESELLSCHAFT FUER LUFTFAHRTFORSCHUNG, BERLIN (BERICHT A)
LG BS-	LILIENTHAL-GESELLSCHAFT FUER LUFTFAHRTFORSCHUNG, BERLIN (BERICHT S)
LGI-	ITALY. COMITATO NAZIONALE PER L'ENERGIA NUCLEARE. LABORATORIO GAS IONIZZATI, FRASCATI
LGI-(YEAR/NUMBER)	ITALY. COMITATO NAZIONALE PER L'ENERGIA NUCLEARE. LABORATORIO GAS IONIZZATI, FRASCATI
LGM-	NORTHROP AIRCRAFT, INC., HAWTHORNE, CALIF. (LANDING GEAR MEMO)
LG/MT-LCR/DR(NUMBER)-	COMPAGNIE FRANCAISE THOMSON HOUSTON-HOTCHKISS BRANDT, PARIS
LGO-	COLUMBIA UNIV., PALISADES, N.Y. LAMONT GEOLOGICAL OBS
LGO-	MIDWEST APPLIED SCIENCE CORP., LAFAYETTE, IND.
LG/PB-LCR/DR(NUMBER)-	COMPAGNIE FRANCAISE THOMSON HOUSTON-HOTCHKISS BRANDT, PARIS
LGR-ER-	LOCKHEED-GEORGIA CO., MARIETTA
LGR-ER/S-	LOCKHEED-GEORGIA CO., MARIETTA
LGR-M/P-	LOCKHEED-GEORGIA CO., MARIETTA
LG TB (NUMBER)/	LILIENTHAL-GESELLSCHAFT FUER LUFTFAHRTFORSCHUNG, BERLIN (TAGUNGSVERICHT)
LG TB-	LILIENTHAL-GESELLSCHAFT FUER LUFTFAHRTFORSCHUNG, BERLIN (TAGUNGSBERICHT)
LH-	KEARFOTT CO., INC., N.Y.C.
LH-	SAAB AIRCRAFT CO., SWEDEN
LHK-	SAAB AIRCRAFT CO., SWEDEN
LHU-	SAAB AIRCRAFT CO., SWEDEN
LI-(6 DIGITS)	AMERICAN REHABILITATION FOUNDATION. INST. FOR INTERDISCIPLINARY STUDIES, MINNEAPOLIS
LI-	LEAR, INC., SANTA MONICA, CALIF.
LI-	LITTON INDUSTRIES, INC. (ALL LOCATIONS)
LI-	WESTERN MICROWAVE LAB., INC., SANTA CLARA, CALIF.
LIBA-	LONG ISLAND BIOLOGICAL ASSN. BIOLOGICAL LAB., COLD SPRING HARBOR, N.Y.
LIBI-	LIBRASCOPE INC.
LIBRARY TRANS-	GT. BRIT. ROYAL AIRCRAFT EST., FARNBOROUGH, HANTS, ENG
LIBRASCOPE-	GENERAL PRECISION SYSTEMS, INC., LIBRASCOPE SYSTEMS DIV., GLENDALE, CALIF.
LIB/TRANS-	AUSTRALIAN ATOMIC ENERGY COMMISSION. RESEARCH ESTABLISHMENT, LUCAS HEIGHTS, NEW SOUTH WALES (TRANSLATION)
LIEGEU-	LIEGE. UNIVERSITE
LIIR-	LOUISVILLE, KY. UNIV. INST. OF INDUSTRIAL RESEARCH
LIM-(YEAR)-/	GENERAL PRECISION, INC. LINK GROUP, BINGHAMTON, N.Y.
LINAC-	RENSSELAER POLYTECHNIC INST., TROY, N.Y.
LINCOLN MANUAL-	MASSACHUSETTS INST. OF TECH., LEXINGTON. LINCOLN LAB.
LINCS-(NUMBER-YEAR)	CENTER FOR APPLIED LINGUISTICS, WASHINGTON, D.C. (LANGUAGE INFORMATION NETWORK)
LIR-	RES. + DEV. BD. COMM. ON ELECTRONICS, WASH., D.C.
LIR-	RESEARCH + DEV. BD. PANEL ON INFRARED, WASH., D.C.
LIR-TR-	MASSACHUSETTS INST. OF TECH., CAMBRIDGE. LAB. FOR INSULATION RESEARCH
LIT-	LITTLE (ARTHUR D.), INC., CAMBRIDGE, MASS.
LIT. SURVEY NO.-	MARTIN CO., DENVER
LIU-	LIEGE. UNIVERSITE. INSTITUT D-ASTROPHYSIQUE
LIVL-	CALIFORNIA. UNIV., LIVERMORE. RADIATION LAB.
LIZ-	RES. + DEV. BD. COMM. ON ELECTRONICS, WASH., D.C.
LJ-(NUMBER)-	SPERRY PHOENIX CO., ARIZ.
LJ-	SPERRY RAND CORP. SPERRY FLIGHT SYSTEMS DIV., PHOENIX, ARIZ.
LJ-	SPERRY UTAH CO., SALT LAKE CITY
LJL(ISCAR)-	IOWA STATE COLL., AMES. INST. FOR ATOMIC RESEARCH (CODE IS AUTHOR'S INITIALS)
LJL-MAC-	IOWA STATE COLL., AMES. INST. FOR ATOMIC RESEARCH (CODE IS AUTHOR'S INITIALS)
LJT-	ENGLISH ELECTRIC CO., LTD.
LK-	GENERAL ELECTRIC CO. INSULATOR DEPT., BALTIMORE
LL-	ARMY TANK-AUTOMOTIVE CENTER, WARREN, MICH. (LAND LOCOMOTION)
LL-	LINDEN LABS., INC., UNIVERSITY PARK, PENNA.
LL-	MASSACHUSETTS INST. OF TECH., CAMBRIDGE. LINCOLN LAB.
LL-	MASSACHUSETTS INST. OF TECH., LEXINGTON. LINCOLN LAB.
LL-	TELEDYNE, INC., ALEXANDRIA, VA.
LLI-	LINDEN LABS., INC., STATE COLLEGE, PENNA.
LL-TN-	MASSACHUSETTS INST. OF TECH., LEXINGTON. LINCOLN LAB.
LM-	(LIGHT METALS) (REF. 2)

LM-	(LIQUID METALS) (REF. 2)
L + M-	ABERDEEN PROVING GROUND, MD.
LM-	FOREIGN ECONOMIC ADMINISTRATION. ENEMY BRANCH. LABOR AND MANPOWER DIVISION
LM-	FRANKLIN INST. LABS. FOR RES. + DEV., PHILADELPHIA
LM-(NUMBER)	GENERAL ELECTRIC CO. LIGHT MILITARY ELECTRONICS DEPT., UTICA, N.Y.
LM-	GT. BRIT. ADVISORY COUNCIL ON SCI. RES. + TECH. DEV. LABOR AND MANPOWER DIVISION
LMA(NUMBER)-	GRUMMAN AEROSPACE CORP. PRODUCT SUPPORT DEPT., BETHPAGE, N.Y.
LMCC-	GT. BRIT. CAPENHURST WORKS, CHES., ENGLAND
LMCC/P-	GT. BRIT. CAPENHURST WORKS, CHES., ENGLAND
LMC-T-	ENGLISH ELECTRIC CO., LTD. GUIDED WEAPONS DIV., LUTON, BEDS., ENGLAND
LMDC-	LONDON MILITARY DOCUMENTS CENTER
LMD/NOS.-	UNITED KINGDOM ATOMIC ENERGY AUTHORITY. RESEARCH GP. ATOMIC ENERGY RES. ESTAB., HARWELL, BERKS, ENGLAND
LME-	GENERAL ELECTRIC CO. AEROSPACE ELECTRONICS DEPT., UTICA, N.Y.
LM-E-(NUMBER)-	GENERAL ELECTRIC CO. AIRCRAFT ENGINE GP., LYNN, MASS.
LMEC-(YEAR)-	ATOMICS INTERNATIONAL. LIQUID METAL ENGINEERING CENTER, CANOGA PARK, CALIF.
LMEC-MEMO-(YEAR)-	ATOMICS INTERNATIONAL. LIQUID METAL ENGINEERING CENTER, CANOGA PARK, CALIF.
LMEC-MR (MONTH/YEAR)	ATOMICS INTERNATIONAL. LIQUID METAL ENGINEERING CENTER, CANOGA PARK, CALIF.
LMEC-TDR-(YEAR)-	ATOMICS INTERNATIONAL. LIQUID METAL ENGINEERING CENTER, CANOGA PARK, CALIF.
LMEC-TPR (MONTH/YEAR)	ATOMICS INTERNATIONAL. LIQUID METAL ENGINEERING CENTER, CANOGA PARK, CALIF.
LMEJ-	GENERAL ELECTRIC CO. ARMAMENT AND CONTROL PRODUCTS SECTION, JOHNSON CITY, N.Y.
LMEJ-	GENERAL ELECTRIC CO. LIGHT MILITARY ELECTRONICS EQUIPMENT DEPT., JOHNSON CITY, N.Y.
LMF-	ITALY. COMITATO NAZIONALE PER L'ENERGIA NUCLEARE. LABORATORI NAZIONALI DI FRASCATI
LMFP/P-	UNITED KINGDOM ATOMIC ENERGY AUTHORITY. RESEARCH GP. ATOMIC ENERGY RES. ESTAB., HARWELL, BERKS, ENGLAND
LMFR-	BROOKHAVEN NATIONAL LAB., UPTON, N.Y.
LMFR/I-	UNITED KINGDOM ATOMIC ENERGY AUTHORITY. RESEARCH GP. ATOMIC ENERGY RES. ESTAB., HARWELL, BERKS, ENGLAND
LMFR/P-	UNITED KINGDOM ATOMIC ENERGY AUTHORITY. RESEARCH GP. ATOMIC ENERGY RES. ESTAB., HARWELL, BERKS, ENGLAND
LMFS/N-	GT. BRIT. ATOMIC EN. RES. ESTAB., HARWELL, BERKS, ENG
LMFS/P-	UNITED KINGDOM ATOMIC ENERGY AUTHORITY. RESEARCH GP. ATOMIC ENERGY RES. ESTAB., HARWELL, BERKS, ENGLAND
LMFT/N-	UNITED KINGDOM ATOMIC ENERGY AUTHORITY. RESEARCH GP. ATOMIC ENERGY RES. ESTAB., HARWELL, BERKS, ENGLAND
LMFT/P-	UNITED KINGDOM ATOMIC ENERGY AUTHORITY. RESEARCH GP. ATOMIC ENERGY RES. ESTAB., HARWELL, BERKS, ENGLAND
LMFT/P-	UNITED KINGDOM ATOMIC ENERGY AUTHORITY. RESEARCH GP. CHEMISTRY DIV., WOOLWICH OUTSTATION, ENGLAND
LMI-(YEAR)-	LOGISTICS MANAGEMENT INST., WASHINGTON, D.C.
LMI-TASK-(YEAR)-	LOGISTICS MANAGEMENT INST., WASHINGTON, D.C.
LML-	LEESONA MOOS LABS. DIV. OF LEESONA CORP., JAMAICA, NY
LML-	LOOKOUT MOUNTAIN LAB., LOS ANGELES
LMO-(NUMBER)-	GRUMMAN AIRCRAFT ENGINEERING CORP., BETHPAGE, N.Y.
LMP-	YALE UNIV., NEW HAVEN. LAB. OF MARINE PHYSICS
LMPR-	STRATEGIC BOMBING SURVEY (LIGHT METALS PLANT RPTS.)
LMSC-	ATOMICS INTERNATIONAL DIV., NORTH AMERICAN AVIATION, INC., CANOGA PARK, CALIF.
LMSC-	KNOLLS ATOMIC POWER LAB. LIQUID METALS SAFETY COMMITTEE, SCHENECTADY, N.Y.
LMSC-	LOCKHEED MISSILES AND SPACE CO., PALO ALTO, CALIF.
LMSC-(NUMBERS)-	LOCKHEED MISSILES AND SPACE CO., PALO ALTO, CALIF.
LMSC-	LOCKHEED MISSILES AND SPACE CO., SUNNYVALE, CALIF.
LMSC-(NUMBERS)	LOCKHEED MISSILES AND SPACE CO., SUNNYVALE, CALIF.
LMSC-	LOCKHEED MISSILES AND SPACE CO., VAN NUYS, CALIF.
LMSC-A-(SIX DIGITS)	LOCKHEED MISSILES AND SPACE CO., PALO ALTO, CALIF. (UNCLASSIFIED AND CONFIDENTIAL REPORTS)
LMSC-A-(SIX DIGITS)	LOCKHEED MISSILES AND SPACE CO., SUNNYVALE, CALIF.
LMSC-B-(SIX DIGITS)	LOCKHEED MISSILES AND SPACE CO., PALO ALTO, CALIF. (SECRET REPORTS)
LMSC-CB(YEAR)-	LOCKHEED MISSILES AND SPACE CO., PALO ALTO, CALIF. (CITATION BIBLIOGRAPHY)
LMSC-CB-(YEAR)-	LOCKHEED MISSILES AND SPACE CO., SUNNYVALE, CALIF. (CITATION BIBLIOGRAPHY)
LMSC-D(NUMBER)	LOCKHEED MISSILES AND SPACE CO., SUNNYVALE, CALIF.
LMSC-EETP-(NUMBER)/	LOCKHEED MISSILES AND SPACE CO., SUNNYVALE, CALIF.
LMSC-HREC-(LETTER)-	LOCKHEED MISSILES AND SPACE CO. HUNTSVILLE RESEARCH AND ENGINEERING CENTER, ALA.
LMSC/HREC-A(NO.S)-	LOCKHEED MISSILES AND SPACE CO., HUNTSVILLE, ALA.
LMSC/HTVL-	LOCKHEED MISSILES AND SPACE CO., HUNTSVILLE, ALA.
LMSC-LS(YEAR)-	LOCKHEED MISSILES AND SPACE CO., PALO ALTO, CALIF. (LITERATURE SEARCH)
LMSC-M-	LOCKHEED MISSILES AND SPACE CO., PALO ALTO, CALIF.
LMSC-N-(NUMBER(S))-	LOCKHEED MISSILES AND SPACE CO., PALO ALTO, CALIF.
LMSC-N-A-(NUMBERS)-	LOCKHEED MISSILES AND SPACE CO., PALO ALTO, CALIF.
LMSC-NSP-(YEAR)-	LOCKHEED MISSILES AND SPACE CO., SUNNYVALE, CALIF.
LMSC-SB-(YEAR)-	LOCKHEED MISSILES AND SPACE CO., SUNNYVALE, CALIF.
LMSC-SRB-(YEAR)-	LOCKHEED MISSILES AND SPACE CO., SUNNYVALE, CALIF.
LMSC-SSD-(NO.)-(YR.)-	LOCKHEED MISSILES AND SPACE CO., SUNNYVALE, CALIF.
LMSC-TR-	ELECTRONICS CORP. OF AMERICA, CAMBRIDGE, MASS.
LMSD-	LOCKHEED MISSILES AND SPACE CO., PALO ALTO, CALIF.
LMSD-SB-(YR)-	LOCKHEED AIRCRAFT CORP. MISSILES AND SPACE DIV., SUNNYVALE, CALIF.
LMSD-SRB-(YEAR)-	LOCKHEED AIRCRAFT CORP. MISSILES AND SPACE DIV., SUNNYVALE, CALIF.
LMSP/P-	GT. BRIT. ATOMIC EN. RES. ESTAB., HARWELL, BERKS, ENG
LMSS-TR-(YEAR)-	CHICAGO. UNIV. LAB. OF MOLECULAR STRUCTURE + SPECTRA
LM/TAB-	MSA RESEARCH CORP., CALLERY, PENNA. (LIQUID METALS TECHNOLOGY ABSTRACT BULLETIN)
LN-	(LABORATORY NOTE) (REF. 2)
LN-	LEEDS AND NORTHRUP CO., PHILADELPHIA.
LN-	NORTHROP AIRCRAFT, INC., HAWTHORNE, CALIF. (LAB. NOTE)
LNF-	ITALY. COMITATO NAZIONALE PER L'ENERGIA NUCLEARE. LABORATORI NAZIONALI DI FRASCATI
LNF-(YEAR/NUMBER)	ITALY. COMITATO NAZIONALE PER L'ENERGIA NUCLEARE. LABORATORI NAZIONALI DI FRASCATI
LNP-	LOCKHEED-GEORGIA CO., MARIETTA (NUCLEAR PRODUCTS)
LNP-ER-	LOCKHEED NUCLEAR PRODUCTS, MARIETTA, GA.
LNP-NM-	LOCKHEED NUCLEAR PRODUCTS, MARIETTA, GA.
LNP-NP-	LOCKHEED NUCLEAR PRODUCTS, MARIETTA, GA.
LNP-NR-	LOCKHEED AIRCRAFT CORP. MISSILE SYSTEMS DIV., PALO ALTO, CALIF.
LNP-NR-	LOCKHEED-GEORGIA CO., MARIETTA (NUCLEAR REPORT)
LNP-NR-	LOCKHEED NUCLEAR PRODUCTS, MARIETTA, GA.
LO-	COAST AND GEODETIC SURVEY, WASH., D.C. (LOCAL CHARTS)
LO-	LOWELL OBSERVATORY, FLAGSTAFF, ARIZ.
LOBUND-AEC RPT. NO.-	UNIV. OF NOTRE DAME, SOUTH BEND, IND. LOBUND INST.
LOC-	LAUNCH OPERATIONS CENTER, COCOA BEACH, FLA.
LOC-(LETTERS)-	LAUNCH OPERATIONS CENTER, COCOA BEACH, FLA.
LOFT-GE-M-	GENERAL ELECTRIC CO. NUCLEAR SYSTEMS PROGRAMS, IDAHO FALLS (LOFT PROGRAM)
LOG-	(LOG RECORD) (REF. 2)
LONDON-	LONDON. UNIV., UNIVERSITY COLL. DEPT. OF PHYSICS
LONG SHOT-	GEOLOGICAL SURVEY, DENVER

LOR-

LOR-	OFFICE OF SCIENTIFIC RES. + DEV. (LIAISON OFF. REF.)
LOST-	COLUMBIA UNIV., ST. DAVID'S, BERMUDA. GEOPHYSICAL FIELD STATION
LOT-	ENGLISH ELECTRIC CO., LTD.
LOU-	LOUISVILLE, KY. UNIV.
LOUU-	LOUVAIN, BELGIUM. UNIVERSITE
LOVELACE-	LOVELACE FOUNDATION FOR MEDICAL EDUCATION AND RESEARCH, ALBUQUERQUE, N. MEX.
LP-	BADGER ORDNANCE WORKS, BARABOO, WIS.
LP-	MASS. INST. OF TECH., CAMBRIDGE. LEXINGTON PROJECT
LP-	NATIONAL RESEARCH COUNCIL OF CANADA, OTTAWA
LP-	OFFICE OF SCIENTIFIC RESEARCH AND DEVELOPMENT
LP-	PARIS. UNIVERSITE, ORSAY. LABORATOIRE DE PHYSIQUE DES PLASMAS
LP.(NO.)	PARIS. UNIV., ORSAY. LABORATOIRE DES HAUTES ENERGIES
LPAM-RI-	COLLEGE DE FRANCE, PARIS. LABORATOIRE DE PHYSIQUE ATOMIQUE ET MOLECULAIRE
LPC-	LOCKHEED PROPULSION CO., REDLANDS, CALIF.
LPC-(NUMBER)(LETTER)	LOCKHEED PROPULSION CO., REDLANDS, CALIF.
LPC-(NUMBER)-S-	LOCKHEED PROPULSION CO., REDLANDS, CALIF.
LPC-MIN.-	LOS ALAMOS SCIENTIFIC LAB., N. MEX.
LPDC-	LIBERTY POWDER DEFENSE CORP., BARABOO, WIS.
LPIA-	JOHNS HOPKINS UNIV., SILVER SPRING, MD. LIQUID PROPELLANT INFORMATION AGENCY
LPIA/D(YR)	JOHNS HOPKINS UNIV., SILVER SPRING, MD. LIQUID PROPELLANT INFORMATION AGENCY
LPN-UM-	MONTREAL. UNIV. LABORATOIRE DE PHYSIQUE NUCLEAIRE
LPS-	CORNELL UNIV., ITHACA, N.Y. LAB. OF PLASMA STUDIES
LPSM-	AEROJET-GENERAL CORP., SACRAMENTO, CALIF.
LPSR-	MASSACHUSETTS INST. OF TECH., LEXINGTON. LINCOLN LAB.
LPTHE-(YEAR/NUMBER)	FRANCE. LABORATOIRE DE PHYSIQUE THEORIQUE ET DES HAUTES ENERGIES, ORSAY
LPTHE-TH-	FRANCE. LABORATOIRE DE PHYSIQUE THEORIQUE ET DES HAUTES ENERGIES, ORSAY
LPTPE-	PARIS. UNIV., ORSAY. LABORATOIRE DE PHYSIQUE THEORIQUE ET PARTICULES ELEMENTAIRES
LR-	(LABORATORY REPORT) (REF. 2)
LR-	(LETTER REPORT) (REF. 2)
LR-	AIR FORCE (LONG RANGE NAVIGATION CHARTS)
LR-	AMES LAB., IOWA
LR-	ARGENTINA. INSTITUTO DE INVESTIGACION AERONAUTICA Y ESPACIAL, BUENOS AIRES
LR-	BUFFALO ELECTRO-CHEMICAL CO., INC.
LR-	CANADA. NATIONAL AERONAUTICAL ESTABLISHMENT, OTTAWA
LR-(2 DIGITS)	GT. BRIT. DEPT. OF SCIENTIFIC AND INDUSTRIAL RESEARCH. WARREN SPRING LAB.
LR-	IOWA STATE COLL., AMES
LR-	LOCKHEED-CALIFORNIA CO., BURBANK
LR-	MARQUARDT CORP., VAN NUYS, CALIF.
LR-	NATIONAL RESEARCH COUNCIL OF CANADA, OTTAWA
LR-	NATL. RES. COUNCIL OF CANADA. NATL. RES. LABS., OTTAWA
LR-	STEVENS INST. OF TECH., HOBOKEN, N.J. DAVIDSON LAB.
LRA-	RES. + DEV. BD. COMM. ON ELECTRONICS, WASH., D.C.
LR-B-	SPERRY GYROSCOPE CO., GREAT NECK, N.Y.
LRBA-(NUMBER)-(YEAR)-EN	FRANCE. LABORATOIRE DE RECHERCHES BALISTIQUES ET AERODYNAMIQUES, VERNON
LRBA-CR-	FRANCE. LABORATOIRE DE RECHERCHES BALISTIQUES ET AERODYNAMIQUES, VERNON
LRBA-E-	FRANCE. LABORATOIRE DE RECHERCHES BALISTIQUES ET AERODYNAMIQUES, VERNON
LRBA-NT-(NUMBER/YR/LTRS)	FRANCE. LABORATOIRE DE RECHERCHES BALISTIQUES ET AERODYNAMIQUES, VERNON
LRBA-NT-E-	FRANCE. LABORATOIRE DE RECHERCHES BALISTIQUES ET AERODYNAMIQUES, VERNON
LRBA-PV-(NUMBER-NO.-LTRS)	FRANCE. LABORATOIRE DE RECHERCHES BALISTIQUES ET AERODYNAMIQUES, VERNON
LRB-IB-(YEAR)-	WISCONSIN. LEGISLATIVE REFERENCE BUREAU, MADISON (INFORMATIONAL BULLETIN)
LRC-	LANGLEY RESEARCH CENTER, LANGLEY FIELD, VA. (REF. 26)
LRC-	TEXAS. UNIV., AUSTIN. LINGUISTICS RESEARCH CENTER
LRC-(YEAR)-(LTRS.)-	TEXAS. UNIV., AUSTIN. LINGUISTICS RESEARCH CENTER
LRD-(NUMBERS)-	AEROJET-GENERAL CORP. LIQUID ROCKET OPERATIONS, SACRAMENTO, CALIF.
LRD-	SPERRY GYROSCOPE CO., GREAT NECK, N.Y.
LRDC-REPRINT-	PITTSBURGH. UNIV. LEARNING RESEARCH + DEV. CENTER
LRDC-TR-	PITTSBURGH. UNIV. LEARNING RESEARCH + DEV. CENTER
LR-E-	BUFFALO ELECTRO-CHEMICAL CO., INC.
LRFSG-	LONG RANGE FUZING STANDARDIZATION GROUP (AEC-DOD), ALBUQUERQUE, N. MEX.
LRG-(YEAR)-B-	AEROSPACE CORP., EL SEGUNDO, CALIF. (LITERATURE RESEARCH GROUP, BIBLIOGRAPHY)
LRG-(YEAR)-T-	AEROSPACE CORP., EL SEGUNDO, CALIF. (LITERATURE RESEARCH GROUP, TRANSLATION)
LRG-	GT. BRIT. DEPT. OF SCI. + IND. RES. DIRECTORATE OF TUBE ALLOYS (ASSIGNED BY MAJOR GEN. L. R. GROVES)
LRI-	LINFIELD RESEARCH INSTITUTE, MC MINNVILLE, ORE.
LRL-	CALIFORNIA RESEARCH AND DEV. CO. LIVERMORE RES. LAB.
LRL LWS-	CALIFORNIA RESEARCH AND DEV. CO. LIVERMORE RES. LAB.
LRL MTA-	CALIFORNIA RESEARCH AND DEV. CO. LIVERMORE RES. LAB.
LRM-	BELL TELEPHONE LABS., INC., N.Y.C.
LRO-	AEROJET-GENERAL CORP. LIQUID ROCKET OPERATIONS, SACRAMENTO, CALIF.
LRO-	AEROJET-GENERAL CORP. LIQUID ROCKET PLANT, SACRAMENTO, CALIF.
LRO-	INDUSTRIAL INSTRUMENTS, INC., CEDAR GROVE, N.J.
LRP-	AEROJET-GENERAL CORP., AZUSA, CALIF.
LRP-	AEROJET-GENERAL CORP. LIQUID ROCKET PLANT, SACRAMENTO, CALIF.
LRP-	SWITZERLAND. LABORATOIRE DE RECHERCHES SUR LA PHYSIQUE DES PLASMAS, LAUSANNE
LRP-(NUMBER/YEAR)	SWITZERLAND. LABORATOIRE DE RECHERCHES SUR LA PHYSIQUE DES PLASMAS, LAUSANNE
LRPGD-	AIR FORCE MISSILE TEST CENTER, PATRICK AFB, FLA.
LRPL-QPR-	PICATINNY ARSENAL. LIQUID ROCKET PROPULSION LAB., DOVER, N.J.
LRR-	SMITH (A.O.) CORP. LONG RANGE RES. LAB., MILWAUKEE
LRSL-	FRANCE. LABORATOIRE DE RECHERCHE TECHNIQUES DE SAINT-LOUIS
LRV/(LETTERS-NUMBER)	UGINE-KUHLMANN. LAB. DE RECHERCHE, VENTHON, FRANCE
LRWE-	AUSTRALIA. LONG RANGE WEAPONS ESTABLISHMENT
LS-	(LITERATURE SEARCH) (REF. 2)
LS-	AEROJET-GENERAL CORP., AZUSA, CALIF.
LS-	CONVAIR, SAN DIEGO, CALIF.
LS-	FOREST SERVICE. LAKE STATES STATION
LS-	GARRETT CORP. AIRESEARCH MFG. DIV., LOS ANGELES
LS-	HUGHES AIRCRAFT CO., CULVER CITY, CALIF.
LS-	ISRAEL. ATOMIC ENERGY COMMISSION, TEL-AVIV
LS-	ISRAEL. ATOMIC ENERGY COMMISSION. SOREQ NUCLEAR RESEARCH CENTER, YAVNE
LSA-	LITERATURE SERVICE ASSOCIATES, BOUND BROOK, N.J. (TRANSLATIONS)
LSB-	LANGUAGE SERVICE BUREAU, CLEVELAND (TRANSLATIONS)
LS-BIB-(YEAR)-	HUGHES AIRCRAFT CO., CULVER CITY, CALIF.
LSC-S-	SYRACUSE UNIV., N.Y. LAB. OF SENSORY COMMUNICATION
LSD-	ABERDEEN PROVING GROUND. LABORATORY SERVICE DIV., MD.

LSI-	LITTON SYSTEMS, INC.
LSI-ASD-	LITTON SYSTEMS, INC. APPLIED SCIENCE DIV., MINNEAPOLIS
LSI/PED-	LEAR SIEGLER, INC. POWER EQUIPMENT DIV., CLEVELAND
LSMC-	LOCKHEED MISSILES AND SPACE CO., PALO ALTO, CALIF.
LSR-	AEROJET-GENERAL CORP., SACRAMENTO, CALIF.
LSR-RM-	STANFORD RESEARCH INST., MENLO PARK, CALIF. (LOGISTIC SYSTEMS RESEARCH)
LSSL-TP-	LITTON SCIENTIFIC SUPPORT LAB., FORT ORD, CALIF.
LST-	ENGLISH ELECTRIC CO., LTD.
LS-TR-	LIFE SCIENCES, INC., FORT WORTH, TEX.
L/STR-	LOCKHEED-CALIFORNIA CO., BURBANK
L/S-U-	GENERAL PRECISION, INC., SUNNYVALE, CALIF.
LSU-	LOUISIANA STATE UNIV., BATON ROUGE
L/T-	BRITISH ELECTRICAL AND ALLIED INDUSTRIES RESEARCH ASSN., LEATHERHEAD, SURREY, ENGLAND
L + T-	LIGHTNING AND TRANSIENTS RESEARCH INST., MINNEAPOLIS
L/T-	LIGHTNING AND TRANSIENTS RESEARCH INST., MINNEAPOLIS
LT-	NATIONAL RESEARCH COUNCIL OF CANADA, OTTAWA
LT-	RAND CORP., SANTA MONICA, CALIF.
LT-(YEAR)-	RAND CORP., SANTA MONICA, CALIF. (TRANSLATION)
LTC-(NUMBER)-	GRUMMAN AIRCRAFT ENGINEERING CORP., BETHPAGE, N.Y.
LTD-	ETHYL CORP. RESEARCH AND DEVELOPMENT DEPT., FERNDALE, MICH. (LABORATORY TECHNICAL DATA)
LTD-TM-	LADRIERE, INC. TECH. DIV., DETROIT
LTEC/PR/	ITALY. COMITATO NAZIONALE PER L'ENERGIA NUCLEARE, ROME
LTIRF-	LOWELL TECHNOLOGICAL INST. RESEARCH FDN., MASS.
LTIRF-(NUMBER)/IP	LOWELL TECHNOLOGICAL INST. RESEARCH FDN., MASS.
LTL-	LOS ALAMOS SCI. LAB., N. MEX. (LITTLE TITLE LIST)
LTP-(NUMBER-NUMBER)	GRUMMAN AIRCRAFT ENGINEERING CORP., BETHPAGE, N.Y.
LTR-HA-	CANADA. NATIONAL AERONAUTICAL ESTABLISHMENT, OTTAWA
LTRI-	LIGHTNING AND TRANSIENTS RESEARCH INST., MINNEAPOLIS
LTRI LR-	LIGHTNING AND TRANSIENTS RESEARCH INST., MINNEAPOLIS
LTRS-RM-	GT. BRIT. LOW TEMPERATURE RESEARCH STATION. RADIATION GROUP, CAMBRIDGE, ENGLAND
LTR-ST-	CANADA. NATIONAL AERONAUTICAL ESTABLISHMENT, OTTAWA
LTV-	LING-TEMCO-VOUGHT, INC., DALLAS
LTV-	LTV AEROSPACE CORP., DALLAS
LTV-(NUMBERS)W/	LTV ELECTROSYSTEMS, INC., GARLAND, TEX.
LTV-	LTV ELECTROSYSTEMS INC., GREENVILLE, TEX.
LTV/A-	LING-TEMCO-VOUGHT, INC. LTV ASTRONAUTICS DIV., DALLAS
LTV/M-	LING-TEMCO-VOUGHT, INC. LTV MILITARY ELECTRONICS DIV., DALLAS
LTV/R-	LING-TEMCO-VOUGHT, INC. LTV RANGE SYSTEMS DIV., DALLAS
LTV/VA-	LING-TEMCO-VOUGHT, INC. LTV VOUGHT AERONAUTICS DIV., DALLAS
LU-	LEHIGH UNIV., BETHLEHEM, PENNA.
LU-	LOUISIANA STATE UNIV., BATON ROUGE
LU-DSR-	LOUISVILLE, KY. UNIV. SCHOOL OF MEDICINE
LU/DSR-(YR)-	LOUISVILLE, KY. UNIV. SCHOOL OF MEDICINE
LU-EE/(YEAR)/	LEEDS, ENGLAND. UNIV. DEPT. OF ELEC. + ELECTRONIC ENG
LU-FEL-	LEHIGH UNIV., BETHLEHEM, PENNA. FRITZ (ENG.) LAB.
LUIP-CR-(YEAR)-	LUND, SWEDEN. UNIV. INST. OF PHYSICS
LUNP-	LUND, SWEDEN. INST. OF TECH. DEPT. OF NUCLEAR PHYSICS (MANY RPTS. ISSUED JOINTLY WITH LUND UNIV.)
LUNP-	LUND, SWEDEN. UNIV. DEPT. OF PHYSICS
LU-QR-	LEHIGH UNIV., BETHLEHEM, PENNA. CHEMISTRY DEPT.
LU-S-	LUND, SWEDEN. UNIV.
LUSY-	LUND, SWEDEN. UNIV. DEPT. OF PHYSICS
LU-TR-	LEHIGH UNIV., BETHLEHEM, PENNA. INST. OF RESEARCH
LUX.N.(NO.)(NO.)	SYNDICAT LUXEMBOURGEOIS POUR L'INDUSTRIE NUCLEAIRE
LV-	EDGERTON, GERMESHAUSEN + GRIER, INC., LAS VEGAS, NEV.
LV-	MARTIN CO., BALTIMORE (LAUNCH VEHICLE REPORT)
LVT-	CONTINENTAL AVIATION AND ENGINEERING CORP., DETROIT
LVT-	ENGLISH ELECTRIC CO., LTD.
LWJ-LJL-MAC-	MIDWESTERN UNIVERSITIES RESEARCH ASSN., URBANA, ILL.
LWL-CR-	ARMY LIMITED WAR LAB., ABERDEEN PROVING GROUND, MD.
LWL-TM-(YEAR)-	ARMY LIMITED WAR LAB., ABERDEEN PROVING GROUND, MD.
LWL-TN-(YEAR)-	ARMY LIMITED WAR LAB., ABERDEEN PROVING GROUND, MD.
LWL-TR-(YEAR)-	ARMY LIMITED WAR LAB., ABERDEEN PROVING GROUND, MD.
LWS-	CALIFORNIA RESEARCH AND DEVELOPMENT CO., BERKELEY
LWS-	CALIFORNIA RESEARCH AND DEVELOPMENT CO., LIVERMORE
LXN/NP-	BRAZIL. LABORATORIO DE ACUSTICA E SONICA, SAO PAULO
LYCEN/	LYON. UNIVERSITE. INSTITUT DE PHYSIQUE NUCLEAIRE.

M-	(MANUAL) (REF. 2)
M-	(MEMORANDUM) (REF. 2)
M-	(MEMORANDUM REPORT) (REF. 2)
M-	(MISCELLANEOUS PROJECT) (REF. 2)
M-	(MONTHLY REPORT) (REF. 2)
M-	(MONTHLY TEST REPORT) (REF. 2)
M-	(TECHNICAL MEMORANDUM) (REF. 2)
M-	AEROJET-GENERAL CORP., AZUSA, CALIF.
M-	AEROJET-GENERAL CORP., SACRAMENTO, CALIF.
M-	AEROJET-GENERAL CORP. LIQUID ROCKET PLANT, SACRAMENTO, CALIF.
M001-	AEROJET NUCLEAR SYSTEMS CO., SACRAMENTO, CALIF.
M-	ALL AMERICAN AIRWAYS, INC. ENGINEERING + RESEARCH DIV., WILMINGTON, DEL. (MISCELLANEOUS REPT.)
M(YEAR)-	AMERICAN OIL CO. RES. AND DEV. DEPT., WHITING, IND.
M-	ATOMIC ENERGY COMMISSION, WASHINGTON, D.C. (REF. 23)
M6/A/(NUMBER/YEAR)	BRITISH STEEL CORP., LONDON
M-	BROOKLYN. POLYTECHNIC INST. MICROWAVE RESEARCH INST.
M1-	CAMBRIDGE CORP.
M-(NUMBER)-C(LETTER)	CHICAGO. UNIV. METALLURGICAL LAB. (REF. 10)
M-	COLES SIGNAL LAB., BELMAR, N.J. (TECH. MEMO.)
M-	COMMERCIAL SOLVENTS CORP., N.Y.C.
M(YR.)-	CONTINENTAL MOTORS CORP. AIRCRAFT ENGINE DIV., MUSKEGON, MICH.
M-(NUMBER)-	DIRECTORATE OF AEROSPACE SAFETY, NORTON AFB, CALIF.
M-	EVANS SIGNAL LAB., BELMAR, N.J. (TECH. MEMO.)
M-	FLORIDA STATE UNIV., TALLAHASSEE
M(NUMBER).-	FRANCE. CENTRE DE MATHEMATIQUES DE L'ECOLE POLYTECHNIQUE, PARIS
M(NO.)-	FRANKFORD ARSENAL, PHILADELPHIA
M62-	FRANKFORD ARSENAL. PITMAN-DUNN LAB., PHILADELPHIA
M-	GARRETT CORP. AIRESEARCH MFG. DIV., LOS ANGELES
M-(NO.)-(NO.)-	GENERAL DYNAMICS/POMONA, CALIF.
M(NO.S)-	GENERAL PRECISION, INC., LITTLE FALLS, N.J.
M(YEAR)-	GIBBS AND COX, INC., N.Y.C.
M-	GT. BRIT. CHEM. DEFENCE EXPTL. STA.,PORTON,WILTS,ENG.
M/-	GT. BRIT. TELECOMMUNICATIONS RESEARCH ESTABLISHMENT, MALVERN, WORCS, ENGLAND
M-	HERMES ELECTRONICS CO., CAMBRIDGE, MASS.
(YEAR)M-	HUGHES RESEARCH LABS DIV. OF HUGHES AIRCRAFT CO., MALIBU, CALIF.
M-	IIT RESEARCH INST., CHICAGO
M(NUMBERS)-	IIT RESEARCH INST., CHICAGO
M-	IIT RESEARCH INST., CHICAGO. TECHNOLOGY CENTER
M-	ITALY. ISTITUTO ELETTRONTECNICO NAZIONALE, TURIN
M40-	JAPAN
M-	KELLEX CORP., N.Y.C.
M-(NUMBER)-(YEAR)-	LOCKHEED MISSILES AND SPACE CO., SUNNYVALE, CALIF.
M-	LOCKHEED MISSILES AND SPACE CO. ELECTRONIC SCIENCES LAB., PALO ALTO, CALIF.
M-	LOCKHEED MISSILES AND SPACE CO. MATERIALS SCIENCES LAB., PALO ALTO, CALIF.
M-	MANHATTAN DISTRICT. RESEARCH CONTROL (SERIES ASSIGNED TO REPORTS FROM VARIOUS AMERICAN SOURCES)
M-(YR.)-	MARTIN CO., DENVER
M-(NUMBER-NUMBER)-	MARTIN CO., DENVER
M-	MARTIN-MARIETTA CORP. DENVER DIV.
M-	MASSACHUSETTS INST. OF TECH.,CAMBRIDGE.RADIATION LAB.
M-	METEOROLOGY INTERNATIONAL, INC., MONTEREY, CALIF.
M720-1-R-	MICHIGAN. UNIV., ANN ARBOR. ENGINEERING RES. INST.
M(YEAR)-	MITRE CORP., BEDFORD, MASS.
M(NUMBER)	MONASH UNIV., CLAYTON, AUSTRALIA
M3-B-	MONSANTO CHEMICAL CO., DAYTON, OHIO
M3-D-	MONSANTO CHEMICAL CO., DAYTON, OHIO
M4-D-	MONSANTO CHEMICAL CO., DAYTON, OHIO
M-	MOTOROLA INC. SEMICONDUCTOR PRODS. DIV.,PHOENIX,ARIZ.
M-	NATIONAL BUREAU OF STANDARDS. BOULDER LABS., COLO.
M-009-	NAVAL CIVIL ENGINEERING RESEARCH AND EVALUATION LAB., PORT HUENEME, CALIF.
M-(PROJECT)-	NAVAL MEDICAL RESEARCH INST., BETHESDA, MD.
M-(YEAR)-	NETHERLANDS. METAALINSTITUUT TNO, DELFT
M(YEAR)-	NETHERLANDS. METAALINSTITUUT TNO, THE HAGUE
M(YEAR)-(NO.)/ADR/	NETHERLANDS. METAALINSTITUUT TNO, THE HAGUE
M-(YEAR)-	NETHERLANDS. TNO. CENTRAAL LABORATORIUM, DELFT
M-	NORTHWESTERN UNIV., EVANSTON, ILL. AERIAL MEASUREMENTS LAB.
M-	OPERATIONS ANALYSIS OFF. (AIR FORCE), WASHINGTON,D.C.
M-52-	ROHM AND HAAS CO., PHILADELPHIA
M-(YEAR)-	ROHM + HAAS CO. REDSTONE ARSENAL RESEARCH DIV., HUNTSVILLE, ALA.
M-1004 (YEAR,-MONTH)	SANDIA CORP., ALBUQUERQUE, N. MEX.
M-	SIGNAL CORPS ENG. LABS.,FT. MONMOUTH, NJ (TECH.MEMO.)
M-	SQUIER SIGNAL LAB., FORT MONMOUTH, N.J. (TECH. MEMO.)
M-	STANFORD UNIV., CALIF. MICROWAVE LAB.
M-	STANFORD UNIV.,CALIF. STANFORD LINEAR ACCELERATOR CTR
M-	STANFORD UNIV., CALIF. W.W. HANSEN LABS. OF PHYSICS
M-	TECHNICAL INFORMATION SERVICE, AEC (SERIES ASSIGNED TO JAPANESE REPORTS ABOUT HIROSHIMA + NAGASAKI)
M-	TECHNICAL INFORMATION SERVICE EXTENSION, AEC
M(YEAR)-	THOMAS (A.S.), INC., WESTWOOD, MASS.
M-	THOMPSON RAMO WOOLDRIDGE INC., LOS ANGELES
M-(NO.)-	UNITED AIRCRAFT CORP., EAST HARTFORD, CONN.
M(YR.)-(NO.)-	WASHINGTON. UNIV., SEATTLE
M-	WESTINGHOUSE ELECTRIC CORP., PITTSBURGH
M(LETTER(S))-	NATIONAL RES.COUNCIL OF CANADA. MONTREAL LAB.(REF.29)
M(LETTER)C-	OFFICE OF SCIENTIFIC RESEARCH AND DEVELOPMENT
MA-	BUREAU OF AERONAUTICS (NAVY). MAINTENANCE DIV.
M/A-	DAYTON, OHIO. UNIV.
MA-	ELECTRO-HYDRAULICS, LTD., WARRINGTON, LANCS, ENGLAND
MA-1-(LETTERS)-	HUGHES AIRCRAFT CO., CULVER CITY, CALIF.
MA-	MICROWAVE ASSOCIATES, INC., BOSTON
MA-	NAVAL AIR DEVELOPMENT CENTER. AVIATION MEDICAL ACCELERATION LAB., JOHNSVILLE, PENNA.
MAA-(YEAR)-	AIR FORCE MATERIALS LAB. MATERIALS APPLICATIONS DIV., WRIGHT-PATTERSON AFB, OHIO
MAA-(LTRS)-	AIR FORCE SYSTEMS COMMAND, WRIGHT-PATTERSON AFB, OHIO
MAAG-	MILITARY ASSISTANCE ADVISORY GROUP, BONN
MAA-TM-(YEAR)-	AIR FORCE MATERIALS LAB. MATERIALS APPLICATIONS DIV., WRIGHT-PATTERSON AFB, OHIO
MAB-(NUMBER-LTR.-NO.)	NATIONAL ACADEMY OF SCIENCES-NATIONAL RES. COUNCIL, MATERIALS ADVISORY BOARD, WASHINGTON, D.C.
MABM-	NAVY (MINE ASSEMBLY BASE MEMORANDUM)
MABS-	MARINE CORPS, WASHINGTON, D.C.
MAC-	MC DONNELL AIRCRAFT CORP., ST. LOUIS
MAC-(NO.)-ALC(NO.)	MC DONNELL AIRCRAFT CORP., ST. LOUIS
MAC-	MARQUARDT CORP., VAN NUYS, CALIF.
MAC-	MASSACHUSETTS INST. OF TECH., CAMBRIDGE
MAC-	MASS. INST. OF TECH., CAMBRIDGE. DEPT. OF METALLURGY

MAC-	MEASUREMENT ANALYSIS CORP., LOS ANGELES
MAC-(YR)-	METALS AND CONTROLS CORP., ATTLEBORO, MASS.
MAC-	MIDWESTERN UNIVERSITIES RESEARCH ASSN., URBANA, ILL.
MAC-(INITIALS)-	MIDWESTERN UNIVERSITIES RESEARCH ASSN., URBANA, ILL.
MAC-E-	MC DONNELL AIRCRAFT CORP., ST. LOUIS
MAC-E(NUMBER)	MC DONNELL ASTRONAUTICS CO., ST. LOUIS
MAC-F-	MC DONNELL ASTRONAUTICS CO., ST. LOUIS
MACHLAB-	NAVAL SHIP RESEARCH AND DEV. CENTER, ANNAPOLIS
MAC-K(NUMBER)-	MC DONNELL ASTRONAUTICS CO., ST. LOUIS
MACL-	MACHLETT LABS., INC., SPRINGDALE, CONN.
MACL-	MACLAREN AND SONS, LTD., LONDON (TRANSLATIONS)
MAC-LB-	MC DONNELL ASTRONAUTICS CO., ST. LOUIS (BIBLIOGRAPHY)
MAC-OT/E-(NO.-NO.-YR.)	MILITARY AIRLIFT COMMAND, SCOTT AFB, ILL.
MAC PLUTO QPR-	MARQUARDT AIRCRAFT CO., VAN NUYS, CALIF.(PLUTO PROJ.)
MAC-PR-	MARQUARDT AIRCRAFT CO., VAN NUYS, CALIF.
MACT-	METALLURGICAL ADVISORY COMMITTEE ON TITANIUM
MAC-TR-	MC DONNELL AIRCRAFT CORP., ST. LOUIS
MAC-TR-(NUMBER)-	MC DONNELL AIRCRAFT CORP., ST. LOUIS
MAC-TR-153-	MARQUARDT AIRCRAFT CO., VAN NUYS, CALIF.
MAC-TR-	MASSACHUSETTS INST. OF TECH., CAMBRIDGE
MAC-TR-	MASSACHUSETTS INST. OF TECH., LEXINGTON. LINCOLN LAB.
MACT-TAD-	MILITARY ASSISTANCE COMMAND. TRAINING AIDS DIV., VIETNAM
MACV-DAR-(LTR.,NO...)	MILITARY ASSISTANCE COMMAND, VIETNAM, APO SAN FRAN.
MACV-SEER-	MILITARY ASSISTANCE COMMAND, VIETNAM. APO SAN FRAN.
MADVT-	BENDIX CORP. BENDIX SYSTEMS DIV., ANN ARBOR, MICH.
MAE-	RESEARCH AND DEVELOPMENT BOARD. PANEL ON MISSILE AUXILIARY EQUIPMENT, WASHINGTON, D.C.
MAEE-F.TN/-	GT. BRIT. MARINE AIRCRAFT EXPERIMENTAL ESTABLISHMENT, FELIXSTOWE, SUFF., ENGLAND
MAF-	GT. BRIT. MINISTRY OF AGRICULTURE AND FISHERIES
MAF B-	GT. BRIT. MINISTRY OF AGRICULTURE AND FISHERIES (BULLETIN)
MAFB-TN-	MAC DILL AFB, FLA.
MAGC-	MAGNAVOX CO., FORT WAYNE
MAGI-	MATHEMATICAL APPLICATIONS GROUP, INC.,WHITE PLAINS,NY
MAGI-MR-	MATHEMATICAL APPLICATIONS GROUP, INC.,WHITE PLAINS,NY
MAG. S.-	CALIFORNIA. UNIV., BERKELEY. RADIATION LAB. (REF. 9)
MAG. SPEC-	CALIFORNIA. UNIV., BERKELEY. RADIATION LAB. (REF. 9)
MAI-	MICROWAVE ASSOCIATES, INC., BOSTON
MAI A-	MARINE ADVISERS, INC., LA JOLLA, CALIF.
MAK-	MARKITE CO., N.Y.C.
MAL-	MALLORY (P.R.) AND CO., INC., INDIANAPOLIS
MALLARD-TD-	AUSTRALIA. DEPT. OF DEFENCE. EDP DIV., CANBERRA
MALLARD-TD-	BRITISH AIRCRAFT CORP. GUIDED WEAPONS DIV., BRISTOL, GLOUCESTER, ENGLAND
MAM-	BUREAU OF ORDNANCE (NAVY) (MAGNETIC ANALYSIS MEMO.)
MAM-(YEAR)-	DEFENSE COMMUNICIATIONS AGENCY, WASHINGTON, D.C.
MAM-	OFFICE OF THE DIRECTOR OF DEFENSE (RES+ENG),WASH.,DC
MAMI-	NAVAL AEROSPACE MEDICAL INST., PENSACOLA, FLA.
MAN-(YEAR)-	OFFICE OF THE DIRECTOR OF DEFENSE (RESEARCH AND ENGINEERING), WASH., D.C.(MANAGEMENT ANALYSIS NOTE)
MANLABS-PR-	MANLABS, INC., CAMBRIDGE, MASS.
MANLABS-TR-	MANLABS, INC., CAMBRIDGE, MASS. (TECHNICAL REPORT)
MANL-TR-	MANUFACTURING LABS., INC., CAMBRIDGE, MASS.
MANPOWER-RB-	MAINE. UNIV., ORONO. MANPOWER RESEARCH PROJECT
MAN-TKA-	MASCHINENFABRIK AUGSBURG-NURNBERG A. G., NUREMBERG
MA-O-	MANHATTAN DISTRICT. LIAISON OFFICE, CHALK RIVER, ONT.
MAP-	GT. BRIT. MINISTRY OF AIRCRAFT PRODUCTION.
MAP-	GT. BRIT. MINISTRY OF SUPPLY
MAPA-	OMAHA-COUNCIL BLUFFS METROPOLITAN AREA PLANNING AGENCY
MAR-	ARMY, WASHINGTON, D.C. (MILITARY ATTACHE REPORT)
MAR-	MARTIN CO. (ALL LOCATIONS)
MAR-	MARTIN-MARIETTA CORP. (ALL LOCATIONS)
MA-R(NUMBER)	MICROWAVE ASSOCIATES, LTD., LUTON, BEDS., ENGLAND
MAR-	NAVAL ORDNANCE LAB., WHITE OAK, MD. (MATHEMATICAL ANALYSIS REQUEST)
MAR-	OFFICE OF THE DIRECTOR OF DEFENSE (RES+ENG),WASH.,DC
MAR-	RUBBER RESEARCH INST. TNO, DELFT, NETHERLANDS
MARA-	MOUNT AUBURN RES. ASSOCIATES INC., CAMBRIDGE, MASS.
MARC-	MARTIN CO., BALTIMORE
MAR CSH-(NUMBERS)-	MARTIN CO., BALTIMORE
MAR-ER-	MARTIN CO., BALTIMORE
MAR M-M-P(YR.)-	MARTIN CO., DENVER
MAR MND-	MARTIN CO., BALTIMORE
MAR-MND-	MARTIN-MARIETTA CORP. NUCLEAR DIV., BALTIMORE
MAR MND-M-	MARTIN CO. NUCLEAR DIV., BALTIMORE
MAR MND-P-	MARTIN CO. NUCLEAR DIV., BALTIMORE
MAR MND-SF-	MARTIN CO. NUCLEAR DIV., BALTIMORE
MAR MND-SR-	MARTIN CO. NUCLEAR DIV., BALTIMORE
MAR OR-	MARTIN CO., ORLANDO, FLA.
MAR P-(YR.)-	MARTIN CO., ORLANDO, FLA.
MARQ-	MARQUARDT CORP., VAN NUYS, CALIF.
MARQUARDT-	MARQUARDT CORP., VAN NUYS, CALIF.
MARR-	NAVAL ORDNANCE LAB., WHITE OAK, MD. (MATHEMATICAL ANALYSIS RESEARCH REQUEST)
MARR-	NAVAL ORDNANCE LAB., WHITE OAK, MD. (MECHANIZED ANALYSIS RESEARCH REQUEST)
MAR S-	MARTIN CO., BALTIMORE
MART-	HUGHES AIRCRAFT CO., CULVER CITY, CALIF. (MOBILE AUTOMATIC RADIATION DETECTOR)
MARTIN-	MARTIN CO., DENVER
MARTIN CR-(NO.)-	MARTIN CO., BALTIMORE
MARTIN-CR-(YR.)-	MARTIN CO., DENVER
MARTIN-CR-	MARTIN CO. CENTER FOR HIGH ENERGY FORMING, DENVER
MARTIN CR-	MARTIN-MARIETTA CORP. DENVER DIV.
MARU-	MARYLAND. UNIV., COLLEGE PARK
MARU-QR-	MARYLAND. UNIV., COLLEGE PARK
MARU-TN-	MARYLAND. UNIV., COLLEGE PARK
MARU TN BN-	MARYLAND. UNIV., COLLEGE PARK. INST. FOR FLUID DYNAMICS AND APPLIED MATHEMATICS
MARU TR-	MARYLAND. UNIV., COLLEGE PARK. DEPT. OF PHYSICS
MARU TR BT-	MARYLAND. UNIV., COLLEGE PARK. INST. FOR FLUID DYNAMICS AND APPLIED MATHEMATICS
MARYLAND U-(MONTH)(YEAR)	MARYLAND. UNIV., COLLEGE PARK
MAS-	ARMY TECH. INTELL. CTR. MEDICAL ANALYSIS SEC., TOKYO
MAS-	LOCKHEED AIRCRAFT CORP., BURBANK, CALIF.
MAS-	MEDICAL ANALYSIS SECTION (AIR FORCE)
MAS-	RES. + DEV. BD. COMM. ON GUIDED MISSILES, WASH., DC
MAS-	RESEARCH AND DEVELOPMENT BOARD. PANEL ON AERO-DYNAMICS AND STRUCTURES, WASHINGTON, D.C.
MASC-(NUMBER)-F	MIDWEST APPLIED SCIENCE CORP., LAFAYETTE, IND.
MASC/N-	GT. BRIT. ATOMIC EN. RES. ESTAB., HARWELL, BERKS, ENG
MASC-RD-	MIDWEST APPLIED SCIENCE CORP., LAFAYETTE, IND.
M/AS/DM/	BRITISH STEEL CORP. MIDLAND GROUP,MOORGATE,YORKS,ENG.
MASSU-	MASSACHUSETTS. UNIV., AMHERST

MA-S/T-
- GT. BRIT. MINISTRY OF AVIATION. TECHNICAL INFORMATION AND LIBRARY SERVICES, LONDON

MASTER PLAN-
- DENVER REGIONAL COUNCIL OF GOVERNMENTS

MA-S/T-MEMO-(NO.)/(YR.)
- GT. BRIT. MINISTRY OF AVIATION. TECHNICAL INFORMATION AND LIBRARY SERVICES, LONDON

MA-S/T-MEMO-
- GT. BRIT. MINISTRY OF CIVIL AVIATION

MAS-TR-
- ROCHESTER, N.Y. UNIV. DEPT. OF MECHANICAL AND AEROSPACE SCIENCES

M-ASTR-IN-(YEAR)-
- GEORGE C. MARSHALL SPACE FLIGHT CTR., HUNTSVILLE, ALA.

MAT-
- CRANFIELD INST. OF TECH., BUCKS, ENGLAND

MATADOR JTG-
- MATADOR JOINT TASK GROUP (REF. 18)

MATERIALS SCIENCE LAB-
- SOUTHERN ILLINOIS UNIV., CARBONDALE. MATERIALS SCIENCE LAB.

MATHEMATICAL NOTE-
- BOEING SCIENTIFIC RESEARCH LABS. MATHEMATICS RESEARCH LAB., SEATTLE

MATH/MEMO(NO.)/WW
- GT. BRIT. TELECOMMUNICATIONS RESEARCH ESTABLISHMENT, MALVERN, WORCS, ENGLAND

MATHREP-
- RENSSELAER POLYTECHNIC INST., TROY, N.Y.

MATHS.MEMO/(NO.)/(LTRS)
- GT. BRIT. ATOMIC ENERGY RESEARCH ESTABLISHMENT. REACTOR PHYS. DIV., HARWELL, BERKS, ENGLAND

MATI-
- NAVY (MINE ASSEMBLY AND TEST INSTRUCTIONS)

MATLAB-
- NAVAL SHIP RESEARCH AND DEV. CENTER, ANNAPOLIS

MATR-
- BEECH AIRCRAFT CORP., WICHITA, KAN. (MISSILE AERO TECHNOLOGY REPORT)

MATS-
- MILITARY AIR TRANSPORT SERVICE, WASHINGTON, D.C.

MATSCIENCE-
- INDIA. INST. OF MATHEMATICAL SCIENCES, MADRAS

MATSCIT PS-
- CALIFORNIA INST. OF TECH., PASADENA. KECK LAB. OF ENGINEERING MATERIALS (POLYMER SCIENCE REPORT)

MATSCIT-PS-(YEAR)-
- CALIFORNIA INST. OF TECH., PASADENA. KECK LAB. OF ENGINEERING MATERIALS (POLYMER SCIENCE REPORT)

MATT-
- PRINCETON UNIV., N.J. PLASMA PHYSICS LAB.

MATT-
- PRINCETON UNIV., N.J. PROJECT MATTERHORN

MATT-BIB-
- PRINCETON UNIV., N.J. PLASMA PHYSICS LAB.

MATT-Q-
- PRINCETON UNIV., N.J. PROJECT MATTERHORN (QUARTERLY)

MATT-TM-
- PRINCETON UNIV., N.J. PROJECT MATTERHORN (TECH. MEMO)

MATT-TRANS.-
- PRINCETON UNIV., N.J. PLASMA PHYSICS LAB.

MAX-
- MAXSON (W.L.) CORP., N.Y.C.

MB-
- CALIFORNIA. UNIV., BERKELEY. RADIATION LAB.

MB-
- GENERAL ELECTRIC CO. RESEARCH LAB., SCHENECTADY, N.Y.

MB-
- MB ASSOCIATES, SAN RAMON, CALIF.

MB-(YEAR)/
- MB ASSOCIATES, SAN RAMON, CALIF.

MB-
- NATIONAL RESEARCH COUNCIL OF CANADA, OTTAWA

MB-(INITIALS)-
- CALIFORNIA. UNIV., BERKELEY. RADIATION LAB.

MBE-
- OFFICE OF SCIENTIFIC RESEARCH AND DEVELOPMENT

MBG-ONR-
- MISSOURI BOTANICAL GARDEN, ST. LOUIS

MBL-
- MARINE BIOLOGICAL LAB., WOODS HOLE, MASS.

MBL-(YEAR)/
- NETHERLANDS. RIJKSVERDEDIGINGSORGANISATIE, TNO. MEDISCH BIOLOGISCH LABORATORIUM, RIJSWIJK

MBM-
- HUGHES AIRCRAFT CO., CULVER CITY, CALIF. (MOBILE BALLISTIC MISSILE)

MB-R-
- CANADA. DEPT. OF MINES AND TECH. SURVEYS. MINES BR.

MB-R-(YEAR)/
- MB ASSOCIATES, SAN RAMON, CALIF.

MC-
- AIR MATERIEL COMMAND, WRIGHT-PATTERSON AFB, OHIO

MC-
- ATOMIC ENERGY OF CANADA LTD. CHALK RIVER PROJ., ONT.

MC-
- MARINE CORPS, WASHINGTON, D.C.

MC-
- MARTIN-MARIETTA CORP. DENVER DIV.

MC (3 DIGITS)
- MAURY CENTER FOR OCEANOGRAPHIC SCIENCE, WASHINGTON,DC

MC-
- MAXSON (W.L.) CORP., N.Y.C.

MC-
- MITHRAS, INC., CAMBRIDGE, MASS.

MC-
- NORTH AMERICAN AVIATION, INC., DOWNEY, CALIF.

MCA-
- GT. BRIT. MINISTRY OF CIVIL AVIATION

MCA-
- MC ALLISTER AND ASSOCIATES, INC. ALBUQUERQUE, N.MEX.

MCA C-
- GT. BRIT. MINISTRY OF CIVIL AVIATION

MCA CA-
- GT. BRIT. MINISTRY OF CIVIL AVIATION

MCAFB-
- COMBAT CREW TRAINING WING (MBOM)(3520TH), MC CONNELL AFB, KAN.

MC-AP-
- GENERAL ELECTRIC CO. AERONAUTIC AND ORDNANCE SYSTEMS DIV., SCHENECTADY, N.Y.

MCA P-
- GT. BRIT. MINISTRY OF CIVIL AVIATION (PUBLICATIONS)

MCAP-
- GT. BRIT. MINISTRY OF CIVIL AVIATION (PUBLICATIONS)

MCASTRO-F(NUMBER)
- MC DONNELL ASTRONAUTICS CO., ST. LOUIS

MCASTRO-G(NO.)
- MC DONNELL ASTRONAUTICS CO., ST. LOUIS

MCB-
- NAVAL TORPEDO STATION, KEYPORT, WASH.

MCC-(NO.)-TR-
- BUREAU OF AERONAUTICS (NAVY)

MCC-1023-TR-
- MATHIESON CHEMICAL CORP., BALTIMORE

MCC-
- MICHIGAN CHEMICAL CORP., ST. LOUIS, MICH.

MCC-
- MONSANTO CHEMICAL CO., EVERETT, MASS.

MCC-
- OLIN MATHIESON CHEMICAL CORP., BALTIMORE

MCC-1023-TR-
- OLIN MATHIESON CHEMICAL CORP., BALTIMORE

MCCC-(YEAR)-
- AIR FORCE LOGISTICS COMMAND. DIRECTORATE OF COST ANALYSIS, WRIGHT-PATTERSON AFB, OHIO

MCCP-
- PANERO (GUY B.) ENGINEERS, N.Y.C.

MCCRONE-
- MC CRONE RESEARCH INST., CHICAGO (FINAL REPORT)

MCD-
- GENERAL ELECTRIC CO. MILITARY COMMUNICATIONS DEPT., OKLAHOMA CITY

MCD-
- MC DONNELL AIRCRAFT CORP., ST. LOUIS

MCDA-
- CONTINENTAL MOTORS CORP., MUSKEGON, MICH.

MCDEC-DB-(NUMBER-YEAR)
- MARINE CORPS DEVELOPMENT AND EDUCATION COMMAND, QUANTICO, VA.

MC DONNELL A-
- MC DONNELL AIRCRAFT CORP., ST. LOUIS

MCDONNELL-A(NO.)
- MC DONNELL AIRCRAFT CORP., ST. LOUIS

MC DONNELL B-
- MC DONNELL AIRCRAFT CORP., ST. LOUIS

MCEB-
- MARINE CORPS SCHOOLS. EQUIPMENT BOARD, QUANTICO, VA.

MC EB TR-
- MARINE CORPS SCHOOLS. EQUIPMENT BOARD, QUANTICO, VA. (TEST REPORTS)

MC EB TR PROJ-
- MARINE CORPS SCHOOLS. EQUIPMENT BOARD, QUANTICO, VA. (TEST REPORTS)

MCER-
- MARTIN CO., BALTIMORE (ENGINEERING REPORT)

MCF-
- FIELD INTELLIGENCE AGENCY, TECH. (MISC. CHEM. FILE)

MCG-
- MARTIN-MARIETTA CORP. DENVER DIV.

MCGU-
- MC GILL UNIV., MONTREAL

MCGU-AMRG-
- MC GILL UNIV., MONTREAL. ARCTIC METEOROLOGY RES. GP.

MCGU-SRI-
- MC GILL UNIV., MONTREAL. SPACE RESEARCH INST.

MCIA-
- AIR MATERIEL COMMAND. ANALYSIS DIV., WRIGHT-PATTERSON AFB, OHIO

MCIC-
- BATTELLE MEMORIAL INST. METALS AND CERAMICS INFORMATION CENTER, COLUMBUS, OHIO

MCKEE-
- MC KEE (ARTHUR G.) AND CO. WESTERN KNAPP ENGINEERING DIV., SAN FRANCISCO

MCL-(NO.)/(NO.)
- AERONAUTICAL SYSTEMS DIV., WRIGHT-PATTERSON AFB, OHIO

MCL-
- FOREIGN TECHNOLOGY DIV., WRIGHT-PATTERSON AFB, OHIO

MCL-(NO.)/(ROMAN NO.)
- FOREIGN TECHNOLOGY DIV., WRIGHT-PATTERSON AFB, OHIO

MCL-
- MC CLELLAN CENTRAL LAB., MC CLELLAN AFB, CALIF.

MCL-
- MC MILLAN LAB., INC., IPSWICH, MASS.

MC/LFB-
- MARINE CORPS, WASHINGTON, D.C.

MCL-TP-
- MC CLELLAN CENTRAL LAB., MC CLELLAN AFB, CALIF.

MCL-TR-
- MC CLELLAN CENTRAL LAB., MC CLELLAN AFB, CALIF.

MCM-
- BUREAU OF ORDNANCE (NAVY) (MAGNETIC CALIBRATION MEMO)

MCM-
- MC MILLAN LAB., INC., MARBLEHEAD, MASS.

MCM-
- NAVAL ORDNANCE LAB., WHITE OAK, MD. (MINE COUNTERMEASURES MEMORANDUM)

MCM-	RES. + DEV. BD. COMM. ON GUIDED MISSILES, WASH., DC
MCM-	RES. + DEV. BD. PANEL ON COUNTERMEASURES, WASH., D.C.
MCML-	MC MILLAN LAB., INC., IPSWICH, MASS.
MCN-	CHEMICAL AND RADIOLOGICAL LABS., ARMY CHEM. CTR., MD.
MCN-	JOINT MATERIEL INTELLIGENCE AGENCY, WASHINGTON, D.C.
MCN-	THIOKOL CHEM. CORP. REACTION MOTORS DIV., DENVILLE, NJ
MCO-(NUMBER)(LETTER)	MARINE CORPS, WASHINGTON, D.C.
3M CO.-	MINNESOTA MINING AND MFG. CO., ST. PAUL
MCOAG RESEARCH CONTRIB-	CENTER FOR NAVAL ANALYSES. MARINE CORPS OPERATIONS ANALYSIS GROUP, WASHINGTON, D.C.
MCOAG-STUDY-	CENTER FOR NAVAL ANALYSES. MARINE CORPS OPERATIONS ANALYSIS GROUP, WASHINGTON, D.C.
MC/ORD-(NUMBER-YEAR)	CANADA. MARITIME COMMAND. OPERATIONAL RESEARCH DIV., HALIFAX
MCP-	MARTIN-MARIETTA CORP. DENVER DIV.
MCPPAF-	DIRECTORATE OF PROCUREMENT + PRODUCTION, WRIGHT-PATTERSON AFB, OHIO
MCR-(YEAR)-	MARTIN CO., DENVER
MCR-(YEAR)-	MARTIN-MARIETTA CORP. DENVER DIV.
MCR-	UNITED NUCLEAR CORP. DEV. DIV., WHITE PLAINS, N.Y.
MCRE-	AIR MATERIEL COMMAND. ENGINEERING DIV., WRIGHT-PATTERSON AFB, OHIO
MCREE-	AIR MATERIEL COMMAND. ENGINEERING DIV. ELECTRONIC SUBDIV., WRIGHT-PATTERSON AFB, OHIO
MCREE-(YEAR)-	AIR MATERIEL COMMAND. ENGINEERING DIV. ELECTRONIC SUBDIV., WRIGHT-PATTERSON AFB, OHIO (MEMO RPTS.)
MCREEXD-	WRIGHT AIR DEVELOPMENT CENTER. AERO MEDICAL LAB., WRIGHT-PATTERSON AFB, OHIO
MCREXA-	AIR MATERIEL COMMAND. ENGINEERING DIV. AIRCRAFT LAB., WRIGHT-PATTERSON AFB, OHIO
MCREXA8-(NUMBERS)-	WRIGHT AIR DEVELOPMENT CENTER. AIRCRAFT LAB., WRIGHT-PATTERSON AFB, OHIO
MCREXD-	AIR MATERIEL COMMAND. ENGINEERING DIV. AERO-MEDICAL LAB., WRIGHT-PATTERSON AFB, OHIO
MCREXE-(NOS.)-	AIR MATERIEL COMMAND. ENGINEERING DIV. EQUIPMENT LAB., WRIGHT-PATTERSON AFB, OHIO
MCREXF-(YR.)-	AIR MATERIEL COMMAND. ENGINEERING DIV. PHOTOGRAPHIC LAB., WRIGHT-PATTERSON AFB, OHIO (MEMO REPT.)
MCREXP-(NO.)-(NO.)	AIR MATERIEL COMMAND. ENGINEERING DIV. POWER PLANT LAB., WRIGHT-PATTERSON AFB, OHIO
MCRFT-	AIR MATERIEL COMMAND. FLIGHT TEST DIV. TEST ENGINEERING SUBDIV., WRIGHT-PATTERSON AFB, OHIO
MCRFT-	DIRECTORATE OF FLIGHT AND ALL-WEATHER TESTING, WRIGHT-PATTERSON AFB, OHIO
MCRL-TP-	OHIO STATE UNIV. RESEARCH FOUNDATION. MAPPING AND CHARTING RESEARCH LAB., COLUMBUS
MC-RN-	MARKITE CO., N.Y.C.
MCRR-	WRIGHT AIR DEVELOPMENT CENTER. AERONAUTICAL RESEARCH LAB., WRIGHT-PATTERSON AFB, OHIO
MCS-	HERCULES POWDER CO., MAGNA, UTAH
MCS-	HERCULES POWDER CO. CHEMICAL PROPULSION DIV. BACCHUS WORKS, MAGNA, UTAH
MCS-	MARINE CORPS SCHOOLS. EQUIPMENT BOARD, QUANTICO, VA.
MCS- 53MC... THRU 56MC...	TECHNICAL IND. INTELL. COMM. MISC. CHEMS. SUB-COMM. AIR MATERIEL COMMAND, WRIGHT-PATTERSON AFB, OHIO (FIRST TWO DIGITS STAND FOR YEAR) (REF. 1)
MCV-	MEDICAL COLL. OF VIRGINIA, RICHMOND
MCV (CONTRACT NO.)	MEDICAL COLL. OF VIRGINIA, RICHMOND
MCW-	MALLINCKRODT CHEMICAL WORKS, ST. LOUIS
MCW-	MALLINCKRODT CHEMICAL WORKS. URANIUM DIV., WELDON SPRING, MO.
MCW-A-	MALLINCKRODT CHEMICAL WORKS, ST. LOUIS
MD-	ASSOCIATION OF AMERICAN RAILROADS. MECHANICAL DIVISION, WASHINGTON, D.C.
MD-	CALIFORNIA. UNIV., BERKELEY
MD-	CANADA. DEPT. OF MINES AND TECH. SURVEYS. MINES BR.
MD-	COAST AND GEODETIC SURVEY, ROCKVILLE, MD. (MAGNETIC DECLINATION)
MD-	GENERAL MOTORS RESEARCH LABS. MECHANICAL DEV. DEPT., WARREN, MICH.
MD-	GT. BRIT. DEPT. OF ATOMIC ENERGY. INDUSTRIAL GROUP H.Q., RISLEY, LANCS, ENGLAND (REF. 36)
MD-	MARYLAND. UNIV., COLLEGE PARK
MD-(NUMBER)	NORTH AMERICAN AVIATION, INC. MISSILE DEVELOPMENT DIV., DOWNEY, CALIF.
MD-	NORTH AMERICAN AVIATION, INC. SPACE AND INFORMATION SYSTEMS DIV., DOWNEY, CALIF.
MD-(NUMBER)-	ROCKETDYNE DIV., NORTH AMERICAN AVIATION, INC., CANOGA PARK, CALIF.
MD-(YEAR)-	ROCKETDYNE DIV., NORTH AMERICAN AVIATION, INC., CANOGA PARK, CALIF.
MD-	STANFORD UNIV., CALIF. DEPT. OF MECHANICAL ENG., THERMOSCIENCES DIV.
MD-	STANFORD UNIV., CALIF. THERMOSCIENCES DIV.
MD-	UTAH. UNIV., SALT LAKE CITY
MDAC-	MC DONNELL DOUGLAS ASTRONAUTICS CO. (ALL LOCATIONS)
MDAC-PAPER-WD-	MC DONNELL DOUGLAS ASTRONAUTICS CO., HUNTINGTON BEACH, CALIF.
MDAC-WD-	MC DONNELL DOUGLAS ASTRONAUTICS CO., HUNTINGTON BEACH, CALIF.
MDA/PMM/(NUMBER/YEAR)	NOTTINGHAM, ENGLAND. UNIV. DEPT. OF GEOGRAPHY (MULTIPLE DISCRIMINANT ANALYSIS)(INITIALS)
MDB-	NAVY (MINE DISPOSAL BULLETIN)
MDC-E(NUMBER)	MC DONNELL DOUGLAS ASTRONAUTICS CO., ST. LOUIS
MDC-G-	MC DONNELL DOUGLAS ASTRONAUTICS CO. (ALL LOCATIONS)
MDC-J(NUMBER)/	DOUGLAS AIRCRAFT CO., INC., LONG BEACH, CALIF.
MDC-J-	MC DONNELL DOUGLAS CORP. DOUGLAS AIRCRAFT DIV., LONG BEACH, CALIF.
MDC-TDR-	AIR FORCE MISSILE DEV. CTR., HOLLOMAN AFB, N. MEX.
MDC-TR-	AIR FORCE MISSILE DEV. CTR., HOLLOMAN AFB, N. MEX.
MDC-TR-(YEAR)-	AIR FORCE MISSILE DEV. CTR., HOLLOMAN AFB, N. MEX.
MDC-TR-(YEAR)-	NEW MEXICO. UNIV., ALBUQUERQUE
MD/D-(NUMBER)/	BRITISH IRON AND STEEL RESEARCH ASSN., LONDON
M/DD.(NUMBER/NUMBER)/	BRITISH STEEL CORP. MIDLAND GROUP, MOORGATE, YORKS, ENG.
MDD-	GT. BRIT. UNDERWATER COUNTERMEASURES + WEAPONS ESTAB.
MDDC-	ATOMIC ENERGY COMMISSION, WASHINGTON, D.C. (REF. 24)
MDE-	WESTINGHOUSE DEFENSE AND SPACE CENTER. SURFACE DIV., BALTIMORE
MDE-	WESTINGHOUSE ELECTRIC CORP., BALTIMORE
MDE-	WESTINGHOUSE ELECTRIC CORP. ENGINEERING OPERATIONS, HUNTSVILLE, ALA.
MDE-	WESTINGHOUSE ELECTRIC CORP. SURFACE DIV., BALTIMORE
MD EA-	CHEMICAL WARFARE SERVICE. MEDICAL DIV., EDGEWOOD ARSENAL, MD. (MEMORANDUM REPORTS)
MDF-	FRIEDMAN (MORRIS D.), INC., WEST CONCORD, MASS.
MDF-	FRIEDMAN (MORRIS D.), INC., WEST NEWTON, MASS. (TRANSLATIONS)
MDFRL-	ARMY MEDICAL RESEARCH LAB., FORT KNOX, KY.
MDG-G-	MC DONNELL DOUGLAS CO., RICHLAND, WASH.
MDI-	MICHIGAN-DYNAMICS, INC. DYNAMIC FILTERS DIV., DETROIT
MDL-	BUREAU OF COMMERCIAL FISHERIES, WASHINGTON, D.C. (MARKET DEVELOPMENT LEAFLETS)
MDL-	MICROWAVE DEVELOPMENT LABS., INC., WALTHAM, MASS.
MDL-	NAVY MINE DEFENSE LAB., PANAMA CITY, FLA.
MDL-	UTAH. UNIV., SALT LAKE CITY
MDL-	UTAH. UNIV., SALT LAKE CITY. MICROWAVE DEVICES LAB.
MDL-Q(NUMBER)	UTAH. UNIV., SALT LAKE CITY. MICROWAVE DEVICES LAB. (QUARTERLY REPORT)
MDL-TP-	NAVY MINE DEFENSE LAB., PANAMA CITY, FLA.
MDL-U-	NAVY MINE DEFENSE LAB., PANAMA CITY, FLA.

MDR-(NO.)-(NO.)-

MDR-(NO.)-(NO.)-	BOEING CO. AERO-SPACE DIV., SEATTLE
MDR-	CHEMICAL CORPS. MEDICAL DIV., ARMY CHEMICAL CTR., MD.
MDR-	CHEMICAL WARFARE SERVICE. MEDICAL DIV., EDGEWOOD ARSENAL, MD.
MDRR-	CHEMICAL CORPS. MEDICAL DIV., ARMY CHEMICAL CENTER, MD. (RESEARCH REPORT)
MDS-	LOCKHEED MISSILES AND SPACE CO., PALO ALTO, CALIF.
MDS-	MARTIN-MARIETTA CORP. DENVER DIV.
MD-S-	RADIO CORP. OF AMERICA, MOORESTOWN, N.J.
MDSR-	CHEMICAL CORPS. MEDICAL DIV., ARMY CHEMICAL CENTER, MD. (SPECIAL REPORT)
MDT-	CALIFORNIA, UNIV., BERKELEY. LAWRENCE RADIATION LAB.
MDTN-	NAVAL ORDNANCE LAB. MECHANICS DIV. (TECHNICAL NOTE)
MD-TR-(NO.)-CC	CANADA. DEPT. OF MINES AND TECH. SURVEYS. MINES BR.
MD-TR-(YEAR)-	MARYLAND. UNIV., COLLEGE PARK
MD-TR-	MARYLAND. UNIV., COLLEGE PARK
MDWG-(YEAR)-	GEORGE C. MARSHALL SPACE FLIGHT CTR.,HUNTSVILLE, ALA.
ME-	ARMY NATICK LABS., MASS.
ME-	COMMITTEE ON MEDICAL RESEARCH (MEDICINE REPORTS)
ME-	GRUMMAN AIRCRAFT ENGINEERING CORP., BETHPAGE, N.Y.
ME-	HUGHES AIRCRAFT CO., NEWPORT BEACH, CALIF. (MICROELECTRONICS)
ME-	ILLINOIS. UNIV., URBANA. DEPT. OF MECHANICAL ENG.
ME-	LUND, SWEDEN. UNIV.
ME-	MARTIN CO., DENVER
ME-	MARTIN-MARIETTA CORP. DENVER DIV.
ME-	MARYLAND. UNIV., COLLEGE PARK
ME-(YEAR)/	MASSACHUSETTS INST. OF TECH., CAMBRIDGE. AERO-ELASTIC AND STRUCTURES RESEARCH LAB.
ME-	NATIONAL RESEARCH COUNCIL OF CANADA. DIVISION OF MECHANICAL ENGINEERING, OTTAWA
ME-	NEW MEXICO. UNIV., ALBUQUERQUE. BUREAU OF ENG. RES.
ME-	ROYAL CANADIAN AIR FORCE. CENTRAL EXPERIMENTAL AND PROVING ESTAB., ROCKCLIFFE, ONT.
ME-	SYRACUSE UNIV., N.Y. RESEARCH INST. DEPT. OF MECHANICAL AND AEROSPACE ENGINEERING
ME(NUMBER)-	SYRACUSE UNIV., N.Y. RESEARCH INST. DEPT. OF MECHANICAL AND AEROSPACE ENGINEERING
ME-(YEAR)-	TENNESSEE. UNIV., KNOXVILLE. COLL. OF ENGINEERING
ME-(NUMBER)-(YEAR)-	TENNESSEE. UNIV., KNOXVILLE. COLL. OF ENGINEERING
ME/A-(YEAR)-	CARLETON UNIV., OTTAWA. DIV. OF AEROTHERMODYNAMICS
MEA-	GT. BRIT. MISSION FOR ECONOMIC AFFAIRS, LONDON
ME/AEROSPACE-	SYRACUSE UNIV., N.Y. RESEARCH INST. DEPT. OF MECHANICAL AND AEROSPACE ENGINEERING
MEC-(NO.)-(NO.)-	MAXSON ELECTRONICS CORP., GREAT RIVER, N.Y.
MEC-	MINNESOTA ELECTRONICS CORP., ST. PAUL
MECAR-	SOCIETE ANONYME BELGE DE MECHANIQUE ET ARMAMENT
MECH-	BUREAU OF RECLAMATION (INTERIOR). MECHANICAL BRANCH, DENVER
MECH. ENG. SPEC.-	CALIFORNIA. UNIV., BERKELEY. RADIATION LAB. (REF. 9)
MECH-GSF-(YEAR)-	AIR FORCE INST. OF TECH., WRIGHT-PATTERSON AFB, OHIO. SCHOOL OF ENGINEERING
MECO-	MECHANICAL EQUIPMENT CO., NEW ORLEANS
ME-CP-(NUMBER)-	GRUMMAN AIRCRAFT ENGINEERING CORP., BETHPAGE, N.Y.
MEC-QTR-NOBSR-	MICROWAVE ELECTRONICS CORP., PALO ALTO, CALIF.
MEC-QTR-NONR-	MICROWAVE ELECTRONICS CORP., PALO ALTO, CALIF.
MED-	AVCO MARINE ELECTRONICS OFFICE, NEW LONDON, CONN.
MED-(YEAR)-	NORTHROP CORP. NORTRONICS DIV., NEEDHAM, MASS. (USED UNTIL 1966 BY MARINE EQUIPMENT DEPT.)
MED-(YEAR)-	OAK RIDGE INST. OF NUCLEAR STUDIES, INC. MEDICAL DIV., TENN.
MED-	QUARTERMASTER RESEARCH AND ENGINEERING COMMAND. MECHANICAL ENGINEERING DIV., NATICK, MASS.
MEDDELELSE-	NORGES TEKNISKE HOEGSKOLE, TRONDHEIM. NORWEGIAN SHIP MODEL EXPERIMENT TANK
MEDDH-	ARMY MEDICAL RESEARCH UNIT, EUROPE (PROGRESS REPORT)
MEDEW-	BROOKE ARMY MEDICAL CENTER, SAN ANTONIO
MEDEW-RS-	BROOKE ARMY MEDICAL CENTER, SAN ANTONIO
MEDF-	TENNESSEE EASTMAN CORP., OAK RIDGE, TENN.
MED TIFR-	ARMY FORCES IN THE PACIFIC (MEDICAL TECHNICAL INTELLIGENCE FIELD REPORT)
MEDUI-	ILLINOIS. UNIV., URBANA
MEDUI-(NO.)-AEC	ILLINOIS. UNIV., URBANA
MEDUI-(NO.)-AF	ILLINOIS. UNIV., URBANA. DEPT. OF MINING AND METALLURGICAL ENGINEERING
MEF-	NAVAL MINE ENGINEERING FACILITY
MEG-	UNION OF SOUTH AFRICA. COUNCIL FOR SCIENTIFIC AND INDUSTRIAL RESEARCH, PRETORIA
MEI-	MELPAR, INC., FALLS CHURCH, VA.
MEL-	DEPARTMENT OF LABOR, WASHINGTON, D.C.
MEL-	MARINE ENGINEERING LAB., ANNAPOLIS
MEL-	MARTIN-MARIETTA CORP. DENVER DIV.
MEL-	MELPAR, INC., ALEXANDRIA, VA.
MEL-	MICROWAVE ENGINEERING LABS., PALO ALTO, CALIF.
MEL-	NAVY MARINE ENGINEERING LAB., ANNAPOLIS
MEL F-PH.2 QPR-	MELPAR, INC., FALLS CHURCH, VA.
MELLON-	MELLON INST., PITTSBURGH
MELLON-IIR-RR-	MELLON INST., PITTSBURGH
MELLON-RR-	MELLON INST., PITTSBURGH (RESEARCH REPORT)
MELLON-TR-	MELLON INST., PITTSBURGH (TECHNICAL REPORT)
MELP-	MELPAR, INC. (ALL LOCATIONS)
MELPAR-	MELPAR, INC., FALLS CHURCH, VA.
MELPAR-QPR-	MELPAR, INC., FALLS CHURCH, VA.
MELPAR-TR-	MELPAR, INC., FALLS CHURCH, VA.
MEL-R + D-	MARINE ENGINEERING LAB., ANNAPOLIS
MEL-TM-	MARINE ENGINEERING LAB., ANNAPOLIS
ME-MF-(NO.)-(NO.)-	GRUMMAN AIRCRAFT ENGINEERING CORP., BETHPAGE, N.Y.
MEMO-	(MEMORANDUM) (REF. 2)
MEMO-	ARGONNE NATIONAL LAB., LEMONT, ILL. (REF. 10)
MEMO-	BALLISTIC RESEARCH LABS.,ABERDEEN PROVING GROUND, MD.
MEMO-	CANADA. DEPT. OF MINES AND TECHNICAL SURVEYS
MEMO-	CARBIDE AND CARBON CHEMICALS CORP. K-25 PLANT, OAK RIDGE, TENN. (REF. 10)
MEMO-	DUKE UNIV., DURHAM, N.C.
MEMO-	GEOLOGICAL SURVEY, WASHINGTON, D.C.
MEMO-	HANFORD ENGINEER WORKS, RICHLAND, WASH.
MEMO-(YEAR)-(MONTH)	MC GILL UNIV., MONTREAL
MEMO-	MICHIGAN. UNIV., ANN ARBOR
MEMO-	MINE SAFETY APPLIANCES CO., CALLERY, PENNA.
MEMO(YR)	NAVAL MEDICAL RESEARCH INST., BETHESDA, MD.
MEMO-(YEAR)	NAVAL SUBMARINE MEDICAL CENTER, GROTON, CONN.
MEMO-	PENNSYLVANIA STATE COLL., STATE COLLEGE. SCHOOL OF MINERAL INDUSTRIES
MEMO-(INITIALS)-	GENERAL ELECTRIC CO., SCHENECTADY, N.Y.
MEMO-(INITIALS)-	KNOLLS ATOMIC POWER LAB., SCHENECTADY, N.Y.
MEMO-EN-(YR)-	GENERAL ELECTRIC CO. RESEARCH LAB., SCHENECTADY, N.Y.
MEMO-M-	AUSTRALIAN ARMY OPERATIONAL RESEARCH GROUP
MEMO-M-	GENERAL ELECTRIC CO. RESEARCH LAB., SCHENECTADY, N.Y.
MEMO M	GT. BRIT. NATIONAL GAS TURBINE ESTABLISHMENT, FARNBOROUGH, HANTS, ENGLAND

MEMO-NEC-	KNOLLS ATOMIC POWER LAB., SCHENECTADY, N.Y. (NUCLEAR ENGINEERING COURSE)
MEMO-NO.-(NO.)-	CALIF. INST. OF TECH., PASADENA. JET PROPULSION LAB.
MEMO. NOLM-	NAVAL ORDNANCE LAB., WHITE OAK, MD.
MEMO REPORT-	PENNSYLVANIA STATE COLL., STATE COLLEGE. SCHOOL OF MINERAL INDUSTRIES
MEMO-SE-PA-	HANFORD WORKS, RICHLAND, WASH.
MEMO-SE-PC-	HANFORD WORKS, RICHLAND, WASH.
ME-MP-(NUMBER)-	GRUMMAN AIRCRAFT ENGINEERING CORP., BETHPAGE, N.Y.
ME-OT-(NUMBER)-	GRUMMAN AIRCRAFT ENGINEERING CORP., BETHPAGE, N.Y.
MEP(YR.-YR.)(INITIALS)-	COLORADO STATE UNIV., FORT COLLINS
MEP-	MICROWAVE ELECTRONICS CORP., PALO ALTO, CALIF.
ME-Q-	DELAWARE. UNIV., NEWARK
ME-Q-	GENERAL ELECTRIC CO., SCHENECTADY, N.Y.
MER(NO.)-(YR.)-(INITLS.)	COLORADO STATE UNIV., FORT COLLINS. COLL. OF ENG.
MERC-TN-	MC GILL UNIV., MONTREAL
MERDL-	ARMY MEDICAL EQUIP. RES. + DEV. LAB., FT. TOTTEN, NY
MERDL-A-	ARMY MEDICAL EQUIP. RES. + DEV. LAB., FT. TOTTEN, NY
MERL-	GT. BRIT. MECHANICAL ENGINEERING RES. LAB., GLASGOW
MERL-	GT. BRIT. NATL. ENG. LAB., EAST KILBRIDE, GLASGOW
MERL-(YEAR)-	MC GILL UNIV., MONTREAL. DEPT. OF MECHANICAL ENG.
MERL-TN-(YEAR/NUMBER)	MC GILL UNIV., MONTREAL. DEPT. OF MECHANICAL ENG.
MERRIMACK-	MERRIMACK COLL., NORTH ANDOVER, MASS.
ME-RT-	STEVENS INST. OF TECH., HOBOKEN, N.J. DEPT. OF MECHANICAL ENGINEERING
MET.-	ARGONNE NATIONAL LAB., ILL.
MET-(MONTH)-(YEAR)	ARMY ELECTRONICS COMMAND. METEOROLOGICAL SUPPORT ACTIVITY, FORT HUACHUCA, ARIZ.
MET-	ATOMIC ENERGY OF CANADA LTD.,CHALK RIVER,ONT. (REF.6)
ME-T-	DELAWARE. UNIV., NEWARK
MET-	GT. BRIT. ROYAL AIRCRAFT EST.,FARNBOROUGH, HANTS, ENG
MET-	SYRACUSE UNIV., N.Y. DEPT. OF CHEMICAL ENGINEERING AND METALLURGY
MET-	SYRACUSE UNIV., N.Y. METALLURGICAL RESEARCH LABS.
MET-	SYRACUSE UNIV., N.Y. RESEARCH INST.
MET.(NO.)(LETTER)	UNITED KINGDOM. AIR MINISTRY. METEOROLOGICAL OFFICE, LONDON
METCUT-	METCUT RESEARCH ASSOCIATES INC., CINCINNATI
MET-E-(NUMBER)-	SYRACUSE UNIV., N.Y. RESEARCH INST. DEPT. OF CHEMICAL ENGINEERING AND METALLURGY
MET-E-(NO.)-(NO.)(LTR.)	SYRACUSE UNIV., N.Y. RESEARCH INST. DEPT. OF CHEMICAL ENGINEERING AND METALLURGY
MET.E-(NO.)-(NO.)-FR	SYRACUSE UNIV., N.Y. RESEARCH INST. DEPT. OF CHEMICAL ENGINEERING AND METALLURGY
MET-E-(NO.)-(NO.)-QP(NO.)	SYRACUSE UNIV., N.Y. RESEARCH INST. DEPT. OF CHEMICAL ENGINEERING AND METALLURGY
METEOR-(NO.)-(NO.)-(NO.)	BELL AIRCRAFT CORP., BUFFALO (PROJECT METEOR)
METEOR-	MASSACHUSETTS INST. OF TECH., CAMBRIDGE. GUIDED MISSILES PROGRAM (PROJECT METEOR)
METEOROLOGY-	MC GILL UNIV., MONTREAL. ARCTIC METEOROLOGY RES. GP.
MET FLUID DYN. RES. LAB.-	MASSACHUSETTS INST. OF TECH., CAMBRIDGE
METGS-	SYRACUSE UNIV., N.Y. METALLURGICAL RESEARCH LABS.
METGS-	SYRACUSE UNIV., N.Y. RESEARCH INST.
MET-I-	ATOMIC ENERGY OF CANADA LTD.,CHALK RIVER,ONT. (REF.6)
METI-	ATOMIC ENERGY OF CANADA LTD. CHALK RIVER PROJECT, ONT. (REF. 6)
ME-TN-	ILLINOIS. UNIV., URBANA. DEPT. OF MECHANICAL ENG.
ME-TN-	ILLINOIS. UNIV., URBANA. ENGINEERING EXPERIMENT STA.
MET-O-	UNITED KINGDOM. AIR MINISTRY. METEOROLOGICAL OFFICE, LONDON
ME-TP-(NUMBER)-	GRUMMAN AIRCRAFT ENGINEERING CORP., BETHPAGE, N.Y.
ME-TR-	ILLINOIS. UNIV., URBANA. DEPT. OF MECHANICAL ENG.
METRI-(NUMBER)-	CLARK, COOPER, FIELD AND WOHL, INC., NEW YORK CITY
METRONICS-	METRONICS ASSOCIATES, INC., PALO ALTO, CALIF.
ME-TR-ORD-1121-	ILLINOIS. UNIV., URBANA
MET-SPT-ACTV-	ARMY ELECTRONICS COMMAND. METEOROLOGICAL SUPPORT ACTIVITY, FORT HUACHUCA, ARIZ.
ME VB-	MESSERSCHMIDT A.G. (VERSUCHSBERICHT)
ME VB-(NO.)-	MESSERSCHMIDT A.G. (VERSUCHSBERICHT)
MEW-	TENNESSEE EASTMAN CORP., OAK RIDGE, TENN.
MEWD-	ARMY SIGNAL MISSILE SUPPORT AGENCY. MISSILE ELECTRONIC WARFARE DIV., WHITE SANDS MISSILE RANGE, NM
ME-WE-(NUMBER)-	GRUMMAN AIRCRAFT ENGINEERING CORP., BETHPAGE, N.Y.
MEXE-	GT. BRIT. MILITARY ENGINEERING EXPERIMENTAL ESTAB.
MF-	AEROJET-GENERAL CORP.,SACRAMENTO,CAL.(MATERIALS RPT)
MF-	AEROJET-GENERAL CORP. LIQUID ROCKET OPERATIONS, SACRAMENTO, CALIF.
MF-	EUROPEAN THEATER OF OPERATIONS (ARMY). ALSOS MISSION (MF FILE)
MF-(PROJECT)	NAVAL MEDICAL RESEARCH INST., BETHESDA, MD.
MF-	NEW YORK. STATE UNIV., SYRACUSE
MF-	NEW YORK UNIV., N.Y.C. INST. OF MATHEMATICAL SCIENCES
MF-	NEW YORK UNIV., N.Y.C. MAGNETO-FLUID DYNAMICS DIV.
MFA-	MERRILL FLOOD AND ASSOCIATES
MFA-TS-	METAL FINISHING ABSTRACTS TRANSLATION SERVICES, TEDDINGTON, MIDDX., ENGLAND
MFDR-	AEROJET-GENERAL CORP.,SACRAMENTO,CAL.(MATERIALS RPT)
MFL-	ARMY BALLISTIC MISSILE AGENCY. MISSILE FIRING LAB., CAPE CANAVERAL, FLA.
MFPG-	MEETING OF THE MECHANICAL FAILURES PREVENTION GROUP, 10TH, PALO ALTO, CALIF.
MFS-	GEORGE C. MARSHALL SPACE FLIGHT CTR.,HUNTSVILLE, ALA.
MFSC-IN-SSL-T-	GEORGE C. MARSHALL SPACE FLIGHT CTR.,HUNTSVILLE, ALA.
MFTD-	RAYTHEON MFG. CO., WALTHAM, MASS.
MFTD-	RAYTHEON MFG. CO. MISSILE FLIGHT TEST DEPT.
MG-	COAST AND GEODETIC SURVEY, ROCKVILLE, MD. (MAGNETOGRAMS)
MG-	GT. BRIT. MINISTRY OF SUPPLY
MG-3-	HUGHES AIRCRAFT CO., CULVER CITY, CALIF.
MG-(2 DIGITS)-	HUGHES AIRCRAFT CO., CULVER CITY, CALIF.
MG/A/	BRITISH STEEL CORP. METALLURGY DIV., LONDON
MG/C/	BRITISH STEEL CORP. METALLURGY DIV., LONDON
MGC-	RES. + DEV. BD. COMM. ON GUIDED MISSILES, WASH., DC
MGCR-	GENERAL ATOMIC DIV.,GEN. DYNAMICS CORP.,SAN DIEGO,CAL
MGCR-M-	GENERAL ATOMIC DIV.,GEN. DYNAMICS CORP.,SAN DIEGO,CAL
MGCR-P-	GENERAL ATOMIC DIV.,GEN. DYNAMICS CORP.,SAN DIEGO,CAL
MGCR-RE-	GENERAL ATOMIC DIV.,GEN. DYNAMICS CORP.,SAN DIEGO,CAL
MGCR-RP-	GENERAL ATOMIC DIV.,GEN. DYNAMICS CORP.,SAN DIEGO,CAL
MG/D-(NUMBER/YEAR)	BRITISH IRON AND STEEL RES. ASSN.,SHEFFIELD,YORKS,ENG
MG/D-(NUMBER/NUMBER)	BRITISH IRON + STEEL RES.ASSN. METALLURGY DIV.,LONDON
MGRT-	NORTH ATLANTIC TREATY ORGANIZATION. STRUCTURES AND MATERIALS PANEL, PARIS
MGTD-(NUMBER)-	LOGICON, INC., REDONDO BEACH, CALIF.
MGX-	MAGNAVOX CO., FORT WAYNE
MH-	CORPS OF ENGINEERS (ARMY).DIV OF MECH. ENG.,WASH.,DC
MH-	METAL HYDRIDES INC., BEVERLY, MASS.
MH-	MINNEAPOLIS-HONEYWELL REGULATOR CO. (ALL LOCATIONS)
MH-	NATIONAL RESEARCH COUNCIL OF CANADA. DIVISION OF MECHANICAL ENGINEERING, OTTAWA
MH-	NEW YORK UNIV., N.Y.C. ATOMIC ENERGY COMMISSION COMPUTING FACILITY. (NOTES ON MAGNETO-HYDRODYNAMICS)

MH-	NEW YORK UNIV., N.Y.C. INST. OF MATHEMATICAL SCIENCES
MH-AD-	MINNEAPOLIS-HONEYWELL REGULATOR CO., MINNEAPOLIS
MH-AERO-	MINNEAPOLIS-HONEYWELL REGULATOR CO., ST. PETERSBURG, FLA
MH-AERO-	MINNEAPOLIS-HONEYWELL REGULATOR CO. AERONAUTICAL DIV. MINNEAPOLIS
MH AERO (NO.) QR-	MINNEAPOLIS-HONEYWELL REGULATOR CO. AERONAUTICAL DIV. MINNEAPOLIS
MH AERO (NO.) TR-	MINNEAPOLIS-HONEYWELL REGULATOR CO. AERONAUTICAL DIV. MINNEAPOLIS
MHBC-	MC GRAW-HILL PUBLISHING CO., INC., N.Y.C.
MHC-	METAL HYDRIDES INC., BEVERLY, MASS.
MHC-	MOUNT HOLYOKE COLL., SOUTH HADLEY, MASS. PSYCHO-PHYSICAL RESEARCH UNIT
MHD-	MHD RESEARCH, INC., NEWPORT BEACH, CALIF.
MHE-	ARMY NATICK LABS. MECHANICAL ENGINEERING DIV., MASS.
MHM-	GEORGE C. MARSHALL SPACE FLIGHT CTR., HUNTSVILLE, ALA.
MH-OD-	MINNEAPOLIS-HONEYWELL REGULATOR CO. ORDNANCE DIV., HOPKINS, MINN.
MH-OD-	MINNEAPOLIS-HONEYWELL REGULATOR CO. ORDNANCE DIV., MONROVIA, CALIF.
MHR-	GEORGE C. MARSHALL SPACE FLIGHT CTR., HUNTSVILLE, ALA.
MHR-	HARVARD UNIV., CAMBRIDGE, MASS. PSYCHO-ACOUSTIC LAB.
MHR-	HONEYWELL, INC. (ALL LOCATIONS)
MHR-	MC GRAW-HILL, INC. F.W. DODGE CO., WRIGHT-PATTERSON AFB, OHIO
MHR-(YEAR)-	MC GRAW-HILL, INC. F.W. DODGE CO., WRIGHT-PATTERSON AFB, OHIO
MHR-	MC GRAW-HILL PUBLISHING CO., INC., N.Y.C.
MHR-	METCUT RESEARCH ASSOCIATES INC., CINCINNATI
MHR AERO (NO.) TR-	MINNEAPOLIS-HONEYWELL REGULATOR CO. AERONAUTICAL DIV. MINNEAPOLIS
MHRC-	MINNEAPOLIS-HONEYWELL REGULATOR CO., MINNEAPOLIS
MHR-MDL-	MINNEAPOLIS-HONEYWELL REGULATOR CO. MISSILE DEVELOPMENT LAB., LOS ANGELES
MHR MDL R-ED-	MINNEAPOLIS-HONEYWELL REGULATOR CO. MISSILE DEVELOPMENT LAB., LOS ANGELES
MHR-SN-	MINNEAPOLIS-HONEYWELL REGULATOR CO. ORDNANCE DIV., HOPKINS, MINN.
MHSCDRC-	GT. BRIT. MINISTRY OF HOME SECURITY. CIVIL DEFENCE RESEARCH COMMITTEE
MH-SM-	MASON AND HANGER-SILAS MASON CO. INC. PANTEX ORDNANCE PLANT, AMARILLO, TEX.
MHSMB-	MASON + HANGER-SILAS MASON CO., INC., BURLINGTON, IOWA
MHSMP-	MASON + HANGER-SILAS MASON CO., INC. PANTEX PLANT, AMARILLO, TEX.
MHSMP-(YEAR)-	MASON + HANGER-SILAS MASON CO., INC. PANTEX PLANT, AMARILLO, TEX.
MHV-	COAST AND GEODETIC SURVEY, ROCKVILLE, MD. (MAGNETOGRAMS AND HOURLY VALUES)
M.I.10-	GT. BRIT. WAR OFFICE
MI-(YEAR)-	MARTIN CO., DENVER
MI-	MARTIN-MARIETTA CORP. DENVER DIV.
MI-	MELLON INST., PITTSBURGH
MI-	MOTOROLA INC. (ALL LOCATIONS)
MI-	NATIONAL RESEARCH COUNCIL OF CANADA. ATOMIC ENERGY PROJECT, CHALK RIVER, ONT. (REF. 29)
MI-	NETHERLANDS. METAALINSTITUUT TNO, THE HAGUE
MIA-	MARBLE INST. OF AMERICA, INC., MOUNT VERNON, N.Y.
MIAMI-U-	MIAMI UNIV., OXFORD, OHIO
MIAPH-AP-(YEAR.NUMBER)	MIAMI. UNIV., CORAL GABLES, FLA. DEPT. OF PHYSICS
MIAPH-OP-(YEAR.NUMBER)	MIAMI. UNIV., CORAL GABLES, FLA. DEPT. OF PHYSICS
MIAPH-PL-	MIAMI. UNIV., CORAL GABLES, FLA. DEPT. OF PHYSICS
MIAPH-PL-(YEAR)-	MIAMI. UNIV., CORAL GABLES, FLA. DEPT. OF PHYSICS
MIAPH-PP-	MIAMI. UNIV., CORAL GABLES, FLA. DEPT. OF PHYSICS
MIAPH-TP-(YEAR.NUMBER)	MIAMI. UNIV., CORAL GABLES, FLA. DEPT. OF PHYSICS
MIC-	COMMITTEE FOR MILITARY INFORMATION CONTROL
MICH-	MICHIGAN. UNIV., ANN ARBOR
MICH-(NUMBER)-(LETTER)	MICHIGAN. UNIV., ANN ARBOR (SQUID PROJECT) (REF. 33)
MICH (NOS.-NO.-LTR.)	MICHIGAN. UNIV., ANN ARBOR
MICH-(NUMBER)-PU	PURDUE UNIV., LAFAYETTE, IND.
MICH-	VIRGINIA. UNIV., CHARLOTTESVILLE
MICH EDG-TR-	MICHIGAN. UNIV., ANN ARBOR. ELECTRONIC DEFENSE GROUP
MICH ERI-(NOS.-NO.-LTR.)	MICHIGAN. UNIV., ANN ARBOR. ENGINEERING RES. INST.
MICH ERI-(YR.-NO.)	MICHIGAN. UNIV., ANN ARBOR. ENGINEERING RES. INST.
MICHIGAN U-(NO.)-(LTR.)-	MICHIGAN. UNIV., ANN ARBOR
MICH M(NOS.) TR-	MICHIGAN. UNIV., ANN ARBOR. ENGINEERING RES. INST.
MICH MX(NOS)-(NOS-NO-LTR)	MICHIGAN. UNIV., ANN ARBOR. ENGINEERING RES. INST.
MICH-U-	MICHIGAN. UNIV., ANN ARBOR
MICHU-	MICHIGAN. UNIV., ANN ARBOR
MICH-U/BAMIRAC-	MICHIGAN. UNIV., ANN ARBOR. BALLISTIC MISSILE RADIATION ANALYSIS CENTER
MICHU-IP-	MICHIGAN. UNIV., ANN ARBOR. COLL. OF ENGINEERING (INDUSTRY PROGRAM)
MICH UNM-	MICHIGAN. UNIV., ANN ARBOR
MICH UNR-	MICHIGAN. UNIV., ANN ARBOR
MICHU-PR-	ATOMIC POWER DEVELOPMENT ASSOCIATES, INC., DETROIT
MICHU-PR-	MICHIGAN. UNIV., ANN ARBOR. ENGINEERING RES. INST.
MICH-U/VESIAC-	MICHIGAN. UNIV., ANN ARBOR. VELA SEISMIC INFORMATION ANALYSIS CENTER
MICH-U-WRL-	MICHIGAN. UNIV., ANN ARBOR. WILLOW RUN LABS.
MICRO (NUMBER)-	EUROPEAN THEATER OF OPERATIONS (ARMY). ALSOS MISSION
MICROCLIMATE-TR-	AGRICULTURAL RESEARCH SERVICE. SOIL AND WATER CONSERVATION RESEARCH DIV., ITHACA, N.Y.
MICRO-NOTES-	NAVAL AMMUNITION DEPOT, CRANE, IND.
MIC U-	MICHIGAN. UNIV., ANN ARBOR
MID-CR-	ARMY MISSILE COMMAND. MISSILE INTELLIGENCE DIRECTORATE, REDSTONE ARSENAL, ALA.
MIE-	MEDICAL + INDUSTRIAL EQUIPMENT CO., LTD.
MIF-	AIR TECH. INTELL. CTR., WRIGHT-PATTERSON AFB, OHIO
MIIR-RR-	MELLON INST., PITTSBURGH
MIIR-TR-	MELLON INST., PITTSBURGH
MIL-(LETTER)-	ARMED FORCES SUPPLY SUPPORT CENTER. STANDARDIZATION DIV., WASHINGTON, D.C. (MILITARY SPECS. + STDS.) (REF. 13)
MIL-(LETTER)-	DEPARTMENT OF DEFENSE, WASHINGTON, D.C. (USED TO IDENTIFY SPECIFICATIONS. THE LETTER FOLLOWING MIL- REPRESENTS THE ARTICLE FOR WHICH THE SPECIFICATIONS ARE GIVEN, E.G. MIL-T-, TENT, TUBE, ETC.) (REF. 13)
MIL-STD-	(MILITARY STANDARD) SEE ALSO MIL-(LETTER)- (REF. 13)
MIL-STD-	MUNITIONS BOARD STANDARDS AGENCY, WASHINGTON, D.C.
MIL-TDR-	ROCKETDYNE DIV., NORTH AMERICAN AVIATION, INC., CANOGA PARK, CALIF.
MIM-	NATIONAL RESEARCH COUNCIL OF CANADA. ATOMIC ENERGY PROJECT, CHALK RIVER, ONT.
MIMEOGRAPHIC SERIES-	PURDUE UNIV., LAFAYETTE, IND. DEPT. OF STATISTICS
MIN-	MINNESOTA. UNIV., MINNEAPOLIS
MIN-DEF-ATT-	UNITED KINGDOM. ARMY PERSONNEL RESEARCH ESTABLISHMENT, BYFLEET, SURREY, ENGLAND
MINN-	MINNESOTA. UNIV., MINNEAPOLIS
MINN EM-	MINNESOTA. UNIV., MINNEAPOLIS. INST. OF TECH.
MINN HTL TR-	MINNESOTA. UNIV., MINNEAPOLIS. HEAT TRANSFER LAB.
MINN RR-	MINNESOTA. UNIV., MINNEAPOLIS. INST. OF TECH.
MINN-SR-	MINNESOTA. UNIV., MINNEAPOLIS. PHYSICAL ELECTRONICS RESEARCH LAB.
MINN-TR-	MINNESOTA. UNIV., MINNEAPOLIS

MINN-TR-AP-	MINNESOTA. UNIV., MINNEAPOLIS (ATMOSPHERIC PHYSICS PROGRAM)
MINN-TR-CR-	MINNESOTA. UNIV., MINNEAPOLIS. COSMIC RAY GROUP
MINNU-	MINNESOTA. UNIV., MINNEAPOLIS
MINS-	MARE ISLAND NAVAL SHIPYARD. RUBBER LAB., CALIF.
MINTECH-T-	UNITED KINGDOM. MINISTRY OF TECHNOLOGY, LONDON
MINU-	MINNESOTA. UNIV., MINNEAPOLIS
MIP-	MANDREL INDUSTRIES, INC., HOUSTON, TEX.
MIPR-	(MONTHLY INTERIM PROGRESS REPORT) (REF. 2)
MIR-	MITRE CORP., BEDFORD, MASS.
MIS-	EUROPEAN THEATER OF OPERATIONS (ARMY). ALSOS MISSION (MIS FILE)
MIS-	MILITARY INTELLIGENCE SERVICE (ARMY), WASHINGTON, DC
MIS-	SUPREME COMMANDER FOR THE ALLIED POWERS. MILITARY INTELLIGENCE SECTION
MIS ATIS-	SUPREME CMDR. FOR THE ALLIED POWERS. MIL. INTELL. SEC GENL. STAFF ALLIED TRANSLATOR + INTERPRETER SEC.
MISC.-	ARMY BIOLOGICAL LABS., FREDERICK, MD.
MISC.-	ATOMIC ENERGY OF CANADA LTD. CHALK RIVER PROJ., ONT.
MISC.-	NATIONAL RESEARCH COUNCIL OF CANADA. ATOMIC ENERGY PROJECT, CHALK RIVER, ONT.
MISCELLANEOUS PAPER-	(MISCELLANEOUS PAPER) (REF. 2)
MISC-SSL-(YEAR)-	GEORGE C. MARSHALL SPACE FLIGHT CTR.,HUNTSVILLE, ALA.
MIS IR-	SUPREME COMMANDER FOR THE ALLIED POWERS. MILITARY INTELLIGENCE SECTION (INTELLIGENCE REPORTS)
MISSC-	MISSISSIPPI STATE COLL., STATE COLLEGE
MISSSU-	MISSISSIPPI STATE UNIV., STATE COLLEGE
MISSU-	MISSISSIPPI. UNIV., UNIVERSITY
MISU-	MISSOURI. UNIV., COLUMBIA
MISU-	MISSOURI. UNIV., ROLLA
MISU-TR-	MISSOURI. UNIV., ROLLA
MIT-	MASSACHUSETTS INST. OF TECH., CAMBRIDGE
MIT-(NUMBER)-(LETTER)	MASSACHUSETTS INST. OF TECH., CAMBRIDGE
MIT-(NUMBER)-M	MASS. INST. OF TECH., CAMBRIDGE (SQUID PROJ.)(REF.33)
MIT-(NUMBER)-P	MASS. INST. OF TECH., CAMBRIDGE (SQUID PROJ.)(REF.33)
MIT-(NUMBER)-R	MASS. INST. OF TECH., CAMBRIDGE (SQUID PROJ.)(REF.33)
MIT-(NUMBER)-T	MASS. INST. OF TECH., CAMBRIDGE (SQUID PROJ.)(REF.33)
MIT-(NO.)-(NO.)-TM-	MASSACHUSETTS INST. OF TECH., CAMBRIDGE. ELECTRONIC SYSTEMS LAB.
MIT-	PADUA UNIVERSITA
MIT-	VIRGINIA. UNIV.,CHARLOTTESVILLE.DEPT.OF AEROSPACE ENG
MIT-(INITIALS)	MASSACHUSETTS INST. OF TECH., CAMBRIDGE
MIT-AE-	MASSACHUSETTS INST. OF TECH., CAMBRIDGE. AERO-ELASTIC AND STRUCTURES RESEARCH LAB.
MIT-AE-	MASSACHUSETTS INST. OF TECH., CAMBRIDGE. DEPT. OF AERONAUTICAL ENG. (AERO-ELASTIC + STRUCTURES RES.)
MIT-AL-	MASS. INST. OF TECH., CAMBRIDGE. ACOUSTICS LAB.
MIT-AL-QR-	MASS. INST. OF TECH., CAMBRIDGE. ACOUSTICS LAB.
MIT-AL-TR-	MASS. INST. OF TECH., CAMBRIDGE. ACOUSTICS LAB.
MIT-AL-TR-	MASS. INST. OF TECH., CAMBRIDGE. AEROPHYSICS LAB.
MIT-AR-(INCL.YEARS)	MASSACHUSETTS INST. OF TECH., CAMBRIDGE
MIT-ASRL-	MASSACHUSETTS INST. OF TECH., CAMBRIDGE. AERO-ELASTIC AND STRUCTURES RESEARCH LAB.
MIT-ASRL-TR25-	MASSACHUSETTS INST. OF TECH., CAMBRIDGE. AERO-ELASTIC AND STRUCTURES RESEARCH LAB.
MIT-ASRL-TR-	MASSACHUSETTS INST. OF TECH., CAMBRIDGE. AERO-ELASTIC AND STRUCTURES RESEARCH LAB.
MIT-CC-PR-	MASS. INST. OF TECH., CAMBRIDGE. COMPUTATION CENTER
MIT-CL-TR-	MASS. INST. OF TECH., CAMBRIDGE. CRYSTALLOGRAPHIC LAB
MIT-CSE-	MASSACHUSETTS INST. OF TECH., CAMBRIDGE. DEPT. OF CIVIL AND SANITARY ENGINEERING
MIT-CSR-T-	MASS. INST. OF TECH., CAMBRIDGE. CENTER FOR SPACE RES
MIT-CSR-TR-(YEAR)-	MASS. INST. OF TECH., CAMBRIDGE. CENTER FOR SPACE RES
MIT-CTP-	MASSACHUSETTS INST. OF TECH., CAMBRIDGE. CENTER FOR THEORETICAL PHYSICS
MIT-DACL-	MASSACHUSETTS INST. OF TECH., CAMBRIDGE. DYNAMIC ANALYSIS AND CONTROL LAB.
MIT-DCE-	MASSACHUSETTS INST. OF TECH., CAMBRIDGE. DEPT. OF CIVIL ENGINEERING
MIT-DCL-	MASS. INST. OF TECH., CAMBRIDGE. DIGITAL COMPUTER LAB
MIT-DCL-SR-	MASS. INST. OF TECH., CAMBRIDGE. DIGITAL COMPUTER LAB
MIT-DFT-	MASSACHUSETTS INST. OF TECH., CAMBRIDGE. DEPT. OF FOOD TECHNOLOGY.
MIT-DIC-	MASSACHUSETTS INST. OF TECH., CAMBRIDGE. DIV. OF INDUSTRIAL COOPERATION
MIT DIC(NOS.)-ER-	MASSACHUSETTS INST. OF TECH., CAMBRIDGE. SERVO-MECHANISMS LAB. (ENGINEERING REPORT)
MIT DIC(NUMBERS)-R-	MASS. INST. OF TECH., CAMBRIDGE. SERVOMECHANISMS LAB.
MIT DIC(NUMBERS)-TM-	MASS. INST. OF TECH., CAMBRIDGE. SERVOMECHANISMS LAB.
MIT-DIC-R-	MASSACHUSETTS INST. OF TECH., CAMBRIDGE. DIV. OF INDUSTRIAL COOPERATION
MIT-DIC-TR-	MASSACHUSETTS INST. OF TECH., CAMBRIDGE. DIV. OF INDUSTRIAL COOPERATION
MIT-DM-	MASS. INST. OF TECH., CAMBRIDGE. DEPT. OF METALLURGY
MIT-DME-	MASS. INST. OF TECH., CAMBRIDGE. DEPT. OF MECH. ENG.
MIT-DMET-	MASSACHUSETTS INST. OF TECH., CAMBRIDGE. DEPT. OF METEOROLOGY
MIT-DME-TR-	MASS. INST. OF TECH., CAMBRIDGE. DEPT. OF MECH. ENG.
MIT-DNA-(YEAR)-	MASSACHUSETTS INST. OF TECH., CAMBRIDGE. DEPT. OF NAVAL ARCHITECTURE AND MARINE ENGINEERING
MIT-DSR-	MASSACHUSETTS INST. OF TECH., CAMBRIDGE
MIT-DSR-TR-	MASSACHUSETTS INST. OF TECH., CAMBRIDGE. DIVISION OF SPONSORED RESEARCH
MIT-EAL-PR-	MASSACHUSETTS INST. OF TECH., CAMBRIDGE. EXPERIMENTAL ASTRONOMY LAB. (PROGRESS REPORT)
MIT-EAL-RE-	MASSACHUSETTS INST. OF TECH., CAMBRIDGE. EXPERIMENTAL ASTRONOMY LAB. (REPORT)
MIT-EAL-RN-	MASSACHUSETTS INST. OF TECH., CAMBRIDGE. EXPERIMENTAL ASTRONOMY LAB. (RESEARCH NOTE)
MIT-EAL-TE-	MASSACHUSETTS INST. OF TECH., CAMBRIDGE. EXPERIMENTAL ASTRONOMY LAB. (THESIS)
MIT-ENIG-	MASSACHUSETTS INST. OF TECH., CAMBRIDGE. ELECTRONIC NUCLEAR INSTRUMENTATION GROUP
MIT-ERL-	MASSACHUSETTS INST. OF TECH., CAMBRIDGE. ELECTRONICS RESEARCH LAB.
MIT-ESL-	MASSACHUSETTS INST. OF TECH., CAMBRIDGE. ELECTRONIC SYSTEMS LAB.
MIT-FCL-(NO.)-R-	MASS. INST. OF TECH., CAMBRIDGE. FLIGHT CONTROL LAB.
MIT-FDRG-(YR)-	MASS. INST. OF TECH.,CAMBRIDGE. FLUID DYNAMICS RES.GP
MIT FLUID DYN.RES.LAB.-	MASS. INST. OF TECH., CAMBRIDGE. FLUID DYNAMICS RESEARCH LAB.
MIT-FML-	MASS. INST. OF TECH., CAMBRIDGE. FLUID MECHANICS LAB.
MIT-FP-	MASSACHUSETTS INST. OF TECH., CAMBRIDGE. FIBERS AND POLYMERS DIV.
MIT-FRL-	MASS. INST. OF TECH., CAMBRIDGE. FUELS RESEARCH LAB.
MIT-FRL-TR-	MASS. INST. OF TECH., CAMBRIDGE. FUELS RESEARCH LAB.
MITG-	MASS. INST. OF TECH., CAMBRIDGE. MINERAL ENG. LAB.
MITG-	MASS. INST. OF TECH., WATERTOWN. MINERAL ENG. LAB.
MITG-A-	MASS. INST. OF TECH., CAMBRIDGE. DEPT. OF METALLURGY
MIT GTL R-	MASS. INST. OF TECH., CAMBRIDGE. GAS TURBINE LAB.
MIT-HL-	MASS. INST. OF TECH., CAMBRIDGE. HYDRODYNAMICS LAB.
MIT-HTL-TR-	MASS. INST. OF TECH., CAMBRIDGE. HEAT TRANSFER LAB.
MIT-HVL-	MASS. INST. OF TECH., CAMBRIDGE. HIGH VOLTAGE LAB.
MIT-IL-	MASS. INST. OF TECH., CAMBRIDGE. INSTRUMENTATION LAB.
MIT-IL-E-	MASSACHUSETTS INST. OF TECH., CAMBRIDGE. INSTRUMENTATION LAB. (ENGINEERING NOTES)

MIT-IL-R-

MIT-IL-R-
    MASSACHUSETTS INST. OF TECH., CAMBRIDGE. INSTRUMENTATION LAB. (REPORTS)
MIT-IL-RR(NO.)R-
    MASS. INST. OF TECH., CAMBRIDGE. INSTRUMENTATION LAB.
MIT-L-
    MASSACHUSETTS INST. OF TECH., CAMBRIDGE
MIT-LAB-TR-
    MASSACHUSETTS INST. OF TECH., CAMBRIDGE. LAB. FOR APPLIED BIOPHYSICS. (TECHNICAL REPORT)
MIT-LCSP-ARR-
    MASSACHUSETTS INST. OF TECH., CAMBRIDGE. LAB OF CHEMICAL AND SOLID-STATE PHYSICS.(ANNUAL RES. REPT)
MIT-LEXP-
    MASS. INST. OF TECH., CAMBRIDGE. LEXINGTON PROJECT
MIT-LIR-
    MASSACHUSETTS INST. OF TECH., CAMBRIDGE. LAB. FOR INSULATION RESEARCH
MIT LIR PR-
    MASSACHUSETTS INST. OF TECH., CAMBRIDGE. LAB. FOR INSULATION RESEARCH
MIT LIR TR-
    MASSACHUSETTS INST. OF TECH., CAMBRIDGE. LAB. FOR INSULATION RESEARCH
MIT-LL-
    MASSACHUSETTS INST. OF TECH., CAMBRIDGE. LINCOLN LAB.
MIT-LL-
    MASSACHUSETTS INST. OF TECH., LEXINGTON. LINCOLN LAB.
MIT-LL-BIB-
    MASSACHUSETTS INST. OF TECH., LEXINGTON. LINCOLN LAB. (BIBLIOGRAPHY)
MIT-LL-GR(NO.)-
    MASSACHUSETTS INST. OF TECH., LEXINGTON. LINCOLN LAB. (GROUP REPORTS)
MIT-LL-QR-(YR)-(MO)-
    MASSACHUSETTS INST. OF TECH., LEXINGTON. LINCOLN LAB. (SOLID STATE RESEARCH QUARTERLY PROGRESS REPORT)
MIT-LL-SSR-
    MASSACHUSETTS INST. OF TECH., LEXINGTON. LINCOLN LAB. (SOLID STATE RESEARCH)
MIT-LL-TM-
    MASSACHUSETTS INST. OF TECH., LEXINGTON. LINCOLN LAB. (TECHNICAL MEMORANDUM)
MIT-LL-TR-
    MASSACHUSETTS INST. OF TECH., LEXINGTON. LINCOLN LAB. (TECHNICAL REPORT)
MIT-LNS-
    MASSACHUSETTS INST. OF TECH., CAMBRIDGE. LAB. FOR NUCLEAR SCIENCE
MIT-LNSE-
    MASSACHUSETTS INST. OF TECH., CAMBRIDGE. LAB. FOR NUCLEAR SCIENCE AND ENGINEERING
MIT-LNSE-PR-
    MASSACHUSETTS INST. OF TECH., CAMBRIDGE. LAB. FOR NUCLEAR SCIENCE AND ENGINEERING (PROGRESS REPORT)
MIT-LNSE-TR-
    MASSACHUSETTS INST. OF TECH., CAMBRIDGE. LAB. FOR NUCLEAR SCIENCE AND ENGINEERING (TECHNICAL REPORT)
MIT-LNS-PR-
    MASSACHUSETTS INST. OF TECH., CAMBRIDGE. LAB. FOR NUCLEAR SCIENCE (PROGRESS REPORT)
MIT-LNS-TR-
    MASSACHUSETTS INST. OF TECH., CAMBRIDGE. LAB. FOR NUCLEAR SCIENCE (TECHNICAL REPORT)
MIT-M-
    MASSACHUSETTS INST. OF TECH., CAMBRIDGE
MIT-METEOR-
    MASSACHUSETTS INST. OF TECH., CAMBRIDGE. GUIDED MISSILES PROGRAM
MIT-ML-(YEAR)
    MASSACHUSETTS INST. OF TECH., CAMBRIDGE. MAGNETOGASDYNAMICS LAB.
MIT-MPD-
    MASS. INST. OF TECH., CAMBRIDGE. METALS PROCESSING DIV
MIT MR-
    MASS. INST. OF TECH., CAMBRIDGE (MEMORANDUM RPTS.)
MIT-MR-
    MASSACHUSETTS INST. OF TECH., CAMBRIDGE.(METEOR REPT.)
MIT-MVT-
    MASS. INST. OF TECH., CAMBRIDGE. MAN-VEHICLE LAB.
MITNE-
    MASSACHUSETTS INST. OF TECH., CAMBRIDGE. DEPT. OF NUCLEAR ENGINEERING
MIT-NE-
    MASSACHUSETTS INST. OF TECH., CAMBRIDGE. DEPT. OF NUCLEAR ENGINEERING
MIT-NSES-
    MASSACHUSETTS INST. OF TECH., CAMBRIDGE
MIT-NSL-
    MASS. INST. OF TECH., CAMBRIDGE. NAVAL SUPERSONIC LAB
MIT-NSL-TR-
    MASSACHUSETTS INST. OF TECH., CAMBRIDGE. NAVAL SUPERSONIC LAB. (TECHNICAL REPORT)
MIT-OR-
    MASS. INST.OF TECH.,OAK RIDGE,TENN. ENG. PRACTICE SCH
MIT-P-
    MASSACHUSETTS INST. OF TECH., CAMBRIDGE
MIT-(P)-
    MASS. INST. OF TECH., CAMBRIDGE. METALLURGICAL PROJ.
MIT-PDC-
    MASSACHUSETTS INST. OF TECH., CAMBRIDGE
MIT-PG-
    MASSACHUSETTS INST. OF TECH., CAMBRIDGE
MIT-PL-
    MASSACHUSETTS INST. OF TECH., CAMBRIDGE
MIT-QPR-
    MASSACHUSETTS INST. OF TECH., CAMBRIDGE
MIT-R(YEAR)-
    MASSACHUSETTS INST. OF TECH., CAMBRIDGE

MIT RAD LAB-
    MASSACHUSETTS INST. OF TECH.,CAMBRIDGE.RADIATION LAB.
MIT RAD LAB (ROMAN NO.)-
    MASSACHUSETTS INST. OF TECH.,CAMBRIDGE.RADIATION LAB.
MIT RAD LAB 483-
    MASSACHUSETTS INST. OF TECH., CAMBRIDGE. RADIATION LABORATORY (RADOME BULLETINS)
MIT RAD LAB (LETTER(S))-
    MASSACHUSETTS INST. OF TECH.,CAMBRIDGE.RADIATION LAB.
MIT-RBR-
    MASSACHUSETTS INST. OF TECH., CAMBRIDGE
MITRE-
    MITRE CORP., BEDFORD, MASS.
MITRE-
    MITRE CORP., LEXINGTON, MASS.
MITRE SR-
    MITRE CORP., LEXINGTON, MASS.
MITRE-TM-
    MITRE CORP., BEDFORD, MASS.
MITRE-W-
    MITRE CORP., BEDFORD, MASS.
MIT-RHFS-
    MASSACHUSETTS INST. OF TECH., SOUTH DARTMOUTH. ROUND HILL FIELD STATION
MIT-RL-
    MASSACHUSETTS INST. OF TECH.,CAMBRIDGE.RADIATION LAB.
MIT-RLE-
    MASSACHUSETTS INST. OF TECH., CAMBRIDGE. RESEARCH LAB. OF ELECTRONICS
MIT-RLE-PTR-
    MASSACHUSETTS INST. OF TECH., CAMBRIDGE. RESEARCH LAB. OF ELECTRONICS (PRELIMINARY TECHNICAL REPORT)
MIT-RLE-QPR-
    MASSACHUSETTS INST. OF TECH., CAMBRIDGE. RESEARCH LAB. OF ELECTRONICS (QUARTERLY PROGRESS REPORT)
MIT RLE TR-
    MASSACHUSETTS INST. OF TECH., CAMBRIDGE. RESEARCH LAB. OF ELECTRONICS
MIT RLMM-
    MASSACHUSETTS INST. OF TECH., CAMBRIDGE. RESEARCH LABORATORY FOR MECHANICS OF MATERIALS
MIT-RL-R-
    MASS. INST. OF TECH.,CAMBRIDGE. RADIATION LAB.(REPT.)
MIT-RM-
    MASSACHUSETTS INST. OF TECH., CAMBRIDGE. DYNAMIC ANALYSIS AND CONTROL LAB.
MIT RMP APR-
    MASSACHUSETTS INST. OF TECH.,CAMBRIDGE. RADIOACTIVITY CENTER (ANNUAL PROGRESS REPORTS)
MIT-RMT-
    MASS. INST. OF TECH., CAMBRIDGE. METALLURGICAL PROJ.
MIT-RR/R(YEAR)-
    MASSACHUSETTS INST. OF TECH., CAMBRIDGE. DEPT. OF CIVIL AND SANITARY ENGINEERING
MIT-S-
    MASSACHUSETTS INST. OF TECH., CAMBRIDGE
MITS-
    MASS. INST. OF TECH., CAMBRIDGE. DEPT. OF METALLURGY
MITS-
    MASSACHUSETTS INST. OF TECH., CAMBRIDGE. RICHARDS MINERAL ENGINEERING LAB.
MIT SL M-
    MASSACHUSETTS INST. OF TECH., CAMBRIDGE. SERVOMECHANISMS LAB. (MEMORANDUM)
MIT SL R-
    MASS. INST. OF TECH., CAMBRIDGE. SERVOMECHANISMS LAB.
MIT-SL-TR-
    MASS. INST. OF TECH., CAMBRIDGE. SPECTROSCOPY LAB.
MIT-SMI-
    MASSACHUSETTS INST. OF TECH., CAMBRIDGE. SOLID-STATE AND MOLECULAR THEORY GROUP
MIT-SMTG-
    MASSACHUSETTS INST. OF TECH., CAMBRIDGE. SOLID-STATE AND MOLECULAR THEORY GROUP
MIT-SMTG-QPR-
    MASSACHUSETTS INST. OF TECH., CAMBRIDGE. SOLID-STATE AND MOLECULAR THEORY GROUP (QUARTERLY PROGRESS RPT)
MIT-SMTG-QR-
    MASSACHUSETTS INST. OF TECH., CAMBRIDGE. SOLID-STATE AND MOLECULAR THEORY GROUP (QUARTERLY PROGRESS RPT)
MIT-SMTG-TR-
    MASSACHUSETTS INST. OF TECH., CAMBRIDGE. SOLID-STATE AND MOLECULAR THEORY GROUP (TECHNICAL REPORT)
MIT-SPL-QTPR-
    MASS. INST. OF TECH., CAMBRIDGE. SPACE PROPULSION LAB
MIT SP TR-
    MASS. INST. OF TECH., CAMBRIDGE. SPECTROSCOPY LAB.
MIT-SSMTG-
    MASSACHUSETTS INST. OF TECH., CAMBRIDGE. SOLID-STATE AND MOLECULAR THEORY GROUP
MIT-SSMTG-PN-
    MASSACHUSETTS INST. OF TECH., CAMBRIDGE. SOLID-STATE AND MOLECULAR THEORY GROUP (PROGRAMMING NOTE)
MIT-SVA-
    MASS. INST. OF TECH., CAMBRIDGE. METALLURGICAL PROJ.
MITT-
    LUDWIG-MAXIMILIANS-UNIVERSITAET, MUNICH. METEOROLOGISCHES INST.
MIT-T-
    MASSACHUSETTS INST. OF TECH., CAMBRIDGE
MIT-TR-
    MASSACHUSETTS INST. OF TECH., CAMBRIDGE
MIT-TTM-
    MASS. INST. OF TECH., CAMBRIDGE. METALLURGICAL PROJ.
MIT-U-
    MASSACHUSETTS INST. OF TECH., CAMBRIDGE
MIT-WCP-
    NUCLEAR METALS, INC., CAMBRIDGE, MASS.
MIT-WTR-
    MASS. INST. OF TECH., CAMBRIDGE. NAVAL SUPERSONIC LAB

MITX-	MASSACHUSETTS INST. OF TECH., CAMBRIDGE
MJ-	WASTE KING CORP., LOS ANGELES
MJL-	MARQUARDT AIRCRAFT CO., VAN NUYS, CALIF.
MJO-	BISSETT-BERMAN CORP., SAN DIEGO, CALIF.
MJS-	PRATT AND WHITNEY AIRCRAFT DIV., UNITED AIRCRAFT CORP., MIDDLETOWN, CONN.
MK-	NATIONAL RESEARCH COUNCIL OF CANADA, OTTAWA
MK-	NATIONAL RESEARCH COUNCIL OF CANADA. DIVISION OF MECHANICAL ENGINEERING, OTTAWA
MKC-	MARKITE CO., N.Y.C.
ML-	AIR FORCE MATERIALS LAB, WRIGHT-PATTERSON AFB, OHIO
ML-	CALIFORNIA. UNIV., BERKELEY. MICROWAVE LAB.
ML-(2 DIGITS)-	HUGHES AIRCRAFT CO. MICROWAVE LAB., CULVER CITY, CALIF.
ML(1 DIGIT)-(LETTERS)-	HUGHES AIRCRAFT CO. MICROWAVE LAB., CULVER CITY, CALIF.
ML-	MIAMI. UNIV., CORAL GABLES, FLA. MARINE LAB.
ML-	MIAMI. UNIV., FLA. RADAR METEOROLOGICAL LAB.
ML-	NAVY UNDERWATER SOUND REFERENCE LAB., MT. LAKES, N.J.
ML-	STANFORD UNIV., CALIF. MICROWAVE LAB.
ML-	TENNESSEE EASTMAN CORP., OAK RIDGE, TENN.
MLCR-	CHEMICAL CORPS MEDICAL LABS., ARMY CHEMICAL CENTER, MD. (CONTRACT REPORT)
MLH-	RESEARCH AND DEVELOPMENT BOARD. PANEL ON LAUNCHING AND HANDLING, WASHINGTON, D.C.
ML-HEPL-	STANFORD UNIV., CALIF. HIGH ENERGY PHYSICS LAB.
MLI-	MALAKER LABS., INC., HIGH BRIDGE, N.J.
MLI-	MAXWELL LABS., INC., SAN DIEGO, CALIF.
MLI-PR-	MANUFACTURING LABS., INC. PHYSICAL METALLURGY DIV., CAMBRIDGE, MASS. (PROGRESS REPORT)
MLI-TR-	MANUFACTURING LABS., INC. PHYSICAL METALLURGY DIV., CAMBRIDGE, MASS. (TECHNICAL REPORT)
ML-LN-	AIR FORCE MATERIALS LAB., WRIGHT-PATTERSON AFB, OHIO
ML-LN-(NUMBER)-	LOWRY AFB MAINTENANCE LAB., COLO.
ML-LOWRY-TM-(YEAR)-	AIR FORCE MAINTENANCE LAB., LOWRY AFB, COLO.
MLM-	MOUND LAB., MIAMISBURG, OHIO
MLM-BC-	MOUND LAB., MIAMISBURG, OHIO
MLM-CF-(YR)-(MO)-	MOUND LAB., MIAMISBURG, OHIO
MLM-M-	MONSANTO CHEMICAL CO., DAYTON, OHIO
MLM-M-	MOUND LAB., MIAMISBURG, OHIO
MLM-OP-	MOUND LAB., MIAMISBURG, OHIO
ML/MPC-	UNITED KINGDOM ATOMIC ENERGY AUTHORITY. RESEARCH GP. ATOMIC ENERGY RES. ESTAB., HARWELL, BERKS, ENGLAND
MLM-REPRINT-	MOUND LAB., MIAMISBURG, OHIO (REPRINT)
MLNS SP-	GT. BRIT. MINISTRY OF LABOUR AND NATIONAL SERVICE. FACTORY DEPARTMENT (SAFETY PAMPHLET)
MLO-(YEAR)-	RADIO CORP. OF AMERICA. WEST COAST DIV., VAN NUYS, CAL.
ML(PR)-	HUGHES AIRCRAFT CO. MICROWAVE LAB., CULVER CITY, CALIF.
MLR-	CHEMICAL CORPS MEDICAL LABS., ARMY CHEMICAL CTR., MD.
MLR-	CHICAGO. UNIV. METALLURGICAL LAB.
MLR-	MAXWELL LABS., INC., SAN DIEGO, CALIF.
MLR-(NO.)-(YEAR)-	NEW YORK UNIV., N.Y.C. DEPT. OF CHEMICAL ENGINEERING
MLR-	WESTINGHOUSE ELECTRIC CORP., PITTSBURGH
MLRA-	LIBRARY OF CONGRESS, WASHINGTON, D.C. (MONTHLY LIST OF RUSSIAN ACCESSIONS)
MLRR-	CHEMICAL CORPS MEDICAL LABS., ARMY CHEMICAL CENTER, MD. (RESEARCH REPORT)
MLRR-AR-	AEROJET-GENERAL CORP. LIQUID ROCKET OPERATIONS, SACRAMENTO, CALIF. (MAINTENANCE, LOGISTICS, RELIABILITY AND READINESS)
MLSR-	CHEMICAL CORPS MEDICAL LABS., ARMY CHEMICAL CENTER, MD. (SPECIAL REPORT)
MLT-	BUREAU OF SHIPS. MATERIALS LAB. (TESTS)
ML-TDR-	AIR FORCE MATERIALS LAB., WRIGHT-PATTERSON AFB, OHIO
ML-TDR-	WRIGHT AIR DEVELOPMENT CENTER. MATERIALS LAB., WRIGHT-PATTERSON AFB, OHIO
ML-TM-	AIR FORCE MATERIALS LAB., WRIGHT-PATTERSON AFB, OHIO
ML-TM-(NUMBER)-	LOWRY AFB MAINTENANCE LAB., COLO.
ML/TN-	MARSHALL LABS., TORRANCE, CALIF.
ML-TR-	AIR FORCE MATERIALS LAB., WRIGHT-PATTERSON AFB, OHIO
MM-	(MAINTENANCE MANUAL) (REF. 2)
MM-	AEROJET-GENERAL CORP., SACRAMENTO, CALIF.
MM-(YR)-	BELL TELEPHONE LABS., INC., N.Y.C.
MM(NO.-NO.-NO.)(M)	BELL TELEPHONE LABS., INC., WHIPPANY, N.J.
MM-	BENDIX CORP. BENDIX PRODUCTS DIV., SOUTH BEND, IND.
MM-	CALIFORNIA. UNIV., BERKELEY. RADIATION LAB.
MM-	FIELD INTELLIGENCE AGENCY, TECHNICAL
MM-	GEORGE C. MARSHALL SPACE FLIGHT CENTER, HUNTSVILLE, ALA. (MARSHALL MEMO)
MM-	HUMAN RESOURCES RESEARCH INSTITUTE, MAXWELL AFB, ALA.
MM-	TENNESSEE EASTMAN CORP., OAK RIDGE, TENN.
M/M-	UNITED KINGDOM ATOMIC ENERGY AUTHORITY. RESEARCH GP. ATOMIC ENERGY RES. ESTAB., HARWELL, BERKS, ENGLAND
MMAB-(NO.)-C	NATL. RES. COUNCIL. MINERALS + METALS ADVISORY BD.
MMAB-(NO.)-M	NATL. RES. COUNCIL. MINERALS + METALS ADVISORY BD.
MMC-	MARTIN-MARIETTA CORP., BALTIMORE
MMC-AR-(YEAR)	MARTIN-MARIETTA CORP., BALTIMORE
MMC-RM-	MARTIN-MARIETTA CORP., BALTIMORE
MME-(YEAR)-	PITTSBURGH. UNIV. METALLURGICAL AND MATERIALS ENG.
(YEAR)-MMEST-	OGDEN AIR MATERIEL AREA. SERVICE ENGINEERING DIV., HILL AFB, UTAH
MML-	DEPARTMENT OF DEFENSE, WASHINGTON, D.C. (GUIDED MISSILE MAILING LIST)
MML-	RES. + DEV. BD. COMM. ON GUIDED MISSILES, WASH., DC
MML-	TECHNION-ISRAEL INST. OF TECH., HAIFA. MATERIAL MECHANICS LAB.
MMM-(NUMBERS)-	GENERAL ELECTRIC CO., PHILADELPHIA
MMM/	MINNESOTA MINING AND MFG. CO., ST. PAUL
MMM-	MINNESOTA MINING AND MFG. CO., ST. PAUL
MMM-	MINNESOTA MINING AND MFG. CO. ISOTOPE POWER LAB., ST. PAUL
MMM-(NUMBER-NUMBER)	MINNESOTA MINING AND MFG. CO. ISOTOPE POWER LAB., ST. PAUL
MMMC-	MINNESOTA MINING AND MFG. CO., ST. PAUL
MMMC-	MINN. MINING + MFG. CO. CENTRAL RES. DEPT., ST. PAUL
MMMC-QPR-	MINNESOTA MINING AND MFG. CO. CENTRAL RESEARCH DEPT., ST. PAUL (QUARTERLY PROGRESS REPORT)
MMMC-RR-	MINN. MINING + MFG. CO. CENTRAL RES. LABS., ST. PAUL
MMPP-	MICHIGAN. UNIV., ANN ARBOR. DEPT. OF NUCLEAR ENG.
MMPP-	MICHIGAN. UNIV., ANN ARBOR. MICHIGAN MEMORIAL-PHOENIX PROJECT
MMPP-	VIRGINIA. UNIV., CHARLOTTESVILLE
MMPP-FRPC-	MASSACHUSETTS INST. OF TECH., CAMBRIDGE. DEPT. OF NUCLEAR ENGINEERING
MMPP-FRPC-	MICHIGAN. UNIV., ANN ARBOR. DEPT. OF NUCLEAR ENG.
MMPP-NPC-	MICHIGAN. UNIV., ANN ARBOR. MICHIGAN MEMORIAL-PHOENIX PROJECT
MMPP-S-	MICHIGAN. UNIV., ANN ARBOR. MICHIGAN MEMORIAL-PHOENIX PROJECT
MMPR-	NORTHROP AIRCRAFT, INC., HAWTHORNE, CALIF. (MONTHLY MISSILE PROGRESS REPORT)
MMR-	AEROJET-GENERAL CORP. SOLID ROCKET PLANT, SACRAMENTO, CALIF. (MINUTEMAN PROGRAM)
MMR-	NETHERLANDS. MINISTRY OF DEFENCE. MATERIEL COUNCIL, THE HAGUE

MMR-(LETTER(S))-
: AEROJET-GENERAL CORP. SOLID ROCKET PLANT, SACRAMENTO, CALIF. (MINUTEMAN PROGRAM)
MMS-
: MC DONNELL AIRCRAFT CORP., ST. LOUIS (MATERIAL SPECS)
MM-TR-
: OGDEN AIR MATERIEL AREA. DIRECTORATE OF MATERIEL MANAGEMENT, HILL AFB, UTAH
MMUM-
: NAVAL ORDNANCE LAB. MINE MODIFICATION UNIT (MEMO.)
MMUR-
: NAVAL ORDNANCE LAB. MINE MODIFICATION UNIT (REPORT)
MN-
: AEROJET-GENERAL CORP., SACRAMENTO, CALIF.
MN-
: COMMITTEE ON MEDICAL RES. (MEDICAL NUTRITION RPTS.)
MN-
: MARKITE CO., N.Y.C.
MN-
: MARTIN-MARIETTA CORP. AEROSPACE DIV., BALTIMORE
MNC-
: MARYLAND. UNIV., COLLEGE PARK. DEPT. OF CHEMISTRY
MND-
: MARTIN CO. NUCLEAR DIV., BALTIMORE
MND-
: MARTIN-MARIETTA CORP. AEROSPACE DIV., BALTIMORE
MND-(NUMBER)-
: MARTIN-MARIETTA CORP. NUCLEAR DIV., BALTIMORE
MND-(LETTER)-
: MARTIN-MARIETTA CORP. NUCLEAR DIV., BALTIMORE
MND-AC-
: MARTIN CO. NUCLEAR DIV., BALTIMORE
MND-ANP-
: MARTIN CO. NUCLEAR DIV., BALTIMORE
MND-C-
: MARTIN CO. NUCLEAR DIV., BALTIMORE
MND-DB-
: MARTIN CO. NUCLEAR DIV., BALTIMORE
MND-E-
: MARTIN CO. NUCLEAR DIV., BALTIMORE
MND-FBR-
: MARTIN CO. NUCLEAR DIV., BALTIMORE (FLUIDIZED BED REACTOR)
MND-LFBR-
: MARTIN CO. NUCLEAR DIV., BALTIMORE (LIQUID FLUIDIZED BED REACTOR)
MND-LIB-
: MARTIN CO., BALTIMORE
MND-M-
: MARTIN CO. NUCLEAR DIV., BALTIMORE
MND-M(NUMBER)(LETTER)-
: MARTIN CO. NUCLEAR DIV., BALTIMORE
MND-MC-
: MARTIN-MARIETTA CORP. NUCLEAR DIV., BALTIMORE
MND-MD-
: MARTIN CO. NUCLEAR DIV., BALTIMORE
MND-MPR-
: MARTIN CO. NUCLEAR DIV., BALTIMORE
MND-P-
: MARTIN CO. NUCLEAR DIV., BALTIMORE
MND-R-
: MARTIN CO. NUCLEAR DIV., BALTIMORE (RADIOISOTOPE HEATER)
MND-ROC-
: MARTIN CO. NUCLEAR DIV., BALTIMORE
MND-RP-
: MARTIN CO. NUCLEAR DIV., BALTIMORE
MND-SF-
: MARTIN CO. NUCLEAR DIV., BALTIMORE
MND-SR-
: MARTIN CO. NUCLEAR DIV., BALTIMORE
MNI-(NUMBER)-
: MARTIN CO., DENVER
MNL-
: ARMY MEDICAL NUTRITION LAB., DENVER.
MN-SW-
: MARTIN CO. NUCLEAR DIV., BALTIMORE
MN-TR-
: MATH AND METRIK, INC., CHESHIRE, CONN.
MNU-
: MINNESOTA. UNIV., MINNEAPOLIS
MO-
: COAST AND GEODETIC SURVEY, ROCKVILLE, MD. (MAGNETIC OBSERVATIONS)
MO-
: RADIO CORP. OF AMERICA. MISSILE AND SURFACE RADAR DIV., MOORESTOWN, N.J.
MO596R-S-
: RADIO CORP. OF AMERICA. MISSILE AND SURFACE RADAR DEPT., MOORESTOWN, N.J.
MO-
: UNITED KINGDOM. AIR MINISTRY. METEOROLOGICAL OFFICE, LONDON
M.O.-
: UNITED KINGDOM. AIR MINISTRY. METEOROLOGICAL OFFICE, LONDON
MO-(2 DIGITS)-
: UPPSALA. UNIV. COSMIC RAY GROUP
MOA-
: UNITED KINGDOM. MINISTRY OF AVIATION, LONDON
MO-BB-S-
: RADIO CORP. OF AMERICA. MISSILE AND SURFACE RADAR DIV., MOORESTOWN, N.J.
MOBBS-
: RADIO CORP. OF AMERICA. MISSILE AND SURFACE RADAR DIV., MOORESTOWN, N.J.
MOC-D-
: MYSTIC OCEANOGRAPHIC CO., CONN.
MOC-S-
: MYSTIC OCEANOGRAPHIC CO., CONN.

(MODEL NO.)(SUBJ.LTR.)-
: FAIRCHILD ENGINE AND AIRPLANE CORP. PILOTLESS PLANE DIV., FARMINGDALE, N.Y. (REF. 1)
(MODEL NUMBER)-(LETTER)-
: CURTISS-WRIGHT CORP. AIRPLANE DIV., COLUMBUS, OHIO (REF. 1)
MOEES-(YEAR)-
: RADIO CORP. OF AMERICA, MOORESTOWN, N.J.
MOEES-(YEAR)-
: RADIO CORP. OF AMERICA. SYSTEMS ENGINEERING EVALUATION AND RESEARCH, MOORESTOWN, N.J.
MOL-
: MARTIN CO., DENVER
MOLEC-
: MOLECULON RESEARCH CORP., CAMBRIDGE, MASS.
MOM-
: UNITED KINGDOM. AIR MINISTRY. METEOROLOGICAL OFFICE, LONDON
MON-
: CLINTON LABS., OAK RIDGE, TENN.
MON-
: EVANS SIGNAL LAB., BELMAR, N.J.
MON-(LETTER)-
: CLINTON LABS., OAK RIDGE, TENN.
MONO-
: FLORIDA. UNIV., GAINESVILLE. REGIONAL REHABILITATION RESEARCH INST.
MONO-
: NAVAL SCHOOL OF AVIATION MEDICINE, PENSACOLA, FLA.
MONO-D-
: OHIO STATE UNIV., COLUMBUS. DISASTER RESEARCH CENTER
MONSANTO-
: MONSANTO CHEMICAL CO. (ALL LOCATIONS)
MONSANTO-200B-QR-
: MONSANTO CHEMICAL CO., EVERETT, MASS.
MONSANTO-QR-
: MONSANTO CHEMICAL CO., EVERETT, MASS.
MONSANTO-TB-PL-
: MONSANTO CHEMICAL CO. ORGANIC CHEM. DIV., ST. LOUIS
MONSANTO-TP-AT-
: MONSANTO CHEMICAL CO. ORGANIC CHEM. DIV., ST. LOUIS
MONTU-
: MONTREAL. UNIV.
MOOR-
: AMERICAN POWER JET CO., MONTCLAIR, N.J.
MOP-
: HUGHES AIRCRAFT CO. SPACE SYSTEMS DIV., EL SEGUNDO, CAL (MISSION OPERATIONS)
MOR-
: (MONTHLY OPERATIONS REPORT) (REF. 2)
MOR-
: GT. BRIT. MILITARY OPERATIONAL RESEARCH
MORI-
: NAVY (MINE OVERHAUL AND REWORK INSTRUCTIONS)
MORP-(YEAR)-
: NATIONAL CENTER FOR RADIOLOGICAL HEALTH, ROCKVILLE, MD
MOS-
: GT. BRIT. MINISTRY OF SUPPLY
MOS-
: GT. BRIT. ROAD RES. LAB., HARMONDSWORTH, MIDDX., ENG.
MO-S-
: RADIO CORP. OF AMERICA. DEFENSE ELECTRONIC PRODUCTS, N.Y.C.
MO-S-
: RADIO CORP. OF AMERICA. MISSILE AND SURFACE RADAR DIV., MOORESTOWN, N.J.
MOS-MONO-(NUMBER)-(NO.)
: GT. BRIT. MINISTRY OF SUPPLY
MOS-TIB GDC-1-
: GT. BRIT. MINISTRY OF SUPPLY. TECHNICAL INFO. BUR.
MOT-
: MOTOROLA INC., PHOENIX, ARIZ.
MOTO-
: MOTOROLA INC., RIVERSIDE, CALIF.
MOTO PR-
: MOTOROLA INC., RIVERSIDE, CALIF.
MOTO RL-
: MOTOROLA INC., RIVERSIDE, CALIF.
MOTO-RL-
: MOTOROLA INC. RIVERSIDE RESEARCH LAB., CALIF.
MOT TR-RL-
: MOTOROLA INC. RIVERSIDE RESEARCH LAB., CALIF.
MOU-IS-
: MISSOURI. UNIV., COLUMBIA. INFORMATION DEPT.
MOU MRG TN-
: MISSOURI. UNIV., COLUMBIA. DEPT. OF MATHEMATICS (MATHEMATICS RESEARCH PROJECT)
MOW-
: GT. BRIT. MINISTRY OF WORKS
MOW (LETTER(S))-
: GT. BRIT. MINISTRY OF WORKS
MOW AL-
: GT. BRIT. MINISTRY OF WORKS (LEAFLET)
MOW EM-
: GT. BRIT. MINISTRY OF WORKS (ECONOMY MEMORANDUM)
MOW NB B-
: GT. BRIT. DEPT. OF SCIENTIFIC AND INDUSTRIAL RESEARCH (NATIONAL BUILDING STUDIES. BULLETIN)
MOW PWB S-
: GT. BRIT. MINISTRY OF WORKS (POSTWAR BLDG. STUDIES)
MP-
: (MISCELLANEOUS PAPER) (REF. 2)
MP-
: (MISCELLANEOUS PUBLICATION) (REF. 2)
MP-
: ARMY ENGINEER WATERWAYS EXPERIMENT STATION, VICKSBURG, MISS.
MP-
: CHRYSLER CORP. MISSILE OPERATIONS

M/P-	CRANFIELD INST. OF TECH., BUCKS, ENGLAND
MP-	DOUGLAS AIRCRAFT CO., INC. MISSILE + SPACE SYSTEMS DIV., NEWPORT BEACH, CALIF.
MP-(YEAR)-	GENERAL DYNAMICS/CONVAIR, SAN DIEGO, CALIF.
MP-	LOS ALAMOS SCIENTIFIC LAB., N. MEX.
MP-1- THRU MP-9-	LOS ALAMOS SCIENTIFIC LAB., N. MEX. (INTERNAL CORRESPONDENCE SERIAL USED BY GROUPS IN MP DIV.)
MP-	MARQUARDT CORP., VAN NUYS, CALIF.
MP-	MEDICAL PHYSICS TRANSLATION FUND, PARK FOREST, ILL.
MP-	NATIONAL RESEARCH COUNCIL OF CANADA. DIVISION OF MECHANICAL ENGINEERING, OTTAWA
MP-	NETHERLANDS. NATIONAAL LUCHT EN RUIMTEVAART-LABORATORIUM, AMSTERDAM
MP-	OFFICE OF CIVIL DEFENSE, WASHINGTON, D.C.
MP-(YEAR)-	UNITED STATES LAKE SURVEY, DETROIT
MP(LETTER(S))-	NATIONAL RES.COUNCIL OF CANADA. MONTREAL LAB.(REF.29)
MPC-	COMBUSTION ENG., INC. NUCLEAR DIV., WINDSOR, CONN.
MPC-	MASSACHUSETTS INST. OF TECH., CAMBRIDGE
MPC-FR-	MARKS POLARIZED CORP., WHITESTONE, N.Y.
MPC-QPR-	MARKS POLARIZED CORP., WHITESTONE, N.Y.
MPD-	GENERAL DYNAMICS/CONVAIR, SAN DIEGO, CALIF.
MPD-	TEXAS. UNIV., AUSTIN. MILITARY PHYSICS DIV.
MPE-	NATIONAL SCIENCE FOUNDATION, WASHINGTON, D.C.(MATHE-MATICAL, PHYSICAL AND ENGINEERING SCIENCES)
MPF-	RES. + DEV. BD. COMM. ON GUIDED MISSILES, WASH., DC
MPF-	RES. + DEV. BD. PANEL ON PROPULSION + FUELS, WASH.,DC
MPG-(YEAR)-	CALIFORNIA. UNIV., LOS ANGELES
MPG-	COLUMBIA UNIV., N.Y.C. MATHEMATICAL PHYSICS GROUP
MPI-	MAX-PLANCK-INSTITUT FUER PHYSIK UND ASTROPHYSIK, MUNICH
MPI-	SWEDEN. MILITARPSYKOLOGISK INSTITUTET, STOCKHOLM
MPIH-(YEAR)/(NO.)/(NO.)	MAX-PLANCK-INSTITUT FUER KERNPHYSIK, HEIDELBERG
MP/IN-	AUSTRALIAN ATOMIC ENERGY COMMISSION. RESEARCH ESTABLISHMENT, LUCAS HEIGHTS, NEW SOUTH WALES
MPI-PA-	MAX-PLANCK-INSTITUT FUER PHYSIK UND ASTROPHYSIK, MUNICH
MPI-PAE/ASTRO-	MAX-PLANCK-INSTITUT FUER PHYSIK UND ASTROPHYSIK, MUNICH
MPI-PAE-EXTRATERP-	MAX-PLANCK-INSTITUT FUER PHYSIK UND ASTROPHYSIK, MUNICH
MPI-PAE/EXTRATERR.-	MAX-PLANCK-INSTITUT FUER PHYSIK UND ASTROPHYSIK, MUNICH
MPI-PAE/P1-(NO.)/(YR.)	MAX-PLANCK-INSTITUT FUER PHYSIK UND ASTROPHYSIK, MUNICH
MPI-PAE/PL-	MAX-PLANCK-INSTITUT FUER PHYSIK UND ASTROPHYSIK, MUNICH
MPL-	AEROJET-GENERAL CORP., SACRAMENTO, CALIF.
MPL-	GENERAL APPLIED SCIENCE LABS., INC., WESTBURY, N.Y.
MPL-	MICROWAVE PHYSICS LAB., MOUNTAIN VIEW, CALIF.
MPL-	SYLVANIA ELECTRIC PRODUCTS INC. MICROWAVE PHYSICS LAB., MOUNTAIN VIEW, CALIF.
MPL-	YORK UNIV., TORONTO. MOLECULAR PSYCHOBIOLOGY LAB.
MPL-C-	SCRIPPS INSTITUTION OF OCEANOGRAPHY. MARINE PHYSICAL LAB., SAN DIEGO, CALIF.
MPL-M-	SYLVANIA ELECTRIC PRODUCTS INC. MICROWAVE PHYSICS LAB., MOUNTAIN VIEW, CALIF.
MPLR-	GEOPHYSICS CORP. OF AMERICA, BEDFORD, MASS.
MPL-S-	SCRIPPS INSTITUTION OF OCEANOGRAPHY. MARINE PHYSICAL LAB., SAN DIEGO, CALIF.
MPL-TM-	SCRIPPS INSTITUTION OF OCEANOGRAPHY. MARINE PHYSICAL LAB., SAN DIEGO, CALIF.
MPL-U-	CALIFORNIA. UNIV., LA JOLLA
MPL-U-	SCRIPPS INSTITUTION OF OCEANOGRAPHY. MARINE PHYSICAL LAB., SAN DIEGO, CALIF.
MPO-	RAYTHEON CO. MICROWAVE POWER OPERATION, BURLINGTON, MASS.
MP.OO-	CURTISS-WRIGHT CORP. WRIGHT AERO. DIV., WOOD-RIDGE,NJ
MPOP-	OFFICE OF THE CHIEF OF NAVAL OPERATIONS, WASHINGTON, D.C. (MASTER PROGRAM OBJECTIVES PLAN)
MPO-PD-	SYLVANIA ELECTRONIC SYSTEMS-EAST, NEEDHAM, MASS.
MPO-SR-(NUMBER)-	SYLVANIA ELECTRIC PRODUCTS, INC. PRODUCT SUPPORT ORGANIZATION, WEST ROXBURY, MASS.
MPO-SR-	SYLVANIA ELECTRONIC SYSTEMS-EAST, NEEDHAM, MASS.
MPP-	ALBUQUERQUE OPERATIONS OFFICE, AEC, N. MEX.
MP PROJ.-	NATIONAL RESEARCH COUNCIL OF CANADA. ATOMIC ENERGY PROJECT, CHALK RIVER, ONT.
MPR-	(MINE PRODUCTION REPORT) (REF. 2)
MPR-	(MONTHLY PROGRESS REPORT) (REF. 2)
MPR-	(MONTHLY PROJECT REPORT) (REF. 2)
MPR-	AEROJET-GENERAL CORP., SACRAMENTO, CALIF.
MPR-	AERONUTRONIC DIV. OF PHILCO CORP., NEWPORT BEACH, CALIF. (MONTHLY PROGRESS REPORT)
MPR-	AIRTECHNOLOGY CORP., CAMBRIDGE, MASS.
MPR-(YEAR)-(MONTH)	AVCO CORP. AVCO-EVERETT RESEARCH LAB., EVERETT, MASS.
MPR(YR)-(MO)	AVCO MFG. CORP. ADVANCED DEV. DIV., STRATFORD, CONN.
MPR(YR)-(MO)	AVCO MFG. CORP. RESEARCH AND ADVANCED DEVELOPMENT DIV., LAWRENCE, MASS.
MPR-(ROMAN NO.)-	COLUMBIA UNIV., N.Y.C. ENGINEERING RESEARCH LABS. (MONTHLY PROGRESS REPORT)
MPR-	DATA CORP., DAYTON, OHIO
MPR-	FAIRCHILD CAMERA AND INSTRUMENT CORP. RECONNAISSANCE SYSTEMS DIV., SYOSSET, N.Y.
MPR-	HEXCEL PRODUCTS, INC., OAKLAND, CALIF.
MPR-	LING-TEMCO-VOUGHT, INC., DALLAS
MPR-	LOCKHEED-CALIFORNIA CO., BURBANK
MPR-	MINNESOTA MINING AND MFG. CO., ST. PAUL
MPR-	NAVAL MINE DEPOT, YORKTOWN, VA.(MINE PRODUCTION RPT.)
MPR-	SPERRY RAND CORP., ST. PAUL
MPR-	TENNESSEE EASTMAN CORP., OAK RIDGE, TENN.
MPR-196-	TENNESSEE EASTMAN CORP., OAK RIDGE, TENN.
MPR-	UTAH STATE UNIV., LOGAN
MPRC-	GT. BRIT. MILITARY PERSONNEL RESEARCH COMMITTEE
MPR-CDC-	CLINTON LABS., OAK RIDGE, TENN.
MPRL-	TEXAS. UNIV., AUSTIN. MILITARY PHYSICS RES. LAB.
MPR-SAT-FE-	GEORGE C. MARSHALL SPACE FLIGHT CTR.,HUNTSVILLE, ALA.
MPS-	CORPS OF ENGINEERS (ARMY). ENGINEER BOARD. MAP PROJECTION SECTION
MPS/DL-INT-(YR.)-	EUROPEAN ORGANIZATION FOR NUCLEAR RESEARCH, GENEVA
MPS/EP-	EUROPEAN ORGANIZATION FOR NUCLEAR RESEARCH, GENEVA
MPS/INT/CO/	EUROPEAN ORGANIZATION FOR NUCLEAR RESEARCH, GENEVA
MPS/INT.DL/B-(YR)-	EUROPEAN ORGANIZATION FOR NUCLEAR RESEARCH, GENEVA
MPS-INT.LIN-	EUROPEAN ORGANIZATION FOR NUCLEAR RESEARCH, GENEVA
MPS-INT.MA-	EUROPEAN ORGANIZATION FOR NUCLEAR RESEARCH, GENEVA
MPS/INT.MU/EP-	EUROPEAN ORGANIZATION FOR NUCLEAR RESEARCH, GENEVA
MPS/SI/INT.DL-	EUROPEAN ORGANIZATION FOR NUCLEAR RESEARCH, GENEVA
M/PZ/	BRITISH STEEL CORP. MIDLAND GROUP,MOORGATE,YORKS,ENG.
MR-	(MEMORANDUM REPORT) (REF. 2)
MR-	(MISCELLANEOUS REPORT) (REF. 2)
MR-	(MONTHLY REPORT) (REF. 2)
MR-	(MONTHLY REVIEW) (REF. 2)
MR-	AEROJET-GENERAL CORP., SACRAMENTO, CALIF.
MR-	AIR FORCE LOGISTICS COMMAND. LOGISTICS SIMULATION CENTER, WRIGHT-PATTERSON AFB, OHIO
MR-	AMERICAN MACHINE AND FOUNDRY CO., CHICAGO

MR-

MR-	BEECH AIRCRAFT CORP., WICHITA, KAN.
MR-(YEAR)-	BUNKER-RAMO CORP. HUMAN ENG. GP., CANOGA PARK, CALIF.
MR-	CALIFORNIA. UNIV., LOS ANGELES
MR-	COMMITTEE ON MEDICAL RESEARCH (MALARIA REPORTS)
MR-	FRANKFORD ARSENAL. PITMAN-DUNN LAB., PHILADELPHIA
MR-	FRANKFORD ARSENAL. PROPELLANT ACTUATED DEVICES DIV., PHILADELPHIA
MR-	FRANKLIN INST., PHILADELPHIA (MONTHLY REVIEW)
MR-(PROJECT)-	FRANKLIN INST. LABS. FOR RES. + DEV., PHILADELPHIA
MR-	GENERAL AMERICAN TRANSPORT. CORP. MRD DIV., NILES, ILL
MR-	GENERAL MOTORS CORP., DETROIT
MR-	MALVESTUTO AERO SPACE CORP., OXNARD, CALIF.
MR-	MARQUARDT CORP. NEWPORT BEACH LAB., CALIF.
MR-	MARQUARDT CORP., VAN NUYS, CALIF.
MR-	MATHEMATICAL APPLICATIONS GROUP, INC., WHITE PLAINS, NY
MR(NO.)-FR-	METEOROLOGY RESEARCH, INC., ALTADENA, CALIF.
MR-	MONTREAL. UNIV.
MR-	NATIONAL BUREAU OF STANDARDS. BOULDER LABS., COLO.
MR-	NATIONAL RESEARCH COUNCIL OF CANADA. ATOMIC ENERGY PROJECT, CHALK RIVER, ONT.
MR-	NAVAL AIR DEVELOPMENT CENTER. AEROSPACE MEDICAL RESEARCH DEPT., JOHNSVILLE, PENNA.
MR-	NAVAL MEDICAL RESEARCH LAB., NEW LONDON, CONN.
M.R.-(NUMBER/YEAR)	NAVAL OCEANOGRAPHIC OFFICE, WASHINGTON, D.C. (MANUSCRIPT REPORT)
MR-	NAVAL POWDER FACTORY. RES.+DEV. DEPT., INDIAN HEAD, MD.
MR-	NAVAL PROPELLANT PLANT. RESEARCH AND DEVELOPMENT DEPT., INDIAN HEAD, MD.
MR-	NAVAL SUBMARINE MEDICAL CENTER, GROTON, CONN.
MR-(YEAR)-	NAVAL SUBMARINE MEDICAL CENTER, GROTON, CONN.
MR-	NETHERLANDS. MATHEMATISCH CENTRUM, AMSTERDAM
MR-	OXFORD UNIV.
MR-	PICATINNY ARSENAL, DOVER, N.J.
MR-	PREFORMED LINE PRODUCTS COMPANY. RESEARCH AND ENGINEERING, CLEVELAND (MEMORANDUM REPORT)
MR-(NO.)-E-	PREFORMED LINE PRODUCTS COMPANY. RESEARCH AND ENGINEERING, CLEVELAND (MEMORANDUM REPORT-EXTERNAL)
M/R 16-	RAYTHEON MFG. CO. MISSILE SYSTEMS DIV., BEDFORD, MASS.
MR-	STANFORD UNIV., CALIF.
MR-	TRW EQUIPMENT LABS. MATERIALS RESEARCH AND DEVELOPMENT DEPT., CLEVELAND
MR-	WAR DEPT., WASHINGTON, D.C. (MOBILIZATION REGULATIONS)
MR-(LETTER)(DATE CODE)	NATIONAL ADVISORY COMMITTEE FOR AERONAUTICS (MEMORANDUM REPORT) (REF. 25)
MR-A-	CONSOLIDATED VULTEE AIRCRAFT CORP., FORT WORTH, TEX.
MRA-	METCUT RESEARCH ASSOCIATES INC., CINCINNATI
MRB-	MONSANTO RESEARCH CORP., DAYTON, OHIO
MRB-	MONSANTO RESEARCH CORP. BOSTON LABS., EVERETT, MASS.
MRB-(NO.)-(LTR(S))-	MONSANTO RESEARCH CORP. BOSTON LABS., EVERETT, MASS.
MRC-	GT. BRIT. MEDICAL RESEARCH COUNCIL, LONDON
MR-C-	ILLINOIS. UNIV., URBANA
MRC-	MAGNETIC RESEARCH CORP., EL SEGUNDO, CALIF.
MRC-	MATERIALS RESEARCH CORP., ORANGEBURG, N.Y.
MRC-	MATERIALS RESEARCH CORP., YONKERS, N.Y.
MRC-	MICROWAVE RADIATION CO., INC., GARDENA, CALIF.
MRC-	MISSION RESEARCH CORP., SANTA BARBARA, CALIF.
MRC-	MOLECULON RESEARCH CORP., CAMBRIDGE, MASS.
MRC-	MONSANTO RESEARCH CORP. (ALL LOCATIONS)
MRC-	WISCONSIN. UNIV., MADISON
MRC-(LTR)(NO.)-(LTR)	MOLECULON RESEARCH CORP., CAMBRIDGE, MASS.
MRC-DA-	MONSANTO RESEARCH CORP., DAYTON, OHIO
MRC IHRB R-	GT. BRIT. MEDICAL RES. COUNCIL. IND. HEALTH RES. BD.
MRC M-	GT. BRIT. MEDICAL RESEARCH COUNCIL, LONDON (MEMO.)
MRC-OLS-	WISCONSIN. UNIV., MADISON. MATHEMATICS RESEARCH CENTER (ORIENTATION LECTURE SERIES)
MRC ORIENTATION LECTURE-	WISCONSIN. UNIV., MADISON. MATHEMATICS RESEARCH CTR.
MRC-R-	MATERIALS RESEARCH CORP., YONKERS, N.Y.
MRC-REPRINT-(YR.)-	PITTSBURGH. UNIV. MANAGEMENT RESEARCH CENTER
MRC SR-	GT. BRIT. MEDICAL RES. COUNCIL, LONDON (SPEC. RPTS.)
MRC-TRS-	WISCONSIN. UNIV., MADISON. MATHEMATICS RESEARCH CTR.
MRC-TS-	WISCONSIN. UNIV., MADISON. MATHEMATICS RESEARCH CENTER (RESEARCH SUMMARY REPORT)
MRC-TSR-	WISCONSIN. UNIV., MADISON. MATHEMATICS RESEARCH CENTER (TECHNICAL SUMMARY REPORT)
MRC-WES-	ARMY ENGINEER WATERWAYS EXPERIMENT STATION, VICKSBURG, MISS.
MRD-	MEDICAL RESEARCH AND DEVELOPMENT BOARD, WASH., D.C.
MRD-	NORTHROP CORP. NORAIR DIV. MANUFACTURING DEPT., HAWTHORNE, CALIF. (MANUFACTURING RESEARCH + DEV.)
MRDC-	METALLURGICAL RES. + DEV. CO., INC., WASHINGTON, D.C.
MRDC-(YEAR)-	THAILAND. MILITARY RES. + DEV. CENTER, BANGKOK
MRDD R-	AEROJET-GENERAL CORP., SACRAMENTO, CALIF.
MRD-GADC-	GENERAL AMERICAN TRANSPORT. CORP. MRD DIV., NILES, ILL
MRDL-(NUMBER-YEAR)	CORPS OF ENGINEERS (ARMY). MISSOURI RIVER DIV. LABS., OMAHA
MRD-LAB-	CORPS OF ENGINEERS (ARMY). MISSOURI RIVER DIV. LABS., OMAHA
MRDL-TR-	CORPS OF ENGINEERS (ARMY). MISSOURI RIVER DIV. LABS., OMAHA
MRDR-	AEROJET-GENERAL CORP., SACRAMENTO, CALIF.
MRD-TE-ERIC-	CORPS OF ENGINEERS (ARMY). MISSOURI RIVER DIV. LABS., OMAHA
MRD-TR-(YEAR)-	CORPS OF ENGINEERS (ARMY). MISSOURI RIVER DIV. LABS., OMAHA
MRD-TR-	CORPS OF ENGINEERS (ARMY). MISSOURI RIVER DIV. LABS., OMAHA
MR+E-(YEAR)-	ARMY NATICK LABS. CLOTHING AND MATERIALS DIV., MASS.
MRE-	CURTISS-WRIGHT CORP. WRIGHT AERO. DIV., WOOD-RIDGE, NJ
MRE-	UNITED KINGDOM. MICROBIOLOGICAL RESEARCH ESTAB., SALISBURY, WILTS, ENGLAND
MRG-	GENERAL DYNAMICS/ASTRONAUTICS, SAN DIEGO, CALIF.
MRI-	BROOKLYN. POLYTECHNIC INST. MICROWAVE RESEARCH INST.
MRI-	CALIFORNIA. UNIV., BERKELEY. MINERALS RESEARCH LAB.
MRI-	LOCKHEED MISSILES AND SPACE CO., SUNNYVALE, CALIF.
MRI-	METEOROLOGY RESEARCH, INC., ALTADENA, CALIF.
MRI(YEAR)-(LETTERS)-	METEOROLOGY RESEARCH, INC., ALTADENA, CALIF.
MRI(YEAR)-FR-	METEOROLOGY RESEARCH, INC., ALTADENA, CALIF.
MRI-(YEAR)-IR	METEOROLOGY RESEARCH, INC., ALTADENA, CALIF.
MRI-(YEAR)-PR-	METEOROLOGY RESEARCH, INC., ALTADENA, CALIF.
MRI-	MIDWEST RESEARCH INST., KANSAS CITY, MO.
MRI-(NO.)-B-QPR-	MIDWEST RESEARCH INST., KANSAS CITY, MO.
MRI-(NO.)-P-QTR-	MIDWEST RESEARCH INST., KANSAS CITY, MO.
MRI-(YEAR)-FR-	SOUTH DAKOTA SCHOOL OF MINES + TECHNOLOGY, RAPID CITY
MRI-	WATERTOWN ARSENAL. ORDNANCE MATERIALS RES. OFF., MASS.
MRI-HOU-(YEAR)-	MECHANICS RESEARCH, INC., HOUSTON, TEX.
MRI-PR-	MIDWEST RESEARCH INST., KANSAS CITY, MO.
MRI-QPR-	MIDWEST RESEARCH INST., KANSAS CITY, MO.
MR-IR-(NUMBER)-	MARQUARDT CORP., VAN NUYS, CALIF.
MRI-R-(NUMBER)R	METEOROLOGY RESEARCH, INC., ALTADENA, CALIF.
MRI-TR-(MONTH/YEAR)	MIDWEST RESEARCH INST., KANSAS CITY, MO.

MRJ-	CURTISS-WRIGHT CORP., WOOD-RIDGE, N.J.
MRJ.00-	CURTISS-WRIGHT CORP. WRIGHT AERO. DIV., WOOD-RIDGE, NJ
MRL-(NUMBER -LTR.-NO.)	MATERIALS RESEARCH LAB., INC., RICHTON PARK, ILL.
MRL-	MISSOURI RESEARCH LABS., INC., ST. LOUIS
MRL-	MULLARD RESEARCH LABS., REIGATE, SURREY, ENGLAND
MRL-	NAVAL MEDICAL RESEARCH LAB., NEW LONDON, CONN.
MRL-	NORTHROP CORP. NORAIR DIV. ENGINEERING DEPT., HAW-THORNE, CALIF. (MATERIALS RESEARCH)
MRL-	WATERTOWN ARSENAL. MATERIALS RESEARCH LAB., MASS.
MRL-(INITIALS)-	CALIFORNIA. UNIV., BERKELEY. RADIATION LAB.
MRL (EA)-	CHEMICAL WARFARE SERVICE. MEDICAL RESEARCH LAB., EDGEWOOD ARSENAL, MD.
MRL-R-	MAGNAVOX CO. RESEARCH LABS., TORRANCE, CALIF.
MRL-TDR-(YEAR)-	AEROSPACE MEDICAL RESEARCH LABS. (6570TH), WRIGHT-PATTERSON AFB, OHIO
MR-N-	CONSOLIDATED VULTEE AIRCRAFT CORP., FORT WORTH, TEX.
MR-N-	CONVAIR, FORT WORTH, TEX.
MR-N-	GENERAL DYNAMICS/FORT WORTH, TEX.
MRO-	(MINE RECOVERY OPERATION)
MRO-(YR.)-(NO.)-(NO.)-	RADIO CORP. OF AMERICA, CAMDEN, N.J.
MRP-	(MONTHLY REPORT OF PROGRESS) (REF. 2)
MRP-	COLUMBIA UNIV., N.Y.C. ENGINEERING RESEARCH LABS.
MRP-	GT. BRIT. AIR MINISTRY. METEOROLOGICAL RES. COMM.
MRP-	OFFICE OF THE CHIEF OF NAVAL OPERATIONS, WASHINGTON, D.C. (MID-RANGE PLAN)
MRP-(NO.)-(NO.)-	REPUBLIC AVIATION CORP., FARMINGDALE, N.Y.
MRR-	(MONTHLY REVIEW REPORT) (REF. 2)
MRR-(YEAR)-	AEROJET-GENERAL CORP., SACRAMENTO, CALIF.
MRR-	MUNICH. TECHNISCHE HOCHSCHULE. INSTITUT FUER MESS- UND REGELUNGSTECHNIK
MR REPORT-	KELLEX CORP., SILVER SPRING, MD. (MATERIALS RESEARCH)
MR-S-	MARQUARDT CORP., VAN NUYS, CALIF.
MR-S-	MARQUARDT CORP. SYSTEMS ENG. DIV., VAN NUYS, CALIF.
MRS-	QUARTERMASTER RESEARCH AND ENGINEERING COMMAND, NATICK, MASS. (METHODS RESEARCH STUDY)
MRT-	GENERAL DYNAMICS/CONVAIR, FORT WORTH, TEX.
MRU-	NAVAL ORDNANCE LAB. MINE RESEARCH UNIT
MS-	(MATERIAL SPECIFICATIONS) (REF. 2)
MS-	(MILITARY SPECIFICATION) (REF. 2)
MS-	(MILITARY STANDARD) (REF. 13)
MS-	AEROJET-GENERAL CORP., SACRAMENTO, CALIF.
MS-	ARMY ENGINEER WATERWAYS EXPERIMENT STATION, VICKSBURG, MISS.
MS-	BOEING CO., SEATTLE
MS-	CANADA. NATIONAL AERONAUTICAL ESTABLISHMENT, OTTAWA
MS-	GT. BRIT. DEPT. OF ATOMIC ENERGY
MS-	GT. BRIT. MINISTRY OF SUPPLY
MS-	GT. BRIT. ROYAL AIRCRAFT EST., FARNBOROUGH, HANTS, ENG
MS-	GT. BRIT. UNDERWATER COUNTERMEASURES + WEAPONS ESTAB.
MS-(YEAR)-	INDUSTRIAL AND ENGINEERING CHEMISTRY. RESEARCH RESULTS SERVICE, WASHINGTON, D.C. (MANUSCRIPT)
MS-	INTERNATIONAL BUSINESS MACHINES CORP. FEDERAL SYSTEMS DIV., BETHESDA, MD.
MS-	LITTON SYSTEMS, INC. DATA SYSTEMS DIV., VAN NUYS, CAL.
MS-	MASSACHUSETTS INST. OF TECH., CAMBRIDGE. NATL. MAGNET LAB
MS-	MASSACHUSETTS INST. OF TECH., LEXINGTON. LINCOLN LAB.
MS-	MITHRAS, INC., CAMBRIDGE, MASS.
MS-	NATIONAL RESEARCH COUNCIL OF CANADA. MONTREAL LAB.
MS-(YEAR)-	PENNSYLVANIA. UNIV., PHILA. MOORE SCH. OF ELEC. ENG.
MS-	QUARTERMASTER RESEARCH AND ENGINEERING COMMAND, NATICK, MASS. (MICRO BIOLOGY SERIES)
MS-(NUMBER)/R1	TECHNION-ISRAEL INST. OF TECH., HAIFA
MS-	VIRGINIA. UNIV., CHARLOTTESVILLE. DEPT. OF MATERIALS SCIENCE
MS-	VIRGINIA. UNIV., CHARLOTTESVILLE. RESEARCH LABS. FOR ENGINEERING SCIENCES
MS-	WHITE SANDS MISSILE RANGE, N. MEX.
MSA-	COAST AND GEODETIC SURVEY, ROCKVILLE, MD. (ABSTRACTS OF EARTHQUAKE REPORTS)
MSA-	MSA RESEARCH CORP., CALLERY, PENNA.
MSA-MEMO-	MINE SAFETY APPLIANCES CO., CALLERY, PENNA.
MS-AP-	GARRETT CORP. AIRESEARCH MFG. DIV., LOS ANGELES
MSA-PR-	MSA RESEARCH CORP., CALLERY, PENNA.
MSAR-	MSA RESEARCH CORP., CALLERY, PENNA.
MSAR-(YEAR)-	MSA RESEARCH CORP., CALLERY, PENNA.
MSAR-(NO.)-(NO.)-	MSA RESEARCH CORP., CALLERY, PENNA.
MSAR-(YEAR)-	MSA RESEARCH CORP., EVANS CITY, PENNA.
MS ARD EM-	GT. BRIT. ARMAMENT RESEARCH DEPT., FORT HALSTEAD, KENT, ENGLAND (ELECTRODEPOSITION MEMO)
MSAR-MEMO-	MSA RESEARCH CORP., CALLERY, PENNA.
MSAR-TR-	MSA RESEARCH CORP., CALLERY, PENNA.
MSASD-	AEROJET-GENERAL CORP., SACRAMENTO, CALIF.
MSA-TM-	AEROJET-GENERAL CORP., SACRAMENTO, CALIF.
MSA-TR-	MINE SAFETY APPLIANCES CO., CALLERY, PENNA.
MSB-	DUKE UNIV., BEAUFORT, N.C. MARINE STATION (BULLETIN)
MSC-	AIR ATTACHE, LONDON
MSC-	CORNELL UNIV., ITHACA, N.Y. MATERIAL SCIENCE CENTER
MSC-	EUROPEAN ORGANIZATION FOR NUCLEAR RESEARCH, GENEVA
MSC-	LEHIGH UNIV., BETHLEHEM, PENNA. MARINE SCIENCE CTR.
MSC-	MANNED SPACECRAFT CENTER, HOUSTON, TEX.
MSC-	MATHEMATICAL SCIENCES CORP., SEATTLE
MSCA-	DOUGLAS AIRCRAFT CO., INC., SANTA MONICA, CALIF.
MSC-BM-MR-	MANNED SPACECRAFT CENTER, HOUSTON, TEX.
MSC-EC-	MAC NEAL-SCHWENDLER CORP., SAN MARINO, CALIF.
MSC-R-	MIAMI SHIPBUILDING CORP., FLA.
MSC-TR-	CORNELL UNIV., ITHACA, N.Y. MATERIAL SCIENCE CENTER
MSD-	GENERAL ELECTRIC CO., PHILADELPHIA
MS-D-	GT. BRIT. MINISTRY OF SUPPLY
MSD-(NUMBER)-TP-	HUGHES AIRCRAFT CO. MISSILE SYSTEMS DIV., CANOGA PARK, CALIF.
MSD(NUMBER)RS	HUGHES AIRCRAFT CO. MISSILE SYSTEMS DIV., CANOGA PARK, CALIF.
MSD-	LOCKHEED AIRCRAFT CORP. MISSILE SYSTEMS DIV., VAN NUYS, CALIF.
MSD-	LOCKHEED MISSILES AND SPACE CO., PALO ALTO, CALIF.
MSD-(NUMBER)-	LTV AEROSPACE CORP. MISSILES AND SPACE DIV., DALLAS
MSD-	REPUBLIC AVIATION CORP., HICKSVILLE, N.Y.
MSD-(NO.)-(NO.)-(NO.)	REPUBLIC AVIATION CORP. MISSILE SYSTEMS DIV., MINEOLA, N.Y.
MSD-	THOMPSON PRODUCTS, INC., CLEVELAND
MS DAE (LETTER(S))-	GT. BRIT. DEPT. OF ATOMIC ENERGY
MSD-ES-	LTV AEROSPACE CORP. MISSILES AND SPACE DIV., DALLAS
MSDH/(YEAR)/	IMPERIAL CHEMICAL INDUSTRIES, LTD., LONDON
MSDH/(NUMBER/NUMBER)	IMPERIAL CHEMICAL INDUSTRIES, LTD. CENTRAL MANAGEMENT SERVICES, WILMSLOW, CHES., ENGLAND
MSD-N-	GENERAL ELECTRIC CO. MISSILE AND SPACE DIV., PHILA.
MSD-P(YEAR)-	HUGHES AIRCRAFT CO. MISSILE SYSTEMS DIV., CULVER CITY, CALIF.
MSD-T-	LTV AEROSPACE CORP. MISSILES AND SPACE DIV., DALLAS

MS DTD B-	GT. BRIT. MINISTRY OF SUPPLY. DEPT. OF TANK DESIGN (BULLETINS)
MS DTD MS-	GT. BRIT. MINISTRY OF SUPPLY. DEPT. OF TANK DESIGN (MATERIAL SPECIFICATIONS)
MS DTD R-	GT. BRIT. MINISTRY OF SUPPLY. DEPT. OF TANK DESIGN (REPORT)
MSD-TM-(NO.)-(NO.)-	LOCKHEED AIRCRAFT CORP. MISSILES + SPACE DIV., PALO ALTO, CALIF.
MSD-TP(NUMBER)	LTV AEROSPACE CORP. MISSILES AND SPACE DIV., DALLAS
MSE-	CALIFORNIA. UNIV., BERKELEY. COLL. OF ENGINEERING
MSEE-	PENNSYLVANIA. UNIV., PHILA. MOORE SCH. OF ELEC. ENG.
MSFC-	GEORGE C. MARSHALL SPACE FLIGHT CTR., HUNTSVILLE, ALA.
MSFC-ASO-IN-	GEORGE C. MARSHALL SPACE FLIGHT CTR., HUNTSVILLE, ALA.
MSFC-LOD-(LETTERS)-	GEORGE C. MARSHALL SPACE FLIGHT CTR., HUNTSVILLE, ALA. (LAUNCH OPERATIONS DIRECTORATE)
MSFC-STD-	DOUGLAS AIRCRAFT CO., INC. MISSILE AND SPACE SYSTEMS DIV., SANTA MONICA, CALIF.
MSFRL-	ARMY MEDICAL RESEARCH LAB., FORT KNOX, KY.
MSH-	MOUNT SINAI HOSPITAL, N.Y.C.
MSI-(NUMBER)A	COAST AND GEODETIC SURVEY, ROCKVILLE, MD. (SEISMOLOGICAL BULLETIN, ANTARCTICA)
MSI-	COAST AND GEODETIC SURVEY, WASHINGTON, D.C.
MSL-	GENERAL MOTORS CORP. MATERIALS AND STRUCTURES LAB., WARREN, MICH.
MSL-(YEAR)-	GENERAL MOTORS TECHNICAL CENTER. MATERIALS AND STRUCTURES LAB., WARREN, MICH.
MSL-	NATIONAL WEATHER SATELLITE CENTER, WASHINGTON, D.C.
MSL-	WEATHER BUREAU, WASHINGTON, D.C.
MS-M-(NUMBER)C	WESTINGHOUSE ELECTRIC CORP. ATOMIC POWER DIV., PITTS.
MSNW-(YEAR)-(NO.)-	MATHEMATICAL SCIENCES NORTHWEST, INC., SEATTLE
MSP-	COAST AND GEODETIC SURVEY, ROCKVILLE, MD. (SEISMOLOGY BULLETIN)
MS-PR-(YEAR)-	KAISER ALUMINUM AND CHEMICAL CORP. DEPT. OF METALLURGICAL RESEARCH, SPOKANE
MSR-	AEROJET-GENERAL CORP., SACRAMENTO, CALIF.
MSR-	BEECH AIRCRAFT CORP., WICHITA, KAN. (MISSILE STRUCTURES REPORT)
MSR-	DENVER. UNIV.
MS-R-	DENVER. UNIV. DENVER RESEARCH INST. DIV. OF MATHEMATICAL SCIENCES
MSR-	PENNSYLVANIA. UNIV., PHILA. MOORE SCH. OF ELEC. ENG.
MSR-	QUARTERMASTER CORPS (MICROBIOLOGICAL SERIES REPORTS)
MSRP-DP-	CALIFORNIA. UNIV., LOS ANGELES. MANAGEMENT SCIENCES RESEARCH PROJECT (DISCUSSION PAPER)
MSRP-RR-	CALIFORNIA. UNIV., LOS ANGELES. MANAGEMENT SCIENCES RESEARCH PROJECT
MSRR-	CARNEGIE INST. OF TECH., PITTSBURGH. MANAGEMENT SCIENCES RESEARCH GROUP
MSS-(NUMBER)-	WESTERN RESERVE ELECTRONICS, INC., CLEVELAND
MSSR-	GRUMMAN AIRCRAFT ENGINEERING CORP. MILITARY SPACE SYSTEMS, BETHPAGE, N.Y.
MST-	MALLORY-SHARON TITANIUM CORP., NILES, OHIO
MST-	NAVAL MINE DEPOT. MINE SERVICE TEST DEPT., YORKTOWN, VA
MSTC-	MALLORY-SHARON TITANIUM CORP., NILES, OHIO
MSTF-	NAVAL MINE DEPOT, YORKTOWN, VA. (MINE SERVICE TEST FINAL)
MSTFR-	NAVAL MINE DEPOT, YORKTOWN, VA. (MINE SERVICE TEST FINAL REPORT)
MS TGS-	GT. BRIT. MINISTRY OF SUPPLY. TECHNICAL COORDINATING COMMITTEE ON GENERAL STORES.
MSTIR-	NAVAL MINE DEPOT, YORKTOWN, VA. (MINE SERVICE TEST INTERIM REPORT)
MSTR-	NAVAL MINE DEPOT, YORKTOWN, VA. (MINE SERVICE TEST REPORT)
MSTS-	MILITARY SEA TRANSPORTATION SERVICE, WASHINGTON, D.C.
MSU-	MICHIGAN STATE UNIV., EAST LANSING
MSUCP-	MICHIGAN STATE UNIV., EAST LANSING (CYCLOTRON PROJ.)
MSU-PR-	MICHIGAN STATE UNIV., EAST LANSING. DEPT. OF CHEMISTRY
MSU-RM-	MICHIGAN STATE UNIV., EAST LANSING. DEPT. OF STATISTICS
MSU-TOR-	MICHIGAN STATE UNIV., EAST LANSING. DEPT. OF PHYSICS (TECHNICAL OPERATING REPORT)
MSU-TR-	MICHIGAN STATE UNIV., EAST LANSING
MSV/-	IMPERIAL CHEMICAL INDUSTRIES, LTD. BILLINGHAM DIV., DURHAM, ENGLAND
MSVD-(YEAR)-SD-	GENERAL ELECTRIC CO. MISSILE AND SPACE VEHICLE DEPT., PHILADELPHIA
MS WM-	GT. BRIT. MINISTRY OF SUPPLY (WELDING MEMORANDA)
MSWP/N-	UNITED KINGDOM ATOMIC ENERGY AUTHORITY. RESEARCH GP. ATOMIC ENERGY RES. ESTAB., HARWELL, BERKS, ENGLAND
MSWP/R-	UNITED KINGDOM ATOMIC ENERGY AUTHORITY. RESEARCH GP. ATOMIC ENERGY RES. ESTAB., HARWELL, BERKS, ENGLAND
MT-	(MISSILE TEST) (REF. 2)
MT-	(MACHINE TRANSLATION) (REF. 2)
MT-	AIR FORCE MISSILE TEST CENTER, PATRICK AFB, FLA.
MT-	ATOMIC ENERGY OF CANADA LTD. CHALK RIVER PROJ., ONT.
MT-	BOELKOW GMBH., OTTOBRUNN BEI MUENCHEN, GERMANY
MT-	CALIFORNIA. UNIV., BERKELEY. LAWRENCE RADIATION LAB.
MT-	CHRYSLER CORP. MISSILE DIV., DETROIT
MT-(NUMBER)-	CONVAIR. ASTRONAUTICS DIV., SAN DIEGO, CALIF.
MT-	DOW CHEMICAL CO. METALLURGICAL LABS., MIDLAND, MICH.
MT-	GT. BRIT. MINISTRY OF SUPPLY
MT-	GRUMMAN AIRCRAFT ENGINEERING CORP., BETHPAGE, N.Y.
MT-(YEAR)-	MARTIN CO., DENVER
MT-	MARTIN-MARIETTA CORP. DENVER DIV.
MT-	NATL. RES. COUNCIL OF CANADA. NATL. RES. LABS., OTTAWA
MT-	NAVAL MISSILE CENTER. MISSILE TEST DEPT., PT.MUGU,CAL.
MTA-	CALIFORNIA RESEARCH AND DEV. CO. LIVERMORE RES. LAB.
MTA-	MAC DONALD (W.S.) CO., INC., CAMBRIDGE, MASS.
MTA-TP-	MACHINE TOOL AUTOMATION INC., SOUTHPORT, CONN.
MTB-	MITSUBISHI HEAVY-INDUSTRIES, LTD., TOKYO
MTC-	AIR FORCE MISSILE TEST CENTER, PATRICK AFB, FLA.
MTC-	NAVAL AIR MISSILE TEST CENTER, POINT MUGU, CALIF.
MTD-	RES. + DEV. BD. COMM. ON GUIDED MISSILES, WASH., DC
MTD-	RES. + DEV. BD. PANEL ON TARGET DRONES, WASH., D.C.
MTDMR-	NAVAL AIR MISSILE TEST CENTER. MISSILE TEST DEPT., POINT MUGU, CALIF. (MEMORANDUM REPORT)
MTDR-	PACIFIC FLEET. MINE TEST DIV. (REPORT)
MTEC-	ATOMIC ENERGY OF CANADA LTD. CHALK RIVER PROJ., ONT.
MTER-	WESTINGHOUSE ELECTRIC CORP., ELMIRA, N.Y.
MTHT-	HOLLOMAN AFB, N. MEX.
MTI-	HERCULES POWDER CO., WILMINGTON, DEL.
MTI-	HERCULES POWDER CO. CHEMICAL PROPULSION DIV. BACCHUS WORKS, MAGNA, UTAH
MTI-	MECHANICAL TECHNOLOGY, INC., LATHAM, N.Y.
MTI-(NO.)TR(NO.)	MECHANICAL TECHNOLOGY, INC., LATHAM, N.Y.
MTL-	UNITED NUCLEAR CORP., WHITE PLAINS, N.Y.
MTLGI-	AIR FORCE MISSILE TEST CENTER, PATRICK AFB, FLA.
MTM-	DOUGLAS AIRCRAFT CO., INC. (MISSILES TECH. MEMO. RPT)
MTMB-	(MILITARY TRAFFIC MANAGEMENT BULLETIN)
MTMR-	(MILITARY TRAFFIC MANAGEMENT REGULATIONS)
MTMTS-	MILITARY TRAFFIC MANAGEMENT AND TERMINAL SERVICE, WASHINGTON, D.C.
MTO-	HERCULES POWDER CO., MAGNA, UTAH

MTO-	HERCULES POWDER CO. CHEMICAL PROPULSION DIV. BACCHUS WORKS, MAGNA, UTAH
MT-OT-(NO.)-(NO.)-	GRUMMAN AIRCRAFT ENGINEERING CORP., BETHPAGE, N.Y.
MTO-TR-(YEAR)-	AIR FORCE LOGISTICS COMMAND. MAINTENANCE TECHNOLOGY OFFICE, TINKER AFB, OKLA.
MTP-	AIR FORCE MISSILE TEST CENTER, PATRICK AFB, FLA.
MTP-(NUMBER-NUMBER)-	ARMY TEST AND EVALUATION COMMAND, ABERDEEN PROVING GROUND, MD.
MTP-	GEORGE C. MARSHALL SPACE FLIGHT CENTER, HUNTSVILLE, ALA. (MARSHALL TECHNICAL PAPER)
MTP-	MITRE CORP., BAILEY'S CROSSROADS, VA.
MTP-	MITRE CORP., BEDFORD, MASS.
MTP-	NATIONAL BUREAU OF STANDARDS, WASHINGTON, D.C.
MTP-	WAR DEPT., WASH., D.C. (MOBILIZATION TRNG. PROGRAM)
MTP-	WORKS PROJECT ADMINISTRATION, WASHINGTON, D.C. (MATHEMATICAL TABLES PROJECT)
MTP-AERO-(YEAR)-	GEORGE C. MARSHALL SPACE FLIGHT CTR.,HUNTSVILLE, ALA.
MTP-ASTR-	GEORGE C. MARSHALL SPACE FLIGHT CTR.,HUNTSVILLE, ALA.
MTP-PVE-	GEORGE C. MARSHALL SPACE FLIGHT CTR.,HUNTSVILLE, ALA.
MTPR-	COLUMBIA UNIV., N.Y.C.
MTP-RP-(YEAR)-	GEORGE C. MARSHALL SPACE FLIGHT CTR.,HUNTSVILLE, ALA.
MTR-	(MATERIALS TESTING REPORT) (REF. 2)
MTR-	BOEING AIRPLANE CO., SEATTLE
MTR-	BOMBARDMENT TRAINING WING (3535TH), MATHER AFB, CALIF
MT-R-	CHRYSLER CORP. MISSILE OPERATIONS
M-TR-	ILLINOIS. UNIV., URBANA
MTR-	MITRE CORP., BEDFORD, MASS.
MTR-	MITRE CORP., LEXINGTON, MASS.
MTR-	PHILLIPS PETROLEUM CO., IDAHO FALLS
MTRI-	RESEARCH AND DEVELOPMENT BOARD, WASHINGTON, D.C. (MISSILE TEST RANGE INSTRUMENTATION)
MTRI-	RES. + DEV. BD. COMM. ON GUIDED MISSILES, WASH., DC
MTRL-	PHILLIPS PETROLEUM CO., IDAHO FALLS
MTRL-	PHILLIPS PETROLEUM CO. ATOMIC EN. DIV., IDAHO FALLS
MTS-	MARINE RESOURCES INC. MARINE TECHNICAL SERVICES DIV., MELVILLE, N.Y.
MTS-	NAVAL GUN FACTORY. METALLURGICAL AND TESTING SECTION, WASHINGTON, D.C.
MTSC-	MIDDLE TENNESSEE STATE COLL., MURFREESBORO
MTSR-	GRAZ. TECHNISCHE HOCHSCHULE
MTT-	AIR FORCE MISSILE TEST CENTER, PATRICK AFB, FLA.
MTT-	RES. + DEV. BD. COMM. ON GUIDED MISSILES, WASH., DC
MTT-	RESEARCH AND DEVELOPMENT BOARD. PANEL ON TEST AND TRAINING EQUIPMENT, WASHINGTON, D.C.
MTTE-	AIR FORCE MISSILE TEST CENTER, PATRICK AFB, FLA.
MT-TOM-	RES. + DEV. BD. COMM. ON MATERIALS, WASHINGTON, D.C.
MTWL-	NETHERLANDS. STICHTING MOEILIJK TOEGANKELIJKE WETENSCHAPPELIJKE LITERATUUR, DELFT
MTX-TM-	AIR FORCE MISSILE TEST CENTER, PATRICK AFB, FLA.
MU-	MARYLAND. UNIV., COLLEGE PARK
MU-	MICHIGAN. UNIV., ANN ARBOR
MU-	MINNESOTA. UNIV., MINNEAPOLIS
MU-	MISSOURI. UNIV., COLUMBIA
MUC-(INITIALS)-	CHICAGO. UNIV. METALLURGICAL LAB.
MUM-	NAVAL ORDNANCE LAB. MINE UNIT (MEMORANDUM)
MU-ML-	MIAMI. UNIV., CORAL GABLES, FLA. MARINE LAB.
MU MRGTN-	MISSOURI. UNIV., COLUMBIA. DEPT. OF MATHEMATICS (MATHEMATICS RESEARCH PROJECT)
MUP-TR-	MARYLAND. UNIV., COLLEGE PARK. DEPT. OF PHYSICS AND ASTRONOMY
M/U/PZ/	BRITISH STEEL CORP. MIDLAND GROUP,MOORGATE,YORKS,ENG.
MUR-	NAVAL ORDNANCE LAB. MINE UNIT (REPORT)
MURA-	MIDWESTERN UNIVERSITIES RESEARCH ASSN., MADISON, WIS.
MURA-	MIDWESTERN UNIVERSITIES RESEARCH ASSN., STOUGHTON,WIS
MURA-	MIDWESTERN UNIVERSITIES RESEARCH ASSN., URBANA, ILL.
MURA-(INITIALS)-	MIDWESTERN UNIVERSITIES RESEARCH ASSN., URBANA, ILL.
MU-RAL-	MINNESOTA. UNIV., MINNEAPOLIS. ROSEMOUNT AERO. LABS.
MUSC-	SOUTH CAROLINA. MEDICAL UNIV., CHARLESTON
MV-(NUMBER)-	MARTIN CO., VANDENBERG AFB, CALIF.
MV-	MARTIN-MARIETTA CORP. DENVER DIV.
MV-(YR.)-	MASSACHUSETTS INST. OF TECH., CAMBRIDGE. MAN-VEHICLE CONTROL LAB. (BEGINNING JULY 1966)
MV-972/-	METROPOLITAN-VICKERS ELECTRICAL CO., LTD., MANCHESTER, LANCS, ENGLAND (PROJECT 972)
MV-	UNITED KINGDOM ATOMIC ENERGY AUTHORITY. RESEARCH GP. ATOMIC ENERGY RES. ESTAB., HARWELL, BERKS, ENGLAND
MVLS-(YEAR)-	MASS. INST. OF TECH., CAMBRIDGE. CENTER FOR SPACE RES
MVS-(NUMBER)-	MARTIN CO., VANDENBERG AFB, CALIF.
MVT-(YR.)-	MASSACHUSETTS INST. OF TECH., CAMBRIDGE. MAN-VEHICLE CONTROL LAB. (THESIS)
MW-	MC GILL UNIV., MONTREAL
MW-	MC GILL UNIV., MONTREAL. MAC DONALD PHYSICS LAB.
MW-	MC GILL UNIV., MONTREAL. STORMY WEATHER GROUP
MW-(YEAR)-	UNITRODE CORP., WATERTOWN, MASS.
MWF-	RES. + DEV. BD. COMM. ON GUIDED MISSILES, WASH., DC
MWF-	RES. + DEV. BD. PANEL ON WARHEADS + FUZES, WASH.,D.C.
MW/HC/(NUMBER)/(YEAR)	BRITISH STEEL CORP., LONDON
MWK-	KELLOGG (M.W.) CO., JERSEY CITY
MWK-RL-(YEAR)-	KELLOGG (M.W.) CO. PETROLEUM AND CHEMICAL RESEARCH DEPT., JERSEY CITY
MWK-SPD-	KELLOGG (M.W.) CO. SPECIAL PROJECTS DEPT.,JERSEY CITY
MWO-	WAR DEPT., WASHINGTON, D.C. (MODIFICATION WORK ORDER)
MWORM-	NAVY (MINE WARFARE OPERATIONAL RESEARCH MEMORANDUM)
MWORR-	NAVY (MINE WARFARE OPERATIONAL RESEARCH REPORT)
MWP(S)/DGD-	UNITED KINGDOM ATOMIC ENERGY AUTHORITY. DEVELOPMENT AND ENGINEERING GROUP, CULCHETH, LANCS, ENGLAND
MWP(S)/P-	UNITED KINGDOM ATOMIC ENERGY AUTHORITY. DEVELOPMENT AND ENGINEERING GROUP, CULCHETH, LANCS, ENGLAND
MW PWB S-	GT. BRIT. MINISTRY OF WORKS (POSTWAR BLDG. STUDIES)
MWR-	BEECH AIRCRAFT CORP., WICHITA, KAN. (MISSILE WEIGHT REPORT)
MWR-	NAVY (MINE WARFARE REPORT)
MWT-	MC GILL UNIV., MONTREAL. STORMY WEATHER GROUP
MX-	(HANDBOOKS AND REPORTS CONCERNING MX COMPONENTS)
MX-(YEAR)-	AIRCO CHEMICAL DIV., CUMBERLAND CHEMICAL CORP., MIDDLESEX, N.J. (USED 1965-)
MX-	GT. BRIT. DIRECTORATE OF MATERIALS AND EXPLOSIVES RESEARCH AND DEVELOPMENT
MXL-(INITIALS)-	TENNESSEE EASTMAN CORP., OAK RIDGE, TENN.
M/Z/	BRITISH STEEL CORP. MIDLAND GROUP,MOORGATE,YORKS,ENG.

N-	(TECHNICAL NOTE) (REF.2)
N-	AIR FORCE CAMBRIDGE RESEARCH CENTER. GEOPHYSICS RESEARCH DIRECTORATE, MASS.
N-	ALL AMERICAN ENGINEERING CO., WILMINGTON, DEL.
N-	ARGONNE NATIONAL LAB., LEMONT, ILL. (REF. 10)
N-(2 DIGITS)-(YEAR)	BRUSSELS. UNIVERSITE. INSTITUT DE MECANIQUE APPLIQUEE
N-	CALIFORNIA INST. OF TECH.,PASADENA. HYDRODYNAMICS LAB
N-	CLINTON LABS., OAK RIDGE, TENN.
N-	DU PONT DE NEMOURS (E.I.) + CO. JACKSON LAB., WILMINGTON, DEL.
N(3 DIGITS)-F	ELECTRO-OPTICAL RESEARCH CO., LOS ANGELES
N-	GT. BRIT. AERONAUTICAL RESEARCH COUNCIL
N-	GT. BRIT. ROYAL AIRCRAFT EST.,FARNBOROUGH, HANTS, ENG
N-	HEWLETT-PACKARD CO. FREQUENCY AND TIME DIV.-EAST, BEVERLY, MASS.
N-	HOLMES AND NARVER, INC. ON-CONTINENT TEST DIV., LAS VEGAS, NEV.
N-	ILLINOIS. UNIV., URBANA. DEPT. OF CIVIL ENGINEERING
N-	INSTITUTE FOR DEFENSE ANALYSES. SCIENCE AND TECHNOLOGY DIV., ARLINGTON, VA.
N-	LOCKHEED MISSILES AND SPACE CO., PALO ALTO, CALIF.
N-	LOS ALAMOS SCIENTIFIC LAB., N. MEX.
N-1- THRU N-7-	LOS ALAMOS SCIENTIFIC LAB., N. MEX. (INTERNAL CORRESPONDENCE SERIAL USED BY N DIV.)
N(YEAR)-	NATIONAL AERONAUTICS AND SPACE ADMINISTRATION, WASHINGTON, D.C. (ACCESSION NUMBER FOLLOWS YEAR)
N-(NOS. ABOVE 30,000)	NATIONAL AERONAUTICS AND SPACE ADM., WASHINGTON, D.C. (ASSIGNED TO MISCELLANEOUS BRITISH, CANADIAN AND U.S. REPORTS)
N-	NAVY, WASHINGTON, D.C.
N(NUMBER)	NEWMARK (N.M.) CONSULTING ENGINEERING SERVICES, URBANA, ILL.
N-(YEAR)H-	NORTH AMERICAN AVIATION, INC. COLUMBUS DIV., OHIO
N(NUMBER)-(YEAR)-	POWER REACTOR AND NUCLEAR FUEL DEVELOPMENT CORP., TOKAI, JAPAN
N(NO.)-R(NO.)	PROTECTION, INC., INGLEWOOD, CALIF.
N-	RUTGERS UNIV., NEW BRUNSWICK, N.J.
N-	SHARP (GEORGE G.) INC., N.Y.C.
N-(NUMBER/NUMBER(S))	SYSTEM DEVELOPMENT CORP., SANTA MONICA, CALIF. (NOTE)
N-	TECHNICAL INFORMATION SERVICE, AEC (SERIES ASSIGNED TO JAPANESE REPORTS ABOUT HIROSHIMA + NAGASAKI)
N-	U.S.S.R. GOSUDARSTVENNYI KOMITET PO ISPOL'ZOVANIYU ATOMNOI ENERGII. INSTITUT TEORETICHESKOI I EKSPERI-MENTAL'NOI FIZIKI, MOSCOW
N(LETTER)C-	OFFICE OF SCIENTIFIC RESEARCH AND DEVELOPMENT
NA-	CALIFORNIA. UNIV., BERKELEY
NA-(YEAR)-	CALIFORNIA. UNIV., BERKELEY. COLL. OF ENGINEERING
NA-	CALIFORNIA. UNIV., BERKELEY. INST. OF ENG. RES.
NA-(YEAR)-	CALIFORNIA. UNIV., BERKELEY. OFFICE OF RES. SERVICES
NA-	EAGLE-PICHER CO., JOPLIN, MO.
NA-(YEAR)-	FEDERAL AVIATION ADMINISTRATION, OKLAHOMA CITY
NA-	KELLETT AIRCRAFT CORP., CAMDEN, N.J.
NA-(YEAR)-	NATIONAL AVIATION FACILITIES EXPTL.CTR.,ATLANTIC CITY
NA-(YEAR)-	NATIONAL BUREAU OF STANDARDS, WASHINGTON, D.C.
NA-	NORTH AMERICAN AVIATION, INC., COLUMBUS, OHIO
NA-	NORTH AMERICAN AVIATION, INC., LOS ANGELES
NA-(YEAR)-	NORTH AMERICAN AVIATION, INC., LOS ANGELES
NA(NUMBER)-	NORTH AMERICAN AVIATION, INC., LOS ANGELES
NA-	NORTH AMERICAN AVIATION, INC. COLUMBUS DIV., OHIO
NA(YEAR)H-	NORTH AMERICAN AVIATION, INC. COLUMBUS DIV., OHIO
NA(NUMBER)-(NOS.)(LTR.)	NORTH AMERICAN AVIATION, INC. ENG. DEPT., LOS ANGELES
NA-(YEAR)-	NORTH AMERICAN ROCKWELL CORP., LOS ANGELES
NA-	NORTHROP AIRCRAFT, INC., HAWTHORNE, CALIF. (NORTHROP ACCIDENT)
NA-(YEAR)-	NORTHROP AIRCRAFT, INC., HAWTHORNE, CALIF.
NA-(NUMBER)-N-	SOCIETE BELGE POUR L'INDUSTRIE NUCLEAIRE, BRUSSELS
NA-(NUMBER)-	SPERRY GYROSCOPE CO., GREAT NECK, N.Y.
NA-(NO.)-(NO.)-(NO.)	SPERRY GYROSCOPE CO. ELECTRONIC TUBE DIV., GREAT NECK, N.Y.
NA-	TELEDYNE INDUSTRIES, INC. GEOTECH DIV., GARLAND, TEX.
NA-(YEAR)-	WESTERN CO. OF NORTH AMERICA, RICHARDSON, TEX.
NAA-	ATOMICS INTERNATIONAL DIV., NORTH AMERICAN AVIATION, INC., CANOGA PARK, CALIF.
NAA-	NORTH AMERICAN AVIATION, INC. (ALL LOCATIONS)
NAA-(YEAR)	NORTH AMERICAN AVIATION, INC. (ALL LOCATIONS)
NAA-(YEAR)MD-	NORTH AMERICAN AVIATION, INC., LOS ANGELES
NAA-(YEAR)LA-	NORTH AMERICAN AVIATION, INC. LOS ANGELES DIV.
NAA-	THOMPSON RAMO WOOLDRIDGE INC. TAPCO DIV., CLEVELAND
NAA-AER-	ATOMICS INTERNATIONAL DIV., NORTH AMERICAN AVIATION, INC., CANOGA PARK, CALIF.
NAA-AER-	NORTH AMERICAN AVIATION, INC. ATOMIC ENERGY RESEARCH DEPT., DOWNEY, CALIF.
NAA-AER-MEMO-	NORTH AMERICAN AVIATION, INC. ATOMIC ENERGY RESEARCH DEPT., DOWNEY, CALIF.
NAA/AI-	ATOMICS INTERNATIONAL DIV., NORTH AMERICAN AVIATION, INC., CANOGA PARK, CALIF.
NAA-AI-	ATOMICS INTERNATIONAL DIV., NORTH AMERICAN AVIATION, INC., CANOGA PARK, CALIF.
NAA-AL-	NORTH AMERICAN AVIATION, INC. AEROPHYSICS LAB., DOWNEY, CALIF.
NAA-AL-	NORTH AMERICAN AVIATION, INC. MISSILE DEVELOPMENT DIV., DOWNEY, CALIF.
NAA-AL-	ROCKETDYNE DIV., NORTH AMERICAN AVIATION, INC., CANOGA PARK, CALIF.
NAA-AL-MEMO-	NORTH AMERICAN AVIATION, INC. AEROPHYSICS LAB., DOWNEY, CALIF.
NAA-C4-	AUTONETICS DIV.,NORTH AMER.AVIATION INC.,ANAHEIM,CAL.
NAA-EM-	NORTH AMERICAN AVIATION, INC. ELECTRO MECHANICAL ENGINEERING DEPT., DOWNEY, CALIF.
NA-AG-	CHICAGO. UNIV.
NAAL-	NORTH AMERICAN AVIATION, INC., LOS ANGELES
NA-AM-	NORTHROP AIRCRAFT, INC., HAWTHORNE, CALIF.
NAA-MD-(YEAR)-	NORTH AMERICAN AVIATION, INC. MISSILE DIV., DOWNEY, CALIF.
NAA-MD-(YEAR)-	ROCKETDYNE DIV., NORTH AMERICAN AVIATION, INC., CANOGA PARK, CALIF.
NAA-NA-(NUMBER(S))-	NORTH AMERICAN AVIATION, INC., LOS ANGELES
NAA-NA-(YR.)H-(NO.)-(NO.)	NORTH AMERICAN AVIATION, INC. COLUMBUS DIV., OHIO
NAA-PC-	NORTH AMERICAN AVIATION, INC., DOWNEY, CALIF.
NAA-PS-	NORTH AMERICAN AVIATION, INC., DOWNEY, CALIF.
NAAR-	NAVAL AIR ATTACHE (REPORT)
NAA-R-	NORTH AMERICAN AVIATION, INC., DOWNEY, CALIF.
NAA-R-	ROCKETDYNE DIV., NORTH AMERICAN AVIATION, INC., CANOGA PARK, CALIF.
NAA-RE-	NORTH AMERICAN AVIATION, INC. (ALL LOCATIONS)
NAA-RF-	NORTH AMERICAN AVIATION, INC., DOWNEY, CALIF.
NAA-RG-	NORTH AMERICAN AVIATION, INC., DOWNEY, CALIF.
NAA-R-G.O.-(NO.)(DATE)	ROCKETDYNE DIV., NORTH AMERICAN AVIATION, INC., CANOGA PARK, CALIF.
NAA-RM-	ROCKETDYNE DIV., NORTH AMERICAN AVIATION, INC., CANOGA PARK, CALIF.
NAA RR(YEAR)-	ROCKETDYNE DIV., NORTH AMERICAN AVIATION, INC., CANOGA PARK, CALIF.
NAAS-	NAVAL AUXILIARY AIR STATION, CORPUS CHRISTI, TEX.
NAASC-TR-	NORTH AMERICAN AVIATION SCIENCE CENTER, THOUSAND OAKS, CALIF.

NAA-SID-	NORTH AMERICAN AVIATION INC. SPACE AND INFORMATION SYSTEMS DIV., DOWNEY, CALIF.
NAA-SPD-P-(NO.)-(YEAR)-	NORTH AMERICAN AVIATION, INC. SPECIAL PROJECTS DIV., LONG BEACH, CALIF.
NAA-SR-	ATOMICS INTERNATIONAL DIV., NORTH AMERICAN AVIATION, INC., CANOGA PARK, CALIF.
NAA-SR-	NORTH AMERICAN AVIATION, INC. (ALL LOCATIONS)
NAA-SR-(YR)-	NORTH AMERICAN AVIATION, INC., DOWNEY, CALIF.
NAA-SR-	RADIO CORP. OF AMERICA, LANCASTER, PENNA.
NAA-SR-	THOMPSON RAMO WOOLDRIDGE INC., CANOGA PARK, CALIF.
NAA-SR-	THOMPSON RAMO WOOLDRIDGE INC., CLEVELAND
NAA-SR-	THOMPSON RAMO WOOLDRIDGE INC. TRW ELECTROMECHANICAL DIV., CLEVELAND
NAA-SRL-	NORTH AMERICAN AVIATION, INC., DOWNEY, CALIF.
NAA-SR-MEMO-	ATOMICS INTERNATIONAL, CANOGA PARK, CALIF.
NAA-SR-MEMO-	ATOMICS INTERNATIONAL DIV., NORTH AMERICAN AVIATION, INC., CANOGA PARK, CALIF.
NAA-SR-MEMO-	NORTH AMERICAN AVIATION, INC., DOWNEY, CALIF.
NAA-SR-MISC.-	NORTH AMERICAN AVIATION, INC., DOWNEY, CALIF.
NAA-SR-MTA-	NORTH AMERICAN AVIATION, INC., DOWNEY, CALIF.
NAA-SR-TDR-	ATOMICS INTERNATIONAL DIV., NORTH AMERICAN AVIATION, INC., CANOGA PARK, CALIF.
NAA-TFD-	NORTH AMERICAN AVIATION, INC., LOS ANGELES
NACA-	NATIONAL ADVISORY COMMITTEE FOR AERONAUTICS (REF. 25)
NACA-(LTRS)-E-(DATE CODE)	AIRCRAFT ENGINE RESEARCH LAB., CLEVELAND (REF. 25)
NACA-(LTRS)-A-(DATE CODE)	AMES AERONAUTICAL LAB., MOFFETT FIELD, CALIF.(REF.25)
NACA-(LTRS)-E-(DATE CODE)	FLIGHT PROPULSION RESEARCH LAB., CLEVELAND (REF. 25)
NACA-(LTRS)-H-(DATE CODE)	HIGH-SPEED FLIGHT STATION, EDWARDS AFB, CAL. (REF.25)
NACA-(LTRS)-L-(DATE CODE)	LANGLEY AERONAUTICAL LAB., LANGLEY FIELD, VA.(REF.25)
NACA-(LTRS)-L-(DATE CODE)	LANGLEY MEMORIAL AERONAUTICAL LAB., LANGLEY FIELD, VA. (REF. 25)
NACA-(LTRS)-E-(DATE CODE)	LEWIS FLIGHT PROPULSION LAB., CLEVELAND (REF. 25)
NACA-AC-	NATIONAL ADVISORY COMMITTEE FOR AERONAUTICS (AIRCRAFT CIRCULAR) (REF. 25)
NACA-ACR-	LEWIS FLIGHT PROPULSION LAB., CLEVELAND
NACA-ACR-(DATE)	NATIONAL ADVISORY COMMITTEE FOR AERONAUTICS (ADVANCE CONFIDENTIAL REPORT) (REF. 25)
NACA-ACR-(LTR)(DATE CODE)	NATIONAL ADVISORY COMMITTEE FOR AERONAUTICS (ADVANCE CONFIDENTIAL REPORT) (REF. 25)
NACA-AR-	NATL. ADVISORY COMM. FOR AERONAUTICS (ANNUAL REPORT)
NACA-ARR-(DATE)	NATIONAL ADVISORY COMMITTEE FOR AERONAUTICS (ADVANCE RESTRICTED REPORT) (REF. 25)
NACA-ARR-(LTR)(DATE CODE)	NATIONAL ADVISORY COMMITTEE FOR AERONAUTICS (ADVANCE RESTRICTED REPORT) (REF. 25)
NACA-CB-(DATE)	NATIONAL ADVISORY COMMITTEE FOR AERONAUTICS (CONFIDENTIAL BULLETIN) (REF. 25)
NACA-CB-(LTR)(DATE CODE)	NATIONAL ADVISORY COMMITTEE FOR AERONAUTICS (CONFIDENTIAL BULLETIN) (REF. 25)
NACA-MR-(DATE)	NATIONAL ADVISORY COMMITTEE FOR AERONAUTICS (MEMORANDUM REPORT) (REF. 25)
NACA-MR-(LTR)(DATE CODE)	NATIONAL ADVISORY COMMITTEE FOR AERONAUTICS (MEMORANDUM REPORT) (REF. 25)
NACAR-	NORTH AMERICAN CARBON, INC., COLUMBUS, OHIO
NACA-RA-(NUMBER)(LETTER)	NATIONAL ADVISORY COMMITTEE FOR AERONAUTICS (RESEARCH ABSTRACTS) (LETTER AT END STANDS FOR SECURITY CLASSIFICATION)
NACA-RB-(DATE)	NATIONAL ADVISORY COMMITTEE FOR AERONAUTICS (RESTRICTED BULLETIN) (REF. 25)
NACA-RB-(LTR)(DATE CODE)	NATIONAL ADVISORY COMMITTEE FOR AERONAUTICS (RESTRICTED BULLETIN) (REF. 25)
NACA-RB-E-	LEWIS FLIGHT PROPULSION LAB., CLEVELAND
NACA-REPORT-	NATIONAL ADVISORY COMM. FOR AERONAUTICS (RPT)(REF.25)
NACA-RM-(LTR)(DATE CODE)	NATIONAL ADVISORY COMMITTEE FOR AERONAUTICS (RESEARCH MEMORANDUM) (REF. 25)
NACA-RM-S-	NATIONAL ADVISORY COMMITTEE FOR AERONAUTICS (RESEARCH MEMORANDUM) (REF. 25)
NACA-RM-SE-	LEWIS FLIGHT PROPULSION LAB., CLEVELAND (RESEARCH MEMORANDUM) (REF. 25)
NACA-RM-SL-	LANGLEY AERONAUTICAL LAB., LANGLEY FIELD, VA. (RESEARCH MEMORANDUM) (REF. 25)
NACA-TM-	NATIONAL ADVISORY COMMITTEE FOR AERONAUTICS (TECHNICAL MEMORANDUM) (REF. 25)
NACA-TN-	NATIONAL ADVISORY COMMITTEE FOR AERONAUTICS (TECHNICAL NOTE) (REF. 25)
NACA-TR-	NATIONAL ADVISORY COMMITTEE FOR AERONAUTICS
NACA-WR-(LETTER)-	NATIONAL ADVISORY COMMITTEE FOR AERONAUTICS (WARTIME REPORT) (REF. 25)
NACA-WRW-	NATIONAL ADVISORY COMMITTEE FOR AERONAUTICS (WARTIME REPORT) (REF. 25)
NAC-CR-MICRO-NOTES-	NAVAL AMMUNITION DEPOT, CRANE, IND.
NAD-	NAVAL AIR DEVELOPMENT CENTER, JOHNSVILLE, PENNA.
NAD-	NAVAL AMMUNITION DEPOT (ALL LOCATIONS)
NAD-AR-	NAVAL AIR DEVELOPMENT CENTER, JOHNSVILLE, PENNA.
NADC-	NAVAL AIR DEVELOPMENT CENTER, JOHNSVILLE, PENNA.
NADC-(LETTERS)-	NAVAL AIR DEVELOPMENT CENTER, JOHNSVILLE, PENNA.
NADC AC-	NAVAL AIR DEV.CTR. AERO. COMPUTER LAB.,JOHNSVILLE,PA.
NADC-AC-	NAVAL AIR DEVELOPMENT CENTER. AEROSPACE CREW EQUIPMENT DEPT., JOHNSVILLE, PENNA.
NADC-AE-	NAVAL AIR DEVELOPMENT CENTER, JOHNSVILLE, PENNA.
NADC-AE-	NAVAL AIR DEVELOPMENT CENTER. AERO ELECTRONIC TECHNOLOGY DEPT., WARMINSTER, PENNA.
NADC-AER-	NAVAL AIR DEVELOPMENT CENTER, JOHNSVILLE, PENNA.
NADC AI-	NAVAL AIR DEV. CTR. AERO. INSTRS. LAB.,JOHNSVILLE,PA.
NADC-AM-	NAVAL AIR DEVELOPMENT CENTER. AERO MECHANICS DEPT., JOHNSVILLE, PENNA.
NADC-AM-	NAVAL AIR DEVELOPMENT CENTER. AERO MECHANICS DEPT., WARMINSTER, PENNA.
NADC-AP-	NAVAL AIR DEVELOPMENT CENTER. AERONAUTICAL PHOTOGRAPHIC EXPERIMENT LAB.
NADC AR-	NAVAL AIR DEVELOPMENT CENTER. AVIATION ARMAMENT LAB., JOHNSVILLE, PENNA.
NADC-ASW-	NAVAL AIR DEVELOPMENT CENTER. ANTI-SUBMARINE WARFARE LAB., JOHNSVILLE, PENNA.
NADC-AW-	NAVAL AIR DEVELOPMENT CENTER. AIR WARFARE RESEARCH DEPT., JOHNSVILLE, PENNA.
NADC-AW-	NAVAL AIR DEVELOPMENT CENTER. ANTI-SUBMARINE WARFARE LAB., JOHNSVILLE, PENNA.
NADC-ED-	NAVAL AIR DEV. CTR. ENG. DEV. LAB., JOHNSVILLE,PENNA.
NADC EL-	NAVAL AIR DEVELOPMENT CENTER. AERONAUTICAL ELECTRONIC AND ELECTRIC LAB., JOHNSVILLE, PENNA.
NADC-MA-	NAVAL AIR DEVELOPMENT CENTER, JOHNSVILLE, PENNA.
NADC-MA-	NAVAL AIR DEVELOPMENT CENTER. AERO MATERIALS DEPT., WARMINSTER, PENNA.
NADC MA-	NAVAL AIR DEVELOPMENT CENTER. AVIATION MEDICAL ACCELERATION LAB., JOHNSVILLE, PENNA.
NADC-MA-L-	NAVAL AIR DEVELOPMENT CENTER. AVIATION MEDICAL ACCELERATION LAB., JOHNSVILLE, PENNA.
NADC-ML-	NAVAL AIR DEVELOPMENT CENTER. AVIATION MEDICAL ACCELERATION LAB., JOHNSVILLE, PENNA.
NADC-MR-	NAVAL AIR DEVELOPMENT CENTER. AEROSPACE MEDICAL RESEARCH DEPT., JOHNSVILLE, PENNA.
NADC-R-	NAVAL AIR DEVELOPMENT CENTER, JOHNSVILLE, PENNA.
NAD-CRANE-	NAVAL AMMUNITION DEPOT, CRANE, IND.
NAD-CRANE-QE/C-(YR)-	NAVAL AMMUNITION DEPOT, CRANE, IND.
NAD-CR-MICRO-NOTES	NAVAL AMMUNITION DEPOT, CRANE, IND.
NAD-CR-QE/C-(YEAR)-	NAVAL AMMUNITION DEPOT, CRANE, IND.
NAD-CR-RDTR-	NAVAL AIR TURBINE TEST STATION, TRENTON
NAD-CR-RDTR-	NAVAL AMMUNITION DEPOT, CRANE, IND.
NAD-CR-ROTR-	NAVAL AMMUNITION DEPOT, CRANE, IND.
NADC-SD-	NAVAL AIR DEVELOPMENT CENTER. SYSTEMS ANALYSIS AND ENGINEERING DEPT., JOHNSVILLE, PENNA.

NADC-ST-

NADC-ST-	NAVAL AIR DEVELOPMENT CENTER. AERO STRUCTURES DEPT., JOHNSVILLE, PENNA.
NADC-SY-	NAVAL AIR DEVELOPMENT CENTER, JOHNSVILLE, PENNA.
NADC-WR-	NAVAL AIR DEVELOPMENT CENTER. AIR WARFARE RESEARCH DEPT., JOHNSVILLE, PENNA.
NAD-EL-	NAVAL AIR DEVELOPMENT CENTER, JOHNSVILLE, PENNA.
NADEVCEN-	NAVAL AIR DEVELOPMENT CENTER, JOHNSVILLE, PENNA.
NADH-	NAVAL AMMUNITION DEPOT, HAWTHORNE, NEV.
NAD/HAW-	NAVAL AMMUNITION DEPOT, HAWTHORNE, NEV.
NAD MATS-	MILITARY AIR TRANSPORT SERVICE. ATLANTIC DIV., WESTOVER AFB, MASS.
NAD-QE-	NAVAL AMMUNITION AND NET DEPOT. QUALITY EVALUATION LAB., SEAL BEACH, CALIF.
NAD-QE/C-(YEAR)-	NAVAL AMMUNITION DEPOT, CRANE, IND.
NAD-QE-OH-	NAVAL AMMUNITION DEPOT. QUALITY EVALUATION LAB., OAHU, HAWAII
NAD-R-	NAVAL AIR DEVELOPMENT CENTER, JOHNSVILLE, PENNA.
NAD-RDTR-	NAVAL AMMUNITION DEPOT, CRANE, IND.
NADS-	NAVAL AIR DEVELOPMENT CENTER, JOHNSVILLE, PENNA.
NAE-	(NOTES ON AVIATION ENGINEERS)
NAE-	NATIONAL ACADEMY OF ENGINEERING, WASHINGTON, D.C.
NAE-(LETTERS)-	CANADA. NATIONAL AERONAUTICAL ESTABLISHMENT, OTTAWA
NAEC-	CANADA. NATIONAL AERONAUTICAL ESTABLISHMENT, OTTAWA
NAEC-ACEL-	NAVAL AIR ENGINEERING CENTER. AEROSPACE CREW EQUIPMENT LAB., PHILADELPHIA
NAEC-AEL-	NAVAL AIR ENG. CTR. AERO. ENGINE LAB., PHILADELPHIA
NAEC-AML-	NAVAL AIR ENGINEERING CENTER. AERONAUTICAL MATERIALS LAB., PHILADELPHIA
NAEC-ASL-	NAVAL AIR ENGINEERING CENTER, PHILADELPHIA
NAEC-ENG-	NAVAL AIR ENGINEERING CTR. ENG. DEPT., PHILADELPHIA
NAEC LR-	CANADA. NATIONAL AERONAUTICAL ESTAB., OTTAWA (LAB.RPT)
NAEC N-	CANADA. NATIONAL AERONAUTICAL ESTAB., OTTAWA (NOTE)
NAEC-NAEL-	NAVAL AIR ENGINEERING CENTER. NAVAL AIR ENGINEERING LAB., PHILADELPHIA
NAEC R-	CANADA. NATIONAL AERONAUTICAL ESTABLISHMENT, OTTAWA
NAEF-ENG-	NAVAL AIR ENG. FACILITY. ENG. DEPT., PHILADELPHIA
NAEL-ENG-	NAVAL AIR ENGINEERING CENTER. NAVAL AIR ENGINEERING LAB., PHILADELPHIA
NAE-LR-	CANADA. NATIONAL AERONAUTICAL ESTAB., OTTAWA (LAB.RPT)
NAE-LTR-UA-	CANADA. NATIONAL AERONAUTICAL ESTABLISHMENT, OTTAWA (UNSTEADY AERODYNAMICS LAB. TECH. REPORT)
NAEM-	TURKEY. AEC. CEKMECE NUCLEAR RESEARCH CENTER, ISTANBUL
NAE-MISC-	CANADA. NATIONAL AERONAUTICAL ESTABLISHMENT, OTTAWA
NAE-MS-	CANADA. NATIONAL AERONAUTICAL ESTABLISHMENT, OTTAWA
NAES-	NAVAL AIR EXPERIMENTAL STATION, PHILADELPHIA
NAES AIL	NAVAL AIR EXPTL. STA. AERO. INSTRS. LAB., PHILA.
NAES AIL (NUMBER)-	NAVAL AIR EXPTL. STA. AERO. INSTRS. LAB., PHILA.
NAES ARRL-	NAVAL AIR EXPERIMENTAL STATION. AERONAUTICAL RADIO AND RADAR LAB., PHILADELPHIA
NAES INSTR-	NAVAL AIR DEV. CTR. AERO. INSTRS. LAB., JOHNSVILLE, PA.
NAESU-	NAVAL AVIATION ENGINEERING SERVICE UNIT
NAF-	NAVAL AIRCRAFT FACTORY, PHILADELPHIA
NAF-	NAVAL AVIONICS FACILITY, INDIANAPOLIS
NAFDU-	GT. BRIT. NAVAL AIR FIGHTING DEVELOPMENT UNIT
NAFI-(NUMBER)-GR	NAVAL AVIONICS FACILITY, INDIANAPOLIS
NAFI-	NAVAL AVIONICS FACILITY, INDIANAPOLIS
NAF IDS-	NAVAL AIRCRAFT FACTORY. INSTRS. DEV. SEC., PHILA.
NAFI-MRR-	NAVAL AVIONICS FACILITY, INDIANAPOLIS
NAFI-MTR-	NAVAL AVIONICS FACILITY, INDIANAPOLIS
NAFI-TP-	NAVAL AVIONICS FACILITY, INDIANAPOLIS
NAFI-TR-	NAVAL AVIONICS FACILITY, INDIANAPOLIS
NAF-MR-	NAVAL AVIONICS FACILITY, INDIANAPOLIS
NAF SPEC (LETTER)-	NAVAL AIRCRAFT FACTORY, PHILADELPHIA (SPECIFICATIONS)
NAGA-TN-	MICHIGAN. UNIV., ANN ARBOR
NA-GM-	NORTHROP AIRCRAFT, INC., HAWTHORNE, CALIF. (GUIDED MISSILES)
NAHRO-PUB-N(NUMBER)	NATIONAL ASSN. OF HOUSING + REDEVELOPMENT OFFICIALS, WASHINGTON, D.C.
NAI-(NO.)PR(MONTH)(YEAR)	NORTH AMERICAN INSTRUMENTS, INC., ALTADENA, CALIF.
NAI-	NORTHROP AIRCRAFT, INC., HAWTHORNE, CALIF.
NAI-(YR)-	NORTHROP AIRCRAFT, INC., HAWTHORNE, CALIF.
NAI-(NUMBERS)-	NORTHROP AIRCRAFT, INC., HAWTHORNE, CALIF.
NAI FTS(NO.)-	NORTHROP AIRCRAFT, INC., HAWTHORNE, CALIF. (FLIGHT TEST PROGRESS REPORTS)
NAI GM-	NORTHROP AIRCRAFT, INC., HAWTHORNE, CALIF.
NAI-LN-	NORTHROP AIRCRAFT, INC., HAWTHORNE, CALIF.
NAILSC-ILS-	NAVAL AVIATION INTEGRATED LOGISTIC SUPPORT CENTER, PATUXENT RIVER, MD.
NAILSC-ILS-IR-	NAVAL AVIATION INTEGRATED LOGISTIC SUPPORT CENTER, PATUXENT RIVER, MD.
NAI-MMPR-	NORTHROP AIRCRAFT, INC., HAWTHORNE, CALIF. (MISSILE MONTHLY PROGRESS REPORT)
NA-L-	COLUMBIA UNIV., NYC. SUBSTITUTE ALLOY MATERIALS LABS.
NAL-(NUMBER)-	NATIONAL ACCELERATOR LAB., BATAVIA, ILL.
NAL-	NATIONAL ACCELERATOR LAB., BATAVIA, ILL.
NAL-	NAVAL AERO. LAB., NAVAL AIR STA., BANANA RIVER, FLA.
NAL-EN-	NATIONAL ACCELERATOR LAB., BATAVIA, ILL.
NAL-FN-	NATIONAL ACCELERATOR LAB., OAK BROOK, ILL.
NAL-THY-	NATIONAL ACCELERATOR LAB., BATAVIA, ILL.
NAL-TM-	NATIONAL ACCELERATOR LAB., OAK BROOK, ILL.
NAL-TN-	INDIA. NATIONAL AERONAUTICAL LAB., BANGALORE
NAL-TR-	JAPAN. NATIONAL AEROSPACE LAB., TOKYO
NAL-TR-	NATIONAL ACCELERATOR LAB., BATAVIA, ILL.
NAL-TT-	INDIA. NATIONAL AERONAUTICAL LAB., BANGALORE
NA-M-	COLUMBIA UNIV., NYC. SUBSTITUTE ALLOY MATERIALS LABS.
NAM-	NAVAL AIR MATERIAL CENTER, PHILADELPHIA
NAM-	OFFICE OF NAVAL RESEARCH (NAVAL ANALYSIS MEMORANDUM)
NAM-AML-	NAVAL AIR MATERIAL CENTER. AERONAUTICAL MATERIALS LAB., PHILADELPHIA
NAMC-	NAVAL AIR MATERIAL CENTER, PHILADELPHIA
NAMC-ACEL-	NAVAL AIR MATERIAL CENTER. AIR CREW EQUIPMENT LAB., PHILADELPHIA
NAMC-AEL-	NAVAL AIR MATERIAL CENTER. AERO. ENGINE LAB., PHILA.
NAMC-AIL-	NAVAL AIR MATERIAL CENTER. AERONAUTICAL INSTRUMENTS LAB., PHILADELPHIA
NAMC AML-	NAVAL AIR EXPTL. STA. AERO. MATERIALS LAB., PHILA.
NAMC-AML-AE-	NAVAL AIR MATERIAL CENTER, PHILADELPHIA
NAMC-AML-RS-	NAVAL AIR MATERIAL CENTER. AERONAUTICAL MATERIALS LAB., PHILADELPHIA
NAMC-APEL-	NAVAL AIR MATERIAL CENTER. AERONAUTICAL PHOTOGRAPHIC EXPERIMENTAL LAB., PHILADELPHIA
NAMC-ASL-	NAVAL AIR MATERIAL CENTER. AERONAUTICAL STRUCTURES LAB., PHILADELPHIA
NA-MCH-	NORTH AMERICAN AVIATION, INC., LOS ANGELES
NAMI-	NAVAL AEROSPACE MEDICAL INST., PENSACOLA, FLA.
NAMI-	NAVAL PERSONNEL RESEARCH ACTIVITY, SAN DIEGO, CALIF.
NAMI-	NAVAL SCHOOL OF AVIATION MEDICINE, PENSACOLA, FLA.
NA/MJM/(NUMBER/YEAR)	NOTTINGHAM, ENGLAND. UNIV. DEPT. OF GEOGRAPHY (NETWORK ANALYSIS)(AUTHOR'S INITIALS)
NAMRI-	NAVAL MEDICAL RESEARCH INST., BETHESDA, MD.
NAMRL-	NAVAL AEROSPACE MEDICAL RESEARCH LAB., PENSACOLA, FLA.

228

NAMRU-	NAVAL MEDICAL RESEARCH UNIT NO...
NAMTC-	NAVAL AIR MISSILE TEST CENTER, POINT MUGU, CALIF.
NAMTC-MR-	NAVAL AIR MISSILE TEST CENTER, POINT MUGU, CALIF.
NAMTC-PR-	NAVAL AIR MISSILE TEST CENTER, POINT MUGU, CALIF.
NAMTC-TM-	NAVAL AIR MISSILE TEST CENTER, POINT MUGU, CALIF.
NAMTC-TR-	NAVAL AIR MISSILE TEST CENTER, POINT MUGU, CALIF.
NAMU-	NAVAL AIR DEVELOPMENT CENTER, JOHNSVILLE, PENNA.
NAND-	NAVAL AMMUNITION + NET DEPOT, SEAL BEACH, CALIF.
NAO-TN-	UNITED KINGDOM. ROYAL GREENWICH OBSERVATORY. NAUTICAL ALMANAC OFFICE, HERSTMONCEUX, SUSSEX
NAOTS-	NAVAL AVIATION ORDNANCE TEST STATION, CHINCOTEAGUE, VA
NAPCA-	NATIONAL AIR POLLUTION CONTROL ADMINISTRATION, WASHINGTON, D.C.
NAPLESU-	NAPLES. UNIVERSITA. ISTITUTO DI FISICA TEORICA
NAPLES-UK-	GT. BRIT. ATOMIC EN. RES. ESTAB., HARWELL, BERKS, ENG
NAPTC-AED-	NAVAL AIR PROPULSION TEST CENTER. AERONAUTICAL ENGINE DEPT., PHILADELPHIA
NAPTC-ATD-	NAVAL AIR PROPULSION TEST CENTER. AERONAUTICAL TURBINE DEPT., TRENTON
NAPTC-ATD-	NAVAL AIR TURBINE TEST STATION, TRENTON
NAR-	CALIFORNIA. UNIV., LOS ANGELES. NUMERICAL ANALYSIS RESEARCH
NAR-	NORTHROP AIRCRAFT, INC., HAWTHORNE, CALIF.
NAR-	OFFICE OF NAVAL RESEARCH (NAVAL ANALYSIS REPORT)
NARBA-	GOVERNMENT-INDUSTRY COMMITTEE ON NARBA PREPARATION
NA-RC-	NORTHROP AIRCRAFT, INC., HAWTHORNE, CALIF.
NARCO-	NATIONAL AERONAUTICAL CORP., AMBLER, PENNA.
NARD-	NORTHROP CORP. NORTRONICS DIV. APPLIED RESEARCH DEPT., NEWBURY PARK, CALIF.
NARD-TR-(NUMBER)-	NORTRONICS. APPLIED RES. DEPT., NEWBURY PARK, CALIF.
NAREC-REF-	NAVAL RESEARCH LAB., WASHINGTON, D.C.
NARF-	GENERAL DYNAMICS/FORT WORTH, TEX.
NARF-(YEAR)-	GENERAL DYNAMICS/FORT WORTH, TEX. (REF. 3)
NARF-	NAVAL AEROSPACE RECOVERY FACILITY, EL CENTRO, CALIF.
NARF-PR-OP-	CONVAIR. NUCLEAR AIRCRAFT RESEARCH FACILITY, FORT WORTH, TEX. (REF. 3)
NARMCO-	NARMCO INDUSTRIES, INC. RES. + DEV. DIV., SAN DIEGO, CAL
NARMCO-	WHITTAKER CORP. NARMCO RESEARCH AND DEVELOPMENT DIV., SAN DIEGO, CALIF.
NARMCO-QR-	NARMCO INDUSTRIES, INC. RES. + DEV. DIV., SAN DIEGO, CAL
NARMCO-STR-	NARMCO INDUSTRIES, INC. RES. + DEV. DIV., SAN DIEGO, CAL
NARS-	NATIONAL ARCHIVES AND RECORDS SERVICE, WASHINGTON, DC
NARTS-	NAVAL AIR ROCKET TEST STATION, DOVER, N.J.
NARTS-	NAVAL AIR ROCKET TEST STATION, LAKE DENMARK, N.J.
NARTS-TN-	NAVAL AIR ROCKET TEST STATION, LAKE DENMARK, N.J.
NAS-	(NAVY AERONAUTICAL SPECIFICATION)
NAS(CONTRACT NO.)-(LTR)-	AEROJET-GENERAL CORP., SACRAMENTO, CALIF. (LETTER INDICATES TYPE OF REPORT OR FREQUENCY)
NAS-	NATIONAL ACADEMY OF SCIENCES
NAS-(CONTRACT NUMBER)-	NATIONAL AERONAUTICS AND SPACE ADM., WASHINGTON, D.C.
NAS-	NAVAL AIR STATION, LAKEHURST, N.J.
NASA-	NATIONAL AERONAUTICS AND SPACE ADM., WASHINGTON, D.C. (REF. 26)
NASA-(LETTERS)-A-	AMES RESEARCH CENTER, MOFFETT FIELD, CALIF. (REF. 26)
NASA-(LETTERS)-H-	HIGH-SPEED FLIGHT STATION, EDWARDS, CALIF. (REF. 26)
NASA-(LETTERS)-L-	LANGLEY RESEARCH CENTER, LANGLEY FIELD, VA. (REF. 26)
NASA-(LETTERS)-E-	LEWIS RESEARCH CENTER, CLEVELAND (REF. 26)
NASA-(LETTERS)-W-	NATIONAL AERONAUTICS AND SPACE ADMINISTRATION. HEADQUARTERS, WASHINGTON, D.C. (REF. 26)
NASA-BIB-	NATIONAL AERONAUTICS AND SPACE ADM., WASHINGTON, D.C.
NASA-C-	NATIONAL AERONAUTICS AND SPACE ADM., WASHINGTON, D.C.
NASA-CASE-(LETTERS)-	NATIONAL AERONAUTICS AND SPACE ADMINISTRATION, WASHINGTON, D.C. (PATENT APPLICATION)
NASA-CASE-(HQN)-	NATIONAL AERONAUTICS AND SPACE ADMINISTRATION, WASHINGTON, D.C. (PATENT APPLICATION)
NASA-CASE-(XMS)-	NATIONAL AERONAUTICS AND SPACE ADMINISTRATION, WASHINGTON, D.C. (PATENT APPLICATION)
NASA-CCN-	NATIONAL AERONAUTICS AND SPACE ADM., WASHINGTON, D.C.
NASA-CF-	NATIONAL AERONAUTICS AND SPACE ADM., WASHINGTON, D.C.
NASA-CR-	NATIONAL AERONAUTICS AND SPACE ADMINISTRATION, WASHINGTON, D.C. (CONTRACT REPORT)
NASA-E-	NATIONAL AERONAUTICS AND SPACE ADMINISTRATION. RES. ADVISORY COMM. ON ELECTRICAL POWER PLANT SYSTEMS, WASHINGTON, D.C.
NASA-EP-	NATIONAL AERONAUTICS AND SPACE ADMINISTRATION. EDUCATIONAL PROGRAMS DIV., WASHINGTON, D.C.
NASA-GODDARD NEWS-	GODDARD SPACE FLIGHT CENTER, GREENBELT, MD.
NASA-KSC-	JOHN F. KENNEDY SPACE CENTER, COCOA BEACH, FLA.
NASA-KSC-DTI-	JOHN F. KENNEDY SPACE CENTER, COCOA BEACH, FLA. (DESIGN TECHNICAL INSTRUCTION)
NASA-KSC-MAB-(NO.-YR.)	JOHN F. KENNEDY SPACE CENTER, COCOA BEACH, FLA.
NASA-KSC-TEST-	JOHN F. KENNEDY SPACE CENTER, COCOA BEACH, FLA.
NASA-KSC-TM-	JOHN F. KENNEDY SPACE CENTER, COCOA BEACH, FLA.
NASA-KSC-TR-	JOHN F. KENNEDY SPACE CENTER, COCOA BEACH, FLA.
NASA-LS-	NATIONAL AERONAUTICS AND SPACE ADMINISTRATION, COLLEGE PARK, MD. (LITERATURE SEARCH NUMBER...)
NASA-M-(DATE LETTER)	NATIONAL AERONAUTICS AND SPACE ADM., WASHINGTON, D.C. (MEMORANDUM) (REF. 26)
NASA-MEMO-(DATE LETTER)	NATIONAL AERONAUTICS AND SPACE ADM., WASHINGTON, D.C. (REF. 26)
NASA-MEMORANDUM(DATE LTR)	NATIONAL AERONAUTICS AND SPACE ADM., WASHINGTON, D.C. (REF. 26)
NASA-MHR-	GEORGE C. MARSHALL SPACE FLIGHT CTR., HUNTSVILLE, ALA.
NASA-MSC-	MANNED SPACECRAFT CENTER, HOUSTON, TEX.
NASA-MSC-G-R-	MANNED SPACECRAFT CENTER, HOUSTON, TEX.
NASA NEWS RELEASE-(YR.)-	NATIONAL AERONAUTICS AND SPACE ADM., WASHINGTON, D.C.
NASA-NPC-	NATIONAL AERONAUTICS AND SPACE ADMINISTRATION, WASHINGTON, D.C. (QUALITY PUBLICATION)
NASA-PA-(NUMBER LETTER)	NATIONAL AERONAUTICS AND SPACE ADM., WASHINGTON, D.C. (PUBLICATION ANNOUNCEMENTS) (LETTER AT END STANDS FOR SECURITY CLASSIFICATION)
NASA-R-	NATIONAL AERONAUTICS AND SPACE ADM., WASHINGTON, D.C.
NAS ARDC COM-...	AIR RES. + DEV. COMMAND, ANDREWS AFB, MD. (COMM.RPT.)
NAS ARDC COM-...	NATIONAL ACADEMY OF SCIENCES
NAS ARDC COM-...	NATIONAL RESEARCH COUNCIL, WASHINGTON, D.C.
NASA-RE-(DATE LETTER)	NATIONAL AERONAUTICS AND SPACE ADM., WASHINGTON, D.C. (REPUBLICATION) (REF. 26)
NASA-REPUBLICATION...	NATIONAL AERONAUTICS AND SPACE ADMINISTRATION, WASHINGTON, D.C. (REF. 26)
NASA-R-ME-IV-S-	NATIONAL AERONAUTICS AND SPACE ADM., WASHINGTON, D.C.
NASA-RP-	NATIONAL AERONAUTICS AND SPACE ADM., WASHINGTON, D.C. (REPRINT SERIES)
NASA-SP-	NATIONAL AERONAUTICS AND SPACE ADMINISTRATION, WASHINGTON, D.C. (SELECTED PUBLICATIONS)
NASA-TM-X-	NATIONAL AERONAUTICS AND SPACE ADMINISTRATION, WASHINGTON, D.C. (TECHNICAL MEMORANDUM) (REF. 26)
NASA-TN-D-	NATIONAL AERONAUTICS AND SPACE ADMINISTRATION, WASHINGTON, D.C. (TECHNICAL NOTE) (REF. 26)
NASA-TPA-INDEX-	NATIONAL AERONAUTICS AND SPACE ADMINISTRATION, WASHINGTON, D.C. (INDEX OF TECHNICAL PUBLICATIONS ANNOUNCEMENTS)
NASA-TP-INDEX-	NATIONAL AERONAUTICS AND SPACE ADMINISTRATION, WASHINGTON, D.C. (INDEX OF TECHNICAL PUBLICATIONS)
NASA-TR-R-	NATIONAL AERONAUTICS AND SPACE ADMINISTRATION, WASHINGTON, D.C. (TECHNICAL REPORT) (REF. 26)
NASA-TT-F-	NATIONAL AERONAUTICS AND SPACE ADMINISTRATION, WASHINGTON, D.C. (TECHNICAL TRANSLATION) (REF. 26)
NA-SB-(NOS.)-(NO.)-	NORTH AMERICAN AVIATION, INC. LOS ANGELES DIV.

NASC-
- NAVAL AIR SYSTEMS COMMAND, WASHINGTON, D.C.

NAS-CR-
- NATIONAL ACADEMY OF SCIENCES

NASC-TRW-(YEAR)-
- TRW SYSTEMS GROUP, REDONDO BEACH, CALIF.

NASL-
- NAVAL APPLIED SCIENCE LAB., BROOKLYN

NASL-9400-(NUMBER)-PR-
- NAVAL APPLIED SCIENCE LAB., BROOKLYN

NASL-(LETTERS)-
- NAVAL APPLIED SCIENCE LAB., BROOKLYN

NASL-IED-(NO.)-TM-
- NAVAL APPLIED SCIENCE LAB., BROOKLYN

NASL-INTEGRATION STDY-
- NAVAL APPLIED SCIENCE LAB., BROOKLYN

NASL-IR-
- NAVAL APPLIED SCIENCE LAB., BROOKLYN

NASL-PR-
- NAVAL APPLIED SCIENCE LAB., BROOKLYN

NASL-TM-
- NAVAL APPLIED SCIENCE LAB., BROOKLYN

NASL-VN-
- NAVAL APPLIED SCIENCE LAB., BROOKLYN

NASL-VNG-
- NAVAL APPLIED SCIENCE LAB., BROOKLYN

NAS-NRC-
- NATIONAL ACADEMY OF SCIENCES + NATIONAL RES. COUNCIL

NAS-NRC/PUB-
- NATIONAL ACADEMY OF SCIENCES

NAS-NRC/PUB-
- NATIONAL RESEARCH COUNCIL, WASHINGTON, D.C.

NAS-NS-
- GHENT. RIJKSUNIVERSITEIT

NAS-NS-
- NATIONAL ACADEMY OF SCIENCES (NUCLEAR SCIENCE SERIES)

NAST-
- BENDIX CORP. BENDIX SYSTEMS DIV., ANN ARBOR, MICH.

NASWF-
- NAVAL AIR SPEC. WEAPONS FACILITY, KIRTLAND AFB, N.MEX

NASWF (LTRS, LTRS)-
- NAVAL AIR SPEC. WEAPONS FACILITY, KIRTLAND AFB, N.MEX

NASWF-(LTRS,LTRS-LTRS)-
- NAVAL AIR SPEC. WEAPONS FACILITY, KIRTLAND AFB, N.MEX

NAT-
- NAVAL AIR TRAINING COMMAND, PENSACOLA, FLA.

NATC-
- NAVAL AIR TEST CENTER, PATUXENT RIVER, MD.

NATC-AT-
- NAVAL AIR TEST CENTER. ARMAMENT TEST DIV., PATUXENT RIVER, MD.

NATC ET-
- NAVAL AIR TEST CENTER. ELECTRONICS TEST DIV., PATUXENT RIVER, MD.

NATC-FT (NUMBER)-
- NAVAL AIR TEST CENTER, PATUXENT RIVER, MD.

NATC-FT-DIR-
- NAVAL AIR TEST CENTER, PATUXENT RIVER, MD.

NATC-ST-IR
- NAVAL AIR TEST CENTER, PATUXENT RIVER, MD.

NATC-TR-
- NAVAL AIR TEST CENTER, PATUXENT RIVER, MD.

NATC-TSD-IR
- NAVAL AIR TEST CENTER, PATUXENT RIVER, MD.

NATC-WST-(NUMBER-YEAR)
- NAVAL AIR TEST CENTER, PATUXENT RIVER, MD.

NATEC-
- NAVAL AIRSHIP TRAINING + EXPTL. COMMAND, LAKEHURST,NJ

NA-TFD-(YEAR)-
- NORTH AMERICAN AVIATION, INC., LOS ANGELES
  (SEE ALSO: AUTONETICS DIV., NORTH AMERICAN AVIATION)

NA-TFD-(NUMBERS)-
- NORTH AMERICAN AVIATION, INC. LOS ANGELES DIV.

NATF-E-
- NAVAL AIR TEST FACILITY. SHIP INSTALLATIONS, LAKEHURST, N.J.

NATF-EN-
- NAVAL AIR TEST FACILITY. SHIP INSTALLATIONS, LAKEHURST, N.J.

NATF-SI-R(NUMBER)
- NAVAL AIR TEST FACILITY. SHIP INSTALLATIONS, LAKEHURST, N.J.

NA-TO-
- NORTH AMERICAN AVIATION, INC. LOS ANGELES DIV.

NATO-AC/
- NORTH ATLANTIC TREATY ORGANIZATION, BRUSSELS

NATO-DC-
- NORTH ATLANTIC TREATY ORGANIZATION. DEFENSE COLLEGE, PARIS

NATS-
- MILITARY AIR TRANSPORT SERVICE, WASHINGTON, D.C.

NATS-NVOO-
- NEVADA OPERATIONS OFFICE, AEC, LAS VEGAS (NEVADA AERIAL TRACKING SYSTEM) (REF. 4)

NATS-PR-(NUMBER)-P-(YR.)
- NEVADA OPERATIONS OFFICE, AEC, LAS VEGAS (NEVADA AERIAL TRACKING SYSTEM) (REF. 4)

NATTS-ATL-
- NAVAL AIR TURBINE TEST STATION, TRENTON

NATTS-ATL-
- NAVAL AIR TURBINE TEST STATION. AERONAUTICAL TURBINE LAB., TRENTON

NATTS-ATL-TN-
- NAVAL AIR TURBINE TEST STATION. AERONAUTICAL TURBINE LAB., TRENTON

NATU-
- NAVAL AIR STATION, QUONSET POINT, R.I.

NATU-
- NAVAL AIRCRAFT TORPEDO UNIT, QUONSET POINT, R.I.

NATU-ITM-(NO.)-(YR.)
- NAVAL AIR STATION, QUONSET POINT, R.I. (INTERNAL TECHNICAL MEMORANDUM)

NAV-
- NAVY

NAV AEL-
- NAVAL AIR EXPERIMENTAL STA. AERO. ENGINE LAB., PHILA.

NAVAER-
- BUREAU OF AERONAUTICS (NAVY)

NAVAER-
- BUREAU OF AERONAUTICS (NAVY) (SPECIFICATIONS)

NAVAER-(YR)-1R-
- OFFICE OF THE CHIEF OF NAVAL OPERATIONS, WASH., D.C.

NAVAER-(YR)-1T-
- OFFICE OF THE CHIEF OF NAVAL OPERATIONS, WASH., D.C.

NAVAER ADR-
- BUREAU OF AERONAUTICS (NAVY) (AVIATION DESIGN RESEARCH REPORTS)

NAVAER ADR (LETTER)-
- BUREAU OF AERONAUTICS (NAVY) (AVIATION DESIGN RESEARCH REPORTS)

NAVAER-AROWA-
- BUREAU OF AERONAUTICS (NAVY) (AROWA PROJECT)

NAVAERORECOVFAC-
- NAVAL AEROSPACE RECOVERY FACILITY, EL CENTRO, CALIF.

NAVAER PAD-
- BUREAU OF AERONAUTICS (NAVY)(PRELIM. AIRCRAFT DESIGN)

NAVAER SB-
- BUREAU OF AERONAUTICS (NAVY) (STRUCTURES BULLETIN)

NAVAER SI-
- BUREAU OF AERONAUTICS (NAVY) (SPERRY INSTRUCTIONS)

NAVAER SM-
- BUREAU OF AERONAUTICS (NAVY) (STRUCTURES MEMO)

NAVAER TN-
- BUREAU OF AERONAUTICS (NAVY) (TECHNICAL NOTES)

NAVAIR-17-20(LETTERS)-
- BUREAU OF NAVAL WEAPONS, WASHINGTON, D.C.

NAVAIR-17-20(LETTERS)-
- BUREAU OF NAVAL WEAPONS REPRESENTATIVE, POMONA, CAL.

NAVAIR-(NO.)-20(LTRS.)-
- NAVAL AIR SYSTEMS COMMAND, WASHINGTON, D.C.

NAVAIR-(NUMBERS,LTRS.)-
- NAVAL AIR SYSTEMS COMMAND, WASHINGTON, D.C.

NAVAIRMATCEN-ASL-
- NAVAL AIR MATERIAL CENTER, PHILADELPHIA

NAVAIRTORPU-TM-
- NAVAL AIRCRAFT TORPEDO UNIT, QUONSET POINT, R.I.

NAV-AS-
- NAVAL AIR STATION, NORTH ISLAND, SAN DIEGO, CALIF.

NAVASC-
- NAVAL AIR SYSTEMS COMMAND, WASHINGTON, D.C.

NAV AS/SD MISC.
- NAVAL AIR STATION, NORTH ISLAND, SAN DIEGO, CALIF.

NAVCERELAB-
- NAVAL CIVIL ENGINEERING RESEARCH AND EVALUATION LAB., PORT HUENEME, CALIF.

NAVCG-
- COAST GUARD, WASHINGTON, D.C.

NAVCOMP-
- OFFICE OF THE COMPTROLLER OF THE NAVY, WASHINGTON, DC

NAVCOSSACT-
- NAVAL COMMAND SYSTEMS SUPPORT ACTIVITY, WASHINGTON,DC

NAVCOSSACT-(NUMBER)-
- NAVAL COMMAND SYSTEMS SUPPORT ACTIVITY, WASHINGTON,DC

NAV-C+RB-
- BUREAU OF CONSTRUCTION AND REPAIR (NAVY)

NAVDOCKS-
- BUREAU OF YARDS AND DOCKS

NAVDOCKS-P-
- BUREAU OF YARDS AND DOCKS

NAVDOCKS TP-(LETTERS)-
- BUREAU OF YARDS AND DOCKS

NAVDOCKS-TP-PL-
- BUREAU OF YARDS AND DOCKS

NAVDOCKS-TP-TE-
- BUREAU OF YARDS AND DOCKS

NAV EES-
- NAVAL ENGINEERING EXPERIMENT STATION, ANNAPOLIS

NAV EES (LETTER(S))-
- NAVAL ENGINEERING EXPERIMENT STATION, ANNAPOLIS

NAVELECSYSCOM-
- NAVAL ELECTRONIC SYSTEMS COMMAND, WASHINGTON, D.C.

NAVENGRXST-
- NAVAL ENGINEERING EXPERIMENT STATION, ANNAPOLIS

NAVEODFAC-
- NAVAL EXPLOSIVE ORDNANCE DISPOSAL FACILITY, INDIAN HEAD, MD.

NAVEODFAC EODR-
- NAVAL EXPLOSIVE ORDNANCE DISPOSAL FACILITY, INDIAN HEAD, MD. (EXPLOSIVE ORDNANCE DISPOSAL REPORT)

NAVEODRAC-
- NAVAL EXPLOSIVE ORDNANCE DISPOSAL FACILITY, INDIAN HEAD, MD.

NAVESC-
- NAVAL ELECTRONIC SYSTEMS COMMAND, WASHINGTON, D.C.

NAVESOS-
- OFFICE OF NAVAL RESEARCH, WASHINGTON, D.C.

NAVEXOS-
- NAVY. EXECUTIVE OFFICE OF THE SECRETARY, WASH., D.C.

NAVEXOS-P-
- MICHIGAN. UNIV., ANN ARBOR

NAVEXOS-P-
- OFFICE OF NAVAL RESEARCH, WASHINGTON, D.C.

NAVFAC-DM-
- NAVAL FACILITIES ENGINEERING COMMAND, WASHINGTON, DC (DESIGN MANUAL)

NAVFEC-
- NAVAL FACILITIES ENGINEERING COMMAND, WASHINGTON, DC

NAV-FWC-
- NAVAL AIR STATION. FLEET WEATHER CTR.,SAN DIEGO, CAL

NAV-GMRWG-	INTER-BUREAU TECHNICAL COMMITTEE (NAVY). GUIDED MISSILE RELAY WORKING GROUP
NAVHO-	HYDROGRAPHIC OFFICE, SUITLAND, MD.
NAV-IBTC-	INTER-BUREAU TECHNICAL COMMITTEE (NAVY), WASH., D.C.
NAV-ILCF-	INTER-LABORATORY COMMITTEE ON FACILITIES (NAVY)
NAVJAP-	(NAVAL ACTIVITIES JAPAN)
NAVMAT-INST-	NAVAL MATERIAL COMMAND, WASHINGTON, D.C.
NAVMC-	MARINE CORPS, WASHINGTON, D.C.
NAVMD-	NAVAL MINE DEPOT, YORKTOWN, VA.
NAVMED-	BUREAU OF MEDICINE AND SURGERY (NAVY)
NAVMED-(NUMBERS)	BUREAU OF MEDICINE AND SURGERY(NAVY), WASHINGTON,D.C.
NAVMED-M-(NUMBERS)	BUREAU OF MEDICINE AND SURGERY(NAVY), WASHINGTON,D.C.
NAVMED-MF(NUMBER)-	NAVAL MEDICAL RESEARCH INST., BETHESDA, MD.
NAVMED-MR(NUMBER)-	NAVAL SUBMARINE MEDICAL CENTER, GROTON, CONN.
NAVMED-P-	BUREAU OF MEDICINE AND SURGERY (NAVY)
NAVMED PROJ X-	BUREAU OF MEDICINE AND SURGERY (NAVY)
NAV-MINS-	MARE ISLAND NAVAL SHIPYARD, CALIF.
NAVML-	NAVAL MATERIAL LAB., BROOKLYN
NAVMRI-	NAVAL MEDICAL RESEARCH INST., BETHESDA, MD.
NAV MRL-	NAVAL MEDICAL RESEARCH LAB., NEW LONDON, CONN.
NAV-NNS-	NORFOLK NAVAL SHIPYARD, PORTSMOUTH, VA.
NAVNUPWRU-	NAVAL NUCLEAR POWER UNIT, FORT BELVOIR, VA.
NAVOBSY-	NAVAL OBSERVATORY, WASHINGTON, D.C.
NAVOCEANO-	NAVAL OCEANOGRAPHIC OFFICE, WASHINGTON, D.C.
NAV-OD-	BUREAU OF ORDNANCE (NAVY)
NAV-OP-	OFFICE OF THE CHIEF OF NAVAL OPERATIONS, WASH., D.C.
NAVOP-	OFFICE OF THE CHIEF OF NAVAL OPERATIONS, WASH., D.C.
NAVORD-	BUREAU OF NAVAL WEAPONS
NAVORD AIS-	BUREAU OF ORDNANCE (NAVY) (AD INTERIM SPECIFICATION)
NAVORD CPR-	BUREAU OF ORDNANCE (NAVY) (COMPUTATION PROJECT RPTS.)
NAVORD ER-	BUREAU OF ORDNANCE (NAVY) (EXPLOSIVES RES. RPT.)
NAVORD ER MEMO-	BUREAU OF ORDNANCE (NAVY) (EXPLOSIVES RES. MEMORANDA)
NAVORD INST-	BUREAU OF ORDNANCE (NAVY) (INSTRUCTION)
NAVORD-NOTS-	NAVAL ORDNANCE TEST STA., INYOKERN (CHINA LAKE), CAL.
NAVORD OCL-	BUREAU OF ORDNANCE (NAVY) (ORDNANCE CIRCULAR LETTER)
NAVORD OD-	BUREAU OF ORDNANCE (NAVY) (ORDNANCE DATA)
NAVORD-OFC-	NAVAL ORDNANCE TEST STA., INYOKERN (CHINA LAKE), CAL.
NAVORD OHI-	BUREAU OF ORDNANCE (NAVY) (ORDNANCE HANDLING INSTRUCTIONS)
NAVORD OMI-	BUREAU OF ORDNANCE (NAVY) (ORDNANCE MODIFICATION INSTRUCTIONS)
NAVORD OP-	BUREAU OF ORDNANCE (NAVY) (ORDNANCE PAMPHLET)
NAVORD OS-	BUREAU OF ORDNANCE (NAVY) (ORDNANCE SPECIFICATION)
NAVORD OS-	NAVAL ORDNANCE LAB., WHITE OAK, MD.
NAVORD OSTD-	BUREAU OF ORDNANCE (NAVY) (ORDNANCE STANDARD)
NAVORD OSTD-	BUREAU OF ORDNANCE (NAVY) (ORDNANCE STD. TECH. DATA)
NAVORD OTI-	BUREAU OF ORDNANCE (NAVY) (ORDNANCE TECHNICAL INSTRUCTIONS)
NAVORD-RDAF-	NAVAL ORDNANCE TEST STA., INYOKERN (CHINA LAKE), CAL.
NAVORD-RRB-	NAVAL ORDNANCE TEST STA., INYOKERN (CHINA LAKE), CAL.
NAVORD SWOP-	JOINT ATOMIC WEAPONS PUBLICATIONS BOARD, ALBUQUERQUE, N. MEX. (REF. 14)
NAVORD-TM-	NAVAL ORDNANCE TEST STA., INYOKERN (CHINA LAKE), CAL.
NAVORD-TN-	BUREAU OF ORDNANCE (NAVY)
NAV PDH-	NAVY (PASSIVE DEFENSE HANDBOOK)
NAVPERS-	BUREAU OF NAVAL PERSONNEL
NAVPERS-PRD-PRM-(YEAR)-	BUREAU OF NAVAL PERSONNEL. PERSONNEL RESEARCH DIV., WASHINGTON, D.C.
NAVPERS-PRD-PRR-(YEAR)-	BUREAU OF NAVAL PERSONNEL. PERSONNEL RESEARCH DIV., WASHINGTON, D.C.
NAVPERS-PRD-PTB-(YEAR)-	BUREAU OF NAVAL PERSONNEL. PERSONNEL RESEARCH DIV., WASHINGTON, D.C.
NAVPERS-PRM-	BUREAU OF NAVAL PERSONNEL. PERSONNEL RESEARCH DIV., WASHINGTON, D.C.
NAVPERS RRP-	BUREAU OF NAVAL PERSONNEL (RESEARCH REPORT PROJS.)
NAVPERS STRP-	BUREAU OF NAVAL PERSONNEL (SELECTION TEST REPORT PROJECTS)
NAV-PGS-	NAVAL POSTGRADUATE SCHOOL, MONTEREY, CALIF.
NAVPHOTOCEN-R/D-	NAVAL PHOTOGRAPHIC CENTER. RESEARCH AND DEVELOPMENT DEPT., WASHINGTON, D.C.
NAVPRO-	NAVAL PLANT REPRESENTATIVE OFFICE, COLUMBUS, OHIO
NAVRECONTECHSUPPCEN-	NAVAL RECONNAISSANCE AND TECHNICAL SUPPORT CENTER, WASHINGTON, D.C.
NAVRETRACOM-	NAVAL RETRAINING COMMAND, SAN DIEGO, CALIF.
NAVSANDA-	BUREAU OF SUPPLIES AND ACCOUNTS (NAVY)
NAVSANDA-	BUREAU OF SUPPLIES + ACCOUNTS (NAVY) (SPECIFICATIONS)
NAVSANDA SPEC-	BUREAU OF SUPPLIES + ACCOUNTS (NAVY) (SPECIFICATIONS)
NAVSEC-	NAVAL SHIP ENGINEERING CTR. PORT HUENEME DIV., CALIF.
NAVSECPHILADIV-A-	NAVAL SHIP ENGINEERING CENTER, PHILADELPHIA
NAVSECPHILADIV-AP-	NAVAL SHIP ENGINEERING CENTER, PHILADELPHIA
NAVSECPHILADIV-B-	NAVAL SHIP ENGINEERING CENTER, PHILADELPHIA
NAVSECPHILADIV-C-	NAVAL SHIP ENGINEERING CENTER, PHILADELPHIA
NAVSECPHILADIV-FB-	NAVAL SHIP ENGINEERING CENTER, PHILADELPHIA
NAVSEC-X-	NAVAL SHIP ENGINEERING CENTER, PHILADELPHIA
NAVSHIPS-	NAVAL SHIP SYSTEMS COMMAND, WASHINGTON, D.C.
NAVSHIPS B-	BUREAU OF SHIPS (BLUDWORTH REPORTS)
NAVSHIPS C-	BUREAU OF SHIPS
NAVSHIPS CP-	BUREAU OF SHIPS (COMPUTATION PROJECT)
NAVSHIPS CPR-	BUREAU OF SHIPS (COMPUTATION PROJECT REPORTS)
NAVSHIPS EL-	BUREAU OF SHIPS. ESSO LAB.
NAVSHIPS ENG-	BUREAU OF SHIPS (ELECTRONICS INSTRUCTION BOOK)
NAVSHIPS I-	BUREAU OF SHIPS (INSTRUCTIONS)
NAVSHIPS IB-	BUREAU OF SHIPS (INSTRUCTION BOOK)
NAVSHIPS IBT-	BUREAU OF SHIPS (INSTRUCTION BOOK, GAROD PART)
NAVSHIPS IC IB-	BUREAU OF SHIPS (INSTRUCTION BOOK)
NAVSHIPS IN-	BUREAU OF SHIPS (INDUSTRIAL NOTES)
NAVSHIPS ITL-	PHILADELPHIA NAVAL SHIPYARD. INDUSTRIAL TEST LAB.
NAVSHIPS MLT-	BUREAU OF SHIPS. MATERIALS LAB. (TESTS)
NAVSHIPS RA-	BUREAU OF SHIPS (RADIO INSTRUCTION BOOK)
NAVSHIPS RE-	BUREAU OF SHIPS (ELECTRONICS INSTRUCTION BOOK)
NAVSHIPS RL-	MARE ISLAND NAVAL SHIPYARD. RUBBER LAB., CALIF.
NAVSHIPS RM-	BUREAU OF SHIPS (RESEARCH MEMORANDA)
NAVSHIPS RM (NUMBER)-	BUREAU OF SHIPS (RESEARCH MEMORANDA)
NAVSHIPS RW-	BUREAU OF SHIPS (ELECTRONICS INSTRUCTION BOOK)
NAVSHIPS SHIPS-	BUREAU OF SHIPS (ELECTRONICS INSTRUCTION BOOK)
NAVSHIPS SPEC-	BUREAU OF SHIPS (SPECIFICATIONS)
NAVSHIPS SSC-	BUREAU OF SHIPS. SHIPS STRUCTURAL COMMITTEE
NAVSHIPS SSS-	BUREAU OF SHIPS (SUBMARINE SIGNAL CO. SPECIFICATIONS)
NAVSHIPS TLR-	BUREAU OF SHIPS. SHIPBUILDING DIVISION. RESEARCH AND STANDARDS BRANCH (TECH. LITERATURE RES. SERIES)
NAVSHIPS UER-	BUREAU OF SHIPS. UNDERWATER EXPLOSION RESEARCH GROUP
NAVSO-	MICHIGAN. UNIV., ANN ARBOR
NAVSPASUR-	NAVAL SPACE SURVEILLANCE SYSTEM, DAHLGREN, VA.
NAVSSC-	NAVAL SUPPLY SYSTEMS COMMAND, WASHINGTON, D.C.
NAVTECHMISJAP-	NAVAL TECHNICAL MISSION TO JAPAN
NAV TEC ICA-	NAVAL TECHNICAL INTELLIGENCE CENTER (ABSTRACT)

NAV TEC MIS EU-
NAVAL TECHNICAL MISSION IN EUROPE
NAV TEC MIS EU LR...-
NAVAL TECHNICAL MISSION IN EUROPE (LETTER REPORTS)
NAV TEC MIS EU TR-
NAVAL TECHNICAL MISSION IN EUROPE
NAV TEC MIS EU TR (NO.)-
NAVAL TECHNICAL MISSION IN EUROPE
NAV TEC MIS JAP-
NAVAL TECHNICAL MISSION TO JAPAN
NAV TEC MIS JAP (LTR.)-
NAVAL TECHNICAL MISSION TO JAPAN
NAV TEC MIS JAP E-
NAVAL TECHNICAL MISSION TO JAPAN (ELECTRONICS RPTS.)
NAV TEC MIS JAP M-
NAVAL TECH. MISSION TO JAPAN (MEDICAL TARGET RPTS.)
NAV TEC MIS JAP O-
NAVAL TECHNICAL MISSION TO JAPAN (ORDNANCE REPORTS)
NAV TEC MIS JAP S-
NAVAL TECHNICAL MISSION TO JAPAN (SUBJECT REPORTS)
NAV TEC MIS JAP X-
NAVAL TECHNICAL MISSION TO JAPAN (MISCELLANEOUS RPTS)
NAV TEC UN EU-
NAVAL TECHNICAL UNIT IN EUROPE
NAV-TME-
NAVAL TECHNICAL MISSION IN EUROPE
NAVTORPSTA-
NAVAL TORPEDO STATION, KEYPORT, WASH.
NAVTRADEVCEN-
NAVAL TRAINING DEVICES CENTER, PORT WASHINGTON, N.Y.
NAVTRADEVCEN-IH-
NAVAL TRAINING DEVICES CENTER, ORLANDO, FLA.
NAVUSL-
NAVY UNDERWATER SOUND LAB., NEW LONDON, CONN.
NAVWAG-
CENTER FOR NAVAL ANALYSES. NAVAL WARFARE ANALYSIS GROUP, WASHINGTON, D.C.
NAVWAG-IRM-
CENTER FOR NAVAL ANALYSES. NAVAL WARFARE ANALYSIS GROUP, WASHINGTON, D.C.
NAVWAG-RESEARCH CONTRIB-
CENTER FOR NAVAL ANALYSES. NAVAL WARFARE ANALYSIS GROUP, WASHINGTON, D.C.
NAVWEARSCHFAC-TP-
NAVY WEATHER RESEARCH FACILITY, NORFOLK, VA.
NAVWEPS-
BUREAU OF NAVAL WEAPONS
NAVWEPSHAC-
BUREAU OF NAVAL WEAPONS. HYDROBALLISTICS ADVISORY COMMITTEE, WASHINGTON, D.C.
NAVWEPS-OD-
BUREAU OF NAVAL WEAPONS
NAVWESS-
NATIONAL AVIATION WEATHER SYSTEMS STUDY, WASH., D.C.
NAVWPNQAO-TR-(YEAR)-
NAVAL WEAPONS QUALITY ASSURANCE OFFICE, WASH., D.C.
NAVY-SPO-
SPECIAL PROJECTS OFFICE (NAVY), WASHINGTON, D.C.
NAVY SWOP-
JOINT ATOMIC WEAPONS PUBLICATIONS BOARD, ALBUQUERQUE, N. MEX. (REF. 14)
NAXSTA-
NAVAL AIR EXPERIMENTAL STATION, PHILADELPHIA
NB-
(NOTEBOOK) (REF. 2)
NB-
HUGHES AIRCRAFT CO., CULVER CITY, CALIF.
NB-(YEAR)-
NORTHROP CORP. NORAIR DIV., HAWTHORNE, CALIF.
NBC-
NATIONAL BATTERY CO., DEPEW, N. Y.
NBC-
NATIONAL BROADCASTING CO., INC., N.Y.C.
NBDC-TB-(NUMBER-YEAR)
BATTELLE MEMORIAL INST. NATIONAL BOMB DATA CENTER, COLUMBUS, OHIO
NBL-
NEW BRUNSWICK LAB., AEC, N.J.
NBR-
NATIONAL BIOMEDICAL RESEARCH FDN., SILVER SPRING, MD.
NBS-(NO. LETTER)-
NATIONAL BUREAU OF STANDARDS, WASHINGTON, D.C.
NBS-
NATIONAL BUREAU OF STANDARDS, WASHINGTON, D.C.
NBS-(LETTER(S))-
NATIONAL BUREAU OF STANDARDS, WASHINGTON, D.C.
NBS-AMS-
NATIONAL BUREAU OF STANDARDS, WASHINGTON, D.C.
NBS-BMS-
NATIONAL BUREAU OF STANDARDS, WASHINGTON, D.C. (BUILDING MATERIALS AND STRUCTURES)
NBS-BSS-
NATIONAL BUREAU OF STANDARDS, WASHINGTON, D.C.
NBS-BSS-
NATIONAL BUREAU OF STANDARDS. BUILDING RESEARCH DIV., WASHINGTON, D.C. (BUILDING SCIENCE SERIES)
NBS-C-
NATIONAL BUREAU OF STANDARDS, WASH., D.C. (CIRCULAR)
NBS-CEL-
NATIONAL BUREAU OF STANDARDS. CRYOGENIC ENGINEERING LABS., BOULDER, COLO.
NBS-CIRC-
NATIONAL BUREAU OF STANDARDS, WASHINGTON, D.C.
NBS-CM-
NATIONAL BUREAU OF STANDARDS. BOULDER LABS., COLO.
NBS-CRPL-
NATIONAL BUREAU OF STANDARDS. BOULDER LABS., COLO.
NBS-CRPL-
NATIONAL BUREAU OF STANDARDS. CENTRAL RADIO PROPA-GATION LAB., WASHINGTON, D.C.

NBS-CRRL-
NATIONAL BUREAU OF STANDARDS. CENTRAL RADIO PROPA-GATION LAB., BOULDER, COLO.
NBS-CS-
NATIONAL BUREAU OF STANDARDS. OFFICE OF COMMODITY STANDARDS
NBS-D-
NATIONAL BUREAU OF STANDARDS, WASHINGTON, D.C.
NBS-FIPS-PUB-(NUMBER)-C
NATIONAL BUREAU OF STANDARDS, WASHINGTON, D.C. (FEDERAL INFORMATION PROCESSING STANDARD)
NBS-HANDBOOK-
NATIONAL BUREAU OF STANDARDS, WASHINGTON, D.C.
NBS LC-
NATIONAL BUREAU OF STANDARDS, WASHINGTON, D.C. (LETTER CIRCULAR)
NBS-LN-(NUMBER)-
NATIONAL BUREAU OF STANDARDS. BOULDER LABS., COLO. (LAB. NOTE)
NBS-LP-
NATIONAL BUREAU OF STANDARDS, WASHINGTON, D.C. (LIST OF PUBLICATIONS)
NBS-M-
NATIONAL BUREAU OF STANDARDS. CRYOGENIC DATA CENTER, BOULDER, COLO.
NBS-MISC.-PUBL.-
NATIONAL BUREAU OF STANDARDS, WASHINGTON, D.C.
NBS-MON-
NATIONAL BUREAU OF STANDARDS, WASHINGTON, D.C.
NBS-MONO-
NATIONAL BUREAU OF STANDARDS, WASHINGTON, D.C.
NBS-MONO-
NATIONAL BUREAU OF STANDARDS. BOULDER LABS., COLO.
NBS-MONOGRAPH-
NATIONAL BUREAU OF STANDARDS, WASHINGTON, D.C.
NBS MONOGRAPH-
NATIONAL BUREAU OF STANDARDS. BOULDER LABS., COLO.
NBS-MR-
NATIONAL BUREAU OF STANDARDS. BOULDER LABS., COLO.
NBS-NOEU-
NATIONAL BUREAU OF STANDARDS. NAVAL ORDNANCE EXPERIMENTAL UNIT, WASHINGTON, D.C.
NBS ODDR-
NATIONAL BUREAU OF STANDARDS. ORDNANCE DEVELOPMENT DIV., WASHINGTON, D.C.
NBS-OSRDB-(YEAR)-
NATIONAL BUREAU OF STANDARDS. OFFICE OF STANDARD REFERENCE DATA, WASH., D.C. (BIBLIOGRAPHIC SERIES)
NBS-P-
NATIONAL BUREAU OF STANDARDS, WASHINGTON, D.C. (PROGRESS REPORT TO AEC)
NBS-PAPER-
NATIONAL BUREAU OF STANDARDS. BOULDER LABS., COLO.
NBS-PM-
NATIONAL BUREAU OF STANDARDS. BOULDER LABS., COLO.
NBS-PR-
NATIONAL BUREAU OF STANDARDS, WASHINGTON, D.C.
NBS-PRL-
NATL. BUR. OF STDS. RADIATION PHYS. LAB., WASH., D.C.
NBS-PROJ.(NO.)-(NO.)-
NATIONAL BUREAU OF STANDARDS, WASHINGTON, D.C.
NBS R-
NATIONAL BUREAU OF STANDARDS, WASHINGTON, D.C.
NBS RP-
NATIONAL BUREAU OF STDS., WASH., D.C. (RES. PAPER)
NBS-SP-
NATIONAL BUREAU OF STANDARDS. BOULDER LABS., COLO.
NBS-SPEC. PUBL.-
NATIONAL BUREAU OF STANDARDS. BOULDER LABS., COLO.
NBS-SPR-R-
NATIONAL BUREAU OF STANDARDS. OFFICE OF COMMODITY STANDARDS
NBS-STR-
NATIONAL BUREAU OF STANDARDS, WASHINGTON, D.C. (SUMMARY TECHNICAL REPORT)
NBS-TN-
NATIONAL BUREAU OF STDS., WASH., D.C. (TECH. NOTE)
NBS-TN-
NATIONAL BUREAU OF STANDARDS. BOULDER LABS., COLO. (TECHNICAL NOTE)
NBS-TR-
NATIONAL BUREAU OF STDS., WASH., D.C. (TECH. REPT.)
NBS TR-
NATIONAL BUREAU OF STDS., WASH., D.C. (TEST REPT.)
NBS-TR-
NATIONAL BUREAU OF STANDARDS. BOULDER LABS., COLO. (TECHNICAL REPORT)
NBS-TRANS-
NATIONAL BUREAU OF STANDARDS. BOULDER LABS., COLO. (TRANSLATION)
NBTL-A-
NAVAL BOILER AND TURBINE LAB., PHILADELPHIA
NBTL-B-
NAVAL BOILER AND TURBINE LAB., PHILADELPHIA
NBTL-P-
NAVAL BOILER AND TURBINE LAB., PHILADELPHIA
NBTL-T-R-
NAVAL BOILER AND TURBINE LAB., PHILADELPHIA
NBTL-T-T-
NAVAL BOILER AND TURBINE LAB., PHILADELPHIA
NC-
FOREST SERVICE. NORTH CENTRAL STATION, ST. PAUL
NC-
GT. BRIT. MINISTRY OF SUPPLY
NC-
NAUGATUCK CHEM. DIV., UNITED STATES RUBBER CO., CONN.
NC-
NORTHROP CAROLINA, INC., ASHEVILLE, N.C.
NC-
NORWAY. INSTITUTT FOR ATOMENERGI, KJELLER

NCA-QPR-(NUMBER)-(YEAR)	NUCLEAR CORP. OF AMERICA. INSTRUMENT AND CONTROL DIV., DENVILLE, N.J.
NCAR-	NATIONAL CENTER FOR ATMOSPHERIC RES., BOULDER, COLO.
NCAR-CT-	NATIONAL CENTER FOR ATMOSPHERIC RES., BOULDER, COLO.
NCAR-MS-	NATIONAL CENTER FOR ATMOSPHERIC RES., BOULDER, COLO.
NCAR-TN-	NATIONAL CENTER FOR ATMOSPHERIC RES., BOULDER, COLO.
NCAR-TN-(YEAR)-	NATIONAL CENTER FOR ATMOSPHERIC RES., BOULDER, COLO.
NCBC-	NICKEL CADMIUM BATTERY CORP., EAST HAMPTON, MASS.
NCC-	BROOKHAVEN NATIONAL LAB., UPTON, N.Y.
NCC-	NATIONAL CARBON CO. RESEARCH LABS., CLEVELAND
NCC-	NORTH CAROLINA STATE COLL., RALEIGH
NCD RP-	NAVAL CLOTHING DEPOT, BROOKLYN (RESEARCH PROJECTS)
NCEL-C-	NAVAL CIVIL ENGINEERING LAB., PORT HUENEME, CALIF. (TECHNICAL COMPILATION)
NCEL-CR-	NAVAL CIVIL ENGINEERING LAB., PORT HUENEME, CALIF. (CONTRACT REPORT)
NCEL-M-	NAVAL CIVIL ENGINEERING LAB., PORT HUENEME, CALIF. (TECHNICAL MEMORANDUM)
NCEL-N-	NAVAL CIVIL ENGINEERING LAB., PORT HUENEME, CALIF. (TECHNICAL NOTE)
NCEL-R-	NAVAL CIVIL ENGINEERING LAB., PORT HUENEME, CALIF. (TECHNICAL REPORT)
NCEL-STIR-	NAVAL CIVIL ENGINEERING LAB., PORT HUENEME, CALIF.
NCEL-TM-	NAVAL CIVIL ENGINEERING LAB., PORT HUENEME, CALIF.
NCEL-TN-	NAVAL CIVIL ENGINEERING LAB., PORT HUENEME, CALIF.
NCEL-TN-N-	NAVAL CIVIL ENGINEERING LAB., PORT HUENEME, CALIF.
NCEL-TR-	NAVAL CIVIL ENGINEERING LAB., PORT HUENEME, CALIF.
NCEL-TR-R-	NAVAL CIVIL ENGINEERING LAB., PORT HUENEME, CALIF.
NCE-NV-	NEWARK COLL. OF ENGINEERING, N.J. DEPT. OF MECHANICAL ENGINEERING
NCER-	NATIONAL CENTER FOR EARTHQUAKE RES., MENLO PARK, CAL.
NCE-REL-	NAVAL CIVIL ENGINEERING RESEARCH AND EVALUATION LAB., PORT HUENEME, CALIF.
NCEREL M-	NAVAL CIVIL ENGINEERING RESEARCH AND EVALUATION LAB., PORT HUENEME, CALIF. (TECHNICAL MEMORANDUM)
NCEREL R-	NAVAL CIVIL ENGINEERING RESEARCH AND EVALUATION LAB., PORT HUENEME, CALIF.
NCEREL TM-	NAVAL CIVIL ENGINEERING RESEARCH AND EVALUATION LAB., PORT HUENEME, CALIF. (TECHNICAL MEMORANDUM)
NCEREL TN-	NAVAL CIVIL ENGINEERING RESEARCH AND EVALUATION LAB., PORT HUENEME, CALIF. (TECHNICAL NOTE)
NCF-	WESTERN ELECTRIC CO., N.Y.C.
NCG-TM-(YEAR)-	ARMY ENGINEER NUCLEAR CRATERING GROUP, LIVERMORE, CAL
NCG-TR-	ARMY ENGINEER NUCLEAR CRATERING GROUP, LIVERMORE, CAL
NCH-	CLEARINGHOUSE FOR FEDERAL SCIENTIFIC AND TECHNICAL INFORMATION, SPRINGFIELD, VA. (NON-CLEARINGHOUSE)
NCHS-RD-	NATIONAL CENTER FOR HEALTH SERVICES RESEARCH AND DEVELOPMENT, ARLINGTON, VA.
NCL-	NORTHROP CORPORATE LABS., HAWTHORNE, CALIF.
NCL-(YEAR)-	NORTHROP CORPORATE LABS., HAWTHORNE, CALIF.
NCL/AE-	GT. BRIT. NATL. CHEMICAL LAB., TEDDINGTON, MIDDX., ENG.
NCL/AE-172	GT. BRIT. NATL. CHEMICAL LAB., TEDDINGTON, MIDDX., ENG.
NCLA-ENG-	CALIFORNIA. UNIV., LOS ANGELES. SCHOOL OF ENGINEERING AND APPLIED SCIENCE
NCL/DEP-	GT. BRIT. NATL. CHEMICAL LAB., TEDDINGTON, MIDDX., ENG.
NCL-S+SN-	NORTHROP CORPORATE LABS., PASADENA, CALIF. (SENSOR AND SIMULATION NOTES)
NCPS-	NATIONAL COMMISSION ON PRODUCT SAFETY, WASHINGTON, DC
NCR-	BELL TELEPHONE LABS., INC., BURLINGTON, N.C.
NCR-	FAIRCHILD ENGINE AND AIRPLANE CORP. NEPA DIV., OAK RIDGE, TENN. (REF. 31)
NCR-	NATIONAL CASH REGISTER CO., DAYTON, OHIO
NCR-ED-	NATIONAL CASH REGISTER CO., HAWTHORNE, CALIF.
NCRE/N-	GT. BRIT. NAVAL CONSTRUCTION RESEARCH ESTABLISHMENT, DUNFERMLINE, FIFE, SCOTLAND
NCRE/R-	GT. BRIT. NAVAL CONSTRUCTION RESEARCH ESTABLISHMENT, DUNFERMLINE, FIFE, SCOTLAND
NCRE/R.498	GT. BRIT. NAVAL CONSTRUCTION RESEARCH ESTABLISHMENT, DUNFERMLINE, FIFE, SCOTLAND
NCRH-	NATIONAL CENTER FOR RADIOLOGICAL HEALTH, ROCKVILLE, MD
NCRH-SIB-	NATIONAL CENTER FOR RADIOLOGICAL HEALTH. STANDARDS AND INTELLIGENCE BRANCH, ROCKVILLE, MD.
NCRL-	NATIONAL CARBON CO. RESEARCH LABS., CLEVELAND
NCRL-BMPR-	NATIONAL CARBON CO. RESEARCH LABS., PARMA, OHIO (BI-MONTHLY PROGRESS REPORT)
NCRL-PR-(NUMBER)-(YEAR)	NATIONAL CARBON CO. RESEARCH LABS., CLEVELAND
NCRL-PR-(NUMBER)-(YEAR)	NATIONAL CARBON CO. RESEARCH LABS., PARMA, OHIO
NCRP-	NATIONAL COUNCIL ON RADIATION PROTECTION AND MEASUREMENTS, WASHINGTON, D.C.
NCRRD-	NATIONAL CASH REGISTER CO., DAYTON, OHIO
NCRRD-(YEAR)-FU-	NATIONAL CASH REGISTER CO. FUNDAMENTAL RESEARCH DEPT., DAYTON, OHIO
NCR-TN-G-	NETHERLANDS. NATIONAAL LUCHT EN RUIMTEVAARTLABORATORIUM, AMSTERDAM
NCRT-PR-	NATIONAL COLLEGE OF RUBBER TECHNOLOGY, LONDON
NCS-	NORTH CAROLINA STATE COLL., RALEIGH
NCSAC-	NUCLEAR CROSS SECTIONS ADVISORY COMMITTEE, AEC
NCSC-	NORTH CAROLINA STATE COLL., RALEIGH
NCSC-	UNITED STATES-EURATOM JOINT RESEARCH + DEV. PROGRAM
NCSH-	PUBLIC HEALTH SERVICE. DIV. OF CHRONIC DISEASES. NATIONAL CLEARINGHOUSE FOR SMOKING + HEALTH
NCSL-	NORTH CAROLINA STATE COLL., RALEIGH. DEPT. OF ENG. RES
NCSU-	NORTH CAROLINA STATE UNIV., RALEIGH
NCSU-QPR-	NORTH CAROLINA STATE UNIV., RALEIGH
NCSU-TR-	NORTH CAROLINA STATE UNIV., RALEIGH. DEPT. OF ENG. RES
NCT-	KNOLLS ATOMIC POWER LAB., SCHENECTADY, N.Y.
NCU-	NORTH CAROLINA. UNIV., CHAPEL HILL
NCU IS MS-	NORTH CAROLINA. UNIV., CHAPEL HILL. INST. OF STATISTICS. (MIMEOGRAPH SERIES)
NCU-MS-	NORTH CAROLINA. UNIV., CHAPEL HILL. INST. OF STATISTICS. (MIMEOGRAPH SERIES)
ND-	BUREAU OF NAVAL PERSONNEL
ND-	ITT FEDERAL ELECTRIC CORP., PARAMUS, N.J.
ND-	NAVAL PERSONNEL RESEARCH ACTIVITY, WASHINGTON, D.C.
ND-(YEAR)-	NAVAL PERSONNEL RESEARCH ACTIVITY, WASHINGTON, D.C.
ND-	PITTSBURGH PLATE GLASS CO. CHEMICAL DIV. NATRIUM PLANT, NEW MARTINSVILLE, W. VA. (NATRIUM DEV.)
NDA-	AMERICAN GAS AND ELECTRIC CO., N.Y.C.
NDA-	DOW CHEMICAL CO., MIDLAND, MICH.
NDA-(NUMBER)-	GENERAL MOTORS CORP., DETROIT
NDA-(NUMBER)-	UNITED NUCLEAR CORP., ELMSFORD, N.Y.
NDA-(NUMBER-LETTERS)-	UNITED NUCLEAR CORP., ELMSFORD, N.Y.
NDA-	UNITED NUCLEAR CORP., WHITE PLAINS, N.Y.
NDA-(NO.)-	UNITED NUCLEAR CORP., WHITE PLAINS, N.Y.
NDA-(NO.-LETTER)-	UNITED NUCLEAR CORP. DEV. DIV., WHITE PLAINS, N.Y.
NDA-(LETTERS)-	UNITED NUCLEAR CORP., ELMSFORD, N.Y.
NDA-(INITIALS)-	UNITED NUCLEAR CORP. DEV. DIV., WHITE PLAINS, N.Y.
NDA-DESE-	UNITED NUCLEAR CORP. DEV. DIV., WHITE PLAINS, N.Y.
NDA-(MEMO)-	UNITED NUCLEAR CORP., ELMSFORD, N.Y.
NDA-MEMO-	UNITED NUCLEAR CORP. DEV. DIV., WHITE PLAINS, N.Y.
NDA-PHYS.-	UNITED NUCLEAR CORP. DEV. DIV., WHITE PLAINS, N.Y.
NDC-	AEROJET-GENERAL CORP. ROCKET ENGINE OPERATIONS-NUCLEAR, SACRAMENTO, CALIF.
NDC-	ATOMIC ENERGY OF CANADA LTD., CHALK RIVER, ONT.

NDC-R/

NDC-R/	NUCLEAR DESIGN AND CONSTRUCTION LTD., WHETSTONE, MIDDX., ENGLAND
NDD-	NUCLEAR METALS, INC., CAMBRIDGE, MASS.
NDEO-	UNITED NUCLEAR CORP. DEV. DIV., WHITE PLAINS, N.Y.
NDL-(YEAR)-(LTR.)-	ARMY NUCLEAR DEFENSE LAB., EDGEWOOD ARSENAL, MD.
NDL-SP-	ARMY NUCLEAR DEFENSE LAB., EDGEWOOD ARSENAL, MD.
NDL-TM-	ARMY NUCLEAR DEFENSE LAB., EDGEWOOD ARSENAL, MD.
NDL-TR-	ARMY NUCLEAR DEFENSE LAB., EDGEWOOD ARSENAL, MD.
NDL-U-FY-(YEAR)	ARMY NUCLEAR DEFENSE LAB., EDGEWOOD ARSENAL, MD.
NDM-	MARTIN CO., BALTIMORE
NDN-	ARGONNE NATIONAL LAB., LEMONT, ILL.
NDN-	CHICAGO. UNIV. INST. FOR NUCLEAR STUDIES
NDN-(YEAR)-	NORTHROP CORP. NORTRONICS DIV. NEEDHAM DEPT., MASS.
NDR-	FORD INSTRUMENT CO., LONG ISLAND CITY, N.Y.
NDRC-	NATIONAL DEFENSE RESEARCH COMMITTEE
NDRC-(1 THRU 19) STR-	NATIONAL DEFENSE RESEARCH COMMITTEE (SUMMARY TECHNICAL REPORT) (REF. 27)
NDRC-(1 THRU 19)-	NATIONAL DEFENSE RESEARCH COMMITTEE (REF. 27)
NDRC-	OFFICE OF SCIENTIFIC RESEARCH AND DEVELOPMENT
NDRC-(LETTER)-	NATIONAL DEFENSE RESEARCH COMMITTEE (REF. 27)
NDRC AES-	NATIONAL DEFENSE RES. COMM. (AIR + EARTH SHOCK RPTS.)
NDRC-AMG-	NATIONAL DEFENSE RESEARCH COMMITTEE. APPLIED MATHEMATICS GROUP (REF. 28)
NDRC AMG-B-	BROWN UNIV., PROVIDENCE. APPLIED MATH. GP. (REF. 28)
NDRC AMG-B-MEMO-	BROWN UNIV., PROVIDENCE. APPLIED MATH. GP. (REF. 28)
NDRC AMG-C-	COLUMBIA UNIV., N.Y.C. APPLIED MATH. GROUP (REF. 28)
NDRC AMG-C-MEMO-	COLUMBIA UNIV., N.Y.C. APPLIED MATH. GROUP (REF. 28)
NDRC AMG-H-	HARVARD UNIV., CAMBRIDGE, MASS. APPLIED MATHEMATICS GROUP (REF. 28)
NDRC AMG-IAS-	INSTITUTE FOR ADVANCED STUDY. APPLIED MATHEMATICS GROUP, PRINCETON, N.J. (REF. 28)
NDRC AMG-N-	NORTHWESTERN UNIV., EVANSTON, ILL. APPLIED MATHEMATICS GROUP (REF. 28)
NDRC AMG-NYU-	NEW YORK UNIV., N.Y.C. APPLIED MATH. GROUP (REF. 28)
NDRC AMG-NYU-MEMO-	NEW YORK UNIV., N.Y.C. APPLIED MATH. GROUP (REF. 28)
NDRC-AMP-	NATIONAL DEFENSE RESEARCH COMMITTEE. APPLIED MATHEMATICS PANEL (REF. 28)
NDRC-AMP-MEMO-	NATIONAL DEFENSE RESEARCH COMMITTEE. APPLIED MATHEMATICS PANEL (REF. 28)
NDRC-AMP-NOTE-	NATIONAL DEFENSE RESEARCH COMMITTEE. APPLIED MATHEMATICS PANEL (REF. 28)
NDRC AOM-	NATIONAL DEFENSE RES. COMM. (ARMOR + ORDNANCE MEMO.)
NDRC AOR-	NATIONAL DEFENSE RES. COMM. (ARMOR + ORDNANCE RPTS.)
NDRC APP-	NATIONAL DEFENSE RES. COMM. (APPLIED PSYCH. PANEL)
NDRC APP-	NATIONAL DEFENSE RESEARCH COMMITTEE. APPLIED PSYCHOLOGY PANEL (REF. 28)
NDRC-DIV.-(1 THRU 19)-	NATIONAL DEFENSE RESEARCH COMMITTEE (REF. 27)
NDRC MEMO-	NATIONAL DEFENSE RES. COMM. (MEMORANDA SERIES)
NDRC PORC-	NATIONAL DEFENSE RESEARCH COMMITTEE (PROJECT REPORTS)
NDRC RS-	NATIONAL DEFENSE RES. COMM. (RPT. TO THE SERVICES)
NDRC RSC CIR-	NATIONAL DEFENSE RES. COMM. (RES. ON SOUND CONTROL)
NDRC RSC IC-	NATIONAL DEFENSE RESEARCH COMMITTEE. (RESEARCH ON SOUND CONTROL. INFORMAL COMMUNICATIONS)
NDRC-STR-	NATIONAL DEFENSE RES. COMM. (SUMMARY TECH. REPTS.)
NDRC-STR-(LETTERS)-	NATIONAL DEFENSE RES. COMM. (SUMMARY TECH. REPTS.)
NDRC WPG-	NATIONAL DEFENSE RES. COMM. (WAVE PROPAGATION GP.)
NDRE-	NORWAY. FORSVARETS FORSKNINGSINSTITUTT, KJELLER
NDRE-	NORWAY. FORSVARETS FORSKNINGSINSTITUTT, OSLO
NDRE-E-	NORWAY. FORSVARETS FORSKNINGSINSTITUTT, KJELLER
NDRE-IR-F-	NORWAY. FORSVARETS FORSKNINGSINSTITUTT, OSLO
NDRE-IR-K-	NORWAY. FORSVARETS FORSKNINGSINSTITUTT, OSLO
NDRE-K-	NORWAY. FORSVARETS FORSKNINGSINSTITUTT, KJELLER
NDRE-S-	NORWAY. FORSVARETS FORSKNINGSINSTITUTT, KJELLER
NDRE-U-	NORWAY. FORSVARETS FORSKNINGSINSTITUTT, KJELLER
NDRE-X-	NORWAY. FORSVARETS FORSKNINGSINSTITUTT, KJELLER
NDRI-(YEAR)-	NAVAL DENTAL RESEARCH INST., GREAT LAKES, ILL.
NDRI-PR-	NAVAL DENTAL RESEARCH INST., GREAT LAKES, ILL.
NDS-TR-	NAVAL DENTAL SCHOOL, BETHESDA, MD.
NDT-	AEROJET-GENERAL CORP., SACRAMENTO, CALIF.
NDU-	UNIVERSITY OF NOTRE DAME, IND.
NDU-TR-	UNIVERSITY OF NOTRE DAME, IND.
NDU-TR-(YEAR)-	UNIVERSITY OF NOTRE DAME, IND.
NE-	FOREST SERVICE. NORTHEASTERN STA., UPPER DARBY, PENNA
NE-	NAVY UNDERWATER SOUND LAB., NEW LONDON, CONN.
NE-	NEW YORK NAVAL SHIPYARD. MATERIAL LAB., BROOKLYN
NE-	PITTSBURGH PLATE GLASS CO. CHEMICAL DIV. NATRIUM PLANT, NEW MARTINSVILLE, W. VA. (NATRIUM ENG.)
NE-	POLAROID CORP., CAMBRIDGE, MASS.
NE-(NUMBER)-(NUMBER)-	VIRGINIA. UNIV., CHARLOTTESVILLE, DIV. OF NUCLEAR ENG
NE-	WHITE SANDS MISSILE RANGE, N. MEX. (NATL. RANGE ENG.)
NEA-	FOSTER WHEELER CORP., N.Y.C.
NEA-	PIONEER SERVICE AND ENGINEERING CO., CHICAGO
NE-AEC-	VIRGINIA. UNIV., CHARLOTTESVILLE. RESEARCH LABS. FOR ENGINEERING SCIENCES
NEAR-TR-	NIELSEN ENGINEERING + RESEARCH, INC., PALO ALTO, CAL.
NEBRASKA STUDY-(YEAR)-	NEBRASKA. UNIV., LINCOLN. DEPT. OF HORTICULTURE AND FORESTRY
NEB-SS-	CANADA. NATIONAL ENERGY BOARD, OTTAWA
NECL-TN-	NAVAL CIVIL ENGINEERING LAB., PORT HUENEME, CALIF.
NECL-TR-	NAVAL CIVIL ENGINEERING LAB., PORT HUENEME, CALIF.
NECPT-	TRW SYSTEMS GROUP. WASHINGTON OPERATIONS, D.C.
NECTP-	DEPT. OF TRANSPORTATION, WASHINGTON, D.C. (NORTHEAST CORRIDOR TRANSPORTATION PROJECT REPORT)
NED-	CORNELL UNIV., ITHACA, N.Y. LAB. OF NUCLEAR STUDIES
NED-	GENERAL MILLS, INC. NUCLEAR EQUIP. DEPT., MINNEAPOLIS
NEDC-	GENERAL ELECTRIC CO. NUCLEAR ENERGY DIV., SAN JOSE, CAL
NEDG-	GENERAL ELECTRIC CO. NUCLEAR ENERGY DIV., SAN JOSE, CAL
NEDO-	GENERAL ELECTRIC CO. NUCLEAR ENERGY DIV., SAN JOSE, CAL
NEDU-RR-	NAVY EXPERIMENTAL DIVING UNIT, WASHINGTON, D.C.
NEEP-	AIR FORCE (NUCLEAR ELECTRONIC EFFECTS PROGRAM)
NEES-	NAVAL ENGINEERING EXPERIMENT STATION, ANNAPOLIS
NEI-	ATOMIC ENERGY OF CANADA LTD., CHALK RIVER, ONT. (REF.6)
NEIC-RR-	POLAND. NUCLEAR ENERGY INFORMATION CENTER, WARSAW
NEL-	NAVY CLOTHING AND TEXTILE OFFICE, BROOKLYN
NEL-	NAVY ELECTRONICS LAB., SAN DIEGO, CALIF.
NEL-	NEW ENGLAND LIME CO., ADAMS, MASS.
NEL-	UNITED KINGDOM. NATIONAL ENGINEERING LAB., GLASGOW
NEL-(LETTER(S))-	NAVY ELECTRONICS LAB., SAN DIEGO, CALIF.
NELC-	NAVAL ELECTRONICS LAB. CENTER, SAN DIEGO, CALIF.
NELC-IR-	NAVAL ELECTRONICS LAB. CENTER, SAN DIEGO, CALIF.
NELC-CR-	NAVAL ELECTRONICS LAB., SAN DIEGO, CALIF.
NELC-TR-	NAVAL ELECTRONICS LAB. CENTER, SAN DIEGO, CALIF.
NEL/DOC-	NAVY ELECTRONICS LAB., SAN DIEGO, CALIF. (DOCUMENT)
NEL-ER-	AMF ATOMICS, INC., GREENWICH, CONN.
NEL-LDR-	GT. BRIT. NATL. ENG. LAB., EAST KILBRIDE, GLASGOW
NELS-	NAVY ELECTRONICS LAB., SAN DIEGO, CALIF.
NELS R-	NAVY ELECTRONICS LAB., SAN DIEGO, CALIF.

NELS S-	NAVY ELECTRONICS LAB., SAN DIEGO, CALIF.(SERIAL RPTS)
NEL/TM-	NAVY ELECTRONICS LAB., SAN DIEGO, CALIF.(TECH. MEMO.)
NEL TR-	NAVY ELECTRONICS LAB., SAN DIEGO, CALIF.(TRANSLATION)
N.E.M.-	CONSOLIDATED VULTEE AIRCRAFT CORP., FORT WORTH, TEX.
NEM-	CONVAIR, FORT WORTH, TEX.
NEMLAB-	NEW ENGLAND MATERIALS LAB., INC., MEDFORD, MASS.
NEODF-	NAVAL EXPLOSIVE ORDNANCE DISPOSAL FACILITY, INDIAN HEAD, MD.
NEPA-	(NUCLEAR ENERGY FOR THE PROPULSION OF AIRCRAFT) (REF. 31)
NEPA-	FAIRCHILD ENGINE AND AIRPLANE CORP. NEPA DIV., OAK RIDGE, TENN. (REF. 31)
NEPA-(NO.)-(LTRS.)-(NO.)	FAIRCHILD ENGINE AND AIRPLANE CORP. NEPA DIV., OAK RIDGE, TENN. (REF. 31)
NEPA-IC-(YEAR)-(MONTH)-	FAIRCHILD ENGINE AND AIRPLANE CORP. NEPA DIV., OAK RIDGE, TENN. (REF. 31)
NEPA-IY-	FAIRCHILD ENGINE AND AIRPLANE CORP. NEPA DIV., OAK RIDGE, TENN. (REF. 31)
NEPA-NCR-	FAIRCHILD ENGINE AND AIRPLANE CORP. NEPA DIV., OAK RIDGE, TENN. (REF. 31)
NEPA-SCR-	FAIRCHILD ENGINE AND AIRPLANE CORP. NEPA DIV., OAK RIDGE, TENN. (REF. 31)
NEPA-SCRM-	FAIRCHILD ENGINE AND AIRPLANE CORP. NEPA DIV., OAK RIDGE, TENN. (REF. 31)
NEPA-SERM-	FAIRCHILD ENGINE AND AIRPLANE CORP. NEPA DIV., OAK RIDGE, TENN. (REF. 31)
NEPA-STRM-	FAIRCHILD ENGINE AND AIRPLANE CORP. NEPA DIV., OAK RIDGE, TENN. (REF. 31)
NER-	BUDD CO., PHILADELPHIA
NERB-	NUCLEAR ENERGY RESEARCH BUREAU, N.Y.C.
NESC-	NATIONAL ENGINEERING SCIENCE CO., PASADENA, CALIF.
NESC-	NATIONAL ENVIRONMENTAL SATELLITE SERVICE, WASH., D.C.
NESC-	NAVAL ELECTRONIC SYSTEMS COMMAND, WASHINGTON, D.C.
NESCO-	NATIONAL ENGINEERING SCIENCE CO., PASADENA, CALIF.
NESCO-P-	NATIONAL ENGINEERING SCIENCE CO., PASADENA, CALIF.
NESCO-S-	NATIONAL ENGINEERING SCIENCE CO., PASADENA, CALIF.
NESCO-SN-	NATIONAL ENGINEERING SCIENCE CO., PASADENA, CALIF.
NESCTM-	NATIONAL ENVIRONMENTAL SATELLITE SERVICE, WASH., D.C.
NESS-	NATIONAL ENVIRONMENTAL SATELLITE SERVICE, WASH., D.C.
NESTEF-(YEAR)-	NAVAL ELECTRONIC SYSTEMS TEST AND EVALUATION FACILITY, PATUXENT RIVER, MD.
NESTEF-(YEAR)-	NAVAL ELECTRONIC SYSTEMS TEST AND EVALUATION FACILITY, ST. INIGOES, MD.
NESTF-	NAVAL ELECTRONIC SYSTEMS TEST AND EVALUATION FACILITY, PATUXENT RIVER, MD.
NET-WB-	DEFENSE ATOMIC SUPPORT AGENCY. FIELD COMMAND, ALBUQUERQUE, N. MEX. (NUCLEAR EMERGENCY TEAM TRAINING)
NEU-	NORTHEASTERN UNIV., BOSTON
NEVIS-	COLUMBIA UNIV., IRVINGTON-ON-HUDSON, N.Y. NEVIS LABS.
NEVU-TR-	NEVADA. UNIV., RENO. DESERT RESEARCH INST.
NEW YORK U-(NO.)-TR-	NEW YORK UNIV., N.Y.C.
NF-	AEROJET GENERAL CORP., AZUSA, CALIF.
NF-	AIR FORCE (NAVIGATIONAL FLIGHT CHARTS)
NF-	BATTELLE MEMORIAL INST., COLUMBUS, OHIO
NF-	CLEVELAND INDUSTRIAL RESEARCH, INC.
NF-	DOW CHEMICAL CO., MIDLAND, MICH.
NF-(NUMBER)Q-(YEAR)	DOW CHEMICAL CO., MIDLAND, MICH.
NF-(NUMBER) Q-(YEAR)	DOW CHEMICAL CO. SCIENTIFIC PROJS. LAB., MIDLAND, MICH
NF-	ENGLISH ELECTRIC CO., LTD.
NF-(YEAR)-	NATIONAL CENTER FOR RADIOLOGICAL HEALTH, ROCKVILLE, MD
NF-	NATIONAL FIREWORKS ORDNANCE CORP., WEST HANOVER, MASS
NF-	OLIN MATHIESON CHEMICAL CORP., BALTIMORE
NF-	OLIN MATHIESON CHEMICAL CORP., NIAGARA FALLS, N.Y.
NF-	REACTION MOTORS, INC., DENVILLE, N.J.
NF-	STANFORD RESEARCH INST., MENLO PARK, CALIF.
NFEC-	NAVAL FACILITIES ENGINEERING COMMAND, WASHINGTON, DC
NFOC-	NATIONAL FIREWORKS ORDNANCE CORP., WEST HANOVER, MASS
NFR-	OLIN MATHIESON CHEMICAL CORP. METALLURGICAL LABS., NEW HAVEN
NFSAIS-	NATIONAL FEDERATION OF SCIENCE ABSTRACTING AND INDEXING SERVICES, PHILADELPHIA
NFSAIS-TR-	NATIONAL FEDERATION OF SCIENCE ABSTRACTING AND INDEXING SERVICES, PHILADELPHIA (TECHNICAL REPORT)
NG(NUMBER)	BOEING CO., SEATTLE
NGC-	NATIONAL GEOPHYSICAL CO., INC., DALLAS
NGF-	NAVAL GUN FACTORY, WASHINGTON, D.C.
NGF MTS-	NAVAL GUN FACTORY. METALLURGICAL AND TESTING SECTION, WASHINGTON, D.C.
NGF MTS TR-	NAVAL GUN FACTORY. METALLURGICAL AND TESTING SECTION, WASHINGTON, D.C. (TEST REPORTS)
NGF OL TR-	NAVAL GUN FACTORY. OPTICAL LAB., WASHINGTON, D.C.
NGF-RR-	EXPERIMENTAL DIVING UNIT (NAVY), WASHINGTON, D.C.
NGF RR-	NAVAL GUN FACTORY, WASHINGTON, D.C. (RESEARCH RPTS.)
NGF-T-	NAVAL GUN FACTORY, WASHINGTON, D.C.
NGF-T-	NAVAL WEAPONS PLANT, WASHINGTON, D.C.
NGG-TR-	ARMY ENGINEER NUCLEAR CRATERING GROUP, LIVERMORE, CAL
NGL-(2 DIGITS-3 DIGITS)-	NATIONAL AERONAUTICS AND SPACE ADMINISTRATION, WASHINGTON, D.C. (CONTRACT NUMBER)
NGTE-M-	GT. BRIT. NATIONAL GAS TURBINE ESTABLISHMENT, FARNBOROUGH, HANTS, ENGLAND
NGTE-NT-	GT. BRIT. NATIONAL GAS TURBINE ESTABLISHMENT, FARNBOROUGH, HANTS, ENGLAND
NGTE-R-	GT. BRIT. NATIONAL GAS TURBINE ESTABLISHMENT, FARNBOROUGH, HANTS, ENGLAND
NH-	(SERIES ASSIGNED TO RPTS. ABOUT NAGASAKI + HIROSHIMA)
NH-	OLIN MATHIESON CHEMICAL CORP., NEW HAVEN
NHA-	NEWMARK, HANSEN AND ASSOCIATES, CAMBRIDGE, MASS.
NHA-	NEWMARK, HANSEN AND ASSOCIATES, URBANA, ILL.
NHA CONSTR. GUIDE-	NEWMARK, HANSEN AND ASSOCIATES, URBANA, ILL.
NHB-	AMES RESEARCH CENTER, MOFFETT FIELD, CALIF.
NHB-(NUMBER.NUMBER)	NATIONAL AERONAUTICS AND SPACE ADMINISTRATION, WASHINGTON, D.C.
NHI-	NATIONAL HEART INSTITUTE, BETHESDA, MD.
NHK-	JAPAN BROADCASTING CORP., TOKYO
NHO-	HYDROGRAPHIC OFFICE, SUITLAND, MD.
NHPC-	NORTH-HOLLAND PUBLISHING CO., AMSTERDAM
NHRL-	NATIONAL HURRICANE RESEARCH LAB., CORAL GABLES, FLA.
NHRP-	NATIONAL HURRICANE RESEARCH PROJECT, WASHINGTON, D.C.
NHS-(NUMBER/YEAR)	GT. BRIT. CENTRAL ELECTRICITY GENERATING BOARD. NUCLEAR HEALTH AND SAFETY DEPT., LONDON
NHU-	NEW HAMPSHIRE. UNIV., DURHAM
N.H. UNIV.-	NEW HAMPSHIRE. UNIV., DURHAM
N.I.-	FRANCE. COMMISSARIAT A L'ENERGIE ATOMIQUE. CENTRE D'ETUDES NUCLEAIRES, GRENOBLE
NI-	ROME. UNIVERSITA. ISTITUTO DI FISICA, G. MARCONI
NIC-TRANS-	ARMY, WASHINGTON, D.C.
NIC-TRANS-	NAVAL SCI. AND TECH. INTELLIGENCE CENTER, WASH., D.C.
NIC-TRANS-	OFFICE OF NAVAL INTELLIGENCE, WASHINGTON, D.C.
NID PG-	GT. BRIT. ADMIRALTY. NAVAL INTELLIGENCE DIRECTOR
NIF-TM-	NORDISK FORSKNINGSINSTITUT FOR MALING OG TRYKFARVER, COPENHAGEN
NIGMS-	NATIONAL INST. OF GEN. MEDICAL SCIENCES, BETHESDA, MD
NIH-	NATIONAL INSTITUTES OF HEALTH, BETHESDA, MD.

NIH-(YEAR-NUMBER)-	NATIONAL INSTITUTES OF HEALTH, BETHESDA, MD.
NIH-ES-	NATIONAL INSTITUTE OF ENVIRONMENTAL HEALTH SCIENCES, RESEARCH TRIANGLE PARK, N.C.
NIIEFA-	U.S.S.R. GOSUDARSTVENNYI KOMITET PO ISPOL'ZOVANIYU ATOMNOI ENERGII. NAUCHNO-ISSLEDOVATEL'SKII INSTITUT ELEKTROFIZICHESKOI APPARATURY, LENINGRAD
NIIP-	GT. BRIT. NATIONAL INST. OF INDUSTRIAL PSYCH., LONDON
NIJS-	YUGOSLAVIA. INSTITUT JOZEF STEFAN, LJUBLJANA
NIJS-P-	YUGOSLAVIA. INSTITUT JOZEF STEFAN, LJUBLJANA
NIJS-R-	YUGOSLAVIA. INSTITUT JOZEF STEFAN, LJUBLJANA
NIKE-X-SR-	ARMY MISSILE TEST AND EVALUATION DIRECTORATE, WHITE SANDS MISSILE RANGE, N. MEX.
NIM-	UNION OF SOUTH AFRICA. NATIONAL INST. FOR METALLURGY, JOHANNESBURG
N-IM-(LETTERS)-	ARMED FORCES SPECIAL WEAPONS PROJECT. FIELD COMMAND, ALBUQUERQUE, N. MEX. (NAVY ASSEMBLYMAN COURSE. INSTRUCTORS MANUAL)
NINDB-	NATIONAL INST. OF NEUROLOGICAL DISEASES AND BLINDNESS, BETHESDA, MD.
NIO-	NAVAL INSPECTOR OF ORDNANCE
NIO-A.(NUMBER)	UNITED KINGDOM. NATIONAL INST. OF OCEANOGRAPHY, GODALMING, SURREY, ENGLAND
NIO-CR/	UNITED KINGDOM. NATIONAL INST. OF OCEANOGRAPHY, GODALMING, SURREY, ENGLAND
NIO-D.(NUMBER)	UNITED KINGDOM. NATIONAL INST. OF OCEANOGRAPHY, GODALMING, SURREY, ENGLAND
NIOT-	UNITED KINGDOM. NATIONAL INST. OF OCEANOGRAPHY, WORMLEY, HERTS, ENGLAND
NIRL-	GT. BRIT. NATL. INST. FOR RES. IN NUCLEAR SCIENCE. RUTHERFORD HIGH ENERGY LAB., HARWELL, BERKS, ENG.
NIRL/LN/	GT. BRIT. NATL. INST. FOR RES. IN NUCLEAR SCIENCE. RUTHERFORD HIGH ENERGY LAB., HARWELL, BERKS, ENG.
NIRL/M/	GT. BRIT. NATIONAL INST. FOR RESEARCH IN NUCLEAR SCIENCE, HARWELL, BERKS, ENGLAND
NIRL/M-	GT. BRIT. NATL. INST. FOR RES. IN NUCLEAR SCIENCE. RUTHERFORD HIGH ENERGY LAB., HARWELL, BERKS, ENG.
NIRL/R/	GT. BRIT. NATL. INST. FOR RES. IN NUCLEAR SCIENCE. RUTHERFORD HIGH ENERGY LAB., HARWELL, BERKS, ENG.
NIRM-	NAVAL WARFARE ANALYSIS GROUP, WASHINGTON, D.C.
NIRNS-LN-	GT. BRIT. NATL. INST. FOR RES. IN NUCLEAR SCIENCE. RUTHERFORD HIGH ENERGY LAB., HARWELL, BERKS, ENG.
NIRNS/R/	GT. BRIT. NATL. INST. FOR RES. IN NUCLEAR SCIENCE. RUTHERFORD HIGH ENERGY LAB., HARWELL, BERKS, ENG.
NIRS-	JAPAN. NATIONAL INST. OF RADIOLOGICAL SCIENCES, CHIBA
NIRS-AR-	JAPAN. NATIONAL INST. OF RADIOLOGICAL SCIENCES, CHIBA
NIRS-PU-	JAPAN. NATIONAL INST. OF RADIOLOGICAL SCIENCES, CHIBA
NIRS-RSD-	JAPAN. NATIONAL INST. OF RADIOLOGICAL SCIENCES, CHIBA
NIS-	CENTRAL INTELLIGENCE AGENCY, WASHINGTON, D.C.
NIT-FTP-AR-	NORGES TEKNISKE HOEGSKOLE, TRONDHEIM
NIU-	OFFICE OF NAVAL INTELLIGENCE, WASHINGTON, D.C.
NJ-	INTERNATIONAL BUSINESS MACHINES CORP., SAN JOSE, CAL.
NJ-	SPERRY ELECTRONIC TUBE DIV., GAINESVILLE, FLA.
NJ-(NUMBER)-	SPERRY GYROSCOPE CO. ELECTRONIC TUBE DIV., GREAT NECK, N.Y.
NJ-	SPERRY RAND CORP., GAINESVILLE, FLA.
NJ CRS-	NEW JERSEY. CERAMIC RESEARCH STATION, NEW BRUNSWICK
NJCRS-	NEW JERSEY. CERAMIC RESEARCH STATION, NEW BRUNSWICK
NL-	NATIONAL DEFENSE RESEARCH COMMITTEE (NEWS LETTER)
NL-	NATIONAL RESEARCH COUNCIL OF CANADA. ATOMIC ENERGY PROJECT, CHALK RIVER, ONT. (REF. 29)
NL-	NORDEN LABS. CORP., WHITE PLAINS, N.Y.
NL-	SANDIA CORP., ALBUQUERQUE, N. MEX.(NOMENCLATURE LIST)
NLABS-TPMR-	MONSANTO RESEARCH CORP., DAYTON, OHIO
NLC-	NORDEN LABS. CORP., WHITE PLAINS, N.Y.
NLCO-	BUREAU OF MINES, NW. ELECTRODEV. LAB., ALBANY, ORE.
NLCO-	NATIONAL LEAD CO. OF OHIO, CINCINNATI
NLCO-TR-	NATIONAL LEAD CO. OF OHIO, CINCINNATI
NLF-(YEAR/NUMBER)	ITALY. COMITATO NAZIONALE PER L'ENERGIA NUCLEARE. LABORATORI NAZIONALI DI FRASCATI
NLL-	NETHERLANDS. NATIONAAL LUCHTVAARTHLABORATORIUM, AMSTERDAM
NLL-AERE-TRANS-	UNITED KINGDOM. NATIONAL LENDING LIBRARY FOR SCIENCE AND TECHNOLOGY, BOSTON SPA, YORKS, ENGLAND
NLL-BRSLC-	UNITED KINGDOM. NATIONAL LENDING LIBRARY FOR SCIENCE AND TECHNOLOGY, BOSTON SPA, YORKS, ENGLAND
NLL-CA-	UNITED KINGDOM. NATIONAL LENDING LIBRARY FOR SCIENCE AND TECHNOLOGY, BOSTON SPA, YORKS, ENGLAND
NLL-CE-TRANS-	UNITED KINGDOM. NATIONAL LENDING LIBRARY FOR SCIENCE AND TECHNOLOGY, BOSTON SPA, YORKS, ENGLAND
NLL-CTO-	UNITED KINGDOM. NATIONAL LENDING LIBRARY FOR SCIENCE AND TECHNOLOGY, BOSTON SPA, YORKS, ENGLAND
NLL-EE-TRANS-	UNITED KINGDOM. NATIONAL LENDING LIBRARY FOR SCIENCE AND TECHNOLOGY, BOSTON SPA, YORKS, ENGLAND
NLL-INC-	UNITED KINGDOM. NATIONAL LENDING LIBRARY FOR SCIENCE AND TECHNOLOGY, BOSTON SPA, YORKS, ENGLAND
NLL-KIRKBY-TRANS-	UNITED KINGDOM. NATIONAL LENDING LIBRARY FOR SCIENCE AND TECHNOLOGY, BOSTON SPA, YORKS, ENGLAND
NLL-LC-	UNITED KINGDOM. NATIONAL LENDING LIBRARY FOR SCIENCE AND TECHNOLOGY, BOSTON SPA, YORKS, ENGLAND
NLL-LIB-COMM-	UNITED KINGDOM. NATIONAL LENDING LIBRARY FOR SCIENCE AND TECHNOLOGY, BOSTON SPA, YORKS, ENGLAND
NLL-LTI-	UNITED KINGDOM. NATIONAL LENDING LIBRARY FOR SCIENCE AND TECHNOLOGY, BOSTON SPA, YORKS, ENGLAND
NLL-M-	UNITED KINGDOM. NATIONAL LENDING LIBRARY FOR SCIENCE AND TECHNOLOGY, BOSTON SPA, YORKS, ENGLAND
NLL-MET-OFF-M-	UNITED KINGDOM. NATIONAL LENDING LIBRARY FOR SCIENCE AND TECHNOLOGY, BOSTON SPA, YORKS, ENGLAND
NLL-MIN-TECH-T-	UNITED KINGDOM. NATIONAL LENDING LIBRARY FOR SCIENCE AND TECHNOLOGY, BOSTON SPA, YORKS, ENGLAND
NLL-MP-	NETHERLANDS. NATIONAAL LUCHTVAARTHLABORATORIUM, AMSTERDAM
NLL-NIOT-	UNITED KINGDOM. NATIONAL LENDING LIBRARY FOR SCIENCE AND TECHNOLOGY, BOSTON SPA, YORKS, ENGLAND
NLL-NSTIC-	UNITED KINGDOM. NATIONAL LENDING LIBRARY FOR SCIENCE AND TECHNOLOGY, BOSTON SPA, YORKS, ENGLAND
NLL-OA-	UNITED KINGDOM. NATIONAL LENDING LIBRARY FOR SCIENCE AND TECHNOLOGY, BOSTON SPA, YORKS, ENGLAND
NLL-PORS-TRANS-	UNITED KINGDOM. NATIONAL LENDING LIBRARY FOR SCIENCE AND TECHNOLOGY, BOSTON SPA, YORKS, ENGLAND
NLL-RAE-TRANS-	UNITED KINGDOM. NATIONAL LENDING LIBRARY FOR SCIENCE AND TECHNOLOGY, BOSTON SPA, YORKS, ENGLAND
NLL-RISLEY-TRANS-	UNITED KINGDOM. NATIONAL LENDING LIBRARY FOR SCIENCE AND TECHNOLOGY, BOSTON SPA, YORKS, ENGLAND
NLL-RRE-TRANS-	UNITED KINGDOM. NATIONAL LENDING LIBRARY FOR SCIENCE AND TECHNOLOGY, BOSTON SPA, YORKS, ENGLAND
NLL-RTS-	UNITED KINGDOM. NATIONAL LENDING LIBRARY FOR SCIENCE AND TECHNOLOGY, BOSTON SPA, YORKS, ENGLAND
NLL-SMRE-TRANS-	UNITED KINGDOM. NATIONAL LENDING LIBRARY FOR SCIENCE AND TECHNOLOGY, BOSTON SPA, YORKS, ENGLAND
NLL-T-	UNITED KINGDOM. NATIONAL LENDING LIBRARY FOR SCIENCE AND TECHNOLOGY, BOSTON SPA, YORKS, ENGLAND
NLL-TRANS-	UNITED KINGDOM. NATIONAL LENDING LIBRARY FOR SCIENCE AND TECHNOLOGY, BOSTON SPA, YORKS, ENGLAND
NLL-WINDSCALE-	UNITED KINGDOM. NATIONAL LENDING LIBRARY FOR SCIENCE AND TECHNOLOGY, BOSTON SPA, YORKS, ENGLAND
NLM-	NATIONAL LIBRARY OF MEDICINE, WASHINGTON, D.C.
NLM L.S.(YEAR)-	NATIONAL LIBRARY OF MEDICINE, WASHINGTON, D.C. (LITERATURE SEARCH)
NLN-	NAVAL ENGINEERING EXPERIMENT STATION, ANNAPOLIS
NLO-	NATIONAL LEAD CO. OF OHIO, CINCINNATI
NLOGM-	CALIFORNIA INST. OF TECH., PASADENA. NAVY LIAISON OFFICE FOR GUIDED MISSILES
NLR-G-	NETHERLANDS. NATIONAAL LUCHT EN RUIMTEVAART-LABORATORIUM, AMSTERDAM
NLR-MEMO-WR-	NETHERLANDS. NATIONAAL LUCHT EN RUIMTEVAART-LABORATORIUM, AMSTERDAM

NLR-MP-	NETHERLANDS. NATIONAAL LUCHT EN RUIMTEVAART-LABORATORIUM, AMSTERDAM
NLR-SR-	NETHERLANDS. NATIONAAL LUCHT EN RUIMTEVAART-LABORATORIUM, AMSTERDAM
NLR-TM-	NETHERLANDS. NATIONAAL LUCHT EN RUIMTEVAART-LABORATORIUM, AMSTERDAM
NLR-TN-	NETHERLANDS. NATIONAAL LUCHT EN RUIMTEVAART-LABORATORIUM, AMSTERDAM
NLR-TN-F-	NETHERLANDS. NATIONAAL LUCHT EN RUIMTEVAART-LABORATORIUM, AMSTERDAM
NLR-TN-G-	NETHERLANDS. NATIONAAL LUCHT EN RUIMTEVAART-LABORATORIUM, AMSTERDAM
NLR-TN-M-	NETHERLANDS. NATIONAAL LUCHT EN RUIMTEVAART-LABORATORIUM, AMSTERDAM
NLR-TN-T-	NETHERLANDS. NATIONAAL LUCHT EN RUIMTEVAART-LABORATORIUM, AMSTERDAM
NLR-TR-	NETHERLANDS. NATIONAAL LUCHT EN RUIMTEVAART-LABORATORIUM, AMSTERDAM
NLR-TR-F-	NETHERLANDS. NATIONAAL LUCHT EN RUIMTEVAART-LABORATORIUM, AMSTERDAM
NLR-TR-S-	NETHERLANDS. NATIONAAL LUCHT EN RUIMTEVAART-LABORATORIUM, AMSTERDAM
NLR-TR-T-	NETHERLANDS. NATIONAAL LUCHT EN RUIMTEVAART-LABORATORIUM, AMSTERDAM
NLR-V-	NETHERLANDS. NATIONAAL LUCHT EN RUIMTEVAART-LABORATORIUM, AMSTERDAM
NM-	(NOTE ON MATERIALS) (REF. 2)
N.M. (PROJECT NUMBER)-	(SEE ENTRIES FILED AS NM-(PROJECT NO.)-)
NM-	LOCKHEED NUCLEAR PRODUCTS, MARIETTA, GA.
NM-	NAVAL MEDICAL FIELD RESEARCH LAB., CAMP LEJEUNE, N.C.
NM-(PROJECT NO.)-	NAVAL MEDICAL RESEARCH INST., BETHESDA, MD.
NM-(PROJECT NO.)-	NAVAL RADIOLOGICAL DEFENSE LAB., SAN FRANCISCO
NM-(PROJECT NO.)-	NAVAL SCHOOL OF AVIATION MEDICINE, PENSACOLA, FLA.
NM-	NORTHROP AIRCRAFT, INC., HAWTHORNE, CALIF. (MISC.)
NMA-	MARTIN CO., BALTIMORE
NMA-(NO.)-(LETTERS)-	NEW MEXICO COLL. OF AGRICULTURE AND MECHANIC ARTS, STATE COLLEGE. PHYSICAL SCIENCE LAB.
NMAB-	NATIONAL RESEARCH COUNCIL. NATIONAL MATERIALS ADVISORY BOARD, WASHINGTON, D.C.
NMAC-	NAVAL MISSILE AND ASTRONAUTICS CENTER, POINT MUGU,CAL
NMAC-TR-(YEAR)-	NAVAL MISSILE AND ASTRONAUTICS CENTER, POINT MUGU,CAL
NMA+M-AER-	NEW MEXICO COLL. OF AGRICULTURE AND MECHANIC ARTS, STATE COLLEGE. PHYSICAL SCIENCE LAB.
NMA-M-CM-	NEW MEXICO COLL. OF AGRICULTURE AND MECHANIC ARTS, STATE COLLEGE. PHYSICAL SCIENCE LAB.
NMA+M-PSL-	NEW MEXICO COLL. OF AGRICULTURE AND MECHANIC ARTS, STATE COLLEGE. PHYSICAL SCIENCE LAB.
NMC-	INTERSTATE ELECTRONICS CORP. NATIONAL MARINE CONSULTANTS DIV., ANAHEIM, CALIF.
NMC-	NAVAL MISSILE CENTER, POINT MUGU, CALIF.
NMC-(LETTERS)-	NAVAL MISSILE CENTER, POINT MUGU, CALIF.
NMC-AER-	NEW MEXICO COLL. OF AGRICULTURE AND MECHANIC ARTS, STATE COLLEGE. PHYSICAL SCIENCE LAB.
NMCL MRR-	NAVAL MINE DEFENSE LAB., PANAMA CITY, FLA.
NMC-MP-	NAVAL MISSILE CENTER, POINT MUGU, CALIF.
NMC-MP-(YEAR)-	NAVAL MISSILE CENTER, POINT MUGU, CALIF.
NMC-MR-TS-	NAVAL MISSILE CENTER, POINT MUGU, CALIF.
NMC-ONR-	INTERSTATE ELECTRONICS CORP. NATIONAL MARINE CONSULTANTS DIV., ANAHEIM, CALIF.
NMCP-	NAVY MATHEMATICAL COMPUTING ADVISORY PANEL, WASH., DC
NMC-PMR-TR-	NAVAL MISSILE CENTER, POINT MUGU, CALIF.
NMCSSC-	NATIONAL MILITARY COMMAND SYSTEM SUPPORT CENTER, WASHINGTON, D.C.
NMCSSC-CSM-	NATIONAL MILITARY COMMAND SYSTEM SUPPORT CENTER, WASHINGTON, D.C.
NMCSSC-CSM-AM-	NATIONAL MILITARY COMMAND SYSTEM SUPPORT CENTER, WASHINGTON, D.C.
NMCSSC-CSM-GD-	NATIONAL MILITARY COMMAND SYSTEM SUPPORT CENTER, WASHINGTON, D.C.
NMCSSC-CSM-OM-	NATIONAL MILITARY COMMAND SYSTEM SUPPORT CENTER, WASHINGTON, D.C.
NMCSSC-CSM-PD-	NATIONAL MILITARY COMMAND SYSTEM SUPPORT CENTER, WASHINGTON, D.C.
NMCSSC-CSM-PSM-	NATIONAL MILITARY COMMAND SYSTEM SUPPORT CENTER, WASHINGTON, D.C.
NMCSSC-CSM-SD-IB-(YEAR)-	NATIONAL MILITARY COMMAND SYSTEM SUPPORT CENTER, WASHINGTON, D.C.
NMCSSC-CSM-UM-	NATIONAL MILITARY COMMAND SYSTEM SUPPORT CENTER, WASHINGTON, D.C.
NMCSSC-DN-	NATIONAL MILITARY COMMAND SYSTEM SUPPORT CENTER, WASHINGTON, D.C.
NMCSSC-SPM-OAM-	NATIONAL MILITARY COMMAND SYSTEM SUPPORT CENTER, WASHINGTON, D.C.
NMCSSC-SPM-OCD-	NATIONAL MILITARY COMMAND SYSTEM SUPPORT CENTER, WASHINGTON, D.C.
NMCSSC-SPM-VTP-	NATIONAL MILITARY COMMAND SYSTEM SUPPORT CENTER, WASHINGTON, D.C.
NMCSSC-TM-	NATIONAL MILITARY COMMAND SYSTEM SUPPORT CENTER, WASHINGTON, D.C.
NMCSSC TR-	NATIONAL MILITARY COMMAND SYSTEM SUPPORT CENTER, WASHINGTON, D.C.
NMC-TM-(YEAR)-	NAVAL MISSILE CENTER, POINT MUGU, CALIF.
NMC-TP-(YEAR)-	NAVAL MISSILE CENTER, POINT MUGU, CALIF.
NMC-TR-	NATIONAL MARINE CONSULTANTS, ANAHEIM, CALIF.
NMC-TR-	NAVAL MISSILE AND ASTRONAUTICS CENTER, POINT MUGU,CAL
NMD-	NAVAL MINE DEPOT, YORKTOWN, VA.
NMDL-	NAVY MINE DEFENSE LAB., PANAMA CITY, FLA.
NMDL TP-	NAVAL MINE DEFENSE LAB., PANAMA CITY, FLA.
NMD/Y-	NAVAL MINE DEPOT, YORKTOWN, VA.
NME-	NATIONAL MILITARY ESTABLISHMENT, WASHINGTON, D.C.
NME/MTRI-	RES. + DEV. BD. COMM. ON GUIDED MISSILES. PANEL ON TEST RANGE PROCEDURES + INSTRUMENTATION, WASH.,D.C.
NME-RS-	NATIONAL MILITARY ESTABLISHMENT. RES. + DEV. BOARD
N.MEX. EE-	NEW MEXICO. UNIV., ALBUQUERQUE
N.MEX. U. TR-EE-	NEW MEXICO. UNIV., ALBUQUERQUE, ENG. EXPERIMENT STA.
NMFRL-	NAVAL MEDICAL FIELD RESEARCH LAB., CAMP LEJEUNE, N.C.
NMFS-CIRC-	NATIONAL MARINE FISHERIES SERVICE, SEATTLE (CIRCULAR)
NMFS-SSRF-	NATIONAL MARINE FISHERIES SERVICE, SEATTLE
NMI-	AMES LAB., IOWA
NMI-	ARGONNE NATIONAL LAB., LEMONT, ILL.
NMI-	ARMOUR RESEARCH FOUNDATION, CHICAGO
NMI-	BATTELLE MEMORIAL INST., COLUMBUS, OHIO
NMI-	BRIDGEPORT BRASS CO., CONN.
NMI-	CALIFORNIA. UNIV., BERKELEY. LAWRENCE RADIATION LAB.
NMI-	DU PONT DE NEMOURS (E.I.) + CO. SAVANNAH RIVER LAB., AIKEN, S.C.
NMI-	GENERAL ELECTRIC CO. AIRCRAFT NUCLEAR PROPULSION DEPT., CINCINNATI
NMI-	GENERAL ELECTRIC CO. NUCLEAR MATERIALS AND PROPULSION OPERATION, CINCINNATI
NMI-	NEW MEXICO INST. OF MINING AND TECH., SOCORRO
NMI-	NUCLEAR MATERIALS AND EQUIPMENT CORP., APOLLO, PENNA.
NMI-	NUCLEAR METALS, INC. (ALL LOCATIONS)
NMI-(NUMBER).(NUMBER)	NUCLEAR METALS, INC., WEST CONCORD, MASS.
NMI-	OAK RIDGE NATIONAL LAB., TENN.
NMI-	TEXTRON INC. NUCLEAR METALS DIV., WEST CONCORD, MASS.
NMI-	WHITTAKER CORP. NUCLEAR METALS DIV., WEST CONCORD, MASS.
NMI-(INITIALS)-	NUCLEAR METALS, INC., CAMBRIDGE, MASS.
NMI-FR-	NUCLEAR METALS, INC., CAMBRIDGE, MASS. (FEASIBILITY REPORT)

NMIMT-	NEW MEXICO INST. OF MINING AND TECH., SOCORRO
NMI-MT-	NEW MEXICO INST. OF MINING AND TECH., SOCORRO. DEPT. OF MINERAL TECHNOLOGY
NMI-TJ-	NUCLEAR METALS, INC., CONCORD, MASS.
NML-(YEAR)-	MASSACHUSETTS INST.OF TECH.,CAMBRIDGE.NATL.MAGNET LAB
NML-	NAVAL MATERIAL LAB., BROOKLYN
NML-(PROJECT NUMBER)-	NAVAL MATERIAL LAB., BROOKLYN
NML-(PROJECT NUMBER)-	NEW YORK NAVAL SHIPYARD. MATERIAL LAB., BROOKLYN
NMMI-	NEW MEXICO MILITARY INST., ROSWELL
NMNRU-(YEAR)-	NAVY MEDICAL NEUROPSYCHIATRIC RESEARCH UNIT, SAN DIEGO, CALIF.
NMP-HTMP-	GENERAL ELECTRIC CO. FLIGHT PROPULSION LAB. DEPT., CINCINNATI
NMP-HTMP-	GENERAL ELECTRIC CO. NUCLEAR MATERIALS AND PROPULSION OPERATION, CINCINNATI
NM PROJECT-	NAVAL MEDICAL RESEARCH INST., BETHESDA, MD.
NMRI-	NAVAL MEDICAL RESEARCH INST., BETHESDA, MD.
NMRI-(LETTERS)-	NAVAL MEDICAL RESEARCH INST., BETHESDA, MD.
NMRI LRS(YR.)-	NAVAL MEDICAL RESEARCH INST., BETHESDA, MD. (LECTURE AND REVIEW SERIES)
NMRI-MR-	NAVAL MEDICAL RESEARCH INST., BETHESDA, MD.
NMRI MR(YR.)-	NAVAL MEDICAL RESEARCH INST., BETHESDA, MD. (MEMORANDUM REPORTS)
NMRI PROJECT-	NAVAL MEDICAL RESEARCH INST., BETHESDA, MD.
NMRI R-	NAVAL MEDICAL RESEARCH INST., BETHESDA, MD.
NMRI RR-	NAVAL MEDICAL RES. INST., BETHESDA, MD.(RES. REPTS.)
NMRL-	NAVAL MEDICAL RESEARCH LAB., NEW LONDON, CONN.
NMS/MN/	NIGERIAN METEOROLOGICAL SERVICE, LAGOS
NMSM/RDD/T-	NEW MEXICO SCHOOL OF MINES, SOCORRO
NMSM-TR-	NEW MEXICO SCHOOL OF MINES. RESEARCH AND DEVELOPMENT DIV., SOCORRO
NM STATE U-(NO)-QPR-	NEW MEXICO STATE UNIV., UNIVERSITY PARK
NM STATE U-FDR-	NEW MEXICO STATE UNIV., UNIVERSITY PARK
NMS/TN/	NIGERIAN METEOROLOGICAL SERVICE, LAGOS
NMS-TRANS-	NAVAL MEDICAL RESEARCH INST., BETHESDA, MD.
NMS-TRANS-	RADIO CORP. OF AMERICA, APPLIED RESEARCH, CAMDEN,N.J.
NMSU-	GT. BRIT. NAVAL MOTION STUDY UNIT
NMSU-	NEW MEXICO STATE UNIV., UNIVERSITY PARK
NMSU-AMC-(NO.)-(NO.)	NEW MEXICO STATE UNIV., UNIVERSITY PARK. PHYSICAL SCIENCE LAB.
NMSU-FDR-	NEW MEXICO STATE UNIV., UNIVERSITY PARK
NMSU-R-	NEW MEXICO STATE UNIV., UNIVERSITY PARK
NMSU-TALOS-	NEW MEXICO STATE UNIV., UNIVERSITY PARK
NMT-T-	NEW MEXICO INST. OF MINING AND TECH., SOCORRO
NMU-	NEW MEXICO. UNIV., ALBUQUERQUE
NMU-EE-	NEW MEXICO. UNIV., ALBUQUERQUE. BUREAU OF ENG. RES.
NMU-QPR-	NEW MEXICO. UNIV., ALBUQUERQUE
NMWTS-	NAVAL MINE WARFARE TEST STATION, SOLOMONS, MD.
NN-	BALLISTIC MISSILE DIV. (AIR FORCE)
NN-	NATIONAL NORTHERN, WEST HANOVER, MASS.
NNC-	NATIONAL NORTHERN CORP., WEST HANOVER, MASS.
NND-	NAVAL NET DEPOT
NND-(YEAR)-	NORTHROP CORP. NORTRONICS DIV. NEEDHAM DEPT., MASS.
NNES-	ATOMIC ENERGY COMMISSION, WASHINGTON, D.C. (NATIONAL NUCLEAR ENERGY SERIES)
NNK-	PRINCETON UNIV., N.J. PALMER PHYSICAL LAB.
NN-M-	NATIONAL NORTHERN, WEST HANOVER, MASS.
NNMC-	NATIONAL NAVAL MEDICAL CENTER, BETHESDA, MD.
NNOEU-	NAVAL NUCLEAR ORDNANCE EVALUATION UNIT, KIRTLAND AFB, N. MEX.
NNOEU VNV	NAVAL NUCLEAR ORDNANCE EVALUATION UNIT, KIRTLAND AFB, N. MEX. (VULNERABILITY NEWS AND VIEWS)
NN-Q-	NATIONAL NORTHERN, WEST HANOVER, MASS.
NNR-	NORTHROP AIRCRAFT, INC., HAWTHORNE, CALIF.
NNRL-	NORTHWESTERN UNIV., EVANSTON, ILL.
NNS-	NORFOLK NAVAL SHIPYARD, PORTSMOUTH, VA.
NNSD-	NEWPORT NEWS SHIPBUILDING AND DRY DOCK CO., VA.
NNSD-NSPS-	NEWPORT NEWS SHIPBUILDING AND DRY DOCK CO., VA.
NNSD-R-	NEWPORT NEWS SHIPBUILDING AND DRY DOCK CO., VA.
NO-	NAVAL OCEANOGRAPHIC OFFICE, WASHINGTON, D.C.
NOAA-	NATIONAL OCEANIC AND ATMOSPHERIC ADM., BOULDER, COLO.
NOAA-CONFR-	NATIONAL OCEAN SURVEY, ROCKVILLE, MD.
NOAA-NWS-	NATIONAL WEATHER SERVICE. OFFICE OF HYDROLOGY, SILVER SPRING, MD.
NOAA/PL-	NATIONAL OCEANIC AND ATMOSPHERIC ADM., BOULDER, COLO.
NOAA-SP-NOS-	NATIONAL OCEAN SURVEY, ROCKVILLE, MD.
NOAA-TM-EDS-	ENVIRONMENTAL DATA SERVICE, SILVER SPRING, MD.
NOAA-TM-EDS-BC-	ENVIRONMENTAL DATA SERVICE, SILVER SPRING, MD. (BIBLIOGRAPHY ON CLIMATE)
NOAA-TM-ERL-AOML-	ATLANTIC OCEANOGRAPHIC AND METEOROLOGICAL LABS., MIAMI, FLA.
NOAA-TM-ERL-APCL-	ATMOSPHERIC PHYSICS AND CHEMISTRY LAB., BOULDER,COLO.
NOAA-TM-ERL-ARL-	AIR RESOURCES LAB., RESEARCH TRIANGLE PARK, N.C.
NOAA-TM-ERL-BOMAP-	ENVIRONMENTAL RESEARCH LABS., ROCKVILLE, MD. (BARBADOS OCEANOGRAPHIC + METEOROLOGICAL ANALYSIS PROJ.)
NOAA-TM-ERL-ESL-	EARTH SCIENCES LABS., BOULDER, COLO.
NOAA-TM-ERL-MMTC-	MARINE MINERALS TECHNOLOGY CENTER, TIBURON, CALIF.
NOAA-TM-ERL-NHRL-	NATIONAL HURRICANE RESEARCH LAB., CORAL GABLES, FLA.
NOAA-TM-ERL-NSSL-	NATIONAL SEVERE STORMS LAB., NORMAN, OKLA.
NOAA-TM-ERL-OD-	ENVIRONMENTAL RESEARCH LABS. OFFICE OF THE DIRECTOR, BOULDER, COLO.
NOAA-TM-ERL-SEL-	SPACE ENVIRONMENT LAB., BOULDER, COLO.
NOAA-TM-ERL-WPL-	WAVE PROPAGATION LAB., BOULDER, COLO.
NOAA-TM-NESS-	NATIONAL ENVIRONMENTAL SATELLITE SERVICE, WASH., D.C.
NOAA-TM-NMS-SOS-	NATIONAL WEATHER SERVICE. SPACE OPERATIONS SUPPORT DIV., SILVER SPRING, MD.
NOAA-TM-NOS-	NATIONAL OCEAN SURVEY, ROCKVILLE, MD.
NOAA-TM-NWS-CR-	NATIONAL WEATHER SERVICE. CENTRAL REGION, KANSAS CITY, MO.
NOAA-TM-NWS-ENG-	NATIONAL WEATHER SERVICE. ENG. DIV., SILVER SPRING,MD
NOAA-TM-NWS-ER-	NATIONAL WEATHER SERVICE. EASTERN REGION, GARDEN CITY, N.Y.
NOAA-TM-NWS-FCST-	NATIONAL WEATHER SERVICE. WEATHER ANALYSIS + PREDICTION DIV.,SILVER SPRING,MD. (NOTES TO FORECASTERS)
NOAA-TM-NWS-HYDRO-	NATIONAL WEATHER SERVICE. OFFICE OF HYDROLOGY, SILVER SPRING, MD.
NOAA-TM-NWS-NMC-	NATIONAL METEOROLOGICAL CENTER, SUITLAND, MD.
NOAA-TM-NWS-PR-	NATIONAL WEATHER SERVICE. PACIFIC REGION, HONOLULU
NOAA-TM-NWS-SOS-	NATIONAL WEATHER SERVICE. SPACE OPERATIONS SUPPORT DIV., SILVER SPRING, MD.
NOAA-TM-NWS-SPDD-	NATIONAL WEATHER SERVICE. SYSTEMS PLANS AND DESIGN DIV., SILVER SPRING, MD.
NOAA-TM-NWS-SR-	NATIONAL WEATHER SERVICE. SOUTHERN REGION, FT. WORTH, TEX.
NOAA-TM-NWS-TDL-	NATIONAL WEATHER SERVICE. TECHNIQUES DEVELOPMENT LAB., SILVER SPRING, MD.
NOAA-TM-NWS-T+EL-	NATIONAL WEATHER SERVICE. TEST AND EVALUATION LAB., STERLING, VA.
NOAA-TM-NWSTM-WR-	NATIONAL WEATHER SERVICE. WESTERN REGION, SALT LAKE CITY
NOAA-TR-ERL-(NO.)-SDL-	SPACE DISTURBANCES LAB., BOULDER, COLO.
NOAA-TR-ERL-(NO.)-WPL-	WAVE PROPAGATION LAB., BOULDER, COLO.
NOAA-TR-NESS-	NATIONAL ENVIRONMENTAL SATELLITE SERVICE, WASH., D.C.

NOAA-TR-NMFS-CIRC-	NATIONAL MARINE FISHERIES SERVICE, SEATTLE (CIRCULAR)
NOAA-TR-NMFS-SSRF-	NATIONAL MARINE FISHERIES SERVICE, SEATTLE (SPECIAL SCIENTIFIC REPORT FISHERIES SERIES)
NOAA-TR-NOS-	NATIONAL OCEAN SURVEY, ROCKVILLE, MD.
NOAA-TR-WB-	NATIONAL WEATHER SERVICE. OFFICE OF HYDROLOGY, SILVER SPRING, MD.
NOAG-RESEARCH CONTRIB-	CENTER FOR NAVAL ANALYSES. NAVAL OBJECTIVES ANALYSIS GROUP, WASHINGTON, D.C.
NOBSR-	WESTINGHOUSE ELECTRIC CORP., BLOOMFIELD, N.J.
NODC-	NATIONAL OCEANOGRAPHIC DATA CENTER, WASHINGTON, D.C.
NOD-ED-	AMBAC INDUSTRIES. ARMA DIV., GARDEN CITY, N.Y.
NODU-	JOHNS HOPKINS UNIV., SILVER SPRING, MD. APPLIED PHYSICS LAB. NAVAL INSPECTOR OF ORDNANCE
NOEU-	NAVAL ORDNANCE EXPERIMENTAL UNIT, WASHINGTON, D.C.
NO.H.-	NAVAL OCEANOGRAPHIC OFFICE. HYDROGRAPHIC SURVEYS DEPT., WASHINGTON, D.C.
NOIL-	UNITED KINGDOM. NAVAL ORDNANCE INSPECTION LAB., CAERWENT, WALES
NOISE-	NEW YORK NAVAL SHIPYARD, BROOKLYN
NOL-	NAVAL ORDNANCE LAB. (ALL LOCATIONS)
NOL-C-	NAVAL ORDNANCE LAB., CORONA, CALIF.
NOLC-	NAVAL ORDNANCE LAB., CORONA, CALIF.
NOL-CORONA-	NAVAL ORDNANCE LAB., CORONA, CALIF.
NOL-CORONA-TM-	NAVAL ORDNANCE LAB., CORONA, CALIF.
NOLC-TM-	NAVAL ORDNANCE LAB., CORONA, CALIF.
NOLH-	NAVAL ORDNANCE LAB., WHITE OAK, MD. (HISTORY)
NOLM-	NAVAL ORDNANCE LAB., WASHINGTON, D.C. (MEMORANDA)
NOLM-	NAVAL ORDNANCE LAB., WHITE OAK, MD. (MEMORANDA)
NOLR-	NAVAL ORDNANCE LAB., WHITE OAK, MD. (REPORTS)
NOLR MUR-	NAVAL ORDNANCE LAB., WHITE OAK, MD. (MINE UNIT RPTS.)
NOLS-	NAVAL ORDNANCE LAB., WHITE OAK, MD. (SUGGESTION)
NOLSS-	NAVAL ORDNANCE LAB., SILVER SPRING, MD.
NOLSS TR-	NAVAL ORDNANCE LAB., SILVER SPRING, MD. (TEST RPTS.)
NOL TASK NO.	NAVAL ORDNANCE LAB., WHITE OAK, MD.
NOLTN-	NAVAL ORDNANCE LAB., WHITE OAK, MD. (TECHNICAL NOTE)
NOLTR-(YEAR)-	NAVAL ORDNANCE LAB., WHITE OAK, MD.
NOL-TR-	NAVAL ORDNANCE LAB., WHITE OAK, MD.
NOLX-	NAVAL ORDNANCE LAB., WHITE OAK, MD. (MISCELLANEOUS PUBLICATION)
NOMEN.-	CALIFORNIA. UNIV., BERKELEY. RADIATION LAB. (REF. 9)
NOO-	NAVAL OCEANOGRAPHIC OFFICE, WASHINGTON, D.C.
NOO-	NEVADA OPERATIONS OFFICE, AEC, LAS VEGAS
NOO-FR-	NAVAL OCEANOGRAPHIC OFFICE, WASHINGTON, D.C.
NOO-HO-(NUMBERS)-	NAVAL OCEANOGRAPHIC OFFICE, WASHINGTON, D.C.
NOO-IM-(YEAR)-	NAVAL OCEANOGRAPHIC OFFICE, WASHINGTON, D.C.
NOO-IMR-M-	NAVAL OCEANOGRAPHIC OFFICE, WASHINGTON, D.C.
NOO-IR-	NAVAL OCEANOGRAPHIC OFFICE, WASHINGTON, D.C.
NOO-IR-H-	NAVAL OCEANOGRAPHIC OFFICE, WASHINGTON, D.C.
NOO-SP-	NAVAL OCEANOGRAPHIC OFFICE, WASHINGTON, D.C.
NOO-TR-	NAVAL OCEANOGRAPHIC OFFICE, WASHINGTON, D.C.
NOO-TRANS-	NAVAL OCEANOGRAPHIC OFFICE, WASHINGTON, D.C.
NOP-	WHITE SANDS MISSILE RANGE, N. MEX. (NORTH OSCURA PEAK)
NOPF-	NAVAL ORDNANCE PLANT, FOREST PARK, ILL.
NOPI-	NAVAL ORDNANCE PLANT, INDIANAPOLIS
NOR-	BUREAU OF MINES. ELECTROTECHNICAL LAB., NORRIS, TENN
NOR-	FOREST SERVICE. NORTHERN STATION, JUNEAU, ALASKA
NOR-	NORTHROP CORP. (ALL LOCATIONS)
NOR-(YEAR)-	NORTHROP CORP. (ALL LOCATIONS)
NOR-	RADIOPLANE. DIV. OF NORTHROP CORP., VAN NUYS, CALIF.
NORA-	NORWAY. INSTITUTT FOR ATOMENERGI, KJELLER
NORAD-	NORTH AMERICAN AIR DEFENSE COMMAND. J-2 SECTION, ENT AFB, COLO.
NORA-MEMO-	NORWAY. INSTITUTT FOR ATOMENERGI, KJELLER
NOR-ARD-(YEAR)-	NORTHROP CORP. NORTRONICS DIV., NEWBURY PARK, CALIF.
NORC-	CHICAGO. UNIV. NATIONAL OPINION RESEARCH CENTER
NORE-	NORWAY. FORSVARETS FORSKNINGSINSTITUUT, KJELLER
NOR-NARD-(YEAR)-	NORTHROP NORTRONICS, NEWBURY PARK, CALIF.
NOR NB(YEAR)-	NORTHROP AIRCRAFT, INC. NORAIR, HAWTHORNE, CALIF.
NOR/NVR-	NORTHROP CORP. VENTURA DIV., NEWBURY PARK, CALIF.
NORT-	NORTHROP CORP. (ALL LOCATIONS)
NORT-(YEAR)-	NORTHROP SYSTEMS LABS., PALOS VERDES, CALIF.
NORT-	NORTRONICS, ANAHEIM, CALIF.
NORT-(YEAR)-	NORTRONICS, HAWTHORNE, CALIF.
NORT-(YEAR)-	NORTRONICS, PALOS VERDES PENINSULA, CALIF.
NORTH CAROLINA SC...	NORTH CAROLINA STATE COLL., RALEIGH
NOS-	NATIONAL OCEAN SURVEY, ROCKVILLE, MD.
NOSC-	NAVAL ORDNANCE SYSTEMS COMMAND, WASHINGTON, D.C.
NOS-IHMR-	NAVAL ORDNANCE STATION, INDIAN HEAD, MD.
NOS-IHSP-	NAVAL ORDNANCE STATION, INDIAN HEAD, MD.
NOS-IHSP-	NAVAL ORDNANCE SYSTEMS COMMAND, WASHINGTON, D.C.
NOS-IHTR-	NAVAL ORDNANCE STATION, INDIAN HEAD, MD.
NOS-TFS-(YEAR)-	JONES + LAUGHLIN STEEL CORP. WIRE ROPE DIV., MUNCY, PA.
NOS-TMR-	NAVAL ORDNANCE STATION, INDIAN HEAD, MD.
NOTAS DE FISICA-	RIO DE JANEIRO. CENTRO BRASILEIRO DE PESQUISAS FISICAS
NOTEBOOK H.E.W.-(NO.)-T	HANFORD WORKS, RICHLAND, WASH.
NOTE NO.-	BRUSSELS. UNIVERSITE. CENTRE DE PHYSIQUE NUCLEAIRE
NOTES BK...	RAYTHEON MFG. CO., WALTHAM, MASS.
NOTE TECHNIQUE-	FRANCE. OFFICE NATIONAL D=ETUDES ET DE RECHERCHES AEROSPATIALES, CHATILLON-SOUS-BAGNEUX
NOTS-	NAVAL ORDNANCE TEST STATION, CHINA LAKE, CALIF.
NOTS-(NUMBER)-	NAVAL ORDNANCE TEST STA., INYOKERN (CHINA LAKE), CAL.
NOTS-	NAVAL ORDNANCE TEST STATION, PASADENA, CALIF.
NOTS-(LETTERS)-	NAVAL ORDNANCE TEST STA., INYOKERN (CHINA LAKE), CAL.
NOTS ADPUB-	NAVAL ORDNANCE TEST STATION, CHINA LAKE, CALIF. (ADMINISTRATIVE PUBLICATION)
NOTS-ASSIGNMENT-	NAVAL ORDNANCE TEST STA., INYOKERN (CHINA LAKE), CAL.
NOTS-DOC-	NAVAL ORDNANCE TEST STA., INYOKERN (CHINA LAKE), CAL.
NOTS/IDP-	NAVAL ORDNANCE TEST STA., INYOKERN (CHINA LAKE), CAL.
NOTS-PM-P-	NAVAL ORDNANCE TEST STA., INYOKERN (CHINA LAKE), CAL.
NOTS REG.-	NAVAL ORDNANCE TEST STA., INYOKERN (CHINA LAKE), CAL.
NOTS T-	NAVAL ORDNANCE TEST STATION, CHINA LAKE, CALIF.
NOTS-TEST REPT.-	NAVAL ORDNANCE TEST STA., INYOKERN (CHINA LAKE), CAL.
NOTS TM-	NAVAL ORDNANCE TEST STATION, CHINA LAKE, CALIF. (TECHNICAL MEMORANDUM)
NOTS-TN-(NUMBERS)-	NAVAL ORDNANCE TEST STATION, CHINA LAKE, CALIF.
NOTS TP-	NAVAL ORDNANCE TEST STATION, CHINA LAKE, CALIF. (TECHNICAL PUBLICATION)
NOTS-TP-	NAVAL ORDNANCE TEST STATION, PASADENA, CALIF.
NOTS TPR-	NAVAL ORDNANCE TEST STATION, CHINA LAKE, CALIF. (TECHNICAL PROGRESS REPORT)
NOW-	PENNA. STATE UNIV., UNIV. PARK. ORDNANCE RES. LAB.
NP-	BUREAU OF AERONAUTICS (NAVY). AIRCRAFT NUCLEAR PROPULSION DIV.
NP-	BUREAU OF NAVAL PERSONNEL
NP-	BURROUGHS CORP., DETROIT
NP-	COMMITTEE ON MEDICAL RESEARCH (NEUROPSYCHIATRY RPTS.)

NP/16/L5/S29/-	NAVAL ENGINEERING EXPERIMENT STATION, ANNAPOLIS
NP-(5 DIGITS)	OFFICE OF INFORMATION SERVICES, AEC, WASHINGTON, D.C. (NON-PROJECT REPORT. INITIALLY ASSIGNED TO ALL NON-AEC REPORTS, BUT LATER USED ONLY FOR THOSE WITHOUT OTHER NUMBERS)
NP-	PITTSBURGH PLATE GLASS CO. CHEMICAL DIV. NATRIUM PLANT, NEW MARTINSVILLE, W. VA.(NATRIUM PRODUCTION)
NPA/	EUROPEAN COUNCIL FOR NUCLEAR RESEARCH, GENEVA
NPA-INT-	EUROPEAN ORGANIZATION FOR NUCLEAR RESEARCH, GENEVA
NPC-	CALIFORNIA. UNIV., BERKELEY. RADIATION LAB.
NPC-(YEAR)/-	GT. BRIT. ATOMIC EN. RES. ESTAB., HARWELL, BERKS, ENG
NPC-	NATIONAL AERONAUTICS AND SPACE ADM., WASHINGTON, D.C.
NPC-	NAVAL PHOTOGRAPHIC CENTER, WASHINGTON, D.C.
NPCC/FEWP/P-	NUCLEAR POWER GROUP, KNUTSFORD, CHES., ENGLAND
NPCC-FEWP/P.-	UNITED KINGDOM ATOMIC ENERGY AUTHORITY. DEVELOPMENT AND ENGINEERING GROUP, SPRINGFIELDS, LANCS. ENG.
NPCC-FEWP/P.	UNITED KINGDOM ATOMIC ENERGY AUTHORITY. REACTOR GP.
NPCC/MWP(G)/	UNITED KINGDOM ATOMIC ENERGY AUTHORITY. INDUSTRIAL GROUP. WINDSCALE WORKS, SELLAFIELD, CUMB., ENGLAND
NPCC/MWP(G)/P.	NUCLEAR POWER GROUP, KNUTSFORD, CHES., ENGLAND
NPCC/MWP/G/P.	UNITED KINGDOM ATOMIC ENERGY AUTHORITY. DEVELOPMENT AND ENGINEERING GROUP, CAPENHURST, CHES., ENGLAND
NPCC/MWP(G)/P.	UNITED KINGDOM ATOMIC ENERGY AUTHORITY. PRODUCTION GP
NPCC/MWP(G)/P.	UNITED KINGDOM ATOMIC ENERGY AUTHORITY. REACTOR GP.
NPCC/MWP/(G)P.	UNITED KINGDOM ATOMIC ENERGY AUTHORITY. RESEARCH GP. ATOMIC ENERGY RES. ESTAB., HARWELL, BERKS, ENGLAND
NPCC-MWP/P.	UNITED KINGDOM ATOMIC ENERGY AUTHORITY. DEVELOPMENT AND ENGINEERING GROUP, CULCHETH, LANCS, ENGLAND
NPCC-MWP-P.	UNITED KINGDOM ATOMIC ENERGY AUTHORITY. REACTOR GROUP, CULCHETH, LANCS, ENGLAND
NPCC-MWP(S)DGA/	UNITED KINGDOM ATOMIC ENERGY AUTHORITY. DEVELOPMENT AND ENGINEERING GROUP, CULCHETH, LANCS, ENGLAND
NPCC-MWP(S)DGD/	UNITED KINGDOM ATOMIC ENERGY AUTHORITY. DEVELOPMENT AND ENGINEERING GROUP, CULCHETH, LANCS, ENGLAND
NPCC-MWP(S)DGD/P.	UNITED KINGDOM ATOMIC ENERGY AUTHORITY. REACTOR GROUP, CULCHETH, LANCS, ENGLAND
NPCC-MWP(S)/P.	UNITED KINGDOM ATOMIC ENERGY AUTHORITY. DEVELOPMENT AND ENGINEERING GROUP, CULCHETH, LANCS, ENGLAND
NPCC-MWP(S)/P.	UNITED KINGDOM ATOMIC ENERGY AUTHORITY. REACTOR GROUP, CULCHETH, LANCS, ENGLAND
NPCC-RKWP/P-	UNITED KINGDOM ATOMIC ENERGY AUTHORITY. REACTOR GP.
NPCC/RPWP-	NUCLEAR POWER GROUP, KNUTSFORD, CHES., ENGLAND
NPCC/RPWP-	UNITED KINGDOM ATOMIC ENERGY AUTHORITY. INDUSTRIAL GROUP H.Q., RISLEY, LANCS, ENGLAND
NPCC/RPWP/N	UNITED KINGDOM ATOMIC ENERGY AUTHORITY. INDUSTRIAL GROUP H.Q., RISLEY, LANCS, ENGLAND
NPCC/RPWP/P.	UNITED KINGDOM ATOMIC ENERGY AUTHORITY. PRODUCTION GP
NPCC/RPWP/P.	UNITED KINGDOM ATOMIC ENERGY AUTHORITY. REACTOR GP.
NP-CC/RPWP/P-	UNITED KINGDOM ATOMIC ENERGY AUTHORITY. RESEARCH GP. ATOMIC ENERGY RES. ESTAB., HARWELL, BERKS, ENGLAND
NPCC/RPWP/P-	UNITED KINGDOM ATOMIC ENERGY AUTHORITY. RESEARCH GP. ATOMIC ENERGY RES. ESTAB., HARWELL, BERKS, ENGLAND
NPC/FEWP/P-	UNITED KINGDOM ATOMIC ENERGY AUTHORITY. INDUSTRIAL GROUP. SPRINGFIELDS WORKS, LANCS, ENGLAND
NPDD-	DETROIT EDISON CO. NUCLEAR POWER DEVELOPMENT DEPT.
NPDT-(YEAR)-	CANADA. HYDRO-ELECTRIC POWER COMMISSION OF ONTARIO, TORONTO
NPEU-	NAVAL PARACHUTE EXPERIMENTAL UNIT, EL CENTRO, CALIF.
NPF-	NAVAL POWDER FACTORY, INDIAN HEAD, MD.
NPF-AR-	NAVAL POWDER FACTORY, INDIAN HEAD, MD.
NPF DIM-	NAVAL POWDER FACTORY. ORDNANCE INVESTIGATION LAB., INDIAN HEAD, MD. (DEMOLITION INVEST. MEMORANDA)
NPF MEMO-	NAVAL POWDER FACTORY, INDIAN HEAD, MD.
NPF-MR-	NAVAL POWDER FACTORY, INDIAN HEAD, MD.
NPF-TR-	NAVAL POWDER FACTORY, INDIAN HEAD, MD.
NPG-	(NEVADA PROVING GROUND)
NPG-(NUMBER)(AECL)	ATOMIC ENERGY OF CANADA LTD. CHALK RIVER PROJ., ONT.
NPG-	NAVAL PROVING GROUND, DAHLGREN, VA.
NPG-(NUMBER)(NAVY)	NAVAL PROVING GROUND, DAHLGREN, VA.
NPG-	NAVAL WEAPONS LAB., DAHLGREN, VA.
NPG-	NEVADA PROVING GROUND
NPG-	NUCLEAR POWER GROUP, CHICAGO
NP/GEN-	GT. BRIT. ATOMIC EN. RES. ESTAB., HARWELL, BERKS, ENG
NPG-LC-	NAVAL PROVING GROUND, DAHLGREN, VA.
NPGR-	NAVAL PROVING GROUND, DAHLGREN, VA.
NPL-	GR. BRIT. NATL. PHYSICAL LAB., TEDDINGTON, MIDDX., ENG
NPL-	TEXAS. UNIV., AUSTIN
NPL/AERO/-	GT. BRIT. NATIONAL PHYSICAL LAB., AERODYNAMICS DIV., TEDDINGTON, MIDDX., ENGLAND
NPL-AERO-AC-	UNITED KINGDOM. NATIONAL PHYSICAL LAB., TEDDINGTON, MIDDX., ENGLAND
NPL-AERO-NOTE-	GT. BRIT. NATL. PHYSICAL LAB., TEDDINGTON, MIDDX., ENG
NPL-AERO-SPECIAL-	UNITED KINGDOM. NATIONAL PHYSICAL LAB., TEDDINGTON, MIDDX., ENGLAND
NPL-AERO-SR-	GT. BRIT. NATL. PHYSICAL LAB., TEDDINGTON, MIDDX., ENG
NPL-CCU-	GT. BRIT. NATL. PHYSICAL LAB., TEDDINGTON, MIDDX., ENG
NPL-COM-SCI-	GT. BRIT. NATL. PHYSICAL LAB., TEDDINGTON, MIDDX., ENG
NPL-DNAM-	UNITED KINGDOM. NATIONAL PHYSICAL LAB. DIV. OF NUMERICAL + APPLIED MATH., TEDDINGTON, MIDDX., ENG.
NPL-IMS-	UNITED KINGDOM. NATIONAL PHYSICAL LAB. DIV. OF INORG. AND METALLIC STRUCTURE, TEDDINGTON, MIDDX., ENG.
NPL-MA-	GT. BRIT. NATL. PHYSICAL LAB., TEDDINGTON, MIDDX., ENG
NPL-MAT-APP-	UNITED KINGDOM. NATIONAL PHYSICAL LAB., TEDDINGTON, MIDDX., ENGLAND
NPL-MC-	GT. BRIT. NATL. PHYSICAL LAB., TEDDINGTON, MIDDX., ENG
NPL-OP-MET-	UNITED KINGDOM. NATIONAL PHYSICAL LAB., TEDDINGTON, MIDDX., ENGLAND
NPL-OP-MET-	UNITED KINGDOM. NATIONAL PHYSICAL LAB. OPTICAL METROLOGY DIV., TEDDINGTON, MIDDX., ENGLAND
NPL-QU-	UNITED KINGDOM. NATIONAL PHYSICAL LAB. QUANTUM METROLOGY DIV., TEDDINGTON, MIDDX., ENGLAND
NPL-SHIP-	UNITED KINGDOM. NATIONAL PHYSICAL LAB., TEDDINGTON, MIDDX., ENGLAND
NPOC/FEWP-	GT. BRIT. SPRINGFIELDS WORKS, LANCS, ENGLAND
NPP-	NAVAL PROPELLANT PLANT, INDIAN HEAD, MD.
NPP-F-	NAVAL PROPELLANT PLANT, INDIAN HEAD, MD.
NPP-MR-	NAVAL PROPELLANT PLANT. RESEARCH AND DEVELOPMENT DEPT., INDIAN HEAD, MD.
NPP/QAS-(YEAR)-	NAVAL PROPELLANT PLANT. QUALITY ASSURANCE DEPT., INDIAN HEAD, MD.
NPP-RP-	NAVAL PROPELLANT PLANT, INDIAN HEAD, MD.
NPP-RR-(YEAR)-	NAVAL PROPELLANT PLANT, INDIAN HEAD, MD.
NPPSA-PRL-	NAVAL PERSONNEL PROGRAM SUPPORT ACTIVITY. PERSONNEL RESEARCH LAB., WASHINGTON, D.C.
NPP-TMR-	NAVAL PROPELLANT PLANT. RESEARCH AND DEVELOPMENT DEPT., INDIAN HEAD, MD.
NPP-TPR-	NAVAL PROPELLANT PLANT, INDIAN HEAD, MD. (TECHNICAL PROGRESS REPORT)
NPP-TR-	NAVAL PROPELLANT PLANT. RESEARCH AND DEVELOPMENT DEPT., INDIAN HEAD, MD.
NP/R-	UNITED KINGDOM ATOMIC ENERGY AUTHORITY. DEVELOPMENT AND ENGINEERING GROUP, CULCHETH, LANCS, ENGLAND
NP/R-	UNITED KINGDOM ATOMIC ENERGY AUTHORITY. REACTOR GROUP, CULCHETH, LANCS, ENGLAND
NPRA-SRR-	NAVAL PERSONNEL RESEARCH ACTIVITY, SAN DIEGO, CALIF.
NPRA-STB-(YEAR)-	NAVAL PERSONNEL RESEARCH ACTIVITY, SAN DIEGO, CALIF.
NPRFA-	NAVAL PERSONNEL RES. FIELD ACTIVITY, SAN DIEGO, CAL.
NPR/P.	UNITED KINGDOM ATOMIC ENERGY AUTHORITY. REACTOR GROUP, CULCHETH, LANCS, ENGLAND
NPS-	NAVAL POSTGRADUATE SCHOOL, MONTEREY, CALIF.

NPS-(NUMBERS,LETTERS)-	NAVAL POSTGRADUATE SCHOOL, MONTEREY, CALIF.
NP-TR-	TECHNICAL INFORMATION SERVICE EXTENSION, AEC. (SERIES ASSIGNED TO NON-AEC UNNUMBERED TRANSLATIONS)
NQR/(YEAR)-	JOHNS HOPKINS UNIV., SILVER SPRING, MD. APPLIED PHYSICS LAB.
NR-	(REPORT) (REF. 2)
NR-(2 DIGITS)	AARHUS UNIV., DENMARK. ROEMER OBSERVATORY
NR-	ADVIESBUREAU DER GENIE, HAGUE
NR-	BURROUGHS CORP., PAOLI, PENNA.
NR-	CHALMERS TEKNISKA HOEGSKOLA, GOETEBORG
N/R-	GT. BRIT. ATOMIC EN. RES. ESTAB., HARWELL, BERKS, ENG
NR-	LOCKHEED-GEORGIA CO., MARIETTA (NUCLEAR REPORT)
NR-	LOCKHEED NUCLEAR PRODUCTS, MARIETTA, GA.
NR-	NATIONAL RESEARCH CORP., CAMBRIDGE, MASS.
NR(YEAR)H-	NORTH AMERICAN ROCKWELL CORP. COLUMBUS DIV., OHIO
NR-	OFFICE OF NAVAL RESEARCH, WASHINGTON, D.C.
NR-	PITTSBURGH PLATE GLASS CO. CHEMICAL DIV. NATRIUM PLANT, NEW MARTINSVILLE, W. VA. (NATRIUM RESEARCH)
NR-	PRINCETON UNIV., N.J. JAMES FORRESTAL RES. CENTER
NR-	STANFORD UNIV., CALIF. MICROWAVE LAB.
NR-	VARIAN ASSOCIATES, BEVERLY, MASS.
NR-	WHITE SANDS MISSILE RANGE, N. MEX. (NATIONAL RANGE)
NR- (INITIALS)-	NAVAL RADIOLOGICAL DEFENSE LAB., SAN FRANCISCO
NRA-	NRA, INC., LONG ISLAND CITY, N.Y.
NRA-	NUCLEAR RESEARCH ASSOCIATES, LONG ISLAND CITY, N.Y.
N-RAD-L-	NAVAL RADIATION LAB., SAN FRANCISCO
N RAD L ADP-	NAVAL RADIATION LAB., SAN FRANCISCO
NRAO-	NATIONAL RADIO ASTRONOMY OBSERVATORY, GREEN BANK, W.VA
NRB-HGR-	DIVISION OF REACTOR DEVELOPMENT, AEC. NAVAL REACTORS BRANCH, WASHINGTON, D.C.
NRC-	ATOMIC ENERGY OF CANADA LTD. CHALK RIVER PROJ., ONT.
NRC-(NUMBERS)-	NATIONAL RESEARCH CORP., CAMBRIDGE, MASS.
NRC-	NATIONAL RESEARCH COUNCIL, WASHINGTON, D.C.
NRC-	NATIONAL RESEARCH COUNCIL OF CANADA, OTTAWA
NRC-(YR)-(NO.)-	PARSONS(C.A.) + CO., LTD., NEWCASTLE-UPON-TYNE, ENG.
NRC ABCC-	ATOMIC BOMB CASUALTY COMMISSION, WASHINGTON, D.C.
NRC AM-	NATIONAL RESEARCH COUNCIL. COMM. ON AVIATION MEDICINE
NRCC-	NATIONAL RESEARCH COUNCIL OF CANADA, OTTAWA
NRCC-	NATIONAL RESEARCH COUNCIL OF CANADA. LIBRARY, OTTAWA (TRANSLATIONS)
NRCC (LETTER(S))-	NATIONAL RESEARCH COUNCIL OF CANADA, OTTAWA
NRCC AEP (LETTER(S))-	NATIONAL RESEARCH COUNCIL OF CANADA, OTTAWA
NRCC AEP CRTEC-	NATIONAL RESEARCH COUNCIL OF CANADA. ATOMIC ENERGY PROJECT, CHALK RIVER, ONT. (TECHNICAL REPORT)
NRC CAO-	NATIONAL RES. COUNCIL. COMM. ON AMPHIBIOUS OPERATIONS
NRC CAP-	NATIONAL RESEARCH COUNCIL. COMM. ON AVIATION PSYCH.
NRCC DBR-	NATIONAL RESEARCH COUNCIL OF CANADA. DIV. OF BUILDING RESEARCH, OTTAWA
NRC CFD IM-	NATIONAL RESEARCH COUNCIL. COMMITTEE ON FORTIFICATION DESIGN (INTERIM MEMORANDA)
NRCC ME-	NATIONAL RESEARCH COUNCIL OF CANADA. DIVISION OF MECHANICAL ENGINEERING, OTTAWA
NRC-CUW-	NATIONAL ACADEMY OF SCIENCES-NATIONAL RES. COUNCIL, COMMITTEE ON UNDERSEA WARFARE, WASHINGTON, D.C.
NRC CUW-	NATIONAL RESEARCH COUNCIL. COMM. ON UNDERSEA WARFARE
NRC CWD-	NATIONAL RESEARCH COUNCIL. COMMITTEE ON WASTE DISPOSAL, WASHINGTON, D.C.
NRC-EC-	NATIONAL RESEARCH COUNCIL OF CANADA, OTTAWA
NRCE/R-	CAMBRIDGE UNIV., ENGLAND
NRC FD-	NATL. RES. COUNCIL. COMM. ON FORTIFICATION DESIGN
NRC FD IM-	NATIONAL RESEARCH COUNCIL. COMMITTEE ON FORTIFICATION DESIGN (INTERIM MEMORANDA)
NRC FD IR-	NATIONAL RESEARCH COUNCIL. COMMITTEE ON FORTIFICATION DESIGN (INTERIM REPORTS)
NRC-FSR-	NATIONAL RESEARCH COUNCIL, WASHINGTON, D.C. (FUNGICIDE SCREENING RPT.)
NRCJ-	KERNFORSCHUNGSANLAGE JUELICH, GERMANY
NRC-LR-	NATL. RES. COUNCIL OF CANADA. NATL. RES. LABS., OTTAWA
NRC MAB-	NATIONAL RESEARCH COUNCIL. MATERIALS ADVISORY BOARD, WASHINGTON, D.C.
NRC-MAC-	NATIONAL ACADEMY OF SCIENCES-NATIONAL RES. COUNCIL, MINE ADVISORY COMMITTEE, WASHINGTON, D.C.
NRC/MD-	NATIONAL RESEARCH CORP. METALS DIV., CAMBRIDGE, MASS.
NRC-ME-	NATIONAL RESEARCH COUNCIL OF CANADA, OTTAWA
NRC-ME-MK-	NATIONAL RESEARCH COUNCIL OF CANADA, OTTAWA
NRC-MMAB-	NATL. RES. COUNCIL. MINERALS + METALS ADVISORY BD.
NRC-MS-	NATIONAL RESEARCH COUNCIL OF CANADA. DIVISION OF MECHANICAL ENGINEERING, OTTAWA
NRC-MT-	NATIONAL RESEARCH COUNCIL OF CANADA. DIVISION OF MECHANICAL ENGINEERING, OTTAWA
NRCN-	ISRAEL. ATOMIC ENERGY COMMISSION. NUCLEAR RESEARCH CENTER-NEGEV, BEERSHEBA
NRCO-	NATIONAL RESEARCH CORP., BOSTON
NRCO-	NATIONAL RESEARCH CORP., CAMBRIDGE, MASS.
NRC P-	NATIONAL RESEARCH COUNCIL, WASH., D.C. (PUBLICATIONS)
NRC PPB IM-	NATIONAL RESEARCH COUNCIL. COMMITTEE ON PASSIVE PROTECTION AGAINST BOMBING (INTERIM MEMORANDA)
NRC PPB IR-	NATIONAL RESEARCH COUNCIL. COMMITTEE ON PASSIVE PROTECTION AGAINST BOMBING (INTERIM REPORTS)
NRC PUBL.	NATIONAL RESEARCH COUNCIL, WASHINGTON, D.C.
NRC/REE-	NATIONAL RESEARCH COUNCIL OF CANADA. RADIO AND ELECTRICAL ENGINEERING DIV., OTTAWA
NRC SSC-	NATIONAL RESEARCH COUNCIL. COMMITTEE ON SHIP STRUCTURE, WASHINGTON, D.C.
NRC-TA-W QR-	NATIONAL RESEARCH CORP. METALS DIV., CAMBRIDGE, MASS. (TANTALUM-TUNGSTEN QUARTERLY REPORT)
NRC-TIS-	NATIONAL RESEARCH COUNCIL OF CANADA. TECHNICAL INFORMATION SERVICE, OTTAWA
NRC-TT-	NATIONAL RESEARCH COUNCIL OF CANADA, OTTAWA (TECHNICAL TRANSLATION)
NRC TWGC-	NATIONAL RESEARCH COUNCIL. COMMITTEE ON TREATMENT OF WAR GAS CASUALTIES
NRD-	WHITTAKER CORP. NARMCO RESEARCH AND DEVELOPMENT DIV., SAN DIEGO, CALIF.
NRDC-	NATIONAL RESEARCH DEVELOPMENT CORP. (GT. BRIT.)
NRDC-	UNITED KINGDOM ATOMIC ENERGY AUTHORITY. RESEARCH GP. ATOMIC ENERGY RES. ESTAB., HARWELL, BERKS, ENGLAND
NRDC (DIV.LETTER)-	NATIONAL DEFENSE RESEARCH COMMITTEE
NRDL-	EUROPEAN-AMERICAN NUCLEAR DATA COMMITTEE
NRDL-	NAVAL RADIOLOGICAL DEFENSE LAB., SAN FRANCISCO
NRDL AD(LETTER)-	NAVAL RADIOLOGICAL DEFENSE LAB., SAN FRAN. (REF. 30)
NRDL-ER-	NAVAL RADIOLOGICAL DEFENSE LAB., SAN FRANCISCO (EVALUATION REPORT)
NRDL IER-	NAVAL RADIOLOGICAL DEFENSE LAB., SAN FRANCISCO (INSTRUMENT EVALUATION REPORTS)
NRDL-P-	NAVAL RADIOLOGICAL DEFENSE LAB., SAN FRANCISCO
NRDL-R+L-	NAVAL RADIOLOGICAL DEFENSE LAB., SAN FRANCISCO (REVIEWS AND LECTURES)
NRDL-TM-	NAVAL RADIOLOGICAL DEFENSE LAB., SAN FRAN. (TECH. MEMO)
NRDL-TR-	NAVAL RADIOLOGICAL DEFENSE LAB., SAN FRAN. (TECH. REPT)
NRDL-TRC-	NAVAL RADIOLOGICAL DEFENSE LAB. CIVIL DEFENSE TECHNICAL GROUP, SAN FRANCISCO
NRD-MJO-	WHITTAKER CORP. NARMCO RESEARCH AND DEVELOPMENT DIV., SAN DIEGO, CALIF.
NRDU-V-	NAVY RESEARCH AND DEVELOPMENT UNIT, VIETNAM
NRE-	CANADA. NAVAL RES. ESTABLISHMENT, DARTMOUTH, N.S.

NRE-(YEAR)/

NRE-(YEAR)/	CANADA. NAVAL RES. ESTABLISHMENT, DARTMOUTH, N.S.
NRE-	NOVA SCOTIA. NAVAL RESEARCH ESTABLISHMENT, DARTMOUTH
NREC-	NATIONAL RESOURCE EVALUATION CENTER, WASHINGTON, D.C.
NREC-(NUMBER)-	NORTHERN RESEARCH AND ENG. CORP., CAMBRIDGE, MASS.
NREC-TM-	NORTH CAROLINA. UNIV., CHAPEL HILL
NREC-TR-	NORTH CAROLINA. UNIV., CHAPEL HILL
NRE-FR/(YEAR)/	CANADA. NAVAL RES. ESTABLISHMENT, DARTMOUTH, N.S.
NRE-ISD-TN	AUSTRALIA. WEAPONS RESEARCH ESTABLISHMENT
NRFA-(LETTERS)-	WESTINGHOUSE ELECTRIC CORP. NAVAL REACTORS FACILITY, IDAHO FALLS
NRFE-(LETTERS)-	WESTINGHOUSE ELECTRIC CORP. NAVAL REACTORS FACILITY, IDAHO FALLS
NRF-ER-	WESTINGHOUSE ELECTRIC CORP. NAVAL REACTORS FACILITY, IDAHO FALLS
NRFS-AT-	WESTINGHOUSE ELECTRIC CORP. NAVAL REACTORS FACILITY, IDAHO FALLS
NRFS-ER-	WESTINGHOUSE ELECTRIC CORP. BETTIS ATOMIC POWER LAB., PITTSBURGH
NRFS-ER-	WESTINGHOUSE ELECTRIC CORP. NAVAL REACTORS FACILITY, IDAHO FALLS
NRFS-PR-	WESTINGHOUSE ELECTRIC CORP. NAVAL REACTORS FACILITY, IDAHO FALLS
NRFTS-(LETTERS)-	WESTINGHOUSE ELECTRIC CORP. NAVAL REACTORS FACILITY, IDAHO FALLS
NRI-	NATIONAL RADIAC, INC., NEWARK, N.J.
NRIC-(YEAR)-	KNOLLS ATOMIC POWER LAB. NAVAL REACTOR INFORMATION CENTER, SCHENECTADY, N.Y.
NR-IR-(NUMBER)-	NORTHROP CORP. NORAIR DIV., HAWTHORNE, CALIF.
NRL-	NAVAL RESEARCH LAB., WASHINGTON, D.C.
NRL-	NEW ZEALAND. NAVAL RESEARCH LAB., AUCKLAND
NRL-	NORTHROP CORP. NORAIR DIV. ENGINEERING DEPT., HAWTHORNE, CALIF.
NRL-(LETTER)-	NAVAL RESEARCH LAB., WASHINGTON, D.C. (WITH REPT. NO. 3500 THE LETTER WAS DROPPED)
NRL A-	NAVAL RESEARCH LAB., WASHINGTON, D.C. (ARMY-NAVY PRECIPITATION STATIC TECHNICAL REPORTS)
NRL-AR-	NEW ZEALAND. NATL. RADIATION LAB., CHRISTCHURCH
NRL-B-	NAVAL RESEARCH LAB., WASHINGTON, D.C.
NRL-BIB-	NAVAL RESEARCH LAB., WASHINGTON, D.C.
NRL-BIBL-	NAVAL RESEARCH LAB., WASHINGTON, D.C.
NRL-BULL-	NAVAL RESEARCH LAB., WASHINGTON, D.C.
NRL-C-	NAVAL RESEARCH LAB., WASHINGTON, D.C.
NRL-(DIV)-(NO./YR)	NAVAL RESEARCH LAB., WASHINGTON, D.C.
NRL-F-	NEW ZEALAND. NATL. RADIATION LAB., CHRISTCHURCH
NRL-F(NO.)-	NEW ZEALAND. NATL. RADIATION LAB., CHRISTCHURCH
NRL-GMRWG-	NAVAL RESEARCH LAB. GUIDED MISSILE RELAY WORKING GROUP, WASHINGTON, D.C.
NRL-H-	NAVAL RESEARCH LAB., WASHINGTON, D.C.
NRLIB-	NAVAL RESEARCH LAB., WASH., D.C. (INSTRUCTION BOOK)
NRL-M-	NAVAL RESEARCH LAB., WASHINGTON, D.C.
NRL-MEMO-	NAVAL RESEARCH LAB., WASHINGTON, D.C.
NRL MR-	NAVAL RESEARCH LAB., WASHINGTON, D.C. (MEMO. RPT.)
NRL-O-	NAVAL RESEARCH LAB., WASHINGTON, D.C.
NRL-P-	NAVAL RESEARCH LAB., WASHINGTON, D.C.
NRL-PR-(MONTH/YEAR)	NAVAL RESEARCH LAB., WASHINGTON, D.C.
NRL-R-	NAVAL RESEARCH LAB., WASHINGTON, D.C.
NRL-RD-	NAVAL RESEARCH LAB., WASHINGTON, D.C.
NRL-S-	NAVAL RESEARCH LAB., WASHINGTON, D.C.
NRL TE-	NAVAL RESEARCH LAB., WASHINGTON, D.C. (TEST RPTS.)
NRL-TER-	NAVAL RESEARCH LAB., WASHINGTON, D.C.
NRL-TEST AND EVALUATION-	NAVAL RESEARCH LAB., WASHINGTON, D.C.
NRL-TM-	NAVAL RESEARCH LAB., WASHINGTON, D.C.
NRL TR-	NAVAL RESEARCH LAB., WASHINGTON, D.C. (TRANSLATIONS)
NRL-TR-(NO.)-U	NETHERLANDS. NATL. AEROSPACE LAB., AMSTERDAM
NRL-TRANS-	NAVAL RESEARCH LAB., WASHINGTON, D.C.
NR/P-	UNITED KINGDOM ATOMIC ENERGY AUTHORITY. WEAPONS GROUP ATOMIC WEAPONS RES. ESTAB., ALDERMASTON, BERKS, ENG.
NRP/ANL-	GT. BRIT. NOISE REDUCTION PANEL
NRP/ANL-	UNITED KINGDOM. NOISE REDUCTION PANEL, LONDON (WORKING PARTY ON ACCEPTABLE NOISE LEVELS)
NRPB-	UNITED KINGDOM. NATIONAL RADIOLOGICAL PROTECTION BOARD, HARWELL, BERKS, ENGLAND
NRR-	NORTHROP AIRCRAFT, INC., HAWTHORNE, CALIF.(RES.REPT.)
NRRI-	WYOMING. UNIV., LARAMIE. NATURAL RESOURCES RESEARCH INSTITUTE
NRRL-	DEPARTMENT OF AGRICULTURE. NORTHERN UTILIZATION RESEARCH BRANCH, PEORIA, ILL.
NRS-	ARMED SERVICES TECHNICAL INFORMATION AGENCY. REFERENCE CENTER, ARLINGTON HALL STATION, ARLINGTON, VA.
NRS-	NEW SOUTH WALES, AUSTRALIA. UNIV., KENSINGTON
NRS-	NEW SOUTH WALES, AUSTRALIA. UNIV., KENSINGTON. DEPT. OF NUCLEAR AND RADIATION CHEMISTRY
NRS-	SUPREME COMMANDER FOR THE ALLIED POWERS. NATURAL RESOURCES SECTION
NRS-(LETTER)(NUMBER)	NUCLEAR RESEARCH SERVICES, INC., DALLAS
NRSL-	NAVY ELECTRONICS LAB., SAN DIEGO, CALIF.
NRS LAB AERO-	NAVY RADIO AND SOUND LAB., SAN DIEGO, CALIF.
NRSL WP-	NAVY RADIO AND SOUND LAB., SAN DIEGO, CALIF. (WAVE PROPAGATION)
NRS-MK-	NATIONAL RESEARCH COUNCIL OF CANADA, OTTAWA
NRSS-	ILLINOIS. UNIV., URBANA (NUCLEAR RADIATION SHIELDING STUDIES)
NRSSG-	AIR FORCE NUCLEAR REACTOR SYSTEM SAFETY GROUP, KIRTLAND AFB, N. MEX.
NRTF-(LETTERS)-	WESTINGHOUSE ELECTRIC CORP. NAVAL REACTORS FACILITY, IDAHO FALLS
NS-	(NUCLEAR SCIENCE) (REF. 2)
NS-	AIR FORCE (SPECIAL AIR NAVIGATION)
NS-	COMMITTEE ON MEDICAL RESEARCH (NEUROSURGERY RPTS.)
NS-	FEDERATION OF AMERICAN SOCIETIES FOR EXPERIMENTAL BIOLOGY, WASHINGTON, D.C.
NS-	IOWA STATE UNIV. OF SCIENCE + TECHNOLOGY, AMES (FILM)
NS-	NEW ZEALAND. DEPT. OF SCIENTIFIC AND INDUSTRIAL RESEARCH. DIV. OF NUCLEAR SCIENCES, LOWER HUTT
NS-	NEW ZEALAND. DEPT. OF SCIENTIFIC AND INDUSTRIAL RESEARCH. DIV. OF NUCLEAR SCIENCES, WELLINGTON
NS-	NORTHROP AIRCRAFT, INC., HAWTHORNE, CALIF. (SPEC.)
NSA-	OFFICE OF INFORMATION SERVICES, AEC, WASHINGTON, D.C. (NUCLEAR SCIENCE ABSTRACTS)
NSAM-	NAVAL SCHOOL OF AVIATION MEDICINE, PENSACOLA, FLA.
NSAM-RR-	NAVAL SCHOOL OF AVIATION MEDICINE, PENSACOLA, FLA.
NSBEO-	NATIONAL SONIC BOOM EVALUATION OFFICE, ARLINGTON, VA.
NSC-(NUMBER)-(NUMBER)	NATIONAL SAFETY COUNCIL, CHICAGO (FILM)
NSC-	NATIONAL SEMICONDUCTOR CORP., DANBURY, CONN.
NSC-CF-R-	MANNED SPACECRAFT CENTER, HOUSTON, TEX.
NSCC MR-	NEW YORK. STATE UNIV. COLL. OF CERAMICS, ALFRED
NSCI-(YEAR)-	CALIFORNIA. UNIV., BERKELEY
NSCI-	STANFORD RESEARCH INST., MENLO PARK, CALIF.
NSCSSC-SPM-CDS-	NATIONAL MILITARY COMMAND SYSTEM SUPPORT CENTER, WASHINGTON, D.C.
NSD-	NAVAL SUPPLY DEPOT
NSDRL/A-(NUMBER)-	NAVAL SHIP RESEARCH AND DEV. LAB., ANNAPOLIS, MD.
NSE-	(NUCLEAR SCIENCE AND ENGINEERING) (REF. 2)
NSEC-(NUMBER)-	INTERNATIONAL CHEMICAL AND NUCLEAR CORP. NUCLEAR SCIENCE DIV., PITTSBURGH

NTSB-AAS-(YEAR)-

NSEC-	NUCLEAR SCIENCE AND ENGINEERING CORP., PITTSBURGH
NSEC-GE-	NUCLEAR SCIENCE AND ENGINEERING CORP., PITTSBURGH
NSEC-T-	NUCLEAR SCIENCE AND ENGINEERING CORP., PITTSBURGH
NSF-	NATIONAL SCIENCE FOUNDATION, WASHINGTON, D.C.
NSF-(YEAR)-	NATIONAL SCIENCE FOUNDATION, WASHINGTON, D.C.
NSFC-IN-SSL-T-	GEORGE C. MARSHALL SPACE FLIGHT CTR., HUNTSVILLE, ALA.
NSF-G-	STANFORD UNIV., CALIF. DEPT. OF MECHANICAL ENG.
NSF-P-	TEXAS. UNIV., AUSTIN. ANTENNAS AND PROPAGATION LAB. (WORK DONE UNDER NATIONAL SCIENCE FOUNDATION GRANT)
NSF-PR-	MASSACHUSETTS. UNIV., AMHERST
NSF-TR-	NATIONAL SCIENCE FOUNDATION, WASH., D.C. (TRANSLATIONS)
NSG-REL-	NUOVA SAN GIORGIO S.P.A., GENOA
NSI-	NUCLEAR SYSTEMS, INC., GARLAND, TEX.
NSJ-	JAPAN. INSTITUTE OF PHYSICAL AND CHEMICAL RES., TOKYO
NSJ-TR-	JAPAN. ATOMIC ENERGY RESEARCH INSTITUTE, TOKYO
NSL-	MASS. INST. OF TECH., CAMBRIDGE. NAVAL SUPERSONIC LAB
NSL-	NATIONAL SCIENTIFIC LABS., INC., WASHINGTON, D.C.
NSL-	NATIONAL SPECTROGRAPHIC LABS., INC., CLEVELAND
NSL-	NORTHROP SPACE LABS., HAWTHORNE, CALIF.
NSL-(YEAR)-	NORTHROP SPACE LABS., HAWTHORNE, CALIF.
NSL-(YEAR)-	NORTHROP SYSTEMS LABS., HAWTHORNE, CALIF.
NSL-	NORTHWESTERN UNIV., EVANSTON, ILL. GAS DYNAMICS LAB.
NSL-E-	NORTHROP SPACE LABS., HAWTHORNE, CALIF.
NSL-LAB-	NATIONAL SPECTROGRAPHIC LABS., INC., CLEVELAND
NSL-TSR-	NATIONAL SPECTROGRAPHIC LABS., INC., CLEVELAND
NSM-	NEW YORK NAVAL SHIPYARD. MATERIAL LAB., BROOKLYN
NSMC-	NAVAL SUBMARINE MEDICAL CENTER, GROTON, CONN.
NSMC-MR-(YEAR)-	NAVAL SUBMARINE MEDICAL CENTER, GROTON, CONN.
NSMSES-	NAVAL SHIP MISSILE SYSTEMS ENG. STA., PORT HUENEME, CAL
NSMW-TP-	NAVAL SCHOOL OF MINE WARFARE, CHARLESTON, S.C.
NSP-(NO.)-	LOCKHEED MISSILES AND SPACE CO., SUNNYVALE, CALIF.
NS-R-(NUMBER)-(NUMBER)-	NAVY ELECTRONICS LAB., SAN DIEGO, CALIF.
NSR-	NEW ZEALAND. DEPT. OF SCIENTIFIC AND INDUSTRIAL RESEARCH. DIV. OF NUCLEAR SCIENCES, LOWER HUTT
NSR-	NEW ZEALAND. DEPT. OF SCIENTIFIC AND INDUSTRIAL RESEARCH. DIV. OF NUCLEAR SCIENCES, LOWER HUTT
NSR-	NORTHROP AIRCRAFT, INC., HAWTHORNE, CALIF. (NUCLEAR SCIENCE REPORT)
NSR-	RASOR (NED S.), DAYTON, OHIO
NSRB-	NATIONAL SECURITY RESOURCES BOARD, WASHINGTON, D.C.
NSRDC-	NAVAL SHIP RESEARCH AND DEV. CENTER, ANNAPOLIS
NSRDC-	NAVAL SHIP RESEARCH AND DEV. CENTER, WASHINGTON, D.C.
NSRDC-TE-P-	NAVAL SHIP RESEARCH AND DEV. CENTER, WASHINGTON, D.C.
NSRDC-TRANS-	NAVAL SHIP RESEARCH AND DEV. CENTER, WASHINGTON, D.C.
NSRDF-	NAVAL SUPPLY RES. AND DEV. FACILITY, BAYONNE, N.J.
NSRDI-W-	NORTH STAR RESEARCH AND DEV. INST., MINNEAPOLIS
NSRDL/A-(NUMBER)-	NAVAL SHIP RESEARCH AND DEV. LAB., ANNAPOLIS, MD.
NSRDL/PC-	NAVAL SHIP RESEARCH AND DEV. LAB., PANAMA CITY, FLA.
NSRDS-NBS-	NATIONAL BUREAU OF STANDARDS. NATIONAL STANDARD REFERENCE DATA SYSTEM, WASHINGTON, D.C.
NSRDS-NBS-	PURDUE UNIV., LAFAYETTE, IND. THERMOPHYSICAL PROPERTIES RESEARCH CENTER
NSS-(YEAR)-	NAVAL PERSONNEL PROGRAM SUPPORT ACTIVITY. PERSONNEL SURVEYS DIV., WASHINGTON, D.C.
NSS-	NEW YORK SHIPBUILDING CORP., CAMDEN, N.J.
NSS-	NORTHROP CORP. NORTRONICS DIV., ANAHEIM, CALIF.
NSSA-R-	NAVY SPACE SYSTEMS ACTIVITY, LOS ANGELES
NSSC-	NAVAL SHIP SYSTEMS COMMAND, WASHINGTON, D.C.
NSSDC-	GODDARD SPACE FLIGHT CENTER, GREENBELT, MD.
NSS/ECO-(LETTER)-	MARITIME ADMIN. OFFICE OF RES. AND DEV., WASHINGTON, D.C. (N.S. SAVANNAH/EXPTL. COMMERCIAL OPERATION)
NSSG-	UNITED KINGDOM ATOMIC ENERGY AUTHORITY. RESEARCH GP. ATOMIC ENERGY RES. ESTAB., HARWELL, BERKS, ENGLAND
NSSG-R-	UNITED KINGDOM ATOMIC ENERGY AUTHORITY. RESEARCH GP. ATOMIC ENERGY RES. ESTAB., HARWELL, BERKS, ENGLAND (NAVAL SECTION SHIELDING GROUP)
NSSL-	NATIONAL SEVERE STORMS LAB., NORMAN, OKLA.
NST-	TECHNICAL INFO. SERVICE, AEC (NUCLEAR SCIENCE + TECH)
NSTIC-	GT. BRIT. NAVAL SCIENTIFIC TECH. INFO. CENTER, LONDON
NSTIC/(NO.)/(YEAR)-	GT. BRIT. NAVAL SCIENTIFIC TECH. INFO. CENTER, LONDON
NSTIC-TRANS-	GT. BRIT. NAVAL SCIENTIFIC TECH. INFO. CENTER, LONDON
NSW-TM-	NAVAL WEAPONS LAB., DAHLGREN, VA.
NSY-	(NAVAL SHIPYARD)
NT-	AIX-MARSEILLES UNIV., FRANCE
NT-	AKADEMIYA NAUK SSSR. RADIOTEKHNICHESKII INST., MOSCOW
NT-(NUMBER)-(YEAR)	BRUSSELS. UNIVERSITE
NT-(NUMBER)/EAS-E/(NO.)	FRANCE. LABORATOIRE DE RECHERCHES BALISTIQUES ET AERODYNAMIQUES, VERNON
NT-(NUMBER)/(YEAR)/EG	FRANCE. LABORATOIRE DE RECHERCHES BALISTIQUES ET AERODYNAMIQUES, VERNON
NT-(NUMBER)/(YEAR)/EN	FRANCE. LABORATOIRE DE RECHERCHES BALISTIQUES ET AERODYNAMIQUES, VERNON
NT-	FRANCE. MINISTERE DE L'AIR, PARIS
NTC-(NUMBER)	NARA TECHNICAL COLL., JAPAN
NTC-	NAVAL TRAINING CENTER, SAN DIEGO, CALIF.
NTD-	AKADEMIYA NAUK SSSR. RADIOTEKHNICHESKII INST., MOSCOW
NT/D1/(YEAR)-	SOCIETE EUROPEENNE D'ETUDE ET D'INVESTIGATION DE SYSTEMS SPATIAUX, COURBEVOIE, FRANCE
NTDC-	NAVAL TRAINING DEVICES CENTER, PORT WASHINGTON, N.Y.
NTDC-TR-	NAVAL TRAINING DEVICES CENTER, PORT WASHINGTON, N.Y.
NTDC TR-	VITRO LABS., SILVER SPRING, MD.
NT/DTA/GI-III/(YR.)-	SOCIETE POUR L'ETUDE ET LA REALISATION D'ENGINS BALISTIQUES, COURBEVOIE, FRANCE
NTH-(YEAR)/	FRANCE. LABORATOIRE DE PHYSIQUE THEORIQUE ET DES HAUTES ENERGIES, NICE
N-TH-(YEAR)/	FRANCE. LABORATOIRE DE PHYSIQUE THEORIQUE ET DES HAUTES ENERGIES, NICE
NTI-	NORTHWESTERN UNIV., EVANSTON, ILL. TECHNOLOGICAL INST.
NTI-(NUMBER)-(LETTER)	NORTHWESTERN UNIV., EVANSTON, ILL. TECHNOLOGICAL INST. (SQUID PROJECT) (REF. 33)
NTIR-	(NONTECHNICAL INTELLIGENCE REPT.) (REF. 2)
NTIR-	ARMY AIR FORCE (NON-TECHNICAL INTELLIGENCE REPORTS)
NTME-	NAVAL TECHNICAL MISSION IN EUROPE
NTMJ-	NAVAL TECHNICAL MISSION TO JAPAN
NTN-	MARTIN-MARIETTA CORP. DENVER DIV.
NTNF-	NORWEGIAN INST. OF TECH., TRONDHEIM. ELECTRONICS RESEARCH LAB.
NTO-(NUMBER)-	CONNECTICUT. UNIV., HARTFORD. DEPT. OF LAB. MEDICINE
NTO-I-	NERVA TEST OPERATIONS, NUCLEAR ROCKET DEVELOPMENT STATION, JACKASS FLATS, NEV.
NTO-R-	NERVA TEST OPERATIONS, NUCLEAR ROCKET DEVELOPMENT STATION, JACKASS FLATS, NEV.
NTO-SOP-	NERVA TEST OPERATIONS, NUCLEAR ROCKET DEVELOPMENT STATION, JACKASS FLATS, NEV.
NTO-T-	NERVA TEST OPERATIONS, NUCLEAR ROCKET DEVELOPMENT STATION, JACKASS FLATS, NEV.
NTS-	GEOLOGICAL SURVEY, DENVER
NTS-	NAVAL TORPEDO STATION, KEYPORT, WASH.
NTS-	NAVAL UNDERWATER ORDNANCE STATION, NEWPORT, R.I.
NTSB-AAR-(YEAR)-	NATIONAL TRANSPORTATION SAFETY BOARD, WASHINGTON, DC
NTSB-AAS-(YEAR)-	NATIONAL TRANSPORTATION SAFETY BOARD. BUREAU OF AVIATION SAFETY, WASHINGTON, D.C.

NTSB-BA-	NATIONAL TRANSPORTATION SAFETY BOARD, WASHINGTON, DC (BRIEFS OF ACCIDENTS)
NTS N CR-	NAVAL TORPEDO STA., NEWPORT, R.I. (CONSECUTIVE RPTS.)
NTU-	NAVAL TECHNICAL UNIT IN EUROPE
NTW-	N. T. W. MISSILE ENGINEERING, INC., LOS ANGELES
NU-	NEBRASKA. UNIV., LINCOLN
NU-	NORTHWESTERN UNIV., EVANSTON, ILL.
NU-	SPERRY RAND CORP., GAINESVILLE, FLA.
NUB-	NEW YORK. STATE UNIV., STONY BROOK
NUB-	NORTHEASTERN UNIV., BOSTON
NUC-	PENNSYLVANIA STATE UNIV., UNIVERSITY PARK
NUC-E-	PENNSYLVANIA STATE UNIV., UNIVERSITY PARK
NUCLEAR-	ORENDA ENGINES LTD., MALTON, ONT.
NUCLEX-	CZECHOSLOVAKIA. SKODA WORKS, PILSEN
NUCOR-	NUCLEAR CORP. OF AMERICA, DENVILLE, N.J.
NUCOR-	SPACE AND MISSILE SYSTEMS ORGANIZATION, LOS ANGELES AIR FORCE STATION, CALIF.
NUC-TN-	NAVAL UNDERSEA RESEARCH AND DEV. CTR., SAN DIEGO, CAL
NUC-TP-	NAVAL UNDERSEA RESEARCH AND DEV. CTR., SAN DIEGO, CAL
NU-GDL-B-	NORTHWESTERN UNIV., EVANSTON, ILL. GAS DYNAMICS LAB.
NU-HYDRO-(NUMBERS)-TS	NEBRASKA. UNIV., LINCOLN. DEPT. OF MECHANICAL ENG.
NUMEC-	NUCLEAR MATERIALS AND EQUIPMENT CORP. (ALL LOCATIONS)
NUMEC-(NUMBERS)-	NUCLEAR MATERIALS AND EQUIPMENT CORP., LEWISTON, N.Y.
NUMEC-P-	NUCLEAR MATERIALS AND EQUIPMENT CORP., APOLLO, PENNA.
NUMEC-TM-P-	NUCLEAR MATERIALS AND EQUIPMENT CORP., APOLLO, PENNA.
NUOS-	NAVAL UNDERWATER ORDNANCE STATION, NEWPORT, R.I.
NUOS-CONSECUTIVE-	NAVAL UNDERWATER ORDNANCE STATION, NEWPORT, R.I.
NUOS-ITN-(NUMBER)-(YR.)	NAVAL UNDERWATER ORDNANCE STATION, NEWPORT, R.I.
NUP-	NIHON UNIV., TOKYO
NUP-(YEAR)-	TOKYO METROPOLITAN UNIV.
NUP-	TOKYO UNIV.
NUP-A-(YEAR-NUMBER)	NIHON UNIV., TOKYO
NURC-	NATIONAL UNION RADIO CORP., HATBORO, PENNA.
NURC-	NATIONAL UNION RADIO CORP., ORANGE, N.J.
NURG/M-	UNITED KINGDOM ATOMIC ENERGY AUTHORITY. RESEARCH GP. ATOMIC ENERGY RES. ESTAB., HARWELL, BERKS, ENGLAND
NUS-	NEW YORK. OFFICE OF ATOMIC DEVELOPMENT, ALBANY
NUS-	NUCLEAR UTILITY SERVICES, INC., WASHINGTON, D.C.
NUS-	NUS CORP., WASHINGTON, D.C.
NUS-	NUS CORP. ENVIRONMENTAL SAFEGUARDS DIV., WASH., D.C.
NUSC-CR-	NAVAL UNDERWATER SYSTEMS CTR., NEWPORT, R.I.
NUSC/NL-	NAVAL UNDERWATER SYSTEMS CTR. NEW LONDON LAB., CONN.
NUSC-TR-	NAVAL UNDERWATER SYSTEMS CTR., NEWPORT, R.I.
NUSL-	NAVY UNDERWATER SOUND LAB., NEW LONDON, CONN.
NUSL-PUB-	NAVY UNDERWATER SOUND LAB., NEW LONDON, CONN.
NUS-TM-ENG-	NUS CORP., WASHINGTON, D.C.
NUS-TM-ES-	NUS CORP., WASHINGTON, D.C.
NUS-TM-S-	NUS CORP., WASHINGTON, D.C.
NUTU-	NEWCASTLE-UPON-TYNE UNIV., NORTHUMBERLAND, ENGLAND
NUWC-TP-	NAVAL UNDERSEA WARFARE CENTER, PASADENA, CALIF.
NUWC-TP-	NAVAL UNDERSEA WARFARE CENTER, SAN DIEGO, CALIF.
NUWS-CR-	NAVAL UNDERWATER WEAPONS RES. + ENG.STA,NEWPORT, R.I.
NUWS-QE/N-(YR.)-(LTRS.)	NAVAL UNDERWATER WEAPONS RES. + ENG.STA,NEWPORT, R.I.
NUWS-TR-	NAVAL UNDERWATER WEAPONS RES. + ENG.STA,NEWPORT, R.I.
NV-	NORTHROP CORP. VENTURA DIV., NEWBURY PARK, CALIF.
NV-	RES. + DEV. BD. COMM. ON NAVIGATION, WASHINGTON, D.C.
NVB-	NORTHROP CORP. DIV. OF NORTHROP VENTURA, NEWBURY PARK, CALIF. (BROCHURES)
NVL-	ARMY ELECTRONICS COMMAND. NIGHT VISION LAB., FORT BELVOIR, VA.
NVO-	NEVADA OPERATIONS OFFICE, AEC, LAS VEGAS (REF. 4)
NVOO-	NEVADA OPERATIONS OFFICE, AEC, LAS VEGAS (REF. 4)
NVP-	NORTHROP CORP. DIV. OF NORTHROP VENTURA, NEWBURY PARK, CALIF. (PROPOSALS)
NVR-	NORTHROP CORP., NEWBURY PARK, CALIF.
NVR-	NORTHROP CORP. VENTURA DIV., NEWBURY PARK, CALIF.
NVT-	NORTHROP CORP. DIV. OF NORTHROP VENTURA, NEWBURY PARK, CALIF. (TECHNICAL LETTERS)
NV-TP-	NORTHROP CORP. DIV. OF NORTHROP VENTURA, NEWBURY PARK, CALIF. (TECHNICAL PAPERS)
NWAG-IRM-	CENTER FOR NAVAL ANALYSES. NAVAL WARFARE ANALYSIS GROUP, WASHINGTON, D.C.
NWAG-RESEARCH CONTRIB-	CENTER FOR NAVAL ANALYSES. NAVAL WARFARE ANALYSIS GROUP, WASHINGTON, D.C.
NWAG-STUDY-	CENTER FOR NAVAL ANALYSES. NAVAL WARFARE ANALYSIS GROUP, WASHINGTON, D.C.
NWCCA-TM-	NAVAL WEAPONS CENTER. CORONA ANNEX, CALIF.
NWCCL-TM-(NUMBER)-	NAVAL WEAPONS CENTER. CORONA LABS., CALIF.
NWCCL-TP-	NAVAL WEAPONS CENTER. CORONA LABS., CALIF.
NWCCL-TR-	NAVAL WEAPONS CENTER. CORONA LABS., CALIF.
NWC-L-(YEAR)-	NATIONAL WATER COMMISSION, ARLINGTON, VA.
NWC-SBS-(YEAR)-	NATIONAL WATER COMMISSION, ARLINGTON, VA.
NWC-TN-	NAVAL WEAPONS CENTER, CHINA LAKE, CALIF.
NWC-TP-	BIRMINGHAM. UNIV., ENGLAND. DEPT. OF IND. METALLURGY
NWC-TP-	NAVAL WEAPONS CENTER, CHINA LAKE, CALIF.
NWE-	OFFICE OF THE CHIEF OF ENGINEERS (ARMY), WASHINGTON, D.C. (NUCLEAR WEAPONS EFFECTS)
NWEF-	NAVAL WEAPONS EVALUATION FACILITY, KIRTLAND AFB, N.M.
NWER-	DEFENSE ATOMIC SUPPORT AGENCY, WASHINGTON, D.C.
NWG-	OFFICE OF THE CHIEF OF NAVAL OPERATIONS. NAVAL WARFARE ANALYSIS GROUP, WASHINGTON, D.C.
NWHL-	NAVAL WEAPONS HANDLING LAB., COLTS NECK, N.J.
NWIP-(NUMBER)-	OFFICE OF THE CHIEF OF NAVAL OPERATIONS, WASH., D.C.
NWL-	NAVAL WEAPONS LAB., DAHLGREN, VA.
NWL-AR-(NUMBER-YEAR)	NAVAL WEAPONS LAB., DAHLGREN, VA.
NWL-AR-W/(NUMBER)-(YEAR)	NAVAL WEAPONS LAB., DAHLGREN, VA.
NWL-CR-	NAVAL WEAPONS LAB., DAHLGREN, VA.
NWL-CR-N(NUMBER)-	NAVAL WEAPONS LAB., DAHLGREN, VA.
NWL-CSR-N(NUMBERS)	NAVAL WEAPONS LAB., DAHLGREN, VA.
NWL-K-(NUMBER)/(YEAR)	NAVAL WEAPONS LAB., DAHLGREN, VA.
NWL-MAL-	NAVAL WEAPONS LAB., DAHLGREN, VA.
NWL-TM K/	NAVAL WEAPONS LAB. COMPUTATIONAL AND ANALYSIS LAB., DAHLGREN, VA. (TECHNICAL MEMO)
NWL-TM-T-	NAVAL WEAPONS LAB., DAHLGREN, VA.
NWL-TM T/	NAVAL WEAPONS LAB. WARHEAD + TERMINAL BALLISTICS LAB., DAHLGREN, VA. (TECHNICAL MEMO)
NWL-TM-W-	NAVAL WEAPONS LAB., DAHLGREN, VA.
NWL-TM W/	NAVAL WEAPONS LAB. WEAPONS DEVELOPMENT + EVALUATION LAB., DAHLGREN, VA. (TECHNICAL MEMO)
NWL-TN-T-(NUMBER-YEAR)	N AL WEAPONS LAB., DAHLGREN, VA.
NWL-TN-W-(NUMBER-YEAR)	NAVAL WEAPONS LAB., DAHLGREN, VA.
NWL-TR-	NAVAL WEAPONS LAB., DAHLGREN, VA.
NWL/W-	NAVAL WEAPONS LAB., DAHLGREN, VA.
NWORM-	MINE WARFARE OPERATIONAL RESEARCH GROUP (NAVY) (NAVAL WARFARE OPERATIONAL RESEARCH MEMORANDUM)
NWP-	(NAVAL WARFARE PUBLICATION)
NWP-	NAVAL WEAPONS PLANT, WASHINGTON, D.C.
NWP-	OFFICE OF THE CHIEF OF NAVAL OPERATIONS, WASH., D.C.

NWP-P-	DU PONT DE NEMOURS (E.I.) + CO. PIGMENTS DEPT., WILMINGTON, DEL.
NWPW-T-(NUMBER)-(YEAR)	NAVAL WEAPONS PLANT, WASHINGTON, D.C.
NWR-	NORTHROP AIRCRAFT, INC., HAWTHORNE, CALIF.
NWRC-	STANFORD RESEARCH INST. NAVAL WARFARE RESEARCH CENTER, MENLO PARK, CALIF.
NWRC-BIB-	STANFORD RESEARCH INST. NAVAL WARFARE RESEARCH CENTER, MENLO PARK, CALIF.
NWRC/LSR-RM-	STANFORD RESEARCH INST. NAVAL WARFARE RESEARCH CENTER, MENLO PARK, CALIF.
NWRC-RM-	STANFORD RESEARCH INST. NAVAL WARFARE RESEARCH CENTER, MENLO PARK, CALIF.
NWRC-TR-	STANFORD RESEARCH INST. NAVAL WARFARE RESEARCH CENTER, MENLO PARK, CALIF.
NWRF-(NO.)-(NO.)-	NAVY WEATHER RESEARCH FACILITY, NORFOLK, VA.
NWRF-	NAVY WEATHER RESEARCH FACILITY, NORFOLK, VA.
NWRF-TP-(NUMBER-NO.)-	NAVY WEATHER RESEARCH FACILITY, NORFOLK, VA.
NWR/H9/(NO.)/-	GT. BRIT. ATOMIC WEAPONS RESEARCH ESTABLISHMENT, ALDERMASTON, BERKS, ENGLAND
NWS-	GT. BRIT. NAVAL WEATHER SERVICE
NWS-	NAVAL WEAPONS STATION, YORKTOWN, VA.
NWS-(NUMBER)/	NAVAL WEATHER SERVICE COMMAND, WASHINGTON, D.C.
NWS-CR-	NATIONAL WEATHER SERVICE. CENTRAL REGION, KANSAS CITY, MO.
NWS-ENG-	NATIONAL WEATHER SERVICE. ENG. DIV., SILVER SPRING, MD
NWS-ER-	NATIONAL WEATHER SERVICE. EASTERN REGION, GARDEN CITY, N.Y.
NWS-FCST-	NATIONAL WEATHER SERVICE. WEATHER ANALYSIS + PREDICTION DIV.,SILVER SPRING,MD. (NOTES TO FORECASTERS)
NWS-HYDRO-	NATIONAL WEATHER SERVICE. OFFICE OF HYDROLOGY, SILVER SPRING, MD.
NWS-MEMO-	GT. BRIT. NAVAL WEATHER SERVICE
NWS-NMC-	NATIONAL METEOROLOGICAL CENTER, SUITLAND, MD.
NWS-PR-	NATIONAL WEATHER SERVICE. PACIFIC REGION, HONOLULU
NWS-R+D-	NAVAL WEAPONS STATION, YORKTOWN, VA.
NWSSC-	AIR FORCE NUCLEAR WEAPON SYSTEM SAFETY GROUP, KIRTLAND AFB, N. MEX.
NWSSG-(YEAR)-	AIR FORCE NUCLEAR WEAPON SYSTEM SAFETY GROUP, KIRTLAND AFB, N. MEX.
NWS-SOS-	NATIONAL WEATHER SERVICE. SPACE OPERATIONS SUPPORT DIV., SILVER SPRING, MD.
NWS-SPDD-	NATIONAL WEATHER SERVICE. SYSTEMS PLANS AND DESIGN DIV., SILVER SPRING, MD.
NWS-SR-	NATIONAL WEATHER SERVICE. SOUTHERN REGION, FT. WORTH, TEX.
NWS-TDL-	NATIONAL WEATHER SERVICE. TECHNIQUES DEVELOPMENT LAB., SILVER SPRING, MD.
NWS-T+EL-	NATIONAL WEATHER SERVICE. TEST AND EVALUATION LAB., STERLING, VA.
NWSTM-WR-	NATIONAL WEATHER SERVICE. WESTERN REGION, SALT LAKE CITY
NWSY-TR-(YEAR)-	NAVAL WEAPONS STATION, YORKTOWN, VA.
NWU-	NORTHWESTERN UNIV., EVANSTON, ILL.
NWU-AR-	NORTHWESTERN UNIV., EVANSTON, ILL.
NWUC-	NORTHEASTERN WOOD UTILIZATION COUNCIL
NWUC B-	NORTHEASTERN WOOD UTILIZATION COUNCIL (BULLETIN)
NWUC WN-	NORTHEASTERN WOOD UTILIZATION COUNCIL (WOOD NOTES)
NWU-TI-TN-	NORTHWESTERN UNIV.,EVANSTON, ILL. TECHNOLOGICAL INST.
NWU-TI-TR-	NORTHWESTERN UNIV.,EVANSTON, ILL. TECHNOLOGICAL INST.
NWU-TR-	NORTHWESTERN UNIV.,EVANSTON. ILL. TECHNOLOGICAL INST.
NY-	OFFICE OF SCIENTIFIC RESEARCH AND DEVELOPMENT
NY-	WESTINGHOUSE ELECTRIC CORP., BALTIMORE
NYDO-	NEW YORK DIRECTED OPERATIONS OFFICE, AEC (REF. 4)
NYEET-	NEW YORK EYE AND EAR INFIRMARY. DEPT. OF ELECTROPHYSIOLOGY, N.Y.C.
NYO-	NEW YORK OPERATIONS OFFICE, AEC (REF. 4)
NYO-GEN-	NEW YORK OPERATIONS OFFICE, AEC
NYOO-	NEW YORK OPERATIONS OFFICE, AEC (REF. 4)
NYSU-	NEW YORK. STATE UNIV. (ALL LOCATIONS)
NYSVC-	CORNELL UNIV., ITHACA, N.Y. STATE VETERINARY COLL.
NYU-	NEW YORK UNIV., N.Y.C.
NYU-(NUMBER)-(LETTER(S))	NEW YORK UNIV., N.Y.C. (SQUID PROJECT) (REF. 33)
NYU-333-	NEW YORK UNIV., N.Y.C. COLL. OF ENGINEERING
NYU-AA-(YEAR)-	NEW YORK UNIV., N.Y.C. DEPT. OF AERONAUTICS AND ASTRONAUTICS
NYU-CC-	NEW YORK. STATE UNIV. COLL. OF CERAMICS, ALFRED
NYU-CX-	NEW YORK UNIV., N.Y.C. INST. OF MATHEMATICAL SCIENCES
NYU-EM-	NEW YORK UNIV., N.Y.C. INST. OF MATHEMATICAL SCIENCES
NYU-IMM-	NEW YORK UNIV., N.Y.C. INST. FOR MATH. + MECHANICS
NYU-IMM-	NEW YORK UNIV., N.Y.C. INST. OF MATHEMATICAL SCIENCES
NYUIMM-	NEW YORK UNIV., N.Y.C. INST. OF MATHEMATICAL SCIENCES
NYU-LPR-QR-	NEW YORK UNIV., N.Y.C. COLL. OF ENGINEERING (LIQUID PROPELLANT REVIEW. QUARTERLY REPORT)
NYU-MME-	NEW YORK UNIV., N.Y.C. INST. OF MATHEMATICAL SCIENCES
NYU-QR-(NO.)(YR/NO.)	NEW YORK UNIV., N.Y.C. COLL. OF ENGINEERING
NYU RR BR-	NEW YORK UNIV., N.Y.C. INSTITUTE OF MATHEMATICAL SCIENCES. (RESEARCH REPORTS)
NYU RR HT-	NEW YORK UNIV., N.Y.C. INSTITUTE OF MATHEMATICAL SCIENCES. (RESEARCH REPORTS)
NYU RR TW-	NEW YORK UNIV., N.Y.C. INSTITUTE OF MATHEMATICAL SCIENCES. (RESEARCH REPORTS)
NYU TM-	NEW YORK UNIV., N.Y.C. (TECHNICAL MEMORANDA)
NYU-TR-	NEW YORK UNIV., N.Y.C.
NZJWCC-	NIKE ZEUS JOINT AEC-DOD WARHEAD COORDINATING COMM. (REF. 15)

O-	LOVELACE FOUNDATION FOR MEDICAL EDUCATION AND RESEARCH, ALBUQUERQUE, N. MEX.
O-(NUMBER(S))/(NUMBER)	TRACERLAB, INC.
O-(NUMBER)/(NUMBER)	UNITED KINGDOM ATOMIC ENERGY AUTHORITY. WEAPONS GROUP ATOMIC WEAPONS RES. ESTAB.,ALDERMASTON, BERKS, ENG.
O(LETTER)C-	OFFICE OF SCIENTIFIC RESEARCH AND DEVELOPMENT
OA-	(OPERATIONS ANALYSIS) (REF. 2)
OA-	AIR FORCE (OPERATIONS ANALYSIS WORKING PAPER)
OA-	ARMY AIR FORCE. OPERATIONS ANALYSIS DIV., WASH., D.C.
OA-	MARTIN CO., ORLANDO, FLA. (TECHNICAL REPORT)
OA-	ODIN ASSOCIATES
OA-	OPERATIONS ANALYSIS DIV. (AIR FORCE), WASHINGTON,D.C.
OAC-	ARMY ORDNANCE AMMUNITION COMMAND, JOLIET, ILL.
OAC/(LETTERS)	ARMY ORDNANCE AMMUNITION COMMAND, JOLIET, ILL.
OACSFOR-OT-RD-(YEAR)-	ADJUTANT GENERAL'S OFF. (ARMY), WASHINGTON, D.C.
OACSFOR-OT-RD-	ASSISTANT CHIEF OF STAFF FOR FORCE DEVELOPMENT (ARMY) WASHINGTON, D.C.
OACSFOR-OT-RD-T(NOS.)	ASSISTANT CHIEF OF STAFF FOR FORCE DEVELOPMENT (ARMY) WASHINGTON, D.C.
OACSFOR-OT-UT-	ASSISTANT CHIEF OF STAFF FOR FORCE DEVELOPMENT (ARMY) WASHINGTON, D.C.
OAD-	OPERATIONS ANALYSIS DIV. (AIR FORCE), WASHINGTON,D.C.
OAD-RM-(NUMBER)-	STANFORD RESEARCH INST. OPERATIONS ANALYSIS DEPT., MENLO PARK, CALIF.
OAD-TN-(NUMBER)-	STANFORD RESEARCH INST. OPERATIONS ANALYSIS DEPT., MENLO PARK, CALIF.
OAG-PRB-TRN-	OFFICE OF THE ADJUTANT GENERAL. PERSONNEL RESEARCH BRANCH, WASHINGTON, D.C.
OAHU-SR-	CALIFORNIA. UNIV., LOS ANGELES. INST. OF GEOPHYSICS
OAI-(YEAR)-	OLSON ASSOCIATES, INC., HUNTINGTON, N.Y.
OAIRA-	AIR ATTACHE, OTTAWA
OAK-	OAK JOINT AEC-DOD WARHEAD COORDINATING COMM. (REF.15)
OAL-	AIR FORCE CAMBRIDGE RESEARCH CENTER. OPERATIONAL APPLICATIONS LAB., WASHINGTON, D.C.
OAL-	CONSOLIDATED VULTEE AIRCRAFT CORP. ORDNANCE AEROPHYSICS LAB., DAINGERFIELD, TEX.
OAL-	GENERAL DYNAMICS/DAINGERFIELD. ORDNANCE AEROPHYSICS LAB., TEX.
OAL-CF-	GENERAL DYNAMICS/DAINGERFIELD. ORDNANCE AEROPHYSICS LAB., TEX.
OA+M-	OKLAHOMA AGRICULTURAL + MECHANICAL COLL., STILLWATER
OA-M-	OKLAHOMA AGRICULTURAL + MECHANICAL COLL., STILLWATER
OAMA-	OPERATIONS ANALYSIS DIV. (AIR FORCE), WASHINGTON,D.C.
OAMA-	AIR MATERIEL AREA, OGDEN, UTAH
OAMA-	OGDEN AIR MATERIEL AREA, HILL AFB, UTAH
OAMA-	OGDEN AIR MATERIEL AREA, OOAMA HILL AFB, UTAH
OAMC-	OKLAHOMA AGRICULTURAL + MECHANICAL COLL., STILLWATER
OANAR-	OFFICE OF NAVAL RESEARCH, LONDON. (OFF. OF THE ASST. NAVAL ATTACHE FOR RESEARCH, LONDON)
OANAR-(NO.)-(YR)	OFFICE OF NAVAL RESEARCH, LONDON
OAP-TN-	STANFORD RESEARCH INST. OPERATIONS ANALYSIS DEPT., MENLO PARK, CALIF.
OAQ-	ECUADOR. OBSERVATORIO ASTRONOMICO, QUITO
OAR-	(OPERATIONS ANALYSIS REPORT) (REF. 2)
OAR-	AIR FORCE LOGISTICS COMMAND,WRIGHT-PATTERSON AFB,OHIO
OAR-	ARMY AIR FORCE. SECOND. OPERATIONS AND TRAINING DIV. (OPERATIONS ANALYSIS REPORT)
OA:R-	DIVISION OF OPERATIONS ANALYSIS AND FORECASTING, AEC, WASHINGTON, D.C.
OAR-	OFFICE OF AEROSPACE RESEARCH (AIR FORCE), WASH., D.C.
OAR-(YEAR)-	OFFICE OF AEROSPACE RESEARCH (AIR FORCE), WASH., D.C.
OA-R-	OFFICE OF AEROSPACE RESEARCH (AIR FORCE), WASH., D.C.
OA-R-	OFFICE OF OPERATIONS ANALYSIS AND PLANNING, AEC
OA-R-	OPERATIONS ANALYSIS DIV. (AIR FORCE), WASHINGTON,D.C.
OAR-	WRIGHT AIR DEVELOPMENT CENTER. AERONAUTICAL RESEARCH LAB., WRIGHT-PATTERSON AFB, OHIO
OAR-BRR-	OFFICE OF AEROSPACE RESEARCH (AIR FORCE), WASH., D.C.
OAR-QI-	OFFICE OF AEROSPACE RESEARCH (AIR FORCE), WASH., D.C. (QUARTERLY INDEX)
OAR-TR-	OFFICE OF AIR RESEARCH, WRIGHT-PATTERSON AFB, OHIO
OASD-	OFF. OF THE ASST. SECY. OF DEFENSE(RES.+ENG.),WASH,DC
OA-SR-	OPERATIONS ANALYSIS DIV. (AIR FORCE), WASHINGTON,D.C. (SUMMARY REPORT)
OATM-	AIR FORCE (OPERATIONS ANALYSIS TECHNICAL MEMORANDUM)
OA-TM-	OPERATIONS ANALYSIS DIV. (AIR FORCE), WASHINGTON,D.C. (TECHNICAL MEMORANDUM)
OAWP-	AIR FORCE LOGISTICS COMMAND,WRIGHT-PATTERSON AFB,OHIO
OA WP-	OPERATIONS ANALYSIS DIV. (AIR FORCE), WASHINGTON,D.C.
OA-WP-	STRATEGIC AIR COMMAND, VANDENBERG AFB, CALIF.
OAX-	RESEARCH AND DEVELOPMENT BOARD. PANEL ON AMMUNITION AND EXPLOSIVES, WASHINGTON, D.C.
OAX-MLH-	RESEARCH AND DEVELOPMENT BOARD, WASHINGTON, D.C.
OB-	GT. BRIT. ORDNANCE BOARD
O + B -	OLD AND BARNES, INC., PASADENA, CALIF.
OB-	OLIVER + BOYD, LTD., EDINBURGH (TRANSLATIONS)
OBE-SP-	OFFICE OF BUSINESS ECONOMICS, WASHINGTON, D.C.
OBI-	GT. BRIT. ORDNANCE BOARD (INVESTIGATION)
OBI-	NATIONAL BUREAU OF STANDARDS. OFFICE OF BASIC INSTRUMENTATION, WASHINGTON, D.C.
O.B. INVESTIGATION NO.	GT. BRIT. ORDNANCE BOARD
OBJ-	(OPERATION BUSTER-JANGLE) (REF. 7)
O.B. PROCEEDINGS NO.	GT. BRIT. ORDNANCE BOARD
OBR-(YEAR)-	BUREAU OF INTERNATIONAL COMMERCE, WASHINGTON, D.C.
OBSERVATION-(YEAR)-	CALIFORNIA INST. OF TECH., PASADENA. OWENS VALLEY RADIO OBSERVATORY
OC-	OCCIDENTAL COLL., LOS ANGELES
OC(LETTERS)-	(OPERATION CROSSROADS) (REF. 7)
OCAFF-	CONTINENTAL ARMY COMMAND, FORT MONROE, VA.
OCAMA-	AIR MATERIEL AREA, OKLAHOMA CITY
OCAMA-	OKLAHOMA CITY AIR MATERIEL AREA, TINKER AFB, OKLA.
OCB-	ORDNANCE CORPS (ARMY) (BULLETIN)
OCCASIONAL REPORT -	GENERAL ELECTRIC CO. RESEARCH LAB., SCHENECTADY, N.Y.
OCCM10-	CHEMICAL CORPS, ARMY CHEMICAL CENTER, MD.
OCD-	BUREAU OF ORDNANCE (NAVY) (ORDNANCE CLASSIFICATION OF DEFECTS)
OCD-	ORDNANCE CORPS (ARMY) (DIRECTIVES)
OCD-	WESTON INSTRUMENTS, INC., NEWARK, N.J.
OCDM-	OFFICE OF CIVIL AND DEFENSE MOBILIZATION, WASH., D.C.
OCD-NP-	OFFICE OF CIVIL DEFENSE, WASHINGTON, D.C. (NATIONAL PLAN APPENDIX SERIES)
OC-DOC-	DIVISION OF CLASSIFICATION, AEC, WASHINGTON, D.C.
OCD-OS-	STANFORD RESEARCH INST., MENLO PARK, CALIF.
OCD-PS-	OFFICE OF CIVIL DEFENSE, WASHINGTON, D.C.
OCD-RR-	OFFICE OF CIVIL DEFENSE, WASHINGTON, D.C.
OCD-TR-	OFFICE OF CIVIL DEFENSE, WASHINGTON, D.C.
OCE-	OFFICE OF THE CHIEF OF ENGINEERS (ARMY), WASH., D.C.
OCEANICS-	OCEANICS, INC., PLAINVIEW, N.Y.
OCE/EIN-	OFFICE OF THE CHIEF OF ENGINEERS (ARMY), WASH., D.C.
OCE-EM-	CORPS OF ENGINEERS (ARMY)
OCE-T-	CORPS OF ENGINEERS (ARMY)
OCE-TR-(YEAR)-	OFFICE OF THE CHIEF OF ENGINEERS (ARMY), WASH., D.C.
OCFC-	OWENS-CORNING FIBERGLAS CORP., N.Y.C.
OCL-	BUREAU OF ORDNANCE (NAVY) (ORDNANCE CIRCULAR LETTER)

OCM-	ORDNANCE COMMITTEE (ARMY) (MEMORANDUM)
OCM-	ORDNANCE DEPT. (ARMY)
OCM-	RESEARCH + DEV. BD. COMM. ON ORDNANCE, WASH., D.C.
OCM-	RESEARCH AND DEVELOPMENT BOARD. PANEL ON ORDNANCE MATERIALS, WASHINGTON, D.C.
OCNE-MN-	OKLAHOMA CITY AIR MATERIEL AREA. DIRECTORATE OF MATERIEL MANAGEMENT, TINKER AFB, OKLA.
OCNO-	OFFICE OF THE CHIEF OF NAVAL OPERATIONS, WASH., D.C.
OCNO ID I-	OFFICE OF THE CHIEF OF NAVAL OPERATIONS. INTELLIGENCE DIVISION (INTELLIGENCE REPORT)
OCO-	OFFICE OF THE CHIEF OF ORDNANCE, WASHINGTON, D.C.
OCPRS-	ATOMIC ENERGY COMMISSION, WASHINGTON, D.C.
OCR-	OFFICE OF COAL RESEARCH, WASHINGTON, D.C.
OCR-RDR-	DYNATECH CORP., CAMBRIDGE, MASS.
OCS-	BUREAU OF RADIOLOGICAL HEALTH. OFFICE OF CRITERIA AND STANDARDS, ROCKVILLE, MD.
OCSA-	OFFICE OF THE CHIEF OF STAFF (ARMY), WASHINGTON, D.C.
OCSC-	BALLISTIC RESEARCH LABS., ABERDEEN PROVING GROUND, MD.
OCSC-(NUMBER)-(YEAR)	ORDNANCE CORPS (ARMY). SHAPED CHARGE RESEARCH AND DEVELOPMENT STEERING AND COORDINATING COMMITTEE
OC-TDR-IX-	JOINT TASK FORCE ONE
OD-	(ORDNANCE DIV.) (REF. 2)
OD-	BUREAU OF ORDNANCE (NAVY) (ORDNANCE DATA)
OD-	NATIONAL BUREAU OF STANDARDS. ORDNANCE DEVELOPMENT DIV., WASHINGTON, D.C.
ODC-	BARNES ENGINEERING CO., STAMFORD, CONN.
ODC-	CANADA. DEPT. OF FORESTRY, OTTAWA
ODC-	OLYMPIC DEVELOPMENT CO., STAMFORD, CONN.
ODDR-	NATIONAL BUREAU OF STANDARDS. ORDNANCE DEVELOPMENT DIV., WASHINGTON, D.C.
ODDRE-	OFFICE OF THE DIRECTOR OF DEFENSE (RES+ENG), WASH., DC
ODDRE-MAM-	OFFICE OF THE DIRECTOR OF DEFENSE (RES+ENG), WASH., DC
ODDRE-MAN-	OFFICE OF THE DIRECTOR OF DEFENSE (RES+ENG), WASH., DC
ODL-	DIAMOND ORDNANCE FUZE LABS., WASHINGTON, D.C.
ODM-	OFFICE OF DEFENSE MOBILIZATION, WASHINGTON, D.C.
OD-OAG-	OFFICE OF SCIENTIFIC RESEARCH AND DEVELOPMENT
ODP-	OFFICE OF SCIENTIFIC RESEARCH AND DEVELOPMENT (ORGANIC DEVELOPMENT PROBLEMS)
ODR-	ORDNANCE DEPT. (ARMY) (REPORT)
OD-SP-	NATIONAL BUREAU OF STANDARDS. ORDNANCE DEVELOPMENT DIV., WASHINGTON, D.C. (SPECIAL PROJECTS)
OD-TB-	NATIONAL BUREAU OF STANDARDS. ORDNANCE DEVELOPMENT DIV., WASHINGTON, D.C. (TOSS BOMBING)
ODTM-	WESTINGHOUSE ELECTRIC CORP. ORDNANCE DEPT., BALTIMORE (TECHNICAL MEMORANDUM)
OE-	DEPARTMENT OF HEALTH, EDUCATION, AND WELFARE. OFFICE OF EDUCATION, WASHINGTON, D.C.
OEA-	OFFICE OF SCIENTIFIC RESEARCH AND DEVELOPMENT
OECD-	ORGANIZATION FOR ECONOMIC COOPERATION AND DEVELOPMENT, PARIS
OED-	NATIONAL BUREAU OF STANDARDS, WASHINGTON, D.C.
OED-RM-	STANFORD RESEARCH INST. OPERATIONAL EVALUATION DEPT., MENLO PARK, CALIF.
OED-RM-(NUMBER)-	STANFORD RESEARCH INST. OPERATIONAL EVALUATION DEPT., MENLO PARK, CALIF.
OEEC-	ORGANIZATION FOR EUROPEAN ECONOMIC COOPERATION, PARIS
OEEC/SCC/	UNITED KINGDOM ATOMIC ENERGY AUTHORITY. RESEARCH GP. ATOMIC ENERGY RES. ESTAB., HARWELL, BERKS, ENGLAND
OEG-	OFFICE OF THE CHIEF OF NAVAL OPERATIONS. OPERATIONS EVALUATION GROUP, WASHINGTON, D.C.
OEG-PROFESSIONAL PAPER-	OFFICE OF THE CHIEF OF NAVAL OPERATIONS. OPERATIONS EVALUATION GROUP, WASHINGTON, D.C.
OEG-R-	OFFICE OF THE CHIEF OF NAVAL OPERATIONS. OPERATIONS EVALUATION GROUP, WASHINGTON, D.C. (REPORTS)
OEG-RC-	OFFICE OF THE CHIEF OF NAVAL OPERATIONS. OPERATIONS EVALUATION GROUP, WASHINGTON, D.C.
OEG-RESEARCH CONTRIB-	CENTER FOR NAVAL ANALYSES. OPERATIONS EVALUATION GROUP, WASHINGTON, D.C.
OEG-S-	OFFICE OF THE CHIEF OF NAVAL OPERATIONS. OPERATIONS EVALUATION GROUP, WASHINGTON, D.C. (STUDIES)
OEG-SR-	OFFICE OF THE CHIEF OF NAVAL OPERATIONS. OPERATIONS EVALUATION GROUP, WASHINGTON, D.C.
OEG-STUDY-	CENTER FOR NAVAL ANALYSES. OPERATIONS EVALUATION GROUP, WASHINGTON, D.C.
OEL-MET-	ORENDA ENGINES LTD., MALTON, ONT.
OEL-NUCLEAR-	ROE(A.V.) CANADA LTD. ORENDA ENGINES DIV., MALTON, ONT.
OEM-	OFFICE OF THE DIRECTOR OF DEFENSE (RES+ENG), WASH., DC
OEO-	OFFICE OF ECONOMIC OPPORTUNITY, WASHINGTON, D.C.
OEO-LN-	OFFICE OF ECONOMIC OPPORTUNITY, WASHINGTON, D.C.
OE-OPPE-(YEAR)-	OFFICE OF EDUCATION, WASHINGTON, D.C.
OEP-	OFFICE OF EMERGENCY PREPAREDNESS, WASHINGTON, D.C.
OER-	BUREAU OF RECLAMATION (INTERIOR). OFFICE OF ENG. REFERENCE, DENVER
OERL-	OFFICER EDUCATION RESEARCH LAB., MAXWELL AFB, ALA.
OERL-TM-	OFFICER EDUCATION RESEARCH LAB., MAXWELL AFB, ALA.
OES-	AIR FORCE. 6TH AIR DIV., MAC DILL AFB, FLA.
OES-	GT. BRIT. ROYAL NAVAL PERSONNEL RESEARCH COMMITTEE
O.F.(NUMBER).(NUMBER)	BUREAU OF ORDNANCE (NAVY)
OFC-	NAVAL ORDNANCE TEST STA., INYOKERN (CHINA LAKE), CAL.
OFC-	RESEARCH + DEV. BD. PANEL ON FIRE CONTROL, WASH., D.C.
OFCM-(YEAR)-	ENVIRONMENTAL SCIENCE SERVICES ADM. OFFICE OF FEDERAL COORDINATOR FOR METEOROLOGICAL SERVICES AND SUPPORTING RESEARCH, WASHINGTON, D.C.
OFFSITE STUDIES-	GEOLOGICAL SURVEY, DENVER
O FILE-	EUROPEAN THEATER OF OPERATIONS (ARMY). ALSOS MISSION (OBERTH FILE)
OFJ-	POLAND. INSTITUTE OF NUCLEAR RESEARCH, WARSAW
OFR-	(OPEN FILE REPORT) (REF. 2)
OFS-	ARMY AIR FORCE. OFFICE OF FLYING SAFETY
OFS FSM-	ARMY AIR FORCE. OFFICE OF FLYING SAFETY (FLYING SAFETY MEMORANDUM)
OFS SAR-A-	ARMY AIR FORCE. OFFICE OF FLYING SAFETY (SPECIAL ACCIDENT REPORT. ANALYSIS)
OFS SAR-SA-	ARMY AIR FORCE. OFFICE OF FLYING SAFETY (SPECIAL ACCIDENT REPORT. SUPPLEMENTAL ANALYSIS)
OFW...	AIR FORCE, WASHINGTON, D.C.
OG-	(OPERATION GREENHOUSE) (REF. 7)
OG-(2 DIGITS)	GODDARD SPACE FLIGHT CENTER, GREENBELT, MD.
OGMC-	REDSTONE ARSENAL, HUNTSVILLE, ALA.
OGMS-LSR-	ARMY ORDNANCE GUIDED MISSILE SCHOOL (LIAISON STATUS REPORT)
OGN-	RESEARCH AND DEV. BD., PANEL ON GUNS., WASH., D.C.
OHI-	BUREAU OF ORDNANCE (NAVY) (ORDNANCE HANDLING INSTRUCTIONS)
OI-	OHIO INJECTOR CO., WADSWORTH
OI-	OWENS-ILLINOIS, INC., TOLEDO
OI-	PALERMO, ITALY. UNIV. ISTITUTO DI TECNOLOGIE MECCANICHE
OIL-	NAVAL POWDER FACTORY. ORDNANCE INVESTIGATION LAB., INDIAN HEAD, MD.
OIM-	BUREAU OF ORDNANCE (NAVY) (ORDNANCE INVESTIGATION MEMORANDUM)
OIM-	NAVAL POWDER FACTORY. ORDNANCE INVESTIGATION LAB., INDIAN HEAD, MD.
OJTA-	AIR TRAINING COMMAND, LOWRY AFB, COLO.
OKLA-	OKLAHOMA. UNIV., NORMAN. RESEARCH INST.
OKLASU-QPR-	OKLAHOMA STATE UNIV., STILLWATER (PROGRESS REPORT)
OKLAU-	OKLAHOMA. UNIV., NORMAN

OKM-

OKM-	GERMANY. OBERKOMMANDO DER KRIEGSMARINE
OKM CPEL T-	GERMANY. OBERKOMMANDO DER KRIEGSMARINE. CHEMISCHES-PHYSIKALISCHES EXPERIMENTELLES LABORATORIUM
OKM MD-	GERMANY. OBERKOMMANDO DER KRIEGSMARINE (M DIV.)
O-KPD-	KNOLLS ATOMIC POWER LAB., SCHENECTADY, N.Y.
OKSU-	OKLAHOMA STATE UNIV., STILLWATER
OL-	COMMITTEE ON MEDICAL RESEARCH (OTOLARYNGOLOGY RPTS.)
OL-	NAVAL GUN FACTORY. OPTICAL DESIGN LAB., WASH., D.C.
OLC-	OPERATIONAL LOGIC CORP., ALEXANDRIA, VA.
OLDC-	OLYMPIC DEVELOPMENT CO., STAMFORD, CONN.
OLIN-	OLIN INDUSTRIES, INC., EAST ALTON, ILL.
OLIN-	OLIN MATHIESON CHEMICAL CORP. CHEMS. DIV., NEW HAVEN
OLIN-NH-	OLIN MATHIESON CHEMICAL CORP., NEW HAVEN
OLIN-QPR-	OLIN MATHIESON CHEMICAL CORP., NEW HAVEN
OLIN-SSR-	OLIN CORP. OLIN RES. CTR. CHEMICALS GP., NEW HAVEN
OLO-	ARMY ORDNANCE LIAISON OFFICE, ALBUQUERQUE, N. MEX.
O-L- THRU O-M-	COLUMBIA UNIV., N.Y.C. SUBSTITUTE ALLOY MATERIALS LAB. (REF. 11)
OL TR-	NAVAL GUN FACTORY. OPTICAL LAB., WASHINGTON, D.C.
OM-	NAVAL ORDNANCE TEST STA., INYOKERN (CHINA LAKE), CAL.
OM-	WHITE SANDS MISSILE RANGE, N. MEX. (ORDNANCE MISSION)
OMB-	BALLISTIC RESEARCH LABS.,ABERDEEN PROVING GROUND, MD.
OMB-	OFFICE OF MANAGEMENT AND BUDGET, WASHINGTON, D.C.
OMC-	OZARK-MAHONING CO., TULSA, OKLA.
OMCC-	OLIN MATHIESON CHEMICAL CORP., BALTIMORE
OMCC-HEF-	OLIN MATHIESON CHEMICAL CORP. ENERGY DIV., NIAGARA FALLS, N.Y.
OMCC-HEF-	OLIN MATHIESON CHEMICAL CORP. HIGH ENERGY FUELS DIV., NIAGARA FALLS, N.Y.
OMCC-MRL-(YEAR)-	OLIN MATHIESON CHEMICAL CORP. METALLURGICAL LABS., NEW HAVEN
OMCC-SCR-	OLIN MATHIESON CHEMICAL CORP. SPECIALTY CHEMICALS RESEARCH ENERGY DIV., NEW HAVEN
OME-	NAVAL ORDNANCE TEST STA., INYOKERN (CHINA LAKE), CAL.
OMEGA-TN-	PALMER (WINSLOW), BABYLON, N.Y.
OMG-	OFFICE OF MILITARY GOVERNMENT FOR GERMANY (U.S.)
OMI-	BUREAU OF ORDNANCE (NAVY) (ORDNANCE MODIFICATION INSTRUCTION)
OML-	BUREAU OF ORDNANCE (NAVY) (ORDNANCE MATERIAL LETTER)
OML-	REDSTONE ARSENAL. ORDNANCE MISSILE LABS., HUNTSVILLE, ALA.
OMM-	WORLD METEOROLOGICAL ORGANIZATION, GENEVA
OMM-(NUMBER)-RP-	WORLD METEOROLOGICAL ORGANIZATION, GENEVA
OMRO-	WATERTOWN ARSENAL. ORDNANCE MATERIALS RES. OFF.,MASS.
OMRO-PUB.-	WATERTOWN ARSENAL. ORDNANCE MATERIALS RES. OFF.,MASS.
OMRO-PUBL-	WATERTOWN ARSENAL. ORDNANCE MATERIALS RES. OFF.,MASS.
OM SM-	CALIFORNIA. UNIV., BERKELEY. RADIATION LAB.
ON-(2 DIGITS)	VIENNA. TECHNISCHE HOCHSCHULE. INST. FUER HOCH-FREQUENZTECHNIK
ONC-	COAST AND GEODETIC SURVEY, WASHINGTON, D.C. (OPERATIONAL NAVIGATION CHART)
ONERA-	FRANCE. OFFICE NATIONAL D'ETUDES ET DE RECHERCHES AEROSPATIALES, PARIS
ONERA-CF-	FRANCE. OFFICE NATIONAL D'ETUDES ET DE RECHERCHES AEROSPATIALES, CHATILLON-SOUS-BAGNEUX
ONERA-DP-	FRANCE. OFFICE NATIONAL D'ETUDES ET DE RECHERCHES AEROSPATIALES, CHATILLON-SOUS-BAGNEUX
ONERA-NOTE TECHNIQUE-	FRANCE. OFFICE NATIONAL D'ETUDES ET DE RECHERCHES AEROSPATIALES, CHATILLON-SOUS-BAGNEUX
ONERA-NT-	FRANCE. OFFICE NATIONAL D'ETUDES ET DE RECHERCHES AEROSPATIALES, PARIS
ONERA-P-	FRANCE. OFFICE NATIONAL D'ETUDES ET DE RECHERCHES AEROSPATIALES, PARIS
ONERA-PUB-	FRANCE. OFFICE NATIONAL D'ETUDES ET DE RECHERCHES AEROSPATIALES, CHATILLON-SOUS-BAGNEUX
ONERA-PUBL-	FRANCE. OFFICE NATIONAL D'ETUDES ET DE RECHERCHES AEROSPATIALES, CHATILLON-SOUS-BAGNEUX
ONERA-PUBL-(NO.)/(YR)	FRANCE. OFFICE NATIONAL D'ETUDES ET DE RECHERCHES AEROSPATIALES, PARIS
ONERA-TN-	FRANCE. OFFICE NATIONAL D'ETUDES ET DE RECHERCHES AEROSPATIALES, CHATILLON-SOUS-BAGNEUX
ONERA-TN-	FRANCE. OFFICE NATIONAL D'ETUDES ET DE RECHERCHES AEROSPATIALES, PARIS
ONERA-TP-	FRANCE. OFFICE NATIONAL D'ETUDES ET DE RECHERCHES AEROSPATIALES, CHATILLON-SOUS-BAGNEUX
ONERA-TP-	FRANCE. OFFICE NATIONAL D'ETUDES ET DE RECHERCHES AEROSPATIALES, PARIS
ONI-	OFFICE OF NAVAL INTELLIGENCE, WASHINGTON, D.C.
ONI-A-	OFFICE OF NAVAL INTELLIGENCE, WASHINGTON, D.C.
ONI-ST-(NUMBER)-(YEAR)	OFFICE OF NAVAL INTELLIGENCE, WASHINGTON, D.C.
ONI-TR-	OFFICE OF NAVAL INTELLIGENCE, WASHINGTON, D.C.
ONI-TRANS-	OFFICE OF NAVAL INTELLIGENCE. TRANSLATION SECTION, WASHINGTON, D.C.
ONM-	OFFICE OF NAVAL MATERIAL. SPEC. PROJS. OFF.,WASH.,DC
ONR-	OFFICE OF NAVAL RESEARCH, WASHINGTON, D.C.
ONR-(YEAR)-	OFFICE OF NAVAL RESEARCH, WASHINGTON, D.C.
ONR(YEAR)-	OFFICE OF NAVAL RESEARCH, WASHINGTON, D.C.
ONR/ACR/(LETTERS OR NOS.)	OFFICE OF NAVAL RESEARCH, WASHINGTON, D.C.
ONR-BOS-	COLORADO. UNIV., BOULDER. ENGINEERING EXPERIMENT STA.
ONR-CR-	OFFICE OF NAVAL RESEARCH, WASHINGTON, D.C.
ONR-H-(YEAR)-	INSTITUTE FOR RESEARCH. DIV. OF PSYCHOBIOLOGY, STATE COLLEGE, PENNA.
ONR-HO(YEAR)-	OFFICE OF NAVAL RESEARCH, WASHINGTON, D.C.
ONRL-	OFFICE OF NAVAL RESEARCH, LONDON
ONRL-(NO.)-(YR)	OFFICE OF NAVAL RESEARCH, LONDON
ONRL A(NUMBER)-(YEAR)	OFFICE OF NAVAL RESEARCH, LONDON
ONRL-C-(NUMBER)-(YEAR)	OFFICE OF NAVAL RESEARCH. LONDON (CONFERENCES)
ONRL ESN-(NUMBER)-	OFFICE OF NAVAL RESEARCH, LONDON (EUROPEAN SCI.NOTES)
ONRL-M-	OFFICE OF NAVAL RESEARCH, LONDON
ONRL-R-	OFFICE OF NAVAL RESEARCH, LONDON
ONRL TR-	OFFICE OF NAVAL RESEARCH, LONDON
ONR-RF-	OFFICE OF NAVAL RESEARCH, WASHINGTON, D.C.
ONR-RM-	OFFICE OF NAVAL RESEARCH, WASHINGTON, D.C.
ONR-RR-	OFFICE OF NAVAL RESEARCH, WASHINGTON, D.C.
ONR SYMPOSIUM-ACR-	OFFICE OF NAVAL RESEARCH, WASHINGTON, D.C.
ONR-TR-	OFFICE OF NAVAL RESEARCH, WASHINGTON, D.C.
ONR-TR-RLT-	OFFICE OF NAVAL RESEARCH, WASHINGTON, D.C.
ON/TN/	IMPERIAL COLLEGE OF SCIENCE AND TECHNOLOGY, LONDON. DEPT. OF MECHANICAL ENGINEERING
OO-	OFFICE OF THE CHIEF OF ORDNANCE, WASHINGTON, D.C.
OOA-	OFFICE OF OPERATIONS ANALYSIS, ENT AFB, COLO.
OOD-ENGINEERING-	PERKIN-ELMER CORP. OPTICAL OPERATIONS DIV., WILTON, CONN.
OOG-	OFFICE OF OIL AND GAS, WASHINGTON, D.C.
OOM-	RESEARCH + DEV. BD. COMM. ON ORDNANCE, WASH., D.C.
OOM-	RESEARCH AND DEVELOPMENT BOARD. PANEL ON ORDNANCE MATERIALS, WASHINGTON, D.C.
OONC-(NUMBER)-(YEAR)	OGDEN AIR MATERIEL AREA. SYSTEM SUPPORT MANAGEMENT DIV., HILL AFB, UTAH
(YEAR)-OONEBT-	OGDEN AIR MATERIEL AREA. DIRECTORATE OF MATERIEL MANAGEMENT, HILL AFB, UTAH
(YEAR)-OONEBT-	OGDEN AIR MATERIEL AREA. SERVICE ENGINEERING DIV., HILL AFB, UTAH
(YEAR)-OONEST-	OGDEN AIR MATERIEL AREA. SERVICE ENGINEERING DIV., HILL AFB, UTAH
OOR-	OFFICE OF ORDNANCE RESEARCH (ARMY), DURHAM, N.C.

OOR-(NUMBER).(NUMBER)	OFFICE OF ORDNANCE RESEARCH (ARMY), DURHAM, N.C.
OOR-R-(NUMBER)-	OFFICE OF ORDNANCE RESEARCH (ARMY), DURHAM, N.C.
OORR-	OFFICE OF ORDNANCE RESEARCH (ARMY), DURHAM, N.C.
OORTM-	OFFICE OF ORDNANCE RESEARCH (ARMY), DURHAM, N.C. (TECHNICAL MEMORANDUM)
OOY-	OGDEN AIR MATERIEL AREA, HILL AFB, UTAH
OOY-P-(YEAR)-	OGDEN AIR MATERIEL AREA, HILL AFB, UTAH
OOY-TR-	OGDEN AIR MATERIEL AREA, HILL AFB, UTAH
OP-	(ORDNANCE PAMPHLET) (REF. 2)
OP-	(OPERATION PLANS) (REF. 2)
OP-	ABERDEEN PROVING GROUND, MD. (ORDNANCE PROGRAM)
OP-	BUREAU OF ORDNANCE (NAVY) (ORDNANCE PAMPHLET)
OP-	ETHYL CORP. RESEARCH AND DEVELOPMENT DEPT., FERNDALE, MICH. (ONE-PAGE REPORTS)
OP-	HUGHES AIRCRAFT CO. COMMUNICATIONS DIV., CULVER CITY, CALIF.
OP/	NAVAL AIR DEV. SQUADRON 5, CHINA LAKE, CALIF.
OP-	NAVAL OPERATIONAL TEST + EVALUATION FORCE, NORFOLK, VA
OP-32-	OFFICE OF NAVAL INTELLIGENCE, WASHINGTON, D.C.
OP/(LETTER(S))...	OPERATIONAL DEV. FORCE, ATLANTIC FLEET, NORFOLK, VA.
OPC-	ORDNANCE DEPT. (ARMY). ORDNANCE PACKAGING + CRATING
OPC AFPS-	ORDNANCE DEPT. ORDNANCE PACKAGING AND CRATING. FIRE CONTROL SUBOFFICE, FRANKFORD ARSENAL, PHILADELPHIA (PACKAGING INSTRUCTIONS)
OPC I-	ORDNANCE DEPT. (ARMY). ORDNANCE PACKAGING + CRATING, DETROIT (PACKAGING INSTRUCTIONS)
OPE-	AIR FORCE
OPER. ANAL.-	CALIFORNIA. UNIV., BERKELEY. RADIATION LAB. (REF. 9)
OPG-	EDGEWOOD ARSENAL, MD. (OPERATIONS RESEARCH)
OPLAN-(NUMBER)-(YEAR)	AIR FORCE SPECIAL WEAPONS CENTER, KIRTLAND AFB, N.M. (OPERATION PLAN)
OPM-	ARMY, WASHINGTON, D.C. (ORDNANCE PROOF MANUAL)
OPNAV-	NAVAL OPERATIONAL TEST + EVALUATION FORCE, NORFOLK, VA
OPNAV-	OFFICE OF THE CHIEF OF NAVAL OPERATIONS, WASH., D.C.
OPP-	OAK RIDGE OPERATIONS OFFICE, AEC. OPERATIONAL PLANNING AND POWER DIV., TENN.
OPPL-	OKLAHOMA AGRICULTURAL + MECHANICAL COLL., STILLWATER OKLAHOMA POWER AND PROPULSION LAB.
OPR-	STRATEGIC BOMBING SURVEY (OIL PLANT REPORTS)
OPRD-	OFFICE OF PRODUCTION RESEARCH AND DEVELOPMENT
OPRD CR-	OFFICE OF PRODUCTION RES. + DEV. (CHLORAL RPTS.)
OPRD QCR-	OFFICE OF PRODUCTION RESEARCH AND DEVELOPMENT (QUALITY CONTROL REPORTS, CARNEGIE INST. OF TECH.)
OPRD RR-	OFFICE OF PRODUCTION RESEARCH AND DEVELOPMENT
OPRD TF-	OFFICE OF PRODUCTION RESEARCH AND DEVELOPMENT
OPSEARCH-	JOHNS HOPKINS UNIV., BETHESDA, MD. OPERATIONS RES.OFF.
OPSR-	HUGHES AIRCRAFT CO. OPTICAL PROPERTIES TECHNICAL INFORMATION CENTER, CULVER CITY, CALIF.
OPTIK-(NUMBER/YEAR)	GERMANY. PHYSIKALISCH-TECHNISCHE BUNDESANSTALT, BRUNSWICK
OPX-	AIR FORCE
OQF-	AIR FORCE LOGISTICS COMMAND. DIRECTORATE OF AEROSPACE FUELS, KELLY AFB, TEX.
OQMG-	QUARTERMASTER CORPS, WASHINGTON, D.C.
OR-	(OPEN REPORT) (REF. 2)
OR-	(OPERATIONAL REQUIREMENT) (REF. 2)
OR-	(OPERATIONS REQUIREMENT) (REF. 2)
OR-	(OPERATIONS RESEARCH) (REF. 2)
OR-	(ORDNANCE REPORT) (REF. 2)
OR-	(OVERALL REPORT) (REF. 2)
OR-	AIR FORCE (OPERATIONS REQUIREMENT)
OR-(YEAR)-	ARMY WEAPONS COMMAND. WEAPONS OPERATIONS RESEARCH OFFICE, ROCK ISLAND, ILL.
OR-	BUREAU OF NAVAL WEAPONS
OR-	CANADA. DEPT. OF NATIONAL DEFENCE, OTTAWA
OR-	ITEK CORP., LEXINGTON, MASS.
OR-	MARTIN CO., ORLANDO, FLA.
OR-	NAVY (OPERATIONAL REQUIREMENT)
OR-	RESEARCH + DEV. BD. COMM. ON ORDNANCE, WASH., D.C.
OR-(YEAR)-	ROCK ISLAND ARSENAL LAB., ILL.
OR-	WASHINGTON UNIV., ST. LOUIS
ORA-(NUMBERS,LETTERS)	MICHIGAN. UNIV., ANN ARBOR. OFFICE OF RESEARCH ADM.
ORA-	MICHIGAN. UNIV., ANN ARBOR. OFFICE OF RESEARCH ADM.
ORA-	OFFICE OF RESEARCH ANALYSES, HOLLOMAN AFB, N. MEX.
ORA-(YEAR)-	OFFICE OF RESEARCH ANALYSES, HOLLOMAN AFB, N. MEX.
ORAU-	OAK RIDGE ASSOCIATED UNIVERSITIES, INC., TENN.
ORB-	CANADA. AIR DEFENCE COMMAND. OPERATIONAL RESEARCH BR.
ORB-	SIGNAL CORPS (ARMY). OPERATIONAL RESEARCH BRANCH
ORC-	CALIFORNIA. UNIV., BERKELEY
ORC-	CALIFORNIA. UNIV., BERKELEY. OPERATIONS RESEARCH CTR.
ORC-(YEAR)-	CALIFORNIA. UNIV., BERKELEY. OPERATIONS RESEARCH CTR.
ORC-(YEAR)-	TORONTO. UNIV. INST. FOR AEROSPACE STUDIES
ORC-	WHITE SANDS MISSILE RANGE. OSCURA RANGE CAMP., N.MEX.
ORCP-	ORDNANCE DEPT. RESEARCH AND DEVELOPMENT CENTER, ABERDEEN PROVING GROUND, MD.
ORD-	CANADA. DEPT. OF NATIONAL DEFENCE. OPERATIONAL RESEARCH DIV., OTTAWA
ORD-	NAVAL ORDNANCE SYSTEMS COMMAND, WASHINGTON, D.C.
ORD-(NUMBER) DMU	OFFICE OF RESEARCH AND DEVELOPMENT (INTERIOR), WASH., D.C. (WATER POLLUTION CONTROL RESEARCH SERIES)
ORD-	ORDNANCE CORPS (ARMY)
ORD-	STANFORD RESEARCH INST., MENLO PARK, CALIF.
ORDAB-	ARMY BALLISTIC MISSILE AGENCY, REDSTONE ARSENAL, ALA.
ORD-AJ-	FAIRCHILD SPACE AND DEFENSE SYSTEMS, SYOSSET, N.Y.
ORDALT-	BUREAU OF ORDNANCE (NAVY) (ORDNANCE ALTERATIONS)
ORD-AP-	FAIRCHILD SPACE AND DEFENSE SYSTEMS, SYOSSET, N.Y.
ORD/APG (LETTERS, NOS.)	ABERDEEN PROVING GROUND, MD.
ORDARK-	ARKANSAS. UNIV., FAYETTEVILLE (ORDNANCE PROJECT)
ORDBB-	PICATINNY ARSENAL. FELTMAN RESEARCH LABS., DOVER, N.J.
ORDBB-DR4-	PICATINNY ARSENAL. ARTILLERY AMMUNITION AND ROCKET DEVELOPMENT LAB., DOVER, N.J.
ORDBB-NR-	PICATINNY ARSENAL. QUALITY ASSURANCE DIV., DOVER, N.J.
ORDBB-T-(NUMBER)-(NO.)	PICATINNY ARSENAL, DOVER, N.J.
ORDBB-TK-	PICATINNY ARSENAL, DOVER, N.J. (TASK GROUP REPORTS)
ORDBB-TK-(NUMBER)-(NO.)	PICATINNY ARSENAL, DOVER, N.J. (TASK GROUP REPORTS)
ORDBB-VC-	PICATINNY ARSENAL, DOVER, N.J.
ORDBG-DPS-LS-	ABERDEEN PROVING GROUND, MD.
ORDBG-OTI-	ORDNANCE TECHNICAL INTELLIGENCE SERVICE, ABERDEEN PROVING GROUND, MD.
ORDBS-	WHITE SANDS PROVING GROUND, N. MEX.
ORDC-ARCHIVE-(NO./NO.)	ORDNANCE DEPT. RESEARCH AND DEV. CTR. FOREIGN DOC. EVALUATION BRANCH, ABERDEEN PROVING GROUND, MD.
ORDCIT-	CALIFORNIA INST. OF TECH., PASADENA (ORDNANCE RESEARCH AND DEVELOPMENT)
ORD-CVS-	AIR FORCE. 12TH, WACO, TEX.
ORD DD PROJ KG-	ORDNANCE DEPT. (ARMY). DEVELOPMENT DIV. (PROJECT KG)
ORDD R TR-	OFFICE OF THE CHIEF OF ORDNANCE. RESEARCH AND DEVELOPMENT DIVISION (SUBOFFICE ROCKET)
ORD EATR-	EDGEWOOD ARSENAL, MD.
ORD FLR-	FRANKFORD ARSENAL, PHILADELPHIA

ORD GMC-
  GENERAL MOTORS CORP. ORDNANCE DEPT.
ORD INFORMAL PAPER-
  CANADA. DEPT. OF NATIONAL DEFENCE. OPERATIONAL RESEARCH DIV., OTTAWA
ORDL-EC-CR-
  CORPS OF ENGINEERS (ARMY), OHIO RIVER DIV. LABS., CINCINNATI
ORDL-EC-TR-
  CORPS OF ENGINEERS (ARMY). OHIO RIVER DIV. LABS., CINCINNATI
ORDL-TM-
  CORPS OF ENGINEERS (ARMY). OHIO RIVER DIV. LABS., CINCINNATI
ORDL-TR-
  CORPS OF ENGINEERS (ARMY). OHIO RIVER DIV. LABS., CINCINNATI
ORDMX-ECM-
  DETROIT ARSENAL, CENTER LINE, MICH.
ORDNANCE BOARD PROC...
  GT. BRIT. ORDNANCE BOARD
ORDNANCE PAMPHLET-
  BUREAU OF ORDNANCE (NAVY)
ORDP-
  ARMY, WASHINGTON, D.C. (ORDNANCE DESIGN PAMPHLET)
ORDP-
  ORDNANCE CORPS (ARMY) (PAMPHLET)
ORDP-
  ORDNANCE CORPS (ARMY), WASHINGTON, D.C.
ORD PATR-
  PICATINNY ARSENAL, DOVER, N.J.
ORD PROJ TB-(NUMBERS)
  BALLISTIC RESEARCH LABS.,ABERDEEN PROVING GROUND, MD.
ORD RAR-
  WATERTOWN ARSENAL, MASS. (ROLLED ARMOR REPORT)
ORD RIA LR-
  ROCK ISLAND ARSENAL LAB., ILL.
ORD-RM-(NUMBER)-
  STANFORD RESEARCH INST. OPERATIONS RESEARCH DEPT., MENLO PARK, CALIF.
ORD SA-
  SPRINGFIELD ARMORY, MASS.
ORD SCEL-
  SIGNAL CORPS ENGINEERING LABS., FORT MONMOUTH, N.J.
ORD SIP-
  ORDNANCE DEPT. (ARMY) (STANDARD INSPECTION PROCEDURE)
ORDSW-
  ARMY ORDNANCE SPECIAL WEAPONS-AMMUNITION COMMAND, DOVER, N.J.
ORD TB-(NUMBERS)
  BALLISTIC RESEARCH LABS.,ABERDEEN PROVING GROUND, MD.
ORDTB-
  ORDNANCE CORPS (ARMY), WASHINGTON, D.C.
ORD TEN SPEC-
  ORDNANCE DEPT. (ARMY) (TENTATIVE SPECIFICATION)
ORD TIR-
  ARMY FORCES IN THE PACIFIC (ORDNANCE TECHNICAL INTELLIGENCE REPORT)
ORD TIR-
  ORDNANCE DEPT. (ARMY) (TECHNICAL INTELLIGENCE REPORT)
ORD-TN-(NUMBER)-
  STANFORD RESEARCH INST. OPERATIONS RESEARCH DEPT., MENLO PARK, CALIF.
ORD-TRART-
  GT. BRIT. ORDNANCE DEPT. TECHNICAL REPRESENTATIVE FOR ARTILLERY, LONDON
ORDTX-
  DIAMOND ORDNANCE FUZE LABS., WASHINGTON, D.C.
ORDTX-
  FRANKFORD ARSENAL. FIRE CONTROL LAB., PHILADELPHIA
ORDTX-
  ORDNANCE CORPS (ARMY), WASHINGTON, D.C.
ORDTX-
  PICATINNY ARSENAL. SAMUEL FELTMAN AMMUNITION LABS., DOVER, N.J.
ORDTX-
  SPRINGFIELD ARMORY. RES. + DEV. DIV., MASS.
ORDWES-
  WESLEYAN UNIV.,WINDSOR LOCKS,CONN. ORDNANCE RES.PROJ.
ORDWES-TN-
  WESLEYAN UNIV.,WINDSOR LOCKS,CONN. ORDNANCE RES.PROJ.
ORDXC-
  ARMY BALLISTIC MISSILE AGENCY, REDSTONE ARSENAL, ALA.
ORDXM-C-(NUMBERS)
  ARMY BALLISTIC MISSILE AGENCY, REDSTONE ARSENAL, ALA.
ORDXMC-C-
  ARMY BALLISTIC MISSILE AGENCY, REDSTONE ARSENAL, ALA.
OREGONU-
  OREGON. UNIV., EUGENE
ORG-
  CANADA. OPERATIONAL RESEARCH GROUP
ORG-
  OFFICE OF NAVAL RES. OPERATIONS RES. GP., WASH., DC
ORG-
  SIGNAL CORPS (ARMY). OPERATIONAL RESEARCH GROUP
ORG/MS-
  INDUSTRIEANLAGEN-BETRIEBSGESELLSCHAFT M.B.H., OTTOBRUNN, GERMANY
ORG NOTE-
  EDGEWOOD ARSENAL, MD. (OPERATIONS RESEARCH)
ORG SPECIAL PUB-
  EDGEWOOD ARSENAL, MD. (OPERATIONS RESEARCH)
OR/HF/(NO./YR.)
  BRITISH IRON AND STEEL RESEARCH ASSN. OPERATIONAL RESEARCH DEPT., LONDON (OPEN REPORT)
ORI-
  OFFICE OF NAVAL RESEARCH, WASHINGTON, D.C.
ORI-
  OFFICE OF RESEARCH AND INVENTIONS (NAVY)
ORI-
  OPERATIONS RESEARCH INC., SILVER SPRING, MD.
ORINS-
  OAK RIDGE INST. OF NUCLEAR STUDIES, INC., TENN.
ORINS-MED-
  OAK RIDGE INST. OF NUCLEAR STUDIES, INC. MEDICAL DIV., TENN.
ORI TR-
  OPERATIONS RESEARCH INC., SILVER SPRING, MD.
ORL-
  PENNA. STATE UNIV., UNIV. PARK. ORDNANCE RES. LAB.
ORL(PSU)TM-
  PENNA. STATE UNIV., UNIV. PARK. ORDNANCE RES. LAB.
ORL/UT-
  TEXAS. UNIV., AUSTIN. OPTICAL RESEARCH LAB.
ORL/UVA-
  VIRGINIA. UNIV., CHARLOTTESVILLE. ORDNANCE RES. LAB.
O.R. MIMEO SER-
  TECHNION-ISRAEL INST. OF TECH., HAIFA. FACULTY OF INDUSTRIAL AND MANAGEMENT ENGINEERING
OR/NF/(NUMBER/YEAR)
  BRITISH STEEL CORP., LONDON
ORNL-
  OAK RIDGE NATIONAL LAB., TENN.
ORNL-AIC-
  OAK RIDGE NATIONAL LAB., TENN.
ORNL-AMPIC-
  OAK RIDGE NATIONAL LAB. ATOMIC AND MOLECULAR PROCESSES INFORMATION CENTER, TENN.
ORNL-CD-
  OAK RIDGE NATIONAL LAB., TENN. (CIVIL DEFENSE)
ORNL-CDC-
  OAK RIDGE NATIONAL LAB., TENN.
ORNL-CF-
  OAK RIDGE NATIONAL LAB., TENN. (CENTRAL FILES NO.)
ORNL CF-(YR. MO.)-
  OAK RIDGE NATIONAL LAB., TENN.
ORNL-CPX-
  OAK RIDGE NATIONAL LAB., TENN.
ORNL-CWS-
  OAK RIDGE NATIONAL LAB., TENN.
ORNL-HUD-
  OAK RIDGE NATIONAL LAB., TENN.
ORNL-IBP-(YEAR)-
  OAK RIDGE NATIONAL LAB., TENN.
ORNL-IIC-
  OAK RIDGE NATIONAL LAB. ISOTOPES INFORMATION CENTER, TENN.
ORNL-LN-
  OAK RIDGE NATIONAL LAB., TENN.
ORNL-MP-
  OAK RIDGE NATIONAL LAB., TENN.
ORNL-MTR-
  OAK RIDGE NATIONAL LAB., TENN.
ORNL-NDIC-
  OAK RIDGE NATIONAL LAB. NUCLEAR DESALINATION INFORMATION CENTER, TENN.
ORNL-NSF-EP-
  OAK RIDGE NATIONAL LAB., TENN.
ORNL-NSIC-
  OAK RIDGE NATIONAL LAB. NUCLEAR SAFETY INFORMATION CENTER, TENN.
ORNL-P-
  OAK RIDGE NATIONAL LAB., TENN.
ORNL-RMIC-
  OAK RIDGE NATIONAL LAB. RESEARCH MATERIALS INFORMATION CENTER, TENN.
ORNL RN-
  OAK RIDGE NATIONAL LAB., TENN. (REPRINT NO.)
ORNL-RSIC-
  OAK RIDGE NATIONAL LAB. RADIATION SHIELDING INFORMATION CENTER, TENN.
ORNL-TM-
  OAK RIDGE NATIONAL LAB., TENN.
ORNL-TR-
  OAK RIDGE NATIONAL LAB., TENN.
ORNO-TM-
  OAK RIDGE NATIONAL LAB., TENN.
ORO-
  JOHNS HOPKINS UNIV.,BETHESDA, MD. OPERATIONS RES.OFF.
ORO-(CONTRACT NO.)-
  OAK RIDGE OPERATIONS OFFICE, AEC, TENN. (REF. 4)
ORO-
  OAK RIDGE OPERATIONS OFFICE, AEC, TENN. (REF. 4)
ORO-JHU-
  JOHNS HOPKINS UNIV., CHEVY CHASE, MD. OPERATIONS RESEARCH OFFICE
ORO-LOG.-
  JOHNS HOPKINS UNIV., CHEVY CHASE, MD. OPERATIONS RESEARCH OFFICE
ORO-R-
  JOHNS HOPKINS UNIV., CHEVY CHASE, MD. OPERATIONS RESEARCH OFFICE
ORO-S-
  JOHNS HOPKINS UNIV., CHEVY CHASE, MD. OPERATIONS RESEARCH OFFICE
ORO-SP-
  JOHNS HOPKINS UNIV.,BETHESDA, MD. OPERATIONS RES.OFF.
ORO-T-
  JOHNS HOPKINS UNIV.,BETHESDA, MD. OPERATIONS RES.OFF.
ORO-TP-
  JOHNS HOPKINS UNIV.,BETHESDA, MD. OPERATIONS RES.OFF.
ORO-TR-
  OAK RIDGE OPERATIONS OFFICE, AEC, TENN. (REF. 4)
ORP/SID (YEAR)-
  ENVIRONMENTAL PROTECTION AGENCY. OFFICE OF RADIATION PROGRAMS, WASH., D.C. (SURVEILLANCE AND INSPECTION)
(NUMBER)-ORR-
  AIR FORCE COMMUNICATIONS SERVICE. OPERATIONS RESEARCH ANALYSIS OFFICE, SCOTT AFB, ILL.

ORR-	OFFICE OF RUBBER RESERVE
ORR CC-	OFFICE OF RUBBER RESERVE. COMPOUNDERS COMMITTEE
ORR CD-	OFFICE OF RUBBER RESERVE (POLYMER DEVELOPMENT)
ORR CR-	OFFICE OF RUBBER RESERVE (POLYMER RESEARCH, DISCRETION GROUP REPORTS)
ORR G-	OFFICE OF RUBBER RESERVE (GENERAL REPORTS)
ORR RM-	OFFICE OF RUBBER RESERVE (RAW MATERIALS REPORTS)
ORR TR-	OFFICE OF RUBBER RESERVE (RECONSTRUCTION FINANCE CORP. TECHNICAL REPORTS)
ORS-	SIGNAL CORPS (ARMY). OPERATIONAL RESEARCH STAFF
ORS/	UNITED KINGDOM ATOMIC ENERGY AUTHORITY. REACTOR GP. DOUNREAY EXPTL. REACTOR ESTAB., CAITHNESS, SCOTLAND
ORS/CC-	GT. BRIT. ROYAL AIR FORCE. COASTAL COMMAND
ORS FT-	GT. BRIT. ROYAL AIRCRAFT EST.,FARNBOROUGH, HANTS, ENG
ORSORT-	OAK RIDGE SCHOOL OF REACTOR TECHNOLOGY, TENN.
ORS(S)-	OPERATIONAL RESEARCH SECTION, SINGAPORE
ORUFE-	OPERATIONAL RESEARCH UNIT (FAR EAST), SINGAPORE
OS-	(OPERATION SANDSTONE) (REF. 7)
OS-	(OPERATION SNAPPER) (REF. 7)
OS-	(ORDNANCE SPECIFICATION) (REF. 2)
OS-	ARMY TECHNICAL INTELLIGENCE CENTER. ORDNANCE (ANALYSIS) SECTION, TOKYO
OS-	BUREAU OF ORDNANCE (NAVY) (ORDNANCE SPECIFICATIONS)
OS-(YEAR-NUMBER)-	GENERAL ELECTRIC CO.ORDNANCE SYSTEMS,PITTSFIELD,MASS.
OS-	ORDNANCE ANALYSIS SECTION (AIR FORCE)
OS-	SANDIA CORP., ALBUQUERQJE, N. MEX.
OSA-	OPTICAL SOCIETY OF AMERICA, CAMBRIDGE, MASS. (TRANSLATIONS)
OSA-	SCIENTIFIC ADVISORY BOARD (AIR FORCE), WASHINGTON, DC
OSAKAU-	OSAKA UNIV.
OSB-	CORPS OF ENGINEERS (ARMY). NUCLEAR POWER FIELD OFFICE, FT. BELVOIR, VA.
OSC-(NO.)-(NO.)-(NO.)-	ARIZONA. UNIV., TUCSON. OPTICAL SCIENCES CENTER
OSC-	OREGON STATE COLL., CORVALLIS
OSI-(YEAR)-	OCEANOGRAPHIC SERVICES, INC., SANTA BARBARA, CALIF.
OSI-SR-	CENTRAL INTELL. AGENCY. OFF. OF SCI. INTELL.,WASH.,DC
OSLO-RIS-	OSLO. UNIVERSITETETSFORLAGETS TRYKNINGSSENTRAL (RADIOBIOLOGY STUDY)
OSP-	CORPS OF ENGINEERS (ARMY). NUCLEAR POWER FIELD OFFICE, FT. BELVOIR, VA.
OSP-	GENERAL DYNAMICS/FORT WORTH, TEX.
OSP-	PETERS (O.S.) CO., WASHINGTON, D.C.
OSR-	AIR FORCE OFFICE OF SCIENTIFIC RESEARCH, WASH., D.C.
OSR-	BUREAU OF ORDNANCE (NAVY) (ORDNANCE STATUS REPORT)
OSR-	DIRECTORATE OF RESEARCH ANALYSES, HOLLOMAN AFB, N.M.
OSR-	GT. BRIT. COMMITTEE ON OVERSEAS SCIENTIFIC RELATIONS
OSR-	NATIONAL RESEARCH COUNCIL OF CANADA. ASSOCIATE COMMITTEE ON SYNTHETIC RUBBER RESEARCH
OSRD-	OFFICE OF SCIENTIFIC RESEARCH AND DEVELOPMENT(REF.32)
OSRD-(NO.)-(NO.)-	OFFICE OF SCIENTIFIC RES. + DEV. (DIV. NO.)(REPT.NO.)
OSRD (LETTERS)-	OFFICE OF SCIENTIFIC RESEARCH AND DEV. (FOR IDENTIFICATION OF CONTRACTORS REPRESENTED BY LETTERS, SEE ENTRIES BEGINNING WITH THOSE LETTERS) (REF. 32)
OSRD ALSOS-	EUROPEAN THEATER OF OPERATIONS (ARMY). ALSOS MISSION
OSRD AMP-	NATIONAL DEFENSE RESEARCH COMMITTEE. APPLIED MATHEMATICS PANEL (REF. 28)
OSRD-AMP-NYU-	NEW YORK UNIV., N.Y.C. APPLIED MATH. GROUP (REF. 28)
OSR/DRA-(YEAR)-	DIRECTORATE OF RESEARCH ANALYSES, HOLLOMAN AFB, N.M.
OSR-QI-	AIR FORCE OFFICE OF SCIENTIFIC RESEARCH, WASH., D.C. (QUARTERLY INDEX)
OSR-SRP-(YEAR/MONTH)	AIR FORCE OFFICE OF SCIENTIFIC RESEARCH, WASH., D.C. (STATUS OF RESEARCH PROPOSALS)
OSR-TN-(YEAR)-	AIR FORCE OFFICE OF SCIENTIFIC RESEARCH, WASH., D.C.
OSR-TR-(YEAR)-	AIR FORCE OFFICE OF SCIENTIFIC RESEARCH, WASH., D.C.
OSS-	OFFICE OF STRATEGIC SERVICES, WASHINGTON, D.C.
OSSC-	ARMY, WASHINGTON, D.C. (ORDNANCE STORAGE AND SHIPMENT CHART)
OST-	OFFICE OF SCIENCE AND TECHNOLOGY, WASHINGTON, D.C.
OSTD-	BUREAU OF ORDNANCE (NAVY) (ORDNANCE STANDARD)
OSTI-	UNITED KINGDOM. OFFICE FOR SCIENTIFIC + TECH. INFO.
OST-ONA-(YEAR)-	DEPT. OF TRANSPORTATION, WASH., D.C. (SYSTEMS DEVELOPMENT + TECHNOLOGY, NOISE ABATEMENT)
OSTRICH-	LOS ALAMOS SCIENTIFIC LAB., N. MEX.
OSU-	OHIO STATE UNIV., COLUMBUS
OSU-(NUMBER)-	OHIO STATE UNIV. RESEARCH FOUNDATION, COLUMBUS
OSU-(NUMBER)-TR-	OHIO STATE UNIV. RESEARCH FOUNDATION, COLUMBUS
OSU-	OREGON STATE UNIV., CORVALLIS
OSU-(LETTERS)-	OHIO STATE UNIV., COLUMBUS
OSU-AL-	OHIO STATE UNIV. RES. FDN. ANTENNA LAB., COLUMBUS
OSU-AS-	OKLAHOMA STATE UNIV., STILLWATER
OSU-DME-(NO.)-SR-(NO.)	OHIO STATE UNIV. RESEARCH FOUNDATION. DEPT. OF METALLURGICAL ENGINEERING, COLUMBUS
OSU-ECL-	OHIO STATE UNIV.,COLUMBUS. ELECTRONIC COMPONENTS LAB.
OSU-MCRL-	OHIO STATE UNIV. RESEARCH FOUNDATION. MAPPING AND CHARTING RESEARCH LAB., COLUMBUS
OSU-QPR-(NUMBER)-	OHIO STATE UNIV. RESEARCH FOUNDATION, COLUMBUS
OSU-RF-(NUMBER)-	OHIO STATE UNIV. RESEARCH FOUNDATION, COLUMBUS
OSURF-	OHIO STATE UNIV. RESEARCH FOUNDATION, COLUMBUS
OSURF MR-	OHIO STATE UNIV. RESEARCH FDN., COLUMBUS (MEMO RPTS.)
OSURF SR-	OHIO STATE UNIV. RESEARCH FDN., COLUMBUS (SPEC.RPTS.)
OSU-SR-	OHIO STATE UNIV., COLUMBUS
OSU-SR-	OHIO STATE UNIV. RES. FDN., COLUMBUS (STATUS REPORT)
OSU-TR-	OHIO STATE UNIV. RESEARCH FOUNDATION, COLUMBUS
OSU-WP-	OKLAHOMA STATE UNIV., STILLWATER. SCHOOL OF ELECTRICAL ENGINEERING
OSWAC-(LETTERS)-	ARMY ORDNANCE SPECIAL WEAPONS-AMMUNITION COMMAND, DOVER, N.J.
OSWD-	OFFICE OF SPECIAL WEAPONS DEVELOPMENTS, FT. BLISS,TEX
OSWDD-	OFFICE OF SPECIAL WEAPONS DEVELOPMENTS, FORT BLISS, TEX. (DOCUMENT)
OSW-PR-	OFFICE OF SALINE WATER, WASHINGTON, D.C.
OSW-RDPR-	OFFICE OF SALINE WATER, WASHINGTON, D.C.
OTA-	ORDNANCE TEST ACTIVITY (ARMY), YUMA, ARIZ.
OTAC-	ARMY TANK-AUTOMOTIVE CENTER, WARREN, MICH. (ORDNANCE TANK-AUTOMOTIVE COMMAND)
OTAC-	DETROIT ARSENAL, CENTER LINE, MICH.
OTAN-TR-	NORTH ATLANTIC TREATY ORGANIZATION. SUBCOMMITTEE ON OCEANOGRAPHIC RESEARCH
OTB-	OFFICE OF SCIENTIFIC RESEARCH AND DEVELOPMENT (ORDNANCE AND TERMINAL BALLISTICS)
OTI-	BUREAU OF ORDNANCE (NAVY) (ORDNANCE TECHNICAL INSTRUCTIONS)
OTI-	OFFICE OF TECHNICAL INFORMATION, AEC
OTI-	OFFICE OF TEST INFORMATION, AEC, LAS VEGAS, NEV.
OTI-(NUMBER)-R-	OPTICS TECHNOLOGY, INC., PALO ALTO, CALIF.
OTIB-	JOINT INTELLIGENCE OBJECTIVES AGENCY (OFFICE, TECHNICAL INTELLIGENCE BRANCH)
OTIE-	OFFICE OF TECHNICAL INFORMATION EXTENSION, AEC
OTIE-AL-(NUMBER)(LETTER)	OFFICE OF TECHNICAL INFORMATION EXTENSION, AEC, OAK RIDGE, TENN. (ACCESSION LIST. LETTER AT END STANDS FOR SECURITY CLASSIFICATION)
OTIO-	ORDNANCE TECHNICAL INTELLIGENCE SERVICE, ABERDEEN PROVING GROUND, MD.
OT/ITS-RR-	GERMANY. INSTITUTE OF AVIATION MEDICINE, FUERSTENFELDBRUCK

O-TM-
         STANDARD TELECOMMUNICATION LABS., HARLOW, ESSEX, ENG.
OTO-(NUMBER)-
         ALBUQUERQUE OPERATIONS OFFICE, AEC, N. MEX.
OTO-
         NEVADA TEST ORGANIZATION. OFF-SITE RADIOLOGICAL
            SAFETY ACTIVITIES, AEC
OTR-
         CLEARINGHOUSE FOR FEDERAL SCIENTIFIC AND TECH. INFO.
            OFFICE OF TECH. RESOURCES, SPRINGFIELD, VA.
OTR-
         COMBINED INTELLIGENCE OBJECTIVES SUB-COMMITTEE
            (ORDNANCE TARGET REPORTS)
OTR-
         NATIONAL BUREAU OF STANDARDS, WASHINGTON, D.C.
OTS-
         OFFICE OF TECHNICAL SERVICES, WASHINGTON, D.C.
OTS-(YEAR)-
         OFFICE OF TECH. SERVICES, WASH., D.C. (TRANSLATION)
OTS CAC-
         OFFICE OF TECH. SERVICES. IND.RES.+DEV.DIV.,WASH.,DC
OTS FB-
         OFFICE OF TECHNICAL SERVICES, WASHINGTON, D.C.
            (FILM BIBLIOGRAPHY)
OTS IR-
         OFFICE OF TECH. SERVICES, WASH.,D.C. (INFORMAL RPT.)
OTS IRDD-
         OFFICE OF TECH. SERVICES. IND.RES.+DEV.DIV.,WASH.,DC
OTS IRDD PROJ-
         OFFICE OF TECHNICAL SERVICES. INDUSTRIAL RESEARCH AND
            DEVELOPMENT DIV., WASH., D.C. (PROJECT REPORTS)
OTS PB-
         OFFICE OF TECH. SERVICES. PUBLICATION BD.,WASH.,D.C.
OTS-PL-
         OFFICE OF TECHNICAL SERVICES, WASH.,D.C. (PRICE LIST)
OTS-SB-
         OFFICE OF TECHNICAL SERVICES, WASHINGTON, D.C.
            (SUBJECT BIBLIOGRAPHY)
OTS TAS-
         OFFICE OF TECHNICAL SERVICES. INVENTIONS AND ENG.
            DIV. TECHNICAL ADVISORY SERVICE, WASHINGTON, D.C.
OU-
         OKLAHOMA. UNIV., NORMAN
OU-
         OREGON. UNIV., EUGENE
OUEL-
         OXFORD UNIV. ENGINEERING LAB.
OU-HEPL-
         OHIO UNIV., ATHENS. HIGH ENERGY PHYSICS LAB.
OU-ITA-
         OSLO. UNIV. INST. OF THEORETICAL ASTROPHYSICS
OUK-
         (OPERATION UPSHOT-KNOTHOLE) (REF. 7)
OU-LNS-
         OSAKA UNIV. LAB. OF NUCLEAR STUDIES
OULNS-(YEAR)-
         OSAKA UNIV. LAB. OF NUCLEAR STUDIES
OUM-
         NAVAL ORDNANCE TEST STATION, INYOKERN (CHINA LAKE),
            CALIF. (ORDNANCE UNDERWATER MEMORANDUM)
OURI-
         OKLAHOMA. UNIV., NORMAN. RESEARCH INST.
OU-TR-
         OKLAHOMA. UNIV., NORMAN. RESEARCH INST. HIGH PRES-
            SURE PHYSICS LAB.
OV-
         AEROSPACE CORP., EL SEGUNDO, CALIF.
OV-LNS-(YEAR)-
         OSAKA UNIV. LAB. OF NUCLEAR STUDIES
(NUMBER)-OWB-
         GENERAL ELECTRIC CO. TUBE DEPT., OWENSBORO, KY.
OWRB-PUB-
         OKLAHOMA WATER RESOURCES BOARD, OKLAHOMA CITY
OWRR-A-(NUMBER-STATE)
         OFFICE OF WATER RESOURCES RES.(INTERIOR), WASH., D.C.
OWRR-B-(NUMBER-STATE)
         OFFICE OF WATER RESOURCES RES.(INTERIOR), WASH., D.C.
OWRR-C-(NO.-NO.-NO.-)
         OFFICE OF WATER RESOURCES RES.(INTERIOR), WASH., D.C.
OWRR-S-(NUMBER-STATE)
         OFFICE OF WATER RESOURCES RES.(INTERIOR), WASH., D.C.
OWRR-W-(NUMBER-STATE)
         OFFICE OF WATER RESOURCES RES.(INTERIOR), WASH., D.C.
OXFORDU-
         OXFORD UNIV.
OXU-
         OXFORD UNIV.

P-	(PAMPHLET) (REF. 2)
P-	(PAPER) (REF. 2)
P-	(PERIODIC PROGRESS REPORT) (REF. 2)
P-	(PREPRINT) (REF. 2)
P-	(PROCESS REPORT) (REF. 2)
P-	(PRODUCTION) (REF. 2)
P-	(PROGRESS REPORT) (REF. 2)
P-	(PROJECT REPORT) (REF. 2)
P-	(PROPOSAL) (REF. 2)
P-	(PUBLICATION) (REF. 2)
P-	(REPRINT) (REF. 2)
P-	AEROJET-GENERAL CORP., AZUSA, CALIF.
P-	AEROJET-GENERAL CORP., SACRAMENTO, CALIF.
P-	AEROSPACE CONTROLS CORP., LOS ANGELES
P-	ARMY FIELD FORCES. BOARD NO. 2, FORT KNOX, KY. (FORMAL PROJECT REPORT)
P-	BELGIUM. CENTRE NATIONAL DE RECHERCHES METALLURGIQUES LIEGE
P-(NO.)-	BRUSH BERYLLIUM CO., CLEVELAND
P-	CALIFORNIA. UNIV., BERKELEY. DEPT. OF PHYSICS
P-(NO.)-C-	CALIFORNIA. UNIV., BERKELEY. RADIATION LAB.
P-	CALIFORNIA INST. OF TECH., PASADENA
P-	COLUMBIA UNIV., N.Y.C.
P-	COLUMBIA UNIV., N.Y.C. ELECTRONICS RESEARCH LABS.
P-(2 DIGITS)-	CORNING GLASS WORKS. ELECTRONICS DIV., RALEIGH, N.C.
P-	DYNAMIC SCIENCE CORP., SOUTH PASADENA, CALIF.
P-	EAST COAST AERONAUTICS, INC., MOUNT VERNON, N.Y. (PROCESS REPORT)
P-	EUROPEAN SPACE RESEARCH ORGANIZATION, PARIS
P-	FEDERAL SCIENTIFIC CORP., N.Y.C.
P-	FRANCE. CENTRE NATIONAL D'ETUDES SPATIALES, PARIS
P-(PROJECT)	FRANKLIN INST. LABS. FOR RES. + DEV., PHILADELPHIA
P-(4 DIGITS)-	FRANKLIN INST. LABS. FOR RES. + DEV., PHILADELPHIA
P-	GENERAL DYNAMICS/ELECTRONICS, ROCHESTER, N.Y.
P(NUMBER)-	GENERAL ELECTRIC CO. FLIGHT PROPULSION DIV., CINCINNATI
P-	HONEYWELL, INC. ORDNANCE DIV., HOPKINS, MINN. (PROPOSAL)
P(YEAR)-	HUGHES AIRCRAFT CO.
P(YEAR)-	HUGHES AIRCRAFT CO., CULVER CITY, CALIF.
P(YEAR)-	HUGHES AIRCRAFT CO. SPACE SYSTEMS DIV., EL SEGUNDO, CAL
P-	INSTITUTE FOR DEFENSE ANALYSES, ARLINGTON, VA.
P-	INTERNUCLEAR CO., INC., CLAYTON, MO.
P-	IOWA. STATE UNIV., IOWA CITY. DEPT. OF PHYSICS AND ASTRONOMY
P-(NUMBER)/(YEAR)	ISRAEL. NEGEV INST. FOR ARID ZONE RESEARCH. DEPT. OF PLANT PHYSIOLOGY, BEERSHEBA
P(NUMBER)(LETTER)-	LOCKHEED-CALIFORNIA CO., BURBANK
P-	LOS ALAMOS SCIENTIFIC LAB., N. MEX.
P-1- THRU P-18-	LOS ALAMOS SCIENTIFIC LAB., N. MEX. (INTERNAL CORRESPONDENCE SERIAL USED BY GROUPS IN P DIV.)
P-	MANHATTAN DISTRICT, OAK RIDGE, TENN. (REF. 23)
P-(NUMBER)-	MARTIN CO., DENVER
P-	MARTIN-MARIETTA CORP. DENVER DIV.
P-	NATIONAL ENG. SCIENCE CO., PASADENA, CAL. (PROPOSALS)
P-	NAVAL RESEARCH LAB., WASHINGTON, D.C.
P-(NO.)-	NEUTRON CROSS SECTION ADVISORY GROUP, AEC
P-	OFFICE OF SCIENTIFIC RESEARCH AND DEVELOPMENT
P-	PARKER APPLIANCE CO., CLEVELAND
P-(NUMBER)-	PHILCO-FORD CORP., BILLINGS, MONT.
P-	PLASMADYNE CORP., SANTA ANA, CALIF.
P-	PRATT AND WHITNEY AIRCRAFT DIV., UNITED AIRCRAFT CORP., HARTFORD, CONN.
P-(NO.)-AEC	RAND CORP., SANTA MONICA, CALIF.
P-(NO.)-(RAND)	RAND CORP., SANTA MONICA, CALIF.
P-(NO.)-ARPA	RAND CORP., SANTA MONICA, CALIF.
P-(4 DIGITS)	RAND CORP., SANTA MONICA, CALIF. (REF. 47)
P(NUMBER)	RAYTHEON CO. SUBMARINE SIGNAL DIV., PORTSMOUTH, R.I.
P-	REDSTONE ARSENAL, HUNTSVILLE, ALA. (QUARTERLY PROGRESS REPORT)
P-(YR)-	ROHM + HAAS CO. REDSTONE ARSENAL RESEARCH DIV., HUNTSVILLE, ALA.
P-	STANFORD UNIV., CALIF. DEPT. OF MECHANICAL ENG.
P-	TETRA TECH, INC., PASADENA, CALIF.
P-	TEXAS. UNIV., AUSTIN. ANTENNAS AND PROPAGATION DIV.
P-	TOWSON LABS., INC., BALTIMORE, MD.
P(YEAR)(MONTH)-	TRAVELERS RESEARCH CENTER, INC., HARTFORD (PROPOSAL)
P-	U.S.S.R. JOINT INST. FOR NUCLEAR RESEARCH, DUBNA
P(NUMBER)-	U.S.S.R. JOINT INST. FOR NUCLEAR RESEARCH, DUBNA
P-3WP/	UNITED KINGDOM ATOMIC ENERGY AUTHORITY. RESEARCH GP. ATOMIC ENERGY RES. ESTAB., HARWELL, BERKS, ENGLAND
P-(LETTER)-	CALIFORNIA RESEARCH AND DEVELOPMENT CO., LIVERMORE
P(LETTER)C-	OFFICE OF SCIENTIFIC RESEARCH AND DEVELOPMENT
PA-	(PUBLICATION ANNOUNCEMENT) (REF. 2)
PA-	AERONAUTICAL CHART AND INFORMATION CENTER, ST. LOUIS
PA-(NO.)-	AIR FORCE, WASHINGTON, D.C.
PA-	AUSTRALIA. NATIONAL STANDARDS LAB., CHIPPENDALE, NEW SOUTH WALES
PA-	GERMANY. ZENTRALE FUER WISSENSCHAFTLICHES BERICHTS-WESEN DER LUFTFAHRTFORSCHUNG, BERLIN
P/A-	GRUMMAN AIRCRAFT ENGINEERING CORP., BETHPAGE, N.Y. (PILOTLESS AIRCRAFT)
PA-	MASSACHUSETTS INST. OF TECH. LEXINGTON. LINCOLN LAB.
PA-	NAVAL AIR DEVELOPMENT CENTER. PILOTLESS AIRCRAFT DEVELOPMENT LAB., JOHNSVILLE, PENNA.
PA-	OFFICE OF THE CHIEF OF NAVAL OPERATIONS. OPERATIONS EVALUATION GROUP, WASH., D.C. (PLANNING ANALYSIS)
PA-	PICATINNY ARSENAL, DOVER, N.J.
PA-(NUMBER)-	SPACE TECHNOLOGY LABS., INC., LOS ANGELES
PAA-	PAN AMERICAN WORLD AIRWAYS, INC., LAS VEGAS, NEV.
PA-AG-	LITTLE (ARTHUR D.), INC., CAMBRIDGE, MASS.
PA-BIB-	PICATINNY ARSENAL. SAMUEL FELTMAN AMMUNITION LABS., DOVER, N.J. (BIBLIOGRAPHY)
PAC-	GT. BRIT. ADVISORY COUNCIL ON SCI. RES. + TECH. DEV.
PAC-	PARSONS-AEROJET CO., LOS ANGELES
PAC-	PARSONS (RALPH M.) CO., PASADENA, CALIF.
PACAF-TEST-(YEAR)-	PACIFIC AIR FORCES, HICKAM AFB, HAWAII
PACER-(NUMBER)-	GODDARD SPACE FLIGHT CENTER, GREENBELT, MD.
PACMIRS-	(PACIFIC MILITARY INTELLIGENCE. RESEARCH SECTION)
PACT-	NAVAL ORDNANCE TEST STA., INYOKERN (CHINA LAKE), CAL.
PAC TR-	PARSONS-AEROJET CO., LOS ANGELES
PAD-	AUSTRALIA. WEAPONS RESEARCH ESTABLISHMENT
PAD-	BUREAU OF AERONAUTICS (NAVY)(PRELIM. AIRCRAFT DESIGN)
PAD-	FRANKFORD ARSENAL, PHILADELPHIA
PAD-	GT. BRIT. FLYING PERSONNEL RESEARCH COMMITTEE
PA-DB-TR-	PICATINNY ARSENAL, DOVER, N.J.
PA-DC-QR-(NOS.)-(YR.)	PICATINNY ARSENAL, DOVER, N.J.
PADL-	NAVAL AIR DEVELOPMENT CENTER. PILOTLESS AIRCRAFT DEVELOPMENT LAB., JOHNSVILLE, PENNA.

PADLOC-   SPERRY GYROSCOPE CO. INFORMATION AND COMMUNICATIONS DIV., GREAT NECK, N.Y.
PA-DL-TR-   PICATINNY ARSENAL. LIQUID ROCKET PROPULSION LAB., DOVER, N.J.
PA-DM-   PICATINNY ARSENAL, DOVER, N.J.
PADOVAU-   PADUA UNIVERSITA
PADR-   ALLIS-CHALMERS MFG. CO., NORWOOD, OHIO
PADS-(MONTH)/(YEAR)   TENNESSEE. UNIV. MEMPHIS COLLEGE OF PHARMACY
PADUAU-   PADUA UNIVERSITA
PAEC-   PHILIPPINES. ATOMIC ENERGY COMMISSION, MANILA
PAEC(A)IN-   PHILIPPINES. ATOMIC ENERGY COMMISSION, MANILA
PAEC/A/PH-   PHILIPPINES. ATOMIC ENERGY COMMISSION, MANILA
PAEC/CHEM-   PAKISTAN. ATOMIC EN. CENTRE. CHEMISTRY DIV., LAHORE
PAEC-CHEM-MISC.-   PAKISTAN. ATOMIC EN. CENTRE. CHEMISTRY DIV., LAHORE
PAEC(C)RE-   PHILIPPINES. ATOMIC ENERGY COMMISSION, MANILA
PAEC(D)   PHILIPPINES. ATOMIC ENERGY COMMISSION, MANILA
PAEC(D)CH-   PHILIPPINES. ATOMIC ENERGY COMMISSION, MANILA
PAEC(D)HP-   PHILIPPINES. ATOMIC ENERGY COMMISSION, MANILA
PAEC(D)IS-   PHILIPPINES. ATOMIC ENERGY COMMISSION, MANILA
PAEC(D)PH-   PHILIPPINES. ATOMIC ENERGY COMMISSION, MANILA
PAEC(D)RE-   PHILIPPINES. ATOMIC ENERGY COMMISSION, MANILA
PAEC-IPA(D)PH-   PHILIPPINES. ATOMIC RES. CENTER, DILIMAN, QUEZON CITY
PAECL/(LTRS.)-   PAKISTAN. ATOMIC ENERGY CENTRE, LAHORE
PAECL/CHEM/MISCELL-   PAKISTAN. ATOMIC ENERGY CENTRE, LAHORE
PAECL/HP-   PAKISTAN. ATOMIC EN. CENTRE.HEALTH PHYSICS DIV.,DACCA
PAEC-RR-   PHILIPPINES. ATOMIC ENERGY COMMISSION, MANILA (REVIEW REPORT)
PAFB-   AIR FORCE MISSILE TEST CENTER, PATRICK AFB, FLA.
PA-FLAM-   PICATINNY ARSENAL, DOVER, N.J.
PA-FREL-   PICATINNY ARSENAL, DOVER, N.J.
PA-FRL-TR-   PICATINNY ARSENAL. FELTMAN RES. + ENG. LABS, DOVER,NJ
PA-FT-   PICATINNY ARSENAL, DOVER, N.J. (FRAGMENTATION TESTS)
PAHO-SC-PUB-   PAN AMERICAN HEALTH ORGANIZATION, WASHINGTON, D.C.
PAI-   AMERICAN INST. OF AERONAUTICS + ASTRONAUTICS,LOS ANG.
PAI-   PROCEDYNE ASSOCIATES, INC., NEW BRUNSWICK, N.J.
PAIN-   OFFICE OF THE CHIEF OF NAVAL OPERATIONS. OPERATIONS EVALUATION GROUP, WASH., D.C. (PLANNING ANALYSIS)
PAL-   AMERICAN INST. OF AERONAUTICS + ASTRONAUTICS,LOS ANG.
PAL-   HARVARD UNIV., CAMBRIDGE, MASS. PSYCHO-ACOUSTIC LAB.
PAL-   PICATINNY ARSENAL, DOVER, N.J. (LIBRARY)
PA-LAB-   PICATINNY ARSENAL, DOVER, N.J.
PALTR-   ISOTOPES, INC. PALO ALTO LABS., CALIF.
PAM-   (PAMPHLET) (REF. 2)
PAM-   ARMY, WASHINGTON, D.C. (PAMPHLET)
PAM-(YEAR)-   COLLEGE DE FRANCE, PARIS. LABORATOIRE DE PHYSIQUE ATOMIQUE ET MOLECULAIRE
PAM-   OFFICE OF THE CHIEF OF NAVAL OPERATIONS. OPERATIONS EVALUATION GROUP, WASH., D.C. (PLANNING ANALYSIS)
PA-M   PICATINNY ARSENAL, DOVER, N.J.
PAMR-   PATTERSON, MOOS RES. DIV.,LEESONA CORP.,JAMAICA, N.Y.
PA-MR-   PICATINNY ARSENAL, DOVER, N.J. (MEMORANDUM REPORTS)
PAN-   POLISH ACADEMY OF MINING AND METALLURGY, KRAKOW.
PAN-   POLISH ACADEMY OF SCIENCES. INST. OF NUCLEAR RESEARCH, WARSAW
PAN-0-   POLISH ACADEMY OF SCIENCES. INST. OF NUCLEAR RESEARCH, WARSAW
PANSDOC-   PAKISTAN NATIONAL SCIENTIFIC AND TECHNICAL DOCUMENTATION CENTRE, KARACHI
PA-ORDBB-   PICATINNY ARSENAL. FELTMAN RESEARCH LABS.,DOVER, N.J.
PA-ORDBB-(LTRS.NO.)-   PICATINNY ARSENAL, DOVER, N.J.
PA-ORDBB-TK-   PICATINNY ARSENAL. FELTMAN RES. + ENG. LABS.,DOVER,NJ
PA-ORDBB-TK-   PICATINNY ARSENAL. SPECIAL WEAPONS GROUP, DOVER, N.J.
PA-PAS-   PICATINNY ARSENAL. AMMUNITION DEV. DIV., DOVER, N.J.
PAPER-   (PAPER) (REF. 2)
PA-PLASTEC-   PICATINNY ARSENAL. PLASTICS TECHNICAL EVALUATION CENTER, DOVER, N.J.
PA-POMM-   PICATINNY ARSENAL, DOVER, N.J.
PAPP-   NORTHERN STATES POWER CO., SIOUX FALLS, S. DAK. (PATHFINDER ATOMIC POWER PLANT)
PA PRS-   PICATINNY ARSENAL. FELTMAN RESEARCH LABS., DOVER, N.J. (PROPELLANT RESEARCH SECTION REPORTS)
PAR-   PACIFIC APPLIED RESEARCH, LOS ANGELES
PAR-   PARSONS (RALPH M.) CO., LOS ANGELES
PAR-   PARSONS (RALPH M.) CO., PASADENA, CALIF.
PA-R-   PICATINNY ARSENAL, DOVER, N.J.
PARISU-   PARIS. UNIVERSITE, ORSAY
PAR MP-   PARSONS (RALPH M.) CO., PASADENA, CALIF. (MISCELLANEOUS PUBLICATIONS)
PAR-RANGE-   PARSONS (RALPH M.) CO., PASADENA, CAL. (PROJ. RANGE)
PARSONS-   PARSONS (RALPH M.) CO., LOS ANGELES
PA/RT-   FRANCE. COMMISSARIAT A L'ENERGIE ATOMIQUE, PARIS
(PART NUMBER)   CLEVELAND PNEUMATIC TOOL CO. (REF. 1)
(PART NUMBER(S))-(LTRS   CLEVELAND PNEUMATIC TOOL CO. (REF. 1)
PAS-(YEAR)   FRANCE. COMMISSARIAT A L'ENERGIE ATOMIQUE. CENTRE D'ETUDES NUCLEAIRES, SACLAY
PAS-   OFFICE OF THE CHIEF OF NAVAL OPERATIONS. OPERATIONS EVALUATION GROUP, WASH., D.C. (PLANNING ANALYSIS)
PAS-(NO.)/(ROMAN NO.)   POLISH ACADEMY OF SCIENCES. INST. OF NUCLEAR RESEARCH, WARSAW
PAS-   RESEARCH AND DEVELOPMENT BOARD. PANEL ON AERODYNAMICS AND STRUCTURES, WASHINGTON, D.C.
PASAP-   DAVID SARNOFF RESEARCH CENTER, PRINCETON, N.J.
PAS-INR-(NO./ROMAN NO.)   POLISH ACADEMY OF SCIENCES. INST. OF NUCLEAR RESEARCH, WARSAW
PATASWDEVGRU-   PATROL ASW DEVELOPMENT GROUP (NAVY), NORFOLK, VA.
PA-TE-   PICATINNY ARSENAL, DOVER, N.J.
PA-TM-   PICATINNY ARSENAL, DOVER, N.J.
PA-TN-   PICATINNY ARSENAL. FELTMAN RES. + ENG. LABS.,DOVER,NJ
PA-TN-   THERM ADVANCED RESEARCH, INC., ITHACA, N.Y.
PA-TN-FRL-TN-   PICATINNY ARSENAL. FELTMAN RESEARCH LABS.,DOVER, N.J.
PA-TR-   PICATINNY ARSENAL, DOVER, N.J.
PATR-   PICATINNY ARSENAL, DOVER, N.J.
PA-TR-FRL-TR-   PICATINNY ARSENAL. FELTMAN RESEARCH LABS.,DOVER, N.J.
PAT SPEC-(YR)   GT. BRIT. DEPT. OF ATOMIC ENERGY
PA-TT-   PICATINNY ARSENAL, DOVER, N.J.
PAU-   NAVAL AIR MISSILE TEST CENTER, POINT MUGU, CALIF.
PAU-   UNITED KINGDOM ATOMIC ENERGY AUTHORITY. PROGRAMMES ANALYSIS UNIT, CHILTON, BERKS, ENGLAND
P.A.U. M-(NUMBER)/(YR.)   UNITED KINGDOM ATOMIC ENERGY AUTHORITY. PROGRAMMES ANALYSIS UNIT, CHILTON, BERKS, ENGLAND
PAU-M-   UNITED KINGDOM ATOMIC ENERGY AUTHORITY. PROGRAMMES ANALYSIS UNIT, CHILTON, BERKS, ENGLAND
PAVCO-   PASTUSHIN AVIATION CORP., LOS ANGELES
PAW-   PETROLEUM ADMINISTRATION FOR WAR
PAWA-   PAN AMERICAN WORLD AIRWAYS, INC., SO. SAN FRANCISCO
PAWA-AROWA-   PAN AMERICAN WORLD AIRWAYS, INC., SO. SAN FRANCISCO
PB-   (PUBLICATION) (REF. 2)
PB-   GERMANY. ZENTRALE FUER WISSENSCHAFTLICHES BERICHTSWESEN DER LUFTFAHRTFORSCHUNG, BERLIN (PRUEFBERICHT)
P/B-   HOLMES AND NARVER, INC., LOS ANGELES

PB-(6 DIGITS)	NATIONAL TECHNICAL INFORMATION SERVICE, SPRINGFIELD, VA. (ACCESSION NUMBER)
PB-	PACKARD-BELL ELECTRONICS CORP. SPACE AND SYSTEMS DIV., NEWBURY PARK, CALIF.
P+B-	PICKARD AND BURNS, INC., NEEDHAM, MASS.
P+B-	PICKARD AND BURNS ELECTRONICS, WALTHAM, MASS.
PB-	QUARTERMASTER RESEARCH AND ENGINEERING COMMAND. PSYCHOLOGY BRANCH, NATICK, MASS.
P-B(NUMBER)	UNITED AIRCRAFT CORP. RES. LABS., EAST HARTFORD, CONN.
PB-	UNIVERSITIES RESEARCH ASSOCIATION, INC., WASH., D.C.
PBD-	ATOMIC ENERGY OF CANADA LTD. CHALK RIVER PROJECT, ONT. (REF. 6)
PBD-	GEOLOGICAL SURVEY, DENVER (PREDICTION BACKGROUND DATA)
PBI-	ATOMIC ENERGY OF CANADA LTD. CHALK RIVER PROJECT, ONT. (REF. 6)
P + B PUB.-	PICKARD AND BURNS, INC., NEEDHAM, MASS.
P/B PUB-	PICKARD AND BURNS ELECTRONICS, WALTHAM, MASS.
P + B PUBL.-	PICKARD AND BURNS ELECTRONICS, WALTHAM, MASS.
P-BR-	GT. BRIT. ATOMIC EN. RES. ESTAB., HARWELL, BERKS, ENG
PBTB-	PARSONS, BRINCKERHOFF-TUDOR-BECHTEL, SAN FRANCISCO
PC-	AIR FORCE (PILOTAGE CHARTS)
PC-	AMES LAB., IOWA
PC-	ARGONNE NATIONAL LAB., LEMONT, ILL.
PC-	CHICAGO. UNIV. METALLURGICAL LAB.
PC-(NUMBER) Q-(YEAR)	DOW CHEMICAL CO. SCIENTIFIC PROJS. LAB., MIDLAND, MICH
PC(NUMBER)(LETTER)(NO.)	FAIRCHILD HILLER CORP. REPUBLIC AVIATION DIV., FARMINGDALE, N.Y.
P/C-	HOLMES AND NARVER, INC., LOS ANGELES
PC-	HYDROGRAPHIC OFFICE, SUITLAND, MD. (PILOT CHARTS)
PC-	NATIONAL RESEARCH COUNCIL OF CANADA. ATOMIC ENERGY PROJECT, CHALK RIVER, ONT.
PC-	OFFICE OF NAVAL RESEARCH (PSYCHOLOGICAL WARFARE AND COLD WAR OPERATIONS)
PC-(YEAR)-	PACIFIC CAR AND FOUNDRY CO., RENTON, WASH.
PC-	PEPPERDINE COLL., LOS ANGELES
PC-	PHILCO CORP., PHILADELPHIA
PC-	PLASMADYNE CORP., SANTA ANA, CALIF.
PC-	PROVIDENCE COLL., R.I.
PC-	PSYCHOLOGICAL CORP., N.Y.C.
PC-	PURDUE UNIV., LAFAYETTE, IND. JET PROPULSION CENTER
PC-	RAYTHEON MFG. CO., WAYLAND, MASS.
PC-	RESEARCH + DEV. BD., WASH., D.C. (PERSONNEL OPERATIONS)
PC-	RESEARCH AND DEVELOPMENT BOARD, WASHINGTON, D.C. (PSYCHOLOGICAL WARFARE AND COLD WAR OPERATIONS)
PCA G-4-	SUPREME HEADQUARTERS, ALLIED EXPEDITIONARY FORCES (PRODUCTION CONTROL AGENCY G-4)
PCC-	CANADA. DEFENCE RESEARCH BOARD. PROJECT COORDINATION CENTRE, OTTAWA
PCC-	PENNSALT CHEMICALS CORP., PHILADELPHIA
PCC-	PITTSBURGH CONSOLIDATED COAL CO.
PCC-	RCA VICTOR DIV., LTD. RESEARCH LABS., MONTREAL
PCC D(NUMBER)-	CANADA. PACIFIC NAVAL LAB., ESQUIMALT, B.C.
PCC-D (NO.)-	RCA VICTOR CO., LTD. RESEARCH LABS., MONTREAL
PCCF-RI-(YEAR/NUMBER)	CLERMONT-FERRAND UNIV., FRANCE. LABORATORIE DE PHYSIQUE CORPUSCULAIRE
PCD-	FAIRCHILD HILLER CORP. REPUBLIC AVIATION DIV., FARMINGDALE, N.Y.
PCD-	REPUBLIC AVIATION CORP., FARMINGDALE, N.Y.
PCD-TR-(YEAR)	FAIRCHILD HILLER CORP. REPUBLIC AVIATION DIV., FARMINGDALE, N.Y.
PCD-TR-(YEAR)-	REPUBLIC AVIATION CORP., FARMINGDALE, N.Y.
PCE-	PAGE COMMUNICATIONS ENGINEERS, INC., WASHINGTON, D.C.
PCE-	ROYAL CANADIAN AIR FORCE. CENTRAL EXPERIMENTAL AND PROVING ESTAB., ROCKCLIFFE, ONT.
PCE-R-(NUMBER)-(NUMBER)	PAGE COMMUNICATIONS ENGINEERS, INC., WASHINGTON, D.C.
PCG-B-	TENNESSEE EASTMAN CORP., OAK RIDGE, TENN.
PC-H-	PHILCO CORP. RESEARCH DIV., PHILADELPHIA
PC-IF-	DOW CHEMICAL CO., MIDLAND, MICH.
PCIR-	GENERAL DYNAMICS/FORT WORTH, TEX.
PCL-	GENERAL DYNAMICS/CONVAIR, SAN DIEGO, CALIF.
PCL-	PHILLIPS CABLES, LTD., BROCKVILLE, ONT.
PC MLP TR-	POMONA COLL., CLAREMONT, CALIF. MILLIKAN LAB. OF PHYS.
PCMTD-	JOINT ARMY-NAVY-AIR FORCE AD HOC PANEL ON PERFORMANCE CALCULATION METHODS AND THERMODYNAMIC DATA
PCO-	NAVAL ORDNANCE LAB., WHITE OAK, MD. (PRODUCTION CHANGE ORDER)
PCP-(YEAR)-	BELL TELEPHONE LABS., INC., WHIPPANY, N.J.
PCR-	FMC CORP., PRINCETON, N.J.
PCR-(YR.)-	PENINSULAR CHEMRESEARCH, INC., GAINESVILLE, FLA.
PCR-	ROCKETDYNE DIV., NORTH AMERICAN AVIATION, INC., CANOGA PARK, CALIF.
PCRD-	ESSO RESEARCH AND ENGINEERING CO. PROCESS RES. DIV., LINDEN, N.J.
PCRD-(YEAR)-	ESSO RESEARCH AND ENGINEERING CO. PROCESS RES. DIV., LINDEN, N.J.
PCRD-MRD-	ESSO RESEARCH AND ENGINEERING CO. PROCESS RES. DIV., LINDEN, N.J.
PCR-QPR-(NUMBER)(YEAR)	PENINSULAR CHEMRESEARCH, INC., GAINESVILLE, FLA.
PC-S-	MEYER (CHARLES A.) AND CO., INC., N.Y.C.
PCS-P/	UNITED KINGDOM ATOMIC ENERGY AUTHORITY. DEVELOPMENT AND ENGINEERING GROUP, CULCHETH, LANCS, ENGLAND
PCT-	GT. BRIT. PENICILLIN CLINICAL TRIALS COMMITTEE
PCTR-	GENERAL DYNAMICS/FORT WORTH, TEX.
PCWP/P-	GT. BRIT. ATOMIC EN. RES. ESTAB., HARWELL, BERKS, ENG
PCWP/P-	UNITED KINGDOM ATOMIC ENERGY AUTHORITY. REACTOR GROUP, CULCHETH, LANCS, ENGLAND
PCZ-	PANAMA CANAL ZONE
PCZ ICSM-	PANAMA CANAL ZONE. DEPT. OF OPERATION AND MAINTENANCE
PD-	(PHYSICS DEPT.) (REF. 2)
PD-	(PRELIMINARY DESIGN) (REF. 2)
PD-	(PRODUCTION DEPT.) (REF. 2)
PD-	(PURCHASE DESCRIPTION) (REF. 2)
PD-	AEROJET-GENERAL CORP., SACRAMENTO, CALIF.
PD-(YEAR)-	AIR REDUCTION SALES CO., UNION, N.J. (USED 1962-63)
PD-	ATOMIC ENERGY OF CANADA LTD., CHALK RIVER, ONT. (REF.6)
PD-	BROOKHAVEN NATIONAL LAB., UPTON, N.Y.
PD-	BUREAU OF AERONAUTICS (NAVY). PRODUCTION DIV.
PD-	BUREAU OF SHIPS (PRELIMINARY DESIGN)
PD-(4 DIGITS)	CANADA. SUFFIELD EXPERIMENTAL STATION, RALSTON, ALBERTA
PD 70.-	CARBIDE AND CARBON CHEMICALS CORP. K-25 PLANT, OAK RIDGE, TENN.
PD-	CHEMICAL WARFARE SERVICE. CHEMICAL DIV. PHYSICAL DEPT., EDGEWOOD ARSENAL, MD.
PD-	COMMITTEE ON MEDICAL RES. (PARKE-DAVIS + CO. RPTS.)
PD-(NUMBER) M-(YEAR)	DOW CHEMICAL CO. SCIENTIFIC PROJS. LAB., MIDLAND, MICH
PD-	NATIONAL RESEARCH COUNCIL OF CANADA. DIV. OF ATOMIC ENERGY, CHALK RIVER, ONT. (REF. 29)
PD-(NO.)-(YR.)	SIMMONDS PRECISION PRODUCTS, INC., TARRYTOWN, N.Y.
PD-	STANDARD OIL DEVELOPMENT CO. PROCESS DIV., N.Y.C.
PDB-	STANFORD UNIV., CALIF. DEPT. OF MECHANICAL ENG.
PDC-	ATOMIC ENERGY OF CANADA LTD., CHALK RIVER, ONT.
PDC-	NATIONAL RESEARCH COUNCIL. PREVENTION OF DETERIORATION CENTER, WASHINGTON, D.C.

PDC PC-

PDCPC-	PHELPS DODGE COPPER PRODUCTS CORP. RESEARCH LABS., YONKERS, N.Y.
PDD-(NO.)-	CARBIDE AND CARBON CHEMICALS CORP. K-25 PLANT, OAK RIDGE, TENN.
PDDTM-	WESTINGHOUSE ELECTRIC CORP., BALTIMORE
PDE-	COAST AND GEODETIC SURVEY, ROCKVILLE, MD. (PRELIMINARY DETERMINATION OF EPICENTER)
PDE-	GENERAL DYNAMICS CORP. ELECTRIC BOAT DIV., GROTON, CONN
PDE-1944/15-	GT. BRIT. MINISTRY OF SUPPLY
PD-EX-	TENNESSEE EASTMAN CORP., OAK RIDGE, TENN.
PDGW-	UNITED KINGDOM. DIRECTORATE OF GUIDED WEAPONS RESEARCH AND DEVELOPMENT, LONDON
PDGW-EMR-	IMPERIAL COLL. OF SCI. + TECH., LONDON. JET RES. LAB.
PDH-	NAVY (PASSIVE DEFENSE HANDBOOK)
PDL-	FRANKFORD ARSENAL. PITMAN-DUNN LAB., PHILADELPHIA
PDL-	NATIONAL RESEARCH COUNCIL. PREVENTION OF DETERIORATION CENTER, WASHINGTON, D.C.
PDM-	GRUMMAN OPERATIONS ANALYSIS GROUP. ADVANCED SYSTEMS SECTION, BETHPAGE, N.Y.
PD-MPR-	TENNESSEE EASTMAN CORP., OAK RIDGE, TENN.
PDN-	STANDARD OIL DEVELOPMENT CO. ESSO LABS., LINDEN, N.J. (PETROLEUM DEVELOPMENT NOTES)
PDO-	BENDIX CORP. KANSAS CITY DIV., MO.
PD-PROC-	TENNESSEE EASTMAN CORP., OAK RIDGE, TENN.
PDR-	AEROJET-GENERAL CORP., SACRAMENTO, CALIF.
PDR-	GRUMMAN OPERATIONS ANALYSIS GROUP. ADVANCED SYSTEMS SECTION, BETHPAGE, N.Y.
PDR-A1-(YEAR)-	GRUMMAN AEROSPACE CORP., BETHPAGE, N.Y.
PDR-ASG-	GRUMMAN AIRCRAFT ENGINEERING CORP., BETHPAGE, N.Y.
PDRC-	NATIONAL DEFENSE RESEARCH COMM. (PROJECT RPTS.)
PDR-N-	GRUMMAN AIRCRAFT ENGINEERING CORP., BETHPAGE, N.Y.
PDS-	(PRELIMINARY DESIGN STUDY) (REF. 2)
PDS-	NORTHROP AIRCRAFT, INC., HAWTHORNE, CALIF. (PRELIMINARY DESIGN STUDY)
PDS-	UNIDYNAMICS/PHOENIX, ARIZ.
PDSA-(NUMBER)-(YEAR)	CONVAIR. ASTRONAUTICS DIV., SAN DIEGO, CALIF.
PD-SPEC-	WESTINGHOUSE ELECTRIC CORP. ATOMIC POWER DIV., PITTS.
PDU-	PURDUE UNIV., LAFAYETTE, IND.
PDU-F-	PURDUE UNIV., LAFAYETTE, IND.
PDU-I-	PURDUE UNIV., LAFAYETTE, IND.
PDU-PR-	PURDUE UNIV., LAFAYETTE, IND.
PDU-TM-	PURDUE UNIV., LAFAYETTE, IND.
PDU-TR-EE-	PURDUE UNIV., LAFAYETTE, IND. SCHOOL OF ELEC. ENG.
PE-(NO.)-	AIR FORCE, WASHINGTON, D.C.
PE-	ATOMIC ENERGY OF CANADA LTD., CHALK RIVER, ONT.
PE-(4 DIGITS)-	GARRETT CORP. AIRESEARCH MFG. CO. OF ARIZONA, PHOENIX
PE-	PERKIN-ELMER CORP., NORWALK, CONN.
PE-	POROLOY EQUIPMENT, INC., VAN NUYS, CALIF.
PE-	VITRO LABS., SILVER SPRING, MD.
PE-	WHITE SANDS MISSILE RANGE, N. MEX. (POST ENGINEER)
PE/A/(NUMBER/YEAR)	BRITISH IRON AND STEEL RESEARCH ASSN. PLANT ENGINEERING AND ENERGY DIV., LONDON
PEB-TM-	VITRO LABS., SILVER SPRING, MD.
PEC-	PERKIN-ELMER CORP., NORWALK, CONN.
PEC-	PHILADELPHIA ELECTRIC CO.
PEC-	POLARAD ELECTRONICS CO., N.Y.C.
PEC-	PRECISION ELECTRONIC COMPONENTS, LTD., TORONTO
PEC-(YEAR)-	PRECISION ELECTRONIC COMPONENTS, LTD., TORONTO
PEC-FL-	PHILADELPHIA ELECTRIC CO.
PEC-MGR-	PHILADELPHIA ELECTRIC CO.
PED-	KELLOGG (M.W.) CO., JERSEY CITY
PED-	PICATINNY ARSENAL, DOVER, N.J.
P + EDD-	AIR REDUCTION CO., INC., MURRAY HILL, N.J.
PEDD-(YEAR)-	AIRCO WELDING PRODUCTS DIV., UNION, N.J. (ENGINEERING AND DEVELOPMENT REPORTS, 1964-)
PE/E/(NUMBER/NUMBER)	BRITISH IRON AND STEEL RESEARCH ASSN. PLANT ENGINEERING AND ENERGY DIV., LONDON
PEEC-	PIERSON ELECTRICAL + ENG. CORP., LOS ANGELES
PE-ER-	PERKIN-ELMER CORP. ELECTRO-OPTICAL DIV., NORWALK, CONN.
PE-ER-	PERKIN-ELMER CORP. OPTICAL GROUP, NORWALK, CONN.
PEGCPR-	KAISER ENGINEERS DIV., HENRY J.KAISER CO., OAKLAND, CAL
PEI-(NUMBERS,LETTERS)-	PHILIPS ELECTRONICS AND PHARMACEUTICAL INDUSTRIES CORP., MOUNT VERNON, N.Y.
PEI-	POROLOY EQUIPMENT, INC., VAN NUYS, CALIF.
PEL-	AEROJET-GENERAL CORP. SOLID ROCKET PLANT, SACRAMENTO, CALIF. (PHOTOELASTIC ANALYSIS)
PEL-	OFFICE OF THE ASST. SECY. OF DEFENSE (RES. + DEV.), TECH. ADVISORY PANEL ON ELECTRONICS, WASH., D.C.
PEL-	UNION OF SOUTH AFRICA. ATOMIC ENERGY BOARD, PRETORIA
PEM-	(PROJECT ENGINEERING MEMORANDUM) (REF. 2)
PEM-	GENERAL ELECTRIC CO. COMMUNICATION PRODUCTS DEPT., LYNCHBURG, VA.
PEM-	RADIO CORP. OF AMERICA. RCA LABS. DIV., PRINCETON, NJ
PENN-	PENNSYLVANIA. UNIV., PHILADELPHIA
PENN-AR-	PENNSYLVANIA. UNIV., PHILADELPHIA
PENNSALT-	PENNSALT CHEMICALS CORP. RES.+ DEV. DEPT., WYNDMOOR, PA
PENNSALT-	PENNSYLVANIA SALT MFG. CO., PHILADELPHIA
PENNSALT-TR-	PENNSALT CHEMICALS CORP. RES.+ DEV. DEPT., WYNDMOOR, PA
PENN U-	PENNSYLVANIA. UNIV., PHILADELPHIA
PENNU-ASR-(YEAR)	PENNSYLVANIA. UNIV., PHILADELPHIA
PENNU-ATR-(YEAR)	PENNSYLVANIA. UNIV., PHILADELPHIA
PENNU-MS-(YEAR)-	PENNSYLVANIA. UNIV., PHILADELPHIA
PENNU-RDR-	PENNSYLVANIA. UNIV., PHILADELPHIA
PENNU-TR-	PENNSYLVANIA. UNIV., PHILADELPHIA
PEO-	(PRODUCTION ENGINEERING ORDER) (REF. 2)
PEPI-	PHILIPS ELECTRONICS AND PHARMACEUTICAL INDUSTRIES CORP., MOUNT VERNON, N.Y.
PEPR-	(PERIODIC PROGRESS REPORT) (REF. 2)
PEPR-	MASSACHUSETTS INST. OF TECH., CAMBRIDGE
PER-(NUMBER)-	GENERAL ELECTRIC CO. MISSILE AND SPACE DIV., PHILA.
PERA-	PRODUCTION ENGINEERING RESEARCH ASSN. OF GT. BRIT. INFO. MGR., MELTON MOWBRAY, LEICS, ENG. (TRANSLATIONS)
PERF-	GT. BRIT. AERONAUTICAL RESEARCH COUNCIL
PERLD ES-D-	SIGNAL CORPS (ARMY). PICTORIAL ENGINEERING AND RESEARCH LAB., LONG ISLAND, N.Y. (DRAWING)
PERSHING-	PERSHING JOINT AEC-DOD WARHEAD COORDINATING COMM. (REF. 15)
PER-T-	ASSOCIATED UNIVERSITIES, INC., N.Y.C.
PES-(LTR.-NO.-NO.-YR.)	SANDIA CORP. PHYSICAL AND ELECTRICAL STANDARDS DEPT., ALBUQUERQUE, N. MEX.
PESCO-	PESCO PRODUCTS DIV., BORG-WARNER CORP., BEDFORD, OHIO
PESCO-ENG. RPT.-	PESCO PRODUCTS DIV., BORG-WARNER CORP., BEDFORD, OHIO
PET-	(PRODUCTION ENVIRONMENTAL TESTS) (REF. 2)
PE-TM-	PERKIN-ELMER CORP. ELECTRO-OPTICAL DIV., NORWALK, CONN.
PE-TR-	PERKIN-ELMER CORP., NORWALK, CONN.
PEU MSEE	PENNSYLVANIA. UNIV., PHILA. MOORE SCH. OF ELEC. ENG.
P-F-	CALIFORNIA RESEARCH AND DEVELOPMENT CO., LIVERMORE
PF-	PFAUDLER CO., ROCHESTER, N.Y.
PF-(YEAR)-	PFAUDLER CO., ROCHESTER, N.Y.
PFC-	BUREAU OF COMMERCIAL FISHERIES, WASHINGTON, D.C. (PROGRESSIVE FISH-CULTURIST)

```
PFC-
        PHILCO-FORD CORP., NEWPORT BEACH, CALIF.
PFL-(NO.)/-
        OFF. OF THE ASST. SECY. OF DEFENSE(RES.+DEV.),WASH,DC
PFR-(YEAR)-
        ITEK CORP. OPTICAL SYSTEMS DIV., LEXINGTON, MASS.
PFRDC/FEWP-
        UNITED KINGDOM ATOMIC ENERGY AUTHORITY. RESEARCH GP.
          ATOMIC ENERGY RES. ESTAB., HARWELL, BERKS, ENGLAND
PFRDC/RPWP/P(YEAR)-
        UNITED KINGDOM ATOMIC ENERGY AUTHORITY. RESEARCH GP.
          ATOMIC ENERGY ESTAB., WINFRITH, DORSET, ENGLAND
PFRDC/RPWP/P(YEAR)-
        UNITED KINGDOM ATOMIC ENERGY AUTHORITY. RESEARCH GP.
          ATOMIC ENERGY RES. ESTAB., HARWELL, BERKS, ENGLAND
PF-RM-(YEAR)-ISI-
        PAN-FAX, INC. INFORMATION SCIENCE INST., GOLETA, CAL.
PFZ-
        RES. + DEV. BD. PANEL ON PROXIMITY FUZES, WASH., D.C.
PG-
        AEROJET-GENERAL CORP., SACRAMENTO, CALIF.
PG-
        FRANCE. COMMISSARIAT A L'ENERGIE ATOMIQUE, PARIS
PG-
        GT. BRIT. ADMIRALTY. NAVAL INTELLIGENCE DEPT.
P/G(NUMBER)-(YEAR)
        HOLMES AND NARVER, INC., LOS ANGELES
PG-
        LINDE AIR PRODUCTS CO., N.Y.C.
PG-
        PRAEGER-KAVANAGH-WATERBURY, ENGINEER-ARCHITECTS,NYC
P+G-
        PROCTOR + GAMBLE DEFENSE CORP.
PG-
        RESEARCH AND DEVELOPMENT BOARD. COMMITTEE ON LONG
          RANGE PROVING GROUND, WASHINGTON, D.C.
PG-
        STANFORD UNIV., CALIF. DEPT. OF MECHANICAL ENG.
PG-
        UNITED KINGDOM ATOMIC ENERGY AUTHORITY. IND. GP.
          DOUNREAY EXPTL. REACTOR ESTAB., CAITHNESS, SCOTLAND
PG-
        UNITED KINGDOM ATOMIC ENERGY AUTHORITY. PRODUCTION
          GROUP, ANNON, SCOTLAND
PG-
        UNITED KINGDOM ATOMIC ENERGY AUTHORITY. PRODUCTION
          GROUP, SPRINGFIELDS, LANCS, ENGLAND
PG-
        UNITED KINGDOM ATOMIC ENERGY AUTHORITY. PRODUCTION
          GROUP. WINDSCALE WORKS, SELLAFIELD, CUMB., ENGLAND
PG-
        UNITED KINGDOM ATOMIC ENERGY AUTHORITY. REACTOR GP.,
          SELLAFIELD, CUMB., ENGLAND
PG-(LTRS)-(NO.)(LTR(S))
        UNITED KINGDOM ATOMIC ENERGY AUTHORITY. PRODUCTION
          GROUP (REF. 39)
PGIB-
        MASSACHUSETTS INST. OF TECH., CAMBRIDGE (POLARIS
          GUIDANCE INFORMATION BOOK)
PG-INF. SER.-
        UNITED KINGDOM ATOMIC ENERGY AUTHORITY. PRODUCTION
          GROUP (REF. 39)
PGIS-(NUMBER)(LETTER(S))
        UNITED KINGDOM ATOMIC ENERGY AUTHORITY. PRODUCTION
          GROUP (REF. 39)
PGL-
        POWER GENERATORS, INC., TRENTON, N.J.
P/GM-
        IMPERIAL CHEMICAL INDUSTRIES, LTD. GENERAL CHEMICAL
          DIV., WIDNES, LANCS, ENGLAND
PGM-(NUMBER)(LETTER(S))
        UNITED KINGDOM ATOMIC ENERGY AUTHORITY. PRODUCTION
          GROUP (REF. 39)
PG-MEMO-
        UNITED KINGDOM ATOMIC ENERGY AUTHORITY. PRODUCTION
          GROUP (REF. 39)
PGN-
        PACIFIC NORTHWEST POWER GROUP, RICHLAND, WASH.
PGR-(NUMBER)(LETTER(S))
        UNITED KINGDOM ATOMIC ENERGY AUTHORITY. PRODUCTION
          GROUP (REF. 39)
PGR-EMC-(YEAR)-
        PACIFIC GEEIA REGION. EMC AND MEASUREMENTS BRANCH,
          SAN FRANCISCO
PG-REPORT-(NUMBER)(CA)
        UNITED KINGDOM ATOMIC ENERGY AUTHORITY. PRODUCTION
          GROUP, CAPENHURST, CHES., ENGLAND (REF. 39)
PG-REPORT-(NUMBER)(R)
        UNITED KINGDOM ATOMIC ENERGY AUTHORITY. PRODUCTION
          GROUP, RISLEY, LANCS, ENGLAND (REF. 39)
PG-REPORT-(NUMBER)(S)
        UNITED KINGDOM ATOMIC ENERGY AUTHORITY. PRODUCTION
          GROUP, SPRINGFIELDS, LANCS, ENGLAND (REF. 39)
PG-REPORT-(NUMBER)(W)
        UNITED KINGDOM ATOMIC ENERGY AUTHORITY. PRODUCTION
          GP. WINDSCALE WORKS, SELLAFIELD,CUMB.,ENG. (REF.39)
PGSAM-
        BANGALORE, INDIA., UNIV. COLL. OF ENGINEERING
PH-
        BUREAU OF AERONAUTICS (NAVY). PHOTOGRAPHIC DIV.
PH-(NO.)-M
        CONVAIR, SAN DIEGO, CALIF.
PH-(NUMBER)Q-(YEAR)
        DOW CHEMICAL CO., MIDLAND, MICH. (PERFLUOROALKYL
          HETEROCYCLIC ELASTOMERS)
PH-
        GENERAL DYNAMICS/CONVAIR, SAN DIEGO, CALIF.
PH-
        GT. BRIT. ROYAL AIRCRAFT EST.,FARNBOROUGH, HANTS, ENG
PH-
        IRAQ. ATOMIC ENERGY COMMISSION, BAGHDAD
PH-
        SWEDEN. KUNGLIGA TEKNISKA HOEGSKOLAN, STOCKHOLM.
          DIV. OF PLASMA PHYSICS
PHARMACOLOGY REPORT-
        ROCHESTER, N.Y. UNIV.
329-PHASE-
        BROOKS + PERKINS, INC. DEFENSE PRODS. DIV., DETROIT
PHEP-
        SOUTHWESTERN RADIOLOGICAL HEALTH LAB., LAS VEGAS,NEV.
          (PUBLIC HEALTH EVALUATION)
PHF-
        MICHIGAN. UNIV., ANN ARBOR. HIGHWAY SAFETY RES. INST.
PHIL-
        PHILCO CORP. (ALL LOCATIONS)
PHILCO-
        PHILCO CORP. (ALL LOCATIONS)
PHILCO-H-
        PHILCO CORP., PHILADELPHIA
PHILCO-R-
        PHILCO CORP., LANSDALE, PENNA.
PHILCO-U-
        AERONUTRONIC DIV. OF PHILCO CORP., NEWPORT BEACH,CAL.
PHL-(YEAR)-
        NETHERLANDS. RIJKSVERDEDIGINGSORGANISATIE, TNO.
          PHYSISCH LABORATORIUM, THE HAGUE
PH.L.(YEAR)-
        NETHERLANDS. RIJKSVERDEDIGINGSORGANISATIE, TNO.
          PHYSISCH LABORATORIUM, THE HAGUE
PHM-(NUMBER-YEAR)
        PHOTOMETRICS, INC., LEXINGTON, MASS.
PH-M-
        PUBLIC HEALTH SERVICE, WASHINGTON, D.C.
PHOTO-
        PHOTOGRAMMETRY, INC., SILVER SPRING, MD.
PHOTOGRAMMETRY-SER-
        ILLINOIS. UNIV., URBANA. DEPT. OF CIVIL ENGINEERING
PHO-TR-
        PHILCO CORP., LANSDALE, PENNA.
PHPC-
        PHILLIPS PETROLEUM CO., BARTLESVILLE, OKLA.
PHS-(NUMBER-YEAR)-
        FOREIGN TECHNOLOGY DIV., WRIGHT-PATTERSON AFB, OHIO
PHS-(NUMBER)-RH-
        NATIONAL CENTER FOR RADIOLOGICAL HEALTH, ROCKVILLE,MD
PHS-
        PUBLIC HEALTH SERVICE, WASHINGTON, D.C.
PHS-PUBL.-(NO.)-WP-
        PUBLIC HEALTH SERVICE, WASHINGTON, D.C.
PHX-
        CANADA. NAVAL RESEARCH ESTABLISHMENT
PHY-
        SYRACUSE UNIV., N.Y. RESEARCH INST.
PHYS/EX-
        GT. BRIT. ADVISORY COUNCIL ON SCI. RES. + TECH. DEV.
PHYS/EX-
        GT. BRIT. ARMAMENT RES. DEPT.,FT.HALSTEAD, KENT, ENG.
PHYSICS-
        SYRACUSE UNIV., N.Y. RESEARCH INST.
PHYSICS NOTE-
        RIO DE JANEIRO.CENTRO BRASILEIRO DE PESQUISAS FISICAS
PHYSICS NOTE NO.-
        RIO DE JANEIRO.CENTRO BRASILEIRO DE PESQUISAS FISICAS
PI-
        CARBIDE AND CARBON CHEMICALS CORP., OAK RIDGE, TENN.
PI-1273-M-
        CORNELL AERONAUTICAL LAB., INC., BUFFALO
PI-
        GENERAL MOTORS CORP. RESEARCH LABS. DIV., DETROIT
          (PHYSICS INSTRUMENTATION)
PI-
        PANAMETRICS, INC., WALTHAM, MASS.
PI-
        PHYSICS INTERNATIONAL CO., SAN LEANDRO, CALIF.
PI-
        POLACOAT INC., CINCINNATI
PI-
        SIGNAL CORPS (ARMY). PLANS AND INTELLIGENCE BRANCH
PI-
        TENNESSEE EASTMAN CORP., OAK RIDGE, TENN.
PIB-
        (PRODUCT IMPROVEMENT BULLETIN) (REF. 2)
PIB-
        BROOKLYN. POLYTECHNIC INST.
PIB-(NUMBER)-(LETTER(S))
        BROOKLYN. POLYTECHNIC INST. (SQUID PROJECT) (REF. 33)
PIB-
        BURROUGHS CORP., DETROIT (PRODUCT IMPROVEMENT BULL.)
PIB-
        FLEET COMPUTER PROGRAMMING CENTER, ATLANTIC,
          VIRGINIA BEACH (PROGRAM INFO. BULLETIN)
PIB AL-
        BROOKLYN. POLYTECHNIC INST.
PIBAL-
        BROOKLYN. POLYTECHNIC INST.
PIBAL-R-
        BROOKLYN. POLYTECHNIC INST.
PIBEE(YEAR)-
        BROOKLYN. POLYTECHNIC INST., FARMINGDALE, N.Y.
PIBEP-
        BROOKLYN. POLYTECHNIC INST., FARMINGDALE, N.Y. DEPT.
          OF ELECTROPHYSICS
PIBEP-(YEAR)-
        BROOKLYN. POLYTECHNIC INST., FARMINGDALE, N.Y. DEPT.
          OF ELECTROPHYSICS
PIB-ISSUE-
        FLEET COMPUTER PROGRAMMING CENTER, ATLANTIC,
          VIRGINIA BEACH (PROGRAM INFO. BULLETIN)
```

PIB-ISSUE-	FLEET COMPUTER PROGRAMMING CENTER, PACIFIC, SAN DIEGO, CALIF.
PIB-MRI-	BROOKLYN. POLYTECHNIC INST. MICROWAVE RESEARCH INST.
PIBMRI-	BROOKLYN. POLYTECHNIC INST. MICROWAVE RESEARCH INST.
PIBMRI-R-	BROOKLYN. POLYTECHNIC INST. MICROWAVE RESEARCH INST.
PIB-PRL-	BROOKLYN. POLYTECHNIC INST. PROPULSION RESEARCH LAB.
PIB R-	BROOKLYN. POLYTECHNIC INST.
PIB-TR-	BROOKLYN. POLYTECHNIC INST.
PIC-	NAVAL PHOTOGRAPHIC INTERPRETATION CENTER, WASH., D.C.
PIC-	PENNSYLVANIA. UNIV., PHILADELPHIA. POWER INFO. CTR.
PIC-ARS-	PICATINNY ARSENAL, DOVER, N.J.
PIC ARS TM-	PICATINNY ARSENAL, DOVER, N.J.
PIC. ARS. TN-	PICATINNY ARSENAL, DOVER, N.J.
PIC. ARS. TR-	PICATINNY ARSENAL, DOVER, N.J.
PIC-BAT-	PENNSYLVANIA. UNIV., PHILADELPHIA. POWER INFO. CTR.
PIC-ELE-MHD-	PENNSYLVANIA. UNIV., PHILADELPHIA. POWER INFO. CTR.
PIC-ELE-TI-	PENNSYLVANIA. UNIV., PHILADELPHIA. POWER INFO. CTR.
PIC-SOL-	PENNSYLVANIA. UNIV., PHILADELPHIA. POWER INFO. CTR.
PIFR-	PHYSICS INTERNATIONAL CO., SAN LEANDRO, CALIF.
PI(INT)-	FRANCE. COMMISSARIAT A L'ENERGIE ATOMIQUE. CENTRE D'ETUDES NUCLEAIRES, GRENOBLE
PIIR-(NUMBER)-(YEAR)	PHYSICS INTERNATIONAL CO., SAN LEANDRO, CALIF.
PILGRIM-	ARMED FORCES SPECIAL WEAPONS PROJECT. WEAPONS EFFECTS PROGRAM, WASH., D.C. (OPERATION PLUMBBOB) (REF. 7)
PIMBRI-R-(NUMBER)-(YEAR)	BROOKLYN. POLYTECHNIC INST. MICROWAVE RESEARCH INST.
PIN-(YEAR)	GENERAL DYNAMICS/CONVAIR, SAN DIEGO, CALIF.
PINSTECH/HP-	PAKISTAN. INST. OF NUCLEAR SCIENCE AND TECHNOLOGY. HEALTH PHYSICS DIV., ISLAMABAD
PINSTECH/PHY-	PAKISTAN. INST. OF NUCLEAR SCIENCE AND TECHNOLOGY, ISLAMABAD (ALL DIVISIONS)
PINSTECH/RIPD-	PAKISTAN. INST. OF NUCLEAR SCIENCE AND TECHNOLOGY. RADIOISOTOPE PRODUCTION DIV., ISLAMABAD
PINSTECH/RO-	PAKISTAN. INST. OF NUCLEAR SCIENCE AND TECHNOLOGY. REACTOR OPERATIONS DIV., ISLAMABAD
PINSTECH/RT-	PAKISTAN. INST. OF NUCLEAR SCIENCE AND TECHNOLOGY. REACTOR THEORY DIV., ISLAMABAD
PI-(NT)-	FRANCE. COMMISSARIAT A L'ENERGIE ATOMIQUE. CENTRE D'ETUDES NUCLEAIRES, GRENOBLE
PI/NT/(NUMBER-NUMBER)	FRANCE. COMMISSARIAT A L'ENERGIE ATOMIQUE. CENTRE D'ETUDES NUCLEAIRES, GRENOBLE
PINT-	PHYSICS INTERNATIONAL CO., BERKELEY, CALIF.
PINT-	PHYSICS INTERNATIONAL CO., SAN LEANDRO, CALIF.
PINT-PIFR-	PHYSICS INTERNATIONAL CO., SAN LEANDRO, CALIF.
PIP/CB-CED-GA-(YEAR)	COOSA VALLEY AREA PLANNING + DEVELOPMENT COMMISSION, ROME, GA.
PIPPA-WA/MEMO-	GT. BRIT. ATOMIC EN. RES. ESTAB., HARWELL, BERKS, ENG
PIPPA-WP/MEMO-	GT. BRIT. DEPT. OF ATOMIC ENERGY. INDUSTRIAL GROUP H.Q., RISLEY, LANCS, ENGLAND
PIPR-	PHYSICS INTERNATIONAL CO., SAN LEANDRO, CALIF.
PIQR-	PHYSICS INTERNATIONAL CO., SAN LEANDRO, CALIF.
PIR-	CANADA. PACIFIC NAVAL LAB., ESQUIMALT, B.C. (PACIFIC INTERIM REPORT)
PIR-(NUMBER)-	GENERAL ELECTRIC CO. MISSILE AND SPACE DIV., PHILA.
PIR-	MARTIN-MARIETTA CORP. DENVER DIV.
PIR-	PHYSICS INTERNATIONAL CO., SAN LEANDRO, CALIF.
PIT-	AEROJET-GENERAL CORP., SACRAMENTO, CALIF.
PITHA-	AACHEN. TECHNISCHE HOCHSCHULE. PHYSIKALISCHES INST.
PI-TN-	PRESEARCH, INC., SILVER SPRING, MD.
PITR-	PHYSICS INTERNATIONAL CO., SAN LEANDRO, CALIF.
PITSR-	PHYSICS INTERNATIONAL CO., SAN LEANDRO, CALIF.
PITT-	PITTSBURGH. UNIV.
PITT-PLFB-	PITTSBURGH. UNIV. PYMATUNING LAB. OF FIELD BIOLOGY
PITT-RR-	PITTSBURGH. UNIV.
PITT-SRCC-	PITTSBURGH. UNIV. SPACE RESEARCH COORDINATION CTR.
PITTU-	PITTSBURGH. UNIV.
PITU-	PITTSBURGH. UNIV.
PITU-SRL-	PITTSBURGH. UNIV. SARAH MELLON SCAIFE RADIATION LAB.
PJ-	ASSOCIATES FOR INTERNATL. RES., INC., CAMBRIDGE, MASS
PJG-A-	TENNESSEE EASTMAN CORP., OAK RIDGE, TENN.
PJM-	(POWER JETS MEMORANDUM) (REF. 2)
PJR-	(POWER JETS REPORT) (REF. 2)
PJR-	POWER JETS, LTD. (GT. BRIT.)
PK-	ARGONNE NATIONAL LAB., ILL.
P-K-	NATIONAL DEFENSE RESEARCH COMMITTEE
PKN-	STANDARD OIL DEVELOPMENT CO., ELIZABETH, N.J.
PL-	(PARTS LIST) (REF. 2)
PL-	CONGRESS, WASHINGTON, D.C. (PUBLIC LAW)
PL-	GOVERNMENT PRINTING OFFICE, WASH., D.C. (PRICE LIST)
PL-	GT. BRIT. DIRECTORATE OF MATERIALS AND EXPLOSIVES RESEARCH AND DEVELOPMENT
PL-	INDIA. NATIONAL PHYSICAL LAB., NEW DELHI
PL-	MARTIN-MARIETTA CORP. DENVER DIV.
PL-	PHILIPS LABS., INC., BRIARCLIFF MANOR, N.Y.
PL-	PHILIPS LABS., INC., IRVINGTON-ON-HUDSON, N.Y.
PL-	STANFORD RES. INST. POULTER LABS., MENLO PARK, CALIF.
PLAC-	UNITED KINGDOM ATOMIC ENERGY AUTHORITY. RESEARCH GP. ATOMIC ENERGY RES. ESTAB., HARWELL, BERKS, ENGLAND
PLANS-(YEAR)-	CANADA. DEFENCE RESEARCH BOARD. CHIEF OF PLANS, OTTAWA
PLAS-	PLASMADYNE CORP., SANTA ANA, CALIF.
PLAS (LTRS.)-	PLASMADYNE CORP., SANTA ANA, CALIF.
PLASTEC-	DE BELL + RICHARDSON, INC., HAZARDVILLE, CONN.
PLASTEC-	PICATINNY ARSENAL. PLASTICS TECHNICAL EVALUATION CENTER, DOVER, N.J.
PLASTEC-NOTE-	PICATINNY ARSENAL. PLASTICS TECHNICAL EVALUATION CENTER, DOVER, N.J.
PLASTEC-TN-	PICATINNY ARSENAL. PLASTICS TECHNICAL EVALUATION CENTER, DOVER, N.J. (TECHNICAL NOTES)
PLASTEC-TR-	PICATINNY ARSENAL. PLASTICS TECHNICAL EVALUATION CENTER, DOVER, N.J. (TECHNICAL REPORTS)
PL-C-TN-	PICATINNY ARSENAL. PYROTECHNICS LAB., DOVER, N.J.
PLESS-	PLESSET (E.H.) ASSOCIATES, INC., LOS ANGELES
PLESS-	PLESSET (E.H.) ASSOCIATES, INC., SANTA MONICA, CALIF.
PLESSET-	PLESSET (E.H.) ASSOCIATES, INC., SANTA MONICA, CALIF.
PLESS TR-	PLESSET (E.H.) ASSOCIATES, INC., LOS ANGELES
PL.GW-	PLESSEY CO., LTD., ILFORD, ESSEX, ENGLAND
PLLRC-(YEAR)-	PUBLIC LAND LAW REVIEW COMMISSION, WASHINGTON, D.C.
PLLRC-STUDY-	PUBLIC LAND LAW REVIEW COMMISSION, WASHINGTON, D.C.
PL/N-	ACADEMIA R.S.R. INSTITUTUL DE FIZICA ATOMICA, BUCHAREST
PLP-	GT. BRIT. DEPT. OF ATOMIC EN. PLANT LOCATION PANEL
PLP-	PLENUM PRESS, INC., N.Y.C. (TRANSLATIONS)
PLR-	HARVARD UNIV., CAMBRIDGE, MASS. PSYCHOLOGICAL LABS.
PLR-	PLASMADYNE CORP., SANTA ANA, CALIF.
PLR-	WASHINGTON. UNIV., SEATTLE. PSYCHOPHYSICS LAB.
PL-TM-(YEAR)-	PERSONNEL RES. LAB. (6570TH), LACKLAND AFB, TEX.
PL-TR-(YEAR)-	PERSONNEL RES. LAB. (6570TH), LACKLAND AFB, TEX.
PL-TR-(NO.)-	STANFORD RES. INST. POULTER LABS., MENLO PARK, CALIF.
PLUTO/IS-	UNITED KINGDOM ATOMIC ENERGY AUTHORITY. RESEARCH GP. ATOMIC ENERGY RES. ESTAB., HARWELL, BERKS, ENGLAND
PLUTO-TN-	CALIFORNIA. UNIV., LIVERMORE. LAWRENCE RADIATION LAB. (PLUTO TECHNICAL NOTE)

PM-	(PROPULSION MEMORANDUM) (REF. 2)
PM-	BUREAU OF AERONAUTICS (NAVY)
PM-	CAMBRIDGE CORP.
PM-	CANADA. DEPT. OF MINES AND TECH. SURVEYS. MINES BR.
PM-	DE HAVILLAND AIRCRAFT CO., LTD. (GT. BRIT.)
PM-	FLORIDA. UNIV., GAINESVILLE. ELECTRON DEVICES LAB.
PM-(NUMBER)	HOEGANAES AB, SWEDEN
PM/(YEAR)/	MONTPELLIER UNIV., FRANCE (PHYSICS AND MATH.)
PM-(NUMBER)	NATIONAL BUREAU OF STANDARDS. BOULDER LABS., COLO.
PM-	NATIONAL RESEARCH COUNCIL OF CANADA. ATOMIC ENERGY PROJECT, CHALK RIVER, ONT.
PM-	OFFICE OF CIVIL DEFENSE, WASHINGTON, D.C.
PM-	OFFICE OF NAVAL MATERIAL, WASH., D.C. (PROJECT MANAGER)
PM-	PRINCETON UNIV., N.J. PROJECT MATTERHORN
PM-	THIOKOL CHEMICAL CORP., BRIGHAM CITY, UTAH
PM(C)-	GT. BRIT. CAPENHURST WORKS, CHES., ENGLAND (REF. 36)
PME-	AEROJET-GENERAL CORP., SACRAMENTO, CALIF.
PMF-(NUMBER)(LETTER)	ATOMIC ENERGY COMMISSION, WASHINGTON, D.C. (RADIOLOGICAL SAFETY)
PM-I-	CANADA. DEPT. OF MINES AND TECH. SURVEYS. MINES BR.
PML-(NO.)-SR-	PARKE MATHEMATICAL LABS., INC., CARLISLE, MASS.
PM-M-	CANADA. DEPT. OF MINES AND TECH. SURVEYS. MINES BR.
PM-MPR-	TENNESSEE EASTMAN CORP., OAK RIDGE, TENN.
PM-MPR-LA-	TENNESSEE EASTMAN CORP., OAK RIDGE, TENN.
PMO-	CALIFORNIA. UNIV., BERKELEY. LAWRENCE RADIATION LAB.
PMO-	WHITE SANDS MISSILE RANGE, N. MEX. (PROVOST MARSHAL OFFICE)
PM-PO-	PRINCETON UNIV., N.J. PROJECT MATTERHORN (PRINT-OUTS)
PMP-TM-	PACIFIC MISSILE RANGE, POINT MUGU, CALIF.
PM-Q-	PRINCETON UNIV., N.J. PROJECT MATTERHORN (QUARTERLY)
PM-R-	CANADA. DEPT. OF MINES AND TECH. SURVEYS. MINES BR.
PM(R)-	GT. BRIT. DEPT. OF ATOMIC ENERGY. INDUSTRIAL GROUP H.Q., RISLEY, LANCS, ENGLAND (REF. 36)
PMR-	GRUMMAN OPERATIONS ANALYSIS GROUP. ADVANCED SYSTEMS SECTION, BETHPAGE, N.Y.
PMR-	MICHIGAN. UNIV., ANN ARBOR
PMR-	OFFICE OF SCIENTIFIC RESEARCH AND DEVELOPMENT
PMR-	PACIFIC MISSILE RANGE, POINT MUGU, CALIF.
PMR-MP-(YR.)-	PACIFIC MISSILE RANGE, POINT MUGU, CALIF.
PMR-MR-(YEAR)-	PACIFIC MISSILE RANGE, POINT MUGU, CALIF.
PMR-TM-	PACIFIC MISSILE RANGE, POINT MUGU, CALIF.
PMR-TN-	PACIFIC MISSILE RANGE, POINT MUGU, CALIF.
PMR-TR-(YEAR)-	PACIFIC MISSILE RANGE, POINT MUGU, CALIF.
P + MS-	A.F. PERSONNEL + TRAINING RES. CTR., LACKLAND AFB, TEX.
PMS-	CIVIL SERVICE COMMISSION, WASHINGTON, D.C. (PERSONNEL MANAGEMENT SERIES)
PM(S)-	GT. BRIT. SPRINGFIELDS WORKS, LANCS, ENGLAND (REF.36)
PM-S-	PRINCETON UNIV., N.J. PROJECT MATTERHORN
PMSRP-	(PHYSICAL AND MATHEMATICAL RESEARCH PAPER) (REF. 2)
PMSRP-	AIR FORCE CAMBRIDGE RESEARCH LABS., BEDFORD, MASS. (PHYSICAL AND MATHEMATICAL RESEARCH PAPER)
PM-T-	CANADA. DEPT. OF MINES AND TECH. SURVEYS. MINES BR.
PMU-	STANFORD RESEARCH INST., MENLO PARK, CALIF.
PM(W)-	GT. BRIT. WINDSCALE WORKS, SELLAFIELD, CUMB., ENGLAND (REF. 36)
PM-WPR-	TENNESSEE EASTMAN CORP., OAK RIDGE, TENN.
PN-	ACADEMIA R.S.R. INSTITUTUL DE FIZICA ATOMICA, BUCHAREST
PN-	GT. BRIT. CHEMICAL DEFENCE EXPERIMENTAL ESTABLISHMENT, PORTON, WILTS, ENGLAND
PN-(NUMBER)-	SCIENCE ENGINEERING ASSOCIATES, SAN MARINO, CALIF.
PNB-	BORDEAUX UNIV. LABORATOIRE DE PHYSIQUE THEORIQUE
PNC-N-	POWER REACTOR AND NUCLEAR FUEL DEVELOPMENT CORP., TOKAI, JAPAN
PNCPU-	POWER REACTOR AND NUCLEAR FUEL DEVELOPMENT CORP., TOKAI, JAPAN
PNCT-	POWER REACTOR AND NUCLEAR FUEL DEVELOPMENT CORP., TOKAI, JAPAN
PNCT-AR-	POWER REACTOR AND NUCLEAR FUEL DEVELOPMENT CORP., TOKAI, JAPAN
PNE-	DIVISION OF APPLIED TECHNOLOGY, AEC, WASHINGTON, D.C. (PEACEFUL USES OF NUCLEAR EXPLOSIONS. PLOWSHARE PROGRAM)
PNE-G-	DIVISION OF APPLIED TECHNOLOGY, AEC, WASHINGTON, D.C. (PROJECT GASBUGGY)
PNE-R-	DIVISION OF APPLIED TECHNOLOGY, AEC, WASHINGTON, D.C. (PROJECT RULISON)
PNG-	AUSTRALIA. BUREAU OF MINERAL RESOURCES, GEOLOGY AND GEOPHYSICS, CANBERRA
PNG-	PACIFIC NORTHWEST POWER GROUP, RICHLAND, WASH.
PNJ-	PRINCETON UNIV., N.J.
PNJ-LA-	PRINCETON UNIV., N.J.
PNL-	CANADA. PACIFIC NAVAL LAB., ESQUIMALT, B.C.
PNL-TM-	CANADA. PACIFIC NAVAL LAB., ESQUIMALT, B.C.
PNM-	HARVARD UNIV., CAMBRIDGE, MASS. PSYCHO-ACOUSTIC LAB.
PNPF-	PIQUA NUCLEAR POWER FACILITY, OHIO
PNR-	HARVARD UNIV., CAMBRIDGE, MASS. PSYCHO-ACOUSTIC LAB. (PSYCHO NAVY RESEARCH)
PNR-	NAVAL RESEARCH LAB., WASHINGTON, D.C.
PNR/DIR-(YEAR)-	FRANCE. COMMISSARIAT A L'ENERGIE ATOMIQUE. CENTRE D'ETUDES NUCLEAIRES, CADARACHE
PNR/EMTR-(YEAR)-	FRANCE. COMMISSARIAT A L'ENERGIE ATOMIQUE. CENTRE D'ETUDES NUCLEAIRES, CADARACHE
PNRO-DEV-	PITTSBURGH NAVAL REACTORS OPERATIONS OFFICE, AEC
PNROO-DEV-	PITTSBURGH NAVAL REACTORS OPERATIONS OFFICE, AEC
PNRO-SMD-	PITTSBURGH NAVAL REACTORS OPERATIONS OFFICE, AEC
PNR/SEPR/	FRANCE. COMMISSARIAT A L'ENERGIE ATOMIQUE
PNR/SETR-R-	FRANCE. COMMISSARIAT A L'ENERGIE ATOMIQUE. CENTRE D'ETUDES NUCLEAIRES, CADARACHE
PNSI-TR-(YEAR)-	POLHEMUS NAVIGATION SCIENCES, INC., BURLINGTON, VT.
PNSI-TR-(NUMBER)-	POLHEMUS NAVIGATION SCIENCES, INC., BURLINGTON, VT.
PNU-	CALIFORNIA. UNIV., LIVERMORE. RADIATION LAB.
PNW-	FOREST SERVICE. PACIFIC NORTHWEST STA., PORTLAND, ORE
PNWL-	BATTELLE-NORTHWEST. PACIFIC NORTHWEST LAB., RICHLAND WASH.
PO-	KELSEY-HAYES CO., UTICA, N.Y.
PO-	KELSEY-HAYES CO. TURBINE PARTS DIV., UTICA, N.Y.
PO-	OFFICE OF NAVAL RESEARCH (PERSONNEL OPERATIONS)
PO-	QUARTERMASTER CORPS (RESEARCH, PURCHASE ORDER)
POC-	POLAROID CORP., CAMBRIDGE, MASS.
POC-A-	GT. BRIT. ATOMIC EN. RES. ESTAB., HARWELL, BERKS, ENG
POC-DATA-	GT. BRIT. ATOMIC EN. RES. ESTAB., HARWELL, BERKS, ENG
POCLM/MEMO-	GT. BRIT. ATOMIC EN. RES. ESTAB., HARWELL, BERKS, ENG
POC/MEM-	UNITED KINGDOM ATOMIC ENERGY AUTHORITY. RESEARCH GP. ATOMIC ENERGY RES. ESTAB., HARWELL, BERKS, ENGLAND
POC/MEMO-	UNITED KINGDOM ATOMIC ENERGY AUTHORITY. RESEARCH GP. ATOMIC ENERGY RES. ESTAB., HARWELL, BERKS, ENGLAND
POCO-PMH-(YEAR)-	CALIFORNIA. UNIV., LOS ANGELES. CENTER FOR THE HEALTH SCIENCES (PHYSIOLOGY OF CHIMPS IN ORBIT)
POC/RIG MANUAL/	UNITED KINGDOM ATOMIC ENERGY AUTHORITY. RESEARCH GP. ATOMIC ENERGY RES. ESTAB., HARWELL, BERKS, ENGLAND
POC/SC-	GT. BRIT. ATOMIC EN. RES. ESTAB., HARWELL, BERKS, ENG
POC/SSC/1-	UNITED KINGDOM ATOMIC ENERGY AUTHORITY. RESEARCH GP. ATOMIC ENERGY RES. ESTAB., HARWELL, BERKS, ENGLAND

POED-

POED-	GT. BRIT. POST OFFICE. ENGINEERING DEPT.
PO-ED/RR-	GT. BRIT. POST OFFICE. ENGINEERING DEPT.
POIR-	(PROJECT OFFICERS INTERIM REPORT) (REF. 7)
POL-	POLAROID CORP., CAMBRIDGE, MASS.
POLC-NC-	POLAROID CORP., CAMBRIDGE, MASS.
POLY-	POLYTECHNIC ENGINEERING CORP., SILVER SPRING, MD.
POM-	OFFICE OF SCIENTIFIC RESEARCH AND DEVELOPMENT
POMM-	ARMY MATERIEL COMMAND, WASHINGTON, D.C.
POMM-	HONEYWELL, INC. ORDNANCE DIV., HOPKINS, MINN.
POP-	MASON + HANGER-SILAS MASON CO., INC. PANTEX PLANT, AMARILLO, TEX.
POR-	(PROJECT OFFICERS REPORT) (REF. 7)
POR-	OFF. OF THE ASST. SECY. OF DEFENSE(RES.+DEV.),WASH,DC
PORTON-	GT. BRIT. CHEMICAL DEFENCE EXPERIMENTAL ESTABLISHMENT, PORTON, WILTS, ENGLAND
PORTON MEMO-	GT. BRIT. CHEMICAL DEFENCE EXPERIMENTAL ESTABLISHMENT, PORTON, WILTS, ENGLAND
PORTON TN-	GT. BRIT. CHEMICAL DEFENCE EXPERIMENTAL ESTABLISHMENT, PORTON, WILTS, ENGLAND
PORTON TP-	GT. BRIT. CHEMICAL DEFENCE EXPERIMENTAL ESTABLISHMENT, PORTON, WILTS, ENGLAND
PP-	(PHYSICAL PROPERTIES) (REF. 2)
PP-	(PROFESSIONAL PAPER) (REF. 2)
PP-	AEROJET-GENERAL CORP., SACRAMENTO, CALIF.
PP-	BUREAU OF AERONAUTICS (NAVY). POWER PLANT DIV.
PP/	CATANIA, ITALY. UNIV. ISTITUTO DI FISICA
PP1-B-	CLINTON ENGINEER WORKS, MANHATTAN DISTRICT, OAK RIDGE, TENN.
PP-(NUMBER)-C	GEOLOGICAL SURVEY, WASHINGTON, D.C.
PP-(NUMBER-YEAR)	GEORGE WASHINGTON UNIV., WASHINGTON, D.C.
PP-	GT. BRIT. AERO. RES. COUNCIL. POWER PLANTS COMM.
PP-	GT. BRIT. ROYAL AIRCRAFT EST.,FARNBOROUGH, HANTS, ENG
P/P-(NUMBER)-(YEAR)	HOLMES AND NARVER, INC., LOS ANGELES
PP-	JOHNS HOPKINS UNIV., SILVER SPRING, MD. PANEL ON PHYSICAL PROPERTIES OF SOLID PROPELLANTS
PP-(NOS.)-	NAVAL AIR MISSILE TEST CENTER, POINT MUGU, CALIF.
PP-	PERGAMON PRESS, N.Y.C. (TRANSLATIONS)
PP-	PROPULSION PRODUCTS, SANTA MONICA, CALIF.
PP-	RADIO CORP. OF AMERICA, BURLINGTON, MASS.
PP-	TENNESSEE EASTMAN CORP., OAK RIDGE, TENN.
PP-(NO.)-(LETTER)-	TENNESSEE EASTMAN CORP., OAK RIDGE, TENN.
PPA-	PRINCETON UNIV., N.J.
PPAD-	PRINCETON-PENNSYLVANIA ACCELERATOR, PRINCETON, N.J.
PPAR-	PRINCETON-PENNSYLVANIA ACCELERATOR, PRINCETON, N.J.
PPB-	NATIONAL RESEARCH COUNCIL. COMMITTEE ON PASSIVE PROTECTION AGAINST BOMBING
PPC-	PARKER PEN CO., JANESVILLE, WIS.
PPC-	PHILLIPS PETROLEUM CO. ATOMIC EN. DIV., IDAHO FALLS
PPC-(NO.)-(YR.)R	PHILLIPS PETROLEUM CO. RESEARCH AND DEVELOPMENT DEPT. BARTLESVILLE, OKLA.
PPC-IDO-	PHILLIPS PETROLEUM CO. ATOMIC EN. DIV., IDAHO FALLS
PPCS-	PHILLIPS PETROLEUM CO. ATOMIC EN. DIV., IDAHO FALLS
PPD-(YEAR)-(LETTER)-	NORTHROP NORTRONICS. PRECISION PRODUCTS DEPT., NORWOOD, MASS.
PPD-	TENNESSEE EASTMAN CORP., OAK RIDGE, TENN.
PPDG-	UNITED KINGDOM ATOMIC ENERGY AUTHORITY. INDUSTRIAL GROUP H.Q., RISLEY, LANCS, ENGLAND
PPDR-	AEROJET-GENERAL CORP., SACRAMENTO, CALIF.
PPD-WPR-	AEROJET-GENERAL CORP. SOLID ROCKET PLANT, SACRAMENTO, CALIF. (FLEET BALLISTIC MISSILE PROGRAM)
PPL-	FAIRCHILD HILLER CORP. REPUBLIC AVIATION DIV. PLASMA PROPULSION LAB., FARMINGDALE, N.Y.
PPL-(YEAR)-	FAIRCHILD HILLER CORP. REPUBLIC AVIATION DIV. PLASMA PROPULSION LAB., FARMINGDALE, N.Y.
P/PL-	HOLMES AND NARVER, INC., LOS ANGELES
P/PL-(NUMBER)-(YEAR)	HOLMES AND NARVER, INC., LOS ANGELES
PPL-	SASKATCHEWAN. UNIV., SASKATOON. PLASMA PHYSICS LAB.
PPL-AF-	PRINCETON UNIV., N.J. PLASMA PHYSICS LAB.
PPL-NP(NUMBER)	PRINCETON UNIV., N.J. PLASMA PHYSICS LAB.
PPL-TR-(YEAR)-	FAIRCHILD HILLER CORP. REPUBLIC AVIATION DIV. PLASMA PROPULSION LAB., FARMINGDALE, N.Y.
PPL-TRANS-	PRINCETON UNIV., N.J. PLASMA PHYSICS LAB.
PPM-	BUREAU OF AERONAUTICS (NAVY) (POWER PLANT MEMORANDUM)
PPM-	HARVARD UNIV., CAMBRIDGE, MASS.
PPN-	UNITED KINGDOM ATOMIC ENERGY AUTHORITY. RESEARCH GP. CULHAM LAB., ABINGDON, BERKS, ENGLAND
PPPM-	NATIONAL RESEARCH COUNCIL OF CANADA. ATOMIC ENERGY PROJECT, CHALK RIVER, ONT.
PPR-	STRATEGIC BOMBING SURVEY (PROPELLANTS PLANT RPTS.)
PPS-	BUREAU OF MINES. BARTLESVILLE PETROLEUM RESEARCH CENTER, OKLA.
PPS-	PICATINNY ARSENAL. PROPULSION AND PROPELLANT SEC., DOVER, N.J.
PPSST-	GEORGE WASHINGTON UNIV., WASHINGTON, D.C.
PPT-	OFFICE OF THE ASSISTANT SECRETARY OF DEFENSE. ADVISORY PANEL ON PERSONNEL AND TRAINING RESEARCH
PPWPR-	GT. BRIT. ATOMIC EN. RES. ESTAB., HARWELL, BERKS, ENG
PQMD-	QUARTERMASTER DEPOT, PHILADELPHIA
PQR/(YEAR)-	JOHNS HOPKINS UNIV., SILVER SPRING, MD. APPLIED PHYSICS LAB.
PQR-	OHIO STATE UNIV., COLUMBUS. CRYOGENIC LAB.
PR-	(PERFORMANCE REQUIREMENT) (REF. 2)
PR-	(PLANT REPORT) (REF. 2)
PR-	(PRELIMINARY REPORT) (REF. 2)
PR-	(PREPRINT) (REF. 2)
PR-	(PROGRESS REPORT) (REF. 2)
PR-	(PROJECT REPORT) (REF. 2)
PR-	ATOMIC ENERGY OF CANADA LTD.,CHALK RIVER,ONT. (REF.6)
PR-	BARDEN CORP., DANBURY, CONN.
PR-(NO.)-(NO.)	CALIFORNIA INST. OF TECH., PASADENA
PR-(YR.)-	DIAMOND ORDNANCE FUZE LABS., WASHINGTON, D.C.
PR-(NUMBER) Q-(YEAR)	DOW CHEMICAL CO. SCIENTIFIC PROJS. LAB., MIDLAND,MICH
PR-	EAGLE-PICHER CO. RESEARCH LABS., MIAMI, OKLA.
PR-(YEAR)-	HARRY DIAMOND LABS., WASHINGTON, D.C.
PR-(NUMBERS,LETTERS)	HESSE-EASTERN DIV., FLIGHTEX FABRICS INC., CAMBRIDGE, MASS.
PR-	HUGHES AIRCRAFT CO., CULVER CITY, CALIF. (PROPOSAL)
PR-(ROMAN NO.)-(NO.LTRS.)	HUGHES AIRCRAFT CO. ENG. DIV., CULVER CITY, CALIF.
PR-(YEAR)-	LOUISVILLE, KY. UNIV. PERFORMANCE RESEARCH LAB.
PR-(NUMBER)-(NUMBER)Q-	MARQUARDT CORP., VAN NUYS, CALIF.
PR-(YEAR)-	MARTIN CO., DENVER
PR-	MARTIN-MARIETTA CORP. DENVER DIV.
PR-	NATIONAL RESEARCH COUNCIL OF CANADA. ATOMIC ENERGY PROJECT, CHALK RIVER, ONT. (REF. 29)
PR-	NAVY (PERFORMANCE REQUIREMENTS)
PR-	OFFICE OF THE ADJUTANT GENERAL, WASHINGTON, D.C.
PR-(NO.)-	OHIO STATE UNIV., COLUMBUS. CRYOGENIC LAB.
PR-	PRINCETON UNIV., N.J.
PR-(NUMBER)-M	PRINCETON UNIV., N.J. (SQUID PROJECT) (REF. 33)
PR-(NUMBER)-M-P	PRINCETON UNIV., N.J. (SQUID PROJECT) (REF. 33)

(PROPOSAL NUMBER)-(LTR)-

PR-(NUMBER)-P
PRINCETON UNIV., N.J. (SQUID PROJECT) (REF. 33)

PR-(NUMBER)-P-R
PRINCETON UNIV., N.J. (SQUID PROJECT) (REF. 33)

PR-(NUMBER)-R
PRINCETON UNIV., N.J. (SQUID PROJECT) (REF. 33)

PR-(NUMBER)-R-P
PRINCETON UNIV., N.J. (SQUID PROJECT) (REF. 33)

PR-
PROCEDYNE ASSOCIATES, INC., NEW BRUNSWICK, N.J.

PR-
QUARTERMASTER RESEARCH AND ENGINEERING COMMAND, NATICK, MASS. (PIONEERING RESEARCH)

PR.(NUMBER)
SYDNEY. UNIV. SCHOOL OF PHYSICS

PR-
TENNESSEE EASTMAN CORP., OAK RIDGE, TENN.

PR (NUMBER)-
UNION CARBIDE CORP. PARMA RESEARCH LAB., OHIO

PR-(LTR(S))-(NO.)-(LTRS)
ATOMIC ENERGY OF CANADA LTD.,CHALK RIVER,ONT. (REF.6)

PR-(LETTERS)-
ATOMIC EN. OF CANADA LTD. CHALK RIVER NUC. LABS., ONT

PR-(LETTER(S))-(LTR(S))-
NATIONAL RESEARCH COUNCIL OF CANADA. ATOMIC ENERGY PROJECT, CHALK RIVER, ONT. (REF. 29)

PRA-
ATOMIC ENERGY OF CANADA LTD. CHALK RIVER PROJ., ONT.

PRA-
NATIONAL RESEARCH COUNCIL OF CANADA, OTTAWA

PRA-
NATIONAL RESEARCH COUNCIL OF CANADA. ATOMIC ENERGY PROJECT, CHALK RIVER, ONT.

PRA-
NAVAL PERSONNEL RESEARCH ACTIVITY, WASHINGTON, D.C.

PRA-
PSYCHOLOGICAL RESEARCH ASSOCIATES, WASHINGTON, D.C.

PR-AEL-
PRINCETON UNIV., N.J. AERONAUTICAL ENGINEERING LAB.

PRAG-
PLYMOUTH ROCK AND GRANITE STATE PROJECT

PR-APPE-
ATOMIC EN. OF CANADA LTD. CHALK RIVER NUC. LABS., ONT

PRA-U(YEAR)-
PENNSYLVANIA RESEARCH ASSOCIATES INC., PHILADELPHIA

PRAW-(YEAR)-
NAVAL PERSONNEL RESEARCH ACTIVITY, WASHINGTON, D.C.

PRAW-RS-(YEAR)-
NAVAL PERSONNEL RESEARCH ACTIVITY, WASHINGTON, D.C.

PR-B-
ATOMIC ENERGY OF CANADA LTD. BIOLOGY AND HEALTH PHYSICS DIV., CHALK RIVER, ONT.

PRB-
OFFICE OF THE ADJUTANT GENERAL, WASHINGTON, D.C.

PR-C-(YR)-EL/S-
CATALYST RESEARCH CORP., BOSTON

PRC-
PLANNING RESEARCH CORP., LOS ANGELES

PRC-
PLANNING RESEARCH CORP., WASHINGTON, D.C.

PRC-
PROPULSION RESEARCH CORP., INGLEWOOD, CALIF.

PRC-
PROPULSION RESEARCH CORP., SANTA MONICA, CALIF.

PRC-D-
PLANNING RESEARCH CORP., WASHINGTON, D.C.

PRCM-
HUGHES AIRCRAFT CO., CULVER CITY, CALIF. (PASSIVE RADIATION COUNTERMEASURES)

PR-CMA-
ATOMIC EN. OF CANADA LTD. CHALK RIVER NUC. LABS., ONT

PR/CONF/(UK)-
UNITED KINGDOM ATOMIC ENERGY AUTHORITY. RESEARCH GP. ATOMIC ENERGY RES. ESTAB., HARWELL, BERKS, ENGLAND

PRC-R-
PITTSBURGH. UNIV.

PRC-R-
PLANNING RESEARCH CORP., LOS ANGELES

PRC-R-
PLANNING RESEARCH CORP., WASHINGTON, D.C.

PRC/TA-
PRC TECHNICAL APPLICATIONS, HUNTSVILLE, ALA.

PRD-(YEAR)-
DEPT. OF AGRICULTURE. PERSONNEL RES. STAFF, WASH.,DC

PRD-(6 DIGITS)-
DOW CHEMICAL CO. ROCKY FLATS DIV., GOLDEN, COLO.

PRD-
GRUMMAN OPERATIONS ANALYSIS GROUP. ADVANCED SYSTEMS SECTION, BETHPAGE, N.Y.

PRD-
POLYTECHNIC RESEARCH AND DEV. CO., INC., BROOKLYN

PRD-
PRD ELECTRONICS, INC., BROOKLYN

PRD-
PURDUE UNIV., LAFAYETTE, IND.

PRDC-
ATOMIC POWER DEVELOPMENT ASSOCIATES, INC., DETROIT

PRDC-
POLYTECHNIC RESEARCH AND DEV. CO., INC., BROOKLYN

PRDC-
POWER REACTOR DEVELOPMENT CO., DETROIT

PRDC-EF-
POWER REACTOR DEVELOPMENT CO. ENRICO FERMI ATOMIC POWER PLANT, DETROIT

PRDC-TR-
POWER REACTOR DEVELOPMENT CO., DETROIT

PRE-
NATIONAL RESEARCH COUNCIL OF CANADA. MONTREAL LAB.

PREAS-
ALLSTATES DESIGN AND DEVELOPMENT CO., INC., TRENTON

PREDC-
GENERAL ELECTRIC CO. AIRCRAFT NUCLEAR PROPULSION DEPT., CINCINNATI (REF. 3)

PRE-TRANS-
GT. BRIT. ROYAL RADAR ESTAB., MALVERN, WORCS, ENGLAND

PREXDC-
GENERAL ELECTRIC CO. AIRCRAFT NUCLEAR PROPULSION PROJECT, CINCINNATI (REF. 3)

PRF-
PURDUE RESEARCH FOUNDATION, LAFAYETTE, IND.

PRI-(NUMBER)X
HUGHES AIRCRAFT CO. SYSTEMS DEVELOPMENT LABS., CULVER CITY, CALIF.

PRI-(NO.)Y
HUGHES AIRCRAFT CO. SYSTEMS DEVELOPMENT LABS., CULVER CITY, CALIF.

PRIN-
PRINCETON UNIV., N.J.

PRIN MATT-
PRINCETON UNIV., N.J. PROJECT MATTERHORN

PRIN-U-
PRINCETON UNIV., N.J.

PRINU-IPR-
PRINCETON UNIV., N.J. (INTERIM PROGRESS REPORTS)

PRINU-QSR-
PRINCETON UNIV., N.J. (QUARTERLY STATUS REPORTS)

PRINU-TR-
PRINCETON UNIV., N.J. (TECHNICAL REPORTS)

PR-IX-
AUBURN RESEARCH FOUNDATION, ALA.

PRL-
DU PONT DE NEMOURS (E.I.) + CO. PARLIN LAB., N.J.

PRL-
PENNSYLVANIA STATE UNIV., UNIVERSITY PARK. PETROLEUM REFINING LAB.

PRL-5.(NUMBER)
PENNSYLVANIA STATE UNIV., UNIVERSITY PARK. PETROLEUM REFINING LAB.

PRL-
PITTSBURGH. UNIV. SARAH MELLON SCAIFE RADIATION LAB.

PRL-
POLAMOLD RESEARCH LABS., INC., SPRINGFIELD, OHIO

PRL-9-
SPACE TECHNOLOGY LABS., INC. PHYSICAL RESEARCH LAB., LOS ANGELES

PRL-LN-
PERSONNEL RES. LAB. (6570TH), LACKLAND AFB, TEX.

PRL-SRM-(YEAR)-
PERSONNEL RES. LAB. (6570TH), LACKLAND AFB, TEX. (STAFF RESEARCH MEMORANDUM)

PRL-TDR-
PERSONNEL RES. LAB. (6570TH), LACKLAND AFB, TEX.

PRL-TDR-(YEAR)-
PERSONNEL RES. LAB. (6570TH), LACKLAND AFB, TEX.

PRL-TM-(YEAR)-
PERSONNEL RES. LAB. (6570TH), LACKLAND AFB, TEX.

PRL-TR-
PERSONNEL RES. LAB. (6570TH), LACKLAND AFB, TEX.

PRL-TR-(YEAR)-
PERSONNEL RES. LAB. (6570TH), LACKLAND AFB, TEX.

PRM-
BUREAU OF NAVAL PERSONNEL. PERSONNEL RESEARCH DIV., WASHINGTON, D.C. (RESEARCH MEMORANUM)

PRM-
PARSONS (RALPH M.) CO., PASADENA, CALIF.

PRM-(YEAR)-
PARSONS (RALPH M.) CO., PASADENA, CALIF.

PRM-
SERENDIPITY ASSOCIATES, CHATSWORTH, CALIF.

PR/MM/(NUMBER-NUMBER)/MT
FRANCE. CENTRE NATIONAL D'ETUDES SPATIALES. DIV. MATHEMATIQUES ET TRAITEMENT, PARIS

PRNC-
BAYLOR UNIV., HOUSTON, TEX. COLL. OF MEDICINE

PRNC-
PUERTO RICO NUCLEAR CENTER, MAYAGUEZ

PRNC-
PUERTO RICO NUCLEAR CENTER, SAN JUAN

PR-NPAP-
ATOMIC ENERGY OF CANADA LTD. NUCLEAR POWER PLANT DIV. TORONTO

PR-NPDP-
ATOMIC ENERGY OF CANADA LTD. NUCLEAR POWER PLANT DIV. TORONTO

PR-NPP-
ATOMIC ENERGY OF CANADA LTD. NUCLEAR POWER PLANT DIV. TORONTO

PRNS-
PUERTO RICO NUCLEAR CENTER, MAYAGUEZ

PROC-
(PROCEEDINGS) (REF. 2)

PROCEDURE DC-
SANDIA CORP. ALO PRIMARY STANDARDS LAB., ALBUQUERQUE, N. MEX. (DIRECT CURRENT PARAMETERS, RESISTANCE)

PROFESSIONAL PAPER-
GEORGE WASHINGTON UNIV., ARLINGTON, VA. HUMAN RESOURCES RESEARCH OFFICE

PROJ-
(PROJECT) (REF. 2)

PROM.-
SWITZERLAND. EIDGENOESSISCHE TECHNISCHE HOCHSCHULE, ZURICH

PR/OP-
UNITED KINGDOM ATOMIC ENERGY AUTHORITY. WEAPONS GROUP ATOMIC WEAPONS RES. ESTAB.,ALDERMASTON, BERKS, ENG.

(PROPOSAL NUMBER)-(LTR)-
CURTISS-WRIGHT CORP. AIRPLANE DIV., COLUMBUS, OHIO (REF. 1)

PRO(TM)-	GT. BRIT. WINDSCALE WORKS, SELLAFIELD, CUMB., ENGLAND (REF. 36)
PROT-SAN.-	ITALY. COMITATO NAZIONALE PER L'ENERGIA NUCLEARE, ROME
PROVE-	PURDUE UNIV., LAFAYETTE, IND. PURDUE RESEARCH ORGANIZATION FOR VULNERABILITY EVALUATIONS
PR-P-	ATOMIC EN. OF CANADA LTD. CHALK RIVER NUC. LABS., ONT
PR-P-(NUMBER)-GP	ATOMIC EN. OF CANADA LTD. CHALK RIVER NUC. LABS., ONT
PRP-(YEAR)-	PHILCO-FORD CORP., NEWPORT BEACH, CALIF.
PRP-	WASHINGTON. UNIV., SEATTLE
PRP/GC/P-	SOCIETE BELGE POUR L'INDUSTRIE NUCLEAIRE, BRUSSELS
PR-PL-	PRINCETON UNIV., N.J. PLASTICS LAB.
PR-PP DEV-	ATOMIC ENERGY OF CANADA LTD. NUCLEAR POWER PLANT DIV. TORONTO
PR-PP ENG-	ATOMIC ENERGY OF CANADA LTD. NUCLEAR POWER PLANT DIV. TORONTO
PR-PPL-	PRINCETON UNIV., N.J. PALMER PHYSICAL LAB.
PRQ-	PHYSICS INTERNATIONAL CO., SAN LEANDRO, CALIF.
PRR-	BUREAU OF NAVAL PERSONNEL. PERSONNEL RESEARCH DIV., WASHINGTON, D.C. (RESEARCH REPORT)
PRR-	GENERAL ELECTRIC CO. MISSILE AND SPACE DIV., PHILA.
PRR-	ONTARIO RESEARCH FOUNDATION, TORONTO
PRR-	OPERATIONS RESEARCH INC., SILVER SPRING, MD.
PRR-(YEAR)-	OPERATIONS RESEARCH INC., SILVER SPRING, MD.
PR-REPORTS-	ATOMIC ENERGY OF CANADA LTD., CHALK RIVER, ONT. (REF.6)
PR-S-	ATOMIC ENERGY OF CANADA LTD., CHALK RIVER, ONT. (REF.6)
PRS-(YEAR)-	DEPT. OF AGRICULTURE. PERSONNEL RES. STAFF, WASH., DC
PRS-	OFFICE OF THE ADJUTANT GENERAL, WASHINGTON, D.C.
PRS-P-	STANFORD UNIV., CALIF. INST. OF POLITICAL STUDIES
PRU-	JOHNS HOPKINS UNIV., BALTIMORE. INST. FOR COOP. RES.
PRU-	PRINCETON UNIV., N.J.
PRU-	PUERTO RICO. UNIV., RIO PIEDRAS
PRUS-	NAVAL PERSONNEL RES. FIELD ACTIVITY, SAN DIEGO, CAL.
PRV-	ITALY. COMITATO NAZIONALE PER L'ENERGIA NUCLEARE, BOLOGNA
PRV-R-(YEAR)(NUMBER)	ITALY. COMITATO NAZIONALE PER L'ENERGIA NUCLEARE, ROME
PR-WAS-	ATOMIC ENERGY OF CANADA LTD. WHITESHELL NUCLEAR RESEARCH ESTABLISHMENT. APPLIED SCIENCE DIV., PINAWA, MANITOBA (PROGRESS REPORT)
PRWRA-	PUERTO RICO WATER RESOURCES AUTHORITY, SAN JUAN
PRWRA-GNEC-	PUERTO RICO WATER RESOURCES AUTHORITY, SAN JUAN
PS-	(PRELIMINARY STUDY) (REF. 2)
PS(4 DIGITS)FR	BRUNSWICK CORP. TECH. PRODUCTS DIV., MUSKEGON, MICH.
P + S-	COLUMBIA UNIV., N.Y.C. COLL. OF PHYSICIANS + SURGEONS
PS/(4 DIGITS)	EUROPEAN ORGANIZATION FOR NUCLEAR RESEARCH, GENEVA
P.S.-	MC DONNELL AIRCRAFT CORP., ST. LOUIS (PROCESS SPECS.)
P+S-	PASS AND SEYMOUR, INC., SYRACUSE, N.Y.
PS-	PRIMARY SOURCES, N.Y.C. (TRANSLATIONS)
PS-	RESEARCH AND DEVELOPMENT BOARD. COMMITTEE ON BASIC PHYSICAL SCIENCES, WASHINGTON, D.C.
PS-	VARO, INC., GARLAND, TEX.
PS/(INITIALS)-	PLANSEE SOCIETY FOR POWDER METALLURGY (SEMINAR REPRINTS)
PS-AP-	WESTINGHOUSE ELECTRIC CORP., PITTSBURGH
PSB-	KERNFORSCHUNGSZENTRUM KARLSRUHE, GERMANY
PSB-	KERNFORSCHUNGSZENTRUM KARLSRUHE, GERMANY
PSBB-	KERNFORSCHUNGSZENTRUM KARLSRUHE, GERMANY
PSB-NB-	KERNFORSCHUNGSZENTRUM KARLSRUHE, GERMANY
PSC-	NATIONAL INSTITUTES OF HEALTH, BETHESDA, MD.
PSC-	PENNSYLVANIA STATE UNIV., UNIVERSITY PARK
PSC- (AERE)	GT. BRIT. ATOMIC EN. RES. ESTAB., HARWELL, BERKS, ENG
PSCC-	PUBLIC SERVICE COMPANY OF COLORADO
PSC EES TP-	PENNSYLVANIA STATE COLL., STATE COLLEGE. ENGINEERING EXPERIMENT STATION (TECHNICAL PAPER)
PSCM-	PULLMAN-STANDARD CAR MFG. CO., HAMMOND, IND.
PSC PRL-	PENNSYLVANIA STATE COLL., STATE COLLEGE. PETROLEUM REFINING LAB.
PSC SMI MR-	PENNSYLVANIA STATE COLL., STATE COLLEGE. SCHOOL OF MINERAL INDUSTRIES (MEMORANDUM REPORTS)
PSC SMI TR-	PENNSYLVANIA STATE COLL., STATE COLLEGE. SCHOOL OF MINERAL INDUSTRIES
PSDC-TR-	PROTECTIVE STRUCTURES DEV. CENTER, FT. BELVOIR, VA.
PSE-	PIONEER SERVICE AND ENGINEERING CO., CHICAGO
PSEA-	PHYSICAL SECURITY EQUIPMENT AGENCY, WASHINGTON, D.C.
PSER-(YEAR)-	GRUMMAN AIRCRAFT ENGINEERING CORP., BETHPAGE, N.Y.
PSI-	PACIFIC SEMICONDUCTORS, INC., CULVER CITY, CALIF.
PSI-(NUMBERS)	PACIFIC SEMICONDUCTORS, INC., CULVER CITY, CALIF.
PSI-(YEAR)-	PHYSICAL STUDIES, INC., DAYTON, OHIO
PSI-	PHYSICAL STUDIES, INC., RENO, NEV.
PSI-	PLANETARY SCIENCES INC., SANTA CLARA, CALIF.
PS/INT.-AR(YEAR)-	EUROPEAN ORGANIZATION FOR NUCLEAR RESEARCH, GENEVA
PSL-	NORTHEASTERN UNIV., BOSTON. PHOTOCHEMISTRY AND SPECTROSCOPY LAB.
PSL/NMSU-	NEW MEXICO STATE UNIV., UNIVERSITY PARK. PHYSICAL SCIENCE LAB.
PSL-PM(NUMBER)	NEW MEXICO STATE UNIV., UNIVERSITY PARK. PHYSICAL SCIENCE LAB.
PSL-PQ(NUMBER)	NEW MEXICO STATE UNIV., UNIVERSITY PARK. PHYSICAL SCIENCE LAB.
PSL-PR(NUMBER)	NEW MEXICO STATE UNIV., UNIVERSITY PARK. PHYSICAL SCIENCE LAB.
PSL-PS(NUMBER)	NEW MEXICO STATE UNIV., UNIVERSITY PARK. PHYSICAL SCIENCE LAB.
PSL-PT(NUMBER)	NEW MEXICO STATE UNIV., UNIVERSITY PARK. PHYSICAL SCIENCE LAB.
PSNS-	PUGET SOUND NAVAL SHIPYARD. MATERIAL LABS., BREMERTON, WASH.
PSR-	(PERIODIC STATUS REPORT) (REF. 2)
PSR-	(PROGRESS SUMMARY REPORT) (REF. 2)
PSR-(YEAR)-	GRUMMAN AIRCRAFT ENGINEERING CORP. PRELIMINARY SYSTEMS SECTION, BETHPAGE, N.Y.
PSR-	PENNSYLVANIA STATE UNIV., UNIVERSITY PARK. DEPT. OF NUCLEAR ENGINEERING
PSRB-PNW-	FOREST SERVICE. PACIFIC NORTHWEST STA., PORTLAND ORE.
PSRP-	(PHYSICAL SCIENCE RESEARCH PAPER) (REF. 2)
PSRP-	AIR FORCE CAMBRIDGE RESEARCH LABS. MICROWAVE PHYSICS LAB., BEDFORD, MASS. (PHYSICAL SCIENCE RES. PAPERS)
PSRP-PNW-	FOREST SERVICE. PACIFIC NORTHWEST STA., PORTLAND ORE.
PSR-PROJECT-	NORTH CAROLINA STATE UNIV., RALEIGH. PHYSICAL SCIENCE RESEARCH
PST-CAT-	NATIONAL SCIENCE FOUNDATION, WASHINGTON, D.C. (PROGRAM FOR SCIENCE TRANSLATION)
PSTR-(NUMBER)-	PENNSYLVANIA STATE UNIV., UNIVERSITY PARK
PSU-	PENNSYLVANIA STATE UNIV., UNIVERSITY PARK
PSU-(NO.)-P	PENNSYLVANIA STATE UNIV., UNIVERSITY PARK
PSUC-	PUGET SOUND UTILITIES COUNCIL
PSU-GI-	PENNSYLVANIA STATE UNIV., UNIVERSITY PARK. GROTH INST.
PSU-IRL-SCI-	PENNSYLVANIA STATE UNIV., UNIVERSITY PARK. IONOSPHERE RESEARCH LAB.
PSU-IR-SR-	PENNSYLVANIA STATE UNIV., UNIVERSITY PARK. IONOSPHERE RESEARCH LAB. (SCIENTIFIC REPORTS)
PSU-NUC-E-	PENNSYLVANIA STATE UNIV., UNIVERSITY PARK
PSU-PR-	PENNSYLVANIA STATE UNIV., UNIVERSITY PARK
PSU-PSR-	PENNSYLVANIA STATE UNIV., UNIVERSITY PARK
PSU-TR-	PENNSYLVANIA STATE UNIV., UNIVERSITY PARK

PSW-	FOREST SERVICE. PACIFIC SOUTHWEST STA., BERKELEY, CAL
PSW-	PACIFIC SOUTHWEST FOREST AND RANGE EXPERIMENT STATION, BERKELEY, CALIF.
PSYC-	PSYCHOLOGICAL CORP., N.Y.C.
PT-	GARRETT CORP. AIRESEARCH MFG. DIV., LOS ANGELES
PT-	HANFORD WORKS, RICHLAND, WASH.
PT-	LOS ALAMOS SCIENTIFIC LAB., N. MEX.
PT-	NATIONAL DEFENSE RESEARCH COMMITTEE
PT(NUMBERS)	NEW MEXICO STATE UNIV., UNIVERSITY PARK
PT-	OFFICE OF SCIENTIFIC RESEARCH AND DEVELOPMENT (PREPARATION AND TESTING OF EXPLOSIVES)
PT-	PRECISION TECHNOLOGY, INC., LIVERMORE, CALIF.
PT-	RAYTHEON CO., BURLINGTON, MASS.
PT-	RAYTHEON CO. MICROWAVE AND POWER TUBE DIV., WALTHAM, MASS.
PT-	STD. OIL + GAS CO. EXPLOSIVES RES. LAB., TULSA, OKLA.
PTB-	BORDEAUX UNIV. LABORATOIRE DE PHYSIQUE THEORIQUE
PTB-	BUREAU OF NAVAL PERSONNEL. PERSONNEL RESEARCH DIV., WASHINGTON, D.C. (TECHNICAL BULLETIN)
PTB-	GERMANY. PHYSIKALISCH-TECHNISCHE BUNDESANSTALT, BRUNSWICK
PTB-(YEAR)-	HRB-SINGER, INC., STATE COLLEGE, PENNA.
PTB-	PERSONNEL RESEARCH CENTER, INC., DETROIT, MICH.
PTB-JB-(NUMBER)/(YEAR)	GERMANY. PHYSIKALISCH-TECHNISCHE BUNDESANSTALT, BRUNSWICK
PTC/P-	UNITED KINGDOM ATOMIC ENERGY AUTHORITY. RESEARCH GP. ATOMIC ENERGY RES. ESTAB., HARWELL, BERKS, ENGLAND
PTDR-	AEROJET-GENERAL CORP. LIQUID ROCKET OPERATIONS, SACRAMENTO, CALIF.
PTEP-	DEPARTMENT OF THE INTERIOR, WASHINGTON, D.C. (POPULATION TRENDS AND ENVIRONMENTAL POLICY)
PTHE-(YEAR/NUMBER)	FRANCE. LABORATOIRE DE PHYSIQUE THEORIQUE ET DES HAUTES ENERGIES, ORSAY
PT-IFT-	TURIN. POLITECNICO. ISTITUTO DI FISICA TECNICA
PTM-	FRANCE. FACULTE DES SCIENCES DE MARSEILLE
PTM-	OFFICE OF SCIENTIFIC RESEARCH AND DEVELOPMENT
PTN-	CALIFORNIA. UNIV., LIVERMORE. LAWRENCE RADIATION LAB. (PLUTO TECHNICAL NOTE)
PTN-	GT. BRIT. CHEMICAL DEFENCE EXPERIMENTAL ESTABLISHMENT, PORTON, WILTS, ENGLAND
PTN-	GT. BRIT. MINISTRY OF SUPPLY
PTN-TU-	GT. BRIT. CHEMICAL DEFENCE EXPERIMENTAL ESTABLISHMENT, PORTON, WILTS, ENGLAND
PTP-	GT. BRIT. CHEMICAL DEFENCE EXPERIMENTAL ESTABLISHMENT, PORTON, WILTS, ENGLAND
PTP-	GT. BRIT. CHEMICAL DEFENCE EXPERIMENTAL ESTABLISHMENT, SALISBURY, WILTS, ENGLAND
PTPR-	NAVAL ORDNANCE TEST STATION, INYOKERN (CHINA LAKE), CALIF. (PRELIMINARY TECHNICAL PROGRESS REPORT)
PTR-	(PRELIMINARY TECHNICAL REPORT) (REF. 2)
PTR-	BERLIN. PHYSIKALISCH-TECHNISCHE REICHSANSTALT
PTR-(NUMBER-NUMBER)	FARRAND OPTICAL CO., INC., N.Y.C.
PTR-(NUMBER)-	GENERAL ELECTRIC CO. MISSILE AND SPACE DIV., PHILA.
PTR-	NAVAL AIR TEST CENTER, PATUXENT RIVER, MD.
PTR-	PHILLIPS PETROLEUM CO. ATOMIC EN. DIV., IDAHO FALLS
PTR-	RADIO CORP. OF AMERICA. RCA LABS. DIV., PRINCETON, NJ
PTRC-	A.F. PERSONNEL + TRAINING RES. CTR.,LACKLAND AFB,TEX.
PU-	NATIONAL RESEARCH COUNCIL OF CANADA. MONTREAL LAB.
PU-	PENNSYLVANIA. UNIV., PHILADELPHIA
PU-	PRINCETON UNIV., N.J.
PU-(NO.)/TR-	PRINCETON UNIV., N.J.
PU-	PURDUE UNIV., LAFAYETTE, IND.
PU-	STANFORD RESEARCH INST., MENLO PARK, CALIF.
P.U.A.E.C.-	PRINCETON UNIV., N.J. DEPT. OF AERONAUTICAL ENG.
PU-AEL-	PRINCETON UNIV., N.J. DEPT. OF AERONAUTICAL ENG.
PU AEL R-	PRINCETON UNIV., N.J. AERONAUTICAL ENGINEERING LAB.
PU-AER-	PRINCETON UNIV., N.J. DEPT. OF AERONAUTICAL ENG.
PU-AMS-	PRINCETON UNIV., N.J. DEPT. OF AEROSPACE AND MECHANICAL SCIENCES
PUB-	(PUBLICATION) (REF. 2)
PUB-	BENDIX CORP. PIONEER-CENTRAL DIV., DAVENPORT, IOWA
PUB-	CALIFORNIA. UNIV., LOS ANGELES. INST. OF GEOPHYSICS
PUB-	CALIFORNIA INST. OF TECH., PASADENA
PUB-	HUGHES AIRCRAFT CO. AEROSPACE GROUP, CULVER CITY,CAL.
PUB-	LITTON SYSTEMS, INC. GUIDANCE AND CONTROLS SYSTEMS DIV., WOODLAND HILLS, CALIF.
PUB-(YEAR)-	MASS. INST. OF TECH., CAMBRIDGE. FLUID MECHANICS LAB.
PUB-(YEAR)-	PENNSYLVANIA STATE UNIV., UNIVERSITY PARK. DEPT. OF GEOLOGY AND GEOPHYSICS
PUB-(LETTER)-	ANACONDA WIRE AND CABLE CO., HASTINGS-ON-HUDSON,N.Y.
PUB-CAT-(YEAR)/E	INTERNATIONAL ATOMIC ENERGY AGENCY, VIENNA
PUBL-	(PUBLICATION) (REF. 2)
PUBL.-	(PUBLICATION) (REF. 2)
PUBL.-	ARINC RESEARCH CORP., WASHINGTON, D.C.
PUBL.-	BENDIX AVIATION CORP. PIONEER-CENTRAL DIV., DAVENPORT, IOWA
PUBL-(YEAR)-	COASTAL PLAINS CENTER FOR MARINE DEVELOPMENT SERVICES, WASHINGTON, D.C.
PUBLICATION 3100-P-	SPERRY FARRAGUT CO. ENG. DEPT., BRISTOL, TENN. (PROPOSAL)
PUBLICATION 3100-R-	SPERRY FARRAGUT CO. ENG. DEPT., BRISTOL, TENN.
PUBLICATION SFCO-B-	SPERRY FARRAGUT CO. ENG. DEPT., BRISTOL, TENN. (BROCHURE)
PUBLICATION SFCO-R-	SPERRY FARRAGUT CO. ENG. DEPT., BRISTOL, TENN.
PUB/UP/(NO.)-	ISRAEL. ATOMIC ENERGY COMMISSION, TEL-AVIV
PUB/UP/R-	ISRAEL. ATOMIC ENERGY COMMISSION, TEL-AVIV
PUC-(NUMBERS)-	OAK RIDGE NATIONAL LAB., TENN.
PUC-	PRINCETON UNIV., N.J.
PUC-(YR.)-(NO.)-	PRINCETON UNIV., N.J. PALMER PHYSICAL LAB.
PUC-MONOGRAFIA-	PONTIFICIA UNIVERSIDADE CATOLICA, RIO DE JANEIRO
PUC/P-	UNITED KINGDOM ATOMIC ENERGY AUTHORITY. RESEARCH GP. ATOMIC ENERGY RES. ESTAB., HARWELL, BERKS, ENGLAND
PUC/RRL-	PACIFIC UNION COLL. ANGWIN, CALIF. RADIATION RES.LAB.
PUF-	MICHIGAN. UNIV., ANN ARBOR. HIGHWAY SAFETY RES. INST.
PU FCR FR-	PRINCETON UNIV., N.J. (FIRE CONTROL RES.,FINAL RPTS.)
PU FCR RR-	PRINCETON UNIV., N.J. (FIRE CONTROL RES., RES. RPTS.)
PU-FLD-	PRINCETON UNIV., N.J. DEPT. OF AEROSPACE AND MECHANICAL SCIENCES
PU PL TR-	PRINCETON UNIV., N.J. PLASTICS LAB.
PUR-	PURDUE RESEARCH FOUNDATION, LAFAYETTE, IND.
PUR-	PURDUE UNIV., LAFAYETTE, IND.
PUR-(NUMBER)-(LETTER(S))	PURDUE UNIV., LAFAYETTE, IND. (SQUID PROJ.) (REF. 33)
PURC-	PRINCETON UNIV., N.J. PALMER PHYSICAL LAB.
PURD (LTR.)-	PURDUE UNIV., LAFAYETTE, IND.
PURD-U-	PURDUE UNIV., LAFAYETTE, IND.
PURDUE-JPC-	PURDUE UNIV., LAFAYETTE, IND. JET PROPULSION CENTER
PURDUE U.-	PURDUE UNIV., LAFAYETTE, IND.
PURU-	PURDUE UNIV., LAFAYETTE, IND.
PU-TR-	PRINCETON UNIV., N.J.
PV-	DIRECTORATE OF INTELLIGENCE (AIR FORCE), WASH., D.C.
PV-	PHYSICAL VULNERABILITY DIV. (AIR FORCE), WASH., D.C.
PVA-	AIR FORCE (PHYSICAL VULNERABILITY ABSTRACTS)

PVIM-	AIR FORCE (PHYSICAL VULNERABILITY INTERIM MEMO.)
PVIM-	OFFICE OF NAVAL INTELLIGENCE. PHYSICAL VULNERABILITY DIV., WASHINGTON, D.C.
PVIM-	PHYSICAL VULNERABILITY DIV. (AIR FORCE), WASH., D.C. (INTERIM MEMORANDUM)
PVM-	TENNESSEE EASTMAN CORP., OAK RIDGE, TENN.
PVTM-	AIR FORCE (PHYSICAL VULNERABILITY TECHNICAL MEMO.)
PVTM-	PHYSICAL VULNERABILITY DIV. (AIR FORCE), WASH., D.C. (TECHNICAL MEMORANDUM)
PW-	PRATT AND WHITNEY AIRCRAFT DIV., UNITED AIRCRAFT CORP., HARTFORD, CONN.
PWA-	PRATT AND WHITNEY AIRCRAFT, EAST HARTFORD, CONN.
PWA-	PRATT AND WHITNEY AIRCRAFT DIV.,UNITED AIRCRAFT CORP. (ALL LOCATIONS)
PWAC-	PRATT AND WHITNEY AIRCRAFT DIV.,UNITED AIRCRAFT CORP. (ALL LOCATIONS)
PWA-FP-(YEAR)-	PRATT AND WHITNEY AIRCRAFT. FLORIDA RESEARCH AND DEVELOPMENT CENTER, WEST PALM BEACH
PWA-FR-	ARMY WEAPONS COMMAND, ROCK ISLAND, ILL.
PWA-FR-	PRATT AND WHITNEY AIRCRAFT DIV., UNITED AIRCRAFT CORP FLORIDA RESEARCH AND DEV. CENTER, WEST PALM BEACH
PWAR-	PRATT AND WHITNEY AIRCRAFT DIV., UNITED AIRCRAFT CORP. CONN. AIRCRAFT NUCLEAR ENGINE LAB.,MIDDLETOWN
PWB S-	GT. BRIT. MINISTRY OF WORKS (POSTWAR BLDG. STUDIES)
PWL-	PRATT AND WHITNEY AIRCRAFT DIV., UNITED AIRCRAFT CORP., HARTFORD, CONN.
PWO-	NAVAL ORDNANCE LAB., WHITE OAK,MD.(PROJECT WHITE OAK)
PWP(P)-	GT. BRIT. ATOMIC EN. RES. ESTAB., HARWELL, BERKS, ENG
PWT-	ARNOLD ENGINEERING DEVELOPMENT CENTER. PROPULSION WIND TUNNEL FACILITY, ARNOLD AIR FORCE STA., TENN.
PX-(NO.)-	AIR FORCE, WASHINGTON, D.C.
PX-	ENGINEERING RESEARCH ASSOCIATES, INC.,WASHINGTON,D.C.
PX-	MASON + HANGER-SILAS MASON CO., INC. PANTEX PLANT, AMARILLO, TEX.
PX-	NEW MEXICO STATE UNIV., UNIVERSITY PARK. PHYSICAL SCIENCE LAB.
PX-	SPERRY RAND CORP., ST. PAUL
PX-	SPERRY RAND CORP. UNIVAC DIV., PHILADELPHIA
PX-	SPERRY RAND CORP. UNIVAC DIV., ST. PAUL, MINN.
PX-	SPERRY RAND CORP. UNIVAC FEDERAL SYSTEMS DIV., SALT LAKE CITY
PYRO-	PYROMET CO., SAN CARLOS, CALIF.
PZ-	NATIONAL RESEARCH COUNCIL OF CANADA, OTTAWA

Q-	(QUARTERLY) (REF. 2)
Q-	BUR. OF MINES. METALLURGICAL DIV., BOULDER CITY, NEV
Q(NUMBER)-	DELAWARE. UNIV., NEWARK. DEPT. OF ELECTRICAL ENG.
Q-	DOW CHEMICAL CO. ROCKY FLATS DIV., DENVER
Q-	FRANKLIN INST., PHILADELPHIA
Q-	FRANKLIN INST. LABS. FOR RES. + DEV., PHILADELPHIA
Q-	GT. BRIT. ORDNANCE BOARD
Q-	NEW YORK. STATE UNIV. COLL. OF CERAMICS, ALFRED
Q-	SANDIA CORP., ALBUQUERQUE, N. MEX.
Q-	SYLVANIA ELECTRONIC SYSTEMS. WALTHAM LABS., MASS.
Q(NUMBER)-	SYLVANIA ELECTRONIC SYSTEMS-EAST, NEEDHAM, MASS.
Q(YEAR-(NUMBER)(LETTER)	SYLVANIA ELECTRONIC SYSTEMS-EAST, NEEDHAM, MASS.
Q-	SYLVANIA ELECTRONIC SYSTEMS-EAST, WALTHAM, MASS.
Q-	SYLVANIA ELECTRONIC SYSTEMS-EAST. APPLIED RESEARCH LABS., WALTHAM, MASS.
QAIL-	SANDIA CORP., ALBUQUERQUE, N. MEX.(QUALITY ASSURANCE INFORMATION LETTER)
QAO-	NAVAL WEAPONS QUALITY ASSURANCE OFFICE, WASH., D.C.
QAR-	AEROJET-GENERAL CORP., SACRAMENTO, CALIF.
QAS/NPP-	NAVAL PROPELLANT PLANT. QUALITY ASSURANCE DEPT., INDIAN HEAD, MD.
Q-B(NO.)-	FRANKLIN INST., PHILADELPHIA
QB-(NUMBER)-(YEAR)	GESELLSCHAFT FUER WELTRAUMFORSCHUNG M.B.H., BAD GODESBERG
QC-	(QUALITY CONTROL) (REF. 2)
QC-	AMF ATOMICS (CANADA) LTD., PORT HOPE, ONT.
QC-	BUREAU OF AERONAUTICS (NAVY) (QUALITY CONTROL)
Q-C(4 DIGITS)-	FRANKLIN INST. RESEARCH LABS., PHILADELPHIA
QC/C-	NAVAL AMMUNITION DEPOT. QUALITY CONTROL LAB., CRANE, IND.
QC/PC-	NAVAL MAGAZINE. QUALITY CONTROL LAB., PORT CHICAGO, CALIF.
QCR-	(QUALITY CONTROL REPORT) (REF. 2)
QCR-	OFFICE OF PRODUCTION RESEARCH AND DEVELOPMENT (QUALITY CONTROL REPORTS, CARNEGIE INST. OF TECH.)
QC/SB-	NAVAL AMMUNITION AND NET DEPOT. QUALITY CONTROL LAB., SEAL BEACH, CALIF.
QCSL-	QUALITY CONTROL SURVEILLANCE LAB.
QDRI-	(QUARTERLY DEVELOPMENT RESEARCH INVESTIGATION)(REF.2)
QDRI-	ARMY, WASHINGTON, D.C. (QUALITATIVE DEVELOPMENT REQUIREMENTS INFORMATION)
QE/B-	NAVAL AMMUNITION DEPOT. QUALITY EVALUATION LAB., BANGOR, WASH.
QE/C-	NAVAL AMMUNITION DEPOT. QUALITY EVALUATION LAB., CRANE, IND.
QE/CO-	NAVAL AMMUNITION DEPOT. QUALITY EVALUATION LAB., CONCORD, CALIF.
QE/CO-(YR.)-	NAVAL AMMUNITION DEPOT. QUALITY EVALUATION LAB., CONCORD, CALIF.
QE/CONCORD-	NAVAL WEAPONS STATION. QUALITY EVALUATION LAB., CONCORD, CALIF.
QE/CO-TEST PROCEDURE-	NAVAL WEAPONS STATION. QUALITY EVALUATION LAB., CONCORD, CALIF.
QE/K-	NAVAL TORPEDO STATION. QUALITY EVALUATION LAB., KEYPORT, WASH.
QEM-(NUMBER-NUMBER-NO.)	AMMUNITION PROCUREMENT AND SUPPLY AGENCY. QUALITY EVALUATION DIV., JOLIET, ILL.
QE/N-(YEAR)-(LETTERS)-	NAVAL UNDERWATER ORDNANCE STATION, NEWPORT, R.I.
QE/N-	NAVAL UNDERWATER ORDNANCE STATION. QUALITY EVALUATION LAB., NEWPORT, R.I.
QE/NPF-	NAVAL POWDER FACTORY. QUALITY EVALUATION LAB., INDIAN HEAD, MD.
QE-OH-	NAVAL AMMUNITION DEPOT. QUALITY EVALUATION LAB., OAHU, HAWAII
QE/OH-	NAVAL AMMUNITION DEPOT. QUALITY EVALUATION LAB., OAHU, HAWAII
QEP-	SANDIA CORP., ALBUQUERQUE, N. MEX.(QUALITY ENG. PROJ)
QE/PC-	NAVAL MAGAZINE. QUALITY EVALUATION LAB., PORT CHICAGO, CALIF.
QE/SB-	NAVAL AMMUNITION AND NET DEPOT. QUALITY EVALUATION LAB., SEAL BEACH, CALIF.
QE/SJ-	NAVAL AMMUNITION DEPOT. QUALITY EVALUATION LAB., PORTSMOUTH (ST. JULIENS CREEK), VA.
QETR-	NAVAL AMMUNITION DEPOT, CRANE, IND.
QE/Y-	NAVAL MINE DEPOT. QUALITY EVALUATION LAB., YORKTOWN, VA.
QF+CI-	QUARTERMASTER FOOD AND CONTAINER INSTITUTE FOR THE ARMED FORCES, CHICAGO
QF/NPP-	NAVAL PROPELLANT PLANT, INDIAN HEAD, MD.
QI-	(QUARTERLY INDEX) (REF. 2)
QITR-	CONDUCTRON CORP., ANN ARBOR, MICH.
QLR-	AERONUTRONIC DIV. OF PHILCO CORP., NEWPORT BEACH, CALIF. (QUARTERLY LETTER REPORT)
QM-	(QUALITY MEMORANDUM) (REF. 2)
QM-	(QUARTERLY MEMORANDUM) (REF. 2)
QM-	CORNELL AERONAUTICAL LAB., INC., BUFFALO
QMAS-	ARMY TECHNICAL INTELLIGENCE CENTER. QUARTERMASTER ANALYSIS SECTION, TOKYO
QMAS-	QUARTERMASTER ANALYSIS SECTION (AIR FORCE)
QMB-	QUARTERMASTER BOARD, CAMP LEE, VA.
QMB PROJ FR-	QUARTERMASTER BD., CAMP LEE, VA. (FIELD SURVEY PROJ.)
QMB PROJ M-	QUARTERMASTER BOARD, CAMP LEE, VA. (MISC. PROJ.)
QMB PROJ S-	QUARTERMASTER BOARD, CAMP LEE, VA. (SURVEY PROJ.)
QMB PROJ T-	QUARTERMASTER BOARD, CAMP LEE, VA. (TEST PROJ.)
QMB-S-	QUARTERMASTER BOARD, CAMP LEE, VA. (SURVEY)
QMB-T-	QUARTERMASTER BOARD, CAMP LEE, VA. (TEST REPORT)
QMBT-	QUARTERMASTER BOARD, CAMP LEE, VA. (TEST)
QMC-	CHEMICAL AND RADIOLOGICAL LABS., ARMY CHEM. CTR., MD.
QMC-	OFFICE OF THE QUARTERMASTER GENERAL
QMC-	QUARTERMASTER CORPS
QMC 17-	QUARTERMASTER CORPS (LEATHER SERIES REPORT)
QMCAL-	QUARTERMASTER FOOD AND CONTAINER INSTITUTE FOR THE ARMED FORCES, CHICAGO
QMC EPS-	OFFICE OF THE QUARTERMASTER GENL. MIL. PLANNING DIV. RES. + DEV. BR. (ENVIRONMENTAL PROTECTION SERIES)
QMC EPS MR-	OFFICE OF THE QUARTERMASTER GENERAL. MILITARY PLANNING DIV. RESEARCH AND DEVELOPMENT BRANCH (ENVIRONMENTAL PROTECTION SERIES, MEMO REPORT)
QMC FCI-	QUARTERMASTER FOOD AND CONTAINER INSTITUTE FOR THE ARMED FORCES, CHICAGO
QMC FCI OS-	QUARTERMASTER FOOD AND CONTAINER INSTITUTE FOR THE ARMED FORCES, CHICAGO (OPERATION STUDIES)
QMC FCI RDB-	QUARTERMASTER FOOD AND CONTAINER INSTITUTE FOR THE ARMED FORCES, CHICAGO (RESEARCH AND DEV. BR.)
QMC HS-	QUARTERMASTER CORPS (HISTORICAL SURVEYS)
QMC LSR-	QUARTERMASTER CORPS (LEATHER SERIES REPORT)
QMC M-	QUARTERMASTER CORPS (MANUAL)
QMC MIT-	QUARTERMASTER CORPS (MASS. INST. OF TECH. PROJECTS)
QMC MSR-	QUARTERMASTER CORPS (MICROBIOLOGICAL SERIES REPORTS)
QMC PROJ R-	QUARTERMASTER CORPS
QMC PROJ S-	QUARTERMASTER CORPS
QMCR PO-	QUARTERMASTER CORPS (RESEARCH, PURCHASE ORDER)
QMC RR BS-	QUARTERMASTER CORPS (RES. RPT., BIOCHEMICAL SERIES)
QMC RR ES-	QUARTERMASTER CORPS (RES. RPT., ENTOMOLOGY SERIES)
QMC RSTR C+P-	QUARTERMASTER CORPS. CHEMICALS AND PLASTICS SECTION. (RESEARCH SERVICE TEST REPORT)
QMC SPEC-	QUARTERMASTER CORPS (SPECIFICATION)

QMC TL BS-	QUARTERMASTER CORPS. TECHNICAL LIBRARY (BIBLIOGRAPHIC SERIES)
QMC TRM-	QUARTERMASTER CORPS (TECHNICAL RECONNAISSANCE MEMO.)
QMC TRR-	QUARTERMASTER CORPS (TENT RESEARCH REPORTS)
QMC TSR-	QUARTERMASTER CORPS (TEXTILE SERIES REPORTS)
QMD JEFFERSONVILLE-	QUARTERMASTER DEPOT, JEFFERSONVILLE, IND.
QMD PHILA-	QUARTERMASTER DEPOT, PHILADELPHIA
QMFCI-	QUARTERMASTER FOOD AND CONTAINER INSTITUTE FOR THE ARMED FORCES, CHICAGO
QMFCIAF-	QUARTERMASTER FOOD AND CONTAINER INSTITUTE FOR THE ARMED FORCES, CHICAGO
QMR+E-	QUARTERMASTER RESEARCH + ENG. COMMAND, NATICK, MASS.
QP-	MERCK SHARP AND DOHME RESEARCH LABS., RAHWAY, N.J.
QPR-	(QUARTERLY PROGRESS REPORT) (REF. 2)
QPR-	FOOD + DRUG ADM. DIV. OF MICROBIOLOGY, CINCINNATI
QPR-	OHIO STATE UNIV., COLUMBUS. CRYOGENIC LAB.
QR-	(QUARTERLY REPORT) (REF. 2)
QR-(NUMBER)-PT-	BERLIN. FREIE UNIVERSITAET. INST. FUER METEOROLOGIE UND GEOPHYSIK
QR-(1 DIGIT)	BERLIN. FREIE UNIVERSITAET. INST. FUER METEOROLOGIE UND GEOPHYSIK
QR-(NUMBER)-	CLEVITE CORP. MECHANICAL RESEARCH DIV., CLEVELAND
QR-	GENERAL ELECTRIC CO. MISSILE AND SPACE VEHICLE DEPT., PHILADELPHIA
QR-	NYTRONICS, INC. CAPACITOR DIV., DARLINGTON, S.C.
QR-	ROCKET POWER, INC. RESEARCH LABS., PASADENA, CALIF.
QR-	SANDIA CORP., ALBUQUERQUE, N. MEX.
QRC-(NO.)-(NO.)(MOD)...	GENERAL ELECTRIC CO. LIGHT MILITARY ELECTRONICS DEPT., UTICA, N.Y.
QRC-	SYLVANIA ELECTRONIC SYSTEMS-WEST, MOUNTAIN VIEW, CAL.
QRC-	TACTICAL AIR WARFARE CENTER, EGLIN AFB, FLA.
QRDC-	QUARTERMASTER RESEARCH + DEV. CTR., NATICK, MASS.
QRDC B-	QUARTERMASTER RES. + DEV. COMMAND, NATICK, MASS.
QRDC-EP-	QUARTERMASTER RESEARCH AND DEVELOPMENT CENTER. ENVIRONMENTAL PROTECTION BRANCH, NATICK, MASS.
QRDC T-	QUARTERMASTER RES. + DEV. COMMAND, NATICK, MASS.
QRDC TSR-	QUARTERMASTER RES. + DEV. COMMAND, NATICK, MASS.
QRDL-	QUARTERMASTER RESEARCH AND DEV. LABS., PHILADELPHIA
QREC-	QUARTERMASTER RESEARCH + ENG. COMMAND, NATICK, MASS.
QR+EC EP-	QUARTERMASTER RESEARCH + ENG. COMMAND, NATICK, MASS.
QREC-EP-	QUARTERMASTER RESEARCH AND ENGINEERING COMMAND. ENVIRONMENTAL PROTECTION BRANCH, NATICK, MASS.
QREC-RER-	QUARTERMASTER RESEARCH + ENG. COMMAND, NATICK, MASS.
QREC/SWPBR-	QUARTERMASTER RESEARCH + ENG. COMMAND, NATICK, MASS.
QREC-T-	QUARTERMASTER RESEARCH + ENG. COMMAND, NATICK, MASS.
QREC/TDL-	QUARTERMASTER RESEARCH + ENG. COMMAND, NATICK, MASS.
QREL-EBR-	QUARTERMASTER RESEARCH AND ENGINEERING COMMAND. CHEMICALS AND PLASTICS DIV., NATICK, MASS. (ELASTOMER BRANCH REPORT)
QRM-	NYTRONICS, INC. CAPACITOR DIV., DARLINGTON, S.C.
QR-STP-(NUMBER/NUMBER)	VIENNA. TECHNISCHE HOCHSCHULE
QSA-H-	ARMY MISSILE COMMAND, REDSTONE ARSENAL, ALA.
QSA-N-	ARMY MISSILE COMMAND, REDSTONE ARSENAL, ALA.
QS-QR-(YEAR)-	ARMY MISSILE COMMAND. QUALITY AND RELIABILITY MANAGEMENT OFFICE, REDSTONE ARSENAL, ALA.
QSR-	(QUARTERLY STATUS REPORT) (REF. 2)
QSR-	(QUARTERLY SUMMARY REPORT) (REF. 2)
QS-TSR-(YEAR)-	ARMY MISSILE COMMAND. QUALITY AND RELIABILITY MANAGEMENT OFFICE, REDSTONE ARSENAL, ALA.
QS-TSR-(YEAR)-	ARMY MISSILE COMMAND. SYSTEMS PERFORMANCE ANALYSIS DIV., REDSTONE ARSENAL, ALA.
QTN-	DENVER. UNIV. DENVER RESEARCH INST.
QTPR-	ION PHYSICS CORP., BURLINGTON, MASS.
QTR-	(QUARTERLY TECHNICAL REPORT) (REF. 2)
QTSR-	BERLIN. FREIE UNIVERSITAET
QTSR-	FULMER RES. INST., LTD., STOKE POGES, BUCKS, ENGLAND
QTSR-	SWEDEN. KUNGLIGA TEKNISKA HOEGSKOLAN, STOCKHOLM
QU-	QUEENS UNIV., BELFAST
QU-LNS-(YEAR)-	OSAKA UNIV. LAB. OF NUCLEAR STUDIES

R-	(FORMAL REPORT) (REF. 2)
R-	(INTERNAL REPORT) (REF. 2)
R-	(REGULATION) (REF. 2)
R-	(REPORT) (REF. 2)
R-	(REPRINT) (REF. 2)
R-	(RESEARCH) (REF. 2)
R-	(RESEARCH REPORT) (REF. 2)
R-	(TECHNICAL REPORT) (REF. 2)
R-(4 DIGITS)-(NO.,LTR)-	AEROJET-GENERAL CORP., SACRAMENTO, CALIF. (4 DIGITS ARE WORK ORDER OR CONTRACT CODE NUMBERS. LETTER MAY SHOW FREQUENCY)
R-	AEROJET-GENERAL CORP. ORDNANCE DIV., DOWNEY, CALIF.
R-	AIR FORCE FLIGHT TEST CENTER, EDWARDS AFB, CALIF.
R(NO.)-	AIRTRON, INC., MORRIS PLAINS, N.J.
R-	AKTIEBOLAGET ATOMENERGI, STOCKHOLM
R-(YEAR)-	ARCON CORP., WAKEFIELD, MASS.
R-(MONTH)-(YEAR)	ARMY COASTAL ENGINEERING RES. CENTER, WASHINGTON, DC
2R-	ARMY ROCKET AND GUIDED MISSILE AGENCY, REDSTONE ARSENAL, ALA.
3R-	ARMY ROCKET AND GUIDED MISSILE AGENCY, REDSTONE ARSENAL, ALA.
R-	BAUSCH AND LOMB OPTICAL CO., ROCHESTER, N.Y.
R-	BELGIUM. CENTRE D'ETUDE DE L'ENERGIE NUCLEAIRE, BRUSSELS
R-	BELGIUM. CENTRE D'ETUDE DE L'ENERGIE NUCLEAIRE, MOL
R-12-	BORG-WARNER CORP. RESEARCH CENTER, DES PLAINES, ILL.
R-(NO.)-(NO.)-	BOWLES ENGINEERING CORP., SILVER SPRING, MD.
R-	BRITISH ALUMINIUM CO., LTD., LONDON
R-(NO.)-(YR)	BRITISH ALUMINIUM CO., LTD., LONDON
R-(NO.)-(NO.)-(YR)	BROOKLYN. POLYTECHNIC INST.
R-(NO.)-(YR)-	BROOKLYN. POLYTECHNIC INST.
R1H RR-	CALIFORNIA. UNIV., BERKELEY. RADIATION LAB. (REF. 9)
R-1-H RR-	CALIFORNIA. UNIV., BERKELEY. RADIATION LAB. (REF. 9)
R1-M-	CALIFORNIA. UNIV., BERKELEY. RADIATION LAB. (REF. 9)
R1-RR-	CALIFORNIA. UNIV., BERKELEY. RADIATION LAB. (REF. 9)
R1-S-	CALIFORNIA. UNIV., BERKELEY. RADIATION LAB. (REF. 9)
R1-SM-	CALIFORNIA. UNIV., BERKELEY. RADIATION LAB. (REF. 9)
R-1-184-S-	CALIFORNIA. UNIV., BERKELEY. RADIATION LAB. (REF. 9)
R1 SPECIAL-	CALIFORNIA. UNIV., BERKELEY. RADIATION LAB. (REF. 9)
R1-W-	CALIFORNIA. UNIV., BERKELEY. RADIATION LAB. (REF. 9)
R-1-184-W-	CALIFORNIA. UNIV., BERKELEY. RADIATION LAB. (REF. 9)
R1-184-W-	CALIFORNIA. UNIV., BERKELEY. RADIATION LAB. (REF. 9)
R(YR)CAP(NO.)	CANADIAN GENERAL ELECTRIC CO., LTD. CIVILIAN ATOMIC POWER DEPT., PETERBOROUGH, ONT.
R (YEAR) CAP-	CANADIAN GENERAL ELECTRIC CO., LTD. COMMERCIAL ATOMIC POWER DIV., PETERBOROUGH, ONT.
R-(3 DIGITS)	CHANDLER EVANS INC. CONTROL SYSTEMS DIV., WEST HARTFORD, CONN.
R-	COLUMBIA UNIV., IRVINGTON-ON-HUDSON, N.Y. NEVIS LABS.
R-4-	COLUMBIA UNIV., NYC. SUBSTITUTE ALLOY MATERIALS LABS.
3-R-	CONVAIR, FORT WORTH, TEX.
R-(NUMBER-YEAR)	DANISH ATOMIC ENERGY COMMISSION. RESEARCH ESTABLISHMENT. ELECTRONICS DEPT., RISOE
R-	DAVID TAYLOR MODEL BASIN, CARDEROCK, MD.
R-(NUMBER)-(YEAR)	ELCON LAB., INC., CAMBRIDGE, MASS.
R-(NUMBER)-(YEAR)	ELCON LAB., INC., SALEM, MASS.
R-(NO.)-	ELDORADO MINING + REFINING LTD. RES.+ DEV.DIV.,OTTAWA
R(4 DIGITS)-F	EON INSTRUMENTATION INC., VAN NUYS, CALIF.
163-R(NUMBER)-	FLADER (FREDRIC) INC., NORTH TONAWANDA, N.Y.
R(NUMBER.NUMBER)	FRANCE. CENTRE DE PHYSIQUE THEORIQUE DE L'ECOLE POLYTECHNIQUE
R-(3 DIGITS)	FRANCIS ASSOCIATES, INC., MARION, MASS.
R-	FRANKFORD ARSENAL, PHILADELPHIA
R.53/-/(DATE)	FULMER RES. INST., LTD., STOKE POGES, BUCKS, ENGLAND
R.(NUMBER/NUMBER)	FULMER RES. INST., LTD., STOKE POGES, BUCKS, ENGLAND
R-12-	GENERAL DEVELOPMENT CORP., ELKTON, MD.
R-(YEAR-NUMBER)-	GENERAL DYNAMICS/ELECTRONICS, SAN DIEGO, CALIF.
R(YEAR)(LETTERS)-	GENERAL ELECTRIC CO. (ALL LOCATIONS) (REF. 46)
R(YR)ELM(NO.S)	GENERAL ELECTRIC CO., PALO ALTO, CALIF.
R(YR)AE-	GENERAL ELECTRIC CO., SCHENECTADY, N.Y.
R-(NUMBER)Q-	GENERAL ELECTRIC CO., SCHENECTADY, N.Y.
R(YR)ELC-	GENERAL ELECTRIC CO. ADVANCED ELECTRONICS CENTER, ITHACA, N.Y.
R(YEAR)FPD-	GENERAL ELECTRIC CO. ADVANCED ENGINE AND TECHNOLOGY DEPT., EVENDALE, OHIO
R(YEAR)FPD(NO.)	GENERAL ELECTRIC CO. ADVANCED TECHNICAL AND DEMONSTRATOR PROGRAMS DEPT., CINCINNATI
R(YR)AO-	GENERAL ELECTRIC CO. AERONAUTIC AND ORDNANCE SYSTEMS DIV., SCHENECTADY, N.Y.
R(YEAR)AEG(NUMBER)	GENERAL ELECTRIC CO. AIRCRAFT ENGINE GROUP, CINCINNATI
R(YR)-(LETTERS)	GENERAL ELECTRIC CO. AIRCRAFT GAS TURBINE DIV., EVENDALE, OHIO
R(YR)AGT-	GENERAL ELECTRIC CO. AIRCRAFT GAS TURBINE DIV., EVENDALE, OHIO
R(YEAR)FPD(NUMBER)-	GENERAL ELECTRIC CO.FLIGHT PROPULSION DIV.,CINCINNATI
R(NUMBER)-	GENERAL ELECTRIC CO. FLIGHT PROPULSION LAB. DEPT., CINCINNATI
R(YR)-GL-	GENERAL ELECTRIC CO. GENERAL ENGINEERING AND CONSULTING LAB., SCHENECTADY, N.Y.
R(YEAR)GL-	GENERAL ELECTRIC CO. GEN. ENG. LAB., SCHENECTADY,N.Y.
R(NO.)(LTRS.)(NO.)	GENERAL ELECTRIC CO. HEAVY MILITARY ELECTRONICS DEPT., SYRACUSE, N.Y.
R(NO.)GML-	GENERAL ELECTRIC CO. LIGHT MILITARY ELECTRONICS DEPT., SCHENECTADY, N.Y.
R(NO.)EML(NO.)G	GENERAL ELECTRIC CO. LIGHT MILITARY ELECTRONICS DEPT., UTICA, N.Y.
R(YEAR)FPD(NUMBER)-	GENERAL ELECTRIC CO. MATERIALS DEVELOPMENT LAB. OPERATION, CINCINNATI
R(YEAR)MCD(NUMBER)	GENERAL ELECTRIC CO. MILITARY COMMUNICATIONS DEPT., OKLAHOMA CITY
R(YEAR)SD(NUMBER)	GENERAL ELECTRIC CO. MISSILE + SPACE DIV., CINCINNATI
R(YEAR)SD-	GENERAL ELECTRIC CO. MISSILE AND SPACE DIV., PHILA.
R-(YEAR)-SD-	GENERAL ELECTRIC CO. MISSILE AND SPACE DIV., PHILA.
R-(YEAR)-ND-	GENERAL ELECTRIC CO. NEUTRON DEVICES DEPT., ST. PETERSBURG, FLA.
R(YEAR)APE-	GENERAL ELECTRIC CO. NUCLEAR ENERGY DIV.,SAN JOSE,CAL (ATOMIC POWER EQUIPMENT)
R-(YEAR)ND(NUMBER)	GENERAL ELECTRIC CO. PINELLAS PENINSULA PLANT, ST. PETERSBURG, FLA.
R(YEAR)ELS-	GENERAL ELECTRIC CO. SEMICONDUCTOR PRODUCTS DEPT., SYRACUSE, N.Y.
R-(YEAR)-SD(NUMBER)	GENERAL ELECTRIC CO. SPACE POWER AND PROPULSION SECTION, CINCINNATI
R(YR)TMP-	GENERAL ELECTRIC CO. TECHNICAL MILITARY PLANNING OPERATION, SANTA BARBARA, CALIF.
R(YR)-TL-	GENERAL ELECTRIC CO. THOMSON LAB., LYNN, MASS.
R(NO.)(LTRS.)-	GENERAL ELECTRIC CO. TUBE DEPT., PALO ALTO, CALIF.
R(YEAR)ELM-(NUMBER)-	GENERAL ELECTRIC CO. TUBE DEPT., PALO ALTO, CALIF.
R(YEAR)ELM(NUMBER)-	GENERAL ELECTRIC CO. TWT BUSINESS SECT.,PALO ALTO,CAL
R-48-	GOODRICH (B.F.) CO., AKRON, OHIO
R-	GT. BRIT. AERO. RES. COUNCIL. ROCKETS SUBCOMM.
R-(NO.)-(YR)	GT. BRIT. MINISTRY OF FUEL AND POWER
R-	ILLINOIS. UNIV., URBANA. CONTROL SYSTEMS LAB.

R-

R-	INSTITUTE FOR DEFENSE ANALYSES, ARLINGTON, VA.
R(NUMBERS)-	LITTON PRECISION PRODUCTS, INC. AIRTRON DIV., MORRIS PLAINS, N.J.
R-	LIBRARY OF CONGRESS. TECH.INFO.DIV.,WASH.,DC (REF.20)
(NUMBER)-R-	LUNDY ELECTRONICS AND SYSTEMS, INC., GLEN HEAD, N.Y.
R-(YR.)-(NO.)	MARTIN CO., DENVER
R-	MARTIN-MARIETTA CORP. DENVER DIV.
R(YR)-	MASSACHUSETTS INST. OF TECH., CAMBRIDGE
R(YEAR)-	MASSACHUSETTS INST. OF TECH., CAMBRIDGE. DEPT. OF CIVIL ENGINEERING
R(NO.)-	MASS. INST. OF TECH., CAMBRIDGE. SCHOOL OF ENG.
R-(YEAR)-	MASS. INST. OF TECH., CAMBRIDGE. SCHOOL OF ENG.
R-	MENASCO MANUFACTURING CO., BURBANK, CALIF.
R(NUMBER)-(YEAR)	MICRONETICS, INC., SAN DIEGO, CALIF.
R-(NO.)-(YR.)	NAVAL AIR DEVELOPMENT CENTER. PILOTLESS AIRCRAFT DEVELOPMENT LAB., JOHNSVILLE, PENNA.
R-	PHILCO CORP., LANSDALE, PENNA.
R-	PRINCETON UNIV., N.J.
R-	RADIOPLANE CO. RECOVERY DIV., VAN NUYS, CALIF.
R-	RAND CORP., SANTA MONICA, CALIF.
R-(NUMBER)-(LETTERS)	RAND CORP., SANTA MONICA, CALIF. (REF. 47)
R-(NUMBER)-AEC	RAND CORP., SANTA MONICA, CALIF. (REF. 47)
R-(NUMBER)-RAND	RAND CORP., SANTA MONICA, CALIF. (REF. 47)
R-(NUMBER)-PR	RAND CORP., SANTA MONICA, CALIF. (REF. 47)
R-(NUMBER)-NYC	RAND INST., N.Y.C.
R-	RAVEN INDUSTRIES, INC., SIOUX FALLS, S.D.
R(NUMBER)-	RAYTHEON CO. SUBMARINE SIGNAL DIV., PORTSMOUTH, R.I.
2R-	REDSTONE ARSENAL. ORDNANCE MISSILE LABS., HUNTSVILLE, ALA.
6R-	REDSTONE ARSENAL. ORDNANCE MISSILE LABS., HUNTSVILLE, ALA.
R-(NUMBER)-	ROCKETDYNE. RESEARCH DEPT., CANOGA PARK, CALIF.
R-(NUMBER)P	ROCKETDYNE DIV., NORTH AMERICAN AVIATION, INC., CANOGA PARK, CALIF.
R-400-	ROCKETDYNE DIV., NORTH AMERICAN AVIATION, INC., CANOGA PARK, CALIF.
R-	SANDERS ASSOCIATES, INC., NASHUA, N.H.
R-	SANDIA CORP., ALBUQUERQUE, N. MEX.
R(YEAR)B-	SOLAR, SAN DIEGO, CALIF.
R-	SPEER CARBON CO., NIAGARA FALLS, N.Y.
R(NUMBER)-	SPERRY FARRAGUT CO., BRISTOL, TENN.
R-	STEVENS INST. OF TECH., HOBOKEN, N.J. DAVIDSON LAB.
R(YEAR)-	TECHNICAL COMMUNICATIONS CORP., LEXINGTON, MASS.
R(NUMBER)-(YEAR)-	TELEDYNE SYSTEMS CO. MICRONETICS DIV., SAN DIEGO,CAL.
R-	TENNESSEE EASTMAN CORP., OAK RIDGE, TENN.
R-(NO.)-	UNITED AIRCRAFT CORP., EAST HARTFORD, CONN.
R-73003-	UNITED AIRCRAFT CORP., EAST HARTFORD, CONN.
R-(NO.)-(NO.)-(LETTER)	WESTINGHOUSE ELECTRIC CORP. RES. LABS., E. PITTSBURGH
R-(NO.)-(LETTER)	WESTINGHOUSE ELECTRIC CORP. RES. LABS., E. PITTSBURGH
R-(NO.)-(NO.)-(LETTER)	WESTINGHOUSE ELECTRIC CORP. RES. LABS., PITTSBURGH
R-(NO.)-(LETTER)	WESTINGHOUSE ELECTRIC CORP. RES. LABS., PITTSBURGH
R(LETTER)C-	OFFICE OF SCIENTIFIC RESEARCH AND DEVELOPMENT
R/A-(1 DIGIT)	ASSOCIATION OF BAY AREA GOVERNMENTS, BERKELEY, CALIF.
RA-	BUREAU OF AERONAUTICS (NAVY). RES. + ANALYSIS OFF.
RA-(NUMBER)-	CANADIAN COMMERCIAL CORP., WASHINGTON, D.C.
RA-	ELECTRO-MECHANICAL RESEARCH, INC., SARASOTA, FLA.
RA-(NUMBER)-EXHIBIT	ELECTRO-MECHANICAL RESEARCH, INC., SARASOTA, FLA.
RA-	LOGICON, INC., SAN PEDRO, CALIF.
RA-	NETHERLANDS. RIJKS INSTITUUT VOOR DE VOLKSGEZONDHEID, UTRECHT
RA-	NETHERLANDS. RIJKS INSTITUUT VOOR DE VOLKSGEZONDHEID. LABORATORIUM VOOR STRALINGSONDERZOEK, BILTHOVEN
RA-	RADD ASSOCIATES, PALO ALTO, CALIF.
RA-	RADFORD ARSENAL, VA.
RA-	RAND CORP., SANTA MONICA, CALIF.
RA-	REDSTONE ARSENAL, ALA.
RA-(NUMBER)(LETTER)-	REDSTONE ARSENAL. ORDNANCE MISSILE LABS., HUNTSVILLE, ALA.
RA-(NUMBER)(LETTER)(NO.)-	REDSTONE ARSENAL. ORDNANCE MISSILE LABS., HUNTSVILLE, ALA.
RAAF-	ROYAL AUSTRALIAN AIR FORCE
RA-AM-	ARMY BALLISTIC MISSILE AGENCY, REDSTONE ARSENAL, ALA.
RAA/R/	NORD-AVIATION, PARIS
RAA-RD-(YEAR)-	NATIONAL WEATHER RECORDS CENTER, ASHEVILLE, N.C.
RAC-	FAIRCHILD HILLER CORP. REPUBLIC AVIATION DIV., FARMINGDALE, N.Y.
RAC (NUMBER)-	REPUBLIC AVIATION CORP., FARMINGDALE, N.Y.
RAC-	RESEARCH ANALYSIS CORP., BETHESDA, MD.
RAC-	RESEARCH ANALYSIS CORP., CHEVY CHASE, MD.
RAC-	RESEARCH ANALYSIS CORP., MC LEAN, VA.
RAC-	RYAN AERO. CO., LINDBERGH FIELD, SAN DIEGO, CALIF.
RAC-CR-	RESEARCH ANALYSIS CORP., MC LEAN, VA.(CLIENT REPORTS)
RAC-D-CR-	RESEARCH ANALYSIS CORP., MC LEAN, VA. (DRAFT CLIENT REPORTS)
RAC-D-TP-	RESEARCH ANALYSIS CORP., MC LEAN, VA. (DRAFT TECHNICAL PAPERS)
RAC-E-	RESEARCH ANALYSIS CORP., MC LEAN, VA.
RAC-EAR-	FAIRCHILD HILLER CORP. REPUBLIC AVIATION DIV., FARMINGDALE, N.Y.
RAC-EARMR-	FAIRCHILD HILLER CORP. REPUBLIC AVIATION DIV., FARMINGDALE, N.Y.
RAC-EASR-	FAIRCHILD HILLER CORP. REPUBLIC AVIATION DIV., FARMINGDALE, N.Y.
RAC-ED-(MODEL DESIGNA.)-	FAIRCHILD HILLER CORP. REPUBLIC AVIATION DIV., FARMINGDALE, N.Y. (ENGINEERING DATA)
RACFO-	ARMY MISSILE TEST AND EVALUATION DIRECTORATE, WHITE SANDS MISSILE RANGE, N. MEX.
RACFO-FP-	RESEARCH AND ANALYSIS CORP. FIELD OFFICE, BANGKOK
RACFO-T-FP-	RESEARCH AND ANALYSIS CORP. FIELD OFFICE, BANGKOK
RAC-FP-	RESEARCH ANALYSIS CORP., MC LEAN, VA. (FIELD PAPERS)
RAC-FR-	RESEARCH ANALYSIS CORP., MC LEAN, VA. (FIELD REPTS.)
RAC-GMD-	REPUBLIC AVIATION CORP. GUIDED MISSILE DIV., HICKSVILLE, N.Y.
RAC GMD (NOS.)-(NO.)-	REPUBLIC AVIATION CORP. GUIDED MISSILE DIV., MINEOLA, N.Y.
RAC-GMD-	REPUBLIC AVIATION CORP. GUIDED MISSILE DIV., N.Y.C.
RAC-GR-	RYAN AERO. CO., LINDBERGH FIELD, SAN DIEGO, CALIF.
RACIC-	BATTELLE MEMORIAL INST. REMOTE AREA CONFLICT INFORMATION CENTER, COLUMBUS, OHIO
RACIC-MR-	BATTELLE MEMORIAL INST. REMOTE AREA CONFLICT INFORMATION CENTER, COLUMBUS, OHIO
RACIC-TR-	BATTELLE MEMORIAL INST. REMOTE AREA CONFLICT INFORMATION CENTER, COLUMBUS, OHIO
RAC MSD (NO.)-(NO.)-	REPUBLIC AVIATION CORP. MISSILE SYSTEMS DIV., MINEOLA, N.Y.
RAC(ORO)-	RESEARCH ANALYSIS CORP., BETHESDA, MD.
RAC(ORO)-SP-	RESEARCH ANALYSIS CORP., BETHESDA, MD.
RAC(ORO)-TP-	RESEARCH ANALYSIS CORP., BETHESDA, MD.
RAC-P-	RESEARCH ANALYSIS CORP., MC LEAN, VA. (PAPERS)
RAC-PPL-	FAIRCHILD HILLER CORP. REPUBLIC AVIATION DIV. PLASMA PROPULSION LAB., FARMINGDALE, N.Y.
RAC-PPL-(YEAR)-	FAIRCHILD HILLER CORP. REPUBLIC AVIATION DIV. PLASMA PROPULSION LAB., FARMINGDALE, N.Y.

RAC-PR-	FAIRCHILD HILLER CORP. REPUBLIC AVIATION DIV., FARMINGDALE, N.Y.
RAC-R-	RESEARCH ANALYSIS CORP., MC LEAN, VA.
RAC-T-	RESEARCH ANALYSIS CORP., MC LEAN, VA.
RAC-TP-	RESEARCH ANALYSIS CORP., BETHESDA, MD.
RAC-TP-	RESEARCH ANALYSIS CORP., MC LEAN, VA.
RAD-	ATOMIC POWER DEVELOPMENT ASSOCIATES, INC., DETROIT
RAD-	AVCO CORP. RES.+ ADVANCED DEV. DIV., WILMINGTON,MASS.
RAD-(NUMBER)-	AVCO CORP. RES.+ ADVANCED DEV. DIV., WILMINGTON,MASS.
RAD-(NO.)-TM-	AVCO MFG. CORP. RESEARCH AND ADVANCED DEVELOPMENT DIV., WILMINGTON, MASS.
R/AD-(NUMBER-NUMBER)	CALIF. INST. OF TECH., PASADENA. JET PROPULSION LAB.
RAD-	DOUGLAS AIRCRAFT CO., INC. PROJECT RAND, SANTA MONICA, CALIF.
RAD-	RADFORD ARSENAL, VA.
RAD-	RAND CORP., SANTA MONICA, CALIF.
RA-DA-	ARMY BALLISTIC MISSILE AGENCY, REDSTONE ARSENAL, ALA.
RADC-	ROME AIR DEVELOPMENT CENTER, GRIFFISS AFB, N.Y.
RADC-AR(NUMBER)-(YEAR)	ROME AIR DEVELOPMENT CENTER, GRIFFISS AFB, N.Y.
RADC-SP-	ROME AIR DEVELOPMENT CENTER, GRIFFISS AFB, N.Y.
RADC-SP-(YEAR)-	ROME AIR DEVELOPMENT CENTER, GRIFFISS AFB, N.Y.
RADC-TDR-	ROME AIR DEVELOPMENT CENTER, GRIFFISS AFB, N.Y. (TECHNICAL DOCUMENTARY REPORT)
RADC-TDR-(YEAR)-	ROME AIR DEVELOPMENT CENTER, GRIFFISS AFB, N.Y.
RADC-TN-	ROME AIR DEVELOPMENT CENTER, GRIFFISS AFB, N.Y.
RADC-TN-(YEAR)-	ROME AIR DEVELOPMENT CENTER, GRIFFISS AFB, N.Y.
RADC-TR-	ROME AIR DEVELOPMENT CENTER, GRIFFISS AFB, N.Y.
RADC-TR-(YEAR)-	ROME AIR DEVELOPMENT CENTER, GRIFFISS AFB, N.Y.
RADET-	DIVISION OF BIOLOGY AND MEDICINE, AEC. RADIATION INSTRUMENTS BRANCH, WASHINGTON, D.C.
RADIO REPORT	GT. BRIT. POST OFFICE. ENGINEERING DEPT.
RAD(L)-	DOUGLAS AIRCRAFT CO., INC. PROJECT RAND, SANTA MONICA, CALIF.
RAD-M-	AVCO CORP. RES.+ ADVANCED DEV. DIV., WILMINGTON,MASS.
RAD-M-	AVCO CORP. RES.+ ADVANCED DEV. DIV., WILMINGTON,MASS.
RADMP-	SOCIETE FRANCAISE DE PHYSIQUE, PARIS
	AVCO MFG. CORP. RESEARCH AND ADVANCED DEVELOPMENT DIV., LAWRENCE, MASS.
RAD-MP-2-(YR.)-	AVCO MFG. CORP. RESEARCH AND ADVANCED DEVELOPMENT DIV., LAWRENCE, MASS.
RAD-SA-(NO.)-	AVCO MFG. CORP. RESEARCH AND ADVANCED DEVELOPMENT DIV., LAWRENCE, MASS.
RAD-SA-	AVCO MFG. CORP. RESEARCH AND ADVANCED DEVELOPMENT DIV., WILMINGTON, MASS.
RAD-SR-(YEAR)-	AVCO CORP. RES.+ ADVANCED DEV. DIV., WILMINGTON,MASS.
RAD-TM-(YEAR)-	AVCO CORP. RES.+ ADVANCED DEV. DIV., WILMINGTON,MASS.
RAD-TR-(YEAR)-	AVCO CORP. RES. AND DEV. DIV., WILMINGTON, MASS.
R/AE	BRITISH COLUMBIA. UNIV., VANCOUVER
R/AE-	CANADA. DEFENCE RESEARCH BOARD, OTTAWA
RAE-	FAIRCHILD ENGINE AND AIRPLANE CORP. (RANGER AIRCRAFT ENGINES)
RAE-	GT. BRIT. ROYAL AIRCRAFT ESTAB., BEDFORD, BEDS.,ENG.
RAE-	GT. BRIT. ROYAL AIRCRAFT EST.,FARNBOROUGH, HANTS, ENG
RAE-AERO-	GT. BRIT. ROYAL AIRCRAFT ESTAB., BEDFORD, BEDS.,ENG.
RAE-AERO-	GT. BRIT. ROYAL AIRCRAFT EST.,FARNBOROUGH, HANTS, ENG
RAE-AERO-TM-	GT. BRIT. ROYAL AIRCRAFT EST.,FARNBOROUGH, HANTS, ENG
RAE-ARM.(NUMBER)	GT. BRIT. ROYAL AIRCRAFT EST.,FARNBOROUGH, HANTS, ENG
RAE-CHEM-	GT. BRIT. ROYAL AIRCRAFT EST.,FARNBOROUGH, HANTS, ENG
RAE-CPM-	GT. BRIT. ROYAL AIRCRAFT EST.,FARNBOROUGH, HANTS, ENG
RAE-EL-	GT. BRIT. ROYAL AIRCRAFT EST.,FARNBOROUGH, HANTS, ENG
RAE-IGT-MEMO-	GT. BRIT. ROYAL AIRCRAFT EST.,FARNBOROUGH, HANTS, ENG
RAE-IM-	GT. BRIT. ROYAL AIRCRAFT EST.,FARNBOROUGH, HANTS, ENG
RAE-LIB-	GT. BRIT. ROYAL AIRCRAFT EST.,FARNBOROUGH, HANTS, ENG
RAE-LIB-BIB-	GT. BRIT. ROYAL AIRCRAFT EST.,FARNBOROUGH, HANTS, ENG
RAE-LIB-BIBL-	GT. BRIT. ROYAL AIRCRAFT EST.,FARNBOROUGH, HANTS, ENG
RAE-LIB-MEMO-	GT. BRIT. ROYAL AIRCRAFT EST.,FARNBOROUGH, HANTS, ENG
RAE-LIB-PRESS-TRANS-	GT. BRIT. ROYAL AIRCRAFT EST.,FARNBOROUGH, HANTS, ENG
RAE-LIB-TRANS-	GT. BRIT. ROYAL AIRCRAFT EST.,FARNBOROUGH, HANTS, ENG
RAE-MET-	GT. BRIT. ROYAL AIRCRAFT EST.,FARNBOROUGH, HANTS, ENG
RAEP-	ROCHESTER, N.Y. UNIV. ATOMIC ENERGY PROJECT
RAE-R+M-	GT. BRIT. ROYAL AIRCRAFT EST.,FARNBOROUGH, HANTS, ENG
RAE-R-MECH.ENG-	GT. BRIT. ROYAL AIRCRAFT EST.,FARNBOROUGH, HANTS, ENG
RAE-R-MET-	GT. BRIT. ROYAL AIRCRAFT EST.,FARNBOROUGH, HANTS, ENG
R-AERO-Y-(NO.-NO.)	GEORGE C. MARSHALL SPACE FLIGHT CTR.,HUNTSVILLE, ALA.
RAE-R.P.D.-	GT. BRIT. ROYAL AIRCRAFT EST.,FARNBOROUGH, HANTS, ENG
RAE-R-STRUCT-	GT. BRIT. ROYAL AIRCRAFT EST.,FARNBOROUGH, HANTS, ENG
RAE-SME-	GT. BRIT. ROYAL AIRCRAFT EST.,FARNBOROUGH, HANTS, ENG
RAE-SN-ARM.(NUMBER)	GT. BRIT. ROYAL AIRCRAFT EST.,FARNBOROUGH, HANTS, ENG
RAE-STRUCT-	GT. BRIT. ROYAL AIRCRAFT EST.,FARNBOROUGH, HANTS, ENG
RAE-STRUCTURES-	GT. BRIT. ROYAL AIRCRAFT EST.,FARNBOROUGH, HANTS, ENG
RAE-TE-	GT. BRIT. ROYAL AIRCRAFT EST.,FARNBOROUGH, HANTS, ENG
RAE-TM-AERO-	UNITED KINGDOM. ROYAL AIRCRAFT ESTAB., FARNBOROUGH, HANTS, ENGLAND
RAE-T/M-ARM-	GT. BRIT. ROYAL AIRCRAFT EST.,FARNBOROUGH, HANTS, ENG
RAE-TM-SPACE-	UNITED KINGDOM. ROYAL AIRCRAFT ESTAB., FARNBOROUGH, HANTS, ENGLAND
RAE-TM-WE-	UNITED KINGDOM. ROYAL AIRCRAFT ESTAB., FARNBOROUGH, HANTS, ENGLAND
RAE-TN-AERO-	GT. BRIT. ROYAL AIRCRAFT ESTAB., BEDFORD, BEDS.,ENG.
RAE-TN-AERO-	GT. BRIT. ROYAL AIRCRAFT EST.,FARNBOROUGH, HANTS, ENG
RAE-TN-ARM-	GT. BRIT. ROYAL AIRCRAFT EST.,FARNBOROUGH, HANTS, ENG
RAE-TN-BL-	GT. BRIT. ROYAL AIRCRAFT EST.,FARNBOROUGH, HANTS, ENG
RAE-TN-CHEM-	GT. BRIT. ROYAL AIRCRAFT EST.,FARNBOROUGH, HANTS, ENG
RAE-TN-CPM-	GT. BRIT. ROYAL AIRCRAFT EST.,FARNBOROUGH, HANTS, ENG
RAE-TN-EL-	GT. BRIT. ROYAL AIRCRAFT EST.,FARNBOROUGH, HANTS, ENG
RAE-TN-GW-	GT. BRIT. ROYAL AIRCRAFT EST.,FARNBOROUGH, HANTS, ENG
RAE-TN-IAP-	GT. BRIT. ROYAL AIRCRAFT EST.,FARNBOROUGH, HANTS, ENG
RAE-TN-INST-	GT. BRIT. ROYAL AIRCRAFT EST.,FARNBOROUGH, HANTS, ENG
RAE-TN-INSTN-	GT. BRIT. ROYAL AIRCRAFT EST.,FARNBOROUGH, HANTS, ENG
RAE-TN-MATH-	GT. BRIT. ROYAL AIRCRAFT EST.,FARNBOROUGH, HANTS, ENG
RAE-TN-ME-	GT. BRIT. ROYAL AIRCRAFT EST.,FARNBOROUGH, HANTS, ENG
RAE-TN-MECH. ENG.-	GT. BRIT. ROYAL AIRCRAFT EST.,FARNBOROUGH, HANTS, ENG
RAE-TN-MET-	GT. BRIT. ROYAL AIRCRAFT EST.,FARNBOROUGH, HANTS, ENG
RAE-TN-MS-	GT. BRIT. ROYAL AIRCRAFT EST.,FARNBOROUGH, HANTS, ENG
RAE-TN-RAD-	GT. BRIT. ROYAL AIRCRAFT EST.,FARNBOROUGH, HANTS, ENG
RAE-TN-RPD-	GT. BRIT. ROYAL AIRCRAFT EST.,FARNBOROUGH, HANTS, ENG
RAE-TN-SPACE-	GT. BRIT. ROYAL AIRCRAFT EST.,FARNBOROUGH, HANTS, ENG
RAE-TN-STRUCT-	GT. BRIT. ROYAL AIRCRAFT EST.,FARNBOROUGH, HANTS, ENG
RAE-TN-WE-	GT. BRIT. ROYAL AIRCRAFT EST.,FARNBOROUGH, HANTS, ENG
RAE-TP-	RESEARCH ANALYSIS CORP., MC LEAN, VA.
RAE-TR-	GT. BRIT. ROCKET PROPULSION EST., WESTCOTT, BUCKS,ENG
RAE-TR-	UNITED KINGDOM. ROYAL AIRCRAFT ESTAB., FARNBOROUGH, HANTS, ENGLAND
RAE-TRANS-	GT. BRIT. ROYAL AIRCRAFT EST.,FARNBOROUGH, HANTS, ENG
RAE-TR-ARM-	GT. BRIT. ROYAL AIRCRAFT EST.,FARNBOROUGH, HANTS, ENG
RAF-ARC-	GT. BRIT. AERONAUTICAL RESEARCH COUNCIL
RAG-	BROWN UNIV., PROVIDENCE. RESEARCH ANALYSIS GROUP
RAG-	REAKTOR A. G., WUERENLINGEN, SWITZERLAND

RAI-	RADIATION APPLICATIONS, INC., LONG ISLAND CITY, N.Y.
RAI-	RADIATION APPLICATIONS, INC., N.Y.C.
RAI-	RADIATION, INC., MELBOURNE, FLA.
RAI-	RADIATION, INC., ORLANDO, FLA.
RAI-	RAI RESEARCH CORP., LONG ISLAND CITY, N.Y.
RAINIER-MESA-	GEOLOGICAL SURVEY, DENVER
RAL-RESEARCH RPT.-	MINNESOTA. UNIV., MINNEAPOLIS. ROSEMOUNT AERO. LABS.
RALRR-	MINNESOTA. UNIV., MINNEAPOLIS. ROSEMOUNT AERO. LABS.
RAL-TM-	ROME AIR DEVELOPMENT CENTER. DIRECTORATE OF AEROSPACE SURVEILLANCE + CONTROL, GRIFFISS AFB, NY
RAMC-	ROYAL ARMY MEDICAL COLL. (GT. BRIT.)
RAND-	RAND CORP., SANTA MONICA, CALIF.
RAND-AR-(YEAR)	RAND CORP., SANTA MONICA, CALIF. (ANNUAL REPORT)
RAND-B-	RAND CORP., SANTA MONICA, CALIF.
R AND C-(YEAR)-	ROWLAND AND CO., HADDONFIELD, N.J.
R AND M-	GT. BRIT. AERONAUTICAL RESEARCH COUNCIL
RAND-P-	RAND CORP., SANTA MONICA, CALIF.
RAND(P)-	RAND CORP., SANTA MONICA, CALIF.
RAND-R-	RAND CORP., SANTA MONICA, CALIF.
RAND-RA-	RAND CORP., SANTA MONICA, CALIF.
RAND-RAOP-	RAND CORP., SANTA MONICA, CALIF.
RAND-RAT-	RAND CORP., SANTA MONICA, CALIF. (TRANSLATION)
RAND RM-	RAND CORP., SANTA MONICA, CALIF.
RAND-S-	RAND CORP., SANTA MONICA, CALIF.
RAND-T-	RAND CORP., SANTA MONICA, CALIF.
RA-OML-	REDSTONE ARSENAL. ORDNANCE MISSILE LABS., HUNTSVILLE, ALA.
RAOP-	RAND CORP., SANTA MONICA, CALIF.
RA-PR-	REDSTONE ARSENAL, HUNTSVILLE, ALA.
RAR-	WATERTOWN ARSENAL, MASS. (ROLLED ARMOR REPORT)
RARA-	RARITAN ARSENAL, METUCHEN, N.J.
RARC-	RADIO RECEPTOR CO. INC., BROOKLYN
RARDE-	GT. BRIT. ROYAL ARMAMENT RESEARCH AND DEVELOPMENT ESTABLISHMENT, FORT HALSTEAD, KENT, ENGLAND
RARDE-B-	GT. BRIT. ROYAL ARMAMENT RESEARCH AND DEVELOPMENT ESTABLISHMENT, FORT HALSTEAD, KENT, ENGLAND
RARDE-E-	GT. BRIT. ROYAL ARMAMENT RESEARCH AND DEVELOPMENT ESTABLISHMENT, FORT HALSTEAD, KENT, ENGLAND
RARDE-MEMO-(NO.)/(YR.)	GT. BRIT. ARMAMENT RESEARCH AND DEVELOPMENT ESTABLISHMENT, FORT HALSTEAD, KENT, ENGLAND
RARDE-MEMO-	GT. BRIT. ROYAL ARMAMENT RESEARCH AND DEVELOPMENT ESTABLISHMENT, SEVENOAKS, KENT, ENGLAND
RARDE-MEMO(B)-	GT. BRIT. ROYAL ARMAMENT RESEARCH AND DEVELOPMENT ESTABLISHMENT, FORT HALSTEAD, KENT, ENGLAND
RARDE-MEMO(P)-	GT. BRIT. ROYAL ARMAMENT RESEARCH AND DEVELOPMENT ESTABLISHMENT, FORT HALSTEAD, KENT, ENGLAND
RARDE-X-(MONTH/YEAR)	GT. BRIT. ROYAL ARMAMENT RESEARCH AND DEVELOPMENT ESTABLISHMENT, FORT HALSTEAD, KENT, ENGLAND
RAS-	GENERAL DYNAMICS CORP. ELECTRIC BOAT DIV., GROTON, CONN
RAS-(LETTER)	TENNESSEE EASTMAN CORP., OAK RIDGE, TENN.
RASA-	ROCHESTER APPLIED SCIENCE ASSOCIATES, INC., N.Y.
RAS-MEMO-	GENERAL DYNAMICS CORP. ELECTRIC BOAT DIV., GROTON, CONN
RASR-	NORTHROP CORP. NORAIR DIV. ENGINEERING DEPT., HAWTHORNE, CALIF. (RESEARCH + ADVANCED SYSTEMS REPT.)
RAS-TM-	ROME AIR DEV. CTR. APPLIED RES. LAB., GRIFFISS AFB,NY
R-ASTR-S-(YEAR)-	GEORGE C. MARSHALL SPACE FLIGHT CTR., HUNTSVILLE, ALA.
RAT-1 THRU RAT-146	RAND CORP., SANTA MONICA, CALIF. (TRANSLATION)
RATM-	RADIO CORP. OF AMERICA, MOORESTOWN, N.J.
RA-TR-	ARMY BALLISTIC MISSILE AGENCY, REDSTONE ARSENAL, ALA.
RATSEC-	ROBERT A. TAFT SANITARY ENGINEERING CTR., CINCINNATI
RAU-TM-	ROME AIR DEVELOPMENT CENTER. DIRECTORATE OF COMMUNICATIONS, GRIFFISS AFB, N.Y.
RAW-TM-	ROME AIR DEVELOPMENT CENTER. ELECTRONICS WARFARE LAB., GRIFFISS AFB, N.Y.
RAY-	RAYTHEON MFG. CO., WALTHAM, MASS.
RAYMC-	RAYTHEON MFG. CO., WALTHAM, MASS.
RAYTHEON-	RAYTHEON CO. (ALL LOCATIONS)
RAYTHEON BR-	RAYTHEON MFG. CO. MISSILE SYSTEMS DIV., BEDFORD, MASS.
RAYTHEON NOTES BK...	RAYTHEON MFG. CO., WALTHAM, MASS.
RB-	(REPORT BIBLIOGRAPHY) (REF. 2)
RB-	(RESEARCH BULLETIN) (REF. 2)
RB-	(RESTRICTED BULLETIN) (REF. 2)
RB-(YEAR)-	AMPEX CORP., REDWOOD CITY, CALIF.
RB-	ARMY COMMAND + GEN. STAFF SCH., FT. LEAVENWORTH, KAN.
RB-	ARMY MEDICAL LAB (5TH), ST. LOUIS
RB-	CLEARINGHOUSE FOR FEDERAL SCIENTIFIC AND TECHNICAL INFORMATION, SPRINGFIELD, VA. (REPORT BIBLIOGRAPHY)
RB-	EASTMAN KODAK CO., ROCHESTER, N.Y. (RESERVE BATTERY)
RB-(YEAR)-	EDUCATIONAL TESTING SERVICE, PRINCETON, N.J.
RB-(NUMBER)-	EG+G, INC., BEDFORD, MASS.
RB-	HONEYWELL, INC., MINNEAPOLIS
RB-(NO.)(LETTER)	KELLEX CORP., N.Y.C.
RB-	MARTIN CO., BALTIMORE
RB-	MARTIN CO. TECHNICAL INFO. CENTER, ORLANDO, FLA.
RB-	PRINCETON UNIV., N.J.
RB-	SPERRY GYROSCOPE CO., LTD., MONTREAL
RB-(NUMBER)-(NUMBER)-	SPERRY GYROSCOPE CO., LTD., MONTREAL
RB-(LETTER)(DATE CODE)	NATIONAL ADVISORY COMMITTEE FOR AERONAUTICS (RESTRICTED BULLETIN) (REF. 25)
RBAR-	NAVAL ORDNANCE TEST STA., INYOKERN (CHINA LAKE), CAL.
RBESO-	BENDIX CORP. RED BANK DIV., EATONTOWN, N.J.
RBH-	RICHARDSON, BELLOWS, HENRY AND CO., INC., N.Y.C.
RB/R-	UNITED KINGDOM ATOMIC ENERGY AUTHORITY. INDUSTRIAL GROUP. CAPENHURST WORKS, CHES., ENGLAND
RB/R-	UNITED KINGDOM ATOMIC ENERGY AUTHORITY. INDUSTRIAL GROUP. WINDSCALE WORKS, SELLAFIELD, CUMB., ENGLAND
R/C-	AEROJET-GENERAL CORP. LIQUID ROCKET PLANT, SACRAMENTO, CALIF.
RC-	AIR FORCE (RADAR CHARTS)
RC-	AIR FORCE AVIONICS LAB. RECONNAISSANCE APPLICATIONS BRANCH, WRIGHT-PATTERSON AFB, OHIO
RC-	ARMY ELECTRONIC PROVING GROUND, FORT HUACHUCA, ARIZ.
RC-	DETROIT CONTROLS CORP., REDWOOD CITY, CALIF.
RC-	FRANCE. COMMISSARIAT A L'ENERGIE ATOMIQUE, PARIS
RC-	GT. BRIT. MINISTRY OF HOME SECURITY
R.C.-	GT. BRIT. ROAD RES. LAB., HARMONDSWORTH, MIDDX., ENG.
RC-	IBM WATSON RESEARCH CENTER, YORKTOWN HEIGHTS, N.Y.
R/C-	LOCKHEED MISSILES AND SPACE CO., SUNNYVALE, CALIF.
RC-	NORTHROP AIRCRAFT, INC., HAWTHORNE, CALIF. (RECORDS CONTROL NUMBER)
RC-	NUCLEAR CORP. OF AMERICA. RES.CHEMS.DIV., BURBANK,CAL
RC-(NO.)(PWAC)	PRATT AND WHITNEY AIRCRAFT DIV., UNITED AIRCRAFT CORP., HARTFORD, CONN.
RC-	R-C SCIENTIFIC INSTRUMENT CO., PLAZA DEL REY, CALIF.
RC-	RADIOPLANE CO., VAN NUYS, CALIF.
RC-	RAND CORP., SANTA MONICA, CALIF.
RC-	RAYTHEON CO. (ALL LOCATIONS)
RC-	RESEARCH CHEMICALS INC., BURBANK, CALIF.
RC-(YEAR)-	ROCKETDYNE DIV., NORTH AMERICAN AVIATION, INC., CANOGA PARK, CALIF.

RC-(NUMBER)-(NUMBER)-	UNITED STATES RUBBER CO. RESEARCH CENTER, WAYNE, N.J.
RC-	WRIGHT STATE UNIV., DAYTON, OHIO
RC-	ZENITH RADIO CORP., CHICAGO
RCA-	AIR FORCE MISSILE TEST CENTER, PATRICK AFB, FLA.
RCA-	RADIO CORP. OF AMERICA (ALL LOCATIONS)
RCA-ATE-MTE-	RADIO CORP. OF AMERICA. DEFENSE ELECTRONIC PRODUCTS, BURLINGTON, MASS. (MULTISYSTEM TEST EQUIPMENT)
RCA-CA-	RADIO CORP. OF AMERICA, CAMDEN, N.J.
RCA-CM-	RADIO CORP. OF AMERICA. RCA LABS.DIV.,ROCKY POINT,NY
RCA-CR-(YEAR)-	RADIO CORP. OF AMERICA (ALL LOCATIONS)
RCA-EM-(YEAR)-(NUMBER)-	RADIO CORP. OF AMERICA. DEFENSE ELECTRONIC PRODUCTS (ALL LOCATIONS) (ENGINEERING MEMORANDUM)
RCA-EM-5001 THRU -5300	RADIO CORP. OF AMERICA. HOME INSTRUMENTS DIV., INDIANAPOLIS (ENGINEERING MEMORANDUM)
RCA-EM-	RCA VICTOR DIV., RADIO CORP. OF AMERICA, CAMDEN, N.J.
RCA ER-	RADIO CORP. OF AMERICA (ALL LOCATIONS) (ENG.REPT.)
RCAF-	ROYAL CANADIAN AIR FORCE
RCAF CEPE-	ROYAL CANADIAN AIR FORCE. CENTRAL EXPERIMENTAL AND PROVING ESTAB., ROCKCLIFFE, ONT.
RCAF EPE-	ROYAL CANADIAN AIR FORCE. CENTRAL EXPERIMENTAL AND PROVING ESTAB., ROCKCLIFFE, ONT.
RCA IB-	RADIO CORP. OF AMERICA (ALL LOCATIONS) (INSTRUC.BK.)
RCAL-	RADIO CORP. OF AMERICA. RCA LABS. DIV., PRINCETON, NJ
RCA-LB-	RADIO CORP. OF AMERICA. RCA LABS. DIV., PRINCETON, NJ (LICENSEE BULLETIN)
RCA-MLPR-	RADIO CORP. OF AMERICA, MOORESTOWN, N.J.
RCA-MO-	RADIO CORP. OF AMERICA (ALL LOCATIONS)
RCA MO-(LTR.)-(LTR.)-	RADIO CORP. OF AMERICA. MISSILE AND SURFACE RADAR DEPT., MOORESTOWN, N.J.
RCA-MPR-	RADIO CORP. OF AMERICA, MOORESTOWN, N.J.
RCA-MSRD-QTR-	RADIO CORP. OF AMERICA. MISSILE AND SURFACE RADAR DIV., MOORESTOWN, N.J.
RCA-NP-	RCA VICTOR DIV., RADIO CORP. OF AMERICA, CAMDEN, N.J.
RCA-PEM-	RADIO CORP. OF AMERICA. RCA LABS. DIV., PRINCETON, NJ
RCA-PTR-	RADIO CORP. OF AMERICA. RCA LABS. DIV., PRINCETON, NJ
RCA-QR-	DAVID SARNOFF RESEARCH CENTER, PRINCETON, N.J.
RCA-QTR-	RADIO CORP. OF AMERICA (ALL LOCATIONS)
RCA RES. REPT.-	RCA VICTOR CO., LTD., MONTREAL
RCA-RES.RPT.-(NO.)-(NO.)-	RCA VICTOR CO., LTD. RESEARCH LABS., MONTREAL
RCA-RR-(NO.)-(NO.)-	RCA VICTOR CO., LTD. RESEARCH LABS., MONTREAL
RCASC-	RCA SERVICE CO., INC., CAMDEN, N.J.
RCA TR-(YEAR)-(NUMBER)-	RADIO CORP. OF AMERICA. DEFENSE ELECTRONIC PRODUCTS (ALL LOCATIONS) (TECHNICAL REPORT)
RCA-TR-2001 THRU -2100	RADIO CORP. OF AMERICA. HOME INSTRUMENTS DIV., INDIANAPOLIS (TECHNICAL REPORT)
RCA-TSP-	RADIO CORP. OF AMERICA, MOORESTOWN, N.J. (TECH.SPEC.)
RCA-VCL-	RCA VICTOR CO., LTD., MONTREAL
RC-BR-	RAYTHEON CO. MISSILE AND SPACE DIV., BEDFORD, MASS.
RC-BR-	RAYTHEON CO. MISSILE SYSTEMS DIV., BEDFORD, MASS.
RCC-	COLUMBIA UNIV., N.Y.C.
RCC-	WHITE SANDS MISSILE RANGE. RANGE COMMANDERS COUNCIL, N. MEX.
R.C.C./M-	UNITED KINGDOM ATOMIC ENERGY AUTHORITY. RESEARCH GP. RADIOCHEMICAL CENTRE, AMERSHAM, BUCKS, ENG.(REF.41)
RCC-M-	UNITED KINGDOM ATOMIC ENERGY AUTHORITY. RESEARCH GP. RADIOCHEMICAL CENTRE, AMERSHAM, BUCKS, ENG.(REF.41)
R.C.C./R-	UNITED KINGDOM ATOMIC ENERGY AUTHORITY. RESEARCH GP. RADIOCHEMICAL CENTRE, AMERSHAM, BUCKS, ENG.(REF.41)
RCC-R-	UNITED KINGDOM ATOMIC ENERGY AUTHORITY. RESEARCH GP. RADIOCHEMICAL CENTRE, AMERSHAM, BUCKS, ENG.(REF.41)
RCE-	ATOMIC ENERGY OF CANADA LTD., CHALK RIVER, ONT.
RCF-	JOHNS HOPKINS UNIV., SILVER SPRING, MD. APPLIED PHYSICS LAB.
RCF-	STANFORD UNIV., CALIF. DEPT. OF MECHANICAL ENG., THERMOSCIENCES DIV.
R.CHEM.P.C./P-	UNITED KINGDOM ATOMIC ENERGY AUTHORITY. RESEARCH GP. ATOMIC ENERGY RES. ESTAB., HARWELL, BERKS, ENGLAND
RCI-	NAVAL ORDNANCE TEST STA., INYOKERN (CHINA LAKE), CAL.
RCI-QR-	RESEARCH CHEMICALS INC., BURBANK, CALIF.
RCL-	GT. BRIT. RADIO COMPONENTS STANDARDISATION COMMITTEE
RCM-	BATTELLE MEMORIAL INST., COLUMBUS, OHIO. (TECHNICAL PROGRESS REVIEWS. REACTOR CORE MATERIALS)
RCN-	NETHERLANDS. REACTOR CENTRUM, THE HAGUE
RCN-	NETHERLANDS. REACTOR CENTRUM, PETTEN
RCN-	NETHERLANDS. REACTOR CENTRUM, THE HAGUE
RCN-	NORWAY. INSTITUTT FOR ATOMENERGI, KJELLER
RCN-INT-	NETHERLANDS. REACTOR CENTRUM, THE HAGUE
RC-QPR-	RAYTHEON CO. RESEARCH DIV., WALTHAM, MASS.
RC-R-	RADIOPLANE CO., VAN NUYS, CALIF.
RCRDE-	ROME AIR DEVELOPMENT CENTER, GRIFFISS AFB, N.Y.
RCRP-	ROME AIR DEVELOPMENT CENTER, GRIFFISS AFB, N.Y.
RC-RPE-(YEAR)-	RADIOPLANE CO., VAN NUYS, CALIF.
RCRSS-	ROME AIR DEVELOPMENT CENTER, GRIFFISS AFB, N.Y.
RCRTC-	ROME AIR DEVELOPMENT CENTER, GRIFFISS AFB, N.Y.
RCRTG-	ROME AIR DEVELOPMENT CENTER, GRIFFISS AFB, N.Y.
RCRTN-	ROME AIR DEVELOPMENT CENTER, GRIFFISS AFB, N.Y.
RCRTR-	ROME AIR DEVELOPMENT CENTER, GRIFFISS AFB, N.Y.
RCS-	ARMY, WASHINGTON, D.C. (REPORTS CONTROL SYMBOL)
RC-S-(YEAR)-	ARMY MISSILE COMMAND, ENGINEERING REQUIREMENTS OFFICE, REDSTONE ARSENAL, ALA.
RCS-	GT. BRIT. RADIO COMPONENTS STANDARDISATION COMMITTEE
RC-S-	RAYTHEON CO. HIGH TEMP. MATERIALS DEPT.,WALTHAM, MASS
RC-S-	RAYTHEON CO. MISSILE SYSTEMS DIV., BEDFORD, MASS.
RC-S-	RAYTHEON CO. RESEARCH DIV., WALTHAM, MASS.
RCSC-	GT. BRIT. ATOMIC EN. RES. ESTAB., HARWELL, BERKS, ENG
RCS-DD-DR+E(M)-	BALLISTIC MISSILE DIV. (AIR FORCE), LOS ANGELES
RCS-DD-SD(M)-	BALLISTIC MISSILE DIV. (AIR FORCE), LOS ANGELES
RCS-MEDDH-	ARMY RESEARCH INST. OF ENVIRONMENTAL MEDICINE, NATICK, MASS.
RCS-OAR-	OFFICE OF AEROSPACE RESEARCH (AIR FORCE), WASH., D.C.
RCS-OSD-	ARMY ELECTRONICS LABS., FORT MONMOUTH, N.J.
RCS-OSD-	GENERAL ELECTRIC CO., OWENSBORO, KY.
RCS-OSD-	RADIO CORP. OF AMERICA, N.Y.C.
RCS-SMIQ-	ARMY MISSILE COMMAND, REDSTONE ARSENAL, ALA.
RCTC-	GT. BRIT. ATOMIC EN. RES. ESTAB., HARWELL, BERKS, ENG
RCTC-	UNITED KINGDOM ATOMIC ENERGY AUTHORITY. RESEARCH GP. ATOMIC ENERGY RES. ESTAB., HARWELL, BERKS, ENGLAND
RCTC/P-	UNITED KINGDOM ATOMIC ENERGY AUTHORITY. RESEARCH GP. ATOMIC ENERGY RES. ESTAB., HARWELL, BERKS, ENGLAND
RC-TM-T-	RAYTHEON CO. RESEARCH DIV., WALTHAM, MASS.
RD-	(RESEARCH AND DEVELOPMENT) (REF. 2)
RD-(YEAR)-	AIRBORNE INSTRUMENTS LAB., INC., DEER PARK, N.Y.
RD-	ALLIS-CHALMERS MFG. CO. ATOMIC ENERGY DIV.,WASH.,D.C.
RD-	ALLISON DIV., GENERAL MOTORS CORP., INDIANAPOLIS
RD-	AMERICAN OPTICAL CO. RESEARCH DIV.,SOUTHBRIDGE, MASS.
RD-	ARGONNE NATIONAL LAB., LEMONT, ILL.
R + D-	BABCOCK + WILCOX CO. (ALL LOCATIONS)
RD-(YEAR)-	BENDIX AVIATION CORP. PACIFIC DIV.,N. HOLLYWOOD, CAL.
R/D-	BRITISH OXYGEN CO., LTD., LONDON
RD-	BROADVIEW RESEARCH CORP., BURLINGAME, CALIF.
R/D-(NUMBER-YEAR)	CALIFORNIA. DIV. OF HIGHWAYS. BUDGET DEPT.

RD

RD-	CHICAGO OPERATIONS OFFICE, AEC
RD-	COLUMBIA UNIV., N.Y.C.
RD-	CONTROL DATA CORP. RESEARCH DIV., MINNEAPOLIS
RD-(NO.-NO.)(MO./YR.)	DEFENSE ATOMIC SUPPORT AGENCY.RADIATION DIV., WASH,DC
RD-	ELDORADO MINING + REFINING LTD. RES.+ DEV.DIV.,OTTAWA
RD-(YEAR)-	ENVIRONMENTAL SCIENCE SERVICES ADM.,ATLANTIC CITY,NJ.
R/D-(4 DIGITS)	FALCON RESEARCH AND DEV. CO. TECHNODYNE DIV., DENVER
RD-(YEAR)	FEDERAL AVIATION ADMINISTRATION. SYSTEMS RESEARCH AND DEV. SERVICE, WASHINGTON, D.C. (RESEARCH + DEV.)
RD-(YEAR)-	GARRETT CORP. AIRESEARCH MFG. DIV., LOS ANGELES
RD-	GENERAL ELECTRIC CO. RESEARCH LAB., SCHENECTADY, N.Y.
RD(NUMBER(S))-	GYRODYNE CO. OF AMERICA, INC., ST. JAMES, N.Y.
RD-	HONEYWELL, INC., MINNEAPOLIS
RD-	INTERCEPTOR WEAPONS SCHOOL, TYNDALL AFB, FLA.
RD-	KELLEX CORP., N.Y.C.
RD-(NUMBER)	KELLOGG (M.W.) CO. RES. + DEV. CTR., PISCATAWAY, N.J.
RD-(YEAR)-	KELLOGG (M.W.) CO. RES. + DEV. CTR., PISCATAWAY, N.J.
RD-(NUMBERS)-	MICRO-RADIONICS, INC., VAN NUYS, CALIF.
RD-(YEAR)-	NATIONAL AVIATION FACILITIES EXPTL.CTR.,ATLANTIC CITY
RD-	NAVAL RESEARCH LAB., WASHINGTON, D.C.
R + D-	NAVAL WEAPONS STATION, YORKTOWN, VA.
RD-	NEW ZEALAND. DEPT. OF SCI.+ IND. RES. RADIO RES. OFF.
RD-	OFF. OF THE ASST. SECY. OF DEFENSE(RES.+DEV.),WASH,DC
RD-(NO.)/-	OFF. OF THE ASST. SECY. OF DEFENSE(RES.+DEV.),WASH,DC
RD-(YEAR)-	PARAMETRICS, INC., WALTHAM, MASS.
RD-	PHILIPS LABS., INC. RESEARCH AND DEVELOPMENT DEPT., IRVINGTON-ON-HUDSON, N.Y.
RD-(NUMBER)-R	REX DIV., GARRETT CORP., LOS ANGELES
RD-	ROCKETDYNE DIV., NORTH AMERICAN AVIATION, INC., CANOGA PARK, CALIF.
RD-(NUMBER)-	SPERRY GYROSCOPE CO. RADIATION DIV., GREAT NECK, NY
RD-	SPERRY RAND CORP. SPERRY GYROSCOPE DIV.,GREAT NECK,NY
RD-	SYLVANIA ELECTRIC PRODUCTS INC., BAYSIDE, N.Y.
RD-(YR)-	WYMAN-GORDON CO., WORCESTER, MASS.
RDA-	DIVISION OF TECHNICAL INFORMATION, AEC (RESEARCH AND DEVELOPMENT ABSTRACTS OF THE USAEC)
RD/AD-	SOCIETE EUROPEENNE DE SEMICONDUCTEURS ET DE MICROELECTRONIQUE, PARIS
RDA-DC-4	VITRO CORP. OF AMERICA, N.Y.C.
RDAF-	NAVAL ORDNANCE TEST STA., INYOKERN (CHINA LAKE), CAL.
RDAP-	NAVAL ORDNANCE TEST STA., INYOKERN (CHINA LAKE), CAL.
RD ARM RI-	GT. BRIT. DIRECTORATE OF ARMAMENT RES. + DEV. (AIR)
RDA-TC-	HANFORD WORKS, RICHLAND, WASH.
RDB-	(AMERICAN RECIPIENTS VERSION OF BRITISH DESIGNATION R.+D.B.-, WHICH SEE ALSO)
RDB-	OFF. OF THE ASST. SECY. OF DEFENSE(RES.+DEV.),WASH,DC
RDB-(NO.)/-	RESEARCH AND DEVELOPMENT BOARD, WASHINGTON, D.C.
RDB(C)-	GT. BRIT. CULCHETH LABS., LANCS, ENGLAND
R.+D.B.(C)(LETTERS)-	GT. BRIT. CULCHETH LABS., LANCS, ENGLAND (REF. 36)
R.+D.B.(CA)(LETTERS)-	UNITED KINGDOM ATOMIC ENERGY AUTHORITY. INDUSTRIAL GROUP. CAPENHURST WORKS, CHES., ENGLAND (REF. 36)
R.+D.B.(CAP.)-	GT. BRIT. CAPENHURST WORKS, CHES., ENGLAND (REF. 36)
RDB(CAP)/R-	GT. BRIT. CAPENHURST WORKS, CHES., ENGLAND
RDB(CAP)-TN-	GT. BRIT. CAPENHURST WORKS, CHES., ENGLAND
RDB(CA)/TN-	UNITED KINGDOM ATOMIC ENERGY AUTHORITY. INDUSTRIAL GROUP. CAPENHURST WORKS, CHES., ENGLAND (REF. 36)
RDB(C)TM-	UNITED KINGDOM ATOMIC ENERGY AUTHORITY. INDUSTRIAL GROUP. CULCHETH LABS., LANCS, ENGLAND
RDB(C)/TN-	UNITED KINGDOM ATOMIC ENERGY AUTHORITY. INDUSTRIAL GROUP. CULCHETH LABS., LANCS, ENGLAND
RDB(C)/TN-	UNITED KINGDOM ATOMIC ENERGY AUTHORITY. METALLURGICAL LABS., DIV. OF ATOMIC ENERGY (PRODUCTION), RESEARCH AND DEVELOPMENT BRANCH, CULCHETH, LANCS, ENGLAND
RDB-EL-(NO.)/-	RESEARCH AND DEVELOPMENT BOARD. AD HOC GROUP ON RELIABILITY OF ELECTRONIC EQUIPMENT, WASH., D. C.
RDB-FRO-	RES. + DEV. BD. COMM. ON FUELS + LUBRICANTS, WASH.,DC
RDB-LAP-	RESEARCH AND DEVELOPMENT BOARD, WASHINGTON, D.C.
RD/B/M-	GT. BRIT. CENTRAL ELECTRICITY GENERATING BOARD. BERKELEY NUCLEAR LABS., BRISTOL, ENGLAND
RDB-MT-(NO.)/-	RESEARCH AND DEVELOPMENT BOARD, WASHINGTON, D.C.
RDB-MTRI-	RESEARCH AND DEVELOPMENT BOARD, WASHINGTON, D.C.
RD/B/N-	GT. BRIT. CENTRAL ELECTRICITY GENERATING BOARD. BERKELEY NUCLEAR LABS., BRISTOL, ENGLAND
RDB-NE-	RADIO CORP. OF AMERICA. RCA LABS. DIV., PRINCETON, NJ
RD/B/R-	GT. BRIT. CENTRAL ELECTRICITY GENERATING BOARD. BERKELEY NUCLEAR LABS., BRISTOL, ENGLAND
RDB(R)/	UNITED KINGDOM ATOMIC ENERGY AUTHORITY. INDUSTRIAL GROUP H.Q., RISLEY, LANCS, ENGLAND
R.+D.B.(R)/(LETTERS)-	UNITED KINGDOM ATOMIC ENERGY AUTHORITY. INDUSTRIAL GROUP H.Q., RISLEY, LANCS, ENGLAND (REF. 36)
RDB(R)/TM-	UNITED KINGDOM ATOMIC ENERGY AUTHORITY. INDUSTRIAL GROUP H.Q., RISLEY, LANCS, ENGLAND
RDB(R)/TN-	UNITED KINGDOM ATOMIC ENERGY AUTHORITY. INDUSTRIAL GROUP H.Q., RISLEY, LANCS, ENGLAND
R.+D.B.(S)(LETTERS)-	UNITED KINGDOM ATOMIC ENERGY AUTHORITY. INDUSTRIAL GROUP. SPRINGFIELDS WORKS, LANCS, ENG. (REF. 36)
RDB(S)/TM-	UNITED KINGDOM ATOMIC ENERGY AUTHORITY. INDUSTRIAL GROUP. SPRINGFIELDS WORKS, LANCS, ENGLAND
RDB(S)/TN-	UNITED KINGDOM ATOMIC ENERGY AUTHORITY. INDUSTRIAL GROUP. SPRINGFIELDS WORKS, LANCS, ENGLAND
RDBT-	BENDIX AVIATION CORP. PACIFIC DIV.,N. HOLLYWOOD, CAL.
R.+D.B.(W)(LETTERS)-	UNITED KINGDOM ATOMIC ENERGY AUTHORITY. INDUSTRIAL GROUP.WINDSCALE WORKS,SELLAFIELD,CUMB.,ENG.(REF.36)
RDB(W)/TM-	UNITED KINGDOM ATOMIC ENERGY AUTHORITY. INDUSTRIAL GROUP. WINDSCALE WORKS, SELLAFIELD, CUMB.,ENGLAND
RDB(W)/TN-	UNITED KINGDOM ATOMIC ENERGY AUTHORITY. INDUSTRIAL GROUP. WINDSCALE WORKS, SELLAFIELD, CUMB., ENGLAND
RDD-	(RESEARCH AND DEVELOPMENT DIV.) (REF. 2)
RDD-	AIR RESEARCH AND DEVELOPMENT COMMAND. DIRECTORATE OF ELECTRONICS, BALTIMORE
RDD-	THOMPSON (H. I.) FIBER GLASS CO., GARDENA, CALIF.
RDD-	WATERVLIET ARSENAL. BENET LABS., N.Y.
RDDR-	NAVAL MINE DEPOT. RESEARCH AND DEVELOPMENT DIV., YORKTOWN, VA. (REPORT)
RDDR-	TURKEY. MINISTRY OF NATIONAL DEFENCE, ANKARA
RDDRT-	WAR DEPT. RESEARCH AND DEV. DIV. (TECH. REPT.)
RDD-TRANS-	UNITED STEEL COMPANIES, LTD., SHEFFIELD, YORKS, ENG.
R.D.EXPLOSIVES REPT.-	GT. BRIT. ADVISORY COUNCIL ON SCI. RES. + TECH. DEV.
RDF-	GT. BRIT. ADVISORY COUNCIL ON SCI. RES. + TECH. DEV.
RDFU/V-	COMBAT DEVELOPMENT AND TEST CENTER, VIETNAM
RDG-	ARGONNE NATIONAL LAB., LEMONT, ILL.
RD(H)/F-	GT. BRIT. MINISTRY OF SUPPLY
RDI-	ATOMIC ENERGY OF CANADA LTD. CHALK RIVER PROJECT, ONT. (REF. 6)
RDMA-	NAVAL ORDNANCE TEST STA., INYOKERN (CHINA LAKE), CAL.
RDMD-	NAVAL ORDNANCE TEST STA., INYOKERN (CHINA LAKE), CAL.
RD-MPR-	TENNESSEE EASTMAN CORP., OAK RIDGE, TENN.
RDMR-	NAVAL ORDNANCE TEST STA., INYOKERN (CHINA LAKE), CAL.
RDO-	TENNESSEE EASTMAN CORP., OAK RIDGE, TENN.
RDO-R-	WRIGHT AIR DEVELOPMENT CENTER. MATERIALS LAB., WRIGHT-PATTERSON AFB, OHIO
RDP-	ATOMIC ENERGY OF CANADA LTD. CHALK RIVER PROJ., ONT.
RD/P-	AUSTRALIA. DEFENCE SCIENTIFIC SERVICE, MELBOURNE
RDP-	COLLINS RADIO CO., CEDAR RAPIDS, IOWA (RESEARCH AND DEVELOPMENT PROPOSAL)

RD-QPR-(YR.)-(NO.)-	REPUBLIC AVIATION CORP., FARMINGDALE, N.Y.
RD-QTR-(YR.)-(NO.)-	REPUBLIC AVIATION CORP., FARMINGDALE, N.Y.
RDR-	(RESEARCH AND DEVELOPMENT REPORT) (REF. 2)
RDR-	(RESEARCH DIVISION REPORT) (REF. 2)
RD-R-	AEROJET-GENERAL CORP., AZUSA, CALIF.
RDR-	FARNSWORTH TELEVISION AND RADIO CORP. RESEARCH DEPT., FORT WAYNE
RDR-	GARRETT CORP. AIRESEARCH MFG. CO. OF ARIZONA, PHOENIX
RDR-(YEAR)-	PENNSYLVANIA. UNIV., PHILADELPHIA. MOORE SCHOOL OF ELECTRICAL ENG. (USED 1946-50 FOR RES. + DEV. RPT.)
RDR-(NUMBER)-	ROCKETDYNE DIV., NORTH AMERICAN AVIATION, INC., CANOGA PARK, CALIF.
RDR-	SOLAR AIRCRAFT CO., SAN DIEGO, CALIF.
R.+D.(R)-	UNITED KINGDOM ATOMIC ENERGY AUTHORITY. INDUSTRIAL GROUP H.Q., RISLEY, LANCS, ENGLAND (REF. 36)
RD(R)-	UNITED KINGDOM ATOMIC ENERGY AUTHORITY. INDUSTRIAL GROUP H.Q., RISLEY, LANCS, ENGLAND
R+D RPT. NO.(MO)-(YR)-	MASON (SILAS) CO., INC., BURLINGTON, IOWA
RDS-	(RESEARCH AND DEVELOPMENT SURVEY) (REF. 2)
RD-S-	BUNKER-RAMO CORP., SILVER SPRING, MD.
RDS-(NO.)(LETTER)	CHEMICAL CORPS SCHOOL, ARMY CHEMICAL CENTER, MD.
RDS-	EVANS SIGNAL LAB., BELMAR, N.J.
RDS-	MINNESOTA MINING AND MFG. CO., ST. PAUL (RESEARCH AND DEVELOPMENT SURVEY)
RDS-	NAVAL RESEARCH LAB., WASHINGTON, D.C.
RDS-	PICATINNY ARSENAL. ROCKET DEV. SEC., DOVER, N.J.
RDSM-	BENDIX AVIATION CORP. PACIFIC DIV.,N. HOLLYWOOD, CAL.
RDT-C-(NUMBER-NUMBER)-T	DIVISION OF REACTOR DEV. + TECH., AEC, WASHINGTON,DC
RDT-E-(NUMBER-NUMBER)-T	DIVISION OF REACTOR DEV. + TECH., AEC, WASHINGTON,DC
RDT-F-(NUMBER-NUMBER)-T	DIVISION OF REACTOR DEV. + TECH., AEC, WASHINGTON,DC
RD-TM-(YEAR)	ARMY MISSILE COMMAND. ADVANCED SYSTEMS LAB., REDSTONE ARSENAL, ALA. (TECHNICAL MEMORANDUM)
RDT-M-(NUMBER)	DIVISION OF REACTOR DEV. + TECH., AEC, WASHINGTON,DC
RD-TN-(YEAR)-	ARMY MISSILE COMMAND. ADVANCED SYSTEMS LAB., REDSTONE ARSENAL, ALA. (TECHNICAL NOTE)
RD-TPR-(YEAR)-	ARMY MISSILE COMMAND. ADVANCED SYSTEMS LAB., REDSTONE ARSENAL, ALA. (TECHNICAL PROGRESS REPORT)
RDTR-	(RESEARCH DIVISION TECHNICAL REPORT) (REF. 2)
RD-TR-(NUMBER)-(YEAR)	ARMY MISSILE COMMAND. ADVANCED SYSTEMS LAB., REDSTONE ARSENAL, ALA.
RD-TR-(YEAR)-	ARMY MISSILE COMMAND. AEROBALLISTICS DIRECTORATE, REDSTONE ARSENAL, ALA. (TECHNICAL REPORT)
RDTR-	INET, INC., LOS ANGELES
RDTR-	NAVAL AMMUNITION DEPOT, CRANE, IND.
RDU-	ORDNANCE RES., INC., FORT WALTON BEACH, FLA.
RDU-	NAVAL ORDNANCE TEST STA., INYOKERN (CHINA LAKE), CAL.
RDUB-	NAVAL ORDNANCE TEST STA., INYOKERN (CHINA LAKE), CAL.
RDUI-	NAVAL ORDNANCE TEST STA., INYOKERN (CHINA LAKE), CAL.
RDX-	GT. BRIT. ADVISORY COUNCIL ON SCI. RES. + TECH. DEV.
R+D/Y-	NAVAL MINE DEPOT, YORKTOWN, VA. (RES. + DEV. REPT.)
RE-(YEAR)-	AIR REDUCTION CO., INC. RESEARCH AND ENGINEERING DEPT., MURRAY HILL, N.J. (USED 1963-)
RE-(YEAR)-(NO.)-CRE-	AIR REDUCTION CO., INC. RESEARCH AND ENGINEERING DEPT., MURRAY HILL, N.J.
RE-	BENDIX CORP. NAVIGATION + CONTROL DIV,.TETERBORO,N.J.
RE-	BUREAU OF SHIPS (ELECTRONICS INSTRUCTION BOOK)
RE-(NUMBER-YEAR)	CULLEN COLL. OF ENGINEERING, HOUSTON, TEX.
RE-(YEAR)-(NUMBER)-	CURTISS-WRIGHT CORP. RESEARCH DIV., CLIFTON, N.J.
RE-(NO.)-	CURTISS-WRIGHT CORP. RES. DIV., QUEHANNA, PENNA.
RE-(YR.)-(NO.)-	CURTISS-WRIGHT CORP. RES. DIV., QUEHANNA, PENNA.
R/E-(YEAR)-	EITEL-MC CULLOUGH, INC., SAN CARLOS, CALIF.
RE-	ELECTRO-MECHANICAL RESEARCH, INC., SARASOTA, FLA.
RE-	GRUMMAN AIRCRAFT ENGINEERING CORP., BETHPAGE, N.Y.
RE-(NO.)-PR-	GRUMMAN AIRCRAFT ENGINEERING CORP., BETHPAGE, N.Y.
RE-(NUMBER)(LETTER)	GRUMMAN AIRCRAFT ENG. CORP. RES. DEPT., BETHPAGE, NY
RE(NO.)-(NO.)-CEJ-	HYCON MFG. CO., MONROVIA, CALIF.
RE-	MASSACHUSETTS INST. OF TECH., CAMBRIDGE
RE-(YEAR)-	UNITED STATES STEEL CORP. APPLIED RESEARCH LAB., MONROEVILLE, PENNA.
REA-	REA (J.B.) CO., LOS ANGELES
REA-	REA (J.B.) CO., SANTA MONICA, CALIF.
REA-	RURAL ELECTRIFICATION ADM. (AGRICULTURE), WASH., D.C.
REAL ESTATE-	CONNECTICUT. UNIV., STORRS. CENTER FOR REAL ESTATE AND URBAN ECONOMIC STUDIES
REB-A-	TENNESSEE EASTMAN CORP., OAK RIDGE, TENN.
REC-	ROSEMOUNT ENGINEERING CO. AIR DATA SENSOR DEPT., MINNEAPOLIS
REC-ERC-(YEAR)-	BUREAU OF RECLAMATION (INTERIOR). ENG + RES. CTR., DENVER
RECI-	BATTELLE MEMORIAL INST. RADIATION EFFECTS INFORMATION CENTER, COLUMBUS, OHIO
REC-OCE-(YEAR)-	BUREAU OF RECLAMATION (INTERIOR). OFFICE OF CHIEF ENGINEER, DENVER
RECON-RC-	RECON, INC., TALLAHASSEE
RECON-TR-	RECON, INC., TALLAHASSEE (TECHNICAL REPORT)
RE-CR-	MONSANTO CHEMICAL CO., DAYTON, OHIO
REC-TF-	RESDEL ENGINEERING CORP., PASADENA, CALIF.
REC-TR-	RESDEL ENGINEERING CORP., PASADENA, CALIF.
R-ED-	HONEYWELL, INC., MINNEAPOLIS
RED-(YEAR)-	KELLOGG (M.W.) CO. RES. + DEV. CTR., PISCATAWAY, N.J.
RED-	REDEL, INC., ANAHEIM, CALIF.
REDAR-RER-	DARLING (R.E.) CO., INC. BETHESDA, MD.
REDEL KPR-	REDEL, INC., ANAHEIM, CALIF.
RED-EPM-	ARGONNE NATIONAL LAB., ILL.
RED-EPM-MEMO-	ARGONNE NATIONAL LAB., ILL.
REEC-	REYNOLDS ELEC. + ENG. CO., INC., MERCURY, NEV.
REED-	REED COLLEGE, PORTLAND, ORE.
REF-	ARMY WEAPONS COMMAND. FUTURE WEAPONS SYSTEMS DIV., ROCK ISLAND, ILL.
REF-(YEAR)-	JOHNS HOPKINS UNIV., ANNAPOLIS. CHESAPEAKE BAY INST.
REF-	OREGON STATE UNIV., CORVALLIS. DEPT. OF OCEANOGRAPHY
REF-	TE CO., SANTA BARBARA, CALIF.
REF-(YEAR)-	TEXAS A + M UNIV., COLLEGE STATION. DEPT. OF METEOROLOGY
REF-(YEAR)-	TEXAS A + M UNIV., COLLEGE STATION. DEPT. OF OCEANOGRAPHY
REF-	WOODS HOLE OCEANOGRAPHIC INSTITUTION, MASS.
REF-(YEAR)-	WOODS HOLE OCEANOGRAPHIC INSTITUTION, MASS.
REF-BIB-	MASSACHUSETTS INST. OF TECH., LEXINGTON. LINCOLN LAB.
REF-GRO-	UNITED KINGDOM ATOMIC ENERGY AUTHORITY. WEAPONS GROUP ATOMIC WEAPONS RES. ESTAB.,ALDERMASTON, BERKS, ENG.
REF-M(YEAR)-	WASHINGTON. UNIV., SEATTLE. DEPT. OF OCEANOGRAPHY
REF. SYM.-...	SANDIA CORP., ALBUQUERQUE, N. MEX.
REF-VF-	ROYAL CANADIAN AIR FORCE
REGIONAL-(NUMBER-NUMBER)	TOLEDO REGIONAL AREA PLAN FOR ACTION, OHIO
REI-	ATOMIC ENERGY OF CANADA LTD. CHALK RIVER PROJECT, ONT. (REF. 6)
REIC-	BATTELLE MEMORIAL INST. RADIATION EFFECTS INFORMATION CENTER, COLUMBUS, OHIO
REIC-ACCESS-	BATTELLE MEMORIAL INST. RADIATION EFFECTS INFORMATION CENTER, COLUMBUS, OHIO (MONTHLY ACCESSION LIST)
REIC ACC LIST-	BATTELLE MEMORIAL INST. RADIATION EFFECTS INFORMATION CENTER, COLUMBUS, OHIO. (ACCESSION LIST)

REIC-AL-
    BATTELLE MEMORIAL INST. RADIATION EFFECTS INFORMATION CENTER, COLUMBUS, OHIO. (ACCESSION LISTS)
REIC-CAL-
    BATTELLE MEMORIAL INST. RADIATION EFFECTS INFORMATION CENTER, COLUMBUS, OHIO. (CLASSIFIED ACCESSION LIST)
REIC-MEMO-
    BATTELLE MEMORIAL INST. RADIATION EFFECTS INFORMATION CENTER, COLUMBUS, OHIO
REIC-TM-
    BATTELLE MEMORIAL INST. RADIATION EFFECTS INFORMATION CENTER, COLUMBUS, OHIO. (TECHNICAL MEMORANDA)
REIC-TR-
    TOHOKU UNIV., SENDAI, JAPAN. RESEARCH INST. OF ELECTRICAL COMMUNICATION
REI-TMD-
    OFFICE OF SCIENTIFIC RESEARCH AND DEVELOPMENT
REL-
    RAYMOND ENGINEERING LAB., INC., MIDDLETOWN, CONN.
REL-
    RESEARCH ENTERPRISES LTD., TORONTO
REL-HFG-
    CHRYSLER CORP. SPACE DIV., NEW ORLEANS
REM-
    BUREAU OF ORDNANCE (NAVY)
REM-
    CONVAIR, FORT WORTH, TEX.
REMC-
    GT. BRIT. RADIO AND ELECTRONICS MEASUREMENTS COMM.
REMC-
    JOINT DEPARTMENTAL RADIO AND ELECTRONICS MEASUREMENTS COMMITTEE
REN-
    GT. BRIT. MINISTRY OF HOME SECURITY
R.E.N.-
    GT. BRIT. MINISTRY OF HOME SECURITY. RESEARCH AND EXPERIMENTS DEPT., LONDON
REP-
    DOW CHEMICAL CO. ROCKY FLATS DIV., GOLDEN, COLO.
REP-(NO.)/(YR.)
    JOHANNESBURG. UNIV. OF THE WITWATERSRAND. DEPT. OF MECHANICAL ENGINEERING
(REPORT NO.)-(PROJ. NO.)
    ECLIPSE-PIONEER DIV., BENDIX AVIATION CORP., TETERBORO, N. J. (REF. 1)
(REPORT NO.)-(CODE LTR)
    MENASCO MANUFACTURING CO., BURBANK, CALIF. (REF. 1)
(REPORT NO.)-(COPY NO.)
    ANTIAIRCRAFT ARTILLERY AND GUIDED MISSILES CENTER, FORT BLISS, TEX. (REF. 1)
REPORT R-
    MASS. INST. OF TECH., CAMBRIDGE. SERVOMECHANISMS LAB.
RE-PR-(YEAR)-
    ARMY MISSILE COMMAND. ELECTROMAGNETICS LAB., REDSTONE ARSENAL, ALA. (PROPOSAL)
REPT-
    (REPORT) (REF. 2)
REPT-(YEAR)-
    CANADA. DEPT. OF MINES AND TECHNICAL SURVEYS, OTTAWA
REPT-
    CRANFIELD INST. OF TECH., BUCKS, ENGLAND
REPT-
    DANISH DEFENCE RESEARCH BOARD, COPENHAGEN
REPT.-
    DANISH METEOROLOGICAL INST., CHARLOTTENLUND
REPT-(NUMBER/NUMBER)SR
    EXPLOSIVE TECHNOLOGY, INC., FAIRFIELD, CALIF.
REPT-
    FINNISH METEOROLOGICAL OFFICE, HELSINKI
REPT-
    FRANCE. DIRECTION DE LA METEOROLOGIE NATIONAL, PARIS
REPT-(NUMBER/YEAR)
    FRAUNHOFER INSTITUT, FREIBURG IM BREISGAU, GERMANY
REPT-(NO.LTR.NO.)-TOPS-
    GENERAL ELECTRIC CO., PHILADELPHIA (THERMOELECTRIC OUTER PLANET SPACECRAFT)
REPT-(YEAR)-HVO-
    GENERAL ELECTRIC CO. SPACE DIV., HUNTSVILLE, ALA.
REPT-(NUMBER)-F
    GERMANY. BUNDESPOST. IONOSPHAREN-INSTITUT, BREISACH
REPT-FB-(NUMBER)-(YEAR)
    FRAUNHOFER-GESELLSCHAFT ZUR FORDERUNG DER ANGEWANDTEN FORSCHUNG E.V., DARMSTADT, GERMANY
REPT-S-(NUMBER-YEAR)
    HELSINKI UNIV.
RER-
    BENDIX CORP. ELECTRODYNAMICS DIV., NO. HOLLYWOOD, CAL
RER-
    GEN. MOTORS CORP. HARRISON RADIATOR DIV., LOCKPORT,NY
RE-S-(YEAR)-
    ARMY MISSILE COMMAND. ELECTROMAGNETICS LAB., REDSTONE ARSENAL, ALA.
RE-S-(YEAR)-
    ARMY SIGNAL MISSILE SUPPORT AGENCY, WHITE SANDS MISSILE RANGE, N. MEX.
RE-S-(YEAR)-
    ARMY TEST AND EVALUATION COMMAND, WHITE SANDS MISSILE RANGE, N. MEX.
RES-
    GRACE (W.R.) AND CO. RESEARCH DIV., CLARKSVILLE, MD.
RES-(YR.)-(NO.)
    GRACE (W.R.) AND CO. RESEARCH DIV., CLARKSVILLE, MD.
RES-
    RCA VICTOR DIV., LTD. RESEARCH LABS., MONTREAL
RE-S-(YEAR)-
    WHITE SANDS MISSILE RANGE. NATIONAL RANGE ENG., N. MEX.
RESD-TR-
    INSTITUTE FOR DEFENSE ANALYSES. RESEARCH AND ENGINEERING SUPPORT DIV., ARLINGTON, VA.

RESEARCH CONTRIBUTION-
    CENTER FOR NAVAL ANALYSES. SYSTEMS EVALUATION GROUP, WASHINGTON, D.C.
RESEARCH CONTRIBUTION-
    NAVAL WARFARE ANALYSIS GROUP, WASHINGTON, D.C.
RESEARCH-CONTRIBUTION-
    OPERATIONS EVALUATION GROUP, WASHINGTON, D.C.
RESEARCH PAPER-
    ARCTIC INSTITUTE OF NORTH AMERICA, WASHINGTON, D.C.
RESEARCH PROGRAM MEETING-
    CALIFORNIA. UNIV., BERKELEY. RADIATION LAB.
RESEARCH STUDY-(NO.-NO.)
    NEBRASKA. STATE DEPT. OF ROADS. DIV. OF MATERIALS AND TESTS, LINCOLN
RE-SR-(YEAR)-
    ARMY MISSILE COMMAND. ELECTROMAGNETICS LAB., REDSTONE ARSENAL, ALA. (SPECIAL REPORT)
RE-TM-(YEAR)-
    ARMY MISSILE COMMAND. ADVANCED SENSORS LAB., REDSTONE ARSENAL, ALA.
RE-TM-(YEAR)-
    ARMY MISSILE COMMAND. ELECTROMAGNETICS LAB., REDSTONE ARSENAL, ALA. (TECHNICAL MEMORANDUM)
RETMA-
    RADIO ELECTRONICS + TELEVISION MFRS. ASSN.,WASH.,DC
RE-TN-(YEAR)-
    ARMY MISSILE COMMAND. ADVANCED SENSORS LAB., REDSTONE ARSENAL, ALA.
RE-TN-(YEAR)-
    ARMY MISSILE COMMAND. ELECTROMAGNETICS LAB., REDSTONE ARSENAL, ALA. (TECHNICAL NOTE)
RE-TR-
    ARMY MISSILE COMMAND. RESEARCH AND ENGINEERING DIRECTORATE, REDSTONE ARSENAL, ALA. (PROVISIONAL)
RE-TSR-
    ARMY MISSILE COMMAND. ADVANCED SENSORS LAB., REDSTONE ARSENAL, ALA.
REVIEW-
    SCHOOL OF AEROSPACE MEDICINE, BROOKS AFB, TEX. (AEROMEDICAL REVIEW)
REXN-
    BUREAU OF ORDNANCE (NAVY)
RF-(NUMBER)-
    ARMY MISSILE COMMAND. FUTURE MISSILE SYSTEMS DIV., REDSTONE ARSENAL, ALA.
RF-(NUMBER)-O
    BOELKOW ENTWICKLUNGEN K.G., MUNICH
RF-
    COAST AND GEODETIC SURVEY, WASHINGTON, D.C. (RADIO FACILITY CHARTS)
RF-(3 DIGITS)
    ELECTRIC + MUSICAL INDUSTRIES, LTD.,HAYES, MIDDX,ENG.
RF-(YEAR)-
    GRUMMAN AIRCRAFT ENGINEERING CORP., BETHPAGE, N.Y.
RF-(NUMBER)-O
    MESSERSCHMITT-BOELKOW G.M.B.H., MUNICH
RF-
    OHIO STATE UNIV., COLUMBUS. SYSTEMS RESEARCH GROUP
RF-(NO.)-AR-(YEAR)-
    OHIO STATE UNIV., COLUMBUS. SYSTEMS RESEARCH GROUP
RF-(NO.)-
    OHIO STATE UNIV. RESEARCH FOUNDATION, COLUMBUS
RF-(NO.)-TR-
    OHIO STATE UNIV. RESEARCH FOUNDATION, COLUMBUS
RFA-
    AKTIEBOLAGET ATOMENERGI, STOCKHOLM
RFB-
    BEERS (ROLAND F.), INC., ALEXANDRIA, VA.
RFC-
    RECONSTRUCTION FINANCE CORP.
RFD-
    DOW CHEMICAL CO. ROCKY FLATS DIV., GOLDEN, COLO.
RFENG-R-(YEAR)-
    GRUMMAN AEROSPACE CORP., BETHPAGE, N.Y.
RFF-
    RESEARCH FLIGHT FACILITY, MIAMI, FLA.
RF/H/(NUMBER/NUMBER)
    ELECTRIC + MUSICAL INDUSTRIES, LTD.,HAYES, MIDDX,ENG.
RFN-
    AKTIEBOLAGET ATOMENERGI, STUDSVIK, SWEDEN
RFP-
    DOW CHEMICAL CO. ROCKY FLATS DIV., DENVER
RFP-
    DOW CHEMICAL CO. ROCKY FLATS DIV., GOLDEN, COLO.
RFP-
    KAMAN AIRCRAFT CORP. NUCLEAR DIV., COLORADO SPRINGS
RFP-TRANS-
    DOW CHEMICAL CO. ROCKY FLATS DIV., GOLDEN, COLO.
RFR-
    AKTIEBOLAGET ATOMENERGI, STOCKHOLM
RFR-
    AKTIEBOLAGET ATOMENERGI, STUDSVIK, SWEDEN.
RFR-
    EUROPEAN THEATER OF OPERATIONS (ARMY). ALSOS MISSION REICHSFORSCHUNGSRAT FILE)
RFS-
    MOTOROLA INC., SCOTTSDALE, ARIZ.
RF-TM-(YEAR)-
    ARMY MISSILE COMMAND. FUTURE MISSILE SYSTEMS DIV., REDSTONE ARSENAL, ALA. (TECHNICAL MEMORANDUM)
RF-TN-(YEAR)-
    ARMY MISSILE COMMAND. FUTURE MISSILE SYSTEMS DIV., REDSTONE ARSENAL, ALA. (TECHNICAL NOTE)
RF/TN/A/
    IMPERIAL COLLEGE OF SCIENCE AND TECHNOLOGY, LONDON. DEPT. OF MECHANICAL ENGINEERING
RF-TR-(YEAR)-
    ARMY MISSILE COMMAND. FUTURE MISSILE SYSTEMS DIV., REDSTONE ARSENAL, ALA. (TECHNICAL REPORT)

RFX-	AKTIEBOLAGET ATOMENERGI, STOCKHOLM
RG-	GEOSCIENCE, INC., CAMBRIDGE, MASS.
RG-(NO.)-(YEAR)-N-	WESTERN CO. OF NORTH AMERICA. RES. DIV., DALLAS
R/GC/-	IMPERIAL CHEMICAL INDUSTRIES, LTD. GENERAL CHEMICAL DIV., WIDNES, LANCS, ENGLAND
RG-CC-	REGULUS I AND REGULUS II COORDINATION COMM. (REF. 15)
RG-CC-(YEAR)-	REGULUS I AND REGULUS II COORDINATION COMM. (REF. 15)
RG-O-	SANDERS ASSOCIATES, INC., NASHUA, N.H.
RG-S-(YEAR)-	ARMY INERTIAL GUIDANCE AND CONTROL LAB. AND CENTER, REDSTONE ARSENAL, ALA.
RG-SR-(YEAR)-	ARMY INERTIAL GUIDANCE AND CONTROL LAB. AND CENTER, REDSTONE ARSENAL, ALA.
RG-TM-(YEAR)-	ARMY INERTIAL GUIDANCE AND CONTROL LAB. AND CENTER, REDSTONE ARSENAL, ALA.
RG-TN-(YEAR)-	ARMY INERTIAL GUIDANCE AND CONTROL LAB. AND CENTER, REDSTONE ARSENAL, ALA.
RG-TR-(YEAR)-	ARMY INERTIAL GUIDANCE AND CONTROL LAB. AND CENTER, REDSTONE ARSENAL, ALA.
RH-	ATOMIC ENERGY COMMISSION, WASH.,DC.(REACTOR HANDBOOK)
RH-	ELECTRIC + MUSICAL INDUSTRIES, LTD.,HAYES, MIDDX,ENG.
RH-	ROCKETDYNE DIV., NORTH AMERICAN AVIATION, INC., CANOGA PARK, CALIF.
R+H-	ROHM AND HAAS CO., PHILADELPHIA
RHC-	ROCKHURST COLL., KANSAS CITY, MO.
RHC-	ROHM AND HAAS CO., PHILADELPHIA
RHCI-	ATOMIC ENERGY OF CANADA LTD., CHALK RIVER, ONT.
RHD-(VOL.)-(NO.)-	PUBLIC HEALTH SERVICE, WASHINGTON, D.C.
RHEEM-	RHEEM MFG. CO., DOWNEY, CALIF.
RHEEM R-	RHEEM MFG. CO. RES. + DEV. LABS., DOWNEY, CALIF.
RHEL/M-	UNITED KINGDOM. RUTHERFORD HIGH ENERGY LAB., CHILTON, DIDCOT, BERKS, ENGLAND
RHEL/R-	UNITED KINGDOM. RUTHERFORD HIGH ENERGY LAB., CHILTON, DIDCOT, BERKS, ENGLAND
RHEL/S-	UNITED KINGDOM. RUTHERFORD HIGH ENERGY LAB., CHILTON, DIDCOT, BERKS, ENGLAND
RHF-(NUMBER-NUMBER)	FRANCE. COMMISSARIAT A L'ENERGIE ATOMIQUE. CENTRE D'ETUDES NUCLEAIRES, GRENOBLE
RHM-	PHILLIPS PETROLEUM CO. ATOMIC EN. DIV., IDAHO FALLS
RHM(YEAR)-	UNITED KINGDOM ATOMIC ENERGY AUTHORITY. INDUSTRIAL GROUP H.Q., RISLEY, LANCS,ENGLAND (REACTOR HAZARDS)
RHMC-	RHEEM MFG. CO., DOWNEY, CALIF.
RHM-Z-(NO.)-(YR.)	ARMY ROCKET AND GUIDED MISSILE AGENCY, REDSTONE ARSENAL, ALA.
RH-TR-(YEAR)-	ARMY MISSILE COMMAND. DEV.DIV., REDSTONE ARSENAL, ALA
RI-	(REPORT OF INVESTIGATION) (REF. 2)
RI-	BUREAU OF MINES (REPORT OF INVESTIGATION)
RI-	HERCULES RESEARCH CENTER, WILMINGTON, DEL. (RESEARCH INVESTIGATION)
RI-	IOWA. STATE UNIV., IOWA CITY. DEPT. OF CHEMISTRY
RI-	IOWA. STATE UNIV., IOWA CITY. DEPT. OF PHYSICS AND ASTRONOMY
RI-	NAVAL MISSILE CENTER. RANGE INSTRUMENTATION DEPT.
RI-	NETHERLANDS. REACTOR INSTITUUT, DELFT
RI-(YEAR)-	PARIS. UNIVERSITE, ORSAY
RI-	PARIS. UNIVERSITE, ORSAY. LABORATOIRE DE L'ACCELERATEUR LINEAIRE
RI-(YEAR)-	PARIS. UNIVERSITE, ORSAY. LABORATOIRE DE L'ACCELERATEUR LINEAIRE
RI-(NUMBER)(MONTH)(YEAR)	RADIATION, INC., MELBOURNE, FLA.
RI-	REDEL, INC., ANAHEIM, CALIF.
RI-	RICE INST., HOUSTON, TEX.
RI-	ROCKEFELLER INST., N.Y.C.
RI-	SUPREME CMDR. FOR THE ALLIED POWERS. RES. + INFO. DIV
RIA-	RESEARCH INTERNATL. ASSOCS., WASH.,D.C.(TRANSLATIONS)
RIA-	ROCK ISLAND ARSENAL, ILL.
RIA-(YR.)-	ROCK ISLAND ARSENAL, ILL.
RIA-(LTRS.)-	ROCK ISLAND ARSENAL, ILL. (INDUSTRIAL PROJ. REPT.)
RIAD-TR-(YEAR)-	MARTIN-MARIETTA CORP. RESEARCH INSTITUTE FOR ADVANCED STUDIES, BALTIMORE
RIA-ES-(YR)-	ROCK ISLAND ARSENAL LAB., ILL.
RIAL-	ROCK ISLAND ARSENAL LAB., ILL.
RIAL (YR.)-	ROCK ISLAND ARSENAL LAB., ILL.
RIALR-	ROCK ISLAND ARSENAL LAB., ILL.
RIAS-	MARTIN-MARIETTA CORP. RESEARCH INSTITUTE FOR ADVANCED STUDIES, BALTIMORE
RIAS-(NUMBER)-	MARTIN-MARIETTA CORP. RESEARCH INSTITUTE FOR ADVANCED STUDIES, BALTIMORE
RIAS-AR(YEAR)	MARTIN-MARIETTA CORP. RESEARCH INSTITUTE FOR ADVANCED STUDIES, BALTIMORE (ANNUAL REPORT)
RIAS-QTR-	MARTIN-MARIETTA CORP. RESEARCH INSTITUTE FOR ADVANCED STUDIES, BALTIMORE
RIAS-TR-(YEAR)-	MARTIN-MARIETTA CORP. RESEARCH INSTITUTE FOR ADVANCED STUDIES, BALTIMORE
RIA-TR-(YEAR)-	ROCK ISLAND ARSENAL LAB., ILL.
RIB-	DIVISION OF BIOLOGY AND MEDICINE, AEC. RADIATION INSTRUMENTS BRANCH, WASHINGTON, D.C.
RIB-IR-	DIVISION OF BIOLOGY AND MEDICINE, AEC. RADIATION INSTRUMENTS BRANCH, WASHINGTON, D.C.
RIB-QR-	DIVISION OF BIOLOGY AND MEDICINE, AEC. RADIATION INSTRUMENTS BRANCH, WASHINGTON, D.C.
RIC-	REEVES INSTRUMENT CORP., N.Y.C.
RIC-	UNITED KINGDOM ATOMIC ENERGY AUTHORITY. RESEARCH GP. ATOMIC ENERGY RES. ESTAB., HARWELL, BERKS, ENGLAND
RICC/	INTERNATIONAL ATOMIC ENERGY AGENCY, VIENNA (CONFERENCE ON THE USE OF RADIOISOTOPES IN PHYSICAL SCIENCES AND INDUSTRY, COPENHAGEN, SEPT., 1960)
RICE-	RICE UNIV., HOUSTON, TEX.
RICE-(NUMBER)-PU	RICE UNIV., HOUSTON, TEX.
RICE-(NUMBER)-P	RICE UNIV., HOUSTON, TEX.
RICE-SR-	RICE UNIV., HOUSTON, TEX. (STATUS REPORT)
RICO-	REEVES INSTRUMENT CORP., N.Y.C.
RID-	SMITHS INDUSTRIES, LTD., LONDON
RID-	WHITE SANDS MISSILE RANGE, N. MEX. (RANGE INSTRUMENTATION DEVELOPMENT)
RID-0-	WHITE SANDS MISSILE RANGE, N. MEX.
RIEC-TR-	TOHOKU UNIV., SENDAI, JAPAN. RESEARCH INST. OF ELECTRICAL COMMUNICATION
RIFP-	KYOTO UNIV. RESEARCH INST. FOR FUNDAMENTAL PHYSICS
RIGO-	NETHERLANDS. RADIOBIOLOGISCH INSTITUUT TNO, RIJSWIJK
RIGO-(YEAR)/	NETHERLANDS. RADIOBIOLOGISCH INSTITUUT TNO, RIJSWIJK
RII-	RAVEN INDUSTRIES, INC., SIOUX FALLS, S.D.
RI-IER-	RADIATION, INC., MELBOURNE, FLA.
RIJNHUIZEN-(YR.)-	NETHERLANDS. FOM-INSTITUUT VOOR PLASMA-FYSICA, JUTPHAAS
RIMR-	ROCKEFELLER INST. FOR MEDICAL RESEARCH, N.Y.C.
RIMT-	GERMANY. REICH INSTITUTE FOR MATERIAL TESTING
RIS-	RESEARCH INFORMATION SERVICE, N.Y.C.
RIS-	RESEARCH INFORMATION SERVICE, N.Y.C. (TRANSLATIONS)
RISLEY-4001 THRU -5000	GT. BRIT. ATOMIC EN. RES. ESTAB., HARWELL, BERKS, ENG (ASSIGNED BY GT. BRIT. DIV. OF ATOMIC ENERGY (PRODUCTION) TO INCOMING REPORTS FROM A.E.R.E.,HARWELL)
RISLEY-5001 THRU -5256	GT. BRIT. DEPT. OF ATOMIC ENERGY. INDUSTRIAL GROUP H.Q., RISLEY, LANCS, ENGLAND (REF. 36)
RISLEY-6001 THRU -6286	GT. BRIT. DEPT. OF ATOMIC ENERGY. INDUSTRIAL GROUP H.Q., RISLEY, LANCS, ENGLAND (REF. 36)
RISLEY-7001 THRU -7008	GT. BRIT. DEPT. OF ATOMIC ENERGY. INDUSTRIAL GROUP H.Q., RISLEY, LANCS, ENGLAND (REF. 36)

RISLEY-8001 THRU -8151
: GT. BRIT. DEPT. OF ATOMIC ENERGY. INDUSTRIAL GROUP H.Q., RISLEY, LANCS, ENGLAND (REF. 36)
RISLEY-1001 THRU -2000
: GT. BRIT. DIV. OF ATOMIC ENERGY (PRODUCTION), RISLEY, LANCS, ENGLAND (ASSIGNED TO MISCELLANEOUS BRITISH DOCUMENTS)
RISLEY-3001 THRU -4000
: GT. BRIT. DIV. OF ATOMIC ENERGY (PRODUCTION), RISLEY, LANCS, ENGLAND (ASSIGNED TO MISCELLANEOUS BRITISH DOCUMENTS)
RISLEY-2001 THRU -3000
: IMPERIAL CHEMICAL INDUSTRIES, LTD. (ASSIGNED BY GT. BRIT. DIV. OF ATOMIC ENERGY (PRODUCTION) TO INCOMING REPORTS FROM IMPERIAL CHEM. INDUSTRIES)
RISLEY-
: UNITED KINGDOM ATOMIC ENERGY AUTHORITY. INDUSTRIAL GROUP H.Q., RISLEY, LANCS, ENGLAND (REF. 36)
RISLEY-TRANS-
: GT. BRIT. DEPT. OF ATOMIC ENERGY. INDUSTRIAL GROUP H.Q., RISLEY, LANCS, ENGLAND
RISO-(NUMBER-YEAR)
: ARMY TEST AND EVALUATION COMMAND, WHITE SANDS MISSILE RANGE, N. MEX.
RISO-
: DANISH ATOMIC ENERGY COMMISSION. RES. ESTAB., RISOE
RISO-
: WHITE SANDS MISSILE RANGE, N. MEX.
RISO-M-
: DANISH ATOMIC ENERGY COMMISSION. RES. ESTAB., RISOE
RITU-
: TEMPLE UNIV., PHILADELPHIA. RESEARCH INST.
RITU-(YEAR)-
: TEMPLE UNIV., PHILADELPHIA. RESEARCH INST.
RIU-
: RHODE ISLAND. UNIV., KINGSTON
RJ-
: BENDIX CORP. BENDIX PRODUCTS DIV., SOUTH BEND, IND.
RJ-
: INTERNATIONAL BUSINESS MACHINES CORP. RESEARCH DIV. SAN JOSE LAB., CALIF.
RJA-
: MARTIN CO. NUCLEAR DIV., BALTIMORE
RJS-
: PICATINNY ARSENAL. ROCKET AND JATO SEC., DOVER, N.J.
RJS-CD-
: TENNESSEE EASTMAN CORP., OAK RIDGE, TENN.
RJS-CD2-
: TENNESSEE EASTMAN CORP., OAK RIDGE, TENN.
RK-
: ARGONNE NATIONAL LAB., LEMONT, ILL.
RK-PR-(YEAR)-
: ARMY MISSILE COMMAND. PROPULSION LAB., REDSTONE ARSENAL, ALA. (PROPOSAL)
RK-SR-
: ARMY MISSILE COMMAND. ARMY PROPULSION LAB. AND CENTER, REDSTONE ARSENAL, ALA.
RK-TM-(YEAR)-
: ARMY MISSILE COMMAND. ARMY PROPULSION LAB. AND CENTER, REDSTONE ARSENAL, ALA.
RK-TN-
: ARMY MISSILE COMMAND. ARMY PROPULSION LAB. AND CENTER, REDSTONE ARSENAL, ALA.
RK-TN-(YEAR)-
: ARMY MISSILE COMMAND. PROPULSION LAB., REDSTONE ARSENAL, ALA. (TECHNICAL NOTE)
RK-TPR-(YEAR)-
: ARMY MISSILE COMMAND. PROPULSION LAB., REDSTONE ARSENAL, ALA. (TECHNICAL PROGRESS REPORT)
RK-TR-(YEAR)-
: ARMY MISSILE COMMAND. ARMY PROPULSION LAB. AND CENTER, REDSTONE ARSENAL, ALA.
RK-TSR-(YEAR)-
: ARMY INERTIAL GUIDANCE AND CONTROL LAB. AND CENTER, REDSTONE ARSENAL, ALA.
RK-TSR-
: ARMY MISSILE COMMAND. ARMY PROPULSION LAB. AND CENTER, REDSTONE ARSENAL, ALA.
RK-TSR-(YEAR)-
: ARMY MISSILE COMMAND. PROPULSION LAB., REDSTONE ARSENAL, ALA. (TECHNICAL STATUS REPORT)
RL-
: (RADIATION LAB.) (REF. 2)
RL-
: (READING LIST) (REF. 2)
RL-
: (RESEARCH LABORATORY) (REF. 2)
RL-1.7- THRU RL-37.9-
: CALIFORNIA. UNIV., BERKELEY. RADIATION LAB. (REF. 9)
65-RL-
: GENERAL ELECTRIC CO. GEN. ENG. LAB., SCHENECTADY, N.Y. (FIRST TWO DIGITS STAND FOR YEAR) (REF. 1)
RL-
: GENERAL ELECTRIC CO. RESEARCH LAB., SCHENECTADY, N.Y.
(YEAR)-RL-(NUMBER)-(LTR)
: GENERAL ELECTRIC CO. RESEARCH LAB., SCHENECTADY, N.Y.
(YEAR)-RL-
: GENERAL ELECTRIC CO. RESEARCH LAB., SCHENECTADY, N.Y. (REF. 1)
RL-(YR.)-
: KELLOGG (M.W.) CO. PETROLEUM AND CHEMICAL RESEARCH DEPT., JERSEY CITY
RL-
: MARE ISLAND NAVAL SHIPYARD. RUBBER LAB., CALIF.
RL-
: MASSACHUSETTS INST. OF TECH., CAMBRIDGE. RADIATION LAB.
RL-
: MOTOROLA INC., RIVERSIDE, CALIF.
RL-
: OREGON STATE UNIV., CORVALLIS
RL-3M-
: STANDARD OIL DEVELOPMENT CO. ESSO LABS., LINDEN, N.J.
RL-
: WASHINGTON. UNIV., SEATTLE. LAB. OF RADIATION BIOLOGY
RLC-(NUMBER)-
: ARGONNE NATIONAL LAB., ILL.
RLC-
: RHODES LEWIS CO., LOS ANGELES
RLC-
: TENNESSEE EASTMAN CORP., OAK RIDGE, TENN.
RL-CL-
: BROWN ENGINEERING CO., INC. RES. LABS., HUNTSVILLE, ALA
RLD-
: BENDIX CORP. RESEARCH LABS. DIV., SOUTHFIELD, MICH.
RL-DG-
: CHRYSLER CORP. MISSILE DIV., DETROIT
RLDP-(YEAR)-
: BENDIX CORP. RESEARCH LABS. DIV., SOUTHFIELD, MICH.
RLE-
: ELLIOTT BROS. RESEARCH LABS., LONDON
RLE-
: MASSACHUSETTS INST. OF TECH., CAMBRIDGE. RESEARCH LAB. OF ELECTRONICS
RLE(LETTERS)-
: HUGHES AIRCRAFT CO., CULVER CITY, CALIF. (RESEARCH)
RLE-TR-
: MASSACHUSETTS INST. OF TECH., CAMBRIDGE. RESEARCH LAB. OF ELECTRONICS
RLF-
: MOTOROLA INC. SYSTEMS RESEARCH LAB., RIVERSIDE, CALIF
RL-FG-
: CHRYSLER CORP. MISSILE DIV., DETROIT
RL-GEN-
: DOUGLAS AIRCRAFT CO., INC., TULSA, OKLA.
RL-GEN-
: HANFORD ATOMIC PRODUCTS OPERATION, RICHLAND, WASH.
RLI-
: ATOMIC ENERGY OF CANADA LTD., CHALK RIVER, ONT. (REF.6)
RLI-
: RAZDOW LABS., INC., NEWARK
RL/JHU/TR-
: JOHNS HOPKINS UNIV., BALTIMORE. RADIATION LAB.
RLK-
: BROWN-FIRTH RESEARCH LABS., SHEFFIELD, YORKS, ENGLAND
RLM-
: GERMANY. REICHSLUFTFAHRTMINISTERIUMS
RLM(LETTERS)-
: HUGHES AIRCRAFT CO., CULVER CITY, CALIF. (RESEARCH)
RL-MISC.-
: CALIFORNIA. UNIV., BERKELEY. RADIATION LAB. (REF. 9)
RLMM-
: MASSACHUSETTS INST. OF TECH., CAMBRIDGE. RESEARCH LABORATORY FOR MECHANICS OF MATERIALS
RL-NPE-
: CALIFORNIA. UNIV., BERKELEY. RADIATION LAB. (REF. 9)
RL-NRD-
: HANFORD ATOMIC PRODUCTS OPERATION, RICHLAND, WASH.
RLO-(CONTRACT NUMBER)-
: RICHLAND OPERATIONS OFFICE, AEC, WASH. (REF. 4)
RLO-(CONTRACT NO.)-T-
: RICHLAND OPERATIONS OFFICE, AEC, WASH.
RLO-
: RICHLAND OPERATIONS OFFICE, AEC, WASH.
RLP/P-
: GT. BRIT. ATOMIC EN. RES. ESTAB., HARWELL, BERKS, ENG
R.L.R.(NUMBER)-
: MASSACHUSETTS INST. OF TECH., CAMBRIDGE. RADIATION LAB.
RL-REA-
: HANFORD ATOMIC PRODUCTS OPERATION, RICHLAND, WASH.
RL-SA-
: HANFORD ATOMIC PRODUCTS OPERATION, RICHLAND, WASH.
RL-SEP-
: HANFORD ATOMIC PRODUCTS OPERATION, RICHLAND, WASH.
RLT-
: PRINCETON UNIV., N.J. FRICK CHEMICAL LAB.
RLT-A
: TENNESSEE EASTMAN CORP., OAK RIDGE, TENN.
RL-TM-(YEAR)-
: ARMY MISSILE COMMAND. GROUND SUPPORT EQUIPMENT LAB., REDSTONE ARSENAL, ALA. (TECHNICAL MEMORANDUM)
RLTM-ARL-
: AIR RESOURCES LAB., LAS VEGAS, NEV.
RLTM-ESL-
: ENVIRONMENTAL SCIENCE SERVICES ADM., BOULDER, COLO.
RLTM-ITS-
: INSTITUTE FOR TELECOMMUNICATION SCIENCES, BOULDER, COLO
RLTM-NSSL-
: NATIONAL SEVERE STORMS LAB., NORMAN, OKLA.
RLTM-POL-
: PACIFIC OCEANOGRAPHIC LAB., SEATTLE (RESEARCH LAB. TECH. MEMO)
RL-TN-(YEAR)-
: ARMY MISSILE COMMAND. ELECTROMAGNETICS LAB., REDSTONE ARSENAL, ALA.
RL-TN-(YEAR)-
: ARMY MISSILE COMMAND. GROUND SUPPORT EQUIPMENT LAB., REDSTONE ARSENAL, ALA. (TECHNICAL NOTE)
RL-TPR-(YEAR)-
: ARMY MISSILE COMMAND. GROUND SUPPORT EQUIPMENT LAB., REDSTONE ARSENAL, ALA. (TECHNICAL PROGRESS REPORT)
RL-TR-(YEAR)-
: ARMY MISSILE COMMAND. GROUND SUPPORT EQUIPMENT LAB., REDSTONE ARSENAL, ALA. (TECHNICAL REPORT)
RM-
: (MEMORANDUM) (REF. 2)
RM-
: (REFERENCE MATERIAL) (REF. 2)
RM-
: (RESEARCH MEMORANDUM) (REF. 2)

RM-	AEROJET-GENERAL CORP., SACRAMENTO, CALIF.
RM-	BUREAU OF SHIPS (RESEARCH MEMORANDA)
RM-	ETHYL CORP. RESEARCH AND DEVELOPMENT DEPT., FERNDALE, MICH. (RESEARCH MEMORANDA)
RM-	FOREST SERVICE. ROCKY MOUNTAIN STA., FT. COLLINS, COLO
RM(YEAR)(LETTERS)-	GENERAL ELECTRIC CO. (ALL LOCATIONS) (REF. 46)
RM-	GENERAL ELECTRIC CO. SPECIALTY TRANSFORMER DEPT., FT. WAYNE, IND.
RM-(NO.)TMP-	GENERAL ELECTRIC CO. TECHNICAL MILITARY PLANNING OPERATION, SANTA BARBARA, CALIF.
RM-(YR.)TMP-	GENERAL ELECTRIC CO. TECHNICAL MILITARY PLANNING OPERATION, SANTA BARBARA, CALIF.
R + M-	GT. BRIT. AERONAUTICAL RESEARCH COUNCIL
RM-	GT. BRIT. AERONAUTICAL RESEARCH COUNCIL
R/M-	GT. BRIT. ATOMIC EN. RES. ESTAB., HARWELL, BERKS, ENG
RM-	GRUMMAN AIRCRAFT ENGINEERING CORP., BETHPAGE, N.Y.
RM-(NUMBER)(LETTER)	GRUNZWEIG UND HARTMANN A.G., LUDWIGSHAFEN AM RHEIN, GERMANY
RM-(YEAR/NUMBER)	LONDON. UNIV.
RM-	MARTIN CO., BALTIMORE
RM-	MICHIGAN. UNIV., ANN ARBOR
RM-	MICHIGAN STATE UNIV., EAST LANSING
RM-	MICHIGAN STATE UNIV., EAST LANSING. DEPT. OF STATISTICS
RM-	NATIONAL RESEARCH COUNCIL OF CANADA. ATOMIC ENERGY PROJECT, CHALK RIVER, ONT.
RM-	OFFICE OF RUBBER RESERVE (RAW MATERIALS REPORTS)
RM-	PHILCO CORP., BLUE BELL, PENNA.
RM-	PRINCETON UNIV., N.J.
RM-	RAND CORP., SANTA MONICA, CALIF.
RM-(NUMBER)-AEC	RAND CORP., SANTA MONICA, CALIF. (REF. 47)
RM-(NUMBER)-ARPA	RAND CORP., SANTA MONICA, CALIF.
RM-(NUMBER)-ESSA	RAND CORP., SANTA MONICA, CALIF.
RM-(NUMBER)-RAND	RAND CORP., SANTA MONICA, CALIF. (REF. 47)
RM-(NUMBER)-PR-	RAND CORP., SANTA MONICA, CALIF.
RM-(NUMBER)-OEO	RAND CORP., SANTA MONICA, CALIF.
RM-(NUMBER)-NRL	RAND CORP., SANTA MONICA, CALIF. (REF. 47)
RM-(NUMBER)-NSF	RAND CORP., WASHINGTON, D.C.
RM-	RESEARCH TRIANGLE INST., DURHAM, N.C.
RM-	ROCKETDYNE DIV., NORTH AMERICAN AVIATION, INC., CANOGA PARK, CALIF.
RM-	STANFORD RESEARCH INST., MENLO PARK, CALIF.
RM-	WESTINGHOUSE ELECTRIC CORP., PITTSBURGH
RM-(YEAR)-	WESTINGHOUSE ELECTRIC CORP. RES. LABS., PITTSBURGH
RM-(LETTER)(DATE CODE)	NATIONAL ADVISORY COMMITTEE FOR AERONAUTICS (RESEARCH MEMORANDUM) (REF. 25)
RMA-	AKTIEBOLAGET ATOMENERGI, STOCKHOLM
RMB-	BROWN UNIV., PROVIDENCE
RMC-	RAYTHEON CO. (ALL LOCATIONS)
RMC-	RHEEM MFG. CO., N.Y.C.
RMC-BR-	RAYTHEON MFG. CO. MISSILE SYSTEMS DIV., BEDFORD, MASS.
RMC-HTM-	RAYTHEON MFG. CO., WALTHAM, MASS. (HIGH TEMPERATURE MATERIAL)
RMC-M/R-	RAYTHEON MFG. CO., WALTHAM, MASS.
RMC-QK-	RAYTHEON MFG. CO., WALTHAM, MASS.
RMC-QPR-	RAYTHEON MFG. CO. RESEARCH DIV., WALTHAM, MASS.
RMC-RD-	RESOURCE MANAGEMENT CORP., INC., BETHESDA, MD.
RMC-S-	RAYTHEON MFG. CO., WALTHAM, MASS.
RMCS-	UNITED KINGDOM. ROYAL MILITARY COLL. OF SCIENCE, SHRIVENHAM. WILTS, ENGLAND
RMC-SO-	RAYTHEON MFG. CO., WALTHAM, MASS.
RMC-UR-	RESOURCE MANAGEMENT CORP., INC., BETHESDA, MD.
RMD-	THIOKOL CHEM. CORP. REACTION MOTORS DIV., DENVILLE, NJ
RMD-(NUMBER)-	THIOKOL CHEM. CORP. REACTION MOTORS DIV., DENVILLE, NJ
RMD-(NUMBER)-(NUMBER)-	THIOKOL CHEM. CORP. REACTION MOTORS DIV., DENVILLE, NJ
RMD-(NUMBER)-(LETTER)	THIOKOL CHEM. CORP. REACTION MOTORS DIV., DENVILLE, NJ
RMD-(NUMBER)-Q-	THIOKOL CHEM. CORP. REACTION MOTORS DIV., DENVILLE, NJ
RME-	DIVISION OF RAW MATERIALS, AEC, N.Y.C. (REF. 4)
RMERC-TR-(YEAR)-	MISSOURI. UNIV., ROLLA. ROCK MECHANICS AND EXPLOSIVES RESEARCH CENTER
RMG-	ROCKY MOUNTAIN NUCLEAR POWER STUDY GROUP, IDAHO FALLS
RMG-T-	WESTERN ONTARIO UNIV., LONDON. RACETRACK MICROTRON GP
RMI-	ATOMIC ENERGY OF CANADA LTD., CHALK RIVER, ONT.
RMI-	REACTION MOTORS, INC., DENVILLE, N.J.
RMI-	REACTION MOTORS, INC., ROCKAWAY, N.J.
RMI-	REACTIVE METALS, INC., ASHTABULA, OHIO
RMI-	REACTIVE METALS, INC., NILES, OHIO
RMI-TPR-	REACTION MOTORS, INC., DENVILLE, N.J.
RML-	REDEL, INC., ANAHEIM, CALIF.
RMM-	AKTIEBOLAGET ATOMENERGI, STOCKHOLM
RMO-	DIVISION OF MATERIALS (EXPLORATION), AEC, N.Y.C.
RMO-	RAW MATERIALS OPERATIONS OFF., AEC, WASH., D.C. (REF.4)
RMOO-	RAW MATERIALS OPERATIONS OFF., AEC, WASH., D.C. (REF.4)
RMP-	PARSONS (RALPH M.) CO., LOS ANGELES
RMP-FR-	PARSONS (RALPH M.) CO., LOS ANGELES
RMP-TR-	PARSONS (RALPH M.) CO., LOS ANGELES
RMR-	AEROJET-GENERAL CORP., SACRAMENTO, CALIF.
RMRC/P-	UNITED KINGDOM ATOMIC ENERGY AUTHORITY. RESEARCH GP. ATOMIC ENERGY RES. ESTAB., HARWELL, BERKS, ENGLAND
RMS-	HUGHES AIRCRAFT CO. RADAR AND MISSILE ELECTRONICS LAB., CULVER CITY, CALIF.
RMS-	NAVAL RESEARCH LAB. RADIO MATERIEL SCHOOL, WASH., D.C.
RM-SHG-	TENNESSEE EASTMAN CORP., OAK RIDGE, TENN.
RM-TMP-	GENERAL ELECTRIC CO. TECHNICAL MILITARY PLANNING OPERATION, SANTA BARBARA, CALIF.
RN-	(RESEARCH NOTE) (REF. 2)
RN-	AEROJET-GENERAL CORP. ROCKET ENGINE OPERATIONS-NUCLEAR, SACRAMENTO, CALIF.
RN-	CONVAIR SCIENTIFIC RESEARCH LAB., SAN DIEGO, CALIF.
RN-	ELECTRO-OPTICAL SYSTEMS, INC., PASADENA, CALIF.
RN-	GT. BRIT. DEPT. OF SCIENTIFIC AND INDUSTRIAL RESEARCH. ROAD RESEARCH LAB.
RN-	GRUMMAN AIRCRAFT ENGINEERING CORP., BETHPAGE, N.Y.
RN-	HELIODYNE CORP., LOS ANGELES
RN(NO.)-	HELIODYNE CORP., LOS ANGELES
RN-	SACRAMENTO PEAK OBSERVATORY, SUNSPOT, N. MEX.
RN-	SUMMERS GYROSCOPE CO., SANTA MONICA, CALIF.
RN-(LETTER(S))-	AEROJET-GENERAL CORP. ROCKET ENGINE OPERATIONS-NUCLEAR, SACRAMENTO, CALIF.
R-NCSC-	NORTH CAROLINA STATE COLL., RALEIGH
RN-DR-	AEROJET-GENERAL CORP., SACRAMENTO, CALIF.
RNI-	DUQUESNE LIGHT CO., SHIPPINGPORT, PENNA.
RNL-	TEXAS. UNIV., AUSTIN
RNM-	GENERAL ELECTRIC CO. FLIGHT PROPULSION LAB. DEPT., CINCINNATI (ROCKET NOZZLE MATERIALS)
RNP-	GT. BRIT. ROYAL NAVAL PERSONNEL RESEARCH COMMITTEE
RNPL-	GT. BRIT. ROYAL NAVAL PHYSIOLOGICAL LAB., ALVERSTOKE
RNPRC-	GT. BRIT. ROYAL NAVAL PERSONNEL RESEARCH COMMITTEE

RN-PSW-	PACIFIC SOUTHWEST FOREST AND RANGE EXPERIMENT STATION, BERKELEY, CALIF.
RN-S-	AEROJET-GENERAL CORP., AZUSA, CALIF.
RN-S-	AEROJET-GENERAL CORP., SACRAMENTO, CALIF.
RN-S-	AEROJET-GENERAL CORP. LIQUID ROCKET OPERATIONS, SACRAMENTO, CALIF.
RNSS-	GT. BRIT. ROYAL NAVAL SCIENTIFIC SERVICE
RN-TM-	AEROJET-GENERAL CORP., SACRAMENTO, CALIF.
RN-TR-	ARMY MISSILE COMMAND, REDSTONE ARSENAL, ALA.
RO-	SANDIA CORP., ALBUQUERQUE, N. MEX.
ROAD-NOTE-	UNITED KINGDOM. ROAD RES. LAB., CROWTHORNE, BERKS, ENG
ROC-	NAVAL ORDNANCE PLANT, INDIANAPOLIS
ROCKETDYNE R-(NUMBER)P	ROCKETDYNE DIV., NORTH AMERICAN AVIATION, INC., CANOGA PARK, CALIF.
ROD-	ROSSFORD ORDNANCE DEPOT, TOLEDO
ROEC-	ARMY COMBAT DEVELOPMENTS COMMAND. EXPERIMENTATION COMMAND. RESEARCH OFFICE, MONTEREY, CALIF.
RO/EERL-(YEAR)/	ENVIRONMENTAL PROTECTION AGENCY. RADIATION OFFICE, ROCKVILLE, MD.
ROE-STS-	GT. BRIT. ROYAL OBSERVATORY, EDINBURGH, SCOTLAND
ROMEU-NI-	ROME. UNIVERSITA. ISTITUTO DI FISICA, G. MARCONI
ROOU-	ROOSEVELT UNIV., CHICAGO
RO-RM-	STANFORD RESEARCH INSTITUTE, FORT ORD, CALIF.
RO-S-(YEAR)-	ARMY TEST AND EVALUATION COMMAND. DEPUTY FOR NATIONAL RANGE OPERATIONS, WHITE SANDS MISSILE RANGE, N. MEX. (RANGE OPERATIONS)
ROSC/P-	GT. BRIT. ATOMIC EN. RES. ESTAB., HARWELL, BERKS, ENG
ROTO(NO.)R/-	GT. BRIT. POST OFFICE. ENGINEERING DEPT.
R-OU-	RESEARCH TRIANGLE INST. OPERATIONS RESEARCH AND ECONOMICS DIV., DURHAM, N.C.
ROVER-	LOS ALAMOS SCIENTIFIC LAB., N. MEX.
ROVER-TN-	CALIFORNIA. UNIV., LIVERMORE. LAWRENCE RADIATION LAB.
RP-	(REFERENCE PUBLICATION) (REF. 2)
RP-	(RESEARCH PAPER) (REF. 2)
RP-	(RESEARCH PROPOSAL) (REF. 2)
RP-	ARCTIC INSTITUTE OF NORTH AMERICA, MONTREAL
RP-	ASSOCIATED SEMICONDUCTOR MANUFACTURERS LTD., SOUTHAMPTON, HANTS, ENGLAND
RP-	CANADA. DEFENCE RESEARCH MEDICAL LABS., TORONTO
RP2-	ENGLISH ELECTRIC VALVE CO., LTD., CHELMSFORD, ESSEX, ENG
RP-(YEAR)-(NUMBER)-	MARTIN-MARIETTA CORP. AEROSPACE DIV., BALTIMORE
RP-	MASSACHUSETTS INST. OF TECH., LEXINGTON. LINCOLN LAB.
RP-	NATIONAL BUREAU OF STANDARDS, WASHINGTON, D.C. (RESEARCH PAPER)
RP-	NATIONAL DEFENSE RESEARCH COMM. (ROCKET PROPELLANTS)
RP-	NAVAL POSTGRADUATE SCHOOL, MONTEREY, CALIF.
RP-(NUMBER)-	NAVAL POWDER FACTORY. RES.+DEV. DEPT., INDIAN HEAD, MD.
RP-	NAVY (REGISTERED PUBLICATION)
RP-	OFFICE OF SCIENTIFIC RESEARCH AND DEVELOPMENT (ROCKET PROPELLANTS)
RP-	RADIOPLANE CO., VAN NUYS, CALIF.
R-P-	RAND CORP., SANTA MONICA, CALIF.
RP-(NUMBER-NUMBER)	STANDARD TELEPHONES AND CABLES, LTD., SIDCUP, KENT, ENG.
RP-(LETTERS)-	AEROJET-GENERAL CORP., SACRAMENTO, CALIF.
RPA-	RAND CORP., SANTA MONICA, CALIF.
RPA-I-(LETTER)-	ARMY SIGNAL RADIO PROPAGATION AGENCY, FT. MONMOUTH, NJ
RPA-TR-	ARMY SIGNAL RADIO PROPAGATION AGENCY, FT. MONMOUTH, NJ
RPC-	COMPAGNIE GENERALE DE TELEGRAPHIE SANS FIL, PUTEAU, FRANCE
RPC-	RADIOPLANE CO., VAN NUYS, CALIF.
RPD-	CANADA. DEPT. OF NATIONAL HEALTH AND WELFARE. RADIATION PROTECTION DIV., OTTAWA
RPD-	GT. BRIT. ROYAL AIRCRAFT EST., FARNBOROUGH, HANTS, ENG
RPD-	NETHERLANDS. RIJKS PSYCHOLOGISCHE DIENST, THE HAGUE
RPD-	ROCKET POWER LABS., GABRIEL CO., PASADENA, CALIF.
RP-DR-	AEROJET-GENERAL CORP., SACRAMENTO, CALIF.
RPE-TM-	GT. BRIT. ROCKET PROPULSION EST., WESTCOTT, BUCKS, ENG
RPE-TN-	GT. BRIT. ROCKET PROPULSION EST., WESTCOTT, BUCKS, ENG
RPE-TR-(YEAR)/	GT. BRIT. ROCKET PROPULSION EST., WESTCOTT, BUCKS, ENG
RPE-TRANS-	GT. BRIT. ROCKET PROPULSION EST., WESTCOTT, BUCKS, ENG
RP-FPL-	FOREST PRODUCTS LAB., MADISON, WIS.
RPI-(NUMBER)(AECL)	ATOMIC ENERGY OF CANADA LTD., CHALK RIVER, ONT.
RPI-	OAK RIDGE NATIONAL LAB., TENN.
RPI-	RENSSELAER POLYTECHNIC INST., TROY, N.Y.
RPI-(NUMBER)-	RENSSELAER POLYTECHNIC INST., TROY, N.Y.
RPIA-	JOHNS HOPKINS UNIV., SILVER SPRING, MD. ROCKET PROPELLANT INFORMATION AGENCY
RPI-(AECL)	ATOMIC ENERGY OF CANADA LTD., CHALK RIVER, ONT.
RPIAL-	RENSSELAER POLYTECHNIC INST., TROY, N.Y. DEPT. OF AERONAUTICAL ENGINEERING
RPIB-	RENSSELAER POLYTECHNIC INST., TROY, N.Y.
RPIEE-	RENSSELAER POLYTECHNIC INST., TROY, N.Y. DEPT. OF ELECTRICAL ENGINEERING
RPI-MATH-	RENSSELAER POLYTECHNIC INST., TROY, N.Y. DEPT. OF MATHEMATICS
RPI-MATH-REP-	RENSSELAER POLYTECHNIC INST., TROY, N.Y.
RPI-MP-	RENSSELAER POLYTECHNIC INST., TROY, N.Y.
RP-INT-	INTERMOUNTAIN FOREST AND RANGE EXPERIMENT STA., OGDEN, UTAH
RPI-PR-	RENSSELAER POLYTECHNIC INST., TROY, N.Y.
RPI-PRL-SSR-	RENSSELAER POLYTECHNIC INST., TROY, NY. PLASMA RES. LAB
RPI-PRL-TR-	RENSSELAER POLYTECHNIC INST., TROY, NY. PLASMA RES. LAB
RPI-QPR-	RENSSELAER POLYTECHNIC INST., TROY, N.Y. WELDING LAB.
RPIS-	INFORMATION TECHNOLOGY LABS., WALTHAM, MASS.
RPI-TN-	RENSSELAER POLYTECHNIC INST., TROY, N.Y.
RPI-TR-	RENSSELAER POLYTECHNIC INST., TROY, N.Y.
RPI TR-AE-	RENSSELAER POLYTECHNIC INST., TROY, N.Y. DEPT. OF AERONAUTICAL ENGINEERING
RPL-	AUSTRALIA. COMMONWEALTH SCI. + IND. RES. ORG. RADIOPHYSICS LAB., SYDNEY
RPL-	CANADA. DEFENCE RES. TELECOMMUNICATIONS ESTAB., OTTAWA
RPL-	NATL. BUR. OF STDS. RADIATION PHYS. LAB., WASH., D.C.
RPL-	UNITED KINGDOM ATOMIC ENERGY AUTHORITY. RESEARCH GP. ATOMIC ENERGY RES. ESTAB., HARWELL, BERKS, ENGLAND
RPL-	WATERTOWN ARSENAL. RODMAN PROCESS LAB., MASS.
RPL-(YR.)-	WESTINGHOUSE ELECTRIC CORP., NEWARK, N.J.
RPL-TDR-	AIR FORCE ROCKET PROPULSION LAB., EDWARDS AFB, CALIF.
RPL-TM-	AIR FORCE ROCKET PROPULSION LAB., EDWARDS AFB, CALIF.
RPL-TM-	GT. BRIT. ROCKET PROPULSION EST., WESTCOTT, BUCKS, ENG
RP/M-	GT. BRIT. ATOMIC EN. RES. ESTAB., HARWELL, BERKS, ENG
RP-MR-	AEROJET-GENERAL CORP., SACRAMENTO, CALIF.
RPM-RFX-	AKTIEBOLAGET ATOMENERGI, STOCKHOLM
RPP-	AUSTRALIA. COMMONWEALTH SCI. + IND. RES. ORG. RADIOPHYSICS LAB., SYDNEY
RP-P-	INSTITUTE FOR DEFENSE ANALYSES, ARLINGTON, VA.
RPP-	TENNESSEE EASTMAN CORP., OAK RIDGE, TENN.
RPP/(LETTER)/(NUMBER)	UNITED KINGDOM. RUTHERFORD HIGH ENERGY LAB., CHILTON, DIDCOT, BERKS, ENGLAND
RPP/H-	UNITED KINGDOM. RUTHERFORD HIGH ENERGY LAB., CHILTON, DIDCOT, BERKS, ENGLAND
RP-P-P-	MARYLAND. UNIV., COLLEGE PARK. COMPUTER SCIENCE CTR.

RP-QR-	AEROJET-GENERAL CORP., SACRAMENTO, CALIF.
R-PR-	AEROJET-GENERAL CORP. SOLID ROCKET PLANT, SACRAMENTO, CALIF. (FLEET BALLISTIC MISSILE PROGRAM)
RPR-	AUSTRALIA. RADIOPHYSICS LAB., CHIPPENDALE, NEW SOUTH WALES
RP/RL-	ROCKET POWER, INC. RESEARCH LABS., PASADENA, CALIF.
RP/RL-QR-(NUMBER)-	ROCKET POWER, INC. RESEARCH LABS., PASADENA, CALIF.
RPS-	RESEARCH + DEV. BOARD. COMM. ON AERONAUTICS, WASH., DC
RPS-	SIGNAL CORPS (ARMY). RADIO PROPAGATION SECTION
RPS-AR-	THAILAND. MINISTRY OF PUBLIC HEALTH. RADIATION PROTECTION SERVICE, BANGKOK
RPSC(WH)P.91/DRAG/P	UNITED KINGDOM ATOMIC ENERGY AUTHORITY. RESEARCH GP. ATOMIC ENERGY ESTAB., WINFRITH, DORSET, ENGLAND
RP-SR-	AEROJET-GENERAL CORP. ROCKET ENGINE OPERATIONS-NUCLEAR, SACRAMENTO, CALIF.
RPSR-	CAMP EVANS SIGNAL LABORATORY. RADIO PROPAGATION BRANCH, BELMAR, N.J.
RPSR-	CAMP EVANS SIGNAL LAB. RADIO PROPAGATION SECTION
RPT-	(REPORT) (REF. 2)
RP/T-	GABRIEL CO. ROCKET POWER/TALCO DIV., MESA, ARIZ.
RP-TM-	AEROJET-GENERAL CORP., SACRAMENTO, CALIF.
RP/TN-	AUSTRALIAN ATOMIC ENERGY COMMISSION. RESEARCH ESTABLISHMENT, LUCAS HEIGHTS, NEW SOUTH WALES
(RPT.NO.)(SUBJ.CODE LTR)	FAIRCHILD CAMERA + INSTR. CORP., JAMAICA, N.Y.(REF. 1)
RP-TR-	AEROJET-GENERAL CORP., SACRAMENTO, CALIF.
RP-TR(NO.)-(YR.)	ARMY BALLISTIC MISSILE AGENCY, REDSTONE ARSENAL, ALA.
RPU-	HOLABIRD SIGNAL DEPOT. RADIO PROPAGATION UNIT, BALTIMORE
RPU-	INDIA. NATIONAL PHYSICAL LAB., NEW DELHI
RPU-	SIGNAL CORPS. RADIO PROPAGATION AGENCY, FORT MONMOUTH, N.J.
RPU-S-	INDIA. NATIONAL PHYSICAL LAB., NEW DELHI
RPU-TR-	ARMY SIGNAL RADIO PROPAGATION AGENCY, FT. MONMOUTH, NJ
RQ(YEAR)EE-	CANADIAN INDUSTRIES, LTD., NEW TORONTO, ONT.
RR-	(RESEARCH REPORT) (REF. 2)
RR-	AIR WAR COLLEGE, MAXWELL AFB, ALA.
RR-(YEAR)-	AMPEX CORP., REDWOOD CITY, CALIF.
RR-	ARMY TROPIC TEST CENTER, FORT CLAYTON, CANAL ZONE
RR-	AVCO CORP. RES.+ ADVANCED DEV. DIV., WILMINGTON, MASS.
RR-(NUMBER).(NUMBER)	BUREAU OF MINES. INTERMOUNTAIN EXPERIMENT STATION, SALT LAKE CITY
37 RR-	CALIFORNIA. UNIV., BERKELEY. RADIATION LAB. (REF. 9)
RR-(NUMBER-YEAR)	FORSCHUNGSINSTITUT FUER HOCHFREQUENZPHYSIK, WERTHOVEN GERMANY
RR-	GT. BRIT. DEPT. OF SCIENTIFIC AND INDUSTRIAL RESEARCH (ROAD RESEARCH BULLETINS)
RR-(NUMBER)-TRI-	HONEYWELL, INC. MILITARY PRODUCTS GP., MINNEAPOLIS
RR-	ILLINOIS. STATE WATER SURVEY, URBANA
RR-(NUMBER)/SMR/	MANCHESTER. UNIV., ENGLAND
RR-(NUMBER)/EKK/	MANCHESTER. UNIV., ENGLAND
RR-(NUMBER)/RM/	MANCHESTER. UNIV., ENGLAND
RR-	MARTIN-MARIETTA CORP. DENVER DIV.
RR-(YEAR)-	QUEEN'S UNIV., KINGSTON, ONTARIO
RR-	RCA LABS., PRINCETON, N.J.
RR-(NUMBER)-(NUMBER)-	RCA VICTOR CO., LTD. RESEARCH LABS., MONTREAL
RR-(NO.)/(INITLS.)-	SHEFFIELD, ENGLAND. UNIV. DEPT. OF PROBABILITY AND STATISTICS
RR-(NO.)-(NO.)-X(NO.)	SOUTHERN RESEARCH INST., BIRMINGHAM, ALA.
RR-	SYLVANIA ELECTRONIC SYSTEMS-EAST, WALTHAM, MASS.
RR-(YEAR)	TECHNIDYNE INC., WEST CHESTER, PA.
RR-	TORONTO. UNIV. DEPT. OF ELECTRICAL ENGINEERING
RR-(NUMBER)-	UNITED STATES LAKE SURVEY, DETROIT
RR-	WATERLOO. UNIV., ONTARIO
RR-	WESTINGHOUSE ELECTRIC CORP. RES. LABS., PITTSBURGH
RR-(YR.)-(NO.)-(NO.)-R	WESTINGHOUSE RESEARCH LABS., PITTSBURGH
RR-(YEAR)	WISCONSIN. UNIV., MADISON. GEOPHYSICAL AND POLAR RESEARCH CENTER
RRA-	RADIATION RESEARCH ASSOCIATES, INC., FORT WORTH, TEX.
RRA-C-	RADIATION RESEARCH ASSOCIATES, INC., FORT WORTH, TEX.
RRAC-	UNITED KINGDOM ATOMIC ENERGY AUTHORITY. RESEARCH GP. ATOMIC ENERGY RES. ESTAB., HARWELL, BERKS, ENGLAND
RRAC/SSC/I-	UNITED KINGDOM ATOMIC ENERGY AUTHORITY. RESEARCH GP. ATOMIC ENERGY RES. ESTAB., HARWELL, BERKS, ENGLAND
RRAM-	NAVAL ORDNANCE TEST STA., INYOKERN (CHINA LAKE), CAL.
RRA-M-	RADIATION RESEARCH ASSOCIATES, INC., FORT WORTH, TEX.
RRA-T(NO.)	RADIATION RESEARCH ASSOCIATES, INC., FORT WORTH, TEX.
RRB-	NAVAL ORDNANCE TEST STA., INYOKERN (CHINA LAKE), CAL.
RRB/C-	RADIO RESEARCH STATION, SLOUGH, BUCKS, ENGLAND
RRC-	ARMY TANK-AUTOMOTIVE CENTER. ADVANCED SYSTEMS AND CONCEPT RESEARCH DIV., WARREN, MICH.
RRC-	INDIA. NATIONAL PHYSICAL LAB., NEW DELHI
RRC-	OFFICE OF SCIENTIFIC RESEARCH AND DEVELOPMENT (RDX AND RELATED COMPOUNDS)
RRC-	RADIATION RESEARCH CORP., N.Y.C.
RRC-	RADIATION RESEARCH CORP., WEST PALM BEACH, FLA.
RRC-(YR.)-	ROCKET RESEARCH CORP., SEATTLE
RRC-	ROYAL RESEARCH CORP., HAYWARD, CALIF.
RRC-A-	INDIA. COUNCIL OF SCIENTIFIC AND IND. RES., NEW DELHI
RRC-A(NO.)-	INDIA. NATIONAL PHYSICAL LAB., NEW DELHI
RRC-B-	INDIA. COUNCIL OF SCIENTIFIC AND IND. RES., NEW DELHI
RR-CER-	AEROJET-GENERAL CORP. LIQUID ROCKET PLANT, SACRAMENTO, CALIF.
RRD/DES. DATA-	UNITED KINGDOM ATOMIC ENERGY AUTHORITY. RESEARCH GP. ATOMIC ENERGY RES. ESTAB., HARWELL, BERKS, ENGLAND
RRDE-	GT. BRIT. RADAR RES. EST., MALVERN, WORCS, ENGLAND
RRE-	GT. BRIT. ROYAL RADAR ESTAB., MALVERN, WORCS, ENGLAND
RR-EERL-	CORNELL UNIV., ITHACA, N.Y. ELECTRICAL ENG. RES. LAB.
RREL-	ARMY COLD REGIONS RES. AND ENG. LAB., HANOVER, N.H.
RRE-MEMO.-	GT. BRIT. ROYAL RADAR ESTAB., MALVERN, WORCS, ENGLAND
RREP-	ROSEN (RAYMOND) ENG. PRODUCTS CO., INC., PHILADELPHIA
RREP-(NO.)(LTRS.)(NOS.)	ROSEN (RAYMOND) ENGINEERING PRODUCTS, INC. TELEMETERING DIV., PHILADELPHIA
RRE-TN-	GT. BRIT. RADAR RES. EST., MALVERN, WORCS, ENGLAND
RRE-TRANS-(NUMBER)	GT. BRIT. ROYAL RADAR ESTAB., MALVERN, WORCS, ENGLAND
RRI-	REED RESEARCH INC., WASHINGTON, D.C.
RRI-P-	RIVERSIDE RESEARCH INST., N.Y.C.
RR-IR-	PENNSYLVANIA STATE UNIV., UNIVERSITY PARK. DEPT. OF CIVIL ENGINEERING
RRI-T-(NO./NO./NO.)-	RIVERSIDE RESEARCH INST., N.Y.C.
RRK-(YEAR)-	HIROSHIMA UNIV., TAKEHARA. RESEARCH INST. FOR THEORETICAL PHYSICS
RRL-(NUMBER)-	CARNEGIE-MELLON UNIV., PITTSBURGH. RADIATION RES. LAB
RRL/	ELLIOTT BROS. LTD., LONDON
RRL-	HARVARD UNIV., CAMBRIDGE, MASS. RADIO RESEARCH LAB.
RRL-	ILLINOIS. UNIV., URBANA. RADIOLOCATION RESEARCH LAB.
RRL-(NUMBER)-(NUMBER)	JOHN CARROLL UNIV., CLEVELAND
RRL-	MELLON INST. RADIATION RESEARCH LABS., PITTSBURGH
RRL-LR-	ROAD RESEARCH LAB., CROWTHORNE, BERKS, ENGLAND
RRL-TR-	ILLINOIS. UNIV., URBANA. RADIOLOCATION RESEARCH LAB.
RRO-	NEW ZEALAND. DEPT. OF SCI.+ IND. RES. RADIO RES. OFF.
RR/OH/104	ROLLS ROYCE, LTD., DERBY, ENGLAND

RRP-   BUREAU OF NAVAL PERSONNEL (RESEARCH REPORT PROJS.)
RRP-   TENNESSEE EASTMAN CORP., OAK RIDGE, TENN.
RRP/P- UNITED KINGDOM ATOMIC ENERGY AUTHORITY. PRODUCTION GROUP. WINDSCALE WORKS, SELLAFIELD, CUMB., ENGLAND
RR-R(YEAR)-   MASSACHUSETTS INST. OF TECH., CAMBRIDGE. DEPT. OF CIVIL ENGINEERING
RR-R-   MASSACHUSETTS INST. OF TECH., CAMBRIDGE. SOIL MECHANICS DIV.
RRSY-(YR.)-   BUREAU OF NAVAL WEAPONS
RR-TM-(YEAR)- ARMY MISSILE COMMAND. PHYSICAL SCIENCES LAB., REDSTONE ARSENAL, ALA. (TECHNICAL MEMORANDUM)
RR-TN-(YEAR)- ARMY MISSILE COMMAND. PHYSICAL SCIENCES LAB., REDSTONE ARSENAL, ALA. (TECHNICAL NOTE)
RR-TR-(YEAR)- ARMY MISSILE COMMAND. PHYSICAL SCIENCES LAB., REDSTONE ARSENAL, ALA. (TECHNICAL REPORT)
RRU-   SPERRY RAND CORP. UNIVAC DIV., PHILADELPHIA
RRU-TR- SPERRY RAND CORP. UNIVAC DIV., PHILADELPHIA
RS-   (RESEARCH STUDY) (REF. 2)
RS-   (RESEARCH SUMMARY) (REF. 2)
RS-   ARECIBO IONOSPHERIC OBSERVATORY, PUERTO RICO
RS-(NUMBER)-O BOELKOW ENTWICKLUNGEN K.G., MUNICH
RS-   BUREAU OF AERONAUTICS (NAVY). RESEARCH DIV.
RS-   COMMITTEE ON MEDICAL RES. (RUSSELL SAGE INST. RPTS.)
RS/(NUMBER)/   CONFERENCE ON PHYSICAL PROBLEMS OF REACTOR SHIELDING
RS-   NATIONAL DEFENSE RES. COMM. (RPT. TO THE SERVICES)
RS-(NUMBER)/(NUMBER)   SANDIA CORP., ALBUQUERQUE, N. MEX.
RS-   SANDIA CORP. LIVERMORE LAB., CALIF. (REFERENCE SYMBOL, ACCESSION NUMBERS)
RS-   SASKATCHEWAN. UNIV., SASKATOON
RS-   SOUTHWEST RESEARCH INST., SAN ANTONIO
RSA-   AKTIEBOLAGET ATOMENERGI, STOCKHOLM
RSAF-   GT. BRIT. ARMAMENT DESIGN EST.,FT.HALSTEAD,KENT,ENG.
RSA/OML-   REDSTONE ARSENAL. ORDNANCE MISSILE LABS., HUNTSVILLE, ALA.
RSC-   AERONUTRONIC DIV. OF PHILCO CORP., NEWPORT BEACH,CAL.
RSC-   ATOMIC ENERGY OF CANADA LTD., CHALK RIVER, ONT.
RSC-(NUMBER)(AERE)   GT. BRIT. ATOMIC EN. RES. ESTAB., HARWELL, BERKS, ENG
RSC-   NATIONAL DEFENSE RES. COMM. (RES. ON SOUND CONTROL)
RSD-   BRITISH EUROPEAN AIRWAYS
RSD-(YEAR)-   GENERAL ELECTRIC CO. MISSILE AND SPACE DIV., PHILA.
RSD-   INDIA. NATIONAL PHYSICAL LAB., NEW DELHI
RSD-(NUMBER)- LOGICON, INC., SAN PEDRO, CALIF.
RSD-(LETTER)(NUMBER)   INDIA. NATIONAL PHYSICAL LAB., NEW DELHI
RSG-   NEW YORK. STATE DEPT. OF HEALTH, ALBANY
RSG-   UNITED KINGDOM ATOMIC ENERGY AUTHORITY. RESEARCH GP. ATOMIC ENERGY RES. ESTAB., HARWELL, BERKS, ENGLAND
RSG-A- NEW YORK. STATE DEPT. OF HEALTH, ALBANY
RSG-C- NEW YORK. STATE DEPT. OF HEALTH, ALBANY
RS/H-   HUGHES AIRCRAFT CO. RES. + DEV. DIV.,CULVER CITY,CAL.
RSIC-   REDSTONE SCIENTIFIC INFO. CTR., REDSTONE ARSENAL, ALA
RSIC-U- REDSTONE SCIENTIFIC INFO. CTR., REDSTONE ARSENAL, ALA
RSL-   SYLVANIA ELECTRONIC SYSTEMS-WEST. RECONNAISSANCE SYSTEMS LABS., MOUNTAIN VIEW, CALIF.
RS/L-   UNITED KINGDOM ATOMIC ENERGY AUTHORITY. REACTOR GP., WINFRITH, DORSET, ENGLAND
RSLR-   STRESAU (R) LAB. INC., SPOONER, WIS.
RS-MC- NEW MEXICO. UNIV., ALBUQUERQUE. TECHNOLOGY APPLICATIONS CENTER
RSPP-   UNITED KINGDOM ATOMIC ENERGY AUTHORITY. AUTHORITY HEALTH AND SAFETY BRANCH, RISLEY, LANCS, ENGLAND
RSRC/S- UNITED KINGDOM ATOMIC ENERGY AUTHORITY. INDUSTRIAL GROUP. SPRINGFIELDS WORKS, LANCS, ENGLAND
RSRI-DP-   REGIONAL SCIENCE RESEARCH INST., PHILADELPHIA
RSRWP/S-   UNITED KINGDOM ATOMIC ENERGY AUTHORITY. RESEARCH GP. ATOMIC ENERGY RES. ESTAB., HARWELL, BERKS, ENGLAND
RSSC-RM-   STANFORD RESEARCH INST. REGIONAL SECURITY STUDIES CENTER, MENLO PARK, CALIF.
RSSC-TN-   STANFORD RESEARCH INST. REGIONAL SECURITY STUDIES CENTER, MENLO PARK, CALIF.
RSSC-TR-   STANFORD RESEARCH INST. REGIONAL SECURITY STUDIES CENTER, MENLO PARK, CALIF.
RST-   HUGHES AIRCRAFT CO. PACKAGE ENG. LAB., TUCSON, ARIZ.
RST-   TECHNICAL INFO. SERVICE, AEC (REACTOR SCIENCE + TECH.
RS-TM-(YEAR)- ARMY MISSILE COMMAND. STRUCTURES AND MECHANICS LAB., REDSTONE ARSENAL, ALA. (TECHNICAL MEMORANDUM)
RS-TN-(YEAR)- ARMY MISSILE COMMAND. STRUCTURES AND MECHANICS LAB., REDSTONE ARSENAL, ALA. (TECHNICAL NOTE)
RSTN-   MARTIN-MARIETTA CORP. DENVER DIV.
RSTP-   NATIONAL INSTITUTES OF HEALTH, BETHESDA, MD.
RS-TR-(YEAR)- ARMY MISSILE COMMAND. STRUCTURES AND MECHANICS LAB., REDSTONE ARSENAL, ALA.
RSU-   RUTGERS UNIV., NEW BRUNSWICK, N.J.
RT-   (REDUCTION TABLES) (REF. 2)
RT-   ARMY ENGINEER REACTORS GP. RESEARCH AND TECHNOLOGY DIV., FORT BELVOIR, VA.
RT-   CONVAIR, SAN DIEGO, CALIF.
RT-   ETHYL CORP. RESEARCH AND DEVELOPMENT DEPT., FERNDALE, MICH. (REFINERY TECHNOLOGY)
RT-   GRISCOM-RUSSELL CO., MASSILLON, OHIO
RT-   LIBRARY OF CONGRESS, WASH.,D.C.(RUSSIAN TRANSLATION)
RT-   TAGGART (ROBERT), INC., FAIRFAX, VA.
RT-   TAGGART (ROBERT), INC., FALLS CHURCH, VA.
RT/A-(NUMBER-MO./YR.) ENTWICKLUNGSRING NORD GMBH, BREMEN
RTB-   BUREAU OF SHIPS (RADAR TECHNICAL BULLETIN)
RTB-   ROCKFORD RESEARCH INST., INC., CAMBRIDGE, MASS.
RT/BIO(YEAR)- ITALY. COMITATO NAZIONALE PER L'ENERGIA NUCLEARE, ROME (BIOLOGY AND AGRICULTURE)
RT-BR-(YEAR)- ARMY MISSILE COMMAND. TEST AND RELIABILITY EVALUATIONS LAB., REDSTONE ARSENAL, ALA. (BROCHURE)
RTCA-   RADIO TECH. COMMISSION FOR AERONAUTICS, WASH., D.C.
RT/CHI(YEAR)- ITALY. COMITATO NAZIONALE PER L'ENERGIA NUCLEARE, ROME (CHEMISTRY)
RT/CNI(YEAR)- ITALY. COMITATO NAZIONALE PER L'ENERGIA NUCLEARE, ISPRA
RTD-   RESEARCH AND TECHNOLOGY DIV., BOLLING AFB, D.C.
RTD-(NUMBER)- RESEARCH AND TECHNOLOGY DIV., BOLLING AFB, D.C.
RTD-   TENNESSEE EASTMAN CORP., OAK RIDGE, TENN.
RTDA-IR-(YEAR)-   RESEARCH AND TECHNOLOGY DIV., WRIGHT-PATTERSON AFB, OHIO
RTD-AT-L- RESEARCH AND TECHNOLOGY DIV. DETACHMENT 4, EGLIN AFB, FLA.
RTD-IR-(NUMBER)-   BUREAU OF MINES. ALBANY METALLURGY RES. CTR., ORE.
RTD-IR- RESEARCH AND TECHNOLOGY DIV., BOLLING AFB, D.C.
RTD-IR- UNION CARBIDE CORP. PLASTICS DIV., BOUND BROOK, N.J.
RTD-LA CLINTON ENGINEER WORKS, MANHATTAN DISTRICT, OAK RIDGE, TENN.
RTD-TB-(NO.)-MMP- RESEARCH AND TECHNOLOGY DIV., BOLLING AFB, D.C.
RTD-TDR-   AIR FORCE MATERIALS LAB., WRIGHT-PATTERSON AFB, OHIO
RTD TDR-63-   AIR FORCE WEAPONS LAB., KIRTLAND AFB, N. MEX. (NUMBERS IN 3000 SERIES ONLY)
RTD-TDR-(YEAR)-   RESEARCH AND TECHNOLOGY DIV., BOLLING AFB, D.C.
RTD-TR- RESEARCH AND TECHNOLOGY DIV., BOLLING AFB, D.C.
RTD-TR-(NUMBER)-   RESEARCH AND TECHNOLOGY DIV., BOLLING AFB, D.C.
RTD (WLA)-TM-(YEAR)-   AIR FORCE WEAPONS LAB., KIRTLAND AFB, N. MEX.
RTD (WLR)-TM-(YEAR)-   AIR FORCE WEAPONS LAB., KIRTLAND AFB, N. MEX.

RT/EC-	ITALY. COMITATO NAZIONALE PER L'ENERGIA NUCLEARE, ROME
RT + E CO.-	RURAL TRANSFORMER AND EQUIPMENT CO., MILWAUKEE
RT/EL(YEAR)-	ITALY. COMITATO NAZIONALE PER L'ENERGIA NUCLEARE, ROME (ELECTRONICS)
RTF-	REDSTONE ARSENAL, ALA. (ROCKET TEST FACILITIES)
RT/FI(YEAR)-	ITALY. COMITATO NAZIONALE PER L'ENERGIA NUCLEARE, ROME (PHYSICS)
RT/FIMA(YEAR)-	ITALY. COMITATO NAZIONALE PER L'ENERGIA NUCLEARE, ROME (PHYSICS AND MATHEMATICS)
RT/GEN(YEAR)-	ITALY. COMITATO NAZIONALE PER L'ENERGIA NUCLEARE, ROME
RT/GEO(YEAR)-	ITALY. COMITATO NAZIONALE PER L'ENERGIA NUCLEARE, ROME (GEOLOGY)
RT/GIU(YEAR)-	ITALY. COMITATO NAZIONALE PER L'ENERGIA NUCLEARE, ROME
RTI-	RESEARCH TRIANGLE INST., DURHAM, N.C.
RTI-	TAGGART (ROBERT), INC., FAIRFAX, VA.
RTI-	TAGGART (ROBERT), INC., FALLS CHURCH, VA.
RTI-CE-(NUMBER)-F	RESEARCH TRIANGLE INST., DURHAM, N.C.
RTI-EU-	RESEARCH TRIANGLE INST., DURHAM, N.C.
RTI/FI(YEAR)-	ITALY. COMITATO NAZIONALE PER L'ENERGIA NUCLEARE, ROME
RTI-FR-OU-(NUMBER)-	RESEARCH TRIANGLE INST. OPERATIONS RESEARCH AND ECONOMICS DIV., DURHAM, N.C.
RTI-GU-	RESEARCH TRIANGLE INST., DURHAM, N.C.
RTI/ING(YEAR)-	ITALY. COMITATO NAZIONALE PER L'ENERGIA NUCLEARE, ROME
RTI-IR-SU-	RESEARCH TRIANGLE INST., DURHAM, N.C.
RT/ING-(YEAR)-	ITALY. COMITATO NAZIONALE PER L'ENERGIA NUCLEARE, ROME
RT/ING(YEAR)-	ITALY. COMITATO NAZIONALE PER L'ENERGIA NUCLEARE, ROME (ENGINEERING)
RTI-OU-	RESEARCH TRIANGLE INST. OPERATIONS RESEARCH AND ECONOMICS DIV., DURHAM, N.C.
RTI-RM-	RESEARCH TRIANGLE INST., DURHAM, N.C.
RTI-RM-OU-(NUMBER)-	RESEARCH TRIANGLE INST. OPERATIONS RESEARCH AND ECONOMICS DIV., DURHAM, N.C.
RTI-R-OU-	RESEARCH TRIANGLE INST. OPERATIONS RESEARCH AND ECONOMICS DIV., DURHAM, N.C.
RTI-RR-SU-	RESEARCH TRIANGLE INST., DURHAM, N.C.
RTI-S-	RESEARCH TRIANGLE INST. STATISTICS RESEARCH DIV., DURHAM, N.C.
RTI-SU-	RESEARCH TRIANGLE INST., DURHAM, N.C.
RTI-TRR-	RESEARCH TRIANGLE INST., DURHAM, N.C.
RTI-TR-SU-(NUMBER)-	RESEARCH TRIANGLE INST., DURHAM, N.C.
RTM-	(RESEARCH TECHNICAL MEMORANDUM) (REF. 2)
RTM-	AEROJET ENGINEERING CORP., AZUSA, CALIF. (RESEARCH TECHNICAL MEMORANDUM)
RTM-	ETHYL CORP. RESEARCH AND DEVELOPMENT DEPT., FERNDALE, MICH. (REFINERY TECHNOLOGY MEMORANDA)
RTM-	OHIO STATE UNIV. RESEARCH FOUNDATION, COLUMBUS
RT/MET(YEAR)	ITALY. COMITATO NAZIONALE PER L'ENERGIA NUCLEARE, ROME (METALLURGY)
RTP-	ATOMICS INTERNATIONAL DIV., NORTH AMERICAN AVIATION, INC., CANOGA PARK, CALIF.
RTP-	GT. BRIT. MINISTRY OF AIRCRAFT PRODUCTION.
RTP-	GT. BRIT. MINISTRY OF SUPPLY (RESEARCH TECHNICAL PUBLICATIONS)
RTP-	NAVAL ORDNANCE LAB., WHITE OAK, MD. (REQUIREMENTS + TEST PROCEDURES)
RT/PROT(YEAR)-	ITALY. COMITATO NAZIONALE PER L'ENERGIA NUCLEARE, ROME (SAFETY AND PROTECTION)
RTP T-	GT. BRIT. MIN. OF AIRCRAFT PRODUCTION (TRANSLATION)
RTP/TIB T-	GT. BRIT. MIN. OF AIRCRAFT PRODUCTION (TRANSLATION)
RTR-	CONSOLIDATED ELECTRODYNAMICS CORP. DATA INSTRUMENTS DIV., PASADENA, CALIF.
RTR-	GIANNINI CONTROLS CORP., PASADENA, CALIF.
RTR-	NAVAL ORDNANCE PLANT, INDIANAPOLIS
RTRC-A-	INDIA. NATIONAL PHYSICAL LAB., NEW DELHI
RTS-	UNITED KINGDOM. NATIONAL LENDING LIBRARY FOR SCIENCE AND TECHNOLOGY, BOSTON SPA, YORKS, ENGLAND
RTS-	WEATHER SQUADRON, 32ND, TYNDALL AFB, FLA.
RTST-(YEAR)-	RESEARCH AND TECHNOLOGY DIV., BOLLING AFB, D.C.
RT-TM-(YEAR)-	ARMY MISSILE COMMAND. TEST AND RELIABILITY EVALUATIONS LAB., REDSTONE ARSENAL, ALA. (TECH. MEMO.)
RT-TN-(YEAR)-	ARMY MISSILE COMMAND. TEST AND RELIABILITY EVALUATIONS LAB., REDSTONE ARSENAL, ALA. (TECH. NOTE)
RT-TP-(YEAR)-	ARMY MISSILE COMMAND. TEST AND RELIABILITY EVALUATIONS LAB., REDSTONE ARSENAL, ALA. (TEST PLAN)
RT-TR-(YEAR)-	ARMY MISSILE COMMAND. TEST AND RELIABILITY EVALUATIONS LAB., REDSTONE ARSENAL, ALA. (TECHNICAL REPT)
RU-	DETROIT CONTROLS CORP., REDWOOD CITY, CALIF.
RU-	GEOSCIENCE, INC., CAMBRIDGE, MASS.
RU-	ROCHESTER, N.Y. UNIV.
RU-	RUTGERS UNIV., NEW BRUNSWICK, N.J.
RU CRS-	NEW JERSEY. CERAMIC RESEARCH STATION, NEW BRUNSWICK
RUFUS-RR-	GEOLOGICAL SURVEY, WASHINGTON, D.C.
RUFUS-RR-	NEVADA OPERATIONS OFFICE, AEC, LAS VEGAS
RULISON-	GEOLOGICAL SURVEY, DENVER
RUP-	TOKYO UNIV.
RURAL SOCIOLOGY-	IOWA STATE UNIV. OF SCIENCE AND TECHNOLOGY, AMES. DEPT. OF SOCIOLOGY AND ANTHROPOLOGY
RUSMD-	ROCHESTER, N.Y. UNIV. SCH. OF MED. AND DENTISTRY
RU-TR-(NUMBER)-MAE-F	RUTGERS UNIV., NEW BRUNSWICK, N.J. DEPT. OF MECHANICAL AND AEROSPACE ENGINEERING
RV-	NETHERLANDS. ROYAL NETHERLANDS AIRCRAFT FACTORIES, FOKKER, AMSTERDAM
RVA(YEAR)-	ITALY. COMITATO NAZIONALE PER L'ENERGIA NUCLEARE, BOLOGNA
RVLTR-	BRIMAR VALVE WORKS, FOOTSCRAY, KENT, ENGLAND
RVP-UTS/(YEAR)/	ITALY. COMITATO NAZIONALE PER L'ENERGIA NUCLEARE, ROME
RVS-	GT. BRIT. ATOMIC EN. RES. ESTAB., HARWELL, BERKS, ENG
RVTM-	BRIMAR VALVE WORKS, FOOTSCRAY, KENT, ENGLAND
RVTP-	LOCKHEED MISSILES AND SPACE CO., SUNNYVALE, CALIF.
RVT-PEC-	ITALY. COMITATO NAZIONALE PER L'ENERGIA NUCLEARE, BOLOGNA
RVT-SEC(YEAR)-	ITALY. COMITATO NAZIONALE PER L'ENERGIA NUCLEARE, BOLOGNA
RVT-SIN(YEAR)-	ITALY. COMITATO NAZIONALE PER L'ENERGIA NUCLEARE, BOLOGNA
RVT-SIR(YEAR)-	ITALY. COMITATO NAZIONALE PER L'ENERGIA NUCLEARE. BOLOGNA
RW-	(RADIOLOGICAL WARFARE) (REF. 2)
RW-	IBM WATSON RESEARCH CENTER, YORKTOWN HEIGHTS, N.Y.
R-W-	RAMO-WOOLDRIDGE CORP., LOS ANGELES
RW-	RAMO-WOOLDRIDGE CORP., LOS ANGELES
RW-(NUMBER).(NUMBER)	RAMO-WOOLDRIDGE CORP. CONTROL SYSTEMS DIV., LOS ANG.
R-W-(NO.)-TR	RAMO-WOOLDRIDGE CORP. GUIDED MISSILE RESEARCH DIV., LOS ANGELES
RW-AL-	RAMO-WOOLDRIDGE CORP., LOS ANGELES (ACCESSION LIST)
R-W-ARL-	RAMO-WOOLDRIDGE CORP. AERO. RES. LAB., LOS ANGELES
RWC-	RAMO WOOLDRIDGE CORP., LOS ANGELES
RWC-	WURLITZER (RUDOLPH) CO., CHICAGO
RWC-AM-	RAMO-WOOLDRIDGE CORP., LOS ANGELES
RWCCPC-(YEAR)-	CITY-COUNTY PLANNING COMMISSION, ROCKFORD, ILL.
RW-DL-	RAMO-WOOLDRIDGE CORP. DENVER LABS.
RW-DL-PR-	RAMO-WOOLDRIDGE CORP. DENVER LABS.
RWD-RL-	RAMO-WOOLDRIDGE DIV., THOMPSON RAMO WOOLDRIDGE CORP., CANOGA PARK, CALIF.
RWD-RLM-	RAMO-WOOLDRIDGE DIV., THOMPSON RAMO WOOLDRIDGE CORP., CANOGA PARK, CALIF.

R/W GM-
: RAMO-WOOLDRIDGE CORP. GUIDED MISSILE RESEARCH DIV., LOS ANGELES

RW-GM(NO.)-
: RAMO-WOOLDRIDGE CORP. GUIDED MISSILE RESEARCH DIV., LOS ANGELES

RW-GMRD-
: RAMO-WOOLDRIDGE CORP. GUIDED MISSILE RESEARCH DIV., LOS ANGELES

RW-GM-TR-(NO.)-
: RAMO-WOOLDRIDGE CORP. GUIDED MISSILE RESEARCH DIV., LOS ANGELES

RW-RL-
: RAMO-WOOLDRIDGE. DIV. OF THOMPSON RAMO WOOLDRIDGE INC., CANOGA PARK, CALIF.

RW-RL-
: THOMPSON RAMO WOOLDRIDGE INC. RESEARCH LAB., CANOGA PARK, CALIF.

RW-WDD-
: RAMO-WOOLDRIDGE CORP., LOS ANGELES

RY-
: ELECTRIC + MUSICAL INDUSTRIES, LTD., HAYES, MIDDX, ENG.

RYAC-
: RYAN AERO. CO., LINDBERGH FIELD, SAN DIEGO, CALIF.

RYAN-
: RYAN AERO. CO., LINDBERGH FIELD, SAN DIEGO, CALIF.

S-	(SCIENTIFIC PROGRESS REPORT) (REF. 2)
S-	(SPECIAL REPORT) (REF. 2)
S-	(SPECIFICATION) (REF. 2)
S-	(STAFF REPORT) (REF. 2)
S(YEAR)-	(SUMMARY) (REF. 2)
S(YEAR)-	AC ELECTRONICS-DEFENSE RES. LABS., SANTA BARBARA, CAL.
S-	AEROJET-GENERAL CORP., AZUSA, CALIF.
S-	AERONUTRONIC DIV. OF PHILCO CORP., NEWPORT BEACH, CALIF. (SECRET)
S-(NUMBER)-D	AIR FORCE
S-	AKTIEBOLAGET ATOMENERGI, STOCKHOLM
S-	AKTIEBOLAGET ATOMENERGI, STUDSVIK, SWEDEN
S-55000 THRU S-55999	ARMOUR RESEARCH FOUNDATION, CHICAGO
S-	BELL AIRCRAFT CORP., BUFFALO
S-(4 DIGITS)	BUREAU OF MINES, PITTSBURGH
S-(NUMBER)-(YEAR)	BUREAU OF ORDNANCE (NAVY)
S-	CALIFORNIA. UNIV., BERKELEY. RADIATION LAB.
S-	CARNEGIE INSTITUTION OF WASHINGTON, D.C.
S-	CHEMICAL CORPS. TECH. COMMAND, ARMY CHEMICAL CTR., MD.
S1C-(NUMBER)-SP-	COMBUSTION ENG., INC. NUCLEAR DIV., WINDSOR, CONN.
S1C-(NUMBER)-CA/	COMBUSTION ENG., INC. REACTOR DEV. DIV., WINDSOR, CONN.
S1C-(NUMBER)-RS-	COMBUSTION ENG., INC. REACTOR DEV. DIV., WINDSOR, CONN.
S1C-(NUMBER)-TM/	COMBUSTION ENG., INC. REACTOR DEV. DIV., WINDSOR, CONN.
S1C/HP/	COMBUSTION ENG., INC. REACTOR DEV. DIV., WINDSOR, CONN.
S1C/MD-	COMBUSTION ENG., INC. REACTOR DEV. DIV., WINDSOR, CONN.
S1C/PP-	COMBUSTION ENG., INC. REACTOR DEV. DIV., WINDSOR, CONN.
S1C/RP-	COMBUSTION ENG., INC. REACTOR DEV. DIV., WINDSOR, CONN.
S-	CONGRESS. SENATE, WASHINGTON, D.C.
S-	DAVID TAYLOR MODEL BASIN, CARDEROCK, MD.
S(NUMBER)-	DENMARK. TEKNISKE HOJSKOLE, LYNGBY
S-(NUMBER)-R-	DENMARK. TEKNISKE HOJSKOLE, LYNGBY
S1-(3 DIGITS)	DREXEL UNIV., PHILADELPHIA
S-	EDGERTON, GERMESHAUSEN + GRIER, INC., SANTA BARBARA, CAL
S-(NUMBER)-R	EG+G, INC., SANTA BARBARA, CALIF.
S-	FRANKFORD ARSENAL. PITMAN-DUNN LAB., PHILADELPHIA
S-(YEAR)-	GENERAL ELECTRIC CO., OWENSBORO, KY.
S-(YEAR)-	GENERAL ELECTRIC CO., SCHENECTADY, N.Y.
S(YEAR)-	GENERAL ELECTRIC CO. MICROWAVE TUBE BUSINESS SECTION, SCHENECTADY, N.Y.
S-(2 DIGITS-YEAR)	HELSINKI UNIV. RADIO LAB.
S(YEAR)-	HONEYWELL, INC. MILITARY AND SPACE SCIENCES DEPT., ARLINGTON, VA.
S4M(NUMBER)/	HUGHES AIRCRAFT CO., CULVER CITY, CALIF.
S-	INSTITUTE FOR DEFENSE ANALYSES, ARLINGTON, VA. (USED BY SEVERAL DIVISIONS)
S-	ITT INDUSTRIAL LABS., FORT WAYNE
S-(NUMBER)-(NUMBER)	JOHNS HOPKINS UNIV., SILVER SPRING, MD. SOLID PROPELLANT INFORMATION AGENCY
S49-	KNOLLS ATOMIC POWER LAB., SCHENECTADY, N.Y.
S(NO.)G-ER-	KNOLLS ATOMIC POWER LAB., SCHENECTADY, N.Y.
S-	KOLLSMAN INSTRUMENT CORP., ELMHURST, N.Y.
S-	LIBRARY OF CONGRESS. TECH. INFO. DIV., WASH., DC (REF.20)
S-	MC GILL UNIV., MONTREAL. MAC DONALD PHYSICS LAB.
S-	MARQUARDT CORP., VAN NUYS, CALIF.
S-	NATIONAL DEFENSE RESEARCH COMMITTEE. URANIUM SECTION
S-(NUMBERS)	NATIONAL ENGINEERING SCIENCE CO., PASADENA, CALIF.
S-(NUMBER)-(NUMBER/NO.)	NAVAL RESEARCH LAB., WASHINGTON, D.C.
S-	NAVY MINE DEFENSE LAB., PANAMA CITY, FLA.
S-(NUMBER)	NETHERLANDS. ROYAL NETHERLANDS AIRCRAFT FACTORIES, FOKKER, AMSTERDAM
S-	OFFICE OF THE CHIEF OF NAVAL OPERATIONS, WASH., D.C.
S-	OULU UNIV., FINLAND
S(YEAR)R-	PARKER AIRCRAFT CO., LOS ANGELES
S-	QUARTERMASTER FOOD AND CONTAINER INSTITUTE FOR THE ARMED FORCES, CHICAGO
S-	RAND CORP., SANTA MONICA, CALIF.
S-(NUMBER)-RAND	RAND CORP., SANTA MONICA, CALIF.
S-1 THRU S-152	RAND CORP., SANTA MONICA, CALIF. (CLASSIFIED PAPER)
S-(NO.)-ARPA	RAND CORP., SANTA MONICA, CALIF.
S-	RAYTHEON CO., WALTHAM, MASS.
S-	RAYTHEON CO. RESEARCH DIV., WALTHAM, MASS.
S-	RESEARCH TRIANGLE INST. STATISTICS RESEARCH DIV., DURHAM, N.C.
S-	ROHM + HAAS CO. REDSTONE ARSENAL RESEARCH DIV., HUNTSVILLE, ALA.
S-19000 THRU S-21999	SCHOOL OF AVIATION MEDICINE, RANDOLPH AFB, TEX.
S-	SHELL DEVELOPMENT CO., EMERYVILLE, CALIF.
S/(3 DIGITS)	SOCIETA RICERCHE IMPIANTI NUCLEARI, SALUGGIA, ITALY
S-	SPINDLETOP RESEARCH, INC., LEXINGTON, KY.
S(YEAR)-	SYLVANIA ELECTRONIC SYSTEMS-EAST, NEEDHAM, MASS.
S-(NUMBER)-	SYLVANIA ELECTRONIC SYSTEMS-EAST. APPLIED RESEARCH LAB., WALTHAM, MASS.
S(NUMBER)-	TEXAS INSTRUMENTS, INC. APPARATUS DIV., DALLAS
S1WS-(LETTER(S))-	WESTINGHOUSE ELECTRIC CORP. ATOMIC POWER DIV., IDAHO FALLS
S(NO.)WS-(LTR/S)-	WESTINGHOUSE ELECTRIC CORP. ATOMIC POWER DIV., IDAHO FALLS
S-(NO.)-W(M)-	WESTINGHOUSE ELECTRIC CORP. BETTIS ATOMIC POWER LAB., PITTSBURGH
SA-	(SEMI-ANNUAL REPT.) (REF. 2)
SA-(YEAR)-(NUMBERS)	AEROSPACE CORP. SAN BERNARDINO OPERATIONS, CALIF.
SA-(YEAR)-	BUNKER-RAMO CORP., CANOGA PARK, CALIF.
SA-	BUREAU OF RECLAMATION (INTERIOR), DENVER
SA-	LIEGE. UNIVERSITE. LABORATOIRE D'AERONAUTIQUE
SA-	RESEARCH AND DEVELOPMENT BOARD, WASHINGTON, D.C. (STRATEGIC AIR OPERATIONS)
SA-	SPRINGFIELD ARMORY, MASS.
SA-	STANLEY AVIATION CORP., DENVER
SAAB-TN-	SAAB AIRCRAFT CO., SWEDEN
SAAB-TN-	SAAB AKTIEBOLAG, LINKIPING, SWEDEN
SAAB TN-	SVENSKA AEROPLANE AKTIEBOLAGET, LINKOPING, SWEDEN
SA/AER-	SUNDSTRAND AVIATION, ROCKFORD, ILL.
SAAP-	SUNFLOWER ARMY AMMUNITION PLANT, LAWRENCE, KANS.
SA/ATR-	SUNDSTRAND AVIATION, ROCKFORD, ILL.
SAB-	SCIENTIFIC ADVISORY BOARD (AIR FORCE), WASHINGTON, DC
SAB-APF-	SANDERS ASSOCIATES, INC., BEDFORD, MASS.
SAB-BCP-	SANDERS ASSOCIATES, INC., BEDFORD, MASS.
SAB-FVY-	SANDERS ASSOCIATES, INC., BEDFORD, MASS.
SAC-	AIR PROVING GROUND CTR. SAC PROJ. OFF., EGLIN AFB, FLA
SAC-	SCHOOL OF AVIATION MEDICINE, BROOKS AFB, TEX.
SAC-	SCHOOL OF AVIATION MEDICINE, RANDOLPH AFB, TEX.
SAC-	STANLEY AVIATION CORP., DENVER
SAC-	STRATEGIC AIR COMMAND, OFFUTT AFB, NEB.
SACLANTCEN-TM-	SACLANT ASW RESEARCH CENTER (NATO), LA SPEZIA, ITALY (ANTI-SUBMARINE WARFARE)
SACLANTCEN-TR-	SACLANT ASW RESEARCH CENTER (NATO), LA SPEZIA, ITALY (ANTI-SUBMARINE WARFARE)

SACM-
: STRATEGIC AIR COMMAND, OFFUTT AFB, NEB.
SAC-OA-
: STRATEGIC AIR COMMAND. OFFICE OF OPERATIONS ANALYSIS, OFFUTT AFB, NEBR.
SACP-
: STRATEGIC AIR COMMAND, OFFUTT AFB, NEB.
SAC-R-
: BELL TELEPHONE LABS., INC., N.Y.C.
SAC TECHNICAL REPT (NO.)-
: STRATEGIC AIR COMMAND, OFFUTT AFB, NEB.
SAC TP-
: STRATEGIC AIR COMMAND, OFFUTT AFB, NEB. (TECHNICAL PAMPHLET)
SAD-(YEAR)/(NO./NO.)
: AUSTRALIA. WEAPONS RESEARCH ESTABLISHMENT, ADELAIDE
SAD-(NUMBER)-T
: ROYAL NORWEGIAN COUNCIL FOR SCI. + IND. RES., BLINDERN
SADD-
: SURFACE ANTI-SUBMARINE DEV. DETACHMENT, KEY WEST, FLA
SADTC-
: NORTH ATLANTIC TREATY ORGANIZATION. SHAPE TECHNICAL CENTRE, THE HAGUE
SADTC-(YEAR)/TM-
: NORTH ATLANTIC TREATY ORGANIZATION. SHAPE TECHNICAL CENTRE, THE HAGUE
SAE-
: SOCIETY OF AUTOMOTIVE ENGINEERS, N.Y.C.
SAEA-
: SOUTHWEST ATOMIC ENERGY ASSOCIATES
SAEA-MPL-
: SOUTHWEST ATOMIC ENERGY ASSOCIATES (SOUTHWEST EXPERIMENTAL FAST OXIDE REACTOR DEVELOPMENT PROGRAM)
SAER-
: FRANCE. COMMISSARIAT A L'ENERGIE ATOMIQUE. CENTRE D'ETUDES NUCLEAIRES, SACLAY
SAF-
: MINNESOTA. UNIV., MINNEAPOLIS. ST. ANTHONY FALLS HYDRAULIC LAB.
SAFB-
: SURVIVAL TRAINING SCHOOL, STEAD AFB, NEV.
SAFEX-
: UNITED KINGDOM ATOMIC ENERGY AUTHORITY. WEAPONS GROUP ATOMIC WEAPONS RES. ESTAB., FOULNESS, ESSEX, ENG.
SAFT-
: SOCIETE DES ACCUMULATEURS FIXES ET DE TRACTION
SAG-
: ANTI-SUBMARINE WARFARE SYSTEMS PROJECT OFFICE. SYSTEMS ANALYSIS GROUP, SILVER SPRING, MD.
SAG-
: CANADA. DEFENCE RESEARCH BOARD, OTTAWA
SAG-
: SCIENTIFIC ADVISORY BOARD (AIR FORCE), WASHINGTON, DC
SAG-MEMO-
: CANADA. DEFENCE RESEARCH BOARD, OTTAWA
SAGR-
: NAVAL ORDNANCE LAB., WHITE OAK, MD.
SA HSE-
: SPRINGFIELD ARMORY, MASS.
SAI-
: SANDERS ASSOCIATES, INC., NASHUA, N.H.
SAI-(YEAR)-
: SCIENCE APPLICATIONS, INC., LA JOLLA, CALIF.
SAI-
: SCIENTIFIC-ATLANTA, INC.
SAI-CC-
: SYSTEMS ASSOCIATES, INC., LONG BEACH, CALIF.
SAIL-
: WEATHER BUREAU. SEA-AIR INTERACTION LAB., WASH., D.C.
SAL-
: SAN-ALOO CLASSIFICATION COORDINATING COMMITTEE, AEC
SAL-
: SASKATCHEWAN. UNIV., SASKATOON. SASKATCHEWAN ACCELERATOR LAB.
SA L-
: SPRINGFIELD ARMORY. ENGINEERING DEPT. LAB., MASS.
SA L (NUMBER(S))-
: SPRINGFIELD ARMORY. ENGINEERING DEPT. LAB., MASS.
SALR-
: STANFORD UNIV., CALIF. AEROSOL LAB.
SAM-
: SCHOOL OF AEROSPACE MEDICINE, BROOKS AFB, TEX.
SAM-
: SCHOOL OF AVIATION MEDICINE, RANDOLPH AFB, TEX.
SAM-(YEAR)-
: SCHOOL OF AVIATION MEDICINE, RANDOLPH AFB, TEX.
SAM-REVIEW-(NO.-YEAR)
: SCHOOL OF AEROSPACE MEDICINE, BROOKS AFB, TEX.
SAMSO-(YEAR)-
: SPACE AND MISSILE SYSTEMS ORGANIZATION, LOS ANGELES AIR FORCE STATION, CALIF.
SAMSO-SMEA-
: SPACE AND MISSILE SYSTEMS ORGANIZATION, LOS ANGELES AIR FORCE STATION, CALIF.
SAMSO-TR-(YEAR)-
: SPACE AND MISSILE SYSTEMS ORGANIZATION, LOS ANGELES AIR FORCE STATION, CALIF.
SAM-TDR-(YEAR)-
: SCHOOL OF AEROSPACE MEDICINE, BROOKS AFB, TEX.
SAMTEC-TR-(YEAR)-
: SPACE AND MISSILE TEST CENTER, VANDENBERG AFB, CAL.
SAM-TR-(YEAR)-
: SCHOOL OF AEROSPACE MED., BROOKS AFB, TEX.(TECH.RPT.)
SAM-TR-
: SCHOOL OF AVIATION MEDICINE, RANDOLPH AFB, TEX.
SAM-TT-(NUMBER)-
: SCHOOL OF AEROSPACE MEDICINE, BROOKS AFB, TEX.
SAM-TT-R-(NUMBER)-
: SCHOOL OF AEROSPACE MEDICINE, BROOKS AFB, TEX.

SAN-(NUMBER)-P-
: ALASKA. UNIV., COLLEGE. INST. OF MARINE SCIENCES
SAN-
: SAN FRANCISCO OPERATIONS OFFICE, AEC (REF. 4)
SAN-(CONTRACT NUMBER)-
: SAN FRANCISCO OPERATIONS OFFICE, AEC (REF. 4)
SAN-
: SANDERS ASSOCIATES, INC., NASHUA, N.H.
SAN-(3 DIGITS)-
: SUNDSTRAND AVIATION, ROCKFORD, ILL.
SAN-BDR-(YEAR)-
: SANDERS ASSOCIATES, INC., NASHUA, N.H.
SAN-BLC-
: SANDERS ASSOCIATES, INC., NASHUA, N.H.
SANCU-
: SANTA CLARA, CALIF. UNIV.
SAND-
: SANDIA CORP., ALBUQUERQUE, N. MEX.
S AND C-
: GT. BRIT. AERONAUTICAL RESEARCH COUNCIL
SANDIA TEST-(NO.)-(NO.)
: SANDIA CORP., ALBUQUERQUE, N. MEX.
SANDSTONE-
: (OPERATION SANDSTONE) (REF. 7)
S AND T MEMO-(NO.)/(YR.)
: GT. BRIT. MINISTRY OF SUPPLY. TECHNICAL INFO. BUR.
SAN-FLG-CR-
: SANDERS ASSOCIATES, INC., BEDFORD, MASS.
SAN-GDJ-(YEAR)-
: SANDERS ASSOCIATES, INC., NASHUA, N.H.
SAN-GKN-(YEAR)-
: SANDERS ASSOCIATES, INC., NASHUA, N.H.
SAN-GOL-
: SANDERS ASSOCIATES, INC., NASHUA, N.H.
SANITARY ENG-
: COLORADO STATE UNIV., FORT COLLINS. DEPT. OF MICROBIOLOGY
SAN-JDK-(YEAR)-
: SANDERS ASSOCIATES, INC., BEDFORD, MASS.
SAN-NDL-(YEAR)-
: SANDERS ASSOCIATES, INC., NASHUA, N.H.
SAN-PAH-
: SANDERS ASSOCIATES, INC., NASHUA, N.H.
SAN-PBD-(YEAR)-
: SANDERS ASSOCIATES, INC., NASHUA, N.H.
SAN-PYC-(YEAR)-
: SANDERS ASSOCIATES, INC. INSTRUMENT DIV., NASHUA, N.H.
SAN-PYD-(YEAR)-
: SANDERS ASSOCIATES, INC. INSTRUMENT DIV., MANCHESTER, N.H.
SAN-TR-
: PASADENA FOUNDATION FOR MEDICAL RESEARCH, CALIF.
SAO-(YEAR)-
: ANTI-SUBMARINE WARFARE SYSTEMS PROJECT OFFICE. SYSTEMS ANALYSIS OFFICE, SILVER SPRING, MD.
SAO-
: SMITHSONIAN ASTROPHYSICAL OBS., CAMBRIDGE, MASS.
SA/ONR-
: SUNDSTRAND AVIATION-DENVER, PACOIMA, CALIF.
SAO SPECIAL REPT.-
: SMITHSONIAN ASTROPHYSICAL OBS., CAMBRIDGE, MASS.
SAO-SR-
: SMITHSONIAN ASTROPHYSICAL OBS., CAMBRIDGE, MASS.
SAO-SR-INDEX-
: SMITHSONIAN ASTROPHYSICAL OBS., CAMBRIDGE, MASS.
SAO-TM-(YEAR)-
: ANTI-SUBMARINE WARFARE SYSTEMS PROJECT OFFICE. SYSTEMS ANALYSIS OFFICE, SILVER SPRING, MD.
SAPR-
: COLUMBIA UNIV., N.Y.C. DEPT. OF CHEMICAL ENGINEERING
SA/PR-
: GT. BRIT. HOME OFFICE, LONDON
SAPR-
: HARVARD UNIV., CAMBRIDGE, MASS.
SAPR-
: MANLABS, INC., CAMBRIDGE, MASS.
SAPR-
: MASSACHUSETTS INST. OF TECH., CAMBRIDGE
SAPR-
: PENNSYLVANIA STATE UNIV., UNIVERSITY PARK
SAPR-
: UNIVERSITY OF SOUTHERN CALIFORNIA, LOS ANGELES
SAPR-
: WOODS HOLE OCEANOGRAPHIC INSTITUTION, MASS.
SAR-
: (SEMI-ANNUAL REPORT) (REF. 2)
SAR-
: ARMY AIR FORCE. OFFICE OF FLYING SAFETY (SPECIAL ACCIDENT REPORT)
SAR-
: ARMY TROPIC TEST CENTER, FORT CLAYTON, CANAL ZONE
SAR-
: ATLANTIC RESEARCH CORP., ALEXANDRIA, VA.
SAR-
: BUREAU OF SHIPS (STATISTICAL ANALYSIS REPORT)
SAR-
: COLORADO. UNIV., BOULDER
SAR-
: ENGELHARD INDUSTRIES, INC., EAST NEWARK, N.J.
SAR-
: FLORIDA. UNIV., GAINESVILLE
SAR-
: FRANCE. SERVICE TECHNIQUE DE L'AIR, PARIS
SAR-
: GENERAL ELECTRIC CO., SYRACUSE, N.Y.
SAR-
: GENERAL PRECISION, INC., GLENDALE, CALIF.
SAR-
: GLOBE-UNION INC., MILWAUKEE

SAR-	KNOLLS ATOMIC POWER LAB., SCHENECTADY, N.Y. (SUBMARINE REACTOR PROJECT)
SAR-	MIAMI. UNIV., CORAL GABLES, FLA.
SAR-	NEW YORK UNIV., N.Y.C.
SAR-	SIKORSKY AIRCRAFT DIV., UNITED AIRCRAFT CORP., STRATFORD, CONN.
SAR-	SYRACUSE UNIV., N.Y. RESEARCH INST.
SAR-	WESTERN MICROWAVE LAB., INC., SANTA CLARA, CALIF.
SAR-	WISCONSIN. UNIV., MADISON
SARC-	SYSTEMS ANALYSIS AND RESEARCH CORP., CAMBRIDGE, MASS.
SAR-G-	FRANCE. COMMISSARIAT A L'ENERGIE ATOMIQUE. CENTRE D'ETUDES NUCLEAIRES, GRENOBLE
SA-RG-	SANDERS ASSOCIATES, INC., NASHUA, N.H.
55SARS-	CONSOLIDATED VULTEE AIRCRAFT CORP., SAN DIEGO, CALIF.
SA-RS-	SANDERS ASSOCIATES, INC., NASHUA, N.H.
SAR SER-	SIKORSKY AIRCRAFT DIV., UNITED AIRCRAFT CORP., STRATFORD, CONN.
SAR.T/	FRANCE. COMMISSARIAT A L'ENERGIE ATOMIQUE. CENTRE D'ETUDES NUCLEAIRES, GRENOBLE
56-SARW-	BEECH AIRCRAFT CORP., WICHITA, KAN.
SAR WTT-	SIKORSKY AIRCRAFT DIV., UNITED AIRCRAFT CORP., STRATFORD, CONN.
SAR X-	SIKORSKY AIRCRAFT DIV., UNITED AIRCRAFT CORP., STRATFORD, CONN.
SAS-	ARMY ENGINEER DISTRICT, SAVANNAH
SAS-(NUMBER)	GRUMMAN AIRCRAFT ENGINEERING CORP., BETHPAGE, N.Y.
SASKU-	SASKATCHEWAN. UNIV., SASKATOON
SASM-(NUMBER)-(YEAR)	JOINT CHIEFS OF STAFF. SPECIAL ASSISTANT FOR STRATEGIC MOBILITY, WASHINGTON, D.C.
SASR-	STANFORD UNIV., CALIF. DEPT. OF AERONAUTICS AND ASTRONAUTICS
SATM-	WESTINGHOUSE ELECTRIC CORP. AEROSPACE DIV., BALTIMORE
SATR-	AMERICAN OPTICAL CO. STAMFORD RESEARCH LABS., SOUTHBRIDGE, MASS.
SATR-	MASSACHUSETTS INST. OF TECH., CAMBRIDGE (SEMIANNUAL TECHNICAL SUMMARY REPORT)
SATR-	MINNESOTA. UNIV., MINNEAPOLIS
SATR-	NEW MEXICO STATE UNIV., UNIVERSITY PARK
SATR-	SAINT LOUIS UNIV.
SA-TR(NUMBER)-	SPRINGFIELD ARMORY, MASS.
SA-TR-	SPRINGFIELD ARMORY, MASS.
SATR-	STANFORD RESEARCH INST., MENLO PARK, CALIF.
SATR-	TEXAS INSTRUMENTS, INC., DALLAS
SATR-(NUMBER)	THIOKOL CHEMICAL CORP. HUNTSVILLE DIV., ALA.
SA-TRI-	SPRINGFIELD ARMORY, MASS.
SA-TRI(NO.)-	SPRINGFIELD ARMORY, MASS.
SATSR-	(SEMI-ANNUAL TECHNICAL SUMMARY REPORT) (REF. 2)
SAVANNA RESEARCH SER-	MC GILL UNIV., MONTREAL. DEPT. OF GEOGRAPHY
SAWC-TDR-	SPECIAL AIR WARFARE CENTER, EGLIN AFB, FLA.
SA WO-	SPRINGFIELD ARMORY. ENGINEERING DEPT., MASS.
SB-	(SERVICE BULLETIN) (REF. 2)
SB-	(SPECIAL BIBLIOGRAPHY) (REF. 2)
SB-	(SUPPLY BULLETIN) (REF. 2)
SB-(YEAR)	ANALYTIC SERVICES, INC., FALLS CHURCH, VA.
SB-(YEAR)-S-	ARMY, WASHINGTON, D.C.
SB-	CLEARINGHOUSE FOR FEDERAL SCIENTIFIC AND TECHNICAL INFORMATION, SPRINGFIELD, VA.
SB-	GENERAL ELECTRIC CO. DASA INFORMATION AND ANALYSIS CENTER, SANTA BARBARA, CALIF.
SB-(YR.)-	LOCKHEED MISSILES AND SPACE CO., SUNNYVALE, CALIF.
SB-(YEAR)-	NATIONAL TRANSPORTATION SAFETY BOARD, WASHINGTON, DC
SB-	NAVY, SEAL BEACH, CALIF.
SB-	OFFICE OF TECHNICAL SERVICES, WASHINGTON, D.C. (SELECTIVE BIBLIOGRAPHY)
SB-	RAND CORP., SANTA MONICA, CALIF.
SBA-	NAVAL ORDNANCE TEST STATION. SCIENCE DEPT. BALLISTICS DIV. AERODYNAMICS SECTION, INYOKERN (CHINA LAKE), CALIF.
SBA-	SMALL BUSINESS ADMINISTRATION, WASHINGTON, D.C.
SBAC-	SOCIETY OF BRITISH AIRCRAFT CONSTRUCTORS
SBAMA-2.5-	AEROJET-GENERAL CORP., SACRAMENTO, CALIF.
SBANP-(YEAR)-	SANTA BARBARA ANALYSIS AND PLANNING CORP., CALIF.
SBE-	NAVAL ORDNANCE TEST STA., INYOKERN (CHINA LAKE), CAL.
SBED-	NAVAL ORDNANCE TEST STA., INYOKERN (CHINA LAKE), CAL.
SBEF-	NAVAL ORDNANCE TEST STATION. SCIENCE DEPT. EXTERIOR BALLISTICS, FINNER GP.,INYOKERN (CHINA LAKE),CALIF.
SBES-	NAVAL ORDNANCE TEST STA., INYOKERN (CHINA LAKE), CAL.
SBNL-TM-(NUMBER-YEAR)	NAVAL SUBMARINE BASE, NEW LONDON, CONN.
SBNL-TR-(NUMBERS)-(YEAR)	NAVAL SUBMARINE BASE, NEW LONDON, CONN.
SBRC-	SANTA BARBARA RESEARCH CENTER, GOLETA, CALIF.
SBS-	STRATEGIC BOMBING SURVEY
SBS AEPR-	STRATEGIC BOMBING SURVEY (AERO ENGINE PLANT RPTS.)
SBS AFPR-	STRATEGIC BOMBING SURVEY (AIR FRAMES PLANT REPORTS)
SBS ASDR-	STRATEGIC BOMBING SURVEY. AERO STUDIES DIV.
SBS CDR-	STRATEGIC BOMBING SURVEY (CIVILIAN DEFENSE REPORTS)
SBS CPR-	STRATEGIC BOMBING SURVEY (COKING PLANT REPORTS)
SBS HIPR-	STRATEGIC BOMBING SURVEY (HEAVY INDUSTRY PLANT RPTS.)
SBS LMPR-	STRATEGIC BOMBING SURVEY (LIGHT METALS PLANT RPTS.)
SBS OPR-	STRATEGIC BOMBING SURVEY (OIL PLANT REPORTS)
SBS P-	STRATEGIC BOMBING SURVEY (PACIFIC REPORTS)
SBS P-I-	STRATEGIC BOMBING SURVEY (PACIFIC INTERROGATION RPTS)
SBS PPR-	STRATEGIC BOMBING SURVEY (PROPELLANTS PLANT RPTS.)
SBS PR-	STRATEGIC BOMBING SURVEY (PLANT REPORTS)
SBS SP-	STRATEGIC BOMBING SURVEY (SPECIAL REPORTS)
SBS UDPR-	STRATEGIC BOMBING SURVEY. UTILITIES DIV.(PLANT RPTS.)
SBW-	OKLAHOMA STATE UNIV., STILLWATER
SC-	CONVAIR, FORT WORTH, TEX.
S-C(NUMBER)-	FRANKLIN INST. RESEARCH LABS., PHILADELPHIA
SC-	GARRETT CORP. AIRESEARCH MFG. DIV., LOS ANGELES
SC-	GT. BRIT. ADVISORY COUNCIL ON SCI. RES. + TECH. DEV.
S.C. 111/	GT. BRIT. AIR MINISTRY. METEOROLOGICAL RES. COMM.
SC-	GT. BRIT. AIR MINISTRY. METEOROLOGICAL RES. COMM.
SC-	GT. BRIT. CAPENHURST WORKS, CHES., ENGLAND (REF. 36)
S.C.-(NUMBER-NUMBER)-	GREENVILLE COUNTY PLANNING COMMISSION, S.C.
SC-(NUMBER-NUMBER)-	GREENVILLE COUNTY PLANNING COMMISSION, S.C.
SC/	INTERNATIONAL CENTRE FOR THEORETICAL PHYSICS, TRIESTE, ITALY
SC-	MASSACHUSETTS INST. OF TECH.,CAMBRIDGE.RADIATION LAB.
SC-	NATIONAL RESEARCH CORP., CAMBRIDGE, MASS.
SC-	OFFICE OF NAVAL RESEARCH (SEA COMBAT OPERATIONS)
SC-	OFFICE OF SCIENTIFIC RES. + DEV. (SHAPED CHARGES)
SC-(NUMBER)-(YEAR)	ORDNANCE CORPS (ARMY). SHAPED CHARGE RESEARCH AND DEVELOPMENT STEERING AND COORDINATING COMMITTEE
SC-	RESEARCH + DEV. BD.,WASH.,D.C.(SEA COMBAT OPERATIONS)
SC-	SANDIA CORP., ALBUQUERQUE, N. MEX.
SC-(NUMBER)-C	SANDIA CORP., ALBUQUERQUE, N. MEX. (CATALOG)
SC-(NUMBER)-CP	SANDIA CORP., ALBUQUERQUE, N. MEX. (CORP. PROCEDURE)
SC-(NUMBER)-M	SANDIA CORP., ALBUQUERQUE, N. MEX.
SC-(NUMBER)-NL	SANDIA CORP., ALBUQUERQUE, N. MEX.(NOMENCLATURE LIST)

SC-(NUMBER)-PL
  SANDIA CORP., ALBUQUERQUE, N. MEX. (PARTS LIST)
SC-(NUMBER)-PR
  SANDIA CORP., ALBUQUERQUE, N. MEX. (PROGRESS REPT.)
SC-(NUMBER)-RC
  SANDIA CORP., ALBUQUERQUE, N. MEX.
SC-(NUMBER)(RR)
  SANDIA CORP., ALBUQUERQUE, N. MEX.
SC-(NUMBER)(SP)
  SANDIA CORP., ALBUQUERQUE, N. MEX. (SPECIFICATION)
SC-(NUMBER)(WD)
  SANDIA CORP., ALBUQUERQUE, N. MEX.
SC-(NUMBER)-TM
  SANDIA CORP., ALBUQUERQUE, N. MEX. (TEST MANUAL)
SC-
  SANDIA CORP. LIVERMORE LAB., CALIF.
SC-(NUMBER)-M
  SANDIA CORP. LIVERMORE LAB., CALIF.
SC-(NUMBER)-TM
  SANDIA CORP. LIVERMORE LAB., CALIF. (TECHNICAL MEMO)
SC-(NUMBER)-TR
  SANDIA CORP. LIVERMORE LAB., CALIF. (TECH. REPORT)
SC-(NUMBER)-WD
  SANDIA CORP. LIVERMORE LAB., CALIF. (WEAPON DATA)
SC-
  SHELL CHEMICAL CORP. INDUSTRIAL CHEMICALS DIV.,N.Y.C.
SC-
  SONOTONE CORP., ELMSFORD, N.Y.
SC-40-0014-
  SOUTH CAROLINA. STATE PLANNING + GRANTS DIV.,COLUMBIA
S-C-
  STROMBERG-CARLSON CORP, ROCHESTER, N.Y.
SC-
  STROMBERG-CARLSON CORP., SAN DIEGO, CALIF.
S-C-
  STROMBERG-CARLSON CORP. DATA PRODUCTS DIV., SAN DIEGO, CALIF.
SCA-
  NAVAL ORDNANCE TEST STA., INYOKERN (CHINA LAKE), CAL.
SCA-
  SERVO CORP. OF AMERICA, HICKSVILLE, N.Y.
SCA-
  SERVO CORP. OF AMERICA, NEW HYDE PARK, N.Y.
SC-(AERE)-
  UNITED KINGDOM ATOMIC ENERGY AUTHORITY. RESEARCH GP. ATOMIC ENERGY RES. ESTAB., HARWELL, BERKS, ENGLAND
SCAL-MISC-
  SANDIA CORP., ALBUQUERQUE, N. MEX.
SCAL-SERIES-
  SANDIA CORP., ALBUQUERQUE, N. MEX.
SC-ANSIC-
  SANDIA LABS., ALBUQUERQUE, N. MEX.
SCAP-
  SUPREME COMMANDER FOR THE ALLIED POWERS
SCAP ESS B-
  SUPREME COMMANDER FOR THE ALLIED POWERS. ECONOMIC AND SCI. SEC. RES. + STATISTICS DIV. (BULL.)
SCAP ESS R-
  SUPREME COMMANDER FOR THE ALLIED POWERS. ECONOMIC AND SCIENTIFIC SECTION
SCAP ESS SR-
  SUPREME COMMANDER FOR THE ALLIED POWERS. ECONOMIC AND SCIENTIFIC SEC. RES. DIV. (SPEC. RPTS.)
SCAP MIS ATIS-
  SUPREME CMDR. FOR THE ALLIED POWERS. MIL. INTELL. SEC GENL. STAFF ALLIED TRANSLATOR + INTERPRETER SEC.
SCAP MIS IR-
  SUPREME COMMANDER FOR THE ALLIED POWERS. MILITARY INTELLIGENCE SECTION (INTELLIGENCE REPORTS)
SCAP NRS-
  SUPREME COMMANDER FOR THE ALLIED POWERS. NATURAL RESOURCES SECTION
SCAP NRS PS-
  SUPREME COMMANDER FOR THE ALLIED POWERS. NATURAL RESOURCES SECTION (PRELIMINARY STUDIES)
SCAP RI-
  SUPREME CMDR. FOR THE ALLIED POWERS. RES. + INFO. DIV
SCAP RI (NO-LTRS-LTR)-
  SUPREME CMDR. FOR THE ALLIED POWERS. RES. + INFO. DIV
SCAP RI (LTR-NO-LTRS)-
  SUPREME CMDR. FOR THE ALLIED POWERS. RES. + INFO. DIV
SCAP SR-
  SUPREME CMDR. FOR THE ALLIED POWERS (SPECIAL REPORTS)
SC-ARPIC-
  SANDIA LABS., ALBUQUERQUE, N. MEX.
SC-B-(YEAR)-
  SANDIA LABS., ALBUQUERQUE, N. MEX. (SELECTIVE BIB.)
SCB-MEMO-(NUMBER)-(YEAR)
  OFFICE OF THE CHIEF OF NAVAL OPERATIONS. SHIP CHARACTERISTICS BOARD, WASHINGTON, D.C.
SCC-
  SHELL CHEMICAL CO., SAN FRANCISCO
SCC-(NUMBER)-R-
  SPEER CARBON CO., NIAGARA FALLS, N.Y.
SCC-
  SPEER CARBON CO. RES. LAB., NIAGARA FALLS, N.Y.
SCC-
  STAUFFER CHEMICAL CO., N.Y.C.
SCC-26-FS(NO.)
  STAUFFER CHEMICAL CO. CHAUNCEY RESEARCH CENTER, N.Y. (FINAL SUMMARY REPORTS)
SCC-26-QPR-
  STAUFFER CHEMICAL CO. CHAUNCEY RESEARCH CENTER, N.Y. (QUARTERLY PROGRESS REPORTS)
SCC-824-Q(NUMBER)
  STAUFFER CHEMICAL CO. CHAUNCEY RESEARCH CENTER, N.Y.
SCC-ANNUAL-TSR-(YR.-YR.)
  STAUFFER CHEMICAL CO. RICHMOND RESEARCH LAB., CALIF.
SCCC-TED-(YEAR)-FR-
  ARMY COMMUNICATIONS-ELECTRONICS ENGINEERING INSTALLATION AGENCY, FORT HUACHUCA, ARIZ.
SCCI-
  FRANCE. COMMISSARIAT A L'ENERGIE ATOMIQUE. CENTRE D'ETUDES NUCLEAIRES, FONTENAY-AUX-ROSES
SC-CR-(YEAR)-
  SANDIA LABS., ALBUQUERQUE, N. MEX.
SCC-RRC-
  STAUFFER CHEMICAL CO. RICHMOND RESEARCH CENTER, CALIF
SCC-RRL-
  STAUFFER CHEMICAL CO. RICHMOND RESEARCH LAB., CALIF.
SCC-TED-
  ARMY STRATEGIC COMMUNICATIONS COMMAND. TEST AND EVALUATION DIRECTORATE, FORT HUACHUCA, ARIZ.
SCC-TSR-
  STAUFFER CHEMICAL CO. WESTERN RESEARCH CENTER, RICHMOND, CALIF.
SCC-WRC-
  STAUFFER CHEMICAL CO. WESTERN RESEARCH CENTER, RICHMOND, CALIF.
SCD-
  ANTIAIRCRAFT ARTILLERY AND GUIDED MISSILE CENTER, FORT BLISS, TEX.
SCD-
  ARMY TANK-AUTOMOTIVE COMMAND. SCIENTIFIC COMPUTER DIV., WARREN, MICH.
SCD-
  STROMBERG-CARLSON DIV., GENERAL DYNAMICS CORP., ROCHESTER, N.Y.
SC-DC-
  GT. BRIT. ATOMIC WEAPONS RESEARCH ESTABLISHMENT, ALDERMASTON, BERKS, ENGLAND
SCDC-
  SANDIA CORP., ALBUQUERQUE, N. MEX.
SC-DC-(YEAR)-
  SANDIA CORP., ALBUQUERQUE, N. MEX.
SC-DC-
  SANDIA LABS., ALBUQUERQUE, N. MEX.
SC-DCR-
  SANDIA CORP., ALBUQUERQUE, N. MEX.
SCDR-
  SANDIA CORP., ALBUQUERQUE, N. MEX. (DEV. REPT.)
SCDR-(NO.)-(YR.)
  SANDIA CORP., ALBUQUERQUE, N. MEX.
SC-DR-
  SANDIA CORP. LIVERMORE LAB., CALIF.
SCDR-(NUMBER)-(YEAR)
  SANDIA CORP. LIVERMORE LAB., CALIF. (DEV. REPORT)
SC-DR-(YEAR)-
  SANDIA LABS., ALBUQUERQUE, N. MEX.
SC-DR-
  SANDIA LABS., ALBUQUERQUE, N. MEX.
SCD-TM-(NO./NO./NO.)
  SYSTEM DEVELOPMENT CORP., SANTA MONICA, CALIF.
SCD-TM-L-(NO./NO./NO.)
  SYSTEM DEVELOPMENT CORP., SANTA MONICA, CALIF.
SC-DW-(YEAR)-
  SANDIA CORP., ALBUQUERQUE, N. MEX.
SCEAG-
  HALLER, RAYMOND, AND BROWN, INC., STATE COLLEGE, PA.
SCEAR/PANEL-
  UNITED NATIONS. SCIENTIFIC COMMITTEE ON THE EFFECTS OF ATOMIC RADIATION
SCEC-
  SOUTHERN CALIFORNIA EDISON CO.
SCEL-
  EVANS SIGNAL LAB., BELMAR, N.J.
SCEL-
  SIGNAL CORPS ENGINEERING LABS., FORT MONMOUTH, N.J.
SCEL-E-
  SIGNAL CORPS ENGINEERING LABS., FORT MONMOUTH, N.J.
SCEL ER-
  SIGNAL CORPS ENG. LABS.,FT. MONMOUTH, N.J.(ENG. RPTS)
SCEL ER E-
  SIGNAL CORPS ENG. LABS.,FT. MONMOUTH, N.J.(ENG. RPTS)
SCEL-ETIB-
  SIGNAL CORPS ENGINEERING LABS., FORT MONMOUTH, N.J. (ELECTRON TUBE INFORMATION BULLETIN)
SCEL-M-
  SIGNAL CORPS ENGINEERING LABS., FORT MONMOUTH, N.J.
SCEL-SSDB-
  SIGNAL CORPS ENGINEERING LABS. SOLID STATE DEVICES BRANCH, FORT MONMOUTH, N.J.
SCEL-T-
  SIGNAL CORPS ENG. LABS.,FT. MONMOUTH, N.J.(TEST RPT.)
SCEL TM-
  SIGNAL CORPS ENG. LABS.,FT. MONMOUTH, N.J.(TECH.MEMO)
SCEL TM M-
  SIGNAL CORPS ENG. LABS.,FT. MONMOUTH, N.J.(TECH MEMO)
SCEL TR-
  SIGNAL CORPS ENG. LABS.,FT. MONMOUTH, N.J.(TEST RPTS)
SCEL TR T-
  SIGNAL CORPS ENG. LABS.,FT. MONMOUTH, N.J.(TEST RPTS)
SCGF-
  AIR FORCE SYSTEMS COMMAND, ANDREWS AFB, MD.
SCI-
  SCIENCE COMMUNICATION, INC., WASHINGTON, D.C.
SCI-
  SPACE CRAFT, INC., HUNTSVILLE, ALA.
SCI-ARIES-
  SCIENCE COMMUNICATION, INC., WASHINGTON, D.C.
SCIENTIFIC-
  (SCIENTIFIC REPORT) (REF. 2)
SCIENTIFIC REPORT
  (SCIENTIFIC REPORT) (REF. 2)
SCI-GI-
  SPACE CRAFT, INC., HUNTSVILLE, ALA.
SCI-RA-
  SPACE CRAFT, INC., HUNTSVILLE, ALA.

SCI-T-(YEAR)-	SANDIA LABS., ALBUQUERQUE, N. MEX. (TRANSLATION)
SCL-	CAPEHART-FARNSWORTH CORP., FORT WAYNE
SCL-	SANDIA CORP. LIBRARY, ALBUQUERQUE, N. MEX. (TRANSLATION)
SCL-	SANDIA CORP. LIVERMORE LAB., CALIF.
SCL-B-	SANDIA CORP., ALBUQUERQUE, N. MEX.
SCL-CR-(YEAR)-	SANDIA CORP. LIVERMORE LAB., CALIF.
SCL-CR-(YEAR)-	SANDIA LABS., LIVERMORE, CALIF.
SCL-DC-(YEAR)-	SANDIA CORP. LIVERMORE LAB., CALIF.
SCL-DC-	SANDIA LABS., LIVERMORE, CALIF.
SCL-DC-(YEAR)-	SANDIA LABS., LIVERMORE, CALIF.
SCL-DR-(YEAR)-	SANDIA LABS., LIVERMORE, CALIF.
SCL-R-	SANDIA CORP. LIVERMORE LAB., CALIF. (REPRINT)
SCL-RR-	SANDIA CORP. LIVERMORE LAB., CALIF.
SCL-T-	SANDIA LABS., ALBUQUERQUE, N. MEX. (TRANSLATION)
SCL-T-	SANDIA CORP. LIVERMORE LAB., CALIF. (TRANSLATION)
SCL-TM-	SANDIA LABS., ALBUQUERQUE, N. MEX.
SCL-TM-	SANDIA LABS., LIVERMORE, CALIF.
SCL-TN-(YEAR)-	SANDIA LABS., LIVERMORE, CALIF.
SCL/TPS/(NO.)(LETTER)	EUROPEAN CONFERENCE ON SATELLITE COMMUNICATIONS. TECHNICAL PLANNING STAFF
SCLV-	SANDIA CORP. LIVERMORE LAB., CALIF.
SCL-WD-	SANDIA CORP. LIVERMORE LAB., CALIF. (WEAPON DATA)
SC-M-(YEAR)-	SANDIA LABS., ALBUQUERQUE, N. MEX.
SCNC-	SYLVANIA ELECTRIC PRODUCTS INC. SYLCOR DIV., BAYSIDE, N.Y.
SCP-	DARTMOUTH COLL., HANOVER, N.H.
SCP-	DARTMOUTH COLL., HANOVER, N.H. THAYER SCHOOL OF ENG.
SC-PC-	CHICAGO. UNIV. METALLURGICAL LAB.
SC-PP-(YEAR)-	NORTH AMERICAN ROCKWELL CORP., THOUSAND OAKS, CALIF.
SC-PR-(YEAR)-	SANDIA LABS., ALBUQUERQUE, N. MEX.
SCPRF-	STRUCTURAL CLAY PRODUCTS RESEARCH FDN., GENEVA, ILL.
SCPRI-	FRANCE. SERVICE CENTRAL DE PROTECTION CONTRE LES RAYONNEMENTS IONISANTS, PARIS
SCPRI(A)-	FRANCE. SERVICE CENTRAL DE PROTECTION CONTRE LES RAYONNEMENTS IONISANTS, PARIS
SCPRI(M)-	FRANCE. SERVICE CENTRAL DE PROTECTION CONTRE LES RAYONNEMENTS IONISANTS, PARIS
SCPRI(RM)	FRANCE. SERVICE CENTRAL DE PROTECTION CONTRE LES RAYONNEMENTS IONISANTS, PARIS
SCPRI(S)-	FRANCE. SERVICE CENTRAL DE PROTECTION CONTRE LES RAYONNEMENTS IONISANTS, PARIS
SCPRI(T)-	FRANCE. SERVICE CENTRAL DE PROTECTION CONTRE LES RAYONNEMENTS IONISANTS, PARIS
SCPUAE-	U.S.S.R. GOSUDARSTVENNYI KOMITET PO ISPOL'ZOVANIYU ATOMNOI ENERGII, MOSCOW
SC-QAA-	SANDIA CORP., ALBUQUERQUE, N. MEX.
SC-QR-	SANDIA LABS., ALBUQUERQUE, N. MEX. (QUARTERLY REPORT)
SCR-	AEROJET-GENERAL CORP., SACRAMENTO, CALIF.
SCR-	AEROJET-GENERAL CORP. LIQUID ROCKET OPERATIONS, SACRAMENTO, CALIF.
SCR-	AEROJET-GENERAL CORP. ORDNANCE DIV., DOWNEY, CALIF.
SCR-	FAIRCHILD ENGINE AND AIRPLANE CORP. NEPA DIV., OAK RIDGE, TENN. (REF. 31)
SC-R-	SANDIA CORP., ALBUQUERQUE, N. MEX.
SCR-	SANDIA CORP., ALBUQUERQUE, N. MEX. (REPORT)
SCR-	SANDIA CORP., ALBUQUERQUE, N. MEX. (REPRINT)
SCR-	SANDIA CORP., ALBUQUERQUE, N. MEX. (RESEARCH)
SCR-	SANDIA CORP. LIVERMORE LAB., CALIF. (REPRINT)
SC-R-(YEAR)-	SANDIA LABS., LIVERMORE, CALIF.
SCR-(NUMBER)-	UNITED AIRCRAFT CORPORATE SYSTEMS CENTER, FARMINGTON, CONN.
SCR-	UNITED AIRCRAFT CORPORATE SYSTEMS CENTER. SUPPORT SYSTEMS GROUP, EL SEGUNDO, CALIF.
SCR-	UNITED AIRCRAFT CORPORATE SYSTEMS CENTER. WEAPONS EFFECTS + SUPPORT SYSTEMS DEPT., EL SEGUNDO, CAL.
SCRB-	SANDIA CORP., ALBUQUERQUE, N. MEX.
SCRB-(YR.)-	SANDIA CORP., ALBUQUERQUE, N. MEX.
SCRL-MONOGRAPH-	SPEECH COMMUNICATIONS RESEARCH LAB., INC., SANTA BARBARA, CALIF.
SCRM-	FAIRCHILD ENGINE AND AIRPLANE CORP. NEPA DIV., OAK RIDGE, TENN. (REF. 31)
SC-RR-	SANDIA CORP. LIVERMORE LAB., CALIF.
SC-RR-(YEAR)-	SANDIA LABS., ALBUQUERQUE, N. MEX.
SCS-	SANDIA CORP., ALBUQUERQUE, N. MEX.
SCS-(LETTER(S))-	UNITED KINGDOM ATOMIC ENERGY AUTHORITY. INDUSTRIAL GROUP. SPRINGFIELDS WORKS, LANCS, ENG. (REF. 36)
SCSES-	SIGNAL CORPS (ARMY) (SCIENTIFIC ELECTRONIC STUDY)
SCS-M-	GT. BRIT. MIN. OF SUPPLY. CHEM. INSPECTORATE, LONDON
SCS-MEMO-	UNITED KINGDOM ATOMIC ENERGY AUTHORITY. INDUSTRIAL GROUP. SPRINGFIELDS WORKS, LANCS, ENGLAND
SCS-R-	GT. BRIT. DIV. OF ATOMIC ENERGY (PRODUCTION), RISLEY, LANCS, ENGLAND
SCS-R-	GT. BRIT. MIN. OF SUPPLY. CHEM. INSPECTORATE, LONDON
SCS-TME-(NUMBER)-	GENERAL ELECTRIC CO. MISSILE AND SPACE DIV., PHILA.
SCT-	SANDIA CORP., ALBUQUERQUE, N. MEX.
SC-T-	SANDIA LABS., ALBUQUERQUE, N. MEX. (TRANSLATIONS)
SCTC-	SIGNAL CORPS TECHNICAL COMMITTEE, WASHINGTON, D.C.
SCTM-	SANDIA CORP., ALBUQUERQUE, N. MEX. (TECHNICAL MEMO.)
SCTM-(NO.)-(YR.)-	SANDIA CORP., ALBUQUERQUE, N. MEX.
SCTM-(NO.)-(YEAR)-(NO.)	SANDIA CORP., ALBUQUERQUE, N. MEX.
SC-TM-(YEAR)-	SANDIA CORP., ALBUQUERQUE, N. MEX.
SCTM-(NUMBER)-(YEAR)	SANDIA CORP. LIVERMORE LAB., CALIF. (TECHNICAL MEMO)
SC-TM-(YEAR)-	SANDIA LABS., ALBUQUERQUE, N. MEX.
SCTM-M-1004 (YR./MO.)	SANDIA CORP., ALBUQUERQUE, N. MEX.
SCTP-(YEAR)-	AIR FORCE SYSTEMS COMMAND, ANDREWS AFB, MD.
SCTR-	NORTH AMERICAN AVIATION SCIENCE CENTER, THOUSAND OAKS, CALIF.
SCT-T-(YEAR)-	SANDIA LABS., LIVERMORE, CALIF.
SCU-	SANTA CLARA, CALIF. UNIV.
SCU-	SOUTH CAROLINA. UNIV., COLUMBIA
SCW-	MIDWESTERN UNIVERSITIES RESEARCH ASSN., URBANA, ILL. (CODE IS AUTHOR'S INITIALS)
SCW-	WASHINGTON. STATE COLL., PULLMAN
SC-WD-(YEAR)-	SANDIA LABS., ALBUQUERQUE, N. MEX.
SCWP-	NAVAL ORDNANCE TEST STA., INYOKERN (CHINA LAKE), CAL.
SD-(NUMBER)-	AEROJET-GENERAL CORP. SPACE DIV., EL MONTE, CALIF.
SD-	BUREAU OF AERONAUTICS (NAVY) (DETAIL SPECIFICATION)
SD-	CHRYSLER CORP., CENTER LINE, MICH.
SD-	FIELD INTELL. AGENCY, TECH. (STAATSDRUCKEREI DOCS.)
SD-	FISH AND WILDLIFE SERVICE (INTERIOR), WASHINGTON, D.C. (STATISTICAL DIGESTS)
(YEAR)SD-	GENERAL ELECTRIC CO. MISSILE AND SPACE VEHICLE DEPT., PHILADELPHIA (REF. 1)
(YEAR)SD-	GENERAL ELECTRIC CO. MISSILE AND SPACE VEHICLE DEPT., SCHENECTADY, N.Y. (REF. 1)
SD-	GOVERNMENT PRINTING OFFICE. SUPERINTENDENT OF DOCUMENTS, WASHINGTON, D.C.
SD-	GT. BRIT. AIR MINISTRY
SD-	LOS ALAMOS SCIENTIFIC LAB., N. MEX.
SD-	NAVAL PERSONNEL RES. FIELD ACTIVITY, SAN DIEGO, CAL.
SD-(YEAR)-(NO.)-(NO.)	NORTH AMERICAN ROCKWELL CORP. SPACE DIV., DOWNEY, CAL
SD-(YEAR)-CS-	NORTH AMERICAN ROCKWELL CORP. SPACE DIV., DOWNEY, CAL (CS IS PROGRAM AREA DESIGNATION)

SD-(YEAR)-SA-
    NORTH AMERICAN ROCKWELL CORP. SPACE DIV., DOWNEY, CAL
    (SA IS PROGRAM AREA DESIGNATION)
SD-(YEAR)-CE-
    NORTH AMERICAN ROCKWELL CORP. SPACE DIV., DOWNEY, CAL
    (CE IS PROGRAM DESIGNATION)
SD-(YEAR)-SH-
    NORTH AMERICAN ROCKWELL CORP. SPACE DIV., DOWNEY, CAL
    (SH IS PROGRAM DESIGNATION)
S/D-
    SPACE/DEFENSE CORP., BIRMINGHAM, MICH.
SD-
    STROMBERG DATAGRAPHICS, INC., SAN DIEGO, CALIF.
S/D-
    UNITED KINGDOM ATOMIC ENERGY AUTHORITY. RESEARCH GP.
    ATOMIC ENERGY RES. ESTAB., HARWELL, BERKS, ENGLAND
SD/A-
    UNITED KINGDOM ATOMIC ENERGY AUTHORITY. DEVELOPMENT
    AND ENGINEERING GROUP, CULCHETH, LANCS, ENGLAND
SD-BA-
    FAIRCHILD CAMERA AND INSTR. CORP., HICKSVILLE, N.Y.
SDC-
    IOWA. STATE UNIV., IOWA CITY
SDC/
    ITT GILFILLAN, INC., LOS ANGELES
SDC-
    NAVAL TRAINING DEVICES CENTER, PORT WASHINGTON, N.Y.
SDC-
    ROCHESTER, N.Y. UNIV.
SDC-
    SHELL DEVELOPMENT CO., EMERYVILLE, CALIF.
SDC-
    SPACE DATA CORP., PHOENIX, ARIZ.
SDC (CONTRACT NO.)-(NO.)-
    SPECIAL DEVICES CENTER, PORT WASHINGTON, N.Y.
SDC-
    SYSTEM DEVELOPMENT CORP., LEXINGTON, MASS.
SDC-
    SYSTEM DEVELOPMENT CORP., SANTA MONICA, CALIF.
SDCO-
    SCHWARZKOPF DEVELOPMENT CORP., YONKERS, N.Y.
SDC-SCIENTIFIC-
    SYSTEM DEVELOPMENT CORP., SANTA MONICA, CALIF.
SDC-SP-
    SYSTEM DEVELOPMENT CORP., SANTA MONICA, CALIF.
SDC-TM-
    SPACE DATA CORP., PHOENIX, ARIZ.
SDC-TM-
    SYSTEM DEVELOPMENT CORP., SANTA MONICA, CALIF.
SDC-TM-BA-(NO./NO./NO.)
    SYSTEM DEVELOPMENT CORP., SANTA MONICA, CALIF.
SDC-TM-L-(NO./NO./NO.)
    SYSTEM DEVELOPMENT CORP., SANTA MONICA, CALIF.
SDC-TM-LX-(L)-(NO./NO.)/
    SYSTEM DEVELOPMENT CORP., SANTA MONICA, CALIF.
SDC-TM-WD-(NUMBER/NO(S).)
    SYSTEM DEVELOPMENT CORP., FALLS CHURCH, VA.
SDC TR-
    SPECIAL DEVICES CENTER, PORT WASHINGTON, N.Y.
SDC TR (NUMBERS)-
    SPECIAL DEVICES CENTER, PORT WASHINGTON, N.Y.
SDD-
    BECHTEL CORP., SAN FRANCISCO
SD-DC-
    MARQUARDT AIRCRAFT CO., VAN NUYS, CALIF.
SDDD-
    NATIONAL RESEARCH COUNCIL OF CANADA. ATOMIC ENERGY
    PROJECT, CHALK RIVER, ONT.
SDE-
    AEROJET-GENERAL CORP., SACRAMENTO, CALIF.
SDES-(YEAR)-
    CHRYSLER CORP. SPACE DIV., NEW ORLEANS
SD/FP-
    GT. BRIT. ADVISORY COUNCIL ON SCI. RES. + TECH. DEV.
    STATIC DETONATION COMM. FRAGMENTATION PANEL
SDI-
    SPECIAL DEVICES, INC., NEWHALL, CALIF.
SDI-
    STRUTHERS-DUNN, INC., PITMAN, N.J.
SDL-
    HUGHES AIRCRAFT CO. SYSTEMS DEVELOPMENT LABS.,
    CULVER CITY, CALIF.
SDL-(NUMBER)-
    NORTH CAROLINA STATE COLL., RALEIGH. SEMICONDUCTOR
    DEVICE LAB.
SDL-
    SPACE DISTURBANCES LAB., BOULDER, COLO.
SDL-
    TELEDYNE, INC. UED EARTH SCIENCES DIV.,ALEXANDRIA,VA.
S/D-LR(YEAR)-
    SPACE/DEFENSE CORP., BIRMINGHAM, MICH.
SDN-
    HUGHES AIRCRAFT CO. (SECRET DOCUMENT NUMBER)
SDN-(NUMBER)-(NUMBER)/
    HUGHES AIRCRAFT CO. SPACE SYSTEMS DIV.,EL SEGUNDO,CAL
SDN-A-
    HUGHES AIRCRAFT CO. AEROSPACE GROUP, CULVER CITY,CAL.
SDN-A-
    HUGHES AIRCRAFT CO. SPACE SYSTEMS DIV.,EL SEGUNDO,CAL
SDO-
    JOHNS HOPKINS UNIV., SILVER SPRING, MD. APPLIED
    PHYSICS LAB.
S-DOC-(NUMBER)-
    CONGRESS. SENATE, WASHINGTON, D.C.
SDP-
    (STAFF DISCUSSION PAPER) (REF. 2)
SDP-(YEAR-NUMBER)-
    GENERAL ELECTRIC CO. SYSTEMS DESIGN AND PROGRAMMING
    UNIT, FALLS CHURCH, VA.
SDP-
    GEORGE WASHINGTON UNIV., WASHINGTON, D.C.
    (STAFF DISCUSSION PAPER)
S/D-P(YEAR)-
    SPACE/DEFENSE CORP., BIRMINGHAM, MICH.
S/D-PO(YEAR)-
    SPACE/DEFENSE CORP., BIRMINGHAM, MICH.
SDR-
    COLLINS RADIO CO., CEDAR RAPIDS, IOWA (SERVICE
    DIVISION REPORT)
SDR-(YR.)-
    DANIEL MANN JOHNSON AND MENDENHALL. SYSTEMS ENG. DIV.
    LOS ANGELES
SDR-
    DEPARTMENT OF DEFENSE, WASHINGTON, D.C. (SYSTEM
    DEVELOPMENT REQUIREMENT)
SDR-
    LOCKHEED AIRCRAFT CORP., BURBANK, CALIF.
SDR-
    LOCKHEED AIRCRAFT SERVICE CO. VEHICLE DEVELOPMENT
    GROUP, ONTARIO, CALIF.
SDR-
    NUCLEAR DEV. CORP. OF AMERICA, WHITE PLAINS, N.Y.
SDR-
    POLYMER CORP., LTD., SARNIA, ONT.
SDS/(NUMBER)-(NUMBER)-
    EUROPEAN SPACE RESEARCH ORGANIZATION. SPACE DOCUMEN-
    TATION SERVICE, ROME
SDSCF-
    SAN DIEGO, CALIF. STATE COLLEGE FOUNDATION
SDS EM-
    SIGNAL CORPS ENGINEERING LABS. SUPPRESSION DEVELOP-
    MENT SECTION, FORT MONMOUTH, N.J. (ENG. MEMO)
S/D-SP(YEAR)-
    SPACE/DEFENSE CORP., BIRMINGHAM, MICH.
SDSU-(YEAR)-
    SOUTH DAKOTA STATE UNIV., BROOKINGS. DEPT. OF
    MECHANICAL ENGINEERING
SDTM-
    GT. BRIT. AIR MINISTRY
S/D-TM(YEAR)-
    SPACE/DEFENSE CORP., BIRMINGHAM, MICH.
SDTM-
    WESTINGHOUSE ELECTRIC CORP., BALTIMORE
SDTM-
    WESTINGHOUSE ELECTRIC CORP. AEROSPACE DIV., BALTIMORE
SD-TR-
    HUGHES AIRCRAFT CO.,NEWPORT BEACH,CAL.(SEMICONDUCTOR)
S/D-TR(YEAR)-
    SPACE/DEFENSE CORP., BIRMINGHAM, MICH.
SE-(YEAR)-(NUMBERS)
    AEROSPACE CORP. SAN BERNARDINO OPERATIONS, CALIF.
SE-(3 DIGITS)
    BROWN ENGINEERING CO., INC. SPACE AND MILITARY
    SYSTEMS DIV., HUNTSVILLE, ALA.
SE-
    BUREAU OF AERONAUTICS (NAVY).SHORE ESTABLISHMENT DIV.
SE-
    FOREST SERVICE. SOUTHEASTERN STA., ASHEVILLE, N.C.
SE-
    GT. BRIT. ADVISORY COUNCIL ON SCI. RES. + TECH. DEV.
SE-
    SHAW AND ESTES, DALLAS
SEA-
    SOCIETE D'ELECTRONIQUE ET D'AUTOMATISME, COURBEVOIE,
    FRANCE
SEAGRANT-
    NATIONAL OCEANIC AND ATMOSPHERIC ADMINISTRATION,
    ROCKVILLE, MD. (SEA GRANT PROGRAM)
SEA GRANT PUB-
    NATIONAL OCEANIC AND ATMOSPHERIC ADMINISTRATION,
    ROCKVILLE, MD. (SEA GRANT PROGRAM)
SEA-PN-
    SCIENCE ENGINEERING ASSOCIATES, SAN MARINO, CALIF.
SEASTAG-
    SOUTHEAST ASIA TREATY ORGANIZATION (STANDARD)
S-E-ASTR-R-
    GEORGE C. MARSHALL SPACE FLIGHT CTR.,HUNTSVILLE, ALA.
S/E-ASTR-S-
    GEORGE C. MARSHALL SPACE FLIGHT CTR.,HUNTSVILLE, ALA.
SEATO-
    SOUTHEAST ASIA TREATY ORGANIZATION
SEC-
    FRANCE. COMMISSARIAT A L'ENERGIE ATOMIQUE. CENTRE
    D'ETUDES NUCLEAIRES, SACLAY
SEC-
    SIERRA ELECTRONIC CORP., SAN CARLOS, CALIF.
SEC-
    SIERRA ELECTRONIC CORP., SAN FRANCISCO
SEC-
    SOUNDRIVE ENGINE CO., LOS ANGELES
SEC-
    SPRAGUE ELECTRICAL CO., NORTH ADAMS, MASS.
SECA-
    FRANCE. COMMISSARIAT A L'ENERGIE ATOMIQUE, CENTRE
    D'ETUDES NUCLEAIRES, CADARACHE
SEC/CA-
    FRANCE. COMMISSARIAT A L'ENERGIE ATOMIQUE. DEPARTE-
    MENT DES ETUDES DE PILES, CADARACHE
(SEC.NO.)(MODEL NO.)(NO.)
    LINK AVIATION, INC., BINGHAMTON, N.Y. (REF. 1)
SECNR-
    FRANCE. COMMISSARIAT A L'ENERGIE ATOMIQUE. CENTRE
    D'ETUDES NUCLEAIRES. SERVICE DES EXPERIENCES
    CRITIQUES A NEUTRONS RAPIDES, CADARACHE
SECO-
    SPACE ELECTRONICS CORP., GLENDALE, CALIF.
SEC-TR-
    ROBERT A. TAFT SANITARY ENGINEERING CTR., CINCINNATI

SED-	STANFORD RESEARCH INST. SYSTEMS EVALUATION DEPT., MENLO PARK, CALIF.
SEDCO-	STRUCTURAL ENGINEERING AND DESIGN CO., LOS ANGELES
SEDR-	MC DONNELL AIRCRAFT CORP., ST. LOUIS (ENGINEERING DEVELOPMENT AND RESEARCH)
SED-RM-	STANFORD RESEARCH INST., MENLO PARK, CALIF.
SEDT-NJ-	SPERRY ELECTRONIC TUBE DIV., GAINESVILLE, FLA.
SEEC-	SPECIALTY ENGINEERING AND ELECTRONICS CO., BROOKLYN
SEEC/IDR(NO.)/CONTRACT	SPECIALTY ENGINEERING AND ELECTRONICS CO., BROOKLYN
SEEO-	UNITED NUCLEAR CORP., WHITE PLAINS, N.Y.
SEFL-(NUMBER)A	SYSTEMS ENGINEERING GROUP, WRIGHT-PATTERSON AFB, OHIO
SEF-TN-(YEAR)-	AIR FORCE SYSTEMS COMMAND, WRIGHT-PATTERSON AFB, OHIO
SEG-	SYSTEMS ENGINEERING GROUP, WRIGHT-PATTERSON AFB, OHIO
SEG RESEARCH CONTRIB-	CENTER FOR NAVAL ANALYSES. SYSTEMS EVALUATION GROUP, WASHINGTON, D.C.
SEG-STUDY-	CENTER FOR NAVAL ANALYSES. SYSTEMS EVALUATION GROUP, ARLINGTON, VA.
SEG-STUDY-	CENTER FOR NAVAL ANALYSES. SYSTEM EVALUATION GROUP, WASHINGTON, D.C.
SEG-TDR-(YEAR)-	SYSTEMS ENGINEERING GROUP, WRIGHT-PATTERSON AFB, OHIO
SEG-TR-(YEAR)-	SYSTEMS ENGINEERING GROUP, WRIGHT-PATTERSON AFB, OHIO
SEI-	STAVID ENGINEERING, INC., PLAINFIELD, N.J.
SEISMIC DATA LAB-	TELEDYNE, INC. UED EARTH SCIENCES DIV., ALEXANDRIA, VA.
SEISMIC DLR-	TELEDYNE, INC. UED EARTH SCIENCES DIV., ALEXANDRIA, VA.
SEL-	ARMY SIGNAL ENGINEERING LABS., FORT MONMOUTH, N.J.
SEL-(YEAR)-	CALIFORNIA. UNIV., BERKELEY. STRUCTURAL ENG. LAB.
SEL-	MICHIGAN. UNIV., ANN ARBOR. SYSTEMS ENGINEERING LAB.
SEL-	SIGNAL CORPS ENGINEERING LABS., FORT MONMOUTH, N.J.
SEL-	STANFORD RESEARCH INST., MENLO PARK, CALIF.
SEL-(YEAR)-	STANFORD UNIV., CALIF. STANFORD ELECTRONICS LABS.
SEL-	STANFORD UNIV., CALIF. STANFORD ELECTRONICS LABS.
SEL TM M-	ARMY SIGNAL ENGINEERING LABS., FORT MONMOUTH, N.J.
SEL-TR-	MICHIGAN. UNIV., ANN ARBOR. SYSTEMS ENGINEERING LAB.
SELWS-	WHITE SANDS MISSILE RANGE, N. MEX.
SEN-IPSA-	EUROPEAN NUCLEAR ENERGY AGENCY, PARIS
SEN/IPSA/(YEAR)/	EUROPEAN SPACE VEHICLE LAUNCHER DEV. ORG., PARIS
SEN-IPSA-	ORGANIZATION FOR ECONOMIC COOPERATION AND DEVELOPMENT, PARIS
SEP-	SYLVANIA ELECTRIC PRODUCTS INC. (ALL LOCATIONS)
SEP-(NUMBER)(P)	SYLVANIA ELECTRIC PRODUCTS INC., BAYSIDE, N.Y.
SEP-(NUMBER)(X)	SYLVANIA ELECTRIC PRODUCTS INC., BAYSIDE, N.Y.
SEP-2-	SYLVANIA ELECTRIC PRODUCTS INC. MISSILE SYSTEMS LABS. WALTHAM, MASS.
SE-PC-	HANFORD WORKS, RICHLAND, WASH.
SEP-CR-	SYLVANIA ELECTRIC PRODUCTS INC., WOBURN, MASS.
SEP-E2-	SYLVANIA ELECTRONIC SYSTEMS. WALTHAM LABS., MASS.
SEP-EDL-	SYLVANIA ELECTRONIC SYSTEMS. ELECTRONIC DEFENSE LABS., MOUNTAIN VIEW, CALIF.
SEP-IR-	SYLVANIA ELECTRIC PRODUCTS INC., TOWANDA, PENNA.
SEP-L-	SYLVANIA ELECTRIC PRODUCTS INC., FLUSHING, N.Y.
SEP-MPL-	SYLVANIA ELECTRIC PRODUCTS INC. MICROWAVE PHYSICS LAB., MOUNTAIN VIEW, CALIF.
SEP-MSL-	SYLVANIA ELECTRIC PRODUCTS INC. MISSILE SYSTEMS LABS. WALTHAM, MASS.
SEP-P-	DU PONT DE NEMOURS (E.I.) + CO. SAVANNAH RIVER PLANT, AIKEN, S.C.
SEP(P)-	SYLVANIA ELECTRIC PRODUCTS INC., BAYSIDE, N.Y.
SEP-SESD-	SYLVANIA ELECTRIC PRODUCTS INC. SYLVANIA ELECTRONIC SYSTEMS DIV., MOUNTAIN VIEW, CALIF.
SEP-SESE-	SYLVANIA ELECTRONIC SYSTEMS-EAST, WALTHAM, MASS.
SEP-TR-(NO.)-(NO.)	SYLVANIA ELECTRIC PRODUCTS INC. MISSILE SYSTEMS LABS. WALTHAM, MASS.
SEP-(X)-	SYLVANIA ELECTRIC PRODUCTS INC., BAYSIDE, N.Y.
SEP-(XD)-	SYLVANIA ELECTRIC PRODUCTS INC., BAYSIDE, N.Y.
SEP-YD-	SYLVANIA ELECTRIC PRODUCTS INC., BAYSIDE, N.Y.
SER-	COMMITTEE ON MEDICAL RESEARCH (SANITARY ENG. RPTS.)
SER-(YEAR)-	ELECTRONIC COMMUNICATIONS, INC., ST. PETERSBURG, FLA.
SER-	ILLINOIS. UNIV., URBANA
SER-	SIKORSKY AIRCRAFT DIV., UNITED AIRCRAFT CORP., BRIDGEPORT, CONN.
SER-	SIKORSKY AIRCRAFT DIV., UNITED AIRCRAFT CORP., STRATFORD, CONN.
SER-	UNITED AIRCRAFT CORP., STRATFORD, CONN.
S/ERA-	SYMPOSIUM ON ELECTRON RING ACCELERATOR, 1968, BERKELEY, CALIF.
SERAI-	BELGIUM. SOCIETE D'ETUDES DE RECHERCHES ET D'APPLICATIONS POUR L'INDUSTRIE, BRUSSELS
SER-B-TOM-	FINLAND. SUOMALAINEN TIEDEAKATEMIA, HELSINKI
SER-D-GEOPHYS.BULL.-	DUBLIN INSTITUTE FOR ADVANCED STUDIES
SEREB-DTA-(NO.)/(NO.)-	SOCIETE POUR L'ETUDE ET LA REALISATION D'ENGINS BALISTIQUES, COURBEVOIE, FRANCE
SERGEANT-JWCC-	SERGEANT JOINT AEC-DOD WARHEAD COORDINATING COMM. (REF. 15)
SERG-TR-	LONG ISLAND UNIV., GREENVALE, N.Y.
SERI-(CONTRACT NO.)-	SOUTHWEST RESEARCH INST. DEPT. OF AEROSPACE PROPULSION RESEARCH, SAN ANTONIO
SERIAL-	GEORGE WASHINGTON UNIV., WASHINGTON, D.C. (LOGISTICS RESEARCH PROJECT)
SERIAL T-	GEORGE WASHINGTON UNIV., WASHINGTON, D.C. (LOGISTICS RESEARCH PROJECT)
SERIAL-TM-DMC-	GEORGE WASHINGTON UNIV., WASHINGTON, D.C. (LOGISTICS RESEARCH PROJECT)
SERIES-(NO.)-ISSUE-	CALIFORNIA. UNIV., BERKELEY. SPACE SCIENCES LAB.
SERL-	CALIFORNIA. UNIV., BERKELEY. SANITARY ENG. RES. LAB.
SERL-TR-M-	GT. BRIT. SERVICES ELECTRONICS RESEARCH LAB., BALDOCK, HERTS, ENGLAND
SERM-	FAIRCHILD ENGINE AND AIRPLANE CORP. NEPA DIV., OAK RIDGE, TENN. (REF. 31)
SES-	CANADA. SUFFIELD EXPERIMENTAL STATION, RALSTON, ALBERTA
SES-(YEAR)-	CANADA. SUFFIELD EXPERIMENTAL STATION, RALSTON, ALBERTA
SES-	CORPS OF ENGINEERS (ARMY). MIL. INTELL. DIV., WASH., DC
SESA PAPER-	FORD MOTOR CO., DEARBORN, MICH.
SES-FIELD EXPERIMENT-	CANADA. SUFFIELD EXPERIMENTAL STATION, RALSTON, ALBERTA
SESM-(YEAR)-	CALIFORNIA. UNIV., BERKELEY. DIV. OF STRUCTURAL ENG. AND STRUCTURAL MECHANICS
SESM-	CALIFORNIA. UNIV., BERKELEY. DIV. OF STRUCTURAL ENG. AND STRUCTURAL MECHANICS
SES-MEMO-(NUMBER)/(YEAR)	CANADA. SUFFIELD EXPERIMENTAL STATION, RALSTON, ALBERTA
SES-SPECIAL PUB-	CANADA. SUFFIELD EXPERIMENTAL STATION, RALSTON, ALBERTA
SES-TN-	CANADA. SUFFIELD EXPERIMENTAL STATION, RALSTON, ALBERTA
SES TP-	CANADA. SUFFIELD EXPERIMENTAL STATION, RALSTON, ALBERTA. (TECHNICAL PAPER)
SESW-(NUMBER)R(NUMBER)-	SYLVANIA ELECTRONIC SYSTEMS-WEST, MOUNTAIN VIEW, CAL.
SESW-	SYLVANIA ELECTRONIC SYSTEMS-WEST, MOUNTAIN VIEW, CAL.
SESW-E(NUMBER)	SYLVANIA ELECTRONIC SYSTEMS-WEST, MOUNTAIN VIEW, CAL.
SESW-M(NUMBER)	SYLVANIA ELECTRONIC SYSTEMS-WEST, MOUNTAIN VIEW, CAL.
SETD-	SPERRY ELECTRONIC TUBE DIV., GAINESVILLE, FLA.
SETD-NJ-	SPERRY RAND CORP. SPERRY ELECTRONIC TUBE DIV., GAINESVILLE, FLA.
SETE-	JOINT MILITARY INDUSTRIAL ELECTRONIC TEST EQUIPMENT SYMPOSIUM
SETE-	NEW YORK UNIV., N.Y.C. SCHOOL OF ENG. AND SCIENCE
SETR-R-	FRANCE. COMMISSARIAT A L'ENERGIE ATOMIQUE. CENTRE D'ETUDES NUCLEAIRES, CADARACHE

SF-	TENNESSEE VALLEY AUTHORITY, OAK RIDGE
SFAL-	PICATINNY ARSENAL. SAMUEL FELTMAN AMMUNITION LABS., DOVER, N.J.
SFAL-MR-	PICATINNY ARSENAL. SAMUEL FELTMAN AMMUNITION LABS., DOVER, N.J. (MONTHLY REPORTS)
SFAL-RM-	PICATINNY ARSENAL. SAMUEL FELTMAN AMMUNITION LABS., DOVER, N.J. (RESEARCH MEMORANDA)
SFAL-TR-	PICATINNY ARSENAL. SAMUEL FELTMAN AMMUNITION LABS., DOVER, N.J. (TECHNICAL REPORTS)
SFCO-R-	SPERRY FARRAGUT CO. ENG. DEPT., BRISTOL, TENN.
SFC-P-	SPERRY FARRAGUT CO., BRISTOL, TENN. (COMPANY TECHNICAL PROPOSAL)
SFC-R-	SPERRY FARRAGUT CO., BRISTOL, TENN. (CO. TECH. RPT.)
SFCSE-AEC-TT(YEAR-NO.)	NATIONAL SCIENCE FOUNDATION, WASHINGTON, D.C. (SPECIAL FOREIGN CURRENCY SCIENCE INFO. PROGRAM)
SFCSI-AEC-TT(YEAR-NO.)	NATIONAL SCIENCE FOUNDATION, WASHINGTON, D.C. (SPECIAL FOREIGN CURRENCY SCIENCE INFO. PROGRAM)
SFCSI-AGR-TT(YEAR-NO.)	NATIONAL SCIENCE FOUNDATION, WASHINGTON, D.C. (SPECIAL FOREIGN CURRENCY SCIENCE INFO. PROGRAM)
SFCSI-COMM-TT(YEAR-NO.)	NATIONAL SCIENCE FOUNDATION, WASHINGTON, D.C. (SPECIAL FOREIGN CURRENCY SCIENCE INFO. PROGRAM)
SFCSI-HEW-TT(YEAR-NO.)	NATIONAL SCIENCE FOUNDATION, WASHINGTON, D.C. (SPECIAL FOREIGN CURRENCY SCIENCE INFO. PROGRAM)
SFCSI-INT-TT(YEAR-NO.)	NATIONAL SCIENCE FOUNDATION, WASHINGTON, D.C. (SPECIAL FOREIGN CURRENCY SCIENCE INFO. PROGRAM)
SFCSI-NASA-TT(YEAR-NO.)	NATIONAL SCIENCE FOUNDATION, WASHINGTON, D.C. (SPECIAL FOREIGN CURRENCY SCIENCE INFO. PROGRAM)
SFCSI-NSF-TT(YEAR-NO.)	NATIONAL SCIENCE FOUNDATION, WASHINGTON, D.C. (SPECIAL FOREIGN CURRENCY SCIENCE INFO. PROGRAM)
SFCSI-SMI-TT(YEAR-NO.)	NATIONAL SCIENCE FOUNDATION, WASHINGTON, D.C. (SPECIAL FOREIGN CURRENCY SCIENCE INFO. PROGRAM)
SFCSI-TVA-TT(YEAR-NO.)	NATIONAL SCIENCE FOUNDATION, WASHINGTON, D.C. (SPECIAL FOREIGN CURRENCY SCIENCE INFO. PROGRAM)
SFD-	SFD LABS., INC., UNION, N.J.
SFD-(NUMBER)-R-	SFD LABS., INC., UNION, N.J.
SFK-AL(NUMBER)L(NUMBER)	SKF INDUSTRIES, INC. RES. LAB., KING OF PRUSSIA, PA.
SFL-	SAN FERNANDO LABS., PACOIMA, CALIF.
SFL-A-	FINLAND. INSTITUTE OF RADIATION PHYSICS, HELSINKI
SFL-QR-	SAN FERNANDO LABS., PACOIMA, CALIF. (QUARTERLY REPORT)
SFO-	SANTA FE OPERATIONS OFFICE, AEC, N. MEX. (REF. 4)
SFPWP/P-	GT. BRIT. CULCHETH LABS., LANCS, ENGLAND
SFQC-	SPERRY FARRAGUT CO. QUALITY CONTROL DEPT., BRISTOL, TENN.
SF/R/	IMPERIAL COLLEGE OF SCIENCE AND TECHNOLOGY, LONDON
SFR-	TECHNICAL INDUSTRIAL INTELL. COMM. (SOLID FUELS RPTS)
SFRL-P(NUMBER)-	TEXAS. UNIV., AUSTIN. STRUCTURES FATIGUE RES. LAB.
SFRL-TR-P(NUMBER)-	TEXAS. UNIV., AUSTIN. STRUCTURES FATIGUE RES. LAB.
SFSCI-NLM-TT(YEAR-NO.)	NATIONAL SCIENCE FOUNDATION, WASHINGTON, D.C. (SPECIAL FOREIGN CURRENCY SCIENCE INFO. PROGRAM)
SF/TN/	IMPERIAL COLLEGE OF SCIENCE AND TECHNOLOGY, LONDON
SF/TN/	IMPERIAL COLLEGE OF SCIENCE AND TECHNOLOGY, LONDON. DEPT. OF MECHANICAL ENGINEERING
SG-	OFFICE OF THE SURGEON GENERAL, WASHINGTON, D.C.
SG-	PHILCO-FORD CORP. SPACE AND RE-ENTRY SYSTEMS DIV., NEWPORT BEACH, CALIF.
SG-(NUMBER)-FR	SPACE-GENERAL, EL MONTE, CALIF.
SG-(NUMBER)FR-	SPACE-GENERAL, EL MONTE, CALIF.
SG-(NUMBER)PR-(NO.)A	SPACE-GENERAL, EL MONTE, CALIF.
SG-(NUMBER)R-	SPACE-GENERAL, EL MONTE, CALIF.
SG-(NUMBER)/SP-	SPACE-GENERAL CORP., LOS ANGELES
SG-(NUMBER)/SR-	SPACE-GENERAL CORP. CENTER FOR RESEARCH AND EDUCATION, LOS ANGELES
SG-(NUMBER)/QR-	SPACE-GENERAL CORP. CENTER FOR RESEARCH AND EDUCATION, LOS ANGELES
SGAE-(YEAR)/	OESTERREICHISCHE STUDIENGESELLSCHAFT FUER ATOMENERGIE G.M.B.H., VIENNA
SGAE-(YEAR/NUMBER)(LTR.)	OESTERREICHISCHE STUDIENGESELLSCHAFT FUER ATOMENERGIE G.M.B.H., VIENNA
SGAE-(LTRS.)-(NO.)/(YR.)	OESTERREICHISCHE STUDIENGESELLSCHAFT FUER ATOMENERGIE G.M.B.H., SEIBERSDORF
SGAE-B-	OESTERREICHISCHE STUDIENGESELLSCHAFT FUER ATOMENERGIE G.M.B.H., VIENNA
SGAE-BL-(NUMBER/YEAR)	OESTERREICHISCHE STUDIENGESELLSCHAFT FUER ATOMENERGIE G.M.B.H. INSTITUT FUER BIOLOGIE UND LANDWIRTSCHAFT, SEIBERSDORF
SGAE-CH-	OESTERREICHISCHE STUDIENGESELLSCHAFT FUER ATOMENERGIE G.M.B.H. INSTITUT FUER CHEMIE, SEIBERSDORF
SGAE-E-(NUMBER/YEAR)	OESTERREICHISCHE STUDIENGESELLSCHAFT FUER ATOMENERGIE G.M.B.H. ELEKTRONIKINSTITUT, SEIBERSDORF
SGAE-G-	OESTERREICHISCHE STUDIENGESELLSCHAFT FUER ATOMENERGIE G.M.B.H., SEIBERSDORF
SGAE-IB-	OESTERREICHISCHE STUDIENGESELLSCHAFT FUER ATOMENERGIE G.M.B.H., VIENNA
SGAE-IB/IA-(NUMBER/YEAR)	OESTERREICHISCHE STUDIENGESELLSCHAFT FUER ATOMENERGIE G.M.B.H. INSTITUT FUER RADIUMFORSCHUNG UND KERN-PHYSIK, SEIBERSDORF
SGAE-LA-(NUMBER/YEAR)	OESTERREICHISCHE STUDIENGESELLSCHAFT FUER ATOMENERGIE G.M.B.H., SIEBERSDORF (LANDWIRTSCHAFTLICHE FORSCHUNGSABTEILUNG)
SGAE-M-	OESTERREICHISCHE STUDIENGESELLSCHAFT FUER ATOMENERGIE G.M.B.H. INSTITUT FUER METALLURGIE, SEIBERSDORF
SGAE-ME-	OESTERREICHISCHE STUDIENGESELLSCHAFT FUER ATOMENERGIE G.M.B.H., VIENNA
SGAE-PH-	OESTERREICHISCHE STUDIENGESELLSCHAFT FUER ATOMENERGIE G.M.B.H. PHYSIK-INSTITUT, SEIBERSDORF
SGAE-R-(NUMBER/YEAR)	OESTERREICHISCHE STUDIENGESELLSCHAFT FUER ATOMENERGIE G.M.B.H. REAKTORINSTITUT, SEIBERSDORF
SGAE-RE-(NUMBER/YEAR)	OESTERREICHISCHE STUDIENGESELLSCHAFT FUER ATOMENERGIE G.M.B.H. INST. FUER REAKTORENTWICKLUNG, SEIBERSDORF
SGAE-RT-(NUMBER/YEAR)	OESTERREICHISCHE STUDIENGESELLSCHAFT FUER ATOMENERGIE G.M.B.H. INST. FUER REAKTORTECHNIK, SEIBERSDORF
SGAE-SS-(NUMBER/YEAR)	OESTERREICHISCHE STUDIENGESELLSCHAFT FUER ATOMENERGIE G.M.B.H. INST. FUER STRAHLENSCHUTZ, SEIBERSDORF
SGC-	AEROJET-GENERAL CORP., SACRAMENTO, CALIF.
SGC-	SPACE-GENERAL CORP., EL MONTE, CALIF.
SGC-(NUMBER)-	SPACE-GENERAL CORP., LOS ANGELES
SGC-(NUMBER)FR-	SPACE-GENERAL CORP. CENTER FOR RESEARCH AND EDUCATION, LOS ANGELES
SGC-	SPERRY GYROSCOPE CO., GREAT NECK, N.Y.
SGC-(NO.)-(NO.)-	SPERRY GYROSCOPE CO., GREAT NECK, N.Y.
SGC/CA-(NO.)-(NO.)-	SPERRY GYROSCOPE CO., GREAT NECK, N.Y.
SGC-FI-	FORD INSTRUMENT CO., LONG ISLAND CITY, N.Y.
SGC-IER-	SPERRY GYROSCOPE CO., GREAT NECK, N.Y. (INTERIM ENGINEERING REPORT)
SGC-P-	SPACE-GENERAL CORP., EL MONTE, CALIF.
SGD-(NUMBER)-	AEROJET-GENERAL CORP., EL MONTE, CALIF.
SGD-(NUMBER)-	SPERRY GYROSCOPE CO., GREAT NECK, N.Y.
SGD-	SPERRY RAND CORP. SPERRY GYROSCOPE DIV., GREAT NECK, NY
SGDG/P-	UNITED KINGDOM ATOMIC ENERGY AUTHORITY. RESEARCH GP. ATOMIC ENERGY RES. ESTAB., HARWELL, BERKS, ENGLAND
SGH-	SIMPSON GUMPERTZ AND HEGER, INC., CAMBRIDGE, MASS.
SGP/P-	GT. BRIT. ATOMIC EN. RES. ESTAB., HARWELL, BERKS, ENG
SGRC/P-	UNITED KINGDOM ATOMIC ENERGY AUTHORITY. RESEARCH GP. ATOMIC ENERGY RES. ESTAB., HARWELL, BERKS, ENGLAND
SGTEB-	AIR FORCE PACKAGING EVALUATION AGENCY, WRIGHT-PATTERSON AFB, OHIO
SG-VAL-	GENERAL ELECTRIC CO. NUCLEAR ENERGY DIV., SAN JOSE, CAL (VALLECITOS REPORT)
SGWG-GAM-72-	AIR RESEARCH AND DEVELOPMENT COMMAND. SPECIAL GAM-72 WORKING GROUP
SH-	BENDIX CORP. BENDIX PRODUCTS DIV., SOUTH BEND, IND.
SH-	COMMITTEE ON MEDICAL RESEARCH (SHELL DEV. CO. RPTS.)
SH-	SHOCK HYDRODYNAMICS, INC., SHERMAN OAKS, CALIF.
SH-(4 DIGITS)-MC	SHOCK HYDRODYNAMICS, INC., SHERMAN OAKS, CALIF.

SHA-	BONN. UNIVERSITAET. INST. FUER PHYSIKALISCHE CHEMIE
SHAEF PCA G-4-	SUPREME HEADQUARTERS, ALLIED EXPEDITIONARY FORCES (PRODUCTION CONTROL AGENCY G-4)
SHANW-	SHANNON AND WILSON INC., SEATTLE
SHG-	SMITH, HINCHMAN AND GRYLLS, INC. AERONAUTICAL ICE RESEARCH LAB., DETROIT
SHI-	SHOCK HYDRODYNAMICS, INC., SHERMAN OAKS, CALIF.
SHIP-	UNITED KINGDOM. NATIONAL PHYSICAL LAB. SHIP DIV., TEDDINGTON, MIDDX., ENGLAND
SHIPS-	NAVAL SHIP SYSTEMS COMMAND, WASHINGTON, D.C.
SHMR-	SYLVANIA ELECTRIC PRODUCTS INC., BOSTON
SHN-	HOLMES AND NARVER, INC., LOS ANGELES
SHO-	SHELL OIL CO., N.Y.C.
SHORT REPORT	METAL HYDRIDES INC. CHEM. RES. LAB., BEVERLY, MASS.
SHSLR-	MICHIGAN STATE UNIV., EAST LANSING. DEPT. OF SPEECH
SI-	BUREAU OF AERONAUTICS (NAVY) (SPERRY INSTRUCTIONS)
SI-	BUREAU OF SHIPS. SHIPS INSTALLATION DIV.
SI-	DUMONT (ALLEN B.) LABS., INC., CLIFTON, N.J.
SI-(YEAR)-	GENERAL PRECISION, INC. LIBRASCOPE DIV., GLENDALE,CAL
SI-	NORWAY. SENTRALINSTITUTT FOR INDUSTRIELL FORSKNING, OSLO
SI-	SMITHSONIAN INSTITUTION, WASHINGTON, D.C.
SI-	SQUIER SIGNAL LAB., FORT MONMOUTH, N.J.
SIARGRAPH-	ROME. UNIVERSITA. SCUOLA DI INGEGNERIA AERONAUTICA
SIB-	NATIONAL CENTER FOR RADIOLOGICAL HEALTH. STANDARDS AND INTELLIGENCE BRANCH, ROCKVILLE, MD.
SIC-	OFFICE OF MANAGEMENT AND BUDGET, WASHINGTON, D.C. (STANDARD INDUSTRIAL CLASSIFICATION)
SICA-	SINTERCAST CORP. OF AMERICA, N.Y.C.
SID-(YEAR)-	NORTH AMERICAN AVIATION, INC. SPACE AND INFORMATION SYSTEMS DIV., CANOGA PARK, CALIF.
SID-	NORTH AMERICAN AVIATION, INC. SPACE AND INFORMATION SYSTEMS DIV., DOWNEY, CALIF.
SID-(YEAR)-	NORTH AMERICAN AVIATION, INC. SPACE AND INFORMATION SYSTEMS DIV., DOWNEY, CALIF.
SID(YEAR)-	NORTH AMERICAN AVIATION, INC. SPACE AND INFORMATION SYSTEMS DIV., EL SEGUNDO, CALIF.
SID-(YEAR)T-	NORTH AMERICAN AVIATION, INC. SPACE AND INFORMATION SYSTEMS DIV., TULSA, OKLA.
SIEC-	SIERRA ELECTRONIC CORP., SAN CARLOS, CALIF.
SIEGLER-	SIEGLER CORP., INGLEWOOD, CALIF.
SIF-	NORWAY. SENTRALINSTITUTT FOR INDUSTRIELL FORSKNING, OSLO
SIG-	AIR FORCE CAMBRIDGE RESEARCH LABS., BEDFORD, MASS.
SIG-	SIGNAL CORPS (ARMY)
SIG-A-	SIGNATRON, INC., LEXINGTON, MASS.
SIG APP DP-	SIGNAL CORPS (ARMY). PICTORIAL ENGINEERING AND RESEARCH LAB., LONG ISLAND, N.Y.
SIGC-	SIGNAL CORPS (ARMY)
SIG CEE-	(CAPTURED ENEMY EQUIPMENT)
SIG CEIS-	SIGNAL INTELLIGENCE SERVICE, 849TH. CAPTURED ENEMY INTELLIGENCE BRANCH
SIG CESE-	SIGNAL CORPS (ARMY) (CAPTURED ENEMY SIGNAL EQUIPMENT)
SIG-CR-	SIGNATRON, INC., LEXINGTON, MASS.
SIG ERL DP-	SIGNAL CORPS (ARMY). PICTORIAL ENGINEERING AND RESEARCH LAB., LONG ISLAND, N.Y. (DIV. REPT.)
SIGFM/EL-XS-3-99-(NO.)-	EVANS SIGNAL LAB. INST. FOR EXPLORATORY RES.,BELMAR, N.J.
SIGIA-	SIGNAL CORPS INTELL. AGENCY (ARMY), WASH.,DC (TRANS.)
SIGIA R-(NUMBER)-(YEAR)	SIGNAL CORPS INTELLIGENCE AGENCY (ARMY), WASH., D.C.
SIG IR-	SIGNAL CORPS (ARMY) (INTELLIGENCE REPORT)
SIG IR RAD-	SIGNAL CORPS (ARMY) (INTELLIGENCE REPORT)
SIG IR RCM-	SIGNAL CORPS (ARMY) (INTELLIGENCE REPORT)
SIG IR SRM-	SIGNAL CORPS (ARMY) (INTELLIGENCE REPORT)
SIG IS T-	SIGNAL CORPS (ARMY). INTELLIGENCE AND SECURITY BOARD
SIG ORB-	SIGNAL CORPS (ARMY). OPERATIONAL RESEARCH BRANCH
SIG ORB (NUMBER)-	SIGNAL CORPS (ARMY). OPERATIONAL RESEARCH BRANCH
SIG ORB (LETTER(S))-	SIGNAL CORPS (ARMY). OPERATIONAL RESEARCH BRANCH
SIG ORG-	SIGNAL CORPS (ARMY). OPERATIONAL RESEARCH GROUP
SIG ORG (LTR(S))-(NO.)-	SIGNAL CORPS (ARMY). OPERATIONAL RESEARCH GROUP
SIG ORS-	SIGNAL CORPS (ARMY). OPERATIONAL RESEARCH STAFF
SIG ORS P-	SIGNAL CORPS (ARMY). OPERATIONAL RESEARCH STAFF
SIG PERLD ES-D-	SIGNAL CORPS (ARMY). PICTORIAL ENGINEERING AND RESEARCH LAB., LONG ISLAND, N.Y. (DRAWING)
SIG PERLD IP-	SIGNAL CORPS (ARMY). PICTORIAL ENGINEERING AND RESEARCH LAB., LONG ISLAND, N.Y.
SIG PI-	SIGNAL CORPS (ARMY). PLANS AND INTELLIGENCE BRANCH
SIG RPS-	SIGNAL CORPS (ARMY). RADIO PROPAGATION SECTION
SIG RPS PR-	SIGNAL CORPS (ARMY). RADIO PROPAGATION SECTION (PROPAGATION REPORT)
SIG RPU-	HOLABIRD SIGNAL DEPOT. RADIO PROPAGATION UNIT, BALTIMORE
SIG RPU TR-	HOLABIRD SIGNAL DEPOT. RADIO PROPAGATION UNIT, BALTIMORE
SIG/SDI-	AMERICAN SOCIETY FOR INFORMATION SCIENCE. SPECIAL INTEREST GROUP ON SELECTIVE DISSEMINATION OF INFORMATION, WASHINGTON, D.C.
SIG T-	SIGNAL CORPS (ARMY) (TRANSLATION)
SIG TEN SPEC-	SIGNAL CORPS (ARMY) (TENTATIVE SPECIFICATION)
SIG-TR-	SIGNATRON, INC., LEXINGTON, MASS.
SIG X	SIGNAL CORPS (ARMY)
SII-QSR-	INTERNATIONAL CONSULTANT SCIENTISTS CORP., BROOKLINE, MASS.
(YEAR)-SIM-(NUMBER/NO.)	GENERAL ELECTRIC CO. APOLLO SUPPORT DEPT., DAYTONA BEACH, FLA.
SIM-	NATIONAL RESEARCH COUNCIL. PACIFIC SCIENCE BOARD
(YEAR)-SIMAR-(NO./NO.)	GENERAL ELECTRIC CO. APOLLO SUPPORT DEPT., DAYTONA BEACH, FLA.
SIM/CAM/(MONTH)/(YEAR)	SIMULMATICS CORP., CAMBRIDGE, MASS.
SIN-	SCHWEIZERISCHES INST. FUER NUKLEARFORSCHUNG, ZURICH
SINP-TH-	SAHA INST. OF NUCLEAR PHYSICS, CALCUTTA
SINP-TH-(YEAR)-	SAHA INST. OF NUCLEAR PHYSICS, CALCUTTA
SINR-E(NUMBER)-	U.S.S.R. JOINT INST. FOR NUCLEAR RESEARCH, DUBNA
SIN-TM-(NUMBER)-(NUMBER)	SCHWEIZERISCHES INST. FUER NUKLEARFORSCHUNG, ZURICH
SIO-	SCRIPPS INSTITUTION OF OCEANOGRAPHY, LA JOLLA, CALIF.
SIO-(YEAR)-	SCRIPPS INSTITUTION OF OCEANOGRAPHY, LA JOLLA, CALIF.
SIO-REF-(YEAR)-	SCRIPPS INSTITUTION OF OCEANOGRAPHY, LA JOLLA, CALIF.
SIO-SEA GRANT-PUB-	SCRIPPS INSTITUTION OF OCEANOGRAPHY, LA JOLLA, CALIF.
SIP-	ORDNANCE DEPT. (ARMY) (STANDARD INSPECTION PROCEDURE)
SIPRE-	SNOW, ICE, + PERMAFROST RES. ESTAB., WILMETTE, ILL.
SIPRE CONFERENCE-	SNOW, ICE, + PERMAFROST RES. ESTAB., WILMETTE, ILL.
SIPRE-RR-	ARMY COLD REGIONS RES. + ENG. LAB., WILMETTE, ILL.
SIPRE-SR-	SNOW, ICE, + PERMAFROST RES. ESTAB., WILMETTE, ILL.
SIPRE TRANS-	SNOW, ICE, + PERMAFROST RESEARCH ESTABLISHMENT, WILMETTE, ILL. (TRANSLATION)
SI-PUB.-	NORWAY. SENTRALINSTITUTT FOR INDUSTRIELL FORSKNING, OSLO
SI PUBL.-	NORWAY. SENTRALINSTITUTT FOR INDUSTRIELL FORSKNING, OSLO
SIR-	KNOLLS ATOMIC POWER LAB., SCHENECTADY, N.Y.
SIR-	LIBRARY OF CONGRESS. AIR INFORMATION DIV., WASH.,D.C.
SIR-	PAN AMERICAN WORLD AIRWAYS, INC. SPECTRUM SIGNATURE PROJECT, SIERRA VISTA, ARIZ.
SISS-	GENERAL ELECTRIC CO. HEAVY MILITARY ELECTRONICS DEPT., SYRACUSE, N.Y.

SISS-
	GENERAL PRECISION, INC. LIBRASCOPE DIV., GLENDALE, CAL
SIT-
	STEVENS INST. OF TECH., HOBOKEN, N.J.
SIT-(NUMBER)-(NUMBER)
	STEVENS INST. OF TECH., HOBOKEN, N.J.
SIT-DL-
	STEVENS INST. OF TECH., HOBOKEN, N.J. DAVIDSON LAB.
SIT ETT-
	STEVENS INST. OF TECH., HOBOKEN, N.J. EXPERIMENTAL TOWING TANK
SIT-ME-RS-
	STEVENS INST. OF TECH., HOBOKEN, N.J. DEPT. OF MECHANICAL ENGINEERING
SIT-ME-RT-
	STEVENS INST. OF TECH., HOBOKEN, N.J. DEPT. OF MECHANICAL ENGINEERING
SIT-OE-(YEAR)-
	STEVENS INST. OF TECH., HOBOKEN, N.J. OCEAN ENG. DEPT.
SIT-P-
	STEVENS INST. OF TECH., HOBOKEN, N.J. DEPT. OF PHYS.
SIT-PML-
	STEVENS INST. OF TECH., HOBOKEN, N.J. POWDER METALLURGY LAB.
SIT-QPR-
	STEVENS INST. OF TECH., HOBOKEN, N.J.
SIT-R-
	STEVENS INST. OF TECH., HOBOKEN, N.J. DAVIDSON LAB.
SIU-
	SOUTHERN ILLINOIS UNIV., CARBONDALE
SJ-(NUMBER)-
	SPERRY MICROWAVE ELECTRONICS CO., CLEARWATER, FLA.
SJC-A-(NUMBER)-(NUMBER)
	JAPAN. ATOMIC ENERGY RESEARCH INSTITUTE, TOKYO
SJC-A-
	TOKYO UNIV.
SJC-A-(YEAR)-
	TOKYO UNIV. INST. FOR NUCLEAR STUDIES
SJC-P-(YEAR)-
	TOKYO UNIV.
SJC-P-(YEAR)-
	TOKYO UNIV. INST. FOR NUCLEAR STUDIES
SJC-T-(YEAR)-
	TOKYO UNIV.
SJC-T-(YEAR)-
	TOKYO UNIV. INST. FOR NUCLEAR STUDIES
SJ-M-(NO.)-(NO.)-
	SPERRY MICROWAVE ELECTRONICS CO., CLEARWATER, FLA.
SJM-
	SPERRY MICROWAVE ELECTRONICS CO., CLEARWATER, FLA.
SJP-
	CHEMICAL WARFARE SERVICE. SAN JOSE PROJECT, CALIF.
SK-
	LEOPOLD-FRANZEN UNIVERSITAET, INNSBRUCK, AUSTRIA
SK-
	SLOAN-KETTERING INST. FOR CANCER RESEARCH, N.Y.C.
SK-(YEAR)-
	STENCEL AERO ENGINEERING CORP., ASHEVILLE, N.C.
SKB-(NUMBER)/M-
	NORWAY. TECH. UNIV., TRONDHEIM
SKF-AL(YR.)L(NO.)
	SKF INDUSTRIES, INC. RES. LAB., KING OF PRUSSIA, PA.
SK MD-
	SLOAN-KETTERING INST. FOR CANCER RESEARCH, N.Y.C.
SKY-
	SKYDYNE, INC., PORT JERVIS, N.Y.
SL-
	ARINC RESEARCH CORP., ANNAPOLIS
SL-
	DOW CHEMICAL CO., MIDLAND, MICH.
SL-
	FISH AND WILDLIFE SERVICE (INTERIOR), WASHINGTON, D.C. (STATISTICAL LISTS)
SL(YEAR)-
	GENERAL DYNAMICS/CONVAIR, SAN DIEGO, CALIF.
SL-
	GENERAL ELECTRIC CO., LTD. APPLIED ELECTRONICS LABS., STANMORE, MIDDX, ENGLAND
SL-
	SANDIA LAB., ALBUQUERQUE, N. MEX.
SL-
	SARGENT AND LUNDY, CHICAGO
SL-
	SIGNAL CORPS (LABORATORIES)
SLA-
	SPECIAL LIBRARIES ASSN. TRANSLATION CENTER, JOHN CRERAR LIBRARY, CHICAGO
SLAC-
	STANFORD UNIV., CALIF. STANFORD LINEAR ACCELERATOR CTR
SLAC-PUB-
	STANFORD UNIV., CALIF. STANFORD LINEAR ACCELERATOR CENTER (SUBMITTED FOR PUBLICATION)
SLAC-TN-(YR.)-
	STANFORD UNIV., CALIF. STANFORD LINEAR ACCELERATOR CENTER (INTERNAL NOTES)
SLAC-TRANS-
	STANFORD UNIV., CALIF. STANFORD LINEAR ACCELERATOR CENTER (TRANSLATION)
SLA-TR-(YEAR)-
	SPECIAL LIBRARIES ASSN. TRANSLATION CENTER, JOHN CRERAR LIBRARY, CHICAGO
SLC-
	DU PONT DE NEMOURS (E.I.) + CO., WILMINGTON, DEL.
SLC-
	DU PONT DE NEMOURS (E.I.) + CO. EASTERN LAB., GIBBSTOWN, N.J.
SLC-
	SYSTEMS LABORATORIES CORP., SHERMAN OAKS, CALIF.
SLF-
	FINLAND. INSTITUTE OF RADIATION PHYSICS, HELSINKI
SLMR-
	BUREAU OF MINES. SALT LAKE EXPERIMENT STATION, SALT LAKE CITY
SLMS-
	SANDIA LAB., ALBUQUERQUE, N. MEX.
SLOE-
	AIR FORCE (SPECIAL LIST OF EQUIPMENT)
SLP-
	MICHIGAN STATE UNIV., EAST LANSING. STATISTICAL LAB.
SLR-
	LOCKHEED AIRCRAFT CORP., BURBANK, CALIF.
SLS-
	UNITED AIRCRAFT CORP., WINDSOR LOCKS, CONN.
SLSR-
	AIR FORCE INST. OF TECH., WRIGHT-PATTERSON AFB, OHIO. SCHOOL OF SYSTEMS AND LOGISTICS
SLTR-(NUMBER-YEAR)
	AIR FORCE INST. OF TECH., WRIGHT-PATTERSON AFB, OHIO. SCHOOL OF SYSTEMS AND LOGISTICS
SLU-
	SAINT LOUIS UNIV.
SLV-1B-
	AIR FORCE SPECIAL WEAPONS CENTER, KIRTLAND AFB, N. MEX. (SLV-1B SPACE PROBE)
SLW/RM/XMG/
	UNITED KINGDOM ATOMIC ENERGY AUTHORITY. INDUSTRIAL GROUP. CAPENHURST WORKS, CHES., ENGLAND
SM-
	(SCIENTIFIC MEMORANDUM) (REF. 2)
SM-
	(SPECIAL MEMORANDUM) (REF. 2)
SM-
	(STAFF MEMORANDUM) (REF. 2)
SM-
	(STRUCTURE MEMORANDUM) (REF. 2)
SM-
	(SUMMARY MEMORANDUM) (REF. 2)
SM-
	AIR FORCE SYSTEMS COMMAND, WRIGHT-PATTERSON AFB, OHIO
SM-
	AIR RESEARCH AND DEVELOPMENT COMMAND, ANDREWS AFB, MD
SM-
	AKADEMIYA NAUK SSSR., MOSCOW
SM-
	ARGONNE NATIONAL LAB., ILL.
SM-
	ATOMIC ENERGY OF CANADA LTD., CHALK RIVER, ONT.
SM-
	AUSTRALIA. AERO. RES. LABS., FISHERMANS BEND, VICTORIA
SM-
	BERN. UNIVERSITAET
SM-
	BOEING CO. AERO-SPACE DIV., SEATTLE
SM-
	BONN. UNIVERSITAET
SM-
	BUREAU OF AERONAUTICS (NAVY) (STRUCTURES MEMORANDUM)
37 SM-
	CALIFORNIA. UNIV., BERKELEY. RADIATION LAB. (REF. 9)
SM-
	CALIFORNIA. UNIV., LA JOLLA
SM-
	CALIFORNIA INST. OF TECH., PASADENA
SM-(NUMBER)/
	CANADA. GEOLOGICAL SURVEY, OTTAWA
SM-(NUMBER)/
	CHICAGO. UNIV.
SM-
	COLUMBIA UNIV., PALISADES, N.Y. LAMONT GEOLOGICAL OBS
SM-
	DANISH ATOMIC ENERGY COMMISSION. RES. ESTAB., RISOE
SM-
	DOUGLAS AIRCRAFT CO., INC.
SM-
	DOUGLAS AIRCRAFT CO., INC., LONG BEACH, CALIF.
SM-
	DOUGLAS AIRCRAFT CO., INC., SANTA MONICA, CALIF.
SM-(NUMBER)-Q-
	DOUGLAS AIRCRAFT CO., INC. ASTROPOWER LAB., NEWPORT BEACH, CALIF.
SM-
	DOUGLAS AIRCRAFT CO., INC. MISSILE AND SPACE SYSTEMS DIV., SANTA MONICA, CALIF.
SM-
	EUROPEAN SPACE DATA CENTRE, DARMSTADT, GERMANY
SM-
	INDIA. ATOMIC ENERGY ESTABLISHMENT, TROMBAY
SM-(NUMBER/NUMBER)
	INTERNATIONAL ATOMIC ENERGY AGENCY, VIENNA
SM-(NUMBER/NUMBER)
	INTERNATIONAL ATOMIC ENERGY AGENCY, VIENNA
SM-
	ISRAEL. ATOMIC ENERGY COMMISSION, TEL-AVIV
SM (NUMBER)/(NUMBER)
	KERNFORSCHUNGSZENTRUM KARLSRUHE, GERMANY
SM-(5 DIGITS)
	MC DONNELL DOUGLAS ASTRONAUTICS CO., HUNTINGTON BEACH, CALIF.
SM-
	MANHATTAN DISTRICT. SAFETY COMMITTEE
SM-(YR.)-
	NEW YORK UNIV., N.Y.C. COLL. OF ENGINEERING
SM-
	OFFICE OF SCIENTIFIC RESEARCH AND DEVELOPMENT
SM-
	RAND CORP., SANTA MONICA, CALIF. (SPECIAL MEMO.)
SM-
	SAFETY COMMITTEE, MANHATTAN DISTRICT
SM-4-
	SANDIA CORP., ALBUQUERQUE, N. MEX.

SM-	SERVOMECHANISMS, INC., HAWTHORNE, CALIF.
SM-	SHOCK HYDRODYNAMICS, INC., SHERMAN OAKS, CALIF.
SM-	UNITED KINGDOM ATOMIC ENERGY AUTHORITY. RESEARCH GP. WANTAGE RESEARCH LABORATORIES, BERKS, ENGLAND
SM-	WESTINGHOUSE ELECTRIC AND MFG. CO., E. PITTSBURGH
SM-(NO.)(WRL)	WESTINGHOUSE ELECTRIC CORP. RES. LABS., E. PITTSBURGH
SM-(NUMBER)(WRL)	WESTINGHOUSE ELECTRIC CORP. RES. LABS., PITTSBURGH
SM-(LETTER)(NUMBER)-6	SANDIA CORP., ALBUQUERQUE, N. MEX.
SM-(LETTER(S))-	SANDIA CORP., ALBUQUERQUE, N. MEX. (READINESS PROGRAM SERIES)
SMAD-	LITTON SYSTEMS, INC., WOODLAND HILLS, CALIF.
SMA-TR-(NUMBER)-	SCIENTIFIC MANAGEMENT ASSOCS., INC., HADDONFIELD, NJ
SMB-(YR.)-	SANDIA CORP. MANUAL BOARD., ALBUQUERQUE, N. MEX.
SM/BE/(NUMBER/YEAR)	BRITISH STEEL CORP. STEELMAKING DIV., LONDON
SMBTI-	SHIP AND MACHINE BUILDING TESTING INST. (KIEL NAVY INST., DEPT. OF SHIP CONSTRUCTION)
SMC-	GT. BRIT. ATOMIC EN. RES. ESTAB., HARWELL, BERKS, ENG
SMC-	MASON (SILAS) CO., INC., BURLINGTON, IOWA
SMC-	SINGER MFG. CO., N.Y.C.
SMCPSTC-(NUMBER)-	SUPPLY AND MAINTENANCE COMMAND (ARMY). PACKAGING STORAGE AND TRANSPORTABILITY CENTER
SMC-R-	SHUFORD-MASSENGILL CORP., LEXINGTON, MASS.
SMC, SMRL-	SUBMARINE MEDICAL RESEARCH LAB., GROTON, CONN.
SMD-	SANDIA CORP., ALBUQUERQUE, N. MEX.
SME-	GT. BRIT. ROYAL AIRCRAFT EST.,FARNBOROUGH, HANTS, ENG
SME-AA-	FAIRCHILD CAMERA AND INSTRUMENT CORP., SYOSSET, N.Y.
SME-AC-	FAIRCHILD SPACE AND DEFENSE SYSTEMS, SYOSSET, N.Y.
SME-AF-	FAIRCHILD SPACE AND DEFENSE SYSTEMS, SYOSSET, N.Y.
SME-AG-	FAIRCHILD SPACE AND DEFENSE SYSTEMS, SYOSSET, N.Y.
SME-AJ-	FAIRCHILD SPACE AND DEFENSE SYSTEMS, SYOSSET, N.Y.
SME-AP-	FAIRCHILD SPACE AND DEFENSE SYSTEMS, SYOSSET, N.Y.
SME-BB-	FAIRCHILD CAMERA AND INSTRUMENT CORP., SYOSSET, N.Y.
SMEC-	SPERRY MICROWAVE ELECTRONICS CO., CLEARWATER, FLA.
SMF-	AMERICAN INST. OF AERONAUTICS + ASTRONAUTICS, N.Y.C.
SMF-	INSTITUTE OF THE AERONAUTICAL SCIENCES, N.Y.C.
SMI-	PENNSYLVANIA STATE COLL., STATE COLLEGE. SCHOOL OF MINERAL INDUSTRIES
SMIR-	SERVOMECHANISMS, INC., GOLETA, CALIF.
SMP-	UNITED AIRCRAFT CORP. HAMILTON STANDARD DIV., WINDSOR LOCKS, CONN.
SM-R-	SHUFORD-MASSENGILL CORP., LEXINGTON, MASS.
SMR-	WHITE SANDS MISSILE RANGE. SMALL MISSILE RANGE, N.MEX
SMRE-	GT. BRIT. SAFETY IN MINES RESEARCH ESTABLISHMENT, BUXTON, DERBY, ENGLAND
SMRE-RR-	GT. BRIT. SAFETY IN MINES RESEARCH ESTABLISHMENT, BUXTON, DERBY, ENGLAND
SMRL-	SUBMARINE MEDICAL RESEARCH LAB., GROTON, CONN.
SMRL-	TEXAS. UNIV., AUSTIN. STRUCTURAL MECHANICS RES. LAB.
SMRL MEMO (YEAR)-	SUBMARINE MEDICAL RESEARCH LAB., GROTON, CONN.
SMRL-MR-(YEAR)-	NAVAL SUBMARINE MEDICAL CENTER. SUBMARINE MEDICAL RESEARCH LAB., GROTON, CONN.
SMRL SP (YEAR)-	SUBMARINE MEDICAL RES. LAB.,GROTON, CONN.(SPEC. RPT.)
SMRL-SR-(YEAR)-	SUBMARINE MEDICAL RESEARCH LAB., GROTON, CONN.
SMRP-	CHICAGO. UNIV. SATELLITE + MESOMETEOROLOGY RES. PROJ.
SMRP-RP-	CHICAGO. UNIV. SATELLITE + MESOMETEOROLOGY RES. PROJ.
SMS-	ARMY SIGNAL MISSILE SUPPORT AGENCY, WHITE SANDS MISSILE RANGE, N. MEX.
S.M.S.-	UNITED KINGDOM. ROYAL NAVAL PERSONNEL RESEARCH COMM., LONDON
SMSA-	ARMY SIGNAL MISSILE SUPPORT AGENCY, WHITE SANDS MISSILE RANGE, N. MEX.
SMSA-OR-LR-	ARMY SIGNAL MISSILE SUPPORT AGENCY. MISSILE METEOROLOGY DIV., WHITE SANDS MISSILE RANGE, N. MEX.
SMSA-MEWD-	ARMY SIGNAL MISSILE SUPPORT AGENCY. MISSILE ELECTRONIC WARFARE DIV., WHITE SANDS MISSILE RANGE, NM
SMSA-PR-	ARMY SIGNAL MISSILE SUPPORT AGENCY, WHITE SANDS MISSILE RANGE, N. MEX. (PROGRESS REPORTS)
SMSA-TM-	ARMY SIGNAL MISSILE SUPPORT AGENCY, WHITE SANDS MISSILE RANGE, N. MEX. (TECHNICAL MEMORANDA)
SMSA-TR-	ARMY SIGNAL MISSILE SUPPORT AGENCY, WHITE SANDS MISSILE RANGE, N. MEX.
SMSD-ABMDA-	BROWN ENGINEERING CO., INC. SPACE AND MILITARY SYSTEMS DIV., HUNTSVILLE, ALA.
SMSD-AMSMI-	BROWN ENGINEERING CO., INC. SPACE AND MILITARY SYSTEMS DIV., HUNTSVILLE, ALA.
SMSRL-	PITTSBURGH. UNIV. SARAH MELLON SCAIFE RADIATION LAB.
SMU-	SOUTHERN METHODIST UNIV., DALLAS
SMUFD-AE-	FORT DETRICK, FREDERICK, MD.
SMUFD-MISC PUB-	FORT DETRICK, FREDERICK, MD.
SMUFD-TECH. MANUSCRIPT-	FORT DETRICK, FREDERICK, MD.
SMUFD-TECHNICAL STUDY-	FORT DETRICK, FREDERICK, MD.
SMUFD-TM-	FORT DETRICK, FREDERICK, MD.
SMUFD-TR-	FORT DETRICK, FREDERICK, MD.
SMUPA-NR-	PICATINNY ARSENAL. NUCLEAR RELIABILITY DIV., DOVER,NJ
SMUPA-TK-	PICATINNY ARSENAL. NUCLEAR WEAPONS GROUP, DOVER, NJ
SMVD-	MULLARD RADIO VALVE CO., LTD.
SMW-	OFFICE OF THE ASSISTANT SECRETARY OF DEFENSE (RES. + ENG.). WORKING GP. ON MICROWAVE TUBES
SN-	(SCIENTIFIC NOTE) (REF. 2)
SN-(3 DIGITS)-F	DYNAMIC SCIENCE CORP., MONROVIA, CALIF.
SN-	JOHNS HOPKINS UNIV., SILVER SPRING, MD. APPLIED PHYSICS LAB.
SN-	NATIONAL ENG. SCIENCE CO., PASADENA, CAL.(CON. RPTS.)
SN-	NATIONAL ENGINEERING SCIENCE CORP., MC LEAN, VA.
SNC-(NUMBER)-	SANDERS NUCLEAR CORP., NASHUA, N.H.
SNDL-	(STANDARD NAVY DISTRIBUTION LIST)
SNE-	FRANCE. SERVICE DE NEUTRONIQUE EXPERIMENTALE
SNE-	SOUTHERN NUCLEAR ENGINEERING, INC., DUNEDIN, FLA.
SNE-(NUMBER)NP	SOUTHERN NUCLEAR ENGINEERING, INC., DUNEDIN, FLA.
SNECMA-	SOCIETE NATIONALE D'ETUDE ET DE CONSTRUCTIONS DE MOTEURS D'AVIATION, PARIS
SNL-	ORDNANCE DEPT. (ARMY) (STANDARD NOMENCLATURE LIST)
SNPO-	SPACE NUCLEAR PROPULSION OFFICE (AEC/NASA), WASH,D.C.
SNW/C/	BRITISH STEEL CORP. SCOTTISH AND NORTH WEST GROUP, MOTHERWELL, SCOTLAND
SNY-LEJ-	SCHENECTADY OPERATIONS OFFICE, AEC, N.Y.
SO-	(SHIPPING ORDER) (REF. 2)
SO-(4 DIGITS)	AMERICAN OPTICAL CO.LASER PRODS.DIV.,SOUTHBRIDGE,MASS
SO-	FOREST SERVICE. SOUTHERN STATION, NEW ORLEANS
SO-	GT. BRIT. CHEM. DEFENCE RES.EST.,SUTTON OAK,LANCS,ENG
SO-	OFFICE OF NAVAL RESEARCH (SUPPLY OPERATIONS)
SO-	SCHENECTADY OPERATIONS OFFICE, AEC, N.Y. (REF. 4)
SO-	SOLAR AIRCRAFT CO., SAN DIEGO, CALIF.
S.O. (5 DIGITS)	TRANS-SONICS, INC., BURLINGTON, MASS
SO-	WESTERN FILTER CO., INC., GARDENA, CALIF.
SOA-	STANDARD OIL DEVELOPMENT CO., ELIZABETH, N.J.
SOA-	STANDARD OIL DEVELOPMENT CO. ESSO LABS., LINDEN, N.J.
SOA-SR-	SMITHSONIAN ASTROPHYSICAL OBS., CAMBRIDGE, MASS.
SOC-	SPECIAL OPERATIONS CENTER, EGLIN AFB, FLA.
SOCAL-	UNIVERSITY OF SOUTHERN CALIFORNIA, LOS ANGELES
SOCU-	SOUTH CAROLINA. UNIV., COLUMBIA

SOD-
STANDARD OIL DEVELOPMENT CO., ELIZABETH, N.J.
SOD-(NO.)-(YR.)
STANDARD OIL DEVELOPMENT CO., N.Y.C.
SOD-
STANDARD OIL DEVELOPMENT CO. ESSO LABS., LINDEN, N.J.
SOD-PKN-
STANDARD OIL DEVELOPMENT CO. ESSO LABS., LINDEN, N.J.
SOF-(NUMBER)A-
AIR FORCE SPECIAL OPERATIONS FORCE, EGLIN AFB, FLA.
SOHIO-
STANDARD OIL CO. (OHIO). RES. + DEV. DEPT., CLEVELAND
SOIL ENGINEERING SER-
CORNELL UNIV., ITHACA, N.Y.
SOIL ENG. RES. SER.-
PRINCETON UNIV.,N.J. DEPT. OF CIVIL + GEOLOGICAL ENG.
SOIL INFORMATION SER-
NORTH CAROLINA STATE UNIV., RALEIGH. DEPT. OF SOIL SCIENCE
SOIL MECHANICS SER-
ILLINOIS. UNIV., URBANA. DEPT. OF CIVIL ENGINEERING
SOILS PUB-
MASSACHUSETTS INST. OF TECH., CAMBRIDGE. DEPT. OF CIVIL ENGINEERING
SO-INT-
SCHENECTADY OPERATIONS OFFICE, AEC, N.Y.
SOLID STATE DEVICE-TR-
MASSACHUSETTS INST. OF TECH., CAMBRIDGE. CENTER FOR MATERIAL SCIENCE AND ENGINEERING
SOM-
OLIN MATHIESON CHEMICAL CORP. NUCLEAR FUEL DIV., NEW HAVEN
SON-
SONOTONE CORP., ELMSFORD, N.Y.
SON-
STANDARD OIL DEVELOPMENT CO., ELIZABETH, N.J.
SOO-
SCHENECTADY OPERATIONS OFFICE, AEC, N.Y. (REF. 4)
SOP-
(STANDARD OPERATING PROCEDURE) (REF. 2)
SOR-(LETTERS-(NO.-NO.)
MARINE CORPS, WASHINGTON, D.C.
SOR-CC-
MARINE CORPS, WASHINGTON, D.C. (SPECIFIC OPERATIONAL REQUIREMENT)
SORI-
SOUTHERN RESEARCH INST., BIRMINGHAM, ALA.
SORI-(NUMBER)-(NUMBER)
SOUTHWEST RESEARCH INST., SAN ANTONIO
SORIN-
SOCIETA RICERCHE IMPIANTI NUCLEARI, SALUGGIA, ITALY
SORIN-M/
SOCIETA RICERCHE IMPIANTI NUCLEARI, SALUGGIA, ITALY
SORIN-T/
SOCIETA RICERCHE IMPIANTI NUCLEARI, SALUGGIA, ITALY
SORI-PR-
SOUTHERN RESEARCH INST., BIRMINGHAM, ALA.
SORL-
SOUTHERN RESEARCH INST., BIRMINGHAM, ALA.
SORO-CINFAC-(NO.-NO.)
AMERICAN UNIV., WASHINGTON, D.C. SPECIAL OPERATIONS RESEARCH OFFICE
SORR-
BUREAU OF ORDNANCE (NAVY) (SUBSURFACE ORDNANCE RESEARCH REPORT)
SOS-FR-
SPACE ORDNANCE SYSTEMS, INC., EL SEGUNDO, CALIF.
SOTS-
SENIOR OBSERVER SECTION, MATHER AFB, CALIF.
SOW-
SUNFLOWER ORDNANCE WORKS, LAWRENCE, KANS.
SP-
(SCIENTIFIC PAPER) (REF. 2)
SP-
(SOLID PROPELLANTS) (REF. 2)
SP-
(SPECIAL PAPER) (REF. 2)
SP-
(SPECIAL PROJECT REPORT) (REF. 2)
SP-
(SPECIAL PROPELLANTS) (REF. 2)
SP-
(SPECIAL PUBLICATION) (REF. 2)
SP-
(SPECIFICATION) (REF. 2)
SP-
(STAFF PAPER) (REF. 2)
SP-
AEROJET-GENERAL CORP., SACRAMENTO, CALIF.
SP-
AIR FORCE (STRATEGIC PLANNING CHARTS)
SP-
BRITISH HYDROMECHANICS RES. ASSN.,CRANFIELD,BEDS,ENG.
SP-
CALLERY CHEMICAL CO., PENNA.
SP-
FORD INSTRUMENT CO., LONG ISLAND CITY, N.Y.
SP-
GENERAL ELECTRIC CO. TEMPO. CENTER FOR ADVANCED STUDIES, SANTA BARBARA, CALIF.
SP-
GOODYEAR AEROSPACE CORP., AKRON, OHIO
SP-5001 THRU -5256
GT. BRIT. DEPT. OF ATOMIC ENERGY. INDUSTRIAL GROUP H.Q., RISLEY, LANCS, ENGLAND (REF. 36)
SP-
GT. BRIT. LEGISLATIVE COUNCIL (SESSIONAL PAPER)
SP-
HAMILTON STANDARD, WINDSOR LOCKS, CONN.
SP-(YEAR)-
LEANDER MC CORMICK OBSERVATORY, CHARLOTTESVILLE, VA.
SP-
NATIONAL DEFENSE RESEARCH COMM. (SPECIAL PROPELLANTS)
SP-(YEAR)-
NAVAL AIR SYSTEMS COMMAND. PROFESSIONAL DEVELOPMENT CENTER, WASHINGTON, D.C.
SP-
NAVAL OCEANOGRAPHIC OFFICE, WASHINGTON, D.C.
SP-
NEW YORK BOTANICAL GARDEN, BRONX
SP-
NORTHROP AIRCRAFT, INC., HAWTHORNE, CALIF. (SPECIAL PROJECTS)
SP-
OFFICE OF SCIENTIFIC RES. + DEV.(SPECIAL PROPELLANTS)
SP-
OLIN MATHIESON CHEMICAL CORP. (SOLID PROPELLANTS)
SP-(YEAR)(NO.)F
PROTEUS, INC., MOUNTAIN LAKES, N.J.
SP-(NUMBER)-
RCA LABS., PRINCETON, N.J.
S + P(NO.)A-
SANDERSON AND PORTER, N.Y.C.
SP-
SMITHSONIAN ASTROPHYSICAL OBS., CAMBRIDGE, MASS.
SP-
SPRAGUE ELECTRICAL CO., NORTH ADAMS, MASS.
SP-
SVERDRUP AND PARCEL, INC., ST. LOUIS
SP-
SYSTEM DEVELOPMENT CORP., SANTA MONICA, CALIF.
SP-(NUMBER/NUMBER(S))
SYSTEM DEVELOPMENT CORP., SANTA MONICA, CALIF. (PAPER)
SP(NUMBER)-(LTR.,NO.)
TEXAS INSTRUMENTS, INC. APPARATUS DIV., DALLAS
SP-(YEAR)-
WESTINGHOUSE ELECTRIC CORP. RES. LABS., PITTSBURGH
SP/(LTRS)/(YR)-
AIR FORCE INST. OF TECH., WRIGHT-PATTERSON AFB, OHIO
SPC-
SPERRY PHOENIX CO., ARIZ.
SPCC-
SHIPS PARTS CONTROL CENTER, MECHANICSBURG, PENNA.
SPC-LJ-(NUMBER)-
SPERRY PHOENIX CO., ARIZ.
SP(CS)-
GT. BRIT. SPRINGFIELDS WORKS, LANCS, ENGLAND (REF.36)
SPD-
BORG-WARNER CORP. SPECIAL PRODUCTS DIV.
SPD-
GENERAL DYNAMICS CORP. ELECTRIC BOAT DIV.,GROTON,CONN
SPD-
INGERSOLL KALAMAZOO DIV., MICH.
SPD-
KELLOGG (M.W.) CO. SPECIAL PROJECTS DEPT.,JERSEY CITY
SPD-
MAGNAVOX CO. SPECIAL PRODUCTS DIV.
SPD/(NUMBER)
SYLVANIA ELECTRONIC SYSTEMS-EAST, NEEDHAM, MASS.
SPDD-
WESTINGHOUSE ELECTRIC CORP., PITTSBURGH
SPDGC-
CORPS OF ENGINEERS (ARMY). SOUTH PACIFIC DIV., SAN FRANCISCO
SPDIR-
INGERSOLL KALAMAZOO DIV., BORG-WARNER CORP., MICH.
SPE-
NATIONAL SCIENCE FOUNDATION. DIV. OF SCIENTIFIC PERSONNEL AND EDUCATION, WASHINGTON, D.C.
SPEC-
(SPECIAL) (REF. 2)
SPEC-
(SPECIFICATION) (REF. 2)
SPEC-
CALIFORNIA. UNIV., BERKELEY. RADIATION LAB. (REF. 9)
SPEC-(NUMBER)/
LOCKHEED MISSILES AND SPACE CO., SUNNYVALE, CALIF.
SPEC-ACC-
AEROJET-GENERAL CORP. LIQUID ROCKET OPERATIONS, SACRAMENTO, CALIF.
SPEC-AIL-
AIRBORNE INSTRUMENTS LAB., INC., DEER PARK, N.Y.
SPEC-D(NUMBER)
HONEYWELL, INC., ST. PETERSBURG, FLA.
SPECDEVCEN-(NO.)-(NO.)-
OFFICE OF NAVAL RES. SPECIAL DEVICES CTR., WASH.,D.C.
SPECDEVCEN-
OFFICE OF NAVAL RES. SPECIAL DEVICES CTR., WASH.,D.C.
SPEC-ES(NUMBER)-P
HONEYWELL, INC., ST. PETERSBURG, FLA.
SPEC-FMS-
HONEYWELL, INC. AERONAUTICAL DIV., ST.PETERSBURG, FLA
SPEC-GS-
SPERRY RAND CORP. UNIVAC DEFENSE SYSTEMS DIV., ST. PAUL
SPECIAL LISTING-
DEPARTMENT OF THE INTERIOR, WASHINGTON, D.C.
SPECIAL PROJECTS-
GEOLOGICAL SURVEY, DENVER
SPECIAL REPORTS-
AIR FORCE CAMBRIDGE RESEARCH LABS., BEDFORD, MASS.
SPECIAL SCIENTIFIC-
AMERICAN SCIENCE AND ENG. INC., CAMBRIDGE, MASS.
SPECIAL STUDIES-
GEOLOGICAL SURVEY, DENVER
SPECIAL TOPIC-
RCA LABS. COMMUNICATIONS RES. LAB., PRINCETON, N.J.

```
SPECIAL TR-
              STANFORD RESEARCH INST., MENLO PARK, CALIF.
SPEC-LA(NUMBERS)-
              NORTH AMERICAN AVIATION, INC. LOS ANGELES DIV.
SPEC-LMSC-
              LOCKHEED MISSILES AND SPACE CO., SUNNYVALE, CALIF.
SPEC-LSP-(NO.)-
              GRUMMAN AIRCRAFT ENGINEERING CORP., BETHPAGE, N.Y.
SPEC-MA(NO.)-
              NORTH AMERICAN ROCKWELL CORP. SPACE DIV., DOWNEY, CAL
SPEC-MIL-(NO.)(NORD)
              BUREAU OF ORDNANCE (NAVY) (SPECIFICATION)
SPEC-MIL-(LTR)-(NO.)(AER)
              BUREAU OF AERONAUTICS (NAVY) (SPECIFICATION)
SPEC-O-
              GENERAL DYNAMICS/CONVAIR, SAN DIEGO, CALIF.
SPEC-RAO-(NUMBER)-
              ROCKETDYNE. CHEMICAL AND MATERIAL SCIENCES DEPT.,
                CANOGA PARK, CALIF.
SPEC-RBO-(NUMBER)-
              ROCKETDYNE. CHEMICAL AND MATERIAL SCIENCES DEPT.,
                CANOGA PARK, CALIF.
SPEC-S-
              AVCO MISSILE SPACE AND ELECTRONICS GROUP. AVCO SPACE
                DIV., WILMINGTON, MASS.
SPEC SEM-
              CALIFORNIA. UNIV., BERKELEY. RADIATION LAB. (REF. 9)
SPEC. SEMI.-
              CALIFORNIA. UNIV., BERKELEY. RADIATION LAB. (SPECIAL
                SEMINAR) (REF. 9)
SPEECH-
              TECHNICAL INFORMATION SERVICE, AEC
SPERRY
              SPERRY GYROSCOPE CO., GREAT NECK, N.Y.
SPERRY-(NO.)-
              SPERRY GYROSCOPE CO., GREAT NECK, N.Y.
SPERRY-(LETTERS)-
              SPERRY GYROSCOPE CO., GREAT NECK, N.Y.
SPG-
              WHITE SANDS PROVING GROUND, N. MEX.
SPGC-
              SPACE-GENERAL CORP., EL MONTE, CALIF.
SP-HB-
              HATHAWAY INSTRUMENT CO., DENVER, COLO.
SPIA-
              JOHNS HOPKINS UNIV., SILVER SPRING, MD. SOLID PROPEL-
                LANT INFORMATION AGENCY
SPIA/A-
              JOHNS HOPKINS UNIV., SILVER SPRING, MD. SOLID PROPEL-
                LANT INFORMATION AGENCY (ABSTRACTS)
SPIA/A-
              JOHNS HOPKINS UNIV., SILVER SPRING, MD. SOLID PROPEL-
                LANT INFORMATION AGENCY (CURRENT ABSTRACTS)
SPIA/LPIA-
              JOHNS HOPKINS UNIV., SILVER SPRING, MD. LIQUID PRO-
                PELLANT INFORMATION AGENCY
SPIA/LPIA-
              JOHNS HOPKINS UNIV., SILVER SPRING, MD. SOLID PROPEL-
                LANT INFORMATION AGENCY
SPIA/M-
              JOHNS HOPKINS UNIV., SILVER SPRING, MD. SOLID PROPEL-
                LANT INFORMATION AGENCY (MANUALS)
SPIA/PP-
              JOHNS HOPKINS UNIV., SILVER SPRING, MD. SOLID PROPEL-
                LANT INFORMATION AGENCY (PHYSICAL PROPERTIES)
SPIA/S-
              JOHNS HOPKINS UNIV., SILVER SPRING, MD. SOLID PROPEL-
                LANT INFORMATION AGENCY
SPIA-S(YR.)-
              JOHNS HOPKINS UNIV., SILVER SPRING, MD. SOLID PROPEL-
                LANT INFORMATION AGENCY
SPIA-SPSP-
              JOHNS HOPKINS UNIV., SILVER SPRING, MD. SOLID PROPEL-
                LANT INFORMATION AGENCY (SOLID PROPELLANT SURVEIL-
                LANCE PANEL)
SPIC-
              OFFICE OF THE DIRECTOR OF DEFENSE (RES) + ENG.)
                AD HOC GROUP ON SOLID PROPELLANT INSTABILITY OF
                COMBUSTION, WASHINGTON, D.C.
SPINST-
              SPECIAL PROJECTS OFFICE (NAVY), WASHINGTON, D.C.
SPIR-
              AIR FORCE (SPECIAL PHOTO-INTERPRETATION REPORT)
SPL-(YEAR)-
              CALIFORNIA RESEARCH CORP. SAN PABLO LAB.
SPL-
              SYRACUSE UNIV., N.Y. RES. CORP. SPEC. PROJS. LAB.
SPL-
              SYRACUSE UNIV., N.Y. SPECIAL PROJECTS LAB.
SPL TR-
              OFFICE OF NAVAL RES. SANDS POINT LAB., PT. WASH.,N.Y.
SPL-TR-
              SYRACUSE UNIV., N.Y. RES. CORP. SPEC. PROJS. LAB.
SPM-
              FRANCE. COMMISSARIAT A L'ENERGIE ATOMIQUE. CENTRE
                D'ETUDES NUCLEAIRES. SERVICE DE PHYSIQUE MATHE-
                MATIQUE, SACLAY
SPM-
              SANDIA CORP., ALBUQUERQUE, N. MEX.(STD. PARTS MANUAL)
SPME-
              SANDIA CORP., ALBUQUERQUE, N. MEX. (ELECTRICAL
                STANDARD PARTS MANUAL)
SPMM-
              SANDIA CORP., ALBUQUERQUE, N. MEX. (MECHANICAL
                STANDARD PARTS MANUAL)
SPNBE-
              FRANCE. COMMISSARIAT A L'ENERGIE ATOMIQUE. CENTRE
                D'ETUDES NUCLEAIRES, SACLAY
SPNO-
              SPACE NUCLEAR PROPULSION OFFICE (AEC/NASA), WASH,D.C.

SP-NOTE-
              NAVY. SPECIAL PROJECTS OFFICE, WASHINGTON, D.C.
SPO-
              PERKIN-ELMER CORP., POMONA, CALIF.
SPOG-
              BRITISH ARMY OF THE RHINE. SPECIAL PROJECTILE OPERA-
                TIONS GROUP
SPP-(YEAR)-
              BENDIX CORP. BENDIX PRODUCTS AEROSPACE DIV., SOUTH
                BEND, IND.
SPR-
              (SPECIAL PROJECT REPORT) (REF. 2)
SPR-
              (SPECIAL REPORT) (REF. 2)
SPR-
              (SUPPLEMENTARY PROGRESS REPORT) (REF. 2)
SPR-
              CALIFORNIA. UNIV., LIVERMORE. LAWRENCE RADIATION LAB.
SPR-
              COMMITTEE ON MEDICAL RESEARCH (SPECIAL REPORTS)
SPR-
              OESTERREICHISCHE STUDIENGESELLSCHAFT FUER ATOMENERGIE
                G.M.B.H. INSTITUT FUER BIOLOGIE UND LANDWIRTSCHAFT,
                SEIBERSDORF (SEIBERSDORF PROJECT REPORT)
SPRA-
              ARMY SIGNAL RADIO PROPAGATION AGENCY, FT. MONMOUTH,NJ
SPRINGFIELDS-(NO.)/
              GT. BRIT. SPRINGFIELDS WORKS, LANCS, ENG. (REPRINTS)
SPRUCE-
              SPRUCE JT. AEC-DOD WARHEAD COORDINATING COMM.(REF.15)
SPS-
              STANDARD PRESSED STEEL CO., JENKINTOWN, PENNA.
SPSD-(NUMBER)-AFS-
              WESTINGHOUSE ELECTRIC CORP. STEAM DIV.,LESTER, PENNA.
SPSP/
              JOHNS HOPKINS UNIV., SILVER SPRING, MD. SOLID PROPEL-
                LANT INFORMATION AGENCY (SOLID PROPELLANT SURVEIL-
                LANCE PANEL)
SPSP-
              JOINT ARMY-NAVY-AIR FORCE SOLID PROPELLANT SUR-
                VEILLANCE PANEL (REF. 13)
SPSR-
              CALIF. INST. OF TECH., PASADENA. JET PROPULSION LAB.
                (SURVEYER PROJECT STATUS REPORT)
SPSTP-
              JOINT ARMY-NAVY-AIR FORCE SOLID PROPELLANT ROCKET
                STATIC TEST PANEL (REF. 13)
SPT-
              FRANCE. COMMISSARIAT A L'ENERGIE ATOMIQUE. CENTRE
                D'ETUDES NUCLEAIRES. SERVICE DE PHYSIQUE
                THEORIQUE, SACLAY
SPT/(INITIALS)-
              FRANCE. COMMISSARIAT A L'ENERGIE ATOMIQUE. CENTRE
                D'ETUDES NUCLEAIRES. SERVICE DE PHYSIQUE
                THEORIQUE, SACLAY
SPT/DOC/(INITIALS)-
              FRANCE. COMMISSARIAT A L'ENERGIE ATOMIQUE. CENTRE
                D'ETUDES NUCLEAIRES. SERVICE DE PHYSIQUE
                THEORIQUE, SACLAY
SP-TP-
              AEROJET-GENERAL CORP., AZUSA, CALIF.
SQ-
              CURTISS-WRIGHT CORP., COLUMBUS, OHIO
SQ-
              CURTISS-WRIGHT CORP. WRIGHT AERO. DIV., WOOD-RIDGE,NJ
SQ-
              SQUIBB INSTITUTE FOR MEDICAL RESEARCH, NEW
                BRUNSWICK, N.J.
SQUID-
              SQUID PROJECT (REF. 33)
SQUID-PR-(NUMBER)-C
              SQUID PROJECT (REF. 33)
SQUID-PR-(NUMBER)-P
              SQUID PROJECT (REF. 33)
SQUID-PSU-(NUMBER)-P
              PENNSYLVANIA STATE UNIV., UNIVERSITY PARK (SQUID
                PROJECT) (REF. 33)
SQUID-PUR-(NUMBER)-M
              PURDUE UNIV., LAFAYETTE, IND. (SQUID PROJ.)(REF. 33)
SQUID-TM-PIB-
              BROOKLYN. POLYTECHNIC INST. (SQUID PROJECT) (REF. 33)
SQUID-TR-AC-(NUMBER)-PU
              PURDUE UNIV., LAFAYETTE, IND. (SQUID PROJ.) (REF. 33)
SQUID-TR-ARC-(NUMBER)-PU
              PURDUE UNIV., LAFAYETTE, IND. (SQUID PROJ.) (REF. 33)
SQUID-TR-CAL-(NO)-PU
              PURDUE UNIV., LAFAYETTE, IND. (SQUID PROJ.) (REF. 33)
SQUID-TR-CCNY-(NO.)-PU
              PURDUE UNIV., LAFAYETTE, IND. (SQUID PROJ.) (REF. 33)
SQUID-TR-CU-(NUMBER)-PU
              PURDUE UNIV., LAFAYETTE, IND. (SQUID PROJ.) (REF. 33)
SQUID-TR-DYN-(NUMBER)-PU
              PURDUE UNIV., LAFAYETTE, IND. (SQUID PROJ.) (REF. 33)
SQUID-TR-GIT-(NO)-PU
              PURDUE UNIV., LAFAYETTE, IND. (SQUID PROJ.) (REF. 33)
SQUID-TR-ILL-(NUMBER-PU
              PURDUE UNIV., LAFAYETTE, IND. (SQUID PROJ.) (REF. 33)
SQUID-TR-MIT-(NO.)-PU
              MASS. INST. OF TECH., CAMBRIDGE (SQUID PROJ.)(REF.33)
SQUID-TR-MIT-(NO)-PU
              PURDUE UNIV., LAFAYETTE, IND. (SQUID PROJ.) (REF. 33)
SQUID-TR-NYU-(NUMBER)P
              PURDUE UNIV., LAFAYETTE, IND. (SQUID PROJ.) (REF. 33)
SQUID-TR-PR-(NUMBER)-PU
              PURDUE UNIV., LAFAYETTE, IND. (SQUID PROJ.) (REF. 33)
SQUID-TR-RICE-(NO)-PU
              PURDUE UNIV., LAFAYETTE, IND. (SQUID PROJ.) (REF. 33)
SQUID-TR-SRI-(NUMBER)-PU
              PURDUE UNIV., LAFAYETTE, IND. (SQUID PROJ.) (REF. 33)
```

SR-	
SR-	(SCIENTIFIC REPORT) (REF. 2)
SR-	(SCIENTIFIC RESEARCH) (REF. 2)
SR-	(SECTION REPORT) (REF. 2)
SR-	(SIMULATION REPORT) (REF. 2)
SR-	(SOUND REPORT) (REF. 2)
SR-	(SPECIAL REGULATION) (REF. 2)
SR-	(SPECIAL REPORT) (REF. 2)
SR-	(STAFF REPORT) (REF. 2)
SR-	(STANDARDIZATION REPORT) (REF. 2)
SR-	(STATUS REPORT) (REF. 2)
SR-	(SUMMARY REPORT) (REF. 2)
SR-	(SUPPLEMENTAL REPORT) (REF. 2)
SR-	AERONUTRONIC SYSTEMS, INC., GLENDALE, CALIF.
SR-	AIR TRANSPORT ASSN. OF AMERICA. SHIPPERS RESEARCH DIV., WASHINGTON, D.C.
SR-	ARMY, WASHINGTON, D.C.
SR-(NUMBER)-(YEAR)	ATLANTIC FLEET. FLEET AIR WINGS, NORFOLK, VA.
S-R-	BABCOCK + WILCOX CO. ATOMIC EN. DIV., LYNCHBURG, VA.
SR-/(YR)	CANADA. DEPT. OF MINES AND RESOURCES. BUREAU OF MINES
SR-	COMMITTEE ON MEDICAL RESEARCH (SHOCK REPORTS)
SR-(YEAR)-	FLORIDA. UNIV., GAINESVILLE. ENGINEERING AND INDUSTRIAL EXPERIMENT STATION
SR-(NUMBER)-(NUMBER)	GEORGIA. UNIV., ATHENS
SR-5001 THRU -5256	GT. BRIT. DEPT. OF ATOMIC ENERGY. INDUSTRIAL GROUP H.Q., RISLEY, LANCS, ENGLAND (REF. 36)
SR-(YEAR)-	INDUSTRIAL COLL. OF THE ARMED FORCES, WASHINGTON,D.C.
SR-(NUMBER)-(NUMBER)	JOHNS HOPKINS UNIV., SILVER SPRING, MD. APPLIED PHYSICS LAB.
SR/(NUMBER)-M/	LOCKHEED MISSILES AND SPACE CO. ADVANCED STRATEGIC PROGRAMS, SUNNYVALE, CALIF.
SR-(YR.)-	MARTIN CO., DENVER
SR-(NUMBER)-	MARTIN CO., DENVER
SR-	MARTIN-MARIETTA CORP. DENVER DIV.
SR-	MITRE CORP., BEDFORD, MASS.
SR-(YEAR)-	NAVAL SUBMARINE MEDICAL CENTER. SUBMARINE MEDICAL RESEARCH LAB., GROTON, CONN.
SR-	OFFICE OF NAVAL RESEARCH (SUPPORTING RESEARCH + DEV.)
SR-	OFFICE OF SCIENTIFIC RESEARCH AND DEVELOPMENT
SR-1-	SAVANNAH RIVER OPERATIONS OFFICE, AEC, AIKEN, S.C.
SR(NO.)-	SPACELABS, INC., VAN NUYS, CALIF.
SR-	STANFORD RESEARCH INST., MENLO PARK, CALIF.
3SR-	SYSTEMS SCIENCE AND SOFTWARE, LA JOLLA, CALIF.
SR-	TENNESSEE EASTMAN CORP., OAK RIDGE, TENN.
SR-	UNITED AIRCRAFT CORP., EAST HARTFORD, CONN.
SR-(NO.)(WRL)	WESTINGHOUSE ELECTRIC CORP. RES. LABS., E. PITTSBURGH
SR-(LETTERS)-	SAVANNAH RIVER OPERATIONS OFFICE, AEC, AIKEN, S.C.
SRA-	SMYTH RESEARCH ASSOCIATES, SAN DIEGO, CALIF.
SRAI-	STANFORD RADIO ASTRONOMY INST., CALIF.
S-(RAND)	RAND CORP., SANTA MONICA, CALIF.
SRARI-	U.S.S.R. SCIENTIFIC RESEARCH ATOMIC REACTOR INST., MELEKESS
SRARI-P-	U.S.S.R. SCIENTIFIC RESEARCH ATOMIC REACTOR INST., MELEKESS
SRB-	LOCKHEED AIRCRAFT CORP. MISSILES AND SPACE DIV., SUNNYVALE, CALIF.
SRB-(YR.)-	LOCKHEED AIRCRAFT CORP. MISSILES AND SPACE DIV., SUNNYVALE, CALIF.
SRC-(NUMBER-LTR.-YR.)-	CASE WESTERN RESERVE UNIV.,CLEVELAND.SYSTEMS RES.CTR.
SRC-(YEAR)-TR-	SYSTEMS RESEARCH CORP., WASHINGTON, D.C.
SRC-A-	CALIFORNIA. UNIV., BERKELEY. SURVEY RESEARCH CENTER
SRCC-	PITTSBURGH. UNIV. SPACE RESEARCH COORDINATION CTR.
SRCC-CR-(YEAR)-	SPERRY RAND RESEARCH CENTER, SUDBURY, MASS.
SRCC-RR-(YEAR)-	SPERRY RAND RESEARCH CENTER, SUDBURY, MASS.
SRC-M-	CALIFORNIA. UNIV., BERKELEY. SURVEY RESEARCH CENTER
SRCR-(YEAR)-	UNION CARBIDE CORP., N.Y.C.
SRCR-(YEAR)-	UNION CARBIDE CORP. LINDE DIV., INDIANAPOLIS
SRCR-	UNION CARBIDE CORP. LINDE DIV. SPEEDWAY LABS., INDIANAPOLIS, IND.
SRC-S-	BUREAU OF MINES. SAFETY RESEARCH CENTER, PITTSBURGH
SRD-	AERONUTRONIC DIV. OF PHILCO CORP., NEWPORT BEACH,CAL.
SRD-	PHILCO-FORD CORP. SPACE AND RE-ENTRY SYSTEMS DIV., NEWPORT BEACH, CALIF.
SRD-	RAYTHEON CO. SYSTEM REQUIREMENTS DEPT., WALTHAM, MASS
SRDA-	ARMY SIGNAL RES. AND DEV. AGENCY, FT. MONMOUTH, N.J.
SRDB-(YR.)-	SANDIA CORP. RES. + DEV. BD., ALBUQUERQUE, N. MEX.
SRDC-	STOCKHEIM RESEARCH AND DEV. CORP., WOODSIDE, N.Y.
SRDE-	GT. BRIT. SIGNALS RESEARCH AND DEVELOPMENT ESTABLISHMENT, CHRISTCHURCH, HANTS, ENGLAND
SRD-G-	PHILCO-FORD CORP. SPACE AND RE-ENTRY SYSTEMS DIV., NEWPORT BEACH, CALIF.
SRDG-	PHILCO-FORD CORP. SPACE AND RE-ENTRY SYSTEMS DIV., NEWPORT BEACH, CALIF.
SRDL-	ARMY SIGNAL RESEARCH + DEV. LAB., FORT MONMOUTH, N.J.
SR+DL R+D SUMMARY...	ARMY SIGNAL RESEARCH AND DEVELOPMENT LAB., FORT MONMOUTH, N.J. (RESEARCH AND DEVELOPMENT SUMMARY)
SRDL-TR-	ARMY SIGNAL RESEARCH + DEV. LAB., FORT MONMOUTH, N.J.
SRD-R-	UNITED KINGDOM ATOMIC ENERGY AUTHORITY. SAFETY AND RELIABILITY DIRECTORATE, RISLEY, LANCS, ENGLAND
SRDS-	FEDERAL AVIATION ADMINISTRATION. SYSTEMS RESEARCH AND DEV. SERVICE, WASHINGTON, D.C. (RESEARCH + DEV.)
SRDS-RD-(YEAR)-	FEDERAL AVIATION ADMINISTRATION. SYSTEMS RESEARCH AND DEV. SERVICE, WASHINGTON, D.C. (RESEARCH + DEV.)
SRE-	GT. BRIT. ROYAL NAVAL SCIENTIFIC SERVICE
SRE/ACS-	GT. BRIT. ADMIRALTY COMPUTING SERVICE
SRE-PR-	FRANCE. COMMISSARIAT A L'ENERGIE ATOMIQUE. CENTRE D'ETUDES NUCLEAIRES, SACLAY
SRES-	ILLINOIS. UNIV., URBANA
SRE/WAVE/	GT. BRIT. ADMIRALTY. DEPT. OF SCI. RES. + EXPERIMENT
SRFB-	NATIONAL RESEARCH COUNCIL OF CANADA. SPACE RESEARCH FACILITIES BRANCH, OTTAWA
SRG-	COLUMBIA UNIV., N.Y.C. STATISTICAL RESEARCH GROUP
SRG-	PHILCO-FORD CORP. SPACE AND RE-ENTRY SYSTEMS DIV., NEWPORT BEACH, CALIF.
SRG-	PRINCETON UNIV., N.J. (STATISTICAL RESEARCH GROUP)
SRG-C-	OFFICE OF SCIENTIFIC RESEARCH AND DEVELOPMENT
SRG MEMO-	COLUMBIA UNIV., N.Y.C. STATISTICAL RESEARCH GROUP
SRG P-	PRINCETON UNIV., N.J. (STATISTICAL RESEARCH GROUP)
SRHL-	SOUTHWESTERN RADIOLOGICAL HEALTH LAB., LAS VEGAS,NEV.
SRI/	GT. BRIT. ATOMIC WEAPONS RESEARCH ESTABLISHMENT, ALDERMASTON, BERKS, ENGLAND
SRI-	ILLINOIS. UNIV., URBANA
SRI-(NUMBER)-	SOUTHERN RESEARCH INST., BIRMINGHAM, ALA.
SRI-	SOUTHWEST RESEARCH INST., SAN ANTONIO
SRI-	STANFORD RESEARCH INST., MENLO PARK, CALIF.
SRI-(NUMBER)	STANFORD RESEARCH INST., MENLO PARK, CALIF.
SRI-(YEAR)	STANFORD RESEARCH INST., MENLO PARK, CALIF.
SRI-(NO.)-P	STANFORD RESEARCH INST., MENLO PARK, CALIF.
SRI-(NUMBER)-(LETTERS)-	STANFORD RESEARCH INST., MENLO PARK, CALIF.
SRI-(NUMBER)-PU	STANFORD RESEARCH INST., MENLO PARK, CALIF.
SRI-	VIRGINIA. UNIV., CHARLOTTESVILLE
SRI-(LTRS.)U-(PROJ. NO.)-	STANFORD RESEARCH INST., MENLO PARK, CALIF. (U DENOTES GOVERNMENT CONTRACT)

SR-IA-
    BUREAU OF NAVAL WEAPONS, WASHINGTON, D.C.
SRIA-
    STANFORD RESEARCH INST., MENLO PARK, CALIF.
SRIA-(NUMBER)-P-(NO.)-
    STANFORD RESEARCH INST., MENLO PARK, CALIF.
SRIA-(NUMBER)-P-
    STANFORD RESEARCH INST., MENLO PARK, CALIF.
SRIA-(NUMBER)-(NUMBER)
    STANFORD RESEARCH INST., MENLO PARK, CALIF.
SR-IB-
    BUREAU OF NAVAL WEAPONS, WASHINGTON, D.C.
SRIB-
    SOUTHERN RESEARCH INST., BIRMINGHAM, ALA.
SRIC-(YEAR)-
    SCIENTIFIC RESEARCH INSTRUMENTS CORP., BALTIMORE
SRI-CU-677-
    STANFORD RESEARCH INST., MENLO PARK, CALIF.
SRI-CU-(NUMBER)-
    STANFORD RESEARCH INST., MENLO PARK, CALIF.
SRI-D-(NUMBER)-
    STANFORD RESEARCH INST., MENLO PARK, CALIF.
SRI-DAC-
    STANFORD RESEARCH INST., MENLO PARK, CALIF.
SRI-GD-(NO.)-FR
    STANFORD RES. INST. POULTER LABS., MENLO PARK, CALIF.
SRI-H-R-
    MC GILL UNIV., MONTREAL. SPACE RESEARCH INST.
SRI-H-TN-
    MC GILL UNIV., MONTREAL. SPACE RESEARCH INST.
SRI-IR-
    STANFORD RESEARCH INST., MENLO PARK, CALIF.
SRI-ITR-
    STANFORD RESEARCH INST., MENLO PARK, CALIF.
SRI-MR-
    STANFORD RESEARCH INST., MENLO PARK, CALIF.
SRI-OAD-TN-(NUMBER)-
    STANFORD RESEARCH INST. OPERATIONS ANALYSIS DEPT.,
    MENLO PARK, CALIF.
SRI-ORD-STR-(NUMBER)-
    STANFORD RESEARCH INST. OPERATIONS RESEARCH DEPT.,
    MENLO PARK, CALIF.
SRI-PAU-(PROJ. NO.)-
    STANFORD RESEARCH INST., MENLO PARK, CALIF.
SRI PGU-
    STANFORD RESEARCH INST., MENLO PARK, CALIF.
SRI-PHU-
    STANFORD RESEARCH INST., MENLO PARK, CALIF.
SRI-PL-
    STANFORD RES. INST. POULTER LABS., MENLO PARK, CALIF.
SRI-PL-TR-(NO.)-(YR.)
    STANFORD RES. INST. POULTER LABS., MENLO PARK, CALIF.
SRI-PR-
    SOUTHERN RESEARCH INST., BIRMINGHAM, ALA.
SRI-PU-
    STANFORD RESEARCH INST., MENLO PARK, CALIF.
SRI-R-
    MC GILL UNIV., MONTREAL. SPACE RESEARCH INST.
SRI-R-
    SPACE RESEARCH INST., INC., NORTH TROY, VT.
SRI-RM-
    STANFORD RESEARCH INST., MENLO PARK, CALIF.
SRI-SCIENTIFIC-
    STANFORD RESEARCH INST., MENLO PARK, CALIF.
SRI-SPECIAL TR-
    STANFORD RESEARCH INST., MENLO PARK, CALIF.
SRI-SR-
    STANFORD RESEARCH INST., MENLO PARK, CALIF.
SRI-SU-(NO.)-(NO.)
    STANFORD RESEARCH INST., MENLO PARK, CALIF.
SRI-SU-1815-
    STANFORD RESEARCH INST., MENLO PARK, CALIF.
SRI-TL-
    MC GILL UNIV., MONTREAL. SPACE RESEARCH INST.
SRI-TM-
    STANFORD RESEARCH INST., MENLO PARK, CALIF.
SRI-TN-(NUMBER)-SEA-
    STANFORD RESEARCH INST., MENLO PARK, CALIF.
SRI-TR-
    STANFORD RESEARCH INST., MENLO PARK, CALIF.
SRI-TR-(NUMBER)-
    STANFORD RESEARCH INST., MENLO PARK, CALIF.
SRI-TR OF (DATE)
    SOUTHERN RESEARCH INST., BIRMINGHAM, ALA.
SRL-
    AIR FORCE ACADEMY, COLORADO SPRINGS. FRANK J. SEILER
    RESEARCH LAB.
SRL-
    G.E.C. ELECTRONICS, LTD. APPLIED ELECTRONICS LABS.,
    STANMORE, MIDDX., ENGLAND
SRL-
    LITTON INDUSTRIES, INC. SPACE RESEARCH LABS.,
    BEVERLY HILLS, CALIF.
SRL-(NUMBER)-FR-(YEAR)-
    MICHIGAN. UNIV., ANN ARBOR. SYSTEMS RESEARCH LAB.
SRL-(YR)-
    NORTH AMERICAN AVIATION, INC., DOWNEY, CALIF.
SRL-
    SYSTEMS RESEARCH LABS., INC., DAYTON, OHIO
SRL-(NUMBER)-
    SYSTEMS RESEARCH LABS., INC., DAYTON, OHIO
SRL-(NUMBER)-A
    SYSTEMS RESEARCH LABS., INC., DAYTON, OHIO
SRL-(LETTERS)-(NO.)-(NO.)
    STANFORD RESEARCH INST., MENLO PARK, CALIF.
SRL-TR-
    AEROSPACE RESEARCH LABS., WRIGHT-PATTERSON AFB, OHIO
SRL-TR-(YEAR)-
    AIR FORCE ACADEMY, COLORADO SPRINGS. FRANK J. SEILER
    RESEARCH LAB.
SRL-TR-(NUMBER)-B
    SYSTEMS RESEARCH LABS., INC., DAYTON, OHIO
SRM-
    (STAFF RESEARCH MEMORANDUM) (REF. 2)
SRM-
    NAVAL PERSONNEL RESEARCH ACTIVITY, SAN DIEGO, CALIF.
    (RESEARCH REPORT)
SRM-(YEAR)-
    NAVAL PERSONNEL RESEARCH ACTIVITY, SAN DIEGO, CALIF.
SRM-
    NORTHWESTERN UNIV., EVANSTON, ILL. TECHNOLOGICAL INST.
SRM-
    STANDARD ROLLING MILLS, INC., BROOKLYN
SRMB-(YR.)-
    SANDIA CORP. ROAD MATERIEL BD., ALBUQUERQUE, N. MEX.
SRM-MEMO-
    UNITED KINGDOM ATOMIC ENERGY AUTHORITY. WEAPONS GROUP
    ATOMIC WEAPONS RES. ESTAB., ALDERMASTON, BERKS, ENG.
SRN-
    SACRAMENTO PEAK OBSERVATORY, SUNSPOT, N. MEX.
SRO-
    AEROJET-GENERAL CORP. SOLID ROCKET PLANT, SACRA-
    MENTO, CALIF.
SRO-(NUMBER)-TM
    AEROJET-GENERAL CORP. SOLID ROCKET PLANT, SACRA-
    MENTO, CALIF.
SRO-
    GT. BRIT. DEPT. OF ATOMIC ENERGY. INDUSTRIAL GROUP
    H.Q., RISLEY, LANCS, ENGLAND (REF. 36)
SRO-
    SAVANNAH RIVER OPERATIONS OFFICE, AEC, AIKEN, S.C.
SRO/ML-
    GT. BRIT. CULCHETH LABS., LANCS, ENGLAND (REF. 36)
SROO-
    SAVANNAH RIVER OPERATIONS OFFICE, AEC, AIKEN, S.C.
SRP-
    AEROJET-GENERAL CORP. SOLID ROCKET PLANT, SACRA-
    MENTO, CALIF.
SRR-
    GT. BRIT. DEPT. OF ATOMIC ENERGY. INDUSTRIAL GROUP
    H.Q., RISLEY, LANCS, ENGLAND (REF. 36)
SRR-
    NAVAL PERSONNEL RESEARCH ACTIVITY, SAN DIEGO, CALIF.
    (RESEARCH REPORT)
SRR-(YEAR)-
    NAVAL PERSONNEL RESEARCH ACTIVITY, SAN DIEGO, CALIF.
SRR-
    NORTHROP AIRCRAFT, INC., HAWTHORNE, CALIF. (STRUC-
    TURES RESEARCH REPORT)
SRRC-(NUMBER/YEAR)
    GT. BRIT. SCOTTISH RES. REACTOR CENTRE, EAST KILBRIDE
SRRC-
    SPERRY RAND RESEARCH CENTER, SUDBURY, MASS.
SRRC-CR-(YEAR)-
    SPERRY RAND RESEARCH CENTER, SUDBURY, MASS.
SRRC-RR-
    SPERRY RAND RESEARCH CENTER, SUDBURY, MASS.
SRRC-RR-(YR)-(NO.)
    SPERRY RAND RESEARCH CENTER, SUDBURY, MASS.
SRRL-
    DEPARTMENT OF AGRICULTURE. SOUTHERN UTILIZATION
    RESEARCH BRANCH, NEW ORLEANS
SRS-
    (STRUCTURAL RESEARCH SERIES) (REF. 2)
SR-S-
    ARMY NATICK LABS. EARTH SCIENCES DIV., MASS.
SRS-
    ILLINOIS. UNIV., URBANA. DEPT. OF CIVIL ENGINEERING
SRS-
    PHILCO-FORD CORP. SPACE AND RE-ENTRY SYSTEMS DIV.,
    NEWPORT BEACH, CALIF.
SRS-
    UNITED KINGDOM ATOMIC ENERGY AUTHORITY. INDUSTRIAL
    GROUP. SPRINGFIELDS WORKS, LANCS, ENG. (REF. 36)
SRS-BNTT-PM-
    PHILCO-FORD CORP. SPACE AND RE-ENTRY SYSTEMS DIV.,
    NEWPORT BEACH, CALIF.
SRSC/P-
    GT. BRIT. SPRINGFIELDS WORKS, LANCS, ENGLAND
SRSI-
    SPACE RECOVERY SYSTEMS, INC. DIV. OF ITEK CORP.,
    EL SEGUNDO, CALIF.
SRSM-
    HUGHES AIRCRAFT CO., CULVER CITY, CALIF. (SPECIAL
    RESEARCH STUDY MEMORANDUM)
SRS(PPL)-
    FAIRCHILD HILLER CORP. REPUBLIC AVIATION DIV.
    PLASMA PROPULSION LAB., FARMINGDALE, N.Y.
SRS-RD-
    FLORIDA. UNIV., GAINESVILLE. REGIONAL REHABILITATION
    RESEARCH INST.
SRS-S-
    PHILCO-FORD CORP. SPACE AND RE-ENTRY SYSTEMS DIV.,
    NEWPORT BEACH, CALIF.
SRS-SG-
    PHILCO-FORD CORP. SPACE AND RE-ENTRY SYSTEMS DIV.,
    NEWPORT BEACH, CALIF.
SRS-SRG-
    PHILCO-FORD CORP. SPACE AND RE-ENTRY SYSTEMS DIV.,
    NEWPORT BEACH, CALIF.
SRS-TM-
    PHILCO-FORD CORP. SPACE AND RE-ENTRY SYSTEMS DIV.,
    PALO ALTO, CALIF.
SRS-TR-
    REPUBLIC AVIATION CORP., FARMINGDALE, N.J.
SRT-(YEAR)-
    SANTA RITA TECHNOLOGY, INC., MENLO PARK, CALIF.
SRTM-
    NATIONAL WEATHER SERVICE. SOUTHERN REGION, FT. WORTH,
    TEX.

```
SR-TN/(NUMBER)-
              NETHERLANDS. NIJVERHEIDSORGANISATIE, TNO, DELFT
SRW-
              GT. BRIT. WINDSCALE WORKS, SELLAFIELD, CUMB., ENGLAND
                 (REF. 36)
SRW-T-
              SAVANNAH RIVER OPERATIONS OFFICE, AEC, AIKEN, S.C.
SS-
              (OPERATION SANDSTONE) (REF. 7)
SS-
              (SPECIAL SERIES) (REF. 2)
SS-
              (SPECIAL STUDY) (REF. 2)
SS-(NUMBER)
              AIR COMMAND AND STAFF COLL., MAXWELL AFB, ALA.
SS-
              FEDERAL AVIATION ADMINISTRATION. OFFICE OF SUPERSONIC
                 TRANSPORT DEVELOPMENT, WASHINGTON, D.C.
SS-
              GARRETT CORP. AIRESEARCH MFG. DIV., LOS ANGELES
SS-
              GT. BRIT. ROYAL NAVAL PERSONNEL RESEARCH COMMITTEE
SS-
              GT. BRIT. UNDERWATER COUNTERMEASURES + WEAPONS ESTAB.
SS-
              HUGHES AIRCRAFT CO. SPACE SYSTEMS DIV.,EL SEGUNDO,CAL
                 (SURVEYOR SPACECRAFT)
SS(NO.)B-T(NO.)-
              LOCKHEED MISSILES AND SPACE CO., PALO ALTO, CALIF.
SS-
              NAVAL TORPEDO STATION, KEYPORT, WASH.
SSA-(NUMBER-YEAR)
              NORD-AVIATION, CHATILLON-SOUS-BAGNEUX, FRANCE
SSA-(NUMBER-YEAR)
              SYMPOSIUM ON SPACE, 1968, VENICE
SSAP-(YEAR)-
              CATHOLIC UNIV. OF AMERICA, WASHINGTON, D.C. DEPT. OF
                 SCIENCE AND APPLIED PHYSICS
SSARL-DPH-(YEAR)-
              SOUTH SHORE ANALYTICAL AND RESEARCH LAB., ISLIP, N.Y.
SSBME-(MONTH/YEAR)
              UNITED KINGDOM ATOMIC ENERGY AUTHORITY. WEAPONS GROUP
                 ATOMIC WEAPONS RES. ESTAB.,ALDERMASTON, BERKS, ENG.
SSC-
              BROWN UNIV., PROVIDENCE
SSC-
              MASSACHUSETTS INST. OF TECH., CAMBRIDGE
SSC-
              NATIONAL RESEARCH COUNCIL. COMMITTEE ON SHIP STRUC-
                 TURE, WASHINGTON, D.C. (SERIES USED FOR REPORTS
                 SPONSORED BY THE COMMITTEE BUT PREPARED BY VARIOUS
                 AGENCIES)
SSC-
              SEISMOGRAPH SERVICE CO., TULSA, OKLA.
SSC-
              SONNTAG SCIENTIFIC CORP., GREENWICH, CONN.
SSCD-
              UNITED KINGDOM ATOMIC ENERGY AUTHORITY. WEAPONS GROUP
                 ATOMIC WEAPONS RES. ESTAB.,ALDERMASTON, BERKS, ENG.
SSC-RM-
              STANFORD RESEARCH INST. STRATEGIC STUDIES CENTER,
                 MENLO PARK, CALIF.
SSD-(YEAR)-
              DUNLAP AND ASSOCIATES, INC., DARIEN, CONN.
SSD-(5 DIGITS)
              HUGHES AIRCRAFT CO. SPACE SYSTEMS DIV.,EL SEGUNDO,CAL
SSD-(YEAR)-
              SPACE SYSTEMS DIV., LOS ANGELES AIR FORCE STATION,
                 CALIF.
SSD-CR-(YEAR)-
              SPACE SYSTEMS DIV. (AIR FORCE), INGLEWOOD, CALIF.
SSDS EM-
              TOMS RIVER SIGNAL LAB. SUPPRESSION SYSTEMS
                 DEVELOPMENT SUBSECTION, N.J. (ENGINEERING MEMO.)
SSD-TDR-(YEAR)-
              SPACE SYSTEMS DIV., LOS ANGELES AIR FORCE STATION,
                 CALIF.
SSD-TR-(YEAR)-
              SPACE SYSTEMS DIV. (AIR FORCE), INGLEWOOD, CALIF.
S(SEISMOLOGY)-
              MASSACHUSETTS INST. OF TECH., CAMBRIDGE. DEPT. OF
                 GEOLOGY AND GEOPHYSICS
SSEL-
              PENNSYLVANIA STATE UNIV., UNIVERSITY PARK. SPACE
                 SCIENCES + ENGINEERING LAB.
S.S.E. MEMO-
              GENERAL ELECTRIC CO. SPACE SCIENCES LAB.,PHILADELPHIA
SSE-T(NUMBER)-
              LOCKHEED MISSILES AND SPACE CO., PALO ALTO, CALIF.
SSF-
              ARMED FORCES SPECIAL WEAPONS PROJECT, WASHINGTON,
                 D.C. (SERVICE STORAGE FACILITIES)
SS-H-
              NATIONAL TRANSPORTATION SAFETY BOARD, WASHINGTON, DC
SSI-
              AKTIEBOLAGET ATOMENERGI, STOCKHOLM
SSI-
              SPACE SCIENCES INC., NATICK, MASS.
SSI-
              SPACE SCIENCES INC., WALTHAM, MASS.
SSI-(NUMBER)-FR-
              SPACE SCIENCES INC., WALTHAM, MASS.
SSI-(NUMBER)-SR-
              SPACE SCIENCES INC., WALTHAM, MASS.
SSI-(3 DIGITS)-
              SPACE SCIENCES INC., WALTHAM, MASS.
SS INFORMAL REPT.-
              GT. BRIT. MINISTRY OF SUPPLY. MINE DESIGN DEPT.
SSI-TR-
              SPACE SCIENCES INC., NATICK, MASS.

SSL-          CALIFORNIA. UNIV., BERKELEY
SSL-          GENERAL ELECTRIC CO. MISSILE AND SPACE DIV., PHILA.
SSL-          SQUIER SIGNAL LAB., FORT MONMOUTH, N.J.
SSL ER-       SQUIER SIGNAL LAB., FORT MONMOUTH, N.J. (ENG. RPTS.)
SSL SI-       SQUIER SIGNAL LAB., FORT MONMOUTH, N.J.
SSL(UC)-      CALIFORNIA. UNIV., BERKELEY. SPACE SCIENCES LAB.
SSME-TN-(MONTH/YEAR)
              UNITED KINGDOM ATOMIC ENERGY AUTHORITY. WEAPONS GROUP
                 ATOMIC WEAPONS RES. ESTAB.,ALDERMASTON, BERKS, ENG.
SSMP-(NUMBER. NO. LTR.)
              ARMY SAFEGUARD SYSTEM COMMAND, HUNTSVILLE, ALA.
SSMP-         ARMY SAFEGUARD SYSTEM COMMAND, HUNTSVILLE, ALA.
                 (SAFEGUARD SYSTEM MASTER PLAN)
SSM-T(YEAR)-
              LOCKHEED MISSILES AND SPACE CO., PALO ALTO, CALIF.
SSM-T(NUMBERS)-
              LOCKHEED MISSILES AND SPACE CO., SUNNYVALE, CALIF.
S/SNDT-       JOHNS HOPKINS UNIV., BALTIMORE (SYMPOSIUM ON SCALE
                 AND NONDESTRUCTIVE TESTING)
SSNR/USA/(YEAR)/
              UNITED KINGDOM ATOMIC ENERGY AUTHORITY. WEAPONS GROUP
                 ATOMIC WEAPONS RES. ESTAB.,ALDERMASTON, BERKS, ENG.
SSN-T(YEAR)-
              LOCKHEED AIRCRAFT CORP. MISSILES AND SPACE DIV.,
                 SUNNYVALE, CALIF.
SS-OL1 (MONTH/YEAR)
              STATISTICS AND REPORTS DIV. (AIR FORCE), WASHINGTON,
                 D.C. (DIRECTORY OF U.S. AIR FORCE ORGANIZATIONS)
SSPA-         SWEDISH STATE SHIPBUILDING EXPERIMENTAL TANK,GOTEBORG
SSPA-PUB-     SWEDISH STATE SHIPBUILDING EXPERIMENTAL TANK
SSPD/USA/     UNITED KINGDOM ATOMIC ENERGY AUTHORITY. WEAPONS GROUP
                 ATOMIC WEAPONS RES. ESTAB.,ALDERMASTON, BERKS, ENG.
SSR-          (SPECIAL SCIENTIFIC REPORT) (REF. 2)
SSR-          MASS. INST. OF TECH., CAMBRIDGE. INSTRUMENTATION LAB.
SS-R-         NATIONAL TRANSPORTATION SAFETY BOARD, WASHINGTON, DC
SSR-          POLYMER CORP., LTD., SARNIA, ONT.
SSR-          SOLID STATE RADIATIONS, INC., LOS ANGELES
SSR-(NUMBER)F-
              SOLID STATE RADIATIONS, INC., LOS ANGELES
SSRC-         PENNSYLVANIA. UNIV., PHILADELPHIA
SSRC-TR-(YR.)-
              HUGHES AIRCRAFT CO. SOLID STATE RESEARCH CENTER,
                 NEWPORT BEACH, CALIF.
SS-R/H-       NATIONAL TRANSPORTATION SAFETY BOARD, WASHINGTON, DC
SSRI-         SOLID STATE RADIATIONS, INC., LOS ANGELES
SSRI-REPRINT-
              HAWAII. UNIV., HONOLULU. SOCIAL SCIENCE RES. INST.
SSS-          BUREAU OF SHIPS (SUBMARINE SIGNAL CO. SPECIFICATIONS)
SSS-          SEA-SPACE SYSTEMS, INC., TORRANCE, CALIF.
SSS-3SR-      SYSTEMS SCIENCE AND SOFTWARE, LA JOLLA, CALIF.
SSS-3SIR-     SYSTEMS SCIENCE AND SOFTWARE, LA JOLLA, CALIF.
SSSR-(LETTER)-
              (SSSR- IS ASSIGNED BY VARIOUS RECIPIENTS AS A PREFIX
                 TO CODES FOR REPORTS ORIGINATING IN THE USSR. FOR
                 EXPLANATION OF THE REMAINDER OF THE CODE, SEE
                 ENTRY UNDER LETTERS FOLLOWING SSSR-)
SST-          FEDERAL AVIATION ADMINISTRATION. OFFICE OF SUPERSONIC
                 TRANSPORT DEVELOPMENT, WASHINGTON, D.C.
SST-          NATIONAL AVIATION FACILITIES EXPTL.CTR.,ATLANTIC CITY
SST/B(NO.)(LTR.)-(NO.)/
              BRITISH AIRCRAFT CORP., FILTON, GLOUCESTER, ENG.
SSTG-         SIEGLER CORP., INGLEWOOD, CALIF.
SSU-          STANFORD RESEARCH INST. SOUTHERN CALIF. LABS.,
                 SO. PASADENA
ST-           (SPECIAL TRANSLATION) (REF. 2)
ST-           ANTIAIRCRAFT ARTILLERY AND GUIDED MISSILE SCHOOL,
                 FORT BLISS, TEX.
ST-(NUMBER)-
              ARMY MILITARY POLICE SCHOOL, FORT GORDON, GA.
ST-(NUMBER)-
              ARMY ORDNANCE CENTER AND SCHOOL, ABERDEEN PROVING
                 GROUND, MD.
ST-(NUMBERS)-
              ARMY SPECIAL WARFARE SCHOOL, FORT BRAGG, N.C.
(NUMBERS)-ST-
              ATLANTIC RESEARCH CORP. MISSILE SYSTEMS DIV.,
                 COSTA MESA, CALIF.
ST-           BRITISH AIRCRAFT CORP. GUIDED WEAPONS DIV.,
                 STEVENAGE, HERTS, ENGLAND
```

ST-	CHEMICAL CORPS SCHOOL, ARMY CHEMICAL CENTER, MD.
ST-	NAVAL AIR TEST CENTER, PATUXENT RIVER, MD.
ST(NO.)-	NAVAL AIR TEST CENTER, PATUXENT RIVER, MD.
ST-	OFFICE OF SCIENTIFIC RESEARCH AND DEVELOPMENT
ST-	ORDNANCE SCHOOL, ABERDEEN PROVING GROUND, MD.
ST-	QUARTERMASTER RESEARCH AND ENGINEERING COMMAND, NATICK, MASS. (SPECIAL TEXT)
STA-	SUPERSONIC TUNNEL ASSN.
ST-AA-(LETTERS)-	NATIONAL AERONAUTICS AND SPACE ADM., WASHINGTON, D.C. (SCIENTIFIC TRANS.)(CATEGORY DESIGNATION)(SUBJECT)
STAC-	STANLEY AVIATION CORP., BUFFALO
ST-AC-(LETTERS)-	NATIONAL AERONAUTICS AND SPACE ADM., WASHINGTON, D.C. (SCIENTIFIC TRANS.)(CATEGORY DESIGNATION)(SUBJECT)
ST-ACH-(LETTERS)-	NATIONAL AERONAUTICS AND SPACE ADM., WASHINGTON, D.C. (SCIENTIFIC TRANS.)(CATEGORY DESIGNATION)(SUBJECT)
ST-AD-(LETTERS)-	NATIONAL AERONAUTICS AND SPACE ADM., WASHINGTON, D.C. (SCIENTIFIC TRANS.)(CATEGORY DESIGNATION)(SUBJECT)
STAG-M-	ARMY STRATEGY AND TACTICS ANALYSIS GP., BETHESDA, MD.
STAGTEN-	ARMY STRATEGY AND TACTICS ANALYSIS GP., BETHESDA, MD.
ST-AI-(LETTERS)-	NATIONAL AERONAUTICS AND SPACE ADM., WASHINGTON, D.C. (SCIENTIFIC TRANS.)(CATEGORY DESIGNATION)(SUBJECT)
ST-AM-(LETTERS)-	NATIONAL AERONAUTICS AND SPACE ADM., WASHINGTON, D.C. (SCIENTIFIC TRANS.)(CATEGORY DESIGNATION)(SUBJECT)
STAN-	STANFORD UNIV., CALIF.
STANAG-	NORTH ATLANTIC TREATY ORGANIZATION. MILITARY AGENCY FOR STANDARDIZATION, PARIS
STAN-CS-	STANFORD UNIV., CALIF. DEPT. OF COMPUTER SCIENCE
STAN-CS-(YEAR)-	STANFORD UNIV., CALIF. DEPT. OF COMPUTER SCIENCE
STAN ML-	STANFORD UNIV., CALIF. MICROWAVE LAB.
STAN SEL TR-(NO.)-	STANFORD UNIV., CALIF. SOLID-STATE ELECTRONICS LAB.
STAN. U./SEL-	STANFORD UNIV., CALIF. STANFORD ELECTRONICS LABS.
STAN. U. TR-	STANFORD UNIV., CALIF. (TECHNICAL REPORTS)
STAR-	NATIONAL AERONAUTICS AND SPACE ADMINISTRATION, WASHINGTON, D.C. (SCIENTIFIC AND TECHNICAL AEROSPACE REPORTS)
STAR-	NAVAL ORDNANCE LAB., WHITE OAK, MD. (STANDARD TEST AUTHORIZATION AND REPORT SYSTEM)
STAR-C-	EMERSON ELECTRIC CO. STAR DEPT., SANTA BARBARA, CAL.
ST-ASD-(LETTERS)-	NATIONAL AERONAUTICS AND SPACE ADM., WASHINGTON, D.C. (SCIENTIFIC TRANS.)(CATEGORY DESIGNATION)(SUBJECT)
STAT.-	ROCHESTER, N.Y. UNIV.
STB-	NAVAL PERSONNEL RESEARCH ACTIVITY, SAN DIEGO, CALIF. (TECHNICAL BULLETIN)
STB-(YEAR)-	NAVAL PERSONNEL RESEARCH ACTIVITY, SAN DIEGO, CALIF.
STC-	NORTH ATLANTIC TREATY ORGANIZATION. SHAPE TECHNICAL CENTRE, THE HAGUE
STC-	SYSTEMS TECHNOLOGY CORP., DALLAS
STC-CR-NSPM-	NORTH ATLANTIC TREATY ORGANIZATION. SHAPE TECHNICAL CENTER, THE HAGUE
ST-CM-(LETTERS)-	NATIONAL AERONAUTICS AND SPACE ADM., WASHINGTON, D.C. (SCIENTIFIC TRANS.)(CATEGORY DESIGNATION)(SUBJECT)
STCO-	AIR TECHNICAL SERVICE COMMAND, WRIGHT-PATTERSON AFB, OHIO (SUPERVISOR TRAINING CONFERENCE OUTLINE)
ST-COM-(LETTERS)-	NATIONAL AERONAUTICS AND SPACE ADM., WASHINGTON, D.C. (SCIENTIFIC TRANS.)(CATEGORY DESIGNATION)(SUBJECT)
ST-CR-(LETTERS)-	NATIONAL AERONAUTICS AND SPACE ADM., WASHINGTON, D.C. (SCIENTIFIC TRANS.)(CATEGORY DESIGNATION)(SUBJECT)
ST-CS-	ARMY MISSILE COMMAND. MISSILE INTELLIGENCE DIRECTORATE, REDSTONE ARSENAL, ALA.
STC-TR-	NORTH ATLANTIC TREATY ORGANIZATION. SHAPE TECHNICAL CENTER, THE HAGUE
STD-	STD RESEARCH CORP., PASADENA, CALIF.
STD-(YEAR)-	STD RESEARCH CORP., PASADENA, CALIF.
S/TD-	SUNDSTRAND TURBO.DIV.OF SUNDSTRAND CORP.,ROCKFORD,ILL
STD-	SUNDSTRAND TURBO DIV., SUNDSTRAND MACHINE TOOL CO., ROCKFORD, ILL.
STD-	WHITE SANDS MISSILE RANGE. SYSTEMS TEST DIV., N. MEX.
ST-EDN-(LETTERS)-	NATIONAL AERONAUTICS AND SPACE ADM., WASHINGTON, D.C. (SCIENTIFIC TRANS.)(CATEGORY DESIGNATION)(SUBJECT)
STELLENBOSCHU-	STELLENBOSCH. UNIV., CAPE PROVINCE, SOUTH AFRICA
ST-EMWP-(LETTERS)-	NATIONAL AERONAUTICS AND SPACE ADM., WASHINGTON, D.C. (SCIENTIFIC TRANS.)(CATEGORY DESIGNATION)(SUBJECT)
ST-ES-(LETTERS)-	NATIONAL AERONAUTICS AND SPACE ADM., WASHINGTON, D.C. (SCIENTIFIC TRANS.)(CATEGORY DESIGNATION)(SUBJECT)
STEWS-ID-	WHITE SANDS MISSILE RANGE, N. MEX.
STEWS-ID-	WHITE SANDS MISSILE RANGE. INSTRUMENTATION R/D DIRECTORATE, N. MEX.
STEWS-RE-I-(YR.)-I	WHITE SANDS MISSILE RANGE. NATIONAL RANGE ENG., N. MEX.
ST-EXP-(LETTERS)-	NATIONAL AERONAUTICS AND SPACE ADM., WASHINGTON, D.C. (SCIENTIFIC TRANS.)(CATEGORY DESIGNATION)(SUBJECT)
STFB-(YR.)-	SANDIA CORP. TECH. FACILITIES BD., ALBUQUERQUE, N.M.
STF/SEF/SVM/SKU-	U.S.S.R. INSTITUT FIZIKI VYSOKIKH ENERGII, SERPUKHOV
STG-	SCHIFFBAUTECHNISCHE GESELLSCHAFT
ST-GC-(LETTERS)-	NATIONAL AERONAUTICS AND SPACE ADM., WASHINGTON, D.C. (SCIENTIFIC TRANS.)(CATEGORY DESIGNATION)(SUBJECT)
ST-GEO-(LETTERS)-	NATIONAL AERONAUTICS AND SPACE ADM., WASHINGTON, D.C. (SCIENTIFIC TRANS.)(CATEGORY DESIGNATION)(SUBJECT)
ST-GM-(LETTERS)-	NATIONAL AERONAUTICS AND SPACE ADM., WASHINGTON, D.C. (SCIENTIFIC TRANS.)(CATEGORY DESIGNATION)(SUBJECT)
STH-	STUTTGART. TECHNISCHE HOCHSCHULE
STI-	SARKES TARZIAN, INC., BLOOMINGTON, IND.
STI-	SYSTEMS TECHNOLOGY, INC., HAWTHORNE, CALIF.
STI-DOC/	INTERNATIONAL ATOMIC ENERGY AGENCY, VIENNA
STI/DOC/10-	INTERNATIONAL ATOMIC ENERGY AGENCY, VIENNA
ST-IGA-(LETTERS)-	NATIONAL AERONAUTICS AND SPACE ADM., WASHINGTON, D.C. (SCIENTIFIC TRANS.)(CATEGORY DESIGNATION)(SUBJECT)
ST-IM-(LETTERS)-	NATIONAL AERONAUTICS AND SPACE ADM., WASHINGTON, D.C. (SCIENTIFIC TRANS.)(CATEGORY DESIGNATION)(SUBJECT)
ST-IMF-(LETTERS)-	NATIONAL AERONAUTICS AND SPACE ADM., WASHINGTON, D.C. (SCIENTIFIC TRANS.)(CATEGORY DESIGNATION)(SUBJECT)
STI/PUB/	INTERNATIONAL ATOMIC ENERGY AGENCY, VIENNA
STI-REP-	INTERNATIONAL ATOMIC ENERGY AGENCY, VIENNA
ST-IS-(LETTERS)-	NATIONAL AERONAUTICS AND SPACE ADM., WASHINGTON, D.C. (SCIENTIFIC TRANS.)(CATEGORY DESIGNATION)(SUBJECT)
STI-TM-(NUMBER)-I-	SYSTEMS TECHNOLOGY, INC., HAWTHORNE, CALIF.
STI-TM-(NUMBER)-A	SYSTEMS TECHNOLOGY, INC., HAWTHORNE, CALIF.
STI-TR-	SYSTEMS TECHNOLOGY, INC., INGLEWOOD, CALIF.
STL-	MAGNA CORP., ANAHEIM, CALIF.
STL-	SANDIA CORP., ALBUQUERQUE, N. MEX. (TITLE LISTS)
STL-(NO.-NO.-LTRS.)-000	SPACE TECHNOLOGY LABS., INC., LOS ANGELES (REF. 42)
STL-	TRW SYSTEMS, REDONDO BEACH, CALIF.
STL-(NO.-NO.-LTRS.)-	TRW SYSTEMS, REDONDO BEACH, CALIF. (REF. 42)
STL-AB-(YEAR)-	SPACE TECHNOLOGY LABS., INC., LOS ANGELES
STL-B-	SPACE TECHNOLOGY LABS., INC., LOS ANGELES
STL-GM-(YR.)-(NO.)-	SPACE TECHNOLOGY LABS., INC., LOS ANGELES
STL GM PTM-(NUMBER)-	SPACE TECHNOLOGY LABS., INC., LOS ANGELES
STL-GM-TN-	SPACE TECHNOLOGY LABS., INC., LOS ANGELES
STL-GM-TR-	SPACE TECHNOLOGY LABS., INC., LOS ANGELES
STL-IBM-V-	RAMO-WOOLDRIDGE CORP. SPACE TECHNOLOGY LABS., LOS ANGELES
STL-II-	MC DONNELL ASTRONAUTICS CO., ST. LOUIS
STLL-	STANDARD TELECOMMUNICATION LABS., HARLOW, ESSEX, ENG.
STL-LN-	SPACE TECHNOLOGY LABS., INC., LOS ANGELES
STL-(MISSILE NAME)-DTO-	SPACE TECHNOLOGY LABS., INC., LOS ANGELES (DETAILED TEST OBJECTIVE)
ST-LMP-(LETTERS)-	NATIONAL AERONAUTICS AND SPACE ADM., WASHINGTON, D.C. (SCIENTIFIC TRANS.)(CATEGORY DESIGNATION)(SUBJECT)
STL-NN-	SPACE TECHNOLOGY LABS., INC., LOS ANGELES

```
STL-OR-
         SPACE TECHNOLOGY LABS., INC., LOS ANGELES
STL-PA-(NO.)-(NO.)
         SPACE TECHNOLOGY LABS., INC., LOS ANGELES
STL-PRL-
         SPACE TECHNOLOGY LABS., INC. PHYSICAL RESEARCH LAB.,
            LOS ANGELES
STL-PRL-
         SPACE TECHNOLOGY LABS., INC. PHYSICAL RESEARCH LAB.,
            REDONDO BEACH, CALIF.
ST-LPS-(LETTERS)-
         NATIONAL AERONAUTICS AND SPACE ADM., WASHINGTON, D.C.
            (SCIENTIFIC TRANS.)(CATEGORY DESIGNATION)(SUBJECT)
STL-QPSR-
         SWEDEN. KUNGLIGA TEKNISKA HOEGSKOLAN, STOCKHOLM.
            TALTRANSMISSIONS-LABORATORIET
STL R-
         AEROJET-GENERAL CORP., SACRAMENTO, CALIF.
STL/TM-(YEAR)-
         SPACE TECHNOLOGY LABS., INC., LOS ANGELES
STL-TN-(YR.)-(NO.)-
         SPACE TECHNOLOGY LABS., INC., LOS ANGELES
STL-TP-
         SPACE TECHNOLOGY LABS., INC., LOS ANGELES
STL-TR-
         SPACE TECHNOLOGY LABS., INC., LOS ANGELES
STL-TR-(YR.)-(NO.)-
         SPACE TECHNOLOGY LABS., INC., LOS ANGELES
STL-TR-
         SPACE TECHNOLOGY LABS., INC., REDONDO BEACH, CALIF.
STM-
         AEROJET-GENERAL CORP. LIQUID ROCKET PLANT, SACRA-
            MENTO, CALIF.
STM-
         AEROJET-GENERAL CORP. SOLID ROCKET PLANT, SACRA-
            MENTO, CALIF. (TECHNICAL MANUAL)
STM.-
         SPERRY MICROWAVE ELECTRONICS CO., CLEARWATER, FLA.
ST-MAT-(LETTERS)-
         NATIONAL AERONAUTICS AND SPACE ADM., WASHINGTON, D.C.
            (SCIENTIFIC TRANS.)(CATEGORY DESIGNATION)(SUBJECT)
S + T-MEMO-
         BRITISH AIRCRAFT CORP., LUTON, BEDS, ENGLAND
S + T-MEMO-
         GT. BRIT. DEPT. OF SCIENTIFIC AND INDUSTRIAL RESEARCH
S+T-MEMO-
         GT. BRIT. MINISTRY OF AVIATION, NOTTINGHAM, ENGLAND
S-T MEMO-
         GT. BRIT. MINISTRY OF SUPPLY. TECHNICAL INFO. BUR.
S + T-MEMO-
         HAWKER SIDDELEY AVIATION, LTD., LONDON
S+T-MEMO-(MO.)/(YR.)
         NORTHAMPTON COLL. OF ADVANCED TECHNOLOGY, LONDON
S/T-MEMO-(NUMBER/YEAR)
         UNITED KINGDOM. MINISTRY OF TECHNOLOGY, LONDON
ST-MHD-(LETTERS)-
         NATIONAL AERONAUTICS AND SPACE ADM., WASHINGTON, D.C.
            (SCIENTIFIC TRANS.)(CATEGORY DESIGNATION)(SUBJECT)
ST-NP-(LETTERS)-
         NATIONAL AERONAUTICS AND SPACE ADM., WASHINGTON, D.C.
            (SCIENTIFIC TRANS.)(CATEGORY DESIGNATION)(SUBJECT)
STO-
         DYNATECH CORP., CAMBRIDGE, MASS.
ST-OA-(LETTERS)-
         NATIONAL AERONAUTICS AND SPACE ADM., WASHINGTON, D.C.
            (SCIENTIFIC TRANS.)(CATEGORY DESIGNATION)(SUBJECT)
STOLLER-
         STOLLER (S.M.) ASSOCIATES, N.Y.C.
ST-PA-(LETTERS)-
         NATIONAL AERONAUTICS AND SPACE ADM., WASHINGTON, D.C.
            (SCIENTIFIC TRANS.)(CATEGORY DESIGNATION)(SUBJECT)
ST-PF-(LETTERS)-
         NATIONAL AERONAUTICS AND SPACE ADM., WASHINGTON, D.C.
            (SCIENTIFIC TRANS.)(CATEGORY DESIGNATION)(SUBJECT)
ST-PHM-(LETTERS)-
         NATIONAL AERONAUTICS AND SPACE ADM., WASHINGTON, D.C.
            (SCIENTIFIC TRANS.)(CATEGORY DESIGNATION)(SUBJECT)
ST-PP-(LETTERS)-
         NATIONAL AERONAUTICS AND SPACE ADM., WASHINGTON, D.C.
            (SCIENTIFIC TRANS.)(CATEGORY DESIGNATION)(SUBJECT)
STPR-
         (SEMI-ANNUAL TECHNICAL PROGRESS REPORT) (REF. 2)
ST-PR-(LETTERS)-
         NATIONAL AERONAUTICS AND SPACE ADM., WASHINGTON, D.C.
            (SCIENTIFIC TRANS.)(CATEGORY DESIGNATION)(SUBJECT)
ST-PS-(LETTERS)-
         NATIONAL AERONAUTICS AND SPACE ADM., WASHINGTON, D.C.
            (SCIENTIFIC TRANS.)(CATEGORY DESIGNATION)(SUBJECT)
STR-
         (SCIENTIFIC TECHNICAL REPORT) (REF. 2)
STR-
         (SUMMARY TECHNICAL REPORT) (REF. 2)
STR-
         AEROJET-GENERAL CORP., SACRAMENTO, CALIF.
STR-
         GENERAL TELEPHONE AND ELECTRONICS LABS. INC. BAY-
            SIDE LABS., N.Y.
STR-
         GT. BRIT. ADMIRALTY TORPEDO EXPTL. EST.,GREENOCK,SCOT
STR-
         MASSACHUSETTS INST. OF TECH., CAMBRIDGE. RESEARCH
            LAB. OF ELECTRONICS
STR-
         NATIONAL BUREAU OF STANDARDS, WASHINGTON, D.C.
STR-
         NATIONAL DEFENSE RESEARCH COMMITTEE
STR-
         OFFICE OF SCIENTIFIC RES. + DEV.(SUMMARY TECH. REPT.)
STR-
         STANFORD RESEARCH INST., MENLO PARK, CALIF.
STR-(YEAR)-
         SYLVANIA ELECTRIC PRODUCTS, INC. HIGH TEMPERATURE
            COMPOSITES LAB., HICKVILLE, N.Y.
STR-
         SYLVANIA ELECTRIC PRODUCTS INC. SYLCOR DIV., BAY-
            SIDE, N.Y.
STR-(YEAR)-(NO.).(NO.)
         SYLVANIA ELECTRIC PRODUCTS INC. SYLCOR DIV., HICKS-
            VILLE, N.Y.
STR-(LETTERS)-
         NATIONAL DEFENSE RESEARCH COMM. (SUMMARY TECH. RPT.)
ST-RA-(LETTERS)-
         NATIONAL AERONAUTICS AND SPACE ADM., WASHINGTON, D.C.
            (SCIENTIFIC TRANS.)(CATEGORY DESIGNATION)(SUBJECT)
STRATCOM-(NOS.)-(LTR.)
         ARMY STRATEGIC COMMUNICATIONS COMMAND, FORT HUACHUCA,
            ARIZ.
STR-DIV.-(1 THRU 19)-
         NATIONAL DEFENSE RESEARCH COMMITTEE (SUMMARY TECH-
            NICAL REPORT) (REF. 27)
ST-RDS-
         CHEMICAL CORPS SCHOOL, ARMY CHEMICAL CENTER, MD.
STRING PROGRAM-
         NEW YORK UNIV., N.Y.C. (LINGUISTIC STRING PROJECT)
STR-IR-
         WESTINGHOUSE ELECTRIC CORP. ATOMIC POWER DIV.,
            IDAHO FALLS
STRM-
         FAIRCHILD ENGINE AND AIRPLANE CORP. NEPA DIV., OAK
            RIDGE, TENN. (REF. 31)
STR-MEMO-
         ARGONNE NATIONAL LAB., LEMONT, ILL.
STRP-
         BUREAU OF NAVAL PERSONNEL (SELECTION TEST REPORT
            PROJECTS)
STR-TOT-
         WESTINGHOUSE ELECTRIC CORP. ATOMIC POWER DIV.,
            IDAHO FALLS
STRUCTURAL RESEARCH SER-
         ILLINOIS. UNIV., URBANA. DEPT. OF CIVIL ENGINEERING
            (STRUCTURAL RESEARCH SERIES)
STRUT.-
         GT. BRIT. AERONAUTICAL RESEARCH COUNCIL, LONDON
STS-
         BABCOCK + WILCOX CO., GALVESTON, TEX.
STS-
         CLEARINGHOUSE FOR FEDERAL SCIENTIFIC AND TECHNICAL
            INFORMATION, SPRINGFIELD, VA.(STATE TECH. SERVICES)
STS-
         SCIENCE TRANSLATION SERVICE
STS-
         SUPERSONIC TRACK SYMPOSIUM
ST-SB-(LETTERS)-
         NATIONAL AERONAUTICS AND SPACE ADM., WASHINGTON, D.C.
            (SCIENTIFIC TRANS.)(CATEGORY DESIGNATION)(SUBJECT)
ST-SP-(LETTERS)-
         NATIONAL AERONAUTICS AND SPACE ADM., WASHINGTON, D.C.
            (SCIENTIFIC TRANS.)(CATEGORY DESIGNATION)(SUBJECT)
ST-SPC-(LETTERS)-
         NATIONAL AERONAUTICS AND SPACE ADM., WASHINGTON, D.C.
            (SCIENTIFIC TRANS.)(CATEGORY DESIGNATION)(SUBJECT)
ST-STR-(LETTERS)-
         NATIONAL AERONAUTICS AND SPACE ADM., WASHINGTON, D.C.
            (SCIENTIFIC TRANS.)(CATEGORY DESIGNATION)(SUBJECT)
STT-
         BRITISH INTELLIGENCE OBJECTIVES SUB-COMMITTEE.
            MILITARY COLL. OF SCIENCE. SCH. OF TANK TECHNOLOGY
STU-(YEAR)-(NUMBER)/U-
         AKTIEBOLAGET ATOMENERGI, STUDSVIK, SWEDEN
STUDY-
         ARMY LOGISTICS MANAGEMENT CENTER, FORT LEE, VA.
STUDY-S-
         INSTITUTE FOR DEFENSE ANALYSES, ARLINGTON, VA.
ST-WP-(LETTERS)-
         NATIONAL AERONAUTICS AND SPACE ADM., WASHINGTON, D.C.
            (SCIENTIFIC TRANS.)(CATEGORY DESIGNATION)(SUBJECT)
SU-
         GOODYEAR AEROSPACE CORP., LITCHFIELD PARK, ARIZ.
SU-
         PENNSYLVANIA. UNIV., PHILADELPHIA. INSTITUTE
            FOR COOPERATIVE RESEARCH (PROJECT SUMMIT)
SU(YEAR)(LETTERS)-
         PENNSYLVANIA. UNIV.,PHILA. INST. FOR COOP. RES.
SU/
         PITTSBURGH. UNIV.
SU-
         RESEARCH TRIANGLE INST., DURHAM, N.C.
SU-
         STANFORD UNIV., CALIF.
SU-(NO.)P(NO.)-
         STANFORD UNIV., CALIF.
SU-(NUMBER)-
         STANFORD UNIV., CALIF.
SU-(NUMBER)-
         SYRACUSE UNIV., N.Y.
SU-1206-
         SYRACUSE UNIV., N.Y.
SU-AA(NUMBER)
         STANFORD UNIV., CALIF. INST. FOR PLASMA RESEARCH
SU-AHT-
         STANFORD UNIV., CALIF. DEPT. OF MECHANICAL ENG.
SU-AMSL-
         STANFORD UNIV., CALIF. APPLIED MATH.+ STATISTICS LAB.
SU-AMSL-TR-
         STANFORD UNIV., CALIF. APPLIED MATH.+ STATISTICS LAB.
SU-AR-
         STANFORD UNIV., CALIF. DEPT. OF AERONAUTICS AND
            ASTRONAUTICS
SUBDEVGRUTWO-(NO.-YR.)
         RAYTHEON CO. MARINE RESEARCH LAB., NEW LONDON, CONN.
```

SUBROC-WJCC-	SUBROC JT. AEC-DOD WARHEAD COORDINATING COMM.(REF.15)
SU-CC-(NUMBER)-(NUMBER)-	STANFORD UNIV., CALIF. COMPUTATION CENTER
SU-CMR-AR-	STANFORD UNIV., CALIF. CENTER FOR MATERIALS RESEARCH
SUCO-EJ-(NUMBER)-	SPERRY UTAH CO., SALT LAKE CITY
SU-CS-	STANFORD UNIV., CALIF. DEPT. OF COMPUTER SCIENCE
SU-DAAR-	STANFORD UNIV., CALIF. DEPT. OF AERONAUTICS AND ASTRONAUTICS
SUDAAR-	STANFORD UNIV., CALIF. DEPT. OF AERONAUTICS AND ASTRONAUTICS
SUDAAR-TR-	STANFORD UNIV., CALIF. DEPT. OF AERONAUTICS AND ASTRONAUTICS
SUDAER-	STANFORD UNIV., CALIF. DEPT. OF AERONAUTICS AND ASTRONAUTICS
SU-DCE-TR-	STANFORD UNIV., CALIF. DEPT. OF CIVIL ENGINEERING
SU-DEM-	STANFORD UNIV., CALIF. DIV. OF ENGINEERING MECHANICS
SU-DEM-TR-	STANFORD UNIV., CALIF. DIV. OF ENGINEERING MECHANICS
SU-DME-TR-	STANFORD UNIV., CALIF. DEPT. OF MECHANICAL ENG.
SU-DMS-(YEAR)-	STANFORD UNIV., CALIF. DEPT. OF MATERIALS SCIENCE
SU-DMS-(YEAR)-R-	STANFORD UNIV., CALIF. DEPT. OF MATERIALS SCIENCE
SU-DMS-(YEAR)-T-	STANFORD UNIV., CALIF. DEPT. OF MATERIALS SCIENCE
SU-EDL-TR-	STANFORD UNIV., CALIF. ELECTRON DEVICES LAB.
SUEL-	SPERRY UTAH ENGINEERING LAB., SALT LAKE CITY
SU-EL-	STANFORD UNIV., CALIF. STANFORD ELECTRONICS LABS.
SUEL-EJ-	SPERRY UTAH ENGINEERING LAB., SALT LAKE CITY
SU-EL-TR-	STANFORD UNIV., CALIF. ELECTRONICS LAB.
SU ERL TR-	STANFORD UNIV., CALIF. ELECTRONICS RESEARCH LAB.
SU-ET-	STANFORD UNIV., CALIF. ELECTRON TUBE LAB.
SUFFIELD-	CANADA. SUFFIELD EXPERIMENTAL STATION, RALSTON, ALBERTA
SUFFIELD FIELD EXPERIMENT	CANADA. SUFFIELD EXPERIMENTAL STATION, RALSTON, ALBERTA
SUFFIELD-MEMO-	CANADA. SUFFIELD EXPERIMENTAL STATION, RALSTON, ALBERTA
SUFFIELD REPT. NO.-	CANADA. SUFFIELD EXPERIMENTAL STATION, RALSTON, ALBERTA
SUFFIELD-SP-	CANADA. SUFFIELD EXPERIMENTAL STATION, RALSTON, ALBERTA
SUFFIELD TECHNICAL PAPER	CANADA. SUFFIELD EXPERIMENTAL STATION, RALSTON, ALBERTA
SUFFIELD-TN-	CANADA. SUFFIELD EXPERIMENTAL STATION, RALSTON, ALBERTA
SUFFIELD TP-	CANADA. SUFFIELD EXPERIMENTAL STATION, RALSTON, ALBERTA
SU-HMT-	STANFORD UNIV., CALIF. THERMOSCIENCES DIV.
SUHP-	STANFORD UNIV., CALIF. HEALTH PHYSICS
SUI-	IOWA. STATE UNIV., IOWA CITY
SUI-(YR)-	IOWA. STATE UNIV., IOWA CITY
SU-IPR-	STANFORD UNIV., CALIF. INST. FOR PLASMA RESEARCH
SUIPR-	STANFORD UNIV., CALIF. INST. FOR PLASMA RESEARCH
SU-IRL-	STANFORD UNIV., CALIF. INSTRUMENTATION RES. LAB.
SUIRP-	STANFORD UNIV., CALIF. PLASMA PHYSICS LAB.
SUL-	KERNFORSCHUNGSANLAGE JUELICH, GERMANY
SUM-	(SUMMARY) (REF. 2)
SUM-	CANADA. DEFENCE RESEARCH BOARD. SCI. INFO. SERVICE
SU-MD-	STANFORD UNIV., CALIF. DEPT. OF MECHANICAL ENG.
SU ME TR-	STANFORD UNIV., CALIF. DEPT. OF MECHANICAL ENG.
SU-ML-	STANFORD UNIV., CALIF. MICROWAVE LAB.
SUMP.-P(MONTH/YEAR)	CALIFORNIA. UNIV., BERKELEY. RADIATION LAB. (REF. 9)
SUMP. SPEC-	CALIFORNIA. UNIV., BERKELEY. RADIATION LAB. (REF. 9)
SUN-	SUNFLOWER ORDNANCE WORKS, LAWRENCE, KANS.
SUNI-	SOUTHERN UNIVERSITIES NUCLEAR INST., FAURE, SOUTH AFRICA
SU-NTL-TR-(NUMBER)-	STANFORD UNIV., CALIF. NUCLEAR TECHNOLOGY LAB.
SUNYBER-	NEW YORK. STATE UNIV., BUFFALO. FACULTY OF ENGINEERING AND APPLIED SCIENCES
SUP-	NAVAL SUPPLY SYSTEMS COMMAND, WASHINGTON, D.C.
SUPAIRCO-	SUPERIOR AIR PRODUCTS, INC., NEWARK, N.J.
SUPDIVE-RR-(NUMBER-YEAR)	BATTELLE MEMORIAL INST., COLUMBUS, OHIO
SUP-M-	ATOMIC ENERGY OF CANADA LTD., CHALK RIVER, ONT.
SU-PPL-	SASKATCHEWAN. UNIV., SASKATOON. PLASMA PHYSICS LAB.
SUPR-	STANFORD UNIV., CALIF. INST. FOR PLASMA RESEARCH
SU-QSR-	STANFORD UNIV., CALIF.
SURASDEVDET-	SURFACE ANTI-SUBMARINE DEV. DETACHMENT, KEY WEST, FLA
SURC-	SYRACUSE UNIV., N.Y. RESEARCH CORP.
SURC-CSL-	SYRACUSE UNIV., N.Y. RESEARCH CORP.
SURC-TR-	SYRACUSE UNIV., N.Y. RESEARCH CORP.
SURI-	SYRACUSE UNIV., N.Y. RESEARCH INST.
SURI-CE-	SYRACUSE UNIV., N.Y. RESEARCH INST.
SURI-CH.E.-	SYRACUSE UNIV., N.Y. RESEARCH INST.
SURI-CHEM.-	SYRACUSE UNIV., N.Y. RESEARCH INST.
SURI-EE-(NUMBER)-	SYRACUSE UNIV., N.Y. RES. INST. ELEC. ENG. DEPT.
SURI-ME-(NUMBER)-	SYRACUSE UNIV., N.Y. RESEARCH INST. DEPT. OF MECHANICAL AND AEROSPACE ENGINEERING
SURI-MET-(NUMBER)-	SYRACUSE UNIV., N.Y. RES. INST. MET. RES. LABS.
SURI-PHYSICS-(NUMBER)-	SYRACUSE UNIV., N.Y. RESEARCH INST.
SU-RSL-TECH-(YEAR)-	STANFORD UNIV., CALIF. REMOTE SENSING LAB.
SU-RSL-TR-(YEAR)-	STANFORD UNIV., CALIF. REMOTE SENSING LAB.
SURVEY IND-(NO.)/(NO.)	WATERTOWN ARSENAL, MASS.
SUS-	COAST AND GEODETIC SURVEY, WASHINGTON, D.C. (SECTIONAL CHARTS, UNITED STATES)
SU-SEL-(YEAR)-	STANFORD UNIV., CALIF. STANFORD ELECTRONICS LABS.
SU-SEL-TR-(NUMBER)-	STANFORD UNIV., CALIF. STANFORD ELECTRONICS LABS.
SU-SLAC-PUB-	STANFORD UNIV., CALIF. STANFORD LINEAR ACCELERATOR CENTER
SU-SRI-	STANFORD RESEARCH INST., MENLO PARK, CALIF.
SU-SSEL-TR-	STANFORD UNIV., CALIF. SOLID-STATE ELECTRONICS LAB.
SUSSEXU-	SUSSEX. UNIV., FALMER, BRIGHTON, ENGLAND
SU-STAN-CS-(YEAR)-	STANFORD UNIV., CALIF. DEPT. OF COMPUTER SCIENCE
SU-STL-TR-	STANFORD UNIV., CALIF. SYSTEMS TECHNIQUES LAB.
SU-TPR-	STANFORD UNIV., CALIF. INST. FOR PLASMA RESEARCH
SU-TR-	STANFORD UNIV., CALIF.
SUU-(NUMBER)A/A	SPECIAL AIR WARFARE CENTER. COMBAT APPLICATIONS GP., EGLIN AFB, FLA.
SU-VRL-	STANFORD UNIV., CALIF. VIBRATION RES. LAB.
SV-(YEAR)-(NUMBERS)	AEROSPACE CORP. VANDENBERG OFFICE
SV-(YEAR)-	BENDIX CORP. BENDIX PRODUCTS AEROSPACE DIV., SOUTH BEND, IND.
SV-	SWEDISH STATE POWER BOARD, STOCKHOLM
SVC-	SAINT VINCENT COLLEGE, LATROBE, PA.
SVETM-(NUMBER)-(NUMBER)	GENERAL ELECTRIC CO. SPACE SCIENCES LAB., PHILADELPHIA (SPACE VEHICLE ENGINEERING TECHNICAL MEMO)
SVHSER-	HAMILTON STANDARD, WINDSOR LOCKS, CONN.
SVLTE-	GT. BRIT. SERVICES VALVE LIFE TEST ESTABLISHMENT
SVM-	BOLT, BERANEK, AND NEWMAN, INC., CAMBRIDGE, MASS.
SVT-	SERVICES VALVE TEST LAB., HASLEMERE, ENGLAND
SW-	ENVIRONMENTAL PROTECTION AGENCY. SOLID WASTE OFFICE, ROCKVILLE, MD.
SW-	GT. BRIT. MINISTRY OF HOME SECURITY
SW-	GT. BRIT. MINISTRY OF HOME SECURITY. INTERDEPARTMENTAL COORDINATING COMMITTEE ON SHOCKWAVES
SW-	OFFICE OF NAVAL RESEARCH (SUBMARINE OPERATIONS)
SWB-	ARMED FORCES SPECIAL WEAPONS PROJECT, WASHINGTON, D.C. (SPECIAL WEAPONS BULLETIN)
SWB-(NO.)-	ARMED FORCES SPECIAL WEAPONS PROJECT, WASHINGTON, D.C.
SWC-	AIR FORCE SPECIAL WEAPONS CENTER, KIRTLAND AFB, N.M.

SWC-(NUMBER)C-	AIR FORCE SPECIAL WEAPONS CENTER, KIRTLAND AFB, N.M.
SWC-(NO.)S-	AIR FORCE SPECIAL WEAPONS CENTER, KIRTLAND AFB, N.M.
SWC-	SPECIAL WEAPONS COMMAND, KIRTLAND AFB, N. MEX.
SWC-	STEWART-WARNER CORP., CHICAGO
SWC-	SWARTHMORE COLL., PENNA.
SWC FTR (YR.)-	AIR FORCE SPECIAL WEAPONS CENTER, KIRTLAND AFB, N.M. (FUNCTIONAL TEST REPORTS)
SWC JFSG (LETTERS)	(JOINT FEASIBILITY STUDY GROUP) (REF. 16)
SWCL-	CALIFORNIA. UNIV., BERKELEY. SEA WATER CONVERSION LAB
SWCL-	CALIFORNIA. UNIV., RICHMOND. SEA WATER CONVERSION LAB
SWC MR WCWGTO2-(YR.)-	AIR FORCE SPECIAL WEAPONS CENTER, KIRTLAND AFB, N.M.
SWC OA WP-	AIR FORCE SPECIAL WEAPONS CENTER. KIRTLAND AFB, N. MEX. (OPERATIONS ANALYSIS. WORKING PAPER)
SWC OAWP-	AIR FORCE SPECIAL WEAPONS CENTER, KIRTLAND AFB, N. MEX. (OPERATIONS ANALYSIS. WORKING PAPER)
SWC OP (NO.)-(YR.)	AIR FORCE SPECIAL WEAPONS CENTER, KIRTLAND AFB, N. MEX. (OPERATIONS ANALYSIS. WORKING PAPER)
SWC-OPLAN-(NUMBER-YEAR)	AIR FORCE SPECIAL WEAPONS CENTER, KIRTLAND AFB, N.M.
SWC-OS-	AIR FORCE SPECIAL WEAPONS CENTER, KIRTLAND AFB, N.M.
SWCP-AR-	DEPARTMENT OF THE INTERIOR, WASHINGTON, D.C. (SALINE WATER CONVERSION)
SWCP-PR-	DEPARTMENT OF THE INTERIOR, WASHINGTON, D.C. (SALINE WATER CONVERSION)
SWC-PROJ.NO.WCO 51-	AIR FORCE SPECIAL WEAPONS CENTER, KIRTLAND AFB, N.M.
SWC PROJ.NO.WCWGTO2-(YR)	AIR FORCE SPECIAL WEAPONS CENTER, KIRTLAND AFB, N.M.
SWCRD-AR-	DEPARTMENT OF AGRICULTURE. SOIL AND WATER CONSERVATION RESEARCH DIV., BELTSVILLE, MD.
SWC SWOTO (YR.)-	AIR FORCE SPECIAL WEAPONS CENTER. 4925TH TEST GROUP (ATOMIC), KIRTLAND AFB, N. MEX.
SWC SWR TM (YR.)-	AIR FORCE SPECIAL WEAPONS CENTER. RESEARCH DIRECTORATE, KIRTLAND AFB, N. MEX. (TECHNICAL MEMORANDUM)
SWC SWV MP-	AIR FORCE SPECIAL WEAPONS CENTER. DEVELOPMENT DIRECTORATE, KIRTLAND AFB, N. MEX. (MISC. PUBLICATION)
SWC SWVN (YR.)-	AIR FORCE SPECIAL WEAPONS CENTER, KIRTLAND AFB, N. MEX. (SAFETY STUDIES)
SWC SWV STS-	AIR FORCE SPECIAL WEAPONS CENTER. DEV. DIRECTORATE, KIRTLAND AFB, N. MEX.(STOCKPILE-TO-TARGET SEQUENCE)
SWC SWV TM (YR.)-	AIR FORCE SPECIAL WEAPONS CENTER. DEVELOPMENT DIRECTORATE, KIRTLAND AFB, N. MEX. (TECHNICAL MEMO.)
SWC-TDR-	AIR FORCE SPECIAL WEAPONS CENTER, KIRTLAND AFB, N. MEX. (TECHNICAL DOCUMENTARY REPORT)
SWC-TDR-(NUMBER)-	AIR FORCE SPECIAL WEAPONS CENTER, KIRTLAND AFB, N. MEX. (TECHNICAL DOCUMENTARY REPORT)
SWC TMR SWV (YR.)-	AIR FORCE SPECIAL WEAPONS CENTER. DEVELOPMENT DIRECTORATE, KIRTLAND AFB, N. MEX. (TECH. MEMO. REPT.)
SWC TN-(YR.)-	AIR FORCE SPECIAL WEAPONS CENTER, KIRTLAND AFB, N. MEX. (TECHNICAL NOTE)
SWC TN SWR (YR.)-	AIR FORCE SPECIAL WEAPONS CENTER. RESEARCH DIRECTORATE, KIRTLAND AFB, N. MEX. (TECHNICAL NOTE)
SWC TN SWW (YR.)-	AIR FORCE SPECIAL WEAPONS CENTER. 4925TH TEST GROUP (ATOMIC), KIRTLAND AFB, N. MEX. (TECHNICAL NOTE)
SWC TR (YR.)-	AIR FORCE SPECIAL WEAPONS CENTER, KIRTLAND AFB, N. MEX. (TECHNICAL REPORT)
SWC-VOL. (NO.). NO.-	AIR FORCE SPECIAL WEAPONS CENTER, KIRTLAND AFB, N. MEX. (SELECTED RECENT DOCUMENTS)
SWC-WCO-	AIR FORCE SPECIAL WEAPONS CENTER, KIRTLAND AFB, N.M.
SWC WCO (YR.)-	AIR FORCE SPECIAL WEAPONS CENTER. 4925TH TEST GROUP (ATOMIC), KIRTLAND AFB, N. MEX.
SWDB-(YR.)-	SANDIA CORP. SPEC. WEAPONS DEV. BD., ALBUQUERQUE, N.M.
SWDG-	SPECIAL WEAPONS DEVELOPMENT GP., KIRTLAND AFB, N.MEX.
SWEL-	SANDIA CORP., ALBUQUERQUE, N. MEX. (SPECIAL WEAPONS EQUIPMENT LIST)
SWET-	(SPECIAL WEAPONS EFFECTS TESTS)
SWF-	NAVAL AIR SPEC. WEAPONS FACILITY, KIRTLAND AFB, N.MEX
SWIR-	ARMED FORCES SPECIAL WEAPONS PROJECT, WASHINGTON, D.C. (SPECIAL WEAPONS INSPECTION REPORT)
SWK-	WESTINGHOUSE ELECTRIC AND MFG. CO., PITTSBURGH
SWND-	STONE + WEBSTER ENGINEERING CORP., N.Y.C.
SW-NIKE-X-	STONE + WEBSTER ENGINEERING CORP., BOSTON
SWODO (YR.)-	AIR FORCE SPECIAL WEAPONS CENTER. 4925TH TEST GROUP (ATOMIC), KIRTLAND AFB, N. MEX.
SWOI-(NUMBER)-	AIR FORCE SPECIAL WEAPONS CENTER. TECHNICAL INFO. DIV., KIRTLAND AFB, N.M. (SELECTED RECENT DOCS.)
SWOP-	JOINT ATOMIC WEAPONS PUBLICATIONS BOARD, ALBUQUERQUE, N. MEX. (REF. 14)
SWOTO-	AIR FORCE SPECIAL WEAPONS CENTER. 4925TH TEST GROUP (ATOMIC), KIRTLAND AFB, N. MEX.
SWOTO (YR.)-	AIR FORCE SPECIAL WEAPONS CENTER. 4925TH TEST GROUP (ATOMIC), KIRTLAND AFB, N. MEX.
SWOTOP (YR.)-	AIR FORCE SPECIAL WEAPONS CENTER. 4925TH TEST GROUP (ATOMIC), KIRTLAND AFB, N. MEX.
SWOV (YEAR)-	NETHERLANDS. INST. FOR ROAD SAFETY RESEARCH, VOORBURG
SWP-	ARMED FORCES SPECIAL WEAPONS PROJECT, WASHINGTON,D.C.
SWPA-	AIR EVALUATION BOARD, SOUTHWEST PACIFIC AREA
SWPCD-	(OPERATION CROSSROADS) (REF. 7)
SWPDV-	ARMED FORCES SPECIAL WEAPONS PROJECT, WASHINGTON,D.C.
SWPFD-	ARMED FORCES SPECIAL WEAPONS PROJECT. WEAPONS EFFECTS DIV., WASH., D.C. (OPERATION WIGWAM) (REF. 7)
SWP/P-	UNITED KINGDOM ATOMIC ENERGY AUTHORITY. INDUSTRIAL GROUP H.Q., RISLEY, LANCS, ENGLAND
SWP/P-	UNITED KINGDOM ATOMIC ENERGY AUTHORITY. INDUSTRIAL GROUP. WINDSCALE WORKS, SELLAFIELD, CUMB., ENGLAND
SWP/P-	UNITED KINGDOM ATOMIC ENERGY AUTHORITY. RESEARCH GP. ATOMIC ENERGY RES. ESTAB., HARWELL, BERKS, ENGLAND
SWPWT-	ARMED FORCES SPECIAL WEAPONS PROJECT, WASHINGTON, D.C. (SPECIAL WEAPONS PROJECT WEAPON TESTS) (REF.7)
SWR-	AIR FORCE SPECIAL WEAPONS CENTER. RESEARCH DIRECTORATE, KIRTLAND AFB, N. MEX.
SWR-	ARMED FORCES SPECIAL WEAPONS PROJECT, WASHINGTON, D.C. (SPECIAL WEAPONS REGULATIONS)
SWRA-(YEAR)-	WATER RESOURCES SCIENTIFIC INFORMATION CENTER (INTERIOR), WASHINGTON, D.C. (SELECTED WATER RESOURCES ABSTRACTS)
SWRHL-	WESTERN ENVIRONMENTAL RESEARCH LAB., LAS VEGAS, NEV.
SWRHL-(NUMBER)R	WESTERN ENVIRONMENTAL RESEARCH LAB., LAS VEGAS, NEV.
SWRI-	SOUTHWEST RESEARCH INST., SAN ANTONIO
SWRI-(CONTRACT NO.)-P-	SOUTHWEST RESEARCH INST., SAN ANTONIO
SWRI-(CONTRACT NO.)-	SOUTHWEST RESEARCH INST., SAN ANTONIO
SWRI-(NUMBER)-(NUMBER)-	SOUTHWEST RESEARCH INST. DEPT. OF INSTRUMENTATION RESEARCH, SAN ANTONIO
SWRI-AR-	SOUTHWEST RESEARCH INST. DEPT. OF AUTOMOTIVE RES., SAN ANTONIO
SWRI-RS-	SOUTHWEST RESEARCH INST., SAN ANTONIO
SWRI-RS-	SOUTHWEST RESEARCH INST. DEPT. OF AEROSPACE PROPULSION RESEARCH, SAN ANTONIO
SWRI-TR-	SOUTHWEST RESEARCH INST. DEPT. OF MECHANICAL SCIENCES, SAN ANTONIO
SWRI/WRAIR/(YR)ANN	STERLING-WINTHROP RESEARCH INST., RENSSELAER, N.Y.
SWRO-	(SPECIAL WEAPONS RETROFIT ORDER)
SWR-TM-	AIR FORCE SPECIAL WEAPONS CENTER. RESEARCH DIRECTORATE, KIRTLAND AFB, N. MEX. (TECHNICAL MEMORANDUM)
SWR-TM-(YR.)-	AIR FORCE SPECIAL WEAPONS CENTER. RESEARCH DIRECTORATE, KIRTLAND AFB, N. MEX. (TECHNICAL MEMORANDUM)
SWR-TN-	AIR FORCE SPECIAL WEAPONS CENTER, KIRTLAND AFB, N.M.
SWSC-	SANDIA CORP., ALBUQUERQUE, N. MEX. (SPECIAL WEAPONS SUPPLY CATALOG)
SWT-	CALIF. INST. OF TECH., PASADENA. JET PROPULSION LAB.
SWT C/R-	GT. BRIT. SPRINGFIELDS WORKS, LANCS, ENGLAND
SWVCS-(NUMBER)-	AIR FORCE SPECIAL WEAPONS CENTER, KIRTLAND AFB, N.M.
SWVN-(NUMBER)-	AIR FORCE SPECIAL WEAPONS CENTER, KIRTLAND AFB, N.M.
SWWVD-	AIR FORCE SPECIAL WEAPONS CENTER, KIRTLAND AFB, N.M.
SX-	HIGH DUTY ALLOYS, LTD., SLOUGH, BUCKS, ENGLAND
SY-	GARRETT CORP. AIRESEARCH MFG. CO. OF ARIZONA, PHOENIX

```
SY-(NUMBER)-R
            GARRETT CORP. AIRESEARCH MFG. CO. OF ARIZONA, PHOENIX
SY-
            NAVAL AIR DEVELOPMENT CENTER. SYSTEMS PROJECT DEPT.,
                JOHNSVILLE, PENNA.
SYDC-
            SYSTEMS DEVELOPMENT CORP., DAYTON, OHIO
SYDNEY-ER.(NUMBER)
            SYDNEY. UNIV. SCHOOL OF PHYSICS
SYDNEY-PR.(NUMBER)
            SYDNEY. UNIV. SCHOOL OF PHYSICS
SYDNEY-TR.(NUMBER)
            SYDNEY. UNIV. SCHOOL OF PHYSICS
SYL-
            SYLVANIA ELECTRIC PRODUCTS INC., TOWANDA, PENNA.
SYLTT-
            SYLVANIA ELECTRIC PRODUCTS INC., TOWANDA, PENNA.
SY-R(NUMBER-YEAR)-
            ARMY WEAPONS COMMAND. SYSTEMS ANALYSIS DIRECTORATE,
                ROCK ISLAND, ILL.
SYRU-
            SYRACUSE UNIV., N.Y.
SYRU (LTRS.)-
            SYRACUSE UNIV., N.Y.
SYSTEM-SI-
            SYSTEM SCIENCES, INC., BETHESDA, MD.
SY-TN(NUMBER-YEAR)
            ARMY WEAPONS COMMAND. SYSTEMS ANALYSIS DIRECTORATE,
                ROCK ISLAND, ILL.
SYU-
            SYRACUSE UNIV., N.Y.
SYU-ME-
            SYRACUSE UNIV., N.Y. RESEARCH INST.
SYU-MET-
            SYRACUSE UNIV., N.Y. DEPT. OF CHEMICAL ENGINEERING
                AND METALLURGY
SZS-
            GERMANY. STAATLICHE ZENTRALE FUER STRAHLENSCHUTZ,
                BERLIN
SZS-(NUMBER)/(YEAR)
            GERMANY. STAATLICHE ZENTRALE FUER STRAHLENSCHUTZ,
                BERLIN
```

T-	
T-	(TECHNICAL MEMORANDUM) (REF. 2)
T-	(TECHNICAL PAPER) (REF. 2)
T-	(TECHNICAL REPORT) (REF. 2)
T-	(TEST REPORT) (REF. 2)
T-	(TRANSLATION) (REF. 2)
T(NUMBER)-TFC-	AEROJET NUCLEAR SYSTEMS CO., SACRAMENTO, CALIF.
T(YR.)-	AEROSPACE TECHNICAL INTELLIGENCE CENTER, WRIGHT-PATTERSON AFB, OHIO
T4-	ALLISON DIV., GENERAL MOTORS CORP., INDIANAPOLIS
T-	ARMY ENGINEER RESEARCH + DEV. LABS., FT. BELVOIR, VA.
T(NUMBER)-(NO.)/(NO.)	AUTONETICS, ANAHEIM, CALIF.
T6-	AUTONETICS DIV., NORTH AMERICAN AVIATION, INC., ANAHEIM, CALIF. (NON-CONTRACTUAL REPORT, 1966)
T-	BENDIX CORP., BALTIMORE
T(NO.)-	BOEING ATLANTIC TEST CENTER, PATRICK AFB, FLA.
T5-	BOEING CO., HUNTSVILLE, ALA. (BOMARC + SATURN)
T206-	BOEING CO., SEATTLE (PROJECT NUMBER, TEST)
T2-	BOEING CO. AERO-SPACE GROUP, SEATTLE
T3-	BOEING CO. AIRPLANE DIV., WICHITA, KAN.
T6-	BOEING CO. COMMERCIAL AIRPLANE DIV., SEATTLE
T7-	BOEING CO. SYSTEMS MANAGEMENT DIV., KENT, WASH.
T4-	BOEING CO. TURBINE DIV., SEATTLE
T8-	BOEING CO. VERTOL DIV., MORTON, PENNA.
T1-	BOEING SCIENTIFIC RESEARCH LABS., SEATTLE (TEST RPT.)
T-	BROOKHAVEN NATIONAL LAB., UPTON, N.Y.
T-(3 DIGITS)	BURROUGHS CORP. ELECTRONIC COMPONENTS DIV., PLAINFIELD, N.J.
T-(NUMBER)-R	CANADA. DEFENCE RESEARCH BOARD, OTTAWA
T-	CANADA. DEFENCE RESEARCH BOARD, OTTAWA
T.(NUMBER).G	CANADA. DEFENCE RESEARCH BOARD. SCIENTIFIC INFORMATION SERVICE (TRANSLATION FROM GERMAN)
T.(NUMBER).I	CANADA. DEFENCE RESEARCH BOARD. SCIENTIFIC INFORMATION SERVICE (TRANSLATION FROM ITALIAN)
T.(NUMBER).J	CANADA. DEFENCE RESEARCH BOARD. SCIENTIFIC INFORMATION SERVICE (TRANSLATION FROM JAPANESE)
T.(NUMBER).R	CANADA. DEFENCE RESEARCH BOARD. SCIENTIFIC INFORMATION SERVICE (TRANSLATION FROM RUSSIAN)
T-(NUMBER)-UKR	CANADA. DIRECTORATE OF SCI. INFO. SERVICES, OTTAWA
T-(NUMBER)R	CANADA. DIRECTORATE OF SCI. INFO. SERVICES, OTTAWA
T-(NUMBER)-C	CANADA. DIRECTORATE OF SCI. INFO. SERVICES, OTTAWA
T-(NUMBER)J	CANADA. DIRECTORATE OF SCI. INFO. SERVICES, OTTAWA
T-	COAST AND GEODETIC SURVEY, ROCKVILLE, MD. (TIDAL BENCH MARKS)
T-	COLES SIGNAL LAB., BELMAR, N.J. (TEST REPORT)
T-(NO.)	COLORADO SCHOOL OF MINES, GOLDEN
T-	COLUMBIA UNIV., N.Y.C. ELECTRONICS RESEARCH LABS.
T-(NUMBER/NUMBER)	COLUMBIA UNIV., N.Y.C. ELECTRONICS RESEARCH LABS.
T-(NUMBER)/(LETTER)	COLUMBIA UNIV., N.Y.C. ELECTRONICS RESEARCH LABS.
T-(NUMBER)-I-(NUMBER)	DEUTSCHER WETTERDIENST, OFFENBACH AM MAIN
T-	DOUGLAS AIRCRAFT CO., INC. PROJECT RAND, SANTA MONICA, CALIF.
T-(NUMBER)-(NO.)Q-(YEAR)	DOW CHEMICAL CO., MIDLAND, MICH. (THERMODYNAMICS)
T-(NUMBER)-Q-(YEAR)-	DOW CHEMICAL CO. SCIENTIFIC PROJS. LAB., MIDLAND, MICH
T-(NUMBER)-Q-(YEAR)-	DOW CHEMICAL CO. THERMAL LAB., MIDLAND, MICH.
T-(NUMBER)-(NO.)Q-(YR)	DOW CHEMICAL CO. THERMAL RESEARCH LAB., MIDLAND, MICH.
T-	DUQUESNE LIGHT CO., SHIPPINGPORT, PENNA.
T-	EVANS SIGNAL LAB., BELMAR, N.J. (TEST REPORT)
T-	FRANKFORD ARSENAL, PHILADELPHIA
T-	FRANKLIN INST. LABS. FOR RES. + DEV., PHILADELPHIA
T-	FRANKLIN INST. RESEARCH LABS., PHILADELPHIA
T-	GEORGE WASHINGTON UNIV., WASHINGTON, D.C.
T9.40-2-	GIRDLER CO. GAS PROCESSES DIV. PROCESS AND DEVELOPMENT RESEARCH LABS., LOUISVILLE, KY.
T-	GT. BRIT. AERONAUTICAL RESEARCH COUNCIL
T-(NO.)/(YR.)	GT. BRIT. ATOMIC WEAPONS RESEARCH ESTABLISHMENT, ALDERMASTON, BERKS, ENGLAND
T(NO.)-	GT. BRIT. FLYING PERSONNEL RESEARCH COMMITTEE
T.-	GT. BRIT. TELECOMMUNICATIONS RESEARCH ESTABLISHMENT, MALVERN, WORCS, ENGLAND
T-	HOLMAN (JOHN F.) AND CO., INC., WASHINGTON, D. C.
20-T-	HOLSTON DEFENSE CORP., KINGSPORT, TENN.
T-	IIT RESEARCH INST., CHICAGO
T-	INSTITUT FRANCO-ALLEMAND DE RECHERCHES, ST. LOUIS, FRANCE
T-	INTER-AMERICAN DEFENSE BOARD, WASHINGTON, D.C.
T-	KAMAN AIRCRAFT CORP., BLOOMFIELD, CONN.
T-	LOS ALAMOS SCIENTIFIC LAB., N. MEX.
T-1- THRU T-9-	LOS ALAMOS SCIENTIFIC LAB., N. MEX. (INTERNAL CORRESPONDENCE SERIAL USED BY GROUPS IN T DIV.)
T(YEAR)-	MASS. INST. OF TECH., CAMBRIDGE. HYDRODYNAMICS LAB.
T-	MASS. INST. OF TECH., CAMBRIDGE. INSTRUMENTATION LAB.
T-	MASSACHUSETTS INST. OF TECH., CAMBRIDGE. SOIL MECHANICS DIV.
T-	MONSANTO CHEMICAL CO., DAYTON, OHIO
T-	MOTOROLA INC., PHOENIX, ARIZ.
T-	NAVY MINE DEFENSE LAB., PANAMA CITY, FLA.
T-	NETHERLANDS. NATIONAAL LUCHT EN RUIMTEVAART-LABORATORIUM, AMSTERDAM
T-	NEW MEXICO INST. OF MINING AND TECH., SOCORRO
T7-(NUMBER)-(NUMBER)	NORTH AMERICAN AVIATION, INC., ANAHEIM, CALIF.
T(NUMBER)-	OGDEN AIR MATERIEL AREA, HILL AFB, UTAH
T-	PLASMADYNE CORP., SANTA ANA, CALIF.
T-	PUGET SOUND NAVAL SHIPYARD. CARR INLET ACOUSTIC RANGE, BREMERTON, WASH.
T-	QUARTERMASTER RESEARCH AND ENGINEERING COMMAND, NATICK, MASS.
T-(NO.)(RAND)	RAND CORP., SANTA MONICA, CALIF.
T-1 THRU T-146	RAND CORP., SANTA MONICA, CALIF. (TRANSLATION)
T-	RAYTHEON CO. RESEARCH DIV., WALTHAM, MASS.
T-	SACRAMENTO PEAK OBSERVATORY, SUNSPOT, N. MEX.
T-	SANDIA CORP., ALBUQUERQUE, N. MEX.
T-	SIGNAL CORPS ENG. LABS.,FT. MONMOUTH, NJ (TEST REPT.)
T-	SQUIER SIGNAL LAB., FORT MONMOUTH, N.J. (TEST REPT.)
T-	SWITZERLAND. EIDGENOESSISCHES FLUGZEUGWERK, EMMEN
T-	TEXACO RESEARCH CENTER, BEACON, N.Y.
T-	TRF, INC., SPRINGFIELD, VA.
T-(YEAR)-	TYCO LABS., INC., WALTHAM, MASS.
T-	WESTINGHOUSE ELECTRIC CORP. ATOMIC POWER DIV., PITTS.
T(NO.)-(NO.)-(YR.)	WHITE SANDS MISSILE RANGE, N. MEX.
T-(LETTER)(NUMBER)-	FRANKLIN INST. RESEARCH LABS., PHILADELPHIA
T(LETTER)C-	OFFICE OF SCIENTIFIC RESEARCH AND DEVELOPMENT
TA-	ALLIED CHEMICAL CORP., MORRISTOWN, N.J.
TA-	ARMSTRONG WHITWORTH (SIR W.G.) AIRCRAFT, LTD.
TA-	ETHYL CORP. RESEARCH AND DEVELOPMENT DEPT., FERNDALE, MICH. (TECHNICAL ANALYSIS)
TA-	FRANKLIN INST., PHILADELPHIA
TA-(NUMBER)-	SPERRY GYROSCOPE CO., GREAT NECK, N.Y.
TA7-(NUMBER)-(NUMBER)-	THIOKOL CHEMICAL CORP., BRIGHAM CITY, UTAH

TA AND D-	COMBAT CREW TRAINING WING (MBOM)(3520TH), MC CONNELL AFB, KAN.
TAAR-	(TARGET AREA ANALYSIS RADAR)
TAB-(YEAR)-	DEFENSE DOCUMENTATION CENTER, ALEXANDRIA, VA. (TECHNICAL ABSTRACT BULLETIN)
TAB-	DIVISION OF BIOLOGY AND MEDICINE, AEC, WASH., D.C.
TAB-	OAK RIDGE NATIONAL LAB., TENN.
TAB-	OAK RIDGE NATIONAL LAB., Y-12 AREA, TENN.
TAB-C-	DEFENSE DOCUMENTATION CENTER, ALEXANDRIA, VA. (TECHNICAL ABSTRACT BULLETIN, CLASSIFIED)
TABL-CONTRIB-	BUREAU OF COMMERCIAL FISHERIES. TROPICAL ATLANTIC BIOLOGICAL LAB., MIAMI, FLA.
TABLELEG-	TABLELEG COMMITTEE (REF. 15)
TAB-R-	DIVISION OF BIOLOGY AND MEDICINE, AEC, WASH., D.C.
TAB-S(YEAR)-	DEFENSE DOCUMENTATION CENTER, ALEXANDRIA, VA. (TECHNICAL ABSTRACT BULLETIN, CLASSIFIED. USED 1962-63)
TAB-U-	DEFENSE DOCUMENTATION CENTER, ALEXANDRIA, VA. (TECHNICAL ABSTRACT BULLETIN,UNCLASSIFIED. USED 1953-64)
TAC-	PETROLEUM ADMINISTRATION FOR WAR. TECH. ADVISORY COMM
TAC-	PETROLEUM INDUSTRY WAR COUNCIL, N.Y.C.
TAC-	SPECIAL AIR WARFARE CENTER. COMBAT APPLICATIONS GP., EGLIN AFB, FLA.
TAC-	TACTICAL AIR COMMAND, LANGLEY AFB, VA.
TAC-	TACTICAL AIR WARFARE CENTER, EGLIN AFB, FLA.
TAC-	TEMCO AIRCRAFT CORP., DALLAS
TAC (LETTER(S))-	PETROLEUM ADMINISTRATION FOR WAR. TECH. ADVISORY COMM
TAC-ER-	TEMCO AIRCRAFT CORP., DALLAS
TAC-OA-M-	TACTICAL AIR COMMAND, LANGLEY AFB, VA. (OPERATIONS ANALYSIS MEMORANDUM)
TAC-OAM-	TACTICAL AIR COMMAND, LANGLEY AFB, VA.
TAC-OA-MEMO-	TACTICAL AIR COMMAND, LANGLEY AFB, VA.
TAC-OA-R-(YEAR)-	TACTICAL AIR COMMAND. OFFICE OF OPERATIONS ANALYSIS, LANGLEY AFB, VA.
TAC-OA-TM-	TACTICAL AIR COMMAND, LANGLEY AFB, VA.
TAC/OA/WP-	TACTICAL AIR COMMAND, LANGLEY AFB, VA. (OPERATIONS ANALYSIS WORKING PAPERS)
TACOM-S-	ARMY TANK-AUTOMOTIVE COMMAND, WARREN, MICH.
TACOM-TR-	ARMY TANK-AUTOMOTIVE COMMAND, WARREN, MICH.
TAC-OPLAN-	TACTICAL AIR COMMAND, LANGLEY AFB, VA.
TACP-	MASSACHUSETTS INST. OF TECH., CAMBRIDGE
TAC-TEST-	TACTICAL AIR COMMAND, LANGLEY AFB, VA.
TAC-TEST-(YEAR)-	TACTICAL AIR WARFARE CENTER, EGLIN AFB, FLA.
TAC-TR-	SPECIAL AIR WARFARE CENTER. COMBAT APPLICATIONS GP., EGLIN AFB, FLA.
TAC-TR-(YEAR)-	TACTICAL AIR COMMAND, LANGLEY AFB, VA.
TAD-	LOS ALAMOS SCIENTIFIC LAB., N. MEX.
TAD-WP-	NATL. BUR. OF STDS. TECHNICAL ANALYSIS DIV., WASH.,DC
TAE-	TECHNION-ISRAEL INST. OF TECH.,HAIFA.DEPT.OF AERO.ENG
TAE-	TECHNION RESEARCH AND DEVELOPMENT FOUNDATION, LTD., HAIFA, ISRAEL
TAEC-D-	TURKEY. ATOMIC ENERGY COMMISSION, ANKARA
TAFB-	INTERCEPTOR WEAPONS SCHOOL, TYNDALL AFB, FLA.
TAGS-	RAYTHEON CO. RADAR AND CONTROL SYSTEMS DEPT., WAYLAND, MASS.
TAGS-(YEAR)-	RAYTHEON CO. RADAR AND CONTROL SYSTEMS DEPT., WAYLAND, MASS.
TAG-TR-	TECHNICAL ANALYSIS GROUP, INC., N.Y.C.
TAIC-	TOKYO ATOMIC INDUSTRY CONFERENCE
TAL-	ARMY BALLISTIC MISSILE AGENCY. TECHNICAL DOCUMENTS LIBRARY, REDSTONE ARSENAL, ALA. (TITLE ANNOUNCEMENT LIST)
TAL-	GERMANY. TECHNISCHE AKADEMIE DER LUFTWAFFE
TALC-OA-TM-	TACTICAL AIRLIFT CENTER. OFFICE OF OPERATION ANALYSIS, POPE AFB, N.C.
TALC-STUDY-DCR-	TACTICAL AIRLIFT CENTER, POPE AFB, N.C.
TAM-	ETHYL CORP. RESEARCH AND DEVELOPMENT DEPT., FERNDALE, MICH. (TECHNICAL ANALYSIS MEMORANDA)
T + AM-	ILLINOIS. UNIV., URBANA. DEPT. OF THEORETICAL AND APPLIED MECHANICS
T/AM-	ILLINOIS. UNIV., URBANA. DEPT. OF THEORETICAL AND APPLIED MECHANICS
T. + A.M.-	ILLINOIS. UNIV., URBANA. ENGINEERING EXPERIMENT STA.
TA+M-	TEXAS A + M UNIV., COLLEGE STATION
TAM-	TEXAS A + M UNIV., COLLEGE STATION. RESEARCH FDN.
TAMM-	ROCKETDYNE DIV., NORTH AMERICAN AVIATION, INC., CANOGA PARK, CALIF.
TAMU-REF-(YEAR)-	TEXAS A + M UNIV., COLLEGE STATION
TAMU-SG-(YEAR)-	TEXAS A + M UNIV., COLLEGE STATION(SEA GRANT PROGRAM)
TAO-	BOEING CO. AERO-SPACE DIV., SEATTLE
TAO-	HUGHES AIRCRAFT CO., WASHINGTON, D. C.(TECH.ANALYSIS)
TA/OST-(YEAR)-	AGENCY FOR INTERNATIONAL DEVELOPMENT. OFFICE OF SCIENCE AND TECHNOLOGY, WASHINGTON, D.C.
TA/OST-AN-(YEAR)-	AGENCY FOR INTERNATIONAL DEVELOPMENT. OFFICE OF SCIENCE AND TECHNOLOGY, WASHINGTON, D.C.
TAO-TR-	HUGHES AIRCRAFT CO., WASHINGTON, D. C.
TAPG-	TANK ARSENAL PROVING GROUND, UTICA, MICH.
TAPI-	ATOMIC ENERGY OF CANADA LTD. NUCLEAR POWER PLANT DIV. TORONTO
TAP-RA-	BUREAU OF NAVAL WEAPONS, WASHINGTON, D.C.
TAR-	THERM, INC. ADVANCE RESEARCH DIV., ITHACA, N.Y.
TAR-	UNITED KINGDOM. FIGHTING VEHICLES RESEARCH AND DEVELOPMENT ESTAB., CHERTSEY, SURREY, ENGLAND
TARC-(NUMBER)-DR(NO.)	TACTICAL AIR RECONNAISSANCE CENTER, SHAW AFB, S.C.
TARC-(NUMBER)-DT(NO.)	TACTICAL AIR RECONNAISSANCE CENTER, SHAW AFB, S.C.
TARC-OA-	TACTICAL AIR RECONNAISSANCE CENTER, SHAW AFB, S.C.
TARL-LAB-NOTE-	PERSONNEL RES. LAB. (6570TH), LACKLAND AFB, TEX. (LABORATORY NOTE)
TARL-LAB NOTE (NUMBER)-	TRAINING AIDS RESEARCH LAB., CHANUTE AFB, ILL.
TARL-LN-(NUMBER)-	PERSONNEL RES. LAB. (6570TH), LACKLAND AFB, TEX. (LABORATORY NOTE)
TARL-TM-(NUMBER)-	TRAINING AIDS RESEARCH LAB., CHANUTE AFB, ILL.
TARS-	AIR FORCE. 5TH. OPERATIONS ANALYSIS OFFICE
TAR-TR-	THERM, INC. ADVANCE RESEARCH DIV., ITHACA, N.Y. (TECHNICAL REPORT)
TAR-TR-	THERM ADVANCED RESEARCH, INC., ITHACA, N.Y.
TAS-	ANTIAIRCRAFT ARTILLERY AND GUIDED MISSILE CENTER, FORT BLISS, TEX.
TAS-(YEAR)-	DEFENSE DOCUMENTATION CENTER, ALEXANDRIA, VA.
TAS-	OFFICE OF TECHNICAL SERVICES. INVENTIONS AND ENG. DIV. TECHNICAL ADVISORY SERVICE, WASHINGTON, D.C.
TASC-TR-(NUMBER)-	ANALYTIC SCIENCES CORP., READING, MASS.
TASK-(YEAR)-	LOGISTICS MANAGEMENT INST., WASHINGTON, D.C.
TASS-(YEAR)-Y-	FIAT S.P.A. DIV. AVIAZIONE, TURIN
TAT-	BENDIX AVIATION CORP. BENDIX PRODUCTS DIV., MISSILES, MISHAWAKA, IND.
TAT-	JOHNS HOPKINS UNIV., SILVER SPRING, MD.
TAUP-(NUMBER)-(YEAR)	TEL-AVIV UNIV., ISRAEL
TAUP-	TEL-AVIV UNIV., ISRAEL
TB-	(TECHNICAL BULLETIN) (REF. 2)
TB-	(TEST BULLETIN) (REF. 2)
TB-(YEAR)-	ANALYTIC SERVICES, INC., FALLS CHURCH, VA.
TB2-	BALLISTIC RESEARCH LABS.,ABERDEEN PROVING GROUND, MD.
TB3-	BALLISTIC RESEARCH LABS.,ABERDEEN PROVING GROUND, MD.
TB-	CANADA. DEPT. OF MINES AND TECHNICAL SURVEYS. MINES BRANCH, OTTAWA

TB-	CORNELL AERONAUTICAL LAB., INC., BUFFALO
TB-	GERMANY. ZENTRALE FUER WISSENSCHAFTLICHES BERICHTSWESEN DER LUFTFAHRTFORSCHUNG, BERLIN (TECHNISCHE BERICHT)
TB(NUMBERS)/TO(NUMBERS)	INSTITUTE FOR TELECOMMUNICATION SCIENCES, BOULDER, COLO
TB-	MITSUBISHI HEAVY-INDUSTRIES, LTD., TOKYO
TB-	TRACERLAB, INC. WESTERN DIV., RICHMOND, CALIF.
TB-	WAR DEPT. (TECHNICAL BULLETIN)
TB CW-	WAR DEPT. (CHEMICAL WARFARE TECHNICAL BULLETIN)
TB ENG-	WAR DEPT. (ENGINEERING TECHNICAL BULLETIN)
TB MED-	WAR DEPT. (MEDICAL TECHNICAL BULLETIN)
TB ORD-	WAR DEPT. (ORDNANCE TECHNICAL BULLETIN)
TB QM-	WAR DEPT. (QUARTERMASTER TECHNICAL BULLETIN)
TB SIG-	WAR DEPT. (SIGNAL TECHNICAL BULLETIN)
TBSIG-	WAR DEPT. (SIGNAL TECHNICAL BULLETIN)
TB TC-	WAR DEPT. TRANSPORTATION CORPS (TECHNICAL BULLETIN)
TBTC-	WAR DEPT. TRANSPORTATION CORPS (TECHNICAL BULLETIN)
TC-	(TECHNICAL COMMUNICATION) (REF. 2)
TC-	(TRANSLATION CODE) (REF. 2)
TC-	AERONAUTICAL CHART AND INFORMATION CENTER, ST. LOUIS
TC-(YEAR)-	ARMED FORCES SPECIAL WEAPONS PROJECT, WASHINGTON, D.C.
TC-	CORNELL AERONAUTICAL LAB., INC., BUFFALO
TC-	OFFICE OF SCIENTIFIC RES. + DEV. (TRACER COMPOSITIONS)
TC-	TARIFF COMMISSION, WASHINGTON, D.C.
TC-	TECHNICAL CAPABILITIES BRANCH (AIR FORCE)
TC-	TELECOMPUTING CORP., LOS ANGELES
TC-	TETRA TECH, INC., PASADENA, CALIF.
TC-	TUFTS UNIV., MEDFORD, MASS.
TC-	WAR DEPT., WASHINGTON, D.C. (TRAINING CIRCULAR)
TCAE-	TELEDYNE CAE, TOLEDO
TCAP/P-	UNITED KINGDOM ATOMIC ENERGY AUTHORITY. RESEARCH GP. ATOMIC ENERGY RES. ESTAB., HARWELL, BERKS, ENGLAND
TCB-	ARMY TRANSPORTATION BOARD, FORT EUSTIS, VA.
TCBR-	AEROJET-GENERAL CORP., SACRAMENTO, CALIF.
TCC-	THIOKOL CHEMICAL CORP. (ALL LOCATIONS)
TCC-(YEAR)-	THIOKOL CHEMICAL CORP., HUNTSVILLE, ALA.
TCC-(NO.)-R(NO.)	THIOKOL CHEMICAL CORP. NUCLEAR DEVELOPMENT CENTER, PARSIPPANY, N.J.
TCC-(NUMBER)-(YEAR)	THIOKOL CHEMICAL CORP. RESEARCH DIV., HUNTSVILLE, ALA
TCC-E(NUMBER)-(YEAR)	THIOKOL CHEMICAL CORP. ELKTON DIV., MD.
TCC-ER-	THIOKOL CHEMICAL CORP., HUNTSVILLE, ALA.
TCC-R(YEAR)-	TECHNICAL COMMUNICATIONS CORP., LEXINGTON, MASS.
TCC-RER-	THIOKOL CHEMICAL CORP., HUNTSVILLE, ALA.
TCC-SR-(NUMBER)-(YEAR)	THIOKOL CHEMICAL CORP. ELKTON DIV., MD.
TCC-TR-	THIOKOL CHEMICAL CORP., HUNTSVILLE, ALA.
TCD-	AEROJET-GENERAL CORP. LIQUID ROCKET PLANT, SACRAMENTO, CALIF.
TCEA-IN-	BELGIUM. TRAINING CENTER FOR EXPERIMENTAL AERODYNAMICS, RHODE-SAINT-GENESE
TCEA-TM-	BELGIUM. TRAINING CENTER FOR EXPERIMENTAL AERODYNAMICS, BRUSSELS
TCED-OP-	AEROJET-GENERAL CORP. SOLID ROCKET OPERATIONS, SACRAMENTO, CALIF.
TCI-	CURTISS-WRIGHT CORP. WRIGHT AERO. DIV., WOOD-RIDGE, NJ
TCI-	TECHNOLOGY FOR COMMUNICATIONS INTERNATIONAL, MOUNTAIN VIEW, CALIF.
TCIR-	CHEMICAL CORPS. TECHNICAL COMMAND, ARMY CHEMICAL CENTER, MD. (INFORMAL REPORT)
TCIR-	CHEMICAL CORPS. TECHNICAL COMMAND, ARMY CHEMICAL CENTER, MD. (TECHNICAL COMMAND INFORMATION REPORTS)
TCI-SFN-	TECHNICAL COMMUNICATIONS, INC., LOS ANGELES
TCM-	AEROJET-GENERAL CORP., SACRAMENTO, CALIF.
TC-PL-	OWENS-CORNING FIBERGLAS CORP. TECHNICAL CENTER, GRANVILLE, OHIO
TCR-	AEROJET-GENERAL CORP., AZUSA, CALIF.
TCR-	AEROJET-GENERAL CORP., SACRAMENTO, CALIF.
TCR-	CHEMICAL CORPS. TECH. COMMAND, ARMY CHEMICAL CTR., MD.
TCREC-	ARMY TRANSPORT. RES. + ENG. COMMAND, FT. EUSTIS, VA.
TCREC-TR-	ARMY TRANSPORTATION RESEARCH COMMAND, FORT EUSTIS, VA
TD-	(TEST DATA) (REF. 2)
TD-	AEROJET-GENERAL CORP., SACRAMENTO, CALIF.
TD-	AMERICAN MACHINE + FOUNDRY CO. TURBO DIV., PACOIMA, CAL.
TD-	BADGER ARMY AMMUNITION PLANT, BARABOO, WIS.
TD-	BUREAU OF AERONAUTICS (NAVY). TECHNICAL DATA DIV.
TD-	BURNDY ENGINEERING CO., INC., N.Y.C. (TEST DATA)
T.D.-(NUMBER)-	FAIRCHILD HILLER CORP. REPUBLIC AVIATION DIV., FARMINGDALE, N.Y.
TD(YR.)-	GENERAL ELECTRIC CO. SPECIAL DEFENSE PROJECTS DEPT., SCHENECTADY, N.Y.
TD-	GENERAL ELECTRIC CO. SPECIALTY CONTROL DEPT., WAYNESBORO, VA.
TD-	HONEYWELL, INC. ORDNANCE DIV., HOPKINS, MINN. (TECH. DOC.)
TD-	ITT FEDERAL LABS., NUTLEY, N.J.
TD-(YEAR)-	ITT FEDERAL LABS., NUTLEY, N.J.
TD-1- THRU TD-6-	LOS ALAMOS SCIENTIFIC LAB., N. MEX. (INTERNAL CORRESPONDENCE SERIAL USED BY GROUPS IN TD DIV.)
TD-100/(NUMBER)	MILITARY ASSISTANCE COMMAND. TRAINING AIDS DIV., VIETNAM
TD-	NAVAL MISSILE CENTER. TARGET DRONE DEPT.
TD-(YEAR)-	NORTH AMERICAN ROCKWELL CORP. TULSA DIV., OKLA.
TD-(YEAR)-	OAK RIDGE NATIONAL LAB., TENN.
TD-	OFFICE OF THE ASSISTANT SECRETARY OF DEFENSE (INSTALLATIONS AND LOGISTICS), WASHINGTON, D.C.
TD-	WAR DEPT. (TRAJECTORY DIAGRAMS)
TDAC-	TROPICAL DETERIORATION ADMINISTRATIVE COMM. (REF. 28)
TDBS-	OFFICE OF THE CHIEF OF ORDNANCE. TECHNICAL DIVISION, BALLISTIC SECTION, WASHINGTON, D.C.
TDC-	GT. BRIT. ATOMIC EN. RES. ESTAB., HARWELL, BERKS, ENG
TDC-	TECHNICAL DEVELOPMENT CORP., CULVER CITY, CALIF.
TDCK-	NETHERLANDS. TECHNISCH DOCUMENTATIE EN INFORMATIE CENTRUM VOOR DE KRIJGSMACHT, THE HAGUE
TDCL-	NETHERLANDS. TECHNISCH DOCUMENTATIE EN INFORMATIE CENTRUM VOOR DE DRIJGSMACHT, THE HAGUE
TDEC-	TECHNICAL DEVELOPMENT + EVALUATION CTR., INDIANAPOLIS
TDFL-	QUARTERMASTER RESEARCH AND DEV. LABS., PHILADELPHIA
T.D. FT-	FAIRCHILD HILLER CORP. REPUBLIC AVIATION DIV., FARMINGDALE, N.Y.
TDI-	AIR FORCE (TARGET DATA INVENTORY)
TDI-	ATOMIC ENERGY OF CANADA LTD. NUCLEAR POWER PLANT DIV. TORONTO
TDIC-	TROPICAL DETERIORATION INFORMATION CENTER
TDL-	TECHNICAL DEVELOPMENT LABS., SAVANNAH (TRANSLATIONS)
TDM-(NUMBER)-	GENERAL ELECTRIC CO. MISSILE AND SPACE DIV., PHILA.
TDM-	NORTHROP AIRCRAFT, INC., HAWTHORNE, CALIF. (THERMODYNAMICS MEMO)
TDM-	PRATT AND WHITNEY AIRCRAFT DIV., UNITED AIRCRAFT CORP., HARTFORD, CONN.
TDM-	TECHNION-ISRAEL INST. OF TECH., HAIFA. DEPT. OF MECHANICS
TDM-(YEAR)-	TECHNION-ISRAEL INST. OF TECH., HAIFA. DEPT. OF MECHANICS
TD-MEMO-	ROYAL AERONAUTICAL SOCIETY. TRANSONIC AERODYNAMICS COMMITTEE, LONDON

TDMR-	CHEMICAL WARFARE SERVICE. TECHNICAL DIV., EDGEWOOD ARSENAL, MD. (MEMORANDUM REPORT)
TDM-W-	AVCO CORP. RES.+ ADVANCED DEV. DIV., WILMINGTON, MASS.
TDNO-	GENERAL DYNAMICS CORP. ELECTRIC BOAT DIV., GROTON, CONN
TDP-	OFFICE OF NAVAL RESEARCH (TECHNICAL DEVELOPMENT PLAN)
TDP-X(NUMBER)	NAVAL ELECTRONIC SYSTEMS COMMAND, SPECIAL COMMUNICATIONS PROJECT OFFICE, WASHINGTON, D.C.
TDR-	(TECHNICAL DATA REPORT) (REF. 2)
TDR-	(TECHNICAL DOCUMENTARY REPORT) (REF. 2)
TDR-	AEROJET-GENERAL CORP., SACRAMENTO, CALIF.
TDR-(CON.NO.)(JOB NO.)-	AEROSPACE CORP., EL SEGUNDO, CALIF. (TECHNICAL DOCUMENTARY REPORT)
TDR-	AEROSPACE CORP., PATRICK AFB, FLA.
TDR-	AEROSPACE CORP., SAN BERNARDINO, CALIF.
TDR-(CON.NO.)A(JOB NO.)-	AEROSPACE CORP. ATLANTIC MISSILE RANGE OFFICE
TDR-(CONTRACT NO)-A(NO.)-	AEROSPACE CORP. EASTERN TEST RANGE OFFICE, PATRICK AFB, FLA.
TDR-(CON.NO.)S(JOB NO.)-	AEROSPACE CORP. SAN BERNARDINO OPERATIONS, CALIF.
TDR-(NUMBERS)TN-	AEROSPACE CORP. SYSTEMS RESEARCH AND PLANNING DIV., EL SEGUNDO, CALIF.
TDR-(YEAR)-	AIR FORCE SPECIAL WEAPONS CENTER, KIRTLAND AFB, N.M.
TDR-	AIR FORCE SYSTEMS COMMAND, EDWARDS AFB, CALIF.
TDR-	ARINC RESEARCH CORP., WASHINGTON, D.C.
TDR-	BENDIX CORP., TETERBORO, N.J.
TDR-	COMMITTEE ON MEDICAL RES. (TROPICAL DISEASES RPTS.)
TD/R/EW/	WESTLAND AIRCRAFT, LTD., YEOVIL, SOMERSET, ENGLAND
TDR-N-	EASTMAN CHEMICAL PRODUCTS, INC., KINGSPORT, TENN.
TDS-	ATOMIC ENERGY OF CANADA LTD. NUCLEAR POWER PLANT DIV. TORONTO (REF. 6)
TDSC-	TROPICAL DETERIORATION STEERING COMMITTEE
TDSI-	ATOMIC ENERGY OF CANADA LTD. NUCLEAR POWER PLANT DIV. TORONTO (REF. 6)
TDU-	GT. BRIT. ROYAL AIR FORCE. TORPEDO DEVELOPMENT UNIT
TDV-	ATOMIC ENERGY OF CANADA LTD. NUCLEAR POWER PLANT DIV. TORONTO
TDV-	COMBUSTION ENGINEERING-SUPERHEATER, INC., N.Y.C.
TD(V)-	COMBUSTION ENGINEERING-SUPERHEATER LTD., MONTREAL
TDVI-	ATOMIC ENERGY OF CANADA LTD. CHALK RIVER PROJECT, ONT. (REF. 6)
TDVI-	ATOMIC ENERGY OF CANADA LTD. NUCLEAR POWER PLANT DIV. TORONTO
TE-(YEAR)-	CALIFORNIA. UNIV. (ALL LOCATIONS)
TE-	NORGES TEKNISKE HOEGSKOLE, TRONDHEIM
TE-(5 DIGITS)	TELEDYNE ELECTRONICS, NEWBURY, CALIF.
TE-	THERMO ELECTRON ENGINEERING CORP., CAMBRIDGE, MASS.
TE-	THERMO ELECTRON ENGINEERING CORP., WALTHAM, MASS.
TE-(NO.)-(YR.)	THERMO ELECTRON ENGINEERING CORP., WALTHAM, MASS.
TE(NO.)-	THERMO ELECTRON ENGINEERING CORP., WALTHAM, MASS.
TE-(NO.-NO.-YEAR)	THERMO ELECTRON ENGINEERING CORP., WALTHAM, MASS.
TE-	TRW SEMICONDUCTORS, INC., LAWNDALE, CALIF.
TE-B-3-	CLINTON ENGINEER WORKS, MANHATTAN DISTRICT, OAK RIDGE, TENN.
TE-B-3-	TENNESSEE EASTMAN CORP., OAK RIDGE, TENN.
TEC-	CANADA. DEPT. OF NATIONAL HEALTH AND WELFARE. RADIATION PROTECTION DIV., OTTAWA
TEC-	CANADA. DEPT. OF TRANSPORT.METEOROLOGICAL BR.,TORONTO
TEC-35-	TENNESSEE EASTMAN CORP., OAK RIDGE, TENN.
TECH-	TECHNOLOGY, INC., DAYTON, OHIO
TECH. MANUSCRIPT-	ARMY BIOLOGICAL LABS., FREDERICK, MD.
TECH MEMO-	(TECHNICAL MEMORANDUM) (REF. 2)
TECH. MEMO.-	SANDIA CORP., ALBUQUERQUE, N. MEX.
TECHNICAL BULL-	WEST VIRGINIA. UNIV.,MORGANTOWN. ENG. EXPERIMENT STA.
TECHNICAL LETTER-	ILLINOIS. STATE WATER SURVEY, URBANA
TECHNICAL PAPER-	(TECHNICAL PAPER) (REF. 2)
TECH REPT-	(TECHNICAL REPORT) (REF. 2)
TEC MIS EU-	NAVAL TECHNICAL MISSION IN EUROPE
TEC MIS JAP-	NAVAL TECHNICAL MISSION TO JAPAN
TECO-	TE CO., SANTA BARBARA, CALIF.
TECO-	TIMBER ENGINEERING CO., WASHINGTON, D.C.
TECOM-(NO.)-(LETTER)	ARMY TEST AND EVALUATION COMMAND, ABERDEEN PROVING GROUND, MD.
TEC-PI-	NATIONAL RESEARCH COUNCIL OF CANADA. ATOMIC ENERGY PROJECT, CHALK RIVER, ONT. (REF. 29)
TECPI-	NATIONAL RESEARCH COUNCIL OF CANADA. ATOMIC ENERGY PROJECT, CHALK RIVER, ONT. (REF. 29)
TEC UN EU-	NAVAL TECHNICAL UNIT IN EUROPE
TED-	NAVY (TEST EXPERIMENTAL AND DEVELOPMENT PROJECTS)
TED BIS-	NAVAL AIR TEST CENTER, PATUXENT RIVER, MD.
TED-NAM-AE-	NAVAL AIR ENGINEERING CENTER. AEROSPACE CREW EQUIPMENT LAB., PHILADELPHIA
TED PTR-	NAVAL AIR TEST CENTER, PATUXENT RIVER, MD.
TED-PTR-(LTRS.)-	NAVAL AIR TEST CENTER, PATUXENT RIVER, MD. (TEST, EXPERIMENTAL AND DEVELOPMENT)
TEE-	GT. BRIT. TORPEDO EXPERIMENTAL ESTABLISHMENT
TEE-	THERMO ELECTRON ENGINEERING CORP., CAMBRIDGE, MASS.
TEE-(NUMBER)-	THERMO ELECTRON ENGINEERING CORP., CAMBRIDGE, MASS.
TEE-	THERMO ELECTRON ENGINEERING CORP., WALTHAM, MASS.
TEE-(NUMBER)-(YEAR)	THERMO ELECTRON ENGINEERING CORP., WALTHAM, MASS.
TEE-(4 DIGITS)-	THERMO ELECTRON ENGINEERING CORP., WALTHAM, MASS.
TEES-	TEXAS A + M UNIV., COLLEGE STATION. ENGINEERING EXPERIMENT STATION
TEES-(NUMBER)	TEXAS A + M UNIV., COLLEGE STATION. ENGINEERING EXPERIMENT STATION
TEI-	GEOLOGICAL SURVEY (ALL LOCATIONS)
TEI-(M)-	GEOLOGICAL SURVEY, WASH., D.C. (TRACE ELEMENTS RPTS.)
TEIR-	GEOLOGICAL SURVEY, WASHINGTON, D.C.
TEI-TM-	TEXACO EXPERIMENT INC., RICHMOND
TEKNISK NOTAT (LTR.)-	NORWAY. FORSVARETS FORSKNINGSINSTITUTT, KJELLER
TEL-	HUGHES AIRCRAFT CO., TUCSON, ARIZ.
TEL-	QUARTERMASTER RESEARCH AND ENGINEERING COMMAND. TEXTILE ENGINEERING LAB., NATICK, MASS.
TEL-	ROYAL CANADIAN AIR FORCE. CENTRAL EXPERIMENTAL AND PROVING ESTAB., ROCKCLIFFE, ONT.
TELC-(NUMBER)-	SUPERIOR CONTINENTAL CORP. RESEARCH AND ENGINEERING CENTER, HICKORY, N.C.
TELEDYNE-SDL-	TELEDYNE SYSTEMS CO., ALEXANDRIA, VA.
TELI-	GEOLOGICAL SURVEY, WASHINGTON, D.C.
TELS-	CANADA. DEFENCE RES. TELECOMMUNICATIONS ESTAB.,OTTAWA
TEM-	GEOLOGICAL SURVEY,WASH.,D.C.(TRACE ELEMENTS MEMO RPT)
TEM-	TENNESSEE EASTMAN CORP., OAK RIDGE, TENN.
TEMCO-	TEMCO AIRCRAFT CORP., DALLAS
TEMCO-	TRANSMITTER EQUIPMENT MFG. CO., INC., N.Y.C.
TEMCO-ER-	TEMCO AIRCRAFT CORP., DALLAS
TEMPLEU-(NO.)(LTRS.)(NO.	TEMPLE UNIV., PHILADELPHIA
TEMR-	GEOLOGICAL SURVEY, WASHINGTON, D.C.
TEMU-	TEMPLE UNIV., PHILADELPHIA
TEMU-RI-	TEMPLE UNIV., PHILADELPHIA. RESEARCH INST.
TENNU-	TENNESSEE. UNIV., KNOXVILLE
TENNU-AES-	TENNESSEE. UNIV., OAK RIDGE. AGRICULTURAL EXPTL. STA.
TENNU-AR-	TENNESSEE. UNIV., KNOXVILLE. DEPT. OF CHEMISTRY
TENNU-PR-	TENNESSEE. UNIV., KNOXVILLE. DEPT. OF CHEMISTRY

TE-PES-

TE-PES-	TENNESSEE EASTMAN CORP., OAK RIDGE, TENN.
TER-	AEROJET-GENERAL CORP., SACRAMENTO, CALIF.
TER-	TELECOM, INC., ARLINGTON, VA.
TER-(3 DIGITS)-	TELECOM, INC., ARLINGTON, VA.
TERRIER CC-	TERRIER COORDINATING COMMITTEE (REF. 15)
TES-(YEAR)-	AIR DEFENSE WEAPONS CENTER, TYNDALL AFB, FLA.
TES-	TEST SQUADRON (OPERATIONAL) (4750TH), TYNDALL AFB,FLA
TE-SEP-	TENNESSEE EASTMAN CORP., OAK RIDGE, TENN.
TE-SES-	TENNESSEE EASTMAN CORP., OAK RIDGE, TENN.
TEST-	HONEYWELL, INC. TEST INSTRUMENTS DIV., ANNAPOLIS, MD.
TEST-	LABORATORY FOR ELECTRONICS, INC., BOSTON
TEST-	THIOKOL CHEMICAL CORP., HUNTSVILLE, ALA.
TEST-A-2901-	GENERAL TESTING LABS., INC., SPRINGFIELD, VA.
TEST-AL-	DAVID TAYLOR MODEL BASIN. AERODYNAMICS LAB., WASH.,DC
TEST-AL-C-	DAVID TAYLOR MODEL BASIN. AERODYNAMICS LAB., WASH.,DC
TEST-B-	DAVID TAYLOR MODEL BASIN. AERODYNAMICS LAB., WASH.,DC
TETRA-P-	TETRA TECH, INC., PASADENA, CALIF.
TEU-	TEMPLE UNIV., PHILADELPHIA
TEU-	TENNESSEE. UNIV., KNOXVILLE
TEUP-(YEAR)-	TOKYO UNIV. OF EDUCATION. DEPT. OF PHYSICS
TEXACO-AF-	TEXACO RESEARCH CENTER, BEACON, N.Y.
TEXI-	TEXAS INSTRUMENTS, INC., DALLAS
TEXU-	TEXAS. UNIV., AUSTIN
TEXU-AR-(YEAR)	TEXAS. UNIV., AUSTIN (ANNUAL REPORT)
TEXU-CPT-	TEXAS. UNIV., AUSTIN
TEXU-DRL-	TEXAS. UNIV., AUSTIN. DEFENSE RESEARCH LAB.
TF-(NO.)-(NO.)P(FILM)	AIR FORCE FILM LIBRARY CENTER, ST. LOUIS
TF-(NUMBER)(LETTER)	AKADEMIYA NAUK SSSR. INSTITUT MATEMATIKI, NOVOSIBIRSK
T-F-	AMERICAN METEOROLOGICAL SOCIETY, BOSTON
T-F-	ELLIOTT BROS. LTD., LONDON
TF-	OFFICE OF PRODUCTION RESEARCH AND DEVELOPMENT
TF-	TAYLOR AND FRANCIS, LTD., LONDON (TRANSLATIONS)
TF-	UPPSALA. UNIV.
T-FC-	AMERICAN METEOROLOGICAL SOCIETY, BOSTON
TFFL-	QUARTERMASTER RESEARCH AND ENGINEERING COMMAND TEXTILE FUNCTIONAL FINISHES LAB., NATICK, MASS.
TFSO-	ARMY BALLISTIC MISSILE AGENCY. TECHNICAL FEASIBILITY STUDIES OFFICE, REDSTONE ARSENAL, ALA.
TG-	(TECHNICAL GENERAL REPORT) (REF. 2)
T-G-	AMERICAN METEOROLOGICAL SOCIETY, BOSTON
TG-	DOW CHEMICAL CO. ROCKY FLATS DIV., GOLDEN, COLO.
TG-	DOW CHEMICAL CO. ROCKY FLATS PLANT, DENVER
TG-	GENERAL DYNAMICS/CONVAIR, SAN DIEGO, CALIF.
TG-	JOHNS HOPKINS UNIV., SILVER SPRING, MD. APPLIED PHYSICS LAB.
TG-(NUMBER)-(NUMBER)	JOHNS HOPKINS UNIV., SILVER SPRING, MD. APPLIED PHYSICS LAB.
TGGM-	GAUTNEY + JONES, CONSULTING ENGINEER, WASHINGTON,D.C.
TH-	COAST AND GEODETIC SURVEY, ROCKVILLE, MD. (TIDAL HARMONIC CONSTANTS)
TH-	DELFT, NETHERLANDS. TECHNISCHE HOGESCHOOL
TH/(YR)-E-	EINDHOVEN, NETHERLANDS. TECHNISCHE HOGESCHOOL
TH-	EUROPEAN ORGANIZATION FOR NUCLEAR RESEARCH, GENEVA
TH/(YEAR)/	FRANCE. COMMISSARIAT A L'ENERGIE ATOMIQUE, SACLAY
TH-	FRANCE. LABORATOIRE DE PHYSIQUE THEORIQUE ET DES HAUTES ENERGIES, ORSAY
(NUMBER)-TH-	GENERAL ELECTRIC CO. ELECTRONICS LAB.,SYRACUSE,N.Y.
TH/(YEAR)/	PARIS. UNIVERSITE, ORSAY. LABORATOIRE DE PHYSIQUE THEORIQUE ET HAUTES ENERGIES
TH-	PARIS. UNIVERSITE, ORSAY. LABORATOIRE DE PHYSIQUE THEORIQUE ET HAUTES ENERGIES
TH-	TOKYO UNIV. INST. FOR NUCLEAR STUDIES
TH-	YESHIVA UNIV., N.Y.C.
TH-A-	DELFT, NETHERLANDS. TECHNISCHE HOGESCHOOL. LABORATORIUM VOOR VOERTUIGTECHNIEK
THAI-AEC-	THAILAND. OFFICE OF ATOMIC ENERGY FOR PEACE, BANGKOK
THD-KR-	DELFT, NETHERLANDS. TECHNISCHE HOGESCHOOL. LABORATORIUM VOOR ENERGIEVOORZIENING EN KERNREACTOREN
THEMIS-(YEAR)-	ILLINOIS INST. OF TECH.,CHICAGO(PROJ. THEMIS)(REF. 48)
THEMIS-(LETTERS)-TR-	OFFICE OF THE DIRECTOR OF DEFENSE (RESEARCH AND ENGINEERING), WASH., D.C. (PROJECT THEMIS) (REF.48)
THEMIS-(LETTERS)-	OFFICE OF THE DIRECTOR OF DEFENSE (RESEARCH AND ENGINEERING), WASH., D.C. (PROJECT THEMIS) (REF.48)
THEMIS-(LETTERS)-T-TR-	OFFICE OF THE DIRECTOR OF DEFENSE (RESEARCH AND ENGINEERING), WASH., D.C. (PROJECT THEMIS) (REF.48)
THEMIS-(LETTERS-YEAR)-	OFFICE OF THE DIRECTOR OF DEFENSE (RESEARCH AND ENGINEERING), WASH., D.C. (PROJECT THEMIS) (REF.48)
THEMIS-(LETTERS)-T-	OFFICE OF THE DIRECTOR OF DEFENSE (RESEARCH AND ENGINEERING), WASH., D.C. (PROJECT THEMIS) (REF.48)
THEMIS-(LETTERS-NO.-YR.)	OFFICE OF THE DIRECTOR OF DEFENSE (RESEARCH AND ENGINEERING), WASH., D.C. (PROJECT THEMIS) (REF.48)
THEMIS-A(YEAR)-	HAWAII. UNIV., HONOLULU (PROJECT THEMIS) (REF. 48)
THEMIS-B(YEAR)-	HAWAII. UNIV., HONOLULU (PROJECT THEMIS) (REF. 48)
THEMIS-KU-RR-	KENTUCKY. UNIV., LEXINGTON (PROJECT THEMIS) (REF. 48)
THEMIS-PROPOSAL-	OFFICE OF THE DIRECTOR OF DEFENSE (RESEARCH AND ENGINEERING), WASH., D.C. (PROJECT THEMIS) (REF.48)
THEMIS-SR-	KANSAS STATE UNIV., MANHATTAN (PROJ. THEMIS)(REF. 48)
THEMIS-TR-	TEXAS A + M UNIV.,COLLEGE STA. (PROJ. THEMIS)(REF.48)
THEMIS-UF-SCIENTIFIC-	FLORIDA. UNIV., GAINESVILLE (PROJECT THEMIS) (REF.48)
THEMIS-UF-TR-	FLORIDA. UNIV., GAINESVILLE (PROJECT THEMIS) (REF.48)
THEMIS-UGA-	GEORGIA. UNIV., ATHENS. DEPT. OF STATISTICS (PROJECT THEMIS) (REF. 48)
THEMIS-UK-RR-	KENTUCKY. UNIV., LEXINGTON (PROJECT THEMIS) (REF. 48)
THEMIS-UTEC-DO-	UTAH. UNIV., SALT LAKE CITY. COLL. OF ENGINEERING
THEO.-P(MONTH/YEAR)	CALIFORNIA. UNIV., BERKELEY. RADIATION LAB. (REF. 9)
THEO.PNU-	CALIFORNIA. UNIV., LIVERMORE. LAWRENCE RADIATION LAB.
THESIS-	(THESIS ISSUED AS A REPORT) (REF. 2)
THI-	THIOKOL CHEMICAL CORP., BRISTOL, PENNA.
THIOKOL (NO.)/(YR.)	THIOKOL CHEMICAL CORP. REDSTONE DIV.,HUNTSVILLE,ALA.
THIOKOL(NUMBER)-(NUMBER)	THIOKOL CHEMICAL CORP. REDSTONE DIV.,HUNTSVILLE,ALA.
THIOKOL C-(NUMBERS)	THIOKOL CHEMICAL CORP. REDSTONE DIV.,HUNTSVILLE,ALA.
THIOKOL C-A-(NUMBERS)	THIOKOL CHEMICAL CORP. REDSTONE DIV.,HUNTSVILLE,ALA.
THIOKOL C-C-(NUMBERS)	THIOKOL CHEMICAL CORP. REDSTONE DIV.,HUNTSVILLE,ALA.
THIOKOL SP-	THIOKOL CHEMICAL CORP. REDSTONE DIV.,HUNTSVILLE,ALA.
THIOKOL TU-(NUMBERS)	THIOKOL CHEMICAL CORP. REDSTONE DIV.,HUNTSVILLE,ALA.
THOR-	FALCON RESEARCH + DEV. CO. THOR DIV.,COCKEYSVILLE,MD.
TH-R-	COLUMBIA UNIV., N.Y.C. DIV. OF WAR RESEARCH
THT-	STANFORD UNIV., CALIF. DEPT. OF MECHANICAL ENG.
THTR-	BROWN, BOVERI/KRUPP REAKTORBAU G.M.B.H., DUSSELDORF (THORIUM HIGH TEMPERATURE REACTOR)
THTR-	KERNFORSCHUNGSANLAGE JUELICH, GERMANY
TI-	(TECHNICAL INSTRUCTION) (REF. 2)
TI-	AMES LAB., IOWA
TI-	BRUSH BERYLLIUM CO., CLEVELAND
TI-	IOWA STATE COLL., AMES
TI-	KNOLLS ATOMIC POWER LAB., SCHENECTADY, N.Y.
TI-	LITTON SYSTEMS, INC. DATA SYSTEMS DIV., VAN NUYS,CAL.
TI-	RES. + DEV. BD. SPEC. COMM. ON TECH. INFO., WASH., DC

TI-	TECHNICAL INFORMATION SERVICE, AEC
TI-	TECHNICAL INFORMATION SERVICE EXTENSION, AEC
TI-	TECHNOLOGY, INC., DAYTON, OHIO
TI-	TECHNOLOGY, INC., SAN ANTONIO
TI-(NUMBER)-(YEAR)-	TEXAS INSTRUMENTS, INC. (ALL LOCATIONS)
TI-	TEXAS INSTRUMENTS, INC., DALLAS
TI-(NUMBER)-LA	TEXAS INSTRUMENTS, INC. DALLAS SERVICES GROUP
TI-(NUMBER)-MMD-	TEXAS INSTRUMENTS, INC. METALLURGICAL MATERIALS DIV., ATTLEBORO, MASS.
TIB-(NUMBER)-(YEAR)	ARMY MISSILE TEST CTR., WHITE SANDS MISSILE RANGE, NM
TIB-	DIVISION OF TECHNICAL INFORMATION, AEC
TIB-	GT. BRIT. MINISTRY OF SUPPLY
TIB-	WHITE SANDS MISSILE RANGE. ELECTRO-MECH. LABS., N.MEX.
TIB/BIB/(U)-	GT. BRIT. MINISTRY OF SUPPLY. TECHNICAL INFO. BUR.
TIB/GDC/	GT. BRIT. MINISTRY OF SUPPLY. TECHNICAL INFO. BUR.
TIB T-	GT. BRIT. MIN. OF AIRCRAFT PRODUCTION (TRANSLATION)
TIB/T-	GT. BRIT. MINISTRY OF SUPPLY. TECHNICAL INFO. BUR.
TIC-	BATTELLE MEMORIAL INST. TRANSDUCER INFORMATION CENTER, COLUMBUS, OHIO
TIC-	TECHNICAL INFORMATION CO. PATENTS DEPT., LONDON (TRANSLATIONS)
TIC-	TECHNOLOGY INTERNATIONAL CORP., BEDFORD, MASS.
TI-C(NUMBER)-(NUMBER)-	TEXAS INSTRUMENTS, INC., DALLAS
TIC-(NUMBER)-(NUMBER)-	TEXAS INSTRUMENTS, INC., DALLAS
TI-C(NUMBER)-	TEXAS INSTRUMENTS, INC. GOVERNMENT PRODS. DIV., DALLAS
TID-	LIBRARY OF CONGRESS. TECH. INFO. DIV., WASH., D.C.
TID-	OFFICE OF INFORMATION SERVICES, AEC, WASHINGTON, D.C. (REF. 5)
TID-	TECHNICAL INFORMATION CENTER, AEC, OAK RIDGE, TENN. (REF. 5)
TIDC PROJ-	TECHNICAL INDUSTRIAL DISARMAMENT COMMITTEE
TID-LS-	TECHNICAL INFORMATION SERVICE EXTENSION, AEC (LITERATURE SEARCH)
TI-DM-	TEXAS INSTRUMENTS, INC. GOVERNMENT PRODS. DIV., DALLAS
TIDM-F(NUMBER)-RM-	AVCO CORP. TELEMETRY AND INSTRUMENTATION DEPT., WILMINGTON, MASS.
TID-OC-	DIVISION OF CLASSIFICATION, AEC, WASHINGTON, D.C.
TIDR-	CHEMICAL WARFARE SERVICE. TECHNICAL INTELLIGENCE DIV., EDGEWOOD ARSENAL, MD.
TID/T-	GT. BRIT. MINISTRY OF SUPPLY. TECHNICAL INFO. BUR.
TIF-	AIR FORCE (TECHNICAL INFORMATION FILE)
TIFR-	INDIA. TATA INST. OF FUNDAMENTAL RESEARCH, BOMBAY
TIH-AC-	AIR TECH. INTELL. CTR., WRIGHT-PATTERSON AFB, OHIO
TI-HB-	TEXAS INSTRUMENTS, INC., DALLAS
TIHR-	TAVISTOCK INST. OF HUMAN RELATIONS, LONDON
TIHR-T(NUMBER)	TAVISTOCK INST. OF HUMAN RELATIONS, LONDON
TIH-WI-	AIR TECH. INTELL. CTR., WRIGHT-PATTERSON AFB, OHIO
TII-	TEMPO INSTRUMENT, INC., HICKSVILLE, N.Y.
TII-	TEXAS INSTRUMENTS, INC., DALLAS
TIIC-	TECHNICAL INDUSTRIAL INTELLIGENCE COMMITTEE
TIIC AS-	TECHNICAL INDUSTRIAL INTELL. COMM. AERO. SUB-COMM.
TIIC C (NO.)-	TECHNICAL INDUSTRIAL INTELL. COMM. (COMMUNICATIONS)
TIIC CS-	TECHNICAL IND. INTELL. COMM. CHEMICALS SUB-COMM.
TIIC MCS-4	TECHNICAL INDUSTRIAL INTELLIGENCE COMMITTEE. MISCELLANEOUS CHEMICALS SUB-COMM. (CHEM 4 RPTS.)
TIIC SFR-	TECHNICAL INDUSTRIAL INTELL. COMM. (SOLID FUELS RPTS)
TIL-	AEROJET-GENERAL CORP. TECHNICAL LIBRARY, SACRAMENTO, CALIF. (REPORT BIBLIOGRAPHY)
TIL-	UNITED KINGDOM. MINISTRY OF AVIATION. TECHNICAL INFORMATION AND LIBRARY SERVICES, LONDON
TIL/BIB/	GT. BRIT. MINISTRY OF SUPPLY. TECHNICAL INFORMATION AND LIBRARY SERVICES
TIL/BIB/(P)/-	GT. BRIT. MINISTRY OF SUPPLY. TECHNICAL INFORMATION AND LIBRARY SERVICES
TIL/BIB/(U)/	GT. BRIT. MINISTRY OF SUPPLY. TECHNICAL INFORMATION AND LIBRARY SERVICES
TIL-BR/	BRITISH AIRCRAFT CORP. GUIDED WEAPONS DIV., STEVENAGE, HERTS, ENGLAND
TIL-BR/	GT. BRIT. MINISTRY OF SUPPLY. TECHNICAL INFORMATION AND LIBRARY SERVICES
TIL-BR-	UNITED KINGDOM. MINISTRY OF TECHNOLOGY. TECHNICAL INFORMATION AND LIBRARY SERVICES, LONDON
TILL-	NAVAL RADIOLOGICAL DEFENSE LAB., SAN FRAN. (INTERNAL LIBRARY ACCESSION LIST OF CLASSIFIED DOCUMENTS)
TIL-LIST-	UNITED KINGDOM. MINISTRY OF TECHNOLOGY. TECHNICAL INFORMATION AND LIBRARY SERVICES, LONDON
TIL/OT-	GT. BRIT. ROYAL RADAR ESTAB., MALVERN, WORCS, ENGLAND
TIL/T.(NUMBER)	GT. BRIT. MINISTRY OF AVIATION. TECHNICAL INFORMATION AND LIBRARY SERVICES, LONDON (TRANSLATION)
TIL/T-	GT. BRIT. MINISTRY OF SUPPLY. TECHNICAL INFORMATION AND LIBRARY SERVICES
TIM-	PRATT AND WHITNEY AIRCRAFT DIV., UNITED AIRCRAFT CORP. (ALL LOCATIONS)
TIO-	ADVANCED RESEARCH PROJECTS AGENCY, WASHINGTON, D.C.
TIO-	INSTITUTE FOR DEFENSE ANALYSES, ARLINGTON, VA.
TIP-	ARMED FORCES TECHNICAL INFORMATION AGENCY. (TECHNICAL INFORMATION PILOT)
TIP-	ARMED SERVICES TECHNICAL INFORMATION AGENCY. REFERENCE CTR., LIBRARY OF CONGRESS, WASH., D.C.
TIP-	DIVISION OF TECHNICAL INFO. EXTENSION, AEC (REF. 5)
TIPI-SPO-(BETIPS)	BENDIX CORP. AEROSPACE SYSTEMS DIV., ANN ARBOR, MICH.
TIR-	(TECHNICAL INTELLIGENCE REPORT) (REF. 2)
TIR-	AIR FORCE (TECHNICAL INTELLIGENCE REPORTS)
TIR-	ORDNANCE DEPT. (ARMY) (TECHNICAL INTELLIGENCE REPORT)
TIR-	PITTSBURGH. UNIV. ARMY MATERIEL RESEARCH STAFF, WASHINGTON, D.C.
TIRL-R-	TUNE INVESTMENTS RESEARCH LABS., HINXTON HALL, CAMBRIDGE, ENGLAND
TIRL-TM-	TUNE INVESTMENTS RESEARCH LABS., HINXTON HALL, CAMBRIDGE, ENGLAND
TIS-	AIR FORCE (TARGET INFORMATION SHEET)
TIS(YEAR)DE(NUMBER)	GENERAL ELECTRIC CO. DIRECT ENERGY CONVERSION OPERATION, WEST LYNN, MASS.
TIS(YEAR)(LETTERS)	GENERAL ELECTRIC CO. TECHNICAL INFORMATION EXCHANGE, SCHENECTADY, N.Y. (TECH. INFO. SERIES) (REF. 46)
TIS-	NATIONAL RESEARCH COUNCIL OF CANADA, OTTAWA
TIS-	TECHNICAL INFORMATION SERVICE, AEC
TIS-	TECHNICAL INFORMATION SERVICE EXTENSION, AEC
TI-S(NUMBER)-	TEXAS INSTRUMENTS, INC., DALLAS
TI-S(NUMBER-NUMBER)-	TEXAS INSTRUMENTS, INC. APPARATUS DIV., DALLAS
TISA-	ARMY ENGINEER DISTRICT, SAVANNAH
TISE-	TECHNICAL INFORMATION SERVICE EXTENSION, AEC, OAK RIDGE, TENN.
TIS-ES-(YR)	AIR TECH. INTELL. CTR., WRIGHT-PATTERSON AFB, OHIO
TI-SP-	TEXAS INSTRUMENTS, INC. SCIENCE SERVICES DIV., DALLAS
TIS-PR-	AIR TECH. INTELL. CTR., WRIGHT-PATTERSON AFB, OHIO
TIS-PR-(YR.)-	AIR TECH. INTELL. CTR., WRIGHT-PATTERSON AFB, OHIO
TIS-R(YEAR)SO-	GENERAL ELECTRIC CO. MISSILE AND SPACE VEHICLE DEPT., PHILADELPHIA
TIS-WI-(YR.)-	AEROSPACE TECHNICAL INTELLIGENCE CENTER, WRIGHT-PATTERSON AFB, OHIO
TI-TR-	TEXAS INSTRUMENTS LTD., BEDFORD, BEDS, ENGLAND
TI-U(NUMBER)-	TEXAS INSTRUMENTS, INC., DALLAS
TJ-(NUMBER)	SPERRY GYROSCOPE CO., GREAT NECK, N.Y.
TK-	TEXAS. UNIV., AUSTIN

TKK-

TKK-	ARGONNE NATIONAL LAB., ILL.
TKK-F-A-	TEKNILLINEN KORKEAKOULU, OTANIEMI, FINLAND
TKS-	BUREAU OF COMMERCIAL FISHERIES, WASHINGTON, D.C. (TEST KITCHEN SERIES)
TL-	(TECHNICAL LETTER) (REF. 2)
TL-	(TECHNICAL LIBRARY) (REF. 2)
TL-	(TITLE LIST) (REF. 2)
TL-(NUMBER)-	AEROJET-GENERAL CORP. LIQUID ROCKET OPERATIONS, SACRAMENTO, CALIF.
T-L-	AMERICAN METEOROLOGICAL SOCIETY, BOSTON
TL-	CALIFORNIA. UNIV., BERKELEY. RADIATION LAB.
TL-	CARBIDE AND CARBON CHEMICALS CORP., OAK RIDGE, TENN.
TL-	GEOLOGICAL SURVEY, DENVER (CRUSTAL STUDIES)
TL-	LOS ALAMOS SCIENTIFIC LAB., N. MEX. (REPORT LIBRARY TITLE LIST)
TL-	NATIONAL RESEARCH COUNCIL OF CANADA. DIV. OF ATOMIC ENERGY, CHALK RIVER, ONT. (REF. 29)
TL-(YEAR)-	NETHERLANDS. TECHNOLOGICAL LAB. RVO-TNO, RIJSWIJK
TL-	TENNESSEE EASTMAN CORP., OAK RIDGE, TENN.
TL-	TRACERLAB, INC., WALTHAM, MASS.
TL-(NO.)TR-	TRACERLAB, INC., WALTHAM, MASS.
TLB-	TRACERLAB. DIV. OF LAB. FOR ELECTRONICS, INC., BOSTON
TLE-(NO.)-	RCA VICTOR DIV., RADIO CORP. OF AMERICA, CAMDEN, N.J.
TLE-	TRACERLAB, INC. EASTERN DIV., WALTHAM, MASS.
TLE-(NO.)(S)-(YR.)-	TRACERLAB, INC. EASTERN DIV., WALTHAM, MASS.
TLI-	SIGNAL CORPS (ARMY) (CAPTURED ENEMY SIGNAL EQUIPMENT) TECHNICAL LIAISON INTELLIGENCE)
TLI-	TYCO LABS., INC., WALTHAM, MASS.
TLI-ER-	TYCO LABS., INC., CLAREMONT, CALIF.
TLM-	(TECHNICAL LIAISON MEMO) (REF. 2)
TLR-	BUREAU OF SHIPS. SHIPBUILDING DIVISION. RESEARCH AND STANDARDS BRANCH (TECH. LITERATURE RES. SERIES)
TLR-	CHICAGO. UNIV. TOXICITY LAB.
TL-TB-(YEAR)-	TRACERLAB. DIV. OF LAB. FOR ELECTRONICS, INC., RICHMOND, CALIF.
TL-UK-	CAMP EVANS SIGNAL LAB., BELMAR, N.J.
TLW-	TRACERLAB. DIV. OF LAB. FOR ELECTRONICS, INC., RICHMOND, CALIF.
TM-	(TECHNICAL MANUAL) (REF. 2)
TM-	(TECHNICAL MEMORANDUM) (REF. 2)
TM-	(TECHNICAL MINUTES) (REF. 2)
TM-	(TECHNICAL MONOGRAPH) (REF. 2)
TM-	(TEST MANUAL) (REF. 2)
TM-	(TRAINING MANUAL) (REF. 2)
TM-(NUMBER)-LRO	AEROJET-GENERAL CORP. LIQUID ROCKET PLANT, SACRAMENTO, CALIF.
TM-(NUMBER)-LRP	AEROJET-GENERAL CORP. LIQUID ROCKET PLANT, SACRAMENTO, CALIF.
TM-(NUMBER)-SRO	AEROJET-GENERAL CORP. SOLID ROCKET PLANT, SACRAMENTO, CALIF.
TM-(NUMBER)-SRP	AEROJET-GENERAL CORP. SOLID ROCKET PLANT, SACRAMENTO, CALIF.
TM-	AEROJET-GENERAL NUCLEONICS, SAN RAMON, CALIF.
TM-	ARGONNE NATIONAL LAB., ILL.
TM-	ARMY, WASHINGTON, D.C. (TECHNICAL MANUAL)
TM-	ARMY COASTAL ENGINEERING RES. CENTER, WASHINGTON, DC
TM-(YEAR)-	AUTONETICS, ANAHEIM, CALIF.
TM(NO.)A(NO.)P(NO.)	BUNKER-RAMO CORP., FORT HUACHUCA, ARIZ.
TM-3-	CHEMICAL WARFARE SERVICE, EDGEWOOD ARSENAL, MD.
TM-(NUMBER)-	CONVAIR, POMONA, CALIF.
TM-5-	CORPS OF ENGINEERS (ARMY)
TM-(NO.)-(NO.)	GENERAL DYNAMICS/POMONA, CALIF.
TM(YEAR)-	GENERAL ELECTRIC CO. MATERIALS DEVELOPMENT LAB. OPERATION, CINCINNATI
TM-	GENERAL ELECTRIC CO. MISSILE AND SPACE DIV., PHILA.
TM-(YEAR)-(NUMBER)-	GENERAL ELECTRIC CO. NUCLEAR MATERIALS AND PROPULSION OPERATION, CINCINNATI
T/M-	GT. BRIT. ATOMIC EN. RES. ESTAB., HARWELL, BERKS, ENG
TM-	GT. BRIT. ATOMIC EN. RES. ESTAB., HARWELL, BERKS, ENG
TM-(YEAR)-	HARRY DIAMOND LABS., WASHINGTON, D.C.
TM-	HUGHES AIRCRAFT CO., CULVER CITY, CALIF.
TM-	INTERSTATE COMMERCE COMMISSION, WASHINGTON, D.C. (TRANSPORT MOBILIZATION)
TM-39-	JOINT ATOMIC WEAPONS PUBLICATIONS BOARD, ALBUQUERQUE, N. MEX. (REF. 14)
TM-	LOS ALAMOS SCIENTIFIC LAB., N. MEX.
TM-	MARTIN-MARIETTA CORP. DENVER DIV.
TM-(NO.)-(NO)-ED	METEOROLOGY INTERNATIONAL, INC., MONTEREY, CALIF.
TM-	MITRE CORP., BEDFORD, MASS.
TM-	NAVAL ORDNANCE TEST STA., INYOKERN (CHINA LAKE), CAL.
TM-9-	ORDNANCE CORPS (ARMY)
TM-	PRATT AND WHITNEY AIRCRAFT, EAST HARTFORD, CONN.
TM-(NO.)-E-	PREFORMED LINE PRODUCTS COMPANY. RESEARCH AND ENG., CLEVELAND (TECHNICAL REPORT-EXTERNAL)
TM-10-	QUARTERMASTER CORPS
TM-(NO.)-(YR.)-	SANDIA CORP., ALBUQUERQUE, N. MEX.
TM-(NUMBER)-(NUMBER)	SCHWEIZERISCHES INST. FUER NUKLEARFORSCHUNG, ZURICH
TM-11-	SIGNAL CORPS (ARMY)
TM-	SQUIER SIGNAL LAB., FORT MONMOUTH, N.J.
TM-(NUMBER/NUMBER(S))	SYSTEM DEVELOPMENT CORP., SANTA MONICA, CALIF.
TM-(NO.)/(NO.)/(NO.)	SYSTEM DEVELOPMENT CORP., SANTA MONICA, CALIF. (TECHNICAL MEMORANDUM)
TM-	TENNESSEE EASTMAN CORP., OAK RIDGE, TENN.
TM-	TEXACO EXPERIMENT INC., RICHMOND
TM-(NUMBER)-(NUMBER)	THOMPSON PRODUCTS, INC., CLEVELAND
TM-(NUMBER)-(NUMBER)	THOMPSON RAMO WOOLDRIDGE INC. TAPCO DIV., CLEVELAND
TMA-	DAVID TAYLOR MODEL BASIN, CARDEROCK, MD.
T-MAS-	ARMY TECH. INTELL. CTR. MEDICAL ANALYSIS SEC., TOKYO
TMB-	DAVID TAYLOR MODEL BASIN, CARDEROCK, MD.
TMB-	EDGERTON, GERMESHAUSEN AND GRIER, INC., BOSTON
TM-B-	UNION CARBIDE CORP. LINDE DIV., TONAWANDA, N.Y.
TMB-AERO-	DAVID TAYLOR MODEL BASIN. AERODYNAMICS LAB., WASH., DC
TMB-C-	DAVID TAYLOR MODEL BASIN, CARDEROCK, MD.
TMB-T-	DAVID TAYLOR MODEL BASIN, CARDEROCK, MD.
TMC-	MARQUARDT CORP., VAN NUYS, CALIF.
TMCA-	TITANIUM METALS CORP. OF AMERICA, N.Y.C.
TM-CH-(NO./NO./NO.)	CHARLOTTE MUNICIPAL INFORMATION SYSTEM PROJECT, N.C.
TMC-S-	MARQUARDT CORP. MATERIALS AND PROCESS DEPT., VAN NUYS, CALIF.
TMC-TM-	NORTH ATLANTIC TREATY ORGANIZATION. SHAPE TECHNICAL CENTER, THE HAGUE
TM-DM-(NO.)-EE-R-	DOUGLAS AIRCRAFT CO., INC. MISSILE AND SPACE SYSTEMS DIV., SANTA MONICA, CALIF.
TMDR-	GENERAL ELECTRIC CO. AIRCRAFT NUCLEAR PROPULSION DEPT., CINCINNATI
TM-DSV4B(LETTERS)-	DOUGLAS AIRCRAFT CO., INC., SANTA MONICA, CALIF.
TM-DSV(NO.)-EF-R-	DOUGLAS AIRCRAFT CO., INC. MISSILE AND SPACE SYSTEMS DIV., SANTA MONICA, CALIF.
TM-DSV(NO.)-PROP-R(NO.)	DOUGLAS AIRCRAFT CO., INC. MISSILE AND SPACE SYSTEMS DIV., SANTA MONICA, CALIF.
TME-	BUREAU OF MINES. METALLURGICAL DIV., TUCSON, ARIZ.

TML-	BATTELLE MEMORIAL INST. TITANIUM METALLURGICAL LAB., COLUMBUS, OHIO
TM(L)-	SYSTEM DEVELOPMENT CORP., SANTA MONICA, CALIF.
TM-L-	SYSTEM DEVELOPMENT CORP., SANTA MONICA, CALIF.
TM-LF2-PROP-R-	DOUGLAS AIRCRAFT CO., INC. SPACE SCIENCES DEPT., SANTA MONICA, CALIF.
TM-LO-	SYSTEM DEVELOPMENT CORP., SANTA MONICA, CALIF.
TM-LX-	SYSTEM DEVELOPMENT CORP., LEXINGTON, MASS.
TM-LX-	SYSTEM DEVELOPMENT CORP., SANTA MONICA, CALIF.
TMON-	(TECHNICAL MONOGRAPH) (REF. 2)
TMP-	GENERAL ELECTRIC CO. TECHNICAL MILITARY PLANNING OPERATION, SANTA BARBARA, CALIF.
(YEAR)TMP-	GENERAL ELECTRIC CO. TEMPO. CENTER FOR ADVANCED STUDIES, SANTA BARBARA, CALIF.
TMR-	(TECHNICAL MEMORANDUM REPORT) (REF. 2)
TMR-	GENERAL ELECTRIC CO. AIRCRAFT NUCLEAR PROPULSION DEPT., CINCINNATI
TMR-	NAVAL PROPELLANT PLANT, INDIAN HEAD, MD.
TMS-	(TECHNICAL MANUSCRIPT) (REF. 2)
TM-SAD-	AUSTRALIA. WEAPONS RESEARCH ESTABLISHMENT, ADELAIDE
TM-SPARTAN-MS-R-	DOUGLAS AIRCRAFT CO., INC. SPACE SCIENCES DEPT., SANTA MONICA, CALIF.
TM-ST-	SWITZERLAND. EIDGENOESSISCHES INSTITUT FUER REAKTORFORSCHUNG, WUERENLINGEN
TM-W-	NAVAL WEAPONS LAB., DAHLGREN, VA.
TM-WD-	SYSTEM DEVELOPMENT CORP., FALLS CHURCH, VA.
TM-X-	AMES RESEARCH CENTER, MOFFETT FIELD, CALIF.
TM-X-53-	GEORGE C. MARSHALL SPACE FLIGHT CENTER, HUNTSVILLE, ALA. (TECHNICAL MEMORANDUM)
TN-	(TECHNICAL NOTE) (REF. 2)
TN-	CANADA. SUFFIELD EXPERIMENTAL STATION, RALSTON, ALBERTA
TN-	FRANKFORD ARSENAL. FIRE CONTROL LAB., PHILADELPHIA
TN-	NATIONAL RESEARCH COUNCIL OF CANADA. ATOMIC ENERGY PROJECT, CHALK RIVER, ONT.
TN-(YR.)-	NAVAL POSTGRADUATE SCHOOL, MONTEREY, CALIF.
TN-	STANFORD RESEARCH INST., MENLO PARK, CALIF.
TN-(NUMBER)-	STANFORD UNIV., CALIF.
TN(NUMBER)-	SYLVANIA ELECTRONIC SYSTEMS-EAST, WALTHAM, MASS.
TN-	VIENNA. TECHNISCHE HOCHSCHULE. INST. OF INORGANIC AND GENERAL CHEMISTRY
TN-(JOB NO.).(NO.-DATE)	VITRO LABS., SILVER SPRING, MD.
TN-(JOB NO.).(TASK NO.)-	VITRO LABS., SILVER SPRING, MD.
TN-AE-	INDIA. NATIONAL AERONAUTICAL LAB., BANGALORE
TN(ALOSU)(NUMBER)-	OHIO STATE UNIV. RESEARCH FOUNDATION, COLUMBUS
TN-AP-(YEAR)-	CHRYSLER CORP., NEW ORLEANS
TN-AST-	BROWN ENGINEERING CO., INC. ADVANCED SYSTEMS AND TECHNOLOGY GROUP, HUNTSVILLE, ALA.
TNB-	(TECHNICAL NEWS BULLETIN) (REF. 2)
TN-BN-	MARYLAND. UNIV., COLLEGE PARK. INST. FOR FLUID DYNAMICS AND APPLIED MATHEMATICS
TNC-	TEXAS NUCLEAR CORP., AUSTIN
TNCC-	TRIPARTITE NUCLEAR CROSS-SECTIONS COMMITTEE (REF. 35)
TNCC(CAN)-	ATOMIC ENERGY OF CANADA LTD. CHALK RIVER PROJ., ONT.
TNCC(CAN)-	TRIPARTITE NUCLEAR CROSS-SECTIONS COMMITTEE (REF. 35)
TNCC(UK)-	TRIPARTITE NUCLEAR CROSS-SECTIONS COMMITTEE (REF. 35)
TNCC/UK-	UNITED KINGDOM ATOMIC ENERGY AUTHORITY. RESEARCH GP. ATOMIC ENERGY RES. ESTAB., HARWELL, BERKS, ENGLAND
TNCC(US)-	TRIPARTITE NUCLEAR CROSS-SECTIONS COMMITTEE (REF. 35)
TN/CS/	MASSACHUSETTS. UNIV., AMHERST. COMPUTER SCIENCE DEPT.
TN-N-	NAVAL CIVIL ENGINEERING LAB., PORT HUENEME, CALIF.
TNN-	TEXAS. UNIV., AUSTIN
TN-N(NUMBER)-(NUMBER)	VITRO LABS., SILVER SPRING, MD.
TNO-	NETHERLANDS. CENTRAAL ORGANISATIE VOOR TOEGEPASTNATURWEIENSCHAPPELIJK ONDERZOEK, THE HAGUE
TN-PAD-	AUSTRALIA. WEAPONS RESEARCH ESTABLISHMENT, SALISBURY
TN-PE-(JOB NO.-LTRS.)	VITRO LABS., SILVER SPRING, MD.
TNPG-	NUCLEAR POWER GROUP, KNUTSFORD, CHES., ENGLAND
TN-PH-	INDIA. NATIONAL AERONAUTICAL LAB., BANGALORE
TN-PT-	RENSSELAER POLYTECHNIC INST., TROY, N.Y. (PROJECT TUBAFLIGHT)
TN-R-	BROWN ENGINEERING CO., INC. RES. LABS., HUNTSVILLE, ALA
TN-R-	NAVAL CIVIL ENGINEERING LAB., PORT HUENEME, CALIF.
TN-S4-	FRANKLIN INST. LABS. FOR RES. + DEV., PHILADELPHIA
TN-SA-	INDIA. NATIONAL AERONAUTICAL LAB., BANGALORE
TN-SAD-	AUSTRALIA. WEAPONS RESEARCH ESTABLISHMENT, SALISBURY
TNSD-R/	TECHNION-ISRAEL INST. OF TECH., HAIFA. DEPT. OF NUCLEAR SCIENCE
TN-SE-	BROWN ENGINEERING CO., INC. RES. LABS., HUNTSVILLE, ALA
TNT-	GT. BRIT. ADVISORY COUNCIL ON SCI. RES. + TECH. DEV.
TNT/RSC	GT. BRIT. ADVISORY COUNCIL ON SCI. RES. + TECH. DEV.
TN-WP-	INDIA. NATIONAL AERONAUTICAL LAB., BANGALORE
TNX-	DU PONT DE NEMOURS (E.I.)+CO. TNX DIV., WILMINGTON, DEL
TNX-	HANFORD WORKS, RICHLAND, WASH.
TNX-PG-	HANFORD WORKS, RICHLAND, WASH.
TNY-	TRANSLATIONS, N.Y.C.
TNY-(NUMBER)(SC)	TRANSLATIONS, INC., N.Y.C.
TO-	(TECHNICAL ORDER) (REF. 2)
T.O.-	(TECHNICAL ORDER) (REF. 2)
TO-	AIR FORCE
T.O.-	AIR FORCE
T.O.11N-	AIR FORCE (TECH ORDER) (REF. 14)
T.O.(NUMBERS)-11N	JOINT ATOMIC WEAPONS PUBLICATIONS BOARD, ALBUQUERQUE, N. MEX. (REF. 14)
TO-	TECHNICAL OPERATIONS, INC. (ALL LOCATIONS)
TO-(LETTER)-(YEAR)-	TECHNICAL OPERATIONS RESEARCH, BURLINGTON, MASS.
TOB-	PUBLIC HEALTH SERVICE. RADIOLOGICAL HEALTH LAB., ROCKVILLE, MD.
TO-B-(NUMBER)-	TECHNICAL OPERATIONS, INC., BURLINGTON, MASS.
TO-B-(YEAR)-	TECHNICAL OPERATIONS, INC., BURLINGTON, MASS.
TOD-(NUMBER)-	AIR FORCE SYSTEMS COMMAND, ANDREWS AFB, MD. (TECHNICAL OBJECTIVE DOCUMENT)
TODD-N-	TODD SHIPYARDS CORP., NUCLEAR DIV., GALVESTON, TEX.
TODD/SML-NSS-	STATE MARINE LINES, INC., YORKTOWN, VA. (N.S. SAVANNAH OPERATIONS)
TODD/SML-NSS-	TODD SHIPYARDS CORP., GALVESTON, TEX.
TOD-SUMMARY GUIDE-(YR)	AIR FORCE SYSTEMS COMMAND, ANDREWS AFB, MD. (TECHNICAL OBJECTIVE DOCUMENT)
TO-EE-	KNOLLS ATOMIC POWER LAB., SCHENECTADY, N.Y.
TOHOKUU-	TOHOKU UNIV., SENDAI, JAPAN
TO/HS/	INTERNATIONAL ATOMIC ENERGY AGENCY, VIENNA
TOI-	TECHNICAL OPERATIONS, INC. (ALL LOCATIONS)
TOI-(YEAR)-	TECHNICAL OPERATIONS, INC. (ALL LOCATIONS)
TOI-B (YEAR)-	TECHNICAL OPERATIONS, INC., BURLINGTON, MASS.
TOI-TR-(YEAR)-	TECHNICAL OPERATIONS, INC. SYSTEM SCIENCES DIV., ALEXANDRIA, VA.
TOM-	TECHNICAL OIL MISSION
TOPOCOM-ID-	ARMY TOPOGRAPHIC COMMAND, WASHINGTON, D.C.
TOPOCOM-TR-	ARMY TOPOGRAPHIC COMMAND, WASHINGTON, D.C.
TOR-	(TECHNICAL OPERATING REPORT) (REF. 2)
TOR-(NUMBERS)OD-	AEROSPACE CORP., EL SEGUNDO, CALIF.

TOR-(CON.NO.)(JOB NO)-

TOR-(CON.NO.)(JOB NO.)-
    AEROSPACE CORP., EL SEGUNDO, CALIF. (TECHNICAL OPERATING REPORT)
TOR-
    AEROSPACE CORP., SAN BERNARDINO, CALIF.
TOR-(CON.NO.)A(JOB NO.)-
    AEROSPACE CORP. ATLANTIC MISSILE RANGE OFFICE
TOR-(CONTRACT NO)-A(NO.)
    AEROSPACE CORP. EASTERN TEST RANGE OFFICE, PATRICK AFB, FLA.
TOR-
    NETHERLANDS. MINISTRY OF ECONOMIC AFFAIRS. MILITARY MISSION TO A.C.C. (TECHNICAL INVESTIGATION RPT.)
TORONTOU-
    TORONTO. UNIV.
TORU-
    TORONTO. UNIV.
TORY II-
    CALIFORNIA. UNIV., LIVERMORE. LAWRENCE RADIATION LAB.
T-OS-
    ARMY TECHNICAL INTELLIGENCE CENTER. ORDNANCE (ANALYSIS) SECTION, TOKYO
TOW-
    HUGHES AIRCRAFT CO. AEROSPACE GROUP, CULVER CITY, CAL.
TOW-
    TECHNICAL OPERATIONS, INC., ARLINGTON, MASS.
TOW-(LETTERS)-
    HUGHES AIRCRAFT CO., CULVER CITY, CALIF.
TP-
    (TECHNICAL PAMPHLET) (REF. 2)
TP-
    (TECHNICAL PAPER) (REF. 2)
TP-
    (TECHNICAL PROBLEM) (REF. 2)
TP-
    (TECHNICAL PUBLICATION) (REF. 2)
TP-
    (TECHNOGRAPHIC PUBLICATION) (REF. 2)
TP-
    (TENTATIVE PAMPHLET) (REF. 2)
TP-
    AEROCHEM RESEARCH LABS., INC., PRINCETON, N.J.
TP-
    AEROJET-GENERAL CORP., SACRAMENTO, CALIF.
TP-(NUMBER)-LRO-
    AEROJET-GENERAL CORP. LIQUID ROCKET OPERATIONS, SACRAMENTO, CALIF.
TP-(NUMBER)-LRO
    AEROJET-GENERAL CORP. LIQUID ROCKET PLANT, SACRAMENTO, CALIF.
TP-(NUMBER)-LRP
    AEROJET-GENERAL CORP. LIQUID ROCKET PLANT, SACRAMENTO, CALIF.
TP-(NUMBER)-SRO
    AEROJET-GENERAL CORP. SOLID ROCKET OPERATIONS, SACRAMENTO, CALIF.
TP-(NUMBER)-SRP
    AEROJET-GENERAL CORP. SOLID ROCKET PLANT, SACRAMENTO, CALIF.
TP-
    FLORIDA. UNIV., GAINESVILLE
TP-
    FRANCE. OFFICE NATIONAL D'ETUDES ET DE RECHERCHES AEROSPATIALES, PARIS
50TP 120-
    GENERAL ELECTRIC CO., SCHENECTADY, N.Y.
TP-
    GENERAL ELECTRIC CO. FLIGHT PROPULSION LAB. DEPT., CINCINNATI
TP-
    GT. BRIT. AERO. RES. COUNCIL. WIND TUNNEL DESIGN COMM
TP-
    GT. BRIT. ATOMIC EN. RES. ESTAB., HARWELL, BERKS, ENG
TP-
    JOINT AIRBORNE TROOP BOARD, FORT BRAGG, N.C.
TP-
    JOINT ATOMIC WEAPONS PUBLICATIONS BOARD, ALBUQUERQUE, N. MEX. (REF. 14)
TP-
    LFE ELECTRONICS, BOSTON
TP(YR)-
    MAGNAVOX CO., FORT WAYNE
TP(YEAR)-
    MAGNAVOX CO. DIFAR ENG. DEPT., FORT WAYNE
TP(NUMBER)
    MAGNAVOX CO. GOVERNMENT AND IND. DIV., URBANA, ILL.
TP-(NUMBER)-
    MSA RESEARCH CORP., CALLERY, PENNA.
TP-
    NORTHROP CORP., NEWBURY PARK, CALIF.
TP-(YEAR)-
    OFFICE OF THE ASSISTANT SECRETARY OF DEFENSE (SYSTEMS ANALYSIS), WASHINGTON, D.C.
TP-
    RADIO CORP. OF AMERICA. ELECTROMAGNETIC AND AVIATION SYSTEMS DIV., VAN NUYS, CAL.
TP(NUMBER)-(LETTER)-
    ROBERTSHAW-FULTON CONTROLS CO., ANAHEIM, CALIF.
TP-
    TEXACO EXPERIMENT INC., RICHMOND
TP-(YEAR AND REPT. NO.)
    TORONTO. UNIV. DEPT. OF MECHANICAL ENGINEERING
TP-(NO.)-(NO.)
    UNITED AIRCRAFT CORP., EAST HARTFORD, CONN.
TP-
    UNITED GLASS BOTTLE MANUFACTURERS, LTD., LONDON
TP-
    UNITED KINGDOM ATOMIC ENERGY AUTHORITY. RESEARCH GP. ATOMIC ENERGY RES. ESTAB., HARWELL, BERKS, ENGLAND

TP-
    UNIVERSITY OF WESTERN ONTARIO, LONDON
TP-(JOB NO.).(TASK NO.)-
    VITRO LABS., SILVER SPRING, MD.
TPA 3-
    GT. BRIT. MINISTRY OF SUPPLY
TPCC-
    BABCOCK + WILCOX CO. ATOMIC EN. DIV., LYNCHBURG, VA.
TPD-
    NETHERLANDS. TECHNISCH PHYSISCHE DIENST TNO-TH, DELFT
TPE-
    WESTINGHOUSE DEFENSE CENTER, BALTIMORE, MD.
TPI-
    ATOMIC ENERGY OF CANADA LTD., CHALK RIVER, ONT.
TPI-
    RADIO CORP. OF AMERICA, VAN NUYS, CALIF.
TPI-
    RADIO CORP. OF AMERICA. WEST COAST DIV., VAN NUYS, CAL.
TPI-
    THOMPSON PRODUCTS, INC., CLEVELAND
TPI-
    TROPICAL PRODUCTS INST., LONDON
TPI-ER-
    THOMPSON PRODUCTS, INC., CLEVELAND
TPJU-(NUMBER/YEAR)
    UNIWERSYTET JAGIELLONSKI, KRAKOW. DEPT. OF THEORETICAL PHYSICS
TPJU-
    UNIWERSYTET JAGIELLONSKI, KRAKOW. DEPT. OF THEORETICAL PHYSICS
TPM-
    AEROJET-GENERAL CORP., SACRAMENTO, CALIF.
TPM-RFR-
    AKTIEBOLAGET ATOMENERGI, STUDSVIK, SWEDEN
TPM-RFX-
    AKTIEBOLAGET ATOMENERGI, STOCKHOLM
TPN-(NUMBER)/(YEAR)
    UNITED KINGDOM ATOMIC ENERGY AUTHORITY. WEAPONS GROUP ATOMIC WEAPONS RES. ESTAB., ALDERMASTON, BERKS, ENG.
TPPD-
    AIR FORCE (TECHNICAL PROGRAM PLANNING DOCUMENT)
TPR-
    (TECHNICAL PROGRESS REPORT) (REF. 2)
TPR-
    (TERRIER PROGRAM REPORT)
TPR-
    AERONUTRONIC DIV. OF PHILCO CORP., NEWPORT BEACH, CALIF. (TECHNICAL PROGRESS REPORT)
TP/R-
    ASSOCIATED ELECTRICAL INDUSTRIES, LTD., MANCHESTER, MANCUNIUM, ENGLAND
TPR-
    FLORIDA. UNIV., GAINESVILLE
TPR-
    GENERAL ELECTRIC CO., CINCINNATI
TPR-
    NAVAL ORDNANCE TEST STATION, CHINA LAKE, CALIF. (TECHNICAL PROGRESS REPORT)
TPR-
    STANFORD RESEARCH INST., MENLO PARK, CALIF.
TPRC-
    PURDUE UNIV., LAFAYETTE, IND. THERMOPHYSICAL PROPERTIES RESEARCH CENTER
TPR-NS-
    OAK RIDGE NATIONAL LAB., TENN.
TPR-PRT-
    GENERAL NUCLEAR ENGINEERING CORP., DUNEDIN, FLA.
TPR-RCM-
    BATTELLE MEMORIAL INST., COLUMBUS, OHIO
TPR-RFT-
    ARGONNE NATIONAL LAB., ILL.
TPR-RM-
    BATTELLE MEMORIAL INST., COLUMBUS, OHIO
TP/S-
    RAYTHEON CO. AUTOMETRIC OPERATION, ALEXANDRIA, VA.
TPT-
    NAVAL AIR TEST CENTER, PATUXENT RIVER, MD.
T QMAS-
    ARMY TECHNICAL INTELLIGENCE CENTER. QUARTERMASTER ANALYSIS SECTION, TOKYO
TR-
    (TECHNICAL REGULATION) (REF. 2)
TR-
    (TECHNICAL REPORT) (REF. 2)
TR-
    (TEST REPORT) (REF. 2)
TR-
    (TEST RUN) (REF. 2)
TR-
    (TRIAL REPORT) (REF. 2)
TR-
    (TRIP REPORT) (REF. 2)
TR-(CON.NO.)(JOB NO.)
    AEROSPACE CORP., EL SEGUNDO, CALIF. (TECH. REPORT)
TR-(CONTRACT NO.)-A(NO.)-
    AEROSPACE CORP. EASTERN TEST RANGE OFFICE, PATRICK AFB, FLA.
TR-
    AIR FORCE, WASHINGTON, D.C.
T-R-
    AMERICAN METEOROLOGICAL SOCIETY, BOSTON (TRANSLATION)
TR-(3 DIGITS)-
    AMERICAN OPTICAL CO. CENTRAL RESEARCH LAB., SOUTHBRIDGE, MASS.
TR-(YEAR)-
    ARMOUR RESEARCH FOUNDATION, CHICAGO
TR-
    ARMY ELECTRONICS LABS., FORT MONMOUTH, N.J.
TR-(YEAR)-(NUMBER)-AD
    ARMY NATICK LABS. AIRDROP ENGINEERING DIV., MASS.

TR-(YEAR)-(NUMBER)-CM	
	ARMY NATICK LABS. CLOTHING AND ORGANIC MATERIALS DIV., MASS.
TR-(YEAR)-(NUMBER)-ES	
	ARMY NATICK LABS. EARTH SCIENCES DIV., MASS.
TR-(YEAR)-(NUMBER)-FD	
	ARMY NATICK LABS. FOOD DIV., MASS.
TR-(YEAR)-(NUMBER)-PR	
	ARMY NATICK LABS. PIONEERING RESEARCH DIV., MASS.
TR-	
	ASTROPHYSICS RESEARCH CORP., LOS ANGELES
TR-(NUMBER)F	
	ASTROSYSTEMS INTERNATIONAL INC., CALDWELL, N.J.
(NUMBERS)-TR-	
	ATLANTIC RESEARCH CORP., DUARTE, CALIF.
(NUMBERS)-TR-	
	ATLANTIC RESEARCH CORP. MISSILE SYSTEMS DIV., COSTA MESA, CALIF.
(NUMBERS)TR-	
	AUERBACH CORP., PHILADELPHIA
(NUMBERS)-TR-	
	AUERBACH CORP., PHILADELPHIA
TR-(4 DIGITS)-(3 DIGITS)	
	AUERBACH CORP. PROFESSIONAL SERVICES GROUP, PHILADELPHIA, PA. (4 DIGITS ARE PROJECT NO., 3 DIGITS ARE PHASE OF PROJECT)
TR-(YR.)-	
	BELL AIRCRAFT CORP., BUFFALO
TR(NO.)-	
	BOLT, BERANEK, AND NEWMAN, INC., CAMBRIDGE, MASS.
TR-	
	CALOROBIC MATERIALS, INC., PEARL RIVER, N.Y.
TR-	
	CARNEGIE-MELLON UNIV., PITTSBURGH
TR-(YEAR)-	
	CASE WESTERN RESERVE UNIV., CLEVELAND
TR-	
	CHICAGO. UNIV.
TR-	
	COMMITTEE ON MEDICAL RESEARCH (TUBERCULOSIS REPORTS)
TR-(YEAR)-	
	CONVAIR. ASTRONAUTICS DIV., SAN DIEGO, CALIF.
TR-	
	DELAWARE. UNIV., NEWARK
TR-(YEAR)-	
	DILWORTH, SECORD, MEAGHER AND ASSOCIATES LTD., TORONTO
TR-	
	EMERSON RADIO AND PHONOGRAPH CORP., N.Y.C.
TR-(1 DIGIT)	
	FLINT TRANSPORTATION AUTHORITY, MICH.
TR-(2 DIGITS)	
	FLORIDA INST. OF TECH., MELBOURNE. LAB. FOR ENVIRONMENTAL AND SOLAR STUDIES
TR-(NUMBER)(LETTER)	
	FLORIDA STATE UNIV., TALLAHASSEE
TR-1-	
	FRANKLIN INST. RESEARCH LABS. OPERATIONS RESEARCH DIV., PHILADELPHIA
TR-	
	GENERAL ELECTRIC CO., SCHENECTADY, N.Y.
TR-(YEAR)-	
	GENERAL MOTORS CORP. DEFENSE RESEARCH LABS., SANTA BARBARA, CALIF.
TR(YEAR)-(NO.)(LTR.)	
	GENERAL MOTORS CORP. DEFENSE RESEARCH LABS., SANTA BARBARA, CALIF.
TR-(NUMBER)-	
	GENERAL TECHNOLOGY CORP., ELGIN, ILL.
TR-(YR)-	
	GENERAL TELEPHONE AND ELECTRONICS LABS. INC. BAYSIDE LABS., N.Y.
T/R-	
	GT. BRIT. ATOMIC EN. RES. ESTAB., HARWELL, BERKS, ENG
TR-	
	HARRY DIAMOND LABS., WASHINGTON, D.C.
TR-	
	HIGH ALTITUDE OBSERVATORY, BOULDER, COLO.
TR-(3 DIGITS)-	
	HYDRONAUTICS, INC., LAUREL, MD.
TR-(YEAR)-(NO.)-(NO.)	
	INFORMATICS, INC., BETHESDA, MD.
TR-(NUMBER)SER-(LETTER)	
	JOHNS HOPKINS UNIV., BALTIMORE. DEPT. OF MECHANICS
TR-(YEAR- 5 DIGITS)-	
	LTV ELECTROSYSTEMS INC. DEPT. OF ADVANCED SYSTEMS DEVELOPMENT, GREENVILLE, TEX.
TR-153-	
	MARQUARDT AIRCRAFT CO., VAN NUYS, CALIF.
TR-(YEAR)-	
	MARTIN-MARIETTA CORP. RESEARCH INSTITUTE FOR ADVANCED STUDIES, BALTIMORE
TR-(YEAR)-	
	MINNESOTA. UNIV., MINNEAPOLIS. DEPT. OF AERONAUTICS AND ENGINEERING SCIENCES
TR-	
	NAVAL TECHNICAL MISSION IN EUROPE
TR-(YEAR)-	
	NEW YORK UNIV., N.Y.C. GEOPHYSICAL SCIENCES LAB.
TR(NUMBER)-(NO.)-(YEAR)-	
	NORTHERN ELECTRIC CO., LTD., OTTAWA
TR-(YEAR)-	
	NORTHWESTERN UNIV., EVANSTON, ILL. INFORMATION PROCESSING AND CONTROL SYSTEMS LAB.
TR-	
	OFFICE OF NAVAL RESEARCH, WASHINGTON, D.C.
TR-(NO.)-	
	OHIO STATE UNIV., COLUMBUS. CRYOGENIC LAB.
TR-(YEAR)-(NO.)-(NO.)	
	OWENS-CORNING FIBERGLAS CORP., TOLEDO, OHIO
TR-	
	PITTSBURGH. UNIV.
TR-(NO.)-E-	
	PREFORMED LINE PRODUCTS COMPANY. RESEARCH AND ENGINEERING, CLEVELAND (TECHNICAL REPORT-EXTERNAL)
TR-(NO.)-ER	
	PREFORMED LINE PRODUCTS CO. RESEARCH AND ENG., CLEVELAND (TECH. REPT. - EXTERNAL, BUT RESTRICTED)
TR-(NUMBER)(LETTER)	
	RADIATION SYSTEMS, INC., ALEXANDRIA, VA.
TR-(NUMBERS)/	
	RHODE ISLAND. UNIV., KINGSTON. DEPT. OF ELEC. ENG.
TR-(NUMBER)-(YEAR)-	
	SERENDIPITY ASSOCIATES, CHATSWORTH, CALIF.
TR-(NUMBER)-(YEAR)-	
	SERENDIPITY ASSOCIATES, LOS ALTOS, CALIF.
TR-(YEAR)-	
	SPACE/DEFENSE CORP., BIRMINGHAM, MICH.
TR-(YR)-	
	SPACE TECHNOLOGY LABS., INC. PHYSICAL RESEARCH LAB., LOS ANGELES
TR-(YR.)-0000-	
	SPACE TECHNOLOGY LABS., INC. PHYSICAL RESEARCH LAB., LOS ANGELES
TR-	
	SPERRY PRODUCTS, INC., DANBURY, CONN.
TR-(NUMBER)-	
	STANFORD RES. INST. POULTER LABS., MENLO PARK, CAL.
TR-(YR.)-(NO.).(NO.)	
	SYLVANIA ELECTRIC PRODUCTS INC. SYLCOR DIV., HICKSVILLE, N.Y.
TR-	
	SYRACUSE UNIV., N.Y. DEPT. OF ELECTRICAL ENGINEERING
TR-	
	SYRACUSE UNIV., N.Y. DEPT. OF PHYSICS
TR-(YEAR)-	
	TECHNICAL OPERATIONS, INC., WASHINGTON, D.C.
TR-(YEAR)-	
	TECHNICAL OPERATIONS, INC. WASHINGTON RESEARCH CENTER, ARLINGTON, VA.
TR-1-	
	TECHNICAL SCIENCES CORP., ELGIN, ILL.
TR-(YEAR)-	
	TECHNIK, INC., JERICHO, N.Y.
TR-	
	TECHNION-ISRAEL INST. OF TECH., HAIFA
TR-(NUMBER)-	
	TEEG RESEARCH, INC., DETROIT
TR-	
	TENNESSEE EASTMAN CORP., OAK RIDGE, TENN.
6TR-	
	TRACERLAB, INC., BOSTON
21TR-	
	TRACERLAB, INC., WALTHAM, MASS.
1TR-	
	TRACERLAB, INC. WESTERN DIV., RICHMOND, CALIF.
5TR-	
	TRACERLAB, INC. WESTERN DIV., RICHMOND, CALIF.
TR-(JOB NO.).(TASK NO.)-	
	VITRO LABS., SILVER SPRING, MD.
TR-	
	WAR DEPT., WASHINGTON, D.C.
TR-(NUMBER)/	
	WISCONSIN. UNIV., MADISON. DEPT. OF MATHEMATICS
TRA-	
	FIBER SCIENCE, INC., GARDENA, CALIF.
TRA-(NUMBER)-(NUMBER)	
	GENERAL ELECTRIC CO., PHILADELPHIA
TR-A-	
	MICHIGAN. OFFICE OF PLANNING COORDINATION
TR-A(YEAR)-	
	NEW YORK UNIV., N.Y.C. DEPT. OF PHYSICS
TR-AC-	
	VIRGINIA. UNIV., CHARLOTTESVILLE
TRACOR-	
	TRACOR, INC., AUSTIN, TEX.
TRACOR-(YEAR)-	
	TRACOR, INC., AUSTIN, TEX.
TRACOR-NYL/(YEAR)-	
	TRACOR, INC., AUSTIN, TEX.
TRACOR-RL/(YEAR)-	
	TRACOR, INC., AUSTIN, TEX.
TRACOR-SD-(YEAR)-	
	TRACOR, INC., SAN DIEGO, CALIF.
TR-AE-	
	CINCINNATI. UNIV. DEPT. OF AEROSPACE ENG.
TR-AE-	
	RENSSELAER POLYTECHNIC INST., TROY, N.Y. AERONAUTICAL ENGINEERING AND ASTRONAUTICS DEPT.
TRAFFIC LAWS COMMENTARY-	
	NATIONAL COMMITTEE ON UNIFORM TRAFFIC LAWS AND ORDINANCES, WASHINGTON, D.C.
TR-AMSMI-RNR-	
	ARMY MISSILE COMMAND, REDSTONE ARSENAL, ALA.
TRANS-	
	(TRANSLATION) (REF. 2)
TRANS-	
	JOHNS HOPKINS UNIV., SILVER SPRING, MD. APPLIED PHYSICS LAB.
TRANS-	
	TRANS-SONICS, INC., BEDFORD, MASS.
TRANS-EMM-(YEAR)-	
	EMMANUEL COLL., BOSTON. ORIENTAL SCIENCE LIBRARY
TRAP-	
	GENERAL ELECTRIC CO. MISSILE AND SPACE DIV., PHILA.
TRAS-	
	TRANS-SONICS, INC., BEDFORD, MASS.
TR-B-	
	MICHIGAN. OFFICE OF PLANNING COORDINATION

T-RC-	AMERICAN METEOROLOGICAL SOCIETY, BOSTON
TRC-	ARMY TRANSPORTATION BOARD, FORT EUSTIS, VA.
TR-C-	NAVAL ACADEMY, ANNAPOLIS. MICHELSON PHYSICS LAB.
TRC-	RAND CORP., SANTA MONICA, CALIF.
TRC-	TEXACO RESEARCH CENTER, BEACON, N.Y.
TRC-	TRAVELERS RESEARCH CENTER, INC., HARTFORD
TRC-	UNION CARBIDE RESEARCH INST., TARRYTOWN, N.Y.
TRC-BR-	UNITED KINGDOM. MINISTRY OF TECHNOLOGY, LONDON
TRCI-	TRAVELERS RESEARCH CENTER, INC., HARTFORD
TR-C+OM-	ARMY NATICK LABS. CLOTHING AND ORGANIC MATERIALS DIV., MASS.
TR-CR-	MINNESOTA. UNIV., MINNEAPOLIS. DEPT. OF PHYSICS
TRC-R-	RAND CORP., SANTA MONICA, CALIF.
TRC-RA-	RAND CORP., SANTA MONICA, CALIF.
TRC-RM-	RAND CORP., SANTA MONICA, CALIF.
TRC-S-	RAND CORP., SANTA MONICA, CALIF.
TRC-TR-	TRIANGLE RESEARCH CORP., ALEXANDRIA, VA.
TR-D-	DELEX SYSTEMS, INC., ARLINGTON, VA.
TR+C-	INTERCEPTOR WEAPONS SCHOOL, TYNDALL AFB, FLA.
TRD-	THIOKOL CHEMICAL CORP. REDSTONE DIV., HUNTSVILLE, ALA.
TR-DA(NUMBER)	PHILCO-FORD CORP. POWER + CONTROL ENGINEERING DEPT., PALO ALTO, CALIF.
TR-DAT-	CHRYSLER CORP. MISSILE DIV., DETROIT
TRDC-	ARMY TRANSPORT. RES. + DEV. COMMAND, FT. EUSTIS, VA.
TRDC-	UNITED KINGDOM ATOMIC ENERGY AUTHORITY. INDUSTRIAL GROUP. SPRINGFIELDS WORKS, LANCS, ENGLAND
TRDC-	UNITED KINGDOM ATOMIC ENERGY AUTHORITY. INDUSTRIAL GROUP. WINDSCALE WORKS, SELLAFIELD, CUMB., ENGLAND
TRDC-	UNITED KINGDOM ATOMIC ENERGY AUTHORITY. RESEARCH GP. ATOMIC ENERGY RES. ESTAB., HARWELL, BERKS, ENGLAND
TRDC/P-	GT. BRIT. CULCHETH LABS., LANCS, ENGLAND
TRDC/P-	GT. BRIT. DEPT. OF ATOMIC ENERGY. INDUSTRIAL GROUP H.Q., RISLEY, LANCS, ENGLAND
TRDC/P-	GT. BRIT. DIV. OF ATOMIC ENERGY (PRODUCTION), RISLEY, LANCS, ENGLAND
TRDC/P-	GT. BRIT. SPRINGFIELDS WORKS, LANCS, ENGLAND
TRDC/T/-	GT. BRIT. ATOMIC EN. RES. ESTAB., HARWELL, BERKS, ENG
TR-DSF-	RENSSELAER POLYTECHNIC INST., TROY, N.Y. DIV. OF SYSTEMS ENGINEERING
TRE-	GT. BRIT. TELECOMMUNICATIONS RESEARCH ESTABLISHMENT, MALVERN, WORCS, ENGLAND
TRE-	UNITED KINGDOM ATOMIC ENERGY AUTHORITY. INDUSTRIAL GROUP. WINDSCALE WORKS, SELLAFIELD, CUMB., ENGLAND
TREC-	ARMY TRANSPORTATION RESEARCH COMMAND, FORT EUSTIS, VA
TR-ECOM-	ARMY ELECTRONICS COMMAND, FORT MONMOUTH, N.J.
TRECOM-	ARMY TRANSPORTATION RESEARCH COMMAND, FORT EUSTIS, VA
TRECOM-TR-(YEAR)-	ARMY TRANSPORTATION RESEARCH COMMAND, FORT EUSTIS, VA
TREC-TR-(YEAR)-	ARMY TRANSPORTATION RESEARCH COMMAND, FORT EUSTIS, VA
TR-EE-	PURDUE UNIV., LAFAYETTE, IND. SCHOOL OF ELEC. ENG.
TR-EE-(YEAR)-	PURDUE UNIV., LAFAYETTE, IND. SCHOOL OF ELEC. ENG.
TRE-M-	GT. BRIT. TELECOMMUNICATIONS RESEARCH ESTABLISHMENT, MALVERN, WORCS, ENGLAND
TRENTON RD-	THIOKOL CHEMICAL CORP. CHEMICAL OPERATIONS, TRENTON
TR-EP-	RENSSELAER POLYTECHNIC INST., TROY, N.Y. DIV. OF ELECTROPHYSICS
TR-ES-	ARMY NATICK LABS. EARTH SCIENCES DIV., MASS.
TRES-	EXOTECH, INC., WASHINGTON, D.C.
TRE-T-	GT. BRIT. TELECOMMUNICATIONS RESEARCH ESTABLISHMENT, MALVERN, WORCS, ENGLAND
TRE-TN-	GT. BRIT. TELECOMMUNICATIONS RESEARCH ESTABLISHMENT, MALVERN, WORCS, ENGLAND
TR-G-	AVCO CORP., TULSA, OKLA.
TRG-(YEAR)-	BABCOCK + WILCOX CO., LYNCHBURG, VA.
TRG-	CONTROL DATA CORP. TRG DIV., MELVILLE, N.Y.
TRG-(NUMBER)-	CONTROL DATA CORP. TRG DIV., MELVILLE, N.Y.
TRG-(NUMBER)-FR	CONTROL DATA CORP. TRG DIV., MELVILLE, N.Y.
TRG-(NUMBER)-TER-	CONTROL DATA CORP. TRG DIV., MELVILLE, N.Y.
TRG-(NUMBER)-TN-(YR)-	CONTROL DATA CORP. TRG DIV., MELVILLE, N.Y.
TRG-(NUMBER)-TR-(NO./YR)	CONTROL DATA CORP. TRG DIV., MELVILLE, N.Y.
TRG-(NUMBER)-USL-(YR)-	CONTROL DATA CORP. TRG DIV., MELVILLE, N.Y.
TRG-	TRG, INC. (ALL LOCATIONS)
TRG-(NUMBER)-FR-	TRG, INC., MELVILLE, N.Y.
TRG-(JOB NUMBER)-QTR-	TRG, INC., MELVILLE, N.Y.
TRG-129-QTR-	TRG, INC., SYOSSET, N.Y.
TRG-(NUMBER)-QTR-	TRG, INC., SYOSSET, N.Y.
TRG-	UNITED KINGDOM ATOMIC ENERGY AUTHORITY. DEVELOPMENT AND ENGINEERING GROUP, SPRINGFIELDS, LANCS, ENGLAND
TRG-	UNITED KINGDOM ATOMIC ENERGY AUTHORITY. INDUSTRIAL GROUP H.Q., RISLEY, LANCS, ENGLAND
TRG-(NUMBER)/(LETTER)/	UNITED KINGDOM ATOMIC ENERGY AUTHORITY. REACTOR GROUP, CULCHETH, LANCS, ENGLAND
TRG-INF. SER.-	UNITED KINGDOM ATOMIC ENERGY AUTHORITY. REACTOR GP.
TRG-MEMO-	UNITED KINGDOM ATOMIC ENERGY AUTHORITY. REACTOR GP.
TRG/OR-	UNITED KINGDOM ATOMIC ENERGY AUTHORITY. REACTOR GP., RISLEY, LANCS, ENGLAND
TRG-REPORT-	GT. BRIT. CENTRAL ELECTRICITY GENERATING BOARD. BERKELEY NUCLEAR LABS., BRISTOL, ENGLAND
TRG-REPORT-	UNITED KINGDOM ATOMIC ENERGY AUTHORITY. REACTOR GP.
TRG-REPORT-	UNITED KINGDOM ATOMIC ENERGY AUTHORITY. RESEARCH GP. ATOMIC ENERGY ESTAB., WINFRITH, DORSET, ENGLAND
TRG-REPORT-	UNITED KINGDOM ATOMIC ENERGY AUTHORITY. RESEARCH GP. ATOMIC EN. RES. EST., HARWELL, BERKS, ENG. (REF.40)
TRG-W-(NUMBER)-F	TRG, INC., MENLO PARK, CALIF.
TR-HT-	JOHN B. PIERCE FOUNDATION, RARITAN, N.J.
TR-HYD-	TEXAS. UNIV., AUSTIN. HYDRAULIC ENGINEERING LAB.
TRI-(YEAR)-	DILWORTH, SECORD, MEAGHER AND ASSOCIATES LTD., TORONTO
TRI-	TEXTILE RESEARCH INST., PRINCETON, N.J.
TRI-(YEAR)-	TRI-UNIVERSITY MESON FACILITY, VANCOUVER, B.C.
TRI-AR-(YEAR)	TRI-UNIVERSITY MESON FACILITY, VANCOUVER, B.C.
TR-IC-	EXOTECH, INC., ROCKVILLE, MD.
TRICO-	UNIVERSITE LOVANIUM DE LEOPOLDVILLE, KINSHASA, DEMOCRATIC REPUBLIC OF CONGO
TRIERU-	TRIER. UNIV., KAISERSLAUTERN, GERMANY
TR-II-	NORTHWESTERN UNIV., EVANSTON, ILL. DEPT. OF PHYSICS
TRIP-(NUMBER-YEAR)	ARMY COMBAT DEVELOPMENTS COMMAND. LIAISON DETACHMENT, SAN FRANCISCO
TR-J-	MICHIGAN. OFFICE OF PLANNING COORDINATION
TRL-	GT. BRIT. ADVISORY COUNCIL ON SCI. RES. + TECH. DEV.
TRL-	TRACERLAB. DIV. OF LAB. FOR ELECTRONICS, INC., BOSTON
TRL-	TRACERLAB. DIV. OF LAB. FOR ELECTRONICS, INC., RICHMOND, CALIF.
TR-LK-	GENERAL ELECTRIC CO. INSULATOR DEPT., BALTIMORE
TRLR-	CHEMICAL WARFARE SERVICE. TOXICOLOGICAL RESEARCH LAB., EDGEWOOD ARSENAL, MD.
TRL TLE-	TRACERLAB, INC. EASTERN DIV. LABS., WALTHAM, MASS.
TRL TLW-	TRACERLAB, INC. WESTERN DIV., RICHMOND, CALIF.
TRM-	QUARTERMASTER CORPS (TEXTILE RECONNAISSANCE MEMO.)
T/RMB-	DAVID TAYLOR MODEL BASIN, CARDEROCK, MD. (TRANSLATION, RUSSIAN MODEL BASIN)
TR-MDL-	UTAH. UNIV., SALT LAKE CITY. MICROWAVE DEVICE AND PHYSICAL ELECTRONICS LAB.
TR-ME-	NEW MEXICO. UNIV., ALBUQUERQUE. BUREAU OF ENG. RES.
TRN-	(TECHNICAL RESEARCH NOTE) (REF. 2)

TRN-	ARMY PERSONNEL RESEARCH OFFICE, WASHINGTON, D.C.
TRN-	CADILLAC GAGE CO. RESEARCH DIV., DETROIT
TRN-	INTERNATIONAL BUSINESS MACHINES CORP., CAMBRIDGE, MASS
TR-N-	RUTGERS UNIV., NEW BRUNSWICK, N.J.
TRN-	TEEG RESEARCH, INC., DETROIT
TR-NTI-	PRINCETON UNIV., N.J. FORRESTAL RESEARCH CENTER
TROC/R-	UNITED KINGDOM ATOMIC ENERGY AUTHORITY. RESEARCH GP. ATOMIC ENERGY RES. ESTAB., HARWELL, BERKS, ENGLAND
TR-ONR-	CALIFORNIA. UNIV., BERKELEY. STATISTICAL LAB.
TR-P-	CALIFORNIA. UNIV., LOS ANGELES. DEPT. OF PHYSICS
TR-PL-	ATLANTIC RESEARCH CORP., ALEXANDRIA, VA.
TR-PL-	LING-TEMCO-VOUGHT, INC., DALLAS
TR-PT-	RENSSELAER POLYTECHNIC INST., TROY, N.Y. (PROJECT TUBAFLIGHT)
TR-Q(NUMBER)-	DELAWARE. UNIV., NEWARK. DEPT. OF ELECTRICAL ENG.
TRR-	(TECHNICAL RESEARCH REPORT) (REF. 2)
TRR-	(THEORETICAL RESEARCH REPORT) (REF. 2)
TRR-	ARMY PERSONNEL RESEARCH OFFICE, WASHINGTON, D.C.
TR-R-	NAVAL CIVIL ENGINEERING LAB., PORT HUENEME, CALIF.
TRR-	QUARTERMASTER CORPS (TEXTILE SERIES REPORTS)
TR-RE-	CHRYSLER CORP. SPACE DIV., NEW ORLEANS
TR-RMG-	WESTERN ONTARIO UNIV., LONDON. RACETRACK MICROTRON GP
TR/RP-	NAVAL POSTGRADUATE SCHOOL, MONTEREY, CALIF.
TR-S-	AERONUTRONIC DIV. OF PHILCO CORP., NEWPORT BEACH, CAL.
TRS-	ARMY ENGINEER RESEARCH + DEV. LABS., FT. BELVOIR, VA.
TRS-	TENNESSEE EASTMAN CORP., OAK RIDGE, TENN.
TRSL SDDS EM-	TOMS RIVER SIGNAL LAB. SUPPRESSION SYSTEMS DEVELOPMENT SUBSECTION, N.J. (ENGINEERING MEMO.)
TRSR-	EXOTECH, INC., WASHINGTON, D.C.
TR-SR-(YEAR)-	EXOTECH, INC., WASHINGTON, D.C.
TRSR-(YEAR)-	EXOTECH, INC., WASHINGTON, D.C.
TR-SRI-	VIRGINIA. UNIV., CHARLOTTESVILLE
TRSWP-MISC-	UNITED KINGDOM ATOMIC ENERGY AUTHORITY. INDUSTRIAL GROUP. CALDER WORKS, CALDERBRIDGE, CUMB., ENGLAND
TRSWP/P-	UNITED KINGDOM ATOMIC ENERGY AUTHORITY. RESEARCH GP. ATOMIC ENERGY RES. ESTAB., HARWELL, BERKS, ENGLAND
TRSWP/R-	GT. BRIT. WINDSCALE WORKS. SELLAFIELD, CUMB., ENGLAND
TRSWP/R-	UNITED KINGDOM ATOMIC ENERGY AUTHORITY. INDUSTRIAL GROUP H.Q., RISLEY, LANCS, ENGLAND
TRSWP/R-	UNITED KINGDOM ATOMIC ENERGY AUTHORITY. INDUSTRIAL GROUP. SPRINGFIELDS WORKS, LANCS, ENGLAND
TR-USC-VACUV-	UNIVERSITY OF SOUTHERN CALIFORNIA, LOS ANGELES. DEPT. OF PHYSICS
TRW-(CONTRACT NO.)-	TRW EQUIPMENT LABS., CLEVELAND
TRW-	TRW INC. (ALL LOCATIONS)
TRW-(NO.-NO.-LTRS.)-	TRW SYSTEMS, REDONDO BEACH, CALIF. (REF. 42)
TRW-(NO.-NO.)-R000	TRW SYSTEMS, REDONDO BEACH, CALIF.
TRW(A)-(NUMBER-NUMBER)	TRW SYSTEMS GROUP, REDONDO BEACH, CALIF.
TRW-ER-	TRW INC., CLEVELAND (ENGINEERING REPORTS)
TR-WIS-2-AEC-	WISCONSIN. UNIV., MADISON. NAVAL RESEARCH LAB.
TRW-R(5 DIGITS-NO.)-R00-	TRW SYSTEMS, REDONDO BEACH, CALIF. (REF. 42)
TRWS-	TRW SEMICONDUCTORS, INC., LAWNDALE, CALIF.
TR-WSTF-	MANNED SPACECRAFT CENTER, HOUSTON, TEX. (WSTF FILTER TEST PROGRAM)
TRW-TM-	TRW INC., CLEVELAND (TECHNICAL MEMORANDA)
TRW-TR-(NUMBER.NUMBER)-	TRW SYSTEMS, REDONDO BEACH, CALIF. (REF. 42)
TRX-	GENERAL TECHNOLOGY CORP., ELGIN, ILL.
TS-	(TECHNICAL STUDY) (REF. 2)
TS-	(TENTATIVE SPECIFICATION) (REF. 2)
TS-	(TEST SUMMARY) (REF. 2)
TS-	AMERICAN CYANAMID CO., BOUND BROOK, N.J.
TS-(YR.)-(NO.)-	BUREAU OF NAVAL WEAPONS
TS-(YR.)-(NO.)	CALIFORNIA. UNIV., BERKELEY. DEPT. OF ENGINEERING
TS-(YEAR)-	CALIFORNIA. UNIV., BERKELEY. DEPT. OF MECH. ENG.
TS-(YEAR)-	CALIFORNIA. UNIV., BERKELEY. THERMAL SYSTEMS DIV.
TS-	ETHYL CORP. RESEARCH AND DEVELOPMENT DEPT., FERNDALE, MICH. (TECHNICAL SERVICE)
TS-	FELTERS CO., BOSTON, MASS.
TS-	NAVAL MISSILE CENTER. TECHNICAL SERVICE DEPT.
TS-	NAVY (TENTATIVE SPECIFICATIONS)
TS-	NAVY (TORPEDO STATION)
TS-	QUARTERMASTER RESEARCH AND ENGINEERING COMMAND, NATICK, MASS. (TEXTILE SERIES)
T SAS-	ARMY TECH. INTELL. CTR. SIGNAL ANALYSIS SEC., TOKYO
TSAWE-	DIRECTORATE OF FLIGHT AND ALL-WEATHER TESTING, WRIGHT-PATTERSON AFB, OHIO
TSB-	(TECHNICAL SERVICE BULLETIN) (REF. 2)
TSB-(YEAR)-	BUREAU OF RADIOLOGICAL HEALTH, ROCKVILLE, MD.
TSB-	NATIONAL CENTER FOR RADIOLOGICAL HEALTH. TECHNICAL SCIENCES BRANCH, ROCKVILLE, MD.
TSB-CONTROL-	ARMY, WASHINGTON, D.C.
TSC-	TELEDYNE SYSTEMS CO., HAWTHORNE, CALIF.
TSC-PD-	TECHNOLOGY SERVICE CORP., SANTA MONICA, CALIF.
TSC-WA-	TERRA-SPACE CORP., MALIBU, CALIF.
TSD-	GENERAL DYNAMICS CORP. ELECTRIC BOAT DIV., GROTON, CONN
TSD-	ROLLS ROYCE, LTD., DERBY, ENGLAND
TSD-TR-(NUMBER-YEAR)	MICHIGAN. DEPT. OF STATE HIGHWAYS. TRAFFIC SAFETY DIV
TSEAA-	AIR TECHNICAL SERVICE COMMAND. ENGINEERING DIV. AERO-MEDICAL LAB., WRIGHT-PATTERSON AFB, OHIO
TSEAC-	AIR TECHNICAL SERVICE COMMAND. ENGINEERING DIV. AIRCRAFT LAB., WRIGHT-PATTERSON AFB, OHIO
TSEAL-	AIR TECHNICAL SERVICE COMMAND. ENGINEERING DIV. AERO-MEDICAL LAB., WRIGHT-PATTERSON AFB, OHIO
TSEAM-	AIR TECHNICAL SERVICE COMMAND. ENGINEERING DIV. MATERIALS LAB., WRIGHT-PATTERSON AFB, OHIO
TSEAM-M-	AIR TECH. SERVICE COMMAND, WRIGHT-PATTERSON AFB, OHIO
TSEAM MR-	AIR TECHNICAL SERVICE COMMAND. ENG. DIV. MATERIALS LAB., WRIGHT-PATTERSON AFB, OHIO (MEMO. RPTS.)
TSEAP-	AIR TECHNICAL SERVICE COMMAND. ENGINEERING DIV. PERSONAL EQUIPMENT LAB., WRIGHT-PATTERSON AFB, OHIO
TSEC-TR-	ROBERT A. TAFT SANITARY ENGINEERING CTR., CINCINNATI
TSEI-	TUNG-SOL ELECTRIC INC., BLOOMFIELD, N.J.
TSELA-	AIR TECHNICAL SERVICE COMMAND. ENG. DIV. ELECTRONICS ADMINISTRATIVE SEC., WRIGHT-PATTERSON AFB, OHIO
TSELC-	AIR TECHNICAL SERVICE COMMAND. ENG. DIV. COMMUNICATION + NAVIGATION LAB., WRIGHT-PATTERSON AFB, OHIO
TSELP-	AIR TECHNICAL SERVICE COMMAND. ENG. DIV. ELECTRONICS PLANS SECTION., WRIGHT-PATTERSON AFB, OHIO
TSELR-	AIR TECHNICAL SERVICE COMMAND. ENGINEERING DIV. ELECTRONICS SUBDIV., WRIGHT-PATTERSON AFB, OHIO
TSELS-	AIR TECHNICAL SERVICE COMMAND. ENGINEERING DIV. SYSTEMS LAB., WRIGHT-PATTERSON AFB, OHIO
TSENG-	AIR MATERIEL COMMAND. ENGINEERING DIV., WRIGHT-PATTERSON AFB, OHIO
TSEPE-	AIR TECHNICAL SERVICE COMMAND. ENGINEERING DIV. EQUIPMENT LAB., WRIGHT-PATTERSON AFB, OHIO
TSEPF-	AIR TECHNICAL SERVICE COMMAND. ENGINEERING DIV. PHOTO LAB., WRIGHT-PATTERSON AFB, OHIO
TSEPL-	AIR TECHNICAL SERVICE COMMAND. ENGINEERING DIV. ENG. PLANS DIV., WRIGHT-PATTERSON AFB, OHIO
TSEPP-	AIR TECHNICAL SERVICE COMMAND. ENGINEERING DIV. POWER PLANT LAB., WRIGHT-PATTERSON AFB, OHIO
TSEPR-	AIR TECHNICAL SERVICE COMMAND. ENGINEERING DIV. PROPELLER LAB., WRIGHT-PATTERSON AFB, OHIO

TSEPS-	AIR TECHNICAL SERVICE COMMAND. ENGINEERING DIV. ARMAMENT LAB., WRIGHT-PATTERSON AFB, OHIO
TSERR-	AIR TECHNICAL SERVICE COMMAND. ENG. DIV. RADIO + RADAR SUBDIV., WRIGHT-PATTERSON AFB, OHIO
TSESA-	AIR TECHNICAL SERVICE COMMAND. ENGINEERING DIV. AIRCRAFT PROJ. SEC., WRIGHT-PATTERSON AFB, OHIO
TSESE-	AIR TECHNICAL SERVICE COMMAND. ENGINEERING DIV. SERVICE ENG. SUBDIV., WRIGHT-PATTERSON AFB, OHIO
TSEST-	AIR TECHNICAL SERVICE COMMAND. ENGINEERING DIV. ENG. STDS. SEC., WRIGHT-PATTERSON AFB, OHIO
TSF-	COMPAGNIES FRANCAISES ASSOCIEES DE TELEGRAPHIE SANS FIL
TSFO-	REDSTONE ARSENAL, HUNTSVILLE, ALA.
TSFTE-	DIRECTORATE OF FLIGHT AND ALL-WEATHER TESTING, WRIGHT-PATTERSON AFB, OHIO
TSI-	TRANS-SONICS, INC., BURLINGTON, MASS.
TSL-	AIR REDUCTION CHEMICAL CO., N.Y.C. (TECHNICAL SALES SERVICE LABORATORY REPORTS)
TSL-	LIBRARY OF CONGRESS. AEROSPACE INFO. DIV., WASH.,D.C.
TSM-	GT. BRIT. DEPT. OF ATOMIC ENERGY. INDUSTRIAL GROUP H.Q., RISLEY, LANCS, ENGLAND (REF. 36)
TSN-	WEIZMANN INST. OF SCIENCE, REHOVOTH, ISRAEL
TSP-	(TECHNICAL SPECIFICATION) (REF. 2)
TSP-	LEWIS RESEARCH CENTER, CLEVELAND
TSR-	(TECHNICAL SUMMARY REPORT) (REF. 2)
TSR-	AMERICAN OPTICAL CO. STAMFORD RESEARCH LABS., SOUTHBRIDGE, MASS.
TSR-	FRANKLIN INST. LABS. FOR RES. + DEV., PHILADELPHIA
TSR-	QUARTERMASTER CORPS (TEXTILE SERIES REPORTS)
TSR-	QUARTERMASTER RESEARCH + DEV. CTR., NATICK, MASS.
TSR-	STANFORD RESEARCH INST., MENLO PARK, CALIF.
TSS-	NAVAL ORDNANCE LAB., WHITE OAK, MD. (TEST SCHEDULE SHEETS)
TSTC-(NUMBER-NUMBER)-	TRI-STATE TRANSPORTATION COMMISSION, N.Y.
TT-	(TECHNICAL TRANSLATION) (REF. 2)
TT-	CANADA. DIRECTORATE OF SCI. INFO. SERVICES, OTTAWA
TT-	FRANKLIN INST. LABS. FOR RES. + DEV., PHILADELPHIA
TT-(YEAR)-	INDIAN NATIONAL SCIENTIFIC DOCUMENTATION CENTER, NEW DELHI
TT-(YEAR)-	ISRAEL PROGRAM FOR SCI. TRANS., LTD., JERUSALEM
TT-	JOHNS HOPKINS UNIV., BALTIMORE
TT-	JOHNS HOPKINS UNIV., SILVER SPRING, MD. APPLIED PHYSICS LAB.
TT-	NATIONAL RESEARCH COUNCIL OF CANADA, OTTAWA (TRANSLATION)
TT-(YEAR)-(5 DIGITS)	NATIONAL TECHNICAL INFORMATION SERVICE, SPRINGFIELD, VA. (ACCESSION NUMBER FOR TRANSLATIONS)
TT-	NAVAL AIR TEST CENTER, PATUXENT RIVER, MD.
TT-(YEAR)-	NAVAL SHIP SYSTEMS COMMAND, WASHINGTON, D.C.
TT-	NAVY, WASHINGTON, D.C.
TT-	RAND CORP., SANTA MONICA, CALIF.
TT-	REDSTONE SCIENTIFIC INFO. CTR., REDSTONE ARSENAL, ALA
TT-	SACRAMENTO PEAK OBSERVATORY, SUNSPOT, N. MEX.
TT-	TETRA TECH, INC., PASADENA, CALIF.
T./T.A. (NUMBER)	LLOYD'S REGISTRY OF SHIPPING. RESEARCH AND TECHNICAL ADVISORY SERVICES DEPT., LONDON
TTB-	AEROJET-GENERAL CORP., SACRAMENTO, CALIF.
TTB/(NUMBER)/	GT. BRIT. ORDNANCE BOARD, LONDON
TTE-	GT. BRIT. TROPICAL TESTING ESTABLISHMENT
TTF-PN-	AEROJET-GENERAL CORP., SACRAMENTO, CALIF.
TTM-	OFFICE OF SCIENTIFIC RES. + DEV.(TORPEDO TECH. MEMO.)
T/TN/	IMPERIAL COLLEGE OF SCIENCE AND TECHNOLOGY, LONDON. DEPT. OF MECHANICAL ENGINEERING
TTP-	JOHNS HOPKINS UNIV., BALTIMORE (TERRIER TEST PROCEDURE)
TTPC-	UNITED KINGDOM ATOMIC ENERGY AUTHORITY
TTR-	(TALOS TEST REPORT)
TTR-	RADIO CORP. OF AMERICA. MISSILE AND SURFACE RADAR DIV., MOORESTOWN, N.J.
TTS/	UNITED KINGDOM ATOMIC ENERGY AUTHORITY. RESEARCH GP. ATOMIC ENERGY RES. ESTAB., HARWELL, BERKS, ENGLAND
TT-TC-	TETRA TECH, INC., PASADENA, CALIF.
TTU-	TENNESSEE TECHNOLOGICAL UNIV., COOKEVILLE
TTU-SOC-	TEXAS TECH UNIV., LUBBOCK. DEPT. OF SOCIOLOGY
T-U-	AMERICAN METEOROLOGICAL SOCIETY, BOSTON
TU-	DOUGLAS AIRCRAFT CO., INC., TULSA, OKLA.
TU-	TEMPLE UNIV., PHILADELPHIA
TU-	TENNESSEE. UNIV., KNOXVILLE
TU-	TEXAS. UNIV., AUSTIN
TUB-	TURCO PRODUCTS, INC., WILMINGTON, CALIF.
TUB-	ARMY, WASHINGTON, D.C.
TUBIK-	BERLIN. TECHNISCHE UNIVERSITAET
TUB-IR-	BERLIN. TECHNISCHE UNIVERSITAET.INST.FUER KERNTECHNIK
TU DRL-	BERLIN. TECHNISCHE UNIV. INST. FUER RAUMFAHRTTECHNIK
TUE-(NUMBER/YEAR)	TEXAS. UNIV., AUSTIN. DEFENSE RESEARCH LAB.
	GERMANY. LABORATORIUM FUER BETRIEBSFESTIGKEIT, DARMSTADT
TUEBINGENU-	TUEBINGEN, GERMANY. UNIV.
TU EERL-	TEXAS. UNIV., AUSTIN. ELECTRICAL ENGINEERING RES.LAB.
TUENS-	TOKYO UNIV. OF EDUCATION. DEPT. OF PHYSICS
TUEP-	TOKYO UNIV.
TUEP-	TOKYO UNIV. OF EDUCATION
TU-ERL-	TEXAS. UNIV., AUSTIN. ENGINEERING RESEARCH LABS.
TUFTS-	TUFTS COLL., MEDFORD, MASS.
TUL-	KARL-MARX-UNIVERSITAET, LEIPZIG. SEKTION PHYSIK
TU-L-	TURCO PRODUCTS, INC., WILMINGTON, CALIF.
TULANEU-TR-	TULANE UNIV., NEW ORLEANS
TULU-	TULANE UNIV., NEW ORLEANS
TU MEMO (NO.)-	TEXAS. UNIV., AUSTIN. MILITARY PHYSICS RES. LAB.
TU MPRL-	TEXAS. UNIV., AUSTIN. MILITARY PHYSICS RES. LAB.
TU MRPL-	TEXAS. UNIV., AUSTIN. MILITARY PHYSICS RES. LAB.
TUN-	TUNG-SOL ELECTRIC INC., BLOOMFIELD, N.J.
TUNL-	TRIANGLE UNIVERSITIES NUCLEAR LAB., DURHAM, N.C.
TU-RI-(LETTERS)-	TEMPLE UNIV., PHILADELPHIA. RESEARCH INST.
TURIN-	TURIN. POLITECNICO. LABORATORIO DI MECCANICA
TU-SMRL-	TEXAS. UNIV., AUSTIN. STRUCTURAL MECHANICS RES. LAB.
T-V-	ANTENNA ENGINEERING LABS., TORRANCE, CALIF.
TV-	ARGONNE NATIONAL LAB., LEMONT, ILL.
TVA-	TENNESSEE VALLEY AUTHORITY, WILSON DAM, ALA.
TVA-P-	TENNESSEE VALLEY AUTHORITY, OAK RIDGE
TVA-P-	TENN. VALLEY AUTH. DIV. OF POWER SUPPLY, CHATTANOOGA
TVC-	GT. BRIT. MINISTRY OF AVIATION.CENTRAL LIBRARY,LONDON
TVC-(LETTERS)-	GT. BRIT. MINISTRY OF AVIATION.CENTRAL LIBRARY,LONDON
TW-	COAST AND GEODETIC SURVEY, ROCKVILLE, MD. (WATER TEMPERATURE)
TW-	FEDERAL CARTRIDGE CORP. TWIN CITIES ORDNANCE PLANT
TW-	GRONINGEN, NETHERLANDS. RIJKSUNIVERSITEIT. MATHEMATISCH INSTITUUT
TW-(3 DIGITS)	NETHERLANDS. MATHEMATISCH CENTRUM. AFDELING TOEGEPASTE WISKUNDE, AMSTERDAM
TW-	THIOKOL CHEMICAL CORP., BRIGHAM CITY, UTAH

TWC-	GEOLOGICAL SURVEY, WASHINGTON, D.C.
TWC-	LOS ALAMOS SCIENTIFIC LAB.,N.M.(TURRET WORKING COMM.)
TWF/R/	IMPERIAL COLLEGE OF SCIENCE AND TECHNOLOGY, LONDON
TWF/R/	IMPERIAL COLLEGE OF SCIENCE AND TECHNOLOGY, LONDON. DEPT. OF MECHANICAL ENGINEERING
TWF/TN/	IMPERIAL COLLEGE OF SCIENCE AND TECHNOLOGY, LONDON
TWG-	LOS ALAMOS SCIENTIFIC LAB., N. MEX.
TWGC-	NATIONAL RESEARCH COUNCIL. COMMITTEE ON TREATMENT OF WAR GAS CASUALTIES
TWP-(YEAR)-	LOCKHEED ELECTRONICS CO. AEROSPACE SYSTEMS DIV., HOUSTON, TEX.
TWP-	THIOKOL CHEMICAL CORP., BRIGHAM CITY, UTAH
TWP-PR-	AIR TECH. INTELL. CTR., WRIGHT-PATTERSON AFB, OHIO
TWP-WI-	AEROSPACE TECHNICAL INTELLIGENCE CENTER, WRIGHT-PATTERSON AFB, OHIO
TWP-WI-(YEAR)-	AEROSPACE TECHNICAL INTELLIGENCE CENTER, WRIGHT-PATTERSON AFB, OHIO
TWR-	NAVAL ORDNANCE LAB., WHITE OAK, MD. (TECHNICAL WORK REQUEST)
TWR-	THIOKOL CHEMICAL CORP. WASATCH DIV.,BRIGHAM CITY,UTAH
TWRC-	ROBERT A. TAFT WATER RESEARCH CENTER, CINCINNATI
TWRC-AWTRL-	ROBERT A. TAFT WATER RESEARCH CENTER. ADVANCED WASTE TREATMENT RESEARCH LAB., CINCINNATI
TYLA-	LIN (T.Y.) AND ASSOCIATES, LOS ANGELES

U-(YEAR)-	AEROJET-GENERAL CORP. LIQUID ROCKET PLANT, SACRAMENTO, CALIF.
U-	AERONUTRONIC DIV. OF PHILCO CORP., NEWPORT BEACH, CALIF. (UNCLASSIFIED)
U-	AERONUTRONIC SYSTEMS, INC., GLENDALE, CALIF.
U-	CAMBRIDGE ACOUSTICAL ASSOCIATES, INC., MASS.
U-	CHICAGO. UNIV. INST. FOR THE STUDY OF METALS
U-	GENERAL DYNAMICS CORP., SAN DIEGO, CALIF.
U(NUMBER)-	GENERAL DYNAMICS CORP. ELECTRIC BOAT DIV.,GROTON,CONN
U(NUMBER)-(YEAR)-	GENERAL DYNAMICS CORP. ELECTRIC BOAT DIV.,GROTON,CONN
U(YEAR)-	GENERAL ELECTRIC CO. NUCLEAR MATERIALS AND PROPULSION OPERATION, CINCINNATI
U(NO.)-	IMPERIAL COLLEGE OF SCIENCE AND TECHNOLOGY, LONDON. DEPT. OF PHYSICS
U-	INSTITUTE FOR DEFENSE ANALYSES, ARLINGTON, VA.
U-	JOHNS HOPKINS UNIV., SILVER SPRING, MD. APPLIED PHYSICS LAB.
U-	LIBRARY OF CONGRESS. TECH.INFO.DIV.,WASH.,DC (REF.20)
U-	NATIONAL BUREAU OF STANDARDS. BOULDER LABS., COLO.
U-	NORWAY. FORSVARETS FORSKNINGSINSTITUTT, OSLO
U(YEAR)-	PENNSYLVANIA RESEARCH ASSOCIATES INC., PHILADELPHIA
U-	PHILCO-FORD CORP., NEWPORT BEACH, CALIF.
U(NO.)-	TEXAS INSTRUMENTS, INC., DALLAS
U-	THIOKOL CHEMICAL CORP., HUNTSVILLE, ALA.
U-(YEAR)-(NUMBER)A	THIOKOL CHEMICAL CORP., HUNTSVILLE, ALA.
U-	THIOKOL CHEMICAL CORP. REDSTONE DIV.,HUNTSVILLE,ALA.
U(LETTER)C-	OFFICE OF SCIENTIFIC RESEARCH AND DEVELOPMENT
UAC-	ARGONNE NATIONAL LAB., LEMONT, ILL.
UAC-	FAIRCHILD ENGINE AND AIRPLANE CORP. NEPA DIV., OAK RIDGE, TENN.
UAC-	UNITED AIRCRAFT CORP., EAST HARTFORD, CONN.
UAC-A-91(4 DIGITS)-	UNITED AIRCRAFT CORP. RES. LABS., EAST HARTFORD,CONN.
UAC-B-91(4 DIGITS)-	UNITED AIRCRAFT CORP. RES. LABS., EAST HARTFORD,CONN.
UAC-C-91(4 DIGITS)-	UNITED AIRCRAFT CORP. RES. LABS., EAST HARTFORD,CONN.
UAC-D-91(4 DIGITS)-	UNITED AIRCRAFT CORP. RES. LABS., EAST HARTFORD,CONN.
UACDCINS-	ARMY COMBAT DEVELOPMENTS COMMAND. INST. OF NUCLEAR STUDIES, FORT BLISS, TEX.
UAC-E-91(4 DIGITS)-	UNITED AIRCRAFT CORP. RES. LABS., EAST HARTFORD,CONN.
UAC-F-91(4 DIGITS)-	UNITED AIRCRAFT CORP. RES. LABS., EAST HARTFORD,CONN.
UACG-(NUMBER)-(YEAR)	MC GILL UNIV., MONTREAL. UPPER ATMOSPHERE CHEMISTRY GROUP
UAC-G-91(4 DIGITS)-	UNITED AIRCRAFT CORP. RES. LABS., EAST HARTFORD,CONN.
UAC-H-91(4 DIGITS)-	UNITED AIRCRAFT CORP. RES. LABS., EAST HARTFORD,CONN.
UAC-HS-	UNITED AIRCRAFT CORP. HAMILTON STANDARD DIV., WINDSOR LOCKS, CONN.
UAC-J-91(4 DIGITS)-	UNITED AIRCRAFT CORP. RES. LABS., EAST HARTFORD,CONN.
UAC-K-91(4 DIGITS)-	UNITED AIRCRAFT CORP. RES. LABS., EAST HARTFORD,CONN.
UACL ER-	UNITED AIRCRAFT OF CANADA LTD., LONGUEUIL, QUEBEC
UACL TN-	UNITED AIRCRAFT OF CANADA LTD., LONGUEUIL, QUEBEC
UAC-M-	UNITED AIRCRAFT CORP., EAST HARTFORD, CONN.
UAC-METEOR-	UNITED AIRCRAFT CORP., EAST HARTFORD, CONN.
UAC M+SS R-	UNITED AIRCRAFT CORP. MISSILES AND SPACE SYSTEMS DIV., EAST HARTFORD, CONN.
UAC-R-	UNITED AIRCRAFT CORP., EAST HARTFORD, CONN.
UAC-R-	UNITED AIRCRAFT CORP. RES. LABS., EAST HARTFORD,CONN.
UACRL-	UNITED AIRCRAFT CORP. RES. LABS., EAST HARTFORD,CONN.
UAC-RL-B-91(4 DIGITS)-	UNITED AIRCRAFT CORP. RES. LABS., EAST HARTFORD,CONN.
UACRL-C-	UNITED AIRCRAFT CORP. RES. LABS., EAST HARTFORD,CONN.
UACRL-D-	UNITED AIRCRAFT CORP. RES. LABS., EAST HARTFORD,CONN.
UACRL-E-	UNITED AIRCRAFT CORP. RES. LABS., EAST HARTFORD,CONN.
UACRL-F(NUMBER)-	UNITED AIRCRAFT CORP., EAST HARTFORD, CONN.
UACRL-G-91(4 DIGITS)-	UNITED AIRCRAFT CORP. RES. LABS., EAST HARTFORD,CONN.
UACRL-J-91(4 DIGITS)-	UNITED AIRCRAFT CORP. RES. LABS., EAST HARTFORD,CONN.
UAC-SA-	SIKORSKY AIRCRAFT DIV., UNITED AIRCRAFT CORP., BRIDGEPORT, CONN.
UAC-SAR-	SIKORSKY AIRCRAFT DIV., UNITED AIRCRAFT CORP., STRATFORD, CONN.
UAC/SCR-	UNITED AIRCRAFT CORPORATE SYSTEMS CENTER, FARMINGTON, CONN.
UAC/SCR-	UNITED AIRCRAFT CORPORATE SYSTEMS CENTER. SUPPORT SYSTEMS GROUP, EL SEGUNDO, CALIF.
UAC-UAR-(LETTER-NUMBER)	UNITED AIRCRAFT CORP. UNITED AIRCRAFT RES. LABS., EAST HARTFORD, CONN.
UAEEL-RR-	ALASKA. UNIV., COLLEGE. ARCTIC ENVIRONMENTAL ENG. LAB
UAE-NPL-	ALBERTA UNIV., EDMONTON. NUCLEAR RESEARCH CENTRE
UAG-	ALASKA. UNIV., COLLEGE. GEOPHYSICAL INST.
UAG-	NATIONAL OCEANIC AND ATMOSPHERIC ADMINISTRATION, BOULDER, COLO. (UPPER ATMOSPHERE GEOPHYSICS)
UAG-	WORLD DATA CENTER A. COORDINATION OFF. NATL. RESEARCH COUNCIL, WASH., D.C. (UPPER ATMOSPHERE GEOPHYSICS)
UAG-C-	ALASKA. UNIV., COLLEGE. GEOPHYSICAL INST.
UAGC-	MC GILL UNIV., MONTREAL. UPPER ATMOSPHERE CHEMISTRY GROUP
UAGIR-	ALASKA, UNIV., COLLEGE
UAG-R-	ALASKA. UNIV., COLLEGE. GEOPHYSICAL INST.
UAHRI-(YEAR)-	ALABAMA. UNIV., HUNTSVILLE. RESEARCH INSTITUTE
UAL METEOROLOGY CIRC-	UNITED AIRLINES INC., CHICAGO
UAPL-	ARKANSAS. UNIV., FAYETTEVILLE. PLASMA LAB.
UAR-	UNITED AIRCRAFT CORP. RES. LABS., EAST HARTFORD,CONN.
UARAEE-	UNITED ARAB REPUBLIC. ATOMIC ENERGY ESTAB., CAIRO
UAR-H(NUMBER)-	UNITED AIRCRAFT CORP. RES. LABS., EAST HARTFORD,CONN.
UARI-	ALABAMA. UNIV., HUNTSVILLE. RESEARCH INSTITUTE
UARI-RR-	ALABAMA. UNIV., HUNTSVILLE. RESEARCH INSTITUTE
UAR-J-	UNITED AIRCRAFT CORP. RES. LABS., EAST HARTFORD,CONN.
UARK-	ARKANSAS. UNIV., FAYETTEVILLE
UARK-	ARKANSAS. UNIV., LITTLE ROCK. SCHOOL OF MEDICINE
UARK-APR (YEAR)	ARKANSAS. UNIV., FAYETTEVILLE (ANNUAL PROGRESS RPT.)
UARL-	UNITED AIRCRAFT CORP. RES. LABS., EAST HARTFORD,CONN.
UARL-D-	UNITED AIRCRAFT CORP. RES. LABS., EAST HARTFORD,CONN.
UARL-F(6 DIGITS)-	UNITED AIRCRAFT CORP. RES. LABS., EAST HARTFORD,CONN.
UARL-G-	UNITED AIRCRAFT CORP. RES. LABS., EAST HARTFORD,CONN.
UARL-H-	UNITED AIRCRAFT CORP. RES. LABS., EAST HARTFORD,CONN.
UARL-J-	UNITED AIRCRAFT CORP. RES. LABS., EAST HARTFORD,CONN.
UARL-UU-(YEAR)-	UTAH. UNIV., SALT LAKE CITY. UPPER AIR RES. LABS.
UBC-	BRITISH COLUMBIA. UNIV., VANCOUVER
UBCAR-	BRITISH COLUMBIA. UNIV., VANCOUVER
UBG-	ADVANCED RESEARCH PROJECTS AGENCY, WASHINGTON, D.C.
UBG-	INSTITUTE FOR DEFENSE ANALYSES, ARLINGTON, VA.
UBIF-	BOLOGNA. UNIVERSITA. ISTITUTO DI FISICA (A. RIGHI)
UBL-	BARKLEY + DEXTER LABS., INC., FITCHBURG, MASS.
UBRC-	GT. BRIT. ARMAMENT RES. DEPT.,FT.HALSTEAD, KENT, ENG.
UC-	CALIFORNIA. UNIV., BERKELEY
UC-	CALIFORNIA. UNIV., LOS ANGELES
UC-(YR.)-	CALIFORNIA. UNIV., LOS ANGELES
UC-	DIVISION OF REACTOR DEVELOPMENT, AEC. NAVAL REACTORS BRANCH, WASHINGTON, D.C.
UC-	ULTRASONIC CORP., CAMBRIDGE, MASS.
UC-	UNION CARBIDE CORP., N.Y.C.
UCAL-	CALIFORNIA. UNIV. (ALL LOCATIONS)

UC-AL-	CALIFORNIA. UNIV., BERKELEY. ANTENNA LAB.
UCAL HE (NO.)-(NO.)-	CALIFORNIA. UNIV., BERKELEY. INST. OF ENG. RES.
UCAL-IER-	CALIFORNIA. UNIV., BERKELEY. INST. OF ENG. RES.
UCAL-IER-	CALIFORNIA. UNIV., RICHMOND. INST. OF ENG. RES.
UCAL IER S.(NO.)-	CALIFORNIA. UNIV., BERKELEY. INST. OF ENG. RES.
UCAL TN DR-	CALIFORNIA. UNIV., BERKELEY
UCB-	CALIFORNIA. UNIV., BERKELEY
UCB-(NUMBER)-P-(NO).-	CALIFORNIA. UNIV., BERKELEY (USED BY MANY DEPTS.)
UCB-(NUMBER)-R	CALIFORNIA. UNIV., BERKELEY (SQUID PROJECT) (REF. 33)
UCB-(YEAR)-	CALIFORNIA. UNIV., BERKELEY
UCB-(CONTRACT NO.)-P-	CALIFORNIA. UNIV., BERKELEY
UCB-ENG-	CALIFORNIA. UNIV., BERKELEY. DEPT. OF CIVIL ENG.
UCC-	ELECTRO METALLURGICAL CO. METALS RESEARCH LAB., NIAGARA FALLS, N.Y.
UCC-	UNION CARBIDE AND CARBON CORP., N.Y.C.
UCC-	UNION CARBIDE CORP. PARMA RESEARCH LAB., OHIO
UCC/AS-	UNION CARBIDE CORP. MATERIALS SYSTEMS DIV., WHITE PLAINS, N.Y.
UC-CD-	CALIFORNIA. UNIV., BERKELEY. INST. OF ENG. RES.
UCC/DSSD-	UNION CARBIDE RESEARCH INST. SPACE SCIENCE AND ENGINEERING GROUP, TARRYTOWN, N.Y.
UCC-TR-C-	UNION CARBIDE CORP. RESEARCH INST., TARRYTOWN, N.Y.
UCD-	CALIFORNIA. UNIV., DAVIS
UCD-(CONTRACT NO.)-P-	CALIFORNIA. UNIV., DAVIS
UCD-(NO.)P-(NO.)-	CALIFORNIA. UNIV., DAVIS. DEPT. OF POMOLOGY
UCD-(NUMBER)-	CALIFORNIA. UNIV., DAVIS. RADIOBIOLOGY LAB.
UCD-CNL-	CALIFORNIA. UNIV., DAVIS. CROCKER NUCLEAR LAB.
UCD-CNL-EX-	CALIFORNIA. UNIV., DAVIS. CROCKER NUCLEAR LAB.
UCD-CR-	IRELAND. NATIONAL UNIV., DUBLIN. UNIV. COLL.
UCDWR-	NAVY ELECTRONICS LAB., SAN DIEGO, CALIF.
UC DWR (LETTER)-	CALIFORNIA. UNIV., SAN DIEGO. DIV. OF WAR RESEARCH
UC DWR I-	CALIFORNIA. UNIV., SAN DIEGO. DIV. OF WAR RESEARCH (INTERNAL RPTS.)
UCH-	CHICAGO. UNIV.
UC-HE-	CALIFORNIA. UNIV., BERKELEY. INST. OF ENG. RES.
UCHI-CC-	CHICAGO. UNIV. METALLURGICAL LAB.
UCHI-CML-	CHICAGO. UNIV. CHICAGO MIDWAY LABS.
UCHI-IAWR-	CHICAGO. UNIV. INST. FOR AIR WEAPONS RESEARCH
UCI-	CALIFORNIA. UNIV., IRVINE
UCI-(NO.)-P-(NO.)-	CALIFORNIA. UNIV., IRVINE
UCID-	CALIFORNIA. UNIV., BERKELEY. LAWRENCE BERKELEY LAB.
UCID-	CALIFORNIA. UNIV., LIVERMORE. LAWRENCE RADIATION LAB.
UCL-	CALIFORNIA. UNIV., LOS ANGELES
UCLA-	CALIFORNIA. UNIV., LOS ANGELES
UCLA-(NUMBER)-P-(NO.)-	CALIFORNIA. UNIV., LOS ANGELES (USED BY MANY DEPTS.)
UCLA-(YEAR)-	CALIFORNIA. UNIV., LOS ANGELES. BIOTECHNOLOGY LAB.
UCLA-(YEAR)-	CALIFORNIA. UNIV., LOS ANGELES. DEPT. OF ENGINEERING
UCLA-(YEAR)/TEP/	CALIFORNIA. UNIV., LOS ANGELES. DEPT. OF PHYSICS
UCLA-(NUMBER)-	CALIFORNIA. UNIV., LOS ANGELES. LAB. OF NUCLEAR MEDICINE + RADIATION BIOLOGY
UCLA-CASE-	CALIFORNIA. UNIV., LOS ANGELES. DEPT. OF ENGINEERING
UCLA-DE-(YEAR)-	CALIFORNIA. UNIV., LOS ANGELES. DEPT. OF ENGINEERING
UCLA-DM-	CALIFORNIA. UNIV., LOS ANGELES. DEPT. OF METEOROLOGY
UCLA-ENG-	CALIFORNIA. UNIV., LOS ANGELES. SCHOOL OF ENGINEERING AND APPLIED SCIENCE
UCLA-IG-	CALIFORNIA. UNIV., LOS ANGELES. INST. OF GEOPHYSICS
UCLA-MPG-	CALIFORNIA. UNIV., LOS ANGELES. DEPT. OF PHYSICS
UCLA-NEL-	CALIFORNIA. UNIV., LOS ANGELES. DEPT. OF ENGINEERING
UCLA-P-	CALIFORNIA. UNIV., LOS ANGELES. DEPT. OF PHYSICS
UCLA-PHONETICS-WP-	CALIFORNIA. UNIV., LOS ANGELES (WORKING PAPERS)
UCLA-TR-	CALIFORNIA. UNIV., LOS ANGELES
UCM-	CALIFORNIA. UNIV., BERKELEY. DEPT. OF ENGINEERING
UCMC-	UNION CARBIDE METALS CO. METALS RESEARCH LABS., NIAGARA FALLS, N.Y.
UCMC-MRL-TR-	UNION CARBIDE METALS CO. METALS RESEARCH LABS., NIAGARA FALLS, N.Y.
UC-MRL-	CALIFORNIA. UNIV., BERKELEY. MINERALS RESEARCH LAB.
UCN-	UNITED STATES-EURATOM JOINT RESEARCH + DEV. PROGRAM
UC-NBL-O-	CALIFORNIA. UNIV., OAKLAND. NAVAL BIOLOGICAL LAB.
UCNC-	UNION CARBIDE NUCLEAR CO. Y-12 PLANT, OAK RIDGE, TENN
UCNC-Y-	UNION CARBIDE NUCLEAR CO. Y-12 PLANT, OAK RIDGE, TENN
UCO-	COLORADO. UNIV., BOULDER
UCO-	COLORADO. UNIV., DENVER. MEDICAL CENTER
UCOL-(NUMBER)-	COLORADO. UNIV., BOULDER
UCOL-(YEAR)-	COLORADO. UNIV., BOULDER
UCOL-P-	COLORADO. UNIV., BOULDER
UCP-	UNITED CARBON PRODUCTS CO., INC., BAY CITY, MICH.
UCPRL-BMPR-	UNION CARBIDE CORP. PARMA RESEARCH LAB., OHIO
UCPRL-RR-	UNION CARBIDE CORP. PARMA RESEARCH LAB., OHIO
UCR-(NO.)-P-(NO.)-B-	CALIFORNIA. UNIV., RICHMOND. INST. OF ENG. RES.
UCR-	CALIFORNIA. UNIV., RIVERSIDE
UCR-(NUMBER)-P-(NO.)-	CALIFORNIA. UNIV., RIVERSIDE (USED BY MANY DEPTS.)
UCRI-	UNION CARBIDE RESEARCH INST., TARRYTOWN, N.Y.
UCRI-TR-C-	UNION CARBIDE RESEARCH INST., TARRYTOWN, N.Y.
UCRL-60-	CALIFORNIA. UNIV., BERKELEY. CROCKER LAB.
UCRL-	CALIFORNIA. UNIV., BERKELEY. LAWRENCE BERKELEY LAB.
UCRL-60-	CALIFORNIA. UNIV., BERKELEY. RADIATION LAB.
UCRL-	CALIFORNIA. UNIV., LIVERMORE. LAWRENCE LIVERMORE LAB.
UCRL-	CALIFORNIA. UNIV., MERCURY, NEV. LAWRENCE RADIATION LAB.
UCRL-BC-	CALIFORNIA. UNIV., BERKELEY. RADIATION LAB. (REF. 9)
UCRL-COC-	CALIFORNIA. UNIV., LIVERMORE. RADIATION LAB.
UCRL-COPD-	CALIFORNIA. UNIV., BERKELEY. RADIATION LAB.
UCRL-COT-	CALIFORNIA. UNIV., BERKELEY. RADIATION LAB.
UCRL-COVA-	CALIFORNIA. UNIV., BERKELEY. RADIATION LAB.
UCRLD-	CALIFORNIA. UNIV., BERKELEY. RADIATION LAB. (SERIES ASSIGNED BY LOS ALAMOS SCIENTIFIC LAB. TO REPORTS RECEIVED FROM UCRL)
UCRL-NB-	CALIFORNIA. UNIV., LIVERMORE. RADIATION LAB.(NOTEBOOK
UCRL-PUB-	CALIFORNIA. UNIV., LIVERMORE. LAWRENCE RADIATION LAB.
UCRL-SPEC-	CALIFORNIA. UNIV., BERKELEY. RADIATION LAB. (REF. 9)
UCRL-TRANS-	CALIFORNIA. UNIV., BERKELEY. RADIATION LAB.
UCR-P(NO.)-	CALIFORNIA. UNIV., RIVERSIDE
UCS-	CALIFORNIA. UNIV., SANTA CRUZ
UCSB-(NUMBER)-P-(NO.)-	CALIFORNIA. UNIV., SANTA BARBARA(USED BY MANY DEPTS.)
UCSB-CRL-	CALIFORNIA. UNIV., SANTA BARBARA
UCSB-ME-(YEAR)-	CALIFORNIA. UNIV., SANTA BARBARA. DEPT. OF MECH. ENG.
UCSC-	UNIVERSAL-CYCLOPS STEEL CORP., BRIDGEVILLE, PENNA.
UCSD-	CALIFORNIA. UNIV., LA JOLLA
UCSD-(NUMBER)-P-(NO.)-	CALIFORNIA. UNIV., LA JOLLA(USED BY MANY DEPTS.)
UCSD-SP-(YEAR)-	CALIFORNIA. UNIV., LA JOLLA. DEPT. OF PHYSICS
UCSESM-	CALIFORNIA. UNIV., BERKELEY. STRUCTURAL ENG. LAB.
UCSF-(NUMBER)-P-(NO.)-	CALIFORNIA. UNIV., SAN FRANCISCO(USED BY MANY DEPTS.)
UCSF-	CALIFORNIA. UNIV., SAN FRANCISCO. SCHOOL OF MEDICINE
UCTL-	CHICAGO. UNIV. AIR FORCE RADIATION LAB.

UD-(NO.)-E-	CORNELL AERONAUTICAL LAB., INC., BUFFALO
UDC-	LUND, SWEDEN. INSTITUTE OF TECHNOLOGY
UDC-	(UNIVERSAL DECIMAL CLASSIFICATION)
UD-FB-	DELAWARE. UNIV., NEWARK
UDPR-	STRATEGIC BOMBING SURVEY. UTILITIES DIV. (PLANT RPTS)
UDRI-PR-(YEAR)-	DAYTON, OHIO. UNIV. RESEARCH INST.
UDRI-TR-(YEAR)-	DAYTON, OHIO. UNIV. RESEARCH INST.
UDT-	UNDERWATER DEMOLITION TEAM NO. 4, INDIAN HEAD, MD.
UE-	NATIONAL DEFENSE RES. COMM. (UNDERWATER EXPLOSIVES)
UE-	OFFICE OF SCIENTIFIC RESEARCH AND DEVELOPMENT (UNDERWATER EXPLOSIVES AND EXPLOSIONS)
UED-	UNITED ELECTRODYNAMICS, INC., PASADENA, CALIF.
UED-ESD-	UNITED ELECTRODYNAMICS, INC. EARTH SCIENCES DIV., ALEXANDRIA, VA.
UED-GBR-	UNITED ELECTRODYNAMICS, INC. EARTH SCIENCES DIV., PASADENA, CALIF.
UED-SDL-	TELEDYNE, INC. UED EARTH SCIENCES DIV.,ALEXANDRIA,VA.
UED-SDLFR-	TELEDYNE INDUSTRIES, INC. EARTH SCIENCES DIV., ALEXANDRIA, VA.
UEM-	WESTINGHOUSE ELECTRIC CORP. UNDERSEAS DIV., BALTIMORE
UER-	UNDERWATER EXPLOSIONS RESEARCH DIV., PORTSMOUTH, VA.
UER-	WESTINGHOUSE ELECTRIC CORP. UNDERSEAS DIV., BALTIMORE
UERD-	UNDERWATER EXPLOSIONS RESEARCH DIV., PORTSMOUTH, VA.
UERD-(NO.)-(YR.)	UNDERWATER EXPLOSIONS RESEARCH DIV., PORTSMOUTH, VA.
UERD-(YR)-	UNDERWATER EXPLOSIONS RESEARCH DIV., PORTSMOUTH, VA.
UERL-	UNDERWATER EXPLOSIVES RESEARCH LAB., WOODS HOLE, MASS
UERL-TECHNICAL MEMO-	UNDERWATER EXPLOSIVES RESEARCH LAB., WOODS HOLE, MASS
UES-	NAVAL ORDNANCE LAB. UNDERWATER ORDNANCE EXPLOSIVE TRAIN SAFETY COMMITTEE
UFCG-	FLORIDA. UNIV., GAINESVILLE
UF-CH-	FLORIDA. UNIV., GAINESVILLE
UFOIRC-(YEAR)-	UFO INFORMATION RETRIEVAL CENTER, RIDERWOOD, MD.
UF/TN/D/	IMPERIAL COLLEGE OF SCIENCE AND TECHNOLOGY, LONDON. DEPT. OF MECHANICAL ENGINEERING
UG-	PHILCO-FORD CORP. SPACE AND RE-ENTRY SYSTEMS DIV., NEWPORT BEACH, CALIF.
UGC-	UNITED GEOPHYSICAL CO., INC., LOS ANGELES
UGC-	UNITED GEOPHYSICAL CORP., PASADENA, CALIF.
UH-	HAWAII. UNIV., HONOLULU
UH-(CONTRACT NO.)-	HAWAII. UNIV., HONOLULU
UH-(CON.NO.)-P-(NO.)-	HAWAII. UNIV., HONOLULU
UH-	NAVY, WASHINGTON, D.C.
UHI-MED-(YEAR)-	HAWAII. UNIV., HONOLULU. DEPT. OF PHYSIOLOGY
UHMET-(YEAR)-	HAWAII. UNIV., HONOLULU. DEPT. OF METEOROLOGY
UI-	ILLINOIS. UNIV., URBANA
UIAL-(YEAR)-	ILLINOIS. UNIV., URBANA. ANTENNA LAB.
UICC-	ILLINOIS. UNIV., CHICAGO
UILU-ENG-	ILLINOIS. UNIV., URBANA
UJ-(NUMBER)/(YEAR)	CESKOSLOVENSKA AKADEMIE VED. USTAV JADERNEHO VYZKUMU, REZ
UJV-	CESKOSLOVENSKA AKADEMIE VED. USTAV JADERNEHO VYZKUMU, REZ
UJV-(NUMBER)-(LETTER)	CESKOSLOVENSKA AKADEMIE VED. USTAV JADERNEHO VYZKUMU, REZ
UK-	BRITISH AIRCRAFT CORP., STEVENAGE, HERTS, ENGLAND
UK-	UNITED KINGDOM. ARMY PERSONNEL RESEARCH ESTABLISHMENT, BYFLEET, SURREY, ENGLAND
U.K.-	UNITED KINGDOM ATOMIC ENERGY AUTHORITY
UK-	UNITED KINGDOM ATOMIC ENERGY AUTHORITY. RESEARCH GP. ATOMIC ENERGY RES. ESTAB., HARWELL, BERKS, ENGLAND
UK-	UNITED KINGDOM STORES AND CLOTHING RESEARCH AND DEVELOPMENT ESTABLISHMENT, COLCHESTER, ESSEX, ENG.
UKAEA-	UNITED KINGDOM ATOMIC ENERGY AUTHORITY
UKAE-CODE-	UNITED KINGDOM ATOMIC ENERGY AUTHORITY. AUTHORITY HEALTH AND SAFETY BRANCH, RISLEY, LANCS, ENGLAND
UK/C/5/	ATOMIC ENERGY OF CANADA LTD. CHALK RIVER PROJ., ONT. (5TH UK-CANADA TECH. CONF., HARWELL, 1956)
UK/C/6/	ATOMIC ENERGY OF CANADA LTD. CHALK RIVER PROJ., ONT. (6TH UK-CANADA TECH. CONF., CHALK RIVER, 1957)
UKE-CR-	ATOMIC ENERGY OF CANADA LTD. CHALK RIVER PROJ., ONT.
UKP-	(OPERATION UPSHOT-KNOTHOLE) (REF. 7)
UK-RN-	UNITED KINGDOM ATOMIC ENERGY AUTHORITY. RESEARCH GP. ATOMIC ENERGY RES. ESTAB., HARWELL, BERKS, ENGLAND (RESEARCH NEWSLETTER)
UK-RN-BE-	UNITED KINGDOM ATOMIC ENERGY AUTHORITY. RESEARCH GP. ATOMIC ENERGY RES. ESTAB., HARWELL, BERKS, ENGLAND (RESEARCH NEWSLETTER, BERYLLIUM)
UK-RN-GC-A-	UNITED KINGDOM ATOMIC ENERGY AUTHORITY. RESEARCH GP. ATOMIC ENERGY RES. ESTAB., HARWELL, BERKS, ENGLAND (RESEARCH NEWSLETTER, GRAPHITE CHEMISTRY)
UK-RN-GC-B-	UNITED KINGDOM ATOMIC ENERGY AUTHORITY. RESEARCH GP. ATOMIC ENERGY RES. ESTAB., HARWELL, BERKS, ENGLAND (RESEARCH NEWSLETTER, GAS COOLANT)
UK-RN-GC-C-	UNITED KINGDOM ATOMIC ENERGY AUTHORITY. RESEARCH GP. ATOMIC ENERGY RES. ESTAB., HARWELL, BERKS, ENGLAND (RESEARCH NEWSLETTER, GRAPHITE PHYSICS)
UK-RN-PLUT-	UNITED KINGDOM ATOMIC ENERGY AUTHORITY. RESEARCH GP. ATOMIC ENERGY RES. ESTAB., HARWELL, BERKS, ENGLAND (RESEARCH NEWSLETTER, PLUTONIUM)
UK-RN-PM-	UNITED KINGDOM ATOMIC ENERGY AUTHORITY. RESEARCH GP. ATOMIC ENERGY RES. ESTAB., HARWELL, BERKS, ENGLAND (RESEARCH NEWSLETTER, PLUTONIUM METALLURGY)
UK-RN-URAN-	UNITED KINGDOM ATOMIC ENERGY AUTHORITY. RESEARCH GP. ATOMIC ENERGY RES. ESTAB., HARWELL, BERKS, ENGLAND (RESEARCH NEWSLETTER, URANIUM)
UL-	LIVERPOOL. UNIV.
ULB-IMA-NT-	BRUSSELS. UNIVERSITE. INSTITUT DE MECANIQUE APPLIQUEE
ULC-	ULTRASONIC CORP., CAMBRIDGE, MASS.
ULDP-	LIVERPOOL. UNIV. DEPT. OF PHYSICS
ULI-TB-	URBAN LAND INST., WASHINGTON, D.C.
ULME/(LETTER. NUMBER)	LIVERPOOL. UNIV. DEPT. OF MECHANICAL ENGINEERING
ULME-B-	LIVERPOOL. UNIV. DEPT. OF MECHANICAL ENGINEERING
ULRAC-MONO-	URBAN LAND RESEARCH ANALYSTS CO., BOSTON
UM-	GERMANY. ZENTRALE FUER WISSENSCHAFTLICHES BERICHTSWESEN DER LUFTFAHRTFORSCHUNG, BERLIN (UNTERSUCHUNGEN UND MITTEILUNGEN)
UM-	MARYLAND. UNIV., COLLEGE PARK
UM-	MICHIGAN. UNIV., ANN ARBOR
UM-	OFFICE OF SCIENTIFIC RESEARCH AND DEVELOPMENT
UMAEC-	MICHIGAN. UNIV., ANN ARBOR
UMAEC (PROJECT)-(YR.)-	MICHIGAN. UNIV., ANN ARBOR
UMC-	UNIVERSAL MATCH CORP., ST. LOUIS
UMC-B(YEAR)-	UNIVERSAL MATCH CORP. ARMAMENT DIV., ST. LOUIS
U-MCREXP-	AIR MATERIEL COMMAND. ENG. DIV. POWER PLANT LAB., WRIGHT-PATTERSON AFB, OHIO (UNSATISFACTORY REPT.)
U MD-	MARYLAND. UNIV., COLLEGE PARK. INST. FOR FLUID DYNAMICS AND APPLIED MATHEMATICS
UM-EDG-	MICHIGAN. UNIV., ANN ARBOR. ELECTRONIC DEFENSE GROUP
UM-ERI-	MICHIGAN. UNIV., ANN ARBOR. ENGINEERING RES. INST.
UMICH-(NUMBER-NUMBER)	MICHIGAN. UNIV., ANN ARBOR. CAVITATION AND MULTIPHASE FLOW LAB.
UMM-	MICHIGAN. UNIV., ANN ARBOR (ALL DEPTS.)(MEMORANDUM)
UMM-MIRO-	MICHIGAN. UNIV., ANN ARBOR. ENG. RES. INST.(PROJ.MIRO
UMM-MX-	MICHIGAN. UNIV., ANN ARBOR. ENGINEERING RES. INST.
UMNE-	ALLIS-CHALMERS MFG. CO. NUCLEAR POWER DEPT., WASHINGTON, D.C.
UMNE-	MARYLAND. UNIV., COLLEGE PARK
UMN/S-	MINNESOTA. UNIV., MINNEAPOLIS

U MO-	MISSOURI. UNIV., COLUMBIA
UM-P-	MELBOURNE. UNIV. SCHOOL OF PHYSICS
UMR-	MICHIGAN. UNIV., ANN ARBOR (REPORTS)
UMR-	MICHIGAN. UNIV., ANN ARBOR. WILLOW RUN RESEARCH CTR.
UMR-	MICHIGAN. UNIV., YPSILANTI. AERONAUTICAL RES. CTR.
UM/RAO-	MICHIGAN. UNIV., ANN ARBOR. RADIO ASTRONOMY OBS.
UMRI-	MICHIGAN. UNIV., ANN ARBOR. RESEARCH INST.
UM-RPP-	OFFICE OF SCIENTIFIC RESEARCH AND DEVELOPMENT
UMS-	MICHIGAN. UNIV., ANN ARBOR. WILLOW RUN RESEARCH CTR.
UMTA-DC-MTD-	URBAN MASS TRANSPORTATION ADMINISTRATION, WASH.,D.C.
UMTA-URT-(NUMBER-YEAR)-	CONSORTIUM OF UNIVERSITIES. URBAN TRANSPORTATION CENTER, WASHINGTON, D.C.
UN-	ATOMIC ENERGY OF CANADA LTD. CHALK RIVER PROJ., ONT.
UN-	MASSACHUSETTS GENERAL HOSPITAL, BOSTON
UN-	NEVADA. UNIV., RENO
UN-	NORTH AMERICAN AVIATION, INC., DOWNEY, CALIF.
UN-	UNITED NATIONS GENERAL ASSEMBLY, N.Y.C.
UNARP-	UNITED KINGDOM ATOMIC ENERGY AUTHORITY. RESEARCH GP. ATOMIC ENERGY RES. ESTAB., HARWELL, BERKS, ENGLAND
UNBA-	BUENOS AIRES. UNIVERSIDAD
UNC-	NORTH CAROLINA. UNIV., CHAPEL HILL
UNC-	UNITED NUCLEAR CORP., ELMSFORD, N.Y.
UNC-	UNITED NUCLEAR CORP., WHITE PLAINS, N.Y.
UNC-(NUMBER)-	UNITED NUCLEAR CORP. RESEARCH AND ENGINEERING CENTER, ELMSFORD, N.Y. (FORMAL R + D REPORT)
UNC-(LETTERS-NUMBER)-	UNITED NUCLEAR CORP. RESEARCH AND ENGINEERING CENTER, ELMSFORD, N.Y. (INTERNAL MEMORANDA REPORT)
UNC-MEMO-	UNITED NUCLEAR CORP. DEV. DIV., WHITE PLAINS, N.Y.
UNC-(MEMO)-	UNITED NUCLEAR CORP. RESEARCH AND ENGINEERING CENTER, ELMSFORD, N.Y. (EXTERNAL MEMORANDA REPORT)
UNC-MTL-	UNITED NUCLEAR CORP. DEV. DIV., WHITE PLAINS, N.Y.
UNC PHYS/MATH-	UNITED NUCLEAR CORP. DEV. DIV., WHITE PLAINS, N.Y.
UNC-PP-	UNITED NUCLEAR CORP. DEV. DIV., WHITE PLAINS, N.Y.
UNC-RD-	UNITED NUCLEAR CORP. DEV. DIV., WHITE PLAINS, N.Y.
UNC-SPLM-	INTERNATIONAL ATOMIC ENERGY AGENCY, VIENNA (SPECIALIST MEETING, MINUTES)
UND-	UNIVERSITY OF NOTRE DAME, SOUTH BEND, IND.
UNDAS-TR-	UNIVERSITY OF NOTRE DAME, IND. DEPT. OF AEROSPACE ENG
UNDEX-	GT. BRIT. ADMIRALTY UNDEX SUB-PANEL
UNDEX-	GT. BRIT. MINISTRY OF HOME SECURITY
UNESCO-	UNITED NATIONS EDUCATIONAL, SCI. + CULTURAL ORG.
UNH-(YEAR)-	NEW HAMPSHIRE. UNIV., DURHAM
UNH-	NEW HAMPSHIRE. UNIV., DURHAM. DEPT. OF PHYSICS
UNH-R-(YEAR)-	NEW HAMPSHIRE. UNIV., DURHAM
UNICHROME-	UNITED CHROMIUM, INC., DETROIT
UNICIV-R-	NEW SOUTH WALES, AUSTRALIA. UNIV., KENSINGTON. DEPT. OF STRUCTURAL MECHANICS
UNICP-FR-	INNSBRUCK UNIV. INST. FOR THEORETICAL PHYSICS
UNICP-SR-	INNSBRUCK UNIV. INST. FOR THEORETICAL PHYSICS
UNILAC-(NUMBER)-(YEAR)	HEIDELBERG UNIV. INSTITUT FUER ANGEWANDTE PHYSIK
UNIV/SU-	STANFORD UNIV., CALIF.
UNIV/UU-	UTAH. UNIV., SALT LAKE CITY
UNM-	NEW MEXICO. UNIV., ALBUQUERQUE
UNM-PR-	NEW MEXICO. UNIV., ALBUQUERQUE. ENG. EXPERIMENT STA.
UNM-PR-EE-	NEW MEXICO. UNIV., ALBUQUERQUE. ENG. EXPERIMENT STA.
UNM-QPR-	NEW MEXICO. UNIV., ALBUQUERQUE. ENG. EXPERIMENT STA.
UNM-TR-	NEW MEXICO. UNIV., ALBUQUERQUE
UNM-TR-EE-	NEW MEXICO. UNIV., ALBUQUERQUE. ENG. EXPERIMENT STA.
UNT-	GT. BRIT. MINISTRY OF SUPPLY. (TRANSLATIONS MADE BY THE GERMAN STAFF OF UNTERLUESS WORK CENTRE)
UN-TR-	NEVADA. UNIV., RENO. MACKAY SCHOOL OF MINES
UNUCOR-	UNITED NUCLEAR CORP. DEV. DIV., WHITE PLAINS, N.Y.
UOC-	CALIFORNIA. UNIV., LOS ANGELES. DEPT. OF CHEMISTRY
U OF C-	CALIFORNIA. UNIV., BERKELEY
U OF IOWA-(YEAR)-	IOWA. STATE UNIV., IOWA CITY
UOPAC-	CALIFORNIA. UNIV., LIVERMORE. LAWRENCE RADIATION LAB.
UOPB-	CALIFORNIA. UNIV., LIVERMORE. LAWRENCE RADIATION LAB.
UOP-DP-	UNIVERSITY OF THE PACIFIC, SAN FRANCISCO. DEPT. OF PHARMACOLOGY
UOPK-	CALIFORNIA. UNIV., LIVERMORE. LAWRENCE RADIATION LAB.
UOPKA-(YEAR)-	CALIFORNIA. UNIV., LIVERMORE. LAWRENCE RADIATION LAB.
UPEE-(YEAR)/	PADUA UNIVERSITA. ISTITUTO DI ELETTROTECNICA E DI ELETTRONICA
UPH(NUMBER)-TR-	PENNSYLVANIA. UNIV., PHILADELPHIA. SCH. OF CHEM. ENG.
UPPSALA-	UPPSALA. UNIV.
UPPSALA-TN-	UPPSALA. UNIV.
UPR-(NUMBER)N	PENNSYLVANIA. UNIV., PHILADELPHIA
UPR-(4 DIGITS)(LETTER)	PENNSYLVANIA. UNIV., PHILADELPHIA. DEPT. OF PHYSICS
UPR-	PUERTO RICO. UNIV., MAYAGUEZ
UPS-	UNITED KINGDOM. ROYAL NAVAL PERSONNEL RESEARCH COMM., LONDON
UR-(NUMBER)/	LEESONA MOOS LABS. DIV. OF LEESONA CORP., GT.NECK, NY
UR-	NAVY (UNSATISFACTORY REPORT)
UR-	ROCHESTER, N.Y. UNIV.
UR-(CONTRACT NO.)-	ROCHESTER, N.Y. UNIV.
UR-(NUMBER)-	ROCHESTER, N.Y. UNIV. DEPT. OF RADIATION BIOLOGY + BIOPHYSICS
URA-	UNIVERSITIES RESEARCH ASSOCIATION, INC., WASH., D.C.
URCRL-	CALIFORNIA. UNIV., BERKELEY. LAWRENCE RADIATION LAB.
URE-TN-WSD-	AUSTRALIA. WEAPONS RESEARCH ESTABLISHMENT, SALISBURY
URI-	RHODE ISLAND. UNIV., KINGSTON
URI-	UNITED RESEARCH INC., CAMBRIDGE, MASS.
URM-	RAND CORP., SANTA MONICA, CALIF.
UR-NSRL-	ROCHESTER, N.Y. UNIV. NUCLEAR STRUCTURE RESEARCH LAB.
UR-NSRL-PR-	ROCHESTER, N.Y. UNIV. NUCLEAR STRUCTURE RESEARCH LAB.
URPA-	ROCHESTER, N.Y. UNIV.
U-RQR/(YEAR)-	JOHNS HOPKINS UNIV., SILVER SPRING, MD. APPLIED PHYSICS LAB. (RES. + DEV. QUARTERLY REPT.)
URR-	UNITED KINGDOM ATOMIC ENERGY AUTHORITY. UNIVERSITIES RESEARCH REACTOR, RISLEY, LANCS, ENGLAND
URS-(NUMBERS)-	AIR FORCE ROCKET PROPULSION LAB., EDWARDS AFB, CALIF.
URS-	UNION CARBIDE CORP. PARMA RESEARCH LAB., OHIO
URS-	URS RESEARCH CO., BURLINGAME, CALIF.
URS-	URS RESEARCH CO., SAN MATEO, CALIF.
URS B(NUMBER)-	UNITED RESEARCH SERVICES, INC., BURLINGAME, CALIF.
URS/CR-	UNITED RESEARCH SERVICES, INC., BURLINGAME, CALIF.
URS-FD-	URS CORP., BURLINGAME, CALIF.
URSI-	CANADIAN NATIONAL COMMITTEE, UNION RADIO-SCIENTIFIQUE INTERNATIONALE, OTTAWA
URSI-	UNITED STATES NATIONAL COMMITTEE, UNION RADIO-SCIENTIFIQUE INTERNATIONALE, WASHINGTON, D.C.
US-	CHALMERS TEKNISKA HOEGSKOLA, GOETEBORG
US-	NAVAL ORDNANCE LAB., WHITE OAK, MD.
US-	SAVANNAH RIVER OPERATIONS OFFICE, AEC, AIKEN, S.C.
USAA-	AIR RES. + DEV. COMMAND. EUROPEAN OFF., BRUSSELS
USAAESW-AB-	ARMY AIRBORNE ELECTRONICS AND SPECIAL WARFARE BOARD, FORT BRAGG, N.C.

USAAESWBD-AB-
    ARMY AIRBORNE ELECTRONICS AND SPECIAL WARFARE BOARD,
    FORT BRAGG, N.C.
USAAESW-CE-
    ARMY AIRBORNE ELECTRONICS AND SPECIAL WARFARE BOARD,
    FORT BRAGG, N.C.
USAAML-
    ARMY AVIATION MATERIAL LABS., FORT EUSTIS, VA.
USAAML-TR-
    ARMY AVIATION MATERIAL LABS., FORT EUSTIS, VA.
USAAMS-STUDY-
    ARMY ARTILLERY AND MISSILE SCHOOL, FORT SILL, OKLA.
USAARL-(YEAR)-
    NAVAL AEROSPACE MEDICAL INST., PENSACOLA, FLA.
USAARTYBD-FA-(NUMBER)-
    ARMY ARTILLERY BOARD, FORT SILL, OKLA.
USAARU-
    ARMY AEROMEDICAL RESEARCH UNIT, FORT RUCKER, ALA.
USAARU-(YEAR)-
    ARMY AEROMEDICAL RESEARCH UNIT, FORT RUCKER, ALA.
USAARU-JR-
    ARMY AEROMEDICAL RESEARCH UNIT, FORT RUCKER, ALA.
USAASTA-(YEAR-NUMBER)
    ARMY AVIATION SYSTEMS TEST ACTIVITY, EDWARDS AFB, CAL
USAATC-
    ARMY ARCTIC TEST CENTER, FORT GREELY, ALASKA
USAAVCOM-TR-(YR-NO-LTR)-
    ARMY AVIATION MATERIEL COMMAND, ST. LOUIS
USAAVLABS-TN-
    ARMY AVIATION MATERIAL LABS., FORT EUSTIS, VA.
USAAVLABS-TR-(YEAR)-
    ARMY AVIATION MATERIAL LABS., FORT EUSTIS, VA.
USAAVLABS-TR-
    HUGHES TOOL CO. AIRCRAFT DIV., CULVER CITY, CALIF.
USAAVLABS-TR-(YEAR)-
    NORTHROP CORP., HAWTHORNE, CALIF.
USAAVNTA-(YEAR)-
    ARMY AVIATION TEST ACTIVITY, EDWARDS AFB, CALIF.
USAAVSCOM-TR-(YR.)-
    ARMY AVIATION SYSTEMS COMMAND, ST. LOUIS
USAAVSCOM-TR-XE-(YR.)-
    ARMY AVIATION SYSTEMS COMMAND, ST. LOUIS
USACDC-(YEAR)-
    ARMY COMBAT DEVELOPMENTS COMMAND, FORT BELVOIR, VA.
USACDCAA-(YEAR)-
    ARMY COMBAT DEVELOPMENTS COMMAND. AVIATION AGENCY,
    FORT RUCKER, ALA.
USACDCARTYA-(YEAR)-
    ARMY COMBAT DEVELOPMENTS COMMAND. ARTILLERY AGENCY,
    FORT SILL, OKLA.
USACDCCAG-(YEAR)-
    ARMY COMBAT DEVELOPMENTS COMMAND. COMBINED ARMS
    AGENCY, FORT LEAVENWORTH, KAN.
USACDCCBRA-
    ARMY COMBAT DEVELOPMENTS COMMAND. CHEMICAL-BIOLOG-
    ICAL-RADIOLOGICAL AGENCY, FORT MC CLELLAND, ALA.
USACDCCEA-(YEAR)-
    ARMY COMBAT DEVELOPMENTS COMMAND. COMMUNICATIONS-
    ELECTRONICS AGENCY, FORT MONMOUTH, N.J.
USACDCCHA-(YEAR)-
    ARMY COMBAT DEVELOPMENTS COMMAND. CHAPLAIN AGENCY,
    FORT LEE, VA.
USACDCEA-(YEAR)-
    ARMY COMBAT DEVELOPMENTS COMMAND. ENGINEER AGENCY,
    FORT BELVOIR, VA.
USACDCEC-
    ARMY COMBAT DEVELOPMENTS COMMAND. EXPERIMENTATION
    CENTER, FORT ORD, CALIF.
USACDCIA-(YEAR)-
    ARMY COMBAT DEVELOPMENTS COMMAND. INFANTRY AGENCY,
    FORT BENNING, GA.
USACDCIAS-
    ARMY COMBAT DEVELOPMENTS COMMAND. INSTITUTE OF AD-
    VANCED STUDIES, CARLISLE BARRACKS, PENNA.
USACDCINS-(MONTH)-(YR.)-
    ARMY COMBAT DEVELOPMENTS COMMAND. INST. OF NUCLEAR
    STUDIES, FORT BLISS, TEXAS
USACDCINTA-
    ARMY INTELLIGENCE CENTER, FORT HOLABIRD, MD.
USACDCJAA-(YEAR)-
    ARMY COMBAT DEVELOPMENTS COMMAND. JUDGE ADVOCATE
    AGENCY, CHARLOTTESVILLE, VA.
USACDCMA-(YEAR)-
    ARMY COMBAT DEVELOPMENTS COMMAND. MAINTENANCE AGENCY,
    ABERDEEN PROVING GROUND, MD.
USACDCMPA-(YEAR)-
    ARMY COMBAT DEVELOPMENTS COMMAND. MILITARY POLICE
    AGENCY, FORT GORDON, GA.
USACDCMSA-(YEAR)-
    ARMY COMBAT DEVELOPMENTS COMMAND. MEDICAL SERVICE
    AGENCY, FORT SAM HOUSTON, TEX.
USACDCNG-
    ARMY COMBAT DEVS. COMMAND. NUCLEAR GP., FT.BLISS,TEX.
USACDC-PAM-(NUMBER)-
    ARMY COMBAT DEVELOPMENTS COMMAND, FORT BELVOIR, VA.
USACRDL-TM-
    ARMY CHEMICAL RES. + DEV. LABS., ARMY CHEM. CTR., MD.
USACRDL-TM-
    ARMY COMBAT DEVS. COMMAND. NUCLEAR GP., FT.BLISS,TEX.
USACSC-
    ARMY COMPUTER SYSTEMS COMMAND, FORT BELVOIR, VA.
USAEEC-
    ARMY ENLISTED EVALUATION CENTER, FORT BENJAMIN
    HARRISON, IND.
USAEHA-
    ARMY ENVIRONMENTAL HYGIENE AGENCY,EDGEWOOD ARSENAL,MD
USAEHA-STUDY-
    ARMY ENVIRONMENTAL HYGIENE AGENCY,EDGEWOOD ARSENAL,MD
USAELRDL-TR-
    ARMY ELECTRONICS RES. + DEV. LAB., FT. MONMOUTH, N.J.

USAEL-TR-
    ARMY ELECTRONICS RES. + DEV. LAB., FT. MONMOUTH, N.J.
USAEL-TR-
    ARMY SIGNAL RESEARCH + DEV. LAB., FORT MONMOUTH, N.J.
USAEPC-SIG-
    ARMY ELECTRONIC PROVING GROUND, FORT HUACHUCA, ARIZ.
USAEPG-
    ARMY ELECTRONIC PROVING GROUND, FORT HUACHUCA, ARIZ.
USAEPG-CR-
    ARMY ELECTRONIC PROVING GROUND, FORT HUACHUCA, ARIZ.
USAEPG-FR-
    ARMY ELECTRONIC PROVING GROUND, FORT HUACHUCA, ARIZ.
USAEPG-LR-
    ARMY ELECTRONIC PROVING GROUND, FORT HUACHUCA, ARIZ.
USAEPG-SIR-
    ARMY ELECTRONIC PROVING GROUND, FORT HUACHUCA, ARIZ.
USAEPG-TP-
    ARMY ELECTRONIC PROVING GROUND, FORT HUACHUCA, ARIZ.
USAERADL-TR-
    ARMY ELECTRONICS RES. + DEV. LAB., FT. MONMOUTH, N.J.
USAERDA-
    ARMY ELECTRONICS RESEARCH AND DEVELOPMENT ACTIVITY,
    WHITE SANDS MISSILE RANGE, N. MEX.
USAERDAA-ELCT-
    ARMY ELECTRONICS RESEARCH AND DEVELOPMENT ACTIVITY,
    FORT HUACHUCA, ARIZ.
USAERDAA-MET-
    ARMY ELECTRONICS RESEARCH AND DEVELOPMENT ACTIVITY,
    FORT HUACHUCA, ARIZ.
USAERDAA-MET-(NO.)-(YR.)
    ARMY ELECTRONICS RESEARCH AND DEVELOPMENT ACTIVITY,
    FORT HUACHUCA, ARIZ.
USAERDL-
    ARMY ENGINEER RESEARCH + DEV. LABS., FT. BELVOIR, VA.
USAERDL-TR-
    ARMY ELECTRONICS RES. + DEV. LAB., FT. MONMOUTH, N.J.
USAETL-(NUMBER)-(LTRS.)
    ARMY ENGINEER TOPOGRAPHIC LABS., FORT BELVOIR, VA.
USAETL-(NUMBER)-TR
    ARMY ENGINEER TOPOGRAPHIC LABS., FORT BELVOIR, VA.
USAETL-RN-
    ARMY ENGINEER TOPOGRAPHIC LABS., FORT BELVOIR, VA.
USAETL-TB-
    ARMY ENGINEER TOPOGRAPHIC LABS., FORT BELVOIR, VA.
USAETL-TR-
    ARMY ENGINEER TOPOGRAPHIC LABS., FORT BELVOIR, VA.
USAF-
    AIR FORCE, WASHINGTON, D.C.
USAFABD-FA-
    ARMY FIELD ARTILLERY BOARD, FORT SILL, OKLA.
USAF-AFM(NUMBER)-
    AIR FORCE, WASHINGTON, D.C. (AIR FORCE MANUAL)
USAFA-TR-(YEAR)-
    AIR FORCE ACADEMY, COLORADO SPRINGS
USAFE-
    AIR FORCES IN EUROPE (U.S.), WIESBADEN, GERMANY
USAFEL-SR-
    AIR FORCE EPIDEMIOLOGICAL LAB., LACKLAND AFB, TEX.
USAF-ETAC-TN-(YEAR)-
    AIR FORCE ENVIRONMENTAL TECHNICAL APPLICATIONS
    CENTER, WASHINGTON, D.C.
USAFIT-TR-
    AIR FORCE INST. OF TECH., WRIGHT-PATTERSON AFB, OHIO
USAF OA R-
    AIR FORCE, WASHINGTON, D.C. (OPERATIONS ANALYSIS)
USAF OA SR-
    AIR FORCE, WASH.,D.C.(OPERATIONS ANALYSIS SPEC. REPT)
USAF OA STM-
    AIR FORCE, WASHINGTON, D.C. (OPERATIONS ANALYSIS
    SPECIAL TECHNICAL MEMORANDUM)
USAF OA TM-
    AIR FORCE, WASH.,D.C.(OPERATIONS ANALYSIS TECH. MEMO)
USAF OA WP-
    AIR FORCE, WASHINGTON, D.C. (OPERATIONS ANALYSIS
    WORKING PAPER)
USAF PV-
    PHYSICAL VULNERABILITY DIV. (AIR FORCE), WASH., D.C.
USAF PVIM-
    PHYSICAL VULNERABILITY DIV. (AIR FORCE), WASH., D.C.
    (INTERIM MEMORANDUM)
USAF PVTM-
    PHYSICAL VULNERABILITY DIV. (AIR FORCE), WASH., D.C.
    (TECHNICAL MEMORANDUM)
USAFSAM-
    SCHOOL OF AVIATION MEDICINE, RANDOLPH AFB, TEX.
USAF-SD-(YEAR)
    AIR FORCE, WASHINGTON, D.C. (STATISTICAL DIGEST)
USAFSOF-(NUMBER)A-
    AIR FORCE SPECIAL OPERATIONS FORCE, EGLIN AFB, FLA.
USAF T.O.-
    AIR FORCE
USAF T.O.11N-
    AIR FORCE (TECH ORDER) (REF. 14)
USAF-TR-(YEAR)-
    AIR FORCE, WASHINGTON, D.C.
USAG-
    OFFICE OF NAVAL RESEARCH, WASHINGTON, D.C.
USAIB-
    ARMY INFANTRY BOARD, FORT BENNING, GA.
USAITAD-
    ARMY INTELLIGENCE THREAT ANALYSIS DETACHMENT,WASH.,DC
USAITAD-(YEAR)-
    MILITARY INTELLIGENCE DETACHMENT (407TH) (STRATEGIC),
    WILLIAMSBURG, VA.
USAJFKSWC-
    JOHN F. KENNEDY CTR. FOR SPECIAL WARFARE,FT. BRAGG,NC
USALMC-ST-
    ARMY LOGISTICS MANAGEMENT CENTER, FORT LEE, VA.
USALMC-STUDY-
    ARMY LOGISTICS MANAGEMENT CENTER, FORT LEE, VA.

USALMC-W-	ARMY LOGISTICS MANAGEMENT CENTER, FORT LEE, VA.
USAMEC-	ARMY MOBILITY EQUIPMENT COMMAND, FORT BELVOIR, VA.
USAMERDC-	ARMY MOBILITY EQUIPMENT COMMAND, FORT BELVOIR, VA.
USAMERDC-	ARMY MOBILITY EQUIPMENT RES. AND DEV. CENTER, FORT BELVOIR, VA.
USAMERDC-M(NO.)-	ARMY MOBILITY EQUIPMENT RES. AND DEV. CENTER, FORT BELVOIR, VA.
USAMERDC-T-	ARMY MOBILITY EQUIPMENT COMMAND, FORT BELVOIR, VA.
USAMERDC-T-(NO.)-	ARMY MOBILITY EQUIPMENT RES. AND DEV. CENTER, FORT BELVOIR, VA.
USAMERDC-TR-	ARMY MOBILITY EQUIPMENT RES. AND DEV. CENTER, FORT BELVOIR, VA.
USAMICOM POMM-	ARMY MISSILE COMMAND, REDSTONE ARSENAL, ALA.
USAMICOM-QSA-(LETTER)-	ARMY MISSILE COMMAND, REDSTONE ARSENAL, ALA.
USAMICOM-U-	ARMY MISSILE COMMAND, REDSTONE ARSENAL, ALA.
USAMRDC-	ARMY MEDICAL RES. AND DEV. COMMAND, WASHINGTON, D.C.
USAMR+DC-	NAVAL RADIOLOGICAL DEFENSE LAB., SAN FRANCISCO
USAMRL-	ARMY MEDICAL RESEARCH LAB., FORT KNOX, KY.
USAMRL-(LETTER)-	ARMY MEDICAL RESEARCH LAB., FORT KNOX, KY.
USAMRNL-	ARMY MEDICAL RESEARCH AND NUTRITION LAB., DENVER
USAMUCOM-(YEAR)-	ARMY MUNITIONS COMMAND, DOVER, N.J.
USAMUCOM-COSTECH-	ARMY MUNITIONS COMMAND, DOVER, N.J.
USA-NLABS-	ARMY NATICK LABS., MASS.
USA-NLABS-FD-(YEAR)-	ARMY NATICK LABS. FOOD DIV., MASS.
USA-NLABS-TN-(YEAR)-	ARMY NATICK LABS. PIONEERING RESEARCH DIV., MASS.
USA-NLABS-TR-	ARMY NATICK LABS., MASS.
USA-NLABS-TR-(YEAR)-	ARMY NATICK LABS., MASS.
USANLABS-TR-(YEAR)-	ARMY NATICK LABS. FOOD LAB., MASS.
USANWSG-	NUCLEAR WEAPONS SURETY GROUP, FORT BELVOIR, VA.
USARADBD-ES-	ARMY AIR DEFENSE BOARD, FORT BLISS, TEX.
USARADBD-GM-	ARMY AIR DEFENSE BOARD, FORT BLISS, TEX.
USARDL-	ARMY SIGNAL RESEARCH + DEV. LAB., FORT MONMOUTH, N.J.
USARDL-TR-	ARMY SIGNAL RESEARCH + DEV. LAB., FORT MONMOUTH, N.J.
USARIEM-TR-(YEAR)-	ARMY RESEARCH INST. OF ENVIRONMENTAL MEDICINE, NATICK, MASS.
USARP-	NATIONAL SCIENCE FOUNDATION, WASHINGTON, D.C. (ANTARCTIC RESEARCH PROGRAM)
USARRL-(YEAR)-	NAVAL AEROSPACE MEDICAL INST., PENSACOLA, FLA.
USASCC-PAM-(NOS.)-	ARMY STRATEGIC COMMUNICATIONS COMMAND, WASHINGTON, DC
USASCC-TM-(NOS.)(NOS.)-	ARMY STRATEGIC COMMUNICATIONS COMMAND, WASHINGTON, DC
USASRDL-TN-	ARMY ELECTRONICS LABS., FORT MONMOUTH, N.J.
USASRDL-TR-	ARMY SIGNAL RESEARCH + DEV. LAB., FORT MONMOUTH, N.J.
USASSA-	ARMY SIGNAL SUPPORT AGENCY
USATEA-(YEAR)-	ARMY TRANSPORTATION ENG. AGENCY, FORT EUSTIS, VA.
USATEA-TR-	ARMY TRANSPORTATION ENG. AGENCY, FORT EUSTIS, VA.
USATRECOM-TR-(YEAR)-	ARMY TRANSPORT. RES. + DEV. COMMAND, FT. EUSTIS, VA.
USATTC-	ARMY TROPIC TEST CENTER, FORT CLAYTON, CANAL ZONE
USBM-	BUREAU OF MINES (ALL LOCATIONS)
USBM-APRL-	BUREAU OF MINES. APPLIED PHYSICS RESEARCH LAB., COLLEGE PARK, MD.
USBM-C-	BUREAU OF MINES
USBM-RC-	BUREAU OF MINES. ALBANY METALLURGY RES. CTR., ORE.
USBM-RI-	BUREAU OF MINES, WASHINGTON, D.C.
USBM-TPR-	BUREAU OF MINES
USBM-TR-	BUREAU OF MINES
USBM-U-	BUREAU OF MINES
USBR-GENERAL-	BUREAU OF RECLAMATION (INTERIOR), DENVER
USC-	UNIVERSITY OF SOUTHERN CALIFORNIA, LOS ANGELES
USC-(NUMBERS)-P-	UNIVERSITY OF SOUTHERN CALIFORNIA, LOS ANGELES
USC-(CONTRACT NO.)-	UNIVERSITY OF SOUTHERN CALIFORNIA, LOS ANGELES. DEPT. OF PHYSICS
USCAE-	PENNSYLVANIA HOSPITAL. UNIT FOR EXPERIMENTAL PSYCHIATRY, PHILADELPHIA
USCAE-	UNIVERSITY OF SOUTHERN CALIFORNIA, LOS ANGELES. DEPT. OF AEROSPACE ENGINEERING
USCAL-	UNIVERSITY OF SOUTHERN CALIFORNIA, LOS ANGELES
USCAL-EC-	UNIVERSITY OF SOUTHERN CALIF., LOS ANGELES. ENG. CTR.
USC-CSPR-	UNIVERSITY OF SOUTHERN CALIF., LOS ANGELES. ENG. CTR.
USCEC-	UNIVERSITY OF SOUTHERN CALIF., LOS ANGELES. ENG. CTR.
USCEC-R-	UNIVERSITY OF SOUTHERN CALIF., LOS ANGELES. ENG. CTR.
USCEE-	UNIVERSITY OF SOUTHERN CALIFORNIA, LOS ANGELES. DEPT. OF ELECTRICAL ENGINEERING
USCEE-	UNIVERSITY OF SOUTHERN CALIF., LOS ANGELES. ENG. CTR.
USCG-	COAST GUARD, WASHINGTON, D.C.
USCG-BULL-	COAST GUARD, WASHINGTON, D.C.
USC-GEOL-(YEAR)-	UNIVERSITY OF SOUTHERN CALIFORNIA, LOS ANGELES. DEPT. OF GEOLOGY
USCG-PROGRAM-ENE-	COAST GUARD. OFFICE OF ENGINEERING, WASH., D.C.
USCISE-	UNIVERSITY OF SOUTHERN CALIFORNIA, LOS ANGELES. DEPT. OF INDUSTRIAL + SYSTEMS ENGINEERING
USCL-T-	UNITED STEEL COMPANIES, LTD., SHEFFIELD, YORKS, ENG.
USCOMM-DC-	DEPARTMENT OF COMMERCE, WASHINGTON, D.C.
USCSC-	CIVIL SERVICE COMMISSION, WASHINGTON, D.C.
USC-TR-(YEAR)-	SOUTH CAROLINA. UNIV., COLUMBIA
USC-VACUV-	UNIVERSITY OF SOUTHERN CALIFORNIA, LOS ANGELES. DEPT. OF PHYSICS
USDA-	DEPARTMENT OF AGRICULTURE, WASHINGTON, D.C.
USDA-ARS-SWC-	AGRICULTURAL RESEARCH SERVICE. SOIL AND WATER CONSERVATION RESEARCH DIV., ITHACA, N.Y.
USFET-	(UNITED STATES FORCES, EUROPEAN THEATER)
USFS-	FOREST SERVICE, WASHINGTON, D.C.
USGA-AMCHITKA-	GEOLOGICAL SURVEY, DENVER
USGA-TD-(YEAR)-	GEOLOGICAL SURVEY. TOPOGRAPHIC DIV., WASHINGTON, D.C.
USGPO-	GOVERNMENT PRINTING OFFICE, WASHINGTON, D.C.
USGS-(NUMBER)-	GEOLOGICAL SURVEY, DENVER
USGS-	GEOLOGICAL SURVEY, WASHINGTON, D.C.
USGS-COMPUTER CONTRIB-	GEOLOGICAL SURVEY, WASHINGTON, D.C.
USGS-DO-(YEAR)-	GEOLOGICAL SURVEY, WASHINGTON, D.C.
USGS-GD-(YEAR)-	GEOLOGICAL SURVEY, WASHINGTON, D.C.
USGS-TD-(YEAR)-	GEOLOGICAL SURVEY. TOPOGRAPHIC DIV., WASHINGTON, D.C.
USGS-TEI-	GEOLOGICAL SURVEY, WASHINGTON, D.C.
USGS-TELI-	GEOLOGICAL SURVEY, WASHINGTON, D.C.
USGS-TEM-	GEOLOGICAL SURVEY, WASHINGTON, D.C.
USGS-WP-	GEOLOGICAL SURVEY, WASHINGTON, D.C.
USI-	UNDERWATER SYSTEMS INC., WHEATON, MD.
USIP-(YEAR)-	STOCKHOLM. UNIV. INST. OF PHYSICS
US-IPR-	SYRACUSE UNIV., N.Y.
USL-	NAVY UNDERWATER SOUND LAB., NEW LONDON, CONN.
USL-	NAVY UNDERWATER SOUND REFERENCE LAB., ORLANDO, FLA.
USL-PUB-	NAVY UNDERWATER SOUND LAB., NEW LONDON, CONN.
USL-SYMPOSIUM-	NAVY UNDERWATER SOUND LAB., NEW LONDON, CONN.
USL-TM-	NAVY UNDERWATER SOUND LAB., NEW LONDON, CONN.
USMA-	MILITARY ACADEMY, WEST POINT, N.Y.
USMC-	UNITED SHOE MACHINERY CORP., BEVERLY, MASS.
USMC-ADO-CC-	MARINE CORPS, WASHINGTON, D.C.
USMC-EDR-CT-	MARINE CORPS, WASHINGTON, D. C.
USMC-EDR-TM-	MARINE CORPS, WASHINGTON, D. C.

USMC-FMFM-	MARINE CORPS, WASHINGTON, D.C.
USMC-GOR-CC-	MARINE CORPS, WASHINGTON, D.C.
USMC-GOR-CT-	MARINE CORPS, WASHINGTON, D. C.
USMC-GOR-FS-	MARINE CORPS, WASHINGTON, D. C.
USMC-GOR-LM-	MARINE CORPS, WASHINGTON, D. C.
USMC-GOR-LO-	MARINE CORPS, WASHINGTON, D. C.
USMC-GOR-NB-	MARINE CORPS, WASHINGTON, D. C.
USMC-ND-	UNITED SHOE MACHINERY CORP., BEVERLY, MASS.
USMC-SOR-(LTR.)(NO.)	MARINE CORPS, WASHINGTON, D.C.
USMC-SOR-CC-	MARINE CORPS, WASHINGTON, D. C.
USMC-SOR-CT-	MARINE CORPS, WASHINGTON, D. C.
USMC-SOR-EW-	MARINE CORPS, WASHINGTON, D. C.
USMC-SOR-IT-	MARINE CORPS, WASHINGTON, D. C.
USMC-SOR-LO-	MARINE CORPS, WASHINGTON, D. C.
USMC-SOR-NB-	MARINE CORPS, WASHINGTON, D. C.
USMC-SOR-TM-	MARINE CORPS, WASHINGTON, D.C.
USNA-	NAVAL ACADEMY, ANNAPOLIS
USNA-E-(YEAR)-	NAVAL ACADEMY, ANNAPOLIS. DEPT. OF ENGINEERING
USNA-S-	NAVAL ACADEMY, ANNAPOLIS
USNA-TSPR-	NAVAL ACADEMY, ANNAPOLIS
USNBD-	NAVY BOMB DISPOSAL SCHOOL, WASHINGTON, D.C.
USN-CR-(YR.)-	MARTIN CO., DENVER
USN DR-	BUREAU OF AERONAUTICS (NAVY)
USNEL-	NAVY ELECTRONICS LAB., SAN DIEGO, CALIF.
USNMI-	(NAVAL MANEUVERING INSTRUCTIONS)
USNMRI-	NAVAL MEDICAL RESEARCH INST., BETHESDA, MD.
USNPGS-	NAVAL POSTGRADUATE SCHOOL, MONTEREY, CALIF.
USNR-	NAVAL RESERVE, WASHINGTON, D.C.
USNRDL-	NAVAL RADIOLOGICAL DEFENSE LAB., SAN FRAN. (REF. 30)
USNRDL-AD-	NAVAL RADIOLOGICAL DEFENSE LAB., SAN FRANCISCO
USNRDL-CP-(YEAR)-	NAVAL RADIOLOGICAL DEFENSE LAB., SAN FRANCISCO
USNRDL-ER-	NAVAL RADIOLOGICAL DEFENSE LAB., SAN FRANCISCO
USNRDL-LR-	NAVAL RADIOLOGICAL DEFENSE LAB., SAN FRANCISCO
USNRDL-P-	NAVAL RADIOLOGICAL DEFENSE LAB., SAN FRANCISCO
USNRDL-REVIEWS/LECTURES-	NAVAL RADIOLOGICAL DEFENSE LAB., SAN FRANCISCO
USNRDL-TM-	NAVAL RADIOLOGICAL DEFENSE LAB., SAN FRANCISCO
USNRDL-TMC-	NAVAL RADIOLOGICAL DEFENSE LAB., SAN FRANCISCO (TECHNICAL MANUAL)
USNRDL-TR-	NAVAL RADIOLOGICAL DEFENSE LAB., SAN FRANCISCO
USNRDL-TRC-	NAVAL RADIOLOGICAL DEFENSE LAB., SAN FRANCISCO
USNRDL-TRL-	NAVAL RADIOLOGICAL DEFENSE LAB., SAN FRANCISCO
USNRDL-TRX-	NAVAL RESEARCH LAB. UNDERWATER SOUND REFERENCE DIV., ORLANDO, FLA.
USNTPS-FTM-	NAVAL TEST PILOT SCHOOL, PATUXENT RIVER, MD.
USNUSL-	NAVY UNDERWATER SOUND LAB., NEW LONDON, CONN.
US-NWSSG-	AIR FORCE NUCLEAR WEAPON SYSTEM SAFETY GROUP, KIRTLAND AFB, N. MEX.
USPP-	UNITED STATES PHOSPHORIC PRODUCTS DIV., TENNESSEE CORP., TAMPA, FLA.
US-PR-	SYRACUSE UNIV., N.Y.
U-SQR/(YEAR)-	JOHNS HOPKINS UNIV., SILVER SPRING, MD. APPLIED PHYSICS LAB. (SPACE QUARTERLY REPORT)
USR-	UNITED RESEARCH SERVICES, INC., BURLINGAME, CALIF.
USRC-	UNITED STATES RUBBER CO., N.Y.C.
USRC-QPR-	UNITED STATES RUBBER CO., N.Y.C.
USRC-SAR-	UNITED STATES RUBBER CO., N.Y.C.
USREQ-	UNITED KINGDOM ATOMIC ENERGY AUTHORITY. WEAPONS GROUP ATOMIC WEAPONS RES. ESTAB.,ALDERMASTON, BERKS. ENG.
USRL-	NAVY UNDERWATER SOUND REFERENCE LAB., ORLANDO, FLA.
USRLG-	NAVY UNDERWATER SOUND REFERENCE LAB., ORLANDO, FLA.
USRL MT. LAKES-	NAVY UNDERWATER SOUND REFERENCE LAB., MT. LAKES, N.J.
USRLO-	NAVY UNDERWATER SOUND REFERENCE LAB., ORLANDO, FLA.
USRL-RR-	NAVY UNDERWATER SOUND REFERENCE LAB., ORLANDO, FLA.
USRL-TRANS-	NAVY UNDERWATER SOUND REFERENCE LAB., ORLANDO, FLA.
USS-	UNITED STATES STEEL CORP. RES. CTR., MONROEVILLE, PA.
USS-	UNITED STATES STEEL CORP. RESEARCH LABS.,KEARNY, N.J.
USS-HDW-IER-	UNITED STATES STEEL CORP. HOMESTEAD DISTRICT WORKS, PITTSBURGH (INTERIM ENGINEERING REPORT)
USTC-	UNITED STATES TESTING CO., HOBOKEN, N.J.
USU-	UTAH STATE UNIV., LOGAN
USUN RS-	UNITED STATES-UNITED NATIONS (REPORT SERIES)
USWD-	BENDIX CORP. ELECTRODYNAMICS DIV., NO. HOLLYWOOD, CAL
UT-	TEXAS. UNIV., AUSTIN
UTAH-	UTAH. UNIV., SALT LAKE CITY
UTAH-PR-	UTAH. UNIV., SALT LAKE CITY
UTAH RR-(NUMBER)-	UTAH. STATE DEPT. OF HIGHWAYS. RESEARCH SECTION
UTAH-SR-	UTAH. UNIV., SALT LAKE CITY. INST. FOR THE STUDY OF RATE PROCESSES
UTAH-TM-	UTAH. UNIV., SALT LAKE CITY. INST. FOR THE STUDY OF RATE PROCESSES
UTAH-TR-	UTAH. UNIV., SALT LAKE CITY (TECHNICAL REPORT)
UTAH U TR-	UTAH. UNIV., SALT LAKE CITY. INST. FOR THE STUDY OF RATE PROCESSES
UT-ARL-	TENNESSEE. UNIV., OAK RIDGE. UT-AEC AGRICULTURAL RESEARCH LAB.
UTC-	UNITED AIRCRAFT CORP., SUNNYVALE, CALIF.
UTC-	UNITED TECHNOLOGY CENTER, SUNNYVALE, CALIF.
UTC-(YEAR)-	UNITED TECHNOLOGY CENTER, SUNNYVALE, CALIF.
UTC-(NUMBER)-ASR-	UNITED TECHNOLOGY CENTER. RESEARCH AND ADVANCED TECHNOLOGY DEPT., SUNNYVALE, CALIF.
UTC-(NO.)-FR	UNITED TECHNOLOGY CENTER. RESEARCH AND ADVANCED TECHNOLOGY DEPT., SUNNYVALE, CALIF.
UTC-(NO.)-QPR(NO.)	UNITED TECHNOLOGY CENTER. RESEARCH AND ADVANCED TECHNOLOGY DEPT., SUNNYVALE, CALIF.
UTC-	UNITED TECHNOLOGY CORP., SUNNYVALE, CALIF.
UTC-(NUMBER)-TSR-	UNITED TECHNOLOGY CORP., SUNNYVALE, CALIF.
UTC-(NUMBER)-QT-	UNITED TECHNOLOGY CORP., SUNNYVALE, CALIF.
UTC-(NUMBER)-QPR-	UNIVAC, BLUE BELL, PENNA.
UTCHE-	TOLEDO. UNIV. DEPT. OF CHEMICAL ENGINEERING
UT-CM-	TEXAS. UNIV., AUSTIN
UTD-	WESTINGHOUSE ELECTRIC CORP., BALTIMORE
UT/DRL-	TEXAS. UNIV., AUSTIN. DEFENSE RESEARCH LAB.
UTEC-DC-(YEAR)-	UTAH. UNIV., SALT LAKE CITY. COLL. OF ENGINEERING
UTEC-DO-	UTAH. UNIV., SALT LAKE CITY. COLL. OF ENGINEERING
UTEC-IMR-(YEAR)-	UTAH. UNIV., SALT LAKE CITY. INST. OF MATERIALS RES.
UTEC-MC-	UTAH. UNIV., SALT LAKE CITY
UTEC-MD-	UTAH. UNIV., SALT LAKE CITY. DEPT. OF ELEC. ENG.
UTEC-MD-(YEAR)-	UTAH. UNIV., SALT LAKE CITY. MICROWAVE DEVICE AND PHYSICAL ELECTRONICS LAB.
UTEC-ME-(YEAR)-	UTAH. UNIV., SALT LAKE CITY. COLL. OF ENG.(MECH. ENG)
UTEC-MR-(YEAR)-	UTAH. UNIV., SALT LAKE CITY. COLL. OF ENGINEERING
UTEC-MSE-(YEAR)-	UTAH. UNIV., SALT LAKE CITY. DIV. OF MATERIALS SCIENCE AND ENGINEERING
UTECS-	TORONTO. UNIV. COMPUTATION CENTRE
UTEC-SI-(YEAR)-	UTAH. UNIV., SALT LAKE CITY. SOLID ROCKET STRUCTURAL INTEGRITY INFORMATION CENTER
UT-EERL-	TEXAS. UNIV., AUSTIN. ELECTRICAL ENGINEERING RES.LAB.
UTG-	ARMED FORCES SPECIAL WEAPONS PROJECT. FIELD COMMAND, ALBUQUERQUE, N.MEX.

UTG/(YR.)-	ARMED FORCES SPECIAL WEAPONS PROJECT. FIELD COMMAND, ALBUQUERQUE, N. MEX.
UT-GSBS-DADA-	TEXAS. UNIV., HOUSTON. GRAD. SCH. OF BIOMEDICAL SCIENCES
UTH-	DELFT, NETHERLANDS. TECHNISCHE HOGESCHOOL
UTIA-	TORONTO. UNIV. INST. OF AEROPHYSICS
UTIA-REV-	TORONTO. UNIV. INST. OF AEROPHYSICS
UTIAS-	CHICAGO. UNIV. INST. FOR AIR WEAPONS RESEARCH
UTIAS-	TORONTO. UNIV. INST. FOR AEROSPACE STUDIES
UTIAS-REV-	TORONTO. UNIV. INST. FOR AEROSPACE STUDIES
UTIAS-TN-	TORONTO. UNIV. INST. FOR AEROSPACE STUDIES
UTIA-TN-	TORONTO. UNIV. INST. OF AEROPHYSICS
UT-MECH-E-	TORONTO. UNIV. DEPT. OF MECHANICAL ENGINEERING
U TSEPP-	AIR TECHNICAL SERVICE COMMAND. ENGINEERING DIV. POWER PLANT LAB., WRIGHT-PATTERSON AFB, OHIO
UTSI-TR-(CON.NO.)-	TENNESSEE. UNIV., TULLAHOMA. SPACE INST.
UTU-	UTAH. UNIV., SALT LAKE CITY
UU-	UTAH. UNIV., SALT LAKE CITY
UU(YEAR)-	UTAH. UNIV., SALT LAKE CITY. UPPER AIR RES. LABS.
UUCD-	UTAH. UNIV., SALT LAKE CITY. DEPT. OF CHEMISTRY
UUIP-	UPPSALA. UNIV. INST. OF PHYSICS
UUIP-TN-	UPPSALA. UNIV. QUANTUM CHEMISTRY GROUP
UU ISR TR-	UTAH. UNIV., SALT LAKE CITY. INST. FOR THE STUDY OF RATE PROCESSES
UUP-	UTAH. UNIV., SALT LAKE CITY
UUT-	UTAH. UNIV., SALT LAKE CITY
UUT-RL-	UTAH. UNIV., SALT LAKE CITY. RADIOBIOLOGY LAB.
UVA-	VIRGINIA. UNIV., CHARLOTTESVILLE
UVA-ORL-	VIRGINIA. UNIV., CHARLOTTESVILLE. ORDNANCE RES. LAB.
UVA/ORL-(NO.)-12RD-59-QPR	VIRGINIA. UNIV., CHARLOTTESVILLE. ORDNANCE RES. LAB.
UVA/ORL-(NO.)-(YR.)-TR-	VIRGINIA. UNIV., CHARLOTTESVILLE. ORDNANCE RES. LAB.
UVAR-	VIRGINIA. UNIV., CHARLOTTESVILLE. SCHOOL OF ENG.
UVA/RLES-EP/	FRANKLIN INST. LABS. FOR RES. + DEV., PHILADELPHIA
UVC-	VIRGINIA. UNIV., CHARLOTTESVILLE
UV-C-	VIRGINIA. UNIV., CHARLOTTESVILLE. ENG. EXPERIMENT STA
UVF-	FEDERAL AVIATION AGENCY. LOS ANGELES AIR ROUTE TRAFFIC CONTROL CENTER, PALMDALE, CALIF.
UV/ORL-	VIRGINIA. UNIV., CHARLOTTESVILLE. ORDNANCE RES. LAB.
UVVVR-	CZECHOSLOVAKIA. INST. FOR RESEARCH, PRODUCTION AND APPLICATION OF RADIOISOTOPES, PRAGUE
UW-	WASHINGTON. UNIV., SEATTLE
UWAL-	WASHINGTON. UNIV., SEATTLE. AERONAUTICAL LAB.
UW-APL-	WASHINGTON. UNIV., SEATTLE. APPLIED PHYSICS LAB.
UWAPL-	WASHINGTON. UNIV., SEATTLE. APPLIED PHYSICS LAB.
UWB-	GT. BRIT. ROYAL NAVAL PERSONNEL RESEARCH COMMITTEE
UWFL-	WASHINGTON. UNIV., SEATTLE. APPLIED FISHERIES LAB.
UWI-	WEST INDIES. UNIV., KINGSTON. DEPT. OF PHYSICS
UWI/CC (NUMBER)	WEST INDIES. UNIV., KINGSTON. COMPUTING CENTER
UWIS-DS-(YEAR)-	WISCONSIN. UNIV., MADISON. DEPT. OF STATISTICS
UWLRB-	WASHINGTON. UNIV., SEATTLE. LAB. OF RADIATION BIOLOGY
UWM-(CON.NO.)-(YEAR)-	WISCONSIN-MILWAUKEE UNIV.
UWM-	WISCONSIN-MILWAUKEE UNIV.
UWNRL-	WISCONSIN. UNIV., MADISON. NAVAL RESEARCH LAB.
UWO-	UNIVERSITY OF WESTERN ONTARIO, LONDON
UWSI-TPR-	UNDERWATER SYSTEMS INC., SILVER SPRING, MD.
UW-TR-	WISCONSIN. UNIV., MADISON

4V-	ADEL PRECISION PRODUCTS CORP., BURBANK, CALIF.
V (YR.)-	CURTISS-WRIGHT CORP. AIRPLANE DIV., COLUMBUS, OHIO
V-	HUGHES AIRCRAFT CO. SYSTEMS DEVELOPMENT LABS., CULVER CITY, CALIF.
V-	LING-TEMCO-VOUGHT, INC., DALLAS
V-	MOTOROLA INC., PHOENIX, ARIZ.
V-(YEAR)	NETHERLANDS. VERFINSTITUUT TNO, DELFT
V-(NUMBER-NUMBER)-OE	ROENTGEN TECHNISCHE DIENST N.V., ROTTERDAM
V-	VEDA, INC., ANN ARBOR, MICH.
V-(NUMBER)U/-	VEDA, INC., ANN ARBOR, MICH.
V-(NO.)-(NO.)-(LETTER)	VITRO LABS., WEST ORANGE, N.J.
VA-	VARIAN ASSOCIATES, PALO ALTO, CALIF.
VA-	VETERANS ADMINISTRATION, WASHINGTON, D.C.
VA-AST-	CHANCE (W.R.) AND ASSOCIATES, ARLINGTON, VA.
VA-AST-	VOUGHT ASTRONAUTICS. DIV. OF CHANCE VOUGHT CORP., DALLAS
VAC-	VERTOL AIRCRAFT CORP., MORTON, PENNA.
VACCO-	VACCO INDUSTRIES, SOUTH EL MONTE, CALIF.
VAC-M-	CALIFORNIA. UNIV., BERKELEY. RADIATION LAB. (REF. 9)
VAC.-P(MONTH/YEAR)	CALIFORNIA. UNIV., BERKELEY. RADIATION LAB. (REF. 9)
VAC. SPEC-	CALIFORNIA. UNIV., BERKELEY. RADIATION LAB. (REF. 9)
VAC-W-	CALIFORNIA. UNIV., BERKELEY. RADIATION LAB. (REF. 9)
VA ENGINEERING RPT.-	VARIAN ASSOCIATES, PALO ALTO, CALIF.
VA-ER-	VARIAN ASSOCIATES, PALO ALTO, CALIF.
VAG-	ALASKA, UNIV., COLLEGE
VAH-	VETERANS ADMINISTRATION HOSPITAL, RICHMOND, VA.
VAL-	GENERAL ELECTRIC CO. NUCLEAR ENERGY DIV.,SAN JOSE,CAL (VALLECITOS REPORT)
VAP-(NUMBER)-(LETTER)-	SOCIETE BELGE POUR L'INDUSTRIE NUCLEAIRE, BRUSSELS
VAR-	VARIAN ASSOCIATES, SAN CARLOS, CALIF.
VAR-ER-	VARIAN ASSOCIATES, PALO ALTO, CALIF.
VARIAN-	VARIAN ASSOCIATES, PALO ALTO, CALIF.
VARL-G-	UNITED AIRCRAFT CORP., EAST HARTFORD, CONN.
VATP/OD-	GARRETT CORP. AIRESEARCH MFG. DIV., LOS ANGELES
VAU-	VIRGINIA. UNIV., CHARLOTTESVILLE
VAU-PPR-	VIRGINIA. UNIV., CHARLOTTESVILLE. DEPT. OF PHYSICS
VBCPC-(YEAR)-	WISCONSIN. STATE DEPT. OF LOCAL AFFAIRS AND DEVELOPMENT, MADISON
VC-	CORNELL AERONAUTICAL LAB., INC., BUFFALO
VCA-	VITRO CORP. OF AMERICA, N.Y.C.
VCA-	VITRO LABS., SILVER SPRING, MD.
VCA-KLX-	VITRO CORP. OF AMERICA, N.Y.C.
VCA-KLX-	VITRO LABS., WEST ORANGE, N.J.
VCAPC-(YEAR)-	WISCONSIN. STATE DEPT. OF LOCAL AFFAIRS AND DEVELOPMENT, MADISON
VCA-TR-	VITRO LABS., SILVER SPRING, MD.
VCLPC-(YEAR)-	WISCONSIN. STATE DEPT. OF LOCAL AFFAIRS AND DEVELOPMENT, MADISON
VC-TR-	VITRO LABS., SILVER SPRING, MD.
VD-	COMMITTEE ON MEDICAL RES. (VENEREAL DISEASES RPTS.)
VDDIT-	AKTIEBOLAGET ATOMENERGI, STOCKHOLM
VDDIT-	AKTIEBOLAGET ATOMENERGI, STUDSVIK, SWEDEN
VDIT-	AKTIEBOLAGET ATOMENERGI, STOCKHOLM
VDIT-(NUMBER).(NUMBER)	AKTIEBOLAGET ATOMENERGI, STOCKHOLM
VE-	CORNELL AERONAUTICAL LAB., INC., BUFFALO
VE-	VOUGHT ELECTRONICS. DIV. OF CHANCE VOUGHT CORP., DALLAS
VEC-(NO.)-RD-	VALUE ENGINEERING CO., ALEXANDRIA, VA.
VEC-	VITRO ENGINEERING CO., N.Y.C.
VEDER-(NUMBER)-(YEAR)-	GRUMMAN AIRCRAFT ENGINEERING CORP., BETHPAGE, N.Y.
VEECO-	VACUUM-ELECTRONIC ENGINEERING CO., BROOKLYN, N.Y.
VER-	CALIFORNIA INST. OF TECH., PASADENA
VER-	VARIAN ASSOCIATES, PALO ALTO, CALIF. (ENG. REPT.)
VERTEX-TR-	VERTEX CORP., KENSINGTON, MD.
VERTEX-TR-(YEAR)-	VERTEX CORP., KENSINGTON, MD.
VERU-	VERMONT. UNIV., BURLINGTON
VESIAC-	MICHIGAN. UNIV., ANN ARBOR. VELA SEISMIC INFORMATION ANALYSIS CENTER
VESIAC BULL-	MICHIGAN. UNIV., ANN ARBOR. VELA SEISMIC INFORMATION ANALYSIS CENTER
VF-(NUMBER)-(LETTER)-	CORNELL AERONAUTICAL LAB., INC., BUFFALO
VFW-H(NUMBER)	VEREINIGTE FLUGTECHNISCHE WERKE G.M.B.H., BREMEN
VG-	(VU-GRAPHS) (REF. 2)
VG-(NUMBER)-(LETTER)-	CORNELL AERONAUTICAL LAB., INC., BUFFALO
VGW/BB-	VICKERS-ARMSTRONGS LTD., LONDON
VGW/RD-	VICKERS-ARMSTRONGS LTD., LONDON
VH-	CORNELL AERONAUTICAL LAB., INC., BUFFALO
VHPC-(YEAR)-	WISCONSIN. STATE DEPT. OF LOCAL AFFAIRS AND DEVELOPMENT, MADISON
VI-	HUGHES AIRCRAFT CO. ENG. DIV., CULVER CITY, CALIF.
VI-	VIRGINIA INST. FOR SCIENTIFIC RESEARCH, RICHMOND
VI-(NO.)-(NO.)-O	VITRO LABS., WEST ORANGE, N.J.
VIC-	VICTOREEN INSTRUMENT CO., CLEVELAND
VICTORIAU-	VICTORIA, B.C. UNIV.
VIDYA-FR-	VIDYA, INC., PALO ALTO, CALIF.
VIDYA-TR-	ITEK CORP. VIDYA DIV., PALO ALTO, CALIF.
VIENNAU-TH/	VIENNA. UNIVERSITAT
VIK-TN-	VON KARMAN INSTITUTE FOR FLUID DYNAMICS, RHODE-SAINT-GENESE, BELGIUM
VIRGINIA U-EE-(NUMBERS)U	VIRGINIA. UNIV., CHARLOTTESVILLE
VISIT NO.-	ATOMIC ENERGY COMMISSION, WASHINGTON, D.C.
VISR-	VIRGINIA INST. FOR SCIENTIFIC RESEARCH, RICHMOND
VISTA-	CALIFORNIA INST. OF TECH., PASADENA
VITRO-	VITRO CORP. OF AMERICA, N.Y.C.
VITRO-(NO.)-(NO.)-(YR.)	VITRO ENGINEERING CO., N.Y.C.
VITRO-	VITRO LABS., SILVER SPRING, MD.
VITRO-	VITRO LABS., WEST ORANGE, N.J.
VITRO-TR-	VITRO CORP. OF AMERICA, N.Y.C.
VKI-CN-	VON KARMAN INSTITUTE FOR FLUID DYNAMICS, RHODE-SAINT-GENESE, BELGIUM
VKI-CR-	VON KARMAN INSTITUTE FOR FLUID DYNAMICS, RHODE-SAINT-GENESE, BELGIUM
VKI-LS-	VON KARMAN INSTITUTE FOR FLUID DYNAMICS, RHODE-SAINT-GENESE, BELGIUM
VKI-TM-	VON KARMAN INSTITUTE FOR FLUID DYNAMICS, RHODE-SAINT-GENESE, BELGIUM
VKI-TN-	VON KARMAN INSTITUTE FOR FLUID DYNAMICS, RHODE-SAINT-GENESE, BELGIUM
VL-	VITRO LABS., SILVER SPRING, MD.
VL-	VITRO LABS., WEST ORANGE, N.J.
VL-(NUMBER)-(NUMBER)-O	VITRO LABS., WEST ORANGE, N.J.
VL-TR-OF (MO./DAY/YR.)	VITRO LABS., WEST ORANGE, N.J.
VML-	REDEL, INC., ANAHEIM, CALIF.
VNIIRT-	U.S.S.R. VSESOYUZNYI NAUCHNO-ISSLEDOVSTELSKII INSTITUT RADIATSIONNOI TEKHNIKI, MOSCOW
VNRR-	(NAVY VOLUNTEER RESEARCH RESERVE UNIT...)
VN+V-	NAVAL NUCLEAR ORDNANCE EVALUATION UNIT, KIRTLAND AFB, N. MEX. (VULNERABILITY NEWS AND VIEWS)

VO-(NO.)-(NO.)-O	VITRO LABS., WEST ORANGE, N.J.
VOY-CR-(YEAR)-	MARTIN CO., DENVER
VP-	CORNELL AERONAUTICAL LAB., INC., BUFFALO
VPC-	TENNESSEE EASTMAN CORP., OAK RIDGE, TENN.
VPENG-	ATOMIC ENERGY OF CANADA LTD., CHALK RIVER, ONT.
VPI-	VIRGINIA POLYTECHNIC INST., BLACKSBURG
VPI-E-	VIRGINIA POLYTECHNIC INST., BLACKSBURG
VPI-E-(YEAR)-	VIRGINIA POLYTECHNIC INST., BLACKSBURG
VPI-QPR-	VIRGINIA POLYTECHNIC INST., BLACKSBURG
VPI-QR-	VIRGINIA POLYTECHNIC INST., BLACKSBURG
VPI-WRRC-	VIRGINIA POLYTECHNIC INST., BLACKSBURG. WATER RESOURCES RESEARCH CENTER
VPN-(YEAR)-	VICTORIA, B.C. UNIV.
VPPR-	CALIF. INST. OF TECH., PASADENA. JET PROPULSION LAB. (VOYAGER PROJECT PROGRESS REPORT)
VP-X-	CANADA. DEPT. OF FORESTRY, VANCOUVER, B.C.
VP-X-	CANADA. FOREST PRODUCTS RES. LAB., VANCOUVER, B.C.
VR-	VEDA, INC., ANN ARBOR, MICH.
VRC-	VEHICLE RESEARCH CORP., PASADENA, CALIF.
VRL TR-	STANFORD UNIV., CALIF. VIBRATION RES. LAB.
VSAF-	TEXAS. UNIV., AUSTIN. APPLIED MECHANICS RESEARCH LAB.
VSC-	AIR FORCE TECHNICAL APPLICATIONS CENTER. VELA SEISMOLOGICAL CENTER, WASHINGTON, D.C.
VSN-	TEXAS. UNIV., AUSTIN. APPLIED MECHANICS RESEARCH LAB.
VT-	CARBIDE AND CARBON CHEMICALS CORP. K-25 PLANT, OAK RIDGE, TENN.
VT(NUMBER)R(NUMBER)	FAIRCHILD HILLER CORP. REPUBLIC AVIATION DIV., FARMINGDALE, N.Y. (VECTORED THRUST)
VT-	TEXAS INSTRUMENTS, INC. SCIENCE SERVICES DIV., DALLAS
VTDC-	RES. + DEV. BD. COMM. ON ELECTRONICS, WASH., D.C.
VTF-	(VT FUZE)
VTFM-	NAVAL ORDNANCE LAB., WHITE OAK, MD. (VT FUZE MEMO.)
VTFR-	NAVAL ORDNANCE LAB., WHITE OAK, MD. (VT FUZE REPORT)
VTH-	DELFT, NETHERLANDS. TECHNISCHE HOGESCHOOL
VTH-	KARLSRUHE, GERMANY. TECHNISCHE HOCHSCHULE
VTLM-	GT. BRIT. SPRINGFIELDS WORKS, LANCS, ENGLAND (REF.36)
VTO/M/-	VICKERS-ARMSTRONGS LTD. WEYBRIDGE WORKS, SURREY, ENG.
VTU-	VERMONT. UNIV., BURLINGTON
VU-	AIR FORCE TECHNICAL APPLICATIONS CENTER. VELA SEISMOLOGICAL CENTER, WASH., D.C. (PROJ. VELA UNIFORM)
VU-	VIRGINIA. UNIV., CHARLOTTESVILLE
VUF-	ADVANCED RESEARCH PROJECTS AGENCY, WASHINGTON, D.C. (SERIES USED FOR REPORTS OF VARIOUS CONTRACTORS CONCERNED WITH THE VELA UNIFORM PROGRAM)
VUP-	ADVANCED RESEARCH PROJECTS AGENCY, WASHINGTON, D.C.
VUP-	EDGERTON, GERMESHAUSEN + GRIER, INC., LAS VEGAS, NEV. (VELA UNIFORM PROJECT)
VUP-	INSTITUTE FOR DEFENSE ANALYSES, ARLINGTON, VA.
VUP-	SANDIA CORP., ALBUQUERQUE, N. MEX.
VUP-	STANFORD RESEARCH INST., MENLO PARK, CALIF. (VELA UNIFORM PROJECT)
VU-TR-	VERMONT. UNIV., BURLINGTON
VWZE-	AIR FORCE SYSTEMS COMMAND, VANDENBERG AFB, CALIF.
VX-(NUMBER)-AEN-	AIR DEVELOPMENT SQUADRON FIVE, CHINA LAKE, CALIF.

W-

W-	AMERICAN POTASH + CHEMICAL CORP., WHITTIER, CALIF.
W7-(NO.)-TNO(NO.)-	ARINC RESEARCH CORP., SANTA ANA, CALIF.
W-	CALIFORNIA. UNIV., BERKELEY
W-	CASE INST. OF TECH., CLEVELAND
W-	CHICAGO. UNIV. METALLURGICAL LAB. (REF. 10)
W-	HUGHES AIRCRAFT CO., LOS ANGELES
W-	HUGHES AIRCRAFT CO. ELECTRON DYNAMICS DIV., TORRANCE, CALIF.
W-(NUMBER)-	HUGHES AIRCRAFT CO. ELECTRON DYNAMICS DIV., TORRANCE, CALIF.
W-	HUGHES AIRCRAFT CO. MICROWAVE TUBE DIV., LOS ANGELES
W-	JOHNS HOPKINS UNIV., BALTIMORE
W-	LOS ALAMOS SCIENTIFIC LAB., N. MEX.
W-(NO.)-RD-	LOS ALAMOS SCIENTIFIC LAB., N. MEX.
W-1- THRU W-11-	LOS ALAMOS SCIENTIFIC LAB., N. MEX. (INTERNAL CORRESPONDENCE SERIAL USED BY GROUPS IN W DIV.)
W-	MITRE CORP., BEDFORD, MASS.
W-	MOTOROLA INC., PHOENIX, ARIZ.
W-	MOTOROLA INC. SEMICONDUCTOR PRODS. DIV., PHOENIX, ARIZ.
W-	NATIONAL RESEARCH COUNCIL OF CANADA, OTTAWA
W-(MONTH)/(YEAR)	NAVAL WEAPONS LAB., DAHLGREN, VA.
W-	NORTH STAR RESEARCH AND DEV. INST., MINNEAPOLIS
W-	OFFICE OF SCIENTIFIC RESEARCH AND DEVELOPMENT
W(NUMBER)-	TRG, INC., MENLO PARK, CALIF.
W-	WAR PRODUCTION BOARD, WASHINGTON, D.C.
W(YEAR)-(5 DIGITS)	WATER RESOURCES SCIENTIFIC INFORMATION CENTER (INTERIOR), WASHINGTON, D.C. (ACCESSION NUMBER)
WA-	ARMY, WASHINGTON, D.C.
WA-(NUMBER)-	ARMY, WASHINGTON, D.C.
WA-	CIVIL SERVICE COMMISSION, WASHINGTON, D.C. (WORK ANNOUNCEMENT)
WA-	OFFICE OF SCIENTIFIC RES. + DEV. LIAISON OFF., LONDON
WA-(NUMBER)	SPERRY GYROSCOPE CO., GREAT NECK, N.Y.
WA-	WATERTOWN ARSENAL, MASS.
WA-	WATERVLIET ARSENAL. BENET LABS., N.Y.
WABC-	WESTINGHOUSE AIR BRAKE CO. UNION SWITCH AND SIGNAL DIV., PITTSBURGH
WAC-	COAST AND GEODETIC SURVEY, WASHINGTON, D.C. (WORLD AERONAUTICAL CHART)
WAC/(NUMBER)/-	GT. BRIT. EXPLOSIVES RESEARCH AND DEVELOPMENT ESTABLISHMENT, WALTHAM ABBEY, ESSEX, ENGLAND
WAC-	WRIGHT AERONAUTICAL CORP., PATERSON, N.J.
WAC-	WRIGHT AERONAUTICAL CORP., WOOD-RIDGE, N.J.
WAC (LETTER)-	WRIGHT AERONAUTICAL CORP., PATERSON, N.J.
W.A.C. SERIAL REPORT-	WRIGHT AERONAUTICAL CORP., WOOD-RIDGE, N.J.
W.A.C. SERIAL REPORT NO.-	CURTISS-WRIGHT CORP. WRIGHT AERO. DIV., WOOD-RIDGE, NJ
WAD-	ARDE ASSOCIATES, NEWARK, N. J.
WAD-	CURTISS-WRIGHT CORP. WRIGHT AERO. DIV., WOOD-RIDGE, NJ
WAD-	GT. BRIT. CHEMICAL INSPECTORATE
WADC-	CURTISS-WRIGHT CORP. WRIGHT AERO. DIV., WOOD-RIDGE, NJ
WADC-	WRIGHT AIR DEV. CENTER, WRIGHT-PATTERSON AFB, OHIO
WADC AEC-	WRIGHT AIR DEV. CENTER, WRIGHT-PATTERSON AFB, OHIO
WADC AF TR-	WRIGHT AIR DEVELOPMENT CENTER, WRIGHT-PATTERSON AFB, OHIO (AIR FORCE TECHNICAL REPORT)
WADC ARL-	WRIGHT AIR DEVELOPMENT CENTER. AIRCRAFT RADIATION LAB., WRIGHT-PATTERSON AFB, OHIO
WADC-MR-	WRIGHT AIR DEV. CENTER, WRIGHT-PATTERSON AFB, OHIO
WADC MR WCEGW R-(NO.)-	WRIGHT AIR DEV. CENTER, WRIGHT-PATTERSON AFB, OHIO
WADC-MR-WCT-	WRIGHT AIR DEV. CENTER, WRIGHT-PATTERSON AFB, OHIO
WADC-PR-	WRIGHT AIR DEV. CENTER, WRIGHT-PATTERSON AFB, OHIO
WADC-PR-(YEAR)-	WRIGHT AIR DEV. CENTER, WRIGHT-PATTERSON AFB, OHIO
WADC-QR-	WRIGHT AIR DEV. CENTER, WRIGHT-PATTERSON AFB, OHIO
WADC-TM-(YEAR)-	WRIGHT AIR DEV. CENTER, WRIGHT-PATTERSON AFB, OHIO
WADC-TN-(YEAR)-	WRIGHT AIR DEV. CENTER, WRIGHT-PATTERSON AFB, OHIO
WADC TN WCLA (YEAR)-	WRIGHT AIR DEV. CENTER, WRIGHT-PATTERSON AFB, OHIO
WADC TN WCLG (YEAR)-	WRIGHT AIR DEV. CENTER, WRIGHT-PATTERSON AFB, OHIO
WADC-TN-WCLR-(YEAR)-	WRIGHT AIR DEVELOPMENT CENTER. AIRCRAFT RADIATION LAB., WRIGHT-PATTERSON AFB, OHIO
WADC-TN-WCLS-	WRIGHT AIR DEVELOPMENT CENTER. AIRCRAFT LAB., WRIGHT-PATTERSON AFB, OHIO
WADC TN WCRR (YEAR)-	WRIGHT AIR DEV. CENTER, WRIGHT-PATTERSON AFB, OHIO
WADC TN WCT (YEAR)-	WRIGHT AIR DEV. CENTER, WRIGHT-PATTERSON AFB, OHIO
WAD-CTR-	CURTISS-WRIGHT CORP. WRIGHT AERO. DIV., WOOD-RIDGE, NJ
WADC-TR-(YEAR)-	WRIGHT AIR DEV. CENTER, WRIGHT-PATTERSON AFB, OHIO
WADC-TR-	WRIGHT AIR DEVELOPMENT CENTER. AEROSPACE MEDICAL LAB., WRIGHT-PATTERSON AFB, OHIO
WADD-	WRIGHT AIR DEV. DIV., WRIGHT-PATTERSON AFB, OHIO
WADD-TDR-(YEAR)-	WRIGHT AIR DEV. DIV., WRIGHT-PATTERSON AFB, OHIO
WADD-TM-(YEAR)-	WRIGHT AIR DEV. DIV., WRIGHT-PATTERSON AFB, OHIO
WADD-TN-	WRIGHT AIR DEV. DIV., WRIGHT-PATTERSON AFB, OHIO
WADD-TN-(YR.)-	WRIGHT AIR DEV. DIV., WRIGHT-PATTERSON AFB, OHIO
WADD-TR-(YR.)-	WRIGHT AIR DEV. DIV., WRIGHT-PATTERSON AFB, OHIO
WADF-	WESTERN AIR DEFENSE FORCE, HAMILTON AFB, CALIF.
WAD-R-	CURTISS-WRIGHT CORP. WRIGHT AERO. DIV., WOOD-RIDGE, NJ
WAD-R(NUMBER)F-	CURTISS-WRIGHT CORP. WRIGHT AERO. DIV., WOOD-RIDGE, NJ
WAD/R-	GT. BRIT. MIN. OF SUPPLY. CHEM. INSPECTORATE, LONDON
WAD-S-	CURTISS-WRIGHT CORP. WRIGHT AERO. DIV., WOOD-RIDGE, NJ
WAD-SST-	CURTISS-WRIGHT CORP. WRIGHT AERO. DIV., WOOD-RIDGE, NJ
WAEC-	NAVAL AIR DEVELOPMENT CENTER, JOHNSVILLE, PENNA.
WAED-	WESTINGHOUSE ELECTRIC CORP. AEROSPACE ELECTRICAL DIV., LIMA, OHIO
WAF-	DELAWARE. UNIV., NEWARK
WAGON-	GEOLOGICAL SURVEY, DENVER
WAHCC-	WAH CHANG CORP. ALBANY DIV., OREGON
WAI-TR-	WEINER ASSOCIATES, INC., BALTIMORE
WAL-	WATERTOWN ARSENAL LAB., MASS.
WAL-(NUMBER)/	WATERTOWN ARSENAL LAB., MASS.
WAL MR-(NUMBER)/	WATERTOWN ARSENAL LAB., MASS.
WAL-MS-	WATERTOWN ARSENAL LAB., MASS.
WALR-	WATERTOWN ARSENAL LAB., MASS.
WALR (NUMBER)/	WATERTOWN ARSENAL LAB., MASS.
WAL-TR-	WATERTOWN ARSENAL LAB., MASS.
WAL-TR-(NUMBER)/	WATERTOWN ARSENAL LAB., MASS.
WAM-	CIVIL SERVICE COMMISSION, WASHINGTON, D.C. (WORK ANNOUNCEMENT)
WANEF-	WESTINGHOUSE ELECTRIC CORP. ASTRONUCLEAR LAB., PITTS.
WANL-	WESTINGHOUSE ELECTRIC CORP. ASTRONUCLEAR LAB., PITTS.
WANL-(NUMBER)-	WESTINGHOUSE ELECTRIC CORP. ASTRONUCLEAR LAB., PITTS.
WANL-MP-	WESTINGHOUSE ELECTRIC CORP. ASTRONUCLEAR LAB., PITTS.
WANL-PD/DDD/-	WESTINGHOUSE ELECTRIC CORP. ASTRONUCLEAR LAB., PITTS.
WANL-PR(AA)-	WESTINGHOUSE ELECTRIC CORP. ASTRONUCLEAR LAB., PITTS.
WANL-PR/CCC/-	WESTINGHOUSE ELECTRIC CORP. ASTRONUCLEAR LAB., PITTS.
WANL-PR-(LL)-	WESTINGHOUSE ELECTRIC CORP. ASTRONUCLEAR LAB., PITTS.
WANL-PR/RRR/-	WESTINGHOUSE ELECTRIC CORP. ASTRONUCLEAR LAB., PITTS.
WANL-PR-(SS)-	WESTINGHOUSE ELECTRIC CORP. ASTRONUCLEAR LAB., PITTS.
WANL-SP-	WESTINGHOUSE ELECTRIC CORP. ASTRONUCLEAR LAB., PITTS.
WANL-TME-	WESTINGHOUSE ELECTRIC CORP. ASTRONUCLEAR LAB., PITTS.

WANL-TNR-	WESTINGHOUSE ELECTRIC CORP. ASTRONUCLEAR LAB., PITTS.
WANL-W-	WESTINGHOUSE ELECTRIC CORP. ASTRONUCLEAR LAB., PITTS.
WA-OMRO-	WATERTOWN ARSENAL. ORDNANCE MATERIALS RES. OFF.,MASS.
WAPD-	WESTINGHOUSE ELECTRIC CORP. BETTIS ATOMIC POWER LAB., PITTSBURGH
WAPD-CTA(LETTERS)-	WESTINGHOUSE ELECTRIC CORP. BETTIS PLANT, PITTSBURGH
WAPD-MRP-	WESTINGHOUSE ELECTRIC CORP. BETTIS ATOMIC POWER LAB., PITTSBURGH
WAPD-PWR-	WESTINGHOUSE ELECTRIC CORP. BETTIS ATOMIC POWER LAB., PITTSBURGH
WAPD-T-	WESTINGHOUSE ELECTRIC CORP. BETTIS ATOMIC POWER LAB., PITTSBURGH
WAPD-TM-	WESTINGHOUSE ELECTRIC CORP. BETTIS ATOMIC POWER LAB., PITTSBURGH
WAPD-TN-	WESTINGHOUSE ELECTRIC CORP. BETTIS ATOMIC POWER LAB., PITTSBURGH
WAPD-TRANS-	WESTINGHOUSE ELECTRIC CORP. BETTIS ATOMIC POWER LAB., PITTSBURGH
WAPT-T-	WESTINGHOUSE ELECTRIC CORP. BETTIS ATOMIC POWER LAB., PITTSBURGH
WAR-	CURTISS-WRIGHT CORP. (WRIGHT AERONAUTICAL REPORT)
WARD-	WATERTOWN ARSENAL, MASS. (RESEARCH AND DEVELOPMENT)
WARD-(NUMBER)-	WESTINGHOUSE ELECTRIC CORP. NUCLEAR ENERGY SYSTEMS. ADVANCED REACTORS DIV., MADISON, PENNA.
WA-RPL-	WATERTOWN ARSENAL. RODMAN PROCESS LAB., MASS.
WARSAWU-	WARSAW. UNIV.
WA-SB-	WATERTOWN ARSENAL LAB., MASS.
WASH-	ATOMIC ENERGY COMMISSION, WASHINGTON, D.C.
W/AT-	ENGLISH ELECTRIC CO., LTD. ATOMIC POWER DIV., WHETSTONE, LEICS, ENGLAND
WATB-	WATERTOWN ARSENAL LAB., MASS. (TECHNICAL BULLETINS)
WATER RESOURCES-	GEOLOGICAL SURVEY. WATER RESOURCES DIV., ROLLA, MO.
WATERVLIET-	WATERVLIET ARSENAL. BENET LABS., N.Y.
WAU-	WASHINGTON UNIV., ST. LOUIS
WAU-TR-	WASHINGTON UNIV., ST. LOUIS
WAW-	CIVIL SERVICE COMMISSION, WASHINGTON, D.C. (WORK ANNOUNCEMENT)
WAYU-	WAYNE UNIV., DETROIT
WB-	(WORK BOOK) (REF. 2)
WB-	WEATHER BUREAU, WASHINGTON, D.C.
WBAN-	AIR FORCE-NAVY WEATHER BUREAU (REF. 13)
WB/BC-	ENVIRONMENTAL DATA SERVICE, SILVER SPRING, MD. (BIBLIOGRAPHY ON CLIMATE)
WB-BM-	ENVIRONMENTAL DATA SERVICE, SILVER SPRING, MD. (BIB)
WBG-TM-	WEATHER BUREAU, WASHINGTON, D.C.
WBI-C-(NUMBER)-BI	BIRKENHEAD (WARREN), INC., SEATTLE
WB-KE-	WAAGNER-BIRO A.G., VIENNA
WBL-	WESTERN BIOLOGICAL LABS., CULVER CITY, CALIF.
WB-NWO-	DEFENSE ATOMIC SUPPORT AGENCY. FIELD COMMAND, ALBUQUERQUE, N.M. (NAVY NUCLEAR WEAPONS OFFICER COURSE)
WB PR-	NATIONAL WEATHER SERVICE. PACIFIC REGION, HONOLULU
WBRS-	WEATHER BUREAU RESEARCH STATION, LAS VEGAS, NEV.
WB SR-	NATIONAL WEATHER SERVICE. SOUTHERN REGION, FT. WORTH, TEX.
WB-T-	ENVIRONMENTAL DATA SERVICE, SILVER SPRING, MD.(TRANS)
WB/TA-	ENVIRONMENTAL DATA SERVICE, SILVER SPRING, MD.
WBTM-CR-	NATIONAL WEATHER SERVICE. CENTRAL REGION, KANSAS CITY, MO.
WBTM-DATAC-	WEATHER BUREAU, SILVER SPRING, MD.
WBTM-EDL-	WEATHER BUREAU, SILVER SPRING, MD.
WBTM-ENG-	NATIONAL WEATHER SERVICE. ENG. DIV., SILVER SPRING,MD
WBTM-ER-	NATIONAL WEATHER SERVICE. EASTERN REGION. GARDEN CITY, N.Y.
WBTM-FCST-	NATIONAL WEATHER SERVICE. WEATHER ANALYSIS + PREDICTION DIV.,SILVER SPRING,MD. (NOTES TO FORECASTERS)
WBTM-HYDRO-	NATIONAL WEATHER SERVICE. OFFICE OF HYDROLOGY, SILVER SPRING, MD.
WBTM-NMC-	NATIONAL METEOROLOGICAL CENTER, SUITLAND, MD.
WBTM-NMC-TM-	NATIONAL METEOROLOGICAL CENTER, SUITLAND, MD.
WBTM-PR-	NATIONAL WEATHER SERVICE. PACIFIC REGION, HONOLULU
WBTM-SOS-	NATIONAL WEATHER SERVICE. SPACE OPERATIONS SUPPORT DIV., SILVER SPRING, MD.
WBTM-SPDD-	NATIONAL WEATHER SERVICE. SYSTEMS PLANS AND DESIGN DIV., SILVER SPRING, MD.
WBTM-SR-	NATIONAL WEATHER SERVICE. SOUTHERN REGION, FT. WORTH, TEX.
WBTM-TDL-	NATIONAL WEATHER SERVICE. TECHNIQUES DEVELOPMENT LAB., SILVER SPRING, MD.
WBTM-T+EL-	NATIONAL WEATHER SERVICE. TEST AND EVALUATION LAB., STERLING, VA.
WBTM-WR-	NATIONAL WEATHER SERVICE. WESTERN REGION, SALT LAKE CITY
WBTM-WR-	WEATHER BUREAU, SALT LAKE CITY
WBTN-FCST-	NATIONAL WEATHER SERVICE. WEATHER ANALYSIS + PREDICTION DIV.,SILVER SPRING,MD. (NOTES TO FORECASTERS)
WB-TP-	WEATHER BUREAU, WASHINGTON, D.C.
WB-TRANS-(YEAR)-	WEATHER BUREAU, WASHINGTON, D.C. (TRANSLATION)
WC-	WELSBACH CORP. OZONE PROCESSES DIV., PHILADELPHIA
WC-	WYANDOTTE CHEMICALS CORP., MICH.
WCA-	BUREAU OF MINES, BRUCETON, PENNA.
WCA-	WAH CHANG CORP. ALBANY DIV., OREGON
WCAP-	CAROLINAS-VIRGINIA NUCLEAR POWER ASSOCIATES, INC., CHARLOTTE, N.C.
WCAP-	UNITED STATES-EURATOM JOINT RESEARCH + DEV. PROGRAM
WCAP-	WESTINGHOUSE ELECTRIC CORP. NUCLEAR ENERGY SYSTEMS, PITTSBURGH (COMMERCIAL ATOMIC POWER)
WCAP-AD-	WESTINGHOUSE ELECTRIC CORP. NUCLEAR ENERGY SYSTEMS, PITTSBURGH
WCC-	WYANDOTTE CHEMICALS CORP., MICH.
WCDC-	WASHINGTON COLL., CHESTERTOWN, MD. DEPT. OF CHEMISTRY
WCE-	DIRECTORATE OF DEVELOPMENT, WRIGHT-PATTERSON AFB,OHIO
WCEE-	WRIGHT AIR DEVELOPMENT CENTER. EQUIPMENT LAB., WRIGHT-PATTERSON AFB, OHIO
WCEGE-	WRIGHT AIR DEVELOPMENT CENTER. WEAPONS GUIDANCE LAB., WRIGHT-PATTERSON AFB, OHIO
WC/ERA-	SYMPOSIUM ON ELECTRON RING ACCELERATOR, 1968, BERKELEY, CALIF.
WCES-	WRIGHT AIR DEVELOPMENT CENTER. ELECTRONIC COMPONENTS LAB., WRIGHT-PATTERSON AFB, OHIO
WCL-	DIRECTORATE OF DEVELOPMENT, WRIGHT-PATTERSON AFB,OHIO
WCLAD-	DIRECTORATE OF DEVELOPMENT, WRIGHT-PATTERSON AFB,OHIO
WCLB-	WRIGHT AIR DEVELOPMENT CENTER. PROPELLER LAB., WRIGHT-PATTERSON AFB, OHIO
WCLC-TN-(YR)-	WRIGHT AIR DEVELOPMENT CENTER. ELECTRONIC COMPONENTS LAB., WRIGHT-PATTERSON AFB, OHIO
WCLD-	WRIGHT AIR DEVELOPMENT CENTER. AERO MEDICAL LAB., WRIGHT-PATTERSON AFB, OHIO
WCLE-	WRIGHT AIR DEVELOPMENT CENTER. EQUIPMENT LAB., WRIGHT-PATTERSON AFB, OHIO
WCLF-	WRIGHT AIR DEVELOPMENT CENTER. AERIAL RECONNAISSANCE LAB., WRIGHT-PATTERSON AFB, OHIO
WCLF-TN-	WRIGHT AIR DEVELOPMENT CENTER. PHOTO RECONNAISSANCE LAB., WRIGHT-PATTERSON AFB, OHIO
WCLG-	WRIGHT AIR DEVELOPMENT CENTER. WEAPONS GUIDANCE LAB., WRIGHT-PATTERSON AFB, OHIO
WCLGS-	WRIGHT AIR DEVELOPMENT CENTER. WEAPONS GUIDANCE LAB., WRIGHT-PATTERSON AFB, OHIO
WCLN-	WRIGHT AIR DEVELOPMENT CENTER. COMMUNICATION AND NAVIGATION LAB., WRIGHT-PATTERSON AFB, OHIO
WCLP-	WRIGHT AIR DEVELOPMENT CENTER. POWER PLANT LAB., WRIGHT-PATTERSON AFB, OHIO

WCLPF- WRIGHT AIR DEVELOPMENT CENTER. POWER PLANT LAB., WRIGHT-PATTERSON AFB, OHIO

WCLPO- WRIGHT AIR DEVELOPMENT CENTER. POWER PLANT LAB., WRIGHT-PATTERSON AFB, OHIO

WCLP-TN-(YR)- WRIGHT AIR DEVELOPMENT CENTER. POWER PLANT LAB., WRIGHT-PATTERSON AFB, OHIO

WCLR- WRIGHT AIR DEVELOPMENT CENTER. AERIAL RECONNAISSANCE LAB., WRIGHT-PATTERSON AFB, OHIO

WCLS- WRIGHT AIR DEVELOPMENT CENTER. AIRCRAFT LAB., WRIGHT-PATTERSON AFB, OHIO

WCLT-TM- WRIGHT AIR DEVELOPMENT CENTER. MATERIALS LAB., WRIGHT-PATTERSON AFB, OHIO

WCM- WRIGHT AIR DEV. CENTER, WRIGHT-PATTERSON AFB, OHIO

WCNE- WRIGHT AIR DEVELOPMENT CENTER. POWER PLANT LAB., WRIGHT-PATTERSON AFB, OHIO

WCNS- WRIGHT AIR DEVELOPMENT CENTER. AIRCRAFT LAB., WRIGHT-PATTERSON AFB, OHIO

WCNSW- WRIGHT AIR DEVELOPMENT CENTER. AIRCRAFT LAB., WRIGHT-PATTERSON AFB, OHIO

WCNSY- WRIGHT AIR DEVELOPMENT CENTER. AIRCRAFT LAB., WRIGHT-PATTERSON AFB, OHIO

WCO-(NUMBER)- AIR FORCE SPECIAL WEAPONS CENTER, KIRTLAND AFB, N.M.

WCOD-(NUMBER)- AIR FORCE SPECIAL WEAPONS CENTER, KIRTLAND AFB, N.M.

WCOWF- DIRECTORATE OF WEAPON SYSTEMS OPERATIONS, WRIGHT-PATTERSON AFB, OHIO

WCOWP- DIRECTORATE OF WEAPON SYSTEMS OPERATIONS, WRIGHT-PATTERSON AFB, OHIO

WCOWS- DIRECTORATE OF WEAPON SYSTEMS OPERATIONS, WRIGHT-PATTERSON AFB, OHIO

WCR- DIRECTORATE OF RESEARCH, WRIGHT-PATTERSON AFB, OHIO

WCRDF- WRIGHT AIR DEVELOPMENT CENTER. AERO MEDICAL LAB., WRIGHT-PATTERSON AFB, OHIO

WCRDM- WRIGHT AIR DEVELOPMENT CENTER. AERO MEDICAL LAB., WRIGHT-PATTERSON AFB, OHIO

WCRDR- WRIGHT AIR DEVELOPMENT CENTER. AERO MEDICAL LAB., WRIGHT-PATTERSON AFB, OHIO

WCRD-TM-(YEAR)- AEROSPACE MEDICAL RESEARCH LABS. (6570TH), WRIGHT-PATTERSON AFB, OHIO

WCRE- WRIGHT AIR DEVELOPMENT CENTER. ELECTRONIC COMPONENTS LAB., WRIGHT-PATTERSON AFB, OHIO

WCRR- WRIGHT AIR DEVELOPMENT CENTER. AERONAUTICAL RESEARCH LAB., WRIGHT-PATTERSON AFB, OHIO

WCRR-TN- WRIGHT AIR DEVELOPMENT CENTER. AERONAUTICAL RESEARCH LAB., WRIGHT-PATTERSON AFB, OHIO

WCRT-(YR)- WRIGHT AIR DEVELOPMENT CENTER. MATERIALS LAB., WRIGHT-PATTERSON AFB, OHIO

WCRTE- WRIGHT AIR DEVELOPMENT CENTER. MATERIALS LAB., WRIGHT-PATTERSON AFB, OHIO

WCRTL-M- WRIGHT AIR DEVELOPMENT CENTER. MATERIALS LAB., WRIGHT-PATTERSON AFB, OHIO

WCRTL-TN- WRIGHT AIR DEVELOPMENT CENTER. MATERIALS LAB., WRIGHT-PATTERSON AFB, OHIO

WCRT-TM- WRIGHT AIR DEVELOPMENT CENTER. MATERIALS LAB., WRIGHT-PATTERSON AFB, OHIO

WCRT-TN-(YR)- WRIGHT AIR DEVELOPMENT CENTER. MATERIALS LAB., WRIGHT-PATTERSON AFB, OHIO

WCS- DIRECTORATE OF WEAPON SYSTEMS OPERATIONS, WRIGHT-PATTERSON AFB, OHIO

WCSDD- DIRECTORATE OF WEAPON SYSTEMS OPERATIONS, WRIGHT-PATTERSON AFB, OHIO

WCSE- DIRECTORATE OF WEAPON SYSTEMS OPERATIONS, WRIGHT-PATTERSON AFB, OHIO

WCSG- DIRECTORATE OF WEAPON SYSTEMS OPERATIONS, WRIGHT-PATTERSON AFB, OHIO

WCSI-(YEAR)-TR- COMMUNICATION SYSTEMS INC., FALLS CHURCH, VA.

WCSP- DIRECTORATE OF WEAPON SYSTEMS OPERATIONS, WRIGHT-PATTERSON AFB, OHIO

WCSPE- DIRECTORATE OF WEAPON SYSTEMS OPERATIONS, WRIGHT-PATTERSON AFB, OHIO

WCSR- DIRECTORATE OF WEAPON SYSTEMS OPERATIONS, WRIGHT-PATTERSON AFB, OHIO

WCSWR- DIRECTORATE OF WEAPON SYSTEMS OPERATIONS, WRIGHT-PATTERSON AFB, OHIO

WCT- DIRECTORATE OF FLIGHT AND ALL-WEATHER TESTING, WRIGHT-PATTERSON AFB, OHIO

53WC... THRU 55WC... WRIGHT AIR DEVELOPMENT CENTER, WRIGHT-PATTERSON AFB, OHIO (FIRST TWO DIGITS STAND FOR YEAR) (REF. 1)

WD- AIR RES.+DEV.COMMAND. WESTERN DEV.DIV.,INGLEWOOD,CAL.

WD-(YR.)- AIR RES.+DEV.COMMAND. WESTERN DEV.DIV.,INGLEWOOD,CAL.

WD- ATOMIC ENERGY OF CANADA LTD. CHALK RIVER PROJECT, ONT. (REF. 6)

WD- BOEING AIRPLANE CO. WICHITA DIV., KAN.

WD-(4 DIGITS) COMPAGNIE GENERALE DE TELEGRAPHIE SANS FIL, PARIS

WD- WAR DEPT.

WDAL- BALLISTIC MISSILE DIV. (AIR FORCE), LOS ANGELES

WDAT- AIR RES.+DEV.COMMAND. WESTERN DEV.DIV.,INGLEWOOD,CAL.

WDC- WASHINGTON DOCUMENT CENTER (ARMY)

WD CCBP- WAR DEPT. COMBINED COMMUNICATIONS BOARD (PROJECTS)

WDD- AIR RES.+DEV.COMMAND. WESTERN DEV.DIV.,INGLEWOOD,CAL.

WDD-(YR.)- AIR RES.+DEV.COMMAND. WESTERN DEV.DIV.,INGLEWOOD,CAL.

WDD- MARTIN-MARIETTA CORP. DENVER DIV.

WDD-AL- AIR RESEARCH AND DEVELOPMENT COMMAND. WESTERN DEVELOPMENT DIV., INGLEWOOD, CALIF. (ACCESSION LISTS)

WDD-DOC.(NO.)- AIR RES.+DEV.COMMAND. WESTERN DEV.DIV.,INGLEWOOD,CAL.

WDD-DOC.(YR.)- AIR RES.+DEV.COMMAND. WESTERN DEV.DIV.,INGLEWOOD,CAL.

WDD-M-MI-(YEAR)- MARTIN CO., DENVER

WDD-M-SR-(NUMBER)- MARTIN CO., DENVER

WD FM- WAR DEPT. (FIELD MANUAL)

WD FT- WAR DEPT. (FIRING TABLE)

WDG-(YR.)- AIR RES.+DEV.COMMAND. WESTERN DEV.DIV.,INGLEWOOD,CAL.

WDGH-(YR.)- AIR RES.+DEV.COMMAND. WESTERN DEV.DIV.,INGLEWOOD,CAL.

WDGH- AIR RES.+DEV.COMMAND. WESTERN DEV.DIV.,INGLEWOOD,CAL.

WDGP-(YR.)- AIR RES.+DEV.COMMAND. WESTERN DEV.DIV.,INGLEWOOD,CAL.

WDGS- ARMY GENERAL STAFF, WASHINGTON, D.C.

WDI- ATOMIC ENERGY OF CANADA LTD., CHALK RIVER, ONT.

WDI- ATOMIC ENERGY OF CANADA LTD. WHITESHELL NUCLEAR RESEARCH ESTAB., PINAWA, MAN.

WDI- BUREAU OF ORDNANCE (NAVY) (WEAPON DATA INDEX)

WDL- PHILCO CORP. WESTERN DEVELOPMENT LABS., PALO ALTO,CAL

WDLPB- BALLISTIC MISSILE DIV. (AIR FORCE), LOS ANGELES

WDLPM- BALLISTIC MISSILE DIV. (AIR FORCE), LOS ANGELES

WDLPR- BALLISTIC MISSILE DIV. (AIR FORCE), LOS ANGELES

WDLPS- SPACE SYSTEMS DIV. (AIR FORCE), INGLEWOOD, CALIF.

WDL/TM- PHILCO CORP. WESTERN DEVELOPMENT LABS., PALO ALTO,CAL (TECHNICAL MANUAL)

WDL/TM- PHILCO CORP. WESTERN DEVELOPMENT LABS., PALO ALTO,CAL (TECHNICAL MEMO)

WDL-TR(NUMBER)- PHILCO-FORD CORP. WESTERN DEV.LABS.,PALO ALTO, CAL.

WD MID SC- WAR DEPT. MILITARY INTELLIGENCE SERVICE

WD MID SS- WAR DEPT. MILITARY INTELLIGENCE DIV. (SPECIAL SERIES)

WD MIS VII- WAR DEPT. MILITARY INTELLIGENCE DIV.

WD MTP- WAR DEPT. (MOBILIZATION TRAINING PROGRAM)

WDP- AIR RES.+DEV.COMMAND, WESTERN DEV.DIV.,INGLEWOOD,CAL.

WD PAM- WAR DEPT. (PAMPHLET)

WDPCR- BALLISTIC MISSILE DIV. (AIR FORCE), LOS ANGELES

WDPP- AIR RES.+DEV.COMMAND. WESTERN DEV.DIV.,INGLEWOOD,CAL.

WDR- BUREAU OF SHIPS (WAR DAMAGE REPORT)

WDR- GENERAL PRECISION SYSTEMS, INC., LINK GP., PALO ALTO, CALIF.

WD SB- WAR DEPT. (SUPPLY BULLETIN)

WDSIT-(YR.)-	AIR RES.+DEV.COMMAND. WESTERN DEV.DIV., INGLEWOOD, CAL.
WDSIT-	AIR RES.+DEV.COMMAND. WESTERN DEV.DIV., INGLEWOOD, CAL.
WDSOT-(YR.)-	AIR RES.+DEV.COMMAND. WESTERN DEV.DIV., INGLEWOOD, CAL.
WDSOT-	AIR RES.+DEV.COMMAND. WESTERN DEV.DIV., INGLEWOOD, CAL.
WDT-	RAMO-WOOLDRIDGE CORP., LOS ANGELES
WD TB-	WAR DEPT. (TECHNICAL BULLETIN)
WD TB CW-	WAR DEPT. (CHEMICAL WARFARE TECHNICAL BULLETIN)
WD TB ENG-	WAR DEPT. (ENGINEERING TECHNICAL BULLETIN)
WD TB MED-	WAR DEPT. (MEDICAL TECHNICAL BULLETIN)
WD TB QM-	WAR DEPT. (QUARTERMASTER TECHNICAL BULLETIN)
WD TB SIG-	WAR DEPT. (SIGNAL TECHNICAL BULLETIN)
WD TB SIG E-	WAR DEPT. (SIGNAL EQUIPMENT TECHNICAL BULLETIN)
WD TB TC-	WAR DEPT. TRANSPORTATION CORPS (TECHNICAL BULLETIN)
WD TD-	WAR DEPT. (TRAJECTORY DIAGRAMS)
WDT-EX-(YEAR)-	BALLISTIC MISSILE DIV. (AIR FORCE), INGLEWOOD, CALIF.
WD-TM-	NATIONAL MILITARY ESTABLISHMENT, WASHINGTON, D.C.
WD TM-	WAR DEPT. (TECHNICAL MANUAL)
WDZAP-	BALLISTIC MISSILE DIV. (AIR FORCE), LOS ANGELES
WE-	WESTERN ELECTRIC CO., N.Y.C.
WEB-	PHILLIPS PETROLEUM CO. ATOMIC ENERGY DIV., IDAHO FALLS (CODE DERIVED FROM AUTHOR'S NAME)
WEBB-(NUMBER)-	WEBB INST. OF NAVAL ARCHITECTURE, GLEN COVE, N.Y.
WEC-	WESTERN ELECTRIC CO., N.Y.C.
WEC-	WESTINGHOUSE ELECTRIC CORP. (ALL LOCATIONS)
WEC-A-	WESTINGHOUSE ELECTRIC CORP. AVIATION GAS TURBINE DIV., KANSAS CITY, MO.
WEC-AAR-	WESTINGHOUSE ELECTRIC CORP. AIR ARM DIV., BALTIMORE
WEC-GO-HS-	WESTINGHOUSE ELECTRIC CORP. DEFENSE + SPACE CENTER, BALTIMORE
WECO-	WESTINGHOUSE ELECTRIC CORP. (ALL LOCATIONS)
WEC-TR-	WESTINGHOUSE ELECTRIC CORP. RES. + DEV. CTR., PITTS.
WEDCOM-	GENERAL ELECTRIC CO., SANTA BARBARA, CALIF.
WEEC-PR-	WEINSCHEL ENGINEERING, KENSINGTON, MD.
WEET-	WESTINGHOUSE ELECTRIC CORP. ELECTRONIC TUBE DIV., ELMIRA, N.Y.
WE-H-	ARMED FORCES SPECIAL WEAPONS PROJECT, WASHINGTON, D.C. (WEAPONS EMPLOYMENT HANDBOOK)
WE IB-	WESTERN ELECTRIC CO., N.Y.C. (INSTRUCTION BOOK)
WE I BUL-	WESTERN ELECTRIC CO., N.Y.C. (INSTRUCTION BULLETIN)
WEIDLINGER-TR-	WEIDLINGER (PAUL) CONSULTANTS, N.Y.C.
WE-M-	ARMED FORCES SPECIAL WEAPONS PROJECT. FIELD COMMAND, ALBUQUERQUE, N. MEX. (WEAPONS EMPLOYMENT MANUAL)
WEMCO-	WESTINGHOUSE ELECTRIC AND MFG. CO., E. PITTSBURGH
WEM IB-	WESTINGHOUSE ELECTRIC AND MANUFACTURING CO., EAST PITTSBURGH (INSTRUCTION BOOK)
WENE-	WESTINGHOUSE ELECTRIC CORP. NUCLEAR ENERGY SYSTEMS, PITTSBURGH
WERG-	WESTON ELECTRICAL INSTRUMENT CORP., NEWARK, N.J.
WERL-	WESTINGHOUSE ELECTRIC CORP. RES. + DEV. CTR., PITTS.
WERL-	WESTINGHOUSE ELECTRIC CORP. RES. LABS., PITTSBURGH (NUMBER USED IS LAST 4 DIGITS OF CONTRACT NO.)
WERL-(CONTRACT NO.)-	WESTINGHOUSE ELECTRIC CORP. RES. LABS., PITTSBURGH
WERL-(NUMBERS)-	WESTINGHOUSE RESEARCH LABS., PITTSBURGH
WERL-(LETTERS)-	WESTINGHOUSE RESEARCH LABS., PITTSBURGH
WERL-ELPLA-	WESTINGHOUSE RESEARCH LABS., PITTSBURGH
WERL-HOLOG-	WESTINGHOUSE RESEARCH LABS., PITTSBURGH
WERL-TAADM-	WESTINGHOUSE RESEARCH LABS., PITTSBURGH
WES-	ARMY ENGINEER WATERWAYS EXPERIMENT STATION, VICKSBURG, MISS.
WES-CR-	ARMY ENGINEER WATERWAYS EXPERIMENT STATION, VICKSBURG, MISS. (CONTRACT REPORT)
WES FR-	ARMY ENGINEER WATERWAYS EXPERIMENT STATION, VICKSBURG, MISS. (FINAL REPORT)
WESI-	WEATHER SERVICES, INC., BOSTON
WES-IR-	ARMY ENGINEER WATERWAYS EXPERIMENT STATION, VICKSBURG, MISS. (INSTRUCTION REPORT)
WES-MISC.PAPER (NO.)-	ARMY ENGINEER WATERWAYS EXPERIMENT STATION, VICKSBURG, MISS.
WES MP-(NUMBER)-	ARMY ENGINEER WATERWAYS EXPERIMENT STATION, VICKSBURG, MISS. (MISCELLANEOUS PAPER)
WES-MP-(LETTER)-(YR.)-	ARMY ENGINEER WATERWAYS EXPERIMENT STATION, VICKSBURG, MISS.
WES-NCG-TR-	ARMY ENGINEER WATERWAYS EXPERIMENT STATION, VICKSBURG, MISS.
WES-RR-	ARMY ENGINEER WATERWAYS EXPERIMENT STATION, VICKSBURG, MISS. (RESEARCH REPORT)
WEST-	WESTCLOX, INC., LA SALLE, ILL.
WEST (NO.)-(NO.)-	WESTINGHOUSE ELECTRIC CORP. AIR ARM DIV., BALTIMORE
WEST AA TN-	WESTINGHOUSE ELECTRIC CORP. AIR ARM DIV., BALTIMORE
WESTERN GEAR-(NUMBER)-	WESTERN GEAR CORP. SYSTEMS MANAGEMENT DIV., LYNWOOD, CALIF.
WESTFIELDC-	WESTFIELD COLL., LONDON
WESTINGHOUSE (NO.)-ML-	WESTINGHOUSE ELECTRIC CORP. (ALL LOCATIONS)
WES-TM-(NUMBER)-	ARMY ENGINEER WATERWAYS EXPERIMENT STATION, VICKSBURG, MISS.
WES-TR-(NUMBER)-	ARMY ENGINEER WATERWAYS EXPERIMENT STATION, VICKSBURG, MISS.
WES-TR-(LETTER)-(YR.)-	ARMY ENGINEER WATERWAYS EXPERIMENT STATION, VICKSBURG, MISS.
WEST WAPD-	WESTINGHOUSE ELECTRIC CORP. BETTIS ATOMIC POWER LAB., PITTSBURGH
WEST WAPD-TM-	WESTINGHOUSE ELECTRIC CORP. BETTIS ATOMIC POWER LAB., PITTSBURGH
WEST-WIAP-	WESTINGHOUSE ELECTRIC CORP. NUCLEAR ENERGY SYSTEMS, PITTSBURGH
WET-(YEAR)-	ARMED FORCES SPECIAL WEAPONS PROJECT. FIELD COMMAND. DIRECTORATE OF WEAPONS EFFECTS TESTS, ALBUQUERQUE, N. MEX. (REF. 7)
WEU-	WESLEYAN UNIV., MIDDLETOWN, CONN.
WF-	MOTOROLA INC., SCOTTSDALE, ARIZ.
WF-	MOTOROLA INC. MILITARY ELECTRONICS DIV., SCOTTSDALE, ARIZ.
WFC-	WAKE FOREST COLL., N.C.
WG-	CIVIL SERVICE COMMISSION, WASHINGTON, D.C. (WORK GRADES)
WG 3-	FORECAST CENTER, ENT AFB, COLO.
WG-	WESTINGHOUSE ELECTRIC CORP., PITTSBURGH
WGC-	WESTERN GEAR CORP., LYNWOOD, CALIF.
WGD-	WESTINGHOUSE ELECTRIC CORP. SURFACE DIV., BALTIMORE
WGLR/DGRR PAPER-	WISSENSCHAFTLICHE GESELLSCHAFT FUER LUFT- UND RAUMFAHRT, COLOGNE
WGP-	AIR WEATHER SERVICE, WASHINGTON, D.C.
WGR-	(WORKING GROUP REPORT) (REF. 2)
(NUMBER)-WH-	BELL TELEPHONE LABS., INC., WHIPPANY, N.J.
WH-	WESTINGHOUSE ELECTRIC CORP. ASTRONUCLEAR LAB., PITTS.
WHAN-FR-	WADCO CORP., RICHLAND, WASH.
WHAN-IR-	WADCO CORP., RICHLAND, WASH.
WHAN-SA-	WADCO CORP., RICHLAND, WASH.
WHC(C)/P-	UNITED KINGDOM ATOMIC ENERGY AUTHORITY. INDUSTRIAL GROUP H.Q., RISLEY, LANCS, ENGLAND
WHC(C)/P-	UNITED KINGDOM ATOMIC ENERGY AUTHORITY. RESEARCH GP. ATOMIC ENERGY RES. ESTAB., HARWELL, BERKS, ENGLAND
WHIRLAJET-	WHIRLAJET, INC., LOS ANGELES
WHITTAKER-D-	WHITTAKER (WM. R) CO., LOS ANGELES
WHO-	WORLD HEALTH ORGANIZATION

WHOI-

WHOI-
　　　WOODS HOLE OCEANOGRAPHIC INSTITUTION, MASS.
WHOI-(YEAR)-
　　　WOODS HOLE OCEANOGRAPHIC INSTITUTION, MASS.
WHOI-CONTRIB-
　　　WOODS HOLE OCEANOGRAPHIC INSTITUTION, MASS.
WHOI-REF-
　　　WOODS HOLE OCEANOGRAPHIC INSTITUTION, MASS.
WHOI-REF-(YEAR)-
　　　WOODS HOLE OCEANOGRAPHIC INSTITUTION, MASS.
WHO-REF-(YR.)-
　　　WOODS HOLE OCEANOGRAPHIC INSTITUTION, MASS.
WHPC-
　　　GT. BRIT. WINDSCALE WORKS, SELLAFIELD, CUMB., ENGLAND
WHTD-
　　　DELFT, NETHERLANDS. TECHNISCHE HOGESCHOOL. LABORATORIUM VOOR TECHNISCHE MECHANICA
WI-(NUMBER-NUMBER-YR.)
　　　NORTH DAKOTA WATER RESOURCES INST., FARGO
WI-
　　　WILMER INSTITUTE
WIAP-
　　　WESTINGHOUSE ELECTRIC CORP. NUCLEAR ENERGY SYSTEMS, PITTSBURGH (INDUSTRIAL ATOMIC POWER)
WIAP-M-
　　　WESTINGHOUSE ELECTRIC CORP. NUCLEAR ENERGY SYSTEMS, PITTSBURGH
WIAP-NL-
　　　WESTINGHOUSE ELECTRIC CORP. NUCLEAR ENERGY SYSTEMS, PITTSBURGH
WIAP-NM-
　　　WESTINGHOUSE ELECTRIC CORP. NUCLEAR ENERGY SYSTEMS, PITTSBURGH
WIAP-P-
　　　WESTINGHOUSE ELECTRIC CORP. NUCLEAR ENERGY SYSTEMS, PITTSBURGH
WIB-
　　　ARMY AIR FORCE. WEATHER INFORMATION BRANCH
WICHU-
　　　WICHITA, KAN. UNIV.
WIC-TCI-
　　　WISCONSIN. UNIV., MADISON. MATHEMATICS RESEARCH CTR.
WIL-
　　　WILLARD STORAGE BATTERY CO., CLEVELAND
WILLOW-(NO.)-(NO.)-(LTR.)
　　　MICHIGAN. UNIV., ANN ARBOR. WILLOW RUN LABS.
WIN-
　　　NATIONAL LEAD CO., INC. RAW MATERIALS DEVELOPMENT LAB., WINCHESTER, MASS.
WINDSCALE-
　　　UNITED KINGDOM ATOMIC ENERGY AUTHORITY. INDUSTRIAL GROUP. WINDSCALE WORKS, SELLAFIELD, CUMB., ENGLAND
WIS-
　　　ARMY AIR FORCE. WEATHER INFORMATION SERVICE
WIS-
　　　WEIZMANN INST. OF SCIENCE, REHOVOTH, ISRAEL
WIS-
　　　WISCONSIN. UNIV., MADISON
WIS-(NUMBER)-(LETTER(S))
　　　WISCONSIN. UNIV., MADISON (SQUID PROJECT) (REF. 33)
WIS-AEC-
　　　WISCONSIN. UNIV., MADISON. THEORETICAL CHEMISTRY LAB.
WIS-AF-
　　　WISCONSIN. UNIV., MADISON. THEORETICAL CHEMISTRY LAB.
WISCO-
　　　WILLAMETTE IRON AND STEEL CO., PORTLAND, ORE.
WISCO-FR-(NO.)-(NO.)
　　　WOOD-IVEY SYSTEMS CORP., WINTER PARK, FLA.
WIS-CRPS-
　　　WISCONSIN. UNIV., MADISON. NUMERICAL ANALYSIS LAB.
WIS-CSTR-
　　　WISCONSIN. UNIV., MADISON. COMPUTER SCIENCES DEPT.
WISCU-
　　　WISCONSIN. UNIV., MADISON
WISK-
　　　UNION OF SOUTH AFRICA. COUNCIL FOR SCIENTIFIC AND INDUSTRIAL RESEARCH, PRETORIA
WIS-MRC-TSR-
　　　WISCONSIN. UNIV., MADISON. MATHEMATICS RESEARCH CTR.
WIS-NSF-
　　　WISCONSIN. UNIV., MADISON. THEORETICAL CHEMISTRY INST
WIS-ONR-
　　　WISCONSIN. UNIV., MADISON. THEORETICAL CHEMISTRY LAB.
WIS-OOR-
　　　WISCONSIN. UNIV., MADISON. THEORETICAL CHEMISTRY LAB.
WIS-QSR-
　　　WISCONSIN. UNIV., MADISON. DEPT. OF PHYSICS
WIS-TCI-
　　　WISCONSIN. UNIV., MADISON. THEORETICAL CHEMISTRY INST
WIS-TR-
　　　WISCONSIN. UNIV., MADISON. DEPT. OF STATISTICS
WISU-
　　　WISCONSIN. UNIV., MADISON
WITWATERSRAND-
　　　UNIVERSITY OF THE WITWATERSRAND, JOHANNESBURG. DEPT. OF MECHANICAL ENGINEERING
WIU-
　　　WICHITA, KAN. UNIV.
WIU-AR-
　　　WICHITA, KAN. UNIV. SCHOOL OF ENG. (AERODYNAMIC RPT.)
WIU-ER-
　　　WICHITA, KAN. UNIV. SCHOOL OF ENGINEERING (ENG. RPT.)
W-J-
　　　WATKINS-JOHNSON CO., PALO ALTO, CALIF.
WJ-(3 DIGITS)
　　　WATKINS-JOHNSON CO., PALO ALTO, CALIF.
WJCC-
　　　WESTERN JOINT COMPUTER CONFERENCE
WKNL-
　　　KIDDE (WALTER) NUCLEAR LABS., INC., GARDEN CITY, N.Y.

332

WL(NUMBER)-TN-(YEAR)-
　　　AIR FORCE WEAPONS LAB., KIRTLAND AFB, N. MEX.
WL-
　　　DIKEWOOD CORP., ALBUQUERQUE, N. MEX.
WL-
　　　WARD LEONARD ELECTRIC CO., MOUNT VERNON, N.Y.
WL-
　　　WESTINGHOUSE ELECTRIC CORP. ATOMIC POWER DIV., PITTS.
WL-
　　　WHEELER LABS., INC., GREAT NECK, N.Y.
WLAD-BL/RP-DCZID/PC-
　　　WISCONSIN. STATE DEPT. OF LOCAL AFFAIRS AND DEVELOPMENT, MADISON (LOCAL + REGIONAL PLANNING)
WLAD-BL/RP-FVCOG-KPC-
　　　WISCONSIN. STATE DEPT. OF LOCAL AFFAIRS AND DEVELOPMENT, MADISON (LOCAL + REGIONAL PLANNING)
WLAD-BL/RP-WPC-(YR.)-
　　　WISCONSIN. STATE DEPT. OF LOCAL AFFAIRS AND DEVELOPMENT, MADISON (LOCAL + REGIONAL PLANNING)
WLAS-
　　　AIR FORCE WEAPONS LAB., KIRTLAND AFB, N. MEX.
W/LAU.(NUMBER)
　　　ENGLISH ELECTRIC CO., LTD. CENTRAL METALLURGICAL LABS., WHETSTONE, LEICS, ENGLAND
WLB-
　　　TENNESSEE EASTMAN CORP., OAK RIDGE, TENN.
WLC-
　　　AIR FORCE CAMBRIDGE RESEARCH CENTER, MASS.
WLC-TM-(YEAR)-
　　　AIR FORCE WEAPONS LAB. CIVIL ENGINEERING DIV., KIRTLAND AFB, N. MEX.
WLC-TN-(YEAR)-
　　　AIR FORCE WEAPONS LAB. CIVIL ENGINEERING DIV., KIRTLAND AFB, N. MEX.
WLC-TN-(YEAR)-
　　　WATER RESOURCES ENGINEERS, INC., WALNUT CREEK, CALIF.
WL-E(NO.)-
　　　SYLVANIA ELECTRONIC SYSTEMS. WALTHAM LABS., MASS.
WLL-
　　　TENNESSEE EASTMAN CORP., OAK RIDGE, TENN.
WLM-
　　　MAXSON (W.L.) CORP., N.Y.C.
WLR-
　　　FOSTER WHEELER CORP. WELDING LAB., N.Y.C.
WLRAF-
　　　PURDUE UNIV., LAFAYETTE, IND. SCHOOL OF MECH. ENG.
WL-TDR-(YEAR)-
　　　AIR FORCE WEAPONS LAB., KIRTLAND AFB, N. MEX.
WL TR-(YEAR)-
　　　AIR FORCE WEAPONS LAB., KIRTLAND AFB, N. MEX.
WM-
　　　(WELDING MEMORANDUM) (REF. 2)
WM-
　　　ATOMIC ENERGY OF CANADA LTD. CHALK RIVER PROJ., ONT.
WM-
　　　CALIFORNIA. UNIV., BERKELEY. RADIATION LAB.
WM-
　　　COLLEGE OF WILLIAM AND MARY, WILLIAMSBURG, VA.
W+M-
　　　COLLEGE OF WILLIAM AND MARY, WILLIAMSBURG, VA.
W/M-(2 DIGITS)
　　　COLLEGE OF WILLIAM AND MARY, WILLIAMSBURG, VA.
WM-
　　　DEFENSE SUPPLY AGENCY, WASHINGTON, D.C.
WM-
　　　GT. BRIT. MINISTRY OF SUPPLY (WELDING MEMORANDA)
WM-
　　　TENNESSEE EASTMAN CORP., OAK RIDGE, TENN.
WMB-
　　　BROBECK (WILLIAM M.) AND ASSOCIATES, BERKELEY, CALIF
WMB-
　　　BROBECK (WILLIAM M.) AND ASSOCIATES, BOULDER, COLO.
WMB/A-
　　　BROBECK (WILLIAM M.) AND ASSOCIATES, BERKELEY, CALIF.
WMC-
　　　WAR METALLURGY COMMITTEE
WMC (LETTER(S))-
　　　WAR METALLURGY COMMITTEE (RESEARCH REPORTS)
WMC METALS-
　　　WAR METALLURGY COMMITTEE (METALS REPORTS)
WMC SP-
　　　WAR METALLURGY COMMITTEE (SPECIAL REPORTS)
WMO-(NUMBER)-TP-
　　　WORLD METEOROLOGICAL ORGANIZATION, GENEVA
WMO-NO-(NUMBER)-TP
　　　WORLD METEOROLOGICAL ORGANIZATION, GENEVA
WMR-
　　　WESTINGHOUSE ELECTRIC CORP. ATOMIC POWER DEPT., PITTS.
WMSI-
　　　CALIFORNIA. UNIV., LOS ANGELES. WESTERN MANAGEMENT SCIENCE INST.
WMSI REPRINT-
　　　CALIFORNIA. UNIV., LOS ANGELES. WESTERN MANAGEMENT SCIENCE INST.
WMSI WORKING PAPER-
　　　CALIFORNIA. UNIV., LOS ANGELES. WESTERN MANAGEMENT SCIENCE INST.
WMSLAW-(YEAR)-
　　　LAWRENCE (WM. S.) AND ASSOCIATES, INC., CHICAGO
WMS-PROJ-
　　　WOLF MANAGEMENT SERVICES, PALO ALTO, CALIF.
WM-TR-
　　　COLLEGE OF WILLIAM AND MARY, WILLIAMSBURG, VA.
WN-
　　　NORTHEASTERN WOOD UTILIZATION COUNCIL (WOOD NOTES)
WNAL-TME-
　　　WESTINGHOUSE ELECTRIC CORP. ASTRONUCLEAR LAB., PITTS.
WNET-
　　　WESTINGHOUSE ELECTRIC CORP. NUCLEAR ENERGY SYSTEMS. NUCLEAR EQUIPMENT DIV., TAMPA, FLA.

WNL-	NATIONAL DEFENSE RESEARCH COMM. (WEEKLY NEWS LETTER)
WNQ-	CURTISS-WRIGHT CORP. RES. DIV., QUEHANNA, PENNA.
WNRE-	ATOMIC ENERGY OF CANADA LTD. WHITESHELL NUCLEAR RESEARCH ESTAB., PINAWA, MAN.
WNRE-(NUMBER)-	WNRE, INC., CHESTERTOWN, MD.
WNY-	WESTERN NEW YORK NUCLEAR RESEARCH CTR., INC., BUFFALO
WNZ-	CURTISS-WRIGHT CORP. RES. DIV., QUEHANNA, PENNA.
W.O.-	AEROJET-GENERAL CORP., SACRAMENTO, CALIF.
WO-	BUREAU OF YARDS AND DOCKS
WO-	FOREST SERVICE, WASHINGTON, D.C. (WASHINGTON OFFICE)
WO-	GT. BRIT. WAR OFFICE
WOC-	WOLLENSAK OPTICAL CO., ROCHESTER, N.Y.
WOLF-	WOLF RESEARCH AND DEVELOPMENT CORP., RIVERDALE, MD.
WO P-	GT. BRIT. WAR OFFICE (PAMPHLET)
WORKING MEMO-	DEFENSE SUPPLY AGENCY, WASHINGTON, D.C.
WORKING MEMO-	LITTLE (ARTHUR D.), INC., CAMBRIDGE, MASS.
WORKING PAPER-	AIR FORCE, WASHINGTON, D.C.
WO S-	GT. BRIT. WAR OFFICE (SPECIFICATION)
WOS-(YEAR)-	NAVAL PERSONNEL RES. + DEV. LAB., WASHINGTON, D.C.
WOU-	UNIVERSITY OF WESTERN ONTARIO, LONDON
WP-	(WORKING PAPER) (REF. 2)
WP-(YEAR)-	AERONAUTICAL SYSTEMS DIV. DEPUTY FOR DEVELOPMENT PLANNING, WRIGHT-PATTERSON AFB, OHIO
WP-	CALIFORNIA. UNIV., BERKELEY
WP-(NUMBER)-(YEAR)-	CARNEGIE-MELLON UNIV., PITTSBURGH
WP-	ENVIRONMENTAL PROTECTION AGENCY. WATER QUALITY OFFICE, ROCKVILLE, MD.
WP-(NUMBER)-AWTR-	ENVIRONMENTAL PROTECTION AGENCY. WATER QUALITY OFFICE, WASHINGTON, D.C. (WASTE TREATMENT)
WP-(NUMBER)-	FEDERAL WATER POLLUTION CONTROL ADMINISTRATION. WATER QUALITY LAB., EDISON, N.J.
WP-(4 DIGITS)	FOSTER-MILLER ASSOCIATES, INC., WALTHAM, MASS.
WP-	INTERNATIONAL ATOMIC ENERGY AGENCY, VIENNA
WP-	KEURING VAN ELECTROTECHNISCHE MATERIALEN, N.V., ARNHEM, NETHERLANDS
WP-	MITRE CORP., BEDFORD, MASS.
WP-	MOTOROLA INC., SCOTTSDALE, ARIZ.
WP-	MOTOROLA INC. MILITARY ELECTRONICS DIV., SCOTTSDALE, ARIZ.
WP-	NAVY ELECTRONICS LAB., SAN DIEGO, CALIF.
WP-	NAVY RADIO AND SOUND LAB., SAN DIEGO, CALIF. (WAVE PROPAGATION NO.)
WPB-	WAR PRODUCTION BOARD, WASHINGTON, D.C.
WPBUATS-	PALM BEACH COUNTY AREA PLANNING BD., WEST PALM BEACH, FLA. (WEST PALM BEACH URBAN AREA TRANSPORT. STUDY)
WPC/P-	UNITED KINGDOM ATOMIC ENERGY AUTHORITY. RESEARCH GP. ATOMIC ENERGY RES. ESTAB., HARWELL, BERKS, ENGLAND
WPD-	AUSTRALIA. WEAPONS RESEARCH ESTABLISHMENT
WPG-	NATIONAL DEFENSE RESEARCH COMM. (WAVE PROPAGATION GP)
WPL-	NAVAL ORDNANCE LAB., WHITE OAK, MD. (WHITE PLAN)
WP-ORD-	FEDERAL WATER POLLUTION CONTROL ADMINISTRATION. WATER QUALITY LAB., EDISON, N.J.
WPR-	ALLEGANY BALLISTICS LAB., CUMBERLAND, MD. (WEEKLY PROGRESS REPORT)
WPR-	GT. BRIT. DEPT. OF SCIENTIFIC AND INDUSTRIAL RESEARCH (WATER POLLUTION RESEARCH TECHNICAL REPORTS)
WPRB-	GT. BRIT. WATER POLLUTION RESEARCH BOARD
WPR/RC-B-	GT. BRIT. WATER POLLUTION RES.LAB.,WATFORD,HERTS,ENG.
WQO-	ENVIRONMENTAL PROTECTION AGENCY. WATER QUALITY OFFICE, ROCKVILLE, MD.
WQR/(YEAR)-	JOHNS HOPKINS UNIV., SILVER SPRING, MD. APPLIED PHYSICS LAB. (WEAPONS QUARTERLY REPORT)
WR-	BROWN UNIV., PROVIDENCE. DIV. OF ENGINEERING
WR-	COMPAGNIE GENERALE DE TELEGRAPHIE SANS FIL, ORSAY, FRANCE
WR-	COMPAGNIE GENERALE DE TELEGRAPHIE SANS FIL, PARIS
WR-	FRANCE. CENTRE DE PHYSIQUE ELECTRONIQUE ET CORPUSCULAIRE
WR-	NAVAL AIR DEVELOPMENT CENTER. AIR WARFARE RESEARCH DEPT., JOHNSVILLE, PENNA.
(NUMBER)-WR-	WESTERN ELECTRIC CO., INC. WINSTON-SALEM, N.C.
WR-	WYLE LABS., INC., HUNTSVILLE, ALA.
WR-(YEAR)-	WYLE LABS., INC., HUNTSVILLE, ALA.
WR-(LETTER)-	NATIONAL ADVISORY COMMITTEE FOR AERONAUTICS (WARTIME REPORT) (REF. 25)
WRA-B-(YEAR-NUMBER)	PUERTO RICO WATER RESOURCES AUTHORITY, SAN JUAN
WR-AF-	JOHNS HOPKINS UNIV., BALTIMORE. CARLYLE BARTON LAB.
WR-AF-	JOHNS HOPKINS UNIV., BALTIMORE. RADIATION LAB.
WRAIR-	WALTER REED ARMY INSTITUTE OF RES., WASHINGTON, D.C.
WRAIR-TR-	WALTER REED ARMY INSTITUTE OF RES., WASHINGTON, D.C.
WRAMC-	WALTER REED ARMY MEDICAL CENTER, WASHINGTON, D.C.
WRC-	CHANCE (W.R.) AND ASSOCIATES, ARLINGTON, VA.
WRC-	WYLE RESEARCH CORP., EL SEGUNDO, CALIF.
WRC-RR-	ILLINOIS. UNIV., URBANA. WATER RESOURCES CENTER
WRC-SM-(YEAR)-	TECHNICAL OPERATIONS, INC. WASHINGTON RESEARCH CENTER, ARLINGTON, VA.
WRC-TR-(YEAR)-	TECHNICAL OPERATIONS, INC. WASHINGTON RESEARCH CENTER, ARLINGTON, VA.
WR(D)-	GT. BRIT. MINISTRY OF SUPPLY. DIRECTORATE OF WEAPONS RESEARCH (DEFENCE)
WRD-	GT. BRIT. MINISTRY OF SUPPLY. DIRECTORATE OF WEAPONS RESEARCH (DEFENCE)
WRE-	AUSTRALIA. WEAPONS RESEARCH ESTABLISHMENT
WREC-	WHITE-RODGERS ELECTRIC CO., ST. LOUIS
WREC-ER-	WHITE-RODGERS ELECTRIC CO., ST. LOUIS
WREC-IER-	WHITE-RODGERS ELECTRIC CO., ST. LOUIS
WRE-CPD-	AUSTRALIA. WEAPONS RESEARCH ESTABLISHMENT
WRE-CPD/T/-	AUSTRALIA. WEAPONS RESEARCH ESTABLISHMENT
WRE-HSA-	AUSTRALIA. WEAPONS RESEARCH ESTABLISHMENT
WRE-ISD-TN-	AUSTRALIA. WEAPONS RESEARCH ESTABLISHMENT
WRE-PAD-	AUSTRALIA. WEAPONS RESEARCH ESTABLISHMENT
WR-ER(NUMBER)	WILLIAMS RESEARCH CORP., WALLED LAKE, MICH.
WRE-SAD	AUSTRALIA. WEAPONS RESEARCH ESTABLISHMENT, SALISBURY
WRE-TM-A-	AUSTRALIA. WEAPONS RESEARCH ESTABLISHMENT, SALISBURY
WRE-TM-CPD/T/	AUSTRALIA. WEAPONS RESEARCH ESTABLISHMENT, SALISBURY
WRE-TM-ED-	AUSTRALIA. WEAPONS RESEARCH ESTABLISHMENT, ADELAIDE
WRE-TM-HSA-	AUSTRALIA. WEAPONS RESEARCH ESTABLISHMENT
WRE-TM-ISD-	AUSTRALIA. WEAPONS RESEARCH ESTABLISHMENT
WRE-TM-PAD-	AUSTRALIA. WEAPONS RESEARCH ESTABLISHMENT. DEPT. OF SUPPLY, ADELAIDE
WRE-TM-TRD-	AUSTRALIA. WEAPONS RESEARCH ESTABLISHMENT
WRE-TM-WSD-	AUSTRALIA. WEAPONS RESEARCH ESTABLISHMENT, ADELAIDE
WRE-TN-APD-	AUSTRALIA. WEAPONS RESEARCH ESTABLISHMENT
WRE-TN-APD-	AUSTRALIA. WEAPONS RESEARCH ESTABLISHMENT, ADELAIDE
WRE-TN-CPD-	AUSTRALIA. WEAPONS RESEARCH ESTABLISHMENT
WRE-TN-CPD-	AUSTRALIA. WEAPONS RESEARCH ESTABLISHMENT, SALISBURY
WRE-TN-DWD-	AUSTRALIA. WEAPONS RESEARCH ESTABLISHMENT, SALISBURY
WRE-TN-EC-	AUSTRALIA. WEAPONS RESEARCH ESTABLISHMENT
WRE-TN-ECD-	AUSTRALIA. WEAPONS RESEARCH ESTABLISHMENT, SALISBURY
WRE-TN-ED-	AUSTRALIA. WEAPONS RESEARCH ESTABLISHMENT, ADELAIDE
WRE-TN-HSA-	AUSTRALIA. WEAPONS RESEARCH ESTABLISHMENT. DEPT. OF SUPPLY, ADELAIDE

WRE-TN-ISD-	AUSTRALIA. WEAPONS RESEARCH ESTABLISHMENT
WRE-TN-PAD-	AUSTRALIA. WEAPONS RESEARCH ESTABLISHMENT
WRE-TN-PD-	AUSTRALIA. WEAPONS RESEARCH ESTABLISHMENT, ADELAIDE
WRE-TN-SAD-	AUSTRALIA. WEAPONS RESEARCH ESTABLISHMENT, ADELAIDE
WRE-TN-TRD-	AUSTRALIA. WEAPONS RESEARCH ESTABLISHMENT
WRE-TN-TSD-	AUSTRALIA. WEAPONS RESEARCH ESTABLISHMENT, ADELAIDE
WRE-TN-TSD-	AUSTRALIA. WEAPONS RESEARCH ESTABLISHMENT, SALISBURY
WRE-TN-WPD-	AUSTRALIA. WEAPONS RESEARCH ESTABLISHMENT
WRE-TN-WSD-	AUSTRALIA. WEAPONS RESEARCH ESTABLISHMENT, SALISBURY
WRE-WPD-	AUSTRALIA. WEAPONS RESEARCH ESTABLISHMENT
WRI-	WINZEN RESEARCH, INC., MINNEAPOLIS
WRL-	MICHIGAN. UNIV., ANN ARBOR. WILLOW RUN LABS.
WRL-	WESTINGHOUSE ELECTRIC CORP. RES. LABS., PITTSBURGH
WRL-IR-	WESTINGHOUSE ELECTRIC CORP. RES. LABS., PITTSBURGH
WRL-ISR-	WESTINGHOUSE ELECTRIC CORP. RES. LABS., PITTSBURGH
WRL-PR-	WESTINGHOUSE ELECTRIC CORP. RES. LABS., PITTSBURGH
WRL-QPR-	WESTINGHOUSE ELECTRIC CORP. RES. LABS., PITTSBURGH
WRL-RR-(YR.)-(NUMBERS)-	WESTINGHOUSE ELECTRIC CORP. RES. LABS., PITTSBURGH
WRL-SP-	WESTINGHOUSE ELECTRIC CORP. RES. LABS., E. PITTSBURGH
WRL-TR-	WESTINGHOUSE ELECTRIC CORP. RES. LABS., PITTSBURGH
WRM-	AMERICAN POTASH + CHEMICAL CORP., WHITTIER, CALIF.
WRM-	NAVAL PERSONNEL PROGRAM SUPPORT ACTIVITY. PERSONNEL RESEARCH LAB., WASHINGTON, D.C. (RESEARCH MEMO.)
WRR-	NAVAL PERSONNEL PROGRAM SUPPORT ACTIVITY. PERSONNEL RESEARCH LAB., WASHINGTON, D.C.
WRRC-BULL-	MINNESOTA. UNIV., MINNEAPOLIS. WATER RESOURCES RESEARCH CENTER
WRRC-PUB-	FLORIDA. UNIV.,GAINESVILLE. WATER RESOURCES RES. CTR.
WRRI-	CLEMSON UNIV., S.C. WATER RESOURCES RESEARCH INST.
WRRI-RR-	NEW MEXICO STATE UNIV., UNIVERSITY PARK. WATER RESOURCES RESEARCH INST.
WRRL-	DEPARTMENT OF AGRICULTURE. WESTERN UTILIZATION RESEARCH BRANCH, ALBANY, CALIF.
WRS-	AIR FORCE (WEAPONS RECOMMENDATION SHEET)
WRSIC-(YEAR)-	WATER RESOURCES SCIENTIFIC INFORMATION CENTER (INTERIOR), WASHINGTON, D.C.
WRU-	WESTERN RESERVE UNIV., CLEVELAND
WRU-URL-TR-	WESTERN RESERVE UNIV.,CLEVELAND. ULTRASONIC RES. LAB.
WS-(CONTRACT NO.)(NO.)-	AEROSPACE CORP., EL SEGUNDO, CALIF. (WORK STATEMENTS)
WS-	HYDROGRAPHIC OFFICE, SUITLAND, MD. (WEATHER SUMMARY)
WS(NO.)-	NAVAL AIR TEST CENTER, PATUXENT RIVER, MD.
WS-	NAVAL ORDNANCE SYSTEMS COMMAND, WASHINGTON, D.C.
WS-	WEATHER BUREAU, ASHEVILLE, N.C.
WS-	WHITE SANDS PROVING GROUND, N. MEX.
WS-ADR-	WHITE SANDS PROVING GROUND. FLIGHT DETERMINATION LAB., N. MEX.
WS-AFC-	ARMY SIGNAL MISSILE SUPPORT AGENCY, WHITE SANDS MISSILE RANGE, N. MEX.
WS-AMTC-TM-	ARMY MISSILE TEST CTR.,WHITE SANDS MISSILE RANGE, NM
WSAO-	NAVAL WEAPON SYSTEMS ANALYSIS OFFICE, QUANTICO, VA.
WSAO-R-	NAVAL WEAPON SYSTEMS ANALYSIS OFFICE, QUANTICO, VA.
WSB-	ARMY AIR FORCE. WEATHER SERVICE (BULLETIN)
WSB-	WILLARD STORAGE BATTERY CO., CLEVELAND
WSC-	WASHINGTON. STATE COLL., PULLMAN
WSC-(INITIALS)-	WASHINGTON. STATE COLL., PULLMAN
WSC-E-	UNITED AIRCRAFT CORPORATE SYSTEMS CENTER. WEATHER SYSTEM CENTER, FARMINGTON, CONN.
WSCI-(YEAR)-	STANFORD RESEARCH INST., MENLO PARK, CALIF.
W+SC-TR-	WARNER AND SWASEY CO. CONTROL INSTR. DIV.,FLUSHING,NY
WSD-	CHICAGO. UNIV.
WS-DRD-N-	WHITE SANDS MISSILE RANGE, N. MEX.
WS-DRDN-	WHITE SANDS MISSILE RANGE, N. MEX.
WSDS-	NORTHROP CORP. NORAIR DIV. ENGINEERING DEPT., HAWTHORNE, CALIF. (WEAPONS SYSTEMS DESIGN STUDY)
WSEG-	WEAPONS SYSTEMS EVALUATION GROUP. INSTITUTE FOR DEFENSE ANALYSES, WASHINGTON, D.C.
WSEG-MEMO-	WEAPONS SYSTEMS EVALUATION GROUP. INSTITUTE FOR DEFENSE ANALYSES, WASHINGTON, D.C.
WSEG RM-	WEAPONS SYSTEMS EVALUATION GROUP. INSTITUTE FOR DEFENSE ANALYSES, WASHINGTON, D.C. (RESEARCH MEMO.)
WS-FDLHR-	WHITE SANDS PROVING GROUND. FLIGHT DETERMINATION LAB., N. MEX.
WS-FDR-	WHITE SANDS MISSILE RANGE, N. MEX.
WS-FTDR-	WHITE SANDS MISSILE RANGE, N. MEX.
WSG-(YEAR)-	WASHINGTON. UNIV., SEATTLE. FISHERIES RESEARCH INST. (WASHINGTON SEA GRANT PROGRAM)
WSI-	WEATHER SERVICES, INC., BOSTON
WSI-R-	GT. BRIT. WINDSCALE WORKS, SELLAFIELD, CUMB., ENGLAND
WSI-WR(NUMBER)	WEATHER SCIENCE, INC., NORMAN, OKLA.
WSL-(LETTER(S))-	UNITED KINGDOM ATOMIC ENERGY AUTHORITY. INDUSTRIAL GROUP.WINDSCALE WORKS,SELLAFIELD,CUMB.,ENG.(REF.36)
WSL-M-	GT. BRIT. WINDSCALE WORKS, SELLAFIELD, CUMB., ENGLAND (REF. 36)
WSL-R-	GT. BRIT. MIN. OF SUPPLY. CHEM. INSPECTORATE, LONDON
WSL-TM-	GT. BRIT. WINDSCALE WORKS, SELLAFIELD, CUMB., ENGLAND (REF. 36)
WSM-	ARMY AIR FORCE. WEATHER SERVICE (MANUAL)
WS-MGP-PR-	WHITE SANDS SIGNAL CORPS AGENCY. WHITE SANDS PROVING GROUND, N. MEX.
WS-(MISSILE NAME)-	WHITE SANDS MISSILE RANGE, N. MEX.
WSMR-	WHITE SANDS MISSILE RANGE, N. MEX.
WSMR/AMTED-	ARMY MISSILE TEST AND EVALUATION DIRECTORATE, WHITE SANDS MISSILE RANGE, N. MEX.
WSMR-AMTED-EMR-	ARMY MISSILE TEST AND EVALUATION DIRECTORATE, WHITE SANDS MISSILE RANGE, N. MEX.
WSMR-AMTED-FDS-(YEAR)-	ARMY MISSILE TEST AND EVALUATION DIRECTORATE, WHITE SANDS MISSILE RANGE, N. MEX.
WSMR-AMTED-HAWK-(YR.)-	ARMY MISSILE TEST AND EVALUATION DIRECTORATE, WHITE SANDS MISSILE RANGE, N. MEX.
WSMR-AMTED-HJ-	ARMY MISSILE TEST AND EVALUATION DIRECTORATE, WHITE SANDS MISSILE RANGE, N. MEX.
WSMR/ARMTE-	ARMY MISSILE TEST AND EVALUATION DIRECTORATE, WHITE SANDS MISSILE RANGE, N. MEX.
WSMR-ARMTE-FDS-(YEAR)-	ARMY MISSILE TEST AND EVALUATION DIRECTORATE, WHITE SANDS MISSILE RANGE, N. MEX.
WSMR-ERDA-	ARMY ELECTRONICS RESEARCH AND DEVELOPMENT ACTIVITY, WHITE SANDS MISSILE RANGE, N. MEX.
WSMR-FTR-	WHITE SANDS MISSILE RANGE, N. MEX. (FINAL TECH. RPT.)
WSMR-MIPR-	ARMY MISSILE TEST CENTER, WHITE SANDS MISSILE RANGE, N. MEX. (MONTHLY INTERIM PROGRESS REPORTS)
WSMR-PR-	ARMY MISSILE TEST CENTER, WHITE SANDS MISSILE RANGE, N. MEX. (MONTHLY INTERIM PROGRESS REPORTS)
WSMR-SR-	ARMY MISSILE TEST CENTER, WHITE SANDS MISSILE RANGE, N. MEX. (SPECIAL REPORTS)
WSMR-TB-	ARMY MISSILE TEST CENTER, WHITE SANDS MISSILE RANGE, N. MEX. (TEST BULLETINS)
WSMR-TM-	ARMY MISSILE TEST CENTER, WHITE SANDS MISSILE RANGE, N. MEX. (TECHNICAL MEMORANDUM)
WSMR-TR-	ARMY MISSILE TEST CENTER, WHITE SANDS MISSILE RANGE, N. MEX. (TECHNICAL REPORT)
WSP-	GEOLOGICAL SURVEY, WASH., D.C. (WATER SUPPLY PAPER)
WS-PDR-	WHITE SANDS PROVING GROUND, N. MEX.
W/SP-DW-	PICATINNY ARSENAL. WARHEADS AND SPECIAL PROJECTS LAB., DOVER, N.J.
WSPG-	WHITE SANDS PROVING GROUND, N. MEX.
WSPG-AFC-	WHITE SANDS PROVING GROUND, N. MEX.

WSPG-ARA-	WHITE SANDS PROVING GROUND, N. MEX.
WSPG-ERTTA-	WHITE SANDS PROVING GROUND, N. MEX. (ELECTRO-MAGNETIC RADIATION THRU THE ATMOSPHERE)
WSPG-MGP-	WHITE SANDS PROVING GROUND, N. MEX. (MISSILE GEOPHYSICS PROGRAM)
WSPG-MIPR-	WHITE SANDS PROVING GROUND, N. MEX. (MONTHLY INTERIM PROGRESS REPORTS)
WSPG-RID-	WHITE SANDS PROVING GROUND. RANGE INSTRUMENTATION DEVELOPMENT DIV., N. MEX.
WSPG-SR-	WHITE SANDS PROVING GROUND, N. MEX. (SPEC. REPTS.)
WSPG-TM-	WHITE SANDS PROVING GROUND, N. MEX. (TECH. MEMOS.)
WSPG-TR-	WHITE SANDS PROVING GROUND, N. MEX. (TECH. REPTS.)
W/SP-TM-	PICATINNY ARSENAL. WARHEADS AND SPECIAL PROJECTS LAB., DOVER, N.J.
WSR-	CURTISS-WRIGHT CORP. WRIGHT AERO. DIV., WOOD-RIDGE, NJ
WSR-	NAVAL PERSONNEL PROGRAM SUPPORT ACTIVITY. PERSONNEL RESEARCH LAB., WASHINGTON, D.C.
WS-RID-	WHITE SANDS PROVING GROUND. RANGE INSTRUMENTATION DEVELOPMENT DIV., N. MEX.
WSS-	WARFARE SYSTEMS SCHOOL, MAXWELL AFB, ALA.
WSSA-	WHITE SANDS SIGNAL CORPS AGENCY, WHITE SANDS MISSILE RANGE, N. MEX.
WSSA-PR-	WHITE SANDS SIGNAL CORPS AGENCY, WHITE SANDS MISSILE RANGE, N. MEX.
WSSA-TM-	WHITE SANDS SIGNAL CORPS AGENCY, WHITE SANDS MISSILE RANGE, N. MEX.
WSSCA-	WHITE SANDS SIGNAL CORPS AGENCY, WHITE SANDS MISSILE RANGE, N. MEX.
WSSCA-PR-	WHITE SANDS SIGNAL CORPS AGENCY, WHITE SANDS PROVING GROUND, N. MEX.
WSSCA-TM-	WHITE SANDS SIGNAL CORPS AGENCY, WHITE SANDS PROVING GROUND, N. MEX.
WSS/CI PAPER (YEAR)-	COMBUSTION INSTITUTE. WESTERN STATES SECTION
WSSCL-	AIR FORCE (WEAPON SYSTEM STOCK CONTROL LIST)
WS-SMSA-	ARMY SIGNAL MISSILE SUPPORT AGENCY, WHITE SANDS MISSILE RANGE, N. MEX.
WS-SR-	ARMY SIGNAL MISSILE SUPPORT AGENCY, WHITE SANDS MISSILE RANGE, N. MEX.
WST-	NAVAL AIR STATION, PATUXENT RIVER, MD.
WS-TB-	ARMY MISSILE TEST CTR., WHITE SANDS MISSILE RANGE, NM
WS-TD-	WHITE SANDS PROVING GROUND, N. MEX.
WST-IR-	NAVAL AIR TEST CENTER, PATUXENT RIVER, MD.
WS-TM-	WHITE SANDS MISSILE RANGE, N. MEX.
WS-TP-	WHITE SANDS PROVING GROUND, N. MEX.
W/S-TR-	WARNER AND SWASEY CO., FLUSHING, N.Y.
W+S-TR-	WARNER AND SWASEY CO. CONTROL INSTR. DIV., FLUSHING, NY
WS-TR-	WHITE SANDS PROVING GROUND, N. MEX.
W-STR-CH-	WESTINGHOUSE ELECTRIC CORP. ATOMIC POWER DIV., PITTS.
W-STR-ER-	WESTINGHOUSE ELECTRIC CORP. ATOMIC POWER DIV., IDAHO FALLS
W-STR-L-	WESTINGHOUSE ELECTRIC CORP. ATOMIC POWER DIV., IDAHO FALLS
W-STR-M-	WESTINGHOUSE ELECTRIC CORP. ATOMIC POWER DIV., IDAHO FALLS
W-STR-TG-	WESTINGHOUSE ELECTRIC CORP. ATOMIC POWER DIV., IDAHO FALLS
W-STR-TS-	WESTINGHOUSE ELECTRIC CORP. ATOMIC POWER DIV., PITTS.
WSU-(YEAR)-	WASHINGTON STATE UNIV., PULLMAN
WSU-	WAYNE STATE UNIV., DETROIT
WSUNRC-	WASHINGTON STATE UNIV., PULLMAN. NUCLEAR RADIATION CTR
WSU-SDL-(YEAR)-	WASHINGTON STATE UNIV., PULLMAN. SHOCK DYNAMICS LAB.
WSU-TR-	WAYNE STATE UNIV., DETROIT
WSWL-	WHITE SANDS MISSILE RANGE. WARHEADS AND SPECIAL WEAPONS LAB., N. MEX.
WT-	(WEAPON TEST REPORT) (REF. 7)
WT-	(WIND TUNNEL) (REF. 2)
WT-	BROWN UNIV., PROVIDENCE. WIND TUNNEL
WTA-	WASHINGTON TECHNOLOGICAL ASSOCS., ROCKVILLE, MD.
WTB-	NAVAL PERSONNEL PROGRAM SUPPORT ACTIVITY. PERSONNEL RESEARCH LAB., WASHINGTON, D.C. (TECHNICAL BULL.)
WTC-	WIND TURBINE CO., WEST CHESTER, PENNA.
WTC/P-	UNITED KINGDOM ATOMIC ENERGY AUTHORITY. INDUSTRIAL GROUP. WINDSCALE WORKS, SELLAFIELD, CUMB., ENGLAND
WTD-	(WIND TUNNEL DATA) (REF. 2)
WTHD-	KARLSRUHE, GERMANY. TECHNISCHE HOCHSCHULE
WTI-	JOINT ATOMIC WEAPONS TECHNICAL INFORMATION GROUP
WTICI-	WIND TUNNEL INSTRUMENT CO., INC., NEWTON, MASS.
WTM-	(WIND TUNNEL MEMORANDUM) (REF. 2)
WTM-	MICHIGAN. UNIV., ANN ARBOR (WIND TUNNEL MEMORANDUM)
WTM-	MICHIGAN. UNIV., ANN ARBOR. ENGINEERING RES. INST.
WTN-	(WIND TUNNEL NOTE) (REF. 2)
WTO-	CORNELL AERONAUTICAL LAB., INC., BUFFALO
WTR-	AMERICAN POTASH + CHEMICAL CORP., WHITTIER, CALIF.
WTR-	MASSACHUSETTS INST. OF TECH., CAMBRIDGE (WIND TUNNEL REPORT)
WTR-	MASS. INST. OF TECH., CAMBRIDGE. NAVAL SUPERSONIC LAB
WTR-	NAVAL ORDNANCE LAB., WHITE OAK, MD. (WIND TUNNEL REQUEST)
WTR-	WESTINGHOUSE ELECTRIC CORP. TESTING REACTOR, WALTZ MILL, PENNA.
WTSC/R-	UNITED KINGDOM ATOMIC ENERGY AUTHORITY. INDUSTRIAL GROUP. WINDSCALE WORKS, SELLAFIELD, CUMB., ENGLAND
WTSMR-	AIR FORCE WESTERN TEST RANGE, VANDENBERG AFB, CALIF.
WU-	WASHINGTON. UNIV., SEATTLE
WU-	WASHINGTON UNIV., ST. LOUIS
WU-	WESLEYAN UNIV., MIDDLETOWN, CONN.
WU-AR-	WICHITA, KAN. UNIV. SCHOOL OF ENGINEERING
WU-REF-M(YEAR)-	WASHINGTON. UNIV., SEATTLE. DEPT. OF OCEANOGRAPHY
WURF-(LTRS.)-	WASHINGTON UNIV., CLAYTON, MO. RESEARCH FOUNDATION
WU-SL-	WASHINGTON UNIV., ST. LOUIS
WUSL-	WASHINGTON UNIV., ST. LOUIS
WUSL TR-	WASHINGTON UNIV., ST. LOUIS (TECHNICAL REPORT)
WUT-	WESTERN UNION TELEGRAPH CO., WATER MILL, N.Y.
WUTC-	WESTERN UNION TELEGRAPH CO., N.Y.C.
WUTC-	WESTERN UNION TELEGRAPH CO., WATER MILL, N.Y.
WVA-	GERMANY. VERSUCHSANSTALT FUER WASSERBAU UND SCHIFF-BAU, BERLIN
WVAL-	WATERVLIET ARSENAL. BENET LABS., N.Y.
WVAL MR-	WATERVLIET ARSENAL. BENET LABS., N.Y.
WVAL RDD-	WATERVLIET ARSENAL. BENET LABS., N.Y.
WVP-	GT. BRIT. ROYAL NAVAL PERSONNEL RESEARCH COMMITTEE
W/V-RR-(YR.)/(NO.)-WD	WHITTENBURG, VAUGHAN ASSOCS., INC., ALEXANDRIA, VA.
W/V-RR-(YR.)/(NO.)-CD	WHITTENBURG, VAUGHAN ASSOCS., INC., ALEXANDRIA, VA.
WVS-	STERLING (WALTER V.) INC., CLAREMONT, CALIF.
WVT-	WATERVLIET ARSENAL. BENET LABS., N.Y.
WVT-ID-	WATERVLIET ARSENAL. BENET LABS., N.Y.
WVT-QA-	WATERVLIET ARSENAL. BENET LABS., N.Y. (QUALITY ASSURANCE)
WVT-RI-	WATERVLIET ARSENAL. BENET LABS., N.Y.
WVT-RR-	WATERVLIET ARSENAL. BENET LABS., N.Y.
WVTRR-	WATERVLIET ARSENAL. BENET LABS., N.Y.
WVT-TR-	WATERVLIET ARSENAL. BENET LABS., N.Y.
WVU-	WEST VIRGINIA. UNIV., MORGANTOWN
WW-	EINDHOVEN, NETHERLANDS. TECHNISCHE HOGESCHOOL

```
WW-(NUMBER)-M-
          EINDHOVEN, NETHERLANDS. TECHNISCHE HOGESCHOOL
WW-(NUMBER)-R-
          EINDHOVEN, NETHERLANDS. TECHNISCHE HOGESCHOOL
WWO-
          EINDHOVEN, NETHERLANDS. TECHNISCHE HOGESCHOOL
WW/R-
          EINDHOVEN, NETHERLANDS. TECHNISCHE HOGESCHOOL
WWR-
          NAVAL PERSONNEL PROGRAM SUPPORT ACTIVITY. PERSONNEL
             RESEARCH LAB., WASHINGTON, D.C.
WWR-(YEAR)-
          OLIN CORP. WINCHESTER-WESTERN RES. DEPT., NEW HAVEN
WX/ERA-
          SYMPOSIUM ON ELECTRON RING ACCELERATOR, 1968,
             BERKELEY, CALIF.
WYANDOTTE (NO.)-QPR-
          WYANDOTTE CHEMICALS CORP., MICH.
WYC-
          WYANDOTTE CHEMICALS CORP., MICH.
WYO NRRI B-
          WYOMING. UNIV., LARAMIE. NATURAL RESOURCES RESEARCH
             INSTITUTE (BULLETIN)
WYOU-
          WYOMING. UNIV., LARAMIE
```

X-(NUMBERS)-(YEAR)-	AMES RESEARCH CENTER, MOFFETT FIELD, CALIF.
X(NUMBER)-(NO.)/(NO.)	AUTONETICS, ANAHEIM, CALIF.
X6-	AUTONETICS DIV., NORTH AMERICAN AVIATION, INC., ANAHEIM, CALIF. (PAPER, REPRINT, OR MONOGRAPH, 1966)
X-	CHICAGO. UNIV. METALLURGICAL LAB. (REF. 10)
X-(NUMBER-YEAR)-	GODDARD SPACE FLIGHT CENTER, GREENBELT, MD.
X-	HUGHES TOOL CO. AIRCRAFT DIV., CULVER CITY, CALIF.
X(YEAR)-(5 DIGITS)	NATIONAL AERONAUTICS AND SPACE ADM., WASHINGTON, D.C.
X-	NATIONAL BERYLLIA CORP., HASKELL, N.J.
X-	NAVAL TECHNICAL MISSION TO JAPAN
X-	NEW YORK. STATE UNIV., SYRACUSE. UPSTATE MEDICAL CTR.
X7-(NUMBER)/(NUMBER)	NORTH AMERICAN AVIATION, INC. OCEAN SYSTEMS OPERATIONS, ANAHEIM, CALIF.
1(LETTER)(NUMBER)-X(LTR)-	COLUMBIA UNIV., N.Y.C. SUBSTITUTE ALLOY MATERIALS LAB. (REF. 11)
1(LETTER)-X(LETTER)-	COLUMBIA UNIV., N.Y.C. SUBSTITUTE ALLOY MATERIALS LAB. (REF. 11)
1-X(LETTER)-	COLUMBIA UNIV., N.Y.C. SUBSTITUTE ALLOY MATERIALS LAB. (REF. 11)
1X(LETTER)-	COLUMBIA UNIV., N.Y.C. SUBSTITUTE ALLOY MATERIALS LAB. (REF. 11)
2(LETTER)(NUMBER)-X(LTR)-	COLUMBIA UNIV., N.Y.C. SUBSTITUTE ALLOY MATERIALS LAB. (REF. 11)
2(LETTER)-X(LETTER)-	COLUMBIA UNIV., N.Y.C. SUBSTITUTE ALLOY MATERIALS LAB. (REF. 11)
2-X(LETTER)-	COLUMBIA UNIV., N.Y.C. SUBSTITUTE ALLOY MATERIALS LAB. (REF. 11)
2X(LETTER)-	COLUMBIA UNIV., N.Y.C. SUBSTITUTE ALLOY MATERIALS LAB. (REF. 11)
3(LETTER)-X(LETTER)-	COLUMBIA UNIV., N.Y.C. SUBSTITUTE ALLOY MATERIALS LAB. (REF. 11)
3-X(LETTER)-	COLUMBIA UNIV., N.Y.C. SUBSTITUTE ALLOY MATERIALS LAB. (REF. 11)
3X(LETTER)-	COLUMBIA UNIV., N.Y.C. SUBSTITUTE ALLOY MATERIALS LAB. (REF. 11)
4-X(LETTER)-	COLUMBIA UNIV., N.Y.C. SUBSTITUTE ALLOY MATERIALS LAB. (REF. 11)
4X(LETTER)-	COLUMBIA UNIV., N.Y.C. SUBSTITUTE ALLOY MATERIALS LAB. (REF. 11)
5(LETTER)(NUMBER)-X(LTR)-	COLUMBIA UNIV., N.Y.C. SUBSTITUTE ALLOY MATERIALS LAB. (REF. 11)
5(LETTER)-X(LETTER)-	COLUMBIA UNIV., N.Y.C. SUBSTITUTE ALLOY MATERIALS LAB. (REF. 11)
5-X(LETTER)-	COLUMBIA UNIV., N.Y.C. SUBSTITUTE ALLOY MATERIALS LAB. (REF. 11)
5X(LETTER)-	COLUMBIA UNIV., N.Y.C. SUBSTITUTE ALLOY MATERIALS LAB. (REF. 11)
6(LETTER)-X(LETTER)-	COLUMBIA UNIV., N.Y.C. SUBSTITUTE ALLOY MATERIALS LAB. (REF. 11)
6-X(LETTER)-	COLUMBIA UNIV., N.Y.C. SUBSTITUTE ALLOY MATERIALS LAB. (REF. 11)
6X(LETTER)-	COLUMBIA UNIV., N.Y.C. SUBSTITUTE ALLOY MATERIALS LAB. (REF. 11)
XA-1-SM-	CALIFORNIA. UNIV., BERKELEY. RADIATION LAB. (REF. 9)
XA-2-SM-	CALIFORNIA. UNIV., BERKELEY. RADIATION LAB. (REF. 9)
XA (NO.)/-	GT. BRIT. ARMAMENT RES. DEPT.,FT.HALSTEAD, KENT, ENG.
XA-	HUGHES AIRCRAFT CO., CULVER CITY, CALIF.
XA-	KELLETT AIRCRAFT CORP., CAMDEN, N.J.
XA-	MSA RESEARCH CORP., CALLERY, PENNA.
XA-2-OD-	TENNESSEE EASTMAN CORP., OAK RIDGE, TENN.
XAC-	COLUMBIA UNIV., NYC. SUBSTITUTE ALLOY MATERIALS LABS.
XAC-	MSA RESEARCH CORP., EVANS CITY, PENNA.
XAC-(LETTER)-	COLUMBIA UNIV., NYC. SUBSTITUTE ALLOY MATERIALS LABS.
XAC-SPEC-	CALIFORNIA. UNIV., BERKELEY. RADIATION LAB. (REF. 9)
XAH-	CALIFORNIA. UNIV., BERKELEY. RADIATION LAB. (REF. 9)
XAH-1-SM-	CALIFORNIA. UNIV., BERKELEY. RADIATION LAB. (REF. 9)
XAH-1-SPEC-	CALIFORNIA. UNIV., BERKELEY. RADIATION LAB. (REF. 9)
XAH-2.39-	CALIFORNIA. UNIV., BERKELEY. RADIATION LAB. (REF. 9)
XAH-2-RR-	CALIFORNIA. UNIV., BERKELEY. RADIATION LAB. (REF. 9)
XAH-2-SM-	CALIFORNIA. UNIV., BERKELEY. RADIATION LAB. (REF. 9)
XAH-SM-	CALIFORNIA. UNIV., BERKELEY. RADIATION LAB. (REF. 9)
XAH-SPEC-	CALIFORNIA. UNIV., BERKELEY. RADIATION LAB. (REF. 9)
XAR-	COLUMBIA UNIV., NYC. SUBSTITUTE ALLOY MATERIALS LABS.
XAR-	GRUMMAN AEROSPACE CORP., BETHPAGE, N.Y.
XAS-	CALIFORNIA. UNIV., BERKELEY. RADIATION LAB. (REF. 9)
XA-SPEC-	CALIFORNIA. UNIV., BERKELEY. RADIATION LAB. (REF. 9)
XB-(NUMBER)/(NUMBER)	NORTH AMERICAN ROCKWELL CORP. OCEAN SYSTEMS OPERATIONS, LONG BEACH, CALIF.
XC-(NUMBER)	ABEX CORP. METALLURGICAL DEPT., MAHWAH, N.J.
XC-	CALIFORNIA. UNIV., BERKELEY. RADIATION LAB. (REF. 9)
XC(2)-	GT. BRIT. ARMAMENT RES. DEPT.,FT.HALSTEAD, KENT, ENG.
XC-	LING-TEMCO-VOUGHT, INC., DALLAS
XCO-	NAVAL ORDNANCE TEST STA., INYOKERN (CHINA LAKE), CAL.
XDC-	GENERAL ELECTRIC CO. AIRCRAFT NUCLEAR PROPULSION DEPT., CINCINNATI
XDC-(YEAR)-(MONTH)-	GENERAL ELECTRIC CO. AIRCRAFT NUCLEAR PROPULSION PROJECT, CINCINNATI (REF. 3)
XDCL-	GENERAL ELECTRIC CO. AIRCRAFT NUCLEAR PROPULSION DEPT., CINCINNATI (REF. 3)
XD-M-	COLUMBIA UNIV., NYC. SUBSTITUTE ALLOY MATERIALS LABS.
XD-R-	COLUMBIA UNIV., NYC. SUBSTITUTE ALLOY MATERIALS LABS.
XG-	NAVAL AIR EXPTL.STA. AERO.MEDICAL EQUIP.LAB., PHILA.
XG-(NUMBER)-(YEAR)	SYSTEMS ANALYSIS AND RESEARCH CORP., BOSTON
XG-(NUMBER)-	SYSTEMS ANALYSIS AND RESEARCH CORP., CAMBRIDGE, MASS.
X-GEAP-	GENERAL ELECTRIC CO. ATOMIC PRODUCTS DIV., CINCINNATI
X-GSG-	HUGHES AIRCRAFT CO. GROUND SYSTEMS GP., FULLERTON,CAL
X-GSG-(LETTER)	HUGHES AIRCRAFT CO. GROUND SYSTEMS GP., FULLERTON,CAL
XI	NATIONAL RESEARCH COUNCIL OF CANADA. ATOMIC ENERGY PROJECT, CHALK RIVER, ONT. (REF. 29)
XL50.1- THRU XL60.9-	CALIFORNIA. UNIV., BERKELEY. RADIATION LAB. (REF. 9)
XL-4.9 THRU XL-58.5-	TENNESSEE EASTMAN CORP., OAK RIDGE, TENN.
XM-	UNITED KINGDOM CHEMICAL ENG. TEAM, CHALK RIVER, ONT.
XMPDC/P-	GT. BRIT. CULCHETH LABS., LANCS, ENGLAND
XPO-	NAVAL ORDNANCE TEST STA., INYOKERN (CHINA LAKE), CAL.
XPR-	NAVAL ORDNANCE TEST STA., INYOKERN (CHINA LAKE), CAL.
XR-	(OPERATION CROSSROADS) (REF. 7)
XR-	ARMY ENGINEER RESEARCH + DEV. LABS., FT. BELVOIR, VA.
X/R-	GT. BRIT. ATOMIC EN. RES. ESTAB., HARWELL, BERKS, ENG
XR-	GT. BRIT. EXPLOSIVES RESEARCH AND DEVELOPMENT ESTABLISHMENT, WALTHAM ABBEY, ESSEX, ENGLAND
XR-	MC MASTER UNIV., HAMILTON, ONT.
XRB-	JOINT TASK FORCE ... (REF. 17)
XRD-	(OPERATION CROSSROADS) (REF. 7)
XRD-	JOINT TASK FORCE ONE
XS-	DAYSTROM, INC. WESTON INSTRUMENTS DIV., NEWARK, N.J.
XS-(NUMBER)	WESTON INSTRUMENTS, INC., NEWARK, N.J.
XTS-	NAVAL ORDNANCE TEST STA., INYOKERN (CHINA LAKE), CAL.
XW-	DEFENSE ATOMIC SUPPORT AGENCY. FIELD COMMAND, ALBUQUERQUE, N. MEX.
XX-(NUMBER-YEAR)-	GODDARD SPACE FLIGHT CENTER, GREENBELT, MD.
XYA-	PFIZER (CHAS.) AND CO. INC., NEW YORK

Y(NUMBER(S))-	GYRODYNE CO. OF AMERICA, INC., ST. JAMES, N.Y.
Y70-(NUMBER)-	GYRODYNE CO. OF AMERICA, INC., ST. JAMES, N.Y.
Y-	LOS ALAMOS SCIENTIFIC LAB., N. MEX. (FILM)
Y-	NAVAL CIVIL ENGINEERING LAB., PORT HUENEME, CALIF.
Y-	OFFICE OF NAVAL INTELLIGENCE, WASHINGTON, D.C.
Y-12-	UNION CARBIDE CORP. Y-12 PLANT, OAK RIDGE, TENN.
Y-	UNION CARBIDE CORP. Y-12 PLANT, OAK RIDGE, TENN.
Y-	YALE UNIV., NEW HAVEN
Y-(LETTER/NUMBER)-	UNION CARBIDE NUCLEAR CO. Y-12 PLANT, OAK RIDGE, TENN
YA-	UNION CARBIDE NUCLEAR CO. Y-12 PLANT, OAK RIDGE, TENN
Y-AE-	UNION CARBIDE CORP. Y-12 PLANT, OAK RIDGE, TENN.
YAEC-	WESTINGHOUSE ELECTRIC CORP. ATOMIC POWER DEPT.,PITTS.
YAEC-	YANKEE ATOMIC ELECTRIC CO., BOSTON
YALE-	YALE UNIV., NEW HAVEN
YALE-(YEAR)-	YALE UNIV., NEW HAVEN
YALE-IR-Y-	YALE UNIV., NEW HAVEN
YALE TR-	YALE UNIV., NEW HAVEN
YALEU-TR-	YALE UNIV., NEW HAVEN
Y-CDC-	UNION CARBIDE CORP. Y-12 PLANT, OAK RIDGE, TENN.
YCP-	SOCIETE NATIONALE D'ETUDE ET DE CONSTRUCTION DE MOTEURS D'AVIATION, PARIS
YD-(YR)-	SYLVANIA ELECTRIC PRODUCTS INC. ATOMIC ENERGY DIV., BAYSIDE, N.Y.
Y-DA-	UNION CARBIDE CORP., OAK RIDGE, TENN.
Y-DA-	UNION CARBIDE CORP. Y-12 PLANT, OAK RIDGE, TENN.
Y-DD-	UNION CARBIDE NUCLEAR CO., OAK RIDGE, TENN.
Y-DE-	UNION CARBIDE CORP., OAK RIDGE, TENN.
YDL-	UNION CARBIDE NUCLEAR CO. Y-12 PLANT, OAK RIDGE, TENN
Y-DM-	YOUNG DEVELOPMENT LABS., INC., ROCKY HILL, N.J.
Y-DP-	UNION CARBIDE CORP. Y-12 PLANT, OAK RIDGE, TENN.
Y-DR-	UNION CARBIDE NUCLEAR CO. Y-12 PLANT, OAK RIDGE, TENN
Y-DR-	UNION CARBIDE CORP., OAK RIDGE, TENN.
YE-(YR)-	UNION CARBIDE CORP. Y-12 PLANT, OAK RIDGE, TENN.
YE-(YR)-	SYLVANIA ELECTRIC PRODUCTS INC. ATOMIC ENERGY DIV., BAYSIDE, N.Y.
Y-E-	SYLVANIA ELECTRIC PRODUCTS INC. METALLURGICAL LABS., BAYSIDE, N.Y.
Y-EB-	UNION CARBIDE CORP. Y-12 PLANT, OAK RIDGE, TENN.
Y-EC-	UNION CARBIDE NUCLEAR CO. Y-12 PLANT, OAK RIDGE, TENN
Y-EF-	UNION CARBIDE CORP. Y-12 PLANT, OAK RIDGE, TENN.
Y-EG-	UNION CARBIDE NUCLEAR CO., OAK RIDGE, TENN.
Y-EH-	UNION CARBIDE CORP. Y-12 PLANT, OAK RIDGE, TENN.
YESHIVAU-	UNION CARBIDE CORP. Y-12 PLANT, OAK RIDGE, TENN.
Y-IA-	YESHIVA UNIV., N.Y.C.
Y-JA-	UNION CARBIDE CORP. Y-12 PLANT, OAK RIDGE, TENN.
Y-KA-	UNION CARBIDE CORP. Y-12 PLANT, OAK RIDGE, TENN.
Y-KB-	UNION CARBIDE CORP. Y-12 PLANT, OAK RIDGE, TENN.
Y-KE-	UNION CARBIDE CORP., OAK RIDGE, TENN.
Y-KG-	UNION CARBIDE NUCLEAR CO. Y-12 PLANT, OAK RIDGE, TENN
Y-MA-	UNION CARBIDE CORP. Y-12 PLANT, OAK RIDGE, TENN.
Y-NA-	UNION CARBIDE CORP. Y-12 PLANT, OAK RIDGE, TENN.
Y-OA-	UNION CARBIDE CORP. Y-12 PLANT, OAK RIDGE, TENN.
Y-PB-	UNION CARBIDE CORP. Y-12 PLANT, OAK RIDGE, TENN.
YPG-	UNION CARBIDE NUCLEAR CO. K-25 PLANT, OAK RIDGE, TENN
Y-SB-	YUMA PROVING GROUND, ARIZ.
YSB-R(NO.-NO.)-	UNION CARBIDE NUCLEAR CO. Y-12 PLANT, OAK RIDGE, TENN
	UNDERWRITERS LABS., INC. MARINE DEPT., WESTWOOD, N.J.
Y-SC-	UNION CARBIDE CORP. Y-12 PLANT, OAK RIDGE, TENN.
YU-	YALE UNIV., NEW HAVEN
YUCCA-(NUMBER)-	GEOLOGICAL SURVEY, DENVER
YU-HL-TR-	YALE UNIV., NEW HAVEN. HAMMOND METALLURGICAL LAB.
YU-HML-	YALE UNIV., NEW HAVEN. HAMMOND METALLURGICAL LAB.
YU-HML-TR-	YALE UNIV., NEW HAVEN. HAMMOND METALLURGICAL LAB.
YUP-	YALE UNIV., NEW HAVEN
YU-SPL-	YALE UNIV., NEW HAVEN. SLOANE PHYSICS LAB.

Z-(2 DIGITS)	CESKOSLOVENSKA AKADEMIE VED, PRAGUE
Z-	CLINTON LABS., OAK RIDGE, TENN.
Z-	EASTMAN KODAK CO., ROCHESTER, N.Y.
Z-	FRANCE. LIGNES TELEGRAPHIQUES ET TELEPHONIQUES, PARIS
Z-	KELLEX CORP., N.Y.C.
Z-	RADAR EVALUATION SQUADRON (4754TH), HILL AFB, UTAH
Z-	SANDIA CORP., ALBUQUERQUE, N. MEX.
ZA-	CONVAIR, SAN DIEGO, CALIF.
ZA-	GT. BRIT. ARMY SUPPLY DIRECTORATE
ZA-	ZAPFFE (CARL A.) AND ASSOCIATES, BALTIMORE
ZAED-	ZENTRALSTELLE FUER ATOMKERNENERGIE-DOKUMENTATION BEIM GMELIN INSTITUT, FRANKFURT AM MAIN
ZAP-	ZAPFFE (CARL A.) AND ASSOCIATES, BALTIMORE
ZB-	CONVAIR, SAN DIEGO, CALIF.
ZB/(NUMBER)/	SUSSEX. UNIV., FALMER, BRIGHTON, ENGLAND
ZBC-	OFFICE OF SCIENTIFIC RESEARCH AND DEVELOPMENT
ZC-	CONVAIR, SAN DIEGO, CALIF.
ZC-	ZIMNEY CORP., MONROVIA, CALIF.
ZC-	ZIMNEY CORP., PASADENA, CALIF.
ZETR-II/P-	UNITED KINGDOM ATOMIC ENERGY AUTHORITY. RESEARCH GP. ATOMIC ENERGY RES. ESTAB., HARWELL, BERKS, ENGLAND
ZFK-	GERMANY. ZENTRALINSTITUT FUER KERNFORSCHUNG, ROSSENDORF BEI DRESDEN
ZFK-	GERMANY. ZENTRALINSTITUT FUER KERNFORSCHUNG, ROSSENDORF BEI DRESDEN
ZFK-DOS-	GERMANY. ZENTRALINSTITUT FUER KERNFORSCHUNG, ROSSENDORF BEI DRESDEN
ZFK-DOS-	GERMANY. ZENTRALINSTITUT FUER KERNPHYSIK, DRESDEN
ZFK-PHA-	GERMANY. ZENTRALINSTITUT FUER KERNFORSCHUNG, ROSSENDORF BEI DRESDEN
ZFK-PHA-	GERMANY. ZENTRALINSTITUT FUER KERNPHYSIK, DRESDEN
ZFK-RCH-	GERMANY. ZENTRALINSTITUT FUER KERNFORSCHUNG, ROSSENDORF BEI DRESDEN
ZFK-RCH-	GERMANY. ZENTRALINSTITUT FUER KERNPHYSIK, DRESDEN
ZFK-RN-	GERMANY. ZENTRALINSTITUT FUER KERNFORSCHUNG, ROSSENDORF BEI DRESDEN
ZFK-RN-	GERMANY. ZENTRALINSTITUT FUER KERNPHYSIK, DRESDEN
ZFK-TPH-	GERMANY. ZENTRALINSTITUT FUER KERNFORSCHUNG, ROSSENDORF BEI DRESDEN
ZFK-TPH-	GERMANY. ZENTRALINSTITUT FUER KERNPHYSIK, DRESDEN
ZFK-WF-	GERMANY. ZENTRALINSTITUT FUER KERNFORSCHUNG, ROSSENDORF BEI DRESDEN
ZFK-WF-	GERMANY. ZENTRALINSTITUT FUER KERNPHYSIK, DRESDEN
ZG-	CONVAIR, SAN DIEGO, CALIF.
ZJ-	CONVAIR, SAN DIEGO, CALIF.
ZJ-	CONVAIR. ASTRONAUTICS DIV., SAN DIEGO, CALIF.
ZJ-(NUMBER)-	GENERAL DYNAMICS/FORT WORTH. AEROSYSTEMS LAB., TEX.
ZJE-	CZECHOSLOVAKIA. SKODA WORKS, PILSEN
ZM-	CONVAIR, SAN DIEGO, CALIF.
ZN-	CONVAIR, SAN DIEGO, CALIF.
ZN-	CONVAIR. ASTRONAUTICS DIV., SAN DIEGO, CALIF.
Z-P(MONTH/YEAR)-	CALIFORNIA. UNIV., BERKELEY. RADIATION LAB. (REF. 9)
ZP-	CONVAIR, SAN DIEGO, CALIF.
ZP-	SANDIA CORP., ALBUQUERQUE, N. MEX.
ZPH-	GENERAL DYNAMICS/ASTRONAUTICS, SAN DIEGO, CALIF.
ZPM-	CONVAIR, SAN DIEGO, CALIF.
ZPR-	ARGONNE NATIONAL LAB., ILL.
ZPR-TM-	ARGONNE NATIONAL LAB., ILL.
ZR-	CONVAIR, SAN DIEGO, CALIF.
ZR-(NO.)-(NO.)-(NO.)	CONVAIR. ASTRONAUTICS DIV., SAN DIEGO, CALIF.
ZRAI-	CONVAIR, SAN DIEGO, CALIF.
ZR-AP-	GENERAL DYNAMICS/CONVAIR, SAN DIEGO, CALIF.
ZRC-	ZENITH RADIO CORP., CHICAGO
Z-SPECIAL-	CALIFORNIA. UNIV., BERKELEY. RADIATION LAB. (REF. 9)
ZTB-	ZATOR CO., CAMBRIDGE, MASS.
ZU-	CONVAIR, SAN DIEGO, CALIF.
ZULU-	ZULU COMM.
ZV-(NUMBER)-0	BOELKOW ENTWICKLUNGEN K.G., MUNICH
ZV-(NUMBER)-0	MESSERSCHMITT-BOELKOW-BLOHM G.M.B.H., MUNICH
ZW-	CONVAIR, SAN DIEGO, CALIF.
ZWB-	GERMANY. ZENTRALE FUER WISSENSCHAFTLICHES BERICHTSWESEN DER LUFTFAHRTFORSCHUNG, BERLIN
ZWB-(LETTERS)-	GERMANY. ZENTRALE FUER WISSENSCHAFTLICHES BERICHTSWESEN DER LUFTFAHRTFORSCHUNG, BERLIN
ZX-	CONVAIR, SAN DIEGO, CALIF.
ZZL-	GENERAL DYNAMICS/CONVAIR, SAN DIEGO, CALIF.

Corporate Entries with Related Report Series Codes

Agency names are arranged alphabetically, word by word. Most abbreviations that are not acronyms file as if they were written out. Within the same agency, the arrangement is:

Agency with no location given
Same Agency (All Locations)
Same Agency, City A.
Same Agency, City B.
Same Agency. Division A, City A.
Same Agency. Division B, City B.
Same Agency. Division C, City A.

In the case of a college or university, the location precedes the name of a department or division.

The Explanatory Notes (p. 14) contain a description of rules for entry, filing, and assigning codes to agencies.

```
A R F PRODUCTS, INC., RATON, N. MEX.                              ARFP-

AACHEN. TECHNISCHE HOCHSCHULE                                     ARL-
AACHEN. TECHNISCHE HOCHSCHULE                                     ASR-
AACHEN. TECHNISCHE HOCHSCHULE                                     HMP-

AACHEN. TECHNISCHE HOCHSCHULE. PHYSIKALISCHES INST.               PITHA-

AAI CORP., COCKEYSVILLE, MD.                                      AAI-
AAI CORP., COCKEYSVILLE, MD.                                      AAI-ER-

AARHUS UNIV., DENMARK. ROEMER OBSERVATORY                         NR-(2 DIGITS)

ABBOTT LABS., NORTH CHICAGO, ILL.                                 AL-

ABERDEEN PROVING GROUND, MD.                                      AD-
ABERDEEN PROVING GROUND, MD.                                      APG-
ABERDEEN PROVING GROUND, MD.                                      APG-AS-
ABERDEEN PROVING GROUND, MD.                                      APG-FR-B-
ABERDEEN PROVING GROUND, MD.                                      APG-FR-M-
ABERDEEN PROVING GROUND, MD.                                      APG-FR-S-
ABERDEEN PROVING GROUND, MD.                                      APG-MISC-
ABERDEEN PROVING GROUND, MD. (ORDNANCE PROGRAM)                   APG OP-
ABERDEEN PROVING GROUND, MD. (ORDNANCE PROGRAM)                   APG OP (NUMBER)/
ABERDEEN PROVING GROUND, MD.                                      APG RPT-
ABERDEEN PROVING GROUND, MD.                                      AS-
ABERDEEN PROVING GROUND, MD.                                      FDEB-
ABERDEEN PROVING GROUND, MD. (FIRING RECORD)                      FR-
ABERDEEN PROVING GROUND, MD.                                      L + M-
ABERDEEN PROVING GROUND, MD. (ORDNANCE PROGRAM)                   OP-
ABERDEEN PROVING GROUND, MD.                                      ORD/APG (LETTERS, NOS.)
ABERDEEN PROVING GROUND, MD.                                      ORDBG-DPS-LS-

ABERDEEN PROVING GROUND. LABORATORY SERVICE DIV., MD.             LSD-

ABERDEEN RESEARCH AND DEVELOPMENT CENTER, ABERDEEN PROVING GROUND, MD.   ARDC-TM-
ABERDEEN RESEARCH AND DEVELOPMENT CENTER, ABERDEEN PROVING GROUND, MD.   ARDC-TR-

ABEX CORP. HYDRODYNAMICS RES. CTR., COLUMBUS, OHIO                HRD-TR-
ABEX CORP. HYDRODYNAMICS RES. CTR., COLUMBUS, OHIO                HY-

ABEX CORP. METALLURGICAL DEPT., MAHWAH, N.J.                      XC-(NUMBER)-

ABT ASSOCIATES, INC., CAMBRIDGE, MASS.                            AA-

AC ELECTRONICS-DEFENSE RES. LABS., SANTA BARBARA,CAL.             AC-DRL-CTN(YEAR)-
AC ELECTRONICS-DEFENSE RES. LABS., SANTA BARBARA,CAL.             AC-DRL-S(YEAR)-
AC ELECTRONICS-DEFENSE RES. LABS., SANTA BARBARA,CAL.             AC-DRL-TR(YEAR)-
AC ELECTRONICS-DEFENSE RES. LABS., SANTA BARBARA,CAL.             GMC-S(YEAR)-
AC ELECTRONICS-DEFENSE RES. LABS., SANTA BARBARA,CAL.             S(YEAR)-

ACADEMIA R.S.R.INSTITUTUL DE FIZICA ATOMICA,BUCHAREST             CRD-
ACADEMIA R.S.R.INSTITUTUL DE FIZICA ATOMICA,BUCHAREST             E.S.R.-
ACADEMIA R.S.R.INSTITUTUL DE FIZICA ATOMICA,BUCHAREST             FN-
```

ACADEMIA R.S.R.INSTITUTUL DE FIZICA ATOMICA,BUCHAREST	FT (NUMBER) (YEAR)
ACADEMIA R.S.R.INSTITUTUL DE FIZICA ATOMICA,BUCHAREST	IFA-
ACADEMIA R.S.R.INSTITUTUL DE FIZICA ATOMICA,BUCHAREST	IFA/A/
ACADEMIA R.S.R.INSTITUTUL DE FIZICA ATOMICA,BUCHAREST	IFA-AL-
ACADEMIA R.S.R.INSTITUTUL DE FIZICA ATOMICA,BUCHAREST	IFA/CO/
ACADEMIA R.S.R.INSTITUTUL DE FIZICA ATOMICA,BUCHAREST	IFA-CR-
ACADEMIA R.S.R.INSTITUTUL DE FIZICA ATOMICA,BUCHAREST	IFA-CRD-
ACADEMIA R.S.R.INSTITUTUL DE FIZICA ATOMICA,BUCHAREST	IFA/DF/
ACADEMIA R.S.R.INSTITUTUL DE FIZICA ATOMICA,BUCHAREST	IFA-EN-
ACADEMIA R.S.R.INSTITUTUL DE FIZICA ATOMICA,BUCHAREST	IFA-ESR-
ACADEMIA R.S.R.INSTITUTUL DE FIZICA ATOMICA,BUCHAREST	IFA-FN-
ACADEMIA R.S.R.INSTITUTUL DE FIZICA ATOMICA,BUCHAREST	IFA-FR-
ACADEMIA R.S.R.INSTITUTUL DE FIZICA ATOMICA,BUCHAREST	IFA/FT/
ACADEMIA R.S.R.INSTITUTUL DE FIZICA ATOMICA,BUCHAREST	IFA-HE-
ACADEMIA R.S.R.INSTITUTUL DE FIZICA ATOMICA,BUCHAREST	IFA-IS-
ACADEMIA R.S.R.INSTITUTUL DE FIZICA ATOMICA,BUCHAREST	IFA-M-
ACADEMIA R.S.R.INSTITUTUL DE FIZICA ATOMICA,BUCHAREST	IFA-MN-
ACADEMIA R.S.R.INSTITUTUL DE FIZICA ATOMICA,BUCHAREST	IFA-MR-
ACADEMIA R.S.R.INSTITUTUL DE FIZICA ATOMICA,BUCHAREST	IFA-NR-
ACADEMIA R.S.R.INSTITUTUL DE FIZICA ATOMICA,BUCHAREST	IFA-NS-
ACADEMIA R.S.R.INSTITUTUL DE FIZICA ATOMICA,BUCHAREST	IFA-PN-
ACADEMIA R.S.R.INSTITUTUL DE FIZICA ATOMICA,BUCHAREST	IFA-TR-
ACADEMIA R.S.R.INSTITUTUL DE FIZICA ATOMICA,BUCHAREST	IRA-
ACADEMIA R.S.R.INSTITUTUL DE FIZICA ATOMICA,BUCHAREST	IS-
ACADEMIA R.S.R.INSTITUTUL DE FIZICA ATOMICA,BUCHAREST	PL/N-
ACADEMIA R.S.R.INSTITUTUL DE FIZICA ATOMICA,BUCHAREST	PN-
ACADEMIC PRESS, INC., N.Y.C. (TRANSLATIONS)	API-
ACADEMY FOR INTERSCIENCE METHODOLOGY, CHICAGO	AIM-
ACADEMY OF NATURAL SCIENCES OF PHILADELPHIA (TRANSLATIONS)	ANSP-
ACCUMETRICS CORP., CAMBRIDGE, MASS.	AC-
ACF INDUSTRIES, INC. (ALL LOCATIONS)	ACF-
ACF INDUSTRIES, INC., ALBUQUERQUE, N. MEX.	ACFI-
ACF INDUSTRIES, INC., ALBUQUERQUE, N. MEX.	ACFI-C-
ACF INDUSTRIES, INC., ALBUQUERQUE, N. MEX.	ACFI-HATE-
ACF INDUSTRIES, INC., ALBUQUERQUE, N. MEX.	AL-(NUMBER)-H
ACF INDUSTRIES, INC., HYATTSVILLE, MD.	ETM-
ACF INDUSTRIES, INC. ALBUQUERQUE DIV., N. MEX.	ACF-SA-
ACF INDUSTRIES, INC. ALBUQUERQUE DIV., N. MEX.	FTR-
ACF INDUSTRIES, INC. AVION DIV., PARAMUS, N.J.	AVION-
ACF INDUSTRIES, INC. ELECTRO-PHYSICS LABS., HYATTSVILLE, MD.	ACF-EPL-
ACORN COMMITTEE (REF. 15)	ACORN-
ACOUSTICA ASSOCIATES, INC., LOS ANGELES	DO-
ACTON LABS., INC., MASS.	ACL-
ACTON LABS., INC., MASS.	ALI-
AD HOC GROUP ON HIGH ALTITUDE DETECTION	AGAAD-
ADAMS-RUSSELL CO., INC., WALTHAM, MASS.	AR-
ADCOM, INC., CAMBRIDGE, MASS.	G-(NUMBER)-(LETTER)-
ADEL PRECISION PRODUCTS CORP., BURBANK, CALIF.(REF.1)	(SUBJECT NO. +/OR LTR.)-
ADEL PRECISION PRODUCTS CORP., BURBANK, CALIF.	4V-
ADELAIDE. UNIV., AUSTRALIA	ADP-
ADELPHI UNIV., GARDEN CITY, N.Y. DEPT. OF GRAD. MATH.	AGM-
ADJUTANT GENERAL'S OFF. (ARMY), WASHINGTON, D.C.	OACSFOR-OT-RD-(YEAR)-
ADMIRAL CORP., CHICAGO	A-
ADMIRAL CORP., CHICAGO	AC(NO.)/(CONTRACT NO.)
ADMIRAL CORP., CHICAGO	ADC-
ADMIRAL CORP., CHICAGO	ADC-SCIENTIFIC RPT.-
ADMIRAL CORP., CHICAGO	ADC-SR-
ADMIRAL CORP., CHICAGO	ADM-
ADMIRAL CORP., CHICAGO (EFFECTS OF NUCLEAR RADIATION. SCIENTIFIC REPORT)	ADM-ENR-SR-
ADMIRAL CORP., CHICAGO	G(NUMBER)
ADVANCE RESEARCH, INC., NEEDHAM HEIGHTS, MASS.	CIRA-
ADVANCED COMPUTER TECHNIQUES CORP, N.Y.C.	ACT-
ADVANCED KINETICS, INC., COSTA MESA, CALIF.	AK-
ADVANCED METALS RESEARCH CORP., BURLINGTON, MASS. (PROGRESS REPORT)	AMR-
ADVANCED METALS RESEARCH CORP., SOMERVILLE, MASS.	AMR-
ADVANCED RESEARCH PROJECTS AGENCY, WASHINGTON, D.C.	ARPA-
ADVANCED RESEARCH PROJECTS AGENCY, WASHINGTON, D.C.	ARPA-AM-
ADVANCED RESEARCH PROJECTS AGENCY, WASHINGTON, D.C.	ARPA-E-
ADVANCED RESEARCH PROJECTS AGENCY, WASHINGTON, D.C.	ARPA-NTDO-(YEAR)-
ADVANCED RESEARCH PROJECTS AGENCY, WASHINGTON, D.C.	ARPA-TIO-(YEAR)-
ADVANCED RESEARCH PROJECTS AGENCY, WASHINGTON, D.C. (SERIES USED FOR REPORTS OF VARIOUS CONTRACTORS CONCERNED WITH THE VELA UNIFORM PROGRAM)	ARPA-VUF-
ADVANCED RESEARCH PROJECTS AGENCY, WASHINGTON, D.C.	TIO-
ADVANCED RESEARCH PROJECTS AGENCY, WASHINGTON, D.C.	UBG-
ADVANCED RESEARCH PROJECTS AGENCY, WASHINGTON, D.C. (SERIES USED FOR REPORTS OF VARIOUS CONTRACTORS CONCERNED WITH THE VELA UNIFORM PROGRAM)	VUF-
ADVANCED RESEARCH PROJECTS AGENCY, WASHINGTON, D.C.	VUP-
ADVANCED SCI. TECHNIQUES RES. ASSOCS., MILFORD, CONN.	ASTRA-(NO.)-E-
ADVANCED SYSTEMS DEVELOPMENT DIV., EL SEGUNDO, CALIF.	ASD-
ADVANCED TECHNOLOGY CORP., TIMONIUM, MD.	ADTEC-

ADVANCED TECHNOLOGY LABS. DIV. OF AMERICAN-STANDARD, MOUNTAIN VIEW, CALIF.	ASATL-
ADVANCED TECHNOLOGY LABS. DIV. OF AMERICAN-STANDARD, MOUNTAIN VIEW, CALIF.	ATL-(NUMBER)A
ADVANCED TECHNOLOGY LABS. DIV. OF AMERICAN-STANDARD, MOUNTAIN VIEW, CALIF.	ATL-A-
ADVANCED TECHNOLOGY LABS. DIV. OF AMERICAN-STANDARD, MOUNTAIN VIEW, CALIF.	ATL-D-
ADVANCED TECHNOLOGY LABS., INC., JERICHO, N.Y.	ATL-TR-
ADVIESBUREAU DER GENIE, HAGUE	NR-
ADVISORY COMMITTEE ON GERMAN WAR DOCUMENTS	GWD-
AERO GEO ASTRO CORP., ALEXANDRIA, VA.	AGA-
AERO GEO ASTRO CORP., ALEXANDRIA, VA.	AGAC-
AERO GEO ASTRO CORP., CORONA, CALIF.	AGAC-
AERO SERVICE CORP., PHILADELPHIA	ASC-
AEROCHEM RESEARCH LABS., INC., PRINCETON, N.J.	AC-
AEROCHEM RESEARCH LABS., INC., PRINCETON, N.J. (PROJECT SQUID)	AC-(NUMBER)-P
AEROCHEM RESEARCH LABS., INC., PRINCETON, N.J.	AC-(NUMBER)-PU
AEROCHEM RESEARCH LABS., INC., PRINCETON, N.J.	ACRL-
AEROCHEM RESEARCH LABS., INC., PRINCETON, N.J.	AEROCHEM-
AEROCHEM RESEARCH LABS., INC., PRINCETON, N.J.	AEROCHEM TN-
AEROCHEM RESEARCH LABS., INC., PRINCETON, N.J.	AEROCHEM TP-
AEROCHEM RESEARCH LABS., INC., PRINCETON, N.J.	ARL-
AEROCHEM RESEARCH LABS., INC., PRINCETON, N.J.	TP-
AERODYNAMISCHE VERSUCHSANSTALT, GOETTINGEN, GERMANY	B-(3 DIGITS)
AEROFLEX LABS., INC., LONG ISLAND CITY, N.Y.	ARMT-
AEROJET DELFT CORP., AZUSA, CALIF.	AEROJET DELFT-
AEROJET ENGINEERING CORP., AZUSA, CALIF. (RESEARCH TECHNICAL MEMORANDUM) (SEE ALSO LATER NAME: AEROJET-GENERAL CORP.)	RTM-
AEROJET-GENERAL CORP. (STATUS REPORT)	AGC-Q-
AEROJET-GENERAL CORP. (ALL LOCATIONS)	AGC-
AEROJET-GENERAL CORP., AZUSA, CALIF. (REF. 1)	(NUMBERS...)
AEROJET-GENERAL CORP., AZUSA, CALIF. (REF. 1)	(WORK ORDER NO.)-
AEROJET-GENERAL CORP., AZUSA, CALIF.	390:64-
AEROJET-GENERAL CORP., AZUSA, CALIF.	AERO-
AEROJET-GENERAL CORP., AZUSA, CALIF.	AEROJET-
AEROJET-GENERAL CORP., AZUSA, CALIF.	AEROJET-L-
AEROJET-GENERAL CORP., AZUSA, CALIF.	AEROJET LRP-
AEROJET-GENERAL CORP., AZUSA, CALIF.	AEROJET TR-
AEROJET-GENERAL CORP., AZUSA, CALIF.	AGC-A-
AEROJET-GENERAL CORP., AZUSA, CALIF.	AGC-AE-
AEROJET-GENERAL CORP., AZUSA, CALIF.	AGC-L-
AEROJET-GENERAL CORP., AZUSA, CALIF.	AGC-RN-
AEROJET-GENERAL CORP., AZUSA, CALIF.(DESIGN CRITERIA)	AGC-RN-S-
AEROJET-GENERAL CORP., AZUSA, CALIF.	AGC SR-
AEROJET-GENERAL CORP., AZUSA, CALIF.	AGC-SY-
AEROJET-GENERAL CORP., AZUSA, CALIF.	AGC-TCR-
AEROJET-GENERAL CORP., AZUSA, CALIF. (TECH. MEMO.)	AGC-TM(NO.)-
AEROJET-GENERAL CORP., AZUSA, CALIF.	AGC-TN-
AEROJET-GENERAL CORP., AZUSA, CALIF.	AJEC-
AEROJET-GENERAL CORP., AZUSA, CALIF.	ARJ-
AEROJET-GENERAL CORP., AZUSA, CALIF.	ARJ-AE-
AEROJET-GENERAL CORP., AZUSA, CALIF.	ARJ-FR-
AEROJET-GENERAL CORP., AZUSA, CALIF.	ARJ-IDO-
AEROJET-GENERAL CORP., AZUSA, CALIF.	ARJ-L-
AEROJET-GENERAL CORP., AZUSA, CALIF.	ARJ-LRP-
AEROJET-GENERAL CORP., AZUSA, CALIF.	ARJ-PR-
AEROJET-GENERAL CORP., AZUSA, CALIF.	ARJ-SR-
AEROJET-GENERAL CORP., AZUSA, CALIF.	ARJ-TM-
AEROJET-GENERAL CORP., AZUSA, CALIF.	L-
AEROJET-GENERAL CORP., AZUSA, CALIF.	LRP-
AEROJET-GENERAL CORP., AZUSA, CALIF.	LS-
AEROJET-GENERAL CORP., AZUSA, CALIF.	M-
AEROJET GENERAL CORP., AZUSA, CALIF.	NF-
AEROJET-GENERAL CORP., AZUSA, CALIF.	P-
AEROJET-GENERAL CORP., AZUSA, CALIF.	RD-R-
AEROJET-GENERAL CORP., AZUSA, CALIF.	RN-S-
AEROJET-GENERAL CORP., AZUSA, CALIF.	S(YEAR)-
AEROJET-GENERAL CORP., AZUSA, CALIF.	SP-TP-
AEROJET-GENERAL CORP., AZUSA, CALIF.	TCR-
AEROJET-GENERAL CORP., EL MONTE, CALIF.	AGC-(NUMBERS)-FR
AEROJET-GENERAL CORP., EL MONTE, CALIF.	AGC-(NUMBERS)-SA
AEROJET-GENERAL CORP., EL MONTE, CALIF.	AGC-(NUMBERS)-SUMMARY
AEROJET-GENERAL CORP., EL MONTE, CALIF.	AGC-SD-
AEROJET-GENERAL CORP., EL MONTE, CALIF.	SGD-(NUMBER)-
AEROJET-GENERAL CORP., SACRAMENTO, CALIF. (REVISION)	ACC-
AEROJET-GENERAL CORP., SACRAMENTO, CALIF.	AGC-(NUMBERS)-(LETTERS)
AEROJET-GENERAL CORP., SACRAMENTO, CALIF.	AGC-CR-
AEROJET-GENERAL CORP., SACRAMENTO, CALIF.	AGC-NJD-
AEROJET-GENERAL CORP., SACRAMENTO, CALIF.	AGC-RM-
AEROJET-GENERAL CORP., SACRAMENTO, CALIF.	AGC-RMR-
AEROJET-GENERAL CORP., SACRAMENTO, CALIF.	AGC-RN-
AEROJET-GENERAL CORP., SACRAMENTO, CALIF.	AGC-RN-Q-
AEROJET-GENERAL CORP., SACRAMENTO, CALIF.	AGC-RN-S-
AEROJET-GENERAL CORP., SACRAMENTO, CALIF. (DEV. REPT)	AGC-RP-DR-
AEROJET-GENERAL CORP., SACRAMENTO, CALIF.	AGC-RP-MR-
AEROJET-GENERAL CORP., SACRAMENTO, CALIF.	AGC-RP-P-
AEROJET-GENERAL CORP., SACRAMENTO, CALIF.	AGC-RP-QR-
AEROJET-GENERAL CORP., SACRAMENTO, CALIF.	AGC-RP-SR-
AEROJET-GENERAL CORP., SACRAMENTO, CALIF.	AGC-RP-TM-
AEROJET-GENERAL CORP., SACRAMENTO, CALIF.	AGC-RP-TR-
AEROJET-GENERAL CORP., SACRAMENTO, CALIF. (SECOND QUARTERLY REPORT)	AGC-SRO-
AEROJET-GENERAL CORP., SACRAMENTO, CALIF.	AGC-TCED-QP-
AEROJET-GENERAL CORP., SACRAMENTO, CALIF.	OOAMA-2.5-
AEROJET-GENERAL CORP., SACRAMENTO, CALIF.	AMDR-
AEROJET-GENERAL CORP., SACRAMENTO, CALIF. (SPECS.)	ATS-
AEROJET-GENERAL CORP., SACRAMENTO, CAL.(TEST PROCEDURE)	CODE IDENT-
AEROJET-GENERAL CORP., SACRAMENTO, CALIF.	CR-(NUMBER)-

AEROJET—GENERAL CORP.

AEROJET-GENERAL CORP., SACRAMENTO, CALIF. (COMPUTING SCIENCES)	CSR-
AEROJET-GENERAL CORP., SACRAMENTO, CALIF.	CTR-
AEROJET-GENERAL CORP., SACRAMENTO, CALIF. (METALLURGICAL INVESTIGATIONS)	DMD-
AEROJET-GENERAL CORP., SACRAMENTO, CALIF.	EAFB-
AEROJET-GENERAL CORP., SACRAMENTO, CALIF.	EL-
AEROJET-GENERAL CORP., SACRAMENTO, CALIF.	EP-
AEROJET-GENERAL CORP., SACRAMENTO, CALIF.	ER-
AEROJET-GENERAL CORP., SACRAMENTO, CALIF.	ESR-
AEROJET-GENERAL CORP., SACRAMENTO, CALIF.	ESTR-
AEROJET-GENERAL CORP., SACRAMENTO, CALIF.	ETDR-
AEROJET-GENERAL CORP., SACRAMENTO, CALIF.	F-
AEROJET-GENERAL CORP., SACRAMENTO, CALIF.	FE-
AEROJET-GENERAL CORP., SACRAMENTO, CALIF.	FOTB-
AEROJET-GENERAL CORP., SACRAMENTO, CALIF.	GSE-
AEROJET-GENERAL CORP., SACRAMENTO, CALIF.	GTP-
AEROJET-GENERAL CORP., SACRAMENTO, CALIF.	GV-
AEROJET-GENERAL CORP., SACRAMENTO, CALIF.	HB-
AEROJET-GENERAL CORP., SACRAMENTO, CALIF.	HTP-
AEROJET-GENERAL CORP., SACRAMENTO, CALIF.	HTR-
AEROJET-GENERAL CORP., SACRAMENTO, CALIF. (INDEPENDENT RESEARCH AND DEVELOPMENT REPORT)	IRA
AEROJET-GENERAL CORP., SACRAMENTO, CALIF. (INDEPENDENT RESEARCH AND DEVELOPMENT REPORT)	IR+D-
AEROJET-GENERAL CORP., SACRAMENTO, CALIF. (INDEPENDENT RESEARCH AND DEVELOPMENT REPORT)	IRP-
AEROJET-GENERAL CORP., SACRAMENTO, CALIF. (INDEPENDENT RESEARCH AND DEVELOPMENT REPORT)	IRWA-
AEROJET-GENERAL CORP., SACRAMENTO, CALIF. (INDEPENDENT RESEARCH AND DEVELOPMENT REPORT)	IWA-
AEROJET-GENERAL CORP., SACRAMENTO, CALIF.	L-
AEROJET-GENERAL CORP., SACRAMENTO, CALIF.	LF-
AEROJET-GENERAL CORP., SACRAMENTO, CALIF.	LPSM-
AEROJET-GENERAL CORP., SACRAMENTO, CALIF.	LSR-
AEROJET-GENERAL CORP., SACRAMENTO, CALIF.	M-
AEROJET-GENERAL CORP., SACRAMENTO, CAL. (MATERIALS RPT)	MF-
AEROJET-GENERAL CORP., SACRAMENTO, CAL. (MATERIALS RPT)	MFDR-
AEROJET-GENERAL CORP., SACRAMENTO, CALIF.	MM-
AEROJET-GENERAL CORP., SACRAMENTO, CALIF.	MN-
AEROJET-GENERAL CORP., SACRAMENTO, CALIF.	MPL-
AEROJET-GENERAL CORP., SACRAMENTO, CALIF.	MPR-
AEROJET-GENERAL CORP., SACRAMENTO, CALIF.	MR-
AEROJET-GENERAL CORP., SACRAMENTO, CALIF.	MRDD R-
AEROJET-GENERAL CORP., SACRAMENTO, CALIF.	MRDR-
AEROJET-GENERAL CORP., SACRAMENTO, CALIF.	MRR-(YEAR)-
AEROJET-GENERAL CORP., SACRAMENTO, CALIF.	MS-
AEROJET-GENERAL CORP., SACRAMENTO, CALIF.	MSASD-
AEROJET-GENERAL CORP., SACRAMENTO, CALIF.	MSA-TM-
AEROJET-GENERAL CORP., SACRAMENTO, CALIF.	MSR-
AEROJET-GENERAL CORP., SACRAMENTO, CALIF. (LETTER INDICATES TYPE OF REPORT OR FREQUENCY)	NAS(CONTRACT NO.)-(LTR)-
AEROJET-GENERAL CORP., SACRAMENTO, CALIF.	NDT-
AEROJET-GENERAL CORP., SACRAMENTO, CALIF.	P-
AEROJET-GENERAL CORP., SACRAMENTO, CALIF.	PD-
AEROJET-GENERAL CORP., SACRAMENTO, CALIF.	PDR-
AEROJET-GENERAL CORP., SACRAMENTO, CALIF.	PG-
AEROJET-GENERAL CORP., SACRAMENTO, CALIF.	PIT-
AEROJET-GENERAL CORP., SACRAMENTO, CALIF.	PME-
AEROJET-GENERAL CORP., SACRAMENTO, CALIF.	PP-
AEROJET-GENERAL CORP., SACRAMENTO, CALIF.	PPDR-
AEROJET-GENERAL CORP., SACRAMENTO, CALIF.	QAR-
AEROJET-GENERAL CORP., SACRAMENTO, CALIF. (4 DIGITS ARE WORK ORDER OR CONTRACT CODE NUMBERS. LETTER MAY SHOW FREQUENCY)	R-(4 DIGITS)-(NO.,LTR)-
AEROJET-GENERAL CORP., SACRAMENTO, CALIF.	RM-
AEROJET-GENERAL CORP., SACRAMENTO, CALIF.	RMR-
AEROJET-GENERAL CORP., SACRAMENTO, CALIF.	RN-DR-
AEROJET-GENERAL CORP., SACRAMENTO, CALIF.	RN-S-
AEROJET-GENERAL CORP., SACRAMENTO, CALIF.	RN-TM-
AEROJET-GENERAL CORP., SACRAMENTO, CALIF.	RP-(LETTERS)-
AEROJET-GENERAL CORP., SACRAMENTO, CALIF.	RP-DR-
AEROJET-GENERAL CORP., SACRAMENTO, CALIF.	RP-MR-
AEROJET-GENERAL CORP., SACRAMENTO, CALIF.	RP-QR-
AEROJET-GENERAL CORP., SACRAMENTO, CALIF.	RP-TM-
AEROJET-GENERAL CORP., SACRAMENTO, CALIF.	RP-TR-
AEROJET-GENERAL CORP., SACRAMENTO, CALIF.	SBAMA-2.5-
AEROJET-GENERAL CORP., SACRAMENTO, CALIF.	SCR-
AEROJET-GENERAL CORP., SACRAMENTO, CALIF.	SDE-
AEROJET-GENERAL CORP., SACRAMENTO, CALIF.	SGC-
AEROJET-GENERAL CORP., SACRAMENTO, CALIF.	SP-
AEROJET-GENERAL CORP., SACRAMENTO, CALIF.	STL R-
AEROJET-GENERAL CORP., SACRAMENTO, CALIF.	STR-
AEROJET-GENERAL CORP., SACRAMENTO, CALIF.	TCBR-
AEROJET-GENERAL CORP., SACRAMENTO, CALIF.	TCM-
AEROJET-GENERAL CORP., SACRAMENTO, CALIF.	TCR-
AEROJET-GENERAL CORP., SACRAMENTO, CALIF.	TD-
AEROJET-GENERAL CORP., SACRAMENTO, CALIF.	TDR-
AEROJET-GENERAL CORP., SACRAMENTO, CALIF.	TER-
AEROJET-GENERAL CORP., SACRAMENTO, CALIF.	TP-
AEROJET-GENERAL CORP., SACRAMENTO, CALIF.	TPM-
AEROJET-GENERAL CORP., SACRAMENTO, CALIF.	TTB-
AEROJET-GENERAL CORP., SACRAMENTO, CALIF.	TTF-PN-
AEROJET-GENERAL CORP., SACRAMENTO, CALIF.	W.O.-
AEROJET-GENERAL CORP. ADVANCED PROPELLANT DEPT., SACRAMENTO, CALIF.	AGC-LR-
AEROJET-GENERAL CORP. ADVANCED PROPELLANT DEPT., SACRAMENTO, CALIF.	AGC-LRP-
AEROJET-GENERAL CORP. AETRON DIV., COVINA, CALIF.	AE-
AEROJET-GENERAL CORP. ASTRIONICS DEPT., AZUSA, CALIF.	AGC-(NUMBER)-(NUMBER)-
AEROJET-GENERAL CORP. EXPERIMENTAL ENGINE DIV., SACRAMENTO, CALIF.	LF-(NUMBERS)-
AEROJET-GENERAL CORP. LIQUID ROCKET OPERATIONS, SACRAMENTO, CALIF.	DM-(NUMBERS)-
AEROJET-GENERAL CORP. LIQUID ROCKET OPERATIONS, SACRAMENTO, CALIF.	DVR-(NUMBERS)-
AEROJET-GENERAL CORP. LIQUID ROCKET OPERATIONS, SACRAMENTO, CALIF.	FSC-(NUMBERS)-
AEROJET-GENERAL CORP. LIQUID ROCKET OPERATIONS, SACRAMENTO, CALIF. (GEMINI PROGRAM)	GEMSIP-
AEROJET-GENERAL CORP. LIQUID ROCKET OPERATIONS, SACRAMENTO, CALIF.	LF-
AEROJET-GENERAL CORP. LIQUID ROCKET OPERATIONS, SACRAMENTO, CALIF.	LRD-(NUMBERS)-
AEROJET-GENERAL CORP. LIQUID ROCKET OPERATIONS, SACRAMENTO, CALIF.	LRO-
AEROJET-GENERAL CORP. LIQUID ROCKET OPERATIONS, SACRAMENTO, CALIF.	MF-
AEROJET-GENERAL CORP. LIQUID ROCKET OPERATIONS, SACRAMENTO, CALIF. (MAINTENANCE, LOGISTICS, RELIABILITY AND READINESS)	MLRR-AR-
AEROJET-GENERAL CORP. LIQUID ROCKET OPERATIONS, SACRAMENTO, CALIF.	PTDR-
AEROJET-GENERAL CORP. LIQUID ROCKET OPERATIONS, SACRAMENTO, CALIF.	RN-S-
AEROJET-GENERAL CORP. LIQUID ROCKET OPERATIONS, SACRAMENTO, CALIF.	SCR-
AEROJET-GENERAL CORP. LIQUID ROCKET OPERATIONS, SACRAMENTO, CALIF.	SPEC-ACC-
AEROJET-GENERAL CORP. LIQUID ROCKET OPERATIONS, SACRAMENTO, CALIF.	TL-(NUMBER)-

AEROJET-GENERAL CORP. LIQUID ROCKET OPERATIONS, SACRAMENTO, CALIF.	TP-(NUMBER)-LRO-
AEROJET-GENERAL CORP. LIQUID ROCKET PLANT, SACRAMENTO, CALIF. (FIRST NO. IS CONTRACT CODE NO.)	A(NO.)-(LTR OR NO.LTR)-
AEROJET-GENERAL CORP. LIQUID ROCKET PLANT, SACRAMENTO, CALIF.	AGC-LRP-
AEROJET-GENERAL CORP. LIQUID ROCKET PLANT, SACRAMENTO, CALIF.	AGC-TM-
AEROJET-GENERAL CORP. LIQUID ROCKET PLANT, SACRAMENTO, CALIF.	AJEC LRP
AEROJET-GENERAL CORP. LIQUID ROCKET PLANT, SACRAMENTO, CALIF. (FINANCIAL MANAGEMENT REPORT)	ASD-2.5-Q-
AEROJET-GENERAL CORP. LIQUID ROCKET PLANT, SACRAMENTO, CALIF.	CPR-
AEROJET-GENERAL CORP. LIQUID ROCKET PLANT, SACRAMENTO, CALIF.	DVR-
AEROJET-GENERAL CORP. LIQUID ROCKET PLANT, SACRAMENTO, CALIF.	DVR-(YEAR)-
AEROJET-GENERAL CORP. LIQUID ROCKET PLANT, SACRAMENTO, CALIF.	ISS-LRP-
AEROJET-GENERAL CORP. LIQUID ROCKET PLANT, SACRAMENTO, CALIF.	LRO-
AEROJET-GENERAL CORP. LIQUID ROCKET PLANT, SACRAMENTO, CALIF.	LRP-
AEROJET-GENERAL CORP. LIQUID ROCKET PLANT, SACRAMENTO, CALIF.	M-
AEROJET-GENERAL CORP. LIQUID ROCKET PLANT, SACRAMENTO, CALIF.	R/C-
AEROJET-GENERAL CORP. LIQUID ROCKET PLANT, SACRAMENTO, CALIF.	RR-CER-
AEROJET-GENERAL CORP. LIQUID ROCKET PLANT, SACRAMENTO, CALIF.	STM-
AEROJET-GENERAL CORP. LIQUID ROCKET PLANT, SACRAMENTO, CALIF.	TCD-
AEROJET-GENERAL CORP. LIQUID ROCKET PLANT, SACRAMENTO, CALIF.	TM-(NUMBER)-LRO
AEROJET-GENERAL CORP. LIQUID ROCKET PLANT, SACRAMENTO, CALIF.	TM-(NUMBER)-LRP
AEROJET-GENERAL CORP. LIQUID ROCKET PLANT, SACRAMENTO, CALIF.	TP-(NUMBER)-LRO
AEROJET-GENERAL CORP. LIQUID ROCKET PLANT, SACRAMENTO, CALIF.	TP-(NUMBER)-LRP
AEROJET-GENERAL CORP. LIQUID ROCKET PLANT, SACRAMENTO, CALIF.	U-(YEAR)-
AEROJET-GENERAL CORP. NUCLEAR ENGINEERING AND MFG. OPERATIONS, SAN RAMON, CALIF.	AN-APCSE-(NUMBERS)-VOL-
AEROJET-GENERAL CORP. NUCLEAR SYSTEMS, DOWNEY, CALIF.	ANWA-PEP-
AEROJET-GENERAL CORP. ORDNANCE DIV., DOWNEY, CALIF.	AGC-(NOS.)-(NOS.)-(NO.)
AEROJET-GENERAL CORP. ORDNANCE DIV., DOWNEY, CALIF.	R-
AEROJET-GENERAL CORP. ORDNANCE DIV., DOWNEY, CALIF.	SCR-
AEROJET-GENERAL CORP. PROPULSION DIV., SACRAMENTO, CALIF.	AGC-SELT-FR-
AEROJET-GENERAL CORP. ROCKET ENGINE OPERATIONS-NUCLEAR, SACRAMENTO, CALIF.	NDC-
AEROJET-GENERAL CORP. ROCKET ENGINE OPERATIONS-NUCLEAR, SACRAMENTO, CALIF.	RN-
AEROJET-GENERAL CORP. ROCKET ENGINE OPERATIONS-NUCLEAR, SACRAMENTO, CALIF.	RN-(LETTER(S))-
AEROJET-GENERAL CORP. ROCKET ENGINE OPERATIONS-NUCLEAR, SACRAMENTO, CALIF.	RP-SR-
AEROJET-GENERAL CORP. SOLID ROCKET OPERATIONS, SACRAMENTO, CALIF.	AGC-TCED-FR-
AEROJET-GENERAL CORP. SOLID ROCKET OPERATIONS, SACRAMENTO, CALIF.	TCED-QP-
AEROJET-GENERAL CORP. SOLID ROCKET OPERATIONS, SACRAMENTO, CALIF.	TP-(NUMBER)-SRO
AEROJET-GENERAL CORP. SOLID ROCKET PLANT, SACRAMENTO, CALIF. (ENVIRONMENTAL TEST MOTOR REPORT)	A2-ETM-
AEROJET-GENERAL CORP. SOLID ROCKET PLANT, SACRAMENTO, CALIF.	AGC-(NUMBERS)-
AEROJET-GENERAL CORP. SOLID ROCKET PLANT, SACRAMENTO, CALIF.	AGC-SRP-
AEROJET-GENERAL CORP. SOLID ROCKET PLANT, SACRAMENTO, CALIF. (FLEET BALLISTIC MISSILE PROGRAM)	B-
AEROJET-GENERAL CORP. SOLID ROCKET PLANT, SACRAMENTO, CALIF (TRAINING MANUAL)	BOOK(NUMBER)
AEROJET-GENERAL CORP. SOLID ROCKET PLANT, SACRAMENTO, CALIF. (MINUTEMAN PROGRAM)	MMR-
AEROJET-GENERAL CORP. SOLID ROCKET PLANT, SACRAMENTO, CALIF. (MINUTEMAN PROGRAM)	MMR-(LETTER(S))-
AEROJET-GENERAL CORP. SOLID ROCKET PLANT, SACRAMENTO, CALIF. (PHOTOELASTIC ANALYSIS)	PEL-
AEROJET-GENERAL CORP. SOLID ROCKET PLANT, SACRAMENTO, CALIF. (FLEET BALLISTIC MISSILE PROGRAM)	PPD-WPR-
AEROJET-GENERAL CORP. SOLID ROCKET PLANT, SACRAMENTO, CALIF. (FLEET BALLISTIC MISSILE PROGRAM)	R-PR-
AEROJET-GENERAL CORP. SOLID ROCKET PLANT, SACRAMENTO, CALIF.	SRO-
AEROJET-GENERAL CORP. SOLID ROCKET PLANT, SACRAMENTO, CALIF.	SRO-(NUMBER)-TM
AEROJET-GENERAL CORP. SOLID ROCKET PLANT, SACRAMENTO, CALIF.	SRP-
AEROJET-GENERAL CORP. SOLID ROCKET PLANT, SACRAMENTO, CALIF. (TECHNICAL MANUAL)	STM-
AEROJET-GENERAL CORP. SOLID ROCKET PLANT, SACRAMENTO, CALIF.	TM-(NUMBER)-SRO
AEROJET-GENERAL CORP. SOLID ROCKET PLANT, SACRAMENTO, CALIF.	TM-(NUMBER)-SRP
AEROJET-GENERAL CORP. SOLID ROCKET PLANT, SACRAMENTO, CALIF.	TP-(NUMBER)-SRP
AEROJET-GENERAL CORP. SPACE DIV., EL MONTE, CALIF.	SD-(NUMBER)-
AEROJET-GENERAL CORP. TECHNICAL LIBRARY, SACRAMENTO, CALIF. (REPORT BIBLIOGRAPHY)	TIL-
AEROJET-GENERAL NUCLEONICS, SAN RAMON, CALIF.	AGC-AN-
AEROJET-GENERAL NUCLEONICS, SAN RAMON, CALIF.	AGN-
AEROJET-GENERAL NUCLEONICS, SAN RAMON, CALIF.	AGN-AN-
AEROJET-GENERAL NUCLEONICS, SAN RAMON, CALIF.	AGN-IDO-
AEROJET-GENERAL NUCLEONICS, SAN RAMON, CALIF.	AGN-RS-
AEROJET-GENERAL NUCLEONICS, SAN RAMON, CALIF.	AGN-TM-
AEROJET-GENERAL NUCLEONICS, SAN RAMON, CALIF.	AGN-TP-
AEROJET-GENERAL NUCLEONICS, SAN RAMON, CALIF.	AN-
AEROJET-GENERAL NUCLEONICS, SAN RAMON, CALIF.	AN-AGCR-
AEROJET-GENERAL NUCLEONICS, SAN RAMON, CALIF.	AN-ENGR-
AEROJET-GENERAL NUCLEONICS, SAN RAMON, CALIF.	AN-FDD-
AEROJET-GENERAL NUCLEONICS, SAN RAMON, CALIF.	AN-IDOP-
AEROJET-GENERAL NUCLEONICS, SAN RAMON, CALIF.	AN-MET-
AEROJET-GENERAL NUCLEONICS, SAN RAMON, CALIF.	APMD-
AEROJET-GENERAL NUCLEONICS, SAN RAMON, CALIF.	EAD-
AEROJET-GENERAL NUCLEONICS, SAN RAMON, CALIF.	GCSD-
AEROJET-GENERAL NUCLEONICS, SAN RAMON, CALIF.	TM-
AEROJET LIQUID ROCKET CO., SACRAMENTO, CALIF.	ALRC-
AEROJET NUCLEAR CO., IDAHO FALLS	ANCR-
AEROJET NUCLEAR SYSTEMS CO., SACRAMENTO, CALIF.	AGC-M001-
AEROJET NUCLEAR SYSTEMS CO., SACRAMENTO, CALIF.	ANSC-
AEROJET NUCLEAR SYSTEMS CO., SACRAMENTO, CALIF.	M001-
AEROJET NUCLEAR SYSTEMS CO., SACRAMENTO, CALIF.	T(NUMBER)-TFC-
AEROJET SOLID PROPULSION CO., SACRAMENTO, CALIF.	ASPC-(NUMBER)-
AEROMEDICAL RES. LAB. (6571ST), HOLLOMAN AFB, N. MEX.	ARL-
AEROMEDICAL RES. LAB. (6571ST), HOLLOMAN AFB, N. MEX.	ARL-TR-(NUMBERS)-
AEROMEDICAL RES. LAB. (6571ST), HOLLOMAN AFB, N. MEX.	ARL-TR-(YEAR)-
AEROMEKANO, STOCKHOLM	AEROMEKANO-
AERONAUTICAL BOARD, WASHINGTON, D.C.	AREC-
AERONAUTICAL CHART AND INFORMATION CENTER, ST. LOUIS	ACIC-
AERONAUTICAL CHART AND INFORMATION CENTER, ST. LOUIS	ACIC-LD-
AERONAUTICAL CHART AND INFORMATION CENTER, ST. LOUIS	ACIC-REF-PUB-
AERONAUTICAL CHART AND INFORMATION CENTER, ST. LOUIS (REFERENCE MATERIAL)	ACIC-RM-
AERONAUTICAL CHART AND INFORMATION CENTER, ST. LOUIS (REFERENCE PUBLICATION)	ACIC-RP-
AERONAUTICAL CHART AND INFORMATION CENTER, ST. LOUIS (SPECIAL PUBLICATION)	ACIC-SP-
AERONAUTICAL CHART AND INFORMATION CENTER, ST. LOUIS	ACIC-STUDY-
AERONAUTICAL CHART AND INFORMATION CENTER, ST. LOUIS	ACIC-STUDY-TC-

AERONAUTICAL CHART AND INFORMATION CENTER (TRANSLATION CODE)

AERONAUTICAL CHART AND INFORMATION CENTER, ST. LOUIS (TRANSLATION CODE)	ACIC-TC-
AERONAUTICAL CHART AND INFORMATION CENTER, ST. LOUIS	ACIC-TECHNICAL PAPER-
AERONAUTICAL CHART AND INFORMATION CENTER, ST. LOUIS (TECHNICAL INSTRUCTION)	ACIC-TI-
AERONAUTICAL CHART AND INFORMATION CENTER, ST. LOUIS (TECHNICAL MANUAL)	ACIC-TM-
AERONAUTICAL CHART AND INFORMATION CENTER, ST. LOUIS (TECHNICAL PAPER)	ACIC-TP-
AERONAUTICAL CHART AND INFORMATION CENTER, ST. LOUIS (TECHNICAL REPORT)	ACIC-TR-
AERONAUTICAL CHART AND INFORMATION CENTER, ST. LOUIS	ARAP-
AERONAUTICAL CHART AND INFORMATION CENTER, ST. LOUIS	ARC-
AERONAUTICAL CHART AND INFORMATION CENTER, ST. LOUIS	ARC-HYP-
AERONAUTICAL CHART AND INFORMATION CENTER, ST. LOUIS	ARC-PERF.-
AERONAUTICAL CHART AND INFORMATION CENTER, ST. LOUIS	ARC-PL-
AERONAUTICAL CHART AND INFORMATION CENTER, ST. LOUIS	ARC-R/M-
AERONAUTICAL CHART AND INFORMATION CENTER, ST. LOUIS	FM-
AERONAUTICAL CHART AND INFORMATION CENTER, ST. LOUIS	PA-
AERONAUTICAL CHART AND INFORMATION CENTER, ST. LOUIS	TC-
AERONAUTICAL ENGINEERING RESEARCH, INC., PASADENA, CAL	AER-
AERONAUTICAL RADIO, INC., WASHINGTON, D.C.	AEEC-
AERONAUTICAL RADIO, INC., WASHINGTON, D.C.	AR-
AERONAUTICAL RADIO, INC., WASHINGTON, D.C.	ARI-
AERONAUTICAL RADIO, INC., WASHINGTON, D.C.	EUCOM-
AERONAUTICAL RESEARCH ASSOCIATES OF PRINCETON, INC., NJ	ARAP-
AERONAUTICAL RESEARCH ASSOCIATES OF PRINCETON, INC., NJ	ARAP-(YEAR)-
AERONAUTICAL RESEARCH ASSOCIATES OF PRINCETON, INC., NJ (PROJECT SQUID)	ARAP-(NUMBER)-P
AERONAUTICAL RESEARCH ASSOCIATES OF PRINCETON, INC., NJ	ARAP-TM-(YEAR)-
AERONAUTICAL RESEARCH FOUNDATION, CAMBRIDGE, MASS.	AERF-
AERONAUTICAL STANDARDS GROUP (NAVY)	AG-
AERONAUTICAL SYSTEMS DIV., WRIGHT-PATTERSON AFB, OHIO	ASB-(YEAR)-
AERONAUTICAL SYSTEMS DIV., WRIGHT-PATTERSON AFB, OHIO	ASB-TM-(YEAR)-
AERONAUTICAL SYSTEMS DIV., WRIGHT-PATTERSON AFB, OHIO	ASC-IR-(NUMBER)-
AERONAUTICAL SYSTEMS DIV., WRIGHT-PATTERSON AFB, OHIO	ASC-TR-
AERONAUTICAL SYSTEMS DIV., WRIGHT-PATTERSON AFB, OHIO	ASD-
AERONAUTICAL SYSTEMS DIV., WRIGHT-PATTERSON AFB, OHIO	ASD-(NUMBER)-
AERONAUTICAL SYSTEMS DIV., WRIGHT-PATTERSON AFB, OHIO	ASD-ASNMH-TM-(YR.)-
AERONAUTICAL SYSTEMS DIV., WRIGHT-PATTERSON AFB, OHIO	ASD-ASTN-
AERONAUTICAL SYSTEMS DIV., WRIGHT-PATTERSON AFB, OHIO (CONTRACT REPORT)	ASD-CR-(YEAR)-
AERONAUTICAL SYSTEMS DIV., WRIGHT-PATTERSON AFB, OHIO	ASD-FR-(YEAR)-(NO.)-
AERONAUTICAL SYSTEMS DIV., WRIGHT-PATTERSON AFB, OHIO	ASD-FTR-(YEAR)-
AERONAUTICAL SYSTEMS DIV., WRIGHT-PATTERSON AFB, OHIO	ASD-IR-
AERONAUTICAL SYSTEMS DIV., WRIGHT-PATTERSON AFB, OHIO	ASD-IR-(NUMBER)-
AERONAUTICAL SYSTEMS DIV., WRIGHT-PATTERSON AFB, OHIO	ASD-MC-
AERONAUTICAL SYSTEMS DIV., WRIGHT-PATTERSON AFB, OHIO	ASD-MP-
AERONAUTICAL SYSTEMS DIV., WRIGHT-PATTERSON AFB, OHIO	ASD-QA-
AERONAUTICAL SYSTEMS DIV., WRIGHT-PATTERSON AFB, OHIO	ASD-TDR-
AERONAUTICAL SYSTEMS DIV., WRIGHT-PATTERSON AFB, OHIO	ASD-TDR-(NUMBER)-
AERONAUTICAL SYSTEMS DIV., WRIGHT-PATTERSON AFB, OHIO	ASD-TDR-(YEAR)-
AERONAUTICAL SYSTEMS DIV., WRIGHT-PATTERSON AFB, OHIO	ASD-TM-(YEAR)-
AERONAUTICAL SYSTEMS DIV., WRIGHT-PATTERSON AFB, OHIO	ASD-TN-(YEAR)-
AERONAUTICAL SYSTEMS DIV., WRIGHT-PATTERSON AFB, OHIO	ASD-TR-
AERONAUTICAL SYSTEMS DIV., WRIGHT-PATTERSON AFB, OHIO	ASD-TR-(NUMBER)-
AERONAUTICAL SYSTEMS DIV., WRIGHT-PATTERSON AFB, OHIO	ASD-TR-(YEAR)-
AERONAUTICAL SYSTEMS DIV., WRIGHT-PATTERSON AFB, OHIO	ASD-TR-(YR.)-(NO)-
AERONAUTICAL SYSTEMS DIV., WRIGHT-PATTERSON AFB, OHIO	ASRCE-TM-
AERONAUTICAL SYSTEMS DIV., WRIGHT-PATTERSON AFB, OHIO	ASTF-
AERONAUTICAL SYSTEMS DIV., WRIGHT-PATTERSON AFB, OHIO	F-
AERONAUTICAL SYSTEMS DIV., WRIGHT-PATTERSON AFB, OHIO	MCL-(NO.)/(NO.)
AERONAUTICAL SYSTEMS DIV. AIMS SYSTEM PROGRAM OFFICE, WRIGHT-PATTERSON AFB, OHIO	DOD-AIMS-(YEAR)-
AERONAUTICAL SYSTEMS DIV. DEPUTY FOR DEVELOPMENT PLANNING, WRIGHT-PATTERSON AFB, OHIO	WP-(YEAR)-
AERONAUTICAL SYSTEMS DIV. DIRECTORATE OF RECONNAISSANCE ENG., WRIGHT-PATTERSON AFB, OHIO	ASNQ-TM-(YEAR)-
AERONAUTICAL SYSTEMS DIV. DIRECTORATE OF PROPULSION AND POWER SUBSYSTEMS, WRIGHT-PATTERSON AFB, OHIO	ASNJ-TN-(YEAR)-
AERONAUTICAL SYSTEMS DIV. FLIGHT DYNAMICS LAB., WRIGHT-PATTERSON AFB, OHIO	ASRMDD-
AERONAUTICAL SYSTEMS DIV. FLIGHT DYNAMICS LAB., WRIGHT-PATTERSON AFB, OHIO	ASRMDD-TM-
AERONAUTICAL SYSTEMS DIV. FLIGHT DYNAMICS LAB., WRIGHT-PATTERSON AFB, OHIO	ASRMDF-TM-
AERONAUTICAL SYSTEMS DIV. FLIGHT TEST ENG. DIV., WRIGHT-PATTERSON AFB, OHIO	ASTDN-(YEAR)-
AERONAUTICAL SYSTEMS DIV. MISSION SIMULATION BRANCH, WRIGHT-PATTERSON AFB, OHIO	ASBES-WP-(YEAR)-
AERONCA MFG. CORP., MIDDLETOWN, OHIO	AERONCA-
AERONCA MFG. CORP., MIDDLETOWN, OHIO	ER-
AERONOMY LAB., BOULDER, COLO.	AL-
AERONOMY LAB., BOULDER, COLO.	ERLTM-AL-
AERONOMY LAB., BOULDER, COLO.	ESSA-TM-ERLTM-AL-
AERONOMY LAB., BOULDER, COLO.	ESSA-TR-ERL-(NO.)-AL-
AERONUTRONIC, NEWPORT BEACH, CALIF.	AL-
AERONUTRONIC, NEWPORT BEACH, CALIF. (SEE ALSO LATER NAME: AERONUTRONIC DIV. OF PHILCO...)	ASA-
AERONUTRONIC DIV., FORD MOTOR CO., LOS ANGELES	AERNU-
AERONUTRONIC DIV., FORD MOTOR CO., LOS ANGELES	AERNU-S-ITR-
AERONUTRONIC DIV., FORD MOTOR CO., LOS ANGELES	AERNU-SRD-
AERONUTRONIC DIV. OF PHILCO CORP., NEWPORT BEACH, CAL.	AERO-
AERONUTRONIC DIV. OF PHILCO CORP., NEWPORT BEACH, CAL.	AERO C-
AERONUTRONIC DIV. OF PHILCO CORP., NEWPORT BEACH, CAL.	AERONUTRONIC-C-
AERONUTRONIC DIV. OF PHILCO CORP., NEWPORT BEACH, CAL.	AERONUTRONIC-S-
AERONUTRONIC DIV. OF PHILCO CORP., NEWPORT BEACH, CAL.	AERONUTRONIC-U-
AERONUTRONIC DIV. OF PHILCO CORP., NEWPORT BEACH, CAL.	AERO S-
AERONUTRONIC DIV. OF PHILCO CORP., NEWPORT BEACH, CAL.	AERO U-
AERONUTRONIC DIV. OF PHILCO CORP., NEWPORT BEACH, CAL.	AMC-(YEAR)-
AERONUTRONIC DIV. OF PHILCO CORP., NEWPORT BEACH, CAL.	ASI-C-
AERONUTRONIC DIV. OF PHILCO CORP., NEWPORT BEACH, CAL.	ASI-S-
AERONUTRONIC DIV. OF PHILCO CORP., NEWPORT BEACH, CAL.	ASI-U-
AERONUTRONIC DIV. OF PHILCO CORP., NEWPORT BEACH, CALIF. (CONFIDENTIAL)	C-
AERONUTRONIC DIV. OF PHILCO CORP., NEWPORT BEACH, CAL.	CTS-
AERONUTRONIC DIV. OF PHILCO CORP., NEWPORT BEACH, CALIF. (FINAL TECHNICAL REPORT)	FTR-
AERONUTRONIC DIV. OF PHILCO CORP., NEWPORT BEACH, CALIF. (MONTHLY PROGRESS REPORT)	MPR-
AERONUTRONIC DIV. OF PHILCO CORP., NEWPORT BEACH, CAL.	PHILCO-U-

AERONUTRONIC DIV. OF PHILCO CORP., NEWPORT BEACH, CALIF. (QUARTERLY LETTER REPORT)	QLR-
AERONUTRONIC DIV. OF PHILCO CORP., NEWPORT BEACH, CAL.	RSC-
AERONUTRONIC DIV. OF PHILCO CORP., NEWPORT BEACH, CALIF. (SECRET)	S-
AERONUTRONIC DIV. OF PHILCO CORP., NEWPORT BEACH, CAL.	SRD-
AERONUTRONIC DIV. OF PHILCO CORP., NEWPORT BEACH, CALIF. (TECHNICAL PROGRESS REPORT)	TPR-
AERONUTRONIC DIV. OF PHILCO CORP., NEWPORT BEACH, CAL.	TR-S-
AERONUTRONIC DIV. OF PHILCO CORP., NEWPORT BEACH, CALIF. (UNCLASSIFIED)	U-
AERONUTRONIC SYSTEMS, INC., GLENDALE, CALIF.	ASI-
AERONUTRONIC SYSTEMS, INC., GLENDALE, CALIF.	ASI AR-
AERONUTRONIC SYSTEMS, INC., GLENDALE, CALIF.	ASI PUB MPR-(YEAR)-
AERONUTRONIC SYSTEMS, INC., GLENDALE, CALIF.	ASI PUB U-
AERONUTRONIC SYSTEMS, INC., GLENDALE, CALIF.	ASI SR-
AERONUTRONIC SYSTEMS, INC., GLENDALE, CALIF.	ASI SRD-
AERONUTRONIC SYSTEMS, INC., GLENDALE, CALIF.	DOC. CONTROL NO.-
AERONUTRONIC SYSTEMS, INC., GLENDALE, CALIF.	SR-
AERONUTRONIC SYSTEMS, INC., GLENDALE, CALIF.	U-
AEROPHYSICS DEVELOPMENT CORP. (REF. 1)	(NUMBERS...)
AEROPHYSICS DEVELOPMENT CORP. (ALL LOCATIONS)	APDC-
AEROPHYSICS DEVELOPMENT CORP., PACIFIC PALISADES, CAL	ADCO-
AEROPHYSICS DEVELOPMENT CORP., SANTA BARBARA, CALIF.	ADCO-
AEROPHYSICS DEVELOPMENT CORP., SANTA BARBARA, CALIF.	ADC-R-
AEROPHYSICS DEVELOPMENT CORP., SANTA BARBARA, CALIF.	ADCSB-
AEROPHYSICS DEVELOPMENT CORP., SANTA BARBARA, CALIF.	CDC-
AEROPHYSICS DEVELOPMENT CORP., SANTA MONICA, CALIF.	ADC-
AEROPHYSICS DEVELOPMENT CORP., SANTA MONICA, CALIF.	ADC-PR-
AEROPHYSICS DEVELOPMENT CORP., SANTA MONICA, CALIF.	ADCSM-
AEROPHYSICS RESEARCH FOUNDATION, GOLETA, CALIF.	AERF-(YR)-(NO)-AROWA
AEROPHYSICS RESEARCH FOUNDATION, GOLETA, CALIF.	APRF-
AEROPHYSICS RESEARCH FOUNDATION, GOLETA, CALIF.	ARF-
AEROPROJECTS, INC., WEST CHESTER, PENNA.	AEPI-
AEROPROJECTS, INC., WEST CHESTER, PENNA.	API-
AEROPROJECTS, INC., WEST CHESTER, PENNA.	API-RR-(YEAR)-
AEROPROJECTS, INC., WEST CHESTER, PENNA.	DP-
AEROSPACE CONTROLS CORP., LOS ANGELES	P-
AEROSPACE CORP., EL SEGUNDO, CALIF. (LIBRARY ACCESSION NUMBER)	A (YEAR)-
AEROSPACE CORP., EL SEGUNDO, CALIF. (ART WORK)	A(NUMBER)AW-
AEROSPACE CORP., EL SEGUNDO, CALIF. (INTEROFFICE NO.)	A(YEAR)-CCC(NUMBER)-
AEROSPACE CORP., EL SEGUNDO, CALIF.	AEROSPACE-TR-
AEROSPACE CORP., EL SEGUNDO, CALIF.	AS-
AEROSPACE CORP., EL SEGUNDO, CALIF.	AS-(YEAR)-(NUMBERS)
AEROSPACE CORP., EL SEGUNDO, CALIF.	ASC-
AEROSPACE CORP., EL SEGUNDO, CALIF.	AS TDR-
AEROSPACE CORP., EL SEGUNDO, CALIF. (AEROSPACE TECHNICAL MEMO)	ATM-(YEAR)(JOB NO.)-
AEROSPACE CORP., EL SEGUNDO, CALIF. (AEROSPACE TECHNICAL NOTE)	ATN-(YEAR)(JOB NO.)-
AEROSPACE CORP., EL SEGUNDO, CALIF. (AEROSPACE TECHNICAL REPORT)	ATR-(YEAR)(JOB NO.)-
AEROSPACE CORP., EL SEGUNDO, CALIF. (ART WORK)	AW-
AEROSPACE CORP., EL SEGUNDO, CALIF. (INTEROFFICE NO.)	CCC(NUMBER)-
AEROSPACE CORP., EL SEGUNDO, CALIF. (CONTRACT STATUS)	CSR-(CON.NO.)(JOB NO.)-
AEROSPACE CORP., EL SEGUNDO, CALIF. (ART WORK)	D(NUMBER)
AEROSPACE CORP., EL SEGUNDO, CALIF.	DCAS-
AEROSPACE CORP., EL SEGUNDO, CALIF.	DCAS-TDR-
AEROSPACE CORP., EL SEGUNDO, CALIF.	DCAS-TR-
AEROSPACE CORP., EL SEGUNDO, CALIF.	EX-(YEAR)-(NUMBERS)
AEROSPACE CORP., EL SEGUNDO, CALIF. (LITERATURE RESEARCH GROUP, BIBLIOGRAPHY)	LRG-(YEAR)-B-
AEROSPACE CORP., EL SEGUNDO, CALIF. (LITERATURE RESEARCH GROUP, TRANSLATION)	LRG-(YEAR)-T-
AEROSPACE CORP., EL SEGUNDO, CALIF.	OV-
AEROSPACE CORP., EL SEGUNDO, CALIF. (TECHNICAL DOCUMENTARY REPORT)	TDR-(CON.NO.)(JOB NO.)-
AEROSPACE CORP., EL SEGUNDO, CALIF.	TDR-(NUMBERS)OD-
AEROSPACE CORP., EL SEGUNDO, CALIF. (TECHNICAL OPERATING REPORT)	TOR-(CON.NO.)(JOB NO.)-
AEROSPACE CORP., EL SEGUNDO, CALIF. (TECH. REPORT)	TR-(CON.NO.)(JOB NO.)
AEROSPACE CORP., EL SEGUNDO, CALIF. (WORK STATEMENTS)	WS-(CONTRACT NO.)(NO.)-
AEROSPACE CORP., PATRICK AFB, FLA.	TDR-
AEROSPACE CORP., SAN BERNARDINO, CALIF.	AEROSPACE-TR-
AEROSPACE CORP., SAN BERNARDINO, CALIF.	TDR-
AEROSPACE CORP., SAN BERNARDINO, CALIF.	TOR-
AEROSPACE CORP. ARMS CONTROL PROGRAM OFFICE, SAN BERNARDINO, CALIF.	66-DINS-
AEROSPACE CORP. ATLANTIC MISSILE RANGE OFFICE	ATM-(YEAR)A(JOB NO.)-
AEROSPACE CORP. ATLANTIC MISSILE RANGE OFFICE	ATN-(YEAR)A(JOB NO.)-
AEROSPACE CORP. ATLANTIC MISSILE RANGE OFFICE	ATR-(YEAR)A(JOB NO.)-
AEROSPACE CORP. ATLANTIC MISSILE RANGE OFFICE	TDR-(CON.NO.)A(JOB NO.)-
AEROSPACE CORP. ATLANTIC MISSILE RANGE OFFICE	TOR-(CON.NO.)A(JOB NO.)-
AEROSPACE CORP. EASTERN TEST RANGE OFFICE, PATRICK AFB, FLA.	ATM-(YEAR)-A(NUMBER)
AEROSPACE CORP. EASTERN TEST RANGE OFFICE, PATRICK AFB, FLA.	ATN-(YEAR)-A(NUMBER)
AEROSPACE CORP. EASTERN TEST RANGE OFFICE, PATRICK AFB, FLA.	ATR-(YEAR)-A(NUMBER)
AEROSPACE CORP. EASTERN TEST RANGE OFFICE, PATRICK AFB, FLA.	TDR-(CONTRACT NO)-A(NO.)-
AEROSPACE CORP. EASTERN TEST RANGE OFFICE, PATRICK AFB, FLA.	TOR-(CONTRACT NO)-A(NO.)
AEROSPACE CORP. EASTERN TEST RANGE OFFICE, PATRICK AFB, FLA.	TR-(CONTRACT NO.)-A(NO.)-
AEROSPACE CORP. SAN BERNARDINO OPERATIONS, CALIF.	ATM-(YEAR)S(JOB NO.)-
AEROSPACE CORP. SAN BERNARDINO OPERATIONS, CALIF.	ATN-(YEAR)S(JOB NO.)-
AEROSPACE CORP. SAN BERNARDINO OPERATIONS, CALIF.	ATR-(YEAR)S(JOB NO.)-
AEROSPACE CORP. SAN BERNARDINO OPERATIONS, CALIF.	CSR-(CON.NO.)S(JOB NO.)-
AEROSPACE CORP. SAN BERNARDINO OPERATIONS, CALIF.	SA-(YEAR)-(NUMBERS)
AEROSPACE CORP. SAN BERNARDINO OPERATIONS, CALIF.	SE-(YEAR)-(NUMBERS)
AEROSPACE CORP. SAN BERNARDINO OPERATIONS, CALIF.	TDR-(CON.NO.)S(JOB NO.)-
AEROSPACE CORP. SYSTEMS RESEARCH AND PLANNING DIV., EL SEGUNDO, CALIF.	TDR-(NUMBERS)TN-
AEROSPACE CORP. VANDENBERG OFFICE	SV-(YEAR)-(NUMBERS)
AEROSPACE INDUSTRIES ASSN., LOS ANGELES	AIA-
AEROSPACE INDUSTRIES ASSN. OF AMERICA, INC., WASHINGTON, D.C.	EETC-
AEROSPACE MEDICAL DIV., BROOKS AFB, TEX.	AMD-
AEROSPACE MEDICAL DIV., BROOKS AFB, TEX.	AMD-CR-(NO)-(NO)-(YR.)-

AEROSPACE MEDICAL DIV.

AEROSPACE MEDICAL DIV., BROOKS AFB, TEX.	AMD-TDR-
AEROSPACE MEDICAL DIV., BROOKS AFB, TEX.	AMD-TR-
AEROSPACE MEDICAL DIV., BROOKS AFB, TEX.	AMD-TR-(YEAR)-
AEROSPACE MEDICAL DIV., BROOKS AFB, TEX.	AMD-TT-
AEROSPACE MEDICAL LAB (CLINICAL), LACKLAND AFB, TEX.	AMLC-TR-(YEAR)-
AEROSPACE MEDICAL RESEARCH LABS. (6570TH), WRIGHT-PATTERSON AFB, OHIO	AMD-AMRL-
AEROSPACE MEDICAL RESEARCH LABS. (6570TH), WRIGHT-PATTERSON AFB, OHIO	AMD-MRL-
AEROSPACE MEDICAL RESEARCH LABS. (6570TH), WRIGHT-PATTERSON AFB, OHIO	AMRL-MEMO-
AEROSPACE MEDICAL RESEARCH LABS. (6570TH), WRIGHT-PATTERSON AFB, OHIO	AMRL-TDR-
AEROSPACE MEDICAL RESEARCH LABS. (6570TH), WRIGHT-PATTERSON AFB, OHIO	AMRL-TR-(YEAR)-
AEROSPACE MEDICAL RESEARCH LABS. (6570TH), WRIGHT-PATTERSON AFB, OHIO	ARML-TR-(YEAR)-
AEROSPACE MEDICAL RESEARCH LABS. (6570TH), WRIGHT-PATTERSON AFB, OHIO	MRL-TDR-(YEAR)-
AEROSPACE MEDICAL RESEARCH LABS. (6570TH), WRIGHT-PATTERSON AFB, OHIO	WCRD-TM-(YEAR)-
AEROSPACE RESEARCH, INC., BOSTON, MASS.	ARI-
AEROSPACE RESEARCH LABS., WRIGHT-PATTERSON AFB, OHIO	ARL-
AEROSPACE RESEARCH LABS., WRIGHT-PATTERSON AFB, OHIO	ARL-(YR)-(NO.)-
AEROSPACE RESEARCH LABS., WRIGHT-PATTERSON AFB, OHIO	ARL-MR-(YEAR)-
AEROSPACE RESEARCH LABS., WRIGHT-PATTERSON AFB, OHIO	ARL-TDR-(YR)-
AEROSPACE RESEARCH LABS., WRIGHT-PATTERSON AFB, OHIO	ARL-TN-(YR)-
AEROSPACE RESEARCH LABS., WRIGHT-PATTERSON AFB, OHIO	ARL-TR-(NO.)(MONTH/YR.)
AEROSPACE RESEARCH LABS., WRIGHT-PATTERSON AFB, OHIO	ARL-TR-(YR.)-
AEROSPACE RESEARCH LABS., WRIGHT-PATTERSON AFB, OHIO	ARL-WSCI-(YEAR)-
AEROSPACE RESEARCH LABS., WRIGHT-PATTERSON AFB, OHIO	SRL-TR-
AEROSPACE TECHNICAL INTELLIGENCE CENTER, WRIGHT-PATTERSON AFB, OHIO	ASTIC-
AEROSPACE TECHNICAL INTELLIGENCE CENTER, WRIGHT-PATTERSON AFB, OHIO	ATIC TIS-(LTRS)(YR)(NO.)
AEROSPACE TECHNICAL INTELLIGENCE CENTER, WRIGHT-PATTERSON AFB, OHIO	T(YR.)-
AEROSPACE TECHNICAL INTELLIGENCE CENTER, WRIGHT-PATTERSON AFB, OHIO	TIS-WI-(YR.)-
AEROSPACE TECHNICAL INTELLIGENCE CENTER, WRIGHT-PATTERSON AFB, OHIO	TWP-WI-
AEROSPACE TECHNICAL INTELLIGENCE CENTER, WRIGHT-PATTERSON AFB, OHIO	TWP-WI-(YEAR)-
AEROTHERM CORP., MOUNTAIN VIEW, CALIF.	AEROTHERM-(YEAR)-
AEROTHERM CORP., MOUNTAIN VIEW, CALIF.	AEROTHERM-UM-(YEAR)-
AETRON-BLUME-ATKINSON, PALO ALTO, CALIF.	ABA-
AGBABIAN-JACOBSEN ASSOCIATES, LOS ANGELES	AJA-
AGBABIAN-JACOBSEN ASSOCIATES, LOS ANGELES	AJA-R-(NUMBER)-(NUMBER)
AGBABIAN-JACOBSEN ASSOCIATES, LOS ANG. (JT. VENTURE)	ATI-AJA-
AGENCY FOR INTERNATIONAL DEVELOPMENT. OFFICE OF PROGRAM AND POLICY COORDINATION, WASHINGTON, D.C.	AID-DP-
AGENCY FOR INTERNATIONAL DEVELOPMENT. OFFICE OF SCIENCE AND TECHNOLOGY, WASHINGTON, D.C.	TA/OST-(YEAR)-
AGENCY FOR INTERNATIONAL DEVELOPMENT. OFFICE OF SCIENCE AND TECHNOLOGY, WASHINGTON, D.C.	TA/OST-AN-(YEAR)-
AGRICULTURAL RESEARCH SERVICE, BELTSVILLE, MD.	AGRIC-
AGRICULTURAL RESEARCH SERVICE, BELTSVILLE, MD.	CR-
AGRICULTURAL RESEARCH SERVICE, BELTSVILLE, MD.	LARS INFORMATION NOTE-
AGRICULTURAL RESEARCH SERVICE. CROP RESEARCH DIV., WASHINGTON, D.C.	CR-ARS-(NUMBER-NUMBER)
AGRICULTURAL RESEARCH SERVICE. SOIL AND WATER CONSERVATION RESEARCH DIV., ITHACA, N.Y.	MICROCLIMATE-IR-
AGRICULTURAL RESEARCH SERVICE. SOIL AND WATER CONSERVATION RESEARCH DIV., ITHACA, N.Y.	USDA-ARS-SWC-
AIR ASSOCIATES, INC., TETERBORO, N.J.	AA-
AIR ATTACHE, LONDON	MSC-
AIR ATTACHE, OTTAWA	OAIRA-
AIR COMMAND AND STAFF COLL., MAXWELL AFB, ALA.	AIRU ACSC-
AIR COMMAND AND STAFF COLL., MAXWELL AFB, ALA.	SS-(NUMBER)-
AIR COORDINATING COMMITTEE, WASHINGTON, D.C.	ACC-
AIR DEFENSE COMMAND, ENT AFB, COLO.	ADC-
AIR DEFENSE COMMAND, ENT AFB, COLO.	ADC M-
AIR DEFENSE COMMAND, ENT AFB, COLO. (OPERATIONS ANALYSIS TECHNICAL MEMORANDUM)	ADC-OA-TM-
AIR DEFENSE COMMAND, ENT AFB, COLO. (OPERATIONS ANALYSIS WORKING PAPER)	ADC-OA-WP-
AIR DEFENSE COMMAND, ENT AFB, COLO.	ADGAS-
AIR DEFENSE WEAPONS CENTER, TYNDALL AFB, FLA.	F-(NUMBER)EC-(NO.)H
AIR DEFENSE WEAPONS CENTER, TYNDALL AFB, FLA.	TES-(YEAR)-
AIR DEVELOPMENT SQUADRON FIVE, CHINA LAKE, CALIF.	VX-(NUMBER)-AEN-
AIR EVALUATION BOARD, SOUTHWEST PACIFIC AREA	SWPA-
AIR FORCE (INTELLIGENCE REPORTS)	AAF IR-
AIR FORCE (NON-TECHNICAL INTELLIGENCE REPORTS)	AAF NTIR-
AIR FORCE (NON-TECHNICAL INTELLIGENCE REPORTS)	AAF NTIR E-
AIR FORCE (TECHNICAL INTELLIGENCE REPORTS)	AAF TIR-
AIR FORCE (TECHNICAL INTELLIGENCE REPORTS)	AAF TIR (LETTER)-
AIR FORCE	AF-
AIR FORCE (FOR EXPLANATION OF THE REMAINDER OF THE CODE, SEE ENTRY UNDER LETTERS FOLLOWING AF-)	AF-(LETTERS)-
AIR FORCE (INTERIM OPERATING INSTRUCTIONS)	AF-IOI (MONTH/YEAR)
AIR FORCE (TECHNICAL INFORMATION FILE)	AFTIF-
AIR FORCE (TECHNICAL ORDER) (REF. 14)	AF-TO-11N-
AIR FORCE (AIR NAVIGATION CHARTS)	AN-
AIR FORCE (APPLIED RESEARCH PLANNING DOCUMENT)	ARPD-
AIR FORCE (SPECIAL AERONAUTICAL CHARTS)	AS-
AIR FORCE (AIRFIELD + SEAPLANE STAS. OF THE WORLD)	ASSOTW-
AIR FORCE (COMMUNICATIONS OPERATING INSTRUCTIONS)	COI-
AIR FORCE	DPS-SA-(NUMBERS)-
AIR FORCE (EQUIPMENT COMPONENT LIST)	ECL-(NO.)-(NO.)-(NO.)
AIR FORCE (FLIGHT CHARTS)	FC-
AIR FORCE (LONG RANGE NAVIGATION CHARTS)	LR-
AIR FORCE (NUCLEAR ELECTRONIC EFFECTS PROGRAM)	NEEP-
AIR FORCE (NAVIGATIONAL FLIGHT CHARTS)	NF-
AIR FORCE (SPECIAL AIR NAVIGATION)	NS-
AIR FORCE (OPERATIONS ANALYSIS WORKING PAPER)	OA-
AIR FORCE (OPERATIONS ANALYSIS TECHNICAL MEMORANDUM)	OATM-
AIR FORCE	OPE-
AIR FORCE	OPX-
AIR FORCE (OPERATIONS REQUIREMENT)	OR-
AIR FORCE (PILOTAGE CHARTS)	PC-

AIR FORCE (PHYSICAL VULNERABILITY ABSTRACTS)	PVA-
AIR FORCE (PHYSICAL VULNERABILITY INTERIM MEMO.)	PVIM-
AIR FORCE (PHYSICAL VULNERABILITY TECHNICAL MEMO.)	PVTM-
AIR FORCE (RADAR CHARTS)	RC-
AIR FORCE	S-(NUMBER)-D
AIR FORCE (SPECIAL LIST OF EQUIPMENT)	SLOE-
AIR FORCE (STRATEGIC PLANNING CHARTS)	SP-
AIR FORCE (SPECIAL PHOTO-INTERPRETATION REPORT)	SPIR-
AIR FORCE (TARGET DATA INVENTORY)	TDI-
AIR FORCE (TECHNICAL INFORMATION FILE)	TIF-
AIR FORCE (TECHNICAL INTELLIGENCE REPORTS)	TIR-
AIR FORCE (TARGET INFORMATION SHEET)	TIS-
AIR FORCE	T.O.-
AIR FORCE	TO-
AIR FORCE (TECH ORDER) (REF. 14)	T.O.11N-
AIR FORCE (TECHNICAL PROGRAM PLANNING DOCUMENT)	TPPD-
AIR FORCE	USAF T.O.-
AIR FORCE (TECH ORDER) (REF. 14)	USAF T.O.11N-
AIR FORCE (WEAPONS RECOMMENDATION SHEET)	WRS-
AIR FORCE (WEAPON SYSTEM STOCK CONTROL LIST)	WSSCL-
AIR FORCE, WASHINGTON, D.C. (REF. 1)	(NUMBERS...)
AIR FORCE, WASHINGTON, D.C.	AAF-
AIR FORCE, WASHINGTON, D.C. (MANUAL)	AF-M-
AIR FORCE, WASHINGTON, D.C. (MANUAL)	AFM-
AIR FORCE, WASHINGTON, D.C. (PAMPHLET)	AFP-
AIR FORCE, WASHINGTON, D.C. (REGULATION)	AFR-
AIR FORCE, WASHINGTON, D.C.	AF-TDR-
AIR FORCE, WASHINGTON, D.C.	AF-TR-
AIR FORCE, WASHINGTON, D.C. (TECHNICAL REPORT)	AFTR-
AIR FORCE, WASHINGTON, D.C. (BOMBING ENCYCLOPEDIA)	BE-
AIR FORCE, WASHINGTON, D.C.	BPE-
AIR FORCE, WASHINGTON, D.C.	BPX-
AIR FORCE, WASHINGTON, D.C.	CSA-
AIR FORCE, WASHINGTON, D.C.	H-
AIR FORCE, WASHINGTON, D.C.	INC/C(NUMBER)/P(NUMBER)
AIR FORCE, WASHINGTON, D.C.	OFW...
AIR FORCE, WASHINGTON, D.C.	PA-(NO.)-
AIR FORCE, WASHINGTON, D.C.	PE-(NO.)-
AIR FORCE, WASHINGTON, D.C.	PX-(NO.)-
AIR FORCE, WASHINGTON, D.C.	TR-
AIR FORCE, WASHINGTON, D.C.	USAF-
AIR FORCE, WASHINGTON, D.C. (AIR FORCE MANUAL)	USAF-AFM(NUMBER)-
AIR FORCE, WASHINGTON, D.C. (OPERATIONS ANALYSIS)	USAF OA R-
AIR FORCE, WASH.,D.C.(OPERATIONS ANALYSIS SPEC. REPT)	USAF OA SR-
AIR FORCE, WASHINGTON, D.C. (OPERATIONS ANALYSIS SPECIAL TECHNICAL MEMORANDUM)	USAF OA STM-
AIR FORCE, WASH.,D.C.(OPERATIONS ANALYSIS TECH. MEMO)	USAF OA TM-
AIR FORCE, WASHINGTON, D.C. (OPERATIONS ANALYSIS WORKING PAPER)	USAF OA WP-
AIR FORCE, WASHINGTON, D.C. (STATISTICAL DIGEST)	USAF-SD-(YEAR)
AIR FORCE, WASHINGTON, D.C.	USAF-TR-(YEAR)-
AIR FORCE, WASHINGTON, D.C.	WORKING PAPER-
AIR FORCE, WRIGHT-PATTERSON AFB, OHIO	AFWP-
AIR FORCE. 5TH. OPERATIONS ANALYSIS OFFICE	TARS-
AIR FORCE. 6TH AIR DIV., MAC DILL AFB, FLA.	DOES-
AIR FORCE. 6TH AIR DIV., MAC DILL AFB, FLA.	OES-
AIR FORCE. 12TH, WACO, TEX.	ORD-CVS-
AIR FORCE. TUSLOG DETACHMENT 36, USAFE, N.Y.C.	DET-36-(YEAR)-
AIR FORCE. TUSLOG DETACHMENT 36, USAFE, N.Y.C.	DET-36-TR-(YEAR)-
AIR FORCE. TUSLOG DETACHMENT 36, USAFE, N.Y.C.	DET-36-PR-(YEAR)-
AIR FORCE ACADEMY, COLORADO SPRINGS	USAFA-TR-(YEAR)-
AIR FORCE ACADEMY, COLORADO SPRINGS. FRANK J. SEILER RESEARCH LAB.	FJSRL-(YEAR)-
AIR FORCE ACADEMY, COLORADO SPRINGS. FRANK J. SEILER RESEARCH LAB.	SRL-
AIR FORCE ACADEMY, COLORADO SPRINGS. FRANK J. SEILER RESEARCH LAB.	SRL-TR-(YEAR)-
AIR FORCE AERO-PROPULSION LAB., WRIGHT-PATTERSON AFB, OHIO	AFAPL-
AIR FORCE AERO-PROPULSION LAB., WRIGHT-PATTERSON AFB, OHIO	AFAPL-CONF-(YEAR)-
AIR FORCE AERO-PROPULSION LAB., WRIGHT-PATTERSON AFB, OHIO	AFAPL-TDR-(NUMBERS)-
AIR FORCE AERO-PROPULSION LAB., WRIGHT-PATTERSON AFB, OHIO	AFAPL-TR-(YEAR)-
AIR FORCE AERO-PROPULSION LAB., WRIGHT-PATTERSON AFB, OHIO	APIT-TM-(YEAR)
AIR FORCE AERO-PROPULSION LAB., WRIGHT-PATTERSON AFB, OHIO	APL-TDR-(YEAR)-
AIR FORCE AERO-PROPULSION LAB., WRIGHT-PATTERSON AFB, OHIO	APL-TR-(YEAR)-
AIR FORCE ARMAMENT CENTER, EGLIN AFB, FLA.	AFAC TM-(YEAR)-
AIR FORCE ARMAMENT CENTER, EGLIN AFB, FLA.	AFAC TN-(YEAR)-
AIR FORCE ARMAMENT CENTER, EGLIN AFB, FLA.	AFAC TR-(YEAR)-
AIR FORCE ARMAMENT LAB., EGLIN AFB, FLA.	AFATL-ATRA-TN-(YEAR)-
AIR FORCE ARMAMENT LAB., EGLIN AFB, FLA.	AFATL-MR-(YEAR)-
AIR FORCE ARMAMENT LAB., EGLIN AFB, FLA.	AFATL-TR-(YEAR)-
AIR FORCE ARMAMENT LAB., EGLIN AFB, FLA.	ATL-
AIR FORCE ARMAMENT LAB., EGLIN AFB, FLA.	ATL-TDR-(YEAR)-
AIR FORCE ARMAMENT LAB., EGLIN AFB, FLA.	ATL-TR-(YEAR)-
AIR FORCE AVIONICS LAB., WRIGHT-PATTERSON AFB, OHIO	AFAL-(NUMBERS)-
AIR FORCE AVIONICS LAB., WRIGHT-PATTERSON AFB, OHIO	AFAL-IR-
AIR FORCE AVIONICS LAB., WRIGHT-PATTERSON AFB, OHIO	AFAL-TDR-(YEAR)-
AIR FORCE AVIONICS LAB., WRIGHT-PATTERSON AFB, OHIO (TECHNICAL PROGRAMS AND CONTACTS)	AFAL-TPC (MONTH/YEAR)
AIR FORCE AVIONICS LAB., WRIGHT-PATTERSON AFB, OHIO	AFAL-TR-(YEAR)-
AIR FORCE AVIONICS LAB., WRIGHT-PATTERSON AFB, OHIO	AL-
AIR FORCE AVIONICS LAB., WRIGHT-PATTERSON AFB, OHIO	AL-TDR-(YEAR)-
AIR FORCE AVIONICS LAB., WRIGHT-PATTERSON AFB, OHIO	AL-TR-(YEAR)-
AIR FORCE AVIONICS LAB., WRIGHT-PATTERSON AFB, OHIO	AVTM-TR-
AIR FORCE AVIONICS LAB. RECONNAISSANCE APPLICATIONS BRANCH, WRIGHT-PATTERSON AFB, OHIO	RC-
AIR FORCE AVIONICS LAB. RESEARCH AND TECHNOLOGY DIV., HOLLOMAN AFB, N. MEX.	AVNI-TR-(YEAR)-
AIR FORCE CAMBRIDGE RESEARCH CENTER, BEDFORD, MASS.	AFCL-
AIR FORCE CAMBRIDGE RESEARCH CENTER, BEDFORD, MASS. (AUTOBAROTROPIC FLOW PROJECT, SCIENTIFIC REPORT)	AFCRC-AFP-SR-
AIR FORCE CAMBRIDGE RESEARCH CENTER, BEDFORD, MASS. (GEOPHYSICAL RESEARCH PAPER)	AFCRC-GRP-
AIR FORCE CAMBRIDGE RESEARCH CENTER, BEDFORD, MASS.	AFCRC-TM-(YEAR)-
AIR FORCE CAMBRIDGE RESEARCH CENTER, BEDFORD, MASS.	AFCRC-TN-(YEAR)-
AIR FORCE CAMBRIDGE RESEARCH CENTER, BEDFORD, MASS.	AFCRC-TR-(YEAR)-

AIR FORCE CAMBRIDGE RESEARCH CENTER

AIR FORCE CAMBRIDGE RESEARCH CENTER, MASS.	AFCRC-
AIR FORCE CAMBRIDGE RESEARCH CENTER, MASS. (AIR FORCE SURVEYS IN GEOPHYSICS)	AFCRC AFSG-
AIR FORCE CAMBRIDGE RESEARCH CENTER, MASS.	CFS-
AIR FORCE CAMBRIDGE RESEARCH CENTER, MASS.	GR-TR-
AIR FORCE CAMBRIDGE RESEARCH CENTER, MASS.	WLC-
(SEE ALSO LATER NAME: AIR FORCE CAMBRIDGE RES. LABS.)	
AIR FORCE CAMBRIDGE RESEARCH CENTER. ELECTRONICS RESEARCH DIRECTORATE, MASS.	AFCRC-E-
AIR FORCE CAMBRIDGE RESEARCH CENTER. ELECTRONICS RESEARCH DIRECTORATE, MASS.	ERD-
AIR FORCE CAMBRIDGE RESEARCH CENTER. GEOPHYSICS RESEARCH DIRECTORATE, BEDFORD, MASS.	GRD-TN-(YEAR)-
AIR FORCE CAMBRIDGE RESEARCH CENTER. GEOPHYSICS RESEARCH DIRECTORATE, BEDFORD, MASS.	GRD-TR-
AIR FORCE CAMBRIDGE RESEARCH CENTER. GEOPHYSICS RESEARCH DIRECTORATE, MASS.	AFCRC-GRD-
AIR FORCE CAMBRIDGE RESEARCH CENTER. GEOPHYSICS RESEARCH DIRECTORATE, MASS.	GEOPHYSICS RES. PAPERS-
AIR FORCE CAMBRIDGE RESEARCH CENTER. GEOPHYSICS RESEARCH DIRECTORATE, MASS.	GRD-
AIR FORCE CAMBRIDGE RESEARCH CENTER. GEOPHYSICS RES. DIRECTORATE, MASS. (GEOPHYSICAL RES. PAPER)	GRP-
AIR FORCE CAMBRIDGE RESEARCH CENTER. GEOPHYSICS RESEARCH DIRECTORATE, MASS.	N-
(SEE ALSO LATER NAME: AIR FORCE CAMBRIDGE RESEARCH LABS. METEOROLOGY LAB., BEDFORD, MASS.)	
AIR FORCE CAMBRIDGE RESEARCH CENTER. OPERATIONAL APPLICATIONS LAB., WASHINGTON, D.C.	OAL-
AIR FORCE CAMBRIDGE RESEARCH LABS., BEDFORD, MASS.	AAF CRL (NUMBER)-
AIR FORCE CAMBRIDGE RESEARCH LABS., BEDFORD, MASS.	AAF CRL E-
AIR FORCE CAMBRIDGE RESEARCH LABS., BEDFORD, MASS.	AAF ERL-
AIR FORCE CAMBRIDGE RESEARCH LABS., BEDFORD, MASS.	AFCRC-TR-(NUMBERS)-
AIR FORCE CAMBRIDGE RESEARCH LABS., BEDFORD, MASS.	AFCRL-
AIR FORCE CAMBRIDGE RESEARCH LABS., BEDFORD, MASS.	AFCRL-(YEAR)-
AIR FORCE CAMBRIDGE RESEARCH LABS., BEDFORD, MASS.	AFCRL-AFSG-
AIR FORCE CAMBRIDGE RESEARCH LABS., BEDFORD, MASS.	AFCRL-AFSIG-
AIR FORCE CAMBRIDGE RESEARCH LABS., BEDFORD, MASS.	AF-CRL-E-
AIR FORCE CAMBRIDGE RESEARCH LABS., BEDFORD, MASS.	AFCRL-ERP-
AIR FORCE CAMBRIDGE RESEARCH LABS., BEDFORD, MASS.	AFCRL-GRD-TR-
AIR FORCE CAMBRIDGE RESEARCH LABS., BEDFORD, MASS. (INSTRUMENTATION PAPER)	AFCRL-IP-
AIR FORCE CAMBRIDGE RESEARCH LABS., BEDFORD, MASS. (PHYSICAL AND MATHEMATICAL RESEARCH PAPER)	AFCRL-PMSRP-
AIR FORCE CAMBRIDGE RESEARCH LABS., BEDFORD, MASS. (PHYSICAL SCIENCE RESEARCH PAPER)	AFCRL-PRSP-
AIR FORCE CAMBRIDGE RESEARCH LABS., BEDFORD, MASS. (PHYSICAL SCIENCE RESEARCH PAPER)	AFCRL-PSRP-
AIR FORCE CAMBRIDGE RESEARCH LABS., BEDFORD, MASS.	AFCRL-SP-
AIR FORCE CAMBRIDGE RESEARCH LABS., BEDFORD, MASS. (SPECIAL REPORT)	AFCRL-SR-
AIR FORCE CAMBRIDGE RESEARCH LABS., BEDFORD, MASS.	AFCRL-TM-
AIR FORCE CAMBRIDGE RESEARCH LABS., BEDFORD, MASS.	AFCRL-TN-(YEAR)-
AIR FORCE CAMBRIDGE RESEARCH LABS., BEDFORD, MASS.	AFCRL-TR-(YEAR)-
AIR FORCE CAMBRIDGE RESEARCH LABS., BEDFORD, MASS. (TRANSLATION)	AFCRL-TRANS-
AIR FORCE CAMBRIDGE RESEARCH LABS., BEDFORD, MASS.	ARCRL-TRANS-
AIR FORCE CAMBRIDGE RESEARCH LABS., BEDFORD, MASS.	CRL-
AIR FORCE CAMBRIDGE RESEARCH LABS., BEDFORD, MASS.	CRO-SR-
AIR FORCE CAMBRIDGE RESEARCH LABS., BEDFORD, MASS.	E-
AIR FORCE CAMBRIDGE RESEARCH LABS., BEDFORD, MASS.	ERD-CRRD-TM-
AIR FORCE CAMBRIDGE RESEARCH LABS., BEDFORD, MASS.	ERD-ORRD-TM-
AIR FORCE CAMBRIDGE RESEARCH LABS., BEDFORD, MASS.	ERL-
AIR FORCE CAMBRIDGE RESEARCH LABS., BEDFORD, MASS.	ERP-
AIR FORCE CAMBRIDGE RESEARCH LABS., BEDFORD, MASS.	ET-
AIR FORCE CAMBRIDGE RESEARCH LABS., BEDFORD, MASS.	FRP-
AIR FORCE CAMBRIDGE RESEARCH LABS., BEDFORD, MASS. (INSTRUMENTATION PAPER)	IP-
AIR FORCE CAMBRIDGE RESEARCH LABS., BEDFORD, MASS. (PHYSICAL AND MATHEMATICAL RESEARCH PAPER)	PMSRP-
AIR FORCE CAMBRIDGE RESEARCH LABS., BEDFORD, MASS.	SIG-
AIR FORCE CAMBRIDGE RESEARCH LABS., BEDFORD, MASS.	SPECIAL REPORTS-
AIR FORCE CAMBRIDGE RESEARCH LABS. ELECTRONICS RESEARCH DIRECTORATE, BEDFORD, MASS.	EERD-CRRC-TM-
AIR FORCE CAMBRIDGE RESEARCH LABS. ELECTRONICS RESEARCH DIRECTORATE, BEDFORD, MASS.	ERD-TN-(YEAR)-
AIR FORCE CAMBRIDGE RESEARCH LABS. METEOROLOGY LAB., BEDFORD, MASS. (AIR FORCE SURVEYS IN GEOPHYSICS)	AFSG-
AIR FORCE CAMBRIDGE RESEARCH LABS. METEOROLOGY LAB., BEDFORD, MASS. (AIR FORCE SURVEYS IN GEOPHYSICS)	AFSIG-
AIR FORCE CAMBRIDGE RESEARCH LABS. MICROWAVE PHYSICS LAB., BEDFORD, MASS. (PHYSICAL SCIENCE RES. PAPERS)	PSRP-
AIR FORCE COMMUNICATIONS SERVICE, SCOTT AFB, ILL.	AFCS-
AIR FORCE COMMUNICATIONS SERVICE, SCOTT AFB, ILL.	AFCS-OPLAN-
AIR FORCE COMMUNICATIONS SERVICE, SCOTT AFB, ILL.	AFCS-SS-
AIR FORCE COMMUNICATIONS SERVICE, SCOTT AFB, ILL.	AFCS-UM-
AIR FORCE COMMUNICATIONS SERVICE. OPERATIONS RESEARCH ANALYSIS OFFICE, SCOTT AFB, ILL.	(NUMBER)-ORR-
AIR FORCE EASTERN TEST RANGE, PATRICK AFB, FLA.	AFETR-
AIR FORCE EASTERN TEST RANGE, PATRICK AFB, FLA.	AFETR-NRTOI-
AIR FORCE EASTERN TEST RANGE, PATRICK AFB, FLA.	AFETR-TR-
AIR FORCE EASTERN TEST RANGE, PATRICK AFB, FLA.	ET-
AIR FORCE EASTERN TEST RANGE, PATRICK AFB, FLA.	ET-(YEAR)-
AIR FORCE ENVIRONMENTAL TECHNICAL APPLICATIONS CENTER, WASHINGTON, D.C.	USAF-ETAC-TN-(YEAR)-
AIR FORCE EPIDEMIOLOGICAL LAB., LACKLAND AFB, TEX.	USAFEL-SR-
AIR FORCE EXPERIMENTAL ROCKET ENGINE TEST STATION, EDWARDS AFB, CALIF.	FTDE-
AIR FORCE FILM LIBRARY CENTER, ST. LOUIS	TF-(NO.)-(NO.)P(FILM)
AIR FORCE FLIGHT DYNAMICS LAB., WRIGHT-PATTERSON AFB, OHIO	AFDL-TR-(YEAR)-
AIR FORCE FLIGHT DYNAMICS LAB., WRIGHT-PATTERSON AFB, OHIO	AFFDL-
AIR FORCE FLIGHT DYNAMICS LAB., WRIGHT-PATTERSON AFB, OHIO	AFFDL-FDCL-TM-(YEAR)-
AIR FORCE FLIGHT DYNAMICS LAB., WRIGHT-PATTERSON AFB, OHIO	AFFDL-FDFM-
AIR FORCE FLIGHT DYNAMICS LAB., WRIGHT-PATTERSON AFB, OHIO	AFFDL-FDFR-TM-
AIR FORCE FLIGHT DYNAMICS LAB., WRIGHT-PATTERSON AFB, OHIO	AFFDL-FDTR-TM-
AIR FORCE FLIGHT DYNAMICS LAB., WRIGHT-PATTERSON AFB, OHIO	AFFDL-TDR-(YEAR)-
AIR FORCE FLIGHT DYNAMICS LAB., WRIGHT-PATTERSON AFB, OHIO	AFFDL-TR-
AIR FORCE FLIGHT DYNAMICS LAB., WRIGHT-PATTERSON AFB, OHIO	ASRMDS-TM-
AIR FORCE FLIGHT DYNAMICS LAB., WRIGHT-PATTERSON AFB, OHIO	FDC/ADPO-TM-(YEAR)-
AIR FORCE FLIGHT DYNAMICS LAB., WRIGHT-PATTERSON AFB, OHIO	FDL-
AIR FORCE FLIGHT DYNAMICS LAB., WRIGHT-PATTERSON AFB, OHIO	FDL-TDR-(YEAR)-
AIR FORCE FLIGHT DYNAMICS LAB., WRIGHT-PATTERSON AFB, OHIO	FDL-TR-(YEAR)-
AIR FORCE FLIGHT DYNAMICS LAB. CONTROL CRITERIA BRANCH, WRIGHT-PATTERSON AFB, OHIO	AFFDC-FDCC-TM-
AIR FORCE FLIGHT DYNAMICS LAB. VEHICLE DYNAMICS DIV., WRIGHT-PATTERSON AFB, OHIO	AFFDL-FDD-(YEAR)-
AIR FORCE FLIGHT TEST CENTER, EDWARDS AFB, CALIF.	AAFTC TN (YR.)-
AIR FORCE FLIGHT TEST CENTER, EDWARDS AFB, CALIF.	AFFTC-SD-
AIR FORCE FLIGHT TEST CENTER, EDWARDS AFB, CALIF.	AFFTC-SP-
AIR FORCE FLIGHT TEST CENTER, EDWARDS AFB, CALIF.	AFFTC-TD-(YEAR)-
AIR FORCE FLIGHT TEST CENTER, EDWARDS AFB, CALIF. (TECHNICAL DOCUMENTARY REPORT)	AFFTC-TDR-

AIR FORCE FLIGHT TEST CENTER, EDWARDS AFB, CALIF.	AFFTC-TIH-
AIR FORCE FLIGHT TEST CENTER, EDWARDS AFB, CALIF.	AFFTC-TM-(YEAR)-
AIR FORCE FLIGHT TEST CENTER, EDWARDS AFB, CALIF.	AFFTC-TN-(YR)-
AIR FORCE FLIGHT TEST CENTER, EDWARDS AFB, CALIF.	AFFTC-TR-
AIR FORCE FLIGHT TEST CENTER, EDWARDS AFB, CALIF.	AFFTC-TR-(YR)-
AIR FORCE FLIGHT TEST CENTER, EDWARDS AFB, CALIF.	FTC-
AIR FORCE FLIGHT TEST CENTER, EDWARDS AFB, CALIF.	FTC-TDR-
AIR FORCE FLIGHT TEST CENTER, EDWARDS AFB, CALIF.	FTC-TR-(YEAR)-
AIR FORCE FLIGHT TEST CENTER, EDWARDS AFB, CALIF.	P-
AIR FORCE GLOBAL WEATHER CENTRAL, OFFUTT AFB, NEB.	AFGWC-TM-
AIR FORCE HUMAN RESOURCES LAB., WRIGHT-PATTERSON AFB, OHIO	AFHRL-TR
AIR FORCE INST. OF TECH., WRIGHT-PATTERSON AFB, OHIO	AF-IT-
AIR FORCE INST. OF TECH., WRIGHT-PATTERSON AFB, OHIO	AFIT-
AIR FORCE INST. OF TECH., WRIGHT-PATTERSON AFB, OHIO (WORK DONE AT COLL. OF AERONAUTICS, CRANFIELD, ENG)	AFIT-CRANFIELD/EE
AIR FORCE INST. OF TECH., WRIGHT-PATTERSON AFB, OHIO	AFIT GAE-
AIR FORCE INST. OF TECH., WRIGHT-PATTERSON AFB, OHIO	AFIT GAS-(YEAR)-
AIR FORCE INST. OF TECH., WRIGHT-PATTERSON AFB, OHIO	AFIT-GNE-
AIR FORCE INST. OF TECH., WRIGHT-PATTERSON AFB, OHIO	GACA-
AIR FORCE INST. OF TECH., WRIGHT-PATTERSON AFB, OHIO	GAE-
AIR FORCE INST. OF TECH., WRIGHT-PATTERSON AFB, OHIO	GA/EE/(YEAR)-
AIR FORCE INST. OF TECH., WRIGHT-PATTERSON AFB, OHIO	GAE/ME/(YEAR)-
AIR FORCE INST. OF TECH., WRIGHT-PATTERSON AFB, OHIO	GAE/PHYS/(YEAR)-
AIR FORCE INST. OF TECH., WRIGHT-PATTERSON AFB, OHIO	GAM-(YEAR)/(LTRS.)/
AIR FORCE INST. OF TECH., WRIGHT-PATTERSON AFB, OHIO	GA/ME/(YEAR)-
AIR FORCE INST. OF TECH., WRIGHT-PATTERSON AFB, OHIO	GA/MECH/(YEAR)-
AIR FORCE INST. OF TECH., WRIGHT-PATTERSON AFB, OHIO	GAO-
AIR FORCE INST. OF TECH., WRIGHT-PATTERSON AFB, OHIO	GA/PH/(YEAR)-
AIR FORCE INST. OF TECH., WRIGHT-PATTERSON AFB, OHIO	GA/PHYS/(YEAR)-
AIR FORCE INST. OF TECH., WRIGHT-PATTERSON AFB, OHIO	GAS-
AIR FORCE INST. OF TECH., WRIGHT-PATTERSON AFB, OHIO	GE/EE/(YEAR)S-
AIR FORCE INST. OF TECH., WRIGHT-PATTERSON AFB, OHIO	GSC-
AIR FORCE INST. OF TECH., WRIGHT-PATTERSON AFB, OHIO	SP/(LTRS)/(YR)-
AIR FORCE INST. OF TECH., WRIGHT-PATTERSON AFB, OHIO	USAFIT-TR-
AIR FORCE INST. OF TECH., WRIGHT-PATTERSON AFB, OHIO. GRADUATE SCHOOL OF BUSINESS (RESEARCH PROJECT REPT)	AFITB-RPR-
AIR FORCE INST. OF TECH., WRIGHT-PATTERSON AFB, OHIO. GRADUATE SCHOOL OF BUSINESS (SEMINAR STUDY)	AFITB-SS-
AIR FORCE INST. OF TECH., WRIGHT-PATTERSON AFB, OHIO. SCHOOL OF ENGINEERING	AFIT-SEP-
AIR FORCE INST. OF TECH., WRIGHT-PATTERSON AFB, OHIO. SCHOOL OF ENGINEERING	AFIT-TR-
AIR FORCE INST. OF TECH., WRIGHT-PATTERSON AFB, OHIO. SCHOOL OF ENGINEERING	DS/EE/(YEAR)-
AIR FORCE INST. OF TECH., WRIGHT-PATTERSON AFB, OHIO. SCHOOL OF ENGINEERING	DS/MC/(YEAR)-
AIR FORCE INST. OF TECH., WRIGHT-PATTERSON AFB, OHIO. SCHOOL OF ENGINEERING (GRADUATE ASTRONAUTICS)	GA/(LTRS.)/(YEAR)-
AIR FORCE INST. OF TECH., WRIGHT-PATTERSON AFB, OHIO. SCHOOL OF ENGINEERING	GAI/(LTRS.)/(YEAR)-
AIR FORCE INST. OF TECH., WRIGHT-PATTERSON AFB, OHIO. SCHOOL OF ENGINEERING (GRADUATE AEROSPACE)	GAM/(LTRS.)/(YEAR)-
AIR FORCE INST. OF TECH., WRIGHT-PATTERSON AFB, OHIO. SCHOOL OF ENGINEERING (GRADUATE AIR WEAPONS)	GAW/(LTRS.)/(YEAR)-
AIR FORCE INST. OF TECH., WRIGHT-PATTERSON AFB, OHIO. SCHOOL OF ENGINEERING (GRADUATE ELECTRONICS)	GE/(LTRS.)/(YEAR)-
AIR FORCE INST. OF TECH., WRIGHT-PATTERSON AFB, OHIO. SCHOOL OF ENGINEERING (GRADUATE GUIDANCE + CONTROL)	GGC/(LTRS.)/(YEAR)-
AIR FORCE INST. OF TECH., WRIGHT-PATTERSON AFB, OHIO. SCHOOL OF ENGINEERING (GRADUATE MECHANICAL ENG.)	GME/(LTRS.)/(YEAR)-
AIR FORCE INST. OF TECH., WRIGHT-PATTERSON AFB, OHIO. SCHOOL OF ENGINEERING (GRADUATE NUCLEAR ENG.)	GNE/(LTRS.)/(YEAR)-
AIR FORCE INST. OF TECH., WRIGHT-PATTERSON AFB, OHIO. SCHOOL OF ENGINEERING	GNE-PH-
AIR FORCE INST. OF TECH., WRIGHT-PATTERSON AFB, OHIO. SCHOOL OF ENGINEERING (GRADUATE NUCLEAR ENG.)	GNE-PHYS/(YEAR)-
AIR FORCE INST. OF TECH., WRIGHT-PATTERSON AFB, OHIO. SCHOOL OF ENGINEERING (GRADUATE RELIABILITY ENG.)	GRE/(LTRS.)/(YEAR)-
AIR FORCE INST. OF TECH., WRIGHT-PATTERSON AFB, OHIO. SCHOOL OF ENGINEERING	GS/(LTRS.)/(YEAR)-
AIR FORCE INST. OF TECH., WRIGHT-PATTERSON AFB, OHIO SCHOOL OF ENGINEERING	GSA/SM/(YEAR)-
AIR FORCE INST. OF TECH., WRIGHT-PATTERSON AFB, OHIO. SCHOOL OF ENGINEERING (GRADUATE SPACE FACILITIES)	GSF/(LTRS.)/(YEAR)-
AIR FORCE INST. OF TECH., WRIGHT-PATTERSON AFB, OHIO. SCHOOL OF ENGINEERING (GRADUATE SYSTEMS MANAGEMENT)	GSM/(LTRS.)/(YEAR)-
AIR FORCE INST. OF TECH., WRIGHT-PATTERSON AFB, OHIO. SCHOOL OF ENGINEERING (GRADUATE SPACE PHYSICS)	GSP/(LTRS.)/(YEAR)-
AIR FORCE INST. OF TECH., WRIGHT-PATTERSON AFB, OHIO. SCHOOL OF ENGINEERING	MECH-GSF-(YEAR)-
AIR FORCE INST. OF TECH., WRIGHT-PATTERSON AFB, OHIO. SCHOOL OF LOGISTICS (ADVANCED LOGISTICS REPORT)	AFITL-ALR-
AIR FORCE INST. OF TECH., WRIGHT-PATTERSON AFB, OHIO. SCHOOL OF LOGISTICS (ADVANCED LOGISTICS REPORT)	AFITL-SRP-
AIR FORCE INST. OF TECH., WRIGHT-PATTERSON AFB, OHIO. SCHOOL OF SYSTEMS AND LOGISTICS	AFIT-SL-
AIR FORCE INST. OF TECH., WRIGHT-PATTERSON AFB, OHIO. SCHOOL OF SYSTEMS AND LOGISTICS	SLSR-
AIR FORCE INST. OF TECH., WRIGHT-PATTERSON AFB, OHIO. SCHOOL OF SYSTEMS AND LOGISTICS	SLTR-(NUMBER-YEAR)
AIR FORCE INSTRUMENT PILOT INSTRUCTOR SCHOOL, RANDOLPH AFB, TEXAS	IPIS-RAFB-TN-(YR.)-
AIR FORCE LOGISTICS COMMAND, WRIGHT-PATTERSON AFB, OHIO	AFLC-(YEAR)-
AIR FORCE LOGISTICS COMMAND, WRIGHT-PATTERSON AFB, OHIO	AFLCP-(NUMBER)-
AIR FORCE LOGISTICS COMMAND, WRIGHT-PATTERSON AFB, OHIO	AFLC-TR-(NUMBER-NUMBER)
AIR FORCE LOGISTICS COMMAND, WRIGHT-PATTERSON AFB, OHIO	GSE/MC/(YEAR)-
AIR FORCE LOGISTICS COMMAND, WRIGHT-PATTERSON AFB, OHIO	OAR-
AIR FORCE LOGISTICS COMMAND, WRIGHT-PATTERSON AFB, OHIO	OAWP-
AIR FORCE LOGISTICS COMMAND. DIRECTORATE OF AEROSPACE FUELS, KELLY AFB, TEX.	OQF-
AIR FORCE LOGISTICS COMMAND. DIRECTORATE OF COST ANALYSIS, WRIGHT-PATTERSON AFB, OHIO	MCCC-(YEAR)-
AIR FORCE LOGISTICS COMMAND. LOGISTICS SIMULATION CENTER, WRIGHT-PATTERSON AFB, OHIO	MR-
AIR FORCE LOGISTICS COMMAND. MAINTENANCE TECHNOLOGY OFFICE, TINKER AFB, OKLA.	MTO-TR-(YEAR)-
AIR FORCE MACHINABILITY DATA CENTER, CINCINNATI	AFMDC-(YEAR)-
AIR FORCE MAINTENANCE LAB., LOWRY AFB, COLO.	ML-LOWRY-TM-(YEAR)-
AIR FORCE MATERIALS LAB., WRIGHT-PATTERSON AFB, OHIO	AFML-
AIR FORCE MATERIALS LAB., WRIGHT-PATTERSON AFB, OHIO (ABSTRACTS OF ACTIVE CONTRACTS)	AFML-AC (MONTH/YEAR)
AIR FORCE MATERIALS LAB., WRIGHT-PATTERSON AFB, OHIO	AFML-IR-
AIR FORCE MATERIALS LAB., WRIGHT-PATTERSON AFB, OHIO	AFML-MAN-(YEAR)-
AIR FORCE MATERIALS LAB., WRIGHT-PATTERSON AFB, OHIO	AFML-RP-(NUMBER)-
AIR FORCE MATERIALS LAB., WRIGHT-PATTERSON AFB, OHIO	AFML-TDR-
AIR FORCE MATERIALS LAB., WRIGHT-PATTERSON AFB, OHIO	AFML-TM-MAN-(YEAR)-
AIR FORCE MATERIALS LAB., WRIGHT-PATTERSON AFB, OHIO	AFML-TR-
AIR FORCE MATERIALS LAB., WRIGHT-PATTERSON AFB, OHIO	AFML-TR-(YEAR)-
AIR FORCE MATERIALS LAB., WRIGHT-PATTERSON AFB, OHIO	AMFL-TR-(YEAR)-
AIR FORCE MATERIALS LAB., WRIGHT-PATTERSON AFB, OHIO	ASTROPOWER-(YEAR)-
AIR FORCE MATERIALS LAB., WRIGHT-PATTERSON AFB, OHIO	IR-(NUMBER)-(NUMBER)
AIR FORCE MATERIALS LAB., WRIGHT-PATTERSON AFB, OHIO	ML-
AIR FORCE MATERIALS LAB., WRIGHT-PATTERSON AFB, OHIO	ML-LN-
AIR FORCE MATERIALS LAB., WRIGHT-PATTERSON AFB, OHIO	ML-TDR-
AIR FORCE MATERIALS LAB., WRIGHT-PATTERSON AFB, OHIO	ML-TM-
AIR FORCE MATERIALS LAB., WRIGHT-PATTERSON AFB, OHIO	ML-TR-
AIR FORCE MATERIALS LAB., WRIGHT-PATTERSON AFB, OHIO	RTD-TDR-
AIR FORCE MATERIALS LAB. MATERIALS APPLICATIONS DIV., WRIGHT-PATTERSON AFB, OHIO	MAA-(YEAR)-

AIR FORCE MATERIALS LAB. MATERIALS APPLICATIONS DIV.

AIR FORCE MATERIALS LAB. MATERIALS APPLICATIONS DIV., WRIGHT-PATTERSON AFB, OHIO	MAA-TM-(YEAR)-
AIR FORCE MISSILE DEV. CTR., HOLLOMAN AFB, N. MEX.	AFMDC-
AIR FORCE MISSILE DEV. CTR., HOLLOMAN AFB, N. MEX.	AFMDC ADJ-(YR)-
AIR FORCE MISSILE DEV. CTR., HOLLOMAN AFB, N. MEX.	AFMDC-DAS-(YEAR)-
AIR FORCE MISSILE DEV. CTR., HOLLOMAN AFB, N. MEX.	AFMDC HDGR-(YR)-
AIR FORCE MISSILE DEV. CTR., HOLLOMAN AFB, N. MEX.	AFMDC MDGR-TM-(YR)-
AIR FORCE MISSILE DEVELOPMENT CENTER, HOLLOMAN AFB, N. MEX. (TECHNICAL DOCUMENTARY REPORT)	AFMDC-TDR-
AIR FORCE MISSILE DEVELOPMENT CENTER, HOLLOMAN AFB, N. MEX. (TEST REPORT)	AFMDC TEST R-
AIR FORCE MISSILE DEVELOPMENT CENTER, HOLLOMAN AFB, N. MEX. (TECHNICAL NOTE)	AFMDC-TN-(YR)-
AIR FORCE MISSILE DEV. CTR., HOLLOMAN AFB, N. MEX.	AFMDC-TR-(YR)-
AIR FORCE MISSILE DEV. CTR., HOLLOMAN AFB, N. MEX.	MDC-TDR-
AIR FORCE MISSILE DEV. CTR., HOLLOMAN AFB, N. MEX.	MDC-TR-
AIR FORCE MISSILE DEV. CTR., HOLLOMAN AFB, N. MEX.	MDC-TR-(YEAR)-
AIR FORCE MISSILE TEST CENTER, PATRICK AFB, FLA.	AFMIC-TR-
AIR FORCE MISSILE TEST CENTER, PATRICK AFB, FLA.	AFMTC-
AIR FORCE MISSILE TEST CENTER, PATRICK AFB, FLA.	AFMTC-FO-
AIR FORCE MISSILE TEST CENTER, PATRICK AFB, FLA.	AFMTC-JUPITER-FTDR-
AIR FORCE MISSILE TEST CENTER, PATRICK AFB, FLA.	AFMTC-MT-(YEAR)-
AIR FORCE MISSILE TEST CENTER, PATRICK AFB, FLA.	AFMTC MTE-TM-(YR)-
AIR FORCE MISSILE TEST CENTER, PATRICK AFB, FLA.	AFMTC-MTO-
AIR FORCE MISSILE TEST CENTER, PATRICK AFB, FLA.	AFMTC-MTW-TM-(YEAR)-
AIR FORCE MISSILE TEST CENTER, PATRICK AFB, FLA.	AFMTC-OD-
AIR FORCE MISSILE TEST CENTER, PATRICK AFB, FLA.	AFMTC-OR-
AIR FORCE MISSILE TEST CENTER, PATRICK AFB, FLA.	AFMTCP-
AIR FORCE MISSILE TEST CENTER, PATRICK AFB, FLA.	AFMTCR-
AIR FORCE MISSILE TEST CENTER, PATRICK AFB, FLA.	AFMTC-REDSTONE-FTDR-
AIR FORCE MISSILE TEST CENTER, PATRICK AFB, FLA.	AFMTC-SNARK-FT-
AIR FORCE MISSILE TEST CENTER, PATRICK AFB, FLA.	AFMTC-SNARK-FTDR-
AIR FORCE MISSILE TEST CENTER, PATRICK AFB, FLA. (TECHNICAL DOCUMENTARY REPORT)	AFMTC-TDR-
AIR FORCE MISSILE TEST CENTER, PATRICK AFB, FLA.	AFMTC TN-(YR)-
AIR FORCE MISSILE TEST CENTER, PATRICK AFB, FLA.	AFMTC-TR-(YR)-
AIR FORCE MISSILE TEST CENTER, PATRICK AFB, FLA.	LRPGD-
AIR FORCE MISSILE TEST CENTER, PATRICK AFB, FLA.	MT-
AIR FORCE MISSILE TEST CENTER, PATRICK AFB, FLA.	MTC-
AIR FORCE MISSILE TEST CENTER, PATRICK AFB, FLA.	MTLGI-
AIR FORCE MISSILE TEST CENTER, PATRICK AFB, FLA.	MTP-
AIR FORCE MISSILE TEST CENTER, PATRICK AFB, FLA.	MTT-
AIR FORCE MISSILE TEST CENTER, PATRICK AFB, FLA.	MTTE-
AIR FORCE MISSILE TEST CENTER, PATRICK AFB, FLA.	MTX-TM-
AIR FORCE MISSILE TEST CENTER, PATRICK AFB, FLA.	PAFB-
AIR FORCE MISSILE TEST CENTER, PATRICK AFB, FLA.	RCA-
AIR FORCE-NAVY CIVIL COMMITTEE (REF. 13)	ANC-
AIR FORCE-NAVY COMMITTEE (REF. 13)	JAFNC-
AIR FORCE-NAVY WEATHER BUREAU (REF. 13)	WBAN-
AIR FORCE NUCLEAR REACTOR SYSTEM SAFETY GROUP, KIRTLAND AFB, N. MEX.	NRSSG-
AIR FORCE NUCLEAR WEAPON SYSTEM SAFETY GROUP, KIRTLAND AFB, N. MEX.	AFC-(YEAR)-SRD-
AIR FORCE NUCLEAR WEAPON SYSTEM SAFETY GROUP, KIRTLAND AFB, N. MEX.	AFSWC-NWSSG-(YEAR)-
AIR FORCE NUCLEAR WEAPON SYSTEM SAFETY GROUP, KIRTLAND AFB, N. MEX.	NWSSC-
AIR FORCE NUCLEAR WEAPON SYSTEM SAFETY GROUP, KIRTLAND AFB, N. MEX.	NWSSG-(YEAR)-
AIR FORCE NUCLEAR WEAPON SYSTEM SAFETY GROUP, KIRTLAND AFB, N. MEX.	US-NWSSG-
AIR FORCE OFFICE OF SCIENTIFIC RESEARCH, ARLINGTON, VA	AFOSR-(YEAR)-
AIR FORCE OFFICE OF SCIENTIFIC RESEARCH, WASH., D.C. (DESIGNATION USED FOR GRANTS)	AF-AFOSR-(NUMBER)-(YEAR)
AIR FORCE OFFICE OF SCIENTIFIC RESEARCH, WASH., D.C.	AFCSR-(YEAR)-(NUMBER)TR
AIR FORCE OFFICE OF SCIENTIFIC RESEARCH, WASH., D.C.	AFORS-(YEAR)-(NUMBER)TR
AIR FORCE OFFICE OF SCIENTIFIC RESEARCH, WASH., D.C.	AFOSR-
AIR FORCE OFFICE OF SCIENTIFIC RESEARCH, WASH., D.C.	AFOSR-(YEAR)-(NUMBER)TR
AIR FORCE OFFICE OF SCIENTIFIC RESEARCH, WASH., D.C.	AFOSR-AFOSR-(YR.)-(NO)TR
AIR FORCE OFFICE OF SCIENTIFIC RESEARCH, WASH., D.C.	AFOSR-J(NUMBER)
AIR FORCE OFFICE OF SCIENTIFIC RESEARCH, WASH., D.C.	AFOSR-O-(NUMBER)TR
AIR FORCE OFFICE OF SCIENTIFIC RESEARCH, WASH., D.C.	AFOSR-SR-(YEAR)-
AIR FORCE OFFICE OF SCIENTIFIC RESEARCH, WASH., D.C.	AFOSR-TD-
AIR FORCE OFFICE OF SCIENTIFIC RESEARCH, WASH., D.C.	AFOSR-TN-
AIR FORCE OFFICE OF SCIENTIFIC RESEARCH, WASH., D.C.	AFOSR-TN-(YEAR)-
AIR FORCE OFFICE OF SCIENTIFIC RESEARCH, WASH., D.C.	AFOSR-TR-
AIR FORCE OFFICE OF SCIENTIFIC RESEARCH, WASH., D.C.	AFOSR-TR-(YEAR)-
AIR FORCE OFFICE OF SCIENTIFIC RESEARCH, WASH., D.C.	AFRO-
AIR FORCE OFFICE OF SCIENTIFIC RESEARCH, WASH., D.C.	AFSOR-(YEAR)-(NUMBER)TR
AIR FORCE OFFICE OF SCIENTIFIC RESEARCH, WASH., D.C.	DAR-
AIR FORCE OFFICE OF SCIENTIFIC RESEARCH, WASH., D.C.	OSR-
AIR FORCE OFFICE OF SCIENTIFIC RESEARCH, WASH., D.C. (QUARTERLY INDEX)	OSR-QI-
AIR FORCE OFFICE OF SCIENTIFIC RESEARCH, WASH., D.C. (STATUS OF RESEARCH PROPOSALS)	OSR-SRP-(YEAR/MONTH)
AIR FORCE OFFICE OF SCIENTIFIC RESEARCH, WASH., D.C.	OSR-TN-(YEAR)-
AIR FORCE OFFICE OF SCIENTIFIC RESEARCH, WASH., D.C.	OSR-TR-(YEAR)-
AIR FORCE OPERATIONAL TEST CENTER, EGLIN AFB, FLA.	AFOTC-
AIR FORCE OPERATIONAL TEST CENTER, EGLIN AFB, FLA.	APC-
AIR FORCE PACKAGING EVALUATION AGENCY, WRIGHT-PATTERSON AFB, OHIO	DSPEA-
AIR FORCE PACKAGING EVALUATION AGENCY, WRIGHT-PATTERSON AFB, OHIO	DSPEB-
AIR FORCE PACKAGING EVALUATION AGENCY, WRIGHT-PATTERSON AFB, OHIO	SGTEB-
A.F. PERSONNEL + TRAINING RES. CTR., LACKLAND AFB, TEX.	AFPTRC-PL-TM-(NO.)-
A.F. PERSONNEL + TRAINING RES. CTR., LACKLAND AFB, TEX.	AFPTRC-TN-(YEAR)-
A.F. PERSONNEL + TRAINING RES. CTR., LACKLAND AFB, TEX.	AFPTRC-TR-(YEAR)-
A.F. PERSONNEL + TRAINING RES. CTR., LACKLAND AFB, TEX.	AO-
A.F. PERSONNEL + TRAINING RES. CTR., LACKLAND AFB, TEX.	P + MS-
A.F. PERSONNEL + TRAINING RES. CTR., LACKLAND AFB, TEX.	PTRC-

(SEE ALSO LATER NAME: PERSONNEL RES. LAB. (6570TH))

AIR FORCE ROCKET PROPULSION LAB., EDWARDS AFB, CALIF.	AFRAL-TR-
AIR FORCE ROCKET PROPULSION LAB., EDWARDS AFB, CALIF.	AFRFL-TR-
AIR FORCE ROCKET PROPULSION LAB., EDWARDS AFB, CALIF.	AFRPL-PR-(YEAR)-
AIR FORCE ROCKET PROPULSION LAB., EDWARDS AFB, CALIF.	AFRPL-TDR-
AIR FORCE ROCKET PROPULSION LAB., EDWARDS AFB, CALIF.	AFRPL-TM-
AIR FORCE ROCKET PROPULSION LAB., EDWARDS AFB, CALIF.	AFRPL-TR-
AIR FORCE ROCKET PROPULSION LAB., EDWARDS AFB, CALIF.	AFRPL-TR-(YEAR)-
AIR FORCE ROCKET PROPULSION LAB., EDWARDS AFB, CALIF.	AFTPL-TR-
AIR FORCE ROCKET PROPULSION LAB., EDWARDS AFB, CALIF.	RPL-TDR-
AIR FORCE ROCKET PROPULSION LAB., EDWARDS AFB, CALIF.	RPL-TM-
AIR FORCE ROCKET PROPULSION LAB., EDWARDS AFB, CALIF.	URS-(NUMBERS)-

AIR FORCE SPECIAL OPERATIONS FORCE, EGLIN AFB, FLA.	SOF-(NUMBER)A-
AIR FORCE SPECIAL OPERATIONS FORCE, EGLIN AFB, FLA.	USAFSOF-(NUMBER)A-
AIR FORCE SPECIAL WEAPONS CENTER, KIRTLAND AFB, N.M.	AFSWC-
AIR FORCE SPECIAL WEAPONS CENTER, KIRTLAND AFB, N. MEX. (FUNCTIONAL TEST REPORT)	AFSWC-FTR-
AIR FORCE SPECIAL WEAPONS CENTER, KIRTLAND AFB, N.M.	AFSWC-KAFB-
AIR FORCE SPECIAL WEAPONS CENTER, KIRTLAND AFB, N. MEX. (OPERATION PLANS)	AFSWC-OP-(YEAR)-
AIR FORCE SPECIAL WEAPONS CENTER, KIRTLAND AFB, N.M.	AFSWC-SWODO-(YEAR)-
AIR FORCE SPECIAL WEAPONS CENTER, KIRTLAND AFB, N.M.	AFSWC-SWOTO-
AIR FORCE SPECIAL WEAPONS CENTER, KIRTLAND AFB, N.M.	AFSWC-SWR-
AIR FORCE SPECIAL WEAPONS CENTER, KIRTLAND AFB, N.M.	AFSWC-SWR-TM-(NUMBER)-
AIR FORCE SPECIAL WEAPONS CENTER, KIRTLAND AFB, N.M.	AFSWC-SWVN-(YEAR)-
AIR FORCE SPECIAL WEAPONS CENTER, KIRTLAND AFB, N. MEX. (TECHNICAL DOCUMENTARY REPORT)	AFSWC-TDR-
AIR FORCE SPECIAL WEAPONS CENTER, KIRTLAND AFB, N. MEX. (TECHNICAL MEMORANDUM)	AFSWC-TM-SWV-(YEAR)-
AIR FORCE SPECIAL WEAPONS CENTER, KIRTLAND AFB, N.M.	AFSWC-TN-(YR)-
AIR FORCE SPECIAL WEAPONS CENTER, KIRTLAND AFB, N.M.	AFSWC-TN-SWR-(YEAR)-
AIR FORCE SPECIAL WEAPONS CENTER, KIRTLAND AFB, N.M.	AFSWC-TN-SWW-(YEAR)-
AIR FORCE SPECIAL WEAPONS CENTER, KIRTLAND AFB, N.M.	AFSWC-TR-(YR)-
AIR FORCE SPECIAL WEAPONS CENTER, KIRTLAND AFB, N.M.	AFSWC-TR-SPEC-EC-
AIR FORCE SPECIAL WEAPONS CENTER, KIRTLAND AFB, N.M.	AFSWC-TR-SWR-(YEAR)-
AIR FORCE SPECIAL WEAPONS CENTER, KIRTLAND AFB, N. MEX. (TEST SUMMARY)	AFSWC-TS-(YEAR)-
AIR FORCE SPECIAL WEAPONS CENTER, KIRTLAND AFB, N.M. (OPERATION PLAN)	OPLAN-(NUMBER)-(YEAR)
AIR FORCE SPECIAL WEAPONS CENTER, KIRTLAND AFB, N. MEX. (SLV-1B SPACE PROBE)	SLV-1B-
AIR FORCE SPECIAL WEAPONS CENTER, KIRTLAND AFB, N.M.	SWC-
AIR FORCE SPECIAL WEAPONS CENTER, KIRTLAND AFB, N.M.	SWC-(NUMBER)C-
AIR FORCE SPECIAL WEAPONS CENTER, KIRTLAND AFB, N.M.	SWC-(NO.)S-
AIR FORCE SPECIAL WEAPONS CENTER, KIRTLAND AFB, N.M. (FUNCTIONAL TEST REPORTS)	SWC FTR (YR.)-
AIR FORCE SPECIAL WEAPONS CENTER, KIRTLAND AFB, N.M.	SWC MR WCWGT02-(YR.)-
AIR FORCE SPECIAL WEAPONS CENTER, KIRTLAND AFB, N. MEX. (OPERATIONS ANALYSIS. WORKING PAPER)	SWC OA WP-
AIR FORCE SPECIAL WEAPONS CENTER, KIRTLAND AFB, N. MEX. (OPERATIONS ANALYSIS. WORKING PAPER)	SWC OAWP-
AIR FORCE SPECIAL WEAPONS CENTER, KIRTLAND AFB, N. MEX. (OPERATIONS ANALYSIS. WORKING PAPER)	SWC OP (NO.)-(YR.)
AIR FORCE SPECIAL WEAPONS CENTER, KIRTLAND AFB, N.M.	SWC-OPLAN-(NUMBER-YEAR)
AIR FORCE SPECIAL WEAPONS CENTER, KIRTLAND AFB, N.M.	SWC-OS-
AIR FORCE SPECIAL WEAPONS CENTER, KIRTLAND AFB, N.M.	SWC-PROJ.NO.WCO 51-
AIR FORCE SPECIAL WEAPONS CENTER, KIRTLAND AFB, N.M.	SWC PROJ.NO.WCWGT02-(YR)
AIR FORCE SPECIAL WEAPONS CENTER, KIRTLAND AFB, N. MEX. (SAFETY STUDIES)	SWC SWVN (YR.)-
AIR FORCE SPECIAL WEAPONS CENTER, KIRTLAND AFB, N. MEX. (TECHNICAL DOCUMENTARY REPORT)	SWC-TDR-
AIR FORCE SPECIAL WEAPONS CENTER, KIRTLAND AFB, N. MEX. (TECHNICAL DOCUMENTARY REPORT)	SWC-TDR-(NUMBER)-
AIR FORCE SPECIAL WEAPONS CENTER, KIRTLAND AFB, N. MEX. (TECHNICAL NOTE)	SWC TN-(YR.)-
AIR FORCE SPECIAL WEAPONS CENTER, KIRTLAND AFB, N. MEX. (TECHNICAL REPORT)	SWC TR (YR.)-
AIR FORCE SPECIAL WEAPONS CENTER, KIRTLAND AFB, N. MEX. (SELECTED RECENT DOCUMENTS)	SWC-VOL. (NO.), NO.-
AIR FORCE SPECIAL WEAPONS CENTER, KIRTLAND AFB, N.M.	SWC-WCO-
AIR FORCE SPECIAL WEAPONS CENTER, KIRTLAND AFB, N.M.	SWR-TN-
AIR FORCE SPECIAL WEAPONS CENTER, KIRTLAND AFB, N.M.	SWVCS-(NUMBER)-
AIR FORCE SPECIAL WEAPONS CENTER, KIRTLAND AFB, N.M.	SWVN-(NUMBER)-
AIR FORCE SPECIAL WEAPONS CENTER, KIRTLAND AFB, N.M.	SWWVD-
AIR FORCE SPECIAL WEAPONS CENTER, KIRTLAND AFB, N.M.	TDR-(YEAR)-
AIR FORCE SPECIAL WEAPONS CENTER, KIRTLAND AFB, N.M.	WCO-(NUMBER)-
AIR FORCE SPECIAL WEAPONS CENTER, KIRTLAND AFB, N.M.	WCOD-(NUMBER)-
AIR FORCE SPECIAL WEAPONS CENTER. 4925TH TEST GROUP (ATOMIC), KIRTLAND AFB, N. MEX.	SWC SWOTO (YR.)-
AIR FORCE SPECIAL WEAPONS CENTER. 4925TH TEST GROUP (ATOMIC), KIRTLAND AFB, N. MEX. (TECHNICAL NOTE)	SWC TN SWW (YR.)-
AIR FORCE SPECIAL WEAPONS CENTER. 4925TH TEST GROUP (ATOMIC), KIRTLAND AFB, N. MEX.	SWC WCO (YR.)-
AIR FORCE SPECIAL WEAPONS CENTER. 4925TH TEST GROUP (ATOMIC), KIRTLAND AFB, N. MEX.	SWODO (YR.)-
AIR FORCE SPECIAL WEAPONS CENTER. 4925TH TEST GROUP (ATOMIC), KIRTLAND AFB, N. MEX.	SWOTO-
AIR FORCE SPECIAL WEAPONS CENTER. 4925TH TEST GROUP (ATOMIC), KIRTLAND AFB, N. MEX.	SWOTO (YR.)-
AIR FORCE SPECIAL WEAPONS CENTER. 4925TH TEST GROUP (ATOMIC), KIRTLAND AFB, N. MEX.	SWOTOP (YR.)-
AIR FORCE SPECIAL WEAPONS CENTER. DEVELOPMENT DIRECTORATE, KIRTLAND AFB, N. MEX. (MISC. PUBLICATION)	SWC SWV MP-
AIR FORCE SPECIAL WEAPONS CENTER. DEV. DIRECTORATE, KIRTLAND AFB, N. MEX.(STOCKPILE-TO-TARGET SEQUENCE)	SWC SWV STS-
AIR FORCE SPECIAL WEAPONS CENTER. DEVELOPMENT DIRECTORATE, KIRTLAND AFB, N. MEX. (TECHNICAL MEMO.)	SWC SWV TM (YR.)-
AIR FORCE SPECIAL WEAPONS CENTER. DEVELOPMENT DIRECTORATE, KIRTLAND AFB, N. MEX. (TECH. MEMO. REPT.)	SWC TMR SWV (YR.)-
AIR FORCE SPECIAL WEAPONS CENTER. RESEARCH DIRECTORATE, KIRTLAND AFB, N. MEX. (TECHNICAL MEMORANDUM)	SWC SWR TM (YR.)-
AIR FORCE SPECIAL WEAPONS CENTER. RESEARCH DIRECTORATE, KIRTLAND AFB, N. MEX. (TECHNICAL NOTE)	SWC TN SWR (YR.)-
AIR FORCE SPECIAL WEAPONS CENTER. RESEARCH DIRECTORATE, KIRTLAND AFB, N. MEX.	SWR-
AIR FORCE SPECIAL WEAPONS CENTER. RESEARCH DIRECTORATE, KIRTLAND AFB, N. MEX. (TECHNICAL MEMORANDUM)	SWR-TM-
AIR FORCE SPECIAL WEAPONS CENTER. RESEARCH DIRECTORATE, KIRTLAND AFB, N. MEX. (TECHNICAL MEMORANDUM)	SWR-TM-(YR.)-
AIR FORCE SPECIAL WEAPONS CENTER. TECHNICAL INFO. DIV., KIRTLAND AFB, N.M. (SELECTED RECENT DOCS.)	SWOI-(NUMBER)-
AIR FORCE SYSTEMS COMMAND, ANDREWS AFB, MD.	AFSC-
AIR FORCE SYSTEMS COMMAND, ANDREWS AFB, MD.	AFSC-(NUMBER)-
AIR FORCE SYSTEMS COMMAND, ANDREWS AFB, MD.	AFSC-CP-(NUMBERS)-(LTR)-
AIR FORCE SYSTEMS COMMAND, ANDREWS AFB, MD.	AFSC-CPN-
AIR FORCE SYSTEMS COMMAND, ANDREWS AFB, MD.	AFSC-HISTORICAL-PUB-
AIR FORCE SYSTEMS COMMAND, ANDREWS AFB, MD.	AFSC-HPS-(YEAR)-
AIR FORCE SYSTEMS COMMAND, ANDREWS AFB, MD.	AFSCM-(NUMBER)-
AIR FORCE SYSTEMS COMMAND, ANDREWS AFB, MD.	AFSC-SPEC-
AIR FORCE SYSTEMS COMMAND, ANDREWS AFB, MD.	AFSC-SPEC-(LTRS.)-
AIR FORCE SYSTEMS COMMAND, ANDREWS AFB, MD.	AFSC-SPEC-DOD-AIMS-(YR)-
AIR FORCE SYSTEMS COMMAND, ANDREWS AFB, MD. (TECH. DOCUMENTARY REPORT)	AFSC-TDR-
AIR FORCE SYSTEMS COMMAND, ANDREWS AFB, MD. (TECHNICAL OBJECTIVE)	AFSC-TOD-(NUMBER)(LTR.)
AIR FORCE SYSTEMS COMMAND, ANDREWS AFB, MD.	AFSC-TR-
AIR FORCE SYSTEMS COMMAND, ANDREWS AFB, MD.	SCGF-
AIR FORCE SYSTEMS COMMAND, ANDREWS AFB, MD.	SCTP-(YEAR)-
AIR FORCE SYSTEMS COMMAND, ANDREWS AFB, MD. (TECHNICAL OBJECTIVE DOCUMENT)	TOD-(NUMBER)-
AIR FORCE SYSTEMS COMMAND, ANDREWS AFB, MD. (TECHNICAL OBJECTIVE DOCUMENT)	TOD-SUMMARY GUIDE-(YR)
AIR FORCE SYSTEMS COMMAND, EDWARDS AFB, CALIF.	TDR-
AIR FORCE SYSTEMS COMMAND, EGLIN AFB, FLA.	AIDR-
AIR FORCE SYSTEMS COMMAND, EGLIN AFB, FLA.	ATL-TDR-
AIR FORCE SYSTEMS COMMAND, VANDENBERG AFB, CALIF.	VWZE-
AIR FORCE SYSTEMS COMMAND, WRIGHT-PATTERSON AFB, OHIO	AGM-
AIR FORCE SYSTEMS COMMAND, WRIGHT-PATTERSON AFB, OHIO	FDTR-TM-(YEAR)-
AIR FORCE SYSTEMS COMMAND, WRIGHT-PATTERSON AFB, OHIO	MAA-(LTRS)-
AIR FORCE SYSTEMS COMMAND, WRIGHT-PATTERSON AFB, OHIO	SEF-TN-(YEAR)-
AIR FORCE SYSTEMS COMMAND, WRIGHT-PATTERSON AFB, OHIO	SM-
AIR FORCE SYSTEMS COMMAND. DIRECTOR OF LABS., WASHINGTON, D.C.	DL-(YEAR)-
AIR FORCE SYSTEMS COMMAND. SPACE SYSTEMS DIV.,LOS ANG	DCASPP-(YEAR)-
AIR FORCE TECHNICAL APPLICATIONS CENTER, WASH., D.C. (LONG RANGE SEISMIC MEASUREMENTS PROJECT)	AFTAC/LRSM-
AIR FORCE TECHNICAL APPLICATIONS CENTER, WASH., D.C.	AFTAC-TR-(YEAR)-
AIR FORCE TECHNICAL APPLICATIONS CENTER, WASH., D.C.	DATDC-

AIR FORCE TECHNICAL APPLICATIONS CENTER VELA SEISMOLOGICAL CENTER

AIR FORCE TECHNICAL APPLICATIONS CENTER. VELA SEISMOLOGICAL CENTER, WASHINGTON, D.C.	VSC-
AIR FORCE TECHNICAL APPLICATIONS CENTER. VELA SEISMOLOGICAL CENTER, WASH., D.C. (PROJ. VELA UNIFORM)	VU-
AIR FORCE WEAPONS LAB., KIRTLAND AFB, N. MEX.	AFWL-
AIR FORCE WEAPONS LAB., KIRTLAND AFB, N. MEX.	AFWLAA-(YEAR)-
AIR FORCE WEAPONS LAB., KIRTLAND AFB, N. MEX.	AFWL CR-(MONTH-DAY-YEAR)
AIR FORCE WEAPONS LAB., KIRTLAND AFB, N. MEX. (ELECTROMAGNETIC PULSE NOTE SERIES)	AFWL-EMP-
AIR FORCE WEAPONS LAB., KIRTLAND AFB, N. MEX.	AFWL-TDR-(YEAR)-
AIR FORCE WEAPONS LAB., KIRTLAND AFB, N. MEX.	AFWL-TM-(YEAR)-
AIR FORCE WEAPONS LAB., KIRTLAND AFB, N. MEX.	AFWL-TR-(YEAR)-
AIR FORCE WEAPONS LAB., KIRTLAND AFB, N. MEX. (ELECTROMAGNETIC PULSE THEORETICAL NOTES)	EMP-2-
AIR FORCE WEAPONS LAB., KIRTLAND AFB, N. MEX. (NUMBERS IN 3000 SERIES ONLY)	RTD TDR-63-
AIR FORCE WEAPONS LAB., KIRTLAND AFB, N. MEX.	RTD (WLA)-TM-(YEAR)-
AIR FORCE WEAPONS LAB., KIRTLAND AFB, N. MEX.	RTD (WLR)-TM-(YEAR)-
AIR FORCE WEAPONS LAB., KIRTLAND AFB, N. MEX.	WL(NUMBER)-TN-(YEAR)-
AIR FORCE WEAPONS LAB., KIRTLAND AFB, N. MEX.	WLAS-
AIR FORCE WEAPONS LAB., KIRTLAND AFB, N. MEX.	WL-TDR-(YEAR)-
AIR FORCE WEAPONS LAB., KIRTLAND AFB, N. MEX.	WL TR-(YEAR)-
AIR FORCE WEAPONS LAB. CIVIL ENGINEERING DIV., KIRTLAND AFB, N. MEX.	WLC-TM-(YEAR)-
AIR FORCE WEAPONS LAB. CIVIL ENGINEERING DIV., KIRTLAND AFB, N. MEX.	WLC-TN-(YEAR)-
AIR FORCE WESTERN TEST RANGE, VANDENBERG AFB, CALIF.	AFWLAW-(YEAR)-
AIR FORCE WESTERN TEST RANGE, VANDENBURG AFB, CALIF.	AFWTR-TR-(YEAR)-
AIR FORCE WESTERN TEST RANGE, VANDENBERG AFB, CALIF.	WTSMR-
AIR FORCES IN EUROPE	AAF EU A-2 P-
AIR FORCES IN EUROPE	EU A-2P-
AIR FORCES IN EUROPE (U.S.), WIESBADEN, GERMANY	ATI-
AIR FORCES IN EUROPE (U.S.), WIESBADEN, GERMANY	USAFE-
AIR MATERIEL AREA, OGDEN, UTAH	OAMA-
AIR MATERIEL AREA, OKLAHOMA CITY	OCAMA-
AIR MATERIEL COMMAND, WRIGHT-PATTERSON AFB, OHIO (AVIATION PSYCHOLOGY PROGRAM)	AAF APP-
AIR MATERIEL COMMAND, WRIGHT-PATTERSON AFB, OHIO	ADM-
AIR MATERIEL COMMAND, WRIGHT-PATTERSON AFB, OHIO	AMC-
AIR MATERIEL COMMAND, WRIGHT-PATTERSON AFB, OHIO	AMC-ARDC-
AIR MATERIEL COMMAND, WRIGHT-PATTERSON AFB, OHIO	AMC-CD-
AIR MATERIEL COMMAND, WRIGHT-PATTERSON AFB, OHIO (EXCERPT REPORTS)	AMC E-
AIR MATERIEL COMMAND, WRIGHT-PATTERSON AFB, OHIO	AMC-IRE-
AIR MATERIEL COMMAND, WRIGHT-PATTERSON AFB, OHIO (MANUAL)	AMCM-
AIR MATERIEL COMMAND, WRIGHT-PATTERSON AFB, OHIO	AMC-MCREOA-
AIR MATERIEL COMMAND, WRIGHT-PATTERSON AFB, OHIO	AMC-MR-EXP-M-51/
AIR MATERIEL COMMAND, WRIGHT-PATTERSON AFB, OHIO	AMC-MR-MCREE-
AIR MATERIEL COMMAND, WRIGHT-PATTERSON AFB, OHIO	AMC-R-
AIR MATERIEL COMMAND, WRIGHT-PATTERSON AFB, OHIO	AMC RPT. CD-
AIR MATERIEL COMMAND, WRIGHT-PATTERSON AFB, OHIO	AMC-TO-
AIR MATERIEL COMMAND, WRIGHT-PATTERSON AFB, OHIO	AMC-TR-
AIR MATERIEL COMMAND, WRIGHT-PATTERSON AFB, OHIO	ATSC-
AIR MATERIEL COMMAND, WRIGHT-PATTERSON AFB, OHIO (EXPERIMENTAL STRESS ANALYSIS)	ESA-
AIR MATERIEL COMMAND, WRIGHT-PATTERSON AFB, OHIO	MC-
AIR MATERIEL COMMAND, WRIGHT-PATTERSON AFB, OHIO (FIRST TWO DIGITS STAND FOR YEAR) (REF. 1)	53MC... THRU 56MC...
AIR MATERIEL COMMAND. AERONAUTICAL ICE RESEARCH LAB.	AMC-AIRL-
AIR MATERIEL COMMAND. AIR DOCUMENTS DIV., WRIGHT-PATTERSON AFB, OHIO (AIR TECH. INTELLIGENCE REVIEW)	AAF T-2 ATIR-
AIR MATERIEL COMMAND. AIR DOCUMENTS DIV., WRIGHT-PATTERSON AFB, OHIO (INTERIM REPORT)	AAF T-2 INT-
AIR MATERIEL COMMAND. AIR DOCUMENTS DIV., WRIGHT-PATTERSON AFB, OHIO (INTELLIGENCE REPORT)	AAF T-2 IR/
AIR MATERIEL COMMAND. AIR DOCUMENTS DIV., WRIGHT-PATTERSON AFB, OHIO	AAF T-2 IRE-
AIR MATERIEL COMMAND. AIR DOCUMENTS DIV., WRIGHT-PATTERSON AFB, OHIO (MICROFILM)	AAF T-2 MICRO-
AIR MATERIEL COMMAND. AIR DOCUMENTS DIV., WRIGHT-PATTERSON AFB, OHIO (SUMMARY REPORT)	AAF T-2 SR/
AIR MATERIEL COMMAND. AIR DOCUMENTS DIV., WRIGHT-PATTERSON AFB, OHIO (TRANSLATION)	AAF T-2 T/
AIR MATERIEL COMMAND. AIR DOCUMENTS DIV., WRIGHT-PATTERSON AFB, OHIO (TECHNICAL REPORT)	AAF T-2 TR-
AIR MATERIEL COMMAND. AIR DOCUMENTS DIV., WRIGHT-PATTERSON AFB, OHIO (AIR TECH. INTELLIGENCE REVIEW)	ATIR-
AIR MATERIEL COMMAND. AIR DOCUMENTS DIV., WRIGHT-PATTERSON AFB, OHIO	IRE-
AIR MATERIEL COMMAND. AIR DOCUMENTS OFFICE, WRIGHT-PATTERSON AFB, OHIO (TRANSLATION)	F-SU-
AIR MATERIEL COMMAND. AIRCRAFT RADIATION LAB., WRIGHT-PATTERSON AFB, OHIO	AMC-ARL-
AIR MATERIEL COMMAND. AIRCRAFT RADIATION LAB., WRIGHT-PATTERSON AFB, OHIO	ARL-
AIR MATERIEL COMMAND. ALL WEATHER FLYING DIV., WRIGHT-PATTERSON AFB, OHIO	AAF AWAF-
AIR MATERIEL COMMAND. ALL-WEATHER FLYING DIV., WRIGHT-PATTERSON AFB, OHIO	AWAF-
AIR MATERIEL COMMAND. ALL-WEATHER FLYING DIV., WRIGHT-PATTERSON AFB, OHIO	AWNW-
AIR MATERIEL COMMAND. ALL-WEATHER FLYING DIV. TRAFFIC AND LANDING BRANCH, WRIGHT-PATTERSON AFB, OHIO	AWTL-
AIR MATERIEL COMMAND. ANALYSIS DIV., WRIGHT-PATTERSON AFB, OHIO	MCIA-
AIR MATERIEL COMMAND. CAMBRIDGE FIELD STATION	AMC-CFS-
AIR MATERIEL COMMAND. CAMBRIDGE FIELD STATION	AMC-ERC-
AIR MATERIEL COMMAND. ELECTRONICS RESEARCH LAB., CAMBRIDGE, MASS.	ERL-E-
AIR MATERIEL COMMAND. ELECTRONICS RESEARCH LAB., WRIGHT-PATTERSON AFB, OHIO	AMC-ERL-
AIR MATERIEL COMMAND. ENGINEERING DIV., WRIGHT-PATTERSON AFB, OHIO	AC-TR-
AIR MATERIEL COMMAND. ENGINEERING DIV., WRIGHT-PATTERSON AFB, OHIO	AMC-AF-TSEAA-(NUMBER)-
AIR MATERIEL COMMAND. ENGINEERING DIV., WRIGHT-PATTERSON AFB, OHIO (MEMORANDUM REPORT)	AMC ED MR-
AIR MATERIEL COMMAND. ENGINEERING DIV., WRIGHT-PATTERSON AFB, OHIO	AMC-MR-MCREXD-
AIR MATERIEL COMMAND. ENGINEERING DIV., WRIGHT-PATTERSON AFB, OHIO	AMC-MR-MCREXE-
AIR MATERIEL COMMAND. ENGINEERING DIV., WRIGHT-PATTERSON AFB, OHIO	AMC-MR-TSEAA-
AIR MATERIEL COMMAND. ENGINEERING DIV., WRIGHT-PATTERSON AFB, OHIO	ENG EB-
AIR MATERIEL COMMAND. ENGINEERING DIV., WRIGHT-PATTERSON AFB, OHIO	MCRE-
AIR MATERIEL COMMAND. ENGINEERING DIV., WRIGHT-PATTERSON AFB, OHIO	TSENG-
AIR MATERIEL COMMAND. ENGINEERING DIV. AERO-MEDICAL LAB., WRIGHT-PATTERSON AFB, OHIO	AAF MCREXD-
AIR MATERIEL COMMAND. ENG. DIV. AERO-MEDICAL LAB., WRIGHT-PATTERSON AFB, OHIO (MEMO. REPT.)	AMC TSEAA (NOS.)(LTRS.)
AIR MATERIEL COMMAND. ENGINEERING DIV. AERO-MEDICAL LAB., WRIGHT-PATTERSON AFB, OHIO	MCREXD-
AIR MATERIEL COMMAND. ENGINEERING DIV. AIRCRAFT LAB., WRIGHT-PATTERSON AFB, OHIO (MEMO. REPT.)	AAF MCREXA-
AIR MATERIEL COMMAND. ENGINEERING DIV. AIRCRAFT LAB., WRIGHT-PATTERSON AFB, OHIO (MEMO. REPT.)	AMC MCREXA-
AIR MATERIEL COMMAND. ENGINEERING DIV. AIRCRAFT LAB., WRIGHT-PATTERSON AFB, OHIO	MCREXA-
AIR MATERIEL COMMAND. ENG. DIV., AIRCRAFT RADIATION LAB., WRIGHT-PATTERSON AFB, OHIO (MEMO. REPT.)	AMC TSELR (NOS.)
AIR MATERIEL COMMAND. ENG. DIV. CLIMATIC PROJECTS SEC., WRIGHT-PATTERSON AFB, OHIO (MEMO. REPT.)	AMC MCREOC-

AIR RESEARCH AND DEVELOPMENT COMMAND

AIR MATERIEL COMMAND. ENGINEERING DIV. ELECTRONIC SUBDIV., WRIGHT-PATTERSON AFB, OHIO (REF. 1)	(CONTRACT NUMBER)-
AIR MATERIEL COMMAND. ENGINEERING DIV. ELECTRONIC SUBDIV., WRIGHT-PATTERSON AFB, OHIO	AAF MCREE-
AIR MATERIEL COMMAND. ENGINEERING DIV. ELECTRONIC SUBDIV., WRIGHT-PATTERSON AFB, OHIO	AMC MCREE-(YR.)-
AIR MATERIEL COMMAND. ENGINEERING DIV. ELECTRONIC SUBDIV., WRIGHT-PATTERSON AFB, OHIO	MCREE-
AIR MATERIEL COMMAND. ENGINEERING DIV. ELECTRONIC SUBDIV., WRIGHT-PATTERSON AFB, OHIO (MEMO RPTS.)	MCREE-(YEAR)-
AIR MATERIEL COMMAND. ENGINEERING DIV. EQUIPMENT LAB., WRIGHT-PATTERSON AFB, OHIO (MEMO. REPT.)	AMC MCREXE-
AIR MATERIEL COMMAND. ENGINEERING DIV. EQUIPMENT LAB., WRIGHT-PATTERSON AFB, OHIO (MEMO. REPT.)	AMC TSEPE-
AIR MATERIEL COMMAND. ENGINEERING DIV. EQUIPMENT LAB., WRIGHT-PATTERSON AFB, OHIO	MCREXE-(NOS.)-
AIR MATERIEL COMMAND. ENGINEERING DIV. PHOTOGRAPHIC LAB., WRIGHT-PATTERSON AFB, OHIO (MEMO REPT.)	MCREXF-(YR.)-
AIR MATERIEL COMMAND. ENGINEERING DIV. POWER PLANT LAB., WRIGHT-PATTERSON AFB, OHIO	MCREXP-(NO.)-(NO.)
AIR MATERIEL COMMAND. ENG. DIV. POWER PLANT LAB., WRIGHT-PATTERSON AFB, OHIO (UNSATISFACTORY REPT.)	U-MCREXP-
AIR MATERIEL COMMAND. ENGINEERING STANDARDS SEC.	EST-
AIR MATERIEL COMMAND. EXPERIMENTAL ENG. SEC.	EXP-
AIR MATERIEL COMMAND. FLIGHT TEST DIV., WRIGHT-PATTERSON AFB, OHIO (MEMO. REPT.)	AMC MCRFTP-
AIR MATERIEL COMMAND. FLIGHT TEST DIV. TEST ENGINEERING SUBDIV., WRIGHT-PATTERSON AFB, OHIO	MCRFT-
AIR MATERIEL COMMAND. GEOPHYSICAL DIVISION, WRIGHT-PATTERSON AFB, OHIO	AMC-GEOD-
AIR MATERIEL COMMAND. ICE RESEARCH BASE	AMC-IRB-
AIR MATERIEL COMMAND. INTELLIGENCE DEPT., WRIGHT-PATTERSON AFB, OHIO (STUDY)	AMC MCI 102-AE-
AIR MATERIEL COMMAND. INTELLIGENCE DEPT., WRIGHT-PATTERSON AFB, OHIO (STUDY)	AMC MCI 102-EL-
AIR MATERIEL COMMAND. INTELLIGENCE DEPT. TECH. ANALYSIS DIV., WRIGHT-PATTERSON AFB, OHIO (REF. 1)	(COMMAND NO.)-(LTRS)...
AIR MATERIEL COMMAND. INTELLIGENCE DIV., WRIGHT-PATTERSON AFB, OHIO (BIBLIOGRAPHY)	AMC F-BB-(NO.)-RE
AIR MATERIEL COMMAND. INTELLIGENCE DIV., WRIGHT-PATTERSON AFB, OHIO (SUMMARY REPORT)	AMC F-SU-(NO.)-ND
AIR MATERIEL COMMAND. INTELLIGENCE DIV., WRIGHT-PATTERSON AFB, OHIO (TECHNICAL REPORT)	AMC F-TR-(NO.)-ND
AIR MATERIEL COMMAND. INTELLIGENCE DIV., WRIGHT-PATTERSON AFB, OHIO (TRANSLATION REPORT)	AMC F-TS-(NO.)-RE
AIR MATERIEL COMMAND. MATERIEL DIV., WRIGHT-PATTERSON AFB, OHIO	AMC AFTR-
AIR MATERIEL COMMAND. SERVICES ENGINEERING SUBDIV., WRIGHT-PATTERSON AFB, OHIO	ESE-
AIR MATERIEL COMMAND. WATSON LABS., RED BANK, N.J.	ENRNT-
AIR NAVIGATION DEVELOPMENT BOARD, WASHINGTON, D.C.	ANDB-
AIR PHOTOGRAPHIC AND CHARTING SERVICE, WRIGHT-PATTERSON AFB, OHIO	APCS-
AIR POLLUTION FOUNDATION, LOS ANGELES	APF-
AIR PRODUCTS, INC., ALLENTOWN, PENNA.	API-
AIR PRODUCTS, INC., ALLENTOWN, PENNA.	API-PR-
AIR PRODUCTS AND CHEMICALS, INC., ALLENTOWN, PENNA.	APCL-
AIR PRODUCTS AND CHEMICALS, INC. RESEARCH AND DEVELOPMENT DEPT., ALLENTOWN, PENNA.	APCI-
AIR PROVING GROUND CENTER, EGLIN AFB, FLA.	APGC-
AIR PROVING GROUND CENTER, EGLIN AFB, FLA.	APGC-AIDR(NUMBER-NUMBER)
AIR PROVING GROUND CENTER, EGLIN AFB, FLA.	APGC-PGY-
AIR PROVING GROUND CENTER, EGLIN AFB, FLA.	APGC-QR-
AIR PROVING GROUND CENTER, EGLIN AFB, FLA.	APGC-SR-(YEAR)-
AIR PROVING GROUND CENTER, EGLIN AFB, FLA. (TECHNICAL DOCUMENTARY REPORT)	APGC-TDR-
AIR PROVING GROUND CENTER, EGLIN AFB, FLA.	APGC-TM-
AIR PROVING GROUND CENTER, EGLIN AFB, FLA.	APGC TN-
AIR PROVING GROUND CENTER, EGLIN AFB, FLA.	APGC-TN-(YR)-
AIR PROVING GROUND CENTER, EGLIN AFB, FLA.	APGC-TR-(YEAR)-
AIR PROVING GROUND CENTER, EGLIN AFB, FLA.	APGC-TR-
AIR PROVING GROUND CTR. SAC PROJ. OFF., EGLIN AFB,FLA	SAC-
AIR PROVING GROUND COMMAND, EGLIN AFB, FLA. (REF. 1)	(BRANCH NO.)-(OTHER NOS.)
AIR PROVING GROUND COMMAND, EGLIN AFB, FLA.	APG-
AIR PROVING GROUND COMMAND, EGLIN AFB, FLA. (PROJECT IDENTIFICATION)	APG/(LTRS.)/
AIR PROVING GROUND COMMAND, EGLIN AFB, FLA.	APGC PROJ-
AIR PROVING GROUND COMMAND, EGLIN AFB, FLA.	APG/CSC/-
AIR PROVING GROUND COMMAND, EGLIN AFB, FLA.	APG/TAT/-
AIR PROVING GROUND COMMAND, KIRTLAND AFB, N. MEX.	DET 1 APGC-
AIR REDUCTION CHEMICAL AND CARBIDE CO., BOUND BROOK, N.J. (USED 1959-64)	(YEAR)-(NUMBER)-BB
AIR REDUCTION CHEMICAL CO.,BOUND BROOK,NJ (1954-60)	BB-
AIR REDUCTION CHEMICAL CO., N.Y.C. (TECHNICAL SALES SERVICE LABORATORY REPORTS)	TSL-
AIR REDUCTION CO., INC., MURRAY HILL, N.J.	P + EDD-
AIR REDUCTION CO., INC. CENTRAL RESEARCH DEPT., MURRAY HILL, N.J. (USED 1961-63)	CRD-(YEAR)-
AIR REDUCTION CO., INC. CENTRAL RESEARCH LABS., MURRAY HILL, N.J.	AIRCO-C-
AIR REDUCTION CO., INC. CENTRAL RESEARCH LABS., MURRAY HILL, N.J.	C-(NUMBER)-
AIR REDUCTION CO., INC. RESEARCH AND ENGINEERING DEPT., MURRAY HILL, N.J. (USED 1963-)	RE-(YEAR)-
AIR REDUCTION CO., INC. RESEARCH AND ENGINEERING DEPT., MURRAY HILL, N.J.	RE-(YEAR)-(NO.)-CRE-
AIR REDUCTION SALES CO., N.Y.C.	ARSC-
AIR REDUCTION SALES CO., UNION, N.J. (USED 1962-63)	PD-(YEAR)-
AIR REDUCTION SALES CO. CRYOGENIC ENG.DEPT.,UNION,NJ	CE-(YEAR)-
AIR RESEARCH AND DEVELOPMENT COMMAND, ANDREWS AFB, MD	ARDC-
AIR RESEARCH AND DEVELOPMENT COMMAND, ANDREWS AFB, MD	ARDC-CATE-
AIR RESEARCH AND DEVELOPMENT COMMAND, ANDREWS AFB, MD	ARDC(RDA-)
AIR RESEARCH AND DEVELOPMENT COMMAND, ANDREWS AFB, MD	ARDC-SR-
AIR RESEARCH AND DEVELOPMENT COMMAND, ANDREWS AFB, MD	ARDC-TN-(YR)-
AIR RESEARCH AND DEVELOPMENT COMMAND, ANDREWS AFB, MD	ARDC-TR-
AIR RESEARCH AND DEVELOPMENT COMMAND, ANDREWS AFB, MD	ARDC-TR-(YR)-
AIR RESEARCH AND DEVELOPMENT COMMAND, ANDREWS AFB, MD	ARPA/TR-
AIR RESEARCH AND DEVELOPMENT COMMAND, ANDREWS AFB, MD	ASTIA AF TPPD-

AIR RES. & DEV. COMMAND

AIR RES. + DEV. COMMAND, ANDREWS AFB, MD. (ASSIGNED TO REPORTS ISSUED BY THE COMMAND AND ITS CONTRACTORS C2- IS 1952, C3- 1953, ETC.)	C2- THRU C9-
AIR RESEARCH AND DEVELOPMENT COMMAND, ANDREWS AFB, MD	C0-
AIR RESEARCH AND DEVELOPMENT COMMAND, ANDREWS AFB, MD	HQARDC-
AIR RES. + DEV. COMMAND, ANDREWS AFB, MD. (COMM.RPT.)	NAS ARDC COM-...
AIR RESEARCH AND DEVELOPMENT COMMAND, ANDREWS AFB, MD	SM-
AIR RESEARCH AND DEVELOPMENT COMMAND, BALTIMORE	AF-ARPD-(NUMBER)(LETTER)
AIR RESEARCH AND DEVELOPMENT COMMAND, BALTIMORE	ARDC C(YR)-
AIR RESEARCH AND DEV. COMMAND, BALTIMORE (MANUAL)	ARDC M-
AIR RESEARCH AND DEVELOPMENT COMMAND, BALTIMORE	ARDCM-
AIR RESEARCH AND DEVELOPMENT COMMAND. DIRECTORATE OF ELECTRONICS, BALTIMORE	RDD-
AIR RESEARCH AND DEVELOPMENT COMMAND. ELECTRONICS RESEARCH DIV., BEDFORD, MASS.	ERD-TN-
AIR RESEARCH AND DEV. COMMAND. EUROPEAN OFFICE	ARDC-EO-
AIR RESEARCH AND DEV. COMMAND. EUROPEAN OFFICE	EOARDC-
AIR RES. + DEV. COMMAND. EUROPEAN OFF., BRUSSELS	USAA-
AIR RESEARCH AND DEVELOPMENT COMMAND. HUMAN FACTORS OPERATIONS RESEARCH LABS., WASHINGTON, D.C.	HFO-
AIR RESEARCH AND DEVELOPMENT COMMAND. HUMAN FACTORS OPERATIONS RESEARCH LABS., WASHINGTON, D.C.	HFORL-
AIR RESEARCH + DEVELOPMENT COMMAND. LIAISON OFFICE, FAR EAST AIR FORCES, TOKYO	ARDCLO-
AIR RESEARCH AND DEVELOPMENT COMMAND. SPECIAL GAM-72 WORKING GROUP	SGWG-GAM-72-
AIR RES.+DEV.COMMAND. WESTERN DEV.DIV.,INGLEWOOD,CAL.	ARDC-CMCC-GM-
AIR RES.+DEV.COMMAND. WESTERN DEV.DIV.,INGLEWOOD,CAL.	ARDC-DAL-
AIR RES.+DEV.COMMAND. WESTERN DEV.DIV.,INGLEWOOD,CAL.	ARDC-RM-
AIR RES.+DEV.COMMAND. WESTERN DEV.DIV.,INGLEWOOD,CAL.	ARDC-WDD-
AIR RES.+DEV.COMMAND. WESTERN DEV.DIV.,INGLEWOOD,CAL.	ARDC-WDOM-(YEAR)-
AIR RES.+DEV.COMMAND. WESTERN DEV.DIV.,INGLEWOOD,CAL.	ARDC-WDPCD-(YEAR)-
AIR RES.+DEV.COMMAND. WESTERN DEV.DIV.,INGLEWOOD,CAL.	ARDC-WDTC-(YEAR)-
AIR RES.+DEV.COMMAND. WESTERN DEV.DIV.,INGLEWOOD,CAL.	WD-
AIR RES.+DEV.COMMAND. WESTERN DEV.DIV.,INGLEWOOD,CAL.	WD-(YR.)-
AIR RES.+DEV.COMMAND. WESTERN DEV.DIV.,INGLEWOOD,CAL.	WDAT-
AIR RES.+DEV.COMMAND. WESTERN DEV.DIV.,INGLEWOOD,CAL.	WDD-
AIR RES.+DEV.COMMAND. WESTERN DEV.DIV.,INGLEWOOD,CAL.	WDD-(YR.)-
AIR RESEARCH AND DEVELOPMENT COMMAND. WESTERN DEVELOPMENT DIV., INGLEWOOD, CALIF. (ACCESSION LISTS)	WDD-AL-
AIR RES.+DEV.COMMAND. WESTERN DEV.DIV.,INGLEWOOD,CAL.	WDD-DOC.(NO.)-
AIR RES.+DEV.COMMAND. WESTERN DEV.DIV.,INGLEWOOD,CAL.	WDD-DOC.(YR.)-
AIR RES.+DEV.COMMAND. WESTERN DEV.DIV.,INGLEWOOD,CAL.	WDG-(YR.)-
AIR RES.+DEV.COMMAND. WESTERN DEV.DIV.,INGLEWOOD,CAL.	WDGH-(YR.)-
AIR RES.+DEV.COMMAND. WESTERN DEV.DIV.,INGLEWOOD,CAL.	WDGH-
AIR RES.+DEV.COMMAND. WESTERN DEV.DIV.,INGLEWOOD,CAL.	WDGP-(YR.)-
AIR RES.+DEV.COMMAND. WESTERN DEV.DIV.,INGLEWOOD,CAL.	WDP-
AIR RES.+DEV.COMMAND. WESTERN DEV.DIV.,INGLEWOOD,CAL.	WDPP-
AIR RES.+DEV.COMMAND. WESTERN DEV.DIV.,INGLEWOOD,CAL.	WDSIT-(YR.)-
AIR RES.+DEV.COMMAND. WESTERN DEV.DIV.,INGLEWOOD,CAL.	WDSIT-
AIR RES.+DEV.COMMAND. WESTERN DEV.DIV.,INGLEWOOD,CAL.	WDSOT-(YR.)-
AIR RES.+DEV.COMMAND. WESTERN DEV.DIV.,INGLEWOOD,CAL.	WDSOT-
(SEE ALSO LATER NAME: BALLISTIC MISSILE DIV (AIR FORCE), INGLEWOOD, CALIF.)	
AIR RESOURCES LAB., LAS VEGAS, NEV.	ARLV-(NUMBER)-(NUMBER)
AIR RESOURCES LAB., LAS VEGAS, NEV.	RLTM-ARL-
AIR RESOURCES LAB., RESEARCH TRIANGLE PARK, N.C.	ERL-ARL-
AIR RESOURCES LAB., RESEARCH TRIANGLE PARK, N.C.	ERLTM-ARL-
AIR RESOURCES LAB., RESEARCH TRIANGLE PARK, N.C.	ESSA-ERLTM-ARL-
AIR RESOURCES LAB., RESEARCH TRIANGLE PARK, N.C.	IOCS-M-ESSA-
AIR RESOURCES LAB., RESEARCH TRIANGLE PARK, N.C.	NOAA-TM-ERL-ARL-
AIR STANDARDIZATION COORDINATING COMMITTEE	ASCC-
AIR TECH. INTELL. CTR., WRIGHT-PATTERSON AFB, OHIO	ATIC-
AIR TECHNICAL INTELLIGENCE CENTER, WRIGHT-PATTERSON AFB, OHIO (STUDY)	ATIC 102-AC-
AIR TECHNICAL INTELLIGENCE CENTER, WRIGHT-PATTERSON AFB, OHIO (STUDY)	ATIC 102-AE-
AIR TECHNICAL INTELLIGENCE CENTER, WRIGHT-PATTERSON AFB, OHIO (STUDY)	ATIC 102-EL-
AIR TECH. INTELL. CTR., WRIGHT-PATTERSON AFB, OHIO	ATIC FE-(NOS.)-(LTRS.)
AIR TECHNICAL INTELLIGENCE CENTER, WRIGHT-PATTERSON AFB, OHIO (TRANSLATION REPORT)	ATIC F-TS-(NO/ROMAN NO.)
AIR TECHNICAL INTELLIGENCE CENTER, WRIGHT-PATTERSON AFB, OHIO (TRANSLATION REPORT)	ATIC F-TS-(NO.)-RE
AIR TECH. INTELL. CTR., WRIGHT-PATTERSON AFB, OHIO	ATIC T-(YR.)-(NO.)
AIR TECH. INTELL. CTR., WRIGHT-PATTERSON AFB, OHIO	ATIC TIH-(LTRS)(YR)(NO.)
AIR TECH. INTELL. CTR., WRIGHT-PATTERSON AFB, OHIO	ATIC TIR-(LTRS)(YR)(NO.)
AIR TECH. INTELL. CTR., WRIGHT-PATTERSON AFB, OHIO	ATIC-TIR-PR-(YEAR)-
AIR TECH. INTELL. CTR., WRIGHT-PATTERSON AFB, OHIO	ATIC-TR-
AIR TECH. INTELL. CTR., WRIGHT-PATTERSON AFB, OHIO	ATIC TR-(LTRS)-(NO.)
AIR TECH. INTELL. CTR., WRIGHT-PATTERSON AFB, OHIO	F-TS-(NUMBER)-RE
AIR TECH. INTELL. CTR., WRIGHT-PATTERSON AFB, OHIO	MIF-
AIR TECH. INTELL. CTR., WRIGHT-PATTERSON AFB, OHIO	TIH-AC-
AIR TECH. INTELL. CTR., WRIGHT-PATTERSON AFB, OHIO	TIH-WI-
AIR TECH. INTELL. CTR., WRIGHT-PATTERSON AFB, OHIO	TIS-ES-(YR)-
AIR TECH. INTELL. CTR., WRIGHT-PATTERSON AFB, OHIO	TIS-PR-
AIR TECH. INTELL. CTR., WRIGHT-PATTERSON AFB, OHIO	TIS-PR-(YR.)-
AIR TECH. INTELL. CTR., WRIGHT-PATTERSON AFB, OHIO	TWP-PR-
(SEE ALSO LATER NAME: AEROSPACE TECH. INTELL. CTR.)	
AIR TECHNICAL INTELLIGENCE SERVICE. FAR EAST COMMAND	ATIS-
AIR TECH. SERVICE COMMAND, WRIGHT-PATTERSON AFB, OHIO	AAF-TSEAM-M-
AIR TECH. SERVICE COMMAND, WRIGHT-PATTERSON AFB, OHIO	ASC-
AIR TECH. SERVICE COMMAND, WRIGHT-PATTERSON AFB, OHIO	ATSC-
AIR TECHNICAL SERVICE COMMAND, WRIGHT-PATTERSON AFB, OHIO (MANUAL)	ATSC-M-
AIR TECHNICAL SERVICE COMMAND, WRIGHT-PATTERSON AFB, OHIO (SUPERVISOR TRAINING CONFERENCE OUTLINE)	ATSC-STCO-
AIR TECHNICAL SERVICE COMMAND, WRIGHT-PATTERSON AFB, OHIO (SUPERVISOR TRAINING CONFERENCE OUTLINE)	STCO-
AIR TECH. SERVICE COMMAND, WRIGHT-PATTERSON AFB, OHIO	TSEAM-M-
AIR TECHNICAL SERVICE COMMAND. ENGINEERING DIV., WRIGHT-PATTERSON AFB, OHIO	AAF ENG-
AIR TECHNICAL SERVICE COMMAND. ENGINEERING DIV. WRIGHT-PATTERSON AFB, OHIO	AAF ENG (NOS.,LETTERS)-
AIR TECHNICAL SERVICE COMMAND. ENGINEERING DIV. WRIGHT-PATTERSON AFB, OHIO	AAF ENG (LETTERS,NOS.)-
AIR TECHNICAL SERVICE COMMAND. ENGINEERING DIV. WRIGHT-PATTERSON AFB, OHIO	AAF TN-
AIR TECHNICAL SERVICE COMMAND. ENGINEERING DIV., WRIGHT-PATTERSON AFB, OHIO	AAF U-ENG-
AIR TECHNICAL SERVICE COMMAND. ENGINEERING DIV., WRIGHT-PATTERSON AFB, OHIO	AAF W-U-ENG-
AIR TECHNICAL SERVICE COMMAND. ENGINEERING DIV. AERO-MEDICAL LAB., WRIGHT-PATTERSON AFB, OHIO	AAF TSEAA-
AIR TECHNICAL SERVICE COMMAND. ENG. DIV. AERO-MEDICAL LAB., WRIGHT-PATTERSON AFB, OHIO (MEMO. RPTS.)	AAF TSEAA MR-
AIR TECHNICAL SERVICE COMMAND. ENGINEERING DIV. AERO-MEDICAL LAB., WRIGHT-PATTERSON AFB, OHIO	AAF TSEAL-
AIR TECHNICAL SERVICE COMMAND. ENGINEERING DIV. AERO-MEDICAL LAB., WRIGHT-PATTERSON AFB, OHIO	AAF TSEAL (NO.,LETTER)-

AIR TECHNICAL SERVICE COMMAND. ENG. DIV. AERO-MEDICAL LAB., WRIGHT-PATTERSON AFB, OHIO (MEMO. RPTS.)	DCRD-
AIR TECHNICAL SERVICE COMMAND. ENG. DIV. AERO-MEDICAL LAB., WRIGHT-PATTERSON AFB, OHIO (MEMO. RPTS.)	DCRDE-
AIR TECHNICAL SERVICE COMMAND. ENGINEERING DIV. AERO-MEDICAL LAB., WRIGHT-PATTERSON AFB, OHIO	TSEAA-
AIR TECHNICAL SERVICE COMMAND. ENGINEERING DIV. AERO-MEDICAL LAB., WRIGHT-PATTERSON AFB, OHIO	TSEAL-
AIR TECHNICAL SERVICE COMMAND. ENGINEERING DIV. AIRCRAFT LAB., WRIGHT-PATTERSON AFB, OHIO	AAF TSEAC-
AIR TECHNICAL SERVICE COMMAND. ENG. DIV. AIRCRAFT LAB., WRIGHT-PATTERSON AFB, OHIO (MEMO. RPTS.)	AAF TSEAC MR-
AIR TECHNICAL SERVICE COMMAND. ENGINEERING DIV. AIRCRAFT LAB., WRIGHT-PATTERSON AFB, OHIO	TSEAC-
AIR TECHNICAL SERVICE COMMAND. ENGINEERING DIV. AIRCRAFT PROJ. SEC., WRIGHT-PATTERSON AFB, OHIO	AAF TSESA-
AIR TECHNICAL SERVICE COMMAND. ENGINEERING DIV. AIRCRAFT PROJ. SEC., WRIGHT-PATTERSON AFB, OHIO	TSESA-
AIR TECHNICAL SERVICE COMMAND. ENGINEERING DIV. ARMAMENT LAB., WRIGHT-PATTERSON AFB, OHIO	AAF TSEPS-
AIR TECHNICAL SERVICE COMMAND. ENGINEERING DIV. ARMAMENT LAB., WRIGHT-PATTERSON AFB, OHIO	TSEPS-
AIR TECHNICAL SERVICE COMMAND. ENGINEERING DIV. CLOTHING BRANCH, WRIGHT-PATTERSON AFB, OHIO	CB-
AIR TECHNICAL SERVICE COMMAND. ENG. DIV. COMMUNICATION + NAVIGATION LAB., WRIGHT-PATTERSON AFB, OHIO	AAF TSELC-
AIR TECHNICAL SERVICE COMMAND. ENG. DIV. COMMUNICATION + NAVIGATION LAB., WRIGHT-PATTERSON AFB, OHIO	TSELC-
AIR TECHNICAL SERVICE COMMAND. ENG. DIV. ELECTRONICS ADMINISTRATIVE SEC., WRIGHT-PATTERSON AFB, OHIO	AAF TSELA-
AIR TECHNICAL SERVICE COMMAND. ENG. DIV. ELECTRONICS ADMINISTRATIVE SEC., WRIGHT-PATTERSON AFB, OHIO	AAF TSELA(NO.,LTR.,NO.)-
AIR TECHNICAL SERVICE COMMAND. ENG. DIV. ELECTRONICS ADMINISTRATIVE SEC., WRIGHT-PATTERSON AFB, OHIO	TSELA-
AIR TECHNICAL SERVICE COMMAND. ENG. DIV. ELECTRONICS PLANS SECTION., WRIGHT-PATTERSON AFB, OHIO	AAF TSELP-
AIR TECHNICAL SERVICE COMMAND. ENG. DIV. ELECTRONICS PLANS SECTION., WRIGHT-PATTERSON AFB, OHIO	TSELP-
AIR TECHNICAL SERVICE COMMAND. ENGINEERING DIV. ELECTRONICS SUBDIV., WRIGHT-PATTERSON AFB, OHIO	AAF TN TSELR-
AIR TECHNICAL SERVICE COMMAND. ENGINEERING DIV. ELECTRONICS SUBDIV., WRIGHT-PATTERSON AFB, OHIO	AAF TSELR-
AIR TECHNICAL SERVICE COMMAND. ENGINEERING DIV. ELECTRONICS SUBDIV., WRIGHT-PATTERSON AFB, OHIO	AAF TSELR ESMR (NO.)-
AIR TECHNICAL SERVICE COMMAND. ENGINEERING DIV. ELECTRONICS SUBDIV., WRIGHT-PATTERSON AFB, OHIO	TSELR-
AIR TECHNICAL SERVICE COMMAND. ENGINEERING DIV. ENG. PLANS DIV., WRIGHT-PATTERSON AFB, OHIO	AAF TSEPL-
AIR TECHNICAL SERVICE COMMAND. ENGINEERING DIV. ENG. PLANS DIV., WRIGHT-PATTERSON AFB, OHIO	AAF U-TSEPL-
AIR TECHNICAL SERVICE COMMAND. ENGINEERING DIV. ENG. PLANS DIV., WRIGHT-PATTERSON AFB, OHIO	AAF W-U-TSEPL-
AIR TECHNICAL SERVICE COMMAND. ENGINEERING DIV. ENG. PLANS DIV., WRIGHT-PATTERSON AFB, OHIO	TSEPL-
AIR TECHNICAL SERVICE COMMAND. ENGINEERING DIV. ENG. STDS. SEC., WRIGHT-PATTERSON AFB, OHIO	AAF EST-
AIR TECHNICAL SERVICE COMMAND. ENGINEERING DIV. ENG. STDS. SEC., WRIGHT-PATTERSON AFB, OHIO	AAF TSEST-
AIR TECHNICAL SERVICE COMMAND. ENGINEERING DIV. ENG. STDS. SEC., WRIGHT-PATTERSON AFB, OHIO	TSEST-
AIR TECHNICAL SERVICE COMMAND. ENGINEERING DIV. EQUIPMENT LAB., WRIGHT-PATTERSON AFB, OHIO	AAF TSEPE-
AIR TECHNICAL SERVICE COMMAND. ENGINEERING DIV. EQUIPMENT LAB., WRIGHT-PATTERSON AFB, OHIO	TSEPE-
AIR TECHNICAL SERVICE COMMAND. ENGINEERING DIV. EXPTL. ENG. SEC., WRIGHT-PATTERSON AFB, OHIO	AAF EXP-
AIR TECHNICAL SERVICE COMMAND. ENGINEERING DIV. EXPTL. ENG. SEC., WRIGHT-PATTERSON AFB, OHIO	AAF EXP-M-(YR-LETTERS)-
AIR TECHNICAL SERVICE COMMAND. ENGINEERING DIV. EXPTL. ENG. SEC., WRIGHT-PATTERSON AFB, OHIO	AAF U-EXP-
AIR TECHNICAL SERVICE COMMAND. ENGINEERING DIV. EXPTL. ENG. SEC., WRIGHT-PATTERSON AFB, OHIO	AAF U-EXP-M-(YR-LTRS)-
AIR TECHNICAL SERVICE COMMAND. ENGINEERING DIV. EXPTL. ENG. SEC., WRIGHT-PATTERSON AFB, OHIO	AAFW-U-EXP-M-(YR)-
AIR TECHNICAL SERVICE COMMAND. ENGINEERING DIV. FLYING CLOTHING BR., WRIGHT-PATTERSON AFB, OHIO	AAF CB-
AIR TECHNICAL SERVICE COMMAND. ENGINEERING DIV. FLYING CLOTHING BR., WRIGHT-PATTERSON AFB, OHIO	AAF FCB-
AIR TECHNICAL SERVICE COMMAND. ENGINEERING DIV. MATERIALS LAB., WRIGHT-PATTERSON AFB, OHIO	AAF TSEAM-
AIR TECHNICAL SERVICE COMMAND. ENG. DIV. MATERIALS LAB., WRIGHT-PATTERSON AFB, OHIO (MEMO. RPTS.)	AAF TSEAM MR-
AIR TECHNICAL SERVICE COMMAND. ENGINEERING DIV. MATERIALS LAB., WRIGHT-PATTERSON AFB, OHIO	TSEAM-
AIR TECHNICAL SERVICE COMMAND. ENG. DIV. MATERIALS LAB., WRIGHT-PATTERSON AFB, OHIO (MEMO. RPTS.)	TSEAM MR-
AIR TECHNICAL SERVICE COMMAND. ENGINEERING DIV. PERSONAL EQUIPMENT LAB., WRIGHT-PATTERSON AFB, OHIO	AAF TSEAP-
AIR TECHNICAL SERVICE COMMAND. ENGINEERING DIV. PERSONAL EQUIPMENT LAB., WRIGHT-PATTERSON AFB, OHIO	TSEAP-
AIR TECHNICAL SERVICE COMMAND. ENGINEERING DIV. PHOTO LAB., WRIGHT-PATTERSON AFB, OHIO	AAF TSEPF-
AIR TECHNICAL SERVICE COMMAND. ENGINEERING DIV. PHOTO LAB., WRIGHT-PATTERSON AFB, OHIO	TSEPF-
AIR TECHNICAL SERVICE COMMAND. ENGINEERING DIV. POWER PLANT LAB., WRIGHT-PATTERSON AFB, OHIO	AAF TSEPP-
AIR TECHNICAL SERVICE COMMAND. ENGINEERING DIV. POWER PLANT LAB., WRIGHT-PATTERSON AFB, OHIO	AAF U-TSEPP-
AIR TECHNICAL SERVICE COMMAND. ENGINEERING DIV. POWER PLANT LAB., WRIGHT-PATTERSON AFB, OHIO	TSEPP-
AIR TECHNICAL SERVICE COMMAND. ENGINEERING DIV. POWER PLANT LAB., WRIGHT-PATTERSON AFB, OHIO	U TSEPP-
AIR TECHNICAL SERVICE COMMAND. ENGINEERING DIV. PROPELLER LAB., WRIGHT-PATTERSON AFB, OHIO	AAF TSEPR-
AIR TECHNICAL SERVICE COMMAND. ENGINEERING DIV. PROPELLER LAB., WRIGHT-PATTERSON AFB, OHIO	TSEPR-
AIR TECHNICAL SERVICE COMMAND. ENG. DIV. RADIO + RADAR SUBDIV., WRIGHT-PATTERSON AFB, OHIO	AAF TSERR-
AIR TECHNICAL SERVICE COMMAND. ENG. DIV. RADIO + RADAR SUBDIV., WRIGHT-PATTERSON AFB, OHIO	TSERR-
AIR TECHNICAL SERVICE COMMAND. ENGINEERING DIV. SERVICE ENG. SUBDIV., WRIGHT-PATTERSON AFB, OHIO	AAF ESE-
AIR TECHNICAL SERVICE COMMAND. ENGINEERING DIV. SERVICE ENG. SUBDIV., WRIGHT-PATTERSON AFB, OHIO	AAF TSESE-
AIR TECHNICAL SERVICE COMMAND. ENGINEERING DIV. SERVICE ENG. SUBDIV., WRIGHT-PATTERSON AFB, OHIO	TSESE-
AIR TECHNICAL SERVICE COMMAND. ENGINEERING DIV. SYSTEMS LAB., WRIGHT-PATTERSON AFB, OHIO	AAF TSELS-
AIR TECHNICAL SERVICE COMMAND. ENGINEERING DIV. SYSTEMS LAB., WRIGHT-PATTERSON AFB, OHIO	TSELS-
AIR TRAINING COMMAND	ATC-
AIR TRAINING COMMAND, LOWRY AFB, COLO.	ATC OJT(LTR.)-(LTRS.)
AIR TRAINING COMMAND, LOWRY AFB, COLO.	OJTA-
AIR TRAINING COMMAND, RANDOLPH AFB, TEXAS	ATC OJT(LTR.)-(LTRS.)
AIR TRAINING COMMAND, SCOTT AFB, ILL.	AFTRC-
AIR TRAINING COMMAND, SCOTT AFB, ILL.	ATRC-
AIR TRANSPORT ASSN. OF AMERICA. SHIPPERS RESEARCH DIV., WASHINGTON, D.C.	SR-
AIR UNIV., MAXWELL AFB, ALA. (REF. 1)	(NUMBERS...)
AIR UNIV., MAXWELL AFB, ALA. (ARCTIC REPORTS)	A-
AIR UNIV., MAXWELL AFB, ALA.	AC+SS-
AIR UNIV., MAXWELL AFB, ALA.	AIRU-
AIR UNIV., MAXWELL AFB, ALA.	AU-THESIS-
AIR UNIV., MAXWELL AFB, ALA. ARCTIC, DESERT, TROPIC INFORMATION CENTER	ADTIC-
AIR UNIV., MAXWELL AFB, ALA. ARCTIC, DESERT, TROPIC INFORMATION CENTER	ADTIC-PUB-G-
AIR UNIV., MAXWELL AFB, ALA. AEROSPACE STUDIES INST.	AU-(NUMBER-NUMBER)-ASI
AIR UNIV., MAXWELL AFB, ALA. RESEARCH STUDIES INST.	AU-
AIR VEHICLE ENVIRONMENTAL RESEARCH TEAM, NATICK, MASS	AE-
AIR VEHICLE ENVIRONMENTAL RESEARCH TEAM, NATICK, MASS	AVERT-

AIR WAR COLLEGE

AIR WAR COLLEGE, MAXWELL AFB, ALA.	AIRU-AWC-
AIR WAR COLLEGE. MAXWELL AFB, ALA.	AWC-
AIR WAR COLLEGE, MAXWELL AFB, ALA.	RR-
AIR WAR COLL., MAXWELL AFB, ALA. EVALUATION STAFF	AIRU-AU-
AIR WAR COLL., MAXWELL AFB, ALA. EVALUATION STAFF	AIRU AU-(NUMBER)-ES
AIR WEATHER SERVICE, MAC DILL AFB., FLA.	BSSR-
AIR WEATHER SERVICE, MAC DILL AFB., FLA.	BS TC-
AIR WEATHER SERVICE, MAC DILL AFB., FLA.	BSTN-
AIR WEATHER SERVICE, SCOTT AFB, ILL.	AWS-
AIR WEATHER SERVICE, SCOTT AFB, ILL.	AWSM-(NUMBER)-
AIR WEATHER SERVICE, SCOTT AFB, ILL.	AWS MANUAL-
AIR WEATHER SERVICE, SCOTT AFB, ILL.	AWSP-(NUMBER)-
AIR WEATHER SERVICE, SCOTT AFB, ILL.	AWS PAM-
AIR WEATHER SERVICE, SCOTT AFB, ILL. (RECURRING PUBLICATION)	AWS-RP-(NOS. MO./YR.)
AIR WEATHER SERVICE, SCOTT AFB, ILL.	AWS SPECIAL STUDY-
AIR WEATHER SERVICE, SCOTT AFB, ILL.	AWS-TR-
AIR WEATHER SERVICE, WASHINGTON, D.C.	AAF AWS-
AIR WEATHER SERVICE, WASHINGTON, D.C.	AAF-AWSM-
AIR WEATHER SERVICE, WASHINGTON, D.C.	AAF-AWSMTR-
AIR WEATHER SERVICE, WASH., D.C. (RUSSIAN REPORT)	AAF AWS R-
AIR WEATHER SERVICE, WASHINGTON, D.C.	AFTSW-
AIR WEATHER SERVICE, WASHINGTON, D.C.	AWS-
AIR WEATHER SERVICE, WASHINGTON, D.C. (BIBLIOGRAPHY)	AWS-BIB-
AIR WEATHER SERVICE, WASHINGTON, D.C. (MEMORANDUM)	AWSM-
AIR WEATHER SERVICE, WASHINGTON, D.C.	AWSM-(NUMBER)-
AIR WEATHER SERVICE, WASHINGTON, D.C.	AWSM-TR-
AIR WEATHER SERVICE, WASHINGTON, D.C.	AWSMTR-
AIR WEATHER SERVICE, WASHINGTON, D.C.	AWSR-
AIR WEATHER SERVICE, WASHINGTON, D.C.	AWS-TR-
AIR WEATHER SERVICE, WASHINGTON, D.C.	WGP-
AIRBORNE INSTRUMENTS LAB., INC., DEER PARK, N.Y.	AIL-
AIRBORNE INSTRUMENTS LAB., INC., DEER PARK, N.Y.	AIL-D/N(NUMBERS)
AIRBORNE INSTRUMENTS LAB., INC., DEER PARK, N.Y.	RD-(YEAR)-
AIRBORNE INSTRUMENTS LAB., INC., DEER PARK, N.Y.	SPEC-AIL-
AIRBORNE INSTRUMENTS LAB., INC., MINEOLA, N.Y.(REF.1)	(PROJECT NUMBER-NO.)-
AIRBORNE INSTRUMENTS LAB., INC., MINEOLA, N.Y.	AIL-
AIRBORNE-QUARTERMASTER RESEARCH + ENG. COMMAND (ARMY)	ABN-
AIRCO CHEMICAL DIV., CUMBERLAND CHEMICAL CORP., MIDDLESEX, N.J. (USED 1965-)	MX-(YEAR)-
AIRCO INDUSTRIAL GASES DIV., JERSEY CITY, NJ (1965-)	IGE-
AIRCO WELDING PRODUCTS DIV., UNION, N.J. (ENGINEERING AND DEVELOPMENT REPORTS, 1964-)	PEDD-(YEAR)-
AIRCRAFT ARMAMENTS, INC., BALTIMORE	AA-
AIRCRAFT ARMAMENTS, INC., COCKEYSVILLE, MD.	AIRA-
AIRCRAFT ARMAMENTS, INC., COCKEYSVILLE, MD. (SEE ALSO LATER NAME: AAI CORP.)	ER-
AIRCRAFT ENGINE RESEARCH LAB., CLEVELAND (REF. 25)	NACA-(LTRS)-E-(DATE CODE)
AIRCRAFT INDUSTRIES ASSOCIATION, WASHINGTON, D.C.	AIA-
AIRCRAFT INDUSTRIES ASSN., AIRCRAFT RESEARCH AND TESTING COMMITTEE, WASHINGTON, D.C.	AIA-ARTC-
AIRCRAFT INDUSTRIES ASSN., AIRCRAFT RESEARCH AND TESTING COMMITTEE, WASHINGTON, D.C.	ARTC-WR-
AIRCRAFT NUCLEAR PROPULSION PROGRAM (REF. 3)	ANP-
AIRCRAFT NUCLEAR PROPULSION PROGRAM (REF. 3)	DC-(YEAR)-(NUMBER)-
AIRCRAFT OBSERVER RESEARCH LAB., MATHER AFB, CALIF.	BON-
AIRCRAFT RESEARCH ASSN., LTD., BEDFORD, BEDS, ENG.	ARA-
AIRCRAFT RESEARCH ASSN., LTD., BEDFORD, BEDS, ENG.	ARA-LIB-TRANS-
AIRCRAFT RESEARCH ASSN., LTD., BEDFORD, BEDS, ENG.	ARA-M.(NO.)/
AIRCRAFT RESEARCH ASSN., LTD., BEDFORD, BEDS, ENG.	ARA-TRANSL-
AIRCRAFT RESEARCH ASSN., LTD., BEDFORD, BEDS., ENG	ARA-WTN-
AIRSHIP EXPERIMENTAL CENTER, LAKEHURST, N.J.	AEC-
AIRTECHNOLOGY CORP., CAMBRIDGE, MASS.	AT-
AIRTECHNOLOGY CORP., CAMBRIDGE, MASS.	MPR-
AIRTRON, INC., LINDEN, N.J.	ETL-
AIRTRON, INC., MORRIS PLAINS, N.J.	R(NO.)-
AIX-MARSEILLES UNIV., FRANCE	NT-
AIX-MARSEILLES UNIV., FRANCE. INSTITUT DE MECANIQUE STATISTIQUE DE LA TURBULENCE (ANNUAL SCI. REPT.)	A.S.R.(NUMBER)
AKADEMIYA NAUK ARMYANSKOI SSR. INSTITUT FIZIKI, EREVAN	EFI-TF-(NUMBER-YEAR)
AKADEMIYA NAUK GRUZINSKOI SSR. FIZIKO-TEKHNICHESKII INSTITUT, SUKHUMI	FTI-(YEAR)-
AKADEMIYA NAUK SSSR., MOSCOW	CN-(NUMBER)/D-
AKADEMIYA NAUK SSSR., MOSCOW	D-(2 DIGITS)
AKADEMIYA NAUK SSSR., MOSCOW	SM-
AKADEMIYA NAUK SSSR. INSTITUT ATOMNOI ENERGII, MOSCOW	IFA-FR-
AKADEMIYA NAUK SSSR. INSTITUT FIZICHESKIKH PROBLEMI, MOSCOW	FEI-
AKADEMIYA NAUK SSSR. INSTITUT FIZIKI, MOSCOW	CTO-
AKADEMIYA NAUK SSSR. INSTITUT FIZIKI, MOSCOW	INP-(NUMBER)/PH
AKADEMIYA NAUK SSSR. INSTITUT MATEMATIKI, NOVOSIBIRSK	TF-(NUMBER)(LETTER)
AKADEMIYA NAUK SSSR. INSTITUT PRIKLADNOI MATEMATIKI, MOSCOW	IPM-
AKADEMIYA NAUK SSSR. INSTITUT TEORETICHESKOI I EKSPERIMENTAL'NOI FIZIKI, MOSCOW	ITEP-
AKADEMIYA NAUK SSSR. INSTITUT YADERNOI FIZIKI, NOVOSIBIRSK	IIF-(NUMBER-YEAR)

AKADEMIYA NAUK SSSR. INSTITUT YADERNOI FIZIKI, NOVOSIBIRSK	IIF-
AKADEMIYA NAUK SSSR. INSTITUT YADERNOI FIZIKI, NOVOSIBIRSK	IIF-LA-
AKADEMIYA NAUK SSSR. INSTITUT YADERNOI FIZIKI, NOVOSIBIRSK	ITF-(YEAR)-
AKADEMIYA NAUK SSSR. INSTITUT YADERNOI FIZIKI, NOVOSIBIRSK	IYAF-(NUMBER)-(YEAR)
AKADEMIYA NAUK SSSR. ORDENA FIZIKO-TECHNICHESKII INSTITUT IMENI A.F. IOFFE, LENINGRAD	FTI-
AKADEMIYA NAUK SSSR. ORDENA LENINA FIZICHESKIYA INST. IMENI P.N. LEBEDEVA	FIPNL-
AKADEMIYA NAUK SSSR. RADIOTEKHNICHESKII INST., MOSCOW	NT-
AKADEMIYA NAUK SSSR. RADIOTEKHNICHESKII INST., MOSCOW	NTD-
AKADEMIYA NAUK UKRAINSKOI SSR. FIZIKO-TEKHNICHESKII INSTITUT, KHARKOV	FTI-(NUMBER)/R-
AKADEMIYA NAUK UKRAINSKOI SSR. FIZIKO-TEKHNICHESKII INSTITUT, KHARKOV	FTI-AN-
AKADEMIYA NAUK UKRAINSKOI SSR. FIZIKO-TEKHNICHESKII INSTITUT, KHARKOV	FTI-LA-
AKADEMIYA NAUK UKRAINSKOI SSR. FIZIKO-TEKHNICHESKII INSTITUT, KHARKOV	ITF-(YEAR)-
AKADEMIYA NAUK UKRAINSKOI SSR. FIZIKO-TEKHNICHESKII INSTITUT, KHARKOV	KFTI-(YEAR/NUMBER)
AKADEMIYA NAUK UKRAINSKOI SSR. FIZIKO-TEKHNICHESKII INSTITUT, KHARKOV	KHFTI-(YEAR/NUMBER)
AKADEMIYA NAUK UKRAINSKOI SSR. INSTITUT TEORETICHESKOI FIZIKI, KIEV	IFA-(YEAR)-
AKADEMIYA NAUK UKRAINSKOI SSR. INSTITUT TEORETICHESKOI FIZIKI, KIEV	INF-(YEAR)-
AKADEMIYA NAUK UKRAINSKOI SSR. INSTITUT TEORETICHESKOI FIZIKI, KIEV	ITF-(YEAR)-
AKADEMIYA NAUK UKRAINSKOI SSR. INSTITUT TEORETICHESKOI FIZIKI, KIEV	ITK-(YEAR)-
AKRON, OHIO. UNIV.	AU-
AKRON, OHIO. UNIV. INST. OF POLYMER SCIENCE	AU-IPS-PR-
AKRON, OHIO. UNIV. INST. OF RUBBER RESEARCH	AU-IRR-FR-
AKRON, OHIO. UNIV. INST. OF RUBBER RESEARCH	AU-IRR-PR-
AKRON, OHIO. UNIV. INST. OF RUBBER RESEARCH	AU-IRR-TR-
AKTIEBOLAGET ATOMENERGI, STOCKHOLM	AE-
AKTIEBOLAGET ATOMENERGI, STOCKHOLM	AEF-
AKTIEBOLAGET ATOMENERGI, STOCKHOLM	AEFI-
AKTIEBOLAGET ATOMENERGI, STOCKHOLM (INELASTIC NEUTRON SCATTERING NEWSLETTER)	AE-INSN-
AKTIEBOLAGET ATOMENERGI, STOCKHOLM	AE-RL-
AKTIEBOLAGET ATOMENERGI, STOCKHOLM	AE-RTG-
AKTIEBOLAGET ATOMENERGI, STOCKHOLM	AES-
AKTIEBOLAGET ATOMENERGI, STOCKHOLM	AE-TR-
AKTIEBOLAGET ATOMENERGI, STOCKHOLM	FRIGG-
AKTIEBOLAGET ATOMENERGI, STOCKHOLM	FTB-
AKTIEBOLAGET ATOMENERGI, STOCKHOLM	FTI-
AKTIEBOLAGET ATOMENERGI, STOCKHOLM	R-
AKTIEBOLAGET ATOMENERGI, STOCKHOLM	RFA-
AKTIEBOLAGET ATOMENERGI, STOCKHOLM	RFR-
AKTIEBOLAGET ATOMENERGI, STOCKHOLM	RFX-
AKTIEBOLAGET ATOMENERGI, STOCKHOLM	RMA-
AKTIEBOLAGET ATOMENERGI, STOCKHOLM	RMM-
AKTIEBOLAGET ATOMENERGI, STOCKHOLM	RPM-RFX-
AKTIEBOLAGET ATOMENERGI, STOCKHOLM	RSA-
AKTIEBOLAGET ATOMENERGI, STOCKHOLM	S-
AKTIEBOLAGET ATOMENERGI, STOCKHOLM	SSI-
AKTIEBOLAGET ATOMENERGI, STOCKHOLM	TPM-RFX-
AKTIEBOLAGET ATOMENERGI, STOCKHOLM	VDDIT-
AKTIEBOLAGET ATOMENERGI, STOCKHOLM	VDIT-
AKTIEBOLAGET ATOMENERGI, STOCKHOLM	VDIT-(NUMBER).(NUMBER)
AKTIEBOLAGET ATOMENERGI, STUDSVIK, SWEDEN	AE-
AKTIEBOLAGET ATOMENERGI, STUDSVIK, SWEDEN	AE-RTL-
AKTIEBOLAGET ATOMENERGI, STUDSVIK, SWEDEN	AE-RTR-
AKTIEBOLAGET ATOMENERGI, STUDSVIK, SWEDEN	AE-SSI-
AKTIEBOLAGET ATOMENERGI, STUDSVIK, SWEDEN	FFN-
AKTIEBOLAGET ATOMENERGI, STUDSVIK, SWEDEN	FTB-
AKTIEBOLAGET ATOMENERGI, STUDSVIK, SWEDEN	RFN-
AKTIEBOLAGET ATOMENERGI, STUDSVIK, SWEDEN.	RFR-
AKTIEBOLAGET ATOMENERGI, STUDSVIK, SWEDEN	S-
AKTIEBOLAGET ATOMENERGI, STUDSVIK, SWEDEN	STU-(YEAR)-(NUMBER)/U-
AKTIEBOLAGET ATOMENERGI, STUDSVIK, SWEDEN	TPM-RFR-
AKTIEBOLAGET ATOMENERGI, STUDSVIK, SWEDEN	VDDIT-
ALABAMA. UNIV., HUNTSVILLE. RESEARCH INSTITUTE	IR-
ALABAMA. UNIV., HUNTSVILLE. RESEARCH INSTITUTE	UAHRI-(YEAR)-
ALABAMA. UNIV., HUNTSVILLE. RESEARCH INSTITUTE	UARI-
ALABAMA. UNIV., HUNTSVILLE. RESEARCH INSTITUTE	UARI-RR-
ALABAMA. UNIV., TUSCALOOSA	ALAU-
ALABAMA. UNIV., UNIVERSITY. BUREAU OF ENG. RESEARCH	BER-(NUMBER-YEAR)
ALABAMA ORDNANCE WORKS, CHILDERSBURG	AOW-
ALABAMA POLYTECHNIC INSTITUTE, AUBURN	API-
ALABAMA STATE HIGHWAY DEPT., MONTGOMERY	HPR-
ALASKA. UNIV., COLLEGE	ALASU-
ALASKA. UNIV., COLLEGE	ALSKU-
ALASKA. UNIV., COLLEGE	UAGIR-
ALASKA. UNIV., COLLEGE	VAG-
ALASKA. UNIV., COLLEGE. ARCTIC ENVIRONMENTAL ENG. LAB	UAEEL-RR-
ALASKA. UNIV., COLLEGE. GEOPHYSICAL INST.(FINAL RPT.)	AU-FR-(YEAR/MONTH)
ALASKA. UNIV., COLLEGE. GEOPHYSICAL INST. (GEOPHYSICAL RESEARCH REPORT)	AU-GRR-
ALASKA. UNIV., COLLEGE. GEOPHYSICAL INST. (INTERIM SCIENTIFIC REPORT)	AU-ISR-
ALASKA. UNIV., COLLEGE. GEOPHYSICAL INST. (QUARTERLY PROGRESS REPORT)	AU-QPR-
ALASKA. UNIV., COLLEGE. GEOPHYSICAL INST. (SUPPLEMENTARY PROGRESS REPORT)	AU-SPR-
ALASKA. UNIV., COLLEGE. GEOPHYSICAL INST. (SCI. RPT.)	AU-SR-
ALASKA. UNIV., COLLEGE. GEOPHYSICAL INST.	UAG-
ALASKA. UNIV., COLLEGE. GEOPHYSICAL INST.	UAG-C-
ALASKA. UNIV., COLLEGE. GEOPHYSICAL INST.	UAG-R-
ALASKA. UNIV., COLLEGE. INST. OF BUSINESS, GOVERNMENT AND RESEARCH	IBEG-
ALASKA. UNIV., COLLEGE. INST. OF MARINE SCIENCES	SAN-(NUMBER)-P-
ALASKA. UNIV., COLLEGE. INST. OF SOCIAL, ECONOMIC AND GOVERNMENT RESEARCH	ISEG-
ALASKA. UNIV., COLLEGE. INST. OF SOCIAL, ECONOMIC AND GOVERNMENT RESEARCH	ISEGR-

ALASKA. UNIV., COLLEGE. INST. OF WATER RESOURCES	IWR-
ALASKAN AIR COMMAND, ELMENDORF AFB	AAC-
ALASKAN AIR COMMAND, ELMENDORF AFB (OPERATIONS ANALYSIS REPORT)	AAC-OAR-
ALASKAN AIR COMMAND, ELMENDORF AFB (OPERATIONS ANALYSIS TECHNICAL MEMORANDUM)	AAC OA TM-
ALASKAN AIR COMMAND, ELMENDORF AFB (OPERATIONS ANALYSIS WORKING PAPER)	AAC OA WP-
ALASKAN AIR COMMAND, LADD AFB	AAC-
ALBERTA UNIV., EDMONTON. NUCLEAR RESEARCH CENTRE	UAE-NPL-
ALBRIGHT + WILSON, LTD., OLDBURY, BIRMINGHAM, ENG.	IT-
ALBUQUERQUE OPERATIONS OFFICE, AEC, N. MEX. (REF.4)	ALO-
ALBUQUERQUE OPERATIONS OFFICE, AEC, N. MEX.	ALO-(NUMBERS)-
ALBUQUERQUE OPERATIONS OFFICE, AEC, N. MEX.	ALO-AEC-CMA-
ALBUQUERQUE OPERATIONS OFFICE, AEC, N. MEX.	AOO-
ALBUQUERQUE OPERATIONS OFFICE, AEC, N. MEX.	COWBOY-OO-
ALBUQUERQUE OPERATIONS OFFICE, AEC, N. MEX.	MPP-
ALBUQUERQUE OPERATIONS OFFICE, AEC, N. MEX.	OTO-(NUMBER)-
ALBUQUERQUE URBAN OBSERVATORY, N. MEX.	ALBU-
ALCO PRODUCTS, INC., FT. BELVOIR, VA.	APAE-
ALCO PRODUCTS, INC., N.Y.C.	APAE-
ALCO PRODUCTS, INC., N.Y.C.	APAE-MEMO-
ALCO PRODUCTS, INC., SCHENECTADY, N.Y. (ATOMIC EN.)	APAE-
ALCO PRODUCTS, INC., SCHENECTADY, N.Y.	APAE-MEMO-
ALCO PRODUCTS, INC., SCHENECTADY, N.Y.	AP-NOTE-
ALFORD (ANDREW) CONSULTING ENGINEER, BOSTON	AACE-
ALFRED UNIV., N.Y.	ALFREDU-
ALFRED UNIV., N.Y.	ALU-
ALIAS BETTY COMMITTEE (REF. 15)	AB-
ALL AMERICAN AIRWAYS, INC. ENGINEERING + RESEARCH DIV., WILMINGTON, DEL. (MISCELLANEOUS REPT.)	M-
ALL AMERICAN ENGINEERING CO., WILMINGTON, DEL.	AAE-N-
ALL AMERICAN ENGINEERING CO., WILMINGTON, DEL.	AA-N-
ALL AMERICAN ENGINEERING CO., WILMINGTON, DEL.	N-
ALLEGANY BALLISTICS LAB., CUMBERLAND, MD. (WEEKLY PROGRESS REPORT)	WPR-
ALLGEMEINE ELEKTRICITAETS-GESELLSCHAFT, FRANKFURT AM MAIN	AEG-
ALLGEMEINE ELEKTRICITAETS-GESELLSCHAFT, FRANKFURT AM MAIN	AEG/E(NUMBER)/
ALLGEMEINE ELEKTRICITAETS-GESELLSCHAFT, FRANKFURT AM MAIN (KERNENERGIEANLAGEN)	KEA-
ALLIED AIR FORCES, CENTRAL EUROPE	AAFCE-
ALLIED AIR FORCES, CENTRAL EUROPE (OPERATIONS ANALYSIS TECHNICAL MEMORANDUM)	AAFCE-OA-TM-
ALLIED AIR FORCES SOUTHERN EUROPE, NAPLES	AIRSOUTH-
ALLIED CHEMICAL CORP., MORRISTOWN, N.J.	ACC-
ALLIED CHEMICAL CORP., MORRISTOWN, N.J.	TA-
ALLIED CHEMICAL CORP., N.Y.C.	ALCC-
ALLIED CHEMICAL CORP., N.Y.C.	GCD-FR-
ALLIED CHEMICAL CORP., N.Y.C.	GCD-QR-
ALLIED CHEM. CORP. GEN. CHEM. DIV.,MORRISTOWN, N.J.	ACC-PR-
ALLIED CHEMICAL CORP. GENERAL CHEMICAL DIV., N.Y.C.	GCD-
ALLIED CHEMICAL CORP. GENERAL CHEMICAL DIV., N.Y.C.	GEN-
ALLIED FORCES ENEMY EQUIPMENT INTELLIGENCE SERVICE	AFEEIS-
ALLIED FORCES ENEMY EQUIPMENT INTELLIGENCE SERVICE	ATSC EEIS
ALLIED FORCES ENEMY EQUIPMENT INTELLIGENCE SERVICE	EEIS-
ALLIED FORCES ENEMY EQUIPMENT INTELLIGENCE SERVICE. CHEMICAL WARFARE SECTION (PRELIMINARY TECH. REPT.)	EEIS CWS PTR-
ALLIED FORCES ENEMY EQUIPMENT INTELLIGENCE SERVICE. INDIA-BURMA THEATER	IB EEIS-
ALLIED FORCES ENEMY EQUIPMENT INTELLIGENCE SERVICE. INDIA-BURMA THEATER. CHEMICAL WARFARE SECTION	CWS-
ALLIED FORCES ENEMY EQUIPMENT INTELL. SERVICE. INDIA-BURMA THEATER. CHEM. WARFARE SEC.(PRELIM.TECH.RPT.)	IB EEIS CWS PTR-
ALLIED FORCES ENEMY EQUIPMENT INTELL. SERVICE. INDIA-BURMA THEATER. CHEM. WARFARE SEC. (TECH. REPT.)	IB EEIS CWS TR-
ALLIED FORCES ENEMY EQUIPMENT INTELLIGENCE SERVICE. INDIA-BURMA THEATER. SIGNAL SECTION. (PRELIM. RPT.)	EEIS SIG PR-
ALLIED FORCES ENEMY EQUIPMENT INTELLIGENCE SERVICE. INDIA-BURMA THEATER. SIGNAL SECTION. (PRELIM. RPT.)	IB EEIS SIG PR-
ALLIED NAVAL FORCES SOUTHERN EUROPE, MALTA	COMNAVSOUTH-OPS-
ALLIED RESEARCH ASSOCIATES, INC., BOSTON	ARAI-
ALLIED RESEARCH ASSOCIATES, INC., BOSTON	ARA-M-
ALLIED RESEARCH ASSOCIATES, INC., CONCORD, MASS.	ARA-
ALLIED RESEARCH ASSOCIATES, INC., CONCORD, MASS.	ARA-F-
ALLIED RESEARCH ASSOCIATES, INC., CONCORD, MASS.	ARA-T-
ALLIED RESEARCH ASSOCIATES, INC., CONCORD, MASS.	ARA-TR-
ALLIED RESEARCH ASSOCIATES, INC. ARADYN DIV., CONCORD, MASS. (VIBRATION ENG. AND TESTING EQUIP.)	DYN-
ALLIS CHALMERS CORP. ATOMIC EN. DIV., BETHESDA, MD.	ACNP-
ALLIS-CHALMERS MFG. CO., MILWAUKEE	ACM-
ALLIS-CHALMERS MFG. CO., NORWOOD, OHIO	PADR-
ALLIS-CHALMERS MFG. CO. ATOMIC ENERGY DIV., MILWAUKEE	KDM-AC-
ALLIS-CHALMERS MFG. CO. ATOMIC ENERGY DIV.,WASH.,D.C. (ELK RIVER REACTOR)	ACNP-ERR-
ALLIS-CHALMERS MFG. CO. ATOMIC ENERGY DIV.,WASH.,D.C.	RD-
ALLIS-CHALMERS MFG. CO. NUCLEAR POWER DEPT., WASHINGTON, D.C.	UMNE-
ALLIS-CHALMERS MFG. CO. RESEARCH DIV., MILWAUKEE	ACR-

ALLIS-CHALMERS MFG. CO. SPACE AND DEFENSE SCIENCES DEPT., MILWAUKEE	ACSD-
ALLIS-CHALMERS MFG. CO. SPACE AND DEFENSE SCIENCES DEPT., MILWAUKEE	ACSDS-
ALLISON DIV., GENERAL MOTORS CORP., CLEVELAND	(NUMBER)-C-(YEAR)
ALLISON DIV., GENERAL MOTORS CORP., INDIANAPOLIS (APPLIED CHEMISTRY SECTION)	ACS-
ALLISON DIV., GENERAL MOTORS CORP., INDIANAPOLIS	ALLISON-EDR-
ALLISON DIV., GENERAL MOTORS CORP., INDIANAPOLIS	EDR-
ALLISON DIV., GENERAL MOTORS CORP., INDIANAPOLIS	GMAD-
ALLISON DIV., GENERAL MOTORS CORP., INDIANAPOLIS	GMC/AD-
ALLISON DIV., GENERAL MOTORS CORP., INDIANAPOLIS	GM EDR-
ALLISON DIV., GENERAL MOTORS CORP., INDIANAPOLIS	RD-
ALLISON DIV., GENERAL MOTORS CORP., INDIANAPOLIS	T4-
ALLMAENNA SVENSKA ELEKTRISKA A.B., VAESTERAES, SWEDEN	KAE-(YEAR)-
ALLOYD CORP., CAMBRIDGE, MASS.	ALLOYD-
ALLSTATES DESIGN AND DEVELOPMENT CO., INC., TRENTON	AS-
ALLSTATES DESIGN AND DEVELOPMENT CO., INC., TRENTON	PREAS-
ALUMINIUM-INDUSTRIE-AKTIEN-GESELLSCHAFT, NEUHAUSEN AM RHEINFALL, SWITZERLAND	AIAG-
ALUMINIUM LABORATORIES LTD., BANBURY, OXFORD., ENGLAND	ALL-
ALUMINUM CO. OF AMERICA, NEW KENSINGTON, PA.	ACA-
ALUMINUM CO. OF AMERICA. ALCOA PROCESS DEVELOPMENT LABS., NEW KENSINGTON, PENNA.	JE-(YR)-
ALUMINUM CO. OF AMERICA. ALCOA RESEARCH LABS., NEW KENSINGTON, PENNA. (PROGRESS REPORT)	ACA-ARL-QPR-
ALUMINUM CO. OF AMERICA. ALCOA RESEARCH LABS., NEW KENSINGTON, PENNA. (TECHNICAL PAPER)	ACA-ARL-TP-
ALUMINUM CO. OF AMERICA. ALCOA RESEARCH LABS., NEW KENSINGTON, PENNA.	ALCOA-
AMBAC INDUSTRIES. ARMA DIV., GARDEN CITY, N.Y.	NOD-ED-
AMCEL PROPULSION CO., ASHEVILLE, N.C.	AMCEL-
AMCEL PROPULSION CO., ASHEVILLE, N.C.	APR-
AMCEL PROPULSION CO., ASHEVILLE, N.C.	ATR-
AMERICAN ASSN. FOR THE ADVANCEMENT OF SCIENCE, WASH, DC	AAAS-
AMERICAN BIOPHYSICS RESEARCH LAB., LANSDALE, PA.	ABRL-
AMERICAN BOSCH ARMA CORP. ARMA DIV., GARDEN CITY, N.Y.	AC-
AMERICAN BOSCH ARMA CORP. ARMA DIV., GARDEN CITY, N.Y.	ARMA-
AMERICAN BOSCH ARMA CORP. ARMA DIV., GARDEN CITY, N.Y.	ARMA-DF-
AMERICAN BOSCH ARMA CORP. ARMA DIV., GARDEN CITY, N.Y.	DAG-
AMERICAN BOSCH ARMA CORP. ARMA DIV., GARDEN CITY, N.Y.	DR-
AMERICAN BOSCH ARMA CORP. ARMA DIV., GARDEN CITY, N.Y.	DS-
AMER. BRAKE SHOE CO. AMER. BRAKEBLOK DIV., DETROIT	ABK-
AMERICAN CAR + FOUNDRY CO., RIVERDALE, MD. (SEE ALSO LATER NAME: ACF INDUSTRIES, INC.)	ERCO-
AMERICAN CAR + FOUNDRY CO. AVION DIV., PARAMUS, N.J.	AIC-
AMERICAN CHAIN AND CABLE CO., INC., DETROIT	CF-
AMERICAN COUNCIL OF COMMERCIAL LABS., WASH., D.C.	ACCL-
AMERICAN CYANAMID CO. (ALL LOCATIONS)	ACCO-
AMERICAN CYANAMID CO., BOUND BROOK, N.J.	TS-
AMERICAN CYANAMID CO., N.Y.C.	ACC-
AMERICAN CYANAMID CO., STAMFORD, CONN. (ANNUAL RPT.)	ACCO-AR-(YEAR)
AMERICAN CYANAMID CO., STAMFORD, CONN. (INTEGRATED SOLID PROPELLANT PROGRAM PROGRESS REPORT)	ACCO-ISPP-PR-
AMERICAN CYANAMID CO., STAMFORD, CONN.	ACCO-PR-
AMERICAN CYANAMID CO., STAMFORD, CONN. (TECH. REPT.)	ACCO-TR-
AMERICAN DOCUMENTATION INSTITUTE, WASHINGTON, D.C.	ADI-
AMERICAN ELECTRIC POWER SERVICE CORP., N.Y.C.	AEPSC-
AMERICAN ELECTRIC POWER SERVICE CORP., N.Y.C.	AEPSCO-
AMERICAN ELECTRO METAL CORP., YONKERS, N.Y.	AEM-
AMERICAN ELECTRO METAL CORP., YONKERS, N.Y.	AEMC-
AMERICAN ELECTRONIC LABS., INC., COLMAR, PENNA.	AEL-
AMERICAN ELECTRONIC LABS., INC., PHILADELPHIA	AEL-
AMERICAN ELECTRONIC LABS., INC., PHILADELPHIA	AEL-TR-
AMERICAN ELECTRONICS, INC., FULLERTON, CALIF.	AEI-
AMERICAN FORCES EUROPEAN THEATER OF OPERATIONS	AFETO-
AMERICAN FORCES EUROPEAN THEATER OF OPERATIONS	ETO-
AMERICAN FORCES EUROPEAN THEATER OF OPERATIONS	ETOUSA ...
AMERICAN FORCES EUROPEAN THEATER OF OPERATIONS. CHIEF ENGINEER	AFETO CE-
AMERICAN FORCES EUROPEAN THEATER OF OPERATIONS. CHIEF SIGNAL OFFICER	AFETO CSO-
AMERICAN FORCES EUROPEAN THEATER OF OPERATIONS. CHIEF SIGNAL OFFICER (FIELD REPORT)	AFETO CSO FR-
AMERICAN FORCES EUROPEAN THEATER OF OPERATIONS. CHIEF SIGNAL OFFICER (INTERROGATION)	AFETO CSO I-
AMERICAN FORCES EUROPEAN THEATER OF OPERATIONS. CHIEF SIGNAL OFFICER (INTELLIGENCE REPORT)	AFETO CSO IR-
AMERICAN GAS AND ELECTRIC CO., N.Y.C.	NDA-
AMERICAN GAS ASSN., INC., N.Y.C.	AGA-
AMERICAN GAS ASSN. TESTING LABS., CLEVELAND (RES. BULL.)	AGA-RB-
AMERICAN GEOGRAPHICAL SOCIETY, N.Y.C. (TRANSLATIONS)	AGS-
AMERICAN GEOGRAPHICAL SOCIETY. DEPT. OF EXPLORATION AND FIELD RESEARCH, N.Y.C.	EP-
AMERICAN GEOLOGICAL INST. TRANSLATIONS OFF., WASH., DC	AGI-

AMERICAN GEOPHYSICAL UNION

AMERICAN GEOPHYSICAL UNION, WASHINGTON, D.C.	AGU-
AMERICAN HELICOPTER CO., INC., LOS ANGELES (REF. 1)	(PROJ. CODE NO.)-
AMERICAN HELICOPTER CO., INC., MANHATTAN BEACH, CALIF	AH-
AMERICAN HELICOPTER CO., INC., MANHATTAN BEACH, CALIF	AHC-
AMERICAN HELICOPTER CO., INC., MANHATTAN BEACH, CALIF	AHCO-
AMERICAN HELICOPTER DIV., FAIRCHILD ENGINE AND AIRPLANE CORP., MESA, ARIZ.	AH-
AMERICAN INST. FOR AEROLOGICAL RES.,PASADENA, CALIF.	AIAR-
AMERICAN INST. FOR RESEARCH, PITTSBURGH	AIR-
AMERICAN INST. FOR RESEARCH, PITTSBURGH	AIR-E(NO.)-(DATE)(LTR.)
AMERICAN INST. FOR RESEARCH, PITTSBURGH	AIR-G(NO.)-(DATE)(LTR.)
(SEE ALSO LATER NAME: AMERICAN INSTITUTES FOR RES.)	
AMERICAN INST. OF AERONAUTICS + ASTRONAUTICS,LOS ANG.	PAI-
AMERICAN INST. OF AERONAUTICS + ASTRONAUTICS,LOS ANG.	PAL-
AMERICAN INST. OF AERONAUTICS AND ASTRONAUTICS, N.Y.C. (REF. 1)	(NUMBER)
AMERICAN INST. OF AERONAUTICS AND ASTRONAUTICS, N.Y.C. (PAPERS)	AIAA-
AMERICAN INST. OF AERONAUTICS AND ASTRONAUTICS, N.Y.C. (CONFERENCE PUBLICATIONS)	AIAA-CP-
AMERICAN INST. OF AERONAUTICS AND ASTRONAUTICS, N.Y.C. (PAPERS)	AIAA-PAPER-(YEAR)-
AMERICAN INST. OF AERONAUTICS AND ASTRONAUTICS, N.Y.C. (CONFERENCE PUBLICATIONS)	CP-
AMERICAN INST. OF AERONAUTICS + ASTRONAUTICS, N.Y.C.	IAS-
AMERICAN INST. OF AERONAUTICS + ASTRONAUTICS, N.Y.C.	SMF-
AMERICAN INST. OF BIOLOGICAL SCIENCES, WASHINGTON, DC	AIBS-
AMERICAN INSTITUTE OF BIOLOGICAL SCIENCES. BIOINSTRUMENTATION ADVISORY COUNCIL, WASHINGTON, D.C.	BIAC INFO. MODULE M(NO.)
AMERICAN INST. OF CROP ECOLOGY, WASHINGTON, D.C.	AICE-
AMERICAN INST. OF ELECTRICAL ENGINEERS, N.Y.C.	AIEE-
AMERICAN INST. OF PHYSICS, N.Y.C. (TRANSLATIONS)	AIP-
AMERICAN INST. OF PHYSICS, N.Y.C.	AIP/45 (LETTER)
AMERICAN INST. OF PHYSICS, N.Y.C. (UNIVERSAL DECIMAL CLASSIFICATION PROJECT)	AIP/UDC-
AMERICAN INST. OF PHYSICS. INFORMATION ANALYSIS AND RETRIEVAL DIV., N.Y.C.	IARD-
AMERICAN INST. OF PHYSICS. INFORMATION ANALYSIS AND RETRIEVAL DIV., N.Y.C.	IARD-(YEAR)-
AMERICAN INST. OF PHYSICS. INFORMATION DIV., N.Y.C.	AIP-ID-
AMERICAN INST. OF PHYSICS. INFORMATION DIV., N.Y.C.	ID-(YEAR)-
AMERICAN INST. OF PHYSICS. SYSTEMS DEV. DIV., N.Y.C.	AIP/SDD-
AMERICAN INSTS. FOR RESEARCH, PITTSBURGH	AIR-(NUMBER-DATE-LTRS.)-
AMERICAN INSTS. FOR RESEARCH, PITTSBURGH	AIR-B(NO.-DATE-LTRS.)-
AMERICAN INSTS. FOR RESEARCH, PITTSBURGH	AIR-C(NO.-DATE-LTRS.)-
AMERICAN INSTS. FOR RESEARCH, PITTSBURGH	AIR-D-
AMERICAN INSTS. FOR RESEARCH, PITTSBURGH	AIR-R(NUMBER)-
AMERICAN INSTS. FOR RESEARCH, PITTSBURGH	AIR-RP(NUMBER)-
AMERICAN INSTS. FOR RESEARCH, PITTSBURGH	GT-(NUMBER)-R
AMERICAN INSTS. FOR RESEARCH, SILVER SPRING, MD.	AIR-E-
AMERICAN INSTS. FOR RESEARCH, SILVER SPRING, MD.	AIR-E(NUMBER)-
AMERICAN INSTS. FOR RESEARCH, SILVER SPRING, MD.	AIR-ES-
AMERICAN INSTS. FOR RESEARCH, SILVER SPRING, MD.	AIR-F-(DATE)-TR
AMERICAN INSTS. FOR RESEARCH. INST. FOR PERFORMANCE TECHNOLOGY, PITTSBURGH	AIR-D(NUMBER)B-
AMERICAN INSTS. FOR RESEARCH IN THE BEHAVIORAL SCIENCES, PALO ALTO, CALIF.	AIR-C28-(DATE)-TR-
AMERICAN INSTRUMENT CO., INC., SILVER SPRING, MD.	AMINCO-
AMERICAN-LA FRANCE-FOAMITE CORP., ELMIRA, N.Y.	ALFCO-
AMERICAN LIBRARY ASSOCIATION, CHICAGO	ALA-
AMERICAN LOCOMOTIVE CO., N.Y.C.	ALCO-
AMERICAN MACHINE AND FOUNDRY CO.	AMF-
AMERICAN MACHINE AND FOUNDRY CO. (INTERIM REPORT)	AMF IR-
AMERICAN MACHINE + FOUNDRY CO.(MONTHLY PROGRESS RPT.)	AMF MPR-
AMERICAN MACHINE AND FOUNDRY CO., ALEXANDRIA, VA.	IPK-
AMERICAN MACHINE AND FOUNDRY CO., CHICAGO	AMFC-
AMERICAN MACHINE AND FOUNDRY CO., CHICAGO	AMF MR-
AMERICAN MACHINE AND FOUNDRY CO., CHICAGO	AMF/MRD TN-
AMERICAN MACHINE AND FOUNDRY CO., CHICAGO	MR-
AMERICAN MACHINE AND FOUNDRY CO., GREENWICH, CONN.	DPEAER-
AMERICAN MACHINE AND FOUNDRY CO., STAMFORD, CONN.	ER-APO-
(SEE ALSO: AMF ATOMICS)	
AMERICAN MACHINE AND FOUNDRY CO. ADVANCED PRODUCTS GROUP, SANTA BARBARA, CALIF.	ER-SB-
AMERICAN MACHINE AND FOUNDRY CO. ADVANCED SYSTEMS LAB., SANTA BARBARA, CALIF.	AMF-ASL-E-(NUMBER)-
AMERICAN MACHINE AND FOUNDRY CO. ADVANCED SYSTEMS LAB., SANTA BARBARA, CALIF.	AMF-G-S-
AMERICAN MACHINE AND FOUNDRY CO. ADVANCED SYSTEMS LAB., SANTA BARBARA, CALIF.	G-S-
AMERICAN MACHINE AND FOUNDRY CO. GENERAL ENGINEERING LAB., GREENWICH, CONN.	AMF-ER-
AMERICAN MACHINE AND FOUNDRY CO. GENERAL ENGINEERING LAB., GREENWICH, CONN.	AMF-G-
AMERICAN MACHINE AND FOUNDRY CO. GENERAL ENGINEERING LAB., GREENWICH, CONN.	AMF-GR-
AMERICAN MACHINE AND FOUNDRY CO. GENERAL ENGINEERING LAB., GREENWICH, CONN.	ER-
AMERICAN MACHINE AND FOUNDRY CO. GENERAL ENGINEERING LAB., GREENWICH, CONN.	FTR-
AMERICAN MACHINE AND FOUNDRY CO. GENERAL ENGINEERING LAB., GREENWICH, CONN.	G-(4 DIGITS)
AMERICAN MACHINE AND FOUNDRY CO. NUCLEAR ENGINEERING LAB., GREENWICH, CONN.	AMF-GR-(NUMBER)-(YEAR)
AMERICAN MACHINE + FOUNDRY CO.TURBO DIV.,PACOIMA,CAL.	AMF/TD-
AMERICAN MACHINE + FOUNDRY CO.TURBO DIV.,PACOIMA,CAL.	TD-
AMERICAN MACHINE AND FOUNDRY CO. TURBO ENGINE DEPT., PACOIMA, CALIF.	AMFC-
AMERICAN MATHEMATICAL SOCIETY, PROVIDENCE	AMS-

AMERICAN MATHEMATICAL SOCIETY, PROVIDENCE(TRANSLATION)	AMST-
AMERICAN MATHEMATICAL SOCIETY, PROVIDENCE	AMS-TR-
AMERICAN METAL PRODUCTS CO. ENGINEERING SCIENCE DIV., ANN ARBOR, MICH.	ES-
AMERICAN METEOROLOGICAL SOCIETY, BOSTON	AM.MS-
AMERICAN METEOROLOGICAL SOCIETY, BOSTON	AMS-
AMERICAN METEOROLOGICAL SOCIETY, BOSTON	T-F-
AMERICAN METEOROLOGICAL SOCIETY, BOSTON	T-FC-
AMERICAN METEOROLOGICAL SOCIETY, BOSTON	T-G-
AMERICAN METEOROLOGICAL SOCIETY, BOSTON	T-L-
AMERICAN METEOROLOGICAL SOCIETY, BOSTON (TRANSLATION)	T-R-
AMERICAN METEOROLOGICAL SOCIETY, BOSTON	T-RC-
AMERICAN METEOROLOGICAL SOCIETY, BOSTON	T-U-
AMERICAN MICRO-SYSTEMS, INC., SANTA CLARA, CALIF.	AMI-(7 DIGITS)
AMERICAN NATIONAL STANDARDS INST., N.Y.C.	ANSI-
AMERICAN NUCLEAR SCIENCE CORP., N.Y.C.	ANSC-
AMERICAN NUCLEAR SOCIETY, HINSDALE, ILL.	ANS-
AMERICAN NUCLEAR SOCIETY, HINSDALE, ILL.	ANS-RPD-
AMERICAN NUCLEAR SOCIETY, HINSDALE, ILL.	ANS-SD-
AMERICAN NUCLEONICS CORP., GLENDALE, CALIF.	ANC-
AMERICAN OIL CO. RES. AND DEV. DEPT., WHITING, IND.	AOC-M(YEAR)-
AMERICAN OIL CO. RES. AND DEV. DEPT., WHITING, IND.	M(YEAR)-
AMERICAN OPTICAL CO. CENTRAL RESEARCH LAB., SOUTHBRIDGE, MASS.	TR-(3 DIGITS)-
AMERICAN OPTICAL CO. J.W. FECKER DIV., PITTSBURGH	DE-
AMERICAN OPTICAL CO. J.W. FECKER DIV., PITTSBURGH	FR-
AMERICAN OPTICAL CO. LASER PRODS. DIV., SOUTHBRIDGE, MASS	SO-(4 DIGITS)
AMERICAN OPTICAL CO. RESEARCH DIV., SOUTHBRIDGE, MASS.	AORC-(NO.)-(NO.)-(N).)-
AMERICAN OPTICAL CO. RESEARCH DIV., SOUTHBRIDGE, MASS.	RD-
AMERICAN OPTICAL CO. STAMFORD RESEARCH LABS., SOUTHBRIDGE, MASS.	SATR-
AMERICAN OPTICAL CO. STAMFORD RESEARCH LABS., SOUTHBRIDGE, MASS.	TSR-
AMERICAN OPTICAL CO. TECHNICAL PRODUCTS GROUP, SOUTHBRIDGE, MASS.	FO-(3 DIGITS)
AMERICAN ORDNANCE ASSOCIATION, WASHINGTON, D.C.	AOA-
AMERICAN PETROLEUM INSTITUTE, N.Y.C.	API-
AMERICAN POTASH + CHEMICAL CORP., WHITTIER, CALIF.	APCC-
AMERICAN POTASH + CHEMICAL CORP., WHITTIER, CALIF.	W-
AMERICAN POTASH + CHEMICAL CORP., WHITTIER, CALIF.	WRM-
AMERICAN POTASH + CHEMICAL CORP., WHITTIER, CALIF.	WTR-
AMERICAN POTASH + CHEM. CORP. RES.DEPT., WHITTIER, CAL.	APCC-QPR-
AMERICAN POWER JET CO., MONTCLAIR, N.J.	APJ-
AMERICAN POWER JET CO., MONTCLAIR, N.J.	MOOR-
AMERICAN POWER JET CO., RIDGEFIELD, N.J.	APJ-
AMERICAN RADIATOR AND STANDARD SANITARY CORP. ATOMIC ENERGY DIV., REDWOOD CITY, CALIF.	ASAE-E-
AMERICAN REHABILITATION FOUNDATION. INST. FOR INTERDISCIPLINARY STUDIES, MINNEAPOLIS	LI-(6 DIGITS)
AMERICAN RESEARCH CORP., FULLERTON, CALIF.	AMRC-
AMERICAN RESEARCH CORP., FULLERTON, CALIF.	ARC-(YEAR)-
AMERICAN ROCKET SOCIETY, N.Y.C.	ARS-
AMERICAN SCIENCE AND ENG. INC., CAMBRIDGE, MASS.	ASE-
AMERICAN SCIENCE AND ENG. INC., CAMBRIDGE, MASS.	ASE TN-
AMERICAN SCIENCE AND ENG. INC., CAMBRIDGE, MASS.	SPECIAL SCIENTIFIC-
AMERICAN SMELTING + REFINING CO., BARBER, N.J.	ASR-
AMERICAN SMELTING + REFINING CO., N.Y.C.	ASARCO-
AMERICAN SOCIETY FOR INFORMATION SCIENCE. SPECIAL INTEREST GROUP ON SELECTIVE DISSEMINATION OF INFORMATION, WASHINGTON, D.C.	SIG/SDI-
AMERICAN SOCIETY FOR TESTING AND MATERIALS, PHILA.	ASTM-
AMERICAN SOCIETY FOR TESTING AND MATERIALS, PHILA.	ASTM-STP-
AMERICAN SOCIETY OF CIVIL ENGINEERS, N.Y.C.	ASCE-
AMERICAN SOCIETY OF MECHANICAL ENGINEERS, N.Y.C.	ASME-
AMERICAN SOCIETY OF TOOL AND MFG. ENGINEERS, DETROIT	ASTME-
AMERICAN SOCIETY OF TOOL ENGINEERS, DETROIT	ASTE-
AMERICAN-STANDARD. ATOMIC EN. DIV., MOUNTAIN VIEW, CAL.	ASAE-
AMERICAN-STANDARD. ATOMIC EN. DIV., MOUNTAIN VIEW, CAL.	ASAE-S-
AMERICAN-STANDARD. ATOMIC EN. DIV., REDWOOD CITY, CAL.	ASAE-
AMERICAN STANDARDS ASSOCIATION, N.Y.C.	ASA-
AMERICAN TURBINE CORP., N.Y.C.	ATC-(YR)-
AMERICAN UNIV., WASHINGTON, D.C. CENTER FOR RESEARCH IN SOCIAL SYSTEMS	CRESS/CINFAC-R-
AMERICAN UNIV., WASHINGTON, D.C. SPECIAL OPERATIONS RESEARCH OFFICE	SORO-CINFAC-(NO.-NO.)
AMERICAN WELDING SOCIETY, N.Y.C.	AWS-
AMES AERONAUTICAL LAB., MOFFETT FIELD, CALIF.(REF.25)	NACA-(LTRS)-A-(DATE CODE)
AMES LAB., IOWA (REF. 10)	CC-

AMES LAB.

AMES LAB., IOWA (REF. 10)	CT-
AMES LAB., IOWA	IS-
AMES LAB., IOWA	ISC-
AMES LAB., IOWA (REACTOR MATERIALS DEVELOPMENT)	IS-RD-
AMES LAB., IOWA	IS-T-
AMES LAB., IOWA	LR-
AMES LAB., IOWA	NMI-
AMES LAB., IOWA	PC-
AMES LAB., IOWA	TI-
AMES RESEARCH CENTER, MOFFETT FIELD, CALIF.	ARC-
AMES RESEARCH CENTER, MOFFETT FIELD, CALIF. (REF. 26)	NASA-(LETTERS)-A-
AMES RESEARCH CENTER, MOFFETT FIELD, CALIF.	NHB-
AMES RESEARCH CENTER, MOFFETT FIELD, CALIF.	TM-X-
AMES RESEARCH CENTER, MOFFETT FIELD, CALIF.	X-(NUMBERS)-(YEAR)-
AMF ATOMICS, INC., GREENWICH, CONN.	AMF-G-
AMF ATOMICS, INC., GREENWICH, CONN.	AMF-Y-
AMF ATOMICS, INC., GREENWICH, CONN.	NEL-ER-
AMF ATOMICS, INC., N.Y.C.	ER-
AMF ATOMICS, INC., YORK, PENNA.	AMR-
AMF ATOMICS (CANADA) LTD., PORT HOPE, ONT.	AMFC-
AMF ATOMICS (CANADA) LTD., PORT HOPE, ONT.	AMFC-(NUMBER)-
AMF ATOMICS (CANADA) LTD., PORT HOPE, ONT.	QC-
AMHERST COLL., MASS.	AC-
AMMAN + WHITNEY, CONSULTING ENGINEERS, N.Y.C.	A+W-
AMMAN + WHITNEY, CONSULTING ENGINEERS, N.Y.C.	AW-
AMMUNITION PROCUREMENT AND SUPPLY AGENCY. QUALITY EVALUATION DIV., JOLIET, ILL.	AER-(NO.)-(NO.)-
AMMUNITION PROCUREMENT AND SUPPLY AGENCY. QUALITY EVALUATION DIV., JOLIET, ILL.	QEM-(NUMBER-NUMBER-NO.)
AMOCO CHEMICALS CORP., SEYMOUR, IND.	AG-
AMP INC. CAPITRON DIV., ELIZABETHTOWN, PENNA.	FOLDER-(NUMBER)-
AMPEX CORP., REDWOOD CITY, CALIF.	AMPEX-
AMPEX CORP., REDWOOD CITY, CAL. (FIELD SERVICE BULL.)	AMPEX FSB-
AMPEX CORP., REDWOOD CITY, CALIF. (RESEARCH BULL.)	AMPEX RB-
AMPEX CORP., REDWOOD CITY, CALIF. (RESEARCH DOC.)	AMPEX RD-
AMPEX CORP., REDWOOD CITY, CALIF. (RES. PROPOSAL)	AMPEX RP-
AMPEX CORP., REDWOOD CITY, CALIF. (TECHNICAL MEMO.)	AMPEX TM-
AMPEX CORP., REDWOOD CITY, CALIF. (TECHNICAL REPT.)	AMPEX TR-
AMPEX CORP., REDWOOD CITY, CALIF.	RB-(YEAR)-
AMPEX CORP., REDWOOD CITY, CALIF.	RR-(YEAR)-
AMPHENOL-BORG ELECTRONICS CORP., CHICAGO	ABEC-
AMPHENOL-BORG ELECTRONICS CORP., CHICAGO	C-
AMSTERDAM. INSTITUUT VOOR KERNPHYSISCH ONDERZOEK	IKO-
AMSTERDAM. INSTITUUT VOOR KERNPHYSISCH ONDERZOEK	IKO-PR-
AMSTERDAM. UNIV.	E-
AMSTERDAM. UNIV.	KR-
ANACONDA WIRE AND CABLE CO., HASTINGS-ON-HUDSON,N.Y.	PUB-(LETTER)-
ANACONDA WIRE AND CABLE CO. MAGNET WIRE RESEARCH, MUSKEGON, MICH.	AWCC-
ANAGRAM CORP., SPRINGFIELD, VA.	AF-AC-(NUMBER)-
ANAGRAM CORP., SPRINGFIELD, VA.	AF(ANAGRAM)-
ANAGRAM CORP., SPRINGFIELD, VA.	DC-AC-(NUMBER)-
ANALYTIC SCIENCES CORP., READING, MASS.	TASC-TR-(NUMBER)-
ANALYTIC SERVICES, INC., BAILEYS CROSSROADS, VA.	AM-
ANALYTIC SERVICES, INC., BAILEYS CROSSROADS, VA.	ANSER-
ANALYTIC SERVICES, INC., BAILEYS CROSSROADS, VA.	DPM-
ANALYTIC SERVICES, INC., FALLS CHURCH, VA.	ANSER-AR-
ANALYTIC SERVICES, INC., FALLS CHURCH, VA.	ANSER-DBN-(YEAR)-
ANALYTIC SERVICES, INC., FALLS CHURCH, VA.	ANSER-RTBN-(YEAR)-
ANALYTIC SERVICES, INC., FALLS CHURCH, VA.	ANSER-SBN-(YEAR)-
ANALYTIC SERVICES, INC., FALLS CHURCH, VA.	ANSER-SDN-(YEAR)-
ANALYTIC SERVICES, INC., FALLS CHURCH, VA.	ANSER-TBN-(YEAR)-
ANALYTIC SERVICES, INC., FALLS CHURCH, VA.	AR-
ANALYTIC SERVICES, INC., FALLS CHURCH, VA.	AR-(YEAR)-
ANALYTIC SERVICES, INC., FALLS CHURCH, VA.	DPM-
ANALYTIC SERVICES, INC., FALLS CHURCH, VA.	DPR-(YEAR)-
ANALYTIC SERVICES, INC., FALLS CHURCH, VA.	SB-(YEAR)
ANALYTIC SERVICES, INC., FALLS CHURCH, VA.	TB-(YEAR)-
ANDERSEN LABS, INC., BLOOMFIELD, CONN.	AL-(NO.-LTR-NO.-LTR)
ANDERSON, GREENWOOD + CO., HOUSTON, TEX.	AGC-FTR-
ANEMOSTAT CORP. OF AMERICA, N.Y.C. (REF. 1)	(NUMBER)
ANSALDO S.P.A., GENOA	GN-
ANSALDO S.P.A., GENOA	ITN/RF-
ANSALDO S.P.A., GENOA	LAB-STU-
ANSCO DIV., GEN. ANILINE + FILM CORP., BINGHAMTON,NY	ANSCO-
ANSCO DIV., GEN. ANILINE + FILM CORP., BINGHAMTON,NY	C-
ANTENNA ENGINEERING LABS., TORRANCE, CALIF.	AEL-
ANTENNA ENGINEERING LABS., TORRANCE, CALIF.	T-V-
ANTI-SUBMARINE WARFARE SYSTEMS PROJECT OFFICE. SYSTEMS ANALYSIS GROUP, SILVER SPRING, MD.	SAG-
ANTI-SUBMARINE WARFARE SYSTEMS PROJECT OFFICE. SYSTEMS ANALYSIS OFFICE, SILVER SPRING, MD.	SAO-(YEAR)-
ANTI-SUBMARINE WARFARE SYSTEMS PROJECT OFFICE. SYSTEMS ANALYSIS OFFICE, SILVER SPRING, MD.	SAO-TM-(YEAR)-
ANTIAIRCRAFT + GUIDED MISSILES BRANCH, FT. BLISS, TEX	AGMB-ST-
ANTIAIRCRAFT ARTILLERY AND GUIDED MISSILE CENTER, FORT BLISS, TEXAS	AA+GM-

	(REPORT NO.)-(COPY NO.)
ANTIAIRCRAFT ARTILLERY AND GUIDED MISSILES CENTER, FORT BLISS, TEX. (REF. 1)	
ANTIAIRCRAFT ARTILLERY AND GUIDED MISSILE CENTER, FORT BLISS, TEX.	SCD-
ANTIAIRCRAFT ARTILLERY AND GUIDED MISSILE CENTER, FORT BLISS, TEX.	TAS-
ANTIAIRCRAFT ARTILLERY AND GUIDED MISSILE SCHOOL, FORT BLISS, TEX.	ST-
ANTIAIRCRAFT COMMAND (ARMY), COLORADO SPRINGS	AA-
ANTIOCH COLL., YELLOW SPRINGS, OHIO	ANTC-
APPLICATION RES. CORP., LOS ANGELES	ARC-C(NUMBER)-(NUMBER)FR
APPLICAZIONI E RICERCHE SCIENTIFICHE, MILAN	ARSN-
APPLIED DATA RESEARCH, INC., PRINCETON, N.J.	ADR-REF-
APPLIED DEVICES CORP., COLLEGE POINT, N.Y.	ADCOR-
APPLIED DEVICES CORP., COLLEGE POINT, N.Y.	EE-
APPLIED MECHANICS REVIEWS, SAN ANTONIO, TEX.	AMR-
APPLIED PARKING TECHNIQUES, INC., RESTON, VA.	APT-
APPLIED PSYCHOLOGY CORP., WASHINGTON, D.C.	APC-
APPLIED RADIATION CORP., WALNUT CREEK, CALIF.	ARCO-AM-
APPLIED SCIENCE CORP., PRINCETON, N.J.	ASC-
APPLIED SCIENCE CORP., PRINCETON, N.J.	ASCOP-
APPLIED SCIENCE CORP., PRINCETON, N.J.	ASCP-ASC-
APPLIED SCIENCE CORP., SANTA PAULA, CALIF.	ASP-
APPLIED TECHNOLOGY ASSOCIATES, INC., RAMSEY, N.J.	ATA-
APPLIED THEORY, INC., LOS ANGELES (JOINT VENTURE)	ATI-AJA-
APPLIED THEORY, INC., LOS ANGELES	ATR-(YEAR)-
APPLIED THEORY INC., SANTA MONICA, CALIF.	ATR-(NUMBERS)
APPROVED MARINE DEVICES CO.	AMD-
ARA, INC., WEST COVINA, CALIF.	ARA-
ARBEITSGEMEINSCHAFT BBC-KRUPP, MANNHEIM, GERMANY	ACC-
ARBEITSGEMEINSCHAFT SATELLITENTRAEGER (ASAT), MUNICH	ASAT-
ARBEITSGEMEINSCHAFT VERSUCHS-REAKTOR GMBH, DUESSELDORF	AVR-
ARCO NUCLEAR CO., LEECHBURG, PENNA.	ARCO-(4 DIGITS)-
ARCON CORP., WAKEFIELD, MASS.	R-(YEAR)-
ARCTIC AEROMEDICAL LAB., FORT WAINWRIGHT, ALASKA	AAL-TDR-(YEAR)-
ARCTIC AEROMEDICAL LAB., FORT WAINWRIGHT, ALASKA	AAL-TR-(YEAR)-
ARCTIC AEROMEDICAL LAB., LADD AFB, ALASKA	AAF AAL-
ARCTIC AEROMEDICAL LAB., LADD AFB, ALASKA	AAF AAL PROJ-
ARCTIC AEROMEDICAL LAB., LADD AFB, ALASKA	AAL-TDR-
ARCTIC AEROMEDICAL LAB., LADD AFB, ALASKA	AAL-TN-(YEAR)-
ARCTIC AEROMEDICAL LAB., LADD AFB, ALASKA	AAL-TR-(YEAR)-
ARCTIC CONSTRUCTION AND FROST EFFECTS LAB., BOSTON	ACFEL-BL-
ARCTIC CONSTRUCTION AND FROST EFFECTS LAB., BOSTON	ACFEL-MISC.PAPER-
ARCTIC CONSTRUCTION AND FROST EFFECTS LAB., BOSTON	ACFEL-TR-
ARCTIC CONSTRUCTION AND FROST EFFECTS LAB., BOSTON	ACFEL-TRANS-
ARCTIC INSTITUTE OF NORTH AMERICA, MONTREAL	AINA-RP-
ARCTIC INSTITUTE OF NORTH AMERICA, MONTREAL	RP-
ARCTIC INSTITUTE OF NORTH AMERICA, WASHINGTON, D.C.	AINA-
ARCTIC INSTITUTE OF NORTH AMERICA, WASHINGTON, D.C.	AINA-TR-
ARCTIC INSTITUTE OF NORTH AMERICA, WASHINGTON, D.C.	RESEARCH PAPER-
ARCTIC RESEARCH LABORATORY, POINT BARROW, ALASKA	ARL-
ARCTIC TEST BOARD (ARMY)	ATB-
ARDE ASSOCIATES, NEWARK, N. J.	ARDE TN-
ARDE ASSOCIATES, NEWARK, N. J.	ARDE TR-
ARDE ASSOCIATES, NEWARK, N. J.	WAD-
ARDE-PORTLAND, INC., PARAMUS, N.J.	AMR-
ARECIBO IONOSPHERIC OBSERVATORY, PUERTO RICO	RS-
ARENBERG ULTRASONIC LAB., JAMAICA PLAIN, MASS.	AUL-
ARGENTINA. COMISION NACIONAL DE ENERGIA ATOMICA, BUENOS AIRES	CNEA-
ARGENTINA. COMISION NACIONAL DE INVESTIGACIONES ESPACIALES, BUENOS AIRES	CNIE-
ARGENTINA. COMISION NACIONAL DE INVESTIGACIONES ESPACIALES, BUENOS AIRES	CNIE-IC-
ARGENTINA. COMISION NACIONAL DE INVESTIGACIONES ESPACIALES, BUENOS AIRES	CNIE-PE-
ARGENTINA. COMISION NACIONAL DE INVESTIGACIONES ESPACIALES, BUENOS AIRES	CNIE-PT-
ARGENTINA. INSTITUTO DE INVESTIGACION AERONAUTICA Y ESPACIAL, BUENOS AIRES	IIAE-INF-(NO.)-(YEAR)
ARGENTINA. INSTITUTO DE INVESTIGACION AERONAUTICA Y ESPACIAL, BUENOS AIRES	IIAE-LR-
ARGENTINA. INSTITUTO DE INVESTIGACION AERONAUTICA Y ESPACIAL, BUENOS AIRES	IIAE-SC-LR-
ARGENTINA. INSTITUTO DE INVESTIGACION AERONAUTICA Y ESPACIAL, BUENOS AIRES	LR-
ARGONNE CANCER RESEARCH HOSPITAL, CHICAGO	ACRH-
ARGONNE CANCER RESEARCH HOSPITAL, CHICAGO	ACRH-1000-
ARGONNE NATIONAL LAB., ILL.	AEC-
ARGONNE NATIONAL LAB., ILL.	ALPR-
ARGONNE NATIONAL LAB., ILL. (NOS. START WITH 4000)	ANL-
ARGONNE NATIONAL LAB., ILL.	ANL-ACL-
ARGONNE NATIONAL LAB., ILL.	ANLAD-
ARGONNE NATIONAL LAB., ILL.	ANL-EBR-

ARGONNE NATIONAL LAB.

ARGONNE NATIONAL LAB., ILL.	ANL/ES-CC-
ARGONNE NATIONAL LAB., ILL.	ANL/ES-CEN-
ARGONNE NATIONAL LAB., ILL.	ANL-ES-RPY-
ARGONNE NATIONAL LAB., ILL. (HIGH ENERGY PARTICLES)	ANL/HEP-
ARGONNE NATIONAL LAB., ILL.	ANL/MET-
ARGONNE NATIONAL LAB., ILL. (MOTION PICTURE FILM)	ANL-MP-(NO.)-(NO.)(FILM)
ARGONNE NATIONAL LAB., ILL.	ANL-RE-
ARGONNE NATIONAL LAB., ILL. (SODIUM TECHNOLOGY)	ANL/ST-
ARGONNE NATIONAL LAB., ILL.	ANL-TR-(YEAR)-
ARGONNE NATIONAL LAB., ILL.	ANL-TRANS-
ARGONNE NATIONAL LAB., ILL.	ANPR-
ARGONNE NATIONAL LAB., ILL.	ATSR-MEMO-
ARGONNE NATIONAL LAB., ILL.	DP-REPORT-
ARGONNE NATIONAL LAB., ILL.	EBWR-TR-
ARGONNE NATIONAL LAB., ILL.	FGF-
ARGONNE NATIONAL LAB., ILL.	FRA-TM-
ARGONNE NATIONAL LAB., ILL.(CODE IS AUTHOR'S INITLS.)	GJB/RJL-
ARGONNE NATIONAL LAB., ILL.	HI-C-MEMO-
ARGONNE NATIONAL LAB., ILL.	MET.-
ARGONNE NATIONAL LAB., ILL.	PK-
ARGONNE NATIONAL LAB., ILL.	RED-EPM-
ARGONNE NATIONAL LAB., ILL.	RED-EPM-MEMO-
ARGONNE NATIONAL LAB., ILL.	RLC-(NUMBER)-
ARGONNE NATIONAL LAB., ILL.	SM-
ARGONNE NATIONAL LAB., ILL.	TKK-
ARGONNE NATIONAL LAB., ILL.	TM-
ARGONNE NATIONAL LAB., ILL.	TPR-RFT-
ARGONNE NATIONAL LAB., ILL.	ZPR-
ARGONNE NATIONAL LAB., ILL.	ZPR-TM-
ARGONNE NATIONAL LAB., IDAHO FALLS (EXPERIMENTAL BREEDER REACTOR)	AN/EBR-
ARGONNE NATIONAL LAB., IDAHO FALLS	ARB-
ARGONNE NATIONAL LAB., LEMONT, ILL.	ADF-
ARGONNE NATIONAL LAB., LEMONT, ILL.	ANL-(INITIALS)-
ARGONNE NATIONAL LAB., LEMONT, ILL.	ANL-(INITLS.)-(INITIALS)-
ARGONNE NATIONAL LAB., LEMONT, ILL.	ANL-NDN-
ARGONNE NATIONAL LAB., LEMONT, ILL. (SUBMARINE THERMAL REACTOR)	ANL-STR-MEMO-
ARGONNE NATIONAL LAB., LEMONT, ILL. (REF. 10)	CA-
ARGONNE NATIONAL LAB., LEMONT, ILL. (REF. 10)	CA-M-
ARGONNE NATIONAL LAB., LEMONT, ILL. (REF. 10)	CA-R-
ARGONNE NATIONAL LAB., LEMONT, ILL. (REF. 10)	CB-
ARGONNE NATIONAL LAB., LEMONT, ILL. (REF. 10)	CC-
ARGONNE NATIONAL LAB., LEMONT, ILL. (REF. 10)	CC-(LETTER(S))-
ARGONNE NATIONAL LAB., LEMONT, ILL. (REF. 10)	CF-
ARGONNE NATIONAL LAB., LEMONT, ILL. (REF. 10)	CH-
ARGONNE NATIONAL LAB., LEMONT, ILL. (REF. 10)	CP-
ARGONNE NATIONAL LAB., LEMONT, ILL. (REF. 10)	CS-
ARGONNE NATIONAL LAB., LEMONT, ILL.	CT-
ARGONNE NATIONAL LAB., LEMONT, ILL. (REF. 10)	DESIGN REPORT NO.-
ARGONNE NATIONAL LAB., LEMONT, ILL.	MEMO-
ARGONNE NATIONAL LAB., LEMONT, ILL. (REF. 10)	N-
ARGONNE NATIONAL LAB., LEMONT, ILL.	NDN-
ARGONNE NATIONAL LAB., LEMONT, ILL.	NMI-
ARGONNE NATIONAL LAB., LEMONT, ILL.	PC-
ARGONNE NATIONAL LAB., LEMONT, ILL.	RD-
ARGONNE NATIONAL LAB., LEMONT, ILL.	RDG-
ARGONNE NATIONAL LAB., LEMONT, ILL.	RK-
ARGONNE NATIONAL LAB., LEMONT, ILL.	STR-MEMO-
ARGONNE NATIONAL LAB., LEMONT, ILL.	TV-
ARGONNE NATIONAL LAB., LEMONT, ILL.	UAC-
(SEE ALSO LATER ADDRESS: ARGONNE NATL. LAB., ILL.)	
ARGUS INFORMATION SERVICE, BALTIMORE (TRANSLATIONS)	AIS-
ARINC RESEARCH CORP., ANNAPOLIS	SL-
ARINC RESEARCH CORP., SANTA ANA, CALIF.	W7-(NO.)-TNO(NO.)-
ARINC RESEARCH CORP., WASHINGTON, D.C.	ARINC-
ARINC RESEARCH CORP., WASHINGTON, D.C.	PUBL.-
ARINC RESEARCH CORP., WASHINGTON, D.C.	TDR-
ARIZONA. UNIV., TUCSON	ARIZ-U-
ARIZONA. UNIV., TUCSON	AZU-
ARIZONA. UNIV., TUCSON	AZU-PR-
ARIZONA. UNIV., TUCSON	AZU-SR-
ARIZONA. UNIV., TUCSON	AZU-TR-
ARIZONA. UNIV., TUCSON	ERDAA-ELCT-
ARIZONA. UNIV., TUCSON. ANALOG/HYBRID COMPUTER LAB.	ACL MEMO-
ARIZONA. UNIV., TUCSON. COMPUTER SCIENCE RESEARCH LAB	CSRL-MEMO-
ARIZONA. UNIV., TUCSON. ENERGY MASS AND MOMENTUM TRANSFER LAB.	EMMT-LAB-TR-
ARIZONA. UNIV., TUCSON. ENGINEERING EXPERIMENT STA.	AR-
ARIZONA. UNIV., TUCSON. ENGINEERING EXPERIMENT STA.	EES-SERIES-
ARIZONA. UNIV., TUCSON. ENGINEERING EXPERIMENT STA.	EES-SR-
ARIZONA. UNIV., TUCSON. INST. OF ATMOSPHERIC PHYS.	AZU-IAP-
ARIZONA. UNIV., TUCSON. LUNAR AND PLANETARY LAB.	COMMUNICATIONS-
ARIZONA. UNIV., TUCSON. OPTICAL SCIENCES CENTER	OSC-(NO.)-(NO.)-(NO.)-
ARIZONA STATE COLL., FLAGSTAFF	ARZ-
ARIZONA STATE COLL., TEMPE	ASC-
ARIZONA STATE UNIV., TEMPE. ENGINEERING RESEARCH CTR.	ERC-FH-(CONTRACT NO.)-
ARIZONA TRANSPORTATION AND TRAFFIC INST., TUCSON	ATTI-
ARKANSAS. UNIV., FAYETTEVILLE	ARK-RE-
ARKANSAS. UNIV., FAYETTEVILLE (ORDNANCE PROJECT)	ORDARK-
ARKANSAS. UNIV., FAYETTEVILLE	UARK-
ARKANSAS. UNIV., FAYETTEVILLE (ANNUAL PROGRESS RPT.)	UARK-APR (YEAR)
ARKANSAS. UNIV., FAYETTEVILLE. PLASMA LAB.	UAPL-

ARKANSAS. UNIV., LITTLE ROCK. SCHOOL OF MEDICINE	ARK-
ARKANSAS. UNIV., LITTLE ROCK. SCHOOL OF MEDICINE	ARK-U-
ARKANSAS. UNIV., LITTLE ROCK. SCHOOL OF MEDICINE	UARK-
ARMAMENT DEV. AND TEST CTR., EGLIN AFB, FLA.	ADTC-TR-(YEAR)-
ARMAMENT SYSTEMS PERSONNEL RES. LAB., LOWRY AFB, COLO	ASPRL-LN-(NUMBERS)
ARMAMENT SYSTEMS PERSONNEL RES. LAB., LOWRY AFB, COLO	ASPRL-TM-
ARMED FORCES MEDICAL POLICY COUNCIL, WASHINGTON, D.C.	AFM-
ARMED FORCES RADIOBIOLOGY RES. INST., BETHESDA, MD.	AFFRI-ARR-
ARMED FORCES RADIOBIOLOGY RES. INST., BETHESDA, MD.	AFFRI-SR-(YEAR)-
ARMED FORCES RADIOBIOLOGY RES. INST., BETHESDA, MD.	AFRII-SR-(YEAR)-
ARMED FORCES RADIOBIOLOGY RES. INST., BETHESDA, MD.	AFRRI-
ARMED FORCES RADIOBIOLOGY RES. INST., BETHESDA, MD. (ANNUAL RESEARCH REPORT)	AFRRI-ARR (MO/YR-MO/YR)
ARMED FORCES RADIOBIOLOGY RES. INST., BETHESDA, MD.	AFRRI-CR-(YEAR)-
ARMED FORCES RADIOBIOLOGY RES. INST., BETHESDA, MD.	AFRRI-SP-(YEAR)-
ARMED FORCES RADIOBIOLOGY RES. INST., BETHESDA, MD.	AFRRI-SR-
ARMED FORCES RADIOBIOLOGY RES. INST., BETHESDA, MD.	AFRRI-SR-(YEAR)-
ARMED FORCES RADIOBIOLOGY RES. INST., BETHESDA, MD.	AFRRI-TN-
ARMED FORCES RADIOBIOLOGY RES. INST., BETHESDA, MD.	AFRRI-TN-(YEAR)-
ARMED FORCES SECURITY AGENCY, WASHINGTON, D.C.	AFSA-
ARMED FORCES SPECIAL WEAPONS PROJECT, WASHINGTON, D.C.	AFSWP-
ARMED FORCES SPECIAL WEAPONS PROJECT, WASHINGTON, D.C.	AFSWP/SWP-
ARMED FORCES SPECIAL WEAPONS PROJECT, WASHINGTON, D.C.	AFSWP/SWPAN-
ARMED FORCES SPECIAL WEAPONS PROJECT, WASHINGTON, D.C. (TECHNICAL LETTER SERIES)	AFSWP TL-
ARMED FORCES SPECIAL WEAPONS PROJECT, WASHINGTON,D.C.	AICBM-
ARMED FORCES SPECIAL WEAPONS PROJECT, WASHINGTON, D.C. (SERVICE STORAGE FACILITIES)	SSF-
ARMED FORCES SPECIAL WEAPONS PROJECT, WASHINGTON, D.C. (SPECIAL WEAPONS BULLETIN)	SWB-
ARMED FORCES SPECIAL WEAPONS PROJECT, WASHINGTON,D.C.	SWB-(NO.)-
ARMED FORCES SPECIAL WEAPONS PROJECT, WASHINGTON, D.C. (SPECIAL WEAPONS INSPECTION REPORT)	SWIR-
ARMED FORCES SPECIAL WEAPONS PROJECT, WASHINGTON,D.C.	SWP-
ARMED FORCES SPECIAL WEAPONS PROJECT, WASHINGTON,D.C.	SWPDV-
ARMED FORCES SPECIAL WEAPONS PROJECT, WASHINGTON, D.C. (SPECIAL WEAPONS PROJECT WEAPON TESTS) (REF.7)	SWPWT-
ARMED FORCES SPECIAL WEAPONS PROJECT, WASHINGTON, D.C. (SPECIAL WEAPONS REGULATIONS)	SWR-
ARMED FORCES SPECIAL WEAPONS PROJECT, WASHINGTON,D.C.	TC-(YEAR)-
ARMED FORCES SPECIAL WEAPONS PROJECT, WASHINGTON, D.C. (WEAPONS EMPLOYMENT HANDBOOK)	WE-H-
ARMED FORCES SPECIAL WEAPONS PROJECT. FIELD COMMAND, ALBUQUERQUE, N. MEX. (WORKBOOK)	AAC-WB-
ARMED FORCES SPECIAL WEAPONS PROJECT. FIELD COMMAND, ALBUQUERQUE, N. MEX. (NOTEBOOK)	ABE-NB-
ARMED FORCES SPECIAL WEAPONS PROJECT. FIELD COMMAND, ALBUQUERQUE, N. MEX. (NOTEBOOK)	ABM-NB-
ARMED FORCES SPECIAL WEAPONS PROJECT. FIELD COMMAND, ALBUQUERQUE, N. MEX. (INSTRUCTORS MANUAL)	ADM-IM-
ARMED FORCES SPECIAL WEAPONS PROJECT. FIELD COMMAND, ALBUQUERQUE, N. MEX. (INSTRUCTORS MANUAL)	AFC-IM-
ARMED FORCES SPECIAL WEAPONS PROJECT. FIELD COMMAND, ALBUQUERQUE, N. MEX. (NOTEBOOK)	AFC-NB-
ARMED FORCES SPECIAL WEAPONS PROJECT. FIELD COMMAND, ALBUQUERQUE, N. MEX. (VU-GRAPHS)	AFC-VG-
ARMED FORCES SPECIAL WEAPONS PROJECT. FIELD COMMAND, ALBUQUERQUE, N. MEX.	AFSWP/FC-
ARMED FORCES SPECIAL WEAPONS PROJECT. FIELD COMMAND, ALBUQUERQUE, N. MEX.	AFSWP/FC- (MO.YR.)-
ARMED FORCES SPECIAL WEAPONS PROJECT. FIELD COMMAND, ALBUQUERQUE, N. MEX. (JOINT TASK GP. RPT.) (REF.18)	AFSWP/FC-JTG-(LTRS-NOS.)
ARMED FORCES SPECIAL WEAPONS PROJECT. FIELD COMMAND, ALBUQUERQUE, N. MEX.	AFSWP-FS-
ARMED FORCES SPECIAL WEAPONS PROJECT. FIELD COMMAND, ALBUQUERQUE, N. MEX.	AFSWP-NW-WB-
ARMED FORCES SPECIAL WEAPONS PROJECT. FIELD COMMAND, ALBUQUERQUE, N. MEX.	AFSWP-SS-(YEAR)-
ARMED FORCES SPECIAL WEAPONS PROJECT. FIELD COMMAND, ALBUQUERQUE, N. MEX.	AFSWP-UTG-
ARMED FORCES SPECIAL WEAPONS PROJECT. FIELD COMMAND, ALBUQUERQUE, N. MEX.	AFSWP-WE-
ARMED FORCES SPECIAL WEAPONS PROJECT. FIELD COMMAND, ALBUQUERQUE, N. MEX.	AFSWP-WU-
ARMED FORCES SPECIAL WEAPONS PROJECT. FIELD COMMAND, ALBUQUERQUE, N. MEX. (INSTRUCTORS MANUAL)	AHE-IM-
ARMED FORCES SPECIAL WEAPONS PROJECT. FIELD COMMAND, ALBUQUERQUE, N. MEX. (NOTEBOOK)	AHE-NB-
ARMED FORCES SPECIAL WEAPONS PROJECT. FIELD COMMAND, ALBUQUERQUE, N. MEX. (VU-GRAPHS)	AHE-VG-
ARMED FORCES SPECIAL WEAPONS PROJECT. FIELD COMMAND, ALBUQUERQUE, N. MEX. (INSTRUCTORS MANUAL)	AHM-IM-
ARMED FORCES SPECIAL WEAPONS PROJECT. FIELD COMMAND, ALBUQUERQUE, N. MEX. (NOTEBOOK)	AHM-NB-
ARMED FORCES SPECIAL WEAPONS PROJECT. FIELD COMMAND, ALBUQUERQUE, N. MEX. (INSTRUCTORS MANUALS)	A-IM-(LETTERS)-
ARMED FORCES SPECIAL WEAPONS PROJECT. FIELD COMMAND, ALBUQUERQUE, N. MEX. (INSTRUCTORS MANUAL)	AKE-IM-
ARMED FORCES SPECIAL WEAPONS PROJECT. FIELD COMMAND, ALBUQUERQUE, N. MEX. (NOTEBOOK)	AKE-NB-
ARMED FORCES SPECIAL WEAPONS PROJECT. FIELD COMMAND, ALBUQUERQUE, N. MEX. (VU-GRAPHS)	AKE-VG-
ARMED FORCES SPECIAL WEAPONS PROJECT. FIELD COMMAND, ALBUQUERQUE, N. MEX. (INSTRUCTORS MANUAL)	AKM-IM-
ARMED FORCES SPECIAL WEAPONS PROJECT. FIELD COMMAND, ALBUQUERQUE, N. MEX. (NOTEBOOK)	AKM-NB-
ARMED FORCES SPECIAL WEAPONS PROJECT. FIELD COMMAND, ALBUQUERQUE, N. MEX. (INSTRUCTORS MANUAL)	ALE-IM-
ARMED FORCES SPECIAL WEAPONS PROJECT. FIELD COMMAND, ALBUQUERQUE, N. MEX. (VU-GRAPHS)	ALE-VG-
ARMED FORCES SPECIAL WEAPONS PROJECT. FIELD COMMAND, ALBUQUERQUE, N. MEX. (WORKBOOK)	ANA-WB-
ARMED FORCES SPECIAL WEAPONS PROJECT. FIELD COMMAND, ALBUQUERQUE, N. MEX. (WORKBOOK)	ANM-WB-
ARMED FORCES SPECIAL WEAPONS PROJECT. FIELD COMMAND, ALBUQUERQUE, N. MEX.	ANWB-(LETTER)-
ARMED FORCES SPECIAL WEAPONS PROJECT. FIELD COMMAND, ALBUQUERQUE, N. MEX. (INSTRUCTORS MANUAL)	APN-IM-
ARMED FORCES SPECIAL WEAPONS PROJECT. FIELD COMMAND, ALBUQUERQUE, N. MEX. (VU-GRAPHS)	APN-VG-
ARMED FORCES SPECIAL WEAPONS PROJECT. FIELD COMMAND, ALBUQUERQUE, N. MEX. (NOTEBOOK)	ARE-NB-
ARMED FORCES SPECIAL WEAPONS PROJECT. FIELD COMMAND, ALBUQUERQUE, N. MEX. (VU-GRAPHS)	ARE-VG-
ARMED FORCES SPECIAL WEAPONS PROJECT. FIELD COMMAND, ALBUQUERQUE, N. MEX. (INSTRUCTORS MANUAL)	ARM-IM-
ARMED FORCES SPECIAL WEAPONS PROJECT. FIELD COMMAND, ALBUQUERQUE, N. MEX. (NOTEBOOK)	ARM-NB-
ARMED FORCES SPECIAL WEAPONS PROJECT. FIELD COMMAND, ALBUQUERQUE, N. MEX. (INSTRUCTORS MANUAL)	ATE-IM-
ARMED FORCES SPECIAL WEAPONS PROJECT. FIELD COMMAND, ALBUQUERQUE, N. MEX. (NOTEBOOK)	ATE-NB-
ARMED FORCES SPECIAL WEAPONS PROJECT. FIELD COMMAND, ALBUQUERQUE, N. MEX. (VU-GRAPHS)	ATE-VG-
ARMED FORCES SPECIAL WEAPONS PROJECT. FIELD COMMAND, ALBUQUERQUE, N. MEX. (INSTRUCTORS MANUAL)	ATM-IM-
ARMED FORCES SPECIAL WEAPONS PROJECT. FIELD COMMAND, ALBUQUERQUE, N. MEX. (NOTEBOOK)	ATM-NB-
ARMED FORCES SPECIAL WEAPONS PROJECT. FIELD COMMAND, ALBUQUERQUE, N. MEX.	AWB-G-
ARMED FORCES SPECIAL WEAPONS PROJECT. FIELD COMMAND, ALBUQUERQUE, N. MEX. (HONEST JOHN, CORPORAL)	AWB-HJC-
ARMED FORCES SPECIAL WEAPONS PROJECT. FIELD COMMAND, ALBUQUERQUE, N. MEX. (ELECTRICAL)	AWB-ME-
ARMED FORCES SPECIAL WEAPONS PROJECT. FIELD COMMAND, ALBUQUERQUE, N. MEX. (MECHANICAL)	AWB-MM-
ARMED FORCES SPECIAL WEAPONS PROJECT. FIELD COMMAND, ALBUQUERQUE, N. MEX. (NIKE HERCULES)	AWB-NH-
ARMED FORCES SPECIAL WEAPONS PROJECT. FIELD COMMAND, ALBUQUERQUE, N. MEX. (REDSTONE)	AWB-R-
ARMED FORCES SPECIAL WEAPONS PROJECT. FIELD COMMAND, ALBUQUERQUE, N. MEX. (INSTRUCTORS MANUAL)	AYE-IM-
ARMED FORCES SPECIAL WEAPONS PROJECT. FIELD COMMAND, ALBUQUERQUE, N. MEX. (NOTEBOOK)	AYE-NB-
ARMED FORCES SPECIAL WEAPONS PROJECT. FIELD COMMAND, ALBUQUERQUE, N. MEX.	CAS-PR-
ARMED FORCES SPECIAL WEAPONS PROJECT. FIELD COMMAND, ALBUQUERQUE, N. MEX.	FC/
ARMED FORCES SPECIAL WEAPONS PROJECT. FIELD COMMAND, ALBUQUERQUE, N. MEX.	FCDR(NUMBER)-(NUMBER)
ARMED FORCES SPECIAL WEAPONS PROJECT. FIELD COMMAND, ALBUQUERQUE, N. MEX.	FCOD (NUMBER(S))-(DATE)
ARMED FORCES SPECIAL WEAPONS PROJECT. FIELD COMMAND, ALBUQUERQUE, N. MEX.	FCPD-
ARMED FORCES SPECIAL WEAPONS PROJECT. FIELD COMMAND, ALBUQUERQUE, N. MEX.	FC-SWTG-
ARMED FORCES SPECIAL WEAPONS PROJECT. FIELD COMMAND, ALBUQUERQUE, N. MEX.	FCTNG(NO./LTR.-NO.,DATE)
ARMED FORCES SPECIAL WEAPONS PROJECT. FIELD COMMAND, ALBUQUERQUE, N. MEX.	FCTS/(YEAR)-
ARMED FORCES SPECIAL WEAPONS PROJECT. FIELD COMMAND, ALBUQUERQUE, N. MEX.	FCTT/(YEAR)-
ARMED FORCES SPECIAL WEAPONS PROJECT. FIELD COMMAND, ALBUQUERQUE, N. MEX.	FCTT-(LETTERS-LETTERS)-
ARMED FORCES SPECIAL WEAPONS PROJECT. FIELD COMMAND, ALBUQUERQUE, N. MEX.	GCE-(LETTERS)-
ARMED FORCES SPECIAL WEAPONS PROJECT. FIELD COMMAND, ALBUQUERQUE, N. MEX. (NOTEBOOK)	GJE-NB-
ARMED FORCES SPECIAL WEAPONS PROJECT. FIELD COMMAND, ALBUQUERQUE, N. MEX. (VU-GRAPHS)	GJE-VG-
ARMED FORCES SPECIAL WEAPONS PROJECT. FIELD COMMAND, ALBUQUERQUE, N. MEX. (INSTRUCTORS MANUAL)	GME-IM-
ARMED FORCES SPECIAL WEAPONS PROJECT. FIELD COMMAND, ALBUQUERQUE, N. MEX. (VU-GRAPHS)	GME-VG-
ARMED FORCES SPECIAL WEAPONS PROJECT. FIELD COMMAND, ALBUQUERQUE, N. MEX. (NOTEBOOK)	GRE-NB-
ARMED FORCES SPECIAL WEAPONS PROJECT. FIELD COMMAND, ALBUQUERQUE, N. MEX. (VU-GRAPHS)	GRE-VG-

ARMED FORCES SPECIAL WEAPONS PROJECT. FIELD COMMAND

ARMED FORCES SPECIAL WEAPONS PROJECT. FIELD COMMAND, ALBUQUERQUE, N. MEX. (NAVY ASSEMBLYMAN COURSE. INSTRUCTORS MANUAL)	N-IM-(LETTERS)-
ARMED FORCES SPECIAL WEAPONS PROJECT. FIELD COMMAND, ALBUQUERQUE, N.MEX.	UTG-
ARMED FORCES SPECIAL WEAPONS PROJECT. FIELD COMMAND, ALBUQUERQUE, N. MEX.	UTG/(YR.)-
ARMED FORCES SPECIAL WEAPONS PROJECT. FIELD COMMAND, ALBUQUERQUE, N. MEX. (WEAPONS EMPLOYMENT MANUAL)	WE-M-
(SEE ALSO LATER NAME: DEFENSE ATOMIC SUPPORT AGENCY. FIELD COMMAND)	
ARMED FORCES SPECIAL WEAPONS PROJECT. FIELD COMMAND. DIRECTORATE OF WEAPONS EFFECTS TESTS, ALBUQUERQUE, N. MEX. (REF. 7)	DWET-
ARMED FORCES SPEC. WEAPONS PROJ. FIELD COMMAND.DIRECTORATE OF WEAPONS EFFECTS TESTS, ALBUQUERQUE,N.MEX.	IOC-
ARMED FORCES SPECIAL WEAPONS PROJECT. FIELD COMMAND. DIRECTORATE OF WEAPONS EFFECTS TESTS, ALBUQUERQUE, N. MEX. (REF. 7)	WET-(YEAR)-
ARMED FORCES SPECIAL WEAPONS PROJECT. FIELD COMMAND. SPECIAL WEAPONS UNIT TRAINING GP., ALBUQUERQUE,N.M.	FCTUG-AFSWP-
ARMED FORCES SPECIAL WEAPONS PROJECT. FIELD COMMAND. TRAINING DIV.,ALBUQUERQUE, N. MEX. (ACCESSION LIST)	FCTD-AL-
ARMED FORCES SPECIAL WEAPONS PROJECT. WEAPONS EFFECTS DIV., WASH., D.C. (OPERATION WIGWAM) (REF. 7)	SWPFD-
ARMED FORCES SPECIAL WEAPONS PROJECT. WEAPONS EFFECTS PROGRAM, WASH., D.C. (OPERATION HARDTACK) (REF. 7)	HARDTACK-
ARMED FORCES SPECIAL WEAPONS PROJECT. WEAPONS EFFECTS PROGRAM, WASH., D.C. (OPERATION PLUMBBOB) (REF. 7)	PILGRIM-
ARMED FORCES STAFF COLLEGE, NORFOLK, VA.	AFSC/NOR-
ARMED FORCES SUPPLY SUPPORT CENTER. STANDARDIZATION DIV., WASHINGTON, D.C. (MILITARY SPECS. + STDS.) (REF. 13)	MIL-(LETTER)-
ARMED FORCES TECHNICAL INFORMATION AGENCY. (TECHNICAL INFORMATION PILOT)	TIP-
ARMED SERVICES ELECTRO STANDARDS AGENCY, FORT MONMOUTH, N.J.	ASESA-
ARMED SERVICES ELECTRO STANDARDS AGENCY, FORT MONMOUTH, N.J.	ASETC-
ARMED SERVICES EXPLOSIVES SAFETY BOARD, WASH., D.C.	ASESB-
ARMED SERVICES PETROLEUM BOARD, WASH., D.C.	ASPB-
ARMED SERVICES TECHNICAL INFO. AGENCY, ARLINGTON, VA.	58AA-
ARMED SERVICES TECHNICAL INFO. AGENCY, ARLINGTON, VA.	56AA-
ARMED SERVICES TECHNICAL INFO. AGENCY, ARLINGTON, VA.	57AA-
ARMED SERVICES TECHNICAL INFO. AGENCY, ARLINGTON, VA.	55AA-
ARMED SERVICES TECHNICAL INFO. AGENCY, ARLINGTON, VA.	54AA-
ARMED SERVICES TECHNICAL INFO. AGENCY, ARLINGTON, VA.	ASTIA
ARMED SERVICES TECHNICAL INFO. AGENCY, ARLINGTON, VA.	ASTIA ARB-
ARMED SERVICES TECHNICAL INFO. AGENCY, ARLINGTON, VA.	ASTIA-CAH-
ARMED SERVICES TECHNICAL INFO. AGENCY, ARLINGTON, VA. (BIBLIOGRAPHICAL TECHNICAL INDEX)	BTI-
(SEE ALSO LATER NAME: DEFENSE DOCUMENTATION CENTER)	
ARMED SERVICES TECHNICAL INFO. AGENCY, DAYTON, OHIO	ASTIA BTI-
ARMED SERVICES TECHNICAL INFORMATION AGENCY. DOCUMENT SERVICE CENTER, DAYTON, OHIO	ATI-
ARMED SERVICES TECHNICAL INFORMATION AGENCY. DOCUMENT SERVICE CENTER, DAYTON, OHIO	DSC-
ARMED SERVICES TECHNICAL INFORMATION AGENCY. DOCUMENT SERVICE CENTER, DAYTON, OHIO	F-TS-
ARMED SERVICES TECHNICAL INFORMATION AGENCY. REFERENCE CENTER, ARLINGTON HALL STATION,ARLINGTON,VA.	NRS-
ARMED SERVICES TECHNICAL INFORMATION AGENCY. REFERENCE CTR., LIBRARY OF CONGRESS, WASH., D.C.	ARC-
ARMED SERVICES TECHNICAL INFORMATION AGENCY. REFERENCE CTR., LIBRARY OF CONGRESS, WASH., D.C.	BTI-ARC-
ARMED SERVICES TECHNICAL INFORMATION AGENCY. REFERENCE CTR., LIBRARY OF CONGRESS, WASH., D.C.	TIP-
(SEE ALSO LATER LOCATION, ALEXANDRIA, VA. AND STILL LATER NAME: DEFENSE DOCUMENTATION CENTER)	
ARMOUR RESEARCH FOUNDATION, CHICAGO (REF. 1)	(NUMBER)
ARMOUR RESEARCH FOUNDATION, CHICAGO	ARF-
ARMOUR RESEARCH FOUNDATION, CHICAGO	ARF-DCM-
ARMOUR RESEARCH FOUNDATION, CHICAGO	ARF-DCP-(YEAR)-
ARMOUR RESEARCH FOUNDATION, CHICAGO	ARF-FR-
ARMOUR RESEARCH FOUNDATION, CHICAGO	ARF-IR-
ARMOUR RESEARCH FOUNDATION, CHICAGO	ARF-ORG-
ARMOUR RESEARCH FOUNDATION, CHICAGO	ARF-PR-
ARMOUR RESEARCH FOUNDATION, CHICAGO	ARF-RP-
ARMOUR RESEARCH FOUNDATION, CHICAGO	ARF-TM-
ARMOUR RESEARCH FOUNDATION, CHICAGO	ARF-TM-(LETTER)-
ARMOUR RESEARCH FOUNDATION, CHICAGO	ARF-TN-(LETTER)-
ARMOUR RESEARCH FOUNDATION, CHICAGO	ARF-TR-
ARMOUR RESEARCH FOUNDATION, CHICAGO	C-
ARMOUR RESEARCH FOUNDATION, CHICAGO	DCM-
ARMOUR RESEARCH FOUNDATION, CHICAGO	DCM-(YEAR)-
ARMOUR RESEARCH FOUNDATION, CHICAGO	DCMR-
ARMOUR RESEARCH FOUNDATION, CHICAGO	DCP-(YEAR)-
ARMOUR RESEARCH FOUNDATION, CHICAGO	G-(3 DIGITS)-PR-
ARMOUR RESEARCH FOUNDATION, CHICAGO	GV-
ARMOUR RESEARCH FOUNDATION, CHICAGO	I-(NO.)-(NO.)-/(NO.)
ARMOUR RESEARCH FOUNDATION, CHICAGO	NMI-
ARMOUR RESEARCH FOUNDATION, CHICAGO	S-55000 THRU S-55999
ARMOUR RESEARCH FOUNDATION, CHICAGO	TR-(YEAR)-
(SEE ALSO SUCCESSOR: IIT RESEARCH FOUNDATION)	
ARMS CONTROL AND DISARMAMENT AGENCY, WASHINGTON, D.C.	ACDA-(NUMBER)-(YEAR)-
ARMS CONTROL AND DISARMAMENT AGENCY, WASHINGTON, D.C.	ACDA/E-
ARMS CONTROL AND DISARMAMENT AGENCY, WASHINGTON, D.C.	ACDA-IR-(NUMBERS)
ARMS CONTROL AND DISARMAMENT AGENCY, WASHINGTON, D.C.	ACDA-PUB-
ARMS CONTROL AND DISARMAMENT AGENCY, WASHINGTON, D.C.	ACDA-RR-(YEAR)-
ARMS CONTROL AND DISARMAMENT AGENCY, WASHINGTON, D.C.	ACDA-ST-
ARMS CONTROL AND DISARMAMENT AGENCY. WEAPONS EVALUATION AND CONTROL BUREAU, WASHINGTON, D.C.	ACDA/WEC-(YEAR)-
ARMS CONTROL AND DISARMAMENT AGENCY. WEAPONS EVALUATION AND CONTROL BUREAU, WASHINGTON, D.C.	ACDA/WEC/FG-(YEAR)-
ARMSTRONG WHITWORTH (SIR W.G.) AIRCRAFT, LTD.	ATR-
ARMSTRONG WHITWORTH (SIR W.G.) AIRCRAFT, LTD.	ETR-
ARMSTRONG WHITWORTH (SIR W.G.) AIRCRAFT, LTD.	TA-
ARMY, VIETNAM	ARV(NUMBERS)
ARMY, WASHINGTON, D.C.	A-
ARMY, WASHINGTON, D.C. (AMMUNITION IDENT. CODE)	AIC-
ARMY, WASHINGTON, D.C. (REGULATION)	AR-
ARMY, WASHINGTON, D.C. (ANNUAL RES. TASK SUMMARY)	ARTS-(YEAR),(VOLUME)
ARMY, WASHINGTON, D.C. (ARMY SUBJECT SCHEDULE)	ASS-
ARMY, WASHINGTON, D.C.	DA-
ARMY, WASHINGTON, D.C. (FOR EXPLANATION OF THE REMAINDER OF THE CODE, SEE ENTRY UNDER LETTERS FOLLOWING DA-)	DA-(LETTERS)-

ARMY, WASHINGTON, D.C. (PAMPHLET)	DAP-
ARMY, WASHINGTON, D.C. (PAMPHLET)	DA-PAM-
ARMY, WASHINGTON D.C. (REORGANIZATION PLAN)	DARPD-(NUMBER)-
ARMY, WASHINGTON, D.C. (SHIPPING DOCUMENT)	DASD-
ARMY, WASHINGTON, D.C. (ENGINEERING ORDER)	EO-
ARMY, WASHINGTON, D.C. (FIELD MANUAL)	FM-
ARMY, WASHINGTON, D.C. (FIRING TABLES)	FT-
ARMY, WASHINGTON, D.C. (FIRING TABLES ROCKETS)	FTR-
ARMY, WASHINGTON, D.C.	H-
ARMY, WASHINGTON, D.C.	I-
ARMY, WASH., D.C. (ATOMIC INCIDENT CONTROL PLAN)	ICP-
ARMY, WASHINGTON, D.C.	J-
ARMY, WASHINGTON, D.C. (MILITARY ATTACHE REPORT)	MAR-
ARMY, WASHINGTON, D.C.	NIC-TRANS-
ARMY, WASHINGTON, D.C. (ORDNANCE PROOF MANUAL)	OPM-
ARMY, WASHINGTON, D.C. (ORDNANCE DESIGN PAMPHLET)	ORDP-
ARMY, WASHINGTON, D.C. (ORDNANCE STORAGE AND SHIPMENT CHART)	OSSC-
ARMY, WASHINGTON, D.C. (PAMPHLET)	PAM-
ARMY, WASHINGTON, D.C. (QUALITATIVE DEVELOPMENT REQUIREMENTS INFORMATION)	QDRI-
ARMY, WASHINGTON, D.C. (REPORTS CONTROL SYMBOL)	RCS-
ARMY, WASHINGTON, D.C.	SB-(YEAR)-S-
ARMY, WASHINGTON, D.C.	SR-
ARMY, WASHINGTON, D.C. (TECHNICAL MANUAL)	TM-
ARMY, WASHINGTON, D.C.	TSB-CONTROL-
ARMY, WASHINGTON, D.C.	TUB-
ARMY, WASHINGTON, D.C.	WA-
ARMY, WASHINGTON, D.C.	WA-(NUMBER)-
ARMY ADVANCED MATERIEL CONCEPTS AGENCY, WASHINGTON,DC (AD HOC WORKING GROUP)	AD HOC-
ARMY ADVANCED MATERIEL CONCEPTS AGENCY, WASHINGTON,DC	AMCA-
ARMY AEROMEDICAL RESEARCH UNIT, FORT RUCKER, ALA.	USAARU-
ARMY AEROMEDICAL RESEARCH UNIT, FORT RUCKER, ALA.	USAARU-(YEAR)-
ARMY AEROMEDICAL RESEARCH UNIT, FORT RUCKER, ALA.	USAARU-JR-
ARMY AIR DEFENSE BOARD	AADB-
ARMY AIR DEFENSE BOARD, FORT BLISS, TEX.	GM-
ARMY AIR DEFENSE BOARD, FORT BLISS, TEX.	USARADBD-ES-
ARMY AIR DEFENSE BOARD, FORT BLISS, TEX.	USARADBD-GM-
ARMY AIR DEFENSE CENTER, FORT BLISS, TEX.	AKBAATC-B-(NO.)-(LTR)-
ARMY AIR DEFENSE SCHOOL, FORT BLISS, TEX.	AADS-
ARMY AIR FORCE (AVIATION PSYCHOLOGY PROGRAM)	AAF APP RR-
ARMY AIR FORCE (BULLETIN)	AAFB-
ARMY AIR FORCE (BOARD PROJECT)	AAF BD PROJ-
ARMY AIR FORCE (MANUAL)	AAF M-
ARMY AIR FORCE (LIAISON MEMORANDUM, SERIES X)	AAF SER. X-LM-
ARMY AIR FORCE (STOCK LIST)	AAF SL-
ARMY AIR FORCE (TRANSLATION)	AAFT-
ARMY AIR FORCE (TECHNICAL ORDER)	AAF TO-
ARMY AIR FORCE (TECHNICAL REPORT)	AAFTR-
ARMY AIR FORCE (AVIATION PSYCHOLOGY PROGRAM)	APP-
ARMY AIR FORCE (NON-TECHNICAL INTELLIGENCE REPORTS)	NTIR-
ARMY AIR FORCE, WASHINGTON, D.C.	AAF P-
ARMY AIR FORCE, WASHINGTON, D.C.	AAF-TR-
ARMY AIR FORCE, WASHINGTON, D.C.	AAF TR (NOS.,LETTERS)-
(SEE ALSO SUCCESSOR: AIR FORCE)	
ARMY AIR FORCE. SECOND. OPERATIONS AND TRAINING DIV. (OPERATIONS ANALYSIS REPORT)	AAF OAR-
ARMY AIR FORCE. SECOND. OPERATIONS AND TRAINING DIV. (OPERATIONS ANALYSIS REPORT)	OAR-
ARMY AIR FORCE. AIR TECHNICAL INTELLIGENCE GROUP	ATIG-
ARMY AIR FORCE. AIRCRAFT RADIATION LAB.	AAF ARL-
ARMY AIR FORCE. AIRCRAFT RADIATION LAB. (ENG. RPT.)	AAF ARL ER-
ARMY AIR FORCE. AIRCRAFT RADIATION LAB. (MEMO. RPT.)	AAF ARL MR-
ARMY AIR FORCE. AIRCRAFT RADIATION LAB. (TEST RPT.)	AAF ARL TR-
ARMY AIR FORCE. DIRECTORATE OF WEATHER	AAF DW-
ARMY AIR FORCE. DIRECTORATE OF WEATHER (SPECIAL SERIES)	AAF DW SS-
ARMY AIR FORCE. DIRECTORATE OF WEATHER	AFTSW-
ARMY AIR FORCE. DIRECTORATE OF WEATHER	DW-
ARMY AIR FORCE. OFFICE OF FLYING SAFETY (FLYING SAFETY MEMORANDUM)	FSM-
ARMY AIR FORCE. OFFICE OF FLYING SAFETY	OFS-
ARMY AIR FORCE. OFFICE OF FLYING SAFETY (FLYING SAFETY MEMORANDUM)	OFS FSM-
ARMY AIR FORCE. OFFICE OF FLYING SAFETY (SPECIAL ACCIDENT REPORT. ANALYSIS)	OFS SAR-A-
ARMY AIR FORCE. OFFICE OF FLYING SAFETY (SPECIAL ACCIDENT REPORT. SUPPLEMENTAL ANALYSIS)	OFS SAR-SA-
ARMY AIR FORCE. OFFICE OF FLYING SAFETY (SPECIAL ACCIDENT REPORT)	SAR-
ARMY AIR FORCE. OFF. OF OPERATIONS ANALYSIS, WASH.,DC	AFOOA-
ARMY AIR FORCE. OPERATIONS ANALYSIS DIV., WASH., D.C.	OA-
ARMY AIR FORCE. WEATHER DIV.	AAF WD-
ARMY AIR FORCE. WEATHER DIV.	AAF-WD-TR-
ARMY AIR FORCE. WEATHER INFORMATION BRANCH	AAF WIB-
ARMY AIR FORCE. WEATHER INFORMATION BRANCH	AFRTW-
ARMY AIR FORCE. WEATHER INFORMATION BRANCH	WIB-
ARMY AIR FORCE. WEATHER INFORMATION SERVICE	AAF WIS-
ARMY AIR FORCE. WEATHER INFORMATION SERVICE	WIS-
ARMY AIR FORCE. WEATHER SERVICE (BULLETIN)	AAF WSB-
ARMY AIR FORCE. WEATHER SERVICE (MANUAL)	AAF WSM-
ARMY AIR FORCE. WEATHER SERVICE (BULLETIN)	WSB-
ARMY AIR FORCE. WEATHER SERVICE (MANUAL)	WSM-
ARMY AIR FORCE BOARD, ORLANDO, FLA.	AAFB-
ARMY AIR FORCES TACTICAL CENTER, ORLANDO, FLA.	AAFTC-
ARMY AIRBORNE ELECTRONICS AND SPECIAL WARFARE BOARD, FORT BRAGG, N.C.	USAAESW-AB-
ARMY AIRBORNE ELECTRONICS AND SPECIAL WARFARE BOARD, FORT BRAGG, N.C.	USAAESWBD-AB-
ARMY AIRBORNE ELECTRONICS AND SPECIAL WARFARE BOARD, FORT BRAGG, N.C.	USAAESW-CE-

ARMY ARCTIC TEST CENTER

ARMY ARCTIC TEST CENTER, FORT GREELY, ALASKA	USAATC-
ARMY ARTILLERY AND MISSILE CENTER, FORT SILL, OKLA.	AAMC-
ARMY ARTILLERY AND MISSILE SCHOOL, FORT SILL, OKLA.	USAAMS-STUDY-
ARMY ARTILLERY BOARD, FORT SILL, OKLA.	USAARTYBD-FA-(NUMBER)-
ARMY ATOMIC WEAPON SYSTEMS SAFETY COMMITTEE	AAWSSC-
ARMY ATTACHE, LONDON	ARMY-AT-LONDON-
ARMY AVIATION MATERIAL LABS., FORT EUSTIS, VA.	AVCIR-
ARMY AVIATION MATERIAL LABS., FORT EUSTIS, VA.	USAAML-
ARMY AVIATION MATERIAL LABS., FORT EUSTIS, VA.	USAAML-TR-
ARMY AVIATION MATERIAL LABS., FORT EUSTIS, VA.	USAAVLABS-TN-
ARMY AVIATION MATERIAL LABS., FORT EUSTIS, VA.	USAAVLABS-TR-(YEAR)-
ARMY AVIATION MATERIEL COMMAND, ST. LOUIS	AVCOM-TR-
ARMY AVIATION MATERIEL COMMAND, ST. LOUIS	USAAVCOM-TR-(YR-NO-LTR)-
ARMY AVIATION SYSTEMS COMMAND, ST. LOUIS	USAAVSCOM-TR-(YR.)-
ARMY AVIATION SYSTEMS COMMAND, ST. LOUIS	USAAVSCOM-TR-XE-(YR.)-
ARMY AVIATION SYSTEMS TEST ACTIVITY, EDWARDS AFB, CAL	USAASTA-(YEAR-NUMBER)
ARMY AVIATION TEST ACTIVITY, EDWARDS AFB, CALIF.	USAAVNTA-(YEAR)-
ARMY BALLISTIC MISSILE AGENCY, HUNTSVILLE, ALA.	AEROBALLISTIC MEMO-
ARMY BALLISTIC MISSILE AGENCY, HUNTSVILLE, ALA.	DG-R-
ARMY BALLISTIC MISSILE AGENCY, HUNTSVILLE, ALA.	DM-R-
ARMY BALLISTIC MISSILE AGENCY, HUNTSVILLE, ALA.	DYR-R-
(SEE ALSO LATER ADDRESS: REDSTONE ARSENAL, ALA.)	
ARMY BALLISTIC MISSILE AGENCY, REDSTONE ARSENAL, ALA.	ABMA-
ARMY BALLISTIC MISSILE AGENCY, REDSTONE ARSENAL, ALA.	ABMA-COMM. MIN./
ARMY BALLISTIC MISSILE AGENCY, REDSTONE ARSENAL, ALA.	ABMA/DCMR TR-
ARMY BALLISTIC MISSILE AGENCY, REDSTONE ARSENAL, ALA.	ABMA-DV-TR-
ARMY BALLISTIC MISSILE AGENCY, REDSTONE ARSENAL, ALA.	ABMA JUPITER TECH BULL-
ARMY BALLISTIC MISSILE AGENCY, REDSTONE ARSENAL, ALA.	ABMA-M-LOD-
ARMY BALLISTIC MISSILE AGENCY, REDSTONE ARSENAL, ALA.	ABMA-MTP-LOD-
ARMY BALLISTIC MISSILE AGENCY, REDSTONE ARSENAL, ALA.	ABMA-OH-(NO.)
ARMY BALLISTIC MISSILE AGENCY, REDSTONE ARSENAL, ALA. (PURCHASE DESCRIPTION)	ABMA PD-H-
ARMY BALLISTIC MISSILE AGENCY, REDSTONE ARSENAL, ALA.	ABMA PD-P-
ARMY BALLISTIC MISSILE AGENCY, REDSTONE ARSENAL, ALA.	ABMA PD-R-
ARMY BALLISTIC MISSILE AGENCY, REDSTONE ARSENAL, ALA.	ABMA PD-S-
ARMY BALLISTIC MISSILE AGENCY, REDSTONE ARSENAL, ALA.	ABMA PD-V-
ARMY BALLISTIC MISSILE AGENCY, REDSTONE ARSENAL, ALA.	ABMA PD-W-
ARMY BALLISTIC MISSILE AGENCY, REDSTONE ARSENAL, ALA.	ABMA-RA-TR-
ARMY BALLISTIC MISSILE AGENCY, REDSTONE ARSENAL, ALA.	ABMA R-DS-
ARMY BALLISTIC MISSILE AGENCY, REDSTONE ARSENAL, ALA.	ABMA-RE-TR-
ARMY BALLISTIC MISSILE AGENCY, REDSTONE ARSENAL, ALA.	ABMA-RG-TR-
ARMY BALLISTIC MISSILE AGENCY, REDSTONE ARSENAL, ALA.	ABMA-RJ-TR-
ARMY BALLISTIC MISSILE AGENCY, REDSTONE ARSENAL, ALA.	ABMA-RP-TR-
ARMY BALLISTIC MISSILE AGENCY, REDSTONE ARSENAL, ALA.	ABMA-RR-R-DS-
ARMY BALLISTIC MISSILE AGENCY, REDSTONE ARSENAL, ALA.	ABMA-RR-TR-
ARMY BALLISTIC MISSILE AGENCY, REDSTONE ARSENAL, ALA.	ABMA-RS-TN-
ARMY BALLISTIC MISSILE AGENCY, REDSTONE ARSENAL, ALA.	ABMA-RS-TR-
ARMY BALLISTIC MISSILE AGENCY, REDSTONE ARSENAL, ALA.	ABMA-RTLA-TM-(YEAR)-...
ARMY BALLISTIC MISSILE AGENCY, REDSTONE ARSENAL, ALA.	ABMA-RT-TR-
ARMY BALLISTIC MISSILE AGENCY, REDSTONE ARSENAL, ALA.	ABMA SP RPT RP-
ARMY BALLISTIC MISSILE AGENCY, REDSTONE ARSENAL, ALA.	ABMA-XMC-
ARMY BALLISTIC MISSILE AGENCY, REDSTONE ARSENAL, ALA.	CAA-
ARMY BALLISTIC MISSILE AGENCY, REDSTONE ARSENAL, ALA. (JUPITER PROGRESS REPORTS)	CR-(YEAR)-J-
ARMY BALLISTIC MISSILE AGENCY, REDSTONE ARSENAL, ALA. (JUPITER PROGRESS REPORTS)	CR-J-(YEAR)-
ARMY BALLISTIC MISSILE AGENCY, REDSTONE ARSENAL, ALA.	DA-TM-(NUMBER)-(NUMBER)
ARMY BALLISTIC MISSILE AGENCY, REDSTONE ARSENAL, ALA. (TECHNICAL NOTE)	DA-TN-
ARMY BALLISTIC MISSILE AGENCY, REDSTONE ARSENAL, ALA.	DA-TR-
ARMY BALLISTIC MISSILE AGENCY, REDSTONE ARSENAL, ALA.	DC-TR-(NUMBER)-(NUMBER)
ARMY BALLISTIC MISSILE AGENCY, REDSTONE ARSENAL, ALA.	DD-DR+E(M)-
ARMY BALLISTIC MISSILE AGENCY, REDSTONE ARSENAL, ALA.	DG-TR-(NUMBER)-(YEAR)
ARMY BALLISTIC MISSILE AGENCY, REDSTONE ARSENAL, ALA.	DIR-TR-(NUMBER)-(YEAR)
ARMY BALLISTIC MISSILE AGENCY, REDSTONE ARSENAL, ALA.	DMCR-TR-(NO.)-(YEAR)
ARMY BALLISTIC MISSILE AGENCY, REDSTONE ARSENAL, ALA.	DSD-TR-(NUMBER)-(YEAR)
ARMY BALLISTIC MISSILE AGENCY, REDSTONE ARSENAL, ALA.	DSF-TN-
ARMY BALLISTIC MISSILE AGENCY, REDSTONE ARSENAL, ALA.	DSF-TR-(NUMBER)-(YEAR)
ARMY BALLISTIC MISSILE AGENCY, REDSTONE ARSENAL, ALA.	DSL-TM-(NUMBER)-(YEAR)
ARMY BALLISTIC MISSILE AGENCY, REDSTONE ARSENAL, ALA.	DSN-TN-(NUMBER)-(YEAR)
ARMY BALLISTIC MISSILE AGENCY, REDSTONE ARSENAL, ALA.	DSN-TR-(NUMBER)-(YEAR)
ARMY BALLISTIC MISSILE AGENCY, REDSTONE ARSENAL, ALA.	DSP-TM-(NUMBER)-(YEAR)
ARMY BALLISTIC MISSILE AGENCY, REDSTONE ARSENAL, ALA.	DSP-TN-(NUMBER)-(YEAR)
ARMY BALLISTIC MISSILE AGENCY, REDSTONE ARSENAL, ALA.	DSP-TR-(NUMBER)-(YEAR)
ARMY BALLISTIC MISSILE AGENCY, REDSTONE ARSENAL, ALA.	DTI-TR-(NUMBER)-(YEAR)
ARMY BALLISTIC MISSILE AGENCY, REDSTONE ARSENAL, ALA.	DT-R-
ARMY BALLISTIC MISSILE AGENCY, REDSTONE ARSENAL, ALA.	DV-TN-(NUMBER)-(YEAR)
ARMY BALLISTIC MISSILE AGENCY, REDSTONE ARSENAL, ALA.	DV-TR-(NUMBER)-(YEAR)
ARMY BALLISTIC MISSILE AGENCY, REDSTONE ARSENAL, ALA. (GUIDANCE AND CONTROL MEMORANDUM)	GCM-
ARMY BALLISTIC MISSILE AGENCY, REDSTONE ARSENAL, ALA.	HSI-(R)-
ARMY BALLISTIC MISSILE AGENCY, REDSTONE ARSENAL, ALA.	IRBB-(NO.)(LTS.)(NO.)
ARMY BALLISTIC MISSILE AGENCY, REDSTONE ARSENAL, ALA.	ORDAB-
ARMY BALLISTIC MISSILE AGENCY, REDSTONE ARSENAL, ALA.	ORDXC-
ARMY BALLISTIC MISSILE AGENCY, REDSTONE ARSENAL, ALA.	ORDXM-C-(NUMBERS)
ARMY BALLISTIC MISSILE AGENCY, REDSTONE ARSENAL, ALA.	ORDXMC-C-
ARMY BALLISTIC MISSILE AGENCY, REDSTONE ARSENAL, ALA.	RA-AM-
ARMY BALLISTIC MISSILE AGENCY, REDSTONE ARSENAL, ALA.	RA-DA-
ARMY BALLISTIC MISSILE AGENCY, REDSTONE ARSENAL, ALA.	RA-TR-
ARMY BALLISTIC MISSILE AGENCY, REDSTONE ARSENAL, ALA.	RP-TR(NO.)-(YR.)
ARMY BALLISTIC MISSILE AGENCY. AEROBALLISTICS LAB., REDSTONE ARSENAL, ALA. (INTERNAL NOTE)	ABMA DA-IN-
ARMY BALLISTIC MISSILE AGENCY. AEROBALLISTICS LAB., REDSTONE ARSENAL, ALA. (MEMORANDUM)	ABMA DA-M-
ARMY BALLISTIC MISSILE AGENCY. AEROBALLISTICS LAB., REDSTONE ARSENAL, ALA. (MEMORANDUM)	ABMA DA-M(NO.)(YR)
ARMY BALLISTIC MISSILE AGENCY. AEROBALLISTICS LAB., REDSTONE ARSENAL, ALA.	ABMA DA MEMO-
ARMY BALLISTIC MISSILE AGENCY. AEROBALLISTICS LAB., REDSTONE ARSENAL, ALA. (REPORT)	ABMA DA-R-
ARMY BALLISTIC MISSILE AGENCY. AEROBALLISTICS LAB., REDSTONE ARSENAL, ALA. (TECHNICAL MEMO)	ABMA DA-TM(NO.)(YR)
ARMY BALLISTIC MISSILE AGENCY. AEROBALLISTICS LAB., REDSTONE ARSENAL, ALA. (TECHNICAL NOTE)	ABMA DA-TN-
ARMY BALLISTIC MISSILE AGENCY. AEROBALLISTICS LAB., REDSTONE ARSENAL, ALA.	ABMA DA-TN(NO.)(YR)
ARMY BALLISTIC MISSILE AGENCY. AEROBALLISTICS LAB., REDSTONE ARSENAL, ALA. (TECHNICAL REPORT)	ABMA DA-TR(NO.)(YR)
ARMY BALLISTIC MISSILE AGENCY. AEROBALLISTICS LAB., REDSTONE ARSENAL, ALA. (WIND TUNNEL NOTES)	ABMA DA-WTN-
ARMY BALLISTIC MISSILE AGENCY. AEROBALLISTICS LAB., REDSTONE ARSENAL, ALA.	ABMA ORDAB-DAA-
ARMY BALLISTIC MISSILE AGENCY. AEROBALLISTICS LAB., REDSTONE ARSENAL, ALA.	ABMA ORDAB-DAE(NO.)(YR)

ARMY BALLISTIC MISSILE AGENCY. AEROBALLISTICS LAB., REDSTONE ARSENAL, ALA.	ABMA ORDAB-DAT-
ARMY BALLISTIC MISSILE AGENCY. COMPUTATION LAB., REDSTONE ARSENAL, ALA. (MEMORANDUM)	ABMA DA-M-
ARMY BALLISTIC MISSILE AGENCY. COMPUTATION LAB., REDSTONE ARSENAL, ALA.	ABMA DCF-TN-
ARMY BALLISTIC MISSILE AGENCY. COMPUTATION LAB., REDSTONE ARSENAL, ALA. (TECHNICAL MEMORANDUM)	ABMA DC-TM(NO.)(YR)
ARMY BALLISTIC MISSILE AGENCY. COMPUTATION LAB., REDSTONE ARSENAL, ALA. (TECHNICAL NOTE)	ABMA DC-TN(NO.)(YR)
ARMY BALLISTIC MISSILE AGENCY. COMPUTATION LAB., REDSTONE ARSENAL, ALA. (TECHNICAL REPORT)	ABMA DC-TR-
ARMY BALLISTIC MISSILE AGENCY. CONTROL OFFICE, REDSTONE ARSENAL, ALA. (CONSOLIDATED REPORT)	ABMA/AFJUPLO CR(NO)(YR)J-
ARMY BALLISTIC MISSILE AGENCY. CONTROL OFFICE, REDSTONE ARSENAL, ALA. (SERIES ENDS WITH CY (FIGURE FOR CALENDAR YEAR)(FIGURE FOR FISCAL YEAR))	ABMA/AFJUPLO CR-J-(QTR.)-
ARMY BALLISTIC MISSILE AGENCY. CONTROL OFFICE, REDSTONE ARSENAL, ALA. (CONSOLIDATED REPORT)	ABMA/AFJUPLO CR-J(NO.YR)-
ARMY BALLISTIC MISSILE AGENCY. CONTROL OFFICE, REDSTONE ARSENAL, ALA. (CONTROL OFFICE PROGRESS RPT.)	ABMA CR-P(YR)-
ARMY BALLISTIC MISSILE AGENCY. CONTROL OFFICE, REDSTONE ARSENAL, ALA. (CONTROL OFFICE REPORT)	ABMA CR-R(NO.)(YR)
ARMY BALLISTIC MISSILE AGENCY. CONTROL OFFICE, REDSTONE ARSENAL, ALA. (CONTROL OFFICE QUARTERLY REPORT ON REDSTONE)	ABMA CR-R-(NO)Q(NO)(YR)
ARMY BALLISTIC MISSILE AGENCY. CONTROL OFFICE, REDSTONE ARSENAL, ALA. (CONTROL OFFICE SATELLITE RPT.)	ABMA CR-S(NO.)(YR)
ARMY BALLISTIC MISSILE AGENCY. CONTROL OFFICE, REDSTONE ARSENAL, ALA. (ORDNANCE ARMY BALLISTIC) (LETTERS ARE OFFICE SYMBOLS)	ABMA ORDAB-CR-
ARMY BALLISTIC MISSILE AGENCY. DEVELOPMENT OPERATIONS OFFICE, REDSTONE ARSENAL, ALA. (REPORT)	ABMA D-R-
ARMY BALLISTIC MISSILE AGENCY. DEVELOPMENT OPERATIONS OFFICE, REDSTONE ARSENAL, ALA. (TECHNICAL NOTE)	ABMA D-TN(NO.)(YR)
ARMY BALLISTIC MISSILE AGENCY. DYNAMIC ANALYSIS BRANCH, REDSTONE ARSENAL, ALA.	DA-
ARMY BALLISTIC MISSILE AGENCY. FABRICATION + ASSEMBLY ENGINEERING LAB., REDSTONE ARSENAL, ALA.	ABMA DFE-TM(NO.)(YR)
ARMY BALLISTIC MISSILE AGENCY. FABRICATION + ASSEMBLY ENGINEERING LAB., REDSTONE ARSENAL, ALA.	ABMA DFE-TN(NO.)(YR)
ARMY BALLISTIC MISSILE AGENCY. FABRICATION + ASSEMBLY ENGINEERING LAB., REDSTONE ARSENAL, ALA. (INTERNAL NOTE)	ABMA DFR-IN(NO.)(YR)
ARMY BALLISTIC MISSILE AGENCY. FABRICATION + ASSEMBLY ENGINEERING LAB., REDSTONE ARSENAL, ALA.	ABMA DFR-TM(NO.)(YR)
ARMY BALLISTIC MISSILE AGENCY. FABRICATION + ASSEMBLY ENGINEERING LAB., REDSTONE ARSENAL, ALA.	ABMA DFR-TN(NO.)(YR)
ARMY BALLISTIC MISSILE AGENCY. FABRICATION + ASSEMBLY ENGINEERING LAB., REDSTONE ARSENAL, ALA.	ABMA DF-TM(NO.)(YR)
ARMY BALLISTIC MISSILE AGENCY. FABRICATION + ASSEMBLY ENGINEERING LAB., REDSTONE ARSENAL, ALA.	ABMA DF-TN(NO.)(YR)
ARMY BALLISTIC MISSILE AGENCY. FABRICATION LAB., REDSTONE ARSENAL, ALA.	ABMA DF-M-
ARMY BALLISTIC MISSILE AGENCY. FIELD OFFICE, LOS ANGELES, CALIF. (TECHNICAL LIAISON MEMO)	ABMA FIELD OFF TLM-
ARMY BALLISTIC MISSILE AGENCY. FIELD SUPPORT DIVISION, REDSTONE ARSENAL, ALA.	ABMA ORDAB-SE-
ARMY BALLISTIC MISSILE AGENCY. FIELD SUPPORT OPERATIONS, REDSTONE ARSENAL, ALA. (MAINTENANCE SERVICE LETTER)	ABMA ORDAB-SM-MSL-
ARMY BALLISTIC MISSILE AGENCY. GUIDANCE AND CONTROL LAB., REDSTONE ARSENAL, ALA. (ACCEPTANCE TEST SPEC.	ABMA ATS-G-
ARMY BALLISTIC MISSILE AGENCY. GUIDANCE AND CONTROL LAB., REDSTONE ARSENAL, ALA.	ABMA DGIC-TN(NO.)(YR)
ARMY BALLISTIC MISSILE AGENCY. GUIDANCE AND CONTROL LAB., REDSTONE ARSENAL, ALA.	ABMA DG-IN(NO.)(YR)
ARMY BALLISTIC MISSILE AGENCY. GUIDANCE AND CONTROL LAB., REDSTONE ARSENAL, ALA.	ABMA DGI-TM(NO.)(YR)
ARMY BALLISTIC MISSILE AGENCY. GUIDANCE AND CONTROL LAB., REDSTONE ARSENAL, ALA. (MEMO)	ABMA DG-M-
ARMY BALLISTIC MISSILE AGENCY. GUIDANCE AND CONTROL LAB., REDSTONE ARSENAL, ALA.	ABMA DG-M(NO.)(YR)
ARMY BALLISTIC MISSILE AGENCY. GUIDANCE AND CONTROL LAB., REDSTONE ARSENAL, ALA.	ABMA DGNA(J)(NO.)(YR)
ARMY BALLISTIC MISSILE AGENCY. GUIDANCE AND CONTROL LAB., REDSTONE ARSENAL, ALA.	ABMA DGNA-(R)(NO.)/(YR)
ARMY BALLISTIC MISSILE AGENCY. GUIDANCE AND CONTROL LAB., REDSTONE ARSENAL, ALA.	ABMA DGNA-(S)(NO.)/(YR)
ARMY BALLISTIC MISSILE AGENCY. GUIDANCE AND CONTROL LAB., REDSTONE ARSENAL, ALA.	ABMA DGNC-(J)(NO.)/(YR)
ARMY BALLISTIC MISSILE AGENCY. GUIDANCE AND CONTROL LAB., REDSTONE ARSENAL, ALA.	ABMA DGNC-(P)(NO.)(YR)
ARMY BALLISTIC MISSILE AGENCY. GUIDANCE AND CONTROL LAB., REDSTONE ARSENAL, ALA.	ABMA DGND-(J)(NO.)(YR)
ARMY BALLISTIC MISSILE AGENCY. GUIDANCE AND CONTROL LAB., REDSTONE ARSENAL, ALA.	ABMA DGNE-(P)(NO.)(YR)
ARMY BALLISTIC MISSILE AGENCY. GUIDANCE AND CONTROL LAB., REDSTONE ARSENAL, ALA.	ABMA DGNE-(S)(NO.)(YR)
ARMY BALLISTIC MISSILE AGENCY. GUIDANCE AND CONTROL LAB., REDSTONE ARSENAL, ALA.	ABMA DGN-(J)(NO.)(YR)
ARMY BALLISTIC MISSILE AGENCY. GUIDANCE AND CONTROL LAB., REDSTONE ARSENAL, ALA.	ABMA DGN-TM(NO.)(YR)
ARMY BALLISTIC MISSILE AGENCY. GUIDANCE AND CONTROL LAB., REDSTONE ARSENAL, ALA.	ABMA DG-R-
ARMY BALLISTIC MISSILE AGENCY. GUIDANCE AND CONTROL LAB., REDSTONE ARSENAL, ALA.	ABMA DGR-TN(NO.)(YR)
ARMY BALLISTIC MISSILE AGENCY. GUIDANCE AND CONTROL LAB., REDSTONE ARSENAL, ALA.	ABMA DG-TM(NO.)(YR)
ARMY BALLISTIC MISSILE AGENCY. GUIDANCE AND CONTROL LAB., REDSTONE ARSENAL, ALA.	ABMA DG-TN(NO.)(YR)
ARMY BALLISTIC MISSILE AGENCY. GUIDANCE AND CONTROL LAB., REDSTONE ARSENAL, ALA.	ABMA DG-TR(NO.)(YR)
ARMY BALLISTIC MISSILE AGENCY. GUIDANCE AND CONTROL LAB., REDSTONE ARSENAL, ALA.	ABMA G+C CONTROL NOTE-
ARMY BALLISTIC MISSILE AGENCY. GUIDANCE AND CONTROL LAB., REDSTONE ARSENAL, ALA.	ABMA G+C DEV RPT-
ARMY BALLISTIC MISSILE AGENCY. GUIDANCE AND CONTROL LAB., REDSTONE ARSENAL, ALA.	ABMA G+C-IN-
ARMY BALLISTIC MISSILE AGENCY. GUIDANCE AND CONTROL LAB., REDSTONE ARSENAL, ALA.	ABMA G+C MEMO-
ARMY BALLISTIC MISSILE AGENCY. GUIDANCE AND CONTROL LAB., REDSTONE ARSENAL, ALA.	ABMA G+C MEMO-GS(SP)-T-
ARMY BALLISTIC MISSILE AGENCY. GUIDANCE AND CONTROL LAB., REDSTONE ARSENAL, ALA.	ABMA G+C-TM-
ARMY BALLISTIC MISSILE AGENCY. GUIDANCE AND CONTROL LAB., REDSTONE ARSENAL, ALA.	ABMA G+C-TN-
ARMY BALLISTIC MISSILE AGENCY. GUIDANCE AND CONTROL LAB., REDSTONE ARSENAL, ALA.	ABMA GS(SP)-T-
ARMY BALLISTIC MISSILE AGENCY. GUIDANCE AND CONTROL LAB., REDSTONE ARSENAL, ALA.	ABMA GS-T-
ARMY BALLISTIC MISSILE AGENCY. LAUNCHING + HANDLING EQUIPMENT LAB., REDSTONE ARSENAL, ALA.	ABMA DLE(NO.)(YR)
ARMY BALLISTIC MISSILE AGENCY. LAUNCHING + HANDLING EQUIPMENT LAB., REDSTONE ARSENAL, ALA.	ABMA DL-M-
ARMY BALLISTIC MISSILE AGENCY. LAUNCHING + HANDLING EQUIPMENT LAB., REDSTONE ARSENAL, ALA.	ABMA DL-TN-
ARMY BALLISTIC MISSILE AGENCY. LAUNCHING + HANDLING EQUIPMENT LAB., REDSTONE ARSENAL, ALA.	ABMA DL-TR(DATE)
ARMY BALLISTIC MISSILE AGENCY. LAUNCHING + HANDLING EQUIPMENT LAB., REDSTONE ARSENAL, ALA.	ABMA TN-DLF(NO.)(YR)
ARMY BALLISTIC MISSILE AGENCY. LAUNCHING + HANDLING EQUIPMENT LAB., REDSTONE ARSENAL, ALA.	ABMA TN-DLG(NO.)(YR)(TN)
ARMY BALLISTIC MISSILE AGENCY. LAUNCHING + HANDLING EQUIPMENT LAB., REDSTONE ARSENAL, ALA.	ABMA TN-DLM(NO.)(YR)
ARMY BALLISTIC MISSILE AGENCY. LAUNCHING + HANDLING EQUIPMENT LAB., REDSTONE ARSENAL, ALA.	ABMA TN-DLP(NO.)(YR)
ARMY BALLISTIC MISSILE AGENCY. LAUNCHING + HANDLING EQUIPMENT LAB., REDSTONE ARSENAL, ALA.	ABMA TP-DLF(NO.)(YR)
ARMY BALLISTIC MISSILE AGENCY. LAUNCHING + HANDLING EQUIPMENT LAB., REDSTONE ARSENAL, ALA.	ABMA TR-DLD(NO.)(YR)
ARMY BALLISTIC MISSILE AGENCY. LAUNCHING + HANDLING EQUIPMENT LAB., REDSTONE ARSENAL, ALA.	ABMA TR-DLE(NO.)(YR)
ARMY BALLISTIC MISSILE AGENCY. LAUNCHING + HANDLING EQUIPMENT LAB., REDSTONE ARSENAL, ALA.	ABMA TR-DLF(NO.)(YR)
ARMY BALLISTIC MISSILE AGENCY. LAUNCHING + HANDLING EQUIPMENT LAB., REDSTONE ARSENAL, ALA.	ABMA TR-DLJ(NO.)(YR)
ARMY BALLISTIC MISSILE AGENCY. LAUNCHING + HANDLING EQUIPMENT LAB., REDSTONE ARSENAL, ALA.	ABMA TR-DLM-
ARMY BALLISTIC MISSILE AGENCY. LAUNCHING + HANDLING EQUIPMENT LAB., REDSTONE ARSENAL, ALA.	ABMA TR-DLM(NO.)(YR)
ARMY BALLISTIC MISSILE AGENCY. LAUNCHING + HANDLING EQUIPMENT LAB., REDSTONE ARSENAL, ALA.	ABMA TR-DLP(NO.)(YR)
ARMY BALLISTIC MISSILE AGENCY. LAUNCHING + HANDLING EQUIPMENT LAB., REDSTONE ARSENAL, ALA.	ABMA TR-DLS(NO.)(YR)
ARMY BALLISTIC MISSILE AGENCY. LAUNCHING + HANDLING EQUIPMENT LAB., REDSTONE ARSENAL, ALA.	ABMA TR-DLT(NO.)(YR)
ARMY BALLISTIC MISSILE AGENCY. LIBRARY, REDSTONE ARSENAL, ALA. (ASSIGNED TO REPORTS NOT IDENTIFIED BY ANY OTHER SERIES CODE)	ABMA HT-
ARMY BALLISTIC MISSILE AGENCY. MISSILE FIRING LAB., CAPE CANAVERAL, FLA.	ABMA DMCR-TR(NO.)(YR)
ARMY BALLISTIC MISSILE AGENCY. MISSILE FIRING LAB., CAPE CANAVERAL, FLA.	ABMA DMG-TR(NO.)(YR)
ARMY BALLISTIC MISSILE AGENCY. MISSILE FIRING LAB., CAPE CANAVERAL, FLA.	ABMA DMMA-TR(NO.)(YR)
ARMY BALLISTIC MISSILE AGENCY. MISSILE FIRING LAB., CAPE CANAVERAL, FLA.	ABMA DMM-TR(NO.)(YR)
ARMY BALLISTIC MISSILE AGENCY. MISSILE FIRING LAB., CAPE CANAVERAL, FLA.	ABMA DMQ-TR(NO.)(YR)
ARMY BALLISTIC MISSILE AGENCY. MISSILE FIRING LAB., CAPE CANAVERAL, FLA.	ABMA DM-R-
ARMY BALLISTIC MISSILE AGENCY. MISSILE FIRING LAB., CAPE CANAVERAL, FLA.	ABMA DM-TM-
ARMY BALLISTIC MISSILE AGENCY. MISSILE FIRING LAB., CAPE CANAVERAL, FLA.	ABMA DM-TM(NO.)(DATE)
ARMY BALLISTIC MISSILE AGENCY. MISSILE FIRING LAB., CAPE CANAVERAL, FLA.	ABMA DM-TR(NO.)(YR)
ARMY BALLISTIC MISSILE AGENCY. MISSILE FIRING LAB., CAPE CANAVERAL, FLA.	ABMA DMTR-TM(NO.)(YR)
ARMY BALLISTIC MISSILE AGENCY. MISSILE FIRING LAB., CAPE CANAVERAL, FLA.	ABMA DMTR-TN(NO.)(YR)
ARMY BALLISTIC MISSILE AGENCY. MISSILE FIRING LAB., CAPE CANAVERAL, FLA.	ABMA DMTR-TR(NO.)(YR)
ARMY BALLISTIC MISSILE AGENCY. MISSILE FIRING LAB., CAPE CANAVERAL, FLA.	ABMA-ORDAB-DM-(LETTERS)-
ARMY BALLISTIC MISSILE AGENCY. MISSILE FIRING LAB., CAPE CANAVERAL, FLA.	MFL-

ARMY BALLISTIC MISSILE AGENCY. MISSILE FIRING LAB.

ARMY BALLISTIC MISSILE AGENCY. MISSILE FIRING LAB., REDSTONE ARSENAL, ALA.	ABMA DMCR-TR(NO.)(YR)
ARMY BALLISTIC MISSILE AGENCY. MISSILE FIRING LAB., REDSTONE ARSENAL, ALA.	ABMA DMG-TR(NO.)(YR)
ARMY BALLISTIC MISSILE AGENCY. MISSILE FIRING LAB., REDSTONE ARSENAL, ALA.	ABMA DMMA-TR(NO.)(YR)
ARMY BALLISTIC MISSILE AGENCY. MISSILE FIRING LAB., REDSTONE ARSENAL, ALA.	ABMA DMM-TR(NO.)(YR)
ARMY BALLISTIC MISSILE AGENCY. MISSILE FIRING LAB., REDSTONE ARSENAL, ALA.	ABMA DMQ-TR(NO.)(YR)
ARMY BALLISTIC MISSILE AGENCY. MISSILE FIRING LAB., REDSTONE ARSENAL, ALA.	ABMA DM-R-
ARMY BALLISTIC MISSILE AGENCY. MISSILE FIRING LAB., REDSTONE ARSENAL, ALA.	ABMA DM-TM-
ARMY BALLISTIC MISSILE AGENCY. MISSILE FIRING LAB., REDSTONE ARSENAL, ALA.	ABMA DM-TM(NO.)(DATE)
ARMY BALLISTIC MISSILE AGENCY. MISSILE FIRING LAB., REDSTONE ARSENAL, ALA.	ABMA DM-TR(NO.)(YR)
ARMY BALLISTIC MISSILE AGENCY. MISSILE FIRING LAB., REDSTONE ARSENAL, ALA.	ABMA DMTR-TM(NO.)(YR)
ARMY BALLISTIC MISSILE AGENCY. MISSILE FIRING LAB., REDSTONE ARSENAL, ALA.	ABMA DMTR-TN(NO.)(YR)
ARMY BALLISTIC MISSILE AGENCY. MISSILE FIRING LAB., REDSTONE ARSENAL, ALA.	ABMA DMTR-TR(NO.)(YR)
ARMY BALLISTIC MISSILE AGENCY. MISSILE FIRING LAB., TITUSVILLE, FLA.	DMCR-TR-(NO.)-(YEAR)
ARMY BALLISTIC MISSILE AGENCY. RESEARCH LAB., REDSTONE ARSENAL, ALA.	ABMA RR-TM(NO.)(YR)
ARMY BALLISTIC MISSILE AGENCY. RESEARCH PROJECTS LAB., REDSTONE ARSENAL, ALA.	ABMA DV-TM(NO.)(DATE)
ARMY BALLISTIC MISSILE AGENCY. RESEARCH PROJECTS LAB., REDSTONE ARSENAL, ALA.	ABMA DV-TN(NO.)(DATE)
ARMY BALLISTIC MISSILE AGENCY. RESEARCH PROJECTS OFFICE, REDSTONE ARSENAL, ALA.	ABMA DV-TN-
ARMY BALLISTIC MISSILE AGENCY. RESEARCH PROJECTS OFFICE, REDSTONE ARSENAL, ALA.	ABMA DV-TR(NO.)(YR)
ARMY BALLISTIC MISSILE AGENCY. RESEARCH PROJECTS OFFICE, REDSTONE ARSENAL, ALA.	ABMA RPO-D-TM-
ARMY BALLISTIC MISSILE AGENCY. RESEARCH PROJECTS OFFICE, REDSTONE ARSENAL, ALA.	ABMA RPO-P-TN-
ARMY BALLISTIC MISSILE AGENCY. RESEARCH PROJECTS OFFICE, REDSTONE ARSENAL, ALA. (RESEARCH BULLETIN)	ABMA RPO-RB-
ARMY BALLISTIC MISSILE AGENCY. RESEARCH PROJECTS OFFICE, REDSTONE ARSENAL, ALA.	ABMA RPO-RS-TN-
ARMY BALLISTIC MISSILE AGENCY. RESEARCH PROJECTS OFFICE, REDSTONE ARSENAL, ALA.	ABMA RPO-T-TN-
ARMY BALLISTIC MISSILE AGENCY. STRUCTURES AND MECHANICS LAB., REDSTONE ARSENAL, ALA.	ABMA DSA-TB-
ARMY BALLISTIC MISSILE AGENCY. STRUCTURES AND MECHANICS LAB., REDSTONE ARSENAL, ALA.	ABMA DSDA-
ARMY BALLISTIC MISSILE AGENCY. STRUCTURES AND MECHANICS LAB., REDSTONE ARSENAL, ALA.	ABMA DSD-TM(NO.)(YR)
ARMY BALLISTIC MISSILE AGENCY. STRUCTURES AND MECHANICS LAB., REDSTONE ARSENAL, ALA.	ABMA DSD-TN(NO.)(YR)
ARMY BALLISTIC MISSILE AGENCY. STRUCTURES AND MECHANICS LAB., REDSTONE ARSENAL, ALA.	ABMA DSD-TR(NO.)(YR)
ARMY BALLISTIC MISSILE AGENCY. STRUCTURES AND MECHANICS LAB., REDSTONE ARSENAL, ALA.	ABMA DSF-TM(NO.)(YR)
ARMY BALLISTIC MISSILE AGENCY. STRUCTURES AND MECHANICS LAB., REDSTONE ARSENAL, ALA.	ABMA DSF-TN(NO.)(YR)
ARMY BALLISTIC MISSILE AGENCY. STRUCTURES AND MECHANICS LAB., REDSTONE ARSENAL, ALA.	ABMA DSF-TR(NO.)(YR)
ARMY BALLISTIC MISSILE AGENCY. STRUCTURES AND MECHANICS LAB., REDSTONE ARSENAL, ALA.	ABMA DS-IN-
ARMY BALLISTIC MISSILE AGENCY. STRUCTURES AND MECHANICS LAB., REDSTONE ARSENAL, ALA.	ABMA DSL-MEMO-
ARMY BALLISTIC MISSILE AGENCY. STRUCTURES AND MECHANICS LAB., REDSTONE ARSENAL, ALA.	ABMA DSL-TB-
ARMY BALLISTIC MISSILE AGENCY. STRUCTURES AND MECHANICS LAB., REDSTONE ARSENAL, ALA.	ABMA DSL-TB(NO.)(YR)
ARMY BALLISTIC MISSILE AGENCY. STRUCTURES AND MECHANICS LAB., REDSTONE ARSENAL, ALA.	ABMA DSL-TM(NO.)(YR)
ARMY BALLISTIC MISSILE AGENCY. STRUCTURES AND MECHANICS LAB., REDSTONE ARSENAL, ALA.	ABMA DSL-TN(NO.)(YR)
ARMY BALLISTIC MISSILE AGENCY. STRUCTURES AND MECHANICS LAB., REDSTONE ARSENAL, ALA.	ABMA DS-M-
ARMY BALLISTIC MISSILE AGENCY. STRUCTURES AND MECHANICS LAB., REDSTONE ARSENAL, ALA.	ABMA DSN-TM(NO.)(YR)
ARMY BALLISTIC MISSILE AGENCY. STRUCTURES AND MECHANICS LAB., REDSTONE ARSENAL, ALA.	ABMA DSN-TN(NO.)(YR)
ARMY BALLISTIC MISSILE AGENCY. STRUCTURES AND MECHANICS LAB., REDSTONE ARSENAL, ALA.	ABMA DSN-TR(NO.)(YR)
ARMY BALLISTIC MISSILE AGENCY. STRUCTURES AND MECHANICS LAB., REDSTONE ARSENAL, ALA.	ABMA DSP-
ARMY BALLISTIC MISSILE AGENCY. STRUCTURES AND MECHANICS LAB., REDSTONE ARSENAL, ALA.	ABMA DSP-IN-
ARMY BALLISTIC MISSILE AGENCY. STRUCTURES AND MECHANICS LAB., REDSTONE ARSENAL, ALA.	ABMA DSP-TB-
ARMY BALLISTIC MISSILE AGENCY. STRUCTURES AND MECHANICS LAB., REDSTONE ARSENAL, ALA.	ABMA DSP-TM(NO.)(YR)
ARMY BALLISTIC MISSILE AGENCY. STRUCTURES AND MECHANICS LAB., REDSTONE ARSENAL, ALA.	ABMA-DSP-TN-
ARMY BALLISTIC MISSILE AGENCY. STRUCTURES AND MECHANICS LAB., REDSTONE ARSENAL, ALA.	ABMA DSP-TN(NO.)(YR)
ARMY BALLISTIC MISSILE AGENCY. STRUCTURES AND MECHANICS LAB., REDSTONE ARSENAL, ALA.	ABMA DSP-TR(NO.)(YR)
ARMY BALLISTIC MISSILE AGENCY. STRUCTURES AND MECHANICS LAB., REDSTONE ARSENAL, ALA.	ABMA DS-R-
ARMY BALLISTIC MISSILE AGENCY. STRUCTURES AND MECHANICS LAB., REDSTONE ARSENAL, ALA.	ABMA DSRP-TM(NO.)(YR)
ARMY BALLISTIC MISSILE AGENCY. STRUCTURES AND MECHANICS LAB., REDSTONE ARSENAL, ALA.	ABMA DSR-TM(NO.)(YR)
ARMY BALLISTIC MISSILE AGENCY. STRUCTURES AND MECHANICS LAB., REDSTONE ARSENAL, ALA.	ABMA DSS-TB-
ARMY BALLISTIC MISSILE AGENCY. STRUCTURES AND MECHANICS LAB., REDSTONE ARSENAL, ALA.	ABMA DS-TB-
ARMY BALLISTIC MISSILE AGENCY. STRUCTURES AND MECHANICS LAB., REDSTONE ARSENAL, ALA.	ABMA DS-TM-
ARMY BALLISTIC MISSILE AGENCY. STRUCTURES AND MECHANICS LAB., REDSTONE ARSENAL, ALA.	ABMA DS-TM(NO.)(YR)
ARMY BALLISTIC MISSILE AGENCY. STRUCTURES AND MECHANICS LAB., REDSTONE ARSENAL, ALA.	ABMA DS-TN-
ARMY BALLISTIC MISSILE AGENCY. STRUCTURES AND MECHANICS LAB., REDSTONE ARSENAL, ALA.	ABMA ORDAB-DSA MEMO-
ARMY BALLISTIC MISSILE AGENCY. STRUCTURES AND MECHANICS LAB., REDSTONE ARSENAL, ALA.	ABMA ORDAB-DSD-
ARMY BALLISTIC MISSILE AGENCY. STRUCTURES AND MECHANICS LAB., REDSTONE ARSENAL, ALA.	ABMA ORDAB-DSDD-
ARMY BALLISTIC MISSILE AGENCY. STRUCTURES AND MECHANICS LAB., REDSTONE ARSENAL, ALA.	ABMA ORDAB-DSDE-
ARMY BALLISTIC MISSILE AGENCY. STRUCTURES AND MECHANICS LAB., REDSTONE ARSENAL, ALA.	ABMA ORDAB-DSDF-
ARMY BALLISTIC MISSILE AGENCY. STRUCTURES AND MECHANICS LAB., REDSTONE ARSENAL, ALA.	ABMA ORDAB-DSDF(NO.)(YR)
ARMY BALLISTIC MISSILE AGENCY. STRUCTURES AND MECHANICS LAB., REDSTONE ARSENAL, ALA.	ABMA ORDAB-DSDG-
ARMY BALLISTIC MISSILE AGENCY. STRUCTURES AND MECHANICS LAB., REDSTONE ARSENAL, ALA.	ABMA ORDAB-DSE-
ARMY BALLISTIC MISSILE AGENCY. STRUCTURES AND MECHANICS LAB., REDSTONE ARSENAL, ALA.	ABMA ORDAB-DSL-
ARMY BALLISTIC MISSILE AGENCY. STRUCTURES AND MECHANICS LAB., REDSTONE ARSENAL, ALA.	ABMA ORDAB-DSLA-MEMO-
ARMY BALLISTIC MISSILE AGENCY. STRUCTURES AND MECHANICS LAB., REDSTONE ARSENAL, ALA.	ABMA ORDAB-DSLB-MEMO-
ARMY BALLISTIC MISSILE AGENCY. STRUCTURES AND MECHANICS LAB., REDSTONE ARSENAL, ALA.	ABMA ORDAB-DSLC-MEMO-
ARMY BALLISTIC MISSILE AGENCY. STRUCTURES AND MECHANICS LAB., REDSTONE ARSENAL, ALA.	ABMA ORDAB-DSLE-MEMO-
ARMY BALLISTIC MISSILE AGENCY. STRUCTURES AND MECHANICS LAB., REDSTONE ARSENAL, ALA.	ABMA ORDAB-DSL MEMO-
ARMY BALLISTIC MISSILE AGENCY. STRUCTURES AND MECHANICS LAB., REDSTONE ARSENAL, ALA.	ABMA ORDAB-DSL-TB(NO)(YR)
ARMY BALLISTIC MISSILE AGENCY. STRUCTURES AND MECHANICS LAB., REDSTONE ARSENAL, ALA.	ABMA ORDAB-DSM-
ARMY BALLISTIC MISSILE AGENCY. STRUCTURES AND MECHANICS LAB., REDSTONE ARSENAL, ALA.	ABMA ORDAB-DSS-TB-
ARMY BALLISTIC MISSILE AGENCY. STRUCTURES AND MECHANICS LAB., REDSTONE ARSENAL, ALA.	ABMA PG-CD-MEMO-
ARMY BALLISTIC MISSILE AGENCY. STRUCTURES AND MECHANICS LAB., REDSTONE ARSENAL, ALA.	ABMA PG-CE-MEMO-
ARMY BALLISTIC MISSILE AGENCY. STRUCTURES AND MECHANICS LAB., REDSTONE ARSENAL, ALA.	ABMA PG-CH-MEMO-
ARMY BALLISTIC MISSILE AGENCY. STRUCTURES AND MECHANICS LAB., REDSTONE ARSENAL, ALA.	ABMA SM/EA MEMO-
ARMY BALLISTIC MISSILE AGENCY. STRUCTURES AND MECHANICS LAB., REDSTONE ARSENAL, ALA.	ABMA S+M-TB-G-
ARMY BALLISTIC MISSILE AGENCY. STRUCTURES AND MECHANICS LAB., REDSTONE ARSENAL, ALA.	ABMA S+M-TDS-
ARMY BALLISTIC MISSILE AGENCY. STRUCTURES AND MECHANICS LAB., REDSTONE ARSENAL, ALA.	ABMA S+M TEST MEMO-
ARMY BALLISTIC MISSILE AGENCY. STRUCTURES AND MECHANICS LAB., REDSTONE ARSENAL, ALA.	ABMA S+M TEST REPORT-
ARMY BALLISTIC MISSILE AGENCY. STRUCTURES AND MECHANICS LAB., REDSTONE ARSENAL, ALA.	ABMA S+M-TM-G-
ARMY BALLISTIC MISSILE AGENCY. STRUCTURES AND MECHANICS LAB., REDSTONE ARSENAL, ALA.	ABMA S+M-TN-B-
ARMY BALLISTIC MISSILE AGENCY. STRUCTURES AND MECHANICS LAB., REDSTONE ARSENAL, ALA.	ABMA S+M-TN-C-
ARMY BALLISTIC MISSILE AGENCY. STRUCTURES AND MECHANICS LAB., REDSTONE ARSENAL, ALA.	ABMA S+M-TN-DDS-
ARMY BALLISTIC MISSILE AGENCY. STRUCTURES AND MECHANICS LAB., REDSTONE ARSENAL, ALA.	ABMA S+M-TN-DSD-
ARMY BALLISTIC MISSILE AGENCY. STRUCTURES AND MECHANICS, REDSTONE ARSENAL, ALA.	ABMA S+M-TN-DSP-
ARMY BALLISTIC MISSILE AGENCY. STRUCTURES AND MECHANICS LAB., REDSTONE ARSENAL, ALA.	ABMA S+M-TN-E-
ARMY BALLISTIC MISSILE AGENCY. STRUCTURES AND MECHANICS LAB., REDSTONE ARSENAL, ALA.	ABMA S+M-TN-G-
ARMY BALLISTIC MISSILE AGENCY. STRUCTURES AND MECHANICS LAB., REDSTONE ARSENAL, ALA.	ABMA S+M-TN-H-
ARMY BALLISTIC MISSILE AGENCY. STRUCTURES AND MECHANICS LAB., REDSTONE ARSENAL, ALA.	ABMA S+M-TN-K-
ARMY BALLISTIC MISSILE AGENCY. STRUCTURES AND MECHANICS LAB., REDSTONE ARSENAL, ALA.	ABMA S+M-TN-M-
ARMY BALLISTIC MISSILE AGENCY. STRUCTURES AND MECHANICS LAB., REDSTONE ARSENAL, ALA.	ABMA S+M-TN-N-
ARMY BALLISTIC MISSILE AGENCY. STRUCTURES AND MECHANICS LAB., REDSTONE ARSENAL, ALA.	ABMA S+M-TN-T-
ARMY BALLISTIC MISSILE AGENCY. STRUCTURES AND MECHANICS, REDSTONE ARSENAL, ALA.	ABMA TN-DSD-
ARMY BALLISTIC MISSILE AGENCY. STRUCTURES AND MECHANICS LAB., REDSTONE ARSENAL, ALA.	ABMA TN-DSF-
ARMY BALLISTIC MISSILE AGENCY. STRUCTURES AND MECHANICS LAB., REDSTONE ARSENAL, ALA.	DS-
ARMY BALLISTIC MISSILE AGENCY. STRUCTURES AND MECHANICS LAB., REDSTONE ARSENAL, ALA.	DSM-
ARMY BALLISTIC MISSILE AGENCY. STRUCTURES AND MECHANICS LAB. STRUCTURAL ANALYSIS SECTION, REDSTONE ARSENAL, ALA.	ABMA DSA-SU MEMO-
ARMY BALLISTIC MISSILE AGENCY. STRUCTURES AND MECHANICS LAB. STRUCTURAL ANALYSIS SECTION, REDSTONE ARSENAL, ALA.	ABMA ORDAB-DSA-DU MEMO-
ARMY BALLISTIC MISSILE AGENCY. SYSTEMS ANALYSIS AND RELIABILITY LAB., REDSTONE ARSENAL, ALA.	ABMA DRMA-TM(NO.)(YR)

ARMY CHEM. CORPS NUCLEAR DEFENSE LAB, ARMY CHEM. CTR

ARMY BALLISTIC MISSILE AGENCY. SYSTEMS ANALYSIS AND RELIABILITY LAB., REDSTONE ARSENAL, ALA.	ABMA DRM-TM(NO.)(YR)
ARMY BALLISTIC MISSILE AGENCY. SYSTEMS ANALYSIS AND RELIABILITY LAB., REDSTONE ARSENAL, ALA.	ABMA DRPG-TM(NO.)(YR)
ARMY BALLISTIC MISSILE AGENCY. SYSTEMS ANALYSIS AND RELIABILITY LAB., REDSTONE ARSENAL, ALA.	ABMA DRP-TM(NO.)(YR)
ARMY BALLISTIC MISSILE AGENCY. SYSTEMS ANALYSIS AND RELIABILITY LAB., REDSTONE ARSENAL, ALA.	ABMA DRR-TM(NO.)(DATE)
ARMY BALLISTIC MISSILE AGENCY. SYSTEMS ANALYSIS AND RELIABILITY LAB., REDSTONE ARSENAL, ALA.	ABMA DRR-TN(NO.)(DATE)
ARMY BALLISTIC MISSILE AGENCY. SYSTEMS ANALYSIS AND RELIABILITY LAB., REDSTONE ARSENAL, ALA.	ABMA DRR-TR(NO.)(DATE)
ARMY BALLISTIC MISSILE AGENCY. SYSTEMS ANALYSIS AND RELIABILITY LAB., REDSTONE ARSENAL, ALA.	ABMA DR-TM(NO.)(YR)
ARMY BALLISTIC MISSILE AGENCY. SYSTEMS ANALYSIS AND RELIABILITY LAB., REDSTONE ARSENAL, ALA.	ABMA DRT-TM(NO.)(DATE)
ARMY BALLISTIC MISSILE AGENCY. SYSTEMS ANALYSIS AND RELIABILITY LAB., REDSTONE ARSENAL, ALA.	ABMA DRT-TN(NO.)(YR)
ARMY BALLISTIC MISSILE AGENCY. SYSTEMS ANALYSIS AND RELIABILITY LAB., REDSTONE ARSENAL, ALA.	ABMA ORDAB-DR-
ARMY BALLISTIC MISSILE AGENCY. SYSTEMS ANALYSIS AND RELIABILITY LAB., REDSTONE ARSENAL, ALA.	ABMA SA+RL MEMO-
ARMY BALLISTIC MISSILE AGENCY. SYSTEMS ANALYSIS AND RELIABILITY LAB., REDSTONE ARSENAL, ALA.	ABMA SA+RL-QR-
ARMY BALLISTIC MISSILE AGENCY. SYSTEMS SUPPORT EQUIPMENT LAB., REDSTONE ARSENAL, ALA.	ABMA DLEC-TN(NO.)(YR)
ARMY BALLISTIC MISSILE AGENCY. SYSTEMS SUPPORT EQUIPMENT LAB., REDSTONE ARSENAL, ALA.	ABMA DLE-TN(NO.)(YR)
ARMY BALLISTIC MISSILE AGENCY. SYSTEMS SUPPORT EQUIPMENT LAB., REDSTONE ARSENAL, ALA.	ABMA DLMD-TN(NO.)(YR)
ARMY BALLISTIC MISSILE AGENCY. SYSTEMS SUPPORT EQUIPMENT LAB., REDSTONE ARSENAL, ALA.	ABMA DLMF-TM(NO.)(YR)
ARMY BALLISTIC MISSILE AGENCY. SYSTEMS SUPPORT EQUIPMENT LAB., REDSTONE ARSENAL, ALA.	ABMA DLMF-TN(NO.)(YR)
ARMY BALLISTIC MISSILE AGENCY. SYSTEMS SUPPORT EQUIPMENT LAB., REDSTONE ARSENAL, ALA.	ABMA DLMF-TR(NO.)(YR)
ARMY BALLISTIC MISSILE AGENCY. SYSTEMS SUPPORT EQUIPMENT LAB., REDSTONE ARSENAL, ALA.	ABMA DLMP-TM(NO.)(YR)
ARMY BALLISTIC MISSILE AGENCY. SYSTEMS SUPPORT EQUIPMENT LAB., REDSTONE ARSENAL, ALA.	ABMA DLMP-TN(NO.)(YR)
ARMY BALLISTIC MISSILE AGENCY. SYSTEMS SUPPORT EQUIPMENT LAB., REDSTONE ARSENAL, ALA.	ABMA DLM-TN(NO.)(YR)
ARMY BALLISTIC MISSILE AGENCY. SYSTEMS SUPPORT EQUIPMENT LAB., REDSTONE ARSENAL, ALA.	ABMA DLMT-TN(NO.)(YR)
ARMY BALLISTIC MISSILE AGENCY. SYSTEMS SUPPORT EQUIPMENT LAB., REDSTONE ARSENAL, ALA.	ABMA DLO-TM(NO.)(YR)
ARMY BALLISTIC MISSILE AGENCY. SYSTEMS SUPPORT EQUIPMENT LAB., REDSTONE ARSENAL, ALA.	ABMA DLS(NO.)(TN)
ARMY BALLISTIC MISSILE AGENCY. SYSTEMS SUPPORT EQUIPMENT LAB., REDSTONE ARSENAL, ALA.	ABMA DLS(NO)(TR)(NO.)(YR)
ARMY BALLISTIC MISSILE AGENCY. SYSTEMS SUPPORT EQUIPMENT LAB., REDSTONE ARSENAL, ALA.	ABMA DLS-TM(NO.)(DATE)
ARMY BALLISTIC MISSILE AGENCY. SYSTEMS SUPPORT EQUIPMENT LAB., REDSTONE ARSENAL, ALA.	ABMA DLS-TN(NO.)(DATE)
ARMY BALLISTIC MISSILE AGENCY. SYSTEMS SUPPORT EQUIPMENT LAB., REDSTONE ARSENAL, ALA.	ABMA DLS-TN(NO.)(YR)
ARMY BALLISTIC MISSILE AGENCY. SYSTEMS SUPPORT EQUIPMENT LAB., REDSTONE ARSENAL, ALA.	ABMA DLS-TR-
ARMY BALLISTIC MISSILE AGENCY. SYSTEMS SUPPORT EQUIPMENT LAB., REDSTONE ARSENAL, ALA.	ABMA DLS-TR(NO.)(YR)
ARMY BALLISTIC MISSILE AGENCY. SYSTEMS SUPPORT EQUIPMENT LAB., REDSTONE ARSENAL, ALA.	ABMA DL-TN(NO.)(YR)
ARMY BALLISTIC MISSILE AGENCY. SYSTEMS SUPPORT EQUIPMENT LAB., REDSTONE ARSENAL, ALA.	ABMA TN-DLE(NO.)(YR)
ARMY BALLISTIC MISSILE AGENCY. SYSTEMS SUPPORT EQUIPMENT LAB., REDSTONE ARSENAL, ALA.	ABMA TN-DLMF(NO.)(YR)
ARMY BALLISTIC MISSILE AGENCY. SYSTEMS SUPPORT EQUIPMENT LAB., REDSTONE ARSENAL, ALA.	ABMA TN-DLS(NO.)(YR)(TN)
ARMY BALLISTIC MISSILE AGENCY. SYSTEMS SUPPORT EQUIPMENT LAB., REDSTONE ARSENAL, ALA.	ABMA TR(NO.)(YR)
ARMY BALLISTIC MISSILE AGENCY. SYSTEMS SUPPORT EQUIPMENT LAB., REDSTONE ARSENAL, ALA.	ABMA TR-DLMF(NO.)(YR)
ARMY BALLISTIC MISSILE AGENCY. SYSTEMS SUPPORT EQUIPMENT LAB., REDSTONE ARSENAL, ALA.	ABMA TR-DLMP(NO.)(YR)
ARMY BALLISTIC MISSILE AGENCY. SYSTEMS SUPPORT EQUIPMENT LAB., REDSTONE ARSENAL, ALA.	ABMA TR-DLMT(NO.)(YR)
ARMY BALLISTIC MISSILE AGENCY. TECHNICAL DOCUMENTS LIBRARY, REDSTONE ARSENAL, ALA.	ABMA TDL BIBLIO-
ARMY BALLISTIC MISSILE AGENCY. TECHNICAL DOCUMENTS LIBRARY, REDSTONE ARSENAL, ALA. (TITLE ANNOUNCEMENT LIST)	ABMA TDL-TAL-
ARMY BALLISTIC MISSILE AGENCY. TECHNICAL DOCUMENTS LIBRARY, REDSTONE ARSENAL, ALA. (TITLE ANNOUNCEMENT LIST)	TAL-
ARMY BALLISTIC MISSILE AGENCY. TECHNICAL FEASIBILITY STUDIES OFFICE, REDSTONE ARSENAL, ALA.	TFSO-
ARMY BALLISTIC MISSILE AGENCY. TECHNICAL LIAISON GROUP OF DOD, REDSTONE ARSENAL, ALA.	ABMA DYR-R-
ARMY BALLISTIC MISSILE AGENCY. TEST EVALUATION AND FIRING LAB., REDSTONE ARSENAL, ALA.	ABMA RT-TM(NO.)(YR)
ARMY BALLISTIC MISSILE AGENCY. TEST LAB., REDSTONE ARSENAL, ALA. (COLD CALIBRATION TEST STAND TEST REPORT)	ABMA CCTS TR(NO)(NO-LTR)
ARMY BALLISTIC MISSILE AGENCY. TEST LAB., REDSTONE ARSENAL, ALA. (CALCULATION NOTES)	ABMA C+F CN-
ARMY BALLISTIC MISSILE AGENCY. TEST LAB., REDSTONE ARSENAL, ALA. (EXPERIMENTAL PROCEDURES)	ABMA C+F EP-
ARMY BALLISTIC MISSILE AGENCY. TEST LAB., REDSTONE ARSENAL, ALA. (FORMULA AND DATA COMPILATIONS)	ABMA C+F F+DC-
ARMY BALLISTIC MISSILE AGENCY. TEST LAB., REDSTONE ARSENAL, ALA. (LABORATORY EXPERIMENTS SUMMARIES)	ABMA C+F LES-
ARMY BALLISTIC MISSILE AGENCY. TEST LAB., REDSTONE ARSENAL, ALA.	ABMA C+F MEMO-
ARMY BALLISTIC MISSILE AGENCY. TEST LAB., REDSTONE ARSENAL, ALA.	ABMA C+F NOTE-
ARMY BALLISTIC MISSILE AGENCY. TEST LAB., REDSTONE ARSENAL, ALA. (TEST EVALUATION PROCEDURES)	ABMA C+F TEP-
ARMY BALLISTIC MISSILE AGENCY. TEST LAB., REDSTONE ARSENAL, ALA.	ABMA DT-EDN-
ARMY BALLISTIC MISSILE AGENCY. TEST LAB., REDSTONE ARSENAL, ALA.	ABMA DT-EM-
ARMY BALLISTIC MISSILE AGENCY. TEST LAB., REDSTONE ARSENAL, ALA.	ABMA DTI-TR(NO.)(YR)
ARMY BALLISTIC MISSILE AGENCY. TEST LAB., REDSTONE ARSENAL, ALA.	ABMA DTM(NO)M1CC
ARMY BALLISTIC MISSILE AGENCY. TEST LAB., REDSTONE ARSENAL, ALA.	ABMA DTM(NO)M1CT
ARMY BALLISTIC MISSILE AGENCY. TEST LAB., REDSTONE ARSENAL, ALA.	ABMA DTM ST-TN(NO.)(YR)
ARMY BALLISTIC MISSILE AGENCY. TEST LAB., REDSTONE ARSENAL, ALA.	ABMA DT-R-
ARMY BALLISTIC MISSILE AGENCY. TEST LAB., REDSTONE ARSENAL, ALA.	ABMA DTR-TM(NO.)(YR)
ARMY BALLISTIC MISSILE AGENCY. TEST LAB., REDSTONE ARSENAL, ALA.	ABMA DTR-TR(NO.)(YR)
ARMY BALLISTIC MISSILE AGENCY. TEST LAB., REDSTONE ARSENAL, ALA.	ABMA DT-TM(NO.)(YR)
ARMY BALLISTIC MISSILE AGENCY. TEST LAB., REDSTONE ARSENAL, ALA.	ABMA DT-TN(NO.)(YR)
ARMY BALLISTIC MISSILE AGENCY. TEST LAB., REDSTONE ARSENAL, ALA.	ABMA DT-TR(NO.)(YR)
ARMY BALLISTIC MISSILE AGENCY. TEST LAB., REDSTONE ARSENAL, ALA.	ABMA ORDAB-DIR-SPU(NO)(YR)
ARMY BALLISTIC MISSILE AGENCY. TEST LAB., REDSTONE ARSENAL, ALA.	ABMA ORDAB-DTM-SPU-
ARMY BALLISTIC MISSILE AGENCY. TEST LAB., REDSTONE ARSENAL, ALA.	ABMA TFS-
ARMY BALLISTIC MISSILE AGENCY. WEAPON SYSTEMS INFORMATION OFFICE, REDSTONE ARSENAL, ALA.	ABMA DI-CC-
ARMY BALLISTIC MISSILE AGENCY. WEAPON SYSTEMS INFORMATION OFFICE, REDSTONE ARSENAL, ALA.	ABMA DIR-TN(NO.)(YR)
ARMY BALLISTIC MISSILE AGENCY. WEAPON SYSTEMS INFORMATION OFFICE, REDSTONE ARSENAL, ALA.	ABMA DIR-TR(NO.)(YR)
ARMY BALLISTIC MISSILE AGENCY. WEAPON SYSTEMS INFORMATION OFFICE, REDSTONE ARSENAL, ALA.	ABMA DI-TM(NO.)(YR)
ARMY BALLISTIC MISSILE AGENCY. WEAPON SYSTEMS INFORMATION OFFICE, REDSTONE ARSENAL, ALA.	ABMA DI-TN(NO.)(YR)
ARMY BALLISTIC MISSILE AGENCY. WEAPON SYSTEMS INFORMATION OFFICE, REDSTONE ARSENAL, ALA.	ABMA DLMT-TM(NO.)(YR)
ARMY BALLISTIC MISSILE AGENCY. WILLOW PROJECT ENGINEERS OFFICE, REDSTONE ARSENAL, ALA.	ABMA DRW-TN(NO.)(YR)
ARMY BEHAVIORAL SCIENCE RES. LAB., WASHINGTON, D.C.	BESRL-TRN-
ARMY BEHAVIORAL SCIENCE RES. LAB., WASHINGTON, D.C.	BESRL-TRR-
ARMY BIOLOGICAL LABS., FREDERICK, MD.	ABL-TR-
ARMY BIOLOGICAL LABS., FREDERICK, MD.	MISC.-
ARMY BIOLOGICAL LABS., FREDERICK, MD.	TECH. MANUSCRIPT-
ARMY BOARD FOR AVIATION ACCIDENT RES.,FORT RUCKER,ALA	HF-
ARMY BOARD FOR AVIATION ACCIDENT RES.,FORT RUCKER,ALA	HF-(YEAR)-
ARMY CHEMICAL CENTER, MD.	EATR-
ARMY CHEM. CTR.,MD.(FOREIGN MATERIEL EVALUATION RPT.)	FMER-
ARMY CHEMICAL CORPS BIOLOGICAL LABS., CAMP DETRICK, MD. (INTERIM REPORTS)	BLIR-
ARMY CHEMICAL CORPS BIOL. LABS., CAMP DETRICK, MD.	CCBL/(LETTERS)-
ARMY CHEMICAL CORPS BIOLOGICAL LABS., CAMP DETRICK, MD. (INTERIM REPORTS)	CC BLIR
ARMY CHEMICAL CORPS BIOLOGICAL LABS., CAMP DETRICK, MD. (SPECIAL REPORTS)	CC BL-SPR-
ARMY CHEMICAL CORPS BIOLOGICAL LABS.,FT. DETRICK, MD.	FD(NUMBER)-
ARMY CHEMICAL CORPS BIOLOGICAL LABS. PHYSICAL DEFENSE DIV., FORT DETRICK, MD.	CCBL-
ARMY CHEM.CORPS NUCLEAR DEFENSE LAB.,ARMY CHEM.CTR,MD	CCNDL-

ARMY CHEMICAL INFORMATION AND DATA SYSTEM

ARMY CHEMICAL INFORMATION AND DATA SYSTEM	CIDS-STATUS-
ARMY CHEMICAL RES. + DEV. LABS., ARMY CHEM. CTR., MD.	CRDL-
ARMY CHEMICAL RES. + DEV. LABS., ARMY CHEM. CTR., MD.	CRDLR-
ARMY CHEMICAL RESEARCH AND DEVELOPMENT LABS., ARMY CHEMICAL CENTER, MD. (SPECIAL PUBLICATIONS)	CRDL-SP-(NUMBER)-
ARMY CHEMICAL RES. + DEV. LABS., ARMY CHEM. CTR., MD.	CRDL-TM-(NUMBER)-
ARMY CHEMICAL RES. + DEV. LABS., ARMY CHEM. CTR., MD.	USACRDL-TM-
ARMY COASTAL ENGINEERING RES. CENTER, WASHINGTON, DC	ACERC-
ARMY COASTAL ENGINEERING RES. CENTER, WASHINGTON, DC	CERC-MISC PAPER-
ARMY COASTAL ENGINEERING RES. CENTER, WASHINGTON, DC	CERC-TM-
ARMY COASTAL ENGINEERING RES. CENTER, WASHINGTON, DC	R-(MONTH)-(YEAR)
ARMY COASTAL ENGINEERING RES. CENTER, WASHINGTON, DC	TM-
ARMY COATING AND CHEMICAL LAB., ABERDEEN PROVING GROUND, MD.	APG-CCL-
ARMY COATING AND CHEMICAL LAB., ABERDEEN PROVING GROUND, MD.	CCL-
ARMY COLD REGIONS RES. AND ENG. LAB., HANOVER, N.H.	APL-UW-
ARMY COLD REGIONS RES. AND ENG. LAB., HANOVER, N.H.	CCREL-SR-
ARMY COLD REGIONS RES. AND ENG. LAB., HANOVER, N.H.	CRREL-
ARMY COLD REGIONS RES. AND ENG. LAB., HANOVER, N.H.	CRREL-CRSE-
ARMY COLD REGIONS RES. AND ENG. LAB., HANOVER, N.H. (SPECIAL REPORT)	CRREL-SR-
ARMY COLD REGIONS RES. AND ENG. LAB., HANOVER, N.H. (TECHNICAL REPORT)	CRREL-TR-
ARMY COLD REGIONS RES. AND ENG. LAB., HANOVER, N.H.	RREL-
ARMY COLD REGIONS RES. + ENG. LAB., WILMETTE, ILL.	CRREL-
ARMY COLD REGIONS RES. + ENG. LAB., WILMETTE, ILL. (RESEARCH REPORT)	CRREL-RR-
ARMY COLD REGIONS RES. + ENG. LAB., WILMETTE, ILL. (SPECIAL REPORT)	CRREL-SR-
ARMY COLD REGIONS RES. + ENG. LAB., WILMETTE, ILL. (TECHNICAL REPORT)	CRREL-TR-
ARMY COLD REGIONS RES. + ENG. LAB., WILMETTE, ILL. (TRANSLATION)	CRREL-TRANS-
ARMY COLD REGIONS RES. + ENG. LAB., WILMETTE, ILL.	SIPRE-RR-
ARMY COMBAT DEVELOPMENTS COMMAND, FORT BELVOIR, VA.	ACN-
ARMY COMBAT DEVELOPMENTS COMMAND, FORT BELVOIR, VA. (COMBAT DEVELOPMENTS OBJECTIVES GUIDE)	CDOG-
ARMY COMBAT DEVELOPMENTS COMMAND, FORT BELVOIR, VA.	USACDC-(YEAR)-
ARMY COMBAT DEVELOPMENTS COMMAND, FORT BELVOIR, VA.	USACDC-PAM-(NUMBER)-
ARMY COMBAT DEVELOPMENTS COMMAND. ARTILLERY AGENCY, FORT SILL, OKLA.	USACDCARTYA-(YEAR)-
ARMY COMBAT DEVELOPMENTS COMMAND. AVIATION AGENCY, FORT RUCKER, ALA.	ACDCAA-
ARMY COMBAT DEVELOPMENTS COMMAND. AVIATION AGENCY, FORT RUCKER, ALA.	USACDCAA-(YEAR)-
ARMY COMBAT DEVELOPMENTS COMMAND. CHAPLAIN AGENCY, FORT LEE, VA.	USACDCCHA-(YEAR)-
ARMY COMBAT DEVELOPMENTS COMMAND. CHEMICAL-BIOLOGICAL-RADIOLOGICAL AGENCY, FORT MC CLELLAN, ALA.	ACDC-CBR-
ARMY COMBAT DEVELOPMENTS COMMAND. CHEMICAL-BIOLOGICAL-RADIOLOGICAL AGENCY, FORT MC CLELLAN, ALA.	CBRSS-
ARMY COMBAT DEVELOPMENTS COMMAND. CHEMICAL-BIOLOGICAL-RADIOLOGICAL AGENCY, FORT MC CLELLAND, ALA.	USACDCCBRA-
ARMY COMBAT DEVELOPMENTS COMMAND. COMBINED ARMS AGENCY, FORT LEAVENWORTH, KAN.	USACDCCAG-(YEAR)-
ARMY COMBAT DEVELOPMENTS COMMAND. COMMUNICATIONS-ELECTRONICS AGENCY, FORT MONMOUTH, N.J.	USACDCCEA-(YEAR)-
ARMY COMBAT DEVELOPMENTS COMMAND. ENGINEER AGENCY, FORT BELVOIR, VA.	USACDCEA-(YEAR)-
ARMY COMBAT DEVELOPMENTS COMMAND. EXPERIMENTATION CENTER, FORT ORD, CALIF.	CDEC-
ARMY COMBAT DEVELOPMENTS COMMAND. EXPERIMENTATION CENTER, FORT ORD, CALIF.	USACDCEC-
ARMY COMBAT DEVELOPMENTS COMMAND. EXPERIMENTATION COMMAND, FORT ORD, CALIF.	CDCEC-
ARMY COMBAT DEVELOPMENTS COMMAND. EXPERIMENTATION COMMAND. RESEARCH OFFICE, MONTEREY, CALIF.	ROEC-
ARMY COMBAT DEVELOPMENTS COMMAND. INFANTRY AGENCY, FORT BENNING, GA.	USACDCIA-(YEAR)-
ARMY COMBAT DEVELOPMENTS COMMAND. INSTITUTE OF ADVANCED STUDIES, CARLISLE BARRACKS, PENNA.	USACDCIAS-
ARMY COMBAT DEVELOPMENTS COMMAND. INST. OF NUCLEAR STUDIES, FORT BLISS, TEX.	UACDCINS-
ARMY COMBAT DEVELOPMENTS COMMAND. INST. OF NUCLEAR STUDIES, FORT BLISS, TEXAS	USACDCINS-(MONTH)-(YR.)
ARMY COMBAT DEVELOPMENTS COMMAND. INST. OF SPECIAL STUDIES, FORT BELVOIR, VA.	AH-
ARMY COMBAT DEVELOPMENTS COMMAND. JUDGE ADVOCATE AGENCY, CHARLOTTESVILLE, VA.	USACDCJAA-(YEAR)-
ARMY COMBAT DEVELOPMENTS COMMAND. LIAISON DETACHMENT, SAN FRANCISCO	TRIP-(NUMBER-YEAR)
ARMY COMBAT DEVELOPMENTS COMMAND. MAINTENANCE AGENCY, ABERDEEN PROVING GROUND, MD.	USACDCMA-(YEAR)-
ARMY COMBAT DEVELOPMENTS COMMAND. MEDICAL SERVICE AGENCY, FORT SAM HOUSTON, TEX.	AMSCD-
ARMY COMBAT DEVELOPMENTS COMMAND. MEDICAL SERVICE AGENCY, FORT SAM HOUSTON, TEX.	USACDCMSA-(YEAR)-
ARMY COMBAT DEVELOPMENTS COMMAND. MILITARY POLICE AGENCY, FORT GORDON, GA.	USACDCMPA-(YEAR)-
ARMY COMBAT DEVS. COMMAND. NUCLEAR GP., FT.BLISS,TEX.	ACDC-NG-
ARMY COMBAT DEVS. COMMAND. NUCLEAR GP., FT. BLISS,TEX	CDC/NG (NUMBERS)
ARMY COMBAT DEVS. COMMAND. NUCLEAR GP., FT.BLISS,TEX.	USACDCNG-
ARMY COMBAT DEVS. COMMAND. NUCLEAR GP., FT.BLISS,TEX.	USACRDL-TM-
ARMY COMBAT DEVELOPMENTS COMMAND. OFFICE OF SPECIAL WEAPONS DEV., FT. BLISS, TEX.	CDC/OSWD (NUMBERS)
ARMY COMBAT DEVELOPMENTS COMMAND. QUARTERMASTER AGENCY, FORT LEE, VA.	ACDCQA-
ARMY COMBAT DEVELOPMENTS COMMAND. SPECIAL WARFARE AND CIVIL AFFAIRS GROUP, FORT BELVOIR, VA.	ACDC-SWCAG-
ARMY COMMAND + GEN. STAFF COLL., FT. LEAVENWORTH,KAN.	CGSC-
ARMY COMMAND + GENERAL STAFF COLL., WASHINGTON, D.C.	FM-
ARMY COMMAND + GEN. STAFF SCH., FT. LEAVENWORTH, KAN.	ACGSC-
ARMY COMMAND + GEN. STAFF SCH., FT. LEAVENWORTH, KAN.	RB-
ARMY COMMUNICATIONS-ELECTRONICS ENGINEERING INSTALLATION AGENCY, FORT HUACHUCA, ARIZ.	SCCC-TED-(YEAR)-FR-
ARMY COMPUTER SYSTEMS COMMAND, FORT BELVOIR, VA.	USACSC-
ARMY CONCEPT TEAM IN VIETNAM, SAN FRANCISCO	ACTIV-
ARMY CONSTRUCTION ENGINEERING RESEARCH LAB., CHAMPAIGN, ILL.	CERL-TR-M-
ARMY ELECTRONIC PROVING GROUND, FORT HUACHUCA, ARIZ.	AEPG-
ARMY ELECTRONIC PROVING GROUND, FORT HUACHUCA, ARIZ.	AEPG-SIG-
ARMY ELECTRONIC PROVING GROUND, FORT HUACHUCA, ARIZ.	AM-

ARMY ELECTRONIC PROVING GROUND, FORT HUACHUCA, ARIZ.	RC-
ARMY ELECTRONIC PROVING GROUND, FORT HUACHUCA, ARIZ.	USAEPC-SIG-
ARMY ELECTRONIC PROVING GROUND, FORT HUACHUCA, ARIZ.	USAEPG-
ARMY ELECTRONIC PROVING GROUND, FORT HUACHUCA, ARIZ.	USAEPG-DR-
ARMY ELECTRONIC PROVING GROUND, FORT HUACHUCA, ARIZ.	USAEPG-FR-
ARMY ELECTRONIC PROVING GROUND, FORT HUACHUCA, ARIZ.	USAEPG-LR-
ARMY ELECTRONIC PROVING GROUND, FORT HUACHUCA, ARIZ.	USAEPG-SIR-
ARMY ELECTRONIC PROVING GROUND, FORT HUACHUCA, ARIZ.	USAEPG-TP-
ARMY ELECTRONIC PROVING GROUND, SPECTRUM SIGNATURE LAB., FORT HUACHUCA, ARIZ.	ASSDV-
ARMY ELECTRONICS COMMAND, FORT MONMOUTH, N.J.	ECOM-
ARMY ELECTRONICS COMMAND, FORT MONMOUTH, N.J.	ECOM-(NUMBER)-
ARMY ELECTRONICS COMMAND, FORT MONMOUTH, N.J.	ECOM-(NUMBER)-F
ARMY ELECTRONICS COMMAND, FORT MONMOUTH, N.J.	ECOM-C-
ARMY ELECTRONICS COMMAND, FORT MONMOUTH, N.J.	ECOM-CATE-
ARMY ELECTRONICS COMMAND, FORT MONMOUTH, N.J.	ECOM-TR-(NUMBER)-
ARMY ELECTRONICS COMMAND, FORT MONMOUTH, N.J.	TR-ECOM-
ARMY ELECTRONICS COMMAND. COMMUNICATIONS/ADP LAB., FORT MONMOUTH, N.J.	CADPL-
ARMY ELECTRONICS COMMAND. METEOROLOGICAL SUPPORT ACTIVITY, FORT HUACHUCA, ARIZ.	MET-(MONTH)-(YEAR)
ARMY ELECTRONICS COMMAND. METEOROLOGICAL SUPPORT ACTIVITY, FORT HUACHUCA, ARIZ.	MET-SPT-ACTV-
ARMY ELECTRONICS COMMAND. NIGHT VISION LAB., FORT BELVOIR, VA.	NVL-
ARMY ELECTRONICS LABS., FORT MONMOUTH, N.J.	AEL-
ARMY ELECTRONICS LABS., FORT MONMOUTH, N.J.	AEL-TR-ECOM-
ARMY ELECTRONICS LABS., FORT MONMOUTH, N.J.	AMSEL-
ARMY ELECTRONICS LABS., FORT MONMOUTH, N.J.	RCS-OSD-
ARMY ELECTRONICS LABS., FORT MONMOUTH, N.J.	TR-
ARMY ELECTRONICS LABS., FORT MONMOUTH, N.J.	USASRDL-TN-
ARMY ELECTRONICS RESEARCH AND DEVELOPMENT ACTIVITY, FORT HUACHUCA, ARIZ.	ERDA-
ARMY ELECTRONICS RESEARCH AND DEVELOPMENT ACTIVITY, FORT HUACHUCA, ARIZ.	ERDAA-ELCT-
ARMY ELECTRONICS RESEARCH AND DEVELOPMENT ACTIVITY, FORT HUACHUCA, ARIZ.	ERDAA-MET-
ARMY ELECTRONICS RESEARCH AND DEVELOPMENT ACTIVITY, FORT HUACHUCA, ARIZ.	USAERDAA-ELCT-
ARMY ELECTRONICS RESEARCH AND DEVELOPMENT ACTIVITY, FORT HUACHUCA, ARIZ.	USAERDAA-MET-
ARMY ELECTRONICS RESEARCH AND DEVELOPMENT ACTIVITY, FORT HUACHUCA, ARIZ.	USAERDAA-MET-(NO.)-(YR.)
ARMY ELECTRONICS RESEARCH AND DEVELOPMENT ACTIVITY, WHITE SANDS MISSILE RANGE, N. MEX.	AERDA-
ARMY ELECTRONICS RESEARCH AND DEVELOPMENT ACTIVITY, WHITE SANDS MISSILE RANGE, N. MEX.	ERDA-
ARMY ELECTRONICS RESEARCH AND DEVELOPMENT ACTIVITY, WHITE SANDS MISSILE RANGE, N. MEX.	ERDAW-
ARMY ELECTRONICS RESEARCH AND DEVELOPMENT ACTIVITY, WHITE SANDS MISSILE RANGE, N. MEX.	USAERDA-
ARMY ELECTRONICS RESEARCH AND DEVELOPMENT ACTIVITY, WHITE SANDS MISSILE RANGE, N. MEX.	WSMR-ERDA-
ARMY ELECTRONICS RES. + DEV. LAB., FT. MONMOUTH, N.J.	AELRDL-
ARMY ELECTRONICS RES. + DEV. LAB., FT. MONMOUTH, N.J.	AELRDL-TR-
ARMY ELECTRONICS RES. + DEV. LAB., FT. MONMOUTH, N.J.	AERDL-T-(NO.)-(YEAR)-
ARMY ELECTRONICS RES. + DEV. LAB., FT. MONMOUTH, N.J.	AERDL-TR-
ARMY ELECTRONICS RES. + DEV. LAB., FT. MONMOUTH, N.J.	USAELRDL-TR-
ARMY ELECTRONICS RES. + DEV. LAB., FT. MONMOUTH, N.J.	USAEL-TR-
ARMY ELECTRONICS RES. + DEV. LAB., FT. MONMOUTH, N.J.	USAERADL-TR-
ARMY ELECTRONICS RES. + DEV. LAB., FT. MONMOUTH, N.J.	USAERDL-TR-
ARMY ENGINEER DISTRICT, SAVANNAH	SAS-
ARMY ENGINEER DISTRICT, SAVANNAH	TISA-
ARMY ENGINEER EXPLOSIVE EXCAVATION RESEARCH OFFICE, LIVERMORE, CALIF.	EERO-TR-
ARMY ENGINEER GEODESY, INTELLIGENCE AND MAPPING RESEARCH AND DEVELOPMENT AGENCY, FORT BELVOIR, VA.	AGIMRDA-
ARMY ENGINEER GEODESY, INTELLIGENCE AND MAPPING RESEARCH AND DEVELOPMENT AGENCY, FORT BELVOIR, VA.	GIMRADA-(NUMBER)-TR
ARMY ENGINEER GEODESY, INTELLIGENCE AND MAPPING RESEARCH AND DEVELOPMENT AGENCY, FORT BELVOIR, VA.	GIMRADA-RN-
ARMY ENGINEER GEODESY, INTELLIGENCE AND MAPPING RESEARCH AND DEVELOPMENT AGENCY, FORT BELVOIR, VA.	GIMRADA-TN-(YEAR)-
ARMY ENGINEER GEODESY, INTELLIGENCE AND MAPPING RESEARCH AND DEVELOPMENT AGENCY, FORT BELVOIR, VA.	GIMRADA-TR-
ARMY ENGINEER NUCLEAR CRATERING GROUP, LIVERMORE, CAL	NCG-TM-(YEAR)-
ARMY ENGINEER NUCLEAR CRATERING GROUP, LIVERMORE, CAL	NCG-TR-
ARMY ENGINEER NUCLEAR CRATERING GROUP, LIVERMORE, CAL	NGG-TR-
ARMY ENGINEER REACTORS GP. ENG. DIV.,FT. BELVOIR, VA.	ED-
ARMY ENGINEER REACTORS GP. RESEARCH AND TECHNOLOGY DIV., FORT BELVOIR, VA.	RT-
ARMY ENGINEER RESEARCH + DEV. LABS., FT. BELVOIR, VA.	AERDL-
ARMY ENGINEER RESEARCH + DEV. LABS., FT. BELVOIR, VA.	AERDL-T-
ARMY ENGINEER RESEARCH + DEV. LABS., FT. BELVOIR, VA.	DC-
ARMY ENGINEER RESEARCH + DEV. LABS., FT. BELVOIR, VA.	EDRL-T-
ARMY ENGINEER RESEARCH + DEV. LABS., FT. BELVOIR, VA.	EPG-
ARMY ENGINEER RESEARCH + DEV. LABS., FT. BELVOIR, VA.	ERDL-
ARMY ENGINEER RESEARCH + DEV. LABS., FT. BELVOIR, VA.	ERDL-(NUMBER)-TR
ARMY ENGINEER RESEARCH + DEV. LABS., FT. BELVOIR, VA.	ERDL R-
ARMY ENGINEER RESEARCH + DEV. LABS., FT. BELVOIR, VA. (TRANSLATION)	ERDL-T-
ARMY ENGINEER RESEARCH + DEV. LABS., FT. BELVOIR, VA.	ERDL-TR-
ARMY ENGINEER RESEARCH + DEV. LABS., FT. BELVOIR, VA.	T-
ARMY ENGINEER RESEARCH + DEV. LABS., FT. BELVOIR, VA.	TRS-
ARMY ENGINEER RESEARCH + DEV. LABS., FT. BELVOIR, VA.	USAERDL-
ARMY ENGINEER RESEARCH + DEV. LABS., FT. BELVOIR, VA.	XR-
ARMY ENGINEER SCHOOL, FORT BELVOIR, VA.	CEDC-
ARMY ENGINEER SCHOOL, FORT BELVOIR, VA.	ES STUDY-
ARMY ENGINEER SCHOOL. COMBAT DEVS. GP., FT.BELVOIR,VA	CECD-
ARMY ENGINEER TOPOGRAPHIC LABS., FORT BELVOIR, VA.	ETL-ETR-
ARMY ENGINEER TOPOGRAPHIC LABS., FORT BELVOIR, VA.	USAETL-(NUMBER)-(LTRS.)
ARMY ENGINEER TOPOGRAPHIC LABS., FORT BELVOIR, VA.	USAETL-(NUMBER)-TR
ARMY ENGINEER TOPOGRAPHIC LABS., FORT BELVOIR, VA.	USAETL-RN-
ARMY ENGINEER TOPOGRAPHIC LABS., FORT BELVOIR, VA.	USAETL-TB-
ARMY ENGINEER TOPOGRAPHIC LABS., FORT BELVOIR, VA.	USAETL-TR-
ARMY ENGINEER WATERWAYS EXPERIMENT STATION,VICKSBURG, MISS.	AEWES-(NUMBER)-
ARMY ENGINEER WATERWAYS EXPERIMENT STATION,VICKSBURG, MISS. (CONTRACT REPORT)	AEWES-CR-(NUMBER)-
ARMY ENGINEER WATERWAYS EXPERIMENT STATION,VICKSBURG, MISS.	AEWES-CR-M-
ARMY ENGINEER WATERWAYS EXPERIMENT STATION,VICKSBURG, MISS.	AEWES-CR-N-(YEAR)-
ARMY ENGINEER WATERWAYS EXPERIMENT STATION,VICKSBURG, MISS.	AEWES-CR-S-
ARMY ENGINEER WATERWAYS EXPERIMENT STATION,VICKSBURG, MISS.	AEWES-INSTRUCTION-N-
ARMY ENGINEER WATERWAYS EXPERIMENT STATION,VICKSBURG, MISS.	AEWES-INSTRUCTION-S-(YR)-
ARMY ENGINEER WATERWAYS EXPERIMENT STATION,VICKSBURG, MISS.	AEWES-M-
ARMY ENGINEER WATERWAYS EXPERIMENT STATION,VICKSBURG, MISS.	AEWES-MISC PAPER-(NO.)-

ARMY ENGINEER WATERWAYS EXPERIMENT STATION

ARMY ENGINEER WATERWAYS EXPERIMENT STATION, VICKSBURG, MISS.	AEWES-MISC. PAPER-M-
ARMY ENGINEER WATERWAYS EXPERIMENT STATION, VICKSBURG, MISS.	AEWES-MISC. PAPER-N-
ARMY ENGINEER WATERWAYS EXPERIMENT STATION, VICKSBURG, MISS.	AEWES-MISC. PAPER-S-
ARMY ENGINEER WATERWAYS EXPERIMENT STATION, VICKSBURG, MISS.	AEWES-N-
ARMY ENGINEER WATERWAYS EXPERIMENT STATION, VICKSBURG, MISS.	AEWES-RR-(LTR.)-(YR.)-
ARMY ENGINEER WATERWAYS EXPERIMENT STATION, VICKSBURG, MISS.	AEWES-S-(YEAR)-
ARMY ENGINEER WATERWAYS EXPERIMENT STATION, VICKSBURG, MISS.	AEWES-TM-(NUMBER)-
ARMY ENGINEER WATERWAYS EXPERIMENT STATION, VICKSBURG, MISS.	AEWES-TR-(NO.)-(NO.)-
ARMY ENGINEER WATERWAYS EXPERIMENT STATION, VICKSBURG, MISS.	AEWES-TR-(NUMBER)-
ARMY ENGINEER WATERWAYS EXPERIMENT STATION, VICKSBURG, MISS.	AEWES-TR-(LTR.)-(YR.)-
ARMY ENGINEER WATERWAYS EXPERIMENT STATION, VICKSBURG, MISS.	AEWES-TRANS-
ARMY ENGINEER WATERWAYS EXPERIMENT STATION, VICKSBURG, MISS.	AEWES-TR-M-
ARMY ENGINEER WATERWAYS EXPERIMENT STATION, VICKSBURG, MISS.	AEWES-TR-N-
ARMY ENGINEER WATERWAYS EXPERIMENT STATION, VICKSBURG, MISS.	AEWES-TR-S-
ARMY ENGINEER WATERWAYS EXPERIMENT STATION, VICKSBURG, MISS.	AEWWS-TR-
ARMY ENGINEER WATERWAYS EXPERIMENT STATION, VICKSBURG, MISS.	CR-
ARMY ENGINEER WATERWAYS EXPERIMENT STATION, VICKSBURG, MISS.	EWES-LIV-
ARMY ENGINEER WATERWAYS EXPERIMENT STATION, VICKSBURG, MISS.	MP-
ARMY ENGINEER WATERWAYS EXPERIMENT STATION, VICKSBURG, MISS.	MRC-WES-
ARMY ENGINEER WATERWAYS EXPERIMENT STATION, VICKSBURG, MISS.	MS-
ARMY ENGINEER WATERWAYS EXPERIMENT STATION, VICKSBURG, MISS.	WES-
ARMY ENGINEER WATERWAYS EXPERIMENT STATION, VICKSBURG, MISS. (CONTRACT REPORT)	WES-CR-
ARMY ENGINEER WATERWAYS EXPERIMENT STATION, VICKSBURG, MISS. (FINAL REPORT)	WES FR-
ARMY ENGINEER WATERWAYS EXPERIMENT STATION, VICKSBURG, MISS. (INSTRUCTION REPORT)	WES-IR-
ARMY ENGINEER WATERWAYS EXPERIMENT STATION, VICKSBURG, MISS.	WES-MISC.PAPER (NO.)-
ARMY ENGINEER WATERWAYS EXPERIMENT STATION, VICKSBURG, MISS. (MISCELLANEOUS PAPER)	WES MP-(NUMBER)-
ARMY ENGINEER WATERWAYS EXPERIMENT STATION, VICKSBURG, MISS.	WES-MP-(LETTER)-(YR.)-
ARMY ENGINEER WATERWAYS EXPERIMENT STATION, VICKSBURG, MISS.	WES-NCG-TR-
ARMY ENGINEER WATERWAYS EXPERIMENT STATION, VICKSBURG, MISS. (RESEARCH REPORT)	WES-RR-
ARMY ENGINEER WATERWAYS EXPERIMENT STATION, VICKSBURG, MISS.	WES-TM-(NUMBER)-
ARMY ENGINEER WATERWAYS EXPERIMENT STATION, VICKSBURG, MISS.	WES-TR-(NUMBER)-
ARMY ENGINEER WATERWAYS EXPERIMENT STATION, VICKSBURG, MISS.	WES-TR-(LETTER)-(YR.)-
ARMY ENLISTED EVALUATION CENTER, FORT BENJAMIN HARRISON, IND.	USAEEC-
ARMY ENVIRONMENTAL HYGIENE AGENCY, EDGEWOOD ARSENAL, MD	USAEHA-
ARMY ENVIRONMENTAL HYGIENE AGENCY, EDGEWOOD ARSENAL, MD	USAEHA-STUDY-
ARMY FIELD ARTILLERY BOARD, FORT SILL, OKLA.	USAFABD-FA-
ARMY FIELD FORCES, FORT BLISS, TEX.	AFF-
ARMY FIELD FORCES, FORT MONROE, VA. (GUIDED MISSILE REPORT)	GM-(NUMBER, YEAR)
ARMY FIELD FORCES. BOARD NO. 2, FORT KNOX, KY. (INFORMAL PROJECT REPORT)	IP-
ARMY FIELD FORCES. BOARD NO. 2, FORT KNOX, KY. (FORMAL PROJECT REPORT)	P-
ARMY FIELD FORCES. BOARD NO. 4, FORT BLISS, TEX. (ANTIAIRCRAFT SERVICE TEST SECTION REPORT)	AA-
ARMY FIELD FORCES. BOARD NO. 4, FORT BLISS, TEX. (GUIDED MISSILE SERVICE TEST SECTION REPORT)	GM-
ARMY FIELD OPERATING COST AGENCY, ALEXANDRIA, VA.	FOCA-OF-
ARMY FIELD OPERATING COST AGENCY, ALEXANDRIA, VA.	FOCA-WS-
ARMY FORCES IN THE PACIFIC	AFPAC-
ARMY FORCES IN THE PACIFIC (ENGINEER, TECHNICAL AND TECHNOLOGICAL SURVEY REPORT)	AFPAC ENG TTS-
ARMY FORCES IN THE PACIFIC (MEDICAL TECHNICAL INTELLIGENCE FIELD REPORT)	AFPAC MED-
ARMY FORCES IN THE PACIFIC (MEDICAL TECHNICAL INTELLIGENCE FIELD REPORT)	AFPAC MED TIFR-
ARMY FORCES IN THE PACIFIC (ORDNANCE TECHNICAL INTELLIGENCE REPORT)	AFPAC ORD-
ARMY FORCES IN THE PACIFIC (ORDNANCE TECHNICAL INTELLIGENCE REPORT)	AFPAC ORD TIR-
ARMY FORCES IN THE PACIFIC (ENGINEER, TECHNICAL AND TECHNOLOGICAL SURVEY REPORT)	ENG TTS-
ARMY FORCES IN THE PACIFIC (MEDICAL TECHNICAL INTELLIGENCE FIELD REPORT)	MED TIFR-
ARMY FORCES IN THE PACIFIC (ORDNANCE TECHNICAL INTELLIGENCE REPORT)	ORD TIR-
ARMY FOREIGN SCIENCE AND TECH. CENTER, WASHINGTON, D.C	FSTC-
ARMY FOREIGN SCIENCE AND TECH. CENTER, WASHINGTON, D.C	FSTC-CS-
ARMY FOREIGN SCIENCE AND TECH. CENTER, WASHINGTON, D.C	FSTC-HF-
ARMY FOREIGN SCIENCE AND TECH. CENTER, WASHINGTON, D.C	FSTC-HT-
ARMY FOREIGN SCIENCE AND TECH. CENTER, WASHINGTON, D.C	FSTC-MP-
ARMY GENERAL STAFF, WASHINGTON, D.C.	WDGS-
ARMY GROUND FORCES (TASK FORCE FRIGID)	AGF-TFF-
ARMY INERTIAL GUIDANCE AND CONTROL LAB. AND CENTER, REDSTONE ARSENAL, ALA.	RG-S-(YEAR)-
ARMY INERTIAL GUIDANCE AND CONTROL LAB. AND CENTER, REDSTONE ARSENAL, ALA.	RG-SR-(YEAR)-
ARMY INERTIAL GUIDANCE AND CONTROL LAB. AND CENTER, REDSTONE ARSENAL, ALA.	RG-TM-(YEAR)-
ARMY INERTIAL GUIDANCE AND CONTROL LAB. AND CENTER, REDSTONE ARSENAL, ALA.	RG-TN-(YEAR)-
ARMY INERTIAL GUIDANCE AND CONTROL LAB. AND CENTER, REDSTONE ARSENAL, ALA.	RG-TR-(YEAR)-
ARMY INERTIAL GUIDANCE AND CONTROL LAB. AND CENTER, REDSTONE ARSENAL, ALA.	RK-TSR-(YEAR)-
ARMY INFANTRY BOARD, FORT BENNING, GA.	USAIB-
ARMY INST. OF PATHOLOGY, WASHINGTON, D.C.	AM-
ARMY INTELLIGENCE CENTER, FORT HOLABIRD, MD.	USACDCINTA-
ARMY INTELLIGENCE THREAT ANALYSIS DETACHMENT, WASH., DC	USAITAD-
ARMY LIBRARY, WASHINGTON, D.C.	ALIB-SPEC. BIB.-
ARMY LIMITED WAR LAB., ABERDEEN PROVING GROUND, MD.	LWL-CR-
ARMY LIMITED WAR LAB., ABERDEEN PROVING GROUND, MD.	LWL-TM-(YEAR)-
ARMY LIMITED WAR LAB., ABERDEEN PROVING GROUND, MD.	LWL-TN-(YEAR)-
ARMY LIMITED WAR LAB., ABERDEEN PROVING GROUND, MD.	LWL-TR-(YEAR)-
ARMY LOGISTICS MANAGEMENT CENTER, FORT LEE, VA.	ALM-(NUMBER-LETTER)
ARMY LOGISTICS MANAGEMENT CENTER, FORT LEE, VA.	STUDY-
ARMY LOGISTICS MANAGEMENT CENTER, FORT LEE, VA.	USALMC-ST-
ARMY LOGISTICS MANAGEMENT CENTER, FORT LEE, VA.	USALMC-STUDY-
ARMY LOGISTICS MANAGEMENT CENTER, FORT LEE, VA.	USALMC-W-
ARMY MAP SERVICE, WASHINGTON, D.C. (REF. 1)	(NUMBERS...)
ARMY MAP SERVICE, WASHINGTON, D.C.	AMS-
ARMY MAP SERVICE, WASHINGTON, D.C.	AMS-GEODETIC MEMO-
ARMY MAP SERVICE, WASHINGTON, D.C. (TECHNICAL BULL.)	AMS-TB-
ARMY MAP SERVICE, WASHINGTON, D.C. (TECH. INSTRUCTION)	AMS-TI-
ARMY MAP SERVICE, WASH., D.C. (TRANSVERSE MERCATOR GRID)	AMS TM-
ARMY MAP SERVICE, WASHINGTON, D.C. (TECHNICAL MANUAL)	AMS-TM-
ARMY MAP SERVICE, WASHINGTON, D.C. (TECHNICAL REPORT)	AMS-TR-
ARMY MAP SERVICE, WASHINGTON, D.C.	EIS-

ARMY MATERIALS + MECHANICS RES. CTR., WATERTOWN, MASS.	AMMRC-CR-
ARMY MATERIALS + MECHANICS RES. CTR., WATERTOWN, MASS.	AMMRC-CR-(YEAR)-
ARMY MATERIALS + MECHANICS RES. CTR., WATERTOWN, MASS.	AMMRC-MS-
ARMY MATERIALS + MECHANICS RES. CTR., WATERTOWN, MASS.	AMMRC-TR-
ARMY MATERIALS RESEARCH AGENCY, WATERTOWN, MASS. (CONTRACT REPORT)	AMRA-CR-(YEAR)-
ARMY MATERIALS RESEARCH AGENCY, WATERTOWN, MASS. (MEMORANDUM REPORT)	AMRA-MR-(YEAR)-
ARMY MATERIALS RESEARCH AGENCY, WATERTOWN, MASS. (MONOGRAPHS)	AMRA-MS-(YEAR)-
ARMY MATERIALS RESEARCH AGENCY, WATERTOWN, MASS. (SPECIAL PUBLICATION)	AMRA-PS-(YEAR)-
ARMY MATERIALS RESEARCH AGENCY, WATERTOWN, MASS. (TECHNICAL REPORT)	AMRA-TR-(YEAR)-
ARMY MATERIALS RESEARCH AGENCY, WATERTOWN, MASS.	CR-(YEAR)-(NUMBER)/
ARMY MATERIALS RES. AGENCY, WATERTOWN ARSENAL, MASS.	AMXMR-P-
ARMY MATERIEL COMMAND, WASHINGTON., D.C.	AMC-
ARMY MATERIEL COMMAND, WASHINGTON, D.C.	AMCP-
ARMY MATERIEL COMMAND, WASHINGTON, D.C.	AMC-PAM-
ARMY MATERIEL COMMAND, WASHINGTON, D.C.	AMC-PAM-(NO.)-(NO.)-
ARMY MATERIEL COMMAND, WASHINGTON, D.C.	AMCR-
ARMY MATERIEL COMMAND, WASHINGTON, D.C.	AMC-STUDY-(NO.)-
ARMY MATERIEL COMMAND, WASHINGTON, D.C.	AMC-TIR-
ARMY MATERIEL COMMAND, WASHINGTON, D.C.	AMC-TIR-(NO.)-(NO.)-
ARMY MATERIEL COMMAND, WASHINGTON, D.C.	POMM-
ARMY MATERIEL SYSTEMS ANALYSIS AGENCY, ABERDEEN PROVING GROUND, MD.	AMSAA-TM-
ARMY MATERIEL SYSTEMS ANALYSIS AGENCY, ABERDEEN PROVING GROUND, MD.	AMSAA-TR-
ARMY MEDICAL EQUIP. RES. + DEV. LAB., FT. TOTTEN, NY	AMERDL-
ARMY MEDICAL EQUIP. RES. + DEV. LAB., FT. TOTTEN, NY	MERDL-
ARMY MEDICAL EQUIP. RES. + DEV. LAB., FT. TOTTEN, NY	MERDL-A-
ARMY MEDICAL LAB (5TH), ST. LOUIS	RB-
ARMY MEDICAL NUTRITION LAB., DENVER	AMNL-
ARMY MEDICAL NUTRITION LAB., DENVER.	MNL-
ARMY MEDICAL RES. AND DEV. COMMAND, WASHINGTON, D.C.	AMRUE-
ARMY MEDICAL RES. AND DEV. COMMAND, WASHINGTON, D.C.	USAMRDC-
ARMY MEDICAL RESEARCH AND NUTRITION LAB., DENVER	AMRNL-
ARMY MEDICAL RESEARCH AND NUTRITION LAB., DENVER	USAMRNL-
ARMY MEDICAL RESEARCH LAB., FORT KNOX, KY.	AMRL-
ARMY MEDICAL RES. LAB., FT. KNOX, KY. (PUBLICATIONS)	AMRL-PUB-
ARMY MEDICAL RESEARCH LAB., FORT KNOX, KY.	AMRL R-
ARMY MEDICAL RESEARCH LAB., FORT KNOX, KY.	MDFRL-
ARMY MEDICAL RESEARCH LAB., FORT KNOX, KY.	MSFRL-
ARMY MEDICAL RESEARCH LAB., FORT KNOX, KY.	USAMRL-
ARMY MEDICAL RESEARCH LAB., FORT KNOX, KY.	USAMRL-(LETTER)-
ARMY MEDICAL RESEARCH UNIT, EUROPE (PROGRESS REPORT)	MEDDH-
ARMY MEDICAL SERVICE. COMBAT DEV. GROUP, WASH., D.C.	AMSCO-
ARMY MEDICAL SERVICE GRADUATE SCHOOL, WASHINGTON, DC	AMSGS-
ARMY MILITARY POLICE SCHOOL, FORT GORDON, GA.	ST-(NUMBER)-
ARMY MISSILE COMMAND, REDSTONE ARSENAL, ALA.	AMC-
ARMY MISSILE COMMAND, REDSTONE ARSENAL, ALA.	AMC-ETO-TM-
ARMY MISSILE COMMAND. REDSTONE ARSENAL, ALA.	AMC/MISS-
ARMY MISSILE COMMAND, REDSTONE ARSENAL, ALA.	AMC-P-
ARMY MISSILE COMMAND, REDSTONE ARSENAL, ALA.	AMC-RA-RE-TR-(YEAR)-
ARMY MISSILE COMMAND, REDSTONE ARSENAL, ALA.	AMC-RA-RG-TR-(YEAR)-
ARMY MISSILE COMMAND, REDSTONE ARSENAL, ALA.	AMC-RA-RN-TR-(YEAR)-
ARMY MISSILE COMMAND, REDSTONE ARSENAL, ALA.	AMC-RA-RR-TR-(YEAR)-
ARMY MISSILE COMMAND, REDSTONE ARSENAL, ALA.	AMC-RA-RS-TR-(YEAR)-
ARMY MISSILE COMMAND, REDSTONE ARSENAL, ALA.	AMC-RA-TIR-
ARMY MISSILE COMMAND, REDSTONE ARSENAL, ALA.	AMC-RK-TR-(YEAR)-
ARMY MISSILE COMMAND, REDSTONE ARSENAL, ALA.	AMC-RP-TN-(YEAR)-
ARMY MISSILE COMMAND, REDSTONE ARSENAL, ALA.	AMC-RS-TM-(YEAR)-
ARMY MISSILE COMMAND, REDSTONE ARSENAL, ALA.	AMC-TIR-
ARMY MISSILE COMMAND, REDSTONE ARSENAL, ALA.	AMC-TR-
ARMY MISSILE COMMAND, REDSTONE ARSENAL, ALA.	EPR-
ARMY MISSILE COMMAND, REDSTONE ARSENAL, ALA.	ETO TM-
ARMY MISSILE COMMAND, REDSTONE ARSENAL, ALA.	QSA-H-
ARMY MISSILE COMMAND, REDSTONE ARSENAL, ALA.	QSA-N-
ARMY MISSILE COMMAND, REDSTONE ARSENAL, ALA.	RCS-SMIQ-
ARMY MISSILE COMMAND, REDSTONE ARSENAL, ALA.	RN-TR-
ARMY MISSILE COMMAND, REDSTONE ARSENAL, ALA.	TR-AMSMI-RNR-
ARMY MISSILE COMMAND, REDSTONE ARSENAL, ALA.	USAMICOM POMM-
ARMY MISSILE COMMAND, REDSTONE ARSENAL, ALA.	USAMICOM-QSA-(LETTER)-
ARMY MISSILE COMMAND, REDSTONE ARSENAL, ALA.	USAMICOM-U-
ARMY MISSILE COMMAND. ADVANCED SENSORS LAB., REDSTONE ARSENAL, ALA.	RE-TM-(YEAR)-
ARMY MISSILE COMMAND. ADVANCED SENSORS LAB., REDSTONE ARSENAL, ALA.	RE-TN-(YEAR)-
ARMY MISSILE COMMAND. ADVANCED SENSORS LAB., REDSTONE ARSENAL, ALA.	RE-TSR-
ARMY MISSILE COMMAND. ADVANCED SYSTEMS LAB., REDSTONE ARSENAL, ALA. (TECHNICAL REPORT)	AMC-RA-RD-TR-(YEAR)-
ARMY MISSILE COMMAND. ADVANCED SYSTEMS LAB., REDSTONE ARSENAL, ALA. (TECHNICAL MEMORANDUM)	AMC-RD-TM-(YEAR)-
ARMY MISSILE COMMAND. ADVANCED SYSTEMS LAB., REDSTONE ARSENAL, ALA. (TECHNICAL REPORT)	AMC-RD-TR-(YEAR)-
ARMY MISSILE COMMAND. ADVANCED SYSTEMS LAB., REDSTONE ARSENAL, ALA. (TECHNICAL MEMORANDUM)	RD-TM-(YEAR)-
ARMY MISSILE COMMAND. ADVANCED SYSTEMS LAB., REDSTONE ARSENAL, ALA. (TECHNICAL NOTE)	RD-TN-(YEAR)-
ARMY MISSILE COMMAND. ADVANCED SYSTEMS LAB., REDSTONE ARSENAL, ALA. (TECHNICAL PROGRESS REPORT)	RD-TPR-(YEAR)-
ARMY MISSILE COMMAND. ADVANCED SYSTEMS LAB., REDSTONE ARSENAL, ALA.	RD-TR-(NUMBER)-(YEAR)
ARMY MISSILE COMMAND. AEROBALLISTICS DIRECTORATE, REDSTONE ARSENAL, ALA. (TECHNICAL REPORT)	RD-TR-(YEAR)-
ARMY MISSILE COMMAND. ARMY INERTIAL GUIDANCE + CONTROL LAB., REDSTONE ARSENAL, ALA. (TECHNICAL MEMO.)	AMC-RG-TM-(YEAR)-
ARMY MISSILE COMMAND. ARMY INERTIAL GUIDANCE + CONTROL LAB., REDSTONE ARSENAL, ALA. (TECHNICAL NOTE)	AMC-RG-TN-(YEAR)-
ARMY MISSILE COMMAND. ARMY INERTIAL GUIDANCE + CONTROL LAB., REDSTONE ARSENAL, ALA. (TECHNICAL REPT.)	AMC-RG-TR-(YEAR)-
ARMY MISSILE COMMAND. ARMY PROPULSION LAB. AND CENTER, REDSTONE ARSENAL, ALA.	RK-SR-
ARMY MISSILE COMMAND. ARMY PROPULSION LAB. AND CENTER, REDSTONE ARSENAL, ALA.	RK-TM-(YEAR)-
ARMY MISSILE COMMAND. ARMY PROPULSION LAB. AND CENTER, REDSTONE ARSENAL, ALA.	RK-TN-
ARMY MISSILE COMMAND. ARMY PROPULSION LAB. AND CENTER, REDSTONE ARSENAL, ALA.	RK-TR-(YEAR)-
ARMY MISSILE COMMAND. ARMY PROPULSION LAB. AND CENTER, REDSTONE ARSENAL, ALA.	RK-TSR-
ARMY MISSILE COMMAND. ARPA DIV., REDSTONE ARSENAL, ALA. (TECHNICAL REPORT)	AMC-RN-TR-(YEAR)-

ARMY MISSILE COMMAND. DEV. DIV., REDSTONE ARSENAL

ARMY MISSILE COMMAND. DEV.DIV., REDSTONE ARSENAL, ALA	RH-TR-(YEAR)-
ARMY MISSILE COMMAND. ELECTROMAGNETICS LAB., REDSTONE ARSENAL, ALA. (TECHNICAL MEMORANDUM)	AMC-RE-TM-(YEAR)-
ARMY MISSILE COMMAND. ELECTROMAGNETICS LAB., REDSTONE ARSENAL, ALA. (TECHNICAL REPORT)	AMC-RE-TR-(YEAR)-
ARMY MISSILE COMMAND. ELECTROMAGNETICS LAB., REDSTONE ARSENAL, ALA. (PROPOSAL)	RE-PR-(YEAR)-
ARMY MISSILE COMMAND. ELECTROMAGNETICS LAB., REDSTONE ARSENAL, ALA.	RE-S-(YEAR)-
ARMY MISSILE COMMAND. ELECTROMAGNETICS LAB., REDSTONE ARSENAL, ALA. (SPECIAL REPORT)	RE-SR-(YEAR)-
ARMY MISSILE COMMAND. ELECTROMAGNETICS LAB., REDSTONE ARSENAL, ALA. (TECHNICAL MEMORANDUM)	RE-TM-(YEAR)-
ARMY MISSILE COMMAND. ELECTROMAGNETICS LAB., REDSTONE ARSENAL, ALA. (TECHNICAL NOTE)	RE-TN-(YEAR)-
ARMY MISSILE COMMAND. ELECTROMAGNETICS LAB., REDSTONE ARSENAL, ALA.	RL-TN-(YEAR)-
ARMY MISSILE COMMAND. ENG. AND QUALITY ASSURANCE DIV. REDSTONE ARSENAL, ALA.	IE-TR-(YEAR)-
ARMY MISSILE COMMAND, ENGINEERING REQUIREMENTS OFFICE, REDSTONE ARSENAL, ALA.	AMC-RC-S-(YEAR)-
ARMY MISSILE COMMAND, ENGINEERING REQUIREMENTS OFFICE, REDSTONE ARSENAL, ALA.	RC-S-(YEAR)-
ARMY MISSILE COMMAND. FUTURE MISSILE SYSTEMS DIV., REDSTONE ARSENAL, ALA. (TECHNICAL REPORT)	AMC-RF-TR-(YEAR)-
ARMY MISSILE COMMAND. FUTURE MISSILE SYSTEMS DIV., REDSTONE ARSENAL, ALA.	RF-(NUMBER)
ARMY MISSILE COMMAND. FUTURE MISSILE SYSTEMS DIV., REDSTONE ARSENAL, ALA. (TECHNICAL MEMORANDUM)	RF-TM-(YEAR)-
ARMY MISSILE COMMAND. FUTURE MISSILE SYSTEMS DIV., REDSTONE ARSENAL, ALA. (TECHNICAL NOTE)	RF-TN-(YEAR)-
ARMY MISSILE COMMAND. FUTURE MISSILE SYSTEMS DIV., REDSTONE ARSENAL, ALA. (TECHNICAL REPORT)	RF-TR-(YEAR)-
ARMY MISSILE COMMAND. GROUND SUPPORT EQUIPMENT LAB., REDSTONE ARSENAL, ALA. (TECHNICAL REPORT)	AMC-RL-TR-(YEAR)-
ARMY MISSILE COMMAND. GROUND SUPPORT EQUIPMENT LAB., REDSTONE ARSENAL, ALA. (TECHNICAL MEMORANDUM)	RL-TM-(YEAR)-
ARMY MISSILE COMMAND. GROUND SUPPORT EQUIPMENT LAB., REDSTONE ARSENAL, ALA. (TECHNICAL NOTE)	RL-TN-(YEAR)-
ARMY MISSILE COMMAND. GROUND SUPPORT EQUIPMENT LAB., REDSTONE ARSENAL, ALA. (TECHNICAL PROGRESS REPORT)	RL-TPR-(YEAR)-
ARMY MISSILE COMMAND. GROUND SUPPORT EQUIPMENT LAB., REDSTONE ARSENAL, ALA. (TECHNICAL REPORT)	RL-TR-(YEAR)-
ARMY MISSILE COMMAND. MISSILE INTELLIGENCE DIRECTORATE, REDSTONE ARSENAL, ALA.	MID-CR-
ARMY MISSILE COMMAND. MISSILE INTELLIGENCE DIRECTORATE, REDSTONE ARSENAL, ALA.	ST-CS-
ARMY MISSILE COMMAND. PHYSICAL SCIENCES LAB.,REDSTONE ARSENAL, ALA. (TECHNICAL MEMORANDUM)	AMC-RR-TM-(YEAR)-
ARMY MISSILE COMMAND. PHYSICAL SCIENCES LAB.,REDSTONE ARSENAL, ALA. (TECHNICAL REPORT)	AMC-RR-TR-(YEAR)-
ARMY MISSILE COMMAND. PHYSICAL SCIENCES LAB.,REDSTONE ARSENAL, ALA. (TECHNICAL MEMORANDUM)	RR-TM-(YEAR)-
ARMY MISSILE COMMAND. PHYSICAL SCIENCES LAB.,REDSTONE ARSENAL, ALA. (TECHNICAL NOTE)	RR-TN-(YEAR)-
ARMY MISSILE COMMAND. PHYSICAL SCIENCES LAB.,REDSTONE ARSENAL, ALA. (TECHNICAL REPORT)	RR-TR-(YEAR)-
ARMY MISSILE COMMAND. PROPULSION LAB., REDSTONE ARSENAL, ALA. (PROPOSAL)	RK-PR-(YEAR)-
ARMY MISSILE COMMAND. PROPULSION LAB., REDSTONE ARSENAL, ALA. (TECHNICAL NOTE)	RK-TN-(YEAR)-
ARMY MISSILE COMMAND. PROPULSION LAB., REDSTONE ARSENAL, ALA. (TECHNICAL PROGRESS REPORT)	RK-TPR-(YEAR)-
ARMY MISSILE COMMAND. PROPULSION LAB., REDSTONE ARSENAL, ALA. (TECHNICAL STATUS REPORT)	RK-TSR-(YEAR)-
ARMY MISSILE COMMAND. QUALITY AND RELIABILITY MANAGEMENT OFFICE, REDSTONE ARSENAL, ALA.	QS-OR-(YEAR)-
ARMY MISSILE COMMAND. QUALITY AND RELIABILITY MANAGEMENT OFFICE, REDSTONE ARSENAL, ALA.	QS-TSR-(YEAR)-
ARMY MISSILE COMMAND. RESEARCH AND ENGINEERING DIRECTORATE, REDSTONE ARSENAL, ALA. (PROVISIONAL)	RE-TR-
ARMY MISSILE COMMAND. STRUCTURES AND MECHANICS LAB., REDSTONE ARSENAL, ALA. (TECHNICAL REPORT)	AMC-RS-TR-(YEAR)-
ARMY MISSILE COMMAND. STRUCTURES AND MECHANICS LAB., REDSTONE ARSENAL, ALA. (TECHNICAL MEMORANDUM)	RS-TM-(YEAR)-
ARMY MISSILE COMMAND. STRUCTURES AND MECHANICS LAB., REDSTONE ARSENAL, ALA. (TECHNICAL NOTE)	RS-TN-(YEAR)-
ARMY MISSILE COMMAND. STRUCTURES AND MECHANICS LAB., REDSTONE ARSENAL, ALA.	RS-TR-(YEAR)-
ARMY MISSILE COMMAND. SYSTEMS ENGINEERING DIV., REDSTONE ARSENAL, ALA.	LCE-TR-(YEAR)-
ARMY MISSILE COMMAND. SYSTEMS PERFORMANCE ANALYSIS DIV., REDSTONE ARSENAL, ALA.	QS-TSR-(YEAR)-
ARMY MISSILE COMMAND. TEST AND RELIABILITY EVALUATIONS LAB., REDSTONE ARSENAL, ALA. (TECHNICAL REPT)	AMC-RT-TR-(YEAR)-
ARMY MISSILE COMMAND. TEST AND RELIABILITY EVALUATIONS LAB., REDSTONE ARSENAL, ALA. (BROCHURE)	RT-BR-(YEAR)-
ARMY MISSILE COMMAND. TEST AND RELIABILITY EVALUATIONS LAB., REDSTONE ARSENAL, ALA. (TECH. MEMO.)	RT-TM-(YEAR)-
ARMY MISSILE COMMAND. TEST AND RELIABILITY EVALUATIONS LAB., REDSTONE ARSENAL, ALA. (TECH. NOTE)	RT-TN-(YEAR)-
ARMY MISSILE COMMAND. TEST AND RELIABILITY EVALUATIONS LAB., REDSTONE ARSENAL, ALA. (TEST PLAN)	RT-TP-(YEAR)-
ARMY MISSILE COMMAND. TEST AND RELIABILITY EVALUATIONS LAB., REDSTONE ARSENAL, ALA. (TECHNICAL REPT)	RT-TR-(YEAR)-
ARMY MISSILE TEST AND EVALUATION DIRECTORATE, WHITE SANDS MISSILE RANGE, N. MEX.	ARMTE-IMPROVED HAWK-E-
ARMY MISSILE TEST AND EVALUATION DIRECTORATE, WHITE SANDS MISSILE RANGE, N. MEX.	ARMTE-IMPROVED HAWK-SE-
ARMY MISSILE TEST AND EVALUATION DIRECTORATE, WHITE SANDS MISSILE RANGE, N. MEX.	ARMTE-TOW-
ARMY MISSILE TEST AND EVALUATION DIRECTORATE, WHITE SANDS MISSILE RANGE, N. MEX.	EMR-
ARMY MISSILE TEST AND EVALUATION DIRECTORATE, WHITE SANDS MISSILE RANGE, N. MEX.	NIKE-X-SR-
ARMY MISSILE TEST AND EVALUATION DIRECTORATE, WHITE SANDS MISSILE RANGE, N. MEX.	RACFO-
ARMY MISSILE TEST AND EVALUATION DIRECTORATE, WHITE SANDS MISSILE RANGE, N. MEX.	WSMR/AMTED-
ARMY MISSILE TEST AND EVALUATION DIRECTORATE, WHITE SANDS MISSILE RANGE, N. MEX.	WSMR-AMTED-EMR-
ARMY MISSILE TEST AND EVALUATION DIRECTORATE, WHITE SANDS MISSILE RANGE, N. MEX.	WSMR-AMTED-FDS-(YEAR)-
ARMY MISSILE TEST AND EVALUATION DIRECTORATE, WHITE SANDS MISSILE RANGE, N. MEX.	WSMR-AMTED-HAWK-(YR.)-
ARMY MISSILE TEST AND EVALUATION DIRECTORATE, WHITE SANDS MISSILE RANGE, N. MEX.	WSMR-AMTED-HJ-
ARMY MISSILE TEST AND EVALUATION DIRECTORATE, WHITE SANDS MISSILE RANGE, N. MEX.	WSMR/ARMTE-
ARMY MISSILE TEST AND EVALUATION DIRECTORATE, WHITE SANDS MISSILE RANGE, N. MEX.	WSMR-ARMTE-FDS-(YEAR)-
ARMY MISSILE TEST AND EVALUATION DIRECTORATE. NUCLEAR WEAPON EFFECTS DIV. WHITE SANDS MISSILE RANGE, N. MEX.	DOSIMETRY-TR-(NO.-NO.)
ARMY MISSILE TEST CTR.,WHITE SANDS MISSILE RANGE, NM	AMTC-
ARMY MISSILE TEST CTR.,WHITE SANDS MISSILE RANGE, NM	AMTC SPEC. RPT.-
ARMY MISSILE TEST CTR.,WHITE SANDS MISSILE RANGE, NM	AMTC-TB-
ARMY MISSILE TEST CTR.,WHITE SANDS MISSILE RANGE, NM	AMTC-TM-
ARMY MISSILE TEST CTR.,WHITE SANDS MISSILE RANGE, NM	AMTC-TR-
ARMY MISSILE TEST CTR.,WHITE SANDS MISSILE RANGE, NM	AMTC-WS-TM-
ARMY MISSILE TEST CTR.,WHITE SANDS MISSILE RANGE, NM	TIB-(NUMBER)-(YEAR)
ARMY MISSILE TEST CTR.,WHITE SANDS MISSILE RANGE, NM	WS-AMTC-TM-
ARMY MISSILE TEST CENTER, WHITE SANDS MISSILE RANGE, N. MEX. (MONTHLY INTERIM PROGRESS REPORTS)	WSMR-MIPR-
ARMY MISSILE TEST CENTER, WHITE SANDS MISSILE RANGE, N. MEX. (MONTHLY INTERIM PROGRESS REPORTS)	WSMR-PR-
ARMY MISSILE TEST CENTER, WHITE SANDS MISSILE RANGE, N. MEX. (SPECIAL REPORTS)	WSMR-SR-
ARMY MISSILE TEST CENTER, WHITE SANDS MISSILE RANGE, N. MEX. (TEST BULLETINS)	WSMR-TB-
ARMY MISSILE TEST CENTER, WHITE SANDS MISSILE RANGE, N. MEX. (TECHNICAL MEMORANDUM)	WSMR-TM-
ARMY MISSILE TEST CENTER, WHITE SANDS MISSILE RANGE, N. MEX. (TECHNICAL REPORT)	WSMR-TR-
ARMY MISSILE TEST CTR.,WHITE SANDS MISSILE RANGE, NM	WS-TB-
ARMY MOBILITY COMMAND. DETROIT ARSENAL	AMC/MOB-
ARMY MOBILITY EQUIPMENT COMMAND, FORT BELVOIR, VA.	USAMEC-
ARMY MOBILITY EQUIPMENT COMMAND, FORT BELVOIR, VA.	USAMERDC-
ARMY MOBILITY EQUIPMENT COMMAND, FORT BELVOIR, VA.	USAMERDC-T-
ARMY MOBILITY EQUIPMENT RES. AND DEV. CENTER, FORT BELVOIR, VA.	AMERDC-
ARMY MOBILITY EQUIPMENT RES. AND DEV. CENTER, FORT BELVOIR, VA.	USAMERDC-
ARMY MOBILITY EQUIPMENT RES. AND DEV. CENTER, FORT BELVOIR, VA.	USAMERDC-M(NO.)-
ARMY MOBILITY EQUIPMENT RES. AND DEV. CENTER, FORT BELVOIR, VA.	USAMERDC-T-(NO.)-
ARMY MOBILITY EQUIPMENT RES. AND DEV. CENTER, FORT BELVOIR, VA.	USAMERDC-TR-
ARMY MUNITIONS COMMAND, DOVER, N.J.	USAMUCOM-(YEAR)-
ARMY MUNITIONS COMMAND, DOVER, N.J.	USAMUCOM-COSTECH-

ARMY MUNITIONS COMMAND. PICATINNY ARSENAL, N.J.	AMC/MUN-
ARMY NATICK LABS., MASS.	ANL-R.C.-
ARMY NATICK LABS., MASS.	CD-
ARMY NATICK LABS., MASS.	(YEAR)-(NUMBER)-CM-
ARMY NATICK LABS., MASS.	EPT-
ARMY NATICK LABS., MASS.	ME-
ARMY NATICK LABS., MASS.	USA-NLABS-
ARMY NATICK LABS., MASS.	USA-NLABS-TR-
ARMY NATICK LABS., MASS.	USA-NLABS-TR-(YEAR)-
ARMY NATICK LABS. AIR DELIVERY EQUIPMENT DIV., MASS.	ADED-
ARMY NATICK LABS. AIRDROP ENGINEERING DIV., MASS.	TR-(YEAR)-(NUMBER)-AD
ARMY NATICK LABS. APPLIED ENTOMOLOGY GROUP, MASS.	AE-
ARMY NATICK LABS. CLOTHING + EQUIP. DEV. BR., MASS.	C/ED-
ARMY NATICK LABS. CLOTHING AND MATERIALS DIV., MASS.	MR+E-(YEAR)-
ARMY NATICK LABS. CLOTHING AND ORGANIC MATERIALS DIV., MASS.	C/ED-
ARMY NATICK LABS. CLOTHING AND ORGANIC MATERIALS DIV., MASS.	C+OM-
ARMY NATICK LABS. CLOTHING AND ORGANIC MATERIALS DIV., MASS.	C/OM-(NUMBER)-
ARMY NATICK LABS. CLOTHING AND ORGANIC MATERIALS DIV., MASS.	C/OM-C/ED-
ARMY NATICK LABS. CLOTHING AND ORGANIC MATERIALS DIV., MASS.	C/OM-MR/E-(YEAR)-
ARMY NATICK LABS. CLOTHING AND ORGANIC MATERIALS DIV., MASS.	TR-(YEAR)-(NUMBER)-CM
ARMY NATICK LABS. CLOTHING AND ORGANIC MATERIALS DIV., MASS.	TR-C+OM-
ARMY NATICK LABS. CLOTHING AND ORGANIC MATERIALS LAB., MASS.	ANL-C+OM-
ARMY NATICK LABS. CLOTHING AND ORGANIC MATERIALS LAB., MASS.	C/OM-
ARMY NATICK LABS. CLOTHING AND ORGANIC MATERIALS LAB., MASS.	C/OM-ER-
ARMY NATICK LABS. CLOTHING AND ORGANIC MATERIALS LAB., MASS.	C/OM-TR-
ARMY NATICK LABS. CLOTHING AND ORGANIC MATERIALS LAB., MASS.	C/OM-TS-
ARMY NATICK LABS. CLOTHING AND PERSONAL LIFE SUPPORT EQUIPMENT LAB., MASS.	ANL-C+PLSEL-
ARMY NATICK LABS. CLOTHING AND PERSONAL LIFE SUPPORT EQUIPMENT LAB., MASS.	C/PLSEL-
ARMY NATICK LABS. CLOTHING AND PERSONAL LIFE SUPPORT EQUIPMENT LAB., MASS.	C/PLSEL-TS-
ARMY NATICK LABS. CLOTHING AND PERSONAL LIFE SUPPORT EQUIPMENT LAB., MASS.	C/PSEL-TS-
ARMY NATICK LABS. EARTH SCIENCES DIV., MASS.	ES-
ARMY NATICK LABS. EARTH SCIENCES DIV., MASS.	FD-
ARMY NATICK LABS. EARTH SCIENCES DIV., MASS.	SR-S-
ARMY NATICK LABS. EARTH SCIENCES DIV., MASS.	TR-(YEAR)-(NUMBER)-ES
ARMY NATICK LABS. EARTH SCIENCES DIV., MASS.	TR-ES-
ARMY NATICK LABS. FOOD DIV., MASS.	FD-
ARMY NATICK LABS. FOOD DIV., MASS.	TR-(YEAR)-(NUMBER)-FD
ARMY NATICK LABS. FOOD DIV., MASS.	USA-NLABS-FD-(YEAR)-
ARMY NATICK LABS. FOOD LAB., MASS.	FL-
ARMY NATICK LABS. FOOD LAB., MASS.	USANLABS-TR-(YEAR)-
ARMY NATICK LABS. MECHANICAL ENGINEERING DIV., MASS.	MHE-
ARMY NATICK LABS. PIONEERING RESEARCH DIV., MASS.	EPR-
ARMY NATICK LABS. PIONEERING RESEARCH DIV., MASS.	TR-(YEAR)-(NUMBER)-PR
ARMY NATICK LABS. PIONEERING RESEARCH DIV., MASS.	USA-NLABS-TN-(YEAR)-
ARMY NATICK LABS. TECHNICAL LIBRARY, MASS.	BIBLIOGRAPHIC-(YEAR)-
ARMY-NAVY (REF. 13)	AN-
ARMY-NAVY (AERONAUTICAL BULLETIN) (REF. 13)	AN AB-
ARMY-NAVY (DRAWING) (REF. 13)	AND-
ARMY-NAVY (REF. 13)	JAN-
ARMY-NAVY (SPECIFICATION) (REF. 13)	JAN-(LETTER)-
ARMY-NAVY (STANDARD) (REF. 13)	JAN-STD-
ARMY-NAVY (WORK MANUAL) (REF. 13)	JAN WM-
ARMY-NAVY-AIR FORCE (COVERS VARIOUS COMMITTEES AND THEIR PUBLICATIONS) (REF. 13)	JANAF-
ARMY-NAVY-AIR FORCE (INTERIM THERMOCHEMICAL TABLES) (REF. 13)	JANAF-ITT-
ARMY-NAVY EXPLOSIVES SAFETY BOARD, WASHINGTON, D.C.	ANESB-
ARMY-NAVY EXPLOSIVES SAFETY BD.,WASH.,DC (TECH.PAPER)	ANESB TP-
ARMY NUCLEAR DEFENSE LAB., EDGEWOOD ARSENAL, MD.	NDL-(YEAR)-(LTR.)-
ARMY NUCLEAR DEFENSE LAB., EDGEWOOD ARSENAL, MD.	NDL-SP-
ARMY NUCLEAR DEFENSE LAB., EDGEWOOD ARSENAL, MD.	NDL-TM-
ARMY NUCLEAR DEFENSE LAB., EDGEWOOD ARSENAL, MD.	NDL-TR-
ARMY NUCLEAR DEFENSE LAB., EDGEWOOD ARSENAL, MD.	NDL-U-FY-(YEAR)
ARMY ORDNANCE AMMUNITION COMMAND, JOLIET, ILL.	OAC-
ARMY ORDNANCE AMMUNITION COMMAND, JOLIET, ILL.	OAC/(LETTERS)
ARMY ORDNANCE CENTER AND SCHOOL, ABERDEEN PROVING GROUND, MD.	ST-(NUMBER)-
ARMY ORDNANCE GUIDED MISSILE SCHOOL	AOGMS-
ARMY ORDNANCE GUIDED MISSILE SCHOOL (LIAISON STATUS REPORT)	OGMS-LSR-
ARMY ORDNANCE LIAISON OFFICE, ALBUQUERQUE, N. MEX.	OLO-
ARMY ORDNANCE MISSILE COMMAND, REDSTONE ARSENAL, ALA.	AOMC-
ARMY ORDNANCE MISSILE COMMAND, REDSTONE ARSENAL, ALA.	AOMC-RA-TR-(YEAR)-
ARMY ORDNANCE MISSILE COMMAND, REDSTONE ARSENAL, ALA.	AOMC-RE-TR-(YEAR)-
ARMY ORDNANCE MISSILE COMMAND, REDSTONE ARSENAL, ALA.	AOMC-RF-TR-(YEAR)-
ARMY ORDNANCE MISSILE COMMAND, REDSTONE ARSENAL, ALA.	AOMC-RM-TR-(YEAR)-
ARMY ORDNANCE MISSILE COMMAND, REDSTONE ARSENAL, ALA.	AOMC-RR-
ARMY ORDNANCE MISSILE COMMAND, REDSTONE ARSENAL, ALA.	AOMC-RR-TR-
ARMY ORDNANCE MISSILE COMMAND, REDSTONE ARSENAL, ALA.	AOMC-RT-TR-(YEAR)-
ARMY ORDNANCE MISSILE COMMAND, REDSTONE ARSENAL, ALA.	CAA-
ARMY ORDNANCE SPECIAL WEAPONS-AMMUNITION COMMAND, DOVER, N.J.	AOSWAC-
ARMY ORDNANCE SPECIAL WEAPONS-AMMUNITION COMMAND, DOVER, N.J.	ORDSW-
ARMY ORDNANCE SPECIAL WEAPONS-AMMUNITION COMMAND, DOVER, N.J.	OSWAC-(LETTERS)-
ARMY ORDNANCE WEAPONS COMMAND, ROCK ISLAND, ILL.	AOWC-
ARMY PERSONNEL RESEARCH OFFICE, WASHINGTON, D.C.	APRO-TRN-
ARMY PERSONNEL RESEARCH OFFICE, WASHINGTON, D.C.	APRO-TRW-
ARMY PERSONNEL RESEARCH OFFICE, WASHINGTON, D.C.	TRN-

ARMY PERSONNEL RESEARCH OFFICE

ARMY PERSONNEL RESEARCH OFFICE, WASHINGTON, D.C.	TRR-
ARMY POLAR RESEARCH AND DEV. CTR., FORT BELVOIR, VA.	APRDC-
ARMY PRODUCTION EQUIPMENT AGENCY, ROCK ISLAND, ILL.	APEA-
ARMY RESEARCH AND DEVELOPMENT GROUP (EUROPE), N.Y.C.	AEDG(E)-E-
ARMY RESEARCH AND DEVELOPMENT GROUP (EUROPE), N.Y.C.	ARDG (E)-
ARMY RESEARCH AND DEVELOPMENT GROUP (EUROPE), N.Y.C.	ARDG(E)-E-
ARMY RESEARCH AND DEVELOPMENT GROUP (EUROPE), N.Y.C.	CRDARE-
ARMY RESEARCH AND DEV. GP. (FAR EAST), SAN FRANCISCO	ARDG (FE)-
ARMY RESEARCH AND DEV. GP. (FAR EAST), SAN FRANCISCO	ARDG(FE)-J-(NO.)-
ARMY RESEARCH INST. OF ENVIRONMENTAL MEDICINE, NATICK, MASS.	RCS-MEDDH-
ARMY RESEARCH INST. OF ENVIRONMENTAL MEDICINE, NATICK, MASS.	USARIEM-TR-(YEAR)-
ARMY RESEARCH OFFICE, DURHAM, N.C.	AORD-(NO.-NO.-LTR.)
ARMY RESEARCH OFFICE, DURHAM, N.C.	AROD-
ARMY RESEARCH OFFICE, DURHAM, N.C.	AROD-(NUMBERONUMBER)-C
ARMY RESEARCH OFFICE, DURHAM, N.C.	AROD-(NO.-NO.-LTR.)
ARMY RESEARCH OFFICE, DURHAM, N.C.	AROD-I-(NUMBERONO.)-RT
ARMY RESEARCH OFFICE, DURHAM, N.C.	ARODR-
ARMY RESEARCH OFFICE, DURHAM, N.C.	AROD-T-(NUMBERONO.)-RT
ARMY RESEARCH OFFICE, WASHINGTON, D.C.	ARO-
ARMY RESEARCH OFFICE, WASHINGTON, D.C.	ARO-ITR-
ARMY ROCKET AND GUIDED MISSILE AGENCY, REDSTONE ARSENAL, ALA.	2A-
ARMY ROCKET AND GUIDED MISSILE AGENCY, REDSTONE ARSENAL, ALA.	ARGMA-
ARMY ROCKET AND GUIDED MISSILE AGENCY, REDSTONE ARSENAL, ALA.	ARGMA-MSP-
ARMY ROCKET AND GUIDED MISSILE AGENCY, REDSTONE ARSENAL, ALA.	ARGMA-PUB-
ARMY ROCKET AND GUIDED MISSILE AGENCY, REDSTONE ARSENAL, ALA.	ARGMA-RCS-...
ARMY ROCKET AND GUIDED MISSILE AGENCY, REDSTONE ARSENAL, ALA.	ARGMA-RHA-
ARMY ROCKET AND GUIDED MISSILE AGENCY, REDSTONE ARSENAL, ALA.	ARGMA-RHM-Z-
ARMY ROCKET AND GUIDED MISSILE AGENCY, REDSTONE ARSENAL, ALA.	ARGMA-TM-
ARMY ROCKET AND GUIDED MISSILE AGENCY, REDSTONE ARSENAL, ALA.	ARGMA-TN-
ARMY ROCKET AND GUIDED MISSILE AGENCY, REDSTONE ARSENAL, ALA.	ARGMA-TR-
ARMY ROCKET AND GUIDED MISSILE AGENCY, REDSTONE ARSENAL, ALA.	2R-
ARMY ROCKET AND GUIDED MISSILE AGENCY, REDSTONE ARSENAL, ALA.	3R-
ARMY ROCKET AND GUIDED MISSILE AGENCY, REDSTONE ARSENAL, ALA.	RHM-Z-(NO.)-(YR.)
ARMY SAFEGUARD SYSTEM COMMAND, HUNTSVILLE, ALA.	ASSC-
ARMY SAFEGUARD SYSTEM COMMAND, HUNTSVILLE, ALA.	SSMP-(NUMBER. NO. LTR.)
ARMY SAFEGUARD SYSTEM COMMAND, HUNTSVILLE, ALA. (SAFEGUARD SYSTEM MASTER PLAN)	SSMP-
ARMY SCIENTIFIC ADVISORY PANEL	CRD-
ARMY SERVICE FORCES	ASF-
ARMY SIGNAL ENGINEERING LABS., FORT MONMOUTH, N.J.	ASEL-
ARMY SIGNAL ENGINEERING LABS., FORT MONMOUTH, N.J.	E-
ARMY SIGNAL ENGINEERING LABS., FORT MONMOUTH, N.J.	SEL-
ARMY SIGNAL ENGINEERING LABS., FORT MONMOUTH, N.J.	SEL TM M-
ARMY SIGNAL MISSILE SUPPORT AGENCY, WHITE SANDS MISSILE RANGE, N. MEX.	ASMSA-
ARMY SIGNAL MISSILE SUPPORT AGENCY, WHITE SANDS MISSILE RANGE, N. MEX. (MISSILE ELECTRONIC WARFARE)	ASMSA-MEW-
ARMY SIGNAL MISSILE SUPPORT AGENCY, WHITE SANDS MISSILE RANGE, N. MEX. (MISSILE METEOROLOGY)	ASMSA-MM-
ARMY SIGNAL MISSILE SUPPORT AGENCY, WHITE SANDS MISSILE RANGE, N. MEX.	RE-S-(YEAR)-
ARMY SIGNAL MISSILE SUPPORT AGENCY, WHITE SANDS MISSILE RANGE, N. MEX.	SMS-
ARMY SIGNAL MISSILE SUPPORT AGENCY, WHITE SANDS MISSILE RANGE, N. MEX.	SMSA-
ARMY SIGNAL MISSILE SUPPORT AGENCY, WHITE SANDS MISSILE RANGE, N. MEX. (PROGRESS REPORTS)	SMSA-PR-
ARMY SIGNAL MISSILE SUPPORT AGENCY, WHITE SANDS MISSILE RANGE, N. MEX. (TECHNICAL MEMORANDA)	SMSA-TM-
ARMY SIGNAL MISSILE SUPPORT AGENCY, WHITE SANDS MISSILE RANGE, N. MEX.	SMSA-TR-
ARMY SIGNAL MISSILE SUPPORT AGENCY, WHITE SANDS MISSILE RANGE, N. MEX.	WS-AFC-
ARMY SIGNAL MISSILE SUPPORT AGENCY, WHITE SANDS MISSILE RANGE, N. MEX.	WS-SMSA-
ARMY SIGNAL MISSILE SUPPORT AGENCY, WHITE SANDS MISSILE RANGE, N. MEX.	WS-SR-
ARMY SIGNAL MISSILE SUPPORT AGENCY. MISSILE ELECTRONIC WARFARE DIV., WHITE SANDS MISSILE RANGE, NM	MEWD-
ARMY SIGNAL MISSILE SUPPORT AGENCY. MISSILE ELECTRONIC WARFARE DIV., WHITE SANDS MISSILE RANGE, NM	SMSA-MEWD-
ARMY SIGNAL MISSILE SUPPORT AGENCY. MISSILE GEOPHYSICS DIV., WHITE SANDS MISSILE RANGE, N. MEX.	ASMSA-MGD-
ARMY SIGNAL MISSILE SUPPORT AGENCY. MISSILE METEOROLOGY DIV., WHITE SANDS MISSILE RANGE, N. MEX.	SMSA-DR-LR-
ARMY SIGNAL RADIO PROPAGATION AGENCY, FT. MONMOUTH, NJ	ASRPA-
ARMY SIGNAL RADIO PROPAGATION AGENCY, FT. MONMOUTH, NJ	RPA-I-(LETTER)-
ARMY SIGNAL RADIO PROPAGATION AGENCY, FT. MONMOUTH, NJ	RPA-TR-
ARMY SIGNAL RADIO PROPAGATION AGENCY, FT. MONMOUTH, NJ	RPU-TR-
ARMY SIGNAL RADIO PROPAGATION AGENCY, FT. MONMOUTH, NJ	SPRA-
ARMY SIGNAL RES. AND DEV. AGENCY, FT. MONMOUTH, N.J.	SRDA-
ARMY SIGNAL RESEARCH + DEV. LAB., FORT MONMOUTH, N.J.	ASRDL-
ARMY SIGNAL RESEARCH + DEV. LAB., FORT MONMOUTH, N.J.	SRDL-
ARMY SIGNAL RESEARCH AND DEVELOPMENT LAB., FORT MONMOUTH, N.J. (RESEARCH AND DEVELOPMENT SUMMARY)	SR+DL R+D SUMMARY...
ARMY SIGNAL RESEARCH + DEV. LAB., FORT MONMOUTH, N.J.	SRDL-TR-
ARMY SIGNAL RESEARCH + DEV. LAB., FORT MONMOUTH, N.J.	USAEL-TR-
ARMY SIGNAL RESEARCH + DEV. LAB., FORT MONMOUTH, N.J.	USARDL-
ARMY SIGNAL RESEARCH + DEV. LAB., FORT MONMOUTH, N.J.	USARDL-TR-
ARMY SIGNAL RESEARCH + DEV. LAB., FORT MONMOUTH, N.J.	USASRDL-TR-
ARMY SIGNAL SUPPORT AGENCY	USASSA-
ARMY SPECIAL WARFARE SCHOOL, FORT BRAGG, N.C.	ST-(NUMBERS)-
ARMY STRATEGIC COMMUNICATIONS COMMAND, FORT HUACHUCA, ARIZ. (TECHNICAL MANUAL)	CCTM-(NUMBER-NUMBER)
ARMY STRATEGIC COMMUNICATIONS COMMAND, FORT HUACHUCA, ARIZ.	STRATCOM-(NOS.)-(LTR.)
ARMY STRATEGIC COMMUNICATIONS COMMAND, WASHINGTON, DC	USASCC-PAM-(NOS.)-
ARMY STRATEGIC COMMUNICATIONS COMMAND, WASHINGTON, DC	USASCC-TM-(NOS.)(NOS.)-
ARMY STRATEGIC COMMUNICATIONS COMMAND. TEST AND EVALUATION DIRECTORATE, FORT HUACHUCA, ARIZ.	SCC-TED-
ARMY STRATEGY AND TACTICS ANALYSIS GP., BETHESDA, MD.	STAG-M-
ARMY STRATEGY AND TACTICS ANALYSIS GP., BETHESDA, MD.	STAGTEN-
ARMY TANK-AUTOMOTIVE CENTER, WARREN, MICH.	ATAC-
ARMY TANK-AUTOMOTIVE CENTER, WARREN, MICH.	ATAC-RRD-
ARMY TANK-AUTOMOTIVE CENTER, WARREN, MICH.	ATAC-TR-

ARMY TANK-AUTOMOTIVE CENTER, WARREN, MICH. (LAND LOCOMOTION)	LL-
ARMY TANK-AUTOMOTIVE CENTER, WARREN, MICH. (ORDNANCE TANK-AUTOMOTIVE COMMAND)	OTAC-
ARMY TANK-AUTOMOTIVE CENTER. ADVANCED SYSTEMS AND CONCEPT RESEARCH DIV., WARREN, MICH.	RRC-
ARMY TANK-AUTOMOTIVE CENTER. FOREIGN TECHNOLOGY OFFICE, WARREN, MICH.	FIO-
ARMY TANK-AUTOMOTIVE COMMAND, WARREN, MICH.	ATACOM-TR-
ARMY TANK-AUTOMOTIVE COMMAND, WARREN, MICH.	TACOM-S-
ARMY TANK-AUTOMOTIVE COMMAND, WARREN, MICH.	TACOM-TR-
ARMY TANK-AUTOMOTIVE COMMAND. SCIENTIFIC COMPUTER DIV., WARREN, MICH.	SCD-
ARMY TECHNICAL INTELLIGENCE CENTER, TOKYO	ATIC-T-
ARMY TECH. INTELL. CTR. MEDICAL ANALYSIS SEC., TOKYO	ATIC-T MAS-
ARMY TECH. INTELL. CTR. MEDICAL ANALYSIS SEC., TOKYO	MAS-
ARMY TECH. INTELL. CTR. MEDICAL ANALYSIS SEC., TOKYO	T-MAS-
ARMY TECHNICAL INTELLIGENCE CENTER. ORDNANCE (ANALYSIS) SECTION, TOKYO	ATIC-T OS-
ARMY TECHNICAL INTELLIGENCE CENTER. ORDNANCE (ANALYSIS) SECTION, TOKYO	OS-
ARMY TECHNICAL INTELLIGENCE CENTER. ORDNANCE (ANALYSIS) SECTION, TOKYO	T-OS-
ARMY TECHNICAL INTELLIGENCE CENTER. QUARTERMASTER ANALYSIS SECTION, TOKYO	ATIC-T QMAS-
ARMY TECHNICAL INTELLIGENCE CENTER. QUARTERMASTER ANALYSIS SECTION, TOKYO	QMAS-
ARMY TECHNICAL INTELLIGENCE CENTER. QUARTERMASTER ANALYSIS SECTION, TOKYO	T QMAS-
ARMY TECH. INTELL. CTR. SIGNAL ANALYSIS SEC., TOKYO	ATIC-T SAS-
ARMY TECH. INTELL. CTR. SIGNAL ANALYSIS SEC., TOKYO	T SAS-
ARMY TEST AND EVALUATION COMMAND, ABERDEEN PROVING GROUND, MD.	MTP-(NUMBER-NUMBER)-
ARMY TEST AND EVALUATION COMMAND, ABERDEEN PROVING GROUND, MD.	TECOM-(NO.)-(LETTER)
ARMY TEST AND EVALUATION COMMAND, WHITE SANDS MISSILE RANGE, N. MEX.	RE-S-(YEAR)-
ARMY TEST AND EVALUATION COMMAND, WHITE SANDS MISSILE RANGE, N. MEX.	RISO-(NUMBER-YEAR)
ARMY TEST AND EVALUATION COMMAND. DEPUTY FOR NATIONAL RANGE OPERATIONS, WHITE SANDS MISSILE RANGE, N. MEX. (RANGE OPERATIONS)	RO-S-(YEAR)-
ARMY TOPOGRAPHIC COMMAND, WASHINGTON, D.C.	TOPOCOM-ID-
ARMY TOPOGRAPHIC COMMAND, WASHINGTON, D.C.	TOPOCOM-TR-
ARMY TRANSPORTATION BOARD, FORT EUSTIS, VA.	TCB-
ARMY TRANSPORTATION BOARD, FORT EUSTIS, VA.	TRC-
ARMY TRANSPORTATION ENG. AGENCY, FORT EUSTIS, VA.	USATEA-(YEAR)-
ARMY TRANSPORTATION ENG. AGENCY, FORT EUSTIS, VA.	USATEA-TR-
ARMY TRANSPORT. RES. + DEV. COMMAND, FT. EUSTIS, VA.	TRDC-
ARMY TRANSPORT. RES. + DEV. COMMAND, FT. EUSTIS, VA.	USATRECOM-TR-(YEAR)-
ARMY TRANSPORT. RES. + ENG. COMMAND, FT. EUSTIS, VA.	TCREC-
ARMY TRANSPORTATION RESEARCH COMMAND, FORT EUSTIS, VA	CRD-
ARMY TRANSPORTATION RESEARCH COMMAND, FORT EUSTIS, VA	TCREC-TR-
ARMY TRANSPORTATION RESEARCH COMMAND, FORT EUSTIS, VA	TREC-
ARMY TRANSPORTATION RESEARCH COMMAND, FORT EUSTIS, VA	TRECOM-
ARMY TRANSPORTATION RESEARCH COMMAND, FORT EUSTIS, VA	TRECOM-TR-(YEAR)-
ARMY TRANSPORTATION RESEARCH COMMAND, FORT EUSTIS, VA	TREC-TR-(YEAR)-
ARMY TROPIC TEST CENTER, FORT CLAYTON, CANAL ZONE	RR-
ARMY TROPIC TEST CENTER, FORT CLAYTON, CANAL ZONE	SAR-
ARMY TROPIC TEST CENTER, FORT CLAYTON, CANAL ZONE	USATTC-
ARMY WAR COLL., CARLISLE BARRACKS, PENNA.	AWC-
ARMY WEAPONS COMMAND, ROCK ISLAND, ILL.	AMC/WPN-
ARMY WEAPONS COMMAND, ROCK ISLAND, ILL.	PWA-FR-
ARMY WEAPONS COMMAND. COST ANALYSIS OFFICE, ROCK ISLAND, ILL.	AMSWE-CPD-
ARMY WEAPONS COMMAND. COST ANALYSIS OFFICE, ROCK ISLAND, ILL.	COSTECH-
ARMY WEAPONS COMMAND. FUTURE WEAPONS SYSTEMS DIV., ROCK ISLAND, ILL.	REF-
ARMY WEAPONS COMMAND. RESEARCH AND ENGINEERING DIRECTORATE, ROCK ISLAND, ILL.	AMSWE-RE-
ARMY WEAPONS COMMAND. SUPPLY AND MAINTENANCE DIRECTORATE, ROCK ISLAND, ILL.	AMSWE-SM-
ARMY WEAPONS COMMAND. SYSTEMS ANALYSIS DIRECTORATE, ROCK ISLAND, ILL.	SY-R(NUMBER-YEAR)-
ARMY WEAPONS COMMAND. SYSTEMS ANALYSIS DIRECTORATE, ROCK ISLAND, ILL.	SY-TN(NUMBER-YEAR)
ARMY WEAPONS COMMAND. WEAPONS OPERATIONS RESEARCH OFFICE, ROCK ISLAND, ILL.	AMSWE-OR-
ARMY WEAPONS COMMAND. WEAPONS OPERATIONS RESEARCH OFFICE, ROCK ISLAND, ILL.	OR-(YEAR)-
ARNOLD ENGINEERING CO., MARENGO, ILL.	ARNEC-
ARNOLD ENGINEERING CO., N.Y.C.	ARNEC-
ARNOLD ENGINEERING DEV. CTR., ARNOLD A.F. STA., TENN.	AECD-TN-(YEAR)-
ARNOLD ENGINEERING DEV. CTR., ARNOLD A.F. STA., TENN.	AECD-TW-(YEAR)-
ARNOLD ENGINEERING DEV. CTR., ARNOLD A.F. STA., TENN.	AEDC-
ARNOLD ENGINEERING DEV. CTR., ARNOLD A.F. STA., TENN.	AEDC-BIB-(YEAR)-
ARNOLD ENGINEERING DEV. CTR., ARNOLD A.F. STA., TENN. (TECHNICAL DOCUMENTARY REPORT)	EDC-TDR-(YEAR)-
ARNOLD ENGINEERING DEV. CTR., ARNOLD A.F. STA., TENN.	AEDC-TM-(YEAR)-
ARNOLD ENGINEERING DEV. CTR., ARNOLD A.F. STA., TENN.	AEDC-TN-(YEAR)-
ARNOLD ENGINEERING DEV. CTR., ARNOLD A.F. STA., TENN.	AEDC-TR-(YEAR)-
ARNOLD ENGINEERING DEVELOPMENT CENTER. GAS DYNAMICS FACILITY, ARNOLD AIR FORCE STATION, TENN.	AEDC-GDF-
ARNOLD ENGINEERING DEVELOPMENT CENTER. GAS DYNAMICS FACILITY, ARNOLD AIR FORCE STATION, TENN.	GDF-
ARNOLD ENGINEERING DEVELOPMENT CENTER. GROUND DEVELOPMENT FORCES, ARNOLD AIR FORCE STATION, TENN.	AEDC-GDF-
ARNOLD ENGINEERING DEVELOPMENT CENTER. PROPULSION WIND TUNNEL FACILITY, ARNOLD AIR FORCE STA., TENN.	PWT-
ARO, INC., ARNOLD AIR FORCE STATION, TENN.	ARO-
ARO EQUIPMENT CORP., BRYAN, OHIO	ARO-
ARTILLERY INFORMATION SERVICE	AIS-

ARTILLERY SCHOOL

ARTILLERY SCHOOL, FORT SILL, OKLA.	FSO-AS-
ASOCIACION VENEZOLANA PARA EL AVANCE DE LA CIENCIA, CARACAS	AVAC-
ASSISTANT CHIEF OF STAFF FOR FORCE DEVELOPMENT (ARMY) WASHINGTON, D.C.	ACSFOR-
ASSISTANT CHIEF OF STAFF FOR FORCE DEVELOPMENT (ARMY) WASHINGTON, D.C.	OACSFOR-OT-RD-
ASSISTANT CHIEF OF STAFF FOR FORCE DEVELOPMENT (ARMY) WASHINGTON, D.C.	OACSFOR-OT-RD-T(NOS.)
ASSISTANT CHIEF OF STAFF FOR FORCE DEVELOPMENT (ARMY) WASHINGTON, D.C.	OACSFOR-OT-UT-
ASSISTANT CHIEF OF STAFF FOR INTELLIGENCE (ARMY), WASHINGTON, D.C.	ACSI-H-
ASSISTANT CHIEF OF STAFF FOR INTELLIGENCE (ARMY), WASHINGTON, D.C.	ACSI-I-
ASSISTANT CHIEF OF STAFF FOR INTELLIGENCE (ARMY), WASHINGTON, D.C.	ACSI-J-
ASSISTANT CHIEF OF STAFF FOR INTELLIGENCE (ARMY), WASHINGTON, D.C.	H-
ASSISTANT CHIEF OF STAFF STUDIES AND ANALYSIS (AIR FORCE), WASHINGTON, D.C.	AFACS/SA-
ASSISTANT CHIEF OF STAFF STUDIES AND ANALYSIS (AIR FORCE), WASHINGTON, D.C.	AFCSA-
ASSOCIATED ELECTRICAL INDUSTRIES, LTD., ALDERMASTON, BERKS., ENGLAND	A-(NO.)(AEI)
ASSOCIATED ELECTRICAL INDUSTRIES, LTD., ALDERMASTON, BERKS, ENGLAND	AEI-
ASSOCIATED ELECTRICAL INDUSTRIES, LTD., ALDERMASTON, BERKS, ENGLAND	AEI-T-
ASSOCIATED ELECTRICAL INDUSTRIES, LTD., ALDERMASTON, BERKS, ENGLAND	BR-A-
ASSOCIATED ELECTRICAL INDUSTRIES, LTD., MANCHESTER, MANCUNIUM, ENGLAND	TP/R-
ASSOCIATED ELECTRICAL INDUSTRIES, LTD. CENTRAL RESEARCH LAB., RUGBY, WARWICK, ENGLAND	AEI-G-
ASSOCIATED ELECTRICAL INDUSTRIES, LTD. RESEARCH LABS. ALDERMASTON, BERKS, ENGLAND	CVD-
ASSOCIATED ELECTRICAL INDUSTRIES, LTD. RESEARCH LABS. RUGBY, WARWICK, ENGLAND	G.-
ASSOCIATED NUCLEONICS, INC., GARDEN CITY, N.Y.	AN-
ASSOCIATED RESEARCH DESIGN, ALBUQUERQUE, N. MEX.	ARD-
ASSOCIATED RESEARCH DESIGN, ALBUQUERQUE, N. MEX.	ARD-(LETTERS)-
ASSOCIATED SEMICONDUCTOR MANUFACTURERS LTD., SOUTHAMPTON, HANTS, ENGLAND	RP-
ASSOCIATED TECHNICAL SERVICES, INC., EAST ORANGE, N.J. (TRANSLATIONS)	ATS-
ASSOCIATED TECHNICAL SERVICES, INC., EAST ORANGE, N.J	ATS-TR-
ASSOCIATED TESTING LABS, INC., WAYNE, N.J.	F(NO.)-
ASSOCIATED UNIVERSITIES, INC., N.Y.C.	ASSOC.U-
ASSOCIATED UNIVERSITIES, INC., N.Y.C.	AU-
ASSOCIATED UNIVERSITIES, INC., N.Y.C.	AUI-
ASSOCIATED UNIVS., INC., N.Y.C. (PROJ. EAST RIVER)	AUI PER-
ASSOCIATED UNIVERSITIES, INC., N.Y.C.	PER-T-
ASSOCIATES FOR INTERNATL. RES., INC., CAMBRIDGE, MASS	PJ-
ASSOCIATION OF AMERICAN RAILROADS. MECHANICAL DIVISION, WASHINGTON, D.C.	MD-
ASSOCIATION OF BAY AREA GOVERNMENTS, BERKELEY, CALIF.	R/A-(1 DIGIT)
ASSN. OF SPEC. LIBRARIES + INFO. BUREAUX, LONDON	ASLIB-
ASTOR JOINT AEC-DOD COORDINATING COMM., SAFETY AND RELIABILITY SUBCOMM.	JCC/ASTOR-
ASTRA, INC., MILFORD, CONN.	ASTRA-
ASTRA, INC., MILFORD, CONN.	ASTRA-(3 DIGITS)-
ASTRA, INC., RALEIGH, N.C.	ASTRA-
ASTRA, INC., RALEIGH, N.C.	ASTRA-(3 DIGITS)-
ASTRA, INC., RALEIGH, N.C.	ASTRA-G-
ASTRO CONSULTANTS, PALM SPRINGS, CALIF.	ACT-TR-(NO.)-(YR.)
ASTRO MET ASSOCIATES INC., CINCINNATI	AMA-
ASTRO MET ASSOCIATES INC., CINCINNATI	ASTRO-MET-
ASTRO RESEARCH CORP., SANTA BARBARA, CALIF.	ARC-LTN-
ASTRO RESEARCH CORP., SANTA BARBARA, CALIF.	ARC-R-
ASTRO-SPACE LABS., INC., HUNTSVILLE, ALA.	ASL-(NUMBERS)-(YR.)-
ASTRO-SPACE LABS., INC., HUNTSVILLE, ALA.	ASL-FE-(YR)-
ASTRO-SPACE LABS., INC., HUNTSVILLE, ALA.	ASL-FR-(YEAR)-
ASTRO TECHNOLOGY CORP., PALO ALTO, CALIF.	ATC-
ASTRODATA, INC., ANAHEIM, CALIF.	ASTRODATA-
ASTROPHYSICS RESEARCH CORP., LOS ANGELES	TR-
ASTROSYSTEMS INTERNATIONAL INC., CALDWELL, N.J.	TR-(NUMBER)F
ATELIERS DE CONSTRUCTIONS ELECTRIQUES DE CHARLEROI, BELGIUM	EN/GI/PS-PW/LP-
ATLANTIC FLEET.ANTISUBMARINE WARFARE FORCE,NORFOLK,VA	ASW MEMO.-
ATLANTIC FLEET.ANTISUBMARINE WARFARE FORCE,NORFOLK,VA	COMASWFORLANT-
ATLANTIC FLEET. FLEET AIR WINGS, NORFOLK, VA.	COMFAIRWINGSLANT-
ATLANTIC FLEET. FLEET AIR WINGS, NORFOLK, VA.	SR-(NUMBER)-(YEAR)
ATLANTIC FLEET. HUNTER KILLER FORCE, N.Y.	(NUMBERS)B(NUMBERS)
ATLANTIC FLEET. HUNTER KILLER FORCE, NORFOLK, VA.	COMHUKFORLAN-
ATLANTIC MISSILE RANGE, PATRICK AFB, FLA.	AMR-
ATLANTIC OCEANOGRAPHIC AND METEOROLOGICAL LABS., MIAMI, FLA.	AOML-
ATLANTIC OCEANOGRAPHIC AND METEOROLOGICAL LABS., MIAMI, FLA.	ERL-AOML-
ATLANTIC OCEANOGRAPHIC AND METEOROLOGICAL LABS., MIAMI, FLA.	ERLTM-ACML-
ATLANTIC OCEANOGRAPHIC AND METEOROLOGICAL LABS., MIAMI, FLA.	ERLTM-AOML-
ATLANTIC OCEANOGRAPHIC AND METEOROLOGICAL LABS., MIAMI, FLA.	ESSA-TM-ERLTM-AOML-
ATLANTIC OCEANOGRAPHIC AND METEOROLOGICAL LABS., MIAMI, FLA.	NOAA-TM-ERL-AOML-
ATLANTIC-PACIFIC INTEROCEANIC CANAL STUDY COMM., WASHINGTON, D.C.	ARLV-
ATLANTIC RESEARCH CORP., ALEXANDRIA, VA.	ARC-
ATLANTIC RESEARCH CORP., ALEXANDRIA, VA. (SQUID PROJECT) (REF. 33)	ARC-(NUMBER)-P

ATOMIC ENERGY OF CANADA LTD.

ATLANTIC RESEARCH CORP., ALEXANDRIA, VA.	ARCAS-
ATLANTIC RESEARCH CORP., ALEXANDRIA, VA.	ARC-QPRS-
ATLANTIC RESEARCH CORP., ALEXANDRIA, VA.	ARC-TRDL-
ATLANTIC RESEARCH CORP., ALEXANDRIA, VA.	ARC-TR-PL-
ATLANTIC RESEARCH CORP., ALEXANDRIA, VA.	ATRC-
ATLANTIC RESEARCH CORP., ALEXANDRIA, VA.	ATRC-SP-QPR-
ATLANTIC RESEARCH CORP., ALEXANDRIA, VA.	J/B-(NUMBERS)-
ATLANTIC RESEARCH CORP., ALEXANDRIA, VA.	SAR-
ATLANTIC RESEARCH CORP., ALEXANDRIA, VA.	TR-PL-
ATLANTIC RESEARCH CORP., DUARTE, CALIF.	(NUMBERS)-TR-
ATLANTIC RESEARCH CORP. MISSILE SYSTEMS DIV., COSTA MESA, CALIF.	(NUMBERS)-EN-
ATLANTIC RESEARCH CORP. MISSILE SYSTEMS DIV., COSTA MESA, CALIF.	(NUMBERS)-ST-
ATLANTIC RESEARCH CORP. MISSILE SYSTEMS DIV., COSTA MESA, CALIF.	(NUMBERS)-TR-
ATLANTIC RES. CORP. SPACE VEHICLES DIV., EL MONTE, CAL	ARC-
ATLANTIC RICHFIELD HANFORD CO., RICHLAND, WASH.	ARH-
ATLANTIC RICHFIELD HANFORD CO., RICHLAND, WASH.	ARH-SA-
ATLAS POWDER CO. (ALL LOCATIONS)	APC-
ATLAS POWDER CO., TAMAQUA, PENNA.	APC-MPR-
ATLAS POWDER CO., WILMINGTON, DEL.	APC-MPR-
ATMOSPHERIC PHYSICS AND CHEMISTRY LAB., BOULDER, COLO.	ERL-(NUMBER)-APCL-
ATMOSPHERIC PHYSICS AND CHEMISTRY LAB., BOULDER, COLO.	ERL-APCL-
ATMOSPHERIC PHYSICS AND CHEMISTRY LAB., BOULDER, COLO.	ESSA-ERL-(NUMBER)-APCL-
ATMOSPHERIC PHYSICS AND CHEMISTRY LAB., BOULDER, COLO.	ESSA-TR-ERL-(NO.)-APCL-
ATMOSPHERIC PHYSICS AND CHEMISTRY LAB., BOULDER, COLO.	ESSA-TR-ERL-APCL-
ATMOSPHERIC PHYSICS AND CHEMISTRY LAB., BOULDER, COLO.	NOAA-TM-ERL-APCL-
ATMOSPHERIC RESEARCH GROUP, ALTADENA, CALIF.	ARG(YR)-FR-
ATMOSPHERIC SCIENCES LAB., WHITE SANDS MISSILE RANGE, NEW MEXICO	DR-
ATMOSPHERIC TURBULENCE AND DIFFUSION LAB., OAK RIDGE, TENN.	ATDL-
ATMOSPHERIC TURBULENCE AND DIFFUSION LAB., OAK RIDGE, TENN.	IERTM-ATDL-
ATOMIC BOMB CASUALTY COMMISSION, HIROSHIMA	ABBC-
ATOMIC BOMB CASUALTY COMMISSION, HIROSHIMA	ABCC-
ATOMIC BOMB CASUALTY COMMISSION, HIROSHIMA	ABCC-(NUMBERS)-(YEAR)
ATOMIC BOMB CASUALTY COMMISSION, HIROSHIMA (ANNUAL RPT)	ABCC-AR-
ATOMIC BOMB CASUALTY COMMISSION, HIROSHIMA	ABCC-MANUAL-(NO.)-(YR.)
ATOMIC BOMB CASUALTY COMMISSION, HIROSHIMA	ABCC-PA-(NUMBER)-(YEAR)
ATOMIC BOMB CASUALTY COMMISSION, HIROSHIMA	ABCC-TR-(NUMBER)-(NO.)
ATOMIC BOMB CASUALTY COMMISSION, HIROSHIMA	ABCC-TR-(NUMBER)-(YEAR)
ATOMIC BOMB CASUALTY COMMISSION, WASHINGTON, D.C.	NRC ABCC-
ATOMIC ELECTRIC PROJECT, ST. LOUIS	AEP-
ATOMIC ENERGY COMMISSION, CHALK RIVER, ONTARIO	AEC-SR-
ATOMIC ENERGY COMMISSION, WASHINGTON, D.C. (REF. 23)	A-
ATOMIC ENERGY COMMISSION, WASHINGTON, D.C.	AEC-
ATOMIC ENERGY COMMISSION, WASHINGTON, D.C. (FOR EXPLANATION OF THE REMAINDER OF THE CODE, SEE ENTRY UNDER LETTERS FOLLOWING AEC-)	AEC-(LETTERS)-
ATOMIC ENERGY COMMISSION, WASHINGTON, D.C. (CONTROLLED THERMONUCLEAR PROGRAM)	AEC-CTR-
ATOMIC ENERGY COMMISSION, WASHINGTON, D.C. (REF. 24)	AECD-
ATOMIC ENERGY COMMISSION, WASHINGTON, D.C. (PROCUREMENT INSTRUCTION)	AECPI-
ATOMIC ENERGY COMMISSION, WASHINGTON, D.C. (PEACEFUL USES OF NUCLEAR EXPLOSIVES)	AEC-PNE-(NUMBER)-
ATOMIC ENERGY COMMISSION, WASHINGTON, D.C. (PROCUREMENT REGULATIONS)	AECPR-
ATOMIC ENERGY COMMISSION, WASH., D.C. (TRANSLATIONS)	AEC-TR-
ATOMIC ENERGY COMMISSION, WASH., D.C. (TRANSLATIONS)	AECTR-
ATOMIC ENERGY COMMISSION, WASHINGTON, D.C.	AEC-WO-
ATOMIC ENERGY COMMISSION, WASHINGTON, D.C. (WEAPON TEST REPORT) (REF. 7)	AEC-WT-
ATOMIC ENERGY COMMISSION, WASHINGTON, D.C. (WEAPON TEST REPORT) (REF. 7)	AECWT-
ATOMIC ENERGY COMMISSION, WASHINGTON, D.C.	CLT-
ATOMIC ENERGY COMMISSION, WASHINGTON, D.C. (REF. 23)	M-
ATOMIC ENERGY COMMISSION, WASHINGTON, D.C. (REF. 24)	MDDC-
ATOMIC ENERGY COMMISSION, WASHINGTON, D.C. (NATIONAL NUCLEAR ENERGY SERIES)	NNES-
ATOMIC ENERGY COMMISSION, WASHINGTON, D.C.	OCPRS-
ATOMIC ENERGY COMMISSION, WASHINGTON, D.C. (RADIOLOGICAL SAFETY)	PMF-(NUMBER)(LETTER)
ATOMIC ENERGY COMMISSION, WASH., DC (REACTOR HANDBOOK)	RH-
ATOMIC ENERGY COMMISSION, WASHINGTON, D.C.	VISIT NO.-
ATOMIC ENERGY COMMISSION, WASHINGTON, D.C.	WASH-
ATOMIC ENERGY COMMISSION, LONDON (INFORMAL SUMMARY)	AEC-LONDON-
ATOMIC ENERGY COMMISSION, ISOTOPE DEV. PROGRAM, WASH, DC	IDP-
ATOMIC ENERGY COMMISSION COMBINED OPERATIONS PLANNING, OAK RIDGE, TENN.	AECOP-
ATOMIC ENERGY OF CANADA LTD. (ALL LOCATIONS)	AECL-
ATOMIC ENERGY OF CANADA LTD., CHALK RIVER, ONT.	AECL-RM-
ATOMIC ENERGY OF CANADA LTD., CHALK RIVER, ONT.	AMF-C-
ATOMIC ENERGY OF CANADA LTD., CHALK RIVER, ONT.	APPE-
ATOMIC ENERGY OF CANADA LTD., CHALK RIVER, ONT.	ARE-
ATOMIC ENERGY OF CANADA LTD., CHALK RIVER, ONT. (ANNUAL SAFETY REPORT)	ASR-
ATOMIC ENERGY OF CANADA LTD., CHALK RIVER, ONT.	BLW-250-NOTE-
ATOMIC ENERGY OF CANADA LTD., CHALK RIVER, ONT.	CAN-
ATOMIC ENERGY OF CANADA LTD., CHALK RIVER, ONT.	CDG-
ATOMIC ENERGY OF CANADA LTD., CHALK RIVER, ONT.	CMP-CF-
ATOMIC ENERGY OF CANADA LTD., CHALK RIVER, ONT.	CMP-FE-
ATOMIC ENERGY OF CANADA LTD., CHALK RIVER, ONT. (REF.6)	CR-(LETTER(S))-
ATOMIC ENERGY OF CANADA LTD., CHALK RIVER, ONT. (REF.6)	CR(LETTER(S))-
ATOMIC ENERGY OF CANADA LTD., CHALK RIVER, ONT.	CRARE-
ATOMIC ENERGY OF CANADA LTD., CHALK RIVER, ONT.	CRDC-
ATOMIC ENERGY OF CANADA LTD., CHALK RIVER, ONT.	CRE-
ATOMIC ENERGY OF CANADA LTD., CHALK RIVER, ONT.	CREL-
ATOMIC ENERGY OF CANADA LTD., CHALK RIVER, ONT.	CRER-
ATOMIC ENERGY OF CANADA LTD., CHALK RIVER, ONT.	CRFD-
ATOMIC ENERGY OF CANADA LTD., CHALK RIVER, ONT.	CRGM-
ATOMIC ENERGY OF CANADA LTD., CHALK RIVER, ONT.	CRGP-
ATOMIC ENERGY OF CANADA LTD., CHALK RIVER, ONT.	CRHP-
ATOMIC ENERGY OF CANADA LTD., CHALK RIVER, ONT.	CRIO-

ATOMIC ENERGY OF CANADA LTD.

ATOMIC ENERGY OF CANADA LTD., CHALK RIVER, ONT.	CRMET-
ATOMIC ENERGY OF CANADA LTD., CHALK RIVER, ONT.	CRNE-
ATOMIC ENERGY OF CANADA LTD., CHALK RIVER, ONT.	CRNP-
ATOMIC ENERGY OF CANADA LTD., CHALK RIVER, ONT.	CRP-
ATOMIC ENERGY OF CANADA LTD., CHALK RIVER, ONT.	CRPL-
ATOMIC ENERGY OF CANADA LTD., CHALK RIVER, ONT.	CRR-
ATOMIC ENERGY OF CANADA LTD., CHALK RIVER, ONT.	CRRCE-
ATOMIC ENERGY OF CANADA LTD., CHALK RIVER, ONT.	CRRD-
ATOMIC ENERGY OF CANADA LTD., CHALK RIVER, ONT.	CRRIS-
ATOMIC ENERGY OF CANADA LTD., CHALK RIVER, ONT.	CRRL-
ATOMIC ENERGY OF CANADA LTD., CHALK RIVER, ONT.	CRRM-
ATOMIC ENERGY OF CANADA LTD., CHALK RIVER, ONT.	CRRP-
ATOMIC ENERGY OF CANADA LTD., CHALK RIVER, ONT.	CRT-
ATOMIC ENERGY OF CANADA LTD., CHALK RIVER, ONT.	DCI-
ATOMIC ENERGY OF CANADA LTD., CHALK RIVER, ONT.	DL-
ATOMIC ENERGY OF CANADA LTD., CHALK RIVER, ONT.	DM-
ATOMIC ENERGY OF CANADA LTD., CHALK RIVER, ONT.	DPNPD-(YEAR)-
ATOMIC ENERGY OF CANADA LTD., CHALK RIVER, ONT.	ELI-
ATOMIC ENERGY OF CANADA LTD., CHALK RIVER, ONT.	ERI-
ATOMIC ENERGY OF CANADA LTD., CHALK RIVER, ONT.	EXP-G-CR-
ATOMIC ENERGY OF CANADA LTD., CHALK RIVER, ONT.	EXP-NPD-
ATOMIC ENERGY OF CANADA LTD., CHALK RIVER, ONT.	EXP-NRU-
ATOMIC ENERGY OF CANADA LTD., CHALK RIVER, ONT.	EXP-NRX-
ATOMIC ENERGY OF CANADA LTD., CHALK RIVER, ONT.	EXPR-
ATOMIC ENERGY OF CANADA LTD., CHALK RIVER, ONT.	FSD/ING-
ATOMIC ENERGY OF CANADA LTD., CHALK RIVER, ONT.	GMI-
ATOMIC ENERGY OF CANADA LTD., CHALK RIVER, ONT.	HRI-
ATOMIC ENERGY OF CANADA LTD.,CHALK RIVER,ONT. (REF.6)	IGI-
ATOMIC ENERGY OF CANADA LTD., CHALK RIVER, ONT.	IGP-
ATOMIC ENERGY OF CANADA LTD.,CHALK RIVER,ONT. (REF.6)	IPD-
ATOMIC ENERGY OF CANADA LTD.,CHALK RIVER,ONT. (REF.6)	MET-
ATOMIC ENERGY OF CANADA LTD.,CHALK RIVER,ONT. (REF.6)	MET-I-
ATOMIC ENERGY OF CANADA LTD., CHALK RIVER, ONT.	NDC-
ATOMIC ENERGY OF CANADA LTD.,CHALK RIVER,ONT. (REF.6)	NEI-
ATOMIC ENERGY OF CANADA LTD.,CHALK RIVER,ONT. (REF.6)	PD-
ATOMIC ENERGY OF CANADA LTD., CHALK RIVER, ONT.	PDB-
ATOMIC ENERGY OF CANADA LTD., CHALK RIVER, ONT.	PE-
ATOMIC ENERGY OF CANADA LTD.,CHALK RIVER,ONT. (REF.6)	PR-
ATOMIC ENERGY OF CANADA LTD.,CHALK RIVER,ONT. (REF.6)	PR-(LTR(S))-(NO.)-(LTRS)
ATOMIC ENERGY OF CANADA LTD.,CHALK RIVER,ONT. (REF.6)	PR-REPORTS-
ATOMIC ENERGY OF CANADA LTD.,CHALK RIVER,ONT. (REF.6)	PR-S-
ATOMIC ENERGY OF CANADA LTD., CHALK RIVER, ONT.	RCE-
ATOMIC ENERGY OF CANADA LTD., CHALK RIVER, ONT.	RHCI-
ATOMIC ENERGY OF CANADA LTD.,CHALK RIVER,ONT. (REF.6)	RLI-
ATOMIC ENERGY OF CANADA LTD., CHALK RIVER, ONT.	RMI-
ATOMIC ENERGY OF CANADA LTD., CHALK RIVER, ONT.	RPI-(NUMBER)(AECL)
ATOMIC ENERGY OF CANADA LTD., CHALK RIVER, ONT.	RPI-(AECL)
ATOMIC ENERGY OF CANADA LTD., CHALK RIVER, ONT.	RSC-
ATOMIC ENERGY OF CANADA LTD., CHALK RIVER, ONT.	SM-
ATOMIC ENERGY OF CANADA LTD., CHALK RIVER, ONT.	SUP-M-
ATOMIC ENERGY OF CANADA LTD., CHALK RIVER, ONT.	TPI-
ATOMIC ENERGY OF CANADA LTD., CHALK RIVER, ONT.	VPENG-
ATOMIC ENERGY OF CANADA LTD., CHALK RIVER, ONT.	WDI-
ATOMIC ENERGY OF CANADA LTD. BIOLOGY AND HEALTH PHYSICS DIV., CHALK RIVER, ONT.	PR-B-
ATOMIC EN. OF CANADA LTD. CHALK RIVER NUC. LABS., ONT	CEI-
ATOMIC EN. OF CANADA LTD. CHALK RIVER NUC. LABS., ONT	CRNL-
ATOMIC EN. OF CANADA LTD. CHALK RIVER NUC. LABS., ONT	GPI-
ATOMIC EN. OF CANADA LTD. CHALK RIVER NUC. LABS., ONT	PR-(LETTERS)-
ATOMIC EN. OF CANADA LTD. CHALK RIVER NUC. LABS., ONT	PR-APPE-
ATOMIC EN. OF CANADA LTD. CHALK RIVER NUC. LABS., ONT	PR-CMA-
ATOMIC EN. OF CANADA LTD. CHALK RIVER NUC. LABS., ONT	PR-P-
ATOMIC EN. OF CANADA LTD. CHALK RIVER NUC. LABS., ONT	PR-P-(NUMBER)-GP
ATOMIC ENERGY OF CANADA LTD. CHALK RIVER PROJECT, ONT. (REF. 1)	(NUMBERS...)
ATOMIC ENERGY OF CANADA LTD. CHALK RIVER PROJ., ONT.	AECL-DR-
ATOMIC ENERGY OF CANADA LTD. CHALK RIVER PROJECT, ONTARIO. (ALUMINUM-NICKEL ALLOY MEETING)	ANAM-
ATOMIC ENERGY OF CANADA LTD. CHALK RIVER PROJECT, ONT. (REF. 6)	B-
ATOMIC ENERGY OF CANADA LTD. CHALK RIVER PROJ., ONT.	BRP-
ATOMIC ENERGY OF CANADA LTD. CHALK RIVER PROJECT, ONT. (REF. 6)	C-
ATOMIC ENERGY OF CANADA LTD. CHALK RIVER PROJECT, ONT. (REF. 6)	CI-
ATOMIC ENERGY OF CANADA LTD. CHALK RIVER PROJ., ONT.	CMP-FF-
ATOMIC ENERGY OF CANADA LTD. CHALK RIVER PROJECT, ONT	CR-B-RHC-
ATOMIC ENERGY OF CANADA LTD. CHALK RIVER PROJ., ONT. (LECTURE)	CRL-
ATOMIC ENERGY OF CANADA LTD. CHALK RIVER PROJECT, ONT. (REF. 6)	CRPR-
ATOMIC ENERGY OF CANADA LTD. CHALK RIVER PROJ., ONT.	DE-
ATOMIC ENERGY OF CANADA LTD. CHALK RIVER PROJ., ONT.	DEV-(LETTER)-(NO.)-
ATOMIC ENERGY OF CANADA LTD. CHALK RIVER PROJ., ONT.	DEV-B-
ATOMIC ENERGY OF CANADA LTD. CHALK RIVER PROJ., ONT.	DEV-C-
ATOMIC ENERGY OF CANADA LTD. CHALK RIVER PROJ., ONT.	DEV-F-
ATOMIC ENERGY OF CANADA LTD. CHALK RIVER PROJ., ONT.	DR-
ATOMIC ENERGY OF CANADA LTD. CHALK RIVER PROJ., ONT.	E-
ATOMIC ENERGY OF CANADA LTD. CHALK RIVER PROJECT, ONT. (REF. 6)	ED-
ATOMIC ENERGY OF CANADA LTD. CHALK RIVER PROJECT, ONT. (REF. 6)	EDI-
ATOMIC ENERGY OF CANADA LTD. CHALK RIVER PROJ., ONT.	EI-
ATOMIC ENERGY OF CANADA LTD. CHALK RIVER PROJECT, ONT. (REF. 6)	ES-
ATOMIC ENERGY OF CANADA LTD. CHALK RIVER PROJECT, ONT. (REF. 6)	FD-
ATOMIC ENERGY OF CANADA LTD. CHALK RIVER PROJ., ONT.	FSP/H-
ATOMIC ENERGY OF CANADA LTD. CHALK RIVER PROJECT, ONT. (REF. 6)	IPI-
ATOMIC ENERGY OF CANADA LTD. CHALK RIVER PROJ., ONT.	J-
ATOMIC ENERGY OF CANADA LTD. CHALK RIVER PROJ., ONT.	JSHP-
ATOMIC ENERGY OF CANADA LTD. CHALK RIVER PROJ., ONT.	LE-
ATOMIC ENERGY OF CANADA LTD. CHALK RIVER PROJECT, ONT. (REF. 6)	MC-
ATOMIC ENERGY OF CANADA LTD. CHALK RIVER PROJ., ONT.	METI-
ATOMIC ENERGY OF CANADA LTD. CHALK RIVER PROJ., ONT.	MISC.-
ATOMIC ENERGY OF CANADA LTD. CHALK RIVER PROJ., ONT.	MT-
ATOMIC ENERGY OF CANADA LTD. CHALK RIVER PROJ., ONT.	MTEC-
ATOMIC ENERGY OF CANADA LTD. CHALK RIVER PROJ., ONT.	NPG-(NUMBER)(AECL)
ATOMIC ENERGY OF CANADA LTD. CHALK RIVER PROJECT, ONT. (REF. 6)	NRC-
ATOMIC ENERGY OF CANADA LTD. CHALK RIVER PROJECT, ONT. (REF. 6)	PBD-
ATOMIC ENERGY OF CANADA LTD. CHALK RIVER PROJ., ONT.	PBI-
ATOMIC ENERGY OF CANADA LTD. CHALK RIVER PROJECT, ONT. (REF. 6)	PRA-
ATOMIC ENERGY OF CANADA LTD. CHALK RIVER PROJ., ONT.	RDI-
ATOMIC ENERGY OF CANADA LTD. CHALK RIVER PROJECT, ONT. (REF. 6)	RDP-
ATOMIC ENERGY OF CANADA LTD. CHALK RIVER PROJECT, ONT. (REF. 6)	REI-
ATOMIC ENERGY OF CANADA LTD. CHALK RIVER PROJ., ONT.	TDVI-
ATOMIC ENERGY OF CANADA LTD. CHALK RIVER PROJ., ONT.	TNCC(CAN)-
ATOMIC ENERGY OF CANADA LTD. CHALK RIVER PROJ., ONT. (5TH UK-CANADA TECH. CONF., HARWELL, 1956)	UK/C/5/
ATOMIC ENERGY OF CANADA LTD. CHALK RIVER PROJ., ONT. (6TH UK-CANADA TECH. CONF., CHALK RIVER, 1957)	UK/C/6/

ATOMIC ENERGY OF CANADA LTD. CHALK RIVER PROJ., ONT.	UKE-CR-
ATOMIC ENERGY OF CANADA LTD. CHALK RIVER PROJ., ONT.	UN-
ATOMIC ENERGY OF CANADA LTD. CHALK RIVER PROJECT, ONT. (REF. 6)	WD-
ATOMIC ENERGY OF CANADA LTD. CHALK RIVER PROJ., ONT.	WM-
ATOMIC ENERGY OF CANADA LTD. COMMERCIAL PRODUCTS DIV., OTTAWA (COMMERCIAL PRODUCTS SCIENCE REPORTS)	CPSR-
ATOMIC ENERGY OF CANADA LTD., DEEP RIVER LAB., ONT.	CRGP-
ATOMIC ENERGY OF CANADA LTD. NUCLEAR POWER PLANT DIV. TORONTO (CANADA-U.S. COOP. ATOMIC POWER PROGRAM)	CUP-
ATOMIC ENERGY OF CANADA LTD. NUCLEAR POWER PLANT DIV. TORONTO	PR-NPAP-
ATOMIC ENERGY OF CANADA LTD. NUCLEAR POWER PLANT DIV. TORONTO	PR-NPDP-
ATOMIC ENERGY OF CANADA LTD. NUCLEAR POWER PLANT DIV. TORONTO	PR-NPP-
ATOMIC ENERGY OF CANADA LTD. NUCLEAR POWER PLANT DIV. TORONTO	PR-PP DEV-
ATOMIC ENERGY OF CANADA LTD. NUCLEAR POWER PLANT DIV. TORONTO	PR-PP ENG-
ATOMIC ENERGY OF CANADA LTD. NUCLEAR POWER PLANT DIV. TORONTO	TAPI-
ATOMIC ENERGY OF CANADA LTD. NUCLEAR POWER PLANT DIV. TORONTO	TDI-
ATOMIC ENERGY OF CANADA LTD. NUCLEAR POWER PLANT DIV. TORONTO (REF. 6)	TDS-
ATOMIC ENERGY OF CANADA LTD. NUCLEAR POWER PLANT DIV. TORONTO (REF. 6)	TDSI-
ATOMIC ENERGY OF CANADA LTD. NUCLEAR POWER PLANT DIV. TORONTO	TDV-
ATOMIC ENERGY OF CANADA LTD. NUCLEAR POWER PLANT DIV. TORONTO	TDVI-
ATOMIC ENERGY OF CANADA LTD. PUBLIC RELATIONS OFFICE, CHALK RIVER, ONT.	AECL-PUB-
ATOMIC ENERGY OF CANADA LTD. WHITESHELL NUCLEAR RESEARCH ESTABLISHMENT, PINAWA, MANITOBA	EXP-WR-
ATOMIC ENERGY OF CANADA LTD. WHITESHELL NUCLEAR RESEARCH ESTAB., PINAWA, MAN.	WDI-
ATOMIC ENERGY OF CANADA LTD. WHITESHELL NUCLEAR RESEARCH ESTAB., PINAWA, MAN.	WNRE-
ATOMIC ENERGY OF CANADA LTD. WHITESHELL NUCLEAR RESEARCH ESTABLISHMENT. APPLIED SCIENCE DIV., PINAWA, MANITOBA (PROGRESS REPORT)	PR-WAS-
ATOMIC INDUSTRIAL FORUM, INC., N.Y.C.	FORUM REPORT-
ATOMIC POWER CONSTRUCTIONS LTD.,SUTTON,SURREY,ENG.	APC/R.-
ATOMIC POWER DEVELOPMENT ASSOCIATES, INC., DETROIT	APDA-
ATOMIC POWER DEVELOPMENT ASSOCIATES, INC., DETROIT	APDA-CFE-
ATOMIC POWER DEVELOPMENT ASSOCIATES, INC., DETROIT (ENRICO FERMI ATOMIC POWER PLANT)	APDA-FAM-(YEAR)-
ATOMIC POWER DEVELOPMENT ASSOCIATES, INC., DETROIT	APDA-NTS-
ATOMIC POWER DEVELOPMENT ASSOCIATES, INC., DETROIT	APDA-SPEC.-
ATOMIC POWER DEVELOPMENT ASSOCIATES, INC., DETROIT	APDA-TM-
ATOMIC POWER DEVELOPMENT ASSOCIATES, INC., DETROIT	APDA-TPI-
ATOMIC POWER DEVELOPMENT ASSOCIATES, INC., DETROIT (JOINT VENTURE)	BMI-APDA-(YEAR)-
ATOMIC POWER DEVELOPMENT ASSOCIATES, INC., DETROIT	FAM-(YEAR)-
ATOMIC POWER DEVELOPMENT ASSOCIATES, INC., DETROIT	MICHU-PR-
ATOMIC POWER DEVELOPMENT ASSOCIATES, INC., DETROIT	PRDC-
ATOMIC POWER DEVELOPMENT ASSOCIATES, INC., DETROIT	RAD-
ATOMICS INTERNATIONAL, CANOGA PARK, CALIF.	AI-(YEAR)-
ATOMICS INTERNATIONAL, CANOGA PARK, CALIF.	AI-AEC-
ATOMICS INTERNATIONAL, CANOGA PARK, CALIF.	AI-AEC-MEMO-
ATOMICS INTERNATIONAL, CANOGA PARK, CALIF. (JOINT VENTURE WITH COMBUSTION ENGINEERING, INC. FOR THE HEAVY WATER ORGANIC COOLED REACTOR)	AI-CE-
ATOMICS INTERNATIONAL, CANOGA PARK, CALIF. (JOINT VENTURE WITH COMBUSTION ENGINEERING, INC. FOR THE HEAVY WATER ORGANIC COOLED REACTOR)	AI-CE-MEMO-
ATOMICS INTERNATIONAL, CANOGA PARK, CALIF. (JOINT VENTURE WITH COMBUSTION ENGINEERING, INC. FOR THE HEAVY WATER ORGANIC COOLED REACTOR)	AI-CE-TDR-
ATOMICS INTERNATIONAL, CANOGA PARK, CALIF.	NAA-SR-MEMO-
ATOMICS INTERNATIONAL. LIQUID METAL ENGINEERING CENTER, CANOGA PARK, CALIF.	LMEC-(YEAR)-
ATOMICS INTERNATIONAL. LIQUID METAL ENGINEERING CENTER, CANOGA PARK, CALIF.	LMEC-MEMO-(YEAR)-
ATOMICS INTERNATIONAL. LIQUID METAL ENGINEERING CENTER, CANOGA PARK, CALIF.	LMEC-MR (MONTH/YEAR)
ATOMICS INTERNATIONAL. LIQUID METAL ENGINEERING CENTER, CANOGA PARK, CALIF.	LMEC-TDR-(YEAR)-
ATOMICS INTERNATIONAL. LIQUID METAL ENGINEERING CENTER, CANOGA PARK, CALIF.	LMEC-TPR (MONTH/YEAR)
ATOMICS INTERNATIONAL DIV., NORTH AMERICAN AVIATION, INC., CANOGA PARK, CALIF.	AI-
ATOMICS INTERNATIONAL DIV., NORTH AMERICAN AVIATION, INC., CANOGA PARK, CALIF.	AI-(YEAR)-MEMO-
ATOMICS INTERNATIONAL DIV., NORTH AMERICAN AVIATION, INC., CANOGA PARK, CALIF.	AI-L-(YEAR)AT(NO.)
ATOMICS INTERNATIONAL DIV., NORTH AMERICAN AVIATION, INC., CANOGA PARK, CALIF.	AI-MEMO-
ATOMICS INTERNATIONAL DIV., NORTH AMERICAN AVIATION, INC., CANOGA PARK, CALIF.	AI NAA-SR-
ATOMICS INTERNATIONAL DIV., NORTH AMERICAN AVIATION, INC., CANOGA PARK, CALIF.	AI-P-
ATOMICS INTERNATIONAL DIV., NORTH AMERICAN AVIATION, INC., CANOGA PARK, CALIF.	AI-PC-
ATOMICS INTERNATIONAL DIV., NORTH AMERICAN AVIATION, INC., CANOGA PARK, CALIF.	AI-TRANS-
ATOMICS INTERNATIONAL DIV., NORTH AMERICAN AVIATION, INC., CANOGA PARK, CALIF.	AL-P-
ATOMICS INTERNATIONAL DIV., NORTH AMERICAN AVIATION, INC., CANOGA PARK, CALIF.	AL-TRANS-
ATOMICS INTERNATIONAL DIV., NORTH AMERICAN AVIATION, INC., CANOGA PARK, CALIF.	LMSC-
ATOMICS INTERNATIONAL DIV., NORTH AMERICAN AVIATION, INC., CANOGA PARK, CALIF.	NAA-
ATOMICS INTERNATIONAL DIV., NORTH AMERICAN AVIATION, INC., CANOGA PARK, CALIF.	NAA-AER-
ATOMICS INTERNATIONAL DIV., NORTH AMERICAN AVIATION, INC., CANOGA PARK, CALIF.	NAA/AI-
ATOMICS INTERNATIONAL DIV., NORTH AMERICAN AVIATION, INC., CANOGA PARK, CALIF.	NAA-AI-
ATOMICS INTERNATIONAL DIV., NORTH AMERICAN AVIATION, INC., CANOGA PARK, CALIF.	NAA-SR-
ATOMICS INTERNATIONAL DIV., NORTH AMERICAN AVIATION, INC., CANOGA PARK, CALIF.	NAA-SR-MEMO-
ATOMICS INTERNATIONAL DIV., NORTH AMERICAN AVIATION, INC., CANOGA PARK, CALIF.	NAA-SR-TDR-
ATOMICS INTERNATIONAL DIV., NORTH AMERICAN AVIATION, INC., CANOGA PARK, CALIF.	RTP-
AUBURN RESEARCH FOUNDATION, ALA.	PR-IX-
AUBURN UNIV., ALA.	FTR-
AUERBACH CORP., PHILADELPHIA	AUER-(NUMBER)-TR-
AUERBACH CORP., PHILADELPHIA	AUER-(NUMBER)-FR-
AUERBACH CORP., PHILADELPHIA	AUERBACH-(NO.)-TR-
AUERBACH CORP., PHILADELPHIA	AUERBACH-PR-
AUERBACH CORP., PHILADELPHIA	(NUMBERS)TR-
AUERBACH CORP., PHILADELPHIA	(NUMBERS)-TR-
AUERBACH CORP. PROFESSIONAL SERVICES GROUP, PHILADELPHIA, PA. (4 DIGITS ARE PROJECT NO., 3 DIGITS ARE PHASE OF PROJECT)	TR-(4 DIGITS)-(3 DIGITS)
AUSTIN RESEARCH ASSOCIATES, INC., TEX.	ARA-
AUSTIN RESEARCH ASSOCIATES, INC., TEX.	ARA-C-
AUSTIN RESEARCH ASSOCIATES, INC., TEX.	ARAI-
AUSTRALIA. AERO. RES. LABS.,FISHERMANS BEND, VICTORIA	ARCC-
AUSTRALIA. AERO. RES. LABS.,FISHERMANS BEND, VICTORIA	ARL-
AUSTRALIA. AERO. RES. LABS.,FISHERMANS BEND, VICTORIA	ARL/A-
AUSTRALIA. AERONAUTICAL RESEARCH LABS., FISHERMANS BEND, VICTORIA (AERODYNAMICS NOTE)	ARL/AERO-N-
AUSTRALIA. AERONAUTICAL RESEARCH LABS., FISHERMANS BEND, VICTORIA (INSTRUMENT NOTE)	ARL/IN-
AUSTRALIA. AERO. RES. LABS.,FISHERMANS BEND, VICTORIA	ARL/ME-
AUSTRALIA. AERO. RES. LABS.,FISHERMANS BEND, VICTORIA	ARL/MET-

AUSTRALIA. AERO. RES. LABS.

AUSTRALIA. AERO. RES. LABS., FISHERMANS BEND, VICTORIA	ARL/SM-
AUSTRALIA. AERO. RES. LABS., FISHERMANS BEND, VICTORIA	AUS ARL/(LETTER)-
AUSTRALIA. AERO. RES. LABS., FISHERMANS BEND, VICTORIA	HSA-
AUSTRALIA. AERO. RES. LABS., FISHERMANS BEND, VICTORIA	SM-
AUSTRALIA. AERONAUTICAL RESEARCH LABS., MELBOURNE	ABL/AERO-
AUSTRALIA. AERONAUTICAL RESEARCH LABS., MELBOURNE	ARL/AERO-
AUSTRALIA. AERONAUTICAL RESEARCH LABS., MELBOURNE	ARL/ENG.FAC.-
AUSTRALIA. AERONAUTICAL RESEARCH LABS., MELBOURNE	ARL/F-
AUSTRALIA. AERONAUTICAL RESEARCH LABS., MELBOURNE	ARL/FL-
AUSTRALIA. AERONAUTICAL RESEARCH LABS., MELBOURNE	ARL/H.G.M.-
AUSTRALIA. AERONAUTICAL RESEARCH LABS., MELBOURNE	ARL/I-
AUSTRALIA. AERONAUTICAL RESEARCH LABS., MELBOURNE	ARL/ME-
AUSTRALIA. AERONAUTICAL RESEARCH LABS., MELBOURNE	ARL/SM-
AUSTRALIA. AIRCRAFT RES. AND DEV. UNIT, LAVERTON	ARDU-
AUSTRALIA. ATOMIC WEAPONS TESTS COMMITTEE, VICTORIA	AWTSC-
AUSTRALIA. BUREAU OF MINERAL RESOURCES, GEOLOGY AND GEOPHYSICS, CANBERRA	PNG-
AUSTRALIA. COMMONWEALTH SCIENTIFIC + IND. RES. ORG.	CSIRO-TRANS-
AUSTRALIA. COMMONWEALTH SCIENTIFIC + IND. RES. ORG.	CSIRO/UI-
AUSTRALIA. COMMONWEALTH SCIENTIFIC AND INDUSTRIAL RES. ORG. DIV. OF FOREST PRODUCTS, MELBOURNE	DFP-TP-
AUSTRALIA. COMMONWEALTH SCI. + IND. RES. ORG. INFO. SERVICE, E. MELBOURNE (TRANSLATIONS)	CSIRO-
AUSTRALIA. COMMONWEALTH SCI. + IND. RES. ORG. RADIOPHYSICS LAB., SYDNEY	RPL-
AUSTRALIA. COMMONWEALTH SCI. + IND. RES. ORG. RADIOPHYSICS LAB., SYDNEY	RPP-
AUSTRALIA. COMMONWEALTH X-RAY AND RADIUM LAB., MELBOURNE	CXRL-
AUSTRALIA. COUNCIL FOR SCIENTIFIC AND INDUSTRIAL RESEARCH. (AERODYNAMICS NOTES)	AUS CSIR DA AN-
AUSTRALIA. COUNCIL FOR SCIENTIFIC AND INDUSTRIAL RESEARCH. (ENGINES NOTE)	AUS CSIR DA EN-
AUSTRALIA. COUNCIL FOR SCIENTIFIC AND INDUSTRIAL RESEARCH. (INSTRUMENTS)	AUS CSIR DA IN-
AUSTRALIA. COUNCIL FOR SCIENTIFIC AND INDUSTRIAL RESEARCH (AERODYNAMICS NOTES)	DA AN-
AUSTRALIA. COUNCIL FOR SCIENTIFIC AND INDUSTRIAL RESEARCH. DIV. OF AERONAUTICS	AUS CSIR-DA-(LETTER)(NO.)
AUSTRALIA. COUNCIL FOR SCIENTIFIC AND INDUSTRIAL RESEARCH. DIVISION OF BUILDING RESEARCH	AUS CSIR BR-
AUSTRALIA. COUNCIL FOR SCIENTIFIC AND INDUSTRIAL RESEARCH. DIVISION OF BUILDING RESEARCH	AUS-CSIR-MP-
AUSTRALIA. COUNCIL FOR SCIENTIFIC AND INDUSTRIAL RESEARCH. DIVISION OF BUILDING RES. (TECH. PAPER)	AUS-MP-TP-
AUSTRALIA. COUNCIL FOR SCIENTIFIC AND INDUSTRIAL RESEARCH. DIVISION OF RADIOPHYSICS	AUS CSIR RP-
AUSTRALIA. COUNCIL FOR SCI. + IND. RES. PHYS. SEC.	AUS CSIR PSS-
AUSTRALIA. COUNCIL FOR SCIENTIFIC AND INDUSTRIAL RES. SEC. OF LUBRICANTS AND BEARINGS. (A SERIES)	AUS CSIR LB A-
AUSTRALIA. COUNCIL FOR SCIENTIFIC AND INDUSTRIAL RES. SEC. OF LUBRICANTS AND BEARINGS. (BEARINGS)	AUS CSIR LB B-
AUSTRALIA. COUNCIL FOR SCIENTIFIC + IND. RES. SEC. OF LUBRICANTS AND BEARINGS. (CUTTING FLUIDS)	AUS CSIR LB CF-
AUSTRALIA. COUNCIL FOR SCI. + IND. RES. SEC. OF LUBRICANTS + BEARINGS. (CUTTING FLUIDS) (INTERIM)	AUS CSIR LB CF I-
AUSTRALIA. COUNCIL FOR SCIENTIFIC AND INDUSTRIAL RESEARCH. SECTION OF LUBRICANTS AND BEARINGS. (CYLINDER AND PISTON WEAR IN AERO ENGINES)	AUS CSIR LB CP-
AUSTRALIA. COUNCIL FOR SCIENTIFIC AND INDUSTRIAL RESEARCH. SEC. OF LUBRICANTS + BEARINGS. (CYLINDER + PISTON WEAR IN AERO ENGINES)(INTERIM)	AUS CSIR LB CP I-
AUSTRALIA. COUNCIL FOR SCIENTIFIC AND IND. RES. SEC. OF LUBRICANTS AND BEARINGS. (EXPLOSIVES)	AUS CSIR LB E-
AUSTRALIA. COUNCIL FOR SCI. + IND. RES. SEC. OF LUBRICANTS + BEARINGS. (LUBRICATION + FRICTION)	AUS CSIR LB LF-
AUSTRALIA. COUNCIL FOR SCI. + IND. RES. SEC. OF LUBRICANTS + BEARINGS. (MUZZLE VELOCITY)	AUS CSIR LB MV-
AUSTRALIA. COUNCIL FOR SCIENTIFIC AND IND. RES. SEC. OF LUBRICANTS + BEARINGS. (PRODUCER GAS)	AUS CSIR LB PG-
AUSTRALIA. COUNCIL FOR SCIENTIFIC AND INDUSTRIAL RESEARCH. SECTION OF TRIBOPHYSICS. (EXPLOSIVES)	AUS CSIR T E-
AUSTRALIA. COUNCIL FOR SCIENTIFIC AND INDUSTRIAL RESEARCH. WOOL TEXTILE RESEARCH LABS.	AUS-WTRL-
AUSTRALIA. DEFENCE SCIENTIFIC SERVICE, MELBOURNE	RD/P-
AUSTRALIA. DEFENCE STANDARDS LABS., MARIBYRNONG	ADSS-
AUSTRALIA. DEFENCE STANDARDS LABS., MARIBYRNONG	DSL-
AUSTRALIA. DEFENCE STANDARDS LABS., MARIBYRNONG	DSL-R-
AUSTRALIA. DEFENCE STANDARDS LABS., MARIBYRNONG	DSL-TM-
AUSTRALIA. DEFENCE STANDARDS LABS., MARIBYRNONG	DSL-TN-
AUSTRALIA. DEFENCE STANDARDS LABS., MARIBYRNONG	DSL-TR-
AUSTRALIA. DEPARTMENT OF CIVIL AVIATION, MELBOURNE	AM-MEMO-
AUSTRALIA. DEPT. OF DEFENCE. EDP DIV., CANBERRA	MALLARD-TD-
AUSTRALIA. DEPARTMENT OF LABOUR AND NATIONAL SERVICE. INDUSTRIAL WELFARE DIVISION	AUS DLNS IWD-
AUSTRALIA. DEPARTMENT OF LABOUR AND NATIONAL SERVICE. INDUSTRIAL WELFARE DIVISION	AUS DLNS IWD TR-
AUSTRALIA. DEPARTMENT OF LABOUR AND NATIONAL SERVICE. INDUSTRIAL WELFARE DIVISION	DLNS IWD-
AUSTRALIA. DEPT. OF SUPPLY. ANTARCTIC DIV., MELBOURNE	ANARE-
AUSTRALIA. JOINT TROPICAL RESEARCH UNIT, INNISFAIL	JTRU-
AUSTRALIA. LONG RANGE WEAPONS ESTABLISHMENT	LRWE-
AUSTRALIA. MUNITIONS SUPPLY LAB., MARIBYRNONG	AUS DM MSL-
AUSTRALIA. MUNITIONS SUPPLY LAB., MARIBYRNONG	AUS DM MSL R-
AUSTRALIA. MUNITIONS SUPPLY LAB., MARIBYRNONG	AUS DM MSL TN-
AUSTRALIA. NATIONAL HEALTH AND MEDICAL RES. COUNCIL	AFV-
AUSTRALIA. NATIONAL STANDARDS LAB., CHIPPENDALE, NEW SOUTH WALES	AUS-NSL-
AUSTRALIA. NATIONAL STANDARDS LAB., CHIPPENDALE, NEW SOUTH WALES	PA-
AUSTRALIA. RADIO PROPAGATION COMMITTEE	ARPC-
AUSTRALIA. RADIOPHYSICS LAB., CHIPPENDALE, NEW SOUTH WALES	RPR-
AUSTRALIA. WEAPONS RESEARCH ESTABLISHMENT	CPD-
AUSTRALIA. WEAPONS RESEARCH ESTABLISHMENT	CPD-(YR)-
AUSTRALIA. WEAPONS RESEARCH ESTABLISHMENT	HSA-
AUSTRALIA. WEAPONS RESEARCH ESTABLISHMENT	ISD-
AUSTRALIA. WEAPONS RESEARCH ESTABLISHMENT	NRE-ISD-TN
AUSTRALIA. WEAPONS RESEARCH ESTABLISHMENT	PAD-
AUSTRALIA. WEAPONS RESEARCH ESTABLISHMENT	WPD-
AUSTRALIA. WEAPONS RESEARCH ESTABLISHMENT	WRE-
AUSTRALIA. WEAPONS RESEARCH ESTABLISHMENT	WRE-CPD-

AVCO CORP.

AUSTRALIA. WEAPONS RESEARCH ESTABLISHMENT	WRE-CPD/T/-
AUSTRALIA. WEAPONS RESEARCH ESTABLISHMENT	WRE-HSA-
AUSTRALIA. WEAPONS RESEARCH ESTABLISHMENT	WRE-ISD-TN-
AUSTRALIA. WEAPONS RESEARCH ESTABLISHMENT	WRE-PAD-
AUSTRALIA. WEAPONS RESEARCH ESTABLISHMENT	WRE-TM-HSA-
AUSTRALIA. WEAPONS RESEARCH ESTABLISHMENT	WRE-TM-ISD-
AUSTRALIA. WEAPONS RESEARCH ESTABLISHMENT	WRE-TM-TRD-
AUSTRALIA. WEAPONS RESEARCH ESTABLISHMENT	WRE-TN-APD-
AUSTRALIA. WEAPONS RESEARCH ESTABLISHMENT	WRE-TN-CPD-
AUSTRALIA. WEAPONS RESEARCH ESTABLISHMENT	WRE-TN-EC-
AUSTRALIA. WEAPONS RESEARCH ESTABLISHMENT	WRE-TN-ISD-
AUSTRALIA. WEAPONS RESEARCH ESTABLISHMENT	WRE-TN-PAD-
AUSTRALIA. WEAPONS RESEARCH ESTABLISHMENT	WRE-TN-TRD-
AUSTRALIA. WEAPONS RESEARCH ESTABLISHMENT	WRE-TN-WPD-
AUSTRALIA. WEAPONS RESEARCH ESTABLISHMENT	WRE-WPD-
AUSTRALIA. WEAPONS RESEARCH ESTABLISHMENT, ADELAIDE	SAD-(YEAR)/(NO./NO.)
AUSTRALIA. WEAPONS RESEARCH ESTABLISHMENT, ADELAIDE	TM-SAD-
AUSTRALIA. WEAPONS RESEARCH ESTABLISHMENT, ADELAIDE	WRE-TM-ED-
AUSTRALIA. WEAPONS RESEARCH ESTABLISHMENT, ADELAIDE	WRE-TM-WSD-
AUSTRALIA. WEAPONS RESEARCH ESTABLISHMENT, ADELAIDE	WRE-TN-APD-
AUSTRALIA. WEAPONS RESEARCH ESTABLISHMENT, ADELAIDE	WRE-TN-ED-
AUSTRALIA. WEAPONS RESEARCH ESTABLISHMENT, ADELAIDE	WRE-TN-PD-
AUSTRALIA. WEAPONS RESEARCH ESTABLISHMENT, ADELAIDE	WRE-TN-SAD-
AUSTRALIA. WEAPONS RESEARCH ESTABLISHMENT, ADELAIDE	WRE-TN-TSD-
AUSTRALIA. WEAPONS RESEARCH ESTABLISHMENT, SALISBURY	APD-
AUSTRALIA. WEAPONS RESEARCH ESTABLISHMENT, SALISBURY	TN-PAD-
AUSTRALIA. WEAPONS RESEARCH ESTABLISHMENT, SALISBURY	TN-SAD-
AUSTRALIA. WEAPONS RESEARCH ESTABLISHMENT, SALISBURY	URE-TN-WSD-
AUSTRALIA. WEAPONS RESEARCH ESTABLISHMENT, SALISBURY	WRE-SAD-
AUSTRALIA. WEAPONS RESEARCH ESTABLISHMENT, SALISBURY	WRE-TM-A-
AUSTRALIA. WEAPONS RESEARCH ESTABLISHMENT, SALISBURY	WRE-TM-CPD/T/
AUSTRALIA. WEAPONS RESEARCH ESTABLISHMENT, SALISBURY	WRE-TN-CPD-
AUSTRALIA. WEAPONS RESEARCH ESTABLISHMENT, SALISBURY	WRE-TN-DWD-
AUSTRALIA. WEAPONS RESEARCH ESTABLISHMENT, SALISBURY	WRE-TN-ECD-
AUSTRALIA. WEAPONS RESEARCH ESTABLISHMENT, SALISBURY	WRE-TN-TSD-
AUSTRALIA. WEAPONS RESEARCH ESTABLISHMENT, SALISBURY	WRE-TN-WSD-
AUSTRALIA. WEAPONS RESEARCH ESTABLISHMENT. DEPT. OF SUPPLY, ADELAIDE	WRE-TM-PAD-
AUSTRALIA. WEAPONS RESEARCH ESTABLISHMENT. DEPT. OF SUPPLY, ADELAIDE	WRE-TN-HSA-
AUSTRALIAN ARMY OPERATIONAL RESEARCH GROUP	MEMO-M-
AUSTRALIAN ATOMIC ENERGY COMMISSION, SYDNEY	AAEC/K-
AUSTRALIAN ATOMIC ENERGY COMMISSION. RESEARCH ESTABLISHMENT, LUCAS HEIGHTS, NEW SOUTH WALES	AAEC-
AUSTRALIAN ATOMIC ENERGY COMMISSION. RESEARCH ESTABLISHMENT, LUCAS HEIGHTS, NEW SOUTH WALES	AAEC/E-
AUSTRALIAN ATOMIC ENERGY COMMISSION. RESEARCH ESTABLISHMENT, LUCAS HEIGHTS, NEW SOUTH WALES	AAEC/LIB/BIB-
AUSTRALIAN ATOMIC ENERGY COMMISSION. RESEARCH ESTABLISHMENT, LUCAS HEIGHTS, NEW SOUTH WALES (LIST OF REPORT PUBLICATIONS)	AAEC-LRP (MONTH/YEAR)
AUSTRALIAN ATOMIC ENERGY COMMISSION. RESEARCH ESTABLISHMENT, LUCAS HEIGHTS, NEW SOUTH WALES	AAEC/M-
AUSTRALIAN ATOMIC ENERGY COMMISSION. RESEARCH ESTABLISHMENT, LUCAS HEIGHTS, NEW SOUTH WALES	AAEC/PM-
AUSTRALIAN ATOMIC ENERGY COMMISSION. RESEARCH ESTABLISHMENT, LUCAS HEIGHTS, NEW SOUTH WALES (PROGRESS REPORT)	AAEC/PR(NUMBER)-(LTR.)
AUSTRALIAN ATOMIC ENERGY COMMISSION. RESEARCH ESTABLISHMENT, LUCAS HEIGHTS, NEW SOUTH WALES	AAEC(SP)/(LTR)(NO.)
AUSTRALIAN ATOMIC ENERGY COMMISSION. RESEARCH ESTABLISHMENT, LUCAS HEIGHTS, NEW SOUTH WALES	AAEC/TM-
AUSTRALIAN ATOMIC ENERGY COMMISSION. RESEARCH ESTABLISHMENT, LUCAS HEIGHTS, NEW SOUTH WALES	AAEC/TRANS.-
AUSTRALIAN ATOMIC ENERGY COMMISSION. RESEARCH ESTABLISHMENT, LUCAS HEIGHTS, NEW SOUTH WALES	AAEC/X-
AUSTRALIAN ATOMIC ENERGY COMMISSION. RESEARCH ESTABLISHMENT, LUCAS HEIGHTS, NEW SOUTH WALES	CE/TN-
AUSTRALIAN ATOMIC ENERGY COMMISSION. RESEARCH ESTABLISHMENT, LUCAS HEIGHTS, NEW SOUTH WALES	ER/TN-
AUSTRALIAN ATOMIC ENERGY COMMISSION. RESEARCH ESTABLISHMENT, LUCAS HEIGHTS, NEW SOUTH WALES	FCD/TN-
AUSTRALIAN ATOMIC ENERGY COMMISSION. RESEARCH ESTABLISHMENT, LUCAS HEIGHTS, NEW SOUTH WALES	FED-
AUSTRALIAN ATOMIC ENERGY COMMISSION. RESEARCH ESTABLISHMENT, LUCAS HEIGHTS, NEW SOUTH WALES (TRANSLATION)	LIB/TRANS-
AUSTRALIAN ATOMIC ENERGY COMMISSION. RESEARCH ESTABLISHMENT, LUCAS HEIGHTS, NEW SOUTH WALES	MP/IN-
AUSTRALIAN ATOMIC ENERGY COMMISSION. RESEARCH ESTABLISHMENT, LUCAS HEIGHTS, NEW SOUTH WALES	RP/TN-
AUSTRALIAN COUNCIL FOR AERONAUTICS, MELBOURNE	ACA-
AUSTRALIAN COUNCIL FOR AERONAUTICS, MELBOURNE	AUS ACA R-
AUSTRALIAN NATIONAL UNIV., CANBERRA	ANU-
AUSTRALIAN NATIONAL UNIV., CANBERRA. DEPT. OF ENGINEERING PHYSICS	EP-RR-
AUSTRALIAN NATIONAL UNIV., CANBERRA. INST. OF ADVANCED STUDIES.	ANU-P-
AUSTRALIAN NATIONAL UNIV., CANBERRA. RESEARCH SCHOOL OF PHYSICAL SCIENCES	ANU/P-
AUTOMATIC DATA FIELD SYSTEMS COMMAND, FT.BELVOIR, VA.	ADFSC-01-(NUMBER-NUMBER)
AUTOMATIC DATA FIELD SYSTEMS COMMAND, FT.BELVOIR, VA.	ADFSC-
AUTOMETRIC CORP., N.Y.C.	CR-(NUMBER)-(NO.,LTRS.)
AUTONETICS, ANAHEIM, CALIF.	C(NUMBER)-(NO.)/(NO.)
AUTONETICS, ANAHEIM, CALIF.	CTR-(NO.)-(NO.)-
AUTONETICS, ANAHEIM, CALIF.	T(NUMBER)-(NO.)/(NO.)
AUTONETICS, ANAHEIM, CALIF.	TM-(YEAR)-
AUTONETICS, ANAHEIM, CALIF.	X(NUMBER)-(NO.)/(NO.)
AUTONETICS, DOWNEY, CALIF.	EM-(NUMBER)-(NUMBER)
AUTONETICS DIV., NORTH AMERICAN AVIATION, INC., ANAHEIM, CALIF. (NOT USED AFTER 1957)	A-
AUTONETICS DIV., NORTH AMERICAN AVIATION, INC., ANAHEIM, CALIF.	BR5-
AUTONETICS DIV., NORTH AMERICAN AVIATION, INC., ANAHEIM, CALIF. (CONTRACTUAL REPORT, 1964)	C4-
AUTONETICS DIV., NORTH AMERICAN AVIATION, INC., ANAHEIM, CALIF. (CONTRACTUAL REPORT, 1965)	C5-
AUTONETICS DIV., NORTH AMERICAN AVIATION, INC., ANAHEIM, CALIF. (CONTRACTUAL REPORT, 1966)	C6-
AUTONETICS DIV., NORTH AMERICAN AVIATION, INC., ANAHEIM, CALIF. (NOT USED AFTER 1963)	EM-
AUTONETICS DIV., NORTH AMER.AVIATION INC.,ANAHEIM,CAL.	NAA-C4-
AUTONETICS DIV., NORTH AMERICAN AVIATION, INC., ANAHEIM, CALIF. (NON-CONTRACTUAL REPORT, 1966)	T6-
AUTONETICS DIV., NORTH AMERICAN AVIATION, INC., ANAHEIM, CALIF. (PAPER, REPRINT, OR MONOGRAPH, 1966)	X6-
AUTONETICS DIV., NORTH AMER. AVIATION INC.,DOWNEY,CAL	AUTO-
AVCO CORP. (ALL LOCATIONS)	AVCO-
AVCO CORP., TULSA, OKLA.	AVCO-TR-
AVCO CORP., TULSA, OKLA.	TR-G-
AVCO CORP., WILMINGTON, MASS.	L(NO.)-

AVCO CORP. APPLIED TECHNOLOGY DIV.

AVCO CORP. APPLIED TECHNOLOGY DIV., LOWELL, MASS.	AVATD-(NO.)-(YEAR)-RR
AVCO CORP. APPLIED TECHNOLOGY DIV., LOWELL, MASS.	AVATD-(NO.)-(YEAR)-CR
AVCO CORP. AVCO-EVERETT RESEARCH LAB., EVERETT, MASS.	AERL-
AVCO CORP. AVCO-EVERETT RESEARCH LAB., EVERETT, MASS.	AERL-(YEAR)-
AVCO CORP. AVCO-EVERETT RESEARCH LAB., EVERETT, MASS.	AERL-AMP-
AVCO CORP. AVCO-EVERETT RESEARCH LAB., EVERETT, MASS. (RESEARCH NOTE)	AERL-RN-
AVCO CORP. AVCO-EVERETT RESEARCH LAB., EVERETT, MASS. (RESEARCH REPORT)	AERL-RR-
AVCO CORP. AVCO-EVERETT RESEARCH LAB., EVERETT, MASS.	AFRD-(NUMBER)-
AVCO CORP. AVCO-EVERETT RESEARCH LAB., EVERETT, MASS.	AMP-
AVCO CORP. AVCO-EVERETT RESEARCH LAB., EVERETT, MASS.	AVCO-AERL-
AVCO CORP. AVCO-EVERETT RESEARCH LAB., EVERETT, MASS.	AVCO-AMP-
AVCO CORP. AVCO-EVERETT RESEARCH LAB., EVERETT, MASS.	AVCO-EVERETT RES. REPT.-
AVCO CORP. AVCO-EVERETT RESEARCH LAB., EVERETT, MASS. (RESEARCH NOTE)	AVCO-RN-
AVCO CORP. AVCO-EVERETT RESEARCH LAB., EVERETT, MASS.	AVCO-RR-
AVCO CORP. AVCO-EVERETT RESEARCH LAB., EVERETT, MASS.	MPR-(YEAR)-(MONTH)
AVCO CORP. AVCO MISSILE, SPACE + ELECTRONICS GROUP, WILMINGTON, MASS.	ARL-(YEAR)-
AVCO CORP. AVCO MISSILE, SPACE + ELECTRONICS GROUP, WILMINGTON, MASS.	AVMSD-(NUMBER)-(YR.)-CR
AVCO CORP. AVCO MISSILE, SPACE + ELECTRONICS GROUP, WILMINGTON, MASS.	ICD-(NUMBER)-(YEAR)-
AVCO CORP. AVCO MISSILE SYSTEMS DIV., WILMINGTON, MASS. (PROFESSIONAL PAPER)	AVMSD-(NUMBER)-(YR.)-PP
AVCO CORP. AVCO MISSILE SYSTEMS DIV., WILMINGTON, MASS. (RESEARCH MEMO)	AVMSD-(NUMBER)-(YR.)-RM
AVCO CORP. AVCO MISSILE SYSTEMS DIV., WILMINGTON, MASS. (RESEARCH REPORT)	AVMSD-(NUMBER)-(YR.)-RR
AVCO CORP. AVCO SPACE SYSTEMS DIV., LOWELL, MASS.	AVSSD-(NUMBER)-(YR.)-CR
AVCO CORP. AVCO SPACE SYSTEMS DIV., WILMINGTON, MASS. (CONTRACT REPORT)	AVSSD-(NUMBER)-(YR.)-CR
AVCO CORP. AVCO SPACE SYSTEMS DIV., WILMINGTON, MASS. (PROFESSIONAL PAPER)	AVSSD-(NUMBER)-(YR.)-PP
AVCO CORP. AVCO SPACE SYSTEMS DIV., WILMINGTON, MASS. (RESEARCH MEMO)	AVSSD-(NUMBER)-(YR.)-RM
AVCO CORP. AVCO SPACE SYSTEMS DIV., WILMINGTON, MASS. (RESEARCH REPORT)	AVSSD-(NUMBER)-(YR.)-RR
AVCO CORP. AVCO SPACE SYSTEMS DIV. AVCO GOVERNMENT PRODUCTS GROUP, LOWELL, MASS.	AVSSD-(NO.)-(YR.)-
AVCO CORP. AVCO SPACE SYSTEMS DIV. AVCO GOVERNMENT PRODUCTS GROUP, LOWELL, MASS.	AVSSD-(LTR)(NO)-EBB-
AVCO CORP. AVCO SYSTEMS DIV. AVCO GOVERNMENT PRODUCTS GROUP, WILMINGTON, MASS.	AVSD-(NUMBER)-(YR.)-RR
AVCO CORP. CROSLEY BROADCASTING CORP., CINCINNATI	AVCO-CR-
AVCO CORP. CROSLEY DIV. ELECTRONIC RES. LAB., BOSTON	EW-
AVCO CORP. ELECTRONICS DIV., CINCINNATI	80045-(LETTERS)-
AVCO CORP. RESEARCH AND ADVANCED DEVELOPMENT DIV., WILMINGTON, MASS. (QUARTERLY REPORT)	AVCO-QR-
AVCO CORP. RES.+ ADVANCED DEV. DIV., WILMINGTON, MASS.	GA-
AVCO CORP. RES.+ ADVANCED DEV. DIV., WILMINGTON, MASS.	L(NO.)-HKA-(YEAR)-
AVCO CORP. RES.+ ADVANCED DEV. DIV., WILMINGTON, MASS.	RAD-
AVCO CORP. RES.+ ADVANCED DEV. DIV., WILMINGTON, MASS.	RAD-(NUMBER)-
AVCO CORP. RES.+ ADVANCED DEV. DIV., WILMINGTON, MASS.	RAD-M-
AVCO CORP. RES.+ ADVANCED DEV. DIV., WILMINGTON, MASS.	RAD-SR-(YEAR)-
AVCO CORP. RES.+ ADVANCED DEV. DIV., WILMINGTON, MASS.	RAD-TM-(YEAR)-
AVCO CORP. RES.+ ADVANCED DEV. DIV., WILMINGTON, MASS.	RR-
AVCO CORP. RES.+ ADVANCED DEV. DIV., WILMINGTON, MASS.	TDM-W-
AVCO CORP. RES. AND DEV. DIV., WILMINGTON, MASS.	RAD-TR-(YEAR)-
AVCO CORP. SPACE SYSTEMS DIV., LOWELL, MASS.	AVSSD-(NUMBER)-(YR.)-PP
AVCO CORP. SYSTEMS DIV., LOWELL, MASS.	AVSD-(NUMBER)-(YR.)-CR
AVCO CORP. TELEMETRY AND INSTRUMENTATION DEPT., WILMINGTON, MASS.	TIDM-F(NUMBER)-RM-
AVCO EVERETT RESEARCH LAB., EVERETT, MASS.	AERL-DSTR-
AVCO MANUFACTURING CORP., EVERETT, MASS.	AVCO-MPR-(YEAR)-
AVCO MANUFACTURING CORP., LAWRENCE, MASS.	AVCO (NO.)-TM(YR.-NO.)
AVCO MANUFACTURING CORP., LAWRENCE, MASS.	AVCO-MPR-(YEAR)-
(SEE ALSO LATER NAME: AVCO CORP.)	
AVCO MANUFACTURING CORP. ADVANCED DEVELOPMENT DIV., STRATFORD, CONN. (REF. 1)	(NUMBERS...)
AVCO MFG. CORP. ADVANCED DEV. DIV., STRATFORD, CONN.	AADD-
AVCO MFG. CORP. ADVANCED DEV. DIV., STRATFORD, CONN.	ADD-
AVCO MFG. CORP. ADVANCED DEV. DIV., STRATFORD, CONN.	AVCO-ADD-
AVCO MFG. CORP. ADVANCED DEV. DIV., STRATFORD, CONN.	MPR(YR)-(MO)
AVCO MFG. CORP. AVCO RESEARCH LAB., EVERETT, MASS.	AERL-BIB-
AVCO MFG. CORP. AVCO RESEARCH LAB., EVERETT, MASS.	ARL-
AVCO MFG. CORP. AVCO RESEARCH LAB., EVERETT, MASS.	AVCO-ARL-
AVCO MFG. CORP. AVCO RESEARCH LAB., EVERETT, MASS.	AVCO-ARL-RR-
AVCO MFG. CORP. AVCO RESEARCH LAB., EVERETT, MASS. (SEMI-ANNUAL REPORT)	AVCO REP SAR (INCL.DATES)
AVCO MFG. CORP. RESEARCH AND ADVANCED DEVELOPMENT DIV., LAWRENCE, MASS.	AVCO AADD (NUMBER)-
AVCO MFG. CORP. RESEARCH AND ADVANCED DEVELOPMENT DIV., LAWRENCE, MASS.	AVCO-RAD-
AVCO MFG. CORP. RESEARCH AND ADVANCED DEVELOPMENT DIV., LAWRENCE, MASS.	AVCO-RAD-(NO.)-
AVCO MFG. CORP. RESEARCH AND ADVANCED DEVELOPMENT DIV., LAWRENCE, MASS.	AVCO-RAD-(NO.)-TM-
AVCO MFG. CORP. RESEARCH AND ADVANCED DEVELOPMENT DIV., LAWRENCE, MASS.	AVCO-RAD-MP-(NO.,YR)-
AVCO MFG. CORP. RESEARCH AND ADVANCED DEVELOPMENT DIV., LAWRENCE, MASS. (SEMI-ANNUAL RPT.)	AVCO-RAD-SA-(NO.)-
AVCO MFG. CORP. RESEARCH AND ADVANCED DEVELOPMENT DIV., LAWRENCE, MASS.	AVCO-RAD-SR-
AVCO MFG. CORP. RESEARCH AND ADVANCED DEVELOPMENT DIV., LAWRENCE, MASS.	AVCO-RAD-TM-
AVCO MFG. CORP. RESEARCH AND ADVANCED DEVELOPMENT DIV., LAWRENCE, MASS.	AVCO-RAD-TR-
AVCO MFG. CORP. RESEARCH AND ADVANCED DEVELOPMENT DIV., LAWRENCE, MASS.	AVCO-SR-(NO.-YR.)-
AVCO MFG. CORP. RESEARCH AND ADVANCED DEVELOPMENT DIV., LAWRENCE, MASS.	AVCO-TR-(NO.-YR.)-
AVCO MFG. CORP. RESEARCH AND ADVANCED DEVELOPMENT DIV., LAWRENCE, MASS.	EEDTR-T-
AVCO MFG. CORP. RESEARCH AND ADVANCED DEVELOPMENT DIV., LAWRENCE, MASS.	MPR(YR)-(MO)
AVCO MFG. CORP. RESEARCH AND ADVANCED DEVELOPMENT DIV., LAWRENCE, MASS.	RADMP-
AVCO MFG. CORP. RESEARCH AND ADVANCED DEVELOPMENT DIV., LAWRENCE, MASS.	RAD-MP-2-(YR.)-
AVCO MFG. CORP. RESEARCH AND ADVANCED DEVELOPMENT DIV., LAWRENCE, MASS.	RAD-SA-(NO.)-
AVCO MFG. CORP. RESEARCH AND ADVANCED DEVELOPMENT DIV., WILMINGTON, MASS. (PROGRESS REPORT)	AVCO-ADD-
AVCO MFG. CORP. RESEARCH AND ADVANCED DEVELOPMENT DIV., WILMINGTON, MASS.	AVCO-RAD-
AVCO MFG. CORP. RESEARCH AND ADVANCED DEVELOPMENT DIV., WILMINGTON, MASS.	AVCO-RAD-MP-(NO.,YR)-
AVCO MFG. CORP. RESEARCH AND ADVANCED DEVELOPMENT DIV., WILMINGTON, MASS.	GRD-TR(YR)-(NO.)-
AVCO MFG. CORP. RESEARCH AND ADVANCED DEVELOPMENT DIV., WILMINGTON, MASS.	RAD-(NO.)-TM-
AVCO MFG. CORP. RESEARCH AND ADVANCED DEVELOPMENT DIV., WILMINGTON, MASS.	RAD-SA-
AVCO MFG. CORP. SYSTEMS MANAGEMENT (CONTRACT REPORT)	AVCSM-(NO.)-(YR.)-CR
AVCO MFG. CORP.SYSTEMS MANAGEMENT(PROFESSIONAL PAPER)	AVCSM-(NO.)-(YR.)-PP
AVCO MFG. CORP. SYSTEMS MANAGEMENT (RESEARCH REPORT)	AVCSM-(NO.)-(YR.)-RM
AVCO MFG. CORP. SYSTEMS MANAGEMENT (RESEARCH REPORT)	AVCSM-(NO.)-(YR.)-RR

AVCO MARINE ELECTRONICS OFFICE, NEW LONDON, CONN.	MED-
AVCO MISSILE SPACE AND ELECTRONICS GROUP. AVCO SPACE DIV., WILMINGTON, MASS.	SPEC-S-
AVIATEST G.M.B.H., DUESSELDORF	BMVTDG-FBWT-(YEAR)-
AVIATION ENGINEERING CORP., WOODSIDE, N.Y.	AVIEN-
AVIATION SAFETY ENG. AND RESEARCH, PHOENIX, ARIZ.	AVSER-(YEAR)-
AVIATION SUPPLY OFFICE, PHILADELPHIA	ASO-
AVIDYNE RESEARCH, INC., BURLINGTON, MASS.	ADR-
AVION ELECTRONICS, INC., PARAMUS, N.J.	AR-
AVION INSTRUMENT CORP., PARAMUS, N.J.	AVIC-
AZTEC SCHOOL OF LANGUAGES, INC., MAYNARD, MASS.	AZT-(YEAR-NUMBER)-FRLL
AZTEC SCHOOL OF LANGUAGES, INC., MAYNARD, MASS.	AZT-(YEAR-NUMBER)-RULL
AZTEC SCHOOL OF LANGUAGES, INC. RESEARCH TRANSLATION DIV., MC LEAN, VA.	AZT-(NO.)-(NO.)-GENRL

BABCOCK & WILCOX CO.

BABCOCK + WILCOX CO. (ALL LOCATIONS)	BAW-(NUMBER)-
BABCOCK + WILCOX CO. (ALL LOCATIONS)	BAW-
BABCOCK + WILCOX CO. (ALL LOCATIONS, LAB. REPORTS)	BW-
BABCOCK + WILCOX CO. (ALL LOCATIONS, LAB. REPORTS)	BW D-
BABCOCK + WILCOX CO. (ALL LOCATIONS, LAB. REPORTS)	BW-RDE-
BABCOCK + WILCOX CO. (ALL LOCATIONS, LAB. REPORTS)	BW RR-
BABCOCK + WILCOX CO. (ALL LOCATIONS)	DC-(YEAR-NUMBER)-
BABCOCK + WILCOX CO. (ALL LOCATIONS)	R + D-
BABCOCK + WILCOX CO.,BARBERTON, OHIO (PROGRESS RPT.)	BAW-PR-
BABCOCK + WILCOX CO., GALVESTON, TEX.	STS-
BABCOCK + WILCOX CO., LYNCHBURG, VA.	TRG-(YEAR)-
BABCOCK + WILCOX CO., N.Y.C.	B AND W-
BABCOCK + WILCOX CO., N.Y.C.	ES-
BABCOCK + WILCOX CO. ATOMIC ENERGY DIV., AKRON, OHIO	BW-AED-
BABCOCK + WILCOX CO. ATOMIC EN. DIV., LYNCHBURG, VA.	ATR-
BABCOCK + WILCOX CO. ATOMIC EN. DIV., LYNCHBURG, VA.	ATRCE-
BABCOCK + WILCOX CO. ATOMIC EN. DIV., LYNCHBURG, VA.	ATR-FE-
BABCOCK + WILCOX CO. ATOMIC EN. DIV., LYNCHBURG, VA. (ADVANCED TEST REACTOR)	ATR-IC-
BABCOCK + WILCOX CO. ATOMIC EN. DIV., LYNCHBURG, VA.	BAW-TM-
BABCOCK + WILCOX CO. ATOMIC EN. DIV., LYNCHBURG, VA.	BAW-TPCC-
BABCOCK + WILCOX CO. ATOMIC EN. DIV., LYNCHBURG, VA.	S-R-
BABCOCK + WILCOX CO. ATOMIC EN. DIV., LYNCHBURG, VA.	TPCC-
BABCOCK + WILCOX CO. RESEARCH CENTER, ALLIANCE, OHIO	BAW-RR-
BABCOCK + WILCOX CO. RESEARCH CENTER, ALLIANCE, OHIO	B/W-
BABCOCK + WILCOX CO. RESEARCH CENTER, ALLIANCE, OHIO	B+W-
BABCOCK ELECTRONICS CORP., COSTA MESA, CALIF.	BEC-
BACON (FREDERICK S.) LABS., WATERTOWN, MASS.	E-
BADGER ARMY AMMUNITION PLANT, BARABOO, WIS.	TD-
BADGER ORDNANCE WORKS, BARABOO, WIS.	LP-
BAIRD ASSOCIATES, INC., CAMBRIDGE, MASS. (REF. 1)	(NUMBER)
BAIRD ASSOCIATES, INC., CAMBRIDGE, MASS.	BA-
BAIRD ASSOCS., INC., CAMBRIDGE, MASS.(FINAL TECH.RPT)	BA-FTR-(YR./MONTH)
BAIRD ASSOCIATES, INC., CAMBRIDGE, MASS.	BAI-
BAIRD ASSOCIATES, INC., CAMBRIDGE, MASS.	BAIR-
BAIRD ASSOCIATES, INC., CAMBRIDGE, MASS.	BAIRD-
BAIRD ASSOCS., INC., CAMBRIDGE, MASS. (QUARTERLY RPT)	BA-QR-
BAIRD ASSOCIATES, INC., CAMBRIDGE, MASS.	J.O.-
BAIRD ATOMIC, INC., BEDFORD, MASS.	(NUMBER)-FR-
BAIRD ATOMIC, INC., CAMBRIDGE, MASS.	(NUMBER)-IR-
BALCO RESEARCH LAB., NEWARK, N.J.	BALCO-
BALDWIN-LIMA-HAMILTON CORP., WALTHAM, MASS.	BLH-TD-(NUMBERS)-
BALDWIN PIANO CO., CINCINNATI	AFAC-
BALLISTIC MISSILE DIV. (AIR FORCE) (ACCESSION LIST)	AF-ARDC-AL-
BALLISTIC MISSILE DIV. (AIR FORCE)	AF-ARDC-BIB-
BALLISTIC MISSILE DIV. (AIR FORCE)	NN-
BALLISTIC MISSILE DIV. (AIR FORCE), INGLEWOOD, CALIF.	AFBMD-
BALLISTIC MISSILE DIV. (AIR FORCE), INGLEWOOD, CALIF.	AFBMD-DWS-TM-
BALLISTIC MISSILE DIV. (AIR FORCE), INGLEWOOD, CALIF.	AFBMD-RCS-DD...-
BALLISTIC MISSILE DIV. (AIR FORCE), INGLEWOOD, CALIF.	AFBMD-TN-(YEAR)-
BALLISTIC MISSILE DIV. (AIR FORCE), INGLEWOOD, CALIF.	AFBMD-TR-(YEAR)-
BALLISTIC MISSILE DIV. (AIR FORCE), INGLEWOOD, CALIF.	AFBMD-WDLPM-
BALLISTIC MISSILE DIV. (AIR FORCE), INGLEWOOD, CALIF.	ARDC-GM-(YEAR)-
BALLISTIC MISSILE DIV. (AIR FORCE), INGLEWOOD, CALIF.	ARDC-WDTP-(YEAR)-
BALLISTIC MISSILE DIV. (AIR FORCE), INGLEWOOD, CALIF.	BMD-
BALLISTIC MISSILE DIV. (AIR FORCE), INGLEWOOD, CALIF.	BMD (LETTER(S))-
BALLISTIC MISSILE DIV. (AIR FORCE), INGLEWOOD, CALIF.	BMD TN-(YEAR)-
BALLISTIC MISSILE DIV. (AIR FORCE), INGLEWOOD, CALIF.	BMD TR-(YEAR)-
BALLISTIC MISSILE DIV. (AIR FORCE), INGLEWOOD, CALIF.	COM-
BALLISTIC MISSILE DIV. (AIR FORCE), INGLEWOOD, CALIF.	GM-(NUMBER)-(NUMBER)-
BALLISTIC MISSILE DIV. (AIR FORCE), INGLEWOOD, CALIF.	GM-TR-
BALLISTIC MISSILE DIV. (AIR FORCE), INGLEWOOD, CALIF.	WDT-EX-(YEAR)-
BALLISTIC MISSILE DIV. (AIR FORCE), LOS ANGELES	RCS-DD-DR+E(M)-
BALLISTIC MISSILE DIV. (AIR FORCE), LOS ANGELES	RCS-DD-SD(M)-
BALLISTIC MISSILE DIV. (AIR FORCE), LOS ANGELES	WDAL-
BALLISTIC MISSILE DIV. (AIR FORCE), LOS ANGELES	WDLPB-
BALLISTIC MISSILE DIV. (AIR FORCE), LOS ANGELES	WDLPM-
BALLISTIC MISSILE DIV. (AIR FORCE), LOS ANGELES	WDLPR-
BALLISTIC MISSILE DIV. (AIR FORCE), LOS ANGELES	WDPCR-
BALLISTIC MISSILE DIV. (AIR FORCE), LOS ANGELES	WDZAP-
BALLISTIC MISSILES COMM. (AIR FORCE), WASHINGTON,D.C.	AFBMC-(LETTERS)-(NUMBER)-
BALLISTIC MISSILES COMM. (AIR FORCE), WASHINGTON,D.C.	AF-WDPC-(YEAR)-
BALLISTIC RESEARCH LABS.,ABERDEEN PROVING GROUND, MD.	AM-
BALLISTIC RESEARCH LABS.,ABERDEEN PROVING GROUND, MD.	APG-BRL-
BALLISTIC RESEARCH LABS.,ABERDEEN PROVING GROUND, MD.	APG-BRLM-
BALLISTIC RESEARCH LABS.,ABERDEEN PROVING GROUND, MD.	APG-BRLR-
BALLISTIC RESEARCH LABS.,ABERDEEN PROVING GROUND, MD.	APG-QPR-
BALLISTIC RESEARCH LABS.,ABERDEEN PROVING GROUND, MD. (TRANSLATIONS OF GERMAN ARCHIVES)	ARCH-
BALLISTIC RESEARCH LABS.,ABERDEEN PROVING GROUND, MD.	BAL-TN-
BALLISTIC RESEARCH LABS.,ABERDEEN PROVING GROUND, MD.	BRL-
BALLISTIC RESEARCH LABS.,ABERDEEN PROVING GROUND, MD.	BRL-AF-(CONTRACT NO.)
BALLISTIC RESEARCH LABS.,ABERDEEN PROVING GROUND, MD.	BRL ANNUAL/
BALLISTIC RESEARCH LABS.,ABERDEEN PROVING GROUND, MD.	BRL BT (NO.,LTRS.)
BALLISTIC RESEARCH LABS.,ABERDEEN PROVING GROUND, MD.	BRL-CR-
BALLISTIC RESEARCH LABS.,ABERDEEN PROVING GROUND, MD.	BRL DATA RPT.-
BALLISTIC RESEARCH LABS.,ABERDEEN PROVING GROUND, MD.	BRL-IBL STATUS-
BALLISTIC RESEARCH LABS.,ABERDEEN PROVING GROUND, MD.	BRL-M-
BALLISTIC RESEARCH LABS.,ABERDEEN PROVING GROUND, MD.	BRLM-
BALLISTIC RESEARCH LABS.,ABERDEEN PROVING GROUND, MD. (MEMORANDUM REPORTS)	BRL-MEMO-
BALLISTIC RESEARCH LABS.,ABERDEEN PROVING GROUND, MD.	

BALLISTIC RESEARCH LABS., ABERDEEN PROVING GROUND, MD. (MEMORANDUM REPORT)	BRL-MR-
BALLISTIC RESEARCH LABS., ABERDEEN PROVING GROUND, MD. (ORDNANCE COMPUTER RES. RPT.)	BRL-OCRR-
BALLISTIC RESEARCH LABS., ABERDEEN PROVING GROUND, MD.	BRL QPR (NO.,LTRS.,NOS.)
BALLISTIC RESEARCH LABS., ABERDEEN PROVING GROUND, MD. (REPORTS)	BRL-R-
BALLISTIC RESEARCH LABS., ABERDEEN PROVING GROUND, MD. (REPORTS)	BRLR-
BALLISTIC RESEARCH LABS., ABERDEEN PROVING GROUND, MD.	BRL-TB-
BALLISTIC RESEARCH LABS., ABERDEEN PROVING GROUND, MD. (TECHNICAL NOTES)	BRL-TN-
BALLISTIC RESEARCH LABS., ABERDEEN PROVING GROUND, MD.	BRL-X-
BALLISTIC RESEARCH LABS., ABERDEEN PROVING GROUND, MD.	BTL-
BALLISTIC RESEARCH LABS., ABERDEEN PROVING GROUND, MD.	MEMO-
BALLISTIC RESEARCH LABS., ABERDEEN PROVING GROUND, MD.	OCSC-
BALLISTIC RESEARCH LABS., ABERDEEN PROVING GROUND, MD.	OMB-
BALLISTIC RESEARCH LABS., ABERDEEN PROVING GROUND, MD.	ORD PROJ TB-(NUMBERS)
BALLISTIC RESEARCH LABS., ABERDEEN PROVING GROUND, MD.	ORD TB-(NUMBERS)
BALLISTIC RESEARCH LABS., ABERDEEN PROVING GROUND, MD.	TB2-
BALLISTIC RESEARCH LABS., ABERDEEN PROVING GROUND, MD.	TB3-
BALLISTIC RESEARCH LABS. EXPLOSION KINETICS BRANCH, ABERDEEN PROVING GROUND, MD.	BRL-DR-
BALLISTIC SYSTEMS DIV. (AIR FORCE), INGLEWOOD, CALIF. (CONTRACT REPORT)	BSD-CR-(YEAR)-
BALLISTIC SYSTEMS DIV., NORTON AFB, CALIF.	AFBSD-TDR-(YEAR)-
BALLISTIC SYSTEMS DIV., NORTON AFB, CALIF.	AFBSD-TN-
BALLISTIC SYSTEMS DIV., NORTON AFB, CALIF.	AFBSD-TR-
BALLISTIC SYSTEMS DIV., NORTON AFB, CALIF.	(YEAR)-BSCPS-
BALLISTIC SYSTEMS DIV., NORTON AFB, CALIF.	BSD-
BALLISTIC SYSTEMS DIV., NORTON AFB, CALIF.	BSD-BSSFR-
BALLISTIC SYSTEMS DIV., NORTON AFB, CALIF.	BSD-EXHIBIT-(NO.)-
BALLISTIC SYSTEMS DIV., NORTON AFB, CALIF.	BSD-TDR-(YEAR)-
BALLISTIC SYSTEMS DIV., NORTON AFB, CALIF.	BSD-TR-(YEAR)-
BANGALORE, INDIA., UNIV. COLL. OF ENGINEERING	PGSAM-
BAR-ILAN UNIVERSITY, RAMAT-GAN, ISRAEL	BAR-ILANU-
BARDEN CORP., DANBURY, CONN.	PR-
BARDWELL AND MC ALISTER, INC., BURBANK, CALIF.	B + M-
BARKLEY + DEXTER LABS., INC., FITCHBURG, MASS.	UBL-
BARNES ENGINEERING CO., STAMFORD, CONN.	BE-
BARNES ENGINEERING CO., STAMFORD, CONN.	BEC-
BARNES ENGINEERING CO., STAMFORD, CONN.	BEC-(NUMBER)-DSR-
BARNES ENGINEERING CO., STAMFORD, CONN.	BEC-TR-
BARNES ENGINEERING CO., STAMFORD, CONN.	ODC-
BARODYNAMICS, INC., GEORGETOWN, COLO.	BARO-
BARODYNAMICS, INC., GEORGETOWN, COLO.	BAROD-(LTRS/LTRS)-
BARRY CONTROLS INC., WATERTOWN, MASS.	BARRY-
BARRY CONTROLS INC., WATERTOWN, MASS.	BCI-
BARRY WRIGHT CORP., WATERTOWN, MASS.	BW-R/D-
BARSS, REITZEL AND ASSOCIATES, INC., CAMBRIDGE, MASS.	BR-OEO-LN-
BARTOL RESEARCH FOUNDATION, SWARTHMORE, PENNA.	BRD-
BARTOL RESEARCH FOUNDATION, SWARTHMORE, PENNA.	BRF-
BARTOL RESEARCH FOUNDATION, SWARTHMORE, PENNA.	FI-BRF-
BASE TEN SYSTEMS, INC., MONMOUTH JUNCTION, N. J.	BTSI-
BASEL. UNIVERSITAET	BASELU-
BASEL. UNIVERSITAET	BMN-
BATELLE-DEFENDER INFO. ANALYSIS CTR., COLUMBUS, OHIO	BDIAC-
BATTELLE MEMORIAL INST., COLUMBUS, OHIO	ACTIAC-
BATTELLE MEMORIAL INST., COLUMBUS, OHIO	APS-
BATTELLE MEMORIAL INST., COLUMBUS, OHIO	BAT-
BATTELLE MEMORIAL INST., COLUMBUS, OHIO	BAT-(NOS.)-(NOS.)-
BATTELLE MEMORIAL INST., COLUMBUS, OHIO	BAT-(NUMBER)A-
BATTELLE MEMORIAL INST., COLUMBUS, OHIO	BAT-G(NUMBER)-
BATTELLE MEMORIAL INST., COLUMBUS, OHIO	BAT-PUB-
BATTELLE MEMORIAL INST., COLUMBUS, OHIO	BAT-RR-
BATTELLE MEMORIAL INST., COLUMBUS, OHIO	BATT-
BATTELLE MEMORIAL INST., COLUMBUS, OHIO	BMI-
BATTELLE MEMORIAL INST., COLUMBUS, OHIO	BMI-(LETTER)-
BATTELLE MEMORIAL INST., COLUMBUS, OHIO	BMI-(INITIALS)-
BATTELLE MEMORIAL INST., COLUMBUS, OHIO	BMI-(INITIALS-LETTER)-
BATTELLE MEMORIAL INST., COLUMBUS, OHIO	BMI-(INITIALS)-(MEMO)-
BATTELLE MEMORIAL INST., COLUMBUS, OHIO (ACCESSION LIST)	BMI-AL-
BATTELLE MEMORIAL INST., COLUMBUS, OHIO (JOINT VENTURE)	BMI-APDA-(YEAR)-
BATTELLE MEMORIAL INST., COLUMBUS, OHIO	BMI-NLVP-TM-
BATTELLE MEMORIAL INST., COLUMBUS, OHIO	BMI-NLVP-TR-
BATTELLE MEMORIAL INST., COLUMBUS, OHIO	BMI-X-
BATTELLE MEMORIAL INST., COLUMBUS, OHIO (REF. 10)	CT-
BATTELLE MEMORIAL INST., COLUMBUS, OHIO	DCIC-(YEAR)-
BATTELLE MEMORIAL INST., COLUMBUS, OHIO	G-(4 DIGITS)-1-
BATTELLE MEMORIAL INST., COLUMBUS, OHIO	NF-
BATTELLE MEMORIAL INST., COLUMBUS, OHIO	NMI-
BATTELLE MEMORIAL INST., COLUMBUS, OHIO. (TECHNICAL PROGRESS REVIEWS. REACTOR CORE MATERIALS)	RCM-
BATTELLE MEMORIAL INST., COLUMBUS, OHIO	SUPDIVE-RR-(NUMBER-YEAR)
BATTELLE MEMORIAL INST., COLUMBUS, OHIO	TPR-RCM-
BATTELLE MEMORIAL INST., COLUMBUS, OHIO	TPR-RM-
BATTELLE MEMORIAL INST., GENEVA	AR-
BATTELLE MEMORIAL INST., WASHINGTON, D.C.	BAT-T(NUMBER)-SR
BATTELLE MEMORIAL INST. DEFENSE CERAMICS INFORMATION CENTER, COLUMBUS, OHIO	DCIC-
BATTELLE MEMORIAL INST. DEFENSE CERAMICS INFORMATION CENTER, COLUMBUS, OHIO (CURRENT AWARENESS BULLETIN)	DCIC-CAB-
BATTELLE MEMORIAL INST. DEFENSE METALS INFORMATION CENTER, COLUMBUS, OHIO	DMIC-
BATTELLE MEMORIAL INST. DEFENSE METALS INFORMATION CENTER, COLUMBUS, OHIO	DMIC-MEMO-
BATTELLE MEMORIAL INST. DEFENSE METALS INFORMATION CENTER, COLUMBUS, OHIO	DMIC-S-
BATTELLE MEMORIAL INST. METALS AND CERAMICS INFORMATION CENTER, COLUMBUS, OHIO	MCIC-

BATTELLE MEMORIAL INST. NATIONAL BOMB DATA CENTER

BATTELLE MEMORIAL INST. NATIONAL BOMB DATA CENTER, COLUMBUS, OHIO	NBDC-TB-(NUMBER-YEAR)
BATTELLE MEMORIAL INST. RADIATION EFFECTS INFORMATION CENTER, COLUMBUS, OHIO	BMI-REIC-
BATTELLE MEMORIAL INST. RADIATION EFFECTS INFORMATION CENTER, COLUMBUS, OHIO	CH-CPR-
BATTELLE MEMORIAL INST. RADIATION EFFECTS INFORMATION CENTER, COLUMBUS, OHIO	RECI-
BATTELLE MEMORIAL INST. RADIATION EFFECTS INFORMATION CENTER, COLUMBUS, OHIO	REIC-
BATTELLE MEMORIAL INST. RADIATION EFFECTS INFORMATION CENTER, COLUMBUS, OHIO (MONTHLY ACCESSION LIST)	REIC-ACCESS-
BATTELLE MEMORIAL INST. RADIATION EFFECTS INFORMATION CENTER, COLUMBUS, OHIO. (ACCESSION LIST)	REIC ACC LIST-
BATTELLE MEMORIAL INST. RADIATION EFFECTS INFORMATION CENTER, COLUMBUS, OHIO. (ACCESSION LISTS)	REIC-AL-
BATTELLE MEMORIAL INST. RADIATION EFFECTS INFORMATION CENTER, COLUMBUS, OHIO. (CLASSIFIED ACCESSION LIST)	REIC-CAL-
BATTELLE MEMORIAL INST. RADIATION EFFECTS INFORMATION CENTER, COLUMBUS, OHIO	REIC-MEMO-
BATTELLE MEMORIAL INST. RADIATION EFFECTS INFORMATION CENTER, COLUMBUS, OHIO. (TECHNICAL MEMORANDA)	REIC-TM-
BATTELLE MEMORIAL INST. REMOTE AREA CONFLICT INFORMATION CENTER, COLUMBUS, OHIO	BAT-171-
BATTELLE MEMORIAL INST. REMOTE AREA CONFLICT INFORMATION CENTER, COLUMBUS, OHIO	RACIC-
BATTELLE MEMORIAL INST. REMOTE AREA CONFLICT INFORMATION CENTER, COLUMBUS, OHIO	RACIC-MR-
BATTELLE MEMORIAL INST. REMOTE AREA CONFLICT INFORMATION CENTER, COLUMBUS, OHIO	RACIC-TR-
BATTELLE MEMORIAL INST. SYSTEMS AND ELECTRONICS DEPT., COLUMBUS, OHIO	BCL-SED-(NUMBER)-
BATTELLE MEMORIAL INST. TITANIUM METALLURGICAL LAB., COLUMBUS, OHIO	BMI-TML-
BATTELLE MEMORIAL INST. TITANIUM METALLURGICAL LAB., COLUMBUS, OHIO	TML-
BATTELLE MEMORIAL INST. TRANSDUCER INFORMATION CENTER, COLUMBUS, OHIO	TIC-
BATTELLE-NORTHWEST. PACIFIC NORTHWEST LAB., RICHLAND, WASH.	BNSA-
BATTELLE-NORTHWEST. PACIFIC NORTHWEST LAB., RICHLAND, WASH.	BNTR-
BATTELLE-NORTHWEST. PACIFIC NORTHWEST LAB., RICHLAND, WASH.	BNWC-
BATTELLE-NORTHWEST. PACIFIC NORTHWEST LAB., RICHLAND, WASH. (FORMAL R + D REPORT)	BNWL-
BATTELLE-NORTHWEST. PACIFIC NORTHWEST LAB., RICHLAND, WASH.	BNWL-B-
BATTELLE-NORTHWEST. PACIFIC NORTHWEST LAB., RICHLAND, WASH.	BNWL-C-
BATTELLE-NORTHWEST. PACIFIC NORTHWEST LAB., RICHLAND, WASH. (INFORMAL R + D REPORT)	BNWL-CC-
BATTELLE-NORTHWEST. PACIFIC NORTHWEST LAB., RICHLAND, WASH. (INVENTION)	BNWL-IR-
BATTELLE-NORTHWEST. PACIFIC NORTHWEST LAB., RICHLAND, WASH. (MANUAL)	BNWL-MA-
BATTELLE-NORTHWEST. PACIFIC NORTHWEST LAB., RICHLAND, WASH. (INTENDED FOR JOURNAL PUBLICATION)	BNWL-SA-
BATTELLE-NORTHWEST. PACIFIC NORTHWEST LAB., RICHLAND, WASH. (TRANSLATION)	BNWL-TR-
BATTELLE-NORTHWEST. PACIFIC NORTHWEST LAB., RICHLAND, WASH.	HW-
BATTELLE-NORTHWEST. PACIFIC NORTHWEST LAB., RICHLAND WASH.	PNWL-
BAUSCH AND LOMB, INC., ROCHESTER, N.Y.	B/L-(NUMBER)-
BAUSCH AND LOMB, INC., ROCHESTER, N.Y.	B/L-(NUMBER)-IR-
BAUSCH AND LOMB OPTICAL CO., ROCHESTER, N.Y.	BL-
BAUSCH AND LOMB OPTICAL CO., ROCHESTER, N.Y.	B-LOC-
BAUSCH AND LOMB OPTICAL CO., ROCHESTER, N.Y.	BLOC-
BAUSCH AND LOMB OPTICAL CO., ROCHESTER, N.Y.	R-
(SEE ALSO LATER NAME: BAUSCH AND LOMB, INC.)	
BAYERISCHE MOTOREN WERKE, A. G., SPANDAU, GERMANY	BMW/ET/(NO.)/(YR)
BAYERISCHE MOTOREN WERKE. ENTWICKLUNG ZENTRALVERWALTUNG	BMW EZV-
BAYERISCHE MOTOREN WERKE. ENTWICKLUNG ZENTRALVERWALTUNG	EZV-
BAYLOR UNIV., HOUSTON, TEX. COLL. OF MEDICINE	ABJ-
BAYLOR UNIV., HOUSTON, TEX. COLL. OF MEDICINE	BAYLOR-
BAYLOR UNIV., HOUSTON, TEX. COLL. OF MEDICINE	BAYLOR-FR/(CONTRACT NO.)
BAYLOR UNIV., HOUSTON, TEX. COLL. OF MEDICINE	BAYLOR-PR/(CONTRACT NO.)
BAYLOR UNIV., HOUSTON, TEX. COLL. OF MEDICINE	BAYLOR UNIV.PR.../MD-(NO)
BAYLOR UNIV., HOUSTON, TEX. COLL. OF MEDICINE	BAYLOR U.PR.../MD-(NO.)
BAYLOR UNIV., HOUSTON, TEX. COLL. OF MEDICINE	PRNC-
BAYLOR UNIV., WACO, TEX.	BAY U-
BECCO CHEMICAL DIV., FOOD MACHINERY AND CHEMICAL CORP., BUFFALO	BECCO-
BECHTEL CORP., SAN FRANCISCO	BC-
BECHTEL CORP., SAN FRANCISCO (JOINT VENTURE)	BCPG-
BECHTEL CORP., SAN FRANCISCO	BCPI-
BECHTEL CORP., SAN FRANCISCO	BC-TOP-
BECHTEL CORP., SAN FRANCISCO	BP-
BECHTEL CORP., SAN FRANCISCO	SDD-
BECKMAN INSTRUMENTS, INC., FULLERTON, CALIF.	BII-TR-
BECKMAN INSTRUMENTS, INC. ADVANCED TECHNOLOGY OPERATIONS, FULLERTON, CALIF.	FR-
BECKMAN INSTRUMENTS, INC. BERKELEY SCIENCE DIV., RICHMOND, CALIF.	BERK-
BEECH AIRCRAFT CORP., BOULDER, COLO.	BR-
BEECH AIRCRAFT CORP., WICHITA, KAN.	BEAC-ER-
BEECH AIRCRAFT CORP., WICHITA, KAN.	BEAD-ER-
BEECH AIRCRAFT CORP., WICHITA, KAN.	BEECH-
BEECH AIRCRAFT CORP., WICHITA, KAN. (DATA MANAGEMENT REPORT)	DMR-
BEECH AIRCRAFT CORP., WICHITA, KAN. (MISSILE AERO TECHNOLOGY REPORT)	MATR-
BEECH AIRCRAFT CORP., WICHITA, KAN.	MR-
BEECH AIRCRAFT CORP., WICHITA, KAN. (MISSILE STRUCTURES REPORT)	MSR-
BEECH AIRCRAFT CORP., WICHITA, KAN. (MISSILE WEIGHT REPORT)	MWR-
BEECH AIRCRAFT CORP., WICHITA, KAN.	56-SARW-
BEECH AIRCRAFT CORP. BOULDER DIV., COLO.	BEECH-ER-
BEECH AIRCRAFT CORP. BOULDER DIV., COLO.	BEECH-TEST REPORT-
BEECHCRAFT RESEARCH AND DEVELOPMENT, INC.	BEECHCRAFT FTRDI-...
BEECHCRAFT RESEARCH AND DEVELOPMENT, INC.	FTRDI-MR(YEAR)-
BEECHCRAFT RESEARCH AND DEV., INC., BOULDER, COLO.	BEECH-
BEECHCRAFT RESEARCH AND DEV., INC., BOULDER, COLO.	BEECH (LETTERS) (LETTERS)
BEECHCRAFT RESEARCH AND DEV., INC., WICHITA, KAN.	BEECHCRAFT-ER-
BEERS (ROLAND F.), INC., ALEXANDRIA, VA.	BEERS-
BEERS (ROLAND F.), INC., ALEXANDRIA, VA.	RFB-
BEGE (J.R.M.) CO., ARLINGTON, MASS.	BEGE-
BEGE (J.R.M.) CO., ARLINGTON, MASS.	BEGE (NUMBERS)-
BELFOUR ENG. CO., SUTTONS BAY, MICH. (INVENTORY REPT)	BEC-IR-
BELFOUR ENG. CO. TECH. INFO. SYSTEMS, SUTTONS BAY, MICH	BELFOUR-IR-

BELFOUR ENG. CO. TECH. INFO. SYSTEMS,SUTTONS BAY,MICH	BELFOUR-PR-
BELGIUM. CENTRE D'ETUDE DE L'ENERGIE NUCLEAIRE, BRUSSELS	BCH/M-
BELGIUM. CENTRE D'ETUDE DE L'ENERGIE NUCLEAIRE, BRUSSELS	BLG-
BELGIUM. CENTRE D'ETUDE DE L'ENERGIE NUCLEAIRE, BRUSSELS	BLG/B-
BELGIUM. CENTRE D'ETUDE DE L'ENERGIE NUCLEAIRE, BRUSSELS	C/(NUMBER)/(YEAR)
BELGIUM. CENTRE D'ETUDE DE L'ENERGIE NUCLEAIRE, BRUSSELS	CEN-R.-
BELGIUM. CENTRE D'ETUDE DE L'ENERGIE NUCLEAIRE, BRUSSELS	D-
BELGIUM. CENTRE D'ETUDE DE L'ENERGIE NUCLEAIRE, BRUSSELS	D/(NUMBER)/
BELGIUM. CENTRE D'ETUDE DE L'ENERGIE NUCLEAIRE, BRUSSELS	R-
BELGIUM. CENTRE D'ETUDE DE L'ENERGIE NUCLEAIRE, MOL	R-
BELGIUM. CENTRE D'ETUDE POUR LES APPLICATIONS DE L'ENERGIE NUCLEAIRE, BRUSSELS	B-
BELGIUM. CENTRE D'ETUDES SOCIALES, BRUSSELS	CES-
BELGIUM. CENTRE DE RECHERCHES SCIENTIFIQUES ET TECHNIQUES DE L'INDUSTRIE DES FABRICATION, BRUSSELS	CRIF-EL-
BELGIUM. CENTRE DE RECHERCHES SCIENTIFIQUES ET TECHNIQUES DE L'INDUSTRIE DES FABRICATION, BRUSSELS	CRIF-MC-
BELGIUM. CENTRE DE RECHERCHES SCIENTIFIQUES ET TECHNIQUES DE L'INDUSTRIE DES FABRICATION, BRUSSELS	CRIF-MT-
BELGIUM. CENTRE DE RECHERCHES SCIENTIFIQUES ET TECHNIQUES DE L'INDUSTRIE DES FABRICATION, BRUSSELS	CRIF-PL-
BELGIUM. CENTRE NATIONAL DE RECHERCHES METALLURGIQUES LIEGE	P-
BELGIUM. FONDS NATIONAL DE LA RECHERCHE SCIENTIFIQUE, BRUSSELS	FNRS-
BELGIUM. INSTITUT D'AERONOMIE SPATIALE DE BELGIQUE, BRUSSELS	A-(NUMBER-YEAR)
BELGIUM. INSTITUT D'AERONOMIE SPATIALE DE BELGIQUE, BRUSSELS	CF-
BELGIUM. INSTITUT INTERUNIVERSITAIRE DES SCIENCES NUCLEAIRES, BRUSSELS	IISN-
BELGIUM. INSTITUT POUR L'ENCOURAGEMENT DE LA RECHERCHE SCIENTIFIQUE DANS L'INDUSTRIE ET L'AGRICULTURE, BRUSSELS	IRSIA-
BELGIUM. ROYAL OBSERVATORY, UCCLE	BULL-(NUMBER)/
BELGIUM. SOCIETE D'ETUDES DE RECHERCHES ET D'APPLICATIONS POUR L'INDUSTRIE, BRUSSELS	BS-
BELGIUM. SOCIETE D'ETUDES DE RECHERCHES ET D'APPLICATIONS POUR L'INDUSTRIE, BRUSSELS	SERAI-
BELGIUM. TRAINING CENTER FOR EXPERIMENTAL AERODYNAMICS, BRUSSELS	TCEA-TM-
BELGIUM. TRAINING CENTER FOR EXPERIMENTAL AERODYNAMICS, RHODE-SAINT-GENESE	TCEA-IN-
BELL AEROSPACE CO. DIV. OF TEXTRON, BUFFALO	A(NUMBER)-
BELL AEROSPACE CO. DIV. OF TEXTRON,BUFFALO(LAB.REPTS)	BA-
BELL AEROSPACE CO. DIV. OF TEXTRON, BUFFALO	BAC-
BELL AEROSPACE CO. DIV. OF TEXTRON, BUFFALO	BA-D-
BELL AEROSPACE CO. DIV. OF TEXTRON,BUFFALO(LAB.REPTS)	BLR-(NUMBER)-
BELL AEROSPACE CO. DIV. OF TEXTRON,BUFFALO(LAB.REPTS)	BLR-
BELL AEROSPACE CO. DIV. OF TEXTRON,ENVIRONMENTAL DATA COLLECTION AND PROCESSING FACILITY, TUCSON, ARIZ.	A(NUMBER)-
BELL AEROSPACE CO. DIV. OF TEXTRON,ENVIRONMENTAL DATA COLLECTION AND PROCESSING FACILITY, TUCSON, ARIZ.	EDCPF-
BELL AEROSYSTEMS CO., BUFFALO	BA-(NUMBER)-
BELL AEROSYSTEMS CO., BUFFALO	BAC-(NUMBER)-(NUMBER)
BELL AEROSYSTEMS CO., BUFFALO	BASC-
BELL AEROSYSTEMS CO., BUFFALO	BELL-(NUMBERS)-
BELL AEROSYSTEMS CO., BUFFALO	BLR-(YEAR)-(NO.)(LTR.)
BELL AEROSYSTEMS CO., BUFFALO	D(NO.)-
BELL AEROSYSTEMS CO., BUFFALO	IMS-(YEAR)-
(SEE ALSO LATER NAME: BELL AEROSPACE CO.)	
BELL AIRCRAFT CORP., BUFFALO (REF. 1)	(NUMBERS...)
BELL AIRCRAFT CORP., BUFFALO	BAC-PR-
BELL AIRCRAFT CORP., BUFFALO (QUARTERLY PROGRESS RPT)	BAC-QR-(NUMBERS)-
BELL AIRCRAFT CORP., BUFFALO (PROJECT METEOR)	BELA-
BELL AIRCRAFT CORP., BUFFALO	BELL-
BELL AIRCRAFT CORP., BUFFALO (PROJECT METEOR)	METEOR-(NO.)-(NO.)-(NO.)
BELL AIRCRAFT CORP., BUFFALO	S-
BELL AIRCRAFT CORP., BUFFALO	TR-(YR.)-
(SEE ALSO LATER NAME: BELL AEROSPACE CO.)	
BELL AND HOWELL CO., CHICAGO	GS-
BELL AND HOWELL RESEARCH LABS., PASADENA, CALIF.	BH-ONR-
BELL HELICOPTER CORP., FORT WORTH, TEX.	BHC-
BELL HELICOPTER CORP., FORT WORTH, TEX.	BHC-TR-
BELL HELICOPTER CORP., FORT WORTH, TEX.	D(NO.)-
BELL TELEPHONE LABS. (ALL LOCATIONS)	BTL-
BELL TELEPHONE LABS., INC., BURLINGTON, N.C.	NCR-
BELL TELEPHONE LABS., INC., MURRAY HILL, N.J.	KN-
BELL TELEPHONE LABS., INC., N.Y.C.	BTL-(NUMBER)-(LETTER)
BELL TELEPHONE LABS., INC., N.Y.C.	BTL-MM-(YEAR)-
BELL TELEPHONE LABS., INC., N.Y.C. (TRANSLATION)	BTL-TRANS-(INITIAL)-
BELL TELEPHONE LABS., INC., N.Y.C.	LRM-
BELL TELEPHONE LABS., INC., N.Y.C.	MM-(YR)-
BELL TELEPHONE LABS., INC., N.Y.C.	SAC-R-
BELL TELEPHONE LABS., INC., WHIPPANY, N.J.	BTL-(NO.)(LTR(S))(NOIS))
BELL TELEPHONE LABS., INC., WHIPPANY, N.J.	BTL MEMO CASE (NO.-NO.)
BELL TELEPHONE LABS., INC., WHIPPANY, N.J.	BTL/(MISSILE NAME)
BELL TELEPHONE LABS., INC., WHIPPANY, N.J.	BTL-PCP-(YEAR)-
BELL TELEPHONE LABS., INC., WHIPPANY, N.J.	BTL-QR-
BELL TELEPHONE LABS., INC., WHIPPANY, N.J.	MM(NO.-NO.-NO.)(M)
BELL TELEPHONE LABS., INC., WHIPPANY, N.J.	PCP-(YEAR)-
BELL TELEPHONE LABS., INC., WHIPPANY, N.J.	(NUMBER)-WH-
BELLARMINE-URSULINE COLL., LOUISVILLE, KY.	BUC-
BELLCOMM, INC., WASHINGTON, D.C.	B(YEAR)-
BELOCK INSTRUMENT CORP., COLLEGE POINT, N.Y.	(NUMBER)-BITR
BELOCK INSTRUMENT CORP., COLLEGE POINT, N.Y.	EE-
BELTONE INST. FOR HEARING RES.,CHICAGO (TRANSLATIONS)	BIHR-

BENDIX AVIATION CORP.

BENDIX AVIATION CORP. (ALL LOCATIONS)	BAC-
BENDIX AVIATION CORP. (FLIGHT TEST PROPOSAL)	FTP-
BENDIX AVIATION CORP.	HT-
BENDIX AVIATION CORP., KANSAS CITY, MO.	KCD-
BENDIX AVIATION CORP., KANSAS CITY, MO.	KCD-E-
(SEE ALSO LATER NAME: BENDIX CORP.)	
BENDIX AVIATION CORP. BENDIX PRODUCTS DIV., MISHAWAKA, IND.	BAC-BPD-
BENDIX AVIATION CORP. BENDIX PRODUCTS DIV., MISHAWAKA, IND.	BAC-BPD-MEMO-
BENDIX AVIATION CORP. BENDIX PRODUCTS DIV., MISHAWAKA, IND.	BPD-
BENDIX AVIATION CORP. BENDIX PRODUCTS DIV., MISSILES, MISHAWAKA, IND.	BEND BDP-
BENDIX AVIATION CORP. BENDIX PRODUCTS DIV., MISSILES, MISHAWAKA, IND.	BEND BDP-M-
BENDIX AVIATION CORP. BENDIX PRODUCTS DIV., MISSILES, MISHAWAKA, IND.	TAT-
BENDIX AVIATION CORP. BENDIX SYSTEMS DIV., ANN ARBOR, MICH.	BAC-BSR-
BENDIX AVIATION CORP. BENDIX SYSTEMS DIV., ANN ARBOR, MICH.	BSD-
BENDIX AVIATION CORP. FRIEZ INSTR. DIV., BALTIMORE	BAC-FI-
BENDIX AVIATION CORP. FRIEZ INSTR. DIV., BALTIMORE	BEND FID-
BENDIX AVIATION CORP. FRIEZ INSTR. DIV., BALTIMORE	BF(NUMBER)-
BENDIX AVIATION CORP. FRIEZ INSTR. DIV., BALTIMORE	FR-
BENDIX AVIATION CORP. KANSAS CITY DIV., MO.	BAC-KC-
BENDIX AVIATION CORP. KANSAS CITY DIV., MO.	BEND-KCD-(LETTERS)-
BENDIX AVIATION CORP. PACIFIC DIV., N. HOLLYWOOD, CAL.	BAC-PD-
BENDIX AVIATION CORP. PACIFIC DIV., N. HOLLYWOOD, CAL.	BAC-RDBT-
BENDIX AVIATION CORP. PACIFIC DIV., N. HOLLYWOOD, CAL.	BEND-(NO.)-
BENDIX AVIATION CORP. PACIFIC DIV., N. HOLLYWOOD, CAL.	DLM-
BENDIX AVIATION CORP. PACIFIC DIV., N. HOLLYWOOD, CAL.	RD-(YEAR)-
BENDIX AVIATION CORP. PACIFIC DIV., N. HOLLYWOOD, CAL.	RDBT-
BENDIX AVIATION CORP. PACIFIC DIV., N. HOLLYWOOD, CAL.	RDSM-
BENDIX AVIATION CORP. PACIFIC DIV. DEVELOPMENT LAB., BURBANK, CALIF.	BAC-P-DLM-
BENDIX AVIATION CORP. PACIFIC DIV., DEVELOPMENT LABS NORTH HOLLYWOOD, CALIF.	BX-
BENDIX AVIATION CORP. PIONEER-CENTRAL DIV., DAVENPORT, IOWA	AMQ-(NO.),LOG.NO.-
BENDIX AVIATION CORP. PIONEER-CENTRAL DIV., DAVENPORT, IOWA	BSC-(NO.)-
BENDIX AVIATION CORP. PIONEER-CENTRAL DIV., DAVENPORT, IOWA	PUBL.-
BENDIX AVIATION CORP. RADIO DIV.	BAC-R-
BENDIX AVIATION CORP. RESEARCH LABS., DETROIT	BAC/RL-
BENDIX AVIATION CORP. RESEARCH LABS., DETROIT	BAC-TM-
BENDIX AVIATION CORP. RESEARCH LABS., DETROIT	BAC-TM-(NO.)-(NO.)-
BENDIX AVIATION CORP. RESEARCH LABS., DETROIT	BEND-R-
BENDIX AVIATION CORP. RESEARCH LABS., DETROIT	DO-
BENDIX AVIATION CORP. RESEARCH LABS., SOUTHFIELD, MICH.	B-
BENDIX AVIATION CORP. RESEARCH LABS. DIV., DETROIT	BAC/RLD-
BENDIX AVIATION CORP. RESEARCH LABS. DIV., SOUTHFIELD, MICH.	AVR-
BENDIX AVIATION CORP. UTICA DIV., N.Y.	BAC-(NO.)-(NO.)U
BENDIX AVIATION CORP. YORK DIV., PENNA.	BEND (LETTERS, DATES)
BENDIX CORP. (ALL LOCATIONS)	BDX-
BENDIX CORP. (ALL LOCATIONS)	BEND-
BENDIX CORP., BALTIMORE	T-
BENDIX CORP., DETROIT	ALSS-TR-
BENDIX CORP., DETROIT	BC-
BENDIX CORP., TETERBORO, N.J.	TDR-
BENDIX CORP. AEROSPACE SYSTEMS DIV., ANN ARBOR, MICH.	BSR-
BENDIX CORP. AEROSPACE SYSTEMS DIV., ANN ARBOR, MICH.	TIPI-SPO-(BETIPS)
BENDIX CORP. AEROSPACE SYSTEMS DIV., MISHAWAKA, IND.	BASD-MO-
BENDIX CORP. BENDIX MISHAWAKA DIV., IND.	BXM-
BENDIX CORP. BENDIX PRODUCTS AEROSPACE DIV., SOUTH BEND, IND.	BPAD-
BENDIX CORP. BENDIX PRODUCTS AEROSPACE DIV., SOUTH BEND, IND.	SPP-(YEAR)-
BENDIX CORP. BENDIX PRODUCTS AEROSPACE DIV., SOUTH BEND, IND.	SV-(YEAR)-
BENDIX CORP. BENDIX PRODUCTS DIV., SOUTH BEND, IND.	MM-
BENDIX CORP. BENDIX PRODUCTS DIV., SOUTH BEND, IND.	RJ-
BENDIX CORP. BENDIX PRODUCTS DIV., SOUTH BEND, IND.	SH-
BENDIX CORP. BENDIX RADIO DIV., BALTIMORE, MD.	BRP-
BENDIX CORP. BENDIX SYSTEMS DIV., ANN ARBOR, MICH.	ADVT-
BENDIX CORP. BENDIX SYSTEMS DIV., ANN ARBOR, MICH.	ATM-
BENDIX CORP. BENDIX SYSTEMS DIV., ANN ARBOR, MICH.	BEND BSC-
BENDIX CORP. BENDIX SYSTEMS DIV., ANN ARBOR, MICH.	BEND-BSD-
BENDIX CORP. BENDIX SYSTEMS DIV., ANN ARBOR, MICH.	BEND-BSR-
BENDIX CORP. BENDIX SYSTEMS DIV., ANN ARBOR, MICH.	BSC-
BENDIX CORP. BENDIX SYSTEMS DIV., ANN ARBOR, MICH.	BSR-
BENDIX CORP. BENDIX SYSTEMS DIV., ANN ARBOR, MICH.	MADVT-
BENDIX CORP. BENDIX SYSTEMS DIV., ANN ARBOR, MICH.	NAST-
BENDIX CORP. COMMUNICATIONS DIV., BALTIMORE	BCD-
BENDIX CORP. ELECTRODYNAMICS DIV., NO. HOLLYWOOD, CAL	RER-
BENDIX CORP. ELECTRODYNAMICS DIV., NO. HOLLYWOOD, CAL	USWD-
BENDIX CORP. ELECTRO-OPTICS DIV., ANN ARBOR, MICH.	EO-TR-
BENDIX CORP. ELECTRO-OPTICS DIV., ANN ARBOR, MICH.	(YEAR)-ETR-
BENDIX CORP. ELECTRO-OPTICS DIV., ANN ARBOR, MICH.	ETR-
BENDIX CORP. ENVIRONMENTAL SCIENCE DIV., BALTIMORE	BF(NUMBER)-
BENDIX CORP. FRIEZ INSTRUMENT DIV., BALTIMORE	EIR-

BENDIX CORP. KANSAS CITY DIV., MO.	BAC-E-
BENDIX CORP. KANSAS CITY DIV., MO.	BWD-TB-
BENDIX CORP. KANSAS CITY DIV., MO.	E-
BENDIX CORP. KANSAS CITY DIV., MO.	KCD-TB-
BENDIX CORP. KANSAS CITY DIV., MO.	PDO-
BENDIX CORP. NAVIGATION + CONTROL DIV., TETERBORO, N.J.	RE-
BENDIX CORP. PACIFIC DIV., NORTH HOLLYWOOD, CALIF.	BEND-DLM-
BENDIX CORP. PIONEER-CENTRAL DIV., DAVENPORT, IOWA	PUB-
BENDIX CORP. RED BANK DIV., EATONTOWN, N.J.	BEND-R-
BENDIX CORP. RED BANK DIV., EATONTOWN, N.J.	RBESO-
BENDIX CORP. RESEARCH LABS. DIV., SOUTHFIELD, MICH.	BC-
BENDIX CORP. RESEARCH LABS. DIV., SOUTHFIELD, MICH.	BC/RL-
BENDIX CORP. RESEARCH LABS. DIV., SOUTHFIELD, MICH.	BC/RLD-
BENDIX CORP. RESEARCH LABS. DIV., SOUTHFIELD, MICH.	BRL-
BENDIX CORP. RESEARCH LABS. DIV., SOUTHFIELD, MICH.	BRLD-
BENDIX CORP. RESEARCH LABS. DIV., SOUTHFIELD, MICH.	RLD-
BENDIX CORP. RESEARCH LABS. DIV., SOUTHFIELD, MICH.	RLDP-(YEAR)-
BENDIX CORP. SCIENCE DIV., BALTIMORE	EIR-
BENDIX CORP. SCINTILLA DIV., SIDNEY, N.Y.	BEND/SD-
BENDIX CORP. SCINTILLA DIV., SIDNEY, N.Y. (RADIO INTERFERENCE GUARD)	BEND/SD TR RIG-
BENDIX CORP. SCINTILLA DIV., SIDNEY, N.Y. (RADIO INTERFERENCE GUARD)	BEND TR RIG-
BENDIX MISSILE SYSTEMS, MISHAWAKA, IND.	BMSD-
BENDIX RADIO CORP. (INSTRUCTION BOOKS)	BRC-IB-
BERGEN. UNIV., DEPT. OF APPLIED MATHEMATICS	BERGEN U-
BERLIN. FREIE UNIVERSITAET	QTSR-
BERLIN. FREIE UNIVERSITAET. INST. FUER METEOROLOGIE UND GEOPHYSIK	QR-(NUMBER)-PT-
BERLIN. FREIE UNIVERSITAET. INST. FUER METEOROLOGIE UND GEOPHYSIK	QR-(1 DIGIT)
BERLIN. HAHN-MEITNER-INSTITUT FUER KERNFORSCHUNG (BERICHT, CHEMIE)	BC-
BERLIN. HAHN-MEITNER-INSTITUT FUER KERNFORSCHUNG	BEW-
BERLIN. HAHN-MEITNER-INSTITUT FUER KERNFORSCHUNG (BERICHT, MATHEMATIK)	BM-
BERLIN. HAHN-MEITNER-INSTITUT FUER KERNFORSCHUNG	B-NDV(NUMBER)
BERLIN. HAHN-MEITNER-INSTITUT FUER KERNFORSCHUNG (BERICHT, PHYSIK)	BP-
BERLIN. HAHN-MEITNER-INSTITUT FUER KERNFORSCHUNG (BERICHT, REAKTOR)	BRP-
BERLIN. HAHN-MEITNER-INSTITUT FUER KERNFORSCHUNG (BERICHT, STRAHLENCHEMIE)	BS-
BERLIN. HAHN-MEITNER-INSTITUT FUER KERNFORSCHUNG (BERICHT, STRAHLENPHYSIK)	BSP-
BERLIN. HAHN-MEITNER-INSTITUT FUER KERNFORSCHUNG	EU-
BERLIN. HAHN-MEITNER-INSTITUT FUER KERNFORSCHUNG	EU-B-
BERLIN. HAHN-MEITNER-INSTITUT FUER KERNFORSCHUNG	HMI-B-
BERLIN. HAHN-MEITNER-INSTITUT FUER KERNFORSCHUNG	IKB-C-B-
BERLIN. PHYSIKALISCH-TECHNISCHE REICHSANSTALT	PTR-
BERLIN. TECHNISCHE UNIVERSITAET	D-
BERLIN. TECHNISCHE UNIVERSITAET	TUB-
BERLIN. TECHNISCHE UNIVERSITAET. INST. FUER KERNTECHNIK	TUBIK-
BERLIN. TECHNISCHE UNIV. INST. FUER RAUMFAHRTTECHNIK	DCA-
BERLIN. TECHNISCHE UNIV. INST. FUER RAUMFAHRTTECHNIK	TUB-IR-
BERMITE POWDER CO., SAUGUS, CALIF.	BPC-
BERN. UNIVERSITAET	SM-
BERYLLIUM CORP., READING, PENNA.	BC-
BERYLLIUM CORP., READING, PENNA.	BERCO-
BETA CORP., RICHMOND, VA.	BETA-
BETA CORP., RICHMOND, VA.	BETA-TR-
BETA INDUSTRIES, INC., DAYTON, OHIO	BII-
BETHLEHEM STEEL CO., PITTSBURGH	BS-S-
BETHLEHEM STEEL CO. SHIPBUILDING DIV., QUINCY, MASS.	BSC-(CONTRACT)-
BETHLEHEM STEEL CO. SHIPBUILDING DIV., QUINCY, MASS.	BSC-(NO.)-
BETHLEHEM STEEL CO. SHIPBUILDING DIV., QUINCY, MASS.	BS-S-
BEUKERS (JOHN M.) LABS., INC., MEDFORD, N.Y.	GC-
BEUKERS (JOHN M.) LABS., INC., STONY BROOK, N.Y.	G(4 DIGITS)-R-
BIO-DYNAMICS, INC., CAMBRIDGE, MASS.	BD-
BIOSCIENCES INFORMATION SERVICE, PHILADELPHIA	BIOSIS-
BIOT AND ARNOLD (CONSULTANTS IN APPLIED MATH. + PHYS)	B+A
BIOT AND ARNOLD (CONSULTANTS IN APPLIED MATH. + PHYS)	FTSR-
BIOTECHNOLOGY, INC., ARLINGTON, VA.	BTI-
BIOTECHNOLOGY, INC., ARLINGTON, VA.	BTI-(YEAR)-
BIRKENHEAD (WARREN), INC., SEATTLE	WBI-C-(NUMBER)-BI
BIRMINGHAM. UNIV., ENGLAND	BIRMINGHAM-
BIRMINGHAM. UNIV., ENGLAND	BIR U-
BIRMINGHAM. UNIV., ENGLAND	FTSR-
BIRMINGHAM. UNIV., ENGLAND. DEPT. OF IND. METALLURGY	NWC-TP-
BISSETT-BERMAN CORP., SAN DIEGO, CALIF.	MJO-
BISSETT-BERMAN CORP., SANTA MONICA, CALIF.	BBC-
BISSETT-BERMAN CORP., SANTA MONICA, CALIF.	C-
BISSETT-BERMAN CORP., SANTA MONICA, CALIF.	C(NO.)-
BISSETT-BERMAN CORP., SANTA MONICA, CALIF.	CDC-

BITUMINOUS COAL RESEARCH, INC.

BITUMINOUS COAL RESEARCH, INC., MONROEVILLE, PA.   BCR-L-

BJORKSTEN RESEARCH LABS., INC., MADISON, WIS.   BJO-
BJORKSTEN RESEARCH LABS., INC., MADISON, WIS.   BJRL-
BJORKSTEN RESEARCH LABS., INC., MADISON, WIS.   BRL-(LETTERS, DATE)
BJORKSTEN RESEARCH LABS., INC., MADISON, WIS.   BRL-N-(NUMBERS)

BLACK AND VEACH CONSULTING ENGINEERS, KANSAS CITY, MO   BLACK AND VEACH-
BLACK AND VEACH CONSULTING ENGINEERS, KANSAS CITY, MO   BVCE MP-
BLACK AND VEACH CONSULTING ENGINEERS, KANSAS CITY, MO   BVCP-

BLASS ANTENNA ELECTRONICS CORP., LONG ISLAND CITY, N.Y.   BAEC-

BLAW-KNOX CO., PITTSBURGH   BKC-

BLAW-KNOX CONSTRUCTION CO., PITTSBURGH   BKC-(NO.)-PE-
BLAW-KNOX CONSTRUCTION CO., PITTSBURGH   D-1-(NO.)S
BLAW-KNOX CONSTRUCTION CO., PITTSBURGH   D-2-(NO.)S
(SEE ALSO LATER NAME: BLAW-KNOX CO.)

BLOCK ENGINEERING INC., CAMBRIDGE, MASS.   BEI-

BLUME (JOHN A.) + ASSOCS. RES. DIV., SAN FRANCISCO   JAB-
BLUME (JOHN A.) + ASSOCS. RES. DIV., SAN FRANCISCO   JAB-(NUMBER)-

BOARD OF ECONOMIC WARFARE, WASH., D.C. (RES. RPTS.)   BEW RR-

BOARD ON GEOGRAPHIC NAMES   BGN-

BOEING AIRPLANE CO., SEATTLE (REF. 1)   (NUMBERS...)
BOEING AIRPLANE CO., SEATTLE   APS 71/RSG-
BOEING AIRPLANE CO., SEATTLE   BAC-
BOEING AIRPLANE CO., SEATTLE   BAC-D-
BOEING AIRPLANE CO., SEATTLE (FLIGHT TEST ANALYSIS REPORT)   BOAC AR FT-
BOEING AIRPLANE CO., SEATTLE (BOMARC PROGRESS RPT.)   BOAC BPR-
BOEING AIRPLANE CO., SEATTLE   BOE-
BOEING AIRPLANE CO., SEATTLE   BOE-D-
BOEING AIRPLANE CO., SEATTLE   BOE-D2-
BOEING AIRPLANE CO., SEATTLE   BOE-D5-
BOEING AIRPLANE CO., SEATTLE   BOE-TD-TOOL DOCUMENT-
BOEING AIRPLANE CO., SEATTLE   BOE-T-TEST-
BOEING AIRPLANE CO., SEATTLE   IM-
BOEING AIRPLANE CO., SEATTLE   MTR-
(SEE ALSO LATER NAME: BOEING CO.)

BOEING AIRPLANE CO. WICHITA DIV., KAN.   BOE-WD-
BOEING AIRPLANE CO. WICHITA DIV., KAN.   WD-

BOEING ATLANTIC TEST CENTER, PATRICK AFB, FLA.   ET(YR.)-
BOEING ATLANTIC TEST CENTER, PATRICK AFB, FLA.   T(NO.)-

BOEING CO., HUNTSVILLE, ALA. (BOMARC + SATURN)   D5-
BOEING CO., HUNTSVILLE, ALA. (BOMARC + SATURN)   T5-

BOEING CO., RENTON, WASH.   D6-

BOEING CO., SEATTLE   B-
BOEING CO., SEATTLE   BAP-(NUMBER)-
BOEING CO., SEATTLE (JOINT VENTURE)   BNJV-(LTR.)-(LTR.)-
BOEING CO., SEATTLE   BOAC-
BOEING CO., SEATTLE (PROJECT NUMBER, DOCUMENT)   D203-
BOEING CO., SEATTLE (PROJECT NUMBER, DOCUMENT)   D180-
BOEING CO., SEATTLE (PROJECT NUMBER, DOCUMENT)   D162-
BOEING CO., SEATTLE   F-
BOEING CO., SEATTLE   MS-
BOEING CO., SEATTLE   NG(NUMBER)
BOEING CO., SEATTLE (PROJECT NUMBER, TEST)   T206-

BOEING CO. AERO-SPACE DIV., SEATTLE   MDR-(NO.)-(NO.)-
BOEING CO. AERO-SPACE DIV., SEATTLE   SM-
BOEING CO. AERO-SPACE DIV., SEATTLE   TAO-

BOEING CO. AERO-SPACE GROUP, SEATTLE   D2-
BOEING CO. AERO-SPACE GROUP, SEATTLE   T2-

BOEING CO. AIRPLANE DIV., WICHITA, KAN.   T3-

BOEING CO. COMMERCIAL AIRPLANE DIV., SEATTLE   D6-
BOEING CO. COMMERCIAL AIRPLANE DIV., SEATTLE   T6-

BOEING CO. SYSTEMS MANAGEMENT DIV., KENT, WASH.   D7-
BOEING CO. SYSTEMS MANAGEMENT DIV., KENT, WASH.   T7-

BOEING CO. TURBINE DIV., SEATTLE   D4-
BOEING CO. TURBINE DIV., SEATTLE   T4-

BOEING CO. VERTOL DIV., MORTON, PENNA.   D8-
BOEING CO. VERTOL DIV., MORTON, PENNA.   T8-

BOEING CO. WICHITA DIV., KAN.   D3-

BOEING SCIENTIFIC RESEARCH LABS., SEATTLE (DOCUMENT)   D1-
BOEING SCIENTIFIC RESEARCH LABS., SEATTLE   D-1-(NUMBER-NUMBER)
BOEING SCIENTIFIC RESEARCH LABS., SEATTLE   DL-(NUMBER-NUMBER)
BOEING SCIENTIFIC RESEARCH LABS., SEATTLE (TEST RPT.)   T1-

BOEING SCIENTIFIC RESEARCH LABS. FLIGHT SCIENCES LAB., SEATTLE   FLIGHT SCI. LAB. REPT.-

BOEING SCIENTIFIC RESEARCH LABS. MATHEMATICS RESEARCH LAB., SEATTLE   MATHEMATICAL NOTE-

BOEING SCIENTIFIC RESEARCH LABS. PLASMA PHYSICS LAB., SEATTLE   DI-(NUMBER-NUMBER)

BOELKOW ENTWICKLUNGEN K.G., MUNICH   BOLDOW-RF-
BOELKOW ENTWICKLUNGEN K.G., MUNICH   DF-(NUMBER)-O
BOELKOW ENTWICKLUNGEN K.G., MUNICH   EP-(NUMBER)-O
BOELKOW ENTWICKLUNGEN K.G., MUNICH   EV-(NUMBER)-O
BOELKOW ENTWICKLUNGEN K.G., MUNICH   FM-(NUMBER)-O
BOELKOW ENTWICKLUNGEN K.G., MUNICH   L-(2 DIGITS)
BOELKOW ENTWICKLUNGEN K.G., MUNICH   RF-(NUMBER)-O
BOELKOW ENTWICKLUNGEN K.G., MUNICH   RS-(NUMBER)-O

BOELKOW ENTWICKLUNGEN K.G., MUNICH	ZV-(NUMBER)-0
BOELKOW ENTWICKLUNGEN K.G. FUTURE PROJS. DEPT.,MUNICH	BOLKOW-DSP-
BOELKOW GMBH., OTTOBRUNN BEI MUENCHEN, GERMANY	MT-
BOHAN (WALTER A.) CO., PARK RIDGE, ILL. (PROJ. FAMOS)	FAMOS-RR-(NUMBER-YEAR)
BOHNA (B.D.) AND CO., INC., SAN FRANCISCO	BOHNA-INV-
BOLOGNA. UNIVERSITA. ISTITUTO DI FISICA (A. RIGHI)	UBIF-
BOLT, BERANEK, AND NEWMAN, INC., CAMBRIDGE, MASS.	BB+N-
BOLT, BERANEK, AND NEWMAN, INC., CAMBRIDGE, MASS.	BBN-
BOLT, BERANEK, AND NEWMAN, INC., CAMBRIDGE, MASS.	BOLT B+N-
BOLT, BERANEK, AND NEWMAN, INC., CAMBRIDGE, MASS.	SVM-
BOLT, BERANEK, AND NEWMAN, INC., CAMBRIDGE, MASS.	TR(NO.)-
BOLT, BERANEK, AND NEWMAN, INC., LOS ANGELES	BBN-
BOLT, BERANEK, AND NEWMAN, INC., LOS ANGELES	FFA-
BOLT, BERANEK, AND NEWMAN, INC., VAN NUYS, CALIF.	DS-(YEAR)-
BOMAC LABS., INC., BEVERLY, MASS.	BL-
BOMAC LABS., INC., BEVERLY, MASS.	BLI-
BOMAC LABS., INC., BEVERLY, MASS.	BOM-
BOMBARDMENT TRAINING WING (3535TH), MATHER AFB, CALIF	MTR-
BOMBRINI PARODI-DELFINO S.P.A., ROME	ELDO/SP/(YEAR)/
BONN. UNIVERSITAET	FR-
BONN. UNIVERSITAET	SM-
BONN. UNIVERSITAET. INST. FUER PHYSIKALISCHE CHEMIE	SHA-
BONN. UNIVERSITAET. PHYSIKALISCHES INSTITUT	BONNU-PI-
BONNEVILLE POWER ADMINISTRATION, PORTLAND, ORE.	BPA-TRANS-
BOOZ, ALLEN AND HAMILTON, CHICAGO	BZAH-
BOOZ-ALLEN APPLIED RESEARCH, INC., ALBUQUERQUE, N.M.	BAARING-(NUMBER)-
BOOZ-ALLEN APPLIED RESEARCH, INC., BETHESDA, MD.	BAAR-
BOOZ-ALLEN APPLIED RESEARCH, INC., BETHESDA, MD.	BAARINC-
BOOZ-ALLEN APPLIED RESEARCH, INC., BETHESDA, MD.	BAARINC-PRO-R-
BOOZ-ALLEN APPLIED RESEARCH, INC., BETHESDA, MD.	BARRINC-
BOOZ-ALLEN APPLIED RESEARCH, INC., CHICAGO	BAARINC-
BOOZ-ALLEN APPLIED RESEARCH, INC., LOS ANGELES	BAARING-(NUMBER)-
BORAX RESEARCH CORP., ANAHEIM, CALIF.	BRC-
BORDEAUX UNIV. LABORATOIRE DE PHYSIQUE THEORIQUE	PNB-
BORDEAUX UNIV. LABORATOIRE DE PHYSIQUE THEORIQUE	PTB-
BORDEN CHEMICAL CO., PHILADELPHIA	BCC-
BORDEN CHEMICAL CO., PHILADELPHIA	BORDEN-ASR (INCL.DATES)
BORDEN CHEMICAL CO., PHILADELPHIA	BORDEN-PR-
BORDEN CHEMICAL CO., PHILADELPHIA	BORDEN-QPR-
BORG-WARNER CORP., KALAMAZOO, MICH.	BWC-
BORG-WARNER CORP. RESEARCH CENTER, DES PLAINES, ILL.	R-12-
BORG-WARNER CORP. ROY C. INGERSOLL RESEARCH CENTER, DES PLAINES, ILL. (REF. 1)	(NUMBER)
BORG-WARNER CORP. ROY C. INGERSOLL RESEARCH CENTER, DES PLAINES, ILL.	BWC-
BORG-WARNER CORP. SPECIAL PRODUCTS DIV.	SPD-
BOSE RESEARCH INST., CALCUTTA	BC-P-
BOSTON COLL., CHESTNUT HILL, MASS. DEPT. OF CHEMISTRY	FRK-
BOSTON INSULATED WIRE AND CABLE	BIW-
BOSTON UNIV.	BOSU-
BOSTON UNIV.	BOSU-QPR-
BOSTON UNIV.	BOSU-TN-
BOSTON UNIV.	BOU-
BOSTON UNIV.	BOU-TN-
BOSTON UNIV.	BU-
BOSTON UNIV. (DRAWING NO.)	BU DWG-(NOS.,LTRS,NOS.)
BOSTON UNIV.	BU ORL TN-
BOSTON UNIV. AREA DEVELOPMENT CENTER	ADC-
BOSTON UNIV. OPTICAL RESEARCH LAB. (REF. 1)	(NUMBER)
BOSTON UNIV. OPTICAL RESEARCH LAB.	BOSU-ORL-PR-
BOSTON UNIV. PHYSICAL RESEARCH LABS.	BU-PRL-TN-
BOURNS/CAI. INC., BARRINGTON, ILL.	CAI-
BOWLES ENGINEERING CORP., SILVER SPRING, MD.	BEC-R-(NUMBER-NO.-YEAR)
BOWLES ENGINEERING CORP., SILVER SPRING, MD.	R-(NO.)-(NO.)-
BOYCE THOMPSON INST. FOR PLANT RES.,INC.,YONKERS,N.Y.	BT-
BOYNTON (A.J.) AND CO., CHICAGO (REF. 1)	(NUMBERS...)
BOYNTON (A.J.) AND CO., CHICAGO	AJBC-
BOYNTON ASSOCIATES, LA CANADA, CALIF.	BOYNTON-
BOYNTON ASSOCIATES, LA CANADA, CALIF.	BOYNTON (LTRS,NOS.)
BRADBERRY (CARROLL E.) AND ASSOCIATES,LOS ALTOS, CAL.	CBA-(NUMBER)-F-
BRADBERRY (CARROLL E.) AND ASSOCIATES,LOS ALTOS, CAL.	HNS-

BRADDOCK, DUNN AND McDONALD, INC.

BRADDOCK, DUNN AND MC DONALD, INC., EL PASO, TEX.	BDM-
BRADDOCK, DUNN AND MC DONALD, INC., EL PASO, TEX.	BDM-(NUMBER-YEAR)-F-
BRADDOCK, DUNN AND MC DONALD, INC., EL PASO, TEX.	BM-(NUMBER-YEAR)-T-
BRADDOCK, DUNN AND MC DONALD, INC., MC LEAN, VA.	BDM-W-
BRANDEIS UNIV., WALTHAM, MASS.	BI-
BRAUN (C. F.) + CO., ALHAMBRA, CALIF.	CFB-
BRAZIL. COMISSAO NACIONAL DE ATIVIDADES ESPACIAS. LABORATORIO DE FISICA ESPACIAL, SAO JOSE DOS CAMPOS	LAFE-
BRAZIL. COMISSAO NACIONAL DE ATIVIDADES ESPACIAS. LABORATORIO DE FISICA ESPACIAL, SAO JOSE DOS CAMPOS	LAFE-SCIENTIFIC-
BRAZIL. COMISSAO NACIONAL DE ATIVIDADES ESPACIAS. LABORATORIO DE FISICA ESPACIAL, SAO JOSE DOS CAMPOS	LAFE-TR-
BRAZIL. COMISSAO NACIONAL DE ENERGIA NUCLEAR	DMP/G-
BRAZIL. COMISSAO NACIONAL DE ENERGIA NUCLEAR	DTC-
BRAZIL. COMISSAO NACIONAL DE ENERGIA NUCLEAR	EBR/PA-
BRAZIL. LABORATORIO DE ACUSTICA E SONICA, SAO PAULO	LXN/NP-
BREWER ENGINEERING LABS., INC., MARION, MASS.	BEL-
BRIDGEPORT BRASS CO., CONN.	BBCO-
BRIDGEPORT BRASS CO., CONN.	BBCO-F-
BRIDGEPORT BRASS CO., CONN.	BRB-
BRIDGEPORT BRASS CO., CONN.	DP-
BRIDGEPORT BRASS CO., CONN.	HAVENS-
BRIDGEPORT BRASS CO., CONN.	NMI-
BRIGHAM YOUNG UNIV., PROVO, UTAH	BYU-
BRIGHAM YOUNG UNIV., PROVO, UTAH	BYU-(NUMBER)-F
BRIGHAM YOUNG UNIV., PROVO, UTAH	BYU-SR-
BRIMAR VALVE WORKS, FOOTSCRAY, KENT, ENGLAND	BVW-
BRIMAR VALVE WORKS, FOOTSCRAY, KENT, ENGLAND	RVLTR-
BRIMAR VALVE WORKS, FOOTSCRAY, KENT, ENGLAND	RVTM-
BRISTOL AEROPLANE CO., LTD., ENGLAND	BAC-
BRISTOL AEROPLANE CO., LTD., ENGLAND	BAC/GW/BRISTOL/TR/102
BRITISH ADMIRALTY DELEGATION, WASHINGTON, D. C.	BAD-
BRITISH AIRCRAFT CORP., BRISTOL, GLOUCESTER, ENG.	ESS/SS-
BRITISH AIRCRAFT CORP., FILTON, GLOUCESTER, ENG.	ESS/ES-
BRITISH AIRCRAFT CORP., FILTON, GLOUCESTER, ENG.	SST/B(NO.)(LTR.)-(NO.)/
BRITISH AIRCRAFT CORP., LUTON, BEDS, ENGLAND	S + T-MEMO-
BRITISH AIRCRAFT CORP., PRESTON, LANCS, ENGLAND	AE-
BRITISH AIRCRAFT CORP., PRESTON, LANCS, ENGLAND	AF/GEN/
BRITISH AIRCRAFT CORP., PRESTON, LANCS, ENGLAND	BAC-AE-
BRITISH AIRCRAFT CORP., STEVENAGE, HERTS, ENGLAND	UK-
BRITISH AIRCRAFT CORP. GUIDED WEAPONS DIV., STEVENAGE, HERTS, ENGLAND	ST-
BRITISH AIRCRAFT CORP. GUIDED WEAPONS DIV., STEVENAGE, HERTS, ENGLAND	TIL-BR/
BRITISH AIRCRAFT CORP. GUIDED WEAPONS DIV., BRISTOL, GLOUCESTER, ENGLAND	MALLARD-TD-
BRITISH ALUMINIUM CO., LTD., LONDON	R-
BRITISH ALUMINIUM CO., LTD., LONDON	R-(NO.)-(YR)
BRITISH ARMY OF THE RHINE. SPECIAL PROJECTILE OPERATIONS GROUP	SPOG-
BRITISH BROADCASTING CORP., KINGSWOOD, SURREY, ENG.	BBC-(YEAR)/
BRITISH COAL UTILISATION RESEARCH ASSOCIATION	CRWP-
BRITISH COLUMBIA. UNIV., VANCOUVER	BCU-
BRITISH COLUMBIA. UNIV., VANCOUVER	R/AE
BRITISH COLUMBIA. UNIV., VANCOUVER	UBC-
BRITISH COLUMBIA. UNIV., VANCOUVER	UBCAR-
BRITISH COLUMBIA RESEARCH COUNCIL, VANCOUVER	CURF-
BRITISH COMMONWEALTH SCIENTIFIC OFFICE, WASH., D.C.	BCSO-
BRITISH COMMONWEALTH SCIENTIFIC OFFICE, WASH., D.C.	BCSO (NO.)-(LETTER)-
BRITISH COMMONWEALTH SCIENTIFIC OFFICE, WASH., D.C.	BCSO SCC-
BRITISH ELECTRICAL AND ALLIED INDUSTRIES RESEARCH ASSN., LEATHERHEAD, SURREY, ENGLAND	BEAIRA-TR-L/T-
BRITISH ELECTRICAL AND ALLIED INDUSTRIES RESEARCH ASSN., LEATHERHEAD, SURREY, ENGLAND	L/T-
BRITISH ELECTRICAL AND ALLIED INDUSTRIES RESEARCH ASSN., LONDON	BR-ERA-G/T-
BRITISH EUROPEAN AIRWAYS	BEA-
BRITISH EUROPEAN AIRWAYS	RSD-
BRITISH HOVERCRAFT CORP. LTD. EXPTL. DEPT., LONDON	BHC-X/O/
BRITISH HYDROMECHANICS RESEARCH ASSN., CRANFIELD, BEDS, ENGLAND (JOINT VENTURE)	BHRA-NEL-JOINT-REPT-
BRITISH HYDROMECHANICS RES. ASSN., CRANFIELD, BEDS, ENG.	BIB-
BRITISH HYDROMECHANICS RES. ASSN., CRANFIELD, BEDS, ENG.	SP-
BRITISH INTELLIGENCE OBJECTIVES SUB-COMMITTEE	BIOS-
BRITISH INTELLIGENCE OBJECTIVES SUB-COMMITTEE	BIOS B-
BRITISH INTELL. OBJECTIVES SUB-COMM. (BAG NUMBER)	BIOS BAG-
BRITISH INTELLIGENCE OBJECTIVES SUB-COMMITTEE	BIOS CPVA-
BRITISH INTELLIGENCE OBJECTIVES SUB-COMM. (DOCUMENTS)	BIOS DOC-
BRITISH INTELL. OBJECTIVES SUB-COMM. (EVALUATION RPT)	BIOS ER-
BRITISH INTELLIGENCE OBJECTIVES SUB-COMMITTEE	BIOS ERA-
BRITISH INTELL. OBJECTIVES SUB-COMM. (FOREIGN DOCS.)	BIOS FD-
BRITISH INTELLIGENCE OBJECTIVES SUB-COMMITTEE (FOREIGN DOCUMENTS, CENTRAL RADIO BUREAU)	BIOS FD CRB-
BRITISH INTELLIGENCE OBJECTIVES SUB-COMMITTEE (MILITARY INTELLIGENCE, REPORTS SECTION)	BIOS FD MIRS-
BRITISH INTELL. OBJECTIVES SUB-COMM. (FINAL RPTS.)	BIOS FR-
BRITISH INTELLIGENCE OBJECTIVES SUB-COMMITTEE (GERMAN BAG ISSUE)	BIOS GBI-
BRITISH INTELLIGENCE OBJECTIVES SUB-COMMITTEE. (INTERIM INVESTIGATION REPORTS)	BIOS II-
BRITISH INTELLIGENCE OBJECTIVES SUB-COMMITTEE. (INTERROGATION REPORTS)	BIOS IR-
BRITISH INTELLIGENCE OBJECTIVES SUB-COMMITTEE. (INTERROGATION REPORTS. SUMMARY)	BIOS IR S-
BRITISH INTELLIGENCE OBJECTIVES SUB-COMMITTEE. (JAPANESE. PRELIMINARY REPORT)	BIOS JAP PR-

BRITISH INTELL. OBJECTIVES SUB-COMM. (MISC. RPTS.)	BIOS M-
BRITISH INTELLIGENCE OBJECTIVES SUB-COMMITTEE	BIOS-MISC/R-
BRITISH INTELL. OBJECTIVES SUB-COMM. (OVERALL RPT.)	BIOS OR-
BRITISH INTELLIGENCE OBJECTIVES SUB-COMMITTEE (OVERSEAS RESEARCH REPORTS)	BIOS ORR-
BRITISH INTELL. OBJECTIVES SUB-COMM. (PRELIM. RPTS.)	BIOS PR-
BRITISH INTELLIGENCE OBJECTIVES SUB-COMMITTEE. (PRELIMINARY REPORTS. LISTS)	BIOS PR L-
BRITISH INTELL. OBJECTIVES SUB-COMM. (FINAL REPORT)	BIOS R-
BRITISH INTELLIGENCE OBJECTIVES SUB-COMMITTEE	BIOS SO-
BRITISH INTELL. OBJECTIVES SUB-COMM. (TRIP NO.)	BIOS TRIP-
BRITISH INTELLIGENCE OBJECTIVES SUB-COMMITTEE	CPVA-
BRITISH INTELLIGENCE OBJECTIVES SUBCOMMITTEE. MILITARY COLL. OF SCIENCE. SCH. OF TANK TECHNOLOGY	ATT (NUMBER) G-
BRITISH INTELLIGENCE OBJECTIVES SUBCOMMITTEE. MILITARY COLL. OF SCIENCE. SCH. OF TANK TECHNOLOGY	BIOS STT-
BRITISH INTELLIGENCE OBJECTIVES SUB-COMMITTEE. MILITARY COLL. OF SCIENCE. SCH. OF TANK TECHNOLOGY	STT-
BRITISH INTELLIGENCE OBJECTIVES SUB-COMMITTEE. TECHNICAL INTELLIGENCE SECTION	BIOS TIS-
BRITISH INTERNAL COMBUSTION ENGINE RESEARCH ASSOCIATION, SLOUGH, BUCKS, ENGLAND	ATR-
BRITISH IRON AND STEEL RESEARCH ASSN., LONDON	MD/D-(NUMBER)/
BRITISH IRON AND STEEL RES. ASSN.,SHEFFIELD,YORKS,ENG	MG/D-(NUMBER/YEAR)
BRITISH IRON + STEEL RES.ASSN. METALLURGY DIV., LONDON	MG/D-(NUMBER/NUMBER)
BRITISH IRON AND STEEL RESEARCH ASSN. OPERATIONAL RESEARCH DEPT., LONDON	BISRA-OR/(NO./YR.)
BRITISH IRON AND STEEL RESEARCH ASSN. OPERATIONAL RESEARCH DEPT., LONDON	BISRA-OR/HF/(NO./YR.)
BRITISH IRON AND STEEL RESEARCH ASSN. OPERATIONAL RESEARCH DEPT., LONDON (OPEN REPORT)	OR/HF/(NO./YR.)
BRITISH IRON AND STEEL RES. ASSN. PHYS. DEPT., LONDON	BISRA-P/(NUMBER/YEAR)
BRITISH IRON AND STEEL RESEARCH ASSN. PLANT ENGINEERING AND ENERGY DIV., LONDON	BISRA-EG/A/(NUMBER/YEAR)
BRITISH IRON AND STEEL RESEARCH ASSN. PLANT ENGINEERING AND ENERGY DIV., LONDON	BISRA-PE/A/(NUMBER/YEAR)
BRITISH IRON AND STEEL RESEARCH ASSN. PLANT ENGINEERING AND ENERGY DIV., LONDON	PE/A/(NUMBER/YEAR
BRITISH IRON AND STEEL RESEARCH ASSN. PLANT ENGINEERING AND ENERGY DIV., LONDON	PE/E/(NUMBER/NUMBER)
BRITISH IRON AND STEEL RES. ASSN. STEELMAKING DIV., LONDON	BISRA-SM/B-
BRITISH IRON AND STEEL RES. ASSN. STEELMAKING DIV., LONDON	BISRA-SM/BE/A/(NO./YR.)
BRITISH JOINT SERVICES MISSION, WASHINGTON, D.C.	BJSM-
BRITISH NON-FERROUS METALS RESEARCH ASSN., LONDON	ATR/
BRITISH NON-FERROUS METALS RESEARCH ASSN., LONDON	BNFMRA-
BRITISH NUCLEAR DESIGN AND CONSTRUCTIONS LTD., WHETSTONE, LEICS, ENGLAND	BNDC/R-
BRITISH OXYGEN CO., LTD., LONDON	R/D-
BRITISH SHIPBUILDING RESEARCH ASSN., ATOMIC ENERGY RESEARCH ESTABLISHMENT, WINFRITH, DORSET, ENGLAND	BSRA/
BRITISH STANDARDS INSTITUTION	BS-
BRITISH STANDARDS INSTITUTION	BSI-BS-
BRITISH STEEL CORP., LONDON	DIS/(NUMBER/YEAR)
BRITISH STEEL CORP., LONDON	M6/A/(NUMBER/YEAR)
BRITISH STEEL CORP., LONDON	MW/HC/(NUMBER)/(YEAR)
BRITISH STEEL CORP., LONDON	OR/NF/(NUMBER/YEAR)
BRITISH STEEL CORP. METALLURGY DIV., LONDON	MG/A/
BRITISH STEEL CORP. METALLURGY DIV., LONDON	MG/C/
BRITISH STEEL CORP. MIDLAND GROUP,MOORGATE,YORKS,ENG.	M/AS/DM/
BRITISH STEEL CORP. MIDLAND GROUP,MOORGATE,YORKS,ENG.	M/DD.(NUMBER/NUMBER)/
BRITISH STEEL CORP. MIDLAND GROUP,MOORGATE,YORKS,ENG.	M/PZ/
BRITISH STEEL CORP. MIDLAND GROUP,MOORGATE,YORKS,ENG.	M/U/PZ/
BRITISH STEEL CORP. MIDLAND GROUP,MOORGATE,YORKS,ENG.	M/Z/
BRITISH STEEL CORP. STEELMAKING DIV., LONDON	SM/BE/(NUMBER/YEAR)
BRITISH STEEL CORP. SCOTTISH AND NORTH WEST GROUP, MOTHERWELL, SCOTLAND	SNW/C/
BRITISH THOMSON-HOUSTON CO.,LTD.RUGBY, WARWICKS,ENG.	L-4002-C-
BRITISH WELDING RESEARCH ASSN., CAMBRIDGE, ENGLAND (TRANSLATIONS)	BWRA-
BRITISH WELDING RESEARCH ASSN., CAMBRIDGE, ENGLAND	C(NUMBER/NUMBER/YEAR)
BRITISH WELDING RESEARCH ASSN., LONDON	BWRA-
BRITISH WELDING RESEARCH ASSN., LONDON	FEI-
BROADVIEW RESEARCH AND DEVELOPMENT, BURLINGAME, CAL.	BRD-
BROADVIEW RESEARCH CORP., BURLINGAME, CALIF.	BRC-
BROADVIEW RESEARCH CORP., BURLINGAME, CALIF.	RD-
BROBECK (WILLIAM M.) AND ASSOCIATES, BERKELEY, CALIF	WMB-
BROBECK (WILLIAM M.) AND ASSOCIATES, BERKELEY, CALIF.	WMB/A-
BROBECK (WILLIAM M.) AND ASSOCIATES, BOULDER, COLO.	WMB-
BROOKE ARMY MEDICAL CENTER, SAN ANTONIO	MEDEW-
BROOKE ARMY MEDICAL CENTER, SAN ANTONIO	MEDEW-RS-
BROOKHAVEN AREA OFFICE, AEC, UPTON, N.Y. (REF. 4)	BHO-(NUMBER)-
BROOKHAVEN NATIONAL LAB., UPTON, N.Y.	ADD-LJL-
BROOKHAVEN NATIONAL LAB., UPTON, N.Y. (ANNUAL REPORT)	AS-
BROOKHAVEN NATIONAL LAB., UPTON, N.Y.	BNL-
BROOKHAVEN NATIONAL LAB., UPTON, N.Y. (AUTHORS INTLS)	BNL-(INITIALS)-
BROOKHAVEN NATIONAL LAB., UPTON, N.Y.	BNL-A
BROOKHAVEN NATIONAL LAB., UPTON, N.Y.	BNL-AS-
BROOKHAVEN NATIONAL LAB., UPTON, N.Y. (CONFERENCES)	BNL-C-
BROOKHAVEN NATIONAL LAB., UPTON, N.Y.	BNL-EDC-
BROOKHAVEN NATIONAL LAB., UPTON, N.Y.	BNL-I-
BROOKHAVEN NATIONAL LAB., UPTON, N.Y.	BNL-L-
BROOKHAVEN NATIONAL LAB., UPTON, N.Y. (LIQUID METAL FUEL REACTORS)	BNL-LMFR-(DATE)
BROOKHAVEN NATIONAL LAB., UPTON, N.Y.	BNL-LOG-
BROOKHAVEN NATIONAL LAB., UPTON, N.Y. (NEUTRON CROSS SECTION EVALUATION GROUP NEWSLETTER)	BNL-N-
BROOKHAVEN NATIONAL LAB., UPTON, N.Y.	BNL-P-
BROOKHAVEN NATIONAL LAB., UPTON, N.Y.	BNL-PD-
BROOKHAVEN NATIONAL LAB., UPTON,N.Y.(PROGRESS REVIEW)	BNL-PR-

BROOKHAVEN NATIONAL LAB.

BROOKHAVEN NATIONAL LAB., UPTON, N.Y. (REACTOR PHYS)	BNL-RP-
BROOKHAVEN NATIONAL LAB., UPTON, N.Y. (SCIENTIFIC PROGRESS REPORT)	BNL-S-
BROOKHAVEN NATIONAL LAB., UPTON, N.Y.	BNL-T-
BROOKHAVEN NATIONAL LAB., UPTON, N.Y. (TRANSLATION)	BNL-TR-
BROOKHAVEN NATIONAL LAB., UPTON, N.Y.	BR-
BROOKHAVEN NATIONAL LAB., UPTON, N.Y.	CRISP-(YEAR)-
BROOKHAVEN NATIONAL LAB., UPTON, N.Y. (CODE IS AUTHOR'S INITIALS)	GTO/JWJ-
BROOKHAVEN NATIONAL LAB., UPTON, N.Y.	IHR-
BROOKHAVEN NATIONAL LAB., UPTON, N.Y.	IS/P-
BROOKHAVEN NATIONAL LAB., UPTON, N.Y. (CODE IS AUTHOR'S INITIALS)	JDK-
BROOKHAVEN NATIONAL LAB., UPTON, N.Y.	JGC/EAK/AVS-
BROOKHAVEN NATIONAL LAB., UPTON, N.Y.	LFR-
BROOKHAVEN NATIONAL LAB., UPTON, N.Y.	LMFR-
BROOKHAVEN NATIONAL LAB., UPTON, N.Y.	NCC-
BROOKHAVEN NATIONAL LAB., UPTON, N.Y.	PD-
BROOKHAVEN NATIONAL LAB., UPTON, N.Y.	T-
BROOKHAVEN NATIONAL LAB. ACCELERATOR DEPT., UPTON, N.Y. (INTERNAL REPORT)	AADD-
BROOKHAVEN NATIONAL LAB. ACCELERATOR DEPT., UPTON, N.Y. (ALTERNATING-GRADIENT SYNCHROTRON)	AGSCD-
BROOKLYN. POLYTECHNIC INST.	BPI-
BROOKLYN. POLYTECHNIC INST.	BTI-
BROOKLYN. POLYTECHNIC INST.	PIB-
BROOKLYN. POLYTECHNIC INST. (SQUID PROJECT) (REF. 33)	PIB-(NUMBER)-(LETTER(S))
BROOKLYN. POLYTECHNIC INST.	PIB AL-
BROOKLYN. POLYTECHNIC INST.	PIBAL-
BROOKLYN. POLYTECHNIC INST.	PIBAL-R-
BROOKLYN. POLYTECHNIC INST.	PIB R-
BROOKLYN. POLYTECHNIC INST.	PIB-TR-
BROOKLYN. POLYTECHNIC INST.	R-(NO.)-(NO.)-(YR)-
BROOKLYN. POLYTECHNIC INST.	R-(NO.)-(YR)-
BROOKLYN. POLYTECHNIC INST. (SQUID PROJECT) (REF. 33)	SQUID-TM-PIB-
BROOKLYN. POLYTECHNIC INST., FARMINGDALE, N.Y.	PIBEE(YEAR)-
BROOKLYN. POLYTECHNIC INST., FARMINGDALE, N.Y. DEPT. OF ELECTROPHYSICS	PIBEP-
BROOKLYN. POLYTECHNIC INST., FARMINGDALE, N.Y. DEPT. OF ELECTROPHYSICS	PIBEP-(YEAR)-
BROOKLYN. POLYTECHNIC INST. MICROWAVE RESEARCH INST.	M-
BROOKLYN. POLYTECHNIC INST. MICROWAVE RESEARCH INST.	MRI-
BROOKLYN. POLYTECHNIC INST. MICROWAVE RESEARCH INST.	PIB-MRI-
BROOKLYN. POLYTECHNIC INST. MICROWAVE RESEARCH INST.	PIBMRI-
BROOKLYN. POLYTECHNIC INST. MICROWAVE RESEARCH INST.	PIBMRI-R-
BROOKLYN. POLYTECHNIC INST. MICROWAVE RESEARCH INST.	PIMBRI-R-(NUMBER)-(YEAR)
BROOKLYN. POLYTECHNIC INST. PROPULSION RESEARCH LAB.	PIB-PRL-
BROOKS + PERKINS, INC. DEFENSE PRODS. DIV., DETROIT	329-PHASE-
BROWN (D.) ASSOCIATES, INC., EAU GALLIE, FLA.	DBA-
BROWN, BOVERI INC., ZURICH-OERLIKON	BBI-
BROWN, BOVERI/KRUPP REAKTORBAU G.M.B.H., DUSSELDORF (THORIUM HIGH TEMPERATURE REACTOR)	THTR-
BROWN ENGINEERING CO., INC., HUNTSVILLE, ALA.	BE-TN-RL-
BROWN ENGINEERING CO., INC., HUNTSVILLE, ALA.	BU-TN-R-
BROWN ENGINEERING CO., INC., HUNTSVILLE, ALA.	CI-SST-(NUMBER-NUMBER)
BROWN ENGINEERING CO., INC. ADVANCED SYSTEMS AND TECHNOLOGY GROUP, HUNTSVILLE, ALA.	TN-AST-
BROWN ENGINEERING CO., INC. INFORMATION AND ELECTRONIC SYSTEMS DIV., HUNTSVILLE, ALA.	IESD-AMC-
BROWN ENGINEERING CO., INC. INFO. SYSTEMS DIV., HUNTSVILLE, ALA.	IS-QRMO-
BROWN ENGINEERING CO., INC. RES. LABS., HUNTSVILLE, ALA	RL-CL-
BROWN ENGINEERING CO., INC. RES. LABS., HUNTSVILLE, ALA	TN-R-
BROWN ENGINEERING CO., INC. RES. LABS., HUNTSVILLE, ALA	TN-SE-
BROWN ENGINEERING CO., INC. SPACE AND MILITARY SYSTEMS DIV., HUNTSVILLE, ALA.	SE-(3 DIGITS)
BROWN ENGINEERING CO., INC. SPACE AND MILITARY SYSTEMS DIV., HUNTSVILLE, ALA.	SMSD-ABMDA-
BROWN ENGINEERING CO., INC. SPACE AND MILITARY SYSTEMS DIV., HUNTSVILLE, ALA.	SMSD-AMSMI-
BROWN-FIRTH RESEARCH LABS., SHEFFIELD, YORKS, ENGLAND	HAR-
BROWN-FIRTH RESEARCH LABS., SHEFFIELD, YORKS, ENGLAND	RLK-
BROWN UNIV., PROVIDENCE	BM-
BROWN UNIV., PROVIDENCE	BRN-
BROWN UNIV., PROVIDENCE (PROJECT SQUID)	BRN-(NUMBER)-P
BROWN UNIV., PROVIDENCE	BROWN-
BROWN UNIV., PROVIDENCE	BROWN-U-
BROWN UNIV., PROVIDENCE	BROWN U. TR-
BROWN UNIV., PROVIDENCE	BT-
BROWN UNIV., PROVIDENCE	BU-
BROWN UNIV., PROVIDENCE	BU-ATR-
BROWN UNIV., PROVIDENCE	BU-B-
BROWN UNIV., PROVIDENCE	BU/NSRDC/(DATE)
BROWN UNIV., PROVIDENCE	BU/NSRDC/
BROWN UNIV., PROVIDENCE (PROJECTS)	BU P-
BROWN UNIV., PROVIDENCE	BU-TR-
BROWN UNIV., PROVIDENCE	C(NO.)-
BROWN UNIV., PROVIDENCE	HALC-TR-
BROWN UNIV., PROVIDENCE	RMB-
BROWN UNIV., PROVIDENCE	SSC-
BROWN UNIV., PROVIDENCE. APPLIED MATH. GP. (REF. 28)	AMG-B-
BROWN UNIV., PROVIDENCE. APPLIED MATH. GP. (REF. 28)	NDRC AMG-B-
BROWN UNIV., PROVIDENCE. APPLIED MATH. GP. (REF. 28)	NDRC AMG-B-MEMO-
BROWN UNIV., PROVIDENCE. CENTER FOR DYNAMICAL SYSTEMS	CDS-LN-(YEAR)-
BROWN UNIV., PROVIDENCE. CENTER FOR DYNAMICAL SYSTEMS	CDS-TR-(YEAR)-
BROWN UNIV., PROVIDENCE. DEPT. OF ENGINEERING	BU-E-
BROWN UNIV., PROVIDENCE. DIV. OF ENGINEERING	AF-
BROWN UNIV., PROVIDENCE. DIV. OF ENGINEERING	BU/DTMB/
BROWN UNIV., PROVIDENCE. DIV. OF ENGINEERING	BU/NSRDC/(NUMBER-YEAR)
BROWN UNIV., PROVIDENCE. DIV. OF ENGINEERING	BU-WT-
BROWN UNIV., PROVIDENCE. DIV. OF ENGINEERING	E-
BROWN UNIV., PROVIDENCE. DIV. OF ENGINEERING	E(NUMBER)-

BROWN UNIV., PROVIDENCE. DIV. OF ENGINEERING	WR-
BROWN UNIV., PROVIDENCE. ENG. MATERIALS RES. LAB.	EMRL-
BROWN UNIV., PROVIDENCE. ENG. MATERIALS RES. LAB.	EMRL-RM-
BROWN UNIV., PROVIDENCE. ENG. MATERIALS RES. LAB.	EMRL-TR-
BROWN UNIV., PROVIDENCE. GRAD. DIV. OF APPLIED MATH.	A11-
BROWN UNIV., PROVIDENCE. GRAD. DIV. OF APPLIED MATH.	A11-S-
BROWN UNIV., PROVIDENCE. GRAD. DIV. OF APPLIED MATH.	A11-T-
BROWN UNIV., PROVIDENCE. GRAD. DIV. OF APPLIED MATH.	A18-
BROWN UNIV., PROVIDENCE. GRAD. DIV. OF APPLIED MATH.	B11-
BROWN UNIV., PROVIDENCE. GRAD. DIV. OF APPLIED MATH.	BU-(LETTER)11-
BROWN UNIV., PROVIDENCE. GRADUATE DIVISION OF APPLIED MATHEMATICS (TRANSLATION)	BU-A9-T-
BROWN UNIV., PROVIDENCE. GRADUATE DIVISION OF APPLIED MATHEMATICS (TRANSLATIONS AND REPORTS)	BU GDAM-
BROWN UNIV., PROVIDENCE. GRAD. DIV. OF APPLIED MATH.	C11-
BROWN UNIV., PROVIDENCE. GRAD. DIV. OF APPLIED MATH.	DA-
BROWN UNIV., PROVIDENCE. GRADUATE DIVISION OF APPLIED MATHEMATICS (TRANSLATIONS AND REPORTS)	GDAM-
BROWN UNIV., PROVIDENCE. GRAD. DIV. OF APPLIED MATH.	GDAM A-9-M-
BROWN UNIV., PROVIDENCE. GRAD. DIV. OF APPLIED MATH.	GDAM A-9-T-
BROWN UNIV., PROVIDENCE. GRAD. DIV. OF APPLIED MATH.	GDAM A-11-
BROWN UNIV., PROVIDENCE. GRADUATE DIVISION OF APPLIED MATHEMATICS (TECHNICAL REPORT)	GDAM TR-
BROWN UNIV., PROVIDENCE. METALS RESEARCH LAB.	BU-MRL-
BROWN UNIV., PROVIDENCE. RESEARCH ANALYSIS GROUP	RAG-
BROWN UNIV., PROVIDENCE. WIND TUNNEL	WT-
BROWNING LABS., INC., WINCHESTER, MASS.	BROL-
BRUBAKER MFG. CO., INC., LOS ANGELES	BMC-
BRUNSWICK CORP. DEFENSE PRODUCTS DIV., MARION, VA.	BR-
BRUNSWICK CORP. TECH. PRODUCTS DIV., MUSKEGON, MICH.	PS(4 DIGITS)FR
BRUSH BERYLLIUM CO., CLEVELAND	BB-
BRUSH BERYLLIUM CO., CLEVELAND	BBC-
BRUSH BERYLLIUM CO., CLEVELAND	BBC-(NO.-NO.)X
BRUSH BERYLLIUM CO., CLEVELAND	BBC-PR-
BRUSH BERYLLIUM CO., CLEVELAND	BBC-TR-
BRUSH BERYLLIUM CO., CLEVELAND	P-(NO.)-
BRUSH BERYLLIUM CO., CLEVELAND	TI-
BRUSH DEVELOPMENT CO., CLEVELAND	BDC-
BRUSH LABS. CO., CLEVELAND	BL-
BRUSH LABS. CO., CLEVELAND	BLC-
BRUSH LABS. CO., CLEVELAND	BLC-PR-
BRUSH LABS. CO., CLEVELAND	BLC-TR-
BRUSSELS. UNIVERSITE	DRME-
BRUSSELS. UNIVERSITE	NT-(NUMBER)-(YEAR)
BRUSSELS. UNIVERSITE. CENTRE DE PHYSIQUE NUCLEAIRE	BULLETIN NOTE NO.-
BRUSSELS. UNIVERSITE. CENTRE DE PHYSIQUE NUCLEAIRE	BULL. NOTE-
BRUSSELS. UNIVERSITE. CENTRE DE PHYSIQUE NUCLEAIRE	NOTE NO.-
BRUSSELS. UNIVERSITE. INSTITUT DE MECANIQUE APPLIQUEE	N-(2 DIGITS)-(YEAR)
BRUSSELS. UNIVERSITE. INSTITUT DE MECANIQUE APPLIQUEE	ULB-IMA-NT-
BRUSSELS. UNIVERSITE. INSTITUT DU CANCER	BUCI-
BRUTCHER (HENRY) TECHNICAL TRANSLATIONS, ALTADENA, CAL	HB-
BRUTCHER (HENRY) TECHNICAL TRANSLATIONS, ALTADENA, CAL	HB-TRANS-
BUENOS AIRES. UNIVERSIDAD	UNBA-
BUDD CO., PHILADELPHIA	NER-
BUDD CO. TESTING LABS., PHILADELPHIA	CTD-
BUFFALO. UNIV.	BUFFU-
BUFFALO. UNIV. (PROGRESS REPORT)	BUFFU-PR-
BUFFALO ELECTRO-CHEMICAL CO., INC.	LR-
BUFFALO ELECTRO-CHEMICAL CO., INC.	LR-E-
BUFFALO FORGE CO.	BFC-
BULOVA RESEARCH + DEV. LABS., INC., WOODSIDE, N.Y.	BRDL-
BULOVA RESEARCH + DEV. LABS., INC., WOODSIDE, N.Y.	BRDL (LETTERS)-
BUNKER-RAMO CORP., CANOGA PARK, CALIF.	SA-(YEAR)-
BUNKER-RAMO CORP., FORT HUACHUCA, ARIZ.	IS(NO.)M(NO.)P(NO.)
BUNKER-RAMO CORP., FORT HUACHUCA, ARIZ.	TM(NO.)A(NO.)P(NO.)
BUNKER-RAMO CORP., SILVER SPRING, MD.	EC-RD-
BUNKER-RAMO CORP., SILVER SPRING, MD.	RD-S-
BUNKER-RAMO CORP. DEFENSE SYSTEMS DIV., CANOGA PARK, CALIF.	D276-5S-
BUNKER-RAMO CORP. DEFENSE SYSTEMS DIV., CANOGA PARK, CALIF.	D(NO.-NO.)S(NUMBER)-
BUNKER-RAMO CORP. DEFENSE SYSTEMS DIV., CANOGA PARK, CALIF.	E324-
BUNKER-RAMO CORP. DEFENSE SYSTEMS DIV., CANOGA PARK, CALIF.	F(NO.-NO.)U(NUMBER)
BUNKER-RAMO CORP. DEFENSE SYSTEMS DIV., CANOGA PARK, CALIF.	G101-
BUNKER-RAMO CORP. DEFENSE SYSTEMS DIV., CANOGA PARK, CALIF.	G(NO.-NO.)U(NUMBER)
BUNKER-RAMO CORP. DEFENSE SYSTEMS DIV., CANOGA PARK, CALIF.	H(NUMBER)-(NO.)(LTR.)-
BUNKER-RAMO CORP. DEFENSE SYSTEMS DIV., SILVER SPRING, MD.	C-(YEAR-NUMBER-NUMBER)-
BUNKER-RAMO CORP. DEFENSE SYSTEMS DIV., SILVER SPRING, MD.	ER-
BUNKER-RAMO CORP. HUMAN ENG. GP., CANOGA PARK, CALIF.	MR-(YEAR)-
BUREAU OF AERONAUTICS (NAVY) (REF. 1)	(NUMBERS...)
BUREAU OF AERONAUTICS (NAVY)	AER-
BUREAU OF AERONAUTICS (NAVY) (AROWA PROJECT)	AROWA-
BUREAU OF AERONAUTICS (NAVY)(ARMAMENT TECHNICAL MEMO)	ATM-
BUREAU OF AERONAUTICS (NAVY)	AW-
BUREAU OF AERONAUTICS (NAVY) (SHIPPING ORDER)	BASO-

BUREAU OF AERONAUTICS (NAVY)

BUREAU OF AERONAUTICS (NAVY)	BUA-
BUREAU OF AERONAUTICS (NAVY)	BUAER-
BUREAU OF AERONAUTICS (NAVY)	BU AER (LETTERS)-
BUREAU OF AERONAUTICS (NAVY) (SPECIFICATIONS)	BUA-XEL-
BUREAU OF AERONAUTICS (NAVY)	BU-DR-
BUREAU OF AERONAUTICS (NAVY)	CCC-(NO.)-TR-
BUREAU OF AERONAUTICS (NAVY)	CGD-
BUREAU OF AERONAUTICS (NAVY)	CO-NAVAER-
BUREAU OF AERONAUTICS (NAVY) (DESIGN RESEARCH)	DR-
BUREAU OF AERONAUTICS (NAVY)	MCC-(NO.)-TR-
BUREAU OF AERONAUTICS (NAVY)	NAVAER-
BUREAU OF AERONAUTICS (NAVY) (SPECIFICATIONS)	NAVAER-
BUREAU OF AERONAUTICS (NAVY) (AVIATION DESIGN RESEARCH REPORTS)	NAVAER ADR-
BUREAU OF AERONAUTICS (NAVY) (AVIATION DESIGN RESEARCH REPORTS)	NAVAER ADR (LETTER)-
BUREAU OF AERONAUTICS (NAVY) (AROWA PROJECT)	NAVAER-AROWA-
BUREAU OF AERONAUTICS (NAVY)(PRELIM. AIRCRAFT DESIGN)	NAVAER PAD-
BUREAU OF AERONAUTICS (NAVY) (STRUCTURES BULLETIN)	NAVAER SB-
BUREAU OF AERONAUTICS (NAVY) (SPERRY INSTRUCTIONS)	NAVAER SI-
BUREAU OF AERONAUTICS (NAVY) (STRUCTURES MEMO)	NAVAER SM-
BUREAU OF AERONAUTICS (NAVY) (TECHNICAL NOTES)	NAVAER TN-
BUREAU OF AERONAUTICS (NAVY)(PRELIM. AIRCRAFT DESIGN)	PAD-
BUREAU OF AERONAUTICS (NAVY)	PM-
BUREAU OF AERONAUTICS (NAVY) (POWER PLANT MEMORANDUM)	PPM-
BUREAU OF AERONAUTICS (NAVY) (QUALITY CONTROL)	QC-
BUREAU OF AERONAUTICS (NAVY) (DETAIL SPECIFICATION)	SD-
BUREAU OF AERONAUTICS (NAVY) (SPERRY INSTRUCTIONS)	SI-
BUREAU OF AERONAUTICS (NAVY) (STRUCTURES MEMORANDUM)	SM-
BUREAU OF AERONAUTICS (NAVY) (SPECIFICATION)	SPEC-MIL-(LTR)-(NO.)(AER)
BUREAU OF AERONAUTICS (NAVY)	USN DR-
(SEE ALSO SUCCESSORS: BUREAU OF NAVAL WEAPONS, AND NAVAL AIR SYSTEMS COMMAND)	
BUREAU OF AERONAUTICS (NAVY). AEROLOGY DIV.	AY-
BUREAU OF AERONAUTICS (NAVY). AIRBORNE EQUIP. DIV.	AE-
BUREAU OF AERONAUTICS (NAVY). AIRCRAFT DESIGN RESEARCH DIV., WASHINGTON, D.C.	ADR-
BUREAU OF AERONAUTICS (NAVY). AIRCRAFT DESIGN RESEARCH DIV., WASHINGTON, D.C.	ADR-M-
BUREAU OF AERONAUTICS (NAVY). AIRCRAFT DESIGN RESEARCH DIV., WASHINGTON, D.C.	ADR-R-
BUREAU OF AERONAUTICS (NAVY). AIRCRAFT DIV.	AC-
BUREAU OF AERONAUTICS (NAVY). AIRCRAFT NUCLEAR PROPULSION DIV.	NP-
BUREAU OF AERONAUTICS (NAVY). AIRFRAME DESIGN DIV.	AD-
BUREAU OF AERONAUTICS (NAVY). AVIONICS DIV.	AD-
BUREAU OF AERONAUTICS (NAVY). AVIONICS DIV.	AV-
BUREAU OF AERONAUTICS (NAVY). COMPUTER DIV.	CM-
BUREAU OF AERONAUTICS (NAVY). EVALUATION DIV.	EV-
BUREAU OF AERONAUTICS (NAVY). GUIDED MISSILE DIV.	GM-
BUREAU OF AERONAUTICS (NAVY). IND. PLANNING DIV.	IP-
BUREAU OF AERONAUTICS (NAVY). MAINTENANCE DIV.	MA-
BUREAU OF AERONAUTICS (NAVY). PHOTOGRAPHIC DIV.	PH-
BUREAU OF AERONAUTICS (NAVY). POWER PLANT DIV.	PP-
BUREAU OF AERONAUTICS (NAVY). PRODUCTION DIV.	PD-
BUREAU OF AERONAUTICS (NAVY). RES. + ANALYSIS OFF.	RA-
BUREAU OF AERONAUTICS (NAVY). RESEARCH DIV.	DR-
BUREAU OF AERONAUTICS (NAVY). RESEARCH DIV.	RS-
BUREAU OF AERONAUTICS (NAVY).SHORE ESTABLISHMENT DIV.	SE-
BUREAU OF AERONAUTICS (NAVY). TECHNICAL DATA DIV.	TD-
BUREAU OF AERONAUTICS REPRESENTATIVE, PASADENA, CAL.	BAR, PASADENA-
BUREAU OF COMMERCIAL FISHERIES, WASHINGTON, D.C. (COMMERCIAL FISHERIES REVIEW)	CFR-
BUREAU OF COMMERCIAL FISHERIES, WASHINGTON, D.C. (CURRENT FISHERIES STATISTICS)	CFS-
BUREAU OF COMMERCIAL FISHERIES, WASHINGTON, D.C. (FISHERY BULLETINS)	FB-
BUREAU OF COMMERCIAL FISHERIES, WASHINGTON, D.C. (FISHERY CIRCULARS)	FC-
BUREAU OF COMMERCIAL FISHERIES, WASHINGTON, D.C. (FISHERY LEAFLETS)	FL-
BUREAU OF COMMERCIAL FISHERIES, WASHINGTON, D.C. (FISH AND WILDLIFE DATA RPT.)	FWS-DR-
BUREAU OF COMMERCIAL FISHERIES, WASHINGTON, D.C. (MARKET DEVELOPMENT LEAFLETS)	MDL-
BUREAU OF COMMERCIAL FISHERIES, WASHINGTON, D.C. (PROGRESSIVE FISH-CULTURIST)	PFC-
BUREAU OF COMMERCIAL FISHERIES, WASHINGTON, D.C. (TEST KITCHEN SERIES)	TKS-
BUREAU OF COMMERCIAL FISHERIES. DIV. OF ECONOMIC RESEARCH, COLLEGE PARK, MD. (MAGNETIC TAPE)	BCF-MT-
BUREAU OF COMMERCIAL FISHERIES. TROPICAL ATLANTIC BIOLOGICAL LAB., MIAMI, FLA.	TABL-CONTRIB-
BUREAU OF CONSTRUCTION AND REPAIR (NAVY)	NAV-C+RB-
BUREAU OF DOMESTIC COMMERCE, WASHINGTON, D.C.	BDC-(YEAR)-TO-
BUREAU OF DOMESTIC COMMERCE, WASHINGTON, D.C.	BDC-QIR-P/P-
BUREAU OF ECONOMIC ANALYSIS, WASHINGTON, D.C. (SURVEY OF CURRENT BUSINESS)	BEA-SCB-(YEAR)-
BUREAU OF ENGINEERING RESEARCH	BER R-
BUREAU OF INTELLIGENCE AND RESEARCH, WASHINGTON, D.C.	BIR (LTRS)-
BUREAU OF INTERNATIONAL COMMERCE, WASHINGTON, D.C. (ECONOMIC TRENDS)	ET-(YEAR)-
BUREAU OF INTERNATIONAL COMMERCE, WASHINGTON, D.C. (INTERNATIONAL MARKETING INFORMATION SERVICE)	IMIS-(YEAR)-
BUREAU OF INTERNATIONAL COMMERCE, WASHINGTON, D.C.	OBR-(YEAR)-
BUREAU OF LABOR STATISTICS, WASHINGTON, D.C.	BLS-
BUREAU OF LABOR STATISTICS, WASHINGTON, D.C.	DOL-BLS-B-
BUREAU OF MEDICINE AND SURGERY (NAVY)	BUMED-
BUREAU OF MEDICINE AND SURGERY (NAVY)	NAVMED-
BUREAU OF MEDICINE AND SURGERY (NAVY)	NAVMED-P-

BUREAU OF MEDICINE AND SURGERY (NAVY)	NAVMED PROJ X-
BUREAU OF MEDICINE AND SURGERY(NAVY), WASHINGTON,D.C.	NAVMED-(NUMBERS)
BUREAU OF MEDICINE AND SURGERY(NAVY), WASHINGTON,D.C.	NAVMED-M-(NUMBERS)
BUREAU OF MINES	BM-
BUREAU OF MINES	BM/ED-TM-
BUREAU OF MINES	BM/ED-TN-
BUREAU OF MINES (INFORMATION CIRCULARS)	BM-IC-
BUREAU OF MINES	BM-OP-(NUMBER-YEAR)
BUREAU OF MINES (REPORTS OF INVESTIGATIONS)	BM-OPEN FILE-(NO./YR.)
BUREAU OF MINES (TRANSLATIONS)	BM-RI-
BUREAU OF MINES	BM-TRANS-
BUREAU OF MINES (OPEN FILE REPT.)	BOM-
BUREAU OF MINES (INFORMATION CIRCULAR)	BUMINES-OFR-(NO.-YR.)
BUREAU OF MINES	I.C.-
BUREAU OF MINES	IC-
BUREAU OF MINES (REPORT OF INVESTIGATION)	INFO. CIRCULAR-
BUREAU OF MINES (ALL LOCATIONS)	RI-
BUREAU OF MINES	USBM-
BUREAU OF MINES	USBM-C-
BUREAU OF MINES	USBM-TPR-
BUREAU OF MINES	USBM-TR-
BUREAU OF MINES	USBM-U-
BUREAU OF MINES, BRUCETON, PENNA.	WCA-
BUREAU OF MINES, PITTSBURGH	BUM-
BUREAU OF MINES, PITTSBURGH (SQUID PROJECT) (REF. 33)	BUM-(NUMBER)-P
BUREAU OF MINES, PITTSBURGH	BUM-PX-
BUREAU OF MINES, PITTSBURGH (SQUID PROJECT) (REF. 33)	BUM-SQUID NO.-
BUREAU OF MINES, PITTSBURGH	BUM-STR-
BUREAU OF MINES, PITTSBURGH	BUM-TPR-
BUREAU OF MINES, PITTSBURGH	S-(4 DIGITS)
BUREAU OF MINES, SALT LAKE CITY	BM-PR-
BUREAU OF MINES, SALT LAKE CITY (RESEARCH REPORT)	BM-RR-
BUREAU OF MINES, WASHINGTON, D.C.	USBM-RI-
BUREAU OF MINES. ALBANY METALLURGY RES. CTR., ORE.	RTD-IR-(NUMBER)-
BUREAU OF MINES. ALBANY METALLURGY RES. CTR., ORE.	USBM-RC-
BUREAU OF MINES. APPLIED PHYSICS RESEARCH LAB., COLLEGE PARK, MD.	APRL-
BUREAU OF MINES. APPLIED PHYSICS RESEARCH LAB., COLLEGE PARK, MD.	APRL-(NUMBER)-
BUREAU OF MINES. APPLIED PHYSICS RESEARCH LAB., COLLEGE PARK, MD.	APRL-E-
BUREAU OF MINES. APPLIED PHYSICS RESEARCH LAB., COLLEGE PARK, MD.	USBM-APRL-
BUREAU OF MINES. BARTLESVILLE PETROLEUM RESEARCH CENTER, OKLA.	BCPR-PPS-
BUREAU OF MINES. BARTLESVILLE PETROLEUM RESEARCH CENTER, OKLA.	BCPR-QPR-
BUREAU OF MINES. BARTLESVILLE PETROLEUM RESEARCH CENTER, OKLA.	BPRC-PPS-
BUREAU OF MINES. BARTLESVILLE PETROLEUM RESEARCH CENTER, OKLA.	BPRC-QPR-
BUREAU OF MINES. BARTLESVILLE PETROLEUM RESEARCH CENTER, OKLA.	PPS-
BUREAU OF MINES. DIV. OF EXPLOSIVES TECHNOLOGY, PITTS	BMET-
BUREAU OF MINES. DIV. OF EXPLOSIVES TECHNOLOGY, PITTS	BMET-S-
BUREAU OF MINES. ELECTROTECHNICAL LAB., NORRIS, TENN.	ETL-
BUREAU OF MINES. ELECTROTECHNICAL LAB., NORRIS, TENN	NOR-
BUREAU OF MINES. EXPLOSIVES AND PHYSICAL SCIENCES DIV., PITTSBURGH	BM-EPSD-PR-
BUREAU OF MINES.EXPLOSIVES RESEARCH CENTER,PITTSBURGH	ERC-
BUREAU OF MINES. EXPLOSIVES RES. LAB., PITTSBURGH	BM-BULL-(NUMBER)
BUREAU OF MINES. EXPLOSIVES RES. LAB., PITTSBURGH	BM-LIV-
BUREAU OF MINES. INTERMOUNTAIN EXPERIMENT STATION, SALT LAKE CITY	RR-(NUMBER).(NUMBER)
BUR. OF MINES. METALLURGICAL DIV., BOULDER CITY, NEV	Q-
BUREAU OF MINES. METALLURGICAL DIV., TUCSON, ARIZ.	TME-
BUREAU OF MINES. NW. ELECTRODEV. LAB., ALBANY, ORE.	BM-B-
BUREAU OF MINES. NW. ELECTRODEV. LAB., ALBANY, ORE.	BM-II-
BUREAU OF MINES. NW. ELECTRODEV. LAB., ALBANY, ORE.	NLCO-
BUREAU OF MINES. OIL SHALE RESEARCH BR.,LARAMIE, WYO.	BM-OSRD-
BUREAU OF MINES. PETROLEUM EXPERIMENT STATION, BARTLESVILLE, OKLA.	BM-PES-
BUREAU OF MINES. ROLLA METALLURGY RESEARCH CTR., MO.	BUM-MPR-
BUREAU OF MINES. ROLLA METALLURGY RESEARCH CTR., MO.	IPR-SP-
BUREAU OF MINES. SALT LAKE EXPERIMENT STATION, SALT LAKE CITY	SLMR-
BUREAU OF MINES. SAFETY RESEARCH CENTER, PITTSBURGH	SRC-S-
BUREAU OF MINES. SPEC. MINERALS INVEST. BR.,WASH.,D.C	BM-SMIB-
BUREAU OF NAVAL PERSONNEL	BUPERS-
BUREAU OF NAVAL PERSONNEL	NAVPERS-
BUREAU OF NAVAL PERSONNEL (RESEARCH REPORT PROJS.)	NAVPERS RRP-
BUREAU OF NAVAL PERSONNEL (SELECTION TEST REPORT PROJECTS)	NAVPERS STRP-
BUREAU OF NAVAL PERSONNEL	ND-
BUREAU OF NAVAL PERSONNEL	NP-
BUREAU OF NAVAL PERSONNEL (RESEARCH REPORT PROJS.)	RRP-
BUREAU OF NAVAL PERSONNEL (SELECTION TEST REPORT PROJECTS)	STRP-
BUREAU OF NAVAL PERSONNEL. PERSONNEL RESEARCH DIV., WASHINGTON, D.C.	NAVPERS-PRD-PRM-(YEAR)-
BUREAU OF NAVAL PERSONNEL. PERSONNEL RESEARCH DIV., WASHINGTON, D.C.	NAVPERS-PRD-PRR-(YEAR)-
BUREAU OF NAVAL PERSONNEL. PERSONNEL RESEARCH DIV., WASHINGTON, D.C.	NAVPERS-PRD-PTB-(YEAR)-
BUREAU OF NAVAL PERSONNEL. PERSONNEL RESEARCH DIV., WASHINGTON, D.C.	NAVPERS-PRM-
BUREAU OF NAVAL PERSONNEL. PERSONNEL RESEARCH DIV., WASHINGTON, D.C. (RESEARCH MEMORANUM)	PRM-
BUREAU OF NAVAL PERSONNEL. PERSONNEL RESEARCH DIV., WASHINGTON, D.C. (RESEARCH REPORT)	PRR-
BUREAU OF NAVAL PERSONNEL. PERSONNEL RESEARCH DIV., WASHINGTON, D.C. (TECHNICAL BULLETIN)	PTB-
BUREAU OF NAVAL WEAPONS	BNW-
BUREAU OF NAVAL WEAPONS	BNW-AL-
BUREAU OF NAVAL WEAPONS	BUWEPS-

BUREAU OF NAVAL WEAPONS

BUREAU OF NAVAL WEAPONS (INSTRUCTION)	BUWEPS INSTR-
BUREAU OF NAVAL WEAPONS	DLI-
BUREAU OF NAVAL WEAPONS	DR-
BUREAU OF NAVAL WEAPONS (EXPLOSIVE ORDNANCE DISPOSAL BULLETIN)	EODB-
BUREAU OF NAVAL WEAPONS	NAVORD-
BUREAU OF NAVAL WEAPONS	NAVWEPS-
BUREAU OF NAVAL WEAPONS	NAVWEPS-OD-
BUREAU OF NAVAL WEAPONS	OR-
BUREAU OF NAVAL WEAPONS	RRSY-(YR.)-
BUREAU OF NAVAL WEAPONS	TS-(YR.)-(NO.)-
BUREAU OF NAVAL WEAPONS, WASHINGTON, D.C.	NAVAIR-17-20(LETTERS)-
BUREAU OF NAVAL WEAPONS, WASHINGTON, D.C.	SR-IA-
BUREAU OF NAVAL WEAPONS, WASHINGTON, D.C.	SR-IB-
BUREAU OF NAVAL WEAPONS, WASHINGTON, D.C.	TAP-RA-
BUREAU OF NAVAL WEAPONS. HYDROBALLISTICS ADVISORY COMMITTEE, WASHINGTON, D.C.	NAVWEPSHAC-
BUREAU OF NAVAL WEAPONS REPRESENTATIVE, POMONA, CAL.	NAVAIR-17-20(LETTERS)-
BUREAU OF NAVAL WEAPONS REPRESENTATIVE, SUNNYVALE, CAL	BUWEPSREP-
BUREAU OF NAVAL WEAPONS REPRESENTATIVE. METROLOGY ENGINEERING CENTER, POMONA, CALIF.	EC-
BUREAU OF NAVAL WEAPONS REPRESENTATIVE. METROLOGY ENGINEERING CENTER, POMONA, CALIF.	ENGINEERING CIRCULAR-
BUREAU OF ORDNANCE (NAVY) (REF. 1)	(NUMBERS...)
BUREAU OF ORDNANCE (NAVY) (ACOUSTIC ANALYSIS MEMO.)	AAM-
BUREAU OF ORDNANCE (NAVY) (ACOUSTIC ANALYSIS REPORT)	AAR-
BUREAU OF ORDNANCE (NAVY) (ACOUSTIC CALIBRATION MEMO)	ACM-
BUREAU OF ORDNANCE (NAVY) (AD INTERIM SPECIFICATION)	AIS-
BUREAU OF ORDNANCE (NAVY) (ACOUSTIC RANGE REPORT)	ARR-
BUREAU OF ORDNANCE (NAVY) (BUMBLEBEE PROJ.) (REF. 8)	BB-
BUREAU OF ORDNANCE (NAVY)	BOI-
BUREAU OF ORDNANCE (NAVY) (SHIPMENT ORDER)	BOSO-
BUREAU OF ORDNANCE (NAVY) (BUMBLEBEE PROJECT)	BR-
BUREAU OF ORDNANCE (NAVY) (BALLISTIC TECHNICAL NOTE)	BTN-
BUREAU OF ORDNANCE (NAVY) (BUMBLEBEE PROJ.) (REF. 8)	BUMBLEBEE-
BUREAU OF ORDNANCE (NAVY)	BUO OP-
BUREAU OF ORDNANCE (NAVY)	BU ORD/
BUREAU OF ORDNANCE (NAVY)	BUORD-
BUREAU OF ORDNANCE (NAVY)	BUORD BM-
BUREAU OF ORDNANCE (NAVY)(BULLETIN OF ORDNANCE INFO.)	BUORD BOI-
BUREAU OF ORDNANCE (NAVY)	BUORD-BOI-(NO.)-(YR.)
BUREAU OF ORDNANCE (NAVY) (LIBRARY WEEKLY BULLETIN)	BUORD DRAWINGS-
BUREAU OF ORDNANCE (NAVY)	BUORD-LWB-(YR.)-
BUREAU OF ORDNANCE (NAVY) (ORDNANCE PAMPHLET)	BUORD (NAVY) SER-(NO.)-SP
BUREAU OF ORDNANCE (NAVY)	BUORD OP-
BUREAU OF ORDNANCE (NAVY)	BUORD REM-
BUREAU OF ORDNANCE (NAVY)	BUORD REXN-
BUREAU OF ORDNANCE (NAVY)	BUORD-S(NUMBER(S))-(YR.)
BUREAU OF ORDNANCE (NAVY)	BUORD-SR-
BUREAU OF ORDNANCE (NAVY) (CAPTURED ENEMY EQUIPMENT)	BUO-REM-
BUREAU OF ORDNANCE (NAVY) (COMPUTATION PROJECT RPTS.)	CEE-
BUREAU OF ORDNANCE (NAVY) (DEGAUSSING MEMORANDUM)	CPR-
BUREAU OF ORDNANCE (NAVY) (DEGAUSSING NOTE)	DGM-
BUREAU OF ORDNANCE (NAVY)(DEGAUSSING STA. INSTRUCTION)	DGN-
BUREAU OF ORDNANCE (NAVY) (DESIGN INSPECTION DATA)	DGSI-
BUREAU OF ORDNANCE (NAVY) (DEGAUSSING MEMORANDUM)	DID-
BUREAU OF ORDNANCE (NAVY) (EXPLOSIVES RES. RPT.)	DM-
BUREAU OF ORDNANCE (NAVY) (EXPLOSIVES RES. MEMO.)	ER-
BUREAU OF ORDNANCE (NAVY) (EXPLOSIVES RESEARCH RPT.)	ERM-
BUREAU OF ORDNANCE (NAVY)	ERR-
BUREAU OF ORDNANCE (NAVY) (FOREIGN ORDNANCE REPORT)	FOIM-
BUREAU OF ORDNANCE (NAVY) (FLASHING + WIPING MEMO.)	FOR-
BUREAU OF ORDNANCE (NAVY) (MAGNETIC ANALYSIS MEMO.)	FWM-
BUREAU OF ORDNANCE (NAVY) (MAGNETIC CALIBRATION MEMO)	MAM-
BUREAU OF ORDNANCE (NAVY)	MCM-
BUREAU OF ORDNANCE (NAVY) (AD INTERIM SPECIFICATION)	NAV-OD-
BUREAU OF ORDNANCE (NAVY) (COMPUTATION PROJECT RPTS.)	NAVORD AIS-
BUREAU OF ORDNANCE (NAVY) (EXPLOSIVES RES. RPT.)	NAVORD CPR-
BUREAU OF ORDNANCE (NAVY) (EXPLOSIVES RES. MEMORANDA)	NAVORD ER-
BUREAU OF ORDNANCE (NAVY) (INSTRUCTION)	NAVORD ER MEMO-
BUREAU OF ORDNANCE (NAVY) (ORDNANCE CIRCULAR LETTER)	NAVORDINST-
BUREAU OF ORDNANCE (NAVY) (ORDNANCE DATA)	NAVORD OCL-
BUREAU OF ORDNANCE (NAVY) (ORDNANCE HANDLING INSTRUCTIONS)	NAVORD OD-
BUREAU OF ORDNANCE (NAVY) (ORDNANCE MODIFICATION INSTRUCTIONS)	NAVORD OHI-
BUREAU OF ORDNANCE (NAVY) (ORDNANCE PAMPHLET)	NAVORD OMI-
BUREAU OF ORDNANCE (NAVY) (ORDNANCE SPECIFICATION)	NAVORD OP-
BUREAU OF ORDNANCE (NAVY) (ORDNANCE STANDARD)	NAVORD OS-
BUREAU OF ORDNANCE (NAVY) (ORDNANCE STD. TECH. DATA)	NAVORD OSTD-
BUREAU OF ORDNANCE (NAVY) (ORDNANCE TECHNICAL INSTRUCTIONS)	NAVORD OSTD-
BUREAU OF ORDNANCE (NAVY)	NAVORD OTI-
BUREAU OF ORDNANCE (NAVY) (ORDNANCE CLASSIFICATION OF DEFECTS)	NAVORD-TN-
BUREAU OF ORDNANCE (NAVY) (ORDNANCE CIRCULAR LETTER)	OCD-
BUREAU OF ORDNANCE (NAVY) (ORDNANCE DATA)	OCL-
BUREAU OF ORDNANCE (NAVY)	OD-
BUREAU OF ORDNANCE (NAVY) (ORDNANCE HANDLING INSTRUCTIONS)	O.F.(NUMBER).(NUMBER)
BUREAU OF ORDNANCE (NAVY) (ORDNANCE INVESTIGATION MEMORANDUM)	OHI-
BUREAU OF ORDNANCE (NAVY) (ORDNANCE MODIFICATION INSTRUCTION)	OIM-
BUREAU OF ORDNANCE (NAVY) (ORDNANCE MATERIAL LETTER)	OMI-
BUREAU OF ORDNANCE (NAVY) (ORDNANCE PAMPHLET)	OML-
BUREAU OF ORDNANCE (NAVY) (ORDNANCE ALTERATIONS)	OP-
BUREAU OF ORDNANCE (NAVY)	ORDALT-
BUREAU OF ORDNANCE (NAVY) (ORDNANCE SPECIFICATIONS)	ORDNANCE PAMPHLET-
BUREAU OF ORDNANCE (NAVY) (ORDNANCE STATUS REPORT)	OS-
BUREAU OF ORDNANCE (NAVY) (ORDNANCE STANDARD)	OSR-
BUREAU OF ORDNANCE (NAVY) (ORDNANCE TECHNICAL INSTRUCTIONS)	OSTD-
BUREAU OF ORDNANCE (NAVY)	OTI-
BUREAU OF ORDNANCE (NAVY)	REM-
BUREAU OF ORDNANCE (NAVY) (SUBSURFACE ORDNANCE RESEARCH REPORT)	REXN-
BUREAU OF ORDNANCE (NAVY) (SPECIFICATION)	S-(NUMBER)-(YEAR)
BUREAU OF ORDNANCE (NAVY) (WEAPON DATA INDEX)	SORR-
(SEE ALSO SUCCESSORS: BUREAU OF NAVAL WEAPONS, AND NAVAL ORDNANCE SYSTEMS COMMAND)	SPEC-MIL-(NO.)(NORD)
	WDI-
BUREAU OF ORDNANCE (NAVY). ASPEN COMMITTEE	ASPEN-
BUREAU OF ORDNANCE (NAVY). EVALUATION + ANALYSIS GP.	EAG-

BUREAU OF ORDNANCE (NAVY). EVALUATION AND ANALYSIS STAFF	BUO-EAG-
BUREAU OF ORDNANCE (NAVY). EVALUATION AND ANALYSIS STAFF	BUO-EAS-
BUREAU OF ORDNANCE (NAVY). EVALUATION AND ANALYSIS STAFF	EAS-
BUREAU OF ORDNANCE (NAVY). EVALUATION AND ANALYSIS STAFF	EAS-TN-
BUREAU OF ORDNANCE (NAVY). HYDROBALLISTICS ADVISORY COMMITTEE, WASHINGTON, D. C.	BOHAC-
BUREAU OF ORDNANCE (NAVY). TECHNICAL LIAISON OFFICE	BOTLO-
BUREAU OF PUBLIC ROADS, WASHINGTON, D.C.	BPR-
BUREAU OF RADIOLOGICAL HEALTH, ROCKVILLE, MD.	BRH/CFS-(YEAR)-
BUREAU OF RADIOLOGICAL HEALTH, ROCKVILLE, MD.	BRH/DBE-(YEAR)-
BUREAU OF RADIOLOGICAL HEALTH, ROCKVILLE, MD.	BRH/OBD-(YEAR)-
BUREAU OF RADIOLOGICAL HEALTH, ROCKVILLE, MD.	BRH/OCS-(YEAR)-
BUREAU OF RADIOLOGICAL HEALTH, ROCKVILLE, MD.	BRH/ORO-(YEAR)-
BUREAU OF RADIOLOGICAL HEALTH, ROCKVILLE, MD.	TSB-(YEAR)-
BUREAU OF RADIOLOGICAL HEALTH. DIV. OF BIOLOGICAL EFFECTS, ROCKVILLE, MD.	DBE-(YEAR)-
BUREAU OF RADIOLOGICAL HEALTH. DIV. OF ELECTRONIC PRODUCTS, ROCKVILLE, MD.	BRH/DEP-(YEAR)-
BUREAU OF RADIOLOGICAL HEALTH. DIV. OF ELECTRONIC PRODUCTS, ROCKVILLE, MD.	DEP-(YEAR)-
BUREAU OF RADIOLOGICAL HEALTH. DIV. OF ENVIRONMENTAL RADIATION, ROCKVILLE, MD.	BRH/DER-(YEAR)-
BUREAU OF RADIOLOGICAL HEALTH. DIV. OF MEDICAL RADIATION EXPOSURE, ROCKVILLE, MD.	BRH/DMRE-(YEAR)-
BUREAU OF RADIOLOGICAL HEALTH. DIV. OF MEDICAL RADIATION EXPOSURE, ROCKVILLE, MD.	DMRE-(YEAR)-
BUREAU OF RADIOLOGICAL HEALTH. NORTHEASTERN RADIOLOGICAL HEALTH LAB., WINCHESTER, MASS.	BRH/NERHL-(YEAR)-
BUREAU OF RADIOLOGICAL HEALTH. OFFICE OF CRITERIA AND STANDARDS, ROCKVILLE, MD.	OCS-
BUREAU OF RADIOLOGICAL HEALTH. SOUTHEASTERN RADIOLOGICAL HEALTH LAB., MONTGOMERY, ALA.	BRH/SERHL-(YEAR)-
BUREAU OF RADIOLOGICAL HEALTH. SOUTHWESTERN RADIOLOGICAL HEALTH LAB., LAS VEGAS, NEV.	BRH/SWRHL-(YEAR)-
BUREAU OF RECLAMATION (INTERIOR), DENVER	ENGINEERING MONO-
BUREAU OF RECLAMATION (INTERIOR), DENVER	SA-
BUREAU OF RECLAMATION (INTERIOR), DENVER	USBR-GENERAL-
BUREAU OF RECLAMATION (INTERIOR). CHEMICAL ENG. BR., DENVER	CHE-
BUREAU OF RECLAMATION (INTERIOR). COMMISSIONERS OFFICE, DENVER (TRANSLATIONS)	BR-
BUREAU OF RECLAMATION (INTERIOR). COMMISSIONERS OFFICE, DENVER	C-
BUREAU OF RECLAMATION (INTERIOR). COMMISSIONERS OFFICE, DENVER	EM-
BUREAU OF RECLAMATION (INTERIOR). COMMISSIONERS OFFICE, DENVER	G-
BUREAU OF RECLAMATION (INTERIOR). COMMISSIONERS OFFICE, DENVER	HYD-
BUREAU OF RECLAMATION (INTERIOR). ENG + RES. CTR., DENVER	REC-ERC-(YEAR)-
BUREAU OF RECLAMATION (INTERIOR). MECHANICAL BRANCH, DENVER	MECH-
BUREAU OF RECLAMATION (INTERIOR). OFFICE OF CHIEF ENGINEER, DENVER	GENERAL-
BUREAU OF RECLAMATION (INTERIOR). OFFICE OF CHIEF ENGINEER, DENVER	REC-OCE-(YEAR)-
BUREAU OF RECLAMATION (INTERIOR). OFFICE OF ENG. REFERENCE, DENVER	OER-
BUREAU OF RETIREMENT, INSURANCE AND OCCUPATIONAL HEALTH (CIVIL SERVICE), WASHINGTON, D.C.	BRI
BUREAU OF SHIPS	BUS-
BUREAU OF SHIPS	BUSHIPS-
BUREAU OF SHIPS	BU SHIPS (NO.-LTR.)-
BUREAU OF SHIPS	BUSHIPS-TM-
BUREAU OF SHIPS (TRANSLATION)	BUSHIPS-TRANS-
BUREAU OF SHIPS (COMPUTATION PROJECT REPORTS)	CPR-
BUREAU OF SHIPS	DTW.6161.DK-
BUREAU OF SHIPS	DTW.GK.MCD-
BUREAU OF SHIPS (INSTRUCTION BOOK, GAROD PART)	IBT-
BUREAU OF SHIPS (INDUSTRIAL NOTES)	IN-
BUREAU OF SHIPS (BLUDWORTH REPORTS)	NAVSHIPS B-
BUREAU OF SHIPS	NAVSHIPS C-
BUREAU OF SHIPS (COMPUTATION PROJECT)	NAVSHIPS CP-
BUREAU OF SHIPS (COMPUTATION PROJECT REPORTS)	NAVSHIPS CPR-
BUREAU OF SHIPS (ELECTRONICS INSTRUCTION BOOK)	NAVSHIPS ENG-
BUREAU OF SHIPS (INSTRUCTIONS)	NAVSHIPS I-
BUREAU OF SHIPS (INSTRUCTION BOOK)	NAVSHIPS IB-
BUREAU OF SHIPS (INSTRUCTION BOOK, GAROD PART)	NAVSHIPS IBT-
BUREAU OF SHIPS (INSTRUCTION BOOK)	NAVSHIPS IC IB-
BUREAU OF SHIPS (INDUSTRIAL NOTES)	NAVSHIPS IN-
BUREAU OF SHIPS (RADIO INSTRUCTION BOOK)	NAVSHIPS RA-
BUREAU OF SHIPS (ELECTRONICS INSTRUCTION BOOK)	NAVSHIPS RE-
BUREAU OF SHIPS (RESEARCH MEMORANDA)	NAVSHIPS RM-
BUREAU OF SHIPS (RESEARCH MEMORANDA)	NAVSHIPS RM (NUMBER)-
BUREAU OF SHIPS (ELECTRONICS INSTRUCTION BOOK)	NAVSHIPS RW-
BUREAU OF SHIPS (ELECTRONICS INSTRUCTION BOOK)	NAVSHIPS SHIPS-
BUREAU OF SHIPS (SPECIFICATIONS)	NAVSHIPS SPEC-
BUREAU OF SHIPS (SUBMARINE SIGNAL CO. SPECIFICATIONS)	NAVSHIPS SSS-
BUREAU OF SHIPS (PRELIMINARY DESIGN)	PD-
BUREAU OF SHIPS (ELECTRONICS INSTRUCTION BOOK)	RE-
BUREAU OF SHIPS (RESEARCH MEMORANDA)	RM-
BUREAU OF SHIPS (RADAR TECHNICAL BULLETIN)	RTB-
BUREAU OF SHIPS (STATISTICAL ANALYSIS REPORT)	SAR-
BUREAU OF SHIPS (SUBMARINE SIGNAL CO. SPECIFICATIONS)	SSS-
BUREAU OF SHIPS (WAR DAMAGE REPORT)	WDR-
(SEE ALSO LATER NAME: NAVAL SHIP SYSTEMS COMMAND)	
BUREAU OF SHIPS. ESSO LAB.	EL-
BUREAU OF SHIPS. ESSO LAB.	NAVSHIPS EL-
BUREAU OF SHIPS. MATERIALS LAB. (TESTS)	MLT-
BUREAU OF SHIPS. MATERIALS LAB. (TESTS)	NAVSHIPS MLT-
BUREAU OF SHIPS. SHIPBUILDING DIVISION. RESEARCH AND STANDARDS BRANCH (TECH. LITERATURE RES. SERIES)	NAVSHIPS TLR-
BUREAU OF SHIPS. SHIPBUILDING DIVISION. RESEARCH AND STANDARDS BRANCH (TECH. LITERATURE RES. SERIES)	TLR-
BUREAU OF SHIPS. SHIPS INSTALLATION DIV.	SI-
BUREAU OF SHIPS. SHIPS STRUCTURAL COMMITTEE	NAVSHIPS SSC-
BUREAU OF SHIPS. UNDERWATER EXPLOSION RESEARCH GROUP	NAVSHIPS UER-

BUREAU OF SOCIAL SCIENCE RESEARCH, INC.

BUREAU OF SOCIAL SCIENCE RESEARCH, INC., WASH., D.C.	BSSR-
BUREAU OF SOLID WASTE MANAGEMENT, ROCKVILLE, MD.	BSWM-SW-
BUREAU OF SUPPLIES AND ACCOUNTS (NAVY)	BUSANDA-
BUREAU OF SUPPLIES AND ACCOUNTS (NAVY)	NAVSANDA-
BUREAU OF SUPPLIES + ACCOUNTS (NAVY) (SPECIFICATIONS)	NAVSANDA-
BUREAU OF SUPPLIES + ACCOUNTS (NAVY) (SPECIFICATIONS)	NAVSANDA SPEC-
BUREAU OF THE BUDGET, WASHINGTON, D.C.	BOB-(NUMBER-NUMBER)
BUREAU OF THE CENSUS, WASHINGTON, D.C.	BD-(YEAR)-
BUREAU OF THE CENSUS, WASHINGTON, D.C.	BDSAF-(NUMBER)(YEAR)
BUREAU OF THE CENSUS, WASHINGTON, D.C.	BW-(YEAR)-
BUREAU OF THE CENSUS, WASHINGTON, D.C. (CURRENT BUSINESS REPORTS)	CB-(YEAR)-
BUREAU OF THE CENSUS, WASHINGTON, D.C.	CDD-(NUMBER-NUMBER)
BUREAU OF YARDS AND DOCKS	BU DOCKS
BUREAU OF YARDS AND DOCKS	BUDOCKS-
BUREAU OF YARDS AND DOCKS	NAVDOCKS-
BUREAU OF YARDS AND DOCKS	NAVDOCKS-P-
BUREAU OF YARDS AND DOCKS	NAVDOCKS TP-(LETTERS)-
BUREAU OF YARDS AND DOCKS	NAVDOCKS-TP-PL-
BUREAU OF YARDS AND DOCKS	NAVDOCKS-TP-TE-
BUREAU OF YARDS AND DOCKS	WO-
BURGESS BATTERY CO., FREEPORT, ILL.	BURGESS (LTRS)-QPR-
BURKE RESEARCH CO., VAN DYKE, MICH.	BURK-
BURNDY CORP., NORWALK, CONN.	BEDR-
BURNDY ENGINEERING CO., INC., N.Y.C. (TEST DATA)	TD-
BURNS AND MC DONNELL ENGINEERING CO., KANSAS CITY, MO.	B + MEC-(NO.)-(NO.,LTR)
BURNS AND ROE, INC., N.Y.C.	BAR-
BURROUGHS CORP. (ALL LOCATIONS)	BURROUGHS(NO.)-MPR-(DATE)
BURROUGHS CORP., DETROIT	BURR-
BURROUGHS CORP., DETROIT	BURR PROD (LETTERS)-
BURROUGHS CORP., DETROIT	NP-
BURROUGHS CORP., DETROIT (PRODUCT IMPROVEMENT BULL.)	PIB-
BURROUGHS CORP., ENCINO, CALIF.	BURR-
BURROUGHS CORP., ENCINO, CALIF.	BURR PROD (LETTERS)-
BURROUGHS CORP., PAOLI, PENNA.	B-(NUMBERS)-
BURROUGHS CORP., PAOLI, PENNA.	BURROUGHS-
BURROUGHS CORP., PAOLI, PENNA.	NR-
BURROUGHS CORP., RANDOR, PENNA.	BURR-(CONTRACT NO.)
BURROUGHS CORP. DEFENSE SPACE AND SPECIAL SYSTEMS GROUP, PAOLI, PA.	F-(YEAR)-
BURROUGHS CORP. ELECTRONIC COMPONENTS DIV., PLAINFIELD, N.J.	T-(3 DIGITS)
BUSINESS AND DEFENSE SERVICES ADM., WASHINGTON, D.C.	BDSA-
BUTLER UNIV., INDIANAPOLIS	BTU-

C-E-I-R, INC., ARLINGTON, VA.	IPT-TM-
C-G ELECTRONICS, INC., ALBUQUERQUE, N. MEX.	CGE-
CADILLAC GAGE CO. RESEARCH DIV., DETROIT	TRN-
CAGLIARI, ITALY. UNIVERSITA. ISTITUTO DE FISICA	LAL-
CALIDYNE CO., WINCHESTER, MASS.	CALY-
CALIFORNIA. DEPARTMENT OF EMPLOYMENT, SACRAMENTO	E-(NUMBER)-
CALIFORNIA. DIV. OF HIGHWAYS. BUDGET DEPT.	R/D-(NUMBER-YEAR)
CALIFORNIA. UNIV. (ALL LOCATIONS)	CALU-
CALIFORNIA. UNIV. (ALL LOCATIONS)	CAL-U-
CALIFORNIA. UNIV. (ALL LOCATIONS)	CAL. U., SERIES-
CALIFORNIA. UNIV. (ALL LOCATIONS)	CAL. U.,SERIES-(NO.ISSUE)
CALIFORNIA. UNIV. (ALL LOCATIONS)	TE-(YEAR)-
CALIFORNIA. UNIV. (ALL LOCATIONS)	UCAL-
CALIFORNIA. UNIV., BERKELEY	AL-
CALIFORNIA. UNIV., BERKELEY	AS-
CALIFORNIA. UNIV., BERKELEY (SQUID PROJECT) (REF. 33)	CAL-(NUMBER)-(LETTER(S))
CALIFORNIA. UNIV., BERKELEY	DP-
CALIFORNIA. UNIV., BERKELEY	DR-
CALIFORNIA. UNIV., BERKELEY	HPS-
CALIFORNIA. UNIV., BERKELEY	MD-
CALIFORNIA. UNIV., BERKELEY	NA-
CALIFORNIA. UNIV., BERKELEY	NSCI-(YEAR)-
CALIFORNIA. UNIV., BERKELEY	ORC-
CALIFORNIA. UNIV., BERKELEY	SSL-
CALIFORNIA. UNIV., BERKELEY	UC-
CALIFORNIA. UNIV., BERKELEY	UCAL TN DR-
CALIFORNIA. UNIV., BERKELEY	UCB-
CALIFORNIA. UNIV., BERKELEY (USED BY MANY DEPTS.)	UCB-(NUMBER)-P-(NO.)-
CALIFORNIA. UNIV., BERKELEY (SQUID PROJECT) (REF. 33)	UCB-(NUMBER)-R
CALIFORNIA. UNIV., BERKELEY	UCB-(YEAR)-
CALIFORNIA. UNIV., BERKELEY	UCB-(CONTRACT NO.)-P-
CALIFORNIA. UNIV., BERKELEY	U OF C-
CALIFORNIA. UNIV., BERKELEY	W-
CALIFORNIA. UNIV., BERKELEY	WP-
CALIFORNIA. UNIV., BERKELEY. ANTENNA LAB.	UC-AL-
CALIFORNIA. UNIV., BERKELEY. COLL. OF ENGINEERING	AS-(YEAR)-
CALIFORNIA. UNIV., BERKELEY. COLL. OF ENGINEERING	MSE-
CALIFORNIA. UNIV., BERKELEY. COLL. OF ENGINEERING	NA-(YEAR)-
CALIFORNIA. UNIV., BERKELEY. CROCKER LAB.	BP-
CALIFORNIA. UNIV., BERKELEY. CROCKER LAB.	UCRL-60-
CALIFORNIA. UNIV., BERKELEY. DEPT. OF CIVIL ENG.	UCB-ENG-
CALIFORNIA. UNIV., BERKELEY. DEPT. OF ELEC. ENG.	JA-
CALIFORNIA. UNIV., BERKELEY. DEPT. OF ENGINEERING	TS-(YR.)-(NO.)
CALIFORNIA. UNIV., BERKELEY. DEPT. OF ENGINEERING	UCM-
CALIFORNIA. UNIV., BERKELEY. DEPT. OF MECH. ENG.	TS-(YEAR)-
CALIFORNIA. UNIV., BERKELEY. DEPT. OF PHYSICS	JCRL-
CALIFORNIA. UNIV., BERKELEY. DEPT. OF PHYSICS	P-
CALIFORNIA. UNIV., BERKELEY. DIV. OF AERONAUTICAL SCIENCES	AS-(YEAR)-
CALIFORNIA. UNIV., BERKELEY. DIV.OF APPLIED MECHANICS	AM-(YEAR)-
CALIFORNIA. UNIV., BERKELEY. DIV. OF ENG. RES.(REF.1)	(NUMBER)
CALIFORNIA. UNIV., BERKELEY. DIV. OF STRUCTURAL ENG. AND STRUCTURAL MECHANICS	SESM-(YEAR)-
CALIFORNIA. UNIV., BERKELEY. DIV. OF STRUCTURAL ENG. AND STRUCTURAL MECHANICS	SESM-
CALIFORNIA. UNIV., BERKELEY. EARTHQUAKE ENG. RES. CTR	EERC-(YEAR)-
CALIFORNIA. UNIV., BERKELEY. ELECTRONICS RES. LAB.	ERL-
CALIFORNIA. UNIV., BERKELEY. ELECTRONICS RES. LAB.	ERL-(YEAR)-
CALIFORNIA. UNIV., BERKELEY. ELECTRONICS RES. LAB.	ERL-M-
CALIFORNIA. UNIV., BERKELEY. ELECTRONICS RES. LAB.	ERL-TR-(NO.)-
CALIFORNIA. UNIV., BERKELEY. ELECTRONICS RES. LAB.	GENIE-P-
CALIFORNIA. UNIV., BERKELEY. FLUID MECHANICS LAB.	HE-116-
CALIFORNIA. UNIV., BERKELEY. HUMAN FACTORS IN TECHNOLOGY RESEARCH GROUP	HFT-
CALIFORNIA. UNIV., BERKELEY. HYDRAULIC ENG. LAB.	HEL-
CALIFORNIA. UNIV., BERKELEY. INST. OF ENG. RES.	AM-(YEAR)-
CALIFORNIA. UNIV., BERKELEY. INST. OF ENG. RES.	CAL. U. HE-
CALIFORNIA. UNIV., BERKELEY. INST. OF ENG. RES.	CAL. U. IER (NUMBERS)
CALIFORNIA. UNIV., BERKELEY. INST. OF ENG. RES.	CDR-P-
CALIFORNIA. UNIV., BERKELEY. INST. OF ENG. RES.	CDRP-
CALIFORNIA. UNIV., BERKELEY. INST. OF ENG. RES.	HE-150-
CALIFORNIA. UNIV., BERKELEY. INST. OF ENG. RES.	IER-
CALIFORNIA. UNIV., BERKELEY. INST. OF ENG. RES.	IER-TR-
CALIFORNIA. UNIV., BERKELEY. INST. OF ENG. RES.	NA-
CALIFORNIA. UNIV., BERKELEY. INST. OF ENG. RES.	UCAL HE (NO.)-(NO.)-
CALIFORNIA. UNIV., BERKELEY. INST. OF ENG. RES.	UCAL-IER-
CALIFORNIA. UNIV., BERKELEY. INST. OF ENG. RES.	UCAL IER S.(NO.)-
CALIFORNIA. UNIV., BERKELEY. INST. OF ENG. RES.	UC-CD-
CALIFORNIA. UNIV., BERKELEY. INST. OF ENG. RES.	UC-HE-
CALIFORNIA. UNIV., BERKELEY. INSTITUTE OF TRANSPORTATION AND TRAFFIC ENGINEERING	ITTE-
CALIFORNIA. UNIV., BERKELEY. LAWRENCE BERKELEY LAB.	LBL-
CALIFORNIA. UNIV., BERKELEY. LAWRENCE BERKELEY LAB.	UCID-
CALIFORNIA. UNIV., BERKELEY. LAWRENCE BERKELEY LAB.	UCRL-
CALIFORNIA. UNIV., BERKELEY. LAWRENCE RADIATION LAB.	BECV-(YEAR-NUMBER)
CALIFORNIA. UNIV., BERKELEY. LAWRENCE RADIATION LAB.	GCTN-

CALIFORNIA. UNIV., LAWRENCE RADIATION LAB.

CALIFORNIA. UNIV., BERKELEY. LAWRENCE RADIATION LAB.	MDT-
CALIFORNIA. UNIV., BERKELEY. LAWRENCE RADIATION LAB.	MT-
CALIFORNIA. UNIV., BERKELEY. LAWRENCE RADIATION LAB.	NMI-
CALIFORNIA. UNIV., BERKELEY. LAWRENCE RADIATION LAB.	PMO-
CALIFORNIA. UNIV., BERKELEY. LAWRENCE RADIATION LAB.	URCRL-
(SEE ALSO LATER NAME: CALIFORNIA. UNIV., BERKELEY. LAWRENCE BERKELEY LAB.)	
CALIFORNIA. UNIV., BERKELEY. MICROWAVE LAB.	ML-
CALIFORNIA. UNIV., BERKELEY. MINERALS RESEARCH LAB.	MRI-
CALIFORNIA. UNIV., BERKELEY. MINERALS RESEARCH LAB.	UC-MRL-
CALIFORNIA. UNIV., BERKELEY. OFFICE OF RES. SERVICES	NA-(YEAR)-
CALIFORNIA. UNIV., BERKELEY. OPERATIONS RESEARCH CTR.	ORC-
CALIFORNIA. UNIV., BERKELEY. OPERATIONS RESEARCH CTR.	ORC-(YEAR)-
CALIFORNIA. UNIV., BERKELEY. RADIATION LAB. (REF. 9)	184-
CALIFORNIA. UNIV., BERKELEY. RADIATION LAB.	AB-
CALIFORNIA. UNIV., BERKELEY. RADIATION LAB. (REF. 9)	ANAL-MISC.
CALIFORNIA. UNIV., BERKELEY. RADIATION LAB.	ARC S-
CALIFORNIA. UNIV., BERKELEY. RADIATION LAB. (REF. 9)	ASSAY-
CALIFORNIA. UNIV., BERKELEY. RADIATION LAB.	BC-
CALIFORNIA. UNIV., BERKELEY. RADIATION LAB. (REF. 9)	BEI-
CALIFORNIA. UNIV., BERKELEY. RADIATION LAB. (REF. 9)	BETA-
CALIFORNIA. UNIV., BERKELEY. RADIATION LAB.	BETA-S-
CALIFORNIA. UNIV., BERKELEY. RADIATION LAB. (REF. 9)	BJM-
CALIFORNIA. UNIV., BERKELEY. RADIATION LAB. (REF. 9)	BP-
CALIFORNIA. UNIV., BERKELEY. RADIATION LAB. (REF. 9)	C-2-184-M-
CALIFORNIA. UNIV., BERKELEY. RADIATION LAB. (REF. 10)	C-2-184-W-
CALIFORNIA. UNIV., BERKELEY. RADIATION LAB. (REF. 10)	CC-
CALIFORNIA. UNIV., BERKELEY. RADIATION LAB.	CH-
CALIFORNIA. UNIV., BERKELEY. RADIATION LAB. (REF. 9)	CHART X-
CALIFORNIA. UNIV., BERKELEY. RADIATION LAB. (REF. 9)	CHEM M-
CALIFORNIA. UNIV., BERKELEY. RADIATION LAB.	CHEM S-
CALIFORNIA. UNIV., BERKELEY. RADIATION LAB.	COC-
CALIFORNIA. UNIV., BERKELEY. RADIATION LAB. (REF. 9)	COI-
CALIFORNIA. UNIV., BERKELEY. RADIATION LAB.	COL. SPEC-
CALIFORNIA. UNIV., BERKELEY. RADIATION LAB.	COPJ-
CALIFORNIA. UNIV., BERKELEY. RADIATION LAB.	COT-
CALIFORNIA. UNIV., BERKELEY. RADIATION LAB.	CVL-
CALIFORNIA. UNIV., BERKELEY. RADIATION LAB. (REF. 9)	D1-RR-
CALIFORNIA. UNIV., BERKELEY. RADIATION LAB. (REF. 9)	D1-S-
CALIFORNIA. UNIV., BERKELEY. RADIATION LAB. (REF. 9)	D1-SM-
CALIFORNIA. UNIV., BERKELEY. RADIATION LAB. (REF. 9)	D1-SPEC-
CALIFORNIA. UNIV., BERKELEY. RADIATION LAB. (REF. 9)	D1M-
CALIFORNIA. UNIV., BERKELEY. RADIATION LAB. (REF. 9)	D1W-
CALIFORNIA. UNIV., BERKELEY. RADIATION LAB. (REF. 9)	EE-
CALIFORNIA. UNIV., BERKELEY. RADIATION LAB. (REF. 9)	EE-SPEC-
CALIFORNIA. UNIV., BERKELEY. RADIATION LAB. (REF. 9)	ELEC. ENG. SPEC.-
CALIFORNIA. UNIV., BERKELEY. RADIATION LAB. (REF. 9)	ENG. SPEC.-
CALIFORNIA. UNIV., BERKELEY. RADIATION LAB. (REF. 9)	E-P(MONTH/YEAR)
CALIFORNIA. UNIV., BERKELEY. RADIATION LAB. (REF. 9)	E-SPEC-
CALIFORNIA. UNIV., BERKELEY. RADIATION LAB. (REF. 9)	FUND.-184 S2-
CALIFORNIA. UNIV., BERKELEY. RADIATION LAB. (REF. 9)	FUND.-(MONTH/YEAR)
CALIFORNIA. UNIV., BERKELEY. RADIATION LAB. (REF. 9)	FUND.-P(MONTH/YEAR)
CALIFORNIA. UNIV., BERKELEY. RADIATION LAB. (REF. 9)	FUND.-S-
CALIFORNIA. UNIV., BERKELEY. RADIATION LAB.	GA-
CALIFORNIA. UNIV., BERKELEY. RADIATION LAB.	HOT SOURCES-
CALIFORNIA. UNIV., BERKELEY. RADIATION LAB. (REF. 9)	ID-SPEC-
CALIFORNIA. UNIV., BERKELEY. RADIATION LAB.	JHW-
CALIFORNIA. UNIV., BERKELEY. RADIATION LAB. (REF. 9)	J-P(NUMBER/YEAR)
CALIFORNIA. UNIV., BERKELEY. RADIATION LAB.	J SPEC-
CALIFORNIA. UNIV., BERKELEY. RADIATION LAB. (REF. 9)	KAPPA-P-
CALIFORNIA. UNIV., BERKELEY. RADIATION LAB. (REF. 9)	K-P(MONTH/YEAR)
CALIFORNIA. UNIV., BERKELEY. RADIATION LAB. (REF. 9)	MAG. S.-
CALIFORNIA. UNIV., BERKELEY. RADIATION LAB. (REF. 9)	MAG. SPEC-
CALIFORNIA. UNIV., BERKELEY. RADIATION LAB.	MB-
CALIFORNIA. UNIV., BERKELEY. RADIATION LAB. (REF. 9)	MB-(INITIALS)-
CALIFORNIA. UNIV., BERKELEY. RADIATION LAB.	MECH. ENG. SPEC.-
CALIFORNIA. UNIV., BERKELEY. RADIATION LAB. (REF. 9)	MM-
CALIFORNIA. UNIV., BERKELEY. RADIATION LAB.	MRL-(INITIALS)-
CALIFORNIA. UNIV., BERKELEY. RADIATION LAB.	NOMEN.-
CALIFORNIA. UNIV., BERKELEY. RADIATION LAB.	NPC-
CALIFORNIA. UNIV., BERKELEY. RADIATION LAB.	OM SM-
CALIFORNIA. UNIV., BERKELEY. RADIATION LAB. (REF. 9)	OPER. ANAL.-
CALIFORNIA. UNIV., BERKELEY. RADIATION LAB.	P-(NO.)-C-
CALIFORNIA. UNIV., BERKELEY. RADIATION LAB. (REF. 9)	R1H RR-
CALIFORNIA. UNIV., BERKELEY. RADIATION LAB. (REF. 9)	R-1-H RR-
CALIFORNIA. UNIV., BERKELEY. RADIATION LAB. (REF. 9)	R1-M-
CALIFORNIA. UNIV., BERKELEY. RADIATION LAB. (REF. 9)	R1-RR-
CALIFORNIA. UNIV., BERKELEY. RADIATION LAB. (REF. 9)	R1-S-
CALIFORNIA. UNIV., BERKELEY. RADIATION LAB. (REF. 9)	R1-SM-
CALIFORNIA. UNIV., BERKELEY. RADIATION LAB. (REF. 9)	R-1-184-S-
CALIFORNIA. UNIV., BERKELEY. RADIATION LAB. (REF. 9)	R1 SPECIAL-
CALIFORNIA. UNIV., BERKELEY. RADIATION LAB. (REF. 9)	R1-W-
CALIFORNIA. UNIV., BERKELEY. RADIATION LAB. (REF. 9)	R-1-184-W-
CALIFORNIA. UNIV., BERKELEY. RADIATION LAB.	R1-184-W-
CALIFORNIA. UNIV., BERKELEY. RADIATION LAB.	RESEARCH PROGRAM MEETING-
CALIFORNIA. UNIV., BERKELEY. RADIATION LAB. (REF. 9)	RL-1.7- THRU RL-37.9-
CALIFORNIA. UNIV., BERKELEY. RADIATION LAB. (REF. 9)	RL-MISC.-
CALIFORNIA. UNIV., BERKELEY. RADIATION LAB. (REF. 9)	RL-NPE-
CALIFORNIA. UNIV., BERKELEY. RADIATION LAB. (REF. 9)	37 RR-
CALIFORNIA. UNIV., BERKELEY. RADIATION LAB.	S-
CALIFORNIA. UNIV., BERKELEY. RADIATION LAB. (REF. 9)	37 SM-
CALIFORNIA. UNIV., BERKELEY. RADIATION LAB. (REF. 9)	SPEC-
CALIFORNIA. UNIV., BERKELEY. RADIATION LAB. (SPECIAL SEMINAR) (REF. 9)	SPEC SEM-
CALIFORNIA. UNIV., BERKELEY. RADIATION LAB. (REF. 9)	SPEC. SEMI.-
CALIFORNIA. UNIV., BERKELEY. RADIATION LAB. (REF. 9)	SUMP.-P(MONTH/YEAR)
CALIFORNIA. UNIV., BERKELEY. RADIATION LAB. (REF. 9)	SUMP. SPEC-
CALIFORNIA. UNIV., BERKELEY. RADIATION LAB.	THEO.-P(MONTH/YEAR)
CALIFORNIA. UNIV., BERKELEY. RADIATION LAB.	TL-
CALIFORNIA. UNIV., BERKELEY. RADIATION LAB. (REF. 9)	UCRL-60-
CALIFORNIA. UNIV., BERKELEY. RADIATION LAB.	UCRL-BC-
CALIFORNIA. UNIV., BERKELEY. RADIATION LAB.	UCRL-COPD-
	UCRL-COT-
	UCRL-COVA-
CALIFORNIA. UNIV., BERKELEY. RADIATION LAB. (SERIES ASSIGNED BY LOS ALAMOS SCIENTIFIC LAB. TO REPORTS RECEIVED FROM UCRL)	UCRLD-
CALIFORNIA. UNIV., BERKELEY. RADIATION LAB. (REF. 9)	UCRL-SPEC-

CALIFORNIA. UNIV., BERKELEY. RADIATION LAB.	UCRL-TRANS-
CALIFORNIA. UNIV., BERKELEY. RADIATION LAB. (REF. 9)	VAC-M-
CALIFORNIA. UNIV., BERKELEY. RADIATION LAB. (REF. 9)	VAC.-P(MONTH/YEAR)
CALIFORNIA. UNIV., BERKELEY. RADIATION LAB. (REF. 9)	VAC. SPEC-
CALIFORNIA. UNIV., BERKELEY. RADIATION LAB. (REF. 9)	VAC-W-
CALIFORNIA. UNIV., BERKELEY. RADIATION LAB.	WM-
CALIFORNIA. UNIV., BERKELEY. RADIATION LAB. (REF. 9)	XA-1-SM-
CALIFORNIA. UNIV., BERKELEY. RADIATION LAB. (REF. 9)	XA-2-SM-
CALIFORNIA. UNIV., BERKELEY. RADIATION LAB. (REF. 9)	XAC-SPEC-
CALIFORNIA. UNIV., BERKELEY. RADIATION LAB. (REF. 9)	XAH-
CALIFORNIA. UNIV., BERKELEY. RADIATION LAB. (REF. 9)	XAH-1-SM-
CALIFORNIA. UNIV., BERKELEY. RADIATION LAB. (REF. 9)	XAH-1-SPEC-
CALIFORNIA. UNIV., BERKELEY. RADIATION LAB. (REF. 9)	XAH-2.39-
CALIFORNIA. UNIV., BERKELEY. RADIATION LAB. (REF. 9)	XAH-2-RR-
CALIFORNIA. UNIV., BERKELEY. RADIATION LAB. (REF. 9)	XAH-2-SM-
CALIFORNIA. UNIV., BERKELEY. RADIATION LAB. (REF. 9)	XAH-SM-
CALIFORNIA. UNIV., BERKELEY. RADIATION LAB. (REF. 9)	XAH-SPEC-
CALIFORNIA. UNIV., BERKELEY. RADIATION LAB. (REF. 9)	XAS-
CALIFORNIA. UNIV., BERKELEY. RADIATION LAB. (REF. 9)	XA-SPEC-
CALIFORNIA. UNIV., BERKELEY. RADIATION LAB. (REF. 9)	XC-
CALIFORNIA. UNIV., BERKELEY. RADIATION LAB. (REF. 9)	XL50.1- THRU XL60.9-
CALIFORNIA. UNIV., BERKELEY. RADIATION LAB. (REF. 9)	Z-P(MONTH/YEAR)-
CALIFORNIA. UNIV., BERKELEY. RADIATION LAB. (REF. 9)	Z-SPECIAL-

(SEE ALSO LATER NAMES UNDER CALIFORNIA. UNIV., BERKELEY: LAWRENCE RADIATION LAB. AND LAWRENCE BERKELEY LAB.)

CALIFORNIA. UNIV., BERKELEY. SANITARY ENG. RES. LAB.	SERL-
CALIFORNIA. UNIV., BERKELEY. SEA WATER CONVERSION LAB	SWCL-
CALIFORNIA. UNIV., BERKELEY. SPACE SCIENCES LAB.	IWP-
CALIFORNIA. UNIV., BERKELEY. SPACE SCIENCES LAB.	SERIES-(NO.)-ISSUE-
CALIFORNIA. UNIV., BERKELEY. SPACE SCIENCES LAB.	SSL(UC)-
CALIFORNIA. UNIV., BERKELEY. STATISTICAL LAB.	TR-ONR-
CALIFORNIA. UNIV., BERKELEY. STRUCTURAL ENG. LAB.	SEL-(YEAR)-
CALIFORNIA. UNIV., BERKELEY. STRUCTURAL ENG. LAB.	UCSESM-
CALIFORNIA. UNIV., BERKELEY. SURVEY RESEARCH CENTER	SRC-A-
CALIFORNIA. UNIV., BERKELEY. SURVEY RESEARCH CENTER	SRC-M-
CALIFORNIA. UNIV., BERKELEY. THERMAL SYSTEMS DIV.	TS-(YEAR)-
CALIFORNIA. UNIV., DAVIS	UCD-
CALIFORNIA. UNIV., DAVIS	UCD-(CONTRACT NO.)-P-
CALIFORNIA. UNIV., DAVIS. CROCKER NUCLEAR LAB.	UCD-CNL-
CALIFORNIA. UNIV., DAVIS. CROCKER NUCLEAR LAB.	UCD-CNL-EX-
CALIFORNIA. UNIV., DAVIS. DEPT. OF POMOLOGY	UCD-(NO.)P-(NO.)-
CALIFORNIA. UNIV., DAVIS. RADIOBIOLOGY LAB.	UCD-(NUMBER)-
CALIFORNIA. UNIV., IRVINE	UCI-
CALIFORNIA. UNIV., IRVINE	UCI-(NO.)-P-(NO.)-
CALIFORNIA. UNIV., LA JOLLA	MPL-U-
CALIFORNIA. UNIV., LA JOLLA	SM-
CALIFORNIA. UNIV., LA JOLLA.	UCSD-
CALIFORNIA. UNIV., LA JOLLA(USED BY MANY DEPTS.)	UCSD-(NUMBER)-P-(NO.)-
CALIFORNIA. UNIV., LA JOLLA. DEPT. OF PHYSICS	UCSD-SP-(YEAR)-
CALIFORNIA. UNIV., LA JOLLA. INST. FOR PURE AND APPLIED PHYSICAL SCIENCES	IPAPS-(YEAR/YEAR)-
CALIFORNIA. UNIV., LA JOLLA. INST. OF MARINE RESOURCES	IMR-TR-
CALIFORNIA. UNIV., LIVERMORE. COLL. OF ENGINEERING	ERL-
CALIFORNIA. UNIV., LIVERMORE. LAWRENCE LIVERMORE LAB.	K-(YEAR)-(3 DIGITS)-
CALIFORNIA. UNIV., LIVERMORE. LAWRENCE LIVERMORE LAB.	UCRL-
CALIFORNIA. UNIV., LIVERMORE. LAWRENCE RADIATION LAB.	ADN-
CALIFORNIA. UNIV., LIVERMORE. LAWRENCE RADIATION LAB.	C7-(NUMBER/-)
CALIFORNIA. UNIV., LIVERMORE. LAWRENCE RADIATION LAB.	CALU/LRL(L)
CALIFORNIA. UNIV., LIVERMORE. LAWRENCE RADIATION LAB.	COC-
CALIFORNIA. UNIV., LIVERMORE. LAWRENCE RADIATION LAB.	COL-
CALIFORNIA. UNIV., LIVERMORE. LAWRENCE RADIATION LAB.	COMW-
CALIFORNIA. UNIV., LIVERMORE. LAWRENCE RADIATION LAB.	COPP-
CALIFORNIA. UNIV., LIVERMORE. LAWRENCE RADIATION LAB.	COVA-
CALIFORNIA. UNIV., LIVERMORE. LAWRENCE RADIATION LAB.	CVL-
CALIFORNIA. UNIV., LIVERMORE. LAWRENCE RADIATION LAB.	ENS-
CALIFORNIA. UNIV., LIVERMORE. LAWRENCE RADIATION LAB.	ENW-
CALIFORNIA. UNIV., LIVERMORE. LAWRENCE RADIATION LAB.	GCTN-
CALIFORNIA. UNIV., LIVERMORE. LAWRENCE RADIATION LAB. (PROJECT HOBO)	HOBO-
CALIFORNIA. UNIV., LIVERMORE. LAWRENCE RADIATION LAB. (PLUTO TECHNICAL NOTE)	PLUTO-TN-
CALIFORNIA. UNIV., LIVERMORE. LAWRENCE RADIATION LAB. (PLUTO TECHNICAL NOTE)	PTN-
CALIFORNIA. UNIV., LIVERMORE. LAWRENCE RADIATION LAB.	ROVER-TN-
CALIFORNIA. UNIV., LIVERMORE. LAWRENCE RADIATION LAB.	SPR-
CALIFORNIA. UNIV., LIVERMORE. LAWRENCE RADIATION LAB.	THEO.PNU-
CALIFORNIA. UNIV., LIVERMORE. LAWRENCE RADIATION LAB.	TORY II-
CALIFORNIA. UNIV., LIVERMORE. LAWRENCE RADIATION LAB.	UCID-
CALIFORNIA. UNIV., LIVERMORE. LAWRENCE RADIATION LAB.	UCRL-PUB-
CALIFORNIA. UNIV., LIVERMORE. LAWRENCE RADIATION LAB.	UOPAC-
CALIFORNIA. UNIV., LIVERMORE. LAWRENCE RADIATION LAB.	UOPB-
CALIFORNIA. UNIV., LIVERMORE. LAWRENCE RADIATION LAB.	UOPK-
CALIFORNIA. UNIV., LIVERMORE. LAWRENCE RADIATION LAB.	UOPKA-(YEAR)-

(SEE ALSO LATER NAME: CALIFORNIA, UNIV., LIVERMORE. LAWRENCE LIVERMORE LAB.)

CALIFORNIA. UNIV., LIVERMORE. LAWRENCE RADIATION LAB. COMPUTER INFORMATION CENTER	CIC-E(NUMBER.NUMBER)-
CALIFORNIA. UNIV., LIVERMORE. LAWRENCE RADIATION LAB. COMPUTER INFORMATION CENTER	CIC-MANUAL-
CALIFORNIA. UNIV., LIVERMORE. LAWRENCE RADIATION LAB. COMPUTER INFORMATION CENTER	CIC-MP-
CALIFORNIA. UNIV., LIVERMORE. LAWRENCE RADIATION LAB. COMPUTER INFORMATION CENTER	CIC-N(NUMBER.NUMBER)-
CALIFORNIA. UNIV., LIVERMORE. LAWRENCE RADIATION LAB. COMPUTER INFORMATION CENTER	CIC-P(NUMBER)-
CALIFORNIA. UNIV., LIVERMORE. LAWRENCE RADIATION LAB. ELECTRONICS ENGINEERING DEPT.	LE-
CALIFORNIA. UNIV., LIVERMORE. RADIATION LAB.	BY-
CALIFORNIA. UNIV., LIVERMORE. RADIATION LAB.	COI-

CALIFORNIA. UNIV., RADIATION LAB.

CALIFORNIA. UNIV., LIVERMORE. RADIATION LAB.	COPB-
CALIFORNIA. UNIV., LIVERMORE. RADIATION LAB.	COPBA-
CALIFORNIA. UNIV., LIVERMORE. RADIATION LAB.	COPD-
CALIFORNIA. UNIV., LIVERMORE. RADIATION LAB.	COPDB-
CALIFORNIA. UNIV., LIVERMORE. RADIATION LAB.	COPJ-
CALIFORNIA. UNIV., LIVERMORE. RADIATION LAB.	COPO-
CALIFORNIA. UNIV., LIVERMORE. RADIATION LAB.	COPV-
CALIFORNIA. UNIV., LIVERMORE. RADIATION LAB.	COT-
CALIFORNIA. UNIV., LIVERMORE. RADIATION LAB.	COTA-
CALIFORNIA. UNIV., LIVERMORE. RADIATION LAB.	CV-(YR)-(MO)-
CALIFORNIA. UNIV., LIVERMORE. RADIATION LAB.	HCRL-
CALIFORNIA. UNIV., LIVERMORE. RADIATION LAB.	LER(NO.)
CALIFORNIA. UNIV., LIVERMORE. RADIATION LAB.	LIVL-
CALIFORNIA. UNIV., LIVERMORE. RADIATION LAB.	PNU-
CALIFORNIA. UNIV., LIVERMORE. RADIATION LAB.	UCRL-COC-
CALIFORNIA. UNIV., LIVERMORE. RADIATION LAB.(NOTEBOOK	UCRL-NB-

(SEE ALSO LATER NAMES UNDER CALIF. UNIV., LIVERMORE. LAWRENCE RADIATION LAB. AND LAWRENCE LIVERMORE LAB.

CALIFORNIA. UNIV., LIVERMORE. RADIATION LAB. ELECTRONIC ENGINEERING DEPT.	JAL-
CALIFORNIA. UNIV., LOS ANGELES	MPG-(YEAR)-
CALIFORNIA. UNIV., LOS ANGELES	MR-
CALIFORNIA. UNIV., LOS ANGELES	UC-
CALIFORNIA. UNIV., LOS ANGELES	UC-(YR.)-
CALIFORNIA. UNIV., LOS ANGELES	UCL-
CALIFORNIA. UNIV., LOS ANGELES	UCLA-
CALIFORNIA. UNIV., LOS ANGELES (USED BY MANY DEPTS.)	UCLA-(NUMBER)-P-(NO.)-
CALIFORNIA. UNIV., LOS ANGELES (WORKING PAPERS)	UCLA-PHONETICS-WP-
CALIFORNIA. UNIV., LOS ANGELES	UCLA-TR-
CALIFORNIA. UNIV., LOS ANGELES. BIOTECHNOLOGY LAB.	UCLA-(YEAR)-
CALIFORNIA. UNIV., LOS ANGELES. CENTER FOR THE HEALTH SCIENCES	AAP-PMH-(YEAR)-
CALIFORNIA. UNIV., LOS ANGELES. CENTER FOR THE HEALTH SCIENCES (PHYSIOLOGY OF CHIMPS IN ORBIT)	POCO-PMH-(YEAR)-
CALIFORNIA. UNIV., LOS ANGELES. DEPT. OF CHEMISTRY	UOC-
CALIFORNIA. UNIV., LOS ANGELES. DEPT. OF ENGINEERING	CLN-
CALIFORNIA. UNIV., LOS ANGELES. DEPT. OF ENGINEERING	CLN AECU-
CALIFORNIA. UNIV., LOS ANGELES. DEPT. OF ENGINEERING	UCLA-(YEAR)-
CALIFORNIA. UNIV., LOS ANGELES. DEPT. OF ENGINEERING	UCLA-CASE-
CALIFORNIA. UNIV., LOS ANGELES. DEPT. OF ENGINEERING	UCLA-DE-(YEAR)-
CALIFORNIA. UNIV., LOS ANGELES. DEPT. OF ENGINEERING	UCLA-NEL-
CALIFORNIA. UNIV., LOS ANGELES. DEPT. OF METEOROLOGY	UCLA-DM-
CALIFORNIA. UNIV., LOS ANGELES. DEPT. OF PHYSICS	TR-P-
CALIFORNIA. UNIV., LOS ANGELES. DEPT. OF PHYSICS	UCLA-(YEAR)/TEP/
CALIFORNIA. UNIV., LOS ANGELES. DEPT. OF PHYSICS	UCLA-MPG-
CALIFORNIA. UNIV., LOS ANGELES. DEPT. OF PHYSICS	UCLA-P-
CALIFORNIA. UNIV., LOS ANGELES. INST. OF GEOPHYSICS	OAHU-SR-
CALIFORNIA. UNIV., LOS ANGELES. INST. OF GEOPHYSICS	PUB-
CALIFORNIA. UNIV., LOS ANGELES. INST. OF GEOPHYSICS	UCLA-IG-
CALIFORNIA. UNIV., LOS ANGELES. LAB. OF NUCLEAR MEDICINE + RADIATION BIOLOGY	UCLA-(NUMBER)-
CALIFORNIA. UNIV., LOS ANGELES. MANAGEMENT SCIENCES RESEARCH PROJECT (DISCUSSION PAPER)	MSRP-DP-
CALIFORNIA. UNIV., LOS ANGELES. MANAGEMENT SCIENCES RESEARCH PROJECT	MSRP-RR-
CALIFORNIA. UNIV., LOS ANGELES. NUMERICAL ANALYSIS RESEARCH	NAR-
CALIFORNIA. UNIV., LOS ANGELES. SCHOOL OF ENGINEERING AND APPLIED SCIENCE	NCLA-ENG-
CALIFORNIA. UNIV., LOS ANGELES. SCHOOL OF ENGINEERING AND APPLIED SCIENCE	UCLA-ENG-
CALIFORNIA. UNIV., LOS ANGELES. WESTERN MANAGEMENT SCIENCE INST.	WMSI-
CALIFORNIA. UNIV., LOS ANGELES. WESTERN MANAGEMENT SCIENCE INST.	WMSI REPRINT-
CALIFORNIA. UNIV., LOS ANGELES. WESTERN MANAGEMENT SCIENCE INST.	WMSI WORKING PAPER-
CALIFORNIA. UNIV., MERCURY, NEV. LAWRENCE RADIATION LAB.	DN-
CALIFORNIA. UNIV., MERCURY, NEV. LAWRENCE RADIATION LAB.	UCRL-
CALIFORNIA. UNIV., OAKLAND. NAVAL BIOLOGICAL LAB.	UC-NBL-O-
CALIFORNIA. UNIV., RICHMOND. INST. OF ENG. RES.	CAL. U./PROJ. CIVIL-
CALIFORNIA. UNIV., RICHMOND. INST. OF ENG. RES.	UCAL-IER-
CALIFORNIA. UNIV., RICHMOND. INST. OF ENG. RES.	UCR-(NO.)-P-(NO.)-B-
CALIFORNIA. UNIV., RICHMOND. SEA WATER CONVERSION LAB	SWCL-
CALIFORNIA. UNIV., RIVERSIDE	CALT-
CALIFORNIA. UNIV., RIVERSIDE	UCR-
CALIFORNIA. UNIV., RIVERSIDE (USED BY MANY DEPTS.)	UCR-(NUMBER)-P-(NO.)-
CALIFORNIA. UNIV., RIVERSIDE	UCR-P(NO.)-

(CALIFORNIA. UNIV., SAN DIEGO. FOR CURRENT ENTRIES, SEE: CALIFORNIA. UNIV., LA JOLLA)

CALIFORNIA. UNIV., SAN DIEGO. DIV. OF WAR RESEARCH	DWR-
CALIFORNIA. UNIV., SAN DIEGO. DIV. OF WAR RESEARCH	UC DWR (LETTER)-
CALIFORNIA. UNIV., SAN DIEGO. DIV. OF WAR RESEARCH (INTERNAL RPTS.)	UC DWR I-
CALIFORNIA. UNIV., SAN FRANCISCO(USED BY MANY DEPTS.)	UCSF-(NUMBER)-P-(NO.)-
CALIFORNIA. UNIV., SAN FRANCISCO. SCHOOL OF MEDICINE	UCSF-
CALIFORNIA. UNIV., SANTA BARBARA(USED BY MANY DEPTS.)	UCSB-(NUMBER)-P-(NO.)-
CALIFORNIA. UNIV., SANTA BARBARA	UCSB-CRL-
CALIFORNIA. UNIV., SANTA BARBARA. DEPT. OF MECH. ENG.	AEN-
CALIFORNIA. UNIV., SANTA BARBARA. DEPT. OF MECH. ENG.	UCSB-ME-(YEAR)-
CALIFORNIA. UNIV., SANTA CRUZ	UCS-
CALIFORNIA ANALYSIS CENTER, INC., SANTA MONICA, CALIF	CACI-
CALIFORNIA INST. OF TECH., PASADENA	BIRC PUBLICATION NO.-
CALIFORNIA INST. OF TECH., PASADENA	CALIT-
CALIFORNIA INST. OF TECH., PASADENA	CALT-
CALIFORNIA INST. OF TECH., PASADENA	CALT-(NUMBER-NUMBER)
CALIFORNIA INST. OF TECH., PASADENA	CAL-TECH-

CALIFORNIA INST. OF TECH., PASADENA	CALTECH-
CALIFORNIA INST. OF TECH., PASADENA	CH-23-R-
CALIFORNIA INST. OF TECH., PASADENA	CIT-
CALIFORNIA INST. OF TECH., PASADENA	CIT-(LETTER(S))-
CALIFORNIA INST. OF TECH., PASADENA	CIT-E-
CALIFORNIA INST. OF TECH., PASADENA	CIT-IGC-
CALIFORNIA INST. OF TECH., PASADENA	CIT IPC-
CALIFORNIA INST. OF TECH., PASADENA	CIT-M-
CALIFORNIA INST. OF TECH., PASADENA (SENSOR AND SIMULATION NOTES)	CIT-S+SN-
CALIFORNIA INST. OF TECH., PASADENA	CIT-TR-
CALIFORNIA INST. OF TECH., PASADENA	CIT-XRR-
CALIFORNIA INST. OF TECH., PASADENA	E-
CALIFORNIA INST. OF TECH., PASADENA	E-(NUMBER).(NUMBER)
CALIFORNIA INST. OF TECH., PASADENA	ENGORD-
CALIFORNIA INST. OF TECH., PASADENA	IGC-
CALIFORNIA INST. OF TECH., PASADENA	IPC-
CALIFORNIA INST. OF TECH., PASADENA	KM-
CALIFORNIA INST. OF TECH., PASADENA (ORDNANCE RESEARCH AND DEVELOPMENT)	ORDCIT-
CALIFORNIA INST. OF TECH., PASADENA	P-
CALIFORNIA INST. OF TECH., PASADENA	PR-(NO.)-(NO.)
CALIFORNIA INST. OF TECH., PASADENA	PUB-
CALIFORNIA INST. OF TECH., PASADENA	SM-
CALIFORNIA INST. OF TECH., PASADENA	VER-
CALIFORNIA INST. OF TECH., PASADENA	VISTA-
CALIFORNIA INST. OF TECH., PASADENA. ANTENNA LAB.	CIT AL TR-
CALIFORNIA INST. OF TECH., PASADENA. CHEMICAL ENGINEERING-POLYMER LAB.	CHECIT-PL-(YEAR)-
CALIFORNIA INST. OF TECH., PASADENA. GATES AND CRELLIN LABS. OF CHEMISTRY (CONTRIBUTION)	CIT-GCLC CON-
CALIFORNIA INST. OF TECH., PASADENA. GATES AND CRELLIN LABS. OF CHEMISTRY (CONTRIBUTION)	GCLC CON-
CALIFORNIA INST. OF TECH., PASADENA. GRADUATE AERONAUTICAL LABS.	GALCIT-
CALIFORNIA INST. OF TECH., PASADENA. GRADUATE AERONAUTICAL LABS. (HYPERSONIC MEMO)	GALCIT-HYPER M-
CALIFORNIA INST. OF TECH., PASADENA. GRADUATE AERONAUTICAL LABS. (JET PROPULSION CENTER TECH. REPT.)	GALCIT JPC TR-
CALIFORNIA INST. OF TECH., PASADENA. GRADUATE AERONAUTICAL LABS. (SOLID MECHANICS)	GALCIT SM TR-
CALIF. INST. OF TECH., PASADENA. GUGGENHEIM AERO. LAB	CIT GAL-
CALIF. INST. OF TECH., PASADENA. GUGGENHEIM AERO. LAB	GAL-
(SEE ALSO LATER NAME: CALIFORNIA INST. OF TECH., PASADENA. GRADUATE AERONAUTICAL LABS.)	
CALIFORNIA INST. OF TECH., PASADENA. GUGGENHEIM JET PROPULSION CENTER	CIT GJPC TR-
CALIFORNIA INST. OF TECH., PASADENA. GUGGENHEIM JET PROPULSION CENTER	GJPC-
CALIF. INST. OF TECH., PASADENA. HYDRAULIC MACH. LAB.	HML-
CALIF. INST. OF TECH., PASADENA. HYDRAULIC MACH. LAB.	HML ND-
CALIFORNIA INST. OF TECH.,PASADENA. HYDRODYNAMICS LAB	CIT-HL-
CALIFORNIA INST. OF TECH.,PASADENA. HYDRODYNAMICS LAB	E-(NUMBER.NUMBER)
CALIFORNIA INST. OF TECH.,PASADENA. HYDRODYNAMICS LAB	N-
CALIFORNIA INST. OF TECH., PASADENA. JET PROPULSION LAB. (REF. 1)	(CONTRACT CODE NO.)-
CALIF. INST. OF TECH., PASADENA. JET PROPULSION LAB.	CAL-TECH/JPL-
CALIF. INST. OF TECH., PASADENA. JET PROPULSION LAB.	CIT-JPL-
CALIFORNIA INST. OF TECH., PASADENA. JET PROPULSION LAB.	CIT JPL ABSTRACTS-
CALIFORNIA INST. OF TECH., PASADENA. JET PROPULSION LAB. (MEMORANDA)	CIT JPL M(NO.)-(NO.)
CALIFORNIA INST. OF TECH., PASADENA. JET PROPULSION LAB. (PROGRESS REPORTS)	CIT JPL PR(NO.)-(NO.)
CALIFORNIA INST. OF TECH., PASADENA. JET PROPULSION LAB. (PUBLICATIONS)	CIT JPL PUB-
CALIFORNIA INST. OF TECH., PASADENA. JET PROPULSION LAB. (REPORTS)	CIT JPL R(NO.)-(NO.)
CALIFORNIA INST. OF TECH., PASADENA. JET PROPULSION LAB. (SECTION REPORTS)	CIT JPL SR(NO.)-
CALIF. INST. OF TECH., PASADENA. JET PROPULSION LAB.	CIT JPL SURVEY-
CALIF. INST. OF TECH., PASADENA. JET PROPULSION LAB.	EP-
CALIF. INST. OF TECH., PASADENA. JET PROPULSION LAB.	EPD-
CALIF. INST. OF TECH., PASADENA. JET PROPULSION LAB.	JPL-
CALIF. INST. OF TECH., PASADENA. JET PROPULSION LAB.	JPL-(PROJECT NO.)-
CALIF. INST. OF TECH., PASADENA. JET PROPULSION LAB.	JPL-(MONTH)(YEAR)-
CALIFORNIA INST. OF TECH., PASADENA. JET PROPULSION LAB. (ASTRONAUTICS INFORMATION)	JPLAI-
CALIFORNIA INST. OF TECH., PASADENA. JET PROPULSION LAB. (ASTRONAUTICS INFORMATION. ABSTRACTS)	JPLAI/ABSTR. VOL.-
CALIFORNIA INST. OF TECH., PASADENA. JET PROPULSION LAB. (ASTRONAUTICS INFORMATION. INDEX)	JPLAI/INDEX/
CALIFORNIA INST. OF TECH., PASADENA. JET PROPULSION LAB. (ASTRONAUTICS INFORMATION. LITERATURE SEARCH)	JPLAI/LS-
CALIF. INST. OF TECH., PASADENA. JET PROPULSION LAB.	JPLAI-TR-
CALIFORNIA INST. OF TECH., PASADENA. JET PROPULSION LAB. (BIBLIOGRAPHY)	JPL-BIB-39-
CALIF. INST. OF TECH., PASADENA. JET PROPULSION LAB. (BIBLIOGRAPHY)	JPL-BIBL-
CALIFORNIA INST. OF TECH., PASADENA. JET PROPULSION LAB. (COMBINED BIMONTHLY SUMMARY)	JPL-CBS-
CALIFORNIA INST. OF TECH., PASADENA. JET PROPULSION LAB. (EXTERNAL PUBLICATION)	JPL-EP-
CALIFORNIA INST. OF TECH., PASADENA. JET PROPULSION LAB. (GUIDED MISSILE SUMMARY)	JPL-GMS-
CALIF. INST. OF TECH., PASADENA. JET PROPULSION LAB.	JPL/JPS-37-
CALIF. INST. OF TECH., PASADENA. JET PROPULSION LAB.	JPL-MEMO-
CALIFORNIA INST. OF TECH., PASADENA. JET PROPULSION LAB. (NUCLEAR PROPULSION SUMMARY)	JPL-NPS-
CALIFORNIA INST. OF TECH., PASADENA. JET PROPULSION LAB. (PUBLICATION)	JPL-P-
CALIFORNIA INST. OF TECH., PASADENA. JET PROPULSION LAB. (PROGRESS REPORT)	JPL-PR-
CALIFORNIA INST. OF TECH., PASADENA. JET PROPULSION LAB. (PROGRESS REPORT)	JPL-PR-(PROJECT NO.)-
CALIF. INST. OF TECH., PASADENA. JET PROPULSION LAB.	JPL PROPOSAL (NO.)-(NO.)
CALIFORNIA INST. OF TECH., PASADENA. JET PROPULSION LAB. (PUBLICATION)	JPL-PUB-
CALIFORNIA INST. OF TECH., PASADENA. JET PROPULSION LAB. (QUARTERLY SUMMARY REPORT)	JPL-QSR-38-
CALIFORNIA INST. OF TECH., PASADENA. JET PROPULSION LAB. (RESEARCH SUMMARY)	JPL-RES. SUMM.-
CALIFORNIA INST. OF TECH., PASADENA. JET PROPULSION LAB. (RESEARCH SUMMARY)	JPL-RS-(PROJECT NO.)-
CALIFORNIA INST. OF TECH., PASADENA. JET PROPULSION LAB. (MATERIAL SPECIFICATION)	JPL-SPEC-(LTR)(NO.)(NO.)
CALIFORNIA INST. OF TECH., PASADENA. JET PROPULSION LAB. (SPACE PROGRAMS SUMMARY)	JPL-SPS-(PROJECT NO.)-
CALIFORNIA INST. OF TECH., PASADENA. JET PROPULSION LAB. (SPACE RESEARCH SUMMARY)	JPL-SRS-
CALIF. INST. OF TECH., PASADENA. JET PROPULSION LAB.	JPL-STUDY MEMO-
CALIF. INST. OF TECH., PASADENA. JET PROPULSION LAB. (SUMMARY)	JPL-SUM-
CALIFORNIA INST. OF TECH., PASADENA. JET PROPULSION LAB. (SUPERSONIC WIND TUNNEL)	JPL-SWT-
CALIF. INST. OF TECH., PASADENA. JET PROPULSION LAB.	JPL-TECH. RELEASE-(PROJ)-
CALIFORNIA INST. OF TECH., PASADENA. JET PROPULSION LAB. (TECHNICAL MEMO)	JPL-TM-(PROJ. NO.)-
CALIFORNIA INST. OF TECH., PASADENA. JET PROPULSION LAB.	JPL-TR-(NUMBER)-
CALIFORNIA INST. OF TECH., PASADENA. JET PROPULSION LAB. (TECHNICAL REPORT)	JPL-TR-(PROJECT NO.)-
CALIF. OF TECH., PASADENA. JET PROPULSION LAB.	LA-
CALIF. INST. OF TECH., PASADENA. JET PROPULSION LAB.	MEMO-NO.-(NO.)-
CALIF. INST. OF TECH., PASADENA. JET PROPULSION LAB.	R/AD-(NUMBER-NUMBER)
CALIF. INST. OF TECH., PASADENA. JET PROPULSION LAB. (SURVEYER PROJECT STATUS REPORT)	SPSR-
CALIF. INST. OF TECH., PASADENA. JET PROPULSION LAB.	SWT-
CALIF. INST. OF TECH., PASADENA. JET PROPULSION LAB. (VOYAGER PROJECT PROGRESS REPORT)	VPPR-
CALIFORNIA INST. OF TECH., PASADENA. KECK LAB. OF ENGINEERING MATERIALS (POLYMER SCIENCE REPORT)	MATSCIT PS-
CALIFORNIA INST. OF TECH., PASADENA. KECK LAB. OF ENGINEERING MATERIALS (POLYMER SCIENCE REPORT)	MATSCIT-PS-(YEAR)-
CALIFORNIA INST. OF TECH., PASADENA. KECK LAB. OF ENVIRONMENTAL HEALTH ENGINEERING	EHE AT CIT-
CALIFORNIA INST. OF TECH., PASADENA. KECK LAB. OF HYDRAULICS AND WATER RESOURCES	KH-R-

CALIFORNIA INST. OF TECH., PASADENA. NAVY LIAISON OFFICE FOR GUIDED MISSILES	NLOGM-
CALIFORNIA INST. OF TECH., PASADENA. OWENS VALLEY RADIO OBSERVATORY	OBSERVATION-(YEAR)-
CALIFORNIA INST. OF TECH., PASADENA. PLASMA LAB.	CIT PL TR-
CALIFORNIA INST. OF TECH., PASADENA. PROJECT CAMEL	K-1.0 THRU K-3.--
CALIFORNIA INST. OF TECH., PASADENA. QUANTUM ELECTRONICS LAB.	CIT QEL TR-
CALIFORNIA INST. OF TECH., PASADENA. QUANTUM ELECTRONICS SOLID STATE	CIT QESS TR-
CALIFORNIA INST. OF TECH., PASADENA. SOLID STATE ELECTRONICS LAB.	CIT SSEL TR-
CALIFORNIA INST. OF TECH., PASADENA. SYNCHROTRON LAB.	CTSL-
CALIFORNIA INST. OF TECH., PASADENA. SYNCHROTRON LAB.	CTSL-IR-
CALIFORNIA INST. OF TECH., PASADENA. W.K.KELLOGG LAB. OF RADIATION	CIT-KRL-TR-(YR/MO)-
CALIFORNIA RESEARCH AND DEVELOPMENT CO., BERKELEY	CRD-T(NUMBER)-
CALIFORNIA RESEARCH AND DEVELOPMENT CO., BERKELEY	LWS-
CALIFORNIA RESEARCH AND DEVELOPMENT CO., LIVERMORE	CRD-
CALIFORNIA RESEARCH AND DEVELOPMENT CO., LIVERMORE	CRD-A(NUMBER)-
CALIFORNIA RESEARCH AND DEVELOPMENT CO., LIVERMORE	CRD-R-
CALIFORNIA RESEARCH AND DEVELOPMENT CO., LIVERMORE	CRD-T(NUMBER)-
CALIFORNIA RESEARCH AND DEVELOPMENT CO., LIVERMORE	D-F-
CALIFORNIA RESEARCH AND DEVELOPMENT CO., LIVERMORE	D-T-
CALIFORNIA RESEARCH AND DEVELOPMENT CO., LIVERMORE	LWS-
CALIFORNIA RESEARCH AND DEVELOPMENT CO., LIVERMORE	P-(LETTER)-
CALIFORNIA RESEARCH AND DEVELOPMENT CO., LIVERMORE	P-F-
CALIFORNIA RESEARCH AND DEV. CO. LIVERMORE RES. LAB.	LRL-
CALIFORNIA RESEARCH AND DEV. CO. LIVERMORE RES. LAB.	LRL LWS-
CALIFORNIA RESEARCH AND DEV. CO. LIVERMORE RES. LAB.	LRL MTA-
CALIFORNIA RESEARCH AND DEV. CO. LIVERMORE RES. LAB.	MTA-
CALIFORNIA RESEARCH CORP. (ALL LOCATIONS)	CR-
CALIFORNIA RESEARCH CORP., RICHMOND	AC-
CALIFORNIA RESEARCH CORP., RICHMOND	CRC-
CALIFORNIA RESEARCH CORP., RICHMOND	CRC-AEC-
CALIFORNIA RESEARCH CORP., RICHMOND	CRC-NEPA-
CALIFORNIA RESEARCH CORP. SAN PABLO LAB.	SPL-(YEAR)-
CALLERY CHEMICAL CO., PENNA.	CALCH-
CALLERY CHEMICAL CO., PENNA.	CCC-
CALLERY CHEMICAL CO., PENNA.	CCC-(CONTRACT)
CALLERY CHEMICAL CO., PENNA.	CCC-(CONTRACT)-(LETTERS)-
CALLERY CHEMICAL CO., PENNA. (ANNUAL SUMMARY RPT.)	CCC-ASR-(DATES)
CALLERY CHEMICAL CO., PENNA. (QUARTERLY REPORT)	CCC-QR-(NUMBER)(YEAR)
CALLERY CHEMICAL CO., PENNA.	D-
CALLERY CHEMICAL CO., PENNA.	D143-(NUMBER)-
CALLERY CHEMICAL CO., PENNA.	SP-
CALOROBIC MATERIALS, INC., PEARL RIVER, N.Y.	TR-
CAMBRIDGE ACOUSTICAL ASSOCIATES, INC., MASS.	C-
CAMBRIDGE ACOUSTICAL ASSOCIATES, INC., MASS.	CAA-U-
CAMBRIDGE ACOUSTICAL ASSOCIATES, INC., MASS.	U-
CAMBRIDGE CORP.	M1-
CAMBRIDGE CORP.	PM-
CAMBRIDGE CORP., BOULDER, COLO.	CAM-
CAMBRIDGE CORP., BOULDER, COLO.	CAMCO-
CAMBRIDGE CORP., BOULDER, COLO.	CAMCO-TM-
CAMBRIDGE CORP., SOMERVILLE, MASS.	CAM-
CAMBRIDGE CORP., SOMERVILLE, MASS.	CAM-(LTRS)-
CAMBRIDGE CORP., SOMERVILLE, MASS.	CAMCO-
CAMBRIDGE CORP., SOMERVILLE, MASS. (ENG. MEMO.)	CAM EM-
CAMBRIDGE ELECTRON ACCELERATOR, MASS.	CEA-(NO.)(M-H)
CAMBRIDGE ELECTRON ACCELERATOR, MASS.	CEAL-
CAMBRIDGE ELECTRON ACCELERATOR, MASS.	CEAL-TM-
CAMBRIDGE ELECTRON ACCELERATOR, MASS. (MASS. INST. OF TECH. + HARVARD UNIV.)	CEA-(M+H)-
CAMBRIDGE LANGUAGE RESEARCH UNIT, ENGLAND	ISU-
CAMBRIDGE UNIV., ENGLAND	ARC-
CAMBRIDGE UNIV., ENGLAND	ARC-T+M-
CAMBRIDGE UNIV., ENGLAND	CAM-
CAMBRIDGE UNIV., ENGLAND	NRCE/R-
CAMBRIDGE UNIV., ENGLAND. CAVENDISH LAB.	CRL-
CAMBRIDGE UNIV., ENGLAND. CAVENDISH LAB.	HEP-(YEAR)-
CAMBRIDGE UNIV., ENGLAND. DEPT. OF APPLIED MATH. AND PHYSICS	CAM-DAMTP-(YEAR)/(NO.)
CAMBRIDGE UNIV., ENGLAND. DEPT. OF APPLIED MATH. AND PHYSICS	DAMTP-(YEAR)/
CAMBRIDGE UNIV., ENGLAND. DEPT. OF ENGINEERING	CUED/A-
CAMBRIDGE UNIV., ENGLAND. DEPT. OF ENGINEERING	CUED/A-TURBO/TR-
CAMBRIDGE UNIV., ENGLAND. DEPT. OF ENGINEERING	CUED/C-
CAMBRIDGE UNIV., ENGLAND. DEPT. OF ENGINEERING	CUED/C-MAT/TR-
CAMBRIDGE UNIV., ENGLAND. DEPT. OF ENGINEERING	CUED/C-STRUCT/TR-
CAMP COLES SIGNAL LAB., BELMAR, N.J.	CCSL-
CAMP COLES SIGNAL LAB., BELMAR, N.J. (ENG. MEMO.)	CCSL EM-
CAMP COLES SIGNAL LAB., BELMAR, N.J. (ENG. REPT.)	CCSL ER-
CAMP COLES SIGNAL LAB., BELMAR, N.J. (MEMO. FILE)	CCSL MF PROJ-
CAMP COLES SIGNAL LAB., BELMAR, N.J. (TECH. MEMO.)	CCSL TM-
CAMP COLES SIGNAL LAB., BELMAR, N.J. (TEST REPT.)	CCSL TR-
CAMP COLES SIGNAL LABORATORY. APPLIED COMMUNICATIONS BRANCH, BELMAR, N.J. (ENGINEERING MEMORANDUM)	ACB EM-
CAMP COLES SIGNAL LABORATORY. APPLIED COMMUNICATIONS BRANCH, BELMAR, N.J. (ENGINEERING MEMORANDUM)	CCSL ACB EM-
CAMP EVANS SIGNAL LAB., BELMAR, N.J.	CESL-
CAMP EVANS SIGNAL LAB., BELMAR, N.J.	CESL (NO.)(LETTER)/

CAMP EVANS SIGNAL LAB., BELMAR, N.J. (SPECIFICATION)	CESL-SPEC-
CAMP EVANS SIGNAL LAB., BELMAR, N.J.	CESL TL-UK-
CAMP EVANS SIGNAL LAB., BELMAR, N.J. (TECH. MEMO.)	CESL TM-
CAMP EVANS SIGNAL LAB., BELMAR, N.J. (TECH. REPT.)	CESL TR-
CAMP EVANS SIGNAL LAB., BELMAR, N.J.	TL-UK-
CAMP EVANS SIGNAL LABORATORY. DEVELOPMENT PLANNING BRANCH, BELMAR, N.J. (ENGINEERING MEMO)	CESL DPB EM-
CAMP EVANS SIGNAL LABORATORY. DEVELOPMENT PLANNING BRANCH, BELMAR, N.J. (ENGINEERING MEMO)	DPB EM-
CAMP EVANS SIGNAL LABORATORY. RADIO PROPAGATION BRANCH, BELMAR, N.J.	CESL RPSR-
CAMP EVANS SIGNAL LABORATORY. RADIO PROPAGATION BRANCH, BELMAR, N.J.	RPSR-
CAMP EVANS SIGNAL LAB. RADIO PROPAGATION SECTION	RPSR-
CANADA. (FOR EXPLANATION OF THE REMAINDER OF THE CODE, SEE ENTRY UNDER LETTERS FOLLOWING CAN-)	CAN-(LETTERS)-
CANADA. AEROSPACE ENGINEERING TEST ESTAB., UPLAND, ONT.	AETE-(YEAR)/
CANADA. AIR DEFENCE COMMAND. OPERATIONAL RESEARCH BR.	ADC/ORB-
CANADA. AIR DEFENCE COMMAND. OPERATIONAL RESEARCH BR.	ORB-
CANADA. ANTI-SUBMARINE WARFARE OPERATIONAL RES. TEAM	ASW/ORT-
CANADA. ARMAMENT RESEARCH AND DEVELOPMENT ESTABLISHMENT, VALCARTIER, QUE. (REF. 1)	(NUMBERS...)
CANADA. ARMAMENT RES. + DEV. EST., VALCARTIER, QUE.	CARDE-(NO.)/(YR)-
CANADA. ARMAMENT RES. + DEV. EST., VALCARTIER, QUE.	CARDE-REPRINT-
CANADA. ARMAMENT RES. + DEV. EST., VALCARTIER, QUE.	CARDE-TM-
CANADA. ARMAMENT RES. + DEV. EST., VALCARTIER, QUE.	CARDE-TM-(NUMBER)/(YEAR)-
CANADA. ARMAMENT RES. + DEV. EST., VALCARTIER, QUE.	CARDE-TN-
CANADA. ARMAMENT RES. + DEV. EST., VALCARTIER, QUE.	CARDE-T.N.-
CANADA. ARMAMENT RES. + DEV. EST., VALCARTIER, QUE.	CARDE-TR-
CANADA. ARMY, OTTAWA (TRAINING MANUAL)	CATM-
CANADA. ARMY EQUIPMENT ENGINEERING ESTABLISHMENT, OTTAWA	AEEE-
CANADA. ARMY OPERATIONAL RESEARCH ESTABLISHMENT	CAORE-
CANADA. ARMY OPERATIONAL RESEARCH ESTABLISHMENT	CAORG-
CANADA. ARMY OPERATIONAL RESEARCH ESTABLISHMENT	CAORT-
CANADA. COMMUNICATIONS RESEARCH CENTRE, OTTAWA	CRC-
CANADA. COMMUNICATIONS RESEARCH CENTRE, OTTAWA	CRC-TN-
CANADA. DEFENCE CHEM. BIOL. + RADIATION LABS., OTTAWA	DCBRE-
CANADA. DEFENCE CHEM. BIOL. + RADIATION LABS., OTTAWA	DCBRL-
CANADA. DEFENCE CHEM. BIOL. + RADIATION LABS., OTTAWA	DCBRL-TN-
CANADA. DEFENCE RESEARCH ANALYSIS ESTAB., OTTAWA	DRAE-
CANADA. DEFENCE RESEARCH BOARD, OTTAWA	CAN-DRB-
CANADA. DEFENCE RESEARCH BOARD, OTTAWA	CD-
CANADA. DEFENCE RESEARCH BOARD, OTTAWA	CDA-
CANADA. DEFENCE RESEARCH BOARD, OTTAWA	DR-
CANADA. DEFENCE RESEARCH BOARD, OTTAWA	DRB-(PROJECT NUMBER)-
CANADA. DEFENCE RESEARCH BOARD, OTTAWA	DRB DR-
CANADA. DEFENCE RESEARCH BOARD, OTTAWA	DRBS-
CANADA. DEFENCE RESEARCH BOARD, OTTAWA	DRB-T-(NUMBER)-R
CANADA. DEFENCE RESEARCH BOARD, OTTAWA	DRB-TELS-
CANADA. DEFENCE RESEARCH BOARD, OTTAWA	DRB-TELS-TN-
CANADA. DEFENCE RESEARCH BOARD, OTTAWA	DRET-
CANADA. DEFENSE RESEARCH BOARD, OTTAWA	DRML-TM-
CANADA. DEFENCE RESEARCH BOARD, OTTAWA	G-
CANADA. DEFENCE RESEARCH BOARD, OTTAWA	HR-
CANADA. DEFENCE RESEARCH BOARD, OTTAWA	R/AE-
CANADA. DEFENCE RESEARCH BOARD, OTTAWA	SAG-
CANADA. DEFENCE RESEARCH BOARD, OTTAWA	SAG-MEMO-
CANADA. DEFENCE RESEARCH BOARD, OTTAWA	T-(NUMBER)-R
CANADA. DEFENCE RESEARCH BOARD, OTTAWA	T-
CANADA. DEFENCE RESEARCH BOARD, CHIEF OF PERSONNEL BRANCH, OTTAWA	CP-
CANADA. DEFENCE RESEARCH BOARD. CHIEF OF PLANS, OTTAWA	PLANS-(YEAR)-
CANADA. DEFENCE RESEARCH BOARD. PROJECT COORDINATION CENTRE, OTTAWA	PCC-
CANADA. DEFENCE RESEARCH BOARD. SCIENTIFIC INFORMATION SERVICE (TRANSLATION)	DRB-T-
CANADA. DEFENCE RESEARCH BOARD. SCIENTIFIC INFORMATION SERVICE (TRANSLATION)	DRBT (NUMBER)(LETTER)
CANADA. DEFENCE RESEARCH BOARD. SCI. INFO. SERVICE	SUM-
CANADA. DEFENCE RESEARCH BOARD. SCIENTIFIC INFORMATION SERVICE (TRANSLATION FROM GERMAN)	T.(NUMBER).G
CANADA. DEFENCE RESEARCH BOARD. SCIENTIFIC INFORMATION SERVICE (TRANSLATION FROM ITALIAN)	T.(NUMBER).I
CANADA. DEFENCE RESEARCH BOARD. SCIENTIFIC INFORMATION SERVICE (TRANSLATION FROM JAPANESE)	T.(NUMBER).J
CANADA. DEFENCE RESEARCH BOARD. SCIENTIFIC INFORMATION SERVICE (TRANSLATION FROM RUSSIAN)	T.(NUMBER).R
CANADA. DEFENCE RESEARCH CHEMICAL LABS., OTTAWA	DRCL-
CANADA. DEFENCE RESEARCH ESTABLISHMENT, OTTAWA	DREO-
CANADA. DEFENCE RESEARCH ESTABLISHMENT, OTTAWA	DREO-TN-(YEAR)-
CANADA. DEFENCE RES. ESTAB. ATLANTIC, DARTMOUTH, NS	DREA-
CANADA. DEFENCE RES. ESTAB. ATLANTIC, DARTMOUTH, NS	DREA-(YEAR)-
CANADA. DEFENCE RES. ESTAB. ATLANTIC, DARTMOUTH, NS	DREA-TN-MATH/(YR)/
CANADA. DEFENCE RES. ESTAB. PACIFIC, VICTORIA, B.C.	DREP-
CANADA. DEFENCE RES. ESTAB. PACIFIC, VICTORIA, B.C.	DREP-TM-
CANADA. DEFENCE RES. ESTAB. SUFFIELD, RALSTON, ALTA.	DRES-
CANADA. DEFENCE RES. ESTAB. SUFFIELD, RALSTON, ALTA.	DRES-MEMO-
CANADA. DEFENCE RES. ESTAB. SUFFIELD, RALSTON, ALTA.	DRES-TN-
CANADA. DEFENCE RES. ESTAB. SUFFIELD, RALSTON, ALTA.	DRES-TP-
CANADA. DEFENCE RES. ESTAB. TORONTO, DOWNSVIEW, ONT.	DRET-
CANADA. DEFENCE RES. ESTAB. TORONTO, DOWNSVIEW, ONT.	DRET-RP-
CANADA. DEFENCE RES. ESTAB. VALCARTIER, QUEBEC	DREV-
CANADA. DEFENCE RES. ESTAB. VALCARTIER, QUEBEC	DREV-R-
CANADA. DEFENCE RES. ESTAB. VALCARTIER, QUEBEC	DREV-REPRINT-
CANADA. DEFENCE RES. ESTAB. VALCARTIER, QUEBEC	DREV-TL-
CANADA. DEFENCE RES. ESTAB. VALCARTIER, QUEBEC	DREV-TN-
CANADA. DEFENCE RESEARCH KINGSTON LAB.	DRKL-

CANADA. DEFENCE RESEARCH MEDICAL LABS.   414

CANADA. DEFENCE RESEARCH MEDICAL LABS., TORONTO	DRET-TN-
CANADA. DEFENCE RESEARCH MEDICAL LABS., TORONTO	DRML-
CANADA. DEFENCE RESEARCH MEDICAL LABS., TORONTO	DRMLRP-
CANADA. DEFENCE RESEARCH MEDICAL LABS., TORONTO	DRML-TN-
CANADA. DEFENCE RESEARCH MEDICAL LABS., TORONTO	EP-
CANADA. DEFENCE RESEARCH MEDICAL LABS., TORONTO	RP-
CANADA. DEFENCE RES. NORTHERN LAB.,FT. CHURCHILL,MAN.	DRNL-
CANADA. DEFENCE RES. NORTHERN LAB.,FT. CHURCHILL,MAN.	DRNL-TP-
CANADA. DEFENCE RES. TELECOMMUNICATIONS ESTAB.,OTTAWA	DRTE-
CANADA. DEFENCE RES. TELECOMMUNICATIONS ESTAB.,OTTAWA	DRTE-(LETTER)-
CANADA. DEFENCE RES. TELECOMMUNICATIONS ESTAB.,OTTAWA	DRTE/EL-
CANADA. DEFENCE RES. TELECOMMUNICATIONS ESTAB.,OTTAWA	DRTE-GEOPHYSICS-
CANADA. DEFENCE RES. TELECOMMUNICATIONS ESTAB.,OTTAWA	DRTE-HAZEN-
CANADA. DEFENCE RES. TELECOMMUNICATIONS ESTAB.,OTTAWA	DRTE-PUBL.-
CANADA. DEFENCE RES. TELECOMMUNICATIONS ESTAB.,OTTAWA	ECRDC PROJECT NO.-
CANADA. DEFENCE RES. TELECOMMUNICATIONS ESTAB.,OTTAWA	EL-
CANADA. DEFENCE RES. TELECOMMUNICATIONS ESTAB.,OTTAWA	RPL-
CANADA. DEFENCE RES. TELECOMMUNICATIONS ESTAB.,OTTAWA	TELS-
CANADA. DEFENCE RES. TELECOMMUNICATIONS ESTABLISHMENT. RADIO PHYSICS LAB., OTTAWA	DRB-REPRINT-
CANADA. DEFENCE SCIENTIFIC INFORMATION SERVICE,OTTAWA	DSIS-B-
CANADA. DEPT. OF FORESTRY, OTTAWA	ODC-
CANADA. DEPT. OF FORESTRY, VANCOUVER, B.C.	VP-X-
CANADA. DEPT. OF HIGHWAYS, DOWNSVIEW	DHO-RRL(YEAR)
CANADA. DEPT. OF MINES AND RESOURCES. BUREAU OF MINES	SR-/(YR)
CANADA. DEPT. OF MINES AND TECHNICAL SURVEYS	ES-
CANADA. DEPT. OF MINES AND TECHNICAL SURVEYS	GRP-(NUMBER)/(YEAR)
CANADA. DEPT. OF MINES AND TECHNICAL SURVEYS	MEMO-
CANADA. DEPT. OF MINES AND TECHNICAL SURVEYS, OTTAWA	REPT-(YEAR)-
CANADA. DEPT. OF MINES AND TECHNICAL SURVEYS. FUELS + MINING PRACTICE DIV., OTTAWA	FMP-(YEAR)/(NUMBER)-MRL
CANADA. DEPT. OF MINES AND TECH. SURVEYS. MINES BR.	CURF-R-
CANADA. DEPT. OF MINES AND TECH. SURVEYS. MINES BR.	GPR-
CANADA. DEPT. OF MINES AND TECH. SURVEYS. MINES BR.	MB-R-
CANADA. DEPT. OF MINES AND TECH. SURVEYS. MINES BR.	MD-
CANADA. DEPT. OF MINES AND TECH. SURVEYS. MINES BR.	MD-TR-(NO.)-CC
CANADA. DEPT. OF MINES AND TECH. SURVEYS. MINES BR.	PM-
CANADA. DEPT. OF MINES AND TECH. SURVEYS. MINES BR.	PM-I-
CANADA. DEPT. OF MINES AND TECH. SURVEYS. MINES BR.	PM-M-
CANADA. DEPT. OF MINES AND TECH. SURVEYS. MINES BR.	PM-R-
CANADA. DEPT. OF MINES AND TECH. SURVEYS. MINES BR.	PM-T-
CANADA. DEPT. OF MINES AND TECHNICAL SURVEYS. MINES BRANCH, OTTAWA	TB-
CANADA. DEPT. OF NATIONAL DEFENCE, OTTAWA	DR-
CANADA. DEPT. OF NATIONAL DEFENCE, OTTAWA	OR-
CANADA. DEPT. OF NATIONAL DEFENCE. OPERATIONAL RESEARCH DIV., OTTAWA	ORD-
CANADA. DEPT. OF NATIONAL DEFENCE. OPERATIONAL RESEARCH DIV., OTTAWA	ORD INFORMAL PAPER-
CANADA. DEPT. OF NATIONAL HEALTH AND WELFARE. RADIATION PROTECTION DIV., OTTAWA	CNHW-
CANADA. DEPT. OF NATIONAL HEALTH AND WELFARE. RADIATION PROTECTION DIV., OTTAWA	CNHW(RP-(NUMBER))
CANADA. DEPT. OF NATIONAL HEALTH AND WELFARE. RADIATION PROTECTION DIV., OTTAWA	RPD-
CANADA. DEPT. OF NATIONAL HEALTH AND WELFARE. RADIATION PROTECTION DIV., OTTAWA	TEC-
CANADA. DEPT. OF TRANSPORT	CIR-
CANADA. DEPT. OF TRANSPORT.METEOROLOGICAL BR.,TORONTO	CIR-
CANADA. DEPT. OF TRANSPORT.METEOROLOGICAL BR.,TORONTO	CMM-
CANADA. DEPT. OF TRANSPORT.METEOROLOGICAL BR.,TORONTO	CMRR-
CANADA. DEPT. OF TRANSPORT.METEOROLOGICAL BR.,TORONTO	ICE-
CANADA. DEPT. OF TRANSPORT.METEOROLOGICAL BR.,TORONTO	TEC-
CANADA. DIRECTORATE OF BIOSCIENCES RESEARCH. ENVIRONMENTAL PROTECTION SECTION, OTTAWA	EP-TR-
CANADA. DIRECTORATE OF ENGINEER DEVELOPMENT	DED-
CANADA. DIRECTORATE OF OPERATIONAL RESEARCH	DOR(N)-
CANADA. DIRECTORATE OF PHYSICAL RESEARCH, OTTAWA	D/PHYS/R-
CANADA. DIRECTORATE OF PHYSICAL RESEARCH, OTTAWA (REPORTS FROM HAZEN CAMP, ELLESMERE ISLAND, N.W.T.)	D-PHYS-R(G)-HAZEN-
CANADA. DIRECTORATE OF PHYSICAL RESEARCH, OTTAWA	D-PHYS-R(G)-MISC-
CANADA. DIRECTORATE OF SCI. INFO. SERVICES, OTTAWA	DSIS-
CANADA. DIRECTORATE OF SCI. INFO. SERVICES, OTTAWA	DSIS-T-(NUMBER)-R
CANADA. DIRECTORATE OF SCI. INFO. SERVICES, OTTAWA	T-(NUMBER)-UKR
CANADA. DIRECTORATE OF SCI. INFO. SERVICES, OTTAWA	T-(NUMBER)R
CANADA. DIRECTORATE OF SCI. INFO. SERVICES, OTTAWA	T-(NUMBER)-C
CANADA. DIRECTORATE OF SCI. INFO. SERVICES, OTTAWA	T-(NUMBER)J
CANADA. DIRECTORATE OF SCI. INFO. SERVICES, OTTAWA	TT-
CANADA. DIRECTORATE OF SCIENTIFIC INTELLIGENCE	DSI-
CANADA. DIRECTORATE OF SCIENTIFIC INTELLIGENCE	DSI-TRANS-
CANADA. FOREST PRODUCTS RES. LAB., VANCOUVER, B.C.	VP-X-
CANADA. GEOLOGICAL SURVEY, OTTAWA	SM-(NUMBER)/
CANADA. HYDRO-ELECTRIC POWER COMMISSION OF ONTARIO	HEPC-
CANADA. HYDRO-ELECTRIC POWER COMMISSION OF ONTARIO, ROLPHTON (DOUGLAS POINT GENERATING STA.MONTHLY RPT)	DPT-(YEAR)-(MONTH)
CANADA. HYDRO-ELECTRIC POWER COMMISSION OF ONTARIO, TORONTO	NPDT-(YEAR)-
CANADA. JOINT SERVICES OPERATIONAL RESEARCH TEAM	JSORT-
CANADA. MARITIME COMMAND. OPERATIONAL RESEARCH DIV., HALIFAX	MC/ORD-(NUMBER-YEAR)
CANADA. NATIONAL AERONAUTICAL ESTABLISHMENT, OTTAWA	LR-
CANADA. NATIONAL AERONAUTICAL ESTABLISHMENT, OTTAWA	LTR-HA-

CANADA. NATIONAL AERONAUTICAL ESTABLISHMENT, OTTAWA	LTR-S1-
CANADA. NATIONAL AERONAUTICAL ESTABLISHMENT, OTTAWA	MS-
CANADA. NATIONAL AERONAUTICAL ESTABLISHMENT, OTTAWA	NAE-(LETTERS)-
CANADA. NATIONAL AERONAUTICAL ESTABLISHMENT, OTTAWA	NAEC-
CANADA. NATIONAL AERONAUTICAL ESTAB., OTTAWA (LAB.RPT)	NAEC LR-
CANADA. NATIONAL AERONAUTICAL ESTAB., OTTAWA (NOTE)	NAEC N-
CANADA. NATIONAL AERONAUTICAL ESTABLISHMENT, OTTAWA	NAEC R-
CANADA. NATIONAL AERONAUTICAL ESTAB., OTTAWA (LAB.RPT)	NAE-LR-
CANADA. NATIONAL AERONAUTICAL ESTABLISHMENT, OTTAWA (UNSTEADY AERODYNAMICS LAB. TECH. REPORT)	NAE-LTR-UA-
CANADA. NATIONAL AERONAUTICAL ESTABLISHMENT, OTTAWA	NAE-MISC-
CANADA. NATIONAL AERONAUTICAL ESTABLISHMENT, OTTAWA	NAE-MS-
CANADA. NATIONAL ENERGY BOARD, OTTAWA	NEB-SS-
CANADA. NAVAL RESEARCH ESTABLISHMENT	PHX-
CANADA. NAVAL RES. ESTABLISHMENT, DARTMOUTH, N.S.	NRE-
CANADA. NAVAL RES. ESTABLISHMENT, DARTMOUTH, N.S.	NRE-(YEAR)/
CANADA. NAVAL RES. ESTABLISHMENT, DARTMOUTH, N.S.	NRE-FR/(YEAR)/
CANADA. OPERATIONAL RESEARCH GROUP	ORG-
CANADA. PACIFIC NAVAL LAB., ESQUIMALT, B.C.	PCC D(NUMBER)-
CANADA. PACIFIC NAVAL LAB., ESQUIMALT, B.C. (PACIFIC INTERIM REPORT)	PIR-
CANADA. PACIFIC NAVAL LAB., ESQUIMALT, B.C.	PNL-
CANADA. PACIFIC NAVAL LAB., ESQUIMALT, B.C.	PNL-TM-
CANADA. SUFFIELD EXPERIMENTAL STATION,RALSTON,ALBERTA	AF-
CANADA. SUFFIELD EXPERIMENTAL STATION,RALSTON,ALBERTA	DRB-S-TP-
CANADA. SUFFIELD EXPERIMENTAL STATION,RALSTON,ALBERTA	F-(YEAR)-9-
CANADA. SUFFIELD EXPERIMENTAL STATION,RALSTON,ALBERTA	PD-(4 DIGITS)
CANADA. SUFFIELD EXPERIMENTAL STATION,RALSTON,ALBERTA	SES-
CANADA. SUFFIELD EXPERIMENTAL STATION,RALSTON,ALBERTA	SES-(YEAR)-
CANADA. SUFFIELD EXPERIMENTAL STATION,RALSTON,ALBERTA	SES-FIELD EXPERIMENT-
CANADA. SUFFIELD EXPERIMENTAL STATION,RALSTON,ALBERTA	SES-MEMO-(NUMBER)/(YEAR)
CANADA. SUFFIELD EXPERIMENTAL STATION,RALSTON,ALBERTA	SES-SPECIAL PUB-
CANADA. SUFFIELD EXPERIMENTAL STATION,RALSTON,ALBERTA	SES-TN-
CANADA. SUFFIELD EXPERIMENTAL STATION, RALSTON, ALBERTA. (TECHNICAL PAPER)	SES TP-
CANADA. SUFFIELD EXPERIMENTAL STATION,RALSTON,ALBERTA	SUFFIELD-
CANADA. SUFFIELD EXPERIMENTAL STATION,RALSTON,ALBERTA	SUFFIELD FIELD EXPERIMENT
CANADA. SUFFIELD EXPERIMENTAL STATION,RALSTON,ALBERTA	SUFFIELD-MEMO-
CANADA. SUFFIELD EXPERIMENTAL STATION,RALSTON,ALBERTA	SUFFIELD REPT. NO.-
CANADA. SUFFIELD EXPERIMENTAL STATION,RALSTON,ALBERTA	SUFFIELD-SP-
CANADA. SUFFIELD EXPERIMENTAL STATION,RALSTON,ALBERTA	SUFFIELD TECHNICAL PAPER
CANADA. SUFFIELD EXPERIMENTAL STATION,RALSTON,ALBERTA	SUFFIELD-TN-
CANADA. SUFFIELD EXPERIMENTAL STATION,RALSTON,ALBERTA	SUFFIELD TP-
CANADA. SUFFIELD EXPERIMENTAL STATION,RALSTON,ALBERTA	TN-
CANADAIR, LTD., MONTREAL	EBR-
CANADAIR LTD. MISSILES AND SYSTEMS DIV., MONTREAL	CAN-MSER-
CANADIAN ADMIRAL CORP., LTD., PORT CREDIT, ONT.	CCC-
CANADIAN ARSENALS, LTD., TORONTO	CA-
CANADIAN ARSENALS, LTD., TORONTO	CAL-
CANADIAN COMMERCIAL CORP., OTTAWA	CCC7PC(NUMBER)-
CANADIAN COMMERCIAL CORP., WASHINGTON, D.C.	RA-(NUMBER)-
CANADIAN FORCES HEADQUARTERS, OTTAWA	CFHQ-TRIAL-
CANADIAN FORCES HEADQUARTERS, OTTAWA	CFP-
CANADIAN GENERAL ELECTRIC CO., LTD., TORONTO	CGE-
CANADIAN GENERAL ELECTRIC CO., LTD. CIVILIAN ATOMIC POWER DIV., PETERBOROUGH, ONT.	DF-(YEAR)-CAP-
CANADIAN GENERAL ELECTRIC CO., LTD. CIVILIAN ATOMIC POWER DEPT., PETERBOROUGH, ONT.	R(YR)CAP(NO.)
CANADIAN GENERAL ELECTRIC CO., LTD. COMMERCIAL ATOMIC POWER DIV., PETERBOROUGH, ONT.	R (YEAR) CAP-
CANADIAN INDUSTRIES, LTD., NEW TORONTO, ONT.	RQ(YEAR)EE-
CANADIAN MARCONI CO., MONTREAL	CMC-TPL-
CANADIAN MARCONI CO. AVIONICS DEPT., MONTREAL	CMC-P-
CANADIAN MILITARY ELECTRONICS STANDARDS AGENCY,OTTAWA	CAMESA-(NUMBER)A
CANADIAN MILITARY ELECTRONICS STANDARDS AGENCY,OTTAWA	CAMESA-L(NUMBER)-
CANADIAN NATIONAL COMMITTEE, UNION RADIO-SCIENTIFIQUE INTERNATIONALE, OTTAWA	URSI-
CANADIAN NUCLEAR ASSN., TORONTO	67-CNA-
CANADIAN OCEANOGRAPHIC DATA CENTRE, OTTAWA	CODE-
CANADIAN PRATT AND WHITNEY AIRCRAFT CO.,LONGUEUIL,QUE	CPWA ER-
CANADIAN PRATT AND WHITNEY AIRCRAFT CO.,LONGUEUIL,QUE	CPWA TN-
CANADIAN RADIO WAVE PROPAGATION COMMITTEE	CRWPC-
CANADIAN STANDARDS ASSOCIATION, OTTAWA	CSA-
CANADIAN WESTINGHOUSE CO., LTD., HAMILTON, ONT.	CR-R+DL-
CANADIAN WESTINGHOUSE CO., LTD. ATOMIC ENERGY DIV., HAMILTON, ONT.	CWAED-
CANADIAN WESTINGHOUSE CO., LTD., RES. + DEV. LABS.	CW-R+DL-
CANOGA CORP., VAN NUYS, CALIF.	CANO-
CAPE TOWN. UNIV.	CAPETOWNU-
CAPEHART-FARNSWORTH CORP., FORT WAYNE (REF. 1)	(NUMBER)
CAPEHART-FARNSWORTH CORP., FORT WAYNE	CFC-
CAPEHART-FARNSWORTH CORP., FORT WAYNE	SCL-
CAPITOL REGIONAL PLANNING AGENCY, HARTFORD, CONN.	CRPA-HOUSING-
CARBIDE AND CARBON CHEMICALS CO., PADUCAH, KY.	KYL-

CARBIDE AND CARBON CHEMICALS CORP.

CARBIDE AND CARBON CHEMICALS CORP., OAK RIDGE, TENN. (REF. 34)	B-(NUMBER.NUMBER.NUMBER)
CARBIDE AND CARBON CHEMICALS CORP., OAK RIDGE, TENN.	BA-
CARBIDE AND CARBON CHEMICALS CORP., OAK RIDGE, TENN.	BB-
CARBIDE AND CARBON CHEMICALS CORP., OAK RIDGE, TENN.	D-
CARBIDE AND CARBON CHEMICALS CORP., OAK RIDGE, TENN.	JC-
CARBIDE AND CARBON CHEMICALS CORP., OAK RIDGE, TENN.	KCI-
CARBIDE AND CARBON CHEMICALS CORP., OAK RIDGE, TENN.	PI-
CARBIDE AND CARBON CHEMICALS CORP., OAK RIDGE, TENN.	TL-
CARBIDE AND CARBON CHEMICALS CORP. K-25 PLANT, OAK RIDGE, TENN. (REF. 1)	1.(NUMBER).-
CARBIDE AND CARBON CHEMICALS CORP. K-25 PLANT, OAK RIDGE, TENN. (REF. 1)	4.(NUMBER).-
CARBIDE AND CARBON CHEMICALS CORP. K-25 PLANT, OAK RIDGE, TENN. (REF. 1)	2.(NUMBER).-
CARBIDE AND CARBON CHEMICALS CORP. K-25 PLANT, OAK RIDGE, TENN. (REF. 1)	3.(NUMBER).-
CARBIDE AND CARBON CHEMICALS CORP. K-25 PLANT, OAK RIDGE, TENN.	A-
CARBIDE AND CARBON CHEMICALS CORP. K-25 PLANT, OAK RIDGE, TENN.	B-1.1-
CARBIDE AND CARBON CHEMICALS CORP. K-25 PLANT, OAK RIDGE, TENN.	B-7.1-
CARBIDE AND CARBON CHEMICALS CORP. K-25 PLANT, OAK RIDGE, TENN. (REF. 10)	C-
CARBIDE AND CARBON CHEMICALS CORP. K-25 PLANT, OAK RIDGE, TENN.	CB-
CARBIDE AND CARBON CHEMICALS CORP. K-25 PLANT, OAK RIDGE, TENN. (REF. 10)	C+C-K-
CARBIDE AND CARBON CHEMICALS CORP. K-25 PLANT, OAK RIDGE, TENN.	CK-
CARBIDE AND CARBON CHEMICALS CORP. K-25 PLANT, OAK RIDGE, TENN. (REF. 10)	CKC-
CARBIDE AND CARBON CHEMICALS CORP. K-25 PLANT, OAK RIDGE, TENN.	CT-
CARBIDE AND CARBON CHEMICALS CORP. K-25 PLANT, OAK RIDGE, TENN.	D3-
CARBIDE AND CARBON CHEMICALS CORP. K-25 PLANT, OAK RIDGE, TENN.	K-
CARBIDE AND CARBON CHEMICALS CORP. K-25 PLANT, OAK RIDGE, TENN.	K-1.2.-THRU K-4.76-
CARBIDE AND CARBON CHEMICALS CORP. K-25 PLANT, OAK RIDGE, TENN.	KA-
CARBIDE AND CARBON CHEMICALS CORP. K-25 PLANT, OAK RIDGE, TENN.	KAE-
CARBIDE AND CARBON CHEMICALS CORP. K-25 PLANT, OAK RIDGE, TENN.	KB-
CARBIDE AND CARBON CHEMICALS CORP. K-25 PLANT, OAK RIDGE, TENN.	KC-
CARBIDE AND CARBON CHEMICALS CORP. K-25 PLANT, OAK RIDGE, TENN.	KD-
CARBIDE AND CARBON CHEMICALS CORP. K-25 PLANT, OAK RIDGE, TENN.	KDD-
CARBIDE AND CARBON CHEMICALS CORP. K-25 PLANT, OAK RIDGE, TENN.	KLO-
CARBIDE AND CARBON CHEMICALS CORP. K-25 PLANT, OAK RIDGE, TENN.	KP-
CARBIDE AND CARBON CHEMICALS CORP. K-25 PLANT, OAK RIDGE, TENN.	KS-
CARBIDE AND CARBON CHEMICALS CORP. K-25 PLANT, OAK RIDGE, TENN.	KU-
CARBIDE AND CARBON CHEMICALS CORP. K-25 PLANT, OAK RIDGE, TENN.	KZ-
CARBIDE AND CARBON CHEMICALS CORP. K-25 PLANT, OAK RIDGE, TENN. (REF. 10)	MEMO-
CARBIDE AND CARBON CHEMICALS CORP. K-25 PLANT, OAK RIDGE, TENN.	PD 70.-
CARBIDE AND CARBON CHEMICALS CORP. K-25 PLANT, OAK RIDGE, TENN.	PDD-(NO.)-
CARBIDE AND CARBON CHEMICALS CORP. K-25 PLANT, OAK RIDGE, TENN.	VT-

(SEE ALSO: OAK RIDGE GASEOUS DIFFUSION PLANT, AND UNION CARBIDE NUCLEAR CO. K-25 PLANT)

CARBIDE AND CARBON CHEMICALS CORP. SUBSTITUTE ALLOY MATERIALS LABS., N.Y.C.	BA-R-
CARBIDE AND CARBON CHEMICALS CORP. SUBSTITUTE ALLOY MATERIALS LABS., N.Y.C.	BD-R-
CARBIDE AND CARBON CHEMICALS CORP. SUBSTITUTE ALLOY MATERIALS LABS., N.Y.C.	CI-R-
CARBIDE AND CARBON CHEMICALS CORP. SUBSTITUTE ALLOY MATERIALS LABS., N.Y.C.	DA-R-
CARBIDE AND CARBON CHEMICALS CORP. SUBSTITUTE ALLOY MATERIALS LABS., N.Y.C.	DL-R-
CARBIDE AND CARBON CHEMICALS CORP. SUBSTITUTE ALLOY MATERIALS LABS., N.Y.C.	D-R-
CARBIDE AND CARBON CHEMICALS CORP. SUBSTITUTE ALLOY MATERIALS LABS., N.Y.C.	EN-R-
CARBIDE AND CARBON CHEMICALS CORP. Y-12 PLANT, OAK RIDGE, TENN.	CEW-TEC-
CARBIDE AND CARBON CHEMICALS CORP. Y-12 PLANT, OAK RIDGE, TENN.	CY-
CARBIDE AND CARBON CHEMICALS CORP. Y-12 PLANT, OAK RIDGE, TENN.	H-(NUMBER).(NO.).(NO.)
CARBORUNDUM CO., NIAGARA FALLS, N.Y.	CARBC-
CARBORUNDUM CO., NIAGARA FALLS, N.Y.	CARBC-BI-MONTHLY-
CARBORUNDUM CO., NIAGARA FALLS, N.Y.	CARBC-FINAL
CARBORUNDUM CO., NIAGARA FALLS, N.Y. (PROGRESS REPT.)	CARBC-PR-
CARBORUNDUM CO., NIAGARA FALLS, N.Y.	CARBC-QR-
CARBORUNDUM CO., NIAGARA FALLS, N.Y.	CARBC-SR-
CARBORUNDUM CO., NIAGARA FALLS, N.Y.	CARBORUNDUM(NO.-NO.)-QPR-
CARBORUNDUM CO., NIAGARA FALLS, N.Y.	CARBORUNDUM-QR-
CARBORUNDUM CO. RES. + DEV. DIV., NIAGARA FALLS, N.Y.	GV-
CARBORUNDUM METALS CO., INC., AKRON, N.Y.	CM-(NUMBER)-D-
CARBORUNDUM METALS CO., INC., AKRON, N.Y.	CMC-U-(YEAR)-
CARBORUNDUM METALS CO., INC., AKRON, N.Y.	CM-U-
CARLETON CONTROLS CORP., EAST AURORA, N.Y.	CCC-P/N-(NO.-NO.-NO.)
CARLETON UNIV., OTTAWA	CARLETONU-
CARLETON UNIV., OTTAWA	E-(YEAR)-
CARLETON UNIV., OTTAWA. DIV. OF AEROTHERMODYNAMICS	ME/A-(YEAR)-
CARNEGIE INST. OF TECH., PITTSBURGH	CARIT-
CARNEGIE INST. OF TECH., PITTSBURGH (REF. 10)	CC-
CARNEGIE INST. OF TECH., PITTSBURGH	CIT-(LETTER(S))-
CARNEGIE INST. OF TECH., PITTSBURGH	CIT-AFBA-
CARNEGIE INST. OF TECH., PITTSBURGH (ORDNANCE)	CIT-ORD-
CARNEGIE INST. OF TECH., PITTSBURGH	CIT-ORD-6D-TR-
CARNEGIE INST. OF TECH., PITTSBURGH	CIT-ORD-(LETTER(S))-
CARNEGIE INST. OF TECH., PITTSBURGH	CRIT-
CARNEGIE INST. OF TECH., PITTSBURGH	EAC-
CARNEGIE INST. OF TECH., PITTSBURGH	EC-

(SEE ALSO LATER NAME: CARNEGIE-MELLON UNIV.)

CARNEGIE INST. OF TECH., PITTSBURGH. COMPUTATION CTR.	CAR-CC-TR-
CARNEGIE INST. OF TECH., PITTSBURGH. DEPT. OF PHYSICS	CAR-(NUMBER)-
CARNEGIE INST. OF TECH., PITTSBURGH. MANAGEMENT SCIENCES RESEARCH GROUP	MSRR-
CARNEGIE INST. OF TECH., PITTSBURGH. METALS RES. LAB.	CAR-17-TR-
CARNEGIE INSTITUTION OF WASHINGTON, D.C.	CARIW-
CARNEGIE INSTITUTION OF WASHINGTON, D.C.	S-
CARNEGIE INSTITUTION OF WASHINGTON, D.C. DEPT. OF TERRESTRIAL MAGNETISM	DTM-
CARNEGIE-MELLON UNIV., PITTSBURGH	CAR-
CARNEGIE-MELLON UNIV., PITTSBURGH	CMU-(NUMBER)-(NUMBER)-
CARNEGIE-MELLON UNIV., PITTSBURGH	CMU/PPL-(MONTH/YEAR)
CARNEGIE-MELLON UNIV., PITTSBURGH	CMU-TR-
CARNEGIE-MELLON UNIV., PITTSBURGH	TR-
CARNEGIE-MELLON UNIV., PITTSBURGH	WP-(NUMBER)-(YEAR)-
CARNEGIE-MELLON UNIV., PITTSBURGH. DEPT. OF ELEC. ENG	A-(NUMBER-NUMBER)

CARNEGIE-MELLON UNIV., PITTSBURGH. RADIATION RES. LAB	RRL-(NUMBER)-
CAROLINAS-VIRGINIA NUCLEAR POWER ASSOCIATES, INC., CHARLOTTE, N.C.	CVNA-
CAROLINAS-VIRGINIA NUCLEAR POWER ASSOCIATES, INC., CHARLOTTE, N.C.	WCAP-
CAROLINAS-VIRGINIA NUCLEAR POWER ASSOCIATES, INC., COLUMBIA, S.C.	CVNA-
CAROLINAS-VIRGINIA NUCLEAR POWER ASSOCIATES, INC., PARR, S.C.	CVNA-
CARTER OBSERVATORY, WELLINGTON, NEW ZEALAND	CO-
CASE INST. OF TECH., CLEVELAND	CAIT-
CASE INST. OF TECH., CLEVELAND	CASE-
CASE INST. OF TECH., CLEVELAND (LIBRARY ACCESSION NO.)	CASE-(YEAR)-
CASE INST. OF TECH., CLEVELAND	CASE (NUMBER)-FR-(DATE)
CASE INST. OF TECH., CLEVELAND	CASE-QSR-
CASE INST. OF TECH., CLEVELAND	CASE-TR-
CASE INST. OF TECH., CLEVELAND	CASIT-
CASE INST. OF TECH., CLEVELAND	CITC-
CASE INST. OF TECH., CLEVELAND	CSI-
CASE INST. OF TECH., CLEVELAND	W-
(SEE ALSO LATER NAME: CASE WESTERN RESERVE UNIV.)	
CASE INST. OF TECH., CLEVELAND. COMPUTING CENTER	CASE-CC-
CASE INST. OF TECH., CLEVELAND. DEPT. OF MET. ENG.	CASE-FR-
CASE INST. OF TECH., CLEVELAND. DIGITAL SYSTEMS ENGINEERING GROUP	CASE-1-(YEAR)-
CASE INST. OF TECH., CLEVELAND. ENG. DESIGN CENTER	EDC-(GROUP NO.)-(YEAR)-
CASE INST. OF TECH., CLEVELAND. FLUID, THERMAL, AND AEROSPACE SCIENCES GROUP	FTAS/TR-(YEAR)-
CASE INST. OF TECH., CLEVELAND. METALS RESEARCH LAB.	CASE-SR-
CASE INST. OF TECH., CLEVELAND. OPERATIONS RES. GP.	CASE-TECH. MEMO.-
CASE WESTERN RESERVE UNIV., CLEVELAND	CWRU-
CASE WESTERN RESERVE UNIV., CLEVELAND	CWRU-TR-
CASE WESTERN RESERVE UNIV., CLEVELAND	TR-(YEAR)-
CASE WESTERN RESERVE UNIV., CLEVELAND. SYSTEMS RES. CTR.	SRC-(NUMBER-LTR.-YR.)-
CATALYST RESEARCH CORP., BALTIMORE	CRC-
CATALYST RESEARCH CORP., BALTIMORE	CRC-DOC.-
CATALYST RESEARCH CORP., BALTIMORE	CRCO-
CATALYST RESEARCH CORP., BALTIMORE	CRC-PR-
CATALYST RESEARCH CORP., BOSTON	PR-C-(YR)-EL/S-
CATALYTIC CONSTRUCTION CO., PHILADELPHIA	CCCO-
CATALYTIC CONSTRUCTION CO., PHILADELPHIA	CCCO-(CONTRACT)-
CATANIA, ITALY. UNIV. ISTITUTO DI FISICA	PP/
CATHOLIC UNIV. OF AMERICA, WASHINGTON, D.C.	CU-
CATHOLIC UNIV. OF AMERICA, WASHINGTON, D.C.	CUA-
CATHOLIC UNIV. OF AMERICA, WASHINGTON, D.C.	CUA-NE-
CATHOLIC UNIV. OF AMERICA, WASHINGTON, D.C.	CUA-PR-
CATHOLIC UNIV. OF AMERICA, WASHINGTON, D.C.	CUA-QPR-
CATHOLIC UNIV. OF AMERICA, WASHINGTON, D.C.	CUA-TR-
CATHOLIC UNIV. OF AMERICA, WASHINGTON, D.C.	CU/F/
CATHOLIC UNIV. OF AMERICA, WASHINGTON, D.C.	CU/P/
CATHOLIC UNIV. OF AMERICA, WASHINGTON, D.C. DEPT. OF SCIENCE AND APPLIED PHYSICS	SSAP-(YEAR)-
CAYWOOD-SCHILLER, ASSOCIATES, CHICAGO	CAY MB-
CAYWOOD SCHILLER, ASSOCIATES, CHICAGO	CSA-
CAYWOOD-SCHILLER, ASSOCIATES, CHICAGO	CSA-(YEAR)-P-
CAYWOOD-SCHILLER, ASSOCIATES, CHICAGO	CSA-(YEAR)-F-
CBS-HYTRON, DANVERS, MASS.	HYTRON-
CBS LABS., STAMFORD, CONN.	CLD-
CBS LABS., STAMFORD, CONN.	(NUMBER)-CLD-
CELANESE CHEMICAL CO. RES. + DEV. DEPT., CLARKWOOD, TEX	CELANESE-QR-
CELESTIAL RESEARCH CORP., SOUTH PASADENA, CALIF.	BD-(NUMBER)-
CELESTIAL RESEARCH CORP., SOUTH PASADENA, CALIF.	CELESCO BD (NUMBER)-
CELESTRON ASSOCIATES, YONKERS, N.Y.	CA-
CENCO EDUCATIONAL FILM CO., CHICAGO	CENCO-
CENTER FOR APPLIED LINGUISTICS, WASHINGTON, D.C.	CALLINGS-(YEAR)-
CENTER FOR APPLIED LINGUISTICS, WASHINGTON, D.C. (LANGUAGE INFORMATION NETWORK)	LINCS-(NUMBER-YEAR)
CENTER FOR NAVAL ANALYSES, ARLINGTON, VA.	CNAOPROFESSIONAL PAPER-
CENTER FOR NAVAL ANALYSES, ARLINGTON, VA.	CNA-PROFESSIONAL PAPER-
CENTER FOR NAVAL ANALYSES, WASHINGTON, D.C.	CNA RESEARCH CONTRIB-
CENTER FOR NAVAL ANALYSES. INST. OF NAVAL STUDIES, CAMBRIDGE, MASS.	INS-
CENTER FOR NAVAL ANALYSES. INST. OF NAVAL STUDIES, CAMBRIDGE, MASS.	INS-RC-
CENTER FOR NAVAL ANALYSES. INST. OF NAVAL STUDIES, CAMBRIDGE, MASS.	INS-RESEARCH CONTRIB-
CENTER FOR NAVAL ANALYSES. INST. OF NAVAL STUDIES, CAMBRIDGE, MASS.	INS-STUDY-
CENTER FOR NAVAL ANALYSES. INST. OF NAVAL STUDIES, WASHINGTON, D.C.	INS-RC-
CENTER FOR NAVAL ANALYSES. INST. OF NAVAL STUDIES, WASHINGTON, D.C.	INS-RESEARCH CONTRIB-
CENTER FOR NAVAL ANALYSES. MARINE CORPS OPERATIONS ANALYSIS GROUP, WASHINGTON, D.C.	MCOAG RESEARCH CONTRIB-
CENTER FOR NAVAL ANALYSES. MARINE CORPS OPERATIONS ANALYSIS GROUP, WASHINGTON, D.C.	MCOAG-STUDY-
CENTER FOR NAVAL ANALYSES. NAVAL OBJECTIVES ANALYSIS GROUP, WASHINGTON, D.C.	NOAG-RESEARCH CONTRIB-
CENTER FOR NAVAL ANALYSES. NAVAL WARFARE ANALYSIS GROUP, WASHINGTON, D.C.	NAVWAG-
CENTER FOR NAVAL ANALYSES. NAVAL WARFARE ANALYSIS GROUP, WASHINGTON, D.C.	NAVWAG-IRM-
CENTER FOR NAVAL ANALYSES. NAVAL WARFARE ANALYSIS GROUP, WASHINGTON, D.C.	NAVWAG-RESEARCH CONTRIB-
CENTER FOR NAVAL ANALYSES. NAVAL WARFARE ANALYSIS GROUP, WASHINGTON, D.C.	NWAG-IRM-

CENTER FOR NAVAL ANALYSES. NAVAL WARFARE ANALYSIS GROUP, WASHINGTON, D.C.	NWAG-RESEARCH CONTRIB-
CENTER FOR NAVAL ANALYSES. NAVAL WARFARE ANALYSIS GROUP, WASHINGTON, D.C.	NWAG-STUDY-
CENTER FOR NAVAL ANALYSES. OPERATIONS EVALUATION GROUP, WASHINGTON, D.C.	OEG-RESEARCH CONTRIB-
CENTER FOR NAVAL ANALYSES. OPERATIONS EVALUATION GROUP, WASHINGTON, D.C.	OEG-STUDY-
CENTER FOR NAVAL ANALYSES. SYSTEMS EVALUATION GROUP, ARLINGTON, VA.	SEG-STUDY-
CENTER FOR NAVAL ANALYSES. SYSTEMS EVALUATION GROUP, WASHINGTON, D.C.	RESEARCH CONTRIBUTION-
CENTER FOR NAVAL ANALYSES. SYSTEMS EVALUATION GROUP, WASHINGTON, D.C.	SEG RESEARCH CONTRIB-
CENTER FOR NAVAL ANALYSES. SYSTEM EVALUATION GROUP, WASHINGTON, D.C.	SEG-STUDY-
CENTER FOR THE ENVIRONMENT AND MAN, INC., HARTFORD, CONN	CEM-(NUMBER)-
CENTRAL AIR DOCUMENTS OFFICE, WRIGHT-PATTERSON AFB, OHIO (ANNOUNCED TECHNICAL INDEX)	ATI-
CENTRAL AIR DOCS. OFF., WRIGHT-PATTERSON AFB, OHIO	CADO-
CENTRAL AIR DOCS. OFF., WRIGHT-PATTERSON AFB, OHIO	F-TR-(NO.)-(LETTERS)
CENTRAL CONNECTICUT REGIONAL PLANNING AGENCY, PLAINVILLE	CCRPA-M-
CENTRAL CONNECTICUT REGIONAL PLANNING AGENCY, PLAINVILLE	CCRPA-SP-
CENTRAL GEEIA REGION, TINKER AFB, OKLAHOMA CITY	CGR-EMC-
CENTRAL INTELLIGENCE AGENCY, WASHINGTON, D.C.	CIA-
CENTRAL INTELLIGENCE AGENCY, WASHINGTON, D.C.	CIA/BGI-CT-(YEAR-NUMBER)
CENTRAL INTELLIGENCE AGENCY, WASHINGTON, D.C.	CIACGI-CD-(YEAR-NUMBER)
CENTRAL INTELLIGENCE AGENCY, WASHINGTON, D.C. (NATIONAL INTELLIGENCE ESTIMATE)	CIA NIE-(DATE)
CENTRAL INTELLIGENCE AGENCY, WASHINGTON, D.C. (NATIONAL INTELLIGENCE SURVEY)	CIA NIS-
CENTRAL INTELLIGENCE AGENCY, WASHINGTON, D.C. (CONSOLIDATED TRANSLATION SURVEY)	CTS-
CENTRAL INTELLIGENCE AGENCY, WASHINGTON, D.C.	NIS-
CENTRAL INTELLIGENCE AGENCY. FOREIGN DOCUMENTS DIV. WASHINGTON, D.C. (CONSOLIDATED TRANSLATION SURVEY)	CIA FDD T-
CENTRAL INTELLIGENCE AGENCY. OFFICE OF COMPUTER SERVICE, WASHINGTON, D.C.	CHIVE/R-
CENTRAL INTELL. AGENCY. OFF. OF SCI. INTELL., WASH., DC	CIA/OSI-(LTR)-(LTRS)(YR)-
CENTRAL INTELL. AGENCY. OFF. OF SCI. INTELL., WASH., DC	CIA/SI-(NUMBER-YEAR)
CENTRAL INTELL. AGENCY. OFF. OF SCI. INTELL., WASH., DC	OSI-SR-
CENTRAL NEW YORK REGIONAL PLANNING AND DEVELOPMENT BOARD, SYRACUSE	INVENTORY-
CENTRAL UTILITIES ATOMIC POWER ASSOCIATES	ANCP-
CENTURY ENGINEERS, INC., BURBANK, CALIF.	CEI-
CERAMIC FINISHING CO., STATE COLLEGE, PENNA.	CFC-TR-
CESKOSLOVENSKA AKADEMIE VED, PRAGUE	IPPCZ-
CESKOSLOVENSKA AKADEMIE VED, PRAGUE	Z-(2 DIGITS)
CESKOSLOVENSKA AKADEMIE VED, USTAV FYZIKY PLAZMATU, PRAGUE.	IPP-
CESKOSLOVENSKA AKADEMIE VED. USTAV JADERNEHO VYZKUMU, REZ	UJ-(NUMBER)/(YEAR)
CESKOSLOVENSKA AKADEMIE VED. USTAV JADERNEHO VYZKUMU, REZ	UJV-
CESKOSLOVENSKA AKADEMIE VED. USTAV JADERNEHO VYZKUMU, REZ (NAME TRANSLATED IS: INST. OF NUCLEAR RESEARCH)	UJV-(NUMBER)-(LETTER)
CGS LABS., INC., STAMFORD, CONN.	CGS-
CGS SCIENTIFIC CORP., WATERTOWN, MASS.	CGS-TR-
CHALLENGER RESEARCH INC., ROCKVILLE, MD.	CRI-TR-(NUMBER)-
CHALMERS TEKNISKA HOEGSKOLA, GOETEBORG	CI-
CHALMERS TEKNISKA HOEGSKOLA, GOETEBORG	CTH-
CHALMERS TEKNISKA HOEGSKOLA, GOETEBORG	CUT-TRANSACTION-
CHALMERS TEKNISKA HOEGSKOLA, GOETEBORG	GIPR-
CHALMERS TEKNISKA HOEGSKOLA, GOETEBORG	LFF-
CHALMERS TEKNISKA HOEGSKOLA, GOETEBORG	NR-
CHALMERS TEKNISKA HOEGSKOLA, GOETEBORG	US-
CHALMERS TEKNISKA HOEGSKOLA, GOETEBORG. DIV. OF SOLID MECHANICS	CUT-AAV-
CHALMERS TEKNISKA HOEGSKOLA, GOETEBORG. INST. FOR REAKTORFYSIK	CTH-RF-
CHAMBERLAIN CORP., WATERLOO, IOWA	CHAM-
CHANCE (W.R.) AND ASSOCIATES, ARLINGTON, VA.	VA-AST-
CHANCE (W.R.) AND ASSOCIATES, ARLINGTON, VA.	WRC-
CHANCE VOUGHT AIRCRAFT, INC., DALLAS	AER-(LTR.)-(NO.)-
CHANCE VOUGHT AIRCRAFT, INC., DALLAS	AER-EOR-
CHANCE VOUGHT AIRCRAFT, INC., DALLAS	AST/EIR-
CHANCE VOUGHT AIRCRAFT, INC., DALLAS	CVA-
CHANCE VOUGHT AIRCRAFT, INC., DALLAS	CVA-E(YEAR)R-
CHANCE VOUGHT AIRCRAFT, INC., DALLAS	CVA-EOR-
CHANCE VOUGHT AIRCRAFT, INC., DALLAS	CVAI-
CHANCE VOUGHT AIRCRAFT, INC., DALLAS	CVA-PR-
CHANCE VOUGHT AIRCRAFT, INC., DALLAS (SEE ALSO LATER NAME: LING-TEMCO-VOUGHT)	FT/F-
CHANCE VOUGHT AIRCRAFT, INC. AERONAUTICS DIV., DALLAS	CVA AER-(LTR-YR-LTR)-
CHANCE VOUGHT AIRCRAFT, INC. ASTRONAUTICS DIV., DALLAS	CVA AST-(LTR-YR-LTR)-
CHANCE VOUGHT CORP., DALLAS	AER-EIR-
CHANCE VOUGHT CORP., DALLAS	CVC-
CHANDLER EVANS INC. CONTROL SYSTEMS DIV., WEST HARTFORD, CONN.	R-(3 DIGITS)
CHARLES RIVER ASSOCIATES, INC., CAMBRIDGE, MASS.	CRA-(NUMBER-YEAR)-
CHARLESTON NAVAL SHIPYARD, S.C.	C-(NUMBER-YEAR)-
CHARLOTTE MUNICIPAL INFORMATION SYSTEM PROJECT, N.C.	TM-CH-(NO./NO./NO.)
CHASE AIRCRAFT CO., INC., WEST TRENTON, N.J.	CAC-
CHASE AIRCRAFT CO., INC., WEST TRENTON, N.J.	CHASE-
CHATHAM COUNTY-SAVANNAH METROPOLITAN PLANNING COMMISSION, GA.	CHATSAV-(YEAR-NUMBER)

CHATHAM ELECTRONICS CORP., LIVINGSTON, N.J.	CHAT-
CHECCHI AND CO., WASHINGTON, D.C.	CHECCHI-
CHEMICAL AND RADIOLOGICAL LABS., ARMY CHEMICAL CENTER MD. (INTERIM REPORT)	CC CRL IR-
CHEMICAL AND RADIOLOGICAL LABS., ARMY CHEM. CTR., MD.	CC-CRLIR-
CHEMICAL AND RADIOLOGICAL LABS., ARMY CHEM. CTR., MD.	CC CRL R-
CHEMICAL AND RADIOLOGICAL LABS., ARMY CHEM. CTR., MD.	CC-CRLR-
CHEMICAL AND RADIOLOGICAL LABS., ARMY CHEM. CTR., MD.	CRL-
CHEMICAL AND RADIOLOGICAL LABS., ARMY CHEMICAL CENTER, MD. (INTERIM REPORTS)	CRLIR-
CHEMICAL AND RADIOLOGICAL LABS., ARMY CHEMICAL CENTER, MD. (REPORTS)	CRLR-
CHEMICAL AND RADIOLOGICAL LABS., ARMY CHEM. CTR., MD.	FMTR-
CHEMICAL AND RADIOLOGICAL LABS., ARMY CHEM. CTR., MD.	MCN-
CHEMICAL AND RADIOLOGICAL LABS., ARMY CHEM. CTR., MD.	QMC-
CHEMICAL CONSTRUCTION CORP., N.Y.C.	CHEMICO-
CHEMICAL CORPS (ARMY)	CMLRE-
CHEMICAL CORPS (ARMY), WASHINGTON, D.C.	CMLCD-
CHEMICAL CORPS, ARMY CHEMICAL CENTER, MD.	CC-
CHEMICAL CORPS, ARMY CHEMICAL CENTER, MD.	CWS-
CHEMICAL CORPS, ARMY CHEMICAL CENTER, MD.	ETF-
CHEMICAL CORPS, ARMY CHEMICAL CENTER, MD.	ETF-(NUMBER)-
CHEMICAL CORPS, ARMY CHEMICAL CENTER, MD.	OCCM10-
CHEMICAL CORPS. ENG. COMMAND, ARMY CHEM. CTR., MD.	EN-
CHEMICAL CORPS. MEDICAL DIV., ARMY CHEMICAL CTR., MD.	CC MD R-
CHEMICAL CORPS. MEDICAL DIV., ARMY CHEMICAL CENTER, MD. (RESEARCH REPORT)	CC MD RR-
CHEMICAL CORPS. MEDICAL DIV., ARMY CHEMICAL CENTER, MD. (SPECIAL REPORT)	CC MD SR-
CHEMICAL CORPS. MEDICAL DIV., ARMY CHEMICAL CENTER, MD. (RESEARCH REPORT)	CMLEM-52-
CHEMICAL CORPS. MEDICAL DIV., ARMY CHEMICAL CTR., MD.	MDR-
CHEMICAL CORPS. MEDICAL DIV., ARMY CHEMICAL CENTER, MD. (RESEARCH REPORT)	MDRR-
CHEMICAL CORPS. MEDICAL DIV., ARMY CHEMICAL CENTER, MD. (SPECIAL REPORT)	MDSR-
CHEMICAL CORPS. METEOROLOGICAL DIV. BIOLOGICAL DEPT., CAMP DETRICK, MD.	BD-
CHEM. CORPS. RES. + ENG. COMMAND, ARMY CHEM. CTR., MD.	ENCTR-
CHEMICAL CORPS. SUPPLY AND PROCUREMENT DIV., ARMY CHEMICAL CENTER, MD.	CMLWD-
CHEMICAL CORPS. TECH. COMMAND, ARMY CHEMICAL CTR., MD.	CCTC-
CHEMICAL CORPS. TECHNICAL COMMAND, ARMY CHEMICAL CENTER, MD. (INFORMAL REPORT)	CC TCIR-
CHEMICAL CORPS. TECH. COMMAND, ARMY CHEMICAL CTR., MD.	CC TCR-
CHEMICAL CORPS. TECH. COMMAND, ARMY CHEMICAL CTR., MD.	CWS TCR-
CHEMICAL CORPS. TECH. COMMAND, ARMY CHEMICAL CTR., MD.	S-
CHEMICAL CORPS. TECHNICAL COMMAND, ARMY CHEMICAL CENTER, MD. (INFORMAL REPORT)	TCIR-
CHEMICAL CORPS. TECHNICAL COMMAND, ARMY CHEMICAL CENTER, MD. (TECHNICAL COMMAND INFORMATION REPORTS)	TCIR-
CHEMICAL CORPS. TECH. COMMAND, ARMY CHEMICAL CTR., MD.	TCR-
CHEMICAL CORPS BIOL. WARFARE LABS., CAMP DETRICK, MD.	BWL-
CHEMICAL CORPS ENG. AGENCY, ARMY CHEM. CTR., MD.	ENASR-
CHEMICAL CORPS ENG. AGENCY, ARMY CHEM. CTR., MD.	ENATR-
CHEMICAL CORPS ENG. AGENCY, ARMY CHEM. CTR., MD.	ENCR-
CHEMICAL CORPS FIELD REQUIREMENTS AGENCY, FORT MCCLELLAN, ALA.	CC-FRA-
CHEMICAL CORPS MEDICAL LABS., ARMY CHEMICAL CTR., MD.	CC-ML-
CHEMICAL CORPS MEDICAL LABS., ARMY CHEMICAL CTR., MD.	CCML-
CHEMICAL CORPS MEDICAL LABS., ARMY CHEM.CTR., MD.(RPTS)	CCMLR-
CHEMICAL CORPS MEDICAL LABS., ARMY CHEMICAL CTR., MD.	CMLRE-ML-
CHEMICAL CORPS MEDICAL LABS., ARMY CHEMICAL CENTER, MD. (CONTRACT REPORT)	MLCR-
CHEMICAL CORPS MEDICAL LABS., ARMY CHEMICAL CTR., MD.	MLR-
CHEMICAL CORPS MEDICAL LABS., ARMY CHEMICAL CENTER, MD. (RESEARCH REPORT)	MLRR-
CHEMICAL CORPS MEDICAL LABS., ARMY CHEMICAL CENTER, MD. (SPECIAL REPORT)	MLSR-
CHEMICAL CORPS SCHOOL, ARMY CHEMICAL CENTER, MD.	CCS-
CHEMICAL CORPS SCHOOL, ARMY CHEMICAL CENTER, MD.	RDS-(NO.)(LETTER)
CHEMICAL CORPS SCHOOL, ARMY CHEMICAL CENTER, MD.	ST-
CHEMICAL CORPS SCHOOL, ARMY CHEMICAL CENTER, MD.	ST-RDS-
CHEMICAL CORPS SCHOOL, FORT MC CLELLAN, ALA.	CCS-
CHEMICAL CORPS SCHOOL, FORT MC CLELLAN, ALA.	CML. C. SCHOOL-
CHEMICAL INST. OF CANADA, OTTAWA	CIC-
CHEMICAL WARFARE LABS., ARMY CHEMICAL CENTER, MD.	CWL-
CHEMICAL WARFARE LABS., ARMY CHEMICAL CENTER, MD.	CWLR-
CHEMICAL WARFARE LABS., ARMY CHEMICAL CENTER, MD. (SPECIAL PUBLICATIONS)	CWLSP-
CHEMICAL WARFARE LABS., ARMY CHEMICAL CENTER, MD. (SPECIAL PUBLICATION)	CWL-SP-
CHEMICAL WARFARE LABS., ARMY CHEMICAL CENTER, MD.	CWL-TM-
CHEMICAL WARFARE SERVICE, EDGEWOOD ARSENAL, MD. (CAPTURED MATERIEL MEMORANDUM REPORT)	CMMR-
CHEMICAL WARFARE SERVICE, EDGEWOOD ARSENAL, MD. (CAPTURED MATERIEL TECHNICAL REPORT)	CMTR-
CHEMICAL WARFARE SERVICE, EDGEWOOD ARSENAL, MD.	CWS-
CHEMICAL WARFARE SERVICE, EDGEWOOD ARSENAL, MD. (CAPTURED MATERIEL MEMORANDUM REPORT)	CWS CMMR-
CHEMICAL WARFARE SERVICE, EDGEWOOD ARSENAL, MD. (CAPTURED MATERIEL TECHNICAL REPORT)	CWS CMTR-
CHEMICAL WARFARE SERVICE, EDGEWOOD ARSENAL, MD. (CAPTURED MATERIEL TECH. RPT. MASS. INST. OF TECH.)	CWS CMTR MIT-
CHEMICAL WARFARE SERVICE, EDGEWOOD ARSENAL, MD. (DRAWING)	CWS DWG-
CHEMICAL WARFARE SERVICE, EDGEWOOD ARSENAL, MD. (FIELD LABORATORY MEMORANDUM)	CWS FLM-
CHEMICAL WARFARE SERVICE, EDGEWOOD ARSENAL, MD. (FIELD LABORATORY MEMORANDUM)	CWS FLM (NO.)-(NO.)-
CHEMICAL WARFARE SERVICE, EDGEWOOD ARSENAL, MD. (FOREIGN MATERIEL TECHNICAL REPORT)	CWS FMTR-
CHEMICAL WARFARE SERVICE, EDGEWOOD ARSENAL, MD. (FOREIGN MATERIEL TECHNICAL REPORT)	CWS FMTR M-
CHEMICAL WARFARE SERVICE, EDGEWOOD ARSENAL, MD. (FOREIGN MATERIEL TECH. RPT. MASS. INST. OF TECH.)	CWS FMTR MIT-
CHEMICAL WARFARE SERVICE, EDGEWOOD ARSENAL, MD. (FRED PROJECT)	CWS FP-
CHEMICAL WARFARE SERVICE, EDGEWOOD ARSENAL, MD. (INSECT AND RODENT CONTROL)	CWS IRC-
CHEMICAL WARFARE SERVICE, EDGEWOOD ARSENAL, MD.	CWS-PCS-
CHEMICAL WARFARE SERVICE, EDGEWOOD ARSENAL, MD. (SPECIFICATION)	CWS SPEC-
CHEMICAL WARFARE SERVICE, EDGEWOOD ARSENAL, MD. (TECHNICAL REPORTS)	EATR-
CHEMICAL WARFARE SERVICE, EDGEWOOD ARSENAL, MD. (FIELD LABORATORY MEMORANDUM)	FLM-
CHEMICAL WARFARE SERVICE, EDGEWOOD ARSENAL, MD. (FOREIGN MATERIEL TECHNICAL REPORT)	FMTR-
CHEMICAL WARFARE SERVICE, EDGEWOOD ARSENAL, MD. (FRED PROJECT)	FP-
CHEMICAL WARFARE SERVICE, EDGEWOOD ARSENAL, MD. (INSECT AND RODENT CONTROL)	IRC-
CHEMICAL WARFARE SERVICE, EDGEWOOD ARSENAL, MD.	TM-3-
CHEMICAL WARFARE SERVICE. CHEMICAL DIV., EDGEWOOD ARSENAL, MD.	EACD-

CHEMICAL WARFARE SERVICE. CHEMICAL DIV. PHYSICAL DEPT., EDGEWOOD ARSENAL

CHEMICAL WARFARE SERVICE. CHEMICAL DIV. PHYSICAL DEPT., EDGEWOOD ARSENAL, MD.	EACD PD-
CHEMICAL WARFARE SERVICE. CHEMICAL DIV. PHYSICAL DEPT., EDGEWOOD ARSENAL, MD.	PD-
CHEMICAL WARFARE SERVICE. (NO.) CHEMICAL LABORATORY CO. (CAPTURED MATERIEL, TECHNICAL REPORTS)	CWS (NO.) CMTR-
CHEMICAL WARFARE SERVICE. (NO.) CHEMICAL LABORATORY CO. (INTELLIGENCE REPORTS)	CWS (NO.) IR-
CHEMICAL WARFARE SERVICE. (NO.) CHEMICAL LAB. CO.	CWS (NO.) TR-
CHEMICAL WARFARE SERVICE. INTELLIGENCE DIV., EDGEWOOD ARSENAL, MD.	CWS IDR-
CHEMICAL WARFARE SERVICE. MEDICAL DIV., EDGEWOOD ARSENAL, MD.	CWS MD (EA)-
CHEMICAL WARFARE SERVICE. MEDICAL DIV., EDGEWOOD ARSENAL, MD. (MEMORANDUM REPORTS)	CWS MD (EA) MR-
CHEMICAL WARFARE SERVICE. MEDICAL DIV., EDGEWOOD ARSENAL, MD.	CWS MDR-
CHEMICAL WARFARE SERVICE. MEDICAL DIV., EDGEWOOD ARSENAL, MD. (MEMORANDUM REPORTS)	MD EA-
CHEMICAL WARFARE SERVICE. MEDICAL DIV., EDGEWOOD ARSENAL, MD.	MDR-
CHEMICAL WARFARE SERVICE. MEDICAL RESEARCH LAB., EDGEWOOD ARSENAL, MD.	CWS MRL (EA)-
CHEMICAL WARFARE SERVICE. MEDICAL RESEARCH LAB., EDGEWOOD ARSENAL, MD.	MRL (EA)-
CHEMICAL WARFARE SERVICE. SAN JOSE PROJECT, CALIF.	SJP-
CHEMICAL WARFARE SERVICE. TECHNICAL COMMAND, EDGEWOOD ARSENAL, MD. (INFORMAL REPORT)	CWS TCIR-
CHEMICAL WARFARE SERVICE. TECHNICAL DIV., EDGEWOOD ARSENAL, MD. (MEMORANDUM REPORT)	CWS TDMR-
CHEMICAL WARFARE SERVICE. TECHNICAL DIV., EDGEWOOD ARSENAL, MD. (MEMORANDA REPORTS)	EATD MR-
CHEMICAL WARFARE SERVICE. TECHNICAL DIV., EDGEWOOD ARSENAL, MD. (MEMORANDUM REPORT)	TDMR-
CHEMICAL WARFARE SERVICE. TECHNICAL INTELLIGENCE DIV., EDGEWOOD ARSENAL, MD.	CWS TIDR-
CHEMICAL WARFARE SERVICE. TECHNICAL INTELLIGENCE DIV., EDGEWOOD ARSENAL, MD.	TIDR-
CHEMICAL WARFARE SERVICE. TOXICOLOGICAL RESEARCH LAB., EDGEWOOD ARSENAL, MD.	CWS TRLR-
CHEMICAL WARFARE SERVICE. TOXICOLOGICAL RESEARCH LAB., EDGEWOOD ARSENAL, MD.	TRLR-
CHEMSTRAND RESEARCH CENTER, INC., DURHAM, N.C.	CHEMSTRAND-QPR-
CHESAPEAKE INSTRUMENT CORP., SHADYSIDE, MD.	CIC-
CHESAPEAKE INSTRUMENT CORP., SHADYSIDE, MD.	CIC-(NUMBER.NUMBER)F
CHICAGO. UNIV.	CH-
CHICAGO. UNIV.	CHI U-
CHICAGO. UNIV.	CH-U-
CHICAGO. UNIV.	CHU-
CHICAGO. UNIV.	CU-
CHICAGO. UNIV.	HL(NO.)-
CHICAGO. UNIV.	LAP-TR-
CHICAGO. UNIV.	NA-AG-
CHICAGO. UNIV.	SM-(NUMBER)/
CHICAGO. UNIV.	TR-
CHICAGO. UNIV.	UCH-
CHICAGO. UNIV.	WSD-
CHICAGO. UNIV. AIR FORCE RADIATION LAB.	CH-RL-
CHICAGO. UNIV. AIR FORCE RADIATION LAB.	CH USAF R-
CHICAGO. UNIV. AIR FORCE RADIATION LAB.	UCTL-
CHICAGO. UNIV. CHICAGO MIDWAY LABS.	CML-
CHICAGO. UNIV. CHICAGO MIDWAY LABS.	CML-(YR)-TN-(PROJ.NO.)-
CHICAGO. UNIV. CHICAGO MIDWAY LABS.	CML-L-
CHICAGO. UNIV. CHICAGO MIDWAY LABS.	CML-M-
CHICAGO. UNIV. CHICAGO MIDWAY LABS.	CML-M-(NUMBER)-
CHICAGO. UNIV. CHICAGO MIDWAY LABS.	CML-SR-M-
CHICAGO. UNIV. CHICAGO MIDWAY LABS.	CML-TN-
CHICAGO. UNIV. CHICAGO MIDWAY LABS.	CML-TN-P-
CHICAGO. UNIV. CHICAGO MIDWAY LABS.	CML-TR-P(NUMBER-NUMBER)
CHICAGO. UNIV. CHICAGO MIDWAY LABS.	UCHI-CML-
CHICAGO. UNIV. CLOUD PHYSICS LAB.	CPL-
CHICAGO. UNIV. DEPT. OF METEOROLOGY	CH-DM-
CHICAGO. UNIV. DEPT. OF METEOROLOGY	CH-IM-
CHICAGO. UNIV. DEPT. OF METEOROLOGY	CHU-IM-
CHICAGO. UNIV. ENRICO FERMI INST. FOR NUCLEAR STUDIES	CHEFINS-(YR)-
CHICAGO. UNIV. ENRICO FERMI INST. FOR NUCLEAR STUDIES	EFI-
CHICAGO. UNIV. ENRICO FERMI INST. FOR NUCLEAR STUDIES	EFI-(YEAR)-
CHICAGO. UNIV. ENRICO FERMI INST. FOR NUCLEAR STUDIES	EFINS-
CHICAGO. UNIV. ENRICO FERMI INST. FOR NUCLEAR STUDIES	EFINS-(NUMBER)-
CHICAGO. UNIV. INST. FOR AIR WEAPONS RESEARCH	IAWR-
CHICAGO. UNIV. INST. FOR AIR WEAPONS RESEARCH	IAWR(YR)-
CHICAGO. UNIV. INST. FOR AIR WEAPONS RESEARCH	UCHI-IAWR-
CHICAGO. UNIV. INST. FOR AIR WEAPONS RESEARCH	UTIAS-
CHICAGO. UNIV. INST. FOR NUCLEAR STUDIES	CH-(NUMBER)(INS)
CHICAGO. UNIV. INST. FOR NUCLEAR STUDIES	CH-(INS)-
CHICAGO. UNIV. INST. FOR NUCLEAR STUDIES	CHU-INS-
CHICAGO. UNIV. INST. FOR NUCLEAR STUDIES	NDN-
CHICAGO. UNIV. INST. FOR SYSTEM RESEARCH	CH-ISR-
CHICAGO. UNIV. INST. FOR THE STUDY OF METALS	U-
CHICAGO. UNIV. LAB. OF MOLECULAR STRUCTURE + SPECTRA	CHU-LMSS-
CHICAGO. UNIV. LAB. OF MOLECULAR STRUCTURE + SPECTRA	LMSS-TR-(YEAR)-
CHICAGO. UNIV. LABS. FOR APPLIED SCIENCES	CHIC U LAS-TR-
CHICAGO. UNIV. LABS. FOR APPLIED SCIENCES	CH LAS TR-
CHICAGO. UNIV. LABS. FOR APPLIED SCIENCES	CH U/LAS TR-
CHICAGO. UNIV. LABS. FOR APPLIED SCIENCES	LAS-
CHICAGO. UNIV. LABS. FOR APPLIED SCIENCES	LAS-L-P-
CHICAGO. UNIV. LABS. FOR APPLIED SCIENCES	LAS-QR-(LTR.)(NO.)-
CHICAGO. UNIV. LABS. FOR APPLIED SCIENCES	LAS-SR-(NUMBER)-(NUMBER)
CHICAGO. UNIV. LABS. FOR APPLIED SCIENCES	LAS-SR-P-
CHICAGO. UNIV. LABS. FOR APPLIED SCIENCES	LAS-TR-(NUMBER)-(NUMBER)
CHICAGO. UNIV. LABS. FOR APPLIED SCIENCES	LAS-TR-(LTR.)(NO.)-
CHICAGO. UNIV. METALLURGICAL LAB. (ASSIGNED TO REPORTS FROM VARIOUS BRITISH SOURCES)	B-
CHICAGO. UNIV. METALLURGICAL LAB. (ASSIGNED TO REPORTS FROM VARIOUS AMERICAN SOURCES)	BETA-
CHICAGO. UNIV. METALLURGICAL LAB.	BF-RATS-
CHICAGO. UNIV. METALLURGICAL LAB.	BK-
CHICAGO. UNIV. METALLURGICAL LAB. (ASSIGNED TO REPORTS FROM VARIOUS BRITISH SOURCES)	BR-

CHICAGO. UNIV. METALLURGICAL LAB.	C-
CHICAGO. UNIV. METALLURGICAL LAB.	CA-
CHICAGO. UNIV. METALLURGICAL LAB. (REF. 10)	CB-
CHICAGO. UNIV. METALLURGICAL LAB. (REF. 10)	CC-
CHICAGO. UNIV. METALLURGICAL LAB.	CE-
CHICAGO. UNIV. METALLURGICAL LAB.	CF-
CHICAGO. UNIV. METALLURGICAL LAB. (ASSIGNED TO REPORTS FROM GREAT BRITAIN) (REF. 10)	CFB-
CHICAGO. UNIV. METALLURGICAL LAB.	CH CC-
CHICAGO. UNIV. METALLURGICAL LAB. (ASSIGNED TO REPORTS FROM VARIOUS AMERICAN SOURCES)	CHEM-S-
CHICAGO. UNIV. METALLURGICAL LAB. (REF. 10)	CHN-
CHICAGO. UNIV. METALLURGICAL LAB. (REF. 10)	CK-
CHICAGO. UNIV. METALLURGICAL LAB. (REF. 10)	CK-(LETTER(S))-
CHICAGO. UNIV. METALLURGICAL LAB. (REF. 10)	CKAN-
CHICAGO. UNIV. METALLURGICAL LAB. (REF. 10)	CKN-
CHICAGO. UNIV. METALLURGICAL LAB. (REF. 10)	CL-
CHICAGO. UNIV. METALLURGICAL LAB. (REF. 10)	CL-(LETTER(S))-
CHICAGO. UNIV. METALLURGICAL LAB.	CM-
CHICAGO. UNIV. METALLURGICAL LAB. (REF. 10)	CN-
CHICAGO. UNIV. METALLURGICAL LAB. (REF. 10)	CN-(LETTER(S))-
CHICAGO. UNIV. METALLURGICAL LAB. (REF. 10)	CNN-
CHICAGO. UNIV. METALLURGICAL LAB. (REF. 10)	CNP-
CHICAGO. UNIV. METALLURGICAL LAB. (REF. 10)	CNS-
CHICAGO. UNIV. METALLURGICAL LAB. (REF. 10)	CNSX-
CHICAGO. UNIV. METALLURGICAL LAB. (REF. 10)	CP-(LETTER(S))-
CHICAGO. UNIV. METALLURGICAL LAB. (ASSIGNED TO REPORTS FROM GREAT BRITAIN) (REF. 10)	CPB-
CHICAGO. UNIV. METALLURGICAL LAB. (REF. 10)	CPN-
CHICAGO. UNIV. METALLURGICAL LAB. (REF. 10)	CPX-
CHICAGO. UNIV. METALLURGICAL LAB.	CS-
CHICAGO. UNIV. METALLURGICAL LAB. (REF. 10)	CS-I-
CHICAGO. UNIV. METALLURGICAL LAB. (REF. 10)	CS-P-
CHICAGO. UNIV. METALLURGICAL LAB. (ASSIGNED TO REPORTS FROM GREAT BRITAIN) (REF. 10)	CTB-
CHICAGO. UNIV. METALLURGICAL LAB. (REF. 10)	CTM-
CHICAGO. UNIV. METALLURGICAL LAB.	CTN-
CHICAGO. UNIV. METALLURGICAL LAB.	CU MLR-
CHICAGO. UNIV. METALLURGICAL LAB. (REF. 10)	G-(NUMBER)-C(LETTER)
CHICAGO. UNIV. METALLURGICAL LAB. (REF. 10)	M-(NUMBER)-C(LETTER)
CHICAGO. UNIV. METALLURGICAL LAB.	MLR-
CHICAGO. UNIV. METALLURGICAL LAB.	MUC-(INITIALS)-
CHICAGO. UNIV. METALLURGICAL LAB.	PC-
CHICAGO. UNIV. METALLURGICAL LAB.	SC-PC-
CHICAGO. UNIV. METALLURGICAL LAB.	UCHI-CC-
CHICAGO. UNIV. METALLURGICAL LAB. (REF. 10)	W-
CHICAGO. UNIV. METALLURGICAL LAB. (REF. 10)	X-
(SEE ALSO LATER NAME: ARGONNE NATIONAL LAB.)	
CHICAGO. UNIV. NATIONAL OPINION RESEARCH CENTER	NORC-
CHICAGO. UNIV. PSYCHOMETRIC LAB.	CU PL-
CHICAGO. UNIV. SATELLITE + MESOMETEOROLOGY RES. PROJ.	SMRP-
CHICAGO. UNIV. SATELLITE + MESOMETEOROLOGY RES. PROJ.	SMRP-RP-
CHICAGO. UNIV. TOXICITY LAB.	CU TLR-
CHICAGO. UNIV. TOXICITY LAB.	TLR-
CHICAGO AERIAL INDUSTRIES, INC. BARRINGTON, ILL.	CAI-
CHICAGO AERIAL INDUSTRIES, INC. BARRINGTON, ILL.	CAI-(NUMBER)-(LETTER)
CHICAGO AERIAL INDUSTRIES, INC. BARRINGTON, ILL.	CAI-(NUMBER)-ITRD-
CHICAGO BRIDGE AND IRON CO., OAK BROOK, ILL.	CBI-
CHICAGO DEVELOPMENT CORP., RIVERDALE, MD.	CDC-
CHICAGO OPERATIONS OFFICE, AEC	CO-
CHICAGO OPERATIONS OFFICE, AEC (REF. 4)	COO-
CHICAGO OPERATIONS OFFICE, AEC (REF. 4)	COO-(CONTRACT NUMBER)-
CHICAGO OPERATIONS OFFICE, AEC (REF. 4)	COOS-(CONTRACT NUMBER)-
CHICAGO OPERATIONS OFFICE, AEC	RD-
CHROMATIC TELEVISION LABS., INC., N.Y.C.	CTL-
CHRYSLER CORP., CENTER LINE, MICH.	AD-
CHRYSLER CORP., CENTER LINE, MICH.	SD-
CHRYSLER CORP., DETROIT	51-(LETTER)-
CHRYSLER CORP., DETROIT	ADB-TN-
CHRYSLER CORP., DETROIT	CC-
CHRYSLER CORP., DETROIT	CC-HEC-D(NO.)
CHRYSLER CORP., DETROIT	CC-PR-
CHRYSLER CORP., DETROIT	CC-TB-
CHRYSLER CORP., DETROIT	CC-TM-AME-M(NUMBER)
CHRYSLER CORP., DETROIT	CC-TR-
CHRYSLER CORP., DETROIT	CHRYSLER-(LETTERS...)
CHRYSLER CORP., DETROIT	CWO-
CHRYSLER CORP., NEW ORLEANS (CHRYSLER IMPROVED NUMERICAL DIFFERENCING ANALYZER)	CINDA-
CHRYSLER CORP., NEW ORLEANS	TN-AP-(YEAR)-
CHRYSLER CORP. AMPLEX DIV., DETROIT	E-
CHRYSLER CORP. DEFENSE OPERATIONS DIV., DETROIT	JTF-
CHRYSLER CORP. ELECTRICAL AND ELECTRONIC ENG. DEPT., HUNTSVILLE, ALA.	HEE-M-(NUMBER-YEAR)
CHRYSLER CORP. ENGINEERING DIV., DETROIT	CC-AD-
CHRYSLER CORP. ENGINEERING DIV., DETROIT	CC ED TR-
CHRYSLER CORP. ENGINEERING DIV., DETROIT	CCED TR-
CHRYSLER CORP. HUNTSVILLE SPACE OPERATION, ALA.	HMD-
CHRYSLER CORP. HUNTSVILLE SPACE OPERATION, ALA.	HMD-R(NUMBER-YEAR)
CHRYSLER CORP. MISSILE DIV.	AME-
CHRYSLER CORP. MISSILE DIV.	CC-AME-(LETTERS)-
CHRYSLER CORP. MISSILE DIV., DETROIT	ADB-
CHRYSLER CORP. MISSILE DIV., DETROIT	BNE-(NUMBER)-
CHRYSLER CORP. MISSILE DIV., DETROIT	BRA-(NUMBERS)-(YEAR)
CHRYSLER CORP. MISSILE DIV., DETROIT	CC-BIB-
CHRYSLER CORP. MISSILE DIV., DETROIT	CCMD-
CHRYSLER CORP. MISSILE DIV., DETROIT	CC-MT-

CHRYSLER CORP. MISSILE DIV.

CHRYSLER CORP. MISSILE DIV., DETROIT	CC-MTC-
CHRYSLER CORP. MISSILE DIV., DETROIT	CC-MT-M(NO.)
CHRYSLER CORP. MISSILE DIV., DETROIT	CC-PR-(NUMBER,LETTER)-
CHRYSLER CORP. MISSILE DIV., DETROIT	CC-RL-
CHRYSLER CORP. MISSILE DIV., DETROIT	CC-TR-(LTRS)-
CHRYSLER CORP. MISSILE DIV., DETROIT	CR-(NUMBER)-
CHRYSLER CORP. MISSILE DIV., DETROIT	CR-D-
CHRYSLER CORP. MISSILE DIV., DETROIT	FRAS-(NUMBERS)-(YEAR)
CHRYSLER CORP. MISSILE DIV., DETROIT	IG-
CHRYSLER CORP. MISSILE DIV., DETROIT	HM-
CHRYSLER CORP. MISSILE DIV., DETROIT	ISC-
CHRYSLER CORP. MISSILE DIV., DETROIT	MT-
CHRYSLER CORP. MISSILE DIV., DETROIT	RL-DG-
CHRYSLER CORP. MISSILE DIV., DETROIT	RL-FG-
CHRYSLER CORP. MISSILE DIV., DETROIT	TR-DAT-
CHRYSLER CORP. MISSILE DIV., HUNTSVILLE, ALA.	HSM-N(NUMBER)-(YEAR)
CHRYSLER CORP. MISSILE OPERATIONS (REF. 1)	(NUMBERS...)
CHRYSLER CORP. MISSILE OPERATIONS (REDSTONE ARTILLERY MISSILE)	AM-R(NO.)-
CHRYSLER CORP. MISSILE OPERATIONS	CC-EL-
CHRYSLER CORP. MISSILE OPERATIONS (WEEKLY ENGINEERING PROGRESS REPORT)	CC-WEPR-
CHRYSLER CORP. MISSILE OPERATIONS	EL-R-
CHRYSLER CORP. MISSILE OPERATIONS	MP-
CHRYSLER CORP. MISSILE OPERATIONS	MT-R-
CHRYSLER CORP. MISSILE OPERATIONS, DETROIT	CHR-
CHRYSLER CORP. MISSILE OPERATIONS. GUIDANCE AND CONTROL DESIGN DEPT.	GC-R(NUMBER)
CHRYSLER CORP. SPACE DIV., HUNTSVILLE, ALA.	HSM-R-
CHRYSLER CORP. SPACE DIV., NEW ORLEANS	REL-HFG-
CHRYSLER CORP. SPACE DIV., NEW ORLEANS	SDES-(YEAR)-
CHRYSLER CORP. SPACE DIV., NEW ORLEANS	TR-RE-
CINCINNATI. UNIV.	AE-
CINCINNATI. UNIV.	CIN-
CINCINNATI. UNIV.	CINU-
CINCINNATI. UNIV.	DAE-
CINCINNATI. UNIV. DEPT. OF AEROSPACE ENG.	TR-AE-
CINCINNATI TEST LABORATORY	CTL-(NO.-LETTER-NO.)
CITIZENS ADVISORY COMMITTEE ON TRANSPORTATION QUALITY, WASHINGTON, D.C.	CACTQ-
CITY COLL., NEW YORK	CCNY-
CITY-COUNTY PLANNING COMMISSION, ROCKFORD, ILL.	RWCCPC-(YEAR)-
CITY UNIV., NEW YORK	CUNY-
CIVIL AEROMEDICAL RESEARCH INSTITUTE, WASHINGTON, D.C. (CODE USED UNTIL 1963)	CARI-
CIVIL AERONAUTICS ADMINISTRATION, WASHINGTON, D.C.	CAA-
CIVIL AERONAUTICS BOARD, WASHINGTON, D.C.	BOSR-(NUMBER)-
CIVIL AERONAUTICS BOARD, WASHINGTON, D.C.	CAB-
CIVIL SERVICE COMMISSION, WASHINGTON, D.C. (BULLETIN)	BTN-
CIVIL SERVICE COMMISSION, WASHINGTON, D.C.	CSC-
CIVIL SERVICE COMMISSION, WASHINGTON, D.C. (HANDBOOK)	CSCHBX-
CIVIL SERVICE COMMISSION, WASHINGTON, D.C. (JOB EMPLOYMENT METHOD)	CSCJEMPS-
CIVIL SERVICE COMMISSION, WASHINGTON, D.C. (POSTER)	CSCP-
CIVIL SERVICE COMMISSION, WASHINGTON, D.C. (FEDERAL EMPLOYEE FACTS)	FEF-
CIVIL SERVICE COMMISSION, WASHINGTON, D.C. (FIELD PERSONNEL MANUAL)	FPM-
CIVIL SERVICE COMMISSION, WASHINGTON, D.C. (POSITION CLASSIFICATION STD.)	GS-
CIVIL SERVICE COMMISSION, WASHINGTON, D.C. (PERSONNEL MANAGEMENT SERIES)	PMS-
CIVIL SERVICE COMMISSION, WASHINGTON, D.C.	USCSC-
CIVIL SERVICE COMMISSION, WASHINGTON, D.C. (WORK ANNOUNCEMENT)	WA-
CIVIL SERVICE COMMISSION, WASHINGTON, D.C. (WORK ANNOUNCEMENT)	WAM-
CIVIL SERVICE COMMISSION, WASHINGTON, D.C. (WORK ANNOUNCEMENT)	WAW-
CIVIL SERVICE COMMISSION, WASHINGTON, D.C. (WORK GRADES)	WG-
CLARK BROS. CO. DEVELOPMENT AND PRODUCTION TEST DEPT., OLEAN, N.Y.	DDR-
CLARK, COOPER, FIELD AND WOHL, INC., NEW YORK CITY	METRI-(NUMBER)-
CLARK UNIV., WORCESTER, MASS.	CLU-
CLARK UNIV., WORCESTER, MASS. JEPPSON LAB.	CLARKU-
CLARKSON COLL. OF TECHNOLOGY, POTSDAM, N.Y.	CPDD-(YEAR)-
CLEARINGHOUSE FOR FEDERAL SCIENTIFIC AND TECHNICAL INFORMATION, SPRINGFIELD, VA.	CFSTI-(YEAR)-
CLEARINGHOUSE FOR FEDERAL SCIENTIFIC AND TECHNICAL INFORMATION, SPRINGFIELD, VA.	CFSTI-BIB-(YEAR)-
CLEARINGHOUSE FOR FEDERAL SCIENTIFIC AND TECHNICAL INFO.,SPRINGFIELD,VA.(FAST ANNOUNCEMENT SERVICE)	FAS-
CLEARINGHOUSE FOR FEDERAL SCIENTIFIC AND TECHNICAL INFORMATION, SPRINGFIELD, VA.(NON-CLEARINGHOUSE)	NCH-
CLEARINGHOUSE FOR FEDERAL SCIENTIFIC AND TECHNICAL INFORMATION, SPRINGFIELD, VA. (REPORT BIBLIOGRAPHY)	RB-
CLEARINGHOUSE FOR FEDERAL SCIENTIFIC AND TECHNICAL INFORMATION, SPRINGFIELD, VA.	SB-
CLEARINGHOUSE FOR FEDERAL SCIENTIFIC AND TECHNICAL INFORMATION, SPRINGFIELD, VA.(STATE TECH. SERVICES)	STS-
(SEE ALSO LATER NAME: NATIONAL TECH. INFO. SERVICE)	
CLEARINGHOUSE FOR FEDERAL SCIENTIFIC AND TECH. INFO. OFFICE OF TECH. RESOURCES, SPRINGFIELD, VA.	OTR-
CLEAVER-HUME PRESS, LTD., LONDON (TRANSLATIONS)	CHP-
CLEMSON UNIV., S.C. ENGINEERING EXPERIMENT STATION	BULL-
CLEMSON UNIV., S.C. WATER RESOURCES RESEARCH INST.	WRRI-
CLERMONT-FERRAND UNIV., FRANCE. LABORATORIE DE PHYSIQUE CORPUSCULAIRE	PCCF-RI-(YEAR/NUMBER)
CLEVELAND DIESEL ENGINE DIV., GEN. MOTORS CORP.,OHIO	CDED-
CLEVELAND INDUSTRIAL RESEARCH, INC.	NF-
CLEVELAND PNEUMATIC TOOL CO. (REF. 1)	(PART NUMBER)
CLEVELAND PNEUMATIC TOOL CO. (REF. 1)	(PART NUMBER(S))-(LTRS)

CLEVITE CORP., CLEVELAND	CLC-
CLEVITE CORP., CLEVELAND	CLEV-
CLEVITE CORP., CLEVELAND	CLEVITE-PR-
CLEVITE CORP., CLEVELAND	CLEVITE-QPR-
CLEVITE CORP., CLEVELAND	CLEVITE-TR-
CLEVITE CORP. MECHANICAL RESEARCH DIV., CLEVELAND	QR-(NUMBER)-
CLEVITE RESEARCH CENTER, CLEVELAND	CRCC-
CLIFTON PRODUCTS, INC., PAINESVILLE, OHIO	CPI-
CLIMAX MOLYBDENUM CO. OF MICH., DETROIT	CMC-
CLINICAL INVESTIGATION CENTER, OAKLAND, CALIF.	CIC-TR-
CLINTON ENGINEER WORKS, MANHATTAN DISTRICT, OAK RIDGE, TENN.	FE-
CLINTON ENGINEER WORKS, MANHATTAN DISTRICT, OAK RIDGE, TENN.	PP1-B-
CLINTON ENGINEER WORKS, MANHATTAN DISTRICT, OAK RIDGE, TENN.	RTD-LA
CLINTON ENGINEER WORKS, MANHATTAN DISTRICT, OAK RIDGE, TENN.	TE-B-3-
CLINTON LABS., OAK RIDGE, TENN.	B-
CLINTON LABS., OAK RIDGE, TENN. (REF. 10)	CC-
CLINTON LABS., OAK RIDGE, TENN.	CC-CJB-
CLINTON LABS., OAK RIDGE, TENN. (REF. 10)	CE-
CLINTON LABS., OAK RIDGE, TENN.	CF-(YR-MO)-
CLINTON LABS., OAK RIDGE, TENN. (REF. 10)	CH-
CLINTON LABS., OAK RIDGE, TENN. (REF. 10)	CL-
CLINTON LABS., OAK RIDGE, TENN.	CL-(INITIALS)-
CLINTON LABS., OAK RIDGE, TENN.	CLM-(INITIALS)-
CLINTON LABS., OAK RIDGE, TENN. (REF. 10)	CLM-IIC-
CLINTON LABS., OAK RIDGE, TENN.	CN-
CLINTON LABS., OAK RIDGE, TENN. (REF. 10)	CNL-
CLINTON LABS., OAK RIDGE, TENN.	CS-
CLINTON LABS., OAK RIDGE, TENN.	CSB-
CLINTON LABS., OAK RIDGE, TENN.	CSN-
CLINTON LABS., OAK RIDGE, TENN.	HP-
CLINTON LABS., OAK RIDGE, TENN.	MON-
CLINTON LABS., OAK RIDGE, TENN.	MON-(LETTER)-
CLINTON LABS., OAK RIDGE, TENN.	MPR-CDC-
CLINTON LABS., OAK RIDGE, TENN.	N-
CLINTON LABS., OAK RIDGE, TENN.	Z-
COAL TAR RESEARCH ASSN., LEEDS, YORKS, ENG. (TRANS.)	CTRA-
COAST AND GEODETIC SURVEY, FREDERICKSBURG, VA.	C/GSDR-
COAST AND GEODETIC SURVEY, LAS VEGAS, NEV.	CGS-C-
COAST AND GEODETIC SURVEY, LAS VEGAS, NEV.	CGS-E-
COAST AND GEODETIC SURVEY, LAS VEGAS, NEV.	CGS-O-
COAST AND GEODETIC SURVEY, LAS VEGAS, NEV.	CGS-P-
COAST AND GEODETIC SURVEY, ROCKVILLE, MD.	C/GS-
COAST AND GEODETIC SURVEY, ROCKVILLE, MD.	CGS-C-
COAST AND GEODETIC SURVEY, ROCKVILLE, MD.	CGS-P-
COAST AND GEODETIC SURVEY, ROCKVILLE, MD. (TECH MEMO)	C/GSTM-
COAST AND GEODETIC SURVEY, ROCKVILLE, MD. (DENSITY OF SEA WATER)	DW-
COAST AND GEODETIC SURVEY, ROCKVILLE, MD. (PRELIMINARY GEODETIC PUBL.)	G-
COAST AND GEODETIC SURVEY, ROCKVILLE, MD. (HOURLY VALUES)	HV-
COAST AND GEODETIC SURVEY, ROCKVILLE, MD. (MAGNETIC DECLINATION)	MD-
COAST AND GEODETIC SURVEY, ROCKVILLE, MD. (MAGNETOGRAMS)	MG-
COAST AND GEODETIC SURVEY, ROCKVILLE, MD. (MAGNETOGRAMS AND HOURLY VALUES)	MHV-
COAST AND GEODETIC SURVEY, ROCKVILLE, MD. (MAGNETIC OBSERVATIONS)	MO-
COAST AND GEODETIC SURVEY, ROCKVILLE, MD. (ABSTRACTS OF EARTHQUAKE REPORTS)	MSA-
COAST AND GEODETIC SURVEY, ROCKVILLE, MD. (SEISMOLOGICAL BULLETIN, ANTARCTICA)	MSI-(NUMBER)A
COAST AND GEODETIC SURVEY, ROCKVILLE, MD. (SEISMOLOGY BULLETIN)	MSP-
COAST AND GEODETIC SURVEY, ROCKVILLE, MD. (PRELIMINARY DETERMINATION OF EPICENTER)	PDE-
COAST AND GEODETIC SURVEY, ROCKVILLE, MD. (TIDAL BENCH MARKS)	T-
COAST AND GEODETIC SURVEY, ROCKVILLE, MD. (TIDAL HARMONIC CONSTANTS)	TH-
COAST AND GEODETIC SURVEY, ROCKVILLE, MD. (WATER TEMPERATURE)	TW-
(SEE ALSO SUCCESSOR: NATIONAL OCEAN SURVEY)	
COAST AND GEODETIC SURVEY, WASHINGTON, D.C. (REF. 1)	(NUMBERS...)
COAST AND GEODETIC SURVEY, WASHINGTON, D.C. (AERONAUTICAL PLANNING CHART)	AP-
COAST AND GEODETIC SURVEY, WASHINGTON, D.C. (BOTTOM SEDIMENT CHART)	BS-
COAST AND GEODETIC SURVEY, WASHINGTON, D.C. (RADIO DIRECTION FINDING CHARTS)	DF-
COAST AND GEODETIC SURVEY, WASH.,D.C. (FLIGHT CHARTS)	FC-
COAST AND GEODETIC SURVEY, WASHINGTON, D.C. (GLOBAL NAVIGATION CHART)	GNC-
COAST AND GEODETIC SURVEY, WASHINGTON, D.C. (INSTRUMENT LANDING SYSTEM CHART)	ILS-
COAST AND GEODETIC SURVEY, WASHINGTON, D.C. (JET NAVIGATION CHART)	JN-
COAST AND GEODETIC SURVEY, WASH., D.C. (LOCAL CHARTS)	LO-
COAST AND GEODETIC SURVEY, WASHINGTON, D.C.	MSI-
COAST AND GEODETIC SURVEY, WASHINGTON, D.C. (OPERATIONAL NAVIGATION CHART)	ONC-
COAST AND GEODETIC SURVEY, WASHINGTON, D.C. (RADIO FACILITY CHARTS)	RF-
COAST AND GEODETIC SURVEY, WASHINGTON, D.C. (SECTIONAL CHARTS, UNITED STATES)	SUS-
COAST AND GEODETIC SURVEY, WASHINGTON, D.C. (WORLD AERONAUTICAL CHART)	WAC-
COAST GUARD (YARD)	CGYD-
COAST GUARD, WASHINGTON, D.C.	CG-
COAST GUARD, WASHINGTON, D.C.	CGRD-
COAST GUARD, WASHINGTON, D.C.	CGTD-
COAST GUARD, WASHINGTON, D.C.	NAVCG-
COAST GUARD, WASHINGTON, D.C.	USCG-
COAST GUARD, WASHINGTON, D.C.	USCG-BULL-
COAST GUARD. OFFICE OF ENGINEERING, WASH., D.C.	USCG-PROGRAM-ENE-
COASTAL PLAINS CENTER FOR MARINE DEVELOPMENT SERVICES, WASHINGTON, D.C.	PUBL-(YEAR)-
COLE (E.K.), LTD.	EKCM-
COLEMAN ENGINEERING CO., INC., LOS ANGELES	COLE-
COLES SIGNAL LAB., BELMAR, N.J.	CSL-
COLES SIGNAL LAB., BELMAR, N.J. (ENGINEERING REPT.)	E-
COLES SIGNAL LAB., BELMAR, N.J. (TECH. MEMO.)	M-
COLES SIGNAL LAB., BELMAR, N.J. (TEST REPORT)	T-
COLLEGE DE FRANCE, PARIS. LABORATOIRE DE PHYSIQUE ATOMIQUE ET MOLECULAIRE	LPAM-RI-

COLLEGE DE FRANCE, LABORATOIRE DE PHYSIQUE ATOMIQUE ET MOLECULAIRE

COLLEGE DE FRANCE, PARIS. LABORATOIRE DE PHYSIQUE ATOMIQUE ET MOLECULAIRE	PAM-(YEAR)-
COLLEGE OF AERONAUTICS, CRANFIELD, BUCKS, ENGLAND	CA-
COLLEGE OF AERONAUTICS, CRANFIELD, BUCKS, ENGLAND	CAC-
COLLEGE OF AERONAUTICS, CRANFIELD, BUCKS, ENGLAND	COA-
COLLEGE OF AERONAUTICS, CRANFIELD, BUCKS, ENGLAND	COA-AERO-
COLLEGE OF AERONAUTICS, CRANFIELD, BUCKS, ENGLAND	COA-E+C-
COLLEGE OF AERONAUTICS, CRANFIELD, BUCKS, ENGLAND	COA-MAT-
COLLEGE OF AERONAUTICS, CRANFIELD, BUCKS, ENGLAND	COA-MATT-
COLLEGE OF AERONAUTICS, CRANFIELD, BUCKS, ENGLAND	COA-M+P-
COLLEGE OF AERONAUTICS, CRANFIELD, BUCKS, ENGLAND	COA-N-
COLLEGE OF AERONAUTICS, CRANFIELD, BUCKS, ENGLAND	COA-N-MAT.-
COLLEGE OF AERONAUTICS, CRANFIELD, BUCKS, ENGLAND	COA-NOTE-MAT-
COLLEGE OF AERONAUTICS, CRANFIELD, BUCKS, ENGLAND	COA-NOTE-M+P-
COLLEGE OF AERONAUTICS, CRANFIELD, BUCKS, ENGLAND	COA-NOTE-M/P-
COLLEGE OF AERONAUTICS, CRANFIELD, BUCKS, ENGLAND. DEPT. OF AERODYNAMICS	COA-NOTE-AERO-
COLLEGE OF AERONAUTICS, CRANFIELD, BUCKS, ENGLAND. DEPT. OF ELECTRICAL AND CONTROL ENGINEERING	COA-NOTE-E/C-
COLLEGE OF WILLIAM AND MARY, WILLIAMSBURG, VA.	WM-
COLLEGE OF WILLIAM AND MARY, WILLIAMSBURG, VA.	W+M-
COLLEGE OF WILLIAM AND MARY, WILLIAMSBURG, VA.	W/M-(2 DIGITS)
COLLEGE OF WILLIAM AND MARY, WILLIAMSBURG, VA.	WM-TR-
COLLINS RADIO CO., CEDAR RAPIDS, IOWA (APOLLO REPORT)	AR-
COLLINS RADIO CO., CEDAR RAPIDS, IOWA (CIVIL AERONAUTICS ADMINISTRATIVE TYPE CERTIFICATE)	CAATC-
COLLINS RADIO CO., CEDAR RAPIDS, IOWA (DEVELOPMENT DESCRIPTION)	CDD-
COLLINS RADIO CO., CEDAR RAPIDS, IOWA (DEV. REPORT)	CDR-
COLLINS RADIO CO., CEDAR RAPIDS, IOWA (DESCRIPTIVE SPECIFICATION)	CDS-
COLLINS RADIO CO., CEDAR RAPIDS, IOWA (DATA SYSTEMS PROPOSAL)	CDSP-
COLLINS RADIO CO., CEDAR RAPIDS, IOWA (ENG. LETTER)	CEL-
COLLINS RADIO CO., CEDAR RAPIDS, IOWA (ENG. ARTICLE)	CEP-
COLLINS RADIO CO., CEDAR RAPIDS, IOWA (ENG. REPORT)	CER-
COLLINS RADIO CO., CEDAR RAPIDS, IOWA	COLL-
COLLINS RADIO CO., CEDAR RAPIDS, IOWA	COLR-
COLLINS RADIO CO., CEDAR RAPIDS, IOWA (PRODUCT DESCRIPTION)	CPD-
COLLINS RADIO CO., CEDAR RAPIDS, IOWA	CR-
COLLINS RADIO CO., CEDAR RAPIDS, IOWA	CRC-
COLLINS RADIO CO., CEDAR RAPIDS, IOWA (RESEARCH RPT.)	CRR-
COLLINS RADIO CO., CEDAR RAPIDS, IOWA (TECHNICAL DESCRIPTION)	CTD-
COLLINS RADIO CO., CEDAR RAPIDS, IOWA (TECH. RPT.)	CTR-
COLLINS RADIO CO., CEDAR RAPIDS, IOWA (RESEARCH AND DEVELOPMENT PROPOSAL)	RDP-
COLLINS RADIO CO., CEDAR RAPIDS, IOWA (SERVICE DIVISION REPORT)	SDR-
COLLINS RADIO CO., DALLAS	CER-D(NUMBER)
COLLINS RADIO CO., DALLAS	IDR-D(NUMBER)-
COLLINS RADIO CO., NEWPORT BEACH, CALIF.	B-
COLOGNE. UNIVERSITAT.	COLOGNEU-
COLOMBIA. INSTITUTO DE ASUNTOS NUCLEARES, BOGOTA	IAN-
COLOMBIA. INSTITUTO DE ASUNTOS NUCLEARES, BOGOTA	IAN-ARI-
COLOMBIA. INSTITUTO DE ASUNTOS NUCLEARES, BOGOTA	IAN-F-
COLOMBIA. INSTITUTO DE ASUNTOS NUCLEARES, BOGOTA	IAN-Q-
COLOMBIA. INSTITUTO DE ASUNTOS NUCLEARES, DIVISION DE RADIOFISICA SANITARIA, BOGOTA	IAN-SR-
COLOMBIA. INSTITUTO DE ASUNTOS NUCLEARES. DIVISION OF MEDICINE AND BIOLOGY, BOGOTA	IAN-B-
COLOMBIA. INSTITUTO DE ASUNTOS NUCLEARES. DIVISION OF REACTORS, BOGOTA	IAN-A-
COLOMBIA. INSTITUTO DE ASUNTOS NUCLEARES. SECCION DE RADIOFISICA SANITARIA, BOGOTA	IAN-RS-ERA-
COLORADO. UNIV., BOULDER	APL-
COLORADO. UNIV., BOULDER	CO-
COLORADO. UNIV., BOULDER	COLU-
COLORADO. UNIV., BOULDER	KIDM-
COLORADO. UNIV., BOULDER	SAR-
COLORADO. UNIV., BOULDER	UCO-
COLORADO. UNIV., BOULDER	UCOL-(NUMBER)-
COLORADO. UNIV., BOULDER	UCOL-(YEAR)-
COLORADO. UNIV., BOULDER	UCOL-P-
COLORADO. UNIV., BOULDER. DEPT. OF MECH. ENG. (REF.1)	(NUMBER)
COLORADO. UNIV., BOULDER. DEPT. OF PHYSICS (REF. 1)	(NUMBER)
COLORADO. UNIV., BOULDER. ENGINEERING EXPERIMENT STA.	CO-29(601)-(NOS...)-
COLORADO. UNIV., BOULDER. ENGINEERING EXPERIMENT STA.	ONR-BOS-
COLORADO. UNIV., CLIMAX. INST. FOR SOLAR TERRESTRIAL RESEARCH	ISTR-TR-
COLORADO. UNIV., DENVER. MEDICAL CENTER	UCO-
COLORADO FUEL + IRON CORP., PUEBLO	CF+I-
COLORADO RESEARCH CORP., BROOMFIELD	COLRC-
COLORADO SCHOOL OF MINES, GOLDEN	CSM-
COLORADO SCHOOL OF MINES, GOLDEN	T-(NO.)
COLORADO SCHOOL OF MINES, GOLDEN. MINING RESEARCH LAB	CSM-MRL-ONR-
COLORADO SCHOOL OF MINES RESEARCH FDN., INC., GOLDEN	CS-
COLORADO SCHOOL OF MINES RESEARCH FDN., INC., GOLDEN	CSM-
COLORADO SCHOOL OF MINES RESEARCH FDN., INC., GOLDEN	CSM-PR-
COLORADO SCHOOL OF MINES RESEARCH FOUNDATION, INC., GOLDEN. (CONTRIBUTIONS)	CSM R(YR.)-
COLORADO STATE UNIV., FORT COLLINS	CER-(NO.)-
COLORADO STATE UNIV., FORT COLLINS	COLOU-
COLORADO STATE UNIV., FORT COLLINS	CSU-
COLORADO STATE UNIV., FORT COLLINS	MEP(YR.-YR.)(INITIALS)-
COLORADO STATE UNIV., FORT COLLINS. COLLABORATIVE RADIOLOGICAL HEALTH LAB.	CRHL-
COLORADO STATE UNIV., FORT COLLINS. COLL. OF ENG.	MER(NO.)-(YR.)-(INITLS.)
COLORADO STATE UNIV., FORT COLLINS. DEPT. OF CIVIL ENGINEERING	CER(YR.-YR.)(INITIALS)-

COLORADO STATE UNIV., FORT COLLINS. DEPT. OF CIVIL ENGINEERING	CER(YEAR)(INITIALS)-
COLORADO STATE UNIV., FORT COLLINS. DEPT. OF CIVIL ENGINEERING	HYDROLOGY PAPERS-
COLORADO STATE UNIV., FORT COLLINS. DEPT. OF MICROBIOLOGY	SANITARY ENG-
COLUMBIA BROADCASTING SYSTEM, N.Y.C.	CBS-
COLUMBIA BROADCASTING SYSTEM, N.Y.C. (ENG. RES.+DEV.)	CBS ERD-
COLUMBIA BROADCASTING SYSTEM, N.Y.C. (ENG. RES.+DEV.)	ERD-
COLUMBIA BROADCASTING SYSTEM, INC. CBS LABORATORIES, STAMFORD, CONN.	CBS-MLR-
COLUMBIA RESEARCH AND DEV. CORP., COLUMBUS, OHIO	CRDC-
COLUMBIA TECHNICAL TRANSLATIONS, WHITE PLAINS, N.Y.	CTT-
COLUMBIA UNIV., DOBBS FERRY, N.Y.	CONTRIB-
COLUMBIA UNIV., DOBBS FERRY, N.Y. HUDSON LABS.	ARTEMIS-
COLUMBIA UNIV., IRVINGTON-ON-HUDSON, N.Y. NEVIS CYCLOTRON LABS.	AEC-CU-
COLUMBIA UNIV., IRVINGTON-ON-HUDSON, N.Y. NEVIS LABS.	CU NEVIS-
COLUMBIA UNIV., IRVINGTON-ON-HUDSON, N.Y. NEVIS LABS.	NEVIS-
COLUMBIA UNIV., IRVINGTON-ON-HUDSON, N.Y. NEVIS LABS.	R-
COLUMBIA UNIV., N.Y.C.	BT-R-
COLUMBIA UNIV., N.Y.C.	CER-
COLUMBIA UNIV., N.Y.C.	COL. SERIAL-
COLUMBIA UNIV., N.Y.C.	COLUMBIA SERIAL--
COLUMBIA UNIV., N.Y.C.	CR-
COLUMBIA UNIV., N.Y.C. (4 DIGITS ARE CONTRACT NO.)	CU-(4 DIGITS)-
COLUMBIA UNIV., N.Y.C.	CU-(NO)-(YR)-(CON)-(DEPT)
COLUMBIA UNIV., N.Y.C.	CU-2-53-AEC-314-CHEM.
COLUMBIA UNIV., N.Y.C. (ALL DEPTS.)	CU-(3 DIGITS)-
COLUMBIA UNIV., N.Y.C.	CU-
COLUMBIA UNIV., N.Y.C.	CUD-
COLUMBIA UNIV., N.Y.C.	CUF-
COLUMBIA UNIV., N.Y.C.	CUG-
COLUMBIA UNIV., N.Y.C.	CUN-
COLUMBIA UNIV., N.Y.C.	CU TR-
COLUMBIA UNIV., N.Y.C. (ALL DEPTS.)	CU-TR-(NO.)/(CON. NO.)
COLUMBIA UNIV., N.Y.C.	CU-TR-(NUMBER-YEAR)
COLUMBIA UNIV., N.Y.C.	DR-
COLUMBIA UNIV., N.Y.C.	ID-M-
COLUMBIA UNIV., N.Y.C.	IEI-
COLUMBIA UNIV., N.Y.C.	MTPR-
COLUMBIA UNIV., N.Y.C.	P-
COLUMBIA UNIV., N.Y.C.	RCC-
COLUMBIA UNIV., N.Y.C.	RD-
COLUMBIA UNIV., N.Y.C. ACOUSTICS LAB.	AL-
COLUMBIA UNIV., N.Y.C. AIRBORNE INSTRUMENTS LAB.	AIL-
COLUMBIA UNIV., N.Y.C. AIRBORNE INSTRUMENTS LAB.	CUN AIL-
COLUMBIA UNIV., N.Y.C. APPLIED MATH. GROUP (REF. 28)	AMG-C-
COLUMBIA UNIV., N.Y.C. APPLIED MATH. GROUP (REF. 28)	AMP RPT.-
COLUMBIA UNIV., N.Y.C. APPLIED MATH. GROUP (REF. 28)	NDRC AMG-C-
COLUMBIA UNIV., N.Y.C. APPLIED MATH. GROUP (REF. 28)	NDRC AMG-C-MEMO-
COLUMBIA UNIV., N.Y.C. CINDA CENTRE	CINDA-
COLUMBIA UNIV., N.Y.C. COLL. OF PHYSICIANS + SURGEONS	P + S-
COLUMBIA UNIV., N.Y.C. COLUMBIA ASTROPHYSICS LAB.	CAL-CONTRIB-
COLUMBIA UNIV., N.Y.C. COLUMBIA RADIATION LAB.	CRL-
COLUMBIA UNIV., N.Y.C. COLUMBIA RADIATION LAB.	CU-(NO.)-(YR)-SC-
COLUMBIA UNIV., N.Y.C. COLUMBIA RADIATION LAB.	CU-(NUMBER-YEAR)-AMC-
COLUMBIA UNIV., N.Y.C. COLUMBIA RADIATION LAB.	CU-CRL-QPR-
COLUMBIA UNIV., N.Y.C. COLUMBIA RADIATION LAB.	CU-PR-
COLUMBIA UNIV., N.Y.C. DEPT. OF CHEMICAL ENGINEERING	SAPR-
COLUMBIA UNIV., N.Y.C. DEPT. OF ELECTRICAL ENG.	CU-(YR.)-(CONTRACT NO.)-
COLUMBIA UNIV., N.Y.C. DEPT. OF ELECTRICAL ENG.	CU-(NO.)-AF-(NO.)-EE
COLUMBIA UNIV., N.Y.C. DEPT. OF ELECTRICAL ENG.	CU-EE-
COLUMBIA UNIV., N.Y.C. DEPT. OF MATHEMATICS	CU-(NUMBER)-ONR-
COLUMBIA UNIV., N.Y.C. DEPT. OF MATHEMATICS	CU-(NO.)-NONR-K-M
COLUMBIA UNIV., N.Y.C. DIV. OF WAR RESEARCH	A-
COLUMBIA UNIV., N.Y.C. DIV. OF WAR RESEARCH	CUDWR-
COLUMBIA UNIV., N.Y.C. DIV. OF WAR RES. (MEMORANDA)	CUN DWR M-
COLUMBIA UNIV., N.Y.C. DIV. OF WAR RES. (MEMORANDA)	CUN DWR M (LTR.,NO./LTR)-
COLUMBIA UNIV., N.Y.C. DIV. OF WAR RESEARCH	CUN DWR R-
COLUMBIA UNIV., N.Y.C. DIV. OF WAR RESEARCH	CUN DWR R (LTR.,NO./LTR)-
COLUMBIA UNIV., N.Y.C. DIV. OF WAR RESEARCH	DWR-
COLUMBIA UNIV., N.Y.C. DIV. OF WAR RESEARCH	IXR-
COLUMBIA UNIV., N.Y.C. DIV. OF WAR RESEARCH	TH-R-
COLUMBIA UNIV., N.Y.C. ELECTRONICS RESEARCH LABS.	CUERL-F/
COLUMBIA UNIV., N.Y.C. ELECTRONICS RESEARCH LABS.	F-
COLUMBIA UNIV., N.Y.C. ELECTRONICS RESEARCH LABS.	P-
COLUMBIA UNIV., N.Y.C. ELECTRONICS RESEARCH LABS.	T-
COLUMBIA UNIV., N.Y.C. ELECTRONICS RESEARCH LABS.	T-(NUMBER/NUMBER)
COLUMBIA UNIV., N.Y.C. ELECTRONICS RESEARCH LABS.	T-(NUMBER)/(LETTER)
COLUMBIA UNIV., N.Y.C. ENGINEERING RESEARCH LABS.	BPR-(ROMAN NO.)-
COLUMBIA UNIV., N.Y.C. ENGINEERING RESEARCH LABS.	IX-QPR-
COLUMBIA UNIV., N.Y.C. ENGINEERING RESEARCH LABS.	IX-TN-
COLUMBIA UNIV., N.Y.C. ENGINEERING RESEARCH LABS. (MONTHLY PROGRESS REPORT)	MPR-(ROMAN NO.)-
COLUMBIA UNIV., N.Y.C. ENGINEERING RESEARCH LABS.	MRP-
COLUMBIA UNIV., N.Y.C. MATHEMATICAL PHYSICS GROUP	MPG-
COLUMBIA UNIV., N.Y.C. PEGRAM NUCLEAR PHYSICS LABS.	CU(PNPL)-
COLUMBIA UNIV., N.Y.C. PUPIN PHYSICS LABS.	DR-

COLUMBIA UNIV., SCHOOL OF ENGINEERING

COLUMBIA UNIV., N.Y.C. SCHOOL OF ENGINEERING	CUSE-
COLUMBIA UNIV., N.Y.C. STATISTICAL RESEARCH GROUP	SRG-
COLUMBIA UNIV., N.Y.C. STATISTICAL RESEARCH GROUP	SRG MEMO-
COLUMBIA UNIV. N.Y.C. SUBSTITUTE ALLOY MATERIALS LAB. (REF. 11)	1-(LETTER)-
COLUMBIA UNIV. N.Y.C. SUBSTITUTE ALLOY MATERIALS LAB. (REF. 11)	1(LETTER)-
COLUMBIA UNIV. N.Y.C. SUBSTITUTE ALLOY MATERIALS LAB. (REF. 11)	1(LETTER)-(LETTER)-
COLUMBIA UNIV. N.Y.C. SUBSTITUTE ALLOY MATERIALS LAB. (REF. 11)	1(LETTER)(NUMBER)-(LTR)-
COLUMBIA UNIV. N.Y.C. SUBSTITUTE ALLOY MATERIALS LAB. (REF. 11)	1(LETTER)(NUMBER)(LTR)-
COLUMBIA UNIV. N.Y.C. SUBSTITUTE ALLOY MATERIALS LAB. (REF. 11)	1(LETTER)(LETTER)-
COLUMBIA UNIV. N.Y.C. SUBSTITUTE ALLOY MATERIALS LAB. (REF. 11)	2-(LETTER)-
COLUMBIA UNIV. N.Y.C. SUBSTITUTE ALLOY MATERIALS LAB. (REF. 11)	2(LETTER)-
COLUMBIA UNIV. N.Y.C. SUBSTITUTE ALLOY MATERIALS LAB. (REF. 11)	2(LETTER)(NUMBER)-(LTR)-
COLUMBIA UNIV. N.Y.C. SUBSTITUTE ALLOY MATERIALS LAB. (REF. 11)	2(LETTER)(NUMBER)(LTR)-
COLUMBIA UNIV. N.Y.C. SUBSTITUTE ALLOY MATERIALS LAB. (REF. 11)	2(LETTER)-(LETTER)-
COLUMBIA UNIV. N.Y.C. SUBSTITUTE ALLOY MATERIALS LAB. (REF. 11)	2(LETTER)(LETTER)-
COLUMBIA UNIV. N.Y.C. SUBSTITUTE ALLOY MATERIALS LAB. (REF. 11)	3-(LETTER)-
COLUMBIA UNIV. N.Y.C. SUBSTITUTE ALLOY MATERIALS LAB. (REF. 11)	3(LETTER)-
COLUMBIA UNIV. N.Y.C. SUBSTITUTE ALLOY MATERIALS LAB. (REF. 11)	3(LETTER)-(LETTER)-
COLUMBIA UNIV. N.Y.C. SUBSTITUTE ALLOY MATERIALS LAB. (REF. 11)	3(LETTER)(LETTER)-
COLUMBIA UNIV. N.Y.C. SUBSTITUTE ALLOY MATERIALS LAB. (REF. 11)	4-(LETTER)-
COLUMBIA UNIV. N.Y.C. SUBSTITUTE ALLOY MATERIALS LAB. (REF. 11)	4(LETTER)-
COLUMBIA UNIV. N.Y.C. SUBSTITUTE ALLOY MATERIALS LAB. (REF. 11)	5-(LETTER)-
COLUMBIA UNIV. N.Y.C. SUBSTITUTE ALLOY MATERIALS LAB. (REF. 11)	5(LETTER)-
COLUMBIA UNIV. N.Y.C. SUBSTITUTE ALLOY MATERIALS LAB. (REF. 11)	5(LETTER)(NUMBER)-(LTR)-
COLUMBIA UNIV. N.Y.C. SUBSTITUTE ALLOY MATERIALS LAB. (REF. 11)	5(LETTER)(NUMBER)(LTR)-
COLUMBIA UNIV. N.Y.C. SUBSTITUTE ALLOY MATERIALS LAB. (REF. 11)	5(LETTER)-(LETTER)-
COLUMBIA UNIV. N.Y.C. SUBSTITUTE ALLOY MATERIALS LAB. (REF. 11)	5(LETTER)(LETTER)-
COLUMBIA UNIV. N.Y.C. SUBSTITUTE ALLOY MATERIALS LAB. (REF. 11)	6-(LETTER)-
COLUMBIA UNIV. N.Y.C. SUBSTITUTE ALLOY MATERIALS LAB. (REF. 11)	6(LETTER)-
COLUMBIA UNIV. N.Y.C. SUBSTITUTE ALLOY MATERIALS LAB. (REF. 11)	6(LETTER)-(LETTER)-
COLUMBIA UNIV. N.Y.C. SUBSTITUTE ALLOY MATERIALS LAB. (REF. 11)	6(LETTER)(LETTER)-
COLUMBIA UNIV. NYC. SUBSTITUTE ALLOY MATERIALS LABS.	AB-M-
COLUMBIA UNIV. NYC. SUBSTITUTE ALLOY MATERIALS LABS.	AB-R-
COLUMBIA UNIV. NYC. SUBSTITUTE ALLOY MATERIALS LABS.	AC-
COLUMBIA UNIV. NYC. SUBSTITUTE ALLOY MATERIALS LABS.	AC-L-
COLUMBIA UNIV. NYC. SUBSTITUTE ALLOY MATERIALS LABS.	AC-M-
COLUMBIA UNIV. NYC. SUBSTITUTE ALLOY MATERIALS LABS.	AC-R-
COLUMBIA UNIV. NYC. SUBSTITUTE ALLOY MATERIALS LABS.	AC-S-
COLUMBIA UNIV. NYC. SUBSTITUTE ALLOY MATERIALS LABS.	AL-
COLUMBIA UNIV. NYC. SUBSTITUTE ALLOY MATERIALS LABS.	AM-
COLUMBIA UNIV. NYC. SUBSTITUTE ALLOY MATERIALS LABS.	AM-R-
COLUMBIA UNIV. NYC. SUBSTITUTE ALLOY MATERIALS LABS.	ANAL-R-
COLUMBIA UNIV. NYC. SUBSTITUTE ALLOY MATERIALS LABS.	A-R-
COLUMBIA UNIV. NYC. SUBSTITUTE ALLOY MATERIALS LABS.	AS-L-
COLUMBIA UNIV. NYC. SUBSTITUTE ALLOY MATERIALS LABS.	AS-M-
COLUMBIA UNIV. NYC. SUBSTITUTE ALLOY MATERIALS LABS.	ASM-
COLUMBIA UNIV. NYC. SUBSTITUTE ALLOY MATERIALS LABS.	AS-R-
COLUMBIA UNIV. NYC. SUBSTITUTE ALLOY MATERIALS LABS.	AXR-
COLUMBIA UNIV. NYC. SUBSTITUTE ALLOY MATERIALS LABS.	BA2-M-
COLUMBIA UNIV. NYC. SUBSTITUTE ALLOY MATERIALS LABS.	BA2-R-
COLUMBIA UNIV. NYC. SUBSTITUTE ALLOY MATERIALS LABS.	BA-M-
COLUMBIA UNIV. NYC. SUBSTITUTE ALLOY MATERIALS LABS.	BA-R-
COLUMBIA UNIV. NYC. SUBSTITUTE ALLOY MATERIALS LABS.	BB-M-
COLUMBIA UNIV. NYC. SUBSTITUTE ALLOY MATERIALS LABS.	BC-M-
COLUMBIA UNIV. NYC. SUBSTITUTE ALLOY MATERIALS LABS.	BC-R-
COLUMBIA UNIV. NYC. SUBSTITUTE ALLOY MATERIALS LABS.	BD-M-
COLUMBIA UNIV. NYC. SUBSTITUTE ALLOY MATERIALS LABS.	BD-R-
COLUMBIA UNIV. NYC. SUBSTITUTE ALLOY MATERIALS LABS.	BE-M-
COLUMBIA UNIV. NYC. SUBSTITUTE ALLOY MATERIALS LABS.	BE-R-
COLUMBIA UNIV. NYC. SUBSTITUTE ALLOY MATERIALS LABS.	BJ-R-
COLUMBIA UNIV. NYC. SUBSTITUTE ALLOY MATERIALS LABS.	BK-R-
COLUMBIA UNIV. NYC. SUBSTITUTE ALLOY MATERIALS LABS.	BL-M-
COLUMBIA UNIV. NYC. SUBSTITUTE ALLOY MATERIALS LABS.	BM-R-
COLUMBIA UNIV. NYC. SUBSTITUTE ALLOY MATERIALS LABS.	BN-
COLUMBIA UNIV. NYC. SUBSTITUTE ALLOY MATERIALS LABS.	BN-R-
COLUMBIA UNIV. NYC. SUBSTITUTE ALLOY MATERIALS LABS.	BR-M-
COLUMBIA UNIV. NYC. SUBSTITUTE ALLOY MATERIALS LABS.	BR-R-
COLUMBIA UNIV. N.Y.C. SUBSTITUTE ALLOY MATERIALS LAB. (REF. 11)	100-BR- THRU 100-ZS-
COLUMBIA UNIV. NYC. SUBSTITUTE ALLOY MATERIALS LABS.	C1-L- THRU C1-S-
COLUMBIA UNIV. NYC. SUBSTITUTE ALLOY MATERIALS LABS.	C2-M-
COLUMBIA UNIV. NYC. SUBSTITUTE ALLOY MATERIALS LABS.	C2-R-
COLUMBIA UNIV. NYC. SUBSTITUTE ALLOY MATERIALS LABS.	C2X-M-
COLUMBIA UNIV. NYC. SUBSTITUTE ALLOY MATERIALS LABS.	CA-L-
COLUMBIA UNIV. NYC. SUBSTITUTE ALLOY MATERIALS LABS.	CA-R-
COLUMBIA UNIV. NYC. SUBSTITUTE ALLOY MATERIALS LABS.	CAX-M-
COLUMBIA UNIV. NYC. SUBSTITUTE ALLOY MATERIALS LABS.	CAX-R-
COLUMBIA UNIV. NYC. SUBSTITUTE ALLOY MATERIALS LABS.	CAX-S-
COLUMBIA UNIV. NYC. SUBSTITUTE ALLOY MATERIALS LABS.	CI-R-
COLUMBIA UNIV. NYC. SUBSTITUTE ALLOY MATERIALS LABS.	CL-R-
COLUMBIA UNIV. NYC. SUBSTITUTE ALLOY MATERIALS LABS.	CPA-
COLUMBIA UNIV. NYC. SUBSTITUTE ALLOY MATERIALS LABS.	CR-R-
COLUMBIA UNIV. NYC. SUBSTITUTE ALLOY MATERIALS LABS.	CT-R-
COLUMBIA UNIV. NYC. SUBSTITUTE ALLOY MATERIALS LABS.	CTR-
COLUMBIA UNIV. NYC. SUBSTITUTE ALLOY MATERIALS LABS.	D1-M-
COLUMBIA UNIV. NYC. SUBSTITUTE ALLOY MATERIALS LABS.	D1B-M-
COLUMBIA UNIV. NYC. SUBSTITUTE ALLOY MATERIALS LABS.	D1FM-
COLUMBIA UNIV. NYC. SUBSTITUTE ALLOY MATERIALS LABS.	D2L-M-
COLUMBIA UNIV. NYC. SUBSTITUTE ALLOY MATERIALS LABS.	D3-M-
COLUMBIA UNIV. NYC. SUBSTITUTE ALLOY MATERIALS LABS.	D3-X-M-
COLUMBIA UNIV. NYC. SUBSTITUTE ALLOY MATERIALS LABS.	DA-L-
COLUMBIA UNIV. NYC. SUBSTITUTE ALLOY MATERIALS LABS.	DA-M-
COLUMBIA UNIV. NYC. SUBSTITUTE ALLOY MATERIALS LABS.	DA-R-
COLUMBIA UNIV. NYC. SUBSTITUTE ALLOY MATERIALS LABS.	DB-M-
COLUMBIA UNIV. NYC. SUBSTITUTE ALLOY MATERIALS LABS.	DB-R-
COLUMBIA UNIV. NYC. SUBSTITUTE ALLOY MATERIALS LABS.	DC-M-
COLUMBIA UNIV. NYC. SUBSTITUTE ALLOY MATERIALS LABS.	DC-R-
COLUMBIA UNIV. NYC. SUBSTITUTE ALLOY MATERIALS LABS.	DD-M-
COLUMBIA UNIV. NYC. SUBSTITUTE ALLOY MATERIALS LABS.	DE-M-
COLUMBIA UNIV. NYC. SUBSTITUTE ALLOY MATERIALS LABS.	DE-R-
COLUMBIA UNIV. NYC. SUBSTITUTE ALLOY MATERIALS LABS.	D-L-
COLUMBIA UNIV. NYC. SUBSTITUTE ALLOY MATERIALS LABS.	D-M-
COLUMBIA UNIV. NYC. SUBSTITUTE ALLOY MATERIALS LABS.	DM-
COLUMBIA UNIV. NYC. SUBSTITUTE ALLOY MATERIALS LABS.	DM-M-
COLUMBIA UNIV. NYC. SUBSTITUTE ALLOY MATERIALS LABS.	D-R-
COLUMBIA UNIV. NYC. SUBSTITUTE ALLOY MATERIALS LABS.	DR-
COLUMBIA UNIV. NYC. SUBSTITUTE ALLOY MATERIALS LABS.	DR-L-
COLUMBIA UNIV. NYC. SUBSTITUTE ALLOY MATERIALS LABS.	DR-M-
COLUMBIA UNIV. NYC. SUBSTITUTE ALLOY MATERIALS LABS.	EA-L-

COLUMBIA UNIV., NYC. SUBSTITUTE ALLOY MATERIALS LABS.	EA-M-
COLUMBIA UNIV., NYC. SUBSTITUTE ALLOY MATERIALS LABS.	EAMX-
COLUMBIA UNIV., NYC. SUBSTITUTE ALLOY MATERIALS LABS.	EA-R-
COLUMBIA UNIV., NYC. SUBSTITUTE ALLOY MATERIALS LABS.	ED-R-
COLUMBIA UNIV., NYC. SUBSTITUTE ALLOY MATERIALS LABS.	EG-M-
COLUMBIA UNIV., NYC. SUBSTITUTE ALLOY MATERIALS LABS.	EG-R-
COLUMBIA UNIV., NYC. SUBSTITUTE ALLOY MATERIALS LABS.	EK-M-
COLUMBIA UNIV., NYC. SUBSTITUTE ALLOY MATERIALS LABS.	EK-R-
COLUMBIA UNIV., NYC. SUBSTITUTE ALLOY MATERIALS LABS.	EL-R-
COLUMBIA UNIV., NYC. SUBSTITUTE ALLOY MATERIALS LABS.	EM-R-
COLUMBIA UNIV., NYC. SUBSTITUTE ALLOY MATERIALS LABS.	EN-R-
COLUMBIA UNIV., NYC. SUBSTITUTE ALLOY MATERIALS LABS.	EO-M-
COLUMBIA UNIV., NYC. SUBSTITUTE ALLOY MATERIALS LABS.	ER-R-
COLUMBIA UNIV., NYC. SUBSTITUTE ALLOY MATERIALS LABS.	ER-X
COLUMBIA UNIV., NYC. SUBSTITUTE ALLOY MATERIALS LABS.	ES-R-
COLUMBIA UNIV., NYC. SUBSTITUTE ALLOY MATERIALS LABS.	EW-M-
COLUMBIA UNIV., NYC. SUBSTITUTE ALLOY MATERIALS LABS.	EW-R-
COLUMBIA UNIV., NYC. SUBSTITUTE ALLOY MATERIALS LABS.	FS-R-
COLUMBIA UNIV., NYC. SUBSTITUTE ALLOY MATERIALS LABS.	GD-
COLUMBIA UNIV., NYC. SUBSTITUTE ALLOY MATERIALS LABS.	KZ-
COLUMBIA UNIV., NYC. SUBSTITUTE ALLOY MATERIALS LABS.	NA-L-
COLUMBIA UNIV., NYC. SUBSTITUTE ALLOY MATERIALS LABS.	NA-M-
COLUMBIA UNIV., N.Y.C. SUBSTITUTE ALLOY MATERIALS LAB. (REF. 11)	O-L- THRU O-M-
COLUMBIA UNIV., NYC. SUBSTITUTE ALLOY MATERIALS LABS.	R-4-
COLUMBIA UNIV., N.Y.C. SUBSTITUTE ALLOY MATERIALS LAB. (REF. 11)	1(LETTER)(NUMBER)-X(LTR)-
COLUMBIA UNIV., N.Y.C. SUBSTITUTE ALLOY MATERIALS LAB. (REF. 11)	1(LETTER)-X(LETTER)-
COLUMBIA UNIV., N.Y.C. SUBSTITUTE ALLOY MATERIALS LAB. (REF. 11)	1-X(LETTER)-
COLUMBIA UNIV., N.Y.C. SUBSTITUTE ALLOY MATERIALS LAB. (REF. 11)	1X(LETTER)-
COLUMBIA UNIV., N.Y.C. SUBSTITUTE ALLOY MATERIALS LAB. (REF. 11)	2(LETTER)(NUMBER)-X(LTR)-
COLUMBIA UNIV., N.Y.C. SUBSTITUTE ALLOY MATERIALS LAB. (REF. 11)	2(LETTER)-X(LETTER)-
COLUMBIA UNIV., N.Y.C. SUBSTITUTE ALLOY MATERIALS LAB. (REF. 11)	2-X(LETTER)-
COLUMBIA UNIV., N.Y.C. SUBSTITUTE ALLOY MATERIALS LAB. (REF. 11)	2X(LETTER)-
COLUMBIA UNIV., N.Y.C. SUBSTITUTE ALLOY MATERIALS LAB. (REF. 11)	3(LETTER)-X(LETTER)-
COLUMBIA UNIV., N.Y.C. SUBSTITUTE ALLOY MATERIALS LAB. (REF. 11)	3-X(LETTER)-
COLUMBIA UNIV., N.Y.C. SUBSTITUTE ALLOY MATERIALS LAB. (REF. 11)	3X(LETTER)-
COLUMBIA UNIV., N.Y.C. SUBSTITUTE ALLOY MATERIALS LAB. (REF. 11)	4-X(LETTER)-
COLUMBIA UNIV., N.Y.C. SUBSTITUTE ALLOY MATERIALS LAB. (REF. 11)	4X(LETTER)-
COLUMBIA UNIV., N.Y.C. SUBSTITUTE ALLOY MATERIALS LAB. (REF. 11)	5(LETTER)(NUMBER)-X(LTR)-
COLUMBIA UNIV., N.Y.C. SUBSTITUTE ALLOY MATERIALS LAB. (REF. 11)	5(LETTER)-X(LETTER)-
COLUMBIA UNIV., N.Y.C. SUBSTITUTE ALLOY MATERIALS LAB. (REF. 11)	5-X(LETTER)-
COLUMBIA UNIV., N.Y.C. SUBSTITUTE ALLOY MATERIALS LAB. (REF. 11)	5X(LETTER)-
COLUMBIA UNIV., N.Y.C. SUBSTITUTE ALLOY MATERIALS LAB. (REF. 11)	6(LETTER)-X(LETTER)-
COLUMBIA UNIV., N.Y.C. SUBSTITUTE ALLOY MATERIALS LAB. (REF. 11)	6-X(LETTER)-
COLUMBIA UNIV., N.Y.C. SUBSTITUTE ALLOY MATERIALS LAB. (REF. 11)	6X(LETTER)-
COLUMBIA UNIV., NYC. SUBSTITUTE ALLOY MATERIALS LABS.	XAC-
COLUMBIA UNIV., NYC. SUBSTITUTE ALLOY MATERIALS LABS.	XAC-(LETTER)-
COLUMBIA UNIV., NYC. SUBSTITUTE ALLOY MATERIALS LABS.	XAR-
COLUMBIA UNIV., NYC. SUBSTITUTE ALLOY MATERIALS LABS.	XD-M-
COLUMBIA UNIV., NYC. SUBSTITUTE ALLOY MATERIALS LABS.	XD-R-
COLUMBIA UNIV., N.Y.C. WAVE PROPAGATION GROUP	CU-WPG-
COLUMBIA UNIV., PALISADES, N.Y. LAMONT-DOHERTY GEOLOGICAL OBSERVATORY	LDGO-
COLUMBIA UNIV., PALISADES, N.Y. LAMONT GEOLOGICAL OBS	CU-LGO-
COLUMBIA UNIV., PALISADES, N.Y. LAMONT GEOLOGICAL OBS	LGO-
COLUMBIA UNIV., PALISADES, N.Y. LAMONT GEOLOGICAL OBS	SM-
COLUMBIA UNIV., ST. DAVID'S, BERMUDA. GEOPHYSICAL FIELD STATION	DD(NUMBER)-(YEAR)
COLUMBIA UNIV., ST. DAVID'S, BERMUDA. GEOPHYSICAL FIELD STATION	LOST-
COMBAT CREW TRAINING WING (MBOM)(3520TH), MC CONNELL AFB, KAN.	MCAFB-
COMBAT CREW TRAINING WING (MBOM)(3520TH), MC CONNELL AFB, KAN.	TA AND D-
COMBAT DEVELOPMENT AND TEST CENTER, VIETNAM	CTDC-V-
COMBAT DEVELOPMENT AND TEST CENTER, VIETNAM	RDFU/V-
COMBINED COMMUNICATIONS BOARD, JOINT CHIEFS OF STAFF, WASHINGTON, D.C.	C/N-
COMBINED INTELLIGENCE OBJECTIVES SUB-COMMITTEE	CIOS-
COMBINED INTELLIGENCE OBJECTIVES SUB-COMMITTEE	CIOS (ROMAN NUMERAL)-
COMBINED INTELLIGENCE OBJECTIVES SUB-COMM. (DOCUMENT)	CIOS DOC-
COMBINED INTELL. OBJECTIVES SUB-COMM.(EVALUATION RPT)	CIOS ER-
COMBINED INTELLIGENCE OBJECTIVES SUB-COMMITTEE	CIOS-ITEM-
COMBINED INTELLIGENCE OBJECTIVES SUB-COMMITTEE. (MILITARY INTELLIGENCE. REPORTS SECTION)	CIOS MIRS (LETTER)-
COMBINED INTELLIGENCE OBJECTIVES SUB-COMMITTEE. (ORDNANCE TARGET REPORTS)	CIOS OTR-
COMBINED INTELLIGENCE OBJECTIVES SUB-COMMITTEE	CIOS-STATUS-
COMBINED INTELLIGENCE OBJECTIVES SUB-COMM. (TARGET)	CIOS T(NO./NO.)
COMBINED INTELLIGENCE OBJECTIVES SUB-COMMITTEE (ORDNANCE TARGET REPORTS)	OTR-
COMBINED INTELLIGENCE OBJECTIVES SUB-COMMITTEE. AIRCRAFT RESOURCES CONTROL OFFICES	ARCO-
COMBUSTION ENGINEERING, INC., N.Y.C.	CENP-
COMBUSTION ENGINEERING, INC., WINDSOR, CONN. (JOINT VENTURE)	CEND-PRWA-
COMBUSTION ENGINEERING, INC., WINDSOR, CONN. (JOINT VENTURE)	CEND/PRWRA-
COMBUSTION ENGINEERING, INC., WINDSOR, CONN.	CENDRD-
COMBUSTION ENGINEERING, INC. MARINE DEPT., WINDSOR, CONN.	C-E-
COMBUSTION ENGINEERING, INC. NAVAL REACTORS DIV., WINDSOR, CONN.	CENRD-
COMBUSTION ENGINEERING, INC. NAVAL REACTORS DIV., WINDSOR, CONN.	CENRD-(NUMBER)-RS-
COMBUSTION ENGINEERING, INC. NAVAL REACTORS DIV., WINDSOR, CONN.	CENRD/HP-
COMBUSTION ENGINEERING, INC. NAVAL REACTORS DIV., WINDSOR, CONN.	CENRD-S1C-
COMBUSTION ENGINEERING, INC. NUCLEAR COMPONENTS ENGINEERING DEPT., CHATTANOOGA	CENC-
COMBUSTION ENG., INC. NUCLEAR DIV., IDAHO FALLS	CEND-
COMBUSTION ENG., INC. NUCLEAR DIV., WINDSOR, CONN.	CEND-
COMBUSTION ENG., INC. NUCLEAR DIV., WINDSOR, CONN.	CEND-(NO.)-MD-(NO.)
COMBUSTION ENG., INC. NUCLEAR DIV., WINDSOR, CONN.	CEND-(NO.)-RS-(NO.)
COMBUSTION ENG., INC. NUCLEAR DIV., WINDSOR, CONN.	CEND/(NUMBER)/TP-
COMBUSTION ENG., INC. NUCLEAR DIV., WINDSOR, CONN.	CEND-S1C-
COMBUSTION ENG., INC. NUCLEAR DIV., WINDSOR, CONN.	MPC-
COMBUSTION ENG., INC. NUCLEAR DIV., WINDSOR, CONN.	S1C-(NUMBER)-SP-
COMBUSTION ENGINEERING, INC. NUCLEAR POWER DEPT., WINDSOR, CONN.	CENPD-
COMBUSTION ENG.,INC. REACTOR DEV. DIV.,WINDSOR,CONN.	CERD-

COMBUSTION ENG., INC. REACTOR DEV. DIV.

COMBUSTION ENG., INC. REACTOR DEV. DIV., WINDSOR, CONN.	CERD-S1C-
COMBUSTION ENG., INC. REACTOR DEV. DIV., WINDSOR, CONN.	S1C-(NUMBER)-CA/
COMBUSTION ENG., INC. REACTOR DEV. DIV., WINDSOR, CONN.	S1C-(NUMBER)-RS-
COMBUSTION ENG., INC. REACTOR DEV. DIV., WINDSOR, CONN.	S1C-(NUMBER)-TM/
COMBUSTION ENG., INC. REACTOR DEV. DIV., WINDSOR, CONN.	S1C/HP/
COMBUSTION ENG., INC. REACTOR DEV. DIV., WINDSOR, CONN.	S1C/MD-
COMBUSTION ENG., INC. REACTOR DEV. DIV., WINDSOR, CONN.	S1C/PP-
COMBUSTION ENG., INC. REACTOR DEV. DIV., WINDSOR, CONN.	S1C/RP-
(SEE ALSO LATER NAME: COMBUSTION ENG., INC., NAVAL REACTORS DIV.)	
COMBUSTION ENGINEERING-SUPERHEATER, INC., N.Y.C.	TDV-
COMBUSTION ENGINEERING-SUPERHEATER LTD., MONTREAL	TD(V)-
COMBUSTION INSTITUTE. WESTERN STATES SECTION	WSS/CI PAPER (YEAR)-
COMITE CONSULTATIF INTERNATIONAL DE RADIOCOMMUNICATIONS, GENEVA	(ROMAN NUMERAL)/
COMMAND AND CONTROL DEVELOPMENT DIV. (AIR FORCE), BEDFORD, MASS.	AFCCDD-
COMMAND AND CONTROL DEVELOPMENT DIV. (AIR FORCE), BEDFORD, MASS.	AFCCDD-TN-(YEAR)-
COMMAND AND CONTROL DEVELOPMENT DIV. (AIR FORCE), BEDFORD, MASS.	AFCCDD-TR-(YEAR)-
COMMERCIAL SOLVENTS CORP., N.Y.C.	M-
COMMERCIAL SOLVENTS CORP., TERRE HAUTE, IND.	CSC-FINAL-
COMMERCIAL SOLVENTS CORP., TERRE HAUTE, IND.	CSC-M-
COMMERCIAL SOLVENTS CORP., TERRE HAUTE, IND.	CSC-Q-
COMMITTEE FOR MILITARY INFORMATION CONTROL	MIC-
COMMITTEE ON AERONAUTICAL (RES. + DEV.) FACILITIES	CAF-
COMMITTEE ON MARINE RESEARCH EDUCATION AND FACILITIES WASHINGTON, D.C.	CMREF-PAMPHLET-
COMMITTEE ON MEDICAL RESEARCH (BEADLE REPORTS)	BE-
COMMITTEE ON MEDICAL RESEARCH. (BRITISH REPORTS)	BR-
COMMITTEE ON MEDICAL RES. (NATL. BUR. OF STDS. RPTS.)	BS-
COMMITTEE ON MEDICAL RES. (BLOOD SUBSTITUTES RPTS.)	BSR-
COMMITTEE ON MEDICAL RES. (CLINICAL INVEST. RPTS.)	CIR-
COMMITTEE ON MEDICAL RESEARCH (CLARK REPORTS)	CL-
COMMITTEE ON MEDICAL RESEARCH, OFFICE OF SCIENTIFIC RESEARCH AND DEVELOPMENT (REF. 32)	CMR-
COMMITTEE ON MEDICAL RESEARCH (ABBOTT LABS. REPORTS)	CMR A-
COMMITTEE ON MEDICAL RESEARCH (BACKMANN REPORTS)	CMR B-
COMMITTEE ON MEDICAL RESEARCH (BEADLE REPORTS)	CMR BE-
COMMITTEE ON MEDICAL RESEARCH. (BRITISH REPORTS)	CMR BR-
COMMITTEE ON MEDICAL RES. (NATL. BUR. OF STDS. RPTS.)	CMR BS-
COMMITTEE ON MEDICAL RES. (BLOOD SUBSTITUTES RPTS.)	CMR BSR-
COMMITTEE ON MEDICAL RESEARCH (BULLETIN)	CMR BULL-
COMMITTEE ON MEDICAL RESEARCH (COGHILL REPORTS)	CMR C-
COMMITTEE ON MEDICAL RES. (CLINICAL INVEST. RPTS.)	CMR CIR-
COMMITTEE ON MEDICAL RESEARCH (CLARK REPORTS)	CMR CL-
COMMITTEE ON MEDICAL RESEARCH (CONVALESCENCE AND REHABILITATION REPORTS)	CMR CR-
COMMITTEE ON MEDICAL RESEARCH (CHEMOTHERAPY REPORTS)	CMR CT-
COMMITTEE ON MEDICAL RESEARCH (CUTTER REPORTS)	CMR CU-
COMMITTEE ON MEDICAL RESEARCH (DU VIGNEAUD REPORTS)	CMR D-
COMMITTEE ON MEDICAL RES. (FOOD + DRUG ADM. RPTS.)	CMR F-
COMMITTEE ON MEDICAL RES. (HEYDEN CHEM. CORP. RPTS.)	CMR H-
COMMITTEE ON MEDICAL RES. (INFECTIOUS DISEASES RPTS.)	CMR ID-
COMMITTEE ON MEDICAL RES. (INFECTED WOUNDS AND BURNS)	CMR IWB-
COMMITTEE ON MEDICAL RESEARCH (JOHNSON REPORTS)	CMR J-
COMMITTEE ON MEDICAL RES. (LILLY RES. LABS. RPTS.)	CMR L-
COMMITTEE ON MEDICAL RESEARCH (MERCK + CO. RPTS.)	CMR M-
COMMITTEE ON MEDICAL RES. (MISSILE CASUALTIES RPTS.)	CMR MC-
COMMITTEE ON MEDICAL RESEARCH (MEDICINE REPORTS)	CMR ME-
COMMITTEE ON MEDICAL RES. (MEDICAL NUTRITION RPTS.)	CMR MN-
COMMITTEE ON MEDICAL RESEARCH (MALARIA REPORTS)	CMR MR-
COMMITTEE ON MEDICAL RES. (MALARIA RPTS. ABSTRACT)	CMR MR A-
COMMITTEE ON MEDICAL RESEARCH (ROCKEFELLER INSTITUTE FOR MEDICAL RESEARCH)	CMR N-
COMMITTEE ON MEDICAL RESEARCH (NEUROPSYCHIATRY RPTS.)	CMR NP-
COMMITTEE ON MEDICAL RESEARCH (NEUROSURGERY REPORTS)	CMR NS-
COMMITTEE ON MEDICAL RESEARCH (OTOLARYNGOLOGY RPTS.)	CMR OL-
COMMITTEE ON MEDICAL RES. (CHAS. PFIZER + CO. RPTS.)	CMR P-
COMMITTEE ON MEDICAL RES. (PARKE-DAVIS + CO. RPTS.)	CMR PD-
COMMITTEE ON MEDICAL RESEARCH (RANDALL REPORTS)	CMR R-
COMMITTEE ON MEDICAL RES. (RUSSELL SAGE INST. RPTS.)	CMR RS-
COMMITTEE ON MEDICAL RESEARCH (SQUIBB INSTITUTE FOR MEDICAL RESEARCH REPORTS)	CMR S-
COMMITTEE ON MEDICAL RESEARCH (SANITARY ENG. RPTS.)	CMR SER-
COMMITTEE ON MEDICAL RES. (SHELL DEV. CO. RPTS.)	CMR SH-
COMMITTEE ON MEDICAL RESEARCH (SPECIAL REPORTS)	CMR SPR-
COMMITTEE ON MEDICAL RESEARCH (SHOCK REPORTS)	CMR SR-
COMMITTEE ON MEDICAL RES. (TROPICAL DISEASES RPTS.)	CMR TDR-
COMMITTEE ON MEDICAL RESEARCH (TUBERCULOSIS REPORTS)	CMR TR-
COMMITTEE ON MEDICAL RESEARCH (UPJOHN CO. REPORTS)	CMR U-
COMMITTEE ON MEDICAL RES. (VENEREAL DISEASES RPTS.)	CMR VD-
COMMITTEE ON MEDICAL RES. (WINTHROP CHEM. CO. RPTS.)	CMR W-
COMMITTEE ON MEDICAL RESEARCH (WOODWARD REPORTS)	CMR WO-
COMMITTEE ON MEDICAL RESEARCH (CONVALESCENCE AND REHABILITATION REPORTS)	CR-
COMMITTEE ON MEDICAL RESEARCH (CHEMOTHERAPY REPORTS)	CT-
COMMITTEE ON MEDICAL RESEARCH (CUTTER REPORTS)	CU-
COMMITTEE ON MEDICAL RES. (INFECTIOUS DISEASES RPTS.)	ID-
COMMITTEE ON MEDICAL RES. (INFECTED WOUNDS AND BURNS)	IWB-
COMMITTEE ON MEDICAL RESEARCH (JOHNSON REPORTS)	J-
COMMITTEE ON MEDICAL RESEARCH (MEDICINE REPORTS)	ME-
COMMITTEE ON MEDICAL RES. (MEDICAL NUTRITION RPTS.)	MN-
COMMITTEE ON MEDICAL RESEARCH (MALARIA REPORTS)	MR-
COMMITTEE ON MEDICAL RESEARCH (NEUROPSYCHIATRY RPTS.)	NP-
COMMITTEE ON MEDICAL RESEARCH (NEUROSURGERY RPTS.)	NS-
COMMITTEE ON MEDICAL RES. (OTOLARYNGOLOGY RPTS.)	OL-
COMMITTEE ON MEDICAL RES. (PARKE-DAVIS + CO. RPTS.)	PD-
COMMITTEE ON MEDICAL RES. (RUSSELL SAGE INST. RPTS.)	RS-
COMMITTEE ON MEDICAL RESEARCH (SANITARY ENG. RPTS.)	SER-
COMMITTEE ON MEDICAL RESEARCH (SHELL DEV. CO. RPTS.)	SH-
COMMITTEE ON MEDICAL RESEARCH (SPECIAL REPORTS)	SPR-
COMMITTEE ON MEDICAL RESEARCH (SHOCK REPORTS)	SR-
COMMITTEE ON MEDICAL RES. (TROPICAL DISEASES RPTS.)	TDR-
COMMITTEE ON MEDICAL RESEARCH (TUBERCULOSIS REPORTS)	TR-
COMMITTEE ON MEDICAL RES. (VENEREAL DISEASES RPTS.)	VD-
COMMITTEE ON MEDICAL RES. COMM. ON AVIATION MED.	CMR CAM-

COMMITTEE ON MEDICAL RESEARCH. COMMITTEE ON THE TREATMENT OF GAS CASUALTIES	CMR CTGC-
COMMITTEE ON MEDICAL RESEARCH. COMMITTEE ON THE TREATMENT OF GAS CASUALTIES	CMR CTGC (LETTER)-
COMMITTEE ON MEDICAL RESEARCH. COMMITTEE ON THE TREATMENT OF GAS CASUALTIES	CTGC-
COMMITTEE ON PENICILLIN SYNTHESIS	CPS-
COMMONWEALTH EDISON CO., CHICAGO (JOINT VENTURE)	CEPS-
COMMUNICATION RESEARCH INST., MIAMI, FLA.	CRI-
COMMUNICATION SATELLITE CORP., WASHINGTON, D.C.	CL-(NUMBER-YEAR)
COMMUNICATION SATELLITE CORP., WASHINGTON, D.C.	COMSAT-
COMMUNICATION SYSTEMS INC., FALLS CHURCH, VA.	C/S-(YEAR)-
COMMUNICATION SYSTEMS INC., FALLS CHURCH, VA.	C/S-R(NUMBER)-
COMMUNICATION SYSTEMS INC., FALLS CHURCH, VA.	WCSI-(YEAR)-TR-
COMMUNICATION SYSTEMS INC., PARAMUS, N.J.	CSI-
COMMUNICATION SYSTEMS INC., PARAMUS, N.J.	CSI-(YEAR)-TR
COMMUNICATIONS AND SYSTEMS, INC., FALLS CHURCH, VA.	C/S-(4 DIGITS)-
COMMUNICATIONS AND SYSTEMS, INC., FALLS CHURCH, VA.	C/S-TR-(NUMBER)-
COMMUNICATIONS AND SYSTEMS, INC., PARAMUS, N.J.	C/S-(YEAR)-TR-
COMMUNICATIONS AND SYSTEMS, INC., PARAMUS, N.J.	C/S-(YEAR-3 DIGITS)-
COMMUNICATIONS RESEARCH LABS., SANTA ANA, CALIF.	CRL-
COMPAGNIE FRANCAISE THOMSON-HOUSTON, BAGNEUX	DRS/(NUMBER-NUMBER)TS
COMPAGNIE FRANCAISE THOMSON-HOUSTON. DIV. ACTIVITIES SOUS-MARINES	D.ASM-(NUMBER)-CL/CM/PT
COMPAGNIE FRANCAISE THOMSON HOUSTON-HOTCHKISS BRANDT, PARIS	CSF-(YEAR)-
COMPAGNIE FRANCAISE THOMSON HOUSTON-HOTCHKISS BRANDT, PARIS	LG/MT-LCR/DR(NUMBER)-
COMPAGNIE FRANCAISE THOMSON HOUSTON-HOTCHKISS BRANDT, PARIS	LG/PB-LCR/DR(NUMBER)-
COMPAGNIE GENERALE DE TELEGRAPHIE SANS FIL, ORSAY, FRANCE	WR-
COMPAGNIE GENERALE DE TELEGRAPHIE SANS FIL, PARIS	DTW-
COMPAGNIE GENERALE DE TELEGRAPHIE SANS FIL, PARIS	EW-(5 DIGITS)
COMPAGNIE GENERALE DE TELEGRAPHIE SANS FIL, PARIS	WD-(4 DIGITS)
COMPAGNIE GENERALE DE TELEGRAPHIE SANS FIL, PARIS	WR-
COMPAGNIE GENERALE DE TELEGRAPHIE SANS FIL, PUTEAU, FRANCE	RPC-
COMPAGNIES FRANCAISES ASSOCIEES DE TELEGRAPHIE SANS FIL	TSF-
COMPTROLLER GENERAL OF THE U.S., WASHINGTON, D.C.	R-
COMPUTER APPLICATIONS, INC., N.Y.C.	CAI-NY-
COMPUTER APPLICATIONS, INC., N.Y.C.	CAI-SED-(NUMBER)-
COMPUTER ASSOCIATES, INC., ARLINGTON, VA.	CA-(NUMBER)-
COMPUTER ASSOCIATES, INC., WAKEFIELD, MASS.	CA-
COMPUTER COMMAND AND CONTROL CO., PHILADELPHIA	CC+CC-
COMPUTER COMMAND AND CONTROL CO., WASHINGTON, D.C.	CCCC-(NUMBER-NUMBER)
COMPUTER RESEARCH CORP., HAWTHORNE, CALIF.	CRC-
COMPUTER SYMBOLIC, INC., ROME, N.Y.	CSI-(YEAR)-
CONCHO VALLEY COUNCIL OF GOVERNMENTS, SAN ANGELO, TEX	CV-(3 DIGITS)
CONCORD RESEARCH CORP., BURLINGTON, MASS.	CRC-(YEAR)-
CONDUCTRON CORP., ANN ARBOR, MICH.	CAA-D(NUMBER)-(NUMBER)-
CONDUCTRON CORP., ANN ARBOR, MICH.	CC-(3 DIGITS)-
CONDUCTRON CORP., ANN ARBOR, MICH.	CC-0(NUMBER)-(NUMBER)-
CONDUCTRON CORP., ANN ARBOR, MICH.	CC-P-
CONDUCTRON CORP., ANN ARBOR, MICH.	COND-
CONDUCTRON CORP., ANN ARBOR, MICH.	D-
CONDUCTRON CORP., ANN ARBOR, MICH.	D(NUMBER)-
CONDUCTRON CORP., ANN ARBOR, MICH.	QITR-
CONESCO INC., CAMBRIDGE, MASS.	CONESCO-
CONF. ON GOVERNMENT MICROCIRCUIT APPLICATIONS, 1969	GOMAC-(YEAR)-
CONFERENCE ON PHYSICAL PROBLEMS OF REACTOR SHIELDING	RS/(NUMBER)/
CONFERENCE ON PLASMA DIAGNOSIS, CULHAM, ENGLAND	IGN/RT-
CONFERENCE ON THE DISCONTINUANCE OF NUCLEAR WEAPON TESTS	GEN/DNT-
CONFERENCE ON THE DISCONTINUANCE OF NUCLEAR WEAPON TESTS	GEN/DNT/(LETTERS)
CONFERENCE ON THE DISCONTINUANCE OF NUCLEAR WEAPON TESTS	GEN/MISC-
CONGRESS, WASHINGTON, D.C. (PUBLIC LAW)	PL-
CONGRESS. HOUSE OF REPRESENTATIVES, WASHINGTON, D.C.	H-DOC-(NUMBER)-
CONGRESS. HOUSE OF REPRESENTATIVES, WASHINGTON, D.C.	HR-
CONGRESS. JOINT COMMITTEE ON PRINTING, WASHINGTON, DC	JCP-
CONGRESS. SENATE, WASHINGTON, D.C.	S-
CONGRESS. SENATE, WASHINGTON, D.C.	S-DOC-(NUMBER)-
CONN (C.G.) LTD., ELKHART, IND.	ER-
CONNECTICUT. DEPT. OF TRANSPORTATION, HARTFORD (CONN. MASTER TRANSPORTATION PLAN)	CMTP-CDOT-CAM-
CONNECTICUT. ENG. EXPERIMENT STA. (BULLETIN)	C EES B-
CONNECTICUT. UNIV., HARTFORD. DEPT. OF LAB. MEDICINE	NTO-(NUMBER)-
CONNECTICUT. UNIV., STORRS	CONN-
CONNECTICUT. UNIV., STORRS. CENTER FOR REAL ESTATE AND URBAN ECONOMIC STUDIES	REAL ESTATE-

CONNECTICUT. UNIV., STORRS. CHEMISTRY DEPT.	CON-
CONNECTICUT. UNIV., STORRS. DEPT. OF CIVIL ENG.	CE-(YEAR)-
CONNECTICUT HARD RUBBER CO., NEW HAVEN	CHR-
CONNECTICUT TELEPHONE + ELECTRIC CORP., MERIDEN	CTE-
CONNECTICUT YANKEE ATOMIC POWER CO., BOSTON	CYAP-
CONNER (MEL) AND ASSOCIATES, INC., TALLAHASSEE, FLA. (JOINT VENTURE)	CK-KPDO-(YEAR)-
CONRAC CORP. NEW JERSEY DIV., FAIRFIELD	CC-
CONSOER, WHITE AND HERSHEY, WASH., D.C.	CWH-
CONSOLIDATED CONTROLS CORP., BETHEL, CONN.	CCC-
CONSOLIDATED CONTROLS CORP., BETHEL, CONN.	CCC-HD-
CONSOLIDATED EDISON CO. OF NEW YORK, INC.	CECNY-
CONSOLIDATED ELECTRODYNAMICS CORP., PASADENA, CALIF.	CEC-
CONSOLIDATED ELECTRODYNAMICS CORP., PASADENA, CALIF.	CEC-M(NUMBER)-
CONSOLIDATED ELECTRODYNAMICS CORP. DATA INSTRUMENTS DIV., PASADENA, CALIF.	RTR-
CONSOLIDATED ENG. TECHNOLOGY CORP., MOUNTAIN VIEW, CAL.	CETEC-FR-
CONSOLIDATED VULTEE AIRCRAFT CORP., FORT WORTH, TEX. (REF. 3)	ANP DOC. (LETTERS)-
CONSOLIDATED VULTEE AIRCRAFT CORP., FORT WORTH, TEX.	CVAC-
CONSOLIDATED VULTEE AIRCRAFT CORP., FORT WORTH, TEX.	MR-A-
CONSOLIDATED VULTEE AIRCRAFT CORP., FORT WORTH, TEX.	MR-N-
CONSOLIDATED VULTEE AIRCRAFT CORP., FORT WORTH, TEX.	N.E.M.-
CONSOLIDATED VULTEE AIRCRAFT CORP., SAN DIEGO, CALIF.	A-ATLAS-
CONSOLIDATED VULTEE AIRCRAFT CORP., SAN DIEGO, CALIF. (REF. 3)	ANP-
CONSOLIDATED VULTEE AIRCRAFT CORP., SAN DIEGO, CALIF.	CD-
CONSOLIDATED VULTEE AIRCRAFT CORP., SAN DIEGO, CALIF.	CVAL-
CONSOLIDATED VULTEE AIRCRAFT CORP., SAN DIEGO, CALIF.	DEVF-
CONSOLIDATED VULTEE AIRCRAFT CORP., SAN DIEGO, CALIF.	55SARS-
(SEE ALSO LATER NAME: GENERAL DYNAMICS/CONVAIR)	
CONSOLIDATED VULTEE AIRCRAFT CORP. FORT WORTH DIV., TEX.	F(LTRS)-
CONSOLIDATED VULTEE AIRCRAFT CORP. FORT WORTH DIV., TEX.	F(LTRS)-(NO.)-
CONSOLIDATED VULTEE AIRCRAFT CORP. ORDNANCE AEROPHYSICS LAB., DAINGERFIELD, TEX. (REF. 1)	(NUMBER)
CONSOLIDATED VULTEE AIRCRAFT CORP. ORDNANCE AEROPHYSICS LAB., DAINGERFIELD, TEX.	CM-
CONSOLIDATED VULTEE AIRCRAFT CORP. ORDNANCE AEROPHYSICS LAB., DAINGERFIELD, TEX.	OAL-
CONSOLIDATED VULTEE AIRCRAFT CORP. SAN DIEGO DIV., CALIF. (REF. 1)	(JOB NUMBER)
CONSORTIUM OF UNIVERSITIES. URBAN TRANSPORTATION CENTER, WASHINGTON, D.C.	UMTA-URT-(NUMBER-YEAR)-
CONSULTANTS BUREAU, INC., N.Y.C. (TRANSLATIONS)	CB-
CONSULTANTS BUREAU, INC., N.Y.C. (TRANSLATION)	CBI-TRANS-
CONSULTANTS CUSTOM TRANSLATIONS, INC., N.Y.C.	CCT-
CONSULTANTS CUSTOM TRANSLATIONS, INC., N.Y.C.	CCT-TRANS-
CONSULTANTS INTERNATIONAL, INC., DALLAS	CII-R(NUMBER)
CONSUMERS PUBLIC POWER DISTRICT, HALLAM, NEB. (MONTHLY RETIREMENT REPORT)	CPPD-MRR-
CONTINENTAL AIR COMMAND, MITCHEL AFB, N.Y.	CAC-
CONTINENTAL ARMY COMMAND, FORT MONROE, VA.	AGF-
CONTINENTAL ARMY COMMAND, FORT MONROE, VA.	ATDEV-
CONTINENTAL ARMY COMMAND, FORT MONROE, VA.	ATING-
CONTINENTAL ARMY COMMAND, FORT MONROE, VA.	BD 1 CN. (NUMBER)-
CONTINENTAL ARMY COMMAND, FORT MONROE, VA.	CAC-ATDEV-
CONTINENTAL ARMY COMMAND, FORT MONROE, VA.	CAC ATSWD-(LTR.-YEAR)-
CONTINENTAL ARMY COMMAND, FORT MONROE, VA.	CAC/CORG-
CONTINENTAL ARMY COMMAND, FORT MONROE, VA.	CAC/SNODGRASS-
CONTINENTAL ARMY COMMAND, FORT MONROE, VA.	CONARC-
CONTINENTAL ARMY COMMAND, FORT MONROE, VA.	CON-CIR.-
CONTINENTAL ARMY COMMAND, FORT MONROE, VA.	OCAFF-
CONTINENTAL ARMY COMMAND. ARCTIC TEST BOARD, FORT GREELY, ALASKA	CAC ATB PROJ.-
CONTINENTAL ARMY COMMAND. ARCTIC TEST BR., SEATTLE	ATB-
CONTINENTAL ARMY COMMAND. ARTILLERY BOARD, FORT SILL, OKLA.	CAC ARTY BD.-
CONTINENTAL ARMY COMMAND. BD. NO. 1, FT. SILL, OKLA.	CAC BD. 1, PROJ.-
CONTINENTAL ARMY COMMAND. BD. NO. 4, FT. BLISS, TEX.	CAC BD. 4-
CONTINENTAL ARMY COMMAND. COMBAT DEVELOPMENTS SECTION FORT MONROE, VA.	CAC-
CONTINENTAL ARMY COMMAND. COMBAT OPERATIONS RESEARCH GROUP, FORT MONROE, VA.	CORG-
CONTINENTAL AVIATION AND ENGINEERING CORP., DETROIT	CAE-
CONTINENTAL AVIATION AND ENGINEERING CORP., DETROIT	LVT-
CONTINENTAL BEARING RESEARCH CORP., N.Y.C.	CBRC-
CONTINENTAL COPPER AND STEEL INDUSTRIES, INC., N.Y.C.	CCSI-
CONTINENTAL ELECTRONICS LTD., BROOKLYN	CEL-
CONTINENTAL MOTORS CORP., MUSKEGON, MICH.	CMC-MCDA-
CONTINENTAL MOTORS CORP., MUSKEGON, MICH.	MCDA-
CONTINENTAL MOTORS CORP. AIRCRAFT ENGINE DIV., MUSKEGON, MICH.	M(YR.)-
CONTINENTAL OIL CO., PONCA CITY, OKLA.	C-
CONTINENTAL OIL CO., PONCA CITY, OKLA.	CONOCO-
CONTROL DATA CORP. (ALL LOCATIONS)	CDC-

CONTROL DATA CORP., BETHESDA, MD.	CDC-A(NUMBER)-
CONTROL DATA CORP., LOS ANGELES	CTDC-TP-
CONTROL DATA CORP. HOWARD RES. DIV., BETHESDA, MD.	CDC/HRD-
CONTROL DATA CORP. RESEARCH DIV., MINNEAPOLIS	RD-
CONTROL DATA CORP. TRG DIV., MELVILLE, N.Y.	TRG-
CONTROL DATA CORP. TRG DIV., MELVILLE, N.Y.	TRG-(NUMBER)-
CONTROL DATA CORP. TRG DIV., MELVILLE, N.Y.	TRG-(NUMBER)-FR
CONTROL DATA CORP. TRG DIV., MELVILLE, N.Y.	TRG-(NUMBER)-IER-
CONTROL DATA CORP. TRG DIV., MELVILLE, N.Y.	TRG-(NUMBER)-TN-(YR)-
CONTROL DATA CORP. TRG DIV., MELVILLE, N.Y.	TRG-(NUMBER)-TR-(NO./YR)
CONTROL DATA CORP. TRG DIV., MELVILLE, N.Y.	TRG-(NUMBER)-USL-(YR)-
CONTROL INSTRUMENT CO., INC., BROOKLYN, N.Y.	DEV-
CONVAIR (ALL LOCATIONS)	CVAC-(LETTERS)-
CONVAIR (ALL LOCATIONS) (FLIGHT TEST REPORT)	FTR-
CONVAIR, FORT WORTH, TEX.	AE(YR)-(NO.)-
CONVAIR, FORT WORTH, TEX.	AEC-(NUMBER)-CVAC
CONVAIR, FORT WORTH, TEX.	CONVAIR-
CONVAIR, FORT WORTH, TEX.	CVC-(LTRS,NOS,LTRS)-
CONVAIR, FORT WORTH, TEX.	CVC CD-
CONVAIR, FORT WORTH, TEX.	CVC FSA (NOS.-NOS.)
CONVAIR, FORT WORTH, TEX.	CVC FSE (NOS.-NOS.)
CONVAIR, FORT WORTH, TEX.	CVC FZA (NOS.-NOS.)
CONVAIR, FORT WORTH, TEX.	CVC FZC (NOS.-NOS.)
CONVAIR, FORT WORTH, TEX.	CVC FZG (NOS.-NOS.)
CONVAIR, FORT WORTH, TEX.	CVC FZK (NOS.-NOS.)
CONVAIR, FORT WORTH, TEX.	CVC FZM (NOS.-NOS.)
CONVAIR, FORT WORTH, TEX.	CVC MRE-
CONVAIR, FORT WORTH, TEX.	CVC MR-N-
CONVAIR, FORT WORTH, TEX.	FPS-
CONVAIR, FORT WORTH, TEX.	FSE-
CONVAIR, FORT WORTH, TEX.	INR-
CONVAIR, FORT WORTH, TEX.	MR-N-
CONVAIR, FORT WORTH, TEX.	NEM-
CONVAIR, FORT WORTH, TEX.	3-R-
CONVAIR, FORT WORTH, TEX.	REM-
CONVAIR, FORT WORTH, TEX.	SC-
CONVAIR, POMONA, CALIF.	CR-(NO.)-(NO.)-(NO.)
CONVAIR, POMONA, CALIF.	CVC CR-(NOS.-NO.)
CONVAIR, POMONA, CALIF.	TM-(NUMBER)-
CONVAIR, SAN DIEGO, CALIF.	CVA-
CONVAIR, SAN DIEGO, CALIF.	CVC-(LTRS,NOS,LTRS)-
CONVAIR, SAN DIEGO, CALIF.	CVC AI-
CONVAIR, SAN DIEGO, CALIF.	CVC AZ (LTR.)-
CONVAIR, SAN DIEGO, CALIF.	CVC DC-W-
CONVAIR, SAN DIEGO, CALIF.	CVC ERR-SD-
CONVAIR, SAN DIEGO, CALIF.	CVC ZA (NO.-NOS.)
CONVAIR, SAN DIEGO, CALIF.	CVC ZC (NO.-NOS.)
CONVAIR, SAN DIEGO, CALIF.	CVC ZD (NO.-NOS.)
CONVAIR, SAN DIEGO, CALIF.	CVC ZJ-
CONVAIR, SAN DIEGO, CALIF.	CVC ZM (NO.-NOS.)
CONVAIR, SAN DIEGO, CALIF.	CVC ZO-P-
CONVAIR, SAN DIEGO, CALIF.	CVC ZP-
CONVAIR, SAN DIEGO, CALIF.	CVC ZPH-
CONVAIR, SAN DIEGO, CALIF.	CVC ZP-M-
CONVAIR, SAN DIEGO, CALIF.	CVC ZR (NOS.-NOS.)
CONVAIR, SAN DIEGO, CALIF.	CVC ZRAI-
CONVAIR, SAN DIEGO, CALIF.	CVC ZRAP-
CONVAIR, SAN DIEGO, CALIF.	CVC ZS-
CONVAIR, SAN DIEGO, CALIF.	CVC ZU-
CONVAIR, SAN DIEGO, CALIF.	LS-
CONVAIR, SAN DIEGO, CALIF.	PH-(NO.)-M
CONVAIR, SAN DIEGO, CALIF.	RT-
CONVAIR, SAN DIEGO, CALIF.	ZA-
CONVAIR, SAN DIEGO, CALIF.	ZB-
CONVAIR, SAN DIEGO, CALIF.	ZC-
CONVAIR, SAN DIEGO, CALIF.	ZG-
CONVAIR, SAN DIEGO, CALIF.	ZJ-
CONVAIR, SAN DIEGO, CALIF.	ZM-
CONVAIR, SAN DIEGO, CALIF.	ZN-
CONVAIR, SAN DIEGO, CALIF.	ZP-
CONVAIR, SAN DIEGO, CALIF.	ZPM-
CONVAIR, SAN DIEGO, CALIF.	ZR-
CONVAIR, SAN DIEGO, CALIF.	ZRAI-
CONVAIR, SAN DIEGO, CALIF.	ZU-
CONVAIR, SAN DIEGO, CALIF.	ZW-
CONVAIR, SAN DIEGO, CALIF.	ZX-
(SEE ALSO LATER NAME: GENERAL DYNAMICS/CONVAIR)	
CONVAIR. ASTRONAUTICS DIV., SAN DIEGO, CALIF.	AZ(LETTER)-
CONVAIR. ASTRONAUTICS DIV., SAN DIEGO, CALIF.	MT-(NUMBER)-
CONVAIR. ASTRONAUTICS DIV., SAN DIEGO, CALIF.	PDSA-(NUMBER)-(YEAR)
CONVAIR. ASTRONAUTICS DIV., SAN DIEGO, CALIF.	TR-(YEAR)-
CONVAIR. ASTRONAUTICS DIV., SAN DIEGO, CALIF.	ZJ-
CONVAIR. ASTRONAUTICS DIV., SAN DIEGO, CALIF.	ZN-
CONVAIR. ASTRONAUTICS DIV., SAN DIEGO, CALIF.	ZR-(NO.)-(NO.)-(NO.)
CONVAIR. NUCLEAR AIRCRAFT RESEARCH FACILITY, FORT WORTH, TEX. (REF. 3)	CVC NARF (NOS.-NOS.-LTR.)
CONVAIR. NUCLEAR AIRCRAFT RESEARCH FACILITY, FORT WORTH, TEX. (REF. 3)	NARF-PR-OP-
CONVAIR. ORDNANCE AEROPHYSICS LAB., DAINGERFIELD,TEX.	CVC OAL/CM-
CONVAIR. ORDNANCE AEROPHYSICS LAB., DAINGERFIELD,TEX.	CVC OAL-R-
CONVAIR-ASTRONAUTICS, SAN DIEGO, CALIF.	AZM-
CONVAIR-ASTRONAUTICS, SAN DIEGO, CALIF.	CVC/A-
CONVAIR-ASTRONAUTICS, SAN DIEGO, CALIF.	EMG-
CONVAIR SCI. RES. LAB., SAN DIEGO, CALIF. (RES. RPT.)	CSRL-RR-
CONVAIR SCIENTIFIC RESEARCH LAB., SAN DIEGO, CALIF.	RN-
COOK ELECTRIC CO., MORTON GROVE, ILL.	CEC-
COOK ELECTRIC CO., MORTON GROVE, ILL.	CEC-PR-

COOK ELECTRIC CO., COOK RESEARCH LABS.

COOK ELECTRIC CO. COOK RESEARCH LABS., SKOKIE, ILL.	COOK FPR-
COOK ELECTRIC CO. TECH-CTR. DIV., MORTON GROVE, ILL.	E-
COOK ELECTRIC CO. TECH-CTR. DIV., MORTON GROVE, ILL.	EO-
COOK ELECTRIC CO. WIRECOM DIV., CHICAGO	C-
COOK RESEARCH LABS., MORTON GROVE, ILL.	CEC CRL-(LETTERS)-
COOK RESEARCH LABS., MORTON GROVE, ILL.	COOK FPR-
COOK RESEARCH LABS., MORTON GROVE, ILL.	CRL-
COOPERATIVE WEAPON DATA INDEXING COMM. (AEC-DOD)	CWDI-
COOPERATIVE WEAPON DATA INDEXING COMM. (AEC-DOD)	CWDIC-
COORDINATING RESEARCH COUNCIL, INC., N.Y.C.	CRC-
COORDINATING RESEARCH COUNCIL, INC., N.Y.C.	CRCI-
COORDINATING RESEARCH COUNCIL, INC., N.Y.C.	CRC-LD-
COORDINATION CENTER REVIEW	CCR-
COORS PORCELAIN CO., GOLDEN, COLO.	CPC-
COORS PORCELAIN CO., GOLDEN, COLO.	CPC-PR-
COOSA VALLEY AREA PLANNING + DEVELOPMENT COMMISSION, ROME, GA.	CFP-CED-GA-(YEAR)
COOSA VALLEY AREA PLANNING + DEVELOPMENT COMMISSION, ROME, GA.	PIP/CB-CED-GA-(YEAR)
COPENHAGEN. UNIVERSITET	COPENHAGENU-
CORNELL AERONAUTICAL LAB., INC., BUFFALO	AA-
CORNELL AERONAUTICAL LAB., INC., BUFFALO	AD-
CORNELL AERONAUTICAL LAB., INC., BUFFALO	AF-(NO.)-(LETTER)
CORNELL AERONAUTICAL LAB., INC., BUFFALO	AM-
CORNELL AERONAUTICAL LAB., INC., BUFFALO	APL-
CORNELL AERONAUTICAL LAB., INC., BUFFALO	BE-
CORNELL AERONAUTICAL LAB., INC., BUFFALO	BE-(NUMBER-LETTER)-
CORNELL AERONAUTICAL LAB., INC., BUFFALO	BUMBLEBEE-
CORNELL AERONAUTICAL LAB., INC., BUFFALO	CA-
CORNELL AERONAUTICAL LAB., INC., BUFFALO	CA-(NUMBER)-P-
CORNELL AERONAUTICAL LAB., INC., BUFFALO	CAF-VJ-
CORNELL AERONAUTICAL LAB., INC., BUFFALO	CAL-
CORNELL AERONAUTICAL LAB., INC., BUFFALO	CAL-AA-
CORNELL AERONAUTICAL LAB., INC., BUFFALO	CAL-AC-
CORNELL AERONAUTICAL LAB., INC., BUFFALO	CAL-AD-
CORNELL AERONAUTICAL LAB., INC., BUFFALO	CAL-AF-
CORNELL AERONAUTICAL LAB., INC., BUFFALO	CAL-AG-
CORNELL AERONAUTICAL LAB., INC., BUFFALO	CAL-AI-
CORNELL AERONAUTICAL LAB., INC., BUFFALO	CAL-AM-
CORNELL AERONAUTICAL LAB., INC., BUFFALO	CAL-AN-
CORNELL AERONAUTICAL LAB., INC., BUFFALO	CAL-BB-
CORNELL AERONAUTICAL LAB., INC., BUFFALO	CAL-BE-
CORNELL AERONAUTICAL LAB., INC., BUFFALO	CAL-BM-
CORNELL AERONAUTICAL LAB., INC., BUFFALO	CAL-C-
CORNELL AERONAUTICAL LAB., INC., BUFFALO	CAL-CA-
CORNELL AERONAUTICAL LAB., INC., BUFFALO	CAL-CM-
CORNELL AERONAUTICAL LAB., INC., BUFFALO	CAL-DF-
CORNELL AERONAUTICAL LAB., INC., BUFFALO	CAL-DM-
CORNELL AERONAUTICAL LAB., INC., BUFFALO	CAL-GI-
CORNELL AERONAUTICAL LAB., INC., BUFFALO	CAL-GM-
CORNELL AERONAUTICAL LAB., INC., BUFFALO	CAL-HF-
CORNELL AERONAUTICAL LAB., INC., BUFFALO	CAL-HM-
CORNELL AERONAUTICAL LAB., INC., BUFFALO	CAL-IG-
CORNELL AERONAUTICAL LAB., INC., BUFFALO	CAL-IH-
CORNELL AERONAUTICAL LAB., INC., BUFFALO	CAL-IM-
CORNELL AERONAUTICAL LAB., INC., BUFFALO	CAL-JA-
CORNELL AERONAUTICAL LAB., INC., BUFFALO	CAL-KA-
CORNELL AERONAUTICAL LAB., INC., BUFFALO	CAL-KB-
CORNELL AERONAUTICAL LAB., INC., BUFFALO	CAL-KC-
CORNELL AERONAUTICAL LAB., INC., BUFFALO	CAL-KD-
CORNELL AERONAUTICAL LAB., INC., BUFFALO	CAL-KF-
CORNELL AERONAUTICAL LAB., INC., BUFFALO	CAL-KM-
CORNELL AERONAUTICAL LAB., INC., BUFFALO	CAL-NM-
CORNELL AERONAUTICAL LAB., INC., BUFFALO	CAL-PI-
CORNELL AERONAUTICAL LAB., INC., BUFFALO	CAL-QM-
CORNELL AERONAUTICAL LAB., INC., BUFFALO	CAL-RA-
CORNELL AERONAUTICAL LAB., INC., BUFFALO	CAL-RM-
CORNELL AERONAUTICAL LAB., INC., BUFFALO	CAL-SA-
CORNELL AERONAUTICAL LAB., INC., BUFFALO	CALT-
CORNELL AERONAUTICAL LAB., INC., BUFFALO	CAL-TB-
CORNELL AERONAUTICAL LAB., INC., BUFFALO	CAL-TG-
CORNELL AERONAUTICAL LAB., INC., BUFFALO	CAL-TM-
CORNELL AERONAUTICAL LAB., INC., BUFFALO	CAL-UA-
CORNELL AERONAUTICAL LAB., INC., BUFFALO	CAL-UB-
CORNELL AERONAUTICAL LAB., INC., BUFFALO	CAL-UF-
CORNELL AERONAUTICAL LAB., INC., BUFFALO	CAL-UM-
CORNELL AERONAUTICAL LAB., INC., BUFFALO	CAL-VB-
CORNELL AERONAUTICAL LAB., INC., BUFFALO	CAL-VC-
CORNELL AERONAUTICAL LAB., INC., BUFFALO	CAL-VE-
CORNELL AERONAUTICAL LAB., INC., BUFFALO	CAL-VF-
CORNELL AERONAUTICAL LAB., INC., BUFFALO	CAL-VG-
CORNELL AERONAUTICAL LAB., INC., BUFFALO	CAL-VH-
CORNELL AERONAUTICAL LAB., INC., BUFFALO	CAL-VJ-
CORNELL AERONAUTICAL LAB., INC., BUFFALO	CAL-VQ-
CORNELL AERONAUTICAL LAB., INC., BUFFALO	CAL-VS-
CORNELL AERONAUTICAL LAB., INC., BUFFALO	CAL-VU-
CORNELL AERONAUTICAL LAB., INC., BUFFALO	CAL-VY-
CORNELL AERONAUTICAL LAB., INC., BUFFALO	CAL-XA-
CORNELL AERONAUTICAL LAB., INC., BUFFALO	CAL-YB-
CORNELL AERONAUTICAL LAB., INC., BUFFALO	CAL-YM-
CORNELL AERONAUTICAL LAB., INC., BUFFALO	DD-(NO.)-(LETTER)-
CORNELL AERONAUTICAL LAB., INC., BUFFALO	DK-
CORNELL AERONAUTICAL LAB., INC., BUFFALO	FDM-
CORNELL AERONAUTICAL LAB., INC., BUFFALO	GI-(NO.)-(LETTER)
CORNELL AERONAUTICAL LAB., INC., BUFFALO	GM-
CORNELL AERONAUTICAL LAB., INC., BUFFALO	HF-
CORNELL AERONAUTICAL LAB., INC., BUFFALO	HFD-PS-(YEAR)-
CORNELL AERONAUTICAL LAB., INC., BUFFALO	HM-
CORNELL AERONAUTICAL LAB., INC., BUFFALO	IG-
CORNELL AERONAUTICAL LAB., INC., BUFFALO	IG-800-S-
CORNELL AERONAUTICAL LAB., INC., BUFFALO	IH-
CORNELL AERONAUTICAL LAB., INC., BUFFALO	IH-(NUMBER)-F-

CORNELL AERONAUTICAL LAB., INC., BUFFALO	IM-(NUMBER)-(LETTER)-
CORNELL AERONAUTICAL LAB., INC., BUFFALO	KA-
CORNELL AERONAUTICAL LAB., INC., BUFFALO	KAI-
CORNELL AERONAUTICAL LAB., INC., BUFFALO	KB-
CORNELL AERONAUTICAL LAB., INC., BUFFALO	KB-826-M-
CORNELL AERONAUTICAL LAB., INC., BUFFALO	KB-935-M-
CORNELL AERONAUTICAL LAB., INC., BUFFALO	KD-
CORNELL AERONAUTICAL LAB., INC., BUFFALO	KF-
CORNELL AERONAUTICAL LAB., INC., BUFFALO	KWP-
CORNELL AERONAUTICAL LAB., INC., BUFFALO	PI-1273-M-
CORNELL AERONAUTICAL LAB., INC., BUFFALO	QM-
CORNELL AERONAUTICAL LAB., INC., BUFFALO	TB-
CORNELL AERONAUTICAL LAB., INC., BUFFALO	TC-
CORNELL AERONAUTICAL LAB., INC., BUFFALO	UD-(NO.)-E-
CORNELL AERONAUTICAL LAB., INC., BUFFALO	VC-
CORNELL AERONAUTICAL LAB., INC., BUFFALO	VE-
CORNELL AERONAUTICAL LAB., INC., BUFFALO	VF-(NUMBER)-(LETTER)-
CORNELL AERONAUTICAL LAB., INC., BUFFALO	VG-(NUMBER)-(LETTER)-
CORNELL AERONAUTICAL LAB., INC., BUFFALO	VH-
CORNELL AERONAUTICAL LAB., INC., BUFFALO	VP-
CORNELL AERONAUTICAL LAB., INC., BUFFALO	WTO-
CORNELL AERONAUTICAL LAB. FLIGHT RES. DEPT., BUFFALO	CAL-FDM-
CORNELL AERONAUTICAL LAB. FLIGHT RES. DEPT., BUFFALO	CAL-FRM-
CORNELL AERONAUTICAL LAB. FLIGHT RES. DEPT., BUFFALO	FRM-
CORNELL-GUGGENHEIM AVIATION SAFETY CENTER, N.Y.C.	AV-CIR-
CORNELL-GUGGENHEIM AVIATION SAFETY CENTER, N.Y.C.	CIR-
CORNELL-SYDNEY UNIV., ITHACA, N.Y. ASTRONOMY CENTER	CSUAC-
CORNELL UNIV., ITHACA, N.Y.	BRN-
CORNELL UNIV., ITHACA, N.Y.	CORN-
CORNELL UNIV., ITHACA, N.Y.	CORNU-
CORNELL UNIV., ITHACA, N.Y. (ALL DEPARTMENTS)	CORNU-TR-
CORNELL UNIV., ITHACA, N.Y.	CORU-
CORNELL UNIV., ITHACA, N.Y. (COGNITIVE SYSTEMS RESEARCH PROGRAM)	CSRP-
CORNELL UNIV., ITHACA, N.Y.	SOIL ENGINEERING SER-
CORNELL UNIV., ITHACA, N.Y. CENTER FOR RADIOPHYSICS AND SPACE RESEARCH	CRSR-
CORNELL UNIV., ITHACA, N.Y. DEPT. OF CHEMISTRY	CU-
CORNELL UNIV., ITHACA, N.Y. DEPT. OF PLANT BREEDING (BIOMETRICS UNIT)	BU-
CORNELL UNIV., ITHACA, N.Y. ELECTRICAL ENG. RES. LAB.	EERL-
CORNELL UNIV., ITHACA, N.Y. ELECTRICAL ENG. RES. LAB.	RR-EERL-
CORNELL UNIV., ITHACA, N.Y. GRAD. SCH. OF AERO. ENG.	CUGSAE-
CORNELL UNIV., ITHACA, N.Y. LAB. OF NUCLEAR STUDIES	CLNS-
CORNELL UNIV., ITHACA, N.Y. LAB. OF NUCLEAR STUDIES	CS-
CORNELL UNIV., ITHACA, N.Y. LAB. OF NUCLEAR STUDIES	NED-
CORNELL UNIV., ITHACA, N.Y. LAB. OF PLASMA STUDIES	LPS-
CORNELL UNIV., ITHACA, N.Y. MATERIAL SCIENCE CENTER	MSC-
CORNELL UNIV., ITHACA, N.Y. MATERIAL SCIENCE CENTER	MSC-TR-
CORNELL UNIV., ITHACA, N.Y. NUCLEAR REACTOR LAB.	CURL-
CORNELL UNIV., ITHACA, N.Y. SCHOOL OF ELECTRICAL ENG.	CORNU-EE-
CORNELL UNIV., ITHACA, N.Y. SCHOOL OF ELECTRICAL ENG.	CUSEE-
CORNELL UNIV., ITHACA, N.Y. SCHOOL OF ELECTRICAL ENG.	EE-
CORNELL UNIV., ITHACA, N.Y. STATE VETERINARY COLL.	NYSVC-
CORNING GLASS WORKS, N.Y.	C-
CORNING GLASS WORKS, N.Y.	CGW-
CORNING GLASS WORKS, N.Y.	IR-
CORNING GLASS WORKS. ELECTRONICS DIV., RALEIGH, N.C.	P-(2 DIGITS)-
CORPS OF ENGINEERS (ARMY)	COE-
CORPS OF ENGINEERS (ARMY)(UNDERGROUND EXPLOSION TEST)	COE-UET-
CORPS OF ENGINEERS (ARMY)	EM-
CORPS OF ENGINEERS (ARMY)	EM-(NO.)-(NO.)-
CORPS OF ENGINEERS (ARMY)	OCE-EM-
CORPS OF ENGINEERS (ARMY)	OCE-T-
CORPS OF ENGINEERS (ARMY)	TM-5-
CORPS OF ENGINEERS (ARMY), FT. BELVOIR, VA.	CE-
CORPS OF ENGINEERS (ARMY), FT. BELVOIR, VA. (ENGINEERING MANUAL)	CE EM-
CORPS OF ENGINEERS (ARMY). BEACH EROSION BD., WASH,DC	ENGINEERING-NOTES-
CORPS OF ENGINEERS (ARMY).DIV OF MECH. ENG.,WASH.,DC	MH-
CORPS OF ENGINEERS (ARMY). ENGINEER BOARD	COE-EB-
CORPS OF ENGINEERS (ARMY). ENGINEER BOARD	EB-
CORPS OF ENGINEERS (ARMY). ENGINEER BOARD	ENG EB-
CORPS OF ENGINEERS (ARMY). ENGINEER BOARD	ENG EB CR-
CORPS OF ENGINEERS (ARMY). ENGINEER BD.(INTERIM RPT.)	ENG EB IR-
CORPS OF ENGINEERS (ARMY). ENGINEER BD. (PROJECT)	ENG EB PROJ (LETTER(S))-
CORPS OF ENGINEERS (ARMY). ENGINEER BOARD	ENG EB R-
CORPS OF ENGINEERS (ARMY). ENGINEER BD.(REGISTER NO.)	ENG EB REG-
CORPS OF ENGINEERS (ARMY). ENGINEER BOARD	ENG EB REPT-
CORPS OF ENGINEERS (ARMY). ENGINEER BD. (ENG. NOTE)	ENG EN-
CORPS OF ENGINEERS (ARMY). ENGINEER BOARD. DESERT TEST BRANCH	DTB REPT-
CORPS OF ENGINEERS (ARMY). ENGINEER BOARD. DESERT TEST BRANCH	ENG EB DTB RPT-
CORPS OF ENGINEERS (ARMY). ENGINEER BOARD. MAP PROJECTION SECTION	ENG EB MPS-
CORPS OF ENGINEERS (ARMY). ENGINEER BOARD. MAP PROJECTION SECTION	MPS-
CORPS OF ENGINEERS (ARMY). MIL. INTELL. DIV.,WASH.,DC	SES-
CORPS OF ENGINEERS (ARMY). MISSOURI RIVER DIV. LABS., OMAHA	COE-MRDL-
CORPS OF ENGINEERS (ARMY). MISSOURI RIVER DIV. LABS., OMAHA	MRDL-(NUMBER-YEAR)
CORPS OF ENGINEERS (ARMY). MISSOURI RIVER DIV. LABS., OMAHA	MRD-LAB-

CORPS OF ENGINEERS (ARMY). MISSOURI RIVER DIV. LABS., OMAHA	MRDL-TR-
CORPS OF ENGINEERS (ARMY). MISSOURI RIVER DIV. LABS., OMAHA	MRD-TE-ERIC-
CORPS OF ENGINEERS (ARMY). MISSOURI RIVER DIV. LABS., OMAHA	MRD-TR-(YEAR)-
CORPS OF ENGINEERS (ARMY). MISSOURI RIVER DIV. LABS., OMAHA	MRD-TR-
CORPS OF ENGINEERS (ARMY). NUCLEAR POWER FIELD OFFICE, FT. BELVOIR, VA.	OSB-
CORPS OF ENGINEERS (ARMY). NUCLEAR POWER FIELD OFFICE, FT. BELVOIR, VA.	OSP-
CORPS OF ENGINEERS (ARMY). OHIO RIVER DIV. LABS., CINCINNATI	COE-ORDL
CORPS OF ENGINEERS (ARMY). OHIO RIVER DIV. LABS., CINCINNATI	ORDL-EC-CR-
CORPS OF ENGINEERS (ARMY). OHIO RIVER DIV. LABS., CINCINNATI	ORDL-EC-TR-
CORPS OF ENGINEERS (ARMY). OHIO RIVER DIV. LABS., CINCINNATI	ORDL-TM-
CORPS OF ENGINEERS (ARMY). OHIO RIVER DIV. LABS., CINCINNATI	ORDL-TR-
CORPS OF ENGINEERS (ARMY). SOUTH PACIFIC DIV., SAN FRANCISCO	SPDGC-
COSMIC, INC., WASHINGTON, D.C.	COSMIC-
COSMODYNE CORP., HAWTHORNE, CALIF.	CE-(NUMBER)-(LETTER)
COURTNEY AND CO., PHILADELPHIA	COURT-
COWLES COMMISSION FOR RESEARCH IN ECONOMICS	CCRE-
CRANFIELD INST. OF TECH., BUCKS, ENGLAND	AERO-
CRANFIELD INST. OF TECH., BUCKS, ENGLAND	MAT-
CRANFIELD INST. OF TECH., BUCKS, ENGLAND	M/P-
CRANFIELD INST. OF TECH., BUCKS, ENGLAND	REPT-
CREARE INC., HANOVER, N.H.	CREARE-TN-N-
CREW RESEARCH LAB., RANDOLPH AFB, TEX.	CRL-
CREW RESEARCH LAB., RANDOLPH AFB, TEX. (LAB. NOTE)	CRL-LN-(YEAR)-
CREW RESEARCH LAB., RANDOLPH AFB, TEX. (TECH. MEMO)	CRL-TM-(YEAR)-
CREW RESEARCH LAB. SURVIVAL FIELD RESEARCH UNIT, STEAD AFB, NEV. (LABORATORY NOTE)	CRL-LN-(YEAR)-
CROP HAIL INSURANCE ACTUARIAL ASSN., CHICAGO	CHIAA-RR-
CROSBY LABS., INC., MINEOLA, N.Y.	CL-
CROSS-MALAKER LABS., INC., MOUNTAINSIDE, N.J.	CML-(NUMBER)-
CROWN ENGINEERING, ALBUQUERQUE, N. MEX.	CE-
CRUCIBLE STEEL CO. OF AMERICA, PITTSBURGH (REF. 1)	(NUMBER-NUMBER-YEAR)
CRUCIBLE STEEL CO. OF AMERICA, PITTSBURGH	CSCO-
CRUCIBLE STEEL CO. OF AMERICA. MIDLAND RES. LAB., PITTSBURGH (INTERIM REPORT)	CSCA-IR-
CUBIC CORP., SAN DIEGO, CALIF.	CUBIC-
CUBIC CORP., SAN DIEGO, CALIF.	CUBIC-FER-
CUBIC CORP., SAN DIEGO, CALIF.	CUBIC-ITR-
CUBIC CORP., SAN DIEGO, CALIF.	FER/(NUMBER)-
CUBIC CORP., SAN DIEGO, CALIF.	FTR/
CULLEN COLL. OF ENGINEERING, HOUSTON, TEX.	RE-(NUMBER-YEAR)
CURTISS-WRIGHT CORP. (ALL LOCATIONS)	C-
CURTISS-WRIGHT CORP.	CI-
CURTISS-WRIGHT CORP.	CURTISS-WRIGHT-
CURTISS-WRIGHT CORP. (ALL LOCATIONS)	CW-
CURTISS-WRIGHT CORP. (ALL LOCATIONS)	CWR-
CURTISS-WRIGHT CORP. (WRIGHT AERONAUTICAL REPORT)	WAR-
CURTISS-WRIGHT CORP., CALDWELL, N.J.	CURTISS-C-
CURTISS-WRIGHT CORP., COLUMBUS, OHIO	SQ-
CURTISS-WRIGHT CORP., WOOD-RIDGE, N.J.	CW-WR-(YEAR)-
CURTISS-WRIGHT CORP., WOOD-RIDGE, N.J.	MRJ-
CURTISS-WRIGHT CORP. AEROPHYSICS DEVELOPMENT DIV., SANTA BARBARA, CALIF.	AEROPHYSICS TR-(YR)-
CURTISS-WRIGHT CORP. AIRPLANE DIV., COLUMBUS, OHIO (REF. 1)	(MODEL NUMBER)-(LETTER)
CURTISS-WRIGHT CORP. AIRPLANE DIV., COLUMBUS, OHIO (REF. 1)	(PROPOSAL NUMBER)-(LTR)-
CURTISS-WRIGHT CORP. AIRPLANE DIV., COLUMBUS, OHIO	V (YR.)-
CURTISS-WRIGHT CORP. CURTISS DIV., CALDWELL, N.J.	CURTISS-
CURTISS-WRIGHT CORP. RESEARCH DIV., CLIFTON, N.J.	CW-R-
CURTISS-WRIGHT CORP. RESEARCH DIV., CLIFTON, N.J.	RE-(YEAR)-(NUMBER)-
CURTISS-WRIGHT CORP. RES. DIV., QUEHANNA, PENNA.	RE-(NO.)-
CURTISS-WRIGHT CORP. RES. DIV., QUEHANNA, PENNA.	RE-(YR.)-(NO.)-
CURTISS-WRIGHT CORP. RES. DIV., QUEHANNA, PENNA.	WNQ-
CURTISS-WRIGHT CORP. RES. DIV., QUEHANNA, PENNA.	WNZ-
CURTISS-WRIGHT CORP. RESEARCH LAB.	CWRL-
CURTISS-WRIGHT CORP. WRIGHT AERO. DIV., WOOD-RIDGE,NJ	CWC-QPR-
CURTISS-WRIGHT CORP. WRIGHT AERO. DIV., WOOD-RIDGE,NJ	CW-MRJ.00-
CURTISS-WRIGHT CORP. WRIGHT AERO. DIV., WOOD-RIDGE,NJ	CW-PR-
CURTISS-WRIGHT CORP. WRIGHT AERO. DIV., WOOD-RIDGE,NJ	D-
CURTISS-WRIGHT CORP. WRIGHT AERO. DIV., WOOD-RIDGE,NJ	MP.00-
CURTISS-WRIGHT CORP. WRIGHT AERO. DIV., WOOD-RIDGE,NJ	MRE-
CURTISS-WRIGHT CORP. WRIGHT AERO. DIV., WOOD-RIDGE,NJ	MRJ.00-
CURTISS-WRIGHT CORP. WRIGHT AERO. DIV., WOOD-RIDGE,NJ	SQ-
CURTISS-WRIGHT CORP. WRIGHT AERO. DIV., WOOD-RIDGE,NJ	TCI-
CURTISS-WRIGHT CORP. WRIGHT AERO. DIV., WOOD-RIDGE,NJ	W.A.C.SERIAL REPORT NO.-
CURTISS-WRIGHT CORP. WRIGHT AERO. DIV., WOOD-RIDGE,NJ	WAD-
CURTISS-WRIGHT CORP. WRIGHT AERO. DIV., WOOD-RIDGE,NJ	WADC-
CURTISS-WRIGHT CORP. WRIGHT AERO. DIV., WOOD-RIDGE,NJ	WAD-CTR-
CURTISS-WRIGHT CORP. WRIGHT AERO. DIV., WOOD-RIDGE,NJ	WAD-R-
CURTISS-WRIGHT CORP. WRIGHT AERO. DIV., WOOD-RIDGE,NJ	WAD-R(NUMBER)F-
CURTISS-WRIGHT CORP. WRIGHT AERO. DIV., WOOD-RIDGE,NJ	WAD-S-
CURTISS-WRIGHT CORP. WRIGHT AERO. DIV., WOOD-RIDGE,NJ	WAD-SST-
CURTISS-WRIGHT CORP. WRIGHT AERO. DIV., WOOD-RIDGE,NJ	WSR-
CUTLER-HAMMER, INC., DEER PARK, N.J. (SEE ALSO: AIRBORNE INSTRUMENTS LAB., INC.)	AIL-(NUMBER)-I-

CZECHOSLOVAKIA. CESKE VYSOKE UCENI TECHNICKE, PRAGUE   B-
CZECHOSLOVAKIA. INST. FOR RESEARCH, PRODUCTION AND APPLICATION OF RADIOISOTOPES, PRAGUE UVVVR-
CZECHOSLOVAKIA. SKODA WORKS, PILSEN   NUCLEX-
CZECHOSLOVAKIA. SKODA WORKS, PILSEN   ZJE-

CZECHOSLOVAKIA. VYZKUMNY USTAV MATEMATICKYCH STROJU, PRAGUE   J-
(NAME TRANSLATED IS: RESEARCH INST. OF MATHEMATICAL MACHINES (COMPUTERS))

DAIMLER-BENZ A.G.

DAIMLER-BENZ A.G., STUTTGART	D-B RPT-
DAIMLER-BENZ A.G., STUTTGART (VERSUCHSBERICHT)	D-B VER-
DAIRYLAND POWER COOPERATIVE, LA CROSSE, WIS.	DPC-(YEAR)-
DAIRYLAND POWER COOPERATIVE, LA CROSSE, WIS.	DPC-(NUMBER-NUMBER)
DALMO VICTOR CO., SAN CARLOS, CALIF.	DVC-
DALY (LEO A.) CO., OMAHA	DALY-
DANA AREA OFFICE, AEC	DPD-(YEAR)-(NUMBER)-(NO.)
DANIEL GUGGENHEIM AIRSHIP INST., AKRON, OHIO	DGAI-
DANIEL MANN JOHNSON AND MENDENHALL. SYSTEMS ENG. DIV. LOS ANGELES	SDR-(YR.)-
DANISH ASSN. FOR IND. DEV. OF ATOMIC ENERGY, HELLERUP	DANATOM-
DANISH ATOMIC ENERGY COMMISSION. RES. ESTAB., RISOE	FRIGG-NOTE-R-
DANISH ATOMIC ENERGY COMMISSION. RES. ESTAB., RISOE	FRIGG-PM-
DANISH ATOMIC ENERGY COMMISSION. RES. ESTAB., RISOE	RISO-
DANISH ATOMIC ENERGY COMMISSION. RES. ESTAB., RISOE	RISO-M-
DANISH ATOMIC ENERGY COMMISSION. RES. ESTAB., RISOE	SM-
DANISH ATOMIC ENERGY COMMISSION. RESEARCH ESTABLISHMENT. ELECTRONICS DEPT., RISOE	R-(NUMBER-YEAR)
DANISH DEFENCE RESEARCH BOARD, COPENHAGEN	F.991.36-
DANISH DEFENCE RESEARCH BOARD, COPENHAGEN	REPT-
DANISH METEOROLOGICAL INST., CHARLOTTENLUND	GR-
DANISH METEOROLOGICAL INST., CHARLOTTENLUND	ISBN-(NO.-NO.-NO.)-
DANISH METEOROLOGICAL INST., CHARLOTTENLUND	REPT.-
DANISH RESEARCH CENTRE FOR APPLIED ELECTRONICS, COPENHAGEN	ECR-
DARLING (R.E.) CO., INC. BETHESDA, MD.	REDAR-RER-
DARTMOUTH COLL., HANOVER, N.H.	DART-
DARTMOUTH COLL., HANOVER, N.H.	DC-
DARTMOUTH COLL., HANOVER, N.H.	SCP-
DARTMOUTH COLL., HANOVER, N.H. DEPT. OF PHYSICS	CART-TR-
DARTMOUTH COLL., HANOVER, N.H. DEPT. OF PHYSICS	DART-TR-
DARTMOUTH COLL., HANOVER, N.H. THAYER SCHOOL OF ENGINEERING (SQUID PROJECT) (REF. 33)	DART-(NUMBER)-(LETTER)
DARTMOUTH COLL., HANOVER, N.H. THAYER SCHOOL OF ENGINEERING (SQUID PROJECT) (REF. 33)	DART TM-
DARTMOUTH COLL., HANOVER, N.H. THAYER SCHOOL OF ENG.	DCW-
DARTMOUTH COLL., HANOVER, N.H. THAYER SCHOOL OF ENG.	SCP-
DATA CORP., DAYTON, OHIO	DFR-(YEAR)-
DATA CORP., DAYTON, OHIO	DTR-(YEAR)-
DATA CORP., DAYTON, OHIO	FR-
DATA CORP., DAYTON, OHIO	FR-(YEAR)-
DATA CORP., DAYTON, OHIO	MPR-
DATA-DESIGN LABS., ONTARIO, CALIF.	DDL-(NUMBER)-
DATA DYNAMICS, INC., LOS ANGELES	DDI-(NO.-NO.-NO.)
DAVID SARNOFF RESEARCH CENTER, PRINCETON, N.J.	DSRC-
DAVID SARNOFF RESEARCH CENTER, PRINCETON, N.J.	DSRC-QPR-
DAVID SARNOFF RESEARCH CENTER, PRINCETON, N.J.	DSRC-SAR-
DAVID SARNOFF RESEARCH CENTER, PRINCETON, N.J.	DSRC-TR-
DAVID SARNOFF RESEARCH CENTER, PRINCETON, N.J.	PASAP-
DAVID SARNOFF RESEARCH CENTER, PRINCETON, N.J.	RCA-QR-
DAVID TAYLOR MODEL BASIN, CARDEROCK, MD. (AERO DATA REPORT)	ADR-
DAVID TAYLOR MODEL BASIN, CARDEROCK, MD.	C-
DAVID TAYLOR MODEL BASIN, CARDEROCK, MD.	DTMB-
DAVID TAYLOR MODEL BASIN, CARDEROCK, MD.	DTMB-AL-
DAVID TAYLOR MODEL BASIN, CARDEROCK, MD.	DTMB-AL-C-
DAVID TAYLOR MODEL BASIN, CARDEROCK, MD.	DTMB-C-
DAVID TAYLOR MODEL BASIN, CARDEROCK, MD.	DTMB-R-
DAVID TAYLOR MODEL BASIN, CARDEROCK, MD.	DTMB-S-
DAVID TAYLOR MODEL BASIN, CARDEROCK, MD.	DTMB-T-
DAVID TAYLOR MODEL BASIN, CARDEROCK, MD. (TRANSLATION)	DTMB-TRANS.-
DAVID TAYLOR MODEL BASIN, CARDEROCK, MD.	DWT-
DAVID TAYLOR MODEL BASIN, CARDEROCK, MD.	DWTMB-
DAVID TAYLOR MODEL BASIN, CARDEROCK, MD.	DWTMB (LETTER)-
DAVID TAYLOR MODEL BASIN, CARDEROCK, MD. (TRANSLATION, RUSSIAN MODEL BASIN)	DWTMB T/RMB-
DAVID TAYLOR MODEL BASIN, CARDEROCK, MD.	EMB-
DAVID TAYLOR MODEL BASIN, CARDEROCK, MD.	R-
DAVID TAYLOR MODEL BASIN, CARDEROCK, MD.	S-
DAVID TAYLOR MODEL BASIN, CARDEROCK, MD.	TMA-
DAVID TAYLOR MODEL BASIN, CARDEROCK, MD.	TMB-
DAVID TAYLOR MODEL BASIN, CARDEROCK, MD.	TMB-C-
DAVID TAYLOR MODEL BASIN, CARDEROCK, MD.	TMB-T-
DAVID TAYLOR MODEL BASIN, CARDEROCK, MD. (TRANSLATION, RUSSIAN MODEL BASIN)	T/RMB-
DAVID TAYLOR MODEL BASIN. AERODYNAMICS LAB., WASH., DC	AERO-
DAVID TAYLOR MODEL BASIN. AERODYNAMICS LAB., WASH., DC	AERO RPT.-
DAVID TAYLOR MODEL BASIN. AERODYNAMICS LAB., WASH., DC	AL-
DAVID TAYLOR MODEL BASIN. AERODYNAMICS LAB., WASH., DC	AL-C-
DAVID TAYLOR MODEL BASIN. AERODYNAMICS LAB., WASH., DC	DTMB-AERO-
DAVID TAYLOR MODEL BASIN. AERODYNAMICS LAB., WASH., DC	TEST-AL-
DAVID TAYLOR MODEL BASIN. AERODYNAMICS LAB., WASH., DC	TEST-AL-C-
DAVID TAYLOR MODEL BASIN. AERODYNAMICS LAB., WASH., DC	TEST-B-
DAVID TAYLOR MODEL BASIN. AERODYNAMICS LAB., WASH., DC	TMB-AERO-
DAVID TAYLOR MODEL BASIN. APPLIED MATHEMATICS LAB., CARDEROCK, MD.	AML-
DAVID TAYLOR MODEL BASIN. APPLIED MATHEMATICS LAB., CARDEROCK, MD.	DTMB-AML-
DAVISON CHEMICAL CORP., BALTIMORE	DCC-
DAYSTROM, INC. MIL. ELECTRONICS DIV., ARCHBALD, PA.	DAYSTROM-
DAYSTROM, INC. WESTON INSTRUMENTS DIV., NEWARK, N.J.	XS-
DAYTON, OHIO. UNIV.	EE-
DAYTON, OHIO. UNIV.	M/A-

DAYTON, OHIO. UNIV. DEPT. OF ELECTRICAL ENG.	EET-
DAYTON, OHIO. UNIV. RESEARCH INST.	UDRI-PR-(YEAR)-
DAYTON, OHIO. UNIV. RESEARCH INST.	UDRI-TR-(YEAR)-
DAYTON ELECTRONIC PRODUCTS CO., INC., OHIO	DEPCO-(LETTER)-
DE BELL + RICHARDSON, INC., HAZARDVILLE, CONN.	DEBELL-(NUMBER)-QPR-
DE BELL + RICHARDSON, INC., HAZARDVILLE, CONN.	PLASTEC-
DE HAVILLAND AIRCRAFT CO., LTD. (GT. BRIT.)	PM-
DE HAVILLAND AIRCRAFT CO., LTD. SPECIAL PRODUCTS AND APPLIED RESEARCH DIV., MALTON, ONT.	DHC-SP-R.(NO.)-(LTR.)
DE HAVILLAND AIRCRAFT OF CANADA LTD.	AEROC-
DE HAVILLAND AIRCRAFT OF CANADA LTD.	DHA-
DE HAVILLAND AIRCRAFT OF CANADA LTD. SPECIAL PRODUCTS AND RESEARCH DIV., MALTON, ONT.	DHC-SP-R.-
DE HAVILLAND PROPELLERS, LTD.	GW-
DECISION SCIENCE, INC., SAN DIEGO, CALIF.	DSI-(NUMBER)-F
DEFENSE ATOMIC SUPPORT AGENCY, WASHINGTON, D.C.	DASA-
DEFENSE ATOMIC SUPPORT AGENCY, WASHINGTON, D.C.	DASA-BIB-
DEFENSE ATOMIC SUPPORT AGENCY, WASHINGTON, D.C.	DASA-EM-
DEFENSE ATOMIC SUPPORT AGENCY, WASHINGTON, D.C.	DASA-ITR-
DEFENSE ATOMIC SUPPORT AGENCY, WASHINGTON, D.C.	DASA-SC-
DEFENSE ATOMIC SUPPORT AGENCY, WASH., DC (TECH. LTR)	DASA TL-
DEFENSE ATOMIC SUPPORT AGENCY, WASHINGTON, D.C.	NWER-
DEFENSE ATOMIC SUPPORT AGENCY. DATA CENTER, SANTA BARBARA, CALIF. (BIBLIOGRAPHY)	DASA-DC-BIB-
DEFENSE ATOMIC SUPPORT AGENCY. FIELD COMMAND, ALBUQUERQUE, N. MEX.	C(NO.)-CR
DEFENSE ATOMIC SUPPORT AGENCY. FIELD COMMAND, ALBUQUERQUE, N. MEX.	C(NO.)-CR-CB(NO.)
DEFENSE ATOMIC SUPPORT AGENCY. FIELD COMMAND, ALBUQUERQUE, N. MEX.	C(NO.)-CR-CN(NO.)
DEFENSE ATOMIC SUPPORT AGENCY. FIELD COMMAND, ALBUQUERQUE, N. MEX.	DASA-
DEFENSE ATOMIC SUPPORT AGENCY. FIELD COMMAND, ALBUQUERQUE, N. MEX.	DASA/FC
DEFENSE ATOMIC SUPPORT AGENCY. FIELD COMMAND, ALBUQUERQUE, N. MEX.	DASA/FC (MO.YR.)-
DEFENSE ATOMIC SUPPORT AGENCY. FIELD COMMAND, ALBUQUERQUE, N. MEX.	FC/
DEFENSE ATOMIC SUPPORT AGENCY. FIELD COMMAND, ALBUQUERQUE, N. MEX.	FC-(8 DIGITS)
DEFENSE ATOMIC SUPPORT AGENCY. FIELD COMMAND, ALBUQUERQUE, N. MEX.	FCCA-
DEFENSE ATOMIC SUPPORT AGENCY. FIELD COMMAND, ALBUQUERQUE, N. MEX.	FCCG-
DEFENSE ATOMIC SUPPORT AGENCY. FIELD COMMAND, ALBUQUERQUE, N. MEX.	FCTG-(NUMBER).(NUMBER)
DEFENSE ATOMIC SUPPORT AGENCY. FIELD COMMAND, ALBUQUERQUE, N. MEX. (WEAPONS TESTS)	FCWT-
DEFENSE ATOMIC SUPPORT AGENCY. FIELD COMMAND, ALBUQUERQUE, N. MEX. (NUCLEAR EMERGENCY TEAM TRAINING)	NET-WB-
DEFENSE ATOMIC SUPPORT AGENCY. FIELD COMMAND, ALBUQUERQUE, N.M. (NAVY NUCLEAR WEAPONS OFFICER COURSE)	WB-NWO-
DEFENSE ATOMIC SUPPORT AGENCY. FIELD COMMAND, ALBUQUERQUE, N. MEX.	XW-
DEFENSE ATOMIC SUPPORT AGENCY.RADIATION DIV., WASH,DC	RD-(NO.-NO.)(MO./YR.)
DEFENSE ATOMIC SUPPORT AGENCY. TEST COMMAND, ALBUQUERQUE, N. MEX.	DASAR-
DEFENSE COMMUNICATIONS AGENCY, WASHINGTON, D.C.	CSI-(YEAR)-TR-
DEFENSE COMMUNICATIONS AGENCY, WASHINGTON, D.C.	CSM-(NO.)(LTR.)-
DEFENSE COMMUNICATIONS AGENCY, WASHINGTON, D.C.	DCA-(NO.)-(LTR.)-
DEFENSE COMMUNICATIONS AGENCY, WASHINGTON, D.C.	DCA-CIR-
DEFENSE COMMUNICATIONS AGENCY, WASHINGTON, D.C.	DCA-TR-
DEFENSE COMMUNCIATIONS AGENCY, WASHINGTON, D.C.	MAM-(YEAR)-
DEFENSE DOCUMENTATION CENTER, ALEXANDRIA, VA.	AD-(6 DIGITS)
DEFENSE DOCUMENTATION CENTER, ALEXANDRIA, VA.	ARB-
DEFENSE DOCUMENTATION CENTER, ALEXANDRIA, VA. (A REPORT BIBLIOGRAPHY, SEARCH CONTROL NUMBER...)	ARB-SC-
DEFENSE DOCUMENTATION CENTER, ALEXANDRIA, VA.	DDC-
DEFENSE DOCUMENTATION CENTER, ALEXANDRIA, VA.	DDC-TAS-(YEAR)-
DEFENSE DOCUMENTATION CENTER, ALEXANDRIA, VA. (TECHNICAL ABSTRACT BULLETIN)	TAB-(YEAR)-
DEFENSE DOCUMENTATION CENTER, ALEXANDRIA, VA. (TECHNICAL ABSTRACT BULLETIN, CLASSIFIED)	TAB-C-
DEFENSE DOCUMENTATION CENTER, ALEXANDRIA, VA. (TECHNICAL ABSTRACT BULLETIN, CLASSIFIED. USED 1962-63)	TAB-S(YEAR)-
DEFENSE DOCUMENTATION CENTER, ALEXANDRIA, VA. (TECHNICAL ABSTRACT BULLETIN,UNCLASSIFIED. USED 1953-64)	TAB-U-
DEFENSE DOCUMENTATION CENTER, ALEXANDRIA, VA.	TAS-(YEAR)-
DEFENSE LOGISTICS SERVICES CENTER, BATTLE CREEK,MICH. (MASTER CROSS REFERENCE LIST)	DLSC-MCRL-PTIC-
DEFENSE LOGISTICS SERVICES CENTER, BATTLE CREEK,MICH. (FEDERAL SUPPLY CODE)	DSA-FSC-
DEFENSE LOGISTICS SERVICES CENTER, BATTLE CREEK,MICH. (DOD ACTIVITY ADDRESS DIRECTORY)	DSAH-
DEFENSE INTELLIGENCE AGENCY, WASHINGTON, D.C.	DIA-
DEFENSE INTELLIGENCE AGENCY, WASHINGTON, D.C.	DIAM-(YEAR)-
DEFENSE INTELLIGENCE AGENCY, WASH., D.C. (LIBRARY)	DIL-
DEFENSE NUCLEAR AGENCY, WASHINGTON, D.C.	DNA-
DEFENSE NUCLEAR AGENCY, WASHINGTON, D.C.	DNA-(NUMBER)-F
DEFENSE NUCLEAR AGENCY, WASHINGTON, D.C.	DNA-(NUMBER)-T
DEFENSE RESEARCH CORP., SANTA BARBARA, CALIF.	CR-(NUMBER)-
DEFENSE RESEARCH CORP., SANTA BARBARA, CALIF.	CSR-
DEFENSE RESEARCH CORP., SANTA BARBARA, CALIF.	DRC-
DEFENSE RESEARCH CORP., SANTA BARBARA, CALIF.	DRC-(YEAR)-
DEFENSE RESEARCH CORP., SANTA BARBARA, CALIF.	DRC-CSR-
DEFENSE RESEARCH CORP., SANTA BARBARA, CALIF.	DRC-CSR-SD-
DEFENSE RESEARCH CORP., SANTA BARBARA, CALIF.	DRC-IMR-
DEFENSE RESEARCH CORP., SANTA BARBARA, CALIF.	DRC-TM-
DEFENSE RESEARCH CORP., SANTA BARBARA, CALIF.	IMR-
DEFENSE SCIENCE BOARD TASK FORCE, WASH., D.C.	DSBTF-
DEFENSE SUPPLY AGENCY, WASHINGTON, D.C.	WM-
DEFENSE SUPPLY AGENCY, WASHINGTON, D.C.	WORKING MEMO-
DEL MAR ENGINEERING LAB., LOS ANGELES	DEL-
DEL MAR ENGINEERING LABS., SANTA MONICA, CALIF.	JA-
DELAWARE. STATE PLANNING OFFICE, DOVER	DEL-SPO-(YEAR)-
DELAWARE. UNIV., NEWARK	DEL-
DELAWARE. UNIV., NEWARK (SQUID PROJECT) (REF. 33)	DEL-(NUMBER)-P
DELAWARE. UNIV., NEWARK	G-
DELAWARE. UNIV., NEWARK	ME-Q-
DELAWARE. UNIV., NEWARK	ME-T-
DELAWARE. UNIV., NEWARK	TR-

DELAWARE. UNIV.

DELAWARE. UNIV., NEWARK	UD-FB-
DELAWARE. UNIV., NEWARK	WAF-
DELAWARE. UNIV., NEWARK. DEPT. OF CHEM.	AFCOMB-
DELAWARE. UNIV., NEWARK. DEPT. OF ELECTRICAL ENG.	Q(NUMBER)-
DELAWARE. UNIV., NEWARK. DEPT. OF ELECTRICAL ENG.	TR-Q(NUMBER)-
DELAWARE VALLEY REGIONAL PLANNING COMMISSION, PHILA.	DVRPC-
DELCO-REMY DIV., GENERAL MOTORS CORP., MUNCIE, IND.	D-R-
DELEX SYSTEMS, INC., ARLINGTON, VA.	TR-D-
DELFT, NETHERLANDS. TECHNISCHE HOGESCHOOL (DELFT MOLTEN SALT PROJECT)	DMSP-(LETTER)-
DELFT, NETHERLANDS. TECHNISCHE HOGESCHOOL	DMSP-G-
DELFT, NETHERLANDS. TECHNISCHE HOGESCHOOL	TH-
DELFT, NETHERLANDS. TECHNISCHE HOGESCHOOL	UTH-
DELFT, NETHERLANDS. TECHNISCHE HOGESCHOOL	VTH-
DELFT, NETHERLANDS. TECHNISCHE HOGESCHOOL. LABORATORIUM VOOR ENERGIEVOORZIENING EN KERNREACTOREN	THD-KR-
DELFT, NETHERLANDS. TECHNISCHE HOGESCHOOL. LABORATORIUM VOOR TECHNISCHE MECHANICA	WHTD-
DELFT, NETHERLANDS. TECHNISCHE HOGESCHOOL. LABORATORIUM VOOR VOERTUIGTECHNIEK	TH-A-
DEMOLITION RESEARCH UNIT (NAVY) (INTERNAL MEMORANDUM)	DRUIM-
DENMARK. AKADEMIET FOR DE TEKNISKE VIDENSKABER	ATS-
DENMARK. FORSVARETS FORSKNINGSRAD, COPENHAGEN	FFR-
DENMARK. FORSVARETS FORSKNINGSRAD, COPENHAGEN	HF-
DENMARK. TECHNICAL UNIV., COPENHAGEN	LFM-R-(YEAR)-
DENMARK. TECHNICAL UNIV., COPENHAGEN. COASTAL ENG.LAB.	CI-
DENMARK. TEKNISKE HOJSKOLE, LYNGBY	D-(3 DIGITS)
DENMARK. TEKNISKE HOJSKOLE, LYNGBY	DTH-
DENMARK. TEKNISKE HOJSKOLE, LYNGBY	IONLAB-P-
DENMARK. TEKNISKE HOJSKOLE, LYNGBY	S(NUMBER)-
DENMARK. TEKNISKE HOJSKOLE, LYNGBY	S-(NUMBER)-R-
DENMARK. TEKNISKE HOJSKOLE, LYNGBY. HYDRO-AND AERODYNAMICS LAB. AERODYNAMICS SECTION	A-
DENMARK. TEKNISKE HOJSKOLE, LYNGBY. HYDRO-AND AERODYNAMICS LAB. HYDRODYNAMICS SECTION	HY-
DENVER. UNIV.	DENVERU-
DENVER. UNIV.	DU-
DENVER. UNIV.	MSR-
DENVER. UNIV. COLL. OF LAW	AR-(1 DIGIT)
DENVER. UNIV. DENVER RESEARCH INST.	DR-
DENVER. UNIV. DENVER RESEARCH INST.	DRI-
DENVER. UNIV. DENVER RESEARCH INST.	DRI-(NUMBER)-
DENVER. UNIV. DENVER RESEARCH INST.	DRI-LAR-
DENVER. UNIV. DENVER RESEARCH INST.	DRI-MP-
DENVER. UNIV. DENVER RESEARCH INST.	DRI-QR-
DENVER. UNIV. DENVER RESEARCH INST.	DU-
DENVER. UNIV. DENVER RESEARCH INST.	DU-DRI-
DENVER. UNIV. DENVER RESEARCH INST.	DU DRI (LETTERS)
DENVER. UNIV. DENVER RESEARCH INST.	DU-ER-
DENVER. UNIV. DENVER RESEARCH INST.	DU-PR-
DENVER. UNIV. DENVER RESEARCH INST.	GV-
DENVER. UNIV. DENVER RESEARCH INST.	LAR-
DENVER. UNIV. DENVER RESEARCH INST.	QTN-
DENVER. UNIV. DENVER RESEARCH INST. DIV. OF MATHEMATICAL SCIENCES	MS-R-
DENVER REGIONAL COUNCIL OF GOVERNMENTS	DRCOG-(YEAR)-
DENVER REGIONAL COUNCIL OF GOVERNMENTS	MASTER PLAN-
DEPARTMENT OF AGRICULTURE, WASHINGTON, D.C.	USDA-
DEPARTMENT OF AGRICULTURE. AGRICULTURAL RESEARCH SERVICE, WASHINGTON, D.C.	ARS-
DEPT. OF AGRICULTURE. CROPS RES. DIV., WASHINGTON, DC	CR-(MONTH)-(YEAR)
DEPT. OF AGRICULTURE. EASTERN REGIONAL RESEARCH LAB., WYNDMOOR, PENNA. (AGRICULTURAL + IND. CHEMISTRY)	AIC-
DEPARTMENT OF AGRICULTURE. EASTERN UTILIZATION RESEARCH BRANCH, WYNDMOOR, PENNA.	ERRL-
DEPARTMENT OF AGRICULTURE. ECONOMIC RESEARCH SERVICE, WASHINGTON, D.C.	ERS-
DEPARTMENT OF AGRICULTURE. NORTHERN UTILIZATION RESEARCH BRANCH, PEORIA, ILL.	NRRL-
DEPT. OF AGRICULTURE. PERSONNEL RES. STAFF, WASH., DC	PRD-(YEAR)-
DEPT. OF AGRICULTURE. PERSONNEL RES. STAFF, WASH., DC	PRS-(YEAR)-
DEPARTMENT OF AGRICULTURE. SOIL AND WATER CONSERVATION RESEARCH DIV., BELTSVILLE, MD.	SWCRD-AR-
DEPARTMENT OF AGRICULTURE. SOUTHERN UTILIZATION RESEARCH BRANCH, NEW ORLEANS	SRRL-
DEPARTMENT OF AGRICULTURE. WESTERN UTILIZATION RESEARCH BRANCH, ALBANY, CALIF.	WRRL-
DEPARTMENT OF COMMERCE, WASH., D.C. (COMMERCIAL STD.)	CS-
DEPARTMENT OF COMMERCE, WASHINGTON, D.C.	USCOMM-DC-
DEPARTMENT OF DEFENSE, WASHINGTON, D.C. (GUIDED MISSILE MAILING LIST)	ANAF/GMML-
DEPARTMENT OF DEFENSE, WASHINGTON, D.C.	DOD-
DEPARTMENT OF DEFENSE, WASH., D.C. (FOR EXPLANATION OF THE REMAINDER OF THE CODE, SEE LTRS AFTER DOD-)	DOD-(LETTERS)-
DEPARTMENT OF DEFENSE, WASHINGTON, D.C. (GUIDED MISSILE MAILING LIST)	GMML-
DEPARTMENT OF DEFENSE, WASHINGTON, D.C. (USED TO IDENTIFY SPECIFICATIONS. THE LETTER FOLLOWING MIL- REPRESENTS THE ARTICLE FOR WHICH THE SPECIFICATIONS ARE GIVEN, E.G. MIL-T-, TENT, TUBE, ETC.) (REF. 13)	MIL-(LETTER)-
DEPARTMENT OF DEFENSE, WASHINGTON, D.C. (GUIDED MISSILE MAILING LIST)	MML-
DEPARTMENT OF DEFENSE, WASHINGTON, D.C. (SYSTEM DEVELOPMENT REQUIREMENT)	SDR-
DEPT. OF DEFENSE. ADVISORY GROUP ON ELECTRON TUBES	AGET-

DEPARTMENT OF DEFENSE. ADVISORY GROUP ON RELIABILITY OF ELECTRONIC EQUIPMENT	AGREE-
DEPARTMENT OF DEFENSE. ADVISORY GROUP ON RELIABILITY OF ELECTRONIC EQUIPMENT	AGREE-(YR)-(MONTH)-
DEPARTMENT OF HEALTH, EDUCATION, AND WELFARE. OFFICE OF EDUCATION, WASHINGTON, D.C.	OE-
DEPARTMENT OF HOUSING AND URBAN DEV., WASHINGTON,D.C.	HUD-MP-
DEPARTMENT OF HOUSING AND URBAN DEVELOPMENT. DIVISION OF INTERNATIONAL AFFAIRS, WASHINGTON, D.C.	IME-
DEPARTMENT OF HOUSING AND URBAN DEVELOPMENT. DIVISION OF INTERNATIONAL AFFAIRS, WASHINGTON, D.C.	LEAFLET-
DEPARTMENT OF JUSTICE. OFFICE OF LAW ENFORCEMENT ASSISTANCE, WASHINGTON, D.C.	LEAA-NI-(YEAR)-
DEPARTMENT OF LABOR, WASHINGTON, D.C.	MEL-
DEPARTMENT OF LABOR. MANPOWER ADM., WASHINGTON, D.C.	DLMA-(CONTRACT NO.)-
DEPT. OF STATE. DIV. OF ACQUISITION AND DISTRIBUTION	IAD-
DEPARTMENT OF STATE. OFFICE OF EXTERNAL RESEARCH	ACD-
DEPARTMENT OF STATE. OFFICE OF FAR EASTERN AFFAIRS, WASHINGTON, D.C.	FE-
DEPT. OF STATE. OFFICE OF RESEARCH AND ANALYSIS FOR THE NEAR EAST AND SOUTH ASIA, WASHINGTON, D.C.	FAR-(5 DIGITS)
DEPARTMENT OF THE INTERIOR, WASHINGTON, D.C.	BIBLIOGRAPHY-
DEPARTMENT OF THE INTERIOR, WASHINGTON, D.C. (POPULATION TRENDS AND ENVIRONMENTAL POLICY)	PTEP-
DEPARTMENT OF THE INTERIOR, WASHINGTON, D.C.	SPECIAL LISTING-
DEPARTMENT OF THE INTERIOR, WASHINGTON, D.C. (SALINE WATER CONVERSION)	SWCP-AR-
DEPARTMENT OF THE INTERIOR, WASHINGTON, D.C. (SALINE WATER CONVERSION)	SWCP-PR-
DEPARTMENT OF THE INTERIOR. CENTRAL LIBRARY, WASHINGTON, D.C. (TRANSLATIONS)	DI-
DEPT. OF TRANSPORTATION, WASHINGTON, D.C.	DOT-CG-
DEPT. OF TRANSPORTATION, WASHINGTON, D.C.	DOT-FRA-OHSGT-
DEPT. OF TRANSPORTATION, WASHINGTON, D.C. (HIGHWAY SAFETY)	DOT-HS-
DEPT. OF TRANSPORTATION, WASH., D.C. (AIRCRAFT NOISE)	DOT/HUD-IANAP-(YEAR)-
DEPT. OF TRANSPORTATION, WASHINGTON, D.C.	DOT/OS-(CONTRACT NO.)-
DEPT. OF TRANSPORTATION, WASHINGTON, D.C. (HIGHWAY SAFETY)	DOTS-HS-
DEPT. OF TRANSPORTATION, WASHINGTON, D.C.	DOT-TD-
DEPT. OF TRANSPORTATION, WASHINGTON, D.C. (NORTHEAST CORRIDOR TRANSPORTATION PROJECT REPORT)	NECTP-
DEPT. OF TRANSPORTATION, WASH., D.C. (SYSTEMS DEVELOPMENT + TECHNOLOGY, NOISE ABATEMENT)	OST-ONA-(YEAR)-
DEPT. OF TRANSPORTATION. LIBRARY SERVICES DIV., WASHINGTON, D.C.	BIBLIOGRAPHIC LIST-
DEPUTY CHIEF OF STAFF, PLANS AND OPERATIONS (AIR FORCE), WASHINGTON, D.C.	AFRDC-DPM-(YEAR)-
DEPUTY CHIEF OF STAFF. PLANS AND OPERATIONS (AIR FORCE), WASHINGTON, D.C.	AFRDC-DPR-(YEAR)-
DEPUTY CHIEF OF STAFF. PLANS AND OPERATIONS (AIR FORCE), WASHINGTON, D.C.	AFXDC-
DEPUTY CHIEF OF STAFF. PLANS AND OPERATIONS (AIR FORCE), WASHINGTON, D.C.	AFXPD-
DEPUTY CHIEF OF STAFF. RESEARCH AND DEVELOPMENT (AIR FORCE), WASHINGTON, D.C.	AFRDC-
DEPUTY CHIEF OF STAFF. RESEARCH AND DEVELOPMENT (AIR FORCE), WASHINGTON, D.C.	AFRDC-DPM-(YEAR)-
DEPUTY CHIEF OF STAFF. RESEARCH AND DEVELOPMENT (AIR FORCE), WASHINGTON, D.C.	AFRDC-DPR-(YEAR)-
DEPUTY CHIEF OF STAFF. RESEARCH AND DEVELOPMENT (AIR FORCE), WASHINGTON, D.C.	AFRDC-DPS-(YEAR)-
DESERET TEST CENTER, FORT DOUGLAS, UTAH	DTC-(YEAR)-
DESERET TEST CENTER, FORT DOUGLAS, UTAH	DTC-
DESERET TEST CENTER, FORT DOUGLAS, UTAH	DTC-(6 DIGITS)R
DESERET TEST CENTER, FORT DOUGLAS, UTAH	DTC-(NUMBER-NUMBER)
DESERET TEST CENTER, FORT DOUGLAS, UTAH	DTC-B-
DESERET TEST CENTER, FORT DOUGLAS, UTAH	DTC-C-
DESERET TEST CENTER, FORT DOUGLAS, UTAH	DTC-DR-E(NO.)-VFC
DESERET TEST CENTER, FORT DOUGLAS, UTAH	DTC-EC-
DESERET TEST CENTER, FORT DOUGLAS, UTAH	DTC-FIRING RECORD-(YR.)-
DESERET TEST CENTER, FORT DOUGLAS, UTAH	DTC-FR-E-
DESERET TEST CENTER, FORT DOUGLAS, UTAH	DTC-IR-E-
DESERET TEST CENTER, FORT DOUGLAS, UTAH	DTC-LR-E-
DESERET TEST CENTER, FORT DOUGLAS, UTAH	DTC-R-E-
DESERET TEST CENTER, FORT DOUGLAS, UTAH	DTC-SPECIAL STUDY-R-
DESERET TEST CENTER, FORT DOUGLAS, UTAH	DTC-SR-
DESERET TEST CENTER, FORT DOUGLAS, UTAH	DTC-TB-
DESERET TEST CENTER, FORT DOUGLAS, UTAH	DTC-TC-
DESERET TEST CENTER, FORT DOUGLAS, UTAH	DTC-TEST-(YEAR)-
DESERET TEST CENTER, FORT DOUGLAS, UTAH	DTC-TJ-
DESERET TEST CENTER. JOINT CONTACT POINT DIV., FORT DOUGLAS, UTAH	JCP-(NO.)-(NO.)
DESIGNERS FOR INDUSTRY, INC., CLEVELAND (REF. 1)	(PROJECT NUMBER)-
DESIGNERS FOR INDUSTRY, INC., CLEVELAND	DFI-
DETROIT ARSENAL, CENTER LINE, MICH.	DA-
DETROIT ARSENAL, CENTER LINE, MICH.	DARS-
DETROIT ARSENAL, CENTER LINE, MICH.	ORDMX-ECM-
DETROIT ARSENAL, CENTER LINE, MICH.	OTAC-
DETROIT CONTROLS CORP., REDWOOD CITY, CALIF.	DCC-
DETROIT CONTROLS CORP., REDWOOD CITY, CALIF.	DETRCC-
DETROIT CONTROLS CORP., REDWOOD CITY, CALIF.	RC-
DETROIT CONTROLS CORP., REDWOOD CITY, CALIF.	RU-
DETROIT CONTROLS CORP. RES. DIV., REDWOOD CITY,CALIF.	DCC-RC-
DETROIT EDISON CO. NUCLEAR POWER DEVELOPMENT DEPT.	NPDD-
DETROIT SIGNAL LABORATORY	DSL-
DETROIT SIGNAL LABORATORY (ASSIGNMENT MEMORANDA)	DSL AM-
DETROIT SIGNAL LABORATORY (ASSIGNMENT MEMORANDA)	DSL AM (LETTER(S))-
DETROIT SIGNAL LABORATORY (ENGINEERING MEMORANDA)	DSL EM-
DETROIT SIGNAL LABORATORY (ENGINEERING MEMORANDA)	DSL EM (LTR.-NO.-LTRS)-
DETROIT TESTING LAB.	DTL-
DEUTSCHE AKADEMIE DER LUFTFAHRTFORSCHUNG, BERLIN	DAL-
DEUTSCHE FORSCHUNGS GESELLSCHAFT	DFG-
DEUTSCHE FORSCHUNGSANSTALT FUER LUFTFAHRT E.V., INSTITUT FUER AERODYNAMIK, BRUNSWICK	DFL-
DEUTSCHE FORSCHUNGSANSTALT FUER SEGELFLUG, MUNICH	DFS-
DEUTSCHE GESELLSCHAFT FUER FLUGWISSENSCHAFTEN, BRUNSWICK	DLR-FB-(YEAR)-

DEUTSCHE GESELLSCHAFT FUER LUFT- UND RAUMFAHRT E.V.

DEUTSCHE GESELLSCHAFT FUER LUFT- UND RAUMFAHRT E.V., COLOGNE	DGLR-(YEAR)-
DEUTSCHE GESELLSCHAFT FUER RAKETENTECHNIK UND RAUMFAHRT, STUTTGART	DGRR/WGLR PAPER-(YR.)-
DEUTSCHE LUFTFAHRTFORSCHUNG (JAHRBUCH)	JDL-
DEUTSCHE LUFTHANSA A.G. (BERICHT)	DLH-
DEUTSCHE VERSUCHSANSTALT FUER LUFT- UND RAUMFAHRT (ALL LOCATIONS) (REPRINT)	DFVLR-SONDERDRUCK-
DEUTSCHE VERSUCHSANSTALT FUER LUFT- UND RAUMFAHRT (ALL LOCATIONS)	DLR-FB-(YEAR)-
DEUTSCHE VERSUCHSANSTALT FUER LUFT- UND RAUMFAHRT (ALL LOCATIONS)	DLR-MITT.-(YEAR)-
DEUTSCHE VERSUCHSANSTALT FUER LUFT- UND RAUMFAHRT (ALL LOCATIONS)	DVL-
DEUTSCHE VERSUCHSANSTALT FUER LUFTFAHRT	ISL-
DEUTSCHER WETTERDIENST, OFFENBACH AM MAIN	DK-(NUMBER.NUMBER.NO.)
DEUTSCHER WETTERDIENST, OFFENBACH AM MAIN	T-(NUMBER)-I-(NUMBER)
DEUTSCHES ELEKTRONEN-SYNCHROTRON, HAMBURG	DESY-
DEUTSCHES ELEKTRONEN-SYNCHROTRON, HAMBURG	DESY-(YEAR)/
DEUTSCHES ELEKTRONEN-SYNCHROTRON, HAMBURG	DESY-A-(NUMBER).(NUMBER)
DEUTSCHES ELEKTRONEN-SYNCHROTRON, HAMBURG	DESY-F-
DEUTSCHES ELEKTRONEN-SYNCHROTRON, HAMBURG	DESY-H-
DEUTSCHES ELEKTRONEN-SYNCHROTRON, HAMBURG	DESY-HERA-
DEUTSCHES ELEKTRONEN-SYNCHROTRON, HAMBURG	DESY-R-
DEUTSCHES ELEKTRONEN-SYNCHROTRON, HAMBURG	DESY-SR-
DEUTSCHES ELEKTRONEN-SYNCHROTRON, HAMBURG	DESY-ST-
DEVELOPMENT + PROOF SERVICES, ABERDEEN PROVING GD.,MD	APG-DPS-
DEVELOPMENT + PROOF SERVICES, ABERDEEN PROVING GD.,MD	D + PS-
DEVELOPMENT + PROOF SERVICES, ABERDEEN PROVING GD.,MD	DPS-
DEVELOPMENT + PROOF SERVICES, ABERDEEN PROVING GD.,MD	DPS-AD-
DEVELOPMENT + PROOF SERVICES, ABERDEEN PROVING GD.,MD	DPS-FR-B-
DEVELOPMENT + PROOF SERVICES, ABERDEEN PROVING GD.,MD	DPS-FR-P-
DEVELOPMENT LABS., INC., LOS ANGELES	DLI-
DEVELOPMENT LABS., INC., LOS ANGELES	DLI(YR)-
DEWEY (G.C.) AND CO., INC., N.Y.C.	DCI-
DEWEY (G.C.) AND CO., INC., N.Y.C.	DCI DATA/ORD-
DEWEY (G.C.) AND CO., INC., N.Y.C.	DCI-R-
DEWEY (G.C.) AND CO., INC., N.Y.C.	DEWEY DATA-
DEWEY (G.C.) AND CO., INC., N.Y.C.	GCD-
DIAMOND ORDNANCE FUZE LABS., WASHINGTON, D.C.	DOFL-
DIAMOND ORDNANCE FUZE LABS., WASHINGTON, D.C.	DOFL-(NUMBER)-(YEAR)
DIAMOND ORDNANCE FUZE LABS., WASHINGTON, D.C.	DOFL-(YEAR)-T-
DIAMOND ORDNANCE FUZE LABS., WASH., D.C. (LIBRARY)	DOFL-LIBRARY-(YEAR)-
DIAMOND ORDNANCE FUZE LABS., WASHINGTON, D.C.	DOFL-PR-
DIAMOND ORDNANCE FUZE LABS., WASHINGTON, D.C.	DOFL-PR-(YEAR)-
DIAMOND ORDNANCE FUZE LABS., WASHINGTON, D.C.	DOFL-R(NO.)-
DIAMOND ORDNANCE FUZE LABS., WASHINGTON, D.C.	DOFL-TL-RM-
DIAMOND ORDNANCE FUZE LABS., WASHINGTON, D.C.	DOFL-TM-(YEAR)-
DIAMOND ORDNANCE FUZE LABS., WASHINGTON, D.C.	ODL-
DIAMOND ORDNANCE FUZE LABS., WASHINGTON, D.C.	ORDTX-
DIAMOND ORDNANCE FUZE LABS., WASHINGTON, D.C.	PR-(YR.)-
(SEE ALSO LATER NAME: HARRY DIAMOND LABS.)	
DIKEWOOD CORP., ALBUQUERQUE, N. MEX.	DC-FR-
DIKEWOOD CORP., ALBUQUERQUE, N. MEX.	DC-SR-
DIKEWOOD CORP., ALBUQUERQUE, N. MEX.	DC-TN-
DIKEWOOD CORP., ALBUQUERQUE, N. MEX.	DC-TR-(NUMBER)
DIKEWOOD CORP., ALBUQUERQUE, N. MEX.	DC-TR-CSAG-
DIKEWOOD CORP., ALBUQUERQUE, N. MEX.	DWC-
DIKEWOOD CORP., ALBUQUERQUE, N. MEX. (FINAL REPORT)	DWC FR-(NO.)-(NO.)
DIKEWOOD CORP., ALBUQUERQUE, N. MEX.	FR-(NUMBER)-(NUMBER)
DIKEWOOD CORP., ALBUQUERQUE, N. MEX.	WL-
DILWORTH, SECORD, MEAGHER AND ASSOCIATES LTD.,TORONTO	DSM-(NUMBER)-(NUMBER)
DILWORTH, SECORD, MEAGHER AND ASSOCIATES LTD.,TORONTO	TR-(YEAR)-
DILWORTH, SECORD, MEAGHER AND ASSOCIATES LTD.,TORONTO	TRI-(YEAR)-
DIRECTOR OF ENG. + IND. SERVICES,EDGEWOOD ARSENAL,MD.	DEIS-
DIRECTOR OF ENG. + IND. SERVICES,EDGEWOOD ARSENAL,MD.	EATR-
DIRECTOR OF SCIENTIFIC RESEARCH. SCI. RES. ESTAB.	DSR SRE (LETTER)-
DIRECTOR RESEARCH PROGRAM PLANNING. SCI. RES. ESTAB.	DRPP SRE-
DIRECTOR RESEARCH PROGRAM PLANNING. SCI. RES. ESTAB.	DRPP SRE (LETTER)-
DIRECTORATE OF AEROSPACE SAFETY, NORTON AFB, CALIF.	M-(NUMBER)-
DIRECTORATE OF AIRFRAME SUBSYSTEMS ENGINEERING, WRIGHT-PATTERSON AFB, OHIO	ASNFS-TM-(YEAR)-
DIRECTORATE OF BIOLOGICAL OPERATIONS, PINE BLUFF ARSENAL, ARK.	DBO-TR-
DIRECTORATE OF COMMUNICATIONS AND ELECTRONICS, ENT AFB, COLO.	ADCRP-(NUMBER)-
DIRECTORATE OF DEVELOPMENT, WRIGHT-PATTERSON AFB,OHIO	WCE-
DIRECTORATE OF DEVELOPMENT, WRIGHT-PATTERSON AFB,OHIO	WCL-
DIRECTORATE OF DEVELOPMENT, WRIGHT-PATTERSON AFB,OHIO	WCLAD-
DIRECTORATE OF FLIGHT AND ALL-WEATHER TESTING, WRIGHT-PATTERSON AFB, OHIO	AWNW-
DIRECTORATE OF FLIGHT AND ALL-WEATHER TESTING, WRIGHT-PATTERSON AFB, OHIO	MCRFT-
DIRECTORATE OF FLIGHT AND ALL-WEATHER TESTING, WRIGHT-PATTERSON AFB, OHIO	TSAWE-
DIRECTORATE OF FLIGHT AND ALL-WEATHER TESTING, WRIGHT-PATTERSON AFB, OHIO	TSFTE-
DIRECTORATE OF FLIGHT AND ALL-WEATHER TESTING, WRIGHT-PATTERSON AFB, OHIO	WCT-
DIRECTORATE OF FLIGHT STANDARDS AND QUALIFICATION RESEARCH (ARMY), ST. LOUIS	FS/Q-TR-
DIRECTORATE OF FLIGHT TEST ENGINEERING, WRIGHT-PATTERSON AFB, OHIO	ASTDV-FTD-(YEAR)-
DIRECTORATE OF FLIGHT TEST ENGINEERING, WRIGHT-PATTERSON AFB, OHIO	ASTDV-FTR-(YEAR)-
DIRECTORATE OF INTELLIGENCE (AIR FORCE), WASH., D.C.	AF-DI-
DIRECTORATE OF INTELLIGENCE (AIR FORCE), WASHINGTON, D.C. (PHYSICAL VULNERABILITY INTERIM MEMORANDUM)	AF-DI-PVIM-
DIRECTORATE OF INTELLIGENCE (AIR FORCE), WASH., D.C.	AF-IR-
DIRECTORATE OF INTELLIGENCE (AIR FORCE), WASH., D.C.	DI/USAF-
DIRECTORATE OF INTELLIGENCE (AIR FORCE), WASH., D.C.	PV-
DIRECTORATE OF MANAGEMENT ANALYSIS (AIR FORCE), WASHINGTON, D.C.	AFAMA-

DIRECTORATE OF NUCLEAR SAFETY, KIRTLAND AFB, N. MEX.	AFRP-
DIRECTORATE OF NUCLEAR SAFETY, KIRTLAND AFB, N. MEX.	DNS-
DIRECTORATE OF PERSONNEL PLANNING (AIR FORCE), WASHINGTON, D.C.	AFPDPL-PR-(YEAR)-
DIRECTORATE OF PROCUREMENT + PRODUCTION, WRIGHT-PATTERSON AFB, OHIO	MCPPAF-
DIRECTORATE OF RESEARCH, WRIGHT-PATTERSON AFB, OHIO	WCR-
DIRECTORATE OF RESEARCH ANALYSES, HOLLOMAN AFB, N.M.	AFOSR/DRA-(YEAR)-
DIRECTORATE OF RESEARCH ANALYSES, HOLLOMAN AFB, N.M.	DRA-
DIRECTORATE OF RESEARCH ANALYSES, HOLLOMAN AFB, N.M.	DRA-(YEAR)-
DIRECTORATE OF RESEARCH ANALYSES, HOLLOMAN AFB, N.M.	OSR-
DIRECTORATE OF RESEARCH ANALYSES, HOLLOMAN AFB, N.M.	OSR/DRA-(YEAR)-
(SEE ALSO LATER NAME: OFFICE OF RESEARCH ANALYSES)	
DIRECTORATE OF STUDIES AND ANALYSIS (AIR FORCE), WASHINGTON, D.C.	AFSCA-E2-
DIRECTORATE OF WEAPON SYSTEMS OPERATIONS, WRIGHT-PATTERSON AFB, OHIO	WCOWF-
DIRECTORATE OF WEAPON SYSTEMS OPERATIONS, WRIGHT-PATTERSON AFB, OHIO	WCOWP-
DIRECTORATE OF WEAPON SYSTEMS OPERATIONS, WRIGHT-PATTERSON AFB, OHIO	WCOWS-
DIRECTORATE OF WEAPON SYSTEMS OPERATIONS, WRIGHT-PATTERSON AFB, OHIO	WCS-
DIRECTORATE OF WEAPON SYSTEMS OPERATIONS, WRIGHT-PATTERSON AFB, OHIO	WCSDD-
DIRECTORATE OF WEAPON SYSTEMS OPERATIONS, WRIGHT-PATTERSON AFB, OHIO	WCSE-
DIRECTORATE OF WEAPON SYSTEMS OPERATIONS, WRIGHT-PATTERSON AFB, OHIO	WCSG-
DIRECTORATE OF WEAPON SYSTEMS OPERATIONS, WRIGHT-PATTERSON AFB, OHIO	WCSP-
DIRECTORATE OF WEAPON SYSTEMS OPERATIONS, WRIGHT-PATTERSON AFB, OHIO	WCSPE-
DIRECTORATE OF WEAPON SYSTEMS OPERATIONS, WRIGHT-PATTERSON AFB, OHIO	WCSR-
DIRECTORATE OF WEAPON SYSTEMS OPERATIONS, WRIGHT-PATTERSON AFB, OHIO	WCSWR-
DIVISION OF APPLIED TECHNOLOGY, AEC, WASHINGTON, D.C. (PEACEFUL USES OF NUCLEAR EXPLOSIONS. PLOWSHARE PROGRAM)	PNE-
DIVISION OF APPLIED TECHNOLOGY, AEC, WASHINGTON, D.C. (PROJECT GASBUGGY)	PNE-G-
DIVISION OF APPLIED TECHNOLOGY, AEC, WASHINGTON, D.C. (PROJECT RULISON)	PNE-R-
DIVISION OF BIOLOGY AND MEDICINE, AEC, WASH., D.C.	DBM-
DIVISION OF BIOLOGY AND MEDICINE, AEC, WASH., D.C.	TAB-
DIVISION OF BIOLOGY AND MEDICINE, AEC, WASH., D.C.	TAB-R-
DIVISION OF BIOLOGY AND MEDICINE. CIVIL EFFECTS TEST OPERATIONS, AEC (SERIES ASSIGNED TO REPORTS PREPARED BY VARIOUS ORGANIZATIONS)	CEX-
DIVISION OF BIOLOGY AND MEDICINE, AEC. RADIATION INSTRUMENTS BRANCH, WASHINGTON, D.C.	DBM-RIB-
DIVISION OF BIOLOGY AND MEDICINE, AEC. RADIATION INSTRUMENTS BRANCH, WASHINGTON, D.C.	RADET-
DIVISION OF BIOLOGY AND MEDICINE, AEC. RADIATION INSTRUMENTS BRANCH, WASHINGTON, D.C.	RIB-
DIVISION OF BIOLOGY AND MEDICINE, AEC. RADIATION INSTRUMENTS BRANCH, WASHINGTON, D.C.	RIB-IR-
DIVISION OF BIOLOGY AND MEDICINE, AEC. RADIATION INSTRUMENTS BRANCH, WASHINGTON, D.C.	RIB-QR-
DIVISION OF CLASSIFICATION, AEC, WASHINGTON, D.C. (CLASSIFICATION GUIDE)	CG-(SUBJ.LTRS)-(EDIT.NO.)
DIVISION OF CLASSIFICATION, AEC, WASHINGTON, D.C.	CG-UF-
DIVISION OF CLASSIFICATION, AEC, WASHINGTON, D.C.	OC-DOC-
DIVISION OF CLASSIFICATION, AEC, WASHINGTON, D.C.	TID-OC-
DIVISION OF INTELLIGENCE, AEC, WASHINGTON, D.C.	AECI-
DIVISION OF MATERIALS (EXPLORATION), AEC, N.Y.C.	RMO-
DIVISION OF OPERATIONS ANALYSIS AND FORECASTING, AEC, WASHINGTON, D.C.	OA:R-
DIVISION OF RAW MATERIALS, AEC, N.Y.C.	(YEAR)-CNA-
DIVISION OF RAW MATERIALS, AEC, N.Y.C. (REF. 4)	RME-
DIVISION OF REACTOR DEVELOPMENT, AEC. NAVAL REACTORS BRANCH, WASHINGTON, D.C.	NRB-HGR-
DIVISION OF REACTOR DEVELOPMENT, AEC. NAVAL REACTORS BRANCH, WASHINGTON, D.C.	UC-
DIVISION OF REACTOR DEV. + TECH., AEC, WASHINGTON,DC	RDT-C-(NUMBER-NUMBER)-T
DIVISION OF REACTOR DEV. + TECH., AEC, WASHINGTON,DC	RDT-E-(NUMBER-NUMBER)-T
DIVISION OF REACTOR DEV. + TECH., AEC, WASHINGTON,DC	RDT-F-(NUMBER-NUMBER)-T
DIVISION OF REACTOR DEV. + TECH., AEC, WASHINGTON,DC	RDT-M-(NUMBER)
DIVISION OF REACTOR LICENSING, AEC, WASHINGTON, D.C.	DOCKET-
DIVISION OF TECHNICAL INFORMATION, AEC (ABSTRACTS OF CLASSIFIED REPORTS)	ACR-
DIVISION OF TECHNICAL INFORMATION, AEC (CLASSIFIED REPORTS FOR CIVILIAN APPLICATIONS)	CRCA-
DIVISION OF TECHNICAL INFORMATION, AEC	DTI-
DIVISION OF TECHNICAL INFORMATION, AEC, WASH., D.C.	EXEP-
DIVISION OF TECHNICAL INFORMATION, AEC (RESEARCH AND DEVELOPMENT ABSTRACTS OF THE USAEC)	RDA-
DIVISION OF TECHNICAL INFORMATION, AEC	TIB-
(SEE ALSO LATER NAME: OFFICE OF INFO. SERVICES, AEC)	
DIVISION OF TECHNICAL INFORMATION EXTENSION, AEC, OAK RIDGE, TENN.	AEC-SP-
DIVISION OF TECHNICAL INFORMATION EXTENSION, AEC, OAK RIDGE, TENN.	CONF-(NUMBER)-
DIVISION OF TECHNICAL INFORMATION EXTENSION, AEC	DTIE-
DIVISION OF TECH. INFO. EXTENSION,AEC(ACCESSION LIST)	DTIE-AL-
DIVISION OF TECHNICAL INFO. EXTENSION, AEC (REF. 5)	TIP-
(SEE ALSO LATER NAME: TECHNICAL INFO. CENTER, AEC)	
DOCUMENTATION INC., WASHINGTON, D.C.	DI-
DOCUMENTATION INC., WASHINGTON, D.C.	DI/
DODCO INC., BLAWENBERG, N.J.	DODCO TR-
DON BOSCO INST. FOR RESEARCH, RAMSEY, N.J.	DBIR-
DORNE AND MARGOLIN, WESTBURY, N.Y.	DM-
DORNIER-WERKE (VERSUCHSBERICHT)	DO VB-
DORTECH, INC., STAMFORD, CONN.	G(4 DIGITS)-A
DOUGLAS AIRCRAFT CO., INC. (ALL LOCATIONS)	DAC-
DOUGLAS AIRCRAFT CO., INC. (ALL LOCATIONS)	DOUGLAS PAPER-
DOUGLAS AIRCRAFT CO., INC. (ARMAMENT GROUP MEMO. RPT)	AG-MR-
DOUGLAS AIRCRAFT CO., INC. (MISSILES TECH. MEMO. RPT)	MTM-
DOUGLAS AIRCRAFT CO., INC.	SM-
DOUGLAS AIRCRAFT CO., INC., CHARLOTTE, N.C.	DAC-CH-
DOUGLAS AIRCRAFT CO., INC., CHARLOTTE, N.C.	E(NO.)-

DOUGLAS AIRCRAFT CO., INC.

DOUGLAS AIRCRAFT CO., INC., EL SEGUNDO, CALIF.	DAC-ES-
DOUGLAS AIRCRAFT CO., INC., EL SEGUNDO, CALIF.	EL-
DOUGLAS AIRCRAFT CO., INC., LONG BEACH, CALIF.	DAC-ENG.PAPER-
DOUGLAS AIRCRAFT CO., INC., LONG BEACH, CALIF.	DAC-LB-
DOUGLAS AIRCRAFT CO., INC., LONG BEACH, CALIF.	ENGINEERING PAPER-
DOUGLAS AIRCRAFT CO., INC., LONG BEACH, CALIF.	LB-
DOUGLAS AIRCRAFT CO., INC., LONG BEACH, CALIF.	MDC-J(NUMBER)/
DOUGLAS AIRCRAFT CO., INC., LONG BEACH, CALIF.	SM-
DOUGLAS AIRCRAFT CO., INC., NEWPORT BEACH, CALIF. (SEMIANNUAL REPORT)	DAC-(NUMBER)-S-
DOUGLAS AIRCRAFT CO., INC., NEWPORT BEACH, CALIF.	LB-
DOUGLAS AIRCRAFT CO., INC., SANTA MONICA, CALIF.	A2-
DOUGLAS AIRCRAFT CO., INC., SANTA MONICA, CALIF.	DAC-ENG.PAPER-
DOUGLAS AIRCRAFT CO., INC., SANTA MONICA, CALIF.	DAC-SM-
DOUGLAS AIRCRAFT CO., INC., SANTA MONICA, CALIF.	DAS-MTM-
DOUGLAS AIRCRAFT CO., INC., SANTA MONICA, CALIF.	DOUGLAS PAPER-
DOUGLAS AIRCRAFT CO., INC., SANTA MONICA, CALIF.	ENGINEERING PAPER-
DOUGLAS AIRCRAFT CO., INC., SANTA MONICA, CALIF.	ENG. PAPER-
DOUGLAS AIRCRAFT CO., INC., SANTA MONICA, CALIF.	ES-
DOUGLAS AIRCRAFT CO., INC., SANTA MONICA, CALIF.	G-
DOUGLAS AIRCRAFT CO., INC., SANTA MONICA, CALIF.	MSCA-
DOUGLAS AIRCRAFT CO., INC., SANTA MONICA, CALIF.	SM-
DOUGLAS AIRCRAFT CO., INC., SANTA MONICA, CALIF.	TM-DSV4B(LETTERS)-
DOUGLAS AIRCRAFT CO., INC., TULSA, OKLA.	DEV-
DOUGLAS AIRCRAFT CO., INC., TULSA, OKLA.	RL-GEN-
DOUGLAS AIRCRAFT CO., INC., TULSA, OKLA.	TU-
DOUGLAS AIRCRAFT CO., INC. ADVANCED RESEARCH LABS., HUNTINGTON BEACH, CALIF.	DARL-
DOUGLAS AIRCRAFT CO., INC. ADVANCED RESEARCH LABS., HUNTINGTON BEACH, CALIF.	DARL-G-
DOUGLAS AIRCRAFT CO., INC. ASTROPOWER LAB., NEWPORT BEACH, CALIF.	DAC-(NUMBER)-(LETTER(S))
DOUGLAS AIRCRAFT CO., INC. ASTROPOWER LAB., NEWPORT BEACH, CALIF.	DAC-(NUMBER)-Q(NUMBER)
DOUGLAS AIRCRAFT CO., INC. ASTROPOWER LAB., NEWPORT BEACH, CALIF.	SM-(NUMBER)-Q-
DOUGLAS AIRCRAFT CO., INC. DEVELOPMENT LABS., SANTA MONICA, CALIF.	DEV-
DOUGLAS AIRCRAFT CO., INC. EL SEGUNDO DIV., CALIF.	ES-
DOUGLAS AIRCRAFT CO., INC. MISSILE AND SPACE SYSTEMS DIV., HUNTINGTON BEACH, CALIF.	DAC-SM-
DOUGLAS AIRCRAFT CO., INC. MISSILE + SPACE SYSTEMS DIV., NEWPORT BEACH, CALIF.	MP-
DOUGLAS AIRCRAFT CO., INC. MISSILE AND SPACE SYSTEMS DIV., SANTA MONICA, CALIF.	DAC R-
DOUGLAS AIRCRAFT CO., INC. MISSILE AND SPACE SYSTEMS DIV., SANTA MONICA, CALIF.	MSFC-STD-
DOUGLAS AIRCRAFT CO., INC. MISSILE AND SPACE SYSTEMS DIV., SANTA MONICA, CALIF.	SM-
DOUGLAS AIRCRAFT CO., INC. MISSILE AND SPACE SYSTEMS DIV., SANTA MONICA, CALIF.	TM-DM-(NO.)-EE-R-
DOUGLAS AIRCRAFT CO., INC. MISSILE AND SPACE SYSTEMS DIV., SANTA MONICA, CALIF.	TM-DSV(NO.)-EF-R-
DOUGLAS AIRCRAFT CO., INC. MISSILE AND SPACE SYSTEMS DIV., SANTA MONICA, CALIF.	TM-DSV(NO.)-PROP-R(NO.)
DOUGLAS AIRCRAFT CO., INC. MISSILES AND SPACE SYSTEMS ENGINEERING, SANTA MONICA, CALIF.	DAC-A(NO.)-(LTRS.)(NO.)
DOUGLAS AIRCRAFT CO., INC. PROJECT RAND, SANTA MONICA, CALIF.	R-
DOUGLAS AIRCRAFT CO., INC. PROJECT RAND, SANTA MONICA, CALIF.	RAD-
DOUGLAS AIRCRAFT CO., INC. PROJECT RAND, SANTA MONICA, CALIF.	RAD(L)-
DOUGLAS AIRCRAFT CO., INC. PROJECT RAND, SANTA MONICA, CALIF.	T-
(SEE ALSO SUCCESSOR: RAND CORP.)	
DOUGLAS AIRCRAFT CO., INC. SANTA MONICA DIV., LONG BEACH, CALIF.	DAC-D(NUMBER)-(NUMBER)
DOUGLAS AIRCRAFT CO., INC. SPACE SCIENCES DEPT., SANTA MONICA, CALIF.	TM-LF2-PROP-R-
DOUGLAS AIRCRAFT CO., INC. SPACE SCIENCES DEPT., SANTA MONICA., CALIF.	TM-SPARTAN-MS-R-
DOUGLAS UNITED NUCLEAR, INC., RICHLAND, WASH.	DUN-
DOUGLAS UNITED NUCLEAR, INC., RICHLAND, WASH.	DUN-AOP-
DOUGLAS UNITED NUCLEAR, INC., RICHLAND, WASH.	DUN-M-
DOUGLAS UNITED NUCLEAR, INC., RICHLAND, WASH.	DUN-SA-
DOUGLAS UNITED NUCLEAR, INC., RICHLAND, WASH.	DUN-TH-
DOUGLAS UNITED NUCLEAR, INC., RICHLAND, WASH.	HW-DWL-
DOUGLAS UNITED NUCLEAR, INC., RICHLAND, WASH.	HWS-
DOW CHEMICAL CO., MIDLAND, MICH. (ALKYL FUELS)	AF-(NUMBER)Q-(YEAR)
DOW CHEMICAL CO., MIDLAND, MICH. (ANTIMALARIAL SYNTHESIS)	AM-(NUMBER)Q-(YEAR)
DOW CHEMICAL CO., MIDLAND, MICH.	AM-(NUMBER)A-(YEAR)
DOW CHEMICAL CO., MIDLAND, MICH. (ADVANCED RES. ON SOLID ROCKET PROPELLANTS. QUARTERLY PROGRESS RPT.)	AR-(NUMBER) Q-(YEAR)
DOW CHEMICAL CO., MIDLAND, MICH. (ADVANCED RESEARCH ON SOLID ROCKET PROPELLANTS)	AR-(NUMBER) S-(YEAR)
DOW CHEMICAL CO., MIDLAND, MICH. (CASTABLE MAGNESIUM)	CM-(NUMBER)-(YEAR)-
DOW CHEMICAL CO., MIDLAND, MICH.	DCC-
DOW CHEMICAL CO., MIDLAND, MICH.	DOW-
DOW CHEMICAL CO., MIDLAND, MICH. (FLUOROANION DISPLACEMENT)	FD-(NUMBER)Q-(YEAR)
DOW CHEMICAL CO., MIDLAND, MICH.	FP-(NUMBER)Q-(YEAR)
DOW CHEMICAL CO., MIDLAND, MICH.	FS-(NUMBER)A-(YEAR)
DOW CHEMICAL CO., MIDLAND, MICH.	FS-(NUMBER)Q-(YEAR)
DOW CHEMICAL CO., MIDLAND, MICH. (GYRO FLOTATION FLUIDS)	GF-(NUMBER)Q-(YEAR)
DOW CHEMICAL CO., MIDLAND, MICH.	HF-
DOW CHEMICAL CO., MIDLAND, MICH. (LMH-1/HYDRAZINE HETEROGENEOUS PROPELLANT DEVELOPMENT)	HZ-(NUMBER)Q-(YEAR)
DOW CHEMICAL CO., MIDLAND, MICH.	NDA-
DOW CHEMICAL CO., MIDLAND, MICH.	NF-
DOW CHEMICAL CO., MIDLAND, MICH.	NF-(NUMBER)Q-(YEAR)
DOW CHEMICAL CO., MIDLAND, MICH.	PC-IF-
DOW CHEMICAL CO., MIDLAND, MICH. (PERFLUOROALKYL HETEROCYCLIC ELASTOMERS)	PH-(NUMBER)Q-(YEAR)
DOW CHEMICAL CO., MIDLAND, MICH.	SL-
DOW CHEMICAL CO., MIDLAND, MICH. (THERMODYNAMICS)	T-(NUMBER)-(NO.)Q-(YEAR)
DOW CHEMICAL CO. METALLURGICAL LABS., MIDLAND, MICH.	MT-
DOW CHEMICAL CO. NUCLEAR RES. LAB., MIDLAND, MICH.	DOW-NR-
DOW CHEMICAL CO. POLYMER RES. LAB., MIDLAND, MICH.	DOW-QPR-
DOW CHEMICAL CO. ROCKY FLATS DIV., DENVER	Q-
DOW CHEMICAL CO. ROCKY FLATS DIV., DENVER	RFP-
DOW CHEMICAL CO. ROCKY FLATS DIV., GOLDEN, COLO.	AMD-
DOW CHEMICAL CO. ROCKY FLATS DIV., GOLDEN, COLO.	CD-(YEAR)-
DOW CHEMICAL CO. ROCKY FLATS DIV., GOLDEN, COLO.	CRDL-
DOW CHEMICAL CO. ROCKY FLATS DIV., GOLDEN, COLO.	DOW-CD(YEAR)-
DOW CHEMICAL CO. ROCKY FLATS DIV., GOLDEN, COLO.	DOW-RFP-

DOW CHEMICAL CO. ROCKY FLATS DIV., GOLDEN, COLO.	PRD-(6 DIGITS)-
DOW CHEMICAL CO. ROCKY FLATS DIV., GOLDEN, COLO.	REP-
DOW CHEMICAL CO. ROCKY FLATS DIV., GOLDEN, COLO.	RFD-
DOW CHEMICAL CO. ROCKY FLATS DIV., GOLDEN, COLO.	RFP-
DOW CHEMICAL CO. ROCKY FLATS DIV., GOLDEN, COLO.	RFP-TRANS-
DOW CHEMICAL CO. ROCKY FLATS DIV., GOLDEN, COLO.	TG-
DOW CHEMICAL CO. ROCKY FLATS PLANT, DENVER	CD-(YEAR)-
DOW CHEMICAL CO. ROCKY FLATS PLANT, DENVER	DOW-CD(YEAR)-
DOW CHEMICAL CO. ROCKY FLATS PLANT, DENVER	DOW-RFP-
DOW CHEMICAL CO. ROCKY FLATS PLANT, DENVER	TG-
DOW CHEMICAL CO. ROCKY FLATS PLANT, GOLDEN, COLO.	CRDL-
DOW CHEMICAL CO. SCIENTIFIC PROJS. LAB., MIDLAND, MICH	EL-(NUMBER) Q-(YEAR)
DOW CHEMICAL CO. SCIENTIFIC PROJECTS LAB., MIDLAND, MICH. (M IS FOR BIMONTHLY)	HF-(NUMBER) M-(YEAR)
DOW CHEMICAL CO. SCIENTIFIC PROJS. LAB., MIDLAND, MICH	NF-(NUMBER) Q-(YEAR)
DOW CHEMICAL CO. SCIENTIFIC PROJS. LAB., MIDLAND, MICH	PC-(NUMBER) Q-(YEAR)
DOW CHEMICAL CO. SCIENTIFIC PROJS. LAB., MIDLAND, MICH	PD-(NUMBER) M-(YEAR)
DOW CHEMICAL CO. SCIENTIFIC PROJS. LAB., MIDLAND, MICH	PR-(NUMBER) Q-(YEAR)
DOW CHEMICAL CO. SCIENTIFIC PROJS. LAB., MIDLAND, MICH	T-(NUMBER)-Q-(YEAR)-
DOW CHEMICAL CO. THERMAL LAB., MIDLAND, MICH.	T-(NUMBER)-Q-(YEAR)-
DOW CHEMICAL CO. THERMAL RESEARCH LAB., MIDLAND, MICH.	T-(NUMBER)-(NO.)Q-(YR)
DOW CHEMICAL-DETROIT EDISON NUCLEAR POWER DEV. PROJ.	DCDE-
DOW CHEMICAL-DETROIT EDISON NUCLEAR POWER DEV. PROJ.	DE-DC-
DREXEL INST. OF TECH., CENTERTON, NJ. LAB. OF CLIMATOLOGY	DIT-PC-
DREXEL INST. OF TECH., PHILADELPHIA	DIT-
DREXEL UNIV., PHILADELPHIA	DREXELU-TR-
DREXEL UNIV., PHILADELPHIA	SI-(3 DIGITS)
DREXEL UNIV., PHILADELPHIA. COMBUSTION KINETICS LAB.	DUCK-
DU PONT DE NEMOURS (E.I.) + CO. (ALL LOCATIONS)	DP-
DU PONT DE NEMOURS (E.I.) + CO. (ALL LOCATIONS)	DU PONT-
DU PONT DE NEMOURS (E.I.) + CO. (ALL LOCATIONS) (FORMAL REPORTS)	DU PONT FOR-
DU PONT DE NEMOURS (E.I.) + CO., GIBBSTOWN, N.J.	ELAB-D-
DU PONT DE NEMOURS (E.I.) + CO., WILMINGTON, DEL.	AD-
DU PONT DE NEMOURS (E.I.) + CO., WILMINGTON, DEL.	BP-
DU PONT DE NEMOURS (E.I.) + CO., WILMINGTON, DEL.	DP-AD-
DU PONT DE NEMOURS (E.I.) + CO., WILMINGTON, DEL.	DPC-
DU PONT DE NEMOURS (E.I.) + CO., WILMINGTON, DEL.	DPD-
DU PONT DE NEMOURS (E.I.) + CO., WILMINGTON, DEL.	DPD-(YR)-
DU PONT DE NEMOURS (E.I.) + CO., WILMINGTON, DEL.	DPE-
DU PONT DE NEMOURS (E.I.) + CO., WILMINGTON, DEL.	DPK-
DU PONT DE NEMOURS (E.I.) + CO., WILMINGTON, DEL.	DPNC-
DU PONT DE NEMOURS (E.I.) + CO., WILMINGTON, DEL.	DPTR-
DU PONT DE NEMOURS (E.I.) + CO., WILMINGTON, DEL.	DPW-
DU PONT DE NEMOURS (E.I.) + CO., WILMINGTON, DEL.	DPWZ-
DU PONT DE NEMOURS (E.I.) + CO., WILMINGTON, DEL.	DUPONT-
DU PONT DE NEMOURS (E.I.) + CO., WILMINGTON, DEL.	EID-BP-
DU PONT DE NEMOURS (E.I.) + CO., WILMINGTON, DEL.	IED-BP-
DU PONT DE NEMOURS (E.I.) + CO., WILMINGTON, DEL.	SLC-
DU PONT DE NEMOURS (E.I.) + CO. ATOMIC ENERGY DIV., TERRE HAUTE, IND.	DPD-
DU PONT DE NEMOURS (E.I.) + CO. ATOMIC ENERGY DIV., WILMINGTON, DEL.	CMX PROGRESS REPORT-
DU PONT DE NEMOURS (E.I.) + CO. ATOMIC ENERGY DIV., WILMINGTON, DEL.	DPDTM-
DU PONT DE NEMOURS (E.I.) + CO. ATOMIC ENERGY DIV., WILMINGTON, DEL.	DPKN-
DU PONT DE NEMOURS (E.I.) + CO. ATOMIC ENERGY DIV., WILMINGTON, DEL.	DPW-DUPONT-
DU PONT DE NEMOURS (E.I.) + CO. ATOMIC ENERGY DIV., WILMINGTON, DEL.	DPXN-
DU PONT DE NEMOURS (E.I.) + CO. EASTERN LAB., GIBBSTOWN, N.J.	DUPONT-ELAB-(LETTER)-
DU PONT DE NEMOURS (E.I.) + CO. EASTERN LAB., GIBBSTOWN, N.J.	ELAB-B-
DU PONT DE NEMOURS (E.I.) + CO. EASTERN LAB., GIBBSTOWN, N.J.	ELAB-D-
DU PONT DE NEMOURS (E.I.) + CO. EASTERN LAB., GIBBSTOWN, N.J.	SLC-
DU PONT DE NEMOURS (E.I.) + CO. ELASTOMER CHEMICALS DEPT., WILMINGTON, DEL.	BL-
DU PONT DE NEMOURS (E.I.) + CO. ELASTOMERS DIV., WILMINGTON, DEL.	BL-(NUMBER) (DATE)
DU PONT DE NEMOURS (E.I.) + CO. ENGINEERING DEPT., WILMINGTON, DEL.	CWC-ED-
DU PONT DE NEMOURS (E.I.) + CO. ENGINEERING DEPT., WILMINGTON, DEL.	DPE-
DU PONT DE NEMOURS (E.I.) + CO. EXPERIMENTAL STATION, WILMINGTON, DEL. (FINAL REPORT)	DP-ESP-FR-
DU PONT DE NEMOURS (E.I.) + CO. EXPERIMENTAL STATION WILMINGTON, DEL.	ESP-(YR)-
DU PONT DE NEMOURS (E.I.) + CO. EXPLOSIVES DEPT., WILMINGTON, DEL.	DPWD-
DU PONT DE NEMOURS (E.I.) + CO. EXPLOSIVES DEPT. EASTERN LAB., GIBBSTOWN, N.J.	DUPONT-ELAB-(LETTER)-
DU PONT DE NEMOURS (E.I.) + CO. EXPLOSIVES DEPT. EASTERN LAB., GIBBSTOWN, N.J.	DUPONT-PR-
DU PONT DE NEMOURS (E.I.) + CO. GRASSELLI CHEMICALS DEPT., WILMINGTON, DEL.	CT-
DU PONT DE NEMOURS (E.I.) + CO. GRASSELLI CHEMICALS DEPT. EXPERIMENTAL LAB., CLEVELAND	GC-
DU PONT DE NEMOURS (E.I.) + CO. JACKSON LAB., WILMINGTON, DEL.	E-59-
DU PONT DE NEMOURS (E.I.) + CO. JACKSON LAB., WILMINGTON, DEL.	JWD-
DU PONT DE NEMOURS (E.I.) + CO. JACKSON LAB., WILMINGTON, DEL.	N-
DU PONT DE NEMOURS (E.I.) + CO. PARLIN LAB., N.J.	DU PONT PRL-
DU PONT DE NEMOURS (E.I.) + CO. PARLIN LAB., N.J.	PRL-
DU PONT DE NEMOURS (E.I.) + CO. PIGMENTS DEPT., WILMINGTON, DEL.	NWP-P-
DU PONT DE NEMOURS (E.I.) + CO. PIONEERING RESEARCH DIV., WILMINGTON, DEL.	DUPONT-QPR-(INCL.MOS/YR)
DU PONT DE NEMOURS (E.I.) + CO. RADIATION PHYSICS LAB., WILMINGTON, DEL. (RESEARCH REPORT)	DUPONT-RR-(YEAR/MONTH)
DU PONT DE NEMOURS (E.I.) + CO. RUBBER CHEMICALS DIV., WILMINGTON, DEL.	BL-(NUMBER) (DATE)
DU PONT DE NEMOURS (E.I.) + CO. SAVANNAH RIVER LAB., AIKEN, S.C.	DESPU-
DU PONT DE NEMOURS (E.I.) + CO. SAVANNAH RIVER LAB., AIKEN, S.C.	DP-MS-(YEAR)-
DU PONT DE NEMOURS (E.I.) + CO. SAVANNAH RIVER LAB., AIKEN, S.C.	DP(NASA)-

DU PONT DE NEMOURS (E.I.) & CO. SAVANNAH RIVER LAB.

DU PONT DE NEMOURS (E.I.) + CO. SAVANNAH RIVER LAB., AIKEN, S.C.	DPSP-
DU PONT DE NEMOURS (E.I.) + CO. SAVANNAH RIVER LAB., AIKEN, S.C.	DPSPU-
DU PONT DE NEMOURS (E.I.) + CO. SAVANNAH RIVER LAB., AIKEN, S.C.	DPST-
DU PONT DE NEMOURS (E.I.) + CO. SAVANNAH RIVER LAB., AIKEN, S.C.	DPST-(YEAR)-
DU PONT DE NEMOURS (E.I.) + CO. SAVANNAH RIVER LAB., AIKEN, S.C.	DPSTM-
DU PONT DE NEMOURS (E.I.) + CO. SAVANNAH RIVER LAB., AIKEN, S.C.	DPSTOM-
DU PONT DE NEMOURS (E.I.) + CO. SAVANNAH RIVER LAB., AIKEN, S.C.	DPWD-
DU PONT DE NEMOURS (E.I.) + CO. SAVANNAH RIVER LAB., AIKEN, S.C.	NMI-
DU PONT DE NEMOURS (E.I.) + CO. SAVANNAH RIVER LAB., AUGUSTA, GA.	DPSP-
DU PONT DE NEMOURS (E.I.) + CO. SAVANNAH RIVER LAB., AUGUSTA, GA. (HEAVY WATER)	DPSP-(YEAR)-(NO.)-(NO.)HW
DU PONT DE NEMOURS (E.I.) + CO. SAVANNAH RIVER LAB., AUGUSTA, GA.	DPSPU-
DU PONT DE NEMOURS (E.I.) + CO. SAVANNAH RIVER LAB., AUGUSTA, GA.	DPSPWD-(YEAR)-
DU PONT DE NEMOURS (E.I.) + CO. SAVANNAH RIVER LAB., AUGUSTA, GA.	DPST-
DU PONT DE NEMOURS (E.I.) + CO. SAVANNAH RIVER LAB., AUGUSTA, GA.	DPSTM-
DU PONT DE NEMOURS (E.I.) + CO. SAVANNAH RIVER LAB., AUGUSTA, GA.	DPSTMWD-
DU PONT DE NEMOURS (E.I.) + CO. SAVANNAH RIVER LAB., AUGUSTA, GA.	DP(WD)-
DU PONT DE NEMOURS (E.I.) + CO. SAVANNAH RIVER PLANT, AIKEN, S.C.	DP-MS-(YEAR)
DU PONT DE NEMOURS (E.I.) + CO. SAVANNAH RIVER PLANT, AIKEN, S.C.	DPSP-
DU PONT DE NEMOURS (E.I.) + CO. SAVANNAH RIVER PLANT, AIKEN, S.C.	DPSPU-(YR.)-(NO.)-
DU PONT DE NEMOURS (E.I.) + CO. SAVANNAH RIVER PLANT, AIKEN, S.C.	SEP-P-
DU PONT DE NEMOURS (E.I.)+CO. TNX DIV., WILMINGTON, DEL	TNX-
DUBLIN INSTITUTE FOR ADVANCED STUDIES	BS-
DUBLIN INSTITUTE FOR ADVANCED STUDIES	SER-D-GEOPHYS.BULL.-
DUBLIN INSTITUTE FOR ADVANCED STUDIES. SCHOOL OF COSMIC PHYSICS	DIAS-
DUGWAY PROVING GROUND, TOOELE, UTAH	BWALR-
DUGWAY PROVING GROUND, TOOELE, UTAH	CW-
DUGWAY PROVING GROUND, TOOELE, UTAH	DPG-(LETTER(S))-
DUGWAY PROVING GROUND, TOOELE, UTAH	DPG-DD-
DUGWAY PROVING GROUND, TOOELE, UTAH	DPGR-
DUGWAY PROVING GROUND, TOOELE, UTAH (SPECIAL REPORTS)	DPG SR-
DUGWAY PROVING GROUND, TOOELE, UTAH	DPGSR-
DUGWAY PROVING GROUND, TOOELE, UTAH	DPG-T(YEAR)-
DUGWAY PROVING GROUND, TOOELE, UTAH	DPG-TEST PLAN-
DUGWAY PROVING GROUND, TOOELE, UTAH (TRIAL RECORD)	DPGTR-
DUGWAY PROVING GROUND, TOOELE, UTAH (TECH. REPORT)	DPG-TR-
DUGWAY PROVING GROUND, TOOELE, UTAH (TRIAL RECORD)	DPG-TR-
DUGWAY PROVING GROUND, TOOELE, UTAH	DPG-TRIAL RECORD-
DUGWAY PROVING GROUND, TOOELE, UTAH	DPG-TR-T(YEAR)-
DUGWAY PROVING GROUND, TOOELE, UTAH	ETDIR-
DUGWAY PROVING GROUND, TOOELE, UTAH	ETLR-
DUKE UNIV., BEAUFORT, N.C. MARINE STATION (BULLETIN)	DU MSB-
DUKE UNIV., BEAUFORT, N.C. MARINE STATION (BULLETIN)	MSB-
DUKE UNIV., DURHAM, N.C.	DKU-
DUKE UNIV., DURHAM, N.C.	DU-
DUKE UNIV., DURHAM, N.C. (INFORMAL COMMUNICATION)	DU IC-
DUKE UNIV., DURHAM, N.C.	DU-TR-
DUKE UNIV., DURHAM, N.C.	MEMO-
DUMONT ELECTRON TUBES, CLIFTON, N.J.	ETC-(NUMBER)-
DUMONT (ALLEN B.) LABS., INC.	C(YR.)-(LTR.)-
DUMONT (ALLEN B.) LABS., INC., CLIFTON, N.J. (REF. 1)	(NUMBER)
DUMONT (ALLEN B.) LABS., INC., CLIFTON, N.J. (REF. 1)	(DECIMAL NUMBERING CODE)
DUMONT (ALLEN B.) LABS., INC., CLIFTON, N.J.	DM-
DUMONT (ALLEN B.) LABS., INC., CLIFTON, N.J.	DMC-
DUMONT (ALLEN B.) LABS., INC., CLIFTON, N.J.	DUML-
DUMONT (ALLEN B.) LABS., INC., CLIFTON, N.J.	DUMONT-
DUMONT (ALLEN B.) LABS., INC., CLIFTON, N.J.	SI-
DUMONT (ALLEN B.) LABS., INC., PASSAIC, N.J.	ABDL-
DUNLAP AND ASSOCIATES, INC., CLIFTON, N.J.	DUN-
DUNLAP AND ASSOCIATES, INC., DARIEN, CONN.	BSD(YEAR)-
DUNLAP AND ASSOCIATES, INC., DARIEN, CONN.	DA-ED-(YEAR)-
DUNLAP AND ASSOCIATES, INC., DARIEN, CONN.	DRD-
DUNLAP AND ASSOCIATES, INC., DARIEN, CONN.	SSD-(YEAR)-
DUNLAP AND ASSOCIATES, INC., STAMFORD, CONN.	DA-
DUNLAP AND ASSOCIATES, INC., STAMFORD, CONN.	DAI-
DUQUESNE LIGHT CO., SHIPPINGPORT, PENNA.	DCLS-
DUQUESNE LIGHT CO., SHIPPINGPORT, PENNA.	DLCS-
DUQUESNE LIGHT CO., SHIPPINGPORT, PENNA.	DL-S-
DUQUESNE LIGHT CO., SHIPPINGPORT, PENNA.	DLWK-
DUQUESNE LIGHT CO., SHIPPINGPORT, PENNA.	RNI-
DUQUESNE LIGHT CO., SHIPPINGPORT, PENNA.	T-
DYNAMIC ELECTRONICS-NEW YORK, INC., GLENDALE, N.Y.	DE-
DYNAMIC ELECTRONICS-NEW YORK, INC., GLENDALE, N.Y.	DENY-
DYNAMIC SCIENCE CORP., MONROVIA, CALIF.	SN-(3 DIGITS)-F
DYNAMIC SCIENCE CORP., SOUTH PASADENA, CALIF.	DSC-SN-
DYNAMIC SCIENCE CORP., SOUTH PASADENA, CALIF.	P-
DYNAMICS CORP. OF AMERICA. MASSA DIV., HINGHAM, MASS.	ER(NUMBERS)C-
DYNAMICS RESEARCH CORP., STONEHAM, MASS.	DRC-E-
DYNAMICS RESEARCH CORP., STONEHAM, MASS.	DRC-R-
DYNAMICS RESEARCH CORP., STONEHAM, MASS.	DRC-R-(NUMBER)C
DYNASCIENCES CORP., BLUE BELL, PENNA.	DCR-
DYNASCIENCES CORP., FORT WASHINGTON, PENNA.	DCR-
DYNATECH CORP., CAMBRIDGE, MASS.	DYNATECH-
DYNATECH CORP., CAMBRIDGE, MASS.	OCR-RDR-
DYNATECH CORP., CAMBRIDGE, MASS.	STO-
DYNELL ELECTRONICS CORP., PLAINVIEW, N.Y.	DY-(NUMBER)-(LETTER)
DYNETICS, INC., MOUNTAIN LAKES, N.J.	J-

E.M.I. COSSOR ELECTRONICS, LTD., DARTMOUTH, N.S.	EMIC-TR-
E.M.I. ELECTRONICS, LTD., HAYES, KENT, ENGLAND	DMP-
E.M.I. ELECTRONICS, LTD., HAYES, KENT, ENGLAND	EMI-DMP-
EAGLE-PICHER CO., JOPLIN, MO.	EP-
EAGLE-PICHER CO., JOPLIN, MO.	EPC-
EAGLE-PICHER CO., JOPLIN, MO.	NA-
EAGLE-PICHER CO. RESEARCH LABS., JOPLIN, MO.	EPRL-
EAGLE-PICHER CO. RESEARCH LABS., MIAMI, OKLA.	EPRL-
EAGLE-PICHER CO. RESEARCH LABS., MIAMI, OKLA.	EPRL-(LETTERS)-
EAGLE-PICHER CO. RESEARCH LABS., MIAMI, OKLA.	PR-
EARTH SCIENCES LABS., BOULDER, COLO.	ERL-ESL-
EARTH SCIENCES LABS., BOULDER, COLO.	ESL-
EARTH SCIENCES LABS., BOULDER, COLO.	ESSA-TR-ERL-(NO.)-ESL-
EARTH SCIENCES LABS., BOULDER, COLO.	ESSA-TR-ESL-(NO.)-ESL-
EARTH SCIENCES LABS., BOULDER, COLO.	NOAA-TM-ERL-ESL-
EARTHQUAKE ENGINEERING RESEARCH INST., SAN FRANCISCO	EERI-
EARTHQUAKE MECHANISM LAB., SAN FRANCISCO	EML-
EARTHQUAKE MECHANISM LAB., SAN FRANCISCO	ESSA-IERTM-EML-
EARTHQUAKE MECHANISM LAB., SAN FRANCISCO	IERTM-EML-
EAST CENTRAL NUCLEAR GROUP	ECNG-
EAST COAST AERONAUTICS, INC., MOUNT VERNON, N.Y. (DESIGN PROPOSAL)	DP-(NUMBER)-
EAST COAST AERONAUTICS, INC., MOUNT VERNON, N.Y. (PROCESS REPORT)	P-
EAST COAST AERONAUTICS, INC., PELHAM MANOR, N.Y.	ECA-
EAST COAST INTERLABORATORY COMMITTEE ON EDITING AND PUBLISHING (NAVY), JOHNSVILLE, PENNA.	ILCEP-EAST MONOGRAPH-
EAST-WEST GATEWAY COORDINATING COUNCIL, ST. LOUIS	EWG-JH-(NO.).(NO.).(NO.)
EAST-WEST GATEWAY COORDINATING COUNCIL, ST. LOUIS	EWG-KW-(NO.).(NO.).(NO.)
EASTERN AIR LINES, ATLANTA	EAL-
EASTERN JOINT COMPUTER CONFERENCE	EJCC-
EASTERN RESEARCH GROUP, N.Y.C.	ERG-
EASTMAN CHEMICAL PRODUCTS, INC., KINGSPORT, TENN. (TECHNICAL DATA REPORT)	ECP-TDR-
EASTMAN CHEMICAL PRODUCTS, INC., KINGSPORT, TENN.	TDR-N-
EASTMAN KODAK CO., ROCHESTER, N.Y.	DOVE
EASTMAN KODAK CO., ROCHESTER, N.Y.	EK-
EASTMAN KODAK CO., ROCHESTER, N.Y.	EK/ARD-ED-
EASTMAN KODAK CO., ROCHESTER, N.Y.	EKC-
EASTMAN KODAK CO., ROCHESTER, N.Y. (MONTHLY LTR. RPT)	EKC-MLR-
EASTMAN KODAK CO., ROCHESTER, N.Y.	EKC-NOD-(LETTERS)-
EASTMAN KODAK CO., ROCHESTER, N.Y.	EKTR-
EASTMAN KODAK CO., ROCHESTER, N.Y. (RESERVE BATTERY)	RB-
EASTMAN KODAK CO., ROCHESTER, N.Y.	Z-
EASTMAN KODAK CO. NAVY ORDNANCE DIV., ROCHESTER, N.Y. (REF. 1)	(NUMBER)
EASTMAN KODAK CO. NAVY ORDNANCE DIV., ROCHESTER, N.Y.	EK-D-
EASTMAN KODAK CO. NAVY ORDNANCE DIV., ROCHESTER, N.Y.	EK/NOD-
EATONTOWN SIGNAL LAB., N.J.	ESL-
EATONTOWN SIGNAL LAB., N.J. (ENGINEERING MEMORANDA)	ESL EM-
EATONTOWN SIGNAL LAB., N.J. (ENGINEERING REPORTS)	ESL ER-
EATONTOWN SIGNAL LAB., N.J.	ESL TM-
EATONTOWN SIGNAL LAB., N.J.	ESL TR T-
EAU CLAIRE ORDNANCE PLANT	EC-
EBASCO SERVICES, INC., N.Y.C.	ESI-
ECLIPSE-PIONEER DIV., BENDIX AVIATION CORP., TETERBORO, N. J. (REF. 1)	(YEAR CODE)(MO.NO.)-(DAY)
ECLIPSE-PIONEER DIV., BENDIX AVIATION CORP., TETERBORO, N. J.	BAC-EP-
ECLIPSE-PIONEER DIV., BENDIX AVIATION CORP., TETERBORO, N. J.	BEND EPD (LTRS)-
ECLIPSE-PIONEER DIV., BENDIX AVIATION CORP., TETERBORO, N. J.	BEND EPD R-
ECLIPSE-PIONEER DIV., BENDIX AVIATION CORP., TETERBORO, N. J.	EDR-
ECLIPSE-PIONEER DIV., BENDIX AVIATION CORP., TETERBORO, N. J. (REF. 1)	(REPORT NO.)-(PROJ. NO.)
ECONOMIC DEVELOPMENT ADMINISTRATION, WASHINGTON, D.C.	EDA-(YEAR)-
ECONOMIC DEVELOPMENT ADMINISTRATION, WASHINGTON, D.C.	EDACOMM-(YEAR)-
ECONOMIC DEVELOPMENT ADMINISTRATION, WASHINGTON, D.C.	EDA/OER-(YEAR)-
ECUADOR. OBSERVATORIO ASTRONOMICO, QUITO	OAQ-
EDGERTON, GERMESHAUSEN AND GRIER, INC., BEDFORD, MASS	EG + G-B-
EDGERTON, GERMESHAUSEN AND GRIER, INC., BOSTON	ARD-(YEAR)/
EDGERTON, GERMESHAUSEN AND GRIER, INC., BOSTON	CLASS. CONTROL NO. LV-
EDGERTON, GERMESHAUSEN AND GRIER, INC., BOSTON	E.G.+G.-
EDGERTON, GERMESHAUSEN AND GRIER, INC., BOSTON	EG+G-
EDGERTON, GERMESHAUSEN AND GRIER, INC., BOSTON	EG+G-(LETTER(S))-
EDGERTON, GERMESHAUSEN AND GRIER, INC., BOSTON	EG+G-B-
EDGERTON, GERMESHAUSEN AND GRIER, INC., BOSTON	EG+G-OUT-
EDGERTON, GERMESHAUSEN AND GRIER, INC., BOSTON (PACIFIC PROVING GROUND)	EG+G-PPG-
EDGERTON, GERMESHAUSEN AND GRIER, INC., BOSTON	TMB-
EDGERTON, GERMESHAUSEN + GRIER, INC., LAS VEGAS, NEV.	EG+G-
EDGERTON, GERMESHAUSEN + GRIER, INC., LAS VEGAS, NEV.	EG+G-L-
EDGERTON, GERMESHAUSEN + GRIER, INC., LAS VEGAS, NEV.	EGG-L-
EDGERTON, GERMESHAUSEN + GRIER, INC., LAS VEGAS, NEV.	EG+G-LV-
EDGERTON, GERMESHAUSEN + GRIER, INC., LAS VEGAS, NEV.	L-
EDGERTON, GERMESHAUSEN + GRIER, INC., LAS VEGAS, NEV.	LV-
EDGERTON, GERMESHAUSEN + GRIER, INC., LAS VEGAS, NEV. (VELA UNIFORM PROJECT)	VUP-
EDGERTON, GERMESHAUSEN + GRIER,INC.,SANTA BARBARA,CAL	ARMS-NVOO-(LTRS)-
EDGERTON, GERMESHAUSEN + GRIER,INC.,SANTA BARBARA,CAL	EASP-(NUMBER)-
EDGERTON, GERMESHAUSEN + GRIER,INC.,SANTA BARBARA,CAL	EA-TM-(NUMBER)-
EDGERTON, GERMESHAUSEN + GRIER,INC.,SANTA BARBARA,CAL	EATM-(NUMBER)-
EDGERTON, GERMESHAUSEN + GRIER,INC.,SANTA BARBARA,CAL	EGG-

EDGERTON, GERMESHAUSEN & GRIER, INC.

EDGERTON, GERMESHAUSEN + GRIER,INC.,SANTA BARBARA,CAL	EG+G-S-(NO.)-MN
EDGERTON, GERMESHAUSEN + GRIER,INC.,SANTA BARBARA,CAL	EG/G-S-(NUMBER)-R
EDGERTON, GERMESHAUSEN + GRIER,INC.,SANTA BARBARA,CAL	S-
(SEE ALSO LATER NAME: EG+G, INC.)	
EDGEWOOD ARSENAL, MD.	EA-(NO.)-
EDGEWOOD ARSENAL, MD.(MALFUNCTION INVESTIGATION)	EA-M1F-A(NUMBER)-
EDGEWOOD ARSENAL, MD.	EA-SP-(NO.)-
EDGEWOOD ARSENAL, MD.	EASP-
EDGEWOOD ARSENAL, MD.	EASP-(NUMBER)-
EDGEWOOD ARSENAL, MD.	EATM-(NUMBER)-
EDGEWOOD ARSENAL, MD.	EA-TN
EDGEWOOD ARSENAL, MD.	EA-TR-
EDGEWOOD ARSENAL, MD.	EATR-
EDGEWOOD ARSENAL, MD.	ESAP-(NUMBER)-
EDGEWOOD ARSENAL, MD. (INCENDIARY EVALUATION PROJ.)	HISTORICAL MONOGRAPH-
EDGEWOOD ARSENAL, MD. (OPERATIONS RESEARCH)	IEPR-
EDGEWOOD ARSENAL, MD.	OPG-
EDGEWOOD ARSENAL, MD. (OPERATIONS RESEARCH)	ORD EATR-
EDGEWOOD ARSENAL, MD. (OPERATIONS RESEARCH)	ORG NOTE-
	ORG SPECIAL PUB-
EDIN COMPANY, INC., WORCESTER, MASS.	EDIN-
EDISON GENERAL ELECTRIC APPLIANCE CO. (TECH. RPTS.)	EGEAC TR-
EDISON (THOMAS A.) INC., ORANGE, N.J.	EDI-
EDO CORP., COLLEGE POINT, N.Y.	EDO-
EDO (CANADA) LTD., CORNWALL, ONTARIO	EDO-XE-
EDUCATIONAL TESTING SERVICE, PRINCETON, N.J.	ETS-
EDUCATIONAL TESTING SERVICE, PRINCETON, N.J.	ETS-RB-(YEAR)-
EDUCATIONAL TESTING SERVICE, PRINCETON, N.J.	RB-(YEAR)-
EDWALD LABORATORIES, INC., RINGWOOD, ILL. (REF. 1)	(NUMBER)
EFFECTS TECHNOLOGY, INC., SANTA BARBARA, CALIF.	ETI-CR-
EG+G, INC. (ALL LOCATIONS)	B-(4 DIGITS)
EG+G, INC. (ALL LOCATIONS)	EGG-
EG+G, INC. (ALL LOCATIONS)	EGG-(NUMBER)-
EG+G, INC. (ALL LOCATIONS)	EG/G-B-
EG+G, INC. (ALL LOCATIONS)	EG/G-S-(NUMBER)-R
EG+G, INC., ALBUQUERQUE, N. MEX.	AL-
EG+G, INC., ALBUQUERQUE, N. MEX.	EG/G-AL-
EG+G, INC., BOSTON	
EG+G, INC., BEDFORD, MASS.	EGG-1183-
EG+G, INC., LAS VEGAS, NEV.	RB-(NUMBER)-
EG+G, INC., LAS VEGAS, NEV.	ARMS-(YEAR.NUMBER)-
EG+G, INC., LAS VEGAS, NEV.	EGG-TM-(LETTER)-
	EGG-TR-(LETTER)-
EG+G, INC., SANTA BARBARA, CALIF. (AERIAL RADIOLOGICAL MONITORING SYSTEM, NEVADA TEST SITE)	ARMA-NVOO-NTS-AR-
EG+G, INC., SANTA BARBARA, CALIF.	S-(NUMBER)-R
EG+G, INC. SYSTEMS DIV.	EGG-B-
EIMAC, SAN CARLOS, CALIF.	EIMAC-TR-(YEAR)-
EIMAC, SAN CARLOS, CALIF.	EIMAC-TR-
EINDHOVEN, NETHERLANDS. TECHNISCHE HOGESCHOOL	TH/(YR)-E-
EINDHOVEN, NETHERLANDS. TECHNISCHE HOGESCHOOL	WW-
EINDHOVEN, NETHERLANDS. TECHNISCHE HOGESCHOOL	WW-(NUMBER)-M-
EINDHOVEN, NETHERLANDS. TECHNISCHE HOGESCHOOL	WW-(NUMBER)-R-
EINDHOVEN, NETHERLANDS. TECHNISCHE HOGESCHOOL	WWO-
EINDHOVEN, NETHERLANDS. TECHNISCHE HOGESCHOOL	WW/R-
EITEL-MC CULLOUGH, INC., SAN CARLOS, CALIF.	CEF-
EITEL-MC CULLOUGH, INC., SAN CARLOS, CALIF.	R/E-(YEAR)-
ELCON LAB., INC., CAMBRIDGE, MASS.	R-(NUMBER)-(YEAR)-
ELCON LAB., INC., SALEM, MASS.	R-(NUMBER)-(YEAR)-
ELDORADO MINING + REFINING LTD., SASKATCHEWAN	AM-
ELDORADO MINING + REFINING LTD. RES.+ DEV.DIV.,OTTAWA	D-(YR)-(NO.)-
ELDORADO MINING + REFINING LTD. RES.+ DEV.DIV.,OTTAWA	R-(NO.)-
ELDORADO MINING + REFINING LTD. RES.+ DEV.DIV.,OTTAWA	RD-
ELECTRIC + MUSICAL INDUSTRIES, LTD.,HAYES, MIDDX,ENG.	RF-(3 DIGITS)
ELECTRIC + MUSICAL INDUSTRIES, LTD.,HAYES, MIDDX,ENG.	RF/H/(NUMBER/NUMBER)
ELECTRIC + MUSICAL INDUSTRIES, LTD.,HAYES, MIDDX,ENG.	RH-
ELECTRIC + MUSICAL INDUSTRIES, LTD.,HAYES, MIDDX,ENG.	RY-
ELECTRIC STORAGE BATTERY CO. CARL F. NORBERG RESEARCH CENTER, YARDLEY, PENNA.	ESB-E-
ELECTRICAL AND ELECTRONIC MEASUREMENT TEST INSTRUMENT CONFERENCE, OTTAWA	EEMTIC-(NO.)-
ELECTRO-CHEMICAL PRODUCTS CORP., EAST ORANGE, N.J.	ECCO-
ELECTRO-ENGINEERING PRODUCTS CO., CHICAGO	EEP-
ELECTRO-HYDRAULICS, LTD., WARRINGTON, LANCS, ENGLAND	MA-
ELECTRO-MECHANICAL RESEARCH, INC., SARASOTA, FLA.	RA-
ELECTRO-MECHANICAL RESEARCH, INC., SARASOTA, FLA.	RA-(NUMBER)-EXHIBIT
ELECTRO-MECHANICAL RESEARCH, INC., SARASOTA, FLA.	RE-
ELECTRO-MECHANICAL RESEARCH, INC. PHOTOELECTRIC DIV., PRINCETON, N.J.	EMR-PD-
ELECTRO-MECHANICS CO., AUSTIN, TEX.	EMCO-
ELECTRO-MECHANICS CO., DALLAS	EMCO-

ELECTRO METALLURGICAL CO. METALS RESEARCH LAB., NIAGARA FALLS, N.Y.	UCC-
ELECTRO-OPTICAL RESEARCH CO., LOS ANGELES	N(3 DIGITS)-F
ELECTRO-OPTICAL SYSTEMS, INC., PASADENA, CALIF.	EOS-
ELECTRO-OPTICAL SYSTEMS, INC., PASADENA, CALIF.	EOS-(NUMBER)-(LETTER)
ELECTRO-OPTICAL SYSTEMS, INC., PASADENA, CALIF.	EOS-(NUMBER)-TDR-
ELECTRO-OPTICAL SYSTEMS, INC., PASADENA, CALIF.	EOSI-
ELECTRO-OPTICAL SYSTEMS, INC., PASADENA, CALIF.	RN-
ELECTRO-VOICE, INC., BUCHANAN, MICH.	EV-
ELECTROCHIMICA CORP., MENLO PARK, CALIF.	ELCA-
ELECTROMAGNETIC COMPATIBILITY ANALYSIS CENTER, ANNAPOLIS, MD.	ECAC-
ELECTROMAGNETIC COMPATIBILITY ANALYSIS CENTER, ANNAPOLIS, MD. (DIRECTORY, RADAR + COMM. EQUIP.)	ECAC-D-
ELECTROMAGNETIC COMPATIBILITY ANALYSIS CENTER, ANNAPOLIS, MD. (FREQUENCY APPLICATION INDEX)	ECAC-FAI-
ELECTROMAGNETIC COMPATIBILITY ANALYSIS CENTER, ANNAPOLIS, MD. (FREQUENCY ALLOCATION LIST)	ECAC-FAL-
ELECTROMAGNETIC COMPATIBILITY ANALYSIS CENTER, ANNAPOLIS, MD. (SPECTRUM SIGNATURE INDEX)	ECAC-I-
ELECTROMAGNETIC COMPATIBILITY ANALYSIS CENTER, ANNAPOLIS, MD. (INTERMEDIATE REPORT)	ECAC-IR-
ELECTROMAGNETIC COMPATIBILITY ANALYSIS CENTER, ANNAPOLIS, MD.	ECAC-IR-J-
ELECTROMAGNETIC COMPATIBILITY ANALYSIS CENTER, ANNAPOLIS, MD.	ECAC-J-
ELECTROMAGNETIC COMPATIBILITY ANALYSIS CENTER, ANNAPOLIS, MD.	ECAC-J-IH-(SS)-
ELECTROMAGNETIC COMPATIBILITY ANALYSIS CENTER, ANNAPOLIS, MD. (ELECTRONIC ENVIRONMENT LISTING)	ECAC-L-
ELECTROMAGNETIC COMPATIBILITY ANALYSIS CENTER, ANNAPOLIS, MD.	ECAC-PTR-
ELECTROMAGNETIC COMPATIBILITY ANALYSIS CENTER, ANNAPOLIS, MD.	ECAC-R-
ELECTROMAGNETIC COMPATIBILITY ANALYSIS CENTER, ANNAPOLIS, MD. (SHORT TERM PROJECT)	ECAC-STP-
ELECTROMAGNETIC COMPATIBILITY ANALYSIS CENTER, ANNAPOLIS, MD. (TECHNICAL DOC. REPT.)	ECAC-TDR-
ELECTROMAGNETIC COMPATIBILITY ANALYSIS CENTER, ANNAPOLIS, MD. (TECHNICAL MEMORANDUM)	ECAC-TM-
ELECTROMAGNETIC COMPATIBILITY ANALYSIS CENTER, ANNAPOLIS, MD. (TECHNICAL NOTE)	ECAC-TN-
ELECTROMAGNETIC COMPATIBILITY ANALYSIS CENTER, ANNAPOLIS, MD. (TECHNICAL REPORT)	ECAC-TR-
ELECTROMAGNETIC SCIENCES, INC., ATLANTA	EMS-J-
ELECTRONA CORP., IRVINGTON, N.J. (QUARTERLY REPORT)	EC-QR-
ELECTRONA CORP., IRVINGTON, N.J.	ELC-
ELECTRONIC ASSOCIATES, INC., LONG BRANCH, N.J.	EA-
ELECTRONIC ASSOCIATES INC., PRINCETON, N.J.	EA-
ELECTRONIC COMMUNICATIONS, INC., ST. PETERSBURG, FLA.	ECI-(NUMBER)-
ELECTRONIC COMMUNICATIONS, INC., ST. PETERSBURG, FLA.	ECI-I-
ELECTRONIC COMMUNICATIONS, INC., ST. PETERSBURG, FLA.	SER-(YEAR)-
ELECTRONIC COMMUNICATIONS INC. RES. DIV., TIMONIUM, MD.	ECI-
ELECTRONIC CONTROL SYSTEMS, INC., LOS ANGELES	ECS-
ELECTRONIC ENGINEERING CO. OF CALIF., LOS ANGELES	EEC-
ELECTRONIC INDUSTRIES ASSN., WASHINGTON, D.C.	EIA-
ELECTRONIC PROPERTIES INFO. CTR., CULVER CITY, CALIF	DS-
ELECTRONIC SPECIALTY CO., LOS ANGELES	ESC-
ELECTRONIC SUPPLY OFFICE, GREAT LAKES, ILL.	ESO-
ELECTRONIC SYSTEMS DIV. (AIR FORCE), BEDFORD, MASS.	AFESD-
ELECTRONIC SYSTEMS DIV. (AIR FORCE), BEDFORD, MASS.	AFESD-TDR-(YEAR)-
ELECTRONIC SYSTEMS DIV. (AIR FORCE), BEDFORD, MASS.	AFESD-TN-(YEAR)-
ELECTRONIC SYSTEMS DIV. (AIR FORCE), BEDFORD, MASS.	ESC-TR-(YEAR)-
ELECTRONIC SYSTEMS DIV. (AIR FORCE), BEDFORD, MASS.	ESD-
ELECTRONIC SYSTEMS DIV. (AIR FORCE), BEDFORD, MASS.	ESD-CR-
ELECTRONIC SYSTEMS DIV. (AIR FORCE), BEDFORD, MASS.	ESD-TD-
ELECTRONIC SYSTEMS DIV. (AIR FORCE), BEDFORD, MASS. (TECHNICAL DOCUMENTARY REPORT)	ESD-TDR-
ELECTRONIC SYSTEMS DIV. (AIR FORCE), BEDFORD, MASS. (TECHNICAL DOCUMENTARY REPORT)	ESD-TDR-(YEAR)-
ELECTRONIC SYSTEMS DIV. (AIR FORCE), BEDFORD, MASS.	ESD-TR-
ELECTRONIC SYSTEMS DIV. (AIR FORCE), BEDFORD, MASS.	ESD-TR-(YEAR)-
ELECTRONIC SYSTEMS DIV. (AIR FORCE), BEDFORD, MASS.	ESYS-
ELECTRONIC SYSTEMS DIV. (AIR FORCE). DECISION SCIENCES LAB., BEDFORD, MASS.	DCBRL-
ELECTRONICS ASSOCIATES, INC., PRINCETON, N.J.	ELAI F/(CONTRACT NO.)
ELECTRONICS CORP. OF AMERICA, CAMBRIDGE, MASS.	ECA-CN-
ELECTRONICS CORP. OF AMERICA, CAMBRIDGE, MASS.	LMSC-TR-
ELECTRONICS MAINTENANCE ENGINEERING CTR., NORFOLK, VA.	EMC-
ELECTRONICS MAINTENANCE ENGINEERING CTR., NORFOLK, VA.	EMEC-
ELECTRONICS PRODUCTION RESOURCES AGENCY, WASH., D.C.	EPRA-
ELECTRONICS RESEARCH, INC., EVANSVILLE, IND.	ERI-
ELECTRONICS RESEARCH, INC., EVANSVILLE, IND.	GAF-
ELECTRONICS RESEARCH, INC. ANTENNA DIV., EVANSVILLE, IND.	GN-
ELECTRONIQUE MARCEL D'ASSAULT, SAINT-CLOUD, FRANCE	EMD-
ELKHART LAKE VILLAGE PLANNING COMMISSION, WIS.	ELVPC-(YEAR)-
ELLIOTT-AUTOMATION SPACE AND ADVANCED MILITARY SYSTEMS LTD., LONDON	ETN-
ELLIOTT BROS. LTD., LONDON	EBL-
ELLIOTT BROS. LTD., LONDON	RRL/
ELLIOTT BROS. LTD., LONDON	T-F-
ELLIOTT BROS. RESEARCH LABS., LONDON	RLE-
ELLIOTT ELECTRONIC TUBES LTD., BOREHAMWOOD, HERTS, ENG	EET-(NUMBER)/PR-
EMERSON AND CUMING, INC., CANTON, MASS.	ER-
EMERSON ELECTRIC CO. STAR DEPT., SANTA BARBARA, CAL.	EMERSON-STAR-C-
EMERSON ELECTRIC CO. STAR DEPT., SANTA BARBARA, CAL.	
EMERSON ELECTRIC MFG. CO., ST. LOUIS	EC-
EMERSON ELECTRIC MFG. CO., ST. LOUIS	EEMC-

EMERSON RADIO AND PHONOGRAPH CORP.

EMERSON RADIO AND PHONOGRAPH CORP., N.Y.C.	TR-
EMERSON RADIO AND PHONOGRAPH CORP., WASHINGTON, D.C.	ERPC-
EMMANUEL COLL., BOSTON	EMM-(YEAR)-
EMMANUEL COLL., BOSTON	E-T-CH-
EMMANUEL COLL., BOSTON	E-T-F-
EMMANUEL COLL., BOSTON	E-T-FC-
EMMANUEL COLL., BOSTON	E-T-G-
EMMANUEL COLL., BOSTON	E-T-J-
EMMANUEL COLL., BOSTON	E-T-R-
EMMANUEL COLL., BOSTON	E-T-RC-
EMMANUEL COLL., BOSTON. ORIENTAL SCIENCE LIBRARY	TRANS-EMM-(YEAR)-
EMORY UNIV., ATLANTA	EMORYU-QR-
EMPIRE DEVICES, INC., BAYSIDE, N.Y.	ED-
EMR-PHOTOELECTRIC, PRINCETON, N.J.	EMR-PE-
EMR-PHOTOELECTRIC, PRINCETON, N.J.	EMR-POA-
ENERGY CONVERSION DEVICES, INC., TROY, MICH.	ECD-
ENGELHARD INDUSTRIES, INC., EAST NEWARK, N.J.	SAR-
ENGINE PROGRAM INFORMATION CENTER, WASHINGTON, D.C.	EPIC-
ENGINEERING COMMAND (ARMY)	ENC-
ENGINEERING ENTERPRISES, INC.	EEI-
ENGINEERING FOUNDATION, N.Y.C. (COLUMN RES. COUNCIL)	EF-CRC-
ENGINEERING INFORMATION SERVICES, PRESTON, LANCS, ENGLAND (TRANSLATIONS)	EIS-
ENGINEERING INST. OF CANADA, MONTREAL	EIC-(YEAR)-MECH-
ENGINEERING PHYSICS CO., ROCKVILLE, MD.	EPC-
ENGINEERING PHYSICS CO., ROCKVILLE, MD.	EPCO-(YEAR)-
ENGINEERING RESEARCH ASSOCIATES, TORONTO	ERA-
ENGINEERING RESEARCH ASSOCIATES, INC.	ERA-TR-
ENGINEERING RESEARCH ASSOCIATES, INC., ST. PAUL	ERA-PR-
ENGINEERING RESEARCH ASSOCIATES, INC., WASHINGTON, D.C.	ERA-
ENGINEERING RESEARCH ASSOCIATES, INC., WASHINGTON, D.C.	ERA-ER-
ENGINEERING RESEARCH ASSOCIATES, INC., WASHINGTON, D.C.	PX-
ENGINEERING-SCIENCE INC., ARCADIA, CALIF.	ES-
ENGLEMAN AND CO. INC., WASHINGTON, D.C.	G-
ENGLISH ELECTRIC CO., LTD.	LAM-
ENGLISH ELECTRIC CO., LTD.	LAT-
ENGLISH ELECTRIC CO., LTD.	LEL-
ENGLISH ELECTRIC CO., LTD.	LEL.T-
ENGLISH ELECTRIC CO., LTD.	LJT-
ENGLISH ELECTRIC CO., LTD.	LOT-
ENGLISH ELECTRIC CO., LTD.	LST-
ENGLISH ELECTRIC CO., LTD.	LVT-
ENGLISH ELECTRIC CO., LTD.	NF-
ENGLISH ELECTRIC CO., LTD. ATOMIC POWER DIV., WHETSTONE, LEICS, ENGLAND	APDR-
ENGLISH ELECTRIC CO., LTD. ATOMIC POWER DIV., WHETSTONE, LEICS, ENGLAND	W/AT-
ENGLISH ELECTRIC CO., LTD. CENTRAL METALLURGICAL LABS., WHETSTONE, LEICS, ENGLAND	W/LAU.(NUMBER)
ENGLISH ELECTRIC CO., LTD. GUIDED WEAPONS DIV., LUTON, BEDS., ENGLAND	LMC-T-
ENGLISH ELECTRIC VALVE CO., LTD., CHELMSFORD, ESSEX, ENG	EEV-
ENGLISH ELECTRIC VALVE CO., LTD., CHELMSFORD, ESSEX, ENG	RP2-
ENJAY LABS., LINDEN, N.J.	ELD-
ENSIGN-BICKFORD CO. RESEARCH DEPT., SIMSBURY, CONN.	EBC-MR-
ENTWICKLUNGSRING NORD GMBH, BREMEN	ERNA-AB-RTT-
ENTWICKLUNGSRING NORD GMBH, BREMEN	RT/A-(NUMBER-MO./YR.)
ENTWICKLUNGSRING SUED GMBH, MUNICH	EWR-(NUMBER)-(YEAR)-
ENTWICKLUNGSRING SUED GMBH, MUNICH	EWR-E-
ENTWICKLUNGSRING SUED GMBH, MUNICH	EWR-SB-(NUMBER)-
ENVIRONMENTAL DATA SERVICE, SILVER SPRING, MD.	EDS-
ENVIRONMENTAL DATA SERVICE, SILVER SPRING, MD.	EDS-BC-
ENVIRONMENTAL DATA SERVICE, SILVER SPRING, MD. (BIB)	EDS-BM-
ENVIRONMENTAL DATA SERVICE, SILVER SPRING, MD.	EDSTM-
ENVIRONMENTAL DATA SERVICE, SILVER SPRING, MD. (BIB)	EDSTM-BC-
ENVIRONMENTAL DATA SERVICE, SILVER SPRING, MD.	ESSA-EDSTM-
ENVIRONMENTAL DATA SERVICE, SILVER SPRING, MD.	ESSA-TM-EDSTM-
ENVIRONMENTAL DATA SERVICE, SILVER SPRING, MD.	ESSA-TR-EDS-
ENVIRONMENTAL DATA SERVICE, SILVER SPRING, MD.	ESSA-WB-T-
ENVIRONMENTAL DATA SERVICE, SILVER SPRING, MD.	NOAA-TM-EDS-
ENVIRONMENTAL DATA SERVICE, SILVER SPRING, MD. (BIBLIOGRAPHY ON CLIMATE)	NOAA-TM-EDS-BC-
ENVIRONMENTAL DATA SERVICE, SILVER SPRING, MD. (BIBLIOGRAPHY ON CLIMATE)	WB/BC-
ENVIRONMENTAL DATA SERVICE, SILVER SPRING, MD. (BIB)	WB-BM-
ENVIRONMENTAL DATA SERVICE, SILVER SPRING, MD. (TRANS)	WB-T-
ENVIRONMENTAL DATA SERVICE, SILVER SPRING, MD.	WB/TA-
ENVIRONMENTAL HEALTH SERVICE, ROCKVILLE, MD.	CPS-(CONTRACT NO.)
ENVIRONMENTAL PROTECTION AGENCY, WASHINGTON, D.C. (CATEGORICAL PROGRAMS)	EPA-CPA-(NUMBER-YEAR)-
ENVIRONMENTAL PROTECTION AGENCY, WASHINGTON, D.C. (SOLID WASTE)	EPA-SW-(NUMBER-LTRS.)-
ENVIRONMENTAL PROTECTION AGENCY. AIR POLLUTION CONTROL OFFICE, ROCKVILLE, MD.	AP-
ENVIRONMENTAL PROTECTION AGENCY. AIR POLLUTION CONTROL OFFICE, ROCKVILLE, MD.	APCO-
ENVIRONMENTAL PROTECTION AGENCY. AIR POLLUTION CONTROL OFFICE, ROCKVILLE, MD.	APID-

ENVIRONMENTAL PROTECTION AGENCY. AIR POLLUTION CONTROL OFFICE, ROCKVILLE, MD. (AIR POLLUTION TECHNICAL DATA)	APTD-
ENVIRONMENTAL PROTECTION AGENCY. AIR POLLUTION CONTROL OFFICE, ROCKVILLE, MD. (AIR POLLUTION TECHNICAL INFORMATION CENTER)	APTIC-
ENVIRONMENTAL PROTECTION AGENCY. OFFICE OF RADIATION PROGRAMS, WASH., D.C. (SURVEILLANCE AND INSPECTION)	ORP/SID (YEAR)-
ENVIRONMENTAL PROTECTION AGENCY. RADIATION OFFICE, ROCKVILLE, MD.	RO/EERL-(YEAR)/
ENVIRONMENTAL PROTECTION AGENCY. SOLID WASTE OFFICE, ROCKVILLE, MD.	SW-
ENVIRONMENTAL PROTECTION AGENCY. WATER QUALITY OFFICE, WASHINGTON, D.C.	EAP-WQO-(NUMBER)-
ENVIRONMENTAL PROTECTION AGENCY. WATER QUALITY OFFICE, WASHINGTON, D.C.	EPA-WQO-(NUMBER)-
ENVIRONMENTAL PROTECTION AGENCY. WATER QUALITY OFFICE, WASHINGTON, D.C.	EPA-WQO-(NO.)-DQH-DATE)
ENVIRONMENTAL PROTECTION AGENCY. WATER QUALITY OFFICE, WASHINGTON, D.C.	EPA-WQO-(NO.)-EBZ-DATE)
ENVIRONMENTAL PROTECTION AGENCY. WATER QUALITY OFFICE, WASHINGTON, D.C.	EPA-WQO-(NO.-LTRS.-DATE)
ENVIRONMENTAL PROTECTION AGENCY. WATER QUALITY OFFICE, ROCKVILLE, MD.	WP-
ENVIRONMENTAL PROTECTION AGENCY. WATER QUALITY OFFICE, WASHINGTON, D.C. (WASTE TREATMENT)	WP-(NUMBER)-AWTR-
ENVIRONMENTAL PROTECTION AGENCY. WATER QUALITY OFFICE, ROCKVILLE, MD.	WQO-
ENVIRONMENTAL RESEARCH CORP., ALEXANDRIA, VA.	ERC-TMR-
ENVIRONMENTAL RESEARCH CORP., LAS VEGAS, NEV.	ERC-
ENVIRONMENTAL RESEARCH LABS., ROCKVILLE, MD. (BARBADOS OCEANOGRAPHIC + METEOROLOGICAL ANALYSIS PROJ.)	ERL-BOMAP-
ENVIRONMENTAL RESEARCH LABS., ROCKVILLE, MD. (BARBADOS OCEANOGRAPHIC + METEOROLOGICAL ANALYSIS PROJ.)	ERLTM-BOMAP-
ENVIRONMENTAL RESEARCH LABS., ROCKVILLE, MD. (BARBADOS OCEANOGRAPHIC + METEOROLOGICAL ANALYSIS PROJ.)	NOAA-TM-ERL-BOMAP-
ENVIRONMENTAL RESEARCH LABS. OFFICE OF THE DIRECTOR, BOULDER, COLO.	ERL-OD-
ENVIRONMENTAL RESEARCH LABS. OFFICE OF THE DIRECTOR, BOULDER, COLO.	ESSA-TR-ERL-(NO.)-OD-
ENVIRONMENTAL RESEARCH LABS. OFFICE OF THE DIRECTOR, BOULDER, COLO.	NOAA-TM-ERL-OD-
(SEE ALSO NAMES OF SEPARATE LABS., AS: SPACE ENVIRONMENT LAB.)	
ENVIRONMENTAL SCIENCE SERVICES ADM., ATLANTIC CITY, NJ.	RD-(YEAR)-
ENVIRONMENTAL SCIENCE SERVICES ADM., BOULDER, COLO.	ERLTM-OD-
ENVIRONMENTAL SCIENCE SERVICES ADM., BOULDER, COLO.	ESSA-ERL-(NO.)-(LTRS.)-
ENVIRONMENTAL SCIENCE SERVICES ADM., BOULDER, COLO.	ESSA-ERL-TM-
ENVIRONMENTAL SCIENCE SERVICES ADM., BOULDER, COLO.	ESSA-TR-ERL-
ENVIRONMENTAL SCIENCE SERVICES ADM., BOULDER, COLO.	RLTM-ESL-
(SEE ALSO SUCCESSOR: NATL. OCEANIC + ATMOS. ADM.)	
ENVIRONMENTAL SCIENCE SERVICES ADM. OFFICE OF ADM. AND TECHNICAL SERVICES, ROCKVILLE, MD. (LIBRARY)	ATSTM-LIB-
ENVIRONMENTAL SCIENCE SERVICES ADM. OFFICE OF FEDERAL COORDINATOR FOR METEOROLOGICAL SERVICES AND SUPPORTING RESEARCH, WASHINGTON, D.C.	OFCM-(YEAR)-
ENVIRONMENTAL TECHNICAL APPLICATIONS CENTER (AIR FORCE), WASHINGTON, D.C.	AF-ETAC-TN-(YEAR)-
ENVIRONMENTAL TECHNICAL APPLICATIONS CENTER (AIR FORCE), WASHINGTON, D.C.	ETAC-IN-
ENVIRONMENTAL TECHNICAL APPLICATIONS CENTER (AIR FORCE), WASHINGTON, D.C.	ETAC-TN-(YEAR)-
EON INSTRUMENTATION INC., VAN NUYS, CALIF.	R(4 DIGITS)-F
ERIE RESISTOR CORP., PENNA.	ERC-
ERIE TECHNOLOGICAL PRODUCTS, INC., PENNA.	AR(NUMBER)
ESL, INC., PALO ALTO, CALIF.	ESLI-TM-
ESSO DEVELOPMENT CO., LTD.	EEL-
ESSO RESEARCH AND ENGINEERING CO., BAYTOWN, TEX.	HUM-
ESSO RESEARCH AND ENGINEERING CO., BAYTOWN, TEX.	HUM-(NUMBER)-P
ESSO RESEARCH AND ENGINEERING CO., LINDEN, N.J. (BUMBLEBEE PROJECT) (REF. 8)	BUMBLEBEE-
ESSO RESEARCH AND ENGINEERING CO., LINDEN, N.J.	ED-
ESSO RESEARCH AND ENGINEERING CO., LINDEN, N.J.	EREC-
ESSO RESEARCH AND ENGINEERING CO., LINDEN, N.J.	ESSO-
ESSO RESEARCH AND ENGINEERING CO., LINDEN, N.J.	ESSO-(YEAR)-
ESSO RESEARCH AND ENGINEERING CO., LINDEN, N.J.	ESSO-GR-(NO.)-FBP-(YR.)
ESSO RESEARCH AND ENGINEERING CO., LINDEN, N.J.	ESSO-GR-(LETTERS)-(YR.)
ESSO RESEARCH AND ENGINEERING CO., LINDEN, N.J.	ESSO-MA-
ESSO RESEARCH AND ENGINEERING CO. GOVERNMENT RESEARCH LAB., LINDEN, N.J.	GR-(NO.)-APD-(YR.)
ESSO RESEARCH AND ENGINEERING CO. GOVERNMENT RESEARCH LAB., LINDEN, N.J. (ADVANCED SOLID PROPELLANTS)	GR-(NUMBER)-ASP-
ESSO RESEARCH AND ENGINEERING CO. GOVERNMENT RESEARCH LAB., LINDEN, N.J. (DEV. CATALYST FOR HYDRAZINE)	GR-(NUMBER)-DCH-
ESSO RESEARCH AND ENG. CO. GOVERNMENT RES. LAB., LINDEN, N.J. (FUNCTIONALITY OF BINDER PREPOLYMERS)	GR-(NUMBER)-FBP-
ESSO RESEARCH AND ENGINEERING CO. GOVERNMENT RESEARCH LAB., LINDEN, N.J. (MAGNETIC GRADIENT MATERIALS)	GR-(NUMBER)-MGM-
ESSO RESEARCH AND ENGINEERING CO. GOVERNMENT RESEARCH LAB., LINDEN, N.J. (SYNTHESIS OF ANTIMALARIALS)	GR-(NUMBER)-SAM-
ESSO RESEARCH AND ENGINEERING CO. GOVERNMENT RESEARCH LAB., LINDEN, N.J. (SYNTHESIS ANTIRADIATION AGENTS)	GR-(NUMBER)-SAR-
ESSO RES. AND ENG. CO. GOVERNMENT RES. LAB., LINDEN, N.J. (SYNTHESIS + EVAL. OF CURING AGENTS)	GR-(NUMBER)-SECA-
ESSO RESEARCH AND ENGINEERING CO. GOVERNMENT RESEARCH LAB., LINDEN, N.J. (THERMALLY STABLE LIQUID FUELS)	GR-(NUMBER)-TSF-
ESSO RESEARCH AND ENGINEERING CO. GOVERNMENT RESEARCH LAB., LINDEN, N.J. (VAPOR PHASE DEPOSITS)	GR-(NUMBER)-VPD-
ESSO RESEARCH AND ENGINEERING CO. GOVERNMENT RESEARCH LAB., LINDEN, N.J.	GR-(NO.)-FRA-(YEAR)
ESSO RESEARCH AND ENGINEERING CO. GOVERNMENT RESEARCH LAB., LINDEN, N.J. (VAPOR DEPOSITS)	GR-(NO.)-VDP-(YEAR)
ESSO RESEARCH AND ENGINEERING CO. GOVERNMENT RESEARCH LAB., LINDEN, N.J.	GR-(NO.)-PAP-(YEAR)
ESSO RESEARCH AND ENGINEERING CO. GOVERNMENT RESEARCH LAB., LINDEN, N.J.	GR-(NO.)-RIR-(YEAR)
ESSO RESEARCH AND ENGINEERING CO. GOVERNMENT RESEARCH LAB., LINDEN, N.J. (ORGANOPHOSPHONATES)	GR-(NO.)-OPS-(YEAR)
ESSO RESEARCH AND ENGINEERING CO. GOVERNMENT RESEARCH LAB., LINDEN, N.J.	GR-(NO.)-MPD-(YEAR)
ESSO RESEARCH AND ENGINEERING CO. PROCESS RES. DIV., LINDEN, N.J.	PCRD-
ESSO RESEARCH AND ENGINEERING CO. PROCESS RES. DIV., LINDEN, N.J.	PCRD-(YEAR)-
ESSO RESEARCH AND ENGINEERING CO. PROCESS RES. DIV., LINDEN, N.J.	PCRD-MRD-
ETHYL CORP., BATON ROUGE, LA.	EC-
ETHYL CORP., FERNDALE, MICH.	GF-
ETHYL CORP., FERNDALE, MICH.	GR-(YEAR)-
ETHYL CORP. RESEARCH AND DEVELOPMENT DEPT., FERNDALE, MICH. (AERONAUTICAL RESEARCH)	AR-
ETHYL CORP. RESEARCH AND DEVELOPMENT DEPT., FERNDALE, MICH. (AERONAUTICAL RESEARCH MEMORANDA)	ARM-
ETHYL CORP. RESEARCH AND DEV. DEPT., FERNDALE, MICH.	EC-PR-
ETHYL CORP. RESEARCH AND DEVELOPMENT DEPT., FERNDALE, MICH. (ENGINEERING RESEARCH)	ER-
ETHYL CORP. RESEARCH AND DEVELOPMENT DEPT., FERNDALE, MICH. (ENGINEERING RESEARCH MEMORANDA)	ERM-
ETHYL CORP. RESEARCH AND DEV. DEPT., FERNDALE, MICH.	ETHYL-RM-
ETHYL CORP. RESEARCH AND DEVELOPMENT DEPT., FERNDALE, MICH. (GENERAL ADMINISTRATION)	GA-
ETHYL CORP. RESEARCH AND DEVELOPMENT DEPT., FERNDALE, MICH. (LABORATORY TECHNICAL DATA)	LTD-
ETHYL CORP. RESEARCH AND DEVELOPMENT DEPT., FERNDALE, MICH. (ONE-PAGE REPORTS)	OP-
ETHYL CORP. RESEARCH AND DEVELOPMENT DEPT., FERNDALE, MICH. (RESEARCH MEMORANDA)	RM-
ETHYL CORP. RESEARCH AND DEVELOPMENT DEPT., FERNDALE, MICH. (REFINERY TECHNOLOGY)	RT-

ETHYL CORP. RESEARCH AND DEVELOPMENT DEPT.

ETHYL CORP. RESEARCH AND DEVELOPMENT DEPT., FERNDALE, MICH. (REFINERY TECHNOLOGY MEMORANDA)	RTM-
ETHYL CORP. RESEARCH AND DEVELOPMENT DEPT., FERNDALE, MICH. (TECHNICAL ANALYSIS)	TA-
ETHYL CORP. RESEARCH AND DEVELOPMENT DEPT., FERNDALE, MICH. (TECHNICAL ANALYSIS MEMORANDA)	TAM-
ETHYL CORP. RESEARCH AND DEVELOPMENT DEPT., FERNDALE, MICH. (TECHNICAL SERVICE)	TS-
EURATOM-BELGIUM FAST REACTOR ASSN.	AS-015-N-
EUREKA WILLIAMS CORP., BLOOMINGTON, ILL.	EWC-
EUROPEAN AMERICAN COMMITTEE ON REACTOR PHYSICS, EUROPEAN NUCLEAR ENERGY AGENCY, PARIS	EACRP-A-
EUROPEAN AMERICAN COMMITTEE ON REACTOR PHYSICS, EUROPEAN NUCLEAR ENERGY AGENCY, PARIS	EACRP-L-
EUROPEAN AMERICAN COMMITTEE ON REACTOR PHYSICS, EUROPEAN NUCLEAR ENERGY AGENCY, PARIS	EACRP-U-
EUROPEAN-AMERICAN NUCLEAR DATA COMMITTEE (REF. 45)	EANDC-(NUMBER)(LETTER)
EUROPEAN-AMERICAN NUCLEAR DATA COMMITTEE (REPORTS FROM CANADIAN PARTICIPANTS)	EANDC (CAN)(NUMBER)(LTR)
EUROPEAN-AMERICAN NUCLEAR DATA COMMITTEE (REPORTS FROM EURATOM MEMBER PARTICIPANTS)	EANDC (E) (NUMBER)(LTR)
EUROPEAN-AMERICAN NUCLEAR DATA COMMITTEE (REPORTS FROM EURATOM MEMBER PARTICIPANTS)	EANDC (EUR)-
EUROPEAN-AMERICAN NUCLEAR DATA COMMITTEE (REPORTS FROM PARTICIPANTS NOT OTHERWISE DESIGNATED)	EANDC (OR) (NUMBER)(LTR)
EUROPEAN-AMERICAN NUCLEAR DATA COMMITTEE (REPORTS FROM UNITED KINGDOM PARTICIPANTS)	EANDC (UK) (NUMBER)(LTR)
EUROPEAN-AMERICAN NUCLEAR DATA COMMITTEE (REPORTS FROM UNITED STATES PARTICIPANTS)	EANDC (US) (NUMBER)(LTR)
EUROPEAN-AMERICAN NUCLEAR DATA COMMITTEE (REPORTS FROM FRENCH PARTICIPANTS)	EUR-(4 DIGITS).F
EUROPEAN-AMERICAN NUCLEAR DATA COMMITTEE (REPORTS FROM GERMAN PARTICIPANTS)	EUR-(4 DIGITS).D
EUROPEAN-AMERICAN NUCLEAR DATA COMMITTEE (REPORTS FROM EURATOM PARTICIPANTS)	EUR-(4 DIGITS).E
EUROPEAN-AMERICAN NUCLEAR DATA COMMITTEE (REPORTS FROM ITALIAN PARTICIPANTS)	EUR-(4 DIGITS).I
EUROPEAN-AMERICAN NUCLEAR DATA COMMITTEE	NRDL-
EUROPEAN-AMERICAN NUCLEAR DATA COMMITTEE. SECRETARIAT, OECD, PARIS (REF. 45)	EANDC-
EUROPEAN-AMERICAN NUCLEAR DATA COMMITTEE. SECRETARIAT, OECD, PARIS (REF. 45)	EANDC(LETTER(S))-
EUROPEAN ATOMIC ENERGY COMMUNITY (ALL LOCATIONS)	EUR-
EUROPEAN ATOMIC ENERGY COMMUNITY (ALL LOCATIONS)	EURATOM-
EUROPEAN ATOMIC ENERGY COMMUNITY	EUR-CEA-
EUROPEAN ATOMIC ENERGY COMMUNITY	GN-
EUROPEAN ATOMIC ENERGY COMMUNITY, BRUSSELS	EUR/C-
EUROPEAN ATOMIC ENERGY COMMUNITY, BRUSSELS	EUR-CEA-FC-
EUROPEAN ATOMIC ENERGY COMMUNITY, BRUSSELS (TRANS.)	EUR-CEA-FC-(NO.)-TR
EUROPEAN ATOMIC ENERGY COMMUNITY. CENTRAL BUREAU FOR NUCLEAR MEASUREMENTS, GEEL, BELGIUM	CBNM-
EUROPEAN ATOMIC ENERGY COMMUNITY. EUROPEAN SCIENTIFIC DATA PROCESSING CENTER, ISPRA, ITALY	CETIS-
EUROPEAN ATOMIC ENERGY COMMUNITY. JOINT NUCLEAR RESEARCH CENTER, ISPRA, ITALY	ISP-
EUROPEAN ATOMIC ENERGY COMMUNITY. JOINT NUCLEAR RESEARCH CENTER, ISPRA, ITALY	ISPRA-
EUROPEAN ATOMIC ENERGY COMMUNITY. JOINT NUCLEAR RESEARCH CENTER, ISPRA, ITALY	JNRC-
EUROPEAN COMPANY FOR THE CHEMICAL PROCESSING OF IRRADIATED FUELS, MOL, BELGIUM	ED-(LETTERS)-(NUMBER)
EUROPEAN COMPANY FOR THE CHEMICAL PROCESSING OF IRRADIATED FUELS, MOL, BELGIUM	EIR-
EUROPEAN COMPANY FOR THE CHEMICAL PROCESSING OF IRRADIATED FUELS, MOL, BELGIUM	ETR-
EUROPEAN COMPANY FOR THE CHEMICAL PROCESSING OF IRRADIATED FUELS, MOL, BELGIUM	EUROCHEMIC-
EUROPEAN CONFERENCE ON SATELLITE COMMUNICATIONS. TECHNICAL PLANNING STAFF	SCL/TPS/(NO.)(LETTER)
EUROPEAN COUNCIL FOR NUCLEAR RESEARCH, GENEVA	BS-
EUROPEAN COUNCIL FOR NUCLEAR RESEARCH, GENEVA	CERN-(NUMBER)/TH.(NUMBER)
EUROPEAN COUNCIL FOR NUCLEAR RESEARCH, GENEVA	CERN/GEN/-
EUROPEAN COUNCIL FOR NUCLEAR RESEARCH, GENEVA	CERN-NPA/(INITLS)-(YEAR)-
EUROPEAN COUNCIL FOR NUCLEAR RESEARCH, GENEVA	CERN-P-
EUROPEAN COUNCIL FOR NUCLEAR RESEARCH, GENEVA	CERN/T/(INITIALS)/-
EUROPEAN COUNCIL FOR NUCLEAR RESEARCH, GENEVA	DI/HP-
EUROPEAN COUNCIL FOR NUCLEAR RESEARCH, GENEVA	NPA/
EUROPEAN NUCLEAR ENERGY AGENCY, PARIS	ENEA-
EUROPEAN NUCLEAR ENERGY AGENCY, PARIS	SEN-IPSA-
EUROPEAN NUCLEAR ENERGY AGENCY. NEUTRON DATA COMPILATION CENTRE, GIF-SUR-YVETTE, FRANCE	CCDN-CI/
EUROPEAN NUCLEAR ENERGY AGENCY. NEUTRON DATA COMPILATION CENTRE, GIF-SUR-YVETTE, FRANCE	CCDN-NW-
EUROPEAN NUCLEAR ENERGY AGENCY. NEUTRON DATA COMPILATION CENTRE, GIF-SUR-YVETTE, FRANCE	CCDN/SYS-
EUROPEAN ORGANIZATION FOR NUCLEAR RESEARCH, GENEVA	AR/INT.SG-
EUROPEAN ORGANIZATION FOR NUCLEAR RESEARCH, GENEVA	CERN-
EUROPEAN ORGANIZATION FOR NUCLEAR RESEARCH, GENEVA	CERN-(YEAR)-
EUROPEAN ORGANIZATION FOR NUCLEAR RESEARCH, GENEVA	CERN-AR-(YEAR)
EUROPEAN ORGANIZATION FOR NUCLEAR RESEARCH, GENEVA	CERN-BIB-
EUROPEAN ORGANIZATION FOR NUCLEAR RESEARCH, GENEVA	CERN-BS-
EUROPEAN ORGANIZATION FOR NUCLEAR RESEARCH, GENEVA	CERN/DD/DH-(YEAR)/(NO.)
EUROPEAN ORGANIZATION FOR NUCLEAR RESEARCH, GENEVA	CERN/DI/HP-
EUROPEAN ORGANIZATION FOR NUCLEAR RESEARCH, GENEVA	CERN/HER/(INITLS)(NO)(YR)
EUROPEAN ORGANIZATION FOR NUCLEAR RESEARCH, GENEVA	CERN/HERA/(YEAR)-
EUROPEAN ORGANIZATION FOR NUCLEAR RESEARCH, GENEVA	CERN-INT-MPS/EP-
EUROPEAN ORGANIZATION FOR NUCLEAR RESEARCH, GENEVA	CERN-ISR-MA/(YEAR)-
EUROPEAN ORGANIZATION FOR NUCLEAR RESEARCH, GENEVA	CERN-MPS-(INITLS)-(YEAR)-
EUROPEAN ORGANIZATION FOR NUCLEAR RESEARCH, GENEVA	CERN/NP/(INITLS)-(YEAR)-
EUROPEAN ORGANIZATION FOR NUCLEAR RESEARCH, GENEVA	CERN/TC/(INITLS)-(YEAR)-
EUROPEAN ORGANIZATION FOR NUCLEAR RESEARCH, GENEVA	CERN/TC/BEBC-(YR.)-
EUROPEAN ORGANIZATION FOR NUCLEAR RESEARCH, GENEVA	CERN-TH-
EUROPEAN ORGANIZATION FOR NUCLEAR RESEARCH, GENEVA	CERN TRANS (YEAR)-
EUROPEAN ORGANIZATION FOR NUCLEAR RESEARCH, GENEVA	DIR/PS/INT.S-
EUROPEAN ORGANIZATION FOR NUCLEAR RESEARCH, GENEVA	FA/WP/(NO.)/-
EUROPEAN ORGANIZATION FOR NUCLEAR RESEARCH, GENEVA	ISR-
EUROPEAN ORGANIZATION FOR NUCLEAR RESEARCH, GENEVA	MPS/DL-INT-(YR.)-
EUROPEAN ORGANIZATION FOR NUCLEAR RESEARCH, GENEVA	MPS/EP-
EUROPEAN ORGANIZATION FOR NUCLEAR RESEARCH, GENEVA	MPS/INT/CO/
EUROPEAN ORGANIZATION FOR NUCLEAR RESEARCH, GENEVA	MPS/INT.DL/B-(YR)-
EUROPEAN ORGANIZATION FOR NUCLEAR RESEARCH, GENEVA	MPS-INT.LIN-
EUROPEAN ORGANIZATION FOR NUCLEAR RESEARCH, GENEVA	MPS-INT.MA-
EUROPEAN ORGANIZATION FOR NUCLEAR RESEARCH, GENEVA	MPS/INT.MU/EP-
EUROPEAN ORGANIZATION FOR NUCLEAR RESEARCH, GENEVA	MPS/SI/INT.DL-
EUROPEAN ORGANIZATION FOR NUCLEAR RESEARCH, GENEVA	MSC-
EUROPEAN ORGANIZATION FOR NUCLEAR RESEARCH, GENEVA	NPA-INT-
EUROPEAN ORGANIZATION FOR NUCLEAR RESEARCH, GENEVA	PS/(4 DIGITS)
EUROPEAN ORGANIZATION FOR NUCLEAR RESEARCH, GENEVA	PS/INT.-AR(YEAR)-
EUROPEAN ORGANIZATION FOR NUCLEAR RESEARCH, GENEVA	TH-
EUROPEAN ORGANIZATION FOR NUCLEAR RESEARCH. EUROPEAN ACCELERATOR STUDY GROUP, GENEVA	CERN/EAG/(NUMBER)
EUROPEAN ORGANIZATION FOR NUCLEAR RESEARCH. EUROPEAN COMMITTEE FOR FUTURE ACCELERATORS, GENEVA	CERN/ECFA-(YEAR)/(NUMBER)
EUROPEAN ORGANIZATION FOR NUCLEAR RESEARCH. EUROPEAN COMMITTEE FOR FUTURE ACCELERATORS, GENEVA (FINAL SECTIONS OF CODE ARE AUTHORS' INITIALS)	CERN/ECFA-(YR)/WG(NO)-

EUROPEAN ORGANIZATION FOR NUCLEAR RESEARCH. EUROPEAN STUDY GROUP ON FUSION, GENEVA	CERN/FSG/(NUMBER)
EUROPEAN ORGANIZATION FOR NUCLEAR RESEARCH. PROTONSYNCHROTRON GROUP, GENEVA	CERN-PS-(INITIALS)-
EUROPEAN SPACE DATA CENTRE, DARMSTADT, GERMANY	CESRO-SN-
EUROPEAN SPACE DATA CENTRE, DARMSTADT, GERMANY	ESDAC-
EUROPEAN SPACE DATA CENTRE, DARMSTADT, GERMANY	ESRO-SM(NUMBER/ESDAC/
EUROPEAN SPACE DATA CENTRE, DARMSTADT, GERMANY	ESRO-SN-(NO.)/ESDAC/
EUROPEAN SPACE DATA CENTRE, DARMSTADT, GERMANY	ESRO-SR-(NO.)/ESDAC/
EUROPEAN SPACE DATA CENTRE, DARMSTADT, GERMANY	SM-
EUROPEAN SPACE RESEARCH INST., FRASCATI, ITALY	ESRIN-IN-
EUROPEAN SPACE RESEARCH LAB., NOORDWIJK, NETHERLANDS	ESRO-SN-(NUMBER)/ESLAB/
EUROPEAN SPACE RESEARCH ORGANIZATION, PARIS	CERS-
EUROPEAN SPACE RESEARCH ORGANIZATION, PARIS	ESRO-
EUROPEAN SPACE RESEARCH ORGANIZATION, PARIS	ESRO-CR-
EUROPEAN SPACE RESEARCH ORGANIZATION, PARIS	ESRO-SM-
EUROPEAN SPACE RESEARCH ORGANIZATION, PARIS	ESRO-SN-
EUROPEAN SPACE RESEARCH ORGANIZATION, PARIS	ESRO-SP-
EUROPEAN SPACE RESEARCH ORGANIZATION, PARIS	ESRO-SR-
EUROPEAN SPACE RESEARCH ORGANIZATION, PARIS	ESRO-TM-
EUROPEAN SPACE RESEARCH ORGANIZATION, PARIS	ESRO-TR-
EUROPEAN SPACE RESEARCH ORGANIZATION, PARIS	EUROSPACE-
EUROPEAN SPACE RESEARCH ORGANIZATION, PARIS	LAS/
EUROPEAN SPACE RESEARCH ORGANIZATION, PARIS	P-
EUROPEAN SPACE RESEARCH ORGANIZATION. SPACE DOCUMENTATION SERVICE, ROME	SDS/(NUMBER)-(NUMBER)-
EUROPEAN SPACE TECHNOLOGY CENTER, DELFT, NETHERLANDS	ESRO-TN-
EUROPEAN SPACE TECHNOLOGY CENTRE, DELFT, NETHERLANDS	ESTEC-
EUROPEAN SPACE TECHNOLOGY CENTRE, DELFT, NETHERLANDS	ESTEC-MT-
EUROPEAN SPACE TECHNOLOGY CENTER, NOORDWIJK, NETHERLANDS	ESRO-TM/(LETTER)/-
EUROPEAN SPACE TECHNOLOGY CENTER, NOORDWIJK, NETHERLANDS	ESRO-TN-(NUMBER)/ESTEC/
EUROPEAN SPACE VEHICLE LAUNCHER DEV. ORG., PARIS	CECLES-
EUROPEAN SPACE VEHICLE LAUNCHER DEV. ORG., PARIS	ELDO-BIB-
EUROPEAN SPACE VEHICLE LAUNCHER DEV. ORG., PARIS	ELDO-PUBL.-
EUROPEAN SPACE VEHICLE LAUNCHER DEV. ORG., PARIS	ELDO-TM-
EUROPEAN SPACE VEHICLE LAUNCHER DEV. ORG., PARIS	ELDO-TM-(LETTER)-
EUROPEAN SPACE VEHICLE LAUNCHER DEV. ORG., PARIS	ELDO-TM/F/
EUROPEAN SPACE VEHICLE LAUNCHER DEV. ORG., PARIS	FP2/ENEA-
EUROPEAN SPACE VEHICLE LAUNCHER DEV. ORG., PARIS	SEN/IPSA/(YEAR)/
EUROPEAN THEATER OF OPERATIONS (LEATHER REPORT)	LETOR-
EUROPEAN THEATER OF OPERATIONS (ARMY). ALSOS MISSION	ALSOS-
EUROPEAN THEATER OF OPERATIONS (ARMY). ALSOS MISSION	ALSOS-DDC-
EUROPEAN THEATER OF OPERATIONS (ARMY). ALSOS MISSION (GERLACH FILE)	ALSOS-G-
EUROPEAN THEATER OF OPERATIONS (ARMY). ALSOS MISSION (HOYT FILE)	ALSOS-H-
EUROPEAN THEATER OF OPERATIONS (ARMY). ALSOS MISSION (HAAGEN FILE)	ALSOS-HA-
EUROPEAN THEATER OF OPERATIONS (ARMY). ALSOS MISSION (KLOTTER FILE)	ALSOS-K-
EUROPEAN THEATER OF OPERATIONS (ARMY). ALSOS MISSION (KLIEWE FILE)	ALSOS-KLI-
EUROPEAN THEATER OF OPERATIONS (ARMY). ALSOS MISSION (MF FILE)	ALSOS-MF-
EUROPEAN THEATER OF OPERATIONS (ARMY). ALSOS MISSION (MIS FILE)	ALSOS MIS-
EUROPEAN THEATER OF OPERATIONS (ARMY). ALSOS MISSION (OBERTH FILE)	ALSOS O-
EUROPEAN THEATER OF OPERATIONS (ARMY). ALSOS MISSION (REICHSFORSCHUNGSRAT FILE)	ALSOS RFR-
EUROPEAN THEATER OF OPERATIONS (ARMY). ALSOS MISSION	DDC-
EUROPEAN THEATER OF OPERATIONS (ARMY). ALSOS MISSION (GERLACH FILE)	G FILE (ALSOS)-
EUROPEAN THEATER OF OPERATIONS (ARMY). ALSOS MISSION (HAAGEN FILE)	HA-
EUROPEAN THEATER OF OPERATIONS (ARMY). ALSOS MISSION (HOYT FILE)	H FILE-
EUROPEAN THEATER OF OPERATIONS (ARMY). ALSOS MISSION (KLIEWE FILE)	KLI-
EUROPEAN THEATER OF OPERATIONS (ARMY). ALSOS MISSION (MF FILE)	MF-
EUROPEAN THEATER OF OPERATIONS (ARMY). ALSOS MISSION	MICRO (NUMBER)-
EUROPEAN THEATER OF OPERATIONS (ARMY). ALSOS MISSION (MIS FILE)	MIS-
EUROPEAN THEATER OF OPERATIONS (ARMY). ALSOS MISSION (OBERTH FILE)	O FILE-
EUROPEAN THEATER OF OPERATIONS (ARMY). ALSOS MISSION	OSRD ALSOS-
EUROPEAN THEATER OF OPERATIONS (ARMY). ALSOS MISSION REICHSFORSCHUNGSRAT FILE)	RFR-
EVANS RESEARCH AND DEVELOPMENT CORP., N.Y.C.	ERD-
EVANS RESEARCH AND DEVELOPMENT CORP., N.Y.C.	ERDC-
EVANS RESEARCH AND DEVELOPMENT CORP., N.Y.C.	ERD-PR-
EVANS RESEARCH AND DEVELOPMENT CORP., N.Y.C.	EVANS-
EVANS SIGNAL LAB., BELMAR, N.J. (ENGINEERING REPT.)	E-
EVANS SIGNAL LAB., BELMAR, N.J.	ESL-
EVANS SIGNAL LAB., BELMAR, N.J. (TECH. MEMO.)	M-
EVANS SIGNAL LAB., BELMAR, N.J.	MON-
EVANS SIGNAL LAB., BELMAR, N.J.	RDS-
EVANS SIGNAL LAB., BELMAR, N.J.	SCEL-
EVANS SIGNAL LAB., BELMAR, N.J. (TEST REPORT)	T-
EVANS SIGNAL LAB. INST. FOR EXPLORATORY RES., BELMAR, N.J.	SIGFM/EL-XS-3-99-(NO.)-
EXOTECH, INC., ALEXANDRIA, VA.	EXOTECH-
EXOTECH, INC., ROCKVILLE, MD.	TR-IC-
EXOTECH, INC., WASHINGTON, D.C.	EXOTECH-TRES-
EXOTECH, INC., WASHINGTON, D.C.	TRES-
EXOTECH, INC., WASHINGTON, D.C.	TRSR-
EXOTECH, INC., WASHINGTON, D.C.	TR-SR-(YEAR)-
EXOTECH, INC., WASHINGTON, D.C.	TRSR-(YEAR)-
EXPERIMENT, INC., RICHMOND	EI-
EXPERIMENT, INC., RICHMOND	EI-TM-
EXPERIMENT, INC., RICHMOND	EI-TP-
EXPERIMENT, INC., RICHMOND (SQUID PROJECT) (REF. 33)	EXP-(NUMBER)-M
EXPERIMENT, INC., RICHMOND (SQUID PROJECT) (REF. 33)	EXP-(NUMBER)-P
EXPERIMENT, INC., RICHMOND	EXP-TP-
EXPERIMENTAL DIVING UNIT (NAVY), WASHINGTON, D.C.	NGF-RR-
EXPLOSIVE ORDNANCE DISPOSAL SCHOOL (NAVY)	EODS-
EXPLOSIVE TECHNOLOGY, INC., FAIRFIELD, CALIF.	REPT-(NUMBER/NUMBER)SR
EXPLOSIVES CORP. OF AMERICA, ISSAQUAH, WASH.	E-(YEAR)-R-
EYE RESEARCH FOUNDATION OF BETHESDA, MD.	ERF-RR-

F-R MACHINE WORKS, INC.

F-R MACHINE WORKS, INC., LONG ISLAND CITY, N.Y.	F-RMW-
FABRIC RESEARCH LABS., INC., DEDHAM, MASS.	FRL-
FACTORY MUTUAL ENGINEERING DIV., BOSTON	FM-
FACTORY MUTUAL RESEARCH CORP., NORWOOD, MASS.	FMRC-
FAIRCHILD CAMERA AND INSTR. CORP., HICKSVILLE, N.Y.	SD-BA-
FAIRCHILD CAMERA AND INSTRUMENT CORP., JAMAICA, N.Y.	FCIC-
FAIRCHILD CAMERA + INSTR. CORP., JAMAICA, N.Y. (REF. 1)	(RPT.NO.)(SUBJ.CODE LTR)
FAIRCHILD CAMERA AND INSTRUMENT CORP., SYOSSET, N.Y.	SME-AA-
FAIRCHILD CAMERA AND INSTRUMENT CORP., SYOSSET, N.Y.	SME-BB-
FAIRCHILD CAMERA AND INSTRUMENT CORP. RECONNAISSANCE SYSTEMS DIV., SYOSSET, N.Y.	MPR-
FAIRCHILD ENGINE AND AIRPLANE CORP.	BCR-
FAIRCHILD ENGINE AND AIRPLANE CORP. (RANGER AIRCRAFT ENGINES)	RAE-
FAIRCHILD ENGINE AND AIRPLANE CORP., HAGERSTOWN, MD. (SEE ALSO LATER NAME: FAIRCHILD HILLER CORP.)	FEAC-
FAIRCHILD ENGINE AND AIRPLANE CORP. FAIRCHILD AIRCRAFT DIV., HAGERSTOWN, MD.	FA-
FAIRCHILD ENGINE AND AIRPLANE CORP. FAIRCHILD AIRCRAFT DIV., HAGERSTOWN, MD. (FLIGHT TEST REPORT)	FT-
FAIRCHILD ENGINE AND AIRPLANE CORP. FAIRCHILD ENGINE DIV., FARMINGDALE, N.Y.	FED-
FAIRCHILD ENGINE AND AIRPLANE CORP. NEPA DIV., OAK RIDGE, TENN.	CDR-
FAIRCHILD ENGINE AND AIRPLANE CORP. NEPA DIV., OAK RIDGE, TENN.	CR-
FAIRCHILD ENGINE AND AIRPLANE CORP. NEPA DIV., OAK RIDGE, TENN.	DC-(YEAR)-(NUMBER)-
FAIRCHILD ENGINE AND AIRPLANE CORP. NEPA DIV., OAK RIDGE, TENN.	DR-
FAIRCHILD ENGINE AND AIRPLANE CORP. NEPA DIV., OAK RIDGE, TENN. (REF. 31)	IC-
FAIRCHILD ENGINE AND AIRPLANE CORP. NEPA DIV., OAK RIDGE, TENN. (REF. 31)	IY-
FAIRCHILD ENGINE AND AIRPLANE CORP. NEPA DIV., OAK RIDGE, TENN. (REF. 31)	NCR-
FAIRCHILD ENGINE AND AIRPLANE CORP. NEPA DIV., OAK RIDGE, TENN. (REF. 31)	NEPA-
FAIRCHILD ENGINE AND AIRPLANE CORP. NEPA DIV., OAK RIDGE, TENN. (REF. 31)	NEPA-(NO.)-(LTRS.)-(NO.)
FAIRCHILD ENGINE AND AIRPLANE CORP. NEPA DIV., OAK RIDGE, TENN. (REF. 31)	NEPA-IC-(YEAR)-(MONTH)-
FAIRCHILD ENGINE AND AIRPLANE CORP. NEPA DIV., OAK RIDGE, TENN. (REF. 31)	NEPA-IY-
FAIRCHILD ENGINE AND AIRPLANE CORP. NEPA DIV., OAK RIDGE, TENN. (REF. 31)	NEPA-NCR-
FAIRCHILD ENGINE AND AIRPLANE CORP. NEPA DIV., OAK RIDGE, TENN. (REF. 31)	NEPA-SCR-
FAIRCHILD ENGINE AND AIRPLANE CORP. NEPA DIV., OAK RIDGE, TENN. (REF. 31)	NEPA-SCRM-
FAIRCHILD ENGINE AND AIRPLANE CORP. NEPA DIV., OAK RIDGE, TENN. (REF. 31)	NEPA-SERM-
FAIRCHILD ENGINE AND AIRPLANE CORP. NEPA DIV., OAK RIDGE, TENN. (REF. 31)	NEPA-STRM-
FAIRCHILD ENGINE AND AIRPLANE CORP. NEPA DIV., OAK RIDGE, TENN. (REF. 31)	SCR-
FAIRCHILD ENGINE AND AIRPLANE CORP. NEPA DIV., OAK RIDGE, TENN. (REF. 31)	SCRM-
FAIRCHILD ENGINE AND AIRPLANE CORP. NEPA DIV., OAK RIDGE, TENN. (REF. 31)	SERM-
FAIRCHILD ENGINE AND AIRPLANE CORP. NEPA DIV., OAK RIDGE, TENN. (REF. 31)	STRM-
FAIRCHILD ENGINE AND AIRPLANE CORP. NEPA DIV., OAK RIDGE, TENN.	UAC-
FAIRCHILD ENGINE AND AIRPLANE CORP. PILOTLESS PLANE DIV., FARMINGDALE, N.Y. (REF. 1)	(MODEL NO.)(SUBJ.LTR.)-
FAIRCHILD GUIDED MISSILES DIV., FARMINGDALE, N.Y.	FGMD-
FAIRCHILD HILLER CORP., GERMANTOWN, MD. (APPLICATIONS TECHNOLOGY SATELLITES)	ATS-(NUMBER-NUMBER)
FAIRCHILD HILLER CORP. AIRCRAFT SERVICE DIV., ST. AUGUSTINE, FLA.	FD(NUMBER)T(NUMBER)
FAIRCHILD HILLER CORP. AIRCRAFT SERVICE DIV., ST. AUGUSTINE, FLA.	FH(5 DIGITS)
FAIRCHILD-HILLER CORP. ELECTRONIC AND INFORMATION SYSTEMS DIV., BLADENSBURG, MD.	FH-EISD-
FAIRCHILD HILLER CORP. REPUBLIC AVIATION DIV., FARMINGDALE, N.Y.	EAR-
FAIRCHILD HILLER CORP. REPUBLIC AVIATION DIV., FARMINGDALE, N.Y.	EDR-
FAIRCHILD HILLER CORP. REPUBLIC AVIATION DIV., FARMINGDALE, N.Y.	ED-SR-
FAIRCHILD HILLER CORP. REPUBLIC AVIATION DIV., FARMINGDALE, N.Y.	EFTMR-
FAIRCHILD HILLER CORP. REPUBLIC AVIATION DIV., FARMINGDALE, N.Y.	EP-
FAIRCHILD HILLER CORP. REPUBLIC AVIATION DIV., FARMINGDALE, N.Y.	ERF-
FAIRCHILD HILLER CORP. REPUBLIC AVIATION DIV., FARMINGDALE, N.Y.	ERS-(NUMBER)-
FAIRCHILD HILLER CORP. REPUBLIC AVIATION DIV., FARMINGDALE, N.Y.	E-SIR-
FAIRCHILD HILLER CORP. REPUBLIC AVIATION DIV., FARMINGDALE, N.Y.	ETL-
FAIRCHILD HILLER CORP. REPUBLIC AVIATION DIV., FARMINGDALE, N.Y.	ETR-
FAIRCHILD HILLER CORP. REPUBLIC AVIATION DIV., FARMINGDALE, N.Y.	EW-
FAIRCHILD HILLER CORP. REPUBLIC AVIATION DIV., FARMINGDALE, N.Y.	FHR-
FAIRCHILD HILLER CORP. REPUBLIC AVIATION DIV., FARMINGDALE, N.Y.	FTMR-
FAIRCHILD HILLER CORP. REPUBLIC AVIATION DIV., FARMINGDALE, N.Y.	PC(NUMBER)(LETTER)(NO.)
FAIRCHILD HILLER CORP. REPUBLIC AVIATION DIV., FARMINGDALE, N.Y.	PCD-
FAIRCHILD HILLER CORP. REPUBLIC AVIATION DIV., FARMINGDALE, N.Y.	PCD-TR-(YEAR)-
FAIRCHILD HILLER CORP. REPUBLIC AVIATION DIV., FARMINGDALE, N.Y.	RAC-
FAIRCHILD HILLER CORP. REPUBLIC AVIATION DIV., FARMINGDALE, N.Y.	RAC-EAR-
FAIRCHILD HILLER CORP. REPUBLIC AVIATION DIV., FARMINGDALE, N.Y.	RAC-EARMR-
FAIRCHILD HILLER CORP. REPUBLIC AVIATION DIV., FARMINGDALE, N.Y.	RAC-EASR-
FAIRCHILD HILLER CORP. REPUBLIC AVIATION DIV., FARMINGDALE, N.Y. (ENGINEERING DATA)	RAC-ED-(MODEL DESIGNA.)-
FAIRCHILD HILLER CORP. REPUBLIC AVIATION DIV., FARMINGDALE, N.Y.	RAC-PR-
FAIRCHILD HILLER CORP. REPUBLIC AVIATION DIV., FARMINGDALE, N.Y.	T.D.-(NUMBER)-
FAIRCHILD HILLER CORP. REPUBLIC AVIATION DIV., FARMINGDALE, N.Y.	T.D. FT-
FAIRCHILD HILLER CORP. REPUBLIC AVIATION DIV., FARMINGDALE, N.Y. (VECTORED THRUST)	VT(NUMBER)R(NUMBER)
FAIRCHILD HILLER CORP. REPUBLIC AVIATION DIV. PLASMA PROPULSION LAB., FARMINGDALE, N.Y.	PPL-
FAIRCHILD HILLER CORP. REPUBLIC AVIATION DIV. PLASMA PROPULSION LAB., FARMINGDALE, N.Y.	PPL-(YEAR)-
FAIRCHILD HILLER CORP. REPUBLIC AVIATION DIV. PLASMA PROPULSION LAB., FARMINGDALE, N.Y.	PPL-TR-(YEAR)-
FAIRCHILD HILLER CORP. REPUBLIC AVIATION DIV. PLASMA PROPULSION LAB., FARMINGDALE, N.Y.	RAC-PPL-
FAIRCHILD HILLER CORP. REPUBLIC AVIATION DIV. PLASMA PROPULSION LAB., FARMINGDALE, N.Y.	RAC-PPL-(YEAR)-
FAIRCHILD HILLER CORP. REPUBLIC AVIATION DIV. PLASMA PROPULSION LAB., FARMINGDALE, N.Y.	SRS(PPL)-
FAIRCHILD RECORDING EQUIP. CORP., WHITESTONE, N.Y.	FRE-
FAIRCHILD SPACE AND DEFENSE SYSTEMS, SYOSSET, N.Y.	ED-AA-
FAIRCHILD SPACE AND DEFENSE SYSTEMS, SYOSSET, N.Y.	ED-AJ-
FAIRCHILD SPACE AND DEFENSE SYSTEMS, SYOSSET, N.Y.	ED-AY-
FAIRCHILD SPACE AND DEFENSE SYSTEMS, SYOSSET, N.Y.	ORD-AJ-
FAIRCHILD SPACE AND DEFENSE SYSTEMS, SYOSSET, N.Y.	ORD-AP-
FAIRCHILD SPACE AND DEFENSE SYSTEMS, SYOSSET, N.Y.	SME-AC-
FAIRCHILD SPACE AND DEFENSE SYSTEMS, SYOSSET, N.Y.	SME-AF-
FAIRCHILD SPACE AND DEFENSE SYSTEMS, SYOSSET, N.Y.	SME-AG-
FAIRCHILD SPACE AND DEFENSE SYSTEMS, SYOSSET, N.Y.	SME-AJ-
FAIRCHILD SPACE AND DEFENSE SYSTEMS, SYOSSET, N.Y.	SME-AP-
FAIRCHILD STRATOS CORP., HAGERSTOWN, MD.	EPMR-

FAIRCHILD STRATOS CORP. AIRCRAFT-MISSILES DIV., HAGERSTOWN, MD.	FS-AMD-
FAIRCHILD STRATOS CORP. AIRCRAFT-MISSILES DIV., HAGERSTOWN, MD.	FWR-
FALCON RESEARCH AND DEVELOPMENT CO. TARGET AND VULNERABILITY ANALYSIS LAB., DENVER	ER-
FALCON RESEARCH AND DEV. CO. TECHNODYNE DIV., DENVER	R/D-(4 DIGITS)
FALCON RESEARCH + DEV. CO. THOR DIV., COCKEYSVILLE, MD.	THOR-
FANSTEEL METALLURGICAL CORP., NORTH CHICAGO, ILL. (REF. 1)	(NUMBERS...)
FAR EAST AIR FORCES, TOKYO	FEAF-
FAR EAST AIR FORCES. OPERATIONS ANALYSIS OFF., TOKYO	FEAF OA-
FAR EAST COMMAND (ARMY)	FEC-
FAR EAST LAND FORCES HQ. OPERATIONAL REQUIREMENTS BRANCH, SINGAPORE	FARELF-(NUMBER/YEAR)
FAR EAST LAND FORCES HQ. OPERATIONAL REQUIREMENTS BRANCH, SINGAPORE	FARELF-MEMO(NO./YR.)
FARADAY TRANSLATIONS, N.Y.C.	FT-
FARNSWORTH ELECTRONICS CO., FORT WAYNE, IND.	FEC-
FARNSWORTH TELEVISION AND RADIO CORP. RESEARCH DEPT., FORT WAYNE	FTRC RDR-
FARNSWORTH TELEVISION AND RADIO CORP. RESEARCH DEPT., FORT WAYNE	RDR-
FARRAND OPTICAL CO., INC., N.Y.C. (REF. 1)	(NUMBER)
FARRAND OPTICAL CO., INC., N.Y.C.	FOC-
FARRAND OPTICAL CO., INC., N.Y.C.	PTR-(NUMBER-NUMBER)
FEDERAL AVIATION ADMINISTRATION, OKLAHOMA CITY	FAA-AM-
FEDERAL AVIATION ADMINISTRATION, OKLAHOMA CITY	NA-(YEAR)-
FEDERAL AVIATION ADMINISTRATION, WASHINGTON, D.C.	FAA-AC-(NUMBER/NUMBER)-
FEDERAL AVIATION ADMINISTRATION, WASHINGTON, D.C. (AVIATION MEDICINE)	FAA-AM-(YEAR)-
FEDERAL AVIATION ADMINISTRATION, WASHINGTON, D.C.	FAA A-RD-
FEDERAL AVIATION ADMINISTRATION, WASHINGTON, D.C.	FAA BIBLIOGRAPHIC LIST-
FEDERAL AVIATION ADMINISTRATION, WASHINGTON, D.C.	FAA-RD-(YEAR)-
FEDERAL AVIATION ADMINISTRATION, WASHINGTON, D.C.	FAA-TDR-
FEDERAL AVIATION ADMINISTRATION, WASHINGTON, D.C.	FS-(NUMBER)-(NUMBER)-
FEDERAL AVIATION ADMINISTRATION, WASHINGTON, D.C.	INC/(LTR.)(NO.)/(LTR.)-
FEDERAL AVIATION ADM. AERONAUTICAL CTR., OKLAHOMA CITY	AC(YEAR)-
FEDERAL AVIATION ADM. AERONAUTICAL CTR., OKLAHOMA CITY	FAA AC(YEAR)-
FEDERAL AVIATION ADM. AIRCRAFT DEV. SERVICE, WASH., DC	ADS-
FEDERAL AVIATION ADM. AIRCRAFT DEV. SERVICE, WASH., DC	FAA-ADS-
FEDERAL AVIATION ADM. AIRCRAFT DEV. SERVICE, WASH., DC	FAA-DS-
FEDERAL AVIATION ADMINISTRATION. FLIGHT STANDARDS SERVICE, WASHINGTON, D.C. (ADVISORY CIRCULAR)	AC-(SUBJECT NUMBER)-
FEDERAL AVIATION ADMINISTRATION. OFFICE OF AVIATION MEDICINE, WASHINGTON, D.C. (CODE USED SINCE 1964)	AM-(YEAR)-
FEDERAL AVIATION ADMINISTRATION. OFFICE OF SUPERSONIC TRANSPORT DEVELOPMENT, WASHINGTON, D.C.	SS-
FEDERAL AVIATION ADMINISTRATION. OFFICE OF SUPERSONIC TRANSPORT DEVELOPMENT, WASHINGTON, D.C.	SST-
FEDERAL AVIATION ADMINISTRATION. SYSTEMS RESEARCH AND DEV. SERVICE, WASHINGTON, D.C. (RESEARCH + DEV.)	FAA-RD-(YEAR)-
FEDERAL AVIATION ADMINISTRATION. SYSTEMS RESEARCH AND DEV. SERVICE, WASHINGTON, D.C. (RESEARCH + DEV.)	RD-(YEAR)-
FEDERAL AVIATION ADMINISTRATION. SYSTEMS RESEARCH AND DEV. SERVICE, WASHINGTON, D.C. (RESEARCH + DEV.)	SRDS-
FEDERAL AVIATION ADMINISTRATION. SYSTEMS RESEARCH AND DEV. SERVICE, WASHINGTON, D.C. (RESEARCH + DEV.)	SRDS-RD-(YEAR)-
FEDERAL AVIATION AGENCY. LOS ANGELES AIR ROUTE TRAFFIC CONTROL CENTER, PALMDALE, CALIF.	UVF-
FEDERAL CARTRIDGE CORP. TWIN CITIES ORDNANCE PLANT	TW-
FEDERAL CIVIL DEFENSE ADMINISTRATION, WASHINGTON, D.C.	AG-
FEDERAL CIVIL DEFENSE ADMINISTRATION, WASHINGTON, D.C.	FCDA-
FEDERAL CIVIL DEFENSE ADMINISTRATION, WASHINGTON, D.C.	FCDA BULL-
FEDERAL CIVIL DEFENSE ADMINISTRATION, WASHINGTON, D.C.	FCDA H-
FEDERAL CIVIL DEFENSE ADMINISTRATION, WASHINGTON, D.C.	FCDA TB-
FEDERAL CIVIL DEFENSE ADMINISTRATION, WASHINGTON, D.C.	FCDA TM-
FEDERAL CIVIL DEFENSE ADMINISTRATION, WASHINGTON, D.C.	FCDA TR-
FEDERAL COMMUNICATIONS COMMISSION, WASHINGTON, D.C.	FCC-
FEDERAL COUNCIL FOR SCIENCE AND TECHNOLOGY. COMMITTEE ON SCIENTIFIC AND TECHNICAL INFORMATION, WASH., DC	COSATI-(YEAR)-
FEDERAL COUNCIL FOR SCIENCE AND TECHNOLOGY. COMMITTEE ON SCIENTIFIC AND TECHNICAL INFORMATION, WASH., DC (SUBJECT CATEGORY LIST)	CSCL-
FEDERAL COUNCIL FOR SCIENCE AND TECHNOLOGY. COMMITTEE ON WATER RESOURCES RESEARCH, WASHINGTON, D.C.	COWRR-PUB-
FEDERAL ELECTRIC CORP., PARAMUS, N.J.	FEC-
FEDERAL HIGHWAY ADMINISTRATION, WASHINGTON, D.C.	FHWA-HP-HSS-(YEAR)-
FEDERAL HIGHWAY ADMINISTRATION. TENN. DIV., NASHVILLE	F-(3 DIGITS)-
FEDERAL-MOGUL CORP. RES. AND DEV. DIV., ANN ARBOR, MICH.	FMRD-
FEDERAL PACIFIC ELECTRIC CO. CORNELL DUBILIER ELECTRONICS DIV., NORWOOD, MASS.	CDE-
FEDERAL RADIATION COUNCIL, WASHINGTON, D.C.	FRC-
FEDERAL SCIENTIFIC CORP., N.Y.C.	FSC-(YEAR)-
FEDERAL SCIENTIFIC CORP., N.Y.C.	FSC-F/
FEDERAL SCIENTIFIC CORP., N.Y.C.	FSC-T-
FEDERAL SCIENTIFIC CORP., N.Y.C.	P-
FEDERAL TELECOMMUNICATION LABS., INC., NUTLEY, N.J.	FTL-5483-QDR-
FEDERAL TELECOMMUNICATION LABS., INC., NUTLEY, N.J.	FTL-TM-
FEDERAL TELEPHONE AND RADIO CORP., CLIFTON, N.J.	FTL-
FEDERAL TRADE COMMISSION, WASHINGTON, D.C.	FTC-
FEDERAL WATER POLLUTION CONTROL ADMINISTRATION, WASH., D.C. (WATER POLLUTION CONTROL RESEARCH SER.)	DAST-
FEDERAL WATER POLLUTION CONTROL ADM., WASHINGTON, DC	FWPCA-(NUMBER-DATE)-
FEDERAL WATER POLLUTION CONTROL ADM., WASHINGTON, DC	FWPCA-(NO.)-DWF-(DATE)
FEDERAL WATER POLLUTION CONTROL ADM., WASHINGTON, DC	FWPCA-(NO.-LTRS.-DATE)

FEDERAL WATER POLLUTION CONTROL ADM.

FEDERAL WATER POLLUTION CONTROL ADM., WASHINGTON, DC	FWPCA-(NO.)-EAG-(DATE)
FEDERAL WATER POLLUTION CONTROL ADM., WASHINGTON, DC (DISPOSAL AND SEWAGE TREATMENT)	FWPCA-DAST-
(SEE ALSO LATER NAME: FEDERAL WATER QUALITY ADM.)	
FEDERAL WATER POLLUTION CONTROL ADMINISTRATION. WATER QUALITY LAB., EDISON, N.J.	WP-(NUMBER)-
FEDERAL WATER POLLUTION CONTROL ADMINISTRATION. WATER QUALITY LAB., EDISON, N.J.	WP-ORD-
FEDERAL WATER QUALITY ADMINISTRATION, WASHINGTON, DC	FWQA-(NUMBER-DATE)
FEDERAL WATER QUALITY ADMINISTRATION, WASHINGTON, DC	FWQA-(NO.)-EAG-(DATE)
FEDERAL WATER QUALITY ADMINISTRATION, WASHINGTON, DC	FWQA-(NO.-LTRS.-DATE)
FEDERAL WATER QUALITY ADMINISTRATION, WASHINGTON, DC	FWQA-(NO.)-DXU-(DATE)
FEDERAL WATER QUALITY ADMINISTRATION, WASHINGTON, DC	FWQO-
(SEE ALSO LATER NAME: ENVIRONMENTAL PROTECTION AGENCY. WATER QUALITY OFFICE)	
FEDERATION OF AMERICAN SOCIETIES FOR EXPERIMENTAL BIOLOGY, WASHINGTON, D.C.	NS-
FEDERATION OF AMERICAN SOCIETIES FOR EXPERIMENTAL BIOLOGY. COUNCIL ON BIOLOGICAL SCIENCES INFORMATION, BETHESDA, MD. (WORKING DOCUMENT)	COBSI-WD-
FELTERS CO., BOSTON, MASS.	TS-
FERGUSON (H.K.) CO., N.Y.C.	HKF-
FERGUSON (H.K.) CO., N.Y.C.	HKF-1942D-
FERGUSON (H.K.) CO. ATOMIC ENERGY DIV., N.Y.C.	HKF-AED-
FERRANTI, LTD. FLIGHT TRIAL DEPT. (INTERNATIONAL AEROSPACE SYMPOSIUM)	EIST/INST/
FIAT. SEZIONE ENERGIA NUCLEARE, TURIN	FIAT-SEN-R-
FIAT. SEZIONE ENERGIA NUCLEARE, TURIN	FN-E-
FIAT S.P.A. DIV. AVIAZIONE, TURIN	TASS-(YEAR)-Y-
FIBER SCIENCE, INC., GARDENA, CALIF.	TRA-
FIELD COMMAND (ARMY)	FC-
FIELD EMISSION CORP., MC MINNVILLE, ORE.	FEC-QPL-
FIELD INTELLIGENCE AGENCY, TECHNICAL	CIRCULAR LETTERS(FIAT)-
FIELD INTELLIGENCE AGENCY, TECH. (CIRCULAR LETTERS)	CL-
FIELD INTELLIGENCE AGENCY, TECHNICAL (CHEM. RPTS.)	CR-
FIELD INTELLIGENCE AGENCY, TECHNICAL	FIAT-
FIELD INTELLIGENCE AGENCY, TECH. (CIRCULAR LETTERS)	FIAT CL-
FIELD INTELLIGENCE AGENCY, TECHNICAL (CHEM. RPTS.)	FIAT CR-
FIELD INTELLIGENCE AGENCY, TECH. (ENEMY PERSONNEL)	FIAT EP-
FIELD INTELLIGENCE AGENCY, TECH. (EVALUATION REPORTS)	FIAT ER-
FIELD INTELLIGENCE AGENCY, TECHNICAL (FINAL RPTS.)	FIAT FR-
FIELD INTELLIGENCE AGENCY, TECHNICAL (INTELL. RPTS.)	FIAT IR-
FIELD INTELLIGENCE AGENCY, TECHNICAL (INTELL. RPTS.)	FIAT IR (LTRS/LTRS)/
FIELD INTELLIGENCE AGENCY, TECH. (MISC. CHEM. FILE)	FIAT MC-
FIELD INTELLIGENCE AGENCY, TECH. (MISC. CHEM. FILE)	FIAT MCF-
FIELD INTELLIGENCE AGENCY, TECH. (DOC. ON MICROFILM)	FIAT MICROFILM-
FIELD INTELLIGENCE AGENCY, TECHNICAL	FIAT MM-
FIELD INTELL. AGENCY, TECH. (STAATSDRUCKEREI DOCS.)	FIAT SD-
FIELD INTELLIGENCE AGENCY, TECHNICAL (TECH. BULLS.)	FIAT TB-
FIELD INTELLIGENCE AGENCY, TECHNICAL (TECH. BULLS.)	FIAT TB T-
FIELD INTELLIGENCE AGENCY, TECH. (MISC. CHEM. FILE)	MCF-
FIELD INTELLIGENCE AGENCY, TECHNICAL	MM-
FIELD INTELL. AGENCY, TECH. (STAATSDRUCKEREI DOCS.)	SD-
FIFTH NAVAL DISTRICT, NORFOLK, VA.	COMFIVE-
FINLAND. INSTITUTE OF RADIATION PHYSICS, HELSINKI	SFL-A-
FINLAND. INSTITUTE OF RADIATION PHYSICS, HELSINKI	SLF-
FINLAND. STATE INST. FOR TECHNICAL RESEARCH, HELSINKI	JULKAISU-
FINLAND. SUOMALAINEN TIEDEAKATEMIA, HELSINKI	SER-B-TOM-
FINNISH ACADEMY OF TECHNICAL SCIENCES, HELSINKI (ACTA POLYTECHNICA SCANDINAVICA. CHEMISTRY)	APS-CH-
FINNISH ACADEMY OF TECHNICAL SCIENCES, HELSINKI (ACTA POLYTECHNICA SCANDINAVICA. CIVIL ENGINEERING)	APS-CI-
FINNISH ACADEMY OF TECHNICAL SCIENCES, HELSINKI (ACTA POLYTECHNICA SCANDINAVICA. ELECTRICAL ENG.)	APS-EI-
FINNISH ACADEMY OF TECHNICAL SCIENCES, HELSINKI (ACTA POLYTECHNICA SCANDINAVICA. ELEC. ENG. SER.)	APS-EL-
FINNISH ACADEMY OF TECHNICAL SCIENCES, HELSINKI (ACTA POLYTECHNICA SCANDINAVICA. MATH. + COMPUTING)	APS-MA-
FINNISH ACADEMY OF TECHNICAL SCIENCES, HELSINKI (ACTA POLYTECHNICA SCANDINAVICA. MECHANICAL ENG.)	APS-ME-
FINNISH ACADEMY OF TECHNICAL SCIENCES, HELSINKI (ACTA POLYTECHNICA SCANDINAVICA. PHYSICS)	APS-PH-
FINNISH ACADEMY OF TECHNICAL SCIENCES, HELSINKI	EI-
FINNISH ACADEMY OF TECHNICAL SCIENCES, HELSINKI (ACTA POLYTECHNICA SCANDINAVICA. ELEC. ENG. SER.)	EL-(2 DIGITS)
FINNISH METEOROLOGICAL OFFICE, HELSINKI	REPT-
FIRENZE UNIV., ITALY. ISTITUTO DI FISICA	FIRENZEU-
FIRESTONE TIRE AND RUBBER CO., AKRON, OHIO	FTRC-
FIRESTONE TIRE AND RUBBER CO. DEFENSE RESEARCH DIV., AKRON, OHIO	DRD-
FIRESTONE TIRE AND RUBBER CO. ENGINEERING LAB., MONTEREY, CALIF.	FTR-
FIRST ATOMIC SHIP TRANSPORT, INC., N.Y.C.	FAST-
FIRTH STERLING, INC., PITTSBURGH	FS-
FIRTH STERLING, INC. AMERICAN ELECTRO METAL DIV., YONKERS, N.Y.	FS-
FISH AND WILDLIFE SERVICE (INTERIOR), WASHINGTON, D.C. (ADMINISTRATIVE REPORTS)	AR-
FISH AND WILDLIFE SERVICE (INTERIOR), WASHINGTON, D.C. (CONSERVATION BULLETINS)	CB-
FISH AND WILDLIFE SERVICE (INTERIOR), WASHINGTON, D.C. (STATISTICAL DIGESTS)	SD-
FISH AND WILDLIFE SERVICE (INTERIOR), WASHINGTON, D.C. (STATISTICAL LISTS)	SL-
FLADER (FREDRIC) INC., NORTH TONAWANDA, N.Y.	FF-115-
FLADER (FREDRIC) INC., NORTH TONAWANDA, N.Y.	FF-180-R-
FLADER (FREDRIC) INC., NORTH TONAWANDA, N.Y.	FFI-
FLADER (FREDRIC) INC., NORTH TONAWANDA, N.Y.	163-R(NUMBER)-
FLEET COMPUTER PROGRAMMING CENTER, ATLANTIC, VIRGINIA BEACH (PROGRAM INFO. BULLETIN)	FCPL-A-PIB-
FLEET COMPUTER PROGRAMMING CENTER, ATLANTIC, VIRGINIA BEACH (PROGRAM INFO. BULLETIN)	PIB-
FLEET COMPUTER PROGRAMMING CENTER, ATLANTIC, VIRGINIA BEACH (PROGRAM INFO. BULLETIN)	PIB-ISSUE-

FLEET COMPUTER PROGRAMMING CENTER, PACIFIC, SAN DIEGO, CALIF. (PROGRAM INFO. BULLETIN)	FCPC-P-PIB-
FLEET COMPUTER PROGRAMMING CENTER, PACIFIC, SAN DIEGO, CALIF.	PIB-ISSUE-
FLEET NUMERICAL WEATHER CENTRAL, MONTEREY, CALIF.	FNWC-TN-
FLEET OPERATIONS CONTROL CENTER, PACIFIC FLEET, SAN FRANCISCO	FOCCPAC-
FLEET OPERATIONS CONTROL CENTER, PACIFIC FLEET, SAN FRANCISCO	FOCCPAC-TN-
FLEET OPERATIONS CONTROL CENTER, PACIFIC FLEET, SAN FRANCISCO	FOCCPAC-TR-
FLEET WORK STUDY GROUP, ATLANTIC, NORFOLK, VA.	FWSGANT-
FLEET WORK STUDY GROUP, ATLANTIC, NORFOLK, VA.	FWSGLANT-
FLEET WORK STUDY GROUP, PACIFIC, SAN DIEGO, CALIF.	FWSGPAC-STUDY-
FLETCHER AVIATION CORP., PASADENA, CALIF. (REF. 1)	(PROJECT NUMBER)-
FLIGHT PROPULSION RESEARCH LAB., CLEVELAND	FPRL-
FLIGHT PROPULSION RESEARCH LAB., CLEVELAND (REF. 25)	NACA-(LTRS)-E-(DATE CODE)
FLIGHT SAFETY FOUNDATION, INC., N.Y.C.	FAA-FS-(YEAR)-
FLIGHT SAFETY FOUNDATION, INC., N.Y.C.	KD-
FLINDERS UNIV., BEDFORD PARK, SOUTH AUSTRALIA (ASTROPHYSICS AND PLASMA THEORY)	FLINDERSU-APT-
FLINDERS UNIV., BEDFORD PARK, SOUTH AUSTRALIA. SCHOOL OF PHYSICAL SCIENCES	FUPH-R-
FLINT (ALBERT D.) ASSOCIATES, INC.	ALFLIN-
FLINT TRANSPORTATION AUTHORITY, MICH.	TR-(1 DIGIT)
FLORENCE. UNIVERSITA	FTR-
FLORIDA. ENGINEERING EXPERIMENT STATION, GAINESVILLE (TECHNICAL PAPER)	F EES TP-
FLORIDA. UNIV., GAINESVILLE	CSL/ONR-
FLORIDA. UNIV., GAINESVILLE	FLAU-
FLORIDA. UNIV., GAINESVILLE (QUANTUM THEORY PROJECT)	FLORIDA-TR-
FLORIDA. UNIV., GAINESVILLE	FLORIDA U-(NO.)-QTR-
FLORIDA. UNIV., GAINESVILLE	FU-
FLORIDA. UNIV., GAINESVILLE	SAR-
FLORIDA. UNIV., GAINESVILLE (PROJECT THEMIS) (REF.48)	THEMIS-UF-SCIENTIFIC-
FLORIDA. UNIV., GAINESVILLE (PROJECT THEMIS) (REF.48)	THEMIS-UF-TR-
FLORIDA. UNIV., GAINESVILLE	TP-
FLORIDA. UNIV., GAINESVILLE	TPR-
FLORIDA. UNIV., GAINESVILLE	UFCG-
FLORIDA. UNIV., GAINESVILLE	UF-CH-
FLORIDA. UNIV., GAINESVILLE. COMMUNICATIONS SCIENCES LAB.	CSL/ONR-
FLORIDA. UNIV., GAINESVILLE. ELECTRON DEVICES LAB.	PM-
FLORIDA. UNIV., GAINESVILLE. ENGINEERING AND INDUSTRIAL EXPERIMENT STATION (REF.1)	(NUMBER)
FLORIDA. UNIV., GAINESVILLE. ENGINEERING AND INDUSTRIAL EXPERIMENT STATION	FU-QPR-(NUMBER)-(YEAR)-
FLORIDA. UNIV., GAINESVILLE. ENGINEERING AND INDUSTRIAL EXPERIMENT STATION	FU-QTR-
FLORIDA. UNIV., GAINESVILLE. ENGINEERING AND INDUSTRIAL EXPERIMENT STATION	SR-(YEAR)-
FLORIDA. UNIV., GAINESVILLE. REGIONAL REHABILITATION RESEARCH INST.	MONO-
FLORIDA. UNIV., GAINESVILLE. REGIONAL REHABILITATION RESEARCH INST.	SRS-RD-
FLORIDA. UNIV.,GAINESVILLE. WATER RESOURCES RES. CTR.	WRRC-PUB-
FLORIDA INST. OF TECH., MELBOURNE. LAB. FOR ENVIRONMENTAL AND SOLAR STUDIES	TR-(2 DIGITS)
FLORIDA STATE UNIV., TALLAHASSEE	FSU-
FLORIDA STATE UNIV., TALLAHASSEE	FSU-SR-
FLORIDA STATE UNIV., TALLAHASSEE	FSU-TR-
FLORIDA STATE UNIV., TALLAHASSEE	IEG-
FLORIDA STATE UNIV., TALLAHASSEE	M-
FLORIDA STATE UNIV., TALLAHASSEE	TR-(NUMBER)(LETTER)
FLORIDA STATE UNIV., TALLAHASSEE. COMPUTER-ASSISTED INSTRUCTION CENTER	CAI-SYSTEMS MEMO-
FLORIDA STATE UNIV., TALLAHASSEE. COMPUTER-ASSISTED INSTRUCTION CENTER	CAI-TM-
FLORIDA STATE UNIV., TALLAHASSEE. COMPUTER-ASSISTED INSTRUCTION CENTER	CAI-TR-
FLORIDA STATE UNIV., TALLAHASSEE. DEPT. OF METEOROLOGY (SCIENTIFIC REPORT)	FSU-SR-
FLORIDA STATE UNIV., TALLAHASSEE. DEPT. OF STATISTICS	FSU-M-
FLORIDA STATE UNIV., TALLAHASSEE. DEPT. OF STATISTICS	FSU-STATISTICS-
FLORIDA STATE UNIV., TALLAHASSEE. GEOPHYSICAL FLUID DYNAMICS INST.	CONTRIB-
FLORIDA STATE UNIV., TALLAHASSEE. HIGH ENERGY PHYSICS LAB.	FSU-HEP-(YEAR)-
FLORIDA STATE UNIV., TALLAHASSEE. INSTITUTE OF MOLECULAR BIOPHYSICS	BSCS-PAMPHLETS-
FLUOR CORP., LTD., LOS ANGELES	FLUOR-
FLUOR CORP., LTD., WHITTIER, CALIF.	FLR-
FMC CORP., BALTIMORE	AR-
FMC CORP., BALTIMORE	FMC-BA-
FMC CORP., PRINCETON, N.J.	FMC-PCR-
FMC CORP., PRINCETON, N.J.	PCR-
FMC CORP. DEFENSE TECHNOLOGY LABS.,SANTA CLARA,CALIF.	DTL-
FMC CORP. DEFENSE TECHNOLOGY LABS.,SANTA CLARA,CALIF.	DTL-(NUMBER)-
FMC CORP. DEFENSE TECHNOLOGY LABS.,SANTA CLARA,CALIF.	FMC-DTL-
FMC CORP. ORDNANCE ENG. DIV., SAN JOSE., CALIF.	FMC-ORD-
FMC CORP. ORGANIC CHEMICALS DIV., BALTIMORE	FMC-BA-(NUMBER)-X
FOCKE-WULF FLUGZEUGBAU G.M.B.H.	GDV-
FOOD + AGRICULTURE ORG. OF THE UNITED NATIONS, ROME	FAO-
FOOD + AGRICULTURE ORG. OF THE UNITED NATIONS, ROME (PANEL ON INTERNATL. INFO. SYSTEM FOR AGRIC...)	FAO/DC/AGRIS-
FOOD + AGRICULTURE ORG. OF THE UNITED NATIONS, ROME	FAO-FID/C-
FOOD + AGRICULTURE ORG. OF THE UNITED NATIONS, ROME	FAO-FID/R-
FOOD + AGRICULTURE ORG. OF THE UNITED NATIONS, ROME	FAO-FRD/T-

FOOD & AGRICULTURE ORG. OF THE UNITED NATIONS

FOOD + AGRICULTURE ORG. OF THE UNITED NATIONS, ROME	FAO-FRM/
FOOD + AGRICULTURE ORG. OF THE UNITED NATIONS, ROME	FAO-FRS/C-
FOOD + AGRICULTURE ORG. OF THE UNITED NATIONS, ROME	FAO-FRS/R-
FOOD + AGRICULTURE ORG. OF THE UNITED NATIONS, ROME	FAO-FRS/T-
FOOD + AGRICULTURE ORG. OF THE UNITED NATIONS, ROME	FAO-FRV/T-
FOOD + DRUG ADM., WASH., D.C. (INTERBUREAU BY-LINES)	CONSECUTIVE-
FOOD + DRUG ADMINISTRATION, WASHINGTON, D.C.	FDL-T-
FOOD + DRUG ADMINISTRATION. CENTER FOR DRUG INFORMATION, ROCKVILLE, MD.	FDA-NDC-(YEAR)-
FOOD + DRUG ADM. DIV. OF MICROBIOLOGY, CINCINNATI	QPR-
FOOD MACHINERY + CHEMICAL CORP., SAN JOSE, CALIF.	CR-CH-
FOOD MACHINERY + CHEMICAL CORP., SAN JOSE, CALIF.	FMC-
FOOTE MINERAL CO., EXTON, PENNA.	FOOTE-
FOOTE MINERAL CO., PHILADELPHIA	FMC-
FORD INSTRUMENT CO., LONG ISLAND CITY, N.Y.	FIC-
FORD INSTRUMENT CO., LONG ISLAND CITY, N.Y.	FICO-
FORD INSTRUMENT CO., LONG ISLAND CITY, N.Y.	NDR-
FORD INSTRUMENT CO., LONG ISLAND CITY, N.Y.	SGC-FI-
FORD INSTRUMENT CO., LONG ISLAND CITY, N.Y.	SP-
(SEE ALSO LATER NAME: SPERRY RAND CORP. FOR INSTR...)	
FORD MOTOR CO., DEARBORN, MICH.	SESA PAPER-
FORDHAM UNIV., N.Y.C.	FOR-
FORDHAM UNIV., N.Y.C.	FORU-
FORDHAM UNIV., N.Y.C.	HRI-
FORECAST CENTER, ENT AFB, COLO.	ADC-FC-TP-
FORECAST CENTER, ENT AFB, COLO.	WG 3-
FOREIGN BROADCAST INFO. SERVICE, WASHINGTON, D.C.	FBIS-
FOREIGN ECONOMIC ADMINISTRATION	FEA-
FOREIGN ECONOMIC ADMINISTRATION. ENEMY BRANCH	EA-
FOREIGN ECONOMIC ADMINISTRATION. ENEMY BRANCH (EXTERNAL ECONOMIC SECURITY STAFF REPORTS)	ES-
FOREIGN ECONOMIC ADMINISTRATION. ENEMY BRANCH	FEA EA-
FOREIGN ECONOMIC ADMINISTRATION. ENEMY BRANCH (EXTERNAL ECONOMIC SECURITY STAFF REPORTS)	FEA ES-
FOREIGN ECONOMIC ADM. ENEMY BR. INDUSTRY DIV.	FEA IND-
FOREIGN ECONOMIC ADM. ENEMY BR. INDUSTRY DIV.	IND-
FOREIGN ECONOMIC ADMINISTRATION. ENEMY BRANCH. LABOR AND MANPOWER DIVISION	FEA LM-
FOREIGN ECONOMIC ADMINISTRATION. ENEMY BRANCH. LABOR AND MANPOWER DIVISION	LM-
FOREIGN ECONOMIC ADMINISTRATION. LIBERATED AREAS BRANCH (ECONOMIC INSTITUTIONS STAFF REPORTS)	EIS-
FOREIGN ECONOMIC ADMINISTRATION. LIBERATED AREAS BRANCH (ECONOMIC INSTITUTIONS STAFF REPORTS)	FEA EIS-
FOREIGN TECHNICAL INTELLIGENCE OFFICE, ABERDEEN PROVING GROUND, MD.	FTIO-(NUMBER)-
FOREIGN TECHNOLOGY DIV., WRIGHT-PATTERSON AFB, OHIO	FTD-(YEAR)-
FOREIGN TECHNOLOGY DIV., WRIGHT-PATTERSON AFB, OHIO	FTD-(7 DIGITS)
FOREIGN TECHNOLOGY DIV., WRIGHT-PATTERSON AFB, OHIO	FTD-(YEAR)-PHE-
FOREIGN TECHNOLOGY DIV., WRIGHT-PATTERSON AFB, OHIO	FTD-CW-(NUMBERS-YEAR)
FOREIGN TECHNOLOGY DIV., WRIGHT-PATTERSON AFB, OHIO	FTD-H-(YEAR)-
FOREIGN TECHNOLOGY DIV., WRIGHT-PATTERSON AFB, OHIO	FTD-HC-(NO.-NO.-YEAR)
FOREIGN TECHNOLOGY DIV., WRIGHT-PATTERSON AFB, OHIO	FTD-HT-(YEAR)-
FOREIGN TECHNOLOGY DIV., WRIGHT-PATTERSON AFB, OHIO (MACHINE TRANSLATION)	FTD-MT-(YEAR)-
FOREIGN TECHNOLOGY DIV., WRIGHT-PATTERSON AFB, OHIO	FTD-MT-(NO.-NO.-YEAR)
FOREIGN TECHNOLOGY DIV., WRIGHT-PATTERSON AFB, OHIO	FTD-PHS-(NUMBER-YEAR)-
FOREIGN TECHNOLOGY DIV., WRIGHT-PATTERSON AFB, OHIO	FTD-PWS-(NUMBER-YEAR)-
FOREIGN TECHNOLOGY DIV., WRIGHT-PATTERSON AFB, OHIO	FTD-PWW-1M-(NUMBER)-
FOREIGN TECHNOLOGY DIV., WRIGHT-PATTERSON AFB, OHIO	FTD-ST-HB-(NUMBERS)-
FOREIGN TECHNOLOGY DIV., WRIGHT-PATTERSON AFB, OHIO	FTD-TT-(YEAR)-
FOREIGN TECHNOLOGY DIV., WRIGHT-PATTERSON AFB, OHIO	FTD-TT-
FOREIGN TECHNOLOGY DIV., WRIGHT-PATTERSON AFB, OHIO	F-TS-
FOREIGN TECHNOLOGY DIV., WRIGHT-PATTERSON AFB, OHIO	MCL-
FOREIGN TECHNOLOGY DIV., WRIGHT-PATTERSON AFB, OHIO	MCL-(NO.)/(ROMAN NO.)
FOREIGN TECHNOLOGY DIV., WRIGHT-PATTERSON AFB, OHIO	PHS-(NUMBER-YEAR)-
FOREST PRODUCTS LAB., MADISON, WIS.	FPL-
FOREST PRODUCTS LAB., MADISON, WIS.	FPL R-
FOREST PRODUCTS LAB., MADISON, WIS.	FPL-TRANS-
FOREST PRODUCTS LAB., MADISON, WIS.	FSPR-FPL-
FOREST PRODUCTS LAB., MADISON, WIS.	FSRB-FPL-
FOREST PRODUCTS LAB., MADISON, WIS.	FSRN-FPL-
FOREST PRODUCTS LAB., MADISON, WIS.	RP-FPL-
FOREST SERVICE, WASHINGTON, D.C.	FS-
FOREST SERVICE, WASHINGTON, D.C.	FS-TP-
FOREST SERVICE, WASHINGTON, D.C.	USFS-
FOREST SERVICE, WASHINGTON, D.C. (WASHINGTON OFFICE)	WO-
FOREST SERVICE. CENTRAL STATES STATION, BEREA, KY.	CS-
FOREST SERVICE. INSTITUTE OF TROPICAL FORESTRY, RIO PIEDRAS, PUERTO RICO	ITF-
FOREST SERVICE. INTERMOUNTAIN STATION, OGDEN, UTAH	FSRP-INT-
FOREST SERVICE. INTERMOUNTAIN STATION, OGDEN, UTAH	INT-
FOREST SERVICE. LAKE STATES STATION	LS-
FOREST SERVICE. NORTH CENTRAL STATION, ST. PAUL	NC-
FOREST SERVICE. NORTHEASTERN STA., UPPER DARBY, PENNA	FSRP-NE-
FOREST SERVICE. NORTHEASTERN STA., UPPER DARBY, PENNA	NE-
FOREST SERVICE. NORTHERN STATION, JUNEAU, ALASKA	NOR-
FOREST SERVICE. PACIFIC NORTHWEST STA., PORTLAND ORE.	FSRP-PNW-
FOREST SERVICE. PACIFIC NORTHWEST STA., PORTLAND, ORE	PNW-
FOREST SERVICE. PACIFIC NORTHWEST STA., PORTLAND ORE.	PSRB-PNW-
FOREST SERVICE. PACIFIC NORTHWEST STA., PORTLAND ORE.	PSRP-PNW-

FOREST SERVICE. PACIFIC SOUTHWEST STA., BERKELEY, CAL	PSW-
FOREST SERVICE. ROCKY MOUNTAIN STA., FT. COLLINS,COLO	FSRP-RM-
FOREST SERVICE. ROCKY MOUNTAIN STA., FT. COLLINS,COLO	RM-
FOREST SERVICE. SOUTHEASTERN STA., ASHEVILLE, N.C.	FSRB-SE-
FOREST SERVICE. SOUTHEASTERN STA., ASHEVILLE, N.C.	FSRN-SE-
FOREST SERVICE. SOUTHEASTERN STA., ASHEVILLE, N.C.	SE-
FOREST SERVICE. SOUTHERN STATION, NEW ORLEANS	SO-
FORMICA CORP., CINCINNATI (REF. 1)	(CONTRACT ORDER NO.)-
FORMICA CORP., CINCINNATI (REF. 1)	(NUMBER)
FORSCHUNGSANSTALT GRAF ZEPPELIN	FGZ-
FORSCHUNGSINSTITUT FUER HOCHFREQUENZPHYSIK, WERTHOVEN GERMANY	RR-(NUMBER-YEAR)
FORSCHUNGSINSTITUT FUER MILITAERISCHE BAUTECHNIK, ZURICH	FMB-(YEAR)-
FORT DETRICK, FREDERICK, MD.	AMXFD-AE-
FORT DETRICK, FREDERICK, MD.	AMXFD-AE-T-
FORT DETRICK, FREDERICK, MD.	SMUFD-AE-
FORT DETRICK, FREDERICK, MD.	SMUFD-MISC PUB-
FORT DETRICK, FREDERICK, MD.	SMUFD-TECH. MANUSCRIPT-
FORT DETRICK, FREDERICK, MD.	SMUFD-TECHNICAL STUDY-
FORT DETRICK, FREDERICK, MD.	SMUFD-TM-
FORT DETRICK, FREDERICK, MD.	SMUFD-TR-
FORT MONMOUTH SIGNAL LAB., N.J.	FMD-
FORT MONMOUTH SIGNAL LAB., N.J.	FMSL-
FORT MONMOUTH SIGNAL LAB., N.J. (ENGINEERING REPT.)	FMSL ER-
FORT MONMOUTH SIGNAL LAB. SUPPRESSION DEVELOPMENT SECTION, N.J. (ENGINEERING MEMORANDUM)	FMSL SDS EM-
FOSTER-MILLER ASSOCIATES, INC., WALTHAM, MASS.	WP-(4 DIGITS)
FOSTER WHEELER CORP., LIVINGSTON, N.J.	FW-RDR-
FOSTER WHEELER CORP., N.Y.C.	FW-
FOSTER WHEELER CORP., N.Y.C.	FW-(YEAR)-(NUMBER)
FOSTER WHEELER CORP., N.Y.C.	NEA-
FOSTER WHEELER CORP. WELDING LAB., N.Y.C.	WLR-
FOUNDATION FOR GLACIER RESEARCH INC., SEATTLE	FFGR-
FRANCE. CENTRE D'ETUDE SPATIALE DES RAYONNEMENTS, TOULOUSE	CESR-(NUMBER)-(NUMBER)
FRANCE. CENTRE D'ETUDES CRYOGENIQUES, SASSENAGE	CEC/SRAP/(INITIALS)-
FRANCE. CENTRE D'ETUDES ET DE RECHERCHES. SERVICE DE PSYCHOLOGIE APPLIQUEE, TOULON	CERPA-ETUDE-
FRANCE. CENTRE D'ETUDES ET DE RECHERCHES EN TECHNOLOGIE SPATIALE, TOULOUSE	CERTS-(NUMBER)-
FRANCE. CENTRE D'ETUDES ET DE RECHERCHES EN TECHNOLOGIE SPATIALE, TOULOUSE	CERTS-NT-(NUMBER)-
FRANCE. CENTRE D'ETUDES ET DE RECHERCHES EN TECHNOLOGIE SPATIALE, TOULOUSE	CERTS-TN-(NUMBER-NUMBER)
FRANCE. CENTRE DE MATHEMATIQUES DE L'ECOLE POLYTECHNIQUE, PARIS	CMEP-M(NUMBER.NUMBER)
FRANCE. CENTRE DE MATHEMATIQUES DE L'ECOLE POLYTECHNIQUE, PARIS	M(NUMBER).-
FRANCE. CENTRE DE PHYSIQUE ELECTRONIQUE ET CORPUSCULAIRE	WR-
FRANCE. CENTRE DE PHYSIQUE THEORIQUE DE L'ECOLE POLYTECHNIQUE	CPTEP-
FRANCE. CENTRE DE PHYSIQUE THEORIQUE DE L'ECOLE POLYTECHNIQUE	CPTEP-NO-A(NO.)-
FRANCE. CENTRE DE PHYSIQUE THEORIQUE DE L'ECOLE POLYTECHNIQUE	R(NUMBER.NUMBER)
FRANCE. CENTRE DE RECHERCHES ET D'EXPERIMENTATION DE GENIE RURAL, ANTONY	BTGR-
FRANCE. CENTRE DE RECHERCHES SCIENTIFIQUES, INDUSTRIELLES ET MARITIMES	CRSIM-
FRANCE. CENTRE NATIONAL D'ETUDES DES TELECOMMUNICATIONS, ISSY LES MOULINEAUX	CNET-
FRANCE. CENTRE NATIONAL D'ETUDES DES TELECOMMUNICATIONS, ISSY LES MOULINEAUX	CNET-NT-EST/APH/
FRANCE. CENTRE NATIONAL D'ETUDES SPATIALES, PARIS	CNES-
FRANCE. CENTRE NATIONAL D'ETUDES SPATIALES, PARIS	CNES-NT-
FRANCE. CENTRE NATIONAL D'ETUDES SPATIALES, PARIS	P-
FRANCE. CENTRE NATIONAL D'ETUDES SPATIALES. DIV. MATHEMATIQUES ET TRAITEMENT, PARIS	PR/MM/(NUMBER-NUMBER)/MT
FRANCE. CENTRE NATIONAL DE LA RECHERCHE SCIENTIFIQUE	CNRS-AR-(YEAR)
FRANCE. CENTRE NATIONAL DE LA RECHERCHE SCIENTIFIQUE, PARIS	CNRS-
FRANCE. CENTRE NATIONAL DE LA RECHERCHE SCIENTIFIQUE, MARSEILLES + CENTRE DE PHYSIQUE THEORIQUE	CNRS-CPT-(YEAR)-P-
FRANCE. CENTRE NATIONAL DE LA RECHERCHE SCIENTIFIQUE. LABORATOIRE D'ASTRONOMIE SPATIALE, MARSEILLES	LAS-(YEAR)-(LETTERS)-
FRANCE. COMMISSARIAT A L'ENERGIE ATOMIQUE (ALL LOCATIONS)	CEA-
FRANCE. COMMISSARIAT A L'ENERGIE ATOMIQUE (ALL LOCATIONS) (BIBLIOGRAPHY)	CEA-BIB-
FRANCE. COMMISSARIAT A L'ENERGIE ATOMIQUE (ALL LOCATIONS) (CONFERENCE PAPERS)	CEA-CONF-
FRANCE. COMMISSARIAT A L'ENERGIE ATOMIQUE (ALL LOCATIONS)	CEA-R-
FRANCE. COMMISSARIAT A L'ENERGIE ATOMIQUE	PNR/SEPR/
FRANCE. COMMISSARIAT A L'ENERGIE ATOMIQUE, PARIS (ANNUAL REPORT)	CEA-AR-(YEAR)-
FRANCE. COMMISSARIAT A L'ENERGIE ATOMIQUE, PARIS	CEA-PA-
FRANCE. COMMISSARIAT A L'ENERGIE ATOMIQUE, PARIS	CEA/PA/RT-
FRANCE. COMMISSARIAT A L'ENERGIE ATOMIQUE, PARIS	CEA-TP-
FRANCE. COMMISSARIAT A L'ENERGIE ATOMIQUE, PARIS (SERIES ASSIGNED BY THE AEC TO TRANSLATIONS RECEIVED FROM CEA)	CEA-TR-
FRANCE. COMMISSARIAT A L'ENERGIE ATOMIQUE, PARIS	CEA-TR-A-
FRANCE. COMMISSARIAT A L'ENERGIE ATOMIQUE, PARIS	CEN-S-PA-
FRANCE. COMMISSARIAT A L'ENERGIE ATOMIQUE, PARIS	CENS/PA/RT-
FRANCE. COMMISSARIAT A L'ENERGIE ATOMIQUE, PARIS	DPC-CPH-
FRANCE. COMMISSARIAT A L'ENERGIE ATOMIQUE, PARIS	DRP/MNF/R/
FRANCE. COMMISSARIAT A L'ENERGIE ATOMIQUE, PARIS	IP-
FRANCE. COMMISSARIAT A L'ENERGIE ATOMIQUE, PARIS	PA/RT-
FRANCE. COMMISSARIAT A L'ENERGIE ATOMIQUE, PARIS	PG-
FRANCE. COMMISSARIAT A L'ENERGIE ATOMIQUE, PARIS	RC-
FRANCE. COMMISSARIAT A L'ENERGIE ATOMIQUE, SACLAY	CEA-TR-R-

FRANCE. COMMISSARIAT A L'ENERGIE ATOMIQUE

FRANCE. COMMISSARIAT A L'ENERGIE ATOMIQUE, SACLAY	CEA-TR-X-
FRANCE. COMMISSARIAT A L'ENERGIE ATOMIQUE, SACLAY	ELDO/SP/(YEAR)/
FRANCE. COMMISSARIAT A L'ENERGIE ATOMIQUE, SACLAY	TH/(YEAR)/
FRANCE. COMMISSARIAT A L'ENERGIE ATOMIQUE. CENTRE D'ETUDES NUCLEAIRES (ALL LOCATIONS)	CEA-N-
FRANCE. COMMISSARIAT A L'ENERGIE ATOMIQUE. CENTRE D'ETUDES NUCLEAIRES (PUBLICATIONS SCIENTIFIQUES)	CEA-PS-
FRANCE. COMMISSARIAT A L'ENERGIE ATOMIQUE. CENTRE D'ETUDES NUCLEAIRES, CADARACHE	DRP/EMTR/CAD.R.(NO.)
FRANCE. COMMISSARIAT A L'ENERGIE ATOMIQUE. CENTRE D'ETUDES NUCLEAIRES, CADARACHE	DRP/ML/FAR R-
FRANCE. COMMISSARIAT A L'ENERGIE ATOMIQUE. CENTRE D'ETUDES NUCLEAIRES, CADARACHE	DRP/PNR/(LETTERS)/
FRANCE. COMMISSARIAT A L'ENERGIE ATOMIQUE. CENTRE D'ETUDES NUCLEAIRES, CADARACHE	DRP/SCR-
FRANCE. COMMISSARIAT A L'ENERGIE ATOMIQUE. CENTRE D'ETUDES NUCLEAIRES, CADARACHE	DRP/SEMTR/
FRANCE. COMMISSARIAT A L'ENERGIE ATOMIQUE. CENTRE D'ETUDES NUCLEAIRES, CADARACHE	DRP/SEMTR/CAD-
FRANCE. COMMISSARIAT A L'ENERGIE ATOMIQUE. CENTRE D'ETUDES NUCLEAIRES, CADARACHE	DRP/SETR-
FRANCE. COMMISSARIAT A L'ENERGIE ATOMIQUE. CENTRE D'ETUDES NUCLEAIRES, CADARACHE	PNR/DIR-(YEAR)-
FRANCE. COMMISSARIAT A L'ENERGIE ATOMIQUE. CENTRE D'ETUDES NUCLEAIRES, CADARACHE	PNR/EMTR-(YEAR)-
FRANCE. COMMISSARIAT A L'ENERGIE ATOMIQUE. CENTRE D'ETUDES NUCLEAIRES, CADARACHE	PNR/SETR-R-
FRANCE. COMMISSARIAT A L'ENERGIE ATOMIQUE. CENTRE D'ETUDES NUCLEAIRES, CADARACHE	SECA-
FRANCE. COMMISSARIAT A L'ENERGIE ATOMIQUE. CENTRE D'ETUDES NUCLEAIRES, CADARACHE	SETR-R-
FRANCE. COMMISSARIAT A L'ENERGIE ATOMIQUE. CENTRE D'ETUDES NUCLEAIRES, FONTENAY-AUX-ROSES	CEN-
FRANCE. COMMISSARIAT A L'ENERGIE ATOMIQUE. CENTRE D'ETUDES NUCLEAIRES, FONTENAY-AUX-ROSES	CENFAR-
FRANCE. COMMISSARIAT A L'ENERGIE ATOMIQUE. CENTRE D'ETUDES NUCLEAIRES, FONTENAY-AUX-ROSES (TRANS.)	DEP-SEPP-
FRANCE. COMMISSARIAT A L'ENERGIE ATOMIQUE. CENTRE D'ETUDES NUCLEAIRES, FONTENAY-AUX-ROSES	DRP/ML/CA ND-
FRANCE. COMMISSARIAT A L'ENERGIE ATOMIQUE. CENTRE D'ETUDES NUCLEAIRES, FONTENAY-AUX-ROSES	DRP/ML/FAR R/(NO.)-
FRANCE. COMMISSARIAT A L'ENERGIE ATOMIQUE. CENTRE D'ETUDES NUCLEAIRES, FONTENAY-AUX-ROSES	SCCI-
FRANCE. COMMISSARIAT A L'ENERGIE ATOMIQUE. CENTRE D'ETUDES NUCLEAIRES, GRENOBLE	CENG-
FRANCE. COMMISSARIAT A L'ENERGIE ATOMIQUE. CENTRE D'ETUDES NUCLEAIRES, GRENOBLE	DEG/SINR-
FRANCE. COMMISSARIAT A L'ENERGIE ATOMIQUE. CENTRE D'ETUDES NUCLEAIRES, GRENOBLE	DR-SAR-G-
FRANCE. COMMISSARIAT A L'ENERGIE ATOMIQUE. CENTRE D'ETUDES NUCLEAIRES, GRENOBLE	INT/EL-
FRANCE. COMMISSARIAT A L'ENERGIE ATOMIQUE. CENTRE D'ETUDES NUCLEAIRES, GRENOBLE	INT/PI (NT)(NO.-NO./YR.)
FRANCE. COMMISSARIAT A L'ENERGIE ATOMIQUE. CENTRE D'ETUDES NUCLEAIRES, GRENOBLE	INT/SPR-
FRANCE. COMMISSARIAT A L'ENERGIE ATOMIQUE. CENTRE D'ETUDES NUCLEAIRES, GRENOBLE	N.I.-
FRANCE. COMMISSARIAT A L'ENERGIE ATOMIQUE. CENTRE D'ETUDES NUCLEAIRES, GRENOBLE	PI(INT)-
FRANCE. COMMISSARIAT A L'ENERGIE ATOMIQUE. CENTRE D'ETUDES NUCLEAIRES, GRENOBLE	PI-(NT)-
FRANCE. COMMISSARIAT A L'ENERGIE ATOMIQUE. CENTRE D'ETUDES NUCLEAIRES, GRENOBLE	PI/NT/(NUMBER-NUMBER)
FRANCE. COMMISSARIAT A L'ENERGIE ATOMIQUE. CENTRE D'ETUDES NUCLEAIRES, GRENOBLE	RHF-(NUMBER-NUMBER)
FRANCE. COMMISSARIAT A L'ENERGIE ATOMIQUE. CENTRE D'ETUDES NUCLEAIRES, GRENOBLE	SAR-G-
FRANCE. COMMISSARIAT A L'ENERGIE ATOMIQUE. CENTRE D'ETUDES NUCLEAIRES, GRENOBLE	SAR.T/
FRANCE. COMMISSARIAT A L'ENERGIE ATOMIQUE. CENTRE D'ETUDES NUCLEAIRES, SACLAY	ASS/EPR-R-
FRANCE. COMMISSARIAT A L'ENERGIE ATOMIQUE. CENTRE D'ETUDES NUCLEAIRES, SACLAY	CEA-A-
FRANCE. COMMISSARIAT A L'ENERGIE ATOMIQUE. CENTRE D'ETUDES NUCLEAIRES, SACLAY	CEA-NOTE-
FRANCE. COMMISSARIAT A L'ENERGIE ATOMIQUE. CENTRE D'ETUDES NUCLEAIRES, SACLAY	CEA-S-
FRANCE. COMMISSARIAT A L'ENERGIE ATOMIQUE. CENTRE D'ETUDES NUCLEAIRES, SACLAY	CEA-TP-
FRANCE. COMMISSARIAT A L'ENERGIE ATOMIQUE. CENTRE D'ETUDES NUCLEAIRES, SACLAY	CEN-R-
FRANCE. COMMISSARIAT A L'ENERGIE ATOMIQUE. CENTRE D'ETUDES NUCLEAIRES, SACLAY	CFA-R-
FRANCE. COMMISSARIAT A L'ENERGIE ATOMIQUE. CENTRE D'ETUDES NUCLEAIRES, SACLAY	DE/SEA-
FRANCE. COMMISSARIAT A L'ENERGIE ATOMIQUE. CENTRE D'ETUDES NUCLEAIRES, SACLAY	DE/SEP/(NUMBER)/
FRANCE. COMMISSARIAT A L'ENERGIE ATOMIQUE. CENTRE D'ETUDES NUCLEAIRES, SACLAY	DM-
FRANCE. COMMISSARIAT A L'ENERGIE ATOMIQUE. CENTRE D'ETUDES NUCLEAIRES, SACLAY	DM-CS-
FRANCE. COMMISSARIAT A L'ENERGIE ATOMIQUE. CENTRE D'ETUDES NUCLEAIRES, SACLAY	DPA.IS/
FRANCE. COMMISSARIAT A L'ENERGIE ATOMIQUE. CENTRE D'ETUDES NUCLEAIRES, SACLAY	DPC.CPH/
FRANCE. COMMISSARIAT A L'ENERGIE ATOMIQUE. CENTRE D'ETUDES NUCLEAIRES, SACLAY	DPC-IS-
FRANCE. COMMISSARIAT A L'ENERGIE ATOMIQUE. CENTRE D'ETUDES NUCLEAIRES, SACLAY	DPC-PCA-
FRANCE. COMMISSARIAT A L'ENERGIE ATOMIQUE. CENTRE D'ETUDES NUCLEAIRES, SACLAY	DPC/PCA/(YEAR)-
FRANCE. COMMISSARIAT A L'ENERGIE ATOMIQUE. CENTRE D'ETUDES NUCLEAIRES, SACLAY	DPH/DOC/(YEAR)-
FRANCE. COMMISSARIAT A L'ENERGIE ATOMIQUE. CENTRE D'ETUDES NUCLEAIRES, SACLAY	DPH-PFC/
FRANCE. COMMISSARIAT A L'ENERGIE ATOMIQUE. CENTRE D'ETUDES NUCLEAIRES, SACLAY	DPH-T/(NO.)-(NO.)
FRANCE. COMMISSARIAT A L'ENERGIE ATOMIQUE. CENTRE D'ETUDES NUCLEAIRES, SACLAY	DPH-T/DOC/
FRANCE. COMMISSARIAT A L'ENERGIE ATOMIQUE. CENTRE D'ETUDES NUCLEAIRES, SACLAY	DRP/EMTR/FAR/R.(NO.)
FRANCE. COMMISSARIAT A L'ENERGIE ATOMIQUE. CENTRE D'ETUDES NUCLEAIRES, SACLAY	DRP/ML/FAR R-
FRANCE. COMMISSARIAT A L'ENERGIE ATOMIQUE. CENTRE D'ETUDES NUCLEAIRES, SACLAY	DRP/SMNF-
FRANCE. COMMISSARIAT A L'ENERGIE ATOMIQUE. CENTRE D'ETUDES NUCLEAIRES, SACLAY	FC.S-
FRANCE. COMMISSARIAT A L'ENERGIE ATOMIQUE. CENTRE D'ETUDES NUCLEAIRES, SACLAY	EN-
FRANCE. COMMISSARIAT A L'ENERGIE ATOMIQUE. CENTRE D'ETUDES NUCLEAIRES, SACLAY (THESIS)	FRNC-TH-
FRANCE. COMMISSARIAT A L'ENERGIE ATOMIQUE. CENTRE D'ETUDES NUCLEAIRES, SACLAY	GEX-R-
FRANCE. COMMISSARIAT A L'ENERGIE ATOMIQUE. CENTRE D'ETUDES NUCLEAIRES, SACLAY	PAS-(YEAR)/
FRANCE. COMMISSARIAT A L'ENERGIE ATOMIQUE. CENTRE D'ETUDES NUCLEAIRES, SACLAY	SAER-
FRANCE. COMMISSARIAT A L'ENERGIE ATOMIQUE. CENTRE D'ETUDES NUCLEAIRES, SACLAY	SEC-
FRANCE. COMMISSARIAT A L'ENERGIE ATOMIQUE. CENTRE D'ETUDES NUCLEAIRES, SACLAY	SPNBE-
FRANCE. COMMISSARIAT A L'ENERGIE ATOMIQUE. CENTRE D'ETUDES NUCLEAIRES, SACLAY	SRE-PR-
FRANCE. COMMISSARIAT A L'ENERGIE ATOMIQUE. CENTRE D'ETUDES NUCLEAIRES. SECTION DE PHYSIQUE ET D'EXPERIMENTATION, SACLAY	DPE/SPE/(YEAR/NUMBER)
FRANCE. COMMISSARIAT A L'ENERGIE ATOMIQUE. CENTRE D'ETUDES NUCLEAIRES. SERVICE DE PHYSIQUE MATHEMATIQUE, SACLAY	SPM-
FRANCE. COMMISSARIAT A L'ENERGIE ATOMIQUE. CENTRE D'ETUDES NUCLEAIRES. SERVICE DE PHYSIQUE THEORIQUE, SACLAY	SPT-
FRANCE. COMMISSARIAT A L'ENERGIE ATOMIQUE. CENTRE D'ETUDES NUCLEAIRES. SERVICE DE PHYSIQUE THEORIQUE, SACLAY	SPT/(INITIALS)-
FRANCE. COMMISSARIAT A L'ENERGIE ATOMIQUE. CENTRE D'ETUDES NUCLEAIRES. SERVICE DE PHYSIQUE THEORIQUE, SACLAY	SPT/DOC/(INITIALS)-
FRANCE. COMMISSARIAT A L'ENERGIE ATOMIQUE. CENTRE D'ETUDES NUCLEAIRES. SERVICE DES EXPERIENCES CRITIQUES A NEUTRONS RAPIDES, CADARACHE	SECNR-
FRANCE. COMMISSARIAT A L'ENERGIE ATOMIQUE. DEPARTEMENT DES ETUDES DE PILES, CADARACHE	SEC/CA-
FRANCE. COMMISSARIAT A L'ENERGIE ATOMIQUE. SERVICE D'ETUDES DES PILES RAPIDES	CEA-ASS/EPR-
FRANCE. DEPARTEMENT D'ETUDES DE RECHERCHES EN TECHNOLOGIE SPATIALE, TOULOUSE	DERTS-NT-(NUMBER)-
FRANCE. DIRECTION DE LA METEOROLOGIE NATIONAL, PARIS	REPT-
FRANCE. FACULTE DES SCIENCES DE MARSEILLE	FDSM-
FRANCE. FACULTE DES SCIENCES DE MARSEILLE	PTM-
FRANCE. FACULTE DES SCIENCES D'ORSAY	FDSD-
FRANCE. GROUPE DE RECHERCHES IONOSPHERIQUES, ISSY-LES-MOULINEAUX	GRI-NT-
FRANCE. GROUPE DE RECHERCHES IONOSPHERIQUES, ISSY-LES-MOULINEAUX	GRI-NTP-
FRANCE. GROUPE DE RECHERCHES IONOSPHERIQUES, PARIS	GRI-NT-
FRANCE. GROUPE DE RECHERCHES IONOSPHERIQUES, PARIS	GRI-NTP-
FRANCE. GROUPE DE RECHERCHES IONOSPHERIQUES, PARIS	GRI-TP-

FRANCE. INSTITUT DE PHYSIQUE NUCLEAIRE, ORSAY	INPO/TH-
FRANCE. INSTITUT DE PHYSIQUE NUCLEAIRE, ORSAY	IPN-
FRANCE. INSTITUT DE PHYSIQUE NUCLEAIRE, ORSAY	IPNO/TH-
FRANCE. INSTITUT DE RECHERCHES DE LA SIDERURGIE, SAINT GERMAIN-EN-LAYE	IRSID-
FRANCE. INSTITUT DE RECHERCHES NUCLEAIRES, STRASBOURG	IRN-
FRANCE. INSTITUT DES HAUTES ETUDES SCIENTIFIQUES, PARIS	IHES-A(NUMBER)-
FRANCE. INSTITUT DES HAUTES ETUDES SCIENTIFIQUES, PARIS	IHES-N-(LETTER)-
FRANCE. INSTITUT NATIONAL DE RECHERCHE CHIMIQUE APPLIQUEE, PARIS	INRCA-QPR-
FRANCE. INSTITUT NATIONAL DE RECHERCHE CHIMIQUE APPLIQUEE, PARIS	IRCHA-
FRANCE. LABORATOIRE CENTRAL DE L'ARMEMENT, ARCUEIL	LCA/MS-RAE-
FRANCE. LABORATOIRE D'AUTOMATIQUE ET DE SES APPLICATIONS SPATIALES, TOULOUSE	LAAS-
FRANCE. LABORATOIRE D'AUTOMATIQUE ET DE SES APPLICATIONS SPATIALES, TOULOUSE	LAAS-PUBL-
FRANCE. LABORATOIRE D'AUTOMATIQUE ET DE SES APPLICATIONS SPATIALES, TOULOUSE	LAAS-STI-
FRANCE. LABORATOIRE D'ETUDES BALISTIQUES DE ST. LOUIS	LDB-
FRANCE. LABORATOIRE DE PHYSIQUE THEORIQUE ET DES HAUTES ENERGIES, NICE	NTH-(YEAR)/
FRANCE. LABORATOIRE DE PHYSIQUE THEORIQUE ET DES HAUTES ENERGIES, NICE	N-TH-(YEAR)/
FRANCE. LABORATOIRE DE PHYSIQUE THEORIQUE ET DES HAUTES ENERGIES, ORSAY	LPTHE-(YEAR/NUMBER)
FRANCE. LABORATOIRE DE PHYSIQUE THEORIQUE ET DES HAUTES ENERGIES, ORSAY	LPTHE-TH-
FRANCE. LABORATOIRE DE PHYSIQUE THEORIQUE ET DES HAUTES ENERGIES, ORSAY	PTHE-(YEAR/NUMBER)
FRANCE. LABORATOIRE DE PHYSIQUE THEORIQUE ET DES HAUTES ENERGIES, ORSAY	TH-
FRANCE. LABORATOIRE DE RECHERCHE TECHNIQUES DE SAINT-LOUIS	LRSL-
FRANCE. LABORATOIRE DE RECHERCHES BALISTIQUES ET AERODYNAMIQUES, VERNON	ELDO-LRBA-TN-E-
FRANCE. LABORATOIRE DE RECHERCHES BALISTIQUES ET AERODYNAMIQUES, VERNON	LRBA-(NUMBER)-(YEAR)-EN
FRANCE. LABORATOIRE DE RECHERCHES BALISTIQUES ET AERODYNAMIQUES, VERNON	LRBA-CR-
FRANCE. LABORATOIRE DE RECHERCHES BALISTIQUES ET AERODYNAMIQUES, VERNON	LRBA-E-
FRANCE. LABORATOIRE DE RECHERCHES BALISTIQUES ET AERODYNAMIQUES, VERNON	LRBA-NT-(NUMBER/YR/LTRS)
FRANCE. LABORATOIRE DE RECHERCHES BALISTIQUES ET AERODYNAMIQUES, VERNON	LRBA-NT-E-
FRANCE. LABORATOIRE DE RECHERCHES BALISTIQUES ET AERODYNAMIQUES, VERNON	LRBA-PV-(NUMBER-NO.-LTRS)
FRANCE. LABORATOIRE DE RECHERCHES BALISTIQUES ET AERODYNAMIQUES, VERNON	NT-(NUMBER)/EAS-E/(NO.)
FRANCE. LABORATOIRE DE RECHERCHES BALISTIQUES ET AERODYNAMIQUES, VERNON	NT-(NUMBER)/(YEAR)/EG
FRANCE. LABORATOIRE DE RECHERCHES BALISTIQUES ET AERODYNAMIQUES, VERNON	NT-(NUMBER)/(YEAR)/EN
FRANCE. LIGNES TELEGRAPHIQUES ET TELEPHONIQUES, PARIS	Z-
FRANCE. MINISTERE DE L'AIR, PARIS	NT-
FRANCE. OFFICE NATIONAL D=ETUDES ET DE RECHERCHES AEROSPATIALES, CHATILLON-SOUS-BAGNEUX	NOTE TECHNIQUE-
FRANCE. OFFICE NATIONAL D'ETUDES ET DE RECHERCHES AEROSPATIALES, CHATILLON-SOUS-BAGNEUX	ONERA-CF-
FRANCE. OFFICE NATIONAL D'ETUDES ET DE RECHERCHES AEROSPATIALES, CHATILLON-SOUS-BAGNEUX	ONERA-DP-
FRANCE. OFFICE NATIONAL D'ETUDES ET DE RECHERCHES AEROSPATIALES, CHATILLON-SOUS-BAGNEUX	ONERA-NOTE TECHNIQUE-
FRANCE. OFFICE NATIONAL D'ETUDES ET DE RECHERCHES AEROSPATIALES, CHATILLON-SOUS-BAGNEUX	ONERA-PUB-
FRANCE. OFFICE NATIONAL D'ETUDES ET DE RECHERCHES AEROSPATIALES, CHATILLON-SOUS-BAGNEUX	ONERA-PUBL.-
FRANCE. OFFICE NATIONAL D'ETUDES ET DE RECHERCHES AEROSPATIALES, CHATILLON-SOUS-BAGNEUX	ONERA-TN-
FRANCE. OFFICE NATIONAL D'ETUDES ET DE RECHERCHES AEROSPATIALES, CHATILLON-SOUS-BAGNEUX	ONERA-TP-
FRANCE. OFFICE NATIONAL D'ETUDES ET DE RECHERCHES AEROSPATIALES, PARIS	ELDO/SP/(YR)/-
FRANCE. OFFICE NATIONAL D'ETUDES ET DE RECHERCHES AEROSPATIALES, PARIS	ONERA-
FRANCE. OFFICE NATIONAL D'ETUDES ET DE RECHERCHES AEROSPATIALES, PARIS	ONERA-NT-
FRANCE. OFFICE NATIONAL D'ETUDES ET DE RECHERCHES AEROSPATIALES, PARIS	ONERA-P-
FRANCE. OFFICE NATIONAL D'ETUDES ET DE RECHERCHES AEROSPATIALES, PARIS	ONERA-PUBL-(NO.)/(YR)
FRANCE. OFFICE NATIONAL D'ETUDES ET DE RECHERCHES AEROSPATIALES, PARIS	ONERA-TN-
FRANCE. OFFICE NATIONAL D'ETUDES ET DE RECHERCHES AEROSPATIALES, PARIS	ONERA-TP-
FRANCE. OFFICE NATIONAL D'ETUDES ET DE RECHERCHES AEROSPATIALES, PARIS	TP-
FRANCE. SERVICE CENTRAL DE PROTECTION CONTRE LES RAYONNEMENTS IONISANTS, PARIS	SCPRI-
FRANCE. SERVICE CENTRAL DE PROTECTION CONTRE LES RAYONNEMENTS IONISANTS, PARIS	SCPRI(A)-
FRANCE. SERVICE CENTRAL DE PROTECTION CONTRE LES RAYONNEMENTS IONISANTS, PARIS	SCPRI(M)-
FRANCE. SERVICE CENTRAL DE PROTECTION CONTRE LES RAYONNEMENTS IONISANTS, PARIS	SCPRI(RM)
FRANCE. SERVICE CENTRAL DE PROTECTION CONTRE LES RAYONNEMENTS IONISANTS, PARIS	SCPRI(S)-
FRANCE. SERVICE CENTRAL DE PROTECTION CONTRE LES RAYONNEMENTS IONISANTS, PARIS	SCPRI(T)-
FRANCE. SERVICE DE NEUTRONIQUE EXPERIMENTALE	SNE-
FRANCE. SERVICE TECHNIQUE DE L'AIR, PARIS	SAR-
FRANCIS ASSOCIATES, INC., MARION, MASS.	R-(3 DIGITS)
FRANKFORD ARSENAL, PHILADELPHIA	A(NO.)-
FRANKFORD ARSENAL, PHILADELPHIA	FA-
FRANKFORD ARSENAL, PHILADELPHIA	FA-A(YEAR)-
FRANKFORD ARSENAL, PHILADELPHIA	FA-FCDD-
FRANKFORD ARSENAL, PHILADELPHIA	FA-IEP-(NUMBERS)-
FRANKFORD ARSENAL, PHILADELPHIA	FA-LDN-
FRANKFORD ARSENAL, PHILADELPHIA	FALR-
FRANKFORD ARSENAL, PHILADELPHIA	FALR MEMO-
FRANKFORD ARSENAL, PHILADELPHIA	FALR R-
FRANKFORD ARSENAL, PHILADELPHIA (STATISTICAL MEMO)	FALR S MEMO-
FRANKFORD ARSENAL, PHILADELPHIA	FALR T-
FRANKFORD ARSENAL, PHILADELPHIA	FA-M(YEAR)-
FRANKFORD ARSENAL, PHILADELPHIA	FA-M(YEAR)-(NUMBER)-
FRANKFORD ARSENAL, PHILADELPHIA (MEMO REPORT)	FA-MR-M-(NO.)(NO.)(NO.)
FRANKFORD ARSENAL, PHILADELPHIA	FA-P(YEAR)-
FRANKFORD ARSENAL, PHILADELPHIA	FA-PP-T-
FRANKFORD ARSENAL, PHILADELPHIA	FA-R-
FRANKFORD ARSENAL, PHILADELPHIA	FA-RF-(YEAR)-
FRANKFORD ARSENAL, PHILADELPHIA	FA-T(NUMBER)-
FRANKFORD ARSENAL, PHILADELPHIA	FA-TDL-
FRANKFORD ARSENAL, PHILADELPHIA	FA-TEST RPT. T-(NUMBERS)
FRANKFORD ARSENAL, PHILADELPHIA	FA-TN-
FRANKFORD ARSENAL, PHILADELPHIA	FA-TN(YEAR)-
FRANKFORD ARSENAL, PHILADELPHIA	M(NO.)-
FRANKFORD ARSENAL, PHILADELPHIA	ORD FLR-
FRANKFORD ARSENAL, PHILADELPHIA	PAD-
FRANKFORD ARSENAL, PHILADELPHIA	R-
FRANKFORD ARSENAL, PHILADELPHIA	T-
FRANKFORD ARSENAL. FIRE CONTROL LAB., PHILADELPHIA	FALR FC-
FRANKFORD ARSENAL. FIRE CONTROL LAB., PHILADELPHIA	FC-
FRANKFORD ARSENAL. FIRE CONTROL LAB., PHILADELPHIA	ORDTX-

FRANKFORD ARSENAL. FIRE CONTROL LAB.

FRANKFORD ARSENAL. FIRE CONTROL LAB., PHILADELPHIA	TN-
FRANKFORD ARSENAL. PITMAN-DUNN LAB., PHILADELPHIA	EA-A(YEAR)-
FRANKFORD ARSENAL. PITMAN-DUNN LAB., PHILADELPHIA	FA-S-
FRANKFORD ARSENAL. PITMAN-DUNN LAB., PHILADELPHIA	FRANKFORD-(LETTERS)-
FRANKFORD ARSENAL. PITMAN-DUNN LAB., PHILADELPHIA	M62-
FRANKFORD ARSENAL. PITMAN-DUNN LAB., PHILADELPHIA	MR-
FRANKFORD ARSENAL. PITMAN-DUNN LAB., PHILADELPHIA	PDL-
FRANKFORD ARSENAL. PITMAN-DUNN LAB., PHILADELPHIA	S-
FRANKFORD ARSENAL. PROPELLANT ACTUATED DEVICES DIV., PHILADELPHIA	MR-
FRANKFURT AM MAIN. UNIVERSITAET	FRANKFURTU-
FRANKFURT AM MAIN. UNIVERSITAET. INSTITUT FUER KERNPHYSIK.	IKF-
FRANKFURT AM MAIN. UNIVERSITAET. INSTITUT FUER KERNPHYSIK	IKF-D-
FRANKLIN GNO CORP., WEST PALM BEACH, FLA.	FGNOC-
FRANKLIN INST., PHILADELPHIA	B(NO.)-T-
FRANKLIN INST., PHILADELPHIA	F-
FRANKLIN INST., PHILADELPHIA	F-(LETTER)(NUMBER)
FRANKLIN INST., PHILADELPHIA	F-B(NO.)-
FRANKLIN INST., PHILADELPHIA	FI-
FRANKLIN INST., PHILADELPHIA	FI-F-
FRANKLIN INST., PHILADELPHIA	FI-P-
FRANKLIN INST., PHILADELPHIA	FI-QR-
FRANKLIN INST., PHILADELPHIA	FI-QT-
FRANKLIN INST., PHILADELPHIA (RESEARCH REPORT)	FI RR-
FRANKLIN INST., PHILADELPHIA	FRAN-
FRANKLIN INST., PHILADELPHIA (MONTHLY REVIEW)	MR-
FRANKLIN INST., PHILADELPHIA	Q-
FRANKLIN INST., PHILADELPHIA	Q-B(NO.)-
FRANKLIN INST., PHILADELPHIA	TA-
FRANKLIN INST. LABS. FOR RES. + DEV., PHILADELPHIA	B-
FRANKLIN INST. LABS. FOR RES. + DEV., PHILADELPHIA	C-
FRANKLIN INST. LABS. FOR RES. + DEV., PHILADELPHIA	CS-
FRANKLIN INST. LABS. FOR RES. + DEV., PHILADELPHIA	F-(PROJECT)-
FRANKLIN INST. LABS. FOR RES. + DEV., PHILADELPHIA	F-A-
FRANKLIN INST. LABS. FOR RES. AND DEV., PHILADELPHIA (ELECTRIC INITIATOR SYMPOSIUM)	F/EIS-
FRANKLIN INST. LABS. FOR RES. + DEV., PHILADELPHIA	FI-B(NUMBER)-T-
FRANKLIN INST. LABS. FOR RES. + DEV., PHILADELPHIA	FI-F-A-
FRANKLIN INST. LABS. FOR RES. + DEV., PHILADELPHIA	FI-FM-
FRANKLIN INST. LABS. FOR RES. + DEV., PHILADELPHIA	FI-FP-
FRANKLIN INST. LABS. FOR RES. + DEV., PHILADELPHIA	FI-I-
FRANKLIN INST. LABS. FOR RES. + DEV., PHILADELPHIA (INTERIM REPORT)	FI-I-A-(NUMBER)-(NUMBER)
FRANKLIN INST. LABS. FOR RES. + DEV., PHILADELPHIA	FIL-
FRANKLIN INST. LABS. FOR RES. + DEV., PHILADELPHIA	FIL-(LTRS)-
FRANKLIN INST. LABS. FOR RES. + DEV., PHILADELPHIA	FIL-F-
FRANKLIN INST. LABS. FOR RES. + DEV., PHILADELPHIA	FIL-I-
FRANKLIN INST. LABS. FOR RES. + DEV., PHILADELPHIA	FIL-MR-
FRANKLIN INST. LABS. FOR RES. + DEV., PHILADELPHIA (MONTHLY REVIEW REPORT)	FI-MR-
FRANKLIN INST. LABS. FOR RES. + DEV., PHILADELPHIA	FI-MU-A-(NUMBER)-(NUMBER)
FRANKLIN INST. LABS. FOR RES. + DEV., PHILADELPHIA	FI-Q-A-(NUMBER)-(NUMBER)
FRANKLIN INST. LABS. FOR RES. + DEV., PHILADELPHIA	FI-Q-B(NUMBER)-
FRANKLIN INST. LABS. FOR RES. + DEV., PHILADELPHIA	FR-
FRANKLIN INST. LABS. FOR RES. + DEV., PHILADELPHIA	FR-(PROJECT)-
FRANKLIN INST. LABS. FOR RES. + DEV., PHILADELPHIA	FRAN (LTR-NO.)
FRANKLIN INST. LABS. FOR RES. + DEV., PHILADELPHIA	FT-(PROJECT)-
FRANKLIN INST. LABS. FOR RES. + DEV., PHILADELPHIA	I-
FRANKLIN INST. LABS. FOR RES. + DEV., PHILADELPHIA	I-A(NUMBER)-
FRANKLIN INST. LABS. FOR RES. + DEV., PHILADELPHIA	I-B(NUMBER)-
FRANKLIN INST. LABS. FOR RES. + DEV., PHILADELPHIA	LM-
FRANKLIN INST. LABS. FOR RES. + DEV., PHILADELPHIA	MR-(PROJECT)-
FRANKLIN INST. LABS. FOR RES. + DEV., PHILADELPHIA	P-(PROJECT)
FRANKLIN INST. LABS. FOR RES. + DEV., PHILADELPHIA	P-(4 DIGITS)-
FRANKLIN INST. LABS. FOR RES. + DEV., PHILADELPHIA	O-
FRANKLIN INST. LABS. FOR RES. + DEV., PHILADELPHIA	T-
FRANKLIN INST. LABS. FOR RES. + DEV., PHILADELPHIA	TN-S4-
FRANKLIN INST. LABS. FOR RES. + DEV., PHILADELPHIA	TSR-
FRANKLIN INST. LABS. FOR RES. + DEV., PHILADELPHIA	TT-
FRANKLIN INST. LABS. FOR RES. + DEV., PHILADELPHIA	UVA/RLES-EP/
FRANKLIN INST. RESEARCH LABS., PHILADELPHIA	C-(4 DIGITS)-F
FRANKLIN INST. RESEARCH LABS., PHILADELPHIA	F-B(NUMBER)
FRANKLIN INST. RESEARCH LABS., PHILADELPHIA	F-C(NUMBER)
FRANKLIN INST. RESEARCH LABS., PHILADELPHIA	FIRL-C(NUMBER)
FRANKLIN INST. RESEARCH LABS., PHILADELPHIA	FIRL-F-B(NUMBER)
FRANKLIN INST. RESEARCH LABS., PHILADELPHIA	FIRL-F-C(NUMBER)
FRANKLIN INST. RESEARCH LABS., PHILADELPHIA	FIRL-I-C(NUMBER)
FRANKLIN INST. RESEARCH LABS., PHILADELPHIA	FI-SA-
FRANKLIN INST. RESEARCH LABS., PHILADELPHIA	I-
FRANKLIN INST. RESEARCH LABS., PHILADELPHIA	Q-C(4 DIGITS)-
FRANKLIN INST. RESEARCH LABS., PHILADELPHIA	S-C(NUMBER)-
FRANKLIN INST. RESEARCH LABS., PHILADELPHIA	T-
FRANKLIN INST. RESEARCH LABS., PHILADELPHIA	T-(LETTER)(NUMBER)-
FRANKLIN INST. RESEARCH LABS. IDRES INFORMATION CENTER, PHILADELPHIA (INDUSTRIAL RESEARCH)	IDRES-
FRANKLIN INST. RESEARCH LABS. OPERATIONS RESEARCH DIV., PHILADELPHIA	TR-1-
FRAUNHOFER-GESELLSCHAFT ZUR FORDERUNG DER ANGEWANDTEN FORSCHUNG E.V., DARMSTADT, GERMANY	FB-(NUMBER)-(YEAR)
FRAUNHOFER-GESELLSCHAFT ZUR FORDERUNG DER ANGEWANDTEN FORSCHUNG E.V., DARMSTADT, GERMANY	REPT-FB-(NUMBER)-(YEAR)
FRAUNHOFER INSTITUT, FREIBURG IM BREISGAU, GERMANY	REPT-(NUMBER/YEAR)
FREDERIC BURK FOUNDATION RESEARCH CTR., SAN FRANCISCO	FBRC-TR-
FREDERICK (CARL L.) AND ASSOCIATES, BETHESDA, MD.	FRED-
FREDERICK RESEARCH CORP., BETHESDA, MD.	C-(NUMBER-NUMBER)-
FREDERICK RESEARCH CORP., BETHESDA, MD.	FRC-(YEAR)-
FREED ELECTRONICS + CONTROLS CORP., N.Y.C.	FRC-
FREIBURG I. B. UNIVERSITAET	FRU-
FRIEDMAN (MORRIS D.), INC., WEST CONCORD, MASS.	MDF-

FRIEDMAN (MORRIS D.), INC., WEST NEWTON, MASS. (TRANSLATIONS)	MDF-
FROST ENGINEERING DEVELOPMENT CORP., DENVER	FEDC-
FROST ENGINEERING DEVELOPMENT CORP., DENVER	FEDC-TR-
FROST ENGINEERING DEVELOPMENT CORP., ENGLEWOOD, COLO.	FROST-(NUMBER)-
FULMER RES. INST., LTD., STOKE POGES, BUCKS, ENGLAND	QTSR-
FULMER RES. INST., LTD., STOKE POGES, BUCKS, ENGLAND	R.53/-/(DATE)
FULMER RES. INST., LTD., STOKE POGES, BUCKS, ENGLAND	R.(NUMBER/NUMBER)
FULTON COUNTY PLANNING DEPT., JOHNSTOWN, N.Y.	FCPD-TBR-(NUMBER)-
FUNDAMENTAL METHODS ASSOCIATES, N.Y.C.	FM-
FURANE PLASTICS, INC., LOS ANGELES	FP-

G.E.C. ELECTRONICS, LTD. APPLIED ELECTRONICS LABS., STANMORE, MIDDX., ENGLAND — SRL-

GABRIEL CO. ROCKET POWER/TALCO DIV., MESA, ARIZ. — RP/T-

GABRIEL LABS., NEEDHAM HEIGHTS, MASS. — GAB-

GAERTNER RESEARCH, INC., STAMFORD, CONN. — GRI-(4 DIGITS)

GARRETT CORP., LOS ANGELES — AAC-

GARRETT CORP. AIRESEARCH MFG. DIV., LOS ANG. (REF. 1)	(YEAR)-
GARRETT CORP. AIRESEARCH MFG. DIV., LOS ANGELES	AE-
GARRETT CORP. AIRESEARCH MFG. DIV., LOS ANGELES	AE-(NUMBER)-R
GARRETT CORP. AIRESEARCH MFG. DIV., LOS ANGELES	AGE-
GARRETT CORP. AIRESEARCH MFG. DIV., LOS ANGELES	AIRESEARCH-FC-
GARRETT CORP. AIRESEARCH MFG. DIV., LOS ANGELES	AIR-K-
GARRETT CORP. AIRESEARCH MFG. DIV., LOS ANGELES	AM-
GARRETT CORP. AIRESEARCH MFG. DIV., LOS ANGELES	AP-
GARRETT CORP. AIRESEARCH MFG. DIV., LOS ANGELES	APS-
GARRETT CORP. AIRESEARCH MFG. DIV., LOS ANGELES	AR-
GARRETT CORP. AIRESEARCH MFG. DIV., LOS ANGELES	ARMC-
GARRETT CORP. AIRESEARCH MFG. DIV., LOS ANGELES	ATP-
GARRETT CORP. AIRESEARCH MFG. DIV., LOS ANGELES	CB-
GARRETT CORP. AIRESEARCH MFG. DIV., LOS ANGELES	CP-
GARRETT CORP. AIRESEARCH MFG. DIV., LOS ANGELES	DATP-
GARRETT CORP. AIRESEARCH MFG. DIV., LOS ANGELES	DS-
GARRETT CORP. AIRESEARCH MFG. DIV., LOS ANGELES	DTD-
GARRETT CORP. AIRESEARCH MFG. DIV., LOS ANGELES	F-
GARRETT CORP. AIRESEARCH MFG. DIV., LOS ANGELES	F-(NUMBER)-(LETTER)
GARRETT CORP. AIRESEARCH MFG. DIV., LOS ANGELES	FC-
GARRETT CORP. AIRESEARCH MFG. DIV., LOS ANGELES (LIBRARY ACCESSION NUMBER)	G(YEAR)-
GARRETT CORP. AIRESEARCH MFG. DIV., LOS ANGELES	GAR-
GARRETT CORP. AIRESEARCH MFG. DIV., LOS ANGELES	GAR/LA-(YEAR)-
GARRETT CORP. AIRESEARCH MFG. DIV., LOS ANGELES	GAR/LA-(LETTERS-YEAR)-
GARRETT CORP. AIRESEARCH MFG. DIV., LOS ANGELES	GAR/LA HT-(YR.-NO./NO.)
GARRETT CORP. AIRESEARCH MFG. DIV., LOS ANGELES	HT-(YEAR)-
GARRETT CORP. AIRESEARCH MFG. DIV., LOS ANGELES	HT-(YEAR)-(NUMBER/NUMBER)
GARRETT CORP. AIRESEARCH MFG. DIV., LOS ANGELES	K-
GARRETT CORP. AIRESEARCH MFG. DIV., LOS ANGELES	L-
GARRETT CORP. AIRESEARCH MFG. DIV., LOS ANGELES	LS-
GARRETT CORP. AIRESEARCH MFG. DIV., LOS ANGELES	M-
GARRETT CORP. AIRESEARCH MFG. DIV., LOS ANGELES	MS-AP-
GARRETT CORP. AIRESEARCH MFG. DIV., LOS ANGELES	PT-
GARRETT CORP. AIRESEARCH MFG. DIV., LOS ANGELES	RD-(YEAR)-
GARRETT CORP. AIRESEARCH MFG. DIV., LOS ANGELES	SC-
GARRETT CORP. AIRESEARCH MFG. DIV., LOS ANGELES	SS-
GARRETT CORP. AIRESEARCH MFG. DIV., LOS ANGELES	VATP/OD-
GARRETT CORP. AIRESEARCH MFG. CO. OF ARIZONA, PHOENIX	AE-
GARRETT CORP. AIRESEARCH MFG. CO. OF ARIZONA, PHOENIX	AIRESEARCH-F-
GARRETT CORP. AIRESEARCH MFG. CO. OF ARIZONA, PHOENIX	AP-
GARRETT CORP. AIRESEARCH MFG. CO. OF ARIZONA, PHOENIX	APD-
GARRETT CORP. AIRESEARCH MFG. CO. OF ARIZONA, PHOENIX	APS-
GARRETT CORP. AIRESEARCH MFG. CO. OF ARIZONA, PHOENIX	ARM-
GARRETT CORP. AIRESEARCH MFG. CO. OF ARIZONA, PHOENIX	CA-
GARRETT CORP. AIRESEARCH MFG. CO. OF ARIZONA, PHOENIX	GT-
GARRETT CORP. AIRESEARCH MFG. CO. OF ARIZONA, PHOENIX	PE-(4 DIGITS)-
GARRETT CORP. AIRESEARCH MFG. CO. OF ARIZONA, PHOENIX	RDR-
GARRETT CORP. AIRESEARCH MFG. CO. OF ARIZONA, PHOENIX	SY-
GARRETT CORP. AIRESEARCH MFG. CO. OF ARIZONA, PHOENIX	SY-(NUMBER)-R
GARRETT CORP. AIRESEARCH MFG. DIV., PHOENIX, ARIZ.	AE-(NUMBER)-D
GARRETT CORP. AIRESEARCH MFG. DIV., PHOENIX, ARIZ.	AP-(NUMBER)-R
GARRETT CORP. AIRESEARCH MFG. DIV., PHOENIX, ARIZ.	APS-(NUMBER)-R
GARRETT CORP. AIRESEARCH MFG. DIV., PHOENIX, ARIZ.	GR-

GARRETT RESEARCH + DEVELOPMENT CO., LA VERNE, CALIF. — GR/D-(YEAR)-

GATES RUBBER CO., DENVER — GRC-
GATES RUBBER CO., DENVER — GRCO-

GAUTNEY + JONES, CONSULTING ENGINEER, WASHINGTON, D.C. — TGGM-

GAUTNEY + JONES COMMUNICATIONS, INC., FALLS CHURCH, VA — G/J-TR-(NUMBER-NUMBER)

GCA CORP. BEDFORD, MASS. (IN-HOUSE RESEARCH) — GCA-(YEAR NUMBER)

GCA CORP. TECHNOLOGY DIV., BEDFORD, MASS. (WORK DONE FOR AIR FORCE) — GCA-TR-(YEAR)-(NO.)-A
GCA CORP. TECHNOLOGY DIV., BEDFORD, MASS. (WORK DONE FOR OTHER AGENCIES) — GCA-TR-(YEAR)-(NO.)-G
GCA CORP. TECHNOLOGY DIV., BEDFORD, MASS. (WORK DONE FOR NASA) — GCA-TR-(YEAR)-(NO.)-N

GCO, INC., ANN ARBOR, MICH. — GCO-(YEAR)-

GENERAL ACCOUNTING OFFICE, WASHINGTON, D.C. — GAO-

GENERAL AMERICAN TRANSPORT. CORP. GAR DIV., NILES, ILL — GATC-
GENERAL AMERICAN TRANSPORT. CORP. GAR DIV., NILES, ILL — GATC-GARD-MR-
GENERAL AMERICAN TRANSPORT. CORP. GAR DIV., NILES, ILL — GATX-

GENERAL AMERICAN TRANSPORTATION CORP. GENERAL AMERICAN RESEARCH DIV., NILES, ILL. — GARD-
GENERAL AMERICAN TRANSPORTATION CORP. GENERAL AMERICAN RESEARCH DIV., NILES, ILL. — GARD-(NUMBER-LETTER)
GENERAL AMERICAN TRANSPORTATION CORP. GENERAL AMERICAN RESEARCH DIV., NILES, ILL. — GARD-(NUMBER-NUMBER-LTR)
GENERAL AMERICAN TRANSPORTATION CORP. GENERAL AMERICAN RESEARCH DIV., NILES, ILL. — GARD-MR-

GENERAL AMERICAN TRANSPORT. CORP. MRD DIV., NILES, ILL — GATC-MRD-
GENERAL AMERICAN TRANSPORT. CORP. MRD DIV., NILES, ILL — GATC-MRD-TN-
GENERAL AMERICAN TRANSPORT. CORP. MRD DIV., NILES, ILL — GATX-MR-
GENERAL AMERICAN TRANSPORT. CORP. MRD DIV., NILES, ILL — MR-
GENERAL AMERICAN TRANSPORT. CORP. MRD DIV., NILES, ILL — MRD-GADC-

GENERAL ANILINE AND FILM CORP. — GAFC (LETTERS)-

GENERAL ANILINE AND FILM CORP., BINGHAMTON, N.Y. — GAF-US-

GENERAL ANILINE AND FILM CORP., EASTON, PENNA. — GA+FC-

GENERAL APPLIED SCIENCE LABS., INC., WESTBURY, N.Y. — D(NUMBER)-
GENERAL APPLIED SCIENCE LABS., INC., WESTBURY, N.Y. — GASL-
GENERAL APPLIED SCIENCE LABS., INC., WESTBURY, N.Y. — GASL-ITR-
GENERAL APPLIED SCIENCE LABS., INC., WESTBURY, N.Y. — GASL-S-

GENERAL APPLIED SCIENCE LABS., INC., WESTBURY, N.Y.	GASL-TR-
GENERAL APPLIED SCIENCE LABS., INC., WESTBURY, N.Y.	GENAPPSL-
GENERAL APPLIED SCIENCE LABS., INC., WESTBURY, N.Y.	GENAPPSL-TR-
GENERAL APPLIED SCIENCE LABS., INC., WESTBURY, N.Y.	MPL-
GENERAL ASTROMETALS CORP., YONKERS, N.Y.	GAC-
GENERAL ATOMIC DIV., GEN. DYNAMICS CORP., SAN DIEGO, CAL	GA-(LETTER(S))-
GENERAL ATOMIC DIV., GEN. DYNAMICS CORP., SAN DIEGO, CAL	GA-C-
GENERAL ATOMIC DIV., GEN. DYNAMICS CORP., SAN DIEGO, CAL	GACD-PR (CONTRACT NO.)
GENERAL ATOMIC DIV., GEN. DYNAMICS CORP., SAN DIEGO, CAL	GACP-
GENERAL ATOMIC DIV., GEN. DYNAMICS CORP., SAN DIEGO, CAL	GADC-
GENERAL ATOMIC DIV., GEN. DYNAMICS CORP., SAN DIEGO, CAL	GA-P-
GENERAL ATOMIC DIV., GEN. DYNAMICS CORP., SAN DIEGO, CAL	MGCR-
GENERAL ATOMIC DIV., GEN. DYNAMICS CORP., SAN DIEGO, CAL	MGCR-M-
GENERAL ATOMIC DIV., GEN. DYNAMICS CORP., SAN DIEGO, CAL	MGCR-P-
GENERAL ATOMIC DIV., GEN. DYNAMICS CORP., SAN DIEGO, CAL	MGCR-RE-
GENERAL ATOMIC DIV., GEN. DYNAMICS CORP., SAN DIEGO, CAL	MGCR-RP-
(SEE ALSO SUCCESSOR: GULF GENERAL ATOMIC...)	
GENERAL ATRONICS CORP., CONSHOHOCKEN, PENNA.	GEATC-
GENERAL ATRONICS CORP., PHILADELPHIA	GAC-
GENERAL DEVELOPMENT CORP., ELKTON, MD.	R-12-
GENERAL DYNAMICS. ADVANCED STUDIES OFFICE, SAN DIEGO, CALIF.	AOK-(YEAR)-
GENERAL DYNAMICS/ASTRONAUTICS, SAN DIEGO, CALIF.	AE-
GENERAL DYNAMICS/ASTRONAUTICS, SAN DIEGO, CALIF.	AE(YR)-(NO.)
GENERAL DYNAMICS/ASTRONAUTICS, SAN DIEGO, CALIF.	AR-592-1-
GENERAL DYNAMICS/ASTRONAUTICS, SAN DIEGO, CALIF.	AXG-(YEAR)-
GENERAL DYNAMICS/ASTRONAUTICS, SAN DIEGO, CALIF.	AY(YEAR)-
GENERAL DYNAMICS/ASTRONAUTICS, SAN DIEGO, CALIF.	BHC-
GENERAL DYNAMICS/ASTRONAUTICS, SAN DIEGO, CALIF.	BMV-
GENERAL DYNAMICS/ASTRONAUTICS, SAN DIEGO, CALIF.	BRJ-
GENERAL DYNAMICS/ASTRONAUTICS, SAN DIEGO, CALIF.	EM-
GENERAL DYNAMICS/ASTRONAUTICS, SAN DIEGO, CALIF.	ERR-AN-
GENERAL DYNAMICS/ASTRONAUTICS, SAN DIEGO, CALIF.	GD/A-
GENERAL DYNAMICS/ASTRONAUTICS, SAN DIEGO, CALIF.	GDA-
GENERAL DYNAMICS/ASTRONAUTICS, SAN DIEGO, CALIF.	GDA/A-(YEAR)-
GENERAL DYNAMICS/ASTRONAUTICS, SAN DIEGO, CALIF.	GDA-AA(YEAR)-
GENERAL DYNAMICS/ASTRONAUTICS, SAN DIEGO, CALIF.	GDA-AE-(YEAR)-(NUMBER)-
GENERAL DYNAMICS/ASTRONAUTICS, SAN DIEGO, CALIF.	GDA/AOC-(YEAR)-
GENERAL DYNAMICS/ASTRONAUTICS, SAN DIEGO, CALIF.	GDA-AOH(YEAR)-
GENERAL DYNAMICS/ASTRONAUTICS, SAN DIEGO, CALIF.	GDA-AOJ(YEAR)-
GENERAL DYNAMICS/ASTRONAUTICS, SAN DIEGO, CALIF.	GDA-AY(YEAR)-
GENERAL DYNAMICS/ASTRONAUTICS, SAN DIEGO, CALIF.	GDA-DBE-
GENERAL DYNAMICS/ASTRONAUTICS, SAN DIEGO, CALIF.	GD/A-DDB-
GENERAL DYNAMICS/ASTRONAUTICS, SAN DIEGO, CALIF.	GD/A-ERR-
GENERAL DYNAMICS/ASTRONAUTICS, SAN DIEGO, CALIF.	GDA-ERR-
GENERAL DYNAMICS/ASTRONAUTICS, SAN DIEGO, CALIF.	GDA-ERR-AN-
GENERAL DYNAMICS/ASTRONAUTICS, SAN DIEGO, CALIF.	GDA-ESG-
GENERAL DYNAMICS/ASTRONAUTICS, SAN DIEGO, CALIF.	GDA-MP-
GENERAL DYNAMICS/ASTRONAUTICS, SAN DIEGO, CALIF.	GDA-REL-R-
GENERAL DYNAMICS/ASTRONAUTICS, SAN DIEGO, CALIF.	GDA-VCP-N-
GENERAL DYNAMICS/ASTRONAUTICS, SAN DIEGO, CALIF.	GDA-VTP-E-
GENERAL DYNAMICS/ASTRONAUTICS, SAN DIEGO, CALIF.	GDA-VTP-F-
GENERAL DYNAMICS/ASTRONAUTICS, SAN DIEGO, CALIF.	GDA-VTP-M-
GENERAL DYNAMICS/ASTRONAUTICS, SAN DIEGO, CALIF.	GDA-VTP-P-
GENERAL DYNAMICS/ASTRONAUTICS, SAN DIEGO, CALIF.	GDA-XM-(NO.)-(NO.)(TN)
GENERAL DYNAMICS/ASTRONAUTICS, SAN DIEGO, CALIF.	GDA-ZDU-(NO.)-(NO.)-TN
GENERAL DYNAMICS/ASTRONAUTICS, SAN DIEGO, CALIF.	GDA-ZP-(NO.)-(NO.)-TN
GENERAL DYNAMICS/ASTRONAUTICS, SAN DIEGO, CALIF.	GDA-ZS-(NUMBER)-
GENERAL DYNAMICS/ASTRONAUTICS, SAN DIEGO, CALIF.	GDA-ZT-(NO.)-(NO.)-TN
GENERAL DYNAMICS/ASTRONAUTICS, SAN DIEGO, CALIF.	MRG-
GENERAL DYNAMICS/ASTRONAUTICS, SAN DIEGO, CALIF.	ZPH-
GENERAL DYNAMICS/CONVAIR, FORT WORTH, TEX.	MRT-
GENERAL DYNAMICS/CONVAIR, SAN DIEGO, CALIF. (REF.1)	27(LETTER)(NUMBERS)
GENERAL DYNAMICS/CONVAIR, SAN DIEGO, CALIF. (REF.1)	55(LETTER)(NUMBERS)
GENERAL DYNAMICS/CONVAIR, SAN DIEGO, CALIF. (REF.1)	69(LETTER)(NUMBERS)
GENERAL DYNAMICS/CONVAIR, SAN DIEGO, CALIF.	AE-(YEAR)-
GENERAL DYNAMICS/CONVAIR, SAN DIEGO, CALIF.	CONVAIR-
GENERAL DYNAMICS/CONVAIR, SAN DIEGO, CALIF.	COWL-TM-
GENERAL DYNAMICS/CONVAIR, SAN DIEGO, CALIF.	CVC-
GENERAL DYNAMICS/CONVAIR, SAN DIEGO, CALIF.	DCM(YEAR)-
GENERAL DYNAMICS/CONVAIR, SAN DIEGO, CALIF.	ERR-AN-
GENERAL DYNAMICS/CONVAIR, SAN DIEGO, CALIF.	ERR-SD-
GENERAL DYNAMICS/CONVAIR, SAN DIEGO, CALIF.	GDA-
GENERAL DYNAMICS/CONVAIR, SAN DIEGO, CALIF.	GD/A-DBG-(NUMBER)-
GENERAL DYNAMICS/CONVAIR, SAN DIEGO, CALIF.	GD/A-DDG-
GENERAL DYNAMICS/CONVAIR, SAN DIEGO, CALIF.	GD/C-
GENERAL DYNAMICS/CONVAIR, SAN DIEGO, CALIF.	GDC-
GENERAL DYNAMICS/CONVAIR, SAN DIEGO, CALIF.	GDC-(NO.)-(LETTER)-
GENERAL DYNAMICS/CONVAIR, SAN DIEGO, CALIF.	GD-C-(YEAR)-
GENERAL DYNAMICS/CONVAIR, SAN DIEGO, CALIF.	GDC-ACX(YEAR)-
GENERAL DYNAMICS/CONVAIR, SAN DIEGO, CALIF.	GDC-AWP-(YEAR)-
GENERAL DYNAMICS/CONVAIR, SAN DIEGO, CALIF.	GDC-AWV(YEAR)-
GENERAL DYNAMICS/CONVAIR, SAN DIEGO, CALIF.	GDC-AZD-(YEAR)-
GENERAL DYNAMICS/CONVAIR, SAN DIEGO, CALIF.	GDC-BFK(YEAR)-
GENERAL DYNAMICS/CONVAIR, SAN DIEGO, CALIF.	GDC-BKM(YEAR)-
GENERAL DYNAMICS/CONVAIR, SAN DIEGO, CALIF.	GDC-BNZ(YEAR)-
GENERAL DYNAMICS/CONVAIR, SAN DIEGO, CALIF.	GD/C-BTD(YEAR)-
GENERAL DYNAMICS/CONVAIR, SAN DIEGO, CALIF.	GDC-BTD(YEAR)-
GENERAL DYNAMICS/CONVAIR, SAN DIEGO, CALIF.	GDC/C-ERR-AN-
GENERAL DYNAMICS/CONVAIR, SAN DIEGO, CALIF.	GD/C-CHB-(YEAR)-
GENERAL DYNAMICS/CONVAIR, SAN DIEGO, CALIF.	GD/C-CT-(NO.)B-(NO.)-
GENERAL DYNAMICS/CONVAIR, SAN DIEGO, CALIF.	GDC-CT-(NUMBER-NUMBER)
GENERAL DYNAMICS/CONVAIR, SAN DIEGO, CALIF.	GD/C-DBB-
GENERAL DYNAMICS/CONVAIR, SAN DIEGO, CALIF.	GD/C-DBE-
GENERAL DYNAMICS/CONVAIR, SAN DIEGO, CALIF.	GD/C-DBE(YEAR)-
GENERAL DYNAMICS/CONVAIR, SAN DIEGO, CALIF.	GDC-DBG(YEAR)-
GENERAL DYNAMICS/CONVAIR, SAN DIEGO, CALIF.	GD/C-DCB-
GENERAL DYNAMICS/CONVAIR, SAN DIEGO, CALIF.	GDC-DCB(YEAR)-
GENERAL DYNAMICS/CONVAIR, SAN DIEGO, CALIF.	GDC-DCD(YEAR)-
GENERAL DYNAMICS/CONVAIR, SAN DIEGO, CALIF.	GDC-DDB(YEAR)-
GENERAL DYNAMICS/CONVAIR, SAN DIEGO, CALIF.	GDC-DDC(YEAR)-
GENERAL DYNAMICS/CONVAIR, SAN DIEGO, CALIF.	GDC-DDE(YEAR)-

GENERAL DYNAMICS/CONVAIR

464

GENERAL DYNAMICS/CONVAIR, SAN DIEGO, CALIF.	GDC-DDG(YEAR)-
GENERAL DYNAMICS/CONVAIR, SAN DIEGO, CALIF.	GDC-DG-G-
GENERAL DYNAMICS/CONVAIR, SAN DIEGO, CALIF.	GDC-ERR-AN-
GENERAL DYNAMICS/CONVAIR, SAN DIEGO, CALIF.	GDC-ERR-SD-
GENERAL DYNAMICS/CONVAIR, SAN DIEGO, CALIF.	GDC-HST-TR-(NO.)-
GENERAL DYNAMICS/CONVAIR, SAN DIEGO, CALIF.	GDC-OR-P-
GENERAL DYNAMICS/CONVAIR, SAN DIEGO, CALIF.	GD/C-SLV-(NO.)-(NO.)-
GENERAL DYNAMICS/CONVAIR, SAN DIEGO, CALIF.	GDC-ZA-(NUMBER)-
GENERAL DYNAMICS/CONVAIR, SAN DIEGO, CALIF.	GDC-ZC-(NUMBER)-
GENERAL DYNAMICS/CONVAIR, SAN DIEGO, CALIF.	GDC-ZJ-(NO.)-(NO.)-TN
GENERAL DYNAMICS/CONVAIR, SAN DIEGO, CALIF.	GDC-ZM-(NO.)-(NO.)TN
GENERAL DYNAMICS/CONVAIR, SAN DIEGO, CALIF.	GDC-ZP-(NO.)-(NO.)-
GENERAL DYNAMICS/CONVAIR, SAN DIEGO, CALIF.	GDC-ZY-(NUMBER)-
GENERAL DYNAMICS/CONVAIR, SAN DIEGO, CALIF.	GDC-ZZC-(YEAR)-
GENERAL DYNAMICS/CONVAIR, SAN DIEGO, CALIF.	GD/O AOK-(YEAR)-
GENERAL DYNAMICS/CONVAIR, SAN DIEGO, CALIF.	GD/PAPER-
GENERAL DYNAMICS/CONVAIR, SAN DIEGO, CALIF.	GD/QPR-
GENERAL DYNAMICS/CONVAIR, SAN DIEGO, CALIF.	GD ZPH-
GENERAL DYNAMICS/CONVAIR, SAN DIEGO, CALIF.	GD ZR-AP-
GENERAL DYNAMICS/CONVAIR, SAN DIEGO, CALIF.	MP-(YEAR)-
GENERAL DYNAMICS/CONVAIR, SAN DIEGO, CALIF.	MPD-
GENERAL DYNAMICS/CONVAIR, SAN DIEGO, CALIF.	PCL-
GENERAL DYNAMICS/CONVAIR, SAN DIEGO, CALIF.	PH-
GENERAL DYNAMICS/CONVAIR, SAN DIEGO, CALIF.	PIN-(YEAR)-
GENERAL DYNAMICS/CONVAIR, SAN DIEGO, CALIF.	SL(YEAR)-
GENERAL DYNAMICS/CONVAIR, SAN DIEGO, CALIF.	SPEC-O-
GENERAL DYNAMICS/CONVAIR, SAN DIEGO, CALIF.	TG-
GENERAL DYNAMICS/CONVAIR, SAN DIEGO, CALIF.	ZR-AP-
GENERAL DYNAMICS/CONVAIR, SAN DIEGO, CALIF.	ZZL-
GENERAL DYNAMICS CORP., SAN DIEGO, CALIF.	U-
GENERAL DYNAMICS CORP. ELECTRIC BOAT DIV., GROTON, CONN	C-
GENERAL DYNAMICS CORP. ELECTRIC BOAT DIV., GROTON, CONN	C(NO.)-
GENERAL DYNAMICS CORP. ELECTRIC BOAT DIV., GROTON, CONN	C(NUMBER)-(YEAR)-
GENERAL DYNAMICS CORP. ELECTRIC BOAT DIV., GROTON, CONN	D(NO.)-
GENERAL DYNAMICS CORP. ELECTRIC BOAT DIV., GROTON, CONN	EBG/S/(NUMBER/NUMBER)
GENERAL DYNAMICS CORP. ELECTRIC BOAT DIV., GROTON, CONN	GD-
GENERAL DYNAMICS CORP. ELECTRIC BOAT DIV., GROTON, CONN	GD/TDNO-
GENERAL DYNAMICS CORP. ELECTRIC BOAT DIV., GROTON, CONN	PDE-
GENERAL DYNAMICS CORP. ELECTRIC BOAT DIV., GROTON, CONN	RAS-
GENERAL DYNAMICS CORP. ELECTRIC BOAT DIV., GROTON, CONN	RAS-MEMO-
GENERAL DYNAMICS CORP. ELECTRIC BOAT DIV., GROTON, CONN	SPD-
GENERAL DYNAMICS CORP. ELECTRIC BOAT DIV., GROTON, CONN	TDNO-
GENERAL DYNAMICS CORP. ELECTRIC BOAT DIV., GROTON, CONN	TSD-
GENERAL DYNAMICS CORP. ELECTRIC BOAT DIV., GROTON, CONN	U(NUMBER)-
GENERAL DYNAMICS CORP. ELECTRIC BOAT DIV., GROTON, CONN	U(NUMBER)-(YEAR)-
GENERAL DYNAMICS CORP. ELECTRIC BOAT DIV., SAN DIEGO, CALIF.	C(NUMBER)-(YEAR)-
GENERAL DYNAMICS CORP. LIQUID CARBONIC DIV., CHICAGO	GD/LCD-
GENERAL DYNAMICS/DAINGERFIELD. ORDNANCE AEROPHYSICS LAB., TEX.	CM-
GENERAL DYNAMICS/DAINGERFIELD. ORDNANCE AEROPHYSICS LAB., TEX.	OAL-
GENERAL DYNAMICS/DAINGERFIELD. ORDNANCE AEROPHYSICS LAB., TEX.	OAL-CF-
GENERAL DYNAMICS/ELECTRONICS, ROCHESTER, N.Y.	GD-
GENERAL DYNAMICS/ELECTRONICS, ROCHESTER, N.Y.	P-
GENERAL DYNAMICS/ELECTRONICS, SAN DIEGO, CALIF.	ER-
GENERAL DYNAMICS/ELECTRONICS, SAN DIEGO, CALIF.	GD-R-
GENERAL DYNAMICS/ELECTRONICS, SAN DIEGO, CALIF.	R-(YEAR-NUMBER)-
GENERAL DYNAMICS/ELECTRONICS. ACOUSTICS DEPT., ROCHESTER, N.Y.	AC-(3 DIGITS-YEAR)
GENERAL DYNAMICS/ELECTRONICS. HYDROACOUSTICS LAB., ROCHESTER, N.Y.	HL-(NUMBER)-(YEAR)
GENERAL DYNAMICS/FORT WORTH, TEX.	ERR-FW-
GENERAL DYNAMICS/FORT WORTH, TEX.	FE-
GENERAL DYNAMICS/FORT WORTH, TEX.	FGT-
GENERAL DYNAMICS/FORT WORTH, TEX.	FMR-(YEAR)-
GENERAL DYNAMICS/FORT WORTH, TEX.	FMS-
GENERAL DYNAMICS/FORT WORTH, TEX.	FPR-
GENERAL DYNAMICS/FORT WORTH, TEX.	FS-
GENERAL DYNAMICS/FORT WORTH, TEX.	FTDM-
GENERAL DYNAMICS/FORT WORTH, TEX.	FZA-
GENERAL DYNAMICS/FORT WORTH, TEX.	FZC-
GENERAL DYNAMICS/FORT WORTH, TEX.	FZD-
GENERAL DYNAMICS/FORT WORTH, TEX.	FZE-
GENERAL DYNAMICS/FORT WORTH, TEX.	FZI-
GENERAL DYNAMICS/FORT WORTH, TEX.	FZK-
GENERAL DYNAMICS/FORT WORTH, TEX.	FZM-
GENERAL DYNAMICS/FORT WORTH, TEX.	FZP-
GENERAL DYNAMICS/FORT WORTH, TEX.	FZS-
GENERAL DYNAMICS/FORT WORTH, TEX.	FZT-
GENERAL DYNAMICS/FORT WORTH, TEX.	GD/FW-
GENERAL DYNAMICS/FORT WORTH, TEX.	GD/FW FZK-
GENERAL DYNAMICS/FORT WORTH, TEX.	GD/FW FZM-
GENERAL DYNAMICS/FORT WORTH, TEX.	GD/FW MR-N-
GENERAL DYNAMICS/FORT WORTH, TEX.	GD FZK-
GENERAL DYNAMICS/FORT WORTH, TEX.	GD MR-N-
GENERAL DYNAMICS/FORT WORTH, TEX.	MR-N-
GENERAL DYNAMICS/FORT WORTH, TEX.	NARF-
GENERAL DYNAMICS/FORT WORTH, TEX. (REF. 3)	NARF-(YEAR)-
GENERAL DYNAMICS/FORT WORTH, TEX.	OSP-
GENERAL DYNAMICS/FORT WORTH, TEX.	PCIR-
GENERAL DYNAMICS/FORT WORTH, TEX.	PCTR-
GENERAL DYNAMICS/FORT WORTH. AEROSYSTEMS LAB., TEX.	ZJ-(NUMBER)-
GENERAL DYNAMICS/FORT WORTH. CONVAIR AEROSPACE DIV., TEX.	GDC-AAX(YEAR)-
GENERAL DYNAMICS/FORT WORTH. NUCLEAR AEROSPACE RESEARCH FACILITY, TEX.	FK-
GENERAL DYNAMICS/FORT WORTH. NUCLEAR AEROSPACE RESEARCH FACILITY, TEX.	FK-(NUMBER)-
GENERAL DYNAMICS/POMONA, CALIF.	CONVAIR-
GENERAL DYNAMICS/POMONA, CALIF.	CR-6-(NO.-NO.-NO.)
GENERAL DYNAMICS/POMONA, CALIF.	ERR-PO-
GENERAL DYNAMICS/POMONA, CALIF.	GDC-ERR-PO-(NO.)-
GENERAL DYNAMICS/POMONA, CALIF.	GDCR-

GENERAL DYNAMICS/POMONA, CALIF.	GD/P-
GENERAL DYNAMICS/POMONA, CALIF.	M-(NO.)-(NO.)-
GENERAL DYNAMICS/POMONA, CALIF.	TM-(NO.)-(NO.)-
GENERAL ELECTRIC CO. (ALL LOCATIONS (REFS. 1 AND 46)	(YEAR)(LETTERS)-
GENERAL ELECTRIC CO. (ALL LOCATIONS)	GE(YEAR)(LETTERS)-
GENERAL ELECTRIC CO. (ALL LOCATIONS)	GE-
GENERAL ELECTRIC CO. (ALL LOCATIONS)(AEC FUNDED WORK)	GEAP-
GENERAL ELECTRIC CO. (ALL LOCATIONS)	GEMP-
GENERAL ELECTRIC CO. (ALL LOCATIONS) (REF. 46)	GE-R(YEAR)(LETTERS)-
GENERAL ELECTRIC CO. (ALL LOCATIONS)	GER(YEAR)(LTRS.)(NOS.)
GENERAL ELECTRIC CO. (ALL LOCATIONS) (REF. 46)	GE-RM(YEAR)(LETTERS)-
GENERAL ELECTRIC CO. (ALL LOCATIONS) (REF. 46)	R(YEAR)(LETTERS)-
GENERAL ELECTRIC CO. (ALL LOCATIONS) (REF. 46)	RM(YEAR)(LETTERS)-
GENERAL ELECTRIC CO., BETHESDA, MD.	DOC.-(YEAR)(LTRS.)(NO.)
GENERAL ELECTRIC CO., BETHESDA, MD.	FR1-
GENERAL ELECTRIC CO., CINCINNATI	FAA-(LETTERS)(YEAR)-
GENERAL ELECTRIC CO., CINCINNATI	TPR-
GENERAL ELECTRIC CO., EVENDALE, OHIO	DF(NO.)AGT(NO.)
GENERAL ELECTRIC CO., ITHACA, N.Y.	IT-
GENERAL ELECTRIC CO., LYNN, MASS.	GE-TM(YEAR)SE-
GENERAL ELECTRIC CO., N.Y.C.	GE-ANPD-DC-
GENERAL ELECTRIC CO., OWENSBORO, KY.	RCS-OSD-
GENERAL ELECTRIC CO., OWENSBORO, KY.	S-(YEAR)-
GENERAL ELECTRIC CO., PALO ALTO, CALIF.	R(YR)ELM(NO.S)-
GENERAL ELECTRIC CO., PHILADELPHIA	ATOM-
GENERAL ELECTRIC CO., PHILADELPHIA	DOC-(YEAR)(LTRS.)(NOS.)
GENERAL ELECTRIC CO., PHILADELPHIA	GEMS-NS-
GENERAL ELECTRIC CO., PHILADELPHIA	GE-RS-(YEAR)-
GENERAL ELECTRIC CO., PHILADELPHIA	MMM-(NUMBERS)
GENERAL ELECTRIC CO., PHILADELPHIA	MSD-
GENERAL ELECTRIC CO., PHILADELPHIA (THERMOELECTRIC OUTER PLANET SPACECRAFT)	REPT-(NO.LTR.NO.)-TOPS-
GENERAL ELECTRIC CO., PHILADELPHIA	TRA-(NUMBER)-(NUMBER)
GENERAL ELECTRIC CO., PITTSFIELD, MASS.	GE/OD-R-
GENERAL ELECTRIC CO., PITTSFIELD, MASS.	HVL-(YEAR)-
GENERAL ELECTRIC CO., SANTA BARBARA, CALIF.	WEDCOM-
GENERAL ELECTRIC CO., SCHENECTADY, N.Y. (DATA FOLDER)	DF-
GENERAL ELECTRIC CO., SCHENECTADY, N.Y.	DF-(YR LAB.)-
GENERAL ELECTRIC CO., SCHENECTADY, N.Y.	DF-(YR)-(LETTER(S))-
GENERAL ELECTRIC CO., SCHENECTADY, N.Y.	EH-
GENERAL ELECTRIC CO., SCHENECTADY, N.Y.	GE-(INITIALS)-
GENERAL ELECTRIC CO., SCHENECTADY, N.Y.	GEA-
GENERAL ELECTRIC CO., SCHENECTADY, N.Y. (INSTRUCTION)	GEI-
GENERAL ELECTRIC CO., SCHENECTADY, N.Y.	GE-IY-
GENERAL ELECTRIC CO., SCHENECTADY, N.Y.	GEJ-
GENERAL ELECTRIC CO., SCHENECTADY, N.Y.	GE-KAK-
GENERAL ELECTRIC CO., SCHENECTADY, N.Y.	GE-LT-
GENERAL ELECTRIC CO., SCHENECTADY, N.Y.	GE-NUC-LIB-
GENERAL ELECTRIC CO., SCHENECTADY, N.Y.	GER-
GENERAL ELECTRIC CO., SCHENECTADY, N.Y.	GES-
GENERAL ELECTRIC CO., SCHENECTADY, N.Y.	GE-SN-
GENERAL ELECTRIC CO., SCHENECTADY, N.Y.	GET-
GENERAL ELECTRIC CO., SCHENECTADY, N.Y.(PROJ. HERMES)	HERMES-
GENERAL ELECTRIC CO., SCHENECTADY, N.Y.	MEMO-(INITIALS)-
GENERAL ELECTRIC CO., SCHENECTADY, N.Y.	ME-Q-
GENERAL ELECTRIC CO., SCHENECTADY, N.Y.	R(YR)AE-
GENERAL ELECTRIC CO., SCHENECTADY, N.Y.	R-(NUMBER)Q-
GENERAL ELECTRIC CO., SCHENECTADY, N.Y.	S-(YEAR)-
GENERAL ELECTRIC CO., SCHENECTADY, N.Y.	50TP 120-
GENERAL ELECTRIC CO., SCHENECTADY, N.Y.	TR-
GENERAL ELECTRIC CO., SUNNYVALE, CALIF.	GE-MPL-
GENERAL ELECTRIC CO., SYRACUSE, N.Y. (AIR WEAPONS CONTROL SYSTEM, COMPUTER PROGRAM INFORMATION)	AWCS-GCP-
GENERAL ELECTRIC CO., SYRACUSE, N.Y.	EH-
GENERAL ELECTRIC CO., SYRACUSE, N.Y.	GACD-
GENERAL ELECTRIC CO., SYRACUSE, N.Y.	SAR-
GENERAL ELECTRIC CO. ADVANCE PRODUCT OPERATION, SAN JOSE, CALIF.	GEAO-
GENERAL ELECTRIC CO. ADVANCE PRODUCT OPERATION, SAN JOSE, CALIF. (CUSTOMER REPORT)	GECD-
GENERAL ELECTRIC CO. ADVANCED ELECTRONICS CENTER, ITHACA, N.Y.	CH-CPR-
GENERAL ELECTRIC CO. ADVANCED ELECTRONICS CENTER, ITHACA, N.Y.	GE R(YEAR)ELC(NO.-NO.)
GENERAL ELECTRIC CO. ADVANCED ELECTRONICS CENTER, ITHACA, N.Y.	R(YR)ELC-
GENERAL ELECTRIC CO. ADVANCED ENGINE AND TECHNOLOGY DEPT., CINCINNATI	DM(YEAR)-
GENERAL ELECTRIC CO. ADVANCED ENGINE AND TECHNOLOGY DEPT., EVENDALE, OHIO	R(YEAR)FPD-
GENERAL ELECTRIC CO. ADVANCED TECHNICAL AND DEMONSTRATOR PROGRAMS DEPT., CINCINNATI	R(YEAR)FPD(NO.)
GENERAL ELECTRIC CO. ADVANCED TECHNOLOGY LABS., SCHENECTADY, N.Y.	ATL-
GENERAL ELECTRIC CO. AERONAUTIC AND ORDNANCE SYSTEMS DIV., SCHENECTADY, N.Y.	GE-GET-
GENERAL ELECTRIC CO. AERONAUTIC AND ORDNANCE SYSTEMS DIV., SCHENECTADY, N.Y. (PROJECT HERMES)	GE-R(YR.)A-
GENERAL ELECTRIC CO. AERONAUTIC AND ORDNANCE SYSTEMS DIV., SCHENECTADY, N.Y.	MC-AP-
GENERAL ELECTRIC CO. AERONAUTIC AND ORDNANCE SYSTEMS DIV., SCHENECTADY, N.Y.	R(YR)AO-
GENERAL ELECTRIC CO. AEROSPACE ELECTRONICS DEPT., UTICA, N.Y.	AED-
GENERAL ELECTRIC CO. AEROSPACE ELECTRONICS DEPT., UTICA, N.Y.	LME-
GENERAL ELECTRIC CO. AIRCRAFT ENGINE GROUP, CINCINNATI	R(YEAR)AEG(NUMBER)
GENERAL ELECTRIC CO. AIRCRAFT ENGINE GP., LYNN, MASS.	LM-E-(NUMBER)-
GENERAL ELECTRIC CO. AIRCRAFT GAS TURBINE DIV.	ENG. PROGRESS RPT.-
GENERAL ELECTRIC CO. AIRCRAFT GAS TURBINE DIV., CINCINNATI	AGT-

GENERAL ELECTRIC CO. AIRCRAFT GAS TURBINE DIV.

466

GENERAL ELECTRIC CO. AIRCRAFT GAS TURBINE DIV., EVENDALE, OHIO	DF-(YR.)(LTRS.)(NO.)-
GENERAL ELECTRIC CO. AIRCRAFT GAS TURBINE DIV., EVENDALE, OHIO	DF-(YEAR)-AGT-
GENERAL ELECTRIC CO. AIRCRAFT GAS TURBINE DIV., EVENDALE, OHIO	R(YR)-(LETTERS)
GENERAL ELECTRIC CO. AIRCRAFT GAS TURBINE DIV., EVENDALE, OHIO	R(YR)AGT-
(SEE ALSO LATER NAME: GENERAL ELECTRIC CO. FLIGHT PROPULSION...)	
GENERAL ELECTRIC CO. AIRCRAFT NUCLEAR PROPULSION DEPT., CINCINNATI (REF. 3)	ANP-
GENERAL ELECTRIC CO. AIRCRAFT NUCLEAR PROPULSION DEPT., CINCINNATI	D-
GENERAL ELECTRIC CO. AIRCRAFT NUCLEAR PROPULSION DEPT., CINCINNATI (REF. 3)	DC-(YEAR)-(NUMBER)-
GENERAL ELECTRIC CO. AIRCRAFT NUCLEAR PROPULSION DEPT., CINCINNATI	DCL-
GENERAL ELECTRIC CO. AIRCRAFT NUCLEAR PROPULSION DEPT., CINCINNATI	DCL-(YR.)-
GENERAL ELECTRIC CO. AIRCRAFT NUCLEAR PROPULSION DEPT., CINCINNATI	DCM-(YR)-(NO.)-
GENERAL ELECTRIC CO. AIRCRAFT NUCLEAR PROPULSION DEPT., CINCINNATI	DCR-
GENERAL ELECTRIC CO. AIRCRAFT NUCLEAR PROPULSION DEPT., CINCINNATI	DS-
GENERAL ELECTRIC CO. AIRCRAFT NUCLEAR PROPULSION DEPT., CINCINNATI (REF. 3)	GE-APEX-
GENERAL ELECTRIC CO. AIRCRAFT NUCLEAR PROPULSION DEPT., CINCINNATI (REF. 3)	GEO-
GENERAL ELECTRIC CO. AIRCRAFT NUCLEAR PROPULSION DEPT., CINCINNATI (REF. 3)	GE-DC-
GENERAL ELECTRIC CO. AIRCRAFT NUCLEAR PROPULSION DEPT., CINCINNATI (REF. 3)	GERD-
GENERAL ELECTRIC CO. AIRCRAFT NUCLEAR PROPULSION DEPT., CINCINNATI (REF. 3)	GE-XDC-
GENERAL ELECTRIC CO. AIRCRAFT NUCLEAR PROPULSION DEPT., CINCINNATI	GV-
GENERAL ELECTRIC CO. AIRCRAFT NUCLEAR PROPULSION DEPT., CINCINNATI	IBM-
GENERAL ELECTRIC CO. AIRCRAFT NUCLEAR PROPULSION DEPT., CINCINNATI	NMI-
GENERAL ELECTRIC CO. AIRCRAFT NUCLEAR PROPULSION DEPT., CINCINNATI (REF. 3)	PREDC-
GENERAL ELECTRIC CO. AIRCRAFT NUCLEAR PROPULSION DEPT., CINCINNATI	TMDR-
GENERAL ELECTRIC CO. AIRCRAFT NUCLEAR PROPULSION DEPT., CINCINNATI	TMR-
GENERAL ELECTRIC CO. AIRCRAFT NUCLEAR PROPULSION DEPT., CINCINNATI	XDC-
GENERAL ELECTRIC CO. AIRCRAFT NUCLEAR PROPULSION DEPT., CINCINNATI (REF. 3)	XDCL-
GENERAL ELECTRIC CO. AIRCRAFT NUCLEAR PROPULSION DEPT., IDAHO FALLS, IDAHO (REF. 3)	DC-(YEAR)-(NUMBER)-
GENERAL ELECTRIC CO. AIRCRAFT NUCLEAR PROPULSION PROJECT, CINCINNATI (REF. 3)	CARP-A-
GENERAL ELECTRIC CO. AIRCRAFT NUCLEAR PROPULSION PROJECT, CINCINNATI	ER-
GENERAL ELECTRIC CO. AIRCRAFT NUCLEAR PROPULSION PROJECT, CINCINNATI (REF. 3)	GE-ANP-
GENERAL ELECTRIC CO. AIRCRAFT NUCLEAR PROPULSION PROJECT, CINCINNATI (ENG. PROGRESS RPT.) (REF. 3)	GE-ANP-APEX-
GENERAL ELECTRIC CO. AIRCRAFT NUCLEAR PROPULSION PROJECT, CINCINNATI (REF. 3)	GEDR-
GENERAL ELECTRIC CO. AIRCRAFT NUCLEAR PROPULSION PROJECT, CINCINNATI (REF. 3)	GEXDC (YEAR)-
GENERAL ELECTRIC CO. AIRCRAFT NUCLEAR PROPULSION PROJECT, CINCINNATI (REF. 3)	GEXDCL (YEAR)-
GENERAL ELECTRIC CO. AIRCRAFT NUCLEAR PROPULSION PROJECT, CINCINNATI (REF. 3)	PREXDC-
GENERAL ELECTRIC CO. AIRCRAFT NUCLEAR PROPULSION PROJECT, CINCINNATI (REF. 3)	XDC-(YEAR)-(MONTH)-
GENERAL ELECTRIC CO. APOLLO SUPPORT DEPT., DAYTONA BEACH, FLA.	DB(YEAR)D(NUMBER)-
GENERAL ELECTRIC CO. APOLLO SUPPORT DEPT., DAYTONA BEACH, FLA.	GE(YEAR)ASD(NUMBER)-
GENERAL ELECTRIC CO. APOLLO SUPPORT DEPT., DAYTONA BEACH, FLA.	(YEAR)-SIM-(NUMBER/NO.)
GENERAL ELECTRIC CO. APOLLO SUPPORT DEPT., DAYTONA BEACH, FLA.	(YEAR)-SIMAR-(NO./NO.)
GENERAL ELECTRIC CO. ARMAMENT AND CONTROL PRODUCTS SECTION, JOHNSON CITY, N.Y.	LMEJ-
GENERAL ELECTRIC CO. ATOMIC PRODUCTS DIV., CINCINNATI	X-GEAP-
GENERAL ELECTRIC CO. ATOMIC PRODUCTS DIV., SCHENECTADY, N.Y.	APE-
GENERAL ELECTRIC CO. AVIONIC CONTROLS DEPT., BINGHAMTON, N.Y.	ACD-
GENERAL ELECTRIC CO. AVIONIC CONTROLS DEPT., BINGHAMTON, N.Y.	JCTR-
GENERAL ELECTRIC CO. CAPACITOR DEPT., HUDSON FALLS, N.Y.	HF-(YEAR)-
GENERAL ELECTRIC CO. COMMUNICATION PRODUCTS DEPT., LYNCHBURG, VA.	CPD-(NUMBER)-
GENERAL ELECTRIC CO. COMMUNICATION PRODUCTS DEPT., LYNCHBURG, VA.	PEM-
GENERAL ELECTRIC CO. DASA DATA CENTER, SANTA BARBARA, CALIF.	B-
GENERAL ELECTRIC CO. DASA DATA CENTER, SANTA BARBARA, CALIF.	GE-DDC-
GENERAL ELECTRIC CO. DASA DATA CENTER, SANTA BARBARA, CALIF.	GE/DDC-SPEC BIB-
GENERAL ELECTRIC CO. DASA DATA CENTER, SANTA BARBARA, CALIF.	GE/DDC-SPEC RPT-
GENERAL ELECTRIC CO. DASA DATA CENTER, SANTA BARBARA, CALIF.	GE/TEMPO-DDC-
GENERAL ELECTRIC CO. DASA INFORMATION AND ANALYSIS CENTER, SANTA BARBARA, CALIF.	DASIAC-B(YEAR)-
GENERAL ELECTRIC CO. DASA INFORMATION AND ANALYSIS CENTER, SANTA BARBARA, CALIF.	DASIAC-SB-
GENERAL ELECTRIC CO. DASA INFORMATION AND ANALYSIS CENTER, SANTA BARBARA, CALIF.	DASIAC-SR-
GENERAL ELECTRIC CO. DASA INFORMATION AND ANALYSIS CENTER, SANTA BARBARA, CALIF. (SPECIAL BIBS.)	GE-DASIAC-SB-
GENERAL ELECTRIC CO. DASA INFORMATION AND ANALYSIS CENTER, SANTA BARBARA, CALIF.	GE-DASIAC-SR-
GENERAL ELECTRIC CO. DASA INFORMATION AND ANALYSIS CENTER, SANTA BARBARA, CALIF.	SB-
GENERAL ELECTRIC CO. DIRECT ENERGY CONVERSION OPERATION, WEST LYNN, MASS.	TIS(YEAR)DE(NUMBER)
GENERAL ELECTRIC CO. ELECTRONICS LAB., SYRACUSE, N.Y.	(NUMBER)-CO-
GENERAL ELECTRIC CO. ELECTRONICS LAB., SYRACUSE, N.Y.	DEL-(NUMBER)-DD-
GENERAL ELECTRIC CO. ELECTRONICS LAB., SYRACUSE, N.Y.	DF(NO.)(LTRS.)-
GENERAL ELECTRIC CO. ELECTRONICS LAB., SYRACUSE, N.Y.	(NUMBER)-GM-
GENERAL ELECTRIC CO. ELECTRONICS LAB., SYRACUSE, N.Y.	(NUMBER)-TH-
GENERAL ELECTRIC CO. FLIGHT PROPULSION DIV., CINCINNATI	GE-R(YEAR)FPD(NO.)
GENERAL ELECTRIC CO. FLIGHT PROPULSION DIV., CINCINNATI	P(NUMBER)-
GENERAL ELECTRIC CO. FLIGHT PROPULSION DIV., CINCINNATI	R(YEAR)FPD(NUMBER)-
GENERAL ELECTRIC CO. FLIGHT PROPULSION LAB DEPT., CINCINNATI	APEX-
GENERAL ELECTRIC CO. FLIGHT PROPULSION LAB. DEPT., CINCINNATI	DM-
GENERAL ELECTRIC CO. FLIGHT PROPULSION LAB. DEPT., CINCINNATI	DM-(YEAR)-
GENERAL ELECTRIC CO. FLIGHT PROPULSION LAB. DEPT., CINCINNATI	GE-GS-
GENERAL ELECTRIC CO. FLIGHT PROPULSION LAB. DEPT., CINCINNATI	GE-NMP-
GENERAL ELECTRIC CO. FLIGHT PROPULSION LAB. DEPT., CINCINNATI	GE-RE-
GENERAL ELECTRIC CO. FLIGHT PROPULSION LAB. DEPT., CINCINNATI	GE-TMS-
GENERAL ELECTRIC CO. FLIGHT PROPULSION LAB. DEPT., CINCINNATI	NMP-HTMP-
GENERAL ELECTRIC CO. FLIGHT PROPULSION LAB. DEPT., CINCINNATI	R(NUMBER)-
GENERAL ELECTRIC CO. FLIGHT PROPULSION LAB. DEPT., CINCINNATI (ROCKET NOZZLE MATERIALS)	RNM-
GENERAL ELECTRIC CO. FLIGHT PROPULSION LAB. DEPT., CINCINNATI	TP-
GENERAL ELECTRIC CO. FLIGHT PROPULSION LAB. DEPT., EVENDALE, OHIO	C-(NUMBER)-C-
GENERAL ELECTRIC CO. FLIGHT PROPULSION LAB. DEPT., EVENDALE, OHIO	GE-C-(NUMBER)-C-(NUMBER)
GENERAL ELECTRIC CO. FORT WAYNE LAB., IND.	FWL-(NUMBER-NUMBER)-
GENERAL ELECTRIC CO. FORT WAYNE LAB., IND.	GE-FWL-
GENERAL ELECTRIC CO. GEN. ENG. + CONSULTING LAB., SCHENECTADY, N.Y. (ADVANCED NUCLEAR SYSTEMS ENG.)	ANSE-
GENERAL ELECTRIC CO. GENERAL ENGINEERING AND CONSULTING LAB., SCHENECTADY, N.Y.	DF-(YR)AO-
GENERAL ELECTRIC CO. GENERAL ENGINEERING AND CONSULTING LAB., SCHENECTADY, N.Y.	EAH-
GENERAL ELECTRIC CO. GENERAL ENGINEERING AND CONSULTING LAB., SCHENECTADY, N.Y.	GEL-
GENERAL ELECTRIC CO. GENERAL ENGINEERING AND CONSULTING LAB., SCHENECTADY, N.Y.	GE-TR-
GENERAL ELECTRIC CO. GENERAL ENGINEERING AND CONSULTING LAB., SCHENECTADY, N.Y.	GI-
GENERAL ELECTRIC CO. GENERAL ENGINEERING AND CONSULTING LAB., SCHENECTADY, N.Y.	GL-
GENERAL ELECTRIC CO. GENERAL ENGINEERING AND CONSULTING LAB., SCHENECTADY, N.Y.	R(YR)-GL-

GENERAL ELECTRIC CO. GEN. ENG. LAB., SCHENECTADY,N.Y.	DF-(YEAR)GL-
GENERAL ELECTRIC CO. GEN. ENG. LAB., SCHENECTADY,N.Y.	ELD-
GENERAL ELECTRIC CO. GEN. ENG. LAB., SCHENECTADY,N.Y.	GE-GEL-
GENERAL ELECTRIC CO. GEN. ENG. LAB., SCHENECTADY,N.Y.	GEL-
GENERAL ELECTRIC CO. GEN. ENG. LAB., SCHENECTADY,N.Y.	GEL-PR-
GENERAL ELECTRIC CO. GEN. ENG. LAB., SCHENECTADY,N.Y.	(YEAR)-GL-
GENERAL ELECTRIC CO. GEN. ENG. LAB., SCHENECTADY,N.Y.	JP-
GENERAL ELECTRIC CO. GEN. ENG. LAB., SCHENECTADY,N.Y.	R(YEAR)GL-
GENERAL ELECTRIC CO. GEN. ENG. LAB., SCHENECTADY,N.Y. (FIRST TWO DIGITS STAND FOR YEAR) (REF. 1)	65-RL-
GENERAL ELECTRIC CO. GUIDED MISSILES DEPT., SCHENECTADY, N.Y. (PROJECT HERMES)	GE-R(YR.)A-
(GENERAL ELECTRIC CO. HANFORD ATOMIC PRODUCTS OPERATION, SEE: HANFORD ATOMIC PRODUCTS OPERATION)	
GENERAL ELECTRIC CO. HEAVY MILITARY ELECTRONIC EQUIPMENT DIV., SYRACUSE, N.Y. (REF. 1)	(NUMBERS...)
GENERAL ELECTRIC CO. HEAVY MILITARY ELECTRONIC EQUIPMENT DIV., SYRACUSE, N.Y.	EHM-
GENERAL ELECTRIC CO. HEAVY MILITARY ELECTRONICS DEPT., SYRACUSE, N.Y.	AWCS-CS-
GENERAL ELECTRIC CO. HEAVY MILITARY ELECTRONICS DEPT., SYRACUSE, N.Y.(CONFORMAL/PLANAR ARRAY SONAR)	CO(NUMBER)-(NUMBER)-
GENERAL ELECTRIC CO. HEAVY MILITARY ELECTRONICS DEPT., SYRACUSE, N.Y.	GE-(NUMBER)-(NUMBER)
GENERAL ELECTRIC CO. HEAVY MILITARY ELECTRONICS DEPT., SYRACUSE, N.Y.	R(NO.)(LTRS.)(NO.)
GENERAL ELECTRIC CO. HEAVY MILITARY ELECTRONICS DEPT., SYRACUSE, N.Y.	SISS-
GENERAL ELECTRIC CO. INSULATOR DEPT., BALTIMORE	LK-
GENERAL ELECTRIC CO. INSULATOR DEPT., BALTIMORE	TR-LK-
GENERAL ELECTRIC CO. JET ENGINE DEPT., CINCINNATI	DF-(YEAR)-AGT-
GENERAL ELECTRIC CO. LIGHT MILITARY ELECTRONICS DEPT., ITHACA, N.Y.	GE-LMED-
GENERAL ELECTRIC CO. LIGHT MILITARY ELECTRONICS DEPT., SCHENECTADY, N.Y.	GE-LM-
GENERAL ELECTRIC CO. LIGHT MILITARY ELECTRONICS DEPT., SCHENECTADY, N.Y.	R(NO.)GML-
GENERAL ELECTRIC CO. LIGHT MILITARY ELECTRONICS DEPT., UTICA, N.Y.	LM-(NUMBER)-
GENERAL ELECTRIC CO. LIGHT MILITARY ELECTRONICS DEPT., UTICA, N.Y.	QRC-(NO.)-(NO.)(MOD)...
GENERAL ELECTRIC CO. LIGHT MILITARY ELECTRONICS DEPT., UTICA, N.Y.	R(NO.)EML(NO.)G
(SEE ALSO LATER NAMEO GENERAL ELECTRIC CO. AEROSPACE ELECTRONICS DEPT.)	
GENERAL ELECTRIC CO. LIGHT MILITARY ELECTRONICS EQUIPMENT DEPT., JOHNSON CITY, N.Y.	LMEJ-
GENERAL ELECTRIC CO. LIGHT MILITARY ELECTRONICS EQUIPMENT DEPT., SCHENECTADY, N.Y.	GE-LMED-
GENERAL ELECTRIC CO. LIGHT MILITARY ELECTRONICS EQUIPMENT DEPT., SCHENECTADY, N.Y.	GE R(YEAR)APS(NO.)(FR)
GENERAL ELECTRIC CO. MALTA TEST STA.,BALLSTON SPA,NY	GE-M-
GENERAL ELECTRIC CO. MARINE TURBINE AND GEAR DEPT., WEST LYNN, MASS.	DF(YEAR)MTG-
GENERAL ELECTRIC CO. MATERIALS AND PROCESSES LAB., SCHENECTADY, N.Y.	DF-
GENERAL ELECTRIC CO. MATERIALS AND PROCESSES LAB., SCHENECTADY, N.Y.	DF(YR)SL-
GENERAL ELECTRIC CO. MATERIALS DEVELOPMENT LAB. OPERATION, CINCINNATI	R(YEAR)FPD(NUMBER)-
GENERAL ELECTRIC CO. MATERIALS DEVELOPMENT LAB. OPERATION, CINCINNATI	TM(YEAR)-
GENERAL ELECTRIC CO. MEDIUM STEAM TURBINE GENERATOR AND GEAR DEPT., WEST LYNN, MASS.	DF-(YR)-MTG-
GENERAL ELECTRIC CO. METALLURGICAL PRODUCTS DEPT., DETROIT (USED 1953-1963)	G.E. (DATE)-CAR-
GENERAL ELECTRIC CO. METALLURGICAL PRODUCTS DEPT., DETROIT (USED 1964-)	G.E. (DATE)-MPD-
GENERAL ELECTRIC CO. MICROWAVE TUBE BUSINESS SECTION, SCHENECTADY, N.Y.	S(YEAR)-
GENERAL ELECTRIC CO. MILITARY COMMUNICATIONS DEPT., OKLAHOMA CITY	MCD-
GENERAL ELECTRIC CO. MILITARY COMMUNICATIONS DEPT., OKLAHOMA CITY	R(YEAR)MCD(NUMBER)
GENERAL ELECTRIC CO. MISSILE AND ORDNANCE SYSTEMS DEPT., PHILA. (AERODYNAMICS FUNDAMENTALS MEMO.)	GE-AFM-
GENERAL ELECTRIC CO. MISSILE AND ORDNANCE SYSTEMS DEPT., PHILADELPHIA	L/A (NO.)-(NO.)-(NO.)
(SEE ALSO LATER NAME: GENERAL ELECTRIC CO. MISSILE AND SPACE VEHICLE DEPT.)	
GENERAL ELECTRIC CO. MISSILE + SPACE DIV., CINCINNATI	G(NO.)(LTRS.)(NO.)
GENERAL ELECTRIC CO. MISSILE + SPACE DIV., CINCINNATI	R(YEAR)SD(NUMBER)
(SEE ALSO LATER NAME: GENERAL ELECTRIC CO. NUCLEAR SYSTEMS PROGRAMS)	
GENERAL ELECTRIC CO. MISSILE AND SPACE DIV., PHILA.	ANSO-
GENERAL ELECTRIC CO. MISSILE AND SPACE DIV., PHILA.	ATPR-(NUMBER-YEAR)-
GENERAL ELECTRIC CO. MISSILE AND SPACE DIV., PHILA.	DIN-(LTRS., NUMBERS)-
GENERAL ELECTRIC CO. MISSILE AND SPACE DIV., PHILA.	DIN-R(YEAR)SD(NO.)
GENERAL ELECTRIC CO. MISSILE AND SPACE DIV., PHILA.	(NUMBER-NO.)-EO-(NO.)-
GENERAL ELECTRIC CO. MISSILE AND SPACE DIV., PHILA.	GE-ANSO-(NUMBER)
GENERAL ELECTRIC CO. MISSILE AND SPACE DIV., PHILA.	GE-D-(YEAR)-
GENERAL ELECTRIC CO. MISSILE AND SPACE DIV., PHILA.	GEMS-
GENERAL ELECTRIC CO. MISSILE AND SPACE DIV., PHILA.	GE-N-
GENERAL ELECTRIC CO. MISSILE AND SPACE DIV., PHILA.	MSD-N-
GENERAL ELECTRIC CO. MISSILE AND SPACE DIV., PHILA.	PER-(NUMBER)-
GENERAL ELECTRIC CO. MISSILE AND SPACE DIV., PHILA.	PIR-(NUMBER)-
GENERAL ELECTRIC CO. MISSILE AND SPACE DIV., PHILA.	PRR-
GENERAL ELECTRIC CO. MISSILE AND SPACE DIV., PHILA.	PTR-(NUMBER)-
GENERAL ELECTRIC CO. MISSILE AND SPACE DIV., PHILA.	R(YEAR)SD-
GENERAL ELECTRIC CO. MISSILE AND SPACE DIV., PHILA.	R-(YEAR)-SD-
GENERAL ELECTRIC CO. MISSILE AND SPACE DIV., PHILA.	RSD-(YEAR)-
GENERAL ELECTRIC CO. MISSILE AND SPACE DIV., PHILA.	SCS-TME-(NUMBER)-
GENERAL ELECTRIC CO. MISSILE AND SPACE DIV., PHILA.	SSL-
GENERAL ELECTRIC CO. MISSILE AND SPACE DIV., PHILA.	TDM-(NUMBER)-
GENERAL ELECTRIC CO. MISSILE AND SPACE DIV., PHILA.	TM-
GENERAL ELECTRIC CO. MISSILE AND SPACE DIV., PHILA.	TRAP-
GENERAL ELECTRIC CO. MISSILE AND SPACE VEHICLE DEPT., PHILADELPHIA	AEROPHYSICS RES. MEMO-
GENERAL ELECTRIC CO. MISSILE AND SPACE VEHICLE DEPT., PHILADELPHIA	GE-(NUMBER-NUMBER)
GENERAL ELECTRIC CO. MISSILE AND SPACE VEHICLE DEPT., PHILADELPHIA	GE(YEAR)SD(NUMBER)
GENERAL ELECTRIC CO. MISSILE AND SPACE VEHICLE DEPT., PHILADELPHIA	GE-ATLAS-
GENERAL ELECTRIC CO. MISSILE AND SPACE VEHICLE DEPT., PHILADELPHIA	GE-DIN-(NO.)-(NO.)QPR
GENERAL ELECTRIC CO. MISSILE AND SPACE VEHICLE DEPT., PHILADELPHIA	GE-MSD-
GENERAL ELECTRIC CO. MISSILE AND SPACE VEHICLE DEPT., PHILADELPHIA	GE-QR-(NUMBER),AEC-(NO.)
GENERAL ELECTRIC CO. MISSILE AND SPACE VEHICLE DEPT., PHILADELPHIA	GE R(YEAR)SD(NUMBER)
GENERAL ELECTRIC CO. MISSILE AND SPACE VEHICLE DEPT., PHILADELPHIA	GE-THOR-
GENERAL ELECTRIC CO. MISSILE AND SPACE VEHICLE DEPT., PHILADELPHIA	L/A (NO.)-(NO.)-(NO.)
GENERAL ELECTRIC CO. MISSILE AND SPACE VEHICLE DEPT., PHILADELPHIA	MSVD-(YEAR)-SD-
GENERAL ELECTRIC CO. MISSILE AND SPACE VEHICLE DEPT., PHILADELPHIA	QR-
GENERAL ELECTRIC CO. MISSILE AND SPACE VEHICLE DEPT., PHILADELPHIA (REF. 1)	(YEAR)SD-
GENERAL ELECTRIC CO. MISSILE AND SPACE VEHICLE DEPT., PHILADELPHIA	TIS-R(YEAR)SD-
GENERAL ELECTRIC CO. MISSILE AND SPACE VEHICLE DEPT., SCHENECTADY, N.Y. (REF. 1)	(YEAR)SD-

GENERAL ELECTRIC CO. NEUTRON DEVICES DEPT.

GENERAL ELECTRIC CO. NEUTRON DEVICES DEPT., ST. PETERSBURG, FLA.	APIO-
GENERAL ELECTRIC CO. NEUTRON DEVICES DEPT., ST. PETERSBURG, FLA.	GEPP-
GENERAL ELECTRIC CO. NEUTRON DEVICES DEPT., ST. PETERSBURG, FLA.	R-(YEAR)-ND-
GENERAL ELECTRIC CO. NUCLEAR ENERGY DIV.,SAN JOSE,CAL (ATOMIC POWER EQUIPMENT)	APED-
GENERAL ELECTRIC CO. NUCLEAR ENERGY DIV.,SAN JOSE,CAL	APIO-
GENERAL ELECTRIC CO. NUCLEAR ENERGY DIV.,SAN JOSE,CAL (FUEL RECOVERY)	FRO-
GENERAL ELECTRIC CO. NUCLEAR ENERGY DIV.,SAN JOSE,CAL (CUSTOMER REPORT)	GECR-
GENERAL ELECTRIC CO. NUCLEAR ENERGY DIV.,SAN JOSE,CAL (IRRADIATION PROCESSING, CUSTOMER REPORT)	GEIC-
GENERAL ELECTRIC CO. NUCLEAR ENERGY DIV.,SAN JOSE,CAL (IRRADIATION PROCESSING, GOVT. CONTRACT WORK)	GEIR-
GENERAL ELECTRIC CO. NUCLEAR ENERGY DIV.,SAN JOSE,CAL	GESR-
GENERAL ELECTRIC CO. NUCLEAR ENERGY DIV.,SAN JOSE,CAL	GEST-
GENERAL ELECTRIC CO. NUCLEAR ENERGY DIV.,SAN JOSE,CAL	LAC-DA-
GENERAL ELECTRIC CO. NUCLEAR ENERGY DIV.,SAN JOSE,CAL	NEDC-
GENERAL ELECTRIC CO. NUCLEAR ENERGY DIV.,SAN JOSE,CAL	NEDG-
GENERAL ELECTRIC CO. NUCLEAR ENERGY DIV.,SAN JOSE,CAL	NEDO-
GENERAL ELECTRIC CO. NUCLEAR ENERGY DIV.,SAN JOSE,CAL (ATOMIC POWER EQUIPMENT)	R(YEAR)APE-
GENERAL ELECTRIC CO. NUCLEAR ENERGY DIV.,SAN JOSE,CAL (VALLECITOS REPORT)	SG-VAL-
GENERAL ELECTRIC CO. NUCLEAR ENERGY DIV.,SAN JOSE,CAL (VALLECITOS REPORT)	VAL-
GENERAL ELECTRIC CO. NUCLEAR MATERIALS AND PROPULSION OPERATION, CINCINNATI (REF. 3)	APEX-
GENERAL ELECTRIC CO. NUCLEAR MATERIALS AND PROPULSION OPERATION, CINCINNATI	GE-P.O.(NUMBER)-
GENERAL ELECTRIC CO. NUCLEAR MATERIALS AND PROPULSION OPERATION, CINCINNATI	GE-TM-
GENERAL ELECTRIC CO. NUCLEAR MATERIALS AND PROPULSION OPERATION, CINCINNATI	NMI-
GENERAL ELECTRIC CO. NUCLEAR MATERIALS AND PROPULSION OPERATION, CINCINNATI	NMP-HTMP-
GENERAL ELECTRIC CO. NUCLEAR MATERIALS AND PROPULSION OPERATION, CINCINNATI	TM-(YEAR)-(NUMBER)-
GENERAL ELECTRIC CO. NUCLEAR MATERIALS AND PROPULSION OPERATION, CINCINNATI	U(YEAR)-
GENERAL ELECTRIC CO. NUCLEAR SYSTEMS PROGRAMS, CINCINNATI	GESP-
GENERAL ELECTRIC CO. NUCLEAR SYSTEMS PROGRAMS, IDAHO FALLS (LOFT PROGRAM)	LOFT-GE-M-
GENERAL ELECTRIC CO. ORDNANCE DEPT., PITTSFIELD, MASS	FDU-(YEAR)-
GENERAL ELECTRIC CO.ORDNANCE SYSTEMS,PITTSFIELD,MASS.	OS-(YEAR-NUMBER)-
GENERAL ELECTRIC CO. PINELLAS PENINSULA PLANT, ST. PETERSBURG, FLA.	GEPP-
GENERAL ELECTRIC CO. PINELLAS PENINSULA PLANT, ST. PETERSBURG, FLA.	R-(YEAR)ND(NUMBER)
GENERAL ELECTRIC CO. POWER TUBE DEPT., SCHENECTADY,NY	GE/PTD(NUMBER)/CON. NO.)
GENERAL ELECTRIC CO. POWER TUBE DEPT., SCHENECTADY,NY	G-ES-
GENERAL ELECTRIC CO. RESEARCH LAB., SCHENECTADY, N.Y.	BPSN-
GENERAL ELECTRIC CO. RESEARCH LAB., SCHENECTADY, N.Y.	DIN-
GENERAL ELECTRIC CO. RESEARCH LAB., SCHENECTADY, N.Y.	(YEAR)-GC-
GENERAL ELECTRIC CO. RESEARCH LAB., SCHENECTADY, N.Y.	GE-(YEAR)-RL-
GENERAL ELECTRIC CO. RESEARCH LAB., SCHENECTADY, N.Y.	GEL-
GENERAL ELECTRIC CO. RESEARCH LAB., SCHENECTADY, N.Y.	GE-MEMO-EN-(YEAR)-
GENERAL ELECTRIC CO. RESEARCH LAB., SCHENECTADY, N.Y.	GEPM-
GENERAL ELECTRIC CO. RESEARCH LAB., SCHENECTADY, N.Y.	GERL-(YEAR)-GC-
GENERAL ELECTRIC CO. RESEARCH LAB., SCHENECTADY, N.Y.	GERL-(LTRS, NOS.)
GENERAL ELECTRIC CO. RESEARCH LAB., SCHENECTADY, N.Y.	GEZR-
GENERAL ELECTRIC CO. RESEARCH LAB., SCHENECTADY, N.Y.	GP-
GENERAL ELECTRIC CO. RESEARCH LAB., SCHENECTADY, N.Y.	MB-
GENERAL ELECTRIC CO. RESEARCH LAB., SCHENECTADY, N.Y.	MEMO-EN-(YR)-
GENERAL ELECTRIC CO. RESEARCH LAB., SCHENECTADY, N.Y.	MEMO-M-
GENERAL ELECTRIC CO. RESEARCH LAB., SCHENECTADY, N.Y.	OCCASIONAL REPORT -
GENERAL ELECTRIC CO. RESEARCH LAB., SCHENECTADY, N.Y.	RD-
GENERAL ELECTRIC CO. RESEARCH LAB., SCHENECTADY, N.Y.	RL-
GENERAL ELECTRIC CO. RESEARCH LAB., SCHENECTADY, N.Y. (REF. 1)	(YEAR)-RL-(NUMBER)-(LTR)
GENERAL ELECTRIC CO. RESEARCH LAB., SCHENECTADY, N.Y.	(YEAR)-RL-
GENERAL ELECTRIC CO. SEMICONDUCTOR PRODUCTS DEPT., SYRACUSE, N.Y.	R-(YEAR)ELS-
GENERAL ELECTRIC CO. SPACE DIV., HUNTSVILLE, ALA.	REPT-(YEAR)-HVO-
GENERAL ELECTRIC CO. SPACE POWER AND PROPULSION SECTION, CINCINNATI	DMS-
GENERAL ELECTRIC CO. SPACE POWER AND PROPULSION SECTION, CINCINNATI	R-(YEAR)-SD(NUMBER)
GENERAL ELECTRIC CO. SPACE SCIENCES LAB.,PHILADELPHIA	S.S.E. MEMO-
GENERAL ELECTRIC CO. SPACE SCIENCES LAB.,PHILADELPHIA (SPACE VEHICLE ENGINEERING TECHNICAL MEMO)	SVETM-(NUMBER)-(NUMBER)
GENERAL ELECTRIC CO. SPEC. DEFENSE PROJS.DEPT.,PHILA.	GE-QPR-
GENERAL ELECTRIC CO. SPEC. DEFENSE PROJS.DEPT.,PHILA.	GE-RS-TM-
GENERAL ELECTRIC CO. SPECIAL DEFENSE PROJECTS DEPT., SCHENECTADY, N.Y. (SEE ALSO LATER NAME: GENERAL ELECTRIC CO. MISSILE AND SPACE VEHICLE DEPT.)	TD(YR.)-
GENERAL ELECTRIC CO. SPECIALTY CONTROL DEPT., WAYNESBORO, VA.	TD-
GENERAL ELECTRIC CO. SPECIALTY TRANSFORMER DEPT., FT. WAYNE, IND.	RM-
GENERAL ELECTRIC CO. SYSTEMS DESIGN AND PROGRAMMING UNIT, FALLS CHURCH, VA.	SDP-(YEAR-NUMBER)-
GENERAL ELECTRIC CO. TECHNICAL INFORMATION EXCHANGE, SCHENECTADY, N.Y. (TECH. INFO. SERIES) (REF. 46)	TIS(YEAR)(LETTERS)-
GENERAL ELECTRIC CO. TECHNICAL MILITARY PLANNING OPERATION, SANTA BARBARA, CALIF.	DSD-
GENERAL ELECTRIC CO. TECHNICAL MILITARY PLANNING OPERATION, SANTA BARBARA, CALIF.	GE-(NUMBERS)SAR-(DATE)
GENERAL ELECTRIC CO. TECHNICAL MILITARY PLANNING OPERATION, SANTA BARBARA, CALIF.	GE R(YEAR)TMP-
GENERAL ELECTRIC CO. TECHNICAL MILITARY PLANNING OPERATION, SANTA BARBARA, CALIF.	GE/TEMP-
GENERAL ELECTRIC CO. TECHNICAL MILITARY PLANNING OPERATION, SANTA BARBARA, CALIF.	GE-TEMP-(YEAR)TMP-
GENERAL ELECTRIC CO. TECHNICAL MILITARY PLANNING OPERATION, SANTA BARBARA, CALIF.	GE/TEMPO-
GENERAL ELECTRIC CO. TECHNICAL MILITARY PLANNING OPERATION, SANTA BARBARA, CALIF.	GE/TEMPO-RM-(YEAR)TMP-
GENERAL ELECTRIC CO. TECHNICAL MILITARY PLANNING OPERATION, SANTA BARBARA, CALIF.	R(YR)TMP-
GENERAL ELECTRIC CO. TECHNICAL MILITARY PLANNING OPERATION, SANTA BARBARA, CALIF.	RM-(NO.)TMP-
GENERAL ELECTRIC CO. TECHNICAL MILITARY PLANNING OPERATION, SANTA BARBARA, CALIF.	RM-(YR.)TMP-
GENERAL ELECTRIC CO. TECHNICAL MILITARY PLANNING OPERATION, SANTA BARBARA, CALIF.	RM-TMP-
GENERAL ELECTRIC CO. TECHNICAL MILITARY PLANNING OPERATION, SANTA BARBARA, CALIF.	TMP-
GENERAL ELECTRIC CO. TEMPO. CENTER FOR ADVANCED STUDIES, SANTA BARBARA, CALIF.	SP-
GENERAL ELECTRIC CO. TEMPO. CENTER FOR ADVANCED STUDIES, SANTA BARBARA, CALIF.	(YEAR)TMP-
GENERAL ELECTRIC CO. THOMSON LAB., LYNN, MASS.	R(YR)-TL-
GENERAL ELECTRIC CO. TUBE DEPT., OWENSBORO, KY.	GE (NUMBER/CONTRACT NO.)
GENERAL ELECTRIC CO. TUBE DEPT., OWENSBORO, KY.	(NUMBER)-OWB-
GENERAL ELECTRIC CO. TUBE DEPT., PALO ALTO, CALIF.	GE/TD QR-
GENERAL ELECTRIC CO. TUBE DEPT., PALO ALTO, CALIF.	R(NO.)(LTRS.)-
GENERAL ELECTRIC CO. TUBE DEPT., PALO ALTO, CALIF.	R(YEAR)ELM-(NUMBER)-

GENERAL ELECTRIC CO. TUBE DEPT., SCHENECTADY, N.Y.	G-E-SD-
GENERAL ELECTRIC CO. TWT BUSINESS SECT.,PALO ALTO,CAL	R(YEAR)ELM(NUMBER)-
GENERAL ELECTRIC CO. X-RAY DEPT., MILWAUKEE	GE-X-
GENERAL ELECTRIC CO. X-RAY DEPT., MILWAUKEE	GEX-
GENERAL ELECTRIC CO., LTD. APPLIED ELECTRONICS LABS., STANMORE, MIDDX, ENGLAND	SL-
GENERAL ELECTRIC CO., LTD. ATOMIC ENERGY DIV., ERITH, KENT, ENGLAND	GEC-
GENERAL ELECTRIC CO., LTD. RESEARCH LABS., WEMBLEY, MIDDX, ENGLAND	BC-
GENERAL ELECTRIC CO., LTD. RESEARCH LABS., WEMBLEY, MIDDX, ENGLAND	GEL RL-
GENERAL ELECTRIC CO., LTD. RESEARCH LABS., WEMBLEY, MIDDX, ENGLAND	GERL-COMM.-
GENERAL ELECTRONIC LABS. INC., BOSTON	E-(NUMBER)-
GENERAL INSTRUMENT CORP., HICKSVILLE, N.Y.	GIC-
GENERAL INSTRUMENT CORP., NEWARK, N.J. (FINAL REPT.)	GIC-
GENERAL INSTRUMENT CORP. CAPACITOR DIV.,DARLINGTON,SC	GI-QRM-
GENERAL INSTRUMENT CORP. ELECTRONIC SYSTEMS DIV., HICKSVILLE, N.Y.	ESD-(NUMBER)-IR-
GENERAL MILLS, INC., MINNEAPOLIS	GM-
GENERAL MILLS, INC., MINNEAPOLIS	GMI-
GENERAL MILLS, INC. ENG. RES. + DEV.DEPT.,MINNEAPOLIS	GMI ERDD-
GENERAL MILLS, INC. NUCLEAR EQUIP. DEPT., MINNEAPOLIS	GM-NED-
GENERAL MILLS, INC. NUCLEAR EQUIP. DEPT., MINNEAPOLIS	NED-
GENERAL MOTORS CORP. (ALL LOCATIONS)	GMC-
GENERAL MOTORS CORP.(ALL LOCATIONS) (MEMORANDUM RPTS)	GMC MR-
(SEE ALSOO ALLISON DIV., GENERAL MOTORS CORP.)	
GENERAL MOTORS CORP., DETROIT	MR-
GENERAL MOTORS CORP., DETROIT	NDA-(NUMBER)-
GENERAL MOTORS CORP., INDIANAPOLIS	EDR-
GENERAL MOTORS CORP., MILWAUKEE	EDF-
GENERAL MOTORS CORP., MILWAUKEE	IV-AG-
GENERAL MOTORS CORP., SANTA BARBARA, CALIF.	CTN(YR)-
GENERAL MOTORS CORP., SANTA BARBARA, CALIF.	GMC-TM-
GENERAL MOTORS CORP. AC ELECTRONICS DIV., MILWAUKEE	EP(YEAR)-
GENERAL MOTORS CORP. AC ELECTRONICS DIV., MILWAUKEE	EP-(4 DIGITS)
GENERAL MOTORS CORP. AC SPARK PLUG DIV., FLINT, MICH.	(YEAR CODE)-(LETTER)-
GENERAL MOTORS CORP. AC SPARK PLUG DIV., FLINT, MICH.	GMAC-
GENERAL MOTORS CORP. AEROPRODUCTS DIV., DAYTON, OHIO	ER-
GENERAL MOTORS CORP. DEFENSE RESEARCH LABS., SANTA BARBARA, CALIF.	CTN-
GENERAL MOTORS CORP. DEFENSE RESEARCH LABS., SANTA BARBARA, CALIF.	GDRL-TR-(YEAR)-
GENERAL MOTORS CORP. DEFENSE RESEARCH LABS., SANTA BARBARA, CALIF.	GMC-TR-(YEAR)-
GENERAL MOTORS CORP. DEFENSE RESEARCH LABS., SANTA BARBARA, CALIF.	GM-DRL-TR(YEAR)-
GENERAL MOTORS CORP. DEFENSE RESEARCH LABS., SANTA BARBARA, CALIF.	GM TR (YEAR)-
GENERAL MOTORS CORP. DEFENSE RESEARCH LABS., SANTA BARBARA, CALIF.	IMR-
GENERAL MOTORS CORP. DEFENSE RESEARCH LABS., SANTA BARBARA, CALIF.	TR-(YEAR)-
GENERAL MOTORS CORP. DEFENSE RESEARCH LABS., SANTA BARBARA, CALIF.	TR(YEAR)-(NO.)(LTR.)
GENERAL MOTORS CORP. DIGITAL COMPUTER LAB., EL SEGUNDO, CALIF.	ACLA-
GENERAL MOTORS CORP. ELECTRO CHEMISTRY DEPT.	GMC EC-
GENERAL MOTORS CORP. ELECTRO-MOTIVE DIV., DETROIT	FR-NPCV-
GEN. MOTORS CORP. HARRISON RADIATOR DIV., LOCKPORT,NY	HRD-AC-(YEAR)-
GEN. MOTORS CORP. HARRISON RADIATOR DIV., LOCKPORT,NY	RER-
GENERAL MOTORS CORP. MATERIALS AND STRUCTURES LAB., WARREN, MICH.	GMC-MSL-
GENERAL MOTORS CORP. MATERIALS AND STRUCTURES LAB., WARREN, MICH.	MSL-
GENERAL MOTORS CORP. ORDNANCE DEPT.	ORD GMC-
GENERAL MOTORS CORP. PROVING GROUND	GMPG-
GENERAL MOTORS CORP. RESEARCH LABS., WARREN, MICH.	GMR-
GENERAL MOTORS CORP. RESEARCH LABS., WARREN, MICH.	GMRL-
GENERAL MOTORS CORP. RESEARCH LABS. DIV., DETROIT (ELECTROCHEMISTRY)	EC-
GENERAL MOTORS CORP. RESEARCH LABS. DIV., DETROIT (PHYSICS INSTRUMENTATION)	GMC PI-
GENERAL MOTORS CORP. RESEARCH LABS. DIV., DETROIT (PHYSICS INSTRUMENTATION)	PI-
GENERAL MOTORS RESEARCH LABS. MECHANICAL DEV. DEPT., WARREN, MICH.	MD-
GENERAL MOTORS TECHNICAL CENTER. MATERIALS AND STRUCTURES LAB., WARREN, MICH.	MSL-(YEAR)-
GENERAL NUCLEAR ENGINEERING CORP., DUNEDIN, FLA.	GNEC-
GENERAL NUCLEAR ENGINEERING CORP., DUNEDIN, FLA.	TPR-PRT-
GENERAL NUCLEONICS CORP., CLAREMONT, CALIF.	GNC-(NUMBER)(LETTER)-
GENERAL OCEANOLOGY, INC., CAMBRIDGE, MASS	GO-(2 DIGITS)
GENERAL POST OFFICE, ENG. DEPT., LONDON	GPO-(2 DIGITS)
GENERAL POST OFFICE. ENG. DEPT., LONDON	GPO-RR-
GENERAL PRECISION, INC., GLENDALE, CALIF.	SAR-
GENERAL PRECISION, INC., LITTLE FALLS, N.J.	ETO-
GENERAL PRECISION, INC., LITTLE FALLS, N.J.	GPI-
GENERAL PRECISION, INC., LITTLE FALLS, N.J.	GPSI-(NUMBER)-
GENERAL PRECISION, INC., LITTLE FALLS, N.J.	M(NO.S)-
GENERAL PRECISION, INC., SUNNYVALE, CALIF.	L/S-U-

GENERAL PRECISION, INC.

GENERAL PRECISION INC., TARRYTOWN, N. J.	GPI-(NO.)-D-
GENERAL PRECISION INC., TARRYTOWN, N. J.	GPI-(NO.)-J-
GENERAL PRECISION, INC., WAYNE, N.J.	GPI-
GENERAL PRECISION, INC. AEROSPACE GP.,LITTLE FALLS,NJ	GPL-A-
GENERAL PRECISION, INC. GPL DIV., PLEASANTVILLE, N.Y.	GPL-
GENERAL PRECISION, INC. GPL DIV., PLEASANTVILLE, N.Y.	GPL-A-
GENERAL PRECISION, INC. KEARFOTT DIV.,LITTLE FALLS,NJ	KEARFOTT-
GENERAL PRECISION, INC. LIBRASCOPE DIV., GLENDALE,CAL	SI-(YEAR)-
GENERAL PRECISION, INC. LIBRASCOPE DIV., GLENDALE,CAL	SISS-
GENERAL PRECISION, INC. LINK GROUP, BINGHAMTON, N.Y.	LIM-(YEAR)/-
GENERAL PRECISION LAB., INC., PLEASANTVILLE, N.Y.	A11-
GENERAL PRECISION SYSTEMS INC., KEARFOTT SYSTEMS INC., WAYNE, N. J.	GPI-(NO. LTR NO. LTR.)
GENERAL PRECISION SYSTEMS, INC., LIBRASCOPE SYSTEMS DIV., GLENDALE, CALIF.	LIBRASCOPE-
GENERAL PRECISION SYSTEMS, INC., LINK GP., PALO ALTO, CALIF.	WDR-
GENERAL RADIO CO., CAMBRIDGE, MASS.	GR-
GENERAL RESEARCH CORP., ARLINGTON, VA.	GRC-CR-(NUMBER)-
GENERAL RESEARCH CORP., SANTA BARBARA, CALIF.	CRC-CAPA-(NUMBER)-
GENERAL RESEARCH CORP., SANTA BARBARA, CALIF.	GRC-
GENERAL RESEARCH CORP., SANTA BARBARA, CALIF.	GRC-CR-(NUMBER)-
GENERAL RESEARCH CORP., SANTA BARBARA, CALIF.	GRC-CR-(NO.)-(NO.)-F-
GENERAL RESEARCH CORP., SANTA BARBARA, CALIF.	GRC-IMR-
GENERAL RESEARCH CORP., SANTA BARBARA, CALIF.	GRC-TM-(NUMBER)/(NUMBER)
GENERAL SERVICES ADMINISTRATION, WASHINGTON, D.C.	GSA-
GENERAL TECHNOLOGIES CORP., ALEXANDRIA, VA.	GTC-
GENERAL TECHNOLOGIES CORP., RESTON, VA.	GTC-
GENERAL TECHNOLOGY CORP., ELGIN, ILL.	TR-(NUMBER)-
GENERAL TECHNOLOGY CORP., ELGIN, ILL.	TRX-
GENERAL TELEPHONE AND ELECTRONICS LABS. INC. BAYSIDE LABS., N.Y.	GTEL-TR-(YEAR)-...
GENERAL TELEPHONE AND ELECTRONICS LABS. INC. BAYSIDE LABS., N.Y.	GT/E-TR-(YEAR-NO.-NO.)
GENERAL TELEPHONE AND ELECTRONICS LABS. INC. BAYSIDE LABS., N.Y.	GTI-
GENERAL TELEPHONE AND ELECTRONICS LABS. INC. BAYSIDE LABS., N.Y.	GTR-
GENERAL TELEPHONE AND ELECTRONICS LABS. INC. BAYSIDE LABS., N.Y.	STR-
GENERAL TELEPHONE AND ELECTRONICS LABS. INC. BAYSIDE LABS., N.Y.	TR-(YR)-
GENERAL TESTING LABS., INC., SPRINGFIELD, VA.	TEST-A-2901-
GENERAL TIME CORP., LA SALLE, ILL.	GTC-
GENERAL TIRE AND RUBBER CO., AKRON, OHIO	GTR-
GENERAL TIRE AND RUBBER CO., AKRON, OHIO	GTRC-
GENERAL TIRE AND RUBBER CO. OF CALIF., VAN NUYS	GEN-
GENERAL TIRE AND RUBBER CO. OF CALIF. RESEARCH AND DEVELOPMENT DIV., PASADENA (REF. 1)	(NUMBER)
GENISTRON INC. APPLIED RES. DIV., COLLEGE PARK, MD.	GENIS-
GENISTRON INC. APPLIED RES. DIV., COLLEGE PARK, MD.	GENISTRON-
GENTEX CORP., CARBONDALE, PENNA.	GTX-(3 DIGITS)
GEODYNAMICS CORP., SANTA BARBARA, CALIF.	CDR-(NUMBER)FR(NUMBER)
GEOLOGICAL SURVEY (ALL LOCATIONS)	TEI-
GEOLOGICAL SURVEY, DENVER	AI-CE-
GEOLOGICAL SURVEY, DENVER	AMCHITKA-
GEOLOGICAL SURVEY, DENVER	AREA-
GEOLOGICAL SURVEY, DENVER	ATLAS-HA-
GEOLOGICAL SURVEY, DENVER	CENTRAL NEVADA-
GEOLOGICAL SURVEY, DENVER (CHIRIKOF ISLAND, ALASKA)	CHIRIKOF-
GEOLOGICAL SURVEY, DENVER	CHRISTMAS TREE-
GEOLOGICAL SURVEY, DENVER (CRUSTAL STUDIES)	CRUSTAL STUDIES-
GEOLOGICAL SURVEY, DENVER	DANNY-BOY-
GEOLOGICAL SURVEY, DENVER	DRIBBLE-
GEOLOGICAL SURVEY, DENVER	FOREIGN APPLICATIONS-
GEOLOGICAL SURVEY, DENVER	GASBUGGY-
GEOLOGICAL SURVEY, DENVER	GEO-SUR-TL-
GEOLOGICAL SURVEY, DENVER	GEOTHERMAL ENERGY-
GEOLOGICAL SURVEY, DENVER	LONG SHOT-
GEOLOGICAL SURVEY, DENVER	NTS-
GEOLOGICAL SURVEY, DENVER	OFFSITE STUDIES-
GEOLOGICAL SURVEY, DENVER(PREDICTION BACKGROUND DATA)	PBD-
GEOLOGICAL SURVEY, DENVER	RAINIER-MESA-
GEOLOGICAL SURVEY, DENVER	RULISON-
GEOLOGICAL SURVEY, DENVER	SPECIAL PROJECTS-
GEOLOGICAL SURVEY, DENVER	SPECIAL STUDIES-
GEOLOGICAL SURVEY, DENVER (CRUSTAL STUDIES)	TL-
GEOLOGICAL SURVEY, DENVER	USGA-AMCHITKA-
GEOLOGICAL SURVEY, DENVER	USGS-(NUMBER)-
GEOLOGICAL SURVEY, DENVER	WAGON-
GEOLOGICAL SURVEY, DENVER	YUCCA-(NUMBER)-
GEOLOGICAL SURVEY, MENLO PARK, CALIF.	ALEUTIAN-
GEOLOGICAL SURVEY, WASHINGTON, D.C.	BULL-(4 DIGITS)-F
GEOLOGICAL SURVEY, WASHINGTON, D.C.	CIRC-
GEOLOGICAL SURVEY, WASHINGTON, D.C.	GS-
GEOLOGICAL SURVEY, WASHINGTON, D.C. (BIBLIOGRAPHY)	GS-B-
GEOLOGICAL SURVEY, WASHINGTON, D.C. (CIRCULAR)	GS-C-
GEOLOGICAL SURVEY, WASHINGTON, D.C. (CIRCULAR)	GSC-
GEOLOGICAL SURVEY, WASHINGTON, D.C. (CIRCULAR)	GS-CIRC-
GEOLOGICAL SURVEY, WASHINGTON, D.C.	GS-P-(NO.)-A

GEOLOGICAL SURVEY, WASH., D.C. (PROFESSIONAL PAPER)	GSPP-
GEOLOGICAL SURVEY, WASH., D.C. (WATER SUPPLY PAPER)	GS WSP-
GEOLOGICAL SURVEY, WASHINGTON, D.C.	INTERAGENCY-
GEOLOGICAL SURVEY, WASHINGTON, D.C. (INTERAGENCY RPT)	IR-USGS-
GEOLOGICAL SURVEY, WASHINGTON, D.C.	MEMO-
GEOLOGICAL SURVEY, WASHINGTON, D.C.	PP-(NUMBER)-C
GEOLOGICAL SURVEY, WASHINGTON, D.C.	RUFUS-RR-
GEOLOGICAL SURVEY, WASH., D.C. (TRACE ELEMENTS RPTS.)	TEI-(M)-
GEOLOGICAL SURVEY, WASHINGTON, D.C.	TEIR-
GEOLOGICAL SURVEY, WASHINGTON, D.C.	TELI-
GEOLOGICAL SURVEY, WASH., D.C. (TRACE ELEMENTS MEMO RPT)	TEM-
GEOLOGICAL SURVEY, WASHINGTON, D.C.	TEMR-
GEOLOGICAL SURVEY, WASHINGTON, D.C.	TWC-
GEOLOGICAL SURVEY, WASHINGTON, D.C.	USGS-
GEOLOGICAL SURVEY, WASHINGTON, D.C.	USGS-COMPUTER CONTRIB-
GEOLOGICAL SURVEY, WASHINGTON, D.C.	USGS-DO-(YEAR)-
GEOLOGICAL SURVEY, WASHINGTON, D.C.	USGS-GD-(YEAR)-
GEOLOGICAL SURVEY, WASHINGTON, D.C.	USGS-TEI-
GEOLOGICAL SURVEY, WASHINGTON, D.C.	USGS-TELI-
GEOLOGICAL SURVEY, WASHINGTON, D.C.	USGS-TEM-
GEOLOGICAL SURVEY, WASHINGTON, D.C.	USGS-WP-
GEOLOGICAL SURVEY, WASH., D.C. (WATER SUPPLY PAPER)	WSP-
GEOLOGICAL SURVEY. TOPOGRAPHIC DIV., WASHINGTON, D.C.	USGA-TD-(YEAR)-
GEOLOGICAL SURVEY. TOPOGRAPHIC DIV., WASHINGTON, D.C.	USGS-TD-(YEAR)-
GEOLOGICAL SURVEY. WATER RESOURCES DIV., ROLLA, MO.	WATER RESOURCES-
GEOMET, INC., ROCKVILLE, MD.	GEOMET-(NUMBER)-
GEONUCLEAR CORP., LAS VEGAS, NEV.	GEONUCLEAR-
GEOPHYSICS CORP. OF AMERICA, BEDFORD, MASS.	GCA-
GEOPHYSICS CORP. OF AMERICA, BEDFORD, MASS.	GCR-TR-(YEAR-NUMBER)-N
GEOPHYSICS CORP. OF AMERICA, BEDFORD, MASS.	MPLR-
(SEE ALSO: GCA CORP.)	
GEORGE C. MARSHALL SPACE FLIGHT CTR., HUNTSVILLE, ALA.	AE(YEAR)-
GEORGE C. MARSHALL SPACE FLIGHT CTR., HUNTSVILLE, ALA.	GCMSFC-
GEORGE C. MARSHALL SPACE FLIGHT CTR., HUNTSVILLE, ALA.	GMSFC-(LETTERS)(NUMBERS)
GEORGE C. MARSHALL SPACE FLIGHT CTR., HUNTSVILLE, ALA.	IN-P+VE-M-(YEAR)-
GEORGE C. MARSHALL SPACE FLIGHT CTR., HUNTSVILLE, ALA.	M-ASTR-IN-(YEAR)-
GEORGE C. MARSHALL SPACE FLIGHT CTR., HUNTSVILLE, ALA.	MDWG-(YEAR)-
GEORGE C. MARSHALL SPACE FLIGHT CTR., HUNTSVILLE, ALA.	MFS-
GEORGE C. MARSHALL SPACE FLIGHT CTR., HUNTSVILLE, ALA.	MFSC-IN-SSL-T-
GEORGE C. MARSHALL SPACE FLIGHT CTR., HUNTSVILLE, ALA.	MHM-
GEORGE C. MARSHALL SPACE FLIGHT CTR., HUNTSVILLE, ALA.	MHR-
GEORGE C. MARSHALL SPACE FLIGHT CTR., HUNTSVILLE, ALA.	MISC-SSL-(YEAR)-
GEORGE C. MARSHALL SPACE FLIGHT CENTER, HUNTSVILLE, ALA. (MARSHALL MEMO)	MM-
GEORGE C. MARSHALL SPACE FLIGHT CTR., HUNTSVILLE, ALA.	MPR-SAT-FE-
GEORGE C. MARSHALL SPACE FLIGHT CTR., HUNTSVILLE, ALA.	MSFC-
GEORGE C. MARSHALL SPACE FLIGHT CTR., HUNTSVILLE, ALA.	MSFC-ASO-IN-
GEORGE C. MARSHALL SPACE FLIGHT CTR., HUNTSVILLE, ALA. (LAUNCH OPERATIONS DIRECTORATE)	MSFC-LOD-(LETTERS)-
GEORGE C. MARSHALL SPACE FLIGHT CENTER, HUNTSVILLE, ALA. (MARSHALL TECHNICAL PAPER)	MTP-
GEORGE C. MARSHALL SPACE FLIGHT CTR., HUNTSVILLE, ALA.	MTP-AERO-(YEAR)-
GEORGE C. MARSHALL SPACE FLIGHT CTR., HUNTSVILLE, ALA.	MTP-ASTR-
GEORGE C. MARSHALL SPACE FLIGHT CTR., HUNTSVILLE, ALA.	MTP-PVE-
GEORGE C. MARSHALL SPACE FLIGHT CTR., HUNTSVILLE, ALA.	MTP-RP-(YEAR)-
GEORGE C. MARSHALL SPACE FLIGHT CTR., HUNTSVILLE, ALA.	NASA-MHR-
GEORGE C. MARSHALL SPACE FLIGHT CTR., HUNTSVILLE, ALA.	NSFC-IN-SSL-T-
GEORGE C. MARSHALL SPACE FLIGHT CTR., HUNTSVILLE, ALA.	R-AERO-Y-(NO.-NO.)
GEORGE C. MARSHALL SPACE FLIGHT CTR., HUNTSVILLE, ALA.	R-ASTR-S-(YEAR)-
GEORGE C. MARSHALL SPACE FLIGHT CTR., HUNTSVILLE, ALA.	S-E-ASTR-R-
GEORGE C. MARSHALL SPACE FLIGHT CTR., HUNTSVILLE, ALA.	S/E-ASTR-S-
GEORGE C. MARSHALL SPACE FLIGHT CENTER, HUNTSVILLE, ALA. (TECHNICAL MEMORANDUM)	TM-X-53-
GEORGE WASHINGTON UNIV., ARLINGTON, VA. HUMAN RESOURCES RESEARCH OFFICE	BASIC RESEARCH-
GEORGE WASHINGTON UNIV., ARLINGTON, VA. HUMAN RESOURCES RESEARCH OFFICE	EXPLORATORY STUDY-
GEORGE WASHINGTON UNIV., ARLINGTON, VA. HUMAN RESOURCES RESEARCH OFFICE	HUMBRO-TR-
GEORGE WASHINGTON UNIV., ARLINGTON, VA. HUMAN RESOURCES RESEARCH OFFICE	HUMRRO-
GEORGE WASHINGTON UNIV., ARLINGTON, VA. HUMAN RESOURCES RESEARCH OFFICE	HUMRRO-RBP-D(NO.-YR.)-
GEORGE WASHINGTON UNIV., ARLINGTON, VA. HUMAN RESOURCES RESEARCH OFFICE	HUMRRO-TR-(YEAR)-
GEORGE WASHINGTON UNIV., ARLINGTON, VA. HUMAN RESOURCES RESEARCH OFFICE	PROFESSIONAL PAPER-
GEORGE WASHINGTON UNIV., WASHINGTON, D.C. (BIOLOGICAL COMMUNICATIONS PROJECT)	BSCP-(NUMBER-NUMBER)
GEORGE WASHINGTON UNIV., WASHINGTON, D.C.	GUPS-MON-
GEORGE WASHINGTON UNIV., WASHINGTON, D.C.	GW-
GEORGE WASHINGTON UNIV., WASHINGTON, D.C. (POLICY STUDIES IN SCIENCE + TECHNOLOGY)	GWPS-MON-
GEORGE WASHINGTON UNIV., WASHINGTON, D.C. (POLICY STUDIES IN SCIENCE + TECHNOLOGY)	GWPS-OP-
GEORGE WASHINGTON UNIV., WASHINGTON, D.C. (POLICY STUDIES IN SCIENCE + TECHNOLOGY)	GWPS-SDP-
GEORGE WASHINGTON UNIV., WASHINGTON, D.C.	GWU-
GEORGE WASHINGTON UNIV., WASHINGTON, D.C.	PP-(NUMBER-YEAR)
GEORGE WASHINGTON UNIV., WASHINGTON, D.C.	PPSST-
GEORGE WASHINGTON UNIV., WASHINGTON, D.C. (STAFF DISCUSSION PAPER)	SDP-
GEORGE WASHINGTON UNIV., WASHINGTON, D.C. (LOGISTICS RESEARCH PROJECT)	SERIAL-
GEORGE WASHINGTON UNIV., WASHINGTON, D.C. (LOGISTICS RESEARCH PROJECT)	SERIAL T-
GEORGE WASHINGTON UNIV., WASHINGTON, D.C. (LOGISTICS RESEARCH PROJECT)	SERIAL-TM-DMC-
GEORGE WASHINGTON UNIV., WASHINGTON, D.C.	T-
GEORGIA. BUREAU OF STATE PLANNING AND COMMUNITY AFFAIRS, ATLANTA	CPRC-(YEAR)-
GEORGIA. UNIV., ATHENS. (CIVIL DEFENSE RESEARCH)	CDR-
GEORGIA. UNIV., ATHENS	GEOU-
GEORGIA. UNIV., ATHENS	GU-
GEORGIA. UNIV., ATHENS	SR-(NUMBER)-(NUMBER)
GEORGIA. UNIV., ATHENS. DEPT. OF STATISTICS (PROJECT THEMIS) (REF. 48)	THEMIS-UGA-
GEORGIA INST. OF TECH., ATLANTA	A-
GEORGIA INST. OF TECH., ATLANTA	AL-ASD-TDR-
GEORGIA INST. OF TECH., ATLANTA	ECAC-TM-
GEORGIA INST. OF TECH., ATLANTA	GIT-
GEORGIA INST. OF TECH., ATLANTA	GIT-A-(NUMBER)-
GEORGIA INST. OF TECH., ATLANTA	G-T-R-
GEORGIA INST. OF TECH., ATLANTA. ELECTRONICS DIV.	GIT-A-(NUMBER)-SR
GEORGIA INST. OF TECH., ATLANTA. ELECTRONICS DIV.	GIT-GTF-
GEORGIA INST. OF TECH., ATLANTA. ELECTRONICS DIV.	GIT-SR-
GEORGIA INST. OF TECH., ATLANTA. ELECTRONICS DIV.	GIT-STR-A-

GEORGIA INST. OF TECH., ATLANTA. ENG. EXPERIMENT STA.	GIT-(LETTERS)-
GEORGIA INST. OF TECH., ATLANTA. ENG. EXPERIMENT STA.	GIT-B-(NUMBER)-
GEORGIA INST. OF TECH., ATLANTA. ENG. EXPERIMENT STA.	GIT/EES-
GEORGIA INST. OF TECH., ATLANTA. ENG. EXPERIMENT STA.	GIT-QR-
GEORGIA INST. OF TECH., ATLANTA. ENG. EXPERIMENT STA.	GIT-TR-
GEORGIA INST. OF TECH., ATLANTA. ENG. EXPERIMENT STA.	GIT-TSR-
GEORGIA INST. OF TECH., ATLANTA. ENG. EXPERIMENT STA.	G-T-G-(YEAR)-
GEORGIA INST. OF TECH., ATLANTA. ENVIRONMENTAL RESOURCES CENTER	ERC-
GEORGIA INST. OF TECH., ATLANTA. HEALTH SYSTEMS RESEARCH CENTER	HSRD-(YEAR)-
GEORGIA INST. OF TECH., ATLANTA. SCHOOL OF AEROSPACE ENGINEERING	GIT-AER-(YEAR)-
GEORGIA INST. OF TECH., ATLANTA. SCHOOL OF CIVIL ENG.	GIT-CE-PW-
GEORGIA INST. OF TECH., ATLANTA. SCHOOL OF ELECTRICAL ENGINEERING	GIT/EES-A(NO.)/T(NO.)
GEORGIA INST. OF TECH., ATLANTA. SCHOOL OF INFORMATION AND COMPUTER SCIENCE	GITIS-(YEAR)-
GEORGIA INST. OF TECH., ATLANTA. SCHOOL OF INFORMATION AND COMPUTER SCIENCE	ITIS-(YEAR)-
GEORGIA MOUNTAINS PLANNING + DEVELOPMENT COMMISSION, GAINESVILLE	GM-P132-(YEAR-NO.-NO.)
GEOSCIENCE, INC., CAMBRIDGE, MASS.	RG-
GEOSCIENCE, INC., CAMBRIDGE, MASS.	RU-
GEOSCIENCE LTD., LA JOLLA, CALIF.	GLR-
GEOSCIENCE LTD., SAN DIEGO, CALIF.	GLR-
GEOSCIENCE LTD., SOLANA BEACH, CALIF.	GLR-
GEOTECHNICAL CORP., GARLAND, TEX.	GEOTECH-
GEOTECHNICAL CORP., GARLAND, TEX.	GEOTECH-TR-
GERMANTOWN LABS., INC., PHILADELPHIA	GL-(YEAR)-
GERMANY. AERODYNAMISCHE VERSUCHSANSTALT, GOETTINGEN	AVA-
GERMANY. AERODYNAMISCHE VERSUCHSANSTALT, GOETTINGEN	AVA-FB-(YEAR)-
GERMANY. AERODYNAMISCHE VERSUCHSANSTALT, GOETTINGEN	AVAG-
GERMANY. AERODYNAMISCHE VERSUCHSANSTALT, GOETTINGEN	AVA-TB-
GERMANY. AERODYNAMISCHE VERSUCHSANSTALT, GOETTINGEN	B-ICAS-PAPER-
GERMANY. BEVOLLMAECHTIGTE FUER HOCHFREQUENZFORSCHUNG	BHF-
GERMANY. BUNDESMINISTERIUM FUER BILDUNG UND WISSENSCHAFT, BONN	BMBW-FBK-(YEAR)-
GERMANY. BUNDESMINISTERIUM FUER BILDUNG UND WISSENSCHAFT, BONN	BMBW-FB-W-(NO.-NO.)
GERMANY. BUNDESMINISTERIUM FUER VERTEIDIGUNG, BONN	BMVG-FBWT-(YEAR)-
GERMANY. BUNDESMINISTERIUM FUER WISSENSCHAFTLICHE FORSCHUNG, BAD GODESBERG	BMWF-
GERMANY. BUNDESMINISTERIUM FUER WISSENSCHAFTLICHE FORSCHUNG, BAD GODESBERG	BMWF-FBK-(YEAR)-
GERMANY. BUNDESMINISTERIUM FUER WISSENSCHAFTLICHE FORSCHUNG, BAD GODESBERG	BMWF-FB-W-(YEAR)-
GERMANY. BUNDESPOST. FERNMELDETECHNISCHES ZENTRALAMT DARMSTADT	FTZ-
GERMANY. BUNDESPOST. FERNMELDETECHNISCHES ZENTRALAMT DARMSTADT	FTZ-A-(NOS.)-(LTRS.)-
GERMANY. BUNDESPOST. IONOSPHAREN-INSTITUT, BREISACH	IIB-
GERMANY. BUNDESPOST. IONOSPHAREN-INSTITUT, BREISACH	REPT-(NUMBER)-F
GERMANY. ERNST-MACH-INSTITUT FREIBURG/BR. DER FRAUNHOFER-GESELLSCHAFT ZUR FORDERUNG DER ANGEWANDTEN FORSCHUNG	EMI/WB-(MO./YR.)
GERMANY. INSTITUT FUER CHEMIE DER TREIBSTOFFE, BERGHAUSEN	ICT-(NUMBER-YEAR)
GERMANY. INSTITUT FUER NEUTRONENPHYSIK UND REAKTORTECHNIK, KARLSRUHE	INR-NR-
GERMANY. INSTITUT FUER THEORETISCHE GASDYNAMIK, AACHEN	DLR-FB-(YEAR)-
GERMANY. INSTITUTE OF AVIATION MEDICINE, FUERSTENFELDBRUCK	OT/ITS-RR-
GERMANY. LABORATORIUM FUER BETRIEBSFESTIGKEIT, DARMSTADT	F-
GERMANY. LABORATORIUM FUER BETRIEBSFESTIGKEIT, DARMSTADT	FB-
GERMANY. LABORATORIUM FUER BETRIEBSFESTIGKEIT, DARMSTADT	LBF-
GERMANY. LABORATORIUM FUER BETRIEBSFESTIGKEIT, DARMSTADT	LBF-S-
GERMANY. LABORATORIUM FUER BETRIEBSFESTIGKEIT, DARMSTADT	TUE-(NUMBER/YEAR)
GERMANY. LUFTWAFFE (LEHRBILDREIHE)	GL L-
GERMANY. OBERKOMMANDO DER KRIEGSMARINE	OKM-
GERMANY. OBERKOMMANDO DER KRIEGSMARINE (M DIV.)	OKM MD-
GERMANY. OBERKOMMANDO DER KRIEGSMARINE. CHEMISCHES-PHYSIKALISCHES EXPERIMENTELLES LABORATORIUM	CPL T-
GERMANY. OBERKOMMANDO DER KRIEGSMARINE. CHEMISCHES-PHYSIKALISCHES EXPERIMENTELLES LABORATORIUM	OKM CPEL T-
GERMANY. PHYSIKALISCH-TECHNISCHE BUNDESANSTALT, BRUNSWICK	FMBR-(NUMBER)/(YEAR)
GERMANY. PHYSIKALISCH-TECHNISCHE BUNDESANSTALT, BRUNSWICK	FMBR-
GERMANY. PHYSIKALISCH-TECHNISCHE BUNDESANSTALT, BRUNSWICK	FMRB-(NUMBER)/(YEAR)
GERMANY. PHYSIKALISCH-TECHNISCHE BUNDESANSTALT, BRUNSWICK	OPTIK-(NUMBER/YEAR)
GERMANY. PHYSIKALISCH-TECHNISCHE BUNDESANSTALT, BRUNSWICK	PTB-
GERMANY. PHYSIKALISCH-TECHNISCHE BUNDESANSTALT, BRUNSWICK	PTB-JB-(NUMBER)/(YEAR)
GERMANY. REICH INSTITUTE FOR MATERIAL TESTING	RIMT-
GERMANY. REICHSLUFTFAHRTMINISTERIUMS	RLM-
GERMANY. SCHOOL FOR THEORETICAL PHYSICS, OBERWOLFACH	AM-
GERMANY. STAATLICHE ZENTRALE FUER STRAHLENSCHUTZ, BERLIN	SZS-
GERMANY. STAATLICHE ZENTRALE FUER STRAHLENSCHUTZ, BERLIN	SZS-(NUMBER)/(YEAR)
GERMANY. TECHNISCHE AKADEMIE DER LUFTWAFFE	TAL-
GERMANY. VERSUCHSANSTALT FUER WASSERBAU UND SCHIFFBAU, BERLIN	WVA-
GERMANY. ZENTRALE FUER WISSENSCHAFTLICHES BERICHTSWESEN DER LUFTFAHRTFORSCHUNG, BERLIN (FORSCHUNGSBERICHT)	FB-
GERMANY. ZENTRALE FUER WISSENSCHAFTLICHES BERICHTSWESEN DER LUFTFAHRTFORSCHUNG, BERLIN (INDUSTRIE-KURZBERICHT)	KB-
GERMANY. ZENTRALE FUER WISSENSCHAFTLICHES BERICHTSWESEN DER LUFTFAHRTFORSCHUNG, BERLIN	PA-

GERMANY. ZENTRALE FUER WISSENSCHAFTLICHES BERICHTSWESEN DER LUFTFAHRTFORSCHUNG, BERLIN (PRUEFBERICHT)	PB-
GERMANY. ZENTRALE FUER WISSENSCHAFTLICHES BERICHTSWESEN DER LUFTFAHRTFORSCHUNG, BERLIN (TECHNISCHE BERICHT)	TB-
GERMANY. ZENTRALE FUER WISSENSCHAFTLICHES BERICHTSWESEN DER LUFTFAHRTFORSCHUNG, BERLIN (UNTERSUCHUNGEN UND MITTEILUNGEN)	UM-
GERMANY. ZENTRALE FUER WISSENSCHAFTLICHES BERICHTSWESEN DER LUFTFAHRTFORSCHUNG, BERLIN	ZWB-
GERMANY. ZENTRALE FUER WISSENSCHAFTLICHES BERICHTSWESEN DER LUFTFAHRTFORSCHUNG, BERLIN	ZWB-(LETTERS)-
GERMANY. ZENTRALINSTITUT FUER KERNFORSCHUNG, ROSSENDORF BEI DRESDEN	ZFK-
GERMANY. ZENTRALINSTITUT FUER KERNFORSCHUNG, ROSSENDORF BEI DRESDEN	ZFK-
GERMANY. ZENTRALINSTITUT FUER KERNFORSCHUNG, ROSSENDORF BEI DRESDEN	ZFK-DOS-
GERMANY. ZENTRALINSTITUT FUER KERNFORSCHUNG, ROSSENDORF BEI DRESDEN	ZFK-PHA-
GERMANY. ZENTRALINSTITUT FUER KERNFORSCHUNG, ROSSENDORF BEI DRESDEN	ZFK-RCH-
GERMANY. ZENTRALINSTITUT FUER KERNFORSCHUNG, ROSSENDORF BEI DRESDEN	ZFK-RN-
GERMANY. ZENTRALINSTITUT FUER KERNFORSCHUNG, ROSSENDORF BEI DRESDEN	ZFK-TPH-
GERMANY. ZENTRALINSTITUT FUER KERNFORSCHUNG, ROSSENDORF BEI DRESDEN	ZFK-WF-
GERMANY. ZENTRALINSTITUT FUER KERNPHYSIK, DRESDEN	KFKI-
GERMANY. ZENTRALINSTITUT FUER KERNPHYSIK, DRESDEN	ZFK-DOS-
GERMANY. ZENTRALINSTITUT FUER KERNPHYSIK, DRESDEN	ZFK-PHA-
GERMANY. ZENTRALINSTITUT FUER KERNPHYSIK, DRESDEN	ZFK-RCH-
GERMANY. ZENTRALINSTITUT FUER KERNPHYSIK, DRESDEN	ZFK-RN-
GERMANY. ZENTRALINSTITUT FUER KERNPHYSIK, DRESDEN	ZFK-TPH-
GERMANY. ZENTRALINSTITUT FUER KERNPHYSIK, DRESDEN	ZFK-WF-
GESELLSCHAFT FUER ANGEWANDTE MATHEMATIK UND MECHANIK	GAMM-
GESELLSCHAFT FUER KERNENERGIEVERWERTUNG IN SCHIFFBAU UND SCHIFFAHRT M.B.H., HAMBURG	EURMAR-
GESELLSCHAFT FUER KERNENERGIEVERWERTUNG IN SCHIFFBAU UND SCHIFFAHRT M.B.H., HAMBURG	GKSS-(YEAR)/E/
GESELLSCHAFT FUER STRAHLENFORSCHUNG M.B.H. INST. FUER STRAHLENSCHUTZ, NEUHERBERG BEI MUNICH.	GSF-
GESELLSCHAFT FUER STRAHLENFORSCHUNG M.B.H. INST. FUER STRAHLENSCHUTZ, NEUHERBERG BEI MUNICH	GSF-T-
GESELLSCHAFT FUER WELTRAUMFORSCHUNG M.B.H., BAD GODESBERG	GFW-TN-(YEAR/NUMBER)
GESELLSCHAFT FUER WELTRAUMFORSCHUNG M.B.H., BAD GODESBERG	QB-(NUMBER)-(YEAR)
GHENT. RIJKSUNIVERSITEIT	NAS-NS-
GHENT RIJKSUNIVERSITEIT. NATUURKUNDIG LABORATORIUM	FR-(YEAR)-(NUMBER)/
GIANNINI (G.M.) AND CO., PASADENA, CALIF.	GIA-
GIANNINI CONTROLS CORP. (ALL LOCATIONS)	GCC-ER-
GIANNINI CONTROLS CORP., DUARTE, CALIF.	ER-
GIANNINI CONTROLS CORP., DUARTE, CALIF.	GCC-
GIANNINI CONTROLS CORP., FAIRFIELD, N.J.	GCC-(YEAR)-(NO.)-
GIANNINI CONTROLS CORP., PASADENA, CALIF.	RTR-
GIANNINI CONTROLS CORP., SANTA ANA, CALIF.	GCC-
GIANNINI CONTROLS CORP. ASTROMECHANICS RESEARCH DIV., MALVERN, PENNA.	ARD-DM-
GIANNINI CONTROLS CORP. ASTROMECHANICS RESEARCH DIV., MALVERN, PENNA.	ARD-FR-
GIANNINI CONTROLS CORP. ASTROMECHANICS RESEARCH DIV., MALVERN, PENNA.	ARD-TR-
GIANNINI CONTROLS CORP. SYSTEMS DIV., SANTA ANA, CAL.	GMGC (LTRS.)-
GIANNINI-IRVINE, COSTA MESA, CALIF.	FR-(3 DIGITS)-(4 DIGITS)
GIANNINI RESEARCH LAB., SANTA ANA, CALIF.	GRC-
GIANNINI RESEARCH LAB., SANTA ANA, CALIF.	GRL-TR-
GIANNINI SCIENTIFIC CORP., SANTA ANA, CALIF.	FR-
GIANNINI SCIENTIFIC CORP., SANTA ANA, CALIF.	FR-(NO.)-(CON.NO.)
GIANNINI SCIENTIFIC CORP., SANTA ANA, CALIF.	GSC-
GIANNOTTI ASSOCS., BELLPORT, N.Y.	GA-(YEAR)-PR-
GIBBS AND COX, INC., N.Y.C.	G + C-
GIBBS AND COX, INC., N.Y.C.	M(YEAR)-
GIFFELS + ROSSETTI, INC., DETROIT	GV-R-
GILBERT ASSOCIATES, INC., WASHINGTON, D.C.	GAI-
GILBERT ASSOCIATES, INC. NUCLEAR ENERGY DEPT., READING, PENNA.	GAI-
GILFILLAN BROS., INC., LOS ANGELES	EDC-1.1.(NUMBER)-C
GILFILLAN BROS., INC., LOS ANGELES	GBI-
GILFILLAN BROS., INC., LOS ANGELES	GIL-
GIRAVIONS DORAND CO., PARIS	DH. (NUMBER)-PH-
GIRDLER CO. GAS PROCESSES DIV. PROCESS AND DEVELOPMENT RESEARCH LABS., LOUISVILLE, KY.	T9.40-2-
GLADDING, MC BEAN AND CO., LOS ANGELES	GMCB-
GLASGOW. UNIV.	GU-
GLASS FIBERS, INC., TOLEDO	GFI-
GLEASON WORKS, ROCHESTER, N.Y.	GW-(4 DIGITS)
GLENCO CORP., METUCHEN, N.J.	GC-
GLOBE-UNION INC., MILWAUKEE	SAR-
GODDARD SPACE FLIGHT CENTER, GREENBELT, MD.	FAR-(NUMBER)-
GODDARD SPACE FLIGHT CENTER, GREENBELT, MD.	GFSC-
GODDARD SPACE FLIGHT CENTER, GREENBELT, MD.	GSFC-TECHNICAL ABSTRACTS-
GODDARD SPACE FLIGHT CENTER, GREENBELT, MD.	NASA-GODDARD NEWS-
GODDARD SPACE FLIGHT CENTER, GREENBELT, MD.	NSSDC-
GODDARD SPACE FLIGHT CENTER, GREENBELT, MD.	OG-(2 DIGITS)
GODDARD SPACE FLIGHT CENTER, GREENBELT, MD.	PACER-(NUMBER)-
GODDARD SPACE FLIGHT CENTER, GREENBELT, MD.	X-(NUMBER-YEAR)
GODDARD SPACE FLIGHT CENTER, GREENBELT, MD.	XX-(NUMBER-YEAR)
GOETTINGEN. GERMANY. UNIV.	ASR-

GOODRICH (B.F.) CO.

GOODRICH (B.F.) CO., AKRON, OHIO	BFG-SR-
GOODRICH (B.F.) CO., AKRON, OHIO	GBF-
GOODRICH (B.F.) CO., AKRON, OHIO	GER-
GOODRICH (B.F.) CO., AKRON, OHIO	R-48-
GOODRICH (B.F.) CO., RIALTO, CALIF.	ABL-
GOODRICH (B.F.) CO. RESEARCH CENTER, BRECKSVILLE, OHIO	BFG-CONTROL-(NO.)-A
GOODRICH (B.F.) CO. RESEARCH CENTER, BRECKSVILLE, OHIO	BFG-QR-
GOODYEAR AEROSPACE CORP., AKRON, OHIO	AAP-
GOODYEAR AEROSPACE CORP., AKRON, OHIO	GAC-
GOODYEAR AEROSPACE CORP., AKRON, OHIO	GER-
GOODYEAR AEROSPACE CORP., AKRON, OHIO	GER-(YEAR)-
GOODYEAR AEROSPACE CORP., AKRON, OHIO	LA-
GOODYEAR AEROSPACE CORP., AKRON, OHIO	SP-
GOODYEAR AEROSPACE CORP., LITCHFIELD PARK, ARIZ.	AL-
GOODYEAR AEROSPACE CORP., LITCHFIELD PARK, ARIZ.	GERA-
GOODYEAR AEROSPACE CORP., LITCHFIELD PARK, ARIZ.	SU-
GOODYEAR AIRCRAFT CORP.	B.P.S.-
GOODYEAR AIRCRAFT CORP., AKRON, OHIO	E.O.-(NUMBER)-
GOODYEAR AIRCRAFT CORP., AKRON, OHIO	GAC-GEK-
GOODYEAR AIRCRAFT CORP., AKRON, OHIO	GAC-GER-
(SEE ALSO LATER NAME: GOODYEAR AEROSPACE CORP.)	
GOODYEAR ATOMIC CORP., PIKETON, OHIO	GAT-
GOODYEAR ATOMIC CORP., PIKETON, OHIO	GAT-P-
GOODYEAR ATOMIC CORP., PIKETON, OHIO	GAT-R-
GOODYEAR ATOMIC CORP., PIKETON, OHIO	GAT-T-
GOODYEAR ATOMIC CORP., PIKETON, OHIO	GAT-Z-
GOODYEAR ATOMIC CORP., PORTSMOUTH, OHIO (REF. 12)	GAT-
GOODYEAR ATOMIC CORP., PORTSMOUTH, OHIO (REF. 12)	GAT-(LETTER(S))-
GOODYEAR ATOMIC CORP., PORTSMOUTH, OHIO	GAT-DM-
GOODYEAR ATOMIC CORP., PORTSMOUTH, OHIO (REF. 12)	GAT-DR-
GOODYEAR ATOMIC CORP., PORTSMOUTH, OHIO	GAT-E-
GOODYEAR ATOMIC CORP., PORTSMOUTH, OHIO (REF. 12)	GAT-L-
GOODYEAR ATOMIC CORP., PORTSMOUTH, OHIO	GAT-OR-
GOODYEAR ATOMIC CORP., PORTSMOUTH, OHIO	GAT-SI-
GOODYEAR TIRE AND RUBBER CO., AKRON, OHIO	GT+R-
GOODYEAR TIRE AND RUBBER CO. AVIATION PRODUCTION DIV., LITCHFIELD PARK, ARIZ.	APR-
GOTHAER WAGGONFABRIK (VERSUCHBERICHT)	GO VB-
GOULD-NATIONAL BATTERIES, INC., DEPEW, N.Y.	GNB-
GOULD-NATIONAL BATTERIES, INC. RES. DIV., MINNEAPOLIS	GNBI-
GOVERNMENT-INDUSTRY COMMITTEE ON NARBA PREPARATION	NARBA-
GOVERNMENT PRINTING OFFICE, WASHINGTON, D.C.	GPO-
GOVERNMENT PRINTING OFFICE, WASH., D.C. (PRICE LIST)	PL-
GOVERNMENT PRINTING OFFICE, WASHINGTON, D.C.	USGPO-
GOVERNMENT PRINTING OFFICE. SUPERINTENDENT OF DOCUMENTS, WASHINGTON, D.C.	SD-
GRACE (W.R.) AND CO. RESEARCH DIV., CLARKSVILLE, MD.	RES-
GRACE (W.R.) AND CO. RESEARCH DIV., CLARKSVILLE, MD.	RES-(YR.)-(NO.)
GRAHAM, CROWLEY AND ASSOCIATES, INC., CHICAGO	GCA-
GRAHAM, SAVAGE, AND ASSOCIATES, INC., JENKINTOWN, PA.	GSA-
GRAND CENTRAL ROCKET CO., REDLANDS, CALIF.	GCR-
GRAND JUNCTION OFFICE, AEC, COLO. (REF. 4)	GJO-(NUMBER)-
GRAND JUNCTION OPERATIONS OFFICE, AEC, COLO. (REF. 4)	GJ-
GRAZ. TECHNISCHE HOCHSCHULE	MTSR-
GRAZ UNIV.	GRAZU-
GT. BRIT. ADMIRALTY	CB-
GT. BRIT. ADMIRALTY. DEPT. OF OPERATIONAL RESEARCH	DOR-
GT. BRIT. ADMIRALTY. DEPT. OF PHYSICAL RESEARCH	DPR/BWS/-
GT. BRIT. ADMIRALTY. DEPT. OF PHYSICAL RESEARCH	DPR/BWS-(YR.)
GT. BRIT. ADMIRALTY. DEPT. OF SCI. RES. + EXPERIMENT	ATR/T-
GT. BRIT. ADMIRALTY. DEPT. OF SCI. RES. + EXPERIMENT	SRE/WAVE/
GT. BRIT. ADMIRALTY. ELECTRICAL ENGINEERING DEPT.	DEE-
GT. BRIT. ADMIRALTY. NAVAL CONSTRUCTION DEPT.	DNC-
GT. BRIT. ADMIRALTY. NAVAL INTELLIGENCE DEPT.	PG-
GT. BRIT. ADMIRALTY. NAVAL INTELLIGENCE DIRECTOR	NID PG-
GT. BRIT. ADMIRALTY CENTRAL METALLURGICAL LAB.	CML-
GT. BRIT. ADMIRALTY CENTRE FOR SCIENTIFIC INFORMATION AND LIAISON, LONDON	ACSIL/-
GT. BRIT. ADMIRALTY CENTRE FOR SCIENTIFIC INFORMATION AND LIAISON, LONDON	ACSIL/AD/-
GT. BRIT. ADMIRALTY CENTRE FOR SCIENTIFIC INFORMATION AND LIAISON, LONDON	ACSIL/ADM/-
GT. BRIT. ADMIRALTY CENTRE FOR SCIENTIFIC INFORMATION AND LIAISON, LONDON	ACSIL/LIBY/-
GT. BRIT. ADMIRALTY CENTRE FOR SCIENTIFIC INFORMATION AND LIAISON, LONDON	ACSIL-TRANS-
GT. BRIT. ADMIRALTY CENTRE FOR SCIENTIFIC INFORMATION AND LIAISON, LONDON	F2-
GT. BRIT. ADMIRALTY COMPUTING SERVICE	ACS-
GT. BRIT. ADMIRALTY COMPUTING SERVICE	SRE/ACS-
GT. BRIT. ADMIRALTY ENGINEERING LAB., WEST DRAYTON, MIDDX, ENGLAND	AEL-
GT. BRIT. ADMIRALTY ENGINEERING LAB., WEST DRAYTON, MIDDX, ENGLAND	AEL-M-

GT. BRIT. ADMIRALTY ENGINEERING LAB., WEST DRAYTON, MIDDX, ENGLAND	AEL-TN-
GT. BRIT. ADMIRALTY EXPERIMENT WORKS, HASLAR, ENGLAND	AEW-(NUMBER)/(YEAR)
GT. BRIT. ADMIRALTY EXPERIMENT WORKS, HASLAR, ENGLAND	AEW-
GT. BRIT. ADMIRALTY EXPERIMENTAL DIVING UNIT, PORTSMOUTH, HANTS, ENGLAND	AEDU-
GT. BRIT. ADMIRALTY FUEL EXPERIMENTAL STATION	AFES-
GT. BRIT. ADMIRALTY GUNNERY ESTABLISHMENT	AGE-
GT. BRIT. ADMIRALTY MATERIALS LAB., POOLE, DORSET, ENG	A/(NO.)(S)
GT. BRIT. ADMIRALTY MATERIALS LAB., POOLE, DORSET, ENG	ACSU/ADM/-
GT. BRIT. ADMIRALTY MATERIALS LAB., POOLE, DORSET, ENG	AML-
GT. BRIT. ADMIRALTY MATERIALS LAB., POOLE, DORSET, ENG	AML/A/
GT. BRIT. ADMIRALTY MATERIALS LAB., POOLE, DORSET, ENG	AML/P/
GT. BRIT. ADMIRALTY MATERIALS LAB., POOLE, DORSET, ENG	HOL-
GT. BRIT. ADMIRALTY OIL LAB. (ALL LOCATIONS)	AOL-
GT. BRIT. ADMIRALTY OIL LAB. TEDDINGTON, MIDDX, ENG.	AOL-TM-
GT. BRIT. ADMIRALTY RES. LAB., TEDDINGTON, MIDDX, ENG	ARL-
GT. BRIT. ADMIRALTY RES. LAB., TEDDINGTON, MIDDX, ENG	ARL/C/R-
GT. BRIT. ADMIRALTY RES. LAB., TEDDINGTON, MIDDX, ENG	ARL/G/N(NO.)
GT. BRIT. ADMIRALTY RES. LAB., TEDDINGTON, MIDDX, ENG	ARL/H/N(NUMBER)
GT. BRIT. ADMIRALTY RES. LAB., TEDDINGTON, MIDDX, ENG	ARL/L/N(NO.)
GT. BRIT. ADMIRALTY RES. LAB., TEDDINGTON, MIDDX, ENG	ARL/L/R(NO.)
GT. BRIT. ADMIRALTY RES. LAB., TEDDINGTON, MIDDX, ENG	ARL/M/N(NO.)
GT. BRIT. ADMIRALTY RES. LAB., TEDDINGTON, MIDDX, ENG	ARL/M/P(NUMBER)-
GT. BRIT. ADMIRALTY RES. LAB., TEDDINGTON, MIDDX, ENG	ARL/M/R(NO.)
GT. BRIT. ADMIRALTY RES. LAB., TEDDINGTON, MIDDX, ENG	ARL/N1/R-
GT. BRIT. ADMIRALTY RES. LAB., TEDDINGTON, MIDDX, ENG	ARL/O/N(NUMBER)
GT. BRIT. ADMIRALTY RES. LAB., TEDDINGTON, MIDDX, ENG	ARL/R1/R-
GT. BRIT. ADMIRALTY RES. LAB., TEDDINGTON, MIDDX, ENG	ARL/R2/R-
GT. BRIT. ADMIRALTY RES. LAB., TEDDINGTON, MIDDX, ENG	ARL/R3/E-
GT. BRIT. ADMIRALTY SIGNAL AND RADAR ESTABLISHMENT, PORTSMOUTH, HANTS, ENGLAND	ASRE-
GT. BRIT. ADMIRALTY SIGNAL AND RADAR ESTABLISHMENT, PORTSMOUTH, HANTS, ENGLAND	GX-
GT. BRIT. ADMIRALTY SIGNAL ESTABLISHMENT	ASE-
GT. BRIT. ADMIRALTY SURFACE WEAPONS ESTAB., PORTSMOUTH, HANTS, ENGLAND	ASWE-LAB NOTE-
GT. BRIT. ADMIRALTY TORPEDO EXPTL. EST., GREENOCK, SCOT	STR-
GT. BRIT. ADMIRALTY UNDEX SUB-PANEL	UNDEX-
GT. BRIT. ADMIRALTY UNDEX WORKS	AUW/TRI-
GT. BRIT. ADVISORY COUNCIL ON SCI. RES. + TECH. DEV.	A.C.-
GT. BRIT. ADVISORY COUNCIL ON SCI. RES. + TECH. DEV.	AC-
GT. BRIT. ADVISORY COUNCIL ON SCI. RES. + TECH. DEV.	CHEM/EX-
GT. BRIT. ADVISORY COUNCIL ON SCI. RES. + TECH. DEV.	HEI-
GT. BRIT. ADVISORY COUNCIL ON SCI. RES. + TECH. DEV.	IB-
GT. BRIT. ADVISORY COUNCIL ON SCI. RES. + TECH. DEV.	LFC-
GT. BRIT. ADVISORY COUNCIL ON SCI. RES. + TECH. DEV.	PAC-
GT. BRIT. ADVISORY COUNCIL ON SCI. RES. + TECH. DEV.	PHYS/EX-
GT. BRIT. ADVISORY COUNCIL ON SCI. RES. + TECH. DEV.	R.D.EXPLOSIVES REPT.-
GT. BRIT. ADVISORY COUNCIL ON SCI. RES. + TECH. DEV.	RDF-
GT. BRIT. ADVISORY COUNCIL ON SCI. RES. + TECH. DEV.	RDX-
GT. BRIT. ADVISORY COUNCIL ON SCI. RES. + TECH. DEV.	SC-
GT. BRIT. ADVISORY COUNCIL ON SCI. RES. + TECH. DEV.	SE-
GT. BRIT. ADVISORY COUNCIL ON SCI. RES. + TECH. DEV.	TNT-
GT. BRIT. ADVISORY COUNCIL ON SCI. RES. + TECH. DEV.	TNT/RSC
GT. BRIT. ADVISORY COUNCIL ON SCI. RES. + TECH. DEV.	TRL-
GT. BRIT. ADVISORY COUNCIL ON SCI. RES. + TECH. DEV. LABOR AND MANPOWER DIVISION	LM-
GT. BRIT. ADVISORY COUNCIL ON SCI. RES. + TECH. DEV. STATIC DETONATION COMM. FRAGMENTATION PANEL	SD/FP-
GT. BRIT. AERONAUTICAL INSPECTION DIRECTORATE, HAREFIELD, MIDDX, ENGLAND	AID-
GT. BRIT. AERONAUTICAL INSPECTION DIRECTORATE, HAREFIELD, MIDDX, ENGLAND	AID/METRO/
GT. BRIT. AERONAUTICAL INSPECTION DIRECTORATE, HAREFIELD, MIDDX, ENGLAND	AID-METRO-
GT. BRIT. AERONAUTICAL RESEARCH COUNCIL	AP-
GT. BRIT. AERONAUTICAL RESEARCH COUNCIL	ARC-
GT. BRIT. AERONAUTICAL RESEARCH COUNCIL	ARC-CP-
GT. BRIT. AERONAUTICAL RES. COUNCIL (CURRENT PAPER)	ARCCP-
GT. BRIT. AERONAUTICAL RESEARCH COUNCIL	ARC-FM-
GT. BRIT. AERONAUTICAL RESEARCH COUNCIL	ARC(GB)-
GT. BRIT. AERONAUTICAL RESEARCH COUNCIL	ARCRL-
GT. BRIT. AERONAUTICAL RESEARCH COUNCIL	ARC-R+M-
GT. BRIT. AERONAUTICAL RESEARCH COUNCIL	ARC RM-
GT. BRIT. AERONAUTICAL RESEARCH COUNCIL	ARC-R/M-
GT. BRIT. AERONAUTICAL RESEARCH COUNCIL	ARC-S+C-
GT. BRIT. AERONAUTICAL RESEARCH COUNCIL	ARC-TN-
GT. BRIT. AERONAUTICAL RESEARCH COUNCIL	ARC TR-
GT. BRIT. AERONAUTICAL RESEARCH COUNCIL	AW-
GT. BRIT. AERONAUTICAL RESEARCH COUNCIL	CF-
GT. BRIT. AERONAUTICAL RESEARCH COUNCIL	CIV-
GT. BRIT. AERONAUTICAL RESEARCH COUNCIL	CP-
GT. BRIT. AERONAUTICAL RESEARCH COUNCIL	HMT-
GT. BRIT. AERONAUTICAL RESEARCH COUNCIL	N-
GT. BRIT. AERONAUTICAL RESEARCH COUNCIL	PERF-
GT. BRIT. AERONAUTICAL RESEARCH COUNCIL	RAF-ARC-
GT. BRIT. AERONAUTICAL RESEARCH COUNCIL	R AND M-
GT. BRIT. AERONAUTICAL RESEARCH COUNCIL	R + M-
GT. BRIT. AERONAUTICAL RESEARCH COUNCIL	RM-
GT. BRIT. AERONAUTICAL RESEARCH COUNCIL	S AND C-
GT. BRIT. AERONAUTICAL RESEARCH COUNCIL	T-
GT. BRIT. AERONAUTICAL RESEARCH COUNCIL, LONDON	ARC-COMP-(FM)-
GT. BRIT. AERONAUTICAL RESEARCH COUNCIL, LONDON	ARC-HMT-
GT. BRIT. AERONAUTICAL RESEARCH COUNCIL, LONDON	ARC-PA-
GT. BRIT. AERONAUTICAL RESEARCH COUNCIL, LONDON	ARC-PERF-
GT. BRIT. AERONAUTICAL RESEARCH COUNCIL, LONDON	ARC-T+M-
GT. BRIT. AERONAUTICAL RESEARCH COUNCIL, LONDON	STRUT.-
GT. BRIT. AERONAUTICAL RESEARCH COUNCIL. ENGINE AERODYNAMICS SUBCOMMITTEE	EA-

GT. BRIT. AERONAUTICAL RESEARCH COUNCIL. ENGINE AERODYNAMICS SUBCOMMITTEE

GT. BRIT. AERONAUTICAL RESEARCH COUNCIL. ENGINE AERODYNAMICS SUBCOMMITTEE	FM-
GT. BRIT. AERO. RES. COUNCIL. FLUID MOTION SUB-COMM.	F.M.-
GT. BRIT. AERO. RES. COUNCIL. FLUID MOTION SUB-COMM.	FM-
GT. BRIT. AERO. RES. COUNCIL. POWER PLANTS COMM.	PP-
GT. BRIT. AERO. RES. COUNCIL. ROCKETS SUBCOMM.	R-
GT. BRIT. AERO. RES. COUNCIL. WIND TUNNEL DESIGN COMM	ARC-TP-
GT. BRIT. AERO. RES. COUNCIL. WIND TUNNEL DESIGN COMM	TP-
GT. BRIT. AEROPLANE AND ARMAMENT EXPERIMENTAL ESTABLISHMENT, BOSCOMBE DOWN, AMESBURY, WILTS, ENGLAND	AAEE/
GT. BRIT. AEROPLANE AND ARMAMENT EXPERIMENTAL ESTABLISHMENT, BOSCOMBE DOWN, AMESBURY, WILTS, ENGLAND	AAEE/ATO/
GT. BRIT. AEROPLANE AND ARMAMENT EXPERIMENTAL ESTABLISHMENT, BOSCOMBE DOWN, AMESBURY, WILTS, ENGLAND	AAEE/INST/
GT. BRIT. AEROPLANE AND ARMAMENT EXPERIMENTAL ESTABLISHMENT, BOSCOMBE DOWN, AMESBURY, WILTS, ENGLAND	A/AEE-MEMO-
GT. BRIT. AEROPLANE AND ARMAMENT EXPERIMENTAL ESTABLISHMENT, BOSCOMBE DOWN, AMESBURY, WILTS, ENGLAND	AAEE/NOTE/
GT. BRIT. AEROPLANE AND ARMAMENT EXPERIMENTAL ESTABLISHMENT, BOSCOMBE DOWN, AMESBURY, WILTS, ENGLAND	AAEE/TECH/
GT. BRIT. AEROPLANE AND ARMAMENT EXPTL. EST. ARMAMENT DIV., BOSCOMBE DOWN, AMESBURY, WILTS, ENGLAND	ADR-(NUMBER/YEAR)
GT. BRIT. AGRICULTURAL RESEARCH COUNCIL. RADIOBIOLOGICAL LAB., GROVE, BERKS, ENGLAND	ARCL-
GT. BRIT. AGRICULTURAL RESEARCH COUNCIL. RADIOBIOLOGICAL LAB., GROVE, BERKS, ENGLAND	ARCRL-
GT. BRIT. AIR DEFENCE RESEARCH AND DEVELOPMENT ESTABLISHMENT, MALVERN, WORCS, ENGLAND	ADRDE-
GT. BRIT. AIR DOCUMENT INTELLIGENCE (CODE LETTER K)	ADIK-
GT. BRIT. AIR INTERCEPTION COMMITTEE	AIC-
GT. BRIT. AIR MINISTRY	AM-
GT. BRIT. AIR MINISTRY (AIR PUBLICATIONS)	AM AP-
GT. BRIT. AIR MINISTRY (AIR PUBLICATIONS)	AN AP-
GT. BRIT. AIR MINISTRY (AIR PUBLICATIONS)	AP-
GT. BRIT. AIR MINISTRY	ASA(O)-
GT. BRIT. AIR MINISTRY	SD-
GT. BRIT. AIR MINISTRY	SDTM-
GT. BRIT. AIR MINISTRY. METEOROLOGICAL RES. COMM.	MRP-
GT. BRIT. AIR MINISTRY. METEOROLOGICAL RES. COMM.	S.C. 111/
GT. BRIT. AIR MINISTRY. METEOROLOGICAL RES. COMM.	SC-
GT. BRIT. AIR SCIENTIFIC INTELLIGENCE (TECH. TRANS.)	ASI-T-
GT. BRIT. AIR SCIENTIFIC INTELLIGENCE (TECH. TRANS.)	GB-ASI-T-
GT. BRIT. AIR/SEA WARFARE DEVELOPMENT UNIT.	ASWDU-
GT. BRIT. APPLIED PSYCH. RES. UNIT, CAMBRIDGE, ENG.	APU-
GT. BRIT. ARMAMENT DESIGN DEPT.,FT.HALSTEAD,KENT,ENG.	ADD-TR-
GT. BRIT. ARMAMENT DESIGN EST.,FT.HALSTEAD,KENT,ENG.	ADE-
GT. BRIT. ARMAMENT DESIGN EST.,FT.HALSTEAD,KENT,ENG.	RSAF-
GT. BRIT. ARMAMENT RESEARCH AND DEVELOPMENT ESTABLISHMENT, FORT HALSTEAD, KENT, ENGLAND	ARD-(NUMBER)/
GT. BRIT. ARMAMENT RESEARCH AND DEVELOPMENT ESTABLISHMENT, FORT HALSTEAD, KENT, ENGLAND	ARDE-
GT. BRIT. ARMAMENT RESEARCH AND DEVELOPMENT ESTABLISHMENT, FORT HALSTEAD, KENT, ENGLAND	ARDE(B)-(NO.)(YR)
GT. BRIT. ARMAMENT RESEARCH AND DEVELOPMENT ESTABLISHMENT, FORT HALSTEAD, KENT, ENGLAND	ARDE(M)-(NO.)(YR)
GT. BRIT. ARMAMENT RESEARCH AND DEVELOPMENT ESTABLISHMENT, FORT HALSTEAD, KENT, ENGLAND	ARDE-MEMO-
GT. BRIT. ARMAMENT RESEARCH AND DEVELOPMENT ESTABLISHMENT, FORT HALSTEAD, KENT, ENGLAND	ARDE-MEMO(B)-(NO.)(YR)
GT. BRIT. ARMAMENT RESEARCH AND DEVELOPMENT ESTABLISHMENT, FORT HALSTEAD, KENT, ENGLAND	ARDE(MX)-(NO.)(YR)
GT. BRIT. ARMAMENT RESEARCH AND DEVELOPMENT ESTABLISHMENT, FORT HALSTEAD, KENT, ENGLAND	ARDE-(N)-(NO.)(YR)
GT. BRIT. ARMAMENT RESEARCH AND DEVELOPMENT ESTABLISHMENT, FORT HALSTEAD, KENT, ENGLAND	ARD EXPLOSIVES REPORT-
GT. BRIT. ARMAMENT RESEARCH AND DEVELOPMENT ESTABLISHMENT, FORT HALSTEAD, KENT, ENGLAND	ARD EXP. RPT.(NUMBER)/
GT. BRIT. ARMAMENT RESEARCH AND DEVELOPMENT ESTABLISHMENT, FORT HALSTEAD, KENT, ENGLAND	AWEC/P-
GT. BRIT. ARMAMENT RESEARCH AND DEVELOPMENT ESTABLISHMENT, FORT HALSTEAD, KENT, ENGLAND	AWEC/P(YEAR)
GT. BRIT. ARMAMENT RESEARCH AND DEVELOPMENT ESTABLISHMENT, FORT HALSTEAD, KENT, ENGLAND	B1/(NUMBER)/(NUMBER)
GT. BRIT. ARMAMENT RESEARCH AND DEVELOPMENT ESTABLISHMENT, FORT HALSTEAD, KENT, ENGLAND	BI(NO.)/(YR)-
GT. BRIT. ARMAMENT RESEARCH AND DEVELOPMENT ESTABLISHMENT, FORT HALSTEAD, KENT, ENGLAND	BRANCH MEMO B-
GT. BRIT. ARMAMENT RESEARCH AND DEVELOPMENT ESTABLISHMENT, FORT HALSTEAD, KENT, ENGLAND	RARDE-MEMO-(NO.)/(YR.)
GT. BRIT. ARMAMENT RESEARCH AND DEVELOPMENT ESTABLISHMENT, HORSHAM, SUSSEX, ENGLAND	ARDE-S-
GT. BRIT. ARMAMENT RES. DEPT.,FT.HALSTEAD, KENT, ENG.	AC-
GT. BRIT. ARMAMENT RES. DEPT.,FT.HALSTEAD, KENT, ENG.	ARDC-TRR-
GT. BRIT. ARMAMENT RESEARCH DEPT., FORT HALSTEAD, KENT, ENGLAND (ELECTRODEPOSITION MEMO)	ARD EM-
GT. BRIT. ARMAMENT RESEARCH DEPT., FORT HALSTEAD, KENT, ENGLAND (THEORETICAL RESEARCH REPORT)	ARD-TRR-
GT. BRIT. ARMAMENT RESEARCH DEPT., FORT HALSTEAD, KENT, ENGLAND (ELECTRODEPOSITION MEMO)	EM-
GT. BRIT. ARMAMENT RESEARCH DEPT., FORT HALSTEAD, KENT, ENGLAND (ELECTRODEPOSITION MEMO)	MS ARD EM-
GT. BRIT. ARMAMENT RES. DEPT.,FT.HALSTEAD, KENT, ENG.	PHYS/EX-
GT. BRIT. ARMAMENT RES. DEPT.,FT.HALSTEAD, KENT, ENG.	UBRC-
GT. BRIT. ARMAMENT RES. DEPT.,FT.HALSTEAD, KENT, ENG.	XA (NO.)/-
GT. BRIT. ARMAMENT RES. DEPT.,FT.HALSTEAD, KENT, ENG.	XC(2)-
GT. BRIT. ARMAMENT RES. EST., FT. HALSTEAD, KENT, ENG	AR-
GT. BRIT. ARMAMENT RES. EST., FT. HALSTEAD, KENT, ENG	ARE-H-(NO.)/(YR)
GT. BRIT. ARMAMENT RES. EST., FT. HALSTEAD, KENT, ENG	ARE-MEMO-
GT. BRIT. ARMAMENT RES. EST., FT. HALSTEAD, KENT, ENG	ARE-MR-
GT. BRIT. ARMAMENT RES. EST., FT. HALSTEAD, KENT, ENG	H.(NUMBER)/(YEAR)
GT. BRIT. ARMAMENT RES. EST., FT. HALSTEAD, KENT, ENG	HER-A(NO.)/(YR)
GT. BRIT. ARMAMENT RES. EST., FT. HALSTEAD, KENT, ENG	HQ-(NUMBER)/(NUMBER)
GT. BRIT. ARMY COUNCIL. ARMY OPERATIONAL RESEARCH GROUP, LONDON	AORG-
GT. BRIT. ARMY OF THE RHINE. SPECIAL PROJECTILE OPERATIONS GROUP	BAOR-
GT. BRIT. ARMY SUPPLY DIRECTORATE	ZA-
GT. BRIT. ATOMIC EN. RES. ESTAB., HARWELL, BERKS, ENG	ACPP/P-
GT. BRIT. ATOMIC EN. RES. ESTAB., HARWELL, BERKS, ENG	AERE-9ER/-
GT. BRIT. ATOMIC EN. RES. ESTAB., HARWELL, BERKS, ENG	AERE-CATALOGUE NO.-
GT. BRIT. ATOMIC EN. RES. ESTAB., HARWELL, BERKS, ENG	AERE-CIRCUS/A-
GT. BRIT. ATOMIC EN. RES. ESTAB., HARWELL, BERKS, ENG	AERE-E.443/N-
GT. BRIT. ATOMIC EN. RES. ESTAB., HARWELL, BERKS, ENG	AERE-FSG-
GT. BRIT. ATOMIC EN. RES. ESTAB., HARWELL, BERKS, ENG	AERE-LEO/N-
GT. BRIT. ATOMIC EN. RES. ESTAB., HARWELL, BERKS, ENG	AERE-LEO-RDC-
GT. BRIT. ATOMIC EN. RES. ESTAB., HARWELL, BERKS, ENG	AERE-LMFP/
GT. BRIT. ATOMIC EN. RES. ESTAB., HARWELL, BERKS, ENG	AERE/MS-
GT. BRIT. ATOMIC EN. RES. ESTAB., HARWELL, BERKS, ENG	AERE-P-(NO.)WP/P-
GT. BRIT. ATOMIC EN. RES. ESTAB., HARWELL, BERKS, ENG	AERE-PIPPA-WP/MEMO-

GT. BRIT. ATOMIC EN. RES. ESTAB., HARWELL, BERKS, ENG	AERE-RCDG-
GT. BRIT. ATOMIC EN. RES. ESTAB., HARWELL, BERKS, ENG	AERE-REV.-
GT. BRIT. ATOMIC EN. RES. ESTAB., HARWELL, BERKS, ENG	AERE-SPAR-
GT. BRIT. ATOMIC EN. RES. ESTAB., HARWELL, BERKS, ENG	AERE-SPEC-
GT. BRIT. ATOMIC EN. RES. ESTAB., HARWELL, BERKS, ENG	AP-
GT. BRIT. ATOMIC EN. RES. ESTAB., HARWELL, BERKS, ENG	ATRS/CONF/
GT. BRIT. ATOMIC EN. RES. ESTAB., HARWELL, BERKS, ENG	BCM/WP-
GT. BRIT. ATOMIC EN. RES. ESTAB., HARWELL, BERKS, ENG	BI-
GT. BRIT. ATOMIC EN. RES. ESTAB., HARWELL, BERKS, ENG	B/L-
GT. BRIT. ATOMIC EN. RES. ESTAB., HARWELL, BERKS, ENG	BLWR/N-
GT. BRIT. ATOMIC EN. RES. ESTAB., HARWELL, BERKS, ENG	BMC/WP-
GT. BRIT. ATOMIC EN. RES. ESTAB., HARWELL, BERKS, ENG	BWR-
GT. BRIT. ATOMIC EN. RES. ESTAB., HARWELL, BERKS, ENG	CAT. NO. 3-7/
GT. BRIT. ATOMIC EN. RES. ESTAB., HARWELL, BERKS, ENG	CCP/(NO.)-
GT. BRIT. ATOMIC EN. RES. ESTAB., HARWELL, BERKS, ENG	CPDC/H-
GT. BRIT. ATOMIC EN. RES. ESTAB., HARWELL, BERKS, ENG	CSPC-(NUMBER)(LETTER)
GT. BRIT. ATOMIC EN. RES. ESTAB., HARWELL, BERKS, ENG	E-
GT. BRIT. ATOMIC EN. RES. ESTAB., HARWELL, BERKS, ENG	E.443/N-
GT. BRIT. ATOMIC EN. RES. ESTAB., HARWELL, BERKS, ENG	ED/448/N-
GT. BRIT. ATOMIC EN. RES. ESTAB., HARWELL, BERKS, ENG	ERG/
GT. BRIT. ATOMIC EN. RES. ESTAB., HARWELL, BERKS, ENG	EXTRA-MURAL RES. NO.-
GT. BRIT. ATOMIC EN. RES. ESTAB., HARWELL, BERKS, ENG	F.72-
GT. BRIT. ATOMIC EN. RES. ESTAB., HARWELL, BERKS, ENG	FPP-
GT. BRIT. ATOMIC EN. RES. ESTAB., HARWELL, BERKS, ENG	FPS-
GT. BRIT. ATOMIC EN. RES. ESTAB., HARWELL, BERKS, ENG	FREWP/P-
GT. BRIT. ATOMIC EN. RES. ESTAB., HARWELL, BERKS, ENG	FRFEWP-
GT. BRIT. ATOMIC EN. RES. ESTAB., HARWELL, BERKS, ENG	FRPC-FEWP-
GT. BRIT. ATOMIC EN. RES. ESTAB., HARWELL, BERKS, ENG	H-
GT. BRIT. ATOMIC EN. RES. ESTAB., HARWELL, BERKS, ENG	HARE/
GT. BRIT. ATOMIC EN. RES. ESTAB., HARWELL, BERKS, ENG	HD-
GT. BRIT. ATOMIC EN. RES. ESTAB., HARWELL, BERKS, ENG	HDS-
GT. BRIT. ATOMIC EN. RES. ESTAB., HARWELL, BERKS, ENG	HER-
GT. BRIT. ATOMIC EN. RES. ESTAB., HARWELL, BERKS, ENG	HGG-
GT. BRIT. ATOMIC EN. RES. ESTAB., HARWELL, BERKS, ENG	HS-
GT. BRIT. ATOMIC EN. RES. ESTAB., HARWELL, BERKS, ENG	HWWP-
GT. BRIT. ATOMIC EN. RES. ESTAB., HARWELL, BERKS, ENG	I-(NO.)-871A-(NO.)/(NO.)
GT. BRIT. ATOMIC EN. RES. ESTAB., HARWELL, BERKS, ENG	IFR/P-
GT. BRIT. ATOMIC EN. RES. ESTAB., HARWELL, BERKS, ENG	JNPO-MWP(G)/P-
GT. BRIT. ATOMIC EN. RES. ESTAB., HARWELL, BERKS, ENG	LEO/A-
GT. BRIT. ATOMIC EN. RES. ESTAB., HARWELL, BERKS, ENG	LEO/N-
GT. BRIT. ATOMIC EN. RES. ESTAB., HARWELL, BERKS, ENG	LEO/RDC-
GT. BRIT. ATOMIC EN. RES. ESTAB., HARWELL, BERKS, ENG	LFEWP-
GT. BRIT. ATOMIC EN. RES. ESTAB., HARWELL, BERKS, ENG	LFLWP-
GT. BRIT. ATOMIC EN. RES. ESTAB., HARWELL, BERKS, ENG	LMFS/N-
GT. BRIT. ATOMIC EN. RES. ESTAB., HARWELL, BERKS, ENG	LMSP/P-
GT. BRIT. ATOMIC EN. RES. ESTAB., HARWELL, BERKS, ENG	MASC/N-
GT. BRIT. ATOMIC EN. RES. ESTAB., HARWELL, BERKS, ENG	NAPLES-UK-
GT. BRIT. ATOMIC EN. RES. ESTAB., HARWELL, BERKS, ENG	NPC-(YEAR)/-
GT. BRIT. ATOMIC EN. RES. ESTAB., HARWELL, BERKS, ENG	NP/GEN-
GT. BRIT. ATOMIC EN. RES. ESTAB., HARWELL, BERKS, ENG	N/R-
GT. BRIT. ATOMIC EN. RES. ESTAB., HARWELL, BERKS, ENG	P-BR-
GT. BRIT. ATOMIC EN. RES. ESTAB., HARWELL, BERKS, ENG	PCWP/P-
GT. BRIT. ATOMIC EN. RES. ESTAB., HARWELL, BERKS, ENG	PIPPA-WA/MEMO-
GT. BRIT. ATOMIC EN. RES. ESTAB., HARWELL, BERKS, ENG	POC-A-
GT. BRIT. ATOMIC EN. RES. ESTAB., HARWELL, BERKS, ENG	POC-DATA-
GT. BRIT. ATOMIC EN. RES. ESTAB., HARWELL, BERKS, ENG	POCLM/MEMO-
GT. BRIT. ATOMIC EN. RES. ESTAB., HARWELL, BERKS, ENG	POC/SC-
GT. BRIT. ATOMIC EN. RES. ESTAB., HARWELL, BERKS, ENG	PPWPR-
GT. BRIT. ATOMIC EN. RES. ESTAB., HARWELL, BERKS, ENG	PSC- (AERE)
GT. BRIT. ATOMIC EN. RES. ESTAB., HARWELL, BERKS, ENG	PWP(P)-
GT. BRIT. ATOMIC EN. RES. ESTAB., HARWELL, BERKS, ENG	RCSC-
GT. BRIT. ATOMIC EN. RES. ESTAB., HARWELL, BERKS, ENG	RCTC-
GT. BRIT. ATOMIC EN. RES. ESTAB., HARWELL, BERKS, ENG (ASSIGNED BY GT. BRIT. DIV. OF ATOMIC ENERGY (PRODUCTION) TO INCOMING REPORTS FROM A.E.R.E.,HARWELL)	RISLEY-4001 THRU -5000
GT. BRIT. ATOMIC EN. RES. ESTAB., HARWELL, BERKS, ENG	RLP/P-
GT. BRIT. ATOMIC EN. RES. ESTAB., HARWELL, BERKS, ENG	R/M-
GT. BRIT. ATOMIC EN. RES. ESTAB., HARWELL, BERKS, ENG	ROSC/P-
GT. BRIT. ATOMIC EN. RES. ESTAB., HARWELL, BERKS, ENG	RP/M-
GT. BRIT. ATOMIC EN. RES. ESTAB., HARWELL, BERKS, ENG	RSC-(NUMBER)(AERE)
GT. BRIT. ATOMIC EN. RES. ESTAB., HARWELL, BERKS, ENG	RVS-
GT. BRIT. ATOMIC EN. RES. ESTAB., HARWELL, BERKS, ENG	SGP/P-
GT. BRIT. ATOMIC EN. RES. ESTAB., HARWELL, BERKS, ENG	SMC-
GT. BRIT. ATOMIC EN. RES. ESTAB., HARWELL, BERKS, ENG	TDC-
GT. BRIT. ATOMIC EN. RES. ESTAB., HARWELL, BERKS, ENG	T/M-
GT. BRIT. ATOMIC EN. RES. ESTAB., HARWELL, BERKS, ENG	TM-
GT. BRIT. ATOMIC EN. RES. ESTAB., HARWELL, BERKS, ENG	TP-
GT. BRIT. ATOMIC EN. RES. ESTAB., HARWELL, BERKS, ENG	T/R-
GT. BRIT. ATOMIC EN. RES. ESTAB., HARWELL, BERKS, ENG	TRDC/T/-
GT. BRIT. ATOMIC EN. RES. ESTAB., HARWELL, BERKS, ENG	X/R-
(SEE ALSO LATER NAME: UNITED KINGDOM ATOMIC ENERGY AUTHORITY. RESEARCH GP. ATOMIC EN. RES. ESTAB.)	
GT. BRIT. ATOMIC ENERGY RESEARCH ESTABLISHMENT. HARWELL POWER COMMITTEE, HARWELL, BERKS, ENGLAND	HPC-
GT. BRIT. ATOMIC ENERGY RESEARCH ESTABLISHMENT. REACTOR PHYS. DIV., HARWELL, BERKS, ENGLAND	MATHS,MEMO/(NO.)/(LTRS)
GT. BRIT. ATOMIC WEAPONS RESEARCH ESTABLISHMENT, ALDERMASTON, BERKS, ENGLAND (REF. 1)	(NUMBERS...)
GT. BRIT. ATOMIC WEAPONS RESEARCH ESTABLISHMENT, ALDERMASTON, BERKS, ENGLAND	AM-
GT. BRIT. ATOMIC WEAPONS RESEARCH ESTABLISHMENT, ALDERMASTON, BERKS, ENGLAND	AWEC/P(YEAR)-
GT. BRIT. ATOMIC WEAPONS RESEARCH ESTABLISHMENT, ALDERMASTON, BERKS, ENGLAND	E-(NUMBER)/(YEAR)
GT. BRIT. ATOMIC WEAPONS RESEARCH ESTABLISHMENT, ALDERMASTON, BERKS, ENGLAND	HER-A-
GT. BRIT. ATOMIC WEAPONS RESEARCH ESTABLISHMENT, ALDERMASTON, BERKS, ENGLAND	HER-H-
GT. BRIT. ATOMIC WEAPONS RESEARCH ESTABLISHMENT, ALDERMASTON, BERKS, ENGLAND	HO/TOTEM-
GT. BRIT. ATOMIC WEAPONS RESEARCH ESTABLISHMENT, ALDERMASTON, BERKS, ENGLAND	NWR/H9/(NO.)/-
GT. BRIT. ATOMIC WEAPONS RESEARCH ESTABLISHMENT, ALDERMASTON, BERKS, ENGLAND	SC-DC-
GT. BRIT. ATOMIC WEAPONS RESEARCH ESTABLISHMENT, ALDERMASTON, BERKS, ENGLAND	SRI/
GT. BRIT. ATOMIC WEAPONS RESEARCH ESTABLISHMENT, ALDERMASTON, BERKS, ENGLAND	T-(NO.)/(YR.)
(SEE ALSO LATER NAME: UNITED KINGDOM ATOMIC ENERGY AUTHORITY. WEAPONS GP. ATOMIC WEAPONS RES. ESTAB.)	
GT. BRIT. BOMBING RESEARCH MISSION	BBRM-
GT. BRIT. BLDG. RES. STA., GARSTON, HERTS, ENGLAND	BRS-
GT. BRIT. BLDG. RES. STA., GARSTON, HERTS, ENGLAND	BRS-DR-
GT. BRIT. BLDG. RES. STA., GARSTON, HERTS, ENGLAND	CP-(NUMBER/YEAR)
GT. BRIT. BLDG. RES. STA., GARSTON, HERTS, ENGLAND	LC-
GT. BRIT. CAPENHURST WORKS, CHES., ENGLAND	CAP-TM-
GT. BRIT. CAPENHURST WORKS, CHES., ENGLAND (REF. 36)	CP/M-
GT. BRIT. CAPENHURST WORKS, CHES., ENGLAND	CSD (CAP.) M-
GT. BRIT. CAPENHURST WORKS, CHES., ENGLAND (REF. 36)	CSD (CAP.) TM-
GT. BRIT. CAPENHURST WORKS, CHES., ENGLAND	CTSC/-

GT. BRIT. CAPENHURST WORKS

GT. BRIT. CAPENHURST WORKS, CHES., ENGLAND	FRDC/P-
GT. BRIT. CAPENHURST WORKS, CHES., ENGLAND	HARD(A)/P-
GT. BRIT. CAPENHURST WORKS, CHES., ENGLAND (REF. 36)	IDLM-
GT. BRIT. CAPENHURST WORKS, CHES., ENG. (INFO. BIB.)	IGRL-IB/CA-
GT. BRIT. CAPENHURST WORKS, CHES., ENGLAND	IPDC/P-
GT. BRIT. CAPENHURST WORKS, CHES., ENGLAND	LMCC-
GT. BRIT. CAPENHURST WORKS, CHES., ENGLAND	LMCC/P-
GT. BRIT. CAPENHURST WORKS, CHES., ENGLAND (REF. 36)	PM(C)-
GT. BRIT. CAPENHURST WORKS, CHES., ENGLAND (REF. 36)	R.+D.B.(CAP.)-
GT. BRIT. CAPENHURST WORKS, CHES., ENGLAND	RDB(CAP)/R-
GT. BRIT. CAPENHURST WORKS, CHES., ENGLAND	RDB(CAP)-TN-
GT. BRIT. CAPENHURST WORKS, CHES., ENGLAND (REF. 36)	SC-

(SEE ALSO LATER NAMES: UNITED KINGDOM ATOMIC ENERGY AUTHORITY. DEV. AND ENG. GP., CAPENHURST... AND U.K.A.E.A. PRODUCTION GROUP, CAPENHURST...)

GT. BRIT. CENTRAL ELECTRICITY GENERATING BOARD, LOND.	CD/
GT. BRIT. CENTRAL ELECTRICITY GENERATING BOARD. BERKELEY NUCLEAR LABS., BRISTOL, ENGLAND	CEGB-RD/B/N-
GT. BRIT. CENTRAL ELECTRICITY GENERATING BOARD. BERKELEY NUCLEAR LABS., BRISTOL, ENGLAND	FEWP/P-
GT. BRIT. CENTRAL ELECTRICITY GENERATING BOARD. BERKELEY NUCLEAR LABS., BRISTOL, ENGLAND	JNPC/SWP/N-
GT. BRIT. CENTRAL ELECTRICITY GENERATING BOARD. BERKELEY NUCLEAR LABS., BRISTOL, ENGLAND	RD/B/M-
GT. BRIT. CENTRAL ELECTRICITY GENERATING BOARD. BERKELEY NUCLEAR LABS., BRISTOL, ENGLAND	RD/B/N-
GT. BRIT. CENTRAL ELECTRICITY GENERATING BOARD. BERKELEY NUCLEAR LABS., BRISTOL, ENGLAND	RD/B/R-
GT. BRIT. CENTRAL ELECTRICITY GENERATING BOARD. BERKELEY NUCLEAR LABS., BRISTOL, ENGLAND	TRG-REPORT-
GT. BRIT. CENTRAL ELECTRICITY GENERATING BOARD. NUCLEAR HEALTH AND SAFETY DEPT., LONDON	NHS-(NUMBER/YEAR)
GT. BRIT. CENTRAL FIGHTER ESTABLISHMENT	CFE-
GT. BRIT. CENTRAL RADIO BUREAU	CRB-
GT. BRIT. CENTRAL SIGNALS ESTABLISHMENT	CSE-
GT. BRIT. CENTRAL SIGNALS ESTABLISHMENT	CSE R-
GT. BRIT. CHEMICAL DEFENCE EXPERIMENTAL ESTABLISHMENT, PORTON, WILTS, ENGLAND	CDEE-
GT. BRIT. CHEMICAL DEFENCE EXPERIMENTAL ESTABLISHMENT, PORTON, WILTS, ENGLAND	CDES-
GT. BRIT. CHEMICAL DEFENCE EXPERIMENTAL ESTABLISHMENT, PORTON, WILTS, ENGLAND	C.D.E.T.N.(NUMBER)
GT. BRIT. CHEMICAL DEFENCE EXPERIMENTAL ESTABLISHMENT, PORTON, WILTS, ENGLAND	CDE-TN-
GT. BRIT. CHEMICAL DEFENCE EXPERIMENTAL ESTABLISHMENT, PORTON, WILTS, ENGLAND	CDE-TP-
GT. BRIT. CHEMICAL DEFENCE EXPERIMENTAL ESTABLISHMENT, PORTON, WILTS, ENGLAND	C.D.E.T.P.(NUMBER)
GT. BRIT. CHEMICAL DEFENCE EXPERIMENTAL ESTABLISHMENT, PORTON, WILTS, ENGLAND	C.D.E.TP(NUMBER)
GT. BRIT. CHEMICAL DEFENCE EXPERIMENTAL ESTABLISHMENT, PORTON, WILTS, ENGLAND	CDL-
GT. BRIT. CHEMICAL DEFENCE EXPERIMENTAL ESTABLISHMENT, PORTON, WILTS, ENGLAND	CTEE PTP-
GT. BRIT. CHEMICAL DEFENCE EXPERIMENTAL ESTABLISHMENT, PORTON, WILTS, ENGLAND	PN-
GT. BRIT. CHEMICAL DEFENCE EXPERIMENTAL ESTABLISHMENT, PORTON, WILTS, ENGLAND	PORTON-
GT. BRIT. CHEMICAL DEFENCE EXPERIMENTAL ESTABLISHMENT, PORTON, WILTS, ENGLAND	PORTON MEMO-
GT. BRIT. CHEMICAL DEFENCE EXPERIMENTAL ESTABLISHMENT, PORTON, WILTS, ENGLAND	PORTON TN-
GT. BRIT. CHEMICAL DEFENCE EXPERIMENTAL ESTABLISHMENT, PORTON, WILTS, ENGLAND	PORTON TP-
GT. BRIT. CHEMICAL DEFENCE EXPERIMENTAL ESTABLISHMENT, PORTON, WILTS, ENGLAND	PTN-
GT. BRIT. CHEMICAL DEFENCE EXPERIMENTAL ESTABLISHMENT, PORTON, WILTS, ENGLAND	PTN-TU-
GT. BRIT. CHEMICAL DEFENCE EXPERIMENTAL ESTABLISHMENT, PORTON, WILTS, ENGLAND	PTP-
GT. BRIT. CHEMICAL DEFENCE EXPERIMENTAL ESTABLISHMENT, SALISBURY, WILTS, ENGLAND	PTP-
GT. BRIT. CHEM. DEFENCE EXPTL. STA.,PORTON,WILTS,ENG.	CS/CJ-
GT. BRIT. CHEM. DEFENCE EXPTL. STA.,PORTON,WILTS,ENG.	CS/TM-
GT. BRIT. CHEM. DEFENCE EXPTL. STA.,PORTON,WILTS,ENG.	M-

(SEE ALSO LATER NAME: GT. BRIT. CHEMICAL DEFENCE EXPERIMENTAL ESTABLISHMENT)

GT. BRIT. CHEM. DEFENCE RES.EST.,SUTTON OAK,LANCS,ENG	CDRE-
GT. BRIT. CHEM. DEFENCE RES.EST.,SUTTON OAK,LANCS,ENG	SO-
GT. BRIT. CHEMICAL INSPECTORATE	WAD-
GT. BRIT. CHEM. RES. LAB., TEDDINGTON, MIDDX., ENG.	CRL-(LETTERS)-
GT. BRIT. CHEM. RES. LAB., TEDDINGTON, MIDDX., ENG.	CRL/AE-
GT. BRIT. CLOTHING + EQUIP. PHYSIOLOGICAL RES. ESTAB.	CEPRE-
GT. BRIT. CLOTHING + EQUIP. PHYSIOLOGICAL RES. ESTAB.	CSEE-
GT. BRIT. COMBINED OPERATIONS HEADQUARTERS	COHQ-
GT. BRIT. COMMITTEE FOR CO-ORDINATION OF CATHODE RAY TUBE DEVELOPMENT	CCRTD-
GT. BRIT. COMMITTEE FOR THE PREVENTION OF CORROSION AND FOULING, LONDON	ACC/U.(NUMBER)/(YEAR)
GT. BRIT. COMMITTEE ON OVERSEAS SCIENTIFIC RELATIONS	OSR-
GT. BRIT. CULCHETH LABS., LANCS, ENGLAND	CULCHETH-
GT. BRIT. CULCHETH LABS., LANCS, ENGLAND	DDC/P-
GT. BRIT. CULCHETH LABS., LANCS, ENGLAND	FRDC/P-
GT. BRIT. CULCHETH LABS., LANCS, ENGLAND	FRFEWP/P-
GT. BRIT. CULCHETH LABS., LANCS, ENGLAND	RDB(C)-
GT. BRIT. CULCHETH LABS., LANCS, ENGLAND (REF. 36)	R.+D.B.(C)(LETTERS)-
GT. BRIT. CULCHETH LABS., LANCS, ENGLAND	SFPWP/P-
GT. BRIT. CULCHETH LABS., LANCS, ENGLAND (REF. 36)	SRO/ML-
GT. BRIT. CULCHETH LABS., LANCS, ENGLAND	TRDC/P-
GT. BRIT. CULCHETH LABS., LANCS, ENGLAND	XMPDC/P-

(SEE ALSO LATER NAMES: UNITED KINGDOM ATOMIC ENERGY AUTHORITY. DEV. AND ENG. GP., CULCHETH... AND U.K.A.E.A. REACTOR GROUP, CULCHETH...)

GT. BRIT. DEPT. OF ATOMIC ENERGY (BRITISH DECLASSIFIED DOCUMENTS, ATOMIC)	BDDA-
GT. BRIT. DEPT. OF ATOMIC ENERGY	BR-
GT. BRIT. DEPT. OF ATOMIC ENERGY	DAE (LETTER)-
GT. BRIT. DEPT. OF ATOMIC ENERGY (REF. 36)	ENQ-
GT. BRIT. DEPT. OF ATOMIC ENERGY	MS-
GT. BRIT. DEPT. OF ATOMIC ENERGY	MS DAE (LETTER(S))-
GT. BRIT. DEPT. OF ATOMIC ENERGY	PAT SPEC-(YR)
GT. BRIT. DEPT. OF ATOMIC ENERGY. INDUSTRIAL GROUP H.Q., RISLEY, LANCS, ENGLAND (REF. 36)	BP-5001 THRU -5256
GT. BRIT. DEPT. OF ATOMIC ENERGY. INDUSTRIAL GROUP H.Q., RISLEY, LANCS, ENGLAND (REF. 36)	CF-5001 THRU -5256
GT. BRIT. DEPT. OF ATOMIC ENERGY. INDUSTRIAL GROUP H.Q., RISLEY, LANCS, ENGLAND	CF-P-
GT. BRIT. DEPT. OF ATOMIC ENERGY. INDUSTRIAL GROUP H.Q., RISLEY, LANCS, ENGLAND (REFS. 1, 36)	(4 DIGITS)
GT. BRIT. DEPT. OF ATOMIC ENERGY. INDUSTRIAL GROUP H.Q., RISLEY, LANCS, ENGLAND (REF. 36)	IG.LIBIB-
GT. BRIT. DEPT. OF ATOMIC ENERGY. INDUSTRIAL GROUP H.Q., RISLEY, LANCS, ENGLAND (INFORMATION BIB.)	IGRL-IB/R-
GT. BRIT. DEPT. OF ATOMIC ENERGY. INDUSTRIAL GROUP H.Q., RISLEY, LANCS, ENGLAND (REF. 36)	MD-
GT. BRIT. DEPT. OF ATOMIC ENERGY. INDUSTRIAL GROUP H.Q., RISLEY, LANCS, ENGLAND	PIPPA-WP/MEMO-
GT. BRIT. DEPT. OF ATOMIC ENERGY. INDUSTRIAL GROUP H.Q., RISLEY, LANCS, ENGLAND	PM(R)-
GT. BRIT. DEPT. OF ATOMIC ENERGY. INDUSTRIAL GROUP H.Q., RISLEY, LANCS, ENGLAND (REF. 36)	RISLEY-5001 THRU -5256
GT. BRIT. DEPT. OF ATOMIC ENERGY. INDUSTRIAL GROUP H.Q., RISLEY, LANCS, ENGLAND (REF. 36)	RISLEY-6001 THRU -6286

GT. BRIT. DEPT. OF ATOMIC ENERGY. INDUSTRIAL GROUP H.Q., RISLEY, LANCS, ENGLAND (REF. 36)	RISLEY-7001 THRU -7008
GT. BRIT. DEPT. OF ATOMIC ENERGY. INDUSTRIAL GROUP H.Q., RISLEY, LANCS, ENGLAND (REF. 36)	RISLEY-8001 THRU -8151
GT. BRIT. DEPT. OF ATOMIC ENERGY. INDUSTRIAL GROUP H.Q., RISLEY, LANCS, ENGLAND	RISLEY-TRANS-
GT. BRIT. DEPT. OF ATOMIC ENERGY. INDUSTRIAL GROUP H.Q., RISLEY, LANCS, ENGLAND (REF. 36)	SP-5001 THRU -5256
GT. BRIT. DEPT. OF ATOMIC ENERGY. INDUSTRIAL GROUP H.Q., RISLEY, LANCS, ENGLAND (REF. 36)	SR-5001 THRU -5256
GT. BRIT. DEPT. OF ATOMIC ENERGY. INDUSTRIAL GROUP H.Q., RISLEY, LANCS, ENGLAND (REF. 36)	SRO-
GT. BRIT. DEPT. OF ATOMIC ENERGY. INDUSTRIAL GROUP H.Q., RISLEY, LANCS, ENGLAND	SRR-
GT. BRIT. DEPT. OF ATOMIC ENERGY. INDUSTRIAL GROUP H.Q., RISLEY, LANCS, ENGLAND (REF. 36)	TRDC/P-
GT. BRIT. DEPT. OF ATOMIC ENERGY. INDUSTRIAL GROUP H.Q., RISLEY, LANCS, ENGLAND	TSM-

(SEE ALSO LATER NAMES: UNITED KINGDOM ATOMIC ENERGY AUTHORITY. DEVELOPMENT AND ENGINEERING GROUP AND U.K.A.E.A. PRODUCTION GROUP)

GT. BRIT. DEPT. OF ATOMIC EN. PLANT LOCATION PANEL	PLP-
GT. BRIT. DEPT. OF FORESTS (PALESTINE REPORTS)	DF P-
GT. BRIT. DEPT. OF SCIENTIFIC AND INDUSTRIAL RESEARCH. (BUILDING RESEARCH BULLETINS)	BRB-
GT. BRIT. DEPT. OF SCIENTIFIC AND INDUSTRIAL RESEARCH. (BUILDING RESEARCH TECHNICAL PAPERS)	BR TP-
GT. BRIT. DEPT. OF SCIENTIFIC AND INDUSTRIAL RES.	CTS-
GT. BRIT. DEPT. OF SCIENTIFIC AND INDUSTRIAL RES.	DSIR-
GT. BRIT. DEPT. OF SCIENTIFIC AND INDUSTRIAL RESEARCH. (BUILDING RESEARCH BULLETINS)	DSIR BRB-
GT. BRIT. DEPT. OF SCIENTIFIC AND INDUSTRIAL RESEARCH. (BUILDING RESEARCH DIGEST)	DSIR BRD-
GT. BRIT. DEPT. OF SCIENTIFIC AND INDUSTRIAL RESEARCH. (BUILDING RESEARCH TECHNICAL PAPERS)	DSIR BR TP-
GT. BRIT. DEPT. OF SCIENTIFIC AND INDUSTRIAL RESEARCH (CHEMISTRY RESEARCH. SPECIAL REPORT)	DSIR CR SR-
GT. BRIT. DEPT. OF SCIENTIFIC AND INDUSTRIAL RESEARCH (FOOD INVESTIGATION. LEAFLET)	DSIR FI L-
GT. BRIT. DEPT. OF SCIENTIFIC AND INDUSTRIAL RESEARCH (FIRE RESEARCH. TECHNICAL PAPER)	DSIR FIRE TP-
GT. BRIT. DEPT. OF SCIENTIFIC AND INDUSTRIAL RESEARCH (FOOD INVESTIGATION. SPECIAL REPORT)	DSIR FI SR-
GT. BRIT. DEPT. OF SCIENTIFIC AND INDUSTRIAL RESEARCH (FOOD INVESTIGATION. TECHNICAL PAPER)	DSIR FI TP-
GT. BRIT. DEPT. OF SCIENTIFIC AND INDUSTRIAL RESEARCH (FOREST PRODUCTS RESEARCH RECORDS)	DSIR FPRR-
GT. BRIT. DEPT. OF SCIENTIFIC AND INDUSTRIAL RESEARCH (FUEL RESEARCH. SURVEY PAPERS)	DSIR FR SP-
GT. BRIT. DEPT. OF SCIENTIFIC AND INDUSTRIAL RESEARCH (FUEL RESEARCH. TECHNICAL PAPERS)	DSIR FR TP-
GT. BRIT. DEPT. OF SCI. + IND. RES. (LEAFLET)	DSIR L-
GT. BRIT. DEPT. OF SCIENTIFIC AND INDUSTRIAL RESEARCH (NATIONAL BUILDING STUDIES, BULLETIN)	DSIR NB B-
GT. BRIT. DEPT. OF SCIENTIFIC AND INDUSTRIAL RESEARCH (NATIONAL BLDG. STUDIES, SPEC. RPT.)	DSIR NB SR-
GT. BRIT. DEPT. OF SCIENTIFIC AND INDUSTRIAL RESEARCH (NATIONAL BLDG. STUDIES, TECH. PAPERS)	DSIR NB TP-
GT. BRIT. DEPT. OF SCIENTIFIC AND INDUSTRIAL RESEARCH (POST WAR BUILDING STUDIES)	DSIR PWBS-
GT. BRIT. DEPT. OF SCIENTIFIC AND INDUSTRIAL RESEARCH (RADIO RESEARCH, SPECIAL REPORT)	DSIR RAD R SR-
GT. BRIT. DEPT. OF SCIENTIFIC AND INDUSTRIAL RESEARCH (ROAD RESEARCH BULLETINS)	DSIR RR-
GT. BRIT. DEPT. OF SCIENTIFIC AND INDUSTRIAL RESEARCH (ROAD RESEARCH SPECIAL REPORTS)	DSIR RR SR-
GT. BRIT. DEPT. OF SCIENTIFIC AND INDUSTRIAL RESEARCH (ROAD RESEARCH TECHNICAL PAPER)	DSIR RR TP-
GT. BRIT. DEPT. OF SCIENTIFIC AND INDUSTRIAL RESEARCH (SPONSORED RESEARCH REPORTS)	DSIR SR-
GT. BRIT. DEPT. OF SCIENTIFIC AND INDUSTRIAL RESEARCH (WATER POLLUTION RESEARCH TECH. RPTS.)	DSIR WPR TP-
GT. BRIT. DEPT. OF SCIENTIFIC AND INDUSTRIAL RESEARCH (FOREST PRODUCTS RESEARCH RECORDS)	FPRR-
GT. BRIT. DEPT. OF SCI. + IND. RES. (FUEL RES.)	FR-
GT. BRIT. DEPT. OF SCIENTIFIC AND INDUSTRIAL RESEARCH (NATIONAL BUILDING STUDIES, BULLETIN)	MOW NB B-
GT. BRIT. DEPT. OF SCIENTIFIC AND INDUSTRIAL RESEARCH (ROAD RESEARCH BULLETINS)	RR-
GT. BRIT. DEPT. OF SCIENTIFIC AND INDUSTRIAL RESEARCH	S + T-MEMO-
GT. BRIT. DEPT. OF SCIENTIFIC AND INDUSTRIAL RESEARCH (WATER POLLUTION RESEARCH TECHNICAL REPORTS)	WPR-
GT. BRIT. DEPT. OF SCI. + IND. RES. DIRECTORATE OF TUBE ALLOYS (ASSIGNED BY MAJOR GEN. L. R. GROVES)	B-LRG-
GT. BRIT. DEPT. OF SCIENTIFIC AND INDUSTRIAL RESEARCH. DIRECTORATE OF TUBE ALLOYS	DSIRLRG-
GT. BRIT. DEPT. OF SCIENTIFIC AND INDUSTRIAL RESEARCH. DIRECTORATE OF TUBE ALLOYS	II-5-
GT. BRIT. DEPT. OF SCI. + IND. RES. DIRECTORATE OF TUBE ALLOYS (ASSIGNED BY MAJOR GEN. L. R. GROVES)	LRG-
GT. BRIT. DEPT. OF SCIENTIFIC AND INDUSTRIAL RESEARCH. FOREST PRODUCTS RESEARCH LAB. (BULLS.)	DSIR FPR-
GT. BRIT. DEPT. OF SCIENTIFIC AND INDUSTRIAL RESEARCH. FOREST PRODUCTS RES. LAB. (LEAFLETS)	DSIR FPR L-
GT. BRIT. DEPT. OF SCIENTIFIC AND INDUSTRIAL RESEARCH. FOREST PRODUCTS RES. LAB. (SPEC. RPTS.)	DSIR FPR SP-
GT. BRIT. DEPT. OF SCIENTIFIC AND INDUSTRIAL RESEARCH. FOREST PRODUCTS RESEARCH LABORATORY	FPR-
GT. BRIT. DEPT. OF SCIENTIFIC AND INDUSTRIAL RESEARCH. ROAD RESEARCH LAB.	RN-
GT. BRIT. DEPT. OF SCIENTIFIC AND INDUSTRIAL RESEARCH. WARREN SPRING LAB.	LR-(2 DIGITS)
GT. BRIT. DEPUTY DIRECTORATE OF TECH. INTELLIGENCE	DDI (TECH.)-
GT. BRIT. DIRECTORATE OF ARMAMENT RES. + DEV. (AIR)	RD ARM RI-
GT. BRIT. DIRECTORATE OF CHEMICAL DEFENCE RES. + DEV.	DCDRD-
GT. BRIT. DIRECTORATE OF CHEMICAL DEFENCE RES. + DEV.	HQ/IT-
GT. BRIT. DIRECTORATE OF CHEMICAL DEFENCE RES. + DEV.	IE-
GT. BRIT. DIRECTORATE OF INTELLIGENCE	ADI(K)-
GT. BRIT. DIRECTORATE OF INTELLIGENCE	AMSIS-
GT. BRIT. DIRECTORATE OF MATERIALS AND EXPLOSIVES RESEARCH AND DEVELOPMENT	DMXRD-
GT. BRIT. DIRECTORATE OF MATERIALS AND EXPLOSIVES RESEARCH AND DEVELOPMENT	MX-
GT. BRIT. DIRECTORATE OF MATERIALS AND EXPLOSIVES RESEARCH AND DEVELOPMENT	PL-
GT. BRIT. DIRECTORATE OF MATERIALS AND STRUCTURES RESEARCH AND DEVELOPMENT	DMSRD-
GT. BRIT. DIRECTORATE OF ORDNANCE FACTORIES (EXPLOSIVES)	DOF(X)-
GT. BRIT. DIRECTORATE OF ORDNANCE FACTORIES (EXPLOSIVES)	H-
GT. BRIT. DIRECTORATE OF SCIENTIFIC INTELLIGENCE. JOINT TECHNICAL INTELLIGENCE COMMITTEE	DSI/JTIC-
GT. BRIT. DIRECTORATE OF WEAPON RESEARCH (DEFENCE)	DWR(D)-
GT. BRIT. DIV. OF ATOMIC ENERGY (PRODUCTION), RISLEY, LANCS, ENGLAND	DDC/P-(4 DIGITS)
GT. BRIT. DIV. OF ATOMIC ENERGY (PRODUCTION), RISLEY, LANCS, ENGLAND (ASSIGNED TO INCOMING MISCELLANEOUS BRITISH REPORTS) (REF. 1)	FPP/P-
GT. BRIT. DIV. OF ATOMIC ENERGY (PRODUCTION), RISLEY, LANCS, ENGLAND	RISLEY-1001 THRU -2000
GT. BRIT. DIV. OF ATOMIC ENERGY (PRODUCTION), RISLEY, LANCS, ENGLAND (ASSIGNED TO MISCELLANEOUS BRITISH DOCUMENTS)	RISLEY-3001 THRU -4000
GT. BRIT. DIV. OF ATOMIC ENERGY (PRODUCTION), RISLEY, LANCS, ENGLAND (ASSIGNED TO MISCELLANEOUS BRITISH DOCUMENTS)	SCS-R-
GT. BRIT. DIV. OF ATOMIC ENERGY (PRODUCTION), RISLEY, LANCS, ENGLAND	TRDC/P-
GT. BRIT. DIV. OF ATOMIC ENERGY (PRODUCTION), RISLEY, LANCS, ENGLAND	

(SEE ALSO LATER NAME: UNITED KINGDOM ATOMIC ENERGY AUTHORITY. PRODUCTION GROUP)

GT. BRIT. ENGINEER-IN-CHIEF OF THE FLEET	E.IN.C.-
GT. BRIT. EXPLOSIVES RESEARCH AND DEVELOPMENT ESTABLISHMENT, WALTHAM ABBEY, ESSEX, ENGLAND	EMR-
GT. BRIT. EXPLOSIVES RESEARCH AND DEVELOPMENT ESTABLISHMENT, WALTHAM ABBEY, ESSEX, ENGLAND	ERDE-(NO.)/EMR/(YR)
GT. BRIT. EXPLOSIVES RESEARCH AND DEVELOPMENT ESTABLISHMENT, WALTHAM ABBEY, ESSEX, ENGLAND	ERDE-(NO.)/M/(YR)
GT. BRIT. EXPLOSIVES RESEARCH AND DEVELOPMENT ESTABLISHMENT, WALTHAM ABBEY, ESSEX, ENGLAND	ERDE-(NO.)/R/(YR)
GT. BRIT. EXPLOSIVES RESEARCH AND DEVELOPMENT ESTABLISHMENT, WALTHAM ABBEY, ESSEX, ENGLAND	ERDE-TM(NO.)/M/(YR.)
GT. BRIT. EXPLOSIVES RESEARCH AND DEVELOPMENT ESTABLISHMENT, WALTHAM ABBEY, ESSEX, ENGLAND	ERDE-TN-
GT. BRIT. EXPLOSIVES RESEARCH AND DEVELOPMENT ESTABLISHMENT, WALTHAM ABBEY, ESSEX, ENGLAND	ERDE-TR-
GT. BRIT. EXPLOSIVES RESEARCH AND DEVELOPMENT ESTABLISHMENT, WALTHAM ABBEY, ESSEX, ENGLAND	WAC/(NUMBER)/-
GT. BRIT. EXPLOSIVES RESEARCH AND DEVELOPMENT ESTABLISHMENT, WALTHAM ABBEY, ESSEX, ENGLAND	XR-

GT. BRIT. FLYING PERSONNEL RESEARCH COMMITTEE

GT. BRIT. FLYING PERSONNEL RESEARCH COMMITTEE	FA-A(YR)-
GT. BRIT. FLYING PERSONNEL RESEARCH COMMITTEE	FA-PAD-
GT. BRIT. FLYING PERSONNEL RESEARCH COMMITTEE	FA-R-
GT. BRIT. FLYING PERSONNEL RESEARCH COMMITTEE	PAD-
GT. BRIT. FLYING PERSONNEL RESEARCH COMMITTEE	T(NO.)-
(SEE ALSO ENTRIES UNDER:UNITED KINGDOM.FLYING PER...)	
GT. BRIT. FORESTRY COMMISSION	FC-
GT. BRIT. FORESTRY COMMISSION (BULLETIN)	FC B-
GT. BRIT. FORESTRY COMMISSION (FOREST OPERATIONS SER)	FC FOS-
GT. BRIT. FORESTRY COMMISSION (FOREST OPERATIONS SER)	FOS-
GT. BRIT. FUNGICIDE + INSECTICIDE RESEARCH CO-ORDINATION SERVICE	IDIC-
GT. BRIT. GEOLOGICAL SURVEY AND MUSEUM, LONDON	GSM-AED-
GT. BRIT. GERMAN DOCUMENT CENTRE	GDC-
GT. BRIT. GUIDED WEAPONS ADVISORY COMMITTEE	GWAC-
GT. BRIT. HALSTEAD EXPLOITING CENTRE	BIOS/GP.2/HEC
GT. BRIT. HALSTEAD EXPLOITING CENTRE	FBM-
GT. BRIT. HER MAJESTY'S STATIONERY OFFICE, LONDON (TRANSLATIONS)	HMSO-
GT. BRIT. HOME OFFICE, LONDON	SA/PR-
GT. BRIT. HOME OFFICE. CIVIL DEFENCE DEPT. SCIENTIFIC ADVISERS BRANCH, LONDON	CD/SA-
GT. BRIT. INTER-SERVICES METALLURGICAL RESEARCH COUNCIL, LONDON	ISMET-
GT. BRIT. JOINT INFRA-RED COMMITTEE	DPR/JIRC-
GT. BRIT. LEGISLATIVE COUNCIL (SESSIONAL PAPER)	LC SP-
GT. BRIT. LEGISLATIVE COUNCIL (SESSIONAL PAPER)	SP-
GT. BRIT. LOW TEMPERATURE RESEARCH STATION. RADIATION GROUP, CAMBRIDGE, ENGLAND	LTRS-RM-
GT. BRIT. MARINE AIRCRAFT EXPERIMENTAL ESTABLISHMENT, FELIXSTOWE, SUFF., ENGLAND	FX/EQ-
GT. BRIT. MARINE AIRCRAFT EXPERIMENTAL ESTABLISHMENT, FELIXSTOWE, SUFF., ENGLAND	MAEE-F.TN/-
GT. BRIT. MECHANICAL ENGINEERING RES. LAB., GLASGOW	MERL-
GT. BRIT. MEDICAL RESEARCH COUNCIL, LONDON	MRC-
GT. BRIT. MEDICAL RESEARCH COUNCIL, LONDON (MEMO.)	MRC M-
GT. BRIT. MEDICAL RES. COUNCIL, LONDON (SPEC. RPTS.)	MRC SR-
GT. BRIT. MEDICAL RES. COUNCIL. IND. HEALTH RES. BD.	MRC IHRB R-
GT. BRIT. MILITARY ENGINEERING EXPERIMENTAL ESTAB.	MEXE-
GT. BRIT. MILITARY OPERATIONAL RESEARCH	MOR-
GT. BRIT. MILITARY PERSONNEL RESEARCH COMMITTEE	BPC-
GT. BRIT. MILITARY PERSONNEL RESEARCH COMMITTEE	MPRC-
GT. BRIT. MINISTRY OF AGRICULTURE AND FISHERIES	MAF-
GT. BRIT. MINISTRY OF AGRICULTURE AND FISHERIES (BULLETIN)	MAF B-
GT. BRIT. MINISTRY OF AIRCRAFT PRODUCTION.	MAP-
GT. BRIT. MINISTRY OF AIRCRAFT PRODUCTION.	RTP-
GT. BRIT. MIN. OF AIRCRAFT PRODUCTION (TRANSLATION)	RTP T-
GT. BRIT. MIN. OF AIRCRAFT PRODUCTION (TRANSLATION)	RTP/TIB T-
GT. BRIT. MIN. OF AIRCRAFT PRODUCTION (TRANSLATION)	TIB T-
GT. BRIT. MINISTRY OF AVIATION, LONDON	JH-
GT. BRIT. MINISTRY OF AVIATION, NOTTINGHAM, ENGLAND	S+T-MEMO-
GT. BRIT. MINISTRY OF AVIATION.CENTRAL LIBRARY,LONDON	TVC-
GT. BRIT. MINISTRY OF AVIATION.CENTRAL LIBRARY,LONDON	TVC-(LETTERS)-
GT. BRIT. MINISTRY OF AVIATION. TECHNICAL INFORMATION AND LIBRARY SERVICES, LONDON	MA-S/T-
GT. BRIT. MINISTRY OF AVIATION. TECHNICAL INFORMATION AND LIBRARY SERVICES, LONDON	MA-S/T-MEMO-(NO.)/(YR.)
GT. BRIT. MINISTRY OF AVIATION. TECHNICAL INFORMATION AND LIBRARY SERVICES, LONDON (TRANSLATION)	TIL/T.(NUMBER)
GT. BRIT. MINISTRY OF CIVIL AVIATION	CA-
GT. BRIT. MINISTRY OF CIVIL AVIATION	MA-S/T-MEMO-
GT. BRIT. MINISTRY OF CIVIL AVIATION	MCA-
GT. BRIT. MINISTRY OF CIVIL AVIATION	MCA C-
GT. BRIT. MINISTRY OF CIVIL AVIATION	MCA CA-
GT. BRIT. MINISTRY OF CIVIL AVIATION (PUBLICATIONS)	MCA P-
GT. BRIT. MINISTRY OF CIVIL AVIATION (PUBLICATIONS)	MCAP-
GT. BRIT. MINISTRY OF DEFENCE, LONDON	ACO/UK/
GT. BRIT. MINISTRY OF DEFENCE, LONDON	DSTI-TRANS-
GT. BRIT. MINISTRY OF DEFENCE, JOINT INTELL.BUR.,LOND	JIB-
GT. BRIT. MINISTRY OF FUEL AND POWER	R-(NO.)-(YR)
GT. BRIT. MINISTRY OF HOME SECURITY	RC-
GT. BRIT. MINISTRY OF HOME SECURITY	REN-
GT. BRIT. MINISTRY OF HOME SECURITY	SW-
GT. BRIT. MINISTRY OF HOME SECURITY	UNDEX-
GT. BRIT. MINISTRY OF HOME SECURITY. CIVIL DEFENCE RESEARCH COMMITTEE	MHSCDRC-
GT. BRIT. MINISTRY OF HOME SECURITY. INTERDEPARTMENTAL COORDINATING COMMITTEE ON SHOCKWAVES	SW-
GT. BRIT. MINISTRY OF HOME SECURITY. RESEARCH AND EXPERIMENTS DEPT., LONDON	R.E.N.-
GT. BRIT. MINISTRY OF LABOUR AND NATIONAL SERVICE. FACTORY DEPARTMENT (SAFETY PAMPHLET)	MLNS SP-
GT. BRIT. MINISTRY OF SUPPLY	AC-
GT. BRIT. MINISTRY OF SUPPLY	ATP-
GT. BRIT. MINISTRY OF SUPPLY	CI/METHOD-EA-
GT. BRIT. MINISTRY OF SUPPLY	CI/METHOD-EB-
GT. BRIT. MINISTRY OF SUPPLY	CI/METHOD-EC-
GT. BRIT. MINISTRY OF SUPPLY	CI/METHOD-ED-

GT. BRIT. MINISTRY OF SUPPLY	CI/METHOD-EE-
GT. BRIT. MINISTRY OF SUPPLY	GLASGOW REPORT-
GT. BRIT. MINISTRY OF SUPPLY	H.(NO.)/(YR)
GT. BRIT. MINISTRY OF SUPPLY	HP/M-
GT. BRIT. MINISTRY OF SUPPLY	LF-
GT. BRIT. MINISTRY OF SUPPLY	MAP-
GT. BRIT. MINISTRY OF SUPPLY	MG-
GT. BRIT. MINISTRY OF SUPPLY	MOS-
GT. BRIT. MINISTRY OF SUPPLY	MOS-MONO-(NUMBER)-(NO.)
GT. BRIT. MINISTRY OF SUPPLY	MS-
GT. BRIT. MINISTRY OF SUPPLY	MS-D-
GT. BRIT. MINISTRY OF SUPPLY (WELDING MEMORANDA)	MS WM-
GT. BRIT. MINISTRY OF SUPPLY	MT-
GT. BRIT. MINISTRY OF SUPPLY	NC-
GT. BRIT. MINISTRY OF SUPPLY	PDE-1944/15-
GT. BRIT. MINISTRY OF SUPPLY	PTN-
GT. BRIT. MINISTRY OF SUPPLY	RD(H)/F-
GT. BRIT. MINISTRY OF SUPPLY (RESEARCH TECHNICAL PUBLICATIONS)	RTP-
GT. BRIT. MINISTRY OF SUPPLY	TIB-
GT. BRIT. MINISTRY OF SUPPLY	TPA 3-
GT. BRIT. MINISTRY OF SUPPLY. (TRANSLATIONS MADE BY THE GERMAN STAFF OF UNTERLUESS WORK CENTRE)	UNT-
GT. BRIT. MINISTRY OF SUPPLY (WELDING MEMORANDA)	WM-
GT. BRIT. MINISTRY OF SUPPLY. AID LABS., HAREFIELD, MIDDX, ENGLAND	AID/MET/-
GT. BRIT. MIN. OF SUPPLY. CHEM. INSPECTORATE, LONDON	CCI-
GT. BRIT. MIN. OF SUPPLY. CHEM. INSPECTORATE, LONDON	CCI-TD-
GT. BRIT. MIN. OF SUPPLY. CHEM. INSPECTORATE, LONDON	CI-
GT. BRIT. MIN. OF SUPPLY. CHEM. INSPECTORATE, LONDON	CI/MEMO-
GT. BRIT. MIN. OF SUPPLY. CHEM. INSPECTORATE, LONDON	CIP/
GT. BRIT. MIN. OF SUPPLY. CHEM. INSPECTORATE, LONDON	CI/R-
GT. BRIT. MIN. OF SUPPLY. CHEM. INSPECTORATE, LONDON	SCS-M-
GT. BRIT. MIN. OF SUPPLY. CHEM. INSPECTORATE, LONDON	SCS-R-
GT. BRIT. MIN. OF SUPPLY. CHEM. INSPECTORATE, LONDON	WAD/R-
GT. BRIT. MIN. OF SUPPLY. CHEM. INSPECTORATE, LONDON	WSL-R-
GT. BRIT. MINISTRY OF SUPPLY. CHEMICAL RESEARCH AND DEVELOPMENT DEPT., WALTHAM ABBEY, ESSEX, ENGLAND	CRDD-
GT. BRIT. MINISTRY OF SUPPLY. DEPT. OF TANK DESIGN	DTD-
GT. BRIT. MINISTRY OF SUPPLY. DEPT. OF TANK DESIGN (BULLETINS)	MS DTD B-
GT. BRIT. MINISTRY OF SUPPLY. DEPT. OF TANK DESIGN (MATERIAL SPECIFICATIONS)	MS DTD MS-
GT. BRIT. MINISTRY OF SUPPLY. DEPT. OF TANK DESIGN (REPORT)	MS DTD R-
GT. BRIT. MINISTRY OF SUPPLY. DIRECTORATE OF WEAPONS RESEARCH (DEFENCE)	WR(D)-
GT. BRIT. MINISTRY OF SUPPLY. DIRECTORATE OF WEAPONS RESEARCH (DEFENCE)	WRD-
GT. BRIT. MINISTRY OF SUPPLY. MATERIALS RESEARCH DIV.(HIGH EXPLOSIVES RES.), FT. HALSTEAD, KENT, ENG	HER-H-
GT. BRIT. MINISTRY OF SUPPLY. MINE DESIGN DEPT.	SS INFORMAL REPT.-
GT. BRIT. MINISTRY OF SUPPLY. TECHNICAL COORDINATING COMMITTEE ON GENERAL STORES.	MS TGS-
GT. BRIT. MINISTRY OF SUPPLY. TECHNICAL INFORMATION AND LIBRARY SERVICES	TIL/BIB/
GT. BRIT. MINISTRY OF SUPPLY. TECHNICAL INFORMATION AND LIBRARY SERVICES	TIL/BIB/(P)/-
GT. BRIT. MINISTRY OF SUPPLY. TECHNICAL INFORMATION AND LIBRARY SERVICES	TIL/BIB/(U)/
GT. BRIT. MINISTRY OF SUPPLY. TECHNICAL INFORMATION AND LIBRARY SERVICES	TIL-BR/
GT. BRIT. MINISTRY OF SUPPLY. TECHNICAL INFORMATION AND LIBRARY SERVICES	TIL/T-
GT. BRIT. MINISTRY OF SUPPLY. TECHNICAL INFO. BUR.	GDC-
GT. BRIT. MINISTRY OF SUPPLY. TECHNICAL INFO. BUR.	MOS-TIB GDC-1-
GT. BRIT. MINISTRY OF SUPPLY. TECHNICAL INFO. BUR.	S AND T MEMO-(NO.)/(YR.)
GT. BRIT. MINISTRY OF SUPPLY. TECHNICAL INFO. BUR.	S-T MEMO-
GT. BRIT. MINISTRY OF SUPPLY. TECHNICAL INFO. BUR.	TIB/BIB/(U)-
GT. BRIT. MINISTRY OF SUPPLY. TECHNICAL INFO. BUR.	TIB/GDC/
GT. BRIT. MINISTRY OF SUPPLY. TECHNICAL INFO. BUR.	TIB/T-
GT. BRIT. MINISTRY OF SUPPLY. TECHNICAL INFO. BUR.	TID/T-
GT. BRIT. MINISTRY OF WORKS	MOW-
GT. BRIT. MINISTRY OF WORKS	MOW (LETTER(S))-
GT. BRIT. MINISTRY OF WORKS (LEAFLET)	MOW AL-
GT. BRIT. MINISTRY OF WORKS (ECONOMY MEMORANDUM)	MOW EM-
GT. BRIT. MINISTRY OF WORKS (POSTWAR BLDG. STUDIES)	MOW PWB S-
GT. BRIT. MINISTRY OF WORKS (POSTWAR BLDG. STUDIES)	MW PWB S-
GT. BRIT. MINISTRY OF WORKS (POSTWAR BLDG. STUDIES)	PWB S-
GT. BRIT. MISSION FOR ECONOMIC AFFAIRS, LONDON	MEA-
GT. BRIT. NATL. CHEMICAL LAB.,TEDDINGTON, MIDDX.,ENG.	HAR-
GT. BRIT. NATL. CHEMICAL LAB.,TEDDINGTON, MIDDX.,ENG.	NCL/AE-
GT. BRIT. NATL. CHEMICAL LAB.,TEDDINGTON, MIDDX.,ENG.	NCL/AE-172
GT. BRIT. NATL. CHEMICAL LAB.,TEDDINGTON, MIDDX.,ENG.	NCL/DEP-
GT. BRIT. NATL. ENG. LAB., EAST KILBRIDE, GLASGOW	MERL-
GT. BRIT. NATL. ENG. LAB., EAST KILBRIDE, GLASGOW	NEL-LDR-
GT. BRIT. NATIONAL GAS TURBINE ESTABLISHMENT, FARNBOROUGH, HANTS, ENGLAND	ARC-
GT. BRIT. NATIONAL GAS TURBINE ESTABLISHMENT, FARNBOROUGH, HANTS, ENGLAND	MEMO M-
GT. BRIT. NATIONAL GAS TURBINE ESTABLISHMENT, FARNBOROUGH, HANTS, ENGLAND	NGTE-M-
GT. BRIT. NATIONAL GAS TURBINE ESTABLISHMENT, FARNBOROUGH, HANTS, ENGLAND	NGTE-NT-
GT. BRIT. NATIONAL GAS TURBINE ESTABLISHMENT, FARNBOROUGH, HANTS, ENGLAND	NGTE-R-
GT. BRIT. NATIONAL INST. FOR RESEARCH IN NUCLEAR SCIENCE, HARWELL, BERKS, ENGLAND	AERE-SC-
GT. BRIT. NATIONAL INST. FOR RESEARCH IN NUCLEAR SCIENCE, HARWELL, BERKS, ENGLAND	NIRL/M/
GT. BRIT. NATL. INST. FOR RES. IN NUCLEAR SCIENCE. RUTHERFORD HIGH ENERGY LAB., HARWELL, BERKS, ENG.	NIRL-
GT. BRIT. NATL. INST. FOR RES. IN NUCLEAR SCIENCE. RUTHERFORD HIGH ENERGY LAB., HARWELL, BERKS, ENG.	NIRL/LN/
GT. BRIT. NATL. INST. FOR RES. IN NUCLEAR SCIENCE. RUTHERFORD HIGH ENERGY LAB., HARWELL, BERKS, ENG.	NIRL/M-
GT. BRIT. NATL. INST. FOR RES. IN NUCLEAR SCIENCE. RUTHERFORD HIGH ENERGY LAB., HARWELL, BERKS, ENG.	NIRL/R/
GT. BRIT. NATL. INST. FOR RES. IN NUCLEAR SCIENCE. RUTHERFORD HIGH ENERGY LAB., HARWELL, BERKS, ENG.	NIRNS-LN-
GT. BRIT. NATL. INST. FOR RES. IN NUCLEAR SCIENCE. RUTHERFORD HIGH ENERGY LAB., HARWELL, BERKS, ENG.	NIRNS/R/
GT. BRIT. NATIONAL INST. OF INDUSTRIAL PSYCH., LONDON	NIIP-
GT. BRIT. NATIONAL PHYSICAL LAB., TEDDINGTON, MIDDX., ENGLAND (REF. 1)	(NUMBERS...)
GT. BRIT. NATL. PHYSICAL LAB., TEDDINGTON, MIDDX.,ENG	BC-
GT. BRIT. NATL. PHYSICAL LAB., TEDDINGTON, MIDDX.,ENG	HT-
GR. BRIT. NATL. PHYSICAL LAB., TEDDINGTON, MIDDX.,ENG	NPL-
GT. BRIT. NATL. PHYSICAL LAB., TEDDINGTON, MIDDX.,ENG	NPL-AERO-NOTE-
GT. BRIT. NATL. PHYSICAL LAB., TEDDINGTON, MIDDX.,ENG	NPL-AERO-SR-
GT. BRIT. NATL. PHYSICAL LAB., TEDDINGTON, MIDDX.,ENG	NPL-CCU-

GT. BRIT. NATL. PHYSICAL LAB., TEDDINGTON, MIDDX., ENG	NPL-COM-SCI-
GT. BRIT. NATL. PHYSICAL LAB., TEDDINGTON, MIDDX., ENG	NPL-MA-
GT. BRIT. NATL. PHYSICAL LAB., TEDDINGTON, MIDDX., ENG	NPL-MC-
GT. BRIT. NATIONAL PHYSICAL LAB., AERODYNAMICS DIV., TEDDINGTON, MIDDX., ENGLAND	NPL/AERO/-
GT. BRIT. NAVAL AIR FIGHTING DEVELOPMENT UNIT	NAFDU-
GT. BRIT. NAVAL CONSTRUCTION RESEARCH ESTABLISHMENT, DUNFERMLINE, FIFE, SCOTLAND	NCRE/N-
GT. BRIT. NAVAL CONSTRUCTION RESEARCH ESTABLISHMENT, DUNFERMLINE, FIFE, SCOTLAND	NCRE/R-
GT. BRIT. NAVAL CONSTRUCTION RESEARCH ESTABLISHMENT, DUNFERMLINE, FIFE, SCOTLAND	NCRE/R.498
GT. BRIT. NAVAL MOTION STUDY UNIT	NMSU-
GT. BRIT. NAVAL SCIENTIFIC TECH. INFO. CENTER, LONDON	NSTIC-
GT. BRIT. NAVAL SCIENTIFIC TECH. INFO. CENTER, LONDON	NSTIC/(NO.)/(YEAR)-
GT. BRIT. NAVAL SCIENTIFIC TECH. INFO. CENTER, LONDON	NSTIC-TRANS-
GT. BRIT. NAVAL WEATHER SERVICE	NWS-
GT. BRIT. NAVAL WEATHER SERVICE	NWS-MEMO-
GT. BRIT. NOISE REDUCTION PANEL	NRP/ANL-
GT. BRIT. ORDNANCE BOARD	OB-
GT. BRIT. ORDNANCE BOARD (INVESTIGATION)	OBI-
GT. BRIT. ORDNANCE BOARD	O.B. INVESTIGATION NO.
GT. BRIT. ORDNANCE BOARD	O.B. PROCEEDINGS NO.
GT. BRIT. ORDNANCE BOARD	ORDNANCE BOARD PROC...
GT. BRIT. ORDNANCE BOARD	O-
GT. BRIT. ORDNANCE BOARD, LONDON	TTB/(NUMBER)/
GT. BRIT. ORDNANCE DEPT. TECHNICAL REPRESENTATIVE FOR ARTILLERY, LONDON	ORD-TRART-
GT. BRIT. PENICILLIN CLINICAL TRIALS COMMITTEE	PCT-
GT. BRIT. POST OFFICE. ENGINEERING DEPT.	POED-
GT. BRIT. POST OFFICE. ENGINEERING DEPT.	PO-ED/RR-
GT. BRIT. POST OFFICE. ENGINEERING DEPT.	RADIO REPORT
GT. BRIT. POST OFFICE. ENGINEERING DEPT.	ROTO(NO.)R/-
GT. BRIT. RADAR RES. EST., MALVERN, WORCS, ENGLAND	RRDE-
GT. BRIT. RADAR RES. EST., MALVERN, WORCS, ENGLAND	RRE-TN-
GT. BRIT. RADIO AND ELECTRONICS MEASUREMENTS COMM.	REMC-
GT. BRIT. RADIO COMPONENTS STANDARDISATION COMMITTEE	RCL-
GT. BRIT. RADIO COMPONENTS STANDARDISATION COMMITTEE	RCS-
GT. BRIT. ROAD RES. LAB., HARMONDSWORTH, MIDDX., ENG.	ARP/(NUMBER)/RJ-
GT. BRIT. ROAD RES. LAB., HARMONDSWORTH, MIDDX., ENG.	RR-
GT. BRIT. ROAD RESEARCH LAB., HARMONDSWORTH, MIDDX., ENGLAND (ROAD NOTE)	DSIR RRL RN-
GT. BRIT. ROAD RESEARCH LAB., HARMONDSWORTH, MIDDX., ENGLAND (TECHNICAL PAPER)	DSIR RRL TP-
GT. BRIT. ROAD RES. LAB., HARMONDSWORTH, MIDDX., ENG.	MOS-
GT. BRIT. ROAD RES. LAB., HARMONDSWORTH, MIDDX., ENG.	R.C.-
GT. BRIT. ROCKET PROPULSION EST., WESTCOTT, BUCKS, ENG	RAE-TR-
GT. BRIT. ROCKET PROPULSION EST., WESTCOTT, BUCKS, ENG	RPE-TM-
GT. BRIT. ROCKET PROPULSION EST., WESTCOTT, BUCKS, ENG	RPE-TN-
GT. BRIT. ROCKET PROPULSION EST., WESTCOTT, BUCKS, ENG	RPE-TR-(YEAR)/
GT. BRIT. ROCKET PROPULSION EST., WESTCOTT, BUCKS, ENG	RPE-TRANS-
GT. BRIT. ROCKET PROPULSION EST., WESTCOTT, BUCKS, ENG	RPL-TM-
GT. BRIT. ROYAL AIR FORCE. BOMBER SUPPORT DEV. UNIT	BSDU-
GT. BRIT. ROYAL A.F. CENTRAL NAVIGATION + CONTROL SCH	CNCS-
GT. BRIT. ROYAL A.F. CENTRAL NAVIGATION + CONTROL SCH	EANS-
GT. BRIT. ROYAL AIR FORCE. COASTAL COMMAND	ORS/CC-
GT. BRIT. ROYAL AIR FORCE. DIRECTOR-GENERAL OF MEDICAL SERVICES	DGMS-
GT. BRIT. ROYAL AIR FORCE. TORPEDO DEVELOPMENT UNIT	TDU-
GT. BRIT. ROYAL AIRCRAFT ESTAB., BEDFORD, BEDS., ENG.	RAE-
GT. BRIT. ROYAL AIRCRAFT ESTAB., BEDFORD, BEDS., ENG.	RAE-AERO-
GT. BRIT. ROYAL AIRCRAFT ESTAB., BEDFORD, BEDS., ENG.	RAE-TN-AERO-
GT. BRIT. ROYAL AIRCRAFT EST., FARNBOROUGH, HANTS, ENG	AERO-
GT. BRIT. ROYAL AIRCRAFT EST., FARNBOROUGH, HANTS, ENG	ATDU-
GT. BRIT. ROYAL AIRCRAFT EST., FARNBOROUGH, HANTS, ENG	AVC-
GT. BRIT. ROYAL AIRCRAFT EST., FARNBOROUGH, HANTS, ENG	BL-
GT. BRIT. ROYAL AIRCRAFT EST., FARNBOROUGH, HANTS, ENGLAND (CIVIL AIRCRAFT AIRWORTHINESS DATA RECORDING PROGRAMME)	CAADRP-TR-
GT. BRIT. ROYAL AIRCRAFT EST., FARNBOROUGH, HANTS, ENG	CE-
GT. BRIT. ROYAL AIRCRAFT EST., FARNBOROUGH, HANTS, ENG	EL-
GT. BRIT. ROYAL AIRCRAFT EST., FARNBOROUGH, HANTS, ENG	GW/
GT. BRIT. ROYAL AIRCRAFT EST., FARNBOROUGH, HANTS, ENG	GW-
GT. BRIT. ROYAL AIRCRAFT EST., FARNBOROUGH, HANTS, ENG	IAP-
GT. BRIT. ROYAL AIRCRAFT EST., FARNBOROUGH, HANTS, ENG	INSTN-
GT. BRIT. ROYAL AIRCRAFT EST., FARNBOROUGH, HANTS, ENG	LIBRARY TRANS-
GT. BRIT. ROYAL AIRCRAFT EST., FARNBOROUGH, HANTS, ENG	MET-
GT. BRIT. ROYAL AIRCRAFT EST., FARNBOROUGH, HANTS, ENG	MS-
GT. BRIT. ROYAL AIRCRAFT EST., FARNBOROUGH, HANTS, ENG	N-
GT. BRIT. ROYAL AIRCRAFT EST., FARNBOROUGH, HANTS, ENG	ORS FT-
GT. BRIT. ROYAL AIRCRAFT EST., FARNBOROUGH, HANTS, ENG	PH-
GT. BRIT. ROYAL AIRCRAFT EST., FARNBOROUGH, HANTS, ENG	PP-
GT. BRIT. ROYAL AIRCRAFT EST., FARNBOROUGH, HANTS, ENG	RAE-
GT. BRIT. ROYAL AIRCRAFT EST., FARNBOROUGH, HANTS, ENG	RAE-AERO-
GT. BRIT. ROYAL AIRCRAFT EST., FARNBOROUGH, HANTS, ENG	RAE-AERO-TM-
GT. BRIT. ROYAL AIRCRAFT EST., FARNBOROUGH, HANTS, ENG	RAE-ARM.(NUMBER)
GT. BRIT. ROYAL AIRCRAFT EST., FARNBOROUGH, HANTS, ENG	RAE-CHEM-
GT. BRIT. ROYAL AIRCRAFT EST., FARNBOROUGH, HANTS, ENG	RAE-CPM-
GT. BRIT. ROYAL AIRCRAFT EST., FARNBOROUGH, HANTS, ENG	RAE-EL-
GT. BRIT. ROYAL AIRCRAFT EST., FARNBOROUGH, HANTS, ENG	RAE-IGT-MEMO-
GT. BRIT. ROYAL AIRCRAFT EST., FARNBOROUGH, HANTS, ENG	RAE-IM-
GT. BRIT. ROYAL AIRCRAFT EST., FARNBOROUGH, HANTS, ENG	RAE-LIB-
GT. BRIT. ROYAL AIRCRAFT EST., FARNBOROUGH, HANTS, ENG	RAE-LIB-BIB-
GT. BRIT. ROYAL AIRCRAFT EST., FARNBOROUGH, HANTS, ENG	RAE-LIB-BIBL-
GT. BRIT. ROYAL AIRCRAFT EST., FARNBOROUGH, HANTS, ENG	RAE-LIB-MEMO-

GT. BRIT. ROYAL AIRCRAFT EST.,FARNBOROUGH, HANTS, ENG	RAE-LIB-PRESS-TRANS-
GT. BRIT. ROYAL AIRCRAFT EST.,FARNBOROUGH, HANTS, ENG	RAE-LIB-TRANS-
GT. BRIT. ROYAL AIRCRAFT EST.,FARNBOROUGH, HANTS, ENG	RAE-MET-
GT. BRIT. ROYAL AIRCRAFT EST.,FARNBOROUGH, HANTS, ENG	RAE-R+M-
GT. BRIT. ROYAL AIRCRAFT EST.,FARNBOROUGH, HANTS, ENG	RAE-R-MECH.ENG-
GT. BRIT. ROYAL AIRCRAFT EST.,FARNBOROUGH, HANTS, ENG	RAE-R-MET-
GT. BRIT. ROYAL AIRCRAFT EST.,FARNBOROUGH, HANTS, ENG	RAE-R.P.D.-
GT. BRIT. ROYAL AIRCRAFT EST.,FARNBOROUGH, HANTS, ENG	RAE-R-STRUCT-
GT. BRIT. ROYAL AIRCRAFT EST.,FARNBOROUGH, HANTS, ENG	RAE-SME-
GT. BRIT. ROYAL AIRCRAFT EST.,FARNBOROUGH, HANTS, ENG	RAE-SN-ARM.(NUMBER)
GT. BRIT. ROYAL AIRCRAFT EST.,FARNBOROUGH, HANTS, ENG	RAE-STRUCT-
GT. BRIT. ROYAL AIRCRAFT EST.,FARNBOROUGH, HANTS, ENG	RAE-STRUCTURES-
GT. BRIT. ROYAL AIRCRAFT EST.,FARNBOROUGH, HANTS, ENG	RAE-TE-
GT. BRIT. ROYAL AIRCRAFT EST.,FARNBOROUGH, HANTS, ENG	RAE-T/M-ARM-
GT. BRIT. ROYAL AIRCRAFT EST.,FARNBOROUGH, HANTS, ENG	RAE-TN-AERO-
GT. BRIT. ROYAL AIRCRAFT EST.,FARNBOROUGH, HANTS, ENG	RAE-TN-ARM-
GT. BRIT. ROYAL AIRCRAFT EST.,FARNBOROUGH, HANTS, ENG	RAE-TN-BL-
GT. BRIT. ROYAL AIRCRAFT EST.,FARNBOROUGH, HANTS, ENG	RAE-TN-CHEM-
GT. BRIT. ROYAL AIRCRAFT EST.,FARNBOROUGH, HANTS, ENG	RAE-TN-CPM-
GT. BRIT. ROYAL AIRCRAFT EST.,FARNBOROUGH, HANTS, ENG	RAE-TN-EL-
GT. BRIT. ROYAL AIRCRAFT EST.,FARNBOROUGH, HANTS, ENG	RAE-TN-GW-
GT. BRIT. ROYAL AIRCRAFT EST.,FARNBOROUGH, HANTS, ENG	RAE-TN-IAP-
GT. BRIT. ROYAL AIRCRAFT EST.,FARNBOROUGH, HANTS, ENG	RAE-TN-INST-
GT. BRIT. ROYAL AIRCRAFT EST.,FARNBOROUGH, HANTS, ENG	RAE-TN-INSTN-
GT. BRIT. ROYAL AIRCRAFT EST.,FARNBOROUGH, HANTS, ENG	RAE-TN-MATH-
GT. BRIT. ROYAL AIRCRAFT EST.,FARNBOROUGH, HANTS, ENG	RAE-TN-ME-
GT. BRIT. ROYAL AIRCRAFT EST.,FARNBOROUGH, HANTS, ENG	RAE-TN-MECH. ENG.-
GT. BRIT. ROYAL AIRCRAFT EST.,FARNBOROUGH, HANTS, ENG	RAE-TN-MET-
GT. BRIT. ROYAL AIRCRAFT EST.,FARNBOROUGH, HANTS, ENG	RAE-TN-MS-
GT. BRIT. ROYAL AIRCRAFT EST.,FARNBOROUGH, HANTS, ENG	RAE-TN-RAD-
GT. BRIT. ROYAL AIRCRAFT EST.,FARNBOROUGH, HANTS, ENG	RAE-TN-RPD-
GT. BRIT. ROYAL AIRCRAFT EST.,FARNBOROUGH, HANTS, ENG	RAE-TN-SPACE-
GT. BRIT. ROYAL AIRCRAFT EST.,FARNBOROUGH, HANTS, ENG	RAE-TN-STRUCT-
GT. BRIT. ROYAL AIRCRAFT EST.,FARNBOROUGH, HANTS, ENG	RAE-TN-WE-
GT. BRIT. ROYAL AIRCRAFT EST.,FARNBOROUGH, HANTS, ENG	RAE-TRANS-
GT. BRIT. ROYAL AIRCRAFT EST.,FARNBOROUGH, HANTS, ENG	RAE-TR-ARM-
GT. BRIT. ROYAL AIRCRAFT EST.,FARNBOROUGH, HANTS, ENG	RPD-
GT. BRIT. ROYAL AIRCRAFT EST.,FARNBOROUGH, HANTS, ENG	SME-
GT. BRIT. ROYAL AIRCRAFT EST. ARMAMENT + INSTR. UNIT	AIEU-
GT. BRIT. ROYAL AIRCRAFT ESTABLISHMENT. BLIND LANDING EXPERIMENTAL UNIT	BLEU-
GT. BRIT. ROYAL ARMAMENT RESEARCH AND DEVELOPMENT ESTABLISHMENT, FORT HALSTEAD, KENT, ENGLAND	RARDE-
GT. BRIT. ROYAL ARMAMENT RESEARCH AND DEVELOPMENT ESTABLISHMENT, FORT HALSTEAD, KENT, ENGLAND	RARDE-B-
GT. BRIT. ROYAL ARMAMENT RESEARCH AND DEVELOPMENT ESTABLISHMENT, FORT HALSTEAD, KENT, ENGLAND	RARDE-E-
GT. BRIT. ROYAL ARMAMENT RESEARCH AND DEVELOPMENT ESTABLISHMENT, FORT HALSTEAD, KENT, ENGLAND	RARDE-MEMO(B)-
GT. BRIT. ROYAL ARMAMENT RESEARCH AND DEVELOPMENT ESTABLISHMENT, FORT HALSTEAD, KENT, ENGLAND	RARDE-MEMO(P)-
GT. BRIT. ROYAL ARMAMENT RESEARCH AND DEVELOPMENT ESTABLISHMENT, FORT HALSTEAD, KENT, ENGLAND	RARDE-X-(MONTH/YEAR)
GT. BRIT. ROYAL ARMAMENT RESEARCH AND DEVELOPMENT ESTABLISHMENT, SEVENOAKS, KENT, ENGLAND	RARDE-MEMO-
GT. BRIT. ROYAL ARMOURED CORPS EQUIPMENT TRIALS WING, WAREHAM, DORSET, ENGLAND	ETW/GT-
GT. BRIT. ROYAL NAVAL PERSONNEL RESEARCH COMMITTEE	CES-
GT. BRIT. ROYAL NAVAL PERSONNEL RESEARCH COMMITTEE	CP-
GT. BRIT. ROYAL NAVAL PERSONNEL RESEARCH COMMITTEE	OES-
GT. BRIT. ROYAL NAVAL PERSONNEL RESEARCH COMMITTEE	RNP-
GT. BRIT. ROYAL NAVAL PERSONNEL RESEARCH COMMITTEE	RNPRC-
GT. BRIT. ROYAL NAVAL PERSONNEL RESEARCH COMMITTEE	SS-
GT. BRIT. ROYAL NAVAL PERSONNEL RESEARCH COMMITTEE	UWB-
GT. BRIT. ROYAL NAVAL PERSONNEL RESEARCH COMMITTEE	WVP-
GT. BRIT. ROYAL NAVAL PHYSIOLOGICAL LAB., ALVERSTOKE	RNPL-
GT. BRIT. ROYAL NAVAL SCIENTIFIC SERVICE	RNSS-
GT. BRIT. ROYAL NAVAL SCIENTIFIC SERVICE	SRE-
GT. BRIT. ROYAL NAVAL SCIENTIFIC SERVICE, LONDON	CVD/QU-
GT. BRIT. ROYAL NAVAL SCIENTIFIC SERVICE, LONDON	CVD/QUI-
GT. BRIT. ROYAL OBSERVATORY, EDINBURGH, SCOTLAND	ROE-STS-
GT. BRIT. ROYAL RADAR ESTAB., MALVERN, WORCS, ENGLAND	PRE-TRANS-
GT. BRIT. ROYAL RADAR ESTAB., MALVERN, WORCS, ENGLAND	RRE-
GT. BRIT. ROYAL RADAR ESTAB., MALVERN, WORCS, ENGLAND	RRE-MEMO.-
GT. BRIT. ROYAL RADAR ESTAB., MALVERN, WORCS, ENGLAND	RRE-TRANS-(NUMBER)
GT. BRIT. ROYAL RADAR ESTAB., MALVERN, WORCS, ENGLAND	TIL/OT-
GT. BRIT. SAFETY IN MINES RESEARCH ESTABLISHMENT, BUXTON, DERBY, ENGLAND	E-
GT. BRIT. SAFETY IN MINES RESEARCH ESTABLISHMENT, BUXTON, DERBY, ENGLAND	SMRE-
GT. BRIT. SAFETY IN MINES RESEARCH ESTABLISHMENT, BUXTON, DERBY, ENGLAND	SMRE-RR-
GT. BRIT. SCOTTISH RES. REACTOR CENTRE, EAST KILBRIDE	SRRC-(NUMBER/YEAR)
GT. BRIT. SERVICES ELECTRONICS RESEARCH LAB., BALDOCK, HERTS, ENGLAND	SERL-TR-M-
GT. BRIT. SERVICES VALVE LIFE TEST ESTABLISHMENT	SVLTE-
GT. BRIT. SIGNALS RESEARCH AND DEVELOPMENT ESTABLISHMENT, CHRISTCHURCH, HANTS, ENGLAND	SRDE-
GT. BRIT. SPRINGFIELDS WORKS, LANCS, ENGLAND	CI-R-
GT. BRIT. SPRINGFIELDS WORKS, LANCS, ENGLAND (REF.36)	CS(SP)-
GT. BRIT. SPRINGFIELDS WORKS, LANCS, ENGLAND (REF.36)	DL(S)(LETTERS)-
GT. BRIT. SPRINGFIELDS WORKS, LANCS, ENGLAND	DL(S)TN-
GT. BRIT. SPRINGFIELDS WORKS, LANCS, ENGLAND	NPOC/FEWP-
GT. BRIT. SPRINGFIELDS WORKS, LANCS, ENGLAND (REF.36)	PM(S)-
GT. BRIT. SPRINGFIELDS WORKS, LANCS, ENGLAND (REF.36)	SP(CS)-
GT. BRIT. SPRINGFIELDS WORKS, LANCS, ENG. (REPRINTS)	SPRINGFIELDS-(NO.)/
GT. BRIT. SPRINGFIELDS WORKS, LANCS, ENGLAND	SRSC/P-
GT. BRIT. SPRINGFIELDS WORKS, LANCS, ENGLAND	SWT C/R-
GT. BRIT. SPRINGFIELDS WORKS, LANCS, ENGLAND	TRDC/P-
GT. BRIT. SPRINGFIELDS WORKS, LANCS, ENGLAND (REF.36)	VTLM-

(SEE ALSO LATER NAMES: UNITED KINGDOM ATOMIC ENERGY AUTHORITY. DEV. AND ENG. GP., SPRINGFIELDS... AND U.K.A.E.A. PRODUCTION GP., SPRINGFIELDS... AND U.K.A.E.A. REACTOR GROUP, SPRINGFIELDS)

GT. BRIT. TELECOMMUNICATIONS RESEARCH ESTABLISHMENT, MALVERN, WORCS, ENGLAND	BBRL-
GT. BRIT. TELECOMMUNICATIONS RESEARCH ESTABLISHMENT, MALVERN, WORCS, ENGLAND	M/-
GT. BRIT. TELECOMMUNICATIONS RESEARCH ESTABLISHMENT, MALVERN, WORCS, ENGLAND	MATH/MEMO(NO.)/WW
GT. BRIT. TELECOMMUNICATIONS RESEARCH ESTABLISHMENT, MALVERN, WORCS, ENGLAND	T.-

GT. BRIT. TELECOMMUNICATIONS RESEARCH ESTABLISHMENT

GT. BRIT. TELECOMMUNICATIONS RESEARCH ESTABLISHMENT, MALVERN, WORCS, ENGLAND	TRE-
GT. BRIT. TELECOMMUNICATIONS RESEARCH ESTABLISHMENT, MALVERN, WORCS, ENGLAND	TRE-M-
GT. BRIT. TELECOMMUNICATIONS RESEARCH ESTABLISHMENT, MALVERN, WORCS, ENGLAND	TRE-T-
GT. BRIT. TELECOMMUNICATIONS RESEARCH ESTABLISHMENT, MALVERN, WORCS, ENGLAND	TRE-TN-
GT. BRIT. TORPEDO EXPERIMENTAL ESTABLISHMENT	TEE-
GT. BRIT. TROPICAL TESTING ESTABLISHMENT	TTE-
GT. BRIT. UNDERWATER COUNTERMEASURES + WEAPONS ESTAB.	AME-
GT. BRIT. UNDERWATER COUNTERMEASURES + WEAPONS ESTAB.	MDD-
GT. BRIT. UNDERWATER COUNTERMEASURES + WEAPONS ESTAB.	MS-
GT. BRIT. UNDERWATER COUNTERMEASURES + WEAPONS ESTAB.	SS-
GT. BRIT. UNDERWATER DETECTION ESTABLISHMENT, PORTLAND, DORSET, ENGLAND	ASEE-
GT. BRIT. UNDERWATER DETECTION ESTABLISHMENT	HMA/SEE-
GT. BRIT. UNDERWATER DETECTION ESTABLISHMENT	HMUDE-
(SEE ALSO SUCCESSOR: UNITED KINGDOM. ADMIRALTY UNDERWATER DETECTION ESTABLISHMENT)	
GT. BRIT. WAR OFFICE	M.I.10-
GT. BRIT. WAR OFFICE	WO-
GT. BRIT. WAR OFFICE (PAMPHLET)	WO P-
GT. BRIT. WAR OFFICE (SPECIFICATION)	WO S-
GT. BRIT. WATER POLLUTION RESEARCH BOARD	WPRB-
GT. BRIT. WATER POLLUTION RES.LAB.,WATFORD,HERTS,ENG.	EBR-
GT. BRIT. WATER POLLUTION RES.LAB.,WATFORD,HERTS,ENG.	WPR/RC-B-
GT. BRIT. WINDSCALE WORKS, SELLAFIELD, CUMB., ENGLAND	CF-
GT. BRIT. WINDSCALE WORKS, SELLAFIELD, CUMB., ENGLAND (REF. 36)	CSD(W)-
GT. BRIT. WINDSCALE WORKS, SELLAFIELD, CUMB., ENGLAND (REF. 36)	D (W) MEMO-
GT. BRIT. WINDSCALE WORKS, SELLAFIELD, CUMB., ENGLAND	FRDC/P-
GT. BRIT. WINDSCALE WORKS, SELLAFIELD, CUMB., ENGLAND	PM(W)-
GT. BRIT. WINDSCALE WORKS, SELLAFIELD, CUMB., ENGLAND (REF. 36)	PRO(TM)-
GT. BRIT. WINDSCALE WORKS, SELLAFIELD, CUMB., ENGLAND (REF. 36)	SRW-
GT. BRIT. WINDSCALE WORKS, SELLAFIELD, CUMB., ENGLAND	TRSWP/R-
GT. BRIT. WINDSCALE WORKS, SELLAFIELD, CUMB., ENGLAND	WHPC-
GT. BRIT. WINDSCALE WORKS, SELLAFIELD, CUMB., ENGLAND	WSI-R-
GT. BRIT. WINDSCALE WORKS, SELLAFIELD, CUMB., ENGLAND	WSL-M-
GT. BRIT. WINDSCALE WORKS, SELLAFIELD, CUMB., ENGLAND (REF. 36)	WSL-TM-
(SEE ALSO LATER NAMES: UNITED KINGDOM ATOMIC ENERGY AUTHORITY. PRODUCTION GP., WINDSCALE... AND U.K.A.E.A. REACTOR GROUP, WINDSCALE...)	
GREAT LAKES RESEARCH CORP., ELIZABETHTON, TENN.	GLRC-
GREEK ATOMIC ENERGY COMMISSION. DEMOKRITOS NUCLEAR RESEARCH CENTRE, ATHENS	DEMO-(YEAR)/
GREENVILLE COUNTY PLANNING COMMISSION, S.C.	S.C.-(NUMBER-NUMBER)-
GREENVILLE COUNTY PLANNING COMMISSION, S.C.	SC-(NUMBER-NUMBER)-
GREY (JERRY), LIVINGSTON, N.J.	GE-
GRISCOM-RUSSELL CO., MASSILLON, OHIO	GRC-
GRISCOM-RUSSELL CO., MASSILLON, OHIO	RT-
GRONINGEN, NETHERLANDS. RIJKSUNIVERSITEIT. MATHEMATISCH INSTITUUT	TW-
GROUND ELECTRONICS ENGINEERING INSTALLATION AGENCY, GRIFFISS AFB, N.Y.	GEEIA-EMC-(YEAR)-
GROUND ELECTRONICS ENGINEERING INSTALLATION AGENCY, GRIFFISS AFB, N.Y.	GEEIA-TR-(YEAR)-
GRUEN APPLIED SCIENCE LABS., INC., WEST HEMPSTEAD, NY	GASL-
GRUMMAN AEROSPACE CORP., BETHPAGE, N.Y.	PDR-AI-(YEAR)-
GRUMMAN AEROSPACE CORP., BETHPAGE, N.Y.	RFENG-R-(YEAR)-
GRUMMAN AEROSPACE CORP., BETHPAGE, N.Y.	XAR-
GRUMMAN AEROSPACE CORP. ADVANCED COMPOSITES GROUP, BETHPAGE, N.Y.	AC-SM-
GRUMMAN AEROSPACE CORP. PRODUCT SUPPORT DEPT., BETHPAGE, N.Y.	LMA(NUMBER)-
GRUMMAN AIRCRAFT ENGINEERING CORP., BETHPAGE, N.Y.	ADN-
GRUMMAN AIRCRAFT ENGINEERING CORP., BETHPAGE, N.Y.	ADR-
GRUMMAN AIRCRAFT ENGINEERING CORP., BETHPAGE, N.Y.	ASR-(NUMBER-NUMBER)
GRUMMAN AIRCRAFT ENGINEERING CORP., BETHPAGE, N.Y.	AV-
GRUMMAN AIRCRAFT ENGINEERING CORP., BETHPAGE, N.Y.	BM-
GRUMMAN AIRCRAFT ENGINEERING CORP., BETHPAGE, N.Y.	DCN-
GRUMMAN AIRCRAFT ENGINEERING CORP., BETHPAGE, N.Y.	FP-
GRUMMAN AIRCRAFT ENGINEERING CORP., BETHPAGE, N.Y.	FSN-AD(NUMBER)-(NUMBER)
GRUMMAN AIRCRAFT ENGINEERING CORP., BETHPAGE, N.Y.	FSR-
GRUMMAN AIRCRAFT ENGINEERING CORP., BETHPAGE, N.Y.	FSR-(LETTER)(NO.)-
GRUMMAN AIRCRAFT ENGINEERING CORP., BETHPAGE, N.Y.	FSR-ST-
GRUMMAN AIRCRAFT ENGINEERING CORP., BETHPAGE, N.Y.	GAEC-
GRUMMAN AIRCRAFT ENGINEERING CORP., BETHPAGE, N.Y.	GAEC-ENG-POD-(YEAR)-
GRUMMAN AIRCRAFT ENGINEERING CORP., BETHPAGE, N.Y.	GAEC-FSR-S(NUMBER)-
GRUMMAN AIRCRAFT ENGINEERING CORP., BETHPAGE, N.Y.	GAEC-PDR-
GRUMMAN AIRCRAFT ENGINEERING CORP., BETHPAGE, N.Y.	GAEC-PDR-AC-(YEAR)-
GRUMMAN AIRCRAFT ENGINEERING CORP., BETHPAGE, N.Y.	GAEC-RE-
GRUMMAN AIRCRAFT ENGINEERING CORP., BETHPAGE, N.Y.	GRD-
GRUMMAN AIRCRAFT ENGINEERING CORP., BETHPAGE, N.Y.	GWT-
GRUMMAN AIRCRAFT ENGINEERING CORP., BETHPAGE, N.Y.	GWTT-
GRUMMAN AIRCRAFT ENGINEERING CORP., BETHPAGE, N.Y.	LMO-(NUMBER)-
GRUMMAN AIRCRAFT ENGINEERING CORP., BETHPAGE, N.Y.	LTC-(NUMBER)-
GRUMMAN AIRCRAFT ENGINEERING CORP., BETHPAGE, N.Y.	LTP-(NUMBER-NUMBER)
GRUMMAN AIRCRAFT ENGINEERING CORP., BETHPAGE, N.Y.	ME-
GRUMMAN AIRCRAFT ENGINEERING CORP., BETHPAGE, N.Y.	ME-CP-(NUMBER)-
GRUMMAN AIRCRAFT ENGINEERING CORP., BETHPAGE, N.Y.	ME-MF-(NO.)-(NO.)-
GRUMMAN AIRCRAFT ENGINEERING CORP., BETHPAGE, N.Y.	ME-MP-(NUMBER)-
GRUMMAN AIRCRAFT ENGINEERING CORP., BETHPAGE, N.Y.	ME-OT-(NUMBER)-
GRUMMAN AIRCRAFT ENGINEERING CORP., BETHPAGE, N.Y.	ME-TP-(NUMBER)-
GRUMMAN AIRCRAFT ENGINEERING CORP., BETHPAGE, N.Y.	ME-WE-(NUMBER)-
GRUMMAN AIRCRAFT ENGINEERING CORP., BETHPAGE, N.Y.	MT-
GRUMMAN AIRCRAFT ENGINEERING CORP., BETHPAGE, N.Y.	MT-OT-(NO.)-(NO.)-
GRUMMAN AIRCRAFT ENGINEERING CORP., BETHPAGE, N.Y. (PILOTLESS AIRCRAFT)	P/A-
GRUMMAN AIRCRAFT ENGINEERING CORP., BETHPAGE, N.Y.	PDR-ASG-
GRUMMAN AIRCRAFT ENGINEERING CORP., BETHPAGE, N.Y.	PDR-N-
GRUMMAN AIRCRAFT ENGINEERING CORP., BETHPAGE, N.Y.	PSER-(YEAR)-
GRUMMAN AIRCRAFT ENGINEERING CORP., BETHPAGE, N.Y.	RE-
GRUMMAN AIRCRAFT ENGINEERING CORP., BETHPAGE, N.Y.	RE-(NO.)-PR-
GRUMMAN AIRCRAFT ENGINEERING CORP., BETHPAGE, N.Y.	RF-(YEAR)-

GRUMMAN AIRCRAFT ENGINEERING CORP., BETHPAGE, N.Y.	RM-
GRUMMAN AIRCRAFT ENGINEERING CORP., BETHPAGE, N.Y.	RN-
GRUMMAN AIRCRAFT ENGINEERING CORP., BETHPAGE, N.Y.	SAS-(NUMBER)
GRUMMAN AIRCRAFT ENGINEERING CORP., BETHPAGE, N.Y.	SPEC-LSP-(NO.)-
GRUMMAN AIRCRAFT ENGINEERING CORP., BETHPAGE, N.Y.	VEDER-(NUMBER)-(YEAR)-
GRUMMAN AIRCRAFT ENGINEERING CORP. MILITARY SPACE SYSTEMS, BETHPAGE, N.Y.	MSSR-
GRUMMAN AIRCRAFT ENGINEERING CORP. PRELIMINARY SYSTEMS SECTION, BETHPAGE, N.Y.	PSR-(YEAR)-
GRUMMAN AIRCRAFT ENG. CORP. RES. DEPT., BETHPAGE, NY	RE-(NUMBER)(LETTER)
GRUMMAN OPERATIONS ANALYSIS GROUP. ADVANCED SYSTEMS SECTION, BETHPAGE, N.Y.	PDM-
GRUMMAN OPERATIONS ANALYSIS GROUP. ADVANCED SYSTEMS SECTION, BETHPAGE, N.Y.	PDR-
GRUMMAN OPERATIONS ANALYSIS GROUP. ADVANCED SYSTEMS SECTION, BETHPAGE, N.Y.	PMR-
GRUMMAN OPERATIONS ANALYSIS GROUP. ADVANCED SYSTEMS SECTION, BETHPAGE, N.Y.	PRD-
GRUMMAN OPERATIONS ANALYSIS/PLANNING RESEARCH, FUTURE SYSTEMS, BETHPAGE, N.Y.	FSR-
GRUNZWEIG UND HARTMANN A.G., LUDWIGSHAFEN AM RHEIN, GERMANY	RM-(NUMBER)(LETTER)
GTE SYLVANIA, INC., NEEDHAM HEIGHTS, MASS.	FR(YEAR-NUMBER)N
GTE SYLVANIA, INC. ELECTRO-OPTICS ORGANIZATION, MOUNTAIN VIEW, CALIF.	EOO-
GULF GENERAL ATOMIC CO., SAN DIEGO, CALIF.	GULF-GA-
GULF GENERAL ATOMIC, INC., SAN DIEGO, CALIF.	GA-
GULF GENERAL ATOMIC, INC., SAN DIEGO, CALIF.	GACD-
GULF GENERAL ATOMIC, INC., SAN DIEGO, CALIF. (HTGR BASE PROGRAM PROGRESS REPORT)	GACD-6135(MONTH-YEAR)
GULF GENERAL ATOMIC, INC., SAN DIEGO, CALIF. (PUBLIC SERVICE CO. OF COLO. PLANT DESIGN+CONSTR)	GACD-6850(MONTH-YEAR)
GULF GENERAL ATOMIC, INC., SAN DIEGO, CALIF. (PUBLIC SERVICE CO. OF COLO. RES. + DEV. PROGRAM)	GACD-6900(MONTH-YEAR)
GULF GENERAL ATOMIC, INC., SAN DIEGO, CALIF.	GACP-
GULF GENERAL ATOMIC, INC., SAN DIEGO, CALIF.	GAMD-
GULF GENERAL ATOMIC, INC., SAN DIEGO, CALIF. (TRANS.)	GAMD-TR-
GULF GENERAL ATOMIC, INC., SAN DIEGO, CALIF.	GA-P-
GULF GENERAL ATOMIC, INC., SAN DIEGO, CALIF. (TRANS.)	GA-TR-
GULF GENERAL ATOMIC, INC., SAN DIEGO, CALIF.	GGA-
GULF RADIATION TECHNOLOGY, SAN DIEGO, CALIF.	GULF-RT-
GULF RADIATION TECHNOLOGY, SAN DIEGO, CALIF.	GULF-RT-A-
GULF RESEARCH AND DEVELOPMENT CO., PITTSBURGH	GRDC-
GULF UNITED NUCLEAR FUELS CORP., ELMSFORD, N.Y.	GU-
GULF UNIVERSITIES RESEARCH SERVICE, WASHINGTON, D.C.	GURC-
GULTON INDUSTRIES, INC. ALKALINE BATTERY DIV., METUCHEN, N.J.	AB-(NUMBER)-
GULTON INDUSTRIES, INC. ENGINEERING MAGNETICS DIV., HAWTHORNE, CALIF.	EM-
GULTON MFG. CORP., METUCHEN, N.J.	GMC-
GULTON MFG. CORP., METUCHEN, N.J.	GUL-
GYRODYNE CO. OF AMERICA, INC., ST. JAMES, N.Y.	RD(NUMBER(S))-
GYRODYNE CO. OF AMERICA, INC., ST. JAMES, N.Y.	Y(NUMBER(S))-
GYRODYNE CO. OF AMERICA, INC., ST. JAMES, N.Y.	Y70-(NUMBER)-

HALLER, RAYMOND, AND BROWN, INC.

	(PROJECT ACCOUNT NUMBER)-
HALLER, RAYMOND, AND BROWN, INC., STATE COLLEGE, PA. (REF. 1)	
HALLER, RAYMOND, AND BROWN, INC., STATE COLLEGE, PA.	HRB-
HALLER, RAYMOND, AND BROWN, INC., STATE COLLEGE, PA.	SCEAG-
HALLICRAFTERS CO., CHICAGO	HALLIC-
HALLICRAFTERS CO., ROLLING MEADOWS, ILL.	HLC-
HAMBURG. UNIVERSITAET. INST. FUER EXPERIMENTAL PHYSIK	EMBW-FBK-
HAMBURGISCHE SCHIFFBAU-VERSUCHSANSTALT	HSVA-
HAMILTON STANDARD, WINDSOR LOCKS, CONN.	HSIR-
HAMILTON STANDARD, WINDSOR LOCKS, CONN.	HSME-
HAMILTON STANDARD, WINDSOR LOCKS, CONN.	HSSP-
HAMILTON STANDARD, WINDSOR LOCKS, CONN.	HSTP-(YEAR)-
HAMILTON STANDARD, WINDSOR LOCKS, CONN.	SP-
HAMILTON STANDARD, WINDSOR LOCKS, CONN.	SVHSER-
HAMILTON STANDARD. ELECTRONICS DEPT., BROAD BROOK, CONN.	HSER-
HAMILTON STANDARD. ELECTRONICS DEPT., WINDSOR LOCKS, CONN.	HSER-
HANDLEY PAGE (READING), LTD., ST. ALBANS, HERTS, ENG.	HPRWT-
HANFORD ATOMIC PRODUCTS OPERATION, RICHLAND, WASH.	BIN-
HANFORD ATOMIC PRODUCTS OPERATION, RICHLAND, WASH.	HAN-
HANFORD ATOMIC PRODUCTS OPERATION, RICHLAND, WASH.	HW-
HANFORD ATOMIC PRODUCTS OPERATION, RICHLAND, WASH.	HWS-
HANFORD ATOMIC PRODUCTS OPERATION, RICHLAND, WASH.	HW-SA-
HANFORD ATOMIC PRODUCTS OPERATION, RICHLAND, WASH.	HW-TR-
HANFORD ATOMIC PRODUCTS OPERATION, RICHLAND, WASH.	RL-GEN-
HANFORD ATOMIC PRODUCTS OPERATION, RICHLAND, WASH.	RL-NRD-
HANFORD ATOMIC PRODUCTS OPERATION, RICHLAND, WASH.	RL-REA-
HANFORD ATOMIC PRODUCTS OPERATION, RICHLAND, WASH.	RL-SA-
HANFORD ATOMIC PRODUCTS OPERATION, RICHLAND, WASH.	RL-SEP-
HANFORD ENGINEER WORKS, RICHLAND, WASH. (ASSIGNED TO INCOMING REPORTS DURING THE PERIOD WHEN HANFORD WAS OPERATED BY E.I. DU PONT DE NEMOURS + CO.)	DUH-
HANFORD ENGINEER WORKS, RICHLAND, WASH. (ASSIGNED TO INCOMING REPORTS DURING THE PERIOD WHEN HANFORD WAS OPERATED BY THE GENERAL ELECTRIC CO.)	GEH-
HANFORD ENGINEER WORKS, RICHLAND, WASH. (REPORTS ISSUED BY THE GENERAL ELECTRIC CO. JULY - OCT. 1947)	HEW-
HANFORD ENGINEER WORKS, RICHLAND, WASH.	MEMO-
HANFORD ENGINEER WORKS. GE NUCLEONICS PROJECT, RICHLAND, WASH.	GE-
HANFORD ENGINEERING DEVELOPMENT LAB., RICHLAND, WASH.	HEDL-SA-
HANFORD ENGINEERING DEVELOPMENT LAB., RICHLAND, WASH.	HEDL-TME-(YEAR)-
HANFORD OCCUPATIONAL HEALTH FDN., RICHLAND, WASH.	HOHF-
HANFORD OCCUPATIONAL HEALTH FOUNDATION. OCCUPATIONAL HYGIENE DEPT., RICHLAND, WASH.	HEHF-
HANFORD OCCUPATIONAL HEALTH FOUNDATION. OCCUPATIONAL HYGIENE DEPT., RICHLAND, WASH.	HOHF-
HANFORD OPERATIONS OFFICE, AEC, WASH.	HAN-
HANFORD OPERATIONS OFFICE, AEC, WASH.	HO-
HANFORD OPERATIONS OFFICE, AEC, WASH. (REF. 4)	HOO-(INITIALS)-
HANFORD OPERATIONS OFFICE, AEC, WASH. (REF. 4)	HOO-
HANFORD WORKS, RICHLAND, WASH.	CTM-K-
HANFORD WORKS, RICHLAND, WASH.	CTNX-
HANFORD WORKS, RICHLAND, WASH. (ASSIGNED TO INCOMING REPORTS)	GEH-
HANFORD WORKS, RICHLAND, WASH.	H-
HANFORD WORKS, RICHLAND, WASH.	HAN-
HANFORD WORKS, RICHLAND, WASH.	HANFORD-
HANFORD WORKS, RICHLAND, WASH.	HOO-(INITIALS)-
HANFORD WORKS, RICHLAND, WASH.	HWN-
HANFORD WORKS, RICHLAND, WASH.	HW OPERATING STANDARDS-
HANFORD WORKS, RICHLAND, WASH.	HWR-
HANFORD WORKS, RICHLAND, WASH.	HW-R-
HANFORD WORKS, RICHLAND, WASH.	HW TECHNICAL MANUALS-
HANFORD WORKS, RICHLAND, WASH.	MEMO-SE-PA-
HANFORD WORKS, RICHLAND, WASH.	MEMO-SE-PC-
HANFORD WORKS, RICHLAND, WASH.	NOTEBOOK H.E.W.-(NO.)-T
HANFORD WORKS, RICHLAND, WASH.	PT-
HANFORD WORKS, RICHLAND, WASH.	RDA-TC-
HANFORD WORKS, RICHLAND, WASH.	SE-PC-
HANFORD WORKS, RICHLAND, WASH.	TNX-
HANFORD WORKS, RICHLAND, WASH.	TNX-PG-
HANFORD WORKS. DESIGN + CONSTRUCTION DIV., RICHLAND, WASH.	HDC-
HANFORD WORKS. DESIGN + CONSTRUCTION DIV., RICHLAND, WASH. (HIGHLY RESTRICTED REPORTS)	HM-
HANFORD WORKS. DESIGN + CONSTRUCTION DIV., RICHLAND, WASH. (ASSIGNED TO INCOMING REPORTS)	INDC-
HANOVER. TECHNISCHE HOCHSCHULE	ABS-THH-
HARBOR PROTECTION PROJECT	HPP-
HARRIS TRANSDUCER CORP., SOUTHBURY, CONN.	HTC-
HARRY DIAMOND LABS., WASH., D.C. (FLUID AMPLIFICATION)	DOFL-FA-
HARRY DIAMOND LABS., WASHINGTON, D.C.	DOFL-IM-
HARRY DIAMOND LABS., WASHINGTON, D.C.	DOFL-TM-
HARRY DIAMOND LABS., WASHINGTON, D.C.	DOFL-TR-
HARRY DIAMOND LABS., WASHINGTON, D.C.	HDL-
HARRY DIAMOND LABS., WASHINGTON, D.C.	HDL-PR-
HARRY DIAMOND LABS., WASHINGTON, D.C.	HDL-RE-
HARRY DIAMOND LABS., WASHINGTON, D.C.	HDL-TM-
HARRY DIAMOND LABS., WASHINGTON, D.C.	HDL-TR-
HARRY DIAMOND LABS., WASHINGTON, D.C.	PR-(YEAR)-
HARRY DIAMOND LABS., WASHINGTON, D.C.	TM-(YEAR)-
HARRY DIAMOND LABS., WASHINGTON, D.C.	TR-
HARSHAW CHEMICAL CO., CLEVELAND	HCC-
HARSHAW CHEMICAL CO. CENTRAL RESEARCH LAB., CLEVELAND	HQ-
HARVARD UNIV., BOSTON. SCHOOL OF PUBLIC HEALTH	HSPH-
HARVARD UNIV., CAMBRIDGE, MASS.	FL-

HARVARD UNIV., CAMBRIDGE, MASS.	HARVARD U.-(LTRS,NOS.)-
HARVARD UNIV., CAMBRIDGE, MASS.	HSRP-
HARVARD UNIV., CAMBRIDGE, MASS.	HU-
HARVARD UNIV., CAMBRIDGE, MASS.	HU FL-
HARVARD UNIV., CAMBRIDGE, MASS. (MANY DEPTS/SCHOOLS)	HUX-(CONTRACT NO.)-
HARVARD UNIV., CAMBRIDGE, MASS.	PPM-
HARVARD UNIV., CAMBRIDGE, MASS.	SAPR-
HARVARD UNIV., CAMBRIDGE, MASS. ACOUSTICS RES. LAB.	ARL-TM-
HARVARD UNIV., CAMBRIDGE, MASS. ACOUSTICS RES. LAB.	HARL-
HARVARD UNIV., CAMBRIDGE, MASS. ACOUSTICS RESEARCH LAB. (TECHNICAL MEMORANDUM)	HU ARL TM-
HARVARD UNIV., CAMBRIDGE, MASS. AMER. BRITISH LAB.	ABL 1045-
HARVARD UNIV., CAMBRIDGE, MASS. AMER. BRITISH LAB.	HU ABL 1045-
HARVARD UNIV., CAMBRIDGE, MASS. AMER. BRITISH LAB.	HU ABL 1045-MR-
HARVARD UNIV., CAMBRIDGE, MASS. AMERICAN BRITISH LAB. (TECHNICAL MEMORANDA)	HU ABL 1045-TM-
HARVARD UNIV., CAMBRIDGE, MASS. APPLIED MATHEMATICS GROUP (REF. 28)	AMG-H-
HARVARD UNIV., CAMBRIDGE, MASS. APPLIED MATHEMATICS GROUP (REF. 28)	NDRC AMG-H-
HARVARD UNIV., CAMBRIDGE, MASS. BLUE HILL METEOROLOGICAL OBSERVATORY	HU-BHMO-
HARVARD UNIV., CAMBRIDGE, MASS. COMPUTATION LAB.	HCL-PR-
HARVARD UNIV., CAMBRIDGE, MASS. CRUFT LAB.	CL-
HARVARD UNIV., CAMBRIDGE, MASS. CRUFT LAB.	CRUFT-TR-
HARVARD UNIV., CAMBRIDGE, MASS. CRUFT LAB.	HU-CL-PR-
HARVARD UNIV., CAMBRIDGE, MASS. CRUFT LAB.	HU CL TR-
HARVARD UNIV., CAMBRIDGE, MASS. CRUFT LAB.	HUTR-
HARVARD UNIV., CAMBRIDGE, MASS. DEPT. OF MATHEMATICS	LA-HU-
HARVARD UNIV., CAMBRIDGE, MASS. DEPT. OF MINERALOGY	HU-DM-
HARVARD UNIV., CAMBRIDGE, MASS. DIVISION OF ENG. AND APPLIED PHYSICS	HP-
HARVARD UNIV., CAMBRIDGE, MASS. DIVISION OF ENG. AND APPLIED PHYSICS (PROGRESS REPORT)	HU-DEAP-PR-
HARVARD UNIV., CAMBRIDGE, MASS. DIVISION OF ENG. AND APPLIED PHYSICS	HU-DEAP-SPR-
HARVARD UNIV., CAMBRIDGE, MASS. DIVISION OF ENG. AND APPLIED PHYSICS	HU-DEAP-TM-
HARVARD UNIV., CAMBRIDGE, MASS. DIVISION OF ENG. AND APPLIED PHYSICS (TECHNICAL REPORT)	HU-DEAP-TR-
HARVARD UNIV., CAMBRIDGE, MASS. DIVISION OF ENG. AND APPLIED PHYSICS	HU-EA-
HARVARD UNIV., CAMBRIDGE, MASS. DUNBAR LAB.	HU-DL-
HARVARD UNIV., CAMBRIDGE, MASS. HARVARD COLL. OBSERVATORY (SOLAR REPORT)	HCO-
HARVARD UNIV., CAMBRIDGE, MASS. HARVARD COLL. OBS.	HCO-SR-
HARVARD UNIV., CAMBRIDGE, MASS. HARVARD COLL. OBS.	HCO-TR-
HARVARD UNIV., CAMBRIDGE, MASS. LYMAN LAB. OF PHYSICS	HU-LLP-TR-
HARVARD UNIV., CAMBRIDGE, MASS. PSYCHO-ACOUSTIC LAB.	HU PAL-
HARVARD UNIV., CAMBRIDGE, MASS. PSYCHO-ACOUSTIC LAB. (INFORMAL COMMUNICATION)	HU PAL IC-
HARVARD UNIV., CAMBRIDGE, MASS. PSYCHO-ACOUSTIC LAB. (INVENTION REPORTS)	HU PAL IR-
HARVARD UNIV., CAMBRIDGE, MASS. PSYCHO-ACOUSTIC LAB.	MHR-
HARVARD UNIV., CAMBRIDGE, MASS. PSYCHO-ACOUSTIC LAB.	PAL-
HARVARD UNIV., CAMBRIDGE, MASS. PSYCHO-ACOUSTIC LAB.	PNM-
HARVARD UNIV., CAMBRIDGE, MASS. PSYCHO-ACOUSTIC LAB. (PSYCHO NAVY RESEARCH)	PNR-
HARVARD UNIV., CAMBRIDGE, MASS. PSYCHOLOGICAL LABS.	PLR-
HARVARD UNIV., CAMBRIDGE, MASS. RADIO RESEARCH LAB.	HU RRL-
HARVARD UNIV., CAMBRIDGE, MASS. RADIO RESEARCH LAB.	HU RRL 411-IB-
HARVARD UNIV., CAMBRIDGE, MASS. RADIO RESEARCH LAB. (MATERIALS TESTING REPORTS)	HU RRL 411-MTR-
HARVARD UNIV., CAMBRIDGE, MASS. RADIO RESEARCH LAB. (TECHNICAL MEMORANDA)	HU RRL 411-TM-
HARVARD UNIV., CAMBRIDGE, MASS. RADIO RESEARCH LAB. (TEST REPORTS)	HU RRL 411-TR-
HARVARD UNIV., CAMBRIDGE, MASS. RADIO RESEARCH LAB. (DISCLOSURE)	HU RRL D-
HARVARD UNIV., CAMBRIDGE, MASS. RADIO RESEARCH LAB. (TENTATIVE SPECIFICATION)	HU RRL TS-
HARVARD UNIV., CAMBRIDGE, MASS. RADIO RESEARCH LAB.	RRL-
HARVARD UNIV., CAMBRIDGE, MASS. UNDERWATER SOUND LAB.	HU USL RPT-
HARVARD UNIV., CAMBRIDGE, MASS. UNDERWATER SOUND LAB.	HU USL RPT (LETTER)-
HARVEY ALUMINUM, INC., TORRANCE, CALIF.	DPWR-
HARVEY ALUMINUM, INC., TORRANCE, CALIF.	HA-
HARVEY ALUMINUM, INC., TORRANCE, CALIF.	HAI-
HARVEY ALUMINUM, INC., TORRANCE, CALIF.	HAI-PR-
HARVEY ENGINEERING LABS. FOR RES. + DEV.,TORRANCE,CAL	HA-
HARVEY MACHINE CO., INC., TORRANCE, CALIF.	HAR-
HARVEY MACHINE CO., INC., TORRANCE, CALIF.	HMC-
HASTINGS INSTRUMENT CO., INC., HAMPTON, VA.	HAST-
HASTINGS INSTRUMENT CO., INC., HAMPTON, VA.	HICO-
HATHAWAY INSTRUMENT CO., DENVER, COLO.	SP-HB-
HAWAII. UNIV., HONOLULU	HAWAIIU-
HAWAII. UNIV., HONOLULU	HAWU-
HAWAII. UNIV., HONOLULU	HEPG-
HAWAII. UNIV., HONOLULU (PROJECT THEMIS) (REF. 48)	THEMIS-A(YEAR)-
HAWAII. UNIV., HONOLULU (PROJECT THEMIS) (REF. 48)	THEMIS-B(YEAR)-
HAWAII. UNIV., HONOLULU	UH-
HAWAII. UNIV., HONOLULU	UH-(CONTRACT NO.)-
HAWAII. UNIV., HONOLULU	UH-(CON.NO.)-P-(NO.)-
HAWAII. UNIV., HONOLULU. DEPT. OF METEOROLOGY	UHMET-(YEAR)-
HAWAII. UNIV., HONOLULU. DEPT. OF PHYSIOLOGY	UHI-MED-(YEAR)-
HAWAII. UNIV., HONOLULU. HAWAII MARINE LAB.	HML-
HAWAII. UNIV., HONOLULU. SOCIAL SCIENCE RES. INST.	SSRI-REPRINT-
HAWAII INST. OF GEOPHYSICS, HONOLULU	HIG-
HAWAII INST. OF GEOPHYSICS, HONOLULU	HIG-(YEAR)-
HAWK JOINT AEC-DOD COORDINATING COMMITTEE	JCC/HAWK-
HAWKER SIDDELEY AVIATION, LTD., LONDON	S + T-MEMO-

HAWKER SIDDELEY DYNAMICS, LTD.

HAWKER SIDDELEY DYNAMICS, LTD., HATFIELD, HERTS, ENG.	HSD-TN-
HAWKER SIDDELEY DYNAMICS, LTD., HATFIELD, HERTS, ENG.	HSD-TP-
HAWKER SIDDELEY DYNAMICS, LTD., STEVENAGE, HERTS, ENG.	ELDO-RFI-
HAWKER SIDDELEY DYNAMICS, LTD., STEVENAGE, HERTS, ENG.	ELDO/SP/(YEAR)/(NUMBER)
HAYES AIRCRAFT CORP., BIRMINGHAM, ALA.	HAYES-ER-
HAYES INTERNATIONAL CORP., BIRMINGHAM, ALA.	HAIC-
HAYES INTERNATIONAL CORP., BIRMINGHAM, ALA.	HIC-
HAYES INTERNATIONAL CORP., HUNTSVILLE, ALA.	HIC-
HAZELTINE CORP. HAZELTINE RES. DIV., LITTLE NECK, NY	HRD-
HAZELTINE ELECTRONICS CORP., LITTLE NECK, N.Y.	9479-
HAZELTINE ELECTRONICS CORP., LITTLE NECK, N.Y.	HEC-
HAZELTINE SERVICE CORP.	HED-
HAZELTINE SERVICE CORP.	HSC-
HAZLETON-NUCLEAR SCIENCE CORP., PALO ALTO, CALIF.	HNS-(CONTRACT NO.)-
HAZLETON-NUCLEAR SCIENCE CORP., PALO ALTO, CALIF.	H-NSC-
HAZLETON-NUCLEAR SCIENCE CORP. EARTH SCIENCES DIV., PALO ALTO, CALIF.	HNSTR-
HAZLETON-NUCLEAR SCIENCE CORP. RADIOCHEMISTRY DEPT., PALO ALTO, CALIF.	HNS-
HEBREW UNIV., JERUSALEM	E-
HEBREW UNIV., JERUSALEM	HEBREWU-
HEBREW UNIV., JERUSALEM	HEBRU-
HEIDELBERG UNIV. INSTITUT FUER ANGEWANDTE PHYSIK	UNILAC-(NUMBER)-(YEAR)
HEINZ (H.J.) CO., PITTSBURGH	HEINZ-
HELIODYNE CORP. (ALL LOCATIONS)	HRN-
HELIODYNE CORP., LOS ANGELES	RN-
HELIODYNE CORP., LOS ANGELES	RN(NO.)-
HELIODYNE CORP., VAN NUYS, CALIF.	HELIODYNE-RR-
HELIODYNE CORP., VAN NUYS, CALIF.	HPD-
HELIODYNE CORP. BSD/BMRS DATA SERVICES AND ANALYSES, NORTON AFB, CALIF.	HRM-
HELSINKI UNIV.	HELSINKIU-(NO./YR.)
HELSINKI UNIV.	REPT-S-(NUMBER-YEAR)
HELSINKI UNIV. RADIO LAB.	S-(2 DIGITS-YEAR)
HERCULES POWDER CO., KENVIL, N.J.	K-(NUMBER)/MR-(NUMBER)-
HERCULES POWDER CO., MAGNA, UTAH	HPC-
HERCULES POWDER CO., MAGNA, UTAH	MCS-
HERCULES POWDER CO., MAGNA, UTAH	MTO-
HERCULES POWDER CO., WILMINGTON, DEL.	C(YEAR)-
HERCULES POWDER CO., WILMINGTON, DEL.	HERCULES MTI-
HERCULES POWDER CO., WILMINGTON, DEL.	HPC-
HERCULES POWDER CO., WILMINGTON, DEL.	HPC-CPD-
HERCULES POWDER CO., WILMINGTON, DEL.	HPC-PRM-
HERCULES POWDER CO., WILMINGTON, DEL.	MTI-
HERCULES POWDER CO. ALLEGANY BALLISTICS LAB., CUMBERLAND, MD.	ABL-
HERCULES POWDER CO. ALLEGANY BALLISTICS LAB., CUMBERLAND, MD.	ABL/AF/QPR-
HERCULES POWDER CO. ALLEGANY BALLISTICS LAB., CUMBERLAND, MD.	ABL/ARPA/QTSR-
HERCULES POWDER CO. ALLEGANY BALLISTICS LAB., CUMBERLAND, MD.	ABL-B-
HERCULES POWDER CO. ALLEGANY BALLISTICS LAB., CUMBERLAND, MD. (FINAL REPORT)	ABL FR
HERCULES POWDER CO. ALLEGANY BALLISTICS LAB., CUMBERLAND, MD. (FINAL REPORT)	ABL FR (LETTER)-
HERCULES POWDER CO. ALLEGANY BALLISTICS LAB., CUMBERLAND, MD.	ABL-LIBRARY-
HERCULES POWDER CO. ALLEGANY BALLISTICS LAB., CUMBERLAND, MD. (MONTHLY PROGRESS REPORT)	ABL/MPR-
HERCULES POWDER CO. ALLEGANY BALLISTICS LAB., CUMBERLAND, MD.	ABL/NASA/QPR-
HERCULES POWDER CO. ALLEGANY BALLISTICS LAB., CUMBERLAND, MD. (QUARTERLY PROGRESS REPORT)	ABL/QPR-
HERCULES POWDER CO. ALLEGANY BALLISTICS LAB., CUMBERLAND, MD.	ABL/R-
HERCULES POWDER CO. ALLEGANY BALLISTICS LAB., CUMBERLAND, MD.	ABL SR-
HERCULES POWDER CO. ALLEGANY BALLISTICS LAB., CUMBERLAND, MD.	ABL-TR-(YEAR)-
HERCULES POWDER CO. ALLEGANY BALLISTICS LAB., CUMBERLAND, MD. (WEEKLY PROGRESS REPT., SUPP.)	ABL WPR SUPP.-
HERCULES POWDER CO. ALLEGANY BALLISTICS LAB., CUMBERLAND, MD.	ABL/X-
HERCULES POWDER CO. ALLEGANY BALLISTICS LAB., CUMBERLAND, MD.	ABL/Z-
HERCULES POWDER CO. CHEMICAL PROPULSION DIV. BACCHUS WORKS, MAGNA, UTAH	APC-
HERCULES POWDER CO. CHEMICAL PROPULSION DIV. BACCHUS WORKS, MAGNA, UTAH	MCS-
HERCULES POWDER CO. CHEMICAL PROPULSION DIV. BACCHUS WORKS, MAGNA, UTAH	MTI-
HERCULES POWDER CO. CHEMICAL PROPULSION DIV. BACCHUS WORKS, MAGNA, UTAH	MTO-
HERCULES RESEARCH CENTER, WILMINGTON, DEL.	HRC-(YEAR)-
HERCULES RESEARCH CENTER, WILMINGTON, DEL.	HRC-
HERCULES RESEARCH CENTER, WILMINGTON, DEL. (RESEARCH INVESTIGATION)	RI-
HERLEC CORP., MILWAUKEE	HC-
HERMES ELECTRONICS CO., CAMBRIDGE, MASS.	M-
HERNER AND CO., WASHINGTON, D.C.	H+C-
HERNER AND CO., WASHINGTON, D.C.	HEC-
HESSE-EASTERN DIV., FLIGHTEX FABRICS INC., CAMBRIDGE, MASS.	PR-(NUMBERS,LETTERS)
HEWLETT-PACKARD CO., COLORADO SPRINGS	HP-AN-
HEWLETT-PACKARD CO., PALO ALTO, CALIF.	HP-
HEWLETT-PACKARD CO., PALO ALTO, CALIF.	HP-AN-
HEWLETT-PACKARD CO., PALO ALTO, CALIF.	HPCO-
HEWLETT-PACKARD CO. FREQUENCY AND TIME DIV.-EAST, BEVERLY, MASS.	N-
HEWLETT-PACKARD LABS., PALO ALTO, CALIF.	GTR-
HEXCEL PRODUCTS, INC., OAKLAND, CALIF.	HEX-

HEXCEL PRODUCTS, INC., OAKLAND, CALIF.   MPR-

HIGH ALTITUDE OBSERVATORY, BOULDER, COLO.   HAO-
HIGH ALTITUDE OBSERVATORY, BOULDER, COLO.   HAO-TR-
HIGH ALTITUDE OBSERVATORY, BOULDER, COLO.   TR-

HIGH ALTITUDE OBSERVATORY, CLIMAX, COLO.   COLU-HAO-
HIGH ALTITUDE OBSERVATORY, CLIMAX, COLO.   HAO-

HIGH DUTY ALLOYS, LTD., SLOUGH, BUCKS, ENGLAND   SX-

HIGH-SPEED FLIGHT STATION, EDWARDS AFB, CAL. (REF.25)   NACA-(LTRS)-H-(DATE CODE)

HIGH-SPEED FLIGHT STATION, EDWARDS, CALIF. (REF. 26)   NASA-(LETTERS)-H-

HIGH TEMPERATURE MATERIALS, INC., BRIGHTON, MASS.   HTM-QPR-

HIGH VOLTAGE ENGINEERING CORP., BURLINGTON, MASS.   HVEC-TN-
HIGH VOLTAGE ENGINEERING CORP., BURLINGTON, MASS.   HVI-

HILGER + WATTS LTD., LONDON   CH-(NUMBER)-(ENGLAND)

HILLER AIRCRAFT CORP., PALO ALTO, CALIF.   APR-
HILLER AIRCRAFT CORP., PALO ALTO, CALIF.   ARD-
HILLER AIRCRAFT CORP., PALO ALTO, CALIF.   ARDT-

HIROSHIMA UNIV., TAKEHARA. RESEARCH INST. FOR THEORETICAL PHYSICS   RRK-(YEAR)-

HITTMAN ASSOCIATES, INC., BALTIMORE   HA-
HITTMAN ASSOCIATES, INC., BALTIMORE   HIT-

HOEGANAES AB, SWEDEN   PM-(NUMBER)-

HOEGANAES SPONGE IRON CORP., RIVERTON, N.J.   HSIC-LAB-(NUMBER)-(YEAR)
HOEGANAES SPONGE IRON CORP., RIVERTON, N.J.   HSIC-PROD HIGH ALLOY-

HOFF RESEARCH AND DEVELOPMENT LABS., INC., CLEVELAND   HRD-

HOFFMAN LABS., INC., LOS ANGELES   HLI-

HOLABIRD SIGNAL DEPOT. RADIO PROPAGATION UNIT, BALTIMORE   RPU-
HOLABIRD SIGNAL DEPOT. RADIO PROPAGATION UNIT, BALTIMORE   SIG RPU-
HOLABIRD SIGNAL DEPOT. RADIO PROPAGATION UNIT, BALTIMORE   SIG RPU TR-

HOLLEY CARBURETOR CO. ELECTRO MECHANICAL DIV.,DETROIT   EM-(YEAR)-

HOLLOMAN AIR DEVELOPMENT CTR., HOLLOMAN AFB, N. MEX.   FDLH-
HOLLOMAN AIR DEVELOPMENT CTR., HOLLOMAN AFB, N. MEX.   FDLHO-
HOLLOMAN AIR DEVELOPMENT CTR., HOLLOMAN AFB, N. MEX.   FDLHR-
HOLLOMAN AIR DEVELOPMENT CTR., HOLLOMAN AFB, N. MEX.   HADC-
HOLLOMAN AIR DEVELOPMENT CTR., HOLLOMAN AFB, N. MEX.   HADC (LETTERS)-
HOLLOMAN AIR DEVELOPMENT CTR., HOLLOMAN AFB, N. MEX.   HADC ADJ-(YR.)-
HOLLOMAN AIR DEVELOPMENT CTR., HOLLOMAN AFB, N. MEX.   HADC HDGR-(YR.)-
HOLLOMAN AIR DEVELOPMENT CTR., HOLLOMAN AFB, N. MEX.   HADC-TR-(YEAR)-
HOLLOMAN AIR DEVELOPMENT CTR., HOLLOMAN AFB, N. MEX.   HDRRM/TM-
HOLLOMAN AIR DEVELOPMENT CTR., HOLLOMAN AFB, N. MEX.   HDT-
HOLLOMAN AIR DEVELOPMENT CTR., HOLLOMAN AFB, N. MEX.   HDTT-
HOLLOMAN AIR DEVELOPMENT CTR., HOLLOMAN AFB, N. MEX.   HDW/TM-

HOLLOMAN AFB, N. MEX.   HADC-TN-
HOLLOMAN AFB, N. MEX.   HAFB-
HOLLOMAN AFB, N. MEX. (HOLLOMAN DEV. RESEARCH RPT.)   HDRR-
HOLLOMAN AFB, N. MEX.   MTHT-

HOLMAN (JOHN F.) AND CO., INC., WASHINGTON, D. C.   T-

HOLMES AND NARVER, INC., LOS ANGELES   E-(NUMBER)-(YEAR)
HOLMES AND NARVER, INC., LOS ANGELES   HN-
HOLMES AND NARVER, INC., LOS ANGELES   HN-(NUMBER) -
HOLMES AND NARVER, INC., LOS ANGELES   H+N-
HOLMES AND NARVER, INC., LOS ANGELES   HN-CEN-
HOLMES AND NARVER, INC., LOS ANGELES   HN-CHN-
HOLMES + NARVER, INC.,LOS ANG. (ENIWETOK PROVING GD.)   HN-E-(NUMBER)-(YEAR)
HOLMES + NARVER, INC.,LOS ANG. (ENIWETOK PROVING GD.)   HN-EPG-(YEAR)/
HOLMES AND NARVER, INC., LOS ANGELES   H+NID-
HOLMES AND NARVER, INC., LOS ANGELES   HN-JOB-
HOLMES + NARVER, INC., LOS ANG. (NEVADA TEST SITE)   HN-N-(NUMBER)-(YEAR)
HOLMES + NARVER, INC., LOS ANG. (NEVADA TEST SITE)   HN-NTS-(YEAR)/
HOLMES AND NARVER, INC., LOS ANG. (PLOWSHARE PROGRAM)   HN-P/(LETTER)-(NO.)-(YR)
HOLMES AND NARVER, INC., LOS ANGELES   HN-SEN-
HOLMES AND NARVER, INC., LOS ANGELES   HN-TDP-
HOLMES AND NARVER, INC., LOS ANGELES   P/B-
HOLMES AND NARVER, INC., LOS ANGELES   P/C-
HOLMES AND NARVER, INC., LOS ANGELES   P/G(NUMBER)-(YEAR)
HOLMES AND NARVER, INC., LOS ANGELES   P/P-(NUMBER)-(YEAR)
HOLMES AND NARVER, INC., LOS ANGELES   P/PL-
HOLMES AND NARVER, INC., LOS ANGELES   P/PL-(NUMBER)-(YEAR)
HOLMES AND NARVER, INC., LOS ANGELES   SHN-

HOLMES AND NARVER, INC. ON-CONTINENT TEST DIV., LAS VEGAS, NEV.   HN-
HOLMES AND NARVER, INC. ON-CONTINENT TEST DIV., LAS VEGAS, NEV.   N-

HOLSTON DEFENSE CORP., KINGSPORT, TENN. (REF. 1)   (NUMBERS...)
HOLSTON DEFENSE CORP., KINGSPORT, TENN.   20-T-

HONEYWELL, INC. (ALL LOCATIONS)   H-
HONEYWELL, INC. (ALL LOCATIONS)   HONEY-
HONEYWELL, INC. (ALL LOCATIONS)   MHR-

HONEYWELL, INC., MINNEAPOLIS   RB-
HONEYWELL, INC., MINNEAPOLIS   RD-
HONEYWELL, INC., MINNEAPOLIS   R-ED-

HONEYWELL, INC., ST. PETERSBURG, FLA.   HONEYWELL-AERO-
HONEYWELL, INC., ST. PETERSBURG, FLA.   SPEC-D(NUMBER)
HONEYWELL, INC., ST. PETERSBURG, FLA.   SPEC-ES(NUMBER)-P

HONEYWELL, INC. AERONAUTICAL DIV., MINNEAPOLIS   AERO-
HONEYWELL, INC. AERONAUTICAL DIV., MINNEAPOLIS   HA-

HONEYWELL, INC. AERONAUTICAL DIV.

HONEYWELL, INC. AERONAUTICAL DIV., ST.PETERSBURG, FLA	SPEC-FMS-
HONEYWELL, INC. CORPORATE PROGRAM CENTER, MINNEAPOLIS	CPC-(NUMBER)-FR-
HONEYWELL, INC. CORPORATE RES. CENTER, HOPKINS, MINN.	HR-
HONEYWELL, INC. CORPORATE RES. CENTER, HOPKINS, MINN.	HR-(YEAR)-
HONEYWELL, INC. MILITARY AND SPACE SCIENCES DEPT., ARLINGTON, VA.	S(YEAR)-
HONEYWELL, INC. MILITARY PRODUCTS GP., MINNEAPOLIS	RR-(NUMBER)-TRI-
HONEYWELL, INC. ORDNANCE DIV., HOPKINS, MINN.	HO-
HONEYWELL, INC. ORDNANCE DIV., HOPKINS, MINN. (PROPOSAL)	P-
HONEYWELL, INC. ORDNANCE DIV., HOPKINS, MINN.	POMM-
HONEYWELL, INC. ORDNANCE DIV., HOPKINS, MINN. (TECH. DOC.)	TD-
HONEYWELL, INC. RADIATION CENTER, BOSTON	HRC-(YR.)-
HONEYWELL, INC. SYSTEMS AND RESEARCH DIV., ST. PAUL	HONEYWELL-(NUMBER)-FRI
HONEYWELL, INC. TEST INSTRUMENTS DIV., ANNAPOLIS, MD.	TEST-
HONEYWELL RESEARCH CENTER, HOPKINS, MINN.	GR-
HONEYWELL RESEARCH CENTER, HOPKINS, MINN.	HRC-TR-
HONEYWELL RESEARCH CENTER, HOPKINS, MINN.(TECH.REPT)	HR-ITR-
HONEYWELL RESEARCH CENTER, HOPKINS, MINN.(TECH.REPT)	HR-TR-
(SEE ALSO LATER NAME:HONEYWELL,INC.CORPORATE RES.CTR)	
HOOKER CHEMICAL CORP., NIAGARA FALLS, N.Y.	HCC-QPR-
HOOKER ELECTROCHEMICAL CO., NIAGARA FALLS, N.Y.	HEC-
HOOKER ELECTROCHEMICAL CO., NIAGARA FALLS, N.Y.	HEC-QPR-
HORIZONS, INC., CLEVELAND	DPWR-
HORIZONS, INC., CLEVELAND	HI-
HORIZONS, INC., CLEVELAND	HZ-
HOUDRY PROCESS AND CHEMICAL CO., LINWOOD, PENNA.	HOUDRY-
HOUDRY PROCESS + CHEMICAL CO., MARCUS HOOK, PENNA.	DABCO-
HOUGHTON (E.F.) + CO., PHILADELPHIA	BMR-
HOUGHTON (E.F.) + CO., PHILADELPHIA (QUARTERLY REPT)	HOUGHTON-QR-
HOUSING AND HOME FINANCE AGENCY, WASHINGTON, D.C.	HHFA-
HOUSING AND HOME FINANCE AGENCY, WASHINGTON, D.C. (RESEARCH PAPER)	HHFA RP-
HOUSING AND HOME FINANCE AGENCY, WASHINGTON, D.C. (TECHNICAL PAPER)	HHFA TP-
HOUSTON RESEARCH INSTITUTE, TEX.	HRI-(5 DIGITS)-
HRB-SINGER, INC., STATE COLLEGE, PENNA.	HRB-
HRB-SINGER, INC., STATE COLLEGE, PENNA.	HRBI-
HRB-SINGER, INC., STATE COLLEGE, PENNA.	HRB/S-
HRB-SINGER, INC., STATE COLLEGE, PENNA.	HRB-SINGER-
HRB-SINGER, INC., STATE COLLEGE, PENNA.	HRB-TP-
HRB-SINGER, INC., STATE COLLEGE, PENNA.	PTB-(YEAR)-
HUDSON INST., INC., CROTON-ON-HUDSON, N.Y.	HH-(NO.)-RR-
HUDSON INST., INC., CROTON-ON-HUDSON, N.Y.	HI-
HUDSON INST., INC., CROTON-ON-HUDSON, N.Y.	HI-(NUMBER)-RR
HUDSON INST., INC., CROTON-ON-HUDSON, N.Y. (DRAFT)	HI-(NUMBER)-D
HUDSON INST., INC., CROTON-ON-HUDSON, N.Y. (BRIEFING NOTES)	HI-BN-
HUDSON INST., INC., CROTON-ON-HUDSON, N.Y. (DISCUSSION PAPER)	HI-DP-
HUDSON INST., INC., CROTON-ON-HUDSON, N.Y. (PAPER)	HI-P-
HUDSON INST., INC., CROTON-ON-HUDSON, N.Y.	HI-PR-
HUDSON INST., INC., CROTON-ON-HUDSON, N.Y. (TRANSLATION)	HI-T-
HUDSON INST., INC., HARMON-ON-HUDSON, N.Y.	HI-
HUDSON INST., INC., HARMON-ON-HUDSON, N.Y.	HI-(NUMBER)-RR
HUGHES AIRCRAFT CO. (ALL LOCATIONS)	HAC-
HUGHES AIRCRAFT CO.	P(YEAR)-
HUGHES AIRCRAFT CO. (SECRET DOCUMENT NUMBER)	SDN-
HUGHES AIRCRAFT CO., CANOGA PARK, CALIF.	HAC-MSD-P(YR)-(NUMBER)
HUGHES AIRCRAFT CO., CULVER CITY, CALIF.	ASG-18-
HUGHES AIRCRAFT CO., CULVER CITY, CALIF.	AX-
HUGHES AIRCRAFT CO., CULVER CITY, CALIF. (BUSINESS MANAGEMENT AND COST PROPOSAL)	BMCP-
HUGHES AIRCRAFT CO., CULVER CITY, CALIF.	CDP-
HUGHES AIRCRAFT CO., CULVER CITY, CALIF. (COCKPIT INSTRUMENT DISPLAYS)	CID-
HUGHES AIRCRAFT CO., CULVER CITY, CALIF.	CM-
HUGHES AIRCRAFT CO., CULVER CITY, CALIF. (DATA SHEET)	DS-
HUGHES AIRCRAFT CO., CULVER CITY, CALIF.(DETAIL SPEC)	DS-
HUGHES AIRCRAFT CO., CULVER CITY, CALIF.	EC-RD-
HUGHES AIRCRAFT CO.,CULVER CITY,CAL.(FIELD SERVICE)	FSSD-SB-
HUGHES AIRCRAFT CO.,CULVER CITY,CAL.(FLIGHT TESTDATA)	FTDS-
HUGHES AIRCRAFT CO.,,CULVER CITY,CAL.(FLIGHT TEST ENG)	FTED-
HUGHES AIRCRAFT CO., CULVER CITY, CALIF.(FLIGHT TEST)	FTID-
HUGHES AIRCRAFT CO., CULVER CITY, CALIF.	GSG-
HUGHES AIRCRAFT CO., CULVER CITY, CALIF.	(YEAR)H-
HUGHES AIRCRAFT CO., CULVER CITY, CALIF.	HA-
HUGHES AIRCRAFT CO., CULVER CITY, CALIF.	HAC-A(NUMBER)
HUGHES AIRCRAFT CO., CULVER CITY, CALIF.	HAC-ASD-
HUGHES AIRCRAFT CO., CULVER CITY, CALIF.	HAC-ASD/H-
HUGHES AIRCRAFT CO., CULVER CITY, CALIF.	HAC-B(NUMBER)
HUGHES AIRCRAFT CO., CULVER CITY, CALIF.	HAC M-
HUGHES AIRCRAFT CO., CULVER CITY, CALIF.	HAC-OP-
HUGHES AIRCRAFT CO., CULVER CITY, CALIF.	HAC-PUB-
HUGHES AIRCRAFT CO., CULVER CITY, CALIF.	HAC-SDL-
HUGHES AIRCRAFT CO., CULVER CITY, CALIF.	HAC SDN (NOS.)/(NOS.)
HUGHES AIRCRAFT CO., CULVER CITY, CALIF. (SPECIAL RESEARCH STUDY MEMORANDUM)	HAC SRSM-
HUGHES AIRCRAFT CO., CULVER CITY, CALIF.	HAC-SSD-
HUGHES AIRCRAFT CO., CULVER CITY, CALIF.	HAC-TM-
HUGHES AIRCRAFT CO., CULVER CITY, CALIF.	HAC-TOW/H-
HUGHES AIRCRAFT CO., CULVER CITY, CALIF. (SCI. RPT.)	HA-SR-
HUGHES AIRCRAFT CO., CULVER CITY, CALIF. (FINISH SPECIFICATION)	HFS-
HUGHES AIRCRAFT CO.,CULVER CITY,CAL.(MATERIAL SPEC.)	HMS-
HUGHES AIRCRAFT CO.,CULVER CITY,CAL.(HUGHES PROCESS)	HP-
HUGHES AIRCRAFT CO., CULVER CITY, CALIF. (HARD POINT DEFENSE INTERCEPTOR)	HPDI-

HUGHES AIRCRAFT CO., CULVER CITY, CALIF. (HARD POINT DEFENSE SYSTEM)	HPDS-(LETTERS)-
HUGHES AIRCRAFT CO., CULVER CITY, CALIF.	HUGHES TM-
HUGHES AIRCRAFT CO., CULVER CITY, CALIF.	LS-
HUGHES AIRCRAFT CO., CULVER CITY, CALIF.	LS-BIB-(YEAR)-
HUGHES AIRCRAFT CO., CULVER CITY, CALIF.	MA-1-(LETTERS)-
HUGHES AIRCRAFT CO., CULVER CITY, CALIF. (MOBILE AUTOMATIC RADIATION DETECTOR)	MART-
HUGHES AIRCRAFT CO., CULVER CITY, CALIF. (MOBILE BALLISTIC MISSILE)	MBM-
HUGHES AIRCRAFT CO., CULVER CITY, CALIF.	MG-3-
HUGHES AIRCRAFT CO., CULVER CITY, CALIF.	MG-(2 DIGITS)-
HUGHES AIRCRAFT CO., CULVER CITY, CALIF.	NB-
HUGHES AIRCRAFT CO., CULVER CITY, CALIF.	P(YEAR)-
HUGHES AIRCRAFT CO., CULVER CITY, CALIF. (PROPOSAL)	PR-
HUGHES AIRCRAFT CO., CULVER CITY, CALIF. (PASSIVE RADIATION COUNTERMEASURES)	PRCM-
HUGHES AIRCRAFT CO., CULVER CITY, CALIF. (RESEARCH)	RLE(LETTERS)-
HUGHES AIRCRAFT CO., CULVER CITY, CALIF. (RESEARCH)	RLM(LETTERS)-
HUGHES AIRCRAFT CO., CULVER CITY, CALIF.	S4M(NUMBER)/
HUGHES AIRCRAFT CO., CULVER CITY, CALIF. (SPECIAL RESEARCH STUDY MEMORANDUM)	SRSM-
HUGHES AIRCRAFT CO., CULVER CITY, CALIF.	TM-
HUGHES AIRCRAFT CO., CULVER CITY, CALIF.	TOW-(LETTERS)-
HUGHES AIRCRAFT CO., CULVER CITY, CALIF.	XA-
HUGHES AIRCRAFT CO., FULLERTON, CALIF.	FR-
HUGHES AIRCRAFT CO., FULLERTON, CALIF.	FR(YEAR)-
HUGHES AIRCRAFT CO., FULLERTON, CALIF.	HAC-FR-
HUGHES AIRCRAFT CO., LOS ANGELES	W-
HUGHES AIRCRAFT CO., NEWPORT BEACH, CALIF. (ELECTRONIC PRODUCTS)	EPD-
HUGHES AIRCRAFT CO., NEWPORT BEACH, CALIF.	HAD F/(CONTRACT NUMBER)
HUGHES AIRCRAFT CO., NEWPORT BEACH, CALIF. (MICROELECTRONICS)	ME-
HUGHES AIRCRAFT CO., NEWPORT BEACH, CAL.(SEMICONDUCTOR)	SD-TR-
HUGHES AIRCRAFT CO., TUCSON, ARIZ.	TEL-
HUGHES AIRCRAFT CO., WASHINGTON, D. C.(TECH.ANALYSIS)	TAO-
HUGHES AIRCRAFT CO., WASHINGTON, D. C.	TAO-TR-
HUGHES AIRCRAFT CO. AERONAUTICAL SYSTEMS DIV., CULVER CITY, CALIF.	ASD-
HUGHES AIRCRAFT CO. AERONAUTICAL SYSTEMS DIV., CULVER CITY, CALIF.	ASD-(NUMBER)(LETTER)
HUGHES AIRCRAFT CO. AERONAUTICAL SYSTEMS DIV., CULVER CITY, CALIF.	HAC-P(YEAR)-
HUGHES AIRCRAFT CO. AEROSPACE ENGINEERING DIV., CULVER CITY, CALIF.	AED-
HUGHES AIRCRAFT CO. AEROSPACE GROUP, CULVER CITY,CAL.	PUB-
HUGHES AIRCRAFT CO. AEROSPACE GROUP, CULVER CITY,CAL.	SDN-A-
HUGHES AIRCRAFT CO. AEROSPACE GROUP, CULVER CITY,CAL.	TOW-
HUGHES AIRCRAFT CO. COMMUNICATIONS DIV., CULVER CITY, CALIF.	OP-
HUGHES AIRCRAFT CO. DEV. LABS., NEWPORT BEACH, CALIF	DL-
HUGHES AIRCRAFT CO. ELECTRON DYNAMICS DIV., TORRANCE, CALIF.	W-
HUGHES AIRCRAFT CO. ELECTRON DYNAMICS DIV., TORRANCE, CALIF.	W-(NUMBER)-
HUGHES AIRCRAFT CO. ELECTRONIC PROPERTIES INFORMATION CENTER, CULVER CITY, CALIF.	EPIC-
HUGHES AIRCRAFT CO. ELECTRONIC PROPERTIES INFORMATION CENTER, CULVER CITY, CALIF.	EPIC-DS-
HUGHES AIRCRAFT CO. ELECTRONIC PROPERTIES INFORMATION CENTER, CULVER CITY, CALIF.	EPIC-IR-
HUGHES AIRCRAFT CO. ELECTRONIC PROPERTIES INFORMATION CENTER, CULVER CITY, CALIF.	EPIC-S-
HUGHES AIRCRAFT CO. ENG. DIV., CULVER CITY, CALIF.	PR-(ROMAN NO.)-(NO.LTRS.)
HUGHES AIRCRAFT CO. ENG. DIV., CULVER CITY, CALIF.	VI-
HUGHES AIRCRAFT CO. GROUND SYSTEMS GP., FULLERTON,CAL	FP-(YEAR)-
HUGHES AIRCRAFT CO. GROUND SYSTEMS GP., FULLERTON, CALIF., (FINAL REPORT)	FR-(YEAR)-(NUMBER)-
HUGHES AIRCRAFT CO. GROUND SYSTEMS GP., FULLERTON,CAL	(YEAR)HF-
HUGHES AIRCRAFT CO. GROUND SYSTEMS GP., FULLERTON,CAL	X-GSG-
HUGHES AIRCRAFT CO. GROUND SYSTEMS GP., FULLERTON,CAL	X-GSG-(LETTER)
HUGHES AIRCRAFT CO. HUGHES INTERNATIONAL, CULVER CITY, CALIF.	HARAS-
HUGHES AIRCRAFT CO. LIBRARY SERVICES, CULVER CITY,CAL	HAC-LS-BIB-
HUGHES AIRCRAFT CO. MICROWAVE LAB.,CULVER CITY,CALIF.	ML-(2 DIGITS)-
HUGHES AIRCRAFT CO. MICROWAVE LAB.,CULVER CITY,CALIF.	ML(1 DIGIT)-(LETTERS)-
HUGHES AIRCRAFT CO. MICROWAVE LAB.,CULVER CITY,CALIF.	ML(PR)-
HUGHES AIRCRAFT CO. MICROWAVE TUBE DIV., LOS ANGELES	W-
HUGHES AIRCRAFT CO. MISSILE SYSTEMS DIV., CANOGA PARK, CALIF.	MSD-(NUMBER)-TP-
HUGHES AIRCRAFT CO. MISSILE SYSTEMS DIV., CANOGA PARK, CALIF.	MSD-(NUMBER)RS
HUGHES AIRCRAFT CO. MISSILE SYSTEMS DIV., CULVER CITY, CALIF.	MSD-P(YEAR)-
HUGHES AIRCRAFT CO. NUCLEAR ELECTRONICS LAB., CULVER CITY, CALIF.	HAC-(NO.).(NO.)/(NO.)
HUGHES AIRCRAFT CO. OPTICAL PROPERTIES TECHNICAL INFORMATION CENTER, CULVER CITY, CALIF.	OPSR-
HUGHES AIRCRAFT CO. PACKAGE ENG. LAB., TUCSON, ARIZ.	RST-
HUGHES AIRCRAFT CO. RADAR AND MISSILE ELECTRONICS LAB., CULVER CITY, CALIF.	RMS-
HUGHES AIRCRAFT CO. RES. + DEV. DIV.,CULVER CITY,CAL.	RS/H-
HUGHES AIRCRAFT CO. RES. LAB., CULVER CITY, CALIF.	HA-RL-
HUGHES AIRCRAFT CO. SOLID STATE RESEARCH CENTER, NEWPORT BEACH, CALIF.	SSRC-TR-(YR.)-
HUGHES AIRCRAFT CO. SPACE SYSTEMS DIV.,EL SEGUNDO,CAL (MISSION OPERATIONS)	MOP-
HUGHES AIRCRAFT CO. SPACE SYSTEMS DIV.,EL SEGUNDO,CAL	P(YEAR)-
HUGHES AIRCRAFT CO. SPACE SYSTEMS DIV.,EL SEGUNDO,CAL	SDN-(NUMBER)-(NUMBER)/
HUGHES AIRCRAFT CO. SPACE SYSTEMS DIV.,EL SEGUNDO,CAL	SDN-A-
HUGHES AIRCRAFT CO. SPACE SYSTEMS DIV.,EL SEGUNDO,CAL (SURVEYOR SPACECRAFT)	SS-
HUGHES AIRCRAFT CO. SPACE SYSTEMS DIV.,EL SEGUNDO,CAL	SSD-(5 DIGITS)
HUGHES AIRCRAFT CO. SYSTEMS DEVELOPMENT LABS., CULVER CITY, CALIF.	HAC-PRI-(NUMBER)Y
HUGHES AIRCRAFT CO. SYSTEMS DEVELOPMENT LABS., CULVER CITY, CALIF.	HAC-PRII-(NUMBER)Y
HUGHES AIRCRAFT CO. SYSTEMS DEVELOPMENT LABS., CULVER CITY, CALIF.	PRI-(NUMBER)X
HUGHES AIRCRAFT CO. SYSTEMS DEVELOPMENT LABS., CULVER CITY, CALIF.	PRI-(NO.)Y
HUGHES AIRCRAFT CO. SYSTEMS DEVELOPMENT LABS., CULVER CITY, CALIF.	SDL-
HUGHES AIRCRAFT CO. SYSTEMS DEVELOPMENT LABS., CULVER CITY, CALIF.	V-

HUGHES AIRCRAFT CO. WEAPON SYSTEMS DEVELOPMENT LABS.

HUGHES AIRCRAFT CO. WEAPON SYSTEMS DEVELOPMENT LABS., CULVER CITY, CALIF.	HA-WSDL-
HUGHES RESEARCH LABS. DIV. OF HUGHES AIRCRAFT CO., MALIBU, CALIF.	HAC-ATR-
HUGHES RESEARCH LABS. DIV. OF HUGHES AIRCRAFT CO., MALIBU, CALIF.	HAC-RR-
HUGHES RESEARCH LABS. DIV. OF HUGHES AIRCRAFT CO., MALIBU, CALIF.	HRL-
HUGHES RESEARCH LABS. DIV. OF HUGHES AIRCRAFT CO., MALIBU, CALIF.	ISR-
HUGHES RESEARCH LABS DIV. OF HUGHES AIRCRAFT CO., MALIBU, CALIF.	(YEAR)M-
HUGHES TOOL CO., HOUSTON, TEXAS	HTC-
HUGHES TOOL CO. AIRCRAFT DIV.,CULVER CITY,CAL.(REF.1)	(PROJ. NO.)-(NO.)-(LTR.)
HUGHES TOOL CO. AIRCRAFT DIV., CULVER CITY, CALIF.	HTC-(YEAR)-
HUGHES TOOL CO. AIRCRAFT DIV., CULVER CITY, CALIF.	HTC-(NO.)-(LETTER(S))-...
HUGHES TOOL CO. AIRCRAFT DIV., CULVER CITY, CALIF.	HTC-1109-(LETTER(S))-
HUGHES TOOL CO. AIRCRAFT DIV., CULVER CITY, CALIF.	HTC-AD-(YEAR)-
HUGHES TOOL CO. AIRCRAFT DIV., CULVER CITY, CALIF.	USAAVLABS-TR-
HUGHES TOOL CO. AIRCRAFT DIV., CULVER CITY, CALIF.	X-
HULL, ENGLAND. UNIV.	HU U-
HUMAN ENGINEERING LABS., ABERDEEN PROVING GROUND, MD.	APG-HEL-...
HUMAN ENGINEERING LABS., ABERDEEN PROVING GROUND, MD.	APG-TM-
HUMAN ENGINEERING LABS., ABERDEEN PROVING GROUND, MD.	HEL-
HUMAN ENGINEERING LABS., ABERDEEN PROVING GROUND, MD.	HEL-BIB-
HUMAN ENGINEERING LABS., ABERDEEN PROVING GROUND, MD.	HEL-S-
HUMAN ENGINEERING LABS., ABERDEEN PROVING GROUND, MD.	HEL STANDARD S-
HUMAN ENGINEERING LABS., ABERDEEN PROVING GROUND, MD.	HEL-TM-
HUMAN ENGINEERING LABS., ABERDEEN PROVING GROUND, MD.	HEL-TN-
HUMAN RESOURCES RESEARCH CENTER, LACKLAND AFB, TEXAS	AAF HRRC-
HUMAN RESOURCES RESEARCH CENTER, LACKLAND AFB, TEXAS	AAF HRRC RB-
HUMAN RESOURCES RESEARCH CENTER, LACKLAND AFB, TEXAS	AAF HRRC RB (NO.)-
HUMAN RESOURCES RESEARCH CENTER, LACKLAND AFB, TEXAS	HRRC-
HUMAN RESOURCES RESEARCH INSTITUTE, MAXWELL AFB, ALA.	AD-
HUMAN RESOURCES RESEARCH INSTITUTE, MAXWELL AFB, ALA.	HR-
HUMAN RESOURCES RESEARCH INSTITUTE, MAXWELL AFB, ALA.	HRRI-
HUMAN RESOURCES RESEARCH INSTITUTE, MAXWELL AFB, ALA.	HRRI-HR-
HUMAN RESOURCES RESEARCH INSTITUTE, MAXWELL AFB, ALA.	MM-
HUMAN RESOURCES RES. LABS., BOLLING AFB, WASH., D.C.	AAF HRRL-
HUMAN RESOURCES RES. LABS., BOLLING AFB, WASH., D.C.	AAF HRRL (NO.)-
HUMAN RESOURCES RES. LABS., BOLLING AFB, WASH., D.C.	HRRL-
HUMAN RESOURCES RES. LABS., BOLLING AFB, WASH., D.C.	HRRL-MR-
HUMAN SCIENCES RESEARCH, INC., ARLINGTON, VA.	HSR RM (YR.)(NO.)(LTR(S))
HUMAN SCIENCES RESEARCH, INC., ARLINGTON, VA.	HSR-RR-(YR.)(NO.)(LTRS)
HUMAN SCIENCES RESEARCH, INC., ARLINGTON, VA.	HSR TN (YR.)(NO.)(LTR(S))
HUMAN SCIENCES RESEARCH, INC., MC LEAN, VA.	HSR-FRM-
HUMAN SCIENCES RESEARCH, INC., MC LEAN, VA.	HSR-RR-
HUMAN SCIENCES RESEARCH, INC., MC LEAN, VA.	HSR-TN-(YEAR)-
HUMBLE OIL AND REFINING CO., BAYTOWN, TEX.	HUM-
HUMBLE OIL AND REFINING CO., BAYTOWN, TEX.	HUM-(NUMBER)-P
HUMPHREYS CORP. TAFA DIV., BOW, N.H.	HC-
HUNGARIAN ACADEMY OF SCIENCES. INST. OF NUCLEAR RESEARCH, DEBRECEN	INRH-
HUNGARIAN TRADING CO.FOR BOOKS AND NEWSPAPERS-KULTURA BUDAPEST (TRANSLATIONS)	HTCBN-
HUYCK CORP. (PROGRESS REPORT)	HUYCK-
HUYCK RESEARCH CENTER, STAMFORD, CONN.	HRC-
HYCON MFG. CO., MONROVIA, CALIF.	HBR-
HYCON MFG. CO., MONROVIA, CALIF.	RE(NO.)-(NO.)-CEJ-
HYCON MFG. CO., PASADENA, CALIF.	HBRA-
HYCON MFG. CO., PASADENA, CALIF.	HYC-
HYCON MFG. CO., PASADENA, CALIF.	HYCON-
HYDRAULIC RESEARCH AND MFG. CO., BURBANK, CALIF. HYDRAULIC RESEARCH LAB.	HR-
HYDRO-ELECTRIC POWER COMMISSION OF ONTARIO, ROLPHTON	E-
HYDRO RESEARCH SCIENCE, SUNNYVALE, CALIF.	HRS-
HYDROCARBON RESEARCH, INC., N.Y.C.	HRI-
HYDROFOIL CORP., ANNAPOLIS, MD.	HM-
HYDROGRAPHIC OFFICE, SUITLAND, MD. (AIR FACILITY DIRECTORIES)	AFD-
HYDROGRAPHIC OFFICE, SUITLAND, MD. (AIR TARGET MATERIALS PROGRAM)	ATMP-
HYDROGRAPHIC OFFICE, SUITLAND,MD.(BATHYMETRIC CHARTS)	BC-
HYDROGRAPHIC OFFICE, SUITLAND, MD.	BC-(NUMBER)N
HYDROGRAPHIC OFFICE, SUITLAND, MD.	H.O.-
HYDROGRAPHIC OFFICE, SUITLAND, MD.	HO-
HYDROGRAPHIC OFFICE, SUITLAND, MD. (DATA SHEET)	HODS-
HYDROGRAPHIC OFFICE, SUITLAND, MD.	NAVHO-
HYDROGRAPHIC OFFICE, SUITLAND, MD.	NHO-
HYDROGRAPHIC OFFICE, SUITLAND, MD. (PILOT CHARTS)	PC-
HYDROGRAPHIC OFFICE, SUITLAND, MD. (WEATHER SUMMARY)	WS-
(SEE ALSO LATER NAME: NAVAL OCEANOGRAPHIC OFFICE)	
HYDRONAUTICS, INC., LAUREL, MD.	FTR-(NUMBER)-
HYDRONAUTICS, INC., LAUREL, MD.	HYDRO-TR-
HYDRONAUTICS, INC., LAUREL, MD.	TR-(3 DIGITS)-
HYDROSPACE RES. CORP., ROCKVILLE, MD.	HRC-TN-
HYDROSPACE RES. CORP., ROCKVILLE, MD.	HRC-TR-
HYDROSPACE RES. CORP., ROCKVILLE, MD.	HYRC-
HYDROSYSTEMS, INC., FARMINGDALE, N.Y.	HI-(YEAR)-
HYMATIC ENGINEERING CO., LTD., REDDITCH, WORCS, ENG.	B-

I.G. FARBENINDUSTRIE A.G.	IG DRAWING-
I.G. FARBENINDUSTRIE A.G.	IG DRAWING (LETTER(S))-
I.G. FARBENINDUSTRIE A.G. KUNSTSTOFFE KOMMISSION	KUKO-
I.T.E. CIRCUIT BREAKER CO., PHILADELPHIA	ITE-
IBM WATSON RESEARCH CENTER, YORKTOWN HEIGHTS, N.Y.	IBM-RC-
IBM WATSON RESEARCH CENTER, YORKTOWN HEIGHTS, N.Y.	IBM-RW-
IBM WATSON RESEARCH CENTER, YORKTOWN HEIGHTS, N.Y.	RC-
IBM WATSON RESEARCH CENTER, YORKTOWN HEIGHTS, N.Y.	RW-
IDAHO. UNIV., MOSCOW	AR-
IDAHO NUCLEAR CORP., IDAHO FALLS	IN-
IDAHO NUCLEAR CORP., IDAHO FALLS	INC-
IDAHO NUCLEAR CORP., IDAHO FALLS	IN-ITR-
IDAHO OPERATIONS OFFICE, AEC, IDAHO FALLS	CI-
IDAHO OPERATIONS OFFICE, AEC, IDAHO FALLS (REF.4)	IDO-
IDAHO OPERATIONS OFFICE. HEALTH AND SAFETY DIV., AEC, IDAHO FALLS	ICF-
IDAMI LANGUAGE RESEARCH SECTION, MILAN	ILRS-T-
IIT RESEARCH INST., ANNAPOLIS, MD.	ECAC-TDR-
IIT RESEARCH INST., ANNAPOLIS, MD.	ECAC-TM-
IIT RESEARCH INST., ANNAPOLIS, MD.	ECAC-TR-
IIT RESEARCH INST., CHICAGO (PROJECT)	ARF-(LETTER)-
IIT RESEARCH INST., CHICAGO	ASR-
IIT RESEARCH INST., CHICAGO	CH-E-
IIT RESEARCH INST., CHICAGO	DOMIIT-
IIT RESEARCH INST., CHICAGO	DOMITT-
IIT RESEARCH INST., CHICAGO	E-
IIT RESEARCH INST., CHICAGO	ECAC-STP-
IIT RESEARCH INST., CHICAGO	G(NUMBER)
IIT RESEARCH INST., CHICAGO	IITRI-
IIT RESEARCH INST., CHICAGO	IITRI-(NUMBER)-P
IIT RESEARCH INST., CHICAGO	IITRI-(NO.-LTR.-NO.-NO.)
IIT RESEARCH INST., CHICAGO	IITRI-A-
IIT RESEARCH INST., CHICAGO	IITRI-B-
IIT RESEARCH INST., CHICAGO	IITRI-C-
IIT RESEARCH INST., CHICAGO	IITRI-G-
IIT RESEARCH INST., CHICAGO	IITRI-J-
IIT RESEARCH INST., CHICAGO	IITRI-L-
IIT RESEARCH INST., CHICAGO	IITRI-M-
IIT RESEARCH INST., CHICAGO	IITRI-N-
IIT RESEARCH INST., CHICAGO	IITRI-PROJ-(LTR.)-
IIT RESEARCH INST., CHICAGO	IITRI-T-
IIT RESEARCH INST., CHICAGO	IITRI-U-
IIT RESEARCH INST., CHICAGO	IITRI-V-
IIT RESEARCH INST., CHICAGO	K-
IIT RESEARCH INST., CHICAGO	M-
IIT RESEARCH INST., CHICAGO	M(NUMBERS)-
IIT RESEARCH INST., CHICAGO	T-
IIT RESEARCH INST., CHICAGO. TECHNOLOGY CENTER	M-
IKOR, INC., WALTHAM, MASS.	I-
ILC INDUSTRIES, INC., DOVER, DEL.	GIER-
ILLINOIS. AGRICULTURAL EXPERIMENT STATION, URBANA	ILU-ES-
ILLINOIS. STATE WATER SURVEY, URBANA	HYDRO-TR-
ILLINOIS. STATE WATER SURVEY, URBANA	ISWS-
ILLINOIS. STATE WATER SURVEY, URBANA	RR-
ILLINOIS. STATE WATER SURVEY, URBANA	TECHNICAL LETTER-
ILLINOIS. UNIV., CHICAGO	UICC-
ILLINOIS. UNIV., CHICAGO. CENTER FOR URBAN STUDIES	DISCUSSION PAPER-
ILLINOIS. UNIV., URBANA	ALR-
ILLINOIS. UNIV., URBANA	AS-
ILLINOIS. UNIV., URBANA	CNM-R-
ILLINOIS. UNIV., URBANA	C-PR-
ILLINOIS. UNIV., URBANA	CSL-R-
ILLINOIS. UNIV., URBANA	ILL-
ILLINOIS. UNIV., URBANA	ILL-(NUMBER)-P
ILLINOIS. UNIV., URBANA (PROJECT SQUID)	ILL-(NUMBER)-PU
ILLINOIS. UNIV., URBANA (PROJECT SQUID)	ILL U-
ILLINOIS. UNIV., URBANA	ILLU-(NUMBER)-
ILLINOIS. UNIV., URBANA	ILL U.(LETTERS)-
ILLINOIS. UNIV., URBANA	ILU-
ILLINOIS. UNIV., URBANA	IU-
ILLINOIS. UNIV., URBANA (STRUCTURAL RESEARCH SERIES)	IU-SRS-
ILLINOIS. UNIV., URBANA	MEDUI-
ILLINOIS. UNIV., URBANA	MEDUI-(NO.)-AEC
ILLINOIS. UNIV., URBANA	ME-TR-ORD-1121-
ILLINOIS. UNIV., URBANA	MR-C-
ILLINOIS. UNIV., URBANA	M-TR-
ILLINOIS. UNIV., URBANA (NUCLEAR RADIATION SHIELDING STUDIES)	NRSS-
ILLINOIS. UNIV., URBANA	SER-
ILLINOIS. UNIV., URBANA	SRES-
ILLINOIS. UNIV., URBANA	SRI-
ILLINOIS. UNIV., URBANA	UI-
ILLINOIS. UNIV., URBANA	UILU-ENG-
ILLINOIS. UNIV., URBANA. ANTENNA LAB.	IU-AL-
ILLINOIS. UNIV., URBANA. ANTENNA LAB.	UIAL-(YEAR)-
ILLINOIS. UNIV., URBANA. BIOLOGICAL COMPUTER LAB.	BCL-
ILLINOIS. UNIV., URBANA. COLL. OF ENGINEERING (STRUCTURAL RESEARCH SERIES)	ILL U.SRS-
ILLINOIS. UNIV., URBANA. COLL. OF ENGINEERING (STRUCTURAL RESEARCH SERIES)	ILL U., STRUC. RES. SER-
ILLINOIS. UNIV., URBANA. COLL. OF ENGINEERING	ILL. U. T + AM-
ILLINOIS. UNIV., URBANA. CONTROL SYSTEMS LAB.	R-

ILLINOIS. UNIV., URBANA. COORDINATED SCIENCE LAB.	I-
ILLINOIS. UNIV., URBANA. COORDINATED SCIENCE LAB.	IU-I-
ILLINOIS. UNIV., URBANA. COORDINATED SCIENCE LAB.	IU-R-
ILLINOIS. UNIV., URBANA. CYCLOTRON LAB.	IU-CL-TR-
ILLINOIS. UNIV., URBANA. DEPT. OF AERONAUTICAL AND ASTRONAUTICAL ENGINEERING	AAE-
ILLINOIS. UNIV., URBANA. DEPARTMENT OF CERAMIC ENG.	DCE-
ILLINOIS. UNIV., URBANA. DEPARTMENT OF CERAMIC ENG.	ILU DCE
ILLINOIS. UNIV., URBANA. DEPARTMENT OF CERAMIC ENG.	ILU DCE MEMO-
ILLINOIS. UNIV., URBANA. DEPARTMENT OF CERAMIC ENG.	ILU DCE MR-
ILLINOIS. UNIV., URBANA. DEPARTMENT OF CERAMIC ENG.	IU DCE-
ILLINOIS. UNIV., URBANA. DEPARTMENT OF CERAMIC ENG.	IU DCE MEMO-
ILLINOIS. UNIV., URBANA. DEPT. OF CIVIL ENGINEERING	HES-
ILLINOIS. UNIV., URBANA. DEPT. OF CIVIL ENGINEERING (CIVIL ENGINEERING STUDIES. STRUCTURAL RES. SERIES)	ILL CES SRS-
ILLINOIS. UNIV., URBANA. DEPT. OF CIVIL ENGINEERING	ILL-DCE-
ILLINOIS. UNIV., URBANA. DEPT. OF CIVIL ENGINEERING	ILL-U-DCE-
ILLINOIS. UNIV., URBANA. DEPT. OF CIVIL ENGINEERING	N-
ILLINOIS. UNIV., URBANA. DEPT. OF CIVIL ENGINEERING	PHOTOGRAMMETRY-SER-
ILLINOIS. UNIV., URBANA. DEPT. OF CIVIL ENGINEERING	SOIL MECHANICS SER-
ILLINOIS. UNIV., URBANA. DEPT. OF CIVIL ENGINEERING (STRUCTURAL RESEARCH SERIES)	SRS-
ILLINOIS. UNIV., URBANA. DEPT. OF CIVIL ENGINEERING	STRUCTURAL RESEARCH SER-
ILLINOIS. UNIV., URBANA. DEPT. OF CIVIL ENGINEERING. CONSTRUCTION RESEARCH LAB.	CRS-
ILLINOIS. UNIV., URBANA. DEPT. OF COMPUTER SCIENCE	DCS-
ILLINOIS. UNIV., URBANA. DEPT. OF COMPUTER SCIENCE	ILLIAC-IV-(NUMBER)-
ILLINOIS. UNIV., URBANA. DEPT. OF COMPUTER SCIENCE	IU-DCS-
ILLINOIS. UNIV., URBANA. DEPT. OF ELECTRICAL ENGINEERING, CHARGED PARTICLE RESEARCH LAB.	CPRL-
ILLINOIS. UNIV., URBANA. DEPT. OF MECHANICAL ENG.	ME-
ILLINOIS. UNIV., URBANA. DEPT. OF MECHANICAL ENG.	ME-TN-
ILLINOIS. UNIV., URBANA. DEPT. OF MECHANICAL ENG.	ME-TR-
ILLINOIS. UNIV., URBANA. DEPT. OF MINING AND METALLURGICAL ENGINEERING	MEDUI-(NO.)-AF
ILLINOIS. UNIV., URBANA. DEPT. OF PHYSICS	ILL-(TH)-(YEAR)-
ILLINOIS. UNIV., URBANA. DEPT. OF THEORETICAL AND APPLIED MECHANICS	IU-T+AM-
ILLINOIS. UNIV., URBANA. DEPT. OF THEORETICAL AND APPLIED MECHANICS	T + AM-
ILLINOIS. UNIV., URBANA. DEPT. OF THEORETICAL AND APPLIED MECHANICS	T/AM-
ILLINOIS. UNIV., URBANA. DIGITAL COMPUTER LAB.	IU-DCL-
ILLINOIS. UNIV., URBANA. ELECTRICAL ENG. RES. LAB.	DF-
ILLINOIS. UNIV., URBANA. ELECTRICAL ENG. RES. LAB.	ILL EERL TN-
ILLINOIS. UNIV., URBANA. ELECTRICAL ENG. RES. LAB.	ILL EERL TR-
ILLINOIS. UNIV., URBANA. ENGINEERING EXPERIMENT STA.	DCP-
ILLINOIS. UNIV., URBANA. ENGINEERING EXPERIMENT STA.	ILL-EES
ILLINOIS. UNIV., URBANA. ENGINEERING EXPERIMENT STATION. (STRUCTURAL RESEARCH SERIES)	ILL EES SRS-
ILLINOIS. UNIV., URBANA. ENGINEERING EXPERIMENT STA.	ILU-EES
ILLINOIS. UNIV., URBANA. ENGINEERING EXPERIMENT STA.	ILU EES TR-
ILLINOIS. UNIV., URBANA. ENGINEERING EXPERIMENT STA.	IU-EES-
ILLINOIS. UNIV., URBANA. ENGINEERING EXPERIMENT STA.	ME-TN-
ILLINOIS. UNIV., URBANA. ENGINEERING EXPERIMENT STA.	T. + A.M.-
ILLINOIS. UNIV., URBANA. HIGHWAY RESEARCH LAB.	HIGHWAY ENGINEERING SER-
ILLINOIS. UNIV., URBANA. RADIOLOCATION RESEARCH LAB.	RRL-
ILLINOIS. UNIV., URBANA. RADIOLOCATION RESEARCH LAB.	RRL-TR-
ILLINOIS. UNIV., URBANA. SMALL HOMES COUNCIL-BUILDING RESEARCH COUNCIL	IU-RR-(YEAR)-
ILLINOIS. UNIV., URBANA. WATER RESOURCES CENTER	WRC-RR-
ILLINOIS INST. OF TECH., CHICAGO (REF. 1)	(PROJECT NUMBER)-
ILLINOIS INST. OF TECH., CHICAGO	IIT-
ILLINOIS INST. OF TECH., CHICAGO (TECHNICAL REPORT)	IIT-TR-
ILLINOIS INST. OF TECH., CHICAGO	ILL-TECH-
ILLINOIS INST. OF TECH., CHICAGO	ILT-
ILLINOIS INST. OF TECH., CHICAGO	ITT-
ILLINOIS INST. OF TECH.,CHICAGO(PROJ. THEMIS)(REF.48) (SEE ALSO: IIT RESEARCH INST.)	THEMIS-(YEAR)-
ILLINOIS INST. OF TECH., CHICAGO. DEPT. OF MATH.	DOMIIT-
ILLINOIS INST. OF TECH., CHICAGO. DEPT. OF MECHANICS	DOMIIT-
ILLINOIS INST. OF TECH., CHICAGO. DEPT. OF MECHANICS	ILL.TECH./DOMIIT-
ILLINOIS INST. OF TECH., CHICAGO. DEPT. OF MEDICINE	IIT-DOMIIT-
ILLINOIS WATER TREATMENT CO., ROCKFORD	ILLCO-
ILLUMINATING ENGINEERING SOCIETY, N.Y.C.	IES-
IMPERIAL CHEMICAL INDUSTRIES, LTD.	BI-
IMPERIAL CHEMICAL INDUSTRIES, LTD. (ASSIGNED BY GT. BRIT. DIV. OF ATOMIC ENERGY (PRODUCTION) TO INCOMING REPORTS FROM IMPERIAL CHEM. INDUSTRIES)	RISLEY-2001 THRU -3000
IMPERIAL CHEMICAL INDUSTRIES, LTD., LONDON	FA/V-
IMPERIAL CHEMICAL INDUSTRIES, LTD., LONDON	FA/V.38-
IMPERIAL CHEMICAL INDUSTRIES, LTD., LONDON	ICI-
IMPERIAL CHEMICAL INDUSTRIES, LTD., LONDON	MSDH/(YEAR)/
IMPERIAL CHEMICAL INDUSTRIES, LTD. BILLINGHAM DIV., DURHAM, ENGLAND	BILLINGHAM RESEARCH...
IMPERIAL CHEMICAL INDUSTRIES, LTD. BILLINGHAM DIV., DURHAM, ENGLAND	MSV/-
IMPERIAL CHEMICAL INDUSTRIES, LTD. CENTRAL MANAGEMENT SERVICES, WILMSLOW, CHES., ENGLAND	MSDH/(NUMBER/NUMBER)
IMPERIAL CHEMICAL INDUSTRIES, LTD. GENERAL CHEMICAL DIV., WIDNES, LANCS, ENGLAND	P/GM-
IMPERIAL CHEMICAL INDUSTRIES, LTD. GENERAL CHEMICAL DIV., WIDNES, LANCS, ENGLAND	R/GC/-
IMPERIAL CHEMICAL INDUSTRIES, LTD. METALS DIV., BIRMINGHAM, WARWICK, ENGLAND	CON/HAR/EMR/
IMPERIAL COLLEGE OF SCIENCE AND TECHNOLOGY, LONDON	DGGW-EMR-(YEAR)-
IMPERIAL COLLEGE OF SCIENCE AND TECHNOLOGY, LONDON	EF/TN/B/

IMPERIAL COLLEGE OF SCIENCE AND TECHNOLOGY, LONDON	EHT/TN/
IMPERIAL COLLEGE OF SCIENCE AND TECHNOLOGY, LONDON	SF/R/
IMPERIAL COLLEGE OF SCIENCE AND TECHNOLOGY, LONDON	SF/TN/
IMPERIAL COLLEGE OF SCIENCE AND TECHNOLOGY, LONDON	TWF/R/
IMPERIAL COLLEGE OF SCIENCE AND TECHNOLOGY, LONDON	TWF/TN/
IMPERIAL COLLEGE OF SCIENCE AND TECHNOLOGY, LONDON. DEPT. OF AERONAUTICS	IC-AERO-(YEAR)-
IMPERIAL COLLEGE OF SCIENCE AND TECHNOLOGY, LONDON. DEPT. OF MECHANICAL ENGINEERING	BL/TN/
IMPERIAL COLLEGE OF SCIENCE AND TECHNOLOGY, LONDON. DEPT. OF MECHANICAL ENGINEERING	BL/TN/(LETTER)/(NUMBER)-
IMPERIAL COLLEGE OF SCIENCE AND TECHNOLOGY, LONDON. DEPT. OF MECHANICAL ENG. (COMBUSTION KINETICS)	CCK/TN/
IMPERIAL COLLEGE OF SCIENCE AND TECHNOLOGY, LONDON. DEPT. OF MECHANICAL ENGINEERING	EF/(LTR.)/(LTR.)/
IMPERIAL COLLEGE OF SCIENCE AND TECHNOLOGY, LONDON. DEPT. OF MECHANICAL ENGINEERING	EF/TN/A/
IMPERIAL COLLEGE OF SCIENCE AND TECHNOLOGY, LONDON. DEPT. OF MECHANICAL ENGINEERING	IC-
IMPERIAL COLLEGE OF SCIENCE AND TECHNOLOGY, LONDON. DEPT. OF MECHANICAL ENGINEERING	ON/TN/
IMPERIAL COLLEGE OF SCIENCE AND TECHNOLOGY, LONDON. DEPT. OF MECHANICAL ENGINEERING	RF/TN/A/
IMPERIAL COLLEGE OF SCIENCE AND TECHNOLOGY, LONDON. DEPT. OF MECHANICAL ENGINEERING	SF/TN/
IMPERIAL COLLEGE OF SCIENCE AND TECHNOLOGY, LONDON. DEPT. OF MECHANICAL ENGINEERING	T/TN/
IMPERIAL COLLEGE OF SCIENCE AND TECHNOLOGY, LONDON. DEPT. OF MECHANICAL ENGINEERING	TWF/R/
IMPERIAL COLLEGE OF SCIENCE AND TECHNOLOGY, LONDON. DEPT. OF MECHANICAL ENGINEERING	UF/TN/D/
IMPERIAL COLLEGE OF SCIENCE AND TECHNOLOGY, LONDON. DEPT. OF PHYSICS	IC-
IMPERIAL COLLEGE OF SCIENCE AND TECHNOLOGY, LONDON. DEPT. OF PHYSICS	U(NO.)-
IMPERIAL COLL. OF SCI. + TECH., LONDON. JET RES. LAB.	JRL-
IMPERIAL COLL. OF SCI. + TECH., LONDON. JET RES. LAB.	PDGW-EMR-
INDIA. AERONAUTICAL RESEARCH COMMISSION, BANGALORE	ARC-TN-
INDIA. ATOMIC ENERGY ESTABLISHMENT, TROMBAY	AEER/HP/SM-
INDIA. ATOMIC ENERGY ESTABLISHMENT, TROMBAY	AEET-
INDIA. ATOMIC ENERGY ESTABLISHMENT, TROMBAY	AEET/RADIOCHEM/
INDIA. ATOMIC ENERGY ESTABLISHMENT, TROMBAY	AEET-RE-
INDIA. ATOMIC ENERGY ESTABLISHMENT, TROMBAY	AEET-RMS-
INDIA. ATOMIC ENERGY ESTABLISHMENT, TROMBAY	AEET-ROD-
INDIA. ATOMIC ENERGY ESTABLISHMENT, TROMBAY	SM-
INDIA. ATOMIC ENERGY ESTABLISHMENT. AIR MONITORING SECTION, TROMBAY	AEET/AM/
INDIA. ATOMIC EN. ESTAB. ANALYTICAL DIV., TROMBAY	AEET/ANAL/
INDIA. ATOMIC EN. ESTAB. CHEMICAL DIV., TROMBAY	AEET/CD/
INDIA. ATOMIC EN. ESTAB. ELECTRONICS DIV., TROMBAY	AEET/ED/SG/
INDIA. ATOMIC EN. ESTAB. HEALTH PHYS. DIV., TROMBAY	AEET/HP/R-
INDIA. ATOMIC EN. ESTAB. HEALTH PHYS. DIV., TROMBAY	AEET/HP/TH/
INDIA. ATOMIC EN. ESTAB. METALLURGY DIV., TROMBAY	AEET/MET/
INDIA. ATOMIC EN. ESTAB. METALLURGY DIV., TROMBAY	AEET/SPEC/
INDIA. ATOMIC EN. ESTAB. NUCLEAR PHYS. DIV., TROMBAY	AEET/NP-
INDIA. ATOMIC EN. ESTAB. TECHNICAL PHYS.DIV.,TROMBAY	AEET/TP/
INDIA. BHABHA ATOMIC RESEARCH CENTRE, BOMBAY	AICS-
INDIA. BHABHA ATOMIC RESEARCH CENTRE, BOMBAY	BARC-
INDIA. BHABHA ATOMIC RESEARCH CENTRE, BOMBAY	BARC/HP/TM-
INDIA. BHABHA ATOMIC RESEARCH CENTRE, BOMBAY	BARC/I-
INDIA. BHABHA ATOMIC RESEARCH CENTRE, BOMBAY	BARC/INF-
INDIA. CENTRAL MECHANICAL ENG. RES. INST., DURGAPUR	A-
INDIA. CENTRAL MECHANICAL ENG. RES. INST., DURGAPUR	B-
INDIA. CENTRAL MECHANICAL ENG. RES. INST., DURGAPUR	CMERI-A(NUMBER)
INDIA. CENTRAL MECHANICAL ENG. RES. INST., DURGAPUR	CMERI-B(NUMBER)
INDIA. CHEMICAL DEFENCE RESEARCH ESTABLISHMENT	CDRE-
INDIA. COUNCIL OF SCIENTIFIC AND IND. RES., NEW DELHI	ARC-TN-
INDIA. COUNCIL OF SCIENTIFIC AND IND. RES., NEW DELHI	ARC-TR-
INDIA. COUNCIL OF SCIENTIFIC AND IND. RES., NEW DELHI	KRC-(LTR.)-
INDIA. COUNCIL OF SCIENTIFIC AND IND. RES., NEW DELHI	RRC-A-
INDIA. COUNCIL OF SCIENTIFIC AND IND. RES., NEW DELHI	RRC-B-
INDIA. DEPT. OF ATOMIC ENERGY, BOMBAY	DAE-AR-
INDIA. INST. OF MATHEMATICAL SCIENCES, MADRAS	MATSCIENCE-
INDIA. NATIONAL AERONAUTICAL LAB., BANGALORE	NAL-TN-
INDIA. NATIONAL AERONAUTICAL LAB., BANGALORE	NAL-TT-
INDIA. NATIONAL AERONAUTICAL LAB., BANGALORE	TN-AE-
INDIA. NATIONAL AERONAUTICAL LAB., BANGALORE	TN-PH-
INDIA. NATIONAL AERONAUTICAL LAB., BANGALORE	TN-SA-
INDIA. NATIONAL AERONAUTICAL LAB., BANGALORE	TN-WP-
INDIA. NATIONAL PHYSICAL LAB., NEW DELHI	KRC-(LETTER)(NUMBER)
INDIA. NATIONAL PHYSICAL LAB., NEW DELHI	PL-
INDIA. NATIONAL PHYSICAL LAB., NEW DELHI	RPU-
INDIA. NATIONAL PHYSICAL LAB., NEW DELHI	RPU-S-
INDIA. NATIONAL PHYSICAL LAB., NEW DELHI	RRC-
INDIA. NATIONAL PHYSICAL LAB., NEW DELHI	RRC-A(NO.)-
INDIA. NATIONAL PHYSICAL LAB., NEW DELHI	RSD-
INDIA. NATIONAL PHYSICAL LAB., NEW DELHI	RSD-(LETTER)(NUMBER)
INDIA. NATIONAL PHYSICAL LAB., NEW DELHI	RTRC-A-
INDIA. TATA INST. OF FUNDAMENTAL RESEARCH, BOMBAY	TIFR-
INDIAN INST. OF SCIENCE, BANGALORE	AE-(NUMBER)(LETTER)
INDIAN INST. OF SCIENCE, BANGALORE	IWTR/
INDIAN NATIONAL SCIENTIFIC DOCUMENTATION CENTER, NEW DELHI	TT-(YEAR)-
INDIANA UNIV., BLOOMINGTON	IND-TR-
INDIANA UNIV., BLOOMINGTON	INDU-
INDIANA UNIV., BLOOMINGTON	INU-
INDIANA UNIV., BLOOMINGTON. CHEMISTRY DEPT.	CHEM (NO.-NO.)
INDUSTRIAL AND ENGINEERING CHEMISTRY. RESEARCH RESULTS SERVICE, WASHINGTON, D.C.	I+EC-
INDUSTRIAL AND ENGINEERING CHEMISTRY. RESEARCH RESULTS SERVICE, WASHINGTON, D.C. (MANUSCRIPT)	MS-(YEAR)-

INDUSTRIAL COLL. OF THE ARMED FORCES, WASHINGTON, D.C.	ICAF-
INDUSTRIAL COLL. OF THE ARMED FORCES, WASHINGTON, D.C.	ICAF-L-(YEAR)-
INDUSTRIAL COLL. OF THE ARMED FORCES, WASHINGTON, D.C.	ICAF-M (YR.)(NO.)
INDUSTRIAL COLL. OF THE ARMED FORCES, WASHINGTON, D.C.	ICAF-TEMPER-
INDUSTRIAL COLL. OF THE ARMED FORCES, WASHINGTON, D.C.	L(YEAR)-
INDUSTRIAL COLL. OF THE ARMED FORCES, WASHINGTON, D.C.	SR-(YEAR)-
INDUSTRIAL ELECTRONICS + TRANSFORMER CO., LOS ANG.	INET-
INDUSTRIAL INSTRUMENTS, INC., CEDAR GROVE, N.J.	LRO-
INDUSTRIAL NUCLEONICS CORP., COLUMBUS, OHIO	INC-
INDUSTRIAL RESEARCH LABS., BALTIMORE	IRL-
INDUSTRIAL SCIENTIFIC CO., N.Y.C.	INDS-
INDUSTRIAL SCIENTIFIC CO., N.Y.C.	ISC-FR-
INDUSTRIAL TECTONICS, INC., COMPTON, CALIF.	ITI-
INDUSTRIEANLAGEN-BETRIEBSGESELLSCHAFT M.B.H., OTTOBRUNN, GERMANY	ORG/MS-
INET, INC., LOS ANGELES	RDTR-
INFORMATICS, INC., BETHESDA, MD.	TR-(YEAR)-(NO.)-(NO.)
INFORMATICS TISCO, INC., RIVERDALE, MD.	ITI-BIB-
INFORMATION DYNAMICS CORP., READING, MASS.	IDC-
INFORMATION GENERAL CORP., PALO ALTO, CALIF.	IGC-PA-(YEAR)-
INFORMATION GENERAL CORP., PALO ALTO, CALIF.	IGG-PA-(YEAR)-
INFORMATION RESEARCH ASSOCIATES, INC., BERKELEY, CAL.	IRA-TR-(NUMBER)-(YEAR)
INFORMATION RESEARCH ASSOCIATES, INC., CAMBRIDGE, MASS	IRA-TN-
INFORMATION RESEARCH ASSOCIATES, INC., LEXINGTON, MASS	IRA-TR-
INFORMATION RESEARCH ASSOCIATES, INC., SAN DIEGO, CAL	IRA-
INFORMATION TECHNOLOGY LABS., WALTHAM, MASS.	RPIS-
INFRARED INFORMATION SYMPOSIA. (PROCEEDINGS)	IRIS SYM-
INGALLS (ARTHUR L.), ANN ARBOR, MICH.	ALI-
INGERSOLL KALAMAZOO DIV., MICH.	IKD-
INGERSOLL KALAMAZOO DIV., MICH.	SPD-
INGERSOLL KALAMAZOO DIV., BORG-WARNER CORP., MICH.	SPDIR-
INLAND TESTING LABS., MORTON GROVE, ILL.	ITL-SR-
INNSBRUCK UNIV. INST. FOR THEORETICAL PHYSICS	UNICP-FR-
INNSBRUCK UNIV. INST. FOR THEORETICAL PHYSICS	UNICP-SR-
INSPECTOR GENERAL OF THE AIR FORCE. DIRECTORATE OF SECURITY POLICE, WASHINGTON, D.C.	AFISP-
INSPECTORATE GENERAL OF AIR (ARMY) (MANUAL)	AIGA-
INSTITUT FRANCO-ALLEMAND DE RECHERCHES, ST. LOUIS, FRANCE	ISL-D-
INSTITUT FRANCO-ALLEMAND DE RECHERCHES, ST. LOUIS, FRANCE	ISL-N-(NUMBER/YEAR)
INSTITUT FRANCO-ALLEMAND DE RECHERCHES, ST. LOUIS, FRANCE	ISL-ST-
INSTITUT FRANCO-ALLEMAND DE RECHERCHES, ST. LOUIS, FRANCE	ISL-T-
INSTITUT FRANCO-ALLEMAND DE RECHERCHES, ST. LOUIS, FRANCE	T-
INSTITUT FUER PLASMAPHYSIK, GMBH, GARCHING, GERMANY	IPP (NUMBER/NUMBER)
INSTITUT FUER PLASMAPHYSIK, GMBH, GARCHING, GERMANY	IPP-
INSTITUT FUER PLASMAPHYSIK, GMBH, GARCHING, GERMANY	IPP-AR-
INSTITUTE FOR ADVANCED STUDY, PRINCETON, N.J.	IAS-
INSTITUTE FOR ADVANCED STUDY, PRINCETON, N.J.	IAS-FR-(MONTH)-(YEAR)
INSTITUTE FOR ADVANCED STUDY. APPLIED MATHEMATICS GROUP, PRINCETON, N.J. (REF. 28)	AMG-IAS-
INSTITUTE FOR ADVANCED STUDY. APPLIED MATHEMATICS GROUP, PRINCETON, N.J. (REF. 28)	NDRC AMG-IAS-
INSTITUTE FOR DEFENSE ANALYSES, ARLINGTON, VA.	ARPA/IDA-
INSTITUTE FOR DEFENSE ANALYSES, ARLINGTON, VA.	ARPA/IDA (LETTER)-
INSTITUTE FOR DEFENSE ANALYSES, ARLINGTON, VA.	IDA-
INSTITUTE FOR DEFENSE ANALYSES, ARLINGTON, VA.	IDA-ARPA-R-(YEAR)-
INSTITUTE FOR DEFENSE ANALYSES, ARLINGTON, VA.	IDA-ARPA-TR-(YEAR)-
INSTITUTE FOR DEFENSE ANALYSES, ARLINGTON, VA.	IDA/HQ-(YEAR)-
INSTITUTE FOR DEFENSE ANALYSES, ARLINGTON, VA.	IDA/HQ-
INSTITUTE FOR DEFENSE ANALYSES, ARLINGTON, VA. (JOURNAL OF DEFENSE RESEARCH)	IDA/HQ-JDR-(YEAR)-
INSTITUTE FOR DEFENSE ANALYSES, ARLINGTON, VA.	IDA-RESG-TR-
INSTITUTE FOR DEFENSE ANALYSES, ARLINGTON, VA. (TECHNICAL NOTE)	IDA-TN-
INSTITUTE FOR DEFENSE ANALYSES, ARLINGTON, VA.	P-
INSTITUTE FOR DEFENSE ANALYSES, ARLINGTON, VA.	R-
INSTITUTE FOR DEFENSE ANALYSES, ARLINGTON, VA.	RP-P-
INSTITUTE FOR DEFENSE ANALYSES, ARLINGTON, VA. (USED BY SEVERAL DIVISIONS)	S-
INSTITUTE FOR DEFENSE ANALYSES, ARLINGTON, VA.	STUDY-S-
INSTITUTE FOR DEFENSE ANALYSES, ARLINGTON, VA.	TIO-
INSTITUTE FOR DEFENSE ANALYSES, ARLINGTON, VA.	U-
INSTITUTE FOR DEFENSE ANALYSES, ARLINGTON, VA.	UBG-
INSTITUTE FOR DEFENSE ANALYSES, ARLINGTON, VA.	VUP-
INSTITUTE FOR DEFENSE ANALYSES. RESEARCH AND ENGINEERING SUPPORT DIV., ARLINGTON, VA.	IDA-R-
INSTITUTE FOR DEFENSE ANALYSES. RESEARCH AND ENGINEERING SUPPORT DIV., ARLINGTON, VA.	IDA-RP-P-
INSTITUTE FOR DEFENSE ANALYSES. RESEARCH AND ENGINEERING SUPPORT DIV., ARLINGTON, VA.	JMDR-
INSTITUTE FOR DEFENSE ANALYSES. RESEARCH AND ENGINEERING SUPPORT DIV., ARLINGTON, VA.	RESD-TR-
INSTITUTE FOR DEFENSE ANALYSES. SCIENCE AND TECHNOLOGY DIV., ARLINGTON, VA.	IDA-SRD-
INSTITUTE FOR DEFENSE ANALYSES. SCIENCE AND TECHNOLOGY DIV., ARLINGTON, VA.	N-
INSTITUTE FOR DEFENSE ANALYSES. WEAPONS SYSTEMS EVALUATION DIV., WASHINGTON, D.C.	IDA-LOG-
INSTITUTE FOR INTERNAL COMBUSTION ENGINES	IICE-
INSTITUTE FOR RESEARCH. DIV. OF PSYCHOBIOLOGY, STATE COLLEGE, PENNA.	ONR-H-(YEAR)-

INSTITUTE FOR TELECOMMUNICATION SCIENCES, BOULDER, COLO	ESSA-TR-ERL-(NO.)-ITS-
INSTITUTE FOR TELECOMMUNICATION SCIENCES, BOULDER, COLO	ITS-
INSTITUTE FOR TELECOMMUNICATION SCIENCES, BOULDER, COLO	RLTM-ITS-
INSTITUTE FOR TELECOMMUNICATION SCIENCES, BOULDER, COLO	TB(NUMBERS)/TO(NUMBERS)
INSTITUTE FOR TELECOMMUNICATION SCIENCES AND AERONOMY, BOULDER, COLO.	CRPL-
INSTITUTE FOR TELECOMMUNICATION SCIENCES AND AERONOMY, BOULDER, COLO.	ESSA-IER-(NO.)-ITSA-
INSTITUTE FOR TELECOMMUNICATION SCIENCES AND AERONOMY, BOULDER, COLO.	IER-(NO.)-ITSA-(SAME NO.)
INSTITUTE FOR TELECOMMUNICATION SCIENCES AND AERONOMY, BOULDER, COLO.	IERTM-ITSA-
INSTITUTE FOR TELECOMMUNICATION SCIENCES AND AERONOMY, BOULDER, COLO.	ITSA-
INSTITUTE OF ELECTRICAL AND ELECTRONICS ENGINEERS, NYC	IEEE-
INSTITUTE OF ELECTRICAL AND ELECTRONICS ENGINEERS, NYC	IEEE-A(NUMBER)-
INSTITUTE OF ELECTRICAL AND ELECTRONICS ENGINEERS, NYC	IEEE PAPER (NO)-PP-(YR)
INSTITUTE OF METEOROLOGY, STOCKHOLM	ESRO-SN-(NUMBER)/ESLAB
INSTITUTE OF PAPER CHEMISTRY, APPLETON, WIS.	IPC-
INSTITUTE OF RADIO ENGINEERS, N.Y.C. (SEE ALSO LATER NAME: INSTITUTE OF ELECTRICAL AND ELECTRONICS ENGINEERS)	IRE-
INSTITUTE OF TEXTILE TECH., CHARLOTTESVILLE, VA.	ITT-R-
INSTITUTE OF THE AERONAUTICAL SCIENCES, N.Y.C. (REF.1)	(NUMBER)
INSTITUTE OF THE AERONAUTICAL SCIENCES, N.Y.C.	SMF-
INSTITUTES FOR ENVIRONMENTAL RESEARCH, BOULDER, COLO.	ESSA-IER-FB-
INSTITUTES FOR ENVIRONMENTAL RESEARCH, BOULDER, COLO.	ESSA-TR-IER-
INSTITUTES FOR ENVIRONMENTAL RESEARCH, BOULDER, COLO.	IER-
INSTRUMENT DEV. LABS., INC., NEEDHAM HEIGHTS, MASS.	IDL-
INSTRUMENT PILOT INSTRUCTOR SCHOOL, RANDOLPH AFB, TEX. (INSTRUMENT EVALUATION)	AF-IEPR-
INSTRUMENT SOCIETY OF AMERICA, PITTSBURGH	ISA-
INSTRUMENT SYSTEMS CORP. TELEPHONICS DIV., HUNTINGTON, N.Y.	IER-
INTELECTRON CORP., N.Y.C.	ER-
INTELLIGENCE DIV. (ARMY), WASHINGTON, D.C. (REF. 1)	(NUMBERS...)
INTELLIGENCE DIV., GENERAL STAFF, WASHINGTON, D.C.	ID GS-
INTER-AMERICAN DEFENSE BOARD, WASHINGTON, D.C.	T-
INTER-BUREAU TECHNICAL COMMITTEE (NAVY), WASH., D.C.	IBTC-
INTER-BUREAU TECHNICAL COMMITTEE (NAVY), WASH., D.C.	NAV-IBTC-
INTER-BUREAU TECHNICAL COMMITTEE (NAVY). GUIDED MISSILE RELAY WORKING GROUP	GMRWG-
INTER-BUREAU TECHNICAL COMMITTEE (NAVY). GUIDED MISSILE RELAY WORKING GROUP	NAV-GMRWG-
INTER-LABORATORY COMMITTEE ON FACILITIES (NAVY)	ILCF-
INTER-LABORATORY COMMITTEE ON FACILITIES (NAVY)	NAV-ILCF-
INTER-POLYMER CORP., PASSAIC, N.J.	IPC-
INTER-RANGE INSTRUMENTATION GROUP	DR AND CWG-
INTER-RANGE INSTRUMENTATION GROUP	IFCG-
INTER-RANGE INSTRUMENTATION GROUP	IRTWG-
INTER-RANGE INSTRUMENTATION GROUP, WASHINGTON, D.C.	DOD IRIG-(YR)
INTER-RANGE INSTRUMENTATION GROUP. METEOROLOGICAL WORKING GROUP, WHITE SANDS MISSILE RANGE, N. MEX.	IRIG-MWG-
INTER-RANGE MISSILE FLIGHT SAFETY GROUP, WHITE SANDS MISSILE RANGE, N. MEX.	IRMFSG-
INTER-RANGE MISSILE GROUND SAFETY GROUP, WHITE SANDS MISSILE RANGE, N. MEX.	IRMGS-
INTER-SERVICE TOPOGRAPHICAL DEPT.	ISTD-
INTER-UNIV SEMINAR ON ARMED FORCES AND SOCIETY, WILMETTE, ILL.	IUSAFS-
INTERAGENCY CHEMICAL ROCKET PROPULSION GROUP	ICRPG-
INTERAGENCY CHEMICAL ROCKET PROPULSION GROUP	ICRPG-ROUND ROBIN-
INTERAGENCY COMMITTEE ON OCEANOGRAPHY, WASH., D.C.	ICO PAMPHLET-
INTERAGENCY DATA EXCHANGE PROGRAM. (SPONSORED BY ARMY, NAVY, AIR FORCE, NASA, AND CANADIAN MILITARY ELECTRONICS STANDARDS AGENCY TO MINIMIZE DUPLICATE TESTS OF PARTS, COMPONENTS AND MATERIALS)	IDEP-
INTERAGENCY MECHANICAL OPERATIONS GP., STEERING COMM.	IMOG-
INTERATOM. INTERNATIONALE ATOMREACTORBAU GMBH, BENSBERG/COLOGNE	INTAT-
INTERCEPTOR PILOT RES. LAB., TYNDALL AFB, FLA.	IPRL-
INTERCEPTOR WEAPONS SCHOOL, TYNDALL AFB, FLA.	RD-
INTERCEPTOR WEAPONS SCHOOL, TYNDALL AFB, FLA.	TAFB-
INTERCEPTOR WEAPONS SCHOOL, TYNDALL AFB, FLA.	TR+D-
INTERDEPARTMENTAL COMMITTEE FOR ATMOSPHERIC SCIENCES, WASHINGTON, D.C.	ICAS-
INTERDEPARTMENTAL COMMITTEE FOR ATMOSPHERIC SCIENCES, WASHINGTON, D.C.	ICAS-PAPER-(YEAR)-
INTERDEPARTMENTAL COMMITTEE FOR THE ACQUISITION OF FOREIGN PUBLICATIONS	IDC-
INTERDEPARTMENTAL COMMITTEE FOR THE ACQUISITION OF FOREIGN PUBLICATIONS (ABSTRACTS ON ENEMY OIL, FUELS AND LUBRICANTS)	IDC F-
INTERDEPARTMENTAL COMMITTEE FOR THE ACQUISITION OF FOREIGN PUBLICATIONS (SUBJECT INDEX)	IDC R-
INTERDEPARTMENTAL COMM. ON SCI. RES. + DEV., WASH., DC	ICSRD-
INTERGOVERNMENTAL OCEANOGRAPHIC COMMISSION, PARIS	IOC-
INTERIM COMMUNICATIONS SATELLITE COMMITTEE	ICSC-(NUMBERS-LTRS.-YR)
INTERLABORATORY COMMITTEE ON EDITING AND PUBLISHING (NAVY), CORONA, CALIF.	ILCEP-
INTERMOUNTAIN FOREST AND RANGE EXPERIMENT STA., OGDEN, UTAH	RP-INT-

INTERMOUNTAIN RES. & ENG. CO., INC.

INTERMOUNTAIN RES.+ENG.CO.,INC.,SALT LAKE CITY,UTAH	IRECO-
INTERNATIONAL AIR TRANSPORT ASSOCIATION, CANADA	IATA-
INTERNATIONAL ASTRONAUTICAL FEDERATION, PARIS	IISL-BIBL.-
INTERNATIONAL ATOMIC ENERGY AGENCY, VIENNA (CONFERENCE PAPER)	CN-(NUMBER/LETTER)-
INTERNATIONAL ATOMIC ENERGY AGENCY, VIENNA	GC(ROMAN NO.)/INF-
INTERNATIONAL ATOMIC ENERGY AGENCY, VIENNA	GC(III)/
INTERNATIONAL ATOMIC ENERGY AGENCY, VIENNA	GEN/PUB/
INTERNATIONAL ATOMIC ENERGY AGENCY, VIENNA	IAEA-
INTERNATIONAL ATOMIC ENERGY AGENCY, VIENNA (CONFERENCE PAPER)	IAEA-CN-(NUMBER/NUMBER)
INTERNATIONAL ATOMIC ENERGY AGENCY, VIENNA	IAEA-NPR-
INTERNATIONAL ATOMIC ENERGY AGENCY, VIENNA	IAEA-R-
INTERNATIONAL ATOMIC ENERGY AGENCY, VIENNA	IAEA-SM-
INTERNATIONAL ATOMIC ENERGY AGENCY, VIENNA	INFCIRC-
INTERNATIONAL ATOMIC ENERGY AGENCY, VIENNA	PUB-CAT-(YEAR)/E
INTERNATIONAL ATOMIC ENERGY AGENCY, VIENNA (CONFERENCE ON THE USE OF RADIOISOTOPES IN PHYSICAL SCIENCES AND INDUSTRY, COPENHAGEN, SEPT., 1960)	RICC/
INTERNATIONAL ATOMIC ENERGY AGENCY, VIENNA	SM-(NUMBER/NUMBER)
INTERNATIONAL ATOMIC ENERGY AGENCY, VIENNA	SM-(NUMBER/NUMBER)
INTERNATIONAL ATOMIC ENERGY AGENCY, VIENNA	STI-DOC/
INTERNATIONAL ATOMIC ENERGY AGENCY, VIENNA	STI/DOC/10-
INTERNATIONAL ATOMIC ENERGY AGENCY, VIENNA	STI/PUB/
INTERNATIONAL ATOMIC ENERGY AGENCY, VIENNA	STI-REP-
INTERNATIONAL ATOMIC ENERGY AGENCY, VIENNA	TO/HS/
INTERNATIONAL ATOMIC ENERGY AGENCY, VIENNA (SPECIALIST MEETING, MINUTES)	UNC-SPLM-
INTERNATIONAL ATOMIC ENERGY AGENCY, VIENNA	WP-
INTERNATIONAL ATOMIC ENERGY AGENCY. INFORMATIONAL NUCLEAR DATA UNIT, VIENNA	INDSWG-
INTERNATIONAL ATOMIC ENERGY AGENCY. INTERNATIONAL NUCLEAR INFORMATION SYSTEM, VIENNA	IAEA-INIS-
INTERNATIONAL ATOMIC ENERGY AGENCY. INTERNATIONAL NUCLEAR INFORMATION SYSTEM, VIENNA	INIS-MF-
INTERNATIONAL ATOMIC ENERGY AGENCY. NUCLEAR DATA UNIT, VIENNA	FVR-
INTERNATIONAL BUSINESS MACHINES CORP. (ALL LOCATIONS)	IBM-(YEAR)-
INTERNATIONAL BUSINESS MACHINES CORP. (ALL LOCATIONS)	IBM-
INTERNATIONAL BUSINESS MACHINES CORP. (ALL LOCATIONS)	IBM-(NO. NO. NO.)
INTERNATIONAL BUSINESS MACHINES CORP. (ALL LOCATIONS)	IBM-TR-
INTERNATIONAL BUSINESS MACHINES CORP., ALBUQUERQUE,NM	GF(NUMBER)-(NUMBER)-
INTERNATIONAL BUSINESS MACHINES CORP., ALBUQUERQUE,NM	IBM-GF(NUMBER)-
INTERNATIONAL BUSINESS MACHINES CORP., CAMBRIDGE,MASS	TRN-
INTERNATIONAL BUSINESS MACHINES CORP., ENDICOTT, N.Y. (INTERIM ENGINEERING REPORT)	IBM IER-(NOS.)
INTERNATIONAL BUSINESS MACHINES CORP.,GAITHERSBURG,MD	ESD-TR-(YEAR)-
INTERNATIONAL BUSINESS MACHINES CORP.,GAITHERSBURG,MD	FSC-(YEAR)-
INTERNATIONAL BUSINESS MACHINES CORP.,GAITHERSBURG,MD	IBM-FSC-(YEAR)-
INTERNATIONAL BUSINESS MACHINES CORP., KINGSTON, N.Y.	IBM ECPX-
INTERNATIONAL BUSINESS MACHINES CORP., OWEGO, N.Y.	(YR.)-(3 DIGIT DEPT. NO.)
INTERNATIONAL BUSINESS MACHINES CORP., OWEGO, N.Y.	IBMC-
INTERNATL. BUSINESS MACHINES CORP.,POUGHKEEPSIE, N.Y.	CD-
INTERNATL. BUSINESS MACHINES CORP.,POUGHKEEPSIE, N.Y.	IBM TN (NO. NO. NO.)
INTERNATIONAL BUSINESS MACHINES CORP., SAN JOSE, CAL.	NJ-
INTERNATIONAL BUSINESS MACHINES CORP., ZURICH	IBM-NZ-
INTERNATIONAL BUSINESS MACHINES CORP. ADVANCED SYSTEMS DEVELOPMENT DIV., YORKTOWN HEIGHTS, N.Y.	ASDD-RC-
INTERNATIONAL BUSINESS MACHINES CORP. CENTER FOR EXPLORATORY STUDIES, CAMBRIDGE, MASS.	CES-(LETTER-NUMBER)
INTERNATIONAL BUSINESS MACHINES CORP. ELECTRONICS SYSTEMS CENTER, OWEGO, N.Y.	IBM-CD-(NO.)-(NO.)-
INTERNATIONAL BUSINESS MACHINES CORP. FEDERAL SYSTEMS DIV., BETHESDA, MD.	MS-
INTERNATIONAL BUSINESS MACHINES CORP. FEDERAL SYSTEMS DIV., GAITHERSBURG, MD.	IBM-CD-
INTERNATIONAL BUSINESS MACHINES CORP. FEDERAL SYSTEMS DIV., GAITHERSBURG, MD.	IBM-CD-(NUMBER)-(NO.)-
INTERNATIONAL BUSINESS MACHINES CORP. FEDERAL SYSTEMS DIV., OWEGO, N.Y.	IBMC-
INTERNATIONAL BUSINESS MACHINES CORP. MILITARY PRODUCTS DIV., KINGSTON, N.Y.	IBM-ECBX-
INTERNATIONAL BUSINESS MACHINES CORP. RESEARCH DIV. SAN JOSE LAB., CALIF.	IBM-RJ-
INTERNATIONAL BUSINESS MACHINES CORP. RESEARCH DIV. SAN JOSE LAB., CALIF.	RJ-
INTERNATIONAL BUSINESS MACHINES CORP. RESEARCH LAB., SAN JOSE, CALIF. (SEE ALSO ENTRIES BEGINNING: IBM...)	IBM-NJ-
INTERNATIONAL CENTRE FOR THEORETICAL PHYSICS, TRIESTE, ITALY	IC/
INTERNATIONAL CENTRE FOR THEORETICAL PHYSICS, TRIESTE, ITALY	IC-(YEAR)-
INTERNATIONAL CENTRE FOR THEORETICAL PHYSICS, TRIESTE, ITALY	IC-AR-(YEAR-YEAR)
INTERNATIONAL CENTRE FOR THEORETICAL PHYSICS, TRIESTE, ITALY	IC/HEPR/(YEAR)/
INTERNATIONAL CENTRE FOR THEORETICAL PHYSICS, TRIESTE, ITALY	ICTP-
INTERNATIONAL CENTRE FOR THEORETICAL PHYSICS, TRIESTE, ITALY	SC/
INTERNATIONAL CHEMICAL AND NUCLEAR CORP. NUCLEAR SCIENCE DIV., PITTSBURGH	NSEC-(NUMBER)-
INTERNATIONAL CIVIL AVIATION ORGANIZATION, MONTREAL	DOC-(4 DIGITS)
INTERNATIONAL CIVIL AVIATION ORGANIZATION, MONTREAL	ICAO-
INTERNATIONAL COMMISSION ON RADIATION UNITS AND MEASUREMENTS, WASHINGTON, D.C.	ICRU-
INTERNATIONAL CONFERENCE ON THE PEACEFUL USES OF ATOMIC ENERGY, 1955, GENEVA	A/CONF.8/P/
INTERNATIONAL CONFERENCE ON THE PEACEFUL USES OF ATOMIC ENERGY, 1958, GENEVA	A/CONF.15/P/
INTERNATIONAL CONFERENCE ON THE PEACEFUL USES OF ATOMIC ENERGY, 1964, GENEVA	A/CONF.28/P/-
INTERNATIONAL CONFERENCE ON THE PEACEFUL USES OF ATOMIC ENERGY, 1971, GENEVA	A/CONF.49/
INTERNATIONAL CONSULTANT SCIENTISTS CORP., BROOKLINE, MASS.	SII-QSR-
INTERNATIONAL ELECTRIC CORP., PARAMUS, N.J.	IEC-

INTERNATIONAL ELECTRONIC RESEARCH CORP., BURBANK, CAL	IERC-
INTERNATIONAL ELECTRONICS ENGINEERING, INC., WASH.,DC	IEEI-
INTERNATIONAL ELECTRONICS MANUFACTURING CO., WASH.,DC	IEMC-
INTERNATIONAL FEDERATION FOR DOCUMENTATION. U.S. NATIONAL COMMITTEE, WASHINGTON, D.C.	FID-
INTERNATIONAL JOINT COMMITTEE ON PSYCHOMETRIC DATA	IJCPD-
INTERNATIONAL MINERALS AND CHEMICAL CORP., CHICAGO	IMCC-
INTERNATIONAL NUCLEAR DATA COMMITTEE, VIENNA	INDC-
INTERNATIONAL NUCLEAR DATA COMMITTEE, VIENNA	INDC(LETTERS)-(NO./LTR.)
INTERNATIONAL NUCLEAR DATA COMMITTEE, VIENNA (PAPER FROM U.S.S.R.)	INDC(CCP)-(NUMBER/LTR.)
INTERNATIONAL NUCLEAR DATA COMMITTEE, VIENNA	INDC(NDS)-
INTERNATIONAL NUCLEAR DATA COMMITTEE, VIENNA (PAPER FROM U.S.A.)	INDC(US)-(NUMBER/LETTER)
INTERNATIONAL ORGANIZATION FOR STANDARDIZATION,GENEVA	ISO-
INTERNATIONAL PUBLIC OPINION RESEARCH, INC., N.Y.C.	IPOR-
INTERNATIONAL RESEARCH + DEVELOPMENT CO. LTD., NEWCASTLE-UPON-TYNE, ENGLAND	IRD-(YEAR)-
INTERNATIONAL RESISTANCE CO., PHILADELPHIA	IRC-
INTERNATIONAL SEISMOLOGICAL CENTRE, EDINBURGH	ISCE-
INTERNATIONAL SILVER CO. TIMES WIRE AND CABLE DIV., WALLINGFORD, CONN.	ENG-(NUMBER)-E-
INTERNATIONAL TELECOMMUNICATION LABS., INC., N.Y.C.	ITL-
INTERNATIONAL TELECOMMUNICATIONS SATELLITE CONSORTIUM (INTELSAT). INTERIM COMMUNICATIONS SATELLITE COMM.	ICSC-
INTERNATIONAL TELECOMMUNICATIONS UNION, GENEVA (INTERNATIONAL FREQUENCY LIST)	IFL-ITU-OTP-
INTERNATIONAL TELEPHONE AND TELEGRAPH CORP., N.Y.C.	ITT-
INTERNATIONAL TELEPHONE + TELEGRAPH CORP.,NUTLEY,N.J. (SEE ALSO ENTRIES BEGINNING: ITT...)	ITT-TR-
INTERNUCLEAR CO., INC., CLAYTON, MO.	IC-(INITIALS)-(YR)-
INTERNUCLEAR CO., INC., CLAYTON, MO.	INT-DGO-
INTERNUCLEAR CO., INC., CLAYTON, MO.	INTERNUC-
INTERNUCLEAR CO., INC., CLAYTON, MO.	INTERNUCLEAR-
INTERNUCLEAR CO., INC., CLAYTON, MO.	P-
INTERSCIENCE PUBLISHERS, INC., N.Y.C. (TRANSLATIONS)	IP-
INTERSTATE COMMERCE COMMISSION, WASHINGTON, D.C.	ICC-
INTERSTATE COMMERCE COMMISSION, WASHINGTON, D.C. (TRANSPORT MOBILIZATION)	TM-
INTERSTATE ELECTRONICS CORP. NATIONAL MARINE CONSULTANTS DIV., ANAHEIM, CALIF.	NMC-
INTERSTATE ELECTRONICS CORP. NATIONAL MARINE CONSULTANTS DIV., ANAHEIM, CALIF.	NMC-ONR-
INTERSTATE ENGINEERING CORP., EL SEGUNDO, CALIF.	IEC-
INTERSTATION SUPERSONIC TRACK CONFERENCE. STRUCTURES WORKING GROUP, HOLLOMAN AFB, N.MEX.	ISTRACON-
INTERUNIVERSITAIR REACTOR INSTITUUT, DELFT	IRI-(NUMBER)-(YEAR)-
ION PHYSICS CORP., BURLINGTON, MASS.	INR-
ION PHYSICS CORP., BURLINGTON, MASS.	ION-
ION PHYSICS CORP., BURLINGTON, MASS.	QTPR-
IONAC CHEMICAL CO., BIRMINGHAM, N.J.	IONAC-
IOWA. STATE UNIV., IOWA CITY	IS-
IOWA. STATE UNIV., IOWA CITY	ISU-
IOWA. STATE UNIV., IOWA CITY	SDC-
IOWA. STATE UNIV., IOWA CITY	SUI-
IOWA. STATE UNIV., IOWA CITY	SUI-(YR)-
IOWA. STATE UNIV., IOWA CITY	U OF IOWA-(YEAR)-
IOWA. STATE UNIV., IOWA CITY. DEPT. OF CHEMISTRY	RI-
IOWA. STATE UNIV., IOWA CITY. DEPT. OF PHYSICS AND ASTRONOMY	IU-
IOWA. STATE UNIV., IOWA CITY. DEPT. OF PHYSICS AND ASTRONOMY	P-
IOWA. STATE UNIV., IOWA CITY. DEPT. OF PHYSICS AND ASTRONOMY	RI-
IOWA. STATE UNIV., IOWA CITY. IOWA INST. OF HYDRAULIC RESEARCH	IIHR-
IOWA. STATE UNIV., IOWA CITY. IOWA INST. OF HYDRAULIC RESEARCH	IOU-IIHR-
IOWA STATE COLL., AMES	ISC-
IOWA STATE COLL., AMES	LR-
IOWA STATE COLL., AMES (SEE ALSO LATER NAME: IOWA STATE UNIV. OF SCIENCE...)	TI-
IOWA STATE COLL., AMES. INST. FOR ATOMIC RESEARCH (CODE IS AUTHOR'S INITIALS)	LJL(ISCAR)-
IOWA STATE COLL., AMES. INST. FOR ATOMIC RESEARCH (CODE IS AUTHOR'S INITIALS)	LJL-MAC-
IOWA STATE COLL., AMES. STATISTICAL LAB.	EP-
IOWA STATE COLL., AMES. STATISTICAL LAB.	ISC-PTR-(NUMBER).(NUMBER)
IOWA STATE COLL., AMES. STATISTICAL LAB. (TECH. REPT)	ISC-TR-
IOWA STATE COLL., AMES. STATISTICAL LAB.	ISC-TR-(NUMBER).(NUMBER)
IOWA STATE UNIV. OF SCIENCE AND TECHNOLOGY, AMES	IS-
IOWA STATE UNIV. OF SCIENCE AND TECHNOLOGY, AMES	IS-PR(MO./YR.-MO./YR.)
IOWA STATE UNIV. OF SCIENCE AND TECHNOLOGY, AMES	IS-T-
IOWA STATE UNIV. OF SCIENCE AND TECHNOLOGY, AMES	ISU-
IOWA STATE UNIV. OF SCIENCE + TECHNOLOGY, AMES (FILM)	NS-
IOWA STATE UNIV. OF SCIENCE AND TECHNOLOGY, AMES. DEPT. OF SOCIOLOGY AND ANTHROPOLOGY	RURAL SOCIOLOGY-
IOWA STATE UNIV. OF SCIENCE AND TECHNOLOGY, AMES. ENGINEERING RESEARCH INST.	ERI-
IOWA STATE UNIV. OF SCIENCE AND TECHNOLOGY, AMES ENGINEERING RESEARCH INST.	ISU-EKI-AMES-
IOWA STATE UNIV. OF SCIENCE AND TECHNOLOGY, AMES. INST. FOR ATOMIC RESEARCH	IS-TRANS-

IOWA STATE UNIV. OF SCIENCE AND TECHNOLOGY, AMES. RARE-EARTH INFORMATION CENTER	IS-RIC-
IOWA STATE UNIV. OF SCIENCE AND TECHNOLOGY, AMES. SOIL RESEARCH LAB.	CONTRIB-(YEAR)-
IOWA UNIVERSITIES MISSION TO PERU	IOWA PERU ES-
IOWA UNIVERSITIES MISSION TO PERU	IOWA PERU SR-
IPSEN INDUSTRIES, INC., ROCKFORD, ILL.	A-
IRAQ. ATOMIC ENERGY COMMISSION, BAGHDAD	PH-
IRCO CORP., N.Y.C.	IRCO R-(NO.)-
IRELAND. NATIONAL UNIV., DUBLIN. UNIV. COLL.	UCD-CR-
IRON AND STEEL INSTITUTE, LONDON (TRANSLATIONS)	BISI-
ISOCHEM, INC., RICHLAND, WASH.	ISO-
ISOCHEM, INC., RICHLAND, WASH.	ISO-SA-
ISOTOPES, INC., PALO ALTO, CALIF.	HNS-
ISOTOPES, INC., WESTWOOD, N.J.	ISO-
ISOTOPES, INC., WESTWOOD, N.J. (PROGRESS REPORT)	ISOTOPES-QR-
ISOTOPES, INC., WESTWOOD, N.J.	IWL-
ISOTOPES, INC. NUCLEAR SYSTEMS DIV., MIDDLE RIVER, MD	INSD-
ISOTOPES, INC. PALO ALTO LABS., CALIF.	PALTR-
ISRAEL. ATOMIC ENERGY COMMISSION (ALL LOCATIONS)	IA-
ISRAEL. ATOMIC ENERGY COMMISSION, TEL-AVIV	LS-
ISRAEL. ATOMIC ENERGY COMMISSION, TEL-AVIV	PUB/UP/(NO.)-
ISRAEL. ATOMIC ENERGY COMMISSION, TEL-AVIV	PUB/UP/R-
ISRAEL. ATOMIC ENERGY COMMISSION, TEL-AVIV	SM-
ISRAEL. ATOMIC ENERGY COMMISSION. NUCLEAR RESEARCH CENTER-NEGEV, BEERSHEBA	NRCN-
ISRAEL. ATOMIC ENERGY COMMISSION. SOREQ NUCLEAR RESEARCH CENTER, YAVNE	LS-
ISRAEL. DEFENSE RESEARCH ESTABLISHMENT	DRE-
ISRAEL. NEGEV INST. FOR ARID ZONE RESEARCH. DEPT. OF GEOBOTANY, BEERSHEBA	G-(NUMBER)/(YEAR)
ISRAEL. NEGEV INST. FOR ARID ZONE RESEARCH. DEPT. OF PLANT PHYSIOLOGY, BEERSHEBA	P-(NUMBER)/(YEAR)
ISRAEL PROGRAM FOR SCI. TRANS., LTD., JERUSALEM	IPST-
ISRAEL PROGRAM FOR SCI. TRANS., LTD., JERUSALEM	IPST-CAT.-
ISRAEL PROGRAM FOR SCI. TRANS., LTD., JERUSALEM	TT-(YEAR)-
	CM-R-
ITALY. CENTRO INFORMAZIONI STUDI ESPERIENZE, MILAN	CISE-
ITALY. CENTRO INFORMAZIONI STUDI ESPERIENZE, MILAN	CISE-E-
ITALY. CENTRO INFORMAZIONI STUDI ESPERIENZE, MILAN	CISE-N-
ITALY. CENTRO INFORMAZIONI STUDI ESPERIENZE, MILAN	CISE-R-
ITALY. CENTRO NAZIONALE DI FISICA DELL ATMOSFERA E METEOROLOGIA, ROME	CENFAM-PV-
ITALY. CENTRO NAZIONALE DI FISICA DELL ATMOSFERA E METEOROLOGIA, ROME	CENFAM-RDP-
ITALY. COMITATO NAZIONALE PER LE RICERCHE NUCLEARI, ROME	CNBIB-
ITALY. COMITATO NAZIONALE PER LE RICERCHE NUCLEARI, ROME	CNC-
ITALY. COMITATO NAZIONALE PER LE RICERCHE NUCLEARI, ROME	CNG-
ITALY. COMITATO NAZIONALE PER LE RICERCHE NUCLEARI, ROME	CNTR-
ITALY. COMITATO NAZIONALE PER L'ENERGIA NUCLEARE, BOLOGNA	ARV (YEAR)-
ITALY. COMITATO NAZIONALE PER L'ENERGIA NUCLEARE, BOLOGNA	CEC-
ITALY. COMITATO NAZIONALE PER L'ENERGIA NUCLEARE, BOLOGNA	CNEA-CEC(YEAR)(NUMBER)
ITALY. COMITATO NAZIONALE PER L'ENERGIA NUCLEARE, BOLOGNA	CNEN-PRV-R-(YEAR)-
ITALY. COMITATO NAZIONALE PER L'ENERGIA NUCLEARE, BOLOGNA	CNEN-RVA-(YEAR)-
ITALY. COMITATO NAZIONALE PER L'ENERGIA NUCLEARE, BOLOGNA	CNEN-RVT-PEC-
ITALY. COMITATO NAZIONALE PER L'ENERGIA NUCLEARE, BOLOGNA	CNEN-RVT-SEC-(YEAR)-
ITALY. COMITATO NAZIONALE PER L'ENERGIA NUCLEARE, BOLOGNA	CNEN-RVT-SIN-(YEAR)-
ITALY. COMITATO NAZIONALE PER L'ENERGIA NUCLEARE, BOLOGNA	CNEN-RVT-SIR-(YEAR)-
ITALY. COMITATO NAZIONALE PER L'ENERGIA NUCLEARE, BOLOGNA	PRV-
ITALY. COMITATO NAZIONALE PER L'ENERGIA NUCLEARE, BOLOGNA	RVA(YEAR)-
ITALY. COMITATO NAZIONALE PER L'ENERGIA NUCLEARE, BOLOGNA	RVT-PEC-
ITALY. COMITATO NAZIONALE PER L'ENERGIA NUCLEARE, BOLOGNA	RVT-SEC(YEAR)-
ITALY. COMITATO NAZIONALE PER L'ENERGIA NUCLEARE, BOLOGNA	RVT-SIN(YEAR)-
ITALY. COMITATO NAZIONALE PER L'ENERGIA NUCLEARE. BOLOGNA	RVT-SIR(YEAR)-
ITALY. COMITATO NAZIONALE PER L'ENERGIA NUCLEARE, ISPRA	CNI-
ITALY. COMITATO NAZIONALE PER L'ENERGIA NUCLEARE, ISPRA	GN-
ITALY. COMITATO NAZIONALE PER L'ENERGIA NUCLEARE, ISPRA	RT/CNI(YEAR)-
ITALY. COMITATO NAZIONALE PER L'ENERGIA NUCLEARE,ROME	BIO/(NUMBER)/(YEAR)
ITALY. COMITATO NAZIONALE PER L'ENERGIA NUCLEARE,ROME	CNEN-
ITALY. COMITATO NAZIONALE PER L'ENERGIA NUCLEARE,ROME	CNEN-RT/B-
ITALY. COMITATO NAZIONALE PER L'ENERGIA NUCLEARE,ROME	CNEN-RT/BIO-
ITALY. COMITATO NAZIONALE PER L'ENERGIA NUCLEARE,ROME	CNEN-RT/EL-(YEAR)-
ITALY. COMITATO NAZIONALE PER L'ENERGIA NUCLEARE,ROME	CNEN-RT/FI-
ITALY. COMITATO NAZIONALE PER L'ENERGIA NUCLEARE,ROME	CNEN-RT/FIMA-(YEAR)-
ITALY. COMITATO NAZIONALE PER L'ENERGIA NUCLEARE,ROME	CNEN-RT/ING-(YEAR)-
ITALY. COMITATO NAZIONALE PER L'ENERGIA NUCLEARE,ROME	CNEN-RT/MET-(YEAR)-
ITALY. COMITATO NAZIONALE PER L'ENERGIA NUCLEARE,ROME	CNEN-RT/PROT-(YEAR)-
ITALY. COMITATO NAZIONALE PER L'ENERGIA NUCLEARE,ROME	DAISE-(YEAR)-
ITALY. COMITATO NAZIONALE PER L'ENERGIA NUCLEARE,ROME	LTEC/PR/
ITALY. COMITATO NAZIONALE PER L'ENERGIA NUCLEARE,ROME	PROT-SAN.-
ITALY. COMITATO NAZIONALE PER L'ENERGIA NUCLEARE,ROME	PRV-R-(YEAR)(NUMBER)
ITALY. COMITATO NAZIONALE PER L'ENERGIA NUCLEARE, ROME (BIOLOGY AND AGRICULTURE)	RT/BIO(YEAR)-
ITALY. COMITATO NAZIONALE PER L'ENERGIA NUCLEARE, ROME (CHEMISTRY)	RT/CHI(YEAR)-
ITALY. COMITATO NAZIONALE PER L'ENERGIA NUCLEARE,ROME	RT/EC-
ITALY. COMITATO NAZIONALE PER L'ENERGIA NUCLEARE, ROME (ELECTRONICS)	RT/EL(YEAR)-
ITALY. COMITATO NAZIONALE PER L'ENERGIA NUCLEARE, ROME (PHYSICS)	RT/FI(YEAR)-
ITALY. COMITATO NAZIONALE PER L'ENERGIA NUCLEARE, ROME (PHYSICS AND MATHEMATICS)	RT/FIMA(YEAR)-
ITALY. COMITATO NAZIONALE PER L'ENERGIA NUCLEARE,ROME	RT/GEN(YEAR)-
ITALY. COMITATO NAZIONALE PER L'ENERGIA NUCLEARE, ROME (GEOLOGY)	RT/GEO(YEAR)-
ITALY. COMITATO NAZIONALE PER L'ENERGIA NUCLEARE,ROME	RT/GIU(YEAR)-
ITALY. COMITATO NAZIONALE PER L'ENERGIA NUCLEARE,ROME	RTI/FI(YEAR)-

ITALY. COMITATO NAZIONALE PER L'ENERGIA NUCLEARE, ROME	RTI/ING(YEAR)-
ITALY. COMITATO NAZIONALE PER L'ENERGIA NUCLEARE, ROME	RT/ING-(YEAR)-
ITALY. COMITATO NAZIONALE PER L'ENERGIA NUCLEARE, ROME (ENGINEERING)	RT/ING(YEAR)-
ITALY. COMITATO NAZIONALE PER L'ENERGIA NUCLEARE, ROME (METALLURGY)	RT/MET(YEAR)-
ITALY. COMITATO NAZIONALE PER L'ENERGIA NUCLEARE, ROME (SAFETY AND PROTECTION)	RT/PROT(YEAR)-
ITALY. COMITATO NAZIONALE PER L'ENERGIA NUCLEARE, ROME	RVP-UTS/(YEAR)/
ITALY. COMITATO NAZIONALE PER L'ENERGIA NUCLEARE. DIVISIONE DI BIOLOGIA E DE PROTEZIONE SANITARIA, ROME	CNB-
ITALY. COMITATO NAZIONALE PER L'ENERGIA NUCLEARE. LABORATORI NAZIONALI DI FRASCATI	CNEN-IEA-
ITALY. COMITATO NAZIONALE PER L'ENERGIA NUCLEARE. LABORATORI NAZIONALI DI FRASCATI	CNEN-IN-
ITALY. COMITATO NAZIONALE PER L'ENERGIA NUCLEARE. LABORATORI NAZIONALI DI FRASCATI	CNF-
ITALY. COMITATO NAZIONALE PER L'ENERGIA NUCLEARE. LABORATORI NAZIONALI DI FRASCATI	CNT-
ITALY. COMITATO NAZIONALE PER L'ENERGIA NUCLEARE. LABORATORI NAZIONALI DI FRASCATI	LMF-
ITALY. COMITATO NAZIONALE PER L'ENERGIA NUCLEARE. LABORATORI NAZIONALI DI FRASCATI	LNF-
ITALY. COMITATO NAZIONALE PER L'ENERGIA NUCLEARE. LABORATORI NAZIONALI DI FRASCATI	LNF-(YEAR/NUMBER)
ITALY. COMITATO NAZIONALE PER L'ENERGIA NUCLEARE. LABORATORI NAZIONALI DI FRASCATI	NLF-(YEAR/NUMBER)
ITALY. COMITATO NAZIONALE PER L'ENERGIA NUCLEARE. LABORATORIO GAS IONIZZATI, FRASCATI	LGI-
ITALY. COMITATO NAZIONALE PER L'ENERGIA NUCLEARE. LABORATORIO GAS IONIZZATI, FRASCATI	LGI-(YEAR/NUMBER)
ITALY. CONSIGLIO NAZIONALE DELLE RICERCHE, ROME	CNR-
ITALY. CONSIGLIO NAZIONALE DELLE RICERCHE, ROME	IDAC-RT-
ITALY. CONSIGLIO NAZIONALE DELLE RICERCHE, ROME	IDAC-TR-
ITALY. CONSIGLIO NAZIONALE DELLE RICERCHE. ISTITUTO DI ACUSTICA, ROME	IDAC-RT-
ITALY. CONSIGLIO NAZIONALE DELLE RICERCHE. ISTITUTO DI FISICA DELL'ATMOSFERA, TURIN	IFA-RDP-
ITALY. CONSIGLIO NAZIONALE DELLE RICERCHE. ISTITUTO DI FISICA DELL'ATMOSFERA, TURIN	IFA-TR-
ITALY. ISTITUTO DE FISICA DELL'ATMOSFERA, ROME	IFA-CP-
ITALY. ISTITUTO DE FISICA DELL'ATMOSFERA, ROME	IFA-SR-
ITALY. ISTITUTO DE FISICA DELL'ATMOSFERA, ROME	IFA-STR-
ITALY. ISTITUTO ELETTRONTECNICO NAZIONALE, TURIN	M-
ITALY. ISTITUTO NAZIONALE DI FISICA NUCLEARE (ALL LOCATIONS)	INFN/BE-(YEAR)/
ITALY. ISTITUTO NAZIONALE DI FISICA NUCLEARE, BOLOGNA	INFN/TC-
ITALY. ISTITUTO NAZIONALE DI FISICA NUCLEARE, FLORENCE	INFN-TC-(YEAR)/
ITALY. ISTITUTO NAZIONALE DI FISICA NUCLEARE, PADUA	IFPTH-(MONTH/YEAR)
ITALY. ISTITUTO NAZIONALE DI FISICA NUCLEARE, PADUA	IFPTK-(MONTH/YEAR)
ITALY. ISTITUTO NAZIONALE DI FISICA NUCLEARE, PADUA	INFN-PD-(YEAR)/
ITALY. ISTITUTO NAZIONALE DI FISICA NUCLEARE, PADUA	INFN/RE-
ITALY. ISTITUTO NAZIONALE DI FISICA NUCLEARE, TRIESTE	INFN/AE-(YEAR)/
ITALY. ISTITUTO NAZIONALE DI ULTRACUSTICA, ROME	FTR-
ITALY. ISTITUTO NAZIONALE DI ULTRACUSTICA, ROME	INUA-RT-N-
ITALY. ISTITUTO NAZIONALE PER LE APPLICAZIONI DEL CALCOLO, ROME	INAC-PUB-
ITALY. ISTITUTO NAZIONALE PER STUDI ED ESPERIENZE DI ARCHITETTURA NAVALE, ROME	INSEAN-QUADERNO-
ITALY. ISTITUTO SUPERIORE DI SANITA, ROME	ISS-
ITALY. ISTITUTO SUPERIORE DI SANITA, ROME	ISS (YEAR/NUMBER)
ITALY. ISTITUTO UNIVERSITARIO NAVALE, NAPLES	ASR-
ITEK CORP., LEXINGTON, MASS.	ITEK-
ITEK CORP., LEXINGTON, MASS.	OR-
ITEK CORP. INFO. SCIENCES LAB., LEXINGTON, MASS.	ISL-
ITEK CORP. OPTICAL SYSTEMS DIV., LEXINGTON, MASS.	PFR-(YEAR)-
ITEK CORP. VIDYA DIV., PALO ALTO, CALIF.	ITEK-
ITEK CORP. VIDYA DIV., PALO ALTO, CALIF.	VIDYA-TR-
ITT COMMUNICATION SYSTEMS, INC., PARAMUS, N.J.	ICS-
ITT COMMUNICATION SYSTEMS, INC., PARAMUS, N.J.	ICS-(NUMBER)-TR-
ITT COMMUNICATION SYSTEMS, INC., PARAMUS, N.J.	ICS-(YR)-TR-(NO.)-
ITT COMMUNICATION SYSTEMS, INC., PARAMUS, N.J.	ITTCS-
ITT DEFENSE COMMUNICATIONS DIV., NUTLEY, N.J.	ITTDC-(YEAR)-
ITT ELECTRO-PHYSICS LABS., INC., HYATTSVILLE, MD.	CB-
ITT ELECTRO-PHYSICS LABS., INC., HYATTSVILLE, MD.	ETM-
ITT ELECTRO-PHYSICS LABS., INC., HYATTSVILLE, MD.	ITT-EPL-
ITT FEDERAL ELECTRIC CORP., PARAMUS, N.J.	FE-
ITT FEDERAL ELECTRIC CORP., PARAMUS, N.J.	ND-
ITT FEDERAL LABS., NUTLEY, N.J.	ITTFL-
IFF FEDERAL LABS., NUTLEY, N.J.	ITTFL-TD-
ITT FEDERAL LABS., NUTLEY, N.J.	TD-
ITT FEDERAL LABS., NUTLEY, N.J.	TD-(YEAR)-
ITT GILFILLAN, INC., LOS ANGELES	GIO-(YEAR)/
ITT GILFILLAN, INC., LOS ANGELES	SDC/
ITT INDUSTRIAL LABS., FORT WAYNE	ITTFL-(YEAR)-
ITT INDUSTRIAL LABS., FORT WAYNE	ITTIL-(YEAR)-
ITT INDUSTRIAL LABS., FORT WAYNE	S-
ITT LABS., NUTLEY, N.J.	ITT-TM-

J. G. ENGINEERING RESEARCH ASSOCIATES, BALTIMORE	JGERA-
JACKSON AND MORELAND, BOSTON	J+M-
JACOBS INSTRUMENT CO., BETHESDA, MD.	JAC-
JAGIELLONIAN UNIV., KRAKOW	INP-(NUMBER)/PH
JANSKY + BAILEY, WASHINGTON, D.C.	JBI-
JAPAN	J-
JAPAN	M-
JAPAN. ATOMIC ENERGY RESEARCH INSTITUTE, TOKYO	JAERI-
JAPAN. ATOMIC ENERGY RESEARCH INSTITUTE, TOKYO	JAERI-MEMO-
JAPAN. ATOMIC ENERGY RESEARCH INSTITUTE, TOKYO	NSJ-TR-
JAPAN. ATOMIC ENERGY RESEARCH INSTITUTE, TOKYO	SJC-A-(NUMBER)-(NUMBER)
JAPAN. CENTRAL AERONAUTICAL RESEARCH INSTITUTE	CARI-
JAPAN. CENTRAL AERONAUTICAL RESEARCH INSTITUTE	CARI (LETTERS)-
JAPAN. INDUSTRIAL SCIENCE + TECHNOLOGY AGENCY, TOKYO	CIRC-
JAPAN. INSTITUTE OF PHYSICAL AND CHEMICAL RES., TOKYO	IEA-
JAPAN. INSTITUTE OF PHYSICAL AND CHEMICAL RES., TOKYO	NSJ-
JAPAN. NATIONAL AEROSPACE LAB., TOKYO	NAL-TR-
JAPAN. NATIONAL INST. OF RADIOLOGICAL SCIENCES, CHIBA	NIRS-
JAPAN. NATIONAL INST. OF RADIOLOGICAL SCIENCES, CHIBA	NIRS-AR-
JAPAN. NATIONAL INST. OF RADIOLOGICAL SCIENCES, CHIBA	NIRS-PU-
JAPAN. NATIONAL INST. OF RADIOLOGICAL SCIENCES, CHIBA	NIRS-RSD-
JAPAN BROADCASTING CORP., TOKYO	NHK-
JAPAN NUCLEAR SHIP DEVELOPMENT AGENCY, TOKYO	JNS-
JEFFERSON PROVING GROUND, MADISON, IND.	JPG-
JERSEY PRODUCTION RESEARCH CO., TULSA	JPR-
JIKEI UNIV., TOKYO. SCHOOL OF MEDICINE	J-
JOHANNESBURG. UNIV. OF THE WITWATERSRAND. DEPT. OF MECHANICAL ENGINEERING	REP-(NO.)/(YR.)
JOHN B. PIERCE FOUNDATION, RARITAN, N.J.	TR-HT-
JOHN CARROLL UNIV., CLEVELAND	JCU-
JOHN CARROLL UNIV., CLEVELAND	RRL-(NUMBER)-(NUMBER)
JOHN CRERAR LIBRARY, CHICAGO	JCL-
JOHN F. KENNEDY CTR. FOR SPECIAL WARFARE, FT. BRAGG, NC	USAJFKSWC-
JOHN F. KENNEDY SPACE CENTER, COCOA BEACH, FLA.	K-IB-(NO.)(NO.)/
JOHN F. KENNEDY SPACE CENTER, COCOA BEACH, FLA. (KENNEDY MANAGEMENT INSTRUCTIONS)	KMI-
JOHN F. KENNEDY SPACE CENTER, COCOA BEACH, FLA.	K-PA-
JOHN F. KENNEDY SPACE CENTER, COCOA BEACH, FLA.	KSC-
JOHN F. KENNEDY SPACE CENTER, COCOA BEACH, FLA.	KSC-(LETTERS)-
JOHN F. KENNEDY SPACE CENTER, COCOA BEACH, FLA. (GENERAL PUBLICATION)	KSC-GP-
JOHN F. KENNEDY SPACE CENTER, COCOA BEACH, FLA. (ADMINISTRATIVE MANAGEMENT)	KSC-K-AM-
JOHN F. KENNEDY SPACE CENTER, COCOA BEACH, FLA.	K-V-
JOHN F. KENNEDY SPACE CENTER, COCOA BEACH, FLA.	NASA-KSC-
JOHN F. KENNEDY SPACE CENTER, COCOA BEACH, FLA. (DESIGN TECHNICAL INSTRUCTION)	NASA-KSC-DTI-
JOHN F. KENNEDY SPACE CENTER, COCOA BEACH, FLA.	NASA-KSC-MAB-(NO.-YR.)
JOHN F. KENNEDY SPACE CENTER, COCOA BEACH, FLA.	NASA-KSC-TEST-
JOHN F. KENNEDY SPACE CENTER, COCOA BEACH, FLA.	NASA-KSC-TM-
JOHN F. KENNEDY SPACE CENTER, COCOA BEACH, FLA.	NASA-KSC-TR-
JOHNS HOPKINS UNIV., ANNAPOLIS. CHESAPEAKE BAY INST.	REF-(YEAR)-
JOHNS HOPKINS UNIV., BALTIMORE	(NUMBERS...)
JOHNS HOPKINS UNIV., BALTIMORE	CBL/64/IMA-
JOHNS HOPKINS UNIV., BALTIMORE	CBL-JHU-
JOHNS HOPKINS UNIV., BALTIMORE	JHU-
JOHNS HOPKINS UNIV., BALTIMORE (SQUID PROJ.)(REF. 33)	JHU-(NUMBER)-(LETTER(S))
JOHNS HOPKINS UNIV., BALTIMORE	JHU(INITIAL)-
JOHNS HOPKINS UNIV., BALTIMORE (SQUID PROJ.)(REF. 33)	JHU-SQUID-
JOHNS HOPKINS UNIV., BALTIMORE	JHU-SR-
JOHNS HOPKINS UNIV., BALTIMORE	JHU-TR-
JOHNS HOPKINS UNIV., BALTIMORE (SYMPOSIUM ON SCALE AND NONDESTRUCTIVE TESTING)	JHUX-
JOHNS HOPKINS UNIV., BALTIMORE	S/SNDT-
JOHNS HOPKINS UNIV., BALTIMORE (TERRIER TEST PROCEDURE)	TT-
JOHNS HOPKINS UNIV., BALTIMORE	TTP-
JOHNS HOPKINS UNIV., BALTIMORE	W-
JOHNS HOPKINS UNIV., BALTIMORE. BALLISTIC ANAL. LAB.	BAL-TM-
JOHNS HOPKINS UNIV., BALTIMORE. BALLISTIC ANAL. LAB.	BAL-TR-
JOHNS HOPKINS UNIV., BALTIMORE. CARLYLE BARTON LAB.	AF-
JOHNS HOPKINS UNIV., BALTIMORE. CARLYLE BARTON LAB.	AF-(NUMBER)-
JOHNS HOPKINS UNIV., BALTIMORE. CARLYLE BARTON LAB.	WR-AF-
JOHNS HOPKINS UNIV., BALTIMORE. CENTER FOR RESEARCH IN SCIENTIFIC COMMUNICATION	JHU-CRSC-
JOHNS HOPKINS UNIV., BALTIMORE. CHESAPEAKE BAY INST.	CBI/JHU-
JOHNS HOPKINS UNIV., BALTIMORE. DEPT. OF AERONAUTICS	BUMBLEBEE-
JOHNS HOPKINS UNIV., BALTIMORE. DEPT. OF AERONAUTICS	CM-
JOHNS HOPKINS UNIV., BALTIMORE. DEPT. OF MECH. ENG.	JHU-DME-
JOHNS HOPKINS UNIV., BALTIMORE. DEPT. OF MECHANICS	TR-(NUMBER)SER-(LETTER)
JOHNS HOPKINS UNIV., BALTIMORE. INST. FOR COOP. RES.	CRL-S-
JOHNS HOPKINS UNIV., BALTIMORE. INST. FOR COOP. RES.	ICR-
JOHNS HOPKINS UNIV., BALTIMORE. INST. FOR COOP. RES.	JHU-ICR-
JOHNS HOPKINS UNIV., BALTIMORE. INST. FOR COOP. RES.	PRU-
JOHNS HOPKINS UNIV., BALTIMORE. PSYCHOLOGICAL LAB.	JHU-PL-

JOHNS HOPKINS UNIV., BALTIMORE. RADIATION LAB.	AF-
JOHNS HOPKINS UNIV., BALTIMORE. RADIATION LAB.	IMA-
JOHNS HOPKINS UNIV., BALTIMORE. RADIATION LAB.	JHU-AF-
JOHNS HOPKINS UNIV., BALTIMORE. RADIATION LAB.	JHU-RL-
JOHNS HOPKINS UNIV., BALTIMORE. RADIATION LAB.	JHU RLAF-
JOHNS HOPKINS UNIV., BALTIMORE. RADIATION LAB.	JHU/RL-TR-AF-
JOHNS HOPKINS UNIV., BALTIMORE. RADIATION LAB.	RL/JHU/TR-
JOHNS HOPKINS UNIV., BALTIMORE. RADIATION LAB.	WR-AF-
JOHNS HOPKINS UNIV., BALTIMORE. SCHOOL OF ENGINEERING	AERO/JHU-
JOHNS HOPKINS UNIV., BALTIMORE. SCHOOL OF ENGINEERING	AERO/JHU-CM-
JOHNS HOPKINS UNIV.,BETHESDA, MD. OPERATIONS RES.OFF.	OPSEARCH-
JOHNS HOPKINS UNIV.,BETHESDA, MD. OPERATIONS RES.OFF.	ORO-
JOHNS HOPKINS UNIV.,BETHESDA, MD. OPERATIONS RES.OFF.	ORO-SP-
JOHNS HOPKINS UNIV.,BETHESDA, MD. OPERATIONS RES.OFF.	ORO-T-
JOHNS HOPKINS UNIV.,BETHESDA, MD. OPERATIONS RES.OFF.	ORO-TP-
(SEE ALSO SUCCESSOR: RESEARCH ANALYSIS CORP.)	
JOHNS HOPKINS UNIV. CHEVY CHASE, MD. OPERATIONS RESEARCH OFFICE	JHU ORO R-
JOHNS HOPKINS UNIV. CHEVY CHASE, MD. OPERATIONS RESEARCH OFFICE. (STAFF MEMORANDA)	JHU ORO S-
JOHNS HOPKINS UNIV. CHEVY CHASE, MD. OPERATIONS RESEARCH OFFICE. (TECHNICAL MEMORANDA)	JHU ORO T-
JOHNS HOPKINS UNIV. CHEVY CHASE, MD. OPERATIONS RESEARCH OFFICE	ORO-JHU-
JOHNS HOPKINS UNIV. CHEVY CHASE, MD. OPERATIONS RESEARCH OFFICE	ORO-LOG.-
JOHNS HOPKINS UNIV. CHEVY CHASE, MD. OPERATIONS RESEARCH OFFICE	ORO-R-
JOHNS HOPKINS UNIV. CHEVY CHASE, MD. OPERATIONS RESEARCH OFFICE	ORO-S-
(SEE ALSO LATER LOCATION, BETHESDA, MD., AND SUCCESSOR: RESEARCH ANALYSIS CORP.)	
JOHNS HOPKINS UNIV., SEABROOK, N.J. (PUBLICATIONS IN CLIMATOLOGY)	JHU-PC-
JOHNS HOPKINS UNIV., SILVER SPRING, MD.	TAT-
JOHNS HOPKINS UNIV., SILVER SPRING, MD. APPLIED PHYSICS LAB.	APL-
JOHNS HOPKINS UNIV., SILVER SPRING, MD. APPLIED PHYSICS LAB.	APL-AQR/(YEAR)-
JOHNS HOPKINS UNIV., SILVER SPRING, MD. APPLIED PHYSICS LAB. (BUMBLEBEE PROJECT) (REF. 8)	APL-BB-
JOHNS HOPKINS UNIV., SILVER SPRING, MD. APPLIED PHYSICS LAB.	APL-BBW-
JOHNS HOPKINS UNIV., SILVER SPRING, MD. APPLIED PHYSICS LAB.	APL-CF-
JOHNS HOPKINS UNIV., SILVER SPRING, MD. APPLIED PHYSICS LAB.	APL-CLB-
JOHNS HOPKINS UNIV., SILVER SPRING, MD. APPLIED PHYSICS LAB.	APL-CM-
JOHNS HOPKINS UNIV., SILVER SPRING, MD. APPLIED PHYSICS LAB.	APL-C-RQR/(YEAR)-
JOHNS HOPKINS UNIV., SILVER SPRING, MD. APPLIED PHYSICS LAB.	APL-DR-
JOHNS HOPKINS UNIV., SILVER SPRING, MD. APPLIED PHYSICS LAB.	APL-E-
JOHNS HOPKINS UNIV., SILVER SPRING, MD. APPLIED PHYSICS LAB.	APL-JHU-
JOHNS HOPKINS UNIV., SILVER SPRING, MD. APPLIED PHYSICS LAB. (BUMBLEBEE PROJECT) (REF. 8)	APL/JHU-BB-
JOHNS HOPKINS UNIV., SILVER SPRING, MD. APPLIED PHYSICS LAB.	APL-JHU-BLP-
JOHNS HOPKINS UNIV., SILVER SPRING, MD. APPLIED PHYSICS LAB.	APL-JHU-BR-
JOHNS HOPKINS UNIV., SILVER SPRING, MD. APPLIED PHYSICS LAB.	APL-JHU-CF-
JOHNS HOPKINS UNIV., SILVER SPRING, MD. APPLIED PHYSICS LAB.	APL-JHU-CLA-RD-
JOHNS HOPKINS UNIV., SILVER SPRING, MD. APPLIED PHYSICS LAB.	APL-JHU-CM-
JOHNS HOPKINS UNIV., SILVER SPRING, MD. APPLIED PHYSICS LAB.	APL-JHU-CPR-
JOHNS HOPKINS UNIV., SILVER SPRING, MD. APPLIED PHYSICS LAB.	APL-JHU-HWL-
JOHNS HOPKINS UNIV., SILVER SPRING, MD. APPLIED PHYSICS LAB.	APL-JHU-PR-
JOHNS HOPKINS UNIV., SILVER SPRING, MD. APPLIED PHYSICS LAB.	APL-JHU-TAT-
JOHNS HOPKINS UNIV., SILVER SPRING, MD. APPLIED PHYSICS LAB. (TRANSPORTATION CONTRACTOR REPORT)	APL-JHU-TCR-
JOHNS HOPKINS UNIV., SILVER SPRING, MD. APPLIED PHYSICS LAB.	APL-JHU-TG-
JOHNS HOPKINS UNIV., SILVER SPRING, MD. APPLIED PHYSICS LAB.	APL-JHU-TPR-
JOHNS HOPKINS UNIV., SILVER SPRING, MD. APPLIED PHYSICS LAB. (TARTAR MISSILE)	APL-JHU-TT-
JOHNS HOPKINS UNIV., SILVER SPRING, MD. APPLIED PHYSICS LAB.	APL-S-
JOHNS HOPKINS UNIV., SILVER SPRING, MD. APPLIED PHYSICS LAB.	APL-SDO-
JOHNS HOPKINS UNIV., SILVER SPRING, MD. APPLIED PHYSICS LAB.	APL-SMS-FS-
JOHNS HOPKINS UNIV., SILVER SPRING, MD. APPLIED PHYSICS LAB.	APL-SQR-(YEAR)-
JOHNS HOPKINS UNIV., SILVER SPRING, MD. APPLIED PHYSICS LAB.	APL-SQR-CAN-(NO.)-(YEAR)
JOHNS HOPKINS UNIV., SILVER SPRING, MD. APPLIED PHYSICS LAB.	APL-SQR-MPCS-
JOHNS HOPKINS UNIV., SILVER SPRING, MD. APPLIED PHYSICS LAB.	APL-TG-
JOHNS HOPKINS UNIV., SILVER SPRING, MD. APPLIED PHYSICS LAB.	APL-TM-
JOHNS HOPKINS UNIV., SILVER SPRING, MD. APPLIED PHYSICS LAB. (TRANSPORTATION PROGRESS REPORT)	APL-TPR-
JOHNS HOPKINS UNIV., SILVER SPRING, MD. APPLIED PHYSICS LAB. (TRANSLATION)	APL-TRANS-
JOHNS HOPKINS UNIV., SILVER SPRING, MD. APPLIED PHYSICS LAB.	APL-U-RQR/(YEAR)-
JOHNS HOPKINS UNIV., SILVER SPRING, MD. APPLIED PHYSICS LAB.	APL-U-SQR-(YEAR)-
JOHNS HOPKINS UNIV., SILVER SPRING, MD. APPLIED PHYSICS LAB.	APL-WQR/(YEAR)-
JOHNS HOPKINS UNIV., SILVER SPRING, MD. APPLIED PHYSICS LAB. (AERONAUTICS DIV. QUARTERLY REPT.)	AQR/(YEAR)-
JOHNS HOPKINS UNIV., SILVER SPRING, MD. APPLIED PHYSICS LAB. (BUMBLEBEE PROJECT) (REF. 8)	BB-
JOHNS HOPKINS UNIV., SILVER SPRING, MD. APPLIED PHYSICS LAB. (BUMBLEBEE PROJECT) (REF. 8)	BUMBLEBEE-
JOHNS HOPKINS UNIV., SILVER SPRING, MD. APPLIED PHYSICS LAB. (BUMBLEBEE PROJECT) (REF. 8)	CF-
JOHNS HOPKINS UNIV., SILVER SPRING, MD. APPLIED PHYSICS LAB.	CLA-
JOHNS HOPKINS UNIV., SILVER SPRING, MD. APPLIED PHYSICS LAB.	CLB-
JOHNS HOPKINS UNIV., SILVER SPRING, MD. APPLIED PHYSICS LAB. (BUMBLEBEE PROJECT) (REF. 8)	CM-
JOHNS HOPKINS UNIV., SILVER SPRING, MD. APPLIED PHYSICS LAB.	CPIA/M2,REV.-
JOHNS HOPKINS UNIV., SILVER SPRING, MD. APPLIED PHYSICS LAB.	CPR-
JOHNS HOPKINS UNIV., SILVER SPRING, MD. APPLIED PHYSICS LAB. (RES. + DEV. QUARTERLY REPT.)	C-RQR/(YEAR)-
JOHNS HOPKINS UNIV., SILVER SPRING, MD. APPLIED PHYSICS LAB. (SPACE QUARTERLY REPORT)	C-SQR/(YEAR)-
JOHNS HOPKINS UNIV., SILVER SPRING, MD. APPLIED PHYSICS LAB.	DQR/(YEAR)-
JOHNS HOPKINS UNIV., SILVER SPRING, MD. APPLIED PHYSICS LAB.	EQR/(YR)-
JOHNS HOPKINS UNIV., SILVER SPRING, MD. APPLIED PHYSICS LAB.	ETR-
JOHNS HOPKINS UNIV., SILVER SPRING, MD. APPLIED PHYSICS LAB.	FIN-
JOHNS HOPKINS UNIV., SILVER SPRING, MD. APPLIED PHYSICS LAB.	FQR/(YEAR)-
JOHNS HOPKINS UNIV., SILVER SPRING, MD. APPLIED PHYSICS LAB.	FR-
JOHNS HOPKINS UNIV., SILVER SPRING, MD. APPLIED PHYSICS LAB.	GQR/(YR)-
JOHNS HOPKINS UNIV., SILVER SPRING, MD. APPLIED PHYSICS LAB.	JHU-APL-
JOHNS HOPKINS UNIV., SILVER SPRING, MD. APPLIED PHYSICS LAB. (BUMBLEBEE PROJECT) (REF. 8)	JHU APL BB-
JOHNS HOPKINS UNIV., SILVER SPRING, MD. APPLIED PHYSICS LAB.	JHU APL CF-
JOHNS HOPKINS UNIV., SILVER SPRING, MD. APPLIED PHYSICS LAB.	JHU APL CM-
JOHNS HOPKINS UNIV., SILVER SPRING, MD. APPLIED PHYSICS LAB.	JHU APL TG-
JOHNS HOPKINS UNIV., SILVER SPRING, MD. APPLIED PHYSICS LAB.	NQR/(YEAR)-
JOHNS HOPKINS UNIV., SILVER SPRING, MD. APPLIED PHYSICS LAB.	PQR/(YEAR)-
JOHNS HOPKINS UNIV., SILVER SPRING, MD. APPLIED PHYSICS LAB.	RCF-
JOHNS HOPKINS UNIV., SILVER SPRING, MD. APPLIED PHYSICS LAB.	SDO-
JOHNS HOPKINS UNIV., SILVER SPRING, MD. APPLIED PHYSICS LAB.	SN-
JOHNS HOPKINS UNIV., SILVER SPRING, MD. APPLIED PHYSICS LAB.	SR-(NUMBER)-(NUMBER)
JOHNS HOPKINS UNIV., SILVER SPRING, MD. APPLIED PHYSICS LAB.	TG-
JOHNS HOPKINS UNIV., SILVER SPRING, MD. APPLIED PHYSICS LAB.	TG-(NUMBER)-(NUMBER)
JOHNS HOPKINS UNIV., SILVER SPRING, MD. APPLIED PHYSICS LAB.	TRANS-
JOHNS HOPKINS UNIV., SILVER SPRING, MD. APPLIED PHYSICS LAB.	TT-
JOHNS HOPKINS UNIV., SILVER SPRING, MD. APPLIED PHYSICS LAB.	U-
JOHNS HOPKINS UNIV., SILVER SPRING, MD. APPLIED PHYSICS LAB. (RES. + DEV. QUARTERLY REPT.)	U-RQR/(YEAR)-
JOHNS HOPKINS UNIV., SILVER SPRING, MD. APPLIED PHYSICS LAB. (SPACE QUARTERLY REPORT)	U-SQR/(YEAR)-
JOHNS HOPKINS UNIV., SILVER SPRING, MD. APPLIED PHYSICS LAB. (WEAPONS QUARTERLY REPORT)	WQR/(YEAR)-
JOHNS HOPKINS UNIV., SILVER SPRING, MD. APPLIED PHYSICS LAB. BUMBLEBEE COMMITTEE (REF. 8)	BBC-

JOHNS HOPKINS UNIV., SILVER SPRING, MD. APPLIED PHYSICS LAB. NAVAL INSPECTOR OF ORDNANCE	NODU-
JOHNS HOPKINS UNIV., SILVER SPRING, MD. CHEMICAL PROPULSION INFORMATION AGENCY	CPA-
JOHNS HOPKINS UNIV., SILVER SPRING, MD. CHEMICAL PROPULSION INFORMATION AGENCY	CPIA-
JOHNS HOPKINS UNIV., SILVER SPRING, MD. CHEMICAL PROPULSION INFORMATION AGENCY	CPIA-LTM-
JOHNS HOPKINS UNIV., SILVER SPRING, MD. CHEMICAL PROPULSION INFORMATION AGENCY	CPIA-PUB.-
JOHNS HOPKINS UNIV., SILVER SPRING, MD. LIQUID PROPELLANT INFORMATION AGENCY	LPIA-
JOHNS HOPKINS UNIV., SILVER SPRING, MD. LIQUID PROPELLANT INFORMATION AGENCY	LPIA/D(YR)
JOHNS HOPKINS UNIV., SILVER SPRING, MD. LIQUID PROPELLANT INFORMATION AGENCY	SPIA/LPIA-
JOHNS HOPKINS UNIV., SILVER SPRING, MD. PANEL ON PHYSICAL PROPERTIES OF SOLID PROPELLANTS	PP-
JOHNS HOPKINS UNIV., SILVER SPRING, MD. ROCKET PROPELLANT INFORMATION AGENCY	RPIA-
JOHNS HOPKINS UNIV., SILVER SPRING, MD. SOLID PROPELLANT INFO. AGENCY.(ABSTRACTS OF MANUSCRIPTS AND INFORMAL REPTS. OF THE JANAF PANEL ON ANAL.CHEMISTRY)	APA-
JOHNS HOPKINS UNIV., SILVER SPRING, MD. SOLID PROPELLANT INFORMATION AGENCY	JHU APL SPIA/(L + R)
JOHNS HOPKINS UNIV., SILVER SPRING, MD. SOLID PROPELLANT INFORMATION AGENCY	S-(NUMBER)-(NUMBER)
JOHNS HOPKINS UNIV., SILVER SPRING, MD. SOLID PROPELLANT INFORMATION AGENCY	SPIA-
JOHNS HOPKINS UNIV., SILVER SPRING, MD. SOLID PROPELLANT INFORMATION AGENCY (ABSTRACTS)	SPIA/A-
JOHNS HOPKINS UNIV., SILVER SPRING, MD. SOLID PROPELLANT INFORMATION AGENCY (CURRENT ABSTRACTS)	SPIA/A-
JOHNS HOPKINS UNIV., SILVER SPRING, MD. SOLID PROPELLANT INFORMATION AGENCY	SPIA/LPIA-
JOHNS HOPKINS UNIV., SILVER SPRING, MD. SOLID PROPELLANT INFORMATION AGENCY (MANUALS)	SPIA/M-
JOHNS HOPKINS UNIV., SILVER SPRING, MD. SOLID PROPELLANT INFORMATION AGENCY (PHYSICAL PROPERTIES)	SPIA/PP-
JOHNS HOPKINS UNIV., SILVER SPRING, MD. SOLID PROPELLANT INFORMATION AGENCY	SPIA/S-
JOHNS HOPKINS UNIV., SILVER SPRING, MD. SOLID PROPELLANT INFORMATION AGENCY	SPIA-S(YR.)-
JOHNS HOPKINS UNIV., SILVER SPRING, MD. SOLID PROPELLANT INFORMATION AGENCY (SOLID PROPELLANT SURVEILLANCE PANEL)	SPIA-SPSP-
JOHNS HOPKINS UNIV., SILVER SPRING, MD. SOLID PROPELLANT INFORMATION AGENCY (SOLID PROPELLANT SURVEILLANCE PANEL)	SPSP/
JOHNSON (ARNE) INGENIEURBUERO, STOCKHOLM	CI-
JOHNSTON (HERRICK L.) INC., COLUMBUS, OHIO	HLJ-
JOHNSTON LABS., INC., BALTIMORE	JLI-(NUMBER)-
JOHNSTON LABS., INC., BALTIMORE	JLI-
JOHNSTON LABS., INC., BALTIMORE	JLI-(NUMBER-NUMBER)-
JOHNSTON LABS., INC., COCKEYSVILLE, MD.	JLI-(NUMBER)-
JOHNSTON (WILLIAM H.) LABS., INC., LAFAYETTE, IND.	JLI-
JOHNSTON (WILLIAM H.) LABS., INC., LAFAYETTE, IND.	JLI-(NO.)-(NO.)-(NO.)
JOINT AIR DEFENSE BOARD, ENT AFB, COLO.	JADB-
JOINT AIRBORNE TROOP BOARD, FORT BRAGG, N.C.	DP-
JOINT AIRBORNE TROOP BOARD, FORT BRAGG, N.C.	JATB-
JOINT AIRBORNE TROOP BOARD, FORT BRAGG, N.C.	TP-
JOINT AIRWORTHINESS COMMITTEE, LONDON	JAC-PAPER-
JOINT ARMY-NAVY-AIR FORCE AD HOC PANEL ON PERFORMANCE CALCULATION METHODS AND THERMODYNAMIC DATA	PCMTD-
JOINT ARMY-NAVY-AIR FORCE FUZE COMM., WASH.,D.C.	JANAF-SERIAL-
JOINT ARMY-NAVY-AIR FORCE IGNITABILITY PANEL (REF.13)	JANAF-IP-
JOINT ARMY-NAVY-AIR FORCE SOLID PROPELLANT GROUP (REF. 13)	JANAF-BUL-
JOINT ARMY-NAVY-AIR FORCE SOLID PROPELLANT ROCKET STATIC TEST PANEL (REF. 13)	SPSTP-
JOINT ARMY-NAVY-AIR FORCE SOLID PROPELLANT SURVEILLANCE PANEL (REF. 13)	SPSP-
JOINT ARMY NAVY AIRCRAFT INSTRUMENTATION RESEARCH PROJECT, WASHINGTON, D.C.	JANAIR-(NUMBER)-(NUMBER)-
JOINT ARMY NAVY AIRCRAFT INSTRUMENTATION RESEARCH PROJECT, WASHINGTON, D.C.	JANAIR-
JOINT ARMY NAVY AIRCRAFT INSTRUMENTATION RESEARCH PROJECT, WASHINGTON, D.C.	JANAIR-D(NUMBER)-
JOINT ARMY NAVY AIRCRAFT INSTRUMENTATION RESEARCH PROJECT, WASHINGTON, D.C.	JANAIR-EL-
JOINT ARMY NAVY AIRCRAFT INSTRUMENTATION RESEARCH PROJECT, WASHINGTON, D.C.	JANAIR-TR-
JOINT ARMY NAVY AIRCRAFT INSTRUMENTATION RESEARCH PROJECT, WASHINGTON, D.C.	JANAIR-TR-D(NUMBERS)-
JOINT ATOMIC ENERGY INTELLIGENCE COMM., WASH. D.C.	JAEIC-
JOINT ATOMIC INFORMATION EXCHANGE GROUP, WASH., D.C.	JAIEG-
JOINT ATOMIC WEAPONS PUBLICATIONS BOARD, ALBUQUERQUE, N. MEX. (REF. 14)	AEC-DASA-TP-
JOINT ATOMIC WEAPONS PUBLICATIONS BOARD, ALBUQUERQUE, N. MEX. (REF. 14)	AEC-TP-
JOINT ATOMIC WEAPONS PUBLICATIONS BOARD, ALBUQUERQUE, N. MEX. (REF. 14)	AF T.O.(NUMBERS)-11N
JOINT ATOMIC WEAPONS PUBLICATIONS BOARD, ALBUQUERQUE, N. MEX. (REF. 14)	AF T.O.-11N-
JOINT ATOMIC WEAPONS PUBLICATIONS BOARD, ALBUQUERQUE, N. MEX. (REF. 14)	AIR FORCE T.O.(NOS.)-11N
JOINT ATOMIC WEAPONS PUBLICATIONS BOARD, ALBUQUERQUE, N. MEX. (REF. 14)	AIR FORCE T.O.-11N-
JOINT ATOMIC WEAPONS PUBLICATIONS BOARD, ALBUQUERQUE, N. MEX. (REF. 14)	ARMY-TM-39-
JOINT ATOMIC WEAPONS PUBLICATIONS BOARD, ALBUQUERQUE, N. MEX. (REF. 14)	DASA-TP-
JOINT ATOMIC WEAPONS PUBLICATIONS BOARD, ALBUQUERQUE, N. MEX. (REF. 14)	JAWPB-
JOINT ATOMIC WEAPONS PUBLICATIONS BOARD, ALBUQUERQUE, N. MEX. (REF. 14)	NAVORD SWOP-
JOINT ATOMIC WEAPONS PUBLICATIONS BOARD, ALBUQUERQUE, N. MEX. (REF. 14)	NAVY SWOP-
JOINT ATOMIC WEAPONS PUBLICATIONS BOARD, ALBUQUERQUE, N. MEX. (REF. 14)	SWOP-
JOINT ATOMIC WEAPONS PUBLICATIONS BOARD, ALBUQUERQUE, N. MEX. (REF. 14)	TM-39-
JOINT ATOMIC WEAPONS PUBLICATIONS BOARD, ALBUQUERQUE, N. MEX. (REF. 14)	T.O.(NUMBERS)-11N
JOINT ATOMIC WEAPONS PUBLICATIONS BOARD, ALBUQUERQUE, N. MEX. (REF. 14)	TP-
JOINT ATOMIC WEAPONS TECHNICAL INFORMATION GROUP	WTI-
JOINT BOARD ON SCIENTIFIC INFORMATION POLICY	JBSIP-
JOINT CHIEFS OF STAFF. SPECIAL ASSISTANT FOR STRATEGIC MOBILITY, WASHINGTON, D.C.	SASM-(NUMBER)-(YEAR)
JOINT COMMITTEE (AEC-DOD) (REF. 15)	JC-(LETTERS)-(YEAR)-
JOINT COMMITTEE ON NEW WEAPONS + EQUIPMENT, WASH., DC	GMC-
JOINT DEPARTMENTAL RADIO AND ELECTRONICS MEASUREMENTS COMMITTEE	REMC-
JOINT DESIGN TEAM (ARMY), WARREN, MICH.	JDT-
JOINT ELECTRONICS INFORMATION AGENCY, WASH., D.C.	JEIA-
JOINT EST. FOR NUCLEAR ENERGY RES., KJELLER, NORWAY	JENER-
JOINT EST. FOR NUCLEAR ENERGY RES., KJELLER, NORWAY	JENER-PUB.-

JOINT INST. FOR LAB. ASTROPHYSICS, BOULDER, COLO.	JILA-
JOINT INTELLIGENCE OBJECTIVES AGENCY (OFFICE, TECHNICAL INTELLIGENCE BRANCH)	OTIB-
JOINT INTELLIGENCE OBJECTIVES AGENCY, WASH., D.C.	BIGS-
JOINT INTELLIGENCE OBJECTIVES AGENCY, WASH., D.C.	JIOA-
JOINT INTELLIGENCE OBJECTIVES AGENCY AND BRITISH INTELLIGENCE OBJECTIVES SUB-COMM. (COMBINED TRIPS)	CIOS TRIP-
JOINT MATERIEL INTELLIGENCE AGENCY, WASHINGTON, D.C.	MCN-
JOINT METEOROLOGICAL COMMITTEE, WASHINGTON, D.C.	JMC-
JOINT METEOROLOGICAL COMMITTEE, WASHINGTON, D.C.	JMC/E-
JOINT METEOROLOGICAL COMMITTEE, WASHINGTON, D.C.	JMC/WC-
JOINT METEOROLOGICAL SATELLITE ADVISORY COMMITTEE	JMSAC-
JOINT MILITARY INDUSTRIAL ELECTRONIC TEST EQUIPMENT SYMPOSIUM	SETE-
JOINT PARACHUTE TEST FACILITY, EL CENTRO, CALIF.	FTL-
JOINT PUBLICATIONS RESEARCH SERVICE, N.Y.C.	GUO-
JOINT PUBLICATIONS RESEARCH SERVICE, N.Y.C.	JPRS-
JOINT PUBLICATIONS RESEARCH SERVICE, N.Y.C.	JPRS-GUO-
JOINT PUBLICATIONS RESEARCH SERVICE, N.Y.C.	JPRS-L-
JOINT PUBLICATIONS RESEARCH SERVICE, N.Y.C.	JPRS(NY)-
JOINT PUBLICATIONS RESEARCH SERVICE, N.Y.C.	JPRS-TR-L-
JOINT SATELLITE STUDIES GROUP	JSSG-
JOINT SERVICES READING PANEL, LONDON	JSRP-
JOINT SPECIAL WEAPONS PUBLICATIONS BOARD, ALBUQUERQUE, N. MEX. (REF. 14)	AEC-AFSWP-TP-
JOINT SPECIAL WEAPONS PUBLICATIONS BOARD, ALBUQUERQUE, N. MEX. (REF. 14)	AFSWP-TP-
JOINT SPEC. WEAPONS PUBLICATIONS BD.,ALBUQUERQUE, N.M	FCAG-
JOINT SPECIAL WEAPONS PUBLICATIONS BOARD, ALBUQUERQUE, N. MEX. (REF. 14)	JSWPB-
(SEE ALSO LATER NAME: JOINT ATOMIC WEAPONS PUBL.BD.)	
JOINT TACTICAL AIR SUPPORT BOARD, FORT BRAGG, N.C.	ASBD-
JOINT TASK FORCE ... (REF. 17)	JF-
JOINT TASK FORCE ... (REF. 17)	JTF-
JOINT TASK FORCE ... (REF. 17)	JTF-(NUMBER)-ATG-
JOINT TASK FORCE ... (REF. 17)	JTF-(NUMBER)-TG-
JOINT TASK FORCE ... (REF. 17)	XRB-
JOINT TASK FORCE, EGLIN AFB, FLA.	JTF-OPLAN-(NUMBER)-
JOINT TASK FORCE ONE	JTFI-
JOINT TASK FORCE ONE	OC-TDR-IX-
JOINT TASK FORCE ONE	XRD-
JOINT TASK FORCE TWO, SANDIA BASE, N.M.	JTF2-
JOINT TASK FORCE TWO, SANDIA BASE, N.M.	JTF2-(NO.)-(LTRS.)-
JOINT TASK FORCE TWO, SANDIA BASE, N.M.	JTF2-OAR-
JOINT TASK FORCE SEVEN, PEARL HARBOR	JTFMC-TP-
JOINT TASK GROUP ... (REF. 18)	JTG-
JOINT TEST + EVALUATION TASK FORCE, MAC DILL AFB, FLA	JTETF-
JOINT WAR GAMES AGENCY, WASHINGTON, D.C.	JWGA-
JOINT WORKING GROUP ... (REF. 19)	JWG-
JOINT WORKING GROUP ... (REF. 19)	JWG (LETTERS OR NUMBERS)-
JOINT WORKING GROUP ... (REF. 19)	JWG (LTRS OR NOS.)-(YR)-
JONES + LAUGHLIN STEEL CORP. WIRE ROPE DIV.,MUNCY,PA.	IH/TFS-(YEAR)-
JONES + LAUGHLIN STEEL CORP. WIRE ROPE DIV.,MUNCY,PA.	NOS-TFS-(YEAR)-
JUNGLE WARFARE SCHOOL. TRAIL AND DEVELOPMENTS WING, JOHORE BAHRU, MALAYSIA	JWS/TD-
JUNIPER COMMITTEE	JUNIPER-
JUPITER NOSE CONE COMMITTEE (REF. 15)	JUPITER-
JUPITER NOSE CONE COMMITTEE (TECH. BULL.) (REF. 15)	JUPITER-TB-

KAISER AEROSPACE AND ELECTRONICS CORP., ELECTRONICS PLANT

KAISER AEROSPACE AND ELECTRONICS CORP., ELECTRONICS PLANT, PALO ALTO, CALIF.	HFR-
KAISER ALUMINUM AND CHEMICAL CORP. DEPT. OF METALLURGICAL RESEARCH, SPOKANE	MS-PR-(YEAR)-
KAISER ENGINEERS DIV., HENRY J.KAISER CO., OAKLAND, CAL	KE-
KAISER ENGINEERS DIV., HENRY J.KAISER CO., OAKLAND, CAL	KE-(YR.)-(NO.)-RE
KAISER ENGINEERS DIV., HENRY J.KAISER CO., OAKLAND, CAL	KE-(YEAR)-
KAISER ENGINEERS DIV., HENRY J.KAISER CO., OAKLAND, CAL	KEAE-
KAISER ENGINEERS DIV., HENRY J.KAISER CO., OAKLAND, CAL	KE-GCPR-
KAISER ENGINEERS DIV., HENRY J.KAISER CO., OAKLAND, CAL	KENPS-(YR)-(NO.)-R
KAISER ENGINEERS DIV., HENRY J.KAISER CO., OAKLAND, CAL	PEGCPR-
KAISER-FRAZER CORP., WILLOW RUN, MICH.	KMC-
KAMAN AIRCRAFT CORP., BLOOMFIELD, CONN.	KAC-
KAMAN AIRCRAFT CORP., BLOOMFIELD, CONN.	KAC-(NO.)-(NO.)-PR-
KAMAN AIRCRAFT CORP., BLOOMFIELD, CONN.	KAC-R-
KAMAN AIRCRAFT CORP., BLOOMFIELD, CONN.	KAMAN-
KAMAN AIRCRAFT CORP., BLOOMFIELD, CONN.	KAMAN-AC-RN-R-(LTR.)(NO.)
KAMAN AIRCRAFT CORP., BLOOMFIELD, CONN.	T-
KAMAN AIRCRAFT CORP., COLORADO SPRINGS	KM-
KAMAN AIRCRAFT CORP. NUCLEAR DIV., COLORADO SPRINGS (SEE ALSO LATER NAME: KAMAN NUCLEAR)	RFP-
KAMAN AVIDYNE, BURLINGTON, MASS.	KA-(YEAR)-
KAMAN AVIDYNE, BURLINGTON, MASS.	KA-AMC-(NUMBER-YEAR)
KAMAN AVIDYNE, BURLINGTON, MASS.	KA-TR-
KAMAN NUCLEAR, COLORADO SPRINGS	KAC KN-(YEAR)-
KAMAN NUCLEAR, COLORADO SPRINGS	KN-(YEAR)-(NUMBER)-(LTR.)
KAMAN NUCLEAR, COLORADO SPRINGS	KN-(NO.)-(YEAR)
KAMAN NUCLEAR, COLORADO SPRINGS	KN-AMC-(YEAR)-
KANSAS. UNIV., LAWRENCE.	KU-
KANSAS. UNIV., LAWRENCE. CENTER FOR RESEARCH IN ENGINEERING SCIENCE	CRES-(NUMBER)-
KANSAS. UNIV., LAWRENCE. CENTER FOR RESEARCH IN ENGINEERING SCIENCE	CRES-REPRINT-
KANSAS. UNIV., LAWRENCE. CENTER FOR RESEARCH IN ENGINEERING SCIENCE	CRES-TR-(NUMBER-NUMBER)
KANSAS. UNIV., LAWRENCE. DEPT. OF MATHEMATICS	KAN AF2-TN-
KANSAS STATE COLL., MANHATTAN	K-
KANSAS STATE COLL., MANHATTAN. ENG. EXPERIMENT STA.	K EE-
KANSAS STATE UNIV., MANHATTAN	KSU-
KANSAS STATE UNIV., MANHATTAN (PROJ. THEMIS)(REF. 48)	THEMIS-SR-
KANSAS STATE UNIV., MANHATTAN. DEPT. OF ELECT. ENG.	EE-TR-
KANSAS STATE UNIV., MANHATTAN. ENG. EXPERIMENT STA.	KEES-SR-
KANSAS STATE UNIV., MANHATTAN. ENG. EXPERIMENT STA.	KSU-SR-
KANSAS STATE UNIV., MANHATTAN. INST. FOR SYSTEMS DESIGN AND OPTIMIZATION	ISDO-
KARL-MARX-UNIVERSATAET, LEIPZIG. SEKTION PHYSIK	TUL-
KARLSRUHE, GERMANY. TECHNISCHE HOCHSCHULE	KVT-
KARLSRUHE, GERMANY. TECHNISCHE HOCHSCHULE	VTH-
KARLSRUHE, GERMANY. TECHNISCHE HOCHSCHULE	WTHD-
KEARFOTT CO., INC., N.Y.C.	LH-
KEIO UNIV., TOKYO. DEPT. OF ANATOMY	J-
KEIO UNIV., TOKYO. SCHOOL OF MEDICINE	J-
KELLETT AIRCRAFT CORP., CAMDEN, N.J.	NA-
KELLETT AIRCRAFT CORP., CAMDEN, N.J.	XA-
KELLETT AIRCRAFT CORP., WILLOW GROVE, PENNA.	KELLETT-(NO. LTR. NO.)-
KELLEX CORP., N.Y.C.	CRK-
KELLEX CORP., N.Y.C.	D-
KELLEX CORP., N.Y.C.	E-(NO.)A
KELLEX CORP., N.Y.C.	KL-
KELLEX CORP., N.Y.C.	KPS-
KELLEX CORP., N.Y.C.	KX-
KELLEX CORP., N.Y.C.	M&O-
KELLEX CORP., N.Y.C.	RB-(NO.)(LETTER)
KELLEX CORP., N.Y.C.	RD-
KELLEX CORP., N.Y.C.	Z-
KELLEX CORP., SILVER SPRING, MD. (MATERIALS RESEARCH) (SEE ALSO LATER NAME: VITRO CORP. OF AMERICA)	MR REPORT-
KELLOGG (M.W.) CO., JERSEY CITY	KEL-
KELLOGG (M.W.) CO., JERSEY CITY	KELL-
KELLOGG (M.W.) CO., JERSEY CITY	KLG-
KELLOGG (M.W.) CO., JERSEY CITY	MWK-
KELLOGG (M.W.) CO., JERSEY CITY	PED-
KELLOGG (M.W.) CO. PETROLEUM AND CHEMICAL RESEARCH DEPT., JERSEY CITY	MWK-RL-(YEAR)-
KELLOGG (M.W.) CO. PETROLEUM AND CHEMICAL RESEARCH DEPT., JERSEY CITY	RL-(YR.)-
KELLOGG (M.W.) CO. RES. + DEV. CTR., PISCATAWAY, N.J.	RD-(NUMBER)-
KELLOGG (M.W.) CO. RES. + DEV. CTR., PISCATAWAY, N.J.	RD-(YEAR)-
KELLOGG (M.W.) CO. RES. + DEV. CTR., PISCATAWAY, N.J.	RED-(YEAR)-
KELLOGG (M.W.) CO. SPECIAL PROJECTS DEPT., JERSEY CITY (RESEARCH AND DEVELOPMENT REPORTS)	KELL RDR SPD-
KELLOGG (M.W.) CO. SPECIAL PROJECTS DEPT., JERSEY CITY	MWK-SPD-
KELLOGG (M.W.) CO. SPECIAL PROJECTS DEPT., JERSEY CITY	SPD-
KELLOGG SWITCHBOARD + SUPPLY CO., CHICAGO.	KSSC-
KELSEY-HAYES CO., UTICA, N.Y.	PO-
KELSEY-HAYES CO. TURBINE PARTS DIV., UTICA, N.Y.	PO-

KELTEC INDUSTRIES INC., CORONA, CALIF.	AGAC-
KENNAMETAL, INC., LATROBE, PENNA.	KENNI-
KENNAMETAL, INC., LATROBE, PENNA.	KI-
KENT STATE UNIV., OHIO. LIQUID CRYSTAL INST.	LCI-
KENTUCKY. UNIV., LEXINGTON	KE-
KENTUCKY. UNIV., LEXINGTON (PROJECT THEMIS) (REF. 48)	THEMIS-KU-RR-
KENTUCKY. UNIV., LEXINGTON (PROJECT THEMIS) (REF. 48)	THEMIS-UK-RR-
KENTUCKY. UNIV., LEXINGTON. ENG. EXPERIMENT STA.	KE EES-
KENTUCKY PROGRAM DEVELOPMENT OFFICE, FRANKFORT (JOINT VENTURE)	CK-KPDO-(YEAR)-
KENYON COLL., GAMBIER, OHIO	KC-
KERNFORSCHUNGSANLAGE JUELICH, GERMANY	BIBL./DOK.-
KERNFORSCHUNGSANLAGE JUELICH, GERMANY	JUB-BIBL-
KERNFORSCHUNGSANLAGE JUELICH, GERMANY	JUEL-(NUMBER)-(LETTERS)
KERNFORSCHUNGSANLAGE JUELICH, GERMANY (LETTERS STAND FOR INSTITUTES, WORKING GROUPS, ETC. BUT DO NOT AFFECT THE NUMBERS, WHICH ARE IN ONE SERIES)	JUL-(NUMBER)-(LETTERS)
KERNFORSCHUNGSANLAGE JUELICH, GERMANY	JUL-(NUMBER)-
KERNFORSCHUNGSANLAGE JUELICH, GERMANY	JUL-(NUMBER)-NP
KERNFORSCHUNGSANLAGE JUELICH, GERMANY	JUL-BIBL-
KERNFORSCHUNGSANLAGE JUELICH, GERMANY	JUL-CONF-
KERNFORSCHUNGSANLAGE JUELICH, GERMANY	KFA-
KERNFORSCHUNGSANLAGE JUELICH, GERMANY	NRCJ-
KERNFORSCHUNGSANLAGE JUELICH, GERMANY	SUL-
KERNFORSCHUNGSANLAGE JUELICH, GERMANY	THTR-
KERNFORSCHUNGSANLAGE JUELICH. INSTITUT FUER BOTANIK UND MIKROBIOLOGIE, GERMANY	JUL-(NUMBER)-BO
KERNFORSCHUNGSANLAGE JUELICH. INSTITUT FUER CHEMISCHE TECHNOLOGIE, GERMANY	JUL-(NUMBER)-CT
KERNFORSCHUNGSANLAGE JUELICH. INSTITUT FUER FESTKOERPER- UND NEUTRONENPHYSIK, GERMANY	JUL-(NUMBER)-FN
KERNFORSCHUNGSANLAGE JUELICH. INSTITUT FUER FESTKOERPERFORSCHUNG, GERMANY	JUL-(NUMBER)-FF
KERNFORSCHUNGSANLAGE JUELICH. INSTITUT FUER KERNPHYSIK, GERMANY	JUL-(NUMBER)-KP
KERNFORSCHUNGSANLAGE JUELICH. INSTITUT FUER MEDIZIN, GERMANY	JUL-(NUMBER)-ME
KERNFORSCHUNGSANLAGE JUELICH. INSTITUT FUER PHYSIKALISCHE CHEMIE	JUL-(NUMBER)-PC
KERNFORSCHUNGSANLAGE JUELICH. INSTITUT FUER PLASMAPHYSIK, GERMANY	JUL-(NUMBER)-PP
KERNFORSCHUNGSANLAGE JUELICH. INSTITUT FUER RADIOCHEMIE, GERMANY	JUL-(NUMBER)-RC
KERNFORSCHUNGSANLAGE JUELICH. INSTITUT FUER REAKTORBAUELEMENTE, GERMANY	JUL-(NUMBER)-RB
KERNFORSCHUNGSANLAGE JUELICH. INSTITUT FUER REAKTORENTWICKLUNG, GERMANY	JUL-(NUMBER)-RG
KERNFORSCHUNGSANLAGE JUELICH. INSTITUT FUER REAKTORWERKSTOFFE, GERMANY	HZ-
KERNFORSCHUNGSANLAGE JUELICH. INSTITUT FUER REAKTORWERKSTOFFE, GERMANY	JUL-(NUMBER)-RW
KERNFORSCHUNGSANLAGE JUELICH. INSTITUT FUER TECHNISCHE PHYSIK, GERMANY	JUL-(NUMBER)-TP
KERNFORSCHUNGSANLAGE JUELICH. ZENTRALABTEILUNG FORSCHUNGSREAKTOREN, GERMANY	JUL-(NUMBER)-RE
KERNFORSCHUNGSANLAGE JUELICH. ZENTRALABTEILUNG STRAHLENSCHUTZ, GERMANY	JUL-(NUMBER)-ST
KERNFORSCHUNGSANLAGE JUELICH. ZENTRALINSTITUT FUER ANGEWANDTE MATHEMATIK, GERMANY	JUL-(NUMBER)-MA
KERNFORSCHUNGSANLAGE JUELICH. ZENTRALINSTITUT FUER REAKTOREXPERIMENTE, GERMANY	JUL-(NUMBER)-RX
KERNFORSCHUNGSANLAGE JUELICH. ZENTRALLABOR FUER CHEMISCHE ANALYSE, GERMANY	JUL-(NUMBER)-CA
KERNFORSCHUNGSANLAG JUELICH. ZENTRALLABOR FUER ELEKTRONIK, GERMANY	JUL-(NUMBER)-ZE
KERNFORSCHUNGSZENTRUM KARLSRUHE, GERMANY (MATERIAL AND SOLID BODY RESEARCH)	IMF-(NUMBER)/(YEAR)
KERNFORSCHUNGSZENTRUM KARLSRUHE, GERMANY	KFK-(NUMBER/YEAR)-
KERNFORSCHUNGSZENTRUM KARLSRUHE, GERMANY	KFK-
KERNFORSCHUNGSZENTRUM KARLSRUHE, GERMANY	KFK-EXT-(NO./YEAR)-
KERNFORSCHUNGSZENTRUM KARLSRUHE, GERMANY(TRANSLATION)	KFK-TR-
KERNFORSCHUNGSZENTRUM KARLSRUHE, GERMANY	PSB-
KERNFORSCHUNGSZENTRUM KARLSRUHE, GERMANY	PSB-
KERNFORSCHUNGSZENTRUM KARLSRUHE, GERMANY	PSBB-
KERNFORSCHUNGSZENTRUM KARLSRUHE, GERMANY	PSB-NB-
KERNFORSCHUNGSZENTRUM KARLSRUHE, GERMANY	SM (NUMBER)/(NUMBER)
KERNFORSCHUNGSZENTRUM KARLSRUHE. INSTITUT FUER ANGEWANDTE KERNPHYSIK, GERMANY	IAK-
KERNFORSCHUNGSZENTRUM KARLSRUHE. INSTITUT FUER ANGEWANDTE KERNPHYSIK, GERMANY	KFK-IAK-(NUMBER/YEAR)-
KERNFORSCHUNGSZENTRUM KARLSRUHE. INSTITUT FUER ANGEWANDTE REAKTORPHYSIK, GERMANY	KFK-IAR-(NUMBER/YEAR)-
KERNFORSCHUNGSZENTRUM KARLSRUHE. INSTITUT FUER EXPERIMENTELLE KERNPHYSIK, GERMANY	KFK-IEK-(NUMBER/YEAR)-
KERNFORSCHUNGSZENTRUM KARLSRUHE. INSTITUT FUER HEISSE CHEMIE, GERMANY	KFK-IHC-(NUMBER/YEAR)-
KERNFORSCHUNGSZENTRUM KARLSRUHE. INSTITUT FUER KERNVERFAHRENSTECHNIK	KEK-
KERNFORSCHUNGSZENTRUM KARLSRUHE. INSTITUT FUER MATERIAL- UND FESTKOERPERFORSCHUNG, GERMANY	KFK-IMF-(NUMBER/YEAR)-
KERNFORSCHUNGSZENTRUM KARLSRUHE. INSTITUT FUER MATERIAL- UND FESTKOERPERFORSCHUNG, GERMANY	KFK-IMFB-(NUMBER/YEAR)-
KERNFORSCHUNGSZENTRUM KARLSRUHE. INSTITUT FUER NEUTRONENPHYSIK UND REAKTORTECHNIK, GERMANY	INR-(NUMBER/YEAR)-
KERNFORSCHUNGSZENTRUM KARLSRUHE. INSTITUT FUER NEUTRONENPHYSIK UND REAKTORTECHNIK, GERMANY	KFK-INR-(NUMBER/YEAR)-
KERNFORSCHUNGSZENTRUM KARLSRUHE. INSTITUT FUER RADIOCHEMIE, GERMANY	IRCH-(NUMBER/YEAR)-
KERNFORSCHUNGSZENTRUM KARLSRUHE. INSTITUT FUER RADIOCHEMIE, GERMANY	KFK-IR-(NUMBER/YEAR)-
KERNFORSCHUNGSZENTRUM KARLSRUHE. INSTITUT FUER REAKTORBAUELEMENTE, GERMANY	IRB-
KERNFORSCHUNGSZENTRUM KARLSRUHE. INSTITUT FUER REAKTORBAUELEMENTE, GERMANY	KFK-IRB-(NUMBER/YEAR)-
KERNFORSCHUNGSZENTRUM KARLSRUHE. INSTITUT FUER REAKTORENTWICKLUNG, GERMANY	KFK-IR-(NUMBER/YEAR)-
KERNKRAFTWERK LINGEN G.M.B.H., GERMANY	KWL-
KERNKRAFTWERK PLANUNGSGESELLSCHAFT M.B.H., VIENNA	KKWP-(NUMBER/YEAR)

KERNKRAFTWERK RWE-BAYERNWERK G.M.B.H., GUNDREMMINGEN, GERMANY	KRB-(YEAR/NUMBER)
KETT CORP., CINCINNATI	KT-
KETTELLE ASSOCIATES, INC., PAOLI, PENNA.	KTM-
KEURING VAN ELECTROTECHNISCHE MATERIALEN, N.V., ARNHEIM, NETHERLANDS	CH-E/INT.R.-
KEURING VAN ELECTROTECHNISCHE MATERIALEN, N.V., ARNHEM, NETHERLANDS	KEMA-
KEURING VAN ELECTROTECHNISCHE MATERIALEN, N.V., ARNHEM, NETHERLANDS	WP-
KEYDATA AND ADAMS ASSOCIATES, INC., WATERTOWN, MASS.	K/AA-
KIDDE (WALTER) NUCLEAR LABS., INC., GARDEN CITY, N.Y.	OLWK-
KIDDE (WALTER) NUCLEAR LABS., INC., GARDEN CITY, N.Y.	WKNL-
KILGORE, INC. (TECHNICAL REPORT)	KTR-
KINETICS CORP., CINCINNATI	GE-
KINETICS CORP., CINCINNATI	KA-
KIP ELECTRONICS CORP., STAMFORD, CONN.	KIP-
KIP ELECTRONICS CORP., STAMFORD, CONN.	KIP-C-
KIP ELECTRONICS CORP., STAMFORD, CONN.	KP-
KIRTLAND AFB, N. MEX. (REF. 1)	(NUMBERS...)
KIRTLAND AFB, N. MEX.	KAFB-
KIRTLAND AFB, N. MEX. (OPERATION PLAN)	KAFB-OP-
KIRTLAND AFB, N. MEX.	KAFB-WCW-
KITT PEAK NATIONAL OBSERVATORY, TUCSON, ARIZ.	KPNO-
KMS INDUSTRIES INC., ANN ARBOR, MICH.	KMS-(NUMBERS)-(NOS.)(LTR)
KMS TECHNOLOGY CENTER, SAN DIEGO, CALIF.	KMSTC-SD-
KNOLLS ATOMIC POWER LAB., SCHENECTADY, N.Y. (REF. 1)	(NUMBERS...)
KNOLLS ATOMIC POWER LAB., SCHENECTADY, N.Y. (CODE IS AUTHOR'S INITIALS)	CJS-
KNOLLS ATOMIC POWER LAB., SCHENECTADY, N.Y.	CVM-
KNOLLS ATOMIC POWER LAB., SCHENECTADY, N.Y.	KAPL-
KNOLLS ATOMIC POWER LAB., SCHENECTADY, N.Y.	KAPL-A-(LETTERS)-
KNOLLS ATOMIC POWER LAB., SCHENECTADY, N.Y. (ADVANCED DEVELOPMENT ACTIVITY)	KAPL-ADA-
KNOLLS ATOMIC POWER LAB., SCHENECTADY, N.Y.	KAPL-ADM-
KNOLLS ATOMIC POWER LAB., SCHENECTADY, N.Y. (ELEMENTS OF SPECIFICATIONS)	KAPL-A-EOS-
KNOLLS ATOMIC POWER LAB., SCHENECTADY, N.Y.	KAPL-A-S(NO.)-(NO.)-
KNOLLS ATOMIC POWER LAB., SCHENECTADY, N.Y. (SUBMARINE INTERMEDIATE REACTOR)	KAPL-A-SIR-
KNOLLS ATOMIC POWER LAB., SCHENECTADY, N.Y.	KAPL-A-SM-
KNOLLS ATOMIC POWER LAB., SCHENECTADY, N.Y.	KAPL-M-
KNOLLS ATOMIC POWER LAB., SCHENECTADY, N.Y.	KAPL-M-(INITIALS)-
KNOLLS ATOMIC POWER LAB., SCHENECTADY, N.Y.	KAPL-M-AME-
KNOLLS ATOMIC POWER LAB., SCHENECTADY, N.Y.	KAPL-M-AMS-
KNOLLS ATOMIC POWER LAB., SCHENECTADY, N.Y. (CONTROL ROD DRIVE MECHANISM STUDY)	KAPL-M-CRDM-
KNOLLS ATOMIC POWER LAB., SCHENECTADY, N.Y.	KAPL-M-CT-
KNOLLS ATOMIC POWER LAB., SCHENECTADY, N.Y. (RADIOACTIVE ACCESSIBILITY REPORTS)	KAPL-M-CTU-
KNOLLS ATOMIC POWER LAB., SCHENECTADY, N.Y. (NAVAL REACTOR D1G)	KAPL-M-D1G-TD-
KNOLLS ATOMIC POWER LAB., SCHENECTADY, N.Y. (NAVAL REACTOR D1G)	KAPL-M-DNA-
KNOLLS ATOMIC POWER LAB., SCHENECTADY, N.Y.	KAPL-M-EDL-
KNOLLS ATOMIC POWER LAB., SCHENECTADY, N.Y.	KAPL-MEMO-(INITIALS)-
KNOLLS ATOMIC POWER LAB., SCHENECTADY, N.Y. (HEALTH PHYSICS)	KAPL-M-HP-
KNOLLS ATOMIC POWER LAB., SCHENECTADY, N.Y. (S3G REACTOR)	KAPL-M-S3G-
KNOLLS ATOMIC POWER LAB., SCHENECTADY, N.Y. (SUBMARINE ADVANCED REACTOR)	KAPL-M-SAR-RES-
KNOLLS ATOMIC POWER LAB., SCHENECTADY, N.Y. (SHIELDING LAB. EXPERIMENT)	KAPL-M-SL-
KNOLLS ATOMIC POWER LAB., SCHENECTADY, N.Y.	KAPL-M-SMS-
KNOLLS ATOMIC POWER LAB., SCHENECTADY, N.Y.	KAPL-M-SR-
KNOLLS ATOMIC POWER LAB., SCHENECTADY, N.Y.	KAPL-M-SSA-
KNOLLS ATOMIC POWER LAB., SCHENECTADY, N.Y. (SYSTEM DESCRIPTION)	KAPL-M-SSD-
KNOLLS ATOMIC POWER LAB., SCHENECTADY, N.Y. (TEST RUN)	KAPL-M-TR-
KNOLLS ATOMIC POWER LAB., SCHENECTADY, N.Y.	KAPL-P-
KNOLLS ATOMIC POWER LAB., SCHENECTADY, N.Y.	KAPL-PC-
KNOLLS ATOMIC POWER LAB., SCHENECTADY, N.Y.	KAPL-RDTR-
KNOLLS ATOMIC POWER LAB., SCHENECTADY, N.Y.	KAPL-S-
KNOLLS ATOMIC POWER LAB., SCHENECTADY, N.Y.	KAPL-SPEC-(NO.)(LETTERS)-
KNOLLS ATOMIC POWER LAB., SCHENECTADY, N.Y.	KAPL-SPEC-(LTRS)-
KNOLLS ATOMIC POWER LAB., SCHENECTADY, N.Y.	KAPL-SSD-
KNOLLS ATOMIC POWER LAB., SCHENECTADY, N.Y.	KAPL-TRANS-
KNOLLS ATOMIC POWER LAB., SCHENECTADY, N.Y.	KF-
KNOLLS ATOMIC POWER LAB., SCHENECTADY, N.Y.	KPE-
KNOLLS ATOMIC POWER LAB., SCHENECTADY, N.Y.	KPM-
KNOLLS ATOMIC POWER LAB., SCHENECTADY, N.Y.	KPP-
KNOLLS ATOMIC POWER LAB., SCHENECTADY, N.Y.	KPTR-
KNOLLS ATOMIC POWER LAB., SCHENECTADY, N.Y. (NUCLEAR ENGINEERING COURSE)	MEMO-(INITIALS)-
KNOLLS ATOMIC POWER LAB., SCHENECTADY, N.Y.	MEMO-NEC-
KNOLLS ATOMIC POWER LAB., SCHENECTADY, N.Y.	NCT-
KNOLLS ATOMIC POWER LAB., SCHENECTADY, N.Y.	O-KPD-
KNOLLS ATOMIC POWER LAB., SCHENECTADY, N.Y.	S49-
KNOLLS ATOMIC POWER LAB., SCHENECTADY, N.Y.	S(NO.)G-ER-
KNOLLS ATOMIC POWER LAB., SCHENECTADY, N.Y. (SUBMARINE REACTOR PROJECT)	SAR-
KNOLLS ATOMIC POWER LAB., SCHENECTADY, N.Y.	SIR-
KNOLLS ATOMIC POWER LAB., SCHENECTADY, N.Y.	TI-
KNOLLS ATOMIC POWER LAB., SCHENECTADY, N.Y.	TO-EE-
KNOLLS ATOMIC POWER LAB. GE NUCLEONICS PROJECT, SCHENECTADY, N.Y.	GE-(INITIALS)-
KNOLLS ATOMIC POWER LAB. GE NUCLEONICS PROJECT, SCHENECTADY, N.Y.	GE-AEP-(YR)-(LETTERS)-
KNOLLS ATOMIC POWER LAB. GE NUCLEONICS PROJECT, SCHENECTADY, N.Y.	GE-KHK-
KNOLLS ATOMIC POWER LAB. GE NUCLEONICS PROJECT, SCHENECTADY, N.Y.	GE-LT-
KNOLLS ATOMIC POWER LAB. LIQUID METALS SAFETY COMMITTEE, SCHENECTADY, N.Y.	LMSC-
KNOLLS ATOMIC POWER LAB. NAVAL REACTOR INFORMATION CENTER, SCHENECTADY, N.Y.	NRIC-(YEAR)-
KNOLLS ATOMIC POWER LAB. REDOX GP., SCHENECTADY, N.Y.	KAPL-M-REDOX-
KODAIKANAL OBSERVATORY, INDIA	BULL-A-
KODAIKANAL OBSERVATORY, INDIA	DGO-(NUMBER-NUMBER)
KOGAKUIN UNIV., TOKYO	KG-
KOLLSMAN INSTRUMENT CO., ELMHURST, N.Y. (REF. 1)	(NUMBER)
KOLLSMAN INSTRUMENT CO., ELMHURST, N.Y.	KIC-
KOLLSMAN INSTRUMENT CO., ELMHURST, N.Y.	KOLL-

KOLLSMAN INSTRUMENT CORP., ELMHURST, N.Y.	KI-
KOLLSMAN INSTRUMENT CORP., ELMHURST, N.Y.	KIC-RD-
KOLLSMAN INSTRUMENT CORP., ELMHURST, N.Y.	S-
KOLLSMAN INSTRUMENT CORP. SYSTEMS MANAGEMENT DIV., ELMHURST, N.Y.	KIC-SYS-TDR-
KOPPERS CO., INC., PITTSBURGH	KC-
KORAD CORP., SANTA MONICA, CALIF.	KC-
KOREA. ATOMIC ENERGY RESEARCH INST., SEOUL	AERI-(LETTERS)-(NO.)-
KYOTO IMPERIAL UNIV. COLLEGE OF AGRICULTURE	KIU-
KYOTO IMPERIAL UNIV. COLL. OF AGRIC. (CHEM. SERIES)	KIU C-
KYOTO IMPERIAL UNIV. COLLEGE OF AGRICULTURE (ENTOMOLOGICAL SERIES)	KIU E-
KYOTO IMPERIAL UNIV. COLLEGE OF AGRICULTURE (GENETICAL SERIES)	KIU G-
KYOTO IMPERIAL UNIV. COLLEGE OF AGRICULTURE (HORTICULTURAL SERIES)	KIU H-
KYOTO IMPERIAL UNIV. COLLEGE OF AGRICULTURE (PHYTOPATHOLOGICAL SERIES)	KIU P-
KYOTO UNIV.	K-TRANS-
KYOTO UNIV. DEPT. OF AERONAUTICAL ENGINEERING	CP-
KYOTO UNIV. DEPT. OF PHYSICS	KUNS-
KYOTO UNIV. RESEARCH INST. FOR FUNDAMENTAL PHYSICS	INS-PT-
KYOTO UNIV. RESEARCH INST. FOR FUNDAMENTAL PHYSICS	RIFP-
KYOTO UNIV. RESEARCH REACTOR INST.	KURRI-TR-
KYUSHU UNIV., FUKUOKA. INST. FOR THEORETICAL PHYSICS	KYUSHU-

LABORATOIRE CENTRAL DES INDUSTRIES ELECTRIQUES

LABORATOIRE CENTRAL DES INDUSTRIES ELECTRIQUES, FONTENAY-AUX-ROSES, FRANCE	LCIE-
LABORATORY FOR ELECTRONICS, INC., BOSTON	LFE-
LABORATORY FOR ELECTRONICS, INC., BOSTON	TEST-
LAB. FOR ELECTRONICS, INC., ELECTRONICS DIV., BOSTON	C-
LABORATORY SERVICE DIV., ABERDEEN PROVING GROUND, MD.	APG LSD-
LACQUER + CHEM. CORP. ALAKA RES. LAB., BROOKLYN	ALAKA-
LACQUER + CHEM. CORP. ALAKA RES. LAB., BROOKLYN	LCC-
LADISH CO., CUDAHY, WIS.	LADCO-
LADRIERE, INC. TECH. DIV., DETROIT	LTD-TM-
LAMBDA CORP., ARLINGTON, VA.	LAMBDA-
LAMONT-DOHERTY GEOLOGICAL OBSERVATORY, PALISADES, NY	LDGO-
LAND-AIR, INC., HOLLOMAN AFB, N. MEX.	LAI-
LAND-AIR, INC., HOLLOMAN AFB, N. MEX.	LAIR-
LAND-AIR, INC. WHITE SANDS-HOLLOMAN DIV., HOLLOMAN AFB, N. MEX.	LAND-AIR-
LANDING AIDS EXPERIMENT STATION, ARCATA, CALIF.	LAES-
LANGLEY AERONAUTICAL LAB., LANGLEY FIELD, VA.(REF.25)	NACA-(LTRS)-L-(DATE CODE)
LANGLEY AERONAUTICAL LAB., LANGLEY FIELD, VA. (RESEARCH MEMORANDUM) (REF. 25)	NACA-RM-SL-
LANGLEY MEMORIAL AERONAUTICAL LAB., LANGLEY FIELD, VA. (REF. 25)	NACA-(LTRS)-L-(DATE CODE)
LANGLEY RESEARCH CENTER, LANGLEY FIELD, VA. (REF. 26)	LRC-
LANGLEY RESEARCH CENTER, LANGLEY FIELD, VA. (REF. 26)	NASA-(LETTERS)-L-
LANGUAGE SERVICE BUREAU, CLEVELAND (TRANSLATIONS)	LSB-
LASER DIODE LABS., INC., METUCHEN, N.J.	LDL-(4 DIGITS)-
LAUNCH OPERATIONS CENTER, COCOA BEACH, FLA.	LOC-
LAUNCH OPERATIONS CENTER, COCOA BEACH, FLA.	LOC-(LETTERS)-
LAWRENCE (WM. S.) AND ASSOCIATES, INC., CHICAGO	WMSLAW-(YEAR)-
LEANDER MC CORMICK OBSERVATORY, CHARLOTTESVILLE, VA.	SP-(YEAR)-
LEAR, INC., SANTA MONICA, CALIF.	LEAR-
LEAR, INC., SANTA MONICA, CALIF.	LEAR ADR-
LEAR, INC., SANTA MONICA, CALIF.	LEAR ER-
LEAR, INC., SANTA MONICA, CALIF.	LI-
LEAR, INC. ELECTRO-MECHANICAL DIV., GRAND RAPIDS	LEAR-ER-GR-
LEAR SIEGLER, INC. (ALL DIVISIONS)	GRR-(NO.-LTR.NO.LTR.)
LEAR SIEGLER, INC., GRAND RAPIDS	GR-
LEAR SIEGLER, INC., GRAND RAPIDS	GRR-
LEAR SIEGLER, INC., GRAND RAPIDS	LEAR SIEGLER-
LEAR SIEGLER, INC. ASTRONAUTICS DIV.,SANTA MONICA,CAL	ADR-
LEAR SIEGLER, INC. INSTRUMENT DIV., GRAND RAPIDS	GRR-(YEAR)-
LEAR SIEGLER, INC. POWER EQUIPMENT DIV., CLEVELAND	LSI/PED-
LEDEX INC., DAYTON, OHIO	LEDEX-
LEEDS, ENGLAND. UNIV. DEPT. OF ELEC. + ELECTRONIC ENG	LU-EE/(YEAR)/
LEEDS AND NORTHRUP CO., PHILADELPHIA.	LN-
LEESONA CORP., PROVIDENCE	LC-
LEESONA MOOS LABS. DIV. OF LEESONA CORP., JAMAICA, NY	LML-
LEESONA MOOS LABS. DIV. OF LEESONA CORP., GT.NECK, NY	UR-(NUMBER)/
LEHIGH UNIV., BETHLEHEM, PENNA.	CAM-(NUMBER)-
LEHIGH UNIV., BETHLEHEM, PENNA.	FEL-(NUMBER)-
LEHIGH UNIV., BETHLEHEM, PENNA.	LEH-
LEHIGH UNIV., BETHLEHEM, PENNA.	LEHIGH-
LEHIGH UNIV., BETHLEHEM, PENNA.	LU-
LEHIGH UNIV., BETHLEHEM, PENNA. CHEMISTRY DEPT.	LU-QR-
LEHIGH UNIV., BETHLEHEM, PENNA. FRITZ LAB.	FEL-
LEHIGH UNIV., BETHLEHEM, PENNA. FRITZ LAB.	FL-
LEHIGH UNIV., BETHLEHEM, PENNA. FRITZ LAB.	FRIT-
LEHIGH UNIV., BETHLEHEM, PENNA. FRITZ LAB.	FRITZ LAB.-
LEHIGH UNIV., BETHLEHEM, PENNA. FRITZ LAB.	LEHIGH PR-
LEHIGH UNIV., BETHLEHEM, PENNA. FRITZ LAB.	LEHIGH TR-
LEHIGH UNIV., BETHLEHEM, PENNA. FRITZ LAB.	LEHIGH U. PR-
LEHIGH UNIV., BETHLEHEM, PENNA. FRITZ (ENG.) LAB.	LU-FEL-
LEHIGH UNIV., BETHLEHEM, PENNA. INST. OF RESEARCH	LEH TR-
LEHIGH UNIV., BETHLEHEM, PENNA. INST. OF RESEARCH	LU-TR-
LEHIGH UNIV., BETHLEHEM, PENNA. MARINE SCIENCE CTR.	MSC-
LEICESTER UNIV., ENGLAND	FTR-
LENKURT ELECTRIC CO., INC., SAN CARLOS, CALIF.	AN/FCC-
LEOPOLD-FRANZEN UNIVERSITAET, INNSBRUCK, AUSTRIA	SK-
LESSELLS AND ASSOCIATES, INC., BOSTON (PROGRESS REPT)	L + AI-PR-
LETTERMAN ARMY INST. OF RESEARCH, SAN FRANCISCO	LAIR-

LEVINTHAL ELECTRONIC PRODUCTS, INC., PALO ALTO. CAL.	LEVINTHAL-
LEWIS FLIGHT PROPULSION LAB., CLEVELAND (REF. 25)	NACA-(LTRS)-E-(DATE CODE)
LEWIS FLIGHT PROPULSION LAB., CLEVELAND	NACA-ACR-
LEWIS FLIGHT PROPULSION LAB., CLEVELAND	NACA-RB-E-
LEWIS FLIGHT PROPULSION LAB., CLEVELAND (RESEARCH MEMORANDUM) (REF. 25)	NACA-RM-SE-
LEWIS RESEARCH CENTER, CLEVELAND	E-
LEWIS RESEARCH CENTER, CLEVELAND (REF. 26)	NASA-(LETTERS)-E-
LEWIS RESEARCH CENTER, CLEVELAND	TSP-
LEXINGTON LABS., INC., CAMBRIDGE, MASS.	C-
LFE ELECTRONICS, BOSTON	TP-
LIBERTY POWDER DEFENSE CORP., BARABOO, WIS.	LPDC-
LIBRARY OF CONGRESS, WASHINGTON, D.C.	LC-
LIBRARY OF CONGRESS, WASHINGTON, D.C. (MONTHLY LIST OF RUSSIAN ACCESSIONS)	MLRA-
LIBRARY OF CONGRESS, WASH.,D.C.(RUSSIAN TRANSLATION)	RT-
LIBRARY OF CONGRESS. AEROSPACE INFO. DIV., WASH.,D.C.	AID-P-
LIBRARY OF CONGRESS. AEROSPACE INFO. DIV., WASH.,D.C.	AID-REPORT-
LIBRARY OF CONGRESS. AEROSPACE INFO. DIV., WASH.,D.C.	AID REPT. B-
LIBRARY OF CONGRESS. AEROSPACE INFO. DIV., WASH.,D.C.	AID REPT. P-
LIBRARY OF CONGRESS. AEROSPACE INFO. DIV., WASH.,D.C.	AID REPT. T-
LIBRARY OF CONGRESS. AEROSPACE INFO. DIV., WASH.,D.C.	AID-T-(YR)-(NO.)-
LIBRARY OF CONGRESS. AEROSPACE INFO. DIV., WASH.,D.C.	AID-T-(YEAR)-
LIBRARY OF CONGRESS. AEROSPACE INFO. DIV., WASH.,D.C.	AID-U-(YEAR)-
LIBRARY OF CONGRESS. AEROSPACE INFO. DIV., WASH.,D.C.	TSL-
LIBRARY OF CONGRESS. AEROSPACE TECHNOLOGY DIV., WASHINGTON, D.C.	AID-U-(YR)-(NO.)-
LIBRARY OF CONGRESS. AEROSPACE TECHNOLOGY DIV., WASHINGTON, D.C.	ATD-
LIBRARY OF CONGRESS. AEROSPACE TECHNOLOGY DIV., WASHINGTON, D.C.	ATD-(YEAR)-
LIBRARY OF CONGRESS. AEROSPACE TECHNOLOGY DIV., WASHINGTON, D.C.	ATD-B-(YEAR-NUMBER)
LIBRARY OF CONGRESS. AEROSPACE TECHNOLOGY DIV., WASHINGTON, D.C.	ATD-P-(YR)-(NO.)
LIBRARY OF CONGRESS. AEROSPACE TECHNOLOGY DIV., WASHINGTON, D.C.	ATD-R-(YEAR)-
LIBRARY OF CONGRESS. AEROSPACE TECHNOLOGY DIV., WASHINGTON, D.C.	ATD-U-(YR)-(NO.)
LIBRARY OF CONGRESS. AIR INFORMATION DIV., WASH.,D.C.	AID-
LIBRARY OF CONGRESS. AIR INFORMATION DIV., WASH.,D.C.	AID-(YR)-(NO.)
LIBRARY OF CONGRESS. AIR INFORMATION DIV., WASH.,D.C.	AID-P-(YEAR)-
LIBRARY OF CONGRESS. AIR INFORMATION DIV., WASH.,D.C.	AID-U-
LIBRARY OF CONGRESS. AIR INFORMATION DIV., WASH.,D.C.	SIR-
LIBRARY OF CONGRESS. LEGISLATIVE REFERENCE SERVICE, WASHINGTON, D.C.	CB-
LIBRARY OF CONGRESS. LEGISLATIVE REFERENCE SERVICE, WASHINGTON, D.C.	F-
LIBRARY OF CONGRESS. NAVY RESEARCH SECTION, WASHINGTON, D.C. (REPORT BIBLIOGRAPHY)	LC-RB-
LIBRARY OF CONGRESS. TECH. INFO. DIV., WASH., D.C.	ARC-
LIBRARY OF CONGRESS. TECH. INFO. DIV., WASH., D.C.	ASTIA ARC-
LIBRARY OF CONGRESS. TECH.INFO.DIV.,WASH.,DC (REF.20)	C-
LIBRARY OF CONGRESS. TECH. INFO. DIV., WASH., D.C.	LC-TID-
LIBRARY OF CONGRESS. TECH.INFO.DIV.,WASH.,DC (REF.20)	R-
LIBRARY OF CONGRESS. TECH.INFO.DIV.,WASH.,DC (REF.20)	S-
LIBRARY OF CONGRESS. TECH. INFO. DIV., WASH., D.C.	TID-
LIBRARY OF CONGRESS. TECH.INFO.DIV.,WASH.,DC (REF.20)	U-
LIBRASCOPE INC.	LIBI-
LIEGE. UNIVERSITE	LIEGEU-
LIEGE. UNIVERSITE. INSTITUT D-ASTROPHYSIQUE	LIU-
LIEGE. UNIVERSITE. LABORATOIRE D'AERONAUTIQUE	SA-
LIFE SCIENCES, INC., FORT WORTH, TEX.	LS-TR-
LIGHTNING AND TRANSIENTS RESEARCH INST., MINNEAPOLIS	DS-(YEAR)-
LIGHTNING AND TRANSIENTS RESEARCH INST., MINNEAPOLIS	L AND T -
LIGHTNING AND TRANSIENTS RESEARCH INST., MINNEAPOLIS	LET-
LIGHTNING AND TRANSIENTS RESEARCH INST., MINNEAPOLIS	L + T-
LIGHTNING AND TRANSIENTS RESEARCH INST., MINNEAPOLIS	L/T-
LIGHTNING AND TRANSIENTS RESEARCH INST., MINNEAPOLIS	LTRI-
LIGHTNING AND TRANSIENTS RESEARCH INST., MINNEAPOLIS	LTRI LR-
LILIENTHAL-GESELLSCHAFT FUER LUFTFAHRTFORSCHUNG, BERLIN (BERICHT A)	BA-
LILIENTHAL-GESELLSCHAFT FUER LUFTFAHRTFORSCHUNG, BERLIN (BERICHT S)	BS-
LILIENTHAL-GESELLSCHAFT FUER LUFTFAHRTFORSCHUNG, BERLIN	LG-
LILIENTHAL-GESELLSCHAFT FUER LUFTFAHRTFORSCHUNG, BERLIN	LGB-
LILIENTHAL-GESELLSCHAFT FUER LUFTFAHRTFORSCHUNG, BERLIN (BERICHT/NACHTRAG)	LG B (NUMBER)/N-
LILIENTHAL-GESELLSCHAFT FUER LUFTFAHRTFORSCHUNG, BERLIN (BERICHT)	LG B-
LILIENTHAL-GESELLSCHAFT FUER LUFTFAHRTFORSCHUNG, BERLIN (VORBERICHT)	LG B (NUMBER) VB-
LILIENTHAL-GESELLSCHAFT FUER LUFTFAHRTFORSCHUNG, BERLIN (BERICHT A)	LG BA (NUMBER)/
LILIENTHAL-GESELLSCHAFT FUER LUFTFAHRTFORSCHUNG, BERLIN (BERICHT A)	LG BA-
LILIENTHAL-GESELLSCHAFT FUER LUFTFAHRTFORSCHUNG, BERLIN (BERICHT S)	LG BS-
LILIENTHAL-GESELLSCHAFT FUER LUFTFAHRTFORSCHUNG, BERLIN (TAGUNGSVERICHT)	LG TB (NUMBER)/
LILIENTHAL-GESELLSCHAFT FUER LUFTFAHRTFORSCHUNG, BERLIN (TAGUNGSBERICHT)	LG TB-
LIN (T.Y.) AND ASSOCIATES, LOS ANGELES	TYLA-
LINDE AIR PRODUCTS CO., N.Y.C.	PG-
LINDE AIR PRODUCTS CO., TONAWANDA, N.Y. (REF. 1)	(NUMBER)
LINDE AIR PRODUCTS CO., TONAWANDA, N.Y.	LAP-
LINDE AIR PRODUCTS CO., TONAWANDA, N.Y.	LAPC-
LINDE CO. RESEARCH AND DEV. LAB., TONAWANDA, N.Y.	LC-TRL-
LINDE CO. SPEEDWAY LABS., INDIANAPOLIS (SEE ALSO: UNION CARBIDE CORP. LINDE DIV.)	LC-SL-
LINDEN LABS., INC., STATE COLLEGE, PENNA.	LLI-
LINDEN LABS., INC., UNIVERSITY PARK, PENNA.	LL-
LINFIELD RESEARCH INSTITUTE, MC MINNVILLE, ORE.	LRI-
LING-TEMCO-VOUGHT, INC., DALLAS	LTV-

LING-TEMCO-VOUGHT, INC.

LING-TEMCO-VOUGHT, INC., DALLAS	MPR-
LING-TEMCO-VOUGHT, INC., DALLAS	TR-PL-
LING-TEMCO-VOUGHT, INC., DALLAS	V-
LING-TEMCO-VOUGHT, INC., DALLAS	XC-
LING-TEMCO-VOUGHT, INC. LTV ASTRONAUTICS DIV., DALLAS	LTV/A-
LING-TEMCO-VOUGHT, INC. LTV MILITARY ELECTRONICS DIV., DALLAS	LTV/M-
LING-TEMCO-VOUGHT, INC. LTV RANGE SYSTEMS DIV., DALLAS	LTV/R-
LING-TEMCO-VOUGHT, INC. LTV VOUGHT AERONAUTICS DIV., DALLAS	LTV/VA-
LINK AVIATION, INC., BINGHAMTON, N.Y.	ER-
LINK AVIATION, INC., BINGHAMTON, N.Y. (REF. 1)	(SEC.NO.)(MODEL NO.)(NO.)
LITERATURE SERVICE ASSOCIATES, BOUND BROOK, N.J. (TRANSLATIONS)	LSA-
LITHIUM CORP. OF AMERICA, INC., BESSEMER CITY, N.C.	LCA-
LITHIUM CORP. OF AMERICA, INC., MINNEAPOLIS	LCA-
LITHIUM CORP. OF AMERICA, INC., N.Y.C.	LCA-
LITTLE (ARTHUR D.), INC., CAMBRIDGE, MASS.	ADL-
LITTLE (ARTHUR D.), INC., CAMBRIDGE, MASS.	ADL-C-
LITTLE (ARTHUR D.), INC., CAMBRIDGE, MASS.	ADL-LOG-
LITTLE (ARTHUR D.), INC., CAMBRIDGE, MASS.	ADL-NCI-
LITTLE (ARTHUR D.), INC., CAMBRIDGE, MASS.	ALI-
LITTLE (ARTHUR D.), INC., CAMBRIDGE, MASS.	ALI-C-
LITTLE (ARTHUR D.), INC., CAMBRIDGE, MASS. (HELIUM)	BULLETIN HEL-
LITTLE (ARTHUR D.), INC., CAMBRIDGE, MASS. (BUMBLEBEE PROJECT) (REF. 8)	BUMBLEBEE-
LITTLE (ARTHUR D.), INC., CAMBRIDGE, MASS.	C-
LITTLE (ARTHUR D.), INC., CAMBRIDGE, MASS.	C-(NUMBER)-
LITTLE (ARTHUR D.), INC., CAMBRIDGE, MASS.	LIT-
LITTLE (ARTHUR D.), INC., CAMBRIDGE, MASS.	PA-AG-
LITTLE (ARTHUR D.), INC., CAMBRIDGE, MASS.	WORKING MEMO-
LITTLETON RESEARCH AND ENGINEERING CORP., MASS.	G-(NUMBER)-
LITTON INDUSTRIES, SAN CARLOS, CALIF.	DDRF-
LITTON INDUSTRIES. AMECOM DIV., SILVER SPRING, MD.	ERR-
LITTON INDUSTRIES. ELECTRON TUBE DIV., SAN CARLOS, CAL	C-(4 DIGITS)
LITTON INDUSTRIES, INC. (ALL LOCATIONS)	LI-
LITTON INDUSTRIES, INC. SPACE RESEARCH LABS., BEVERLY HILLS, CALIF.	SRL-
LITTON PRECISION PRODUCTS, INC. ADVANCED DEVICE LABS. CANOGA PARK, CALIF.	ADL-
LITTON PRECISION PRODUCTS, INC. AIRTRON DIV., MORRIS PLAINS, N.J.	R(NUMBERS)-
LITTON SCIENTIFIC SUPPORT LAB., FORT ORD, CALIF.	LSSL-TP-
LITTON SYSTEMS, INC.	LSI-
LITTON SYSTEMS, INC., BEVERLY HILLS, CALIF.	BH-
LITTON SYSTEMS, INC., WALTHAM, MASS.	DS-
LITTON SYSTEMS, INC., WOODLAND HILLS, CALIF.	HADES-R-
LITTON SYSTEMS, INC., WOODLAND HILLS, CALIF.	SMAD-
LITTON SYSTEMS, INC. APPLIED SCIENCE DIV., MINNEAPOLIS	ASD-
LITTON SYSTEMS, INC. APPLIED SCIENCE DIV., MINNEAPOLIS	LSI-ASD-
LITTON SYSTEMS, INC. DATA SYSTEMS DIV., VAN NUYS, CAL.	MS-
LITTON SYSTEMS, INC. DATA SYSTEMS DIV., VAN NUYS, CAL.	TI-
LITTON SYSTEMS, INC. GUIDANCE AND CONTROL SYSTEMS DIV., WOODLAND HILLS, CALIF.	AQ(NUMBER)D(YEAR)
LITTON SYSTEMS, INC. GUIDANCE AND CONTROL SYSTEMS DIV., WOODLAND HILLS, CALIF.	AQ(NUMBER)C(YEAR)
LITTON SYSTEMS, INC. GUIDANCE AND CONTROL SYSTEMS DIV., WOODLAND HILLS, CALIF.	G/CSD-AK(NUMBER)
LITTON SYSTEMS, INC. GUIDANCE AND CONTROL SYSTEMS DIV., WOODLAND HILLS, CALIF.	G/CSD-AQ(NUMBER)
LITTON SYSTEMS, INC. GUIDANCE AND CONTROL SYSTEMS DIV., WOODLAND HILLS, CALIF.	G/CSD-AY(NUMBER)
LITTON SYSTEMS, INC. GUIDANCE AND CONTROL SYSTEMS DIV., WOODLAND HILLS, CALIF.	G/CSD-R-
LITTON SYSTEMS, INC. GUIDANCE AND CONTROL SYSTEMS DIV., WOODLAND HILLS, CALIF.	G/CSD-TR-
LITTON SYSTEMS, INC. GUIDANCE AND CONTROLS SYSTEMS DIV., WOODLAND HILLS, CALIF.	PUB-
LITTON SYSTEMS, INC. GUIDANCE SYSTEMS LAB., WOODLAND HILLS, CALIF.	AQ(NUMBER)M(YEAR)
LIVERPOOL. UNIV.	UL-
LIVERPOOL. UNIV. DEPT. OF BUILDING SCIENCE	BS/A/(NUMBER-NUMBER)
LIVERPOOL. UNIV. DEPT. OF MECHANICAL ENGINEERING	ULME/(LETTER. NUMBER)
LIVERPOOL. UNIV. DEPT. OF MECHANICAL ENGINEERING	ULME-B-
LIVERPOOL. UNIV. DEPT. OF PHYSICS	ULDP-
LLOYD'S REGISTRY OF SHIPPING. RESEARCH AND TECHNICAL ADVISORY SERVICES DEPT., LONDON	T./T.A. (NUMBER)
LOCKHEED AIRCRAFT CORP. (ALL LOCATIONS) (DEVELOPMENT PLANNING REPORT)	DPR/
LOCKHEED AIRCRAFT CORP. (ALL LOCATIONS) (INTRADEPARTMENTAL COMMUNICATION)	IAD-
LOCKHEED AIRCRAFT CORP. (ALL LOCATIONS)	LAC-
LOCKHEED AIRCRAFT CORP. (ALL LOCATIONS) (DEVELOPMENT PLANNING REPORT)	LAC DPR-
LOCKHEED AIRCRAFT CORP., BURBANK, CALIF. (REF. 1)	(NUMBER)
LOCKHEED AIRCRAFT CORP., BURBANK, CALIF.	ERM-
LOCKHEED AIRCRAFT CORP., BURBANK, CALIF. (DEVELOPMENT PLANNING TECHNICAL MEMORANDA)	LAC-DPTM-
LOCKHEED AIRCRAFT CORP., BURBANK, CALIF.	MAS-
LOCKHEED AIRCRAFT CORP., BURBANK, CALIF.	SDR-
LOCKHEED AIRCRAFT CORP., BURBANK, CALIF.	SLR-
(SEE ALSO LATER NAME: LOCKHEED-CALIFORNIA CO.)	
LOCKHEED AIRCRAFT CORP., MARIETTA, GA.	C/(NO.)
LOCKHEED AIRCRAFT CORP., MARIETTA, GA. (SEMI-ANNUAL RADIATION EFFECTS SYMPOSIA)	LAC ANP RE SYM SA-
(SEE ALSO LATER NAME: LOCKHEED-GEORGIA CO.)	

LOCKHEED AIRCRAFT CORP. MISSILE SYSTEMS DIV., PALO ALTO, CALIF.	LAC MSD-
LOCKHEED AIRCRAFT CORP. MISSILE SYSTEMS DIV., PALO ALTO, CALIF.	LNP-NR-
LOCKHEED AIRCRAFT CORP. MISSILE SYSTEMS DIV., VAN NUYS, CALIF.	LAC MSD-
LOCKHEED AIRCRAFT CORP. MISSILE SYSTEMS DIV., VAN NUYS, CALIF. (SEE ALSO LATER NAME: LOCKHEED MISSILES AND SPACE CO)	MSD-
LOCKHEED AIRCRAFT CORP. MISSILES + SPACE DIV., PALO ALTO, CALIF.	MSD-TM-(NO.)-(NO.)-
LOCKHEED AIRCRAFT CORP. MISSILES AND SPACE DIV., SUNNYVALE, CALIF. (SPECIAL BIBLIOGRAPHY)	LAC-SB-(YEAR)-
LOCKHEED AIRCRAFT CORP. MISSILES AND SPACE DIV., SUNNYVALE, CALIF.	LAC-SRB-(YEAR)-
LOCKHEED AIRCRAFT CORP. MISSILES AND SPACE DIV., SUNNYVALE, CALIF.	LMSD-SB-(YR)-
LOCKHEED AIRCRAFT CORP. MISSILES AND SPACE DIV., SUNNYVALE, CALIF.	LMSD-SRB-(YEAR)-
LOCKHEED AIRCRAFT CORP. MISSILES AND SPACE DIV., SUNNYVALE, CALIF.	SRB-
LOCKHEED AIRCRAFT CORP. MISSILES AND SPACE DIV., SUNNYVALE, CALIF.	SRB-(YR.)-
LOCKHEED AIRCRAFT CORP. MISSILES AND SPACE DIV., SUNNYVALE, CALIF. (SEE ALSO LATER NAME: LOCKHEED MISSILES AND SPACE CO)	SSN-T(YEAR)-
LOCKHEED AIRCRAFT CORP. STRUCTURAL RESEARCH LAB., MARIETTA, GA.	ER-
LOCKHEED AIRCRAFT SERVICE CO. VEHICLE DEVELOPMENT GROUP, ONTARIO, CALIF.	SDR-
LOCKHEED ELECTRONICS CO., PLAINFIELD, N.J.	LEC-
LOCKHEED ELECTRONICS CO., PLAINFIELD, N.J.	LEC-(NUMBER)-(NUMBER)-
LOCKHEED ELECTRONICS CO., WHITE SANDS MISSILE RANGE, N. MEX.	LEC-TR-(YEAR)-
LOCKHEED ELECTRONICS CO. AEROSPACE SYSTEMS DIV., HOUSTON, TEX.	LEC/HASD-
LOCKHEED ELECTRONICS CO. AEROSPACE SYSTEMS DIV., HOUSTON, TEX.	TWP-(YEAR)-
LOCKHEED ELECTRONICS CO. MILITARY SYSTEMS, PLAINFIELD, N.J.	EC-RD-
LOCKHEED-CALIFORNIA CO., BURBANK	LAC-(NUMBER)-LFL-T-
LOCKHEED-CALIFORNIA CO., BURBANK	LAC-CMRI-
LOCKHEED-CALIFORNIA CO., BURBANK	LAC-LR-
LOCKHEED-CALIFORNIA CO., BURBANK	LAC-OEA/SST-
LOCKHEED-CALIFORNIA CO., BURBANK	LAC-SST/
LOCKHEED-CALIFORNIA CO., BURBANK	LR-
LOCKHEED-CALIFORNIA CO., BURBANK	L/STR-
LOCKHEED-CALIFORNIA CO., BURBANK	MPR-
LOCKHEED-CALIFORNIA CO., BURBANK	P(NUMBER)(LETTER)-
LOCKHEED-GEORGIA CO., MARIETTA	ER-
LOCKHEED-GEORGIA CO., MARIETTA (ENGINEERING REPORTS)	GELAC-ER-
LOCKHEED-GEORGIA CO., MARIETTA	GMRI-
LOCKHEED-GEORGIA CO., MARIETTA (ENGINEERING REPORTS)	LAC-ER-
LOCKHEED-GEORGIA CO., MARIETTA (NUCLEAR REPORT)	LAC-NR-
LOCKHEED-GEORGIA CO., MARIETTA	LAC-ORD-
LOCKHEED-GEORGIA CO., MARIETTA (CONTRACT REPORT)	LG(NO. LTRS. NO.)
LOCKHEED-GEORGIA CO., MARIETTA	LGR-ER-
LOCKHEED-GEORGIA CO., MARIETTA	LGR-ER/S-
LOCKHEED-GEORGIA CO., MARIETTA	LGR-M/P-
LOCKHEED-GEORGIA CO., MARIETTA (NUCLEAR PRODUCTS)	LNP-
LOCKHEED-GEORGIA CO., MARIETTA (NUCLEAR REPORT)	LNP-NR-
LOCKHEED-GEORGIA CO., MARIETTA (NUCLEAR REPORT)	NR-
LOCKHEED MISSILES AND SPACE CO., HUNTSVILLE, ALA.	HTVL/LMSC-
LOCKHEED MISSILES AND SPACE CO., HUNTSVILLE, ALA.	LMSC/HREC-A(NO.S)-
LOCKHEED MISSILES AND SPACE CO., HUNTSVILLE, ALA.	LMSC/HTVL-
LOCKHEED MISSILES AND SPACE CO., PALO ALTO, CALIF.	LAC-LMSC-
LOCKHEED MISSILES AND SPACE CO., PALO ALTO, CALIF.	LMSC-
LOCKHEED MISSILES AND SPACE CO., PALO ALTO, CALIF.	LMSC-(NUMBERS)-
LOCKHEED MISSILES AND SPACE CO., PALO ALTO, CALIF. (UNCLASSIFIED AND CONFIDENTIAL REPORTS)	LMSC-A-(SIX DIGITS)
LOCKHEED MISSILES AND SPACE CO., PALO ALTO, CALIF. (SECRET REPORTS)	LMSC-B-(SIX DIGITS)
LOCKHEED MISSILES AND SPACE CO., PALO ALTO, CALIF. (CITATION BIBLIOGRAPHY)	LMSC-CB(YEAR)-
LOCKHEED MISSILES AND SPACE CO., PALO ALTO, CALIF. (LITERATURE SEARCH)	LMSC-LS(YEAR)-
LOCKHEED MISSILES AND SPACE CO., PALO ALTO, CALIF.	LMSC-M-
LOCKHEED MISSILES AND SPACE CO., PALO ALTO, CALIF.	LMSC-N-(NUMBER(S))-
LOCKHEED MISSILES AND SPACE CO., PALO ALTO, CALIF.	LMSC-N-A-(NUMBERS)-
LOCKHEED MISSILES AND SPACE CO., PALO ALTO, CALIF.	LMSD-
LOCKHEED MISSILES AND SPACE CO., PALO ALTO, CALIF.	LSMC-
LOCKHEED MISSILES AND SPACE CO., PALO ALTO, CALIF.	MDS-
LOCKHEED MISSILES AND SPACE CO., PALO ALTO, CALIF.	MSD-
LOCKHEED MISSILES AND SPACE CO., PALO ALTO, CALIF.	N-
LOCKHEED MISSILES AND SPACE CO., PALO ALTO, CALIF.	SS(NO.)B-T(NO.)-
LOCKHEED MISSILES AND SPACE CO., PALO ALTO, CALIF.	SSE-T(NUMBER)-
LOCKHEED MISSILES AND SPACE CO., PALO ALTO, CALIF.	SSM-T(YEAR)-
LOCKHEED MISSILES AND SPACE CO., SUNNYVALE, CALIF.	B-
LOCKHEED MISSILES AND SPACE CO., SUNNYVALE, CALIF. (CITATION BIBLIOGRAPHY)	CB-(YEAR)-
LOCKHEED MISSILES AND SPACE CO., SUNNYVALE, CALIF.	LAC-LMSD-
LOCKHEED MISSILES AND SPACE CO., SUNNYVALE, CALIF.	LMSC-
LOCKHEED MISSILES AND SPACE CO., SUNNYVALE, CALIF.	LMSC-(NUMBERS)
LOCKHEED MISSILES AND SPACE CO., SUNNYVALE, CALIF.	LMSC-A-(SIX DIGITS)
LOCKHEED MISSILES AND SPACE CO., SUNNYVALE, CALIF. (CITATION BIBLIOGRAPHY)	LMSC-CB-(YEAR)-
LOCKHEED MISSILES AND SPACE CO., SUNNYVALE, CALIF.	LMSC-D(NUMBER)
LOCKHEED MISSILES AND SPACE CO., SUNNYVALE, CALIF.	LMSC-EETP-(NUMBER)/
LOCKHEED MISSILES AND SPACE CO., SUNNYVALE, CALIF.	LMSC-NSP-(YEAR)-
LOCKHEED MISSILES AND SPACE CO., SUNNYVALE, CALIF.	LMSC-SB-(YEAR)-
LOCKHEED MISSILES AND SPACE CO., SUNNYVALE, CALIF.	LMSC-SRB-(YEAR)-
LOCKHEED MISSILES AND SPACE CO., SUNNYVALE, CALIF.	LMSC-SSD-(NO.)-(YR.)-
LOCKHEED MISSILES AND SPACE CO., SUNNYVALE, CALIF.	M
LOCKHEED MISSILES AND SPACE CO., SUNNYVALE, CALIF.	MRI-
LOCKHEED MISSILES AND SPACE CO., SUNNYVALE, CALIF.	NSP-(NO.)-
LOCKHEED MISSILES AND SPACE CO., SUNNYVALE, CALIF.	R/C-
LOCKHEED MISSILES AND SPACE CO., SUNNYVALE, CALIF.	RVTP-
LOCKHEED MISSILES AND SPACE CO., SUNNYVALE, CALIF.	SB-(YR.)-
LOCKHEED MISSILES AND SPACE CO., SUNNYVALE, CALIF.	SPEC-(NUMBER)/
LOCKHEED MISSILES AND SPACE CO., SUNNYVALE, CALIF.	SPEC-LMSC-
LOCKHEED MISSILES AND SPACE CO., SUNNYVALE, CALIF.	SSM-T(NUMBERS)-
LOCKHEED MISSILES AND SPACE CO., VAN NUYS, CALIF.	LMSC-
LOCKHEED MISSILES AND SPACE CO. ADVANCED STRATEGIC PROGRAMS, SUNNYVALE, CALIF.	SR/(NUMBER)-M/
LOCKHEED MISSILES AND SPACE CO. ELECTRONIC SCIENCES LAB., PALO ALTO, CALIF.	M-(NUMBER)-(YEAR)-
LOCKHEED MISSILES AND SPACE CO. HUNTSVILLE RESEARCH AND ENGINEERING CENTER, ALA.	HREC-
LOCKHEED MISSILES AND SPACE CO. HUNTSVILLE RESEARCH AND ENGINEERING CENTER, ALA.	LMSC/HREC-(LETTER)-

LOCKHEED MISSLES AND SPACE CO. MATERIALS SCIENCES LAB.

LOCKHEED MISSILES AND SPACE CO. MATERIALS SCIENCES LAB., PALO ALTO, CALIF.	M-
LOCKHEED MISSILES AND SPACE CO. RESEARCH DIV., PALO ALTO, CALIF.	L-(NUMBER)-(YEAR)-
LOCKHEED NUCLEAR PRODUCTS, MARIETTA, GA.	LAC-NR-
LOCKHEED NUCLEAR PRODUCTS, MARIETTA, GA.	LNP-ER-
LOCKHEED NUCLEAR PRODUCTS, MARIETTA, GA.	LNP-NM-
LOCKHEED NUCLEAR PRODUCTS, MARIETTA, GA.	LNP-NP-
LOCKHEED NUCLEAR PRODUCTS, MARIETTA, GA.	LNP-NR-
LOCKHEED NUCLEAR PRODUCTS, MARIETTA, GA.	NM-
LOCKHEED NUCLEAR PRODUCTS, MARIETTA, GA.	NR-
LOCKHEED NUCLEAR PRODUCTS. GEORGIA DIV., MARIETTA	ER-
LOCKHEED PROPULSION CO., REDLANDS, CALIF.	LPC-
LOCKHEED PROPULSION CO., REDLANDS, CALIF.	LPC-(NUMBER)(LETTER)
LOCKHEED PROPULSION CO., REDLANDS, CALIF.	LPC-(NUMBER)-S-
LOCKLAND AIRCRAFT REACTORS OPERATIONS OFFICE, AEC (REF. 4)	LAR-
LOCKLAND AREA OFFICE, AEC (REF. 4)	LAO-
LOCKLAND AREA OFFICE, AEC	LAO-A-
LOGICON, INC., REDONDO BEACH, CALIF.	ATS-(NUMBER)-
LOGICON, INC., REDONDO BEACH, CALIF.	CS-(NUMBER)-R(NUMBER)
LOGICON, INC., REDONDO BEACH, CALIF.	MGTD-(NUMBER)-
LOGICON, INC., SAN PEDRO, CALIF.	FGE-
LOGICON, INC., SAN PEDRO, CALIF.	RA-
LOGICON, INC., SAN PEDRO, CALIF.	RSD-(NUMBER)-
LOGISTICS MANAGEMENT INST., WASHINGTON, D.C.	LMI-(YEAR)-
LOGISTICS MANAGEMENT INST., WASHINGTON, D.C.	LMI-TASK-(YEAR)-
LOGISTICS MANAGEMENT INST., WASHINGTON, D.C.	TASK-(YEAR)-
LONDON. CITY UNIV. DEPT. OF AERONAUTICS	AERO.(YEAR)/
LONDON. UNIV.	ARC-
LONDON. UNIV.	FTR-
LONDON. UNIV.	RM-(YEAR/NUMBER)
LONDON. UNIV., UNIVERSITY COLL. DEPT. OF PHYSICS	LONDON-
LONDON MILITARY DOCUMENTS CENTER	LMDC-
LONG BEACH STATE COLLEGE, CALIF.	LBSC-
LONG ISLAND BIOLOGICAL ASSN. BIOLOGICAL LAB., COLD SPRING HARBOR, N.Y.	LIBA-
LONG ISLAND UNIV., GREENVALE, N.Y.	SERG-TR-
LONG RANGE FUZING STANDARDIZATION GROUP (AEC-DOD), ALBUQUERQUE, N. MEX.	LRFSG-
LOOKOUT MOUNTAIN LAB., LOS ANGELES	LML-
LOS ALAMOS NUCLEAR CORP., N. MEX.	LANC-R-
LOS ALAMOS SCIENTIFIC LAB., N. MEX. (REF. 1)	(NUMBERS...)
LOS ALAMOS SCIENTIFIC LAB., N. MEX. (INTERNAL CORRESPONDENCE SERIAL USED BY GROUPS IN A DIV.)	A-1- THRU A-3-
LOS ALAMOS SCIENTIFIC LAB., N. MEX.	ADWD-
LOS ALAMOS SCIENTIFIC LAB., N. MEX. (ASSIGNED BEFORE 1950 TO MISCELLANEOUS AMERICAN REPORTS) (REF. 22)	AM-
LOS ALAMOS SCIENTIFIC LAB., N. MEX.	AN-
LOS ALAMOS SCIENTIFIC LAB., N. MEX.	AW-
LOS ALAMOS SCIENTIFIC LAB., N. MEX. (ASSIGNED BEFORE 1950 TO MISC. BRITISH AND CANADIAN REPORTS)(REF.22)	BM-
LOS ALAMOS SCIENTIFIC LAB., N. MEX. (INTERNAL CORRESPONDENCE SERIAL USED BY GROUPS IN C DIV.)	C-1- THRU C-8-
LOS ALAMOS SCIENTIFIC LAB., N. MEX.	CMB-
LOS ALAMOS SCIENTIFIC LAB., N. MEX. (INTERNAL CORRESPONDENCE SERIAL USED BY GROUPS IN CMB DIV.)	CMB-1- THRU CMB-14-
LOS ALAMOS SCIENTIFIC LAB., N. MEX.	CMF-
LOS ALAMOS SCIENTIFIC LAB., N. MEX. (INTERNAL CORRESPONDENCE SERIAL USED BY GROUPS IN CMF DIV., WHICH LATER WAS INCORPORATED INTO CMB DIV.)	CMF-1- THRU CMF-13-
LOS ALAMOS SCIENTIFIC LAB., N. MEX.	CMR-
LOS ALAMOS SCIENTIFIC LAB., N. MEX. (INTERNAL CORRESPONDENCE SERIAL USED BY GROUPS IN CMR DIV., WHICH LATER SPLIT INTO CMB AND CMF DIVS.)	CMR-1- THRU CMR-13-
LOS ALAMOS SCIENTIFIC LAB., N. MEX.	CMR-AE-
LOS ALAMOS SCIENTIFIC LAB., N. MEX.	CMR-DO-GS-
LOS ALAMOS SCIENTIFIC LAB., N. MEX.	CMR-DO-TECH-
LOS ALAMOS SCIENTIFIC LAB., N. MEX.	CMR-M-
LOS ALAMOS SCIENTIFIC LAB., N. MEX.	CMR-TA-
LOS ALAMOS SCIENTIFIC LAB., N. MEX. (INTERNAL CORRESPONDENCE SERIAL USED BY GROUPS IN CNC DIV.)	CNC-1- THRU CNC-11-
LOS ALAMOS SCIENTIFIC LAB., N. MEX. (BIBLIOGRAPHIES)	D-BIB-
LOS ALAMOS SCIENTIFIC LAB., N. MEX.	DCL-
LOS ALAMOS SCIENTIFIC LAB., N. MEX.	D-DIV-
LOS ALAMOS SCIENTIFIC LAB., N. MEX.	D-DOC-
LOS ALAMOS SCIENTIFIC LAB., N. MEX.	DIR-
LOS ALAMOS SCIENTIFIC LAB., N. MEX.	DIRX-
LOS ALAMOS SCIENTIFIC LAB., N. MEX.	D-LIB-
LOS ALAMOS SCIENTIFIC LAB., N. MEX.	D-RL-
LOS ALAMOS SCIENTIFIC LAB., N. MEX. (INTERNAL CORRESPONDENCE SERIAL USED BY GROUPS IN E DIV.)	E-1- THRU E-5-
LOS ALAMOS SCIENTIFIC LAB., N. MEX.	GMX-
LOS ALAMOS SCIENTIFIC LAB., N. MEX. (INTERNAL CORRESPONDENCE SERIAL USED BY GROUPS IN GMX DIV.)	GMX-1- THRU GMX-11-
LOS ALAMOS SCIENTIFIC LAB., N. MEX.	H(NO.)MR(YR.)-
LOS ALAMOS SCIENTIFIC LAB., N. MEX. (INTERNAL CORRESPONDENCE SERIAL USED BY GROUPS IN H DIV.)	H-1- THRU H-9-
LOS ALAMOS SCIENTIFIC LAB., N. MEX. (BIBLIOGRAPHY)	ISD-BIB-
LOS ALAMOS SCIENTIFIC LAB., N. MEX.	J-
LOS ALAMOS SCIENTIFIC LAB., N. MEX. (INTERNAL CORRESPONDENCE SERIAL USED BY GROUPS IN J DIV.)	J-1- THRU J-18-
LOS ALAMOS SCIENTIFIC LAB., N. MEX.	JB-
LOS ALAMOS SCIENTIFIC LAB., N. MEX.	JO-
LOS ALAMOS SCIENTIFIC LAB., N. MEX.	K-
LOS ALAMOS SCIENTIFIC LAB., N. MEX. (INTERNAL CORRESPONDENCE SERIAL USED BY GROUPS IN K DIV.)	K-1- THRU K-4-
LOS ALAMOS SCIENTIFIC LAB., N. MEX.	KIWI-TNT-
LOS ALAMOS SCIENTIFIC LAB., N. MEX. (INTERNAL CORRESPONDENCE SERIAL USED BY GROUPS IN L DIV.)	L-1- THRU L-5-
LOS ALAMOS SCIENTIFIC LAB., N. MEX. (REF. 21)	LA-
LOS ALAMOS SCIENTIFIC LAB., N. MEX. (REF. 21)	LA-(NUMBER)-MS
LOS ALAMOS SCIENTIFIC LAB., N.M. (PROCEDURES MANUAL)	LA-(NUMBER)-SOP
LOS ALAMOS SCIENTIFIC LAB., N. MEX. (WORK DONE AT NEW MEXICO. UNIV., ALBUQUERQUE)	LA-(NO.)-UNM
LOS ALAMOS SCIENTIFIC LAB., N. MEX. (PROGRESS RPT.)	LA-(NUMBER)-PR
LOS ALAMOS SCIENTIFIC LAB., N. MEX. (TRANSLATION)	LA-(NUMBER)-TR
LOS ALAMOS SCIENTIFIC LAB., N. MEX. (BIBLIOGRAPHY)	LA-(NUMBER)-BIB
LOS ALAMOS SCIENTIFIC LAB., N. MEX. (RPT. ABSTRACTS)	LAA-
LOS ALAMOS SCIENTIFIC LAB., N. MEX.	LAB-(LETTERS)-

LOS ALAMOS SCIENTIFIC LAB., N. MEX.	LA-DC-
LOS ALAMOS SCIENTIFIC LAB., N. MEX. (DECLASSIFICATION CODE)	LADC-
LOS ALAMOS SCIENTIFIC LAB., N. MEX. (CODE USED 1972-)	LA-DC-(YEAR)-
LOS ALAMOS SCIENTIFIC LAB., N. MEX.	LA-DIR-
LOS ALAMOS SCIENTIFIC LAB., N. MEX.	LA-GMX-
LOS ALAMOS SCIENTIFIC LAB., N. MEX.	LA-J-
LOS ALAMOS SCIENTIFIC LAB., N. MEX. (REF. 21)	LAMD-
LOS ALAMOS SCIENTIFIC LAB., N. MEX.	LAMS-
LOS ALAMOS SCIENTIFIC LAB., N. MEX.	LASL-
LOS ALAMOS SCIENTIFIC LAB., N. MEX. (PROPOSAL)	LASL-P-
LOS ALAMOS SCIENTIFIC LAB., N. MEX. (TRANSLATION) (CODE USED 1964-)	LA-TR-(YEAR)-
LOS ALAMOS SCIENTIFIC LAB., N. MEX. (TRANSLATION) (CODE USED BEFORE 1964)	LA-TR-
LOS ALAMOS SCIENTIFIC LAB., N. MEX.	LPC-MIN.-
LOS ALAMOS SCI. LAB., N. MEX. (LITTLE TITLE LIST)	LTL-
LOS ALAMOS SCIENTIFIC LAB., N. MEX.	MP-
LOS ALAMOS SCIENTIFIC LAB., N. MEX. (INTERNAL CORRESPONDENCE SERIAL USED BY GROUPS IN MP DIV.)	MP-1- THRU MP-9-
LOS ALAMOS SCIENTIFIC LAB., N. MEX.	N-
LOS ALAMOS SCIENTIFIC LAB., N. MEX. (INTERNAL CORRESPONDENCE SERIAL USED BY GROUPS IN N DIV.)	N-1- THRU N-7-
LOS ALAMOS SCIENTIFIC LAB., N. MEX.	OSTRICH-
LOS ALAMOS SCIENTIFIC LAB., N. MEX.	P-
LOS ALAMOS SCIENTIFIC LAB., N. MEX. (INTERNAL CORRESPONDENCE SERIAL USED BY GROUPS IN P DIV.)	P-1- THRU P-18-
LOS ALAMOS SCIENTIFIC LAB., N. MEX.	PT-
LOS ALAMOS SCIENTIFIC LAB., N. MEX.	ROVER-
LOS ALAMOS SCIENTIFIC LAB., N. MEX.	SD-
LOS ALAMOS SCIENTIFIC LAB., N. MEX.	T-
LOS ALAMOS SCIENTIFIC LAB., N. MEX. (INTERNAL CORRESPONDENCE SERIAL USED BY GROUPS IN T DIV.)	T-1- THRU T-9-
LOS ALAMOS SCIENTIFIC LAB., N. MEX.	TAD-
LOS ALAMOS SCIENTIFIC LAB., N. MEX. (INTERNAL CORRESPONDENCE SERIAL USED BY GROUPS IN TD DIV.)	TD-1- THRU TD-6-
LOS ALAMOS SCIENTIFIC LAB., N. MEX. (REPORT LIBRARY TITLE LIST)	TL-
LOS ALAMOS SCIENTIFIC LAB., N. MEX.	TM-
LOS ALAMOS SCIENTIFIC LAB.,N.M.(TURRET WORKING COMM.)	TWC-
LOS ALAMOS SCIENTIFIC LAB., N. MEX.	TWG-
LOS ALAMOS SCIENTIFIC LAB., N. MEX.	W-
LOS ALAMOS SCIENTIFIC LAB., N. MEX.	W-(NO.)-RD-
LOS ALAMOS SCIENTIFIC LAB., N. MEX. (INTERNAL CORRESPONDENCE SERIAL USED BY GROUPS IN W DIV.)	W-1- THRU W-11-
LOS ALAMOS SCIENTIFIC LAB., N. MEX. (FILM)	Y-
LOS ANGELES REGIONAL TECH. INFO. USERS COUNCIL	LARTIUC-
LOUISIANA STATE UNIV., BATON ROUGE (JOINT VENTURE)	BULL-GT-
LOUISIANA STATE UNIV., BATON ROUGE	LSU-
LOUISIANA STATE UNIV., BATON ROUGE	LU-
LOUISIANA STATE UNIV., BATON ROUGE. DEPT. OF AGRICULTURAL ECONOMICS AND AGRIBUSINESS	DAE-
LOUISIANA WATER RESOURCES RESEARCH INST., BATON ROUGE (JOINT VENTURE)	BULL-GT-
LOUISVILLE, KY. UNIV.	APR-
LOUISVILLE, KY. UNIV.	LOU-
LOUISVILLE, KY. UNIV. INST. OF INDUSTRIAL RES.(REF.1)	(NUMBER)
LOUISVILLE, KY. UNIV. INST. OF INDUSTRIAL RESEARCH	LIIR-
LOUISVILLE, KY. UNIV. PERFORMANCE RESEARCH LAB.	ITR-(YEAR)-
LOUISVILLE, KY. UNIV. PERFORMANCE RESEARCH LAB.	PR-(YEAR)-
LOUISVILLE, KY. UNIV. SCHOOL OF MEDICINE	LU-DSR-
LOUISVILLE, KY. UNIV. SCHOOL OF MEDICINE	LU/DSR-(YR)-
LOUVAIN, BELGIUM. UNIVERSITE	LOUU-
LOUVAIN, BELGIUM. UNIVERSITE. CENTRE DE PHYSIQUE	CPNL-WI-
LOVELACE FOUNDATION FOR MEDICAL EDUCATION AND RESEARCH, ALBUQUERQUE, N. MEX.	LF-
LOVELACE FOUNDATION FOR MEDICAL EDUCATION AND RESEARCH, ALBUQUERQUE, N. MEX.	LFMER-
LOVELACE FOUNDATION FOR MEDICAL EDUCATION AND RESEARCH, ALBUQUERQUE, N. MEX. (INTERNAL REPORTS)	LFMR-
LOVELACE FOUNDATION FOR MEDICAL EDUCATION AND RESEARCH, ALBUQUERQUE, N. MEX. (TRANSLATIONS)	LF-TR-
LOVELACE FOUNDATION FOR MEDICAL EDUCATION AND RESEARCH, ALBUQUERQUE, N. MEX.	LOVELACE-
LOVELACE FOUNDATION FOR MEDICAL EDUCATION AND RESEARCH, ALBUQUERQUE, N. MEX.	O-
LOWELL OBSERVATORY, FLAGSTAFF, ARIZ.	LO-
LOWELL TECHNOLOGICAL INST. RESEARCH FDN., MASS.	A(YEAR)-
LOWELL TECHNOLOGICAL INST. RESEARCH FDN., MASS.	LTIRF-
LOWELL TECHNOLOGICAL INST. RESEARCH FDN., MASS.	LTIRF-(NUMBER)/IP
LOWRY AFB MAINTENANCE LAB., COLO.	ML-LN-(NUMBER)-
LOWRY AFB MAINTENANCE LAB., COLO.	ML-TM-(NUMBER)-
LTV AEROSPACE CORP., DALLAS	LTV-
LTV AEROSPACE CORP. MISSILES AND SPACE DIV., DALLAS	MSD-(NUMBER)-
LTV AEROSPACE CORP. MISSILES AND SPACE DIV., DALLAS	MSD-ES-
LTV AEROSPACE CORP. MISSILES AND SPACE DIV., DALLAS	MSD-T-
LTV AEROSPACE CORP. MISSILES AND SPACE DIV., DALLAS	MSD-TP(NUMBER)-
LTV ELECTROSYSTEMS, INC., DALLAS	G(NO.).(NO.).(NO.)(LTR.)
LTV ELECTROSYSTEMS, INC., GARLAND, TEX.	LTV-(NUMBERS;W/
LTV ELECTROSYSTEMS INC., GREENVILLE, TEX.	G8406.(NUMBER).(NUMBER)
LTV ELECTROSYSTEMS INC., GREENVILLE, TEX.	G(NO.-NO.-NO.)B
LTV ELECTROSYSTEMS INC., GREENVILLE, TEX.	LTV-
LTV ELECTROSYSTEMS INC. DEPT. OF ADVANCED SYSTEMS DEVELOPMENT, GREENVILLE, TEX.	TR-(YEAR- 5 DIGITS)-
LUDWIG-MAXIMILIANS-UNIVERSITAET, MUNICH. METEOROLOGISCHES INST.	MITT-
LUFTFAHRTFORSCHUNG, MUNICH	LFM-
LUFTFAHRTFORSCHUNGSANSTALT HERMANN GOERING	LFA-
LUFTWAFFENAMT, PORZ, GERMANY (GEOPHYSIKALISCHER BERATUNGSDIENST DER BUNDESWEHR)	GEOPHYBDBW-FM-
LUFTWAFFENAMT, PORZ, GERMANY (GEOPHYSIKALISCHER BERATUNGSDIENST DER BUNDESWEHR)	GEOPHYSBDBW-FM/
LUND, SWEDEN. INSTITUTE OF TECHNOLOGY	UDC-
LUND, SWEDEN. INST. OF TECH. DEPT. OF NUCLEAR PHYSICS (MANY RPTS. ISSUED JOINTLY WITH LUND UNIV.)	LUNP-
LUND, SWEDEN. UNIV.	LU-S-

LUND, SWEDEN. UNIV.

LUND, SWEDEN. UNIV.	ME-
LUND, SWEDEN. UNIV. DEPT. OF PHYSICS	LUNP-
LUND, SWEDEN. UNIV. DEPT. OF PHYSICS	LUSY-
LUND, SWEDEN. UNIV. INST. OF PHYSICS	LUIP-CR-(YEAR)-
LUNDBERG (BO), BROMMA, SWEDEN	BL-(3 DIGITS)
LUNDY ELECTRONICS AND SYSTEMS, INC., GLEN HEAD, N.Y.	(NUMBER)-R-
LURIA-COURNAND, INC., HAVRE DE GRACE, MD.	ER-
LYON. UNIVERSITE. INSTITUT DE PHYSIQUE NUCLEAIRE.	LYCEN/

M.C. MANUFACTURING CO., LAKE ORION, MICH. (REF. 1)	(NUMBER)
MC ALLISTER AND ASSOCIATES, INC. ALBUQUERQUE, N.MEX.	MCA-
MC CLELLAN CENTRAL LAB., MC CLELLAN AFB, CALIF.	MCL-
MC CLELLAN CENTRAL LAB., MC CLELLAN AFB, CALIF.	MCL-TP-
MC CLELLAN CENTRAL LAB., MC CLELLAN AFB, CALIF.	MCL-TR-
MC CRONE RESEARCH INST., CHICAGO (FINAL REPORT)	MCCRONE-
MAC DILL AFB, FLA.	MAFB-TN-
MAC DONALD (W.S.) CO., INC., CAMBRIDGE, MASS.	MTA-
MC DONNELL AIRCRAFT CORP., ST. LOUIS (REF. 1)	(NUMBER)
MC DONNELL AIRCRAFT CORP., ST. LOUIS	A-59-
MC DONNELL AIRCRAFT CORP., ST. LOUIS	A001 THRU A999
MC DONNELL AIRCRAFT CORP., ST. LOUIS	B001 THRU B999
MC DONNELL AIRCRAFT CORP., ST. LOUIS	B(NUMBER)-(NUMBER)
MC DONNELL AIRCRAFT CORP., ST. LOUIS	BNW-
MC DONNELL AIRCRAFT CORP., ST. LOUIS	E001 THRU E999
MC DONNELL AIRCRAFT CORP., ST. LOUIS	E(NUMBER)
MC DONNELL AIRCRAFT CORP., ST. LOUIS	F001 THRU F999
MC DONNELL AIRCRAFT CORP., ST. LOUIS (FLIGHT TEST RESEARCH MEMORANDUM)	FTRM-
MC DONNELL AIRCRAFT CORP., ST. LOUIS	G(NUMBER)
MC DONNELL AIRCRAFT CORP., ST. LOUIS	GMS-
MC DONNELL AIRCRAFT CORP., ST. LOUIS (LIBRARY BIB.)	LB-
MC DONNELL AIRCRAFT CORP., ST. LOUIS	MAC-
MC DONNELL AIRCRAFT CORP., ST. LOUIS	MAC-(NO.)-ALC(NO.)
MC DONNELL AIRCRAFT CORP., ST. LOUIS	MAC-E-
MC DONNELL AIRCRAFT CORP., ST. LOUIS	MAC-TR-
MC DONNELL AIRCRAFT CORP., ST. LOUIS	MAC-TR-(NUMBER)-
MC DONNELL AIRCRAFT CORP., ST. LOUIS	MCD-
MC DONNELL AIRCRAFT CORP., ST. LOUIS	MC DONNELL A-
MC DONNELL AIRCRAFT CORP., ST. LOUIS	MCDONNELL-A(NO.)
MC DONNELL AIRCRAFT CORP., ST. LOUIS	MC DONNELL B-
MC DONNELL AIRCRAFT CORP., ST. LOUIS (MATERIAL SPECS)	MMS-
MC DONNELL AIRCRAFT CORP., ST. LOUIS (PROCESS SPECS.)	P.S.-
MC DONNELL AIRCRAFT CORP., ST. LOUIS (ENGINEERING DEVELOPMENT AND RESEARCH)	SEDR-
MC DONNELL ASTRONAUTICS CO., ST. LOUIS	F(NUMBER)
MC DONNELL ASTRONAUTICS CO., ST. LOUIS	MAC-E(NUMBER)
MC DONNELL ASTRONAUTICS CO., ST. LOUIS	MAC-F-
MC DONNELL ASTRONAUTICS CO., ST. LOUIS	MAC-K(NUMBER)-
MC DONNELL ASTRONAUTICS CO., ST. LOUIS (BIBLIOGRAPHY)	MAC-LB-
MC DONNELL ASTRONAUTICS CO., ST. LOUIS	MCASTRO-F(NUMBER)
MC DONNELL ASTRONAUTICS CO., ST. LOUIS	MCASTRO-G(NO.)
MC DONNELL ASTRONAUTICS CO., ST. LOUIS	STL-II-
MC DONNELL DOUGLAS ASTRONAUTICS CO. (ALL LOCATIONS)	DAC-
MC DONNELL DOUGLAS ASTRONAUTICS CO. (ALL LOCATIONS)	DAC-PAPER-
MC DONNELL DOUGLAS ASTRONAUTICS CO. (ALL LOCATIONS)	MDAC-
MC DONNELL DOUGLAS ASTRONAUTICS CO. (ALL LOCATIONS)	MDC-G-
MC DONNELL DOUGLAS ASTRONAUTICS CO., HUNTINGTON BEACH, CALIF.	MDAC-PAPER-WD-
MC DONNELL DOUGLAS ASTRONAUTICS CO., HUNTINGTON BEACH, CALIF.	MDAC-WD-
MC DONNELL DOUGLAS ASTRONAUTICS CO., HUNTINGTON BEACH, CALIF.	SM-(5 DIGITS)
MC DONNELL DOUGLAS ASTRONAUTICS CO., ST. LOUIS	MDC-E(NUMBER)
MC DONNELL DOUGLAS CO., RICHLAND, WASH.	DWDL-(NUMBER-NUMBER)
MC DONNELL DOUGLAS CO., RICHLAND, WASH.	MDG-G-
MC DONNELL-DOUGLAS CO., SANTA MONICA, CALIF.	DAC-
MC DONNELL-DOUGLAS CO., SANTA MONICA, CALIF.	DOUGLAS PAPER-
MC DONNELL DOUGLAS CORP., ST. LOUIS	G(NUMBER)
MC DONNELL DOUGLAS CORP. DOUGLAS AIRCRAFT DIV., LONG BEACH, CALIF.	DAD-
MC DONNELL DOUGLAS CORP. DOUGLAS AIRCRAFT DIV., LONG BEACH, CALIF.	MDC-J-
MC GILL UNIV., MONTREAL (REF. 1)	(NUMBERS...)
MC GILL UNIV., MONTREAL	AE-
MC GILL UNIV., MONTREAL	EIC-
MC GILL UNIV., MONTREAL	MCGU-
MC GILL UNIV., MONTREAL	MEMO-(YEAR)-(MONTH)
MC GILL UNIV., MONTREAL	MERC-TN-
MC GILL UNIV., MONTREAL	MW-
MC GILL UNIV., MONTREAL. ARCTIC METEOROLOGY RES. GP.	MCGU-AMRG-
MC GILL UNIV., MONTREAL. ARCTIC METEOROLOGY RES. GP.	METEOROLOGY-
MC GILL UNIV., MONTREAL. AVIATION MEDICAL RES. UNIT	AMRU-R-(NUMBER-NUMBER)
MC GILL UNIV., MONTREAL. DEPT. OF CHEMISTRY	CE-(NO.)-PR-
MC GILL UNIV., MONTREAL. DEPT. OF GEOGRAPHY	SAVANNA RESEARCH SER-
MC GILL UNIV., MONTREAL. DEPT. OF MECHANICAL ENG.	MERL-(YEAR)-
MC GILL UNIV., MONTREAL. DEPT. OF MECHANICAL ENG.	MERL-TN-(YEAR/NUMBER)
MC GILL UNIV., MONTREAL. EATON ELECTRONICS RES. LAB.	EERL-
MC GILL UNIV., MONTREAL. MAC DONALD PHYSICS LAB.	G-
MC GILL UNIV., MONTREAL. MAC DONALD PHYSICS LAB.	MW-
MC GILL UNIV., MONTREAL. MAC DONALD PHYSICS LAB.	S-
MC GILL UNIV., MONTREAL. SPACE RESEARCH INST.	MCGU-SRI-
MC GILL UNIV., MONTREAL. SPACE RESEARCH INST.	SRI-H-R-
MC GILL UNIV., MONTREAL. SPACE RESEARCH INST.	SRI-H-TN-
MC GILL UNIV., MONTREAL. SPACE RESEARCH INST.	SRI-R-
MC GILL UNIV., MONTREAL. SPACE RESEARCH INST.	SRI-TL-
MC GILL UNIV., MONTREAL. STORMY WEATHER GROUP	MW-
MC GILL UNIV., MONTREAL. STORMY WEATHER GROUP	MWT-
MC GILL UNIV., MONTREAL. UPPER ATMOSPHERE CHEMISTRY GROUP	UACG-(NUMBER)-(YEAR)
MC GILL UNIV., MONTREAL. UPPER ATMOSPHERE CHEMISTRY GROUP	UAGC-
MC GRAW-HILL, INC. F.W. DODGE CO., WRIGHT-PATTERSON AFB, OHIO	MHR-

MC GRAW-HILL, INC. F.W. DODGE CO.

MC GRAW-HILL, INC. F.W. DODGE CO., WRIGHT-PATTERSON AFB, OHIO	MHR-(YEAR)-
MC GRAW-HILL PUBLISHING CO., INC., N.Y.C.	MHBC-
MC GRAW-HILL PUBLISHING CO., INC., N.Y.C.	MHR-
MACHINE TOOL AUTOMATION INC., SOUTHPORT, CONN.	MTA-TP-
MACHLETT LABS., INC., SPRINGDALE, CONN.	MACL-
MC KEE (ARTHUR G.) AND CO. WESTERN KNAPP ENGINEERING DIV., SAN FRANCISCO	MCKEE-
MACLAREN AND SONS, LTD., LONDON (TRANSLATIONS)	MACL-
MC MASTER UNIV., HAMILTON, ONT.	DR-
MC MASTER UNIV., HAMILTON, ONT.	XR-
MC MILLAN LAB., INC., IPSWICH, MASS.	MCL-
MC MILLAN LAB., INC., IPSWICH, MASS.	MCML-
MC MILLAN LAB., INC., MARBLEHEAD, MASS.	MCM-
MAC NEAL-SCHWENDLER CORP., SAN MARINO, CALIF.	MSC-EC-
MAGNA CORP., ANAHEIM, CALIF.	STL-
MAGNAVOX CO., FORT WAYNE	MAGC-
MAGNAVOX CO., FORT WAYNE	MGX-
MAGNAVOX CO., FORT WAYNE	TP(YR)-
MAGNAVOX CO. DIFAR ENG. DEPT., FORT WAYNE	TP(YEAR)-
MAGNAVOX CO. GOVERNMENT AND IND. DIV., URBANA, ILL.	TP(NUMBER)
MAGNAVOX CO. RESEARCH LABS., TORRANCE, CALIF.	MRL-R-
MAGNAVOX CO. SPECIAL PRODUCTS DIV.	SPD-
MAGNETIC RESEARCH CORP., EL SEGUNDO, CALIF.	MRC-
MAGYAR TUDOMANYOS AKADEMIA KOZPONTI FIZIKAI KUTATO INTEZETE, BUDAPEST	KFKI-
MAGYAR TUDOMANYOS AKADEMIA KOZPONTI FIZIKAI KUTATO INTEZETE, BUDAPEST	KFKI-(YEAR)-(NO.)-(LTRS)
MAGYAR TUDOMANYOS AKADEMIA KOZPONTI FIZIKAI KUTATO INTEZETE, BUDAPEST	KFKI-(NUMBER)/(LETTER)
MAGYAR TUDOMANYOS AKADEMIA KOZPONTI FIZIKAI KUTATO INTEZETE, BUDAPEST	KFKI-(YEAR)-(NUMBER)
MAGYAR TUDOMANYOS AKADEMIA KOZPONTI FIZIKAI KUTATO INTEZETE, BUDAPEST	KFKL-(NO.)/(YR)-
MAINE. UNIV., ORONO. MANPOWER RESEARCH PROJECT	MANPOWER-RB-
MALAKER LABS., INC., HIGH BRIDGE, N.J.	CM-
MALAKER LABS., INC., HIGH BRIDGE, N.J.	MLI-
MALLINCKRODT CHEMICAL WORKS, ST. LOUIS	DP-
MALLINCKRODT CHEMICAL WORKS, ST. LOUIS	MCW-
MALLINCKRODT CHEMICAL WORKS, ST. LOUIS	MCW-A-
MALLINCKRODT CHEMICAL WORKS. URANIUM DIV., WELDON SPRING, MO.	MCW-
MALLORY (P.R.) AND CO., INC., INDIANAPOLIS	ENG.-
MALLORY (P.R.) AND CO., INC., INDIANAPOLIS	MAL-
MALLORY-SHARON TITANIUM CORP., NILES, OHIO	MST-
MALLORY-SHARON TITANIUM CORP., NILES, OHIO	MSTC-
MALVESTUTO AERO SPACE CORP., OXNARD, CALIF.	MR-
MANCHESTER. UNIV., ENGLAND	RR-(NUMBER)/SMR/
MANCHESTER. UNIV., ENGLAND	RR-(NUMBER)/EKK/
MANCHESTER. UNIV., ENGLAND	RR-(NUMBER)/RM/
MANDREL INDUSTRIES, INC., HOUSTON, TEX.	MIP-
MANHATTAN DISTRICT, OAK RIDGE, TENN. (REF. 23) (SEE ALSO LATER NAME: ATOMIC ENERGY COMMISSION)	P-
MANHATTAN DISTRICT. LIAISON OFFICE, CHALK RIVER, ONT.	MA-O-
MANHATTAN DISTRICT. RESEARCH CONTROL (SERIES ASSIGNED TO REPORTS FROM VARIOUS AMERICAN SOURCES)	A-
MANHATTAN DISTRICT. RESEARCH CONTROL (SERIES ASSIGNED TO REPORTS FROM VARIOUS AMERICAN SOURCES)	/A/M-
MANHATTAN DISTRICT. RESEARCH CONTROL (SERIES ASSIGNED TO REPORTS FROM VARIOUS AMERICAN SOURCES)	M-
MANHATTAN DISTRICT. SAFETY COMMITTEE	SM-
MANLABS, INC., CAMBRIDGE, MASS.	MANLABS-PR-
MANLABS, INC., CAMBRIDGE, MASS. (TECHNICAL REPORT)	MANLABS-TR-
MANLABS, INC., CAMBRIDGE, MASS.	SAPR-
MANNED SPACECRAFT CENTER, HOUSTON, TEX.	APO-
MANNED SPACECRAFT CENTER, HOUSTON, TEX. (FLIGHT SAFETY INFORMATION BULLETIN)	FSIB-
MANNED SPACECRAFT CENTER, HOUSTON, TEX.	MSC-
MANNED SPACECRAFT CENTER, HOUSTON, TEX.	MSC-BM-MR-
MANNED SPACECRAFT CENTER, HOUSTON, TEX.	NASA-MSC-
MANNED SPACECRAFT CENTER, HOUSTON, TEX.	NASA-MSC-G-R-
MANNED SPACECRAFT CENTER, HOUSTON, TEX.	NSC-CF-R-
MANNED SPACECRAFT CENTER, HOUSTON, TEX. (WSTF FILTER TEST PROGRAM)	TR-WSTF-
MANNED SPACECRAFT CENTER. APOLLO SPECIAL PROJECT OFFICE, HOUSTON, TEX.	ASPO-
MANUFACTURING LABS., INC., CAMBRIDGE, MASS.	MANL-TR-
MANUFACTURING LABS., INC. PHYSICAL METALLURGY DIV., CAMBRIDGE, MASS. (PROGRESS REPORT)	MLI-PR-
MANUFACTURING LABS., INC. PHYSICAL METALLURGY DIV., CAMBRIDGE, MASS. (TECHNICAL REPORT)	MLI-TR-
MARBLE INST. OF AMERICA, INC., MOUNT VERNON, N.Y.	MIA-
MARCHETTI (J.W.), INC., NATICK, MASS.	G-
MARE ISLAND NAVAL SHIPYARD, CALIF.	NAV-MINS-
MARE ISLAND NAVAL SHIPYARD. RUBBER LAB., CALIF.	MINS-
MARE ISLAND NAVAL SHIPYARD. RUBBER LAB., CALIF.	NAVSHIPS RL-
MARE ISLAND NAVAL SHIPYARD. RUBBER LAB., CALIF.	RL-

MARINE ADVISERS, INC., LA JOLLA, CALIF.	MAI A-
MARINE BIOLOGICAL LAB., WOODS HOLE, MASS.	MBL-
MARINE CORPS, WASHINGTON, D.C.	EDR-CT-(NO.-NO.-NO.)
MARINE CORPS, WASHINGTON, D.C.	EDR-IT-(NO.-NO.-NO.)
MARINE CORPS, WASHINGTON, D.C.	FMFM-
MARINE CORPS, WASHINGTON, D.C.	GOR-(LETTERS)-
MARINE CORPS, WASHINGTON, D.C.	GOR-CC-
MARINE CORPS, WASHINGTON, D.C.	MABS-
MARINE CORPS, WASHINGTON, D.C.	MC-
MARINE CORPS, WASHINGTON, D.C.	MC/LFB-
MARINE CORPS, WASHINGTON, D.C.	MCO-(NUMBER)(LETTER)
MARINE CORPS, WASHINGTON, D.C.	NAVMC-
MARINE CORPS, WASHINGTON, D.C.	SOR-(LETTERS-(NO.-NO.)
MARINE CORPS, WASHINGTON, D.C. (SPECIFIC OPERATIONAL REQUIREMENT)	SOR-CC-
MARINE CORPS, WASHINGTON, D. C.	USMC-ADO-CC-
MARINE CORPS, WASHINGTON, D. C.	USMC-EDR-CT-
MARINE CORPS, WASHINGTON, D. C.	USMC-EDR-TM-
MARINE CORPS, WASHINGTON, D.C.	USMC-FMFM-
MARINE CORPS, WASHINGTON, D. C.	USMC-GOR-CC-
MARINE CORPS, WASHINGTON, D. C.	USMC-GOR-CT-
MARINE CORPS, WASHINGTON, D. C.	USMC-GOR-FS-
MARINE CORPS, WASHINGTON, D. C.	USMC-GOR-LM-
MARINE CORPS, WASHINGTON, D. C.	USMC-GOR-LO-
MARINE CORPS, WASHINGTON, D. C.	USMC-GOR-NB-
MARINE CORPS, WASHINGTON, D. C.	USMC-SOR-(LTR.)(NO.)
MARINE CORPS, WASHINGTON, D. C.	USMC-SOR-CC-
MARINE CORPS, WASHINGTON, D. C.	USMC-SOR-CT-
MARINE CORPS, WASHINGTON, D. C.	USMC-SOR-EW-
MARINE CORPS, WASHINGTON, D. C.	USMC-SOR-IT-
MARINE CORPS, WASHINGTON, D. C.	USMC-SOR-LO-
MARINE CORPS, WASHINGTON, D. C.	USMC-SOR-NB-
MARINE CORPS, WASHINGTON, D.C.	USMC-SOR-TM-
MARINE CORPS DEVELOPMENT AND EDUCATION COMMAND, QUANTICO, VA.	MCDEC-DB-(NUMBER-YEAR)
MARINE CORPS LANDING FORCE DEV. CENTER, QUANTICO, VA.	DB-(NUMBER)-(YEAR)
MARINE CORPS LANDING FORCE DEV. CENTER, QUANTICO, VA.	GOR-CC-
MARINE CORPS SCHOOLS. EQUIPMENT BOARD, QUANTICO, VA.	EB-
MARINE CORPS SCHOOLS. EQUIPMENT BOARD, QUANTICO, VA.	MCEB-
MARINE CORPS SCHOOLS. EQUIPMENT BOARD, QUANTICO, VA. (TEST REPORTS)	MC EB TR-
MARINE CORPS SCHOOLS. EQUIPMENT BOARD, QUANTICO, VA. (TEST REPORTS)	MC EB TR PROJ-
MARINE CORPS SCHOOLS. EQUIPMENT BOARD, QUANTICO, VA.	MCS-
MARINE ENGINEERING LAB., ANNAPOLIS	MEL-
MARINE ENGINEERING LAB., ANNAPOLIS	MEL-R + D-
MARINE ENGINEERING LAB., ANNAPOLIS	MEL-TM-
MARINE MINERALS TECHNOLOGY CENTER, TIBURON, CALIF.	ERL-MMTC-
MARINE MINERALS TECHNOLOGY CENTER, TIBURON, CALIF.	NOAA-TM-ERL-MMTC-
MARINE RESOURCES INC. MARINE TECHNICAL SERVICES DIV., MELVILLE, N.Y.	MTS-
MARITIME ADMIN. OFFICE OF RES. AND DEV., WASHINGTON, D.C. (N.S. SAVANNAH/EXPTL. COMMERCIAL OPERATION)	NSS/ECO-(LETTER)-
MARKITE CO., N.Y.C.	MAK-
MARKITE CO., N.Y.C.	MC-RN-
MARKITE CO., N.Y.C.	MKC-
MARKITE CO., N.Y.C.	MN-
MARKS POLARIZED CORP., WHITESTONE, N.Y.	MPC-FR-
MARKS POLARIZED CORP., WHITESTONE, N.Y.	MPC-QPR-
MARQUARDT AIRCRAFT CO., VAN NUYS, CALIF. (REF. 1)	(NUMBER)
MARQUARDT AIRCRAFT CO., VAN NUYS, CALIF.	E.O.-(NUMBER)-
MARQUARDT AIRCRAFT CO., VAN NUYS, CALIF.(PLUTO PROJ.)	MAC PLUTO QPR-
MARQUARDT AIRCRAFT CO., VAN NUYS, CALIF.	MAC-PR-
MARQUARDT AIRCRAFT CO., VAN NUYS, CALIF.	MAC-TR-153-
MARQUARDT AIRCRAFT CO., VAN NUYS, CALIF.	MJL-
MARQUARDT AIRCRAFT CO., VAN NUYS, CALIF.	SD-DC-
MARQUARDT AIRCRAFT CO., VAN NUYS, CALIF.	TR-153-
MARQUARDT AIRCRAFT CO. ASTRO SCIENCES GP,VAN NUYS,CAL (SEE ALSO LATER NAME: MARQUARDT CORP.)	ASG-TM-(YEAR)-
MARQUARDT CORP. NEWPORT BEACH LAB., CALIF.	MR-
MARQUARDT CORP., VAN NUYS, CALIF.	LR-
MARQUARDT CORP., VAN NUYS, CALIF.	MAC-
MARQUARDT CORP., VAN NUYS, CALIF.	MARQ-
MARQUARDT CORP., VAN NUYS, CALIF.	MARQUARDT-
MARQUARDT CORP., VAN NUYS, CALIF.	MP-
MARQUARDT CORP., VAN NUYS, CALIF.	MR-
MARQUARDT CORP., VAN NUYS, CALIF.	MR-IR-(NUMBER)-
MARQUARDT CORP., VAN NUYS, CALIF.	MR-S-
MARQUARDT CORP., VAN NUYS, CALIF.	PR-(NUMBER)-(NUMBER)Q-
MARQUARDT CORP., VAN NUYS, CALIF.	S-
MARQUARDT CORP., VAN NUYS, CALIF.	TMC-
MARQUARDT CORP. FACILITIES ENG., VAN NUYS, CALIF.	FE-(NUMBER)-
MARQUARDT CORP. MATERIALS AND PROCESS DEPT., VAN NUYS, CALIF.	TMC-S-
MARQUARDT CORP. SYSTEMS ENG. DIV., VAN NUYS, CALIF.	MR-S-
MARSHALL LABS., TORRANCE, CALIF.	ML/TN-
MARTIN CO. (ALL LOCATIONS)	MAR-
MARTIN CO., BALTIMORE	CCI-
MARTIN CO., BALTIMORE	CR-
MARTIN CO., BALTIMORE	CR-(NO.)-
MARTIN CO., BALTIMORE	DPR-(NUMBER)-(NUMBER)
MARTIN CO., BALTIMORE	DSR-S-
MARTIN CO., BALTIMORE	GLM-
MARTIN CO., BALTIMORE	GLM-(LETTERS)-
MARTIN CO., BALTIMORE	GLM-B-
MARTIN CO., BALTIMORE	GLM-ER-

MARTIN CO.

MARTIN CO., BALTIMORE (LAUNCH VEHICLE REPORT)	LV-
MARTIN CO., BALTIMORE	MARC-
MARTIN CO., BALTIMORE	MAR CSH-(NUMBERS)-
MARTIN CO., BALTIMORE	MAR-ER-
MARTIN CO., BALTIMORE	MAR MND-
MARTIN CO., BALTIMORE	MAR S-
MARTIN CO., BALTIMORE	MARTIN CR-(NO.)-
MARTIN CO., BALTIMORE (ENGINEERING REPORT)	MCER-
MARTIN CO., BALTIMORE	MND-LIB-
MARTIN CO., BALTIMORE	NDM-
MARTIN CO., BALTIMORE	NMA-
MARTIN CO., BALTIMORE	RB-
MARTIN CO., BALTIMORE	RM-
MARTIN CO., DENVER	CR-(YEAR)-
MARTIN CO., DENVER (FINAL REPORT)	CR-(YEAR)-(NO.)-F-
MARTIN CO., DENVER	CR-(YEAR)-(NO.)-PT-
MARTIN CO., DENVER (CONTRACT REPORT)	DA-CR-(YR.)-
MARTIN CO., DENVER (CONTRACT REPORT)	FTC-CR-(YR.)-
MARTIN CO., DENVER	IFS-
MARTIN CO., DENVER (INTERFACE SPECIFICATION)	IFS-MOL-EFT-
MARTIN CO., DENVER	IR-(YR.)-
MARTIN CO., DENVER	LIT. SURVEY NO.-
MARTIN CO., DENVER	M-(YR.)-
MARTIN CO., DENVER	M-(NUMBER-NUMBER)-
MARTIN CO., DENVER	MAR M-M-P(YR.)-
MARTIN CO., DENVER	MARTIN-
MARTIN CO., DENVER	MARTIN-CR-(YR.)-
MARTIN CO., DENVER	MCR-(YEAR)-
MARTIN CO., DENVER	ME-
MARTIN CO., DENVER	MI-(YEAR)-
MARTIN CO., DENVER	MNI-(NUMBER)-
MARTIN CO., DENVER	MOL-
MARTIN CO., DENVER	MT-(YEAR)-
MARTIN CO., DENVER	P-(NUMBER)-
MARTIN CO., DENVER	PR-(YEAR)-
MARTIN CO., DENVER	R-(YR.)-(NO.)
MARTIN CO., DENVER	SR-(YR.)-
MARTIN CO., DENVER	SR-(NUMBER)-
MARTIN CO., DENVER	USN-CR-(YR.)-
MARTIN CO., DENVER	VOY-CR-(YEAR)-
MARTIN CO., DENVER	WDD-M-MI-(YEAR)-
MARTIN CO., DENVER	WDD-M-SR-(NUMBER)-
MARTIN CO., ORLANDO, FLA.	FTC-CR-
MARTIN CO., ORLANDO, FLA.	MAR OR-
MARTIN CO., ORLANDO, FLA.	MAR P-(YR.)-
MARTIN CO., ORLANDO, FLA. (TECHNICAL REPORT)	OA-
MARTIN CO., ORLANDO, FLA.	OR-
MARTIN CO., VANDENBERG AFB, CALIF.	MV-(NUMBER)-
MARTIN CO., VANDENBERG AFB, CALIF.	MVS-(NUMBER)-
MARTIN CO. CENTER FOR HIGH ENERGY FORMING, DENVER	MARTIN-CR-
MARTIN CO. ELECTRONIC SYSTEMS AND PRODUCT DIV., BALTIMORE	ER-(NUMBER-NUMBER)-
MARTIN CO. NUCLEAR DIV., BALTIMORE	MAR MND-M-
MARTIN CO. NUCLEAR DIV., BALTIMORE	MAR MND-P-
MARTIN CO. NUCLEAR DIV., BALTIMORE	MAR MND-SF-
MARTIN CO. NUCLEAR DIV., BALTIMORE	MAR MND-SR-
MARTIN CO. NUCLEAR DIV., BALTIMORE	MND-
MARTIN CO. NUCLEAR DIV., BALTIMORE	MND-AC-
MARTIN CO. NUCLEAR DIV., BALTIMORE	MND-ANP-
MARTIN CO. NUCLEAR DIV., BALTIMORE	MND-C-
MARTIN CO. NUCLEAR DIV., BALTIMORE	MND-DB-
MARTIN CO. NUCLEAR DIV., BALTIMORE	MND-E-
MARTIN CO. NUCLEAR DIV., BALTIMORE (FLUIDIZED BED REACTOR)	MND-FBR-
MARTIN CO. NUCLEAR DIV., BALTIMORE (LIQUID FLUIDIZED BED REACTOR)	MND-LFBR-
MARTIN CO. NUCLEAR DIV., BALTIMORE	MND-M-
MARTIN CO. NUCLEAR DIV., BALTIMORE	MND-M(NUMBER)(LETTER)-
MARTIN CO. NUCLEAR DIV., BALTIMORE	MND-MD-
MARTIN CO. NUCLEAR DIV., BALTIMORE	MND-MPR-
MARTIN CO. NUCLEAR DIV., BALTIMORE	MND-P-
MARTIN CO. NUCLEAR DIV., BALTIMORE (RADIOISOTOPE HEATER)	MND-R-
MARTIN CO. NUCLEAR DIV., BALTIMORE	MND-ROC-
MARTIN CO. NUCLEAR DIV., BALTIMORE	MND-RP-
MARTIN CO. NUCLEAR DIV., BALTIMORE	MND-SF-
MARTIN CO. NUCLEAR DIV., BALTIMORE	MND-SR-
MARTIN CO. NUCLEAR DIV., BALTIMORE	MN-SW-
MARTIN CO. NUCLEAR DIV., BALTIMORE	RJA-
MARTIN CO. TECHNICAL INFO. CENTER, ORLANDO, FLA.	RB-
MARTIN-HUBBARD CORP., BOSTON (REF. 1)	(PROJECT ACCOUNT NO.)-
MARTIN-MARIETTA CORP. (ALL LOCATIONS)	MAR-
MARTIN-MARIETTA CORP., BALTIMORE	ER-
MARTIN-MARIETTA CORP., BALTIMORE	MMC-
MARTIN-MARIETTA CORP., BALTIMORE	MMC-AR-(YEAR)
MARTIN-MARIETTA CORP., BALTIMORE	MMC-RM-
MARTIN-MARIETTA CORP. AEROSPACE DIV., BALTIMORE	MN-
MARTIN-MARIETTA CORP. AEROSPACE DIV., BALTIMORE	MND-
MARTIN-MARIETTA CORP. AEROSPACE DIV., BALTIMORE	RP-(YEAR)-(NUMBER)-
MARTIN-MARIETTA CORP. DENVER DIV.	BMS-TII-
MARTIN-MARIETTA CORP. DENVER DIV.	CR-
MARTIN-MARIETTA CORP. DENVER DIV.	DEN-
MARTIN-MARIETTA CORP. DENVER DIV.	DM-
MARTIN-MARIETTA CORP. DENVER DIV.	DS-
MARTIN-MARIETTA CORP. DENVER DIV.	ED-
MARTIN-MARIETTA CORP. DENVER DIV.	ER-
MARTIN-MARIETTA CORP. DENVER DIV.	IR-
MARTIN-MARIETTA CORP. DENVER DIV.	M-
MARTIN-MARIETTA CORP. DENVER DIV.	MARTIN CR-
MARTIN-MARIETTA CORP. DENVER DIV.	MC-
MARTIN-MARIETTA CORP. DENVER DIV.	MCG-
MARTIN-MARIETTA CORP. DENVER DIV.	MCP-

MARTIN-MARIETTA CORP. DENVER DIV.	MCR-(YEAR)-
MARTIN-MARIETTA CORP. DENVER DIV.	MDS-
MARTIN-MARIETTA CORP. DENVER DIV.	ME-
MARTIN-MARIETTA CORP. DENVER DIV.	MEL-
MARTIN-MARIETTA CORP. DENVER DIV.	MI-
MARTIN-MARIETTA CORP. DENVER DIV.	MT-
MARTIN-MARIETTA CORP. DENVER DIV.	MV-
MARTIN-MARIETTA CORP. DENVER DIV.	NTN-
MARTIN-MARIETTA CORP. DENVER DIV.	P-
MARTIN-MARIETTA CORP. DENVER DIV.	PIR-
MARTIN-MARIETTA CORP. DENVER DIV.	PL-
MARTIN-MARIETTA CORP. DENVER DIV.	PR-
MARTIN-MARIETTA CORP. DENVER DIV.	R-
MARTIN-MARIETTA CORP. DENVER DIV.	RR-
MARTIN-MARIETTA CORP. DENVER DIV.	RSTN-
MARTIN-MARIETTA CORP. DENVER DIV.	SR-
MARTIN-MARIETTA CORP. DENVER DIV.	TM-
MARTIN-MARIETTA CORP. DENVER DIV.	WDD-
MARTIN-MARIETTA CORP. NUCLEAR DIV., BALTIMORE	MAR-MND-
MARTIN-MARIETTA CORP. NUCLEAR DIV., BALTIMORE	MND-(NUMBER)-
MARTIN-MARIETTA CORP. NUCLEAR DIV., BALTIMORE	MND-(LETTER)-
MARTIN-MARIETTA CORP. NUCLEAR DIV., BALTIMORE	MND-MC-
MARTIN-MARIETTA CORP. RESEARCH INSTITUTE FOR ADVANCED STUDIES, BALTIMORE	RIAD-TR-(YEAR)-
MARTIN-MARIETTA CORP. RESEARCH INSTITUTE FOR ADVANCED STUDIES, BALTIMORE	RIAS-
MARTIN-MARIETTA CORP. RESEARCH INSTITUTE FOR ADVANCED STUDIES, BALTIMORE	RIAS-(NUMBER)-
MARTIN-MARIETTA CORP. RESEARCH INSTITUTE FOR ADVANCED STUDIES, BALTIMORE (ANNUAL REPORT)	RIAS-AR(YEAR)
MARTIN-MARIETTA CORP. RESEARCH INSTITUTE FOR ADVANCED STUDIES, BALTIMORE	RIAS-QTR-
MARTIN-MARIETTA CORP. RESEARCH INSTITUTE FOR ADVANCED STUDIES, BALTIMORE	RIAS-TR-(YEAR)-
	TR-(YEAR)-
MARYLAND. DEPT. OF RESEARCH AND EDUCATION. CHESAPEAKE BIOLOGICAL LAB.	CBL-
MARYLAND. UNIV., COLLEGE PARK	BT-
MARYLAND. UNIV., COLLEGE PARK	MARU-
MARYLAND. UNIV., COLLEGE PARK	MARU-QR-
MARYLAND. UNIV., COLLEGE PARK	MARU-TN-
MARYLAND. UNIV., COLLEGE PARK	MARYLAND U-(MONTH)(YEAR)
MARYLAND. UNIV., COLLEGE PARK	MD-
MARYLAND. UNIV., COLLEGE PARK	MD-TR-(YEAR)-
MARYLAND. UNIV., COLLEGE PARK	MD-TR-
MARYLAND. UNIV., COLLEGE PARK	ME-
MARYLAND. UNIV., COLLEGE PARK	MU-
MARYLAND. UNIV., COLLEGE PARK	UM-
MARYLAND. UNIV., COLLEGE PARK	UMNE-
MARYLAND. UNIV., COLLEGE PARK. COMPUTER SCIENCE CTR.	CSC-TR-(YEAR)-
MARYLAND. UNIV., COLLEGE PARK. COMPUTER SCIENCE CTR.	RP-P-P-
MARYLAND. UNIV., COLLEGE PARK. DEPT. OF AEROSPACE ENGINEERING	AERO-(YEAR)-
MARYLAND. UNIV., COLLEGE PARK. DEPT. OF CHEMISTRY	MNC-
MARYLAND. UNIV., COLLEGE PARK. DEPT. OF PHYSICS	MARU TR-
MARYLAND. UNIV., COLLEGE PARK. DEPT. OF PHYSICS AND ASTRONOMY	AEC-ORO-2504-
MARYLAND. UNIV., COLLEGE PARK. DEPT. OF PHYSICS AND ASTRONOMY	MUP-TR-
MARYLAND. UNIV., COLLEGE PARK. INST. FOR FLUID DYNAMICS AND APPLIED MATHEMATICS	BN-
MARYLAND. UNIV., COLLEGE PARK. INST. FOR FLUID DYNAMICS AND APPLIED MATHEMATICS	MARU TN BN-
MARYLAND. UNIV., COLLEGE PARK. INST. FOR FLUID DYNAMICS AND APPLIED MATHEMATICS	MARU TR BT-
MARYLAND. UNIV., COLLEGE PARK. INST. FOR FLUID DYNAMICS AND APPLIED MATHEMATICS	TN-BN-
MARYLAND. UNIV., COLLEGE PARK. INST. FOR FLUID DYNAMICS AND APPLIED MATHEMATICS	U MD-
MARYLAND. UNIV.,COLLEGE PARK.INST. OF MOLECULAR PHYS.	IMP-
MARYLAND. UNIV.,COLLEGE PARK.INST. OF MOLECULAR PHYS.	IMP-AEC-
MARYLAND. UNIV.,COLLEGE PARK.INST. OF MOLECULAR PHYS.	IMP-NASA-
MARYLAND. UNIV.,COLLEGE PARK.INST. OF MOLECULAR PHYS.	IMP-ONR-
MARYLAND. UNIV.,COLLEGE PARK.INST. OF MOLECULAR PHYS.	IMP-OSR-
MASCHINENFABRIK AUGSBURG-NURNBERG A. G., NUREMBERG	MAN-TKA-
MASON + HANGER-SILAS MASON CO., INC., BURLINGTON,IOWA	MHSMB-
MASON AND HANGER-SILAS MASON CO. INC. PANTEX ORDNANCE PLANT, AMARILLO, TEX.	CMR-
MASON AND HANGER-SILAS MASON CO. INC. PANTEX ORDNANCE PLANT, AMARILLO, TEX.	MH-SM-
MASON + HANGER-SILAS MASON CO., INC. PANTEX PLANT, AMARILLO, TEX.	MHSMP-
MASON + HANGER-SILAS MASON CO., INC. PANTEX PLANT, AMARILLO, TEX.	MHSMP-(YEAR)-
MASON + HANGER-SILAS MASON CO., INC. PANTEX PLANT, AMARILLO, TEX.	POP-
MASON + HANGER-SILAS MASON CO., INC. PANTEX PLANT, AMARILLO, TEX.	PX-
MASON (SILAS) CO., INC., BURLINGTON, IOWA	R+D RPT. NO.(MO)-(YR)-
MASON (SILAS) CO., INC., BURLINGTON, IOWA	SMC-
MASSACHUSETTS. UNIV., AMHERST	MASSU-
MASSACHUSETTS. UNIV., AMHERST	NSF-PR-
MASSACHUSETTS. UNIV., AMHERST. COMPUTER SCIENCE DEPT.	TN/CS/
MASSACHUSETTS. UNIV., AMHERST. SCHOOL OF ENGINEERING	EVE-(NO.)-(YEAR)-
MASSACHUSETTS. UNIV., AMHERST. WATER RESOURCES RESEARCH CENTER	COMPLETION-(YEAR)-
MASSACHUSETTS COMPUTER ASSOCIATES, INC., WAKEFIELD	CA-(NUMBER)-
MASSACHUSETTS GENERAL HOSPITAL, BOSTON	UN-
MASSACHUSETTS INST. OF TECH., CAMBRIDGE	A(NUMBER)
MASSACHUSETTS INST. OF TECH., CAMBRIDGE	BSR-
MASSACHUSETTS INST. OF TECH., CAMBRIDGE (REF. 10)	CT-
MASSACHUSETTS INST. OF TECH., CAMBRIDGE	D3-
MASSACHUSETTS INST. OF TECH., CAMBRIDGE	D.I.C.-
MASSACHUSETTS INST. OF TECH., CAMBRIDGE	E-
MASSACHUSETTS INST. OF TECH., CAMBRIDGE	GAG-
MASSACHUSETTS INST. OF TECH., CAMBRIDGE	GE-E-
MASSACHUSETTS INST. OF TECH., CAMBRIDGE	MAC-
MASSACHUSETTS INST. OF TECH., CAMBRIDGE	MAC-TR-
MASSACHUSETTS INST. OF TECH., CAMBRIDGE	MET FLUID DYN. RES. LAB.-

MASSACHUSETTS INST. OF TECH.

MASSACHUSETTS INST. OF TECH., CAMBRIDGE	MIT-
MASSACHUSETTS INST. OF TECH., CAMBRIDGE	MIT-(NUMBER)-(LETTER)
MASS. INST. OF TECH., CAMBRIDGE (SQUID PROJ.)(REF.33)	MIT-(NUMBER)-M
MASS. INST. OF TECH., CAMBRIDGE (SQUID PROJ.)(REF.33)	MIT-(NUMBER)-P
MASS. INST. OF TECH., CAMBRIDGE (SQUID PROJ.)(REF.33)	MIT-(NUMBER)-R
MASS. INST. OF TECH., CAMBRIDGE (SQUID PROJ.)(REF.33)	MIT-(NUMBER)-T
MASSACHUSETTS INST. OF TECH., CAMBRIDGE	MIT-(INITIALS)
MASSACHUSETTS INST. OF TECH., CAMBRIDGE	MIT-AR-((INCL.YEARS)
MASSACHUSETTS INST. OF TECH., CAMBRIDGE	MIT-DSR-
MASSACHUSETTS INST. OF TECH., CAMBRIDGE	MIT-L-
MASSACHUSETTS INST. OF TECH., CAMBRIDGE	MIT-M-
MASS. INST. OF TECH., CAMBRIDGE (MEMORANDUM RPTS.)	MIT MR-
MASSACHUSETTS INST. OF TECH.,CAMBRIDGE.(METEOR REPT.)	MIT-MR-
MASSACHUSETTS INST. OF TECH., CAMBRIDGE	MIT-NSES-
MASSACHUSETTS INST. OF TECH., CAMBRIDGE	MIT-P-
MASSACHUSETTS INST. OF TECH., CAMBRIDGE	MIT-PDC-
MASSACHUSETTS INST. OF TECH., CAMBRIDGE	MIT-PG-
MASSACHUSETTS INST. OF TECH., CAMBRIDGE	MIT-PL-
MASSACHUSETTS INST. OF TECH., CAMBRIDGE	MIT-QPR-
MASSACHUSETTS INST. OF TECH., CAMBRIDGE	MIT-R(YEAR)-
MASSACHUSETTS INST. OF TECH., CAMBRIDGE	MIT-RBR-
MASSACHUSETTS INST. OF TECH., CAMBRIDGE	MIT-S-
MASSACHUSETTS INST. OF TECH., CAMBRIDGE	MIT-T-
MASSACHUSETTS INST. OF TECH., CAMBRIDGE	MIT-TR-
MASSACHUSETTS INST. OF TECH., CAMBRIDGE	MIT-U-
MASSACHUSETTS INST. OF TECH., CAMBRIDGE	MITX-
MASSACHUSETTS INST. OF TECH., CAMBRIDGE	MPC-
MASSACHUSETTS INST. OF TECH., CAMBRIDGE	PEPR-
MASSACHUSETTS INST. OF TECH., CAMBRIDGE (POLARIS GUIDANCE INFORMATION BOOK)	PGIB-
MASSACHUSETTS INST. OF TECH., CAMBRIDGE	R(YR)-
MASSACHUSETTS INST. OF TECH., CAMBRIDGE	RE-
MASSACHUSETTS INST. OF TECH., CAMBRIDGE (SEMIANNUAL TECHNICAL SUMMARY REPORT)	SAPR-
MASSACHUSETTS INST. OF TECH., CAMBRIDGE	SATR-
MASS. INST. OF TECH., CAMBRIDGE (SQUID PROJ.)(REF.33)	SQUID-TR-MIT-(NO.)-PU
MASSACHUSETTS INST. OF TECH., CAMBRIDGE	SSC-
MASSACHUSETTS INST. OF TECH., CAMBRIDGE	TACP-
MASSACHUSETTS INST. OF TECH., CAMBRIDGE (WIND TUNNEL REPORT)	WTR-
MASSACHUSETTS INST. OF TECH., CAMBRIDGE. ACOUSTICS AND VIBRATION LAB.	A/V-(5 DIGITS)-
MASS. INST. OF TECH., CAMBRIDGE. ACOUSTICS LAB.	MIT-AL-
MASS. INST. OF TECH., CAMBRIDGE. ACOUSTICS LAB.	MIT-AL-QR-
MASS. INST. OF TECH., CAMBRIDGE. ACOUSTICS LAB.	MIT-AL-TR-
MASSACHUSETTS INST. OF TECH., CAMBRIDGE. AERO-ELASTIC AND STRUCTURES RESEARCH LAB.	ASRL-
MASSACHUSETTS INST. OF TECH., CAMBRIDGE. AERO-ELASTIC AND STRUCTURES RESEARCH LAB.	ASRL-LR-(NUMBER)-
MASSACHUSETTS INST. OF TECH., CAMBRIDGE. AERO-ELASTIC AND STRUCTURES RESEARCH LAB.	ASRL-TR-
MASSACHUSETTS INST. OF TECH., CAMBRIDGE. AERO-ELASTIC AND STRUCTURES RESEARCH LAB.	ME-(YEAR)/
MASSACHUSETTS INST. OF TECH., CAMBRIDGE. AERO-ELASTIC AND STRUCTURES RESEARCH LAB.	MIT-AE-
MASSACHUSETTS INST. OF TECH., CAMBRIDGE. AERO-ELASTIC AND STRUCTURES RESEARCH LAB.	MIT-ASRL-
MASSACHUSETTS INST. OF TECH., CAMBRIDGE. AERO-ELASTIC AND STRUCTURES RESEARCH LAB.	MIT-ASRL-TR25-
MASSACHUSETTS INST. OF TECH., CAMBRIDGE. AERO-ELASTIC AND STRUCTURES RESEARCH LAB.	MIT-ASRL-TR-
MASS. INST. OF TECH., CAMBRIDGE. AEROPHYSICS LAB.	MIT-AL-TR-
MASSACHUSETTS INST. OF TECH., CAMBRIDGE. CENTER FOR INTERNATIONAL STUDIES	B/68-
MASSACHUSETTS INST. OF TECH., CAMBRIDGE. CENTER FOR INTERNATIONAL STUDIES	C/64-
MASSACHUSETTS INST. OF TECH., CAMBRIDGE. CENTER FOR INTERNATIONAL STUDIES	C/65-
MASSACHUSETTS INST. OF TECH., CAMBRIDGE. CENTER FOR INTERNATIONAL STUDIES	C/69-
MASSACHUSETTS INST. OF TECH., CAMBRIDGE. CENTER FOR INTERNATIONAL STUDIES	D/(YEAR)-
MASSACHUSETTS INST. OF TECH., CAMBRIDGE. CENTER FOR MATERIAL SCIENCE AND ENGINEERING	SOLID STATE DEVICE-TR-
MASSACHUSETTS INST. OF TECH., CAMBRIDGE. CENTER FOR SPACE RESEARCH	CSR-P-
MASSACHUSETTS INST. OF TECH., CAMBRIDGE. CENTER FOR SPACE RESEARCH	CSR-P-(YEAR)-
MASSACHUSETTS INST. OF TECH., CAMBRIDGE. CENTER FOR SPACE RESEARCH (PROGRESS REPORT)	CSR-PR-(NUMBERS)-
MASSACHUSETTS INST. OF TECH., CAMBRIDGE. CENTER FOR SPACE RESEARCH (THESIS)	CSR-T-(YR.)-
MASSACHUSETTS INST. OF TECH., CAMBRIDGE. CENTER FOR SPACE RESEARCH (TECHNICAL NOTE)	CSR-TN-(YR.)-
MASSACHUSETTS INST. OF TECH., CAMBRIDGE. CENTER FOR SPACE RESEARCH (TECHNICAL REPORT)	CSR-TR-(YR.)-
MASS. INST. OF TECH., CAMBRIDGE. CENTER FOR SPACE RES	MIT-CSR-T-
MASS. INST. OF TECH., CAMBRIDGE. CENTER FOR SPACE RES	MIT-CSR-TR-(YEAR)-
MASS. INST. OF TECH., CAMBRIDGE. CENTER FOR SPACE RES	MVLS-(YEAR)-
MASSACHUSETTS INST. OF TECH., CAMBRIDGE. CENTER FOR THEORETICAL PHYSICS	CPT-
MASSACHUSETTS INST. OF TECH., CAMBRIDGE. CENTER FOR THEORETICAL PHYSICS	CTP-
MASSACHUSETTS INST. OF TECH., CAMBRIDGE. CENTER FOR THEORETICAL PHYSICS	MIT-CTP-
MASSACHUSETTS INST. OF TECH., CAMBRIDGE. CHARLES STARK DRAPER LAB.	E-(4 DIGITS)
MASS. INST. OF TECH., CAMBRIDGE. COMPUTATION CENTER	MIT-CC-PR-
MASS. INST. OF TECH., CAMBRIDGE. CRYSTALLOGRAPHIC LAB	MIT-CL-TR-
MASSACHUSETTS INST. OF TECH., CAMBRIDGE. DEPT. OF AERONAUTICAL ENG. (AERO-ELASTIC + STRUCTURES RES.)	MIT-AE-
MASSACHUSETTS INST. OF TECH., CAMBRIDGE. DEPT. OF CIVIL AND SANITARY ENGINEERING	MIT-CSE-
MASSACHUSETTS INST. OF TECH., CAMBRIDGE. DEPT. OF CIVIL AND SANITARY ENGINEERING	MIT-RR/R(YEAR)-
MASSACHUSETTS INST. OF TECH., CAMBRIDGE. DEPT. OF CIVIL ENGINEERING	MIT-DCE-
MASSACHUSETTS INST. OF TECH., CAMBRIDGE. DEPT. OF CIVIL ENGINEERING	R(YEAR)-
MASSACHUSETTS INST. OF TECH., CAMBRIDGE. DEPT. OF CIVIL ENGINEERING	RR-R(YEAR)-
MASSACHUSETTS INST. OF TECH., CAMBRIDGE. DEPT. OF CIVIL ENGINEERING	SOILS PUB-
MASSACHUSETTS INST. OF TECH., CAMBRIDGE. DEPT. OF FOOD TECHNOLOGY.	MIT-DFT-
MASSACHUSETTS INST. OF TECH., CAMBRIDGE. DEPT. OF GEOLOGY AND GEOPHYSICS	S(SEISMOLOGY)-
MASS. INST. OF TECH., CAMBRIDGE. DEPT. OF MECH. ENG.	MIT-DME-
MASS. INST. OF TECH., CAMBRIDGE. DEPT. OF MECH. ENG.	MIT-DME-TR-
MASS. INST. OF TECH., CAMBRIDGE. DEPT. OF METALLURGY	MAC-
MASS. INST. OF TECH., CAMBRIDGE. DEPT. OF METALLURGY	MIT-DM-
MASS. INST. OF TECH., CAMBRIDGE. DEPT. OF METALLURGY	MITG-A-
MASS. INST. OF TECH., CAMBRIDGE. DEPT. OF METALLURGY	MITS-
MASSACHUSETTS INST. OF TECH., CAMBRIDGE. DEPT. OF METEOROLOGY	MIT-DMET-
MASSACHUSETTS INST. OF TECH., CAMBRIDGE. DEPT. OF NAVAL ARCHITECTURE AND MARINE ENGINEERING	MIT-DNA-(YEAR)-
MASSACHUSETTS INST. OF TECH., CAMBRIDGE. DEPT. OF NUCLEAR ENGINEERING	MITNE-

MASSACHUSETTS INST. OF TECH., CAMBRIDGE. DEPT. OF NUCLEAR ENGINEERING	MIT-NE-
MASSACHUSETTS INST. OF TECH., CAMBRIDGE. DEPT. OF NUCLEAR ENGINEERING	MMPP-FRPC-
MASSACHUSETTS INST. OF TECH., CAMBRIDGE. DIGITAL COMPUTER LAB. (REF. 1)	(NUMBERS...)
MASS. INST. OF TECH., CAMBRIDGE. DIGITAL COMPUTER LAB	DCL-
MASS. INST. OF TECH., CAMBRIDGE. DIGITAL COMPUTER LAB	MIT-DCL-
MASS. INST. OF TECH., CAMBRIDGE. DIGITAL COMPUTER LAB	MIT-DCL-SR-
MASSACHUSETTS INST. OF TECH., CAMBRIDGE. DIV. OF INDUSTRIAL COOPERATION	DIC-
MASSACHUSETTS INST. OF TECH., CAMBRIDGE. DIV. OF INDUSTRIAL COOPERATION	DIC-(LETTER(S))-
MASSACHUSETTS INST. OF TECH., CAMBRIDGE. DIV. OF INDUSTRIAL COOPERATION	MIT-DIC-
MASSACHUSETTS INST. OF TECH., CAMBRIDGE. DIV. OF INDUSTRIAL COOPERATION	MIT-DIC-R-
MASSACHUSETTS INST. OF TECH., CAMBRIDGE. DIV. OF INDUSTRIAL COOPERATION	MIT-DIC-TR-
MASSACHUSETTS INST. OF TECH., CAMBRIDGE. DIVISION OF SPONSORED RESEARCH	DSR-
MASSACHUSETTS INST. OF TECH., CAMBRIDGE. DIVISION OF SPONSORED RESEARCH	DSR-(NUMBER-NUMBER)
MASSACHUSETTS INST. OF TECH., CAMBRIDGE. DIVISION OF SPONSORED RESEARCH	MIT-DSR-TR-
MASSACHUSETTS INST. OF TECH., CAMBRIDGE. DYNAMIC ANALYSIS AND CONTROL LAB.	DACL-
MASSACHUSETTS INST. OF TECH., CAMBRIDGE. DYNAMIC ANALYSIS AND CONTROL LAB.	DACL-RM-
MASSACHUSETTS INST. OF TECH., CAMBRIDGE. DYNAMIC ANALYSIS AND CONTROL LAB.	MIT-DACL-
MASSACHUSETTS INST. OF TECH., CAMBRIDGE. DYNAMIC ANALYSIS AND CONTROL LAB.	MIT-RM-
MASSACHUSETTS INST. OF TECH., CAMBRIDGE. ELECTRONIC NUCLEAR INSTRUMENTATION GROUP	MIT-ENIG-
MASSACHUSETTS INST. OF TECH., CAMBRIDGE. ELECTRONIC SYSTEMS LAB.	ESL-
MASSACHUSETTS INST. OF TECH., CAMBRIDGE. ELECTRONIC SYSTEMS LAB.	ESL-FR-
MASSACHUSETTS INST. OF TECH., CAMBRIDGE. ELECTRONIC SYSTEMS LAB.	ESL-IR-
MASSACHUSETTS INST. OF TECH., CAMBRIDGE. ELECTRONIC SYSTEMS LAB.	ESL-P-
MASSACHUSETTS INST. OF TECH., CAMBRIDGE. ELECTRONIC SYSTEMS LAB.	ESL-QR-
MASSACHUSETTS INST. OF TECH., CAMBRIDGE. ELECTRONIC SYSTEMS LAB.	ESL-R-
MASSACHUSETTS INST. OF TECH., CAMBRIDGE. ELECTRONIC SYSTEMS LAB.	ESL-SR-
MASSACHUSETTS INST. OF TECH., CAMBRIDGE. ELECTRONIC SYSTEMS LAB.	ESL-TM-
MASSACHUSETTS INST. OF TECH., CAMBRIDGE. ELECTRONIC SYSTEMS LAB.	MIT-(NO.)-(NO.)-TM-
MASSACHUSETTS INST. OF TECH., CAMBRIDGE. ELECTRONIC SYSTEMS LAB.	MIT-ESL-
MASSACHUSETTS INST. OF TECH., CAMBRIDGE. ELECTRONICS RESEARCH LAB.	MIT-ERL-
MASSACHUSETTS INST. OF TECH., CAMBRIDGE. ENGINEERING PROJECTS LAB.	EPL-
MASSACHUSETTS INST. OF TECH., CAMBRIDGE. EXPERIMENTAL ASTRONOMY LAB. (PROGRESS REPORT)	MIT-EAL-PR-
MASSACHUSETTS INST. OF TECH., CAMBRIDGE. EXPERIMENTAL ASTRONOMY LAB. (REPORT)	MIT-EAL-RE-
MASSACHUSETTS INST. OF TECH., CAMBRIDGE. EXPERIMENTAL ASTRONOMY LAB. (RESEARCH NOTE)	MIT-EAL-RN-
MASSACHUSETTS INST. OF TECH., CAMBRIDGE. EXPERIMENTAL ASTRONOMY LAB. (THESIS)	MIT-EAL-TE-
MASSACHUSETTS INST. OF TECH., CAMBRIDGE. FIBERS AND POLYMERS DIV.	FP-
MASSACHUSETTS INST. OF TECH., CAMBRIDGE. FIBERS AND POLYMERS DIV.	MIT-FP-
MASS. INST. OF TECH., CAMBRIDGE. FLIGHT CONTROL LAB.	FCL-
MASS. INST. OF TECH., CAMBRIDGE. FLIGHT CONTROL LAB.	MIT-FCL-(NO.)-R-
MASS. INST. OF TECH., CAMBRIDGE. FLIGHT TRANSPORTATION LAB.	FTL-R-
MASS. INST. OF TECH.,CAMBRIDGE. FLUID DYNAMICS RES.GP	MIT-FDRG-(YR)-
MASS. INST. OF TECH., CAMBRIDGE. FLUID DYNAMICS RESEARCH LAB.	MIT FLUID DYN.RES.LAB.-
MASS. INST. OF TECH., CAMBRIDGE. FLUID MECHANICS LAB.	FML-(YEAR)-
MASS. INST. OF TECH., CAMBRIDGE. FLUID MECHANICS LAB.	FML-PUBL-(YEAR-NUMBER)
MASS. INST. OF TECH., CAMBRIDGE. FLUID MECHANICS LAB.	MIT-FML-
MASS. INST. OF TECH., CAMBRIDGE. FLUID MECHANICS LAB.	PUB-(YEAR)-
MASS. INST. OF TECH., CAMBRIDGE. FUELS RESEARCH LAB.	FRL/MIT-TR-
MASS. INST. OF TECH., CAMBRIDGE. FUELS RESEARCH LAB.	MIT-FRL-
MASS. INST. OF TECH., CAMBRIDGE. FUELS RESEARCH LAB.	MIT-FRL-TR-
MASS. INST. OF TECH., CAMBRIDGE. GAS TURBINE LAB.	MIT GTL R-
MASSACHUSETTS INST. OF TECH., CAMBRIDGE. GUIDED MISSILES PROGRAM (PROJECT METEOR)	METEOR-
MASSACHUSETTS INST. OF TECH., CAMBRIDGE. GUIDED MISSILES PROGRAM	MIT-METEOR-
MASS. INST. OF TECH., CAMBRIDGE. HEAT TRANSFER LAB.	MIT-HTL-TR-
MASS. INST. OF TECH., CAMBRIDGE. HIGH VOLTAGE LAB.	MIT-HVL-
MASS. INST. OF TECH., CAMBRIDGE. HYDRODYNAMICS LAB.	MIT-HL-
MASS. INST. OF TECH., CAMBRIDGE. HYDRODYNAMICS LAB.	T(YEAR)-
MASS. INST. OF TECH., CAMBRIDGE. INSTRUMENTATION LAB.	ARG-
MASS. INST. OF TECH., CAMBRIDGE. INSTRUMENTATION LAB.	E-
MASS. INST. OF TECH., CAMBRIDGE. INSTRUMENTATION LAB.	MIT-IL-
MASSACHUSETTS INST. OF TECH., CAMBRIDGE. INSTRUMENTATION LAB. (ENGINEERING NOTES)	MIT-IL-E-
MASSACHUSETTS INST. OF TECH., CAMBRIDGE. INSTRUMENTATION LAB. (REPORTS)	MIT-IL-R-
MASS. INST. OF TECH., CAMBRIDGE. INSTRUMENTATION LAB.	MIT-IL-RR(NO.)R-
MASS. INST. OF TECH., CAMBRIDGE. INSTRUMENTATION LAB.	SSR-
MASS. INST. OF TECH., CAMBRIDGE. INSTRUMENTATION LAB.	T-
MASSACHUSETTS INST. OF TECH., CAMBRIDGE. LAB. FOR APPLIED BIOPHYSICS. (TECHNICAL REPORT)	MIT-LAB-TR-
MASSACHUSETTS INST. OF TECH., CAMBRIDGE. LAB. FOR INSULATION RESEARCH	LIR-TR-
MASSACHUSETTS INST. OF TECH., CAMBRIDGE. LAB. FOR INSULATION RESEARCH	MIT-LIR-
MASSACHUSETTS INST. OF TECH., CAMBRIDGE. LAB. FOR INSULATION RESEARCH	MIT LIR PR-
MASSACHUSETTS INST. OF TECH., CAMBRIDGE. LAB. FOR INSULATION RESEARCH	MIT LIR TR-
MASSACHUSETTS INST. OF TECH., CAMBRIDGE. LAB. FOR NUCLEAR SCIENCE	MIT-LNS-
MASSACHUSETTS INST. OF TECH., CAMBRIDGE. LAB. FOR NUCLEAR SCIENCE (PROGRESS REPORT)	MIT-LNS-PR-
MASSACHUSETTS INST. OF TECH., CAMBRIDGE. LAB. FOR NUCLEAR SCIENCE (TECHNICAL REPORT)	MIT-LNS-TR-
MASSACHUSETTS INST. OF TECH., CAMBRIDGE. LAB. FOR NUCLEAR SCIENCE AND ENGINEERING	MIT-LNSE-
MASSACHUSETTS INST. OF TECH., CAMBRIDGE. LAB. FOR NUCLEAR SCIENCE AND ENGINEERING (PROGRESS REPORT)	MIT-LNSE-PR-
MASSACHUSETTS INST. OF TECH., CAMBRIDGE. LAB. FOR NUCLEAR SCIENCE AND ENGINEERING (TECHNICAL REPORT)	MIT-LNSE-TR-
MASSACHUSETTS INST. OF TECH., CAMBRIDGE. LAB OF CHEMICAL AND SOLID-STATE PHYSICS.(ANNUAL RES. REPT)	MIT-LCSP-ARR-
MASS. INST. OF TECH., CAMBRIDGE. LEXINGTON PROJECT	LEXP-
MASS. INST. OF TECH., CAMBRIDGE. LEXINGTON PROJECT	LP-
MASS. INST. OF TECH., CAMBRIDGE. LEXINGTON PROJECT	MIT-LEXP-
MASSACHUSETTS INST. OF TECH., CAMBRIDGE. LINCOLN LAB.	ACP-
MASSACHUSETTS INST. OF TECH., CAMBRIDGE. LINCOLN LAB.	LL-

MASSACHUSETTS INST. OF TECH., LINCOLN LAB.

MASSACHUSETTS INST. OF TECH., CAMBRIDGE. LINCOLN LAB.	MIT-LL-
MASSACHUSETTS INST. OF TECH., CAMBRIDGE. MAGNETOGASDYNAMICS LAB.	MIT-ML-(YEAR)
MASSACHUSETTS INST. OF TECH., CAMBRIDGE. MAN-VEHICLE CONTROL LAB. (BEGINNING JULY 1966)	MV-(YR.)-
MASSACHUSETTS INST. OF TECH., CAMBRIDGE. MAN-VEHICLE CONTROL LAB. (THESIS)	MVT-(YR.)-
MASS. INST. OF TECH., CAMBRIDGE. MAN-VEHICLE LAB.	MIT-MVT-
MASS. INST. OF TECH., CAMBRIDGE. METALLURGICAL PROJ.	MIT-(P)-
MASS. INST. OF TECH., CAMBRIDGE. METALLURGICAL PROJ.	MIT-RMT-
MASS. INST. OF TECH., CAMBRIDGE. METALLURGICAL PROJ.	MIT-SVA-
MASS. INST. OF TECH., CAMBRIDGE. METALLURGICAL PROJ.	MIT-TTM-
MASS. INST. OF TECH., CAMBRIDGE. METALS PROCESSING DIV	MIT-MPD-
MASS. INST. OF TECH., CAMBRIDGE. MINERAL ENG. LAB.	MITG-
MASSACHUSETTS INST. OF TECH., CAMBRIDGE. NATL. MAGNET LAB	JP-
MASSACHUSETTS INST. OF TECH., CAMBRIDGE. NATL. MAGNET LAB	MS-
MASSACHUSETTS INST. OF TECH., CAMBRIDGE. NATL. MAGNET LAB	NML-(YEAR)-
MASS. INST. OF TECH., CAMBRIDGE. NAVAL SUPERSONIC LAB	MIT-NSL-
MASSACHUSETTS INST. OF TECH., CAMBRIDGE. NAVAL SUPERSONIC LAB. (TECHNICAL REPORT)	MIT-NSL-TR-
MASS. INST. OF TECH., CAMBRIDGE. NAVAL SUPERSONIC LAB	MIT-WTR-
MASS. INST. OF TECH., CAMBRIDGE. NAVAL SUPERSONIC LAB	NSL-
MASS. INST. OF TECH., CAMBRIDGE. NAVAL SUPERSONIC LAB	WTR-
MASSACHUSETTS INST. OF TECH., CAMBRIDGE. RADIATION LAB.	41-
MASSACHUSETTS INST. OF TECH., CAMBRIDGE. RADIATION LAB.	42-
MASSACHUSETTS INST. OF TECH., CAMBRIDGE. RADIATION LAB.	43-
MASSACHUSETTS INST. OF TECH., CAMBRIDGE. RADIATION LAB.	45-
MASSACHUSETTS INST. OF TECH., CAMBRIDGE. RADIATION LAB.	52-
MASSACHUSETTS INST. OF TECH., CAMBRIDGE. RADIATION LAB.	53-
MASSACHUSETTS INST. OF TECH., CAMBRIDGE. RADIATION LAB.	53/(MO.)/(DAY)/(YR.)
MASSACHUSETTS INST. OF TECH., CAMBRIDGE. RADIATION LAB.	M-
MASSACHUSETTS INST. OF TECH., CAMBRIDGE. RADIATION LAB.	MIT RAD LAB-
MASSACHUSETTS INST. OF TECH., CAMBRIDGE. RADIATION LAB.	MIT RAD LAB (ROMAN NO.)-
MASSACHUSETTS INST. OF TECH., CAMBRIDGE. RADIATION LABORATORY (RADOME BULLETINS)	MIT RAD LAB 483-
MASSACHUSETTS INST. OF TECH., CAMBRIDGE. RADIATION LAB.	MIT RAD LAB (LETTER(S))-
MASSACHUSETTS INST. OF TECH., CAMBRIDGE. RADIATION LAB.	MIT-RL-
MASS. INST. OF TECH., CAMBRIDGE. RADIATION LAB. (REPT.)	MIT-RL-R-
MASSACHUSETTS INST. OF TECH., CAMBRIDGE. RADIATION LAB.	RL-
MASSACHUSETTS INST. OF TECH., CAMBRIDGE. RADIATION LAB.	R.L.R.(NUMBER)-
MASSACHUSETTS INST. OF TECH., CAMBRIDGE. RADIATION LAB.	SC-
MASSACHUSETTS INST. OF TECH., CAMBRIDGE. RADIOACTIVITY CENTER (ANNUAL PROGRESS REPORTS)	MIT RMP APR-
MASSACHUSETTS INST. OF TECH., CAMBRIDGE. RESEARCH LABORATORY FOR MECHANICS OF MATERIALS	MIT RLMM-
MASSACHUSETTS INST. OF TECH., CAMBRIDGE. RESEARCH LABORATORY FOR MECHANICS OF MATERIALS	RLMM-
MASSACHUSETTS INST. OF TECH., CAMBRIDGE. RESEARCH LAB. OF ELECTRONICS	JA-
MASSACHUSETTS INST. OF TECH., CAMBRIDGE. RESEARCH LAB. OF ELECTRONICS	MIT-RLE-
MASSACHUSETTS INST. OF TECH., CAMBRIDGE. RESEARCH LAB. OF ELECTRONICS (PRELIMINARY TECHNICAL REPORT)	MIT-RLE-PTR-
MASSACHUSETTS INST. OF TECH., CAMBRIDGE. RESEARCH LAB. OF ELECTRONICS (QUARTERLY PROGRESS REPORT)	MIT-RLE-QPR-
MASSACHUSETTS INST. OF TECH., CAMBRIDGE. RESEARCH LAB. OF ELECTRONICS	MIT RLE TR-
MASSACHUSETTS INST. OF TECH., CAMBRIDGE. RESEARCH LAB. OF ELECTRONICS	RLE-
MASSACHUSETTS INST. OF TECH., CAMBRIDGE. RESEARCH LAB. OF ELECTRONICS	RLE-TR-
MASSACHUSETTS INST. OF TECH., CAMBRIDGE. RESEARCH LAB. OF ELECTRONICS	STR-
MASSACHUSETTS INST. OF TECH., CAMBRIDGE. RICHARDS MINERAL ENGINEERING LAB.	MITS-
MASS. INST. OF TECH., CAMBRIDGE. SCHOOL OF ENG.	R(NO.)-
MASS. INST. OF TECH., CAMBRIDGE. SCHOOL OF ENG.	R-(YEAR)-
MASS. INST. OF TECH., CAMBRIDGE. SERVOMECHANISMS LAB.	E-
MASSACHUSETTS INST. OF TECH., CAMBRIDGE. SERVOMECHANISMS LAB. (ENGINEERING REPORT)	MIT DIC(NOS.)-ER-
MASS. INST. OF TECH., CAMBRIDGE. SERVOMECHANISMS LAB.	MIT DIC(NUMBERS)-R-
MASS. INST. OF TECH., CAMBRIDGE. SERVOMECHANISMS LAB.	MIT DIC(NUMBERS)-TM-
MASSACHUSETTS INST. OF TECH., CAMBRIDGE. SERVOMECHANISMS LAB. (MEMORANDUM)	MIT SL M-
MASS. INST. OF TECH., CAMBRIDGE. SERVOMECHANISMS LAB.	MIT SL R-
MASS. INST. OF TECH., CAMBRIDGE. SERVOMECHANISMS LAB.	REPORT R-
MASSACHUSETTS INST. OF TECH., CAMBRIDGE. SOIL MECHANICS DIV.	AWEWS-
MASSACHUSETTS INST. OF TECH., CAMBRIDGE. SOIL MECHANICS DIV.	RR-R-
MASSACHUSETTS INST. OF TECH., CAMBRIDGE. SOIL MECHANICS DIV.	T-
MASSACHUSETTS INST. OF TECH., CAMBRIDGE. SOLID-STATE AND MOLECULAR THEORY GROUP	MIT-SMI-
MASSACHUSETTS INST. OF TECH., CAMBRIDGE. SOLID-STATE AND MOLECULAR THEORY GROUP	MIT-SMTG-
MASSACHUSETTS INST. OF TECH., CAMBRIDGE. SOLID-STATE AND MOLECULAR THEORY GROUP (QUARTERLY PROGRESS RPT)	MIT-SMTG-QPR-
MASSACHUSETTS INST. OF TECH., CAMBRIDGE. SOLID-STATE AND MOLECULAR THEORY GROUP (QUARTERLY PROGRESS RPT)	MIT-SMTG-QR-
MASSACHUSETTS INST. OF TECH., CAMBRIDGE. SOLID-STATE AND MOLECULAR THEORY GROUP (TECHNICAL REPORT)	MIT-SMTG-TR-
MASSACHUSETTS INST. OF TECH., CAMBRIDGE. SOLID-STATE AND MOLECULAR THEORY GROUP	MIT-SSMTG-
MASSACHUSETTS INST. OF TECH., CAMBRIDGE. SOLID-STATE AND MOLECULAR THEORY GROUP (PROGRAMMING NOTE)	MIT-SSMTG-PN-
MASS. INST. OF TECH., CAMBRIDGE. SPACE PROPULSION LAB	MIT-SPL-QTPR-
MASS. INST. OF TECH., CAMBRIDGE. SPECTROSCOPY LAB.	MIT-SL-TR-
MASS. INST. OF TECH., CAMBRIDGE. SPECTROSCOPY LAB.	MIT SP TR-
MASSACHUSETTS INST. OF TECH., LEXINGTON. LINCOLN LAB. (REF. 1)	(NUMBERS...)
MASSACHUSETTS INST. OF TECH., LEXINGTON. LINCOLN LAB.	AZT-(YEAR)-
MASSACHUSETTS INST. OF TECH., LEXINGTON. LINCOLN LAB.	DOR-
MASSACHUSETTS INST. OF TECH., LEXINGTON. LINCOLN LAB.	JA-
MASSACHUSETTS INST. OF TECH., LEXINGTON. LINCOLN LAB.	LINCOLN MANUAL-
MASSACHUSETTS INST. OF TECH., LEXINGTON. LINCOLN LAB.	LL-
MASSACHUSETTS INST. OF TECH., LEXINGTON. LINCOLN LAB.	LL-TN-
MASSACHUSETTS INST. OF TECH., LEXINGTON. LINCOLN LAB.	LPSR-
MASSACHUSETTS INST. OF TECH., LEXINGTON. LINCOLN LAB.	MAC-TR-
MASSACHUSETTS INST. OF TECH., LEXINGTON. LINCOLN LAB.	MIT-LL-
MASSACHUSETTS INST. OF TECH., LEXINGTON. LINCOLN LAB. (BIBLIOGRAPHY)	MIT-LL-BIB-
MASSACHUSETTS INST. OF TECH., LEXINGTON. LINCOLN LAB. (GROUP REPORTS)	MIT-LL-GR(NO.)-
MASSACHUSETTS INST. OF TECH., LEXINGTON. LINCOLN LAB. (SOLID STATE RESEARCH QUARTERLY PROGRESS REPORT)	MIT-LL-QR-(YR)-(MO)-
MASSACHUSETTS INST. OF TECH., LEXINGTON. LINCOLN LAB. (SOLID STATE RESEARCH)	MIT-LL-SSR-
MASSACHUSETTS INST. OF TECH., LEXINGTON. LINCOLN LAB. (TECHNICAL MEMORANDUM)	MIT-LL-TM-
MASSACHUSETTS INST. OF TECH., LEXINGTON. LINCOLN LAB. (TECHNICAL REPORT)	MIT-LL-TR-
MASSACHUSETTS INST. OF TECH., LEXINGTON. LINCOLN LAB.	MS-
MASSACHUSETTS INST. OF TECH. LEXINGTON. LINCOLN LAB.	PA-
MASSACHUSETTS INST. OF TECH., LEXINGTON. LINCOLN LAB.	REF-BIB-
MASSACHUSETTS INST. OF TECH., LEXINGTON. LINCOLN LAB.	RP-

MASS. INST.OF TECH.,OAK RIDGE,TENN. ENG. PRACTICE SCH	EPS-K-
MASS. INST.OF TECH.,OAK RIDGE,TENN. ENG. PRACTICE SCH	EPS-X-
MASS. INST.OF TECH.,OAK RIDGE,TENN. ENG. PRACTICE SCH	EPS-Y-
MASS. INST.OF TECH.,OAK RIDGE,TENN. ENG. PRACTICE SCH	K-T-
MASS. INST.OF TECH.,OAK RIDGE,TENN. ENG. PRACTICE SCH	KT-
MASS. INST.OF TECH.,OAK RIDGE,TENN. ENG. PRACTICE SCH	MIT-OR-
MASSACHUSETTS INST. OF TECH., SOUTH DARTMOUTH. ROUND HILL FIELD STATION	MIT-RHFS-
MASS. INST. OF TECH., WATERTOWN. MINERAL ENG. LAB.	MITG-
MATADOR JOINT TASK GROUP (REF. 18)	MATADOR JTG-
MATERIALS RESEARCH CORP., ORANGEBURG, N.Y.	MRC-
MATERIALS RESEARCH CORP., YONKERS, N.Y.	MRC-
MATERIALS RESEARCH CORP., YONKERS, N.Y.	MRC-R-
MATERIALS RESEARCH LAB., INC., RICHTON PARK, ILL.	MRL-(NUMBER -LTR.-NO.)
MATERIEL TEST DIRECTORATE, ABERDEEN PROVING GD., MD.	APG-MT-
MATERIEL TEST DIRECTORATE, ABERDEEN PROVING GD., MD.	DPS-
MATH AND METRIK, INC., CHESHIRE, CONN.	MN-TR-
MATHEMATICA, PRINCETON, N.J.	F-
MATHEMATICA, INC., PRINCETON, N.J.	(NUMBER)-FP-
MATHEMATICAL APPLICATIONS GROUP, INC.,WHITE PLAINS,NY	MAGI-
MATHEMATICAL APPLICATIONS GROUP, INC.,WHITE PLAINS,NY	MAGI-MR-
MATHEMATICAL APPLICATIONS GROUP, INC.,WHITE PLAINS,NY	MR-
MATHEMATICAL SCIENCES CORP., SEATTLE	MSC-
MATHEMATICAL SCIENCES NORTHWEST, INC., SEATTLE	MSNW-(YEAR)-(NO.)-
MATHIESON CHEMICAL CORP., BALTIMORE	CLAB-(NUMBER)-(NUMBER)-
MATHIESON CHEMICAL CORP., BALTIMORE	MCC-1023-TR-
MATSUSHITA RES. INST. TOKYO, INC., KAWASAKI, JAPAN	J-(NUMBER)-
MAURER (J.A.), INC., LONG ISLAND, N.Y.	GOP-
MAURY CENTER FOR OCEANOGRAPHIC SCIENCE, WASHINGTON,DC	MC (3 DIGITS)
MAX-PLANCK-INSTITUT FUER KERNPHYSIK, HEIDELBERG	AR-
MAX-PLANCK-INSTITUT FUER KERNPHYSIK, HEIDELBERG	MPIH-(YEAR)/(NO.)/(NO.)
MAX-PLANCK-INSTITUT FUER PHYSIK UND ASTROPHYSIK, MUNICH	MPI-
MAX-PLANCK-INSTITUT FUER PHYSIK UND ASTROPHYSIK, MUNICH	MPI-PA-
MAX-PLANCK-INSTITUT FUER PHYSIK UND ASTROPHYSIK, MUNICH	MPI-PAE/ASTRO-
MAX-PLANCK-INSTITUT FUER PHYSIK UND ASTROPHYSIK, MUNICH	MPI-PAE-EXTRATERP-
MAX-PLANCK-INSTITUT FUER PHYSIK UND ASTROPHYSIK, MUNICH	MPI-PAE/EXTRATERR.-
MAX-PLANCK-INSTITUT FUER PHYSIK UND ASTROPHYSIK, MUNICH	MPI-PAE/PI-(NO.)/(YR.)
MAX-PLANCK-INSTITUT FUER PHYSIK UND ASTROPHYSIK, MUNICH	MPI-PAE/PL-
MAX-PLANCK-INSTITUT FUER STROEMUNGSFORSCHUNG, GOETTINGEN	KWIS-
MAXSON (W.L.) CORP., N.Y.C. (REF. 1)	(WORK ORDER NUMBER)-
MAXSON (W.L.) CORP., N.Y.C.	MAX-
MAXSON (W.L.) CORP., N.Y.C.	MC-
MAXSON (W.L.) CORP., N.Y.C.	WLM-
MAXSON ELECTRONICS CORP., GREAT RIVER, N.Y.	MEC-(NO.)-(NO.)-
MAXWELL LABS., INC., SAN DIEGO, CALIF.	MLI-
MAXWELL LABS., INC., SAN DIEGO, CALIF.	MLR-
MB ASSOCIATES, SAN RAMON, CALIF.	MB-
MB ASSOCIATES, SAN RAMON, CALIF.	MB-(YEAR)/
MB ASSOCIATES, SAN RAMON, CALIF.	MB-R-(YEAR)/
MEASUREMENT ANALYSIS CORP., LOS ANGELES	MAC-
MECHANICAL EQUIPMENT CO., NEW ORLEANS	MECO-
MECHANICAL TECHNOLOGY, INC., LATHAM, N.Y.	MTI-
MECHANICAL TECHNOLOGY, INC., LATHAM, N.Y.	MTI-(NO.)TR(NO.)
MECHANICS RESEARCH, INC., EL SEGUNDO, CALIF.	DA-(YEAR)-
MECHANICS RESEARCH, INC., HOUSTON, TEX.	MRI-HOU-(YEAR)-
MEDICAL ANALYSIS SECTION (AIR FORCE)	MAS-
MEDICAL + INDUSTRIAL EQUIPMENT CO., LTD.	MIE-
MEDICAL COLL. OF VIRGINIA, RICHMOND	MCV-
MEDICAL COLL. OF VIRGINIA, RICHMOND	MCV (CONTRACT NO.)
MEDICAL PHYSICS TRANSLATION FUND, PARK FOREST, ILL.	MP-
MEDICAL RESEARCH AND DEVELOPMENT BOARD, WASH., D.C.	MRD-
MEETING OF THE MECHANICAL FAILURES PREVENTION GROUP, 10TH, PALO ALTO, CALIF.	MFPG-
MELBOURNE. UNIV. SCHOOL OF PHYSICS	UM-P-
MELLON INST., PITTSBURGH	MELLON-
MELLON INST., PITTSBURGH	MELLON-IIR-RR-
MELLON INST., PITTSBURGH (RESEARCH REPORT)	MELLON-RR-
MELLON INST., PITTSBURGH (TECHNICAL REPORT)	MELLON-TR-
MELLON INST., PITTSBURGH	MI-
MELLON INST., PITTSBURGH	MIIR-RR-
MELLON INST., PITTSBURGH	MIIR-TR-
MELLON INST. RADIATION RESEARCH LABS., PITTSBURGH	RRL-
MELPAR, INC. (ALL LOCATIONS)	MELP-

MELPAR, INC.

MELPAR, INC., ALEXANDRIA, VA.	MEL-
MELPAR, INC., FALLS CHURCH, VA.	FPL-TDR-
MELPAR, INC., FALLS CHURCH, VA.	GS-
MELPAR, INC., FALLS CHURCH, VA.	MEI-
MELPAR, INC., FALLS CHURCH, VA.	MEL F-PH.2 QPR-
MELPAR, INC., FALLS CHURCH, VA.	MELPAR-
MELPAR, INC., FALLS CHURCH, VA.	MELPAR-QPR-
MELPAR, INC., FALLS CHURCH, VA.	MELPAR-TR-
MENASCO MANUFACTURING CO., BURBANK, CALIF.	A(NO.)-
MENASCO MANUFACTURING CO., BURBANK, CALIF.	R-
MENASCO MANUFACTURING CO., BURBANK, CALIF. (REF. 1)	(REPORT NO.)-(CODE LTR)
MERCK SHARP AND DOHME RESEARCH LABS., RAHWAY, N.J.	QP-
MERRILL FLOOD AND ASSOCIATES	MFA-
MERRIMACK COLL., NORTH ANDOVER, MASS.	MERRIMACK-
MESSERSCHMIDT A.G. (VERSUCHSBERICHT)	ME VB-
MESSERSCHMIDT A.G. (VERSUCHSBERICHT)	ME VB-(NO.)-
MESSERSCHMITT-BOELKOW-BLOHM G.M.B.H., MUNICH	ZV-(NUMBER)-O
MESSERSCHMITT-BOELKOW G.M.B.H., MUNICH	RF-(NUMBER)-O
METAL FINISHING ABSTRACTS TRANSLATION SERVICES, TEDDINGTON, MIDDX., ENGLAND	MFA-TS-
METAL HYDRIDES INC., BEVERLY, MASS.	MH-
METAL HYDRIDES INC., BEVERLY, MASS.	MHC-
METAL HYDRIDES INC. CHEM. RES. LAB., BEVERLY, MASS.	SHORT REPORT
METALLURGICAL ADVISORY COMMITTEE ON TITANIUM	INFO. BULL. T-
METALLURGICAL ADVISORY COMMITTEE ON TITANIUM	MACT-
METALLURGICAL RES. + DEV. CO., INC., WASHINGTON, D.C.	MRDC-
METALS AND CONTROLS CORP., ATTLEBORO, MASS.	FSM-
METALS AND CONTROLS CORP., ATTLEBORO, MASS.	MAC-(YR)-
METCUT RESEARCH ASSOCIATES INC., CINCINNATI	METCUT-
METCUT RESEARCH ASSOCIATES INC., CINCINNATI	MHR-
METCUT RESEARCH ASSOCIATES INC., CINCINNATI	MRA-
METEOROLOGY INTERNAT'L,INC.,MONTEREY,CAL.(PROJ.FAMOS)	FAMOS-RR-(NUMBER-YEAR)
METEOROLOGY INTERNATIONAL, INC., MONTEREY, CALIF.	M-
METEOROLOGY INTERNATIONAL, INC., MONTEREY, CALIF.	TM-(NO.)-(NO)-ED
METEOROLOGY RESEARCH, INC., ALTADENA, CALIF.	MR(NO.)-FR-
METEOROLOGY RESEARCH, INC., ALTADENA, CALIF.	MRI-
METEOROLOGY RESEARCH, INC., ALTADENA, CALIF.	MRI(YEAR)-(LETTERS)-
METEOROLOGY RESEARCH, INC., ALTADENA, CALIF.	MRI(YEAR)-FR-
METEOROLOGY RESEARCH, INC., ALTADENA, CALIF.	MRI-(YEAR)-IR
METEOROLOGY RESEARCH, INC., ALTADENA, CALIF.	MRI-(YEAR)-PR-
METEOROLOGY RESEARCH, INC., ALTADENA, CALIF.	MRI-R-(NUMBER)R
METRONICS ASSOCIATES, INC., PALO ALTO, CALIF.	METRONICS-
METROPOLITAN-VICKERS ELECTRICAL CO., LTD., MANCHESTER, LANCS, ENGLAND (PROJECT 972)	MV-972/-
MEXICO. COMISION NACIONAL DE ENERGIA NUCLEAR. CENTRO NUCLEAR DE MEXICO, MEXICO CITY (CONF. PROCEEDINGS)	CNM-R-
MEXICO. INSTITUTO POLITECNICO NACIONAL. CENTRO DE INVESTIGACION Y DE ESTUDIOS AVANZADOS, MEXICO CITY	CIEA-
MEYER (CHARLES A.) AND CO., INC., N.Y.C.	PC-S-
MHD RESEARCH, INC., NEWPORT BEACH, CALIF.	MHD-
MIAMI. UNIV., CORAL GABLES, FLA.	SAR-
MIAMI. UNIV., CORAL GABLES, FLA. CENTER FOR THEORETICAL STUDIES	CST-PA-(YEAR)-
MIAMI. UNIV., CORAL GABLES, FLA. CENTER FOR THEORETICAL STUDIES	CTS-B-
MIAMI. UNIV., CORAL GABLES, FLA. CENTER FOR THEORETICAL STUDIES	CTS-B-(YEAR)-
MIAMI. UNIV., CORAL GABLES, FLA. CENTER FOR THEORETICAL STUDIES	CTS-H-(YEAR)-
MIAMI. UNIV., CORAL GABLES, FLA. CENTER FOR THEORETICAL STUDIES	CTS-HE-(YEAR)-
MIAMI. UNIV., CORAL GABLES, FLA. CENTER FOR THEORETICAL STUDIES	CTS-LN-(YEAR)-
MIAMI. UNIV., CORAL GABLES, FLA. CENTER FOR THEORETICAL STUDIES	CTS-PA-(YR.)-
MIAMI. UNIV., CORAL GABLES, FLA. CENTER FOR THEORETICAL STUDIES	CTS-PHIL. S-(YEAR)-
MIAMI. UNIV., CORAL GABLES, FLA. CENTER FOR THEORETICAL STUDIES	CTS-QED-(YEAR)-
MIAMI. UNIV., CORAL GABLES, FLA. CENTER FOR THEORETICAL STUDIES	CTS-SSP-(YEAR)-
MIAMI. UNIV., CORAL GABLES, FLA. CENTER FOR THEORETICAL STUDIES	CTS-TC-(YEAR)-
MIAMI. UNIV., CORAL GABLES, FLA. CENTER FOR THEORETICAL STUDIES	CTS-T-PHYS-(YEAR)-
MIAMI. UNIV., CORAL GABLES, FLA. DEPT. OF PHYSICS	MIAPH-AP-(YEAR.NUMBER)
MIAMI. UNIV., CORAL GABLES, FLA. DEPT. OF PHYSICS	MIAPH-OP-(YEAR.NUMBER)
MIAMI. UNIV., CORAL GABLES, FLA. DEPT. OF PHYSICS	MIAPH-PL-
MIAMI. UNIV., CORAL GABLES, FLA. DEPT. OF PHYSICS	MIAPH-PL-(YEAR)-
MIAMI. UNIV., CORAL GABLES, FLA. DEPT. OF PHYSICS	MIAPH-PP-
MIAMI. UNIV., CORAL GABLES, FLA. DEPT. OF PHYSICS	MIAPH-TP-(YEAR.NUMBER)
MIAMI. UNIV., CORAL GABLES, FLA. MARINE LAB.	ML-
MIAMI. UNIV., CORAL GABLES, FLA. MARINE LAB.	MU-ML-
MIAMI. UNIV., FLA. RADAR METEOROLOGICAL LAB.	ML-
MIAMI SHIPBUILDING CORP., FLA.	MSC-R-
MIAMI UNIV., OXFORD, OHIO	MIAMI-U-
MICHIGAN. DEPT. OF STATE HIGHWAYS. TRAFFIC SAFETY DIV	TSD-TR-(NUMBER-YEAR)
MICHIGAN. OFFICE OF PLANNING COORDINATION	TR-A-
MICHIGAN. OFFICE OF PLANNING COORDINATION	TR-B-
MICHIGAN. OFFICE OF PLANNING COORDINATION	TR-J-
MICHIGAN. UNIV., ANN ARBOR	BMR-
MICHIGAN. UNIV., ANN ARBOR	EMD-
MICHIGAN. UNIV., ANN ARBOR	GRP-

MICHIGAN. UNIV., ANN ARBOR	IN-R-QUAL-(YEAR)-
MICHIGAN. UNIV., ANN ARBOR	ISR-
MICHIGAN. UNIV., ANN ARBOR	MEMO-
MICHIGAN. UNIV., ANN ARBOR	MICH-
MICHIGAN. UNIV., ANN ARBOR (SQUID PROJECT) (REF. 33)	MICH-(NUMBER)-(LETTER)
MICHIGAN. UNIV., ANN ARBOR	MICH (NOS.-NO.-LTR.)
MICHIGAN. UNIV., ANN ARBOR	MICHIGAN U-(NO.)-(LTR.)-
MICHIGAN. UNIV., ANN ARBOR	MICH-U-
MICHIGAN. UNIV., ANN ARBOR	MICHU-
MICHIGAN. UNIV., ANN ARBOR	MICH UNM-
MICHIGAN. UNIV., ANN ARBOR	MICH UNR-
MICHIGAN. UNIV., ANN ARBOR	MIC U-
MICHIGAN. UNIV., ANN ARBOR	MU-
MICHIGAN. UNIV., ANN ARBOR	NACA-TN-
MICHIGAN. UNIV., ANN ARBOR	NAVEXOS-P-
MICHIGAN. UNIV., ANN ARBOR	NAVSO-
MICHIGAN. UNIV., ANN ARBOR	PMR-
MICHIGAN. UNIV., ANN ARBOR	RM-
MICHIGAN. UNIV., ANN ARBOR	UM-
MICHIGAN. UNIV., ANN ARBOR	UMAEC-
MICHIGAN. UNIV., ANN ARBOR	UMAEC (PROJECT)-(YR.)-
MICHIGAN. UNIV., ANN ARBOR	UMM-
MICHIGAN. UNIV., ANN ARBOR	UMR-
MICHIGAN. UNIV., ANN ARBOR (ALL DEPTS.)(MEMORANDUM)	WTM-
MICHIGAN. UNIV., ANN ARBOR (REPORTS)	
MICHIGAN. UNIV., ANN ARBOR (WIND TUNNEL MEMORANDUM)	
MICHIGAN. UNIV., ANN ARBOR. BALLISTIC MISSILE RADIATION ANALYSIS CENTER	BAMIRAC-
MICHIGAN. UNIV., ANN ARBOR. BALLISTIC MISSILE RADIATION ANALYSIS CENTER	MICH-U/BAMIRAC-
MICHIGAN. UNIV., ANN ARBOR. CAVITATION AND MULTIPHASE FLOW LAB.	UMICH-(NUMBER-NUMBER)
MICHIGAN. UNIV., ANN ARBOR. COLL. OF ENGINEERING	IP-
MICHIGAN. UNIV., ANN ARBOR. COLL. OF ENGINEERING	IRIA-
MICHIGAN. UNIV., ANN ARBOR. COLL. OF ENGINEERING (INDUSTRY PROGRAM)	MICHU-IP-
MICHIGAN. UNIV., ANN ARBOR.COMMUNICATION SCIENCES LAB	CSL-
MICHIGAN. UNIV., ANN ARBOR. DEPT. OF AERONAUTICAL AND ASTRONAUTICAL ENGINEERING	CM-
MICHIGAN. UNIV., ANN ARBOR. DEPT. OF NUCLEAR ENG.	MMPP-
MICHIGAN. UNIV., ANN ARBOR. DEPT. OF NUCLEAR ENG.	MMPP-FRPC-
MICHIGAN. UNIV., ANN ARBOR. ELECTRON PHYSICS LAB.	EPL-TR-
MICHIGAN. UNIV., ANN ARBOR. ELECTRONIC DEFENSE GROUP	MICH EDG-TR-
MICHIGAN. UNIV., ANN ARBOR. ELECTRONIC DEFENSE GROUP	UM-EDG-
MICHIGAN. UNIV., ANN ARBOR. ENG. RES. INST. (REF. 1)	(NUMBERS...)
MICHIGAN. UNIV., ANN ARBOR. ENGINEERING RES. INST.	AIR-
MICHIGAN. UNIV., ANN ARBOR. ENGINEERING RES. INST.	CT-
MICHIGAN. UNIV., ANN ARBOR. ENGINEERING RES. INST.	EDG-
MICHIGAN. UNIV., ANN ARBOR. ENGINEERING RES. INST.	ERI-
MICHIGAN. UNIV., ANN ARBOR. ENGINEERING RES. INST.	ERI-(NUMBER)-(NO.)-T
MICHIGAN. UNIV., ANN ARBOR. ENGINEERING RES. INST.	ERI/MICH-
MICHIGAN. UNIV., ANN ARBOR. ENGINEERING RES. INST.	HAO-
MICHIGAN. UNIV., ANN ARBOR. ENGINEERING RES. INST.	M720-1-R-
MICHIGAN. UNIV., ANN ARBOR. ENGINEERING RES. INST.	MICH ERI-(NOS.-NO.-LTR.)
MICHIGAN. UNIV., ANN ARBOR. ENGINEERING RES. INST.	MICH ERI-(YR.-NO.)
MICHIGAN. UNIV., ANN ARBOR. ENGINEERING RES. INST.	MICH M(NOS.) TR-
MICHIGAN. UNIV., ANN ARBOR. ENGINEERING RES. INST.	MICH MX(NOS)-(NOS-NO-LTR)
MICHIGAN. UNIV., ANN ARBOR. ENGINEERING RES. INST.	MICHU-PR-
MICHIGAN. UNIV., ANN ARBOR. ENGINEERING RES. INST.	UM-ERI-
MICHIGAN. UNIV., ANN ARBOR. ENG. RES. INST.(PROJ.MIRO	UMM-MIRO-
MICHIGAN. UNIV., ANN ARBOR. ENGINEERING RES. INST.	UMM-MX-
MICHIGAN. UNIV., ANN ARBOR. ENGINEERING RES. INST.	WTM-
MICHIGAN. UNIV., ANN ARBOR. HIGHWAY SAFETY RES. INST.	HSRI-B(NUMBER)-
MICHIGAN. UNIV., ANN ARBOR. HIGHWAY SAFETY RES. INST.	PHF-
MICHIGAN. UNIV., ANN ARBOR. HIGHWAY SAFETY RES. INST.	PUF-
MICHIGAN. UNIV., ANN ARBOR. INFORMATION SYSTEMS LAB.	ISL-
MICHIGAN. UNIV., ANN ARBOR. MICHIGAN MEMORIALPHOENIX PROJECT	MMPP-
MICHIGAN. UNIV., ANN ARBOR. MICHIGAN MEMORIALPHOENIX PROJECT	MMPP-NPC-
MICHIGAN. UNIV., ANN ARBOR. MICHIGAN MEMORIALPHOENIX PROJECT	MMPP-S-
MICHIGAN. UNIV., ANN ARBOR. OFFICE OF RESEARCH ADM.	BUMBLEBEE-
MICHIGAN. UNIV., ANN ARBOR. OFFICE OF RESEARCH ADM.	ORA-(NUMBERS,LETTERS)
MICHIGAN. UNIV., ANN ARBOR. OFFICE OF RESEARCH ADM.	ORA-
MICHIGAN. UNIV., ANN ARBOR. RADIO ASTRONOMY OBS.	UM/RAO-
MICHIGAN. UNIV., ANN ARBOR. RESEARCH INST.	UMRI-
MICHIGAN. UNIV., ANN ARBOR. SCHOOL OF PUBLIC HEALTH	DPWR-
MICHIGAN. UNIV., ANN ARBOR. SYSTEMS ENGINEERING LAB.	SEL-
MICHIGAN. UNIV., ANN ARBOR. SYSTEMS ENGINEERING LAB.	SEL-TR-
MICHIGAN. UNIV., ANN ARBOR. SYSTEMS RESEARCH LAB.	SRL-(NUMBER)-FR-(YEAR)-
MICHIGAN. UNIV., ANN ARBOR. VELA SEISMIC INFORMATION ANALYSIS CENTER	MICH-U/VESIAC-
MICHIGAN. UNIV., ANN ARBOR. VELA SEISMIC INFORMATION ANALYSIS CENTER	VESIAC-
MICHIGAN. UNIV., ANN ARBOR. VELA SEISMIC INFORMATION ANALYSIS CENTER	VESIAC BULL-
MICHIGAN. UNIV., ANN ARBOR. WILLOW RUN LABS.	MICH-U-WRL-
MICHIGAN. UNIV., ANN ARBOR. WILLOW RUN LABS.	WILLOW-(NO.)-(NO.)-(LTR.)
MICHIGAN. UNIV., ANN ARBOR. WILLOW RUN LABS.	WRL-
MICHIGAN. UNIV., ANN ARBOR. WILLOW RUN RESEARCH CTR.	UMR-
MICHIGAN. UNIV., ANN ARBOR. WILLOW RUN RESEARCH CTR.	UMS-
MICHIGAN. UNIV., YPSILANTI. AERONAUTICAL RES. CTR.	UMR-
MICHIGAN. UNIV., YPSILANTI. WILLOW RUN RESEARCH CTR.	EMD-
MICHIGAN. UNIV., YPSILANTI. WILLOW RUN RESEARCH CTR.	EMF-
MICHIGAN. UNIV., YPSILANTI. WILLOW RUN RESEARCH CTR.	EMG-
MICHIGAN. UNIV., YPSILANTI. WILLOW RUN RESEARCH CTR.	FMV-
MICHIGAN CHEMICAL CORP., ST. LOUIS, MICH.	MCC-

MICHIGAN-DYNAMICS, INC. DYNAMIC FILTERS DIV.

MICHIGAN-DYNAMICS, INC. DYNAMIC FILTERS DIV., DETROIT	MDI-
MICHIGAN STATE UNIV., EAST LANSING	MSU-
MICHIGAN STATE UNIV., EAST LANSING (CYCLOTRON PROJ.)	MSUCP-
MICHIGAN STATE UNIV., EAST LANSING	MSU-TR-
MICHIGAN STATE UNIV., EAST LANSING	RM-
MICHIGAN STATE UNIV., EAST LANSING. COLLEGE OF COMMUNICATION ARTS	BG-
MICHIGAN STATE UNIV., EAST LANSING. DEPT. OF CHEMISTRY	MSU-PR-
MICHIGAN STATE UNIV., EAST LANSING. DEPT. OF PHYSICS (TECHNICAL OPERATING REPORT)	MSU-TOR-
MICHIGAN STATE UNIV., EAST LANSING. DEPT. OF SPEECH	SHSLR-
MICHIGAN STATE UNIV., EAST LANSING. DEPT. OF STATISTICS	JG-
MICHIGAN STATE UNIV., EAST LANSING. DEPT. OF STATISTICS	MSU-RM-
MICHIGAN STATE UNIV., EAST LANSING. DEPT. OF STATISTICS	RM-
MICHIGAN STATE UNIV., EAST LANSING. DIV. OF ENGINEERING RESEARCH	INTERIM SCIENTIFIC-
MICHIGAN STATE UNIV., EAST LANSING. STATISTICAL LAB.	SLP-
MICRO-RADIONICS, INC., VAN NUYS, CALIF.	RD-(NUMBERS)-
MICRONETICS, INC., SAN DIEGO, CALIF.	R(NUMBER)-(YEAR)
MICROWAVE ASSOCIATES, INC., BOSTON	MA-
MICROWAVE ASSOCIATES, INC., BOSTON	MAI-
MICROWAVE ASSOCIATES, LTD., LUTON, BEDS., ENGLAND	MA-R(NUMBER)
MICROWAVE DEVELOPMENT LABS., INC., WALTHAM, MASS.	MDL-
MICROWAVE ELECTRONICS CORP., PALO ALTO, CALIF.	MEC-QTR-NOBSR-
MICROWAVE ELECTRONICS CORP., PALO ALTO, CALIF.	MEC-QTR-NONR-
MICROWAVE ELECTRONICS CORP., PALO ALTO, CALIF.	MEP-
MICROWAVE ENGINEERING LABS., PALO ALTO, CALIF.	MEL-
MICROWAVE PHYSICS LAB., MOUNTAIN VIEW, CALIF.	MPL-
MICROWAVE RADIATION CO., INC., GARDENA, CALIF.	MRC-
MIDDLE TENNESSEE STATE COLL., MURFREESBORO	MTSC-
MIDLAND-WRIGHT, KANSAS CITY, KANS.	CR-/XM-
MIDWEST APPLIED SCIENCE CORP., LAFAYETTE, IND.	LGO-
MIDWEST APPLIED SCIENCE CORP., LAFAYETTE, IND.	MASC-(NUMBER)-F
MIDWEST APPLIED SCIENCE CORP., LAFAYETTE, IND.	MASC-RD-
MIDWEST ENGINEERING DEVELOPMENT CO., INC., KANSAS CITY, MO. (REF. 1)	(NUMBER)
MIDWEST RESEARCH INST., KANSAS CITY, MO.	MRI-
MIDWEST RESEARCH INST., KANSAS CITY, MO.	MRI-(NO.)-B-QPR-
MIDWEST RESEARCH INST., KANSAS CITY, MO.	MRI-(NO.)-P-QTR-
MIDWEST RESEARCH INST., KANSAS CITY, MO.	MRI-PR-
MIDWEST RESEARCH INST., KANSAS CITY, MO.	MRI-QPR-
MIDWEST RESEARCH INST., KANSAS CITY, MO.	MRI-TR-(MONTH/YEAR)
MIDWESTERN UNIVERSITIES RESEARCH ASSN., MADISON, WIS.	MURA-
MIDWESTERN UNIVERSITIES RESEARCH ASSN., STOUGHTON, WIS	MURA-
MIDWESTERN UNIVERSITIES RESEARCH ASSN., URBANA, ILL.	AMS-
MIDWESTERN UNIVERSITIES RESEARCH ASSN., URBANA, ILL.	AMS-MURA/-
MIDWESTERN UNIVERSITIES RESEARCH ASSN., URBANA, ILL.	ESA-MURA-
MIDWESTERN UNIVERSITIES RESEARCH ASSN., URBANA, ILL.	KMT-MAC-
MIDWESTERN UNIVERSITIES RESEARCH ASSN., URBANA, ILL.	KRS-MURA-
MIDWESTERN UNIVERSITIES RESEARCH ASSN., URBANA, ILL.	LWJ-LJL-MAC-
MIDWESTERN UNIVERSITIES RESEARCH ASSN., URBANA, ILL.	MAC-
MIDWESTERN UNIVERSITIES RESEARCH ASSN., URBANA, ILL.	MAC-(INITIALS)-
MIDWESTERN UNIVERSITIES RESEARCH ASSN., URBANA, ILL.	MURA-
MIDWESTERN UNIVERSITIES RESEARCH ASSN., URBANA, ILL.	MURA-(INITIALS)-
MIDWESTERN UNIVERSITIES RESEARCH ASSN., URBANA, ILL. (CODE IS AUTHOR'S INITIALS)	SCW-
MILAN. UNIV. INSTITUTE OF PHARMACOLOGY	HER/AP-(NO.)-(YEAR)
MILAN. UNIV. ISTITUTO DI FISICA	IFUM-(NUMBER)/FT(RI)
MILITARY ACADEMY, WEST POINT, N.Y.	USMA-
MILITARY ACADEMY, WEST POINT, NY. DEPT. OF ELECTRICITY	ELEC-(YEAR)-
MILITARY AIR TRANSPORT SERVICE, WASHINGTON, D.C.	MATS-
MILITARY AIR TRANSPORT SERVICE, WASHINGTON, D.C.	NATS-
MILITARY AIR TRANSPORT SERVICE. ATLANTIC DIV., WESTOVER AFB, MASS.	NAD MATS-
MILITARY AIRLIFT COMMAND, SCOTT AFB, ILL.	MAC-OT/E-(NO.-NO.-YR.)
MILITARY ASSISTANCE ADVISORY GROUP, BONN	MAAG-
MILITARY ASSISTANCE COMMAND, VIETNAM, APO SAN FRAN.	MACV-DAR-(LTR.,NO...)
MILITARY ASSISTANCE COMMAND, VIETNAM. APO SAN FRAN.	MACV-SEER-
MILITARY ASSISTANCE COMMAND. TRAINING AIDS DIV., VIETNAM	MACT-TAD-
MILITARY ASSISTANCE COMMAND. TRAINING AIDS DIV., VIETNAM	TD-100/(NUMBER)
MILITARY INTELLIGENCE DETACHMENT (407TH) (STRATEGIC), WILLIAMSBURG, VA.	USAITAD-(YEAR)-
MILITARY INTELLIGENCE SERVICE (ARMY), WASHINGTON, DC	MIS-
MILITARY LIAISON COMMITTEE, WASHINGTON, D.C.	DOD MLC-
MILITARY SEA TRANSPORTATION SERVICE, WASHINGTON, D.C.	MSTS-
MILITARY TRAFFIC MANAGEMENT AND TERMINAL SERVICE, WASHINGTON, D.C.	MTMTS-

MINAS GERAIS UNIV., BELO HORIZONTE, BRAZIL. DIVISAO DE ENGENHARIA DE REATORES	DER-
MINE SAFETY APPLIANCES CO., CALLERY, PENNA.	C-
MINE SAFETY APPLIANCES CO., CALLERY, PENNA.	MEMO-
MINE SAFETY APPLIANCES CO., CALLERY, PENNA.	MSA-MEMO-
MINE SAFETY APPLIANCES CO., CALLERY, PENNA.	MSA-TR-
(SEE ALSO SUBSIDIARY, MSA RESEARCH CORP.)	
MINE WARFARE OPERATIONAL RESEARCH GROUP (NAVY) (NAVAL WARFARE OPERATIONAL RESEARCH MEMORANDUM)	NWORM-
MINNEAPOLIS-HONEYWELL REGULATOR CO. (ALL LOCATIONS)	MH-
MINNEAPOLIS-HONEYWELL REGULATOR CO., DUARTE, CALIF.	ASROC-D-
MINNEAPOLIS-HONEYWELL REGULATOR CO., MINNEAPOLIS	AD-
MINNEAPOLIS-HONEYWELL REGULATOR CO., MINNEAPOLIS	ASROC-
MINNEAPOLIS-HONEYWELL REGULATOR CO., MINNEAPOLIS	HRC-
MINNEAPOLIS-HONEYWELL REGULATOR CO., MINNEAPOLIS	MH-AD-
MINNEAPOLIS-HONEYWELL REGULATOR CO., MINNEAPOLIS	MHRC-
MINNEAPOLIS-HONEYWELL REGULATOR CO., MONROVIA, CALIF	ASROC-
MINNEAPOLIS-HONEYWELL REGULATOR CO., ST.PETERSBURG,FLA	CCN-
MINNEAPOLIS-HONEYWELL REGULATOR CO., ST.PETERSBURG,FLA	MH-AERO-
(SEE ALSO LATER NAME: HONEYWELL, INC.)	
MINNEAPOLIS-HONEYWELL REGULATOR CO. AERONAUTICAL DIV. MINNEAPOLIS	MH-AERO-
MINNEAPOLIS-HONEYWELL REGULATOR CO. AERONAUTICAL DIV. MINNEAPOLIS	MH AERO (NO.) QR-
MINNEAPOLIS-HONEYWELL REGULATOR CO. AERONAUTICAL DIV. MINNEAPOLIS	MH AERO (NO.) TR-
MINNEAPOLIS-HONEYWELL REGULATOR CO. AERONAUTICAL DIV. MINNEAPOLIS	MHR AERO (NO.) TR-
MINNEAPOLIS-HONEYWELL REGULATOR CO. MISSILE DEVELOPMENT LAB., LOS ANGELES	MHR-MDL-
MINNEAPOLIS-HONEYWELL REGULATOR CO. MISSILE DEVELOPMENT LAB., LOS ANGELES	MHR MDL R-ED-
MINNEAPOLIS-HONEYWELL REGULATOR CO. ORDNANCE DIV., DUARTE, CALIF.	D-
MINNEAPOLIS-HONEYWELL REGULATOR CO. ORDNANCE DIV., HOPKINS, MINN.	MH-OD-
MINNEAPOLIS-HONEYWELL REGULATOR CO. ORDNANCE DIV., HOPKINS, MINN.	MHR-SN-
MINNEAPOLIS-HONEYWELL REGULATOR CO. ORDNANCE DIV., MONROVIA, CALIF.	MH-OD-
MINNESOTA. DEPT. OF HIGHWAYS. OFFICE OF MATERIALS	INVESTIGATION-
MINNESOTA. UNIV., DULUTH	AEB-
MINNESOTA. UNIV., MINNEAPOLIS	LADC-
MINNESOTA. UNIV., MINNEAPOLIS	MIN-
MINNESOTA. UNIV., MINNEAPOLIS	MINN-
MINNESOTA. UNIV., MINNEAPOLIS	MINN-TR-
MINNESOTA. UNIV., MINNEAPOLIS (ATMOSPHERIC PHYSICS PROGRAM)	MINN-TR-AP-
MINNESOTA. UNIV., MINNEAPOLIS	MINNU-
MINNESOTA. UNIV., MINNEAPOLIS	MINU-
MINNESOTA. UNIV., MINNEAPOLIS	MNU-
MINNESOTA. UNIV., MINNEAPOLIS	MU-
MINNESOTA. UNIV., MINNEAPOLIS	SATR-
MINNESOTA. UNIV., MINNEAPOLIS	UMN/S-
MINNESOTA. UNIV., MINNEAPOLIS. COSMIC RAY GROUP	CR-
MINNESOTA. UNIV., MINNEAPOLIS. COSMIC RAY GROUP	MINN-TR-CR-
MINNESOTA. UNIV., MINNEAPOLIS. DEPT. OF AERONAUTICAL ENGINEERING (REF. 1)	(NUMBER)
MINNESOTA. UNIV., MINNEAPOLIS. DEPT. OF AERONAUTICS AND ENGINEERING SCIENCES	TR-(YEAR)-
MINNESOTA. UNIV., MINNEAPOLIS. DEPT. OF PHYSICS	TR-CR-
MINNESOTA. UNIV., MINNEAPOLIS. HEAT TRANSFER LAB.	HTL-TR-
MINNESOTA. UNIV., MINNEAPOLIS. HEAT TRANSFER LAB.	HTL-TR-(YEAR)-
MINNESOTA. UNIV., MINNEAPOLIS. HEAT TRANSFER LAB.	MINN HTL TR-
MINNESOTA. UNIV., MINNEAPOLIS. INST. OF TECH.	MINN EM-
MINNESOTA. UNIV., MINNEAPOLIS. INST. OF TECH.	MINN RR-
MINNESOTA. UNIV., MINNEAPOLIS. PHYSICAL ELECTRONICS RESEARCH LAB.	MINN-SR-
MINNESOTA. UNIV., MINNEAPOLIS. ROSEMOUNT AERO. LABS.	MU-RAL-
MINNESOTA. UNIV., MINNEAPOLIS. ROSEMOUNT AERO. LABS.	RAL-RESEARCH RPT.-
MINNESOTA. UNIV., MINNEAPOLIS. ROSEMOUNT AERO. LABS.	RALRR-
MINNESOTA. UNIV., MINNEAPOLIS. SCHOOL OF PHYSICS	AP-
MINNESOTA. UNIV., MINNEAPOLIS. ST. ANTHONY FALLS HYDRAULIC LAB.	SAF-
MINNESOTA. UNIV., MINNEAPOLIS. WATER RESOURCES RESEARCH CENTER	WRRC-BULL-
MINNESOTA ELECTRONICS CORP., ST. PAUL	MEC-
MINNESOTA MINING AND MFG. CO., ST. PAUL (REF. 1)	(NUMBERS...)
MINNESOTA MINING AND MFG. CO., ST. PAUL	EO-
MINNESOTA MINING AND MFG. CO., ST. PAUL	3M CO.-
MINNESOTA MINING AND MFG. CO., ST. PAUL	MMM/
MINNESOTA MINING AND MFG. CO., ST. PAUL	MMM-
MINNESOTA MINING AND MFG. CO., ST. PAUL	MMMC-
MINNESOTA MINING AND MFG. CO., ST. PAUL	MPR-
MINNESOTA MINING AND MFG. CO., ST. PAUL (RESEARCH AND DEVELOPMENT SURVEY)	RDS-
MINN. MINING + MFG. CO. CENTRAL RES. DEPT., ST. PAUL	MMMC-
MINNESOTA MINING AND MFG. CO. CENTRAL RESEARCH DEPT., ST. PAUL. (QUARTERLY PROGRESS REPORT)	MMMC-QPR-
MINN. MINING + MFG. CO. CENTRAL RES. LABS., ST. PAUL	MMMC-RR-
MINNESOTA MINING AND MFG. CO. DEFENSE PRODUCTS DEPT., ST. PAUL	DPD-
MINNESOTA MINING AND MFG. CO. ISOTOPE POWER LAB., ST. PAUL	MMM-
MINNESOTA MINING AND MFG. CO. ISOTOPE POWER LAB., ST. PAUL	MMM-(NUMBER-NUMBER)
MISSION RESEARCH CORP., SANTA BARBARA, CALIF.	MRC-
MISSISSIPPI. UNIV., UNIVERSITY	MISSU-

MISSISSIPPI STATE COLL., STATE COLLEGE	MISSC-
MISSISSIPPI STATE UNIV., STATE COLLEGE	MISSSU-
MISSISSIPPI STATE UNIV., STATE COLLEGE. AGRICULTURAL EXPT. STA.(AEROPHYSICS + AEROSPACE ENG. RES. REPT.)	AAERR-
MISSISSIPPI STATE UNIV., STATE COLLEGE. AGRICULTURAL EXPT. STA.(AEROPHYSICS + AEROSPACE ENG. RES. REPT.)	AAERR-(NO.),JOURNAL-
MISSISSIPPI STATE UNIV., STATE COLLEGE. DEPT. OF AEROPHYSICS	AEROPHYSICS-RR-
MISSOURI. UNIV., COLUMBIA	E-
MISSOURI. UNIV., COLUMBIA	EP-
MISSOURI. UNIV., COLUMBIA	MISU-
MISSOURI. UNIV., COLUMBIA	MU-
MISSOURI. UNIV., COLUMBIA	U MO-
MISSOURI. UNIV., COLUMBIA. DEPT. OF MATHEMATICS (MATHEMATICS RESEARCH PROJECT)	MOU MRG TN-
MISSOURI. UNIV., COLUMBIA. DEPT. OF MATHEMATICS (MATHEMATICS RESEARCH PROJECT)	MU MRGTN-
MISSOURI. UNIV., COLUMBIA. INFORMATION DEPT.	MOU-IS-
MISSOURI. UNIV., ROLLA	MISU-
MISSOURI. UNIV., ROLLA	MISU-TR-
MISSOURI. UNIV., ROLLA. ROCK MECHANICS AND EXPLOSIVES RESEARCH CENTER	RMERC-TR-(YEAR)-
MISSOURI BOTANICAL GARDEN, ST. LOUIS	MBG-ONR-
MISSOURI RESEARCH LABS., INC., ST. LOUIS	MRL-
MITHRAS, INC., CAMBRIDGE, MASS.	MC-
MITHRAS, INC., CAMBRIDGE, MASS.	MS-
MITRE CORP., BAILEY'S CROSSROADS, VA.	MTP-
MITRE CORP., BEDFORD, MASS.	M(YEAR)-
MITRE CORP., BEDFORD, MASS.	MIR-
MITRE CORP., BEDFORD, MASS.	MITRE-
MITRE CORP., BEDFORD, MASS.	MITRE-TM-
MITRE CORP., BEDFORD, MASS.	MITRE-W-
MITRE CORP., BEDFORD, MASS.	MTP-
MITRE CORP., BEDFORD, MASS.	MTR-
MITRE CORP., BEDFORD, MASS.	SR-
MITRE CORP., BEDFORD, MASS.	TM-
MITRE CORP., BEDFORD, MASS.	W-
MITRE CORP., BEDFORD, MASS.	WP-
MITRE CORP., LEXINGTON, MASS.	MITRE-
MITRE CORP., LEXINGTON, MASS.	MITRE SR-
MITRE CORP., LEXINGTON, MASS.	MTR-
MITSUBISHI HEAVY-INDUSTRIES, LTD., TOKYO	MTB-
MITSUBISHI HEAVY-INDUSTRIES, LTD., TOKYO	TB-
MOLECULON RESEARCH CORP., CAMBRIDGE, MASS.	MOLEC-
MOLECULON RESEARCH CORP., CAMBRIDGE, MASS.	MRC-
MOLECULON RESEARCH CORP., CAMBRIDGE, MASS.	MRC-(LTR)(NO.)-(LTR)
MONASH UNIV., CLAYTON, AUSTRALIA	M(NUMBER)
MONMOUTH COUNTY PLANNING BOARD, N.J.	BASIC STUDIES-
MONSANTO CHEMICAL CO. (ALL LOCATIONS)	MONSANTO-
MONSANTO CHEMICAL CO., DAYTON, OHIO	FINAL REPORT-
MONSANTO CHEMICAL CO., DAYTON, OHIO	M3-B-
MONSANTO CHEMICAL CO., DAYTON, OHIO	M3-D-
MONSANTO CHEMICAL CO., DAYTON, OHIO	M4-D-
MONSANTO CHEMICAL CO., DAYTON, OHIO	MLM-M-
MONSANTO CHEMICAL CO., DAYTON, OHIO	RE-CR-
MONSANTO CHEMICAL CO., DAYTON, OHIO	T-
MONSANTO CHEMICAL CO., EVERETT, MASS.	MCC-
MONSANTO CHEMICAL CO., EVERETT, MASS.	MONSANTO-2008-QR-
MONSANTO CHEMICAL CO., EVERETT, MASS.	MONSANTO-QR-
MONSANTO CHEMICAL CO. ORGANIC CHEM. DIV., ST. LOUIS	MONSANTO-TB-PL-
MONSANTO CHEMICAL CO. ORGANIC CHEM. DIV., ST. LOUIS	MONSANTO-TP-AT-
MONSANTO RESEARCH CORP. (ALL LOCATIONS)	MRC-
MONSANTO RESEARCH CORP., DAYTON, OHIO	MRB-
MONSANTO RESEARCH CORP., DAYTON, OHIO	MRC-DA-
MONSANTO RESEARCH CORP., DAYTON, OHIO	NLABS-TPMR-
MONSANTO RESEARCH CORP., ST. LOUIS, MO.	HPC-(YEAR)-
MONSANTO RESEARCH CORP. BOSTON LABS., EVERETT, MASS.	MRB-
MONSANTO RESEARCH CORP. BOSTON LABS., EVERETT, MASS.	MRB-(NO.)-(LTR(S))-
MONTANA STATE UNIV., BOZEMAN. ELECTRONICS RES. LAB.	ERL-(NUMBER)-
MONTPELLIER UNIV., FRANCE (PHYSICS AND MATH.)	PM/(YEAR)/
MONTREAL. UNIV.	MONTU-
MONTREAL. UNIV.	MR-
MONTREAL. UNIV. LABORATOIRE DE PHYSIQUE NUCLEAIRE	LPN-UM-
MOTOROLA INC. (ALL LOCATIONS)	MI-
MOTOROLA INC., PHOENIX, ARIZ.	J-
MOTOROLA INC., PHOENIX, ARIZ.	MOT-
MOTOROLA INC., PHOENIX, ARIZ.	T-
MOTOROLA INC., PHOENIX, ARIZ.	V-
MOTOROLA INC., PHOENIX, ARIZ.	W-
MOTOROLA INC., RIVERSIDE, CALIF.	MOTO-
MOTOROLA INC., RIVERSIDE, CALIF.	MOTO PR-
MOTOROLA INC., RIVERSIDE, CALIF.	MOTO RL-
MOTOROLA INC., RIVERSIDE, CALIF.	RL-

MOTOROLA INC., SCOTTSDALE, ARIZ.	RFS-
MOTOROLA INC., SCOTTSDALE, ARIZ.	WF-
MOTOROLA INC., SCOTTSDALE, ARIZ.	WP-
MOTOROLA INC. AEROSPACE CENTER, PHOENIX, ARIZ.	ACP-(NUMBER)-
MOTOROLA INC. MILITARY ELECTRONICS DIV., SCOTTSDALE, ARIZ.	ESP-
MOTOROLA INC. MILITARY ELECTRONICS DIV., SCOTTSDALE, ARIZ.	WF-
MOTOROLA INC. MILITARY ELECTRONICS DIV., SCOTTSDALE, ARIZ.	WP-
MOTOROLA INC. RIVERSIDE RESEARCH LAB., CALIF.	MOTO-RL-
MOTOROLA INC. RIVERSIDE RESEARCH LAB., CALIF.	MOT TR-RL-
MOTOROLA INC. SEMICONDUCTOR PRODS. DIV., PHOENIX, ARIZ.	D-
MOTOROLA INC. SEMICONDUCTOR PRODS. DIV., PHOENIX, ARIZ.	G-
MOTOROLA INC. SEMICONDUCTOR PRODS. DIV., PHOENIX, ARIZ.	M-
MOTOROLA INC. SEMICONDUCTOR PRODS. DIV., PHOENIX, ARIZ.	W-
MOTOROLA INC. SYSTEMS RESEARCH LAB., RIVERSIDE, CALIF	RLF-
MOUND LAB., MIAMISBURG, OHIO (REF. 1)	(YEAR CODE)-(MONTH NO.)-
MOUND LAB., MIAMISBURG, OHIO	BC-
MOUND LAB., MIAMISBURG, OHIO	CF-(YEAR)-(MONTH)-
MOUND LAB., MIAMISBURG, OHIO	MLM-
MOUND LAB., MIAMISBURG, OHIO	MLM-BC-
MOUND LAB., MIAMISBURG, OHIO	MLM-CF-(YR)-(MO)-
MOUND LAB., MIAMISBURG, OHIO	MLM-M-
MOUND LAB., MIAMISBURG, OHIO	MLM-OP-
MOUND LAB., MIAMISBURG, OHIO (REPRINT)	MLM-REPRINT-
MOUNT AUBURN RES. ASSOCIATES INC., CAMBRIDGE, MASS.	MARA-
MOUNT HOLYOKE COLL., SOUTH HADLEY, MASS. PSYCHOPHYSICAL RESEARCH UNIT	MHC-
MOUNT SINAI HOSPITAL, N.Y.C.	MSH-
MSA RESEARCH CORP., CALLERY, PENNA.	D143-(NUMBER)-
MSA RESEARCH CORP., CALLERY, PENNA. (LIQUID METALS TECHNOLOGY ABSTRACT BULLETIN)	LM/TAB-
MSA RESEARCH CORP., CALLERY, PENNA.	MSA-
MSA RESEARCH CORP., CALLERY, PENNA.	MSA-PR-
MSA RESEARCH CORP., CALLERY, PENNA.	MSAR-
MSA RESEARCH CORP., CALLERY, PENNA.	MSAR-(YEAR)-
MSA RESEARCH CORP., CALLERY, PENNA.	MSAR-(NO.)-(NO.)-
MSA RESEARCH CORP., CALLERY, PENNA.	MSAR-MEMO-
MSA RESEARCH CORP., CALLERY, PENNA.	MSAR-TR-
MSA RESEARCH CORP., CALLERY, PENNA.	TP-(NUMBER)-
MSA RESEARCH CORP., CALLERY, PENNA.	XA-
MSA RESEARCH CORP., EVANS CITY, PENNA.	MSAR-(YEAR)-
MSA RESEARCH CORP., EVANS CITY, PENNA.	XAC-
MULLARD, LTD. CENTRAL MATERIALS LAB., MITCHAM, SURREY, ENGLAND	CML-
MULLARD RADIO VALVE CO., LTD.	DVT-
MULLARD RADIO VALVE CO., LTD.	SMVD-
MULLARD RESEARCH LABS., REIGATE, SURREY, ENGLAND	MRL-
MUNICH. TECHNISCHE HOCHSCHULE	ASR-
MUNICH. TECHNISCHE HOCHSCHULE. INSTITUT FUER MESS- UND REGELUNGSTECHNIK	MRR-
MUNICH. TECHNISCHE HOCHSCHULE. PHYSIKALISCHES INSTITUT	FRM-
MUNITIONS BOARD. AIRCRAFT COMMITTEE, WASH., D.C.	ANC-
MUNITIONS BOARD STANDARDS AGENCY, WASHINGTON, D.C.	JAN-STD-
MUNITIONS BOARD STANDARDS AGENCY, WASHINGTON, D.C.	MIL-STD-
MYSTIC OCEANOGRAPHIC CO., CONN.	MOC-D-
MYSTIC OCEANOGRAPHIC CO., CONN.	MOC-S-

N.T.W. MISSILE ENGINEERING, INC.

N. T. W. MISSILE ENGINEERING, INC., LOS ANGELES	NTW-
NAGOYA, JAPAN. UNIV. INST. OF PLASMA PHYSICS	IPPJ-
NAGOYA, JAPAN. UNIV. INST. OF PLASMA PHYSICS	IPPJ-DT-
NAGOYA, JAPAN. UNIV. INST. OF PLASMA PHYSICS	IPPJ-T-
NAPLES. UNIVERSITA. ISTITUTO DI AERODINAMICA	IA-
NAPLES. UNIVERSITA. ISTITUTO DI AERONAUTICA	A.I. REPORT NO.-
NAPLES. UNIVERSITA. ISTITUTO DI FISICA TEORICA	AM-
NAPLES. UNIVERSITA. ISTITUTO DI FISICA TEORICA	NAPLESU-
NARA MEDICAL COLL., KASHIHARA CITY, JAPAN. DEPT. OF PHYSIOLOGY	J-
NARA TECHNICAL COLL., JAPAN	NTC-(NUMBER)-
NARMCO INDUSTRIES,INC. RES. + DEV. DIV.,SAN DIEGO,CAL	NARMCO-
NARMCO INDUSTRIES,INC. RES. + DEV. DIV.,SAN DIEGO,CAL	NARMCO-QR-
NARMCO INDUSTRIES,INC. RES. + DEV. DIV.,SAN DIEGO,CAL	NARMCO-STR-
NATIONAL ACADEMY OF ENGINEERING, WASHINGTON, D.C.	NAE-
NATIONAL ACADEMY OF SCIENCES (INTERNATIONAL GEOPHYSICAL YEAR BULLETIN)	IGY BULLETIN-
NATIONAL ACADEMY OF SCIENCES (INTERNATIONAL GEOPHYSICAL YEAR. GENERAL REPORT SERIES)	IGY GRS-
NATIONAL ACADEMY OF SCIENCES (INTERNATIONAL GEOPHYSICAL YEAR. SATELLITE REPORT SERIES)	IGY SRS-
NATIONAL ACADEMY OF SCIENCES	NAS-
NATIONAL ACADEMY OF SCIENCES	NAS ARDC COM-...
NATIONAL ACADEMY OF SCIENCES	NAS-CR-
NATIONAL ACADEMY OF SCIENCES	NAS-NRC/PUB-
NATIONAL ACADEMY OF SCIENCES (NUCLEAR SCIENCE SERIES)	NAS-NS-
NATIONAL ACADEMY OF SCIENCES. COMMITTEE ON SENSORY DEVICES	CSD-
NATIONAL ACADEMY OF SCIENCES + NATIONAL RES. COUNCIL	NAS-NRC-
NATIONAL ACADEMY OF SCIENCES-NATIONAL RES. COUNCIL, COMMITTEE ON UNDERSEA WARFARE, WASHINGTON, D.C.	NRC-CUW-
NATIONAL ACADEMY OF SCIENCES-NATIONAL RES. COUNCIL, MATERIALS ADVISORY BOARD, WASHINGTON, D.C.	MAB-(NUMBER-LTR.-NO.)
NATIONAL ACADEMY OF SCIENCES-NATIONAL RES. COUNCIL, MINE ADVISORY COMMITTEE, WASHINGTON, D.C.	NRC-MAC-
NATIONAL ACADEMY OF SCIENCES-NATIONAL RES. COUNCIL, PACIFIC SCIENCE BOARD, WASHINGTON, D.C.	ATOLL RESEARCH BULL-
NATIONAL ACCELERATOR LAB., BATAVIA, ILL.	NAL-(NUMBER)-
NATIONAL ACCELERATOR LAB., BATAVIA, ILL.	NAL-
NATIONAL ACCELERATOR LAB., BATAVIA, ILL.	NAL-EN-
NATIONAL ACCELERATOR LAB., BATAVIA, ILL.	NAL-THY-
NATIONAL ACCELERATOR LAB., BATAVIA, ILL.	NAL-TR-
NATIONAL ACCELERATOR LAB., OAK BROOK, ILL.	FN-
NATIONAL ACCELERATOR LAB., OAK BROOK, ILL.	NAL-FN-
NATIONAL ACCELERATOR LAB., OAK BROOK, ILL. (SEE ALSO LATER ADDRESS, BATAVIA, ILL.)	NAL-TM-
NATIONAL ADVISORY COMMITTEE FOR AERONAUTICS (AIRCRAFT CIRCULAR) (REF. 25)	AC-
NATIONAL ADVISORY COMMITTEE FOR AERONAUTICS (ADVANCE CONFIDENTIAL REPORT) (REF. 25)	ACR-(LETTER)(DATE CODE)
NATIONAL ADVISORY COMMITTEE FOR AERONAUTICS (ADVANCE RESTRICTED REPORT) (REF. 25)	ARR-(LETTER)(DATE CODE)
NATIONAL ADVISORY COMMITTEE FOR AERONAUTICS (CONFIDENTIAL BULLETIN)	CB-
NATIONAL ADVISORY COMMITTEE FOR AERONAUTICS (CONFIDENTIAL BULLETIN) (REF. 25)	CB-(LETTER)(DATE CODE)
NATIONAL ADVISORY COMMITTEE FOR AERONAUTICS (MEMORANDUM REPORT) (REF. 25)	MR-(LETTER)(DATE CODE)
NATIONAL ADVISORY COMMITTEE FOR AERONAUTICS (REF. 25)	NACA-
NATIONAL ADVISORY COMMITTEE FOR AERONAUTICS (AIRCRAFT CIRCULAR) (REF. 25)	NACA-AC-
NATIONAL ADVISORY COMMITTEE FOR AERONAUTICS (ADVANCE CONFIDENTIAL REPORT) (REF. 25)	NACA-ACR-(DATE)
NATIONAL ADVISORY COMMITTEE FOR AERONAUTICS (ADVANCE CONFIDENTIAL REPORT) (REF. 25)	NACA-ACR-(LTR)(DATE CODE)
NATL. ADVISORY COMM. FOR AERONAUTICS (ANNUAL REPORT)	NACA-AR-
NATIONAL ADVISORY COMMITTEE FOR AERONAUTICS (ADVANCE RESTRICTED REPORT) (REF. 25)	NACA-ARR-(DATE)
NATIONAL ADVISORY COMMITTEE FOR AERONAUTICS (ADVANCE RESTRICTED REPORT) (REF. 25)	NACA-ARR-(LTR)(DATE CODE)
NATIONAL ADVISORY COMMITTEE FOR AERONAUTICS (CONFIDENTIAL BULLETIN) (REF. 25)	NACA-CB-(DATE)
NATIONAL ADVISORY COMMITTEE FOR AERONAUTICS (CONFIDENTIAL BULLETIN) (REF. 25)	NACA-CB-(LTR)(DATE CODE)
NATIONAL ADVISORY COMMITTEE FOR AERONAUTICS (MEMORANDUM REPORT) (REF. 25)	NACA-MR-(DATE)
NATIONAL ADVISORY COMMITTEE FOR AERONAUTICS (MEMORANDUM REPORT) (REF. 25)	NACA-MR-(LTR)(DATE CODE)
NATIONAL ADVISORY COMMITTEE FOR AERONAUTICS (RESEARCH ABSTRACTS) (LETTER AT END STANDS FOR SECURITY CLASSIFICATION)	NACA-RA-(NUMBER)(LETTER)
NATIONAL ADVISORY COMMITTEE FOR AERONAUTICS (RESTRICTED BULLETIN) (REF. 25)	NACA-RB-(DATE)
NATIONAL ADVISORY COMMITTEE FOR AERONAUTICS (RESTRICTED BULLETIN) (REF. 25)	NACA-RB-(LTR)(DATE CODE)
NATIONAL ADVISORY COMM. FOR AERONAUTICS (RPT)(REF.25)	NACA-REPORT-
NATIONAL ADVISORY COMMITTEE FOR AERONAUTICS (RESEARCH MEMORANDUM) (REF. 25)	NACA-RM-(LTR)(DATE CODE)
NATIONAL ADVISORY COMMITTEE FOR AERONAUTICS (RESEARCH MEMORANDUM) (REF. 25)	NACA-RM-S-
NATIONAL ADVISORY COMMITTEE FOR AERONAUTICS (TECHNICAL MEMORANDUM) (REF. 25)	NACA-TM-
NATIONAL ADVISORY COMMITTEE FOR AERONAUTICS (TECHNICAL NOTE) (REF. 25)	NACA-TN-
NATIONAL ADVISORY COMMITTEE FOR AERONAUTICS	NACA-TR-
NATIONAL ADVISORY COMMITTEE FOR AERONAUTICS (WARTIME REPORT) (REF. 25)	NACA-WR-(LETTER)-
NATIONAL ADVISORY COMMITTEE FOR AERONAUTICS (WARTIME REPORT) (REF. 25)	NACA-WRW-
NATIONAL ADVISORY COMMITTEE FOR AERONAUTICS (RESTRICTED BULLETIN) (REF. 25)	RB-(LETTER)(DATE CODE)
NATIONAL ADVISORY COMMITTEE FOR AERONAUTICS (RESEARCH MEMORANDUM) (REF. 25)	RM-(LETTER)(DATE CODE)
NATIONAL ADVISORY COMMITTEE FOR AERONAUTICS (WARTIME REPORT) (REF. 25)	WR-(LETTER)-
(SEE ALSO LATER NAME: NATIONAL AERONAUTICS AND SPACE ADMINISTRATION)	
NATIONAL AERONAUTICAL CORP., AMBLER, PENNA.	NARCO-
NATIONAL AERONAUTICS AND SPACE ADMINISTRATION, COLLEGE PARK, MD. (LITERATURE SEARCH NUMBER...)	NASA-LS-
NATIONAL AERONAUTICS AND SPACE ADM., WASHINGTON, D.C. (TRANSLATOR'S INITIALS) (CATEGORY) (SUBJECT)	AB-CR-(LETTERS)-
NATIONAL AERONAUTICS AND SPACE ADM., WASHINGTON, D.C. (TRANSLATOR'S INITIALS) (CATEGORY)	AB-NP-
NATIONAL AERONAUTICS AND SPACE ADM., WASHINGTON, D.C. (TRANSLATION)	GFW-AB-(YEAR)/
NATIONAL AERONAUTICS AND SPACE ADMINISTRATION, WASHINGTON, D.C. (ACCESSION NUMBER FOLLOWS YEAR)	N(YEAR)-
NATIONAL AERONAUTICS AND SPACE ADM., WASHINGTON, D.C. (ASSIGNED TO MISCELLANEOUS BRITISH, CANADIAN AND U.S. REPORTS)	N-(NOS. ABOVE 30,000)
NATIONAL AERONAUTICS AND SPACE ADM., WASHINGTON, D.C.	NAS-(CONTRACT NUMBER)-
NATIONAL AERONAUTICS AND SPACE ADM., WASHINGTON, D.C. (REF. 26)	NASA-
NATIONAL AERONAUTICS AND SPACE ADM., WASHINGTON, D.C.	NASA-BIB-
NATIONAL AERONAUTICS AND SPACE ADM., WASHINGTON, D.C.	NASA-C-
NATIONAL AERONAUTICS AND SPACE ADMINISTRATION, WASHINGTON, D.C. (PATENT APPLICATION)	NASA-CASE-(LETTERS)-
NATIONAL AERONAUTICS AND SPACE ADMINISTRATION, WASHINGTON, D.C. (PATENT APPLICATION)	NASA-CASE-(HQN)-
NATIONAL AERONAUTICS AND SPACE ADMINISTRATION, WASHINGTON, D.C. (PATENT APPLICATION)	NASA-CASE-(XMS)-
NATIONAL AERONAUTICS AND SPACE ADM., WASHINGTON, D.C.	NASA-CCN-
NATIONAL AERONAUTICS AND SPACE ADM., WASHINGTON, D.C.	NASA-CF-
NATIONAL AERONAUTICS AND SPACE ADMINISTRATION, WASHINGTON, D.C. (CONTRACT REPORT)	NASA-CR-
NATIONAL AERONAUTICS AND SPACE ADM., WASHINGTON, D.C. (MEMORANDUM) (REF. 26)	NASA-M-(DATE LETTER)

NATIONAL BUREAU OF STANDARDS

NATIONAL AERONAUTICS AND SPACE ADM., WASHINGTON, D.C. (REF. 26)	NASA-MEMO-(DATE LETTER)
NATIONAL AERONAUTICS AND SPACE ADM., WASHINGTON, D.C. (REF. 26)	NASA-MEMORANDUM(DATE LTR)
NATIONAL AERONAUTICS AND SPACE ADM., WASHINGTON, D.C.	NASA NEWS RELEASE-(YR.)-
NATIONAL AERONAUTICS AND SPACE ADMINISTRATION, WASHINGTON, D.C. (QUALITY PUBLICATION)	NASA-NPC-
NATIONAL AERONAUTICS AND SPACE ADM., WASHINGTON, D.C. (PUBLICATION ANNOUNCEMENTS) (LETTER AT END STANDS FOR SECURITY CLASSIFICATION)	NASA-PA-(NUMBER LETTER)
NATIONAL AERONAUTICS AND SPACE ADM., WASHINGTON, D.C.	NASA-R-
NATIONAL AERONAUTICS AND SPACE ADM., WASHINGTON, D.C. (REPUBLICATION) (REF. 26)	NASA-RE-(DATE LETTER)
NATIONAL AERONAUTICS AND SPACE ADMINISTRATION, WASHINGTON, D.C. (REF. 26)	NASA-REPUBLICATION...
NATIONAL AERONAUTICS AND SPACE ADM., WASHINGTON, D.C.	NASA-R-ME-IV-S-
NATIONAL AERONAUTICS AND SPACE ADM., WASHINGTON, D.C. (REPRINT SERIES)	NASA-RP-
NATIONAL AERONAUTICS AND SPACE ADMINISTRATION, WASHINGTON, D.C. (SELECTED PUBLICATIONS)	NASA-SP-
NATIONAL AERONAUTICS AND SPACE ADM., WASHINGTON, D.C. (TECHNICAL MEMORANDUM) (REF. 26)	NASA-TM-X-
NATIONAL AERONAUTICS AND SPACE ADMINISTRATION, WASHINGTON, D.C. (TECHNICAL NOTE) (REF. 26)	NASA-TN-D-
NATIONAL AERONAUTICS AND SPACE ADMINISTRATION, WASHINGTON, D.C.(INDEX OF TECHNICAL PUBLICATIONS ANNOUNCEMENTS)	NASA-TPA-INDEX-
NATIONAL AERONAUTICS AND SPACE ADMINISTRATION, WASHINGTON, D.C. (INDEX OF TECHNICAL PUBLICATIONS)	NASA-TP-INDEX-
NATIONAL AERONAUTICS AND SPACE ADMINISTRATION, WASHINGTON, D.C. (TECHNICAL REPORT) (REF. 26)	NASA-TR-R-
NATIONAL AERONAUTICS AND SPACE ADMINISTRATION, WASHINGTON, D.C. (TECHNICAL TRANSLATION) (REF. 26)	NASA-TT-F-
NATIONAL AERONAUTICS AND SPACE ADMINISTRATION, WASHINGTON, D.C. (CONTRACT NUMBER)	NGL-(2 DIGITS-3 DIGITS)-
NATIONAL AERONAUTICS AND SPACE ADMINISTRATION, WASHINGTON, D.C.	NHB-(NUMBER.NUMBER)
NATIONAL AERONAUTICS AND SPACE ADM., WASHINGTON, D.C.	NPC-
NATIONAL AERONAUTICS AND SPACE ADM., WASHINGTON, D.C. (SCIENTIFIC TRANS.)(CATEGORY DESIGNATION)(SUBJECT)	ST-AA-(LETTERS)-
NATIONAL AERONAUTICS AND SPACE ADM., WASHINGTON, D.C. (SCIENTIFIC TRANS.)(CATEGORY DESIGNATION)(SUBJECT)	ST-AC-(LETTERS)-
NATIONAL AERONAUTICS AND SPACE ADM., WASHINGTON, D.C. (SCIENTIFIC TRANS.)(CATEGORY DESIGNATION)(SUBJECT)	ST-ACH-(LETTERS)-
NATIONAL AERONAUTICS AND SPACE ADM., WASHINGTON, D.C. (SCIENTIFIC TRANS.)(CATEGORY DESIGNATION)(SUBJECT)	ST-AD-(LETTERS)-
NATIONAL AERONAUTICS AND SPACE ADM., WASHINGTON, D.C. (SCIENTIFIC TRANS.)(CATEGORY DESIGNATION)(SUBJECT)	ST-AI-(LETTERS)-
NATIONAL AERONAUTICS AND SPACE ADM., WASHINGTON, D.C. (SCIENTIFIC TRANS.)(CATEGORY DESIGNATION)(SUBJECT)	ST-AM-(LETTERS)-
NATIONAL AERONAUTICS AND SPACE ADMINISTRATION, WASHINGTON, D.C. (SCIENTIFIC AND TECHNICAL AEROSPACE REPORTS)	STAR-
NATIONAL AERONAUTICS AND SPACE ADM., WASHINGTON, D.C. (SCIENTIFIC TRANS.)(CATEGORY DESIGNATION)(SUBJECT)	ST-ASD-(LETTERS)-
NATIONAL AERONAUTICS AND SPACE ADM., WASHINGTON, D.C. (SCIENTIFIC TRANS.)(CATEGORY DESIGNATION)(SUBJECT)	ST-CM-(LETTERS)-
NATIONAL AERONAUTICS AND SPACE ADM., WASHINGTON, D.C. (SCIENTIFIC TRANS.)(CATEGORY DESIGNATION)(SUBJECT)	ST-COM-(LETTERS)-
NATIONAL AERONAUTICS AND SPACE ADM., WASHINGTON, D.C. (SCIENTIFIC TRANS.)(CATEGORY DESIGNATION)(SUBJECT)	ST-CR-(LETTERS)-
NATIONAL AERONAUTICS AND SPACE ADM., WASHINGTON, D.C. (SCIENTIFIC TRANS.)(CATEGORY DESIGNATION)(SUBJECT)	ST-EDN-(LETTERS)-
NATIONAL AERONAUTICS AND SPACE ADM., WASHINGTON, D.C. (SCIENTIFIC TRANS.)(CATEGORY DESIGNATION)(SUBJECT)	ST-EMWP-(LETTERS)-
NATIONAL AERONAUTICS AND SPACE ADM., WASHINGTON, D.C. (SCIENTIFIC TRANS.)(CATEGORY DESIGNATION)(SUBJECT)	ST-ES-(LETTERS)-
NATIONAL AERONAUTICS AND SPACE ADM., WASHINGTON, D.C. (SCIENTIFIC TRANS.)(CATEGORY DESIGNATION)(SUBJECT)	ST-EXP-(LETTERS)-
NATIONAL AERONAUTICS AND SPACE ADM., WASHINGTON, D.C. (SCIENTIFIC TRANS.)(CATEGORY DESIGNATION)(SUBJECT)	ST-GC-(LETTERS)-
NATIONAL AERONAUTICS AND SPACE ADM., WASHINGTON, D.C. (SCIENTIFIC TRANS.)(CATEGORY DESIGNATION)(SUBJECT)	ST-GEO-(LETTERS)-
NATIONAL AERONAUTICS AND SPACE ADM., WASHINGTON, D.C. (SCIENTIFIC TRANS.)(CATEGORY DESIGNATION)(SUBJECT)	ST-GM-(LETTERS)-
NATIONAL AERONAUTICS AND SPACE ADM., WASHINGTON, D.C. (SCIENTIFIC TRANS.)(CATEGORY DESIGNATION)(SUBJECT)	ST-IGA-(LETTERS)-
NATIONAL AERONAUTICS AND SPACE ADM., WASHINGTON, D.C. (SCIENTIFIC TRANS.)(CATEGORY DESIGNATION)(SUBJECT)	ST-IM-(LETTERS)-
NATIONAL AERONAUTICS AND SPACE ADM., WASHINGTON, D.C. (SCIENTIFIC TRANS.)(CATEGORY DESIGNATION)(SUBJECT)	ST-IMF-(LETTERS)-
NATIONAL AERONAUTICS AND SPACE ADM., WASHINGTON, D.C. (SCIENTIFIC TRANS.)(CATEGORY DESIGNATION)(SUBJECT)	ST-IS-(LETTERS)-
NATIONAL AERONAUTICS AND SPACE ADM., WASHINGTON, D.C. (SCIENTIFIC TRANS.)(CATEGORY DESIGNATION)(SUBJECT)	ST-LMP-(LETTERS)-
NATIONAL AERONAUTICS AND SPACE ADM., WASHINGTON, D.C. (SCIENTIFIC TRANS.)(CATEGORY DESIGNATION)(SUBJECT)	ST-LPS-(LETTERS)-
NATIONAL AERONAUTICS AND SPACE ADM., WASHINGTON, D.C. (SCIENTIFIC TRANS.)(CATEGORY DESIGNATION)(SUBJECT)	ST-MAT-(LETTERS)-
NATIONAL AERONAUTICS AND SPACE ADM., WASHINGTON, D.C. (SCIENTIFIC TRANS.)(CATEGORY DESIGNATION)(SUBJECT)	ST-MHD-(LETTERS)-
NATIONAL AERONAUTICS AND SPACE ADM., WASHINGTON, D.C. (SCIENTIFIC TRANS.)(CATEGORY DESIGNATION)(SUBJECT)	ST-NP-(LETTERS)-
NATIONAL AERONAUTICS AND SPACE ADM., WASHINGTON, D.C. (SCIENTIFIC TRANS.)(CATEGORY DESIGNATION)(SUBJECT)	ST-OA-(LETTERS)-
NATIONAL AERONAUTICS AND SPACE ADM., WASHINGTON, D.C. (SCIENTIFIC TRANS.)(CATEGORY DESIGNATION)(SUBJECT)	ST-PA-(LETTERS)-
NATIONAL AERONAUTICS AND SPACE ADM., WASHINGTON, D.C. (SCIENTIFIC TRANS.)(CATEGORY DESIGNATION)(SUBJECT)	ST-PF-(LETTERS)-
NATIONAL AERONAUTICS AND SPACE ADM., WASHINGTON, D.C. (SCIENTIFIC TRANS.)(CATEGORY DESIGNATION)(SUBJECT)	ST-PHM-(LETTERS)-
NATIONAL AERONAUTICS AND SPACE ADM., WASHINGTON, D.C. (SCIENTIFIC TRANS.)(CATEGORY DESIGNATION)(SUBJECT)	ST-PP-(LETTERS)-
NATIONAL AERONAUTICS AND SPACE ADM., WASHINGTON, D.C. (SCIENTIFIC TRANS.)(CATEGORY DESIGNATION)(SUBJECT)	ST-PR-(LETTERS)-
NATIONAL AERONAUTICS AND SPACE ADM., WASHINGTON, D.C. (SCIENTIFIC TRANS.)(CATEGORY DESIGNATION)(SUBJECT)	ST-PS-(LETTERS)-
NATIONAL AERONAUTICS AND SPACE ADM., WASHINGTON, D.C. (SCIENTIFIC TRANS.)(CATEGORY DESIGNATION)(SUBJECT)	ST-RA-(LETTERS)-
NATIONAL AERONAUTICS AND SPACE ADM., WASHINGTON, D.C. (SCIENTIFIC TRANS.)(CATEGORY DESIGNATION)(SUBJECT)	ST-SB-(LETTERS)-
NATIONAL AERONAUTICS AND SPACE ADM., WASHINGTON, D.C. (SCIENTIFIC TRANS.)(CATEGORY DESIGNATION)(SUBJECT)	ST-SP-(LETTERS)-
NATIONAL AERONAUTICS AND SPACE ADM., WASHINGTON, D.C. (SCIENTIFIC TRANS.)(CATEGORY DESIGNATION)(SUBJECT)	ST-SPC-(LETTERS)-
NATIONAL AERONAUTICS AND SPACE ADM., WASHINGTON, D.C. (SCIENTIFIC TRANS.)(CATEGORY DESIGNATION)(SUBJECT)	ST-STR-(LETTERS)-
NATIONAL AERONAUTICS AND SPACE ADM., WASHINGTON, D.C. (SCIENTIFIC TRANS.)(CATEGORY DESIGNATION)(SUBJECT)	ST-WP-(LETTERS)-
NATIONAL AERONAUTICS AND SPACE ADM., WASHINGTON, D.C.	X(YEAR)-(5 DIGITS)
NATIONAL AERONAUTICS AND SPACE ADMINISTRATION. EDUCATIONAL PROGRAMS DIV., WASHINGTON, D.C.	NASA-EP-
NATIONAL AERONAUTICS AND SPACE ADMINISTRATION. HEADQUARTERS, WASHINGTON, D.C. (REF. 26)	NASA-(LETTERS)-W-
NATIONAL AERONAUTICS AND SPACE ADMINISTRATION. RES. ADVISORY COMM. ON ELECTRICAL POWER PLANT SYSTEMS, WASHINGTON, D.C.	NASA-E-
NATIONAL AIR POLLUTION CONTROL ADMINISTRATION, WASHINGTON, D.C. (SEE ALSO LATER NAME: ENVIRONMENTAL PROTECTION AGENCY. AIR POLLUTION CONTROL OFFICE)	NAPCA-
NATIONAL ARCHIVES AND RECORDS SERVICE, WASHINGTON, DC (CODE OF FEDERAL REGULATIONS)	CFR-
NATIONAL ARCHIVES AND RECORDS SERVICE, WASHINGTON, DC	NARS-
NATIONAL ASSN. OF HOUSING + REDEVELOPMENT OFFICIALS, WASHINGTON, D.C.	NAHRO-PUB-N(NUMBER)
NATIONAL AVIATION FACILITIES EXPTL.CTR.,ATLANTIC CITY	ADS-
NATIONAL AVIATION FACILITIES EXPTL.CTR.,ATLANTIC CITY	FAA-ADS-
NATIONAL AVIATION FACILITIES EXPTL.CTR.,ATLANTIC CITY	FAA-D-
NATIONAL AVIATION FACILITIES EXPTL.CTR.,ATLANTIC CITY	FAA-NA-(YEAR)-
NATIONAL AVIATION FACILITIES EXPTL.CTR.,ATLANTIC CITY	NA-(YEAR)-
NATIONAL AVIATION FACILITIES EXPTL.CTR.,ATLANTIC CITY	RD-(YEAR)-
NATIONAL AVIATION FACILITIES EXPTL.CTR.,ATLANTIC CITY	SST-
NATIONAL AVIATION WEATHER SYSTEMS STUDY, WASH., D.C.	NAVWESS-
NATIONAL BATTERY CO., DEPEW, N. Y.	NBC-
NATIONAL BERYLLIA CORP., HASKELL, N.J.	X-
NATIONAL BIOMEDICAL RESEARCH FDN., SILVER SPRING, MD.	NBR-
NATIONAL BROADCASTING CO., INC., N.Y.C.	NBC-
NATIONAL BUREAU OF STANDARDS, WASH., D.C. (REF. 1)	(NUMBERS...)
NATIONAL BUREAU OF STANDARDS, WASHINGTON, D.C.	1B-
NATIONAL BUREAU OF STANDARDS, WASHINGTON, D.C.	BMS-
NATIONAL BUREAU OF STANDARDS, WASHINGTON, D.C.	2C-
NATIONAL BUREAU OF STANDARDS, WASHINGTON, D.C.	CJR-RM-
NATIONAL BUREAU OF STANDARDS, WASHINGTON, D.C.	CT-
NATIONAL BUREAU OF STANDARDS, WASHINGTON, D.C.	D-
NATIONAL BUREAU OF STANDARDS, WASHINGTON, D.C.	INA-
NATIONAL BUREAU OF STANDARDS, WASHINGTON, D.C.	INC/C(NO.)/P(NO.)
NATIONAL BUREAU OF STANDARDS, WASHINGTON, D.C. (KINGFISHER TECH. RPT.)	KT-
NATIONAL BUREAU OF STANDARDS, WASHINGTON, D.C. (LETTER CIRCULAR)	LC-
NATIONAL BUREAU OF STANDARDS, WASHINGTON, D.C.	MTP-

NATIONAL BUREAU OF STANDARDS

NATIONAL BUREAU OF STANDARDS, WASHINGTON, D.C.	NA-(YEAR)-
NATIONAL BUREAU OF STANDARDS, WASHINGTON, D.C.	NBS-(NO. LETTER)-
NATIONAL BUREAU OF STANDARDS, WASHINGTON, D.C.	NBS-
NATIONAL BUREAU OF STANDARDS, WASHINGTON, D.C.	NBS-(LETTER(S))-
NATIONAL BUREAU OF STANDARDS, WASHINGTON, D.C.	NBS-AMS-
NATIONAL BUREAU OF STANDARDS, WASHINGTON, D.C. (BUILDING MATERIALS AND STRUCTURES)	NBS-BMS-
NATIONAL BUREAU OF STANDARDS, WASHINGTON, D.C.	NBS-BSS-
NATIONAL BUREAU OF STANDARDS, WASH., D.C. (CIRCULAR)	NBS-C-
NATIONAL BUREAU OF STANDARDS, WASHINGTON, D.C.	NBS-CIRC-
NATIONAL BUREAU OF STANDARDS, WASHINGTON, D.C.	NBS-D-
NATIONAL BUREAU OF STANDARDS, WASHINGTON, D.C. (FEDERAL INFORMATION PROCESSING STANDARD)	NBS-FIPS-PUB-(NUMBER)-C
NATIONAL BUREAU OF STANDARDS, WASHINGTON, D.C.	NBS-HANDBOOK-
NATIONAL BUREAU OF STANDARDS, WASHINGTON, D.C. (LETTER CIRCULAR)	NBS LC-
NATIONAL BUREAU OF STANDARDS, WASHINGTON, D.C. (LIST OF PUBLICATIONS)	NBS-LP-
NATIONAL BUREAU OF STANDARDS, WASHINGTON, D.C.	NBS-MISC.-PUBL.-
NATIONAL BUREAU OF STANDARDS, WASHINGTON, D.C.	NBS-MON-
NATIONAL BUREAU OF STANDARDS, WASHINGTON, D.C.	NBS-MONO-
NATIONAL BUREAU OF STANDARDS, WASHINGTON, D.C.	NBS-MONOGRAPH-
NATIONAL BUREAU OF STANDARDS, WASHINGTON, D.C. (PROGRESS REPORT TO AEC)	NBS-P-
NATIONAL BUREAU OF STANDARDS, WASHINGTON, D.C.	NBS-PR-
NATIONAL BUREAU OF STANDARDS, WASHINGTON, D.C.	NBS-PROJ.(NO.)-(NO.)-
NATIONAL BUREAU OF STANDARDS, WASHINGTON, D.C.	NBS R-
NATIONAL BUREAU OF STDS., WASH., D.C. (RES. PAPER)	NBS RP-
NATIONAL BUREAU OF STANDARDS, WASHINGTON, D.C. (SUMMARY TECHNICAL REPORT)	NBS-STR-
NATIONAL BUREAU OF STDS., WASH., D.C. (TECH. NOTE)	NBS-TN-
NATIONAL BUREAU OF STDS., WASH., D.C. (TECH. REPT.)	NBS-TR-
NATIONAL BUREAU OF STDS., WASH., D.C. (TEST REPT.)	NBS TR-
NATIONAL BUREAU OF STANDARDS, WASHINGTON, D.C.	OED-
NATIONAL BUREAU OF STANDARDS, WASHINGTON, D.C.	OTR-
NATIONAL BUREAU OF STANDARDS, WASHINGTON, D.C. (RESEARCH PAPER)	RP-
NATIONAL BUREAU OF STANDARDS, WASHINGTON, D.C.	STR-
NATIONAL BUREAU OF STANDARDS. BOULDER LABS., COLO.	CM-
NATIONAL BUREAU OF STANDARDS. BOULDER LABS., COLO.	CRPL-F-
NATIONAL BUREAU OF STANDARDS. BOULDER LABS., COLO.	ITSA-TM-
NATIONAL BUREAU OF STANDARDS. BOULDER LABS., COLO.	M-
NATIONAL BUREAU OF STANDARDS. BOULDER LABS., COLO.	MR-
NATIONAL BUREAU OF STANDARDS. BOULDER LABS., COLO.	NBS-CM-
NATIONAL BUREAU OF STANDARDS. BOULDER LABS., COLO.	NBS-CRPL-
NATIONAL BUREAU OF STANDARDS. BOULDER LABS., COLO. (LAB. NOTE)	NBS-LN-(NUMBER)-
NATIONAL BUREAU OF STANDARDS. BOULDER LABS., COLO.	NBS-MONO-
NATIONAL BUREAU OF STANDARDS. BOULDER LABS., COLO.	NBS MONOGRAPH-
NATIONAL BUREAU OF STANDARDS. BOULDER LABS., COLO.	NBS-MR-
NATIONAL BUREAU OF STANDARDS. BOULDER LABS., COLO.	NBS-PAPER-
NATIONAL BUREAU OF STANDARDS. BOULDER LABS., COLO.	NBS-PM-
NATIONAL BUREAU OF STANDARDS. BOULDER LABS., COLO.	NBS-SP-
NATIONAL BUREAU OF STANDARDS. BOULDER LABS., COLO.	NBS-SPEC. PUBL.-
NATIONAL BUREAU OF STANDARDS. BOULDER LABS., COLO. (TECHNICAL NOTE)	NBS-TN-
NATIONAL BUREAU OF STANDARDS. BOULDER LABS., COLO. (TECHNICAL REPORT)	NBS-TR-
NATIONAL BUREAU OF STANDARDS. BOULDER LABS., COLO. (TRANSLATION)	NBS-TRANS-
NATIONAL BUREAU OF STANDARDS. BOULDER LABS., COLO.	PM-(NUMBER)-
NATIONAL BUREAU OF STANDARDS. BOULDER LABS., COLO.	U-
NATIONAL BUREAU OF STANDARDS. BUILDING RESEARCH DIV., WASHINGTON, D.C. (BUILDING SCIENCE SERIES)	BSS-
NATIONAL BUREAU OF STANDARDS. BUILDING RESEARCH DIV., WASHINGTON, D.C. (BUILDING SCIENCE SERIES)	NBS-BSS-
NATIONAL BUREAU OF STANDARDS. CENTRAL RADIO PROPAGATION LAB., BOULDER, COLO.	CRPL-
NATIONAL BUREAU OF STANDARDS. CENTRAL RADIO PROPAGATION LAB., BOULDER, COLO.	IRPL-
NATIONAL BUREAU OF STANDARDS. CENTRAL RADIO PROPAGATION LAB., BOULDER, COLO.	NBS-CRRL-
NATIONAL BUREAU OF STANDARDS. CENTRAL RADIO PROPAGATION LAB., WASHINGTON, D.C.	CRL-
NATIONAL BUREAU OF STANDARDS. CENTRAL RADIO PROPAGATION LAB., WASHINGTON, D.C.	NBS-CRPL-
NATIONAL BUREAU OF STANDARDS. CORONA LABS., CALIF.	KINGFISHER-
NATIONAL BUREAU OF STANDARDS. CRYOGENIC DATA CENTER, BOULDER, COLO.	NBS-M-
NATIONAL BUREAU OF STANDARDS. CRYOGENIC ENGINEERING LABS., BOULDER, COLO.	NBS-CEL-
NATIONAL BUREAU OF STANDARDS. DIFFUSION IN METALS DATA CENTER, WASHINGTON, D.C.	DMDC-
NATIONAL BUREAU OF STANDARDS. INTERSERVICE RADIO PROPAGATION LAB., WASH., D.C. (USED 1941-48)	IRPL-
NATIONAL BUREAU OF STANDARDS. NATIONAL STANDARD REFERENCE DATA SYSTEM, WASHINGTON, D.C.	NSRDS-NBS-
NATIONAL BUREAU OF STANDARDS. NAVAL ORDNANCE EXPERIMENTAL UNIT, WASHINGTON, D.C.	NBS-NOEU-
NATIONAL BUREAU OF STANDARDS. OFFICE OF BASIC INSTRUMENTATION, WASHINGTON, D.C.	OBI-
NATIONAL BUREAU OF STANDARDS. OFFICE OF COMMODITY STANDARDS	NBS-CS-
NATIONAL BUREAU OF STANDARDS. OFFICE OF COMMODITY STANDARDS	NBS-SPR-R-
NATIONAL BUREAU OF STANDARDS. OFFICE OF INFORMATION PROCESSING STANDARDS, WASHINGTON, D.C.	FIPS-PUB-
NATIONAL BUREAU OF STANDARDS. OFFICE OF STANDARD REFERENCE DATA, WASH., D.C. (BIBLIOGRAPHIC SERIES)	NBS-OSRDB-(YEAR)-
NATIONAL BUREAU OF STANDARDS. ORDNANCE DEVELOPMENT DIV., WASHINGTON, D.C.	NBS ODDR-
NATIONAL BUREAU OF STANDARDS. ORDNANCE DEVELOPMENT DIV., WASHINGTON, D.C.	OD-
NATIONAL BUREAU OF STANDARDS. ORDNANCE DEVELOPMENT DIV., WASHINGTON, D.C.	ODDR-
NATIONAL BUREAU OF STANDARDS. ORDNANCE DEVELOPMENT DIV., WASHINGTON, D.C. (SPECIAL PROJECTS)	OD-SP-
NATIONAL BUREAU OF STANDARDS. ORDNANCE DEVELOPMENT DIV., WASHINGTON, D.C. (TOSS BOMBING)	OD-TB-

(SEE ALSO LATER NAMES: DIAMOND ORDNANCE FUZE LABS. AND HARRY DIAMOND LABS.)

NATL. BUR. OF STDS. RADIATION PHYS. LAB., WASH., D.C.	NBS-PRL-
NATL. BUR. OF STDS. RADIATION PHYS. LAB., WASH., D.C.	RPL-
NATL. BUR. OF STDS. TECHNICAL ANALYSIS DIV., WASH., DC	TAD-WP-
NATIONAL CARBON CO. RESEARCH LABS., CLEVELAND	NCC-
NATIONAL CARBON CO. RESEARCH LABS., CLEVELAND	NCRL-
NATIONAL CARBON CO. RESEARCH LABS., CLEVELAND	NCRL-PR-(NUMBER)-(YEAR)
NATIONAL CARBON CO. RESEARCH LABS., PARMA, OHIO (BI-MONTHLY PROGRESS REPORT)	NCRL-BMPR-
NATIONAL CARBON CO. RESEARCH LABS., PARMA, OHIO	NCRL-PR-(NUMBER)-(YEAR)
NATIONAL CASH REGISTER CO., DAYTON, OHIO	NCR-
NATIONAL CASH REGISTER CO., DAYTON, OHIO	NCRRD-
NATIONAL CASH REGISTER CO., HAWTHORNE, CALIF.	NCR-ED-

NATIONAL CASH REGISTER CO. FUNDAMENTAL RESEARCH DEPT., DAYTON, OHIO	NCRRD-(YEAR)-FU-
NATIONAL CENTER FOR ATMOSPHERIC RES., BOULDER, COLO.	NCAR-
NATIONAL CENTER FOR ATMOSPHERIC RES., BOULDER, COLO.	NCAR-CT-
NATIONAL CENTER FOR ATMOSPHERIC RES., BOULDER, COLO.	NCAR-MS-
NATIONAL CENTER FOR ATMOSPHERIC RES., BOULDER, COLO.	NCAR-TN-
NATIONAL CENTER FOR ATMOSPHERIC RES., BOULDER, COLO.	NCAR-TN-(YEAR)-
NATIONAL CENTER FOR EARTHQUAKE RES., MENLO PARK, CAL.	NCER-
NATIONAL CENTER FOR HEALTH SERVICES RESEARCH AND DEVELOPMENT, ARLINGTON, VA.	NCHS-RD-
NATIONAL CENTER FOR RADIOLOGICAL HEALTH, ROCKVILLE, MD	MORP-(YEAR)-
NATIONAL CENTER FOR RADIOLOGICAL HEALTH, ROCKVILLE, MD	NCRH-
NATIONAL CENTER FOR RADIOLOGICAL HEALTH, ROCKVILLE, MD	NF-(YEAR)-
NATIONAL CENTER FOR RADIOLOGICAL HEALTH, ROCKVILLE, MD	PHS-(NUMBER)-RH-
NATIONAL CENTER FOR RADIOLOGICAL HEALTH. STANDARDS AND INTELLIGENCE BRANCH, ROCKVILLE, MD.	NCRH-SIB-
NATIONAL CENTER FOR RADIOLOGICAL HEALTH. STANDARDS AND INTELLIGENCE BRANCH, ROCKVILLE, MD.	SIB-
NATIONAL CENTER FOR RADIOLOGICAL HEALTH. TECHNICAL SCIENCES BRANCH, ROCKVILLE, MD.	TSB-
NATIONAL COLLEGE OF RUBBER TECHNOLOGY, LONDON	NCRT-PR-
NATIONAL COMMISSION ON PRODUCT SAFETY, WASHINGTON, DC	NCPS-
NATIONAL COMMITTEE ON UNIFORM TRAFFIC LAWS AND ORDINANCES, WASHINGTON, D.C.	TRAFFIC LAWS COMMENTARY-
NATIONAL COUNCIL ON RADIATION PROTECTION AND MEASUREMENTS, WASHINGTON, D.C.	NCRP-
NATIONAL DEFENSE RESEARCH COMMITTEE (REF. 27)	A-
NATIONAL DEFENSE RESEARCH COMMITTEE (REF. 27)	A-(NUMBER)(LETTER)
NATIONAL DEFENSE RES. COMM. (AIR AND EARTH SHOCK)	AES-
NATIONAL DEFENSE RES. COMM. (ARMOR + ORDNANCE MEMO.)	AOM-
NATIONAL DEFENSE RES. COMM. (ARMOR + ORDNANCE REPORT)	AOR-
NATIONAL DEFENSE RESEARCH COMMITTEE (REF. 27)	B-
NATIONAL DEFENSE RESEARCH COMMITTEE	CF-
NATIONAL DEFENSE RES. COMM. (CONFIDENTIAL NEWS LETTER)	CNL-
NATIONAL DEFENSE RESEARCH COMMITTEE	I-K-
NATIONAL DEFENSE RESEARCH COMMITTEE (JET PROPULSION)	JP-
NATIONAL DEFENSE RESEARCH COMMITTEE	NDRC-
NATIONAL DEFENSE RESEARCH COMMITTEE (SUMMARY TECHNICAL REPORT) (REF. 27)	NDRC-(1 THRU 19) STR-
NATIONAL DEFENSE RESEARCH COMMITTEE (REF. 27)	NDRC-(1 THRU 19)-
NATIONAL DEFENSE RESEARCH COMMITTEE (REF. 27)	NDRC-(LETTER)-
NATIONAL DEFENSE RES. COMM. (AIR + EARTH SHOCK RPTS.)	NDRC AES-
NATIONAL DEFENSE RES. COMM. (ARMOR + ORDNANCE MEMO.)	NDRC AOM-
NATIONAL DEFENSE RES. COMM. (ARMOR + ORDNANCE RPTS.)	NDRC AOR-
NATIONAL DEFENSE RES. COMM. (APPLIED PSYCH. PANEL)	NDRC APP-
NATIONAL DEFENSE RESEARCH COMMITTEE (REF. 27)	NDRC-DIV.-(1 THRU 19)-
NATIONAL DEFENSE RES. COMM. (MEMORANDA SERIES)	NDRC MEMO-
NATIONAL DEFENSE RESEARCH COMMITTEE (PROJECT REPORTS)	NDRC PDRC-
NATIONAL DEFENSE RES. COMM. (RPT. TO THE SERVICES)	NDRC RS-
NATIONAL DEFENSE RES. COMM. (RES. ON SOUND CONTROL)	NDRC RSC CIR-
NATIONAL DEFENSE RESEARCH COMMITTEE. (RESEARCH ON SOUND CONTROL. INFORMAL COMMUNICATIONS)	NDRC RSC IC-
NATIONAL DEFENSE RES. COMM. (SUMMARY TECH. REPTS.)	NDRC-STR-
NATIONAL DEFENSE RES. COMM. (SUMMARY TECH. REPTS.)	NDRC-STR-(LETTERS)-
NATIONAL DEFENSE RES. COMM. (WAVE PROPAGATION GP.)	NDRC WPG-
NATIONAL DEFENSE RESEARCH COMMITTEE (NEWS LETTER)	NL-
NATIONAL DEFENSE RESEARCH COMMITTEE	NRDC (DIV.LETTER)-
NATIONAL DEFENSE RESEARCH COMM. (PROJECT RPTS.)	PDRC-
NATIONAL DEFENSE RESEARCH COMMITTEE	P-K-
NATIONAL DEFENSE RESEARCH COMMITTEE	PT-
NATIONAL DEFENSE RESEARCH COMM. (ROCKET PROPELLANTS)	RP-
NATIONAL DEFENSE RES. COMM. (RPT. TO THE SERVICES)	RS-
NATIONAL DEFENSE RES. COMM. (RES. ON SOUND CONTROL)	RSC-
NATIONAL DEFENSE RESEARCH COMM. (SPECIAL PROPELLANTS)	SP-
NATIONAL DEFENSE RESEARCH COMMITTEE	STR-
NATIONAL DEFENSE RESEARCH COMM. (SUMMARY TECH. RPT.)	STR-(LETTERS)-
NATIONAL DEFENSE RESEARCH COMMITTEE (SUMMARY TECHNICAL REPORT) (REF. 27)	STR-DIV.-(1 THRU 19)-
NATIONAL DEFENSE RES. COMM. (UNDERWATER EXPLOSIVES)	UE-
NATIONAL DEFENSE RESEARCH COMM. (WEEKLY NEWS LETTER)	WNL-
NATIONAL DEFENSE RESEARCH COMM. (WAVE PROPAGATION GP)	WPG-
NATIONAL DEFENSE RESEARCH COMMITTEE. APPLIED MATHEMATICS GROUP (REF. 28)	AMG-
NATIONAL DEFENSE RESEARCH COMMITTEE. APPLIED MATHEMATICS GROUP (REF. 28)	NDRC-AMG-
NATIONAL DEFENSE RESEARCH COMMITTEE. APPLIED MATHEMATICS PANEL (REF. 28)	AM-
NATIONAL DEFENSE RESEARCH COMMITTEE. APPLIED MATHEMATICS PANEL (REF. 28)	AMP-
NATIONAL DEFENSE RESEARCH COMMITTEE. APPLIED MATHEMATICS PANEL (REF. 28)	AMP-MEMO-
NATIONAL DEFENSE RESEARCH COMMITTEE. APPLIED MATHEMATICS PANEL (REF. 28)	AMP-NOTE-
NATIONAL DEFENSE RESEARCH COMMITTEE. APPLIED MATHEMATICS PANEL (REF. 28)	AMP-RPT.-
NATIONAL DEFENSE RESEARCH COMMITTEE. APPLIED MATHEMATICS PANEL (REF. 28)	NDRC-AMP-
NATIONAL DEFENSE RESEARCH COMMITTEE. APPLIED MATHEMATICS PANEL (REF. 28)	NDRC-AMP-MEMO-
NATIONAL DEFENSE RESEARCH COMMITTEE. APPLIED MATHEMATICS PANEL (REF. 28)	NDRC-AMP-NOTE-
NATIONAL DEFENSE RESEARCH COMMITTEE. APPLIED MATHEMATICS PANEL (REF. 28)	OSRD AMP-
NATIONAL DEFENSE RESEARCH COMMITTEE. APPLIED PSYCHOLOGY PANEL (REF. 28)	APP-
NATIONAL DEFENSE RESEARCH COMMITTEE. APPLIED PSYCHOLOGY PANEL (REF. 28)	NDRC APP-
NATIONAL DEFENSE RESEARCH COMMITTEE. URANIUM SECTION	S-
NATIONAL ENGINEERING SCIENCE CO., PASADENA, CALIF.	NESC-
NATIONAL ENGINEERING SCIENCE CO., PASADENA, CALIF.	NESCO-
NATIONAL ENGINEERING SCIENCE CO., PASADENA, CALIF.	NESCO-P-
NATIONAL ENGINEERING SCIENCE CO., PASADENA, CALIF.	NESCO-S-
NATIONAL ENGINEERING SCIENCE CO., PASADENA, CALIF.	NESCO-SN-
NATIONAL ENG. SCIENCE CO., PASADENA, CAL. (PROPOSALS)	P-
NATIONAL ENGINEERING SCIENCE CO., PASADENA, CALIF.	S-(NUMBERS)
NATIONAL ENG. SCIENCE CO., PASADENA, CAL.(CON. RPTS.)	SN-
NATIONAL ENGINEERING SCIENCE CORP., MC LEAN, VA.	SN-
NATIONAL ENVIRONMENTAL SATELLITE SERVICE, WASH., D.C.	ESSA-TM-NESCTM-
NATIONAL ENVIRONMENTAL SATELLITE SERVICE, WASH., D.C.	ESSA-TR-NESC-
NATIONAL ENVIRONMENTAL SATELLITE SERVICE, WASH., D.C.	NESC-
NATIONAL ENVIRONMENTAL SATELLITE SERVICE, WASH., D.C.	NESCTM-
NATIONAL ENVIRONMENTAL SATELLITE SERVICE, WASH., D.C.	NESS-
NATIONAL ENVIRONMENTAL SATELLITE SERVICE, WASH., D.C.	NOAA-TM-NESS-
NATIONAL ENVIRONMENTAL SATELLITE SERVICE, WASH., D.C.	NOAA-TR-NESS-

NATIONAL FEDERATION OF SCIENCE ABSTRACTING AND INDEXING SERVICES

NATIONAL FEDERATION OF SCIENCE ABSTRACTING AND INDEXING SERVICES, PHILADELPHIA	NFSAIS-
NATIONAL FEDERATION OF SCIENCE ABSTRACTING AND INDEXING SERVICES, PHILADELPHIA (TECHNICAL REPORT)	NFSAIS-TR-
NATIONAL FIREWORKS ORDNANCE CORP., WEST HANOVER, MASS	NF-
NATIONAL FIREWORKS ORDNANCE CORP., WEST HANOVER, MASS	NFOC-
NATIONAL GEOPHYSICAL CO., INC., DALLAS	NGC-
NATIONAL HEART INSTITUTE, BETHESDA, MD.	NHI-
NATIONAL HURRICANE RESEARCH LAB., CORAL GABLES, FLA.	IERTM-NHRL-
NATIONAL HURRICANE RESEARCH LAB., CORAL GABLES, FLA.	NHRL-
NATIONAL HURRICANE RESEARCH LAB., CORAL GABLES, FLA.	NOAA-TM-ERL-NHRL-
NATIONAL HURRICANE RESEARCH LAB., MIAMI, FLA.	ESSA-TM-ERLTM-NHRL-
NATIONAL HURRICANE RESEARCH PROJECT, WASHINGTON, D.C.	NHRP-
NATIONAL INSTITUTE OF ENVIRONMENTAL HEALTH SCIENCES, RESEARCH TRIANGLE PARK, N.C.	ENVIRONMENTAL REVIEW-
NATIONAL INSTITUTE OF ENVIRONMENTAL HEALTH SCIENCES, RESEARCH TRIANGLE PARK, N.C.	NIH-ES-
NATIONAL INST. OF GEN. MEDICAL SCIENCES, BETHESDA, MD	NIGMS-
NATIONAL INST. OF NEUROLOGICAL DISEASES AND BLINDNESS, BETHESDA, MD.	NINDB-
NATIONAL INSTITUTES OF HEALTH, BETHESDA, MD.	NIH-
NATIONAL INSTITUTES OF HEALTH, BETHESDA, MD.	NIH-(YEAR-NUMBER)-
NATIONAL INSTITUTES OF HEALTH, BETHESDA, MD.	PSC-
NATIONAL INSTITUTES OF HEALTH, BETHESDA, MD.	RSTP-
NATIONAL INSTITUTES OF HEALTH. INFORMATION EXCHANGE GROUP NUMBER- BETHESDA, MD.	IEG-(NO.) SCI. MEMO-
NATIONAL LEAD CO., INC. RAW MATERIALS DEVELOPMENT LAB., WINCHESTER, MASS.	WIN-
NATIONAL LEAD CO. OF OHIO, CINCINNATI (FEED MATERIALS PRODUCTION CENTER REPORTS)	FMPC-
NATIONAL LEAD CO. OF OHIO, CINCINNATI	GEM-
NATIONAL LEAD CO. OF OHIO, CINCINNATI	NLCO-
NATIONAL LEAD CO. OF OHIO, CINCINNATI	NLCO-TR-
NATIONAL LEAD CO. OF OHIO, CINCINNATI	NLO-
NATIONAL LIBRARY OF MEDICINE, WASHINGTON, D.C.	NLM-
NATIONAL LIBRARY OF MEDICINE, WASHINGTON, D.C. (LITERATURE SEARCH)	NLM L.S.(YEAR)-
NATIONAL MARINE CONSULTANTS, ANAHEIM, CALIF.	NMC-TR-
NATIONAL MARINE FISHERIES SERVICE, SEATTLE (CIRCULAR)	NMFS-CIRC-
NATIONAL MARINE FISHERIES SERVICE, SEATTLE	NMFS-SSRF-
NATIONAL MARINE FISHERIES SERVICE, SEATTLE (CIRCULAR)	NOAA-TR-NMFS-CIRC-
NATIONAL MARINE FISHERIES SERVICE, SEATTLE (SPECIAL SCIENTIFIC REPORT FISHERIES SERIES)	NOAA-TR-NMFS-SSRF-
NATIONAL METEOROLOGICAL CENTER, SUITLAND, MD.	ESSA-TM-WBTM-NMC-
NATIONAL METEOROLOGICAL CENTER, SUITLAND, MD.	NOAA-TM-NWS-NMC-
NATIONAL METEOROLOGICAL CENTER, SUITLAND, MD.	NWS-NMC-
NATIONAL METEOROLOGICAL CENTER, SUITLAND, MD.	WBTM-NMC-
NATIONAL METEOROLOGICAL CENTER, SUITLAND, MD.	WBTM-NMC-TM-
NATIONAL MILITARY COMMAND SYSTEM SUPPORT CENTER, WASHINGTON, D.C.	CPM-GD/SD-
NATIONAL MILITARY COMMAND SYSTEM SUPPORT CENTER, WASHINGTON, D.C. (COMPUTER SYSTEMS MANUAL)	CSM-UG-(NO.)-(YR.)-
NATIONAL MILITARY COMMAND SYSTEM SUPPORT CENTER, WASHINGTON, D.C.	NMCSSC-
NATIONAL MILITARY COMMAND SYSTEM SUPPORT CENTER, WASHINGTON, D.C.	NMCSSC-CSM-
NATIONAL MILITARY COMMAND SYSTEM SUPPORT CENTER, WASHINGTON, D.C.	NMCSSC-CSM-AM-
NATIONAL MILITARY COMMAND SYSTEM SUPPORT CENTER, WASHINGTON, D.C.	NMCSSC-CSM-GD-
NATIONAL MILITARY COMMAND SYSTEM SUPPORT CENTER, WASHINGTON, D.C.	NMCSSC-CSM-OM-
NATIONAL MILITARY COMMAND SYSTEM SUPPORT CENTER, WASHINGTON, D.C.	NMCSSC-CSM-PD-
NATIONAL MILITARY COMMAND SYSTEM SUPPORT CENTER, WASHINGTON, D.C.	NMCSSC-CSM-PSM-
NATIONAL MILITARY COMMAND SYSTEM SUPPORT CENTER, WASHINGTON, D.C.	NMCSSC-CSM-SD-IB-(YEAR)-
NATIONAL MILITARY COMMAND SYSTEM SUPPORT CENTER, WASHINGTON, D.C.	NMCSSC-CSM-UM-
NATIONAL MILITARY COMMAND SYSTEM SUPPORT CENTER, WASHINGTON, D.C.	NMCSSC-DN-
NATIONAL MILITARY COMMAND SYSTEM SUPPORT CENTER, WASHINGTON, D.C.	NMCSSC-SPM-OAM-
NATIONAL MILITARY COMMAND SYSTEM SUPPORT CENTER, WASHINGTON, D.C.	NMCSSC-SPM-OCD-
NATIONAL MILITARY COMMAND SYSTEM SUPPORT CENTER, WASHINGTON, D.C.	NMCSSC-SPM-VTP-
NATIONAL MILITARY COMMAND SYSTEM SUPPORT CENTER, WASHINGTON, D.C.	NMCSSC-TM-
NATIONAL MILITARY COMMAND SYSTEM SUPPORT CENTER, WASHINGTON, D.C.	NMCSSC TR-
NATIONAL MILITARY COMMAND SYSTEM SUPPORT CENTER, WASHINGTON, D.C.	NSCSSC-SPM-CDS-
NATIONAL MILITARY ESTABLISHMENT, WASHINGTON, D.C.	NME-
NATIONAL MILITARY ESTABLISHMENT, WASHINGTON, D.C.	WD-TM-
NATIONAL MILITARY ESTABLISHMENT. RES. + DEV. BOARD	NME-RS-
NATIONAL NAVAL MEDICAL CENTER, BETHESDA, MD.	NNMC-
NATIONAL NEUTRON CROSS SECTIONS CENTER, BROOKHAVEN NATL. LAB., N.Y. (EVALUATED NUCLEAR DATA FILE)	ENDF-
NATIONAL NORTHERN, WEST HANOVER, MASS.	NN-
NATIONAL NORTHERN, WEST HANOVER, MASS.	NN-M-
NATIONAL NORTHERN, WEST HANOVER, MASS.	NN-Q-
NATIONAL NORTHERN CORP., WEST HANOVER, MASS.	NNC-
NATIONAL OCEAN SURVEY, ROCKVILLE, MD.	C+GS-
NATIONAL OCEAN SURVEY, ROCKVILLE, MD.	CGS-(NUMBER)-
NATIONAL OCEAN SURVEY, ROCKVILLE, MD.	C+GSTM-
NATIONAL OCEAN SURVEY, ROCKVILLE, MD.	ESSA-TR-C+GS
NATIONAL OCEAN SURVEY, ROCKVILLE, MD.	NOAA-CONFR-
NATIONAL OCEAN SURVEY, ROCKVILLE, MD.	NOAA-SP-NOS-
NATIONAL OCEAN SURVEY, ROCKVILLE, MD.	NOAA-TM-NOS-
NATIONAL OCEAN SURVEY, ROCKVILLE, MD.	NOAA-TR-NOS-
NATIONAL OCEAN SURVEY, ROCKVILLE, MD.	NOS-
NATIONAL OCEANIC AND ATMOSPHERIC ADMINISTRATION, ROCKVILLE, MD. (SEA GRANT PROGRAM)	SEAGRANT-
NATIONAL OCEANIC AND ATMOSPHERIC ADMINISTRATION, ROCKVILLE, MD. (SEA GRANT PROGRAM)	SEA GRANT PUB-
NATIONAL OCEANIC AND ATMOSPHERIC ADM., BOULDER, COLO.	NOAA-
NATIONAL OCEANIC AND ATMOSPHERIC ADM., BOULDER, COLO.	NOAA/PL-
NATIONAL OCEANIC AND ATMOSPHERIC ADMINISTRATION, BOULDER, COLO. (UPPER ATMOSPHERE GEOPHYSICS) (SEE ALSO NAMES OF SEPARATE LABS. AND CENTERS AS: NATIONAL HURRICANE RESEARCH LAB.)	UAG-
NATIONAL OCEANOGRAPHIC DATA CENTER, WASHINGTON, D.C.	G-
NATIONAL OCEANOGRAPHIC DATA CENTER, WASHINGTON, D.C.	NODC-

NATIONAL PARK SERVICE. DIV. OF HISTORY, WASH., D.C.	FNP-HH-(YEAR)-
NATIONAL RADIAC, INC., NEWARK, N.J.	NRI-
NATIONAL RADIO ASTRONOMY OBSERVATORY, GREEN BANK, W.VA	NRAO-
NATIONAL REACTOR TESTING STATION, IDAHO FALLS	IN-
NATIONAL RESEARCH CORP., BOSTON	NRCO-
NATIONAL RESEARCH CORP., CAMBRIDGE, MASS.	NR-
NATIONAL RESEARCH CORP., CAMBRIDGE, MASS.	NRC-(NUMBERS)-
NATIONAL RESEARCH CORP., CAMBRIDGE, MASS.	NRCO-
NATIONAL RESEARCH CORP., CAMBRIDGE, MASS.	SC-
NATIONAL RESEARCH CORP. METALS DIV., CAMBRIDGE, MASS.	NRC/MD-
NATIONAL RESEARCH CORP. METALS DIV., CAMBRIDGE, MASS. (TANTALUM-TUNGSTEN QUARTERLY REPORT)	NRC-TA-W QR-
NATIONAL RESEARCH COUNCIL, WASHINGTON, D.C.	NAS ARDC COM-...
NATIONAL RESEARCH COUNCIL, WASHINGTON, D.C.	NAS-NRC/PUB-
NATIONAL RESEARCH COUNCIL, WASHINGTON, D.C.	NRC-
NATIONAL RESEARCH COUNCIL, WASHINGTON, D.C. (FUNGICIDE SCREENING RPT.)	NRC-FSR-
NATIONAL RESEARCH COUNCIL, WASH., D.C. (PUBLICATIONS)	NRC P-
NATIONAL RESEARCH COUNCIL, WASHINGTON, D.C.	NRC PUBL.
NATIONAL RESEARCH COUNCIL. BUILDING RESEARCH ADVISORY BOARD (CONFERENCE REPORT)	BRAB CR-
NATL. RES. COUNCIL. CHEM.-BIOLOGICAL COORDINATION CTR.	CBCC-
NATIONAL RES. COUNCIL. COMM. ON AMPHIBIOUS OPERATIONS	CAO-
NATIONAL RES. COUNCIL. COMM. ON AMPHIBIOUS OPERATIONS	NRC CAO-
NATIONAL RESEARCH COUNCIL. COMM. ON AVIATION MEDICINE	CAM-
NATIONAL RESEARCH COUNCIL. COMM. ON AVIATION MEDICINE	NRC AM-
NATIONAL RESEARCH COUNCIL. COMM. ON AVIATION PSYCH.	NRC CAP-
NATL. RES. COUNCIL. COMM. ON FORTIFICATION DESIGN	FD-
NATIONAL RESEARCH COUNCIL. COMMITTEE ON FORTIFICATION DESIGN (INTERIM MEMORANDA)	NRC CFD IM-
NATL. RES. COUNCIL. COMM. ON FORTIFICATION DESIGN	NRC FD-
NATIONAL RESEARCH COUNCIL. COMMITTEE ON FORTIFICATION DESIGN (INTERIM MEMORANDA)	NRC FD IM-
NATIONAL RESEARCH COUNCIL. COMMITTEE ON FORTIFICATION DESIGN (INTERIM REPORTS)	NRC FD IR-
NATIONAL RESEARCH COUNCIL. COMMITTEE ON HEARING AND BIO-ACOUSTICS, WASHINGTON, D.C.	CHABA-
NATIONAL RESEARCH COUNCIL. COMMITTEE ON PASSIVE PROTECTION AGAINST BOMBING (INTERIM MEMORANDA)	NRC PPB IM-
NATIONAL RESEARCH COUNCIL. COMMITTEE ON PASSIVE PROTECTION AGAINST BOMBING (INTERIM REPORTS)	NRC PPB IR-
NATIONAL RESEARCH COUNCIL. COMMITTEE ON PASSIVE PROTECTION AGAINST BOMBING	PPB-
NATIONAL RESEARCH COUNCIL. COMMITTEE ON POLLUTION ABATEMENT AND CONTROL, WASHINGTON, D.C.	COPAC-
NATIONAL RESEARCH COUNCIL. COMMITTEE ON SHIP STRUCTURE, WASHINGTON, D.C.	NRC SSC-
NATIONAL RESEARCH COUNCIL. COMMITTEE ON SHIP STRUCTURE, WASHINGTON, D.C. (SERIES USED FOR REPORTS SPONSORED BY THE COMMITTEE BUT PREPARED BY VARIOUS AGENCIES)	SSC-
NATIONAL RESEARCH COUNCIL. COMMITTEE ON TREATMENT OF WAR GAS CASUALTIES	NRC TWGC-
NATIONAL RESEARCH COUNCIL. COMMITTEE ON TREATMENT OF WAR GAS CASUALTIES	TWGC-
NATIONAL RESEARCH COUNCIL. COMM. ON UNDERSEA WARFARE	CUW-
NATIONAL RESEARCH COUNCIL. COMM. ON UNDERSEA WARFARE	NRC CUW-
NATIONAL RESEARCH COUNCIL. COMMITTEE ON WASTE DISPOSAL, WASHINGTON, D.C.	NRC CWD-
NATIONAL RESEARCH COUNCIL. HIGHWAY RESEARCH BOARD	HRB-
NATIONAL RES. COUNCIL. HIGHWAY RES. BD. (BIBLIOGRAPHY)	HRB-B-
NATIONAL RESEARCH COUNCIL. HIGHWAY RES. BD. (BULLETIN)	HRB-BUL-
NATIONAL RES. COUNCIL. HIGHWAY RES. BD. (RES. RPT.)	HRB-RR-
NATIONAL RES. COUNCIL. HIGHWAY RES. BD. (SPEC. RPT.)	HRB-SR-
NATIONAL RESEARCH COUNCIL. INSECT CONTROL COMMITTEE	ICC-
NATIONAL RESEARCH COUNCIL. INSECT CONTROL COMMITTEE (ABSTRACT BULLETIN)	ICC AB-
NATIONAL RESEARCH COUNCIL. MATERIALS ADVISORY BOARD, WASHINGTON, D.C.	NRC MAB-
NATL. RES. COUNCIL. MINERALS + METALS ADVISORY BD.	MMAB-(NO.)-C
NATL. RES. COUNCIL. MINERALS + METALS ADVISORY BD.	MMAB-(NO.)-M
NATL. RES. COUNCIL. MINERALS + METALS ADVISORY BD.	NRC-MMAB-
NATIONAL RESEARCH COUNCIL. NATIONAL MATERIALS ADVISORY BOARD, WASHINGTON, D.C.	NMAB-
NATIONAL RESEARCH COUNCIL. PACIFIC SCIENCE BOARD	CIMA-
NATIONAL RESEARCH COUNCIL. PACIFIC SCIENCE BOARD	SIM-
NATIONAL RESEARCH COUNCIL. PREVENTION OF DETERIORATION CENTER, WASHINGTON, D.C.	PDC-
NATIONAL RESEARCH COUNCIL. PREVENTION OF DETERIORATION CENTER, WASHINGTON, D.C.	PDL-
NATIONAL RESEARCH COUNCIL OF CANADA, OTTAWA	AERONAUTICAL LR-
NATIONAL RESEARCH COUNCIL OF CANADA, OTTAWA	CM(NUMBER)-S-
NATIONAL RESEARCH COUNCIL OF CANADA, OTTAWA	CNRC-
NATIONAL RESEARCH COUNCIL OF CANADA, OTTAWA	DME/NAE-(YEAR)-/(NO.)/
NATIONAL RESEARCH COUNCIL OF CANADA, OTTAWA	IS-
NATIONAL RESEARCH COUNCIL OF CANADA, OTTAWA	LP-
NATIONAL RESEARCH COUNCIL OF CANADA, OTTAWA	LR-
NATIONAL RESEARCH COUNCIL OF CANADA, OTTAWA	LT-
NATIONAL RESEARCH COUNCIL OF CANADA, OTTAWA	MB-
NATIONAL RESEARCH COUNCIL OF CANADA, OTTAWA	MK-
NATIONAL RESEARCH COUNCIL OF CANADA, OTTAWA	NRC-
NATIONAL RESEARCH COUNCIL OF CANADA, OTTAWA	NRCC-
NATIONAL RESEARCH COUNCIL OF CANADA, OTTAWA	NRCC (LETTER(S))-
NATIONAL RESEARCH COUNCIL OF CANADA, OTTAWA	NRCC AEP (LETTER(S))-
NATIONAL RESEARCH COUNCIL OF CANADA, OTTAWA	NRC-EC-
NATIONAL RESEARCH COUNCIL OF CANADA, OTTAWA	NRC-ME-
NATIONAL RESEARCH COUNCIL OF CANADA, OTTAWA	NRC-ME-MK-
NATIONAL RESEARCH COUNCIL OF CANADA, OTTAWA (TECHNICAL TRANSLATION)	NRC-TT-
NATIONAL RESEARCH COUNCIL OF CANADA, OTTAWA	NRS-MK-
NATIONAL RESEARCH COUNCIL OF CANADA, OTTAWA	PRA-
NATIONAL RESEARCH COUNCIL OF CANADA, OTTAWA	PZ-
NATIONAL RESEARCH COUNCIL OF CANADA, OTTAWA	TIS-
NATIONAL RESEARCH COUNCIL OF CANADA, OTTAWA (TRANSLATION)	TT-
NATIONAL RESEARCH COUNCIL OF CANADA, OTTAWA	W-

NATIONAL RESEARCH COUNCIL OF CANADA. ASSOCIATE COMM. ON AVIATION MEDICAL RESEARCH

NATIONAL RESEARCH COUNCIL OF CANADA. ASSOCIATE COMM. ON AVIATION MEDICAL RESEARCH	ACAMR-
NATL. RES. COUNCIL OF CANADA. ASSOCIATE COMM. ON AVIATION MEDICAL RES. NO. 1 CLINICAL INVESTIGATION UNIT (RCAF) (LETTERED REPORTS)	ACAMR NO. 1 CIU-
NATL. RES. COUNCIL OF CANADA. ASSOCIATE COMM. ON AVIATION MEDICAL RES. NO. 1 CLINICAL INVESTIGATION UNIT (RCAF)	1 CIU-
NATIONAL RESEARCH COUNCIL OF CANADA. ASSOCIATE COMMITTEE ON SYNTHETIC RUBBER RESEARCH	OSR-
NATIONAL RESEARCH COUNCIL OF CANADA. ATOMIC ENERGY PROJECT, CHALK RIVER, ONT.	AECL GPI-
NATIONAL RESEARCH COUNCIL OF CANADA. ATOMIC ENERGY PROJECT, CHALK RIVER, ONT.	AEPC-
NATIONAL RESEARCH COUNCIL OF CANADA. ATOMIC ENERGY PROJECT, CHALK RIVER, ONT.	AEP CRTEC-
NATIONAL RESEARCH COUNCIL OF CANADA. ATOMIC ENERGY PROJECT, CHALK RIVER, ONT.	CC-(NO.)(AEPC)-
NATIONAL RESEARCH COUNCIL OF CANADA. ATOMIC ENERGY PROJECT, CHALK RIVER, ONT.	CC-(NO.)-H-
NATIONAL RESEARCH COUNCIL OF CANADA. ATOMIC ENERGY PROJECT, CHALK RIVER, ONT.	CCI-
NATIONAL RESEARCH COUNCIL OF CANADA. ATOMIC ENERGY PROJECT, CHALK RIVER, ONT.	CD-
NATIONAL RESEARCH COUNCIL OF CANADA. ATOMIC ENERGY PROJECT, CHALK RIVER, ONT.	CLE-
NATIONAL RESEARCH COUNCIL OF CANADA. ATOMIC ENERGY PROJECT, CHALK RIVER, ONT.	CPP-
NATIONAL RESEARCH COUNCIL OF CANADA. ATOMIC ENERGY PROJECT, CHALK RIVER, ONT.	CRA-
NATIONAL RESEARCH COUNCIL OF CANADA. ATOMIC ENERGY PROJECT, CHALK RIVER, ONT.	CRB-
NATIONAL RESEARCH COUNCIL OF CANADA. ATOMIC ENERGY PROJECT, CHALK RIVER, ONT.	CRC-
NATIONAL RESEARCH COUNCIL OF CANADA. ATOMIC ENERGY PROJECT, CHALK RIVER, ONT.	CRCE-
NATIONAL RESEARCH COUNCIL OF CANADA. ATOMIC ENERGY PROJECT, CHALK RIVER, ONT.	CRD-
NATIONAL RESEARCH COUNCIL OF CANADA. ATOMIC ENERGY PROJECT, CHALK RIVER, ONT.	CRHR-
NATIONAL RESEARCH COUNCIL OF CANADA. ATOMIC ENERGY PROJECT, CHALK RIVER, ONT.	CRIB-
NATIONAL RESEARCH COUNCIL OF CANADA. ATOMIC ENERGY PROJECT, CHALK RIVER, ONT. (REF. 29)	CRL-
NATIONAL RESEARCH COUNCIL OF CANADA. ATOMIC ENERGY PROJECT, CHALK RIVER, ONT.	CRM-
NATIONAL RESEARCH COUNCIL OF CANADA. ATOMIC ENERGY PROJECT, CHALK RIVER, ONT. (REF. 29)	CRP(LETTER(S))-
NATIONAL RESEARCH COUNCIL OF CANADA. ATOMIC ENERGY PROJECT, CHALK RIVER, ONT.	CRPC-
NATIONAL RESEARCH COUNCIL OF CANADA. ATOMIC ENERGY PROJECT, CHALK RIVER, ONT.	CRPE-
NATIONAL RESEARCH COUNCIL OF CANADA. ATOMIC ENERGY PROJECT, CHALK RIVER, ONT.	CRPG-
NATIONAL RESEARCH COUNCIL OF CANADA. ATOMIC ENERGY PROJECT, CHALK RIVER, ONT.	CRPP-
NATIONAL RESEARCH COUNCIL OF CANADA. ATOMIC ENERGY PROJECT, CHALK RIVER, ONT.	CR-PRG-
NATIONAL RESEARCH COUNCIL OF CANADA. ATOMIC ENERGY PROJECT, CHALK RIVER, ONT.	CR-PRG-(NO.)-TEC.-
NATIONAL RESEARCH COUNCIL OF CANADA. ATOMIC ENERGY PROJECT, CHALK RIVER, ONT.	CRPX-
NATIONAL RESEARCH COUNCIL OF CANADA. ATOMIC ENERGY PROJECT, CHALK RIVER, ONT.	CR-RHC-
NATIONAL RESEARCH COUNCIL OF CANADA. ATOMIC ENERGY PROJECT, CHALK RIVER, ONT.	CRTEC-
NATIONAL RESEARCH COUNCIL OF CANADA. ATOMIC ENERGY PROJECT, CHALK RIVER, ONT.	CRX-
NATIONAL RESEARCH COUNCIL OF CANADA. ATOMIC ENERGY PROJECT, CHALK RIVER, ONT.	CTSG-
NATIONAL RESEARCH COUNCIL OF CANADA. ATOMIC ENERGY PROJECT, CHALK RIVER, ONT.	DE-
NATIONAL RESEARCH COUNCIL OF CANADA. ATOMIC ENERGY PROJECT, CHALK RIVER, ONT.	DR-
NATIONAL RESEARCH COUNCIL OF CANADA. ATOMIC ENERGY PROJECT, CHALK RIVER, ONT.	E-(INITIALS)-
NATIONAL RESEARCH COUNCIL OF CANADA. ATOMIC ENERGY PROJECT, CHALK RIVER, ONT. (REF.29)	EI-
NATIONAL RESEARCH COUNCIL OF CANADA. ATOMIC ENERGY PROJECT, CHALK RIVER, ONT.	EM-
NATIONAL RESEARCH COUNCIL OF CANADA. ATOMIC ENERGY PROJECT, CHALK RIVER, ONT. (REF. 29)	GPI-
NATIONAL RESEARCH COUNCIL OF CANADA. ATOMIC ENERGY PROJECT, CHALK RIVER, ONT. (REF. 29)	HI-
NATIONAL RESEARCH COUNCIL OF CANADA. ATOMIC ENERGY PROJECT, CHALK RIVER, ONT.	HRAC-
NATIONAL RESEARCH COUNCIL OF CANADA. ATOMIC ENERGY PROJECT, CHALK RIVER, ONT. (REF. 29)	HRI-
NATIONAL RESEARCH COUNCIL OF CANADA. ATOMIC ENERGY PROJECT, CHALK RIVER, ONT.	IM-
NATIONAL RESEARCH COUNCIL OF CANADA. ATOMIC ENERGY PROJECT, CHALK RIVER, ONT.	LC-
NATIONAL RESEARCH COUNCIL OF CANADA. ATOMIC ENERGY PROJECT, CHALK RIVER, ONT.	LE-
NATIONAL RESEARCH COUNCIL OF CANADA. ATOMIC ENERGY PROJECT, CHALK RIVER, ONT. (REF. 29)	MI-
NATIONAL RESEARCH COUNCIL OF CANADA. ATOMIC ENERGY PROJECT, CHALK RIVER, ONT.	MIM-
NATIONAL RESEARCH COUNCIL OF CANADA. ATOMIC ENERGY PROJECT, CHALK RIVER, ONT.	MISC.-
NATIONAL RESEARCH COUNCIL OF CANADA. ATOMIC ENERGY PROJECT, CHALK RIVER, ONT.	MP PROJ.-
NATIONAL RESEARCH COUNCIL OF CANADA. ATOMIC ENERGY PROJECT, CHALK RIVER, ONT.	MR-
NATIONAL RESEARCH COUNCIL OF CANADA. ATOMIC ENERGY PROJECT, CHALK RIVER, ONT. (REF. 29)	NL-
NATIONAL RESEARCH COUNCIL OF CANADA. ATOMIC ENERGY PROJECT, CHALK RIVER, ONT. (TECHNICAL REPORT)	NRCC AEP CRTEC-
NATIONAL RESEARCH COUNCIL OF CANADA. ATOMIC ENERGY PROJECT, CHALK RIVER, ONT.	PC-
NATIONAL RESEARCH COUNCIL OF CANADA. ATOMIC ENERGY PROJECT, CHALK RIVER, ONT.	PM-
NATIONAL RESEARCH COUNCIL OF CANADA. ATOMIC ENERGY PROJECT, CHALK RIVER, ONT.	PPPM-
NATIONAL RESEARCH COUNCIL OF CANADA. ATOMIC ENERGY PROJECT, CHALK RIVER, ONT. (REF. 29)	PR-
NATIONAL RESEARCH COUNCIL OF CANADA. ATOMIC ENERGY PROJECT, CHALK RIVER, ONT. (REF. 29)	PR-(LETTER(S))-(LTR(S))-
NATIONAL RESEARCH COUNCIL OF CANADA. ATOMIC ENERGY PROJECT, CHALK RIVER, ONT.	PRA-
NATIONAL RESEARCH COUNCIL OF CANADA. ATOMIC ENERGY PROJECT, CHALK RIVER, ONT.	RM-
NATIONAL RESEARCH COUNCIL OF CANADA. ATOMIC ENERGY PROJECT, CHALK RIVER, ONT.	SDDD-
NATIONAL RESEARCH COUNCIL OF CANADA. ATOMIC ENERGY PROJECT, CHALK RIVER, ONT. (REF. 29)	TEC-PI-
NATIONAL RESEARCH COUNCIL OF CANADA. ATOMIC ENERGY PROJECT, CHALK RIVER, ONT. (REF. 29)	TECPI-
NATIONAL RESEARCH COUNCIL OF CANADA. ATOMIC ENERGY PROJECT, CHALK RIVER, ONT.	TN-
NATIONAL RESEARCH COUNCIL OF CANADA. ATOMIC ENERGY PROJECT, CHALK RIVER, ONT. (REF. 29)	XI-

(SEE ALSO LATER NAME: ATOMIC ENERGY OF CANADA LTD.)

NATIONAL RESEARCH COUNCIL OF CANADA. DIV. OF APPLIED PHYSICS, OTTAWA	APR-
NATIONAL RESEARCH COUNCIL OF CANADA. DIV. OF APPLIED PHYSICS, OTTAWA	APXNR-
NATIONAL RESEARCH COUNCIL OF CANADA. DIV. OF ATOMIC ENERGY, CHALK RIVER, ONT.	D-
NATIONAL RESEARCH COUNCIL OF CANADA. DIV. OF ATOMIC ENERGY, CHALK RIVER, ONT.	D-3-
NATIONAL RESEARCH COUNCIL OF CANADA. DIV. OF ATOMIC ENERGY, CHALK RIVER, ONT.	FSD-
NATIONAL RESEARCH COUNCIL OF CANADA. DIV. OF ATOMIC ENERGY, CHALK RIVER, ONT. (REF. 29)	PD-
NATIONAL RESEARCH COUNCIL OF CANADA. DIV. OF ATOMIC ENERGY, CHALK RIVER, ONT. (REF. 29)	TL-

(SEE ALSO LATER NAME: ATOMIC ENERGY OF CANADA LTD.)

NATIONAL RESEARCH COUNCIL OF CANADA. DIV. OF BUILDING RESEARCH, OTTAWA (CBA FRENCH EDITION)	ACC-
NATIONAL RESEARCH COUNCIL OF CANADA. DIV. OF BUILDING RESEARCH, OTTAWA (BIBLIOGRAPHIES)	BIB-
NATIONAL RESEARCH COUNCIL OF CANADA. DIV. OF BUILDING RESEARCH, OTTAWA (BLDG. RES. NEWS)	BRN-
NATIONAL RESEARCH COUNCIL OF CANADA. DIV. OF BUILDING RESEARCH, OTTAWA (CAN. BLDG. ABSTRACTS)	CBA-
NATIONAL RESEARCH COUNCIL OF CANADA. DIV. OF BUILDING RESEARCH, OTTAWA (CAN. BLDG. DIGEST)	CBD-
NATIONAL RESEARCH COUNCIL OF CANADA. DIV. OF BUILDING RESEARCH, OTTAWA (CBD FRENCH EDITION)	CBDF-
NATIONAL RESEARCH COUNCIL OF CANADA. DIV. OF BUILDING RESEARCH, OTTAWA	DBR-
NATIONAL RESEARCH COUNCIL OF CANADA. DIV. OF BUILDING RESEARCH, OTTAWA (FIRE RESEARCH NOTES)	FRNOTE-
NATIONAL RESEARCH COUNCIL OF CANADA. DIV. OF BUILDING RESEARCH, OTTAWA (HOUSING NOTES)	HN-
NATIONAL RESEARCH COUNCIL OF CANADA. DIV. OF BUILDING RESEARCH, OTTAWA	NRCC DBR-
NATIONAL RESEARCH COUNCIL OF CANADA, DIVISION OF MECHANICAL ENGINEERING, OTTAWA	DME-ME-
NATIONAL RESEARCH COUNCIL OF CANADA, DIVISION OF MECHANICAL ENGINEERING, OTTAWA	DME-MER-
NATIONAL RESEARCH COUNCIL OF CANADA, DIVISION OF MECHANICAL ENGINEERING, OTTAWA	DME-MET-
NATIONAL RESEARCH COUNCIL OF CANADA, DIVISION OF MECHANICAL ENGINEERING, OTTAWA	DME-MH-
NATIONAL RESEARCH COUNCIL OF CANADA, DIVISION OF MECHANICAL ENGINEERING, OTTAWA	DME-MK-
NATIONAL RESEARCH COUNCIL OF CANADA, DIVISION OF MECHANICAL ENGINEERING, OTTAWA	DME-ML-
NATIONAL RESEARCH COUNCIL OF CANADA, DIVISION OF MECHANICAL ENGINEERING, OTTAWA	DME-MP-
NATIONAL RESEARCH COUNCIL OF CANADA, DIVISION OF MECHANICAL ENGINEERING, OTTAWA	DME/NAE-
NATIONAL RESEARCH COUNCIL OF CANADA, DIVISION OF MECHANICAL ENGINEERING, OTTAWA	ME-
NATIONAL RESEARCH COUNCIL OF CANADA, DIVISION OF MECHANICAL ENGINEERING, OTTAWA	MH-
NATIONAL RESEARCH COUNCIL OF CANADA, DIVISION OF MECHANICAL ENGINEERING, OTTAWA	MK-
NATIONAL RESEARCH COUNCIL OF CANADA, DIVISION OF MECHANICAL ENGINEERING, OTTAWA	MP-
NATIONAL RESEARCH COUNCIL OF CANADA, DIVISION OF MECHANICAL ENGINEERING, OTTAWA	NRCC ME-
NATIONAL RESEARCH COUNCIL OF CANADA, DIVISION OF MECHANICAL ENGINEERING, OTTAWA	NRC-MS-
NATIONAL RESEARCH COUNCIL OF CANADA, DIVISION OF MECHANICAL ENGINEERING, OTTAWA	NRC-MT-

NATIONAL RESEARCH COUNCIL OF CANADA. LIBRARY, OTTAWA (TRANSLATIONS)	NRCC-
NATIONAL RESEARCH COUNCIL OF CANADA. LOW TEMPERATURE SECTION, OTTAWA	DME-MISC.-
NATIONAL RES.COUNCIL OF CANADA. MONTREAL LAB.(REF.29)	EL-
NATIONAL RES.COUNCIL OF CANADA. MONTREAL LAB.(REF.29)	M(LETTER(S))-
NATIONAL RES.COUNCIL OF CANADA. MONTREAL LAB.(REF.29)	MP(LETTER(S))-
NATIONAL RESEARCH COUNCIL OF CANADA. MONTREAL LAB.	MS-
NATIONAL RESEARCH COUNCIL OF CANADA. MONTREAL LAB.	PRE-
NATIONAL RESEARCH COUNCIL OF CANADA. MONTREAL LAB.	PU-
(SEE ALSO LATER NAME: ATOMIC ENERGY OF CANADA LTD.)	
NATL. RES. COUNCIL OF CANADA. NATL. RES. LABS., OTTAWA	C-(NO.-YR)S-
NATL. RES. COUNCIL OF CANADA. NATL. RES. LABS., OTTAWA	LR-
NATL. RES. COUNCIL OF CANADA. NATL. RES. LABS., OTTAWA	MT-
NATL. RES. COUNCIL OF CANADA. NATL. RES. LABS., OTTAWA	NRC-LR-
NATIONAL RESEARCH COUNCIL OF CANADA. RADIO AND ELECTRICAL ENGINEERING DIV., OTTAWA	ERA-
NATIONAL RESEARCH COUNCIL OF CANADA. RADIO AND ELECTRICAL ENGINEERING DIV., OTTAWA	ERB-
NATIONAL RESEARCH COUNCIL OF CANADA. RADIO AND ELECTRICAL ENGINEERING DIV., OTTAWA	NRC/REE-
NATIONAL RESEARCH COUNCIL OF CANADA. SPACE RESEARCH FACILITIES BRANCH, OTTAWA	SRFB-
NATIONAL RESEARCH COUNCIL OF CANADA. TECHNICAL INFORMATION SERVICE, OTTAWA	NRC-TIS-
NATIONAL RESEARCH DEVELOPMENT CORP. (GT. BRIT.)	NRDC-
NATIONAL RESOURCE ANALYSIS CENTER, WASH., D.C.	IST-
NATIONAL RESOURCE EVALUATION CENTER, WASHINGTON, D.C.	NREC-
NATIONAL SAFETY COUNCIL, CHICAGO (FILM)	NSC-(NUMBER)-(NUMBER)
NATIONAL SCIENCE FOUNDATION, WASHINGTON, D.C.(MATHEMATICAL, PHYSICAL AND ENGINEERING SCIENCES)	MPE-
NATIONAL SCIENCE FOUNDATION, WASHINGTON, D.C.	NSF-
NATIONAL SCIENCE FOUNDATION, WASHINGTON, D.C.	NSF-(YEAR)-
NATIONAL SCIENCE FOUNDATION, WASH.,D.C. (TRANSLATIONS)	NSF-TR-
NATIONAL SCIENCE FOUNDATION, WASHINGTON, D.C. (PROGRAM FOR SCIENCE TRANSLATION)	PST-CAT-
NATIONAL SCIENCE FOUNDATION, WASHINGTON, D.C. (SPECIAL FOREIGN CURRENCY SCIENCE INFO. PROGRAM)	SFCSE-AEC-TT(YEAR-NO.)
NATIONAL SCIENCE FOUNDATION, WASHINGTON, D.C. (SPECIAL FOREIGN CURRENCY SCIENCE INFO. PROGRAM)	SFCSI-AEC-TT(YEAR-NO.)
NATIONAL SCIENCE FOUNDATION, WASHINGTON, D.C. (SPECIAL FOREIGN CURRENCY SCIENCE INFO. PROGRAM)	SFCSI-AGR-TT(YEAR-NO.)
NATIONAL SCIENCE FOUNDATION, WASHINGTON, D.C. (SPECIAL FOREIGN CURRENCY SCIENCE INFO. PROGRAM)	SFCSI-COMM-TT(YEAR-NO.)
NATIONAL SCIENCE FOUNDATION, WASHINGTON, D.C. (SPECIAL FOREIGN CURRENCY SCIENCE INFO. PROGRAM)	SFCSI-HEW-TT(YEAR-NO.)
NATIONAL SCIENCE FOUNDATION, WASHINGTON, D.C. (SPECIAL FOREIGN CURRENCY SCIENCE INFO. PROGRAM)	SFCSI-INT-TT(YEAR-NO.)
NATIONAL SCIENCE FOUNDATION, WASHINGTON, D.C. (SPECIAL FOREIGN CURRENCY SCIENCE INFO. PROGRAM)	SFCSI-NASA-TT(YEAR-NO.)
NATIONAL SCIENCE FOUNDATION, WASHINGTON, D.C. (SPECIAL FOREIGN CURRENCY SCIENCE INFO. PROGRAM)	SFCSI-NSF-TT(YEAR-NO.)
NATIONAL SCIENCE FOUNDATION, WASHINGTON, D.C. (SPECIAL FOREIGN CURRENCY SCIENCE INFO. PROGRAM)	SFCSI-SMI-TT(YEAR-NO.)
NATIONAL SCIENCE FOUNDATION, WASHINGTON, D.C. (SPECIAL FOREIGN CURRENCY SCIENCE INFO. PROGRAM)	SFCSI-TVA-TT(YEAR-NO.)
NATIONAL SCIENCE FOUNDATION, WASHINGTON, D.C. (SPECIAL FOREIGN CURRENCY SCIENCE INFO. PROGRAM)	SFSCI-NLM-TT(YEAR-NO.)
NATIONAL SCIENCE FOUNDATION, WASHINGTON, D.C. (ANTARCTIC RESEARCH PROGRAM)	USARP-
NATIONAL SCIENCE FOUNDATION. DIV. OF SCIENTIFIC PERSONNEL AND EDUCATION, WASHINGTON, D.C.	SPE-
NATIONAL SCIENTIFIC LABS., INC., WASHINGTON, D.C.	NSL-
NATIONAL SECURITY AGENCY, WASH., D.C. (SPEC. REPT.)	COSA-SR-(YR.)
NATIONAL SECURITY RESOURCES BOARD, WASHINGTON, D.C.	NSRB-
NATIONAL SEMICONDUCTOR CORP., DANBURY, CONN.	NSC-
NATIONAL SEVERE STORMS LAB., NORMAN, OKLA.	ERL-NSSL-
NATIONAL SEVERE STORMS LAB., NORMAN, OKLA.	ERLTM-NSSL-
NATIONAL SEVERE STORMS LAB., NORMAN, OKLA.	ESSA-TM-ERLTM-NSSL-
NATIONAL SEVERE STORMS LAB., NORMAN, OKLA.	IERTM-NSSL-
NATIONAL SEVERE STORMS LAB., NORMAN, OKLA.	NOAA-TM-ERL-NSSL-
NATIONAL SEVERE STORMS LAB., NORMAN, OKLA.	NSSL-
NATIONAL SEVERE STORMS LAB., NORMAN, OKLA.	RLTM-NSSL-
NATIONAL SONIC BOOM EVALUATION OFFICE, ARLINGTON, VA.	NSBEO-
NATIONAL SPECTROGRAPHIC LABS., INC., CLEVELAND	NSL-
NATIONAL SPECTROGRAPHIC LABS., INC., CLEVELAND	NSL-LAB-
NATIONAL SPECTROGRAPHIC LABS., INC., CLEVELAND	NSL-TSR-
NATIONAL TECHNICAL INFORMATION SERVICE, SPRINGFIELD, VA. (ACCESSION NUMBER FOR REPORTS FROM OTHER COMMERCE DEPT. AGENCIES. USED 1971-)	COM-(YEAR)-(5 DIGITS)
NATIONAL TECHNICAL INFORMATION SERVICE, SPRINGFIELD, VA. (ACCESSION NUMBER)	PB-(6 DIGITS)
NATIONAL TECHNICAL INFORMATION SERVICE, SPRINGFIELD, VA. (ACCESSION NUMBER FOR TRANSLATIONS)	TT-(YEAR)-(5 DIGITS)
NATIONAL TRANSPORTATION SAFETY BOARD, WASHINGTON, DC	FILE-(NUMBER)-
NATIONAL TRANSPORTATION SAFETY BOARD, WASHINGTON, DC	NTSB-AAR-(YEAR)-
NATIONAL TRANSPORTATION SAFETY BOARD, WASHINGTON, DC (BRIEFS OF ACCIDENTS)	NTSB-BA-
NATIONAL TRANSPORTATION SAFETY BOARD, WASHINGTON, DC	SB-(YEAR)-
NATIONAL TRANSPORTATION SAFETY BOARD, WASHINGTON, DC	SS-H-
NATIONAL TRANSPORTATION SAFETY BOARD, WASHINGTON, DC	SS-R-
NATIONAL TRANSPORTATION SAFETY BOARD, WASHINGTON, DC	SS-R/H-
NATIONAL TRANSPORTATION SAFETY BOARD. BUREAU OF AVIATION SAFETY, WASHINGTON, D.C.	NTSB-AAS-(YEAR)-
NATIONAL UNION RADIO CORP., HATBORO, PENNA.	NURC-
NATIONAL UNION RADIO CORP., ORANGE, N.J.	NURC-
NATIONAL WATER COMMISSION, ARLINGTON, VA.	NWC-L-(YEAR)-
NATIONAL WATER COMMISSION, ARLINGTON, VA.	NWC-SBS-(YEAR)-
NATIONAL WEATHER RECORDS CENTER, ASHEVILLE, N.C.	ESD-TM-
NATIONAL WEATHER RECORDS CENTER, ASHEVILLE, N.C.	RAA-RD-(YEAR)-
NATIONAL WEATHER SATELLITE CENTER, WASHINGTON, D.C.	MSL-
NATIONAL WEATHER SERVICE. CENTRAL REGION, KANSAS CITY, MO.	ESSA-TM-CRTM-
NATIONAL WEATHER SERVICE. CENTRAL REGION, KANSAS CITY, MO.	ESSA-TM-WBTM-CR-
NATIONAL WEATHER SERVICE. CENTRAL REGION, KANSAS CITY, MO.	NOAA-TM-NWS-CR-
NATIONAL WEATHER SERVICE. CENTRAL REGION, KANSAS CITY, MO.	NWS-CR-
NATIONAL WEATHER SERVICE. CENTRAL REGION, KANSAS CITY, MO.	WBTM-CR-
NATIONAL WEATHER SERVICE. EASTERN REGION, GARDEN CITY, N.Y.	ERTM-
NATIONAL WEATHER SERVICE. EASTERN REGION, GARDEN CITY, N.Y.	ESSA-WBTM-ER-

NATIONAL WEATHER SERVICE. EASTERN REGION

NATIONAL WEATHER SERVICE. EASTERN REGION, GARDEN CITY, N.Y.	NOAA-TM-NWS-ER-
NATIONAL WEATHER SERVICE. EASTERN REGION, GARDEN CITY, N.Y.	NWS-ER-
NATIONAL WEATHER SERVICE. EASTERN REGION. GARDEN CITY, N.Y.	WBTM-ER-
NATIONAL WEATHER SERVICE. ENG. DIV., SILVER SPRING, MD	ESSA-TM-WBTM-ENG-
NATIONAL WEATHER SERVICE. ENG. DIV., SILVER SPRING, MD	NOAA-TM-NWS-ENG-
NATIONAL WEATHER SERVICE. ENG. DIV., SILVER SPRING, MD	NWS-ENG-
NATIONAL WEATHER SERVICE. ENG. DIV., SILVER SPRING, MD	WBTM-ENG-
NATIONAL WEATHER SERVICE. OFFICE OF HYDROLOGY, SILVER SPRING, MD.	ESSA-TR-WB-
NATIONAL WEATHER SERVICE. OFFICE OF HYDROLOGY, SILVER SPRING, MD.	NOAA-NWS-
NATIONAL WEATHER SERVICE. OFFICE OF HYDROLOGY, SILVER SPRING, MD.	NOAA-TM-NWS-HYDRO-
NATIONAL WEATHER SERVICE. OFFICE OF HYDROLOGY, SILVER SPRING, MD.	NOAA-TR-WB-
NATIONAL WEATHER SERVICE. OFFICE OF HYDROLOGY, SILVER SPRING, MD.	NWS-HYDRO-
NATIONAL WEATHER SERVICE. OFFICE OF HYDROLOGY, SILVER SPRING, MD.	WBTM-HYDRO-
NATIONAL WEATHER SERVICE. PACIFIC REGION, HONOLULU	ESSA-TM-WBTM-PR-
NATIONAL WEATHER SERVICE. PACIFIC REGION, HONOLULU	NOAA-TM-NWS-PR-
NATIONAL WEATHER SERVICE. PACIFIC REGION, HONOLULU	NWS-PR-
NATIONAL WEATHER SERVICE. PACIFIC REGION, HONOLULU	WB PR-
NATIONAL WEATHER SERVICE. PACIFIC REGION, HONOLULU	WBTM-PR-
NATIONAL WEATHER SERVICE. SOUTHERN REGION, FT. WORTH, TEX.	ESSA-TM-SRTM-
NATIONAL WEATHER SERVICE. SOUTHERN REGION, FT. WORTH, TEX.	ESSA-WBTM-SR-
NATIONAL WEATHER SERVICE. SOUTHERN REGION, FT. WORTH, TEX.	NOAA-TM-NWS-SR-
NATIONAL WEATHER SERVICE. SOUTHERN REGION, FT. WORTH, TEX.	NWS-SR-
NATIONAL WEATHER SERVICE. SOUTHERN REGION, FT. WORTH, TEX.	SRTM-
NATIONAL WEATHER SERVICE. SOUTHERN REGION, FT. WORTH, TEX.	WB SR-
NATIONAL WEATHER SERVICE. SOUTHERN REGION, FT. WORTH, TEX.	WBTM-SR-
NATIONAL WEATHER SERVICE. SPACE OPERATIONS SUPPORT DIV., SILVER SPRING, MD.	ESSA-TM-WBTM-SOS-
NATIONAL WEATHER SERVICE. SPACE OPERATIONS SUPPORT DIV., SILVER SPRING, MD.	NOAA-TM-NMS-SOS-
NATIONAL WEATHER SERVICE. SPACE OPERATIONS SUPPORT DIV., SILVER SPRING, MD.	NOAA-TM-NWS-SOS-
NATIONAL WEATHER SERVICE. SPACE OPERATIONS SUPPORT DIV., SILVER SPRING, MD.	NWS-SOS-
NATIONAL WEATHER SERVICE. SPACE OPERATIONS SUPPORT DIV., SILVER SPRING, MD.	WBTM-SOS-
NATIONAL WEATHER SERVICE. SYSTEMS PLANS AND DESIGN DIV., SILVER SPRING, MD.	ESSA-WBTM-SPDD-
NATIONAL WEATHER SERVICE. SYSTEMS PLANS AND DESIGN DIV., SILVER SPRING, MD.	NOAA-TM-NWS-SPDD-
NATIONAL WEATHER SERVICE. SYSTEMS PLANS AND DESIGN DIV., SILVER SPRING, MD.	NWS-SPDD-
NATIONAL WEATHER SERVICE. SYSTEMS PLANS AND DESIGN DIV., SILVER SPRING, MD.	WBTM-SPDD-
NATIONAL WEATHER SERVICE. TECHNIQUES DEVELOPMENT LAB., SILVER SPRING, MD.	ESSA-TM-WBTM-TDL-
NATIONAL WEATHER SERVICE. TECHNIQUES DEVELOPMENT LAB., SILVER SPRING, MD.	ESSA-WBTM-TDL-
NATIONAL WEATHER SERVICE. TECHNIQUES DEVELOPMENT LAB., SILVER SPRING, MD.	NOAA-TM-NWS-TDL-
NATIONAL WEATHER SERVICE. TECHNIQUES DEVELOPMENT LAB., SILVER SPRING, MD.	NWS-TDL-
NATIONAL WEATHER SERVICE. TECHNIQUES DEVELOPMENT LAB., SILVER SPRING, MD.	WBTM-TDL-
NATIONAL WEATHER SERVICE. TEST AND EVALUATION LAB., STERLING, VA.	ESSA-WBTM-T+EL-
NATIONAL WEATHER SERVICE. TEST AND EVALUATION LAB., STERLING, VA.	NOAA-TM-NWS-T+EL-
NATIONAL WEATHER SERVICE. TEST AND EVALUATION LAB., STERLING, VA.	NWS-T+EL-
NATIONAL WEATHER SERVICE. TEST AND EVALUATION LAB., STERLING, VA.	WBTM-T+EL-
NATIONAL WEATHER SERVICE. WEATHER ANALYSIS + PREDICTION DIV., SILVER SPRING, MD. (NOTES TO FORECASTERS)	ESSA-TM-WBTM-FCST-
NATIONAL WEATHER SERVICE. WEATHER ANALYSIS + PREDICTION DIV., SILVER SPRING, MD. (NOTES TO FORECASTERS)	NOAA-TM-NWS-FCST-
NATIONAL WEATHER SERVICE. WEATHER ANALYSIS + PREDICTION DIV., SILVER SPRING, MD. (NOTES TO FORECASTERS)	NWS-FCST-
NATIONAL WEATHER SERVICE. WEATHER ANALYSIS + PREDICTION DIV., SILVER SPRING, MD. (NOTES TO FORECASTERS)	WBTM-FCST-
NATIONAL WEATHER SERVICE. WEATHER ANALYSIS + PREDICTION DIV., SILVER SPRING, MD. (NOTES TO FORECASTERS)	WBTN-FCST-
NATIONAL WEATHER SERVICE. WESTERN REGION, SALT LAKE CITY	ESSA-TM-WBTM-WR-
NATIONAL WEATHER SERVICE. WESTERN REGION, SALT LAKE CITY	ESSA-TM-WRTM-
NATIONAL WEATHER SERVICE. WESTERN REGION, SALT LAKE CITY	ESSA-WBTM-WR-
NATIONAL WEATHER SERVICE. WESTERN REGION, SALT LAKE CITY	NOAA-TM-NWSTM-WR-
NATIONAL WEATHER SERVICE. WESTERN REGION, SALT LAKE CITY	NWSTM-WR-
NATIONAL WEATHER SERVICE. WESTERN REGION, SALT LAKE CITY	WBTM-WR-
NAUGATUCK CHEM. DIV., UNITED STATES RUBBER CO., CONN.	NC-
NAVAL ACADEMY, ANNAPOLIS	E-(YEAR)-
NAVAL ACADEMY, ANNAPOLIS	USNA-
NAVAL ACADEMY, ANNAPOLIS	USNA-S-
NAVAL ACADEMY, ANNAPOLIS	USNA-TSPR-
NAVAL ACADEMY, ANNAPOLIS. DEPT. OF ENGINEERING	USNA-E-(YEAR)-
NAVAL ACADEMY, ANNAPOLIS. MICHELSON PHYSICS LAB.	TR-C-
NAVAL AERO. LAB., NAVAL AIR STA., BANANA RIVER, FLA.	NAL-
NAVAL AERONAUTICAL ROCKET LAB., LAKE DENMARK, DOVER, N.J. (REF. 1)	(NUMBER)-(YEAR)
NAVAL AEROSPACE MEDICAL INST., PENSACOLA, FLA.	MAMI-
NAVAL AEROSPACE MEDICAL INST., PENSACOLA, FLA.	NAMI-
NAVAL AEROSPACE MEDICAL INST., PENSACOLA, FLA.	USAARL-(YEAR)-
NAVAL AEROSPACE MEDICAL INST., PENSACOLA, FLA.	USARRL-(YEAR)-
NAVAL AEROSPACE MEDICAL RESEARCH LAB., PENSACOLA, FLA.	NAMRL-
NAVAL AEROSPACE RECOVERY FACILITY, EL CENTRO, CALIF.	NARF-
NAVAL AEROSPACE RECOVERY FACILITY, EL CENTRO, CALIF.	NAVAERORECOVFAC-
NAVAL AIR ADVANCED TRNG. COMMAND, JACKSONVILLE, FLA.	CNAADTRA-
NAVAL AIR ATTACHE (REPORT)	NAAR-
NAVAL AIR DEVELOPMENT CENTER, JOHNSVILLE, PENNA.	ADC-
NAVAL AIR DEVELOPMENT CENTER, JOHNSVILLE, PENNA.	ADS-
NAVAL AIR DEVELOPMENT CENTER, JOHNSVILLE, PENNA.	NAD-
NAVAL AIR DEVELOPMENT CENTER, JOHNSVILLE, PENNA.	NAD-AR-
NAVAL AIR DEVELOPMENT CENTER, JOHNSVILLE, PENNA.	NADC-
NAVAL AIR DEVELOPMENT CENTER, JOHNSVILLE, PENNA.	NADC-(LETTERS)-
NAVAL AIR DEVELOPMENT CENTER, JOHNSVILLE, PENNA.	NADC-AE-
NAVAL AIR DEVELOPMENT CENTER, JOHNSVILLE, PENNA.	NADC-AER-
NAVAL AIR DEVELOPMENT CENTER, JOHNSVILLE, PENNA.	NADC-MA-
NAVAL AIR DEVELOPMENT CENTER, JOHNSVILLE, PENNA.	NADC-R-
NAVAL AIR DEVELOPMENT CENTER, JOHNSVILLE, PENNA.	NADC-SY-
NAVAL AIR DEVELOPMENT CENTER, JOHNSVILLE, PENNA.	NAD-EL-
NAVAL AIR DEVELOPMENT CENTER, JOHNSVILLE, PENNA.	NADEVCEN-
NAVAL AIR DEVELOPMENT CENTER, JOHNSVILLE, PENNA.	NAD-R-
NAVAL AIR DEVELOPMENT CENTER, JOHNSVILLE, PENNA.	NADS-
NAVAL AIR DEVELOPMENT CENTER, JOHNSVILLE, PENNA.	NAMU-

NAVAL AIR DEVELOPMENT CENTER, JOHNSVILLE, PENNA.	WAEC-
NAVAL AIR DEVELOPMENT CENTER. AERO ELECTRONIC TECHNOLOGY DEPT., WARMINSTER, PENNA.	NADC-AE-
NAVAL AIR DEVELOPMENT CENTER. AERO ELECTRONICS TECHNOLOGY DEPT., JOHNSVILLE, PENNA.	AE-
NAVAL AIR DEVELOPMENT CENTER. AERO MATERIALS DEPT., WARMINSTER, PENNA.	NADC-MA-
NAVAL AIR DEVELOPMENT CENTER. AERO MECHANICS DEPT., JOHNSVILLE, PENNA.	AM-
NAVAL AIR DEVELOPMENT CENTER. AERO MECHANICS DEPT., JOHNSVILLE, PENNA.	NADC-AM-
NAVAL AIR DEVELOPMENT CENTER. AERO MECHANICS DEPT., WARMINSTER, PENNA.	NADC-AM-
NAVAL AIR DEVELOPMENT CENTER. AERO STRUCTURES DEPT., JOHNSVILLE, PENNA.	NADC-ST-
NAVAL AIR DEV.CTR. AERO. COMPUTER LAB., JOHNSVILLE, PA.	AC-
NAVAL AIR DEV.CTR. AERO. COMPUTER LAB., JOHNSVILLE, PA.	AER-REP-EL-3-
NAVAL AIR DEV.CTR. AERO. COMPUTER LAB., JOHNSVILLE, PA.	NADC AC-
NAVAL AIR DEVELOPMENT CENTER. AERONAUTICAL ELECTRONIC AND ELECTRIC LAB., JOHNSVILLE, PENNA.	ADC-EL-
NAVAL AIR DEVELOPMENT CENTER. AERONAUTICAL ELECTRONIC AND ELECTRIC LAB., JOHNSVILLE, PENNA.	ADC-EL-T-
NAVAL AIR DEVELOPMENT CENTER. AERONAUTICAL ELECTRONIC AND ELECTRIC LAB., JOHNSVILLE, PENNA.	ADS-EL-
NAVAL AIR DEVELOPMENT CENTER. AERONAUTICAL ELECTRONIC AND ELECTRIC LAB., JOHNSVILLE, PENNA.	AEEL-
NAVAL AIR DEVELOPMENT CENTER. AERONAUTICAL ELECTRONIC AND ELECTRIC LAB., JOHNSVILLE, PENNA.	EL-
NAVAL AIR DEVELOPMENT CENTER. AERONAUTICAL ELECTRONIC AND ELECTRIC LAB., JOHNSVILLE, PENNA.	NADC EL-
NAVAL AIR DEV. CTR. AERO. INSTRS. LAB., JOHNSVILLE, PA.	AI-
NAVAL AIR DEV. CTR. AERO. INSTRS. LAB., JOHNSVILLE, PA.	IDS-
NAVAL AIR DEV. CTR. AERO. INSTRS. LAB., JOHNSVILLE, PA.	INSTR-
NAVAL AIR DEV. CTR. AERO. INSTRS. LAB., JOHNSVILLE, PA.	NADC AI-
NAVAL AIR DEV. CTR. AERO. INSTRS. LAB., JOHNSVILLE, PA.	NAES INSTR-
NAVAL AIR DEVELOPMENT CENTER. AERONAUTICAL PHOTOGRAPHIC EXPERIMENT LAB.	AP-
NAVAL AIR DEVELOPMENT CENTER. AERONAUTICAL PHOTOGRAPHIC EXPERIMENT LAB.	NADC-AP-
NAVAL AIR DEVELOPMENT CENTER. AEROSPACE CREW EQUIPMENT DEPT., JOHNSVILLE, PENNA.	NADC-AC-
NAVAL AIR DEVELOPMENT CENTER. AEROSPACE MEDICAL RESEARCH DEPT., JOHNSVILLE, PENNA.	MR-
NAVAL AIR DEVELOPMENT CENTER. AEROSPACE MEDICAL RESEARCH DEPT., JOHNSVILLE, PENNA.	NADC-MR-
NAVAL AIR DEVELOPMENT CENTER. AIR WARFARE RESEARCH DEPT., JOHNSVILLE, PENNA. (USED AFTER JULY 1965)	AW-
NAVAL AIR DEVELOPMENT CENTER. AIR WARFARE RESEARCH DEPT., JOHNSVILLE, PENNA. (MEMORANDUM)	AWRDM-
NAVAL AIR DEVELOPMENT CENTER. AIR WARFARE RESEARCH DEPT., JOHNSVILLE, PENNA.	NADC-AW-
NAVAL AIR DEVELOPMENT CENTER. AIR WARFARE RESEARCH DEPT., JOHNSVILLE, PENNA.	NADC-WR-
NAVAL AIR DEVELOPMENT CENTER. AIR WARFARE RESEARCH DEPT., JOHNSVILLE, PENNA.	WR-
NAVAL AIR DEVELOPMENT CENTER. ANTI-SUBMARINE WARFARE LAB., JOHNSVILLE, PENNA. (USED BEFORE JULY 1965)	AW-
NAVAL AIR DEVELOPMENT CENTER. ANTI-SUBMARINE WARFARE LAB., JOHNSVILLE, PENNA.	NADC-ASW-
NAVAL AIR DEVELOPMENT CENTER. ANTI-SUBMARINE WARFARE LAB., JOHNSVILLE, PENNA.	NADC-AW-
NAVAL AIR DEVELOPMENT CENTER. AVIATION ARMAMENT LAB., JOHNSVILLE, PENNA.	AAL-
NAVAL AIR DEVELOPMENT CENTER. AVIATION ARMAMENT LAB., JOHNSVILLE, PENNA.	AR-
NAVAL AIR DEVELOPMENT CENTER. AVIATION ARMAMENT LAB., JOHNSVILLE, PENNA.	NADC AR-
NAVAL AIR DEVELOPMENT CENTER. AVIATION MEDICAL ACCELERATION LAB., JOHNSVILLE, PENNA.	MA-
NAVAL AIR DEVELOPMENT CENTER. AVIATION MEDICAL ACCELERATION LAB., JOHNSVILLE, PENNA.	NADC MA-
NAVAL AIR DEVELOPMENT CENTER. AVIATION MEDICAL ACCELERATION LAB., JOHNSVILLE, PENNA.	NADC-MA-L-
NAVAL AIR DEVELOPMENT CENTER. AVIATION MEDICAL ACCELERATION LAB., JOHNSVILLE, PENNA.	NADC-ML-
NAVAL AIR DEV. CTR. ENG. DEV. LAB., JOHNSVILLE, PENNA.	NADC-ED-
NAVAL AIR DEVELOPMENT CENTER. PILOTLESS AIRCRAFT DEVELOPMENT LAB., JOHNSVILLE, PENNA.	PA-
NAVAL AIR DEVELOPMENT CENTER. PILOTLESS AIRCRAFT DEVELOPMENT LAB., JOHNSVILLE, PENNA.	PADL-
NAVAL AIR DEVELOPMENT CENTER. PILOTLESS AIRCRAFT DEVELOPMENT LAB., JOHNSVILLE, PENNA.	R-(NO.)-(YR.)
NAVAL AIR DEVELOPMENT CENTER. SYSTEMS ANALYSIS AND ENGINEERING DEPT., JOHNSVILLE, PENNA.	NADC-SD-
NAVAL AIR DEVELOPMENT CENTER. SYSTEMS PROJECT DEPT., JOHNSVILLE, PENNA.	SY-
NAVAL AIR DEV. SQUADRON 5, CHINA LAKE, CALIF.	OP/
NAVAL AIR ENGINEERING CENTER, PHILADELPHIA	NAEC-ASL-
NAVAL AIR ENG. CTR. AERO. ENGINE LAB., PHILADELPHIA	NAEC-AEL-
NAVAL AIR ENGINEERING CENTER. AERONAUTICAL MATERIALS LAB., PHILADELPHIA	AML-
NAVAL AIR ENGINEERING CENTER. AERONAUTICAL MATERIALS LAB., PHILADELPHIA	NAEC-AML-
NAVAL AIR ENGINEERING CENTER. AEROSPACE CREW EQUIPMENT LAB., PHILADELPHIA	NAEC-ACEL-
NAVAL AIR ENGINEERING CENTER. AEROSPACE CREW EQUIPMENT LAB., PHILADELPHIA	TED-NAM-AE-
NAVAL AIR ENGINEERING CTR. ENG. DEPT., PHILADELPHIA	NAEC-ENG-
NAVAL AIR ENGINEERING CENTER. NAVAL AIR ENGINEERING LAB., PHILADELPHIA	NAEC-NAEL-
NAVAL AIR ENGINEERING CENTER. NAVAL AIR ENGINEERING LAB., PHILADELPHIA	NAEL-ENG-
NAVAL AIR ENG. FACILITY. ENG. DEPT., PHILADELPHIA	NAEF-ENG-
NAVAL AIR EXPERIMENTAL STATION, PHILADELPHIA	NAES-
NAVAL AIR EXPERIMENTAL STATION, PHILADELPHIA	NAXSTA-
NAVAL AIR EXPERIMENTAL STA. AERO. ENGINE LAB., PHILA.	AEL-
NAVAL AIR EXPERIMENTAL STA. AERO. ENGINE LAB., PHILA.	F23-
NAVAL AIR EXPERIMENTAL STA. AERO. ENGINE LAB., PHILA.	NAV AEL-
NAVAL AIR EXPTL. STA. AERO. INSTRS. LAB., PHILA.	AIL-
NAVAL AIR EXPTL. STA. AERO. INSTRS. LAB., PHILA.	NAES AIL-
NAVAL AIR EXPTL. STA. AERO. INSTRS. LAB., PHILA.	NAES AIL (NUMBER)-
NAVAL AIR EXPTL. STA. AERO. MATERIALS LAB., PHILA.	AML-NAM-
NAVAL AIR EXPTL. STA. AERO. MATERIALS LAB., PHILA.	AML-NAM-AE-
NAVAL AIR EXPTL. STA. AERO. MATERIALS LAB., PHILA.	NAMC AML-
NAVAL AIR EXPTL.STA. AERO.MEDICAL EQUIP.LAB., PHILA.	XG-
NAVAL AIR EXPERIMENTAL STATION. AERONAUTICAL PHOTOGRAPHIC EXPERIMENTAL LAB., PHILADELPHIA	APEL-
NAVAL AIR EXPERIMENTAL STATION. AERONAUTICAL PHOTOGRAPHIC EXPERIMENTAL LAB., PHILADELPHIA	APEL-(NO.)-(YR.)
NAVAL AIR EXPERIMENTAL STATION. AERONAUTICAL RADIO AND RADAR LAB., PHILADELPHIA	ARRL-
NAVAL AIR EXPERIMENTAL STATION. AERONAUTICAL RADIO AND RADAR LAB., PHILADELPHIA	NAES ARRL-

NAVAL AIR EXPTL. STA. AERO. STRUCTURES LAB.

NAVAL AIR EXPTL. STA. AERO. STRUCTURES LAB., PHILA.	ASL-
NAVAL AIR EXPTL. STA. AERO. STRUCTURES LAB., PHILA.	ASL-NAM-
NAVAL AIR EXPTL. STA. AERO. STRUCTURES LAB., PHILA.	ASL-NAM-AD-
NAVAL AIR MATERIAL CENTER, PHILADELPHIA	NAM-
NAVAL AIR MATERIAL CENTER, PHILADELPHIA	NAMC-
NAVAL AIR MATERIAL CENTER, PHILADELPHIA	NAMC-AML-AE-
NAVAL AIR MATERIAL CENTER, PHILADELPHIA	NAVAIRMATCEN-ASL-
(SEE ALSO LATER NAME: NAVAL AIR ENGINEERING CTR.)	
NAVAL AIR MATERIAL CENTER. AERO. ENGINE LAB., PHILA.	NAMC-AEL-
NAVAL AIR MATERIAL CENTER. AERONAUTICAL INSTRUMENTS LAB., PHILADELPHIA	NAMC-AIL-
NAVAL AIR MATERIAL CENTER. AERONAUTICAL MATERIALS LAB., PHILADELPHIA	NAM-AML-
NAVAL AIR MATERIAL CENTER. AERONAUTICAL MATERIALS LAB., PHILADELPHIA	NAMC-AML-RS-
NAVAL AIR MATERIAL CENTER. AERONAUTICAL PHOTOGRAPHIC EXPERIMENTAL LAB., PHILADELPHIA	NAMC-APEL-
NAVAL AIR MATERIAL CENTER. AERONAUTICAL STRUCTURES LAB., PHILADELPHIA	NAMC-ASL-
NAVAL AIR MATERIAL CENTER. AIR CREW EQUIPMENT LAB., PHILADELPHIA	NAMC-ACEL-
NAVAL AIR MISSILE TEST CENTER, POINT MUGU, CALIF.	GSR-
NAVAL AIR MISSILE TEST CENTER, POINT MUGU, CALIF.	MTC-
NAVAL AIR MISSILE TEST CENTER, POINT MUGU, CALIF.	NAMTC-
NAVAL AIR MISSILE TEST CENTER, POINT MUGU, CALIF.	NAMTC-MR-
NAVAL AIR MISSILE TEST CENTER, POINT MUGU, CALIF.	NAMTC-PR-
NAVAL AIR MISSILE TEST CENTER, POINT MUGU, CALIF.	NAMTC-TM-
NAVAL AIR MISSILE TEST CENTER, POINT MUGU, CALIF.	NAMTC-TR-
NAVAL AIR MISSILE TEST CENTER, POINT MUGU, CALIF.	PAU-
NAVAL AIR MISSILE TEST CENTER, POINT MUGU, CALIF.	PP-(NOS.)-
NAVAL AIR MISSILE TEST CENTER. COMPONENT TEST DEPT., POINT MUGU, CALIF.	CTD-
NAVAL AIR MISSILE TEST CENTER. LAUNCHER DIV., POINT MUGU, CALIF. (MEMORANDUM REPORT)	LDMR-
NAVAL AIR MISSILE TEST CENTER. MISSILE TEST DEPT., POINT MUGU, CALIF. (MEMORANDUM REPORT)	MTDMR-
NAVAL AIR PROPULSION TEST CENTER, PHILADELPHIA	DS-(YEAR)-
NAVAL AIR PROPULSION TEST CENTER. AERONAUTICAL ENGINE DEPT., PHILADELPHIA	NAPTC-AED-
NAVAL AIR PROPULSION TEST CENTER. AERONAUTICAL TURBINE DEPT., TRENTON	NAPTC-ATD-
NAVAL AIR ROCKET TEST STATION, DOVER, N.J.	NARTS-
NAVAL AIR ROCKET TEST STATION, LAKE DENMARK, N.J.	ARTS-
NAVAL AIR ROCKET TEST STATION, LAKE DENMARK, N.J.	NARTS-
NAVAL AIR ROCKET TEST STATION, LAKE DENMARK, N.J.	NARTS-TN-
NAVAL AIR SPEC. WEAPONS FACILITY, KIRTLAND AFB, N.MEX	NASWF-
NAVAL AIR SPEC. WEAPONS FACILITY, KIRTLAND AFB, N.MEX	NASWF (LTRS, LTRS)-
NAVAL AIR SPEC. WEAPONS FACILITY, KIRTLAND AFB, N.MEX	NASWF-(LTRS,LTRS-LTRS)-
NAVAL AIR SPEC. WEAPONS FACILITY, KIRTLAND AFB, N.MEX	SWF-
NAVAL AIR STATION, EL CENTRO, CALIF.	ELC-
NAVAL AIR STATION, LAKEHURST, N.J.	NAS-
NAVAL AIR STATION, NORTH ISLAND, SAN DIEGO, CALIF.	NAV-AS-
NAVAL AIR STATION, NORTH ISLAND, SAN DIEGO, CALIF.	NAV AS/SD MISC.
NAVAL AIR STATION, PATUXENT RIVER, MD.	WST-
NAVAL AIR STATION, PENSACOLA, FLA.	DA-
NAVAL AIR STATION, QUONSET POINT, R.I.	NATU-
NAVAL AIR STATION, QUONSET POINT, R.I. (INTERNAL TECHNICAL MEMORANDUM)	NATU-ITM-(NO.)-(YR.)
NAVAL AIR STATION. ELECTRONICS STDS. LAB.,NORFOLK, VA	ESL-
NAVAL AIR STATION. FLEET WEATHER CTR.,SAN DIEGO, CAL	NAV-FWC-
NAVAL AIR SYSTEMS COMMAND, WASHINGTON, D.C.	AIR-
NAVAL AIR SYSTEMS COMMAND, WASHINGTON, D.C.	NASC-
NAVAL AIR SYSTEMS COMMAND, WASHINGTON, D.C.	NAVAIR-(NO.)-20(LTRS.)-
NAVAL AIR SYSTEMS COMMAND, WASHINGTON, D.C.	NAVAIR-(NUMBERS,LTRS.)-
NAVAL AIR SYSTEMS COMMAND, WASHINGTON, D.C.	NAVASC-
NAVAL AIR SYSTEMS COMMAND. ADVANCED SYSTEMS DIV., WASHINGTON, D.C.	A-(NUMBER-YEAR)-
NAVAL AIR SYSTEMS COMMAND. PROFESSIONAL DEVELOPMENT CENTER, WASHINGTON, D.C.	SP-(YEAR)-
NAVAL AIR TECHNICAL TRAINING COMMAND, MEMPHIS	CNATT-
NAVAL AIR TEST CENTER, PATUXENT RIVER, MD.	AT-
NAVAL AIR TEST CENTER, PATUXENT RIVER, MD.	CT-
NAVAL AIR TEST CENTER, PATUXENT RIVER, MD.	FT-
NAVAL AIR TEST CENTER, PATUXENT RIVER, MD.	FT31-
NAVAL AIR TEST CENTER, PATUXENT RIVER, MD.	FT-TM-(NO.)-(YEAR)
NAVAL AIR TEST CENTER, PATUXENT RIVER, MD.	NATC-
NAVAL AIR TEST CENTER, PATUXENT RIVER, MD.	NATC-FT (NUMBER)-
NAVAL AIR TEST CENTER, PATUXENT RIVER, MD.	NATC-FT-DIR-
NAVAL AIR TEST CENTER, PATUXENT RIVER, MD.	NATC-ST-IR
NAVAL AIR TEST CENTER, PATUXENT RIVER, MD.	NATC-TR-
NAVAL AIR TEST CENTER, PATUXENT RIVER, MD.	NATC-TSD-IR
NAVAL AIR TEST CENTER, PATUXENT RIVER, MD.	NATC-WST-(NUMBER-YEAR)
NAVAL AIR TEST CENTER, PATUXENT RIVER, MD.	PTR-
NAVAL AIR TEST CENTER, PATUXENT RIVER, MD.	ST-
NAVAL AIR TEST CENTER, PATUXENT RIVER, MD.	ST(NO.)-
NAVAL AIR TEST CENTER, PATUXENT RIVER, MD.	TED BIS-
NAVAL AIR TEST CENTER, PATUXENT RIVER, MD.	TED PTR-
NAVAL AIR TEST CENTER, PATUXENT RIVER, MD. (TEST, EXPERIMENTAL AND DEVELOPMENT)	TED-PTR-(LTRS.)-
NAVAL AIR TEST CENTER, PATUXENT RIVER, MD.	TPT-
NAVAL AIR TEST CENTER, PATUXENT RIVER, MD.	TT-
NAVAL AIR TEST CENTER, PATUXENT RIVER, MD.	WS(NO.)-
NAVAL AIR TEST CENTER, PATUXENT RIVER, MD.	WST-IR-
NAVAL AIR TEST CENTER. ARMAMENT TEST DIV., PATUXENT RIVER, MD.	NATC-AT-

NAVAL AIR TEST CENTER. ELECTRONICS TEST DIV., PATUXENT RIVER, MD.	ET-
NAVAL AIR TEST CENTER. ELECTRONICS TEST DIV., PATUXENT RIVER, MD.	NATC ET-
NAVAL AIR TEST FACILITY. SHIP INSTALLATIONS, LAKEHURST, N.J.	NATF-E-
NAVAL AIR TEST FACILITY. SHIP INSTALLATIONS, LAKEHURST, N.J.	NATF-EN-
NAVAL AIR TEST FACILITY. SHIP INSTALLATIONS, LAKEHURST, N.J.	NATF-SI-R(NUMBER)
NAVAL AIR TRAINING COMMAND, PENSACOLA, FLA.	CNATRA-
NAVAL AIR TRAINING COMMAND, PENSACOLA, FLA.	NAT-
NAVAL AIR TURBINE TEST STATION, TRENTON	A-
NAVAL AIR TURBINE TEST STATION, TRENTON	ATTS-
NAVAL AIR TURBINE TEST STATION, TRENTON	NAD-CR-RDTR-
NAVAL AIR TURBINE TEST STATION, TRENTON	NAPTC-ATD-
NAVAL AIR TURBINE TEST STATION, TRENTON	NATTS-ATL-
NAVAL AIR TURBINE TEST STATION. AERONAUTICAL TURBINE LAB., TRENTON	NATTS-ATL-
NAVAL AIR TURBINE TEST STATION. AERONAUTICAL TURBINE LAB., TRENTON	NATTS-ATL-TN-
NAVAL AIRCRAFT FACTORY, PHILADELPHIA	NAF-
NAVAL AIRCRAFT FACTORY, PHILADELPHIA (SPECIFICATIONS)	NAF SPEC (LETTER)-
NAVAL AIRCRAFT FACTORY. INSTRS. DEV. SEC., PHILA.	IDS-
NAVAL AIRCRAFT FACTORY. INSTRS. DEV. SEC., PHILA.	NAF IDS-
NAVAL AIRCRAFT TORPEDO UNIT, QUONSET POINT, R.I.	NATU-
NAVAL AIRCRAFT TORPEDO UNIT, QUONSET POINT, R.I.	NAVAIRTORPU-TM-
NAVAL AIRSHIP TRAINING + EXPTL. COMMAND, LAKEHURST, NJ	NATEC-
NAVAL AMMUNITION + NET DEPOT, SEAL BEACH, CALIF.	NAND-
NAVAL AMMUNITION AND NET DEPOT. QUALITY CONTROL LAB., SEAL BEACH, CALIF.	QC/SB-
NAVAL AMMUNITION AND NET DEPOT. QUALITY EVALUATION LAB., SEAL BEACH, CALIF.	NAD-QE-
NAVAL AMMUNITION AND NET DEPOT. QUALITY EVALUATION LAB., SEAL BEACH, CALIF.	QE/SB-
NAVAL AMMUNITION DEPOT (ALL LOCATIONS)	NAD-
NAVAL AMMUNITION DEPOT, CRANE, IND.	MICRO-NOTES-
NAVAL AMMUNITION DEPOT, CRANE, IND.	NAC-CR-MICRO-NOTES-
NAVAL AMMUNITION DEPOT, CRANE, IND.	NAD-CRANE-
NAVAL AMMUNITION DEPOT, CRANE, IND.	NAD-CRANE-QE/C-(YR)-
NAVAL AMMUNITION DEPOT, CRANE, IND.	NAD-CR-MICRO-NOTES
NAVAL AMMUNITION DEPOT, CRANE, IND.	NAD-CR-QE/C-(YEAR)-
NAVAL AMMUNITION DEPOT, CRANE, IND.	NAD-CR-RDTR-
NAVAL AMMUNITION DEPOT, CRANE, IND.	NAD-CR-RDTR-
NAVAL AMMUNITION DEPOT, CRANE, IND.	NAD-QE/C-(YEAR)-
NAVAL AMMUNITION DEPOT, CRANE, IND.	NAD-RDTR-
NAVAL AMMUNITION DEPOT, CRANE, IND.	QETR-
NAVAL AMMUNITION DEPOT, CRANE, IND.	RDTR-
NAVAL AMMUNITION DEPOT, FALLBROOK, CALIF.	FB-
NAVAL AMMUNITION DEPOT, HAWTHORNE, NEV.	NADH-
NAVAL AMMUNITION DEPOT, HAWTHORNE, NEV.	NAD/HAW-
NAVAL AMMUNITION DEPOT. QUALITY CONTROL LAB., CRANE, IND.	QC/C-
NAVAL AMMUNITION DEPOT. QUALITY EVALUATION LAB., BANGOR, WASH.	QE/B-
NAVAL AMMUNITION DEPOT. QUALITY EVALUATION LAB., CONCORD, CALIF.	QE/CO-
NAVAL AMMUNITION DEPOT. QUALITY EVALUATION LAB., CONCORD, CALIF.	QE/CO-(YR.)-
NAVAL AMMUNITION DEPOT. QUALITY EVALUATION LAB., CRANE, IND.	QE/C-
NAVAL AMMUNITION DEPOT. QUALITY EVALUATION LAB., OAHU, HAWAII	NAD-QE-OH-
NAVAL AMMUNITION DEPOT. QUALITY EVALUATION LAB., OAHU, HAWAII	QE-OH-
NAVAL AMMUNITION DEPOT. QUALITY EVALUATION LAB., OAHU, HAWAII	QE/OH-
NAVAL AMMUNITION DEPOT. QUALITY EVALUATION LAB., PORTSMOUTH (ST. JULIENS CREEK), VA.	QE/SJ-
NAVAL APPLIED SCIENCE LAB., BROOKLYN	IRP-
NAVAL APPLIED SCIENCE LAB., BROOKLYN	NASL-
NAVAL APPLIED SCIENCE LAB., BROOKLYN	NASL-9400-(NUMBER)-PR-
NAVAL APPLIED SCIENCE LAB., BROOKLYN	NASL-(LETTERS)-
NAVAL APPLIED SCIENCE LAB., BROOKLYN	NASL-IED-(NO.)-TM-
NAVAL APPLIED SCIENCE LAB., BROOKLYN	NASL-INTEGRATION STDY-
NAVAL APPLIED SCIENCE LAB., BROOKLYN	NASL-IR-
NAVAL APPLIED SCIENCE LAB., BROOKLYN	NASL-PR-
NAVAL APPLIED SCIENCE LAB., BROOKLYN	NASL-TM-
NAVAL APPLIED SCIENCE LAB., BROOKLYN	NASL-VN-
NAVAL APPLIED SCIENCE LAB., BROOKLYN	NASL-VNG-
NAVAL AUXILIARY AIR STATION, CORPUS CHRISTI, TEX.	NAAS-
NAVAL AVIATION ENGINEERING SERVICE UNIT	NAESU-
NAVAL AVIATION INTEGRATED LOGISTIC SUPPORT CENTER, PATUXENT RIVER, MD.	NAILSC-ILS-
NAVAL AVIATION INTEGRATED LOGISTIC SUPPORT CENTER, PATUXENT RIVER, MD.	NAILSC-ILS-IR-
NAVAL AVIATION ORDNANCE TEST STATION, CHINCOTEAGUE, VA	NAOTS-
NAVAL AVIONICS FACILITY, INDIANAPOLIS	NAF-
NAVAL AVIONICS FACILITY, INDIANAPOLIS	NAFI-(NUMBER)-GR
NAVAL AVIONICS FACILITY, INDIANAPOLIS	NAFI-
NAVAL AVIONICS FACILITY, INDIANAPOLIS	NAFI-MRR-
NAVAL AVIONICS FACILITY, INDIANAPOLIS	NAFI-MTR-
NAVAL AVIONICS FACILITY, INDIANAPOLIS	NAFI-TP-
NAVAL AVIONICS FACILITY, INDIANAPOLIS	NAFI-TR-
NAVAL AVIONICS FACILITY, INDIANAPOLIS	NAF-MR-
NAVAL BOILER AND TURBINE LAB., PHILADELPHIA	NBTL-A-
NAVAL BOILER AND TURBINE LAB., PHILADELPHIA	NBTL-B-
NAVAL BOILER AND TURBINE LAB., PHILADELPHIA	NBTL-P-
NAVAL BOILER AND TURBINE LAB., PHILADELPHIA	NBTL-T-R-
NAVAL BOILER AND TURBINE LAB., PHILADELPHIA	NBTL-T-T-
NAVAL CENTRAL TORPEDO OFFICE	CTO-

NAVAL CIVIL ENGINEERING LAB., PORT HUENEME, CALIF. (TECHNICAL COMPILATION)	NCEL-C-
NAVAL CIVIL ENGINEERING LAB., PORT HUENEME, CALIF. (CONTRACT REPORT)	NCEL-CR-
NAVAL CIVIL ENGINEERING LAB., PORT HUENEME, CALIF. (TECHNICAL MEMORANDUM)	NCEL-M-
NAVAL CIVIL ENGINEERING LAB., PORT HUENEME, CALIF. (TECHNICAL NOTE)	NCEL-N-
NAVAL CIVIL ENGINEERING LAB., PORT HUENEME, CALIF. (TECHNICAL REPORT)	NCEL-R-
NAVAL CIVIL ENGINEERING LAB., PORT HUENEME, CALIF.	NCEL-STIR-
NAVAL CIVIL ENGINEERING LAB., PORT HUENEME, CALIF.	NCEL-TM-
NAVAL CIVIL ENGINEERING LAB., PORT HUENEME, CALIF.	NCEL-TN-
NAVAL CIVIL ENGINEERING LAB., PORT HUENEME, CALIF.	NCEL-TN-N-
NAVAL CIVIL ENGINEERING LAB., PORT HUENEME, CALIF.	NCEL-TR-
NAVAL CIVIL ENGINEERING LAB., PORT HUENEME, CALIF.	NCEL-TR-R-
NAVAL CIVIL ENGINEERING LAB., PORT HUENEME, CALIF.	NECL-TN-
NAVAL CIVIL ENGINEERING LAB., PORT HUENEME, CALIF.	NECL-TR-
NAVAL CIVIL ENGINEERING LAB., PORT HUENEME, CALIF.	TN-N-
NAVAL CIVIL ENGINEERING LAB., PORT HUENEME, CALIF.	TN-R-
NAVAL CIVIL ENGINEERING LAB., PORT HUENEME, CALIF.	TR-R-
NAVAL CIVIL ENGINEERING LAB., PORT HUENEME, CALIF.	Y-
NAVAL CIVIL ENGINEERING RESEARCH AND EVALUATION LAB., PORT HUENEME, CALIF.	M-009-
NAVAL CIVIL ENGINEERING RESEARCH AND EVALUATION LAB., PORT HUENEME, CALIF.	NAVCERELAB-
NAVAL CIVIL ENGINEERING RESEARCH AND EVALUATION LAB., PORT HUENEME, CALIF.	NCE-REL-
NAVAL CIVIL ENGINEERING RESEARCH AND EVALUATION LAB., PORT HUENEME, CALIF. (TECHNICAL MEMORANDUM)	NCEREL M-
NAVAL CIVIL ENGINEERING RESEARCH AND EVALUATION LAB., PORT HUENEME, CALIF.	NCEREL R-
NAVAL CIVIL ENGINEERING RESEARCH AND EVALUATION LAB., PORT HUENEME, CALIF. (TECHNICAL MEMORANDUM)	NCEREL TM-
NAVAL CIVIL ENGINEERING RESEARCH AND EVALUATION LAB., PORT HUENEME, CALIF. (TECHNICAL NOTE)	NCEREL TN-
NAVAL CLOTHING DEPOT, BROOKLYN (RESEARCH PROJECTS)	NCD RP-
NAVAL COMMAND SYSTEMS SUPPORT ACTIVITY, WASHINGTON, DC	COSSACT-
NAVAL COMMAND SYSTEMS SUPPORT ACTIVITY, WASHINGTON, DC	NAVCOSSACT-
NAVAL COMMAND SYSTEMS SUPPORT ACTIVITY, WASHINGTON, DC	NAVCOSSACT-(NUMBER)-
NAVAL DENTAL RESEARCH INST., GREAT LAKES, ILL.	NDRI-(YEAR)-
NAVAL DENTAL RESEARCH INST., GREAT LAKES, ILL.	NDRI-PR-
NAVAL DENTAL SCHOOL, BETHESDA, MD.	NDS-TR-
NAVAL ELECTRONIC SYSTEMS COMMAND, WASHINGTON, D.C.	ELEX-
NAVAL ELECTRONIC SYSTEMS COMMAND, WASHINGTON, D.C.	NAVELECSYSCOM-
NAVAL ELECTRONIC SYSTEMS COMMAND, WASHINGTON, D.C.	NAVESC-
NAVAL ELECTRONIC SYSTEMS COMMAND, WASHINGTON, D.C.	NESC-
NAVAL ELECTRONIC SYSTEMS COMMAND, SPECIAL COMMUNICATIONS PROJECT OFFICE, WASHINGTON, D.C.	TDP-X(NUMBER)
NAVAL ELECTRONIC SYSTEMS TEST AND EVALUATION FACILITY, PATUXENT RIVER, MD.	NESTEF-(YEAR)-
NAVAL ELECTRONIC SYSTEMS TEST AND EVALUATION FACILITY, PATUXENT RIVER, MD.	NESTF-
NAVAL ELECTRONIC SYSTEMS TEST AND EVALUATION FACILITY, ST. INIGOES, MD.	NESTEF-(YEAR)-
NAVAL ELECTRONICS LAB., SAN DIEGO, CALIF.	NEL-CR-
NAVAL ELECTRONICS LAB. CENTER, SAN DIEGO, CALIF.	NELC-
NAVAL ELECTRONICS LAB. CENTER, SAN DIEGO, CALIF.	NELC-IR-
NAVAL ELECTRONICS LAB. CENTER, SAN DIEGO, CALIF.	NELC-TR-
NAVAL ENGINEERING EXPERIMENT STATION, ANNAPOLIS	EES-
NAVAL ENGINEERING EXPERIMENT STATION, ANNAPOLIS	EES-(NO. LETTER)-
NAVAL ENGINEERING EXPERIMENT STATION, ANNAPOLIS	EES-B-
NAVAL ENGINEERING EXPERIMENT STATION, ANNAPOLIS	EES-C-
NAVAL ENGINEERING EXPERIMENT STATION, ANNAPOLIS	NAV EES-
NAVAL ENGINEERING EXPERIMENT STATION, ANNAPOLIS	NAV EES (LETTER(S))-
NAVAL ENGINEERING EXPERIMENT STATION, ANNAPOLIS	NAVENGRXST-
NAVAL ENGINEERING EXPERIMENT STATION, ANNAPOLIS	NEES-
NAVAL ENGINEERING EXPERIMENT STATION, ANNAPOLIS	NLN-
NAVAL ENGINEERING EXPERIMENT STATION, ANNAPOLIS	NP/16/L5/S29/-
NAVAL EXPLOSIVE ORDNANCE DISPOSAL FACILITY, INDIAN HEAD, MD. (ADVANCED TECHNICAL INFORMATION LETTER)	ATIL-
NAVAL EXPLOSIVE ORDNANCE DISPOSAL FACILITY, INDIAN HEAD, MD. (EXPLOSIVE ORDNANCE DISPOSAL LETTER)	EODL-
NAVAL EXPLOSIVE ORDNANCE DISPOSAL FACILITY, INDIAN HEAD, MD. (EXPLOSIVE ORDNANCE DISPOSAL REPORT)	EODR-
NAVAL EXPLOSIVE ORDNANCE DISPOSAL FACILITY, INDIAN HEAD, MD. (EMERGENCY SAFING PROCEDURE)	ESP-
NAVAL EXPLOSIVE ORDNANCE DISPOSAL FACILITY, INDIAN HEAD, MD. (FOREIGN ORDNANCE INFORMATION)	FOI-
NAVAL EXPLOSIVE ORDNANCE DISPOSAL FACILITY, INDIAN HEAD, MD.	NAVEODFAC-
NAVAL EXPLOSIVE ORDNANCE DISPOSAL FACILITY, INDIAN HEAD, MD. (EXPLOSIVE ORDNANCE DISPOSAL REPORT)	NAVEODFAC EODR-
NAVAL EXPLOSIVE ORDNANCE DISPOSAL FACILITY, INDIAN HEAD, MD.	NAVEODRAC-
NAVAL EXPLOSIVE ORDNANCE DISPOSAL FACILITY, INDIAN HEAD, MD.	NEODF-
NAVAL FACILITIES ENGINEERING COMMAND, WASHINGTON, DC	FAC-
NAVAL FACILITIES ENGINEERING COMMAND, WASHINGTON, DC (DESIGN MANUAL)	NAVFAC-DM-
NAVAL FACILITIES ENGINEERING COMMAND, WASHINGTON, DC	NAVFEC-
NAVAL FACILITIES ENGINEERING COMMAND, WASHINGTON, DC	NFEC-
NAVAL FLEET MISSILE SYSTEMS, CORONA, CALIF.	E7-
NAVAL FORCES IN EUROPE, U.S.	COMNAVEU-
NAVAL FORCES IN GERMANY, U.S.	COMNAVFORGER-
NAVAL FORCES IN THE FAR EAST, U.S.	COMNAVFE-
NAVAL GUN FACTORY, WASHINGTON, D.C.	NGF-
NAVAL GUN FACTORY, WASHINGTON, D.C. (RESEARCH RPTS.)	NGF RR-
NAVAL GUN FACTORY, WASHINGTON, D.C.	NGF-T-
NAVAL GUN FACTORY. METALLURGICAL AND TESTING SECTION, WASHINGTON, D.C.	MTS-
NAVAL GUN FACTORY. METALLURGICAL AND TESTING SECTION, WASHINGTON, D.C.	NGF MTS-
NAVAL GUN FACTORY. METALLURGICAL AND TESTING SECTION, WASHINGTON, D.C. (TEST REPORTS)	NGF MTS TR-
NAVAL GUN FACTORY. OPTICAL DESIGN LAB., WASH., D.C.	OL-
NAVAL GUN FACTORY. OPTICAL LAB., WASHINGTON, D.C.	NGF OL TR-
NAVAL GUN FACTORY. OPTICAL LAB., WASHINGTON, D.C.	OL TR-
NAVAL INSPECTOR OF ORDNANCE	NIO-
NAVAL MAGAZINE. QUALITY CONTROL LAB., PORT CHICAGO, CALIF.	QC/PC-
NAVAL MAGAZINE. QUALITY EVALUATION LAB., PORT CHICAGO, CALIF.	QE/PC-
NAVAL MATERIAL COMMAND, WASHINGTON, D.C.	NAVMAT-INST-

NAVAL OCEANOGRAPHIC OFFICE

NAVAL MATERIAL LAB., BROOKLYN	NAVML-
NAVAL MATERIAL LAB., BROOKLYN	NML-
NAVAL MATERIAL LAB., BROOKLYN	NML-(PROJECT NUMBER)-
NAVAL MEDICAL FIELD RESEARCH LAB., CAMP LEJEUNE, N.C.	NM-
NAVAL MEDICAL FIELD RESEARCH LAB., CAMP LEJEUNE, N.C.	NMFRL-
NAVAL MEDICAL RESEARCH INST., BETHESDA, MD.	M-(PROJECT)-
NAVAL MEDICAL RESEARCH INST., BETHESDA, MD.	MEMO(YR)-
NAVAL MEDICAL RESEARCH INST., BETHESDA, MD.	MF-(PROJECT)-
NAVAL MEDICAL RESEARCH INST., BETHESDA, MD.	NAMRI-
NAVAL MEDICAL RESEARCH INST., BETHESDA, MD.	NAVMED-MF(NUMBER)-
NAVAL MEDICAL RESEARCH INST., BETHESDA, MD.	NAVMRI-
NAVAL MEDICAL RESEARCH INST., BETHESDA, MD.	NM-(PROJECT NO.)-
NAVAL MEDICAL RESEARCH INST., BETHESDA, MD.	NM PROJECT-
NAVAL MEDICAL RESEARCH INST., BETHESDA, MD.	NMRI-
NAVAL MEDICAL RESEARCH INST., BETHESDA, MD.	NMRI-(LETTERS)-
NAVAL MEDICAL RESEARCH INST., BETHESDA, MD. (LECTURE AND REVIEW SERIES)	NMRI LRS(YR.)-
NAVAL MEDICAL RESEARCH INST., BETHESDA, MD.	NMRI-MR-
NAVAL MEDICAL RESEARCH INST., BETHESDA, MD. (MEMORANDUM REPORTS)	NMRI MR(YR.)-
NAVAL MEDICAL RESEARCH INST., BETHESDA, MD.	NMRI PROJECT-
NAVAL MEDICAL RESEARCH INST., BETHESDA, MD.	NMRI R-
NAVAL MEDICAL RES. INST., BETHESDA, MD.(RES. REPTS.)	NMRI RR-
NAVAL MEDICAL RESEARCH INST., BETHESDA, MD.	NMS-TRANS-
NAVAL MEDICAL RESEARCH INST., BETHESDA, MD.	USNMRI-
NAVAL MEDICAL RESEARCH LAB., NEW LONDON, CONN.	MR-
NAVAL MEDICAL RESEARCH LAB., NEW LONDON, CONN.	MRL-
NAVAL MEDICAL RESEARCH LAB., NEW LONDON, CONN.	NAV MRL-
NAVAL MEDICAL RESEARCH LAB., NEW LONDON, CONN.	NMRL-
NAVAL MEDICAL RESEARCH UNIT NO...	NAMRU-
NAVAL MINE DEFENSE LAB., PANAMA CITY, FLA.	NMCL MRR-
NAVAL MINE DEFENSE LAB., PANAMA CITY, FLA.	NMDL TP-
NAVAL MINE DEPOT, YORKTOWN, VA.(MINE PRODUCTION RPT.)	MPR-
NAVAL MINE DEPOT, YORKTOWN, VA. (MINE SERVICE TEST FINAL)	MSTF-
NAVAL MINE DEPOT, YORKTOWN, VA. (MINE SERVICE TEST FINAL REPORT)	MSTFR-
NAVAL MINE DEPOT, YORKTOWN, VA. (MINE SERVICE TEST INTERIM REPORT)	MSTIR-
NAVAL MINE DEPOT, YORKTOWN, VA. (MINE SERVICE TEST REPORT)	MSTR-
NAVAL MINE DEPOT, YORKTOWN, VA.	NAVMD-
NAVAL MINE DEPOT, YORKTOWN, VA.	NMD-
NAVAL MINE DEPOT, YORKTOWN, VA.	NMD/Y-
NAVAL MINE DEPOT, YORKTOWN, VA. (RES. + DEV. REPT.)	R+D/Y-
NAVAL MINE DEPOT. MINE SERVICE TEST DEPT.,YORKTOWN,VA	MST-
NAVAL MINE DEPOT. QUALITY EVALUATION LAB., YORKTOWN, VA.	QE/Y-
NAVAL MINE DEPOT. RESEARCH AND DEVELOPMENT DIV., YORKTOWN, VA. (REPORT)	RDDR-
NAVAL MINE ENGINEERING FACILITY	MEF-
NAVAL MINE WARFARE TEST STATION, SOLOMONS, MD.	NMWTS-
NAVAL MISSILE AND ASTRONAUTICS CENTER, POINT MUGU,CAL	NMAC-
NAVAL MISSILE AND ASTRONAUTICS CENTER, POINT MUGU,CAL	NMAC-TR-(YEAR)-
NAVAL MISSILE AND ASTRONAUTICS CENTER, POINT MUGU,CAL	NMC-TR-
NAVAL MISSILE CENTER (CENTER COMMAND)	CEN-
NAVAL MISSILE CENTER (COMPONENT TEST)	CT-
NAVAL MISSILE CENTER, POINT MUGU, CALIF.	COMNAVMISCEN-
NAVAL MISSILE CENTER, POINT MUGU, CALIF.	NMC-
NAVAL MISSILE CENTER, POINT MUGU, CALIF.	NMC-(LETTERS)-
NAVAL MISSILE CENTER, POINT MUGU, CALIF.	NMC-MP-
NAVAL MISSILE CENTER, POINT MUGU, CALIF.	NMC-MP-(YEAR)-
NAVAL MISSILE CENTER, POINT MUGU, CALIF.	NMC-MR-TS-
NAVAL MISSILE CENTER, POINT MUGU, CALIF.	NMC-PMR-TR-
NAVAL MISSILE CENTER, POINT MUGU, CALIF.	NMC-TM-(YEAR)-
NAVAL MISSILE CENTER, POINT MUGU, CALIF.	NMC-TP-(YEAR)-
NAVAL MISSILE CENTER. DIRECTOR OF TESTS.	DT-
NAVAL MISSILE CENTER. LABORATORY EVALUATION DEPT., POINT MUGU, CALIF.	LE-
NAVAL MISSILE CENTER. MISSILE TEST DEPT.,PT.MUGU,CAL.	MT-
NAVAL MISSILE CENTER. RANGE INSTRUMENTATION DEPT.	RI-
NAVAL MISSILE CENTER. TARGET DRONE DEPT.	TD-
NAVAL MISSILE CENTER. TECHNICAL SERVICE DEPT.	TS-
NAVAL NET DEPOT	NND-
NAVAL NUCLEAR ORDNANCE EVALUATION UNIT, KIRTLAND AFB, N. MEX.	NNOEU-
NAVAL NUCLEAR ORDNANCE EVALUATION UNIT, KIRTLAND AFB, N. MEX. (VULNERABILITY NEWS AND VIEWS)	NNOEU VNV
NAVAL NUCLEAR ORDNANCE EVALUATION UNIT, KIRTLAND AFB, N. MEX. (VULNERABILITY NEWS AND VIEWS)	VN+V-
NAVAL NUCLEAR POWER UNIT, FORT BELVOIR, VA.	NAVNUPWRU-
NAVAL OBSERVATORY, WASHINGTON, D.C.	NAVOBSY-
NAVAL OCEANOGRAPHIC OFFICE, WASHINGTON, D.C. (HYDROGRAPHIC OFFICE MISCELLANEOUS)	H.O. MISC.-
NAVAL OCEANOGRAPHIC OFFICE, WASHINGTON, D.C. (HYDROGRAPHIC OFFICE PUBLICATION)	H.O. PUB.-
NAVAL OCEANOGRAPHIC OFFICE, WASHINGTON, D.C. (INFORMAL MANUSCRIPT REPORT)	I.M.R.-(NUMBER/YEAR)
NAVAL OCEANOGRAPHIC OFFICE, WASHINGTON, D.C. (INFORMAL MANUSCRIPT REPORTS)	IMR-
NAVAL OCEANOGRAPHIC OFFICE, WASHINGTON, D.C. (INFORMAL OCEANOGRAPHIC REPORT)	I.O.M.-(NUMBER/YEAR)
NAVAL OCEANOGRAPHIC OFFICE, WASHINGTON, D.C. (INFORMAL OCEANOGRAPHIC REPORT)	IOM-
NAVAL OCEANOGRAPHIC OFFICE, WASHINGTON, D.C. (SUPERSEDED IMR AND IOM SERIES)	IR-
NAVAL OCEANOGRAPHIC OFFICE, WASHINGTON, D.C.	IR-(YEAR)-
NAVAL OCEANOGRAPHIC OFFICE, WASHINGTON, D.C. (MANUSCRIPT REPORT)	M.R.-(NUMBER/YEAR)
NAVAL OCEANOGRAPHIC OFFICE, WASHINGTON, D.C.	NAVOCEANO-
NAVAL OCEANOGRAPHIC OFFICE, WASHINGTON, D.C.	NO-
NAVAL OCEANOGRAPHIC OFFICE, WASHINGTON, D.C.	NOO-
NAVAL OCEANOGRAPHIC OFFICE, WASHINGTON, D.C.	NOO-FR-
NAVAL OCEANOGRAPHIC OFFICE, WASHINGTON, D.C.	NOO-HO-(NUMBERS)-
NAVAL OCEANOGRAPHIC OFFICE, WASHINGTON, D.C.	NOO-IM-(YEAR)-

NAVAL OCEANOGRAPHIC OFFICE

NAVAL OCEANOGRAPHIC OFFICE, WASHINGTON, D.C.	NOO-IMR-M-
NAVAL OCEANOGRAPHIC OFFICE, WASHINGTON, D.C.	NOO-IR-
NAVAL OCEANOGRAPHIC OFFICE, WASHINGTON, D.C.	NOO-IR-H-
NAVAL OCEANOGRAPHIC OFFICE, WASHINGTON, D.C.	NOO-SP-
NAVAL OCEANOGRAPHIC OFFICE, WASHINGTON, D.C.	NOO-TR-
NAVAL OCEANOGRAPHIC OFFICE, WASHINGTON, D.C.	NOO-TRANS-
NAVAL OCEANOGRAPHIC OFFICE, WASHINGTON, D.C.	SP-
NAVAL OCEANOGRAPHIC OFFICE. ANTISUBMARINE WARFARE ENVIRONMENTAL PREDICTION SERVICE, WASHINGTON, D.C.	ASWEPS-
NAVAL OCEANOGRAPHIC OFFICE. HYDROGRAPHIC SURVEYS DEPT., WASHINGTON, D.C.	H-
NAVAL OCEANOGRAPHIC OFFICE. HYDROGRAPHIC SURVEYS DEPT., WASHINGTON, D.C.	H-(NUMBER)-(YEAR)
NAVAL OCEANOGRAPHIC OFFICE. HYDROGRAPHIC SURVEYS DEPT., WASHINGTON, D.C.	NO.H.-
NAVAL OPERATIONAL TEST + EVALUATION FORCE, NORFOLK,VA	OP-
NAVAL OPERATIONAL TEST + EVALUATION FORCE, NORFOLK,VA	OPNAV-
NAVAL ORDNANCE EXPERIMENTAL UNIT, WASHINGTON, D.C.	NOEU-
NAVAL ORDNANCE LAB. (ALL LOCATIONS)	NOL-
NAVAL ORDNANCE LAB., CORONA, CALIF.	IED-G(NUMBER)
NAVAL ORDNANCE LAB., CORONA, CALIF.	NOL-C-
NAVAL ORDNANCE LAB., CORONA, CALIF.	NOLC-
NAVAL ORDNANCE LAB., CORONA, CALIF.	NOL-CORONA-
NAVAL ORDNANCE LAB., CORONA, CALIF.	NOL-CORONA-TM-
NAVAL ORDNANCE LAB., CORONA, CALIF.	NOLC-TM-
NAVAL ORDNANCE LAB., SILVER SPRING, MD.	AERODYNAMICS RES. REPT.-
NAVAL ORDNANCE LAB., SILVER SPRING, MD.	NOLSS-
NAVAL ORDNANCE LAB., SILVER SPRING, MD. (TEST RPTS.)	NOLSS IR-
NAVAL ORDNANCE LAB., WASHINGTON, D.C. (MEMORANDA)	NOLM-
NAVAL ORDNANCE LAB., WHITE OAK, MD. (ACTUATION DATA COMMUNICATION)	ADC-
NAVAL ORDNANCE LAB., WHITE OAK, MD.	AEROBALLISTICS RES. RPT-
NAVAL ORDNANCE LAB., WHITE OAK, MD. (AEROBALLISTICS RESEARCH REPORT)	ARR-
NAVAL ORDNANCE LAB., WHITE OAK, MD.	BALLISTICS RES. REPT.-
NAVAL ORDNANCE LAB., WHITE OAK, MD.	BRR-
NAVAL ORDNANCE LAB., WHITE OAK, MD. (DESIGN ANALYSIS MEMORANDUM)	DAM-
NAVAL ORDNANCE LAB., WHITE OAK, MD.	LC-PR-
NAVAL ORDNANCE LAB., WHITE OAK, MD. (MATHEMATICAL ANALYSIS REQUEST)	MAR-
NAVAL ORDNANCE LAB., WHITE OAK, MD. (MATHEMATICAL ANALYSIS RESEARCH REQUEST)	MARR-
NAVAL ORDNANCE LAB., WHITE OAK, MD. (MECHANIZED ANALYSIS RESEARCH REQUEST)	MARR-
NAVAL ORDNANCE LAB., WHITE OAK, MD. (MINE COUNTERMEASURES MEMORANDUM)	MCM-
NAVAL ORDNANCE LAB., WHITE OAK, MD.	MEMO. NOLM-
NAVAL ORDNANCE LAB., WHITE OAK, MD.	NAVORD OS-
NAVAL ORDNANCE LAB., WHITE OAK, MD. (HISTORY)	NOLH-
NAVAL ORDNANCE LAB., WHITE OAK, MD. (MEMORANDA)	NOLM-
NAVAL ORDNANCE LAB., WHITE OAK, MD. (REPORTS)	NOLR-
NAVAL ORDNANCE LAB., WHITE OAK, MD. (MINE UNIT RPTS.)	NOLR MUR-
NAVAL ORDNANCE LAB., WHITE OAK, MD. (SUGGESTION)	NOLS-
NAVAL ORDNANCE LAB., WHITE OAK, MD.	NOL TASK NO.
NAVAL ORDNANCE LAB., WHITE OAK, MD. (TECHNICAL NOTE)	NOLTN-
NAVAL ORDNANCE LAB., WHITE OAK, MD.	NOLTR-(YEAR)-
NAVAL ORDNANCE LAB., WHITE OAK, MD.	NOL-TR-
NAVAL ORDNANCE LAB., WHITE OAK, MD. (MISCELLANEOUS PUBLICATION)	NOLX-
NAVAL ORDNANCE LAB., WHITE OAK, MD. (PRODUCTION CHANGE ORDER)	PCO-
NAVAL ORDNANCE LAB., WHITE OAK,MD.(PROJECT WHITE OAK)	PWO-
NAVAL ORDNANCE LAB., WHITE OAK, MD. (REQUIREMENTS + TEST PROCEDURES)	RTP-
NAVAL ORDNANCE LAB., WHITE OAK, MD.	SAGR-
NAVAL ORDNANCE LAB., WHITE OAK, MD. (STANDARD TEST AUTHORIZATION AND REPORT SYSTEM)	STAR-
NAVAL ORDNANCE LAB., WHITE OAK, MD. (TEST SCHEDULE SHEETS)	TSS-
NAVAL ORDNANCE LAB., WHITE OAK, MD. (TECHNICAL WORK REQUEST)	TWR-
NAVAL ORDNANCE LAB., WHITE OAK, MD.	US-
NAVAL ORDNANCE LAB., WHITE OAK, MD. (VT FUZE MEMO.)	VTFM-
NAVAL ORDNANCE LAB., WHITE OAK, MD. (VT FUZE REPORT)	VTFR-
NAVAL ORDNANCE LAB., WHITE OAK, MD. (WHITE PLAN)	WPL-
NAVAL ORDNANCE LAB., WHITE OAK, MD. (WIND TUNNEL REQUEST)	WTR-
NAVAL ORDNANCE LAB. DEGAUSSING RANGE (DATA SHEET)	DG-RD-
NAVAL ORDNANCE LAB. ELECTRICAL EVALUATION DIV. (COMMUNICATION)	EEDC-
NAVAL ORDNANCE LAB. INFLUENCE DIV. (COMMUNICATION)	IDC-
NAVAL ORDNANCE LAB. KENSINGTON MAGNETIC LAB. (COMMUNICATION)	KMLC-
NAVAL ORDNANCE LAB. MECHANICS DIV. (TECHNICAL NOTE)	MDTN-
NAVAL ORDNANCE LAB. MINE MODIFICATION UNIT (MEMO.)	MMUM-
NAVAL ORDNANCE LAB. MINE MODIFICATION UNIT (REPORT)	MMUR-
NAVAL ORDNANCE LAB. MINE RESEARCH UNIT	MRU-
NAVAL ORDNANCE LAB. MINE UNIT (MEMORANDUM)	MUM-
NAVAL ORDNANCE LAB. MINE UNIT (REPORT)	MUR-
NAVAL ORDNANCE LAB. UNDERWATER ORDNANCE EXPLOSIVE TRAIN SAFETY COMMITTEE	UES-
NAVAL ORDNANCE PLANT, FOREST PARK, ILL.	NOPF-
NAVAL ORDNANCE PLANT, INDIANAPOLIS	NOPI-
NAVAL ORDNANCE PLANT, INDIANAPOLIS	ROC-
NAVAL ORDNANCE PLANT, INDIANAPOLIS	RTR-
NAVAL ORDNANCE STATION, INDIAN HEAD, MD.	NOS-IHMR-
NAVAL ORDNANCE STATION, INDIAN HEAD, MD.	NOS-IHSP-
NAVAL ORDNANCE STATION, INDIAN HEAD, MD.	NOS-IHTR-
NAVAL ORDNANCE STATION, INDIAN HEAD, MD.	NOS-TMR-
NAVAL ORDNANCE STATION. QUALITY ASSURANCE DEPT., INDIAN HEAD, MD.	IH/QAS-(YEAR)-
NAVAL ORDNANCE STATION. RESEARCH AND DEVELOPMENT DEPT., INDIAN HEAD, MD.	IH/RE-(YEAR)-
NAVAL ORDNANCE STATION. TEST AND EVALUATION DEPT., INDIAN HEAD, MD.	IH/TFS-
NAVAL ORDNANCE SYSTEMS COMMAND, WASHINGTON, D.C.	NOSC-
NAVAL ORDNANCE SYSTEMS COMMAND, WASHINGTON, D.C.	NOS-IHSP-
NAVAL ORDNANCE SYSTEMS COMMAND, WASHINGTON, D.C.	ORD-

NAVAL ORDNANCE SYSTEMS COMMAND, WASHINGTON, D.C.	WS-
NAVAL ORDNANCE TEST STATION, CHINA LAKE, CALIF. (ADMINISTRATIVE PUBLICATION)	ADPUB-
NAVAL ORDNANCE TEST STATION, CHINA LAKE, CALIF. (INTERNAL DISTRIBUTION PUBLICATION)	IDP-
NAVAL ORDNANCE TEST STATION, CHINA LAKE, CALIF.	NOTS-
NAVAL ORDNANCE TEST STATION, CHINA LAKE, CALIF. (ADMINISTRATIVE PUBLICATION)	NOTS ADPUB-
NAVAL ORDNANCE TEST STATION, CHINA LAKE, CALIF.	NOTS T-
NAVAL ORDNANCE TEST STATION, CHINA LAKE, CALIF. (TECHNICAL MEMORANDUM)	NOTS TM-
NAVAL ORDNANCE TEST STATION, CHINA LAKE, CALIF.	NOTS-TN-(NUMBERS)-
NAVAL ORDNANCE TEST STATION, CHINA LAKE, CALIF. (TECHNICAL PUBLICATION)	NOTS TP-
NAVAL ORDNANCE TEST STATION, CHINA LAKE, CALIF. (TECHNICAL PROGRESS REPORT)	NOTS TPR-
NAVAL ORDNANCE TEST STATION, CHINA LAKE, CALIF. (TECHNICAL PROGRESS REPORT)	TPR-
NAVAL ORDNANCE TEST STATION, INYOKERN (CHINA LAKE), CALIF. (REF. 1)	(NUMBERS...)
NAVAL ORDNANCE TEST STATION, INYOKERN (CHINA LAKE), CALIF. (ANNUAL TECHNICAL PROGRESS REPORT)	ATPR-
NAVAL ORDNANCE TEST STA., INYOKERN (CHINA LAKE), CAL.	CSRD-
NAVAL ORDNANCE TEST STA., INYOKERN (CHINA LAKE), CAL.	CSRD-I(RI)(MO./YR.)-
NAVAL ORDNANCE TEST STA., INYOKERN (CHINA LAKE), CAL.	ECTF-
NAVAL ORDNANCE TEST STATION, INYOKERN (CHINA LAKE), CALIF. (TERRIER EVALUATION PROGRAM)	ETP-
NAVAL ORDNANCE TEST STA., INYOKERN (CHINA LAKE), CAL.	NAVORD-NOTS-
NAVAL ORDNANCE TEST STA., INYOKERN (CHINA LAKE), CAL.	NAVORD-OFC-
NAVAL ORDNANCE TEST STA., INYOKERN (CHINA LAKE), CAL.	NAVORD-RDAF-
NAVAL ORDNANCE TEST STA., INYOKERN (CHINA LAKE), CAL.	NAVORD-RRB-
NAVAL ORDNANCE TEST STA., INYOKERN (CHINA LAKE), CAL.	NAVORD-TM-
NAVAL ORDNANCE TEST STA., INYOKERN (CHINA LAKE), CAL.	NOTS-(NUMBER)-
NAVAL ORDNANCE TEST STA., INYOKERN (CHINA LAKE), CAL.	NOTS-(LETTERS)-
NAVAL ORDNANCE TEST STA., INYOKERN (CHINA LAKE), CAL.	NOTS-ASSIGNMENT-
NAVAL ORDNANCE TEST STA., INYOKERN (CHINA LAKE), CAL.	NOTS-DOC.
NAVAL ORDNANCE TEST STA., INYOKERN (CHINA LAKE), CAL.	NOTS/IDP-
NAVAL ORDNANCE TEST STA., INYOKERN (CHINA LAKE), CAL.	NOTS-PM-P-
NAVAL ORDNANCE TEST STA., INYOKERN (CHINA LAKE), CAL.	NOTS REG.-
NAVAL ORDNANCE TEST STA., INYOKERN (CHINA LAKE), CAL.	NOTS-TEST REPT.-
NAVAL ORDNANCE TEST STA., INYOKERN (CHINA LAKE), CAL.	OFC-
NAVAL ORDNANCE TEST STA., INYOKERN (CHINA LAKE), CAL.	OM-
NAVAL ORDNANCE TEST STA., INYOKERN (CHINA LAKE), CAL.	OME-
NAVAL ORDNANCE TEST STATION, INYOKERN (CHINA LAKE), CALIF. (ORDNANCE UNDERWATER MEMORANDUM)	OUM-
NAVAL ORDNANCE TEST STA., INYOKERN (CHINA LAKE), CAL.	PACT-
NAVAL ORDNANCE TEST STATION, INYOKERN (CHINA LAKE), CALIF. (PRELIMINARY TECHNICAL PROGRESS REPORT)	PTPR-
NAVAL ORDNANCE TEST STA., INYOKERN (CHINA LAKE), CAL.	RBAR-
NAVAL ORDNANCE TEST STA., INYOKERN (CHINA LAKE), CAL.	RCI-
NAVAL ORDNANCE TEST STA., INYOKERN (CHINA LAKE), CAL.	RDAF-
NAVAL ORDNANCE TEST STA., INYOKERN (CHINA LAKE), CAL.	RDAP-
NAVAL ORDNANCE TEST STA., INYOKERN (CHINA LAKE), CAL.	RDMA-
NAVAL ORDNANCE TEST STA., INYOKERN (CHINA LAKE), CAL.	RDMD-
NAVAL ORDNANCE TEST STA., INYOKERN (CHINA LAKE), CAL.	RDMR-
NAVAL ORDNANCE TEST STA., INYOKERN (CHINA LAKE), CAL.	RDU-
NAVAL ORDNANCE TEST STA., INYOKERN (CHINA LAKE), CAL.	RDUB-
NAVAL ORDNANCE TEST STA., INYOKERN (CHINA LAKE), CAL.	RDUI-
NAVAL ORDNANCE TEST STA., INYOKERN (CHINA LAKE), CAL.	RRAM-
NAVAL ORDNANCE TEST STA., INYOKERN (CHINA LAKE), CAL.	RRB-
NAVAL ORDNANCE TEST STA., INYOKERN (CHINA LAKE), CAL.	SBE-
NAVAL ORDNANCE TEST STA., INYOKERN (CHINA LAKE), CAL.	SBED-
NAVAL ORDNANCE TEST STA., INYOKERN (CHINA LAKE), CAL.	SBES-
NAVAL ORDNANCE TEST STA., INYOKERN (CHINA LAKE), CAL.	SCA-
NAVAL ORDNANCE TEST STA., INYOKERN (CHINA LAKE), CAL.	SCWP-
NAVAL ORDNANCE TEST STA., INYOKERN (CHINA LAKE), CAL.	TM-
NAVAL ORDNANCE TEST STA., INYOKERN (CHINA LAKE), CAL.	XCO-
NAVAL ORDNANCE TEST STA., INYOKERN (CHINA LAKE), CAL.	XPO-
NAVAL ORDNANCE TEST STA., INYOKERN (CHINA LAKE), CAL.	XPR-
NAVAL ORDNANCE TEST STA., INYOKERN (CHINA LAKE), CAL.	XTS-
(SEE ALSO LATER NAME: NAVAL ORDNANCE TEST STATION, CHINA LAKE, CALIF.)	
NAVAL ORDNANCE TEST STATION, PASADENA, CALIF.	NOTS-
NAVAL ORDNANCE TEST STATION, PASADENA, CALIF.	NOTS-TP-
NAVAL ORDNANCE TEST STATION. SCIENCE DEPT. BALLISTICS DIV. AERODYNAMICS SECTION, INYOKERN (CHINA LAKE), CALIF.	SBA-
NAVAL ORDNANCE TEST STATION. SCIENCE DEPT. EXTERIOR BALLISTICS, FINNER GP., INYOKERN (CHINA LAKE), CALIF.	SBEF-
NAVAL PARACHUTE EXPERIMENTAL UNIT, EL CENTRO, CALIF.	NPEU-
NAVAL PERSONNEL PROGRAM SUPPORT ACTIVITY. PERSONNEL RESEARCH LAB., WASHINGTON, D.C.	NPPSA-PRL-
NAVAL PERSONNEL PROGRAM SUPPORT ACTIVITY. PERSONNEL RESEARCH LAB., WASHINGTON, D.C. (RESEARCH MEMO.)	WRM-
NAVAL PERSONNEL PROGRAM SUPPORT ACTIVITY. PERSONNEL RESEARCH LAB., WASHINGTON, D.C.	WRR-
NAVAL PERSONNEL PROGRAM SUPPORT ACTIVITY. PERSONNEL RESEARCH LAB., WASHINGTON, D.C.	WSR-
NAVAL PERSONNEL PROGRAM SUPPORT ACTIVITY. PERSONNEL RESEARCH LAB., WASHINGTON, D.C. (TECHNICAL BULL.)	WTB-
NAVAL PERSONNEL PROGRAM SUPPORT ACTIVITY. PERSONNEL RESEARCH LAB., WASHINGTON, D.C.	WWR-
NAVAL PERSONNEL PROGRAM SUPPORT ACTIVITY. PERSONNEL SURVEYS DIV., WASHINGTON, D.C.	NSS-(YEAR)-
NAVAL PERSONNEL RESEARCH ACTIVITY, SAN DIEGO, CALIF.	NAMI-
NAVAL PERSONNEL RESEARCH ACTIVITY, SAN DIEGO, CALIF.	NPRA-SRR-
NAVAL PERSONNEL RESEARCH ACTIVITY, SAN DIEGO, CALIF.	NPRA-STB-(YEAR)-
NAVAL PERSONNEL RESEARCH ACTIVITY, SAN DIEGO, CALIF. (RESEARCH REPORT)	SRM-
NAVAL PERSONNEL RESEARCH ACTIVITY, SAN DIEGO, CALIF.	SRM-(YEAR)-
NAVAL PERSONNEL RESEARCH ACTIVITY, SAN DIEGO, CALIF. (RESEARCH REPORT)	SRR-
NAVAL PERSONNEL RESEARCH ACTIVITY, SAN DIEGO, CALIF.	SRR-(YEAR)-
NAVAL PERSONNEL RESEARCH ACTIVITY, SAN DIEGO, CALIF. (TECHNICAL BULLETIN)	STB-
NAVAL PERSONNEL RESEARCH ACTIVITY, SAN DIEGO, CALIF.	STB-(YEAR)-
NAVAL PERSONNEL RESEARCH ACTIVITY, WASHINGTON, D.C.	ND-
NAVAL PERSONNEL RESEARCH ACTIVITY, WASHINGTON, D.C.	ND-(YEAR)-
NAVAL PERSONNEL RESEARCH ACTIVITY, WASHINGTON, D.C.	PRA-
NAVAL PERSONNEL RESEARCH ACTIVITY, WASHINGTON, D.C.	PRAW-(YEAR)-
NAVAL PERSONNEL RESEARCH ACTIVITY, WASHINGTON, D.C.	PRAW-RS-(YEAR)-
NAVAL PERSONNEL RES. + DEV. LAB., WASHINGTON, D.C.	WOS-(YEAR)-
NAVAL PERSONNEL RES. FIELD ACTIVITY, SAN DIEGO, CAL.	BUPERS TB-
NAVAL PERSONNEL RES. FIELD ACTIVITY, SAN DIEGO, CAL.	NPRFA-
NAVAL PERSONNEL RES. FIELD ACTIVITY, SAN DIEGO, CAL.	PRUS-
NAVAL PERSONNEL RES. FIELD ACTIVITY, SAN DIEGO, CAL.	SD-
(SEE ALSO LATER NAME: NAVAL PERSONNEL RES. ACTIVITY)	
NAVAL PHOTOGRAPHIC CENTER, WASHINGTON, D.C.	NPC-
NAVAL PHOTOGRAPHIC CENTER. RESEARCH AND DEVELOPMENT DEPT., WASHINGTON, D.C.	NAVPHOTOCEN-R/D-
NAVAL PHOTOGRAPHIC INTERPRETATION CENTER, WASH., D.C.	PIC-

NAVAL PLANT REPRESENTATIVE OFFICE

NAVAL PLANT REPRESENTATIVE OFFICE, COLUMBUS, OHIO	NAVPRO-
NAVAL POSTGRADUATE SCHOOL, MONTEREY, CALIF.	NAV-PGS-
NAVAL POSTGRADUATE SCHOOL, MONTEREY, CALIF.	NPS-
NAVAL POSTGRADUATE SCHOOL, MONTEREY, CALIF.	NPS-(NUMBERS,LETTERS)-
NAVAL POSTGRADUATE SCHOOL, MONTEREY, CALIF.	RP-
NAVAL POSTGRADUATE SCHOOL, MONTEREY, CALIF.	TN-(YR.)-
NAVAL POSTGRADUATE SCHOOL, MONTEREY, CALIF.	TR/RP-
NAVAL POSTGRADUATE SCHOOL, MONTEREY, CALIF.	USNPGS-
NAVAL POWDER FACTORY, INDIAN HEAD, MD. (EXPLOSIVES INVESTIGATION MEMORANDUM)	EIM-
NAVAL POWDER FACTORY, INDIAN HEAD, MD.	NPF-
NAVAL POWDER FACTORY, INDIAN HEAD, MD.	NPF-AR-
NAVAL POWDER FACTORY, INDIAN HEAD, MD.	NPF MEMO-
NAVAL POWDER FACTORY, INDIAN HEAD, MD.	NPF-MR-
NAVAL POWDER FACTORY, INDIAN HEAD, MD.	NPF-TR-
NAVAL POWDER FACTORY. ORDNANCE INVESTIGATION LAB., INDIAN HEAD, MD. (DEMOLITION INVEST. MEMORANDA)	DIM-
NAVAL POWDER FACTORY. ORDNANCE INVESTIGATION LAB., INDIAN HEAD, MD. (DEMOLITION INVEST. MEMORANDA)	NPF DIM-
NAVAL POWDER FACTORY. ORDNANCE INVESTIGATION LAB., INDIAN HEAD, MD.	OIL-
NAVAL POWDER FACTORY. ORDNANCE INVESTIGATION LAB., INDIAN HEAD, MD.	OIM-
NAVAL POWDER FACTORY. QUALITY EVALUATION LAB., INDIAN HEAD, MD.	QE/NPF-
NAVAL POWDER FACTORY. RES.+DEV. DEPT.,INDIAN HEAD,MD.	MR-
NAVAL POWDER FACTORY. RES.+DEV. DEPT.,INDIAN HEAD,MD.	RP-(NUMBER)-
NAVAL PROPELLANT PLANT, INDIAN HEAD, MD.	F-
NAVAL PROPELLANT PLANT, INDIAN HEAD, MD.	NPP-
NAVAL PROPELLANT PLANT, INDIAN HEAD, MD.	NPP-F-
NAVAL PROPELLANT PLANT, INDIAN HEAD, MD.	NPP-RP-
NAVAL PROPELLANT PLANT, INDIAN HEAD, MD.	NPP-RR-(YEAR)-
NAVAL PROPELLANT PLANT, INDIAN HEAD, MD. (TECHNICAL PROGRESS REPORT)	NPP-TPR-
NAVAL PROPELLANT PLANT, INDIAN HEAD, MD.	QF/NPP-
NAVAL PROPELLANT PLANT, INDIAN HEAD, MD.	TMR-
NAVAL PROPELLANT PLANT. QUALITY ASSURANCE DEPT., INDIAN HEAD, MD.	NPP/QAS-(YEAR)-
NAVAL PROPELLANT PLANT. QUALITY ASSURANCE DEPT., INDIAN HEAD, MD.	QAS/NPP-
NAVAL PROPELLANT PLANT. RESEARCH AND DEVELOPMENT DEPT., INDIAN HEAD, MD.	MR-
NAVAL PROPELLANT PLANT. RESEARCH AND DEVELOPMENT DEPT., INDIAN HEAD, MD.	NPP-MR-
NAVAL PROPELLANT PLANT. RESEARCH AND DEVELOPMENT DEPT., INDIAN HEAD, MD.	NPP-TMR-
NAVAL PROPELLANT PLANT. RESEARCH AND DEVELOPMENT DEPT., INDIAN HEAD, MD.	NPP-TR-
NAVAL PROVING GROUND, DAHLGREN, VA.	NPG-
NAVAL PROVING GROUND, DAHLGREN, VA.	NPG-(NUMBER)(NAVY)
NAVAL PROVING GROUND, DAHLGREN, VA.	NPG-LC-
NAVAL PROVING GROUND, DAHLGREN, VA.	NPGR-
NAVAL RADIATION LAB., SAN FRANCISCO	ADP-
NAVAL RADIATION LAB., SAN FRANCISCO	N-RAD-L-
NAVAL RADIATION LAB., SAN FRANCISCO	N RAD L ADP-
NAVAL RADIOLOGICAL DEFENSE LAB., SAN FRANCISCO(REF.1)	(NUMBERS...)
NAVAL RADIOLOGICAL DEFENSE LAB., SAN FRAN. (REF. 30)	AD-(NUMBER)-(LETTER)
NAVAL RADIOLOGICAL DEFENSE LAB., SAN FRAN. (REF. 30)	AD(LETTER)-
NAVAL RADIOLOGICAL DEFENSE LAB., SAN FRANCISCO	ADO/(NOS.)-SER-(LTR.)
NAVAL RADIOLOGICAL DEFENSE LAB., SAN FRANCISCO	ASWP ITR-
NAVAL RADIOLOGICAL DEFENSE LAB., SAN FRANCISCO	BUMED-(NO.)-(NO.)-
NAVAL RADIOLOGICAL DEFENSE LAB., SAN FRANCISCO	ER-
NAVAL RADIOLOGICAL DEFENSE LAB., SAN FRANCISCO	FBFRC-TR-
NAVAL RADIOLOGICAL DEFENSE LAB., SAN FRANCISCO	NM-(PROJECT NO.)-
NAVAL RADIOLOGICAL DEFENSE LAB., SAN FRANCISCO	NR- (INITIALS)-
NAVAL RADIOLOGICAL DEFENSE LAB., SAN FRANCISCO	NRDL-
NAVAL RADIOLOGICAL DEFENSE LAB., SAN FRAN. (REF. 30)	NRDL AD(LETTER)-
NAVAL RADIOLOGICAL DEFENSE LAB., SAN FRANCISCO (EVALUATION REPORT)	NRDL-ER-
NAVAL RADIOLOGICAL DEFENSE LAB., SAN FRANCISCO (INSTRUMENT EVALUATION REPORTS)	NRDL IER-
NAVAL RADIOLOGICAL DEFENSE LAB., SAN FRANCISCO	NRDL-P-
NAVAL RADIOLOGICAL DEFENSE LAB., SAN FRANCISCO (REVIEWS AND LECTURES)	NRDL-R+L-
NAVAL RADIOLOGICAL DEFENSE LAB.,SAN FRAN.(TECH. MEMO)	NRDL-TM-
NAVAL RADIOLOGICAL DEFENSE LAB.,SAN FRAN.(TECH. REPT)	NRDL-TR-
NAVAL RADIOLOGICAL DEFENSE LAB., SAN FRAN. (INTERNAL LIBRARY ACCESSION LIST OF CLASSIFIED DOCUMENTS)	TILL-
NAVAL RADIOLOGICAL DEFENSE LAB., SAN FRANCISCO	USAMR+DC-
NAVAL RADIOLOGICAL DEFENSE LAB., SAN FRAN. (REF. 30)	USNRDL-
NAVAL RADIOLOGICAL DEFENSE LAB., SAN FRANCISCO	USNRDL-AD-
NAVAL RADIOLOGICAL DEFENSE LAB., SAN FRANCISCO	USNRDL-CP-(YEAR)-
NAVAL RADIOLOGICAL DEFENSE LAB., SAN FRANCISCO	USNRDL-ER-
NAVAL RADIOLOGICAL DEFENSE LAB., SAN FRANCISCO	USNRDL-LR-
NAVAL RADIOLOGICAL DEFENSE LAB., SAN FRANCISCO	USNRDL-P-
NAVAL RADIOLOGICAL DEFENSE LAB., SAN FRANCISCO	USNRDL-REVIEWS/LECTURES-
NAVAL RADIOLOGICAL DEFENSE LAB., SAN FRANCISCO (TECHNICAL MANUAL)	USNRDL-TM-
NAVAL RADIOLOGICAL DEFENSE LAB., SAN FRANCISCO	USNRDL-TMC-
NAVAL RADIOLOGICAL DEFENSE LAB., SAN FRANCISCO	USNRDL-TR-
NAVAL RADIOLOGICAL DEFENSE LAB., SAN FRANCISCO	USNRDL-TRC-
NAVAL RADIOLOGICAL DEFENSE LAB., SAN FRANCISCO	USNRDL-TRL-
NAVAL RADIOLOGICAL DEFENSE LAB. CIVIL DEFENSE TECHNICAL GROUP, SAN FRANCISCO	NRDL-TRC-
NAVAL RECONNAISSANCE AND TECHNICAL SUPPORT CENTER, WASHINGTON, D.C.	NAVRECONTECHSUPPCEN-
NAVAL RESEARCH LAB., WASHINGTON, D.C. (REF. 1)	(NUMBERS...)
NAVAL RESEARCH LAB., WASHINGTON, D.C.	AICBM-
NAVAL RESEARCH LAB., WASHINGTON, D.C.	AMB-
NAVAL RESEARCH LAB., WASHINGTON, D.C.	C-
NAVAL RESEARCH LAB., WASHINGTON, D.C.	CM-
NAVAL RESEARCH LAB., WASHINGTON, D.C.	HYDRA-SUMMARY-
NAVAL RESEARCH LAB., WASHINGTON, D.C.	NAREC-REF-
NAVAL RESEARCH LAB., WASHINGTON, D.C.	NRL-
NAVAL RESEARCH LAB., WASHINGTON, D.C. (WITH REPT. NO. 3500 THE LETTER WAS DROPPED)	NRL-(LETTER)-
NAVAL RESEARCH LAB., WASHINGTON, D.C. (ARMY-NAVY PRECIPITATION STATIC TECHNICAL REPORTS)	NRL A-
NAVAL RESEARCH LAB., WASHINGTON, D.C.	NRL-B-
NAVAL RESEARCH LAB., WASHINGTON, D.C.	NRL-BIB-
NAVAL RESEARCH LAB., WASHINGTON, D.C.	NRL-BIBL-
NAVAL RESEARCH LAB., WASHINGTON, D.C.	NRL-BULL-
NAVAL RESEARCH LAB., WASHINGTON, D.C.	NRL-C-
NAVAL RESEARCH LAB., WASHINGTON, D.C.	NRL-(DIV)-(NO./YR)
NAVAL RESEARCH LAB., WASHINGTON, D.C.	NRL-H-
NAVAL RESEARCH LAB., WASH., D.C. (INSTRUCTION BOOK)	NRLIB-
NAVAL RESEARCH LAB., WASHINGTON, D.C.	NRL-M-
NAVAL RESEARCH LAB., WASHINGTON, D.C.	NRL-MEMO-

NAVAL RESEARCH LAB., WASHINGTON, D.C. (MEMO. RPT.)	NRL MR-
NAVAL RESEARCH LAB., WASHINGTON, D.C.	NRL-O-
NAVAL RESEARCH LAB., WASHINGTON, D.C.	NRL-P-
NAVAL RESEARCH LAB., WASHINGTON, D.C.	NRL-PR-(MONTH/YEAR)
NAVAL RESEARCH LAB., WASHINGTON, D.C.	NRL-R-
NAVAL RESEARCH LAB., WASHINGTON, D.C.	NRL-RD-
NAVAL RESEARCH LAB., WASHINGTON, D.C.	NRL-S-
NAVAL RESEARCH LAB., WASHINGTON, D.C. (TEST RPTS.)	NRL TE-
NAVAL RESEARCH LAB., WASHINGTON, D.C.	NRL-TER-
NAVAL RESEARCH LAB., WASHINGTON, D.C.	NRL-TEST AND EVALUATION-
NAVAL RESEARCH LAB., WASHINGTON, D.C.	NRL-TM-
NAVAL RESEARCH LAB., WASHINGTON, D.C. (TRANSLATIONS)	NRL TR-
NAVAL RESEARCH LAB., WASHINGTON, D.C.	NRL-TRANS-
NAVAL RESEARCH LAB., WASHINGTON, D.C.	P-
NAVAL RESEARCH LAB., WASHINGTON, D.C.	PNR-
NAVAL RESEARCH LAB., WASHINGTON, D.C.	RD-
NAVAL RESEARCH LAB., WASHINGTON, D.C.	RDS-
NAVAL RESEARCH LAB., WASHINGTON, D.C.	S-(NUMBER)-(NUMBER/NO.)
NAVAL RES. LAB. COMBINED COMMUNICATIONS BD., WASH., DC	CCB-
NAVAL RESEARCH LAB. COMBINED RESEARCH GP., WASH., D.C.	CRG-
NAVAL RESEARCH LAB. CYCLOTRON BRANCH, WASHINGTON, DC	CYCLOTRON-
NAVAL RESEARCH LAB. GUIDED MISSILE RELAY WORKING GROUP, WASHINGTON, D.C.	NRL-GMRWG-
NAVAL RESEARCH LAB. RADIO MATERIEL SCHOOL, WASH., D.C.	RMS-
NAVAL RESEARCH LAB. UNDERWATER SOUND REFERENCE DIV., ORLANDO, FLA.	USNRDL-TRX-
NAVAL RESERVE, WASHINGTON, D.C.	USNR-
NAVAL RETRAINING COMMAND, SAN DIEGO, CALIF.	NAVRETRACOM-
NAVAL SCHOOL OF AVIATION MEDICINE, PENSACOLA, FLA.	MONO-
NAVAL SCHOOL OF AVIATION MEDICINE, PENSACOLA, FLA.	NAMI-
NAVAL SCHOOL OF AVIATION MEDICINE, PENSACOLA, FLA.	NM-(PROJECT NO.)-
NAVAL SCHOOL OF AVIATION MEDICINE, PENSACOLA, FLA.	NSAM-
NAVAL SCHOOL OF AVIATION MEDICINE, PENSACOLA, FLA.	NSAM-RR-
NAVAL SCHOOL OF MINE WARFARE, CHARLESTON, S.C.	NSMW-TP-
NAVAL SCI. AND TECH. INTELLIGENCE CENTER, WASH., D.C.	NIC-TRANS-
NAVAL SHIP ENGINEERING CENTER, PHILADELPHIA	NAVSECPHILADIV-A-
NAVAL SHIP ENGINEERING CENTER, PHILADELPHIA	NAVSECPHILADIV-AP-
NAVAL SHIP ENGINEERING CENTER, PHILADELPHIA	NAVSECPHILADIV-B-
NAVAL SHIP ENGINEERING CENTER, PHILADELPHIA	NAVSECPHILADIV-C-
NAVAL SHIP ENGINEERING CENTER, PHILADELPHIA	NAVSECPHILADIV-FB-
NAVAL SHIP ENGINEERING CENTER, PHILADELPHIA	NAVSEC-X-
NAVAL SHIP ENGINEERING CTR. PORT HUENEME DIV., CALIF.	NAVSEC-
NAVAL SHIP MISSILE SYSTEMS ENG. STA., PORT HUENEME, CAL	NSMSES-
NAVAL SHIP RESEARCH AND DEV. CENTER, ANNAPOLIS	ANNADIV-
NAVAL SHIP RESEARCH AND DEV. CENTER, ANNAPOLIS	ELECLAB-
NAVAL SHIP RESEARCH AND DEV. CENTER, ANNAPOLIS	ELECLAB-(NO.)/(YEAR)
NAVAL SHIP RESEARCH AND DEV. CENTER, ANNAPOLIS	MACHLAB-
NAVAL SHIP RESEARCH AND DEV. CENTER, ANNAPOLIS	MATLAB-
NAVAL SHIP RESEARCH AND DEV. CENTER, ANNAPOLIS	NSRDC-
NAVAL SHIP RESEARCH AND DEV. CENTER, WASHINGTON, D.C.	NSRDC-
NAVAL SHIP RESEARCH AND DEV. CENTER, WASHINGTON, D.C.	NSRDC-TE-P-
NAVAL SHIP RESEARCH AND DEV. CENTER, WASHINGTON, D.C.	NSRDC-TRANS-
NAVAL SHIP RESEARCH AND DEVELOPMENT CENTER. AERODYNAMICS LAB., WASHINGTON, D.C.	AL-
NAVAL SHIP RESEARCH AND DEVELOPMENT CENTER. DEPT. OF ACOUSTICS AND VIBRATION, WASHINGTON, D.C.	DAV-(NUMBER)-
NAVAL SHIP RESEARCH AND DEV. LAB., ANNAPOLIS, MD.	NSDRL/A-(NUMBER)-
NAVAL SHIP RESEARCH AND DEV. LAB., ANNAPOLIS, MD.	NSRDL/A-(NUMBER)-
NAVAL SHIP RESEARCH AND DEV. LAB., PANAMA CITY, FLA.	NSRDL/PC-
NAVAL SHIP SYSTEMS COMMAND, WASHINGTON, D.C.	NAVSHIPS-
NAVAL SHIP SYSTEMS COMMAND, WASHINGTON, D.C.	NSSC-
NAVAL SHIP SYSTEMS COMMAND, WASHINGTON, D.C.	SHIPS-
NAVAL SHIP SYSTEMS COMMAND, WASHINGTON, D.C.	TT-(YEAR)-
NAVAL SPACE SURVEILLANCE SYSTEM, DAHLGREN, VA.	NAVSPASUR-
NAVAL SUBMARINE BASE, NEW LONDON, CONN.	SBNL-TM-(NUMBER-YEAR)
NAVAL SUBMARINE BASE, NEW LONDON, CONN.	SBNL-TR-(NUMBERS)-(YEAR)
NAVAL SUBMARINE MEDICAL CENTER, GROTON, CONN.	MEMO-(YEAR)-
NAVAL SUBMARINE MEDICAL CENTER, GROTON, CONN.	MR-
NAVAL SUBMARINE MEDICAL CENTER, GROTON, CONN.	MR-(YEAR)-
NAVAL SUBMARINE MEDICAL CENTER, GROTON, CONN.	NAVMED-MR(NUMBER)-
NAVAL SUBMARINE MEDICAL CENTER, GROTON, CONN.	NSMC-
NAVAL SUBMARINE MEDICAL CENTER, GROTON, CONN.	NSMC-MR-(YEAR)-
NAVAL SUBMARINE MEDICAL CENTER. SUBMARINE MEDICAL RESEARCH LAB., GROTON, CONN.	SMRL-MR-(YEAR)-
NAVAL SUBMARINE MEDICAL CENTER. SUBMARINE MEDICAL RESEARCH LAB., GROTON, CONN.	SR-(YEAR)-
NAVAL SUPPLY DEPOT	NSD-
NAVAL SUPPLY DEPOT, MECHANICSBURG, PENNA.	ALRAND-(NUMBERS)(LTR.)
NAVAL SUPPLY RES. AND DEV. FACILITY, BAYONNE, N.J.	NSRDF-
NAVAL SUPPLY SYSTEMS COMMAND, WASHINGTON, D.C.	NAVSSC-
NAVAL SUPPLY SYSTEMS COMMAND, WASHINGTON, D.C.	SUP-
NAVAL TECHNICAL INTELLIGENCE CENTER (ABSTRACT)	NAV TEC ICA-
NAVAL TECHNICAL MISSION IN EUROPE	NAV TEC MIS EU-
NAVAL TECHNICAL MISSION IN EUROPE (LETTER REPORTS)	NAV TEC MIS EU LR-...
NAVAL TECHNICAL MISSION IN EUROPE	NAV TEC MIS EU TR-

NAVAL TECHNICAL MISSION IN EUROPE

NAVAL TECHNICAL MISSION IN EUROPE	NAV TEC MIS EU TR (NO.)-
NAVAL TECHNICAL MISSION IN EUROPE	NAV-TME-
NAVAL TECHNICAL MISSION IN EUROPE	NTME-
NAVAL TECHNICAL MISSION IN EUROPE	TEC MIS EU-
NAVAL TECHNICAL MISSION IN EUROPE	TR-
NAVAL TECHNICAL MISSION TO JAPAN	NAVTECHMISJAP-
NAVAL TECHNICAL MISSION TO JAPAN	NAV TEC MIS JAP-
NAVAL TECHNICAL MISSION TO JAPAN	NAV TEC MIS JAP (LTR.)-
NAVAL TECHNICAL MISSION TO JAPAN (ELECTRONICS RPTS.)	NAV TEC MIS JAP E-
NAVAL TECH. MISSION TO JAPAN (MEDICAL TARGET RPTS.)	NAV TEC MIS JAP M-
NAVAL TECHNICAL MISSION TO JAPAN (ORDNANCE REPORTS)	NAV TEC MIS JAP O-
NAVAL TECHNICAL MISSION TO JAPAN (SUBJECT REPORTS)	NAV TEC MIS JAP S-
NAVAL TECHNICAL MISSION TO JAPAN (MISCELLANEOUS RPTS)	NAV TEC MIS JAP X-
NAVAL TECHNICAL MISSION TO JAPAN	NTMJ-
NAVAL TECHNICAL MISSION TO JAPAN	TEC MIS JAP-
NAVAL TECHNICAL MISSION TO JAPAN	X-
NAVAL TECHNICAL UNIT IN EUROPE	NAV TEC UN EU-
NAVAL TECHNICAL UNIT IN EUROPE	NTU-
NAVAL TECHNICAL UNIT IN EUROPE	TEC UN EU-
NAVAL TEST PILOT SCHOOL, PATUXENT RIVER, MD.	USNTPS-FTM-
NAVAL TORPEDO STATION, KEYPORT, WASH.	AG(SS)-
NAVAL TORPEDO STATION, KEYPORT, WASH.	B-
NAVAL TORPEDO STATION, KEYPORT, WASH.	CG-
NAVAL TORPEDO STATION, KEYPORT, WASH.	DD-
NAVAL TORPEDO STATION, KEYPORT, WASH.	DDE-
NAVAL TORPEDO STATION, KEYPORT, WASH.	DLG-
NAVAL TORPEDO STATION, KEYPORT, WASH.	MCB-
NAVAL TORPEDO STATION, KEYPORT, WASH.	NAVTORPSTA-
NAVAL TORPEDO STATION, KEYPORT, WASH.	NTS-
NAVAL TORPEDO STATION, KEYPORT, WASH.	SS-
NAVAL TORPEDO STA., NEWPORT, R.I. (CONSECUTIVE RPTS.)	NTS N CR-
NAVAL TORPEDO STATION. QUALITY EVALUATION LAB., KEYPORT, WASH.	QE/K-
NAVAL TRAINING CENTER, SAN DIEGO, CALIF.	NTC-
NAVAL TRAINING DEVICES CENTER, ORLANDO, FLA.	NAVTRADEVCEN-IH-
NAVAL TRAINING DEVICES CENTER, PORT WASHINGTON, N.Y.	NAVTRADEVCEN-
NAVAL TRAINING DEVICES CENTER, PORT WASHINGTON, N.Y.	NTDC-
NAVAL TRAINING DEVICES CENTER, PORT WASHINGTON, N.Y.	NTDC-TR-
NAVAL TRAINING DEVICES CENTER, PORT WASHINGTON, N.Y.	SDC-
NAVAL UNDERSEA RESEARCH AND DEV. CTR., SAN DIEGO, CAL	NUC-TN-
NAVAL UNDERSEA RESEARCH AND DEV. CTR., SAN DIEGO, CAL	NUC-TP-
NAVAL UNDERSEA WARFARE CENTER, PASADENA, CALIF.	NUWC-TP-
NAVAL UNDERSEA WARFARE CENTER, SAN DIEGO, CALIF.	NUWC-TP-
NAVAL UNDERWATER ORDNANCE STATION, NEWPORT, R.I.	NTS-
NAVAL UNDERWATER ORDNANCE STATION, NEWPORT, R.I.	NUOS-
NAVAL UNDERWATER ORDNANCE STATION, NEWPORT, R.I.	NUOS-CONSECUTIVE-
NAVAL UNDERWATER ORDNANCE STATION, NEWPORT, R.I.	NUOS-ITN-(NUMBER)-(YR.)
NAVAL UNDERWATER ORDNANCE STATION, NEWPORT, R.I.	QE/N-(YEAR)-(LETTERS)-
NAVAL UNDERWATER ORDNANCE STATION. QUALITY EVALUATION LAB., NEWPORT, R.I.	QE/N-
NAVAL UNDERWATER SYSTEMS CTR. NEW LONDON LAB., CONN.	NUSC/NL-
NAVAL UNDERWATER SYSTEMS CTR., NEWPORT, R.I.	NUSC-CR-
NAVAL UNDERWATER SYSTEMS CTR., NEWPORT, R.I.	NUSC-TR-
NAVAL UNDERWATER WEAPONS RES. + ENG.STA,NEWPORT, R.I.	NUWS-CR-
NAVAL UNDERWATER WEAPONS RES. + ENG.STA,NEWPORT, R.I.	NUWS-QE/N-(YR.)-(LTRS.)
NAVAL UNDERWATER WEAPONS RES. + ENG.STA,NEWPORT, R.I.	NUWS-TR-
NAVAL WARFARE ANALYSIS GROUP, WASHINGTON, D.C.	NIRM-
NAVAL WARFARE ANALYSIS GROUP, WASHINGTON, D.C.	RESEARCH CONTRIBUTION-
NAVAL WEAPON SYSTEMS ANALYSIS OFFICE, QUANTICO, VA.	WSAO-
NAVAL WEAPON SYSTEMS ANALYSIS OFFICE, QUANTICO, VA.	WSAO-R-
NAVAL WEAPONS CENTER, CHINA LAKE, CALIF.	NWC-TN-
NAVAL WEAPONS CENTER, CHINA LAKE, CALIF.	NWC-TP-
NAVAL WEAPONS CENTER. CORONA ANNEX, CALIF.	NWCCA-TM-
NAVAL WEAPONS CENTER. CORONA LABS., CALIF.	NWCCL-TM-(NUMBER)-
NAVAL WEAPONS CENTER. CORONA LABS., CALIF.	NWCCL-TP-
NAVAL WEAPONS CENTER. CORONA LABS., CALIF.	NWCCL-TR-
NAVAL WEAPONS EVALUATION FACILITY, KIRTLAND AFB, N.M.	K-
NAVAL WEAPONS EVALUATION FACILITY, KIRTLAND AFB, N.M.	NWEF-
NAVAL WEAPONS HANDLING LAB., COLTS NECK, N.J.	NWHL-
NAVAL WEAPONS LAB., DAHLGREN, VA.	K-
NAVAL WEAPONS LAB., DAHLGREN, VA.	NPG-
NAVAL WEAPONS LAB., DAHLGREN, VA.	NSW-TM-
NAVAL WEAPONS LAB., DAHLGREN, VA.	NWL-
NAVAL WEAPONS LAB., DAHLGREN, VA.	NWL-AR-(NUMBER-YEAR)
NAVAL WEAPONS LAB., DAHLGREN, VA.	NWL-AR-W/(NUMBER)-(YEAR)
NAVAL WEAPONS LAB., DAHLGREN, VA.	NWL-CR-
NAVAL WEAPONS LAB., DAHLGREN, VA.	NWL-CR-N(NUMBER)-
NAVAL WEAPONS LAB., DAHLGREN, VA.	NWL-CSR-N(NUMBERS)
NAVAL WEAPONS LAB., DAHLGREN, VA.	NWL-K-(NUMBER)/(YEAR)
NAVAL WEAPONS LAB., DAHLGREN, VA.	NWL-MAL-
NAVAL WEAPONS LAB., DAHLGREN, VA.	NWL-TM-T-
NAVAL WEAPONS LAB., DAHLGREN, VA.	NWL-TM-W-
N AL WEAPONS LAB., DAHLGREN, VA.	NWL-TN-T-(NUMBER-YEAR)
NAVAL WEAPONS LAB., DAHLGREN, VA.	NWL-TN-W-(NUMBER-YEAR)
NAVAL WEAPONS LAB., DAHLGREN, VA.	NWL-TR-
NAVAL WEAPONS LAB., DAHLGREN, VA.	NWL/W-
NAVAL WEAPONS LAB., DAHLGREN, VA.	TM-W-

NAVAL WEAPONS LAB., DAHLGREN, VA.	W-(MONTH)/(YEAR)
NAVAL WEAPONS LAB. COMPUTATIONAL AND ANALYSIS LAB., DAHLGREN, VA. (TECHNICAL MEMO)	NWL-TM K/
NAVAL WEAPONS LAB. WARHEAD + TERMINAL BALLISTICS LAB., DAHLGREN, VA. (TECHNICAL MEMO)	NWL-TM T/
NAVAL WEAPONS LAB. WEAPONS DEVELOPMENT + EVALUATION LAB., DAHLGREN, VA. (TECHNICAL MEMO)	NWL-TM W/
NAVAL WEAPONS PLANT, WASHINGTON, D.C.	NGF-T-
NAVAL WEAPONS PLANT, WASHINGTON, D.C.	NWP-
NAVAL WEAPONS PLANT, WASHINGTON, D.C.	NWPW-T-(NUMBER)-(YEAR)
NAVAL WEAPONS QUALITY ASSURANCE OFFICE, WASH., D.C.	NAVWPNQAO-TR-(YEAR)-
NAVAL WEAPONS QUALITY ASSURANCE OFFICE, WASH., D.C.	QAO-
NAVAL WEAPONS STATION, YORKTOWN, VA.	NWS-
NAVAL WEAPONS STATION, YORKTOWN, VA.	NWS-R+D-
NAVAL WEAPONS STATION, YORKTOWN, VA.	NWSY-TR-(YEAR)-
NAVAL WEAPONS STATION, YORKTOWN, VA.	R + D-
NAVAL WEAPONS STATION. QUALITY EVALUATION LAB., CONCORD, CALIF.	QE/CONCORD-
NAVAL WEAPONS STATION. QUALITY EVALUATION LAB., CONCORD, CALIF.	QE/CO-TEST PROCEDURE-
NAVAL WEATHER SERVICE COMMAND, WASHINGTON, D.C.	NWS-(NUMBER)/
NAVY (ANTI-AIRCRAFT STUDY)	AAS-
NAVY (AVIATION CIRCULAR LETTER)	ACL-
NAVY (CONSOLIDATED STOCK STATUS REPORT)	CSSR-
NAVY (DESTROYER TACTICAL BULLETIN)	DTB-
NAVY (ELECTRONICS INFORMATION BULLETIN)	EIB-
NAVY (ELECTRONIC MAINTENANCE BOOK)	EMB-
NAVY (FAST DEPLOYMENT LOGISTIC PROJECT)	FDL-
NAVY (FLEET TRAINING PAMPHLET)	FTP-
NAVY (FLEET TRAINING PUBLICATION)	FTP-
NAVY (WORK IMPROVEMENT PROGRAM, GROUP IV-B. SUPERVISORY CONFERENCE REPORTS)	G IVB-
NAVY (ILLUSTRATED PARTS BREAKDOWN)	IPB-
NAVY (MINE ASSEMBLY BASE MEMORANDUM)	MABM-
NAVY (MINE ASSEMBLY AND TEST INSTRUCTIONS)	MATI-
NAVY (MINE DISPOSAL BULLETIN)	MDB-
NAVY (MINE OVERHAUL AND REWORK INSTRUCTIONS)	MORI-
NAVY (MINE WARFARE OPERATIONAL RESEARCH MEMORANDUM)	MWORM-
NAVY (MINE WARFARE OPERATIONAL RESEARCH REPORT)	MWORR-
NAVY (MINE WARFARE REPORT)	MWR-
NAVY	NAV-
NAVY (PASSIVE DEFENSE HANDBOOK)	NAV PDH-
NAVY (OPERATIONAL REQUIREMENT)	OR-
NAVY (PASSIVE DEFENSE HANDBOOK)	PDH-
NAVY (PERFORMANCE REQUIREMENTS)	PR-
NAVY (REGISTERED PUBLICATION)	RP-
NAVY (TEST EXPERIMENTAL AND DEVELOPMENT PROJECTS)	TED-
NAVY (TENTATIVE SPECIFICATIONS)	TS-
NAVY (TORPEDO STATION)	TS-
NAVY (UNSATISFACTORY REPORT)	UR-
NAVY, SEAL BEACH, CALIF.	SB-
NAVY, WASHINGTON, D.C. (ALLIED TACTICAL PUBLICATION)	ATP-
NAVY, WASHINGTON, D.C. (ATOMIC UNDERWATER WEAPONS)	AUW-
NAVY, WASHINGTON, D.C.	ENGINEERING REPORT AD-
NAVY, WASHINGTON, D.C.	N-
NAVY, WASHINGTON, D.C.	TT-
NAVY, WASHINGTON, D.C.	UH-
NAVY. EXECUTIVE OFFICE OF THE SECRETARY, WASH., D.C.	NAVEXOS-
NAVY. SPECIAL PROJECTS OFFICE, WASHINGTON, D.C.	SP-NOTE-
NAVY BOMB DISPOSAL SCHOOL, WASHINGTON, D.C.	USNBD-
NAVY CLOTHING AND TEXTILE OFFICE, BROOKLYN	NEL-
NAVY ELECTRONICS LAB., SAN DIEGO, CALIF.	NEL-
NAVY ELECTRONICS LAB., SAN DIEGO, CALIF.	NEL-(LETTER(S))-
NAVY ELECTRONICS LAB., SAN DIEGO, CALIF. (DOCUMENT)	NEL/DOC-
NAVY ELECTRONICS LAB., SAN DIEGO, CALIF.	NELS-
NAVY ELECTRONICS LAB., SAN DIEGO, CALIF.	NELS R-
NAVY ELECTRONICS LAB., SAN DIEGO, CALIF.(SERIAL RPTS)	NELS S-
NAVY ELECTRONICS LAB., SAN DIEGO, CALIF.(TECH. MEMO.)	NEL/TM-
NAVY ELECTRONICS LAB., SAN DIEGO, CALIF.(TRANSLATION)	NEL TR-
NAVY ELECTRONICS LAB., SAN DIEGO, CALIF.	NRSL-
NAVY ELECTRONICS LAB., SAN DIEGO, CALIF.	NS-R-(NUMBER)-(NUMBER)-
NAVY ELECTRONICS LAB., SAN DIEGO, CALIF.	UCDWR-
NAVY ELECTRONICS LAB., SAN DIEGO, CALIF.	USNEL-
NAVY ELECTRONICS LAB., SAN DIEGO, CALIF.	WP-
(SEE ALSO LATER FORM OF NAMEO NAVAL ELECTRONICS LAB. CENTER, SAN DIEGO, CALIF.)	
NAVY EXPERIMENTAL DIVING UNIT, WASHINGTON, D.C.	NEDU-RR-
NAVY MARINE ENGINEERING LAB., ANNAPOLIS	MEL-
NAVY MATHEMATICAL COMPUTING ADVISORY PANEL, WASH., DC	NMCP-
NAVY MEDICAL NEUROPSYCHIATRIC RESEARCH UNIT, SAN DIEGO, CALIF.	NMNRU-(YEAR)-
NAVY MINE DEFENSE LAB., PANAMA CITY, FLA.	MDL-
NAVY MINE DEFENSE LAB., PANAMA CITY, FLA.	MDL-TP-
NAVY MINE DEFENSE LAB., PANAMA CITY, FLA.	MDL-U-
NAVY MINE DEFENSE LAB., PANAMA CITY, FLA.	NMDL-
NAVY MINE DEFENSE LAB., PANAMA CITY, FLA.	S-
NAVY MINE DEFENSE LAB., PANAMA CITY, FLA.	T-
NAVY OCEANOGRAPHIC INSTRUMENTATION CTR., WASH., D.C.	IFS-(YEAR)(NUMBER)
NAVY RADIO AND SOUND LAB., SAN DIEGO, CALIF.	AERO-
NAVY RADIO AND SOUND LAB., SAN DIEGO, CALIF.	NRS LAB AERO-
NAVY RADIO AND SOUND LAB., SAN DIEGO, CALIF. (WAVE PROPAGATION)	NRSL WP-
NAVY RADIO AND SOUND LAB., SAN DIEGO, CALIF. (WAVE PROPAGATION NO.)	WP-
NAVY RESEARCH AND DEVELOPMENT UNIT, VIETNAM	NRDU-V-

NAVY SPACE SYSTEMS ACTIVITY

NAVY SPACE SYSTEMS ACTIVITY, LOS ANGELES	NSSA-R-
NAVY UNDERWATER SOUND LAB., NEW LONDON, CONN.	NAVUSL-
NAVY UNDERWATER SOUND LAB., NEW LONDON, CONN.	NE-
NAVY UNDERWATER SOUND LAB., NEW LONDON, CONN.	NUSL-
NAVY UNDERWATER SOUND LAB., NEW LONDON, CONN.	NUSL-PUB-
NAVY UNDERWATER SOUND LAB., NEW LONDON, CONN.	USL-
NAVY UNDERWATER SOUND LAB., NEW LONDON, CONN.	USL-PUB-
NAVY UNDERWATER SOUND LAB., NEW LONDON, CONN.	USL-SYMPOSIUM-
NAVY UNDERWATER SOUND LAB., NEW LONDON, CONN.	USL-TM-
NAVY UNDERWATER SOUND LAB., NEW LONDON, CONN.	USNUSL-
NAVY UNDERWATER SOUND REFERENCE LAB., MT. LAKES, N.J.	ML-
NAVY UNDERWATER SOUND REFERENCE LAB., MT. LAKES, N.J.	USRL MT. LAKES-
NAVY UNDERWATER SOUND REFERENCE LAB., ORLANDO, FLA.	USL-
NAVY UNDERWATER SOUND REFERENCE LAB., ORLANDO, FLA.	USRL-
NAVY UNDERWATER SOUND REFERENCE LAB., ORLANDO, FLA.	USRLG-
NAVY UNDERWATER SOUND REFERENCE LAB., ORLANDO, FLA.	USRLO-
NAVY UNDERWATER SOUND REFERENCE LAB., ORLANDO, FLA.	USRL-RR-
NAVY UNDERWATER SOUND REFERENCE LAB., ORLANDO, FLA.	USRL-TRANS-
NAVY WEATHER RESEARCH FACILITY, NORFOLK, VA.	NAVWEARSCHFAC-TP-
NAVY WEATHER RESEARCH FACILITY, NORFOLK, VA.	NWRF-(NO.)-(NO.)-
NAVY WEATHER RESEARCH FACILITY, NORFOLK, VA.	NWRF-
NAVY WEATHER RESEARCH FACILITY, NORFOLK, VA.	NWRF-TP-(NUMBER-NO.)-
NCS COMPUTING CORP., ROME, N.Y.	CS-(YEAR)-
NEBRASKA. STATE DEPT. OF ROADS. DIV. OF MATERIALS AND TESTS, LINCOLN	RESEARCH STUDY-(NO.-NO.)
NEBRASKA. UNIV., LINCOLN	NU-
NEBRASKA. UNIV., LINCOLN. DEPT. OF HORTICULTURE AND FORESTRY	HORTICULTURE-PR-
NEBRASKA. UNIV., LINCOLN. DEPT. OF HORTICULTURE AND FORESTRY	NEBRASKA STUDY-(YEAR)-
NEBRASKA. UNIV., LINCOLN. DEPT. OF MECHANICAL ENG.	NU-HYDRO-(NUMBERS)-TS
NERVA TEST OPERATIONS, NUCLEAR ROCKET DEVELOPMENT STATION, JACKASS FLATS, NEV.	NTO-I-
NERVA TEST OPERATIONS, NUCLEAR ROCKET DEVELOPMENT STATION, JACKASS FLATS, NEV.	NTO-R-
NERVA TEST OPERATIONS, NUCLEAR ROCKET DEVELOPMENT STATION, JACKASS FLATS, NEV.	NTO-SOP-
NERVA TEST OPERATIONS, NUCLEAR ROCKET DEVELOPMENT STATION, JACKASS FLATS, NEV.	NTO-T-
NETHERLANDS. ADVIESBUREAU DER GENIE, HAGUE	ABG-
NETHERLANDS. CENTRAAL INSTITUUT VOOR VOEDINGSONDERZOEK TNO, ZEIST	A-(YEAR)/KL/
NETHERLANDS. CENTRAAL INSTITUUT VOOR VOEDINGSONDERZOEK TNO, ZEIST	CIVO-
NETHERLANDS. CENTRAAL INSTITUUT VOOR VOEDINGSONDERZOEK TNO, ZEIST	CIVO-R-
NETHERLANDS. CENTRAAL ORGANISATIE VOOR TOEGEPASTNATURWETENSCHAPPELIJK ONDERZOEK, THE HAGUE (NAME TRANSLATED IS: NETHERLANDS ORGANIZATION FOR APPLIED SCIENTIFIC RESEARCH)	TNO-
NETHERLANDS. DIRECTIE OVERHEIDS-PERSONEELSBELEID, THE HAGUE	DIR.O.P.-(YEAR)/
NETHERLANDS. FDN. FOR FUNDAMENTAL RES. OF MATTER	FOM-
NETHERLANDS. FOM-INSTITUUT VOOR PLASMA-FYSICA, JUTPHAAS	RIJNHUIZEN-(YR.)-
NETHERLANDS. HOUTINSTITUUT TNO, DELFT	H-(YEAR-NUMBER)
NETHERLANDS. INSTITUTE FOR NUCLEAR PHYSICS RESEARCH, AMSTERDAM	INTERIKO-(YEAR/NUMBER)
NETHERLANDS. INSTITUTE FOR NUCLEAR PHYSICS RESEARCH, AMSTERDAM	KE-
NETHERLANDS. INST. FOR ROAD SAFETY RESEARCH, VOORBURG	SWOV (YEAR)-
NETHERLANDS. INSTITUUT TNO VOOR VERPAKKINGEN, DELFT	IVA-MEDD-
NETHERLANDS. INSTITUUT TNO VOOR VERPAKKINGEN, DELFT	IVV-
NETHERLANDS. INSTITUUT VOOR TOEPASSING VAN ATOOMENERGIE IN DE LANDBOUW, WAGENINGEN (NAME TRANSLATED IS: INSTITUTE FOR ATOMIC SCIENCES IN AGRICULTURE)	ITAL-
NETHERLANDS. INSTITUUT VOOR TEXTIELREINIGING TNO, DELFT	CB-
NETHERLANDS. INSTITUUT VOOR TEXTIELREINIGING TNO, DELFT	CS-
NETHERLANDS. INSTITUUT VOOR ZINTUIGFYSIOLOGIE RVO-TNO, SOESTERBERG	IZF-
NETHERLANDS. MATHEMATISCH CENTRUM, AMSTERDAM	MR-
NETHERLANDS. MATHEMATISCH CENTRUM. AFDELING TOEGEPASTE WISKUNDE, AMSTERDAM	TW-(3 DIGITS)
NETHERLANDS. METAALINSTITUUT TNO, DELFT	C-(YEAR)-
NETHERLANDS. METAALINSTITUUT TNO, DELFT	M-(YEAR)-
NETHERLANDS. METAALINSTITUUT TNO, THE HAGUE	M(YEAR)-
NETHERLANDS. METAALINSTITUUT TNO, THE HAGUE	M(YEAR)-(NO.)/ADR/
NETHERLANDS. METAALINSTITUUT TNO, THE HAGUE	MI-
NETHERLANDS. MINISTRY OF DEFENCE. MATERIEL COUNCIL, THE HAGUE	MMR-
NETHERLANDS. MINISTRY OF ECONOMIC AFFAIRS. MILITARY MISSION TO A.C.C. (TECHNICAL INVESTIGATION RPT.)	TOR-
NETHERLANDS. NATIONAAL LUCHT EN RUIMTEVAARTLABORATORIUM, AMSTERDAM	ELDO-M-/(YEAR)/
NETHERLANDS. NATIONAAL LUCHT EN RUIMTEVAARTLABORATORIUM, AMSTERDAM	FF-
NETHERLANDS. NATIONAAL LUCHT EN RUIMTEVAARTLABORATORIUM, AMSTERDAM	G-
NETHERLANDS. NATIONAAL LUCHT EN RUIMTEVAARTLABORATORIUM, AMSTERDAM	MP-
NETHERLANDS. NATIONAAL LUCHT EN RUIMTEVAARTLABORATORIUM, AMSTERDAM	NCR-TN-G-
NETHERLANDS. NATIONAAL LUCHT EN RUIMTEVAARTLABORATORIUM, AMSTERDAM	NLR-G-
NETHERLANDS. NATIONAAL LUCHT EN RUIMTEVAARTLABORATORIUM, AMSTERDAM	NLR-MEMO-WR-
NETHERLANDS. NATIONAAL LUCHT EN RUIMTEVAARTLABORATORIUM, AMSTERDAM	NLR-MP-
NETHERLANDS. NATIONAAL LUCHT EN RUIMTEVAARTLABORATORIUM, AMSTERDAM	NLR-SR-
NETHERLANDS. NATIONAAL LUCHT EN RUIMTEVAARTLABORATORIUM, AMSTERDAM	NLR-TM-
NETHERLANDS. NATIONAAL LUCHT EN RUIMTEVAARTLABORATORIUM, AMSTERDAM	NLR-TN-
NETHERLANDS. NATIONAAL LUCHT EN RUIMTEVAARTLABORATORIUM, AMSTERDAM	NLR-TN-F-
NETHERLANDS. NATIONAAL LUCHT EN RUIMTEVAARTLABORATORIUM, AMSTERDAM	NLR-TN-G-
NETHERLANDS. NATIONAAL LUCHT EN RUIMTEVAARTLABORATORIUM, AMSTERDAM	NLR-TN-M-
NETHERLANDS. NATIONAAL LUCHT EN RUIMTEVAARTLABORATORIUM, AMSTERDAM	NLR-TN-T-
NETHERLANDS. NATIONAAL LUCHT EN RUIMTEVAARTLABORATORIUM, AMSTERDAM	NLR-TR-
NETHERLANDS. NATIONAAL LUCHT EN RUIMTEVAARTLABORATORIUM, AMSTERDAM	NLR-TR-F-
NETHERLANDS. NATIONAAL LUCHT EN RUIMTEVAARTLABORATORIUM, AMSTERDAM	NLR-TR-S-
NETHERLANDS. NATIONAAL LUCHT EN RUIMTEVAARTLABORATORIUM, AMSTERDAM	NLR-TR-T-

NETHERLANDS. NATIONAAL LUCHT EN RUIMTEVAARTLABORATORIUM, AMSTERDAM	NLR-V-
NETHERLANDS. NATIONAAL LUCHT EN RUIMTEVAARTLABORATORIUM, AMSTERDAM (NAME TRANSLATED IS: NATIONAL AERO- AND ASTRONAUTICAL RESEARCH INSTITUTE)	T-
NETHERLANDS. NATIONAAL LUCHTVAARTHLABORATORIUM, AMSTERDAM	NLL-
NETHERLANDS. NATIONAAL LUCHTVAARTHLABORATORIUM, AMSTERDAM (NAME TRANSLATED ISO NATL. AERONAUTICAL RES. LAB.)	NLL-MP-
NETHERLANDS. NATL. AEROSPACE LAB., AMSTERDAM	F-
NETHERLANDS. NATL. AEROSPACE LAB., AMSTERDAM	NRL-TR-(NO.)-U
NETHERLANDS. NIJVERHEIDSORGANISATIE, TNO, DELFT	SR-TN/(NUMBER)-
NETHERLANDS. RADIOBIOLOGISCH INSTITUUT TNO, RIJSWIJK	RIGO-
NETHERLANDS. RADIOBIOLOGISCH INSTITUUT TNO, RIJSWIJK	RIGO-(YEAR)/
NETHERLANDS. REACTOR CENTRUM, PETTEN	KR-
NETHERLANDS. REACTOR CENTRUM, PETTEN	RCN-
NETHERLANDS. REACTOR CENTRUM, THE HAGUE	RCM-
NETHERLANDS. REACTOR CENTRUM, THE HAGUE	RCN-
NETHERLANDS. REACTOR CENTRUM, THE HAGUE	RCN-INT-
NETHERLANDS. REACTOR INSTITUUT, DELFT	RI-
NETHERLANDS. RIJKS INSTITUUT VOOR DE VOLKSGEZONDHEID, UTRECHT	RA-
NETHERLANDS. RIJKS INSTITUUT VOOR DE VOLKSGEZONDHEID. LABORATORIUM VOOR STRALINGSONDERZOEK, BILTHOVEN	RA-
NETHERLANDS. RIJKS PSYCHOLOGISCHE DIENST, THE HAGUE	RPD-
NETHERLANDS. RIJKSVERDEDIGINGSORGANISATIE, TNO. CHEMICAL LAB., RIJSWIJK	CL-(YEAR)-
NETHERLANDS. RIJKSVERDEDIGINGSORGANISATIE, TNO. MEDISCH BIOLOGISCH LABORATORIUM, RIJSWIJK	MBL-(YEAR)/
NETHERLANDS. RIJKSVERDEDIGINGSORGANISATIE, TNO. PHYSISCH LABORATORIUM, THE HAGUE	PHL-(YEAR)-
NETHERLANDS. RIJKSVERDEDIGINGSORGANISATIE, TNO. PHYSISCH LABORATORIUM, THE HAGUE	PH.L.(YEAR)-
NETHERLANDS. ROYAL NETHERLANDS AIRCRAFT FACTORIES, FOKKER, AMSTERDAM	RV-
NETHERLANDS. ROYAL NETHERLANDS AIRCRAFT FACTORIES, FOKKER, AMSTERDAM	S-(NUMBER)
NETHERLANDS. STAATSMIJNEN IN LIMBURG, GELEEN	CL-V-
NETHERLANDS. STICHTING MOEILIJK TOEGANKELIJKE WETENSCHAPPELIJKE LITERATUUR, DELFT (NAME TRANSLATED IS: FOUNDATION FOR SCIENTIFIC LITERATURE DIFFICULT OF ACCESS)	MTWL-
NETHERLANDS. TECHNISCH DOCUMENTATIE EN INFORMATIE CENTRUM VOOR DE KRIJGSMACHT, THE HAGUE	IDCK-
NETHERLANDS. TECHNISCH DOCUMENTATIE EN INFORMATIE CENTRUM VOOR DE KRIJGSMACHT, THE HAGUE	TDCK-
NETHERLANDS. TECHNISCH DOCUMENTATIE EN INFORMATIE CENTRUM VOOR DE DRIJGSMACHT, THE HAGUE	TDCL-
NETHERLANDS. TECHNISCH PHYSISCHE DIENST TNO-TH, DELFT	TPD-
NETHERLANDS. TECHNOLOGICAL LAB. RVO-TNO, RIJSWIJK	TL-(YEAR)-
NETHERLANDS. TNO. CENTRAAL LABORATORIUM, DELFT	CL-
NETHERLANDS. TNO. CENTRAAL LABORATORIUM, DELFT	CL-(YEAR)-
NETHERLANDS. TNO. CENTRAAL LABORATORIUM, DELFT	M-(YEAR)-
NETHERLANDS. VERFINSTITUUT TNO, DELFT	V-(YEAR)-
NEUTRON CROSS SECTION ADVISORY GROUP, AEC	P-(NO.)-
NEVADA. UNIV., RENO	UN-
NEVADA. UNIV., RENO. DESERT RESEARCH INST.	NEVU-TR-
NEVADA. UNIV., RENO. MACKAY SCHOOL OF MINES	UN-TR-
NEVADA OPERATIONS OFFICE, AEC, LAS VEGAS (NEVADA AERIAL TRACKING SYSTEM) (REF. 4)	NATS-NVOO-
NEVADA OPERATIONS OFFICE, AEC, LAS VEGAS (NEVADA AERIAL TRACKING SYSTEM) (REF. 4)	NATS-PR-(NUMBER)-P-(YR.)
NEVADA OPERATIONS OFFICE, AEC, LAS VEGAS	NOO-
NEVADA OPERATIONS OFFICE, AEC, LAS VEGAS	NVO-
NEVADA OPERATIONS OFFICE, AEC, LAS VEGAS (REF. 4)	NVOO-
NEVADA OPERATIONS OFFICE, AEC, LAS VEGAS (REF. 4)	RUFUS-RR-
NEVADA OPERATIONS OFFICE, AEC, LAS VEGAS	
NEVADA PROVING GROUND	NPG-
NEVADA TEST ORGANIZATION. CIVIL EFFECTS TEST GP., AEC	CETG-
NEVADA TEST ORGANIZATION. OFF-SITE RADIOLOGICAL SAFETY ACTIVITIES, AEC	OTO-
NEW BRUNSWICK LAB., AEC, N.J.	NBL-
NEW ENGLAND LIME CO., ADAMS, MASS.	NEL-
NEW ENGLAND MATERIALS LAB., INC., MEDFORD, MASS.	NEMLAB-
NEW HAMPSHIRE. UNIV., DURHAM	AR-
NEW HAMPSHIRE. UNIV., DURHAM	NHU-
NEW HAMPSHIRE. UNIV., DURHAM	N.H. UNIV.-
NEW HAMPSHIRE. UNIV., DURHAM	UNH-(YEAR)-
NEW HAMPSHIRE. UNIV., DURHAM	UNH-R-(YEAR)-
NEW HAMPSHIRE. UNIV., DURHAM. DEPT. OF PHYSICS	UNH-
NEW JERSEY. CERAMIC RESEARCH STATION, NEW BRUNSWICK	CRS-
NEW JERSEY. CERAMIC RESEARCH STATION, NEW BRUNSWICK	NJ CRS-
NEW JERSEY. CERAMIC RESEARCH STATION, NEW BRUNSWICK	NJCRS-
NEW JERSEY. CERAMIC RESEARCH STATION, NEW BRUNSWICK	RU CRS-
NEW MEXICO. UNIV., ALBUQUERQUE	MDC-TR-(YEAR)-
NEW MEXICO. UNIV., ALBUQUERQUE	N.MEX. EE-
NEW MEXICO. UNIV., ALBUQUERQUE	NMU-
NEW MEXICO. UNIV., ALBUQUERQUE	NMU-QPR-
NEW MEXICO. UNIV., ALBUQUERQUE	UNM-
NEW MEXICO. UNIV., ALBUQUERQUE	UNM-TR-
NEW MEXICO. UNIV., ALBUQUERQUE. BUREAU OF ENG. RES.	CE-(2 DIGITS)(YEAR)-
NEW MEXICO. UNIV., ALBUQUERQUE. BUREAU OF ENG. RES.	EE-
NEW MEXICO. UNIV., ALBUQUERQUE. BUREAU OF ENG. RES.	EE-(NUMBER)/(YEAR)/

NEW MEXICO. UNIV., ALBUQUERQUE. BUREAU OF ENG. RES.	EE-(3 DIGITS)(YEAR)DC-
NEW MEXICO. UNIV., ALBUQUERQUE. BUREAU OF ENG. RES.	EE-(3 DIGITS)(YEAR)ONR-
NEW MEXICO. UNIV., ALBUQUERQUE. BUREAU OF ENG. RES.	EE-(3 DIGITS)(YEAR)HAFB-
NEW MEXICO. UNIV., ALBUQUERQUE. BUREAU OF ENG. RES.	ME-
NEW MEXICO. UNIV., ALBUQUERQUE. BUREAU OF ENG. RES.	NMU-EE-
NEW MEXICO. UNIV., ALBUQUERQUE. BUREAU OF ENG. RES.	TR-ME-
NEW MEXICO. UNIV., ALBUQUERQUE. ENG. EXPERIMENT STA.	EE-
NEW MEXICO. UNIV., ALBUQUERQUE. ENG. EXPERIMENT STA.	N.MEX. U. TR-EE-
NEW MEXICO. UNIV., ALBUQUERQUE. ENG. EXPERIMENT STA.	UNM-PR-
NEW MEXICO. UNIV., ALBUQUERQUE. ENG. EXPERIMENT STA.	UNM-PR-EE-
NEW MEXICO. UNIV., ALBUQUERQUE. ENG. EXPERIMENT STA.	UNM-QPR-
NEW MEXICO. UNIV., ALBUQUERQUE. ENG. EXPERIMENT STA.	UNM-TR-EE-
NEW MEXICO. UNIV., ALBUQUERQUE. TECHNOLOGY APPLICATIONS CENTER	RS-MC-
NEW MEXICO COLL. OF AGRICULTURE AND MECHANIC ARTS, STATE COLLEGE	AEX-
NEW MEXICO COLL. OF AGRICULTURE AND MECHANIC ARTS, STATE COLLEGE. PHYSICAL SCIENCE LAB.	AER-
NEW MEXICO COLL. OF AGRICULTURE AND MECHANIC ARTS, STATE COLLEGE. PHYSICAL SCIENCE LAB.	NMA-(NO.)-(LETTERS)-
NEW MEXICO COLL. OF AGRICULTURE AND MECHANIC ARTS, STATE COLLEGE. PHYSICAL SCIENCE LAB.	NMA+M-AER-
NEW MEXICO COLL. OF AGRICULTURE AND MECHANIC ARTS, STATE COLLEGE. PHYSICAL SCIENCE LAB.	NMA-M-CM-
NEW MEXICO COLL. OF AGRICULTURE AND MECHANIC ARTS, STATE COLLEGE. PHYSICAL SCIENCE LAB.	NMA+M-PSL-
NEW MEXICO COLL. OF AGRICULTURE AND MECHANIC ARTS, STATE COLLEGE. PHYSICAL SCIENCE LAB.	NMC-AER-
NEW MEXICO INST. OF MINING AND TECH., SOCORRO	NMI-
NEW MEXICO INST. OF MINING AND TECH., SOCORRO	NMIMT-
NEW MEXICO INST. OF MINING AND TECH., SOCORRO	NMT-T-
NEW MEXICO INST. OF MINING AND TECH., SOCORRO	T-
NEW MEXICO INST. OF MINING AND TECH., SOCORRO. DEPT. OF MINERAL TECHNOLOGY	NMI-MT-
NEW MEXICO MILITARY INST., ROSWELL	NMMI-
NEW MEXICO SCHOOL OF MINES, SOCORRO	NMSM/RDD/T-
NEW MEXICO SCHOOL OF MINES. RESEARCH AND DEVELOPMENT DIV., SOCORRO	NMSM-TR-
NEW MEXICO STATE UNIV., UNIVERSITY PARK	CR-
NEW MEXICO STATE UNIV., UNIVERSITY PARK	NM STATE U-(NO)-QPR-
NEW MEXICO STATE UNIV., UNIVERSITY PARK	NM STATE U-FDR-
NEW MEXICO STATE UNIV., UNIVERSITY PARK	NMSU-
NEW MEXICO STATE UNIV., UNIVERSITY PARK	NMSU-FDR-
NEW MEXICO STATE UNIV., UNIVERSITY PARK	NMSU-R-
NEW MEXICO STATE UNIV., UNIVERSITY PARK	NMSU-TALOS-
NEW MEXICO STATE UNIV., UNIVERSITY PARK	PT(NUMBERS)
NEW MEXICO STATE UNIV., UNIVERSITY PARK	SATR-
NEW MEXICO STATE UNIV., UNIVERSITY PARK. PHYSICAL SCIENCE LAB.	AMC-
NEW MEXICO STATE UNIV., UNIVERSITY PARK. PHYSICAL SCIENCE LAB.	NMSU-AMC-(NO.)-(NO.)
NEW MEXICO STATE UNIV., UNIVERSITY PARK. PHYSICAL SCIENCE LAB.	PSL/NMSU-
NEW MEXICO STATE UNIV., UNIVERSITY PARK. PHYSICAL SCIENCE LAB.	PSL-PM(NUMBER)
NEW MEXICO STATE UNIV., UNIVERSITY PARK. PHYSICAL SCIENCE LAB.	PSL-PQ(NUMBER)
NEW MEXICO STATE UNIV., UNIVERSITY PARK. PHYSICAL SCIENCE LAB.	PSL-PR(NUMBER)
NEW MEXICO STATE UNIV., UNIVERSITY PARK. PHYSICAL SCIENCE LAB.	PSL-PS(NUMBER)
NEW MEXICO STATE UNIV., UNIVERSITY PARK. PHYSICAL SCIENCE LAB.	PSL-PT(NUMBER)
NEW MEXICO STATE UNIV., UNIVERSITY PARK. PHYSICAL SCIENCE LAB.	PX-
NEW MEXICO STATE UNIV., UNIVERSITY PARK. WATER RESOURCES RESEARCH INST.	WRRI-RR-
NEW SOUTH WALES, AUSTRALIA. UNIV., KENSINGTON	NRS-
NEW SOUTH WALES, AUSTRALIA. UNIV., KENSINGTON. DEPT. OF NUCLEAR AND RADIATION CHEMISTRY	NRS-
NEW SOUTH WALES, AUSTRALIA. UNIV., KENSINGTON. DEPT. OF STRUCTURAL MECHANICS	UNICIV-R-
NEW SOUTH WALES, AUSTRALIA. UNIV., KENSINGTON. INST. OF NUCLEAR ENGINEERING	INE/M-
NEW YORK. OFFICE OF ATOMIC DEVELOPMENT, ALBANY	NUS-
NEW YORK. STATE DEPT. OF HEALTH, ALBANY	EXEP-
NEW YORK. STATE DEPT. OF HEALTH, ALBANY	RSG-
NEW YORK. STATE DEPT. OF HEALTH, ALBANY	RSG-A-
NEW YORK. STATE DEPT. OF HEALTH, ALBANY	RSG-C-
NEW YORK. STATE UNIV. (ALL LOCATIONS)	NYSU-
NEW YORK. STATE UNIV., BUFFALO. FACULTY OF ENGINEERING AND APPLIED SCIENCES	SUNYBER-
NEW YORK. STATE UNIV., STONY BROOK	NUB-
NEW YORK. STATE UNIV., SYRACUSE	MF-
NEW YORK. STATE UNIV., SYRACUSE. UPSTATE MEDICAL CTR.	X-
NEW YORK. STATE UNIV. COLL. OF CERAMICS, ALFRED	NSCC MR-
NEW YORK. STATE UNIV. COLL. OF CERAMICS, ALFRED	NYU-CC-
NEW YORK. STATE UNIV. COLL. OF CERAMICS, ALFRED	Q-
NEW YORK BOTANICAL GARDEN, BRONX	SP-
NEW YORK DIRECTED OPERATIONS OFFICE, AEC (REF. 4)	NYDO-
NEW YORK EYE AND EAR INFIRMARY, N.Y.C.	FA-R-
NEW YORK EYE AND EAR INFIRMARY. DEPT. OF ELECTROPHYSIOLOGY, N.Y.C.	NYEEI-
NEW YORK NAVAL SHIPYARD, BROOKLYN	NOISE-
NEW YORK NAVAL SHIPYARD. MATERIAL LAB., BROOKLYN	NE-
NEW YORK NAVAL SHIPYARD. MATERIAL LAB., BROOKLYN	NML-(PROJECT NUMBER)-
NEW YORK NAVAL SHIPYARD. MATERIAL LAB., BROOKLYN	NSM-
NEW YORK OPERATIONS OFFICE, AEC	AMES-
NEW YORK OPERATIONS OFFICE, AEC (REF. 4)	NYO-
NEW YORK OPERATIONS OFFICE, AEC	NYO-GEN-
NEW YORK OPERATIONS OFFICE, AEC (REF. 4)	NYOO-
NEW YORK OPERATIONS OFFICE, AEC. HEALTH + SAFETY LAB.	HASD-
NEW YORK OPERATIONS OFFICE, AEC. HEALTH + SAFETY LAB. (REF. 4)	HASL-

NEW YORK OPERATIONS OFFICE, AEC. HEALTH + SAFETY LAB.	HASL-S-
NEW YORK OPERATIONS OFFICE, AEC. HEALTH + SAFETY LAB.	HASL-SC-
NEW YORK OPERATIONS OFFICE, AEC. HEALTH + SAFETY LAB.	HASL-TM-
NEW YORK SHIPBUILDING CORP., CAMDEN, N.J.	NSS-
NEW YORK UNIV., N.Y.C. (TECHNICAL PROGRESS REPORT)	AEC-
NEW YORK UNIV., N.Y.C.	ALG-
NEW YORK UNIV., N.Y.C.	BR-
NEW YORK UNIV., N.Y.C.	EM-
NEW YORK UNIV., N.Y.C.	EMP-
NEW YORK UNIV., N.Y.C.	FR-
NEW YORK UNIV., N.Y.C.	NEW YORK U-(NO.)-TR-
NEW YORK UNIV., N.Y.C.	NYU-
NEW YORK UNIV., N.Y.C. (SQUID PROJECT) (REF. 33)	NYU-(NUMBER)-(LETTER(S))
NEW YORK UNIV., N.Y.C. (TECHNICAL MEMORANDA)	NYU TM-
NEW YORK UNIV., N.Y.C.	NYU-TR-
NEW YORK UNIV., N.Y.C.	SAR-
NEW YORK UNIV., N.Y.C. (LINGUISTIC STRING PROJECT)	STRING PROGRAM-
NEW YORK UNIV., N.Y.C. APPLIED MATH. GROUP (REF. 28)	AMG-NYU-
NEW YORK UNIV., N.Y.C. APPLIED MATH. GROUP (REF. 28)	AMP-MEMO-
NEW YORK UNIV., N.Y.C. APPLIED MATH. GROUP (REF. 28)	NDRC AMG-NYU-
NEW YORK UNIV., N.Y.C. APPLIED MATH. GROUP (REF. 28)	NDRC AMG-NYU-MEMO-
NEW YORK UNIV., N.Y.C. APPLIED MATH. GROUP (REF. 28)	OSRD-AMP-NYU-
NEW YORK UNIV., N.Y.C. ATOMIC ENERGY COMMISSION COMPUTING FACILITY. (NOTES ON MAGNETO-HYDRODYNAMICS)	MH-
NEW YORK UNIV., N.Y.C. COLL. OF ENGINEERING (REF. 1)	(NUMBERS...)
NEW YORK UNIV., N.Y.C. COLL. OF ENGINEERING	AM-
NEW YORK UNIV., N.Y.C. COLL. OF ENGINEERING	NYU-333-
NEW YORK UNIV., N.Y.C. COLL. OF ENGINEERING (LIQUID PROPELLANT REVIEW. QUARTERLY REPORT)	NYU-LPR-QR-
NEW YORK UNIV., N.Y.C. COLL. OF ENGINEERING	NYU-QR-(NO.)(YR/NO.)
NEW YORK UNIV., N.Y.C. COLL. OF ENGINEERING	SM-(YR.)-
NEW YORK UNIV., N.Y.C. COURANT INSTITUTE OF MATHEMATICAL SCIENCES	EL-
NEW YORK UNIV., N.Y.C. COURANT INSTITUTE OF MATHEMATICAL SCIENCES	EM-
NEW YORK UNIV., N.Y.C. DEPT. OF AERONAUTICS AND ASTRONAUTICS	NYU-AA-(YEAR)-
NEW YORK UNIV., N.Y.C. DEPT. OF CHEMICAL ENGINEERING	MLR-(NO.)-(YEAR)-
NEW YORK UNIV., N.Y.C. DEPT. OF PHYSICS	TR-A(YEAR)-
NEW YORK UNIV., N.Y.C. GEOPHYSICAL SCIENCES LAB.	GSL-TR-(YEAR)-
NEW YORK UNIV., N.Y.C. GEOPHYSICAL SCIENCES LAB.	TR-(YEAR)-
NEW YORK UNIV., N.Y.C. INST. FOR MATH. + MECHANICS	ACC-
NEW YORK UNIV., N.Y.C. INST. FOR MATH. + MECHANICS	IMM-NYU-
NEW YORK UNIV., N.Y.C. INST. FOR MATH. + MECHANICS	NYU-IMM-
NEW YORK UNIV., N.Y.C. INST. OF MATHEMATICAL SCIENCES	BR-
NEW YORK UNIV., N.Y.C. INST. OF MATHEMATICAL SCIENCES	CX-
NEW YORK UNIV., N.Y.C. INST. OF MATHEMATICAL SCIENCES	IMM-
NEW YORK UNIV., N.Y.C. INST. OF MATHEMATICAL SCIENCES	IMM-NYU-
NEW YORK UNIV., N.Y.C. INST. OF MATHEMATICAL SCIENCES	MF-
NEW YORK UNIV., N.Y.C. INST. OF MATHEMATICAL SCIENCES	MH-
NEW YORK UNIV., N.Y.C. INST. OF MATHEMATICAL SCIENCES	NYU-CX-
NEW YORK UNIV., N.Y.C. INST. OF MATHEMATICAL SCIENCES	NYU-EM-
NEW YORK UNIV., N.Y.C. INST. OF MATHEMATICAL SCIENCES	NYU-IMM-
NEW YORK UNIV., N.Y.C. INST. OF MATHEMATICAL SCIENCES	NYUIMM-
NEW YORK UNIV., N.Y.C. INST. OF MATHEMATICAL SCIENCES	NYU-MME-
NEW YORK UNIV., N.Y.C. INSTITUTE OF MATHEMATICAL SCIENCES. (RESEARCH REPORTS)	NYU RR BR-
NEW YORK UNIV., N.Y.C. INSTITUTE OF MATHEMATICAL SCIENCES. (RESEARCH REPORTS)	NYU RR HT-
NEW YORK UNIV., N.Y.C. INSTITUTE OF MATHEMATICAL SCIENCES. (RESEARCH REPORTS)	NYU RR TW-
NEW YORK UNIV., N.Y.C. MAGNETO-FLUID DYNAMICS DIV.	MF-
NEW YORK UNIV., N.Y.C. SCHOOL OF ENG. AND SCIENCE	SETE-
NEW YORK UNIV., N.Y.C. WASHINGTON SQUARE COLL. OF ARTS AND SCIENCE	EM-
NEW ZEALAND. DEPT. OF SCIENTIFIC AND INDUSTRIAL RESEARCH. DIV. OF NUCLEAR SCIENCES, LOWER HUTT	A.E.C.-
NEW ZEALAND. DEPT. OF SCIENTIFIC AND INDUSTRIAL RESEARCH. DIV. OF NUCLEAR SCIENCES, LOWER HUTT	NS-
NEW ZEALAND. DEPT. OF SCIENTIFIC AND INDUSTRIAL RESEARCH. DIV. OF NUCLEAR SCIENCES, LOWER HUTT	NS-R-
NEW ZEALAND. DEPT. OF SCIENTIFIC AND INDUSTRIAL RESEARCH. DIV. OF NUCLEAR SCIENCES, LOWER HUTT	NSR-
NEW ZEALAND. DEPT. OF SCIENTIFIC AND INDUSTRIAL RESEARCH. DIV. OF NUCLEAR SCIENCES, WELLINGTON	NS-
NEW ZEALAND. DEPT. OF SCIENTIFIC AND INDUSTRIAL RESEARCH. PHYSICS AND ENGINEERING LAB., WELLINGTON	DSIR-TN-
NEW ZEALAND. DEPT. OF SCI.+ IND. RES. RADIO RES. OFF.	DSIR-
NEW ZEALAND. DEPT. OF SCI.+ IND. RES. RADIO RES. OFF.	RD-
NEW ZEALAND. DEPT. OF SCI.+ IND. RES. RADIO RES. OFF.	RRO-
NEW ZEALAND. DOMINION LAB., WELLINGTON	CE-
NEW ZEALAND. DOMINION X-RAY + RADIUM LAB., CHRISTCHURCH	DXRL-
NEW ZEALAND. DOMINION X-RAY + RADIUM LAB., CHRISTCHURCH	DXRL-(LTRS)-
NEW ZEALAND. INST. OF NUCLEAR SCIENCES, LOWER HUTT	INS-
NEW ZEALAND. INST. OF NUCLEAR SCIENCES, LOWER HUTT	INS-R-
NEW ZEALAND. NATL. RADIATION LAB., CHRISTCHURCH	NRL-AR-
NEW ZEALAND. NATL. RADIATION LAB., CHRISTCHURCH	NRL-F-
NEW ZEALAND. NATL. RADIATION LAB., CHRISTCHURCH	NRL-F(NO.)-
NEW ZEALAND. NAVAL RESEARCH LAB., AUCKLAND	NRL-
NEWARK COLL. OF ENGINEERING, N.J. DEPT. OF MECHANICAL ENGINEERING	NCE-NV-
NEWCASTLE-UPON-TYNE UNIV., NORTHUMBERLAND, ENGLAND	NUTU-
NEWMARK (N.M.) CONSULTING ENGINEERING SERVICES, URBANA, ILL.	N(NUMBER)
NEWMARK, HANSEN AND ASSOCIATES, CAMBRIDGE, MASS.	NHA-
NEWMARK, HANSEN AND ASSOCIATES, URBANA, ILL.	NHA-
NEWMARK, HANSEN AND ASSOCIATES, URBANA, ILL.	NHA CONSTR. GUIDE-

NEWPORT NEWS SHIPBUILDING AND DRY DOCK CO.

NEWPORT NEWS SHIPBUILDING AND DRY DOCK CO., VA.	NNSD-
NEWPORT NEWS SHIPBUILDING AND DRY DOCK CO., VA.	NNSD-NSPS-
NEWPORT NEWS SHIPBUILDING AND DRY DOCK CO., VA.	NNSD-R-
NICKEL CADMIUM BATTERY CORP., EAST HAMPTON, MASS.	NCBC-
NIELSEN ENGINEERING + RESEARCH, INC., PALO ALTO, CAL.	NEAR-TR-
NIGERIAN METEOROLOGICAL SERVICE, LAGOS	NMS/MN/
NIGERIAN METEOROLOGICAL SERVICE, LAGOS	NMS/TN/
NIHON UNIV., TOKYO	INS-
NIHON UNIV., TOKYO	NUP-
NIHON UNIV., TOKYO	NUP-A-(YEAR-NUMBER)
NIKE-HERCULES SAFETY SUBCOMM. OF THE JOINT COMM. ON NIKE-HERCULES WARHEAD INSTALLATIONS	JC/NHWI-
NIKE ZEUS JOINT AEC-DOD WARHEAD COORDINATING COMM. (REF. 15)	NZJWCC-
NORD-AVIATION, CHATILLON-SOUS-BAGNEUX, FRANCE	SSA-(NUMBER-YEAR)
NORD-AVIATION, PARIS	RAA/R/
NORDEN LABS. CORP., WHITE PLAINS, N.Y.	NL-
NORDEN LABS. CORP., WHITE PLAINS, N.Y.	NLC-
NORDISK FORSKNINGSINSTITUT FOR MALING OG TRYKFARVER, COPENHAGEN	NIF-TM-
NORFOLK NAVAL SHIPYARD, PORTSMOUTH, VA.	NAV-NNS-
NORFOLK NAVAL SHIPYARD, PORTSMOUTH, VA.	NNS-
NORGES TEKNISKE HOEGSKOLE, TRONDHEIM	AE-
NORGES TEKNISKE HOEGSKOLE, TRONDHEIM	ELAB-AE-
NORGES TEKNISKE HOEGSKOLE, TRONDHEIM	ELAB-IR-
NORGES TEKNISKE HOEGSKOLE, TRONDHEIM	ELAB-IT-
NORGES TEKNISKE HOEGSKOLE, TRONDHEIM	ELAB-TE-
NORGES TEKNISKE HOEGSKOLE, TRONDHEIM	ELAB-TL-
NORGES TEKNISKE HOEGSKOLE, TRONDHEIM	ELAB-TR-
NORGES TEKNISKE HOEGSKOLE, TRONDHEIM	ELAB-VT-
NORGES TEKNISKE HOEGSKOLE, TRONDHEIM	ILEA-
NORGES TEKNISKE HOEGSKOLE, TRONDHEIM	LEA-TR-
NORGES TEKNISKE HOEGSKOLE, TRONDHEIM	NIT-FTP-AR-
NORGES TEKNISKE HOEGSKOLE, TRONDHEIM	TE-
NORGES TEKNISKE HOEGSKOLE, TRONDHEIM. NORWEGIAN SHIP MODEL EXPERIMENT TANK	MEDDELELSE-
NORTH AMERICAN AIR DEFENSE COMMAND. J-2 SECTION, ENT AFB, COLO.	NORAD-
NORTH AMERICAN AVIATION, INC. (ALL LOCATIONS)	NAA-
NORTH AMERICAN AVIATION, INC. (ALL LOCATIONS)	NAA-(YEAR)-
NORTH AMERICAN AVIATION, INC. (ALL LOCATIONS)	NAA-RE-
NORTH AMERICAN AVIATION, INC. (ALL LOCATIONS)	NAA-SR-
NORTH AMERICAN AVIATION, INC., ANAHEIM, CALIF.	C6-(YEAR)/
NORTH AMERICAN AVIATION, INC., ANAHEIM, CALIF.	T7-(NUMBER)-(NUMBER)
NORTH AMERICAN AVIATION, INC., COLUMBUS, OHIO	NA-
NORTH AMERICAN AVIATION, INC., DOWNEY, CALIF.	AER-(YR)-(LETTER)-
NORTH AMERICAN AVIATION, INC., DOWNEY, CALIF.	CR-
NORTH AMERICAN AVIATION, INC., DOWNEY, CALIF.	MC-
NORTH AMERICAN AVIATION, INC., DOWNEY, CALIF.	NAA-PC-
NORTH AMERICAN AVIATION, INC., DOWNEY, CALIF.	NAA-PS-
NORTH AMERICAN AVIATION, INC., DOWNEY, CALIF.	NAA-R-
NORTH AMERICAN AVIATION, INC., DOWNEY, CALIF.	NAA-RF-
NORTH AMERICAN AVIATION, INC., DOWNEY, CALIF.	NAA-RG-
NORTH AMERICAN AVIATION, INC., DOWNEY, CALIF.	NAA-SR-(YR)-
NORTH AMERICAN AVIATION, INC., DOWNEY, CALIF.	NAA-SRL-
NORTH AMERICAN AVIATION, INC., DOWNEY, CALIF.	NAA-SR-MEMO-
NORTH AMERICAN AVIATION, INC., DOWNEY, CALIF.	NAA-SR-MISC.-
NORTH AMERICAN AVIATION, INC., DOWNEY, CALIF.	NAA-SR-MTA-
NORTH AMERICAN AVIATION, INC., DOWNEY, CALIF.	SRL-(YR)-
NORTH AMERICAN AVIATION, INC., DOWNEY, CALIF.	UN-
NORTH AMERICAN AVIATION, INC., LOS ANG.(JT. VENTURE)	BNJV-(LTR.)-(LTR.)-
NORTH AMERICAN AVIATION, INC., LOS ANGELES	CD-
NORTH AMERICAN AVIATION, INC., LOS ANGELES	D-
NORTH AMERICAN AVIATION, INC., LOS ANGELES	NA-
NORTH AMERICAN AVIATION, INC., LOS ANGELES	NA-(YEAR)-
NORTH AMERICAN AVIATION, INC., LOS ANGELES	NA(NUMBER)-
NORTH AMERICAN AVIATION, INC., LOS ANGELES	NAA-(YEAR)MD-
NORTH AMERICAN AVIATION, INC., LOS ANGELES	NAAL-
NORTH AMERICAN AVIATION, INC., LOS ANGELES	NAA-NA-(NUMBER(S))-
NORTH AMERICAN AVIATION, INC., LOS ANGELES	NAA-TFD-
NORTH AMERICAN AVIATION, INC., LOS ANGELES	NA-MCH-
NORTH AMERICAN AVIATION, INC., LOS ANGELES (SEE ALSO: AUTONETICS DIV., NORTH AMERICAN AVIATION)	NA-TFD-(YEAR)-
NORTH AMER. AVIATION, INC. AEROPHYSICS DEPT. (NAVAHO)	AL-1900-
NORTH AMERICAN AVIATION, INC. AEROPHYSICS LAB., DOWNEY, CALIF.	AL-
NORTH AMERICAN AVIATION, INC. AEROPHYSICS LAB., DOWNEY, CALIF. (MEMO)	ALM-
NORTH AMERICAN AVIATION, INC. AEROPHYSICS LAB., DOWNEY, CALIF.	NAA-AL-
NORTH AMERICAN AVIATION, INC. AEROPHYSICS LAB., DOWNEY, CALIF.	NAA-AL-MEMO-
NORTH AMERICAN AVIATION, INC. ATOMIC ENERGY RESEARCH DEPT., DOWNEY, CALIF.	NAA-AER-
NORTH AMERICAN AVIATION, INC. ATOMIC ENERGY RESEARCH DEPT., DOWNEY, CALIF.	NAA-AER-MEMO-
NORTH AMERICAN AVIATION, INC. COLUMBUS DIV., OHIO	N-(YEAR)H-
NORTH AMERICAN AVIATION, INC. COLUMBUS DIV., OHIO	NA-
NORTH AMERICAN AVIATION, INC. COLUMBUS DIV., OHIO	NA(YEAR)H-
NORTH AMERICAN AVIATION, INC. COLUMBUS DIV., OHIO	NAA-NA-(YR.)H-(NO.)-(NO.)
NORTH AMERICAN AVIATION, INC. ELECTRO MECHANICAL ENGINEERING DEPT., DOWNEY, CALIF.	NAA-EM-
NORTH AMERICAN AVIATION, INC. ENG. DEPT., LOS ANGELES	NA(NUMBER)-(NOS.)(LTR.)
NORTH AMERICAN AVIATION, INC. LOS ANGELES DIV.	LA(NUMBERS)-(NOS.)(LTR)
NORTH AMERICAN AVIATION, INC. LOS ANGELES DIV.	NAA-(YEAR)LA-
NORTH AMERICAN AVIATION, INC. LOS ANGELES DIV.	NA-SB-(NOS.)-(NO.)-
NORTH AMERICAN AVIATION, INC. LOS ANGELES DIV.	NA-TFD-(NUMBERS)-

NORTH AMERICAN AVIATION, INC. LOS ANGELES DIV.	NA-T0-
NORTH AMERICAN AVIATION, INC. LOS ANGELES DIV.	SPEC-LA(NUMBERS)-
NORTH AMERICAN AVIATION, INC. MISSILE DEVELOPMENT DIV., DOWNEY, CALIF.	MD-(NUMBER)-
NORTH AMERICAN AVIATION, INC. MISSILE DEVELOPMENT DIV., DOWNEY, CALIF.	NAA-AL-
NORTH AMERICAN AVIATION, INC. MISSILE DIV., DOWNEY, CALIF.	NAA-MD-(YEAR)-
NORTH AMERICAN AVIATION, INC. OCEAN SYSTEMS OPERATIONS, ANAHEIM, CALIF.	C-
NORTH AMERICAN AVIATION, INC. OCEAN SYSTEMS OPERATIONS, ANAHEIM, CALIF.	X7-(NUMBER)/(NUMBER)
NORTH AMERICAN AVIATION, INC. SPACE AND INFORMATION SYSTEMS DIV., CANOGA PARK, CALIF.	SID-(YEAR)-
NORTH AMERICAN AVIATION, INC. SPACE AND INFORMATION SYSTEMS DIV., DOWNEY, CALIF.	MD-
NORTH AMERICAN AVIATION INC. SPACE AND INFORMATION SYSTEMS DIV., DOWNEY, CALIF.	NAA-SID-
NORTH AMERICAN AVIATION, INC. SPACE AND INFORMATION SYSTEMS DIV., DOWNEY, CALIF.	SID-
NORTH AMERICAN AVIATION, INC. SPACE AND INFORMATION SYSTEMS DIV., DOWNEY, CALIF.	SID-(YEAR)-
NORTH AMERICAN AVIATION, INC. SPACE AND INFORMATION SYSTEMS DIV., EL SEGUNDO, CALIF.	SID(YEAR)-
NORTH AMERICAN AVIATION, INC. SPACE AND INFORMATION SYSTEMS DIV., TULSA, OKLA.	SID-(YEAR)T-
NORTH AMERICAN AVIATION, INC. SPECIAL PROJECTS DIV., LONG BEACH, CALIF.	NAA-SPD-P-(NO.)-(YEAR)-
NORTH AMERICAN AVIATION SCIENCE CENTER, THOUSAND OAKS, CALIF.	NAASC-TR-
NORTH AMERICAN AVIATION SCIENCE CENTER, THOUSAND OAKS, CALIF.	SCTR-
NORTH AMERICAN CARBON, INC., COLUMBUS, OHIO	NACAR-
NORTH AMERICAN INSTRUMENTS, INC., ALTADENA, CALIF.	NAI-(NO.)PR(MONTH)(YEAR)
NORTH AMERICAN ROCKWELL CORP., LOS ANGELES	NA-(YEAR)-
NORTH AMERICAN ROCKWELL CORP., THOUSAND OAKS, CALIF.	SC-PP-(YEAR)-
NORTH AMERICAN ROCKWELL CORP. AUTONETICS DIV., ANAHEIM, CALIF.	C9-(4 DIGITS./
NORTH AMERICAN ROCKWELL CORP. COLUMBUS DIV., OHIO	NR(YEAR)H-
NORTH AMERICAN ROCKWELL CORP. OCEAN SYSTEMS OPERATIONS, LONG BEACH, CALIF.	XB-(NUMBER)/(NUMBER)
NORTH AMERICAN ROCKWELL CORP. SPACE DIV., DOWNEY, CAL	SD-(YEAR)-(NO.)-(NO.)
NORTH AMERICAN ROCKWELL CORP. SPACE DIV., DOWNEY, CAL (CS IS PROGRAM AREA DESIGNATION)	SD-(YEAR)-CS-
NORTH AMERICAN ROCKWELL CORP. SPACE DIV., DOWNEY, CAL (SA IS PROGRAM AREA DESIGNATION)	SD-(YEAR)-SA-
NORTH AMERICAN ROCKWELL CORP. SPACE DIV., DOWNEY, CAL (CE IS PROGRAM DESIGNATION)	SD-(YEAR)-CE-
NORTH AMERICAN ROCKWELL CORP. SPACE DIV., DOWNEY, CAL (SH IS PROGRAM DESIGNATION)	SD-(YEAR)-SH-
NORTH AMERICAN ROCKWELL CORP. SPACE DIV., DOWNEY, CAL	SPEC-MA(NO.)-
NORTH AMERICAN ROCKWELL CORP. TULSA DIV., OKLA.	TD-(YEAR)-
NORTH ATLANTIC TREATY ORGANIZATION, BRUSSELS	NATO-AC/
NORTH ATLANTIC TREATY ORGANIZATION. ADVISORY GROUP FOR AEROSPACE RESEARCH AND DEVELOPMENT, PARIS	AG-
NORTH ATLANTIC TREATY ORGANIZATION. ADVISORY GROUP FOR AEROSPACE RESEARCH AND DEVELOPMENT, PARIS	AGARD-
NORTH ATLANTIC TREATY ORGANIZATION. ADVISORY GROUP FOR AEROSPACE RESEARCH AND DEVELOPMENT, PARIS	AGARD-AC/
NORTH ATLANTIC TREATY ORGANIZATION. ADVISORY GROUP FOR AEROSPACE RESEARCH AND DEVELOPMENT, PARIS	AGARD-ADVISORY-
NORTH ATLANTIC TREATY ORGANIZATION. ADVISORY GROUP FOR AEROSPACE RESEARCH AND DEVELOPMENT, PARIS	AGARD-AG-(NUMBER)-(YEAR)
NORTH ATLANTIC TREATY ORGANIZATION. ADVISORY GROUP FOR AEROSPACE RESEARCH AND DEVELOPMENT, PARIS	AGARD-AR-(NUMBER)-(YR.)
NORTH ATLANTIC TREATY ORGANIZATION. ADVISORY GROUP FOR AEROSPACE RESEARCH AND DEVELOPMENT, PARIS	AGARD-BIB-
NORTH ATLANTIC TREATY ORGANIZATION. ADVISORY GROUP FOR AEROSPACE RESEARCH AND DEVELOPMENT, PARIS	AGARD-BULL-
NORTH ATLANTIC TREATY ORGANIZATION. ADVISORY GROUP FOR AEROSPACE RESEARCH AND DEVELOPMENT, PARIS	AGARD-CP-
NORTH ATLANTIC TREATY ORGANIZATION. ADVISORY GROUP FOR AEROSPACE RESEARCH AND DEVELOPMENT, PARIS	AGARD-ES-
NORTH ATLANTIC TREATY ORGANIZATION. ADVISORY GROUP FOR AEROSPACE RESEARCH AND DEVELOPMENT, PARIS	AGARD-IB-(YEAR)/
NORTH ATLANTIC TREATY ORGANIZATION. ADVISORY GROUP FOR AEROSPACE RESEARCH AND DEVELOPMENT, PARIS	AGARD-IND-(NO./NO.)-(YR)
NORTH ATLANTIC TREATY ORGANIZATION. ADVISORY GROUP FOR AEROSPACE RESEARCH AND DEVELOPMENT, PARIS	AGARD-LS-
NORTH ATLANTIC TREATY ORGANIZATION. ADVISORY GROUP FOR AEROSPACE RESEARCH AND DEVELOPMENT, PARIS	AGARD-MAN-(NUMBER)-(YR.)
NORTH ATLANTIC TREATY ORGANIZATION. ADVISORY GROUP FOR AEROSPACE RESEARCH AND DEVELOPMENT, PARIS	AGARDOGRAPH-
NORTH ATLANTIC TREATY ORGANIZATION. ADVISORY GROUP FOR AEROSPACE RESEARCH AND DEVELOPMENT, PARIS	AGARD-OGRAPH-
NORTH ATLANTIC TREATY ORGANIZATION. ADVISORY GROUP FOR AEROSPACE RESEARCH AND DEVELOPMENT, PARIS	AGARD-R-(NUMBER)-(YEAR)
NORTH ATLANTIC TREATY ORGANIZATION. ADVISORY GROUP FOR AEROSPACE RESEARCH AND DEVELOPMENT, PARIS	AGARD-SPEC-
NORTH ATLANTIC TREATY ORGANIZATION. DEFENSE COLLEGE, PARIS	NATO-DC-
NORTH ATLANTIC TREATY ORGANIZATION. MILITARY AGENCY FOR STANDARDIZATION, PARIS	AAP-
NORTH ATLANTIC TREATY ORGANIZATION. MILITARY AGENCY FOR STANDARDIZATION, PARIS	STANAG-
NORTH ATLANTIC TREATY ORGANIZATION. SHAPE TECHNICAL CENTRE, THE HAGUE	SADTC-
NORTH ATLANTIC TREATY ORGANIZATION. SHAPE TECHNICAL CENTRE, THE HAGUE	SADTC-(YEAR)/TM-
NORTH ATLANTIC TREATY ORGANIZATION. SHAPE TECHNICAL CENTRE, THE HAGUE	STC-
NORTH ATLANTIC TREATY ORGANIZATION. SHAPE TECHNICAL CENTER, THE HAGUE	STC-CR-NSPM-
NORTH ATLANTIC TREATY ORGANIZATION. SHAPE TECHNICAL CENTER, THE HAGUE	STC-TR-
NORTH ATLANTIC TREATY ORGANIZATION. SHAPE TECHNICAL CENTER, THE HAGUE	TMC-TM-
NORTH ATLANTIC TREATY ORGANIZATION. STRUCTURES AND MATERIALS PANEL, PARIS	MGRT-
NORTH ATLANTIC TREATY ORGANIZATION. SUBCOMMITTEE ON OCEANOGRAPHIC RESEARCH	OTAN-TR-
NORTH CAROLINA. UNIV., CHAPEL HILL	NCU-
NORTH CAROLINA. UNIV., CHAPEL HILL	NREC-TM-
NORTH CAROLINA. UNIV., CHAPEL HILL	NREC-TR-
NORTH CAROLINA. UNIV., CHAPEL HILL	UNC-
NORTH CAROLINA. UNIV., CHAPEL HILL.DEPT.OF STATISTICS	INST. OF STATISTICS-MS-
NORTH CAROLINA. UNIV., CHAPEL HILL. INST. OF STATISTICS. (MIMEOGRAPH SERIES)	NCU IS MS-
NORTH CAROLINA. UNIV., CHAPEL HILL. INST. OF STATISTICS. (MIMEOGRAPH SERIES)	NCU-MS-
NORTH CAROLINA STATE COLL., RALEIGH	NCC-
NORTH CAROLINA STATE COLL., RALEIGH	NCS-
NORTH CAROLINA STATE COLL., RALEIGH	NCSC-
NORTH CAROLINA STATE COLL., RALEIGH	NORTH CAROLINA SC...
NORTH CAROLINA STATE COLL., RALEIGH	R-NCSC-
NORTH CAROLINA STATE COLL., RALEIGH. DEPT. OF ENG.RES	NCSL-
NORTH CAROLINA STATE UNIV., RALEIGH	NCSU-
NORTH CAROLINA STATE UNIV., RALEIGH	NCSU-QPR-
NORTH CAROLINA STATE UNIV., RALEIGH. CENTER FOR ACOUSTICAL STUDIES	CAS-

NORTH CAROLINA STATE UNIV. DEPT. OF ENG. RES.

NORTH CAROLINA STATE UNIV., RALEIGH. DEPT. OF ENG. RES	NCSU-TR-
NORTH CAROLINA STATE UNIV., RALEIGH. DEPT. OF SOIL SCIENCE	SOIL INFORMATION SER-
NORTH CAROLINA STATE UNIV., RALEIGH. PHYSICAL SCIENCE RESEARCH	PSR-PROJECT-
NORTH CAROLINA STATE COLL., RALEIGH. SEMICONDUCTOR DEVICE LAB.	SDL-(NUMBER)-
NORTH DAKOTA WATER RESOURCES INST., FARGO	WI-(NUMBER-NUMBER-YR.)
NORTH-HOLLAND PUBLISHING CO., AMSTERDAM	NHPC-
NORTH STAR RESEARCH AND DEV. INST., MINNEAPOLIS	NSRDI-W-
NORTH STAR RESEARCH AND DEV. INST., MINNEAPOLIS	W-
NORTHAMPTON COLL. OF ADVANCED TECHNOLOGY, LONDON	S+T-MEMO-(MO.)/(YR.)
NORTHEASTERN FOREST EXPERIMENT STATION, UPPER DARBY, PENNA. (FOREST SERVICE RESEARCH NOTE)	FSRN-NE-
NORTHEASTERN FOREST EXP. STA., UPPER DARBY, PENNA.	FSRP-NE-
NORTHEASTERN UNIV., BOSTON	NEU-
NORTHEASTERN UNIV., BOSTON	NUB-
NORTHEASTERN UNIV., BOSTON. PHOTOCHEMISTRY AND SPECTROSCOPY LAB.	PSL-
NORTHEASTERN WOOD UTILIZATION COUNCIL	NWUC-
NORTHEASTERN WOOD UTILIZATION COUNCIL (BULLETIN)	NWUC B-
NORTHEASTERN WOOD UTILIZATION COUNCIL (WOOD NOTES)	NWUC WN-
NORTHEASTERN WOOD UTILIZATION COUNCIL (WOOD NOTES)	WN-
NORTHERN ELECTRIC CO., LTD., OTTAWA	TR(NUMBER)-(NO.)-(YEAR)-
NORTHERN FOREST FIRE LAB., MISSOULA, MONT.	INT-
NORTHERN MICHIGAN UNIV., MARQUETTE. DEPT. OF PHYSICS	LAKE SUPERIOR STUDIES-
NORTHERN RESEARCH AND ENG. CORP., CAMBRIDGE, MASS.	NREC-(NUMBER)-
NORTHERN STATES POWER CO., SIOUX FALLS, S. DAK. (PATHFINDER ATOMIC POWER PLANT)	PAPP-
NORTHROP AIRCRAFT, INC., HAWTHORNE, CALIF. (AERODYNAMICS MEMO)	AM-
NORTHROP AIRCRAFT, INC., HAWTHORNE, CALIF. (ARMAMENT)	ARM-
NORTHROP AIRCRAFT, INC., HAWTHORNE, CALIF. (BOUNDARY LAYER CONTROL)	BLC-
NORTHROP AIRCRAFT, INC., HAWTHORNE, CALIF.	BTG-
NORTHROP AIRCRAFT, INC., HAWTHORNE, CALIF. (DYNAMICS)	DYN-
NORTHROP AIRCRAFT, INC., HAWTHORNE, CALIF. (ENGINEERING TEST REPORT)	ETR-
NORTHROP AIRCRAFT, INC., HAWTHORNE, CAL.(FLIGHT TEST)	FT-
NORTHROP AIRCRAFT, INC., HAWTHORNE, CALIF. (FLIGHT TEST SUMMARY)	FTS-
NORTHROP AIRCRAFT, INC., HAWTHORNE, CALIF. (GUIDED MISSILE)	GM-
NORTHROP AIRCRAFT, INC., HAWTHORNE, CALIF. (GUIDED MISSILE FLIGHT TEST)	GMFT-
NORTHROP AIRCRAFT, INC., HAWTHORNE, CALIF.	ID-(NUMBERS)-
NORTHROP AIRCRAFT, INC., HAWTHORNE, CALIF. (LANDING GEAR MEMO)	LGM-
NORTHROP AIRCRAFT, INC., HAWTHORNE, CALIF.(LAB.NOTE)	LN-
NORTHROP AIRCRAFT, INC., HAWTHORNE, CALIF. (MONTHLY MISSILE PROGRESS REPORT)	MMPR-
NORTHROP AIRCRAFT, INC., HAWTHORNE, CALIF. (NORTHROP ACCIDENT)	NA-
NORTHROP AIRCRAFT, INC., HAWTHORNE, CALIF.	NA-(YEAR)-
NORTHROP AIRCRAFT, INC., HAWTHORNE, CALIF.	NA-AM-
NORTHROP AIRCRAFT, INC., HAWTHORNE, CALIF. (GUIDED MISSILES)	NA-GM-
NORTHROP AIRCRAFT, INC., HAWTHORNE, CALIF.	NAI-
NORTHROP AIRCRAFT, INC., HAWTHORNE, CALIF.	NAI-(YR)-
NORTHROP AIRCRAFT, INC., HAWTHORNE, CALIF. (FLIGHT TEST PROGRESS REPORTS)	NAI-(NUMBERS)-
NORTHROP AIRCRAFT, INC., HAWTHORNE, CALIF.	NAI FTS(NO.)-
NORTHROP AIRCRAFT, INC., HAWTHORNE, CALIF.	NAI GM-
NORTHROP AIRCRAFT, INC., HAWTHORNE, CALIF. (MISSILE MONTHLY PROGRESS REPORT)	NAI-LN-
NORTHROP AIRCRAFT, INC., HAWTHORNE, CALIF.	NAI-MMPR-
NORTHROP AIRCRAFT, INC., HAWTHORNE, CALIF.	NAR-
NORTHROP AIRCRAFT, INC., HAWTHORNE, CALIF.	NA-RC-
NORTHROP AIRCRAFT, INC., HAWTHORNE, CALIF. (MISC.)	NM-
NORTHROP AIRCRAFT, INC., HAWTHORNE, CALIF.	NNR-
NORTHROP AIRCRAFT, INC., HAWTHORNE, CALIF.(RES.REPT.)	NRR-
NORTHROP AIRCRAFT, INC., HAWTHORNE, CALIF. (SPEC.)	NS-
NORTHROP AIRCRAFT, INC., HAWTHORNE, CALIF. (NUCLEAR SCIENCE REPORT)	NSR-
NORTHROP AIRCRAFT, INC., HAWTHORNE, CALIF.	NWR-
NORTHROP AIRCRAFT, INC., HAWTHORNE, CALIF. (PRELIMINARY DESIGN STUDY)	PDS-
NORTHROP AIRCRAFT, INC., HAWTHORNE, CALIF. (RECORDS CONTROL NUMBER)	RC-
NORTHROP AIRCRAFT, INC., HAWTHORNE, CALIF. (SPECIAL PROJECTS)	SP-
NORTHROP AIRCRAFT, INC., HAWTHORNE, CALIF. (STRUCTURES RESEARCH REPORT)	SRR-
NORTHROP AIRCRAFT, INC., HAWTHORNE, CALIF. (THERMODYNAMICS MEMO)	TDM-
(SEE ALSO LATER NAME: NORTHROP CORP.)	
NORTHROP AIRCRAFT, INC. NORAIR, HAWTHORNE, CALIF.	NOR NB(YEAR)-
NORTHROP CAROLINA, INC., ASHEVILLE, N.C.	APR-
NORTHROP CAROLINA, INC., ASHEVILLE, N.C.	ATR-
NORTHROP CAROLINA, INC., ASHEVILLE, N.C.	NC-
NORTHROP CORPORATE LABS., HAWTHORNE, CALIF.	NCL-
NORTHROP CORPORATE LABS., HAWTHORNE, CALIF.	NCL-(YEAR)-
NORTHROP CORPORATE LABS., PASADENA, CALIF. (SENSOR AND SIMULATION NOTES)	NCL-S+SN-
NORTHROP CORP. (ALL LOCATIONS)	NOR-
NORTHROP CORP. (ALL LOCATIONS)	NOR-(YEAR)-
NORTHROP CORP. (ALL LOCATIONS)	NORT-
NORTHROP CORP., HAWTHORNE, CALIF.	USAAVLABS-TR-(YEAR)-
NORTHROP CORP., NEWBURY PARK, CALIF.	NVR-
NORTHROP CORP., NEWBURY PARK, CALIF.	TP-
NORTHROP CORP. DIV. OF NORTHROP VENTURA, NEWBURY PARK, CALIF. (BROCHURES)	NVB-
NORTHROP CORP. DIV. OF NORTHROP VENTURA, NEWBURY PARK, CALIF. (PROPOSALS)	NVP-
NORTHROP CORP. DIV. OF NORTHROP VENTURA, NEWBURY PARK, CALIF. (TECHNICAL LETTERS)	NVT-
NORTHROP CORP. DIV. OF NORTHROP VENTURA, NEWBURY PARK, CALIF. (TECHNICAL PAPERS)	NV-TP-
NORTHROP CORP. NORAIR DIV., HAWTHORNE, CALIF.	NB-(YEAR)-
NORTHROP CORP. NORAIR DIV., HAWTHORNE, CALIF.	NR-IR-(NUMBER)-
NORTHROP CORP. NORAIR DIV. ENGINEERING DEPT., HAWTHORNE, CALIF. (MATERIALS RESEARCH)	MRL-
NORTHROP CORP. NORAIR DIV. ENGINEERING DEPT., HAWTHORNE, CALIF.	NRL-

NORTHROP CORP. NORAIR DIV. ENGINEERING DEPT., HAWTHORNE, CALIF. (RESEARCH + ADVANCED SYSTEMS REPT.)	RASR-
NORTHROP CORP. NORAIR DIV. ENGINEERING DEPT., HAWTHORNE, CALIF. (WEAPONS SYSTEMS DESIGN STUDY)	WSDS-
NORTHROP CORP. NORAIR DIV. MANUFACTURING DEPT., HAWTHORNE, CALIF. (MANUFACTURING RESEARCH + DEV.)	MRD-
NORTHROP CORP. NORTHROP SPACE LABS., RESEARCH AND ANALYSIS SECTION, HUNTSVILLE, ALA.	E(NUMBER)-
NORTHROP CORP. NORTRONICS DIV., ANAHEIM, CALIF.	NSS-
NORTHROP CORP. NORTRONICS DIV., NEEDHAM, MASS. (USED UNTIL 1966 BY MARINE EQUIPMENT DEPT.)	MED-(YEAR)-
NORTHROP CORP. NORTRONICS DIV., NEWBURY PARK, CALIF.	NOR-ARD-(YEAR)-
NORTHROP CORP. NORTRONICS DIV. APPLIED RESEARCH DEPT., NEWBURY PARK, CALIF.	ARD-(YEAR)-
NORTHROP CORP. NORTRONICS DIV. APPLIED RESEARCH DEPT., NEWBURY PARK, CALIF.	NARD-
NORTHROP CORP. NORTRONICS DIV. NEEDHAM DEPT., MASS.	NDN-(YEAR)-
NORTHROP CORP. NORTRONICS DIV. NEEDHAM DEPT., MASS.	NND-(YEAR)-
NORTHROP CORP. VENTURA DIV., NEWBURY PARK, CALIF.	NOR/NVR-
NORTHROP CORP. VENTURA DIV., NEWBURY PARK, CALIF.	NV-
NORTHROP CORP. VENTURA DIV., NEWBURY PARK, CALIF.	NVR-
NORTHROP NORTRONICS, NEWBURY PARK, CALIF.	NOR-NARD-(YEAR)-
NORTHROP NORTRONICS. PRECISION PRODUCTS DEPT., NORWOOD, MASS.	PPD-(YEAR)-(LETTER)-
NORTHROP SPACE LABS.,HAWTHORNE,CAL.(ASTRO SCIENCES)	ASG-TM-(YEAR)-
NORTHROP SPACE LABS.,HAWTHORNE,CAL.(ASTRO SCIENCES)	ASL-
NORTHROP SPACE LABS.,HAWTHORNE,CAL.(ASTRO SCIENCES)	ASRL-TM-(YEAR)-
NORTHROP SPACE LABS., HAWTHORNE, CALIF.	NSL-
NORTHROP SPACE LABS., HAWTHORNE, CALIF.	NSL-(YEAR)-
NORTHROP SPACE LABS., HAWTHORNE, CALIF.	NSL-E-
NORTHROP SYSTEMS LABS., HAWTHORNE, CALIF.	NSL-(YEAR)-
NORTHROP SYSTEMS LABS., PALOS VERDES, CALIF.	NORT-(YEAR)-
NORTHWEST AIRLINES, INC. AERONAUTICAL ICE RESEARCH LABORATORY, MINNEAPOLIS	AIRL-
NORTHWESTERN UNIV., EVANSTON, ILL.	AR-
NORTHWESTERN UNIV., EVANSTON, ILL.	LA-
NORTHWESTERN UNIV., EVANSTON, ILL.	NNRL-
NORTHWESTERN UNIV., EVANSTON, ILL.	NU-
NORTHWESTERN UNIV., EVANSTON, ILL.	NWU-
NORTHWESTERN UNIV., EVANSTON, ILL.	NWU-AR-
NORTHWESTERN UNIV., EVANSTON, ILL. AERIAL MEASUREMENTS LAB.	M-
NORTHWESTERN UNIV., EVANSTON, ILL. APPLIED MATHEMATICS GROUP (REF. 28)	AMG-N-
NORTHWESTERN UNIV., EVANSTON, ILL. APPLIED MATHEMATICS GROUP (REF. 28)	NDRC AMG-N-
NORTHWESTERN UNIV., EVANSTON, ILL. DEPT. OF PHYSICS	TR-II-
NORTHWESTERN UNIV., EVANSTON, ILL. GAS DYNAMICS LAB.	NSL-
NORTHWESTERN UNIV., EVANSTON, ILL. GAS DYNAMICS LAB.	NU-GDL-B-
NORTHWESTERN UNIV., EVANSTON, ILL. INFORMATION PROCESSING AND CONTROL SYSTEMS LAB.	IPAC-(YEAR)-
NORTHWESTERN UNIV., EVANSTON, ILL. INFORMATION PROCESSING AND CONTROL SYSTEMS LAB.	TR-(YEAR)-
NORTHWESTERN UNIV.,EVANSTON, ILL. TECHNOLOGICAL INST.	NTI-
NORTHWESTERN UNIV.,EVANSTON, ILL. TECHNOLOGICAL INST. (SQUID PROJECT) (REF. 33)	NTI-(NUMBER)-(LETTER)
NORTHWESTERN UNIV.,EVANSTON, ILL. TECHNOLOGICAL INST.	NWU-TI-TN-
NORTHWESTERN UNIV.,EVANSTON, ILL. TECHNOLOGICAL INST.	NWU-TI-TR-
NORTHWESTERN UNIV.,EVANSTON, ILL. TECHNOLOGICAL INST.	NWU-TR-
NORTHWESTERN UNIV.,EVANSTON, ILL. TECHNOLOGICAL INST.	SRM-
NORTON CO., WORCESTER, MASS.	BXR-
NORTRONICS, ANAHEIM, CALIF.	NORT-
NORTRONICS, HAWTHORNE, CALIF.	NORT-(YEAR)-
NORTRONICS. PALOS VERDES PENINSULA, CALIF.	NORT-(YEAR)-
NORTRONICS. APPLIED RES. DEPT., NEWBURY PARK, CALIF.	ARD-(YEAR-NUMBER)(LTR.)
NORTRONICS. APPLIED RES. DEPT., NEWBURY PARK, CALIF.	EC-RD-
NORTRONICS. APPLIED RES. DEPT., NEWBURY PARK, CALIF.	NARD-TR-(NUMBER)-
NORWAY. FORSVARETS FORSKNINGSINSTITUTT, BERGEN	FFIR-IR-R-
NORWAY. FORSVARETS FORSKNINGSINSTITUTT, KJELLER	E-
NORWAY. FORSVARETS FORSKNINGSINSTITUTT, KJELLER	F-
NORWAY. FORSVARETS FORSKNINGSINSTITUTT, KJELLER	FFIF-F-
NORWAY. FORSVARETS FORSKNINGSINSTITUTT, KJELLER	FFIF-INTERN RAPPORT F-
NORWAY. FORSVARETS FORSKNINGSINSTITUTT, KJELLER	FFIF-IR-E-
NORWAY. FORSVARETS FORSKNINGSINSTITUTT, KJELLER	FFIF-IR-F-
NORWAY. FORSVARETS FORSKNINGSINSTITUTT, KJELLER	FFIF-IR-S-
NORWAY. FORSVARETS FORSKNINGSINSTITUTT, KJELLER	FFIF TEKNISK NOTAT F-
NORWAY. FORSVARETS FORSKNINGSINSTITUTT, KJELLER	FFIK-
NORWAY. FORSVARETS FORSKNINGSINSTITUTT, KJELLER	FFIK-IR-K-
NORWAY. FORSVARETS FORSKNINGSINSTITUTT, KJELLER	FFIK TEKNISK-NOTAT-K-
NORWAY. FORSVARETS FORSKNINGSINSTITUTT, KJELLER	FF-IR-K-
NORWAY. FORSVARETS FORSKNINGSINSTITUTT, KJELLER	FFIS-
NORWAY. FORSVARETS FORSKNINGSINSTITUTT, KJELLER	FFIS-INTERN RAPPORT F-
NORWAY. FORSVARETS FORSKNINGSINSTITUTT, KJELLER	FFIS-IR-S-
NORWAY. FORSVARETS FORSKNINGSINSTITUTT, KJELLER	FFIS-RAPPORT S-
NORWAY. FORSVARETS FORSKNINGSINSTITUTT, KJELLER	FFIU-TEKNISK U-
NORWAY. FORSVARETS FORSKNINGSINSTITUTT, KJELLER	INTERN RAPPORT-(LTR.)-
NORWAY. FORSVARETS FORSKNINGSINSTITUTT, KJELLER	NDRE-
NORWAY. FORSVARETS FORSKNINGSINSTITUTT, KJELLER	NDRE-E-
NORWAY. FORSVARETS FORSKNINGSINSTITUTT, KJELLER	NDRE-K-
NORWAY. FORSVARETS FORSKNINGSINSTITUTT, KJELLER	NDRE-S-
NORWAY. FORSVARETS FORSKNINGSINSTITUTT, KJELLER	NDRE-U-
NORWAY. FORSVARETS FORSKNINGSINSTITUTT, KJELLER	NDRE-X-
NORWAY. FORSVARETS FORSKNINGSINSTITUUT, KJELLER	NORE-
NORWAY. FORSVARETS FORSKNINGSINSTITUTT, KJELLER	TEKNISK NOTAT (LTR.)-
(NAME TRANSLATED IS NORWEGIAN DEFENSE RESEARCH ESTAB)	
NORWAY. FORSVARETS FORSKNINGSINSTITUTT, OSLO	FFI/F-

NORWAY. FORSVARETS FORSKNINGSINSTITUTT, OSLO	FFIF-IR-F-
NORWAY. FORSVARETS FORSKNINGSINSTITUTT, OSLO	FFIU-TEKNISK-NOTATU-
NORWAY. FORSVARETS FORSKNINGSINSTITUTT, OSLO	KIR-
NORWAY. FORSVARETS FORSKNINGSINSTITUTT, OSLO	KIR-(NO.)/(YR.)
NORWAY. FORSVARETS FORSKNINGSINSTITUTT, OSLO	NDRE-
NORWAY. FORSVARETS FORSKNINGSINSTITUTT, OSLO	NDRE-IR-F-
NORWAY. FORSVARETS FORSKNINGSINSTITUTT, OSLO	NDRE-IR-K-
NORWAY. FORSVARETS FORSKNINGSINSTITUTT, OSLO	U-
NORWAY. INSTITUTT FOR ATOMENERGI, HALDEN	HRP-
NORWAY. INSTITUTT FOR ATOMENERGI, HALDEN	KAE-(YEAR)-
NORWAY. INSTITUTT FOR ATOMENERGI, KJELLER	KIR-
NORWAY. INSTITUTT FOR ATOMENERGI, KJELLER	KIR-N-
NORWAY. INSTITUTT FOR ATOMENERGI, KJELLER	KR-
NORWAY. INSTITUTT FOR ATOMENERGI, KJELLER	NC-
NORWAY. INSTITUTT FOR ATOMENERGI, KJELLER	NORA-
NORWAY. INSTITUTT FOR ATOMENERGI, KJELLER	NORA-MEMO-
NORWAY. INSTITUTT FOR ATOMENERGI, KJELLER	RCN-
NORWAY. INSTITUTT FOR ATOMENERGI. HALDEN REACTOR PROJECT, LILLESTROM	HPR-
NORWAY. SENTRALINSTITUTT FOR INDUSTRIELL FORSKNING, OSLO	CIIR-
NORWAY. SENTRALINSTITUTT FOR INDUSTRIELL FORSKNING, OSLO	CIIR-PUB-
NORWAY. SENTRALINSTITUTT FOR INDUSTRIELL FORSKNING, OSLO	CIIR-SI PUBL.-
NORWAY. SENTRALINSTITUTT FOR INDUSTRIELL FORSKNING, OSLO	CIIR-ST.-
NORWAY. SENTRALINSTITUTT FOR INDUSTRIELL FORSKNING, OSLO	SI-
NORWAY. SENTRALINSTITUTT FOR INDUSTRIELL FORSKNING, OSLO	SIF-
NORWAY. SENTRALINSTITUTT FOR INDUSTRIELL FORSKNING, OSLO	SI-PUB.-
NORWAY. SENTRALINSTITUTT FOR INDUSTRIELL FORSKNING, OSLO	SI PUBL.-
NORWAY. TECH. UNIV., TRONDHEIM	SKB-(NUMBER)/M-
NORWEGIAN DEFENCE CONSTRUCTION SERVICE. OFFICE OF TEST AND DEVELOPMENT, OSLO	FORTIFIKATORISK-NOTAT-
NORWEGIAN INST. OF TECH., TRONDHEIM. ELECTRONICS RESEARCH LAB.	NTNF-
NOTTINGHAM, ENGLAND. UNIV. DEPT. OF GEOGRAPHY (CLUSTER ANALYSIS)(AUTHOR'S INITIALS)	CA/PMM/(NUMBER/YEAR)
NOTTINGHAM, ENGLAND. UNIV. DEPT. OF GEOGRAPHY (CURVE FITTINGS)(AUTHOR'S INITIALS)	CF/JAD/(NUMBER/YEAR)
NOTTINGHAM, ENGLAND. UNIV. DEPT. OF GEOGRAPHY (CORRELATION + REGRESSION)(AUTHOR'S INITIALS)	CRP/RT/(NUMBER/YEAR)
NOTTINGHAM, ENGLAND. UNIV. DEPT. OF GEOGRAPHY (DATA PROCESSING PACKAGE)(AUTHOR'S INITIALS)	DPP/MJM/(NUMBER/YEAR)
NOTTINGHAM, ENGLAND. UNIV. DEPT. OF GEOGRAPHY (MULTIPLE DISCRIMINANT ANALYSIS)(INITIALS)	MDA/PMM/(NUMBER/YEAR)
NOTTINGHAM, ENGLAND. UNIV. DEPT. OF GEOGRAPHY (NETWORK ANALYSIS)(AUTHOR'S INITIALS)	NA/MJM/(NUMBER/YEAR)
NOVA SCOTIA. NAVAL RESEARCH ESTABLISHMENT, DARTMOUTH	NRE-
NRA, INC., LONG ISLAND CITY, N.Y.	NRA-
NUCLEAR CORP. OF AMERICA, DENVILLE, N.J.	NUCOR-
NUCLEAR CORP. OF AMERICA. INSTRUMENT AND CONTROL DIV., DENVILLE, N.J.	NCA-QPR-(NUMBER)-(YEAR)
NUCLEAR CORP. OF AMERICA. RES.CHEMS.DIV., BURBANK,CAL	RC-
NUCLEAR CROSS SECTIONS ADVISORY COMMITTEE, AEC	NCSAC-
NUCLEAR DESIGN AND CONSTRUCTION LTD., WHETSTONE, MIDDX., ENGLAND	NDC-R/
NUCLEAR DEV.CORP. OF AMERICA,WHITE PLAINS,N.Y.(REF.1)	(NUMBERS...)
NUCLEAR DEV. CORP. OF AMERICA, WHITE PLAINS, N.Y.	CIS-
NUCLEAR DEV. CORP. OF AMERICA, WHITE PLAINS, N.Y.	SDR-
NUCLEAR ENERGY RESEARCH BUREAU, N.Y.C.	NERB-
NUCLEAR MATERIALS AND EQUIPMENT CORP. (ALL LOCATIONS)	NUMEC-
NUCLEAR MATERIALS AND EQUIPMENT CORP., APOLLO, PENNA.	NMI-
NUCLEAR MATERIALS AND EQUIPMENT CORP., APOLLO, PENNA.	NUMEC-P-
NUCLEAR MATERIALS AND EQUIPMENT CORP., APOLLO, PENNA.	NUMEC-TM-P-
NUCLEAR MATERIALS AND EQUIPMENT CORP., LEWISTON, N.Y.	NUMEC-(NUMBERS)-
NUCLEAR METALS, INC. (ALL LOCATIONS)	NMI-
NUCLEAR METALS, INC., CAMBRIDGE, MASS.	MIT-WCP-
NUCLEAR METALS, INC., CAMBRIDGE, MASS.	NDD-
NUCLEAR METALS, INC., CAMBRIDGE, MASS.	NMI-(INITIALS)-
NUCLEAR METALS, INC., CAMBRIDGE, MASS. (FEASIBILITY REPORT)	NMI-FR-
NUCLEAR METALS, INC., CONCORD, MASS.	DP-
NUCLEAR METALS, INC., CONCORD, MASS.	NMI-TJ-
NUCLEAR METALS, INC., WEST CONCORD, MASS.	NMI-(NUMBER).(NUMBER)
NUCLEAR POWER GROUP, CHICAGO	NPG-
NUCLEAR POWER GROUP, KNUTSFORD, CHES., ENGLAND	NPCC/FEWP/P-
NUCLEAR POWER GROUP, KNUTSFORD, CHES., ENGLAND	NPCC/MWP(G)/P.
NUCLEAR POWER GROUP, KNUTSFORD, CHES., ENGLAND	NPCC/RPWP-
NUCLEAR POWER GROUP, KNUTSFORD, CHES., ENGLAND	TNPG-
NUCLEAR PRODUCTS-ERCO DIV., ACF INDUSTRIES, INC., WASHINGTON, D.C.	ACF-ERR-
NUCLEAR PRODUCTS-ERCO DIV., ACF INDUSTRIES, INC., WASHINGTON, D.C.	ACF-GCPR-
NUCLEAR PRODUCTS-ERCO DIV., ACF INDUSTRIES, INC., WASHINGTON, D.C.	KE-ACF-GCPR-
NUCLEAR RESEARCH ASSOCIATES, LONG ISLAND CITY, N.Y.	NRA-
NUCLEAR RESEARCH SERVICES, INC., DALLAS	NRS-(LETTER)(NUMBER)
NUCLEAR SCIENCE AND ENGINEERING CORP., PITTSBURGH	NSEC-
NUCLEAR SCIENCE AND ENGINEERING CORP., PITTSBURGH	NSEC-GE-
NUCLEAR SCIENCE AND ENGINEERING CORP., PITTSBURGH	NSEC-T-
NUCLEAR SYSTEMS, INC., GARLAND, TEX.	NSI-
NUCLEAR UTILITY SERVICES, INC., WASHINGTON, D.C.	NUS-
NUCLEAR UTILITY SERVICES, INC.,CONSULTEC DIV.,WASH,DC	CONSULTEC-
NUCLEAR WEAPON SYSTEMS SAFETY GROUP, KIRTLAND AFB, NM	AFINS-

NUCLEAR WEAPONS SURETY GROUP, FORT BELVOIR, VA.	USANWSG-
NUOVA SAN GIORGIO S.P.A., GENOA	NSG-REL-
NUS CORP., WASHINGTON, D.C.	NUS-
NUS CORP., WASHINGTON, D.C.	NUS-TM-ENG-
NUS CORP., WASHINGTON, D.C.	NUS-TM-ES-
NUS CORP., WASHINGTON, D.C.	NUS-TM-S-
NUS CORP. ENVIRONMENTAL SAFEGUARDS DIV., WASH., D.C.	NUS-
NUTRILITE PRODUCTS, INC., BUENA PARK, CALIF.(BIOL.)	B-
NYTRONICS, INC. CAPACITOR DIV., DARLINGTON, S.C.	QR-
NYTRONICS, INC. CAPACITOR DIV., DARLINGTON, S.C.	QRM-

OAK JOINT AEC-DOD WARHEAD COORDINATING COMM. (REF.15)	OAK-
OAK RIDGE ASSOCIATED UNIVERSITIES, INC., TENN.	ORAU-
OAK RIDGE GASEOUS DIFFUSION PLANT, TENN.	K-
OAK RIDGE GASEOUS DIFFUSION PLANT, TENN.	K-C-
OAK RIDGE GASEOUS DIFFUSION PLANT, TENN.	KC-
OAK RIDGE GASEOUS DIFFUSION PLANT, TENN.	K-D-
OAK RIDGE GASEOUS DIFFUSION PLANT, TENN.	K-DP-
OAK RIDGE GASEOUS DIFFUSION PLANT, TENN.	KFM-
OAK RIDGE GASEOUS DIFFUSION PLANT, TENN.	K-L-
OAK RIDGE GASEOUS DIFFUSION PLANT, TENN.	KL-
OAK RIDGE GASEOUS DIFFUSION PLANT, TENN.	KLD-
OAK RIDGE GASEOUS DIFFUSION PLANT, TENN.	KLI-
OAK RIDGE GASEOUS DIFFUSION PLANT, TENN.	K-M-
OAK RIDGE GASEOUS DIFFUSION PLANT, TENN.	K-OA-
OAK RIDGE GASEOUS DIFFUSION PLANT, TENN.	KOA-
OAK RIDGE GASEOUS DIFFUSION PLANT, TENN.	K-P-
OAK RIDGE GASEOUS DIFFUSION PLANT, TENN.	KSA-
OAK RIDGE GASEOUS DIFFUSION PLANT, TENN.	K-TL-
OAK RIDGE GASEOUS DIFFUSION PLANT, TENN.	K-TRANS-
OAK RIDGE GASEOUS DIFFUSION PLANT, TENN.	KX-
OAK RIDGE INST. OF NUCLEAR STUDIES, INC., TENN.	ORINS-
OAK RIDGE INST. OF NUCLEAR STUDIES, INC. MEDICAL DIV., TENN.	MED-(YEAR)-
OAK RIDGE INST. OF NUCLEAR STUDIES, INC. MEDICAL DIV., TENN.	ORINS-MED-
OAK RIDGE NATIONAL LAB., TENN. (REF. 3)	ANP-
OAK RIDGE NATIONAL LAB., TENN.	AR-
OAK RIDGE NATIONAL LAB., TENN.	ARC-
OAK RIDGE NATIONAL LAB., TENN.	CD-
OAK RIDGE NATIONAL LAB., TENN.	CF-(YR-MO)-
OAK RIDGE NATIONAL LAB., TENN.	CL-P-
OAK RIDGE NATIONAL LAB., TENN. (EASTERN DECIDUOUS FOREST BIOME)	EDFB-IBP-(YEAR)-
OAK RIDGE NATIONAL LAB., TENN.	H-
OAK RIDGE NATIONAL LAB., TENN. (HEAVY SECTION STEEL TECHNOLOGY PROGRAM)	HSSTP-TR-
OAK RIDGE NATIONAL LAB., TENN.	ISOTOPE CATALOG-
OAK RIDGE NATIONAL LAB., TENN.	K-
OAK RIDGE NATIONAL LAB., TENN.	K-L-
OAK RIDGE NATIONAL LAB., TENN.	NMI-
OAK RIDGE NATIONAL LAB., TENN.	ORNL-
OAK RIDGE NATIONAL LAB., TENN.	ORNL-AIC-
OAK RIDGE NATIONAL LAB., TENN. (CIVIL DEFENSE)	ORNL-CD-
OAK RIDGE NATIONAL LAB., TENN.	ORNL-CDC-
OAK RIDGE NATIONAL LAB., TENN. (CENTRAL FILES NO.)	ORNL-CF-
OAK RIDGE NATIONAL LAB., TENN.	ORNL CF-(YR. MO.)-
OAK RIDGE NATIONAL LAB., TENN.	ORNL-CPX-
OAK RIDGE NATIONAL LAB., TENN.	ORNL-CWS-
OAK RIDGE NATIONAL LAB., TENN.	ORNL-HUD-
OAK RIDGE NATIONAL LAB., TENN.	ORNL-IBP-(YEAR)-
OAK RIDGE NATIONAL LAB., TENN.	ORNL-LN-
OAK RIDGE NATIONAL LAB., TENN.	ORNL-MP-
OAK RIDGE NATIONAL LAB., TENN.	ORNL-MTR-
OAK RIDGE NATIONAL LAB., TENN.	ORNL-NSF-EP-
OAK RIDGE NATIONAL LAB., TENN.	ORNL-P-
OAK RIDGE NATIONAL LAB., TENN. (REPRINT NO.)	ORNL RN-
OAK RIDGE NATIONAL LAB., TENN.	ORNL-TM-
OAK RIDGE NATIONAL LAB., TENN.	ORNL-TR-
OAK RIDGE NATIONAL LAB., TENN.	ORNO-TM-
OAK RIDGE NATIONAL LAB., TENN.	PUC-(NUMBERS)-
OAK RIDGE NATIONAL LAB., TENN.	RPI-
OAK RIDGE NATIONAL LAB., TENN.	TAB-
OAK RIDGE NATIONAL LAB., TENN.	TD-(YEAR)-
OAK RIDGE NATIONAL LAB., TENN.	TPR-NS-
OAK RIDGE NATIONAL LAB. ATOMIC AND MOLECULAR PROCESSES INFORMATION CENTER, TENN.	ORNL-AMPIC-
OAK RIDGE NATIONAL LAB. ISOTOPES INFORMATION CENTER, TENN.	ORNL-IIC-
OAK RIDGE NATIONAL LAB. NUCLEAR DESALINATION INFORMATION CENTER, TENN.	ORNL-NDIC-
OAK RIDGE NATIONAL LAB. NUCLEAR SAFETY INFORMATION CENTER, TENN.	ORNL-NSIC-
OAK RIDGE NATIONAL LAB. RADIATION SHIELDING INFORMATION CENTER, TENN.	ORNL-RSIC-
OAK RIDGE NATIONAL LAB. RESEARCH MATERIALS INFORMATION CENTER, TENN.	ORNL-RMIC-
OAK RIDGE NATIONAL LAB., Y-12 AREA, TENN.	TAB-
OAK RIDGE OPERATIONS OFFICE, AEC, TENN. (REF. 4)	AEC-ORO-(CONTRACT NO.)-
OAK RIDGE OPERATIONS OFFICE, AEC, TENN. (REF. 4)	ORO-(CONTRACT NO.)-
OAK RIDGE OPERATIONS OFFICE, AEC, TENN. (REF. 4)	ORO-
OAK RIDGE OPERATIONS OFFICE, AEC, TENN. (REF. 4)	ORO-TR-
OAK RIDGE OPERATIONS OFFICE, AEC. ISOTOPES DIV., TENN.	IDA-
OAK RIDGE OPERATIONS OFFICE, AEC. ISOTOPES DIV., TENN.	IDB-
OAK RIDGE OPERATIONS OFFICE, AEC. ISOTOPES DIV., TENN.	IDC-
OAK RIDGE OPERATIONS OFFICE, AEC. ISOTOPES DIV., TENN.	IDD-
OAK RIDGE OPERATIONS OFFICE, AEC. ISOTOPES DIV., TENN.	IDE-
OAK RIDGE OPERATIONS OFFICE, AEC. OPERATIONAL PLANNING AND POWER DIV., TENN.	OPP-
OAK RIDGE SCHOOL OF REACTOR TECHNOLOGY, TENN.	CF-(YEAR)-(MONTH)-
OAK RIDGE SCHOOL OF REACTOR TECHNOLOGY, TENN.	ORSORT-
OCCIDENTAL COLL., LOS ANGELES	OC-
OCEANICS, INC., PLAINVIEW, N.Y.	OVL-
OCEANICS, INC., PLAINVIEW, N.Y.	OCEANICS-
OCEANOGRAPHIC SERVICES, INC., SANTA BARBARA, CALIF.	OSI-(YEAR)-
ODIN ASSOCIATES	OA-
OESTERREICHISCHE STUDIENGESELLSCHAFT FUER ATOMENERGIE G.M.B.H., SEIBERSDORF	SGAE-(LTRS.)-(NO.)/(YR.)
OESTERREICHISCHE STUDIENGESELLSCHAFT FUER ATOMENERGIE G.M.B.H., SEIBERSDORF	SGAE-G-
OESTERREICHISCHE STUDIENGESELLSCHAFT FUER ATOMENERGIE G.M.B.H., SIEBERSDORF (LANDWIRTSCHAFTLICHE FORSCHUNGSABTEILUNG)	SGAE-LA-(NUMBER/YEAR)

OESTERREICHISCHE STUDIENGESELLSCHAFT FUER ATOMENERGIE G.M.B.H., VIENNA	SGAE-(YEAR)/
OESTERREICHISCHE STUDIENGESELLSCHAFT FUER ATOMENERGIE G.M.B.H., VIENNA	SGAE-(YEAR/NUMBER)(LTR.)
OESTERREICHISCHE STUDIENGESELLSCHAFT FUER ATOMENERGIE G.M.B.H., VIENNA	SGAE-B-
OESTERREICHISCHE STUDIENGESELLSCHAFT FUER ATOMENERGIE G.M.B.H., VIENNA	SGAE-IB-
OESTERREICHISCHE STUDIENGESELLSCHAFT FUER ATOMENERGIE G.M.B.H., VIENNA	SGAE-ME-
OESTERREICHISCHE STUDIENGESELLSCHAFT FUER ATOMENERGIE G.M.B.H. ELEKTRONIKINSTITUT, SEIBERSDORF	SGAE-E-(NUMBER/YEAR)
OESTERREICHISCHE STUDIENGESELLSCHAFT FUER ATOMENERGIE G.M.B.H. INSTITUT FUER BIOLOGIE UND LANDWIRTSCHAFT, SEIBERSDORF	SGAE-BL-(NUMBER/YEAR)
OESTERREICHISCHE STUDIENGESELLSCHAFT FUER ATOMENERGIE G.M.B.H. INSTITUT FUER BIOLOGIE UND LANDWIRTSCHAFT, SEIBERSDORF (SEIBERSDORF PROJECT REPORT)	SPR-
OESTERREICHISCHE STUDIENGESELLSCHAFT FUER ATOMENERGIE G.M.B.H. INSTITUT FUER CHEMIE, SEIBERSDORF	SGAE-CH-
OESTERREICHISCHE STUDIENGESELLSCHAFT FUER ATOMENERGIE G.M.B.H. INSTITUT FUER METALLURGIE, SEIBERSDORF	SGAE-M-
OESTERREICHISCHE STUDIENGESELLSCHAFT FUER ATOMENERGIE G.M.B.H. INSTITUT FUER RADIUMFORSCHUNG UND KERNPHYSIK, SEIBERSDORF	SGAE-IB/IA-(NUMBER/YEAR)
OESTERREICHISCHE STUDIENGESELLSCHAFT FUER ATOMENERGIE G.M.B.H. INST. FUER REAKTORENTWICKLUNG, SEIBERSDORF	SGAE-RE-(NUMBER/YEAR)
OESTERREICHISCHE STUDIENGESELLSCHAFT FUER ATOMENERGIE G.M.B.H. INST. FUER REAKTORTECHNIK, SEIBERSDORF	SGAE-RT-(NUMBER/YEAR)
OESTERREICHISCHE STUDIENGESELLSCHAFT FUER ATOMENERGIE G.M.B.H. INST. FUER STRAHLENSCHUTZ, SEIBERSDORF	SGAE-SS-(NUMBER/YEAR)
OESTERREICHISCHE STUDIENGESELLSCHAFT FUER ATOMENERGIE G.M.B.H. PHYSIK-INSTITUT, SEIBERSDORF	SGAE-PH-
OESTERREICHISCHE STUDIENGESELLSCHAFT FUER ATOMENERGIE G.M.B.H. REAKTORINSTITUT, SEIBERSDORF	SGAE-R-(NUMBER/YEAR)
OFFICE FOR ATOMIC ENERGY (AIR FORCE), WASHINGTON, D.C.	AFOAT-1-
OFFICE OF AEROSPACE RESEARCH (AIR FORCE). EUROPEAN OFFICE	AF-EOAR-
OFFICE OF AEROSPACE RESEARCH (AIR FORCE), WASH., D.C. (QUARTERLY INDEX)	AFOAR-QI-(MO.-MO./YR.)
OFFICE OF AEROSPACE RESEARCH (AIR FORCE), WASH., D.C. (STATUS OF RESEARCH PROPOSALS)	AFOAR-SRP-(YEAR/MONTH)
OFFICE OF AEROSPACE RESEARCH (AIR FORCE), WASH., D.C. (DEVELOPMENT PLANNING REPORT)	AFRDC-DPR-
OFFICE OF AEROSPACE RESEARCH (AIR FORCE), WASH., D.C. (DEVELOPMENT PLANNING STUDY)	AFRDC-DPS-
OFFICE OF AEROSPACE RESEARCH (AIR FORCE), WASH., D.C.	OAR-
OFFICE OF AEROSPACE RESEARCH (AIR FORCE), WASH., D.C.	OAR-(YEAR)-
OFFICE OF AEROSPACE RESEARCH (AIR FORCE), WASH., D.C.	OAR-BRR-
OFFICE OF AEROSPACE RESEARCH (AIR FORCE), WASH., D.C. (QUARTERLY INDEX)	OAR-QI-
OFFICE OF AEROSPACE RESEARCH (AIR FORCE), WASH., D.C.	RCS-OAR-
OFFICE OF AIR RESEARCH, WRIGHT-PATTERSON AFB, OHIO	AAF OAR-
OFFICE OF AIR RESEARCH, WRIGHT-PATTERSON AFB, OHIO	AAF-OAR-TR-
OFFICE OF AIR RESEARCH, WRIGHT-PATTERSON AFB, OHIO	OAR-TR-
OFFICE OF BUSINESS ECONOMICS, WASHINGTON, D.C.	OBE-SP-
OFFICE OF CIVIL DEFENSE, WASHINGTON, D.C.	FG-E-
OFFICE OF CIVIL DEFENSE, WASHINGTON, D.C. (HOME FALLOUT PROTECTION SURVEY)	HFPS-
OFFICE OF CIVIL DEFENSE, WASHINGTON, D.C.	MP-
OFFICE OF CIVIL DEFENSE, WASHINGTON, D.C. (NATIONAL PLAN APPENDIX SERIES)	OCD-NP-
OFFICE OF CIVIL DEFENSE, WASHINGTON, D.C.	OCD-PS-
OFFICE OF CIVIL DEFENSE, WASHINGTON, D.C.	OCD-RR-
OFFICE OF CIVIL DEFENSE, WASHINGTON, D.C.	OCD-TR-
OFFICE OF CIVIL DEFENSE, WASHINGTON, D.C.	PM-
(SEE ALSO LATER NAME: OFF. OF EMERGENCY PREPAREDNESS)	
OFFICE OF CIVIL AND DEFENSE MOBILIZATION, WASH., D.C.	OCDM-
(SEE ALSO LATER NAME: OFF. OF EMERGENCY PREPAREDNESS)	
OFFICE OF COAL RESEARCH, WASHINGTON, D.C.	OCR-
OFFICE OF DEFENSE MOBILIZATION, WASHINGTON, D.C.	ODM-
(SEE ALSO LATER NAME: OFF. OF EMERGENCY PREPAREDNESS)	
OFFICE OF ECONOMIC OPPORTUNITY, WASHINGTON, D.C.	OEO-
OFFICE OF ECONOMIC OPPORTUNITY, WASHINGTON, D.C.	OEO-LN-
OFFICE OF EDUCATION, WASHINGTON, D.C.	OE-OPPE-(YEAR)-
OFFICE OF EDUCATION. EDUCATIONAL RESOURCES INFORMATION CENTER, WASHINGTON, D.C. (ERIC DOC.)	ED-(6 DIGITS)
OFFICE OF EDUCATION. EDUCATIONAL RESOURCES INFORMATION CENTER, WASHINGTON, D.C. (ERIC JNL.)	EJ-(6 DIGITS)
OFFICE OF EDUCATION. EDUCATIONAL RESOURCES INFORMATION CENTER, WASHINGTON, D.C.	ERIC-
OFFICE OF EMERGENCY PREPAREDNESS, WASHINGTON, D.C. (DEFENSE MOBILIZATION ORDER)	DMO-
OFFICE OF EMERGENCY PREPAREDNESS, WASHINGTON, D.C. (DEFENSE MANPOWER POLICY)	DMP-
OFFICE OF EMERGENCY PREPAREDNESS, WASHINGTON, D.C. (GENERAL ADMINISTRATIVE ORDER)	GAO-
OFFICE OF EMERGENCY PREPAREDNESS, WASHINGTON, D.C.	OEP-
OFFICE OF INFORMATION SERVICES, AEC, WASHINGTON, D.C. (NON-PROJECT REPORT. INITIALLY ASSIGNED TO ALL NON-AEC REPORTS, BUT LATER USED ONLY FOR THOSE WITHOUT OTHER NUMBERS)	NP-(5 DIGITS)
OFFICE OF INFORMATION SERVICES, AEC, WASHINGTON, D.C. (NUCLEAR SCIENCE ABSTRACTS)	NSA-
OFFICE OF INFORMATION SERVICES, AEC, WASHINGTON, D.C. (REF. 5)	TID-
OFFICE OF MANAGEMENT AND BUDGET, WASHINGTON, D.C.	OMB-
OFFICE OF MANAGEMENT AND BUDGET, WASHINGTON, D.C. (STANDARD INDUSTRIAL CLASSIFICATION)	SIC-
OFFICE OF MILITARY ATTACHE (AIR FORCE), LONDON	AF-OMA-
OFFICE OF MILITARY GOVERNMENT FOR GERMANY (U.S.)	OMG-
OFFICE OF NAVAL INTELLIGENCE, WASHINGTON, D.C.	DNI-
OFFICE OF NAVAL INTELLIGENCE, WASHINGTON, D.C.	NIC-TRANS-
OFFICE OF NAVAL INTELLIGENCE, WASHINGTON, D.C.	NIU-
OFFICE OF NAVAL INTELLIGENCE, WASHINGTON, D.C.	ONI-
OFFICE OF NAVAL INTELLIGENCE, WASHINGTON, D.C.	ONI-A-
OFFICE OF NAVAL INTELLIGENCE, WASHINGTON, D.C.	ONI-ST-(NUMBER)-(YEAR)
OFFICE OF NAVAL INTELLIGENCE, WASHINGTON, D.C.	ONI-TR-
OFFICE OF NAVAL INTELLIGENCE, WASHINGTON, D.C.	OP-32-
OFFICE OF NAVAL INTELLIGENCE, WASHINGTON, D.C.	Y-
OFFICE OF NAVAL INTELLIGENCE. PHYSICAL VULNERABILITY DIV., WASHINGTON, D.C.	PVIM-
OFFICE OF NAVAL INTELLIGENCE. TRANSLATION SECTION, WASHINGTON, D.C.	ONI-TRANS-
OFFICE OF NAVAL MATERIAL, WASH., D.C. (PROJECT MANAGER)	PM-

OFFICE OF NAVAL MATERIAL. SPEC. PROJS. OFF.

OFFICE OF NAVAL MATERIAL. SPEC. PROJS. OFF.,WASH.,DC	ONM-
OFFICE OF NAVAL RESEARCH (AIR DEFENSE OPERATIONS)	AD-
OFFICE OF NAVAL RESEARCH (AIRBORNE LANDING OPERATIONS	AL-
OFFICE OF NAVAL RESEARCH (ANTI-SUBMARINE OPERATIONS)	AS-
OFFICE OF NAVAL RESEARCH (ATOMIC WARFARE OPERATIONS)	AW-
OFFICE OF NAVAL RESEARCH (BASIC RESEARCH)	BR-
OFFICE OF NAVAL RESEARCH (BIOL. WARFARE OPERATIONS)	BW-
OFFICE OF NAVAL RESEARCH	C-
OFFICE OF NAVAL RES. (COMBAT AIR SUPPORT OPERATIONS)	CA-
OFFICE OF NAVAL RESEARCH (NAVAL ANALYSIS MEMORANDUM)	NAM-
OFFICE OF NAVAL RESEARCH (NAVAL ANALYSIS REPORT)	NAR-
OFFICE OF NAVAL RESEARCH (PSYCHOLOGICAL WARFARE AND COLD WAR OPERATIONS)	PC-
OFFICE OF NAVAL RESEARCH (PERSONNEL OPERATIONS)	PO-
OFFICE OF NAVAL RESEARCH (SEA COMBAT OPERATIONS)	SC-
OFFICE OF NAVAL RESEARCH (SUPPLY OPERATIONS)	SO-
OFFICE OF NAVAL RESEARCH (SUPPORTING RESEARCH + DEV.)	SR-
OFFICE OF NAVAL RESEARCH (SUBMARINE OPERATIONS)	SW-
OFFICE OF NAVAL RESEARCH (TECHNICAL DEVELOPMENT PLAN)	TDP-
OFFICE OF NAVAL RESEARCH, LONDON	D-
OFFICE OF NAVAL RESEARCH, LONDON	ESN-
OFFICE OF NAVAL RESEARCH, LONDON. (OFF. OF THE ASST. NAVAL ATTACHE FOR RESEARCH, LONDON)	OANAR-
OFFICE OF NAVAL RESEARCH, LONDON	OANAR-(NO.)-(YR)
OFFICE OF NAVAL RESEARCH, LONDON	ONRL-
OFFICE OF NAVAL RESEARCH, LONDON	ONRL-(NO.)-(YR)
OFFICE OF NAVAL RESEARCH, LONDON	ONRL A(NUMBER)-(YEAR)
OFFICE OF NAVAL RESEARCH, LONDON (CONFERENCES)	ONRL-C-(NUMBER)-(YEAR)
OFFICE OF NAVAL RESEARCH, LONDON (EUROPEAN SCI.NOTES)	ONRL ESN-(NUMBER)-
OFFICE OF NAVAL RESEARCH, LONDON	ONRL-M-
OFFICE OF NAVAL RESEARCH, LONDON	ONRL-R-
OFFICE OF NAVAL RESEARCH, LONDON	ONRL TR-
OFFICE OF NAVAL RESEARCH, WASH., D.C. (SERIES FOR PROCEEDINGS OF MEETINGS, CONFERENCES AND SYMPOSIA SPONSORED BY THE OFFICE OF NAVAL RESEARCH)	ACR-
OFFICE OF NAVAL RESEARCH, WASHINGTON, D.C.	AMP NO. NR-
OFFICE OF NAVAL RESEARCH, WASHINGTON, D.C.	NAVEXOS-
OFFICE OF NAVAL RESEARCH, WASHINGTON, D.C.	NAVEXOS-P-
OFFICE OF NAVAL RESEARCH, WASHINGTON, D.C.	NR-
OFFICE OF NAVAL RESEARCH, WASHINGTON, D.C.	ONR-
OFFICE OF NAVAL RESEARCH, WASHINGTON, D.C.	ONR-(YEAR)-
OFFICE OF NAVAL RESEARCH, WASHINGTON, D.C.	ONR(YEAR)-
OFFICE OF NAVAL RESEARCH, WASHINGTON, D.C.	ONR/ACR/(LETTERS OR NOS.)
OFFICE OF NAVAL RESEARCH, WASHINGTON, D.C.	ONR-CR-
OFFICE OF NAVAL RESEARCH, WASHINGTON, D.C.	ONR-HO(YEAR)-
OFFICE OF NAVAL RESEARCH, WASHINGTON, D.C.	ONR-RF-
OFFICE OF NAVAL RESEARCH, WASHINGTON, D.C.	ONR-RM-
OFFICE OF NAVAL RESEARCH, WASHINGTON, D.C.	ONR-RR-
OFFICE OF NAVAL RESEARCH, WASHINGTON, D.C.	ONR SYMPOSIUM-ACR-
OFFICE OF NAVAL RESEARCH, WASHINGTON, D.C.	ONR-TR-
OFFICE OF NAVAL RESEARCH, WASHINGTON, D.C.	ONR-TR-RLT-
OFFICE OF NAVAL RESEARCH, WASHINGTON, D.C.	ORI-
OFFICE OF NAVAL RESEARCH, WASHINGTON, D.C.	TR-
OFFICE OF NAVAL RESEARCH, WASHINGTON, D.C.	USAG-
OFFICE OF NAVAL RESEARCH. GROUP PSYCHOLOGY BRANCH	GP-
OFFICE OF NAVAL RESEARCH. NAVAL ANALYSIS GROUP, WASHINGTON, D.C.	ACR/NAM-
OFFICE OF NAVAL RES. OPERATIONS RES. GP., WASH., DC	ORG-
OFFICE OF NAVAL RES. SANDS POINT LAB., PT. WASH.,N.Y.	SPL TR-
OFFICE OF NAVAL RES. SPECIAL DEVICES CTR., WASH.,D.C.	SPECDEVCEN-(NO.)-(NO.)-
OFFICE OF NAVAL RES. SPECIAL DEVICES CTR., WASH.,D.C.	SPECDEVCEN-
OFFICE OF OIL AND GAS, WASHINGTON, D.C.	OOG-
OFFICE OF OPERATIONS ANALYSIS, ENT AFB, COLO.	OOA-
OFFICE OF OPERATIONS ANALYSIS AND PLANNING, AEC	OA-R-
OFFICE OF ORDNANCE RESEARCH (ARMY), DURHAM, N.C.	CSGLD-
OFFICE OF ORDNANCE RESEARCH (ARMY), DURHAM, N.C.	CSPRD-
OFFICE OF ORDNANCE RESEARCH (ARMY), DURHAM, N.C.	OOR-
OFFICE OF ORDNANCE RESEARCH (ARMY), DURHAM, N.C.	OOR-(NUMBER).(NUMBER)
OFFICE OF ORDNANCE RESEARCH (ARMY), DURHAM, N.C.	OOR-R-(NUMBER)-
OFFICE OF ORDNANCE RESEARCH (ARMY), DURHAM, N.C.	OORR-
OFFICE OF ORDNANCE RESEARCH (ARMY), DURHAM, N.C. (TECHNICAL MEMORANDUM)	OORTM-
OFFICE OF PRODUCTION RESEARCH AND DEVELOPMENT	OPRD-
OFFICE OF PRODUCTION RES. + DEV. (CHLORAL RPTS.)	OPRD CR-
OFFICE OF PRODUCTION RESEARCH AND DEVELOPMENT (QUALITY CONTROL REPORTS, CARNEGIE INST. OF TECH.)	OPRD QCR-
OFFICE OF PRODUCTION RESEARCH AND DEVELOPMENT	OPRD RR-
OFFICE OF PRODUCTION RESEARCH AND DEVELOPMENT	OPRD TF-
OFFICE OF PRODUCTION RESEARCH AND DEVELOPMENT (QUALITY CONTROL REPORTS, CARNEGIE INST. OF TECH.)	QCR-
OFFICE OF PRODUCTION RESEARCH AND DEVELOPMENT	TF-
OFFICE OF RESEARCH ANALYSES, HOLLOMAN AFB, N. MEX.	ORA-
OFFICE OF RESEARCH ANALYSES, HOLLOMAN AFB, N. MEX.	ORA-(YEAR)-
OFFICE OF RESEARCH AND DEVELOPMENT (INTERIOR), WASH., D.C. (WATER POLLUTION CONTROL RESEARCH SERIES)	ORD-(NUMBER) DMU
OFFICE OF RESEARCH AND INVENTIONS (NAVY)	ORI-
OFFICE OF RUBBER RESERVE (POLYMER DEVELOPMENT)	CD-
OFFICE OF RUBBER RESERVE (POLYMER RESEARCH, DISCRETION GROUP REPORTS)	CR-
OFFICE OF RUBBER RESERVE	ORR-
OFFICE OF RUBBER RESERVE (POLYMER DEVELOPMENT)	ORR CD-
OFFICE OF RUBBER RESERVE (POLYMER RESEARCH, DISCRETION GROUP REPORTS)	ORR CR-
OFFICE OF RUBBER RESERVE (GENERAL REPORTS)	ORR G-
OFFICE OF RUBBER RESERVE (RAW MATERIALS REPORTS)	ORR RM-
OFFICE OF RUBBER RESERVE (RECONSTRUCTION FINANCE CORP. TECHNICAL REPORTS)	ORR TR-
OFFICE OF RUBBER RESERVE (RAW MATERIALS REPORTS)	RM-
OFFICE OF RUBBER RESERVE. COMPOUNDERS COMMITTEE	CC-
OFFICE OF RUBBER RESERVE. COMPOUNDERS COMMITTEE	ORR CC-
OFFICE OF SALINE WATER, WASHINGTON, D.C.	OSW-PR-
OFFICE OF SALINE WATER, WASHINGTON, D.C.	OSW-RDPR-

OFFICE OF SCIENCE AND TECHNOLOGY, WASHINGTON, D.C.	OST-
OFFICE OF SCIENTIFIC RESEARCH AND DEVELOPMENT (ASSIGNED TO REPORTS FROM GREAT BRITAIN)	B-
OFFICE OF SCI. RES. + DEV. (CONTROLLED FRAGMENTATION)	CF-
OFFICE OF SCIENTIFIC RESEARCH AND DEVELOPMENT	CZC-
OFFICE OF SCIENTIFIC RESEARCH AND DEVELOPMENT	DDC-
OFFICE OF SCI. RES. + DEV.(DETONATION+FRAGMENTATION)	DF-
OFFICE OF SCIENTIFIC RESEARCH AND DEVELOPMENT (DETONATION, FRAGMENTATION AND AIR BLAST)	DFA-
OFFICE OF SCIENTIFIC RESEARCH AND DEVELOPMENT	ERD-
OFFICE OF SCIENTIFIC RESEARCH AND DEVELOPMENT (EFFECTS OF WEAPONS ON TARGETS)	EWT-
OFFICE OF SCIENTIFIC RESEARCH AND DEVELOPMENT (EFFECTS OF WEAPONS ON TARGETS)	EWT-(NUMBER)(LETTER)
OFFICE OF SCIENTIFIC RESEARCH AND DEVELOPMENT	F-
OFFICE OF SCIENTIFIC RESEARCH AND DEVELOPMENT (FUNDAMENTAL STUDY OF EXPLOSIVES)	FS-
OFFICE OF SCIENTIFIC RESEARCH AND DEVELOPMENT	I(LETTER)C-
OFFICE OF SCIENTIFIC RESEARCH AND DEVELOPMENT	J-
OFFICE OF SCIENTIFIC RESEARCH AND DEVELOPMENT	J(LETTER)C-
OFFICE OF SCIENTIFIC RES. + DEV. (JET PROPULSION)	JP-
OFFICE OF SCIENTIFIC RESEARCH AND DEVELOPMENT	K-
OFFICE OF SCIENTIFIC RESEARCH AND DEVELOPMENT	L(LETTER)C-
OFFICE OF SCIENTIFIC RES. + DEV. (LIAISON OFF. REF.)	LOR-
OFFICE OF SCIENTIFIC RESEARCH AND DEVELOPMENT	LP-
OFFICE OF SCIENTIFIC RESEARCH AND DEVELOPMENT	M(LETTER)C-
OFFICE OF SCIENTIFIC RESEARCH AND DEVELOPMENT	MBE-
OFFICE OF SCIENTIFIC RESEARCH AND DEVELOPMENT	N(LETTER)C-
OFFICE OF SCIENTIFIC RESEARCH AND DEVELOPMENT	NDRC-
OFFICE OF SCIENTIFIC RESEARCH AND DEVELOPMENT	NY-
OFFICE OF SCIENTIFIC RESEARCH AND DEVELOPMENT	O(LETTER)C-
OFFICE OF SCIENTIFIC RESEARCH AND DEVELOPMENT	OD-OAG-
OFFICE OF SCIENTIFIC RESEARCH AND DEVELOPMENT (ORGANIC DEVELOPMENT PROBLEMS)	ODP-
OFFICE OF SCIENTIFIC RESEARCH AND DEVELOPMENT	OEA-
OFFICE OF SCIENTIFIC RESEARCH AND DEVELOPMENT(REF.32)	OSRD-
OFFICE OF SCIENTIFIC RES. + DEV. (DIV. NO.)(REPT.NO.)	OSRD-(NO.)-(NO.)-
OFFICE OF SCIENTIFIC RESEARCH AND DEV. (FOR IDENTIFICATION OF CONTRACTORS REPRESENTED BY LETTERS, SEE ENTRIES BEGINNING WITH THOSE LETTERS) (REF. 32)	OSRD (LETTERS)-
OFFICE OF SCIENTIFIC RESEARCH AND DEVELOPMENT (ORDNANCE AND TERMINAL BALLISTICS)	OTB-
OFFICE OF SCIENTIFIC RESEARCH AND DEVELOPMENT	P-
OFFICE OF SCIENTIFIC RESEARCH AND DEVELOPMENT	P(LETTER)C-
OFFICE OF SCIENTIFIC RESEARCH AND DEVELOPMENT	PMR-
OFFICE OF SCIENTIFIC RESEARCH AND DEVELOPMENT	POM-
OFFICE OF SCIENTIFIC RESEARCH AND DEVELOPMENT (PREPARATION AND TESTING OF EXPLOSIVES)	PT-
OFFICE OF SCIENTIFIC RESEARCH AND DEVELOPMENT	PTM-
OFFICE OF SCIENTIFIC RESEARCH AND DEVELOPMENT	R(LETTER)C-
OFFICE OF SCIENTIFIC RESEARCH AND DEVELOPMENT	REI-TMD-
OFFICE OF SCIENTIFIC RESEARCH AND DEVELOPMENT (ROCKET PROPELLANTS)	RP-
OFFICE OF SCIENTIFIC RESEARCH AND DEVELOPMENT (RDX AND RELATED COMPOUNDS)	RRC-
OFFICE OF SCIENTIFIC RES. + DEV. (SHAPED CHARGES)	SC-
OFFICE OF SCIENTIFIC RESEARCH AND DEVELOPMENT	SM-
OFFICE OF SCIENTIFIC RES. + DEV.(SPECIAL PROPELLANTS)	SP-
OFFICE OF SCIENTIFIC RESEARCH AND DEVELOPMENT	SR-
OFFICE OF SCIENTIFIC RESEARCH AND DEVELOPMENT	SRG-C-
OFFICE OF SCIENTIFIC RESEARCH AND DEVELOPMENT	ST-
OFFICE OF SCIENTIFIC RES. + DEV.(SUMMARY TECH. REPT.)	STR-
OFFICE OF SCIENTIFIC RESEARCH AND DEVELOPMENT	T(LETTER)C-
OFFICE OF SCIENTIFIC RES. + DEV.(TRACER COMPOSITIONS)	TC-
OFFICE OF SCIENTIFIC RES. + DEV.(TORPEDO TECH. MEMO.)	TTM-
OFFICE OF SCIENTIFIC RESEARCH AND DEVELOPMENT	U(LETTER)C-
OFFICE OF SCIENTIFIC RESEARCH AND DEVELOPMENT (UNDERWATER EXPLOSIVES AND EXPLOSIONS)	UE-
OFFICE OF SCIENTIFIC RESEARCH AND DEVELOPMENT	UM-
OFFICE OF SCIENTIFIC RESEARCH AND DEVELOPMENT	UM-RPP-
OFFICE OF SCIENTIFIC RESEARCH AND DEVELOPMENT	W-
OFFICE OF SCIENTIFIC RESEARCH AND DEVELOPMENT	ZBC-
OFFICE OF SCIENTIFIC RES. + DEV. LIAISON OFF., LONDON	WA-
OFFICE OF SPECIAL WEAPONS DEVELOPMENTS, FT. BLISS,TEX	CAC-
OFFICE OF SPECIAL WEAPONS DEVELOPMENTS, FT. BLISS,TEX	CAC/OSWD-
OFFICE OF SPECIAL WEAPONS DEVELOPMENTS, FORT BLISS, TEX. (DOCUMENT)	CAC-OSWDD-
OFFICE OF SPECIAL WEAPONS DEVELOPMENTS, FT. BLISS,TEX	IL-(NUMBER)-
OFFICE OF SPECIAL WEAPONS DEVELOPMENTS, FT. BLISS,TEX	OSWD-
OFFICE OF SPECIAL WEAPONS DEVELOPMENTS, FORT BLISS, TEX. (DOCUMENT)	OSWDD-
OFFICE OF STRATEGIC SERVICES, WASHINGTON, D.C.	OSS-
OFFICE OF TECHNICAL INFORMATION, AEC (SEE ALSO LATER NAMES: DIV. OF TECH. INFO., AEC, AND OFFICE OF INFO. SERVICES, AEC)	OTI-
OFFICE OF TECHNICAL INFORMATION EXTENSION, AEC	OTIE-
OFFICE OF TECHNICAL INFORMATION EXTENSION, AEC, OAK RIDGE, TENN. (ACCESSION LIST. LETTER AT END STANDS FOR SECURITY CLASSIFICATION) (SEE ALSO LATER NAMES: DIV. OF TECH. INFO. EXTENSION, AEC, AND TECH. INFO. CENTER, AEC)	OTIE-AL-(NUMBER)(LETTER)
OFFICE OF TECH.SERVICES,WASH.,DC.(BOOK EXPLOITATIONS)	BE-
OFFICE OF TECHNICAL SERVICES, WASHINGTON, D.C. (CATALOG OF TECHNICAL REPORTS)	CTR-
OFFICE OF TECHNICAL SERVICES, WASHINGTON, D.C.	IB-
OFFICE OF TECHNICAL SERVICES, WASHINGTON, D.C.	OTS-
OFFICE OF TECH. SERVICES, WASH., D.C. (TRANSLATION)	OTS-(YEAR)-
OFFICE OF TECHNICAL SERVICES, WASHINGTON, D.C. (FILM BIBLIOGRAPHY)	OTS FB-
OFFICE OF TECH. SERVICES, WASH.,D.C. (INFORMAL RPT.)	OTS IR-
OFFICE OF TECHNICAL SERVICES, WASH.,D.C. (PRICE LIST)	OTS-PL-
OFFICE OF TECHNICAL SERVICES, WASHINGTON, D.C. (SUBJECT BIBLIOGRAPHY)	OTS-SB-
OFFICE OF TECHNICAL SERVICES, WASHINGTON, D.C. (SELECTIVE BIBLIOGRAPHY) (SEE ALSO LATER NAMES: CLEARINGHOUSE FOR FEDERAL SCI. + TECH. INFO. AND NATL. TECH. INFO. SERVICE)	SB-
OFFICE OF TECHNICAL SERVICES. INDUSTRIAL RESEARCH AND DEVELOPMENT DIV., WASH., D.C. (PROJ. CONTRACT CAC-)	CAC (NUMBER)-
OFFICE OF TECH. SERVICES. IND.RES.+DEV.DIV.,WASH.,DC	IRDD-
OFFICE OF TECH. SERVICES. IND.RES.+DEV.DIV.,WASH.,DC	OTS CAC-
OFFICE OF TECH. SERVICES. IND.RES.+DEV.DIV.,WASH.,DC	OTS IRDD-
OFFICE OF TECHNICAL SERVICES. INDUSTRIAL RESEARCH AND DEVELOPMENT DIV., WASH., D.C. (PROJECT REPORTS)	OTS IRDD PROJ-
OFFICE OF TECHNICAL SERVICES. INVENTIONS AND ENG. DIV. TECHNICAL ADVISORY SERVICE, WASHINGTON, D.C.	OTS TAS-
OFFICE OF TECHNICAL SERVICES. INVENTIONS AND ENG. DIV. TECHNICAL ADVISORY SERVICE, WASHINGTON, D.C.	TAS-
OFFICE OF TECH. SERVICES. PUBLICATION BD.,WASH.,D.C.	OTS PB-
OFFICE OF TEST INFORMATION, AEC, LAS VEGAS, NEV.	OTI-
OFFICE OF THE ADJUTANT GENERAL, WASHINGTON, D.C.	AGO-
OFFICE OF THE ADJUTANT GENERAL, WASHINGTON, D.C.	PR-
OFFICE OF THE ADJUTANT GENERAL, WASHINGTON, D.C.	PRB-
OFFICE OF THE ADJUTANT GENERAL, WASHINGTON, D.C.	PRS-

OFFICE OF THE ADJUTANT GENERAL. PERSONNEL RESEARCH BRANCH, WASHINGTON, D.C.   OAG-PRB-TRN-

OFFICE OF THE ASSISTANT SECRETARY OF DEFENSE. ADVISORY PANEL ON PERSONNEL AND TRAINING RESEARCH   PPT-

OFFICE OF THE ASSISTANT SECRETARY OF DEFENSE (INSTALLATIONS AND LOGISTICS), WASHINGTON, D.C.   H-
OFFICE OF THE ASSISTANT SECRETARY OF DEFENSE (INSTALLATIONS AND LOGISTICS), WASHINGTON, D.C.   TD-

OFFICE OF THE ASSISTANT SECRETARY OF DEFENSE (SYSTEMS ANALYSIS), WASHINGTON, D.C.   TP-(YEAR)-

OFF. OF THE ASST. SECY. OF DEFENSE(RES.+DEV.),WASH,DC   CAF-
OFF. OF THE ASST. SECY. OF DEFENSE(RES.+DEV.),WASH,DC   CES-
OFF. OF THE ASST. SECY. OF DEFENSE(RES.+DEV.),WASH,DC   DOD/R AND E-
OFF. OF THE ASST. SECY. OF DEFENSE(RES.+DEV.),WASH,DC   JRDB-
OFF. OF THE ASST. SECY. OF DEFENSE(RES.+DEV.),WASH,DC   LET-
OFF. OF THE ASST. SECY. OF DEFENSE(RES.+DEV.),WASH,DC   PFL-(NO.)/-
OFF. OF THE ASST. SECY. OF DEFENSE(RES.+DEV.),WASH,DC   POR-
OFF. OF THE ASST. SECY. OF DEFENSE(RES.+DEV.),WASH,DC   RD-
OFF. OF THE ASST. SECY. OF DEFENSE(RES.+DEV.),WASH,DC   RD-(NO.)/-
OFF. OF THE ASST. SECY. OF DEFENSE(RES.+DEV.),WASH,DC   RDB-

OFFICE OF THE ASSISTANT SECRETARY OF DEFENSE (RES. AND DEV.) ADVISORY GROUP ON ELECTRON TUBES   GET-
OFFICE OF THE ASSISTANT SECRETARY OF DEFENSE (RES. AND DEV.) ADVISORY GROUP ON ELECTRON TUBES   GET/P-
OFFICE OF THE ASSISTANT SECRETARY OF DEFENSE (RESEARCH AND DEVELOPMENT) ADVISORY GROUP ON ELECTRON   GET-REL-
   TUBES. (MILITARY RELIABLE TUBE PROGRAM)

OFFICE OF THE ASSISTANT SECRETARY OF DEFENSE (RES. + ENG.),WASH.,D.C. ADVISORY GP. ON ELECTRONIC PARTS   GEP-

OFFICE OF THE ASST. SECY. OF DEFENSE (RES. + DEV.) RES. + DEV. COORDINATING COMM. ON GEN. SCIENCES   CGS-

OFFICE OF THE ASST. SECY. OF DEFENSE (RES. + DEV.), TECH. ADVISORY PANEL ON ELECTRONICS, WASH., D.C.   PEL-

OFF. OF THE ASST. SECY. OF DEFENSE(RES.+ENG.),WASH,DC   OASD-

OFFICE OF THE ASSISTANT SECRETARY OF DEFENSE (RES. + ENG.). WORKING GP. ON MICROWAVE TUBES   SMW-

OFFICE OF THE ASST. SECY. OF INTERIOR, WATER AND POWER DEV. OFFICE OF SALINE WATER, WASH., D.C.   AM-

OFFICE OF THE CHIEF OF ENGINEERS (ARMY), WASH., D.C.   ATLIS-
OFFICE OF THE CHIEF OF ENGINEERS (ARMY), WASH., D.C.   EIN-
OFFICE OF THE CHIEF OF ENGINEERS (ARMY), WASH., D.C.   ER-(NO.)-(NO.)-
OFFICE OF THE CHIEF OF ENGINEERS (ARMY), WASHINGTON, D.C. (NUCLEAR WEAPONS EFFECTS)   NWE-
OFFICE OF THE CHIEF OF ENGINEERS (ARMY), WASH., D.C.   OCE-
OFFICE OF THE CHIEF OF ENGINEERS (ARMY), WASH., D.C.   OCE/EIN-
OFFICE OF THE CHIEF OF ENGINEERS (ARMY), WASH., D.C.   OCE-TR-(YEAR)-

OFFICE OF THE CHIEF OF ENGINEERS (ARMY). BEACH EROSION BOARD   BEB-
OFFICE OF THE CHIEF OF ENGINEERS (ARMY). BEACH EROSION BOARD (TECHNICAL MEMORANDUM)   ENG BEB TM-

OFFICE OF THE CHIEF OF NAVAL OPERATIONS, WASH., D.C.   CNO-
OFFICE OF THE CHIEF OF NAVAL OPERATIONS, WASHINGTON, D.C. (DEVELOPMENT CHARACTERISTIC)   DC-
OFFICE OF THE CHIEF OF NAVAL OPERATIONS, WASH., D.C.   DCNO-
OFFICE OF THE CHIEF OF NAVAL OPERATIONS, WASH., D.C.   DIO,(LTRS)(NO)-(LTR)-(YR)
OFFICE OF THE CHIEF OF NAVAL OPERATIONS, WASHINGTON, D.C. (MASTER PROGRAM OBJECTIVES PLAN)   MPOP-
OFFICE OF THE CHIEF OF NAVAL OPERATIONS, WASHINGTON, D.C. (MID-RANGE PLAN)   MRP-
OFFICE OF THE CHIEF OF NAVAL OPERATIONS, WASH., D.C.   NAVAER-(YR)-1R-
OFFICE OF THE CHIEF OF NAVAL OPERATIONS, WASH., D.C.   NAVAER-(YR)-1T-
OFFICE OF THE CHIEF OF NAVAL OPERATIONS, WASH., D.C.   NAV-OP-
OFFICE OF THE CHIEF OF NAVAL OPERATIONS, WASH., D.C.   NAVOP-
OFFICE OF THE CHIEF OF NAVAL OPERATIONS, WASH., D.C.   NWIP-(NUMBER)-
OFFICE OF THE CHIEF OF NAVAL OPERATIONS, WASH., D.C.   NWP-
OFFICE OF THE CHIEF OF NAVAL OPERATIONS, WASH., D.C.   OCNO-
OFFICE OF THE CHIEF OF NAVAL OPERATIONS, WASH., D.C.   OPNAV-
OFFICE OF THE CHIEF OF NAVAL OPERATIONS, WASH., D.C.   S-

OFFICE OF THE CHIEF OF NAVAL OPERATIONS. INTELL. DIV   ID-
OFFICE OF THE CHIEF OF NAVAL OPERATIONS. INTELLIGENCE DIVISION (INTELLIGENCE REPORT)   OCNO ID I-

OFFICE OF THE CHIEF OF NAVAL OPERATIONS. NAVAL WARFARE ANALYSIS GROUP, WASHINGTON, D.C.   NWG-

OFFICE OF THE CHIEF OF NAVAL OPERATIONS. OPERATIONS EVALUATION GROUP, WASHINGTON, D.C.   IRM-
OFFICE OF THE CHIEF OF NAVAL OPERATIONS. OPERATIONS EVALUATION GROUP, WASHINGTON, D.C.   OEG-
OFFICE OF THE CHIEF OF NAVAL OPERATIONS. OPERATIONS EVALUATION GROUP, WASHINGTON, D.C.   OEG-PROFESSIONAL PAPER-
OFFICE OF THE CHIEF OF NAVAL OPERATIONS. OPERATIONS EVALUATION GROUP, WASHINGTON, D.C. (REPORTS)   OEG-R-
OFFICE OF THE CHIEF OF NAVAL OPERATIONS. OPERATIONS EVALUATION GROUP, WASHINGTON, D.C.   OEG-RC-
OFFICE OF THE CHIEF OF NAVAL OPERATIONS. OPERATIONS EVALUATION GROUP, WASHINGTON, D.C. (STUDIES)   OEG-S-
OFFICE OF THE CHIEF OF NAVAL OPERATIONS. OPERATIONS EVALUATION GROUP, WASHINGTON, D.C.   OEG-SR-
OFFICE OF THE CHIEF OF NAVAL OPERATIONS. OPERATIONS EVALUATION GROUP, WASH., D.C. (PLANNING ANALYSIS)   PA-
OFFICE OF THE CHIEF OF NAVAL OPERATIONS. OPERATIONS EVALUATION GROUP, WASH., D.C. (PLANNING ANALYSIS)   PAIN-
OFFICE OF THE CHIEF OF NAVAL OPERATIONS. OPERATIONS EVALUATION GROUP, WASH., D.C. (PLANNING ANALYSIS)   PAM-
OFFICE OF THE CHIEF OF NAVAL OPERATIONS. OPERATIONS EVALUATION GROUP, WASH., D.C. (PLANNING ANALYSIS)   PAS-

OFFICE OF THE CHIEF OF NAVAL OPERATIONS. SHIP CHARACTERISTICS BOARD, WASHINGTON, D.C.   SCB-MEMO-(NUMBER)-(YEAR)

OFFICE OF THE CHIEF OF ORDNANCE, WASHINGTON, D.C.   AAWSSC-
OFFICE OF THE CHIEF OF ORDNANCE, WASHINGTON, D.C.   OCO-
OFFICE OF THE CHIEF OF ORDNANCE, WASHINGTON, D.C.   OO-

OFFICE OF THE CHIEF OF ORDNANCE. RESEARCH AND DEVELOPMENT DIVISION (SUBOFFICE ROCKET)   ORDD R TR-

OFFICE OF THE CHIEF OF ORDNANCE. TECHNICAL DIVISION, BALLISTIC SECTION, WASHINGTON, D.C.   TDBS-

OFFICE OF THE CHIEF OF RESEARCH AND DEVELOPMENT, ARMY RESEARCH OFFICE, WASHINGTON, D.C.   CSCRD-

OFFICE OF THE CHIEF OF STAFF (ARMY), WASHINGTON, D.C.   OCSA-

OFFICE OF THE COMPTROLLER OF THE ARMY. DIRECTOR OF MANAGEMENT, WASHINGTON, D.C.   COMPT-M(MEFA)-
OFFICE OF THE COMPTROLLER OF THE ARMY. DIRECTOR OF MANAGEMENT, WASHINGTON, D.C.   COMPT-M(MEI)-MPL-

OFFICE OF THE COMPTROLLER OF THE NAVY, WASHINGTON, DC   NAVCOMP-

OFFICE OF THE DEPUTY INSPECTOR GENERAL, NORTON AFB, CALIF.   DIG-

OFFICE OF THE DIRECTOR OF DEFENSE (RES+ENG),WASH.,DC   AGAAD-
OFFICE OF THE DIRECTOR OF DEFENSE (RES+ENG),WASH.,DC   BULLETIN NO.-
OFFICE OF THE DIRECTOR OF DEFENSE (RES+ENG),WASH,DC   DDR + E-
OFFICE OF THE DIRECTOR OF DEFENSE (RES+ENG),WASH.,DC   DOD/R AND E-
OFFICE OF THE DIRECTOR OF DEFENSE (RES+ENG),WASH.,DC   FL (NUMBER)/
OFFICE OF THE DIRECTOR OF DEFENSE (RES+ENG),WASH.,DC   MAM-
OFFICE OF THE DIRECTOR OF DEFENSE (RESEARCH AND ENGINEERING), WASH., D.C.(MANAGEMENT ANALYSIS NOTE)   MAN-(YEAR)-
OFFICE OF THE DIRECTOR OF DEFENSE (RES+ENG),WASH.,DC   MAR-

OFFICE OF THE DIRECTOR OF DEFENSE (RES+ENG),WASH.,DC	ODDRE-
OFFICE OF THE DIRECTOR OF DEFENSE (RES+ENG),WASH.,DC	ODDRE-MAM-
OFFICE OF THE DIRECTOR OF DEFENSE (RES+ENG),WASH.,DC	ODDRE-MAN-
OFFICE OF THE DIRECTOR OF DEFENSE (RES+ENG),WASH.,DC	OEM-
OFFICE OF THE DIRECTOR OF DEFENSE (RESEARCH AND ENGINEERING), WASH., D.C. (PROJECT THEMIS) (REF.48)	THEMIS-(LETTERS)-TR-
OFFICE OF THE DIRECTOR OF DEFENSE (RESEARCH AND ENGINEERING), WASH., D.C. (PROJECT THEMIS) (REF.48)	THEMIS-(LETTERS)-
OFFICE OF THE DIRECTOR OF DEFENSE (RESEARCH AND ENGINEERING), WASH., D.C. (PROJECT THEMIS) (REF.48)	THEMIS-(LETTERS)-T-TR-
OFFICE OF THE DIRECTOR OF DEFENSE (RESEARCH AND ENGINEERING), WASH., D.C. (PROJECT THEMIS) (REF.48)	THEMIS-(LETTERS-YEAR)-
OFFICE OF THE DIRECTOR OF DEFENSE (RESEARCH AND ENGINEERING), WASH., D.C. (PROJECT THEMIS) (REF.48)	THEMIS-(LETTERS)-T-
OFFICE OF THE DIRECTOR OF DEFENSE (RESEARCH AND ENGINEERING), WASH., D.C. (PROJECT THEMIS) (REF.48)	THEMIS-(LETTERS-NO.-YR.)
OFFICE OF THE DIRECTOR OF DEFENSE (RESEARCH AND ENGINEERING), WASH., D.C. (PROJECT THEMIS) (REF.48)	THEMIS-PROPOSAL-
OFFICE OF THE DIRECTOR OF DEFENSE (RES) + ENG.) AD HOC GROUP ON SOLID PROPELLANT INSTABILITY OF COMBUSTION, WASHINGTON, D.C.	SPIC-
OFFICE OF THE DIRECTOR OF DEFENSE (RES. + ENG.) ADVISORY GROUP ON ELECTRON DEVICES, WASH., D.C.	GED-0-
OFFICE OF THE DIRECTOR OF DEFENSE (RES. + ENG.) ADVISORY GROUP ON ELECTRONIC PARTS AND ELECTRON TUBES (ELECTRONIC COMPONENT PARTS)	ECP-
OFFICE OF THE DIRECTOR OF DEFENSE (RES. + ENG.) ADVISORY GROUP ON HIGH ALTITUDE DETECTION, WASH.,D.C.	ADHAD-
OFFICE OF THE DIRECTOR OF DEFENSE (RES. + ENG.) ADVISORY GROUP ON HIGH ALTITUDE DETECTION, WASH.,D.C.	AGHAD-
OFFICE OF THE INSPECTOR GENERAL, NORTON AFB, CALIF.	AFIAS-STUDY-
OFFICE OF THE JOINT CHIEFS OF STAFF, WASH., D.C.	ACP-
OFFICE OF THE JOINT CHIEFS OF STAFF, WASH., D.C. (JOINT ARMY-NAVY-AIR FORCE PUBLICATION) (REF. 13)	JANAP-
OFFICE OF THE JOINT CHIEFS OF STAFF, WASH., D.C. (JOINT ARMY-NAVY PUBLICATION) (REF. 13)	JANP-
OFFICE OF THE JOINT CHIEFS OF STAFF, WASH., D.C.	JCB-
OFFICE OF THE JOINT CHIEFS OF STAFF, WASH., D.C.	JCS-
OFFICE OF THE JOINT CHIEFS OF STAFF, WASH., D.C.	J/N-
OFFICE OF THE JUDGE ADVOCATE GEN. (NAVY) WASH., D.C.	JAG-
OFFICE OF THE QUARTERMASTER GENERAL	QMC-
OFFICE OF THE QUARTERMASTER GENL. MIL. PLANNING DIV. RES. + DEV. BR. (ENVIRONMENTAL PROTECTION SERIES)	EPS-
OFFICE OF THE QUARTERMASTER GENL. MIL. PLANNING DIV. RES. + DEV. BR. (ENVIRONMENTAL PROTECTION SERIES)	QMC EPS-
OFFICE OF THE QUARTERMASTER GENERAL. MILITARY PLANNING DIV. RESEARCH AND DEVELOPMENT BRANCH (ENVIRONMENTAL PROTECTION SERIES, MEMO REPORT)	QMC EPS MR-
OFFICE OF THE SECRETARY OF DEFENSE. DIRECTOR OF GUIDED MISSILES, WASHINGTON, D.C.	DOD D/GM-
OFFICE OF THE SURGEON GENERAL, WASHINGTON, D.C.	SG-
OFFICE OF WATER RESOURCES RES.(INTERIOR), WASH., D.C.	OWRR-A-(NUMBER-STATE)
OFFICE OF WATER RESOURCES RES.(INTERIOR), WASH., D.C.	OWRR-B-(NUMBER-STATE)
OFFICE OF WATER RESOURCES RES.(INTERIOR), WASH., D.C.	OWRR-C-(NO.-NO.-NO.-)
OFFICE OF WATER RESOURCES RES.(INTERIOR), WASH., D.C.	OWRR-S-(NUMBER-STATE)
OFFICE OF WATER RESOURCES RES.(INTERIOR), WASH., D.C.	OWRR-W-(NUMBER-STATE)
OFFICER EDUCATION RESEARCH LAB., MAXWELL AFB, ALA.	OERL-
OFFICER EDUCATION RESEARCH LAB., MAXWELL AFB, ALA.	OERL-TM-
OGDEN AIR MATERIEL AREA, HILL AFB, UTAH	OAMA-
OGDEN AIR MATERIEL AREA, HILL AFB, UTAH	OOY-
OGDEN AIR MATERIEL AREA, HILL AFB, UTAH	OOY-P-(YEAR)-
OGDEN AIR MATERIEL AREA, HILL AFB, UTAH	OOY-TR-
OGDEN AIR MATERIEL AREA, HILL AFB, UTAH	T(NUMBER)-
OGDEN AIR MATERIEL AREA, OOAMA HILL AFB, UTAH	OAMA-
OGDEN AIR MATERIEL AREA. DIRECTORATE OF MATERIEL MANAGEMENT, HILL AFB, UTAH	MM-TR-
OGDEN AIR MATERIEL AREA. DIRECTORATE OF MATERIEL MANAGEMENT, HILL AFB, UTAH	(YEAR)-OONEBT-
OGDEN AIR MATERIEL AREA. SERVICE ENGINEERING DIV., HILL AFB, UTAH	(YEAR)-MMEST-
OGDEN AIR MATERIEL AREA. SERVICE ENGINEERING DIV., HILL AFB, UTAH	(YEAR)-OONEBT-
OGDEN AIR MATERIEL AREA. SERVICE ENGINEERING DIV., HILL AFB, UTAH	(YEAR)-OONEST-
OGDEN AIR MATERIEL AREA. SYSTEM SUPPORT MANAGEMENT DIV., HILL AFB, UTAH	OONC-(NUMBER)-(YEAR)
OHIO INJECTOR CO., WADSWORTH	OI-
OHIO PROPULSION LAB., WRIGHT-PATTERSON AFB	ASRMPR-TM-
OHIO STATE UNIV., COLUMBUS	OSU-
OHIO STATE UNIV., COLUMBUS	OSU-(LETTERS)-
OHIO STATE UNIV., COLUMBUS	OSU-SR-
OHIO STATE UNIV., COLUMBUS. COMPUTER AND INFORMATION SCIENCE RESEARCH CENTER	CISRC-RT-(YEAR)-
OHIO STATE UNIV., COLUMBUS. COMPUTER AND INFORMATION SCIENCE RESEARCH CENTER	CISRC-TR-(YEAR)-
OHIO STATE UNIV., COLUMBUS. COMPUTER AND INFORMATION SCIENCE RESEARCH CENTER.	CSIRC-TR-(YEAR)-
OHIO STATE UNIV., COLUMBUS. CRYOGENIC LAB.	PQR-
OHIO STATE UNIV., COLUMBUS. CRYOGENIC LAB.	PR-(NO.)-
OHIO STATE UNIV., COLUMBUS. CRYOGENIC LAB.	QPR-
OHIO STATE UNIV., COLUMBUS. CRYOGENIC LAB.	TR-(NO.)-
OHIO STATE UNIV., COLUMBUS. DEPT. OF GEODETIC SCIENCE	DGS-
OHIO STATE UNIV., COLUMBUS. DISASTER RESEARCH CENTER	DRC-
OHIO STATE UNIV., COLUMBUS. DISASTER RESEARCH CENTER	DRC-MONOGRAPH-
OHIO STATE UNIV., COLUMBUS. DISASTER RESEARCH CENTER	DRC-SER-
OHIO STATE UNIV., COLUMBUS. DISASTER RESEARCH CENTER	DRC-TR-
OHIO STATE UNIV., COLUMBUS. DISASTER RESEARCH CENTER	MONO-D-
OHIO STATE UNIV.,COLUMBUS. ELECTRONIC COMPONENTS LAB.	OSU-ECL-
OHIO STATE UNIV., COLUMBUS. SYSTEMS RESEARCH GROUP	RF-
OHIO STATE UNIV., COLUMBUS. SYSTEMS RESEARCH GROUP	RF-(NO.)-AR-(YEAR)-
OHIO STATE UNIV. RESEARCH FOUNDATION, COLUMBUS	ASP-TDR-
OHIO STATE UNIV. RESEARCH FOUNDATION, COLUMBUS	IGPC-
OHIO STATE UNIV. RESEARCH FOUNDATION, COLUMBUS	OSU-(NUMBER)-
OHIO STATE UNIV. RESEARCH FOUNDATION, COLUMBUS	OSU-(NUMBER)-TR-
OHIO STATE UNIV. RESEARCH FOUNDATION, COLUMBUS	OSU-QPR-(NUMBER)-
OHIO STATE UNIV. RESEARCH FOUNDATION, COLUMBUS	OSU-RF-(NUMBER)-
OHIO STATE UNIV. RESEARCH FOUNDATION, COLUMBUS	OSURF-
OHIO STATE UNIV. RESEARCH FOUNDATION, COLUMBUS	OSURF MR-
OHIO STATE UNIV. RESEARCH FDN., COLUMBUS (MEMO RPTS.)	OSURF SR-
OHIO STATE UNIV. RESEARCH FDN., COLUMBUS (SPEC.RPTS.)	

OHIO STATE UNIV. RES. FDN.

OHIO STATE UNIV. RES. FDN., COLUMBUS (STATUS REPORT)	OSU-SR-
OHIO STATE UNIV. RESEARCH FOUNDATION, COLUMBUS	OSU-TR-
OHIO STATE UNIV. RESEARCH FOUNDATION, COLUMBUS	RF-(NO.)-
OHIO STATE UNIV. RESEARCH FOUNDATION, COLUMBUS	RF-(NO.)-TR-
OHIO STATE UNIV. RESEARCH FOUNDATION, COLUMBUS	RTM-
OHIO STATE UNIV. RESEARCH FOUNDATION, COLUMBUS	TN(ALOSU)(NUMBER)-
OHIO STATE UNIV. RES. FDN. ANTENNA LAB., COLUMBUS	OSU-AL-
OHIO STATE UNIV. RESEARCH FOUNDATION. DEPT. OF METALLURGICAL ENGINEERING, COLUMBUS	OSU-DME-(NO.)-SR-(NO.)
OHIO STATE UNIV. RESEARCH FOUNDATION. MAPPING AND CHARTING RESEARCH LAB., COLUMBUS	MCRL-TP-
OHIO STATE UNIV. RESEARCH FOUNDATION. MAPPING AND CHARTING RESEARCH LAB., COLUMBUS	OSU-MCRL-
OHIO UNIV., ATHENS	ASR-
OHIO UNIV., ATHENS	EER-
OHIO UNIV., ATHENS. CENTER FOR INTERNATIONAL STUDIES	AFRICA-
OHIO UNIV., ATHENS. DEPT. OF ELECTRICAL ENGINEERING	EER-(NUMBER-NUMBER)
OHIO UNIV., ATHENS. ENGINEERING EXPERIMENT STATION	EES-(NUMBER)B-
OHIO UNIV., ATHENS. ENGINEERING EXPERIMENT STATION	EES-(NUMBER-NUMBER)
OHIO UNIV., ATHENS. HIGH ENERGY PHYSICS LAB.	OU-HEPL-
OKLAHOMA. STATE DEPT. OF INSTITUTIONS, SOCIAL AND REHABILITATION SERVICES, OKLAHOMA CITY	DSSM-
OKLAHOMA. UNIV., NORMAN	OKLAU-
OKLAHOMA. UNIV., NORMAN	OU-
OKLAHOMA. UNIV., NORMAN. RESEARCH INST.	OKLA-
OKLAHOMA. UNIV., NORMAN. RESEARCH INST.	OURI-
OKLAHOMA. UNIV., NORMAN. RESEARCH INST. ATMOSPHERIC RESEARCH LAB.	ARL-
OKLAHOMA. UNIV., NORMAN. RESEARCH INST. HIGH PRESSURE PHYSICS LAB.	OU-TR-
OKLAHOMA AGRICULTURAL + MECHANICAL COLL., STILLWATER	OA+M-
OKLAHOMA AGRICULTURAL + MECHANICAL COLL., STILLWATER	OAMC-
OKLAHOMA AGRICULTURAL + MECHANICAL COLL., STILLWATER OKLAHOMA POWER AND PROPULSION LAB.	OPPL-
OKLAHOMA CITY AIR MATERIEL AREA, TINKER AFB, OKLA.	OCAMA-
OKLAHOMA CITY AIR MATERIEL AREA. DIRECTORATE OF MATERIEL MANAGEMENT, TINKER AFB, OKLA.	OCNE-MN-
OKLAHOMA STATE UNIV., STILLWATER (PROGRESS REPORT)	OKLASU-QPR-
OKLAHOMA STATE UNIV., STILLWATER	OKSU-
OKLAHOMA STATE UNIV., STILLWATER	OSU-AS-
OKLAHOMA STATE UNIV., STILLWATER	SBW-
OKLAHOMA STATE UNIV., STILLWATER. SCHOOL OF ELECTRICAL ENGINEERING	OSU-WP-
OKLAHOMA WATER RESOURCES BOARD, OKLAHOMA CITY	OWRB-PUB-
OLD AND BARNES, INC., PASADENA, CALIF.	O + B -
OLIN CORP. OLIN RES. CTR. CHEMICALS GP., NEW HAVEN	OLIN-SSR--
OLIN CORP. WINCHESTER-WESTERN RES. DEPT., NEW HAVEN	WWR-(YEAR)-
OLIN INDUSTRIES, INC., EAST ALTON, ILL.	OLIN-
OLIN MATHIESON CHEMICAL CORP. (SOLID PROPELLANTS)	SP-
OLIN MATHIESON CHEMICAL CORP., BALTIMORE	MCC-
OLIN MATHIESON CHEMICAL CORP., BALTIMORE	MCC-1023-TR-
OLIN MATHIESON CHEMICAL CORP., BALTIMORE	NF-
OLIN MATHIESON CHEMICAL CORP., BALTIMORE	OMCC-
OLIN MATHIESON CHEMICAL CORP., EAST ALTON, ILL.	EA-
OLIN MATHIESON CHEMICAL CORP., NEW HAVEN	NH-
OLIN MATHIESON CHEMICAL CORP., NEW HAVEN	OLIN-NH-
OLIN MATHIESON CHEMICAL CORP., NEW HAVEN	OLIN-QPR-
OLIN MATHIESON CHEMICAL CORP., NIAGARA FALLS, N.Y.	NF-
OLIN MATHIESON CHEMICAL CORP. CHEMS. DIV., NEW HAVEN	OLIN-
OLIN MATHIESON CHEMICAL CORP. ENERGY DIV., NIAGARA FALLS, N.Y.	OMCC-HEF-
OLIN MATHIESON CHEMICAL CORP. HIGH ENERGY FUELS DIV., NIAGARA FALLS, N.Y.	OMCC-HEF-
OLIN MATHIESON CHEMICAL CORP. METALLURGICAL LABS., NEW HAVEN	NFR-
OLIN MATHIESON CHEMICAL CORP. METALLURGICAL LABS., NEW HAVEN	OMCC-MRL-(YEAR)-
OLIN MATHIESON CHEMICAL CORP. NUCLEAR FUEL DIV., NEW HAVEN	SOM-
OLIN MATHIESON CHEMICAL CORP. SPECIALTY CHEMICALS RESEARCH ENERGY DIV., NEW HAVEN	OMCC-SCR-
OLIVER + BOYD, LTD., EDINBURGH (TRANSLATIONS)	OB-
OLSON ASSOCIATES, INC., HUNTINGTON, N.Y.	OAI-(YEAR)-
OLYMPIC DEVELOPMENT CO., STAMFORD, CONN.	ODC-
OLYMPIC DEVELOPMENT CO., STAMFORD, CONN.	OLDC-
OMAHA-COUNCIL BLUFFS METROPOLITAN AREA PLANNING AGENCY	MAPA-
ONTARIO. HYDRO-ELECTRIC POWER COMMISSION, TORONTO	DND64-
ONTARIO RESEARCH FOUNDATION, TORONTO	PRR-
OPERATIONAL DEV. FORCE, ATLANTIC FLEET, NORFOLK, VA.	COMOPDEVFOR-SER-
OPERATIONAL DEV. FORCE, ATLANTIC FLEET, NORFOLK, VA.	OP/(LETTER(S))...
OPERATIONAL LOGIC CORP., ALEXANDRIA, VA.	OLC-

OPERATIONAL RESEARCH SECTION, SINGAPORE	ORS(S)-
OPERATIONAL RESEARCH UNIT (FAR EAST), SINGAPORE	ORUFE-
OPERATIONS ANALYSIS DIV. (AIR FORCE), WASHINGTON, D.C.	AFOA-MEMO-(YEAR)-
OPERATIONS ANALYSIS DIV. (AIR FORCE), WASHINGTON, D.C.	AFOA-TN-(YEAR)-
OPERATIONS ANALYSIS DIV. (AIR FORCE), WASHINGTON, D.C.	AF-OAD-
OPERATIONS ANALYSIS DIV. (AIR FORCE), WASHINGTON, D.C.	AF OA R-
OPERATIONS ANALYSIS DIV. (AIR FORCE), WASHINGTON, D.C.	AF OA STM-
OPERATIONS ANALYSIS DIV. (AIR FORCE), WASHINGTON, D.C.	AF OA WP-
OPERATIONS ANALYSIS DIV. (AIR FORCE), WASHINGTON, D.C.	AFOA-S-
OPERATIONS ANALYSIS DIV. (AIR FORCE), WASHINGTON, D.C.	OA-
OPERATIONS ANALYSIS DIV. (AIR FORCE), WASHINGTON, D.C.	OAD-
OPERATIONS ANALYSIS DIV. (AIR FORCE), WASHINGTON, D.C.	OA-M-
OPERATIONS ANALYSIS DIV. (AIR FORCE), WASHINGTON, D.C.	OA-R-
OPERATIONS ANALYSIS DIV. (AIR FORCE), WASHINGTON, D.C. (SUMMARY REPORT)	OA-SR-
OPERATIONS ANALYSIS DIV. (AIR FORCE), WASHINGTON, D.C. (TECHNICAL MEMORANDUM)	OA-TM-
OPERATIONS ANALYSIS DIV. (AIR FORCE), WASHINGTON, D.C.	OA WP-
OPERATIONS ANALYSIS OFF. (AIR FORCE), WASHINGTON, D.C.	GMOA-
OPERATIONS ANALYSIS OFF. (AIR FORCE), WASHINGTON, D.C.	M-
OPERATIONS EVALUATION GROUP, WASHINGTON, D.C.	RESEARCH-CONTRIBUTION-
OPERATIONS RESEARCH INC., SILVER SPRING, MD.	ORI-
OPERATIONS RESEARCH INC., SILVER SPRING, MD.	ORI TR-
OPERATIONS RESEARCH INC., SILVER SPRING, MD.	PRR-
OPERATIONS RESEARCH INC., SILVER SPRING, MD.	PRR-(YEAR)-
OPTICAL SOCIETY OF AMERICA, CAMBRIDGE, MASS. (TRANSLATIONS)	OSA-
OPTICS TECHNOLOGY, INC., PALO ALTO, CALIF.	OTI-(NUMBER)-R-
ORDNANCE ANALYSIS SECTION (AIR FORCE)	OS-
ORDNANCE COMMITTEE (ARMY) (MEMORANDUM)	OCM-
ORDNANCE CORPS (ARMY) (BULLETIN)	OCB-
ORDNANCE CORPS (ARMY) (DIRECTIVES)	OCD-
ORDNANCE CORPS (ARMY)	ORD-
ORDNANCE CORPS (ARMY) (PAMPHLET)	ORDP-
ORDNANCE CORPS (ARMY)	TM-9-
ORDNANCE CORPS (ARMY), WASHINGTON, D.C.	ORDP-
ORDNANCE CORPS (ARMY), WASHINGTON, D.C.	ORDTB-
ORDNANCE CORPS (ARMY), WASHINGTON, D.C.	ORDTX-
ORDNANCE CORPS (ARMY). SHAPED CHARGE RESEARCH AND DEVELOPMENT STEERING AND COORDINATING COMMITTEE	OCSC-(NUMBER)-(YEAR)
ORDNANCE CORPS (ARMY). SHAPED CHARGE RESEARCH AND DEVELOPMENT STEERING AND COORDINATING COMMITTEE	SC-(NUMBER)-(YEAR)
ORDNANCE DEPT. (ARMY) (CATALOG)	ASF-CAT-ORD-
ORDNANCE DEPT. (ARMY)	OCM-
ORDNANCE DEPT. (ARMY) (REPORT)	ODR-
ORDNANCE DEPT. (ARMY) (STANDARD INSPECTION PROCEDURE)	ORD SIP-
ORDNANCE DEPT. (ARMY) (TENTATIVE SPECIFICATION)	ORD TEN SPEC-
ORDNANCE DEPT. (ARMY) (TECHNICAL INTELLIGENCE REPORT)	ORD TIR-
ORDNANCE DEPT. (ARMY) (STANDARD INSPECTION PROCEDURE)	SIP-
ORDNANCE DEPT. (ARMY) (STANDARD NOMENCLATURE LIST)	SNL-
ORDNANCE DEPT. (ARMY) (TECHNICAL INTELLIGENCE REPORT)	TIR-
(SEE ALSO LATER NAME: ORDNANCE CORPS (ARMY))	
ORDNANCE DEPT. (ARMY). DEVELOPMENT DIV. (PROJECT KG)	DD PROJ KG-
ORDNANCE DEPT. (ARMY). DEVELOPMENT DIV. (PROJECT KG)	KG-PROJECT-
ORDNANCE DEPT. (ARMY). DEVELOPMENT DIV. (PROJECT KG)	ORD DD PROJ KG-
ORDNANCE DEPT. (ARMY). ORDNANCE PACKAGING + CRATING	OPC-
ORDNANCE DEPT. (ARMY). ORDNANCE PACKAGING + CRATING, DETROIT (PACKAGING INSTRUCTIONS)	OPC I-
ORDNANCE DEPT. ORDNANCE PACKAGING AND CRATING. FIRE CONTROL SUBOFFICE, FRANKFORD ARSENAL, PHILADELPHIA (PACKAGING INSTRUCTIONS)	AFPS-
ORDNANCE DEPT. ORDNANCE PACKAGING AND CRATING. FIRE CONTROL SUBOFFICE, FRANKFORD ARSENAL, PHILADELPHIA (PACKAGING INSTRUCTIONS)	OPC AFPS-
ORDNANCE DEPT. RESEARCH AND DEVELOPMENT CENTER, ABERDEEN PROVING GROUND, MD.	APG PROJ-REPT-
ORDNANCE DEPT. RESEARCH AND DEVELOPMENT CENTER, ABERDEEN PROVING GROUND, MD.	ORCP-
ORDNANCE DEPT. RESEARCH AND DEV. CTR. FOREIGN DOC. EVALUATION BRANCH, ABERDEEN PROVING GROUND, MD.	ORDC-ARCHIVE-(NO./NO.)
ORDNANCE RES., INC., FORT WALTON BEACH, FLA.	RDTR-
ORDNANCE SCHOOL, ABERDEEN PROVING GROUND, MD.	APG-OS-
ORDNANCE SCHOOL, ABERDEEN PROVING GROUND, MD.	ST-
ORDNANCE TECHNICAL INTELLIGENCE SERVICE, ABERDEEN PROVING GROUND, MD.	ORDBG-OTI-
ORDNANCE TECHNICAL INTELLIGENCE SERVICE, ABERDEEN PROVING GROUND, MD.	OTIO-
ORDNANCE TEST ACTIVITY (ARMY), YUMA, ARIZ.	OTA-
OREGON. UNIV., EUGENE	OREGONU-
OREGON. UNIV., EUGENE	OU-
OREGON METALLURGICAL CORP., ALBANY	HW-OREMET-
OREGON STATE COLL., CORVALLIS	OSC-
OREGON STATE UNIV., CORVALLIS	C-(YEAR)-
OREGON STATE UNIV., CORVALLIS	DR-
OREGON STATE UNIV., CORVALLIS	OSU-
OREGON STATE UNIV., CORVALLIS	RL-
OREGON STATE UNIV., CORVALLIS. COMPUTER CENTER	C-(YEAR)-
OREGON STATE UNIV., CORVALLIS. COMPUTER CENTER	CC-(YEAR)-
OREGON STATE UNIV., CORVALLIS. DEPT. OF OCEANOGRAPHY	DATA-
OREGON STATE UNIV., CORVALLIS. DEPT. OF OCEANOGRAPHY	REF-
ORENDA ENGINES LTD., MALTON, ONT.	NUCLEAR-
ORENDA ENGINES LTD., MALTON, ONT.	OEL-MET-

ORGANIZATION FOR ECONOMIC COOPERATION AND DEVELOPMENT

ORGANIZATION FOR ECONOMIC COOPERATION AND DEVELOPMENT, PARIS	OECD-
ORGANIZATION FOR ECONOMIC COOPERATION AND DEVELOPMENT, PARIS	SEN-IPSA-
ORGANIZATION FOR ECONOMIC COOPERATION AND DEVELOPMENT. DIRECTORATE OF SCIENTIFIC AFFAIRS, PARIS	DAS-CSI-(LETTER/YR.NO.)
ORGANIZATION FOR EUROPEAN ECONOMIC COOPERATION, PARIS	DPD-
ORGANIZATION FOR EUROPEAN ECONOMIC COOPERATION, PARIS	OEEC-
(SEE ALSO SUCCESSOR: ORGANIZATION FOR ECONOMIC COOPERATION AND DEVELOPMENT)	
OSAKA UNIV.	INS-TCA-
OSAKA UNIV.	OSAKAU-
OSAKA UNIV. LAB. OF NUCLEAR STUDIES	CU-LNS-(YEAR)-
OSAKA UNIV. LAB. OF NUCLEAR STUDIES	OU-LNS-
OSAKA UNIV. LAB. OF NUCLEAR STUDIES	OULNS-(YEAR)-
OSAKA UNIV. LAB. OF NUCLEAR STUDIES	OV-LNS-(YEAR)-
OSAKA UNIV. LAB. OF NUCLEAR STUDIES	QU-LNS-(YEAR)-
OSLO. UNIVERSITETETSFORLAGETS TRYKNINGSSENTRAL (RADIOBIOLOGY STUDY)	OSLO-RIS-
OSLO. UNIV. INST. OF THEORETICAL ASTROPHYSICS	OU-ITA-
OULU UNIV., FINLAND	S-
OWENS-CORNING FIBERGLAS CORP., N.Y.C.	OCFC-
OWENS-CORNING FIBERGLAS CORP., TOLEDO, OHIO	TR-(YEAR)-(NO.)-(NO.)
OWENS-CORNING FIBERGLAS CORP. TECHNICAL CENTER. GRANVILLE, OHIO	TC-PL-
OWENS-ILLINOIS, INC., TOLEDO	OI-
OXFORD UNIV.	E-
OXFORD UNIV.	LFEN.(NUMBER)/BETA
OXFORD UNIV.	MR-
OXFORD UNIV.	OXFORDU-
OXFORD UNIV.	OXU-
OXFORD UNIV. ENGINEERING LAB.	OUEL-
OXFORD UNIV., NUFFIELD, ENGLAND, DEPT. OF ORTHOPAEDIC SURGERY	E-
OZARK-MAHONING CO., TULSA, OKLA.	OMC-

PACIFIC AIR FORCES, HICKAM AFB, HAWAII	PACAF-TEST-(YEAR)-
PACIFIC APPLIED RESEARCH, LOS ANGELES	PAR-
PACIFIC CAR AND FOUNDRY CO., RENTON, WASH.	PC-(YEAR)-
PACIFIC FLEET. MINE TEST DIV. (REPORT)	MTDR-
PACIFIC GAS AND ELECTRIC CO., SAN FRANCISCO (JOINT VENTURE)	BCPG-
PACIFIC GAS AND ELECTRIC CO., SAN FRANCISCO	BP-
PACIFIC GEEIA REGION. EMC AND MEASUREMENTS BRANCH, SAN FRANCISCO	PGR-EMC-(YEAR)-
PACIFIC MISSILE RANGE, POINT MUGU, CALIF.	PMP-TM-
PACIFIC MISSILE RANGE, POINT MUGU, CALIF.	PMR-
PACIFIC MISSILE RANGE, POINT MUGU, CALIF.	PMR-MP-(YR.)-
PACIFIC MISSILE RANGE, POINT MUGU, CALIF.	PMR-MR-(YEAR)-
PACIFIC MISSILE RANGE, POINT MUGU, CALIF.	PMR-TM-
PACIFIC MISSILE RANGE, POINT MUGU, CALIF.	PMR-TN-
PACIFIC MISSILE RANGE, POINT MUGU, CALIF.	PMR-TR-(YEAR)-
PACIFIC NORTHWEST POWER GROUP, RICHLAND, WASH.	PGN-
PACIFIC NORTHWEST POWER GROUP, RICHLAND, WASH.	PNG-
PACIFIC OCEANOGRAPHIC LAB., SEATTLE	ESSA-TR-ERL-(NO.)-POL-
PACIFIC OCEANOGRAPHIC LAB., SEATTLE (RESEARCH LAB. TECH. MEMO)	RLTM-POL-
PACIFIC SEMICONDUCTORS, INC., CULVER CITY, CALIF.	PSI-
PACIFIC SEMICONDUCTORS, INC., CULVER CITY, CALIF.	PSI-(NUMBERS)
PACIFIC SOUTHWEST FOREST AND RANGE EXPERIMENT STATION, BERKELEY, CALIF. (FOREST SERVICE RESEARCH PAPER)	FSPR-PSW-
PACIFIC SOUTHWEST FOREST AND RANGE EXPERIMENT STATION, BERKELEY, CALIF.	PSW-
PACIFIC SOUTHWEST FOREST AND RANGE EXPERIMENT STATION, BERKELEY, CALIF.	RN-PSW-
PACIFIC UNION COLL. ANGWIN, CALIF. RADIATION RES.LAB.	PUC/RRL-
PACKARD-BELL ELECTRONICS CORP. SPACE AND SYSTEMS DIV., NEWBURY PARK, CALIF.	PB-
PACKARD MOTOR CAR CO., DETROIT (DESIGN STUDY ON TURBOJET ENGINES)	DD-
PADUA UNIVERSITA	MIT-
PADUA UNIVERSITA	PADOVAU-
PADUA UNIVERSITA	PADUAU-
PADUA UNIVERSITA. ISTITUTO DI ELETTROTECNICA E DI ELETTRONICA	UPEE-(YEAR)/
PADUCAH AREA OFFICE, AEC, KY. (REF. 4)	KY-
PADUCAH GASEOUS DIFFUSION PLANT, KY.	KY-
PADUCAH GASEOUS DIFFUSION PLANT, KY.	KY-(LETTER)-
PADUCAH GASEOUS DIFFUSION PLANT, KY.	KY-D-
PADUCAH GASEOUS DIFFUSION PLANT, KY.	KY-L-
PAGE COMMUNICATIONS ENGINEERS, INC., WASHINGTON, D.C.	PCE-
PAGE COMMUNICATIONS ENGINEERS, INC., WASHINGTON, D.C.	PCE-R-(NUMBER)-(NUMBER)
PAKISTAN. ATOMIC ENERGY CENTRE, DACCA	AECD/EL-
PAKISTAN. ATOMIC ENERGY CENTRE. DACCA	AECD/MISC-
PAKISTAN. ATOMIC ENERGY CENTRE, LAHORE	AECL-HP-
PAKISTAN. ATOMIC ENERGY CENTRE, LAHORE	PAECL/(LTRS.)-
PAKISTAN. ATOMIC ENERGY CENTRE, LAHORE	PAECL/CHEM/MISCELL-
PAKISTAN. ATOMIC EN. CENTRE. AGRICULTURE DIV., DACCA	AECD-AG-
PAKISTAN. ATOMIC ENERGY CENTRE. CHEMISTRY DIV., DACCA	AECD/CH/
PAKISTAN. ATOMIC EN. CENTRE. CHEMISTRY DIV., LAHORE	PAEC/CHEM-
PAKISTAN. ATOMIC EN. CENTRE. CHEMISTRY DIV., LAHORE	PAEC-CHEM-MISC.-
PAKISTAN. ATOMIC ENERGY CENTRE. ENG. DIV., LAHORE	ENGG/HT-
PAKISTAN. ATOMIC EN. CENTRE. EXPTL.PHYSICS DIV.,DACCA	AECD-EP-
PAKISTAN. ATOMIC EN. CENTRE. EXPTL.PHYSICS DIV.,DACCA	AECD-TP-
PAKISTAN. ATOMIC EN. CENTRE.HEALTH PHYSICS DIV.,DACCA	AECD/HP/
PAKISTAN. ATOMIC EN. CENTRE.HEALTH PHYSICS DIV.,DACCA	PAECL/HP-
PAKISTAN. ATOMIC ENERGY CENTRE. LIBRARY DIV., LAHORE	AECL-PAK/LIB-
PAKISTAN. ATOMIC EN. CENTRE. RADIOBIOLOGY DIV., DACCA	AECD/RB-
PAKISTAN. ATOMIC EN. CENTRE.RADIOBIOLOGY DIV.,LAHORE	AECL-PAK/RB-
PAKISTAN. ATOMIC ENERGY COMMISSION, LAHORE	AECL-PAK-
PAKISTAN. INST. OF NUCLEAR SCIENCE AND TECHNOLOGY, ISLAMABAD (ALL DIVISIONS)	PINSTECH/PHY-
PAKISTAN. INST. OF NUCLEAR SCIENCE AND TECHNOLOGY. HEALTH PHYSICS DIV., ISLAMABAD	PINSTECH/HP-
PAKISTAN. INST. OF NUCLEAR SCIENCE AND TECHNOLOGY. RADIOISOTOPE PRODUCTION DIV., ISLAMABAD	PINSTECH/RIPD-
PAKISTAN. INST. OF NUCLEAR SCIENCE AND TECHNOLOGY. REACTOR OPERATIONS DIV., ISLAMABAD	PINSTECH/RO-
PAKISTAN. INST. OF NUCLEAR SCIENCE AND TECHNOLOGY. REACTOR THEORY DIV., ISLAMABAD	PINSTECH/RT-
PAKISTAN NATIONAL SCIENTIFIC AND TECHNICAL DOCUMENTATION CENTRE, KARACHI	PANSDOC-
PALERMO, ITALY. UNIV. ISTITUTO DI TECNOLOGIE MECCANICHE	OI-
PALM BEACH COUNTY AREA PLANNING BD., WEST PALM BEACH, FLA. (WEST PALM BEACH URBAN AREA TRANSPORT. STUDY)	WPBUATS-
PALMER (WINSLOW), BABYLON, N.Y.	OMEGA-TN-
PAN AMERICAN HEALTH ORGANIZATION, WASHINGTON, D.C.	PAHO-SC-PUB-
PAN AMERICAN WORLD AIRWAYS, INC., LAS VEGAS, NEV.	PAA-
PAN AMERICAN WORLD AIRWAYS, INC., PATRICK AFB, FLA.	ETV-TM-(YR)-(NO)-TM-

PAN AMERICAN WORLD AIRWAYS, INC.

PAN AMERICAN WORLD AIRWAYS, INC., SO. SAN FRANCISCO	PAWA-
PAN AMERICAN WORLD AIRWAYS, INC., SO. SAN FRANCISCO	PAWA-AROWA-
PAN AMERICAN WORLD AIRWAYS, INC. GUIDED MISSILES RANGE DIV., PATRICK AFB, FLA.	ETV-TM-(YEAR)-
PAN AMERICAN WORLD AIRWAYS, INC. SPECTRUM SIGNATURE PROJECT, SIERRA VISTA, ARIZ.	SIR-
PAN-FAX, INC. INFORMATION SCIENCE INST., GOLETA, CAL.	PF-RM-(YEAR)-ISI-
PANAMA CANAL CO., WASHINGTON, D.C.	ICSM (NUMBER)-P-
PANAMA CANAL CO., WASHINGTON, D.C.	ICS MEMO-
PANAMA CANAL CO., WASHINGTON, D.C.	ICS MEMO (NUMBER)-P-
PANAMA CANAL ZONE	PCZ-
PANAMA CANAL ZONE. DEPT. OF OPERATION AND MAINTENANCE	PCZ ICSM-
PANAMETRICS, INC., WALTHAM, MASS.	PI-
PANERO (GUY B.) ENGINEERS, N.Y.C.	MCCP-
PARAMETRICS, INC., WALTHAM, MASS. (SEE ALSO SUCCESSORO PANAMETRICS, INC.)	RD-(YEAR)-
PARIS. UNIVERSITE, ORSAY	LAL-RI-
PARIS. UNIVERSITE, ORSAY	PARISU-
PARIS. UNIVERSITE, ORSAY	RI-(YEAR)-
PARIS. UNIVERSITE, ORSAY. ECOLE NORMALE SUPERIEURE	AM-
PARIS. UNIV., ORSAY. INSTITUT DE PHYSIQUE NUCLEAIRE	IPNO/LA-
PARIS. UNIV., ORSAY. INSTITUT DE PHYSIQUE NUCLEAIRE	IPNO/TH-
PARIS. UNIV., ORSAY. INSTITUT DE PHYSIQUE NUCLEAIRE	IRPO-(YEAR)(LETTER)-
PARIS. UNIVERSITE, ORSAY. INSTITUT DU RADIUM	IRPO-(YR.)E-
PARIS. UNIVERSITE, ORSAY. INSTITUT DU RADIUM	IRPO-
PARIS. UNIVERSITE, ORSAY. LABORATOIRE DE L'ACCELERATEUR LINEAIRE	LAL-
PARIS. UNIVERSITE, ORSAY. LABORATOIRE DE L'ACCELERATEUR LINEAIRE	LAL-RT-(NUMBER)-(YEAR)
PARIS. UNIVERSITE, ORSAY. LABORATOIRE DE L'ACCELERATEUR LINEAIRE	RI-
PARIS. UNIVERSITE, ORSAY. LABORATOIRE DE L'ACCELERATEUR LINEAIRE	RI-(YEAR)-
PARIS. UNIVERSITE, ORSAY. LABORATOIRE DE PHYSIQUE DES PLASMAS	LP-
PARIS. UNIVERSITE, ORSAY. LABORATOIRE DE PHYSIQUE THEORIQUE ET HAUTES ENERGIES	TH/(YEAR)/
PARIS. UNIVERSITE, ORSAY. LABORATOIRE DE PHYSIQUE THEORIQUE ET HAUTES ENERGIES	TH-
PARIS. UNIV., ORSAY. LABORATOIRE DE PHYSIQUE THEORIQUE ET PARTICULES ELEMENTAIRES	LPTPE-
PARIS. UNIV., ORSAY. LABORATOIRE DES HAUTES ENERGIES	LP.(NO.)
PARKE MATHEMATICAL LABS., INC., CARLISLE, MASS.	PML-(NO.)-SR-
PARKER AIRCRAFT CO., LOS ANGELES	EER-
PARKER AIRCRAFT CO., LOS ANGELES	S(YEAR)R-
PARKER APPLIANCE CO., CLEVELAND	P-
PARKER PEN CO., JANESVILLE, WIS.	PPC-
PARMA, ITALY. UNIVERSITA. INST. OF HUMAN PHYSIOLOGY	IHP-
PARSONS-AEROJET CO., LOS ANGELES	PAC-
PARSONS-AEROJET CO., LOS ANGELES	PAC TR-
PARSONS(C.A.) + CO., LTD.,NEWCASTLE-UPON-TYNE, ENG.	NRC-(YR)-(NO.)-
PARSONS (RALPH M.) CO., LOS ANGELES	PAR-
PARSONS (RALPH M.) CO., LOS ANGELES	PARSONS-
PARSONS (RALPH M.) CO., LOS ANGELES	RMP-
PARSONS (RALPH M.) CO., LOS ANGELES	RMP-FR-
PARSONS (RALPH M.) CO., LOS ANGELES	RMP-TR-
PARSONS (RALPH M.) CO., PASADENA, CALIF.	BRL-(NO.-LTR.,NO.)
PARSONS (RALPH M.) CO., PASADENA, CALIF.	PAC-
PARSONS (RALPH M.) CO., PASADENA, CALIF.	PAR-
PARSONS (RALPH M.) CO., PASADENA, CALIF. (MISCELLANEOUS PUBLICATIONS)	PAR MP-
PARSONS (RALPH M.) CO., PASADENA, CAL. (PROJ. RANGE)	PAR-RANGE-
PARSONS (RALPH M.) CO., PASADENA, CALIF.	PRM-
PARSONS (RALPH M.) CO., PASADENA, CALIF.	PRM-(YEAR)-
PARSONS, BRINCKERHOFF-TUDOR-BECHTEL, SAN FRANCISCO	PBTB-
PASADENA FOUNDATION FOR MEDICAL RESEARCH, CALIF.	SAN-TR-
PASS AND SEYMOUR, INC., SYRACUSE, N.Y.	P+S-
PASTUSHIN AVIATION CORP., LOS ANGELES	PAVCO-
PATROL ASW DEVELOPMENT GROUP (NAVY), NORFOLK, VA.	PATASWDEVGRU-
PATTERSON, MOOS RES. DIV.,LEESONA CORP.,JAMAICA, N.Y.	LC-
PATTERSON, MOOS RES. DIV.,LEESONA CORP.,JAMAICA, N.Y.	PAMR-
PEAT MARWICK AND LIVINGSTON CO., WASHINGTON, D.C.	CFE/GFE POLICY STUDY-
PENINSULAR CHEMRESEARCH, INC., GAINESVILLE, FLA.	PCR-(YR.)-
PENINSULAR CHEMRESEARCH, INC., GAINESVILLE, FLA.	PCR-QPR-(NUMBER)(YEAR)
PENNSALT CHEMICALS CORP., PHILADELPHIA	PCC-
PENNSALT CHEMICALS CORP. RES.+ DEV. DEPT.,WYNDMOOR,PA	PENNSALT-
PENNSALT CHEMICALS CORP. RES.+ DEV. DEPT.,WYNDMOOR,PA	PENNSALT-TR-
PENNSYLVANIA. UNIV., PHILADELPHIA	AEL-
PENNSYLVANIA. UNIV., PHILADELPHIA	AR-
PENNSYLVANIA. UNIV., PHILADELPHIA	LAP-
PENNSYLVANIA. UNIV., PHILADELPHIA	PENN-
PENNSYLVANIA. UNIV., PHILADELPHIA	PENN-AR-
PENNSYLVANIA. UNIV., PHILADELPHIA	PENN U-

PERKIN-ELMER CORP. ELECTRO-OPTICAL DIV.

PENNSYLVANIA. UNIV., PHILADELPHIA	PENNU-ASR-(YEAR)
PENNSYLVANIA. UNIV., PHILADELPHIA	PENNU-ATR-(YEAR)
PENNSYLVANIA. UNIV., PHILADELPHIA	PENNU-MS-(YEAR)-
PENNSYLVANIA. UNIV., PHILADELPHIA	PENNU-RDR-
PENNSYLVANIA. UNIV., PHILADELPHIA	PENNU-TR-
PENNSYLVANIA. UNIV., PHILADELPHIA	PU-
PENNSYLVANIA. UNIV., PHILADELPHIA	SSRC-
PENNSYLVANIA. UNIV., PHILADELPHIA	UPR-(NUMBER)N
PENNSYLVANIA. UNIV., PHILADELPHIA. DEPT. OF PHYSICS	UPR-(4 DIGITS)(LETTER)
PENNSYLVANIA. UNIV., PHILA. INST. FOR COOP. RES.	BI (NUMBER) CR-
PENNSYLVANIA. UNIV., PHILA. INST. FOR COOP. RES.	CIDS-
PENNSYLVANIA. UNIV., PHILA. INST. FOR COOP. RES.	GR-
PENNSYLVANIA. UNIV., PHILA. INST. FOR COOP. RES.	ICR-
PENNSYLVANIA. UNIV., PHILADELPHIA. INSTITUTE FOR COOPERATIVE RESEARCH (PROJECT SUMMIT)	SU-
PENNSYLVANIA. UNIV., PHILA. INST. FOR COOP. RES.	SU(YEAR)(LETTERS)-
PENNSYLVANIA. UNIV., PHILADELPHIA. INSTITUTE FOR DIRECT ENERGY CONVERSION	INDEC-
PENNSYLVANIA. UNIV., PHILADELPHIA. INSTITUTE FOR DIRECT ENERGY CONVERSION	INDEC-SR-
PENNSYLVANIA. UNIV., PHILA. MOORE SCH. OF ELEC. ENG.	MS-(YEAR)-
PENNSYLVANIA. UNIV., PHILA. MOORE SCH. OF ELEC. ENG.	MSEE-
PENNSYLVANIA. UNIV., PHILA. MOORE SCH. OF ELEC. ENG.	MSR-
PENNSYLVANIA. UNIV., PHILA. MOORE SCH. OF ELEC. ENG.	PEU MSEE-
PENNSYLVANIA. UNIV., PHILADELPHIA. MOORE SCHOOL OF ELECTRICAL ENG. (USED 1946-50 FOR RES. + DEV. RPT.)	RDR-(YEAR)-
PENNSYLVANIA. UNIV., PHILADELPHIA. POWER INFO. CTR.	PIC-
PENNSYLVANIA. UNIV., PHILADELPHIA. POWER INFO. CTR.	PIC-BAT-
PENNSYLVANIA. UNIV., PHILADELPHIA. POWER INFO. CTR.	PIC-ELE-MHD-
PENNSYLVANIA. UNIV., PHILADELPHIA. POWER INFO. CTR.	PIC-ELE-TI-
PENNSYLVANIA. UNIV., PHILADELPHIA. POWER INFO. CTR.	PIC-SOL-
PENNSYLVANIA. UNIV., PHILADELPHIA. SCH. OF CHEM. ENG.	UPH(NUMBER)-TR-
PENNSYLVANIA HOSPITAL. UNIT FOR EXPERIMENTAL PSYCHIATRY, PHILADELPHIA	USCAE-
PENNSYLVANIA RESEARCH ASSOCIATES INC., PHILADELPHIA	EC-SR-
PENNSYLVANIA RESEARCH ASSOCIATES INC., PHILADELPHIA	PRA-U(YEAR)-
PENNSYLVANIA RESEARCH ASSOCIATES INC., PHILADELPHIA	U(YEAR)-
PENNSYLVANIA SALT MFG. CO., PHILADELPHIA	PENNSALT-
PENNSYLVANIA STATE COLL., STATE COLLEGE. ENGINEERING EXPERIMENT STATION (TECHNICAL PAPER)	PSC EES TP-
PENNSYLVANIA STATE COLL., STATE COLLEGE. PETROLEUM REFINING LAB.	PSC PRL-
PENNSYLVANIA STATE COLL., STATE COLLEGE. SCHOOL OF MINERAL INDUSTRIES	MEMO-
PENNSYLVANIA STATE COLL., STATE COLLEGE. SCHOOL OF MINERAL INDUSTRIES	MEMO REPORT-
PENNSYLVANIA STATE COLL., STATE COLLEGE. SCHOOL OF MINERAL INDUSTRIES	PSC SMI MR-
PENNSYLVANIA STATE COLL., STATE COLLEGE. SCHOOL OF MINERAL INDUSTRIES (MEMORANDUM REPORTS)	PSC SMI TR-
PENNSYLVANIA STATE COLL., STATE COLLEGE. SCHOOL OF MINERAL INDUSTRIES	SMI-
PENNSYLVANIA STATE UNIV., UNIVERSITY PARK	B-
PENNSYLVANIA STATE UNIV., UNIVERSITY PARK	NUC-
PENNSYLVANIA STATE UNIV., UNIVERSITY PARK	NUC-E-
PENNSYLVANIA STATE UNIV., UNIVERSITY PARK	PSC-
PENNSYLVANIA STATE UNIV., UNIVERSITY PARK	PSTR-(NUMBER)-
PENNSYLVANIA STATE UNIV., UNIVERSITY PARK	PSU-
PENNSYLVANIA STATE UNIV., UNIVERSITY PARK	PSU-(NO.)-P
PENNSYLVANIA STATE UNIV., UNIVERSITY PARK	PSU-NUC-E-
PENNSYLVANIA STATE UNIV., UNIVERSITY PARK	PSU-PR-
PENNSYLVANIA STATE UNIV., UNIVERSITY PARK	PSU-PSR-
PENNSYLVANIA STATE UNIV., UNIVERSITY PARK	PSU-TR-
PENNSYLVANIA STATE UNIV., UNIVERSITY PARK	SAPR-
PENNSYLVANIA STATE UNIV., UNIVERSITY PARK (SQUID PROJECT) (REF. 33)	SQUID-PSU-(NUMBER)-P
PENNSYLVANIA STATE UNIV., UNIVERSITY PARK. COLL. OF ENGINEERING	ENGINEERING RESEARCH-
PENNSYLVANIA STATE UNIV., UNIVERSITY PARK. COMPUTER AIDED DESIGN AND SIMULATION LAB.	CAD-LAB-(YEAR-NUMBER)
PENNSYLVANIA STATE UNIV., UNIVERSITY PARK. COMPUTER ASSISTED INSTRUCTION LAB.	CAI-R-
PENNSYLVANIA STATE UNIV., UNIVERSITY PARK. DEPT. OF CIVIL ENGINEERING	RR-IR-
PENNSYLVANIA STATE UNIV., UNIVERSITY PARK. DEPT. OF FUEL SCIENCE	FS(YEAR)-
PENNSYLVANIA STATE UNIV., UNIVERSITY PARK. DEPT. OF GEOLOGY AND GEOPHYSICS	PUB-(YEAR)-
PENNSYLVANIA STATE UNIV., UNIVERSITY PARK. DEPT. OF MECHANICAL ENGINEERING	JOINT ROAD FRICTION-
PENNSYLVANIA STATE UNIV., UNIVERSITY PARK. DEPT. OF NUCLEAR ENGINEERING	PSR-
PENNSYLVANIA STATE UNIV., UNIVERSITY PARK. GROTH INST.	GI-
PENNSYLVANIA STATE UNIV., UNIVERSITY PARK. GROTH INST.	PSU-GI-
PENNSYLVANIA STATE UNIV., UNIVERSITY PARK. IONOSPHERE RESEARCH LAB.	PSU-IRL-SCI-
PENNSYLVANIA STATE UNIV., UNIVERSITY PARK. IONOSPHERE RESEARCH LAB. (SCIENTIFIC REPORTS)	PSU-IR-SR-
PENNA. STATE UNIV., UNIV. PARK. ORDNANCE RES. LAB.	D-
PENNA. STATE UNIV., UNIV. PARK. ORDNANCE RES. LAB.	NOW-
PENNA. STATE UNIV., UNIV. PARK. ORDNANCE RES. LAB.	ORL-
PENNA. STATE UNIV., UNIV. PARK. ORDNANCE RES. LAB.	ORL(PSU)TM-
PENNSYLVANIA STATE UNIV., UNIVERSITY PARK. PETROLEUM REFINING LAB.	PRL-
PENNSYLVANIA STATE UNIV., UNIVERSITY PARK. PETROLEUM REFINING LAB.	PRL-5.(NUMBER)
PENNSYLVANIA STATE UNIV., UNIVERSITY PARK. SPACE SCIENCES + ENGINEERING LAB.	SSEL-
PEPPERDINE COLL., LOS ANGELES	PC-
PERGAMON PRESS, N.Y.C. (TRANSLATIONS)	PP-
PERKIN-ELMER CORP., NORWALK, CONN.	PE-
PERKIN-ELMER CORP., NORWALK, CONN.	PEC-
PERKIN-ELMER CORP., NORWALK, CONN.	PE-TR-
PERKIN-ELMER CORP., POMONA, CALIF.	SPO-
PERKIN-ELMER CORP. ELECTRO-OPTICAL DIV., NORWALK, CONN.	PE-ER-

PERKIN-ELMER CORP. ELECTRO-OPTICAL DIV.

PERKIN-ELMER CORP. ELECTRO-OPTICAL DIV., NORWALK, CONN.	PE-TM-
PERKIN-ELMER CORP. OPTICAL GROUP, NORWALK, CONN.	PE-ER-
PERKIN-ELMER CORP. OPTICAL OPERATIONS DIV., WILTON, CONN.	OOD-ENGINEERING-
PERSHING JOINT AEC-DOD WARHEAD COORDINATING COMM. (REF. 15)	PERSHING-
PERSONNEL RESEARCH CENTER, INC., DETROIT, MICH.	PTB-
PERSONNEL RES. LAB. (6570TH), LACKLAND AFB, TEX.	ASD-TR-(YEAR)-
PERSONNEL RES. LAB. (6570TH), LACKLAND AFB, TEX.	PL-TM-(YEAR)-
PERSONNEL RES. LAB. (6570TH), LACKLAND AFB, TEX.	PL-TR-(YEAR)-
PERSONNEL RES. LAB. (6570TH), LACKLAND AFB, TEX.	PRL-LN-
PERSONNEL RES. LAB. (6570TH), LACKLAND AFB, TEX. (STAFF RESEARCH MEMORANDUM)	PRL-SRM-(YEAR)-
PERSONNEL RES. LAB. (6570TH), LACKLAND AFB, TEX.	PRL-TDR-
PERSONNEL RES. LAB. (6570TH), LACKLAND AFB, TEX.	PRL-TDR-(YEAR)-
PERSONNEL RES. LAB. (6570TH), LACKLAND AFB, TEX.	PRL-TM-(YEAR)-
PERSONNEL RES. LAB. (6570TH), LACKLAND AFB, TEX.	PRL-TR-
PERSONNEL RES. LAB. (6570TH), LACKLAND AFB, TEX.	PRL-TR-(YEAR)-
PERSONNEL RES. LAB. (6570TH), LACKLAND AFB, TEX. (LABORATORY NOTE)	TARL-LAB-NOTE-
PERSONNEL RES. LAB. (6570TH), LACKLAND AFB, TEX. (LABORATORY NOTE)	TARL-LN-(NUMBER)-
PESCO PRODUCTS DIV., BORG-WARNER CORP., BEDFORD, OHIO	PESCO-
PESCO PRODUCTS DIV., BORG-WARNER CORP., BEDFORD, OHIO	PESCO-ENG. RPT.-
PETERS (O.S.) CO., WASHINGTON, D.C.	OSP-
PETROLEUM ADMINISTRATION FOR WAR	PAW-
PETROLEUM ADMINISTRATION FOR WAR. TECH. ADVISORY COMM	TAC-
PETROLEUM ADMINISTRATION FOR WAR. TECH. ADVISORY COMM	TAC (LETTER(S))-
PETROLEUM INDUSTRY WAR COUNCIL, N.Y.C.	TAC-
PFAUDLER CO., ROCHESTER, N.Y.	PF-
PFAUDLER CO., ROCHESTER, N.Y.	PF-(YEAR)-
PFIZER (CHAS.) AND CO. INC., NEW YORK	XYA-
PHELPS DODGE COPPER PRODUCTS CORP. RESEARCH LABS., YONKERS, N.Y.	PDCPC-
PHILADELPHIA ELECTRIC CO.	PEC-
PHILADELPHIA ELECTRIC CO.	PEC-FL-
PHILADELPHIA ELECTRIC CO.	PEC-MOR-
PHILADELPHIA NAVAL SHIPYARD. INDUSTRIAL TEST LAB.	ITL-
PHILADELPHIA NAVAL SHIPYARD. INDUSTRIAL TEST LAB.	NAVSHIPS ITL-
PHILCO CORP. (ALL LOCATIONS)	PHIL-
PHILCO CORP. (ALL LOCATIONS)	PHILCO-
PHILCO CORP., BLUE BELL, PENNA.	A(NO.)-
PHILCO CORP., BLUE BELL, PENNA.	AO(NUMBER)-(LETTER)
PHILCO CORP., BLUE BELL, PENNA.	B(NO.)-
PHILCO CORP., BLUE BELL, PENNA.	RM-
PHILCO CORP., LANSDALE, PENNA.	H-
PHILCO CORP., LANSDALE, PENNA.	PHILCO-R-
PHILCO CORP., LANSDALE, PENNA.	PHO-TR-
PHILCO CORP., LANSDALE, PENNA.	R-
PHILCO CORP., PHILADELPHIA	H-
PHILCO CORP., PHILADELPHIA	PC-
PHILCO CORP., PHILADELPHIA	PHILCO-H-
(SEE ALSO: AERONUTRONIC DIV..., AND LATER NAME: PHILCO-FORD CORP.)	
PHILCO CORP. COMMUNICATIONS+ELECTRONICS DIV., PHILA.	EMC-(YEAR)-TR-
PHILCO CORP. GOVERNMENT AND INDUSTRIAL DIV., PHILA.	GI-(NUMBER)-(NUMBER)-
PHILCO CORP. RESEARCH DIV., PHILADELPHIA	PC-H-
PHILCO CORP. WESTERN DEVELOPMENT LABS., PALO ALTO, CAL	WDL-
PHILCO CORP. WESTERN DEVELOPMENT LABS., PALO ALTO, CAL (TECHNICAL MANUAL)	WDL/TM-
PHILCO CORP. WESTERN DEVELOPMENT LABS., PALO ALTO, CAL (TECHNICAL MEMO)	WDL/TM-
PHILCO-FORD CORP., BILLINGS, MONT.	P-(NUMBER)-
PHILCO-FORD CORP., BLUE BELL, PENNA.	B(3 DIGITS)-
PHILCO-FORD CORP., NEWPORT BEACH, CALIF.	PFC-
PHILCO-FORD CORP., NEWPORT BEACH, CALIF.	PRP-(YEAR)-
PHILCO-FORD CORP., NEWPORT BEACH, CALIF.	U-
PHILCO-FORD CORP. COMMUNICATIONS AND ELECTRONICS DIV., WILLOW GROVE, PA.	HX(NUMBER)-
PHILCO-FORD CORP. POWER + CONTROL ENGINEERING DEPT., PALO ALTO, CALIF.	TR-DA(NUMBER)
PHILCO-FORD CORP. SPACE AND RE-ENTRY SYSTEMS DIV., NEWPORT BEACH, CALIF.	CG-
PHILCO-FORD CORP. SPACE AND RE-ENTRY SYSTEMS DIV., NEWPORT BEACH, CALIF.	DR-
PHILCO-FORD CORP. SPACE AND RE-ENTRY SYSTEMS DIV., NEWPORT BEACH, CALIF.	SG-
PHILCO-FORD CORP. SPACE AND RE-ENTRY SYSTEMS DIV., NEWPORT BEACH, CALIF.	SRD-
PHILCO-FORD CORP. SPACE AND RE-ENTRY SYSTEMS DIV., NEWPORT BEACH, CALIF.	SRD-G-
PHILCO-FORD CORP. SPACE AND RE-ENTRY SYSTEMS DIV., NEWPORT BEACH, CALIF.	SRDG-
PHILCO-FORD CORP. SPACE AND RE-ENTRY SYSTEMS DIV., NEWPORT BEACH, CALIF.	SRG-
PHILCO-FORD CORP. SPACE AND RE-ENTRY SYSTEMS DIV., NEWPORT BEACH, CALIF.	SRS-
PHILCO-FORD CORP. SPACE AND RE-ENTRY SYSTEMS DIV., NEWPORT BEACH, CALIF.	SRS-BNTT-PM-
PHILCO-FORD CORP. SPACE AND RE-ENTRY SYSTEMS DIV., NEWPORT BEACH, CALIF.	SRS-S-
PHILCO-FORD CORP. SPACE AND RE-ENTRY SYSTEMS DIV., NEWPORT BEACH, CALIF.	SRS-SG-
PHILCO-FORD CORP. SPACE AND RE-ENTRY SYSTEMS DIV., NEWPORT BEACH, CALIF.	SRS-SRG-
PHILCO-FORD CORP. SPACE AND RE-ENTRY SYSTEMS DIV., NEWPORT BEACH, CALIF.	UG-
PHILCO-FORD CORP. SPACE AND RE-ENTRY SYSTEMS DIV., PALO ALTO, CALIF.	SRS-TM-
PHILCO-FORD CORP. WESTERN DEV.LABS., PALO ALTO, CAL.	WDL-TR(NUMBER)-
PHILIPPINES. ATOMIC ENERGY COMMISSION, MANILA	PAEC-
PHILIPPINES. ATOMIC ENERGY COMMISSION, MANILA	PAEC(A)IN-
PHILIPPINES. ATOMIC ENERGY COMMISSION, MANILA	PAEC/A/PH-

PHILIPPINES. ATOMIC ENERGY COMMISSION, MANILA	PAEC(C)RE-
PHILIPPINES. ATOMIC ENERGY COMMISSION, MANILA	PAEC(D)
PHILIPPINES. ATOMIC ENERGY COMMISSION, MANILA	PAEC(D)CH-
PHILIPPINES. ATOMIC ENERGY COMMISSION, MANILA	PAEC(D)HP-
PHILIPPINES. ATOMIC ENERGY COMMISSION, MANILA	PAEC(D)IS-
PHILIPPINES. ATOMIC ENERGY COMMISSION, MANILA	PAEC(D)PH-
PHILIPPINES. ATOMIC ENERGY COMMISSION, MANILA	PAEC(D)RE-
PHILIPPINES. ATOMIC ENERGY COMMISSION, MANILA (REVIEW REPORT)	PAEC-RR-
PHILIPPINES. ATOMIC RES. CENTER, DILIMAN, QUEZON CITY	PAEC-IPA(D)PH-
PHILIPS ELECTRONICS AND PHARMACEUTICAL INDUSTRIES CORP., MOUNT VERNON, N.Y.	PEI-(NUMBERS,LETTERS)-
PHILIPS ELECTRONICS AND PHARMACEUTICAL INDUSTRIES CORP., MOUNT VERNON, N.Y.	PEPI-
PHILIPS LABS., INC., BRIARCLIFF MANOR, N.Y.	PL-
PHILIPS LABS., INC., IRVINGTON-ON-HUDSON, N.Y.	PL-
PHILIPS LABS., INC. RESEARCH AND DEVELOPMENT DEPT., IRVINGTON-ON-HUDSON, N.Y.	RD-
PHILLIPS CABLES, LTD., BROCKVILLE, ONT.	PCL-
PHILLIPS PETROLEUM CO., BARTLESVILLE, OKLA.	PHPC-
PHILLIPS PETROLEUM CO., IDAHO FALLS	MTR-
PHILLIPS PETROLEUM CO., IDAHO FALLS	MTRL-
PHILLIPS PETROLEUM CO. ARCO CHEM. PLANT, IDAHO FALLS	ARCO-
PHILLIPS PETROLEUM CO. ATOMIC EN. DIV., IDAHO FALLS	CPP-(YR)-(NO.)-(PPC)
PHILLIPS PETROLEUM CO. ATOMIC EN. DIV., IDAHO FALLS	CPP-(NO.)-(PPC)
PHILLIPS PETROLEUM CO. ATOMIC EN. DIV., IDAHO FALLS	CPP-(YEAR)-
PHILLIPS PETROLEUM CO. ATOMIC EN. DIV., IDAHO FALLS	CPP-
PHILLIPS PETROLEUM CO. ATOMIC ENERGY DIV., IDAHO FALLS (CODE IS AUTHOR'S INITIALS)	HUF-
PHILLIPS PETROLEUM CO. ATOMIC EN. DIV., IDAHO FALLS	LEY-
PHILLIPS PETROLEUM CO. ATOMIC EN. DIV., IDAHO FALLS	MTRL-
PHILLIPS PETROLEUM CO. ATOMIC EN. DIV., IDAHO FALLS	PPC-
PHILLIPS PETROLEUM CO. ATOMIC EN. DIV., IDAHO FALLS	PPC-IDO-
PHILLIPS PETROLEUM CO. ATOMIC EN. DIV., IDAHO FALLS	PPCS-
PHILLIPS PETROLEUM CO. ATOMIC EN. DIV., IDAHO FALLS	PTR-
PHILLIPS PETROLEUM CO. ATOMIC EN. DIV., IDAHO FALLS	RHM-
PHILLIPS PETROLEUM CO. ATOMIC ENERGY DIV., IDAHO FALLS (CODE DERIVED FROM AUTHOR'S NAME)	WEB-
PHILLIPS PETROLEUM CO. RESEARCH AND DEVELOPMENT DEPT. BARTLESVILLE, OKLA.	PPC-(NO.)-(YR.)R
PHOTOCIRCUITS CORP., GLEN COVE, N.Y.	FR-
PHOTOGRAMMETRY, INC., SILVER SPRING, MD.	PHOTO-
PHOTOMETRICS, INC., LEXINGTON, MASS.	PHM-(NUMBER-YEAR)
PHYSICAL SECURITY EQUIPMENT AGENCY, WASHINGTON, D.C.	PSEA-
PHYSICAL STUDIES, INC., DAYTON, OHIO	PSI-(YEAR)-
PHYSICAL STUDIES, INC., RENO, NEV.	PSI-
PHYSICAL VULNERABILITY DIV. (AIR FORCE), WASH., D.C.	PV-
PHYSICAL VULNERABILITY DIV. (AIR FORCE), WASH., D.C. (INTERIM MEMORANDUM)	PVIM-
PHYSICAL VULNERABILITY DIV. (AIR FORCE), WASH., D.C. (TECHNICAL MEMORANDUM)	PVTM-
PHYSICAL VULNERABILITY DIV. (AIR FORCE), WASH., D.C.	USAF PV-
PHYSICAL VULNERABILITY DIV. (AIR FORCE), WASH., D.C. (INTERIM MEMORANDUM)	USAF PVIM-
PHYSICAL VULNERABILITY DIV. (AIR FORCE), WASH., D.C. (TECHNICAL MEMORANDUM)	USAF PVTM-
PHYSICS INTERNATIONAL CO., BERKELEY, CALIF.	PINT-
PHYSICS INTERNATIONAL CO., SAN LEANDRO, CALIF.	PI-
PHYSICS INTERNATIONAL CO., SAN LEANDRO, CALIF.	PIFR-
PHYSICS INTERNATIONAL CO., SAN LEANDRO, CALIF.	PIIR-(NUMBER)-(YEAR)
PHYSICS INTERNATIONAL CO., SAN LEANDRO, CALIF.	PINT-
PHYSICS INTERNATIONAL CO., SAN LEANDRO, CALIF.	PINT-PIFR-
PHYSICS INTERNATIONAL CO., SAN LEANDRO, CALIF.	PIPR-
PHYSICS INTERNATIONAL CO., SAN LEANDRO, CALIF.	PIQR-
PHYSICS INTERNATIONAL CO., SAN LEANDRO, CALIF.	PIR-
PHYSICS INTERNATIONAL CO., SAN LEANDRO, CALIF.	PITR-
PHYSICS INTERNATIONAL CO., SAN LEANDRO, CALIF.	PITSR-
PHYSICS INTERNATIONAL CO., SAN LEANDRO, CALIF.	PRQ-
PICATINNY ARSENAL, DOVER, N.J.	DB-TR-
PICATINNY ARSENAL, DOVER, N.J.	FR-
PICATINNY ARSENAL, DOVER, N.J. (FRAGMENTATION TEST RECORD)	FTR-
PICATINNY ARSENAL, DOVER, N.J.	IED-
PICATINNY ARSENAL, DOVER, N.J.	MR-
PICATINNY ARSENAL, DOVER, N.J.	ORDBB-T-(NUMBER)-(NO.)
PICATINNY ARSENAL, DOVER, N.J. (TASK GROUP REPORTS)	ORDBB-TK-
PICATINNY ARSENAL, DOVER, N.J. (TASK GROUP REPORTS)	ORDBB-TK-(NUMBER)-(NO.)
PICATINNY ARSENAL, DOVER, N.J.	ORDBB-VC-
PICATINNY ARSENAL, DOVER, N.J.	ORD PATR-
PICATINNY ARSENAL, DOVER, N.J.	PA-
PICATINNY ARSENAL, DOVER, N.J.	PA-DB-TR-
PICATINNY ARSENAL, DOVER, N.J.	PA-DC-QR-(NOS.)-(YR.)
PICATINNY ARSENAL, DOVER, N.J.	PA-DM-
PICATINNY ARSENAL, DOVER, N.J.	PA-FLAM-
PICATINNY ARSENAL, DOVER, N.J.	PA-FREL-
PICATINNY ARSENAL, DOVER, N.J.	PA-FT-
PICATINNY ARSENAL, DOVER, N.J. (FRAGMENTATION TESTS)	PAL-
PICATINNY ARSENAL, DOVER, N.J. (LIBRARY)	PA-LAB-
PICATINNY ARSENAL, DOVER, N.J.	PA-M
PICATINNY ARSENAL, DOVER, N.J. (MEMORANDUM REPORTS)	PA-MR-
PICATINNY ARSENAL, DOVER, N.J.	PA-ORDBB-(LTRS.NO.)-
PICATINNY ARSENAL, DOVER, N.J.	PA-POMM-
PICATINNY ARSENAL, DOVER, N.J.	PA-R-
PICATINNY ARSENAL, DOVER, N.J.	PA-TE-
PICATINNY ARSENAL, DOVER, N.J.	PA-TM-
PICATINNY ARSENAL, DOVER, N.J.	PA-TR-
PICATINNY ARSENAL, DOVER, N.J.	PATR-
PICATINNY ARSENAL, DOVER, N.J.	PA-TT-
PICATINNY ARSENAL, DOVER, N.J.	PED-
PICATINNY ARSENAL, DOVER, N.J.	PIC-ARS-
PICATINNY ARSENAL, DOVER, N.J.	PIC ARS TM-

PICATINNY ARSENAL

PICATINNY ARSENAL, DOVER, N.J.	PIC. ARS. TN-
PICATINNY ARSENAL, DOVER, N.J.	PIC. ARS. TR-
PICATINNY ARSENAL. AMMUNITION DEV. DIV., DOVER, N.J.	PA-PAS-
PICATINNY ARSENAL. AMMUNITION ENG. LAB., DOVER, N.J.	DC-QR-
PICATINNY ARSENAL. AMMUNITION GROUP, DOVER, N.J.	DW-
PICATINNY ARSENAL. ARTILLERY AMMUNITION AND ROCKET DEVELOPMENT LAB., DOVER, N.J.	ORDBB-DR4-
PICATINNY ARSENAL. DATA PROCESSING SYSTEMS OFFICE, DOVER, N.J.	DPSO-
PICATINNY ARSENAL. DATA PROCESSING SYSTEMS OFFICE, DOVER, N.J.	DPSO-IR-
PICATINNY ARSENAL. DATA PROCESSING SYSTEMS OFFICE, DOVER, N.J.	DPSO-TM-
PICATINNY ARSENAL. ENGINEERING ANALYSIS AND SPECIAL AMMUNITION SECTION, DOVER, N.J.	EAS-
PICATINNY ARSENAL. FELTMAN RES. + ENG. LABS.,DOVER,NJ	FREL-
PICATINNY ARSENAL. FELTMAN RES. + ENG. LABS.,DOVER,NJ	FREL-LS-
PICATINNY ARSENAL. FELTMAN RES. + ENG. LABS, DOVER,NJ	PA-FRL-TR-
PICATINNY ARSENAL. FELTMAN RES. + ENG. LABS.,DOVER,NJ	PA-ORDBB-TK-
PICATINNY ARSENAL. FELTMAN RES. + ENG. LABS.,DOVER,NJ	PA-TN-
(SEE ALSO LATER NAME: PICATINNY...FELTMAN RES. LABS.	
PICATINNY ARSENAL. FELTMAN RESEARCH LABS.,DOVER, N.J.	FRL-
PICATINNY ARSENAL. FELTMAN RESEARCH LABS., DOVER, N.J. (TECHNICAL MEMORANDA)	FRL-TM-
PICATINNY ARSENAL. FELTMAN RESEARCH LABS., DOVER, N.J. (TECHNICAL NOTES)	FRL-TN-
PICATINNY ARSENAL. FELTMAN RESEARCH LABS., DOVER, N.J. (TECHNICAL REPORTS)	FRL-TR-
PICATINNY ARSENAL. FELTMAN RESEARCH LABS.,DOVER, N.J.	ORDBB-
PICATINNY ARSENAL. FELTMAN RESEARCH LABS.,DOVER, N.J.	PA-ORDBB-
PICATINNY ARSENAL. FELTMAN RESEARCH LABS., DOVER, N.J. (PROPELLANT RESEARCH SECTION REPORTS)	PA PRS-
PICATINNY ARSENAL. FELTMAN RESEARCH LABS.,DOVER, N.J.	PA-TN-FRL-TN-
PICATINNY ARSENAL. FELTMAN RESEARCH LABS.,DOVER, N.J.	PA-TR-FRL-TR-
PICATINNY ARSENAL. INDUSTRIAL ENG. DIV., DOVER, N.J.	DB-TR-
PICATINNY ARSENAL. INDUSTRIAL ENG. DIV., DOVER, N.J.	DC-TR-(NUMBER)-(NUMBER)
PICATINNY ARSENAL. LIQUID ROCKET PROPULSION LAB., DOVER, N.J.	DL-TR-
PICATINNY ARSENAL. LIQUID ROCKET PROPULSION LAB., DOVER, N.J.	LRPL-QPR-
PICATINNY ARSENAL. LIQUID ROCKET PROPULSION LAB., DOVER, N.J.	PA-DL-TR-
PICATINNY ARSENAL. NUCLEAR RELIABILITY DIV., DOVER,NJ	SMUPA-NR-
PICATINNY ARSENAL. NUCLEAR WEAPONS GROUP, DOVER, NJ	SMUPA-TK-
PICATINNY ARSENAL. OFFICE OF THE PROJECT MANAGER FOR BOMBS AND RELATED COMPONENTS, DOVER, N.J.	AMCPM-BR-TR-
PICATINNY ARSENAL. PLASTICS TECHNICAL EVALUATION CENTER, DOVER, N.J.	PA-PLASTEC-
PICATINNY ARSENAL. PLASTICS TECHNICAL EVALUATION CENTER, DOVER, N.J.	PLASTEC-
PICATINNY ARSENAL. PLASTICS TECHNICAL EVALUATION CENTER, DOVER, N.J.	PLASTEC-NOTE-
PICATINNY ARSENAL. PLASTICS TECHNICAL EVALUATION CENTER, DOVER, N.J. (TECHNICAL NOTES)	PLASTEC-TN-
PICATINNY ARSENAL. PLASTICS TECHNICAL EVALUATION CENTER, DOVER, N.J. (TECHNICAL REPORTS)	PLASTEC-TR-
PICATINNY ARSENAL. PROPULSION AND PROPELLANT SEC., DOVER, N.J.	PPS-
PICATINNY ARSENAL. PYROTECHNICS LAB., DOVER, N.J.	PL-C-TN-
PICATINNY ARSENAL. QUALITY ASSURANCE DIV., DOVER,N.J.	ORDBB-NR-
PICATINNY ARSENAL. ROCKET AND JATO SEC., DOVER, N.J.	RJS-
PICATINNY ARSENAL. ROCKET DEV. SEC., DOVER, N.J.	RDS-
PICATINNY ARSENAL. WARHEADS AND SPECIAL PROJECTS LAB., DOVER, N.J.	DW-
PICATINNY ARSENAL. WARHEADS AND SPECIAL PROJECTS LAB., DOVER, N.J.	W/SP-DW-
PICATINNY ARSENAL. SAMUEL FELTMAN AMMUNITION LABS., DOVER, N.J.	ORDTX-
PICATINNY ARSENAL. SAMUEL FELTMAN AMMUNITION LABS., DOVER, N.J. (BIBLIOGRAPHY)	PA-BIB-
PICATINNY ARSENAL. SAMUEL FELTMAN AMMUNITION LABS., DOVER, N.J.	SFAL-
PICATINNY ARSENAL. SAMUEL FELTMAN AMMUNITION LABS., DOVER, N.J. (MONTHLY REPORTS)	SFAL-MR-
PICATINNY ARSENAL. SAMUEL FELTMAN AMMUNITION LABS., DOVER, N.J. (RESEARCH MEMORANDA)	SFAL-RM-
PICATINNY ARSENAL. SAMUEL FELTMAN AMMUNITION LABS., DOVER, N.J. (TECHNICAL REPORTS)	SFAL-TR-
(SEE ALSO LATER NAME: PICATINNY...FELTMAN RES. LABS.	
PICATINNY ARSENAL. SPECIAL WEAPONS GROUP, DOVER, N.J.	PA-ORDBB-TK-
PICATINNY ARSENAL. WARHEADS AND SPECIAL PROJECTS LAB., DOVER, N.J.	W/SP-TM-
PICKARD AND BURNS, INC., NEEDHAM, MASS.	P+B-
PICKARD AND BURNS, INC., NEEDHAM, MASS.	P + B PUB.-
PICKARD AND BURNS ELECTRONICS, WALTHAM, MASS.	P+B-
PICKARD AND BURNS ELECTRONICS, WALTHAM, MASS.	P/B PUB-
PICKARD AND BURNS ELECTRONICS, WALTHAM, MASS.	P + B PUBL.-
PIERSON ELECTRICAL + ENG. CORP., LOS ANGELES	PEEC-
PIONEER SERVICE AND ENGINEERING CO., CHICAGO	NEA-
PIONEER SERVICE AND ENGINEERING CO., CHICAGO	PSE-
PIQUA NUCLEAR POWER FACILITY, OHIO	PNPF-
PITTSBURGH. UNIV.	PITT-
PITTSBURGH. UNIV.	PITT-RR-
PITTSBURGH. UNIV.	PITTU-
PITTSBURGH. UNIV.	PITU-
PITTSBURGH. UNIV.	PRC-R-
PITTSBURGH. UNIV.	SU/
PITTSBURGH. UNIV.	TR-
PITTSBURGH. UNIV. ARMY MATERIEL RESEARCH STAFF, WASHINGTON, D.C.	TIR-
PITTSBURGH. UNIV. ATOMIC AND PLASMA PHYSICS LAB.	APP-
PITTSBURGH. UNIV. LEARNING RESEARCH + DEV. CENTER	LRDC-REPRINT-
PITTSBURGH. UNIV. LEARNING RESEARCH + DEV. CENTER	LRDC-TR-
PITTSBURGH. UNIV. MANAGEMENT RESEARCH CENTER	MRC-REPRINT-(YR.)-
PITTSBURGH. UNIV. METALLURGICAL AND MATERIALS ENG.	MME-(YEAR)-

PITTSBURGH. UNIV. PYMATUNING LAB. OF FIELD BIOLOGY	PITT-PLFB-
PITTSBURGH. UNIV. SARAH MELLON SCAIFE RADIATION LAB.	PITU-SRL-
PITTSBURGH. UNIV. SARAH MELLON SCAIFE RADIATION LAB.	PRL-
PITTSBURGH. UNIV. SARAH MELLON SCAIFE RADIATION LAB.	SMSRL-
PITTSBURGH. UNIV. SPACE RESEARCH COORDINATION CTR.	PITT-SRCC-
PITTSBURGH. UNIV. SPACE RESEARCH COORDINATION CTR.	SRCC-
PITTSBURGH CONSOLIDATED COAL CO.	PCC-
PITTSBURGH NAVAL REACTORS OPERATIONS OFFICE, AEC	PNRO-DEV-
PITTSBURGH NAVAL REACTORS OPERATIONS OFFICE, AEC	PNROO-DEV-
PITTSBURGH NAVAL REACTORS OPERATIONS OFFICE, AEC	PNRO-SMD-
PITTSBURGH PLATE GLASS CO. CHEMICAL DIV. NATRIUM PLANT, NEW MARTINSVILLE, W. VA. (NATRIUM DEV.)	ND-
PITTSBURGH PLATE GLASS CO. CHEMICAL DIV. NATRIUM PLANT, NEW MARTINSVILLE, W. VA. (NATRIUM ENG.)	NE-
PITTSBURGH PLATE GLASS CO. CHEMICAL DIV. NATRIUM PLANT, NEW MARTINSVILLE, W. VA. (NATRIUM PRODUCTION)	NP-
PITTSBURGH PLATE GLASS CO. CHEMICAL DIV. NATRIUM PLANT, NEW MARTINSVILLE, W. VA. (NATRIUM RESEARCH)	NR-
PLANETARY SCIENCES INC., SANTA CLARA, CALIF.	PSI-
PLANNING RESEARCH CORP., LOS ANGELES	FTM-
PLANNING RESEARCH CORP., LOS ANGELES	PRC-
PLANNING RESEARCH CORP., LOS ANGELES	PRC-R-
PLANNING RESEARCH CORP., WASHINGTON, D.C.	PRC-
PLANNING RESEARCH CORP., WASHINGTON, D.C.	PRC-D-
PLANNING RESEARCH CORP., WASHINGTON, D.C.	PRC-R-
PLANSEE SOCIETY FOR POWDER METALLURGY (SEMINAR REPRINTS)	PS/(INITIALS)-
PLASMADYNE CORP., SANTA ANA, CALIF.	FR(NO.)-
PLASMADYNE CORP., SANTA ANA, CALIF.	GRC-
PLASMADYNE CORP., SANTA ANA, CALIF.	P-
PLASMADYNE CORP., SANTA ANA, CALIF.	PC-
PLASMADYNE CORP., SANTA ANA, CALIF.	PLAS-
PLASMADYNE CORP., SANTA ANA, CALIF.	PLAS (LTRS.)-
PLASMADYNE CORP., SANTA ANA, CALIF.	PLR-
PLASMADYNE CORP., SANTA ANA, CALIF.	T-
PLENUM PRESS, INC., N.Y.C. (TRANSLATIONS)	PLP-
PLESSET (E.H.) ASSOCIATES, INC., LOS ANGELES	C(NO.)-
PLESSET (E.H.) ASSOCIATES, INC., LOS ANGELES	PLESS-
PLESSET (E.H.) ASSOCIATES, INC., LOS ANGELES	PLESS TR-
PLESSET (E.H.) ASSOCIATES, INC., SANTA MONICA, CALIF	C-(NUMBER)-(YEAR)-
PLESSET (E.H.) ASSOCIATES, INC., SANTA MONICA, CALIF	EHP-(LETTER NUMBERS...)
PLESSET (E.H.) ASSOCIATES, INC., SANTA MONICA, CALIF	EHPA-(LETTER NUMBERS...)
PLESSET (E.H.) ASSOCIATES, INC., SANTA MONICA, CALIF.	PLESS-
PLESSET (E.H.) ASSOCIATES, INC., SANTA MONICA, CALIF.	PLESSET-
PLESSEY CO., LTD., ILFORD, ESSEX, ENGLAND	PL.GW-
PLYMOUTH ROCK AND GRANITE STATE PROJECT	PRAG-
POLACOAT INC., CINCINNATI	PI-
POLAMOLD RESEARCH LABS., INC., SPRINGFIELD, OHIO	PRL-
POLAND. CENTRAL LAB. FOR RADIOLOGICAL PROTECT.,WARSAW	CLOR-(NUMBER)/D
POLAND. CENTRAL LAB. FOR RADIOLOGICAL PROTECT.,WARSAW	CLOR-
POLAND. CENTRAL LAB. FOR RADIOLOGICAL PROTECT.,WARSAW	CLOR-FONTON-
POLAND. CENTRAL LAB. FOR RADIOLOGICAL PROTECT.,WARSAW	CLOR-FOTON-
POLAND. CENTRAL LAB. FOR RADIOLOGICAL PROTECT.,WARSAW	CLOR-GUM-
POLAND. CENTRAL LAB. FOR RADIOLOGICAL PROTECT.,WARSAW	CLOR-I-
POLAND. CENTRAL LAB. FOR RADIOLOGICAL PROTECT.,WARSAW	CLOR-IBJ-
POLAND. CENTRAL LAB. FOR RADIOLOGICAL PROTECT.,WARSAW	CLOR-IO-
(POLISH NAME: CENTRALNE LABORATORIUM OCHRONY RADIOLOGICZNEJ)	
POLAND. CENTRALNY INSTYTUT DOKUMENTACJI NAUKOWOTECHNICZEJ, WARSAW (TRANSLATIONS)	CIDNT-
POLAND. INSTITUTE OF NUCLEAR PHYSICS, KRAKOW	IFJ-
POLAND. INSTITUTE OF NUCLEAR PHYSICS, KRAKOW	INP-(NUMBER)/PS
POLAND. INSTITUTE OF NUCLEAR PHYSICS, KRAKOW	INP-
POLAND. INSTITUTE OF NUCLEAR PHYSICS, KRAKOW	INP-(NUMBER)/PL/PH
POLAND. INSTITUTE OF NUCLEAR PHYSICS, KRAKOW	INP-(NUMBER)-PL
POLAND. INSTITUTE OF NUCLEAR PHYSICS, KRAKOW	INP-(NUMBER)-C
(POLISH NAME: INSTYTUT FIZYKI JADROWEJ)	
POLAND. INSTITUTE OF NUCLEAR RESEARCH, WARSAW	IBJ-(NO./ROMAN NO./LTR.)
POLAND. INSTITUTE OF NUCLEAR RESEARCH, WARSAW	INR-
POLAND. INSTITUTE OF NUCLEAR RESEARCH, WARSAW	INR-(ROMAN NO.)
POLAND. INSTITUTE OF NUCLEAR RESEARCH, WARSAW	INR-(NO./ROMAN NO./LTR.)
POLAND. INSTITUTE OF NUCLEAR RESEARCH, WARSAW	INR-LP-(YEAR)-
POLAND. INSTITUTE OF NUCLEAR RESEARCH, WARSAW	INR-P-
POLAND. INSTITUTE OF NUCLEAR RESEARCH, WARSAW	INR-P-(NO./ROMAN NO./LTR)
POLAND. INSTITUTE OF NUCLEAR RESEARCH, WARSAW	INR-P-(NUMBER/NO.)/PL
POLAND. INSTITUTE OF NUCLEAR RESEARCH, WARSAW	INR-PT-
POLAND. INSTITUTE OF NUCLEAR RESEARCH, WARSAW	OFJ-
(POLISH NAME: INSTYTUT BADAN JADROWYCH)	
POLAND. INST. OF NUCLEAR TECHNIQUES, KRAKOW	INP-(NUMBER)/PS
POLAND. INST. OF NUCLEAR TECHNIQUES, KRAKOW	INT-(NUMBER)/PS
POLAND. INST. OF NUCLEAR TECHNIQUES, KRAKOW	ITJ-
POLAND. INST. OF NUCLEAR TECHNIQUES, KRAKOW	ITJ-(NUMBER)/(LTR(S))
POLAND. NUCLEAR ENERGY INFORMATION CENTER, WARSAW	BPR-(NUMBER)-
POLAND. NUCLEAR ENERGY INFORMATION CENTER, WARSAW	NEIC-RR-
POLARAD ELECTRONICS CO., N.Y.C.	PEC-
POLARIS MARK 2 RE-ENTRY BODY COORD. COMM.	CC/POLARIS-
POLAROID CORP., CAMBRIDGE, MASS.	NE-
POLAROID CORP., CAMBRIDGE, MASS.	POC-
POLAROID CORP., CAMBRIDGE, MASS.	POL-
POLAROID CORP., CAMBRIDGE, MASS.	POLC-NC-

POLHEMUS NAVIGATION SCIENCES, INC.

POLHEMUS NAVIGATION SCIENCES, INC., BURLINGTON, VT.	PNSI-TR-(YEAR)-
POLHEMUS NAVIGATION SCIENCES, INC., BURLINGTON, VT.	PNSI-TR-(NUMBER)-
POLISH ACADEMY OF MINING AND METALLURGY, KRAKOW	PAN-
POLISH ACADEMY OF SCIENCES. INST. OF NUCLEAR RESEARCH, WARSAW	PAN-
POLISH ACADEMY OF SCIENCES. INST. OF NUCLEAR RESEARCH, WARSAW	PAN-0-
POLISH ACADEMY OF SCIENCES. INST. OF NUCLEAR RESEARCH, WARSAW	PAS-(NO.)/(ROMAN NO.)
POLISH ACADEMY OF SCIENCES. INST. OF NUCLEAR RESEARCH, WARSAW	PAS-INR-(NO./ROMAN NO.)
(FROM 1963- THE INST. PUBLS. SHOW NO CONNECTION WITH THE ACADEMY. SEE: POLAND. INST. OF NUCLEAR RES.)	
POLYMER CORP., LTD., SARNIA, ONT.	SDR-
POLYMER CORP., LTD., SARNIA, ONT.	SSR-
POLYTECHNIC ENGINEERING CORP., SILVER SPRING, MD.	POLY-
POLYTECHNIC RESEARCH AND DEV. CO., INC., BROOKLYN	PRD-
POLYTECHNIC RESEARCH AND DEV. CO., INC., BROOKLYN	PRDC-
POMONA COLL.,CLAREMONT, CALIF. MILLIKAN LAB. OF PHYS.	PC MLP TR-
PONTIFICIA UNIVERSIDADE CATOLICA, RIO DE JANEIRO	PUC-MONOGRAFIA-
POROLOY EQUIPMENT, INC., VAN NUYS, CALIF.	PE-
POROLOY EQUIPMENT, INC., VAN NUYS, CALIF.	PEI-
PORTSMOUTH NAVAL SHIPYARD. ELEC. TESTING LAB., N.H.	ETL-
PORTSMOUTH NAVAL SHIPYARD. ELEC. TESTING LAB., N.H.	ETL-N-
PORTUGAL. JUNTA DE ENERGIA NUCLEAR. LABORATORIO DE FISICA E ENGENHARIA NUCLEARES, SACAVEM	DOC.-
PORTUGAL. JUNTA DE ENERGIA NUCLEAR. LABORATORIO DE FISICA E ENGENHARIA NUCLEARES, SACAVEM	DOC-LFEN-
PORTUGAL. JUNTA DE ENERGIA NUCLEAR. LABORATORIO DE FISICA E ENGENHARIA NUCLEARES, SACAVEM.	LFEN-
PORTUGAL. JUNTA DE ENERGIA NUCLEAR. LABORATORIO DE FISICA E ENGENHARIA NUCLEARES, SACAVEM	LFEN-NI.-
PORTUGAL. JUNTA DE INVESTIGACOES DO ULTRAMAR, LISBON	JIU-
POWER GENERATORS, INC., TRENTON, N.J.	PGL-
POWER JETS, LTD. (GT. BRIT.)	PJR-
POWER REACTOR AND NUCLEAR FUEL DEVELOPMENT CORP., TOKAI, JAPAN	N(NUMBER)-(YEAR)-
POWER REACTOR AND NUCLEAR FUEL DEVELOPMENT CORP., TOKAI, JAPAN	PNC-N-
POWER REACTOR AND NUCLEAR FUEL DEVELOPMENT CORP., TOKAI, JAPAN	PNCPU-
POWER REACTOR AND NUCLEAR FUEL DEVELOPMENT CORP., TOKAI, JAPAN	PNCT-
POWER REACTOR AND NUCLEAR FUEL DEVELOPMENT CORP., TOKAI, JAPAN	PNCT-AR-
POWER REACTOR DEVELOPMENT CO., DETROIT	PRDC-
POWER REACTOR DEVELOPMENT CO., DETROIT	PRDC-TR-
POWER REACTOR DEVELOPMENT CO. ENRICO FERMI ATOMIC POWER PLANT, DETROIT	EF-
POWER REACTOR DEVELOPMENT CO. ENRICO FERMI ATOMIC POWER PLANT, DETROIT	PRDC-EF-
PRAEGER-KAVANAGH-WATERBURY, ENGINEER-ARCHITECTS,NYC	PG-
PRATT AND WHITNEY AIRCRAFT, EAST HARTFORD, CONN.	PWA-
PRATT AND WHITNEY AIRCRAFT, EAST HARTFORD, CONN.	TM-
PRATT AND WHITNEY AIRCRAFT. CONNECTICUT AIRCRAFT NUCLEAR ENGINE LAB., MIDDLETOWN	CANEL-
PRATT AND WHITNEY AIRCRAFT. FLORIDA RESEARCH AND DEVELOPMENT CENTER, WEST PALM BEACH	PWA-FP-(YEAR)-
PRATT AND WHITNEY AIRCRAFT DIV.,UNITED AIRCRAFT CORP. (ALL LOCATIONS)	ENG. PROGRESS RPT.-
PRATT AND WHITNEY AIRCRAFT DIV.,UNITED AIRCRAFT CORP. (ALL LOCATIONS)	PWA-
PRATT AND WHITNEY AIRCRAFT DIV.,UNITED AIRCRAFT CORP. (ALL LOCATIONS)	PWAC-
PRATT AND WHITNEY AIRCRAFT DIV.,UNITED AIRCRAFT CORP. (ALL LOCATIONS)	TIM-
PRATT AND WHITNEY AIRCRAFT DIV., UNITED AIRCRAFT CORP., HARTFORD, CONN.	D-
PRATT AND WHITNEY AIRCRAFT DIV., UNITED AIRCRAFT CORP., HARTFORD, CONN.	FSK-
PRATT AND WHITNEY AIRCRAFT DIV., UNITED AIRCRAFT CORP., HARTFORD, CONN.	FXM-
PRATT AND WHITNEY AIRCRAFT DIV., UNITED AIRCRAFT CORP., HARTFORD, CONN.	FXR-
PRATT AND WHITNEY AIRCRAFT DIV., UNITED AIRCRAFT CORP., HARTFORD, CONN.	P-
PRATT AND WHITNEY AIRCRAFT DIV., UNITED AIRCRAFT CORP., HARTFORD, CONN.	PW-
PRATT AND WHITNEY AIRCRAFT DIV., UNITED AIRCRAFT CORP., HARTFORD, CONN.	PWL-
PRATT AND WHITNEY AIRCRAFT DIV., UNITED AIRCRAFT CORP., HARTFORD, CONN.	RC-(NO.)(PWAC)
PRATT AND WHITNEY AIRCRAFT DIV., UNITED AIRCRAFT CORP., HARTFORD, CONN.	TDM-
PRATT AND WHITNEY AIRCRAFT DIV., UNITED AIRCRAFT CORP., MIDDLETOWN, CONN.	AERL-
PRATT AND WHITNEY AIRCRAFT DIV., UNITED AIRCRAFT CORP., MIDDLETOWN, CONN.	MJS-
PRATT AND WHITNEY AIRCRAFT DIV., UNITED AIRCRAFT CORP. CONN. AIRCRAFT NUCLEAR ENGINE LAB.,MIDDLETOWN (REF. 3)	CNLM-
PRATT AND WHITNEY AIRCRAFT DIV., UNITED AIRCRAFT CORP. CONN. AIRCRAFT NUCLEAR ENGINE LAB.,MIDDLETOWN	PWAR-
PRATT AND WHITNEY AIRCRAFT DIV., UNITED AIRCRAFT CORP FLORIDA RESEARCH AND DEV. CENTER, WEST PALM BEACH	PWA-FR-
PRC TECHNICAL APPLICATIONS, HUNTSVILLE, ALA.	PRC/TA-
PRD ELECTRONICS, INC., BROOKLYN	PRD-
PRECISION ELECTRONIC COMPONENTS, LTD., TORONTO	PEC-
PRECISION ELECTRONIC COMPONENTS, LTD., TORONTO	PEC-(YEAR)-
PRECISION TECHNOLOGY, INC., LIVERMORE, CALIF.	COEP-
PRECISION TECHNOLOGY, INC., LIVERMORE, CALIF.	PT-
PREFORMED LINE PRODUCTS COMPANY. RESEARCH AND ENGINEERING, CLEVELAND (FIELD INSTRUMENTATION)	FI-
PREFORMED LINE PRODUCTS COMPANY. RESEARCH AND ENGINEERING, CLEVELAND (MEMORANDUM REPORT)	MR-
PREFORMED LINE PRODUCTS COMPANY. RESEARCH AND ENGINEERING, CLEVELAND (MEMORANDUM REPORT-EXTERNAL)	MR-(NO.)-E-
PREFORMED LINE PRODUCTS COMPANY. RESEARCH AND ENG., CLEVELAND (TECHNICAL REPORT-EXTERNAL)	TM-(NO.)-E-
PREFORMED LINE PRODUCTS COMPANY. RESEARCH AND ENGINEERING, CLEVELAND (TECHNICAL REPORT-EXTERNAL)	TR-(NO.)-E-
PREFORMED LINE PRODUCTS CO. RESEARCH AND ENG., CLEVELAND (TECH. REPT. - EXTERNAL,BUT RESTRICTED)	TR-(NO.)-ER
PRESEARCH, INC., SILVER SPRING, MD.	PI-TN-
PREWITT AIRCRAFT CO., CLIFTON HEIGHTS, PENNA. (REF.1)	(PROJECT NO.)-(SUBJ.NO.)-
PRIMARY SOURCES, N.Y.C. (TRANSLATIONS)	PS-
PRINCETON-PENNSYLVANIA ACCELERATOR, PRINCETON, N.J.	PPAD-
PRINCETON-PENNSYLVANIA ACCELERATOR, PRINCETON, N.J.	PPAR-

PUBLIC HEALTH SERVICE

PRINCETON UNIV., N.J.	F-20-M-
PRINCETON UNIV., N.J.	F-20-P-
PRINCETON UNIV., N.J.	FLD-
PRINCETON UNIV., N.J.	PNJ-
PRINCETON UNIV., N.J.	PNJ-LA-
PRINCETON UNIV., N.J.	PPA-
PRINCETON UNIV., N.J.	PR-
PRINCETON UNIV., N.J. (SQUID PROJECT) (REF. 33)	PR-(NUMBER)-M
PRINCETON UNIV., N.J. (SQUID PROJECT) (REF. 33)	PR-(NUMBER)-M-P
PRINCETON UNIV., N.J. (SQUID PROJECT) (REF. 33)	PR-(NUMBER)-P
PRINCETON UNIV., N.J. (SQUID PROJECT) (REF. 33)	PR-(NUMBER)-P-R
PRINCETON UNIV., N.J. (SQUID PROJECT) (REF. 33)	PR-(NUMBER)-R
PRINCETON UNIV., N.J. (SQUID PROJECT) (REF. 33)	PR-(NUMBER)-R-P
PRINCETON UNIV., N.J.	PRIN-
PRINCETON UNIV., N.J.	PRIN-U-
PRINCETON UNIV., N.J. (INTERIM PROGRESS REPORTS)	PRINU-IPR-
PRINCETON UNIV., N.J. (QUARTERLY STATUS REPORTS)	PRINU-QSR-
PRINCETON UNIV., N.J. (TECHNICAL REPORTS)	PRINU-TR-
PRINCETON UNIV., N.J.	PRU-
PRINCETON UNIV., N.J.	PU-
PRINCETON UNIV., N.J.	PU-(NO.)/TR-
PRINCETON UNIV., N.J.	PUC-
PRINCETON UNIV., N.J. (FIRE CONTROL RES., FINAL RPTS.)	PU FCR FR-
PRINCETON UNIV., N.J. (FIRE CONTROL RES., RES. RPTS.)	PU FCR RR-
PRINCETON UNIV., N.J.	PU-TR-
PRINCETON UNIV., N.J.	R-
PRINCETON UNIV., N.J.	RB-
PRINCETON UNIV., N.J.	RM-
PRINCETON UNIV., N.J. (STATISTICAL RESEARCH GROUP)	SRG-
PRINCETON UNIV., N.J. (STATISTICAL RESEARCH GROUP)	SRG P-
PRINCETON UNIV., N.J. AERONAUTICAL ENGINEERING LAB.	AEL-
PRINCETON UNIV., N.J. AERONAUTICAL ENGINEERING LAB.	PR-AEL-
PRINCETON UNIV., N.J. AERONAUTICAL ENGINEERING LAB.	PU AEL R-
PRINCETON UNIV., N.J. DEPT. OF AERONAUTICAL ENG.	AER-
PRINCETON UNIV., N.J. DEPT. OF AERONAUTICAL ENG.	AERO. ENG. LAB. NO.-
PRINCETON UNIV., N.J. DEPT. OF AERONAUTICAL ENG.	P.U.A.E.D.-
PRINCETON UNIV., N.J. DEPT. OF AERONAUTICAL ENG.	PU-AEL-
PRINCETON UNIV., N.J. DEPT. OF AERONAUTICAL ENG.	PU-AER-
PRINCETON UNIV., N.J. DEPT. OF AEROSPACE AND MECHANICAL SCIENCES	AMS-
PRINCETON UNIV., N.J. DEPT. OF AEROSPACE AND MECHANICAL SCIENCES	AMS-T-
PRINCETON UNIV., N.J. DEPT. OF AEROSPACE AND MECHANICAL SCIENCES	PU-AMS-
PRINCETON UNIV., N.J. DEPT. OF AEROSPACE AND MECHANICAL SCIENCES	PU-FLD-
PRINCETON UNIV., N.J. DEPT. OF CIVIL + GEOLOGICAL ENG.	GEOLOGICAL ENGINEERING-
PRINCETON UNIV., N.J. DEPT. OF CIVIL + GEOLOGICAL ENG.	SOIL ENG. RES. SER.-
PRINCETON UNIV., N.J. FORRESTAL RESEARCH CENTER	TR-NTI-
PRINCETON UNIV., N.J. FRICK CHEMICAL LAB.	RLT-
PRINCETON UNIV., N.J. GUGGENHEIM LABS. FOR THE AEROSPACE PROPULSION SCIENCES	AMS-
PRINCETON UNIV., N.J. JAMES FORRESTAL RES. CENTER	NR-
PRINCETON UNIV., N.J. PALMER PHYSICAL LAB.	NNK-
PRINCETON UNIV., N.J. PALMER PHYSICAL LAB.	PR-PPL-
PRINCETON UNIV., N.J. PALMER PHYSICAL LAB.	PUC-(YR.)-(NO.)-
PRINCETON UNIV., N.J. PALMER PHYSICAL LAB.	PURC-
PRINCETON UNIV., N.J. PLASMA PHYSICS LAB.	MATT-
PRINCETON UNIV., N.J. PLASMA PHYSICS LAB.	MATT-BIB-
PRINCETON UNIV., N.J. PLASMA PHYSICS LAB.	MATT-TRANS.-
PRINCETON UNIV., N.J. PLASMA PHYSICS LAB.	PPL-AF-
PRINCETON UNIV., N.J. PLASMA PHYSICS LAB.	PPL-NP(NUMBER)
PRINCETON UNIV., N.J. PLASMA PHYSICS LAB.	PPL-TRANS-
PRINCETON UNIV., N.J. PLASTICS LAB.	PR-PL-
PRINCETON UNIV., N.J. PLASTICS LAB.	PU PL TR-
PRINCETON UNIV., N.J. PROJECT MATTERHORN	MATT-
PRINCETON UNIV., N.J. PROJECT MATTERHORN (QUARTERLY)	MATT-Q-
PRINCETON UNIV., N.J. PROJECT MATTERHORN (TECH. MEMO)	MATT-TM-
PRINCETON UNIV., N.J. PROJECT MATTERHORN	PM-
PRINCETON UNIV., N.J. PROJECT MATTERHORN (PRINT-OUTS)	PM-PO-
PRINCETON UNIV., N.J. PROJECT MATTERHORN (QUARTERLY)	PM-Q-
PRINCETON UNIV., N.J. PROJECT MATTERHORN	PM-S-
PRINCETON UNIV., N.J. PROJECT MATTERHORN	PRIN MATT-
(SEE ALSO LATER NAME: PRINCETON UNIV., N.J. PLASMA PHYSICS LAB.)	
PROCEDYNE ASSOCIATES, INC., NEW BRUNSWICK, N.J.	PAI-
PROCEDYNE ASSOCIATES, INC., NEW BRUNSWICK, N.J.	PR-
PROCTOR + GAMBLE DEFENSE CORP.	P+G-
PRODUCTION ENGINEERING RESEARCH ASSN. OF GT. BRIT. INFO. MGR., MELTON MOWBRAY, LEICS, ENG. (TRANSLATIONS)	PERA-
PROPULSION PRODUCTS, SANTA MONICA, CALIF.	PP-
PROPULSION RESEARCH CORP., INGLEWOOD, CALIF.	PRC-
PROPULSION RESEARCH CORP., SANTA MONICA, CALIF.	PRC-
PROTECTION, INC., INGLEWOOD, CALIF.	N(NO.)-R(NO.)
PROTECTIVE STRUCTURES DEV. CENTER, FT. BELVOIR, VA.	PSDC-TR-
PROTEUS, INC., MOUNTAIN LAKES, N.J.	SP-(YEAR)(NO.)F
PROVIDENCE COLL., R.I.	PC-
PSYCHOLOGICAL CORP., N.Y.C.	PC-
PSYCHOLOGICAL CORP., N.Y.C.	PSYC-
PSYCHOLOGICAL RESEARCH ASSOCIATES, WASHINGTON, D.C.	PRA-
PUBLIC HEALTH SERVICE, WASHINGTON, D.C.	PH-M-
PUBLIC HEALTH SERVICE, WASHINGTON, D.C.	PHS-

PUBLIC HEALTH SERVICE

PUBLIC HEALTH SERVICE, WASHINGTON, D.C.	PHS-PUBL.-(NO.)-WP-
PUBLIC HEALTH SERVICE, WASHINGTON, D.C.	RHD-(VOL.)-(NO.)-
PUBLIC HEALTH SERVICE. DIV. OF CHRONIC DISEASES. NATIONAL CLEARINGHOUSE FOR SMOKING + HEALTH	NCSH-
PUBLIC HEALTH SERVICE. DIV. OF RADIOLOGICAL HEALTH, WASH., D.C.(INTERLAB TECHNICAL ADVISORY COMM.)	ITAC-
PUBLIC HEALTH SERVICE. RADIOLOGICAL HEALTH LAB., ROCKVILLE, MD.	TOB-
PUBLIC LAND LAW REVIEW COMMISSION, WASHINGTON, D.C.	PLLRC-(YEAR)-
PUBLIC LAND LAW REVIEW COMMISSION, WASHINGTON, D.C.	PLLRC-STUDY-
PUBLIC SERVICE COMPANY OF COLORADO	PSCC-
PUBLIC SERVICE OF NORTHERN ILL., CHICAGO (JT.VENTURE)	CEPS-
PUERTO RICO. UNIV., MAYAGUEZ	UPR-
PUERTO RICO. UNIV., RIO PIEDRAS	PRU-
PUERTO RICO. UNIV., RIO PIEDRAS. COSMIC RAY LAB.	BULL. NO.-
PUERTO RICO NUCLEAR CENTER, MAYAGUEZ	PRNC-
PUERTO RICO NUCLEAR CENTER, MAYAGUEZ	PRNS-
PUERTO RICO NUCLEAR CENTER, SAN JUAN	PRNC-
PUERTO RICO WATER RESOURCES AUTHORITY, SAN JUAN (JOINT VENTURE)	CEND-PRWA-
PUERTO RICO WATER RESOURCES AUTHORITY, SAN JUAN (JOINT VENTURE)	CEND/PRWRA-
PUERTO RICO WATER RESOURCES AUTHORITY, SAN JUAN	PRWRA-
PUERTO RICO WATER RESOURCES AUTHORITY, SAN JUAN	PRWRA-GNEC-
PUERTO RICO WATER RESOURCES AUTHORITY, SAN JUAN	WRA-B-(YEAR-NUMBER)
PUGET SOUND NAVAL SHIPYARD. CARR INLET ACOUSTIC RANGE, BREMERTON, WASH.	T-
PUGET SOUND NAVAL SHIPYARD. MATERIAL LABS.,BREMERTON, WASH.	PSNS-
PUGET SOUND NAVAL SHIPYARD. QUALITY ASSURANCE DIV., BREMERTON, WASH.	E-
PUGET SOUND UTILITIES COUNCIL	PSUC-
PULLMAN-STANDARD CAR MFG. CO., HAMMOND, IND.	PSCM-
PURDUE RESEARCH FOUNDATION, LAFAYETTE, IND.	A-59-
PURDUE RESEARCH FOUNDATION, LAFAYETTE, IND.	FMTR-
PURDUE RESEARCH FOUNDATION, LAFAYETTE, IND.	PRF-
PURDUE RESEARCH FOUNDATION, LAFAYETTE, IND.	PUR-
PURDUE UNIV., LAFAYETTE, IND.	AC-(NUMBER)-PU
PURDUE UNIV., LAFAYETTE, IND.	AEES-
PURDUE UNIV., LAFAYETTE, IND.	ARC-(NUMBER)-PU
PURDUE UNIV., LAFAYETTE, IND.	F-(YEAR)-
PURDUE UNIV., LAFAYETTE, IND.	FMTR-(YEAR)-
PURDUE UNIV., LAFAYETTE, IND.	GIT-(NUMBER)-PU
PURDUE UNIV., LAFAYETTE, IND.	HTGDL-
PURDUE UNIV., LAFAYETTE, IND.	I-(YEAR)-
PURDUE UNIV., LAFAYETTE, IND.	MICH-(NUMBER)-PU
PURDUE UNIV., LAFAYETTE, IND.	PDU-
PURDUE UNIV., LAFAYETTE, IND.	PDU-F-
PURDUE UNIV., LAFAYETTE, IND.	PDU-I-
PURDUE UNIV., LAFAYETTE, IND.	PDU-PR-
PURDUE UNIV., LAFAYETTE, IND.	PDU-TM-
PURDUE UNIV., LAFAYETTE, IND.	PRD-
PURDUE UNIV., LAFAYETTE, IND.	PU-
PURDUE UNIV., LAFAYETTE, IND.	PUR-
PURDUE UNIV., LAFAYETTE, IND. (SQUID PROJ.) (REF. 33)	PUR-(NUMBER)-(LETTER(S))
PURDUE UNIV., LAFAYETTE, IND.	PURD (LTR.)-
PURDUE UNIV., LAFAYETTE, IND.	PURD-U-
PURDUE UNIV., LAFAYETTE, IND.	PURDUE U.-
PURDUE UNIV., LAFAYETTE, IND.	PURU-
PURDUE UNIV., LAFAYETTE, IND. (SQUID PROJ.)(REF. 33)	SQUID-PUR-(NUMBER)-M
PURDUE UNIV., LAFAYETTE, IND. (SQUID PROJ.) (REF. 33)	SQUID-TR-AC-(NUMBER)-PU
PURDUE UNIV., LAFAYETTE, IND. (SQUID PROJ.) (REF. 33)	SQUID-TR-ARC-(NUMBER)-PU
PURDUE UNIV., LAFAYETTE, IND. (SQUID PROJ.) (REF. 33)	SQUID-TR-CAL-(NO)-PU
PURDUE UNIV., LAFAYETTE, IND. (SQUID PROJ.) (REF. 33)	SQUID-TR-CCNY-(NO.)-PU
PURDUE UNIV., LAFAYETTE, IND. (SQUID PROJ.) (REF. 33)	SQUID-TR-CU-(NUMBER)-PU
PURDUE UNIV., LAFAYETTE, IND. (SQUID PROJ.) (REF. 33)	SQUID-TR-DYN-(NUMBER)-PU
PURDUE UNIV., LAFAYETTE, IND. (SQUID PROJ.) (REF. 33)	SQUID-TR-GIT-(NO)-PU
PURDUE UNIV., LAFAYETTE, IND. (SQUID PROJ.) (REF. 33)	SQUID-TR-ILL-(NUMBER-PU
PURDUE UNIV., LAFAYETTE, IND. (SQUID PROJ.) (REF. 33)	SQUID-TR-MIT-(NO)-PU
PURDUE UNIV., LAFAYETTE, IND. (SQUID PROJ.) (REF. 33)	SQUID-TR-NYU-(NUMBER)P
PURDUE UNIV., LAFAYETTE, IND. (SQUID PROJ.) (REF. 33)	SQUID-TR-PR-(NUMBER)-PU
PURDUE UNIV., LAFAYETTE, IND. (SQUID PROJ.) (REF. 33)	SQUID-TR-RICE-(NO)-PU
PURDUE UNIV., LAFAYETTE, IND. (SQUID PROJ.) (REF. 33)	SQUID-TR-SRI-(NUMBER)-PU
PURDUE UNIV.,LAFAYETTE,IND. DEPT. OF COMPUTER SCIENCE	CSD-TR-
PURDUE UNIV., LAFAYETTE, IND. DEPT. OF STATISTICS	MIMEOGRAPHIC SERIES-
PURDUE UNIV., LAFAYETTE, IND. JET PROPULSION CENTER	JPC-
PURDUE UNIV., LAFAYETTE, IND. JET PROPULSION CENTER	PC-
PURDUE UNIV., LAFAYETTE, IND. JET PROPULSION CENTER	PURDUE-JPC-
PURDUE UNIV., LAFAYETTE, IND. JET PROPULSION CENTER, SCHOOL OF MECHANICAL ENGINEERING (PROJECT SQUID)	DYN-
PURDUE UNIV., LAFAYETTE,IND. JOINT HIGHWAY RES. PROJ.	JHRP-
PURDUE UNIV., LAFAYETTE, IND. PURDUE RESEARCH ORGANIZATION FOR VULNERABILITY EVALUATIONS	PROVE-
PURDUE UNIV., LAFAYETTE, IND. SCHOOL OF AERONAUTICAL AND ENGINEERING SCIENCES	A + ES-(NO.)-(NO.)-
PURDUE UNIV., LAFAYETTE, IND. SCHOOL OF AERONAUTICAL AND ENGINEERING SCIENCES	AES-
PURDUE UNIV., LAFAYETTE, IND. SCHOOL OF AERONAUTICAL AND ENGINEERING SCIENCES	A + ES-(YEAR)-
PURDUE UNIV., LAFAYETTE, IND. SCHOOL OF AERO. ENG.	A-59-
PURDUE UNIV., LAFAYETTE, IND. SCHOOL OF AERONAUTICS	A-(NOS.)(PU)
PURDUE UNIV., LAFAYETTE, IND. SCHOOL OF AERONAUTICS, ASTRONAUTICS AND ENGINEERING SCIENCES	AA/ES-
PURDUE UNIV., LAFAYETTE, IND. SCHOOL OF AERONAUTICS, ASTRONAUTICS AND ENGINEERING SCIENCES	AA+ES-(YEAR)-
PURDUE UNIV., LAFAYETTE, IND. SCHOOL OF AERONAUTICS, ASTRONAUTICS AND ENGINEERING SCIENCES	A/ES-

PURDUE UNIV., LAFAYETTE, IND. SCHOOL OF AERONAUTICS, ASTRONAUTICS AND ENGINEERING SCIENCES	A/ES-(YEAR)-
PURDUE UNIV., LAFAYETTE, IND. SCHOOL OF ELEC. ENG.	PDU-TR-EE-
PURDUE UNIV., LAFAYETTE, IND. SCHOOL OF ELEC. ENG.	TR-EE-
PURDUE UNIV., LAFAYETTE, IND. SCHOOL OF ELEC. ENG.	TR-EE-(YEAR)-
PURDUE UNIV., LAFAYETTE, IND. SCHOOL OF MECH. ENG.	WLRAF-
PURDUE UNIV., LAFAYETTE, IND. THERMOPHYSICAL PROPERTIES RESEARCH CENTER	NSRDS-NBS-
PURDUE UNIV., LAFAYETTE, IND. THERMOPHYSICAL PROPERTIES RESEARCH CENTER	TPRC-
PYROGENICS, INC., WOODSIDE, N.Y.	J-(NO.)(LTR.)-
PYROMET CO., SAN CARLOS, CALIF.	PYRO-

QUALITY CONTROL SURVEILLANCE LAB.

QUALITY CONTROL SURVEILLANCE LAB.	QCSL-
QUARTERMASTER ANALYSIS SECTION (AIR FORCE)	QMAS-
QUARTERMASTER BOARD, CAMP LEE, VA.	QMB-
QUARTERMASTER BD., CAMP LEE, VA. (FIELD SURVEY PROJ.)	QMB PROJ FR-
QUARTERMASTER BOARD, CAMP LEE, VA. (MISC. PROJ.)	QMB PROJ M-
QUARTERMASTER BOARD, CAMP LEE, VA. (SURVEY PROJ.)	QMB PROJ S-
QUARTERMASTER BOARD, CAMP LEE, VA. (TEST PROJ.)	QMB PROJ T-
QUARTERMASTER BOARD, CAMP LEE, VA. (SURVEY)	QMB-S-
QUARTERMASTER BOARD, CAMP LEE, VA. (TEST REPORT)	QMB-T-
QUARTERMASTER BOARD, CAMP LEE, VA. (TEST)	QMBT-
QUARTERMASTER CORPS (HISTORICAL SURVEYS)	HS-
QUARTERMASTER CORPS (LEATHER SERIES REPORT)	LETOR-
QUARTERMASTER CORPS (MICROBIOLOGICAL SERIES REPORTS)	MSR-
QUARTERMASTER CORPS (RESEARCH, PURCHASE ORDER)	PO-
QUARTERMASTER CORPS	QMC-
QUARTERMASTER CORPS (LEATHER SERIES REPORT)	QMC 17-
QUARTERMASTER CORPS (HISTORICAL SURVEYS)	QMC HS-
QUARTERMASTER CORPS (LEATHER SERIES REPORT)	QMC LSR-
QUARTERMASTER CORPS (MANUAL)	QMC M-
QUARTERMASTER CORPS (MASS. INST. OF TECH. PROJECTS)	QMC MIT-
QUARTERMASTER CORPS (MICROBIOLOGICAL SERIES REPORTS)	QMC MSR-
QUARTERMASTER CORPS	QMC PROJ R-
QUARTERMASTER CORPS	QMC PROJ S-
QUARTERMASTER CORPS (RESEARCH, PURCHASE ORDER)	QMCR PO-
QUARTERMASTER CORPS (RES. RPT., BIOCHEMICAL SERIES)	QMC RR BS-
QUARTERMASTER CORPS (RES. RPT., ENTOMOLOGY SERIES)	QMC RR ES-
QUARTERMASTER CORPS (SPECIFICATION)	QMC SPEC-
QUARTERMASTER CORPS (TECHNICAL RECONNAISSANCE MEMO.)	QMC TRM-
QUARTERMASTER CORPS (TENT RESEARCH REPORTS)	QMC TRR-
QUARTERMASTER CORPS (TEXTILE SERIES REPORTS)	QMC TSR-
QUARTERMASTER CORPS	TM-10-
QUARTERMASTER CORPS (TEXTILE RECONNAISSANCE MEMO.)	TRM-
QUARTERMASTER CORPS (TEXTILE SERIES REPORTS)	TRR-
QUARTERMASTER CORPS (TEXTILE SERIES REPORTS)	TSR-
QUARTERMASTER CORPS, WASHINGTON, D.C.	OQMG-
QUARTERMASTER CORPS. CHEMICALS AND PLASTICS SECTION. (RESEARCH SERVICE TEST REPORT)	QMC RSTR C+P-
QUARTERMASTER CORPS. TECHNICAL LIBRARY (BIBLIOGRAPHIC SERIES)	QMC TL BS-
QUARTERMASTER DEPOT, JEFFERSONVILLE, IND.	JQMD-
QUARTERMASTER DEPOT, JEFFERSONVILLE, IND.	QMD JEFFERSONVILLE-
QUARTERMASTER DEPOT, PHILADELPHIA	PQMD-
QUARTERMASTER DEPOT, PHILADELPHIA	QMD PHILA-
QUARTERMASTER FOOD AND CONTAINER INSTITUTE FOR THE ARMED FORCES, CHICAGO	FCI-
QUARTERMASTER FOOD AND CONTAINER INSTITUTE FOR THE ARMED FORCES, CHICAGO	QF+CI-
QUARTERMASTER FOOD AND CONTAINER INSTITUTE FOR THE ARMED FORCES, CHICAGO	QMCAL-
QUARTERMASTER FOOD AND CONTAINER INSTITUTE FOR THE ARMED FORCES, CHICAGO	QMC FCI-
QUARTERMASTER FOOD AND CONTAINER INSTITUTE FOR THE ARMED FORCES, CHICAGO (OPERATION STUDIES)	QMC FCI OS-
QUARTERMASTER FOOD AND CONTAINER INSTITUTE FOR THE ARMED FORCES, CHICAGO (RESEARCH AND DEV. BR.)	QMC FCI RDB-
QUARTERMASTER FOOD AND CONTAINER INSTITUTE FOR THE ARMED FORCES, CHICAGO	QMFCI-
QUARTERMASTER FOOD AND CONTAINER INSTITUTE FOR THE ARMED FORCES, CHICAGO	QMFCIAF-
QUARTERMASTER FOOD AND CONTAINER INSTITUTE FOR THE ARMED FORCES, CHICAGO	S-
QUARTERMASTER RESEARCH + DEV. CTR., NATICK, MASS.	EP-
QUARTERMASTER RESEARCH + DEV. CTR., NATICK, MASS.	QRDC-
QUARTERMASTER RESEARCH + DEV. CTR., NATICK, MASS.	TSR-
QUARTERMASTER RESEARCH AND DEVELOPMENT CENTER. CLIMATIC RESEARCH LAB., LAWRENCE, MASS.	CRL-
QUARTERMASTER RESEARCH AND DEVELOPMENT CENTER. CLIMATIC RESEARCH LAB., LAWRENCE, MASS. (MEMO RPTS.)	CRL M-
QUARTERMASTER RESEARCH AND DEVELOPMENT CENTER. CLIMATIC RESEARCH LAB., LAWRENCE, MASS.	CRL R (NUMBER(S))-
QUARTERMASTER RESEARCH AND DEVELOPMENT CENTER. CLIMATIC RESEARCH LAB., LAWRENCE, MASS.	CRLR-
QUARTERMASTER RESEARCH AND DEVELOPMENT CENTER. CLIMATIC RESEARCH LAB., LAWRENCE, MASS.	CRL R (LETTER(S))-
QUARTERMASTER RESEARCH AND DEVELOPMENT CENTER. ENVIRONMENTAL PROTECTION BRANCH, NATICK, MASS.	QRDC-EP-
QUARTERMASTER RES. + DEV. COMMAND, NATICK, MASS.	QRDC B-
QUARTERMASTER RES. + DEV. COMMAND, NATICK, MASS.	QRDC T-
QUARTERMASTER RES. + DEV. COMMAND, NATICK, MASS.	QRDC TSR-
QUARTERMASTER RESEARCH AND DEV. LABS., PHILADELPHIA	BIB-SER-
QUARTERMASTER RESEARCH AND DEV. LABS., PHILADELPHIA	ORDL-
QUARTERMASTER RESEARCH AND DEV. LABS., PHILADELPHIA	TDFL-
QUARTERMASTER RESEARCH AND ENGINEERING COMMAND, NATICK, MASS. (ANALYTICAL SERIES)	AS-
QUARTERMASTER RESEARCH AND ENGINEERING COMMAND, NATICK, MASS. (BIOPHYSICS)	BP-
QUARTERMASTER RESEARCH AND ENGINEERING COMMAND, NATICK, MASS. (CHEMICALS AND PLASTICS)	CP-
QUARTERMASTER RESEARCH AND ENGINEERING COMMAND, NATICK, MASS. (ENVIRONMENTAL ANALYSIS)	EA-
QUARTERMASTER RESEARCH AND ENGINEERING COMMAND, NATICK, MASS. (EXPLOITATION OF FOREIGN MATERIEL)	EFM-
QUARTERMASTER RESEARCH AND ENGINEERING COMMAND, NATICK, MASS. (ENVIRONMENT PROTECTION)	EP-
QUARTERMASTER RESEARCH AND ENGINEERING COMMAND, NATICK, MASS. (ENGINEERING RESEARCH STUDY)	ERS-
QUARTERMASTER RESEARCH AND ENG. COMMAND, NATICK, MASS., (FIRE, WATER, WEATHER, MILDEW RESISTANT)	FWWMR-
QUARTERMASTER RESEARCH AND ENGINEERING COMMAND, NATICK, MASS. (HIGH POLYMER)	HP-
QUARTERMASTER RESEARCH AND ENGINEERING COMMAND, NATICK, MASS. (METHODS RESEARCH STUDY)	MRS-
QUARTERMASTER RESEARCH AND ENGINEERING COMMAND, NATICK, MASS. (MICRO BIOLOGY SERIES)	MS-
QUARTERMASTER RESEARCH AND ENGINEERING COMMAND, NATICK, MASS. (PIONEERING RESEARCH)	PR-
QUARTERMASTER RESEARCH + ENG. COMMAND, NATICK, MASS.	QMR+E-
QUARTERMASTER RESEARCH + ENG. COMMAND, NATICK, MASS.	QREC-
QUARTERMASTER RESEARCH + ENG. COMMAND, NATICK, MASS.	QR+EC EP-
QUARTERMASTER RESEARCH + ENG. COMMAND, NATICK, MASS.	QREC-RER-
QUARTERMASTER RESEARCH + ENG. COMMAND, NATICK, MASS.	QREC/SWPBR-
QUARTERMASTER RESEARCH + ENG. COMMAND, NATICK, MASS.	QREC-T-
QUARTERMASTER RESEARCH + ENG. COMMAND, NATICK, MASS.	QREC/TDL-
QUARTERMASTER RESEARCH AND ENGINEERING COMMAND, NATICK, MASS. (SPECIAL TEXT)	ST-
QUARTERMASTER RESEARCH AND ENGINEERING COMMAND, NATICK, MASS.	T-
QUARTERMASTER RESEARCH AND ENGINEERING COMMAND, NATICK, MASS. (TEXTILE SERIES)	TS-
QUARTERMASTER RESEARCH AND ENGINEERING COMMAND. CHEMICALS AND PLASTICS DIV., NATICK, MASS. (ELASTOMER BRANCH REPORT)	QREL-EBR-
QUARTERMASTER RESEARCH AND ENGINEERING COMMAND. ENVIRONMENTAL PROTECTION BRANCH, NATICK, MASS.	QREC-EP-
QUARTERMASTER RESEARCH AND ENGINEERING COMMAND. MECHANICAL ENGINEERING DIV., NATICK, MASS.	MED-

QUARTERMASTER RESEARCH AND ENGINEERING COMMAND. PSYCHOLOGY BRANCH, NATICK, MASS.	PB-
QUARTERMASTER RESEARCH AND ENGINEERING COMMAND. TEXTILE ENGINEERING LAB., NATICK, MASS.	TEL-
QUARTERMASTER RESEARCH AND ENGINEERING COMMAND TEXTILE FUNCTIONAL FINISHES LAB., NATICK, MASS.	TFFL-
QUARTERMASTER RESEARCH AND ENGINEERING FIELD EVALUATION AGENCY, FORT LEE, VA.	FEA-
QUARTERMASTER RESEARCH AND ENGINEERING FIELD EVALUATION AGENCY, FORT LEE, VA.	FEA-ABN-
QUEEN'S UNIV., KINGSTON, ONTARIO	RR-(YEAR)-
QUEENS UNIV., BELFAST	CVD/QU-
QUEENS UNIV., BELFAST	QU-

R-C SCIENTIFIC INSTRUMENT CO.

R-C SCIENTIFIC INSTRUMENT CO., PLAZA DEL REY, CALIF.	RC-
RADAR EVALUATION SQUADRON (4754TH), HILL AFB, UTAH	C-
RADAR EVALUATION SQUADRON (4754TH), HILL AFB, UTAH	Z-
RADD ASSOCIATES, PALO ALTO, CALIF.	RA-
RADFORD ARSENAL, VA.	RA-
RADFORD ARSENAL, VA.	RAD-
RADIATION APPLICATIONS, INC., LONG ISLAND CITY, N.Y.	RAI-
RADIATION APPLICATIONS, INC., N.Y.C.	RAI-
RADIATION, INC., MELBOURNE, FLA.	RAI-
RADIATION, INC., MELBOURNE, FLA.	RI-(NUMBER)(MONTH)(YEAR)
RADIATION, INC., MELBOURNE, FLA.	RI-IER-
RADIATION, INC., ORLANDO, FLA.	RAI-
RADIATION RESEARCH ASSOCIATES, INC., FORT WORTH, TEX.	RRA-
RADIATION RESEARCH ASSOCIATES, INC., FORT WORTH, TEX.	RRA-C-
RADIATION RESEARCH ASSOCIATES, INC., FORT WORTH, TEX.	RRA-M-
RADIATION RESEARCH ASSOCIATES, INC., FORT WORTH, TEX.	RRA-T(NO.)
RADIATION RESEARCH CORP., N.Y.C.	RRC-
RADIATION RESEARCH CORP., WEST PALM BEACH, FLA.	RRC-
RADIATION SYSTEMS, INC., ALEXANDRIA, VA.	TR-(NUMBER)(LETTER)
RADIO CORP. OF AMERICA (ALL LOCATIONS)	CR-(YEAR)-
RADIO CORP. OF AMERICA (ALL LOCATIONS)	CR-
RADIO CORP. OF AMERICA (ALL LOCATIONS)	RCA-
RADIO CORP. OF AMERICA (ALL LOCATIONS)	RCA-CR-(YEAR)-
RADIO CORP. OF AMERICA (ALL LOCATIONS) (ENG.REPT.)	RCA ER-
RADIO CORP. OF AMERICA (ALL LOCATIONS) (INSTRUC.BK.)	RCA IB-
RADIO CORP. OF AMERICA (ALL LOCATIONS)	RCA-MO-
RADIO CORP. OF AMERICA (ALL LOCATIONS)	RCA-QTR-
RADIO CORP. OF AMERICA, BURLINGTON, MASS.	PP-
RADIO CORP. OF AMERICA, CAMDEN, N.J.	A-
RADIO CORP. OF AMERICA, CAMDEN, N.J.	MRO-(YR.)-(NO.)-(NO.)-
RADIO CORP. OF AMERICA, CAMDEN, N.J.	RCA-CA-
RADIO CORP. OF AMERICA, LANCASTER, PENNA.	AOO-
RADIO CORP. OF AMERICA, LANCASTER, PENNA.	NAA-SR-
RADIO CORP. OF AMERICA, MOORESTOWN, N.J.	DAMP-TM-
RADIO CORP. OF AMERICA, MOORESTOWN, N.J.	MD-S-
RADIO CORP. OF AMERICA, MOORESTOWN, N.J.	MOEES-(YEAR)-
RADIO CORP. OF AMERICA, MOORESTOWN, N.J.	RATM-
RADIO CORP. OF AMERICA, MOORESTOWN, N.J.	RCA-MLPR-
RADIO CORP. OF AMERICA, MOORESTOWN, N.J.	RCA-MPR-
RADIO CORP. OF AMERICA, MOORESTOWN, N.J. (TECH.SPEC.)	RCA-TSP-
RADIO CORP. OF AMERICA, N.Y.C.	RCS-OSD-
RADIO CORP. OF AMERICA, VAN NUYS, CALIF.	TPI-
RADIO CORP. OF AMERICA. AEROSPACE SYSTEMS DIV., BURLINGTON, MASS.	ATE-L-
RADIO CORP. OF AMERICA. AEROSPACE SYSTEMS DIV., BURLINGTON, MASS.	ATE-MS-
RADIO CORP. OF AMERICA. AEROSPACE SYSTEMS DIV., BURLINGTON, MASS.	ATE-MTE-L-
RADIO CORP. OF AMERICA, APPLIED RESEARCH, CAMDEN,N.J.	NMS-TRANS-
RADIO CORP. OF AMERICA. ASTRO-ELECTRONICS DIV., PRINCETON, N.J.	AED-R-
RADIO CORP. OF AMERICA. DEFENSE ELECTRONIC PRODUCTS (ALL LOCATIONS) (ENGINEERING MEMORANDUM)	RCA-EM-(YEAR)-(NUMBER)-
RADIO CORP. OF AMERICA. DEFENSE ELECTRONIC PRODUCTS (ALL LOCATIONS) (TECHNICAL REPORT)	RCA TR-(YEAR)-(NUMBER)-
RADIO CORP. OF AMERICA. DEFENSE ELECTRONIC PRODUCTS, BURLINGTON, MASS. (MULTISYSTEM TEST EQUIPMENT)	ATE-MTE-
RADIO CORP. OF AMERICA. DEFENSE ELECTRONIC PRODUCTS, BURLINGTON, MASS. (MULTISYSTEM TEST EQUIPMENT)	RCA-ATE-MTE-
RADIO CORP. OF AMERICA. DEFENSE ELECTRONIC PRODUCTS, N.Y.C.	MO-S-
RADIO CORP. OF AMERICA. ELECTROMAGNETIC AND AVIATION SYSTEMS DIV., VAN NUYS, CAL.	TP-
RADIO CORP. OF AMERICA. HOME INSTRUMENTS DIV., INDIANAPOLIS (ENGINEERING MEMORANDUM)	RCA-EM-5001 THRU -5300
RADIO CORP. OF AMERICA. HOME INSTRUMENTS DIV., INDIANAPOLIS (TECHNICAL REPORT)	RCA-TR-2001 THRU -2100
RADIO CORP. OF AMERICA. MISSILE AND SURFACE RADAR DIV., MOORESTOWN, N.J.	MO-
RADIO CORP. OF AMERICA. MISSILE AND SURFACE RADAR DEPT., MOORESTOWN, N.J.	MO596R-S-
RADIO CORP. OF AMERICA. MISSILE AND SURFACE RADAR DIV., MOORESTOWN, N.J.	MO-BB-S-
RADIO CORP. OF AMERICA. MISSILE AND SURFACE RADAR DIV., MOORESTOWN, N.J.	MOBBS-
RADIO CORP. OF AMERICA. MISSILE AND SURFACE RADAR DIV., MOORESTOWN, N.J.	MO-S-
RADIO CORP. OF AMERICA. MISSILE AND SURFACE RADAR DEPT., MOORESTOWN, N.J.	RCA MO-(LTR.)-(LTR.)-
RADIO CORP. OF AMERICA. MISSILE AND SURFACE RADAR DIV., MOORESTOWN, N.J.	RCA-MSRD-QTR-
RADIO CORP. OF AMERICA. MISSILE AND SURFACE RADAR DIV., MOORESTOWN, N.J.	TTR-
RADIO CORP. OF AMERICA. RCA LABS. DIV., PRINCETON, NJ	C-
RADIO CORP. OF AMERICA. RCA LABS. DIV., PRINCETON, NJ	CM-
RADIO CORP. OF AMERICA. RCA LABS. DIV., PRINCETON, NJ	LB-
RADIO CORP. OF AMERICA. RCA LABS. DIV., PRINCETON, NJ	PEM-
RADIO CORP. OF AMERICA. RCA LABS. DIV., PRINCETON, NJ	PTR-
RADIO CORP. OF AMERICA. RCA LABS. DIV., PRINCETON, NJ	RCAL-
RADIO CORP. OF AMERICA. RCA LABS. DIV., PRINCETON, NJ (LICENSEE BULLETIN)	RCA-LB-
RADIO CORP. OF AMERICA. RCA LABS. DIV., PRINCETON, NJ	RCA-PEM-
RADIO CORP. OF AMERICA. RCA LABS. DIV., PRINCETON, NJ	RCA-PTR-
RADIO CORP. OF AMERICA. RCA LABS. DIV., PRINCETON, NJ	ROB-NE-
(SEE ALSO: DAVID SARNOFF RESEARCH CENTER)	
RADIO CORP. OF AMERICA. RCA LABS.DIV.,ROCKY POINT,NY	RCA-CM-
RADIO CORP. OF AMERICA. SYSTEMS ENGINEERING EVALUATION AND RESEARCH, MOORESTOWN, N.J.	MOEES-(YEAR)-
RADIO CORP. OF AMERICA. WEST COAST DIV.,VAN NUYS,CAL.	MLO-(YEAR)-
RADIO CORP. OF AMERICA. WEST COAST DIV.,VAN NUYS,CAL.	TPI-

RADIO ELECTRONICS + TELEVISION MFRS. ASSN., WASH., DC	RETMA-
RADIO RECEPTOR CO. INC., BROOKLYN	RARC-
RADIO RESEARCH STATION, SLOUGH, BUCKS, ENGLAND	RRB/C-
RADIO TECH. COMMISSION FOR AERONAUTICS, WASH., D.C.	DO-(3 DIGITS)
RADIO TECH. COMMISSION FOR AERONAUTICS, WASH., D.C.	RTCA-
RADIOPLANE. DIV. OF NORTHROP CORP., VAN NUYS, CALIF.	NOR-
RADIOPLANE CO., VAN NUYS, CALIF.	RC-
RADIOPLANE CO., VAN NUYS, CALIF.	RC-R-
RADIOPLANE CO., VAN NUYS, CALIF.	RC-RPE-(YEAR)-
RADIOPLANE CO., VAN NUYS, CALIF.	RP-
RADIOPLANE CO., VAN NUYS, CALIF.	RPC-
RADIOPLANE CO. RECOVERY DIV., VAN NUYS, CALIF.	R-
RAI RESEARCH CORP., LONG ISLAND CITY, N.Y.	RAI-
RAMO-WOOLDRIDGE. DIV. OF THOMPSON RAMO WOOLDRIDGE INC., CANOGA PARK, CALIF.	RW-RL-
RAMO-WOOLDRIDGE CORP., LOS ANGELES	AM-(NO.)-
RAMO-WOOLDRIDGE CORP., LOS ANGELES	AP-
RAMO-WOOLDRIDGE CORP., LOS ANGELES	CMCC-
RAMO-WOOLDRIDGE CORP., LOS ANGELES	CMCC-GM-
RAMO-WOOLDRIDGE CORP., LOS ANGELES	R-W-
RAMO-WOOLDRIDGE CORP., LOS ANGELES	RW-
RAMO-WOOLDRIDGE CORP., LOS ANGELES (ACCESSION LIST)	RW-AL-
RAMO WOOLDRIDGE CORP., LOS ANGELES	RWC-
RAMO-WOOLDRIDGE CORP., LOS ANGELES	RWC-AM-
RAMO-WOOLDRIDGE CORP., LOS ANGELES	RW-WDD-
RAMO-WOOLDRIDGE CORP., LOS ANGELES	WDT-
RAMO-WOOLDRIDGE CORP. AERO. RES. LAB., LOS ANGELES	ARL-
RAMO-WOOLDRIDGE CORP. AERO. RES. LAB., LOS ANGELES	R-W-ARL-
RAMO-WOOLDRIDGE CORP. CONTROL SYSTEMS DIV., LOS ANG.	RW-(NUMBER).(NUMBER)
RAMO-WOOLDRIDGE CORP. DENVER LABS.	RW-DL-
RAMO-WOOLDRIDGE CORP. DENVER LABS.	RW-DL-PR-
RAMO-WOOLDRIDGE CORP. ELECTRONIC RES. LAB., LOS ANG.	ERL-
RAMO-WOOLDRIDGE CORP. ELECTRONIC RES. LAB., LOS ANG.	FRL-LM-
RAMO-WOOLDRIDGE CORP. ELECTRONIC RES. LAB., LOS ANG.	GM-TR-(NUMBER)-
RAMO-WOOLDRIDGE CORP. GUIDED MISSILE RESEARCH DIV., LOS ANGELES	D-(YEAR)-
RAMO-WOOLDRIDGE CORP. GUIDED MISSILE RESEARCH DIV., LOS ANGELES	GM-
RAMO-WOOLDRIDGE CORP. GUIDED MISSILE RESEARCH DIV., LOS ANGELES	GM-(NUMBER.NUMBER)-
RAMO-WOOLDRIDGE CORP. GUIDED MISSILE RESEARCH DIV., LOS ANGELES	GMRD-R/W-
RAMO-WOOLDRIDGE CORP. GUIDED MISSILE RESEARCH DIV., LOS ANGELES	GM-TR-
RAMO-WOOLDRIDGE CORP. GUIDED MISSILE RESEARCH DIV., LOS ANGELES	R-W-(NO.)-TR
RAMO-WOOLDRIDGE CORP. GUIDED MISSILE RESEARCH DIV., LOS ANGELES	R/W GM-
RAMO-WOOLDRIDGE CORP. GUIDED MISSILE RESEARCH DIV., LOS ANGELES	RW-GM(NO.)-
RAMO-WOOLDRIDGE CORP. GUIDED MISSILE RESEARCH DIV., LOS ANGELES	RW-GMRD-
RAMO-WOOLDRIDGE CORP. GUIDED MISSILE RESEARCH DIV., LOS ANGELES	RW-GM-TR-(NO.)-
RAMO-WOOLDRIDGE CORP. SPACE TECHNOLOGY LABS., LOS ANGELES	STL-IBM-V.
RAMO-WOOLDRIDGE DIV., THOMPSON RAMO WOOLDRIDGE CORP., CANOGA PARK, CALIF.	RWD-RL-
RAMO-WOOLDRIDGE DIV., THOMPSON RAMO WOOLDRIDGE CORP., CANOGA PARK, CALIF.	RWD-RLM-
RAND CORP., SANTA MONICA, CALIF.	AM-
RAND CORP., SANTA MONICA, CALIF. (BRIEFING)	B-
RAND CORP., SANTA MONICA, CALIF.	CSM-
RAND CORP., SANTA MONICA, CALIF.	D-
RAND CORP., SANTA MONICA, CALIF. (REF. 47)	D-(NUMBER)-(LETTERS)
RAND CORP., SANTA MONICA, CALIF.	D(L)-(NUMBER)-AEC
RAND CORP., SANTA MONICA, CALIF.	LT-
RAND CORP., SANTA MONICA, CALIF. (TRANSLATION)	LT-(YEAR)-
RAND CORP., SANTA MONICA, CALIF.	P-(NO.)-AEC
RAND CORP., SANTA MONICA, CALIF.	P-(NO.)-(RAND)
RAND CORP., SANTA MONICA, CALIF.	P-(NO.)-ARPA
RAND CORP., SANTA MONICA, CALIF. (REF. 47)	P-(4 DIGITS)
RAND CORP., SANTA MONICA, CALIF.	R-
RAND CORP., SANTA MONICA, CALIF. (REF. 47)	R-(NUMBER)-(LETTERS)
RAND CORP., SANTA MONICA, CALIF. (REF. 47)	R-(NUMBER)-AEC
RAND CORP., SANTA MONICA, CALIF. (REF. 47)	R-(NUMBER)-RAND
RAND CORP., SANTA MONICA, CALIF. (REF. 47)	R-(NUMBER)-PR
RAND CORP., SANTA MONICA, CALIF.	RA-
RAND CORP., SANTA MONICA, CALIF.	RAD-
RAND CORP., SANTA MONICA, CALIF.	RAND-
RAND CORP., SANTA MONICA, CALIF. (ANNUAL REPORT)	RAND-AR-(YEAR)
RAND CORP., SANTA MONICA, CALIF.	RAND-B-
RAND CORP., SANTA MONICA, CALIF.	RAND-P-
RAND CORP., SANTA MONICA, CALIF.	RAND(P)-
RAND CORP., SANTA MONICA, CALIF.	RAND-R-
RAND CORP., SANTA MONICA, CALIF.	RAND-RA-
RAND CORP., SANTA MONICA, CALIF.	RAND-RAOP-
RAND CORP., SANTA MONICA, CALIF. (TRANSLATION)	RAND-RAT-
RAND CORP., SANTA MONICA, CALIF.	RAND RM-
RAND CORP., SANTA MONICA, CALIF.	RAND-S-
RAND CORP., SANTA MONICA, CALIF.	RAND-T-
RAND CORP., SANTA MONICA, CALIF. (TRANSLATION)	RAOP-
RAND CORP., SANTA MONICA, CALIF.	RAT-1 THRU RAT-146
RAND CORP., SANTA MONICA, CALIF. (REF. 47)	RC-
RAND CORP., SANTA MONICA, CALIF.	RM-
RAND CORP., SANTA MONICA, CALIF.	RM-(NUMBER)-AEC
RAND CORP., SANTA MONICA, CALIF. (REF. 47)	RM-(NUMBER)-ARPA
RAND CORP., SANTA MONICA, CALIF.	RM-(NUMBER)-ESSA
RAND CORP., SANTA MONICA, CALIF.	RM-(NUMBER)-RAND
RAND CORP., SANTA MONICA, CALIF. (REF. 47)	RM-(NUMBER)-PR-
RAND CORP., SANTA MONICA, CALIF.	RM-(NUMBER)-OEO
RAND CORP., SANTA MONICA, CALIF.	RM-(NUMBER)-NRL
RAND CORP., SANTA MONICA, CALIF.	R-P-
RAND CORP., SANTA MONICA, CALIF.	RPA-
RAND CORP., SANTA MONICA, CALIF.	S-
RAND CORP., SANTA MONICA, CALIF.	S-(NUMBER)-RAND
RAND CORP., SANTA MONICA, CALIF. (CLASSIFIED PAPER)	S-1 THRU S-152

RAND CORP.

RAND CORP., SANTA MONICA, CALIF.	S-(NO.)-ARPA
RAND CORP., SANTA MONICA, CALIF.	SB-
RAND CORP., SANTA MONICA, CALIF. (SPECIAL MEMO.)	SM-
RAND CORP., SANTA MONICA, CALIF.	S-(RAND)
RAND CORP., SANTA MONICA, CALIF.	T-(NO.)(RAND)
RAND CORP., SANTA MONICA, CALIF. (TRANSLATION)	T-1 THRU T-146
RAND CORP., SANTA MONICA, CALIF.	TRC-
RAND CORP., SANTA MONICA, CALIF.	TRC-R-
RAND CORP., SANTA MONICA, CALIF.	TRC-RA-
RAND CORP., SANTA MONICA, CALIF.	TRC-RM-
RAND CORP., SANTA MONICA, CALIF.	TRC-S-
RAND CORP., SANTA MONICA, CALIF.	TT-
RAND CORP., SANTA MONICA, CALIF.	URM-
RAND CORP., WASHINGTON, D.C.	RM-(NUMBER)-NSF
RAND INST., N.Y.C.	R-(NUMBER)-NYC
RARITAN ARSENAL, METUCHEN, N.J.	RARA-
RASOR (NED S.), DAYTON, OHIO	NSR-
RAULAND CORP., CHICAGO	DC-
RAVEN INDUSTRIES, INC., SIOUX FALLS, S.D.	R-
RAVEN INDUSTRIES, INC., SIOUX FALLS, S.D.	RII-
RAW MATERIALS OPERATIONS OFF.,AEC, WASH.,D.C.(REF.4)	RMO-
RAW MATERIALS OPERATIONS OFF.,AEC, WASH.,D.C.(REF.4)	RMOO-
RAYMOND ENGINEERING LAB., INC., MIDDLETOWN, CONN.	REL-
RAYTHEON CO. (ALL LOCATIONS)	BR-
RAYTHEON CO. (ALL LOCATIONS)	RAYTHEON-
RAYTHEON CO. (ALL LOCATIONS)	RC-
RAYTHEON CO. (ALL LOCATIONS)	RMC-
RAYTHEON CO., ANDOVER, MASS.	ANO(NUMBER)
RAYTHEON CO., ANDOVER, MASS.	ARG-
RAYTHEON CO., BURLINGTON, MASS.	PT-
RAYTHEON CO., NORWOOD, MASS.	AMR-
RAYTHEON CO., NORWOOD, MASS.	CDP-TR-
RAYTHEON CO., WALTHAM, MASS.	S-
RAYTHEON CO. AUTOMETRIC OPERATION, ALEXANDRIA, VA.	FR-(YEAR)-
RAYTHEON CO. AUTOMETRIC OPERATION, ALEXANDRIA, VA.	TP/S-
RAYTHEON CO. EQUIPMENT DIV. HDQTRS., WALTHAM, MASS.	EDH-
RAYTHEON CO. HIGH TEMP. MATERIALS DEPT.,WALTHAM, MASS	RC-S-
RAYTHEON CO. MARINE RESEARCH LAB., NEW LONDON, CONN.	SUBDEVGRUTWO-(NO.-YR.)
RAYTHEON CO. MICROWAVE AND POWER TUBE DIV., WALTHAM, MASS.	PT-
RAYTHEON CO. MICROWAVE POWER OPERATION, BURLINGTON, MASS.	MPO-
RAYTHEON CO. MISSILE AND SPACE DIV., BEDFORD, MASS.	RC-BR-
RAYTHEON CO. MISSILE SYSTEMS DIV., BEDFORD, MASS.	BR-S-
RAYTHEON CO. MISSILE SYSTEMS DIV., BEDFORD, MASS.	RC-BR-
RAYTHEON CO. MISSILE SYSTEMS DIV., BEDFORD, MASS.	RC-S-
RAYTHEON CO. RADAR AND CONTROL SYSTEMS DEPT., WAYLAND, MASS.	TAGS-
RAYTHEON CO. RADAR AND CONTROL SYSTEMS DEPT., WAYLAND, MASS.	TAGS-(YEAR)-
RAYTHEON CO. RESEARCH DIV., WALTHAM, MASS.	RC-QPR-
RAYTHEON CO. RESEARCH DIV., WALTHAM, MASS.	RC-S-
RAYTHEON CO. RESEARCH DIV., WALTHAM, MASS.	RC-TM-T-
RAYTHEON CO. RESEARCH DIV., WALTHAM, MASS.	S-
RAYTHEON CO. RESEARCH DIV., WALTHAM, MASS.	T-
RAYTHEON CO. SPACE + INFO. SYSTEMS DIV., SUDBURY,MASS	FR-(YEAR)-
RAYTHEON CO. SUBMARINE SIGNAL DIV., PORTSMOUTH, R.I.	P(NUMBER)-
RAYTHEON CO. SUBMARINE SIGNAL DIV., PORTSMOUTH, R.I.	R(NUMBER)-
RAYTHEON CO. SYSTEM REQUIREMENTS DEPT., WALTHAM, MASS	SRD-
RAYTHEON MFG. CO., WALTHAM, MASS. (REF. 1)	(NUMBER)
RAYTHEON MFG. CO., WALTHAM, MASS.	MFTD-
RAYTHEON MFG. CO., WALTHAM, MASS.	NOTES BK...
RAYTHEON MFG. CO., WAYLAND, MASS.	PC-
RAYTHEON MFG. CO., WALTHAM, MASS.	RAY-
RAYTHEON MFG. CO., WALTHAM, MASS.	RAYMC-
RAYTHEON MFG. CO., WALTHAM, MASS.	RAYTHEON NOTES BK...
RAYTHEON MFG. CO., WALTHAM, MASS. (HIGH TEMPERATURE MATERIAL)	RMC-HTM-
RAYTHEON MFG. CO., WALTHAM, MASS.	RMC-M/R-
RAYTHEON MFG. CO., WALTHAM, MASS.	RMC-QK-
RAYTHEON MFG. CO., WALTHAM, MASS.	RMC-S-
RAYTHEON MFG. CO., WALTHAM, MASS.	RMC-SO-
(SEE ALSO LATER NAME: RAYTHEON CO.)	
RAYTHEON MFG. CO. MISSILE FLIGHT TEST DEPT.	MFTD-
RAYTHEON MFG. CO. MISSILE SYSTEMS DIV.,BEDFORD, MASS.	M/R 16-
RAYTHEON MFG. CO. MISSILE SYSTEMS DIV.,BEDFORD, MASS.	RAYTHEON BR-
RAYTHEON MFG. CO. MISSILE SYSTEMS DIV.,BEDFORD, MASS.	RMC-BR-
RAYTHEON MFG. CO. RESEARCH DIV., WALTHAM, MASS.	RMC-QPR-
RAZDOW LABS., INC., NEWARK	RLI-
RCA LABS., PRINCETON, N.J.	RR-
RCA LABS., PRINCETON, N.J.	SP-(NUMBER)-
RCA LABS. COMMUNICATIONS RES. LAB., PRINCETON, N.J.	SPECIAL TOPIC-

RCA SERVICE CO., INC., CAMDEN, N.J.	AES-
RCA SERVICE CO., INC., CAMDEN, N.J.	RCASC-
RCA SERVICE CO., INC., PATRICK AFB, FLA.	ETR-TR-
RCA VICTOR CO., LTD., MONTREAL	RCA RES. REPT.-
RCA VICTOR CO., LTD., MONTREAL	RCA-VCL-
RCA VICTOR CO., LTD. RESEARCH LABS., MONTREAL	PCC-D (NO.)-
RCA VICTOR CO., LTD. RESEARCH LABS., MONTREAL	RCA-RES.RPT.-(NO.)-(NO.)-
RCA VICTOR CO., LTD. RESEARCH LABS., MONTREAL	RCA-RR-(NO.)-(NO.)-
RCA VICTOR CO., LTD. RESEARCH LABS., MONTREAL	RR-(NUMBER)-(NUMBER)-
RCA VICTOR DIV., LTD. RESEARCH LABS., MONTREAL	PCC-
RCA VICTOR DIV., LTD. RESEARCH LABS., MONTREAL	RES-
RCA VICTOR DIV., RADIO CORP. OF AMERICA, CAMDEN, N.J.	ASG-(CONTRACT NO.)
RCA VICTOR DIV., RADIO CORP. OF AMERICA, CAMDEN, N.J.	EXPERIMENT-
RCA VICTOR DIV., RADIO CORP. OF AMERICA, CAMDEN, N.J.	RCA-EM-
RCA VICTOR DIV., RADIO CORP. OF AMERICA, CAMDEN, N.J.	RCA-NP-
RCA VICTOR DIV., RADIO CORP. OF AMERICA, CAMDEN, N.J.	TLE-(NO.)-
REA (J.B.) CO., LOS ANGELES	REA-
REA (J.B.) CO., SANTA MONICA, CALIF.	REA-
REACTION MOTORS, INC., DENVILLE, N.J.	NF-
REACTION MOTORS, INC., DENVILLE, N.J.	RMI-
REACTION MOTORS, INC., DENVILLE, N.J.	RMI-TPR-
REACTION MOTORS, INC., ROCKAWAY, N.J.	RMI-
REACTIVE METALS, INC., ASHTABULA, OHIO	RMI-
REACTIVE METALS, INC., NILES, OHIO	RMI-
REACTIVE METALS, INC., SEYMOUR, CONN.	BRB-
READING, ENGLAND. UNIV.	CVD/RU-
REAKTOR A. G., WUERENLINGEN, SWITZERLAND	RAG-
RECON. INC., TALLAHASSEE	RECON-RC-
RECON. INC., TALLAHASSEE (TECHNICAL REPORT)	RECON-TR-
RECONSTRUCTION FINANCE CORP.	RFC-
REDEL, INC., ANAHEIM, CALIF.	APR-
REDEL, INC., ANAHEIM, CALIF.	KPR-
REDEL, INC., ANAHEIM, CALIF.	RED-
REDEL, INC., ANAHEIM, CALIF.	REDEL KPR-
REDEL, INC., ANAHEIM, CALIF.	RI-
REDEL, INC., ANAHEIM, CALIF.	RML-
REDEL, INC., ANAHEIM, CALIF.	VML-
REDSTONE ARSENAL, ALA.	RA-
REDSTONE ARSENAL, ALA. (ROCKET TEST FACILITIES)	RTF-
REDSTONE ARSENAL, HUNTSVILLE, ALA. (INTERIM AND FINAL REPORT)	A-(NO.LTR.)
REDSTONE ARSENAL, HUNTSVILLE, ALA.	OGMC-
REDSTONE ARSENAL, HUNTSVILLE, ALA. (QUARTERLY PROGRESS REPORT)	P-
REDSTONE ARSENAL, HUNTSVILLE, ALA.	RA-PR-
REDSTONE ARSENAL, HUNTSVILLE, ALA.	TSFO-
REDSTONE ARSENAL. ORDNANCE MISSILE LABS., HUNTSVILLE, ALA.	2A-
REDSTONE ARSENAL. ORDNANCE MISSILE LABS., HUNTSVILLE, ALA.	OML-
REDSTONE ARSENAL. ORDNANCE MISSILE LABS., HUNTSVILLE, ALA.	2R-
REDSTONE ARSENAL. ORDNANCE MISSILE LABS., HUNTSVILLE, ALA.	6R-
REDSTONE ARSENAL. ORDNANCE MISSILE LABS., HUNTSVILLE, ALA.	RA-(NUMBER)(LETTER)-
REDSTONE ARSENAL. ORDNANCE MISSILE LABS., HUNTSVILLE, ALA.	RA-(NUMBER)(LETTER)(NO.)-
REDSTONE ARSENAL. ORDNANCE MISSILE LABS., HUNTSVILLE, ALA.	RA-OML-
REDSTONE ARSENAL. ORDNANCE MISSILE LABS., HUNTSVILLE, ALA.	RSA/OML-
REDSTONE SCIENTIFIC INFO. CTR., REDSTONE ARSENAL, ALA	AMC-RSIC-
REDSTONE SCIENTIFIC INFO. CTR., REDSTONE ARSENAL, ALA	RSIC-
REDSTONE SCIENTIFIC INFO. CTR., REDSTONE ARSENAL, ALA	RSIC-U-
REDSTONE SCIENTIFIC INFO. CTR., REDSTONE ARSENAL, ALA	TT-
REED COLLEGE, PORTLAND, ORE.	REED-
REED RESEARCH INC., WASHINGTON, D.C.	RRI-
REEVES INSTRUMENT CORP., N.Y.C. (REF. 1)	(NUMBER)
REEVES INSTRUMENT CORP., N.Y.C.	RIC-
REEVES INSTRUMENT CORP., N.Y.C.	RICO-
REGIONAL SCIENCE RESEARCH INST., PHILADELPHIA	RSRI-DP-
REGULUS I AND REGULUS II COORDINATION COMM. (REF. 15)	RG-CC-
REGULUS I AND REGULUS II COORDINATION COMM. (REF. 15)	RG-CC-(YEAR)-
REMINGTON ARMS CO., INC., BRIDGEPORT, CONN.	AB-
RENSSELAER POLYTECHNIC INST., TROY, N.Y.	LINAC-
RENSSELAER POLYTECHNIC INST., TROY, N.Y.	MATHREP-
RENSSELAER POLYTECHNIC INST., TROY, N.Y.	RPI-
RENSSELAER POLYTECHNIC INST., TROY, N.Y.	RPI-(NUMBER)-
RENSSELAER POLYTECHNIC INST., TROY, N.Y.	RPIB-
RENSSELAER POLYTECHNIC INST., TROY, N.Y.	RPI-MATH-REP-
RENSSELAER POLYTECHNIC INST., TROY, N.Y.	RPI-MP-
RENSSELAER POLYTECHNIC INST., TROY, N.Y.	RPI-PR-
RENSSELAER POLYTECHNIC INST., TROY, N.Y.	RPI-TN-
RENSSELAER POLYTECHNIC INST., TROY, N.Y.	RPI-TR-
RENSSELAER POLYTECHNIC INST., TROY, N.Y. (PROJECT TUBAFLIGHT)	TN-PT-
RENSSELAER POLYTECHNIC INST., TROY, N.Y. (PROJECT TUBAFLIGHT)	TR-PT-
RENSSELAER POLYTECHNIC INST., TROY, N.Y. AERONAUTICAL ENGINEERING AND ASTRONAUTICS DEPT.	TR-AE-
RENSSELAER POLYTECHNIC INST., TROY, N.Y. DEPT. OF AERONAUTICAL ENGINEERING	RPIAL-
RENSSELAER POLYTECHNIC INST., TROY, N.Y. DEPT. OF AERONAUTICAL ENGINEERING	RPI TR-AE-

RENSSELAER POLYTECHNIC INST., TROY, N.Y. DEPT. OF ELECTRICAL ENGINEERING	RPIEE-
RENSSELAER POLYTECHNIC INST., TROY, N.Y. DEPT. OF MATHEMATICS	RPI-MATH-
RENSSELAER POLYTECHNIC INST., TROY, N.Y. DIV. OF ELECTROPHYSICS	TR-EP-
RENSSELAER POLYTECHNIC INST., TROY, N.Y. DIV. OF SYSTEMS ENGINEERING	TR-DSF-
RENSSELAER POLYTECHNIC INST.,TROY,NY. PLASMA RES. LAB	RPI-PRL-SSR-
RENSSELAER POLYTECHNIC INST.,TROY,NY. PLASMA RES. LAB	RPI-PRL-TR-
RENSSELAER POLYTECHNIC INST.,TROY, N.Y. WELDING LAB.	RPI-QPR-
REPUBLIC AVIATION CORP., FARMINGDALE, N.Y.	AVD-(NUMBER)-
REPUBLIC AVIATION CORP., FARMINGDALE, N.Y.	C-
REPUBLIC AVIATION CORP., FARMINGDALE, N.Y.	ESRD-(NO.)-
REPUBLIC AVIATION CORP., FARMINGDALE, N.Y.	MRP-(NO.)-(NO.)-
REPUBLIC AVIATION CORP., HICKSVILLE, N.Y.	MSD-
REPUBLIC AVIATION CORP., FARMINGDALE, N.Y.	PCD-
REPUBLIC AVIATION CORP., FARMINGDALE, N.Y.	PCD-TR-(YEAR)-
REPUBLIC AVIATION CORP., FARMINGDALE, N.Y.	RAC (NUMBER)-
REPUBLIC AVIATION CORP., FARMINGDALE, N.Y.	RD-QPR-(YR.)-(NO.)-
REPUBLIC AVIATION CORP., FARMINGDALE, N.Y.	RD-QTR-(YR.)-(NO.)-
REPUBLIC AVIATION CORP., FARMINGDALE, N.J.	SRS-TR-
(SEE ALSO LATER NAME: FAIRCHILD HILLER CORP. REPUBLIC AVIATION DIV.)	
REPUBLIC AVIATION CORP. APPLIED RESEARCH AND DEVELOPMENT MATERIALS LAB., FARMINGDALE, N.Y.	ARD (NUMBER)-
REPUBLIC AVIATION CORP. GUIDED MISSILE DIV., HICKSVILLE, N.Y.	GMD-
REPUBLIC AVIATION CORP. GUIDED MISSILE DIV., HICKSVILLE, N.Y.	RAC-GMD-
REPUBLIC AVIATION CORP. GUIDED MISSILE DIV., MINEOLA, N.Y.	GMD-(NUMBER)-
REPUBLIC AVIATION CORP. GUIDED MISSILE DIV., MINEOLA, N.Y.	RAC GMD (NOS.)-(NO.)-
REPUBLIC AVIATION CORP. GUIDED MISSILE DIV., N.Y.C.	RAC-GMD-
REPUBLIC AVIATION CORP. MISSILE SYSTEMS DIV., MINEOLA, N.Y.	MSD-(NO.)-(NO.)-(NO.)
REPUBLIC AVIATION CORP. MISSILE SYSTEMS DIV., MINEOLA, N.Y.	RAC MSD (NO.)-(NO.)-
RESDEL ENGINEERING CORP., PASADENA, CALIF.	REC-TF-
RESDEL ENGINEERING CORP., PASADENA, CALIF.	REC-TR-
RESEARCH ANALYSIS CORP., BETHESDA, MD.	RAC-
RESEARCH ANALYSIS CORP., BETHESDA, MD.	RAC(ORO)-
RESEARCH ANALYSIS CORP., BETHESDA, MD.	RAC(ORO)-SP-
RESEARCH ANALYSIS CORP., BETHESDA, MD.	RAC(ORO)-TP-
RESEARCH ANALYSIS CORP., BETHESDA, MD.	RAC-TP-
RESEARCH ANALYSIS CORP., CHEVY CHASE, MD.	RAC-
RESEARCH ANALYSIS CORP., MC LEAN, VA.	RAC-
RESEARCH ANALYSIS CORP., MC LEAN, VA.(CLIENT REPORTS)	RAC-CR-
RESEARCH ANALYSIS CORP., MC LEAN, VA. (DRAFT CLIENT REPORTS)	RAC-D-CR-
RESEARCH ANALYSIS CORP., MC LEAN, VA. (DRAFT TECHNICAL PAPERS)	RAC-D-TP-
RESEARCH ANALYSIS CORP., MC LEAN, VA.	RAC-E-
RESEARCH ANALYSIS CORP., MC LEAN, VA. (FIELD PAPERS)	RAC-FP-
RESEARCH ANALYSIS CORP., MC LEAN, VA. (FIELD REPTS.)	RAC-FR-
RESEARCH ANALYSIS CORP., MC LEAN, VA. (PAPERS)	RAC-P-
RESEARCH ANALYSIS CORP., MC LEAN, VA.	RAC-R-
RESEARCH ANALYSIS CORP., MC LEAN, VA.	RAC-T-
RESEARCH ANALYSIS CORP., MC LEAN, VA.	RAC-TP-
RESEARCH ANALYSIS CORP., MC LEAN, VA.	RAE-TP-
RESEARCH AND ANALYSIS CORP. FIELD OFFICE, BANGKOK	RACFO-FP-
RESEARCH AND ANALYSIS CORP. FIELD OFFICE, BANGKOK	RACFO-T-FP-
RESEARCH + DEV. BD.,WASH.,DC (AIR DEFENCE OPERATIONS)	AD-
RES. + DEV. BD.,WASH.,DC (AIRBORNE LANDING OPERATIONS	AL-
RES. + DEV. BD., WASH., D.C. (AMPHIBIOUS OPERATIONS)	AO-
RES. + DEV. BD.,WASH.,DC.(ANTI-SUBMARINE OPERATIONS)	AS-
RES. + DEV. BD.,WASH.,DC(ATOMIC WARFARE OPERATIONS)	AW-
RES. + DEV. BD.,WASH.,D.C. (BIOL. WARFARE OPERATIONS)	BW-
RESEARCH AND DEVELOPMENT BOARD, WASHINGTON, D.C. (COMBAT AIR SUPPORT OPERATIONS)	CA-
RES. + DEV. BD.,WASH.,DC. (CHEM. WARFARE OPERATIONS)	CW-
RESEARCH AND DEVELOPMENT BOARD, WASHINGTON, D.C.	DOD RSB-
RESEARCH AND DEVELOPMENT BOARD, WASHINGTON, D.C. (LANDING COMBAT OPERATIONS)	LC-
RESEARCH AND DEVELOPMENT BOARD, WASHINGTON, D.C. (MISSILE TEST RANGE INSTRUMENTATION)	MTRI-
RESEARCH AND DEVELOPMENT BOARD, WASHINGTON, D.C.	OAX-MLH-
RESEARCH + DEV. BD.,WASH.,D.C. (PERSONNEL OPERATIONS)	PC-
RESEARCH AND DEVELOPMENT BOARD, WASHINGTON, D.C. (PSYCHOLOGICAL WARFARE AND COLD WAR OPERATIONS)	PC-
RESEARCH AND DEVELOPMENT BOARD, WASHINGTON, D.C.	RDB-(NO.)/-
RESEARCH AND DEVELOPMENT BOARD, WASHINGTON, D.C.	RDB-LAP-
RESEARCH AND DEVELOPMENT BOARD, WASHINGTON, D.C.	RDB-MT-(NO.)/-
RESEARCH AND DEVELOPMENT BOARD, WASHINGTON, D.C.	RDB-MTRI-
RESEARCH AND DEVELOPMENT BOARD, WASHINGTON, D.C. (STRATEGIC AIR OPERATIONS)	SA-
RESEARCH + DEV. BD.,WASH.,D.C.(SEA COMBAT OPERATIONS)	SC-
RESEARCH AND DEVELOPMENT BOARD. AD HOC COMMITTEE ON SCIENTIFIC AND SYNTHETIC ANALYSIS, WASHINGTON, D.C.	ADO-
RESEARCH AND DEVELOPMENT BOARD. AD HOC GROUP ON RELIABILITY OF ELECTRONIC EQUIPMENT, WASH., D. C.	RDB-EL-(NO.)/-
RESEARCH + DEV. BOARD. COMM. ON AERONAUTICS,WASH.,DC	AR-
RESEARCH + DEV. BOARD. COMM. ON AERONAUTICS,WASH.,DC	RPS-
RESEARCH AND DEVELOPMENT BOARD. COMMITTEE ON BASIC PHYSICAL SCIENCES, WASHINGTON, D.C.	PS-
RES. + DEV. BD. COMM. ON ELECTRONICS, WASH., D.C.	EL-
RES. + DEV. BD. COMM. ON ELECTRONICS, WASH., D.C.	EL (NUMBER)/
RES. + DEV. BD. COMM. ON ELECTRONICS, WASH., D.C.	LAP-
RES. + DEV. BD. COMM. ON ELECTRONICS, WASH., D.C.	LCM-
RES. + DEV. BD. COMM. ON ELECTRONICS, WASH., D.C.	LCS-
RES. + DEV. BD. COMM. ON ELECTRONICS, WASH., D.C.	LEA-
RES. + DEV. BD. COMM. ON ELECTRONICS, WASH., D.C.	LIR-
RES. + DEV. BD. COMM. ON ELECTRONICS, WASH., D.C.	LIZ-
RES. + DEV. BD. COMM. ON ELECTRONICS, WASH., D.C.	LRA-
RES. + DEV. BD. COMM. ON ELECTRONICS, WASH., D.C.	VTDC-
RES. + DEV. BD. COMM. ON FUELS + LUBRICANTS, WASH.,DC	FAP-
RES. + DEV. BD. COMM. ON FUELS + LUBRICANTS, WASH.,DC	FL-
RES. + DEV. BD. COMM. ON FUELS + LUBRICANTS, WASH.,DC	FPL-

RES. + DEV. BD. COMM. ON FUELS + LUBRICANTS, WASH.,DC	RDB-FRO-
RESEARCH + DEV. BD. COMM. ON GEN. SCIENCES, WASH.,DC.	CGS-
RESEARCH AND DEVELOPMENT BOARD. COMMITTEE ON GEOGRAPHICAL EXPLORATION, WASHINGTON, D.C.	GE-
RES.+ DEV.BD. COMM. ON GEOPHYSICAL SCIENCES, WASH.,DC	CGS-
RES.+ DEV.BD. COMM. ON GEOPHYSICAL SCIENCES, WASH.,DC	GS-
RES.+ DEV.BD. COMM.ON GEOPHYSICS + GEOGRAPHY, WASH,DC	GCG-
RES.+ DEV.BD. COMM.ON GEOPHYSICS + GEOGRAPHY, WASH,DC	GG-
RES.+ DEV.BD. COMM.ON GEOPHYSICS + GEOGRAPHY, WASH,DC	GHY-
RES.+ DEV.BD. COMM.ON GEOPHYSICS + GEOGRAPHY, WASH,DC	GRT-
RES. + DEV. BD. COMM. ON GUIDED MISSILES, WASH., DC	GM-
RES. + DEV. BD. COMM. ON GUIDED MISSILES, WASH., DC	HTD-GM-
RES. + DEV. BD. COMM. ON GUIDED MISSILES, WASH., DC	MAS-
RES. + DEV. BD. COMM. ON GUIDED MISSILES, WASH., DC	MCM-
RES. + DEV. BD. COMM. ON GUIDED MISSILES, WASH., DC	MGC-
RES. + DEV. BD. COMM. ON GUIDED MISSILES, WASH., DC	MML-
RES. + DEV. BD. COMM. ON GUIDED MISSILES, WASH., DC	MPF-
RES. + DEV. BD. COMM. ON GUIDED MISSILES, WASH., DC	MTD-
RES. + DEV. BD. COMM. ON GUIDED MISSILES, WASH., DC	MTRI-
RES. + DEV. BD. COMM. ON GUIDED MISSILES, WASH., DC	MTT-
RES. + DEV. BD. COMM. ON GUIDED MISSILES, WASH., DC	MWF-
RES. + DEV. BD. COMM. ON GUIDED MISSILES. PANEL ON TEST RANGE PROCEDURES + INSTRUMENTATION, WASH.,D.C.	NME/MTRI-
RES. + DEV. BD. COMM. ON HUMAN RESOURCES, WASH., D.C.	HBM-
RES. + DEV. BD. COMM. ON HUMAN RESOURCES, WASH., D.C.	HPS-
RES. + DEV. BD. COMM. ON HUMAN RESOURCES, WASH., D.C.	HR-
RES. + DEV. BD. COMM. ON HUMAN RESOURCES, WASH., D.C.	HR-HML-
RES. + DEV. BD. COMM. ON HUMAN RESOURCES, WASH., D.C.	HR-HTD-
RESEARCH AND DEVELOPMENT BOARD. COMMITTEE ON LONG RANGE PROVING GROUND, WASHINGTON, D.C.	PG-
RES. + DEV. BD. COMM. ON MATERIALS, WASHINGTON, D.C.	MT-TOM-
RES. + DEV. BD. COMM. ON NAVIGATION, WASHINGTON, D.C.	NV-
RESEARCH + DEV. BD. COMM. ON ORDNANCE, WASH., D.C.	OCM-
RESEARCH + DEV. BD. COMM. ON ORDNANCE, WASH., D.C.	OOM-
RESEARCH + DEV. BD. COMM. ON ORDNANCE, WASH., D.C.	OR-
RESEARCH + DEV. BD. PANEL ON ACOUSTICS, WASH., D.C.	LEA-
RESEARCH AND DEVELOPMENT BOARD. PANEL ON AERODYNAMICS AND STRUCTURES, WASHINGTON, D.C.	MAS-
RESEARCH AND DEVELOPMENT BOARD. PANEL ON AERODYNAMICS AND STRUCTURES, WASHINGTON, D.C.	PAS-
RESEARCH AND DEVELOPMENT BOARD. PANEL ON AMMUNITION AND EXPLOSIVES, WASHINGTON, D.C.	OAX-
RESEARCH AND DEVELOPMENT BOARD. PANEL ON ANTENNAS AND PROPAGATION, WASHINGTON, D.C.	LAP-
RES. + DEV. BD. PANEL ON COMMUNICATIONS, WASH., D.C.	LCM-
RESEARCH + DEV. BD. PANEL ON COMPONENTS, WASH., D.C.	LCP-
RES. + DEV. BD. PANEL ON COUNTERMEASURES, WASH., D.C.	MCM-
RES. + DEV. BD. PANEL ON ELECTRON TUBES, WASH., D.C.	LET-
RESEARCH + DEV. BD. PANEL ON FIRE CONTROL, WASH.,D.C.	OFC-
RESEARCH AND DEVELOPMENT BOARD. PANEL ON FUELS REQUIRING OXIDIZERS, WASHINGTON, D.C.	FRO-
RESEARCH AND DEV. BD., PANEL ON GUNS., WASH., D.C.	OGN-
RESEARCH AND DEVELOPMENT BOARD. PANEL ON HUMAN ENGINEERING AND PSYCHOPHYSIOLOGY, WASHINGTON, D.C.	HPS-
RESEARCH + DEV. BD. PANEL ON INFRARED, WASH., D.C.	LIR-
RESEARCH AND DEVELOPMENT BOARD. PANEL ON LAUNCHING AND HANDLING, WASHINGTON, D.C.	MLH-
RES. + DEV. BD. PANEL ON MANPOWER, WASHINGTON, D.C.	HMP-
RESEARCH AND DEVELOPMENT BOARD. PANEL ON MISSILE AUXILIARY EQUIPMENT, WASHINGTON, D.C.	MAE-
RESEARCH AND DEVELOPMENT BOARD. PANEL ON ORDNANCE MATERIALS, WASHINGTON, D.C.	OCM-
RESEARCH AND DEVELOPMENT BOARD. PANEL ON ORDNANCE MATERIALS, WASHINGTON, D.C.	OOM-
RES. + DEV. BD. PANEL ON PROPULSION + FUELS, WASH.,DC	MPF-
RES. + DEV. BD. PANEL ON PROXIMITY FUZES, WASH., D.C.	PFZ-
RES. + DEV. BD. PANEL ON TARGET DRONES, WASH., D.C.	MTD-
RESEARCH AND DEVELOPMENT BOARD. PANEL ON TEST AND TRAINING EQUIPMENT, WASHINGTON, D.C.	MTT-
RES. + DEV. BD. PANEL ON WARHEADS + FUZES, WASH.,D.C.	MWF-
RES. + DEV. BD. SPEC. COMM. ON TECH. INFO., WASH., DC	TI-
RESEARCH AND TECHNOLOGY DIV., BOLLING AFB, D.C.	RTD-
RESEARCH AND TECHNOLOGY DIV., BOLLING AFB, D.C.	RTD-(NUMBER)-
RESEARCH AND TECHNOLOGY DIV., BOLLING AFB, D.C.	RTD-IR-
RESEARCH AND TECHNOLOGY DIV., BOLLING AFB, D.C.	RTD-TB-(NO.)-MMP-
RESEARCH AND TECHNOLOGY DIV., BOLLING AFB, D.C.	RTD-TDR-(YEAR)-
RESEARCH AND TECHNOLOGY DIV., BOLLING AFB, D.C.	RTD-TR-
RESEARCH AND TECHNOLOGY DIV., BOLLING AFB, D.C.	RTD-TR-(NUMBER)-
RESEARCH AND TECHNOLOGY DIV., BOLLING AFB, D.C.	RTST-(YEAR)-
RESEARCH AND TECHNOLOGY DIV., WRIGHT-PATTERSON AFB, OHIO	RTDA-IR-(YEAR)-
RESEARCH AND TECHNOLOGY DIV. DETACHMENT 4, EGLIN AFB, FLA.	RTD-AT-L-
RESEARCH CHEMICALS INC., BURBANK, CALIF.	RC-
RESEARCH CHEMICALS INC., BURBANK, CALIF.	RCI-QR-
RESEARCH DIV. (AIR FORCE), WASHINGTON, D.C. (QUARTERLY INDEX)	AFRD-QI-(YEAR/MO.-MO.)
RESEARCH DIV. (AIR FORCE), WASHINGTON, D.C. (STATUS OF RESEARCH PROPOSALS)	AFRD-SRP-(YEAR/MONTH)

RESEARCH DIV. (AIR FORCE), WASHINGTON, D.C. (QUARTERLY INDEX)	ARDC-QI-(YEAR/MO.-MO.)
RESEARCH ENTERPRISES LTD., TORONTO	REL-
RESEARCH FLIGHT FACILITY, MIAMI, FLA.	RFF-
RESEARCH INFORMATION SERVICE, N.Y.C.	RIS-
RESEARCH INFORMATION SERVICE, N.Y.C. (TRANSLATIONS)	RIS-
RESEARCH INTERNATL. ASSOCS., WASH.,D.C.(TRANSLATIONS)	RIA-
RESEARCH TRIANGLE INST., DURHAM, N.C.	EU-
RESEARCH TRIANGLE INST., DURHAM, N.C.	RM-
RESEARCH TRIANGLE INST., DURHAM, N.C.	RTI-
RESEARCH TRIANGLE INST., DURHAM, N.C.	RTI-CE-(NUMBER)-F
RESEARCH TRIANGLE INST., DURHAM, N.C.	RTI-EU-
RESEARCH TRIANGLE INST., DURHAM, N.C.	RTI-OU-
RESEARCH TRIANGLE INST., DURHAM, N.C.	RTI-IR-SU-
RESEARCH TRIANGLE INST., DURHAM, N.C.	RTI-RM-
RESEARCH TRIANGLE INST., DURHAM, N.C.	RTI-RR-SU-
RESEARCH TRIANGLE INST., DURHAM, N.C.	RTI-SU-
RESEARCH TRIANGLE INST., DURHAM, N.C.	RTI-TRR-
RESEARCH TRIANGLE INST., DURHAM, N.C.	RTI-TR-SU-(NUMBER)-
RESEARCH TRIANGLE INST., DURHAM, N.C.	SU-
RESEARCH TRIANGLE INST. OPERATIONS RESEARCH AND ECONOMICS DIV., DURHAM, N.C.	R-OU-
RESEARCH TRIANGLE INST. OPERATIONS RESEARCH AND ECONOMICS DIV., DURHAM, N.C.	RTI-FR-OU-(NUMBER)-
RESEARCH TRIANGLE INST. OPERATIONS RESEARCH AND ECONOMICS DIV., DURHAM, N.C.	RTI-OU-
RESEARCH TRIANGLE INST. OPERATIONS RESEARCH AND ECONOMICS DIV., DURHAM, N.C.	RTI-RM-OU-(NUMBER)-
RESEARCH TRIANGLE INST. OPERATIONS RESEARCH AND ECONOMICS DIV., DURHAM, N.C.	RTI-R-OU-
RESEARCH TRIANGLE INST. STATISTICS RESEARCH DIV., DURHAM, N.C.	RTI-S-
RESEARCH TRIANGLE INST. STATISTICS RESEARCH DIV., DURHAM, N.C.	S-
RESOURCE MANAGEMENT CORP., INC., BETHESDA, MD.	RMC-RD-
RESOURCE MANAGEMENT CORP., INC., BETHESDA, MD.	RMC-UR-
REX DIV., GARRETT CORP., LOS ANGELES	RD-(NUMBER)-R
REYNOLDS ELEC. + ENG. CO., INC., MERCURY, NEV.	REEC-
RHEEM MFG. CO., DOWNEY, CALIF.	RHEEM-
RHEEM MFG. CO., DOWNEY, CALIF.	RHMC-
RHEEM MFG. CO., N.Y.C.	RMC-
RHEEM MFG. CO. RES. + DEV. LABS., DOWNEY, CALIF.	RHEEM R-
RHODE ISLAND. UNIV., KINGSTON	RIU-
RHODE ISLAND. UNIV., KINGSTON	URI-
RHODE ISLAND. UNIV., KINGSTON. DEPT. OF ELEC. ENG.	TR-(NUMBERS)/
RHODE ISLAND. UNIV.,KINGSTON. DIV. OF ENG. RES. + DEV	ENGINEERING BULL-
RHODE ISLAND STATE COLL., KINGSTON. DEPT. OF CHEMISTRY (REF. 1)	(NUMBER)
RHODES LEWIS CO., LOS ANGELES	RLC-
RICE INST., HOUSTON, TEX.	RI-
RICE UNIV., HOUSTON, TEX.	AAR-
RICE UNIV., HOUSTON, TEX.	RICE-
RICE UNIV., HOUSTON, TEX.	RICE-(NUMBER)-PU
RICE UNIV., HOUSTON, TEX.	RICE-(NUMBER)-P
RICE UNIV., HOUSTON, TEX. (STATUS REPORT)	RICE-SR-
RICHARDSON, BELLOWS, HENRY AND CO., INC., N.Y.C.	RBH-
RICHLAND OPERATIONS OFFICE, AEC, WASH. (REF. 4)	RLO-(CONTRACT NUMBER)-
RICHLAND OPERATIONS OFFICE, AEC, WASH.	RLO-(CONTRACT NO.)-T-
RICHLAND OPERATIONS OFFICE, AEC, WASH.	RLO-
RIKKYO UNIV.,YOKOSUKA,JAPAN. INST. FOR ATOMIC ENERGY	IAERU-
RIO DE JANEIRO.CENTRO BRASILEIRO DE PESQUISAS FISICAS	NOTAS DE FISICA-
RIO DE JANEIRO.CENTRO BRASILEIRO DE PESQUISAS FISICAS	PHYSICS NOTE-
RIO DE JANEIRO.CENTRO BRASILEIRO DE PESQUISAS FISICAS	PHYSICS NOTE NO.-
RIVERSIDE RESEARCH INST., N.Y.C.	RRI-P-
RIVERSIDE RESEARCH INST., N.Y.C.	RRI-T-(NO./NO./NO.)-
ROAD RESEARCH LAB., CROWTHORNE, BERKS, ENGLAND	RRL-LR-
ROBERT A. TAFT SANITARY ENGINEERING CTR., CINCINNATI	RATSEC-
ROBERT A. TAFT SANITARY ENGINEERING CTR., CINCINNATI	SEC-TR-
ROBERT A. TAFT SANITARY ENGINEERING CTR., CINCINNATI	TSEC-TR-
ROBERT A. TAFT WATER RESEARCH CENTER, CINCINNATI	TWRC-
ROBERT A. TAFT WATER RESEARCH CENTER. ADVANCED WASTE TREATMENT RESEARCH LAB., CINCINNATI	AWTR-
ROBERT A. TAFT WATER RESEARCH CENTER. ADVANCED WASTE TREATMENT RESEARCH LAB., CINCINNATI	TWRC-AWTRL-
ROBERTSHAW-FULTON CONTROLS CO., ANAHEIM, CALIF.	TP(NUMBER)-(LETTER)-
ROCHESTER, N.Y. UNIV.	II-188-
ROCHESTER, N.Y. UNIV.	II-189-
ROCHESTER, N.Y. UNIV.	PHARMACOLOGY REPORT-
ROCHESTER, N.Y. UNIV.	RU-
ROCHESTER, N.Y. UNIV.	SDC-
ROCHESTER, N.Y. UNIV.	STAT.-
ROCHESTER, N.Y. UNIV.	UR-
ROCHESTER, N.Y. UNIV.	UR-(CONTRACT NO.)-
ROCHESTER, N.Y. UNIV.	URPA-
ROCHESTER, N.Y. UNIV. ATOMIC ENERGY PROJECT	RAEP-
ROCHESTER, N.Y. UNIV. CENTER FOR SYSTEM SCIENCE	CSS-(YEAR)-
ROCHESTER, N.Y. UNIV. DEPT. OF MECHANICAL AND AEROSPACE SCIENCES	MAS-TR-

ROCHESTER, N.Y. UNIV. DEPT. OF RADIATION BIOLOGY + BIOPHYSICS	UR-(NUMBER)-
ROCHESTER, N.Y. UNIV. INST. OF OPTICS (REF. 1)	(NUMBERS...)
ROCHESTER, N.Y. UNIV. NUCLEAR STRUCTURE RESEARCH LAB.	UR-NSRL-
ROCHESTER, N.Y. UNIV. NUCLEAR STRUCTURE RESEARCH LAB.	UR-NSRL-PR-
ROCHESTER, N.Y. UNIV. SCH. OF MED. AND DENTISTRY	RUSMD-
ROCHESTER APPLIED SCIENCE ASSOCIATES, INC., N.Y.	RASA-
ROCK ISLAND ARSENAL, ILL.	RIA-
ROCK ISLAND ARSENAL, ILL.	RIA-(YR.)-
ROCK ISLAND ARSENAL, ILL. (INDUSTRIAL PROJ. REPT.)	RIA-(LTRS.)-
ROCK ISLAND ARSENAL LAB., ILL.	ES-
ROCK ISLAND ARSENAL LAB., ILL.	OR-(YEAR)-
ROCK ISLAND ARSENAL LAB., ILL.	ORD RIA LR-
ROCK ISLAND ARSENAL LAB., ILL.	RIA-ES-(YR)-
ROCK ISLAND ARSENAL LAB., ILL.	RIAL-
ROCK ISLAND ARSENAL LAB., ILL.	RIAL (YR.)-
ROCK ISLAND ARSENAL LAB., ILL.	RIALR-
ROCK ISLAND ARSENAL LAB., ILL.	RIA-TR-(YEAR)-
ROCKEFELLER INST., N.Y.C.	RI-
ROCKEFELLER INST. FOR MEDICAL RESEARCH, N.Y.C.	RIMR-
ROCKET POWER, INC. RESEARCH LABS., PASADENA, CALIF.	QR-
ROCKET POWER, INC. RESEARCH LABS., PASADENA, CALIF.	RP/RL-
ROCKET POWER, INC. RESEARCH LABS., PASADENA, CALIF.	RP/RL-QR-(NUMBER)-
ROCKET POWER LABS., GABRIEL CO., PASADENA, CALIF.	RPD-
ROCKET PROFICIENCY CENTER (AIR FORCE), YUMA, ARIZ.	ADC/YAFB/RF-
ROCKET RESEARCH CORP., SEATTLE	RRC-(YR.)-
ROCKETDYNE, CANOGA PARK, CALIF.	IES-(NUMBER)-
ROCKETDYNE. CHEMICAL AND MATERIAL SCIENCES DEPT., CANOGA PARK, CALIF.	SPEC-RAO-(NUMBER)-
ROCKETDYNE. CHEMICAL AND MATERIAL SCIENCES DEPT., CANOGA PARK, CALIF.	SPEC-RBO-(NUMBER)-
ROCKETDYNE. RESEARCH DEPT., CANOGA PARK, CALIF.	R-(NUMBER)-
ROCKETDYNE DIV., NORTH AMERICAN AVIATION, INC., CANOGA PARK, CALIF. (REF. 1)	(NUMBERS...)
ROCKETDYNE DIV., NORTH AMERICAN AVIATION, INC., CANOGA PARK, CALIF.	AER-
ROCKETDYNE DIV., NORTH AMERICAN AVIATION, INC., CANOGA PARK, CALIF.	AL-
ROCKETDYNE DIV., NORTH AMERICAN AVIATION, INC., CANOGA PARK, CALIF.	BC-(NUMBER)-
ROCKETDYNE DIV., NORTH AMERICAN AVIATION, INC., CANOGA PARK, CALIF.	BCI-
ROCKETDYNE DIV., NORTH AMERICAN AVIATION, INC., CANOGA PARK, CALIF.	CER-(NUMBER)-
ROCKETDYNE DIV., NORTH AMERICAN AVIATION, INC., CANOGA PARK, CALIF.	G.O.-
ROCKETDYNE DIV., NORTH AMERICAN AVIATION, INC., CANOGA PARK, CALIF.	HRD-
ROCKETDYNE DIV., NORTH AMERICAN AVIATION, INC., CANOGA PARK, CALIF.	MD-(NUMBER)-
ROCKETDYNE DIV., NORTH AMERICAN AVIATION, INC., CANOGA PARK, CALIF.	MD-(YEAR)-
ROCKETDYNE DIV., NORTH AMERICAN AVIATION, INC., CANOGA PARK, CALIF.	MIL-TDR-
ROCKETDYNE DIV., NORTH AMERICAN AVIATION, INC., CANOGA PARK, CALIF.	NAA-AL-
ROCKETDYNE DIV., NORTH AMERICAN AVIATION, INC., CANOGA PARK, CALIF.	NAA-MD-(YEAR)-
ROCKETDYNE DIV., NORTH AMERICAN AVIATION, INC., CANOGA PARK, CALIF.	NAA-R-
ROCKETDYNE DIV., NORTH AMERICAN AVIATION, INC., CANOGA PARK, CALIF.	NAA-R-G.O.-(NO.)(DATE)
ROCKETDYNE DIV., NORTH AMERICAN AVIATION, INC., CANOGA PARK, CALIF.	NAA-RM-
ROCKETDYNE DIV., NORTH AMERICAN AVIATION, INC., CANOGA PARK, CALIF.	NAA RR(YEAR)-
ROCKETDYNE DIV., NORTH AMERICAN AVIATION, INC., CANOGA PARK, CALIF.	PCR-
ROCKETDYNE DIV., NORTH AMERICAN AVIATION, INC., CANOGA PARK, CALIF.	R-(NUMBER)P
ROCKETDYNE DIV., NORTH AMERICAN AVIATION, INC., CANOGA PARK, CALIF.	R-400-
ROCKETDYNE DIV., NORTH AMERICAN AVIATION, INC., CANOGA PARK, CALIF.	RC-(YEAR)-
ROCKETDYNE DIV., NORTH AMERICAN AVIATION, INC., CANOGA PARK, CALIF.	RD-
ROCKETDYNE DIV., NORTH AMERICAN AVIATION, INC., CANOGA PARK, CALIF.	RDR-(NUMBER)-
ROCKETDYNE DIV., NORTH AMERICAN AVIATION, INC., CANOGA PARK, CALIF.	RH-
ROCKETDYNE DIV., NORTH AMERICAN AVIATION, INC., CANOGA PARK, CALIF.	RM-
ROCKETDYNE DIV., NORTH AMERICAN AVIATION, INC., CANOGA PARK, CALIF.	ROCKETDYNE R-(NUMBER)P
ROCKETDYNE DIV., NORTH AMERICAN AVIATION, INC., CANOGA PARK, CALIF.	TAMM-
ROCKFORD RESEARCH INST., INC., CAMBRIDGE, MASS.	RTB-
ROCKHURST COLL., KANSAS CITY, MO.	RHC-
ROCKY MOUNTAIN NUCLEAR POWER STUDY GROUP, IDAHO FALLS	RMG-
ROE(A.V.) CANADA LTD. ORENDA ENGINES DIV., MALTON, ONT.	OEL-NUCLEAR-
ROENTGEN TECHNISCHE DIENST N.V., ROTTERDAM	V-(NUMBER-NUMBER)-OE
ROHM AND HAAS CO., PHILADELPHIA	IEL-
ROHM AND HAAS CO., PHILADELPHIA	M-52-
ROHM AND HAAS CO., PHILADELPHIA	R+H-
ROHM AND HAAS CO., PHILADELPHIA	RHC-
ROHM + HAAS CO. REDSTONE ARSENAL RESEARCH DIV., HUNTSVILLE, ALA.	AF-
ROHM + HAAS CO. REDSTONE ARSENAL RESEARCH DIV., HUNTSVILLE, ALA.	M-(YEAR)-
ROHM + HAAS CO. REDSTONE ARSENAL RESEARCH DIV., HUNTSVILLE, ALA.	P-(YR)-
ROHM + HAAS CO. REDSTONE ARSENAL RESEARCH DIV., HUNTSVILLE, ALA.	S-
ROLLS ROYCE, LTD., DERBY, ENGLAND	INC-
ROLLS ROYCE, LTD., DERBY, ENGLAND	RR/OH/104
ROLLS ROYCE, LTD., DERBY, ENGLAND	TSD-
ROME. UNIVERSITA. ISTITUTO DI FISICA, G. MARCONI	NI-
ROME. UNIVERSITA. ISTITUTO DI FISICA, G. MARCONI	ROMEU-NI-
ROME. UNIVERSITA. SCUOLA DI INGEGNERIA AERONAUTICA	SIARGRAPH-
ROME AIR DEVELOPMENT CENTER, GRIFFISS AFB, N.Y.	GAFB-
ROME AIR DEVELOPMENT CENTER, GRIFFISS AFB, N.Y.	RADC-
ROME AIR DEVELOPMENT CENTER, GRIFFISS AFB, N.Y.	RADC-AR(NUMBER)-(YEAR)
ROME AIR DEVELOPMENT CENTER, GRIFFISS AFB, N.Y.	RADC-SP-
ROME AIR DEVELOPMENT CENTER, GRIFFISS AFB, N.Y.	RADC-SP-(YEAR)-
ROME AIR DEVELOPMENT CENTER, GRIFFISS AFB, N.Y. (TECHNICAL DOCUMENTARY REPORT)	RADC-TDR-
ROME AIR DEVELOPMENT CENTER, GRIFFISS AFB, N.Y.	RADC-TDR-(YEAR)-
ROME AIR DEVELOPMENT CENTER, GRIFFISS AFB, N.Y.	RADC-TN-

ROME AIR DEVELOPMENT CENTER

ROME AIR DEVELOPMENT CENTER, GRIFFISS AFB, N.Y.	RADC-TN-(YEAR)-
ROME AIR DEVELOPMENT CENTER, GRIFFISS AFB, N.Y.	RADC-TR-
ROME AIR DEVELOPMENT CENTER, GRIFFISS AFB, N.Y.	RADC-TR-(YEAR)-
ROME AIR DEVELOPMENT CENTER, GRIFFISS AFB, N.Y.	RCRDE-
ROME AIR DEVELOPMENT CENTER, GRIFFISS AFB, N.Y.	RCRP-
ROME AIR DEVELOPMENT CENTER, GRIFFISS AFB, N.Y.	RCRSS-
ROME AIR DEVELOPMENT CENTER, GRIFFISS AFB, N.Y.	RCRTC-
ROME AIR DEVELOPMENT CENTER, GRIFFISS AFB, N.Y.	RCRTG-
ROME AIR DEVELOPMENT CENTER, GRIFFISS AFB, N.Y.	RCRTN-
ROME AIR DEVELOPMENT CENTER, GRIFFISS AFB, N.Y.	RCRTR-
ROME AIR DEV. CTR. APPLIED RES. LAB., GRIFFISS AFB,NY	RAS-TM-
ROME AIR DEVELOPMENT CENTER. DIRECTORATE OF AEROSPACE SURVEILLANCE + CONTROL, GRIFFISS AFB, NY	RAL-TM-
ROME AIR DEVELOPMENT CENTER. DIRECTORATE OF COMMUNICATIONS, GRIFFISS AFB, N.Y.	RAU-TM-
ROME AIR DEVELOPMENT CENTER. ELECTRONICS WARFARE LAB., GRIFFISS AFB, N.Y.	RAW-TM-
ROOSEVELT UNIV., CHICAGO	ROOU-
ROSEMOUNT ENGINEERING CO. AIR DATA SENSOR DEPT., MINNEAPOLIS	REC-
ROSEN (RAYMOND) ENG. PRODUCTS CO., INC., PHILADELPHIA	RREP-
ROSEN (RAYMOND) ENGINEERING PRODUCTS, INC. TELEMETERING DIV., PHILADELPHIA	RREP-(NO.)(LTRS.)(NOS.)
ROSSFORD ORDNANCE DEPOT, TOLEDO	ROD-
ROWLAND AND CO., HADDONFIELD, N.J.	R AND C-(YEAR)-
ROYAL AERONAUTICAL SOCIETY. TRANSONIC AERODYNAMICS COMMITTEE, LONDON	TD-MEMO-
ROYAL ARMY MEDICAL COLL. (GT. BRIT.)	RAMC-
ROYAL AUSTRALIAN AIR FORCE	RAAF-
ROYAL CANADIAN AIR FORCE	RCAF-
ROYAL CANADIAN AIR FORCE	REF-VF-
ROYAL CANADIAN AIR FORCE. CENTRAL EXPERIMENTAL AND PROVING ESTAB., ROCKCLIFFE, ONT.	ARM-
ROYAL CANADIAN AIR FORCE. CENTRAL EXPERIMENTAL AND PROVING ESTAB., ROCKCLIFFE, ONT.	CE AND PE-
ROYAL CANADIAN AIR FORCE. CENTRAL EXPERIMENTAL AND PROVING ESTAB., ROCKCLIFFE, ONT.	CEPE-
ROYAL CANADIAN AIR FORCE. CENTRAL EXPERIMENTAL AND PROVING ESTAB., ROCKCLIFFE, ONT.	EI-
ROYAL CANADIAN AIR FORCE. CENTRAL EXPERIMENTAL AND PROVING ESTAB., ROCKCLIFFE, ONT.	EPE-
ROYAL CANADIAN AIR FORCE. CENTRAL EXPERIMENTAL AND PROVING ESTAB., ROCKCLIFFE, ONT.	GE-
ROYAL CANADIAN AIR FORCE. CENTRAL EXPERIMENTAL AND PROVING ESTAB., ROCKCLIFFE, ONT.	ME-
ROYAL CANADIAN AIR FORCE. CENTRAL EXPERIMENTAL AND PROVING ESTAB., ROCKCLIFFE, ONT.	PCE-
ROYAL CANADIAN AIR FORCE. CENTRAL EXPERIMENTAL AND PROVING ESTAB., ROCKCLIFFE, ONT.	RCAF CEPE-
ROYAL CANADIAN AIR FORCE. CENTRAL EXPERIMENTAL AND PROVING ESTAB., ROCKCLIFFE, ONT.	RCAF EPE-
ROYAL CANADIAN AIR FORCE. CENTRAL EXPERIMENTAL AND PROVING ESTAB., ROCKCLIFFE, ONT.	TEL-
ROYAL CANADIAN AIR FORCE. INST. OF AVIATION MEDICINE	IAM-
ROYAL CANADIAN A.F. OPERATIONAL RES. SEC., OTTAWA	CAOPS/ORS-
ROYAL CANADIAN A.F. OPERATIONAL RES. SEC., OTTAWA	EP-
ROYAL NORWEGIAN COUNCIL FOR SCI. + IND. RES.,BLINDERN	SAD-(NUMBER)-T
ROYAL ORDNANCE FACTORY, CHORLEY, LANCS, ENGLAND	DCROF-(NUMBER)-
ROYAL RESEARCH CORP., HAYWARD, CALIF.	RRC-
RUBBER RESEARCH INST. TNO, DELFT, NETHERLANDS	MAR-
RUGBY COLL. OF ENG. TECHNOLOGY, WARWICKS, ENGLAND	DED-(NUMBER)-MCB-
RURAL COOPERATIVE POWER ASSN., ELK RIVER, MINN.	JAT-(YEAR)-
RURAL ELECTRIFICATION ADM. (AGRICULTURE), WASH., D.C.	REA-
RURAL TRANSFORMER AND EQUIPMENT CO., MILWAUKEE	RT + E CO.-
RUTGERS UNIV., NEW BRUNSWICK, N.J.	N-
RUTGERS UNIV., NEW BRUNSWICK, N.J.	RSU-
RUTGERS UNIV., NEW BRUNSWICK, N.J.	RU-
RUTGERS UNIV., NEW BRUNSWICK, N.J.	TR-N-
RUTGERS UNIV., NEW BRUNSWICK, N.J. DEPT. OF MECHANICAL AND AEROSPACE ENGINEERING	RU-TR-(NUMBER)-MAE-F
RUTGERS UNIV., NEW BRUNSWICK, N.J. SCH. OF CHEMISTRY	A-
RYAN AERO. CO., LINDBERGH FIELD, SAN DIEGO, CALIF.	G-(NO.)-(YR)
RYAN AERO. CO., LINDBERGH FIELD, SAN DIEGO, CALIF.	RAC-
RYAN AERO. CO., LINDBERGH FIELD, SAN DIEGO, CALIF.	RAC-GR-
RYAN AERO. CO., LINDBERGH FIELD, SAN DIEGO, CALIF.	RYAC-
RYAN AERO. CO., LINDBERGH FIELD, SAN DIEGO, CALIF.	RYAN-

SAAB AIRCRAFT CO., SWEDEN	LH-
SAAB AIRCRAFT CO., SWEDEN	LHK-
SAAB AIRCRAFT CO., SWEDEN	LHU-
SAAB AIRCRAFT CO., SWEDEN	SAAB-TN-
SAAB AKTIEBOLAG, LINKIPING, SWEDEN	SAAB-TN-
SACLANT ASW RESEARCH CENTER (NATO), LA SPEZIA, ITALY (ANTI-SUBMARINE WARFARE)	SACLANTCEN-TM-
SACLANT ASW RESEARCH CENTER (NATO), LA SPEZIA, ITALY (ANTI-SUBMARINE WARFARE)	SACLANTCEN-TR-
SACRAMENTO PEAK OBSERVATORY, SUNSPOT, N. MEX.	RN-
SACRAMENTO PEAK OBSERVATORY, SUNSPOT, N. MEX.	SRN-
SACRAMENTO PEAK OBSERVATORY, SUNSPOT, N. MEX.	T-
SACRAMENTO PEAK OBSERVATORY, SUNSPOT, N. MEX.	TT-
SAFETY COMMITTEE, MANHATTAN DISTRICT	SM-
SAHA INST. OF NUCLEAR PHYSICS, CALCUTTA	SINP-TH-
SAHA INST. OF NUCLEAR PHYSICS, CALCUTTA	SINP-TH-(YEAR)-
SAIGON UNIV., SOUTH VIETNAM. DEPT. OF PHYSICS	KHVL-
SAINT LOUIS UNIV.	SATR-
SAINT LOUIS UNIV.	SLU-
SAINT VINCENT COLLEGE, LATROBE, PA.	SVC-
SAN-ALOO CLASSIFICATION COORDINATING COMMITTEE, AEC	SAL-
SAN ANTONIO AIR MATERIEL AREA, TEX.	AFPS-SAAMA-
SAN DIEGO, CALIF. STATE COLLEGE FOUNDATION	SDSCF-
SAN DIEGO COUNTY COMPREHENSIVE PLANNING ORG., CALIF.	JOB-
SAN FERNANDO LABS., PACOIMA, CALIF.	SFL-
SAN FERNANDO LABS., PACOIMA, CALIF.(QUARTERLY REPORT)	SFL-QR-
SAN FRANCISCO BAY NAVAL SHIPYARD. HUNTER'S POINT DIV.	HPNS-TR-(NUMBER-YEAR)
SAN FRANCISCO BAY NAVAL SHIPYARD. HUNTER'S POINT DIV.	HUNTERS PT-TR-(NUMBER-YR)
SAN FRANCISCO OPERATIONS OFFICE, AEC (REF. 4)	SAN-
SAN FRANCISCO OPERATIONS OFFICE, AEC (REF. 4)	SAN-(CONTRACT NUMBER)-
SANBORN CO., WALTHAM, MASS. (INSTRUCTION MANUAL)	IM-(MODEL NUMBER)-
SANDERS ASSOCIATES, INC., BEDFORD, MASS.	APF-CR-
SANDERS ASSOCIATES, INC., BEDFORD, MASS.	BCP-CR-
SANDERS ASSOCIATES, INC., BEDFORD, MASS.	JAJ-CR-
SANDERS ASSOCIATES, INC., BEDFORD, MASS.	JHJ-CR-
SANDERS ASSOCIATES, INC., BEDFORD, MASS.	SAB-APF-
SANDERS ASSOCIATES, INC., BEDFORD, MASS.	SAB-BCP-
SANDERS ASSOCIATES, INC., BEDFORD, MASS.	SAB-FVY-
SANDERS ASSOCIATES, INC., BEDFORD, MASS.	SAN-FLG-CR-
SANDERS ASSOCIATES, INC., BEDFORD, MASS.	SAN-JDK-(YEAR)-
SANDERS ASSOCIATES, INC., NASHUA, N.H.	ISCS-
SANDERS ASSOCIATES, INC., NASHUA, N.H.	R-
SANDERS ASSOCIATES, INC., NASHUA, N.H.	RG-O-
SANDERS ASSOCIATES, INC., NASHUA, N.H.	SAI-
SANDERS ASSOCIATES, INC., NASHUA, N.H.	SAN-
SANDERS ASSOCIATES, INC., NASHUA, N.H.	SAN-BDR-(YEAR)-
SANDERS ASSOCIATES, INC., NASHUA, N.H.	SAN-BLC-
SANDERS ASSOCIATES, INC., NASHUA, N.H.	SAN-GDJ-(YEAR)-
SANDERS ASSOCIATES, INC., NASHUA, N.H.	SAN-GKN-(YEAR)-
SANDERS ASSOCIATES, INC., NASHUA, N.H.	SAN-GOL-
SANDERS ASSOCIATES, INC., NASHUA, N.H.	SAN-NDL-(YEAR)-
SANDERS ASSOCIATES, INC., NASHUA, N.H.	SAN-PAH-
SANDERS ASSOCIATES, INC., NASHUA, N.H.	SAN-PBO-(YEAR)-
SANDERS ASSOCIATES, INC., NASHUA, N.H.	SA-RG-
SANDERS ASSOCIATES, INC., NASHUA, N.H.	SA-RS-
SANDERS ASSOCIATES, INC. INSTRUMENT DIV., MANCHESTER, N.H.	SAN-PYD-(YEAR)-
SANDERS ASSOCIATES, INC. INSTRUMENT DIV., NASHUA, N.H.	SAN-PYC-(YEAR)-
SANDERS NUCLEAR CORP., NASHUA, N.H.	SNC-(NUMBER)-
SANDERSON AND PORTER, N.Y.C.	S + P(NO.)A-
SANDIA CORP., ALBUQUERQUE, N. MEX. (REF. 1)	(NUMBERS...)
SANDIA CORP., ALBUQUERQUE, N. MEX. (TECH. MANUAL)	AEC-DASA-
SANDIA CORP., ALBUQUERQUE, N. MEX. (MATADOR)	AMG-M-(NUMBER)-
SANDIA CORP., ALBUQUERQUE, N. MEX.	BEJG-
SANDIA CORP., ALBUQUERQUE, N. MEX. (BASE SPARES ALLOWANCE LIST)	BSAL-
SANDIA CORP., ALBUQUERQUE, N. MEX.	CASE 400.- THRU 800.-
SANDIA CORP., ALBUQUERQUE, N. MEX. (COMMERCIAL PACKAGING SPECIFICATION)	CPS-
SANDIA CORP., ALBUQUERQUE, N. MEX.	DC-
SANDIA CORP., ALBUQUERQUE, N. MEX. (DESIGN COORDINATION HANDBOOK)	DCH-(LETTERS)-(NUMBER)
SANDIA CORP., ALBUQUERQUE, N. MEX. (DESIGN MANUAL)	DM-
SANDIA CORP., ALBUQUERQUE, N. MEX.	HWT-
SANDIA CORP., ALBUQUERQUE, N. MEX. (INSPECTION METHODS INSTRUCTION)	IMI-(NUMBER)-
SANDIA CORP., ALBUQUERQUE, N. MEX.	M-1004 (YEAR,-MONTH)
SANDIA CORP., ALBUQUERQUE, N. MEX.(NOMENCLATURE LIST)	NL-
SANDIA CORP., ALBUQUERQUE, N. MEX.	OS-
SANDIA CORP., ALBUQUERQUE, N. MEX.(QUALITY ASSURANCE INFORMATION LETTER)	Q-
SANDIA CORP., ALBUQUERQUE, N. MEX.(QUALITY ENG. PROJ)	QAIL-
SANDIA CORP., ALBUQUERQUE, N. MEX.	QEP-
SANDIA CORP., ALBUQUERQUE, N. MEX.	QR-
SANDIA CORP., ALBUQUERQUE, N. MEX.	R-
SANDIA CORP., ALBUQUERQUE, N. MEX.	REF. SYM.-...
SANDIA CORP., ALBUQUERQUE, N. MEX.	RO-
SANDIA CORP., ALBUQUERQUE, N. MEX.	RS-(NUMBER)/(NUMBER)
SANDIA CORP., ALBUQUERQUE, N. MEX.	SAND-
SANDIA CORP., ALBUQUERQUE, N. MEX.	SANDIA TEST-(NO.)-(NO.)
SANDIA CORP., ALBUQUERQUE, N. MEX.	SC-
SANDIA CORP., ALBUQUERQUE, N. MEX. (CATALOG)	SC-(NUMBER)-C
SANDIA CORP., ALBUQUERQUE, N. MEX. (CORP. PROCEDURE)	SC-(NUMBER)-CP
SANDIA CORP., ALBUQUERQUE, N. MEX.	SC-(NUMBER)-M
SANDIA CORP., ALBUQUERQUE, N. MEX.(NOMENCLATURE LIST)	SC-(NUMBER)-NL

SANDIA CORP.

SANDIA CORP., ALBUQUERQUE, N. MEX. (PARTS LIST)	SC-(NUMBER)-PL
SANDIA CORP., ALBUQUERQUE, N. MEX. (PROGRESS REPT.)	SC-(NUMBER)-PR
SANDIA CORP., ALBUQUERQUE, N. MEX.	SC-(NUMBER)-RC
SANDIA CORP., ALBUQUERQUE, N. MEX.	SC-(NUMBER)(RR)
SANDIA CORP., ALBUQUERQUE, N. MEX. (SPECIFICATION)	SC-(NUMBER)(SP)
SANDIA CORP., ALBUQUERQUE, N. MEX.	SC-(NUMBER)(WD)
SANDIA CORP., ALBUQUERQUE, N. MEX. (TEST MANUAL)	SC-(NUMBER)-TM
SANDIA CORP., ALBUQUERQUE, N. MEX.	SCAL-MISC-
SANDIA CORP., ALBUQUERQUE, N. MEX.	SCAL-SERIES-
SANDIA CORP., ALBUQUERQUE, N. MEX.	SCDC-
SANDIA CORP., ALBUQUERQUE, N. MEX.	SC-DC-(YEAR)-
SANDIA CORP., ALBUQUERQUE, N. MEX.	SC-DCR-
SANDIA CORP., ALBUQUERQUE, N. MEX. (DEV. REPT.)	SCDR-
SANDIA CORP., ALBUQUERQUE, N. MEX.	SCDR-(NO.)-(YR.)
SANDIA CORP., ALBUQUERQUE, N. MEX.	SC-DW-(YEAR)-
SANDIA CORP., ALBUQUERQUE, N. MEX.	SCL-B-
SANDIA CORP., ALBUQUERQUE, N. MEX.	SC-QAA-
SANDIA CORP., ALBUQUERQUE, N. MEX. (REPORT)	SC-R-
SANDIA CORP., ALBUQUERQUE, N. MEX. (REPRINT)	SCR-
SANDIA CORP., ALBUQUERQUE, N. MEX. (RESEARCH)	SCR-
SANDIA CORP., ALBUQUERQUE, N. MEX.	SCR-
SANDIA CORP., ALBUQUERQUE, N. MEX.	SCRB-
SANDIA CORP., ALBUQUERQUE, N. MEX.	SCRR-(YR.)-
SANDIA CORP., ALBUQUERQUE, N. MEX.	SCS-
SANDIA CORP., ALBUQUERQUE, N. MEX.	SCT-
SANDIA CORP., ALBUQUERQUE, N. MEX. (TECHNICAL MEMO.)	SCTM-
SANDIA CORP., ALBUQUERQUE, N. MEX.	SCTM-(NO.)-(YR.)
SANDIA CORP., ALBUQUERQUE, N. MEX.	SCTM-(NO.)-(YEAR)-(NO.)
SANDIA CORP., ALBUQUERQUE, N. MEX.	SC-TM-(YEAR)-
SANDIA CORP., ALBUQUERQUE, N. MEX.	SCTM-M-1004 (YR./MO.)
SANDIA CORP., ALBUQUERQUE, N. MEX.	SM-4-
SANDIA CORP., ALBUQUERQUE, N. MEX. (READINESS PROGRAM SERIES)	SM-(LETTER)(NUMBER)-6
SANDIA CORP., ALBUQUERQUE, N. MEX.	SM-(LETTER(S))-
SANDIA CORP., ALBUQUERQUE, N. MEX.(STD. PARTS MANUAL)	SMD-
SANDIA CORP., ALBUQUERQUE, N. MEX. (ELECTRICAL STANDARD PARTS MANUAL)	SPM-
SANDIA CORP., ALBUQUERQUE, N. MEX. (MECHANICAL STANDARD PARTS MANUAL)	SPME-
SANDIA CORP., ALBUQUERQUE, N. MEX. (TITLE LISTS)	SPMM-
SANDIA CORP., ALBUQUERQUE, N. MEX. (SPECIAL WEAPONS EQUIPMENT LIST)	STL-
SANDIA CORP., ALBUQUERQUE, N. MEX. (SPECIAL WEAPONS SUPPLY CATALOG)	SWEL-
SANDIA CORP., ALBUQUERQUE, N. MEX.	SWSC-
SANDIA CORP., ALBUQUERQUE, N. MEX.	T-
SANDIA CORP., ALBUQUERQUE, N. MEX.	TECH. MEMO.-
SANDIA CORP., ALBUQUERQUE, N. MEX.	TM-(NO.)-(YR.)-
SANDIA CORP., ALBUQUERQUE, N. MEX.	VUP-
SANDIA CORP., ALBUQUERQUE, N. MEX.	Z-
SANDIA CORP., ALBUQUERQUE, N. MEX.	ZP-
(SEE ALSO LATER NAME:SANDIA LABS., ALBUQUERQUE, N.MEX.)	
SANDIA CORP. AD HOC WORKING GROUP, ALBUQUERQUE, N.M.	AWG-
SANDIA CORP. ALO PRIMARY STANDARDS LAB., ALBUQUERQUE, N. MEX. (DIRECT CURRENT PARAMETERS, RESISTANCE)	DC-
SANDIA CORP. ALO PRIMARY STANDARDS LAB., ALBUQUERQUE, N. MEX. (DIRECT CURRENT PARAMETERS, RESISTANCE)	PROCEDURE DC-
SANDIA CORP. LIBRARY, ALBUQUERQUE, N.MEX.(TRANSLATION)	SCL-
SANDIA CORP. LIVERMORE LAB., CALIF. (REFERENCE SYMBOL, ACCESSION NUMBERS)	RS-
SANDIA CORP. LIVERMORE LAB., CALIF.	SC-
SANDIA CORP. LIVERMORE LAB., CALIF.	SC-(NUMBER)-M
SANDIA CORP. LIVERMORE LAB., CALIF. (TECHNICAL MEMO)	SC-(NUMBER)-TM
SANDIA CORP. LIVERMORE LAB., CALIF. (TECH. REPORT)	SC-(NUMBER)-TR
SANDIA CORP. LIVERMORE LAB., CALIF. (WEAPON DATA)	SC-(NUMBER)-WD
SANDIA CORP. LIVERMORE LAB., CALIF.	SC-DR-
SANDIA CORP. LIVERMORE LAB., CALIF. (DEV. REPORT)	SCDR-(NUMBER)-(YEAR)
SANDIA CORP. LIVERMORE LAB., CALIF.	SCL-
SANDIA CORP. LIVERMORE LAB., CALIF.	SCL-CR-(YEAR)-
SANDIA CORP. LIVERMORE LAB., CALIF.	SCL-DC-(YEAR)-
SANDIA CORP. LIVERMORE LAB., CALIF. (REPRINT)	SCL-R-
SANDIA CORP. LIVERMORE LAB., CALIF.	SCLV-
SANDIA CORP. LIVERMORE LAB., CALIF. (WEAPON DATA)	SCL-WD-
SANDIA CORP. LIVERMORE LAB., CALIF. (REPRINT)	SCR-
SANDIA CORP. LIVERMORE LAB., CALIF.	SC-RR-
SANDIA CORP. LIVERMORE LAB., CALIF. (TECHNICAL MEMO)	SCTM-(NUMBER)-(YEAR)
(SEE ALSO LATER NAME: SANDIA LABS., LIVERMORE, CALIF)	
SANDIA CORP. MANUAL BOARD., ALBUQUERQUE, N. MEX.	SMB-(YR.)-
SANDIA CORP. PHYSICAL AND ELECTRICAL STANDARDS DEPT., ALBUQUERQUE, N. MEX.	PES-(LTR.-NO.-NO.-YR.)
SANDIA CORP. QUALITY ASSURANCE DEPT., ALBUQUERQUE, NM	AEC-ALO-DOC.
SANDIA CORP. QUALITY ASSURANCE DEPT., ALBUQUERQUE, NM	AEC-ALO-QAP'S
SANDIA CORP. RES. + DEV. BD., ALBUQUERQUE, N. MEX.	SRDB-(YR.)-
SANDIA CORP. ROAD MATERIEL BD., ALBUQUERQUE, N. MEX.	SRMB-(YR.)-
SANDIA CORP. SPEC. WEAPONS DEV. BD., ALBUQUERQUE, N.M.	SWDB-(YR.)-
SANDIA CORP. SPECIAL WEAPONS DEVELOPMENT BOARD. GUIDED MISSILE COMMITTEE, ALBUQUERQUE, N. MEX.	GMC-
SANDIA CORP. TECH. FACILITIES BD., ALBUQUERQUE, N.M.	STFB-(YR.)-
SANDIA LAB., ALBUQUERQUE, N. MEX.	SL-
SANDIA LAB., ALBUQUERQUE, N. MEX.	SLMS-
SANDIA LABS., ALBUQUERQUE, N. MEX.	SC-ANSIC-
SANDIA LABS., ALBUQUERQUE, N. MEX.	SC-ARPIC-
SANDIA LABS., ALBUQUERQUE, N. MEX. (SELECTIVE BIB.)	SC-B-(YEAR)-
SANDIA LABS., ALBUQUERQUE, N. MEX.	SC-CR-(YEAR)-
SANDIA LABS., ALBUQUERQUE, N. MEX.	SC-DC-
SANDIA LABS., ALBUQUERQUE, N. MEX.	SC-DR-(YEAR)-
SANDIA LABS., ALBUQUERQUE, N. MEX.	SC-DR-
SANDIA LABS., ALBUQUERQUE, N. MEX. (TRANSLATION)	SCI-T-(YEAR)-
SANDIA LABS., ALBUQUERQUE, N. MEX. (TRANSLATION)	SCL-T-
SANDIA LABS., ALBUQUERQUE, N. MEX.	SCL-TM-
SANDIA LABS., ALBUQUERQUE, N. MEX.	SC-M-(YEAR)-
SANDIA LABS., ALBUQUERQUE, N. MEX.	SC-PR-(YEAR)-
SANDIA LABS., ALBUQUERQUE, N. MEX. (QUARTERLY REPORT)	SC-QR-
SANDIA LABS., ALBUQUERQUE, N. MEX.	SC-RR-(YEAR)-
SANDIA LABS., ALBUQUERQUE, N. MEX. (TRANSLATIONS)	SC-T-
SANDIA LABS., ALBUQUERQUE, N. MEX.	SC-TM-(YEAR)-

SANDIA LABS., ALBUQUERQUE, N. MEX.	SC-WD-(YEAR)-
SANDIA LABS., LIVERMORE, CALIF.	SCL-CR-(YEAR)-
SANDIA LABS., LIVERMORE, CALIF.	SCL-DC-
SANDIA LABS., LIVERMORE, CALIF.	SCL-DC-(YEAR)-
SANDIA LABS., LIVERMORE, CALIF.	SCL-DR-(YEAR)-
SANDIA LABS., LIVERMORE, CALIF.	SCL-RR-
SANDIA LABS., LIVERMORE, CALIF. (TRANSLATION)	SCL-T-
SANDIA LABS., LIVERMORE, CALIF.	SCL-TM-
SANDIA LABS., LIVERMORE, CALIF.	SCL-TN-(YEAR)-
SANDIA LABS., LIVERMORE, CALIF.	SC-R-(YEAR)-
SANDIA LABS., LIVERMORE, CALIF.	SCT-T-(YEAR)-
SANTA BARBARA ANALYSIS AND PLANNING CORP., CALIF.	SBANP-(YEAR)-
SANTA BARBARA RESEARCH CENTER, GOLETA, CALIF.	SBRC-
SANTA CLARA, CALIF. UNIV.	SANCU-
SANTA CLARA, CALIF. UNIV.	SCU-
SANTA FE OPERATIONS OFFICE, AEC, N. MEX. (REF. 4)	SFO-
SANTA RITA TECHNOLOGY, INC., MENLO PARK, CALIF.	SRT-(YEAR)-
SAO PAULO, BRAZIL. UNIVERSIDADE	CMN-
SAO PAULO, BRAZIL. UNIVERSIDADE. CENTER OF STUDIES IN CELESTIAL MECHANICS	CEMC-IMP-USP-(YR/YR-NO.)
SAO PAULO, BRAZIL. UNIV. CENTRE DE MEDICINA NUCLEAR	CMN-PUB-(NUMBER/YEAR)
SAO PAULO, BRAZIL. UNIVERSIDADE. INSTITUTO DE ENERGIA ATOMICA	CNEN-IEA-
SAO PAULO, BRAZIL. UNIVERSIDADE. INSTITUTO DE ENERGIA ATOMICA	IEA-
SAO PAULO, BRAZIL. UNIVERSIDADE. INSTITUTO DE ENERGIA ATOMICA	IEA-INF-
SAO PAULO, BRAZIL. UNIVERSIDADE. INSTITUTO DE ENERGIA ATOMICA	IEA/RQ-
SARGENT AND LUNDY, CHICAGO	SL-
SARKES TARZIAN, INC., BLOOMINGTON, IND.	STI-
SASKATCHEWAN. UNIV., SASKATOON	AR-
SASKATCHEWAN. UNIV., SASKATOON	DR-
SASKATCHEWAN. UNIV., SASKATOON	RS-
SASKATCHEWAN. UNIV., SASKATOON	SASKU-
SASKATCHEWAN. UNIV., SASKATOON. PLASMA PHYSICS LAB.	PPL-
SASKATCHEWAN. UNIV., SASKATOON. PLASMA PHYSICS LAB.	SU-PPL-
SASKATCHEWAN. UNIV., SASKATOON. SASKATCHEWAN ACCELERATOR LAB.	SAL-
SAVANNAH RIVER OPERATIONS OFFICE, AEC, AIKEN, S.C.	SR-1-
SAVANNAH RIVER OPERATIONS OFFICE, AEC, AIKEN, S.C.	SR-(LETTERS)-
SAVANNAH RIVER OPERATIONS OFFICE, AEC, AIKEN, S.C.	SRO-
SAVANNAH RIVER OPERATIONS OFFICE, AEC, AIKEN, S.C.	SROO-
SAVANNAH RIVER OPERATIONS OFFICE, AEC, AIKEN, S.C.	SRW-T-
SAVANNAH RIVER OPERATIONS OFFICE, AEC, AIKEN, S.C.	US-
SCHENECTADY OPERATIONS OFFICE, AEC, N.Y.	SNY-LEJ-
SCHENECTADY OPERATIONS OFFICE, AEC, N.Y.	SO-
SCHENECTADY OPERATIONS OFFICE, AEC, N.Y. (REF. 4)	SO-INT-
SCHENECTADY OPERATIONS OFFICE, AEC, N.Y. (REF. 4)	SOO-
SCHIFFBAUTECHNISCHE GESELLSCHAFT	STG-
SCHJELDAHL (G.T.) CO., NORTHFIELD, MINN.	C-
SCHOOL OF AEROSPACE MEDICINE, BROOKS AFB, TEX. (AEROMEDICAL REVIEW)	REVIEW-
SCHOOL OF AEROSPACE MEDICINE, BROOKS AFB, TEX.	SAM-
SCHOOL OF AEROSPACE MEDICINE, BROOKS AFB, TEX.	SAM-REVIEW-(NO.-YEAR)
SCHOOL OF AEROSPACE MEDICINE, BROOKS AFB, TEX.	SAM-TDR-(YEAR)-
SCHOOL OF AEROSPACE MEDICINE, BROOKS AFB, TEX.	SAM-TR-(YEAR)-
SCHOOL OF AEROSPACE MED., BROOKS AFB, TEX.(TECH.RPT.)	SAM-TT-(NUMBER)-
SCHOOL OF AEROSPACE MEDICINE, BROOKS AFB, TEX.	SAM-TT-R-(NUMBER)-
SCHOOL OF AEROSPACE MEDICINE, BROOKS AFB, TEX.	
SCHOOL OF AVIATION MEDICINE, BROOKS AFB, TEX.	AF-SAM-(YEAR)-
SCHOOL OF AVIATION MEDICINE, BROOKS AFB, TEX. (TECHNICAL DOCUMENTARY REPORT)	AF-SAM-TDR-
SCHOOL OF AVIATION MEDICINE, BROOKS AFB, TEX.	SAC-
SCHOOL OF AVIATION MEDICINE, RANDOLPH AFB, TEX.	AAF SAM-
SCHOOL OF AVIATION MEDICINE, RANDOLPH AFB, TEX.	AAF SAM PROJ-
SCHOOL OF AVIATION MEDICINE, RANDOLPH AFB, TEX.	AF-SAM-
SCHOOL OF AVIATION MEDICINE, RANDOLPH AFB, TEX.	AF-SAM (PROJ. NO.)
SCHOOL OF AVIATION MEDICINE, RANDOLPH AFB, TEX.	AF-SAM-(YR)-
SCHOOL OF AVIATION MEDICINE, RANDOLPH AFB, TEX.	AIRU AM-
SCHOOL OF AVIATION MEDICINE, RANDOLPH AFB, TEX.	AIRU AM SRU-
SCHOOL OF AVIATION MEDICINE, RANDOLPH AFB, TEX. (AIRCRAFT NUCLEAR PROPULSION BIOMEDICAL SUMMARY)	AIRU-ANPBP-
SCHOOL OF AVIATION MEDICINE, RANDOLPH AFB, TEX.	CSA-(YEAR)-
SCHOOL OF AVIATION MEDICINE, RANDOLPH AFB, TEX.	S-19000 THRU S-21999
SCHOOL OF AVIATION MEDICINE, RANDOLPH AFB, TEX.	SAC-
SCHOOL OF AVIATION MEDICINE, RANDOLPH AFB, TEX.	SAM-
SCHOOL OF AVIATION MEDICINE, RANDOLPH AFB, TEX.	SAM-(YEAR)-
SCHOOL OF AVIATION MEDICINE, RANDOLPH AFB, TEX.	SAM-TR-
SCHOOL OF AVIATION MEDICINE, RANDOLPH AFB, TEX.	USAFSAM-
SCHWABLAND (GEORGE A.), ALEXANDRIA, VA.	GAS-
SCHWARZKOPF DEVELOPMENT CORP., YONKERS, N.Y.	SDCO-
SCHWEIZERISCHES INST. FUER NUKLEARFORSCHUNG, ZURICH	SIN-
SCHWEIZERISCHES INST. FUER NUKLEARFORSCHUNG, ZURICH	SIN-TM-(NUMBER)-(NUMBER)
SCHWEIZERISCHES INST. FUER NUKLEARFORSCHUNG, ZURICH	TM-(NUMBER)-(NUMBER)
SCIENCE APPLICATIONS, INC., LA JOLLA, CALIF.	SAI-(YEAR)-
SCIENCE COMMUNICATION, INC., WASHINGTON, D.C.	SCI-
SCIENCE COMMUNICATION, INC., WASHINGTON, D.C.	SCI-ARIES-
SCIENCE ENGINEERING ASSOCIATES, SAN MARINO, CALIF.	PN-(NUMBER)-
SCIENCE ENGINEERING ASSOCIATES, SAN MARINO, CALIF.	SEA-PN-
SCIENCE TRANSLATION SERVICE	STS-

SCIENTIFIC ADVISORY BOARD (AIR FORCE)

SCIENTIFIC ADVISORY BOARD (AIR FORCE), WASHINGTON, DC	OSA-
SCIENTIFIC ADVISORY BOARD (AIR FORCE), WASHINGTON, DC	SAB-
SCIENTIFIC ADVISORY BOARD (AIR FORCE), WASHINGTON, DC	SAG-
SCIENTIFIC-ATLANTA, INC.	J(NUMBER)-FR-
SCIENTIFIC-ATLANTA, INC.	J(NUMBER)-QR-
SCIENTIFIC-ATLANTA, INC.	SAI-
SCIENTIFIC MANAGEMENT ASSOCS., INC., HADDONFIELD, NJ	SMA-TR-(NUMBER)-
SCIENTIFIC RESEARCH INSTRUMENTS CORP., BALTIMORE	SRIC-(YEAR)-
SCIENTIFIC TRANSLATION SERVICE, ANN ARBOR, MICH.	K-TRANS-
SCRIPPS INSTITUTION OF OCEANOGRAPHY, LA JOLLA, CALIF.	IMP-TR-
SCRIPPS INSTITUTION OF OCEANOGRAPHY, LA JOLLA, CALIF.	SIO-
SCRIPPS INSTITUTION OF OCEANOGRAPHY, LA JOLLA, CALIF.	SIO-(YEAR)-
SCRIPPS INSTITUTION OF OCEANOGRAPHY, LA JOLLA, CALIF.	SIO-REF-(YEAR)-
SCRIPPS INSTITUTION OF OCEANOGRAPHY, LA JOLLA, CALIF.	SIO-SEA GRANT-PUB-
SCRIPPS INSTITUTION OF OCEANOGRAPHY. APPLIED OCEANOGRAPHY GROUP, LA JOLLA, CALIF.	AOG-
SCRIPPS INSTITUTION OF OCEANOGRAPHY. APPLIED OCEANOGRAPHY GROUP, LA JOLLA, CALIF.	AOG-S-
SCRIPPS INSTITUTION OF OCEANOGRAPHY. APPLIED OCEANOGRAPHY GROUP, LA JOLLA, CALIF.	AOG-U-
SCRIPPS INSTITUTION OF OCEANOGRAPHY. MARINE PHYSICAL LAB., SAN DIEGO, CALIF.	MPL-C-
SCRIPPS INSTITUTION OF OCEANOGRAPHY. MARINE PHYSICAL LAB., SAN DIEGO, CALIF.	MPL-S-
SCRIPPS INSTITUTION OF OCEANOGRAPHY. MARINE PHYSICAL LAB., SAN DIEGO, CALIF.	MPL-TM-
SCRIPPS INSTITUTION OF OCEANOGRAPHY. MARINE PHYSICAL LAB., SAN DIEGO, CALIF.	MPL-U-
SEA-SPACE SYSTEMS, INC., TORRANCE, CALIF.	SSS-
SEISMOGRAPH SERVICE CO., TULSA, OKLA.	SSC-
SELENIA S.P.A., ROME	ELDO-RFI-
SELENIA S.P.A., ROME	ELDO-RFI-GA-
SELENIA S.P.A., ROME	ELDO-TR-
SENIOR OBSERVER SECTION, MATHER AFB, CALIF.	SOTS-
SERCK RADIATORS, LTD., BIRMINGHAM, WAR., ENG.	K-
SERENDIPITY ASSOCIATES, CHATSWORTH, CALIF.	PRM-
SERENDIPITY ASSOCIATES, CHATSWORTH, CALIF.	TR-(NUMBER)-(YEAR)-
SERENDIPITY ASSOCIATES, LOS ALTOS, CALIF.	TR-(NUMBER)-(YEAR)-
SERGEANT JOINT AEC-DOD WARHEAD COORDINATING COMM.	JCC/SRGNT-
SERGEANT JOINT AEC-DOD WARHEAD COORDINATING COMM. (REF. 15)	SERGEANT-JWCC-
SERVICES VALVE TEST LAB., HASLEMERE, ENGLAND	SVT-
SERVO CORP. OF AMERICA, HICKSVILLE, N.Y.	SCA-
SERVO CORP. OF AMERICA, NEW HYDE PARK, N.Y.	SCA-
SERVOMECHANISMS, INC., GOLETA, CALIF.	SMIR-
SERVOMECHANISMS, INC., HAWTHORNE, CALIF.	SM-
SFD LABS., INC., UNION, N.J.	SFD-
SFD LABS., INC., UNION, N.J.	SFD-(NUMBER)-R-
SHANNON AND WILSON INC., SEATTLE	SHANW-
SHARP (GEORGE G.) INC., N.Y.C.	N-
SHARPLES CORP., PHILADELPHIA	59A-HSB-
SHAW AND ESTES, DALLAS	SE-
SHEFFIELD, ENGLAND. UNIV.	GSR-
SHEFFIELD, ENGLAND. UNIV. DEPT. OF FUEL TECHNOLOGY AND CHEMICAL ENGINEERING	HIC-
SHEFFIELD, ENGLAND. UNIV. DEPT. OF FUEL TECHNOLOGY AND CHEMICAL ENGINEERING	HIC-(NUMBER)-APP
SHEFFIELD, ENGLAND. UNIV. DEPT. OF PROBABILITY AND STATISTICS	RR-(NO.)/(INITLS.)-
SHELL CHEMICAL CO., SAN FRANCISCO	SCC-
SHELL CHEMICAL CORP. INDUSTRIAL CHEMICALS DIV.,N.Y.C.	SC-
SHELL DEVELOPMENT CO., EMERYVILLE, CALIF.	EMS-
SHELL DEVELOPMENT CO., EMERYVILLE, CALIF.	S-
SHELL DEVELOPMENT CO., EMERYVILLE, CALIF.	SDC-
SHELL OIL CO., N.Y.C. (REF. 1)	(NUMBER)
SHELL OIL CO., N.Y.C.	SHO-
SHIP AND MACHINE BUILDING TESTING INST. (KIEL NAVY INST., DEPT. OF SHIP CONSTRUCTION)	SMBTI-
SHIPS PARTS CONTROL CENTER, MECHANICSBURG, PENNA.	SPCC-
SHOCK AND VIBRATION INFORMATION CENTER (DEFENSE), WASHINGTON, D.C.	BULL-(NUMBER)-PT-
SHOCK HYDRODYNAMICS, INC., SHERMAN OAKS, CALIF.	SH-
SHOCK HYDRODYNAMICS, INC., SHERMAN OAKS, CALIF.	SH-(4 DIGITS)-MC
SHOCK HYDRODYNAMICS, INC., SHERMAN OAKS, CALIF.	SHI-
SHOCK HYDRODYNAMICS, INC., SHERMAN OAKS, CALIF.	SM-
SHORT BROS. AND HARLAND, LTD.	GW-
SHUFORD-MASSENGILL CORP., LEXINGTON, MASS.	SMC-R-
SHUFORD-MASSENGILL CORP., LEXINGTON, MASS.	SM-R-
SIEGLER CORP., INGLEWOOD, CALIF.	SIEGLER-
SIEGLER CORP., INGLEWOOD, CALIF.	SSTG-
SIERRA ELECTRONIC CORP., SAN CARLOS, CALIF.	SEC-
SIERRA ELECTRONIC CORP., SAN CARLOS, CALIF.	SIEC-

SIERRA ELECTRONIC CORP., SAN FRANCISCO	SEC-
SIGNAL CORPS (ARMY) (MANUAL)	ASF-M-SIG-
SIGNAL CORPS (ARMY) (CAPTURED ENEMY SIGNAL EQUIPMENT)	CESE-
SIGNAL CORPS (ARMY) (CAPTURED ENEMY SIGNAL EQUIPMENT. TECHNICAL LIAISON INTELLIGENCE)	CESE TLI-
SIGNAL CORPS (ARMY) (INFRARED REPORT)	IFR-
SIGNAL CORPS (ARMY) (SCIENTIFIC ELECTRONIC STUDY)	SCSES-
SIGNAL CORPS (ARMY)	SIG-
SIGNAL CORPS (ARMY)	SIGC-
SIGNAL CORPS (ARMY) (CAPTURED ENEMY SIGNAL EQUIPMENT)	SIG CESE-
SIGNAL CORPS (ARMY) (INTELLIGENCE REPORT)	SIG IR-
SIGNAL CORPS (ARMY) (INTELLIGENCE REPORT)	SIG IR RAD-
SIGNAL CORPS (ARMY) (INTELLIGENCE REPORT)	SIG IR RCM-
SIGNAL CORPS (ARMY) (INTELLIGENCE REPORT)	SIG IR SRM-
SIGNAL CORPS (ARMY) (TRANSLATION)	SIG T-
SIGNAL CORPS (ARMY) (TENTATIVE SPECIFICATION)	SIG TEN SPEC-
SIGNAL CORPS (ARMY)	SIG X
SIGNAL CORPS (LABORATORIES)	SL-
SIGNAL CORPS (ARMY) (CAPTURED ENEMY SIGNAL EQUIPMENT) TECHNICAL LIAISON INTELLIGENCE)	TLI-
SIGNAL CORPS (ARMY)	TM-11-
SIGNAL CORPS (ARMY). BATTERY DEVELOPMENT SECTION	BD-
SIGNAL CORPS (ARMY). ENEMY EQUIPMENT INTELLIGENCE DIV. CHINA-BURMA-INDIA THEATER (PRELIMINARY REPORT)	CBI EEIS SIG PR-
SIGNAL CORPS (ARMY). ENEMY EQUIPMENT INTELLIGENCE DIV. CHINA-BURMA-INDIA THEATER	EEIS-
SIGNAL CORPS (ARMY). ENEMY EQUIPMENT INTELLIGENCE DIV. CHINA THEATER (PRELIMINARY REPORT)	CHINA EEIS SIG IC PR-
SIGNAL CORPS (ARMY). INTELLIGENCE AND SECURITY BOARD	SIG IS T-
SIGNAL CORPS (ARMY). OPERATIONAL RESEARCH BRANCH	ORB-
SIGNAL CORPS (ARMY). OPERATIONAL RESEARCH BRANCH	SIG ORB-
SIGNAL CORPS (ARMY). OPERATIONAL RESEARCH BRANCH	SIG ORB (NUMBER)-
SIGNAL CORPS (ARMY). OPERATIONAL RESEARCH BRANCH	SIG ORB (LETTER(S))-
SIGNAL CORPS (ARMY). OPERATIONAL RESEARCH GROUP	ORG-
SIGNAL CORPS (ARMY). OPERATIONAL RESEARCH GROUP	SIG ORG-
SIGNAL CORPS (ARMY). OPERATIONAL RESEARCH GROUP	SIG ORG (LTR(S))-(NO.)-
SIGNAL CORPS (ARMY). OPERATIONAL RESEARCH STAFF	ORS-
SIGNAL CORPS (ARMY). OPERATIONAL RESEARCH STAFF	SIG ORS-
SIGNAL CORPS (ARMY). OPERATIONAL RESEARCH STAFF	SIG ORS P-
SIGNAL CORPS (ARMY). PICTORIAL ENGINEERING AND RESEARCH LAB., LONG ISLAND, N.Y.	APP DP-
SIGNAL CORPS (ARMY). PICTORIAL ENGINEERING AND RESEARCH LAB., LONG ISLAND, N.Y. (DIV. REPT.)	ERL DP-
SIGNAL CORPS (ARMY). PICTORIAL ENGINEERING AND RESEARCH LAB., LONG ISLAND, N.Y. (DRAWING)	PERLD ES-D-
SIGNAL CORPS (ARMY). PICTORIAL ENGINEERING AND RESEARCH LAB., LONG ISLAND, N.Y.	SIG APP DP-
SIGNAL CORPS (ARMY). PICTORIAL ENGINEERING AND RESEARCH LAB., LONG ISLAND, N.Y. (DIV. REPT.)	SIG ERL DP-
SIGNAL CORPS (ARMY). PICTORIAL ENGINEERING AND RESEARCH LAB., LONG ISLAND, N.Y. (DRAWING)	SIG PERLD ES-D-
SIGNAL CORPS (ARMY). PICTORIAL ENGINEERING AND RESEARCH LAB., LONG ISLAND, N.Y.	SIG PERLD IP-
SIGNAL CORPS (ARMY). PLANS AND INTELLIGENCE BRANCH	PI-
SIGNAL CORPS (ARMY). PLANS AND INTELLIGENCE BRANCH	SIG PI-
SIGNAL CORPS (ARMY). RADIO PROPAGATION SECTION	RPS-
SIGNAL CORPS (ARMY). RADIO PROPAGATION SECTION	SIG RPS-
SIGNAL CORPS (ARMY). RADIO PROPAGATION SECTION (PROPAGATION REPORT)	SIG RPS PR-
SIGNAL CORPS INTELL. AGENCY (ARMY), WASH.,DC (TRANS.)	SIGIA-
SIGNAL CORPS INTELLIGENCE AGENCY (ARMY), WASH., D.C.	SIGIA R-(NUMBER)-(YEAR)
SIGNAL CORPS. RADIO PROPAGATION AGENCY, FORT MONMOUTH, N.J.	RPU-
SIGNAL CORPS ENG. LABS.,FT. MONMOUTH, NJ (ENG. REPT.)	E-
SIGNAL CORPS ENG. LABS.,FT.MONMOUTH,N.J.(FISCAL YEAR)	FY-
SIGNAL CORPS ENG. LABS.,FT. MONMOUTH, NJ (TECH.MEMO)	M-
SIGNAL CORPS ENGINEERING LABS., FORT MONMOUTH, N.J.	ORD SCEL-
SIGNAL CORPS ENGINEERING LABS., FORT MONMOUTH, N.J.	SCEL-
SIGNAL CORPS ENGINEERING LABS., FORT MONMOUTH, N.J.	SCEL-E-
SIGNAL CORPS ENG. LABS.,FT. MONMOUTH, N.J.(ENG. RPTS)	SCEL ER-
SIGNAL CORPS ENG. LABS.,FT. MONMOUTH, N.J.(ENG. RPTS)	SCEL ER E-
SIGNAL CORPS ENGINEERING LABS., FORT MONMOUTH, N.J. (ELECTRON TUBE INFORMATION BULLETIN)	SCEL-ETIB-
SIGNAL CORPS ENGINEERING LABS., FORT MONMOUTH, N.J.	SCEL-M-
SIGNAL CORPS ENG. LABS.,FT. MONMOUTH, N.J.(TEST RPT.)	SCEL-T-
SIGNAL CORPS ENG. LABS.,FT. MONMOUTH, N.J.(TECH.MEMO)	SCEL TM-
SIGNAL CORPS ENG. LABS.,FT. MONMOUTH, N.J.(TECH MEMO)	SCEL TM M-
SIGNAL CORPS ENG. LABS.,FT. MONMOUTH, N.J.(TEST RPTS)	SCEL TR-
SIGNAL CORPS ENG. LABS.,FT. MONMOUTH, N.J.(TEST RPTS)	SCEL TR T-
SIGNAL CORPS ENGINEERING LABS., FORT MONMOUTH, N.J.	SEL-
SIGNAL CORPS ENG. LABS.,FT. MONMOUTH, NJ (TEST REPT.)	T-
SIGNAL CORPS ENGINEERING LABS. SOLID STATE DEVICES BRANCH, FORT MONMOUTH, N.J.	SCEL-SSDB-
SIGNAL CORPS ENGINEERING LABS. SUPPRESSION DEVELOPMENT SECTION, FORT MONMOUTH, N.J. (ENG. MEMO)	SDS EM-
SIGNAL CORPS TECHNICAL COMMITTEE, WASHINGTON, D.C.	SCTC-
SIGNAL INTELLIGENCE SERVICE, 849TH. CAPTURED ENEMY INTELLIGENCE BRANCH	CEIS-
SIGNAL INTELLIGENCE SERVICE, 849TH. CAPTURED ENEMY INTELLIGENCE BRANCH	SIG CEIS-
SIGNATRON, INC., LEXINGTON, MASS.	SIG-A-
SIGNATRON, INC., LEXINGTON, MASS.	SIG-CR-
SIGNATRON, INC., LEXINGTON, MASS.	SIG-TR-
SIKORSKY AIRCRAFT DIV., UNITED AIRCRAFT CORP., BRIDGEPORT, CONN.	SER-
SIKORSKY AIRCRAFT DIV., UNITED AIRCRAFT CORP., BRIDGEPORT, CONN.	UAC-SA-
SIKORSKY AIRCRAFT DIV., UNITED AIRCRAFT CORP., STRATFORD, CONN.	SAR-
SIKORSKY AIRCRAFT DIV., UNITED AIRCRAFT CORP., STRATFORD, CONN.	SAR SER-
SIKORSKY AIRCRAFT DIV., UNITED AIRCRAFT CORP., STRATFORD, CONN.	SAR WTT-
SIKORSKY AIRCRAFT DIV., UNITED AIRCRAFT CORP., STRATFORD, CONN.	SAR X-
SIKORSKY AIRCRAFT DIV., UNITED AIRCRAFT CORP., STRATFORD, CONN.	SER-
SIKORSKY AIRCRAFT DIV., UNITED AIRCRAFT CORP., STRATFORD, CONN.	UAC-SAR-
SIMMONDS PRECISION PRODUCTS, INC., TARRYTOWN, N.Y.	PD-
SIMPSON GUMPERTZ AND HEGER, INC., CAMBRIDGE, MASS.	SGH-
SIMULMATICS CORP., CAMBRIDGE, MASS.	SIM/CAM/(MONTH)/(YEAR)

SINGER-GENERAL PRECISION, INC.

SINGER-GENERAL PRECISION, INC. KEARFOTT DIV., LITTLE FALLS, N.J.	B(8 DIGITS)
SINGER MFG. CO., N.Y.C.	SMC-
SINTERCAST CORP. OF AMERICA, N.Y.C.	SICA-
SKF INDUSTRIES, INC. RES. LAB., KING OF PRUSSIA, PA.	AL(YEAR)L(NO.)
SKF INDUSTRIES, INC. RES. LAB., KING OF PRUSSIA, PA.	AL-(YEAR)Q-
SKF INDUSTRIES, INC. RES. LAB., KING OF PRUSSIA, PA.	AL(YEAR)T(NO.)
SKF INDUSTRIES, INC. RES. LAB., KING OF PRUSSIA, PA.	SFK-AL(NUMBER)L(NUMBER)
SKF INDUSTRIES, INC. RES. LAB., KING OF PRUSSIA, PA.	SKF-AL(YR.)L(NO.)
SKYDYNE, INC., PORT JERVIS, N.Y.	SKY-
SLOAN-KETTERING INST. FOR CANCER RESEARCH, N.Y.C.	SK-
SLOAN-KETTERING INST. FOR CANCER RESEARCH, N.Y.C.	SK MD-
SMALL BUSINESS ADMINISTRATION, WASHINGTON, D.C.	SBA-
SMITH (A.O.) CORP., MILWAUKEE	AD-
SMITH (A.O.) CORP., MILWAUKEE	ERL-
SMITH (A.O.) CORP. ADMINISTRATION, RESEARCH AND ENGINEERING LABS., MILWAUKEE (HEAT EXCHANGERS)	ARL-HE-
SMITH (A.O.) CORP. CERAMIC RES. LAB., MILWAUKEE	CRL-HTC-
SMITH (A.O.) CORP. CERAMIC RES. LAB., MILWAUKEE	CRL-HTC-TOPICAL-
SMITH (A.O.) CORP. LONG RANGE RES. LAB., MILWAUKEE	LRR-
SMITH (S.) AND SONS, LTD., LONDON	GW/SS-
SMITH, HINCHMAN AND GRYLLS, INC. AERONAUTICAL ICE RESEARCH LAB., DETROIT	SHG-
SMITH, HINCHMAN + GRYLLS, INC. AERONAUTICAL ICING RESEARCH LAB., YPSILANTI, MICH.	AIRL-
SMITHS INDUSTRIES, LTD., LONDON	RID-
SMITHSONIAN ASTROPHYSICAL OBS., CAMBRIDGE, MASS.	AERONOMY-
SMITHSONIAN ASTROPHYSICAL OBS., CAMBRIDGE, MASS.	SAO-
SMITHSONIAN ASTROPHYSICAL OBS., CAMBRIDGE, MASS.	SAO SPECIAL REPT.-
SMITHSONIAN ASTROPHYSICAL OBS., CAMBRIDGE, MASS.	SAO-SR-
SMITHSONIAN ASTROPHYSICAL OBS., CAMBRIDGE, MASS.	SAO-SR-INDEX-
SMITHSONIAN ASTROPHYSICAL OBS., CAMBRIDGE, MASS.	SOA-SR-
SMITHSONIAN ASTROPHYSICAL OBS., CAMBRIDGE, MASS.	SP-
SMITHSONIAN INSTITUTION, WASHINGTON, D.C.	SI-
SMYTH RESEARCH ASSOCIATES, SAN DIEGO, CALIF.	SRA-
SNIA VISCOSA, COLLEFERRO, ITALY	GSR/CSE/
SNOW, ICE, + PERMAFROST RES. ESTAB., WILMETTE, ILL.	SIPRE-
SNOW, ICE, + PERMAFROST RES. ESTAB., WILMETTE, ILL.	SIPRE CONFERENCE-
SNOW, ICE, + PERMAFROST RES. ESTAB., WILMETTE, ILL.	SIPRE-SR-
SNOW, ICE, + PERMAFROST RESEARCH ESTABLISHMENT, WILMETTE, ILL. (TRANSLATION) (SEE ALSO: ARMY COLD REGIONS RES. + ENG. LAB.)	SIPRE TRANS-
SOCIETA RICERCHE IMPIANTI NUCLEARI, SALUGGIA, ITALY	S/(3 DIGITS)
SOCIETA RICERCHE IMPIANTI NUCLEARI, SALUGGIA, ITALY	SORIN-
SOCIETA RICERCHE IMPIANTI NUCLEARI, SALUGGIA, ITALY	SORIN-M/
SOCIETA RICERCHE IMPIANTI NUCLEARI, SALUGGIA, ITALY	SORIN-T/
SOCIETE ANONYME BELGE DE MECHANIQUE ET ARMAMENT	MECAR-
SOCIETE BELGE POUR L'INDUSTRIE NUCLEAIRE, BRUSSELS	BN-(4 DIGITS)-
SOCIETE BELGE POUR L'INDUSTRIE NUCLEAIRE, BRUSSELS	EF-(NO.)-(LTRS.)-
SOCIETE BELGE POUR L'INDUSTRIE NUCLEAIRE, BRUSSELS	NA-(NUMBER)-N-
SOCIETE BELGE POUR L'INDUSTRIE NUCLEAIRE, BRUSSELS	PRP/GC/P-
SOCIETE BELGE POUR L'INDUSTRIE NUCLEAIRE, BRUSSELS	VAP-(NUMBER)-(LETTER)-
SOCIETE D'ELECTRONIQUE ET D'AUTOMATISME, COURBEVOIE, FRANCE	SEA-
SOCIETE D'ETUDE DE LA PROPULSION PAR REACTION, VILLEJUIF, FRANCE	ELDO/SP/(YEAR)/
SOCIETE DES ACCUMULATEURS FIXES ET DE TRACTION	SAFT-
SOCIETE EUROPEENNE DE SEMICONDUCTEURS ET DE MICROELECTRONIQUE, PARIS	RD/AD-
SOCIETE EUROPEENNE D'ETUDE ET D'INVESTIGATION DE SYSTEMS SPATIAUX, COURBEVOIE, FRANCE	NT/D1/(YEAR)/
SOCIETE FRANCAISE DE PHYSIQUE, PARIS	RAD-M-
SOCIETE L'AIR LIQUIDE, SASSENAGE, FRANCE	ELDO/SP/(YEAR)/
SOCIETE NATIONALE D'ETUDE ET DE CONSTRUCTIONS DE MOTEURS D'AVIATION, PARIS	SNECMA-
SOCIETE NATIONALE D'ETUDE ET DE CONSTRUCTION DE MOTEURS D'AVIATION, PARIS	YCP-
SOCIETE NATIONALE D'ETUDE ET DE CONSTRUCTIONS DE MOTEURS D'AVIATION. DIV. ATOMIQUE, SURESNES, FRANCE	EE/(YEAR)/
SOCIETE POUR L'ETUDE ET LA REALISATION D'ENGINS BALISTIQUES, COURBEVOIE, FRANCE	CDTA-(NUMBER)-
SOCIETE POUR L'ETUDE ET LA REALISATION D'ENGINS BALISTIQUES, COURBEVOIE, FRANCE	DTA/(NUMBER)/
SOCIETE POUR L'ETUDE ET LA REALISATION D'ENGINS BALISTIQUES, COURBEVOIE, FRANCE	ELDO/SP/(YEAR)/
SOCIETE POUR L'ETUDE ET LA REALISATION D'ENGINS BALISTIQUES, COURBEVOIE, FRANCE	NT/DTA/GI-III/(YR.)-
SOCIETE POUR L'ETUDE ET LA REALISATION D'ENGINS BALISTIQUES, COURBEVOIE, FRANCE	SEREB-DTA-(NO.)/(NO.)-
SOCIETY OF AUTOMOTIVE ENGINEERS, N.Y.C.	AIR-
SOCIETY OF AUTOMOTIVE ENGINEERS, N.Y.C.	ARP-
SOCIETY OF AUTOMOTIVE ENGINEERS, N.Y.C.	INC/C(NO.)/P(NO.)
SOCIETY OF AUTOMOTIVE ENGINEERS, N.Y.C.	SAE-
SOCIETY OF BRITISH AIRCRAFT CONSTRUCTORS	SBAC-
SOLAR, SAN DIEGO, CALIF.	ER-
SOLAR, SAN DIEGO, CALIF.	ER-(NUMBER)-
SOLAR, SAN DIEGO, CALIF.	R(YEAR)B-
SOLAR AIRCRAFT CO., SAN DIEGO, CALIF. (REF. 1)	(NUMBER)
SOLAR AIRCRAFT CO., SAN DIEGO, CALIF.	RDR-
SOLAR AIRCRAFT CO., SAN DIEGO, CALIF.	SO-

SOLID STATE RADIATIONS, INC., LOS ANGELES	SSR-
SOLID STATE RADIATIONS, INC., LOS ANGELES	SSR-(NUMBER)F-
SOLID STATE RADIATIONS, INC., LOS ANGELES	SSRI-
SONNTAG SCIENTIFIC CORP., GREENWICH, CONN.	SSC-
SONOTONE CORP., ELMSFORD, N.Y.	SC-
SONOTONE CORP., ELMSFORD, N.Y.	SON-
SOUNDRIVE ENGINE CO., LOS ANGELES	SEC-
SOUTH CAROLINA. MEDICAL UNIV., CHARLESTON	MUSC-
SOUTH CAROLINA. STATE PLANNING + GRANTS DIV., COLUMBIA	SC-40-0014-
SOUTH CAROLINA. UNIV., COLUMBIA	SCU-
SOUTH CAROLINA. UNIV., COLUMBIA	SOCU-
SOUTH CAROLINA. UNIV., COLUMBIA	USC-TR-(YEAR)-
SOUTH DAKOTA SCHOOL OF MINES + TECHNOLOGY, RAPID CITY	MRI-(YEAR)-FR-
SOUTH DAKOTA STATE UNIV., BROOKINGS. DEPT. OF MECHANICAL ENGINEERING	SDSU-(YEAR)-
SOUTH SHORE ANALYTICAL AND RESEARCH LAB., ISLIP, N.Y.	SSARL-DPH-(YEAR)-
SOUTHAMPTON, ENGLAND. UNIV., DEPT. OF AERONAUTICS AND ASTRONAUTICS	AASU-
SOUTHAMPTON, ENGLAND. UNIV., DEPT. OF AERONAUTICS AND ASTRONAUTICS	AASU-TN-
SOUTHAMPTON, ENGLAND. UNIV. INSTITUTE OF SOUND AND VIBRATION RESEARCH	ISAV-MEMO-
SOUTHAMPTON, ENGLAND. UNIV. INSTITUTE OF SOUND AND VIBRATION RESEARCH	ISVR-
SOUTHAMPTON, ENGLAND. UNIV. INSTITUTE OF SOUND AND VIBRATION RESEARCH	ISVR-TR-
SOUTHEAST ASIA TREATY ORGANIZATION (STANDARD)	SEASTAG-
SOUTHEAST ASIA TREATY ORGANIZATION	SEATO-
SOUTHEASTERN FOREST EXPERIMENT STATION, ASHEVILLE, NC	HICKORY TF-
SOUTHERN CALIFORNIA EDISON CO.	AI-
SOUTHERN CALIFORNIA EDISON CO.	SCEC-
SOUTHERN COOPERATIVE WIND TUNNEL, PASADENA, CALIF.	CWT-
SOUTHERN ILLINOIS UNIV., CARBONDALE	SIU-
SOUTHERN ILLINOIS UNIV., CARBONDALE. MATERIALS SCIENCE LAB.	MATERIALS SCIENCE LAB-
SOUTHERN METHODIST UNIV., DALLAS	SMU-
SOUTHERN NUCLEAR ENGINEERING, INC., DUNEDIN, FLA.	SNE-
SOUTHERN NUCLEAR ENGINEERING, INC., DUNEDIN, FLA.	SNE-(NUMBER)NP
SOUTHERN RESEARCH INST., BIRMINGHAM, ALA. (REF. 1)	(NUMBERS...)
SOUTHERN RESEARCH INST., BIRMINGHAM, ALA.	RR-(NO.)-(NO.)-X(NO.)
SOUTHERN RESEARCH INST., BIRMINGHAM, ALA.	SORI-
SOUTHERN RESEARCH INST., BIRMINGHAM, ALA.	SORI-PR-
SOUTHERN RESEARCH INST., BIRMINGHAM, ALA.	SORL-
SOUTHERN RESEARCH INST., BIRMINGHAM, ALA.	SRI-(NUMBER)-
SOUTHERN RESEARCH INST., BIRMINGHAM, ALA.	SRIB-
SOUTHERN RESEARCH INST., BIRMINGHAM, ALA.	SRI-PR-
SOUTHERN RESEARCH INST., BIRMINGHAM, ALA.	SRI-TR OF (DATE)
SOUTHERN UNIVERSITIES NUCLEAR INST., FAURE, SOUTH AFRICA	SUNI-
SOUTHWEST ATOMIC ENERGY ASSOCIATES	SAEA-
SOUTHWEST ATOMIC ENERGY ASSOCIATES (SOUTHWEST EXPERIMENTAL FAST OXIDE REACTOR DEVELOPMENT PROGRAM)	SAEA-MPL-
SOUTHWEST CENTER FOR ADVANCED STUDIES, DALLAS	DASS-(YEAR)-
SOUTHWEST RESEARCH INST., SAN ANTONIO	RS-
SOUTHWEST RESEARCH INST., SAN ANTONIO	SORI-(NUMBER)-(NUMBER)
SOUTHWEST RESEARCH INST., SAN ANTONIO	SRI-
SOUTHWEST RESEARCH INST., SAN ANTONIO	SWRI-
SOUTHWEST RESEARCH INST., SAN ANTONIO	SWRI-(CONTRACT NO.)-P-
SOUTHWEST RESEARCH INST., SAN ANTONIO	SWRI-(CONTRACT NO.)-
SOUTHWEST RESEARCH INST., SAN ANTONIO	SWRI-RS-
SOUTHWEST RESEARCH INST. ARMY FUELS AND LUBRICANTS RESEARCH LAB., SAN ANTONIO	FLRL-
SOUTHWEST RESEARCH INST. DEPT. OF AEROSPACE PROPULSION RESEARCH, SAN ANTONIO	SERI-(CONTRACT NO.)-
SOUTHWEST RESEARCH INST. DEPT. OF AEROSPACE PROPULSION RESEARCH, SAN ANTONIO	SWRI-RS-
SOUTHWEST RESEARCH INST. DEPT. OF AUTOMOTIVE RES., SAN ANTONIO	AR-
SOUTHWEST RESEARCH INST. DEPT. OF AUTOMOTIVE RES., SAN ANTONIO	SWRI-AR-
SOUTHWEST RESEARCH INST. DEPT. OF INSTRUMENTATION RESEARCH, SAN ANTONIO	SWRI-(NUMBER)-(NUMBER)-
SOUTHWEST RESEARCH INST. DEPT. OF MECHANICAL SCIENCES, SAN ANTONIO	SWRI-TR-
SOUTHWESTERN RADIOLOGICAL HEALTH LAB., LAS VEGAS, NEV. (PUBLIC HEALTH EVALUATION)	PHEP-
SOUTHWESTERN RADIOLOGICAL HEALTH LAB., LAS VEGAS, NEV. (SEE ALSO LATER NAME: WESTERN ENVIRONMENTAL RES.LAB.)	SRHL-
SPACE AND MISSILE SYSTEMS ORGANIZATION, LOS ANGELES AIR FORCE STATION, CALIF.	NUCOR-
SPACE AND MISSILE SYSTEMS ORGANIZATION, LOS ANGELES AIR FORCE STATION, CALIF.	SAMSO-(YEAR)-
SPACE AND MISSILE SYSTEMS ORGANIZATION, LOS ANGELES AIR FORCE STATION, CALIF.	SAMSO-SMEA-
SPACE AND MISSILE SYSTEMS ORGANIZATION, LOS ANGELES AIR FORCE STATION, CALIF.	SAMSO-TR-(YEAR)-
SPACE AND MISSILE TEST CENTER, VANDENBERG AFB, CAL.	SAMTEC-TR-(YEAR)-
SPACE CRAFT, INC., HUNTSVILLE, ALA.	SCI-
SPACE CRAFT, INC., HUNTSVILLE, ALA.	SCI-GI-
SPACE CRAFT, INC., HUNTSVILLE, ALA.	SCI-RA-
SPACE DATA CORP., PHOENIX, ARIZ.	SDC-
SPACE DATA CORP., PHOENIX, ARIZ.	SDC-TM-
SPACE/DEFENSE CORP., BIRMINGHAM, MICH.	S/D-
SPACE/DEFENSE CORP., BIRMINGHAM, MICH.	S/D-LR(YEAR)-
SPACE/DEFENSE CORP., BIRMINGHAM, MICH.	S/D-P(YEAR)-
SPACE/DEFENSE CORP., BIRMINGHAM, MICH.	S/D-PO(YEAR)-

SPACE/DEFENSE CORP.

SPACE/DEFENSE CORP., BIRMINGHAM, MICH.	S/D-SP(YEAR)-
SPACE/DEFENSE CORP., BIRMINGHAM, MICH.	S/D-TM(YEAR)-
SPACE/DEFENSE CORP., BIRMINGHAM, MICH.	S/D-TR(YEAR)-
SPACE/DEFENSE CORP., BIRMINGHAM, MICH.	TR-(YEAR)-
SPACE DISTURBANCES LAB., BOULDER, COLO.	ERLTM-SDL-
SPACE DISTURBANCES LAB., BOULDER, COLO.	ESSA-SDL-
SPACE DISTURBANCES LAB., BOULDER, COLO.	ESSA-TR-ERL-(NO.)-SDL-
SPACE DISTURBANCES LAB., BOULDER, COLO.	NOAA-TR-ERL-(NO.)-SDL-
SPACE DISTURBANCES LAB., BOULDER, COLO.	SDL-
SPACE ELECTRONICS CORP., GLENDALE, CALIF.	SECO-
SPACE ENVIRONMENT LAB., BOULDER, COLO.	ERL-SEL-
SPACE ENVIRONMENT LAB., BOULDER, COLO.	NOAA-TM-ERL-SEL-
SPACE-GENERAL, EL MONTE, CALIF.	SG-(NUMBER)-FR
SPACE-GENERAL, EL MONTE, CALIF.	SG-(NUMBER)FR-
SPACE-GENERAL, EL MONTE, CALIF.	SG-(NUMBER)PR-(NO.)A
SPACE-GENERAL, EL MONTE, CALIF.	SG-(NUMBER)R-
SPACE-GENERAL CORP., EL MONTE, CALIF.	SGC-
SPACE-GENERAL CORP., EL MONTE, CALIF.	SGC-P-
SPACE-GENERAL CORP., EL MONTE, CALIF.	SPGC-
SPACE-GENERAL CORP., LOS ANGELES	SG-(NUMBER)/SP-
SPACE-GENERAL CORP., LOS ANGELES	SGC-(NUMBER)-
SPACE-GENERAL CORP. CENTER FOR RESEARCH AND EDUCATION, LOS ANGELES	SG-(NUMBER)/SR-
SPACE-GENERAL CORP. CENTER FOR RESEARCH AND EDUCATION, LOS ANGELES	SG-(NUMBER)/QR-
SPACE-GENERAL CORP. CENTER FOR RESEARCH AND EDUCATION, LOS ANGELES	SGC-(NUMBER)FR-
SPACE NUCLEAR PROPULSION OFFICE (AEC/NASA), WASH,D.C.	SNPO-
SPACE NUCLEAR PROPULSION OFFICE (AEC/NASA), WASH,D.C.	SPNO-
SPACE ORDNANCE SYSTEMS, INC., EL SEGUNDO, CALIF.	SOS-FR-
SPACE RECOVERY SYSTEMS, INC. DIV. OF ITEK CORP., EL SEGUNDO, CALIF.	SRSI-
SPACE RESEARCH INST., INC., NORTH TROY, VT.	SRI-R-
SPACE SCIENCES INC., NATICK, MASS.	SSI-
SPACE SCIENCES INC., NATICK, MASS.	SSI-TR-
SPACE SCIENCES INC., WALTHAM, MASS.	SSI-
SPACE SCIENCES INC., WALTHAM, MASS.	SSI-(NUMBER)-FR-
SPACE SCIENCES INC., WALTHAM, MASS.	SSI-(NUMBER)-SR-
SPACE SCIENCES INC., WALTHAM, MASS.	SSI-(3 DIGITS)-
SPACE SYSTEMS DIV. (AIR FORCE), INGLEWOOD, CALIF.	SSD-CR-(YEAR)-
SPACE SYSTEMS DIV. (AIR FORCE), INGLEWOOD, CALIF.	SSD-TR-(YEAR)-
SPACE SYSTEMS DIV. (AIR FORCE), INGLEWOOD, CALIF.	WDLPS-
SPACE SYSTEMS DIV., LOS ANGELES AIR FORCE STATION, CALIF.	SSD-(YEAR)-
SPACE SYSTEMS DIV., LOS ANGELES AIR FORCE STATION, CALIF.	SSD-TDR-(YEAR)-
SPACE TECHNOLOGY LABS., INC., EL SEGUNDO, CALIF.	GM-TR-
SPACE TECHNOLOGY LABS., INC., LOS ANGELES	AS(YEAR)-V(NUMBER)-(NO.)
SPACE TECHNOLOGY LABS., INC., LOS ANGELES	EM-(NO.)-(NO.)
SPACE TECHNOLOGY LABS., INC., LOS ANGELES	GM-
SPACE TECHNOLOGY LABS., INC., LOS ANGELES	GM-(NUMBER)-(NUMBER)-
SPACE TECHNOLOGY LABS., INC., LOS ANGELES	PA-(NUMBER)-
SPACE TECHNOLOGY LABS., INC., LOS ANGELES (REF. 42)	STL-(NO.-NO.-LTRS.)-000
SPACE TECHNOLOGY LABS., INC., LOS ANGELES	STL/AB-(YEAR)-
SPACE TECHNOLOGY LABS., INC., LOS ANGELES	STL-B-
SPACE TECHNOLOGY LABS., INC., LOS ANGELES	STL-GM-(YR.)-(NO.)-
SPACE TECHNOLOGY LABS., INC., LOS ANGELES	STL GM PTM-(NUMBER)-
SPACE TECHNOLOGY LABS., INC., LOS ANGELES	STL-GM-TN-
SPACE TECHNOLOGY LABS., INC., LOS ANGELES	STL-GM-TR-
SPACE TECHNOLOGY LABS., INC., LOS ANGELES	STL-LN-
SPACE TECHNOLOGY LABS., INC., LOS ANGELES (DETAILED TEST OBJECTIVE)	STL-(MISSILE NAME)-DTO-
SPACE TECHNOLOGY LABS., INC., LOS ANGELES	STL-NN-
SPACE TECHNOLOGY LABS., INC., LOS ANGELES	STL-OR-
SPACE TECHNOLOGY LABS., INC., LOS ANGELES	STL-PA-(NO.)-(NO.)
SPACE TECHNOLOGY LABS., INC., LOS ANGELES	STL/TM-(YEAR)-
SPACE TECHNOLOGY LABS., INC., LOS ANGELES	STL-TN-(YR.)-(NO.)-
SPACE TECHNOLOGY LABS., INC., LOS ANGELES	STL-TP-
SPACE TECHNOLOGY LABS., INC., LOS ANGELES	STL-TR-
SPACE TECHNOLOGY LABS., INC., LOS ANGELES	STL-TR-(YR.)-(NO.)-
SPACE TECHNOLOGY LABS., INC., REDONDO BEACH, CALIF.	CGM-
SPACE TECHNOLOGY LABS., INC., REDONDO BEACH, CALIF.	STL-TR-
(SEE ALSO LATER NAME: TRW SYSTEMS)	
SPACE TECHNOLOGY LABS., INC. ELECTRONIC LAB,LOS ANG.	GM-TM-
SPACE TECHNOLOGY LABS., INC. PHYSICAL RESEARCH LAB., LOS ANGELES	GM-TR-
SPACE TECHNOLOGY LABS., INC. PHYSICAL RESEARCH LAB., LOS ANGELES	PRL-9-
SPACE TECHNOLOGY LABS., INC. PHYSICAL RESEARCH LAB., LOS ANGELES	STL-PRL-
SPACE TECHNOLOGY LABS., INC. PHYSICAL RESEARCH LAB., LOS ANGELES	TR-(YR)-
SPACE TECHNOLOGY LABS., INC. PHYSICAL RESEARCH LAB., LOS ANGELES	TR-(YR.)-0000-
SPACE TECHNOLOGY LABS., INC. PHYSICAL RESEARCH LAB., REDONDO BEACH, CALIF.	STL-PRL-
SPACELABS, INC., VAN NUYS, CALIF.	SR(NO.)-
SPAIN. INSTITUTO NACIONAL DE TECNICA AERONAUTICA, MADRID	INTA-
SPAIN. JUNTA DE ENERGIA NUCLEAR, MADRID	IS-(NUMBER)/I-
SPAIN. JUNTA DE ENERGIA NUCLEAR, MADRID	JEN-
SPAIN. JUNTA DE ENERGIA NUCLEAR, MADRID	LFEN.-
SPAIN. JUNTA DE ENERGIA NUCLEAR. DIRECCION DE QUIMICA E ISOTOPES, MADRID.	JEN-(NO.)-DQ/I-
SPAIN. JUNTA DE ENERGIA NUCLEAR. DIVISION DE FISICA, MADRID	JEN-(NO.)-IFIC/I-
SPAIN. JUNTA DE ENERGIA NUCLEAR. DIVISION DE FISICA, MADRID	JEN-(NO.)-DF/I-
SPAIN. JUNTA DE ENERGIA NUCLEAR. DIVISION DE MATERIALES, MADRID	JEN-(NO.)-DMA/I-

SPAIN. JUNTA DE ENERGIA NUCLEAR. DIVISION DE METALLURGIA, MADRID	JEN-(NO.)-DME/I-
SPECIAL AIR WARFARE CENTER, EGLIN AFB, FLA.	SAWC-TDR-
SPECIAL AIR WARFARE CENTER. COMBAT APPLICATIONS GP., EGLIN AFB, FLA.	SUU-(NUMBER)A/A
SPECIAL AIR WARFARE CENTER. COMBAT APPLICATIONS GP., EGLIN AFB, FLA.	TAC-
SPECIAL AIR WARFARE CENTER. COMBAT APPLICATIONS GP., EGLIN AFB, FLA.	TAC-TR-
SPECIAL DEVICES, INC., NEWHALL, CALIF.	SDI-
SPECIAL DEVICES CENTER, PORT WASHINGTON, N.Y.	SDC (CONTRACT NO.)-(NO.)-
SPECIAL DEVICES CENTER, PORT WASHINGTON, N.Y.	SDC TR-
SPECIAL DEVICES CENTER, PORT WASHINGTON, N.Y.	SDC TR (NUMBERS)-
SPECIAL LIBRARIES ASSN. TRANSLATION CENTER, JOHN CRERAR LIBRARY, CHICAGO	SLA-
SPECIAL LIBRARIES ASSN. TRANSLATION CENTER, JOHN CRERAR LIBRARY, CHICAGO	SLA-TR-(YEAR)-
SPECIAL OPERATIONS CENTER, EGLIN AFB, FLA.	SOC-
SPECIAL PROJECTS OFFICE (NAVY), WASHINGTON, D.C.	NAVY-SPO-
SPECIAL PROJECTS OFFICE (NAVY), WASHINGTON, D.C.	SPINST-
SPECIAL WEAPONS COMMAND, KIRTLAND AFB, N. MEX.	SWC-
SPECIAL WEAPONS DEVELOPMENT GP., KIRTLAND AFB, N.MEX.	SWDG-
SPECIALTY ENGINEERING AND ELECTRONICS CO., BROOKLYN	SEEC-
SPECIALTY ENGINEERING AND ELECTRONICS CO., BROOKLYN	SEEC/IDR(NO.)/CONTRACT
SPEECH COMMUNICATIONS RESEARCH LAB., INC., SANTA BARBARA, CALIF.	SCRL-MONOGRAPH-
SPEER CARBON CO., NIAGARA FALLS, N.Y.	R-
SPEER CARBON CO., NIAGARA FALLS, N.Y.	SCC-(NUMBER)-R-
SPEER CARBON CO. RES. LAB., NIAGARA FALLS, N.Y.	SCC-
SPERRY ELECTRONIC TUBE DIV., GAINESVILLE, FLA.	NJ-
SPERRY ELECTRONIC TUBE DIV., GAINESVILLE, FLA.	SEDT-NJ-
SPERRY ELECTRONIC TUBE DIV., GAINESVILLE, FLA.	SETD-
SPERRY FARRAGUT CO., BRISTOL, TENN.	R(NUMBER)-
SPERRY FARRAGUT CO., BRISTOL, TENN. (COMPANY TECHNICAL PROPOSAL)	SFC-P-
SPERRY FARRAGUT CO., BRISTOL, TENN. (CO. TECH. RPT.)	SFC-R-
SPERRY FARRAGUT CO. ENG. DEPT., BRISTOL, TENN. (PROPOSAL)	PUBLICATION 3100-P-
SPERRY FARRAGUT CO. ENG. DEPT., BRISTOL, TENN.	PUBLICATION 3100-R-
SPERRY FARRAGUT CO. ENG. DEPT., BRISTOL, TENN. (BROCHURE)	PUBLICATION SFCO-B-
SPERRY FARRAGUT CO. ENG. DEPT., BRISTOL, TENN.	PUBLICATION SFCO-R-
SPERRY FARRAGUT CO. ENG. DEPT., BRISTOL, TENN.	SFCO-R-
SPERRY FARRAGUT CO. QUALITY CONTROL DEPT., BRISTOL, TENN.	SFQC-
SPERRY GYROSCOPE CO., GREAT NECK, N.Y.	A-
SPERRY GYROSCOPE CO., GREAT NECK, N.Y.	AJ-(NUMBER)-
SPERRY GYROSCOPE CO., GREAT NECK, N.Y.	AK-(NUMBER)-
SPERRY GYROSCOPE CO., GREAT NECK, N.Y.	AR-(NUMBER)-
SPERRY GYROSCOPE CO., GREAT NECK, N.Y.	B-
SPERRY GYROSCOPE CO., GREAT NECK, N.Y.	CA-(NUMBER)-
SPERRY GYROSCOPE CO., GREAT NECK, N.Y.	CJ-(NUMBER)-
SPERRY GYROSCOPE CO., GREAT NECK, N.Y.	DA-(NUMBER)-
SPERRY GYROSCOPE CO., GREAT NECK, N.Y. (REF. 1)	(DEPT. NUMBER)-
SPERRY GYROSCOPE CO., GREAT NECK, N.Y.	EB-(NUMBER)-
SPERRY GYROSCOPE CO., GREAT NECK, N.Y.	GJ-
SPERRY GYROSCOPE CO., GREAT NECK, N.Y.	GK-(NO.)-(NO.)-(NO.)-BK-
SPERRY GYROSCOPE CO., GREAT NECK, N.Y.	H-
SPERRY GYROSCOPE CO., GREAT NECK, N.Y.	HB-(NUMBER)-
SPERRY GYROSCOPE CO., GREAT NECK, N.Y.	LR-B-
SPERRY GYROSCOPE CO., GREAT NECK, N.Y.	LRD-
SPERRY GYROSCOPE CO., GREAT NECK, N.Y.	NA-(NUMBER)-
SPERRY GYROSCOPE CO., GREAT NECK, N.Y.	SGC-
SPERRY GYROSCOPE CO., GREAT NECK, N.Y.	SGC-(NO.)-(NO.)-
SPERRY GYROSCOPE CO., GREAT NECK, N.Y.	SGC/CA-(NO.)-(NO.)-
SPERRY GYROSCOPE CO., GREAT NECK, N.Y. (INTERIM ENGINEERING REPORT)	SGC-IER-
SPERRY GYROSCOPE CO., GREAT NECK, N.Y.	SGD-(NUMBER)-
SPERRY GYROSCOPE CO., GREAT NECK, N.Y.	SPERRY-
SPERRY GYROSCOPE CO., GREAT NECK, N.Y.	SPERRY-(NO.)-
SPERRY GYROSCOPE CO., GREAT NECK, N.Y.	SPERRY-(LETTERS)-
SPERRY GYROSCOPE CO., GREAT NECK, N.Y.	TA-(NUMBER)-
SPERRY GYROSCOPE CO., GREAT NECK, N.Y.	TJ-(NUMBER)-
SPERRY GYROSCOPE CO., GREAT NECK, N.Y.	WA-(NUMBER)-
SPERRY GYROSCOPE CO., SYOSSET, N.Y.	GA-(NUMBER)-
SPERRY GYROSCOPE CO., SYOSSET, N.Y.	GJ-(NUMBER)-
SPERRY GYROSCOPE CO. AIR ARMAMENT DIV., GT.NECK,N.Y.	CA-
SPERRY GYROSCOPE CO. ELECTRONIC TUBE DIV., GREAT NECK, N.Y.	NA-(NO.)-(NO.)-(NO.)
SPERRY GYROSCOPE CO. FLECTRONIC TUBE DIV., GREAT NECK, N.Y.	NJ-(NUMBER)-
SPERRY GYROSCOPE CO. INERTIAL DIV., GREAT NECK, N.Y.	CA-(NUMBER)-
SPERRY GYROSCOPE CO. INFORMATION AND COMMUNICATIONS DIV., GREAT NECK, N.Y.	PADLOC-
SPERRY GYROSCOPE CO. RADIATION DIV., GREAT NECK, NY	RD-(NUMBER)-
SPERRY GYROSCOPE CO., LTD., MONTREAL	RB-
SPERRY GYROSCOPE CO., LTD., MONTREAL	RB-(NUMBER)-(NUMBER)-
SPERRY MICROWAVE ELECTRONICS CO., CLEARWATER, FLA.	SJ-(NUMBER)-
SPERRY MICROWAVE ELECTRONICS CO., CLEARWATER, FLA.	SJ-M-(NO.)-(NO.)-
SPERRY MICROWAVE ELECTRONICS CO., CLEARWATER, FLA.	SJM-
SPERRY MICROWAVE ELECTRONICS CO., CLEARWATER, FLA.	SMEC-
SPERRY MICROWAVE ELECTRONICS CO., CLEARWATER, FLA.	STM.-
SPERRY PHOENIX CO., ARIZ.	LJ-(NUMBER)-
SPERRY PHOENIX CO., ARIZ.	SPC-
SPERRY PHOENIX CO., ARIZ.	SPC-LJ-(NUMBER)-
SPERRY PIEDMONT CO., CHARLOTTESVILLE, VA.	JA-(NUMBER)-

SPERRY PRODUCTS, INC.

SPERRY PRODUCTS, INC., DANBURY, CONN.	TR-
SPERRY RAND CORP., GAINESVILLE, FLA.	NJ-
SPERRY RAND CORP., GAINESVILLE, FLA.	NU-
SPERRY RAND CORP., GREAT NECK, N.Y.	GJ-(NUMBER)-
SPERRY RAND CORP., ST. PAUL	MPR-
SPERRY RAND CORP., ST. PAUL	PX-
SPERRY RAND CORP. FORD INSTRUMENT DIV., LONG ISLAND CITY, N.Y.	CA-
SPERRY RAND CORP. FORD INSTRUMENT DIV., LONG ISLAND CITY, N.Y.	FD-
SPERRY RAND CORP. SPERRY ELECTRO OPTICS, GT.NECK,N.Y.	AB-
SPERRY RAND CORP. SPERRY ELECTRONIC TUBE DIV., GAINESVILLE, FLA.	SETD-NJ-
SPERRY RAND CORP. SPERRY GYROSCOPE DIV.,GREAT NECK,NY	RD-
SPERRY RAND CORP. SPERRY GYROSCOPE DIV.,GREAT NECK,NY	SGD-
SPERRY RAND CORP. SPERRY FLIGHT SYSTEMS DIV., PHOENIX, ARIZ.	LJ-
SPERRY RAND CORP. SPERRY SYSTEMS MANAGEMENT DIV., GREAT NECK, N.Y.	GB-(4 DIGITS)-
SPERRY RAND CORP. SPERRY SYSTEMS MANAGEMENT DIV., GREAT NECK, N.Y.	GK-
SPERRY RAND CORP. UNIVAC DEFENSE SYSTEMS DIV., ST. PAUL	SPEC-GS-
SPERRY RAND CORP. UNIVAC DIV., PHILADELPHIA	EF-
SPERRY RAND CORP. UNIVAC DIV., PHILADELPHIA	PX-
SPERRY RAND CORP. UNIVAC DIV., PHILADELPHIA	RRU-
SPERRY RAND CORP. UNIVAC DIV., PHILADELPHIA	RRU-TR-
SPERRY RAND CORP. UNIVAC DIV., ST. PAUL, MINN.	PX-
SPERRY RAND CORP. UNIVAC FEDERAL SYSTEMS DIV., ST. PAUL, MINN.	EF-
SPERRY RAND CORP. UNIVAC FEDERAL SYSTEMS DIV., SALT LAKE CITY	PX-
SPERRY RAND RESEARCH CENTER, SUDBURY, MASS.	SRCC-CR-(YEAR)-
SPERRY RAND RESEARCH CENTER, SUDBURY, MASS.	SRCC-RR-(YEAR)-
SPERRY RAND RESEARCH CENTER, SUDBURY, MASS.	SRRC-
SPERRY RAND RESEARCH CENTER, SUDBURY, MASS.	SRRC-CR-(YEAR)-
SPERRY RAND RESEARCH CENTER, SUDBURY, MASS.	SRRC-RR-
SPERRY RAND RESEARCH CENTER, SUDBURY, MASS.	SRRC-RR-(YR)-(NO.)
SPERRY UTAH CO., SALT LAKE CITY	EJ-(NUMBER)-
SPERRY UTAH CO., SALT LAKE CITY	FJ-
SPERRY UTAH CO., SALT LAKE CITY	LJ-
SPERRY UTAH CO., SALT LAKE CITY	SUCO-EJ-(NUMBER)-
SPERRY UTAH ENGINEERING LAB., SALT LAKE CITY	SUEL-
SPERRY UTAH ENGINEERING LAB., SALT LAKE CITY	SUEL-EJ-
SPINDLETOP RESEARCH, INC., LEXINGTON, KY.	DOC-B-
SPINDLETOP RESEARCH, INC., LEXINGTON, KY.	DOC-C-
SPINDLETOP RESEARCH, INC., LEXINGTON, KY.	DOC-G-
SPINDLETOP RESEARCH, INC., LEXINGTON, KY.	DOC-S-
SPINDLETOP RESEARCH, INC., LEXINGTON, KY.	S-
SPRAGUE ELECTRIC CO., NORTH ADAMS, MASS.	A-
SPRAGUE ELECTRICAL CO., NORTH ADAMS, MASS.	SEC-
SPRAGUE ELECTRICAL CO., NORTH ADAMS, MASS.	SP-
SPRINGFIELD ARMORY, MASS.	HSE-
SPRINGFIELD ARMORY, MASS.	ORD SA-
SPRINGFIELD ARMORY, MASS.	SA-
SPRINGFIELD ARMORY, MASS.	SA HSE-
SPRINGFIELD ARMORY, MASS.	SA-TR(NUMBER)-
SPRINGFIELD ARMORY, MASS.	SA-TR-
SPRINGFIELD ARMORY, MASS.	SA-TRI-
SPRINGFIELD ARMORY, MASS.	SA-TRI(NO.)-
SPRINGFIELD ARMORY. ENGINEERING DEPT., MASS.	SA WO-
SPRINGFIELD ARMORY. ENGINEERING DEPT. LAB., MASS.	SA L-
SPRINGFIELD ARMORY. ENGINEERING DEPT. LAB., MASS.	SA L (NUMBER(S))-
SPRINGFIELD ARMORY. RES. + DEV. DIV., MASS.	ORDTX-
SPRUCE JT. AEC-DOD WARHEAD COORDINATING COMM.(REF.15)	SPRUCE-
SQUIBB INSTITUTE FOR MEDICAL RESEARCH, NEW BRUNSWICK, N.J.	SQ-
SQUID PROJECT (REF. 33)	SQUID-
SQUID PROJECT (REF. 33)	SQUID-PR-(NUMBER)-C
SQUID PROJECT (REF. 33)	SQUID-PR-(NUMBER)-P
SQUIER SIGNAL LAB., FORT MONMOUTH, N.J. (ENG. REPT.)	E-
SQUIER SIGNAL LAB., FORT MONMOUTH, N.J. (TECH. MEMO.)	M-
SQUIER SIGNAL LAB., FORT MONMOUTH, N.J.	SI-
SQUIER SIGNAL LAB., FORT MONMOUTH, N.J.	SSL-
SQUIER SIGNAL LAB., FORT MONMOUTH, N.J. (ENG. RPTS.)	SSL ER-
SQUIER SIGNAL LAB., FORT MONMOUTH, N.J.	SSL SI-
SQUIER SIGNAL LAB., FORT MONMOUTH, N.J. (TEST REPT.)	T-
SQUIER SIGNAL LAB., FORT MONMOUTH, N.J.	TM-
STACK GAS PROBLEM WORKING GROUP, AEC	AIRCO-MISC.-
STD. OIL + GAS CO. EXPLOSIVES RES. LAB., TULSA, OKLA.	PT-
STANDARD OIL CO. OF INDIANA, CHICAGO	F-(YR.)-
STANDARD OIL CO. (OHIO). RES. + DEV. DEPT., CLEVELAND	SOHIO-
STANDARD OIL DEVELOPMENT CO., ELIZABETH, N.J.	PKN-
STANDARD OIL DEVELOPMENT CO., ELIZABETH, N.J.	SOA-
STANDARD OIL DEVELOPMENT CO., ELIZABETH, N.J.	SOD-
STANDARD OIL DEVELOPMENT CO., ELIZABETH, N.J.	SON-

STANDARD OIL DEVELOPMENT CO., N.Y.C.	SOD-(NO.)-(YR.)
STANDARD OIL DEVELOPMENT CO. ESSO LABS., LINDEN, N.J. (BUMBLEBEE PROJECT) (REF. 8)	BUMBLEBEE-
STANDARD OIL DEVELOPMENT CO. ESSO LABS., LINDEN, N.J. (PETROLEUM DEVELOPMENT NOTES)	PDN-
STANDARD OIL DEVELOPMENT CO. ESSO LABS., LINDEN, N.J.	RL-3M-
STANDARD OIL DEVELOPMENT CO. ESSO LABS., LINDEN, N.J.	SOA-
STANDARD OIL DEVELOPMENT CO. ESSO LABS., LINDEN, N.J.	SOD-
STANDARD OIL DEVELOPMENT CO. ESSO LABS., LINDEN, N.J.	SOD-PKN-
STANDARD OIL DEVELOPMENT CO. PROCESS DIV., N.Y.C.	PD-(NO.)-(YR.)
STANDARD PRESSED STEEL CO., JENKINTOWN, PENNA.	SPS-
STANDARD ROLLING MILLS, INC., BROOKLYN	SRM-
STANDARD TELECOMMUNICATION LABS., HARLOW, ESSEX, ENG.	O-TM-
STANDARD TELECOMMUNICATION LABS., HARLOW, ESSEX, ENG.	STLL-
STANDARD TELEPHONES AND CABLES, LTD., SIDCUP, KENT, ENG.	RP-(NUMBER-NUMBER)
STANFORD RADIO ASTRONOMY INST., CALIF.	SRAI-
STANFORD RESEARCH INSTITUTE, FORT ORD, CALIF.	RO-RM-
STANFORD RESEARCH INST., MENLO PARK, CALIF.	AR-
STANFORD RESEARCH INST., MENLO PARK, CALIF.	EQ-(YEAR)-
STANFORD RESEARCH INST., MENLO PARK, CALIF.	FMU-
STANFORD RESEARCH INST., MENLO PARK, CALIF.	GO-
STANFORD RESEARCH INST., MENLO PARK, CALIF.	IPR-
STANFORD RESEARCH INST., MENLO PARK, CALIF. (LOGISTIC SYSTEMS RESEARCH)	LSR-RM-
STANFORD RESEARCH INST., MENLO PARK, CALIF.	NF-
STANFORD RESEARCH INST., MENLO PARK, CALIF.	NSCI-
STANFORD RESEARCH INST., MENLO PARK, CALIF.	OCD-OS-
STANFORD RESEARCH INST., MENLO PARK, CALIF.	ORD-
STANFORD RESEARCH INST., MENLO PARK, CALIF.	PMU-
STANFORD RESEARCH INST., MENLO PARK, CALIF.	PU-
STANFORD RESEARCH INST., MENLO PARK, CALIF.	RM-
STANFORD RESEARCH INST., MENLO PARK, CALIF.	SATR-
STANFORD RESEARCH INST., MENLO PARK, CALIF.	SED-RM-
STANFORD RESEARCH INST., MENLO PARK, CALIF.	SEL-
STANFORD RESEARCH INST., MENLO PARK, CALIF.	SPECIAL TR-
STANFORD RESEARCH INST., MENLO PARK, CALIF.	SR-
STANFORD RESEARCH INST., MENLO PARK, CALIF.	SRI-
STANFORD RESEARCH INST., MENLO PARK, CALIF.	SRI-(NUMBER)-
STANFORD RESEARCH INST., MENLO PARK, CALIF.	SRI-(YEAR)-
STANFORD RESEARCH INST., MENLO PARK, CALIF.	SRI-(NO.)-P
STANFORD RESEARCH INST., MENLO PARK, CALIF.	SRI-(NUMBER)-(LETTERS)-
STANFORD RESEARCH INST., MENLO PARK, CALIF.	SRI-(NUMBER)-PU
STANFORD RESEARCH INST., MENLO PARK, CALIF. (U DENOTES GOVERNMENT CONTRACT)	SRI-(LTRS.)U-(PROJ. NO.)-
STANFORD RESEARCH INST., MENLO PARK, CALIF.	SRIA-
STANFORD RESEARCH INST., MENLO PARK, CALIF.	SRIA-(NUMBER)-P-(NO.)-
STANFORD RESEARCH INST., MENLO PARK, CALIF.	SRIA-(NUMBER)-P-
STANFORD RESEARCH INST., MENLO PARK, CALIF.	SRIA-(NUMBER)-(NUMBER)
STANFORD RESEARCH INST., MENLO PARK, CALIF.	SRI-CU-677-
STANFORD RESEARCH INST., MENLO PARK, CALIF.	SRI-CU-(NUMBER)-
STANFORD RESEARCH INST., MENLO PARK, CALIF.	SRI-D-(NUMBER)-
STANFORD RESEARCH INST., MENLO PARK, CALIF.	SRI-DAC-
STANFORD RESEARCH INST., MENLO PARK, CALIF.	SRI-IR-
STANFORD RESEARCH INST., MENLO PARK, CALIF.	SRI-ITR-
STANFORD RESEARCH INST., MENLO PARK, CALIF.	SRI-MR-
STANFORD RESEARCH INST., MENLO PARK, CALIF.	SRI-PAU-(PROJ. NO.)-
STANFORD RESEARCH INST., MENLO PARK, CALIF.	SRI PGU-
STANFORD RESEARCH INST., MENLO PARK, CALIF.	SRI-PHU-
STANFORD RESEARCH INST., MENLO PARK, CALIF.	SRI-PU-
STANFORD RESEARCH INST., MENLO PARK, CALIF.	SRI-RM-
STANFORD RESEARCH INST., MENLO PARK, CALIF.	SRI-SCIENTIFIC-
STANFORD RESEARCH INST., MENLO PARK, CALIF.	SRI-SPECIAL TR-
STANFORD RESEARCH INST., MENLO PARK, CALIF.	SRI-SR-
STANFORD RESEARCH INST., MENLO PARK, CALIF.	SRI-SU-(NO.)-(NO.)
STANFORD RESEARCH INST., MENLO PARK, CALIF.	SRI-SU-1815-
STANFORD RESEARCH INST., MENLO PARK, CALIF.	SRI-TM-
STANFORD RESEARCH INST., MENLO PARK, CALIF.	SRI-TN-(NUMBER)-SEA-
STANFORD RESEARCH INST., MENLO PARK, CALIF.	SRI-TR-
STANFORD RESEARCH INST., MENLO PARK, CALIF.	SRI-TR-(NUMBER)-
STANFORD RESEARCH INST., MENLO PARK, CALIF.	SRL-(LETTERS)-(NO.)-(NO.)
STANFORD RESEARCH INST., MENLO PARK, CALIF.	STR-
STANFORD RESEARCH INST., MENLO PARK, CALIF.	SU-SRI-
STANFORD RESEARCH INST., MENLO PARK, CALIF.	TN-
STANFORD RESEARCH INST., MENLO PARK, CALIF.	TPR-
STANFORD RESEARCH INST., MENLO PARK, CALIF.	TSR-
STANFORD RESEARCH INST., MENLO PARK, CALIF. (VELA UNIFORM PROJECT)	VUP-
STANFORD RESEARCH INST., MENLO PARK, CALIF.	WSCI-(YEAR)-
STANFORD RESEARCH INST. EARTH SCIENCES DEPT., MENLO PARK, CALIF.	ESD-TN-
STANFORD RESEARCH INST. NAVAL WARFARE RESEARCH CENTER, MENLO PARK, CALIF.	NWRC-
STANFORD RESEARCH INST. NAVAL WARFARE RESEARCH CENTER, MENLO PARK, CALIF.	NWRC-BIB-
STANFORD RESEARCH INST. NAVAL WARFARE RESEARCH CENTER, MENLO PARK, CALIF.	NWRC/LSR-RM-
STANFORD RESEARCH INST. NAVAL WARFARE RESEARCH CENTER, MENLO PARK, CALIF.	NWRC-RM-
STANFORD RESEARCH INST. NAVAL WARFARE RESEARCH CENTER, MENLO PARK, CALIF.	NWRC-TR-
STANFORD RESEARCH INST. OPERATIONAL EVALUATION DEPT., MENLO PARK, CALIF.	OED-RM-
STANFORD RESEARCH INST. OPERATIONAL EVALUATION DEPT., MENLO PARK, CALIF.	OED-RM-(NUMBER)-
STANFORD RESEARCH INST. OPERATIONS ANALYSIS DEPT., MENLO PARK, CALIF.	OAD-RM-(NUMBER)-
STANFORD RESEARCH INST. OPERATIONS ANALYSIS DEPT., MENLO PARK, CALIF.	OAD-TN-(NUMBER)-
STANFORD RESEARCH INST. OPERATIONS ANALYSIS DEPT., MENLO PARK, CALIF.	OAP-TN-
STANFORD RESEARCH INST. OPERATIONS ANALYSIS DEPT., MENLO PARK, CALIF.	SRI-OAD-TN-(NUMBER)-
STANFORD RESEARCH INST. OPERATIONS RESEARCH DEPT., MENLO PARK, CALIF.	ORD-RM-(NUMBER)-
STANFORD RESEARCH INST. OPERATIONS RESEARCH DEPT., MENLO PARK, CALIF.	ORD-TN-(NUMBER)-
STANFORD RESEARCH INST. OPERATIONS RESEARCH DEPT., MENLO PARK, CALIF.	SRI-ORD-STR-(NUMBER)-
STANFORD RES. INST. POULTER LABS., MENLO PARK, CALIF.	PL-
STANFORD RES. INST. POULTER LABS., MENLO PARK, CALIF.	PL-TR-(NO.)-
STANFORD RES. INST. POULTER LABS., MENLO PARK, CALIF.	SRI-GD-(NO.)-FR
STANFORD RES. INST. POULTER LABS., MENLO PARK, CALIF.	SRI-PL-
STANFORD RES. INST. POULTER LABS., MENLO PARK, CALIF.	SRI-PL-TR-(NO.)-(YR.)
STANFORD RES. INST. POULTER LABS., MENLO PARK, CAL.	TR-(NUMBER)-

STANFORD RESEARCH INST. REGIONAL SECURITY STUDIES CENTER

STANFORD RESEARCH INST. REGIONAL SECURITY STUDIES CENTER, MENLO PARK, CALIF.	RSSC-RM-
STANFORD RESEARCH INST. REGIONAL SECURITY STUDIES CENTER, MENLO PARK, CALIF.	RSSC-TN-
STANFORD RESEARCH INST. REGIONAL SECURITY STUDIES CENTER, MENLO PARK, CALIF.	RSSC-TR-
STANFORD RESEARCH INST. SOUTHERN CALIF. LABS., SO. PASADENA	SSU-
STANFORD RESEARCH INST. STRATEGIC STUDIES CENTER, MENLO PARK, CALIF.	SSC-RM-
STANFORD RESEARCH INST. SYSTEMS EVALUATION DEPT., MENLO PARK, CALIF.	SED-
STANFORD UNIV., CALIF.	ESRP-TN-
STANFORD UNIV., CALIF.	MR-
STANFORD UNIV., CALIF.	STAN-
STANFORD UNIV., CALIF. (TECHNICAL REPORTS)	STAN. U. TR-
STANFORD UNIV., CALIF.	SU-
STANFORD UNIV., CALIF.	SU-(NO.)P(NO.)-
STANFORD UNIV., CALIF.	SU-(NUMBER)-
STANFORD UNIV., CALIF.	SU-QSR-
STANFORD UNIV., CALIF.	SU-TR-
STANFORD UNIV., CALIF.	TN-(NUMBER)-
STANFORD UNIV., CALIF.	UNIV/SU-
STANFORD UNIV., CALIF. AEROSOL LAB.	SALR-
STANFORD UNIV., CALIF. APPLIED MATH.+ STATISTICS LAB.	SU-AMSL-
STANFORD UNIV., CALIF. APPLIED MATH.+ STATISTICS LAB.	SU-AMSL-TR-
STANFORD UNIV., CALIF. BIOPHYSICS LAB.	BL-
STANFORD UNIV., CALIF. CENTER FOR MATERIALS RESEARCH	SU-CMR-AR-
STANFORD UNIV., CALIF. COMPUTATION CENTER	SU-CC-(NUMBER)-(NUMBER)-
STANFORD UNIV., CALIF. DEPT. OF AERONAUTICS AND ASTRONAUTICS	SASR-
STANFORD UNIV., CALIF. DEPT. OF AERONAUTICS AND ASTRONAUTICS	SU-AR-
STANFORD UNIV., CALIF. DEPT. OF AERONAUTICS AND ASTRONAUTICS	SU-DAAR-
STANFORD UNIV., CALIF. DEPT. OF AERONAUTICS AND ASTRONAUTICS	SUDAAR-
STANFORD UNIV., CALIF. DEPT. OF AERONAUTICS AND ASTRONAUTICS	SUDAAR-TR-
STANFORD UNIV., CALIF. DEPT. OF AERONAUTICS AND ASTRONAUTICS	SUDAER-
STANFORD UNIV., CALIF. DEPT. OF CIVIL ENGINEERING	SU-DCE-TR-
STANFORD UNIV., CALIF. DEPT. OF COMPUTER SCIENCE	CS-
STANFORD UNIV., CALIF. DEPT. OF COMPUTER SCIENCE	CS-TR-
STANFORD UNIV., CALIF. DEPT. OF COMPUTER SCIENCE	STAN-CS-
STANFORD UNIV., CALIF. DEPT. OF COMPUTER SCIENCE	STAN-CS-(YEAR)-
STANFORD UNIV., CALIF. DEPT. OF COMPUTER SCIENCE	SU-CS-
STANFORD UNIV., CALIF. DEPT. OF COMPUTER SCIENCE	SU-STAN-CS-(YEAR)-
STANFORD UNIV., CALIF. DEPT. OF GENETICS	IRL-
STANFORD UNIV., CALIF. DEPT. OF MATERIALS SCIENCE	DMS-(YEAR)-
STANFORD UNIV., CALIF. DEPT. OF MATERIALS SCIENCE	DMS-
STANFORD UNIV., CALIF. DEPT. OF MATERIALS SCIENCE	SU-DMS-(YEAR)-
STANFORD UNIV., CALIF. DEPT. OF MATERIALS SCIENCE	SU-DMS-(YEAR)-R-
STANFORD UNIV., CALIF. DEPT. OF MATERIALS SCIENCE	SU-DMS-(YEAR)-T-
STANFORD UNIV., CALIF. DEPT. OF MECHANICAL ENG.	FM-
STANFORD UNIV., CALIF. DEPT. OF MECHANICAL ENG.	FP-
STANFORD UNIV., CALIF. DEPT. OF MECHANICAL ENG.	HS-
STANFORD UNIV., CALIF. DEPT. OF MECHANICAL ENG.	LG-
STANFORD UNIV., CALIF. DEPT. OF MECHANICAL ENG.	NSF-G-
STANFORD UNIV., CALIF. DEPT. OF MECHANICAL ENG.	P-
STANFORD UNIV., CALIF. DEPT. OF MECHANICAL ENG.	PD-
STANFORD UNIV., CALIF. DEPT. OF MECHANICAL ENG.	PG-
STANFORD UNIV., CALIF. DEPT. OF MECHANICAL ENG.	SU-AHT-
STANFORD UNIV., CALIF. DEPT. OF MECHANICAL ENG.	SU-DME-TR-
STANFORD UNIV., CALIF. DEPT. OF MECHANICAL ENG.	SU-MD-
STANFORD UNIV., CALIF. DEPT. OF MECHANICAL ENG.	SU ME TR-
STANFORD UNIV., CALIF. DEPT. OF MECHANICAL ENG.	THT-
STANFORD UNIV., CALIF. DEPT. OF MECHANICAL ENG., THERMOSCIENCES DIV.	AHT-
STANFORD UNIV., CALIF. DEPT. OF MECHANICAL ENG., THERMOSCIENCES DIV.	IT-
STANFORD UNIV., CALIF. DEPT. OF MECHANICAL ENG., THERMOSCIENCES DIV.	MD-
STANFORD UNIV., CALIF. DEPT. OF MECHANICAL ENG., THERMOSCIENCES DIV.	RCF-
STANFORD UNIV., CALIF. DIV. OF ENGINEERING MECHANICS	SU-DEM-
STANFORD UNIV., CALIF. DIV. OF ENGINEERING MECHANICS	SU-DEM-TR-
STANFORD UNIV., CALIF. ELECTRON DEVICES LAB.	SU-EDL-TR-
STANFORD UNIV., CALIF. ELECTRON TUBE LAB.	SU-ET-
STANFORD UNIV., CALIF. ELECTRONICS LAB.	SU-EL-TR-
STANFORD UNIV., CALIF. ELECTRONICS RESEARCH LAB.	SU ERL TR-
STANFORD UNIV., CALIF. HEALTH PHYSICS	SUHP-
STANFORD UNIV., CALIF. HIGH ENERGY PHYSICS LAB.	HEPL-
STANFORD UNIV., CALIF. HIGH ENERGY PHYSICS LAB.	ML-HEPL-
STANFORD UNIV., CALIF. INST. FOR PLASMA RESEARCH	SU-AA(NUMBER)
STANFORD UNIV., CALIF. INST. FOR PLASMA RESEARCH	SU-IPR-
STANFORD UNIV., CALIF. INST. FOR PLASMA RESEARCH	SUIPR-
STANFORD UNIV., CALIF. INST. FOR PLASMA RESEARCH	SUPR-
STANFORD UNIV., CALIF. INST. FOR PLASMA RESEARCH	SU-TPR-
STANFORD UNIV., CALIF. INST. IN COMPUTER COORDINATED SYSTEMS	CCS-
STANFORD UNIV., CALIF. INST. IN ENG. ECONOMIC SYSTEMS	DPS-
STANFORD UNIV., CALIF. INST. OF POLITICAL STUDIES	PRS-P-
STANFORD UNIV., CALIF. INST. OF THEORETICAL PHYSICS	ITP-
STANFORD UNIV., CALIF. INSTRUMENTATION RES. LAB.	IRL-
STANFORD UNIV., CALIF. INSTRUMENTATION RES. LAB.	SU-IRL-
STANFORD UNIV., CALIF. MICROWAVE LAB.	C-

STANFORD UNIV., CALIF. MICROWAVE LAB.	M-
STANFORD UNIV., CALIF. MICROWAVE LAB.	ML-
STANFORD UNIV., CALIF. MICROWAVE LAB.	NR-
STANFORD UNIV., CALIF. MICROWAVE LAB.	STAN ML-
STANFORD UNIV., CALIF. MICROWAVE LAB.	SU-ML-
STANFORD UNIV., CALIF. NUCLEAR TECHNOLOGY LAB.	SU-NTL-TR-(NUMBER)-
STANFORD UNIV., CALIF. PLASMA PHYSICS LAB.	SUIRP-
STANFORD UNIV., CALIF. PROJECT IN ENGINEERING ECONOMIC PLANNING	EEP-
STANFORD UNIV., CALIF. REMOTE SENSING LAB.	SU-RSL-TECH-(YEAR)-
STANFORD UNIV., CALIF. REMOTE SENSING LAB.	SU-RSL-TR-(YEAR)-
STANFORD UNIV., CALIF. SOLID-STATE ELECTRONICS LAB.	STAN SEL TR-(NO.)-
STANFORD UNIV., CALIF. SOLID-STATE ELECTRONICS LAB.	SU-SSEL-TR-
STANFORD UNIV., CALIF. STANFORD ELECTRONICS LABS.	SEL-(YEAR)-
STANFORD UNIV., CALIF. STANFORD ELECTRONICS LABS.	SEL-
STANFORD UNIV., CALIF. STANFORD ELECTRONICS LABS.	STAN. U./SEL-
STANFORD UNIV., CALIF. STANFORD ELECTRONICS LABS.	SU-EL-
STANFORD UNIV., CALIF. STANFORD ELECTRONICS LABS.	SU-SEL-(YEAR)-
STANFORD UNIV., CALIF. STANFORD ELECTRONICS LABS.	SU-SEL-TR-(NUMBER)-
STANFORD UNIV., CALIF. STANFORD LINEAR ACCELERATOR CENTER	CPT-
STANFORD UNIV., CALIF. STANFORD LINEAR ACCELERATOR CENTER (HYDROGEN BUBBLE CHAMBER GROUP)	HBC-
STANFORD UNIV.,CALIF. STANFORD LINEAR ACCELERATOR CTR	M-
STANFORD UNIV.,CALIF. STANFORD LINEAR ACCELERATOR CTR	SLAC-
STANFORD UNIV., CALIF. STANFORD LINEAR ACCELERATOR CENTER (SUBMITTED FOR PUBLICATION)	SLAC-PUB-
STANFORD UNIV., CALIF. STANFORD LINEAR ACCELERATOR CENTER (INTERNAL NOTES)	SLAC-TN-(YR.)-
STANFORD UNIV., CALIF. STANFORD LINEAR ACCELERATOR CENTER (TRANSLATION)	SLAC-TRANS-
STANFORD UNIV., CALIF. STANFORD LINEAR ACCELERATOR CENTER	SU-SLAC-PUB-
STANFORD UNIV., CALIF. SYSTEMS TECHNIQUES LAB.	SU-STL-TR-
STANFORD UNIV., CALIF. THERMOSCIENCES DIV.	MD-
STANFORD UNIV., CALIF. THERMOSCIENCES DIV.	SU-HMT-
STANFORD UNIV., CALIF. VIBRATION RES. LAB.	SU-VRL-
STANFORD UNIV., CALIF. VIBRATION RES. LAB.	VRL TR-
STANFORD UNIV., CALIF. W.W. HANSEN LABS. OF PHYSICS	M-
STANLEY AVIATION CORP., BUFFALO	STAC-
STANLEY AVIATION CORP., DENVER	CS-
STANLEY AVIATION CORP., DENVER	SA-
STANLEY AVIATION CORP., DENVER	SAC-
STATE MARINE LINES, INC., YORKTOWN, VA. (N.S. SAVANNAH OPERATIONS)	TODD/SML-NSS-
STATISTICS AND REPORTS DIV. (AIR FORCE), WASHINGTON, D.C. (DIRECTORY OF U.S. AIR FORCE ORGANIZATIONS)	SS-OL1 (MONTH/YEAR)
STAUFFER CHEMICAL CO., N.Y.C.	SCC-
STAUFFER CHEMICAL CO. CHAUNCEY RESEARCH CENTER, N.Y.	CH-
STAUFFER CHEMICAL CO. CHAUNCEY RESEARCH CENTER, N.Y. (FINAL SUMMARY REPORTS)	SCC-26-FS(NO.)
STAUFFER CHEMICAL CO. CHAUNCEY RESEARCH CENTER, N.Y. (QUARTERLY PROGRESS REPORTS)	SCC-26-QPR-
STAUFFER CHEMICAL CO. CHAUNCEY RESEARCH CENTER, N.Y.	SCC-824-Q(NUMBER)
STAUFFER CHEMICAL CO. RICHMOND RESEARCH CENTER, CALIF	SCC-RRC-
STAUFFER CHEMICAL CO. RICHMOND RESEARCH LAB., CALIF.	SCC-ANNUAL-TSR-(YR.-YR.)
STAUFFER CHEMICAL CO. RICHMOND RESEARCH LAB., CALIF.	SCC-RRL-
STAUFFER CHEMICAL CO. WESTERN RESEARCH CENTER, RICHMOND, CALIF.	SCC-TSR-
STAUFFER CHEMICAL CO. WESTERN RESEARCH CENTER, RICHMOND, CALIF.	SCC-WRC-
STAVID ENGINEERING, INC., PLAINFIELD, N.J.	SEI-
STD RESEARCH CORP., PASADENA, CALIF.	STD-
STD RESEARCH CORP., PASADENA, CALIF.	STD-(YEAR)-
STELLENBOSCH. UNIV., CAPE PROVINCE, SOUTH AFRICA	STELLENBOSCHU-
STENCEL AERO ENGINEERING CORP., ASHEVILLE, N.C.	SK-(YEAR)-
STERLING (WALTER V.) INC., CLAREMONT, CALIF.	WVS-
STERLING-WINTHROP RESEARCH INST., RENSSELAER, N.Y.	SWRI/WRAIR/(YR)ANN
STEVENS INST. OF TECH., HOBOKEN, N.J.	SIT-
STEVENS INST. OF TECH., HOBOKEN, N.J.	SIT-(NUMBER)-(NUMBER)
STEVENS INST. OF TECH., HOBOKEN, N.J.	SIT-QPR-
STEVENS INST. OF TECH., HOBOKEN, N.J. DAVIDSON LAB.	LR-
STEVENS INST. OF TECH., HOBOKEN, N.J. DAVIDSON LAB.	R-
STEVENS INST. OF TECH., HOBOKEN, N.J. DAVIDSON LAB.	SIT-DL-
STEVENS INST. OF TECH., HOBOKEN, N.J. DAVIDSON LAB.	SIT-R-
STEVENS INST. OF TECH., HOBOKEN, N.J. DEPT. OF MECHANICAL ENGINEERING	ME-RT-
STEVENS INST. OF TECH., HOBOKEN, N.J. DEPT. OF MECHANICAL ENGINEERING	SIT-ME-RS-
STEVENS INST. OF TECH., HOBOKEN, N.J. DEPT. OF MECHANICAL ENGINEERING	SIT-ME-RT-
STEVENS INST. OF TECH., HOBOKEN, N.J. DEPT. OF METALLURGY	C-
STEVENS INST. OF TECH., HOBOKEN, N.J. DEPT. OF PHYS.	SIT-P-
STEVENS INST. OF TECH., HOBOKEN, N.J. EXPERIMENTAL TOWING TANK (REF. 1)	(NUMBER)
STEVENS INST. OF TECH., HOBOKEN, N.J. EXPERIMENTAL TOWING TANK	ETT-
STEVENS INST. OF TECH., HOBOKEN, N.J. EXPERIMENTAL TOWING TANK	SIT ETT-
STEVENS INST. OF TECH., HOBOKEN, N.J. OCEAN ENG. DEPT.	SIT-OE-(YEAR)-
STEVENS INST. OF TECH., HOBOKEN, N.J. POWDER METALLURGY LAB.	SIT-PML-
STEWART-WARNER CORP., CHICAGO	SWC-
STEWART-WARNER CORP. SOUTH WIND DIV., INDPLS. (REF.1)	(NUMBER)

STOCHKEIM RESEARCH AND DEV. CORP.

STOCHHEIM RESEARCH AND DEV. CORP., WOODSIDE, N.Y.	SRDC-
STOCKHOLM. ARBETSMEDICINSKA INSTITUTET	AI-REPORT-
STOCKHOLM. UNIV. INST. OF PHYSICS	USIP-(YEAR)-
STOLLER (S.M.) ASSOCIATES, N.Y.C.	STOLLER-
STONE + WEBSTER ENGINEERING CORP., BOSTON	SW-NIKE-X-
STONE + WEBSTER ENGINEERING CORP., N.Y.C.	AE-
STONE + WEBSTER ENGINEERING CORP., N.Y.C.	SWND-
STRASBOURG. UNIVERSITE	LAL-
STRATEGIC AIR COMMAND, OFFUTT AFB, NEB.	AF-SAC-
STRATEGIC AIR COMMAND, OFFUTT AFB, NEB.	SAC-
STRATEGIC AIR COMMAND, OFFUTT AFB, NEB.	SACM-
STRATEGIC AIR COMMAND, OFFUTT AFB, NEB.	SACP-
STRATEGIC AIR COMMAND, OFFUTT AFB, NEB.	SAC TECHNICAL REPT (NO.)-
STRATEGIC AIR COMMAND, OFFUTT AFB, NEB. (TECHNICAL PAMPHLET)	SAC TP-
STRATEGIC AIR COMMAND, VANDENBERG AFB, CALIF.	OA-WP-
STRATEGIC AIR COMMAND. OFFICE OF OPERATIONS ANALYSIS, OFFUTT AFB, NEBR.	SAC-OA-
STRATEGIC BOMBING SURVEY (AERO ENGINE PLANT REPORTS)	AEPR-
STRATEGIC BOMBING SURVEY (AIR FRAMES PLANT REPORTS)	AFPR-
STRATEGIC BOMBING SURVEY (CIVILIAN DEFENSE REPORTS)	CDR-
STRATEGIC BOMBING SURVEY (COKING PLANT REPORTS)	CPR-
STRATEGIC BOMBING SURVEY (EUROPE)	EW-
STRATEGIC BOMBING SURVEY (HEAVY INDUSTRY PLANT RPTS.)	HIPR-
STRATEGIC BOMBING SURVEY (LIGHT METALS PLANT RPTS.)	LMPR-
STRATEGIC BOMBING SURVEY (OIL PLANT REPORTS)	OPR-
STRATEGIC BOMBING SURVEY (PROPELLANTS PLANT RPTS.)	PPR-
STRATEGIC BOMBING SURVEY	SBS-
STRATEGIC BOMBING SURVEY (AERO ENGINE PLANT RPTS.)	SBS AEPR-
STRATEGIC BOMBING SURVEY (AIR FRAMES PLANT REPORTS)	SBS AFPR-
STRATEGIC BOMBING SURVEY (CIVILIAN DEFENSE REPORTS)	SBS CDR-
STRATEGIC BOMBING SURVEY (COKING PLANT REPORTS)	SBS CPR-
STRATEGIC BOMBING SURVEY (HEAVY INDUSTRY PLANT RPTS.)	SBS HIPR-
STRATEGIC BOMBING SURVEY (LIGHT METALS PLANT RPTS.)	SBS LMPR-
STRATEGIC BOMBING SURVEY (OIL PLANT REPORTS)	SBS OPR-
STRATEGIC BOMBING SURVEY (PACIFIC REPORTS)	SBS P-
STRATEGIC BOMBING SURVEY (PACIFIC INTERROGATION RPTS)	SBS P-I-
STRATEGIC BOMBING SURVEY (PROPELLANTS PLANT RPTS.)	SBS PPR-
STRATEGIC BOMBING SURVEY (PLANT REPORTS)	SBS PR-
STRATEGIC BOMBING SURVEY (SPECIAL REPORTS)	SBS SP-
STRATEGIC BOMBING SURVEY. AERO STUDIES DIV.	ASDR-
STRATEGIC BOMBING SURVEY. AERO STUDIES DIV.	SBS ASDR-
STRATEGIC BOMBING SURVEY. UTILITIES DIV.(PLANT RPTS.)	SBS UDPR-
STRATEGIC BOMBING SURVEY. UTILITIES DIV. (PLANT RPTS)	UDPR-
STRESAU (R) LAB. INC., SPOONER, WIS.	RSLR-
STROMBERG-CARLSON CORP, ROCHESTER, N.Y.	S-C-
STROMBERG-CARLSON CORP., SAN DIEGO, CALIF.	SC-
STROMBERG-CARLSON CORP. DATA PRODUCTS DIV., SAN DIEGO, CALIF.	S-C-
STROMBERG-CARLSON DIV., GENERAL DYNAMICS CORP., ROCHESTER, N.Y.	SCD-
STROMBERG DATAGRAPHICS, INC., SAN DIEGO, CALIF.	SD-
STRUCTURAL CLAY PRODUCTS RESEARCH FDN., GENEVA, ILL.	SCPRF-
STRUCTURAL ENGINEERING AND DESIGN CO., LOS ANGELES	SEDCO-
STRUTHERS-DUNN, INC., PITMAN, N.J.	SDI-
STUTTGART. TECHNISCHE HOCHSCHULE	STH-
STUTTGART. TECHNISCHE HOCHSCHULE. FORSCHUNGSINSTITUT FUER KRAFTFAHRWESEN UND FAHRZEUNGMOTOREN	FKFS RPT-
STUTTGART. TECHNISCHE HOCHSCHULE. INSTITUT FUER AERODYNAMIK UND GASDYNAMIK	AGRR-TB-
STUTTGART. TECHNISCHE HOCHSCHULE. INSTITUT FUER HOCHTEMPERATURFORSCHUNG	IFH-(NUMBER)-
STUTTGART. TECHNISCHE HOCHSCHULE. INSTITUT FUER HOCHTEMPERATURFORSCHUNG	IHT-
STUTTGART. TECHNISCHE HOCHSCHULE. INSTITUT FUER HOCHTEMPERATURFORSCHUNG	IHT-BERICHT-
STUTTGART. TECHNISCHE HOCHSCHULE. INSTITUT FUER PLASMAFORSCHUNG	IFP-(NUMBER)-
STUTTGART. UNIVERSITAET. INST. FUER PLASMAFORSCHUNG	IPF-(YEAR)-
SUBMARINE MEDICAL RESEARCH LAB., GROTON, CONN.	SMC, SMRL-
SUBMARINE MEDICAL RESEARCH LAB., GROTON, CONN.	SMRL-
SUBMARINE MEDICAL RESEARCH LAB., GROTON, CONN.	SMRL MEMO (YEAR)-
SUBMARINE MEDICAL RES. LAB.,GROTON, CONN.(SPEC. RPT.)	SMRL SP (YEAR)-
SUBMARINE MEDICAL RESEARCH LAB., GROTON, CONN.	SMRL-SR-(YEAR)-
SUBROC JT. AEC-DOD WARHEAD COORDINATING COMM.(REF.15)	SUBROC-WJCC-
SUD-AVIATION, PARIS	DTS/ES-
SUMMERS GYROSCOPE CO., SANTA MONICA, CALIF.	RN-
SUNDSTRAND AVIATION, DENVER	CDRD-(YEAR)-
SUNDSTRAND AVIATION, ROCKFORD, ILL.	SA/AER-
SUNDSTRAND AVIATION, ROCKFORD, ILL.	SA/ATR-
SUNDSTRAND AVIATION, ROCKFORD, ILL.	SAN-(3 DIGITS)-
SUNDSTRAND AVIATION-DENVER, PACOIMA, CALIF.	SA/ONR-
SUNDSTRAND TURBO.DIV.OF SUNDSTRAND CORP.,ROCKFORD,ILL	AMF-TD-
SUNDSTRAND TURBO.DIV.OF SUNDSTRAND CORP.,ROCKFORD,ILL	S/TD-

SUNDSTRAND TURBO DIV., SUNDSTRAND MACHINE TOOL CO., ROCKFORD, ILL.	STD-
SUNFLOWER ARMY AMMUNITION PLANT, LAWRENCE, KANS.	SAAP-
SUNFLOWER ORDNANCE WORKS, LAWRENCE, KANS.	SOW-
SUNFLOWER ORDNANCE WORKS, LAWRENCE, KANS.	SUN-
SUPERIOR AIR PRODUCTS, INC., NEWARK, N.J.	SUPAIRCO-
SUPERIOR CONTINENTAL CORP. RESEARCH AND ENGINEERING CENTER, HICKORY, N.C.	TELC-(NUMBER)-
SUPERSONIC TRACK SYMPOSIUM	STS-
SUPERSONIC TUNNEL ASSN.	STA-
SUPPLY AND MAINTENANCE COMMAND (ARMY). PACKAGING STORAGE AND TRANSPORTABILITY CENTER	SMCPSTC-(NUMBER)-
SUPREME COMMANDER FOR THE ALLIED POWERS	SCAP-
SUPREME CMDR. FOR THE ALLIED POWERS (SPECIAL REPORTS)	SCAP SR-
SUPREME COMMANDER FOR THE ALLIED POWERS. ECONOMIC AND SCIENTIFIC SECTION	SCAP ESS R-
SUPREME COMMANDER FOR THE ALLIED POWERS. ECONOMIC AND SCI. SEC. RES. + STATISTICS DIV. (BULL.)	ESS B-
SUPREME COMMANDER FOR THE ALLIED POWERS. ECONOMIC AND SCI. SEC. RES. + STATISTICS DIV. (BULL.)	SCAP ESS B-
SUPREME COMMANDER FOR THE ALLIED POWERS. ECONOMIC AND SCIENTIFIC SEC. RES. DIV. (SPEC. RPTS.)	ESS SR-
SUPREME COMMANDER FOR THE ALLIED POWERS. ECONOMIC AND SCIENTIFIC SEC. RES. DIV. (SPEC. RPTS.)	SCAP ESS SR-
SUPREME COMMANDER FOR THE ALLIED POWERS. MILITARY INTELLIGENCE SECTION	MIS-
SUPREME COMMANDER FOR THE ALLIED POWERS. MILITARY INTELLIGENCE SECTION (INTELLIGENCE REPORTS)	MIS IR-
SUPREME COMMANDER FOR THE ALLIED POWERS. MILITARY INTELLIGENCE SECTION (INTELLIGENCE REPORTS)	SCAP MIS IR-
SUPREME CMDR. FOR THE ALLIED POWERS. MIL. INTELL. SEC GENL. STAFF ALLIED TRANSLATOR + INTERPRETER SEC.	ATIS-
SUPREME CMDR. FOR THE ALLIED POWERS. MIL. INTELL. SEC GENL. STAFF ALLIED TRANSLATOR + INTERPRETER SEC.	MIS ATIS-
SUPREME CMDR. FOR THE ALLIED POWERS. MIL. INTELL. SEC GENL. STAFF ALLIED TRANSLATOR + INTERPRETER SEC.	SCAP MIS ATIS-
SUPREME COMMANDER FOR THE ALLIED POWERS. NATURAL RESOURCES SECTION	NRS-
SUPREME COMMANDER FOR THE ALLIED POWERS. NATURAL RESOURCES SECTION	SCAP NRS-
SUPREME COMMANDER FOR THE ALLIED POWERS. NATURAL RESOURCES SECTION (PRELIMINARY STUDIES)	SCAP NRS PS-
SUPREME CMDR. FOR THE ALLIED POWERS. RES. + INFO. DIV	RI-
SUPREME CMDR. FOR THE ALLIED POWERS. RES. + INFO. DIV	SCAP RI-
SUPREME CMDR. FOR THE ALLIED POWERS. RES. + INFO. DIV	SCAP RI (NO-LTRS-LTR)-
SUPREME CMDR. FOR THE ALLIED POWERS. RES. + INFO. DIV	SCAP RI (LTR-NO-LTRS)-
SUPREME HEADQUARTERS, ALLIED EXPEDITIONARY FORCES (PRODUCTION CONTROL AGENCY G-4)	PCA G-4-
SUPREME HEADQUARTERS, ALLIED EXPEDITIONARY FORCES (PRODUCTION CONTROL AGENCY G-4)	SHAEF PCA G-4-
SURFACE ANTI-SUBMARINE DEV. DETACHMENT, KEY WEST, FLA	SADD-
SURFACE ANTI-SUBMARINE DEV. DETACHMENT, KEY WEST, FLA	SURASDEVDET-
SURVIVAL TRAINING SCHOOL, STEAD AFB, NEV.	SAFB-
SUSSEX. UNIV., FALMER, BRIGHTON, ENGLAND	SUSSEXU-
SUSSEX. UNIV., FALMER, BRIGHTON, ENGLAND	ZB/(NUMBER)/
SVENSKA AEROPLANE AKTIEBOLAGET, LINKOPING, SWEDEN	SAAB TN-
SVERDRUP AND PARCEL, INC., ST. LOUIS	SP-
SWARTHMORE COLL., PENNA.	SWC-
SWEDEN. ATOMKOMMITTEN, STOCKHOLM	IPV-
SWEDEN. FLYGTEKNISKA FOERSOEKSANSTALTEN, STOCKHOLM	ARIS-
SWEDEN. FLYGTEKNISKA FOERSOEKSANSTALTEN, STOCKHOLM	FFA-
SWEDEN. FLYGTEKNISKA FOERSOEKSANSTALTEN, STOCKHOLM	FFA-AU-
SWEDEN. FLYGTEKNISKA FOERSOEKSANSTALTEN, STOCKHOLM	FFAP-
SWEDEN. FLYGTEKNISKA FOERSOEKSANSTALTEN, STOCKHOLM	FFA-R-
SWEDEN. FLYGTEKNISKA FOERSOEKSANSTALTEN, STOCKHOLM	HE-
(NAME TRANSLATED IS AERONAUTICAL RESEARCH INST.)	
SWEDEN. FOERSKNINGRADENS LABORATORIUM, STUDSVIK	LFF-
SWEDEN. FOERSVARETS FORSKNINGSANSTALT, STOCKHOLM	A-
SWEDEN. FOERSVARETS FORSKNINGSANSTALT, STOCKHOLM	A-(4 DIGITS)-
SWEDEN. FOERSVARETS FORSKNINGSANSTALT, STOCKHOLM	B-
SWEDEN. FOERSVARETS FORSKNINGSANSTALT, STOCKHOLM	C-
SWEDEN. FOERSVARETS FORSKNINGSANSTALT, STOCKHOLM	FAO-P-D-
SWEDEN. FOERSVARETS FORSKNINGSANSTALT, STOCKHOLM	FOA-
SWEDEN. FOERSVARETS FORSKNINGSANSTALT, STOCKHOLM	FOA-A-
SWEDEN. FOERSVARETS FORSKNINGSANSTALT, STOCKHOLM	FOA-P-
SWEDEN. FOERSVARETS FORSKNINGSANSTALT, STOCKHOLM	FOA-P-C-
SWEDEN. FOERSVARETS FORSKNINGSANSTALT, STOCKHOLM	LF-
(NAME TRANSLATED IS RES. INST. OF NATIONAL DEFENSE)	
SWEDEN. FOERSVARETS TELETEKNISKA LAB., STOCKHOLM	FTL-A-A(NUMBER-NUMBER)
SWEDEN. INGENIORVETENSKAPSAKADEMIEN, STOCKHOLM	IVA-
SWEDEN. INST. FOR BUILDING RESEARCH, STOCKHOLM	BYGGFORSKNINGEN-
SWEDEN. KIRUNA GEOFYSISKA OBSERVATORIUM	K60-TR-(YEAR)-
SWEDEN. KIRUNA GEOFYSISKA OBSERVATORIUM	KGO-
SWEDEN. KUNGLIGA TEKNISKA HOEGSKOLAN, STOCKHOLM	AL-
SWEDEN. KUNGLIGA TEKNISKA HOEGSKOLAN, STOCKHOLM	FI-
SWEDEN. KUNGLIGA TEKNISKA HOEGSKOLAN, STOCKHOLM	QTSR-
SWEDEN. KUNGLIGA TEKNISKA HOEGSKOLAN, STOCKHOLM. DIV. OF PLASMA PHYSICS	PH-
SWEDEN. KUNGLIGA TEKNISKA HOEGSKOLAN, STOCKHOLM. FLYGTEKNISKA LABORATINET	FL-
SWEDEN. KUNGLIGA TEKNISKA HOEGSKOLAN, STOCKHOLM. INSTITUTIONEN FOER FLYGTEKNIK	AERO TN-
SWEDEN. KUNGLIGA TEKNISKA HOEGSKOLAN, STOCKHOLM. INSTITUTIONEN FOER FLYGTEKNIK	KTH-AERO TN-
SWEDEN. KUNGLIGA TEKNISKA HOEGSKOLAN, STOCKHOLM. TALTRANSMISSIONS-LABORATORIET	STL-QPSR-
SWEDEN. MILITARPSYKOLOGISK INSTITUTET, STOCKHOLM	MPI-

SWEDEN. SCHOOL OF EDUCATION, DEPT. OF EDUCATIONAL AND PSYCHOLOGICAL RESEARCH

SWEDEN. SCHOOL OF EDUCATION, MALMO. DEPT. OF EDUCATIONAL AND PSYCHOLOGICAL RESEARCH	DIDAKOMETRY-
SWEDISH DETONIC RESEARCH FOUNDATION, STOCKHOLM	DI-(YEAR)-
SWEDISH DETONIC RESEARCH FOUNDATION, STOCKHOLM	DL-(YEAR)-
SWEDISH STATE POWER BOARD, STOCKHOLM	SV-
SWEDISH STATE SHIPBUILDING EXPERIMENTAL TANK	SSPA-PUB-
SWEDISH STATE SHIPBUILDING EXPERIMENTAL TANK, GOTEBORG	SSPA-
SWITZERLAND. EIDGENOESSISCHES INSTITUT FUER REAKTORFORSCHUNG, WUERENLINGEN.	AN-ST-
SWITZERLAND. EIDGENOESSISCHES INSTITUT FUER REAKTORFORSCHUNG, WUERENLINGEN	EIR-
SWITZERLAND. EIDGENOESSISCHES INSTITUT FUER REAKTORFORSCHUNG, WUERENLINGEN	EIR-TM-
SWITZERLAND. EIDGENOESSISCHES INSTITUT FUER REAKTORFORSCHUNG, WUERENLINGEN (NAME TRANSLATED IS: FEDERAL INST. FOR REACTOR RES.)	TM-ST-
SWITZERLAND. EIDGENOESSISCHE TECHNISCHE HOCHSCHULE, ZURICH	AFIF-
SWITZERLAND. EIDGENOESSISCHE TECHNISCHE HOCHSCHULE, ZURICH	DISS-
SWITZERLAND. EIDGENOESSISCHE TECHNISCHE HOCHSCHULE, ZURICH	DISS-NR-
SWITZERLAND. EIDGENOESSISCHE TECHNISCHE HOCHSCHULE, ZURICH	ETH-
SWITZERLAND. EIDGENOESSISCHE TECHNISCHE HOCHSCHULE, ZURICH	PROM.-
SWITZERLAND. EIDGENOESSISCHE TECHNISCHE HOCHSCHULE, ZURICH. LABORATORIUM FUER ATOMOSPHAERNPHYSIK	LAPETH-
SWITZERLAND. EIDGENOESSISCHES FLUGZEUGWERK, EMMEN	T-
SWITZERLAND. LABORATOIRE DE RECHERCHES SUR LA PHYSIQUE DES PLASMAS, LAUSANNE	IRP-(NUMBER/YEAR)
SWITZERLAND. LABORATOIRE DE RECHERCHES SUR LA PHYSIQUE DES PLASMAS, LAUSANNE	LRP-
SWITZERLAND. LABORATOIRE DE RECHERCHES SUR LA PHYSIQUE DES PLASMAS, LAUSANNE	LRP-(NUMBER/YEAR)
SYDNEY. UNIV. SCHOOL OF PHYSICS	ER.(NUMBER)
SYDNEY. UNIV. SCHOOL OF PHYSICS	PR.(NUMBER)
SYDNEY. UNIV. SCHOOL OF PHYSICS	SYDNEY-ER.(NUMBER)
SYDNEY. UNIV. SCHOOL OF PHYSICS	SYDNEY-PR.(NUMBER)
SYDNEY. UNIV. SCHOOL OF PHYSICS	SYDNEY-TR.(NUMBER)
SYLVANIA-CORNING NUCLEAR CORP., BAYSIDE, N.Y.	DCF-
SYLVANIA-CORNING NUCLEAR CORP., HICKSVILLE, N.Y.	DCF-(NUMBER)-CH
SYLVANIA ELECTRIC PRODUCTS INC. (ALL LOCATIONS)	SEP-
SYLVANIA ELECTRIC PRODUCTS INC., BAYSIDE, N.Y.	B2/(NUMBER)/(YEAR)-
SYLVANIA ELECTRIC PRODUCTS, INC., BAYSIDE, N.Y.	BA-
SYLVANIA ELECTRIC PRODUCTS INC., BAYSIDE, N.Y.	BI(NO.)/(YR)-
SYLVANIA ELECTRIC PRODUCTS INC., BAYSIDE, N.Y.	BL/(NUMBER)/(YEAR)-
SYLVANIA ELECTRIC PRODUCTS INC., BAYSIDE, N.Y.	CR-31-555-
SYLVANIA ELECTRIC PRODUCTS INC., BAYSIDE, N.Y.	DCF-(NUMBER)-H
SYLVANIA ELECTRIC PRODUCTS INC., BAYSIDE, N.Y.	RD-
SYLVANIA ELECTRIC PRODUCTS INC., BAYSIDE, N.Y.	SEP-(NUMBER)(P)
SYLVANIA ELECTRIC PRODUCTS INC., BAYSIDE, N.Y.	SEP-(NUMBER)(X)
SYLVANIA ELECTRIC PRODUCTS INC., BAYSIDE, N.Y.	SEP(P)-
SYLVANIA ELECTRIC PRODUCTS INC., BAYSIDE, N.Y.	SEP-(X)-
SYLVANIA ELECTRIC PRODUCTS INC., BAYSIDE, N.Y.	SEP-(XD)-
SYLVANIA ELECTRIC PRODUCTS INC., BAYSIDE, N.Y.	SEP-YD-
SYLVANIA ELECTRIC PRODUCTS INC., BOSTON	SHMR-
SYLVANIA ELECTRIC PRODUCTS INC., FLUSHING, N.Y.	SEP-L-
SYLVANIA ELECTRIC PRODUCTS INC., TOWANDA, PENNA.	SEP-IR-
SYLVANIA ELECTRIC PRODUCTS INC., TOWANDA, PENNA.	SYL-
SYLVANIA ELECTRIC PRODUCTS INC., TOWANDA, PENNA.	SYLTT-
SYLVANIA ELECTRIC PRODUCTS INC., WALTHAM, MASS.	F-
SYLVANIA ELECTRIC PRODUCTS INC., WALTHAM, MASS.	F(NO.)-
SYLVANIA ELECTRIC PRODUCTS INC., WOBURN, MASS.	BW-
SYLVANIA ELECTRIC PRODUCTS INC., WOBURN, MASS.	SEP-CR-
SYLVANIA ELECTRIC PRODUCTS INC. ATOMIC ENERGY DIV., BAYSIDE, N.Y.	YD-(YR)-
SYLVANIA ELECTRIC PRODUCTS INC. ATOMIC ENERGY DIV., BAYSIDE, N.Y.	YE-(YR)-
SYLVANIA ELECTRIC PRODUCTS INC. ELECTRONICS DIV., BOSTON	CR31-(NUMBER)-
SYLVANIA ELECTRIC PRODUCTS, INC. HIGH TEMPERATURE COMPOSITES LAB., HICKVILLE, N.Y.	STR-(YEAR)-
SYLVANIA ELECTRIC PRODUCTS INC. METALLURGICAL LABS., BAYSIDE, N.Y.	YE-(YR)-
SYLVANIA ELECTRIC PRODUCTS INC. MICROWAVE PHYSICS LAB., MOUNTAIN VIEW, CALIF.	MPL-
SYLVANIA ELECTRIC PRODUCTS INC. MICROWAVE PHYSICS LAB., MOUNTAIN VIEW, CALIF.	MPL-N-
SYLVANIA ELECTRIC PRODUCTS INC. MICROWAVE PHYSICS LAB., MOUNTAIN VIEW, CALIF.	SEP-MPL-
SYLVANIA ELECTRIC PRODUCTS INC. MISSILE SYSTEMS LABS. WALTHAM, MASS.	SEP-2-
SYLVANIA ELECTRIC PRODUCTS INC. MISSILE SYSTEMS LABS. WALTHAM, MASS.	SEP-MSL-
SYLVANIA ELECTRIC PRODUCTS INC. MISSILE SYSTEMS LABS. WALTHAM, MASS.	SEP-TR-(NO.)-(NO.)
SYLVANIA ELECTRIC PRODUCTS INC. PRODUCT DEVELOPMENT LABS., FLUSHING, N.Y.	B(NO.)/(NO.)/(YR.)-
SYLVANIA ELECTRIC PRODUCTS, INC. PRODUCT SUPPORT ORGANIZATION, WEST ROXBURY, MASS.	MPO-SR-(NUMBER)-
SYLVANIA ELECTRIC PRODUCTS INC. RADIO TUBE DIV., N.Y.C. (REF. 1)	(NUMBER)
SYLVANIA ELECTRIC PRODUCTS INC. SYLCOR DIV., BAYSIDE, N.Y.	SCNC-
SYLVANIA ELECTRIC PRODUCTS INC. SYLCOR DIV., BAYSIDE, N.Y.	STR-
SYLVANIA ELECTRIC PRODUCTS INC. SYLCOR DIV., HICKSVILLE, N.Y.	STR-(YEAR)-(NO.).(NO.)
SYLVANIA ELECTRIC PRODUCTS INC. SYLCOR DIV., HICKSVILLE, N.Y.	TR-(YR.)-(NO.).(NO.)
SYLVANIA ELECTRIC PRODUCTS INC. SYLVANIA ELECTRONIC SYSTEMS DIV., MOUNTAIN VIEW, CALIF.	SEP-SESD-
SYLVANIA ELECTRONIC SYSTEMS, NEEDHAM, MASS.	DEP-
SYLVANIA ELECTRONIC SYSTEMS. ELECTRONIC DEFENSE LABS., MOUNTAIN VIEW, CALIF.	EDL-
SYLVANIA ELECTRONIC SYSTEMS. ELECTRONIC DEFENSE LABS., MOUNTAIN VIEW, CALIF.	EDL-E(NUMBER)
SYLVANIA ELECTRONIC SYSTEMS. ELECTRONIC DEFENSE LABS., MOUNTAIN VIEW, CALIF.	EDL-G(NUMBER)
SYLVANIA ELECTRONIC SYSTEMS. ELECTRONIC DEFENSE LABS., MOUNTAIN VIEW, CALIF.	EDL-M(NUMBER)
SYLVANIA ELECTRONIC SYSTEMS. ELECTRONIC DEFENSE LABS., MOUNTAIN VIEW, CALIF.	SEP-EDL-

SYLVANIA ELECTRONIC SYSTEMS. WALTHAM LABS., MASS.	F-
SYLVANIA ELECTRONIC SYSTEMS. WALTHAM LABS., MASS.	Q-
SYLVANIA ELECTRONIC SYSTEMS. WALTHAM LABS., MASS.	SEP-E2-
SYLVANIA ELECTRONIC SYSTEMS. WALTHAM LABS., MASS.	WL-E(NO.)-
SYLVANIA ELECTRONIC SYSTEMS-CENTRAL, WILLIAMSVILLE, NY	C(NUMBER)-
SYLVANIA ELECTRONIC SYSTEMS-EAST, NEEDHAM, MASS.	F(NUMBER)
SYLVANIA ELECTRONIC SYSTEMS-EAST, NEEDHAM, MASS.	FR(YEAR)-
SYLVANIA ELECTRONIC SYSTEMS-EAST, NEEDHAM, MASS.	MPO-PD-
SYLVANIA ELECTRONIC SYSTEMS-EAST, NEEDHAM, MASS.	MPO-SR-
SYLVANIA ELECTRONIC SYSTEMS-EAST, NEEDHAM, MASS.	Q(NUMBER)-
SYLVANIA ELECTRONIC SYSTEMS-EAST, NEEDHAM, MASS.	O(YEAR-(NUMBER)(LETTER)
SYLVANIA ELECTRONIC SYSTEMS-EAST, NEEDHAM, MASS.	S(YEAR)-
SYLVANIA ELECTRONIC SYSTEMS-EAST, NEEDHAM, MASS.	SPD/(NUMBER)-
SYLVANIA ELECTRONIC SYSTEMS-EAST, WALTHAM, MASS.	F-(NUMBER)-
SYLVANIA ELECTRONIC SYSTEMS-EAST, WALTHAM, MASS.	Q-
SYLVANIA ELECTRONIC SYSTEMS-EAST, WALTHAM, MASS.	RR-
SYLVANIA ELECTRONIC SYSTEMS-EAST, WALTHAM, MASS.	SEP-SESE-
SYLVANIA ELECTRONIC SYSTEMS-EAST, WALTHAM, MASS.	TN(NUMBER)-
SYLVANIA ELECTRONIC SYSTEMS-EAST. ADVANCED SYSTEMS PLANNING, WALTHAM, MASS.	ASPM-
SYLVANIA ELECTRONIC SYSTEMS-EAST. APPLIED RESEARCH LAB., WALTHAM, MASS.	F-(NUMBER)-
SYLVANIA ELECTRONIC SYSTEMS-EAST. APPLIED RESEARCH LAB., WALTHAM, MASS.	I-(NUMBER)-
SYLVANIA ELECTRONIC SYSTEMS-EAST. APPLIED RESEARCH LABS., WALTHAM, MASS.	Q-
SYLVANIA ELECTRONIC SYSTEMS-EAST. APPLIED RESEARCH LAB., WALTHAM, MASS.	S-(NUMBER)-
SYLVANIA ELECTRONIC SYSTEMS-WEST, MOUNTAIN VIEW, CAL.	CTR-(NUMBER)-
SYLVANIA ELECTRONIC SYSTEMS-WEST, MOUNTAIN VIEW, CAL.	QRC-
SYLVANIA ELECTRONIC SYSTEMS-WEST, MOUNTAIN VIEW, CAL.	SESW-(NUMBER)R(NUMBER)-
SYLVANIA ELECTRONIC SYSTEMS-WEST, MOUNTAIN VIEW, CAL.	SESW-
SYLVANIA ELECTRONIC SYSTEMS-WEST, MOUNTAIN VIEW, CAL.	SESW-E(NUMBER)
SYLVANIA ELECTRONIC SYSTEMS-WEST, MOUNTAIN VIEW, CAL.	SESW-M(NUMBER)
SYLVANIA ELECTRONIC SYSTEMS-WEST. ELECTRONIC DEFENSE LABS., MOUNTAIN VIEW, CALIF.	EDL-(LETTER)(NUMBER)
SYLVANIA ELECTRONIC SYSTEMS-WEST. ELECTRONIC DEFENSE LABS., MOUNTAIN VIEW, CALIF.	EDL-CD-(NUMBER)-M
SYLVANIA ELECTRONIC SYSTEMS-WEST. RECONNAISSANCE SYSTEMS LABS., MOUNTAIN VIEW, CALIF.	RSL-
SYMPOSIUM ON ELECTRON RING ACCELERATOR, 1968, BERKELEY, CALIF.	ERAC-
SYMPOSIUM ON ELECTRON RING ACCELERATOR, 1968, BERKELEY, CALIF.	S/ERA-
SYMPOSIUM ON ELECTRON RING ACCELERATOR, 1968, BERKELEY, CALIF.	WC/ERA-
SYMPOSIUM ON ELECTRON RING ACCELERATOR, 1968, BERKELEY, CALIF.	WX/ERA-
SYMPOSIUM ON SPACE, 1968, VENICE	SSA-(NUMBER-YEAR)
SYNDICAT LUXEMBOURGEOIS POUR L'INDUSTRIE NUCLEAIRE	LUX.N.(NO.)(NO.)
SYRACUSE UNIV., N.Y.	SU-(NUMBER)-
SYRACUSE UNIV., N.Y.	SU-1206-
SYRACUSE UNIV., N.Y.	SYRU-
SYRACUSE UNIV., N.Y.	SYRU (LTRS.)-
SYRACUSE UNIV., N.Y.	SYU-
SYRACUSE UNIV., N.Y.	US-IPR-
SYRACUSE UNIV., N.Y.	US-PR-
SYRACUSE UNIV., N.Y. DEPT. OF CHEMICAL ENGINEERING AND METALLURGY	MET-
SYRACUSE UNIV., N.Y. DEPT. OF CHEMICAL ENGINEERING AND METALLURGY	SYU-MET-
SYRACUSE UNIV., N.Y. DEPT. OF ELECTRICAL ENGINEERING	EE-(NUMBER)-
SYRACUSE UNIV., N.Y. DEPT. OF ELECTRICAL ENGINEERING	EE(NUMBER)-(NUMBER)F(NO.)
SYRACUSE UNIV., N.Y. DEPT. OF ELECTRICAL ENGINEERING	EE(NUMBER-(NUMBER)-T-
SYRACUSE UNIV., N.Y. DEPT. OF ELECTRICAL ENGINEERING	EE(4 DIGITS)-(YEAR)-
SYRACUSE UNIV., N.Y. DEPT. OF ELECTRICAL ENGINEERING	TR-
SYRACUSE UNIV., N.Y. DEPT. OF PHYSICS	TR-
SYRACUSE UNIV., N.Y. LAB. OF SENSORY COMMUNICATION	LSC-S-
SYRACUSE UNIV., N.Y. METALLURGICAL RESEARCH LABS.	MET-
SYRACUSE UNIV., N.Y. METALLURGICAL RESEARCH LABS.	METGS-
SYRACUSE UNIV., N.Y. RESEARCH CORP.	CSL-
SYRACUSE UNIV., N.Y. RESEARCH CORP.	CSL-(YEAR)-
SYRACUSE UNIV., N.Y. RESEARCH CORP.	DSD-R-
SYRACUSE UNIV., N.Y. RESEARCH CORP.	DSD-TM-
SYRACUSE UNIV., N.Y. RESEARCH CORP.	SURC-
SYRACUSE UNIV., N.Y. RESEARCH CORP.	SURC-CSL-
SYRACUSE UNIV., N.Y. RESEARCH CORP.	SURC-TR-
SYRACUSE UNIV., N.Y. RES. CORP. DEFENSE SYSTEMS DIV.	DSD-R-
SYRACUSE UNIV., N.Y. RES. CORP. DEFENSE SYSTEMS DIV.	DSD-TM-
SYRACUSE UNIV., N.Y. RES. CORP. DEFENSE SYSTEMS LAB.	DSL-R-
SYRACUSE UNIV., N.Y. RES. CORP. SPEC. PROJS. LAB.	SPL-
SYRACUSE UNIV., N.Y. RES. CORP. SPEC. PROJS. LAB.	SPL-TR-
SYRACUSE UNIV., N.Y. RESEARCH INST.	ARI-
SYRACUSE UNIV., N.Y. RESEARCH INST.	CHE (NO.-NO.-LTR.-NO.)
SYRACUSE UNIV., N.Y. RESEARCH INST.	MET-
SYRACUSE UNIV., N.Y. RESEARCH INST.	METGS-
SYRACUSE UNIV., N.Y. RESEARCH INST.	PHY-
SYRACUSE UNIV., N.Y. RESEARCH INST.	PHYSICS-
SYRACUSE UNIV., N.Y. RESEARCH INST.	SAR-
SYRACUSE UNIV., N.Y. RESEARCH INST.	SURI-
SYRACUSE UNIV., N.Y. RESEARCH INST.	SURI-CE-
SYRACUSE UNIV., N.Y. RESEARCH INST.	SURI-CH.E.-
SYRACUSE UNIV., N.Y. RESEARCH INST.	SURI-CHEM.-
SYRACUSE UNIV., N.Y. RESEARCH INST.	SURI-PHYSICS-(NUMBER)-
SYRACUSE UNIV., N.Y. RESEARCH INST.	SYU-ME-
SYRACUSE UNIV., N.Y. RESEARCH INST. DEFENSE SYSTEMS LAB.	DSL-R-
SYRACUSE UNIV., N.Y. RESEARCH INST. DEPT. OF CHEMICAL ENGINEERING AND METALLURGY	MET-E-(NUMBER)-
SYRACUSE UNIV., N.Y. RESEARCH INST. DEPT. OF CHEMICAL ENGINEERING AND METALLURGY	MET-E-(NO.)-(NO.)(LTR.)
SYRACUSE UNIV., N.Y. RESEARCH INST. DEPT. OF CHEMICAL ENGINEERING AND METALLURGY	MET-E-(NO.)-(NO.)-FR
SYRACUSE UNIV., N.Y. RESEARCH INST. DEPT. OF CHEMICAL ENGINEERING AND METALLURGY	MET-E-(NO.)-(NO.)-QP(NO.)

SYRACUSE UNIV., N. Y. RES. INST. DEPT. OF CIVIL ENG.

SYRACUSE UNIV., N.Y. RES. INST. DEPT. OF CIVIL ENG.	CE-
SYRACUSE UNIV., N.Y. RESEARCH INST. DEPT. OF MECHANICAL AND AEROSPACE ENGINEERING	ME-
SYRACUSE UNIV., N.Y. RESEARCH INST. DEPT. OF MECHANICAL AND AEROSPACE ENGINEERING	ME(NUMBER)-
SYRACUSE UNIV., N.Y. RESEARCH INST. DEPT. OF MECHANICAL AND AEROSPACE ENGINEERING	ME/AEROSPACE-
SYRACUSE UNIV., N.Y. RESEARCH INST. DEPT. OF MECHANICAL AND AEROSPACE ENGINEERING	SURI-ME-(NUMBER)-
SYRACUSE UNIV., N.Y. RES. INST. ELEC. ENG. DEPT.	SURI-EE-(NUMBER)-
SYRACUSE UNIV., N.Y. RES. INST. ELECTRONICS RES. LAB.	ERL-
SYRACUSE UNIV., N.Y. RES. INST. MET. RES. LABS.	SURI-MET-(NUMBER)-
SYRACUSE UNIV., N.Y. SPECIAL PROJECTS LAB.	SPL-
SYSTEM DEVELOPMENT CORP. (LTRS. GIVE LOCATION)	BR-(LTRS.-NO./NO(S).)
SYSTEM DEVELOPMENT CORP. (LTRS. GIVE LOCATION)	BREA-(LTRS.-NO./NO(S).)
SYSTEM DEVELOPMENT CORP. (LTRS. GIVE LOCATION)	BRG-(LTRS.-NO./NO(S).)
SYSTEM DEVELOPMENT CORP. (LTRS. GIVE LOCATION)	BRIA-(LTRS.-NO./NO(S).)
SYSTEM DEVELOPMENT CORP. (LTRS. GIVE LOCATION)	BRT-(LTRS.-NO./NO(S).)
SYSTEM DEVELOPMENT CORP., FALLS CHURCH, VA.	SDC-TM-WD-(NUMBER/NO(S).)
SYSTEM DEVELOPMENT CORP., FALLS CHURCH, VA.	TM-WD-
SYSTEM DEVELOPMENT CORP., LEXINGTON, MASS.	SDC-
SYSTEM DEVELOPMENT CORP., LEXINGTON, MASS.	TM-LX-
SYSTEM DEVELOPMENT CORP., SANTA MONICA, CALIF. (BROCHURE)	BR-(NUMBER/NUMBER(S))
SYSTEM DEVELOPMENT CORP., SANTA MONICA, CALIF. (BROCHURE, EXTERNAL ADMINISTRATION)	BREA-(NUMBER/NUMBER(S))
SYSTEM DEVELOPMENT CORP., SANTA MONICA, CALIF. (BROCHURE, GENERAL)	BRG-(NUMBER/NUMBER(S))
SYSTEM DEVELOPMENT CORP., SANTA MONICA, CALIF. (BROCHURE, INTERNAL ADMINISTRATION)	BRIA-(NUMBER/NUMBER(S))
SYSTEM DEVELOPMENT CORP., SANTA MONICA, CALIF. (BROCHURE, TECHNICAL)	BRT-(NUMBER/NUMBER(S))
SYSTEM DEVELOPMENT CORP., SANTA MONICA, CALIF. (DOCUMENT)	D-(NUMBER/NUMBER(S))
SYSTEM DEVELOPMENT CORP., SANTA MONICA, CALIF. (FIELD NOTE)	FN-(NUMBER/NUMBER(S))
SYSTEM DEVELOPMENT CORP., SANTA MONICA, CALIF. (NOTE)	N-(NUMBER/NUMBER(S))
SYSTEM DEVELOPMENT CORP., SANTA MONICA, CALIF.	SCD-TM-(NO./NO./NO.)
SYSTEM DEVELOPMENT CORP., SANTA MONICA, CALIF.	SCD-TM-L-(NO./NO./NO.)
SYSTEM DEVELOPMENT CORP., SANTA MONICA, CALIF.	SDC-
SYSTEM DEVELOPMENT CORP., SANTA MONICA, CALIF.	SDC-SCIENTIFIC-
SYSTEM DEVELOPMENT CORP., SANTA MONICA, CALIF.	SDC-SP-
SYSTEM DEVELOPMENT CORP., SANTA MONICA, CALIF.	SDC-TM-
SYSTEM DEVELOPMENT CORP., SANTA MONICA, CALIF.	SDC-TM-BA-(NO./NO./NO.)
SYSTEM DEVELOPMENT CORP., SANTA MONICA, CALIF.	SDC-TM-L-(NO./NO./NO.)
SYSTEM DEVELOPMENT CORP., SANTA MONICA, CALIF.	SDC-TM-LX-(L)-(NO./NO.)/
SYSTEM DEVELOPMENT CORP., SANTA MONICA, CALIF.	SP-
SYSTEM DEVELOPMENT CORP., SANTA MONICA, CALIF. (PAPER)	SP-(NUMBER/NUMBER(S))
SYSTEM DEVELOPMENT CORP., SANTA MONICA, CALIF.	TM-(NUMBER/NUMBER(S))
SYSTEM DEVELOPMENT CORP., SANTA MONICA, CALIF. (TECHNICAL MEMORANDUM)	TM-(NO.)/(NO.)/(NO.)
SYSTEM DEVELOPMENT CORP., SANTA MONICA, CALIF.	TM(L)-
SYSTEM DEVELOPMENT CORP., SANTA MONICA, CALIF.	TM-L-
SYSTEM DEVELOPMENT CORP., SANTA MONICA, CALIF.	TM-LO-
SYSTEM DEVELOPMENT CORP., SANTA MONICA, CALIF.	TM-LX-
SYSTEM SCIENCES, INC., BETHESDA, MD.	SYSTEM-SI-
SYSTEMS ANALYSIS AND RESEARCH CORP., BOSTON	XG-(NUMBER)-(YEAR)
SYSTEMS ANALYSIS AND RESEARCH CORP., CAMBRIDGE, MASS.	SARC-
SYSTEMS ANALYSIS AND RESEARCH CORP., CAMBRIDGE, MASS.	XG-(NUMBER)-
SYSTEMS ASSOCIATES, INC., LONG BEACH, CALIF.	SAI-CC-
SYSTEMS DEVELOPMENT CORP., DAYTON, OHIO	SYDC-
SYSTEMS ENGINEERING GROUP, WRIGHT-PATTERSON AFB, OHIO	SEFL-(NUMBER)A
SYSTEMS ENGINEERING GROUP, WRIGHT-PATTERSON AFB, OHIO	SEG-
SYSTEMS ENGINEERING GROUP, WRIGHT-PATTERSON AFB, OHIO	SEG-TDR-(YEAR)-
SYSTEMS ENGINEERING GROUP, WRIGHT-PATTERSON AFB, OHIO	SEG-TR-(YEAR)-
SYSTEMS LABORATORIES CORP., SHERMAN OAKS, CALIF.	SLC-
SYSTEMS RESEARCH CORP., WASHINGTON, D.C.	SRC-(YEAR)-TR-
SYSTEMS RESEARCH LABS., INC., DAYTON, OHIO	SRL-
SYSTEMS RESEARCH LABS., INC., DAYTON, OHIO	SRL-(NUMBER)-
SYSTEMS RESEARCH LABS., INC., DAYTON, OHIO	SRL-(NUMBER)-A
SYSTEMS RESEARCH LABS., INC., DAYTON, OHIO	SRL-TR-(NUMBER)-B
SYSTEMS SCIENCE AND SOFTWARE, LA JOLLA, CALIF.	3SR-
SYSTEMS SCIENCE AND SOFTWARE, LA JOLLA, CALIF.	SSS-3SR-
SYSTEMS SCIENCE AND SOFTWARE, LA JOLLA, CALIF.	SSS-3SIR-
SYSTEMS TECHNOLOGY CORP., DALLAS	STC-
SYSTEMS TECHNOLOGY, INC., HAWTHORNE, CALIF.	STI-
SYSTEMS TECHNOLOGY, INC., HAWTHORNE, CALIF.	STI-TM-(NUMBER)-I-
SYSTEMS TECHNOLOGY, INC., HAWTHORNE, CALIF.	STI-TM-(NUMBER)-A
SYSTEMS TECHNOLOGY, INC., INGLEWOOD, CALIF.	STI-TR-

TABLELEG COMMITTEE (REF. 15)   TABLELEG-

TACTICAL AIR COMMAND, LANGLEY AFB, VA.   AFTAC-
TACTICAL AIR COMMAND, LANGLEY AFB, VA.   B-
TACTICAL AIR COMMAND, LANGLEY AFB, VA.   CDS-
TACTICAL AIR COMMAND, LANGLEY AFB, VA.   TAC-
TACTICAL AIR COMMAND, LANGLEY AFB, VA. (OPERATIONS ANALYSIS MEMORANDUM)   TAC-OA-M-
TACTICAL AIR COMMAND, LANGLEY AFB, VA.   TAC-OAM-
TACTICAL AIR COMMAND, LANGLEY AFB, VA.   TAC-OA-MEMO-
TACTICAL AIR COMMAND, LANGLEY AFB, VA.   TAC-OA-TM-
TACTICAL AIR COMMAND, LANGLEY AFB, VA. (OPERATIONS ANALYSIS WORKING PAPERS)   TAC/OA/WP-
TACTICAL AIR COMMAND, LANGLEY AFB, VA.   TAC-OPLAN-
TACTICAL AIR COMMAND, LANGLEY AFB, VA.   TAC-TEST-
TACTICAL AIR COMMAND, LANGLEY AFB, VA.   TAC-TR-(YEAR)-

TACTICAL AIR COMMAND. OFFICE OF OPERATIONS ANALYSIS, LANGLEY AFB, VA.   TAC-OA-R-(YEAR)-

TACTICAL AIR RECONNAISSANCE CENTER, SHAW AFB, S.C.   TARC-(NUMBER)-DR(NO.)
TACTICAL AIR RECONNAISSANCE CENTER, SHAW AFB, S.C.   TARC-(NUMBER)-DT(NO.)
TACTICAL AIR RECONNAISSANCE CENTER, SHAW AFB, S.C.   TARC-OA-

TACTICAL AIR WARFARE CENTER, EGLIN AFB, FLA.   AN/GPA-
TACTICAL AIR WARFARE CENTER, EGLIN AFB, FLA.   ORC-
TACTICAL AIR WARFARE CENTER, EGLIN AFB, FLA.   TAC-
TACTICAL AIR WARFARE CENTER, EGLIN AFB, FLA.   TAC-TEST-(YEAR)-

TACTICAL AIRLIFT CENTER, POPE AFB, N.C.   TALC-STUDY-DCR-

TACTICAL AIRLIFT CENTER. OFFICE OF OPERATION ANALYSIS, POPE AFB, N.C.   TALC-OA-TM-

TAGGART (ROBERT), INC., FAIRFAX, VA.   RT-
TAGGART (ROBERT), INC., FAIRFAX, VA.   RTI-

TAGGART (ROBERT), INC., FALLS CHURCH, VA.   RT-
TAGGART (ROBERT), INC., FALLS CHURCH, VA.   RTI-

TANK ARSENAL PROVING GROUND, UTICA, MICH.   TAPG-

TARIFF COMMISSION, WASHINGTON, D.C.   TC-

TASK GROUP (NO.), FLEET POST OFFICE, SAN FRANCISCO (COMMANDER TASK GROUP)   CTG-

TASK GROUP DELTA, NORFOLK, VA.   DELTA-

TAVISTOCK INST. OF HUMAN RELATIONS, LONDON   TIHR-
TAVISTOCK INST. OF HUMAN RELATIONS, LONDON   TIHR-T(NUMBER)

TAYLOR AND FRANCIS, LTD., LONDON (TRANSLATIONS)   TF-

TE CO., SANTA BARBARA, CALIF.   E-
TE CO., SANTA BARBARA, CALIF.   REF-
TE CO., SANTA BARBARA, CALIF.   TECO-

TECHNE, LTD., CAMBRIDGE, ENGLAND   J-

TECHNICAL ANALYSIS GROUP, INC., N.Y.C.   TAG-TR-

TECHNICAL CAPABILITIES BRANCH (AIR FORCE)   TC-

TECHNICAL COMMUNICATIONS CORP., LEXINGTON, MASS.   R(YEAR)-
TECHNICAL COMMUNICATIONS CORP., LEXINGTON, MASS.   TCC-R(YEAR)-

TECHNICAL COMMUNICATIONS, INC., LOS ANGELES   TCI-SFN-

TECHNICAL COOPERATION PROGRAM, AEC   AEC-43/-

TECHNICAL DEVELOPMENT + EVALUATION CTR., INDIANAPOLIS   TDEC-

TECHNICAL DEVELOPMENT CORP., CULVER CITY, CALIF.   TDC-

TECHNICAL DEVELOPMENT LABS., SAVANNAH (TRANSLATIONS)   TDL-

TECHNICAL INDUSTRIAL DISARMAMENT COMMITTEE   TIDC PROJ-

TECHNICAL INDUSTRIAL INTELL. COMM. (COMMUNICATIONS)   C-(NO.)-
TECHNICAL INDUSTRIAL INTELL. COMM. (COMMUNICATIONS)   JIOA C-
TECHNICAL INDUSTRIAL INTELL. COMM. (SOLID FUELS RPTS)   SFR-
TECHNICAL INDUSTRIAL INTELLIGENCE COMMITTEE   TIIC-
TECHNICAL INDUSTRIAL INTELL. COMM. (COMMUNICATIONS)   TIIC C (NO.)-
TECHNICAL INDUSTRIAL INTELL. COMM. (SOLID FUELS RPTS)   TIIC SFR-

TECHNICAL INDUSTRIAL INTELL. COMM. AERO. SUB-COMM.   AS-
TECHNICAL INDUSTRIAL INTELL. COMM. AERO. SUB-COMM.   TIIC AS-

TECHNICAL IND. INTELL. COMM. CHEMICALS SUB-COMM.   CS-
TECHNICAL IND. INTELL. COMM. CHEMICALS SUB-COMM.   TIIC CS-

TECHNICAL IND. INTELL. COMM. MISC. CHEMS. SUB-COMM.   MCS-
TECHNICAL INDUSTRIAL INTELLIGENCE COMMITTEE. MISCELLANEOUS CHEMICALS SUB-COMM. (CHEM 4 RPTS.)   TIIC MCS-4

TECHNICAL INFORMATION CENTER, AEC, OAK RIDGE, TENN. (ASSIGNED TO CONF. PAPERS. FIRST 2 DIGITS ARE YEAR,   CONF-(6 DIGITS)-
   NEXT 2 ARE MONTH, LAST 2 ARE SERIAL NUMBER OF CONF. SERIAL NO. OF INDIVIDUAL PAPER MAY FOLLOW DASH)
TECHNICAL INFORMATION CENTER, AEC, OAK RIDGE, TENN. (ASSIGNED TO REPORTS FROM THE U.S.-JAPANESE FAST   JAPFNR-
   REACTOR EXCHANGE PROGRAM)
TECHNICAL INFORMATION CENTER, AEC, OAK RIDGE, TENN. (REF. 5)   TID-

TECHNICAL INFORMATION CO. PATENTS DEPT., LONDON (TRANSLATIONS)   TIC-

TECHNICAL INFO. SERVICE, AEC (UNCLASSIFIED RPTS.)   AECU-
TECHNICAL INFORMATION SERVICE, AEC (CIVILIAN APPLICATION PROGRAM ENGINEERING DRAWINGS)   CAPE-
TECHNICAL INFORMATION SERVICE, AEC (SERIES ASSIGNED TO JAPANESE REPORTS ABOUT HIROSHIMA + NAGASAKI) J-
TECHNICAL INFORMATION SERVICE, AEC (JOURNAL OF METALLURGY AND CERAMICS)   JMC-
TECHNICAL INFORMATION SERVICE, AEC (SERIES ASSIGNED TO JAPANESE REPORTS ABOUT HIROSHIMA + NAGASAKI) M-
TECHNICAL INFORMATION SERVICE, AEC (SERIES ASSIGNED TO JAPANESE REPORTS ABOUT HIROSHIMA + NAGASAKI) N-
TECHNICAL INFO. SERVICE, AEC (NUCLEAR SCIENCE + TECH)   NST-
TECHNICAL INFO. SERVICE, AEC (REACTOR SCIENCE + TECH.   RST-
TECHNICAL INFORMATION SERVICE, AEC   SPEECH-
TECHNICAL INFORMATION SERVICE, AEC   TI-
TECHNICAL INFORMATION SERVICE, AEC   TIS-
(SEE ALSO LATER NAMES: DIVISION OF TECHNICAL INFORMATION, AEC, AND OFFICE OF INFORMATION SERVICES, AEC)

TECHNICAL INFORMATION SERVICE EXTENSION, AEC

TECHNICAL INFORMATION SERVICE EXTENSION, AEC, OAK RIDGE, TENN. (FOREIGN WEAPONS EFFECTS - ASSIGNED TO UNITED KINGDOM AND CANADIAN REPORTS RELEASED TO THE U. S.)	FWE-
TECHNICAL INFORMATION SERVICE EXTENSION, AEC	M-
TECHNICAL INFORMATION SERVICE EXTENSION, AEC. (SERIES ASSIGNED TO NON-AEC UNNUMBERED TRANSLATIONS)	NP-TR-
TECHNICAL INFORMATION SERVICE EXTENSION, AEC	TI-
TECHNICAL INFORMATION SERVICE EXTENSION, AEC (LITERATURE SEARCH)	TID-LS-
TECHNICAL INFORMATION SERVICE EXTENSION, AEC	TIS-
TECHNICAL INFORMATION SERVICE EXTENSION, AEC, OAK RIDGE, TENN.	TISE-
(SEE ALSO LATER NAMES: DIVISION OF TECHNICAL INFORMATION EXTENSION, AEC, AND TECH. INFO. CENTER, AEC)	
TECHNICAL OIL MISSION	TOM-
TECHNICAL OPERATIONS, INC. (ALL LOCATIONS)	TO-
TECHNICAL OPERATIONS, INC. (ALL LOCATIONS)	TOI-
TECHNICAL OPERATIONS, INC. (ALL LOCATIONS)	TOI-(YEAR)-
TECHNICAL OPERATIONS, INC., ARLINGTON, MASS.	CP-
TECHNICAL OPERATIONS, INC., ARLINGTON, MASS.	TOW-
TECHNICAL OPERATIONS, INC., BURLINGTON, MASS.	TO-B-(NUMBER)-
TECHNICAL OPERATIONS, INC., BURLINGTON, MASS.	TO-B-(YEAR)-
TECHNICAL OPERATIONS, INC., BURLINGTON, MASS.	TOI-B (YEAR)-
TECHNICAL OPERATIONS, INC., WASHINGTON, D.C.	TR-(YEAR)-
TECHNICAL OPERATIONS, INC. COMBAT OPERATIONS RESEARCH GROUP, FORT BELVOIR, VA.	CORG-M-
TECHNICAL OPERATIONS, INC. COMBAT OPERATIONS RESEARCH GROUP, FORT BELVOIR, VA.	CORG-R-
TECHNICAL OPERATIONS, INC. COMBAT OPERATIONS RESEARCH GROUP, FORT BELVOIR, VA.	CORG-SP-
TECHNICAL OPERATIONS, INC. COMBAT OPERATIONS RESEARCH GROUP, FORT BELVOIR, VA.	CORP-SP-
TECHNICAL OPERATIONS, INC. COMBAT OPERATIONS RESEARCH GROUP, FORT MONROE, VA.	CORG-(LTRS.)-
TECHNICAL OPERATIONS, INC. SYSTEM SCIENCES DIV., ALEXANDRIA, VA.	TOI-TR-(YEAR)-
TECHNICAL OPERATIONS, INC. WASHINGTON RESEARCH CENTER, ARLINGTON, VA.	TR-(YEAR)-
TECHNICAL OPERATIONS, INC. WASHINGTON RESEARCH CENTER, ARLINGTON, VA.	WRC-SM-(YEAR)-
TECHNICAL OPERATIONS, INC. WASHINGTON RESEARCH CENTER, ARLINGTON, VA.	WRC-TR-(YEAR)-
TECHNICAL OPERATIONS RESEARCH, BURLINGTON, MASS.	TO-(LETTER)-(YEAR)-
TECHNICAL SCIENCES CORP., ELGIN, ILL.	TR-1-
TECHNIDYNE INC., WEST CHESTER, PA.	RR-(YEAR)
TECHNIK, INC., JERICHO, N.Y.	TR-(YEAR)-
TECHNION-ISRAEL INST. OF TECH., HAIFA	AE-
TECHNION-ISRAEL INST. OF TECH., HAIFA	MS-(NUMBER)/R1
TECHNION-ISRAEL INST. OF TECH., HAIFA	TR-
TECHNION-ISRAEL INST. OF TECH., HAIFA. DEPT. OF AERO.ENG	TAE-
TECHNION-ISRAEL INST. OF TECH., HAIFA. DEPT. OF MECHANICS	TDM-
TECHNION-ISRAEL INST. OF TECH., HAIFA. DEPT. OF MECHANICS	TDM-(YEAR)-
TECHNION-ISRAEL INST. OF TECH., HAIFA. DEPT. OF NUCLEAR SCIENCE	TNSD-R/
TECHNION-ISRAEL INST. OF TECH., HAIFA. FACULTY OF INDUSTRIAL AND MANAGEMENT ENGINEERING	O.R. MIMEO SER-
TECHNION-ISRAEL INST. OF TECH., HAIFA. MATERIAL MECHANICS LAB.	MML-
TECHNION RESEARCH AND DEVELOPMENT FOUNDATION, LTD., HAIFA, ISRAEL	TAE-
TECHNISCHER UEBERWACHUNGS-VEREIN E.V., COLOGNE. INSTITUT FUER REAKTORSICHERHEIT	IRS-(LETTER)-
TECHNISCHER UEBERWACHUNGS-VEREIN E.V., COLOGNE. INSTITUT FUER REAKTORSICHERHEIT	IRS-T-
TECHNOLOGY FOR COMMUNICATIONS INTERNATIONAL, MOUNTAIN VIEW, CALIF.	TCI-
TECHNOLOGY, INC., DAYTON, OHIO	TECH-
TECHNOLOGY, INC., DAYTON, OHIO	TI-
TECHNOLOGY, INC., SAN ANTONIO	TI-
TECHNOLOGY INTERNATIONAL CORP., BEDFORD, MASS.	TIC-
TECHNOLOGY SERVICE CORP., SANTA MONICA, CALIF.	TSC-PD-
TEEG RESEARCH, INC., DETROIT	TR-(NUMBER)-
TEEG RESEARCH, INC., DETROIT	TRN-
TEKNILLINEN KORKEAKOULU, OTANIEMI, FINLAND (NAME TRANSLATED IS: TECHNICAL UNIVERSITY)	TKK-F-A-
TEL-AVIV UNIV., ISRAEL	TAUP-(NUMBER)-(YEAR)
TEL-AVIV UNIV., ISRAEL	TAUP-
TELECOM, INC., ARLINGTON, VA.	TER-
TELECOM, INC., ARLINGTON, VA.	TER-(3 DIGITS)-
TELECOMPUTING CORP., LOS ANGELES	TC-
TELEDYNE BROWN ENGINEERING. BALLISTIC MISSILE DEFENSE DIV., HUNTSVILLE, ALA.	BMDD-SAISC-
TELEDYNE CAE, TOLEDO	TCAE-
TELEDYNE CONTINENTAL MOTORS, WARREN, MICH. (DESIGN AND DEVELOPMENT)	D-(3 DIGITS)
TELEDYNE ELECTRONICS, NEWBURY, CALIF.	TE-(5 DIGITS)
TELEDYNE, INC., ALEXANDRIA, VA.	LL-
TELEDYNE, INC. UED EARTH SCIENCES DIV., ALEXANDRIA, VA.	SDL-
TELEDYNE, INC. UED EARTH SCIENCES DIV., ALEXANDRIA, VA.	SEISMIC DATA LAB-
TELEDYNE, INC. UED EARTH SCIENCES DIV., ALEXANDRIA, VA.	SEISMIC DLR-
TELEDYNE, INC. UED EARTH SCIENCES DIV., ALEXANDRIA, VA.	UED-SDL-
TELEDYNE INDUSTRIES, INC. EARTH SCIENCES DIV., ALEXANDRIA, VA.	UED-SDLFR-
TELEDYNE INDUSTRIES, INC. GEOTECH DIV., GARLAND, TEX.	NA-

TELEDYNE SYSTEMS CO., ALEXANDRIA, VA.	TELEDYNE-SDL-
TELEDYNE SYSTEMS CO., HAWTHORNE, CALIF.	TSC-
TELEDYNE SYSTEMS CO. MICRONETICS DIV., SAN DIEGO, CAL.	R(NUMBER)-(YEAR)-
TEMCO AIRCRAFT CORP., DALLAS (REF. 1)	(NUMBERS...)
TEMCO AIRCRAFT CORP., DALLAS	TAC-
TEMCO AIRCRAFT CORP., DALLAS	TAC-ER-
TEMCO AIRCRAFT CORP., DALLAS	TEMCO-
TEMCO AIRCRAFT CORP., DALLAS	TEMCO-ER-
TEMPLE UNIV., PHILADELPHIA	APR-
TEMPLE UNIV., PHILADELPHIA	TEMPLEU-(NO.)(LTRS.)(NO.
TEMPLE UNIV., PHILADELPHIA	TEMU-
TEMPLE UNIV., PHILADELPHIA	TEU-
TEMPLE UNIV., PHILADELPHIA	TU-
TEMPLE UNIV., PHILADELPHIA. RESEARCH INST.	RITU-
TEMPLE UNIV., PHILADELPHIA. RESEARCH INST.	RITU-(YEAR)-
TEMPLE UNIV., PHILADELPHIA. RESEARCH INST.	TEMU-RI-
TEMPLE UNIV., PHILADELPHIA. RESEARCH INST.	TU-RI-(LETTERS)-
TEMPO INSTRUMENT, INC., HICKSVILLE, N.Y.	TII-
TENNESSEE. UNIV., KNOXVILLE	TENNU-
TENNESSEE. UNIV., KNOXVILLE	TEU-
TENNESSEE. UNIV., KNOXVILLE	TU-
TENNESSEE. UNIV., KNOXVILLE. COLL. OF ENGINEERING	ME-(YEAR)-
TENNESSEE. UNIV., KNOXVILLE. COLL. OF ENGINEERING	ME-(NUMBER)-(YEAR)-
TENNESSEE. UNIV., KNOXVILLE. DEPT. OF CHEMISTRY	TENNU-AR-
TENNESSEE. UNIV., KNOXVILLE. DEPT. OF CHEMISTRY	TENNU-PR-
TENNESSEE. UNIV. MEMPHIS COLLEGE OF PHARMACY	PADS-(MONTH)/(YEAR)
TENNESSEE. UNIV., OAK RIDGE. AGRICULTURAL EXPTL. STA.	TENNU-AES-
TENNESSEE. UNIV., OAK RIDGE. UT-AEC AGRICULTURAL RESEARCH LAB.	UT-ARL-
TENNESSEE. UNIV., TULLAHOMA. SPACE INST.	UTSI-TR-(CON.NO.)-
TENNESSEE EASTMAN CORP., OAK RIDGE, TENN.	A-
TENNESSEE EASTMAN CORP., OAK RIDGE, TENN. (REF. 34)	A-0.600- THRU A-7.390-
TENNESSEE EASTMAN CORP., OAK RIDGE, TENN.	A-2TD-
TENNESSEE EASTMAN CORP., OAK RIDGE, TENN.	AB-
TENNESSEE EASTMAN CORP., OAK RIDGE, TENN.	AC-
TENNESSEE EASTMAN CORP., OAK RIDGE, TENN.	AHD-B3-
TENNESSEE EASTMAN CORP., OAK RIDGE, TENN.	AHM-13-
TENNESSEE EASTMAN CORP., OAK RIDGE, TENN.	AJM-
TENNESSEE EASTMAN CORP., OAK RIDGE, TENN.	ALM-SL-
TENNESSEE EASTMAN CORP., OAK RIDGE, TENN. (REF. 34)	B-0.341- THRU B-7.470-
TENNESSEE EASTMAN CORP., OAK RIDGE, TENN.	BCD-MHR-
TENNESSEE EASTMAN CORP., OAK RIDGE, TENN.	BETA CD-
TENNESSEE EASTMAN CORP., OAK RIDGE, TENN.	BETA CONF-
TENNESSEE EASTMAN CORP., OAK RIDGE, TENN.	BOS-
TENNESSEE EASTMAN CORP., OAK RIDGE, TENN.	BOS-P-
TENNESSEE EASTMAN CORP., OAK RIDGE, TENN.	BPI-
TENNESSEE EASTMAN CORP., OAK RIDGE, TENN.	BTG-
TENNESSEE EASTMAN CORP., OAK RIDGE, TENN. (REF. 34)	C-0.100- THRU C-5.385-
TENNESSEE EASTMAN CORP., OAK RIDGE, TENN.	CD-
TENNESSEE EASTMAN CORP., OAK RIDGE, TENN.	CD-(NO.)-(INITIALS)-
TENNESSEE EASTMAN CORP., OAK RIDGE, TENN.	CD-(INITIALS)-
TENNESSEE EASTMAN CORP., OAK RIDGE, TENN.	CD-AC-
TENNESSEE EASTMAN CORP., OAK RIDGE, TENN.	CD-BETA-S-
TENNESSEE EASTMAN CORP., OAK RIDGE, TENN.	CDC-
TENNESSEE EASTMAN CORP., OAK RIDGE, TENN.	CD-CHEM-S-
TENNESSEE EASTMAN CORP., OAK RIDGE, TENN.	CD-CONF-
TENNESSEE EASTMAN CORP., OAK RIDGE, TENN.	CD-F-
TENNESSEE EASTMAN CORP., OAK RIDGE, TENN.	CD-GS-
TENNESSEE EASTMAN CORP., OAK RIDGE, TENN.	CD-MPR-
TENNESSEE EASTMAN CORP., OAK RIDGE, TENN.	CD-MPR-191-
TENNESSEE EASTMAN CORP., OAK RIDGE, TENN.	CD-MPR-(INITIALS)-
TENNESSEE EASTMAN CORP., OAK RIDGE, TENN.	CD-PPO-
TENNESSEE EASTMAN CORP., OAK RIDGE, TENN.	CD-SP-
TENNESSEE EASTMAN CORP., OAK RIDGE, TENN.	CG-
TENNESSEE EASTMAN CORP., OAK RIDGE, TENN.	CHEM-
TENNESSEE EASTMAN CORP., OAK RIDGE, TENN.	CHEM CD-
TENNESSEE EASTMAN CORP., OAK RIDGE, TENN.	CHEM-S-
TENNESSEE EASTMAN CORP., OAK RIDGE, TENN.	CO-
TENNESSEE EASTMAN CORP., OAK RIDGE, TENN.	COM-
TENNESSEE EASTMAN CORP., OAK RIDGE, TENN.	COUNTING-
TENNESSEE EASTMAN CORP., OAK RIDGE, TENN.	CPK-
TENNESSEE EASTMAN CORP., OAK RIDGE, TENN.	CPK-E-
TENNESSEE EASTMAN CORP., OAK RIDGE, TENN.	CPKE-
TENNESSEE EASTMAN CORP., OAK RIDGE, TENN. (REF. 10)	CS-
TENNESSEE EASTMAN CORP., OAK RIDGE, TENN. (REF. 34)	D-1.210- THRU D-7.440-
TENNESSEE EASTMAN CORP., OAK RIDGE, TENN.	D SPECIAL-
TENNESSEE EASTMAN CORP., OAK RIDGE, TENN.	DWS-PM-
TENNESSEE EASTMAN CORP., OAK RIDGE, TENN. (REF. 34)	E-0.100- THRU E-5.430-
TENNESSEE EASTMAN CORP., OAK RIDGE, TENN.	ENG-(NO. INITIAL(S))
TENNESSEE EASTMAN CORP., OAK RIDGE, TENN.	ER-A-
TENNESSEE EASTMAN CORP., OAK RIDGE, TENN.	EVM-
TENNESSEE EASTMAN CORP., OAK RIDGE, TENN.	EX-
TENNESSEE EASTMAN CORP., OAK RIDGE, TENN. (REF. 34)	F-3.20- THRU F-4.40-
TENNESSEE EASTMAN CORP., OAK RIDGE, TENN.	FTH-
TENNESSEE EASTMAN CORP., OAK RIDGE, TENN. (REF. 34)	G-1.133- THRU G-3.200-
TENNESSEE EASTMAN CORP., OAK RIDGE, TENN.	GCD-MPR-
TENNESSEE EASTMAN CORP., OAK RIDGE, TENN. (REF. 34)	H-0.100- THRU H-10.430-
TENNESSEE EASTMAN CORP., OAK RIDGE, TENN.	H-WRG-
TENNESSEE EASTMAN CORP., OAK RIDGE, TENN.	JCH-
TENNESSEE EASTMAN CORP., OAK RIDGE, TENN.	JDT-S
TENNESSEE EASTMAN CORP., OAK RIDGE, TENN.	JM-
TENNESSEE EASTMAN CORP., OAK RIDGE, TENN.	JMG-CD-
TENNESSEE EASTMAN CORP., OAK RIDGE, TENN.	JPBA-
TENNESSEE EASTMAN CORP., OAK RIDGE, TENN.	JRC-
TENNESSEE EASTMAN CORP., OAK RIDGE, TENN.	JWM-
TENNESSEE EASTMAN CORP., OAK RIDGE, TENN.	MEDF-
TENNESSEE EASTMAN CORP., OAK RIDGE, TENN.	MEW-

TENNESSEE EASTMAN CORP.

TENNESSEE EASTMAN CORP., OAK RIDGE, TENN.	ML-
TENNESSEE EASTMAN CORP., OAK RIDGE, TENN.	MM-
TENNESSEE EASTMAN CORP., OAK RIDGE, TENN.	MPR-
TENNESSEE EASTMAN CORP., OAK RIDGE, TENN.	MPR-196-
TENNESSEE EASTMAN CORP., OAK RIDGE, TENN.	MXL-(INITIALS)-
TENNESSEE EASTMAN CORP., OAK RIDGE, TENN.	PCG-B-
TENNESSEE EASTMAN CORP., OAK RIDGE, TENN.	PD-EX-
TENNESSEE EASTMAN CORP., OAK RIDGE, TENN.	PD-MPR-
TENNESSEE EASTMAN CORP., OAK RIDGE, TENN.	PD-PROC-
TENNESSEE EASTMAN CORP., OAK RIDGE, TENN.	PI-
TENNESSEE EASTMAN CORP., OAK RIDGE, TENN.	PJG-A-
TENNESSEE EASTMAN CORP., OAK RIDGE, TENN.	PM-MPR-
TENNESSEE EASTMAN CORP., OAK RIDGE, TENN.	PM-MPR-LA-
TENNESSEE EASTMAN CORP., OAK RIDGE, TENN.	PM-WPR-
TENNESSEE EASTMAN CORP., OAK RIDGE, TENN.	PP-
TENNESSEE EASTMAN CORP., OAK RIDGE, TENN.	PP-(NO.)-(LETTER)-
TENNESSEE EASTMAN CORP., OAK RIDGE, TENN.	PPD-
TENNESSEE EASTMAN CORP., OAK RIDGE, TENN.	PR-
TENNESSEE EASTMAN CORP., OAK RIDGE, TENN.	PVM-
TENNESSEE EASTMAN CORP., OAK RIDGE, TENN.	R-
TENNESSEE EASTMAN CORP., OAK RIDGE, TENN.	RAS-(LETTER)
TENNESSEE EASTMAN CORP., OAK RIDGE, TENN.	RD-MPR-
TENNESSEE EASTMAN CORP., OAK RIDGE, TENN.	RDO-
TENNESSEE EASTMAN CORP., OAK RIDGE, TENN.	REB-A-
TENNESSEE EASTMAN CORP., OAK RIDGE, TENN.	RJS-CD-
TENNESSEE EASTMAN CORP., OAK RIDGE, TENN.	RJS-CD2-
TENNESSEE EASTMAN CORP., OAK RIDGE, TENN.	RLC-
TENNESSEE EASTMAN CORP., OAK RIDGE, TENN.	RLT-A
TENNESSEE EASTMAN CORP., OAK RIDGE, TENN.	RM-SHG-
TENNESSEE EASTMAN CORP., OAK RIDGE, TENN.	RPP-
TENNESSEE EASTMAN CORP., OAK RIDGE, TENN.	RRP-
TENNESSEE EASTMAN CORP., OAK RIDGE, TENN.	RTD-
TENNESSEE EASTMAN CORP., OAK RIDGE, TENN.	SR-
TENNESSEE EASTMAN CORP., OAK RIDGE, TENN.	TE-B-3-
TENNESSEE EASTMAN CORP., OAK RIDGE, TENN.	TEC-35-
TENNESSEE EASTMAN CORP., OAK RIDGE, TENN.	TEM-
TENNESSEE EASTMAN CORP., OAK RIDGE, TENN.	TE-PES-
TENNESSEE EASTMAN CORP., OAK RIDGE, TENN.	TE-SEP-
TENNESSEE EASTMAN CORP., OAK RIDGE, TENN.	TE-SES-
TENNESSEE EASTMAN CORP., OAK RIDGE, TENN.	TL-
TENNESSEE EASTMAN CORP., OAK RIDGE, TENN.	TM-
TENNESSEE EASTMAN CORP., OAK RIDGE, TENN.	TR-
TENNESSEE EASTMAN CORP., OAK RIDGE, TENN.	TRS-
TENNESSEE EASTMAN CORP., OAK RIDGE, TENN.	VPC-
TENNESSEE EASTMAN CORP., OAK RIDGE, TENN.	WLB-
TENNESSEE EASTMAN CORP., OAK RIDGE, TENN.	WLL-
TENNESSEE EASTMAN CORP., OAK RIDGE, TENN.	WM-
TENNESSEE EASTMAN CORP., OAK RIDGE, TENN.	XA-2-OD-
TENNESSEE EASTMAN CORP., OAK RIDGE, TENN.	XL-4.9 THRU XL-58.5-
TENNESSEE TECHNOLOGICAL UNIV., COOKEVILLE	TTU-
TENNESSEE VALLEY AUTHORITY, OAK RIDGE (CHEMICAL ENGINEERING REPRINT)	CD-
TENNESSEE VALLEY AUTHORITY, OAK RIDGE (CHEM.ENG.BULL)	CEB-
TENNESSEE VALLEY AUTHORITY, OAK RIDGE (CHEM.ENG.RPT.)	CER-
TENNESSEE VALLEY AUTHORITY, OAK RIDGE	EGCR-
TENNESSEE VALLEY AUTHORITY, OAK RIDGE	EGCR-STUDY-
TENNESSEE VALLEY AUTHORITY, OAK RIDGE	SF-
TENNESSEE VALLEY AUTHORITY, OAK RIDGE	TVA-P-
TENNESSEE VALLEY AUTHORITY, WILSON DAM, ALA.	TVA-
TENN. VALLEY AUTH. DIV. OF POWER SUPPLY, CHATTANOOGA	GCRD-
TENN. VALLEY AUTH. DIV. OF POWER SUPPLY, CHATTANOOGA	TVA-P-
TERRA-SPACE CORP., MALIBU, CALIF.	TSC-WA-
TERRIER COORDINATING COMMITTEE (REF. 15)	TERRIER CC-
TERRIER COORD. COMM., VULNERABILITY TEST SUBCOMM.	CC/TERRIER-
TEST SQUADRON (OPERATIONAL) (4750TH), TYNDALL AFB,FLA	TES-
TEST WING (DEV.)(6555TH), PATRICK AFB, FLA.	ATW-
TEST WING (DEV.)(6555TH), PATRICK AFB, FLA.	DWS-TM-(YEAR)-
TETRA TECH, INC., PASADENA, CALIF.	P-
TETRA TECH, INC., PASADENA, CALIF.	TC-
TETRA TECH, INC., PASADENA, CALIF.	TETRA-P-
TETRA TECH, INC., PASADENA, CALIF.	TT-
TETRA TECH, INC., PASADENA, CALIF.	TT-TC-
TEXACO EXPERIMENT INC., RICHMOND	BUMBLEBEE-
TEXACO EXPERIMENT INC., RICHMOND	EXP-
TEXACO EXPERIMENT INC., RICHMOND	TEI-TM-
TEXACO EXPERIMENT INC., RICHMOND	TM-
TEXACO EXPERIMENT INC., RICHMOND	TP-
TEXACO RESEARCH CENTER, BEACON, N.Y.	BLR-AF-
TEXACO RESEARCH CENTER, BEACON, N.Y.	T-
TEXACO RESEARCH CENTER, BEACON, N.Y.	TEXACO-AF-
TEXACO RESEARCH CENTER, BEACON, N.Y.	TRC-
TEXAS. UNIV., AUSTIN	DRL-TM-
TEXAS. UNIV., AUSTIN	HYD-
TEXAS. UNIV., AUSTIN	NPL-
TEXAS. UNIV., AUSTIN	RNL-
TEXAS. UNIV., AUSTIN	TEXU-
TEXAS. UNIV., AUSTIN (ANNUAL REPORT)	TEXU-AR-(YEAR)
TEXAS. UNIV., AUSTIN	TEXU-CPT-
TEXAS. UNIV., AUSTIN	TK-
TEXAS. UNIV., AUSTIN	TNN-
TEXAS. UNIV., AUSTIN	TU-
TEXAS. UNIV., AUSTIN	UT-
TEXAS. UNIV., AUSTIN	UT-CM-
TEXAS. UNIV., AUSTIN. ANTENNAS AND PROPAGATION DIV.	P-
TEXAS. UNIV., AUSTIN. ANTENNAS AND PROPAGATION LAB. (WORK DONE UNDER NATIONAL SCIENCE FOUNDATION GRANT)	NSF-P-

TEXAS. UNIV., AUSTIN. APPLIED MECHANICS RESEARCH LAB.	VSAF-
TEXAS. UNIV., AUSTIN. APPLIED MECHANICS RESEARCH LAB.	VSN-
TEXAS. UNIV., AUSTIN. CENTER FOR CYBERNETIC STUDIES	CS-(2 DIGITS)
TEXAS. UNIV., AUSTIN. CENTER FOR PARTICLE THEORY	AEC-
TEXAS. UNIV., AUSTIN. CENTER FOR PARTICLE THEORY	CPT-
TEXAS. UNIV., AUSTIN. CENTER FOR PLASMA PHYSICS AND THERMONUCLEAR RESEARCH	CPPT-
TEXAS. UNIV., AUSTIN. CENTER FOR RESEARCH IN WATER RESOURCES	CRWR-
TEXAS. UNIV., AUSTIN. COMMUNICATIONS THEORY AND SYSTEMS RESEARCH LAB.	CTSRL-(NUMBER-NUMBER)
TEXAS. UNIV., AUSTIN. DEFENSE RESEARCH LAB.	BUMBLEBEE-
TEXAS. UNIV., AUSTIN. DEFENSE RESEARCH LAB.	DRL-
TEXAS. UNIV., AUSTIN. DEFENSE RESEARCH LAB. (ACOUSTICS REPORT)	DRL-A-
TEXAS. UNIV., AUSTIN. DEFENSE RESEARCH LAB.	DRL-TM-
TEXAS. UNIV., AUSTIN. DEFENSE RESEARCH LAB.	DRL-TR-
TEXAS. UNIV., AUSTIN. DEFENSE RESEARCH LAB.	DRL/UT-
TEXAS. UNIV., AUSTIN. DEFENSE RESEARCH LAB. (TRANS.)	DRL/UT T-
TEXAS. UNIV., AUSTIN. DEFENSE RESEARCH LAB.	TEXU-DRL-
TEXAS. UNIV., AUSTIN. DEFENSE RESEARCH LAB.	TU DRL-
TEXAS. UNIV., AUSTIN. DEFENSE RESEARCH LAB.	UT/DRL-
TEXAS. UNIV., AUSTIN. ELECTRICAL ENGINEERING RES.LAB.	CM-
TEXAS. UNIV., AUSTIN. ELECTRICAL ENGINEERING RES.LAB.	EERL-
TEXAS. UNIV., AUSTIN. ELECTRICAL ENGINEERING RES.LAB.	EERL-TR-(YEAR)-
TEXAS. UNIV., AUSTIN. ELECTRICAL ENGINEERING RES.LAB.	TU EERL-
TEXAS. UNIV., AUSTIN. ELECTRICAL ENGINEERING RES.LAB.	UT-EERL-
TEXAS. UNIV., AUSTIN. ELECTRONICS RESEARCH CENTER	JSEP-TR-
TEXAS. UNIV., AUSTIN. ENGINEERING MECHANICS RES. LAB.	EMRL-
TEXAS. UNIV., AUSTIN. ENGINEERING MECHANICS RES. LAB.	EMRL-RM-
TEXAS. UNIV., AUSTIN. ENGINEERING MECHANICS RES. LAB.	EMRL-TR-
TEXAS. UNIV., AUSTIN. ENGINEERING RESEARCH LABS.	TU-ERL-
TEXAS. UNIV., AUSTIN. ENVIRONMENTAL HEALTH ENGINEERING RESEARCH LAB.	EHE-
TEXAS. UNIV., AUSTIN. ENVIRONMENTAL HEALTH ENGINEERING RESEARCH LAB.	EHE-(NUMBER)-
TEXAS. UNIV., AUSTIN. HYDRAULIC ENGINEERING LAB.	TR-HYD-
TEXAS. UNIV., AUSTIN. LABS. FOR ELECTRONICS AND RELATED SCIENCE RESEARCH	JSEP-TR-
TEXAS. UNIV., AUSTIN. LINGUISTICS RESEARCH CENTER	LRC-
TEXAS. UNIV., AUSTIN. LINGUISTICS RESEARCH CENTER	LRC-(YEAR)-(LTRS.)-
TEXAS. UNIV., AUSTIN. MILITARY PHYSICS DIV.	MPD-
TEXAS. UNIV., AUSTIN. MILITARY PHYSICS RES. LAB.	MPRL-
TEXAS. UNIV., AUSTIN. MILITARY PHYSICS RES. LAB.	TU MEMO (NO.)-
TEXAS. UNIV., AUSTIN. MILITARY PHYSICS RES. LAB.	TU MPRL-
TEXAS. UNIV., AUSTIN. MILITARY PHYSICS RES. LAB.	TU MRPL-
TEXAS. UNIV., AUSTIN. OPTICAL RESEARCH LAB.	ORL/UT-
TEXAS. UNIV., AUSTIN. STRUCTURAL MECHANICS RES. LAB.	SMRL-
TEXAS. UNIV., AUSTIN. STRUCTURAL MECHANICS RES. LAB.	TU-SMRL-
TEXAS. UNIV., AUSTIN. STRUCTURES FATIGUE RES. LAB.	SFRL-P(NUMBER)-
TEXAS. UNIV., AUSTIN. STRUCTURES FATIGUE RES. LAB.	SFRL-TR-P(NUMBER)-
TEXAS. UNIV., HOUSTON. GRAD. SCH. OF BIOMEDICAL SCIENCES	UT-GSBS-DADA-
TEXAS A + M UNIV., COLLEGE STATION	AMCT-
TEXAS A + M UNIV., COLLEGE STATION	TA+M-
TEXAS A + M UNIV., COLLEGE STATION	TAMU-REF-(YEAR)-
TEXAS A + M UNIV., COLLEGE STATION(SEA GRANT PROGRAM)	TAMU-SG-(YEAR)-
TEXAS A + M UNIV.,COLLEGE STA. (PROJ. THEMIS)(REF.48)	THEMIS-TR-
TEXAS A + M UNIV., COLLEGE STATION. DEPT. OF METEOROLOGY	REF-(YEAR)-
TEXAS A + M UNIV., COLLEGE STATION. DEPT. OF OCEANOGRAPHY	A/M-REF-(YEAR)-IT
TEXAS A + M UNIV., COLLEGE STATION. DEPT. OF OCEANOGRAPHY	REF-(YEAR)-
TEXAS A + M UNIV., COLLEGE STATION. ENGINEERING EXPERIMENT STATION	TEES-
TEXAS A + M UNIV., COLLEGE STATION. ENGINEERING EXPERIMENT STATION	TEES-(NUMBER)-
TEXAS A + M UNIV., COLLEGE STATION. RESEARCH FDN.	A/M-REF-
TEXAS A + M UNIV., COLLEGE STATION. RESEARCH FDN.	TAM-
TEXAS INSTRUMENTS, INC. (ALL LOCATIONS)	TI-(NUMBER)-(YEAR)-
TEXAS INSTRUMENTS, INC., DALLAS	C(NO.)-
TEXAS INSTRUMENTS, INC., DALLAS	SATR-
TEXAS INSTRUMENTS, INC., DALLAS	TEXI-
TEXAS INSTRUMENTS, INC., DALLAS	TI-
TEXAS INSTRUMENTS, INC., DALLAS	TI-C(NUMBER)-(NUMBER)-
TEXAS INSTRUMENTS, INC., DALLAS	TIC-(NUMBER)-(NUMBER)-
TEXAS INSTRUMENTS, INC., DALLAS	TI-HB-
TEXAS INSTRUMENTS, INC., DALLAS	TII-
TEXAS INSTRUMENTS, INC., DALLAS	TI-S(NUMBER)-
TEXAS INSTRUMENTS, INC., DALLAS	TI-U(NUMBER)-
TEXAS INSTRUMENTS, INC., DALLAS	U(NO.)-
TEXAS INSTRUMENTS, INC. APPARATUS DIV., DALLAS	C(NUMBER)-(NUMBER)-
TEXAS INSTRUMENTS, INC. APPARATUS DIV., DALLAS	S(NUMBER)-
TEXAS INSTRUMENTS, INC. APPARATUS DIV., DALLAS	SP(NUMBER)-(LTR.,NO.)
TEXAS INSTRUMENTS, INC. APPARATUS DIV., DALLAS	TI-S(NUMBER-NUMBER)-
TEXAS INSTRUMENTS, INC. DALLAS SERVICES GROUP	TI-(NUMBER)-LA
TEXAS INSTRUMENTS, INC. ELECTRONIC AND OPTICAL SYSTEMS DEPT., DALLAS	DM-(YEAR)/(NO.-NO.)
TEXAS INSTRUMENTS, INC. GOVERNMENT PRODS. DIV.,DALLAS	TI-C(NUMBER)-
TEXAS INSTRUMENTS, INC. GOVERNMENT PRODS. DIV.,DALLAS	TI-DM-
TEXAS INSTRUMENTS, INC. METALLURGICAL MATERIALS DIV., ATTLEBORO, MASS.	TI-(NUMBER)-MMD-

TEXAS INSTRUMENTS, INC. SCIENCE SERVICES DIV., DALLAS	TI-SP-
TEXAS INSTRUMENTS, INC. SCIENCE SERVICES DIV., DALLAS	VT-
TEXAS INSTRUMENTS LTD., BEDFORD, BEDS, ENGLAND	TI-TR-
TEXAS NUCLEAR CORP., AUSTIN	TNC-
TEXAS TECH UNIV., LUBBOCK. DEPT. OF SOCIOLOGY	TTU-SOC-
TEXTILE RESEARCH INST., PRINCETON, N.J.	TRI-
TEXTRON INC. DALMO VICTOR CO., SAN CARLOS, CALIF.	DVC-
TEXTRON INC. DALMO VICTOR CO., SAN CARLOS, CALIF.	DVC-
TEXTRON INC. NUCLEAR METALS DIV., WEST CONCORD, MASS.	NMI-
THAILAND. MILITARY RES. + DEV. CENTER, BANGKOK	MRDC-(YEAR)-
THAILAND. MINISTRY OF PUBLIC HEALTH. RADIATION PROTECTION SERVICE, BANGKOK	RPS-AR-
THAILAND. OFFICE OF ATOMIC ENERGY FOR PEACE, BANGKOK	THAI-AEC-
THERM, INC. ADVANCE RESEARCH DIV., ITHACA, N.Y.	TAR-
THERM, INC. ADVANCE RESEARCH DIV., ITHACA, N.Y. (TECHNICAL REPORT)	TAR-TR-
THERM ADVANCED RESEARCH, INC., ITHACA, N.Y.	PA-TN-
THERM ADVANCED RESEARCH, INC., ITHACA, N.Y.	TAR-TR-
THERMAL SYNDICATE LTD., WALLSEND, NORTHUMBERLAND, ENG	CVD/(NUMBER)/(YEAR)
THERMO ELECTRON ENGINEERING CORP., CAMBRIDGE, MASS.	TE-
THERMO ELECTRON ENGINEERING CORP., CAMBRIDGE, MASS.	TEE-
THERMO ELECTRON ENGINEERING CORP., CAMBRIDGE, MASS.	TEE-(NUMBER)-
THERMO ELECTRON ENGINEERING CORP., WALTHAM, MASS.	TE-
THERMO ELECTRON ENGINEERING CORP., WALTHAM, MASS.	TE-(NO.)-(YR.)
THERMO ELECTRON ENGINEERING CORP., WALTHAM, MASS.	TE(NO.)-
THERMO ELECTRON ENGINEERING CORP., WALTHAM, MASS.	TE-(NO.-NO.-YEAR)
THERMO ELECTRON ENGINEERING CORP., WALTHAM, MASS.	TEE-
THERMO ELECTRON ENGINEERING CORP., WALTHAM, MASS.	TEE-(NUMBER)-(YEAR)
THERMO ELECTRON ENGINEERING CORP., WALTHAM, MASS.	TEE-(4 DIGITS)-
THIOKOL CHEMICAL CORP. (ALL LOCATIONS)	TCC-
THIOKOL CHEMICAL CORP., BRIGHAM CITY, UTAH	PM-
THIOKOL CHEMICAL CORP., BRIGHAM CITY, UTAH	TA7-(NUMBER)-(NUMBER)-
THIOKOL CHEMICAL CORP., BRIGHAM CITY, UTAH	TW-
THIOKOL CHEMICAL CORP., BRIGHAM CITY, UTAH	TWP-
THIOKOL CHEMICAL CORP., BRISTOL, PENNA.	THI-
THIOKOL CHEMICAL CORP., HUNTSVILLE, ALA.	C-(YEAR)-
THIOKOL CHEMICAL CORP., HUNTSVILLE, ALA.	LD-
THIOKOL CHEMICAL CORP., HUNTSVILLE, ALA.	TCC-(YEAR)-
THIOKOL CHEMICAL CORP., HUNTSVILLE, ALA.	TCC-ER-
THIOKOL CHEMICAL CORP., HUNTSVILLE, ALA.	TCC-RER-
THIOKOL CHEMICAL CORP., HUNTSVILLE, ALA.	TCC-TR-
THIOKOL CHEMICAL CORP., HUNTSVILLE, ALA.	TEST-
THIOKOL CHEMICAL CORP., HUNTSVILLE, ALA.	U-
THIOKOL CHEMICAL CORP., HUNTSVILLE, ALA.	U-(YEAR)-(NUMBER)A
THIOKOL CHEMICAL CORP. CHEMICAL OPERATIONS, TRENTON	TRENTON RD-
THIOKOL CHEMICAL CORP. ELKTON DIV., MD.	E(NUMBER)-(YEAR)
THIOKOL CHEMICAL CORP. ELKTON DIV., MD.	TCC-E(NUMBER)-(YEAR)
THIOKOL CHEMICAL CORP. ELKTON DIV., MD.	TCC-SR-(NUMBER)-(YEAR)
THIOKOL CHEMICAL CORP. HUNTSVILLE DIV., ALA.	SATR-(NUMBER)
THIOKOL CHEMICAL CORP. NUCLEAR DEVELOPMENT CENTER, PARSIPPANY, N.J.	TCC-(NO.)-R(NO.)
THIOKOL CHEM. CORP. REACTION MOTORS DIV., DENVILLE, NJ	(NUMBER)-IR-
THIOKOL CHEM. CORP. REACTION MOTORS DIV., DENVILLE, NJ	MCN-
THIOKOL CHEM. CORP. REACTION MOTORS DIV., DENVILLE, NJ	RMD-
THIOKOL CHEM. CORP. REACTION MOTORS DIV., DENVILLE, NJ	RMD-(NUMBER)-
THIOKOL CHEM. CORP. REACTION MOTORS DIV., DENVILLE, NJ	RMD-(NUMBER)-(NUMBER)-
THIOKOL CHEM. CORP. REACTION MOTORS DIV., DENVILLE, NJ	RMD-(NUMBER)-(LETTER)
THIOKOL CHEM. CORP. REACTION MOTORS DIV., DENVILLE, NJ	RMD-(NUMBER)-Q-
THIOKOL CHEMICAL CORP. REDSTONE DIV., HUNTSVILLE, ALA.	C-
THIOKOL CHEMICAL CORP. REDSTONE DIV., HUNTSVILLE, ALA.	THIOKOL (NO.)/(YR.)
THIOKOL CHEMICAL CORP. REDSTONE DIV., HUNTSVILLE, ALA.	THIOKOL(NUMBER)-(NUMBER)
THIOKOL CHEMICAL CORP. REDSTONE DIV., HUNTSVILLE, ALA.	THIOKOL C-(NUMBERS)
THIOKOL CHEMICAL CORP. REDSTONE DIV., HUNTSVILLE, ALA.	THIOKOL C-A-(NUMBERS)
THIOKOL CHEMICAL CORP. REDSTONE DIV., HUNTSVILLE, ALA.	THIOKOL C-C-(NUMBERS)
THIOKOL CHEMICAL CORP. REDSTONE DIV., HUNTSVILLE, ALA.	THIOKOL SP-
THIOKOL CHEMICAL CORP. REDSTONE DIV., HUNTSVILLE, ALA.	THIOKOL TU-(NUMBERS)
THIOKOL CHEMICAL CORP. REDSTONE DIV., HUNTSVILLE, ALA.	TRD-
THIOKOL CHEMICAL CORP. REDSTONE DIV., HUNTSVILLE, ALA.	U-
THIOKOL CHEMICAL CORP. RESEARCH DIV., HUNTSVILLE, ALA	TCC-(NUMBER)-(YEAR)
THIOKOL CHEMICAL CORP. WASATCH DIV., BRIGHAM CITY, UTAH	TWR-
THOMAS (A.S.), INC., WESTWOOD, MASS.	M(YEAR)-
THOMPSON (H. I.) FIBER GLASS CO., GARDENA, CALIF.	RDD-
THOMPSON PRODUCTS, INC., CLEVELAND	ING ER-
THOMPSON PRODUCTS, INC., CLEVELAND	JER-
THOMPSON PRODUCTS, INC., CLEVELAND	MSD-
THOMPSON PRODUCTS, INC., CLEVELAND	TM-(NUMBER)-(NUMBER)
THOMPSON PRODUCTS, INC., CLEVELAND	TPI-
THOMPSON PRODUCTS, INC., CLEVELAND	TPI-ER-
THOMPSON PRODUCTS, INC. ANTENNA RESEARCH LAB, COLUMBUS, OHIO	ARL-
THOMPSON PRODUCTS, INC. JET DIV. (ENGINEERING REPORT)	JER-

THOMPSON RAMO WOOLDRIDGE INC., CANOGA PARK, CALIF.	NAA-SR-
THOMPSON RAMO WOOLDRIDGE INC., CLEVELAND	ER-
THOMPSON RAMO WOOLDRIDGE INC., CLEVELAND	NAA-SR-
THOMPSON RAMO WOOLDRIDGE INC., LOS ANGELES	C20-OU(NUMBER)-
THOMPSON RAMO WOOLDRIDGE INC., LOS ANGELES	GM-TM-
THOMPSON RAMO WOOLDRIDGE INC., LOS ANGELES	M-
(SEE ALSO LATER NAME: TRW INC.)	
THOMPSON RAMO WOOLDRIDGE INC. RESEARCH LAB., CANOGA PARK, CALIF.	RW-RL-
THOMPSON RAMO WOOLDRIDGE INC. TAPCO DIV., CLEVELAND	ER-
THOMPSON RAMO WOOLDRIDGE INC. TAPCO DIV., CLEVELAND	NAA-
THOMPSON RAMO WOOLDRIDGE INC. TAPCO DIV., CLEVELAND	TM-(NUMBER)-(NUMBER)
THOMPSON RAMO WOOLDRIDGE INC. TRW ELECTROMECHANICAL DIV., CLEVELAND	NAA-SR-
THORN-AEI RADIO VALVES AND TUBES, LTD., ROCHESTER,ENG	BTR-
TIMBER ENGINEERING CO., WASHINGTON, D.C.	TECO-
TITANIUM METALS CORP. OF AMERICA, N.Y.C.	TMCA-
TODD SHIPYARDS CORP., GALVESTON, TEX.	TODD/SML-NSS-
TODD SHIPYARDS CORP., NUCLEAR DIV., GALVESTON, TEX.	TODD-N-
TOHOKU UNIV., SENDAI, JAPAN	TOHOKUU-
TOHOKU UNIV., SENDAI, JAPAN. RESEARCH INST. OF ELECTRICAL COMMUNICATION	REIC-TR-
TOHOKU UNIV., SENDAI, JAPAN. RESEARCH INST. OF ELECTRICAL COMMUNICATION	RIEC-TR-
TOKYO ATOMIC INDUSTRY CONFERENCE	TAIC-
TOKYO IMPERIAL UNIV. PATHOLOGICAL INST.	C-(NO.)-B-
TOKYO METROPOLITAN UNIV.	NUP-(YEAR)-
TOKYO UNIV.	INST-
TOKYO UNIV.	ISAS-
TOKYO UNIV.	J-
TOKYO UNIV.	NUP-
TOKYO UNIV.	RUP-
TOKYO UNIV.	SJC-A-
TOKYO UNIV.	SJC-P-(YEAR)-
TOKYO UNIV.	SJC-T-(YEAR)-
TOKYO UNIV.	TUEP-
TOKYO UNIV. INST. FOR NUCLEAR STUDIES	INS-
TOKYO UNIV. INST. FOR NUCLEAR STUDIES	INSJ-
TOKYO UNIV. INST. FOR NUCLEAR STUDIES	INS-PT-
TOKYO UNIV. INST. FOR NUCLEAR STUDIES	INS-TCA-
TOKYO UNIV. INST. FOR NUCLEAR STUDIES	INS-TH-
TOKYO UNIV. INST. FOR NUCLEAR STUDIES	INS-TL-
TOKYO UNIV. INST. FOR NUCLEAR STUDIES	SJC-A-(YEAR)-
TOKYO UNIV. INST. FOR NUCLEAR STUDIES	SJC-P-(YEAR)-
TOKYO UNIV. INST. FOR NUCLEAR STUDIES	SJC-T-(YEAR)-
TOKYO UNIV. INST. FOR NUCLEAR STUDIES	TH-
TOKYO UNIV. INST. FOR SOLID STATE PHYSICS	ISSP-A-
TOKYO UNIV. OF EDUCATION	TUEP-
TOKYO UNIV. OF EDUCATION. DEPT. OF PHYSICS	TEUP-(YEAR)-
TOKYO UNIV. OF EDUCATION. DEPT. OF PHYSICS	TUENS-
TOLEDO. UNIV. DEPT. OF CHEMICAL ENGINEERING	UTCHE-
TOLEDO REGIONAL AREA PLAN FOR ACTION, OHIO	REGIONAL-(NUMBER.NUMBER)
TOMS RIVER SIGNAL LAB. SUPPRESSION SYSTEMS DEVELOPMENT SUBSECTION, N.J. (ENGINEERING MEMO.)	SSDS EM-
TOMS RIVER SIGNAL LAB. SUPPRESSION SYSTEMS DEVELOPMENT SUBSECTION, N.J. (ENGINEERING MEMO.)	TRSL SDDS EM-
TORONTO. UNIV.	CI-
TORONTO. UNIV.	TORONTOU-
TORONTO. UNIV.	TORU-
TORONTO. UNIV. COMPUTATION CENTRE	UTECS-
TORONTO. UNIV. DEPT. OF ELECTRICAL ENGINEERING	RR-
TORONTO. UNIV. DEPT. OF MECHANICAL ENGINEERING	TP-(YEAR AND REPT. NO.)
TORONTO. UNIV. DEPT. OF MECHANICAL ENGINEERING	UT-MECH-E-
TORONTO. UNIV. INST. FOR AEROSPACE STUDIES	ORC-(YEAR)-
TORONTO. UNIV. INST. FOR AEROSPACE STUDIES	UTIAS-
TORONTO. UNIV. INST. FOR AEROSPACE STUDIES	UTIAS-REV-
TORONTO. UNIV. INST. FOR AEROSPACE STUDIES	UTIAS-TN-
TORONTO. UNIV. INST. OF AEROPHYSICS	CAN UTIA (NUMBER)
TORONTO. UNIV. INST. OF AEROPHYSICS	UTIA-
TORONTO. UNIV. INST. OF AEROPHYSICS	UTIA-REV-
TORONTO. UNIV. INST. OF AEROPHYSICS	UTIA-TN-
TOWSON LABS., INC., BALTIMORE, MD.	P-
TRACERLAB, INC.	I-(NO.)/(NO.)
TRACERLAB, INC.	I-(NO.)-(NO.)/(NO.)
TRACERLAB, INC.	O-(NUMBER(S))/(NUMBER)
TRACERLAB, INC., BOSTON (REF. 1)	(NUMBERS...)
TRACERLAB, INC., BOSTON	6TR-
TRACERLAB, INC., WALTHAM, MASS. (REF. 1)	(NUMBERS...)
TRACERLAB, INC., WALTHAM, MASS.	TL-
TRACERLAB, INC., WALTHAM, MASS.	TL-(NO.)TR-
TRACERLAB, INC., WALTHAM, MASS.	21TR-
TRACERLAB, INC. EASTERN DIV., WALTHAM, MASS.	TLE-

TRACERLAB, INC. EASTERN DIV.

TRACERLAB, INC. EASTERN DIV., WALTHAM, MASS.	TLE-(NO.)(S)-(YR.)-
TRACERLAB, INC. EASTERN DIV. LABS., WALTHAM, MASS.	TRL TLE-
TRACERLAB, INC. WESTERN DIV., RICHMOND, CAL. (REF. 1)	(NUMBERS...)
TRACERLAB, INC. WESTERN DIV., RICHMOND, CALIF.	TB-
TRACERLAB, INC. WESTERN DIV., RICHMOND, CALIF.	1TR-
TRACERLAB, INC. WESTERN DIV., RICHMOND, CALIF.	5TR-
TRACERLAB, INC. WESTERN DIV., RICHMOND, CALIF.	TRL TLW-
(SEE ALSO LATER NAME: TRACERLAB. DIV. OF LAB. FOR ELECTRONICS)	
TRACERLAB. DIV. OF LAB. FOR ELECTRONICS, INC., BOSTON	TLB-
TRACERLAB. DIV. OF LAB. FOR ELECTRONICS, INC., BOSTON	TRL-
TRACERLAB. DIV. OF LAB. FOR ELECTRONICS, INC., RICHMOND, CALIF.	TL-TB-(YEAR)-
TRACERLAB. DIV. OF LAB. FOR ELECTRONICS, INC., RICHMOND, CALIF.	TLW-
TRACERLAB. DIV. OF LAB. FOR ELECTRONICS, INC., RICHMOND, CALIF.	TRL-
TRACOR, INC., AUSTIN, TEX.	TRACOR-
TRACOR, INC., AUSTIN, TEX.	TRACOR-(YEAR)-
TRACOR, INC., AUSTIN, TEX.	TRACOR-NYL/(YEAR)-
TRACOR, INC., AUSTIN, TEX.	TRACOR-RL/(YEAR)-
TRACOR, INC., SAN DIEGO, CALIF.	TRACOR-SD-(YEAR)-
TRAINING AIDS RESEARCH LAB., CHANUTE AFB, ILL.	TARL-LAB NOTE (NUMBER)-
TRAINING AIDS RESEARCH LAB., CHANUTE AFB, ILL.	TARL-TM-(NUMBER)-
TRANS-SONICS, INC., BEDFORD, MASS.	TRANS-
TRANS-SONICS, INC., BEDFORD, MASS.	TRAS-
TRANS-SONICS, INC., BURLINGTON, MASS	S.O. (5 DIGITS)
TRANS-SONICS, INC., BURLINGTON, MASS.	TSI-
TRANSCONTINENTAL + WESTERN AIR, INC., KANSAS CITY, MO. (REF. 1)	(NUMBER)
TRANSLATIONS, N.Y.C.	TNY-
TRANSLATIONS, INC., N.Y.C.	TNY-(NUMBER)(SC)
TRANSMITTER EQUIPMENT MFG. CO., INC., N.Y.C.	TEMCO-
TRANSPORTATION CORPS (ARMY) (CATALOG)	ASF-CAT-TC-
TRAVELERS RESEARCH CENTER, INC., HARTFORD (NUMBER MAY BE 7000-7999. IT IS CONTRACT NO.)(REF.1)	7000-
TRAVELERS RESEARCH CENTER, INC., HARTFORD (PROPOSAL)	P(YEAR)(MONTH)-
TRAVELERS RESEARCH CENTER, INC., HARTFORD	TRC-
TRAVELERS RESEARCH CENTER, INC., HARTFORD	TRCI-
TRAVENOL LABS., INC., MORTON GROVE, ILL.	BX-(NUMBER-YEAR)
TREFILERIES ET LAMINOIRS DU HAVRE. CENTRE DE RECHERCHES, ANTONY, FRANCE	CDRA-
TRF, INC., SPRINGFIELD, VA.	T-
TRG, INC. (ALL LOCATIONS)	TRG-
TRG, INC., MELVILLE, N.Y.	FSC-
TRG, INC., MELVILLE, N.Y.	TRG-(NUMBER)-FR-
TRG, INC., MELVILLE, N.Y.	TRG-(JOB NUMBER)-QTR-
TRG, INC., MENLO PARK, CALIF.	TRG-W-(NUMBER)-F
TRG, INC., MENLO PARK, CALIF.	W(NUMBER)-
TRG, INC., SYOSSET, N.Y.	TRG-129-QTR-
TRG, INC., SYOSSET, N.Y.	TRG-(NUMBER)-QTR-
TRI-STATE TRANSPORTATION COMMISSION, N.Y.	TSTC-(NUMBER-NUMBER)-
TRI-UNIVERSITY MESON FACILITY, VANCOUVER, B.C.	TRI-(YEAR)-
TRI-UNIVERSITY MESON FACILITY, VANCOUVER, B.C.	TRI-AR-(YEAR)
TRIANGLE RESEARCH CORP., ALEXANDRIA, VA.	TRC-TR-
TRIANGLE UNIVERSITIES NUCLEAR LAB., DURHAM, N.C.	TUNL-
TRIER. UNIV., KAISERSLAUTERN, GERMANY	TRIERU-
TRIPARTITE NUCLEAR CROSS-SECTIONS COMMITTEE (REF. 35)	TNCC-
TRIPARTITE NUCLEAR CROSS-SECTIONS COMMITTEE (REF. 35)	TNCC(CAN)-
TRIPARTITE NUCLEAR CROSS-SECTIONS COMMITTEE (REF. 35)	TNCC(UK)-
TRIPARTITE NUCLEAR CROSS-SECTIONS COMMITTEE (REF. 35)	TNCC(US)-
TROPICAL DETERIORATION ADMINISTRATIVE COMM. (REF. 28)	TDAC-
TROPICAL DETERIORATION INFORMATION CENTER	TDIC-
TROPICAL DETERIORATION STEERING COMMITTEE	TDSC-
TROPICAL PRODUCTS INST., LONDON	TPI-
TRW EQUIPMENT LABS., CLEVELAND	TRW-(CONTRACT NO.)-
TRW EQUIPMENT LABS. MATERIALS RESEARCH AND DEVELOPMENT DEPT., CLEVELAND	MR-
TRW INC. (ALL LOCATIONS)	TRW-
TRW INC., CLEVELAND (ENGINEERING REPORTS)	TRW-ER-
TRW INC., CLEVELAND (TECHNICAL MEMORANDA)	TRW-TM-
TRW SEMICONDUCTORS, INC., LAWNDALE, CALIF.	TE-
TRW SEMICONDUCTORS, INC., LAWNDALE, CALIF.	TRWS-
TRW SYSTEMS, REDONDO BEACH, CALIF. (REFS. 1, 42)	(4 DIGITS-NO.-LTRS.)-
TRW SYSTEMS, REDONDO BEACH, CALIF. (REFS. 1, 42)	(5 DIGITS-NO.-LTRS.)-
TRW SYSTEMS, REDONDO BEACH, CALIF.	STL-
TRW SYSTEMS, REDONDO BEACH, CALIF. (REF. 42)	STL-(NO.-NO.-LTRS.)-
TRW SYSTEMS, REDONDO BEACH, CALIF. (REF. 42)	TRW-(NO.-NO.-LTRS.)-
TRW SYSTEMS, REDONDO BEACH, CALIF.	TRW-(NO.-NO.)-R00O
TRW SYSTEMS, REDONDO BEACH, CALIF. (REF. 42)	TRW-R(5 DIGITS-NO.)-R00-

TRW SYSTEMS, REDONDO BEACH, CALIF. (REF. 42)	TRW-TR-(NUMBER.NUMBER)-
TRW SYSTEMS. ENG. MECHANICS LAB., REDONDO BEACH, CAL	EM-
TRW SYSTEMS GROUP, REDONDO BEACH, CALIF.	NASC-TRW-(YEAR)-
TRW SYSTEMS GROUP, REDONDO BEACH, CALIF.	TRW(A)-(NUMBER-NUMBER)
TRW SYSTEMS GROUP. WASHINGTON OPERATIONS, D.C.	NECPT-
TUEBINGEN, GERMANY. UNIV.	TUEBINGENU-
TUFTS COLL., MEDFORD, MASS.	TUFTS-
TUFTS UNIV., MEDFORD, MASS.	HEIAS-
TUFTS UNIV., MEDFORD, MASS.	TC-
TULANE UNIV., NEW ORLEANS	TULANEU-TR-
TULANE UNIV., NEW ORLEANS	TULU-
TUNE INVESTMENTS RESEARCH LABS., HINXTON HALL, CAMBRIDGE, ENGLAND	TIRL-R-
TUNE INVESTMENTS RESEARCH LABS., HINXTON HALL, CAMBRIDGE, ENGLAND	TIRL-TM-
TUNG-SOL ELECTRIC INC., BLOOMFIELD, N.J.	TSEI-
TUNG-SOL ELECTRIC INC., BLOOMFIELD, N.J.	TUN-
TURCO PRODUCTS, INC., WILMINGTON, CALIF.	TU-
TURCO PRODUCTS, INC., WILMINGTON, CALIF.	TU-L-
TURIN. POLITECNICO. ISTITUTO DI FISICA TECNICA	PT-IFT-
TURIN. POLITECNICO. LABORATORIO DI MECCANICA	TURIN-
TURKEY. ATOMIC ENERGY COMMISSION, ANKARA	TAEC-D-
TURKEY. AEC. CEKMECE NUCLEAR RESEARCH CENTER,ISTANBUL	CNAEM-
TURKEY. AEC. CEKMECE NUCLEAR RESEARCH CENTER,ISTANBUL	CNAM-
TURKEY. AEC. CEKMECE NUCLEAR RESEARCH CENTER,ISTANBUL	NAEM-
TURKEY. MINISTRY OF NATIONAL DEFENCE, ANKARA	RDDR-
TYCO LABS., INC., CLAREMONT, CALIF.	ER-
TYCO LABS., INC., CLAREMONT, CALIF.	TLI-ER-
TYCO LABS., INC., WALTHAM, MASS.	T-(YEAR)-
TYCO LABS., INC., WALTHAM, MASS.	TLI-
TYPHON JOINT AEC-DOD COORDINATING COMM.	JCC/TYPHON-

U.S.S.R. CENTRAL AERO-HYDRODYNAMICAL INST.	CAHI-
U.S.S.R. EREVAN INST. OF PHYSICS	EFI-ME-(MONTH/YEAR)
U.S.S.R. EREVAN INST. OF PHYSICS	EFI-TF-(MONTH/YEAR)
U.S.S.R. GOSUDARSTVENNYI KOMITET PO ISPOL'ZOVANIYU ATOMNOI ENERGII, MOSCOW	ITEF-
U.S.S.R. GOSUDARSTVENNYI KOMITET PO ISPOL'ZOVANIYU ATOMNOI ENERGII, MOSCOW (NAME TRANSLATED IS: STATE COMMITTEE FOR THE UTILIZATION OF ATOMIC ENERGY)	SCPUAE-
U.S.S.R. GOSUDARSTVENNYI KOMITET PO ISPOL'ZOVANIYU ATOMNOI ENERGII.FIZIKO-ENERGETICHESKII INST,OBNINSK	FEI-
U.S.S.R. GOSUDARSTVENNYI KOMITET PO ISPOL'ZOVANIYU ATOMNOI ENERGII. FIZIKO-TEKHNICHESKII INST.,SUKHUMI	FTI-
U.S.S.R. GOSUDARSTVENNYI KOMITET PO ISPOL'ZOVANIYU ATOMNOI ENERGII. INSTITUT ATOMNOI ENERGII, MOSCOW	IAE-
U.S.S.R. GOSUDARSTVENNYI KOMITET PO ISPOL'ZOVANIYU ATOMNOI ENERGII. INSTITUT TEORETICHESKOI I EKSPERI-MENTAL'NOI FIZIKI, MOSCOW	N-
U.S.S.R. GOSUDARSTVENNYI KOMITET PO ISPOL'ZOVANIYU ATOMNOI ENERGII. NAUCHNO-ISSLEDOVATEL'SKII INSTITUT ELEKTROFIZICHESKOI APPARATURY, LENINGRAD	NIIEFA-
U.S.S.R. INSTITUT FIZIKI VYSOKIKH ENERGII, SERPUKHOV	IFVE-(LETTERS)-
U.S.S.R. INSTITUT FIZIKI VYSOKIKH ENERGII, SERPUKHOV	IFVE-INZH-(YEAR)-
U.S.S.R. INSTITUT FIZIKI VYSOKIKH ENERGII, SERPUKHOV	IFVE-OP-(YEAR)-
U.S.S.R. INSTITUT FIZIKI VYSOKIKH ENERGII, SERPUKHOV	IFVE-OP-(YEAR)-(NO.)-K
U.S.S.R. INSTITUT FIZIKI VYSOKIKH ENERGII, SERPUKHOV	IFVE-SEF-(YEAR)-
U.S.S.R. INSTITUT FIZIKI VYSOKIKH ENERGII, SERPUKHOV	IFVE-SEF-(YEAR)-(NO.)-K
U.S.S.R. INSTITUT FIZIKI VYSOKIKH ENERGII, SERPUKHOV	IFVE-SEF/(LTRS.)-(YR)-
U.S.S.R. INSTITUT FIZIKI VYSOKIKH ENERGII, SERPUKHOV	IFVE-SKR-(YEAR)-(NO.)-
U.S.S.R. INSTITUT FIZIKI VYSOKIKH ENERGII, SERPUKHOV	IFVE-SKU-(YEAR)-
U.S.S.R. INSTITUT FIZIKI VYSOKIKH ENERGII, SERPUKHOV	IFVE-SPK/(LTRS.)-(YR)-
U.S.S.R. INSTITUT FIZIKI VYSOKIKH ENERGII, SERPUKHOV	IFVE-STF-(YEAR)-(NO.)-
U.S.S.R. INSTITUT FIZIKI VYSOKIKH ENERGII, SERPUKHOV	IFVE-SVM-(YEAR)-
U.S.S.R. INSTITUT FIZIKI VYSOKIKH ENERGII, SERPUKHOV	STF/SEF/SVM/SKU-
U.S.S.R. JOINT INST. FOR NUCLEAR RESEARCH, DUBNA	DINR-E-
U.S.S.R. JOINT INST. FOR NUCLEAR RESEARCH, DUBNA	DINR-P-
U.S.S.R. JOINT INST. FOR NUCLEAR RESEARCH, DUBNA	DINR-R-
U.S.S.R. JOINT INST. FOR NUCLEAR RESEARCH, DUBNA	E-
U.S.S.R. JOINT INST. FOR NUCLEAR RESEARCH, DUBNA	JINR-(LETTER)(NUMBER)-
U.S.S.R. JOINT INST. FOR NUCLEAR RESEARCH, DUBNA	JINR-D-
U.S.S.R. JOINT INST. FOR NUCLEAR RESEARCH, DUBNA	JINR-DC-
U.S.S.R. JOINT INST. FOR NUCLEAR RESEARCH, DUBNA	JINR-E-
U.S.S.R. JOINT INST. FOR NUCLEAR RESEARCH, DUBNA	JINR-E(NUMBER)-
U.S.S.R. JOINT INST. FOR NUCLEAR RESEARCH, DUBNA	JINR-P-
U.S.S.R. JOINT INST. FOR NUCLEAR RESEARCH, DUBNA	JINR-P(NUMBER)-
U.S.S.R. JOINT INST. FOR NUCLEAR RESEARCH, DUBNA	P-
U.S.S.R. JOINT INST. FOR NUCLEAR RESEARCH, DUBNA	P(NUMBER)-
U.S.S.R. JOINT INST. FOR NUCLEAR RESEARCH, DUBNA	SINR-E(NUMBER)-
U.S.S.R. SCIENTIFIC RESEARCH ATOMIC REACTOR INST., MELEKESS	SRARI-
U.S.S.R. SCIENTIFIC RESEARCH ATOMIC REACTOR INST., MELEKESS	SRARI-P-
U.S.S.R. VSESOYUZNYI NAUCHNO-ISSLEDOVSTELSKII INSTITUT RADIATSIONNOI TEKHNIKI, MOSCOW	VNIIRT-
UFO INFORMATION RETRIEVAL CENTER, RIDERWOOD, MD.	UFOIRC-(YEAR)-
UGINE-KUHLMANN. LAB. DE RECHERCHE, VENTHON, FRANCE	LRV/(LETTERS-NUMBER)
ULTRASONIC CORP., CAMBRIDGE, MASS.	UC-
ULTRASONIC CORP., CAMBRIDGE, MASS.	ULC-
UNDERWATER DEMOLITION TEAM NO. 4, INDIAN HEAD, MD.	UDT-
UNDERWATER EXPLOSIONS RESEARCH DIV., PORTSMOUTH, VA. (REF. 1)	(NUMBERS...)
UNDERWATER EXPLOSIONS RESEARCH DIV., PORTSMOUTH, VA.	C-
UNDERWATER EXPLOSIONS RESEARCH DIV., PORTSMOUTH, VA.	UER-
UNDERWATER EXPLOSIONS RESEARCH DIV., PORTSMOUTH, VA.	UERD-
UNDERWATER EXPLOSIONS RESEARCH DIV., PORTSMOUTH, VA.	UERD-(NO.)-(YR.)
UNDERWATER EXPLOSIONS RESEARCH DIV., PORTSMOUTH, VA.	UERD-(YR)-
UNDERWATER EXPLOSIVES RESEARCH LAB., WOODS HOLE, MASS	UERL-
UNDERWATER EXPLOSIVES RESEARCH LAB., WOODS HOLE, MASS	UERL-TECHNICAL MEMO-
UNDERWATER SYSTEMS INC., SILVER SPRING, MD.	UWSI-TPR-
UNDERWATER SYSTEMS INC., WHEATON, MD.	USI-
UNDERWRITERS LABS., INC., N.Y.C.	FU-
UNDERWRITERS LABS., INC. MARINE DEPT., WESTWOOD, N.J.	YSB-R(NO.-NO.)-
UNIDYNAMICS/PHOENIX, ARIZ.	DTR-
UNIDYNAMICS/PHOENIX, ARIZ.	PDS-
UNIDYNAMICS/PHOENIX. DIV. OF UNIVERSAL MATCH CORP., ARIZ.	D(YEAR)-(NUMBER)
UNION CARBIDE AND CARBON CORP., N.Y.C.	UCC-
UNION CARBIDE CORP., N.Y.C.	SRCR-(YEAR)-
UNION CARBIDE CORP., N.Y.C.	UC-
UNION CARBIDE CORP., OAK RIDGE, TENN.	K-
UNION CARBIDE CORP., OAK RIDGE, TENN.	Y-DA-
UNION CARBIDE CORP., OAK RIDGE, TENN.	Y-DD-
UNION CARBIDE CORP., OAK RIDGE, TENN.	Y-DR-
UNION CARBIDE CORP., OAK RIDGE, TENN.	Y-KB-
UNION CARBIDE CORP. LINDE DIV., INDIANAPOLIS	L-
UNION CARBIDE CORP. LINDE DIV., INDIANAPOLIS	SRCR-(YEAR)-
UNION CARBIDE CORP. LINDE DIV., NEWARK, N.J.	L-
UNION CARBIDE CORP. LINDE DIV., TONAWANDA, N.Y.	ASR-
UNION CARBIDE CORP. LINDE DIV., TONAWANDA, N.Y.	TM-B-
UNION CARBIDE CORP. LINDE DIV. SPEEDWAY LABS., INDIANAPOLIS, IND.	SRCR-
UNION CARBIDE CORP. MATERIALS SYSTEMS DIV., WHITE PLAINS, N.Y.	UCC/AS-

UNION CARBIDE CORP. PARMA RESEARCH LAB., OHIO	C-
UNION CARBIDE CORP. PARMA RESEARCH LAB., OHIO	PR (NUMBER)-
UNION CARBIDE CORP. PARMA RESEARCH LAB., OHIO	UCC-
UNION CARBIDE CORP. PARMA RESEARCH LAB., OHIO	UCPRL-BMPR-
UNION CARBIDE CORP. PARMA RESEARCH LAB., OHIO	UCPRL-RR-
UNION CARBIDE CORP. PARMA RESEARCH LAB., OHIO	URS-
UNION CARBIDE CORP. PLASTICS DIV., BOUND BROOK, N.J.	RTD-IR-
UNION CARBIDE CORP. RESEARCH INST., TARRYTOWN, N.Y.	UCC-TR-C-
UNION CARBIDE CORP. Y-12 PLANT, OAK RIDGE, TENN.	Y-12-
UNION CARBIDE CORP. Y-12 PLANT, OAK RIDGE, TENN.	Y-
UNION CARBIDE CORP. Y-12 PLANT, OAK RIDGE, TENN.	Y-AE-
UNION CARBIDE CORP. Y-12 PLANT, OAK RIDGE, TENN.	Y-CDC-
UNION CARBIDE CORP. Y-12 PLANT, OAK RIDGE, TENN.	Y-DA-
UNION CARBIDE CORP. Y-12 PLANT, OAK RIDGE, TENN.	Y-DM-
UNION CARBIDE CORP. Y-12 PLANT, OAK RIDGE, TENN.	Y-DR-
UNION CARBIDE CORP. Y-12 PLANT, OAK RIDGE, TENN.	Y-E-
UNION CARBIDE CORP. Y-12 PLANT, OAK RIDGE, TENN.	Y-EC-
UNION CARBIDE CORP. Y-12 PLANT, OAK RIDGE, TENN.	Y-EG-
UNION CARBIDE CORP. Y-12 PLANT, OAK RIDGE, TENN.	Y-EH-
UNION CARBIDE CORP. Y-12 PLANT, OAK RIDGE, TENN.	Y-IA-
UNION CARBIDE CORP. Y-12 PLANT, OAK RIDGE, TENN.	Y-JA-
UNION CARBIDE CORP. Y-12 PLANT, OAK RIDGE, TENN.	Y-KA-
UNION CARBIDE CORP. Y-12 PLANT, OAK RIDGE, TENN.	Y-KG-
UNION CARBIDE CORP. Y-12 PLANT, OAK RIDGE, TENN.	Y-MA-
UNION CARBIDE CORP. Y-12 PLANT, OAK RIDGE, TENN.	Y-NA-
UNION CARBIDE CORP. Y-12 PLANT, OAK RIDGE, TENN.	Y-OA-
UNION CARBIDE CORP. Y-12 PLANT, OAK RIDGE, TENN.	Y-SC-
UNION CARBIDE METALS CO. METALS RESEARCH LABS., NIAGARA FALLS, N.Y.	UCMC-
UNION CARBIDE METALS CO. METALS RESEARCH LABS., NIAGARA FALLS, N.Y.	UCMC-MRL-TR-
UNION CARBIDE NUCLEAR CO., OAK RIDGE, TENN.	K-
UNION CARBIDE NUCLEAR CO., OAK RIDGE, TENN.	KT-
UNION CARBIDE NUCLEAR CO., OAK RIDGE, TENN.	Y-DA-
UNION CARBIDE NUCLEAR CO., OAK RIDGE, TENN.	Y-EF-
UNION CARBIDE NUCLEAR CO., PADUCAH, KY.	KY-F-
UNION CARBIDE NUCLEAR CO. COMPUTING TECHNOLOGY CENTER, OAK RIDGE, TENN.	CTC-
UNION CARBIDE NUCLEAR CO. COMPUTING TECHNOLOGY CENTER, OAK RIDGE, TENN.	CTC-INF-
UNION CARBIDE NUCLEAR CO. K-25 PLANT, OAK RIDGE, TENN	KB-
UNION CARBIDE NUCLEAR CO. K-25 PLANT, OAK RIDGE, TENN	KD-
UNION CARBIDE NUCLEAR CO. K-25 PLANT, OAK RIDGE, TENN	Y-PB-
UNION CARBIDE NUCLEAR CO. PADUCAH PLANT, KY.	KY-
UNION CARBIDE NUCLEAR CO. PADUCAH PLANT, KY.	KYB-
UNION CARBIDE NUCLEAR CO. PADUCAH PLANT, KY.	KYM-
UNION CARBIDE NUCLEAR CO. Y-12 PLANT, OAK RIDGE, TENN	C+C-Y-
UNION CARBIDE NUCLEAR CO. Y-12 PLANT, OAK RIDGE, TENN	UCNC-
UNION CARBIDE NUCLEAR CO. Y-12 PLANT, OAK RIDGE, TENN	UCNC-Y-
UNION CARBIDE NUCLEAR CO. Y-12 PLANT, OAK RIDGE, TENN	Y-(LETTER/NUMBER)-
UNION CARBIDE NUCLEAR CO. Y-12 PLANT, OAK RIDGE, TENN	YA-
UNION CARBIDE NUCLEAR CO. Y-12 PLANT, OAK RIDGE, TENN	Y-DE-
UNION CARBIDE NUCLEAR CO. Y-12 PLANT, OAK RIDGE, TENN	Y-DP-
UNION CARBIDE NUCLEAR CO. Y-12 PLANT, OAK RIDGE, TENN	Y-EB-
UNION CARBIDE NUCLEAR CO. Y-12 PLANT, OAK RIDGE, TENN	Y-KE-
UNION CARBIDE NUCLEAR CO. Y-12 PLANT, OAK RIDGE, TENN	Y-SB-
(SEE ALSO LATER NAME: UNION CARBIDE CORP. Y-12 PLANT)	
UNION CARBIDE RESEARCH INST., TARRYTOWN, N.Y.	C-
UNION CARBIDE RESEARCH INST., TARRYTOWN, N.Y. (INST. RESEARCH MEMORANDUM)	IRM-NO.-
UNION CARBIDE RESEARCH INST., TARRYTOWN, N.Y. (INST. RESEARCH SUMMARY)	IRS-NO.-
UNION CARBIDE RESEARCH INST., TARRYTOWN, N.Y.	TRC-
UNION CARBIDE RESEARCH INST., TARRYTOWN, N.Y.	UCRI-
UNION CARBIDE RESEARCH INST., TARRYTOWN, N.Y.	UCRI-TR-C-
UNION CARBIDE RESEARCH INST. SPACE SCIENCE AND ENGINEERING GROUP, TARRYTOWN, N.Y.	UCC/DSSD-
UNION OF SOUTH AFRICA. ATOMIC ENERGY BOARD, PRETORIA	G-
UNION OF SOUTH AFRICA. ATOMIC ENERGY BOARD, PRETORIA	PEL-
UNION OF SOUTH AFRICA. COUNCIL FOR SCIENTIFIC AND INDUSTRIAL RESEARCH, PRETORIA	AR-
UNION OF SOUTH AFRICA. COUNCIL FOR SCIENTIFIC AND INDUSTRIAL RESEARCH, PRETORIA	CHEM-(YEAR)
UNION OF SOUTH AFRICA. COUNCIL FOR SCIENTIFIC AND INDUSTRIAL RESEARCH, PRETORIA	CSIR-
UNION OF SOUTH AFRICA. COUNCIL FOR SCIENTIFIC AND INDUSTRIAL RESEARCH, PRETORIA	CSIR-AR-
UNION OF SOUTH AFRICA. COUNCIL FOR SCIENTIFIC AND INDUSTRIAL RESEARCH, PRETORIA	CSIR-FIS-
UNION OF SOUTH AFRICA. COUNCIL FOR SCIENTIFIC AND INDUSTRIAL RESEARCH, PRETORIA	CSIR-TEL-
UNION OF SOUTH AFRICA. COUNCIL FOR SCIENTIFIC AND INDUSTRIAL RESEARCH, PRETORIA	FIS-
UNION OF SOUTH AFRICA. COUNCIL FOR SCIENTIFIC AND INDUSTRIAL RESEARCH, PRETORIA	MEG-
UNION OF SOUTH AFRICA. COUNCIL FOR SCIENTIFIC AND INDUSTRIAL RESEARCH, PRETORIA	WISK-
UNION OF SOUTH AFRICA. GOVERNMENT METALLURGICAL LAB., JOHANNESBURG	GML-
UNION OF SOUTH AFRICA. MAGNETIC OBS., HERMANUS	A.-
UNION OF SOUTH AFRICA. MAGNETIC OBS., HERMANUS	B-
UNION OF SOUTH AFRICA. NATIONAL INST. FOR METALLURGY, JOHANNESBURG	NIM-
UNION OF SOUTH AFRICA. NATIONAL INST. FOR TELECOMMUNICATIONS RES. DEEP SPACE INSTRUMENTATION LAB., HARTEBEESTHOEK	DSIF-(NO.)/OPS/(NO.)/
UNITED AIRCRAFT CORPORATE SYSTEMS CENTER, FARMINGTON, CONN.	SCR-(NUMBER)-
UNITED AIRCRAFT CORPORATE SYSTEMS CENTER, FARMINGTON, CONN.	UAC/SCR-
UNITED AIRCRAFT CORPORATE SYSTEMS CENTER. SUPPORT SYSTEMS GROUP, EL SEGUNDO, CALIF.	SCR-
UNITED AIRCRAFT CORPORATE SYSTEMS CENTER. SUPPORT SYSTEMS GROUP, EL SEGUNDO, CALIF.	UAC/SCR-
UNITED AIRCRAFT CORPORATE SYSTEMS CENTER. WEAPONS EFFECTS + SUPPORT SYSTEMS DEPT., EL SEGUNDO, CAL.	SCR-
UNITED AIRCRAFT CORPORATE SYSTEMS CENTER. WEATHER SYSTEM CENTER, FARMINGTON, CONN.	WSC-E-
UNITED AIRCRAFT CORP., EAST HARTFORD, CONN.	A-
UNITED AIRCRAFT CORP., EAST HARTFORD, CONN.	C-
UNITED AIRCRAFT CORP., EAST HARTFORD, CONN.	C(NO.)-
UNITED AIRCRAFT CORP., EAST HARTFORD, CONN.	D(NO.)-

UNITED AIRCRAFT CORP.

UNITED AIRCRAFT CORP., EAST HARTFORD, CONN.	G-
UNITED AIRCRAFT CORP., EAST HARTFORD, CONN.	J-
UNITED AIRCRAFT CORP., EAST HARTFORD, CONN.	M-(NO.)-
UNITED AIRCRAFT CORP., EAST HARTFORD, CONN.	R-(NO.)-
UNITED AIRCRAFT CORP., EAST HARTFORD, CONN.	R-73003-
UNITED AIRCRAFT CORP., EAST HARTFORD, CONN.	SR-
UNITED AIRCRAFT CORP., EAST HARTFORD, CONN.	TP-(NO.)-(NO.)
UNITED AIRCRAFT CORP., EAST HARTFORD, CONN.	UAC-
UNITED AIRCRAFT CORP., EAST HARTFORD, CONN.	UAC-M-
UNITED AIRCRAFT CORP., EAST HARTFORD, CONN.	UAC-METEOR-
UNITED AIRCRAFT CORP., EAST HARTFORD, CONN.	UAC-R-
UNITED AIRCRAFT CORP., EAST HARTFORD, CONN.	UACRL-F(NUMBER)-
UNITED AIRCRAFT CORP., EAST HARTFORD, CONN.	VARL-G-
UNITED AIRCRAFT CORP., STRATFORD, CONN.	SER-
UNITED AIRCRAFT CORP., SUNNYVALE, CALIF.	UTC-
UNITED AIRCRAFT CORP., WINDSOR LOCKS, CONN.	EBR-
UNITED AIRCRAFT CORP., WINDSOR LOCKS, CONN.	SLS-
(SEE ALSO PRATT AND WHITNEY AIRCRAFT DIV... AND SIKORSKY AIRCRAFT DIV...)	
UNITED AIRCRAFT CORP. HAMILTON STANDARD DIV., WINDSOR LOCKS, CONN.	EP-
UNITED AIRCRAFT CORP. HAMILTON STANDARD DIV., WINDSOR LOCKS, CONN.	HS-
UNITED AIRCRAFT CORP. HAMILTON STANDARD DIV., WINDSOR LOCKS, CONN.	HSP-
UNITED AIRCRAFT CORP. HAMILTON STANDARD DIV., WINDSOR LOCKS, CONN.	SMP-
UNITED AIRCRAFT CORP. HAMILTON STANDARD DIV., WINDSOR LOCKS, CONN.	UAC-HS-
(SEE ALSO LATER NAMED HAMILTON STANDARD...)	
UNITED AIRCRAFT CORP. HAMILTON STANDARD PROPELLERS DIV., EAST HARTFORD, CONN. (REF. 1)	(NUMBER)
UNITED AIRCRAFT CORP. MISSILES AND SPACE SYSTEMS DIV., EAST HARTFORD, CONN.	UAC M+SS R-
UNITED AIRCRAFT CORP. RES. LABS., EAST HARTFORD, CONN.	B-
UNITED AIRCRAFT CORP. RES. LABS., EAST HARTFORD, CONN.	C-
UNITED AIRCRAFT CORP. RES. LABS., EAST HARTFORD, CONN.	D(NUMBERS)-
UNITED AIRCRAFT CORP. RES. LABS., EAST HARTFORD, CONN.	E(NUMBERS)-
UNITED AIRCRAFT CORP. RES. LABS., EAST HARTFORD, CONN.	F-(NUMBER)-
UNITED AIRCRAFT CORP. RES. LABS., EAST HARTFORD, CONN.	G-91(4 DIGITS)-
UNITED AIRCRAFT CORP. RES. LABS., EAST HARTFORD, CONN.	H-(NUMBER)-
UNITED AIRCRAFT CORP. RES. LABS., EAST HARTFORD, CONN.	K-(NUMBER)-
UNITED AIRCRAFT CORP. RES. LABS., EAST HARTFORD, CONN.	P-B(NUMBER)
UNITED AIRCRAFT CORP. RES. LABS., EAST HARTFORD, CONN.	UAC-A-91(4 DIGITS)-
UNITED AIRCRAFT CORP. RES. LABS., EAST HARTFORD, CONN.	UAC-B-91(4 DIGITS)-
UNITED AIRCRAFT CORP. RES. LABS., EAST HARTFORD, CONN.	UAC-C-91(4 DIGITS)-
UNITED AIRCRAFT CORP. RES. LABS., EAST HARTFORD, CONN.	UAC-D-91(4 DIGITS)-
UNITED AIRCRAFT CORP. RES. LABS., EAST HARTFORD, CONN.	UAC-E-91(4 DIGITS)-
UNITED AIRCRAFT CORP. RES. LABS., EAST HARTFORD, CONN.	UAC-F-91(4 DIGITS)-
UNITED AIRCRAFT CORP. RES. LABS., EAST HARTFORD, CONN.	UAC-G-91(4 DIGITS)-
UNITED AIRCRAFT CORP. RES. LABS., EAST HARTFORD, CONN.	UAC-H-91(4 DIGITS)-
UNITED AIRCRAFT CORP. RES. LABS., EAST HARTFORD, CONN.	UAC-J-91(4 DIGITS)-
UNITED AIRCRAFT CORP. RES. LABS., EAST HARTFORD, CONN.	UAC-K-91(4 DIGITS)-
UNITED AIRCRAFT CORP. RES. LABS., EAST HARTFORD, CONN.	UAC-R-
UNITED AIRCRAFT CORP. RES. LABS., EAST HARTFORD, CONN.	UACRL-
UNITED AIRCRAFT CORP. RES. LABS., EAST HARTFORD, CONN.	UAC-RL-B-91(4 DIGITS)-
UNITED AIRCRAFT CORP. RES. LABS., EAST HARTFORD, CONN.	UACRL-C-
UNITED AIRCRAFT CORP. RES. LABS., EAST HARTFORD, CONN.	UACRL-D-
UNITED AIRCRAFT CORP. RES. LABS., EAST HARTFORD, CONN.	UACRL-E-
UNITED AIRCRAFT CORP. RES. LABS., EAST HARTFORD, CONN.	UACRL-G-91(4 DIGITS)-
UNITED AIRCRAFT CORP. RES. LABS., EAST HARTFORD, CONN.	UACRL-J-91(4 DIGITS)-
UNITED AIRCRAFT CORP. RES. LABS., EAST HARTFORD, CONN.	UAR-
UNITED AIRCRAFT CORP. RES. LABS., EAST HARTFORD, CONN.	UAR-H(NUMBER)
UNITED AIRCRAFT CORP. RES. LABS., EAST HARTFORD, CONN.	UAR-J-
UNITED AIRCRAFT CORP. RES. LABS., EAST HARTFORD, CONN.	UARL-
UNITED AIRCRAFT CORP. RES. LABS., EAST HARTFORD, CONN.	UARL-D-
UNITED AIRCRAFT CORP. RES. LABS., EAST HARTFORD, CONN.	UARL-F(6 DIGITS)-
UNITED AIRCRAFT CORP. RES. LABS., EAST HARTFORD, CONN.	UARL-G-
UNITED AIRCRAFT CORP. RES. LABS., EAST HARTFORD, CONN.	UARL-H-
UNITED AIRCRAFT CORP. RES. LABS., EAST HARTFORD, CONN.	UARL-J-
UNITED AIRCRAFT CORP. UNITED AIRCRAFT RES. LABS., EAST HARTFORD, CONN.	UAC-UAR-(LETTER-NUMBER)
UNITED AIRCRAFT OF CANADA LTD., LONGUEUIL, QUEBEC	UACL ER-
UNITED AIRCRAFT OF CANADA LTD., LONGUEUIL, QUEBEC	UACL TN-
UNITED AIRLINES INC., CHICAGO	UAL METEOROLOGY CIRC-
UNITED ARAB REPUBLIC. ATOMIC ENERGY ESTAB., CAIRO	UARAEE-
UNITED CARBON PRODUCTS CO., INC., BAY CITY, MICH.	UCP-
UNITED CHROMIUM, INC., DETROIT	UNICHROME-
UNITED CONTROL CORP., SOUTH EL MONTE, CALIF.	DS-(YEAR)-
UNITED ELECTRODYNAMICS, INC., PASADENA, CALIF.	UED-
UNITED ELECTRODYNAMICS, INC. DATA ANALYSIS AND TECHNIQUE DEVELOPMENT CENTER, ALEXANDRIA, VA.	DATDC-
UNITED ELECTRODYNAMICS, INC. EARTH SCIENCES DIV., ALEXANDRIA, VA.	UED-ESD-
UNITED ELECTRODYNAMICS, INC. EARTH SCIENCES DIV., PASADENA, CALIF.	GBR-
UNITED ELECTRODYNAMICS, INC. EARTH SCIENCES DIV., PASADENA, CALIF.	UED-GBR-
UNITED GEOPHYSICAL CO., INC., LOS ANGELES	UGC-
UNITED GEOPHYSICAL CORP., PASADENA, CALIF.	UGC-
UNITED GLASS BOTTLE MANUFACTURERS, LTD., LONDON	TP-
UNITED KINGDOM. ADMIRALITY CORROSION COMMISSION, LONDON	ACC-
UNITED KINGDOM. ADMIRALTY CORROSION COMMISSION, LONDON	ACC/H-
UNITED KINGDOM. ADMIRALTY EXPERIMENTAL DIVING UNIT, PORTSMOUTH, HANTS, ENGLAND	AEDU-
UNITED KINGDOM. ADMIRALTY MARINE ENGINEERING ESTABLISHMENT, GOSPORT, HANTS, ENGLAND	AMEE-
UNITED KINGDOM. ADMIRALTY MARINE ENGINEERING ESTABLISHMENT, GOSPORT, HANTS, ENGLAND	AMEE-TM-(NUMBER/YEAR)
UNITED KINGDOM. ADMIRALTY SURFACE WEAPONS ESTABLISHMENT, PORTSMOUTH, HANTS, ENGLAND	ASWE-
UNITED KINGDOM. ADMIRALTY SURFACE WEAPONS ESTABLISHMENT, PORTSMOUTH, HANTS, ENGLAND	ASWE LAB NOTE-XP-(YR.)-

UNITED KINGDOM. ADMIRALTY UNDERWATER WEAPONS ESTABLISHMENT, PORTLAND, DORSET, ENGLAND	AUWE-
UNITED KINGDOM. ADMIRALTY UNDERWATER WEAPONS ESTABLISHMENT, PORTLAND, DORSET, ENGLAND	AUWE-TN-
UNITED KINGDOM. AERONAUTICAL INSPECTION DIRECTORATE, HAREFIELD, MIDDX., ENGLAND	AID/D.(NUMBER/YEAR)
UNITED KINGDOM. AERONAUTICAL INSPECTION DIRECTORATE, HAREFIELD, MIDDX., ENGLAND	AID/DEV/(NUMBER/YEAR)
UNITED KINGDOM. AERONAUTICAL QUALITY ASSURANCE DIRECTORATE, HAREFIELD, MIDDX., ENGLAND	A.Q.D./D(NUMBER/YEAR)
UNITED KINGDOM. AERONAUTICAL QUALITY ASSURANCE DIRECTORATE, UXBRIDGE, MIDDX., ENGLAND	AQD/Y/
UNITED KINGDOM. AIR MINISTRY. METEOROLOGICAL OFFICE, LONDON	AM MO-
UNITED KINGDOM. AIR MINISTRY. METEOROLOGICAL OFFICE, LONDON	JMRP-
UNITED KINGDOM. AIR MINISTRY. METEOROLOGICAL OFFICE, LONDON	MET.(NO.)(LETTER)
UNITED KINGDOM. AIR MINISTRY. METEOROLOGICAL OFFICE, LONDON	MET-O-
UNITED KINGDOM. AIR MINISTRY. METEOROLOGICAL OFFICE, LONDON	MO-
UNITED KINGDOM. AIR MINISTRY. METEOROLOGICAL OFFICE, LONDON	M.O.-
UNITED KINGDOM. AIR MINISTRY. METEOROLOGICAL OFFICE, LONDON	MOM-
UNITED KINGDOM. AIR REGISTRATION BOARD, REIGATE, SURREY, ENGLAND	ARB-TN-
UNITED KINGDOM. AIR TRAFFIC CONTROL EVALUATION UNIT, BOURNEMOUTH, HANTS, ENGLAND	ACTEU-
UNITED KINGDOM. AIR TRAFFIC CONTROL EVALUATION UNIT, BOURNEMOUTH, HANTS, ENGLAND	ATCEU-
UNITED KINGDOM. ARMY AIR TRANSPORT TRAINING AND DEVELOPMENT CENTER, OLD SARUM, WILTS, ENGLAND	AATDC-
UNITED KINGDOM. ARMY OPERATIONAL RESEARCH ESTABLISHMENT, BYFLEET, SURREY, ENGLAND	AORE-
UNITED KINGDOM. ARMY PERSONNEL RESEARCH COMM., LONDON	APRC-
UNITED KINGDOM. ARMY PERSONNEL RESEARCH ESTABLISHMENT, BYFLEET, SURREY, ENGLAND	APRE-
UNITED KINGDOM. ARMY PERSONNEL RESEARCH ESTABLISHMENT, BYFLEET, SURREY, ENGLAND	APRE-RM-Q/
UNITED KINGDOM. ARMY PERSONNEL RESEARCH ESTABLISHMENT, BYFLEET, SURREY, ENGLAND	APRE-TRIAL-MEMO-
UNITED KINGDOM. ARMY PERSONNEL RESEARCH ESTABLISHMENT, BYFLEET, SURREY, ENGLAND	MIN-DEF-ATT-
UNITED KINGDOM. ARMY PERSONNEL RESEARCH ESTABLISHMENT, BYFLEET, SURREY, ENGLAND	UK-
UNITED KINGDOM. ARMY WORK STUDY GROUP, GUILDFORD, SURREY, ENGLAND	AWSG-(NUMBER)/(YEAR)
UNITED KINGDOM. CENTRAL DOCKYARD LAB., PORTSMOUTH, HANTS, ENGLAND	C.D.L.ACC/N-
UNITED KINGDOM. BOARD OF TRADE, LONDON	CAP-
UNITED KINGDOM. COMMONWEALTH AND EMPIRE RADIO FOR CIVIL AVIATION, LONDON	CERCA-
UNITED KINGDOM. DARESBURY NUCLEAR PHYSICS LAB., CHESHIRE, ENGLAND	DNPL-
UNITED KINGDOM. DARESBURY NUCLEAR PHYSICS LAB., CHESHIRE, ENGLAND	DNPL-AR-(YEAR)
UNITED KINGDOM. DARESBURY NUCLEAR PHYSICS LAB., CHESHIRE, ENGLAND	DNPL/P-
UNITED KINGDOM. DARESBURY NUCLEAR PHYSICS LAB., CHESHIRE, ENGLAND	DNPL/R-
UNITED KINGDOM. DIRECTORATE OF GUIDED WEAPONS RESEARCH AND DEVELOPMENT, LONDON	DGGW-
UNITED KINGDOM. DIRECTORATE OF GUIDED WEAPONS RESEARCH AND DEVELOPMENT, LONDON	DGWP-
UNITED KINGDOM. DIRECTORATE OF GUIDED WEAPONS RESEARCH AND DEVELOPMENT, LONDON	DGWRD-
UNITED KINGDOM. DIRECTORATE OF GUIDED WEAPONS RESEARCH AND DEVELOPMENT, LONDON	PDGW-
UNITED KINGDOM. DIRECTORATE OF MATERIALS RESEARCH AND DEVELOPMENT, LONDON	CD-MAT-
UNITED KINGDOM. DIRECTORATE OF MATERIALS RESEARCH AND DEVELOPMENT, LONDON	D-MAT-
UNITED KINGDOM. DIRECTORATE OF MATERIALS RESEARCH AND DEVELOPMENT, LONDON	D-MAT-AVIATION-
UNITED KINGDOM. ELECTRICITY COUNCIL RESEARCH CENTRE, CAPENHURST, BERKS, ENGLAND	ECRC/
UNITED KINGDOM. ELECTRICITY COUNCIL RESEARCH CENTRE, CAPENHURST, BERKS, ENGLAND	ECRC/M-
UNITED KINGDOM. ELECTRICITY COUNCIL RESEARCH CENTRE, CAPENHURST, BERKS, ENGLAND	ECRC/N-
UNITED KINGDOM. ELECTRICITY COUNCIL RESEARCH CENTRE, CAPENHURST, BERKS, ENGLAND	ECRC-R-
UNITED KINGDOM. FIGHTING VEHICLES RESEARCH AND DEVELOPMENT ESTAB., CHERTSEY, SURREY, ENGLAND	AFV-
UNITED KINGDOM. FIGHTING VEHICLES RESEARCH AND DEVELOPMENT ESTAB., CHERTSEY, SURREY, ENGLAND	FT-
UNITED KINGDOM. FIGHTING VEHICLES RESEARCH AND DEVELOPMENT ESTAB., CHERTSEY, SURREY, ENGLAND	FVDD-
UNITED KINGDOM. FIGHTING VEHICLES RESEARCH AND DEVELOPMENT ESTAB., CHERTSEY, SURREY, ENGLAND	FVPE-
UNITED KINGDOM. FIGHTING VEHICLES RESEARCH AND DEVELOPMENT ESTAB., CHERTSEY, SURREY, ENGLAND	FVRDE-
UNITED KINGDOM. FIGHTING VEHICLES RESEARCH AND DEVELOPMENT ESTAB., CHERTSEY, SURREY, ENGLAND	FVRDE-(LETTERS)-
UNITED KINGDOM. FIGHTING VEHICLES RESEARCH AND DEVELOPMENT ESTAB., CHERTSEY, SURREY, ENGLAND	TAR-
UNITED KINGDOM. FLYING PERSONNEL RES. COMM., LONDON	FPRC/(NUMBERS)/A
UNITED KINGDOM. FLYING PERSONNEL RES. COMM., LONDON	FPRC/
UNITED KINGDOM. FLYING PERSONNEL RES. COMM., LONDON	FPRC/(NUMBERS)/B
UNITED KINGDOM. FLYING PERSONNEL RES. COMM., LONDON	FPRC/AM-
UNITED KINGDOM. FLYING PERSONNEL RES. COMM., LONDON	FPRC/MEMO-
UNITED KINGDOM. MICROBIOLOGICAL RESEARCH ESTAB., SALISBURY, WILTS, ENGLAND	MRE-
UNITED KINGDOM. MINISTRY OF AGRICULTURE, FISHERIES, AND FOOD. FISHERIES RADIOBIOLOGICAL LAB., LOWESTOFT E. SUFFOLK, ENGLAND	FRL-
UNITED KINGDOM. MINISTRY OF AVIATION, LONDON	MOA-
UNITED KINGDOM. MINISTRY OF AVIATION. TECHNICAL INFORMATION AND LIBRARY SERVICES, LONDON	TIL-
UNITED KINGDOM. MINISTRY OF TECHNOLOGY, LONDON	MINTECH-T-
UNITED KINGDOM. MINISTRY OF TECHNOLOGY, LONDON	S/T-MEMO-(NUMBER/YEAR)
UNITED KINGDOM. MINISTRY OF TECHNOLOGY, LONDON	TRC-BR-
UNITED KINGDOM. MINISTRY OF TECHNOLOGY. HYDRAULICS RESEARCH STATION, WALLINGFORD, BERKS, ENGLAND	INT-
UNITED KINGDOM. MINISTRY OF TECHNOLOGY. TECHNICAL INFORMATION AND LIBRARY SERVICES, LONDON	TIL-BR-
UNITED KINGDOM. MINISTRY OF TECHNOLOGY. TECHNICAL INFORMATION AND LIBRARY SERVICES, LONDON	TIL-LIST-
UNITED KINGDOM. NATIONAL ENGINEERING LAB., GLASGOW (JOINT VENTURE)	BHRA-NEL-JOINT-REPT-
UNITED KINGDOM. NATIONAL ENGINEERING LAB., GLASGOW	E(NUMBER)
UNITED KINGDOM. NATIONAL ENGINEERING LAB., GLASGOW	NEL-
UNITED KINGDOM. NATIONAL INST. OF OCEANOGRAPHY, GODALMING, SURREY, ENGLAND	NIO-A.(NUMBER)
UNITED KINGDOM. NATIONAL INST. OF OCEANOGRAPHY, GODALMING, SURREY, ENGLAND	NIO-CR/
UNITED KINGDOM. NATIONAL INST. OF OCEANOGRAPHY, GODALMING, SURREY, ENGLAND	NIO-D.(NUMBER)
UNITED KINGDOM. NATIONAL INST. OF OCEANOGRAPHY, WORMLEY, HERTS, ENGLAND	NIOT-
UNITED KINGDOM. NATIONAL LENDING LIBRARY FOR SCIENCE AND TECHNOLOGY, BOSTON SPA, YORKS, ENGLAND	NLL-AERE-TRANS-
UNITED KINGDOM. NATIONAL LENDING LIBRARY FOR SCIENCE AND TECHNOLOGY, BOSTON SPA, YORKS, ENGLAND	NLL-BRSLC-
UNITED KINGDOM. NATIONAL LENDING LIBRARY FOR SCIENCE AND TECHNOLOGY, BOSTON SPA, YORKS, ENGLAND	NLL-CA-
UNITED KINGDOM. NATIONAL LENDING LIBRARY FOR SCIENCE AND TECHNOLOGY, BOSTON SPA, YORKS, ENGLAND	NLL-CE-TRANS-
UNITED KINGDOM. NATIONAL LENDING LIBRARY FOR SCIENCE AND TECHNOLOGY, BOSTON SPA, YORKS, ENGLAND	NLL-CTO-
UNITED KINGDOM. NATIONAL LENDING LIBRARY FOR SCIENCE AND TECHNOLOGY, BOSTON SPA, YORKS, ENGLAND	NLL-EE-TRANS-

UNITED KINGDOM. NATIONAL LENDING LIBRARY FOR SCIENCE AND TECHNOLOGY

UNITED KINGDOM. NATIONAL LENDING LIBRARY FOR SCIENCE AND TECHNOLOGY, BOSTON SPA, YORKS, ENGLAND	NLL-INC-
UNITED KINGDOM. NATIONAL LENDING LIBRARY FOR SCIENCE AND TECHNOLOGY, BOSTON SPA, YORKS, ENGLAND	NLL-KIRKBY-TRANS-
UNITED KINGDOM. NATIONAL LENDING LIBRARY FOR SCIENCE AND TECHNOLOGY, BOSTON SPA, YORKS, ENGLAND	NLL-LC-
UNITED KINGDOM. NATIONAL LENDING LIBRARY FOR SCIENCE AND TECHNOLOGY, BOSTON SPA, YORKS, ENGLAND	NLL-LIB-COMM-
UNITED KINGDOM. NATIONAL LENDING LIBRARY FOR SCIENCE AND TECHNOLOGY, BOSTON SPA, YORKS, ENGLAND	NLL-LTI-
UNITED KINGDOM. NATIONAL LENDING LIBRARY FOR SCIENCE AND TECHNOLOGY, BOSTON SPA, YORKS, ENGLAND	NLL-M-
UNITED KINGDOM. NATIONAL LENDING LIBRARY FOR SCIENCE AND TECHNOLOGY, BOSTON SPA, YORKS, ENGLAND	NLL-MET-OFF-M-
UNITED KINGDOM. NATIONAL LENDING LIBRARY FOR SCIENCE AND TECHNOLOGY, BOSTON SPA, YORKS, ENGLAND	NLL-MIN-TECH-T-
UNITED KINGDOM. NATIONAL LENDING LIBRARY FOR SCIENCE AND TECHNOLOGY, BOSTON SPA, YORKS, ENGLAND	NLL-NIOT-
UNITED KINGDOM. NATIONAL LENDING LIBRARY FOR SCIENCE AND TECHNOLOGY, BOSTON SPA, YORKS, ENGLAND	NLL-NSTIC-
UNITED KINGDOM. NATIONAL LENDING LIBRARY FOR SCIENCE AND TECHNOLOGY, BOSTON SPA, YORKS, ENGLAND	NLL-OA-
UNITED KINGDOM. NATIONAL LENDING LIBRARY FOR SCIENCE AND TECHNOLOGY, BOSTON SPA, YORKS, ENGLAND	NLL-PORS-TRANS-
UNITED KINGDOM. NATIONAL LENDING LIBRARY FOR SCIENCE AND TECHNOLOGY, BOSTON SPA, YORKS, ENGLAND	NLL-RAE-TRANS-
UNITED KINGDOM. NATIONAL LENDING LIBRARY FOR SCIENCE AND TECHNOLOGY, BOSTON SPA, YORKS, ENGLAND	NLL-RISLEY-TRANS-
UNITED KINGDOM. NATIONAL LENDING LIBRARY FOR SCIENCE AND TECHNOLOGY, BOSTON SPA, YORKS, ENGLAND	NLL-RRE-TRANS-
UNITED KINGDOM. NATIONAL LENDING LIBRARY FOR SCIENCE AND TECHNOLOGY, BOSTON SPA, YORKS, ENGLAND	NLL-RTS-
UNITED KINGDOM. NATIONAL LENDING LIBRARY FOR SCIENCE AND TECHNOLOGY, BOSTON SPA, YORKS, ENGLAND	NLL-SMRE-TRANS-
UNITED KINGDOM. NATIONAL LENDING LIBRARY FOR SCIENCE AND TECHNOLOGY, BOSTON SPA, YORKS, ENGLAND	NLL-T-
UNITED KINGDOM. NATIONAL LENDING LIBRARY FOR SCIENCE AND TECHNOLOGY, BOSTON SPA, YORKS, ENGLAND	NLL-TRANS-
UNITED KINGDOM. NATIONAL LENDING LIBRARY FOR SCIENCE AND TECHNOLOGY, BOSTON SPA, YORKS, ENGLAND	NLL-WINDSCALE-
UNITED KINGDOM. NATIONAL LENDING LIBRARY FOR SCIENCE AND TECHNOLOGY, BOSTON SPA, YORKS, ENGLAND	RTS-
UNITED KINGDOM. NATIONAL PHYSICAL LAB., TEDDINGTON, MIDDX., ENGLAND	NPL-AERO-AC-
UNITED KINGDOM. NATIONAL PHYSICAL LAB., TEDDINGTON, MIDDX., ENGLAND	NPL-AERO-SPECIAL-
UNITED KINGDOM. NATIONAL PHYSICAL LAB., TEDDINGTON, MIDDX., ENGLAND	NPL-MAT-APP-
UNITED KINGDOM. NATIONAL PHYSICAL LAB., TEDDINGTON, MIDDX., ENGLAND	NPL-OP-MET-
UNITED KINGDOM. NATIONAL PHYSICAL LAB., TEDDINGTON, MIDDX., ENGLAND	NPL-SHIP-
UNITED KINGDOM. NATIONAL PHYSICAL LAB. DIV. OF COMPUTER SCIENCE, TEDDINGTON, MIDDX., ENGLAND	COM-SCI-
UNITED KINGDOM. NATIONAL PHYSICAL LAB. DIV. OF INORG. AND METALLIC STRUCTURE, TEDDINGTON, MIDDX., ENG.	IMS-
UNITED KINGDOM. NATIONAL PHYSICAL LAB. DIV. OF INORG. AND METALLIC STRUCTURE, TEDDINGTON, MIDDX., ENG.	NPL-IMS-
UNITED KINGDOM. NATIONAL PHYSICAL LAB. DIV. OF NUMERICAL + APPLIED MATH., TEDDINGTON, MIDDX., ENG.	NPL-DNAM-
UNITED KINGDOM. NATIONAL PHYSICAL LAB. OPTICAL METROLOGY DIV., TEDDINGTON, MIDDX., ENGLAND	NPL-OP-MET-
UNITED KINGDOM. NATIONAL PHYSICAL LAB. QUANTUM METROLOGY DIV., TEDDINGTON, MIDDX., ENGLAND	NPL-QU-
UNITED KINGDOM. NATIONAL PHYSICAL LAB. SHIP DIV., TEDDINGTON, MIDDX., ENGLAND	SHIP-
UNITED KINGDOM. NATIONAL RADIOLOGICAL PROTECTION BOARD, HARWELL, BERKS, ENGLAND	NRPB-
UNITED KINGDOM. NAVAL ORDNANCE INSPECTION LAB., CAERWENT, WALES	NOIL-
UNITED KINGDOM. NOISE REDUCTION PANEL, LONDON (WORKING PARTY ON ACCEPTABLE NOISE LEVELS)	NRP/ANL-
UNITED KINGDOM. OFFICE FOR SCIENTIFIC + TECH. INFO.	OSTI-
UNITED KINGDOM. ROAD RES. LAB., CROWTHORNE, BERKS, ENG	ROAD-NOTE-
UNITED KINGDOM. ROYAL AIRCRAFT ESTAB., FARNBOROUGH, HANTS, ENGLAND	RAE-TM-AERO-
UNITED KINGDOM. ROYAL AIRCRAFT ESTAB., FARNBOROUGH, HANTS, ENGLAND	RAE-TM-SPACE-
UNITED KINGDOM. ROYAL AIRCRAFT ESTAB., FARNBOROUGH, HANTS, ENGLAND	RAE-TM-WE-
UNITED KINGDOM. ROYAL AIRCRAFT ESTAB., FARNBOROUGH, HANTS, ENGLAND	RAE-TR-
UNITED KINGDOM. ROYAL GREENWICH OBSERVATORY. NAUTICAL ALMANAC OFFICE, HERSTMONCEUX, SUSSEX	NAO-TN-
UNITED KINGDOM. ROYAL MILITARY COLL. OF SCIENCE, SHRIVENHAM, WILTS, ENGLAND	C-
UNITED KINGDOM. ROYAL MILITARY COLL. OF SCIENCE, SHRIVENHAM, WILTS, ENGLAND	RMCS-
UNITED KINGDOM. ROYAL NAVAL PERSONNEL RESEARCH COMM., LONDON	S.M.S.-
UNITED KINGDOM. ROYAL NAVAL PERSONNEL RESEARCH COMM., LONDON	UPS-
UNITED KINGDOM. RUTHERFORD HIGH ENERGY LAB., CHILTON, DIDCOT, BERKS, ENGLAND	HEP/MISC/
UNITED KINGDOM. RUTHERFORD HIGH ENERGY LAB., CHILTON, DIDCOT, BERKS, ENGLAND	RHEL/M-
UNITED KINGDOM. RUTHERFORD HIGH ENERGY LAB., CHILTON, DIDCOT, BERKS, ENGLAND	RHEL/R-
UNITED KINGDOM. RUTHERFORD HIGH ENERGY LAB., CHILTON, DIDCOT, BERKS, ENGLAND	RHEL/S-
UNITED KINGDOM. RUTHERFORD HIGH ENERGY LAB., CHILTON, DIDCOT, BERKS, ENGLAND	RPP/(LETTER)/(NUMBER)
UNITED KINGDOM. RUTHERFORD HIGH ENERGY LAB., CHILTON, DIDCOT, BERKS, ENGLAND	RPP/H-
UNITED KINGDOM. SCIENCE RESEARCH COUNCIL. ASTROPHYSICS RES. UNIT, ABINGDON, BERKS, ENGLAND	ARU-R(NUMBER)
UNITED KINGDOM ATOMIC ENERGY AUTHORITY	AERE-CMND-
UNITED KINGDOM ATOMIC ENERGY AUTHORITY (HOMOGENEOUS AQUEOUS REACTOR DESIGN)	AERE-HARD(C)P-
UNITED KINGDOM ATOMIC ENERGY AUTHORITY	TTPC-
UNITED KINGDOM ATOMIC ENERGY AUTHORITY	U.K.-
UNITED KINGDOM ATOMIC ENERGY AUTHORITY	UKAEA-
UNITED KINGDOM ATOMIC ENERGY AUTHORITY, LONDON	DPR/INF/
UNITED KINGDOM ATOMIC ENERGY AUTHORITY, LONDON	HL(YEAR/NUMBER)
UNITED KINGDOM ATOMIC ENERGY AUTHORITY. ATOMIC ENERGY RESEARCH ESTABLISHMENT, HARWELL, BERKS, ENGLAND	DPC.IS/
UNITED KINGDOM ATOMIC ENERGY AUTHORITY. AUTHORITY HEALTH AND SAFETY BRANCH, HARWELL, BERKS, ENGLAND	AHSB(A)-R-
UNITED KINGDOM ATOMIC ENERGY AUTHORITY. AUTHORITY HEALTH AND SAFETY BRANCH, LONDON	AHSB(A)-R-
UNITED KINGDOM ATOMIC ENERGY AUTHORITY. AUTHORITY HEALTH AND SAFETY BRANCH, LONDON	AHSB/RP/-R-
UNITED KINGDOM ATOMIC ENERGY AUTHORITY. AUTHORITY HEALTH AND SAFETY BRANCH, RISLEY, LANCS, ENGLAND	AHSB-
UNITED KINGDOM ATOMIC ENERGY AUTHORITY. AUTHORITY HEALTH AND SAFETY BRANCH, RISLEY, LANCS, ENGLAND	AHSB-MEMO-
UNITED KINGDOM ATOMIC ENERGY AUTHORITY. AUTHORITY HEALTH AND SAFETY BRANCH, RISLEY, LANCS, ENGLAND	AHSB-REPORT-
UNITED KINGDOM ATOMIC ENERGY AUTHORITY. AUTHORITY HEALTH AND SAFETY BRANCH, RISLEY, LANCS, ENGLAND	AHSB (S) (LETTER)-
UNITED KINGDOM ATOMIC ENERGY AUTHORITY. AUTHORITY HEALTH AND SAFETY BRANCH, RISLEY, LANCS, ENGLAND	ASB(S)R-
UNITED KINGDOM ATOMIC ENERGY AUTHORITY. AUTHORITY HEALTH AND SAFETY BRANCH, RISLEY, LANCS, ENGLAND	DPR/INF/
UNITED KINGDOM ATOMIC ENERGY AUTHORITY. AUTHORITY HEALTH AND SAFETY BRANCH, RISLEY, LANCS, ENGLAND	HS/CR-
UNITED KINGDOM ATOMIC ENERGY AUTHORITY. AUTHORITY HEALTH AND SAFETY BRANCH, RISLEY, LANCS, ENGLAND	RSPP-
UNITED KINGDOM ATOMIC ENERGY AUTHORITY. AUTHORITY HEALTH AND SAFETY BRANCH, RISLEY, LANCS, ENGLAND	UKAE-CODE-
UNITED KINGDOM ATOMIC ENERGY AUTHORITY. AUTHORITY HEALTH AND SAFETY BRANCH. RADIOLOGICAL PROTECTION DIV., HARWELL, BERKS, ENGLAND	AHSB(RP)-
UNITED KINGDOM ATOMIC ENERGY AUTHORITY. AUTHORITY HEALTH AND SAFETY BRANCH. RADIOLOGICAL PROTECTION DIV., HARWELL, BERKS, ENGLAND	AHSB (RP) (LETTER)-
UNITED KINGDOM ATOMIC ENERGY AUTHORITY. AUTHORITY HEALTH AND SAFETY BRANCH. RADIOLOGICAL PROTECTION DIV., HARWELL, BERKS, ENGLAND	AHSB(RP)R-
UNITED KINGDOM ATOMIC ENERGY AUTHORITY. AUTHORITY HEALTH AND SAFETY BRANCH. RADIOLOGICAL PROTECTION DIV., HARWELL, BERKS, ENGLAND	CTO/
UNITED KINGDOM ATOMIC ENERGY AUTHORITY. AUTHORITY HEALTH AND SAFETY BRANCH. SAFEGUARDS DIV., LONDON	AHSB (S) R-

UNITED KINGDOM ATOMIC ENERGY AUTHORITY. CULHAM LAB., ABINGDON, BERKS, ENGLAND	CLM-
UNITED KINGDOM ATOMIC ENERGY AUTHORITY. CULHAM LAB., ABINGDON, BERKS, ENGLAND	CLM-BIB-
UNITED KINGDOM ATOMIC ENERGY AUTHORITY. CULHAM LAB., ABINGDON, BERKS, ENGLAND	CLM-L-
UNITED KINGDOM ATOMIC ENERGY AUTHORITY. CULHAM LAB., ABINGDON, BERKS, ENGLAND	CLM-LM-(NUMBER/YEAR)
UNITED KINGDOM ATOMIC ENERGY AUTHORITY. CULHAM LAB., ABINGDON, BERKS, ENGLAND	CLM-M-
UNITED KINGDOM ATOMIC ENERGY AUTHORITY. CULHAM LAB., ABINGDON, BERKS, ENGLAND	CLM-PR-
UNITED KINGDOM ATOMIC ENERGY AUTHORITY. CULHAM LAB., ABINGDON, BERKS, ENGLAND	CLM-R-
UNITED KINGDOM ATOMIC ENERGY AUTHORITY. CULHAM LAB., ABINGDON, BERKS, ENGLAND (TRANSLATIONS)	CLM-TRANS-
UNITED KINGDOM ATOMIC ENERGY AUTHORITY. CULHAM LAB., ABINGDON, BERKS, ENGLAND	CTO-
UNITED KINGDOM ATOMIC ENERGY AUTHORITY. DEVELOPMENT AND ENGINEERING GROUP (REF. 39)	DEG-(LTRS)-(NO.)(LTR(S))
UNITED KINGDOM ATOMIC ENERGY AUTHORITY. DEVELOPMENT AND ENGINEERING GROUP (REF. 39)	DEGIS-(NUMBER)(LTR(S))
UNITED KINGDOM ATOMIC ENERGY AUTHORITY. DEVELOPMENT AND ENGINEERING GROUP (REF. 39)	DEGM-(NUMBER)(LETTER(S))
UNITED KINGDOM ATOMIC ENERGY AUTHORITY. DEVELOPMENT AND ENGINEERING GROUP (REF. 39)	DEGR-(NUMBER)(LETTER(S))
UNITED KINGDOM ATOMIC ENERGY AUTHORITY. DEVELOPMENT AND ENG. GROUP, CAPENHURST, CHES., ENG. (REF. 39)	DEG-(LETTERS)-(NO.)(CA)
UNITED KINGDOM ATOMIC ENERGY AUTHORITY. DEVELOPMENT AND ENGINEERING GROUP, CAPENHURST, CHES., ENGLAND	IAE-
UNITED KINGDOM ATOMIC ENERGY AUTHORITY. DEVELOPMENT AND ENGINEERING GROUP, CAPENHURST, CHES., ENGLAND	IGR-(LETTER(S))/CA-
UNITED KINGDOM ATOMIC ENERGY AUTHORITY. DEVELOPMENT AND ENGINEERING GROUP, CAPENHURST, CHES., ENGLAND	NPCC/MWP/G/P.
UNITED KINGDOM ATOMIC ENERGY AUTHORITY. DEVELOPMENT AND ENGINEERING GROUP, CULCHETH, LANCS, ENGLAND	MWP(S)/DGD.
UNITED KINGDOM ATOMIC ENERGY AUTHORITY. DEVELOPMENT AND ENGINEERING GROUP, CULCHETH, LANCS, ENGLAND	MWP(S)/P.
UNITED KINGDOM ATOMIC ENERGY AUTHORITY. DEVELOPMENT AND ENGINEERING GROUP, CULCHETH, LANCS, ENGLAND	NPCC-MWP/P.
UNITED KINGDOM ATOMIC ENERGY AUTHORITY. DEVELOPMENT AND ENGINEERING GROUP, CULCHETH, LANCS, ENGLAND	NPCC-MWP(S)DGA/
UNITED KINGDOM ATOMIC ENERGY AUTHORITY. DEVELOPMENT AND ENGINEERING GROUP, CULCHETH, LANCS, ENGLAND	NPCC-MWP(S)DGD/
UNITED KINGDOM ATOMIC ENERGY AUTHORITY. DEVELOPMENT AND ENGINEERING GROUP, CULCHETH, LANCS, ENGLAND	NPCC-MWP(S)/P.
UNITED KINGDOM ATOMIC ENERGY AUTHORITY. DEVELOPMENT AND ENGINEERING GROUP, CULCHETH, LANCS, ENGLAND	NP/R-
UNITED KINGDOM ATOMIC ENERGY AUTHORITY. DEVELOPMENT AND ENGINEERING GROUP, CULCHETH, LANCS, ENGLAND	PCS-P/
UNITED KINGDOM ATOMIC ENERGY AUTHORITY. DEVELOPMENT AND ENGINEERING GROUP, CULCHETH, LANCS, ENGLAND	SD/A-
UNITED KINGDOM ATOMIC ENERGY AUTHORITY. DEVELOPMENT AND ENGINEERING GROUP, RISLEY, LANCS, ENGLAND	DEG-
UNITED KINGDOM ATOMIC ENERGY AUTHORITY. DEVELOPMENT AND ENG. GROUP, RISLEY, LANCS, ENGLAND (REF. 39)	DEG-(LETTERS)-(NO.)(R)
UNITED KINGDOM ATOMIC ENERGY AUTHORITY. DEVELOPMENT AND ENGINEERING GROUP, RISLEY, LANCS, ENGLAND	DEG-INF-SER-
UNITED KINGDOM ATOMIC ENERGY AUTHORITY. DEVELOPMENT AND ENGINEERING GROUP, RISLEY, LANCS, ENGLAND	DEG-MEMO-
UNITED KINGDOM ATOMIC ENERGY AUTHORITY. DEVELOPMENT AND ENGINEERING GROUP, RISLEY, LANCS, ENGLAND	DEG-REPORT-
UNITED KINGDOM ATOMIC ENERGY AUTHORITY. DEVELOPMENT AND ENGINEERING GROUP, RISLEY, LANCS, ENGLAND	DPC-CPH-
UNITED KINGDOM ATOMIC ENERGY AUTHORITY. DEVELOPMENT AND ENGINEERING GROUP, RISLEY, LANCS, ENGLAND	EAES-U.K.-
UNITED KINGDOM ATOMIC ENERGY AUTHORITY. DEVELOPMENT AND ENGINEERING GROUP, SPRINGFIELDS, LANCS, ENG.	NPCC-FEWP/P.-
UNITED KINGDOM ATOMIC ENERGY AUTHORITY. DEVELOPMENT AND ENGINEERING GROUP, SPRINGFIELDS, LANCS, ENGLAND	TRG-
UNITED KINGDOM ATOMIC ENERGY AUTHORITY. ENGINEERING GROUP, RISLEY, LANCS., ENGLAND	EG-REPORT-
UNITED KINGDOM ATOMIC ENERGY AUTHORITY. INDUSTRIAL GROUP (COMMITTEES) (REF. 37)	IGC-(LETTER(S))/(LTR(S))-
UNITED KINGDOM ATOMIC ENERGY AUTHORITY. INDUSTRIAL GROUP (REF. 38)	IG-INF.SER-(NO.)LTR/LTR)
UNITED KINGDOM ATOMIC ENERGY AUTHORITY. INDUSTRIAL GROUP (REF. 38)	IGIS-(NUMBER)(LTR(S)/LTR)
UNITED KINGDOM ATOMIC ENERGY AUTHORITY. INDUSTRIAL GROUP (REF. 38)	IGM-(NUMBER)(LTR(S)/LTR)
UNITED KINGDOM ATOMIC ENERGY AUTHORITY. INDUSTRIAL GROUP (REF. 38)	IG-MEMO-(NO.)(LTR(S)/LTR)
UNITED KINGDOM ATOMIC ENERGY AUTHORITY. INDUSTRIAL GROUP (PROGRESS REPORT) (REF. 37)	IGM-PR-
UNITED KINGDOM ATOMIC ENERGY AUTHORITY. INDUSTRIAL GROUP (REF. 37)	IGO-(LETTER(S))/(LTR(S))-
UNITED KINGDOM ATOMIC ENERGY AUTHORITY. INDUSTRIAL GROUP (REF. 38)	IGR-(NUMBER)(LTR(S)/LTR)
UNITED KINGDOM ATOMIC ENERGY AUTHORITY. INDUSTRIAL GROUP (REF. 37)	IGR-(LETTER(S))/(LTR(S))-
UNITED KINGDOM ATOMIC ENERGY AUTHORITY. INDUSTRIAL GROUP (REF. 38)	IG-REPORT-(NO.)(LTR/LTR)
UNITED KINGDOM ATOMIC ENERGY AUTHORITY. INDUSTRIAL GROUP (REF. 37)	IGRL-(LTR(S))/(LTR(S))-
UNITED KINGDOM ATOMIC ENERGY AUTHORITY. INDUSTRIAL GROUP (REF. 37)	IGS-(LETTER(S))/(LTR(S))-
UNITED KINGDOM ATOMIC ENERGY AUTHORITY. INDUSTRIAL GROUP H.Q., RISLEY, LANCS, ENGLAND	AGR/USAEC/
UNITED KINGDOM ATOMIC ENERGY AUTHORITY. INDUSTRIAL GROUP H.Q., RISLEY, LANCS, ENGLAND (REF. 36)	CF-
UNITED KINGDOM ATOMIC ENERGY AUTHORITY. INDUSTRIAL GROUP H.Q., RISLEY, LANCS, ENGLAND	EMC/P-
UNITED KINGDOM ATOMIC ENERGY AUTHORITY. INDUSTRIAL GROUP H.Q., RISLEY, LANCS, ENGLAND	FPP-
UNITED KINGDOM ATOMIC ENERGY AUTHORITY. INDUSTRIAL GROUP H.Q., RISLEY, LANCS, ENGLAND	FRDC/P-
UNITED KINGDOM ATOMIC ENERGY AUTHORITY. INDUSTRIAL GROUP H.Q., RISLEY, LANCS, ENGLAND (REF. 37)	IGC-(LETTER(S))/(LTR(S))-
UNITED KINGDOM ATOMIC ENERGY AUTHORITY. INDUSTRIAL GROUP H.Q., RISLEY, LANCS, ENGLAND (REF. 37)	IGE-(LETTER(S))-
UNITED KINGDOM ATOMIC ENERGY AUTHORITY. INDUSTRIAL GROUP H.Q., RISLEY, LANCS, ENGLAND (REF. 37)	IGI-(LETTER(S))-
UNITED KINGDOM ATOMIC ENERGY AUTHORITY. INDUSTRIAL GROUP H.Q., RISLEY, LANCS, ENGLAND (REF. 38)	IGM-(NUMBER)(RD/R)
UNITED KINGDOM ATOMIC ENERGY AUTHORITY. INDUSTRIAL GROUP H.Q., RISLEY, LANCS, ENGLAND	I.G. MEMO. (NO.)-(RD/R)
UNITED KINGDOM ATOMIC ENERGY AUTHORITY. INDUSTRIAL GROUP H.Q., RISLEY, LANCS, ENGLAND (REF. 37)	IGO-(LETTER(S))/R-
UNITED KINGDOM ATOMIC ENERGY AUTHORITY. INDUSTRIAL GROUP H.Q., RISLEY, LANCS, ENGLAND (REF. 38)	IGR-(NUMBER)(LTR(S)/R)
UNITED KINGDOM ATOMIC ENERGY AUTHORITY. INDUSTRIAL GROUP H.Q., RISLEY, LANCS, ENGLAND (REF. 37)	IGR-(LETTER(S))/R-
UNITED KINGDOM ATOMIC ENERGY AUTHORITY. INDUSTRIAL GROUP H.Q., RISLEY, LANCS, ENGLAND (REF. 37)	IGRL-(LETTER(S))/R-
UNITED KINGDOM ATOMIC ENERGY AUTHORITY. INDUSTRIAL GROUP H.Q., RISLEY, LANCS, ENGLAND	IGRL-T/W-
UNITED KINGDOM ATOMIC ENERGY AUTHORITY. INDUSTRIAL GROUP H.Q., RISLEY, LANCS, ENGLAND (REF. 37)	IGS-(LETTER(S))/R-
UNITED KINGDOM ATOMIC ENERGY AUTHORITY. INDUSTRIAL GROUP H.Q., RISLEY, LANCS, ENGLAND (REF. 37)	IGT-(LETTER(S))-
UNITED KINGDOM ATOMIC ENERGY AUTHORITY. INDUSTRIAL GROUP H.Q., RISLEY, LANCS, ENGLAND	LEO-RDC/P-
UNITED KINGDOM ATOMIC ENERGY AUTHORITY. INDUSTRIAL GROUP H.Q., RISLEY, LANCS, ENGLAND	NPCC/RPWP-
UNITED KINGDOM ATOMIC ENERGY AUTHORITY. INDUSTRIAL GROUP H.Q., RISLEY, LANCS, ENGLAND	NPCC/RPWP/N
UNITED KINGDOM ATOMIC ENERGY AUTHORITY. INDUSTRIAL GROUP H.Q., RISLEY, LANCS, ENGLAND	PPDG-
UNITED KINGDOM ATOMIC ENERGY AUTHORITY. INDUSTRIAL GROUP H.Q., RISLEY, LANCS, ENGLAND	RDB(R)/
UNITED KINGDOM ATOMIC ENERGY AUTHORITY. INDUSTRIAL GROUP H.Q., RISLEY, LANCS, ENGLAND (REF. 36)	R.+D.B.(R)/(LETTERS)
UNITED KINGDOM ATOMIC ENERGY AUTHORITY. INDUSTRIAL GROUP H.Q., RISLEY, LANCS, ENGLAND	RDB(R)/TM-
UNITED KINGDOM ATOMIC ENERGY AUTHORITY. INDUSTRIAL GROUP H.Q., RISLEY, LANCS, ENGLAND	RDB(R)/TN-
UNITED KINGDOM ATOMIC ENERGY AUTHORITY. INDUSTRIAL GROUP H.Q., RISLEY, LANCS, ENGLAND (REF. 36)	R.+D.(R)-
UNITED KINGDOM ATOMIC ENERGY AUTHORITY. INDUSTRIAL GROUP H.Q., RISLEY, LANCS, ENGLAND	RD(R)-
UNITED KINGDOM ATOMIC ENERGY AUTHORITY. INDUSTRIAL GROUP H.Q., RISLEY, LANCS, ENGLAND (REACTOR HAZARDS)	RHM(YEAR)-
UNITED KINGDOM ATOMIC ENERGY AUTHORITY. INDUSTRIAL GROUP H.Q., RISLEY, LANCS, ENGLAND (REF. 36)	RISLEY-
UNITED KINGDOM ATOMIC ENERGY AUTHORITY. INDUSTRIAL GROUP H.Q., RISLEY, LANCS, ENGLAND	SWP/P-
UNITED KINGDOM ATOMIC ENERGY AUTHORITY. INDUSTRIAL GROUP H.Q., RISLEY, LANCS, ENGLAND	TRG-
UNITED KINGDOM ATOMIC ENERGY AUTHORITY. INDUSTRIAL GROUP H.Q., RISLEY, LANCS, ENGLAND	TRSWP/R-
UNITED KINGDOM ATOMIC ENERGY AUTHORITY. INDUSTRIAL GROUP H.Q., RISLEY, LANCS, ENGLAND	WHC(C)/P-
UNITED KINGDOM ATOMIC ENERGY AUTHORITY. INDUSTRIAL GROUP H.Q. LIBRARY, RISLEY, LANCS, ENG. (REF. 36)	IG-TRANS-(R)-
UNITED KINGDOM ATOMIC ENERGY AUTHORITY. INDUSTRIAL GROUP. CALDER WORKS, CALDERBRIDGE, CUMB., ENGLAND	IGM-(LETTER(S))/CR-
UNITED KINGDOM ATOMIC ENERGY AUTHORITY. INDUSTRIAL GROUP. CALDER WORKS, CALDERBRIDGE, CUMB., ENGLAND	IGO-(LETTER(S))/CR-
UNITED KINGDOM ATOMIC ENERGY AUTHORITY. INDUSTRIAL GROUP. CALDER WORKS, CALDERBRIDGE, CUMB., ENGLAND	TRSWP-MISC-
UNITED KINGDOM ATOMIC ENERGY AUTHORITY. INDUSTRIAL GROUP. CAPENHURST WORKS, CHES., ENGLAND	CAPENHURST-
UNITED KINGDOM ATOMIC ENERGY AUTHORITY. INDUSTRIAL GROUP. CAPENHURST WORKS, CHES., ENGLAND	CTSC/P-
UNITED KINGDOM ATOMIC ENERGY AUTHORITY. INDUSTRIAL GROUP. CAPENHURST WORKS, CHES., ENGLAND	DDC/P-
UNITED KINGDOM ATOMIC ENERGY AUTHORITY. INDUSTRIAL GROUP. CAPENHURST WORKS, CHES., ENGLAND	HRWP/(NO.)B-
UNITED KINGDOM ATOMIC ENERGY AUTHORITY. INDUSTRIAL GROUP. CAPENHURST WORKS, CHES., ENGLAND (REF. 37)	IGO-(LETTER(S))/CA-
UNITED KINGDOM ATOMIC ENERGY AUTHORITY. INDUSTRIAL GROUP. CAPENHURST WORKS, CHES., ENGLAND (REF. 38)	IGR-(NUMBER)(LTR(S)/CA)
UNITED KINGDOM ATOMIC ENERGY AUTHORITY. INDUSTRIAL GROUP. CAPENHURST WORKS, CHES., ENGLAND	I.G.REPORT (NO.)-(RD/CA)
UNITED KINGDOM ATOMIC ENERGY AUTHORITY. INDUSTRIAL GROUP. CAPENHURST WORKS, CHES., ENGLAND (REF. 37)	IGRL-(LETTER(S))/CA-
UNITED KINGDOM ATOMIC ENERGY AUTHORITY. INDUSTRIAL GROUP. CAPENHURST WORKS, CHES., ENGLAND	RB/R-
UNITED KINGDOM ATOMIC ENERGY AUTHORITY. INDUSTRIAL GROUP. CAPENHURST WORKS, CHES., ENGLAND (REF. 36)	R.+D.B.(CA)(LETTERS)-
UNITED KINGDOM ATOMIC ENERGY AUTHORITY. INDUSTRIAL GROUP. CAPENHURST WORKS, CHES., ENGLAND (REF. 36)	RDB(CA)/TN-
UNITED KINGDOM ATOMIC ENERGY AUTHORITY. INDUSTRIAL GROUP. CAPENHURST WORKS, CHES., ENGLAND	SLW/RM/XMG/
UNITED KINGDOM ATOMIC ENERGY AUTHORITY. INDUSTRIAL GROUP. CHAPELCROSS WORKS, DUMFRIESSHIRE, SCOTLAND	IGO-(LETTER(S))/CC-

UNITED KINGDOM ATOMIC ENERGY AUTHORITY. INDUSTRIAL GROUP

UNITED KINGDOM ATOMIC ENERGY AUTHORITY. INDUSTRIAL GROUP. CULCHETH LABS., LANCS, ENGLAND	GCM/UK/
UNITED KINGDOM ATOMIC ENERGY AUTHORITY. INDUSTRIAL GROUP. CULCHETH LABS., LANCS, ENGLAND	IC/P-
UNITED KINGDOM ATOMIC ENERGY AUTHORITY. INDUSTRIAL GROUP. CULCHETH LABS., LANCS, ENGLAND (REF. 37)	IGR-(LETTER(S))/C-
UNITED KINGDOM ATOMIC ENERGY AUTHORITY. INDUSTRIAL GROUP. CULCHETH LABS., LANCS, ENGLAND	RDB(C)TM-
UNITED KINGDOM ATOMIC ENERGY AUTHORITY. INDUSTRIAL GROUP. CULCHETH LABS., LANCS, ENGLAND	RDB(C)/TN-
UNITED KINGDOM ATOMIC ENERGY AUTHORITY. IND. GP. DOUNREAY EXPTL. REACTOR ESTAB., CAITHNESS, SCOTLAND (FAST REACTOR NEWSLETTER)	DERE-FRN-(MONTH/YEAR)
UNITED KINGDOM ATOMIC ENERGY AUTHORITY. IND. GP. DOUNREAY EXPTL.REACTOR EST.,CAITHNESS,SCOT.(REF.37)	IG-
UNITED KINGDOM ATOMIC ENERGY AUTHORITY. IND. GP. DOUNREAY EXPTL.REACTOR EST.,CAITHNESS,SCOT.(REF.37)	IGD-(LETTER(S))-
UNITED KINGDOM ATOMIC ENERGY AUTHORITY. IND. GP. DOUNREAY EXPTL.REACTOR EST.,CAITHNESS,SCOT.(REF.38)	IGM-(NUMBER)(D)
UNITED KINGDOM ATOMIC ENERGY AUTHORITY. IND. GP. DOUNREAY EXPTL. REACTOR ESTAB., CAITHNESS, SCOTLAND	PG-
UNITED KINGDOM ATOMIC ENERGY AUTHORITY. INDUSTRIAL GROUP. DOUNREAY WORKS, CAITHNESS, SCOT. (REF. 37)	IGR-(LETTER(S))/D-
UNITED KINGDOM ATOMIC ENERGY AUTHORITY. INDUSTRIAL GROUP. SPRINGFIELDS WORKS, LANCS, ENG.	ARDC/P/-
UNITED KINGDOM ATOMIC ENERGY AUTHORITY. INDUSTRIAL GROUP. SPRINGFIELDS WORKS, LANCS, ENG. (REF. 38)	IGR-(NUMBER)(LTR(S)/S)
UNITED KINGDOM ATOMIC ENERGY AUTHORITY. INDUSTRIAL GROUP. SPRINGFIELDS WORKS, LANCS, ENG. (REF. 37)	IGR-(LETTER(S))/S-
UNITED KINGDOM ATOMIC ENERGY AUTHORITY. INDUSTRIAL GROUP. SPRINGFIELDS WORKS, LANCS, ENGLAND	NPC/FEWP/P-
UNITED KINGDOM ATOMIC ENERGY AUTHORITY. INDUSTRIAL GROUP. SPRINGFIELDS WORKS, LANCS, ENG. (REF. 36)	R.+D.B.(S)(LETTERS)-
UNITED KINGDOM ATOMIC ENERGY AUTHORITY. INDUSTRIAL GROUP. SPRINGFIELDS WORKS, LANCS, ENGLAND	RDB(S)/TM-
UNITED KINGDOM ATOMIC ENERGY AUTHORITY. INDUSTRIAL GROUP. SPRINGFIELDS WORKS, LANCS, ENGLAND	RDB(S)/TN-
UNITED KINGDOM ATOMIC ENERGY AUTHORITY. INDUSTRIAL GROUP. SPRINGFIELDS WORKS, LANCS, ENGLAND	RSRC/S-
UNITED KINGDOM ATOMIC ENERGY AUTHORITY. INDUSTRIAL GROUP. SPRINGFIELDS WORKS, LANCS, ENG. (REF. 36)	SCS-(LETTER(S))-
UNITED KINGDOM ATOMIC ENERGY AUTHORITY. INDUSTRIAL GROUP. SPRINGFIELDS WORKS, LANCS, ENGLAND	SCS-MEMO-
UNITED KINGDOM ATOMIC ENERGY AUTHORITY. INDUSTRIAL GROUP. SPRINGFIELDS WORKS, LANCS, ENG. (REF. 36)	SRS-
UNITED KINGDOM ATOMIC ENERGY AUTHORITY. INDUSTRIAL GROUP. SPRINGFIELDS WORKS, LANCS, ENGLAND	TRDC-
UNITED KINGDOM ATOMIC ENERGY AUTHORITY. INDUSTRIAL GROUP. SPRINGFIELDS WORKS, LANCS, ENGLAND	TRSWP/R-
UNITED KINGDOM ATOMIC ENERGY AUTHORITY. INDUSTRIAL GROUP.WINDSCALE WORKS,SELLAFIELD,CUMB.,ENG.(REF.38)	IGR-(NUMBER)(LTR(S)/W)
UNITED KINGDOM ATOMIC ENERGY AUTHORITY. INDUSTRIAL GROUP.WINDSCALE WORKS,SELLAFIELD,CUMB.,ENG.(REF.37)	IGR-(LETTER(S))/W-
UNITED KINGDOM ATOMIC ENERGY AUTHORITY. INDUSTRIAL GROUP. WINDSCALE WORKS, SELLAFIELD, CUMB., ENGLAND	NPCC/MWP(G)/
UNITED KINGDOM ATOMIC ENERGY AUTHORITY. INDUSTRIAL GROUP. WINDSCALE WORKS, SELLAFIELD, CUMB., ENGLAND	RB/R-
UNITED KINGDOM ATOMIC ENERGY AUTHORITY. INDUSTRIAL GROUP.WINDSCALE WORKS,SELLAFIELD,CUMB.,ENG.(REF.36)	R.+D.B.(W)(LETTERS)-
UNITED KINGDOM ATOMIC ENERGY AUTHORITY. INDUSTRIAL GROUP. WINDSCALE WORKS, SELLAFIELD, CUMB.,ENGLAND	RDB(W)/TM-
UNITED KINGDOM ATOMIC ENERGY AUTHORITY. INDUSTRIAL GROUP. WINDSCALE WORKS, SELLAFIELD, CUMB., ENGLAND	RDB(W)/TN-
UNITED KINGDOM ATOMIC ENERGY AUTHORITY. INDUSTRIAL GROUP. WINDSCALE WORKS, SELLAFIELD, CUMB., ENGLAND	SWP/P-
UNITED KINGDOM ATOMIC ENERGY AUTHORITY. INDUSTRIAL GROUP. WINDSCALE WORKS, SELLAFIELD, CUMB., ENGLAND	TRDC-
UNITED KINGDOM ATOMIC ENERGY AUTHORITY. INDUSTRIAL GROUP. WINDSCALE WORKS, SELLAFIELD, CUMB., ENGLAND	TRE-
UNITED KINGDOM ATOMIC ENERGY AUTHORITY. INDUSTRIAL GROUP. WINDSCALE WORKS, SELLAFIELD, CUMB., ENGLAND	WINDSCALE-
UNITED KINGDOM ATOMIC ENERGY AUTHORITY. INDUSTRIAL GROUP.WINDSCALE WORKS,SELLAFIELD,CUMB.,ENG.(REF.36)	WSL-(LETTER(S))-
UNITED KINGDOM ATOMIC ENERGY AUTHORITY. INDUSTRIAL GROUP. WINDSCALE WORKS, SELLAFIELD, CUMB., ENGLAND	WTC/P-
UNITED KINGDOM ATOMIC ENERGY AUTHORITY. INDUSTRIAL GROUP. WINDSCALE WORKS, SELLAFIELD, CUMB., ENGLAND	WTSC/R-
(SEE ALSO LATER NAMES: UNITED KINGDOM ATOMIC ENERGY AUTHORITY. DEVELOPMENT AND ENGINEERING GROUP AND U.K.A.E.A. PRODUCTION GROUP)	
UNITED KINGDOM ATOMIC ENERGY AUTHORITY. METALLURGICAL LABS., DIV. OF ATOMIC ENERGY (PRODUCTION), RESEARCH AND DEVELOPMENT BRANCH, CULCHETH, LANCS, ENGLAND	RDB(C)/TN-
UNITED KINGDOM ATOMIC ENERGY AUTHORITY. PRODUCTION GP	NPCC/MWP(G)/P.
UNITED KINGDOM ATOMIC ENERGY AUTHORITY. PRODUCTION GP	NPCC/RPWP/P.
UNITED KINGDOM ATOMIC ENERGY AUTHORITY. PRODUCTION GROUP (REF. 39)	PG-(LTRS)-(NO.)(LTR(S))
UNITED KINGDOM ATOMIC ENERGY AUTHORITY. PRODUCTION GROUP (REF. 39)	PG-INF. SER.-
UNITED KINGDOM ATOMIC ENERGY AUTHORITY. PRODUCTION GROUP (REF. 39)	PGIS-(NUMBER)(LETTER(S))
UNITED KINGDOM ATOMIC ENERGY AUTHORITY. PRODUCTION GROUP (REF. 39)	PGM-(NUMBER)(LETTER(S))
UNITED KINGDOM ATOMIC ENERGY AUTHORITY. PRODUCTION GROUP (REF. 39)	PG-MEMO-
UNITED KINGDOM ATOMIC ENERGY AUTHORITY. PRODUCTION GROUP (REF. 39)	PGR-(NUMBER)(LETTER(S))
UNITED KINGDOM ATOMIC ENERGY AUTHORITY. PRODUCTION GROUP, ANNON, SCOTLAND	PG-
UNITED KINGDOM ATOMIC ENERGY AUTHORITY. PRODUCTION GROUP, CAPENHURST, CHES., ENGLAND (REF. 39)	PG-REPORT-(NUMBER)(CA)
UNITED KINGDOM ATOMIC ENERGY AUTHORITY. PRODUCTION GROUP, RISLEY, LANCS, ENGLAND (REF. 39)	PG-REPORT-(NUMBER)(R)
UNITED KINGDOM ATOMIC ENERGY AUTHORITY. PRODUCTION GROUP, SPRINGFIELDS, LANCS, ENGLAND	IGD-(LETTER(S))/S-
UNITED KINGDOM ATOMIC ENERGY AUTHORITY. PRODUCTION GROUP, SPRINGFIELDS, LANCS, ENGLAND	PG-
UNITED KINGDOM ATOMIC ENERGY AUTHORITY. PRODUCTION GROUP, SPRINGFIELDS, LANCS, ENGLAND (REF. 39)	PG-REPORT-(NUMBER)(S)
UNITED KINGDOM ATOMIC ENERGY AUTHORITY. PRODUCTION GROUP. WINDSCALE WORKS, SELLAFIELD, CUMB., ENGLAND	AGR/OC/P.
UNITED KINGDOM ATOMIC ENERGY AUTHORITY. PRODUCTION GROUP. WINDSCALE WORKS, SELLAFIELD, CUMB., ENGLAND	C/N-
UNITED KINGDOM ATOMIC ENERGY AUTHORITY. PRODUCTION GROUP. WINDSCALE WORKS, SELLAFIELD, CUMB., ENGLAND	DC/P-
UNITED KINGDOM ATOMIC ENERGY AUTHORITY. PRODUCTION GROUP. WINDSCALE WORKS, SELLAFIELD, CUMB., ENGLAND	PG-
UNITED KINGDOM ATOMIC ENERGY AUTHORITY. PRODUCTION GP. WINDSCALE WORKS, SELLAFIELD,CUMB.,ENG. (REF.39)	PG-REPORT-(NUMBER)(W)
UNITED KINGDOM ATOMIC ENERGY AUTHORITY. PRODUCTION GROUP. WINDSCALE WORKS, SELLAFIELD, CUMB., ENGLAND	RRP/P-
UNITED KINGDOM ATOMIC EN. AUTH., PRODUCTION GP. CHEMICAL SERVICES DEPT., WINDSCALE, SELLAFIELD,ENG	IGC-ARDC/P-
UNITED KINGDOM ATOMIC EN. AUTH., PRODUCTION GP. CHEMICAL SERVICES DEPT., WINDSCALE, SELLAFIELD, ENG	IGD-(LETTER(S))/W-
UNITED KINGDOM ATOMIC ENERGY AUTHORITY. PROGRAMMES ANALYSIS UNIT, CHILTON, BERKS, ENGLAND	PAU-
UNITED KINGDOM ATOMIC ENERGY AUTHORITY. PROGRAMMES ANALYSIS UNIT, CHILTON, BERKS, ENGLAND	P.A.U. M-(NUMBER)/(YR.)
UNITED KINGDOM ATOMIC ENERGY AUTHORITY. PROGRAMMES ANALYSIS UNIT, CHILTON, BERKS, ENGLAND	PAU-M-
UNITED KINGDOM ATOMIC ENERGY AUTHORITY. REACTOR GP.	NPCC-FEWP/P.
UNITED KINGDOM ATOMIC ENERGY AUTHORITY. REACTOR GP.	NPCC/MWP(G)/P.
UNITED KINGDOM ATOMIC ENERGY AUTHORITY. REACTOR GP.	NPCC-RKWP/P.
UNITED KINGDOM ATOMIC ENERGY AUTHORITY. REACTOR GP.	NPCC/RPWP/P.
UNITED KINGDOM ATOMIC ENERGY AUTHORITY. REACTOR GP.	TRG-INF. SER.-
UNITED KINGDOM ATOMIC ENERGY AUTHORITY. REACTOR GP.	TRG-MEMO-
UNITED KINGDOM ATOMIC ENERGY AUTHORITY. REACTOR GP.	TRG-REPORT-
UNITED KINGDOM ATOMIC ENERGY AUTHORITY. REACTOR GROUP, CULCHETH, LANCS, ENGLAND	GCM/UK/A-
UNITED KINGDOM ATOMIC ENERGY AUTHORITY. REACTOR GROUP, CULCHETH, LANCS, ENGLAND	HWR-FEMWG/P-(YEAR)-
UNITED KINGDOM ATOMIC ENERGY AUTHORITY. REACTOR GROUP, CULCHETH, LANCS, ENGLAND	JNPC-MWP(G)/P-
UNITED KINGDOM ATOMIC ENERGY AUTHORITY. REACTOR GROUP, CULCHETH, LANCS, ENGLAND	JNPC-MWP-SSG/P(YR)-
UNITED KINGDOM ATOMIC ENERGY AUTHORITY. REACTOR GROUP, CULCHETH, LANCS, ENGLAND	JNPC-SSG/P(YEAR)-
UNITED KINGDOM ATOMIC ENERGY AUTHORITY. REACTOR GROUP, CULCHETH, LANCS, ENGLAND	NPCC-MWP-P.
UNITED KINGDOM ATOMIC ENERGY AUTHORITY. REACTOR GROUP, CULCHETH, LANCS, ENGLAND	NPCC-MWP(S)DGD/P.
UNITED KINGDOM ATOMIC ENERGY AUTHORITY. REACTOR GROUP, CULCHETH, LANCS, ENGLAND	NPCC-MWP(S)/P.
UNITED KINGDOM ATOMIC ENERGY AUTHORITY. REACTOR GROUP, CULCHETH, LANCS, ENGLAND	NP/R-
UNITED KINGDOM ATOMIC ENERGY AUTHORITY. REACTOR GROUP, CULCHETH, LANCS, ENGLAND	NPR/P.
UNITED KINGDOM ATOMIC ENERGY AUTHORITY. REACTOR GROUP, CULCHETH, LANCS, ENGLAND	PCWP/P-
UNITED KINGDOM ATOMIC ENERGY AUTHORITY. REACTOR GROUP, CULCHETH, LANCS, ENGLAND	TRG-(NUMBER)/(LETTER)/
UNITED KINGDOM ATOMIC ENERGY AUTHORITY. REACTOR GP., RISLEY,LANCS,ENG.(ADVANCED GAS COOLED REACTOR)	AGR-FESG-P(YEAR)-
UNITED KINGDOM ATOMIC ENERGY AUTHORITY. REACTOR GP., RISLEY, LANCS, ENGLAND	CON/R-
UNITED KINGDOM ATOMIC ENERGY AUTHORITY. REACTOR GP., RISLEY, LANCS, ENGLAND (FAST REACTOR NEWSLETTER)	FR/N-(YEAR)(NUMBER)
UNITED KINGDOM ATOMIC ENERGY AUTHORITY. REACTOR GP., RISLEY, LANCS, ENGLAND	FRPC/RSWP/
UNITED KINGDOM ATOMIC ENERGY AUTHORITY. REACTOR GP., RISLEY, LANCS, ENGLAND	FRPMC/RSWP/
UNITED KINGDOM ATOMIC ENERGY AUTHORITY. REACTOR GP., RISLEY, LANCS, ENGLAND	JNPC-RKWP-
UNITED KINGDOM ATOMIC ENERGY AUTHORITY. REACTOR GP., RISLEY, LANCS, ENGLAND	JNPC-SWP/P-

UNITED KINGDOM ATOMIC ENERGY AUTHORITY. REACTOR GP., RISLEY, LANCS, ENGLAND — TRG/OR-
UNITED KINGDOM ATOMIC ENERGY AUTHORITY. REACTOR GP., SELLAFIELD, CUMB., ENGLAND — PG-
UNITED KINGDOM ATOMIC ENERGY AUTHORITY. REACTOR GP., SPRINGFIELDS, LANCS, ENGLAND — FRPA-FEWP/
UNITED KINGDOM ATOMIC ENERGY AUTHORITY. REACTOR GP., WINDSCALE, CUMB., ENGLAND — JNPC-FEWP(YEAR)/P-
UNITED KINGDOM ATOMIC ENERGY AUTHORITY. REACTOR GP., WINFRITH, DORSET, ENGLAND — DPCPWP/P.-
UNITED KINGDOM ATOMIC ENERGY AUTHORITY. REACTOR GP., WINFRITH, DORSET, ENGLAND — RS/L-
UNITED KINGDOM ATOMIC ENERGY AUTHORITY. REACTOR GP. CONTROL AND INSTRUMENTATION DIV., ATOMIC ENERGY ESTAB., WINFRITH, DORSET, ENGLAND — JNPC/RPWP/NFISG/P35-
UNITED KINGDOM ATOMIC ENERGY AUTHORITY. REACTOR GP. DOUNREAY EXPTL. REACTOR ESTAB., CAITHNESS, SCOTLAND — ORS/
UNITED KINGDOM ATOMIC ENERGY AUTHORITY. RESEARCH GP. (SERIES ASSIGNED TO REPORTS ON DRAGON PROJECT OF ORGANIZATION FOR ECONOMIC COOPERATION AND DEV.) — DP REPORT-

UNITED KINGDOM ATOMIC ENERGY AUTHORITY. RESEARCH GP. (ALL LOCATIONS) (REF. 40) — AERE-M-
UNITED KINGDOM ATOMIC ENERGY AUTHORITY. RESEARCH GP. (ALL LOCATIONS) (REF. 40) — AERE-R-

UNITED KINGDOM ATOMIC ENERGY AUTHORITY. RESEARCH GP. ATOMIC ENERGY ESTAB., WINFRITH, DORSET, ENGLAND — AEEW-
UNITED KINGDOM ATOMIC ENERGY AUTHORITY. RESEARCH GP. ATOMIC ENERGY ESTAB., WINFRITH, DORSET, ENGLAND — AEEW-AR-
UNITED KINGDOM ATOMIC ENERGY AUTHORITY. RESEARCH GP. ATOMIC ENERGY ESTAB., WINFRITH, DORSET, ENGLAND — AEEW-M-
UNITED KINGDOM ATOMIC ENERGY AUTHORITY. RESEARCH GP. ATOMIC ENERGY ESTAB., WINFRITH, DORSET, ENGLAND — AEEW-R-
UNITED KINGDOM ATOMIC ENERGY AUTHORITY. RESEARCH GP. ATOMIC ENERGY ESTAB., WINFRITH, DORSET, ENGLAND — AEEW-TRANS-
UNITED KINGDOM ATOMIC ENERGY AUTHORITY. RESEARCH GP. ATOMIC ENERGY ESTAB., WINFRITH, DORSET, ENGLAND — DP-
UNITED KINGDOM ATOMIC ENERGY AUTHORITY. RESEARCH GP. ATOMIC ENERGY ESTAB., WINFRITH, DORSET, ENGLAND — DP-MEMO-
UNITED KINGDOM ATOMIC ENERGY AUTHORITY. RESEARCH GP. ATOMIC ENERGY ESTAB., WINFRITH, DORSET, ENGLAND — DPRDD-
UNITED KINGDOM ATOMIC ENERGY AUTHORITY. RESEARCH GP. ATOMIC ENERGY ESTAB., WINFRITH, DORSET, ENGLAND — D.P. REPORT-
 (SERIES ASSIGNED TO REPORTS ON DRAGON PROJECT OF ORGANIZATION FOR ECONOMIC COOPERATION AND DEV.)
UNITED KINGDOM ATOMIC ENERGY AUTHORITY. RESEARCH GP. ATOMIC ENERGY ESTAB., WINFRITH, DORSET, ENGLAND — JNPC/NFISG/-
UNITED KINGDOM ATOMIC ENERGY AUTHORITY. RESEARCH GP. ATOMIC ENERGY ESTAB., WINFRITH, DORSET, ENGLAND — PFRDC/RPWP/P(YEAR)-
UNITED KINGDOM ATOMIC ENERGY AUTHORITY. RESEARCH GP. ATOMIC ENERGY ESTAB., WINFRITH, DORSET, ENGLAND — RPSC(WH)P.91/DRAG/P
UNITED KINGDOM ATOMIC ENERGY AUTHORITY. RESEARCH GP. ATOMIC ENERGY ESTAB., WINFRITH, DORSET, ENGLAND — TRG-REPORT-

UNITED KINGDOM ATOMIC ENERGY AUTHORITY. RESEARCH GP. ATOMIC ENERGY RES. ESTAB., HARWELL, BERKS, ENGLAND — AAEC/ARC/P-
UNITED KINGDOM ATOMIC ENERGY AUTHORITY. RESEARCH GP. ATOMIC ENERGY RES. ESTAB., HARWELL, BERKS, ENGLAND — ACPP-
UNITED KINGDOM ATOMIC ENERGY AUTHORITY. RESEARCH GP. ATOMIC EN. RES. EST., HARWELL, BERKS, ENG. (REF.40) — A.E.R.E.-(LETTERS)-
UNITED KINGDOM ATOMIC ENERGY AUTHORITY. RESEARCH GP. ATOMIC EN. RES. EST., HARWELL, BERKS, ENG. (REF.40) — A.E.R.E.-(LTRS)/(LTRS)-
UNITED KINGDOM ATOMIC ENERGY AUTHORITY. RESEARCH GP. ATOMIC ENERGY RES. ESTAB., HARWELL, BERKS, ENGLAND — AERE-AM-
UNITED KINGDOM ATOMIC ENERGY AUTHORITY. RESEARCH GP. ATOMIC ENERGY RES. ESTAB., HARWELL, BERKS, ENGLAND — AERE-A/R-
UNITED KINGDOM ATOMIC ENERGY AUTHORITY. RESEARCH GP. ATOMIC EN. RES. EST., HARWELL, BERKS, ENG. (REF.40) — AERE-BIB-
UNITED KINGDOM ATOMIC ENERGY AUTHORITY. RESEARCH GP. ATOMIC ENERGY RES. ESTAB., HARWELL, BERKS, ENGLAND — AERE/EMR/PR-
UNITED KINGDOM ATOMIC ENERGY AUTHORITY. RESEARCH GP. ATOMIC ENERGY RES. ESTAB., HARWELL, BERKS, ENGLAND — AERE-L-
UNITED KINGDOM ATOMIC ENERGY AUTHORITY. RESEARCH GP. ATOMIC EN. RES. EST., HARWELL, BERKS, ENG. (REF.40) — AERE-NP-R-
UNITED KINGDOM ATOMIC ENERGY AUTHORITY. RESEARCH GP. ATOMIC ENERGY RES. ESTAB., HARWELL, BERKS, ENGLAND — AERE-P-(NUMBER)/P-
UNITED KINGDOM ATOMIC ENERGY AUTHORITY. RESEARCH GP. ATOMIC ENERGY RES. ESTAB., HARWELL, BERKS, ENGLAND — AERE-PGEC/L-
UNITED KINGDOM ATOMIC ENERGY AUTHORITY. RESEARCH GP. ATOMIC ENERGY RES. ESTAB., HARWELL, BERKS, ENGLAND — AERE-PR/HPM-.
UNITED KINGDOM ATOMIC ENERGY AUTHORITY. RESEARCH GP. ATOMIC ENERGY RES. ESTAB., HARWELL, BERKS, ENGLAND — AERE-PR/SSP-
UNITED KINGDOM ATOMIC EN. AUTH. RES. GP. ATOMIC EN. RES. EST., HARWELL, BERKS, ENG. (READING LIST) — AERE-RL-
UNITED KINGDOM ATOMIC ENERGY AUTHORITY. RESEARCH GP. ATOMIC EN. RES. EST., HARWELL, BERKS, ENG. (REF.40) — AERE-RP/R-
UNITED KINGDOM ATOMIC ENERGY AUTHORITY. RESEARCH GROUP. ATOMIC ENERGY RESEARCH ESTABLISHMENT, HARWELL, BERKS, ENG. (TRIPARTITE CONFERENCE) — AERE-TC-(YEAR)
UNITED KINGDOM ATOMIC EN. AUTH. RES. GP. ATOMIC EN. RES. ESTAB., HARWELL, BERKS, ENG. (TRANSLATIONS) — AERE-TRANS-
UNITED KINGDOM ATOMIC ENERGY AUTHORITY. RESEARCH GP. ATOMIC ENERGY RES. ESTAB., HARWELL, BERKS, ENGLAND — AERE-W/AT-
UNITED KINGDOM ATOMIC ENERGY AUTHORITY. RESEARCH GP. ATOMIC ENERGY RES. ESTAB., HARWELL, BERKS, ENGLAND — AERE-X/
UNITED KINGDOM ATOMIC ENERGY AUTHORITY. RESEARCH GP. ATOMIC ENERGY RES. ESTAB., HARWELL, BERKS, ENGLAND — AERE-Z/(LETTERS)-
UNITED KINGDOM ATOMIC ENERGY AUTHORITY. RESEARCH GP. ATOMIC ENERGY RES. ESTAB., HARWELL, BERKS, ENGLAND — AERO/CONF./
UNITED KINGDOM ATOMIC ENERGY AUTHORITY. RESEARCH GP. ATOMIC ENERGY RES. ESTAB., HARWELL, BERKS, ENGLAND — AGRC/PWP/P-
UNITED KINGDOM ATOMIC ENERGY AUTHORITY. RESEARCH GP. ATOMIC ENERGY RES. ESTAB., HARWELL, BERKS, ENGLAND — AHSB/RP/P-
UNITED KINGDOM ATOMIC ENERGY AUTHORITY. RESEARCH GP. ATOMIC ENERGY RES. ESTAB., HARWELL, BERKS, ENGLAND — AIC/P-
UNITED KINGDOM ATOMIC ENERGY AUTHORITY. RESEARCH GP. ATOMIC ENERGY RES. ESTAB., HARWELL, BERKS, ENGLAND — BPWP/P-
UNITED KINGDOM ATOMIC ENERGY AUTHORITY. RESEARCH GP. ATOMIC ENERGY RES. ESTAB., HARWELL, BERKS, ENGLAND — CPDC-P-
UNITED KINGDOM ATOMIC ENERGY AUTHORITY. RESEARCH GP. ATOMIC ENERGY RES. ESTAB., HARWELL, BERKS, ENGLAND — CPN/
UNITED KINGDOM ATOMIC ENERGY AUTHORITY. RESEARCH GP. ATOMIC ENERGY RES. ESTAB., HARWELL, BERKS, ENGLAND — CRTC/R-
UNITED KINGDOM ATOMIC ENERGY AUTHORITY. RESEARCH GP. ATOMIC ENERGY RES. ESTAB., HARWELL, BERKS, ENGLAND — CSC/P-
UNITED KINGDOM ATOMIC ENERGY AUTHORITY. RESEARCH GP. ATOMIC ENERGY RES. ESTAB., HARWELL, BERKS, ENGLAND — DCPWP-P-
UNITED KINGDOM ATOMIC ENERGY AUTHORITY. RESEARCH GP. ATOMIC ENERGY RES. ESTAB., HARWELL, BERKS, ENGLAND — D/L-
UNITED KINGDOM ATOMIC ENERGY AUTHORITY. RESEARCH GP. ATOMIC ENERGY RES. ESTAB., HARWELL, BERKS, ENGLAND — DPC.P-
UNITED KINGDOM ATOMIC ENERGY AUTHORITY. RESEARCH GP. ATOMIC ENERGY RES. ESTAB., HARWELL, BERKS, ENGLAND — DPED-
UNITED KINGDOM ATOMIC ENERGY AUTHORITY. RESEARCH GP. ATOMIC ENERGY RES. ESTAB., HARWELL, BERKS, ENGLAND — DRSWP/R-
UNITED KINGDOM ATOMIC ENERGY AUTHORITY. RESEARCH GP. ATOMIC ENERGY RES. ESTAB., HARWELL, BERKS, ENGLAND — DSU/P-
UNITED KINGDOM ATOMIC ENERGY AUTHORITY. RESEARCH GP. ATOMIC ENERGY RES. ESTAB., HARWELL, BERKS, ENGLAND — DW(YR)(NO.)
UNITED KINGDOM ATOMIC ENERGY AUTHORITY. RESEARCH GP. ATOMIC ENERGY RES. ESTAB., HARWELL, BERKS, ENGLAND — FFP-
UNITED KINGDOM ATOMIC ENERGY AUTHORITY. RESEARCH GP. ATOMIC ENERGY RES. ESTAB., HARWELL, BERKS, ENGLAND — FPSD/P-
UNITED KINGDOM ATOMIC ENERGY AUTHORITY. RESEARCH GP. ATOMIC ENERGY RES. ESTAB., HARWELL, BERKS, ENGLAND — FRDC/P-
UNITED KINGDOM ATOMIC ENERGY AUTHORITY. RESEARCH GP. ATOMIC ENERGY RES. ESTAB., HARWELL, BERKS, ENGLAND — FRDWP/P-
UNITED KINGDOM ATOMIC ENERGY AUTHORITY. RESEARCH GP. ATOMIC ENERGY RES. ESTAB., HARWELL, BERKS, ENGLAND — FRFEWP/P-
UNITED KINGDOM ATOMIC ENERGY AUTHORITY. RESEARCH GP. ATOMIC ENERGY RES. ESTAB., HARWELL, BERKS, ENGLAND — FROPC/P-
UNITED KINGDOM ATOMIC ENERGY AUTHORITY. RESEARCH GP. ATOMIC ENERGY RES. ESTAB., HARWELL, BERKS, ENGLAND — FROP-C/R-
UNITED KINGDOM ATOMIC ENERGY AUTHORITY. RESEARCH GP. ATOMIC ENERGY RES. ESTAB., HARWELL, BERKS, ENGLAND — FRWP-
UNITED KINGDOM ATOMIC ENERGY AUTHORITY. RESEARCH GP. ATOMIC ENERGY RES. ESTAB., HARWELL, BERKS, ENGLAND — FSG-
UNITED KINGDOM ATOMIC ENERGY AUTHORITY. RESEARCH GP. ATOMIC ENERGY RES. ESTAB., HARWELL, BERKS, ENGLAND — GCHW-(LETTERS)-
UNITED KINGDOM ATOMIC ENERGY AUTHORITY. RESEARCH GP. ATOMIC ENERGY RES. ESTAB., HARWELL, BERKS, ENGLAND — GCHW/MEMO-
UNITED KINGDOM ATOMIC ENERGY AUTHORITY. RESEARCH GP. ATOMIC ENERGY RES. ESTAB., HARWELL, BERKS, ENGLAND — GCHW-NOTE-
UNITED KINGDOM ATOMIC ENERGY AUTHORITY. RESEARCH GP. ATOMIC ENERGY RES. ESTAB., HARWELL, BERKS, ENGLAND — GCM/UK-
UNITED KINGDOM ATOMIC ENERGY AUTHORITY. RESEARCH GP. ATOMIC ENERGY RES. ESTAB., HARWELL, BERKS, ENGLAND — GCM/UK/B-
UNITED KINGDOM ATOMIC ENERGY AUTHORITY. RESEARCH GP. ATOMIC ENERGY RES. ESTAB., HARWELL, BERKS, ENGLAND — GRDC/P-
UNITED KINGDOM ATOMIC ENERGY AUTHORITY. RESEARCH GP. ATOMIC ENERGY RES. ESTAB., HARWELL, BERKS, ENGLAND — HAP-
UNITED KINGDOM ATOMIC ENERGY AUTHORITY. RESEARCH GP. ATOMIC ENERGY RES. ESTAB., HARWELL, BERKS, ENGLAND — HAR-
UNITED KINGDOM ATOMIC ENERGY AUTHORITY. RESEARCH GP. ATOMIC ENERGY RES. ESTAB., HARWELL, BERKS, ENGLAND — HARD(LETTER)/(LETTER)-
UNITED KINGDOM ATOMIC ENERGY AUTHORITY. RESEARCH GP. ATOMIC ENERGY RES. ESTAB., HARWELL, BERKS, ENGLAND — HARD-(A)/P-
UNITED KINGDOM ATOMIC ENERGY AUTHORITY. RESEARCH GP. ATOMIC ENERGY RES. ESTAB., HARWELL, BERKS, ENGLAND — HARD(B)/P-
UNITED KINGDOM ATOMIC ENERGY AUTHORITY. RESEARCH GP. ATOMIC ENERGY RES. ESTAB., HARWELL, BERKS, ENGLAND — HARD(C)/P-
UNITED KINGDOM ATOMIC ENERGY AUTHORITY. RESEARCH GP. ATOMIC ENERGY RES. ESTAB., HARWELL, BERKS, ENGLAND — HARD/DATA-
UNITED KINGDOM ATOMIC ENERGY AUTHORITY. RESEARCH GP. ATOMIC ENERGY RES. ESTAB., HARWELL, BERKS, ENGLAND — HARD/P-
UNITED KINGDOM ATOMIC ENERGY AUTHORITY. RESEARCH GP. ATOMIC ENERGY RES. ESTAB., HARWELL, BERKS, ENGLAND — HCCC/P-
UNITED KINGDOM ATOMIC ENERGY AUTHORITY. RESEARCH GP. ATOMIC ENERGY RES. ESTAB., HARWELL, BERKS, ENGLAND — HCP/P-
UNITED KINGDOM ATOMIC ENERGY AUTHORITY. RESEARCH GP. ATOMIC ENERGY RES. ESTAB., HARWELL, BERKS, ENGLAND — HGTC-(LETTERS)/
UNITED KINGDOM ATOMIC ENERGY AUTHORITY. RESEARCH GP. ATOMIC ENERGY RES. ESTAB., HARWELL, BERKS, ENGLAND — HIM/CONF-
UNITED KINGDOM ATOMIC ENERGY AUTHORITY. RESEARCH GP. ATOMIC ENERGY RES. ESTAB., HARWELL, BERKS, ENGLAND — HL-(YEAR/NUMBER)
UNITED KINGDOM ATOMIC ENERGY AUTHORITY. RESEARCH GP. ATOMIC ENERGY RES. ESTAB., HARWELL, BERKS, ENGLAND — H/PAN-P-
UNITED KINGDOM ATOMIC ENERGY AUTHORITY. RESEARCH GP. ATOMIC ENERGY RES. ESTAB., HARWELL, BERKS, ENGLAND — HP/GEN-
UNITED KINGDOM ATOMIC ENERGY AUTHORITY. RESEARCH GP. ATOMIC ENERGY RES. ESTAB., HARWELL, BERKS, ENGLAND — HRGP/P-
UNITED KINGDOM ATOMIC ENERGY AUTHORITY. RESEARCH GP. ATOMIC ENERGY RES. ESTAB., HARWELL, BERKS, ENGLAND — HRSCC/P-
UNITED KINGDOM ATOMIC ENERGY AUTHORITY. RESEARCH GP. ATOMIC ENERGY RES. ESTAB., HARWELL, BERKS, ENGLAND — HRSCC/S-
UNITED KINGDOM ATOMIC ENERGY AUTHORITY. RESEARCH GP. ATOMIC ENERGY RES. ESTAB., HARWELL, BERKS, ENGLAND — HRSP/S-
UNITED KINGDOM ATOMIC ENERGY AUTHORITY. RESEARCH GP. ATOMIC ENERGY RES. ESTAB., HARWELL, BERKS, ENGLAND — HTGC/CSC/P.(NUMBER)

UNITED KINGDOM ATOMIC ENERGY AUTHORITY. RESEARCH GP.

UNITED KINGDOM ATOMIC ENERGY AUTHORITY. RESEARCH GP. ATOMIC ENERGY RES. ESTAB., HARWELL, BERKS, ENGLAND	HTGC-CWP/MEMO-
UNITED KINGDOM ATOMIC ENERGY AUTHORITY. RESEARCH GP. ATOMIC ENERGY RES. ESTAB., HARWELL, BERKS, ENGLAND	HTGC-FEWP/MEMO-
UNITED KINGDOM ATOMIC ENERGY AUTHORITY. RESEARCH GP. ATOMIC ENERGY RES. ESTAB., HARWELL, BERKS, ENGLAND	HTGC-FPWP/P-
UNITED KINGDOM ATOMIC ENERGY AUTHORITY. RESEARCH GP. ATOMIC ENERGY RES. ESTAB., HARWELL, BERKS, ENGLAND	HTGC-FRWP/M
UNITED KINGDOM ATOMIC ENERGY AUTHORITY. RESEARCH GP. ATOMIC ENERGY RES. ESTAB., HARWELL, BERKS, ENGLAND	HTGC-FRWP/MEMO-
UNITED KINGDOM ATOMIC ENERGY AUTHORITY. RESEARCH GP. ATOMIC ENERGY RES. ESTAB., HARWELL, BERKS, ENGLAND	HTGC-MEMO-
UNITED KINGDOM ATOMIC ENERGY AUTHORITY. RESEARCH GP. ATOMIC ENERGY RES. ESTAB., HARWELL, BERKS, ENGLAND	HTGC-MT/MEMO-
UNITED KINGDOM ATOMIC ENERGY AUTHORITY. RESEARCH GP. ATOMIC ENERGY RES. ESTAB., HARWELL, BERKS, ENGLAND	HTGC/P-
UNITED KINGDOM ATOMIC ENERGY AUTHORITY. RESEARCH GP. ATOMIC ENERGY RES. ESTAB., HARWELL, BERKS, ENGLAND	HTR/(LETTERS)/P-
UNITED KINGDOM ATOMIC ENERGY AUTHORITY. RESEARCH GP. ATOMIC ENERGY RES. ESTAB., HARWELL, BERKS, ENGLAND	HTR/CWP/P-
UNITED KINGDOM ATOMIC ENERGY AUTHORITY. RESEARCH GP. ATOMIC ENERGY RES. ESTAB., HARWELL, BERKS, ENGLAND	HTRDC/-
UNITED KINGDOM ATOMIC ENERGY AUTHORITY. RESEARCH GP. ATOMIC ENERGY RES. ESTAB., HARWELL, BERKS, ENGLAND	HTRDC/P-
UNITED KINGDOM ATOMIC ENERGY AUTHORITY. RESEARCH GP. ATOMIC ENERGY RES. ESTAB., HARWELL, BERKS, ENGLAND	HTR/FEWP-MEMO-
UNITED KINGDOM ATOMIC ENERGY AUTHORITY. RESEARCH GP. ATOMIC ENERGY RES. ESTAB., HARWELL, BERKS, ENGLAND	HTR-FEWP/P.
UNITED KINGDOM ATOMIC ENERGY AUTHORITY. RESEARCH GP. ATOMIC ENERGY RES. ESTAB., HARWELL, BERKS, ENGLAND	HTR/PWP/MEMO-
UNITED KINGDOM ATOMIC ENERGY AUTHORITY. RESEARCH GP. ATOMIC ENERGY RES. ESTAB., HARWELL, BERKS, ENGLAND	HTR/PWP/P-
UNITED KINGDOM ATOMIC ENERGY AUTHORITY. RESEARCH GP. ATOMIC ENERGY RES. ESTAB., HARWELL, BERKS, ENGLAND	LB/G-
UNITED KINGDOM ATOMIC ENERGY AUTHORITY. RESEARCH GP. ATOMIC ENERGY RES. ESTAB., HARWELL, BERKS, ENGLAND	LMD/NOS.-
UNITED KINGDOM ATOMIC ENERGY AUTHORITY. RESEARCH GP. ATOMIC ENERGY RES. ESTAB., HARWELL, BERKS, ENGLAND	LMFP/P-
UNITED KINGDOM ATOMIC ENERGY AUTHORITY. RESEARCH GP. ATOMIC ENERGY RES. ESTAB., HARWELL, BERKS, ENGLAND	LMFR/I-
UNITED KINGDOM ATOMIC ENERGY AUTHORITY. RESEARCH GP. ATOMIC ENERGY RES. ESTAB., HARWELL, BERKS, ENGLAND	LMFR/P-
UNITED KINGDOM ATOMIC ENERGY AUTHORITY. RESEARCH GP. ATOMIC ENERGY RES. ESTAB., HARWELL, BERKS, ENGLAND	LMFS/P-
UNITED KINGDOM ATOMIC ENERGY AUTHORITY. RESEARCH GP. ATOMIC ENERGY RES. ESTAB., HARWELL, BERKS, ENGLAND	LMFT/N-
UNITED KINGDOM ATOMIC ENERGY AUTHORITY. RESEARCH GP. ATOMIC ENERGY RES. ESTAB., HARWELL, BERKS, ENGLAND	LMFT/P-
UNITED KINGDOM ATOMIC ENERGY AUTHORITY. RESEARCH GP. ATOMIC ENERGY RES. ESTAB., HARWELL, BERKS, ENGLAND	ML/MPC-
UNITED KINGDOM ATOMIC ENERGY AUTHORITY. RESEARCH GP. ATOMIC ENERGY RES. ESTAB., HARWELL, BERKS, ENGLAND	M/M-
UNITED KINGDOM ATOMIC ENERGY AUTHORITY. RESEARCH GP. ATOMIC ENERGY RES. ESTAB., HARWELL, BERKS, ENGLAND	MSWP/N-
UNITED KINGDOM ATOMIC ENERGY AUTHORITY. RESEARCH GP. ATOMIC ENERGY RES. ESTAB., HARWELL, BERKS, ENGLAND	MSWP/R-
UNITED KINGDOM ATOMIC ENERGY AUTHORITY. RESEARCH GP. ATOMIC ENERGY RES. ESTAB., HARWELL, BERKS, ENGLAND	MV-
UNITED KINGDOM ATOMIC ENERGY AUTHORITY. RESEARCH GP. ATOMIC ENERGY RES. ESTAB., HARWELL, BERKS, ENGLAND	NPCC/MWP/(G)P.
UNITED KINGDOM ATOMIC ENERGY AUTHORITY. RESEARCH GP. ATOMIC ENERGY RES. ESTAB., HARWELL, BERKS, ENGLAND	NP-CC/RPWP/P-
UNITED KINGDOM ATOMIC ENERGY AUTHORITY. RESEARCH GP. ATOMIC ENERGY RES. ESTAB., HARWELL, BERKS, ENGLAND	NPCC/RPWP/P-
UNITED KINGDOM ATOMIC ENERGY AUTHORITY. RESEARCH GP. ATOMIC ENERGY RES. ESTAB., HARWELL, BERKS, ENGLAND	NRDC-
UNITED KINGDOM ATOMIC ENERGY AUTHORITY. RESEARCH GP. ATOMIC ENERGY RES. ESTAB., HARWELL, BERKS, ENGLAND	NSSG-
UNITED KINGDOM ATOMIC ENERGY AUTHORITY. RESEARCH GP. ATOMIC ENERGY RES. ESTAB., HARWELL, BERKS, ENGLAND (NAVAL SECTION SHIELDING GROUP)	NSSG-R-
UNITED KINGDOM ATOMIC ENERGY AUTHORITY. RESEARCH GP. ATOMIC ENERGY RES. ESTAB., HARWELL, BERKS, ENGLAND	NURG/M-
UNITED KINGDOM ATOMIC ENERGY AUTHORITY. RESEARCH GP. ATOMIC ENERGY RES. ESTAB., HARWELL, BERKS, ENGLAND	OEEC/SCC/
UNITED KINGDOM ATOMIC ENERGY AUTHORITY. RESEARCH GP. ATOMIC ENERGY RES. ESTAB., HARWELL, BERKS, ENGLAND	P-3WP/
UNITED KINGDOM ATOMIC ENERGY AUTHORITY. RESEARCH GP. ATOMIC ENERGY RES. ESTAB., HARWELL, BERKS, ENGLAND	PFRDC/FEWP-
UNITED KINGDOM ATOMIC ENERGY AUTHORITY. RESEARCH GP. ATOMIC ENERGY RES. ESTAB., HARWELL, BERKS, ENGLAND	PFRDC/RPWP/P(YEAR)-
UNITED KINGDOM ATOMIC ENERGY AUTHORITY. RESEARCH GP. ATOMIC ENERGY RES. ESTAB., HARWELL, BERKS, ENGLAND	PLAC-
UNITED KINGDOM ATOMIC ENERGY AUTHORITY. RESEARCH GP. ATOMIC ENERGY RES. ESTAB., HARWELL, BERKS, ENGLAND	PLUTO/IS-
UNITED KINGDOM ATOMIC ENERGY AUTHORITY. RESEARCH GP. ATOMIC ENERGY RES. ESTAB., HARWELL, BERKS, ENGLAND	POC/MEM-
UNITED KINGDOM ATOMIC ENERGY AUTHORITY. RESEARCH GP. ATOMIC ENERGY RES. ESTAB., HARWELL, BERKS, ENGLAND	POC/MEMO-
UNITED KINGDOM ATOMIC ENERGY AUTHORITY. RESEARCH GP. ATOMIC ENERGY RES. ESTAB., HARWELL, BERKS, ENGLAND	POC/RIG MANUAL/
UNITED KINGDOM ATOMIC ENERGY AUTHORITY. RESEARCH GP. ATOMIC ENERGY RES. ESTAB., HARWELL, BERKS, ENGLAND	POC/SSC/I-
UNITED KINGDOM ATOMIC ENERGY AUTHORITY. RESEARCH GP. ATOMIC ENERGY RES. ESTAB., HARWELL, BERKS, ENGLAND	PR/CONF/(UK)-
UNITED KINGDOM ATOMIC ENERGY AUTHORITY. RESEARCH GP. ATOMIC ENERGY RES. ESTAB., HARWELL, BERKS, ENGLAND	PTC/P-
UNITED KINGDOM ATOMIC ENERGY AUTHORITY. RESEARCH GP. ATOMIC ENERGY RES. ESTAB., HARWELL, BERKS, ENGLAND	PUC/P-
UNITED KINGDOM ATOMIC ENERGY AUTHORITY. RESEARCH GP. ATOMIC ENERGY RES. ESTAB., HARWELL, BERKS, ENGLAND	R.CHEM.P.C./P-
UNITED KINGDOM ATOMIC ENERGY AUTHORITY. RESEARCH GP. ATOMIC ENERGY RES. ESTAB., HARWELL, BERKS, ENGLAND	RCTC-
UNITED KINGDOM ATOMIC ENERGY AUTHORITY. RESEARCH GP. ATOMIC ENERGY RES. ESTAB., HARWELL, BERKS, ENGLAND	RCTC/P-
UNITED KINGDOM ATOMIC ENERGY AUTHORITY. RESEARCH GP. ATOMIC ENERGY RES. ESTAB., HARWELL, BERKS, ENGLAND	RIC-
UNITED KINGDOM ATOMIC ENERGY AUTHORITY. RESEARCH GP. ATOMIC ENERGY RES. ESTAB., HARWELL, BERKS, ENGLAND	RMRC/P-
UNITED KINGDOM ATOMIC ENERGY AUTHORITY. RESEARCH GP. ATOMIC ENERGY RES. ESTAB., HARWELL, BERKS, ENGLAND	RPL-
UNITED KINGDOM ATOMIC ENERGY AUTHORITY. RESEARCH GP. ATOMIC ENERGY RES. ESTAB., HARWELL, BERKS, ENGLAND	RRAC-
UNITED KINGDOM ATOMIC ENERGY AUTHORITY. RESEARCH GP. ATOMIC ENERGY RES. ESTAB., HARWELL, BERKS, ENGLAND	RRAC/SSC/I-
UNITED KINGDOM ATOMIC ENERGY AUTHORITY. RESEARCH GP. ATOMIC ENERGY RES. ESTAB., HARWELL, BERKS, ENGLAND	RRD/DES. DATA-
UNITED KINGDOM ATOMIC ENERGY AUTHORITY. RESEARCH GP. ATOMIC ENERGY RES. ESTAB., HARWELL, BERKS, ENGLAND	RSG-
UNITED KINGDOM ATOMIC ENERGY AUTHORITY. RESEARCH GP. ATOMIC ENERGY RES. ESTAB., HARWELL, BERKS, ENGLAND	RSRWP/S-
UNITED KINGDOM ATOMIC ENERGY AUTHORITY. RESEARCH GP. ATOMIC ENERGY RES. ESTAB., HARWELL, BERKS, ENGLAND	SC-(AERE)-
UNITED KINGDOM ATOMIC ENERGY AUTHORITY. RESEARCH GP. ATOMIC ENERGY RES. ESTAB., HARWELL, BERKS, ENGLAND	S/D-
UNITED KINGDOM ATOMIC ENERGY AUTHORITY. RESEARCH GP. ATOMIC ENERGY RES. ESTAB., HARWELL, BERKS, ENGLAND	SGDG/P-
UNITED KINGDOM ATOMIC ENERGY AUTHORITY. RESEARCH GP. ATOMIC ENERGY RES. ESTAB., HARWELL, BERKS, ENGLAND	SGRC/P-
UNITED KINGDOM ATOMIC ENERGY AUTHORITY. RESEARCH GP. ATOMIC ENERGY RES. ESTAB., HARWELL, BERKS, ENGLAND	SWP/P-
UNITED KINGDOM ATOMIC ENERGY AUTHORITY. RESEARCH GP. ATOMIC ENERGY RES. ESTAB., HARWELL, BERKS, ENGLAND	TCAP/P-
UNITED KINGDOM ATOMIC ENERGY AUTHORITY. RESEARCH GP. ATOMIC ENERGY RES. ESTAB., HARWELL, BERKS, ENGLAND	TNCC/UK-
UNITED KINGDOM ATOMIC ENERGY AUTHORITY. RESEARCH GP. ATOMIC ENERGY RES. ESTAB., HARWELL, BERKS, ENGLAND	TP-
UNITED KINGDOM ATOMIC ENERGY AUTHORITY. RESEARCH GP. ATOMIC ENERGY RES. ESTAB., HARWELL, BERKS, ENGLAND	TRDC-
UNITED KINGDOM ATOMIC ENERGY AUTHORITY. RESEARCH GP. ATOMIC EN. RES. EST., HARWELL, BERKS, ENG. (REF.40)	TRG-REPORT-
UNITED KINGDOM ATOMIC ENERGY AUTHORITY. RESEARCH GP. ATOMIC ENERGY RES. ESTAB., HARWELL, BERKS, ENGLAND	TROC/R-
UNITED KINGDOM ATOMIC ENERGY AUTHORITY. RESEARCH GP. ATOMIC ENERGY RES. ESTAB., HARWELL, BERKS, ENGLAND	TRSWP/P-
UNITED KINGDOM ATOMIC ENERGY AUTHORITY. RESEARCH GP. ATOMIC ENERGY RES. ESTAB., HARWELL, BERKS, ENGLAND	TTS/
UNITED KINGDOM ATOMIC ENERGY AUTHORITY. RESEARCH GP. ATOMIC ENERGY RES. ESTAB., HARWELL, BERKS, ENGLAND	UK-
UNITED KINGDOM ATOMIC ENERGY AUTHORITY. RESEARCH GP. ATOMIC ENERGY RES. ESTAB., HARWELL, BERKS, ENGLAND (RESEARCH NEWSLETTER)	UK-RN-
UNITED KINGDOM ATOMIC ENERGY AUTHORITY. RESEARCH GP. ATOMIC ENERGY RES. ESTAB., HARWELL, BERKS, ENGLAND (RESEARCH NEWSLETTER, BERYLLIUM)	UK-RN-BE-
UNITED KINGDOM ATOMIC ENERGY AUTHORITY. RESEARCH GP. ATOMIC ENERGY RES. ESTAB., HARWELL, BERKS, ENGLAND (RESEARCH NEWSLETTER, GRAPHITE CHEMISTRY)	UK-RN-GC-A-
UNITED KINGDOM ATOMIC ENERGY AUTHORITY. RESEARCH GP. ATOMIC ENERGY RES. ESTAB., HARWELL, BERKS, ENGLAND (RESEARCH NEWSLETTER, GAS COOLANT)	UK-RN-GC-B-
UNITED KINGDOM ATOMIC ENERGY AUTHORITY. RESEARCH GP. ATOMIC ENERGY RES. ESTAB., HARWELL, BERKS, ENGLAND (RESEARCH NEWSLETTER, GRAPHITE PHYSICS)	UK-RN-GC-C-
UNITED KINGDOM ATOMIC ENERGY AUTHORITY. RESEARCH GP. ATOMIC ENERGY RES. ESTAB., HARWELL, BERKS, ENGLAND (RESEARCH NEWSLETTER, PLUTONIUM)	UK-RN-PLUT-
UNITED KINGDOM ATOMIC ENERGY AUTHORITY. RESEARCH GP. ATOMIC ENERGY RES. ESTAB., HARWELL, BERKS, ENGLAND (RESEARCH NEWSLETTER, PLUTONIUM METALLURGY)	UK-RN-PM-
UNITED KINGDOM ATOMIC ENERGY AUTHORITY. RESEARCH GP. ATOMIC ENERGY RES. ESTAB., HARWELL, BERKS, ENGLAND (RESEARCH NEWSLETTER, URANIUM)	UK-RN-URAN-
UNITED KINGDOM ATOMIC ENERGY AUTHORITY. RESEARCH GP. ATOMIC ENERGY RES. ESTAB., HARWELL, BERKS, ENGLAND	UNARP-
UNITED KINGDOM ATOMIC ENERGY AUTHORITY. RESEARCH GP. ATOMIC ENERGY RES. ESTAB., HARWELL, BERKS, ENGLAND	WHC(C)/P-
UNITED KINGDOM ATOMIC ENERGY AUTHORITY. RESEARCH GP. ATOMIC ENERGY RES. ESTAB., HARWELL, BERKS, ENGLAND	WPC/P-
UNITED KINGDOM ATOMIC ENERGY AUTHORITY. RESEARCH GP. ATOMIC ENERGY RES. ESTAB., HARWELL, BERKS, ENGLAND	ZETR-II/P-
UNITED KINGDOM ATOMIC EN. AUTH. RES. GP. CHEM. ENG. DIV. ATOMIC ENERGY RES. EST., HARWELL, BERKS, ENG.	AERE-CE/(LETTERS)-
UNITED KINGDOM ATOMIC EN. AUTH. RES. GP. CHEMISTRY DIV. ATOMIC ENERGY RES. EST., HARWELL, BERKS, ENG.	AERE-C/(LETTERS)-
UNITED KINGDOM ATOMIC EN. AUTH. RES. GROUP. CHEMISTRY DIV., CHATHAM OUTSTATION, KENT, ENGLAND	AERE-A/M-
UNITED KINGDOM ATOMIC ENERGY AUTHORITY. RESEARCH GP. CHEMISTRY DIV., WOOLWICH OUTSTATION, ENGLAND	AERE-A/M-
UNITED KINGDOM ATOMIC ENERGY AUTHORITY. RESEARCH GP. CHEMISTRY DIV., WOOLWICH OUTSTATION, ENGLAND	LMFT/P-
UNITED KINGDOM ATOMIC ENERGY AUTHORITY. RESEARCH GP. CULHAM LAB., ABINGDON, BERKS, ENGLAND	CLM-R-
UNITED KINGDOM ATOMIC ENERGY AUTHORITY. RESEARCH GP. CULHAM LAB., ABINGDON, BERKS, ENGLAND	CLM-AR-
UNITED KINGDOM ATOMIC ENERGY AUTHORITY. RESEARCH GP. CULHAM LAB., ABINGDON, BERKS, ENGLAND	CLM-BIB-
UNITED KINGDOM ATOMIC ENERGY AUTHORITY. RESEARCH GP. CULHAM LAB., ABINGDON, BERKS, ENGLAND	CLM-L-
UNITED KINGDOM ATOMIC ENERGY AUTHORITY. RESEARCH GP. CULHAM LAB., ABINGDON, BERKS, ENGLAND	CLM-M-

UNITED KINGDOM ATOMIC ENERGY AUTHORITY. RESEARCH GP. CULHAM LAB., ABINGDON, BERKS, ENGLAND (PREPRINT)	CLM-P-
UNITED KINGDOM ATOMIC ENERGY AUTHORITY. RESEARCH GP. CULHAM LAB., ABINGDON, BERKS, ENGLAND	CLM-PR-
UNITED KINGDOM ATOMIC ENERGY AUTHORITY. RESEARCH GP. CULHAM LAB., ABINGDON, BERKS, ENGLAND	CLM-R-
UNITED KINGDOM ATOMIC ENERGY AUTHORITY. RESEARCH GP. CULHAM LAB., ABINGDON, BERKS, ENGLAND	PPN-
UNITED KINGDOM ATOMIC EN. AUTH. RES. GP. DIRECTORS OFF. ATOMIC ENERGY RES. EST., HARWELL, BERKS, ENG.	AERE-D/(LETTERS)-
UNITED KINGDOM ATOMIC EN. AUTH. RES. GP. ELECTRONICS DIV. ATOMIC ENERGY RES. EST., HARWELL, BERKS, ENG.	AERE-EL/(LETTERS)-
UNITED KINGDOM ATOMIC EN. AUTH. RES. GP. ENGINEERING DIV. ATOMIC ENERGY RES. EST., HARWELL, BERKS, ENG.	AERE-E/(LETTERS)-
UNITED KINGDOM ATOMIC EN. AUTH. RES. GP. ENG.SERVICES DIV. ATOMIC ENERGY RES. EST., HARWELL, BERKS, ENG.	AERE-ES/(LETTERS)-
UNITED KINGDOM ATOMIC EN. AUTH. RES. GP. GEN. PHYSICS DIV. ATOMIC ENERGY RES. EST., HARWELL, BERKS, ENG.	AERE-G/(LETTERS)-
UNITED KINGDOM ATOMIC EN. AUTH. RES. GP. GEN. PHYSICS DIV. ATOMIC ENERGY RES. EST., HARWELL, BERKS, ENG.	AERE-GP/(LETTERS)-
UNITED KINGDOM ATOMIC EN. AUTH. RES. GP. HEALTH DIV. ATOMIC ENERGY RES. EST., HARWELL, BERKS, ENG.	AERE-H/(LETTERS)-
UNITED KINGDOM ATOMIC EN. AUTH. RES. GP. HEALTH PHYS. DIV. ATOMIC ENERGY RES. EST., HARWELL, BERKS, ENG.	AERE-HP/(LETTERS)-
UNITED KINGDOM ATOMIC EN. AUTH. RES. GP. ISOTOPES DIV. ATOMIC ENERGY RES. EST., HARWELL, BERKS, ENG.	AERE-I/(LETTERS)-
UNITED KINGDOM ATOMIC EN. AUTH. RES. GP. MEDICAL DIV. ATOMIC ENERGY RES. EST., HARWELL, BERKS, ENG.	AERE-MED/(LETTERS)-
UNITED KINGDOM ATOMIC ENERGY AUTHORITY. RESEARCH GROUP. MEDICAL RES. COUNCIL. RADIOBIOLOGICAL RES. UNIT. ATOMIC EN. RES. EST., HARWELL, BERKS, ENG.	AERE-MRC/(LETTERS)-
UNITED KINGDOM ATOMIC EN. AUTH. RES. GP. METALLURGY DIV. ATOMIC ENERGY RES. EST., HARWELL, BERKS, ENG.	AERE-M/(LETTERS)-
UNITED KINGDOM ATOMIC EN. AUTH. RES. GP. NUCLEAR PHYS. DIV. ATOMIC EN. RES. EST., HARWELL,BERKS,ENG.	AERE-N/(LETTERS)-
UNITED KINGDOM ATOMIC EN. AUTH. RES. GP. NUCLEAR PHYS. DIV. ATOMIC EN. RES. EST., HARWELL,BERKS,ENG.	AERE-NP/(LETTERS)-
UNITED KINGDOM ATOMIC EN. AUTH. RES. GP. NUCLEAR PHYS. DIV. ATOMIC EN. RES. EST., HARWELL,BERKS,ENG.	AERE-PR/NP-
UNITED KINGDOM ATOMIC ENERGY AUTHORITY. RESEARCH GP. RADIOCHEMICAL CENTRE, AMERSHAM, BUCKS, ENG.(REF.41)	R.C.C./M-
UNITED KINGDOM ATOMIC ENERGY AUTHORITY. RESEARCH GP. RADIOCHEMICAL CENTRE, AMERSHAM, BUCKS, ENG.(REF.41)	RCC-M-
UNITED KINGDOM ATOMIC ENERGY AUTHORITY. RESEARCH GP. RADIOCHEMICAL CENTRE, AMERSHAM, BUCKS, ENG.(REF.41)	R.C.C./R-
UNITED KINGDOM ATOMIC ENERGY AUTHORITY. RESEARCH GP. RADIOCHEMICAL CENTRE, AMERSHAM, BUCKS, ENG.(REF.41)	RCC-R-
UNITED KINGDOM ATOMIC EN. AUTH. RES. GP. REACTOR DIV. ATOMIC ENERGY RES. EST., HARWELL, BERKS, ENG.	AERE-R/(LETTERS)-
UNITED KINGDOM ATOMIC EN. AUTH. RES. GP. REACTOR ENG. DIV. ATOMIC EN. RES. EST., HARWELL,BERKS,ENG.	AERE-RE/(LETTERS)-
UNITED KINGDOM ATOMIC EN. AUTH. RES. GP. REACTOR PHYS. DIV. ATOMIC EN. RES. EST., HARWELL,BERKS,ENG.	AERE-R/(LETTERS)-
UNITED KINGDOM ATOMIC EN. AUTH. RES. GP. REACTOR PHYS. DIV. ATOMIC EN. RES. EST., HARWELL,BERKS,ENG.	AERE-RP/(LETTERS)-
UNITED KINGDOM ATOMIC EN. AUTH. RES. GP. REACTOR SCH. ATOMIC ENERGY RES. EST., HARWELL, BERKS, ENG.	AERE-RS/(LETTERS)-
UNITED KINGDOM ATOMIC EN. AUTH. RES. GP. THEORETICAL PHYS. DIV. ATOMIC EN. RES. EST., HARWELL,BERKS,ENG.	AERE-T/(LETTERS)-
UNITED KINGDOM ATOMIC ENERGY AUTHORITY. RESEARCH GP. WANTAGE RESEARCH LABORATORIES, BERKS, ENGLAND	SM-
UNITED KINGDOM ATOMIC ENERGY AUTHORITY. SAFETY AND RELIABILITY DIRECTORATE, RISLEY, LANCS, ENGLAND	SRD-R-
UNITED KINGDOM ATOMIC ENERGY AUTHORITY. SAFETY COMMITTEE, HARWELL,BERKS, ENGLAND	AERE-SC-
UNITED KINGDOM ATOMIC ENERGY AUTHORITY. UNIVERSITIES RESEARCH REACTOR, RISLEY, LANCS, ENGLAND	URR-
UNITED KINGDOM ATOMIC ENERGY AUTHORITY. WEAPONS GROUP ATOMIC WEAPONS RES. ESTAB.,ALDERMASTON, BERKS, ENG.	ARE-(NUMBER)/(YEAR)
UNITED KINGDOM ATOMIC ENERGY AUTHORITY. WEAPONS GROUP ATOMIC WEAPONS RES. ESTAB.,ALDERMASTON, BERKS, ENG.	ARL/R4/C-
UNITED KINGDOM ATOMIC ENERGY AUTHORITY. WEAPONS GROUP ATOMIC WEAPONS RES. ESTAB.,ALDERMASTON, BERKS, ENG.	AWRE-
UNITED KINGDOM ATOMIC ENERGY AUTHORITY. WEAPONS GROUP ATOMIC WEAPONS RES. ESTAB.,ALDERMASTON, BERKS, ENG. (DRAWING)	AWRE-DWG-HR/(LTR)(NO.)
UNITED KINGDOM ATOMIC ENERGY AUTHORITY. WEAPONS GROUP ATOMIC WEAPONS RES. ESTAB.,ALDERMASTON, BERKS, ENG.	AWRE-E(NO.)/(YR.)
UNITED KINGDOM ATOMIC ENERGY AUTHORITY. WEAPONS GROUP ATOMIC WEAPONS RES. ESTAB.,ALDERMASTON, BERKS, ENG.	AWRE-EIVR-
UNITED KINGDOM ATOMIC ENERGY AUTHORITY. WEAPONS GROUP ATOMIC WEAPONS RES. ESTAB.,ALDERMASTON, BERKS, ENG.	AWRE-ERN-
UNITED KINGDOM ATOMIC ENERGY AUTHORITY. WEAPONS GROUP ATOMIC WEAPONS RES. ESTAB.,ALDERMASTON, BERKS, ENG.	AWRE-ERN-(NO.)/(YR.)
UNITED KINGDOM ATOMIC ENERGY AUTHORITY. WEAPONS GROUP ATOMIC WEAPONS RES. ESTAB.,ALDERMASTON, BERKS, ENG.	AWRE-FMP/IMOG/
UNITED KINGDOM ATOMIC ENERGY AUTHORITY. WEAPONS GROUP ATOMIC WEAPONS RES. ESTAB.,ALDERMASTON, BERKS, ENG.	AWRE-GRO/
UNITED KINGDOM ATOMIC ENERGY AUTHORITY. WEAPONS GROUP ATOMIC WEAPONS RES. ESTAB.,ALDERMASTON, BERKS, ENG.	AWRE-HR-
UNITED KINGDOM ATOMIC ENERGY AUTHORITY. WEAPONS GROUP ATOMIC WEAPONS RES. ESTAB.,ALDERMASTON, BERKS, ENG.	AWRE-INSP/
UNITED KINGDOM ATOMIC ENERGY AUTHORITY. WEAPONS GROUP ATOMIC WEAPONS RES. ESTAB.,ALDERMASTON, BERKS, ENG.	AWRE-J(NO.)/(NO.)-
UNITED KINGDOM ATOMIC ENERGY AUTHORITY. WEAPONS GROUP ATOMIC WEAPONS RES. ESTAB.,ALDERMASTON, BERKS, ENG.	AWRE-J(NO.)/(NO.)-UK/
UNITED KINGDOM ATOMIC ENERGY AUTHORITY. WEAPONS GROUP ATOMIC WEAPONS RES. ESTAB.,ALDERMASTON, BERKS, ENG.	AWRE/LIB/BIB/
UNITED KINGDOM ATOMIC ENERGY AUTHORITY. WEAPONS GROUP ATOMIC WEAPONS RES. ESTAB.,ALDERMASTON, BERKS, ENG.	AWRE-NR-(NO.)/(YR.)
UNITED KINGDOM ATOMIC ENERGY AUTHORITY. WEAPONS GROUP ATOMIC WEAPONS RES. ESTAB.,ALDERMASTON, BERKS, ENG.	AWRE-NR/A-(NO.)/(YR.)
UNITED KINGDOM ATOMIC ENERGY AUTHORITY. WEAPONS GROUP ATOMIC WEAPONS RES. ESTAB.,ALDERMASTON, BERKS, ENG.	AWRE-NR/C-(NO.)/(YR.)
UNITED KINGDOM ATOMIC ENERGY AUTHORITY. WEAPONS GROUP ATOMIC WEAPONS RES. ESTAB.,ALDERMASTON, BERKS, ENG.	AWRE-NR/P-(NO.)/(YR.)
UNITED KINGDOM ATOMIC ENERGY AUTHORITY. WEAPONS GROUP ATOMIC WEAPONS RES. ESTAB.,ALDERMASTON, BERKS, ENG.	AWRE-O-
UNITED KINGDOM ATOMIC ENERGY AUTHORITY. WEAPONS GROUP ATOMIC WEAPONS RES. ESTAB.,ALDERMASTON, BERKS, ENG.	AWRE-O-(NO.)/(YR.)
UNITED KINGDOM ATOMIC ENERGY AUTHORITY. WEAPONS GROUP ATOMIC WEAPONS RES. ESTAB.,ALDERMASTON, BERKS, ENG.	AWRE-PAPER-A.-
UNITED KINGDOM ATOMIC ENERGY AUTHORITY. WEAPONS GROUP ATOMIC WEAPONS RES. ESTAB.,ALDERMASTON, BERKS, ENG.	AWRE-R-(NO.)/(YR.)
UNITED KINGDOM ATOMIC ENERGY AUTHORITY. WEAPONS GROUP ATOMIC WEAPONS RES. ESTAB.,ALDERMASTON, BERKS, ENG.	AWRE-SLE-(NO.)/(YR.)
UNITED KINGDOM ATOMIC ENERGY AUTHORITY. WEAPONS GROUP ATOMIC WEAPONS RES. ESTAB.,ALDERMASTON, BERKS, ENG.	AWRE-SSBME-(NO.)/(YR.)
UNITED KINGDOM ATOMIC ENERGY AUTHORITY. WEAPONS GROUP ATOMIC WEAPONS RES. ESTAB.,ALDERMASTON, BERKS, ENG.	AWRE-SSME-TN-(NO.)/(YR.)
UNITED KINGDOM ATOMIC ENERGY AUTHORITY. WEAPONS GROUP ATOMIC WEAPONS RES. ESTAB.,ALDERMASTON, BERKS, ENG.	AWRE-SSPD/USA/
UNITED KINGDOM ATOMIC ENERGY AUTHORITY. WEAPONS GROUP ATOMIC WEAPONS RES. ESTAB.,ALDERMASTON, BERKS, ENG.	AWRE-SWAN-(NO.)/(NO.)
UNITED KINGDOM ATOMIC ENERGY AUTHORITY. WEAPONS GROUP ATOMIC WEAPONS RES. ESTAB.,ALDERMASTON, BERKS, ENG.	AWRE-T-(NO.)/(YR.)
UNITED KINGDOM ATOMIC ENERGY AUTHORITY. WEAPONS GROUP ATOMIC WEAPONS RES. ESTAB.,ALDERMASTON, BERKS, ENG.	AWRE-TERN-(NO.)/(YR.)
UNITED KINGDOM ATOMIC ENERGY AUTHORITY. WEAPONS GROUP. ATOMIC WEAPONS RESEARCH ESTABLISHMENT, ALDERMASTON, BERKS, ENGLAND (THEORETICAL PHYSICS NOTE)	AWRE-TPN-
UNITED KINGDOM ATOMIC ENERGY AUTHORITY. WEAPONS GROUP ATOMIC WEAPONS RES. ESTAB.,ALDERMASTON, BERKS, ENGLAND (THEORETICAL PHYSICS NOTE)	AWRE-TPN-(NO.)/(YR.)
UNITED KINGDOM ATOMIC ENERGY AUTHORITY. WEAPONS GROUP ATOMIC WEAPONS RES. ESTAB.,ALDERMASTON, BERKS, ENG.	AWRE-TRANS-
UNITED KINGDOM ATOMIC ENERGY AUTHORITY. WEAPONS GROUP ATOMIC WEAPONS RES. ESTAB.,ALDERMASTON, BERKS, ENG.	CFRSWP/P(YEAR)(NUMBER)
UNITED KINGDOM ATOMIC ENERGY AUTHORITY. WEAPONS GROUP ATOMIC WEAPONS RES. ESTAB.,ALDERMASTON, BERKS, ENG.	CNR/PR/
UNITED KINGDOM ATOMIC ENERGY AUTHORITY. WEAPONS GROUP ATOMIC WEAPONS RES. ESTAB.,ALDERMASTON, BERKS, ENG.	COS-
UNITED KINGDOM ATOMIC ENERGY AUTHORITY. WEAPONS GROUP ATOMIC WEAPONS RES. ESTAB.,ALDERMASTON, BERKS, ENG.	CWD/
UNITED KINGDOM ATOMIC ENERGY AUTHORITY. WEAPONS GROUP ATOMIC WEAPONS RES. ESTAB.,ALDERMASTON, BERKS, ENG.	EIVR-
UNITED KINGDOM ATOMIC ENERGY AUTHORITY. WEAPONS GROUP ATOMIC WEAPONS RES. ESTAB.,ALDERMASTON, BERKS, ENG. (EXPLOSIVES RESEARCH NOTE)	FRN-(NUMBER)/(YEAR)
UNITED KINGDOM ATOMIC ENERGY AUTHORITY. WEAPONS GROUP ATOMIC WEAPONS RES. ESTAB.,ALDERMASTON, BERKS, ENG. (FINE MACHINING PANEL)	FMP/IMOG/
UNITED KINGDOM ATOMIC ENERGY AUTHORITY. WEAPONS GROUP ATOMIC WEAPONS RES. ESTAB.,ALDERMASTON, BERKS, ENG.	GRO/(NUMBER)/(NUMBER)/
UNITED KINGDOM ATOMIC ENERGY AUTHORITY. WEAPONS GROUP ATOMIC WEAPONS RES. ESTAB.,ALDERMASTON, BERKS, ENG.	NR/P-
UNITED KINGDOM ATOMIC ENERGY AUTHORITY. WEAPONS GROUP ATOMIC WEAPONS RES. ESTAB.,ALDERMASTON, BERKS, ENG.	O-(NUMBER)/(NUMBER)
UNITED KINGDOM ATOMIC ENERGY AUTHORITY. WEAPONS GROUP ATOMIC WEAPONS RES. ESTAB.,ALDERMASTON, BERKS, ENG.	PR/OP-
UNITED KINGDOM ATOMIC ENERGY AUTHORITY. WEAPONS GROUP ATOMIC WEAPONS RES. ESTAB.,ALDERMASTON, BERKS, ENG.	REF-GRO-
UNITED KINGDOM ATOMIC ENERGY AUTHORITY. WEAPONS GROUP ATOMIC WEAPONS RES. ESTAB.,ALDERMASTON, BERKS, ENG.	SRM-MEMO-
UNITED KINGDOM ATOMIC ENERGY AUTHORITY. WEAPONS GROUP ATOMIC WEAPONS RES. ESTAB.,ALDERMASTON, BERKS, ENG.	SSBME-(MONTH/YEAR)
UNITED KINGDOM ATOMIC ENERGY AUTHORITY. WEAPONS GROUP ATOMIC WEAPONS RES. ESTAB.,ALDERMASTON, BERKS, ENG.	SSCD-

UNITED KINGDOM ATOMIC ENERGY AUTHORITY. WEAPONS GROUP ATOMIC WEAPONS RES. ESTAB.,ALDERMASTON, BERKS, ENG.	SSME-TN-(MONTH/YEAR)
UNITED KINGDOM ATOMIC ENERGY AUTHORITY. WEAPONS GROUP ATOMIC WEAPONS RES. ESTAB.,ALDERMASTON, BERKS, ENG.	SSNR/USA/(YEAR)/
UNITED KINGDOM ATOMIC ENERGY AUTHORITY. WEAPONS GROUP ATOMIC WEAPONS RES. ESTAB.,ALDERMASTON, BERKS, ENG.	SSPD/USA/
UNITED KINGDOM ATOMIC ENERGY AUTHORITY. WEAPONS GROUP ATOMIC WEAPONS RES. ESTAB.,ALDERMASTON, BERKS, ENG.	TPN-(NUMBER)/(YEAR)
UNITED KINGDOM ATOMIC ENERGY AUTHORITY. WEAPONS GROUP ATOMIC WEAPONS RES. ESTAB.,ALDERMASTON, BERKS, ENG.	USREQ-
UNITED KINGDOM ATOMIC ENERGY AUTHORITY. WEAPONS GROUP ATOMIC WEAPONS RES. ESTAB., FOULNESS, ESSEX, ENG.	SAFEX-
UNITED KINGDOM CHEMICAL ENG. TEAM, CHALK RIVER, ONT.	XM-
UNITED KINGDOM STORES AND CLOTHING RESEARCH AND DEVELOPMENT ESTABLISHMENT, COLCHESTER, ESSEX, ENG.	UK-
UNITED NATIONS. SCIENTIFIC COMMITTEE ON THE EFFECTS OF ATOMIC RADIATION	A/AC.82/INF-
UNITED NATIONS. SCIENTIFIC COMMITTEE ON THE EFFECTS OF ATOMIC RADIATION	SCEAR/PANEL-
UNITED NATIONS. SECRETARIAT	A/AC.82/G/1.
UNITED NATIONS ATOMIC ENERGY COMMISSION	AEC/INF/-
UNITED NATIONS EDUCATIONAL, SCI. + CULTURAL ORG.	UNESCO-
UNITED NATIONS GENERAL ASSEMBLY, N.Y.C.	A/AC.82/G/L.-
UNITED NATIONS GENERAL ASSEMBLY, N.Y.C.	A/AC.82/G/R.-
UNITED NATIONS GENERAL ASSEMBLY, N.Y.C.	A/AC.82/R.-
UNITED NATIONS GENERAL ASSEMBLY, N.Y.C.	A/RES/
UNITED NATIONS GENERAL ASSEMBLY, N.Y.C.	UN-
UNITED NUCLEAR CORP., ELMSFORD, N.Y.	NDA-(NUMBER)-
UNITED NUCLEAR CORP., ELMSFORD, N.Y.	NDA-(NUMBER-LETTERS)-
UNITED NUCLEAR CORP., ELMSFORD, N.Y.	NDA-(LETTERS)-
UNITED NUCLEAR CORP., ELMSFORD, N.Y.	NDA-(MEMO)-
UNITED NUCLEAR CORP., ELMSFORD, N.Y.	UNC-
UNITED NUCLEAR CORP., WHITE PLAINS, N.Y.	MTL-
UNITED NUCLEAR CORP., WHITE PLAINS, N.Y.	NDA-
UNITED NUCLEAR CORP., WHITE PLAINS, N.Y.	NDA-(NO.)-
UNITED NUCLEAR CORP., WHITE PLAINS, N.Y.	SEEO-
UNITED NUCLEAR CORP., WHITE PLAINS, N.Y.	UNC-
UNITED NUCLEAR CORP. DEV. DIV., WHITE PLAINS, N.Y.	MCR-
UNITED NUCLEAR CORP. DEV. DIV., WHITE PLAINS, N.Y.	NDA-(NO.-LETTER)-
UNITED NUCLEAR CORP. DEV. DIV., WHITE PLAINS, N.Y.	NDA-((INITIALS)-
UNITED NUCLEAR CORP. DEV. DIV., WHITE PLAINS, N.Y.	NDA-DESE-
UNITED NUCLEAR CORP. DEV. DIV., WHITE PLAINS, N.Y.	NDA-MEMO-
UNITED NUCLEAR CORP. DEV. DIV., WHITE PLAINS, N.Y.	NDA-PHYS.-
UNITED NUCLEAR CORP. DEV. DIV., WHITE PLAINS, N.Y.	NDEO-
UNITED NUCLEAR CORP. DEV. DIV., WHITE PLAINS, N.Y.	UNC-MEMO-
UNITED NUCLEAR CORP. DEV. DIV., WHITE PLAINS, N.Y.	UNC-MTL-
UNITED NUCLEAR CORP. DEV. DIV., WHITE PLAINS, N.Y.	UNC PHYS/MATH-
UNITED NUCLEAR CORP. DEV. DIV., WHITE PLAINS, N.Y.	UNC-PP-
UNITED NUCLEAR CORP. DEV. DIV., WHITE PLAINS, N.Y.	UNC-RD-
UNITED NUCLEAR CORP. DEV. DIV., WHITE PLAINS, N.Y.	UNUCOR-
UNITED NUCLEAR CORP. RESEARCH AND ENGINEERING CENTER, ELMSFORD, N.Y. (FORMAL R + D REPORT)	UNC-(NUMBER)-
UNITED NUCLEAR CORP. RESEARCH AND ENGINEERING CENTER, ELMSFORD, N.Y. (INTERNAL MEMORANDA REPORT)	UNC-(LETTERS-NUMBER)-
UNITED NUCLEAR CORP. RESEARCH AND ENGINEERING CENTER, ELMSFORD, N.Y. (EXTERNAL MEMORANDA REPORT)	UNC-(MEMO)-
UNITED RESEARCH INC., CAMBRIDGE, MASS.	URI-
UNITED RESEARCH SERVICES, INC., BURLINGAME, CALIF.	URS B(NUMBER)-
UNITED RESEARCH SERVICES, INC., BURLINGAME, CALIF.	URS/CR-
UNITED RESEARCH SERVICES, INC., BURLINGAME, CALIF.	USR-
(SEE ALSO LATER NAME: URS RESEARCH CO.)	
UNITED SHOE MACHINERY CORP., BEVERLY, MASS.	USMC-
UNITED SHOE MACHINERY CORP., BEVERLY, MASS.	USMC-ND-
UNITED STATES CONSULATE GENERAL. SCIENCE OFFICE, FRANKFURT AM MAIN (GERMAN SCIENCE BULLETIN)	GSB-
UNITED STATES EMBASSY, PARIS	AE, PARIS-
UNITED STATES-EURATOM FAST REACTOR EXCHANGE PROGRAM	EURFNR-
UNITED STATES-EURATOM JOINT RESEARCH + DEV. PROGRAM	CEND-
UNITED STATES-EURATOM JOINT RESEARCH + DEV. PROGRAM	EURAEC-
UNITED STATES-EURATOM JOINT RESEARCH + DEV. PROGRAM	NCSC-
UNITED STATES-EURATOM JOINT RESEARCH + DEV. PROGRAM	UCN-
UNITED STATES-EURATOM JOINT RESEARCH + DEV. PROGRAM	WCAP-
UNITED STATES LAKE SURVEY, DETROIT	BD-
UNITED STATES LAKE SURVEY, DETROIT	BD-(NUMBER)-
UNITED STATES LAKE SURVEY, DETROIT	MP-(YEAR)-
UNITED STATES LAKE SURVEY, DETROIT	RR-(NUMBER)-
UNITED STATES LEGATION, BERNE, SWITZERLAND	ALBS-
UNITED STATES NATIONAL COMMITTEE, UNION RADIOSCIENTIFIQUE INTERNATIONALE, WASHINGTON, D.C.	URSI-
UNITED STATES PHOSPHORIC PRODUCTS DIV., TENNESSEE CORP., TAMPA, FLA.	USPP-
UNITED STATES RUBBER CO., N.Y.C.	USRC-
UNITED STATES RUBBER CO., N.Y.C.	USRC-QPR-
UNITED STATES RUBBER CO., N.Y.C.	USRC-SAR-
UNITED STATES RUBBER CO. RESEARCH CENTER, WAYNE, N.J.	RC-(NUMBER)-(NUMBER)-
UNITED STATES STEEL CORP. APPLIED RESEARCH LAB., MONROEVILLE, PENNA.	ARL-B-
UNITED STATES STEEL CORP. APPLIED RESEARCH LAB., MONROEVILLE, PENNA.	ARL-S-
UNITED STATES STEEL CORP. APPLIED RESEARCH LAB., MONROEVILLE, PENNA.	RE-(YEAR)-
UNITED STATES STEEL CORP. HOMESTEAD DISTRICT WORKS, PITTSBURGH (INTERIM ENGINEERING REPORT)	USS-HDW-IER-
UNITED STATES STEEL CORP. RES. CTR., MONROEVILLE, PA.	USS-
UNITED STATES STEEL CORP. RESEARCH LABS., KEARNY, N.J.	USS-
UNITED STATES TESTING CO., HOBOKEN, N.J.	USTC-
UNITED STATES UNDERSEAS CABLE CORP., WASHINGTON, D.C.	ETR-

UNITED STATES-UNITED KINGDOM JOINT WORKING GROUP	JOWOG-(NUMBER)/PR-
UNITED STATES-UNITED KINGDOM JOINT WORKING GROUP	JOWOG-(NUMBER)/(NUMBER)
UNITED STATES-UNITED NATIONS (REPORT SERIES)	USUN RS-
UNITED STEEL COMPANIES, LTD., SHEFFIELD, YORKS, ENG.	RDD-TRANS-
UNITED STEEL COMPANIES, LTD., SHEFFIELD, YORKS, ENG.	USCL-T-
UNITED TECHNOLOGY CENTER, SUNNYVALE, CALIF.	ER-UTC-(YEAR)-
UNITED TECHNOLOGY CENTER, SUNNYVALE, CALIF.	UTC-
UNITED TECHNOLOGY CENTER, SUNNYVALE, CALIF.	UTC-(YEAR)-
UNITED TECHNOLOGY CENTER. RESEARCH AND ADVANCED TECHNOLOGY DEPT., SUNNYVALE, CALIF.	UTC-(NUMBER)-ASR-
UNITED TECHNOLOGY CENTER. RESEARCH AND ADVANCED TECHNOLOGY DEPT., SUNNYVALE, CALIF.	UTC-(NO.)-FR
UNITED TECHNOLOGY CENTER. RESEARCH AND ADVANCED TECHNOLOGY DEPT., SUNNYVALE, CALIF.	UTC-(NO.)-QPR(NO.)
UNITED TECHNOLOGY CORP., SUNNYVALE, CALIF.	ER-UTC-
UNITED TECHNOLOGY CORP., SUNNYVALE, CALIF.	UTC-
UNITED TECHNOLOGY CORP., SUNNYVALE, CALIF.	UTC-(NUMBER)-TSR-
UNITED TECHNOLOGY CORP., SUNNYVALE, CALIF.	UTC-(NUMBER)-QT-
UNITRODE CORP., WATERTOWN, MASS.	MW-(YEAR)-
UNIVAC, BLUE BELL, PENNA.	UTC-(NUMBER)-QPR-
UNIVAC, ST. PAUL, MINN.	EE-
UNIVERSAL-CYCLOPS STEEL CORP., BRIDGEVILLE, PENNA.	UCSC-
(UNIVERSAL DECIMAL CLASSIFICATION)	UDC-
UNIVERSAL MATCH CORP., ST. LOUIS	UMC-
UNIVERSAL MATCH CORP. ARMAMENT DIV., ST. LOUIS	UMC-B(YEAR)-
UNIVERSIDAD MAYOR DE SAN ANDRES, LA PAZ, BOLIVIA	CUADERNO-
UNIVERSIDADE MACKENZIE, SAO PAULO, BRAZIL. CENTRO DE RADIO-ASTRONOMIA E ASTROFISICA	CRAAM-
UNIVERSIDADE MACKENZIE, SAO PAULO, BRAZIL. GRUPO DE RADIO-ASTRONOMIA	GRAM-
UNIVERSITATEA DIN TIMISOARA, ROUMANIA	F.T.-
UNIVERSITE LOVANIUM DE LEOPOLDVILLE, KINSHASA, DEMOCRATIC REPUBLIC OF CONGO	TRICO-
UNIVERSITIES RESEARCH ASSOCIATION, INC., WASH., D.C.	PB-
UNIVERSITIES RESEARCH ASSOCIATION, INC., WASH., D.C.	URA-
UNIV. COLL. OF SOUTH WALES AND MONMOUTHSHIRE, CARDIFF	ATR-
UNIVERSITY OF NOTRE DAME, IND.	NDU-
UNIVERSITY OF NOTRE DAME, IND.	NDU-TR-
UNIVERSITY OF NOTRE DAME, IND.	NDU-TR-(YEAR)-
UNIVERSITY OF NOTRE DAME, IND. DEPT. OF AEROSPACE ENG	L-
UNIVERSITY OF NOTRE DAME, IND. DEPT. OF AEROSPACE ENG	UNDAS-TR-
UNIVERSITY OF NOTRE DAME, SOUTH BEND, IND.	UND-
UNIV. OF NOTRE DAME, SOUTH BEND, IND. LOBUND INST.	L-
UNIV. OF NOTRE DAME, SOUTH BEND, IND. LOBUND INST.	LOBUND-AEC RPT. NO.-
UNIVERSITY OF SOUTHERN CALIFORNIA, LOS ANG. (REF. 1)	(CONTRACT CODE NO.)-(NOS)
UNIVERSITY OF SOUTHERN CALIFORNIA, LOS ANGELES	EE-
UNIVERSITY OF SOUTHERN CALIFORNIA, LOS ANGELES	SAPR-
UNIVERSITY OF SOUTHERN CALIFORNIA, LOS ANGELES	SOCAL-
UNIVERSITY OF SOUTHERN CALIFORNIA, LOS ANGELES	USC-
UNIVERSITY OF SOUTHERN CALIFORNIA, LOS ANGELES	USC-(NUMBERS)-P-
UNIVERSITY OF SOUTHERN CALIFORNIA, LOS ANGELES	USCAL-
UNIVERSITY OF SOUTHERN CALIFORNIA, LOS ANGELES. DEPT. OF AEROSPACE ENGINEERING	USCAE-
UNIVERSITY OF SOUTHERN CALIFORNIA, LOS ANGELES. DEPT. OF ELECTRICAL ENGINEERING	USCEE-
UNIVERSITY OF SOUTHERN CALIFORNIA, LOS ANGELES. DEPT. OF GEOLOGY	USC-GEOL-(YEAR)-
UNIVERSITY OF SOUTHERN CALIFORNIA, LOS ANGELES. DEPT. OF INDUSTRIAL + SYSTEMS ENGINEERING	USCISE-
UNIVERSITY OF SOUTHERN CALIFORNIA, LOS ANGELES. DEPT. OF PHYSICS	TR-USC-VACUV-
UNIVERSITY OF SOUTHERN CALIFORNIA, LOS ANGELES. DEPT. OF PHYSICS	USC-(CONTRACT NO.)-
UNIVERSITY OF SOUTHERN CALIFORNIA, LOS ANGELES. DEPT. OF PHYSICS	USC-VACUV-
UNIVERSITY OF SOUTHERN CALIF., LOS ANGELES. ENG. CTR.	USCAL-EC-
UNIVERSITY OF SOUTHERN CALIF., LOS ANGELES. ENG. CTR.	USC-CSPR-
UNIVERSITY OF SOUTHERN CALIF., LOS ANGELES. ENG. CTR.	USCEC-
UNIVERSITY OF SOUTHERN CALIF., LOS ANGELES. ENG. CTR.	USCEC-R-
UNIVERSITY OF SOUTHERN CALIF., LOS ANGELES. ENG. CTR.	USCEE-
UNIVERSITY OF THE PACIFIC, SAN FRANCISCO. DEPT. OF PHARMACOLOGY	UOP-DP-
UNIVERSITY OF WESTERN ONTARIO, LONDON	TP-
UNIVERSITY OF WESTERN ONTARIO, LONDON	UWO-
UNIVERSITY OF WESTERN ONTARIO, LONDON	WOU-
UNIVERSITY OF THE WITWATERSRAND, JOHANNESBURG. DEPT. OF MECHANICAL ENGINEERING	WITWATERSRAND-
UNIWERSYTET JAGIELLONSKI, KRAKOW. DEPT. OF THEORETICAL PHYSICS	TPJU-(NUMBER/YEAR)
UNIWERSYTET JAGIELLONSKI, KRAKOW. DEPT. OF THEORETICAL PHYSICS	TPJU-
UPPSALA. UNIV.	TF-
UPPSALA. UNIV.	UPPSALA-
UPPSALA. UNIV.	UPPSALA-TN-
UPPSALA. UNIV. COSMIC RAY GROUP	MO-(2 DIGITS)
UPPSALA. UNIV. INST. OF PHYSICS	UUIP-
UPPSALA. UNIV. QUANTUM CHEMISTRY GROUP	UUIP-TN-
URBAN LAND INST., WASHINGTON, D.C.	ULI-TB-

URBAN LAND RESEARCH ANALYSTS CO.

URBAN LAND RESEARCH ANALYSTS CO., BOSTON	ULRAC-MONO-
URBAN MASS TRANSPORTATION ADMINISTRATION, WASH.,D.C.	UMTA-DC-MTD-
URS CORP., BURLINGAME, CALIF.	URS-FD-
URS RESEARCH CO., BURLINGAME, CALIF.	URS-
URS RESEARCH CO., SAN MATEO, CALIF.	URS-
UTAH. STATE DEPT. OF HIGHWAYS. RESEARCH SECTION	UTAH RR-(NUMBER)-
UTAH. UNIV., SALT LAKE CITY	EC-
UTAH. UNIV., SALT LAKE CITY	MD-
UTAH. UNIV., SALT LAKE CITY	MDL-
UTAH. UNIV., SALT LAKE CITY	UNIV/UU-
UTAH. UNIV., SALT LAKE CITY	UTAH-
UTAH. UNIV., SALT LAKE CITY	UTAH-PR-
UTAH. UNIV., SALT LAKE CITY (TECHNICAL REPORT)	UTAH-TR-
UTAH. UNIV., SALT LAKE CITY	UTEC-MC-
UTAH. UNIV., SALT LAKE CITY	UTU-
UTAH. UNIV., SALT LAKE CITY	UU-
UTAH. UNIV., SALT LAKE CITY	UUP-
UTAH. UNIV., SALT LAKE CITY	UUT-
UTAH. UNIV., SALT LAKE CITY. COLL. OF ENGINEERING	THEMIS-UTEC-DO-
UTAH. UNIV., SALT LAKE CITY. COLL. OF ENGINEERING	UTEC-DC-(YEAR)-
UTAH. UNIV., SALT LAKE CITY. COLL. OF ENGINEERING	UTEC-DO-
UTAH. UNIV., SALT LAKE CITY. COLL. OF ENG.(MECH. ENG)	UTEC-ME-(YEAR)-
UTAH. UNIV., SALT LAKE CITY. COLL. OF ENGINEERING	UTEC-MR-(YEAR)-
UTAH. UNIV., SALT LAKE CITY. DEPT. OF CHEMISTRY	UUCD-
UTAH. UNIV., SALT LAKE CITY. DEPT. OF ELEC. ENG.	UTEC-MD-
UTAH. UNIV., SALT LAKE CITY. DIV. OF MATERIALS SCIENCE AND ENGINEERING	UTEC-MSE-(YEAR)-
UTAH. UNIV., SALT LAKE CITY. ECOLOGY AND EPIZOOLOGY RESEARCH GROUP	ECOLOGY AND EPIZOOLOGY-
UTAH. UNIV., SALT LAKE CITY. INST. FOR THE STUDY OF RATE PROCESSES	UTAH-SR-
UTAH. UNIV., SALT LAKE CITY. INST. FOR THE STUDY OF RATE PROCESSES	UTAH-TM-
UTAH. UNIV., SALT LAKE CITY. INST. FOR THE STUDY OF RATE PROCESSES	UTAH U TR-
UTAH. UNIV., SALT LAKE CITY. INST. FOR THE STUDY OF RATE PROCESSES	UU ISR TR-
UTAH. UNIV., SALT LAKE CITY. INST. OF MATERIALS RES.	UTEC-IMR-(YEAR)-
UTAH. UNIV., SALT LAKE CITY. INST. OF METALS AND EXPLOSIVES RESEARCH	IMER-CONTROL-(NO.)-
UTAH. UNIV., SALT LAKE CITY. MICROWAVE DEVICE AND PHYSICAL ELECTRONICS LAB.	TR-MDL-
UTAH. UNIV., SALT LAKE CITY. MICROWAVE DEVICE AND PHYSICAL ELECTRONICS LAB.	UTEC-MD-(YEAR)-
UTAH. UNIV., SALT LAKE CITY. MICROWAVE DEVICES LAB.	MDL-
UTAH. UNIV., SALT LAKE CITY. MICROWAVE DEVICES LAB. (QUARTERLY REPORT)	MDL-Q(NUMBER)
UTAH. UNIV., SALT LAKE CITY. RADIOBIOLOGY LAB.	UUT-RL-
UTAH. UNIV., SALT LAKE CITY. SOLID ROCKET STRUCTURAL INTEGRITY INFORMATION CENTER	UTEC-SI-(YEAR)-
UTAH. UNIV., SALT LAKE CITY. UPPER AIR RES. LABS.	UARL-UU-(YEAR)-
UTAH. UNIV., SALT LAKE CITY. UPPER AIR RES. LABS.	UU(YEAR)-
UTAH CENTER FOR WATER RESOURCES RESEARCH, LOGAN	CWRR-(NUMBER-NUMBER)
UTAH STATE UNIV., LOGAN	MPR-
UTAH STATE UNIV., LOGAN	USU-

VACCO INDUSTRIES, SOUTH EL MONTE, CALIF.	VACCO-
VACUUM-ELECTRONIC ENGINEERING CO., BROOKLYN, N.Y.	VEECO-
VALUE ENGINEERING CO., ALEXANDRIA, VA.	VEC-(NO.)-RD-
VANDERBILT UNIV., NASHVILLE	APR-
VARIAN ASSOCIATES, BEVERLY, MASS.	NR-
VARIAN ASSOCIATES, PALO ALTO, CALIF.	VA-
VARIAN ASSOCIATES, PALO ALTO, CALIF.	VA ENGINEERING RPT.-
VARIAN ASSOCIATES, PALO ALTO, CALIF.	VA-ER-
VARIAN ASSOCIATES, PALO ALTO, CALIF.	VAR-ER-
VARIAN ASSOCIATES, PALO ALTO, CALIF.	VARIAN-
VARIAN ASSOCIATES, PALO ALTO, CALIF. (ENG. REPT.)	VER-
VARIAN ASSOCIATES, SAN CARLOS, CALIF.	VAR-
VARIAN ASSOCIATES. CALIFORNIA AVENUE OPERATIONS, PALO ALTO, CALIF.	(YEAR)-CAO-
VARIAN ASSOCIATES. CALIFORNIA AVENUE OPERATIONS, PALO ALTO, CALIF.	(YEAR)-CAO-(NUMBER)-
VARIAN ASSOCIATES. EIMAC DIV., SAN CARLOS, CALIF.	EIMAC-
VARIAN ASSOCIATES. EIMAC DIV., SAN CARLOS, CALIF.	EIMAC-QR-
VARO, INC., GARLAND, TEX.	EP-F(NUMBER)(LTR.)-(YR)-
VARO, INC., GARLAND, TEX.	EP-F(NUMBER)/I/
VARO, INC., GARLAND, TEX.	PS-
VEDA, INC., ANN ARBOR, MICH.	V-
VEDA, INC., ANN ARBOR, MICH.	V-(NUMBER)U/-
VEDA, INC., ANN ARBOR, MICH.	VR-
VEHICLE RESEARCH CORP., PASADENA, CALIF.	VRC-
VEREINIGTE FLUGTECHNISCHE WERKE G.M.B.H., BREMEN	VFW-H(NUMBER)
VERMONT. UNIV., BURLINGTON	VERU-
VERMONT. UNIV., BURLINGTON	VTU-
VERMONT. UNIV., BURLINGTON	VU-TR-
VERTEX CORP., KENSINGTON, MD.	VERTEX-TR-
VERTEX CORP., KENSINGTON, MD.	VERTEX-TR-(YEAR)-
VERTOL AIRCRAFT CORP., MORTON, PENNA.	VAC-
VETERANS ADMINISTRATION, WASHINGTON, D.C.	IB-
VETERANS ADMINISTRATION, WASHINGTON, D.C.	VA-
VETERANS ADMINISTRATION. PROSTHETIC AND SENSORY AIDS SERVICE, WASHINGTON, D.C.	BPR-
VETERANS ADMINISTRATION HOSPITAL, RICHMOND, VA.	VAH-
VICKERS-ARMSTRONGS LTD., LONDON	VGW/BB-
VICKERS-ARMSTRONGS LTD., LONDON	VGW/RD-
VICKERS-ARMSTRONGS LTD. WEYBRIDGE WORKS, SURREY, ENG.	VTO/M/-
VICTOREEN INSTRUMENT CO., CLEVELAND	VIC-
VICTORIA, B.C. UNIV.	VICTORIAU-
VICTORIA, B.C. UNIV.	VPN-(YEAR)-
VIDYA, INC., PALO ALTO, CALIF.	VIDYA-FR-
VIENNA. TECHNISCHE HOCHSCHULE	AR-STP-
VIENNA. TECHNISCHE HOCHSCHULE	HL-
VIENNA. TECHNISCHE HOCHSCHULE	QR-STP-(NUMBER/NUMBER)
VIENNA. TECHNISCHE HOCHSCHULE. INST. FUER HOCHFREQUENZTECHNIK	ON-(2 DIGITS)
VIENNA. TECHNISCHE HOCHSCHULE. INST. OF INORGANIC AND GENERAL CHEMISTRY	TN-
VIENNA. UNIVERSITAT	VIENNAU-TH/
VIRGINIA. UNIV., CHARLOTTESVILLE	A-
VIRGINIA. UNIV., CHARLOTTESVILLE	AC-
VIRGINIA. UNIV., CHARLOTTESVILLE	ARC-
VIRGINIA. UNIV., CHARLOTTESVILLE	CAL-
VIRGINIA. UNIV., CHARLOTTESVILLE	DEL-
VIRGINIA. UNIV., CHARLOTTESVILLE	EE-
VIRGINIA. UNIV., CHARLOTTESVILLE	HUM-
VIRGINIA. UNIV., CHARLOTTESVILLE	JHU-
VIRGINIA. UNIV., CHARLOTTESVILLE	MICH-
VIRGINIA. UNIV., CHARLOTTESVILLE	MMPP-
VIRGINIA. UNIV., CHARLOTTESVILLE	SRI-
VIRGINIA. UNIV., CHARLOTTESVILLE	TR-AC-
VIRGINIA. UNIV., CHARLOTTESVILLE	TR-SRI-
VIRGINIA. UNIV., CHARLOTTESVILLE	UVA-
VIRGINIA. UNIV., CHARLOTTESVILLE	UVC-
VIRGINIA. UNIV., CHARLOTTESVILLE	VAU-
VIRGINIA. UNIV., CHARLOTTESVILLE	VIRGINIA U-EE-(NUMBERS)U
VIRGINIA. UNIV., CHARLOTTESVILLE	VU-
VIRGINIA. UNIV., CHARLOTTESVILLE. COBB CHEMICAL LAB.	CCL-
VIRGINIA. UNIV., CHARLOTTESVILLE. DEPT. OF AEROSPACE ENG	MIT-
VIRGINIA. UNIV., CHARLOTTESVILLE. DEPT. OF AEROSPACE ENGINEERING AND ENGINEERING PHYSICS	AEEP-(NO.)-(NO.)-(YR.)
VIRGINIA. UNIV., CHARLOTTESVILLE. DEPT. OF MATERIALS SCIENCE	MS-
VIRGINIA. UNIV., CHARLOTTESVILLE. DEPT. OF PHYSICS	VAU-PPR-
VIRGINIA. UNIV., CHARLOTTESVILLE. DIV. OF NUCLEAR ENG	NE-(NUMBER)-(NUMBER)-
VIRGINIA. UNIV., CHARLOTTESVILLE. ENG. EXPERIMENT STA	UV-C-
VIRGINIA. UNIV., CHARLOTTESVILLE. INDUSTRIAL RESEARCH AND DEVELOPMENT CENTER	IRDC-

VIRGINIA. UNIV., CHARLOTTESVILLE. ORDNANCE RES. LAB.	ORL/UVA-
VIRGINIA. UNIV., CHARLOTTESVILLE. ORDNANCE RES. LAB.	UVA-ORL-
VIRGINIA. UNIV., CHARLOTTESVILLE. ORDNANCE RES. LAB.	UVA/ORL-(NO.)-12RD-59-QPR
VIRGINIA. UNIV., CHARLOTTESVILLE. ORDNANCE RES. LAB.	UVA/ORL-(NO.)-(YR.)-TR-
VIRGINIA. UNIV., CHARLOTTESVILLE. ORDNANCE RES. LAB.	UV/ORL-
VIRGINIA. UNIV., CHARLOTTESVILLE. RESEARCH LABS. FOR ENGINEERING SCIENCES	AST-(NUMBER-NUMBER)-
VIRGINIA. UNIV., CHARLOTTESVILLE. RESEARCH LABS. FOR ENGINEERING SCIENCES	CE-
VIRGINIA. UNIV., CHARLOTTESVILLE. RESEARCH LABS. FOR ENGINEERING SCIENCES	EE-(NO.)-(NO.)-(YR.LTR.)
VIRGINIA. UNIV., CHARLOTTESVILLE. RESEARCH LABS. FOR ENGINEERING SCIENCES	EME-(NO-NO)-(YR.CLASS)-
VIRGINIA. UNIV., CHARLOTTESVILLE. RESEARCH LABS. FOR ENGINEERING SCIENCES	EMI-
VIRGINIA. UNIV., CHARLOTTESVILLE. RESEARCH LABS. FOR ENGINEERING SCIENCES	EP-
VIRGINIA. UNIV., CHARLOTTESVILLE. RESEARCH LABS. FOR ENGINEERING SCIENCES	MS-
VIRGINIA. UNIV., CHARLOTTESVILLE. RESEARCH LABS. FOR ENGINEERING SCIENCES	NE-AEC-
VIRGINIA. UNIV., CHARLOTTESVILLE. SCHOOL OF ENG.	UVAR-
VIRGINIA INST. FOR SCIENTIFIC RESEARCH, RICHMOND	VI-
VIRGINIA INST. FOR SCIENTIFIC RESEARCH, RICHMOND	VISR-
VIRGINIA POLYTECHNIC INST., BLACKSBURG	VPI-
VIRGINIA POLYTECHNIC INST., BLACKSBURG	VPI-E-
VIRGINIA POLYTECHNIC INST., BLACKSBURG	VPI-E-(YEAR)-
VIRGINIA POLYTECHNIC INST., BLACKSBURG	VPI-QPR-
VIRGINIA POLYTECHNIC INST., BLACKSBURG	VPI-QR-
VIRGINIA POLYTECHNIC INST., BLACKSBURG. DEPT. OF ENGINEERING MECHANICS	DEMVPI-(NUMBER-NUMBER)
VIRGINIA POLYTECHNIC INST., BLACKSBURG. WATER RESOURCES RESEARCH CENTER	VPI-WRRC-
VITRO CORP. OF AMERICA, N.Y.C.	KLX-
VITRO CORP. OF AMERICA, N.Y.C.	RDA-DC-4
VITRO CORP. OF AMERICA, N.Y.C.	VCA-
VITRO CORP. OF AMERICA, N.Y.C.	VCA-KLX-
VITRO CORP. OF AMERICA, N.Y.C.	VITRO-
VITRO CORP. OF AMERICA, N.Y.C.	VITRO-TR-
VITRO CORP. OF AMERICA. ATOMIC POWER STUDY GROUP, NYC	KLXS-
VITRO ENGINEERING CO., N.Y.C.	VEC-
VITRO ENGINEERING CO., N.Y.C.	VITRO-(NO.)-(NO.)-(YR.)
VITRO LABS., SILVER SPRING, MD.	BR-(JOB NO.).(TASK NO.)-
VITRO LABS., SILVER SPRING, MD.	NTDC TR-
VITRO LABS., SILVER SPRING, MD.	PE-
VITRO LABS., SILVER SPRING, MD.	PEB-TM-
VITRO LABS., SILVER SPRING, MD.	TN-(JOB NO.).(NO.-DATE)
VITRO LABS., SILVER SPRING, MD.	TN-(JOB NO.).(TASK NO.)
VITRO LABS., SILVER SPRING, MD.	TN-N(NUMBER)-(NUMBER)
VITRO LABS., SILVER SPRING, MD.	TN-PE-(JOB NO.-LTRS.)
VITRO LABS., SILVER SPRING, MD.	TP-(JOB NO.).(TASK NO.)-
VITRO LABS., SILVER SPRING, MD.	TR-(JOB NO.).(TASK NO.)-
VITRO LABS., SILVER SPRING, MD.	VCA-
VITRO LABS., SILVER SPRING, MD.	VCA-TR-
VITRO LABS., SILVER SPRING, MD.	VC-TR-
VITRO LABS., SILVER SPRING, MD.	VITRO-
VITRO LABS., SILVER SPRING, MD.	VL-
VITRO LABS., WEST ORANGE, N.J.	KLX-
VITRO LABS., WEST ORANGE, N.J.	V-(NO.)-(NO.)-(LETTER)
VITRO LABS., WEST ORANGE, N.J.	VCA-KLX-
VITRO LABS., WEST ORANGE, N.J.	VI-(NO.)-(NO.)-O
VITRO LABS., WEST ORANGE, N.J.	VITRO-
VITRO LABS., WEST ORANGE, N.J.	VL-
VITRO LABS., WEST ORANGE, N.J.	VL-(NUMBER)-(NUMBER)-O
VITRO LABS., WEST ORANGE, N.J.	VL-TR-OF (MO./DAY/YR.)
VITRO LABS., WEST ORANGE, N.J.	VO-(NO.)-(NO.)-O
VON KARMAN INSTITUTE FOR FLUID DYNAMICS, RHODESAINT-GENESE, BELGIUM	ARL-
VON KARMAN INSTITUTE FOR FLUID DYNAMICS, RHODESAINT-GENESE, BELGIUM	IN-
VON KARMAN INSTITUTE FOR FLUID DYNAMICS, RHODESAINT-GENESE, BELGIUM	VIK-TN-
VON KARMAN INSTITUTE FOR FLUID DYNAMICS, RHODESAINT-GENESE, BELGIUM	VKI-CN-
VON KARMAN INSTITUTE FOR FLUID DYNAMICS, RHODESAINT-GENESE, BELGIUM	VKI-CR-
VON KARMAN INSTITUTE FOR FLUID DYNAMICS, RHODESAINT-GENESE, BELGIUM	VKI-LS-
VON KARMAN INSTITUTE FOR FLUID DYNAMICS, RHODESAINT-GENESE, BELGIUM	VKI-TM-
VON KARMAN INSTITUTE FOR FLUID DYNAMICS, RHODESAINT-GENESE, BELGIUM	VKI-TN-
VOORHEES (ALAN M.) AND ASSOCIATES, INC., MC LEAN, VA.	AMV-R-
VOUGHT AERONAUTICS. DIV. OF CHANCE VOUGHT CORP., DALLAS	E(NO.)(LETTER)-
VOUGHT ASTRONAUTICS. DIV. OF CHANCE VOUGHT CORP., DALLAS	CVA-AST-
VOUGHT ASTRONAUTICS. DIV. OF CHANCE VOUGHT CORP., DALLAS	VA-AST-
VOUGHT ELECTRONICS. DIV. OF CHANCE VOUGHT CORP., DALLAS	CVA-VE-
VOUGHT ELECTRONICS. DIV. OF CHANCE VOUGHT CORP., DALLAS	VE-
VOUGHT RANGE SYSTEMS. DIV. OF CHANCE VOUGHT CORP., DALLAS	CVA-VRS-

WAAGNER-BIRO A.G., VIENNA	WB-KE-
WADCO CORP., RICHLAND, WASH.	WHAN-FR-
WADCO CORP., RICHLAND, WASH.	WHAN-IR-
WADCO CORP., RICHLAND, WASH.	WHAN-SA-
WAH CHANG CORP. ALBANY DIV., OREGON	WAHCC-
WAH CHANG CORP. ALBANY DIV., OREGON	WCA-
WAKE FOREST COLL., N.C.	WFC-
WALTER REED ARMY INSTITUTE OF RES., WASHINGTON, D.C.	WRAIR-
WALTER REED ARMY INSTITUTE OF RES., WASHINGTON, D.C.	WRAIR-TR-
WALTER REED ARMY MEDICAL CENTER, WASHINGTON, D.C.	WRAMC-
WAR DEPT., WASHINGTON, D.C.(MOBILIZATION REGULATIONS)	MR-
WAR DEPT., WASH., D.C. (MOBILIZATION TRNG. PROGRAM)	MTP-
WAR DEPT., WASHINGTON, D.C. (MODIFICATION WORK ORDER)	MWO-
WAR DEPT. (TECHNICAL BULLETIN)	TB-
WAR DEPT. (CHEMICAL WARFARE TECHNICAL BULLETIN)	TB CW-
WAR DEPT. (ENGINEERING TECHNICAL BULLETIN)	TB ENG-
WAR DEPT. (MEDICAL TECHNICAL BULLETIN)	TB MED-
WAR DEPT. (ORDNANCE TECHNICAL BULLETIN)	TB ORD-
WAR DEPT. (QUARTERMASTER TECHNICAL BULLETIN)	TB QM-
WAR DEPT. (SIGNAL TECHNICAL BULLETIN)	TB SIG-
WAR DEPT. (SIGNAL TECHNICAL BULLETIN)	TBSIG-
WAR DEPT., WASHINGTON, D.C. (TRAINING CIRCULAR)	TC-
WAR DEPT. (TRAJECTORY DIAGRAMS)	TD-
WAR DEPT., WASHINGTON, D.C.	TR-
WAR DEPT.	WD-
WAR DEPT. (FIELD MANUAL)	WD FM-
WAR DEPT. (FIRING TABLE)	WD FT-
WAR DEPT. (MOBILIZATION TRAINING PROGRAM)	WD MTP-
WAR DEPT. (PAMPHLET)	WD PAM-
WAR DEPT. (SUPPLY BULLETIN)	WD SB-
WAR DEPT. (TECHNICAL BULLETIN)	WD TB-
WAR DEPT. (CHEMICAL WARFARE TECHNICAL BULLETIN)	WD TB CW-
WAR DEPT. (ENGINEERING TECHNICAL BULLETIN)	WD TB ENG-
WAR DEPT. (MEDICAL TECHNICAL BULLETIN)	WD TB MED-
WAR DEPT. (QUARTERMASTER TECHNICAL BULLETIN)	WD TB QM-
WAR DEPT. (SIGNAL TECHNICAL BULLETIN)	WD TB SIG-
WAR DEPT. (SIGNAL EQUIPMENT TECHNICAL BULLETIN)	WD TB SIG E-
WAR DEPT. (TRAJECTORY DIAGRAMS)	WD TD-
WAR DEPT. (TECHNICAL MANUAL)	WD TM-
WAR DEPT. COMBINED COMMUNICATIONS BOARD (PROJECTS)	CCBP-
WAR DEPT. COMBINED COMMUNICATIONS BOARD (PUPLICATIONS)	CCBP-
WAR DEPT. COMBINED COMMUNICATIONS BOARD (PROJECTS)	WD CCBP-
WAR DEPT. MILITARY INTELLIGENCE DIV. (SPECIAL SERIES)	WD MID SS-
WAR DEPT. MILITARY INTELLIGENCE DIV.	WD MIS VII-
WAR DEPT. MILITARY INTELLIGENCE SERVICE	WD MID SC-
WAR DEPT. RESEARCH AND DEV. DIV. (TECH. REPT.)	RDDRT-
WAR DEPT. TRANSPORTATION CORPS (TECHNICAL BULLETIN)	TB TC-
WAR DEPT. TRANSPORTATION CORPS (TECHNICAL BULLETIN)	TBTC-
WAR DEPT. TRANSPORTATION CORPS (TECHNICAL BULLETIN)	WD TB TC-
WAR METALLURGY COMMITTEE	WMC-
WAR METALLURGY COMMITTEE (RESEARCH REPORTS)	WMC (LETTER(S))-
WAR METALLURGY COMMITTEE (METALS REPORTS)	WMC METALS-
WAR METALLURGY COMMITTEE (SPECIAL REPORTS)	WMC SP-
WAR PRODUCTION BOARD, WASHINGTON, D.C.	ATI-
WAR PRODUCTION BOARD, WASHINGTON, D.C.	W-
WAR PRODUCTION BOARD, WASHINGTON, D.C.	WPB-
WARD LEONARD ELECTRIC CO., MOUNT VERNON, N.Y.	WL-
WARFARE SYSTEMS SCHOOL, MAXWELL AFB, ALA.	WSS-
WARNER AND SWASEY CO., FLUSHING, N.Y.	W/S-TR-
WARNER AND SWASEY CO. CONTROL INSTR. DIV.,FLUSHING,NY	W+SC-TR-
WARNER AND SWASEY CO. CONTROL INSTR. DIV.,FLUSHING,NY	W+S-TR-
WARSAW. UNIV.	WARSAWU-
WARSAW. UNIV. INST. OF THEORETICAL PHYSICS	IFT-
WASHINGTON DOCUMENT CENTER (ARMY)	WDC-
WASHINGTON. STATE COLL., PULLMAN	SCW-
WASHINGTON. STATE COLL., PULLMAN	WSC-
WASHINGTON. STATE COLL., PULLMAN	WSC-(INITIALS)-
WASHINGTON. UNIV., SEATTLE	M(YR.)-(NO.)-
WASHINGTON. UNIV., SEATTLE	PRP-
WASHINGTON. UNIV., SEATTLE	UW-
WASHINGTON. UNIV., SEATTLE	WU-
WASHINGTON. UNIV., SEATTLE. AERONAUTICAL LAB.	UWAL-
WASHINGTON. UNIV., SEATTLE. APPLIED FISHERIES LAB.	UWFL-
WASHINGTON. UNIV., SEATTLE. APPLIED PHYSICS LAB.	APL-UW-
WASHINGTON. UNIV., SEATTLE. APPLIED PHYSICS LAB.	APL UW/TE-
WASHINGTON. UNIV., SEATTLE. APPLIED PHYSICS LAB.	UW-APL-
WASHINGTON. UNIV., SEATTLE. APPLIED PHYSICS LAB.	UWAPL-
WASHINGTON. UNIV., SEATTLE. DEPT. OF ATMOSPHERIC SCIENCES	CONTRIB-
WASHINGTON. UNIV., SEATTLE. DEPT. OF ELECTRICAL ENG.	EE-
WASHINGTON. UNIV., SEATTLE. DEPT. OF OCEANOGRAPHY	REF-M(YEAR)-
WASHINGTON. UNIV., SEATTLE. DEPT. OF OCEANOGRAPHY	WU-REF-M(YEAR)-

WASHINGTON. UNIV. FISHERIES RESEARCH INST.

WASHINGTON. UNIV., SEATTLE. FISHERIES RESEARCH INST. (WASHINGTON SEA GRANT PROGRAM)	WSG-(YEAR)-
WASHINGTON. UNIV., SEATTLE. LAB. OF RADIATION BIOLOGY	RL-
WASHINGTON. UNIV., SEATTLE. LAB. OF RADIATION BIOLOGY	UWLRB-
WASHINGTON. UNIV., SEATTLE. PSYCHOPHYSICS LAB.	PLR-
WASHINGTON COLL., CHESTERTOWN, MD. DEPT. OF CHEMISTRY	WCDC-
WASHINGTON STATE UNIV., PULLMAN	WSU-(YEAR)-
WASHINGTON STATE UNIV., PULLMAN. NUCLEAR RADIATION CTR	WSUNRC-
WASHINGTON STATE UNIV., PULLMAN. SHOCK DYNAMICS LAB.	WSU-SDL-(YEAR)-
WASHINGTON TECHNOLOGICAL ASSOCS., ROCKVILLE, MD.	WTA-
WASHINGTON UNIV., CLAYTON, MO. RESEARCH FOUNDATION	WURF-(LTRS.)-
WASHINGTON UNIV., ST. LOUIS	DC-(YEAR)-(NUMBER)-
WASHINGTON UNIV., ST. LOUIS	OR-
WASHINGTON UNIV., ST. LOUIS	WAU-
WASHINGTON UNIV., ST. LOUIS	WAU-TR-
WASHINGTON UNIV., ST. LOUIS	WU-
WASHINGTON UNIV., ST. LOUIS	WU-SL-
WASHINGTON UNIV., ST. LOUIS	WUSL-
WASHINGTON UNIV., ST. LOUIS (TECHNICAL REPORT)	WUSL TR-
WASHINGTON UNIV., ST. LOUIS. CONTROL SYSTEMS SCIENCE AND ENGINEERING	CSSE-
WASTE KING CORP., LOS ANGELES	MJ-
WATER RESOURCES ENGINEERS, INC., WALNUT CREEK, CALIF.	WLC-TN-(YEAR)-
WATER RESOURCES SCIENTIFIC INFORMATION CENTER (INTERIOR), WASHINGTON, D.C. (SELECTED WATER RESOURCES ABSTRACTS)	SWRA-(YEAR)-
WATER RESOURCES SCIENTIFIC INFORMATION CENTER (INTERIOR), WASHINGTON, D.C. (ACCESSION NUMBER)	W(YEAR)-(5 DIGITS)
WATER RESOURCES SCIENTIFIC INFORMATION CENTER (INTERIOR), WASHINGTON, D.C.	WRSIC-(YEAR)-
WATERLOO. UNIV., ONTARIO	BR-
WATERLOO. UNIV., ONTARIO	RR-
WATERTOWN ARSENAL, MASS. (ROLLED ARMOR REPORT)	ORD RAR-
WATERTOWN ARSENAL, MASS. (ROLLED ARMOR REPORT)	RAR-
WATERTOWN ARSENAL, MASS.	SURVEY IND-(NO.)/(NO.)
WATERTOWN ARSENAL, MASS.	WA-
WATERTOWN ARSENAL, MASS. (RESEARCH AND DEVELOPMENT)	WARD-
WATERTOWN ARSENAL. MATERIALS RESEARCH LAB., MASS.	MRL-
WATERTOWN ARSENAL. ORDNANCE MATERIALS RES. OFF., MASS.	MRI-
WATERTOWN ARSENAL. ORDNANCE MATERIALS RES. OFF., MASS.	OMRO-
WATERTOWN ARSENAL. ORDNANCE MATERIALS RES. OFF., MASS.	OMRO-PUB.-
WATERTOWN ARSENAL. ORDNANCE MATERIALS RES. OFF., MASS.	OMRO-PUBL-
WATERTOWN ARSENAL. ORDNANCE MATERIALS RES. OFF., MASS.	WA-OMRO-
WATERTOWN ARSENAL. RODMAN PROCESS LAB., MASS.	RPL-
WATERTOWN ARSENAL. RODMAN PROCESS LAB., MASS.	WA-RPL-
WATERTOWN ARSENAL LAB., MASS. (CAST ARMOR REPORTS)	CAR-
WATERTOWN ARSENAL LAB., MASS.	H-
WATERTOWN ARSENAL LAB., MASS.	WAL-
WATERTOWN ARSENAL LAB., MASS.	WAL-(NUMBER)/
WATERTOWN ARSENAL LAB., MASS.	WAL MR-(NUMBER)/
WATERTOWN ARSENAL LAB., MASS.	WAL-MS-
WATERTOWN ARSENAL LAB., MASS.	WALR-
WATERTOWN ARSENAL LAB., MASS.	WALR (NUMBER)/
WATERTOWN ARSENAL LAB., MASS.	WAL-TR-
WATERTOWN ARSENAL LAB., MASS.	WAL-TR-(NUMBER)/
WATERTOWN ARSENAL LAB., MASS.	WA-SB-
WATERTOWN ARSENAL LAB., MASS. (TECHNICAL BULLETINS)	WATB-
WATERVLIET ARSENAL. BENET LABS., N.Y.	RDD-
WATERVLIET ARSENAL. BENET LABS., N.Y.	WA-
WATERVLIET ARSENAL. BENET LABS., N.Y.	WATERVLIET-
WATERVLIET ARSENAL. BENET LABS., N.Y.	WVAL-
WATERVLIET ARSENAL. BENET LABS., N.Y.	WVAL MR-
WATERVLIET ARSENAL. BENET LABS., N.Y.	WVAL RDD-
WATERVLIET ARSENAL. BENET LABS., N.Y.	WVT-
WATERVLIET ARSENAL. BENET LABS., N.Y.	WVT-ID-
WATERVLIET ARSENAL. BENET LABS., N.Y. (QUALITY ASSURANCE)	WVT-QA-
WATERVLIET ARSENAL. BENET LABS., N.Y.	WVT-RI-
WATERVLIET ARSENAL. BENET LABS., N.Y.	WVT-RR-
WATERVLIET ARSENAL. BENET LABS., N.Y.	WVTRR-
WATERVLIET ARSENAL. BENET LABS., N.Y.	WVT-TR-
WATKINS-JOHNSON CO., PALO ALTO, CALIF.	W-J-
WATKINS-JOHNSON CO., PALO ALTO, CALIF.	WJ-(3 DIGITS)
WAVE PROPAGATION LAB., BOULDER, COLO.	ERL-(NUMBER)-WPL-
WAVE PROPAGATION LAB., BOULDER, COLO.	ERL-WPL-
WAVE PROPAGATION LAB., BOULDER, COLO.	ESSA-TR-ERL-(NO.)-WPL-
WAVE PROPAGATION LAB., BOULDER, COLO.	NOAA-TM-ERL-WPL-
WAVE PROPAGATION LAB., BOULDER, COLO.	NOAA-TR-ERL-(NO.)-WPL-
WAYNE STATE UNIV., DETROIT	WSU-
WAYNE STATE UNIV., DETROIT	WSU-TR-
WAYNE UNIV., DETROIT	INT.-KRS-MURA-
WAYNE UNIV., DETROIT	WAYU-
WEAPONS SYSTEMS EVALUATION GROUP. INSTITUTE FOR DEFENSE ANALYSES, WASHINGTON, D.C.	WSEG-
WEAPONS SYSTEMS EVALUATION GROUP. INSTITUTE FOR DEFENSE ANALYSES, WASHINGTON, D.C.	WSEG-MEMO-
WEAPONS SYSTEMS EVALUATION GROUP. INSTITUTE FOR DEFENSE ANALYSES, WASHINGTON, D.C. (RESEARCH MEMO.)	WSEG RM-
WEATHER BUREAU, ASHEVILLE, N.C.	WS-
WEATHER BUREAU, SALT LAKE CITY	WBTM-WR-
WEATHER BUREAU, SILVER SPRING, MD.	ESSA-TR-WB-

WEATHER BUREAU, SILVER SPRING, MD.	WBTM-DATAC-
WEATHER BUREAU, SILVER SPRING, MD.	WBTM-EDL-
WEATHER BUREAU, WASHINGTON, D.C.	MSL-
WEATHER BUREAU, WASHINGTON, D.C.	WB-
WEATHER BUREAU, WASHINGTON, D.C.	WBG-TM-
WEATHER BUREAU, WASHINGTON, D.C.	WB-TP-
WEATHER BUREAU, WASHINGTON, D.C. (TRANSLATION)	WB-TRANS-(YEAR)-
(SEE ALSO SUCCESSOR: NATIONAL WEATHER SERVICE AND ITS: ENVIRONMENTAL DATA SERVICE)	
WEATHER BUREAU. SEA-AIR INTERACTION LAB., WASH., D.C.	SAIL-
WEATHER BUREAU RESEARCH STATION, LAS VEGAS, NEV.	WBRS-
WEATHER SCIENCE, INC., NORMAN, OKLA.	WSI-WR(NUMBER)
WEATHER SERVICES, INC., BOSTON	WESI-
WEATHER SERVICES, INC., BOSTON	WSI-
WEATHER SQUADRON, 32ND, TYNDALL AFB, FLA.	RTS-
WEATHER WING (1ST), SAN FRANCISCO	1WW-SPECIAL STUDY-
WEBB INST. OF NAVAL ARCHITECTURE, GLEN COVE, N.Y.	WEBB-(NUMBER)-
WEBER AIRCRAFT CORP., BURBANK, CALIF.	ACR-(YEAR)-
WEIDLINGER (PAUL) CONSULTANTS, N.Y.C.	WEIDLINGER-TR-
WEINER ASSOCIATES, INC., BALTIMORE	WAI-TR-
WEINSCHEL ENGINEERING, KENSINGTON, MD.	WEEC-PR-
WEIZMANN INST. OF SCIENCE, REHOVOTH, ISRAEL	TSN-
WEIZMANN INST. OF SCIENCE, REHOVOTH, ISRAEL	WIS-
WELSBACH CORP. OZONE PROCESSES DIV., PHILADELPHIA	WC-
WESLEYAN UNIV., MIDDLETOWN, CONN.	WEU-
WESLEYAN UNIV., MIDDLETOWN, CONN.	WU-
WESLEYAN UNIV.,WINDSOR LOCKS,CONN. ORDNANCE RES.PROJ.	ORDWES-
WESLEYAN UNIV.,WINDSOR LOCKS,CONN. ORDNANCE RES.PROJ.	ORDWES-TN-
WEST INDIES. UNIV., KINGSTON. COMPUTING CENTER	UWI/CC (NUMBER)
WEST INDIES. UNIV., KINGSTON. DEPT. OF PHYSICS	UWI-
WEST VIRGINIA. UNIV., MORGANTOWN	WVU-
WEST VIRGINIA. UNIV.,MORGANTOWN. ENG. EXPERIMENT STA.	TECHNICAL BULL-
WEST VIRGINIA. UNIV., MORGANTOWN. MEDICAL CENTER	OSSC-S-
WEST VIRGINIA. UNIV.,MORGANTOWN. WATER RESEARCH INST.	JOURNAL-
WESTCLOX, INC., LA SALLE, ILL.	WEST-
WESTERN AIR DEFENSE FORCE, HAMILTON AFB, CALIF.	WADF-
WESTERN BIOLOGICAL LABS., CULVER CITY, CALIF.	WBL-
WESTERN CO. OF NORTH AMERICA, RICHARDSON, TEX.	NA-(YEAR)-
WESTERN CO. OF NORTH AMERICA. RES. DIV., DALLAS	RG-(NO.)-(YEAR)-N-
WESTERN ELECTRIC CO., N.Y.C.	NCF-
WESTERN ELECTRIC CO., N.Y.C.	WE-
WESTERN ELECTRIC CO., N.Y.C.	WEC-
WESTERN ELECTRIC CO., N.Y.C. (INSTRUCTION BOOK)	WE IB-
WESTERN ELECTRIC CO., N.Y.C. (INSTRUCTION BULLETIN)	WE I BUL-
WESTERN ELECTRIC CO., INC. WINSTON-SALEM, N.C.	(NUMBER)-WR-
WESTERN ENVIRONMENTAL RESEARCH LAB., LAS VEGAS, NEV.	SWRHL-
WESTERN ENVIRONMENTAL RESEARCH LAB., LAS VEGAS, NEV.	SWRHL-(NUMBER)R
WESTERN FILTER CO., INC., GARDENA, CALIF.	SO-
WESTERN GEAR CORP., LYNWOOD, CALIF.	WGC-
WESTERN GEAR CORP. SYSTEMS MANAGEMENT DIV., LYNWOOD, CALIF.	WESTERN GEAR-(NUMBER)-
WESTERN JOINT COMPUTER CONFERENCE	WJCC-
WESTERN MICROWAVE LAB., INC., SANTA CLARA, CALIF.	LI-
WESTERN MICROWAVE LAB., INC., SANTA CLARA, CALIF.	SAR-
WESTERN NEW YORK NUCLEAR RESEARCH CTR., INC., BUFFALO	WNY-
WESTERN ONTARIO UNIV., LONDON. BOUNDARY LAYER WIND TUNNEL LAB.	BLWT-(NUMBER-YEAR)
WESTERN ONTARIO UNIV., LONDON. RACETRACK MICROTRON GP	RMG-T-
WESTERN ONTARIO UNIV., LONDON. RACETRACK MICROTRON GP	TR-RMG-
WESTERN RESERVE ELECTRONICS, INC., CLEVELAND	MSS-(NUMBER)-
WESTERN RESERVE UNIV., CLEVELAND	CSL-TR-
WESTERN RESERVE UNIV., CLEVELAND	WRU-
WESTERN RESERVE UNIV.,CLEVELAND. ULTRASONIC RES. LAB.	WRU-URL-TR-
WESTERN UNION TELEGRAPH CO., N.Y.C.	WUTC-
WESTERN UNION TELEGRAPH CO., WATER MILL, N.Y.	WUT-
WESTERN UNION TELEGRAPH CO., WATER MILL, N.Y.	WUTC-
WESTFIELD COLL., LONDON	WESTFIELDC-
WESTINGHOUSE AIR BRAKE CO. UNION SWITCH AND SIGNAL DIV., PITTSBURGH	WABC-

WESTINGHOUSE DEFENSE CENTER

WESTINGHOUSE DEFENSE CENTER, BALTIMORE, MD.	TPE-
WESTINGHOUSE DEFENSE AND SPACE CENTER. SURFACE DIV., BALTIMORE	MDE-
WESTINGHOUSE ELECTRIC AND MFG. CO., E. PITTSBURGH	SM-
WESTINGHOUSE ELECTRIC AND MFG. CO., E. PITTSBURGH	WEMCO-
WESTINGHOUSE ELECTRIC AND MANUFACTURING CO., EAST PITTSBURGH (INSTRUCTION BOOK)	WEM IB-
WESTINGHOUSE ELECTRIC AND MFG. CO., PITTSBURGH	SWK-
WESTINGHOUSE ELECTRIC CORP. (BETTIS TECHNICAL REVIEW)	BT-
WESTINGHOUSE ELECTRIC CORP. (ALL LOCATIONS)	WEC-
WESTINGHOUSE ELECTRIC CORP. (ALL LOCATIONS)	WECO-
WESTINGHOUSE ELECTRIC CORP. (ALL LOCATIONS)	WESTINGHOUSE (NO.)-ML-
WESTINGHOUSE ELECTRIC CORP., BALTIMORE	A-
WESTINGHOUSE ELECTRIC CORP., BALTIMORE	AA-
WESTINGHOUSE ELECTRIC CORP., BALTIMORE	EE-
WESTINGHOUSE ELECTRIC CORP., BALTIMORE	G.O.-
WESTINGHOUSE ELECTRIC CORP., BALTIMORE	MDE-
WESTINGHOUSE ELECTRIC CORP., BALTIMORE	NY-
WESTINGHOUSE ELECTRIC CORP., BALTIMORE	PDDTM-
WESTINGHOUSE ELECTRIC CORP., BALTIMORE	SDTM-
WESTINGHOUSE ELECTRIC CORP., BALTIMORE	UTD-
WESTINGHOUSE ELECTRIC CORP., BLOOMFIELD, N.J.	NOBSR-
WESTINGHOUSE ELECTRIC CORP., EAST PITTSBURGH, PENNA.	GM-
WESTINGHOUSE ELECTRIC CORP., ELMIRA, N.Y.	MTER-
WESTINGHOUSE ELECTRIC CORP., NEWARK, N.J.	RPL-(YR.)-
WESTINGHOUSE ELECTRIC CORP., PITTSBURGH	AE-
WESTINGHOUSE ELECTRIC CORP., PITTSBURGH	(YR.)-(NO.)EO-RADIO-
WESTINGHOUSE ELECTRIC CORP., PITTSBURGH	GSE-
WESTINGHOUSE ELECTRIC CORP., PITTSBURGH	M-
WESTINGHOUSE ELECTRIC CORP., PITTSBURGH	MLR-
WESTINGHOUSE ELECTRIC CORP., PITTSBURGH	PS-AP-
WESTINGHOUSE ELECTRIC CORP., PITTSBURGH	RM-
WESTINGHOUSE ELECTRIC CORP., PITTSBURGH	SPDD-
WESTINGHOUSE ELECTRIC CORP., PITTSBURGH	WG-
WESTINGHOUSE ELECTRIC CORP. ADVANCED DESIGN AND DEVELOPMENT DEPT., KANSAS CITY, MO.	ADD-
WESTINGHOUSE ELECTRIC CORP. ADVANCED DESIGN AND DEVELOPMENT DEPT., KANSAS CITY, MO.	ADD-(YR)-
WESTINGHOUSE ELECTRIC CORP. ADVANCED SYSTEMS ENGINEERING, SUNNYVALE, CALIF.	ASE-
WESTINGHOUSE ELECTRIC CORP. AEROSPACE DIV., BALTIMORE	AAD-
WESTINGHOUSE ELECTRIC CORP. AEROSPACE DIV., BALTIMORE	CQ-
WESTINGHOUSE ELECTRIC CORP. AEROSPACE DIV., BALTIMORE	SATM-
WESTINGHOUSE ELECTRIC CORP. AEROSPACE DIV., BALTIMORE	SDTM-
WESTINGHOUSE ELECTRIC CORP. AEROSPACE ELECTRICAL DIV., LIMA, OHIO	D-
WESTINGHOUSE ELECTRIC CORP. AEROSPACE ELECTRICAL DIV., LIMA, OHIO	WAED-
WESTINGHOUSE ELECTRIC CORP. AEROSPACE TEST LAB., BALTIMORE	ATL-
WESTINGHOUSE ELECTRIC CORP. AIR ARM DIV., BALTIMORE	WEC-AAR-
WESTINGHOUSE ELECTRIC CORP. AIR ARM DIV., BALTIMORE	WEST (NO.)-(NO.)-
WESTINGHOUSE ELECTRIC CORP. AIR ARM DIV., BALTIMORE	WEST AA TN-
WESTINGHOUSE ELECTRIC CORP. ASTRONUCLEAR LAB., PITTS.	EE-
WESTINGHOUSE ELECTRIC CORP. ASTRONUCLEAR LAB., PITTS.	WANEF-
WESTINGHOUSE ELECTRIC CORP. ASTRONUCLEAR LAB., PITTS.	WANL-
WESTINGHOUSE ELECTRIC CORP. ASTRONUCLEAR LAB., PITTS.	WANL-(NUMBER)-
WESTINGHOUSE ELECTRIC CORP. ASTRONUCLEAR LAB., PITTS.	WANL-MP-
WESTINGHOUSE ELECTRIC CORP. ASTRONUCLEAR LAB., PITTS.	WANL-PD/DDD/-
WESTINGHOUSE ELECTRIC CORP. ASTRONUCLEAR LAB., PITTS.	WANL-PR(AA)-
WESTINGHOUSE ELECTRIC CORP. ASTRONUCLEAR LAB., PITTS.	WANL-PR/CCC/-
WESTINGHOUSE ELECTRIC CORP. ASTRONUCLEAR LAB., PITTS.	WANL-PR-(LL)-
WESTINGHOUSE ELECTRIC CORP. ASTRONUCLEAR LAB., PITTS.	WANL-PR/RRR/-
WESTINGHOUSE ELECTRIC CORP. ASTRONUCLEAR LAB., PITTS.	WANL-PR-(SS)-
WESTINGHOUSE ELECTRIC CORP. ASTRONUCLEAR LAB., PITTS.	WANL-SP-
WESTINGHOUSE ELECTRIC CORP. ASTRONUCLEAR LAB., PITTS.	WANL-TME-
WESTINGHOUSE ELECTRIC CORP. ASTRONUCLEAR LAB., PITTS.	WANL-TNR-
WESTINGHOUSE ELECTRIC CORP. ASTRONUCLEAR LAB., PITTS.	WANL-W-
WESTINGHOUSE ELECTRIC CORP. ASTRONUCLEAR LAB., PITTS.	WH-
WESTINGHOUSE ELECTRIC CORP. ASTRONUCLEAR LAB., PITTS.	WNAL-TME-
WESTINGHOUSE ELECTRIC CORP. ATOMIC EQUIP. DIV., CHESWICK, PENNA.	EM-
WESTINGHOUSE ELECTRIC CORP. ATOMIC POWER DEPT., PITTS.	BP-AT-
WESTINGHOUSE ELECTRIC CORP. ATOMIC POWER DEPT., PITTS.	WMR-
WESTINGHOUSE ELECTRIC CORP. ATOMIC POWER DEPT., PITTS.	YAEC-
(SEE ALSO LATER NAME: WESTINGHOUSE ELECTRIC CORP. NUCLEAR ENERGY SYSTEMS)	
WESTINGHOUSE ELECTRIC CORP. ATOMIC POWER DIV., IDAHO FALLS	S1WS-(LETTER(S))-
WESTINGHOUSE ELECTRIC CORP. ATOMIC POWER DIV., IDAHO FALLS	S(NO.)WS-(LTR/S)-
WESTINGHOUSE ELECTRIC CORP. ATOMIC POWER DIV., IDAHO FALLS	STR-IR-
WESTINGHOUSE ELECTRIC CORP. ATOMIC POWER DIV., IDAHO FALLS	STR-TOT-
WESTINGHOUSE ELECTRIC CORP. ATOMIC POWER DIV., IDAHO FALLS	W-STR-ER-
WESTINGHOUSE ELECTRIC CORP. ATOMIC POWER DIV., IDAHO FALLS	W-STR-L-
WESTINGHOUSE ELECTRIC CORP. ATOMIC POWER DIV., IDAHO FALLS	W-STR-M-
WESTINGHOUSE ELECTRIC CORP. ATOMIC POWER DIV., IDAHO FALLS	W-STR-TG-
WESTINGHOUSE ELECTRIC CORP. ATOMIC POWER DIV., PITTS.	AP-
WESTINGHOUSE ELECTRIC CORP. ATOMIC POWER DIV., PITTS.	APD-
WESTINGHOUSE ELECTRIC CORP. ATOMIC POWER DIV., PITTS.	APIC-(NO.)SM-
WESTINGHOUSE ELECTRIC CORP. ATOMIC POWER DIV., PITTS.	FR-(NUMBER)-(LETTER)
WESTINGHOUSE ELECTRIC CORP. ATOMIC POWER DIV., PITTS.	MS-M-(NUMBER)C
WESTINGHOUSE ELECTRIC CORP. ATOMIC POWER DIV., PITTS.	PD-SPEC-
WESTINGHOUSE ELECTRIC CORP. ATOMIC POWER DIV., PITTS.	T-
WESTINGHOUSE ELECTRIC CORP. ATOMIC POWER DIV., PITTS.	WL-
WESTINGHOUSE ELECTRIC CORP. ATOMIC POWER DIV., PITTS.	W-STR-CH-
WESTINGHOUSE ELECTRIC CORP. ATOMIC POWER DIV., PITTS.	W-STR-TS-
(SEE ALSO LATER NAME: WESTINGHOUSE ELECTRIC CORP. BETTIS ATOMIC POWER LAB.)	
WESTINGHOUSE ELECTRIC CORP. AVIATION GAS TURBINE DIV., KANSAS CITY, MO.	A-(NO.)(WEC)

WESTINGHOUSE ELECTRIC CORP. AVIATION GAS TURBINE DIV., KANSAS CITY, MO.	WEC-A-
WESTINGHOUSE ELECTRIC CORP. AVIATION GAS TURBINE DIV., PHILADELPHIA	A-
WESTINGHOUSE ELECTRIC CORP. BETTIS ATOMIC POWER LAB., PITTSBURGH	AP-
WESTINGHOUSE ELECTRIC CORP. BETTIS ATOMIC POWER LAB., PITTSBURGH	BETTIS DESIGNATION-
WESTINGHOUSE ELECTRIC CORP. BETTIS ATOMIC POWER LAB., PITTSBURGH	F-C-
WESTINGHOUSE ELECTRIC CORP. BETTIS ATOMIC POWER LAB., PITTSBURGH	NRFS-ER-
WESTINGHOUSE ELECTRIC CORP. BETTIS ATOMIC POWER LAB., PITTSBURGH	S-(NO.)-W(M)-
WESTINGHOUSE ELECTRIC CORP. BETTIS ATOMIC POWER LAB., PITTSBURGH	WAPD-
WESTINGHOUSE ELECTRIC CORP. BETTIS ATOMIC POWER LAB., PITTSBURGH	WAPD-MRP-
WESTINGHOUSE ELECTRIC CORP. BETTIS ATOMIC POWER LAB., PITTSBURGH	WAPD-PWR-
WESTINGHOUSE ELECTRIC CORP. BETTIS ATOMIC POWER LAB., PITTSBURGH	WAPD-T-
WESTINGHOUSE ELECTRIC CORP. BETTIS ATOMIC POWER LAB., PITTSBURGH	WAPD-TM-
WESTINGHOUSE ELECTRIC CORP. BETTIS ATOMIC POWER LAB., PITTSBURGH	WAPD-TN-
WESTINGHOUSE ELECTRIC CORP. BETTIS ATOMIC POWER LAB., PITTSBURGH	WAPD-TRANS-
WESTINGHOUSE ELECTRIC CORP. BETTIS ATOMIC POWER LAB., PITTSBURGH	WAPT-T-
WESTINGHOUSE ELECTRIC CORP. BETTIS ATOMIC POWER LAB., PITTSBURGH	WEST WAPD-
WESTINGHOUSE ELECTRIC CORP. BETTIS ATOMIC POWER LAB., PITTSBURGH	WEST WAPD-TM-
WESTINGHOUSE ELECTRIC CORP. BETTIS PLANT, PITTSBURGH	FR-(NUMBER)-(LETTER)
WESTINGHOUSE ELECTRIC CORP. BETTIS PLANT, PITTSBURGH	FR-(NO.)-B
WESTINGHOUSE ELECTRIC CORP. BETTIS PLANT, PITTSBURGH	WAPD-CTA(LETTERS)-
(SEE ALSO LATER NAME: WESTINGHOUSE ELECTRIC CORP. BETTIS ATOMIC POWER LAB.)	
WESTINGHOUSE ELECTRIC CORP. DECO COMMUNICATIONS DEPT., LEESBURG, VA.	DECO-(NUMBER)-F
WESTINGHOUSE ELECTRIC CORP. DEFENSE + SPACE CENTER, BALTIMORE	WEC-GO-HS-
WESTINGHOUSE ELECTRIC CORP. DEVELOPMENT ENG. DEPT., LESTER, PENNA.	EC-
WESTINGHOUSE ELECTRIC CORP. ELECTRONIC TUBE DIV., BALTIMORE	C-(NUMBER)(LETTER)-
WESTINGHOUSE ELECTRIC CORP. ELECTRONIC TUBE DIV., ELMIRA, N.Y.	WEET-
WESTINGHOUSE ELECTRIC CORP. ENGINEERING OPERATIONS, HUNTSVILLE, ALA.	MDE-
WESTINGHOUSE ELECTRIC CORP. HUNTSVILLE ENGINEERING OPERATIONS, ALA.	HEO-(NUMBER)-
WESTINGHOUSE ELECTRIC CORP. NAVAL REACTORS FACILITY, IDAHO FALLS	NRFA-(LETTERS)-
WESTINGHOUSE ELECTRIC CORP. NAVAL REACTORS FACILITY, IDAHO FALLS	NRFE-(LETTERS)-
WESTINGHOUSE ELECTRIC CORP. NAVAL REACTORS FACILITY, IDAHO FALLS	NRF-ER-
WESTINGHOUSE ELECTRIC CORP. NAVAL REACTORS FACILITY, IDAHO FALLS	NRFS-AT-
WESTINGHOUSE ELECTRIC CORP. NAVAL REACTORS FACILITY, IDAHO FALLS	NRFS-ER-
WESTINGHOUSE ELECTRIC CORP. NAVAL REACTORS FACILITY, IDAHO FALLS	NRFS-PR-
WESTINGHOUSE ELECTRIC CORP. NAVAL REACTORS FACILITY, IDAHO FALLS	NRFTS-(LETTERS)-
WESTINGHOUSE ELECTRIC CORP. NAVAL REACTORS FACILITY, IDAHO FALLS	NRTF-(LETTERS)-
WESTINGHOUSE ELECTRIC CORP. NUCLEAR ENERGY SYSTEMS, PITTSBURGH (COMMERCIAL ATOMIC POWER)	WCAP-
WESTINGHOUSE ELECTRIC CORP. NUCLEAR ENERGY SYSTEMS, PITTSBURGH	WCAP-AD-
WESTINGHOUSE ELECTRIC CORP. NUCLEAR ENERGY SYSTEMS, PITTSBURGH	WENE-
WESTINGHOUSE ELECTRIC CORP. NUCLEAR ENERGY SYSTEMS, PITTSBURGH	WEST-WIAP-
WESTINGHOUSE ELECTRIC CORP. NUCLEAR ENERGY SYSTEMS, PITTSBURGH (INDUSTRIAL ATOMIC POWER)	WIAP-
WESTINGHOUSE ELECTRIC CORP. NUCLEAR ENERGY SYSTEMS, PITTSBURGH	WIAP-M-
WESTINGHOUSE ELECTRIC CORP. NUCLEAR ENERGY SYSTEMS, PITTSBURGH	WIAP-NL-
WESTINGHOUSE ELECTRIC CORP. NUCLEAR ENERGY SYSTEMS, PITTSBURGH	WIAP-NM-
WESTINGHOUSE ELECTRIC CORP. NUCLEAR ENERGY SYSTEMS, PITTSBURGH	WIAP-P-
WESTINGHOUSE ELECTRIC CORP. NUCLEAR ENERGY SYSTEMS. ADVANCED REACTORS DIV., MADISON, PENNA.	WARD-(NUMBER)-
WESTINGHOUSE ELECTRIC CORP. NUCLEAR ENERGY SYSTEMS. NUCLEAR EQUIPMENT DIV., TAMPA, FLA.	WNET-
WESTINGHOUSE ELECTRIC CORP. ORDNANCE DEPT., BALTIMORE (TECHNICAL MEMORANDUM)	ODTM-
WESTINGHOUSE ELECTRIC CORP. RES. + DEV. CTR., PITTS.	(YR.)-9F(NO.)-MIOPT-R
WESTINGHOUSE ELECTRIC CORP. RES. + DEV. CTR., PITTS.	WEC-TR-
WESTINGHOUSE ELECTRIC CORP. RES. + DEV. CTR., PITTS.	WERL-
WESTINGHOUSE ELECTRIC CORP. RES. LABS., E. PITTSBURGH	GR-
WESTINGHOUSE ELECTRIC CORP. RES. LABS., E. PITTSBURGH	R-(NO.)-(NO.)-(LETTER)
WESTINGHOUSE ELECTRIC CORP. RES. LABS., E. PITTSBURGH	R-(NO.)-(LETTER)
WESTINGHOUSE ELECTRIC CORP. RES. LABS., E. PITTSBURGH	SM-(NO.)(WRL)
WESTINGHOUSE ELECTRIC CORP. RES. LABS., E. PITTSBURGH	SR-(NO.)(WRL)
WESTINGHOUSE ELECTRIC CORP. RES. LABS., E. PITTSBURGH	WRL-SP-
WESTINGHOUSE ELECTRIC CORP. RES. LABS., PITTSBURGH	CCR-
WESTINGHOUSE ELECTRIC CORP. RES. LABS., PITTSBURGH	R-(NO.)-(NO.)-(LETTER)
WESTINGHOUSE ELECTRIC CORP. RES. LABS., PITTSBURGH	R-(NO.)-(LETTER)
WESTINGHOUSE ELECTRIC CORP. RES. LABS., PITTSBURGH	RM-(YEAR)-
WESTINGHOUSE ELECTRIC CORP. RES. LABS., PITTSBURGH	RR-
WESTINGHOUSE ELECTRIC CORP. RES. LABS., PITTSBURGH	SM-(NUMBER)(WRL)
WESTINGHOUSE ELECTRIC CORP. RES. LABS., PITTSBURGH	SP-(YEAR)-
WESTINGHOUSE ELECTRIC CORP. RES. LABS., PITTSBURGH (NUMBER USED IS LAST 4 DIGITS OF CONTRACT NO.)	WERL-
WESTINGHOUSE ELECTRIC CORP. RES. LABS., PITTSBURGH	WERL-(CONTRACT NO.)-
WESTINGHOUSE ELECTRIC CORP. RES. LABS., PITTSBURGH	WRL-
WESTINGHOUSE ELECTRIC CORP. RES. LABS., PITTSBURGH	WRL-IR-
WESTINGHOUSE ELECTRIC CORP. RES. LABS., PITTSBURGH	WRL-ISR-
WESTINGHOUSE ELECTRIC CORP. RES. LABS., PITTSBURGH	WRL-PR-
WESTINGHOUSE ELECTRIC CORP. RES. LABS., PITTSBURGH	WRL-QPR-
WESTINGHOUSE ELECTRIC CORP. RES. LABS., PITTSBURGH	WRL-RR-(YR.)-(NUMBERS)-
WESTINGHOUSE ELECTRIC CORP. RES. LABS., PITTSBURGH	WRL-TR-
WESTINGHOUSE ELECTRIC CORP. SPECIAL PRODUCTS DEVELOPMENT DIV., PITTSBURGH (REF. 1)	(NUMBER)
WESTINGHOUSE ELECTRIC CORP. STEAM DIV., LESTER, PENNA.	EC-
WESTINGHOUSE ELECTRIC CORP. STEAM DIV., LESTER, PENNA.	SPSD-(NUMBER)-AFS-
WESTINGHOUSE ELECTRIC CORP. SURFACE DIV., BALTIMORE	MDE-
WESTINGHOUSE ELECTRIC CORP. SURFACE DIV., BALTIMORE	WGD-
WESTINGHOUSE ELECTRIC CORP. TESTING REACTOR, WALTZ MILL, PENNA.	WTR-
WESTINGHOUSE ELECTRIC CORP. UNDERSEAS DIV., BALTIMORE	UEM-
WESTINGHOUSE ELECTRIC CORP. UNDERSEAS DIV., BALTIMORE	UER-
WESTINGHOUSE RESEARCH LABS., PITTSBURGH	(YR.)-9F(NO.)-(NO.)-
WESTINGHOUSE RESEARCH LABS., PITTSBURGH	(YR.)-9F(NO.)-WAVES-P
WESTINGHOUSE RESEARCH LABS., PITTSBURGH	RR-(YR.)-(NO.)-(NO.)-R
WESTINGHOUSE RESEARCH LABS., PITTSBURGH	WERL-(NUMBERS)-
WESTINGHOUSE RESEARCH LABS., PITTSBURGH	WERL-(LETTERS)-

WESTINGHOUSE RESEARCH LABS.

WESTINGHOUSE RESEARCH LABS., PITTSBURGH	WERL-ELPLA-
WESTINGHOUSE RESEARCH LABS., PITTSBURGH	WERL-HOLOG-
WESTINGHOUSE RESEARCH LABS., PITTSBURGH	WERL-TAADM-
WESTINGHOUSE RESEARCH LABS. ATOMIC AND MOLECULAR SCIENCES, PITTSBURGH	(YR.)-(NO.)E2-GASES-P
WESTINGHOUSE RESEARCH LABS. INSULATION AND CHEM. TECHNOLOGY DEPT., PITTSBURGH	(YR.)-9B5-LUBER-R(NO.)
WESTINGHOUSE RESEARCH LABS. QUANTUM ELECTRONICS DEPT., PITTSBURGH	(YR.)-9C1-ARCSO-R(NO.)
WESTLAND AIRCRAFT, LTD., YEOVIL, SOMERSET, ENGLAND	TD/R/EW/
WESTON ELECTRICAL INSTRUMENT CORP., NEWARK, N.J.	WERG-
WESTON INSTRUMENTS, INC., NEWARK, N.J.	OCD-
WESTON INSTRUMENTS, INC., NEWARK, N.J.	XS-(NUMBER)-
WHEELER LABS., INC., GREAT NECK, N.Y.	WL-
WHIRLAJET, INC., LOS ANGELES	WHIRLAJET-
WHIRLPOOL CORP. EVANSVILLE ORDNANCE DEPT., IND.	EV-
WHITE-RODGERS ELECTRIC CO., ST. LOUIS	WREC-
WHITE-RODGERS ELECTRIC CO., ST. LOUIS	WREC-ER-
WHITE-RODGERS ELECTRIC CO., ST. LOUIS	WREC-IER-
WHITE SANDS MISSILE RANGE, N. MEX.	DR-
WHITE SANDS MISSILE RANGE, N. MEX.	DR/CWG-
WHITE SANDS MISSILE RANGE, N. MEX.	DRD-N-
WHITE SANDS MISSILE RANGE, N. MEX.	DRDN-
WHITE SANDS MISSILE RANGE, N. MEX. (INTEGRATED RANGE MISSION)	IRM-
WHITE SANDS MISSILE RANGE, N. MEX.	MS-
WHITE SANDS MISSILE RANGE, N. MEX. (NATL. RANGE ENG.)	NE-
WHITE SANDS MISSILE RANGE, N. MEX. (NORTH OSCURA PEAK)	NOP-
WHITE SANDS MISSILE RANGE, N. MEX. (NATIONAL RANGE)	NR-
WHITE SANDS MISSILE RANGE, N. MEX. (ORDNANCE MISSION)	OM-
WHITE SANDS MISSILE RANGE, N. MEX. (POST ENGINEER)	PE-
WHITE SANDS MISSILE RANGE, N. MEX. (PROVOST MARSHAL OFFICE)	PMO-
WHITE SANDS MISSILE RANGE, N. MEX. (RANGE INSTRUMENTATION DEVELOPMENT)	RID-
WHITE SANDS MISSILE RANGE, N. MEX.	RID-O-
WHITE SANDS MISSILE RANGE, N. MEX.	RISO-
WHITE SANDS MISSILE RANGE, N. MEX.	SELWS-
WHITE SANDS MISSILE RANGE, N. MEX.	STEWS-ID-
WHITE SANDS MISSILE RANGE, N. MEX.	T(NO.)-(NO.)-(YR.)
WHITE SANDS MISSILE RANGE, N. MEX.	WS-DRD-N-
WHITE SANDS MISSILE RANGE, N. MEX.	WS-DRDN-
WHITE SANDS MISSILE RANGE, N. MEX.	WS-FDR-
WHITE SANDS MISSILE RANGE, N. MEX.	WS-FTDR-
WHITE SANDS MISSILE RANGE, N. MEX.	WS-(MISSILE NAME)-
WHITE SANDS MISSILE RANGE, N. MEX.	WSMR-
WHITE SANDS MISSILE RANGE, N. MEX. (FINAL TECH. RPT.)	WSMR-FTR-
WHITE SANDS MISSILE RANGE, N. MEX.	WS-TM-
WHITE SANDS MISSILE RANGE. ATMOSPHERIC SCIENCES LAB., N. MEX. (DATA REPORT)	ASL/DR-
WHITE SANDS MISSILE RANGE. DATA ANALYSIS DIRECTORATE, N. MEX.	DAD-
WHITE SANDS MISSILE RANGE. DATA COLLECTION DIRECTORATE, N. MEX. (ENGINEERING REPORT)	DCD-ER-
WHITE SANDS MISSILE RANGE. ELECTRO-MECH. LABS., N.MEX.	EML-
WHITE SANDS MISSILE RANGE. ELECTRO-MECH. LABS., N.MEX.	TIB-
WHITE SANDS MISSILE RANGE. ELECTRONICS RESEARCH AND DEVELOPMENT ACTIVITY, N. MEX.	ERDA-
WHITE SANDS MISSILE RANGE. FLIGHT SIMULATION LAB., N.M.	FSL-
WHITE SANDS MISSILE RANGE. FREQUENCY MANAGEMENT DIV., N. MEX.	FMD-
WHITE SANDS MISSILE RANGE. FREQUENCY MANAGEMENT DIV., N. MEX.	FMD-CD-(YR.)-S-(NO.)-R
WHITE SANDS MISSILE RANGE. FREQUENCY MANAGEMENT DIV., N. MEX.	FMD-TM-
WHITE SANDS MISSILE RANGE. INSTRUMENTATION R/D DIRECTORATE, N. MEX.	STEWS-ID-
WHITE SANDS MISSILE RANGE. INTER-RANGE INSTRUMENTATION GROUP, N. MEX.	IRIG-(NUMBER)-(YEAR)
WHITE SANDS MISSILE RANGE. INTER-RANGE INSTRUMENTATION GROUP, N. MEX.	IRIG-
WHITE SANDS MISSILE RANGE. NATIONAL RANGE ENG., N. MEX.	RE-S-(YEAR)-
WHITE SANDS MISSILE RANGE. NATIONAL RANGE ENG., N. MEX.	STEWS-RE-I-(YR.)-I
WHITE SANDS MISSILE RANGE. OSCURA RANGE CAMP., N.MEX.	ORC-
WHITE SANDS MISSILE RANGE. RANGE COMMANDERS COUNCIL, N. MEX.	RCC-
WHITE SANDS MISSILE RANGE. SMALL MISSILE RANGE, N.MEX	SMR-
WHITE SANDS MISSILE RANGE. SYSTEMS TEST DIV., N. MEX.	STD-
WHITE SANDS MISSILE RANGE. WARHEADS AND SPECIAL WEAPONS LAB., N. MEX.	WSWL-
WHITE SANDS PROVING GROUND, N. MEX.	ORDBS-
WHITE SANDS PROVING GROUND, N. MEX.	SPG-
WHITE SANDS PROVING GROUND, N. MEX.	WS-
WHITE SANDS PROVING GROUND, N. MEX.	WS-PDR-
WHITE SANDS PROVING GROUND, N. MEX.	WSPG-
WHITE SANDS PROVING GROUND, N. MEX.	WSPG-AFC-
WHITE SANDS PROVING GROUND, N. MEX.	WSPG-ARA-
WHITE SANDS PROVING GROUND, N. MEX. (ELECTROMAGNETIC RADIATION THRU THE ATMOSPHERE)	WSPG-ERTTA-
WHITE SANDS PROVING GROUND, N. MEX. (MISSILE GEOPHYSICS PROGRAM)	WSPG-MGP-
WHITE SANDS PROVING GROUND, N. MEX. (MONTHLY INTERIM PROGRESS REPORTS)	WSPG-MIPR-
WHITE SANDS PROVING GROUND, N. MEX. (SPEC. REPTS.)	WSPG-SR-
WHITE SANDS PROVING GROUND, N. MEX. (TECH. MEMOS.)	WSPG-TM-
WHITE SANDS PROVING GROUND, N. MEX. (TECH. REPTS.)	WSPG-TR-
WHITE SANDS PROVING GROUND, N. MEX.	WS-TD-
WHITE SANDS PROVING GROUND, N. MEX.	WS-TP-
WHITE SANDS PROVING GROUND, N. MEX.	WS-TR-
(SEE ALSO LATER NAME: WHITE SANDS MISSILE RANGE)	
WHITE SANDS PROVING GD. ELECTRONIC WARFARE DIV., N.MEX	EW-
WHITE SANDS PROVING GROUND. FLIGHT DETERMINATION LAB., N. MEX.	WS-ADR-

WHITE SANDS PROVING GROUND. FLIGHT DETERMINATION LAB., N. MEX.	WS-FDLHR-
WHITE SANDS PROVING GROUND. RANGE INSTRUMENTATION DEVELOPMENT DIV., N. MEX.	WSPG-RID-
WHITE SANDS PROVING GROUND. RANGE INSTRUMENTATION DEVELOPMENT DIV., N. MEX.	WS-RID-
WHITE SANDS SIGNAL AGENCY. OFFICE OF THE AREA FREQUENCY COORDINATOR, WHITE SANDS MISSILE RANGE,N.M.	AFC-(LTRS)-(DAY-MO.-YR)
WHITE SANDS SIGNAL CORPS AGENCY, WHITE SANDS MISSILE RANGE, N. MEX.	WSSA-
WHITE SANDS SIGNAL CORPS AGENCY, WHITE SANDS MISSILE RANGE, N. MEX.	WSSA-PR-
WHITE SANDS SIGNAL CORPS AGENCY, WHITE SANDS MISSILE RANGE, N. MEX.	WSSA-TM-
WHITE SANDS SIGNAL CORPS AGENCY, WHITE SANDS MISSILE RANGE, N. MEX.	WSSCA-
WHITE SANDS SIGNAL CORPS AGENCY, WHITE SANDS PROVING GROUND, N. MEX.	WS-MGP-PR-
WHITE SANDS SIGNAL CORPS AGENCY, WHITE SANDS PROVING GROUND, N. MEX.	WSSCA-PR-
WHITE SANDS SIGNAL CORPS AGENCY, WHITE SANDS PROVING GROUND, N. MEX.	WSSCA-TM-
WHITTAKER (WM. R) CO., LOS ANGELES	WHITTAKER-D-
WHITTAKER CORP. NARMCO RESEARCH AND DEVELOPMENT DIV., SAN DIEGO, CALIF.	NARMCO-
WHITTAKER CORP. NARMCO RESEARCH AND DEVELOPMENT DIV., SAN DIEGO, CALIF.	NRD-
WHITTAKER CORP. NARMCO RESEARCH AND DEVELOPMENT DIV., SAN DIEGO, CALIF.	NRD-MJO-
WHITTAKER CORP. NUCLEAR METALS DIV., WEST CONCORD, MASS.	NM1-
WHITTENBURG, VAUGHAN ASSOCS., INC., ALEXANDRIA, VA.	W/V-RR-(YR.)/(NO.)-WD
WHITTENBURG, VAUGHAN ASSOCS., INC., ALEXANDRIA, VA.	W/V-RR-(YR.)/(NO.)-CD
WICHITA, KAN. UNIV.	WICHU-
WICHITA, KAN. UNIV.	WIU-
WICHITA, KAN. UNIV. SCHOOL OF ENG. (AERODYNAMIC RPT.)	WIU-AR-
WICHITA, KAN. UNIV. SCHOOL OF ENGINEERING (ENG. RPT.)	WIU-ER-
WICHITA, KAN. UNIV. SCHOOL OF ENGINEERING	WU-AR-
WICHITA STATE UNIV., KANS. DEPT OF AERONAUTICAL ENG.	AR-(YEAR)-
WILLAMETTE IRON AND STEEL CO., PORTLAND, ORE.	WISCO-
WILLARD STORAGE BATTERY CO., CLEVELAND	WIL-
WILLARD STORAGE BATTERY CO., CLEVELAND	WSB-
WILLIAMS RESEARCH CORP., WALLED LAKE, MICH.	WR-ER(NUMBER)
WILMER INSTITUTE	WI-
WIND TUNNEL INSTRUMENT CO., INC., NEWTON, MASS.	WTICI-
WIND TURBINE CO., WEST CHESTER, PENNA.	WTC-
WINZEN RESEARCH, INC., MINNEAPOLIS	WRI-
WISCONSIN. BUREAU OF STATE PLANNING, MADISON (INFORMATION SYSTEM)	BSP-IS-(NUMBER-NUMBER)
WISCONSIN. LEGISLATIVE REFERENCE BUREAU, MADISON (INFORMATIONAL BULLETIN)	LRB-IB-(YEAR)-
WISCONSIN. STATE DEPT. OF LOCAL AFFAIRS AND DEVELOPMENT, MADISON	VBCPC-(YEAR)-
WISCONSIN. STATE DEPT. OF LOCAL AFFAIRS AND DEVELOPMENT, MADISON	VCAPC-(YEAR)-
WISCONSIN. STATE DEPT. OF LOCAL AFFAIRS AND DEVELOPMENT, MADISON	VCLPC-(YEAR)-
WISCONSIN. STATE DEPT. OF LOCAL AFFAIRS AND DEVELOPMENT, MADISON	VHPC-(YEAR)-
WISCONSIN. STATE DEPT. OF LOCAL AFFAIRS AND DEVELOPMENT, MADISON (LOCAL + REGIONAL PLANNING)	WLAD-BL/RP-DCZID/PC-
WISCONSIN. STATE DEPT. OF LOCAL AFFAIRS AND DEVELOPMENT, MADISON (LOCAL + REGIONAL PLANNING)	WLAD-BL/RP-FVCOG-KPC-
WISCONSIN. STATE DEPT. OF LOCAL AFFAIRS AND DEVELOPMENT, MADISON (LOCAL + REGIONAL PLANNING)	WLAD-BL/RP-WPC-(YR.)-
WISCONSIN. UNIV., MADISON	C/N-
WISCONSIN. UNIV., MADISON	FD-
WISCONSIN. UNIV., MADISON	HG-
WISCONSIN. UNIV., MADISON	MRC-
WISCONSIN. UNIV., MADISON	SAR-
WISCONSIN. UNIV., MADISON	UW-TR-
WISCONSIN. UNIV., MADISON	WIS-
WISCONSIN. UNIV., MADISON (SQUID PROJECT) (REF. 33)	WIS-(NUMBER)-(LETTER(S))
WISCONSIN. UNIV., MADISON	WISCU-
WISCONSIN. UNIV., MADISON	WISU-
WISCONSIN. UNIV., MADISON. COMPUTER SCIENCES DEPT.	WIS-CSTR-
WISCONSIN. UNIV., MADISON. DEPT. OF CHEMISTRY	CF-1498-A THRU -1509
WISCONSIN. UNIV., MADISON. DEPT. OF MATHEMATICS	TR-(NUMBER)/
WISCONSIN. UNIV., MADISON. DEPT. OF PHYSICS	WIS-OSR-
WISCONSIN. UNIV., MADISON. DEPT. OF STATISTICS	UWIS-DS-(YEAR)-
WISCONSIN. UNIV., MADISON. DEPT. OF STATISTICS	WIS-TR-
WISCONSIN. UNIV., MADISON. GEOPHYSICAL AND POLAR RESEARCH CENTER	RR-(YEAR)-
WISCONSIN. UNIV., MADISON. MATHEMATICS RESEARCH CENTER (ORIENTATION LECTURE SERIES)	MRC-OLS-
WISCONSIN. UNIV., MADISON. MATHEMATICS RESEARCH CTR.	MRC ORIENTATION LECTURE-
WISCONSIN. UNIV., MADISON. MATHEMATICS RESEARCH CTR.	MRC-TRS-
WISCONSIN. UNIV., MADISON. MATHEMATICS RESEARCH CENTER (RESEARCH SUMMARY REPORT)	MRC-TS-
WISCONSIN. UNIV., MADISON. MATHEMATICS RESEARCH CENTER (TECHNICAL SUMMARY REPORT)	MRC-TSR-
WISCONSIN. UNIV., MADISON. MATHEMATICS RESEARCH CTR.	WIC-TCI-
WISCONSIN. UNIV., MADISON. MATHEMATICS RESEARCH CTR.	WIS-MRC-TSR-
WISCONSIN. UNIV., MADISON. NAVAL RESEARCH LAB.	CM-
WISCONSIN. UNIV., MADISON. NAVAL RESEARCH LAB.	TR-WIS-2-AEC-
WISCONSIN. UNIV., MADISON. NAVAL RESEARCH LAB.	UWNRL-
WISCONSIN. UNIV., MADISON. NUMERICAL ANALYSIS LAB.	WIS-CRPS-
WISCONSIN. UNIV., MADISON. THEORETICAL CHEMISTRY INST	WIS-NSF-
WISCONSIN. UNIV., MADISON. THEORETICAL CHEMISTRY INST	WIS-TCI-
WISCONSIN. UNIV., MADISON. THEORETICAL CHEMISTRY LAB.	WIS-AEC-
WISCONSIN. UNIV., MADISON. THEORETICAL CHEMISTRY LAB.	WIS-AF-
WISCONSIN. UNIV., MADISON. THEORETICAL CHEMISTRY LAB.	WIS-ONR-
WISCONSIN. UNIV., MADISON. THEORETICAL CHEMISTRY LAB.	WIS-OOR-
WISCONSIN-MILWAUKEE UNIV.	UWM-(CON.NO.)-(YEAR)-

WISCONSIN-MILWAUKEE UNIV.

WISCONSIN-MILWAUKEE UNIV.	UWM-
WISSENSCHAFTLICHE GESELLSCHAFT FUER LUFT- UND RAUMFAHRT, COLOGNE	DLR-MITT-(YEAR)-
WISSENSCHAFTLICHE GESELLSCHAFT FUER LUFT- UND RAUMFAHRT, COLOGNE	WGLR/DGRR PAPER-
WNRE, INC., CHESTERTOWN, MD.	WNRE-(NUMBER)-
WOLF MANAGEMENT SERVICES, PALO ALTO, CALIF.	WMS-PROJ-
WOLF RESEARCH AND DEVELOPMENT CORP., RIVERDALE, MD.	WOLF-
WOLLENSAK OPTICAL CO., ROCHESTER, N.Y.	WOC-
WOOD-IVEY SYSTEMS CORP., WINTER PARK, FLA.	WISCO-FR-(NO.)-(NO.)
WOODS HOLE OCEANOGRAPHIC INSTITUTION, MASS.	REF-
WOODS HOLE OCEANOGRAPHIC INSTITUTION, MASS.	REF-(YEAR)-
WOODS HOLE OCEANOGRAPHIC INSTITUTION, MASS.	SAPR-
WOODS HOLE OCEANOGRAPHIC INSTITUTION, MASS.	WHOI-
WOODS HOLE OCEANOGRAPHIC INSTITUTION, MASS.	WHOI-(YEAR)-
WOODS HOLE OCEANOGRAPHIC INSTITUTION, MASS.	WHOI-CONTRIB-
WOODS HOLE OCEANOGRAPHIC INSTITUTION, MASS.	WHOI-REF-
WOODS HOLE OCEANOGRAPHIC INSTITUTION, MASS.	WHOI-REF-(YEAR)-
WOODS HOLE OCEANOGRAPHIC INSTITUTION, MASS.	WHO-REF-(YR.)-
WORKS PROJECT ADMINISTRATION, WASHINGTON, D.C. (MATHEMATICAL TABLES PROJECT)	MTP-
WORLD DATA CENTER A. COORDINATION OFF. NATL. RESEARCH COUNCIL, WASH., D.C. (UPPER ATMOSPHERE GEOPHYSICS)	UAG-
WORLD HEALTH ORGANIZATION	WHO-
WORLD HEALTH ORGANIZATION. INSTITUTO DE NUTRICION DE CENTRO AMERICA Y PANAMA, GUATEMALA CITY	INCAP-L-
WORLD HEALTH ORGANIZATION. REGIONAL OFFICE FOR EUROPE, COPENHAGEN	EURO-
WORLD METEOROLOGICAL ORGANIZATION, GENEVA	OMM-
WORLD METEOROLOGICAL ORGANIZATION, GENEVA	OMM-(NUMBER)-RP-
WORLD METEOROLOGICAL ORGANIZATION, GENEVA	WMO-(NUMBER)-TP-
WORLD METEOROLOGICAL ORGANIZATION, GENEVA	WMO-NO-(NUMBER)-TP
WRIGHT AERONAUTICAL CORP., PATERSON, N.J.	WAC-
WRIGHT AERONAUTICAL CORP., PATERSON, N.J.	WAC (LETTER)-
WRIGHT AERONAUTICAL CORP., WOOD-RIDGE, N.J.	WAC-
WRIGHT AERONAUTICAL CORP., WOOD-RIDGE, N.J.	W.A.C. SERIAL REPORT-
WRIGHT AIR DEV. CENTER, WRIGHT-PATTERSON AFB, OHIO	AMC-WC-
WRIGHT AIR DEV. CENTER, WRIGHT-PATTERSON AFB, OHIO	AMC-WCEG-
WRIGHT AIR DEV. CENTER, WRIGHT-PATTERSON AFB, OHIO	AMC-WCNSW-
WRIGHT AIR DEV. CENTER, WRIGHT-PATTERSON AFB, OHIO	AMC-WCRR-
WRIGHT AIR DEV. CENTER, WRIGHT-PATTERSON AFB, OHIO	AMC-WCSG-(YEAR)-
WRIGHT AIR DEV. CENTER, WRIGHT-PATTERSON AFB, OHIO	WADC-
WRIGHT AIR DEV. CENTER, WRIGHT-PATTERSON AFB, OHIO	WADC AEC-
WRIGHT AIR DEVELOPMENT CENTER, WRIGHT-PATTERSON AFB, OHIO (AIR FORCE TECHNICAL REPORT)	WADC AF TR-
WRIGHT AIR DEV. CENTER, WRIGHT-PATTERSON AFB, OHIO	WADC-MR-
WRIGHT AIR DEV. CENTER, WRIGHT-PATTERSON AFB, OHIO	WADC MR WCEGW R-(NO.)-
WRIGHT AIR DEV. CENTER, WRIGHT-PATTERSON AFB, OHIO	WADC-MR-WCT-
WRIGHT AIR DEV. CENTER, WRIGHT-PATTERSON AFB, OHIO	WADC-PR-
WRIGHT AIR DEV. CENTER, WRIGHT-PATTERSON AFB, OHIO	WADC-PR-(YEAR)-
WRIGHT AIR DEV. CENTER, WRIGHT-PATTERSON AFB, OHIO	WADC-QR-
WRIGHT AIR DEV. CENTER, WRIGHT-PATTERSON AFB, OHIO	WADC-TM-(YEAR)-
WRIGHT AIR DEV. CENTER, WRIGHT-PATTERSON AFB, OHIO	WADC-TN-(YEAR)-
WRIGHT AIR DEV. CENTER, WRIGHT-PATTERSON AFB, OHIO	WADC TN WCLA (YEAR)-
WRIGHT AIR DEV. CENTER, WRIGHT-PATTERSON AFB, OHIO	WADC TN WCLG (YEAR)-
WRIGHT AIR DEV. CENTER, WRIGHT-PATTERSON AFB, OHIO	WADC TN WCRR (YEAR)-
WRIGHT AIR DEV. CENTER, WRIGHT-PATTERSON AFB, OHIO	WADC TN WCT (YEAR)-
WRIGHT AIR DEV. CENTER, WRIGHT-PATTERSON AFB, OHIO	WADC-TR-(YEAR)-
WRIGHT AIR DEV. CENTER, WRIGHT-PATTERSON AFB, OHIO	WCM-
WRIGHT AIR DEVELOPMENT CENTER, WRIGHT-PATTERSON AFB, OHIO (FIRST TWO DIGITS STAND FOR YEAR) (REF. 1)	53WC... THRU 55WC...
WRIGHT AIR DEVELOPMENT CENTER. AERIAL RECONNAISSANCE LAB., WRIGHT-PATTERSON AFB, OHIO	DU-
WRIGHT AIR DEVELOPMENT CENTER. AERIAL RECONNAISSANCE LAB., WRIGHT-PATTERSON AFB, OHIO	WCLF-
WRIGHT AIR DEVELOPMENT CENTER. AERIAL RECONNAISSANCE LAB., WRIGHT-PATTERSON AFB, OHIO	WCLR-
WRIGHT AIR DEVELOPMENT CENTER. AERO MEDICAL LAB., WRIGHT-PATTERSON AFB, OHIO	DCRDM-
WRIGHT AIR DEVELOPMENT CENTER. AERO MEDICAL LAB., WRIGHT-PATTERSON AFB, OHIO	MCREEXD-
WRIGHT AIR DEVELOPMENT CENTER. AERO MEDICAL LAB., WRIGHT-PATTERSON AFB, OHIO	WCLD-
WRIGHT AIR DEVELOPMENT CENTER. AERO MEDICAL LAB., WRIGHT-PATTERSON AFB, OHIO	WCRDF-
WRIGHT AIR DEVELOPMENT CENTER. AERO MEDICAL LAB., WRIGHT-PATTERSON AFB, OHIO	WCRDM-
WRIGHT AIR DEVELOPMENT CENTER. AERO MEDICAL LAB., WRIGHT-PATTERSON AFB, OHIO	WCRDR-
WRIGHT AIR DEVELOPMENT CENTER. AERONAUTICAL RESEARCH LAB., WRIGHT-PATTERSON AFB, OHIO	MCRR-
WRIGHT AIR DEVELOPMENT CENTER. AERONAUTICAL RESEARCH LAB., WRIGHT-PATTERSON AFB, OHIO	OAR-
WRIGHT AIR DEVELOPMENT CENTER. AERONAUTICAL RESEARCH LAB., WRIGHT-PATTERSON AFB, OHIO	WCRR-
WRIGHT AIR DEVELOPMENT CENTER. AERONAUTICAL RESEARCH LAB., WRIGHT-PATTERSON AFB, OHIO	WCRR-TN-
WRIGHT AIR DEVELOPMENT CENTER. AEROSPACE MEDICAL LAB., WRIGHT-PATTERSON AFB, OHIO	WADC-TR-
WRIGHT AIR DEVELOPMENT CENTER. AIRCRAFT LAB., WRIGHT-PATTERSON AFB, OHIO	MCREXA8-(NUMBERS)-
WRIGHT AIR DEVELOPMENT CENTER. AIRCRAFT LAB., WRIGHT-PATTERSON AFB, OHIO	WADC-TN-WCLS-
WRIGHT AIR DEVELOPMENT CENTER. AIRCRAFT LAB., WRIGHT-PATTERSON AFB, OHIO	WCLS-
WRIGHT AIR DEVELOPMENT CENTER. AIRCRAFT LAB., WRIGHT-PATTERSON AFB, OHIO	WCNS-
WRIGHT AIR DEVELOPMENT CENTER. AIRCRAFT LAB., WRIGHT-PATTERSON AFB, OHIO	WCNSW-
WRIGHT AIR DEVELOPMENT CENTER. AIRCRAFT LAB., WRIGHT-PATTERSON AFB, OHIO	WCNSY-
WRIGHT AIR DEVELOPMENT CENTER. AIRCRAFT RADIATION LAB., WRIGHT-PATTERSON AFB, OHIO	WADC ARL-
WRIGHT AIR DEVELOPMENT CENTER. AIRCRAFT RADIATION LAB., WRIGHT-PATTERSON AFB, OHIO	WADC-TN-WCLR-(YEAR)-
WRIGHT AIR DEVELOPMENT CENTER. COMMUNICATION AND NAVIGATION LAB., WRIGHT-PATTERSON AFB, OHIO	WCLN-
WRIGHT AIR DEVELOPMENT CENTER. ELECTRONIC COMPONENTS LAB., WRIGHT-PATTERSON AFB, OHIO	WCES-
WRIGHT AIR DEVELOPMENT CENTER. ELECTRONIC COMPONENTS LAB., WRIGHT-PATTERSON AFB, OHIO	WCLC-TN-(YR)-
WRIGHT AIR DEVELOPMENT CENTER. ELECTRONIC COMPONENTS LAB., WRIGHT-PATTERSON AFB, OHIO	WCRE-
WRIGHT AIR DEVELOPMENT CENTER. EQUIPMENT LAB., WRIGHT-PATTERSON AFB, OHIO	WCEE-
WRIGHT AIR DEVELOPMENT CENTER. EQUIPMENT LAB., WRIGHT-PATTERSON AFB, OHIO	WCLE-
WRIGHT AIR DEVELOPMENT CENTER. MATERIALS LAB., WRIGHT-PATTERSON AFB, OHIO	DCRTT-
WRIGHT AIR DEVELOPMENT CENTER. MATERIALS LAB., WRIGHT-PATTERSON AFB, OHIO	ML-TDR-
WRIGHT AIR DEVELOPMENT CENTER. MATERIALS LAB., WRIGHT-PATTERSON AFB, OHIO	RDO-R-

WRIGHT AIR DEVELOPMENT CENTER. MATERIALS LAB., WRIGHT-PATTERSON AFB, OHIO	WCLT-TM-
WRIGHT AIR DEVELOPMENT CENTER. MATERIALS LAB., WRIGHT-PATTERSON AFB, OHIO	WCRT-(YR)-
WRIGHT AIR DEVELOPMENT CENTER. MATERIALS LAB., WRIGHT-PATTERSON AFB, OHIO	WCRTE-
WRIGHT AIR DEVELOPMENT CENTER. MATERIALS LAB., WRIGHT-PATTERSON AFB, OHIO	WCRTL-M-
WRIGHT AIR DEVELOPMENT CENTER. MATERIALS LAB., WRIGHT-PATTERSON AFB, OHIO	WCRTL-TN-
WRIGHT AIR DEVELOPMENT CENTER. MATERIALS LAB., WRIGHT-PATTERSON AFB, OHIO	WCRT-TM-
WRIGHT AIR DEVELOPMENT CENTER. MATERIALS LAB., WRIGHT-PATTERSON AFB, OHIO	WCRT-TN-(YR)-
WRIGHT AIR DEVELOPMENT CENTER. PHOTO RECONNAISSANCE LAB., WRIGHT-PATTERSON AFB, OHIO	WCLF-TN-
WRIGHT AIR DEVELOPMENT CENTER. POWER PLANT LAB., WRIGHT-PATTERSON AFB, OHIO	WCLP-
WRIGHT AIR DEVELOPMENT CENTER. POWER PLANT LAB., WRIGHT-PATTERSON AFB, OHIO	WCLPF-
WRIGHT AIR DEVELOPMENT CENTER. POWER PLANT LAB., WRIGHT-PATTERSON AFB, OHIO	WCLPO-
WRIGHT AIR DEVELOPMENT CENTER. POWER PLANT LAB., WRIGHT-PATTERSON AFB, OHIO	WCLP-TN-(YR)-
WRIGHT AIR DEVELOPMENT CENTER. POWER PLANT LAB., WRIGHT-PATTERSON AFB, OHIO	WCNE-
WRIGHT AIR DEVELOPMENT CENTER. PROPELLER LAB., WRIGHT-PATTERSON AFB, OHIO	WCLB-
WRIGHT AIR DEVELOPMENT CENTER. WEAPONS GUIDANCE LAB., WRIGHT-PATTERSON AFB, OHIO	WCEGE-
WRIGHT AIR DEVELOPMENT CENTER. WEAPONS GUIDANCE LAB., WRIGHT-PATTERSON AFB, OHIO	WCLG-
WRIGHT AIR DEVELOPMENT CENTER. WEAPONS GUIDANCE LAB., WRIGHT-PATTERSON AFB, OHIO	WCLGS-
WRIGHT AIR DEV. DIV., WRIGHT-PATTERSON AFB, OHIO	ARTC/R + D-
WRIGHT AIR DEV. DIV., WRIGHT-PATTERSON AFB, OHIO	WADD-
WRIGHT AIR DEV. DIV., WRIGHT-PATTERSON AFB, OHIO	WADD-TDR-(YEAR)-
WRIGHT AIR DEV. DIV., WRIGHT-PATTERSON AFB, OHIO	WADD-TM-(YEAR)-
WRIGHT AIR DEV. DIV., WRIGHT-PATTERSON AFB, OHIO	WADD-TN-
WRIGHT AIR DEV. DIV., WRIGHT-PATTERSON AFB, OHIO	WADD-TN-(YR.)-
WRIGHT AIR DEV. DIV., WRIGHT-PATTERSON AFB, OHIO	WADD-TR-(YR.)-
WRIGHT STATE UNIV., DAYTON, OHIO	RC-
WURLITZER (RUDOLPH) CO., CHICAGO	RWC-
WYANDOTTE CHEMICALS CORP., MICH.	WC-
WYANDOTTE CHEMICALS CORP., MICH.	WCC-
WYANDOTTE CHEMICALS CORP., MICH.	WYANDOTTE (NO.)-QPR-
WYANDOTTE CHEMICALS CORP., MICH.	WYC-
WYLE LABS., INC., HUNTSVILLE, ALA.	WR-
WYLE LABS., INC., HUNTSVILLE, ALA.	WR-(YEAR)-
WYLE RESEARCH CORP., EL SEGUNDO, CALIF.	WRC-
WYMAN-GORDON CO., WORCESTER, MASS.	RD-(YR)-
WYOMING. UNIV., LARAMIE	WYOU-
WYOMING. UNIV., LARAMIE. DIV. OF BUSINESS AND ECONOMIC RESEARCH	ECONOMIC PLANNING SER-
WYOMING. UNIV., LARAMIE. NATURAL RESOURCES RESEARCH INSTITUTE	NRRI-
WYOMING. UNIV., LARAMIE. NATURAL RESOURCES RESEARCH INSTITUTE (BULLETIN)	WYO NRRI B-

YALE UNIV.

YALE UNIV., NEW HAVEN	CT-
YALE UNIV., NEW HAVEN	Y-
YALE UNIV., NEW HAVEN	YALE-
YALE UNIV., NEW HAVEN	YALE-(YEAR)-
YALE UNIV., NEW HAVEN	YALE-IR-Y-
YALE UNIV., NEW HAVEN	YALE TR
YALE UNIV., NEW HAVEN	YALEU-TR-
YALE UNIV., NEW HAVEN	YU-
YALE UNIV., NEW HAVEN	YUP-
YALE UNIV., NEW HAVEN. COWLES FOUNDATION FOR RESEARCH IN ECONOMICS	COWLES FOUNDATION PAPER-
YALE UNIV., NEW HAVEN. COWLES FOUNDATION FOR RESEARCH IN ECONOMICS	FOUNDATION PAPER-
YALE UNIV., NEW HAVEN. DUNHAM LAB.	CT-
YALE UNIV., NEW HAVEN. EDWARDS STREET LAB.	ESL-
YALE UNIV., NEW HAVEN. EDWARDS STREET LAB.	HPP-
YALE UNIV., NEW HAVEN. HAMMOND METALLURGICAL LAB.	YU-HL-TR-
YALE UNIV., NEW HAVEN. HAMMOND METALLURGICAL LAB.	YU-HML-
YALE UNIV., NEW HAVEN. HAMMOND METALLURGICAL LAB.	YU-HML-TR-
YALE UNIV., NEW HAVEN. LAB. OF MARINE PHYSICS	LMP-
YALE UNIV., NEW HAVEN. OBSERVATORY	ARL-(YEAR)-
YALE UNIV., NEW HAVEN. SLOANE PHYSICS LAB.	YU-SPL-
YANKEE ATOMIC ELECTRIC CO., BOSTON	YAEC-
YESHIVA UNIV., N.Y.C.	TH-
YESHIVA UNIV., N.Y.C.	YESHIVAU-
YORK UNIV., TORONTO. MOLECULAR PSYCHOBIOLOGY LAB.	MPL-
YOUNG DEVELOPMENT LABS., INC., ROCKY HILL, N.J.	YDL-
YUGOSLAVIA. INSTITUT JOZEF STEFAN, LJUBLJANA	IJS-
YUGOSLAVIA. INSTITUT JOZEF STEFAN, LJUBLJANA	NIJS-
YUGOSLAVIA. INSTITUT JOZEF STEFAN, LJUBLJANA	NIJS-P-
YUGOSLAVIA. INSTITUT JOZEF STEFAN, LJUBLJANA	NIJS-R-
YUGOSLAVIA. INSTITUT RUDJER BOSKOVIC, ZAGREB	IRB-TP-(NUMBER)-(YEAR)
YUGOSLAVIA. INSTITUT ZA NUKLEARNE NAUKE BORIS KIDRIC, BELGRADE	BKI-
YUGOSLAVIA. INSTITUT ZA NUKLEARNE NAUKE BORIS KIDRIC, BELGRADE	IBK-
YUMA PROVING GROUND, ARIZ.	YPG-

ZAPFFE (CARL A.) AND ASSOCIATES, BALTIMORE — ZA-
ZAPFFE (CARL A.) AND ASSOCIATES, BALTIMORE — ZAP-

ZATOR CO., CAMBRIDGE, MASS. — ZTB-

ZENITH RADIO CORP., CHICAGO — RC-
ZENITH RADIO CORP., CHICAGO — ZRC-

ZENTRALSTELLE FUER ATOMKERNENERGIE-DOKUMENTATION BEIM GMELIN INSTITUT, FRANKFURT AM MAIN — AEC-C-
ZENTRALSTELLE FUER ATOMKERNENERGIE-DOKUMENTATION BEIM GMELIN INSTITUT, FRANKFURT AM MAIN — AED-
ZENTRALSTELLE FUER ATOMKERNENERGIE-DOKUMENTATION BEIM GMELIN INSTITUT, FRANKFURT AM MAIN — AED-A-
ZENTRALSTELLE FUER ATOMKERNENERGIE-DOKUMENTATION BEIM GMELIN INSTITUT, FRANKFURT AM MAIN — AED-AB-(YEAR)-
ZENTRALSTELLE FUER ATOMKERNENERGIE-DOKUMENTATION BEIM GMELIN INSTITUT, FRANKFURT AM MAIN — AED-BRD-
ZENTRALSTELLE FUER ATOMKERNENERGIE-DOKUMENTATION BEIM GMELIN INSTITUT, FRANKFURT AM MAIN — AED-BRD-C-
ZENTRALSTELLE FUER ATOMKERNENERGIE-DOKUMENTATION BEIM GMELIN INSTITUT, FRANKFURT AM MAIN — AED-CONF.(YEAR)-
ZENTRALSTELLE FUER ATOMKERNENERGIE-DOKUMENTATION BEIM GMELIN INSTITUT, FRANKFURT AM MAIN (THESIS) — AED-D-
ZENTRALSTELLE FUER ATOMKERNENERGIE-DOKUMENTATION BEIM GMELIN INSTITUT, FRANKFURT AM MAIN — AED-M(NUMBER)
ZENTRALSTELLE FUER ATOMKERNENERGIE-DOKUMENTATION BEIM GMELIN INSTITUT, FRANKFURT AM MAIN — ZAED

ZIMNEY CORP., MONROVIA, CALIF. — ZC-

ZIMNEY CORP., PASADENA, CALIF. — ZC-

ZULU COMM. — ZULU-

ZURICH. UNIVERSITAET — ASR-

Z
6945
A2
D5
1973

NOV 15 1974

RAYMOND H. FOGLER LIBRARY
DATE DUE

BOOKS ARE SUBJECT TO
RECALL AFTER TWO WEEKS